ENCYCLOPEDIA OF SPACE AND ASTRONOMY

JOSEPH A. ANGELO, JR.

An imprint of Infobase Publishing

To the memory of my parents, Rose and Joseph Angelo, whose sacrifices, love, and dedication to family taught me what is most important in life and showed me how to see the hand of God in all things, large and small, in this beautiful universe.

Encyclopedia of Space and Astronomy

Facts On File, Inc.
An imprint of Infobase Publishing
132 West 31st Street
New York NY 10001

Library of Congress Cataloging-in-Publication Data

Angelo, Joseph A.
Encyclopedia of space and astronomy / Joseph A. Angelo, Jr.
p. cm.
Includes bibliographical references and index.
ISBN 0-8160-5330-8 (hardcover)
1. Space astronomy—Encyclopedias. 2. Astronomy—Encyclopedias. I. Title.
QB136.A55 2006
520'.3—dc222004030800

Text design by Joan M. Toro
Cover design by Cathy Rincon
Illustrations by Richard Garratt

Printed in the United States of America

VB Hermitage 10 9 8 7 6 5 4 3 2

This book is printed on acid-free paper.

CONTENTS

Appendixes:

ACKNOWLEDGMENTS

I wish to publicly acknowledge the generous support of the National Aeronautics and Space Administration (NASA), the United States Air Force (USAF), the U.S. Geological Survey (USGS), the National Reconnaissance Office (NRO), and the European Space Agency (Washington, D.C., office) during the preparation of this book. Special thanks are also extended to the editorial staff at Facts On File, particularly my editor, Frank K. Darmstadt. The staff at the Evans Library of Florida Tech again provided valuable research support. Finally, without the help of my wife, Joan, the manuscript and illustrations for this book would never have survived three consecutive hurricanes during the summer of 2004 and then emerge from chaotic piles of hastily moved boxes to become a workable document.

INTRODUCTION

The *Encyclopedia of Space and Astronomy* introduces the exciting relationship between modern astronomy and space technology. The book also examines the technical, social, and philosophical influences this important combination of science and technology exerts on our global civilization. With the start of the space age in 1957, scientists gained the ability to place sophisticated observatories in outer space. Since such orbiting astronomical facilities operate above the masking limitations imposed by our planet's atmosphere, they can collect scientific data in regions of the electromagnetic spectrum previously unavailable to astronomers who could only look at the heavens from the surface of Earth.

Data from orbiting astronomical observatories, as well as from an armada of planetary exploration spacecraft, have completely transformed observational astronomy, astrophysics, and cosmology. Space technology has expanded our view of the universe. Scientists now enjoy meeting the universe face-to-face and harvesting enormous quantities of interesting new scientific data from all the information-rich portions of the electromagnetic spectrum. Some spacecraft are designed to investigate mysterious cosmic rays—those very tiny, but extremely energetic, pieces of galactic and extragalactic material that provide tantalizing clues about extremely violent cosmic processes, such as exploding stars and colliding galaxies. Other spacecraft have explored the solar system.

Through developments in space technology, planetary bodies within the solar system are no longer the unreachable points of wandering light that intrigued ancient stargazers. Today, as a result of flybys, orbital reconnaissance missions, and even landings by a variety of interesting robot spacecraft, these mysterious celestial objects have become familiar worlds. The marvel of space technology has also allowed 12 human beings to walk on the surface of another world (the Moon) and to return safely to Earth with hand-selected collections of rock and soil samples for detailed investigation by fellow scientists.

The major entries and special essays within this book highlight the synthesis of modern astronomy and space technology. This fortuitous union of science and technology has created an explosion in knowledge amounting to the start of a second scientific revolution—similar, but even more consequential, than the first. The first scientific revolution began in the 16th century, when a few bold astronomers used their pioneering observations and mathematics to challenge the long-cherished philosophical position that Earth was the stationary center of the universe.

This book serves as a guide and introduction to the fundamental concepts, basic principles, famous and less-known people, major events, and impact of astronomy and space technology. The collection of biographical entries, some brief and others a bit more extensive, allows the reader to discover firsthand the genius, sacrifice, visionary brilliance, and hard work of the men and women who established modern astronomy and/or brought about the age of space. Special essays address a variety of interesting, intellectually stimulating topics and should help high school and college students better understand, appreciate, and even participate in the great space age revolution that embraces our global civilization. The general reader will find major entries, such as *rocket, telescope,* and *spacecraft,* very useful as introductory treatments of complicated subjects. Major entries are prepared in a simple, easy to read style and are generously complemented by illustrations, photographs, and visionary artist renderings. The numerous supporting entries throughout the book serve as concise capsules of basic information. An extensive network of cross-references assists the reader in further pursuing a particular scientific topic or technical theme. Entries that describe the significance of past, current, and future space activities generally include a balanced combination of visual (graphic) and written material. Images from contemporary space missions take the reader beyond Earth and provide a firsthand view of some of the most interesting celestial objects in our solar system.

Many major entries provide compact technical discussions concerning basic concepts considered fundamental in understanding modern astronomy and space technology. While mathematics is certainly important in astronomy and space technology, the book avoids an oppressive use of formulas and equations. Such detailed mathematical treatments are considered more appropriate for specialized textbooks and are beyond the scope of this introductory work. Other entries summarize contemporary astronomical knowledge or discuss the results of important space exploration missions. Because of the rapidly changing nature of contemporary space missions, the reader is encouraged to pursue updates through many of the excellent Internet sites suggested in Appendix II. Still other entries, such as *horizon mission methodology* and *starship,* are designed to stimulate intellectual curiosity by challenging the reader to think "outside the box," or, more appropriately, "beyond this planet." These entries pose intriguing questions, introduce scientifically based speculations, and suggest some of the anticipated consequences of future space missions. The search for extraterrestrial life is another important example. Right now the subject of extraterrestrial life resides in the nebulous buffer zone between science fiction and highly speculative science, but, through advanced space exploration projects, these discussions could easily become the factual centerpiece of contemporary scientific investigation.

The confirmed discovery of extraterrestrial life, extinct or existent—no matter how humble in form—on Mars or perhaps on one of the intriguing Galilean moons of Jupiter could force a major revision in the chauvinistic planetary viewpoint human beings have tacitly embraced for centuries. This deeply embedded terrestrial chauvinism suggests that planet Earth and, by extrapolation, the universe were created primarily for human benefit. Perhaps this is so, but the discovery of life elsewhere could shatter such a myopic viewpoint and encourage a wholesale reevaluation of long-standing philosophical questions such as "Who are we as a species?" and "What is our role in the cosmic scheme of things?"

The somewhat speculative approach taken by certain entries is necessary if the book is to properly project the potential impact of space exploration. What has occurred in the four or so decades of space exploration is just the tip of the intellectual iceberg. Space missions planned for the next few decades promise to accelerate the pace and excitement of the contemporary scientific revolution triggered by the union of space technology and astronomy.

Special effort has been made to provide not only easily understood technical entries, but also entries that are set in an appropriate scientific, social, and/or philosophical context. This approach extends the information content of the book to a wider audience—an audience that includes those who want to become scientists and engineers as well as those who plan to pursue other careers but desire to understand how astronomy and space technology affect their lives.

Most early civilizations paid close attention to the sky. Naked eye astronomy, the most ancient and widely practiced of the physical sciences, provided social and political cohesion to early societies and often became an integral part of religious customs. To many ancient peoples the easily observed and recorded cyclic movement of the Sun, Moon, planets (only five are visible without the astronomical telescope), and stars provided a certain degree of order and stability in their lives. After all, these ancient peoples did not possess the highly accurate personal timepieces that tend to control our modern lives. In fact, people living in today's schedule-dominated, fast-paced societies often become totally detached from the natural diurnal cycle that closely regulated the lifestyles and activities of ancient peoples. As citizens of a 24/7-wired world that is always "open for business," it is sometimes difficult for us to appreciate how the heavens and nature touched every aspect of a person's life in ancient societies. For these early peoples stargazing, myth, religion, and astrology often combined to form what scientists now refer to as "ancient astronomy." So it should not come as too great a surprise to us that these peoples often deified the Sun, the Moon, and other important celestial objects. Their lives were interwoven with and became dependent on the reasonably predictable motions of such commonly observed but apparently unreachable celestial objects.

Looking back in history from the vantage point of 21st-century science, some of us might be tempted to ridicule such ancient activities. How could the early Egyptians or the Aztecs deify the Sun? But are we that much different with respect to how celestial objects control our lives? As a result of space technology our global civilization has become highly interwoven with and dependent on *human-made celestial objects*. In addition to the revolution in modern astronomy, spacecraft now support global communications, are an essential component of national defense, monitor weather conditions, provide valuable scientific data about the Earth as a complex, highly interactive system, and help travelers find their way on land, on sea, and in the air. A number of special entries and companion essays provide introductory discussions on just how much space technology affects many aspects of modern life.

The scientific method—that is, the practice of science as a form of natural philosophy or an organized way of looking at and explaining the world and how things work—emerged in the 16th and 17th centuries in western Europe. Telescope-assisted astronomical observations encouraged Galileo Galilei, Johannes Kepler, Sir Isaac Newton, and other pioneering thinkers to overthrow two millennia of geocentric (Earth-centered) cosmology and replace it with a more scientifically sound, heliocentric view of the solar system. Observational astronomy gave rise to the first scientific revolution and led to the practice of organized science (the scientific method)—one of the greatest contributions of Western civilization to the human race.

Today the combination of astronomy and space technology is encouraging scientists to revisit some of humankind's most important and long-pondered philosophical questions: Who are we? Where did we come from? Where are we going? Are we alone in this vast universe? Future space missions will define the cosmic philosophy of an emerging solar system civilization. As the detailed exploration of distant worlds leads to presently unimaginable discoveries, space technology and modern astronomy will enable us to learn more about our role and place as an intelligent species in a vast and beautiful universe.

This book is not just a carefully prepared collection of technical facts. It also serves as a special guide for those readers who want to discover how astronomy and space technology are making the universe both a destination and a destiny for the human race. For example, the entry on *nucleosynthesis* and the essay "We Are Made of Stardust" will help readers discover that the biogenic elements in their bodies, such as carbon, nitrogen, oxygen, and phosphorous, came from ancient stars that exploded and hurled their core materials into the interstellar void long before the solar system formed. Throughout most of human history, people considered the universe a place apart from life on Earth. Through modern astronomy and space technology we can now proclaim that "We have met the universe and it is US." So come share the excitement, the vision, and the intellectual accomplishments of the astronomers and space technologists who helped humankind reach for the stars.

—Joseph A. Angelo, Jr.
Cape Canaveral, 2005

ENTRIES A–Z

A Symbol for RELATIVE MASS NUMBER.

Abell, George O. (1927–1983) American *Astronomer* George O. Abell is best known for his investigation and classification of galactic clusters. Using photographic plates from the Palomar Observatory Sky Survey (POSS) made with the 1.2-m (48-inch) Schmidt Telescope at the Palomar Observatory in California, he characterized more than 2,700 Abell clusters of galaxies. In 1958 he summarized this work in a book now known as the *Abell Catalogue.*

See also ABELL CLUSTER; GALACTIC CLUSTER; PALOMAR OBSERVATORY; PALOMAR OBSERVATORY SKY SURVEY (POSS).

Abell cluster A rich (high-concentration density) cluster of galaxies as characterized by the American astronomer GEORGE O. ABELL (1927–83). In 1958 Abell produced a catalog describing over 2,700 of such high-density galactic clusters using photographic data from the Palomar Observatory. Abell required that each cluster of galaxies satisfy certain criteria before he included them in this catalog. His selection criteria included population (each Abell cluster had to contain 50 or more galaxies) and high-concentration density (richness). Abell also characterized such rich galactic clusters by their appearance—listing them as either regular or irregular.

See also PALOMAR OBSERVATORY.

aberration 1. In optics, a specific deviation from a perfect image, such as spherical aberration, astigmatism, coma, curvature of field, and distortion. For example, spherical aberration results in a point of light (that is, a point image) appearing as a circular disk at the focal point of a spherical LENS or curved MIRROR. This occurs because the focal points of light rays far from the optical axis are different from the focal points of light rays passing through or near the center of the lens or mirror. Light rays passing near the center of a spherical lens or curved mirror have longer focal lengths than do those passing near the edges of the optical system. 2. In astronomy, the apparent angular displacement of the position of a celestial body in the direction of motion of the observer. This effect is caused by the combination of the velocity of the observer and the velocity of light (c).

See also ABERRATION OF STARLIGHT.

This *Chandra X-Ray Observatory* image acquired on May 25, 2000, of the galactic cluster Abell 2104 revealed six bright X-ray sources that are associated with supermassive black holes in red galaxies in the cluster. Since red galaxies are thought to be composed primarily of older stars and to contain little gas, the observation came as a surprise to astronomers. Abell 2104 is about 2 billion light-years away. *(Courtesy of NASA/CXC/P. Martini, et al.)*

aberration of starlight The apparent angular displacement of the position of a celestial body in the direction of motion of the observer. This effect is caused by the combination of the velocity of the observer and the velocity of light (c). An observer on Earth would have Earth's orbital velocity around the sun (V_{Earth}), which is approximately 30 km/s. As a result of this effect in the course of a year, the light from a fixed star appears to move in a small ellipse around its mean position on the celestial sphere. The British astronomer JAMES BRADLEY (1693–1762) discovered this phenomenon in 1728.

abiotic Not involving living things; not produced by living organisms.

See also LIFE IN THE UNIVERSE.

ablation A form of mass transfer cooling that involves the removal of a special surface material (called an *ablative material*) from a body, such as a reentry vehicle, a planetary probe, or a reusable aerospace vehicle, by melting, vaporization, sublimation, chipping, or other erosive process, due to the aerodynamic heating effects of moving through a planetary atmosphere at very high speed. The rate at which ablation occurs is a function of the reentering body's passage through the aerothermal environment, a high-temperature environment caused by atmospheric friction. Ablation is also a function of other factors including (1) the amount of thermal energy (i.e., heat) needed to raise the temperature of the ablative material to the ablative temperature (including phase change), (2) the head-blocking action created by boundary layer thickening due to the injection of mass, and (3) the thermal energy dissipated into the interior of the body by CONDUCTION at the ablative temperature. To promote maximum thermal protection, the ablative material should not easily conduct heat into the reentry body. To minimize mass loss during the ablative cooling process, the ablative material also should have a high value for its effective heat of ablation—a thermophysical property that describes the efficiency with which thermal energy (in joules [J]) is removed per unit mass lost or "ablated" (in kilograms [kg]). Contemporary fiberglass resin ablative materials can achieve more than 10^7 J/kg thermal energy removal efficiencies through sublimation processes during reentry. Ablative cooling generally is considered to be the least mass-intensive approach to reentry vehicle thermal protection. However, these mass savings are achieved at the expense of heat shield (and possibly reentry vehicle) reusability.

ablative material A special material designed to provide thermal protection to a reentry body traveling at hypersonic speed in a planetary atmosphere. Ablative materials are used on the surfaces of reentry vehicles and planetary probes to absorb thermal energy (i.e., heat) by removing mass. This mass loss process prevents heat transfer to the rest of the vehicle and maintains the temperatures of the vehicle's structure and interior (including crew compartment for missions involving humans or other living creatures) within acceptable levels. Ablative materials absorb thermal energy by increasing in temperature and then by undergoing changes in their physical state through melting, vaporization, or sublimation. This absorbed thermal energy is then dissipated from the vehicle's surface by a loss of mass (generally in either the liquid or vapor phase) during high-speed flow interactions with the atmosphere. The departing ablative material also can block part of the aerodynamic heat transfer to the remaining surface material in a manner similar to transpiration cooling. Modern fiberglass resin compound ablative materials can achieve more than 10 million joules (J) of effective thermal energy transfer per kilogram (kg) of mass removed.

abort 1. To cancel, cut short, or break off an action, operation, or procedure with an aircraft, space vehicle, or the like, especially because of equipment failure. For example, the lunar landing mission was *aborted* during the *Apollo 13* flight. 2. In defense, failure to accomplish a military mission for any reason other than enemy action. The abort may occur at any point from the initiation of an operation to arrival at the target or destination. 3. An aircraft, space vehicle, or planetary probe that aborts. 4. An act or instance of aborting. 5. To cancel or cut short a flight after it has been launched.

A malfunction in the first stage of the Vanguard launch vehicle caused the vehicle to lose thrust after just two seconds, aborting the mission on December 6, 1957. The catastrophic destruction of this rocket vehicle and its small scientific satellite temporarily shattered American hopes of effectively responding to the successful launches of two different Sputnik satellites by the Soviet Union at the start of the space age in late 1957. ***(Courtesy of U.S. Navy)***

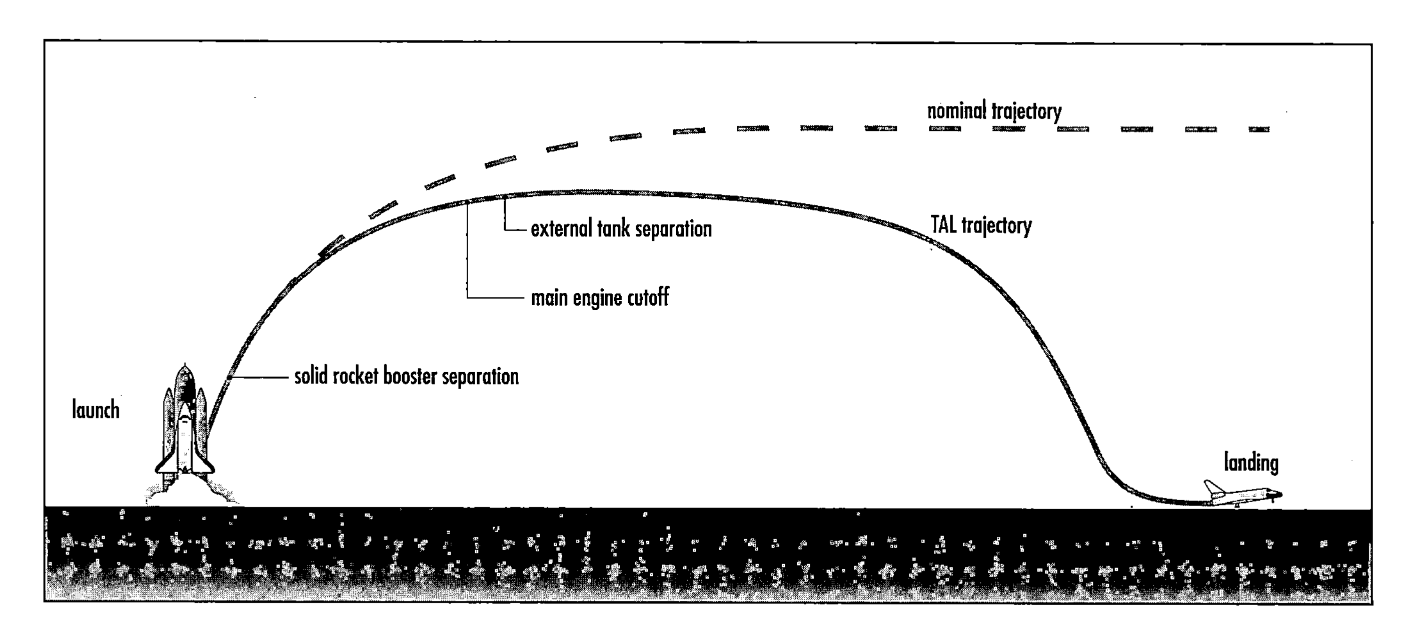

Transatlantic landing abort mode for the space shuttle

abort modes (Space Transportation System) Selection of an ascent abort mode may become necessary if there is a failure that affects space shuttle vehicle performance, such as the failure of a main engine or the orbital maneuvering system (OMS). Other failures requiring early termination of a space shuttle flight, such as a cabin leak, could require the crew to select an abort mode.

There are two basic types of ascent abort modes for space shuttle missions: *intact aborts* and *contingency aborts.* Intact aborts are designed to provide a safe return of the Orbiter vehicle and its crew to a planned landing site. Contingency aborts are designed to permit flight crew survival following more severe failures when an intact abort is not possible. A contingency abort would generally result in a ditch operation.

There are four types of intact aborts: Abort-to-Orbit (ATO), Abort-Once-Around (AOA), Transatlantic-Landing (TAL), and Return-to-Launch-Site (RTLS).

The *Abort-To-Orbit* (ATO) mode is designed to allow the Orbiter vehicle to achieve a temporary orbit that is lower than the nominal mission orbit. This abort mode requires less performance and provides time to evaluate problems and then choose either an early deorbit maneuver or an OMS thrusting maneuver to raise the orbit and continue the mission.

The *Abort-Once-Around* (AOA) mode is designed to allow the Orbiter vehicle to fly once around Earth and make a normal entry and landing. This abort mode generally involves two OMS thrusting sequences, with the second sequence being a deorbit maneuver. The atmospheric entry sequence would be similar to a normal mission entry.

The *Transatlantic-Landing* (TAL) mode is designed to permit an intact landing on the other side of the Atlantic Ocean at emergency landing sites in either Morocco, the Gambia, or Spain. This abort mode results in a ballistic trajectory that does not require an OMS burn.

Finally, the *Return-to-Launch-Site* (RTLS) mode involves flying downrange to dissipate propellant and then turning around under power to return the Orbiter vehicle, its crew, and its payload directly to a landing at or near the KENNEDY SPACE CENTER (KSC).

The type of failure (e.g., loss of one main engine) and the time of the failure determine which type of abort mode would be selected. For example, if the problem is a space shuttle main engine (SSME) failure, the flight crew and mission control center (MCC) (at the NASA JOHNSON SPACE CENTER in Houston, Texas) would select the best option available at the time when the main engine fails.

See also SPACE TRANSPORTATION SYSTEM.

absentee ratio With respect to a hypothetical constellation of orbiting space weapon platforms, the ratio of the number of platforms not in position to participate in a battle to the number that are. In contemporary space defense studies typical absentee ratios for postulated kinetic energy weapon systems were approximately 10 to 30, depending on the orbits of the space platforms and the hypothesized space battle scenarios.

See also BALLISTIC MISSILE DEFENSE (BMD).

absolute magnitude (symbol: M) The measure of the brightness (or apparent magnitude) that a star would have if it were hypothetically located at a reference distance of 10 parsecs (10 pc), about 32.6 light-years, from the Sun.

See also APPARENT MAGNITUDE; STAR.

absolute temperature Temperature value relative to absolute zero, which corresponds to 0 K, or –273.15°C (after the Swedish astronomer ANDERS CELSIUS.) In international system (SI) units, the absolute temperature values are expressed in kelvins (K), a unit named in honor of the Scottish physicist

BARON WILLIAM THOMSON KELVIN. In the traditional engineering unit system, absolute temperature values are expressed in degrees Rankine (R), named after the Scottish engineer William Rankine (1820–72).

In SI units the *absolute temperature scale* is the Kelvin scale. By international agreement a reference value of 273.16 K has been assigned to the triple point of water. The Celsius (formerly centigrade) temperature scale (symbol °C) is a *relative temperature scale* that is related to the absolute Kelvin scale by the formula

$$T_C = T_K - 273.16$$

The Celsius scale was originally developed such that 1°C = 1 K, but using the ice point of water (273.15 K) to establish 0°C. However, from decisions made at the International Practical Temperature Scale Agreements of 1968, the triple point of water was established as 273.16 K, or 0.01°C, consequently shifting the zero point of the Celsius scale slightly so that now 0 K actually corresponds to –273.15°C.

A second absolute temperature scale that sometimes appears in American aerospace engineering activities is the Rankine scale. In this case the triple point for water is fixed by international agreement at 491.69°R. The Fahrenheit scale (°F) is a relative temperature scale related to the absolute Rankine scale by the formula

$$T_F = T_R - 459.67$$

Similarly, the Fahrenheit relative temperature scale was developed such that 1°F = 1°R, but because by international agreement the triple point of water (491.69°R) represents 32.02°F on the Fahrenheit relative temperature scale, 0°F now corresponds to 459.67°R.

Note that the international scientific community uses SI units exclusively and that the proper symbol for kelvins is simply K (*without* the symbol °).

absolute zero The temperature at which molecular motion vanishes and an object has no thermal energy (or heat). From thermodynamics, absolute zero is the lowest possible temperature, namely zero kelvin (0 K).

See also ABSOLUTE TEMPERATURE.

absorbed dose (symbol: D) When ionizing radiation passes through matter some of its energy is imparted to the matter. The amount of energy absorbed per unit mass of irradiated material is called the absorbed dose. The traditional unit of absorbed dose is the rad (an acronym for *r*adiation *a*bsorbed *d*ose), while the SI unit for absorbed dose is called the gray (Gy). One gray is defined as one joule (J) of energy deposited per kilogram (kg) of irradiated matter, while one rad is defined as 100 ergs of energy deposited per gram of irradiated matter.

The traditional and SI units of absorbed dose are related as follows:

$$100 \text{ rad} = 1 \text{ Gy}$$

The absorbed dose is often loosely referred to in radiation protection activities as "dose" (although this use is neither precise nor correct) and is also frequently confused with the term DOSE EQUIVALENT (H).

See also IONIZING RADIATION; RADIATION SICKNESS.

absorptance (absorptivity) In heat transfer by thermal radiation the absorptance (commonly used symbol: α) of a body is defined as the ratio of the incident radiant energy absorbed by the body to the total radiant energy falling upon the body. For the special case of an ideal blackbody, all radiant energy incident upon this blackbody is absorbed, regardless of the wavelength or direction. Therefore, the absorptance for a blackbody has a value of unity, that is, $\alpha_{\text{blackbody}} = 1$. All other real-world solid objects have an absorptance of less than 1. *Compare with* REFLECTANCE and TRANSMITTANCE.

absorption line A gap, dip, or dark line occurring in a spectrum caused by the absorption of electromagnetic radiation (radiant energy) at a specific wavelength by an absorbing substance, such as a planetary atmosphere or a monatomic gas.

See also ABSORPTION SPECTRUM.

absorption spectrum The array of absorption lines and bands that results from the passage of electromagnetic radiation (i.e., radiant energy) from a continuously emitting high-temperature source through a selectively absorbing medium that is cooler than the source. The absorption spectrum is characteristic of the absorbing medium, just as an emission spectrum is characteristic of the radiating source.

An absorption spectrum formed by a monatomic gas (e.g., helium) exhibits discrete dark lines, while one formed by a polyatomic gas (e.g., carbon dioxide, CO_2) exhibits ordered arrays (bands) of dark lines that appear regularly spaced and very close together. This type of absorption is often referred to as line absorption. Line spectra occur because the atoms of the absorbing gas are making transitions between specific energy levels. In contrast, the absorption spectrum formed by a selectively absorbing liquid or solid is generally continuous in nature, that is, there is a continuous wavelength region over which radiation is absorbed.

See also ELECTROMAGNETIC SPECTRUM.

Abu'l Wafa (940–998) Arab *Mathematician, Astronomer* This early Arab astronomer developed spherical trigonometry and worked in the Baghdad Observatory, constructed by Muslim prince Sharaf al-Dawla. Specifically, he introduced the use of tangent and cotangent functions in his astronomical activities, which included careful observations of solstices and equinoxes.

abundance of elements (in the universe) Stellar spectra provide an estimate of the cosmic abundance of elements as a percentage of the total mass of the universe. The 10 most common elements are hydrogen (H) at 73.5 percent of the total mass, helium (He) at 24.9 percent, oxygen (O) at 0.7 percent, carbon (C) at 0.3 percent, iron at 0.15 percent, neon (Ne) at 0.12 percent, nitrogen (N) at 0.10 percent, silicon (Si) at 0.07 percent, magnesium (Mg) at 0.05 percent, and sulfur (S) at 0.04 percent.

We Are Made of Stardust

Throughout most of human history people considered themselves and the planet they lived on as being apart from the rest of the universe. After all, the heavens were clearly unreachable and therefore had to remain the abode of the deities found in the numerous mythologies that enriched ancient civilizations. It is only with the rise of modern astronomy and space technology that scientists have been able to properly investigate the chemical evolution of the universe. And the results are nothing short of amazing.

While songwriters and poets often suggest that a loved one is made of stardust, modern scientists have shown us that this is *not* just a fanciful artistic expression. It is quite literally true. All of us are made of stardust! Thanks to a variety of astrophysical phenomena, including ancient stellar explosions that took place long before the solar system formed, the chemical elements enriching our world and supporting life came from the stars. This essay provides a brief introduction to the cosmic connection of the chemical elements.

The chemical elements, such as carbon (C), oxygen (O), and calcium (Ca), are all around us and are part of us. Furthermore, the composition of planet Earth and the chemical processes that govern life within our planet's biosphere are rooted in these chemical elements. To acknowledge the relationship between the chemical elements and life, scientists have given a special name to the group of chemical elements they consider essential for all living systems—whether here on Earth, possibly elsewhere in the solar system, or perhaps on habitable planets around other stars. Scientists refer to this special group of life-sustaining chemical elements as the *biogenic elements.*

Biologists focus their studies on life as it occurs on Earth in its many varied and interesting forms. Exobiologists extend basic concepts about terrestrial carbon-based life in order to create their speculations about the possible characteristics of life beyond Earth's biosphere. When considering the biogenic elements, scientists usually place primary emphasis on the elements hydrogen (H), carbon, nitrogen (N), oxygen, sulfur (S), and phosphorous (P). The chemical compounds of major interest are those normally associated with water (H_2O) and with other organic chemicals, in which carbon bonds with itself or with other biogenic elements. There are also several "life-essential" inorganic chemical elements, including iron (Fe), magnesium (Mg), calcium, sodium (Na), potassium (K), and chlorine (Cl).

All the natural chemical elements found here on Earth and elsewhere in the universe have their ultimate origins in cosmic events. Since different elements come from different events, the elements that make up life itself reflect a variety of astrophysical phenomena that have taken place in the universe. For example, the hydrogen found in water and hydrocarbon molecules formed just a few moments after the big bang event that started the universe. Carbon, the element considered the basis for all terrestrial life, formed in small stars. Elements such as calcium and iron formed in the interiors of large stars. Heavier elements with atomic numbers beyond iron, such as silver (Ag) and gold (Au), formed in the tremendous explosive releases of supernovae. Certain light elements, such as lithium (Li), beryllium (Be), and boron (B), resulted from energetic cosmic ray interactions with other atoms, including the hydrogen and helium nuclei found in interstellar space.

Following the big bang explosion the early universe contained the primordial mixture of energy and matter that evolved into all the forms of energy and matter we observe in the universe today. For example, about 100 seconds after the big bang the temperature of this expanding mixture of matter and energy fell to approximately one billion degrees kelvin (K)—"cool" enough that neutrons and protons began to stick to each other during certain collisions and form light nuclei, such as deuterium and lithium. When the universe was about three minutes old 95 percent of the atoms were hydrogen, 5 percent were helium, and there were only trace amounts of lithium. At the time, these three elements were the only ones that existed.

As the universe continued to expand and cool, the early atoms (mostly hydrogen and a small amount of helium) began to gather through gravitational attraction into very large clouds of gas. For millions of years these giant gas clouds were the only matter in the universe, because neither stars nor planets had yet formed. Then, about 200 million years after the big bang, the first stars began to shine, and the creation of important new chemical elements started in their thermonuclear furnaces.

Stars form when giant clouds of mostly hydrogen gas, perhaps light-years across, begin to contract under their own gravity. Over millions of years various clumps of hydrogen gas would eventually collect into a giant ball of gas that was hundreds of thousands of times more massive than Earth. As the giant gas ball continued to contract under its own gravitational influence, an enormous pressure arose in its interior. Consistant with the laws of physics, the increase in pressure at the center of this "protostar" was accompanied by an increase in temperature. Then, when the center reached a minimum temperature of about 15 million degrees kelvin, the hydrogen nuclei in the center of the contracting gas ball moved fast enough that when they collided these light (low mass) atomic nuclei would undergo fusion. This is the very moment when a new star is born.

The process of nuclear fusion releases a great amount of energy at the center of the star. Once thermonuclear burning begins in a star's core, the internal energy release counteracts the continued contraction of stellar mass by gravity. The ball of gas becomes stable—as the inward pull of gravity exactly balances the outward radiant pressure from thermonuclear fusion reactions in the core. Ultimately, the energy released in fusion flows upward to the star's outer surface, and the new star "shines."

Stars come in a variety of sizes, ranging from about a 10th to 60 (or more) times the mass of our own star, the Sun. It was not until the mid-1930s that astrophysicists began to recognize how the process of nuclear fusion takes place in the interiors of all normal stars and fuels their enormous radiant energy outputs. Scientists use the term *nucleosynthesis* to describe the complex process of how different size stars create different elements through nuclear fusion reactions.

Astrophysicists and astronomers consider stars less than about five times the mass of the Sun medium and small sized stars. The production of elements in stars within this mass range is similar. Small and medium sized stars also share a similar fate at the end of life. At birth small stars begin their stellar life by fusing hydrogen into helium in their cores. This process generally continues for billions of

(continues)

We Are Made of Stardust *(continued)*

years, until there is no longer enough hydrogen in a particular stellar core to fuse into helium. Once hydrogen burning stops so does the release of the thermonuclear energy that produced the radiant pressure, which counteracted the relentless inward attraction of gravity. At this point in its life a small star begins to collapse inward. Gravitational contraction causes an increase in temperature and pressure. As a consequence, any hydrogen remaining in the star's middle layers soon becomes hot enough to undergo thermonuclear fusion into helium in a "shell" around the dying star's core. The release of fusion energy in this shell enlarges the star's outer layers, causing the star to expand far beyond its previous dimensions. This expansion process cools the outer layers of the star, transforming them from brilliant white hot or bright yellow in color to a shade of dull glowing red. Quite understandably, astronomers call a star at this point in its life cycle a *red giant*.

Gravitational attraction continues to make the small star collapse until the pressure in its core reaches a temperature of about 100 million degrees kelvin (K). This very high temperature is sufficient to allow the thermonuclear fusion of helium into carbon. The fusion of helium into carbon now releases enough energy to prevent further gravitational collapse—at least until the helium runs out. This stepwise process continues until oxygen is fused. When there is no more material to fuse at the progressively increasing high-temperature conditions within the collapsing core, gravity again exerts its relentless attractive influence on matter. This time, however, the heat released during gravitational collapse causes the outer layers of the small star to blow off, creating an expanding symmetrical cloud of material that astronomers call a planetary nebula. This expanding cloud may contain up to ten percent of the small or medium size star's mass. The explosive blow-off process is very important because it disperses into space the elements created in the small star's core by nucleosynthesis.

The final collapse that causes the small star to eject a planetary nebula also liberates thermal energy, but this time the energy release is not enough to fuse other elements. So the remaining core material continues to collapse until all the atoms are crushed together and only the repulsive force between the electrons counteracts gravity's relentless pull. Astronomers refer to this type of condensed matter as a degenerate star and give the final compact object a special name—the white dwarf star. The white dwarf star represents the final phase in the evolution of most low-mass stars, including our Sun.

If the white dwarf star is a member of a binary star system, its intense gravity might pull some gas away from the outer regions of the companion (normal) star. When this happens the intense gravity of the white dwarf causes the inflowing new gas to rapidly reach very high temperatures, and a sudden explosion occurs. Astronomers call this event a nova. The nova explosion can make a white dwarf appear up to 10,000 times brighter for a short period of time. Thermonuclear fusion reactions that take place during the nova explosion also create new elements, such as carbon, oxygen, nitrogen, and neon. These elements are then dispersed into space.

In some very rare cases a white dwarf might undergo a gigantic explosion that astrophysicists call a Type 1a supernova. This happens when a white dwarf is part of a binary star system and pulls too much matter from its stellar companion. Suddenly, the compact star can no longer support the additional mass, and even the repulsive pressure of electrons in crushed atoms can no longer prevent further gravitational collapse. This new wave of gravitational collapse heats the helium and carbon nuclei in a white dwarf and causes them to fuse into nickel, cobalt, and iron. However, the thermonuclear burning now occurs so fast that the white dwarf completely explodes. During this rare occurrence, nothing is left behind. All the elements created by nucleosynthesis during the lifetime of the small star now scatter into space as a result of this spectacular supernova detonation.

Large stars have more than five times the mass of our Sun. These stars begin their lives in pretty much the same way as small stars—by fusing hydrogen into helium. However, because of their size, large stars burn faster and hotter, generally fusing all the hydrogen in their cores into helium in less than 1 billion years. Once the hydrogen in the large star's core is fused into helium, it becomes a red supergiant—a stellar object similar to the red giant star previously mentioned, only larger. However, unlike a red giant, the much larger red supergiant star has enough mass to produce much higher core temperatures as a result of gravitational contraction. A red supergiant fuses helium into carbon, carbon and helium into oxygen, and even two carbon nuclei into magnesium. Thus, through a combination of intricate nucleosynthesis reactions the supergiant star forms progressively heavier elements up to and including the element iron. Astrophysicists suggest that the red supergiant has an onionlike structure, with different elements being fused at different temperatures in layers around the core. The process of convection brings these elements from the star's interior to near its surface, where strong stellar winds then disperse them into space.

Thermonuclear fusion continues in a red supergiant star until the element iron is formed. Iron is the most stable of all the elements. So elements lighter than (below) iron on the periodic table generally emit energy when joined or fused in thermonuclear reactions, while elements heavier than (above) iron on the periodic table emit energy only when their nuclei split or fission. So where did the elements more massive than iron come from? Astrophysicists tell us that neutron capture is one way the more massive elements form. Neutron capture occurs when a free neutron (one outside an atomic nucleus) collides with an atomic nucleus and "sticks." This capture process changes the nature of the compound nucleus, which is often radioactive and undergoes decay, thereby creating a different element with a new atomic number.

While neutron capture can take place in the interior of a star, it is during a supernova explosion that many of the heavier elements, such as iodine, xenon, gold, and the majority of the naturally occurring radioactive elements, are formed by rapid neutron capture reactions.

Let us briefly examine what happens when a large star goes supernova. The red supergiant eventually produces the element iron in its intensely hot core. However, because of nuclear stability phenomena, iron is the last chemical element formed in nucleosynthesis. When fusion begins to fill the core of a red supergiant star with iron, thermonuclear energy release in the large star's interior decreases. Because of this decline, the star no longer has the internal radiant pressure to resist the attractive force of gravity, and so the red supergiant begins to collapse. Suddenly, this gravitational collapse causes the core temperature to rise to more than 100 billion degrees kelvin, smashing the electrons and protons in each iron atom together to form neutrons. The force of gravity now draws this massive collection of neutrons incredibly close together. For about a second the neutrons fall very fast toward the center of the star. Then they smash into each other and suddenly stop. This sudden stop causes the neutrons to recoil violently, and an explosive shockwave travels outward from the highly compressed core. As this shockwave travels from the core, it heats and accelerates the outer layers of material of the red supergiant star. The traveling shockwave causes the majority of the large star's mass to be blown off into space. Astrophysicists call this enormous explosion a Type II supernova.

A supernova will often release (for a brief moment) enough energy to outshine an entire galaxy. Since supernovae explosions scatter elements made within red supergiant stars far out into space, they are one of the most important ways the chemical elements disperse in the universe. Just before the outer material is driven off into space, the tremendous force of the supernova explosion provides nuclear conditions that support rapid capture of neutrons. Rapid neutron capture reactions transform elements in the outer layers of the red supergiant star into radioactive isotopes that decay into elements heavier than iron.

This essay could only provide a very brief glimpse of our cosmic connection to the chemical elements. But the next time you look up at the stars on a clear night, just remember that you, all other persons, and everything in our beautiful world is made of stardust.

accelerated life test(s) The series of test procedures for a spacecraft or aerospace system that approximate in a relatively short period of time the deteriorating effects and possible failures that might be encountered under normal, long-term space mission conditions. Accelerated life tests help aerospace engineers detect critical design flaws and material incompatibilities (for example, excessive wear or friction) that eventually might affect the performance of a spacecraft component or subsystem over its anticipated operational lifetime.

See also LIFE CYCLE.

acceleration (usual symbol: a) The rate at which the velocity of an object changes with time. Acceleration is a vector quantity and has the physical dimensions of length per unit time to the second power (for example, meters per second per second, or m/s^2).

See also NEWTON'S LAWS OF MOTION.

acceleration of gravity The local acceleration due to gravity on or near the surface of a planet. On Earth, the acceleration due to gravity (g) of a free-falling object has the standard value of 9.80665 m/s^2 by international agreement. According to legend, the famous Italian scientist GALILEO GALILEI simultaneously dropped a large and small cannonball from the top of the Tower of Pisa to investigate the acceleration of gravity. As he anticipated, each object fell to the ground in exactly the same amount of time (neglecting air resistance)—despite the difference in their masses. During the Apollo Project astronauts repeated a similar experiment on the surface of the Moon, dramatically demonstrating the universality of physical laws and honoring Galileo's scientific genius. It was Galileo's pioneering investigation of the physics of free fall that helped SIR ISAAC NEWTON unlock the secrets of motion of the mechanical universe.

accelerator A device used in nuclear physics for increasing the velocity and energy of charged elementary particles, such as electrons and protons, through the application of electromagnetic forces.

See also ELEMENTARY PARTICLE(S).

accelerometer An instrument that measures acceleration or gravitational forces capable of imparting acceleration. Frequently used on space vehicles to assist in guidance and navigation and on planetary probes to support scientific data collection.

acceptance test(s) In the aerospace industry, the required formal tests conducted to demonstrate the acceptability of a unit, component, or system for delivery. These tests demonstrate performance to purchase specification requirements and serve as quality-control screens to detect deficiencies of workmanship and materials.

accretion The gradual accumulation of small particles, gas, and dust into larger material bodies. Accretion takes place when small particles collide and stick together, forming bigger clumps of matter. Once these larger masses reach sufficient size, gravitational attraction helps accelerate the accretion process. Here on Earth relatively tiny snowflakes start their journey to the ground from the belly of some cold, gray winter cloud. As they float down, they often collide with other snowflakes, sticking together and growing into large clusters of flakes. By the time the original snowflakes reach the ground, they have often accreted into beautiful, giant flakes—much to the delight of people who can watch the falling snow from the comfort of a warm, cozy mountain lodge. Astrophysicists generally associate the process of accretion in outer space with a swirling disk of small particles, dust, and gas. There are many types of accreting celestial objects, including protostars, protoplanets, X-ray binary star systems, and black holes. For example, in the early stages of stellar formation matter begins to collect into a nebula, a giant interstellar cloud of gas and dust. Eventually protostars form under the influence of accretion and gravitational contraction in regions of the nebula that contain enough mass. When a new star forms, small quantities of residual matter in the outer regions of the swirling disk of protostar material may begin to collect by accretion and then through gravitational contraction and condensation grow into one or more planets.

See also ACCRETION DISK; BLACK HOLE; PROTOPLANET; PROTOSTAR; X-RAY BINARY STAR SYSTEM.

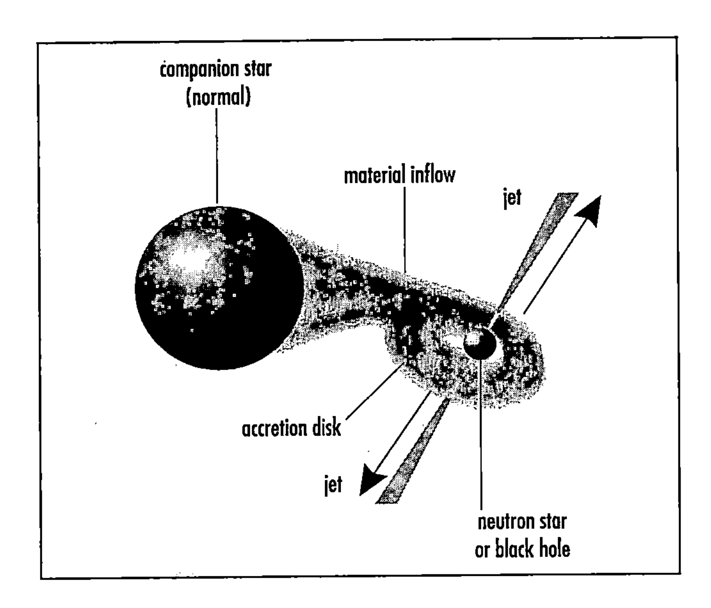

The inflow of material from a companion star creating an accretion disk around a neutron star or black hole

accretion disk The whirling disk of inflowing (or infalling) material from a normal stellar companion that develops around a massive compact body, such as a neutron star or a black hole. The conservation of angular momentum shapes this disk, which is often accompanied by a pair of very high-speed material jets departing in opposite directions perpendicular to the plane of the disk.

See also ACCRETION.

Achilles The first asteroid of the Trojan group discovered. The German astronomer MAXIMILLIAN (MAX) WOLF found the 115-kilometer-diameter minor planet in 1906. This asteroid is located at the L_4 Lagrangian point 60° ahead of Jupiter and characterized by the following orbital parameters: a period of 11.77 years, an inclination of 10.3 degrees, an aphelion of 5.95 A.U., and a perihelion of 4.40 A.U. Also referred to as Asteroid-588 or Achilles-588.

See also ASTEROID; LAGRANGIAN LIBRATION POINTS; TROJAN GROUP.

achondrite A class of stony meteorite that generally does not contain chondrules—silicate spheres embedded in a smooth matrix.

See also MARTIAN METEORITES; METEORITE.

achromatic lens A compound lens, generally containing two or more optical components, designed to correct for chromatic aberration. Chromatic aberration is an undesirable condition (a blurring color fringe around the image) that arises when a simple converging lens focuses different colors of incident visible light (say blue and red) at different points along its principal axis. This happens because the index of refraction of the material from which the lens is constructed often varies with the wavelength of the incident light. For example, while passing through a converging lens, a ray of violet light (shortest wavelength visible light) will refract (bend) more than a ray of red light (longest wavelength visible light). So the ray of violet light will cross the principal axis closer to the lens than will the ray of red light. In other words, the focal length for violet light is shorter than that for red light. Physicists call the phenomenon when beams of light of different colors refract by different amounts *dispersion.* Lens makers have attempted to overcome this physical problem by using the compound lens—that is, a converging lens and diverging lens in tandem. This arrangement often brings different colors more nearly to the same focal point. The word *achromatic* comes from the Greek *achromotos,* which means "without color." High-quality optical systems, such as expensive cameras, use achromatic lenses.

See also LENS.

Acidalia Planitia A distinctive surface feature on Mars. More than 2,600 kilometers in diameter, the Acidalia Planitia is the most prominent dark marking in the northern hemisphere of the Red Planet.

See also LATIN SPACE DESIGNATIONS; MARS.

acoustic absorber An array of acoustic resonators distributed along the wall of a combustion chamber, designed to prevent oscillatory combustion by increasing damping in the rocket engine system.

See also ROCKET.

acquisition 1. The process of locating the orbit of a satellite or the trajectory of a space probe so that mission control personnel can track the object and collect its telemetry data. 2. The process of pointing an antenna or telescope so that it is properly oriented to gather tracking or telemetry data from a satellite or space probe. 3. The process of searching for and detecting a potentially threatening object in space. An acquisition sensor is designed to search a large area of space and to distinguish potential targets from other objects against the background of space.

acronym A word formed from the first letters of a name, such as HST—which means the *Hubble Space Telescope.* It is also a word formed by combining the initial parts of a series of words, such as lidar—which means *l*ight *d*etection *a*nd *r*anging. Acronyms are frequently used in the aerospace industry, space technology, and astronomy.

activation analysis A method for identifying and measuring chemical elements in a sample of material. The sample is first made radioactive by bombarding (i.e., by irradiating) it with neutrons, protons, or other nuclear particles. The newly formed radioactive atoms in the sample then experience radioactive decay, giving off characteristic nuclear radiations, such as gamma rays of specific energy levels, that reveal what kinds of atoms are present and possibly how many.

active control The automatic activation of various control functions and equipment onboard an aerospace vehicle or satellite. For example, to achieve active attitude control of a satellite, the satellite's current attitude is measured automatically and compared with a reference or desired value. Any significant difference between the satellite's current attitude and the reference or desired attitude produces an error signal, which is then used to initiate appropriate corrective maneuvers by onboard actuators. Since both the measurements and the automatically imposed corrective maneuvers will not be perfect, the active control cycle usually continues through a number of iterations until the difference between the satellite's actual and desired attitude is within preselected, tolerable limits.

active discrimination In ballistic missile defense (BMD), the illumination of a potential target with electromagnetic radiation in order to determine from the characteristics of the reflected radiation whether this object is a genuine threat object (e.g., an enemy reentry vehicle or postboost vehicle) or just a decoy. Radar and laser radar systems are examples of active discrimination devices.

See also BALLISTIC MISSILE DEFENSE (BMD).

active galactic nucleus (AGN) The central region of a distant active galaxy that appears to be a point-like source of intense X-ray or gamma ray emissions. Astrophysicists speculate that the AGN is caused by the presence of a centrally located, supermassive black hole accreting nearby matter.

See also ACTIVE GALAXIES.

active galaxies (AGs) A galaxy is a system of stars, gas, and dust bound together by their mutual gravity. A typical galaxy has billions of stars, and some galaxies even have trillions of stars.

Although galaxies come in many different shapes, the basic structure is the same: a dense core of stars called the *galactic nucleus* surrounded by other stars and gas. Normally, the core of an elliptical or disk galaxy is small, relatively faint, and composed of older, redder stars. However, in some galaxies the core is intensely bright—shining with a power level equivalent to trillions of sunlike stars and easily outshining the combined light from all of the rest of the stars in that galaxy. Astronomers call a galaxy that emits such tremendous amounts of energy an *active galaxy* (AG), and they call the center of an active galaxy the *active galactic nucleus* (AGN). Despite the fact that active galaxies are actually quite rare, because they are so bright they can be observed at great distances—even across the entire visible universe.

Scientists currently believe that at the center of these bright galaxies lies a supermassive black hole, an incredibly massive object that contains the masses of millions or perhaps billions of stars the size of our Sun. As matter falls toward a supermassive black hole, the material forms an accretion disk, a flattened disk of gravitationally trapped material swirling around the black hole. Friction and magnetic forces inside the accretion disk heat the material to millions of kelvins, and it glows brightly nearly all the way across the electromagnetic spectrum, from radio waves to X-rays. Although our home galaxy, the Milky Way, has a central supermassive black hole, it is not an active galaxy. For reasons that astrophysicists cannot currently explain, the black hole at the center of our galaxy is inactive, or quiescent, as are most present-day galaxies.

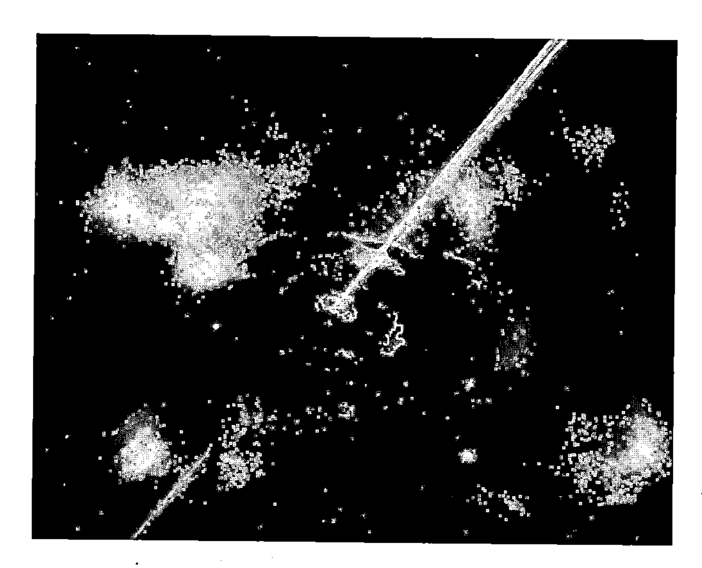

An artist's rendering of an active galaxy with jets ***(Courtesy of NASA)***

Although the physics underlying the phenomenon is not well understood, scientists know that in some cases the accretion disk of an active galaxy focuses long jets of matter that streak away from the AGN at speeds near the speed of light. The jets are highly collimated (meaning they retain their narrow focus over vast distances) and are emitted in a direction perpendicular to the accretion disk. Eventually these jets slow to a stop due to friction with gas well outside the galaxy, forming giant clouds of matter that radiate strongly at radio wavelength. In addition, surrounding the accretion disk is a large torus (donut-shaped cloud) of molecular material. When viewed from certain angles, this torus can obscure observations of the accretion disk surrounding the supermassive black hole.

There are many types of active galaxies. Initially, when astronomers were first studying active galaxies, they thought that the different types of AGs were fundamentally different celestial objects. Now many (but not all) astronomers and astrophysicists generally accept the *unified model* of AGs. This means that most or all AGs are actually just different versions of the same object. Many of the apparent differences between types of AGs are due to viewing them at different orientations with respect to the accretion disk or due to observing them in different wavelength regions of the electromagnetic (EM) spectrum (such as the radio frequency, visible, and X-ray portions of the EM spectrum).

Basically, the unified model of AGs suggests that the type of AG astronomers see depends on the way they see it. If they saw the accretion disk and gas torus edge on, they called the AG a *radio galaxy*. The torus of cool gas and dust blocks most of the visible, ultraviolet, and X-ray radiation from the intensely hot inflowing material as it approaches the event horizon of the supermassive black hole or as it swirls nearby in the accretion disk. As a consequence, the most obvious observable features are the radio-wave-emitting jets and giant lobes well outside the AG.

If the disk is tipped slightly to our line of sight, we can see higher-energy (shorter-wavelength) electromagnetic radiation from the accretion disk inside the gas torus in addition to the lower-energy (longer-wavelength). Astronomers call this type of active galaxy a *Seyfert galaxy,* named after the American astronomer CARL SEYFERT, who first cataloged these galaxies in 1943. A Seyfert galaxy looks very much like a normal galaxy but with a very bright core (AG nucleus) and may be giving off high-energy photons such as X-rays.

If the active galaxy is very far away from Earth, astronomers may observe the core (AGN) as a starlike object even if the fainter surrounding galaxy is undetected. In this case they call the AG a *quasar,* which is scientific shorthand for quasi-stellar radio source—so named because the first such objects detected appeared to be starlike through a telescope but, unlike regular stars, emitted copious quantities of radio waves. The Dutch-American astronomer MAARTEN SCHMIDT discovered the first quasar (called 3C 273) in 1963. This quasar is an AG a very great distance away and receding from us at more than 90 percent of the speed of light. Quasars are among the most distant, and therefore youngest, extragalactic objects astronomers can observe.

If the AG is tipped 90 degrees with respect to observers on Earth, astronomers would be looking straight down ajet from the AG. They call this type of object a *blazar.* The first blazar detected was a BL LAC OBJECT. But in the late 1920s they mistakenly classified this type of extragalactic object as a variable star because of its change in visual brightness. It was not until the 1970s that astronomers recognized the extragalactic nature of this interesting class of objects. More recently, using advanced space-based observatories, such as the *Compton Gamma Ray Observatory* (CGRO), astronomers have detected very energetic gamma ray emissions from blazars.

In summary, the basic components of an AG are a supermassive black hole core, an accretion disk surrounding this core, and a torus of gas and dust. In some but not all cases, there are also a pair of highly focused jets of energy and matter. The type of AG astronomers see depends on the viewing angle at which they observe a particular AG. The generally accepted unified model of AGs includes blazars, quasars, radio galaxies, and Seyfert galaxies.

See also ACTIVE GALACTIC NUCLEUS; BLAZAR; GALAXY; QUASARS, RADIO GALAXY; SEYFERT GALAXY.

active homing guidance A system of homing guidance wherein the missile carries within itself both the source for illuminating the target and the receiver for detecting the signal reflected by the target. Active homing guidance systems also can be used to assist space systems in rendezvous and docking operations.

active microwave instrument A microwave instrument, such as a radar altimeter, that provides its own source of illumination. For example, by measuring the radar returns from the ocean or sea, a radar altimeter on a spacecraft can be used to deduce wave height, which is an indirect measure of surface wind speed.

See also REMOTE SENSING.

active remote sensing A remote sensing technique in which the sensor supplies its own source of electromagnetic radiation to illuminate a target. A synthetic aperture radar (SAR) system is an example.

See also REMOTE SENSING; SYNTHETIC APERTURE RADAR.

active satellite A satellite that transmits a signal, in contrast to a passive (dormant) satellite.

active sensor A sensor that illuminates a target, producing return secondary radiation that is then detected in order to track and possibly identify the target. A lidar is an example of an active sensor.

See also LIDAR; RADAR IMAGING; REMOTE SENSING.

active Sun The name scientists have given to the collection of dynamic solar phenomena, including sunspots, solar flares, and prominences, associated with intense variations in the Sun's magnetic activity. *Compare with* QUIET SUN.

active tracking system A system that requires the addition of a transponder or transmitter on board a space vehicle or missile to repeat, transmit, or retransmit information to the tracking equipment.

actuator A servomechanism that supplies and transmits energy for the operation of other mechanisms, systems, or process control equipment.

See also ROBOTICS IN SPACE.

acute radiation syndrome (ARS) The acute organic disorder that follows exposure to relatively severe doses of ionizing radiation. A person will initially experience nausea, diarrhea, or blood cell changes. In the later stages loss of hair, hemorrhaging, and possibly death can take place. *Radiation dose equivalent* values of about 4.5 to 5 sievert (450 to 500 rem) will prove fatal to 50 percent of the exposed individuals in a large general population. Also called *radiation sickness.*

Adams, John Couch (1819–1892) British *Astronomer* John Couch Adams is cocredited with the mathematical discovery of Neptune. From 1843 to 1845 he investigated irregularities in the orbit of Uranus and predicted the existence of a planet beyond. However, his work was essentially ignored until the French astronomer URBAIN JEAN JOSEPH LEVERRIER made similar calculations that enabled the German astronomer JOHANN GOTTFRIED GALLE to discover Neptune on September 23, 1846. Couch was a professor of astronomy and also the director of the Cambridge Observatory.

See also NEPTUNE.

Adams, Walter Sydney (1876–1956) American *Astronomer* Walter Sydney Adams specialized in stellar spectroscopic studies and codeveloped the important technique called spectroscopic parallax for determining stellar distances. In 1915 his spectral studies of Sirius B led to the discovery of the first white dwarf star. From 1923 to 1946 he served as the director of the Mount Wilson Observatory in California.

Adams was born on December 20, 1876, in the village of Kessab near Antioch in northern Syria. His parents were American missionaries to the part of Syria then under Turkish rule as part of the former Ottoman Empire. By 1885 his parents completed their missionary work, and the family returned to New Hampshire. In Dartmouth College he earned a reputation as a brilliant undergraduate student. As a result of his courses with Professor Edwin B. Frost (1866–1935), Adams selected a career in astronomy. In 1898 he graduated from Dartmouth College and then followed his mentor, Professor Frost, to the Yerkes Observatory, operated by the University of Chicago. While learning spectroscopic methods at the observatory, Adams also continued his formal studies in astronomy at the University of Chicago, where he obtained a graduate degree in 1900.

The American astronomer GEORGE ELLERY HALE founded Yerkes Observatory in 1897, and its 1-meter (40-inch) refractor telescope was the world's largest. As a young graduate, Adams had the unique opportunity to work with Hale as that famous scientist established a new department devoted to stellar spectroscopy. This experience cultivated Adams's lifelong interest in stellar spectroscopy.

From 1900 to 1901 he studied in Munich, Germany, under several eminent German astronomers, including KARL SCHWARZSCHILD. Upon his return to the United States, Adams worked at the Yerkes Observatory on a program that measured the radial velocities of early-type stars. In 1904 he accompanied Hale to the newly established Mount Wilson Observatory on Mount Wilson in the San Gabriel Mountains about 30 kilometers northwest of Los Angeles, California. Adams became assistant director of this observatory in 1913 and served in that capacity until 1923, when he succeeded Hale as director.

Adams married his first wife, Lillian Wickham, in 1910. After her death 10 years later he married Adeline L. Miller in 1922, and the couple had two sons, Edmund M. and John F. From 1923 until his retirement in 1946 Walter Adams served as the director of the Mount Wilson Observatory. Following his retirement on January 1, 1946, he continued his astronomical activities at the Hale Solar Laboratory in Pasadena, California.

At Mount Wilson Adams was closely involved in the design, construction, and operation of the observatory's initial 1.5-meter (60-inch) reflecting telescope and then the newer 2.5-meter (100-inch) reflector that came on line in 1917. Starting in 1914 he collaborated with the German astronomer Arnold Kohlschütter (1883–1969) in developing a method of establishing the surface temperature, luminosity, and distance of stars from their spectral data. In particular, Adams showed how it was possible for astronomers to distinguish between a dwarf star and a giant star simply from their spectral data. As defined by astronomers, a dwarf star is any main sequence star, while a giant star is a highly luminous one that has departed the main sequence toward the end of its life and swollen significantly in size. Giant stars typically have diameters from 5 to 25 times the diameter of the Sun and luminosities that range from tens to hundreds of times the luminosity of the Sun. Adams showed that it was possible to determine the luminosity of a star from its spectrum. This allowed him to introduce the important method of *spectroscopic parallax,* whereby the luminosity deduced from a star's spectrum is then used to estimate its distance. Astronomers have estimated the distance of many thousands with this important method.

Adams is perhaps best known for his work involving his work with Sirius B, the massive but small companion to Sirius, the Dog Star—brightest star in the sky after the Sun. In 1844 the German mathematician and astronomer FRIEDRICH BESSEL first showed that Sirius must have a companion, and he even estimated that its mass must be about the same as that of the Sun. Then, in 1862 the American optician ALVAN CLARK made the first telescopic observation of Sirius B, sometimes a called "the Pup." Their preliminary work set the stage for Adams to make his great discovery.

Adams obtained the spectrum of Sirius B in 1915. This was a very difficult task because of the brightness of its stellar companion, Sirius. The spectral data indicated that the small star was considerably hotter than the Sun. A skilled astronomer, Adams immediately realized that such a hot celestial object, just eight light-years distant, could remain invisible to the naked eye only if it were very much smaller than the Sun. Sirius B is actually slightly less than the size of Earth. Assuming all his observations and reasoning were true, Adams reached the following important conclusion: Sirius B must have an extremely high density, possibly approaching 1 million times the density of water.

Thus, Adams made astronomical history by identifying the first white dwarf, a small, dense object that represents the end product of stellar evolution for all but the most massive stars. A golf-ball-sized chunk taken from the central region of a white dwarf star would have a mass of about 35,000 kilograms—that is, 35 metric tons, or some 15 fully equipped sport utility vehicles (SUVs) neatly squeezed into the palm of your hand.

Almost a decade later Adams parlayed his identification of this very dense white dwarf star. He assumed a compact object like Sirius B should possess a very strong gravitational field. He further reasoned that according to ALBERT EINSTEIN's general relativity theory, Sirius B's strong gravitational field should redshift any light it emitted. Between 1924 and 1925 he successfully performed difficult spectroscopic measurements that detected the anticipated slight redshift of the star's light. This work provided other scientists independent observational evidence that Einstein's general relativity theory was valid.

In 1932 Adams conducted spectroscopic investigations of the Venusian atmosphere, showing that it was rich in carbon dioxide (CO_2). He retired as director of the Mount Wilson Observatory in 1946 but continued his research at the Hale Laboratory. He died in Pasadena, California, on May 11, 1956.

adapter skirt A flange or extension on a launch vehicle stage or spacecraft section that provides a means of fitting on another stage or section.

adaptive optics Optical systems that can be modified (such as by adjusting the shape of a mirror) to compensate for distortions. An example is the use of information from a beam of light passing through the atmosphere to compensate for the distortion experienced by another beam of light on its passage through the atmosphere. Adaptive optics systems are used in observational astronomy to eliminate the "twinkling" of stars and in ballistic missile defense to reduce the dispersive effect of the atmosphere on laser beam weapons. At visible and near-infrared wavelengths, the angular resolution of Earth-based telescopes with apertures greater than 10 to 20 centimeters is limited by turbulence in Earth's atmosphere rather than by the inherent, diffraction-limited image size of the system. Large telescopes are often equipped with adaptive optics systems to compensate for atmospheric turbulence effects, enabling these systems to achieve imaging on scales that approach the diffraction limit. Adaptive optics systems continuously measure the wave front errors resulting from atmospheric turbulence. Then, using a pointlike reference source situated above the distorting layers of Earth's atmosphere, compensation is achieved by rapidly adjusting a deformable optical element located in or near a pupil plane of the optical system.

For example, in the adaptive optics system built for the 2.54-meter (100 inch) telescope at the Mount Wilson Observatory, the incoming light reflected from the telescope mirror is divided into several hundred smaller beams or regions. Looking at the beam of light from a star, the system sees hundreds of separate beams that are going in different directions because of the effects of Earth's atmosphere. The electron circuits in the system compute the bent shape of a deformable mirror surface that would straighten out the separate beams so that they are all going in the same direction. Then a signal is sent to the deformable mirror to change its shape in accordance with these electronic signals. Simply stated, in an adaptive optics system a crooked beam of light hits a crooked mirror and a straight beam of light is reflected.

See also KECK OBSERVATORY; MIRROR; MOUNT WILSON OBSERVATORY; TELESCOPE.

adiabatic process A change of state (condition) of a thermodynamic system in which there is no heat transfer across the boundaries of the system. For example, an adiabatic compression process results in "warming" (i.e., raising the internal energy level of) a working fluid, while an adiabatic expansion process results in "cooling" (i.e., decreasing the internal energy level of) a working fluid.

Adonis The approximately one-kilometer-diameter Apollo group asteroid discovered in 1936 by the Belgian astronomer Eugène Joseph Delporte, when the minor planet passed within 2.2 million kilometers (0.015 astronomical unit [AU]) of Earth. A period of 2.57 years, an inclination of two degrees, an aphelion of 3.3 AU, and a perihelion of 0.5 AU characterize the asteroid's orbit around the Sun. Following its discovery in 1936, Adonis was not observed again until 1977. Some astronomers speculate that this celestial object might actually be an inactive comet nucleus that is associated with the minor meteor showers called the Capricornids and the Sagittariids. Also called Asteroid 2101.

See also APOLLO GROUP; ASTEROID; METEOR SHOWER.

Adrastea A small (about 20 kilometers in diameter) moon of Jupiter that orbits the giant planet at a distance of 129,000 kilometers. It was discovered in 1979 as a result of the *Voyager 2* spacecraft flyby. The orbit of this tiny, irregularly shaped (26-km × 20-km × 16-km) satellite has a period of approximately 0.30 day and an inclination of zero degrees. Adrastea is one of four minor Jovian moons whose orbits lie inside the orbit of Io, the innermost of the Galilean moons. The other small Jovian moons are Amalthea, Metis, and Thebe. Like Metis, Adrastea has an orbit that lies inside the synchronous orbit radius of Jupiter. This means that the moon rotates around the giant planet faster than the planet rotates on its axis. As a physical consequence, both of the two tiny moons will eventually experience orbital decay and fall into the giant planet. Adrastea and Metis also orbit inside Jupiter's main ring, causing astronomers to suspect that these tiny moons are the source of the material in the ring.

See also AMALTHEA; JUPITER; METIS; THEBE.

Advanced Composition Explorer **(ACE)** The primary purpose of NASA's *Advanced Composition Explorer* (ACE) scientific spacecraft is to determine and compare the isotopic and elemental composition of several distinct samples of matter, including the solar corona, the interplanetary medium, the local interstellar medium, and galactic matter. Earth is constantly bombarded with a stream of accelerated nuclear particles that arrive not only from the Sun but also from interstellar and galactic sources. The ACE spacecraft carries six high-resolution spectrometers that measure the elemental, isotopic, and ionic charge state composition of nuclei with atomic numbers ranging from hydrogen (H) ($Z = 1$) to nickel (Ni) ($Z = 28$) and with energies ranging from solar wind energies (about 1 keV/nucleon) to galactic cosmic energies (about 500 MeV/nucleon). The ACE spacecraft also carries three monitoring instruments that sample low-energy particles of solar origin.

The ACE spacecraft is 1.6 meters across and 1 meter high, not including the four solar arrays and magnetometer booms attached to two of the solar panels. At launch the spacecraft had a mass of 785 kilograms, including 189 kilograms of hydrazine fuel for orbit insertion and orbit maintenance. The solar arrays generate about 500 watts of electric power. The spacecraft spins at a rate of five revolutions per minute (5 rpm), with the spin axis generally pointed along the Earth-Sun line. Most of the spacecraft's scientific instruments are on the top (sunward) deck.

On August 25, 1997, a Delta II rocket successfully launched the ACE spacecraft from Cape Canaveral Air Force Station in Florida. In order to get away from the effects of Earth's magnetic field, the ACE spacecraft then used its onboard propulsion system to travel almost 1.5 million kilometers away from Earth and reached its operational orbit at the Earth-Sun libration point (L_1). By operating at the L_1 libration point, the ACE spacecraft stays in a relatively constant position with respect to Earth as Earth revolves around the Sun. From a vantage point that is approximately 1/100th of the distance from Earth to the Sun, the ACE spacecraft performs measurements over a wide range of energy and nuclear mass—under all solar wind flow conditions and during both large and small particle events, including solar flares. In addition to scientific observations, the ACE mission also provides real-time solar wind measurements to the National Oceanic and Atmospheric Administration (NOAA) for use in forecasting space weather. When reporting space weather ACE can provide an advance warning (about one hour) of geomagnetic storms that could overload power grids, disrupt communications on Earth, and represent a hazard to astronauts.

As previously mentioned, the ACE spacecraft operates at the L_1 libration point, which is a point of Earth-Sun gravitational equilibrium about 1.5 million kilometers from Earth and 148.5 million kilometers from the Sun. This operational location provides the ACE spacecraft a prime view of the Sun and the galactic regions beyond. The spacecraft has enough hydrazine propellant on board to maintain its orbit at the L_1 libration point until about the year 2019.

See also EXPLORER SPACECRAFT; LAGRANGIAN LIBRATION POINTS; SPACE WEATHER.

Advanced Earth Observation Satellite **(ADEOS)** The ADEOS-I spacecraft (called *Midori-I* in Japan) was the first international space platform dedicated to Earth environmental research and is managed by the Japanese Aerospace Exploration Agency (JAXA), formerly called the National Space Development Agency of Japan (NASDA). The environmental satellite was launched on August 17, 1996, into an 830-km Sun-synchronous orbit around Earth by an H-II expendable rocket from the Tanegashima Space Center in Japan. ADEOS-I operated until June 1997, when it experienced an operational malfunction in orbit.

In December 2002 Japan launched the *Advanced Earth Observing Satellite-II* (ADEOS-II) as the successor of ADEOS-I. The ADEOS-II (called *Midori-II* in Japan) was carried into an approximate 810-km altitude polar orbit by an H-II expendable launch vehicle. This spacecraft is equipped with two principal sensors: the advanced microwave scanning radiometer (AMSR), which, day and night, observes geographical parameters related to water, and the global imager (GLI), which extensively and accurately observes oceans, land, and clouds. On October 25, 2003, a major anomaly was detected on the satellite while aerospace personnel were sending commands to analyze its status. As a result of the subsequent investigation by Japanese aerospace personnel and based on the fact that they have not reestablished any further communications with the satellite, JAXA officials have concluded that there is only an extremely small probability of restoring the ADEOS-II to operational status. However, JAXA officials will continue to send commands to the disabled satellite and to investigate its deteriorating condition in order to better clarify the cause of the operational anomaly and to prevent its recurrence in future satellite programs.

The primary application of the ADEOS (Midori) spacecraft is to monitor global environmental changes, such as maritime meteorological conditions, atmospheric ozone, and gases that promote global warming.

See also EARTH SYSTEM SCIENCE; GLOBAL CHANGE; JAPANESE AEROSPACE EXPLORATION AGENCY.

Advanced Satellite for Cosmology and Astrophysics **(ASCA)** ASCA was Japan's fourth cosmic X-ray astronomy mission and the second for which the United States provided part of the scientific payload. The satellite was successfully launched on February 20, 1993, and then operated successfully until July 15, 2000, when mission managers placed the spacecraft into a safe-hold mode. After seven and a half years of scientific observations, the satellite reentered on March 2, 2001. The spacecraft carried four large-area X-ray telescopes and was the first X-ray astronomy mission to combine imaging capability with a broad pass filter, good spectral resolution, and a large effective detection area. ASCA was also the first scientific satellite to use charge-coupled device (CCD) detectors for X-ray astronomy. Data from this spacecraft provided spectroscopy from interacting binary star systems and examined the abundance of heavy elements in clusters of galaxies, consistent with Type II supernova origin.

See also X-RAY ASTRONOMY.

Advanced X-Ray Astrophysics Facility **(AXAF)** *See* CHANDRA X-RAY OBSERVATORY (CXO).

Aegis A surface ship–launched missile used by the U.S. Navy as part of a totally integrated shipboard weapon system that combines computers, radars, and missiles to provide a defense umbrella for surface shipping. The system is capable of automatically detecting, tracking, and destroying a hostile air-launched, sea-launched, or land-launched weapon. In the late 1960s the U.S. Navy developed the Advanced Surface Missile System (ASMS) and then renamed this system Aegis after the shield of the god Zeus in Greek mythology. The Aegis system provides effective defense against antiship cruise missiles and human-crewed enemy aircraft in all environmental conditions.

aeolian Pertaining to, carried by, or caused by the wind; for example, aeolian sand dunes on the surface of Mars. Also spelled eolian.

aero- A prefix that means of or pertaining to the air, the atmosphere, aircraft, or flight through the atmosphere of a planet.

aeroassist The use of the thin upper regions of a planet's atmosphere to provide the lift or drag needed to maneuver a

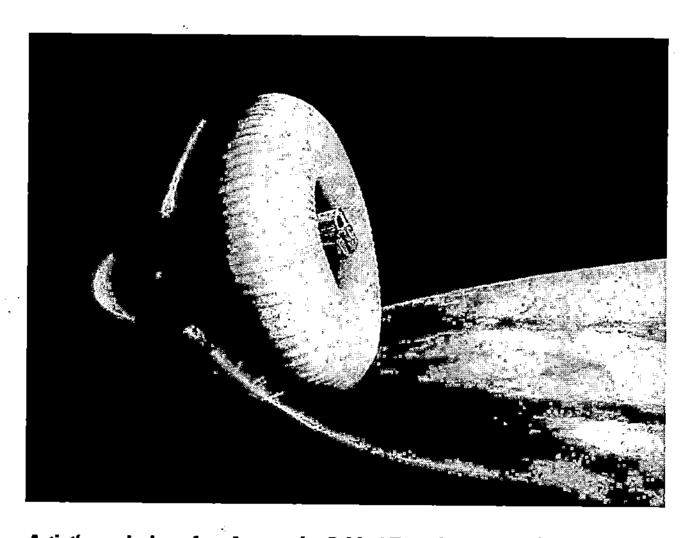

Artist's rendering of an Aeroassist Orbital Transfer Vehicle (AOTV) ***(Courtesy of NASA)***

spacecraft. Near a planet with a sensible atmosphere, aeroassist allows a spacecraft to change direction or to slow down without expending propellant from the control rocket system.

See also AEROBRAKING.

aerobraking The use of a specially designed spacecraft structure to deflect rarefied (very low-density) airflow around a spacecraft, thereby supporting aeroassist maneuvers in the vicinity of a planet. Such maneuvers reduce the spacecraft's need to perform large propulsive burns when making orbital changes near a planet. In 1993 NASA's Magellan Mission became the first planetary exploration spacecraft to use aerobraking as a means of changing its orbit around the target planet (Venus).

See also AEROASSIST; *MAGELLAN* MISSION.

aerocapture The use of a planet's atmosphere to slow down a spacecraft. Like aerobraking, aerocapture is part of a special family of aeroassist technologies that support more robust science missions to the most distant planets of the solar system. An aerocapture vehicle approaching a planet on a hyperbolic trajectory is captured into orbit as it passes through the atmosphere, without the use of rocket propellant by an on-board propulsion system. Aerospace engineers suggest that this "fuel-free" capture method will reduce the typical mass of an interplanetary robotic spacecraft by half or greater, allowing them to design a smaller, less expensive space vehicle, yet one better equipped to perform extensive long-term science at its destination. To experience aerocapture, a spacecraft needs adequate drag to slow its speed and some protection from aerodynamic heating. Aerospace engineers fulfill these two functions in a variety of ways. One technique involves a traditional blunt, rigid aeroshell, as was successfully used on the *Mars Pathfinder* spacecraft's entry and descent in 1997. Engineers are also considering the use of a lighter, inflatable aeroshell, as well as the employment of a large, trailing *ballute*—a combination parachute and balloon that is made of durable thin material and is stowed behind the space vehicle for deployment. In addition to the exploration of Mars with more sophisticated aerocapture technology spacecraft, engineers are considering the use of this technology for a potential science orbiter mission to Neptune and an advanced explorer mission to Saturn's moon Titan.

See also AEROASSIST; AEROBRAKING; *MARS PATHFINDER*; NEPTUNE; TITAN.

aerodynamic force The lift (L) and/or drag (D) forces exerted by a moving gas upon a body completely immersed in it. Lift acts in a direction normal to the flight path, while drag acts in a direction parallel and opposite to the flight path. (*See* figure.) The aerodynamic forces acting upon a body flying through the atmosphere of a planet are dependent on the velocity of the object, its geometric characteristics (i.e., size and shape), and the thermophysical properties of the atmosphere (e.g., temperature and density) at flight altitude. For low flight speeds, lift and drag are determined primarily by the angle of attack (α). At high (i.e., supersonic) speeds, these forces become a function of both the angle of attack and the Mach number (M). For a typical ballistic missile, the angle of attack is usually very low (i.e., $\alpha < 1°$).

See also AIRFOIL; REENTRY VEHICLE.

aerodynamic heating Frictional surface heating experienced by an aerospace vehicle, space system, or reentry vehicle as it enters the upper regions of a planetary atmosphere at very high velocities. Peak aerodynamic heating generally occurs in stagnation point regions on the object, such as on

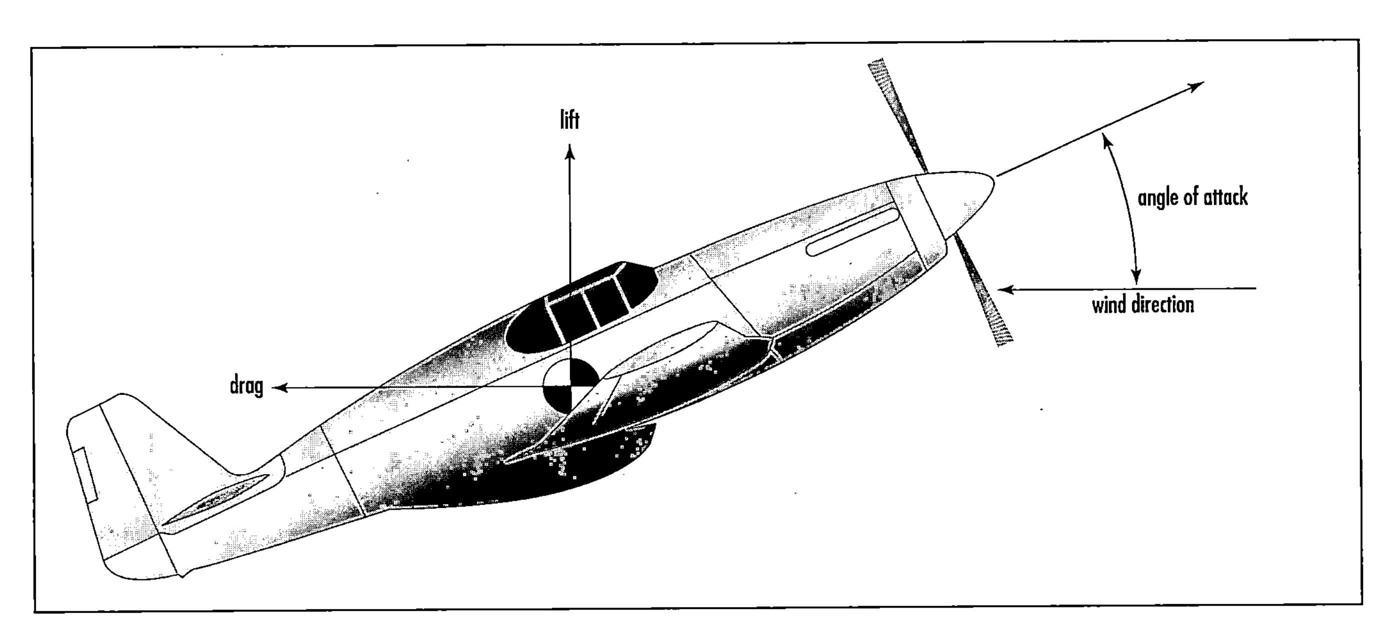

Aerodynamic forces (lift and drag) acting on aircraft flying at a certain angle of attack

the leading edge of a wing or on the blunt surfaces of a nose cone. Special thermal protection is needed to prevent structural damage or destruction. NASA's space shuttle Orbiter vehicle, for example, uses thermal protection tiles to survive the intense aerodynamic heating environment that occurs during reentry and landing.

See also ABLATION.

aerodynamic missile A missile that uses aerodynamic forces to maintain its flight path, generally employing propulsion guidance.

See also BALLISTIC MISSILE; CRUISE MISSILE; GUIDED MISSILE.

aerodynamic skip An atmospheric entry abort caused by entering a planet's atmosphere at too shallow an angle. Much like a stone skipping across the surface of a pond, this condition results in a trajectory back out into space rather than downward toward the planet's surface.

aerodynamic throat area The effective flow area of a nozzle's throat; generally, the effective flow area is less than the geometric flow area because the flow is not uniform.

See also NOZZLE.

aerodynamic vehicle A craft that has lifting and control surfaces to provide stability, control, and maneuverability while flying through a planet's atmosphere. For aerodynamic missiles, the flight profile both during and after thrust depends primarily upon aerodynamic forces. A glider or an airplane is capable of flight only within a sensible atmosphere, and such vehicles rely on aerodynamic forces to keep them aloft.

aerogel A silicon-based solid with a porous, spongelike structure in which 99.8 percent of the volume is empty space. By comparison, aerogel is 1,000 times less dense than glass, another silicon-based solid. Discovered in the 1930s by a researcher at Stanford University in California, material scientists at NASA's Jet Propulsion Laboratory (JPL) altered the original recipe to develop an important new aerogel for space exploration. The JPL-made aerogel approaches the density of air, but is a durable solid that easily survives launch and space environments. The space age aerogel is a smoky blue-colored substance that has many unusual physical properties and can withstand extreme temperatures. One important feature is the fact that the JPL-made aerogel can provide 39 times more thermal insulation than the best fiberglass insulation. JPL engineers used the aerogel to insulate the electronics box on the *Mars Pathfinder* rover, which explored the Red Planet in 1997. They also used the material to thermally insulate the batteries on the 2003 Mars Exploration Rovers, *Spirit* and *Opportunity.*

Perhaps the most innovative space technology application of aerogel to date involved the *Stardust* spacecraft and its mission to bring back samples of primordial extraterrestrial materials. Specifically, aerospace engineers fitted the aerogel aboard the *Stardust* spacecraft into a tennis racket–shaped collector grid. One side of the collector (called the A side) faced the tiny high-speed particles from Comet Wild, while the reverse (or B side) encountered streams of interstellar dust at various points in the spacecraft's trajectory.

As a tiny piece of comet dust (typically much smaller than a grain of sand) hits the aerogel, it buries itself in the special material and creates a carrot-shaped track up to 200 times its own length. The aerogel gradually slows each high-speed particle down and brings it to a stop without damaging the sample or altering its shape and chemical composition. The JPL-made aerogel on the *Stardust* spacecraft is less dense at the impact face where the particle first encounters the material. Then, it presents a gradually increasing density. So, as a high-speed dust particle burrows deeper into the material, it slows down gently and eventually comes to rest. The gradually increasing density concept is similar to the progressive lens concept used in some eyeglasses.

See also MARS EXPLORATION ROVER MISSION; *MARS PATHFINDER;* STARDUST MISSION.

aeronautics The science of flight within the atmosphere of a planet; the engineering principles associated with the design and construction of craft for flight within the atmosphere; the theory and practice of operating craft within the atmosphere. *Compare with* ASTRONAUTICS.

aerosol A very small dust particle or droplet of liquid (other than water or ice) in a planet's atmosphere, ranging in size from about 0.001 micrometer (μm) to larger than 100 micrometers (μm) in radius. Terrestrial aerosols include smoke, dust, haze, and fumes. They are important in Earth's atmosphere as nucleation sites for the condensation of water droplets and ice crystals, as participants in various chemical cycles, and as absorbers and scatterers of solar radiation. Aerosols influence Earth's radiation budget (i.e., the overall balance of incoming versus outgoing radiant energy), which in turn influences the climate on the surface of the planet.

See also EARTH RADIATION BUDGET; GLOBAL CHANGE.

aerospace A term derived from *aero*nautics and *space* meaning of or pertaining to Earth's atmospheric envelope and outer space beyond it. These two separate physical entities are taken as a single realm for activities involving launch, guidance, control, and recovery of vehicles and systems that can travel through and function in both physical regions. For example, NASA's space shuttle *Orbiter* is called an aerospace vehicle because it operates both in the atmosphere and in outer space.

See also AERONAUTICS; ASTRONAUTICS.

aerospace ground equipment (AGE) All the support and test equipment needed on Earth's surface to make an aerospace system or spacecraft function properly during in its intended space mission.

See also CAPE CANAVERAL AIR FORCE STATION; KENNEDY SPACE CENTER; LAUNCH SITE.

aerospace medicine The branch of medical science that deals with the effects of flight on the human body. The treatment of space sickness (space adaptation syndrome) falls within this field.

See also SPACE SICKNESS.

aerospace vehicle A vehicle capable of operating both within Earth's sensible (measurable) atmosphere and in outer space. The space shuttle *Orbiter* vehicle is an example.

See also SPACE TRANSPORTATION SYSTEM.

aerospike nozzle A rocket nozzle design that allows combustion to occur around the periphery of a spike (or center plug). The thrust-producing hot exhaust flow is then shaped and adjusted by the ambient (atmospheric) pressure. Sometimes called a *plug nozzle* or a *spike nozzle.*

See also NOZZLE.

aerozine A liquid rocket fuel consisting of a mixture of hydrazine (N_2H_4) and unsymmetrical dimethylhydrazine (acronym: UDMH), which has the chemical formula $(CH_3)_2NNH_2$.

See also ROCKET.

afterbody Any companion body (usually jettisoned, expended hardware) that trails a spacecraft following launch and contributes to the space debris problem. Any expended portion of a launch vehicle or rocket that enters Earth's atmosphere unprotected behind a returning nose cone or space capsule that is protected against the aerodynamic heating. Finally, it is any unprotected, discarded portion of a space probe or spacecraft that trails behind the protected probe or lander spacecraft as either enters a planet's atmosphere to accomplish a mission.

See also SPACE DEBRIS.

Agena A versatile, upper stage rocket that supported numerous American military and civilian space missions in the 1960s and 1970s. One special feature of this liquid propellant system was its in-space engine restart capability. The U.S. Air Force originally developed the Agena for use in combination with Thor or Atlas rocket first stages. Agena A, the first version of this upper stage, was followed by Agena B, which had a larger fuel capacity and engines that could restart in space. The later Agena D was standardized to provide a launch vehicle for a variety of military and NASA payloads. For example, NASA used the Atlas-Agena vehicles to launch large Earth-orbiting satellites as well as lunar and interplanetary space probes; Thor-Agena vehicles launched scientific satellites, such as the *Orbiting Geophysical Observatory* (OGO), and applications satellites, such as Nimbus meteorological satellites. In the Gemini Project, the Agena D vehicle, modified to suit specialized requirements of space rendezvous and docking maneuvers, became the Gemini Agena Target Vehicle (GATV).

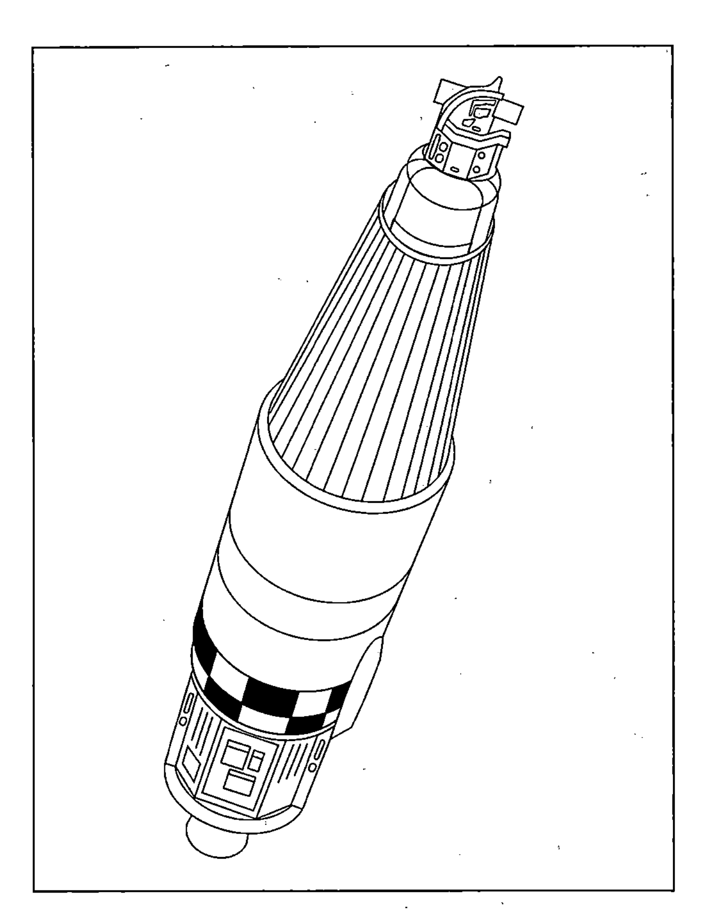

Agena, an upper stage vehicle configured for the SNAPSHOT mission flown by the U.S. government in 1965. During this mission the Agena successfully placed its payload, an experimental SNAP-10A space nuclear reactor (the small, wastebasket-sized structure shown in the upper portion of the drawing), into a near-circular, 1,300-kilometer-altitude orbit. Once at this orbit, the reactor was commanded to start up and operated at approximately 500 watts (electric) for 45 days.

age of the Earth Planet Earth is very old, about 4.5 billion years or more, according to recent estimates. Most of the evidence for an ancient Earth is contained in the rocks that form the planet's crust. The rock layers themselves, like pages in a long and complicated history, record the surface-shaping events of the past, and buried within them are traces of life—that is, the plants and animals that evolved from organic structures, existing perhaps 3 billion years ago. Also contained in once molten rocks are radioactive elements whose isotopes provide Earth with an atomic clock. Within these rocks parent isotopes decay at a predictable rate to form daughter isotopes. By determining the relative amounts of parent and daughter isotopes, the age of these rocks can be calculated. Therefore, the results of studies of rock layers (*stratigraphy*) and of fossils (*paleontology*), coupled with the ages of certain rocks as measured by atomic clocks (*geochronology*), indicate a very old Earth.

Up until now scientists have not found a way to determine the exact age of Earth directly from terrestrial rocks because our planet's oldest rocks have been recycled and destroyed by the process of plate tectonics. If there are any of Earth's primordial rocks left in their original state, scientists have not yet found them. Nevertheless, scientists have been able to determine the probable age of the solar system and to calculate an age for Earth by assuming that Earth and the rest of the solid bodies in the solar system formed at the same time and are, therefore, of the same age.

The ages of Earth and Moon rocks and of meteorites are measured by the decay of long-lived radioactive isotopes of elements that occur naturally in rocks and minerals and that decay with half-lives of 700 million to more than 100 billion years into stable isotopes of other elements. Scientists use these dating techniques, which are firmly grounded in physics and are known collectively as *radiometric dating,* to measure the last time that the rock being dated was either melted or disturbed sufficiently to rehomogenize its radioactive elements. Ancient rocks exceeding 3.5 billion years in age are found on all of the planet's continents. The oldest rocks on Earth found so far are the Acasta Gneisses in northwestern Canada near Great Slave Lake (4.03 billion years old) and the Isua Supracrustal rocks in West Greenland (about 3.7 to 3.8 billion years old), but well-studied rocks nearly as old are also found in the Minnesota River Valley and northern Michigan (some 3.5–3.7 billion years old), in Swaziland (about 3.4–3.5 billion years old), and in Western Australia (approximately 3.4–3.6 billion years old). Scientists have dated these ancient rocks using a number of radiometric methods, and the consistency of the results gives scientists confidence that the estimated ages are correct to within a few percent. An interesting feature of these ancient rocks is that they are not from any sort of "primordial crust" but are lava flows and sediments deposited in shallow water, an indication that Earth history began well before these rocks were deposited. In Western Australia single zircon crystals found in younger sedimentary rocks have radiometric ages of as much as 4.3 billion years, making these tiny crystals the oldest materials to be found on Earth so far. Scientists have not yet found the source rocks for these zircon crystals.

The ages measured for Earth's oldest rocks and oldest crystals show that our planet is at least 4.3 billion years in age but do not reveal the exact age of Earth's formation. The best age for Earth is currently estimated as 4.54 billion years. This value is based on old, presumed single-stage leads (Pb) coupled with the Pb ratios in troilite from iron meteorites, specifically the Canyon Diablo meteorite. In addition, mineral grains (zircon) with uranium-lead (U-Pb) ages of 4.4 billion years have recently been reported from sedimentary rocks found in west-central Australia.

The Moon is a more primitive planet than Earth because it has not been disturbed by plate tectonics. Therefore, some of the Moon's more ancient rocks are more plentiful. The six American Apollo Project human missions and three Russian Luna robotic spacecraft missions returned only a small number of lunar rocks to Earth. The returned lunar rocks vary greatly in age, an indication of their different ages of formation and their subsequent histories. The oldest dated Moon rocks, however, have ages between 4.4 and 4.5 billion years and provide a minimum age for the formation of our nearest planetary neighbor.

Thousands of meteorites, which are fragments of asteroids that fall to Earth, have been recovered. These primitive extraterrestrial objects provide the best ages for the time of formation of the solar system. There are more than 70 meteorites of different types whose ages have been measured using radiometric dating techniques. The results show that the meteorites, and therefore the solar system, formed between 4.53 and 4.58 billion years ago. The best age for Earth comes not from dating individual rocks but by considering Earth and meteorites as part of the same evolving system in which the isotopic composition of lead, specifically the ratio of lead-207 to lead-206, changes over time owing to the decay of radioactive uranium-235 and uranium-238, respectively. Scientists have used this approach to determine the time required for the isotopes in Earth's oldest lead ores, of which there are only a few, to evolve from its primordial composition, as measured in uranium-free phases of iron meteorites, to its compositions at the time these lead ores separated from their mantle reservoirs.

According to scientists at the U.S. Geological Survey (USGS), these calculations result in an age for Earth and meteorites, and therefore the solar system, of 4.54 billion years with an uncertainty of less than 1 percent. To be precise, this age represents the last time that lead isotopes were homogeneous throughout the inner solar system and the time that lead and uranium were incorporated into the solid bodies of the solar system. The age of 4.54 billion years found for the solar system and Earth is consistent with current estimates of 11 to 13 billion years for the age of the Milky Way galaxy (based on the stage of evolution of globular cluster stars) and the age of 10 to 15 billion years for the age of the universe (based on the recession of distant galaxies).

age of the Moon In astronomy, the elapsed time, usually expressed in days, since the last new Moon.

See also PHASES OF THE MOON.

age of the universe One way scientists obtain an estimate for the minimum age of the universe is by directly determining the age of the oldest objects in our galaxy. They then use this information to constrain various cosmological models. Astronomers reason, for example, that the oldest objects in our galaxy have to be at least as old as the universe—since these objects are part of the universe. They also postulate that the expansion of the universe is a function of the Hubble constant (H_0), the average density of the universe, and possibly something called the cosmological constant.

At present scientists base their best estimate for the age of the oldest stars in the galaxy on the absolute magnitude of the *main-sequence turn-off point* in globular clusters. Astronomers define the turn-off point as the point on the Hertzsprung-Russell (H-R) diagram at which stars leave the main sequence and enter a more evolved phase. The turn-off point of a globular cluster is considered the most accurate means of determining its age. Globular clusters are Population II systems, meaning all the stars found within them are relatively old. About 150 globular clusters are known in the Milky Way galaxy. The distribution and other physical characteristics of globular clusters imply that they formed early in the life of the galaxy. Globular clusters have a measured age of about 13 billion years, so scientists postulate that the universe is at least that old.

Astronomers also use the rate of recession of distant galaxies to estimate the age of the universe. Current cosmological models postulate that the universe has been expanding ever since the big bang event. By measuring the Hubble constant, scientists can tell how fast the universe is expanding today. In the simplest cosmological model, astronomers assume that the rate of expansion of the universe is unchanging with time. This model implies a linear relationship between the distance to a galaxy and its velocity of recession. The inverse of the Hubble constant—the quantity $1/H_0$, which is also called the *Hubble time*—then provides an estimate of the age of the universe. The current measured value of the Hubble constant is about 22 kilometers per second per million light years (km/s-Mly). So, within the linear (or uniform) rate expansion model, the Hubble time suggests that the universe is about 13.6 billion years old.

If the rate of expansion of the universe has actually slowed down over time, then the universe could be somewhat (but not much) younger than 13.6 billion years. But the suggestion of a universe much younger than 13.6 billion years creates a dilemma for scientists, because they have measured globular clusters that are 13 billion years old. Clearly, these ancient astronomical objects in our own galaxy cannot be older than the universe itself.

On the other hand, if the rate of expansion of the universe has actually increased over time, then the universe could be somewhat older than the 13.6 billion years implied by Hubble time. Recent astronomical observations of Type I supernova suggest that the expansion of the universe is actually accelerating. Based on such observations, scientists now suggest that the universe is about 15 billion years old. Why is the expansion of the universe speeding up? At present astrophysicists and cosmologists are not really sure. However, some suggest the phenomenon is due to the influence of a not yet completely understood concept called *cosmological constant*—a correction factor first suggested by ALBERT EINSTEIN to help explain how the influence of gravity might be opposed in a static universe.

Distilling all of these hypotheses, extrapolations, and observations down to a manageable piece of information, scientists generally consider the age of the universe to lie between 13 and 15 billion years.

See also AGE OF THE EARTH; BIG BANG; COSMOLOGICAL CONSTANT; COSMOLOGY; HUBBLE CONSTANT; UNIVERSE.

agglutinate(s) Common type of particle found on the Moon, consisting of small rock, mineral, and glass fragments impact-bonded together with glass.

airborne astronomy To complement space-based astronomical observations at infrared (IR) wavelengths, the National Aeronautics and Space Administration (NASA) uses airborne observatories—that is, infrared telescopes carried by airplanes that reach altitudes above most of the infrared radiation–absorbing water vapor in Earth's atmosphere. The first such observatory was a modified Lear jet, the second a modified C-141A Starlifter jet transport. NASA called the C-141A airborne observatory the Kuiper Airborne Observatory (KAO) in honor of the Dutch-American astronomer GERARD PETER KUIPER. The KAO's 0.9-meter aperture reflector was a Cassegrain telescope designed primarily for observations in the 1 to 500 micrometer (μm) wavelength spectral range. This aircraft collected infrared images of the universe from 1974 until its retirement in 1995.

In 2005 NASA in cooperation with the German Space Agency (DLR) will resume airborne infrared astronomy measurements using a new aircraft called the Stratospheric Observatory for Infrared Astronomy (SOFIA). NASA Ames Research Center (Moffett Field, California) serves as the home base for this refurbished and modified Boeing 747SP aircraft. SOFIA's 2.5-meter telescope will collect infrared radiation from celestial objects at wavelengths of 50 to 200 micrometers (μm). The aircraft will fly in the stratosphere at an altitude of more than 12.2-kilometers, placing its infrared telescope above 95 percent of the water vapor in Earth's atmosphere.

Turbulence is a major concern for airborne astronomical observations. Drawing upon operational experience with the Kuiper Airborne Observatory, engineers are designing SOFIA's telescope control system with an advanced jitter control system to avoid blurry images. One of the advantages of an airborne infrared observatory is that it can fly anywhere in the world on relatively short notice to take advantage of a special scientific opportunity such as the sudden appearance of a supernova. Also, if an instrument fails on SOFIA, the aircraft can simply fly back to its home base, where technicians can remove the defective instrument and plug in a new one. One of SOFIA's missions is to serve as a flying test bed for new infrared sensor technologies before these instruments are flown in space on an astronomical satellite.

air-breathing missile A missile with an engine requiring the intake of air for combustion of its fuel, as in a ramjet or turbojet. It can be contrasted with the rocket-engine missile, which carries its own oxidizer and can operate beyond the atmosphere.

See also CRUISE MISSILE; GUIDED MISSILE.

aircraft A vehicle designed to be supported by the air, either by the dynamic action of the air upon the surfaces of the vehicle or else by its own buoyancy. Aircraft include fixed-wing airplanes, gliders, helicopters, free and captive balloons, and airships.

airfoil A wing, fin, or canard designed to provide an aerodynamic force (e.g., lift) when it moves through the air (on Earth) or through the sensible atmosphere of a planet (such as Mars or Venus) or of Titan, the largest moon of Saturn.

airframe The assembled structural and aerodynamic components of an aircraft or rocket vehicle that support the different systems and subsystems integral to the vehicle. The term *airframe,* a carryover from aviation technology, is still appropriate for rocket vehicles, since a major function of the airframe is performed during the rocket's flight within the atmosphere.

airglow A faint general luminosity in Earth's sky, most apparent at night. As a relatively steady (visible) radiant emission from the upper atmosphere, it is distinguished from the sporadic emission of aurora. Airglow is a chemiluminescence that occurs during recombination of ionized atmospheric atoms and molecules that have collided with

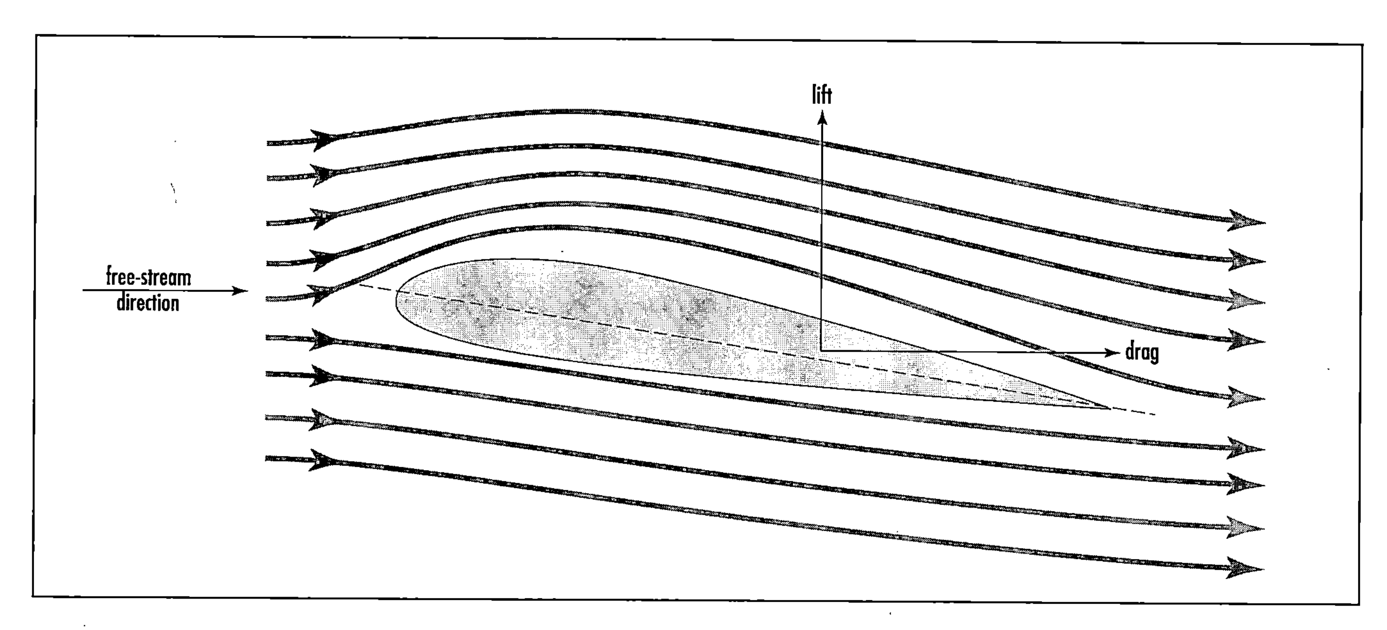

Basic forces generated by an airfoil in an airstream

high-energy particles and electromagnetic radiation, primarily from the sun. Airglow includes emissions from the oxygen molecule (O_2), the nitrogen molecule (N_2), the hydroxyl radical (OH), atomic oxygen (O), and sodium (Na).

See also ATMOSPHERE; AURORA.

air-launched cruise missile (ALCM) The U.S. Air Force developed the AGM-86B/C as a guided, air-launched, surface attack cruise missile. This small, subsonic winged missile is powered by a turbofan jet engine. Launched from aircraft, such as the B-52 bomber, it can fly complicated routes to a target through the use of modern information technology, such as interaction between its inertial navigation computer system and the Global Positioning System (GPS). *Compare with* BALLISTIC MISSILE.

airlock A small chamber with "airtight" doors that is capable of being pressurized and depressurized. The airlock serves as a passageway for crew members and equipment between places at different pressure levels—for example, between a spacecraft's pressurized crew cabin and outer space.

air-to-air missile (AAM) A missile launched from an airborne vehicle at a target above the surface of Earth (i.e., at a target in the air). One example is the Advance Medium Range, Air-to-Air Missile (AMRAAM), also called the AIM-120, developed by the U.S. Air Force and Navy in conjunction with several NATO allies. AMRAAM serves as a new generation follow-on to the Sparrow (AIM-7) air-to-air missile. The AIM-120 missile is faster, smaller, and lighter and has improved capabilities against low-altitude targets.

See also CRUISE MISSILE; GUIDED MISSILE.

air-to-ground missile (AGM) *See* AIR-TO-SURFACE MISSILE.

air-to-surface missile (ASM) A missile launched from an airborne platform at a surface target. Sometimes referred to as an AGM, or air-to-ground missile. One example is the AGM-130A missile, equipped with a television and imaging infrared seeker that allows an airborne weapon system officer to observe and/or steer the missile as it attacks a ground target. Developed by the U.S. Air Force, the AGM-130A supports high- and low-altitude precision strikes at relatively safe standoff ranges.

See also AIR LAUNCHED CRUISE MISSILE; CRUISE MISSILE; GUIDED MISSILE.

air-to-underwater missile (AUM) A missile launched from an airborne vehicle toward an underwater target, such as a submarine.

See also CRUISE MISSILE; GUIDED MISSILE.

Airy, Sir George Biddell (1801–1892) British *Astronomer* Sir George Airy modernized the Royal Greenwich Observatory (RGO) and served as England's seventh Astronomer Royal from 1835 to 1881. Despite many professional accomplishments, he is often remembered for his failure to recognize the value of the theoretical work of JOHN COUCH ADAMS, thereby delaying the discovery of Neptune until 1846.

Airy disk In optics, the intensity distribution around a circular aperture caused by diffraction, such that a point source of light becomes a bright, disklike image. Because of the phenomenon of diffraction, even a telescope with perfect optics cannot form a pointlike image of a star. Instead, an optical system with a circular aperture forms an image that consists of a central disk, called the Airy disk after GEORGE BIDDELL AIRY, who first investigated the phenomenon in 1834, surrounded by several faint diffraction rings. The diameter of the Airy disk characterizes the resolving power (angular resolution) limits of an optical telescope.

See also TELESCOPE.

Aitken, Robert Grant (1864–1951) American *Astronomer* Robert Aitken worked at Lick Observatory in California and specialized in binary star systems. In 1932 he published *New General Catalog of Double Stars,* a seminal work that contains measurements of more than 17,000 double stars and is often called the *Aitken Double Star Catalog (ADS).*

albedo The ratio of the electromagnetic radiation (such as visible light) reflected by the surface of a nonluminous object to the total amount of electromagnetic radiation incident upon that surface. The albedo is usually expressed as a percentage. For example, the planetary albedo of Earth is about 30 percent. This means that approximately 30 percent of the total solar radiation falling upon Earth is reflected back to outer space. With respect to the Moon, a high albedo indicates a light-colored region, while a low albedo indicates a dark region. Lunar maria ("seas") have low albedo; lunar highlands have high albedo.

Aldebaran The brightest (alpha) star in the constellation Taurus. Aldebaran is a K-spectral class red giant located about 60 light-years from Earth. With a surface temperature under 4,000 kelvins, the star has a distinctive orange color. Early Arab astronomers named it "Aldebaran"—which means "the Follower" (in Arabic)—because the star appeared to follow the Seven Sisters (Pleiades) star cluster across the sky. This star is positioned just in front of the Hyades star cluster, tempting amateur stargazers to incorrectly include it in the cluster. However, Aldebaran only lies along the line of sight and is not a member of the Hyades star cluster, which is much farther away, at a distance of about 150 light-years.

Aldrin, Edwin E. "Buzz" (1930–) American *Astronaut, U.S. Air Force Officer* Edwin ("Buzz") Aldrin served as the lunar module pilot during the National Aeronautics and Space Administration's *Apollo 11* mission to the Moon. This mission took place from July 16 to 24, 1969, and involved the first landing expedition. During this historic mission, Aldrin followed NEIL A. ARMSTRONG on July 20 and became the second human being to step onto the lunar surface.

On July 20, 1969, the American astronaut Edwin "Buzz" Aldrin became the second human being to walk on the Moon. He accomplished this historic feat as a member of NASA's *Apollo 11* lunar landing mission. *(Courtesy of NASA)*

Alfonso X of Castile (1221–1284) Spanish *Monarch, Astronomer* Alfonso X of Castile was the scholarly Spanish monarch who ordered the compilation of revised planetary tables based on Arab astronomy (primarily *The Almagest* of PTOLEMY), but also updated its contents with astronomical observations made at Toledo, Spain. As a result of his actions, *The Alfonsine Tables* were completed in 1272 and served medieval astronomers for more than three centuries.

Alfvén, Hannes Olof Gösta (1908–1995) Swedish *Physicist, Space Scientist, Cosmologist* Hannes Alfvén developed the theory of magnetohydrodynamics (MHD)—the branch of physics that helps astrophysicists understand sunspot formation. MHD also helps scientists understand the magnetic field–plasma interactions (now called Alfvén waves in his honor) taking place in the outer regions of the Sun and other stars. For this pioneering work and its applications to many areas of plasma physics, Alfvén shared the 1970 Nobel Prize in physics.

Alfvén was born in Norrköping, Sweden, on May 30, 1908. He enrolled at the University of Uppsala in 1926 and obtained a doctor of philosophy (Ph.D.) degree in 1934. That same year he received an appointment to lecture in physics at the University of Uppsala. He married Kerstin Maria Erikson in 1935, and the couple had a total of five children, one son and four daughters.

In 1937 Alfvén joined the Nobel Institute for Physics in Stockholm as a research physicist. Starting in 1940 he held a number of positions at the Royal Institute of Technology in Stockholm. He served as an appointed professor in the theory of electricity from 1940 to 1945, a professor of electronics from 1945 to 1963, and a professor of plasma physics from 1963 to 1967. In 1967 he came to the United States and joined the University of California at San Diego as a visiting professor of physics until retiring in 1991. In 1970 he shared the Nobel Prize in physics with the French physicist Louis Néel (1904–2000). Alfvén received his half of the prize for his fundamental work and discoveries in magnetohydrodynamics (MHD) and its numerous applications in various areas of plasma physics.

Alfvén is best known for his pioneering plasma physics research that led to the creation of the field of magnetohydrodynamics, the branch of physics concerned with the interactions between plasma and a magnetic field. Plasma is an electrically neutral gaseous mixture of positive and negative ions. Physicists often refer to plasma as the fourth state of matter, because the plasma behaves quite differently than do solids, liquids, or gases. In the 1930s Alfvén went against conventional beliefs in classical electromagnetic theory. Going against all "conventional wisdom," he boldly proposed a theory of sunspot formation centered on his hypothesis that under certain physical conditions a plasma can become locked, or "frozen," in a magnetic field and then participate in its wave motion. He built upon this hypothesis and further suggested in 1942 that under certain conditions, as found, for example, in the Sun's atmosphere, waves (now called "Alfvén waves") can propagate through plasma influenced by magnetic fields. Today solar physicists believe that at least a portion of the energy heating the solar corona is due to action of Alfvén waves propagating from the outer layers of the Sun. Much of his early work on magnetohydrodynamics and plasma physics is collected in the books *Cosmical Electrodynamics* published in 1948 and *Cosmical Electrodynamics: Fundamental Principles,* coauthored in 1963 with Swedish physicist Carl-Gunne Fälthammar (b. 1931).

In the 1950s Alfvén again proved to be a scientific rebel when he disagreed with both steady state cosmology and big bang cosmology. Instead, he suggested an alternate cosmology based on the principles of plasma physics. He used his 1956 book, *On the Origin of the Solar System,* to introduce his hypothesis that the larger bodies in the solar system were formed by the condensation of small particles from primordial magnetic plasma. He further pursued such electrodynamic cosmology concepts in his 1975 book, *On the Evolution of the Solar System,* which he authored with Gustaf Arrhenius. He also developed an interesting model of the early universe in which he assumed the presence of a huge spherical cloud containing equal amounts of matter and antimatter. This theoretical model, sometimes referred to as *Alfvén's antimatter cosmology model,* is not currently supported by observational evidence of large quantities of annihilation radiation—the anticipated by-product when galaxies and antigalaxies collide. So while contemporary astrophysicists and astronomers generally disagree with his theories in cosmology, Alfvén's work in magnetohydrodynamics and plasma physics remains of prime importance to scientists engaged in controlled thermonuclear fusion research.

After retiring in 1991 from his academic positions at the University of California at San Diego and the Royal Institute of Technology in Stockholm, Alfvén enjoyed his remaining years by "commuting" on a semiannual basis between homes in the San Diego area and Stockholm. He died in Stockholm on April 2, 1995, and is best-remembered as the Nobel laureate theoretical physicist who founded the important field of magnetohydrodynamics.

Alfvén wave In magnetohydrodynamics, a transverse wave that occurs in a region that contains a magnetic field and plasma. As first suggested by HANNES ALFVÉN in the 1930s, under certain physical conditions a plasma can become locked, or "frozen," in a magnetic field and then participate in its wave motion.

Algol The beta star in the constellation Perseus. Called the "Demon Star" by Arab astronomers, Algol was the first eclipsing binary to be discovered. Although previous astronomers had reported variations in Algol's brightness, the British astronomer JOHN GOODRICKE was the first astronomer to determine the period (2.867 days). In 1782 Goodricke also suggested that Algol's periodic behavior might be due to a dark, not visible stellar companion that regularly passed in front of (eclipsed) its visible companion. Modern astronomers appreciate the correctness of Goodricke's daring hypothesis and treat Algol is an eclipsing binary star system in which the darker companion star (called Algol B) periodically cuts off the light of the brighter star (called Algol A), making the brightness of Algol A appear variable. Algol A is a bright main sequence star with a mass of approximately 3.7 solar masses, while its less massive (0.8 solar mass) companion, Algol B, is an evolved star that has become a subgiant. This multiple star system also contains a third more distant star called Algol C, which orbits both Algol A and Algol B in 1.86 years.

In addition to being the first known eclipsing binary, Algol gives rise to what astronomers call the *Algol paradox.* Generally, the greater the mass of a star, the shorter its lifetime, as it consumes more nuclear fuel and burns brighter. Yet, here, the more massive star (Algol A) is still on the main sequence, while the less massive companion (Algol B) has evolved into a dying subgiant. Astronomers resolve this paradox by suggesting that since the stars are so close together (separated by less than 0.1 astronomical unit), as the giant star (Algol B) grew large, tidal effects caused by the presence of Algol A caused a transfer of mass directly onto Algol A from Algol B.

See also BINARY STAR SYSTEM; STAR; SUBGIANT.

algorithm A special mathematical procedure or rule for solving a particular type of problem.

alien life form (ALF) A general, though at present hypothetical, expression for extraterrestrial life, especially life that exhibits some degree of intelligence.

See also EXOBIOLOGY; EXTREMOPHILE.

Almagest, The The Arabic name (meaning "the greatest") for the collection of ancient Greek astronomical and mathematical knowledge written by PTOLEMY in about 150 C.E. and translated by Arab astronomers in about 820. This compendium included the 48 ancient Greek constellations upon which today's astronomers base the modern system of constellations.

Alouette spacecraft Small scientific spacecraft designed to study the ionosphere that were built by Canada and launched by the United States in the 1960s. *Alouette 1* was placed into a 996-km by 1,032-km, 80.5° inclination polar orbit by a Thrust-Augmented Thor–Agena-B launch vehicle combination on September 28, 1962 (local time), from Vandenberg Air Force Base in California. With the launch of *Alouette 1,* Canada became the third nation (behind the Soviet Union and the United States) to have a national satellite in orbit around Earth. The spacecraft was a small ionospheric observatory instrumented with an ionospheric sounder, a very low frequency (VLF) receiver, an energetic particle detector, and a cosmic noise experiment. The satellite was spin-stabilized at approximately 1.4 revolutions per minute (rpm) after antenna extension. About 500 days following launch, the spacecraft's spin rate slowed more than anticipated, to a rate of just 0.6 rpm, when the satellite's spin stabilization failed. From that point on attitude information could be deduced only from a single magnetometer and temperature measurements on the upper and lower heat shields. This early scientific satellite, like others of the time, did not carry a tape recorder, so data was available only when the spacecraft's orbital track passed telemetry stations located in areas near Hawaii, Singapore, Australia, Europe, and Central Africa. In September 1972 Canadian ground controllers terminated operation of the spacecraft. The success of *Alouette 1* led not only to the follow-on international scientific program, but more importantly led to the establishment of the Canadian space industry.

Alouette 2 was launched on November 28, 1965 (local time), by a Thrust-Augmented Thor–Agena B rocket configuration from Vandenberg AFB and placed into a 505-km by 2,987-km polar orbit with an inclination of 79.8°. Like its predecessor, this small ionospheric observatory functioned successfully for almost 10 years. Routine operations were terminated in July 1975, although the spacecraft was successfully reactivated on November 28–29, 1975, in order to obtain data on its 10th anniversary in orbit.

Success of the Alouette spacecraft encouraged development and launch of the International Satellite for Ionospheric Studies (ISIS) spacecraft. *ISIS-1* was launched in 1969 and *ISIS-2* in 1971. The ISIS satellites were used until 1984, when the program was concluded.

See also CANADIAN SPACE AGENCY (CSA).

Alpha Centauri The closest star system, about 4.3 light-years away. It is actually a triple-star system, with two stars orbiting around each other and a third star, called Proxima Centauri, revolving around the pair at some distance. In various celestial configurations, Proxima Centauri becomes the closest known star to our solar system—approximately 4.2 light-years away.

alpha particle (symbol: α) A positively charged atomic particle emitted by certain radionuclides. It consists of two neutrons and two protons bound together and is identical to the nucleus of a helium-4 ($^{4}_{2}He$) atom. Alpha particles are the least penetrating of the three common types of ionizing radiation (alpha particle, beta particle, and gamma ray). They are stopped easily by materials such as a thin sheet of paper or even a few centimeters of air.

Alpha Regio A high plateau on Venus about 1,300 kilometers across.

See also VENUS.

alpha (α) star Within the system of stellar nomenclature introduced in 1603 by the German astronomer JOHANN BAYER (1572–1625), the brightest star in a constellation.

Alpher, Ralph Asher (1921–) American *Physicist* Ralph Alpher is the theoretical physicist who collaborated with GEORGE GAMOW (and other scientists) in 1948 to develop the big bang theory of the origin of the universe. He also extended nuclear transmutation theory (nucleosynthesis) within a very hot, early fireball to predict the existence of a now-cooled, residual cosmic microwave background.

See also BIG BANG (THEORY); COSMIC MICROWAVE BACKGROUND; NUCLEOSYNTHESIS.

altazimuth mounting A telescope mounting that has one axis pointing to the zenith.

altimeter An instrument for measuring the height (altitude) above a planet's surface, generally reported relative to a common planetary reference point, such as sea level on Earth. A barometric altimeter is an instrument that displays the height of an aircraft or aerospace vehicle above a specified pressure datum; a radio/radar altimeter is an instrument that displays the distance between an aircraft or spacecraft and the surface vertically below as determined by a reflected radio/radar transmission.

altitude 1. (astronomy) The angle between an observer's horizon and a target celestial body. The altitude is 0° if the

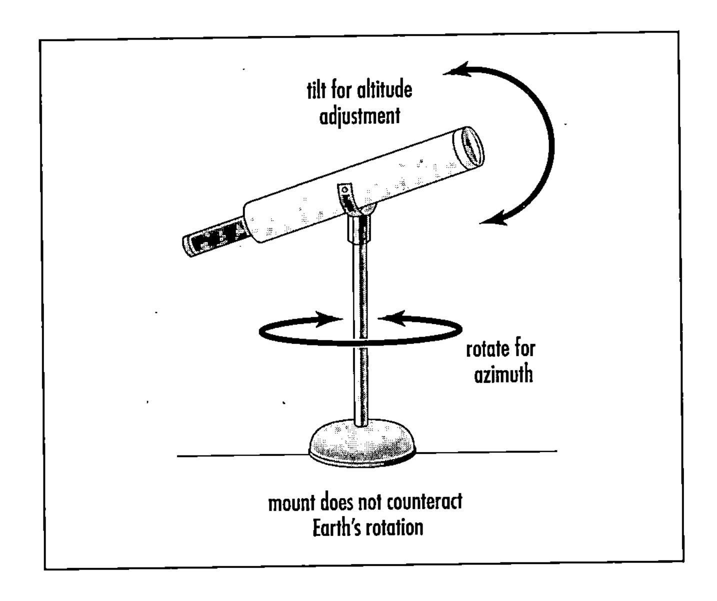

Altazimuth mounting

object is on the horizon, and 90° if the object is at zenith (that is, directly overhead). 2. (spacecraft) In space vehicle navigation, the height above the mean surface of the reference celestial body. Note that the *distance* of a space vehicle or spacecraft from the reference celestial body is taken as the distance from the center of the object.

altitude chamber A chamber within which the air pressure and temperature can be adjusted to simulate different conditions that occur at different altitudes; used for crew training, experimentation, and aerospace system testing. Also called *hypobaric chamber.*

altitude engine Rocket engine that is designed to operate at high-altitude conditions.

Alvarez, Luis Walter (1911–1988) American *Physicist* Luis Walter Alvarez, a Nobel Prize–winning physicist, collaborated with his son, Walter (a geologist), to rock the scientific community in 1980 by proposing an extraterrestrial catastrophe theory that is sometimes called the "Alvarez hypothesis." This popular theory suggests that a large asteroid struck Earth some 65 million years ago, causing a mass extinction of life, including the dinosaurs. Alvarez supported the hypothesis by gathering interesting geologic evidence—namely, the discovery of a worldwide enrichment of iridium in the thin sediment layer between the Cretaceous and Tertiary periods. The unusually high elemental abundance of iridium has been attributed to the impact of an extraterrestrial mineral source. The Alvarez hypothesis has raised a great deal of scientific interest in cosmic impacts, both as a way to possibly explain the disappearance of the dinosaurs and as a future threat to planet Earth that might be avoidable through innovations in space technology.

Luis Alvarez was born in San Francisco, California, on June 13, 1911. His family moved from San Francisco to Rochester, Minnesota, in 1925 so his father could join the staff at the Mayo Clinic. During his high school years the young Alvarez spent his summers developing experimental skills by working as an apprentice in the Mayo Clinic's instrument shop. He initially enrolled in chemistry at the University of Chicago but quickly embraced physics, especially experimental physics, with an enthusiasm and passion that remained for a lifetime. In rapid succession he earned his bachelor's degree (1932), master's degree (1934), and doctor of philosophy degree (1936) in physics at the University of Chicago. He married Geraldine Smithwick in 1936, and they had two children, a son (Walter) and a daughter (Jean). Two decades later, in 1958, Luis Alvarez married his second wife, Janet L. Landis, with whom he had two other children, a son (Donald) and a daughter (Helen).

After obtaining his doctorate in physics from the University of Chicago in 1936, Alvarez began his long professional association with the University of California at Berkeley. Only World War II disrupted this affiliation. For example, from 1940 to 1943 Alvarez conducted special wartime radar research at the Massachusetts Institute of Technology in Boston. From 1944 to 1945 Alvarez was part of the atomic bomb team at Los Alamos Laboratory in New Mexico. He played a key role in the development of the first plutonium implosion weapon. Specifically, as part of his Manhattan Project responsibilities, Alvarez had the difficult task of developing a reliable, high-speed way to multipoint detonate the chemical high explosive used to symmetrically squeeze the bomb's plutonium core into a supercritical mass. Then, during the world's first atomic bomb test (called Trinity) on July 16, 1945, near Alamogordo, New Mexico, Alvarez flew overhead as a scientific observer. He was the only witness to precisely sketch the first atomic debris cloud as part of his report. He also served as a scientific observer onboard one of the escort aircraft that accompanied the *Enola Gay,* the American B-29 bomber that dropped the first atomic weapon on Hiroshima, Japan, on August 6, 1945.

After World War II Alvarez returned to the University of California at Berkeley and served as a professor of physics from 1945 until his retirement in 1978. His brilliant career as an experimental physicist involved many interesting discoveries that go well beyond the scope of this entry. Before describing his major contribution to modern astronomy, however, it is appropriate to briefly discuss the experimental work that earned him the Nobel Prize in physics in 1968. Simply stated, Alvarez helped start the great elementary particle stampede that began in the early 1960s. He did this by transforming the liquid hydrogen bubble chamber into a large, enormously powerful research instrument of modern high-energy physics. His innovative work allowed teams of researchers at Berkeley and elsewhere to detect and identify many new species of very short-lived subnuclear particles. This opened the way for the development of the quark model in modern nuclear physics. When an elementary particle passes through the chamber's liquid hydrogen (kept at the very low temperature of –250°C), the cryogenic fluid is warmed to the boiling point along the track that the particle leaves. The tiny telltale trail of bubbles is photographed and then computer analyzed by Alvarez's device. Nuclear physicists examine the bubble track photographs in order to extract new information about whichever member of the "nuclear particle zoo" they have just captured. Alvarez's large liquid-hydrogen bubble chamber came into operation in March 1959 and almost immediately led to the discovery of many interesting new elementary particles. In an analogy quite appropriate here, Alvarez's hydrogen bubble chamber did for elementary particle physics what the telescope did for observational astronomy.

Just before his retirement from the University of California, Alvarez became intrigued by a very unusual geologic phenomenon. In 1977 his son Walter (b. 1940), a geologist, showed him an interesting rock specimen, which the younger Alvarez had collected from a site near Gubbio, a medieval Italian town in the Apennine Mountains. The rock sample was about 65 million years old and consisted of two layers of limestone: one from the Cretaceous period (symbol K—after the German word *Kreide* for Cretaceous) and the other from the Tertiary period (symbol T). A thin, approximately one-centimeter-thick clay strip separated the two limestone layers. According to geologic history, as this layered rock specimen formed eons ago, the dinosaurs flourished and then mysteriously passed into extinction. Perhaps the thin clay strip contained information that might answer the question: Why did the great dinosaurs suddenly disappear?

The father and son scientific team carefully examined the rock and were puzzled by the presence of a very high concentration of iridium in the peculiar sedimentary clay. Here on Earth iridium is quite rare, and typically no more than about 0.03 parts per billion are found in the planet's crust. Soon geologists discovered this same iridium enhancement (sometimes called the "iridium anomaly") in other places around the world in the same thin sedimentary layer (called the KT boundary) that was laid down about 65 million years ago—the suspected time of a great mass extinction. Since iridium is quite rare in Earth's crust and more abundant in other solar system bodies, the Alvarez team postulated that a large asteroid—possibly one about 10 kilometers or more in diameter—had struck prehistoric Earth.

This cosmic collision would have caused an environmental catastrophe throughout the planet. The Alvarez team reasoned that such a giant asteroid would largely vaporize while passing through Earth's atmosphere, spreading a dense cloud of dust particles including large quantities of extraterrestrial iridium atoms uniformly around the globe. They further speculated that after the impact of this killer asteroid, a dense cloud of ejected dirt, dust, and debris would encircle the globe for many years, blocking photosynthesis and destroying the food chains upon which many ancient animals depended.

When Luis Alvarez published this hypothesis in 1980, it created a stir in the scientific community. In fact, the Nobel laureate spent much of the final decade of his life explaining and defending his extraterrestrial catastrophe theory. Despite the geophysical evidence of a global iridium anomaly in the thin sedimentary clay layer at the KT boundary, many geologists and paleontologists preferred other explanations concerning the mass extinction that occurred on Earth about 65 million years ago. While still controversial, the Alvarez hypothesis about a killer asteroid emerged from the 1980s as the most popular reason to explain why the dinosaurs disappeared.

Shortly after Alvarez's death, two very interesting scientific events took place that gave additional support to his extraterrestrial catastrophe theory. First, in the early 1990s a 180-kilometer-diameter ring structure, called Chicxulub, was identified from geophysical data collected in the Yucatán region of Mexico. The Chicxulub crater has been age-dated at 65 million years. The impact of a 10-kilometer-diameter asteroid could have created this very large crater, as well as causing enormous tidal waves. Second, a wayward comet called Shoemaker-Levy 9 slammed into Jupiter in July 1994. Scientists using a variety of space-based and Earth-based observatories studied how the giant planet's atmosphere convulsed after getting hit by a runaway "cosmic train" consisting of 20-kilometer-diameter or so chunks of this fragmented comet that plowed into the southern hemisphere of Jupiter. The comet's fragments deposited the energy equivalent of about 40 million megatons of trinitrotoluene (TNT), and their staccato impacts sent huge plumes of hot material shooting more than 1,000 kilometers above the visible Jovian cloud tops.

About a year before his death in Berkeley, California, on August 31, 1988, Alvarez published a colorful account of his life in the autobiography entitled *Alvarez, Adventures of a Physicist* (1987). In addition to his 1968 Nobel Prize in physics, he received numerous other awards, including the Collier Trophy in Aviation (1946) and the National Medal of Science, personally presented to him in 1964 by President Lyndon Johnson.

Amalthea The largest of Jupiter's smaller moons. The American astronomer EDWARD EMERSON BARNARD discovered Amalthea on September 9, 1892, using the 0.91-meter (36-inch) refractor telescope at the Lick Observatory. This irregularly shaped (270-km × 170-km × 150-km) moon is one of four tiny inner moons that orbit Jupiter inside the orbit of Io. Amalthea has a mean distance from Jupiter of 181,300 kilometers, an orbital period of 0.498 days, and a rotation period of 0.498 days. This means that Amalthea is in synchronous rotation and keeps the same blunt end always pointing toward Jupiter. Amalthea's orbit around Jupiter has an eccentricity of 0.003 and an inclination of 0.40 degrees.

The tiny moon is heavily cratered and has two prominent craters. The larger crater, called Pan, is 90 kilometers long and lies in the small moon's northern hemisphere; the other large crater, called Gaea, is 75 kilometers long and straddles the moon's south pole. There are also two mountains on Amalthea, Mons Ida and Mons Lyctos. Amalthea is a dark object (0.06 visual albedo) and reddish in color. In fact, astronomers consider Amalthea the reddest object in the solar system. The moon's red color is most likely caused by sulfur from Io spiraling down to Jupiter and impacting on Amalthea.

The small moon has an estimated mass of approximately 7.2×10^{18} kilograms, an average density of about 1.8 g/cm^3, a surface gravity ranging between 0.054 and 0.083 m/s^2, and an escape velocity of 0.084 km/s. Amalthea has a striking resemblance to Phobos, the potato-shaped tiny Martian moon. However, Amalthea is 10 times larger than Phobos and orbits its parent planet in a much more inhospitable environment. Being so close to Jupiter (*see* figure), Amalthea is exposed to the giant planet's intense radiation field and experiences high doses of the energetic ions, protons, and electrons present in the Jovian magnetosphere. In addition, Amalthea is bombarded with high-velocity micrometeorites, as well as heavy ions—predominantly sulfur, oxygen, and sodium diffusing away from Io.

In Greek mythology Amalthea was the nymph who nursed the infant Jupiter with goat's milk. The discovery of Amalthea in 1892 represented the first discovery of a Jovian moon since 1610, when GALILEO GALILEI ushered in the age of telescopic astronomy with his discovery of Jupiter's four large moons, Callisto, Europa, Ganymede, and Io. By interesting historic coincidence, Amalthea was also the last moon of any planet discovered by direct visual observation. All new moons discovered since then have been found through the use of photography, including imagery collected by scientific spacecraft such as the *Voyager* and *Galileo* spacecraft.

See also JUPITER.

Ambartsumian, Viktor Amazaspovich (1908–1996) Armenian *Astrophysicist* Viktor A. Ambartsumian founded the Byurakan Astrophysical Observatory in 1946 on Mount

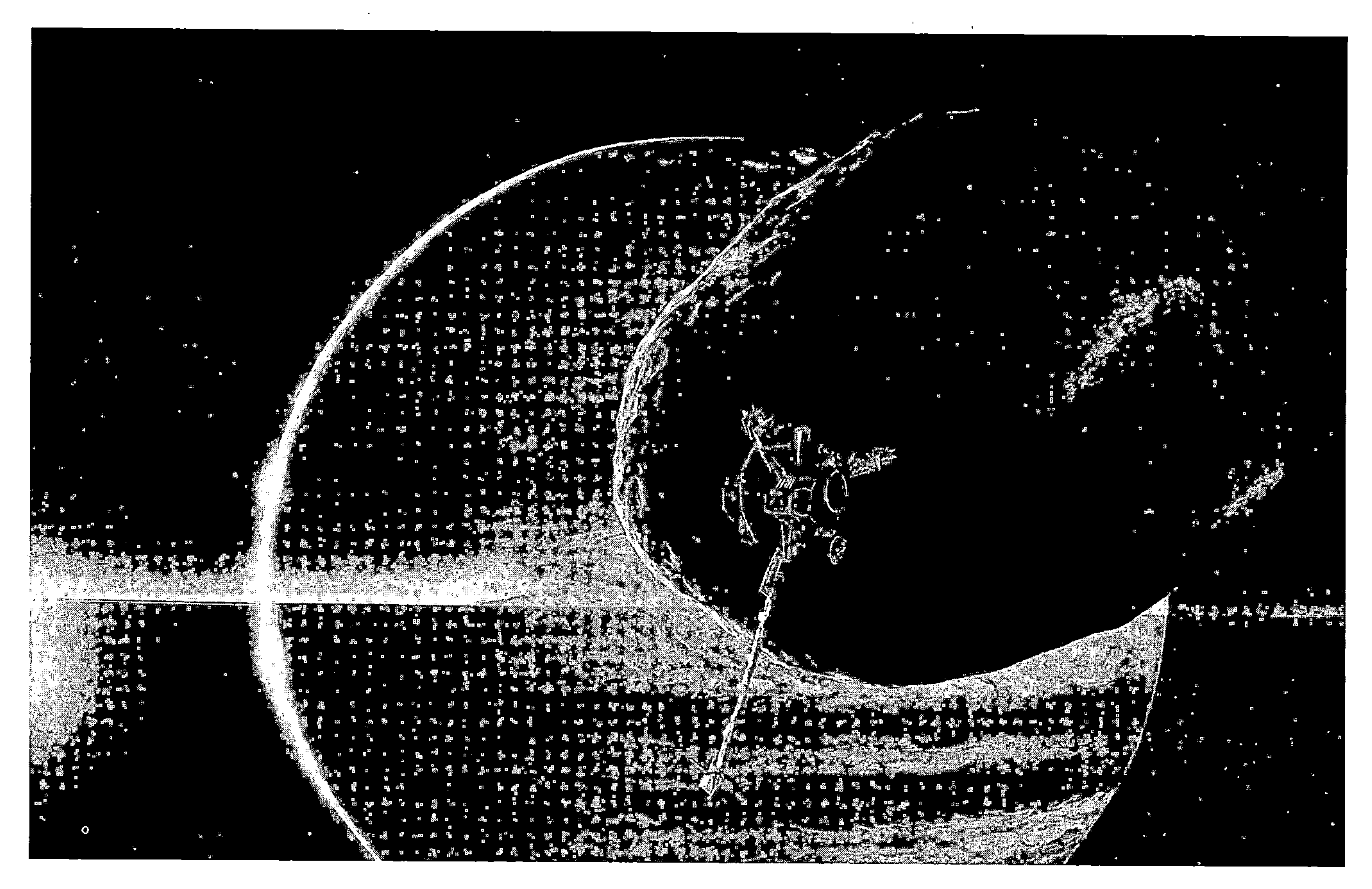

This artist's rendering depicts NASA's *Galileo* spacecraft as it performed a very close flyby of Jupiter's tiny inner moon Amalthea in November 2002. *(Courtesy of NASA)*

Aragatz near Yerevan, Armenia. This facility served as one of the major astronomical observatories in the Soviet Union. Considered the father of Russian theoretical astrophysics, he introduced the idea of stellar association into astronomy in 1947. His major theoretical contributions to astrophysics involved the origin and evolution of stars, the nature of interstellar matter, and phenomena associated with active galactic nuclei.

Ambartsumian was born on September 18, 1908, in Tbilisi, Georgia (then part of the czarist Russian Empire). His father was a distinguished Armenian philologist (that is, teacher of classical Greco-Roman literature), so as a young man he was exposed to the value of intense intellectual activity. When Ambartsumian was just 11 years old, he wrote the first two of his many astronomical papers, "The New Sixteen-Year Period for Sunspots" and "Description of Nebulae in Connection with the Hypothesis on the Origin of the Universe." His father quickly recognized his son's mathematical talents and aptitude for physics and encouraged him to pursue higher education in St. Petersburg (then called Leningrad), Russia.

In 1925 Ambartsumian enrolled at the University of Leningrad. Before receiving his degree in physics in 1928, he published 10 papers on astrophysics while still an undergraduate student. From 1928 to 1931 he did his graduate work in astrophysics at the Pulkovo Astronomical Observatory. This historic observatory near St. Petersburg was founded in the 1830s by the German astronomer FRIEDRICH GEORG WILHELM STRUVE and served as a major observatory for the Soviet Academy of Sciences until its destruction during World War II. During the short period from 1928 to 1930, Ambartsumian published 22 astrophysics papers in various journals while still only a graduate student at Pulkovo Observatory. The papers signified the emergence of his broad theoretical research interest, and his scientific activities began to attract official attention and recognition within the former Soviet government. At age 26 Ambartsumian became a professor at the University of Leningrad, where he soon organized and headed the first department of astrophysics in the Soviet Union. His pioneering academic activities and numerous scientific publications in the field eventually earned him the title of "father of Soviet theoretical astrophysics."

He served as a member of the faculty at the University of Leningrad from 1931 to 1943 and was elected to the Soviet Academy of Sciences in 1939. During the disruptive days of World War II, Ambartsumian was drawn by his heritage and returned to Armenia (then a republic within the former Soviet Union). There, in 1943 he assumed a teaching position at Yerevan State University in the capital city of the Republic of Armenia. That year he also became a member of the Armenian Academy of Sciences. In 1946 he organized the development and construction of the Byurakan Astronomical Observatory. Located on Mount Aragatz near Yerevan, this facility served as one of the major astronomical observatories in the former Soviet Union.

Ambartsumian served as the director of the Byurakan Astrophysical Observatory and resumed his activities investigating the evolution and nature of star systems. In 1947 he introduced the important new concept of stellar association, the very loose groupings of young stars of similar spectral type that commonly occur in the gas- and dust-rich regions of the Milky Way galaxy's spiral arms. He was the first to suggest the notion that interstellar matter occurs in the form of clouds. Some of his most important work took place in 1955, when he proposed that certain galactic radio sources indicated the occurrence of violent explosions at the centers of these particular galaxies. The English translation of his influential textbook, *Theoretical Astrophysics,* appeared in 1958 and became standard reading for astronomers-in-training around the globe.

The eminent Armenian astrophysicist died at the Byurakan Observatory on August 12, 1996. His fellow scientists had publicly acknowledged his technical contributions and leadership by electing him president of the International Astronomical Union (1961–64) and president of the International Council of Scientific Unions (1970–74). He is best remembered for his empirical approach to complex astrophysical problems dealing with the origin and evolution of stars and galaxies. He summarized his most important work in the paper "On Some Trends in the Development of Astrophysics" in *Annual Review of Astronomy and Astrophysics* 18, 1–13 (1980).

ambient The surroundings, especially of or pertaining to the environment around an aircraft, aerospace vehicle, or other aerospace object (e.g., nose cone or planetary probe). The natural conditions found in a given environment, such as the ambient pressure or ambient temperature on the surface of a planet. For example, a planetary probe on the surface of Venus must function in an infernolike environment where the ambient temperature is about 480°C (753 K).

Amici, Giovanni Battista (1786–1863) Italian *Astronomer, Optician* As an optician with an interest in astronomy, Giovanni Amici developed better mirrors for reflecting telescopes. While serving as astronomer to the grand duke of Tuscany, he used his improved telescopes to make more refined observations of the Jovian moons and of selected binary star systems.

See also BINARY STAR SYSTEM; TELESCOPE.

amino acid An acid containing the amino (NH_2) group, a group of molecules necessary for life. More than 80 amino acids are presently known, but only some 20 occur naturally in living organisms, where they serve as the building blocks of proteins. On Earth many microorganisms and plants can synthesize amino acids from simple inorganic compounds. However, terrestrial animals (including human beings) must rely on their diet to provide adequate supplies of amino acids.

Amino acids have been synthesized nonbiologically under conditions that simulate those that may have existed on the primitive Earth, followed by the synthesis of most of the biologically important molecules. Amino acids and other biologically significant organic substances have also been found to occur naturally in meteorites and are not considered to have been produced by living organisms.

See also EXOBIOLOGY; LIFE IN THE UNIVERSE; METEOROIDS.

Amor group A collection of near-Earth asteroids that cross the orbit of Mars but do not cross the orbit of Earth. This asteroid group acquired its name from the 1-kilometer-diameter Amor asteroid, discovered in 1932 by the Belgian astronomer EUGÈNE DELPORTE.

See also ASTEROID; NEAR-EARTH ASTEROID.

amorphotoi Term used by the early Greek astronomers to describe the spaces in the night sky populated by dim stars between the prominent groups of stars making up the ancient constellations. It is Greek for "unformed."

ampere (symbol: A) The SI unit of electric current, defined as the constant current that—if maintained in two straight parallel conductors of infinite length, of negligible circular cross sections, and placed 1 meter apart in a vacuum—would produce a force between these conductors equal to 2×10^{-7} newtons per meter of length. The unit is named after the French physicist André M. Ampère (1775–1836).

amplitude Generally, the maximum value of the displacement of a wave or other periodic phenomenon from a reference (average) position. Specifically, in astronomy and astrophysics, the overall range of brightness (from maximum magnitude to minimum magnitude) of a variable star.

amplitude modulation (AM) In telemetry and communications, a form of modulation in which the amplitude of the carrier wave is varied, or "modulated," about its unmodulated value. The amount of modulation is proportional to the amplitude of the signal wave. The frequency of the carrier wave is kept constant. *Compare with* FREQUENCY MODULATION.

Ananke The small moon of Jupiter discovered in 1951 by the American astronomer SETH BARNES NICHOLSON (1891–1963). Ananke has a diameter of 30 kilometers and a mass of about 3.8×10^{16} kilograms. It travels around the giant planet at a distance of 21,200,000 kilometers in a retrograde orbit. A period of 631 days, an eccentricity of 0.169, and an inclination of 147 degrees characterize the small moon's orbit. The Jovian moons Carme, Pasiphae, and Sinope have similarly unusual retrograde orbits. Astronomers sometimes refer to Ananke as Jupiter XII.

Anaxagoras (ca. 500–ca. 428 B.C.E.) Greek *Philosopher, Early Cosmologist* Anaxagoras was an ancient Greek thinker who speculated that the Sun was really a huge, incandescent rock, that the planets and stars were flaming rocks, and that the Moon had materials similar to Earth, including possibly inhabitants. For these bold hypotheses, the Athenian authorities charged him with "religious impiety" and banished him from the city. His work in early cosmology included the concept that "mind" (Greek *nous,* νους) formed the material objects of the universe. Perhaps Anaxagoras was simply well ahead of his time, because one of the most interesting areas of contemporary physics is the investigation of possible linkages between human consciousness and the material universe.

See also ANTHROPHIC PRINCIPLE; COSMOLOGY.

Anaximander (ca. 610–ca. 546 B.C.E.) Greek *Philosopher, Astronomer* Anaximander was the earliest Hellenistic thinker to propose a systematic worldview. Within his cosmology a stationary, cylindrically shaped Earth floated freely in space at the center of the universe. Recognizing that the heavens appeared to revolve around the North Star (Polaris), he used a complete sphere to locate celestial bodies (that is, stars and planets) in the night sky. He also proposed *apeiron* (Greek for "unlimited" or "infinite") as the source element for all material objects—a formless, imperceptible substance from which all things originate and back into which all things return.

See also COSMOLOGY; UNIVERSE.

ancient astronaut theory A contemporary hypothesis that Earth was visited in the past by intelligent alien beings. Although *unproven* in the context of traditional scientific investigation, this hypothesis has given rise to many popular stories, books, and motion pictures—all speculations that seek to link such phenomena as the legends from ancient civilizations concerning superhuman creatures, unresolved mysteries of the past, and unidentified flying objects (UFOs) with extraterrestrial sources.

ancient astronomical instruments Ancient societies, including the Babylonians, Greeks, Egyptians, Chinese, and Mayans, took stargazing seriously. Within the limits of naked eye astronomy, they used a variety of basic observational techniques and later very simple instruments to define, monitor, and predict the motion of celestial objects. Prior to the invention of the telescope, the most elaborate astronomical instruments were constructed out of metal and wood. Even societies that did not develop sophisticated metal-working skills were able to make reasonably accurate astronomical observations by using other clever techniques, such as the construction and careful alignment of buildings. As revealed by archaeoastronomy, the alignment of many special ancient buildings or monuments corresponded to the locally observed position of the Sun at equinox or solstice. Other ancient structures, such as the great Egyptian pyramids, were aligned to highlight the appearance of certain conspicuous (bright) stars.

Many ancient astronomers used *horizon intercepts*—the alignment of a rising or setting celestial object with some feature on the distant horizons as observed from a special location on Earth's surface—to track the motions of the Sun, Moon, and planets. These observations allowed them to develop early calendars, especially calendars based on the phases of the Moon or Earth's annual movement around the Sun. Pretelescopic astronomers also used *heliacal risings* as a date-keeping device. A heliacal rising is the rising of a star or planet at or just prior to sunrise, such that the celestial object is observable in the morning sky. For example, the ancient Egyptians used the heliacal rising of Sirius (the Dog star) to predict the annual flooding of the Nile River.

In addition to naked eye observations involving the use of special buildings, horizon intercepts, and heliacal risings, pretelescopic astronomers used some or all of these early instruments to monitor the heavens: the armillary sphere, the astrolabe, the cross staff, the quadrant, the sundial, and the triquetrum.

The ancient Greeks developed instruments such as the *armillary sphere* to track the movement of objects in the plane of the celestial equator against the annual motion of the Sun. This basic device consisted of a set of graduated rings (called armillaries) that represented important circles on the celestial sphere, such as the horizon, the celestial equator, the ecliptic, and the meridian. These rings formed a skeletal celestial sphere. A movable sighting arrangement allowed early astronomers to observe a celestial object and then read off its position (coordinates) using the markings on the relevant circles.

The *cross staff* was another tool widely used by astronomers and navigators before the invention of the telescope. The device consisted of a main staff with a perpendicular crosspiece attached at its middle to the staff and able to slide up and down along it. Reportedly invented by Rabbi Levi ben Gershon (1288–1344), a Jewish scholar who lived in southern France, the cross staff allowed medieval astronomers to measure the angle between the directions of two stars. Similar older instruments for this purpose existed and were used by early astronomers such as HIPPARCHOS and PTOLEMY, but none of these early instruments were as portable or flexible as Gershon's cross staff. This device also proved eminently suitable for navigation across the sea. For example, a ship's officer could use the cross staff to measure the elevation angle of the noontime Sun above the horizon, a measurement that allowed the officer to then estimate the ship's latitude.

Ptolemy and other Greek astronomers used the quadrant, a graduated quarter of a circle constructed to allow an observer to measure the altitude of celestial objects above the horizon. An astronomer would sight a target celestial object along one arm of the quadrant and then read off its elevation from a scale (from 0 to 90 degrees) with the help of a plumb line suspended from the center of the quarter circle. With this arrangement a celestial object just on the horizon would have an elevation of 0 degrees, while an object at zenith (directly overhead) would have an elevation of 90 degrees. The ancient Greeks adopted the Babylonian division of the circle into 360 degrees and as early as 300 B.C.E. were able to measure and describe the heavens with remarkable precision. From antiquity through the Renaissance, the quadrant served as one of the key instruments of pretelescope astronomy. For example, the most celebrated astronomical quadrant of the European Renaissance was TYCHO BRAHE's 2-meter-radius mural quadrant, which he constructed in 1582.

Unquestionably, the *astrolabe* is the most famous ancient astronomical instrument. Essentially an early mechanical computer, this versatile device helped astronomers and navigators solve problems concerning time and the position of the Sun and stars. The planispheric astrolabe was the most popular design. A *planisphere* is a two-dimensional map projection that is centered on the northern or southern pole of the celestial sphere. Generally, a planisphere illustrates the major stars of the constellations and has a movable overlay capable of showing these stars at a particular time for a particular terrestrial latitude. The planispheric astrolabe had the celestial sphere projected onto the equatorial plane. The instrument, usually made of brass, had a ring so it could easily hang suspended in a vertical plane. It also had a movable sighting device (called the alidade) that pivoted at the center of the disk. The typical astrolabe often came with several removable sky map plates to accommodate use at various terrestrial latitudes. Although dating back to antiquity, Arab astronomers refined the astrolabe and transformed it into a very versatile observing and computing instrument.

Another basic and commonly used ancient astronomical instrument was the *sundial*. In its simplest form, the sundial indicates local solar time by the position of the Sun's shadow as cast by the gnomon.

Finally, the *triquetrum* was a medieval instrument used to measure elevation angles. It consisted of a calibrated sighting device (called the alidade) supported between vertical and horizontal spars. Astronomers calibrated the horizontal spar in degrees.

For thousands of years astronomy, the oldest form of organized science, remained highly interwoven with ancient cultures and religions. Then, long before the arrival of the telescope, early instruments such as the armillary sphere, the quadrant, the sundial, the astrolabe, and the triquetrum provided stargazers with a consistent means of monitoring and recording the motions of familiar celestial objects. The early data sets developed with these instruments allowed astronomers to start modeling the universe and to begin inquiring about its composition and extent. Starting with Ptolemy and continuing through Renaissance Europe, these early instruments supported the emergence of a more sophisticated form of astronomy and, in so doing, helped set the stage for the arrival of the telescope and the scientific revolution in 17th-century Europe.

See also ARCHAEOLOGICAL ASTRONOMY; ASTROLABE; QUADRANT.

ancient constellations The collection of approximately 50 constellations drawn up by ancient astronomers and recorded by PTOLEMY, including such familiar constellations as the signs of the zodiac, Ursa Major (the Great Bear), Boötes (the Herdsman), and Orion (the Hunter).

See also ZODIAC.

Anders, William A. (1933–) American *Astronaut, U.S. Air Force Officer* William Anders was a member of NASA's *Apollo 8* mission (December 21–27, 1968), the first human flight to the vicinity of the Moon. Along with FRANK BORMAN and JAMES A. LOVELL, JR., Anders flew to the Moon in the Apollo spacecraft, completed 10 orbital revolutions of the Moon, and then returned safely to Earth.

androgynous interface A nonpolar interface; one that physically can join with another of the same design; literally, having both male and female characteristics.

Andromeda galaxy The Great Spiral Galaxy (or M31) in the constellation of Andromeda, about 2.2 million light-years (670 kiloparsecs) away. It is the most distant object visible to the naked eye and is the closest spiral galaxy to the Milky Way galaxy.

See also GALAXY; MILKY WAY GALAXY.

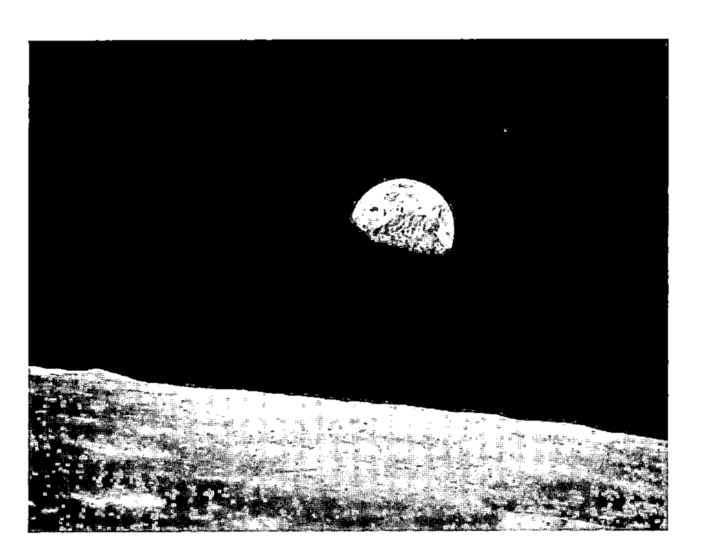

This inspirational view of the "rising" Earth greeted astronaut William Anders and his *Apollo 8* mission crewmates, astronauts Frank Borman and James A. Lovell, Jr., as they came from behind the Moon after the lunar orbit insertion burn (December 1968). *(Courtesy of NASA)*

anechoic chamber A test enclosure specially designed for experiments in acoustics. The interior walls of the chamber are covered with special materials (typically sound-absorbing, pyramid-shaped surfaces) that absorb sufficiently well the sound incident upon the walls, thereby creating an essentially "sound-free" condition in the frequency range(s) of interest.

angle The inclination of two intersecting lines to each other, measured by the arc of a circle intercepted between the two lines forming the angle. There are many types of angles. An *acute angle* is less than 90°; a *right angle* is precisely 90°; an *obtuse angle* is greater than 90° but less than 180°; and a *straight angle* is 180°.

angle of attack The angle (commonly used symbol: α) between a reference line fixed with respect to an airframe and a line in the direction of movement of the body.

See also AERODYNAMIC FORCE.

angle of incidence The angle at which a ray of light (or other type of electromagnetic radiation) impinges on a surface. This angle is usually measured between the direction of propagation and a perpendicular to the surface at the point of incidence and, therefore, has a value lying between 0° and 90°. The figure shows a ray of light impinging upon a plane mirror. In this case the angle of incidence equals the angle of reflection.

angle of reflection The angle at which a reflected ray of light (or other type of electromagnetic radiation) leaves a reflecting surface. This angle is usually measured between the direction of the outgoing ray and a perpendicular to the sur-

The European Space Agency's fully deployed ERS-1 spacecraft in a large anechoic chamber at the Interspace Test Facility in Toulouse, France (1990). ERS-1 synthetic aperture radar (SAR) images, together with the data from other onboard instruments, supported the worldwide scientific community from 1991 to 2000. ***(Courtesy of S. Vermeer–European Space Agency)***

face at the point of reflection. For a plane mirror the angle of reflection is equal to the angle of incidence.

angstrom (symbol: Å) A unit of length used to indicate the wavelength of electromagnetic radiation in the visible, near-infrared, and near-ultraviolet portions of the electromagnetic spectrum. Named after the Swedish physicist ANDERS JONAS ÅNGSTRÖM (1814–74), who quantitatively described the Sun's spectrum in 1868. One angstrom equals 0.1 nanometer (10^{-10} m).

Ångström, Anders Jonas (1814–1874) Swedish *Physicist, Astronomer* Anders Ångström was a Swedish physicist and solar astronomer who performed pioneering spectral studies of the Sun. In 1862 he discovered that hydrogen was present in the solar atmosphere and went on to publish a detailed map of the Sun's spectrum, covering other elements present as well. A special unit of wavelength, the angstrom (symbol Å), now honors his accomplishments in spectroscopy and astronomy.

Ångström was born in Lögdö, Sweden, on August 13, 1814. He studied physics and astronomy and graduated in 1839 with a doctorate from the University of Uppsala. This university was founded in 1477 and is the oldest of the Scandinavian universities. Upon graduation Ångström joined the university faculty as a lecturer in physics and astronomy. For more than three decades he remained at this institution, serving it in a variety of academic and research positions. For example, in 1843 he became an astronomical observer at the famous Uppsala Observatory, the observatory founded in 1741 by ANDERS CELSIUS (1701–44). In 1858 Ångström became chair of the physics department and remained a professor in that department for the remainder of his life.

He performed important research in heat transfer, spectroscopy, and solar astronomy. With respect to his contributions in heat transfer phenomena, Ångström developed a method to measure thermal conductivity by showing that it was proportional to electrical conductivity. He was also was one of the 19th-century pioneers of spectroscopy. Ångström observed that an electrical spark produces two superimposed spectra. One spectrum is associated with the metal of the electrode generating the spark, while the other spectrum originates from the gas through which the spark passes.

Ångström applied LEONHARD EULER's resonance theorem to his experimentally derived atomic spectra data and discovered an important principle of spectral analysis. In his paper "Optiska Undersökningar" (Optical investigations), which he presented to the Swedish Academy in 1853, Ångström reported that an incandescent (hot) gas emits light at precisely the same wavelength as it absorbs light when it is cooled. This finding represents Ångström's finest research work in spectroscopy, and his results anticipated the spectroscopic discoveries of GUSTAV KIRCHHOFF that led to the subsequent formulation of Kirchhoff's laws of radiation. Ångström was also able to demonstrate the composite nature of the visible spectra of various metal alloys.

His laboratory activities at the University of Uppsala also provided Ångström the "hands-on" experience in the

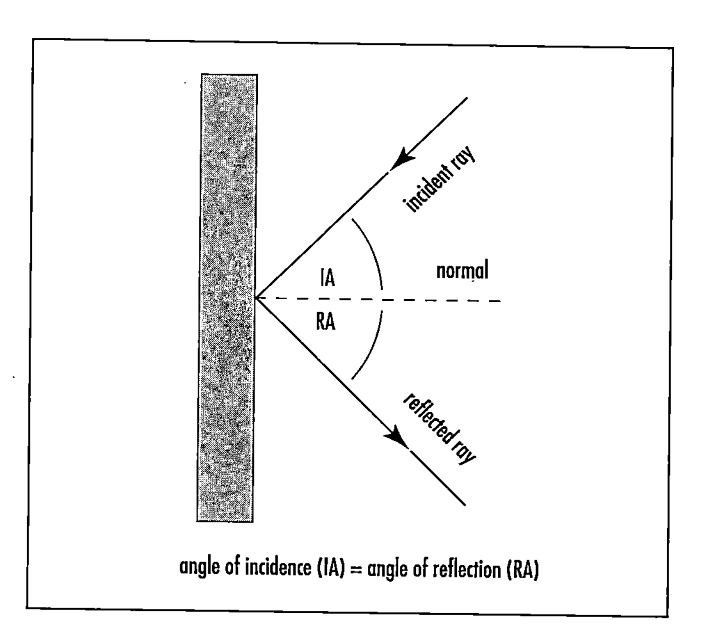

The relationship between the angle of incidence (IA) and the angle of reflection (RA) for a ray of light incident upon a plane mirror

emerging field of spectroscopy necessary to accomplish his pioneering observational work in solar astronomy. By 1862 Ångström's initial spectroscopic investigations of the solar spectrum enabled him to announce his discovery that the Sun's atmosphere contained hydrogen. In 1868 he published *Recherches sur le spectre solaire* (Researches on the solar spectrum), his famous atlas of the solar spectrum, containing his careful measurements of approximately 1,000 Fraunhofer lines. Unlike other pioneering spectroscopists such as ROBERT BUNSEN and Gustav Kirchhoff, who used an arbitrary measure, Ångström precisely measured the corresponding wavelengths in units equal to one-ten-millionth of a meter.

Ångström's map of the solar spectrum served as a standard of reference for astronomers for nearly two decades. In 1905 the international scientific community honored his contributions by naming the unit of wavelength he used the "angstrom"—where one angstrom (symbol Å) corresponds to a length of 10^{-10} meter. Physicists, spectroscopists, and microscopists use the angstrom when they discuss the visible light portion of the electromagnetic spectrum. The human eye is sensitive to electromagnetic radiation with wavelengths between 4×10^{-7} and 7×10^{-7} meter. These numbers are very small, so scientists often find it more convenient to use the angstrom in their technical discussions. For example, the range of human vision can now be expressed as ranging between 4,000 and 7,000 angstroms (Å).

In 1867 Ångström became the first scientist to examine the spectrum of the aurora borealis (northern lights). Because of this pioneering work, his name is sometimes associated with the aurora's characteristic bright yellow-green light. He was a member of the Royal Swedish Academy (Stockholm) and the Royal Academy of Sciences of Uppsala. In 1870 Ångström was elected a fellow of the Royal Society in London, from which society he received the prestigious Rumford Medal in 1872. He died in Uppsala on June 21, 1874.

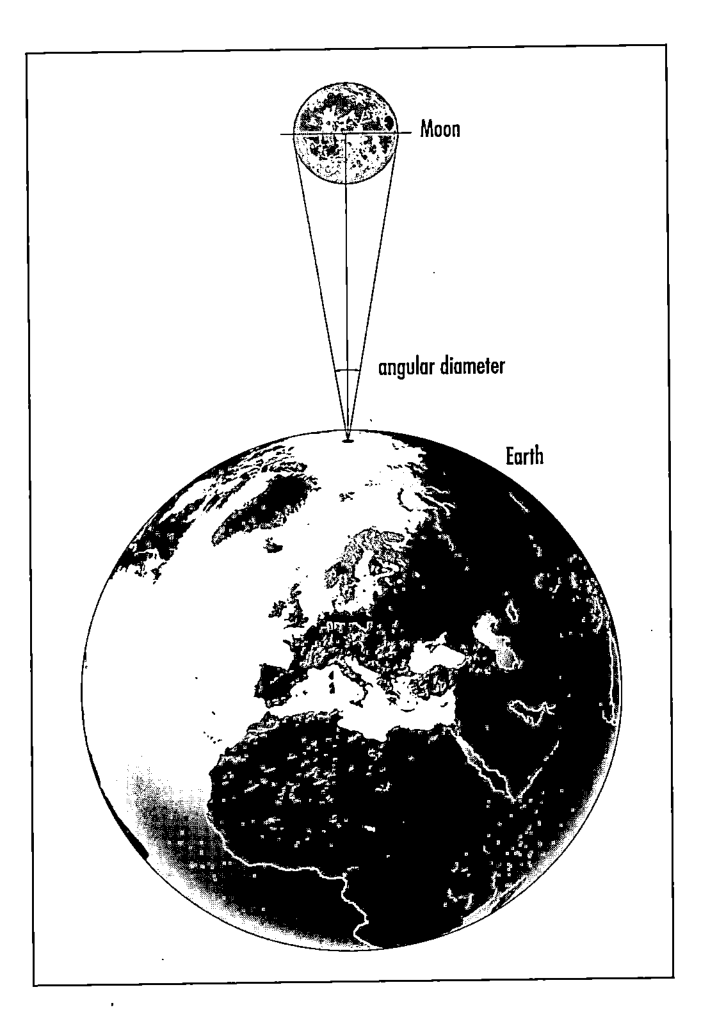

Angular diameter

angular acceleration (symbol: α) The time rate of change of angular velocity (ω).

angular diameter The angle formed by the lines projected from a common point to the opposite sides of a body. Astronomers use the angular diameter to express the apparent diameter of a celestial body in angular units (that is, degrees, minutes, and seconds of arc).

angular distance The apparent angular spacing between two objects in the sky, expressed in degrees, minutes, and seconds of arc.

angular frequency (symbol: ω) The frequency of a periodic quantity expressed as angular velocity in radians per second. It is equal to the frequency (in hertz, or cycles per second) times 2π radians per cycle.

angular measure Units of angle generally expressed in terms of degrees (°), arc minutes (′), and arc seconds (″), where 1° of angle equals 60′, and 1′ equals 60″. A full circle contains 360°, or 2π radians. In the case of the Keplerian element called the right ascension of the ascending node (Ω), angular measure is expressed in terms of hours, minutes, and seconds. This particular orbital element describes the rotation of the orbital plane from the reference axis and helps uniquely specify the position and path of a satellite (natural or human-made) in its orbit as a function of time.

See also KEPLERIAN ELEMENTS; RADIAN.

angular momentum (symbol: L) A measure of an object's tendency to continue rotating at a particular rate around a certain axis. It is defined as the product of the angular velocity (ω) of the object and its moment of inertia (I) about the axis of rotation, that is, $L = I\,\omega$.

angular resolution The angle between two points that can just be distinguished by a detector and a collimator. The normal human eye has an angular resolution of about 1 arc minute (1′). A collimator is a device (often cylindrical in form) that produces a certain size beam of particles or light.

angular separation *See* ANGULAR DISTANCE.

angular velocity (symbol: ω) The change of angle per unit of time; usually expressed in radians per second.

***Anik* spacecraft** In the Inuit language, *anik* means "little brother." This is the name given to an evolving family of highly successful Canadian communications satellites. In 1969 a combined effort of the Canadian government and various Canadian broadcast organizations created Telecast Canada. The role of Telesat Canada was to create a series of geostationary communications satellites that could carry radio, television, and telephone messages between the people in the remote northern portions of Canada and the more heavily populated southern portions. Such satellite-based telecommunication services were viewed as promoting a stronger feeling of fraternity and brotherhood throughout all regions of the country, and therefore the name Anik.

The first three Anik communications satellites—called *Anik A1, A2* and *A3*—were identical. Hughes Aircraft in California (now a part of the Boeing Company) constructed these satellites, although some of their components were made by various Canadian firms. *Anik A1* was launched into geostationary orbit on November 9, 1972 (local time), by an expendable Delta rocket from Cape Canaveral Air Force Station in Florida. Activation of the *Anik 1* satellite made Canada the first country in the world to have a domestic commercial geostationary satellite system. In 1982 *Anik A1* was retired and replaced by *Anik D1,* a more advanced member of this trail-blazing communications satellite family.

Similarly, a Delta rocket successfully launched the *Anik A2* satellite from Cape Canaveral on April 20, 1973 (local time). It also provided telecommunications services to domestic users throughout Canada until retirement in 1982. The third of the initial Anik spacecraft, *Anik A3,* was successfully launched by a Delta rocket from Cape Canaveral on May 7, 1975 (local time), and supported telecommunications services to Canadian markets until retirement in 1983.

Anik B1 was launched by a Delta rocket from Cape Canaveral on December 15, 1978 (local time). This spacecraft provided the world's first commercial service in the 14/12 gigahertz (GHz) frequency band and revolutionized satellite communications by demonstrating the feasibility of the direct broadcast satellite concept. The successful operation of *Anik B1* showed that future communications satellites could serve very small Earth stations that could easily be erected almost anywhere. Telecommunication experiments with *Anik B1* also revealed that less power than anticipated would be required to provide direct broadcast service to remote areas. Telesat Canada decommissioned *Anik B1* in 1986.

On July 17, 2004 (local time), an Ariane 5 rocket launched the *Anik F2* satellite from the Kourou Launch Complex in French Guiana. The Anik F series was introduced in 1998 and is the first generation of Anik satellites to use the Boeing 702 spacecraft, a large spacecraft (7.3 meters by 3.8 meters by 3.4 meters) that uses a xenon ion propulsion system for orbit maintenance. *Anik F2* operates at the geostationary orbital slot of 111.1 degrees west longitude and provides broadband Internet, distance learning, and telemedicine to rural areas of the United States and Canada.

See also CANADIAN SPACE AGENCY; COMMUNICATIONS SATELLITE; DIRECT BROADCAST SATELLITE.

anisotropic Exhibiting different properties along axes in different directions; an anisotropic radiator would, for example, emit different amounts of radiation in different directions as compared to an isotropic radiator, which would emit radiation uniformly in all directions.

annihilation radiation Upon collision, the conversion of an atomic particle and its corresponding antiparticle into pure electromagnetic energy. For example, when an electron (e^-) and positron (e^+) collide, the minimum annihilation radiation released consists of a pair of gamma rays, each of approximately 0.511 million electron volts (MeV) energy.

annual parallax (symbol: π) The parallax of a star that results from the change in the position of a reference observing point during Earth's annual revolution around the Sun. It is the maximum angular displacement of the star that occurs when the star-Sun-Earth angle is 90° (as illustrated in the figure). Also called the *heliocentric parallax.*

anomalistic period The time interval between two successive perigee passages of a satellite in orbit about its primary body. For example, the term *anomalistic month* defines the

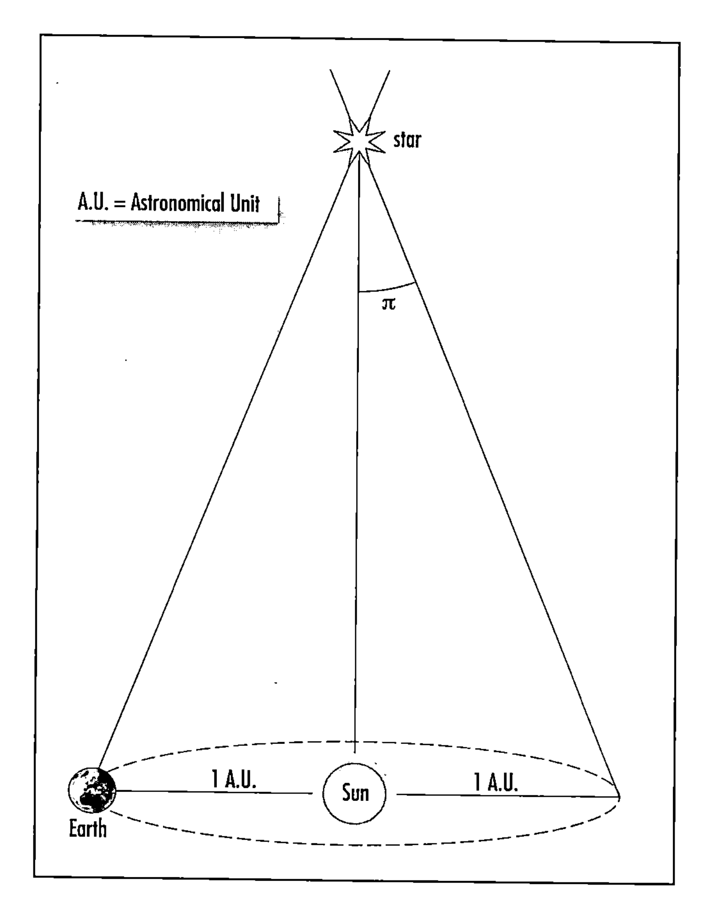

Annual parallax

mean time interval between successive passages of the Moon through its closest point to Earth (perigee), about 27.555 days.

anomalistic year *See* YEAR.

anomaly 1. (aerospace operations) A deviation from the normal or anticipated result. 2. (astronomy) The angle used to define the position (at a particular time) of a celestial object, such as a planet or artificial satellite in an elliptical orbit about its primary body. The *true anomaly* of a planet is the angle (in the direction of the planet's motion) between the point of closest approach (i.e., the perihelion), the focus (the Sun), and the planet's current orbital position.

Antarctic meteorites Meteorites from the Antarctic continent are a relatively recent resource for study of the material formed early in the solar system. Scientists think most of these meteorites come from asteroids, but some may have originated on larger planets. In 1969 Japanese scientists first discovered concentrations of meteorites in Antarctica. Most of these meteorites have fallen onto the ice sheet in the last million years. They seem to be concentrated in places where the flowing ice, acting as a conveyor belt, runs into an obstacle and is then worn away, leaving behind the meteorites.

A NASA geochemist puts his gear aboard a snowmobile for an Antarctic field trip in search of meteorites ca. January 1980. *(Courtesy of NASA)*

Compared with meteorites collected in more temperate regions on Earth, the Antarctic meteorites are relatively well preserved. This large collection of meteorites provides scientists a better understanding of the abundance of meteorite types in the solar system and how meteorites relate to asteroids and comets. Among the Antarctic meteorites are pieces blasted off the Moon, and probably Mars, by impacts. Because meteorites in space absorb and record cosmic radiation, the time elapsed since the meteorite impacted Earth can be determined from laboratory studies. The elapsed time since fall, or terrestrial residence age, of a meteorite represents additional information that might be useful in environmental studies of Antarctic ice sheets.

For American scientists the collection and curation of Antarctic meteorites is a cooperative effort among NASA, the National Science Foundation (NSF), and the Smithsonian Institution. NSF science teams camping on the ice collect the meteorites. (*See* figure.) Since 1977 the still-frozen meteorites have been returned to the NASA Johnson Space Center (JSC) in Houston, Texas, for curation and distribution. Some of the specimens are forwarded to the Smithsonian Institution, but JSC scientists curate more than 4,000 meteorites for more than 250 scientists worldwide and eagerly await the arrival of the several hundred new meteorites each year.

Antares The alpha star in the constellation Scorpius, this huge red supergiant is about 520 light-years away from Earth. Because of its reddish color, the early Greek astronomers named it Antares, meaning opposite or rival of Mars. Modern astronomers estimate that the colossal star has a diameter about 400 times that of the Sun and is some 10,000 times more luminous. Antares is the dominant member of a binary star system. Its much smaller and hotter dwarf companion orbits the supergiant with a period of approximately 900 years.

antenna A device used to detect, collect, or transmit radio waves. A radio telescope is a large receiving antenna, while many spacecraft have both a directional antenna and an omnidirectional antenna to transmit (downlink) telemetry and to receive (uplink) instructions.

See also TELEMETRY.

antenna array A group of antennas coupled together into a system to obtain directional effects or to increase sensitivity.

See also VERY LARGE ARRAY.

anthropic principle An interesting, though highly controversial, hypothesis in modern cosmology that suggests that the universe evolved after the big bang in just the right way so that life, especially intelligent life, could develop. The proponents of this hypothesis contend that the fundamental physical constants of the universe actually support the existence of life and (eventually) the emergence of conscious intelligence—including, of course, human beings. The advocates of this hypothesis further suggest that with just a slight change in the value of any of these fundamental physical constants, the universe would have evolved very differently after the big bang.

For example, if the force of gravitation were weaker than it is, expansion of matter after the big bang would have been much more rapid, and the development of stars, planets, and galaxies from extremely sparse (nonaccreting) nebular materials could not have occurred: no stars, no planets, no development of life (as we know it)! If, on the other hand, the force of gravitation were stronger than it is, the expansion of primordial material would have been sluggish and retarded, encouraging a total gravitational collapse (i.e., the big crunch) long before the development of stars and planets: again, no stars, no planets, no life!

Opponents of this hypothesis suggest that the values of the fundamental physical constants are just a coincidence. In any event, the anthropic principle represents a lively subject for philosophical debate and speculation. Until, however, the hypothesis can actually be tested, it must remain out of the mainstream of demonstrable science.

See also BIG BANG; BIG CRUNCH; COSMOLOGY; EXTRATERRESTRIAL LIFE CHAUVINISMS; LIFE IN THE UNIVERSE; PRINCIPLE OF MEDIOCRITY; UNIVERSE.

antiballistic missile system (ABM) A missile system designed to intercept and destroy a strategic offensive ballistic missile or its deployed reentry vehicles.

See also ANTIMISSILE MISSILE; BALLISTIC MISSILE DEFENSE.

antimatter Matter in which the ordinary nuclear particles (such as electrons, protons, and neutrons) have been replaced by their corresponding antiparticles (that is, positrons, antiprotons, antineutrons, etc.) For example, an antimatter hydrogen atom (or *antihydrogen*) would consist of a negatively charged antiproton as the nucleus surrounded by a positively charged orbiting positron.

Normal matter and antimatter mutually annihilate each other upon contact and are converted into pure energy, called annihilation radiation. Although extremely small quantities of antimatter (primarily antihydrogen) have been produced in laboratories, significant quantities of antimatter have yet to be produced or even observed elsewhere in the universe.

An interesting variety of matter-antimatter propulsion schemes (sometimes referred to as *photon rockets*) has been suggested for interstellar travel. Should we be able to manufacture and contain significant quantities (e.g., milligrams to kilograms) of antimatter (especially antihydrogen) in the 21st century, such photon rocket concepts might represent a possible way of propelling robot probes to neighboring star systems. One challenging technical problem for future aerospace engineers would be the safe storage of antimatter "propellant" in a normal matter spacecraft. Another engineering challenge would be properly shielding the interstellar probe's payload from exposure to the penetrating, harmful annihilation radiation released when normal matter and antimatter collide. Also called *mirror matter.*

See also ANNIHILATION RADIATION; ANTIPARTICLE; STARSHIP.

antimatter cosmology A cosmological model proposed by the Swedish scientists HANNES ALFVÉN and Oskar Benjamin Klein (1894–1977) as an alternate to the big bang model. In their model the scientists assumed that early universe consisted of a huge, spherical cloud, called a metagalaxy, containing equal amounts of matter and antimatter. As this cloud collapsed under the influence of gravity, its density increased, and a condition was reached in which matter and antimatter collided, producing large quantities of annihilation radiation. The radiation pressure from the annihilation process caused the universe to stop collapsing and to expand. In time, clouds of matter and antimatter formed into equivalent numbers of galaxies and antigalaxies. An *antigalaxy* is a galaxy composed of antimatter.

There are many technical difficulties with the Alfvén-Klein cosmological model. For example, no observational evidence has been found to date indicating the presence of large quantities of antimatter in the universe. If the postulated antigalaxies actually existed, large quantities of annihilation radiation (gamma rays) would certainly be emitted at the interface points between the matter and antimatter regions of the universe.

See also ANNIHILATION RADIATION; ANTIMATTER; ASTROPHYSICS; BIG BANG (THEORY); COSMOLOGY.

antimissile missile (AMM) A MISSILE launched against a hostile missile in flight.

See also BALLISTIC MISSILE DEFENSE; SPRINT.

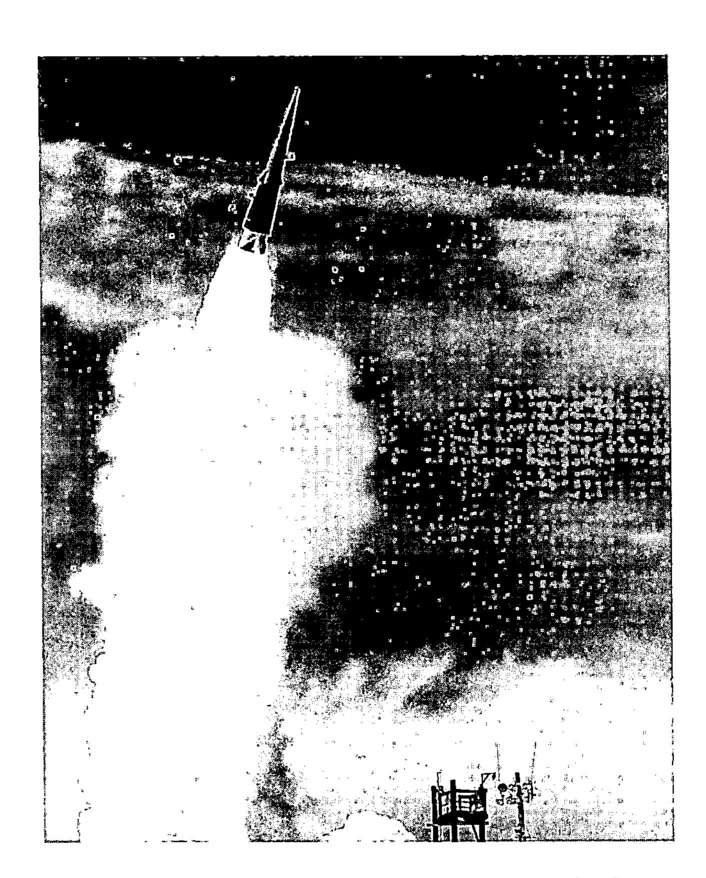

An early U.S. antimissile missile called the Sprint being tested at the White Sands Missile Range in 1967 ***(Courtesy of the U.S. Army/White Sands Missile Range)***

antiparticle Every elementary particle has a corresponding (or hypothetical) antiparticle, which has equal mass but opposite electrical charge (or other property, as in the case of the neutron and antineutron). The antiparticle of the electron is the positron; of the proton, the antiproton; and so on. However, the photon is its own antiparticle. When a particle and its corresponding antiparticle collide, their masses are converted into energy in a process called *annihilation.* For example, when an electron and positron collide, the two particles disappear and are converted into annihilation radiation consisting of two gamma ray photons, each with a characteristic energy level of 0.511 million electron volts (MeV).

See also ANNIHILATION RADIATION; ANTIMATTER; ELEMENTARY PARTICLE(S); GAMMA RAYS.

antiradiation missile (ARM) A missile that homes passively on a radiation source, such as an enemy radar signal.

antisatellite spacecraft (ASAT) A spacecraft designed to destroy other satellites in space. An ASAT spacecraft could be deployed in space disguised as a peaceful satellite that quietly lurks as a secret hunter/killer satellite, awaiting instructions to track and attack its prey. In-orbit testing of ASAT systems could litter near-Earth space with large quantities of space debris.

antislosh baffle A device provided in a propellant tank of a liquid fuel rocket to dampen unwanted liquid motion, or sloshing, during flight. This device can take many forms, including a flat ring, truncated cone, and vane.

See also ROCKET.

antisubmarine rocket (ASROC) A surface ship–launched, rocket-propelled, nuclear depth charge or homing torpedo.

This artist's rendering depicts the on-orbit testing of a Soviet antisatellite weapon system. In the 1980s the Soviet Union tested and operated an orbital antisatellite weapon, a killer spacecraft designed to destroy space targets by orbiting near them and then releasing a multipellet blast. The cloud of high-speed pellets would destroy the target spacecraft. *(Courtesy of the U.S. Department of Defense/Defense Intelligence Agency; artist Ronald C. Wittmann, 1986)*

antivortex baffles Assemblies installed in the propellant tanks of liquid fuel rockets to prevent gas from entering the rocket engine(s). The baffles minimize the rotating action as the propellants flow out the bottom of the tanks. Without these baffles, the rotating propellants would create a vortex similar to a whirlpool in a bathtub drain.

See also ROCKET.

apareon That point on a Mars-centered orbit at which a satellite or spacecraft is at its greatest distance from the planet. The term is analogous to the term APOGEE.

apastron 1. The point in a body's orbit around a star at which the object is at a maximum distance from the star. 2. Also, that point in the orbit of one member of a binary star system at which the stars are farthest apart. *Compare with* PERIASTRON.

aperture 1. The opening in front of a telescope, camera, or other optical instrument through which light passes. 2. The diameter of the objective of a telescope or other optical instrument. 3. Concerning a unidirectional antenna, that portion of a plane surface near the antenna, perpendicular to the direction of maximum radiation, through which the major portion of the electromagnetic radiation passes.

aperture synthesis A resolution-improving technique in radio astronomy that uses a variable aperture radio interferometer to mimic the full dish size of a huge radio telescope.

See also RADAR ASTRONOMY; RADAR IMAGING; RADIO ASTRONOMY.

apex The direction in the sky (celestial sphere) toward which the Sun and its system of planets appear to be moving relative to the local stars. Also called the *solar apex,* it is located in the constellation of Hercules.

aphelion The point in an object's orbit around the Sun that is most distant from the Sun. *Compare with* PERIHELION.

Aphrodite Terra A large, fractured highland region near the equator of Venus.

See also BETA REGIO; ISHTAR TERRA; VENUS.

apoapsis That point in an orbit farthest from the orbited central body or center of attraction.

See also ORBITS OF OBJECTS IN SPACE.

apocynthion That point in the lunar orbit of a spacecraft or satellite launched from Earth that is farthest from the Moon. *Compare with* APOLUNE.

apogee 1. In general, the point at which a missile trajectory or a satellite orbit is farthest from the center of the gravitational field of the controlling body or bodies. 2. Specifically, the point that is farthest from Earth in a geocentric orbit. The term is applied to both the orbit of the Moon around Earth as well as to the orbits of artificial satellites around Earth. At apogee, the velocity of the orbiting object is at a minimum.

To enlarge or "circularize" the orbit, a spacecraft's thruster (i.e., the apogee motor) is fired at apogee, giving the space vehicle and its payload necessary increase in velocity. Opposite of PERIGEE. 3. The highest altitude reached by a sounding rocket launched from Earth's surface.

See also ORBITS OF OBJECTS IN SPACE.

apogee motor A solid propellant rocket motor that is attached to a spacecraft and fired when the deployed spacecraft is at the apogee of an initial (relatively low-altitude) parking orbit around Earth. This firing establishes a new orbit farther from Earth or permits the spacecraft to achieve escape velocity. Also called *apogee kick motor* and *apogee rocket.*

Apollo group A collection of near-Earth asteroids that have perihelion distances of 1.017 astronomical units or less, taking them across the orbit of Earth around the Sun. This group acquired its name from the first asteroid to be discovered, Apollo, in 1932 by the German astronomer KARL REINMUTH.

See also ASTEROID; NEAR-EARTH ASTEROID.

Apollo Lunar Surface Experiments Package (ALSEP) Scientific devices and equipment placed on the Moon by the Apollo Project astronauts and left there to transmit data back to Earth. Experiments included the study of meteorite impacts, lunar surface characteristics, seismic activity on the Moon, solar wind interaction, and analysis of the very tenuous lunar atmosphere.

See also APOLLO PROJECT.

Apollonius of Perga (ca. 262–ca. 190 B.C.E.) Greek *Mathematician* Apollonius of Perga developed the theory of conic sections (including the circle, ellipse, parabola, and hyperbola) that allowed JOHANNES KEPLER and SIR ISAAC NEWTON to accurately describe the motion of celestial bodies in the solar system. The ancient Greek mathematician also created the key mathematical treatment that enabled HIPPARCHUS OF NICAEA and (later) PTOLEMY to promote a geocentric epicycle theory of planetary motion.

See also CONIC SECTIONS; EPICYCLE.

Apollo Project On May 25, 1961, President JOHN F. KENNEDY proposed that the United States establish as a national goal a human lunar landing and safe return to Earth by the end of the decade. In response to this presidential initiative, the National Aeronautics and Space Administration (NASA) instituted the Apollo Project (which was preceded by the Mercury and Gemini Projects). A giant new rocket, the Saturn V, would be needed to send astronauts and their equipment safely to the lunar surface. The Saturn V first stage used a cluster of five F-1 engines that generated some 33,360,000 newtons of thrust. The second stage used a cluster of five J-2 engines that developed a combined thrust of 4.4 million newtons. The third stage of this colossal "moon rocket" was powered by a single J-2 engine with 889,600 newtons of thrust capability.

From October 1961 through April 1968, a variety of unmanned flight tests of the Saturn 1B and V launch vehicles and the Apollo spacecraft/lunar module were conducted successfully. These unmanned tests included the *Apollo 4, 5,* and *6* missions (November 1967 to April 1968)—missions that involved Earth orbital trajectories.

On January 27, 1967, tragedy struck the U.S. space program when a fire erupted inside an Apollo spacecraft during ground testing at Complex 34, Cape Canaveral Air Force Station, Florida. This flash fire resulted in the deaths of astronauts VIRGIL (GUS) I. GRISSOM, EDWARD H. WHITE II, and ROGER B. CHAFFEE. As a result of this accident, major modifications had to be made to the Apollo spacecraft prior to its first crewed mission in space.

Liftoff of the first crewed Apollo launch, *Apollo 7,* took place on October 11, 1968. A Saturn 1B vehicle left Complex 34 carrying astronauts WALTER SCHIRRA, DONN EISELE, and R. WALTER CUNNINGHAM on an 11-day mission that tested the Apollo spacecraft's performance in Earth orbit. All subsequent Apollo missions were flown from Complex 39 at the NASA Kennedy Space Center, which is adjacent to Cape Canaveral.

Apollo 8, with astronauts FRANK BORMAN, WILLIAM A. ANDERS, and JAMES A. LOVELL, JR., lifted off on December 21, 1968. This flight involved the first voyage by human beings from Earth to the vicinity of another planetary body. In 147 hours *Apollo 8* took its crew on a faultless 800,000-kilometer space journey—a historic flight that included 10 lunar orbits. Exciting lunar and Earth photography as well as live television broadcasts from the Apollo capsule highlighted this pioneering mission.

Apollo 9 took place in March 1969 and was the first *all-up* flight of the Apollo/Saturn V space vehicle configuration—that is, all lunar mission equipment was flown together for the first time, including a flight article lunar module instead of the lunar module test article used in the *Apollo 8* mission. While in Earth orbit the *Apollo 9* astronauts demonstrated key docking and rendezvous maneuvers between the Apollo spacecraft and the lunar module.

The *Apollo 10* mission in May 1969 successfully completed the second human flight that orbited the Moon. In a dress rehearsal for the actual lunar landing mission (which occurred in *Apollo 11*), the *Apollo 10* astronauts came within 14.5 kilometers of the lunar surface and spent nearly 62 hours (31 revolutions) in lunar orbit.

The *Apollo 11* mission achieved the national goal set by President Kennedy in 1961—namely landing human beings on the surface of the Moon and returning them safely to Earth within the decade of the 1960s. On July 20, 1969, astronauts NEIL A. ARMSTRONG and EDWIN E. "BUZZ" ALDRIN flew the lunar module to the surface of the Moon, touching down safely in the Sea of Tranquility. While Armstrong and then Aldrin became the first two persons to walk on another world, astronaut MICHAEL COLLINS, the command module pilot, orbited above.

Four months later, in November 1969, the crew of *Apollo 12* repeated the journey to the lunar surface, this time landing at and exploring a site in the Ocean of Storms. Their lunar surface activities included a visit to the landing site of the *Surveyor 3* spacecraft.

Apollo 13 was launched on April 11, 1970. Its crew was scheduled to land on the Fra Mauro upland area of the Moon.

Moon Bases and the Third Millennium

We came, we saw . . . we left. Unfortunately, a little more than three decades after the magnificent Apollo Project, the incredible sights and sounds associated with placing human footsteps on the lunar surface have begun to fade in our collective memory. The beginning of the third millennium provides a new opportunity to make the Moon not only a home away from home, but also humankind's portal to the solar system and beyond.

To appreciate the true significance of a return to the Moon and the establishment of a permanent lunar base (perhaps as early as 2020), we need to exercise a strategic vision of the future that is unconstrained by the short-term (five years or less) thinking. Short-term thinking often encumbers corporate and government decision makers, who function in bureaucracies configured to preserve the status quo. Embracing a millennial timeframe, this essay boldly examines how recently proposed lunar base developments could alter the course of human history. As we look across the next thousand years of human development, one important technology trend becomes glaringly apparent—long before our descendants celebrate the arrival of the fourth millennium (the year 3,001, to be precise), space technology will have become the major engine of economic growth, not only for planet Earth but for our solar system civilization.

Reviewing the visionary aerospace literature of the 20th century, we encounter many exciting scenarios for human development beyond the boundaries of Earth. Many of these proposals build upon the basic "open world" philosophy that outer space and its resources are an integral part of all future terrestrial economic growth and social development.

The modern liquid propellant chemical rocket is the dream machine that enabled space travel and the emergence of an open world philosophy. People and their robotic exploring machines can now reach places inaccessible to all previous human generations. Taken to the physical extremes of the solar system, this open world philosophy stimulates a strategic vision containing future space-related activities such as the creation of a diverse family of city-state planetoids throughout heliocentric space and the successful terraforming of Mars and Venus. However, such "blue sky" ideas should not come as too much of a surprise. The concept of custom-made worlds abounds in the space literature and even extends far back into the 19th century. The term *space literature* is used here because science fiction has an uncanny way of becoming science fact. The visionary American rocket scientist Robert H. Goddard stated: "It is difficult to say what is impossible, for the dream of yesterday is the hope of today and the reality of tomorrow."

At the beginning of the 20th century another space travel visionary, the Russian schoolteacher Konstantin Tsiolkovsky, boldly predicted the construction of miniature worlds in space, complete with closed life support systems. In the mid-1970s the German-American rocket engineer Krafft Ehricke introduced the Androcell concept, a collection of large, human-made miniworlds sprinkled throughout the solar system. Through creativity and engineering, each Androcell would represent a unique space-based habitat that used extraterrestrial resources far more efficiently than the naturally formed planets. By the year 3001, living in various human-made worlds that offer multiple-gravity levels (ranging from almost 0 g to about 1 g) could become a pleasant choice for a large portion of the human race.

However fantastic such future space developments may now appear, their technical and economic heritage will stem from one common, easily understood and envisioned technical ancestor—the permanent Moon base. The term *Moon base* (or *lunar base*) as used here includes an entire spectrum of concepts ranging from an early habitable "miner's shack" cobbled together with space station–like modules to a deluxe, multifunctional, self-sufficient, well-populated settlement of more than 10,000 people who live on (or just below) the lunar surface.

A variety of interesting lunar surface base (LSB) studies were performed prior to, during, and just after NASA's politically driven Apollo Project (1967–72). The earlier of these Moon base studies tended to be a bit more optimistic in their assumptions concerning extraterrestrial resources. For example, many speculated that water might be available in underground reservoirs of lunar ice. However, detailed studies of the Apollo samples soon revealed the anhydrous nature of the returned lunar rocks. As a result, post-Apollo lunar base scenarios logically focused on the much more expensive process of hauling hydrogen and/or water from Earth.

Then, at the end of the 20th century, the lunar resource equation shifted dramatically once again—this time in favor of permanent habitation. On January 6, 1998, NASA launched a modest, low-budget robot spacecraft called the *Lunar Prospector.* Among other duties, its instruments were to investigate hints of lunar ice deposits in the Moon's permanently shadowed polar regions. By March 1998 the *Lunar Prospector* mission team announced that the spacecraft had discovered the presence of somewhere between 10 and 300 million tons of water ice scattered inside the craters of the lunar poles, both northern and southern. While these estimates are very exciting, they also remain controversial within the scientific community. So the first order of business for any new robotic spacecraft mission to the Moon should be a careful global assessment of lunar resources, especially a confirmation of the presence of water ice.

In January 2004 NASA received a presidential mandate to initiate a series of robotic spacecraft missions to the Moon by not later than 2008 in order to prepare for and support future human exploration activities there. The confirmed presence of large quantities of water ice would make the Moon the most important piece of extraterrestrial real estate in the entire solar system.

The Apollo Project—that daring cold war era space technology demonstration feat proposed by President John F. Kennedy in May 1961—was unquestionably a tremendous technical achievement, perhaps the greatest in all human history. However, this project was not undertaken in response to a long-term national goal to support the permanent human occupancy of space. Rather, as its chief technical architect, the famous German-American rocket scientist Wernher von Braun, was fond of noting this daring feat was primarily "a large rocket building contest" with the former Soviet Union. Von Braun felt confident (and was proven correct by history) that his team of aerospace engineers could beat the Soviets in building enormously large, Moon-mission class expendable rockets. Any serious consideration of permanent Moon bases was simply tossed aside during the politically charged, schedule-driven superpower race to the Moon of the 1960s.

Consequently, the Apollo Project was never planned as part of a long-term American commitment to the permanent occupancy of cislunar space. The expensive, expendable Apollo era space hardware won the "race to the Moon," but 30 years later little direct space technology benefit remains from that schedule-driven design approach. For example, the monstrous Saturn V launch vehicle is as extinct as the dinosaurs.

Today, most federal bureaucracies (including NASA) reflect the impatience of a fast-paced American society that demands and thrives on instant gratification. Nobody wants to wait for the "water to boil." So, in the absence of the compelling geopolitical threats that existed during the cold war, how can we expect the majority of American political leaders and their constituents to get excited about expensive space projects that will have a significant impact only after decades, if not centuries? In a somewhat humorous historic analogy, one must doubt whether even a founding-father president, such as Thomas Jefferson, would be able to "sell" an extraterrestrial version of his 1803 Louisiana Purchase to members of the present day U.S. Congress or the people whom they represent. Daily job stress, the fear of terrorism, the obsession with consumptive materialism, and concerns about health care and unemployment cause transitory mental distractions that prevent millions of people from discovering and appreciating the "long-term" role that space technology (including permanent lunar bases) is now playing and will continue to play in the future of the human race.

Despite such day-to-day distractions, there is one important fact that we should never forget: Space technology now allows us to expand into the universe. This off-planet expansion process creates an extraterrestrial manifest destiny for the entire human family. We will venture into space not as Americans, Russians, Italians, or Japanese, but as "terrans"—"star sailors" from Earth who set forth upon the cosmic ocean in search of our primordial roots. The permanent Moon base catalyzes that important vision by creating a human presence in the cosmos that lies a significant distance beyond the boundaries of Earth.

So when we return to the Moon later this century, it will not be for a brief moment of scientific inquiry, as happened during Apollo. Rather, we will go as permanent inhabitants of a new world. We will build surface bases from which to complete lunar exploration, establish science and technology centers that take advantage of the special properties of the lunar environment, and begin to harvest in situ resources (particularly mineral and suspected water ice deposits) in support of the creation of humankind's solar system civilization.

In the first stage of an overall lunar development scenario, men and women, along with their smart robotic companions, will return to the Moon (possibly between 2015 and 2020) to conduct more extensive site explorations and resource evaluations. These efforts, supported by precursor robotic missions that will begin in about 2008, will pave the way for the first permanent lunar base. Many interesting lunar base applications, both scientific and commercial, have been suggested since the Apollo Project. Some of these concepts include:

- a lunar scientific laboratory complex,
- a lunar commercial and industrial complex in support of space-based manufacturing,
- an astrophysical observatory for solar system and deep space surveillance,
- a training site and assembly point for human expeditions to Mars,
- a nuclear waste repository for spent space nuclear power plants,
- a rapid response complex for a planetary defense system that protects Earth from menacing short-warning comets and rogue asteroids,
- an exobiology laboratory and extraterrestrial sample quarantine facility,
- a studio complex for extraterrestrial entertainment industries that use virtual reality and telepresence to entertain millions of people on Earth, and
- the site of innovative political, social, and cultural developments that help rejuvenate humankind's sense of purpose in the cosmic scheme of things.

As surface activities expand and mature, the initial lunar base will grow into a permanent settlement of about 1,000 inhabitants. During this period lunar mining operations (including water ice harvesting) should expand to satisfy growing demands for "moon products" throughout cislunar space. With the rise of highly automated lunar agriculture, the Moon could also become our "extraterrestrial breadbasket," providing the majority of all food products consumed by the space-faring portion of the human race. In-flight meals for Martian travelers could easily be "prepared with pride on the Moon." As our solar system civilization grows, the difference in surface gravity between the Moon and the Earth provides a major economic advantage for lunar-made products delivered to destinations and customers throughout cislunar space and beyond. Long-term space logistics studies should include the role and impact of a fully functional, highly automated lunar spaceport. By the mid-21st century outsourcing the assembly and delivery of replacement spacecraft to lunar based companies may be the most economical approach to maintaining the armada of space systems that support a completely wired global civilization here on Earth.

By skillfully using in-situ lunar resources, the early permanent lunar settlement will grow, expand, and then replicate itself in a manner similar to the way the original European settlements in the New World took hold and then expanded during the 16th and 17th centuries. At some point in the late 21st century, the lunar population should reach about 500,000 people, a number that some social scientists suggest is the critical mass for achieving political diversity and economic self-sufficiency. This event would correspond to a very special milestone in human history. Why? From that moment forward the human race would exist in two distinct, separate, and independent planetary niches. We would be "terran" and "nonterran" (or extraterrestrial).

Throughout human history frontiers have generally provided the physical conditions and inspiration needed to stimulate technical, social, and cultural progress. The knowledge that human beings permanently dwell upon a clearly "visible" nearby world may also trigger a major renaissance back on Earth. This process would be a home planet–wide awakening of the human spirit that leads to the creation of new wealth, the search for new knowledge, and the pur-

(continues)

Moon Bases and the Third Millennium *(continued)*

suit of beauty in many new forms of artistic expression. These tangible and intangible benefits will gradually accrue as a natural part of the creation and evolutionary development of the Moon base.

Our present civilization—as the first to permanently venture into cislunar space and to construct lunar bases—will long be admired, not only for its great technical accomplishments, but also for its innovative intellectual and cultural achievements. It is not too great a speculation to further suggest here that the descendants of these lunar settlers will go on to become first the interplanetary and then the interstellar portions of the human race. The true significance of the permanent lunar surface base is that it makes the Moon our gateway to the universe.

The late Krafft Ehricke most eloquently described this connection in 1984 at an international lunar base conference in Washington, D.C. He proclaimed to a standing-room-only audience of engineers, scientists, and technical writers that "The Creator of our universe wanted human beings to become space travelers. We were given a Moon that was just far enough away to require the development of sophisticated space technologies, yet close enough to allow us to be successful on our first concentrated attempt."

However, a rupture of the service module oxygen tank on April 13 caused a life-threatening power failure of the command and service module electrical system. Using their lunar module (named *Aquarius*) as a "lifeboat," astronauts JAMES A. LOVELL, JR., JOHN L. SWIGERT, JR., and FRED W. HAISE, JR., skillfully maneuvered their disabled spacecraft around the Moon and came back to Earth on a "free-return" trajectory. Just prior to reentry into Earth's atmosphere, they abandoned the severely damaged service module and the "lifesaving" lunar module and rode the Apollo command module to a safe splashdown and recovery in the Pacific Ocean on April 17, 1970.

Apollo 14 was retargeted to accomplish the mission planned for *Apollo 13*. The *Apollo 14* spacecraft was launched on January 31, 1971, and on February 5 the lunar module successfully touched down on the Fra Mauro formation, just 18.3 meters from the targeted point. The astronauts carried out numerous experiments and conducted successful exploration and specimen collections in the lunar highlands. The return flight to Earth was normal, and the *Apollo 14* command module splashed down in the South Pacific on February 9, 1971.

The fourth successful lunar landing mission, *Apollo 15*, was launched on July 26, 1971. The astronauts landed at the Hadley Apennine site. During their extended (approximately 67-hour) stay on the lunar surface, they used the electric-powered lunar rover vehicle for the first time. Approximately 77 kilograms of surface samples were collected, and extensive scientific experiments were performed during this mission. Splashdown in the Pacific Ocean took place on August 7, 1971.

Apollo 16, the fifth lunar landing mission, was launched on April 16, 1972. The astronauts explored the Descartes highlands and returned approximately 93.5 kilograms of

Apollo 11 **astronaut Edwin E. "Buzz" Aldrin descends the ladder of the lunar module and becomes the second human to walk on the Moon on July 20, 1969.** ***(Courtesy of NASA)***

Apollo 12 **astronauts Charles Conrad, Jr., and Alan L. Bean on the lunar surface in November 1969** ***(Courtesy of NASA)***

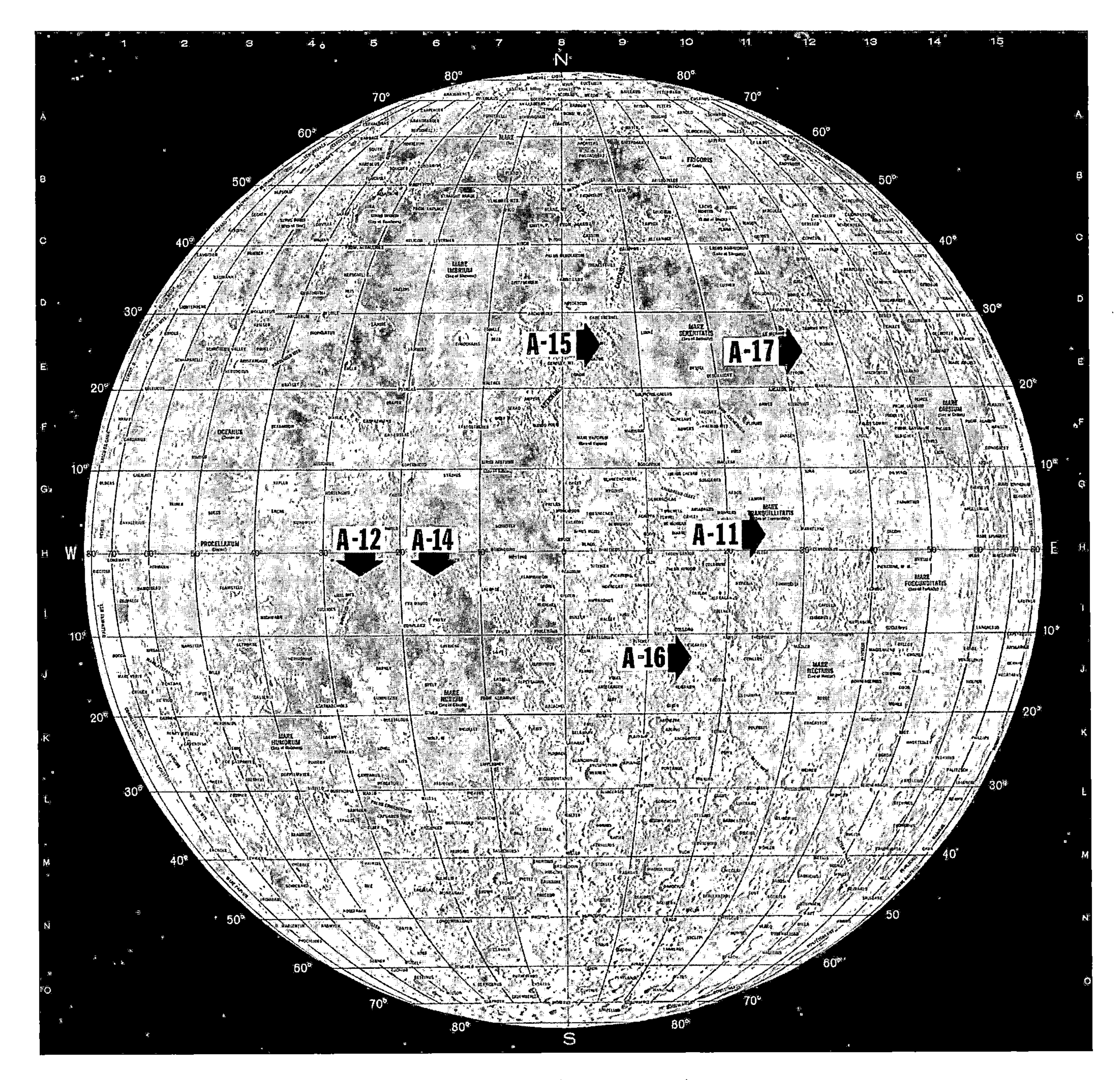

View of the Moon's nearside with the Apollo Project landing sites marked ***(Courtesy of NASA)***

Moon rocks and soil samples. A lunar rover vehicle was again used to assist in surface exploration activities, and the liftoff of the lunar module from the Moon's surface was recorded for the first time with the help of a television camera teleoperated (that is, remotely controlled) from Earth.

The final Apollo mission to the Moon, *Apollo 17,* was launched on December 7, 1972. During this 12-day mission the astronauts explored the Taurus-Littrow landing site, emplaced geophysical instruments, and collected 108 kilograms of lunar soil and rock samples. Astronauts EUGENE A. CERNAN and HARRISON SCHMITT were the last human beings to walk on the surface of the Moon in the 20th century.

The Apollo Project and its exciting missions to the Moon represent one of the greatest triumphs of both human technology and the human spirit. The Apollo Project included numerous uncrewed test missions and 12 crewed missions: three Earth orbiting missions (*Apollo 7, Apollo 9,* and the APOLLO-SOYUZ TEST PROJECT), two lunar orbiting missions

Apollo Project Summary

Spacecraft Name	Crew	Date	Flight time (hours, minutes, seconds)	Revolutions	Remarks
Apollo 7	Walter H. Schirra Donn Eisele Walter Cunningham	10/11–22/68	260:08:45	163	First crewed Apollo flight demonstrated the spacecraft, crew, and support elements. All performed as required.
Apollo 8	Frank Borman James A. Lovell, Jr. William Anders	12/21–27/68	147:00:41	10 rev. of Moon	History's first crewed flight to the vicinity of another celestial body.
Apollo 9	James A. McDivitt David R. Scott Russell L. Schweikart	3/3–13/69	241:00:53	151	First all-up crewed Apollo flight (with Saturn V and command, service, and lunar modules). First Apollo extravehicular activity. First docking of command service module with lunar module (LM).
Apollo 10	Thomas P. Stafford John W. Young Eugene A. Cernan	5/18–26/69	192:03:23	31 rev. of Moon	Apollo LM descended to within 14.5 km of Moon and later rejoined command service module. First rehearsal in lunar environment.
Apollo 11	Neil A. Armstrong Michael Collins Edwin E. Aldrin	7/16–24/69	195:18:35	30 rev. of Moon	First landing of a person on the Moon. Total stay time: 21 hr., 36 min.
Apollo 12	Charles Conrad, Jr. Richard F. Gordon, Jr Alan L. Bean	11/14–24/69	244:36:25	45 rev. of Moon	Second crewed exploration of the Moon. Total stay time: 31 hr., 31 min.
Apollo 13	James A. Lovell, Jr. John L. Swigert, Jr. Fred W. Haise, Jr.	4/11–17/70	142:54:41	—	Mission aborted because of service module oxygen tank failure.
Apollo 14	Alan B. Shepard, Jr. Stuart A. Roosa Edgar D. Mitchell	1/31–2/9/71	216:01:59	34 rev. of Moon	First crewed landing in and exploration of lunar highlands. Total stay time: 33 hr., 31 min.
Apollo 15	David R. Scott Alfred M. Worden James B. Irwin	7/26–8/7/71	295:11:53	74 rev. of Moon	First use of lunar roving vehicle. Total stay time: 66 hr., 55 min.
Apollo 16	John W. Young Thomas K. Mattingly II Charles M. Duke, Jr.	4/16–27/72	265:51:05	64 rev. of Moon	First use of remote-controlled television camera to record liftoff of the lunar module ascent stage from the lunar surface. Total stay time: 71 hr., 2 min.
Apollo 17	Eugene A. Cernan Ronald E. Evans Harrison H. Schmitt	12/7–19/72	301:51:59	75 rev. of Moon	Last crewed lunar landing and exploration of the Moon in the Apollo Project returned 110 kg of lunar samples to Earth. Total stay time: 75 hr.

Source: NASA.

(*Apollo 8* and *10*), a lunar swingby (*Apollo 13*), and six lunar landing missions (*Apollo 11, 12, 14, 15, 16,* and *17*). The table on page 40 provides a summary of this magnificent project and the figure shows the locations visited on the Moon's surface. Overall, 12 human beings (sometimes called the "Moon walkers") performed extravehicular activities (EVAs) on the lunar surface during the Apollo Project. As part of the *Apollo 11* mission, astronauts Armstrong and Aldrin landed in the Sea of Tranquility (0.67 degrees N latitude, 23.5 degrees E longitude). During the *Apollo 12* mission, astronauts ALAN L. BEAN and CHARLES (PETE) CONRAD, JR. explored the Moon's surface in the Ocean of Storms (3.0 degrees S latitude, 23.4 degrees W longitude). As part of the *Apollo 14* mission, astronauts ALAN B. SHEPARD, JR., and EDGAR DEAN MITCHELL landed at Fra Mauro (3.7 degrees S latitude, 17.5 degrees W longitude). In the *Apollo 15* mission astronauts DAVID R. SCOTT and JAMES B. IRWIN explored the Moon's Hadley-Apennine region (26.1 degrees N latitude, 3.7 degrees E longitude) and used an electric-powered lunar rover vehicle during their EVA. As part of the *Apollo 16* mission astronauts JOHN W. YOUNG and CHARLES M. DUKE, JR., landed at Descartes (9.0 degrees S latitude, 15.5 degrees E longitude) and explored the region with the assistance of an electric-powered lunar rover. During the *Apollo 17* mission astronauts Eugene Cernan and Harrison Schmitt explored the Moon's Taurus-Littrow region (20.2 N latitude, 30.8 E longitude) and collected 110 kilograms of rock samples with the assistance of an electric-powered lunar rover.

Their EVA marked the end of human lunar exploration in the 20th century.

See also GEMINI PROJECT; MERCURY PROJECT; MOON; RANGER PROJECT; SURVEYOR PROJECT.

Apollo-Soyuz Test Project (ASTP) The joint United States–Soviet Union space mission that took place in July 1975. The mission centered around the rendezvous and docking of the *Apollo 18* spacecraft (three astronaut crew: THOMAS P. STAFFORD, Vance Brand, and DEKE SLAYTON, JR.) and the *Soyuz 19* spacecraft (two cosmonaut crew: ALEXEI LEONOV and Valeriy Kubasov).

Both the *Soyuz 19* and *Apollo 18* spacecraft were launched on July 15, 1975; the US *Apollo 18* spacecraft lifted off approximately seven and a half hours after the Russians launched the *Soyuz 19* spacecraft. The Russian cosmonauts maneuvered their *Soyuz 19* spacecraft to the planned orbit for docking at an altitude of 222 kilometers over Europe. The *Apollo 18* astronauts then completed the rendezvous sequence, eventually docking with the Soyuz spacecraft on July 17, 1975. During the next two days the crews accomplished four transfer operations between the two spacecraft and completed five scheduled experiments. Following the first undocking, a joint solar eclipse experiment was performed. Then, the *Apollo 18* spacecraft accomplished a second successful docking, this time with the *Soyuz 19* apparatus locking the two spacecraft together. The final undocking occurred on July 19. The two spacecraft moved to a station-keeping distance, and a joint ultraviolet absorption experiment was performed involving a complicated series of orbital maneuvers. Afterward, the *Apollo 18* spacecraft entered a separate orbit, and both the *Soyuz 19* and *Apollo 18* crews conducted unilateral activities. The *Soyuz 19* landed safely on July 21 after six mission days, and the *Apollo 18* flight successfully concluded on July 24, 1975, nine days after launch. The primary objectives of this first international human-crewed mission were met, including rendezvous, docking, crew transfer, and control center–crew interaction.

See also APOLLO PROJECT; SOYUZ SPACECRAFT.

Apollo Telescope Mount (ATM) The sophisticated telescope mount on NASA's *Skylab* space station. From May 1973 until February 1974 this device supported solar astronomy instruments that operated primarily in the ultraviolet and X-ray portions of the electromagnetic spectrum.

See also SKYLAB.

apolune That point in an orbit around the Moon of a spacecraft launched from the lunar surface that is farthest from the Moon. *Compare with* PERILUNE.

aposelene The farthest point in an orbit around the Moon.

See also APOCYNTHION; APOLUNE.

apparent In astronomy, observed. True values are reduced from apparent (observed) values by eliminating factors such as refraction and flight time that can affect the observation.

apparent diameter The observed diameter (but not necessarily the actual diameter) of a celestial body. Usually expressed in degrees, minutes, and seconds of arc. Also called angular diameter.

apparent magnitude (symbol: m) The brightness of a star (or other celestial body) as measured by an observer on Earth. Its value depends on the star's intrinsic brightness (luminosity), how far away it is, and how much of its light has been absorbed by the intervening interstellar medium.

See also ABSOLUTE MAGNITUDE; MAGNITUDE.

apparent motion The observed motion of a heavenly body across the celestial sphere, assuming that Earth is at the center of the celestial sphere and is standing still (stationary).

apparition The period during which a celestial body within the solar system is visible to an observer on Earth; for example, the apparition of a periodic comet or the morning apparition of the planet Venus. This term is not generally used in astronomy to describe the observation of objects that are visible regularly, such as the Sun and the fixed stars.

Applications Technology Satellite (ATS) From the mid-1960s to the mid-1970s NASA sponsored the Applications Technology Satellites series of six spacecraft that explored and flight tested a wide variety of new technologies for use in future communications, meteorological, and navigation satellites. Some of the important space technology areas investigated during this program included spin stabilization, gravity gradient stabilization, complex synchronous orbit maneuvers, and a number of communications experiments. The ATS flights also investigated the space environment in geostationary orbit. Although intended primarily as engineering testbeds, the ATS satellites also collected and transmitted meteorological data and served (on occasion) as communications satellites. The first five ATS satellites in the series shared many common design features, while the sixth spacecraft represented a new design.

ATS-1 was launched from Cape Canaveral Air Force Station in Florida on December 7, 1966, by an Atlas-Agena D rocket vehicle combination. The mission of this prototype weather satellite was to test techniques of satellite orbit and motion at geostationary orbit and to transmit meteorological imagery and data to ground stations. Controllers deactivated the *ATS-1* spacecraft on December 1, 1978. During its 12-year lifetime *ATS-1* explored the geostationary environment and performed several C-band communications experiments, including the transmission of educational and health-related programs to rural areas of the United States and several countries in the Pacific Basin. *ATS-1* also provided meteorologists with the first full-Earth cloud cover images.

The *ATS-2* satellite had a mission similar to that of its predecessor, but a launch vehicle failure (Atlas-Agena D configuration from Cape Canaveral) on April 6, 1967, placed the spacecraft in an undesirable orbit around Earth. Atmospheric torques caused by the spacecraft's improper and unintended orbit eventually overcame the ability of the satellite's gravity gradient stabilization system, so the spacecraft began to slowly tumble. Although *ATS-2* remained functional, controllers deactivated the distressed spacecraft six months after its launch because they were receiving only a limited amount of useful data from the tumbling satellite.

Slightly larger, but similar in design to *ATS-1,* NASA's *ATS-3* spacecraft was successfully launched on November 5, 1967, by an Atlas-Agena D rocket configuration from Cape Canaveral. The goals of the *ATS-3* flight included the investigation of spin stabilization techniques and VHF and C-band communications experiments. In addition to fulfilling its primary space technology demonstration mission, *ATS-3* provided regular telecommunications service to sites in the Pacific Basin and Antarctica, provided emergency communications links during the 1987 Mexican earthquake and the Mount Saint Helens volcanic eruption disaster, and supported the Apollo Project Moon landings. *ATS-3* also advanced weather satellite technology by providing the first color images of Earth from space, as well as regular cloud cover images for meteorological studies. On December 1, 1978, ground controllers deactivated the spacecraft.

On August 10, 1967, an Atlas-Centaur rocket vehicle attempted to send the *ATS-4* satellite to geostationary orbit from Cape Canaveral. However, when the Centaur rocket failed to ignite, ground controllers deactivated the satellite about 61 minutes after the launch. Without the Centaur stage burn, the satellite's orbit was simply too low, and atmospheric drag eventually caused it to reenter Earth's atmosphere and disintegrate on October 17, 1968.

The mission of the *ATS-5* satellite was to evaluate gravity gradient stabilization and new imaging techniques for meteorological data retrieval. On August 12, 1969, an Atlas-Centaur rocket vehicle successfully lifted *ATS-5* into space. However, following the firing of the satellite's apogee kick motor, *ATS-5* went into an unplanned flat spin. The space vehicle recovered and began spinning about the proper axis, but in a direction opposite to the planned direction. As a result, the spacecraft's gravity gradient booms could not be deployed, and some of its experiments (such as the gravity gradient stabilization and meteorological imagery acquisition demonstrations) could not function. The *ATS-5* spacecraft did perform a limited number of other experiments before ground controllers boosted it above geostationary orbit at the end of the mission.

A Titan 3C rocket successfully launched NASA's *ATS-6* satellite from Cape Canaveral on May 30, 1974. In addition to accomplishing its technology demonstration mission, *ATS-6* became the world's first educational satellite. During its five-year life *ATS-6* transmitted educational programs to India, to rural regions of the United States, and to other countries. This satellite also performed air traffic control tests and direct broadcast television experiments, and helped demonstrate the concept of satellite-assisted search and rescue. *ATS-6* also played a major role in relaying signals to the Johnson Space Center in Houston, Texas, during the Apollo-Soyuz Test Project. At the end of its mission ground controllers boosted the spacecraft above geostationary orbit.

approach The maneuvers of a spacecraft or aerospace vehicle from its normal orbital position (sometimes called its station-keeping position) toward another orbiting object (e.g., a satellite, space station, or spacecraft) for the purpose of conducting rendezvous and docking or capture operations.

appulse 1. The apparent near approach of one celestial body to another. For example, the apparent close approach (on the celestial sphere) of a planet or asteroid to a star, without the occurrence of an occultation. 2. A penumbral eclipse of the Moon.

apsis (plural: apsides) In celestial mechanics, either of the two orbital points nearest (*periapsis*) or farthest (*apoapsis*) from the center of gravitational attraction. The apsides are called the perigee and apogee for an orbit around Earth and the perihelion and aphelion for an orbit around the Sun. The straight line connecting these two points is called the *line of apsides* and represents the major axis of an elliptical orbit.

See also ORBITS OF OBJECTS IN SPACE.

***Aqua* (spacecraft)** A NASA-sponsored advanced Earth-observing satellite (EOS) placed into polar orbit by a Delta II expendable rocket from Vandenberg Air Force Base, California, on May 4, 2002. The primary role of *Aqua,* as its name implies (i.e., Latin for "water"), is to gather information about changes in ocean circulation and how clouds and surface water processes affect Earth's climate. Equipped with six state-of-the-art remote sensing instruments, the satellite is collecting data on global precipitation, evaporation, and the cycling of water on a planetary basis. The spacecraft's data include information on water vapor and clouds in the atmosphere, precipitation from the atmosphere, soil wetness on the land, glacial ice on the land, sea ice in the oceans, snow cover on both land and sea ice, and surface waters throughout the world's oceans, bays, and lakes. Such information is helping scientists improve the quantification of the global water cycle and examine such issues as whether the hydrologic cycle (that is, cycling of water through the Earth system) is possibly accelerating.

Aqua is a joint project among the United States, Japan, and Brazil. The United States provided the spacecraft and the following four instruments: the atmospheric infrared sounder (AIRS), the clouds and Earth's radiant energy system (CERES), the moderate resolution imaging spectroradiometer (MODIS), and the advanced microwave sounding unit (AMSU). Japan provided the advanced microwave scanning radiometer for EOS (AMSR-E), and the Brazilian Institute for Space provided the humidity sounder for Brazil (HSB). Overall management of the *Aqua* mission takes place at NASA's Goddard Space Flight Center in Greenbelt, Maryland.

See also EARTH SYSTEM SCIENCE; GLOBAL CHANGE; REMOTE SENSING; *TERRA* SPACECRAFT.

Arago, Dominique-François (1786–1853) French *Scientist, Statesman* Dominique-François Arago developed instruments to study the polarization of light. He had a special interest in polarized light from comets and determined through careful observation that Comet Halley (1835 passage) and other comets were not self-luminous. His compendium, *Popular Astronomy,* extended scientific education to a large portion of the European middle class.

archaeological astronomy Scientific investigation concerning the astronomical significance of ancient structures and sites. Many early peoples looked up at the sky and made up stories about what they saw but could not physically explain, so archaeoastronomy also includes the study of the celestial lore, mythologies, and the primitive cosmologies (world views) of ancient cultures.

Prehistoric cave paintings (some up to 30 millennia old) provide a lasting testament that early peoples engaged in stargazing and incorporated such astronomical observations in their cultures. Early peoples also carved astronomical symbols in stones (petroglyphs) now found at ancient ceremonial locations and ruins. The figure depicts two ancient Native American petroglyphs of astronomical significance. Photograph A (on the left) was taken in the Cave of Life of the Anasazi Indians in Arizona's Petrified Forest. The solar markings form a ceremonial calendar related to the winter solstice. Photograph B (on the right) shows a petroglyph made by the Hohokam Indians who once lived in what is now the Painted Rocks State Park in Arizona. This petroglyph also contains solar markings that appear to form a prehistoric astronomical calendar, depicting the summer solstice, equinox, and winter solstice.

Many of the great monuments and ceremonial structures of ancient civilizations have alignments with astronomical significance. One of the oldest astronomical observatories is Stonehenge. During travel to Greece and Egypt in the early 1890s, SIR JOSEPH NORMAN LOCKYER noticed how many ancient temples had their foundations aligned along an east-west axis, a consistent alignment that suggested to him some astronomical significance with respect to the rising and setting Sun. To pursue this interesting hypothesis, Lockyer then visited Karnack, one of the great temples of ancient Egypt. He discussed the hypothesis in his 1894 book *The Dawn of Astronomy.* This book is often regarded as the beginning of archaeoastronomy. As part of his efforts, Lockyer studied studied Stonehenge, an ancient site located in southern England. However, he could not accurately determine the site's construction date. As a result, he could not confidently project the solar calendar back to a sufficiently precise moment in history that would reveal how the curious circular ring of large vertical stones topped by capstones might be connected to some astro-

Shown here are two ancient Native American petroglyphs of astronomical significance. Anasazi Indians who lived in Arizona's Petrified Forest made the petroglyph shown in the photograph on the left. The solar markings form a ceremonial calendar related to the winter solstice. Similarly, Hohokam Indians who lived in what is now Painted Rocks State Park in Arizona made the petroglyph appearing in the photograph on the right. This petroglyph also contains solar markings considered to form an astronomical calendar for the summer solstice, equinox, and winter solstice. ***(Courtesy of NASA)***

nomical practice of the ancient Britons. Lockyer's visionary work clearly anticipated the results of modern studies of Stonehenge—results that suggest the site could have served as an ancient astronomical calendar around 2,000 B.C.E.

The Egyptians and the Maya both used the alignment of structures to assist in astronomical observations and the construction of calendars. Modern astronomers have discovered that the Great Pyramid at Giza, Egypt, has a significant astronomical alignment, as do certain Maya structures such as those found at Uxmal in the Yucatán, Mexico. Mayan astronomers were particularly interested in times (called "zenial passages") when the Sun crossed over certain latitudes in Central America. The Maya were also greatly interested in the planet Venus and treated the planet with as much importance as the Sun. The Maya had a good knowledge of astronomy and were able to calculate planetary movements and eclipses over millennia.

For many ancient peoples the motion of the Moon, the Sun, and the planets and the appearance of certain constellations of stars served as natural calendars that helped regulate daily life. Since these celestial bodies were beyond physical reach or understanding, various mythologies emerged along with native astronomies. Within ancient cultures the sky became the home of the gods, and the Moon and Sun were often deified.

While no anthropologist really knows what the earliest human beings thought about the sky, the culture of the Australian Aborigines, which has been passed down for more than 40,000 years through the use of legends, dances, and songs, gives us a glimpse of how these early people interpreted the Sun, Moon, and stars. The Aboriginal culture is the world's oldest and most long-lived, and the Aboriginal view of the cosmos involves a close interrelationship between people, nature, and sky. Fundamental to their ancient culture is the concept of "the Dreaming"—a distant past when the spirit ancestors created the world. Aboriginal legends, dance, and songs express how in the distant past the spirit ancestors created the natural world and entwined people into a close relationship with the sky and with nature. Within the Aboriginal culture the Sun is regarded as a woman. She awakes in her camp in the east each day and lights a torch that she then carries across the sky. In contrast, Aborigines consider the Moon male, and, because of the coincidental association of the lunar cycle with the female menstrual cycle, they linked the Moon with fertility and consequently gave it a great magical status. These ancient people also regarded a solar eclipse as the male Moon uniting with the female Sun.

For the ancient Egyptians, Ra (also called Re) was regarded as the all-powerful sun god who created the world and sailed across the sky each day. As a sign of his power, an Egyptian pharaoh would use the title "son of Ra." Within Greek mythology Apollo was the god of the Sun and his twin sister, Artemis (Diana in Roman mythology), the Moon goddess.

From prehistory astronomical observations have played a major role in the evolution of human cultures. Archaeoastronomy helps link our contemporary knowledge of the heavens with the way our distant ancestors viewed the sky and interpreted the mysterious objects they saw.

arc-jet engine An electric rocket engine in which the propellant gas is heated by passing through an electric arc. In general, the cathode is located axially in the arc-jet engine, and the anode is formed into a combined plenum chamber/constrictor/nozzle configuration. The main problems associated with arc-jet engines are electrode erosion and low overall efficiency. Electrode erosion occurs as a result of the intense heating experienced by the electrodes and can seriously limit the thruster lifetime. The electrode erosion problem can be reduced by careful design of the electrodes. Overall, arc-jet engine efficiency is dominated by "frozen-flow" losses, which are the result of dissociation and ionization of the propellant. These frozen-flow losses are much more difficult to reduce since they are dependent on the thermodynamics of the flow and heating processes. A high-power arc-jet engine with exhaust velocity values between 8×10^3 and 2×10^4 meters per second is an attractive option for propelling an orbital transfer vehicle.

See also ELECTRIC PROPULSION.

arc minute One sixtieth (1/60th) of a degree of angle. This unit of angle is associated with precise measurements of motions and positions of celestial objects as occurs in the science of astrometry.

$$1° = 60 \text{ arc min} = 60'$$

See also ARC SECOND; ASTROMETRY.

arc second One/three thousand six hundredth (1/3,600th) of a degree of angle. This unit of angle is associated with very precise measurements of stellar motions and positions in the science of astrometry.

$$1'(\text{arc min}) = 60 \text{ arc sec} = 60''$$

See also ARC MINUTE; ASTROMETRY.

Arecibo Interstellar Message To help inaugurate the powerful radio/radar telescope of the Arecibo Observatory in the tropical jungles of Puerto Rico, an interstellar message of friendship was beamed to the fringes of the Milky Way galaxy. On November 16, 1974, this interstellar radio signal was transmitted toward the Great Cluster in Hercules (Messier 13, or M13, for short), which lies about 25,000 light-years away from Earth. The globular cluster M13 contains about 300,000 stars within a radius of approximately 18 light-years.

This transmission, often called the Arecibo Interstellar Message, was made at the 2,380-megahertz (MHz) radio frequency with a 10 hertz (Hz) bandwidth. The average effective radiated power was 3×10^{12} watts (3 terawatts [TW]) in the direction of transmission. The signal is considered to be the strongest radio signal yet beamed into space by our planetary civilization. Perhaps 25,000 years from now a radio telescope operated by members of an intelligent alien civilization somewhere in the M13 cluster will receive and decode this interesting signal. If they do, they will learn that intelligent life has evolved here on Earth.

The Arecibo Interstellar Message of 1974 consisted of 1,679 consecutive characters. It was written in a binary format—that is, only two different characters were used. In binary notation the two different characters are denoted as "0" and "1." In the actual transmission, each character was represented by one of two specific radio frequencies, and the message was transmitted by shifting the frequency of the Arecibo Observatory's radio transmitter between these two radio frequencies in accordance with the plan of the message.

The message itself was constructed by the staff of the National Astronomy and Ionosphere Center (NAIC). It can be decoded by breaking up the message into 73 consecutive groups of 23 characters each and then arranging these groups in sequence one under the other. The numbers 73 and 23 are prime numbers. Their use should facilitate the discovery by any alien civilization receiving the message that the above format is the right way to interpret the message. The figure shows the decoded message: The first character transmitted (or received) is located in the upper right-hand corner.

This message describes some of the characteristics of terrestrial life that the scientific staff at NAIC felt would be of particular interest and technical relevance to an extraterrestrial civilization. The NAIC staff interpretation of the interstellar message is as follows.

The Arecibo message begins with a "lesson" that describes the number system being used. This number system is the binary system, in which numbers are written in powers of 2 rather than of 10, as in the decimal system used in everyday life. NAIC staff scientists believe that the binary system is one of the simplest number systems and is particularly easy to code in a simple message. Written across the top of the message (from right to left) are the numbers 1 through 10 in binary notation. Each number is marked with a *number label*—that is, a single character, which denotes the start of a number.

The next block of information in the message occurs just below the numbers. It is recognizable as five numbers. From right to left these numbers are 1, 6, 7, 8, and 15. This otherwise unlikely sequence of numbers should eventually be interpreted as the atomic numbers of the elements hydrogen, carbon, nitrogen, oxygen, and phosphorus.

Next in the message are 12 groups on lines 12 through 30 that are similar groups of five numbers. Each of these groups represents the chemical formula of a molecule or radical. The numbers from right to left in each case provide the number of atoms of hydrogen, carbon, nitrogen, oxygen, and phosphorus, respectively, that are present in the molecule or radical.

Since the limitations of the message did not permit a description of the physical structure of the radicals and molecules, the simple chemical formulas do not define in all cases the precise identity of the radical or molecule. However, these structures are arranged as they are organized within the macromolecule described in the message. Intelligent alien organic chemists somewhere in the M13 cluster should eventually be able to arrive at a unique solution for the molecular structures being described in the message.

The most specific of these structures, and perhaps the one that should point the way to correctly interpreting the others, is the molecular structure that appears four times on lines 17 through 20 and lines 27 through 30. This is a structure containing one phosphorus atom and four oxygen atoms, the well-known phosphate group. The outer structures on lines 12 through 15 and lines 22 through 25 give the formula for a sugar molecule, deoxyribose. The two sugar molecules on lines 12 through 15 have between them two structures: the chemical formulas for thymine (left structure) and adenine (right structure). Similarly, the molecules between the sugar molecules on lines 22 through 25 are guanine (on the left) and cytosine (on the right).

The macromolecule, or overall chemical structure, is that of deoxyribonucleic acid (DNA). The DNA molecule contains the genetic information that controls the form, living processes, and behavior of all terrestrial life. This structure is actually wound as a double helix, as depicted in lines 32 through 46 of the message. The complexity and degree of development of intelligent life on Earth is described by the number of characters in the genetic code, that is, by the number of adenine-thymine and guanine-cytosine combinations in the DNA molecule. The fact that there are some 4 billion such pairs in human DNA is illustrated in the message by the number given in the center of double helix between lines 27 and 43. Note that the number

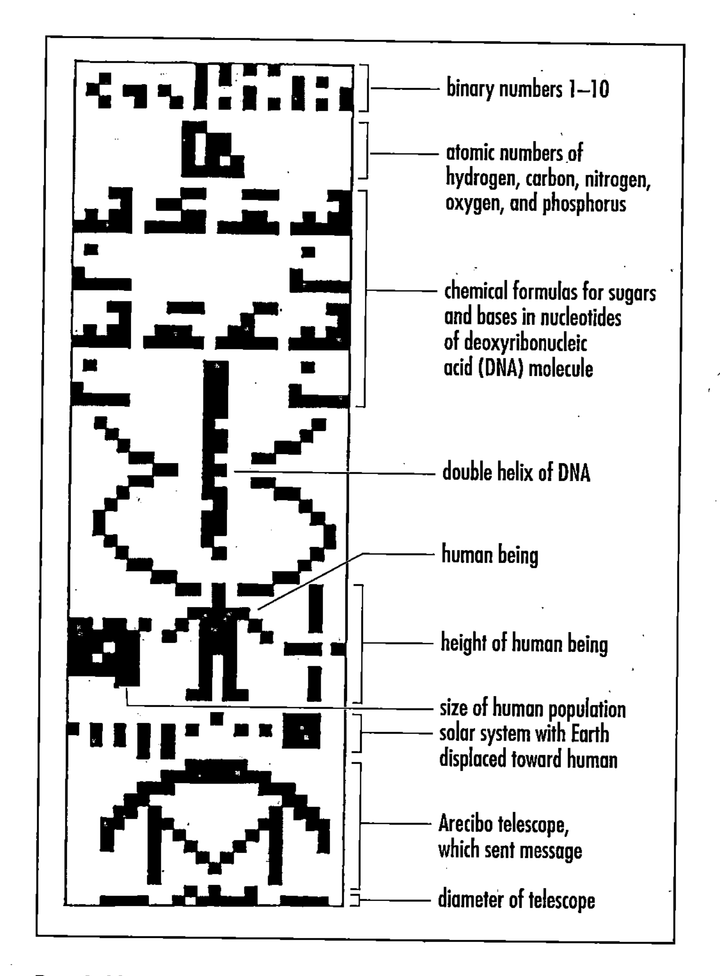

Decoded form of the Arecibo Message of 1974 ***(Courtesy of Frank D. Drake and the staff of the National Astronomy and Ionosphere Center, operated by Cornell University under contract with the National Science Foundation)***

label is used here to establish this portion of the message as a number and to show where the number begins.

The double helix leads to the "head" in a crude sketch of a human being. The scientists who composed the message hoped that this would indicate connections among the DNA molecule, the size of the helix, and the presence of an "intelligent" creature. To the right of the sketch of a human being is a line that extends from the head to the feet of the "message human." This line is accompanied by the number 14. This portion of the message is intended to convey the fact that the "creature" drawn is 14 units of length in size. The only possible unit of length associated with the message is the wavelength of the transmission, namely 12.6 centimeters. This makes the creature in the message 176 centimeters, or about five feet nine inches, tall. To the left of the human being is a number, approximately 4 billion. This number represents the approximate human population on planet Earth when the message was transmitted.

Below the sketch of the human being is a representation of our solar system. The Sun is at the right, followed by nine planets with some coarse representation of relative sizes. The third planet, Earth, is displaced to indicate that there is something special about it. In fact, it is displaced toward the drawing of the human being, who is centered on it. An extraterrestial scientist in pondering this message should recognize that Earth is the home of the intelligent creatures who sent it.

Below the solar system and centered on the third planet is an image of a telescope. The concept of "telescope" is described by showing a device that directs rays to a point. The mathematical curve leading to such a diversion of paths is crudely indicated. The telescope is not upside down, but rather "up" with respect to the symbol for the planet Earth.

At the very end of the message, the size of the telescope is indicated. Here, it is both the size of the largest radio telescope on Earth and also the size of the telescope that sent the message (namely, the radio telescope at the Arecibo Observatory). It is shown as 2,430 wavelengths across, or roughly 305 meters. No one, of course, expects an alien civilization to have the same unit system we use on Earth, but physical quantities, such as the wavelength of transmission, provide a common reference frame.

This interstellar message was transmitted at a rate of 10 characters per second, and it took 169 seconds to transmit the entire information package. It is interesting to realize that just one minute after completion of transmission, the interstellar greetings passed the orbit of Mars. After 35 minutes the message passed the orbit of Jupiter, and after 71 minutes it silently crossed the orbit of Saturn. Some 5 hours and 20 minutes after transmission the message passed the orbit of Pluto, leaving the solar system and entering "interstellar space." It will be detectable by telescopes anywhere in our galaxy of approximately the same size and capability as the Arecibo facility that sent it.

See also ARECIBO OBSERVATORY; INTERSTELLAR COMMUNICATION AND CONTACT.

Arecibo Observatory The world's largest radio/radar telescope, with a 305-m-diameter dish. It is located in a large, bowl-shaped natural depression in the tropical jungle of Puerto Rico. (*See* figure below.) The Arecibo Observatory is the main observing instrument of the National Astronomy and Ionosphere Center (NAIC), a national center for radio and radar astronomy and ionospheric physics operated by Cornell University under contract with the National Science Foundation. The observatory operates on a continuous basis, 24 hours a day every day, providing observing time and logistical support to visiting scientists. When the giant telescope operates as a radio wave receiver, it can listen for signals from celestial objects at the farthest reaches of the universe. As a radar transmitter/receiver, it assists astronomers and planetary scientists by bouncing signals off the Moon, off nearby planets and their satellites, off asteroids, and even off layers of Earth's ionosphere.

The Arecibo Observatory has made many contributions to astronomy and astrophysics. In 1965 the facility (operating as a radar transmitter/receiver) determined that the rotation rate of the planet Mercury is 59 days rather than the previously estimated 88 days. In 1974 the facility (operating as a radio wave receiver) supported the discovery of the first binary pulsar system. This discovery led to an important confirmation of ALBERT EINSTEIN's theory of general relativity and earned the American physicists RUSSELL A. HULSE and JOSEPH H. TAYLOR, JR., the 1993 Nobel Prize in physics. In the early 1990s astronomers used the facility to discover extrasolar planets in orbit around the rapidly rotating pulsar B1257+12.

In May 2000 astronomers used the Arecibo Observatory as a radar transmitter/receiver to collect the first-ever radar images of a main belt asteroid, named 216 Kleopatra. Kleopatra is a large, dog-bone shaped minor planet about 217 kilometers long and 94 kilometers wide. Discovered in 1880, the exact

The Arecibo Observatory ***(Courtesy of NASA)***

shape of Kleopatra was unknown until then. Astronomers used the telescope to bounce radar signals off Kleopatra. Then, with sophisticated computer analysis techniques, the scientists decoded the echoes, transformed them into images, and assembled a computer model of the asteroid's shape. This activity was made possible because the Arecibo telescope underwent major upgrades in the 1990s—improvements that dramatically increased its sensitivity and made feasible the radar imaging of more distant objects in the solar system.

The Arecibo Observatory is also uniquely suited to search for signals from extraterrestrial life by focusing on thousands of star systems in the 1,000 MHz to 3,000 MHz range. To date, no such signals have been found.

See also ARECIBO INTERSTELLAR MESSAGE; RADAR ASTRONOMY; RADIO ASTRONOMY; SEARCH FOR EXTRATERRESTRIAL INTELLIGENCE (SETI).

areocentric With Mars as the center.

areodesy The branch of science that determines by careful observation and precise measurement the size and shape of the planet Mars and the exact position of points and features on the surface of the planet. *Compare with* GEODESY.

areography The study of the surface features of Mars; the "geography" of Mars.

Ares The planet Mars. This name for the fourth planet in our solar system is derived from the ancient Greek word Ares (Αρες), their name for the mythological god of war (who was called Mars by the Romans). Today the term *Ares* is seldom used for Mars except in combined forms, such as *areocentric* and *areodesy*.

Argelander, Friedrich Wilhelm August (1799–1875) German *Astronomer* Friedrich Argelander investigated variable stars and compiled a major telescopic (but prephotography) survey of all the stars in the Northern Hemisphere brighter than the 9th magnitude. From 1859 to 1862 he published the four-volume *Bonn Survey (Bonner Durchmusterung)*—an important work that contains more than 324,000 stars.

He was born on March 22, 1799, in the Baltic port of Memel, East Prussia (now Klaipeda, Lithuania). His father was a wealthy Finnish merchant and his mother a German. Argelander studied at the University of Königsberg in Prussia, where he was one of FRIEDRICH BESSEL's most outstanding students. In 1820 he decided to pursue astronomy as a career when he became Bessel's assistant at the Königsberg Observatory. Argelander received a Ph.D. degree in astronomy in 1822 from the University of Königsberg. His doctoral dissertation involved a critical review of the celestial observations made by JOHN FLAMSTEED. Argelander's academic research interest in assessing the observational quality of earlier star catalogs so influenced his later professional activities that the hard-working astronomer would eventually develop his own great catalog of northern hemisphere stars.

In 1823 Bessel's letter of recommendation helped Argelander secure a position as an observer at the newly established Turku (Åbo) Observatory in southwestern Finland (then an autonomous grand duchy within the Imperial Russian Empire). There the young astronomer poured his energies into the study of stellar motions. Unfortunately, a great fire in September 1827 totally destroyed Turku, the former capital of Finland, halting Argelander's work at the observatory. Following the great fire the entire university community moved from Turku to the new Finnish capital at Helsinki.

The university promoted Argelander to the rank of professor of astronomy in 1828 and also gave him the task of designing and constructing a new observatory in Helsinki. Argelander found a suitable site on a hill south of the city, and by 1832 construction of the new observatory was completed. The beautiful and versatile new observatory in Helsinki served as the model for the Pulkovo Observatory constructed by FRIEDRICH GEORG WILHELM STRUVE near St. Petersburg for use as the major observatory of the Imperial Russian Empire. Argelander summarized his work on stellar motions in the 1837 book *About the Proper Motion of the Solar System.*

His work in Helsinki ended in 1837, when Bonn University in his native Prussia offered him a professorship in astronomy that he could not refuse. The offer included construction of a new observatory at Bonn, financed by the German crown prince, Friedrich Wilhelm IV (1795–1861), who became king in 1840. Argelander was a personal friend of Friedrich Wilhelm IV, having offered the crown prince refuge in his own home in Memel following Napoleon's defeat of the Prussian Army in 1806.

The new observatory in Bonn was inaugurated in 1845. From then on Argelander devoted himself to the development and publication of his famous star catalog, the *Bonner Durchmusterung (Bonn Survey)*, published between 1859 and 1862. Argelander and his assistants worked hard measuring the position and brightness of 324,198 stars in order to compile the largest and most comprehensive star catalog ever produced without the assistance of photography. This enormous work listed all the stars observable in the Northern Hemisphere down to the 9th magnitude. In 1863 Argelander founded the Astronomische Gesellschaft (Astronomical Society), whose mission was to continue his work by developing a complete celestial survey using input by observers throughout Europe.

Friedrich Argelander died in Bonn on February 17, 1875. His assistant and successor, Eduard Schönfeld (1828–91), extended Argelander's astronomical legacy into the skies of the Southern Hemisphere by adding another 133,659 stars. Schönfeld's own efforts ended in 1886, but other astronomers continued Argelander's quest. In 1914 the *CÓRDOBA DURCHMUSTERUNG (Córdoba Survey)* appeared; it contained the positions of 578,802 Southern Hemisphere stars measured down to the 10th magnitude, as mapped from the Cordoba Observatory in Argentina. This effort completed the huge systematic (preastrophotography) survey of stars begun by Bessel and his hard-working assistant Argelander nearly a century before.

argument of periapsis The argument (angular distance) of periapsis from the ascending node.

See also ORBITS OF OBJECTS IN SPACE.

Ariane The Ariane family of launch vehicles evolved from a European desire, first expressed in the early 1960s, to achieve and maintain an independent access to space. Early manifestation of the efforts that ultimately resulted in the creation of Arianespace (the international company that now markets the Ariane launch vehicles) included France's Diamant launch vehicle program (with operations at Hammaguir, Algeria, in the Sahara Desert) and the Europa launch vehicle program, which operated in the Woomera Range in Australia before moving to Kourou, French Guiana, in 1970. These early efforts eventually yielded the first Ariane flight on December 24, 1979. That mission, called L01, was followed by 10 more Ariane-1 flights over the next six years. The initial Ariane vehicle family (Ariane-1 through Ariane-4) centered on a three-stage launch vehicle configuration with evolving capabilities. The Ariane-1 vehicle gave way in 1984 to the more powerful Ariane-2 and Ariane-3 vehicle configurations. These configurations, in turn, were replaced in June 1988 with the successful launch of the Ariane-4 vehicle, a launcher that has been called Europe's "space workhorse." Ariane-4 vehicles are designed to orbit satellites with a total mass value of up to 4,700 kg. The various launch vehicle versions differ according to the number and type of strap-on boosters and the size of the fairings.

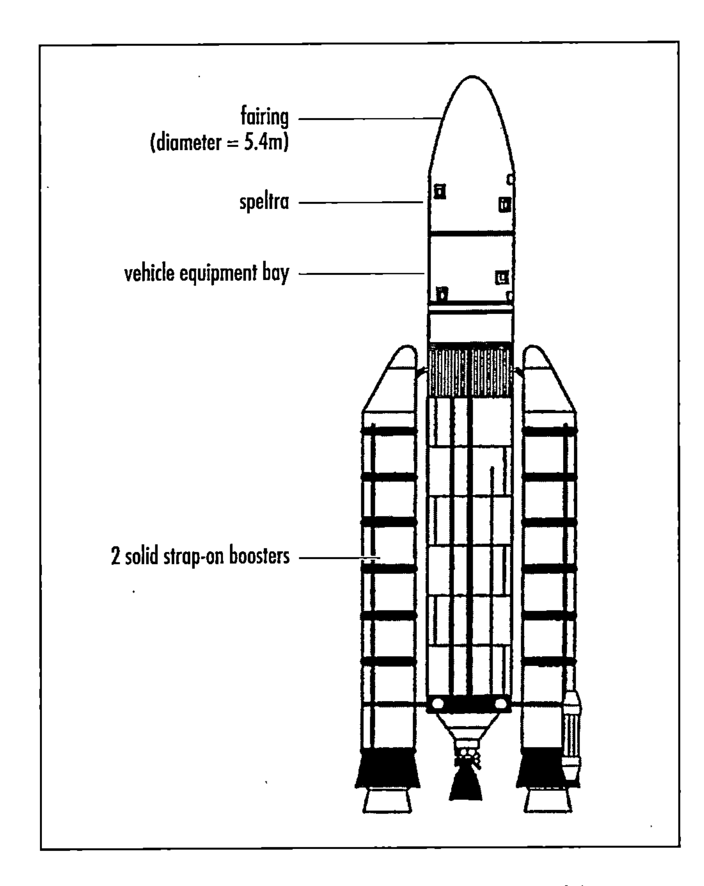

The general features of the Ariane-5 "two-stage" launch vehicle, consisting of a powerful liquid hydrogen/liquid oxygen–fueled Vulcain main engine and two strap-on solid propellant rockets

On October 30, 1997, a new, more powerful launcher, called Ariane-5, joined the Ariane family, when its second qualification flight (called V502 or Ariane 502) successfully took place at the Guiana Space Center in Kourou. At 27 minutes into this flight the *Maqsat-H* and *Maqsat-B* platforms, carrying instruments to analyze the new launch vehicle's performance, and the *TeamSat* technology satellite were injected into orbit.

The Ariane-5 launch vehicle is an advanced "two-stage" system, consisting of a powerful liquid hydrogen/liquid oxygen–fueled main engine (called the Vulcain) and two strap-on solid propellant rockets. It is capable of placing a satellite payload of 5,900 kg (dual satellite launch) or 6,800 kg (single satellite launch) into geostationary transfer orbit (GTO) and approximately 20,000 kg into low-Earth orbit (LEO).

However, the development of the Ariane-5 has not been without some technical challenges. On June 4, 1996, the maiden flight of the Ariane-5 launch vehicle ended in a spectacular failure. Approximately 40 seconds after initiation of the flight sequence, at an altitude of about 3,700 m, a guidance system error caused the vehicle to suddenly veer off its flight path, break up, and explode—as the errant vehicle properly self-destructed. Then, on December 12, 2002, approximately three minutes after lift-off, the more powerful (10-tonne) Ariane-5 vehicle involved in Flight 157 experienced another catastrophic failure. A leak in the Vulcain 2 nozzle's cooling circuit during ascent appears to have produced a nozzle failure that then led to a loss of control over the launch vehicle's trajectory.

With the retirement of the Ariane-4 system, the continuously improving Ariane-5 propulsion system now serves as the main launch vehicle for the European Community. The Ariane-5 launch vehicle can place a wide variety of payloads into low-Earth orbit, into geostationary Earth orbit (GEO), and on interplanetary trajectories.

Ariane vehicles are launched from Kourou, French Guiana, in South America. Kourou was chosen in part because it is close to Earth's equator, which makes the site ideal for launching most satellites. Following the first firing of a Diamant rocket by CNES (the French Space Agency) in 1970, Europe decided to use Kourou for its Europa launch vehicles. In 1975 the European Space Agency (ESA) took over the existing European facilities at the Guiana Space Center to build the ELA-1 complex (the Ensemble de Lancement Ariane, or Ariane Launch Complex) for its Ariane-1, 2, and 3 launch vehicles. The ESA next created the ELA-2 complex for the Ariane-4 vehicle. The Ariane-5 vehicle is launched from ELA-3, a modern launch complex whose facilities spread over 21 square km. In all, the European spaceport at Kourou covers 960 square km and has a workforce of more than 1,000 persons.

See also CENTRE NATIONAL D'ÉTUDES SPATIALES (CNES); EUROPEAN SPACE AGENCY; KOUROU LAUNCH COMPLEX; LAUNCH VEHICLE.

Ariel Discovered in 1851 by the British astronomer WILLIAM LASSELL, this moon of Uranus is named after the mischievous airy spirit in William Shakespeare's play *The Tempest.* Ariel has a diameter of 1158 km and a mass of 1.3×10^{21} kg. Like

Titania but smaller, the bright moon is a mixture of water ice and rock. Its cratered surface has a system of deep interconnected rift valleys (chasmata) that are hundreds of kilometers long. Ariel is in synchronous orbit and travels around Uranus in 2.52 days at a distance of 191,020 km.

See also URANUS.

Ariel spacecraft The family of scientific satellites developed by the United Kingdom and launched and operated in cooperation with the United States (NASA) between 1962 and 1979. The first four satellites, called *Ariel 1, 2, 3,* and *4,* focused on investigating Earth's ionosphere. *Ariel 5* was dedicated to X-ray astronomy, and *Ariel 6* supported both cosmic ray research and X-ray astronomy.

Ariel 1, the United Kingdom's first Earth-orbiting satellite, was launched by a Thor-Delta rocket from Cape Canaveral Air Force Station in Florida on April 26, 1962. The satellite went into a 389-km (pergiee) by 1,214-km (apogee) orbit around Earth and had a period of approximately 101 minutes, an inclination of 53.9 degrees, and an eccentricity of 0.0574. The small 62-kg cylindrically shaped spacecraft had a diameter of 0.58-meter and a height of 0.22-meter. *Ariel 1* carried a tape recorder and instrumentation for one cosmic ray, two solar emission, and three ionosphere experiments. Following launch, this early scientific satellite operated nominally until July 9, 1962—except for the solar Lyman–alpha emission experiment, which failed upon launch. From July 9 to September 8, 1962, the spacecraft had only limited operation. Then, spacecraft controllers operated *Ariel 1* again from August 25, 1964, to November 9, 1964, in order to obtain scientific data concurrently with the operation of NASA's *Explorer 20* satellite.

Ariel 2 was launched by a Scout rocket from NASA's Wallops Flight Facility on March 27, 1964. The spacecraft carried British experiments to measure galactic radio noise. *Ariel 2* traveled around Earth in a 180-km (apogee) by 843-km (perigee) orbit with a period of 101.3 minutes, an inclination of 51.6 degrees, and an eccentricity of 0.048.

Ariel 3 was launched on May 5, 1967, by a Scout rocket from Vandenberg Air Force Base (AFB) in California. The small scientific satellite traveled around Earth in a 497-km (perigee) by 608-km (apogee) polar orbit with a period of 95.7 minutes, an inclination of 80.2 degrees, and an eccentricity of 0.008. Instruments on board *Ariel 3* extended earlier British investigations of the ionosphere and near-Earth space. Controllers turned off the satellite in September 1969, and it decayed from orbit and disintegrated in Earth's upper atmosphere on December 14, 1970.

Ariel 4 was launched on December 11, 1971, by a Scout rocket from Vandenberg AFB. The small scientific satellite traveled around Earth in a 480-km (perigee) by 590-km (apogee) polar orbit with a period of 95.1 minutes, an inclination of 83 degrees, and an eccentricity of 0.008. British scientists designed the scientific payload of *Ariel 4* to investigate the interactions among electromagnetic waves, plasmas, and energetic particles present in Earth's upper ionosphere. The spacecraft had a design lifetime of one year.

The scientific instruments on board *Ariel 5* supported X-ray astronomy. A Scout rocket carried the satellite into orbit on October 15, 1974, from the San Marco launch platform in the Indian Ocean off the coast of Kenya, Africa. *Ariel 5* traveled around Earth every 95 minutes in an almost circular (512-km by 557-km) low-inclination (2.9 degree) orbit. Data were stored on board the spacecraft in a core storage and then dumped to ground stations once per orbit. The *Ariel 5* satellite monitored the sky with six different instruments (5 British and 1 American) from October 15, 1974, to March 14, 1980, when it reentered Earth's atmosphere.

Ariel 6 was the final satellite in this series of scientific spacecraft supporting a United Kingdom–United States collaborative space research program. A Scout rocket launched *Ariel 6* from NASA's Wallops Island Flight Facility on June 3, 1979 (local time), and placed the spacecraft into a near circular 625-km-altitude, 55-degree-inclination orbit. The objective of this mission was to perform scientific observations in support of high-energy astrophysics, primarily cosmic ray and X-ray measurements. *Ariel 6* continued to provide data until February 1982.

Aristarchus of Samos (ca. 320–ca. 250 B.C.E.) Greek *Mathematician, Astronomer* Aristarchus was the first Greek astronomer to suggest that Earth not only revolved on its axis, but also traveled around the Sun along with the other known planets. Unfortunately, he was severely criticized for his bold hypothesis of a moving Earth. At the time Greek society favored the teachings of ARISTOTLE and other, who advocated a "geocentric cosmology" with an immovable Earth at the center of the universe. Almost 18 centuries would pass before NICHOLAS COPERNICUS revived heliocentric (Sun-centered) cosmology.

Aristotle (384–322 B.C.E.) Greek *Philosopher* Aristotle was one of the most influential thinkers in the development of Western civilization. He endorsed and embellished the geocentric (Earth-centered) cosmology of EUDOXUS OF CNIDUS and championed a worldview that dominated astronomy for almost 2,000 years. In Aristotelian cosmology a nonmoving Earth was surrounded by a system of 49 concentric transparent ("crystal") spheres, each helping to account for the motion of all visible celestial bodies. The outermost sphere contained the fixed stars. PTOLEMY later replaced Aristotle's spheres with a system of epicycles. Modified but unchallenged, Aristotelian cosmology dominated Western thinking for almost two millennia, until finally surrendering to the heliocentric (Sun-centered) cosmology of NICHOLAS COPERNICUS, JOHANNES KEPLER, and GALILEO GALILEI.

Arizona meteorite crater One of Earth's most famous and best-preserved meteorite impact craters. It is approximately 175 meters deep and 1,200 meters across and is located in northern Arizona about 55 kilometers east of Flagstaff. This impact crater formed about 50,000 years ago when a large (possibly 50- to 70-meter-diameter) iron meteorite crashed into the desert surface at an estimated speed of approximately 16 km/s. Also called Meteor Crater or the Barringer Crater after the American mining engineer Daniel Barringer (1860–1929), who investigated the crater and was the first to suggest its extraterrestrial origin as the result of a massive meteorite strike.

See also CRATER; IMPACT CRATER; METEOROIDS.

Armstrong, Neil A. (1930–) American *Astronaut* Neil Armstrong is the American astronaut and X-15 test pilot who served as the commander for NASA's *Apollo 11* lunar landing mission in July 1969. As he became the first human being to set foot on the Moon (July 20, 1969), he uttered these historic words: "That's one small step for a man, one giant leap for mankind."

See also APOLLO PROJECT.

Arp, Halton Christian (1927–) American *Astronomer* The astronomer Harlton Arp published the *Atlas of Peculiar Galaxies* in 1966. He also proposed a controversial theory concerning redshift phenomena associated with distant quasars.

See also GALAXY; QUASARS; REDSHIFT.

Arrest, Heinrich Louis d' (1822–1875) German *Astronomer* Heinrich d'Arrest worked as an astronomer at the Berlin Observatory, Germany, and the Copenhagen Observatory, Denmark. While working at the Copenhagen Observatory he discovered numerous celestial objects, including asteroids, comets, and nebulas. While working at the Berlin Observatory he participated with the German astronomer JOHANN GALLE in the 1846 telescopic search that led to the discovery of Neptune.

See also ASTEROID; COMET; NEPTUNE.

Apollo 11 **astronaut Neil Armstrong, the first human being to walk on the Moon** ***(Courtesy of NASA)***

Arrhenius, Svante August (1859–1927) Swedish *Chemist, Exobiologist* Years ahead of his time, Svante A. Arrhenius was the pioneering physical chemist who won the 1903 Nobel Prize in chemistry for a brilliant idea that his conservative doctoral dissertation committee barely approved in 1884. His wide-ranging talents anticipated such space age scientific disciplines as planetary science and exobiology. In 1895 he became the first scientist to formally associate the presence of "heat trapping" gases, such as carbon dioxide, in a planet's atmosphere with the greenhouse effect. Then, early in the 20th century he caused another scientific commotion when he boldly speculated about how life might spread from planet to planet and might even be abundant throughout the universe.

Arrhenius was born on February 19, 1859, in the town of Vik, Sweden. His father was a land surveyor employed by the University of Uppsala and responsible for managing the university's estates at Vik, where Svante Arrhenius was born. His uncle was a well-respected professor of botany and rector of the Agricultural High School near Uppsala.

In 1860 his family moved to Uppsala. While a student at the cathedral school in Uppsala, the young Arrhenius demonstrated his aptitude for arithmetical calculations and developed a great interest in mathematics and physics. Upon graduation from the cathedral school in 1876, he entered the University of Uppsala, where he studied mathematics, chemistry, and physics. He earned a bachelor's degree from the university in 1878 and continued there for an additional three years as a graduate student. However, Arrhenius encountered professors at the University of Uppsala who dwelled in too conservative a technical environment and would not support his innovative doctoral research topic involving the electrical conductivity of solutions. So in 1881 he went to Stockholm to perform his dissertation research in absentia under Professor Eric Edlund at the Physical Institute of the Swedish Academy of Sciences.

Responding to the more favorable research environment with Edlund in Stockholm, Arrhenius pursued his scientific quest to answer the mystery in chemistry of why a solution of salt water conducts electricity when neither salt nor water by themselves do. His brilliant hunch was "ions"—that is, electrolytes when dissolved in water split or dissociate into electrically opposite positive and negative ions. In 1884 Arrhenius presented this scientific breakthrough in his thesis, "Investigations on the galvanic conductivity of electrolytes." However, the revolutionary nature of his ionic theory simply overwhelmed the orthodox-thinking reviewers on his doctoral committee at the University of Uppsala. They just barely passed him by giving his thesis the equivalent of a "blackball" fourth-class rank, declaring his work was "not without merit."

Undeterred, Arrhenius accepted his doctoral degree, continued to promote his new ionic theory, visited other innovative thinkers throughout Europe, and explored new areas of science that intrigued him. For example, in 1887 he worked with LUDWIG BOLTZMANN in Graz, Austria. Then, in 1891

Arrhenius accepted a position as a lecturer in physics at the Technical University in Stockholm (Stockholms Högskola). There, in 1894, he met and married his first wife, Sofia Rudeck, his student and assistant. The following year he received a promotion to professor in physics, and the newly married couple had a son, Olev Wilhelm. His first marriage was only a brief one and ended in divorce in 1896.

The Nobel Prize committee viewed the quality of Arrhenius's pioneering work in ionic theory quite differently than did his doctoral committee. They awarded Arrhenius the 1903 Nobel Prize in chemistry "in recognition of the extraordinary services he has rendered to the advancement of chemistry by his electrolytic theory of dissociation." At the awards ceremony in Stockholm he met another free-spirited genius, Marie Curie (1867–1934), who shared the 1903 Nobel Prize in physics with her husband, Pierre (1859–1906), and A. Henri Becquerel (1852–1908) for the codiscovery of radioactivity. Certainly, the year 1903 was an interesting and challenging one for the members of the Nobel Prize selection committee. Arrhenius's discovery shattered conventional wisdom in both physics and chemistry, while Marie Curie's pioneering radiochemistry discoveries forced the committee to include her as a recipient, shattering male dominance in science and making her the first woman to receive the prestigious award.

In 1905 Arrhenius retired from his professorship in physics and accepted a position as the director of the newly created Nobel Institute of Physical Chemistry in Stockholm. This position was expressly tailored for Arrhenius by the Swedish Academy of Sciences to accommodate his wide-ranging technical interests. That same year he married his second wife, Maria Johansson, who eventually bore him two daughters and a son.

Soon a large number of collaborators came to the Nobel Institute of Physical Chemistry from all over Sweden and numerous other countries. The institute's creative environment allowed Arrhenius to spread his many ideas far and wide. Throughout his life he took a very lively interest in various branches of physics and chemistry and published many influential books, including *Textbook of Theoretical Electrochemistry* (1900), *Textbook of Cosmic Physics* (1903), *Theories of Chemistry* (1906), and *Theories of Solutions* (1912).

In 1895 Arrhenius boldly ventured into the fields of climatology, geophysics, and planetary science when he presented an interesting paper to the Stockholm Physical Society: "On the influence of carbonic acid in the air upon the temperature of the ground." This visionary paper anticipated by decades contemporary concerns about the greenhouse effect and the rising carbon dioxide (carbonic acid) content in Earth's atmosphere. In the article Arrhenius argued that variations in trace atmospheric constituents, especially carbon dioxide, could greatly influence Earth's overall heat (energy) budget.

During the next 10 years Arrhenius continued his pioneering work on the effects of carbon dioxide on climate, including his concern about rising levels of anthropogenic (human-caused) carbon dioxide emissions. He summarized his major thoughts on the issue in the 1903 textbook *Lehrbuch der kosmichen Physik* (Textbook of cosmic physics)—an interesting work that anticipated scientific disciplines (namely, planetary science and Earth system science) that did not yet exist.

A few years later Arrhenius published the first of several of his popular technical books: *Worlds in the Making* (1908). In this book he described the "hot-house theory" (now called the greenhouse effect) of the atmosphere. He was especially interested in explaining how high-latitude temperature changes could promote the onset of ice ages and interglacial periods. But in *Worlds in the Making* he also introduced his *panspermia* hypothesis—a bold speculation that life could be spread through outer space from planet to planet or even from star system to star system by the diffusion of spores, bacteria, or other microorganisms.

In 1901 he was elected to the Swedish Academy of Sciences despite lingering academic opposition in Sweden to his internationally recognized achievements in physical chemistry. The Royal Society of London awarded him the Davy Medal in 1902 and elected him a foreign member in 1911. That same year, during a visit to the United States, he received the first Willard Gibbs Medal from the American Chemical Society. Finally, in 1914 the British Chemical Society presented him with its prestigious Faraday Medal. He remained intellectually active as the director of the Nobel Institute of Physical Chemistry until his death in Stockholm on October 2, 1927.

See also EARTH RADIATION BUDGET; EARTH SYSTEM SCIENCE; GREENHOUSE EFFECT; PANSPERMIA.

artificial gravity Simulated gravity conditions established within a spacecraft, space station, or space settlement. Rotating the space system about an axis creates this condition, since the centrifugal force generated by the rotation produces effects similar to the force of gravity within the vehicle. The Russian astronautics pioneer KONSTANTIN TSIOLKOVSKY first suggested this technique at the start of the 20th century.

artificial intelligence (AI) A term commonly taken to mean the study of thinking and perceiving as general information-processing functions, or the science of machine intelligence (MI). In the past few decades computer systems have been programmed to diagnose diseases; prove theorems; analyze electronic circuits; play complex games such as chess, poker, and backgammon; solve differential equations; assemble mechanical equipment using robotic manipulator arms and end effectors (the "hands" at the end of the manipulator arms); pilot uncrewed vehicles across complex terrestrial terrain, as well as through the vast reaches of interplanetary space; analyze the structure of complex organic molecules; understand human speech patterns; and even write other computer programs. All of these computer-accomplished functions require a degree of "intelligence" similar to mental activities performed by the human brain. Someday, a general theory of intelligence may emerge from the current efforts of scientists and engineers who are now engaged in the field of artificial intelligence. This general theory would help guide the design and development of even "smarter" robot spacecraft and exploratory probes, allowing us to more fully explore and use the resources that await us throughout the solar system and beyond.

Artificial intelligence generally includes a number of elements or subdisciplines. Some of these are planning and problem solving; perception; natural language; expert systems; automation, teleoperation, and robotics; distributed data management; and cognition and learning. Each of these A1 subdisciplines will now be discussed briefly.

All artificial intelligence involves elements of planning and problem solving. The problem-solving function implies a wide range of tasks, including decision making, optimization, dynamic resource allocation, and many other calculations and logical operations.

Perception is the process of obtaining data from one or more sensors and processing or analyzing these data to assist in making some subsequent decision or taking some subsequent action. The basic problem in perception is to extract from a large amount of (remotely) sensed data some feature or characteristic that then permits object identification.

One of the most challenging problems in the evolution of the digital computer has been the communication that must occur between the human operator and the machine. The human operator would like to use an everyday, or natural, language to gain access to the computer system. The process of communication between machines and people is very complex and frequently requires sophisticated computer hardware and software.

An expert system permits the scientific or technical expertise of a particular human being to be stored in a computer for subsequent use by other human beings who have not had the equivalent professional or technical experience. These expert systems have been developed for use in such diverse fields as medical diagnosis, mineral exploration, and mathematical problem solving. To create such an expert system, a team of software specialists will collaborate with a scientific expert to construct a computer-based interactive dialogue system that is capable, at least to some extent, of making the expert's professional knowledge and experience available to other individuals. In this case the computer, or "thinking machine," not only stores the scientific (or profes-

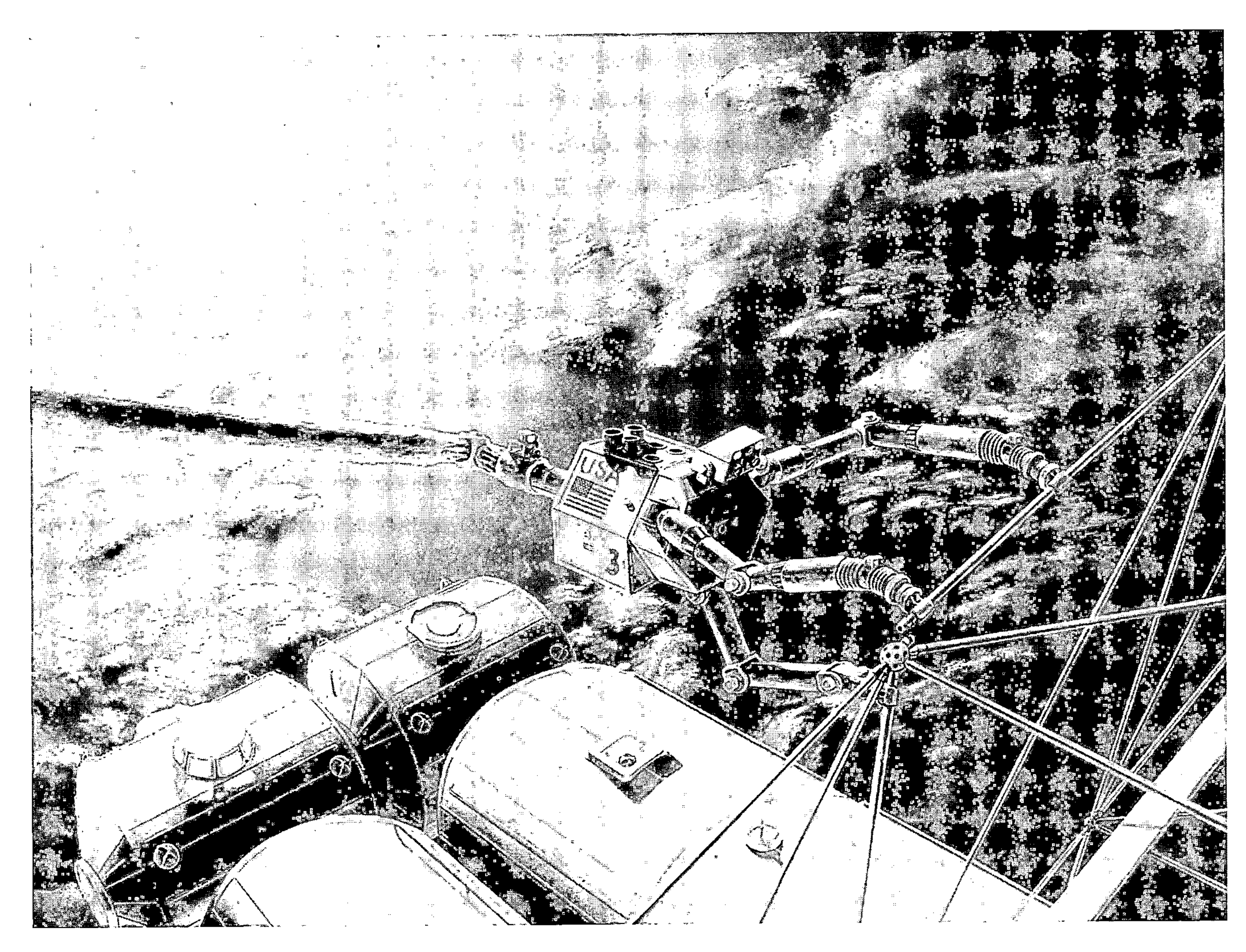

Future space robots like the one shown in this artist's rendering will need the ability to learn and possibly even understand as they encounter new situations and are forced to change their originally planned mode of operation. ***(Artist's rendering courtesy of NASA)***

sional) expertise of one human being, but also permits ready access to this valuable knowledge base because of its artificial intelligence, which guides other human users.

Automatic devices are those that operate without direct human control. NASA has used many such automated smart machines to explore alien worlds. For example, the two Viking landers placed on the Martian surface in 1976 represent one of the early great triumphs of robotic space exploration. After separation from the Viking orbiter spacecraft, the lander (protected by an aeroshell) descended into the thin Martian atmosphere at speeds of approximately 16,000 kilometers per hour. It was slowed down by aerodynamic drag until its aeroshell was discarded. Each robotic lander spacecraft slowed down further by releasing a parachute and then achieved a gentle landing by automatically firing retrorockets. Both Viking landers successfully accomplished the entire soft landing sequence automatically without any direct human intervention or guidance.

Teleoperation implies that a human operator is in remote control of a mechanical system. Control signals can be sent by means of "hardwire" (if the device under control is nearby) or via electromagnetic signals (for example, laser or radio frequency) if the robot system is some distance away. NASA's *Pathfinder* mission to the surface of Mars in 1997 successfully demonstrated teleoperation of a minirobot rover at planetary distances. The highly successful *Mars Pathfinder* mission consisted of a stationary "lander" spacecraft and a small "surface rover" called *Sojourner*. This six-wheeled minirobot rover vehicle was actually controlled (or "teleoperated") by the Earth-based flight team at the Jet Propulsion Laboratory (JPL) in Pasadena, California. The human operators used images of the Martian surface obtained by both the rover and the lander systems. These interplanetary teleoperations required that the rover be capable of some semiautonomous operation, since there was a time delay of the signals that averaged between 10 and 15 minutes depending on the relative position of Earth and Mars over the courses of the mission. This rover had a hazard avoidance system, and surface movement was performed very slowly. The 2003 Mars Exploration Rovers *Spirit* and *Opportunity* provided even more sophisticated and rewarding teleoperation experiences at interplanetary distances as they successfully scampered across different portions of the Red Planet in 2004.

Of course, in dealing with the great distances in interplanetary exploration, a situation is eventually reached when electromagnetic wave transmission cannot accommodate effective "real-time" control. When the device to be controlled on an alien world is many light-minutes or even light-hours away and when actions or discoveries require split-second decisions, teleoperation must yield to increasing levels of autonomous, machine intelligence–dependent robotic operation.

Robot devices are computer-controlled mechanical systems that are capable of manipulating or controlling other machine devices, such as end effectors. Robots may be mobile or fixed in place and either fully automatic or teleoperated.

Large quantities of data are frequently involved in the operation of automatic robotic devices. The field of distributed data management is concerned with ways of organizing cooperation among independent, but mutually interacting, databases.

In the field of artificial intelligence, the concept of cognition and learning refers to the development of a machine intelligence that can deal with new facts, unexpected events, and even contradictory information. Today's smart machines handle new data by means of preprogrammed methods or logical steps. Tomorrow's "smarter" machines will need the ability to learn, possibly even to understand, as they encounter new situations and are forced to change their mode of operation.

Perhaps late in the 21st century, as the field of artificial intelligence sufficiently matures, we will send fully automatic robot probes on interstellar voyages. Each very smart interstellar probe will have to be capable of independently examining a new star system for suitable extrasolar planets and, if successful in locating one, beginning the search for extraterrestrial life. Meanwhile, back on Earth, scientists will have to wait for its electromagnetic signals to travel light-years across the interstellar void, eventually informing its human builders that the extraterrestrial exploration plan has been successfully accomplished.

See also MARS EXPLORATION ROVER MISSION; *MARS PATHFINDER;* ROBOTICS IN SPACE; VIKING PROJECT.

artificial satellite A human-made object, such as a spacecraft, placed in orbit around Earth or other celestial body. *Sputnik 1* was the first artificial satellite to be placed in orbit around Earth.

Aryabhata India's first satellite, named after the fifth-century Indian mathematician Aryabhata the Elder (476–550). The spacecraft was launched by the Soviet Union on April 19, 1975, from Kapustin Yar. The satellite traveled around Earth in an orbit that had a period of 96.3 minutes, an apogee of 619 kilometers, and a perigee of 563 kilometers, at an inclination of 50.7 degrees. *Aryabhata* was built by the Indian Space Research Organization (ISRO) to conduct experiments in X-ray astronomy and solar physics. The spacecraft was a 26-sided polygon 1.4 meters in diameter. All faces of this polygon (except the top and bottom) were covered with solar cells. A power failure halted experiments after four days in orbit. All signals from the spacecraft were lost after five days of operation. The satellite reentered the Earth's atmosphere on February 11, 1992.

ascending node That point in the orbit of a celestial body when it travels from south to north across a reference plane, such as the equatorial plane of the celestial sphere or the plane of the ecliptic. Also called the *northbound node. Compare with* DESCENDING NODE.

AsiaSat 3 A communications satellite launched on December 24, 1997, by a Russian Proton-K rocket from the Baikonur Cosmodrome in Kazakhstan. Originally owned by Asia Satellite Telecommunications Co. Ltd. (AsiaSat) of Hong Kong, People's Republic of China, this communications satellite was intended for use primarily for television distribution and telecommunications services throughout Asia, the Middle East, and the Pacific Basin (including Australia). Specifically, it has multiple spot beams to provide telecommunications services to selected geographic areas.

The body-stabilized satellite was 26.2 meters tip-to-tip along the axis of the solar arrays and 10 meters across the axis of the antennas. The bus was essentially a cube, roughly four meters on a side. Power to the spacecraft was generated using two sun-tracking, four-panel solar wings covered with gallium-arsenide (Ga-As) solar cells, which provided up to 9,900 watts of electric power. A 29-cell nickel-hydrogen (Ni-H) battery provided electric power to the spacecraft during eclipse operations. For station-keeping the spacecraft used a bipropellant propulsion system consisting of 12 conventional bipropellant thrusters. Two 2.72-meter-diameter antennas were mounted on opposing sides of the bus, perpendicular to the axis along which the solar arrays were mounted. One of these antennas operated in C-band, the other in Ku-band. A 1.3-m-diameter reflector operating in the Ku-band provided focused area coverage. A one-meter-diameter Ku-band steerable spot-beam antenna provided the spacecraft with the ability to direct five-degree coverage of any area on Earth's surface visible to the spacecraft from its orbital location. Both of these antennas were mounted on the nadir side of the spacecraft.

Although the satellite reached geosynchronous altitude after launch, a malfunction in the fourth propulsion stage resulted in a dysfunctional orbit. Following its failure to achieve the proper operational orbit, the manufacturer (Hughes Global Services) purchased the spacecraft back from the insurance companies and renamed it *HGS 1*. Ground controllers then maneuvered the *HGS 1* spacecraft into two successive flybys of the Moon—innovative operations that eventually placed the wayward communications satellite into a useful geosynchronous orbit. This was the first time aerospace ground controllers used orbital mechanics maneuvers around the Moon to salvage a commercial satellite.

See also BAIKONOUR COSMODROME; COMMUNICATIONS SATELLITE.

A stars Stars of spectral type A that appear blue-white or white in color and have surface temperatures ranging between approximately 7,500 and 11,000 K. Altair, Sirius, and Vega are examples.

See also SPECTRAL CLASSIFICATION; STAR.

asteroid A small, solid, rocky body without atmosphere that orbits the Sun independent of a planet. The vast majority of asteroids, which are also called minor planets, have orbits that congregate in the asteroid belt or the main belt, a vast doughnut-shaped region of heliocentric space located between the orbits of Mars and Jupiter. The asteroid belt extends from approximately 2 to 4 astronomical units (AU) (or about 300 million to 600 million kilometers from the Sun) and probably contains millions of asteroids ranging widely in size from Ceres—which at about 940 km in diameter is about one-quarter the diameter of our Moon—to solid bodies that are less than 1 km across. There are more than 20,000 numbered asteroids.

NASA's *Galileo* spacecraft was the first to observe an asteroid close up, flying past main belt asteroids Gaspra and Ida in 1991 and 1993, respectively. Gaspra (19 km long) and Ida (58 km long) proved to be irregularly shaped objects, rather like potatoes, riddled with craters and fractures. The *Galileo* spacecraft also discovered that Ida had its own moon, a tiny body called Dactyl in orbit around its parent asteroid. Astronomers suggest Dactyl may be a fragment from past collisions in the asteroid belt.

NASA's Near-Earth Asteroid Rendezvous (NEAR) mission was the first scientific mission dedicated to the exploration of an asteroid. The NEAR *Shoemaker* spacecraft caught up with the asteroid Eros in February 2000 and orbited the small body for a year, studying its surface, orbit, mass, composition, and magnetic field. Then, in February 2001 mission controllers guided the spacecraft to the first-ever landing on an asteroid.

Scientists currently believe that asteroids are the primordial material that was prevented by Jupiter's strong gravity from accreting (accumulating) into a planet-sized body when the solar system was born about 4.6 billion years ago. It is estimated that the total mass of all the asteroids (if assembled together) would create a celestial body about 1,500 kilometers in diameter—an object less than half the size of the Moon.

Known asteroids range in size from about 940 kilometers in average diameter (Ceres, the first asteroid discovered) down to "pebbles" a few centimeters in diameter. There are 16 asteroids that have diameters of 240 kilometers or more. The majority of main belt asteroids follow slightly elliptical, stable orbits, revolving around the Sun in the same direction as Earth and the other planets and taking from three to six years to complete a full circuit of the Sun.

Minor planet 1, Ceres, is the largest main belt asteroid, with a diameter that is some 960 by 932 kilometers. The Italian astronomer GIUSEPPI PIAZZI discovered Ceres in 1801. This type-C asteroid has a rotation period of 9.075 hours and an orbital period (around the Sun) of 4.6 years. The German astronomer HEINRICH W. OLBERS discovered minor planet 2, Pallas, in 1802. This large asteroid has irregular dimensions

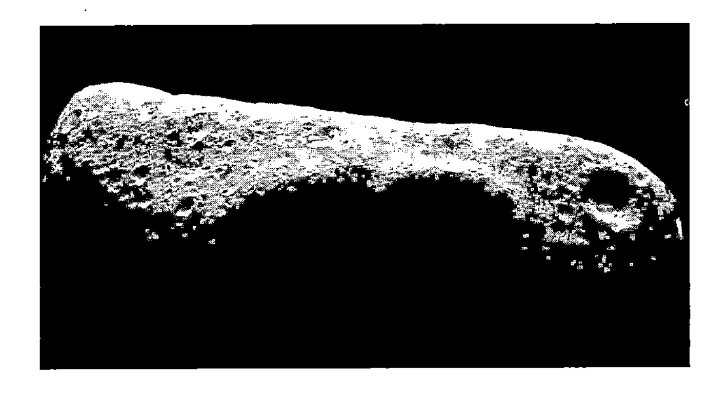

A mosaic image of the southern hemisphere of the asteroid Eros taken by NASA's *NEAR Shoemaker* spacecraft on November 30, 2000. The image provides a long-distance view of the 33-km-long asteroid and its cratered terrain. The direction south is to the top of the image. *(Courtesy of NASA and Johns Hopkins University Applied Physics Laboratory)*

(570 km by 525 km by 482 km) and a rotation period of 7.81 hours. Its orbital period is 4.61 years. Minor planet 3, Juno, was discovered in 1804 by the German astronomer Karl Harding (1765–1834). Juno is a type-S asteroid with a diameter of 240 kilometers and a rotation period of 7.21 hours. Its orbital period is 4.36 years. Minor planet 4, Vesta, was the fourth asteroid discovered. German astronomer Heinrich W. Olbers made the discovery in 1807. Vesta has a diameter of approximately 530 kilometers and is the third-largest asteroid (behind Ceres and Pallas). This interesting minor planet has a rotation period of 5.34 hours and an orbital period of 3.63 years. With an average albedo of 0.38, it is the only asteroid bright enough to be observed by the naked eye. Observations in 1997 by the *Hubble Space Telescope* (HST) revealed surface features that suggest a complex geological history, including lava flows and impact basins. Especially noteworthy, the HST instruments found an enormous crater that formed about a billion years ago. Scientists point out that while Vesta is an uncommon asteroid type, meteorites having the same composition as Vesta have been found on Earth. This leads to the interesting speculation that such Vestalike meteorites may actually be remnants from the ancient collision that created the minor planet's giant crater. Future exploration by robotic spacecraft should help solve this mystery.

Our understanding of asteroids has been derived from three main sources: Earth-based remote sensing, robot spacecraft flybys and landings (including the *Galileo* and NEAR-*Shoemaker* missions), and laboratory analyses of meteorites that are assumed to be related to asteroids. Scientists classify asteroids according to the shape and slope of their reflectance spectra. An elaborate characterization system of 14 classes has evolved, including C (0.03–0.09 albedo), B (0.04–0.08 albedo), F (0.03–0.06 albedo), G (0.05–0.09 albedo), P (0.02–0.06 albedo), D (0.02–0.05 albedo), T (0.04–0.11 albedo), S (0.10–0.22 albedo), M (0.10–0.18 albedo), E (0.25–0.60 albedo), A (0.13–0.40 albedo), Q (moderate albedo), R (moderately high albedo), and V (moderately high albedo). The last four classes (namely, A, Q, R and V) are quite rare types of asteroids, generally with just one or a few asteroids falling into any of these classes. The term *albedo* refers to an object's reflectivity, or intrinsic brightness. A white, perfectly reflecting surface has an albedo of 1.0, while a black, perfectly absorbing surface has an albedo of 0. However, albedo value, while quite useful, does not by itself uniquely establish the class of a particular asteroid.

The majority of asteroids fall into the following three classes or categories: C-type, S-type, and M-type. The C-type, or carbonaceous, asteroids appear to be covered with a dark, carbon-based material and are assumed to be similar to carbonaceous chondrite meteorites. This class includes more than 75 percent of known asteroids. C-type asteroids are very dark and have albedos ranging from 0.03 to 0.09. They are predominantly located in the outer regions of the main belt.

The S-type, or silicaceous, asteroids account for about 17 percent of the known asteroids. They are relatively bright, with an albedo ranging between 0.10 and 0.22. S-type asteroids dominate the inner portion of the main belt. This type of asteroid is thought to be similar to silicaceous or stony-iron meteorites. Of particular interest in the development of an extraterrestrial civilization is the fact that S-class asteroids (especially the more readily accessible Earth-crossing ones) may contain up to 30 percent free metals—that is, alloys of iron, nickel, and cobalt, along with high concentrations of precious metals in the platinum group.

The M-type, or metallic, asteroids are relatively bright, with an albedo ranging from 0.10 to 0.18. These asteroids are thought to be the remaining metallic cores of very small, differentiated planetoids that have been stripped of their crusts through collisions with other asteroids. M-type minor planets are found in the middle region of the main belt.

To date there is no general scientific consensus on the origin or nature of the various types of asteroids and their relationship to different types of meteorites. The most common meteorites, known as ordinary chondrites, are composed of small grains of rock and appear to be relatively unchanged since the solar system formed. Stony-iron meteorites, on the other hand, appear to be remnants of larger bodies that were once melted so that the heavier metals and lighter rocks separated into different layers. Space scientists have also hypothesized that some of the dark (that is, low-albedo) asteroids whose perihelia (orbital points nearest the Sun) fall within the orbit of Mars may actually be extinct cometary nuclei.

Earth-approaching and Earth-crossing asteroids (ECAs) are of special interest from both the perspective of extraterrestrial resource opportunities as well the possibility that one of these near-Earth objects (NEOs) could undergo a catastrophic collision with Earth. A "killer asteroid," a kilometer or greater in diameter, impacting Earth would contain enough kinetic energy to trigger an extinction level event (ELE) similar to the violent cosmic collision that appears to have happened 65 million years ago, wiping out the dinosaurs and many other forms of life.

These inner solar system asteroids, or near-Earth asteroids (NEAs), fall into three groups. Each group was named for a prominent asteroid within it, namely Amor (after minor planet 1,221 Amor), Apollo (after minor planet 1,862 Apollo), and Aten (after minor planet 2,062 Aten).

Amor group asteroids have perihelia between 1.017 astronomical units (AU) (Earth's aphelion) and 1.3 AU (an arbitrarily chosen distance) and cross the orbit of Mars. Since these asteroids approach Earth (but do not cross its orbit), they are sometimes called Earth-approaching asteroids, or Earth-grazing asteroids (Earth grazers). Minor planet 1,221 Amor was discovered in 1932 by the Belgian astronomer EUGÈNE J. DELPORTE. This asteroid has a diameter of approximately 1 kilometer and an orbital period of 2.66 years. Minor planet 433 Eros is a member of the Amor group. It is an S-type asteroid with an orbital period of 1.76 years. Eros is irregularly shaped (33 by 13 by 13 km) and has a rotation period of 5.270 hours. The semimajor axis of its orbit around the Sun is 1.458 AU.

The Apollo group of asteroids have semimajor axes greater than 1.0 AU and aphelion distances less than 1.017 AU—taking them across the orbit of Earth. Minor planet 1,862 Apollo (the group's namesake) was discovered in 1932 by the German astronomer KARL REINMUTH. This S-type Earth-crossing asteroid has a diameter of about 1.6 kilometers and an orbital period of 1.81 years. After passing within

0.07 AU (~10.5 million km) of Earth in 1932, this minor became "lost in space" until 1973, when it was rediscovered. The largest member of the Apollo group is minor planet 1,866 Sisyphus. This ECA has a diameter of about 8 kilometers and an orbital period of 2.6 years.

The Aten group of asteroids also cross Earth's orbit but have semimajor axes less than 1.0 AU and aphelion distances greater than 0.983 AU (Earth's perihelion distance). Consequently, members of this group have an average distance from the Sun that is less than that of Earth. Minor planet 2,062 Aten was discovered in 1976 by the American astronomer Eleanor Kay Helin. It is an S-type asteroid and has a diameter of approximately 1 kilometer and an orbital period of 0.95 years. The largest member of this Earth-crossing group is called minor planet 3,753 (it has yet to be given a formal name). It has a diameter of about 3 kilometers, an orbital period of approximately 1 year, and travels in a complex orbit around the Sun.

Near-Earth asteroids are a dynamically young population of planetary bodies, whose orbits evolve on 100-million-year time scales because of collisions and gravitational interactions with the Sun and the terrestrial planets. The largest NEA discovered to date is minor planet 1,036 Ganymed, an S-type member of the Amor group. Ganymed has a diameter of about 41 kilometers. While less than 300 NEAs have been identified, current estimates suggest that there are more than 1,000 NEAs large enough (i.e., 1-kilometer diameter or more) to threaten Earth in this century and beyond.

No discussion of the minor planets would be complete without at least a brief mention of the Trojan asteroid group and the interesting asteroid/comet object Chiron. The Trojan asteroids form two groups of minor planets that are clustered near the two stable Lagrangian points of Jupiter's orbit, that is, 60° leading [L_4 point] and 60° following [L_5] the giant planet at a mean distance of 5.2 AU from the Sun. The first Trojan asteroid, minor planet 588 Achilles was discovered in 1906 at the L_4 Lagrangian point—60° ahead of Jupiter. More than 200 Trojan asteroids are now known. As part of astronomical tradition, major members of this group of asteroids have been named after the mythical heros of the Trojan War.

Asteroid 2060 Chiron was discovered in 1977 by the American astronomer Charles Kowal. With a perihelion distance of just 8.46 AU and an aphelion distance of about 18.9 AU, this unusual object lies well beyond the main asteroid belt and follows a highly chaotic, eccentric orbit that goes from within the orbit of Saturn out to the orbit of Uranus. Chiron's diameter is currently estimated at between 150 and 210 kilometers, and it has a rotation rate of approximately 5.9 hours. This enigmatic object has an orbital period of 50.7 years. Chiron, which is also called Comet 95P, is especially unusual because it has a detectable coma, indicating it is (or behaves like) a cometary body. However, it is also more than 50,000 times the characteristic volume of a typical comet's nucleus, possessing a size more commensurate with a large asteroid—which is what astronomers initially assumed it to be. Furthermore, its curious orbit is unstable (chaotic) on time scales of a million years, indicating that it has not been in its present orbit very long (from a cosmic time scale perspective). Chiron is actually the first of several such objects with similar orbits and properties discovered to date. In recognition of their dual comet-asteroid nature, astronomers call these peculiar objects Centaurs, after the race of half-human/half-horse beings found in Greek mythology. In fact, Chiron is named after the wisest of the Centaurs, the one who was the tutor of both Achilles and Hercules. Astronomers currently believe that Centaurs may be objects that have escaped from the Kuiper Belt, a disk of distant, icy planetesimals, or frozen comet nuclei, that orbit the Sun beyond Neptune. Kuiper Belt objects are also referred to as trans-Neptunian objects, or TNOs.

See also ASTEROID DEFENSE SYSTEM; ASTEROID MINING; EXTRATERRESTRIAL CATASTROPHE THEORY; EXTRATERRESTRIAL RESOURCES; GALILEO PROJECT; KUIPER BELT; NEAR EARTH ASTEROID RENDEZVOUS MISSION; ROSETTA MISSION.

asteroid belt The region of outer space between the orbits of Mars and Jupiter that contains the great majority of the asteroids. These minor planets (or planetoids) have orbital periods lying between three and six years and travel around the Sun at distances of between 2.2 to 3.3 astronomical units. Also called the *main belt.*

See also ASTEROID.

asteroid defense system Impacts by Earth-approaching asteroids and comets, often collectively referred to as near-Earth objects, or NEOs, pose a significant hazard to life and to property. Although the annual probability of Earth being struck by a *large* asteroid or comet (i.e., one with a diameter greater than 1 kilometer) is extremely small, the consequences of such a cosmic collision would be catastrophic on a global scale.

Scientists recommend that we carefully assess the true nature of this extraterrestrial threat and then prepare to use advances in space technology to deal with it as necessary in the 21st century and beyond. During the third United Nations conference on the exploration and peaceful uses of outer space held in Vienna, Austria, in July 1999, the International Astronomical Union (IAU) officially endorsed the Torino Impact Hazard Scale. The table on page 57 presents this risk assessment scale, which was developed to help the experts communicate to the public the potential danger of a threatening near-Earth object. On this scale a value 0 or 1 means virtually no chance of impact or hazard, while a value of 10 means a certain global catastrophe. This particular risk assessment scale was first introduced at an IAU workshop in Torino and was subsequently named for that Italian city. Similarly, in 2002 at another international meeting, risk assessment specialists introduced a second, more sophisticated impact hazard scale, subsequently called the Palermo Scale.

The Torino Scale is designed to communicate to the public the risk associated with a future Earth approach by an asteroid or comet. The Palermo Scale, which has integer values from 0 to 10, takes into consideration the predicted impact energy of the event as well as its likelihood of actually happening (that is, the event's impact probability). The Palermo Scale is used by specialists in the field to quantify in more detail the level of concern warranted for a future potential impact. Much of the utility of the more sophisticated Palermo Scale lies in its ability to more carefully assess the risk posed

Torino Impact Hazard Scale
(Assessing Asteroid and Comet Impact Hazard Predictions in the 21st Century)

Event Category	Scale	Significance and Potential Consequences
Events Having No Likely Consequences (White Zone on colored chart)	0	The likelihood of a collision is zero, or well below the chance that a random object of the same size will strike Earth within next few decades. This designation also applies to any small object that, in the event of a collision, is unlikely to reach Earth's surface intact.
Events Meriting Careful Monitoring (Green Zone)	1	The chance of collision is extremely unlikely, about the same as a random object of the same size striking Earth within the next few decades.
Events Meriting Concern (Yellow Zone)	2	A somewhat close, but not unusual encounter. Collision is very unlikely.
	3	A close encounter, with a 1% or greater chance of a collision capable of causing localized destruction.
	4	A close encounter, with 1% or greater chance of a collision capable of causing regional devastation.
Threatening Events (Orange Zone)	5	A close encounter, with a significant threat of a collision capable of causing regional devastation.
	6	A close encounter, with a significant threat of a collision capable of causing a global catastrophe.
	7	A close encounter, with an extremely significant threat of a collision capable of causing a global catastrophe.
Certain Collisions (Red Zone)	8	A collision capable of causing localized destruction. Such events occur somewhere on Earth between once per 50 years and once per 1000 years.
	9	A collision capable of causing regional devastation. Such events occur between once per 1000 years and once per 100,000 years.
	10	A collision capable of causing a global climatic catastrophe. Such events occur once per 100,000 years, or less often.

Source: NASA (July 1999).

by less threatening Torino Scale 0 events—events that actually constitute nearly all the potential impact hazards detected to date. Since the Palermo Scale is continuous and it depends upon the number of years until the potential impact, there is no convenient conversion between the two scales. In general, however, if an event rises above the background level, it will achieve a value greater than zero on both the Torino and Palermo Impact Hazard Scales. For the purposes of this discussion, however, the Torino Scale is more than adequate and far less cluttered with the mathematical details of probabilistic risk assessment.

Earth resides in a swarm of comets and asteroids that can and do impact its surface. In fact, the entire solar system contains a long-lived population of asteroids and comets, some fraction of which are perturbed into orbits that then cross the orbits of Earth and other planets. Since the discovery of the first Earth-crossing asteroid (called 1,862 Apollo) in 1932, alarming evidence concerning the true population of "killer asteroids" and "doomsday comets" in near-Earth space has accumulated. Continued improvements in telescopic search techniques over the past several decades have resulted in the discovery of dozens of near-Earth asteroids and short-period comets each year. Here, by arbitrary convention, a *short-period comet* refers to a comet that takes 20 years or less to travel around the Sun; while a *long-period comet* takes more than 20 years. Finally, the violent crash of Comet Shoemaker-Levy 9 into Jupiter during the summer of 1994 helped move the issue of "asteroid defense" from one of mere technical speculation to one of focused discussion within the international scientific community.

In general, an asteroid defense system would perform two principal functions: surveillance and threat mitigation. The surveillance function incorporates optical and radar tracking systems to continuously monitor space for threatening NEOs. Should a large potential impactor be detected, specific planetary defense activities would be determined by the amount of warning time given and the level of technology available. With sufficient warning time a killer impactor might either be deflected or sufficiently disrupted to save Earth.

Mitigation techniques greatly depend on the amount of warning time and are usually divided into two basic categories: techniques that physically destroy the threatening NEO by fragmenting it and techniques that deflect the threatening NEO by changing its velocity a small amount. For the first half of this century high-yield (megaton class) nuclear explosives appear to be the tool of choice to achieve either deflection or disruption of a large impactor. Interceptions far from Earth, made possible by a good surveillance system, are much more desirable than interceptions near the Earth-Moon system, because fewer explosions would be needed.

For example, a 1 megaton nuclear charge exploded on the surface of a 300-meter-diameter asteroid will most likely deflect this object if the explosion occurs when the asteroid is quite far away from Earth (say about 1 AU distant.) However, it would take hundreds of gigatons (10^9) of explosive yield to change the velocity of a 10-km-diameter asteroid by about 10 meters per second when the object is only two weeks from impacting Earth.

By the middle of this century a variety of advanced space systems should be available to nudge threatening objects into harmless orbits. Focused nuclear detonations, high-thrust nuclear thermal rockets, low- (but continuous) thrust electric propulsion vehicles, and even mass-driver propulsion systems that use chunks of the object as reaction mass could be ready to deflect a threatening NEO.

However, should deflection prove unsuccessful or perhaps impractical (due to the warning time available), then physical destruction or fragmentation of the impactor would be required to save the planet. Numerous megaton-class nuclear detonations could be used to shatter the approaching object into smaller, less dangerous pieces. A small asteroid (perhaps 10 to 100 meters in diameter) might also be maneuvered into the path of the killer impactor, causing a violent collision that shatters both objects. If such impactor fragmentation is attempted, it is essential that all the major debris fragments miss Earth; otherwise, the defense system merely changes a cosmic cannon ball into a celestial cluster bomb.

Contemporary studies suggest that the threshold diameter for an "extinction-level" NEO lies between 1 and 2 kilometers. Smaller impactors (with diameters from hundreds of meters down to tens of meters) can cause severe regional or local damage, but are not considered a global threat. However, the impact of a 10- to 15-km-diameter asteroid would most certainly cause massive extinctions. At the point of impact there would be intense shock wave heating and fires, tremendous earthquakes, giant tidal waves (if the killer impactor crashed into a watery region), hurricane-force winds, and hundreds of billions of tons of debris thrown into the atmosphere. As a giant cloud of dust spread across the planet, months of darkness and much cooler temperatures would result. In addition to millions of immediate casualties (depending on where the impactor hit), most global food crops would eventually be destroyed, and modern civilization would collapse.

The primary NEO threat is associated with Earth-crossing asteroids. About 2,000 large asteroids are believed to reside in near-Earth space, although less than 200 have been detected so far. Between 25 percent and 50 percent of these objects will eventually impact Earth, but the average time interval between such impacts is quite long—typically more than 100,000 years.

The largest currently known Earth-crossing asteroid is 1,866 Sisyphus. It has a diameter of approximately 8 km, making it only slightly smaller than the 10-km-diameter impactor believed to have hit Earth about 65 million years ago. That event, sometimes called the Cretaceous/Tertiary Impact, is thought to have caused the extinction of the dinosaurs and up to 75 percent of the other prehistoric species.

Short-period comets constitute only about 1 percent of the asteroidal NEO hazard. However, long-period comets, many of which are not detected until they enter the inner solar system for the first time, represent the second-most-important impact hazard. While their numbers amount to only a few percent of the asteroid impacts, these long-period comets approach Earth with greater speeds and therefore higher kinetic energy in proportion to their mass. It is now estimated that as many as 25 percent of the objects reaching Earth with energies in excess of 100,000 megatons (of equivalent trinitrotoluene [TNT] explosive yield) have been long-period comets. On average, one such comet passes between Earth and the Moon every century, and one strikes Earth about every million years.

Since long-period comets do not pass near Earth frequently, it is not really possible to obtain an accurate census of such objects. Each must be detected on its initial approach to the inner solar system. Fortunately, comets are much brighter than asteroids of the same size as a result of the outgassing of volatile materials stimulated by solar heating. Comets in the size range of interest (i.e., greater than 1 kilometer in diameter) will generally be visible to the telescopes of a planetary defense system by the time they reach the outer asteroid belt (about 500 million kilometers distant), providing several months of warning before they approach Earth. However, the short time span available for observation will result in less well-determined orbits and, therefore, greater uncertainty as to whether a planetary impact is likely. As a result, we can expect a greater potential for "false alarms" with threatening long-period comets than with Earth-crossing asteroids.

At present no *known* NEO has an orbit that will lead to a collision with Earth during the first half of this century. Scientists further anticipate that the vast majority of "to be discovered" NEOs (both asteroids and short-period comets) will likewise pose no immediate threat to our planet. Even if an Earth-crossing asteroid is discovered to be an impactor, it will typically make hundreds or at least tens of moderately close (near-miss) passes before it represents any real danger to our planet. Under these circumstances, there would be ample time (perhaps several decades) for developing an appropriate response. However, the lead time to counter a threat from a newly discovered long-period comet could be much less, perhaps a year or so. As a "planetary insurance policy," we might decide to deploy the key components of the planetary defense system in space by the middle of this century so these components can respond promptly to any short-notice NEO threat.

See also ASTEROID; ASTEROID MINING; COMET; EXTRATERRESTRIAL CATASTROPHE THEORY; IMPACT CRATER; METEOROIDS; TUNGUSKA EVENT.

asteroid mining The asteroids, especially Earth-crossing asteroids (ECAs), can serve as "extraterrestrial warehouses" from which future space workers may extract certain critical materials needed in space-based industrial activities and in the construction and operation of large space platforms and habitats. On various types (or classes) of asteroids scientists expect to find such important materials as water (trapped), organic compounds, and essential metals, including iron, nickel, cobalt, and the platinum group.

The "threat" of Earth-crossing asteroids has recently been acknowledged in both the scientific literature and popular media. However, beyond deflecting or fragmenting an Earth-threatening ECA there may be some great economic advantage in "capturing" such an asteroid and parking it in a safe Earth-Moon system orbit. In addition to the scientific and space technology "lessons learned" in such a capture mission, many economic benefits would also be gained by harvesting (mining) the asteroid's natural resources. Contemporary studies suggest that large-scale mining operations involving a single 1- to 2-kilometer-diameter S-type or M-type ECA could net trillions (10^{12}) of dollars in nickel, cobalt, iron, and precious platinum group metals. These revenues would help offset the overall cost of running a planetary asteroid defense system and of conducting a special mitigation (defense) mission against a particularly threatening ECA.

The stable Lagrangian libration points (L_4 and L_5) within cislunar space have been suggested as suitable "parking locations" for a captured ECA. Once parked in such a "safe" stable orbit, the captured ECA (*see* figure) might be harvested not only for its minerals but could also serve as a natural space platform from which to mount interplanetary expeditions or on which to conduct large-scale space construction operations (such as the development of satellite power stations).

Therefore, with the evolution of visionary asteroid mining strategies later in this century, threatening ECAs might actually represent opportunistic resources that could play a

An artist's concept of an asteroid mining operation ca. 2025 ***(Artist's rendering courtesy of NASA)***

major role in the evolution of a permanent human civilization in space.

See also ASTEROID; ASTEROID DEFENSE SYSTEM; EXTRATERRESTRIAL RESOURCES.

asteroid negation system (ANS) A proposed planetary defense system capable of intercepting any (natural) space object (i.e., an asteroid or comet) that was determined to be a threat to Earth. The system would accomplish this negation task in sufficient time to deflect the space object from its collision course with Earth or else fragment it into smaller pieces that would not pose a major planetary threat. Deflection and/or fragmentation could be accomplished with a variety of means, ranging from the detonation of nuclear explosives to the use of high specific impulse thrusters, kinetic energy projectiles, or directed energy devices.

See also ASTEROID DEFENSE SYSTEM.

astro- A prefix that means "star" or (by extension) outer space or celestial; for example, astronaut, astronautics, and astrophysics.

astrobiology The search for and study of living organisms found on celestial bodies beyond Earth.

See also EXOBIOLOGY.

astrobleme A geological structure (often eroded) produced by the hypervelocity impact of a meteoroid, comet, or asteroid.

astrochimp(s) The nickname given to the primates used during the early U.S. space program. In the Mercury Project (the first U.S. crewed space program), these astrochimps were used to test space capsule and launch vehicle system hardware prior to its commitment to human flight. For example, on January 31, 1961, a 17-kg chimpanzee named Ham was launched from Cape Canaveral Air Force Station by a Redstone rocket on a suborbital flight test of the Mercury Project spacecraft. During this mission the propulsion system developed more thrust than planned, and the simian space traveler experienced overacceleration. After recovery Ham appeared to be in good physical condition. (*See* figure.) However, when shown another space capsule, his reactions made it quite clear to his handlers that this particular astrochimp wanted no further role in the space program.

Astrochimp Ham reaches out for an apple after his successful suborbital rocket flight from Cape Canaveral on January 31, 1961. ***(Courtesy of NASA)***

On November 29, 1961, another astrochimp, a 17-kg chimpanzee named Enos, was launched successfully from Cape Canaveral by an Atlas rocket on the final orbital qualification flight test of the Mercury Project spacecraft. During the flight Enos performed psychomotor duties and upon recovery was found to be in excellent physical condition.

See also MERCURY PROJECT.

astrodynamics The application of celestial mechanics, propulsion system theory, and related fields of science and engineering to the problem of carefully planning and directing the trajectory of a space vehicle.

See also ORBITS OF OBJECTS IN SPACE.

astroengineering Incredible feats of engineering and technology involving the energy and material resources of an entire star system or several star systems. The detection of such astroengineering projects would be a positive indication of the presence of a Type II, or even Type III, extraterrestrial civilization in our galaxy. On example of an astroengineering project would be the creation of a *Dyson sphere*—a cluster of structures and habitats made by an intelligent alien species to encircle their native star and effectively intercept all of its radiant energy. *Compare with* PLANETARY ENGINEERING.

See also DYSON SPHERE; EXTRATERRESTRIAL CIVILIZATIONS.

astrolabe An ancient multipurpose astronomical instrument that provided early astronomers with a two-dimensional model of the celestial sphere with which to measure the position of celestial objects, measure the time of night or the time of year, and determine the altitude of any celestial object above the horizon. Historically, the astrolabe was an elaborately inscribed brass disk that functioned as the equivalent of a mechanical "lap-top" astronomical computer. The earliest known astrolabes appeared around 300 to 200 B.C.E. and were possibly used by the famous Greek astronomer HIPPARCHUS. From ancient Greece to the European Middle Ages, astronomers continuously refined the astrolabe and gradually transformed the device into a complex portable instrument that became indispensable in the practice of pretelescope astronomy. Arab astronomers contributed some of the most important improvements to this instrument, turning it into the most widely used multipurpose device of early astronomy.

The most important part of the traditional astrolabe was a circular brass plate typically about 15 centimeters or so in diameter with a convenient ring at the top by which the instrument could be suspended and hang perfectly vertical. Engraved upon the backside of the brass disk were several circles hosting convenient gradations, such as 360 degrees,

365 days, 12 months, and so forth. In time these backside engravings could also accommodate trigonometric calculations. The front of the brass disk also had several engraved circles. The outer circle usually had 24 divisions, corresponding to the hours in a day. Another circle was divided to serve as a calendar, using the constellations of the zodiac as a convenient reference. The celestial pole (northern or southern) was engraved in the central part of the brass front, as were lines representing the tropics and the equator.

Astronomers would place another disk (called the sky disk) on the front of the astrolabe in such a manner that they could rotate this removable sky disk. The thin brass sky disk had many opening cut into it. These cuts were precisely arranged so an astronomer could see the engraved front of the brass disk that made up the body of the astrolabe. By carefully adjusting each sky disk, early astronomers could determine the visible parts of the sky and the altitude of different celestial bodies for various times of the year and for different latitudes. The astrolabe also had a pivotal pointer or ruler attached to the center of the backside of the central brass plate. Early astronomers would suspend the astrolabe by its ring. Once the device was hanging in a proper vertical position, they would align the pointer with some target celestial object and then read its altitude from an appropriate measurement scale engraved on the backside of the instrument.

See also ANCIENT ASTRONOMICAL INSTRUMENTS.

astrology Attempt by many early astronomers to forecast future events on Earth by observing and interpreting the relative positions of the fixed stars, the Sun, the known planets, and the Moon. Such mystical stargazing was a common activity in most ancient societies and was enthusiastically practiced in western Europe up through the 17th century. For example, at the dawn of the scientific revolution GALILEO GALILEI taught a required university course on medical astrology, and JOHANNES KEPLER earned a living as a court astrologer. Astrology still lingers in modern society, especially in the form of daily horoscopes found in many newspapers. The popular pseudoscience of astrology is based on the unfounded and unscientific hypothesis that the motion of celestial bodies controls and influences human lives and terrestrial events. At the bare minimum, this erroneous hypothesis denies the existence of human free will.

See also ZODIAC.

astrometric binary A binary star system in which irregularities in the proper motion (wobbling) of a visible star imply the presence of an undetected companion.

See also BINARY STAR SYSTEM.

astrometry The branch of astronomy that is concerned with the very precise measurement of the motion and position of celestical objects; a subset or branch of astronomy. It is often divided into two major categories: *global astrometry* (addressing positions and motions over large areas of the sky) and *small-field astrometry* (dealing with relative positions and motions that are measured within the area observed by a telescope—that is, within the instantaneous field of view of the telescope).

Scientists involved in the practice of astrometry are often concerned with proper motion, nutation, and precession, phenomena that can cause the positions of celestial bodies to change over time. Astronomers participate in *photographic astrometry* when they accurately determine the positions of planets, stars, or other celestial objects with respect to the positions of reference stars as seen on photographic plates or high-resolution digital images.

Measuring the trigonometric parallax for a star provides the only completely reliable way of measuring distances in the local universe. Ground-based astrometric measurements are limited by several factors, including fluctuations in the atmosphere, limited sky coverage per observing site, and instrument flexure. Space-based astrometry helps overcome these disadvantages and supports global astrometry with many inherent advantages versus pointed (small-field) observations from the ground. The European Space Agency's *Hipparcos* satellite was the first space-based astrometry mission and demonstrated that milliarcsecond accuracy is achievable by means of a continuously scanning satellite that observes two directions simultaneously.

See also HIPPARCOS SPACECRAFT; NUTATION; PRECESSION; PROPER MOTION.

astronaut Within the American space program, a person who travels in outer space; a person who flies in an aerospace vehicle to an altitude of more than 80 km. The word comes from a combination of two ancient Greek words that literally mean "star" (*astro*) and "sailor or traveler" (*naut*). *Compare with* COSMONAUT.

astronautics The branch of engineering science dealing with spaceflight and the design and operation of space vehicles.

Astronomer Royal The honorary title created in 1675 by King Charles II and given to a prominent English astronomer. Up until 1971 the Astronomer Royal also served as the director of the Royal Greenwich Observatory. JOHN FLAMSTEED was the first English astronomer to hold this position. He served from 1675 until his death in 1719.

astronomical unit (AU) A convenient unit of distance defined as the semimajor axis of Earth's orbit around the Sun. One AU, the average distance between Earth and the Sun, is equal to approximately 149.6×10^6 km, 499.01 light-seconds.

Astronomische Nederlandse Satellite (ANS) The ANS spacecraft was launched by an expendable Scout rocket from the Western Test Range at Vandenberg Air Force Base in California on August 30, 1974. Following a successful launch the scientific spacecraft entered a polar orbit around Earth and operated until June 14, 1977. ANS was a collaborative effort between the Netherlands and the United States. The Universities of Groningen and Utrecht in the Netherlands provided the ultraviolet and the soft X-ray experiments, while NASA provided the hard X-ray experiment, the spacecraft, and the launch vehicle. The scientific instrument package on this satellite discovered X-ray bursts and made the initial detection of X-ray flares from the flare star UV Ceti.

See also FLARE STAR; X-RAY ASTRONOMY.

astronomy Branch of science that deals with celestial bodies and studies their size, composition, position, origin, and dynamic behavior. Astronomy dates back to antiquity and represents one of humankind's oldest forms of science. Prior to the use of the telescope in the early 17th century, the practice of astronomy was generally limited to naked eye observations, the use of ancient instruments (such as the astrolabe and the cross staff), and the application of progressively more sophisticated forms of mathematics. The arrival of the telescope gave birth to optical astronomy—an influential discipline that greatly stimulated the scientific revolution and accelerated the blossoming of modern science.

The arrival of the space age in 1957 allowed astronomers to observe the universe across all portions of the electromagnetic spectrum and to visit the previously unreachable, alien worlds in our solar system. Today modern astronomy has many branches and subdisciplines, including astrometry, gamma-ray astronomy, X-ray astronomy, ultraviolet astronomy, visual (optical) astronomy, infrared astronomy, microwave astronomy, radar astronomy, and radio astronomy. Closely related scientific fields include high-energy astrophysics, cosmic-ray physics, neutrino physics, solar physics, condensed-matter physics, space science, planetary geology, exobiology, and cosmology.

See also ASTROPHYSICS; COSMOLOGY.

astrophotography The use of photographic techniques to create images of celestial bodies. Astronomers are now replacing light-sensitive photographic emulsions with charged-coupled devices (CCDs) to create digital images in the visible, infrared, and ultraviolet portions of the electromagnetic spectrum. The application of photography in observational astronomy began in the middle part of the 19th century (*see* figure) and has exerted great influence on the field ever since.

This is the oldest close-up picture of the Moon on record. The picture was made in March 1851 and was originally a crystalotype print from a dauguerrotype photograph. The words *crystalotype* and *dauguerrotype* describe some of the early-19th-century photographic processes. The crystalotype is a sunlight-formed picture on glass, while the dauguerrotype photographic process involved the use of a mirrorlike polished silver surface. Since the dauguerrotype photographic process was a positive-only process, it did not support reproduction of the original image. *(Courtesy of NASA)*

astrophysics Astronomy addressed fundamental questions that have occupied humans since our primitive beginnings. What is the nature of the universe? How did it begin, how is it evolving, and what will be its eventual fate? As important as these questions are, there is another motive for astronomical studies. Since the 17th century, when SIR ISAAC NEWTON's studies of celestial mechanics helped him to formulate the three basic laws of motion and the universal law of gravitation, the sciences of astronomy and physics have become intertwined. *Astrophysics* can be defined as the study of the nature and physics of stars and star systems. It provides the theoretical framework for understanding astronomical observations. At times astrophysics can be used to predict phenomena before they have even been observed by astronomers, such as black holes. The laboratory of outer space makes it possible to investigate large-scale physical processes that cannot be duplicated in a terrestrial laboratory. Although the immediate, tangible benefits to humankind from progress in astrophysics cannot easily be measured or predicted, the opportunity to extend our understanding of the workings of the universe is really an integral part of the rise of our civilization.

Role of Modern Astrophysics

Today astrophysics has within its reach the ability to bring about one of the greatest scientific achievements ever—a unified understanding of the total evolutionary scheme of the universe. This remarkable revolution in astrophysics is happening now as a result of the confluence of two streams of technical development: remote sensing and spaceflight. Through the science of remote sensing we have acquired sensitive instruments capable of detecting and analyzing radiation across the whole range of the electromagnetic (EM) spectrum. Spaceflight lets astrophysicists place sophisticated remote sensing instruments above Earth's atmosphere. The wavelengths transmitted through the interstellar medium and arriving in the vicinity of near-Earth space are spread over approximately 24 decades of the spectrum. (A decade is a group, series, or power of 10.) However, the majority of this interesting EM radiation never reaches the surface of Earth because the terrestrial atmosphere effectively blocks such radiation across most of the spectrum. It should be remembered that the visible and infrared atmospheric windows occupy a spectral slice whose width is roughly one decade. Ground-based radio observatories can detect stellar radiation over a spectral range that adds about five more decades to the range of observable frequencies, but the remaining 18 decades of the spectrum are still blocked and are effectively invisible to astrophysicists on Earth's surface. Consequently, information that can be gathered by observers at the bottom of Earth's atmosphere represents only a small fraction of the total amount of information available concerning extraterrestrial objects. Sophisticated remote sensing instruments placed above Earth's atmosphere are now capable of sensing electromagnetic radiation over nearly the entire spectrum, and these instruments are rapidly changing our picture of the cosmos.

For example, we previously thought that the interstellar medium was a fairly uniform collection of gas and dust, but space-borne ultraviolet telescopes have shown us that its structure is very inhomogenous and complex. There are newly discovered components of the interstellar medium, such as extremely hot gas that is probably heated by shock waves from exploding stars. In fact, there is a great deal of interstellar pushing and shoving going on. Matter gathers and cools in some places because matter elsewhere is heated and dispersed. Besides discovering the existence of the very hot gas, the orbiting astronomical observatories have discovered two potential sources of the gas: the intense stellar winds that boil off hot stars and the rarer but more violent blasts of matter from exploding supernovas. In addition, X-ray and gamma-ray astronomy have contributed substantially to the discovery that the universe is not relatively serene and unchanging as previously imagined, but is actually dominated by the routine occurrence of incredibly violent events.

And this series of remarkable new discoveries is just beginning. Future astrophysics missions will provide access to the full range of the electromagnetic spectrum at increased angular and spectral resolution. They will support experimentation in key areas of physics, especially relativity and gravitational physics. Out of these exciting discoveries, perhaps, will emerge the scientific pillars for constructing a mid-21st-century civilization based on technologies unimaginable in the framework of contemporary physics.

The Tools of Astrophysics

Virtually all the information we receive about celestial objects comes to us through observation of electromagnetic radiation. Cosmic ray particles are an obvious and important exception, as are extraterrestrial material samples that have been returned to Earth (for example, lunar rocks). Each portion of the electromagnetic spectrum carries unique information about the physical conditions and processes in the universe. Infrared radiation reveals the presence of thermal emission from relatively cool objects; ultraviolet and extreme ultraviolet radiation may indicate thermal emission from very hot objects. Various types of violent events can lead to the production of X-rays and gamma rays.

Although EM radiation varies over many decades of energy and wavelength, the basic principles of measurement are quite common to all regions of the spectrum. The fundamental techniques used in astrophysics can be classified as imaging, spectrometry, photometry and polarimetry. Imaging provides basic information about the distribution of material in a celestial object, its overall structure, and, in some cases, its physical nature. Spectrometry is a measure of radiation intensity as a function of wavelength. It provides information on nuclear, atomic, and molecular phenomena occurring in and around the extraterrestrial object under observation. Photometry involves measuring radiation intensity as a function of time. It provides information about the time variations of physical processes within and around celestial objects, as well as their absolute intensities. Finally, polarimetry is a measurement of radiation intensity as a function of polarization angle. It provides information on ionized particles rotating in strong magnetic fields.

An incredible look deep into the universe. This view of nearly 10,000 galaxies is the deepest visible-light image of the cosmos. Called the Hubble Ultra Deep Field, this galaxy-studded view collected by the Hubble Space Telescope in 2004 represents a deep core sample of the universe, cutting across billions of light-years. The snapshot includes galaxies of various ages, sizes, shapes, and colors. Some of the galaxies shown here existed when the universe was just 800 million years old and are among the most distant known. *(Courtesy of NASA/STScI)*

High-Energy Astrophysics

High-energy astrophysics encompasses the study of extraterrestrial X-rays, gamma rays, and energetic cosmic ray particles. Prior to space-based high-energy astrophysics, scientists believed that violent processes involving high-energy emissions were rare in stellar and galactic evolution. Now, because of studies of extraterrestrial X-rays and gamma rays, we know that such processes are quite common rather than exceptional. The observation of X-ray emissions has been very valuable in the study of high-energy events, such as mass transfer in binary star systems, interaction of supernova remnants with interstellar gas, and quasars (whose energy source is presently unknown but believed to involve matter accreting to [falling into] a black hole). It is thought that gamma rays might be the missing link in understanding the physics of interesting high-energy objects such as pulsars and black holes. The study of cosmic ray particles provides important information about the physics of nucleosynthesis and about the interactions of particles and strong magnetic fields. High-energy phenomena that are suspected sources of cosmic rays include supernovas, pulsars, radio galaxies, and quasars.

X-Ray Astronomy

X-ray astronomy is the most advanced of the three high-energy astrophysics disciplines. Space-based X-ray observatories, such as NASA's *Chandra X-Ray Observatory,* increase our understanding in the following areas: (1) stellar structure and evolution, including binary star systems, supernova remnants, pulsar and plasma effects, and relativity effects in intense gravitational fields; (2) large-scale galactic phenomena, including interstellar media and soft X-ray mapping of local galaxies; (3) the nature of active galaxies, including spectral characteristics and the time variation of X-ray emissions from the nuclear or central regions of such galaxies; and (4) rich clusters of galaxies, including X-ray background radiation and cosmology modeling.

Gamma-Ray Astronomy

Gamma rays consist of extremely energetic photons (that is, energies greater than 10^5 electron volts [eV]) and result from physical processes different than those associated with X-rays. The processes associated with gamma ray emissions in astrophysics include: (1) the decay of radioactive nuclei; (2) cosmic ray interactions; (3) curvature radiation in extremely strong magnetic fields; and (4) matter-antimatter annihilation. Gamma-ray astronomy reveals the explosive, high-energy processes associated with such celestial phenomena as supernovas, exploding galaxies and quasars, pulsars, and black holes.

Gamma-ray astronomy is especially significant because the gamma rays being observed can travel across our entire galaxy and even across most of the universe without suffering appreciable alteration or absorption. Therefore, these energetic gamma rays reach our solar system with the same characteristics, including directional and temporal features, as they started with at their sources, possibly many light-years distant and deep within regions or celestial objects opaque to other wavelengths. Consequently, gamma-ray astronomy provides information on extraterrestrial phenomena not observable at any other wavelength in the electromagnetic spectrum and on spectacularly energetic events that may have occurred far back in the evolutionary history of the universe.

Cosmic-Ray Astronomy

Cosmic rays are extremely energetic particles that extend in energy from one million (10^6) electron volts (eV) to more than 10^{20} eV and range in composition from hydrogen (atomic number $Z = 1$) to a predicted atomic number of $Z = 114$. This composition also includes small percentages of electrons, positrons, and possibly antiprotons. Cosmic-ray astronomy provides information on the origin of the elements (nucleosynthetic processes) and the physics of particles at ultrahigh-energy levels. Such information addresses astrophysical questions concerning the nature of stellar explosions and the effects of cosmic rays on star formation and galactic structure and stability.

Optical Astronomy (Hubble Space Telescope)

Astronomical work in a number of areas has greatly benefited from large, high-resolution optical systems that have operated or are now operating outside Earth's atmosphere. Some of these interesting areas include investigation of the interstellar medium, detailed study of quasars and black holes, observation of binary X-ray sources and accretion disks, extragalactic astronomy, and observational cosmology. The *Hubble Space Telescope* (HST) constitutes the very heart of NASA's spaceborne ultraviolet/optical astronomy program at the beginning of the 21st century. Launched in 1990 and repaired and refurbished in orbit by space shuttle crews, the HST's continued ability to cover a wide range of wavelengths, to provide fine angular resolution, and to detect faint sources makes it one of the most powerful and important astronomical instruments ever built. Within a decade the *James Webb Space Telescope* (JWST) will begin operation and provide scientists with even more amazing views of the universe.

Ultraviolet Astronomy

Another interesting area of astrophysics involves the extreme ultraviolet (EUV) region of the electromagnetic spectrum. The interstellar medium is highly absorbent at EUV wavelengths (100 to 1,000 angstroms [Å]). EUV data gathered from space-based instruments, such as those placed on NASA's *Extreme Ultraviolet Explorer* (EUVE), are being used to confirm and refine contemporary theories of the late stages of stellar evolution, to analyze the effects of EUV radiation on the interstellar medium, and to map the distribution of matter in our solar neighborhood.

Infrared Astronomy

Infrared (IR) astronomy involves studies of the electromagnetic (EM) spectrum from 1 to 100 micrometers wavelength, while radio astronomy involves wavelengths greater than 100 micrometers. (A micrometer is one millionth [10^{-6}] of a meter.) Infrared radiation is emitted by all classes of "cool" objects (stars, planets, ionized gas and dust regions, and galaxies) and the cosmic background radiation. Most emissions from objects with temperatures ranging from about 3 to 2,000 kelvins are in the infrared region of the spectrum. In order of decreasing wavelength, the sources of infrared and microwave (radio) radiation are: (1) galactic synchrotron radiation; (2) galactic thermal bremsstrahlung radiation in regions of ionized hydrogen; (3) the cosmic background radiation; (4) 15-kelvins cool galactic dust and 100-kelvins stellar-heated galactic dust; (5) infrared galaxies and primeval galaxies; (6) 300-kelvins interplanetary dust; and (7) 3,000-kelvins starlight. Advanced space-based IR observatories, such as NASA's *Spitzer Space Telescope,* are revolutionizing how astrophysicists and astronomers perceive the universe.

The *Spitzer Space Telescope* is the fourth and final element in NASA's family of orbiting great observatories, which includes the *Hubble Space Telescope,* the *Compton Gamma Ray Observatory,* and the *Chandra X-Ray Observatory.* These great space-based observatories give astrophysicists and astronomers an orbiting "toolbox" that provides multiwavelength studies of the universe. The *Spitzer Space Telescope* joined this spectacular suite of orbiting instruments in August 2003. The sophisticated IR observatory provides a fresh vantage point on processes that up to now have remained mostly in the dark, such as the formation of galaxies, stars, and planets. Within our Milky Way galaxy infrared astronomy helps scientists discover and characterize dust discs around nearby stars, thought to be the signposts of planetary system formation. The *Spitzer Space Telescope,* for example, is now provid-

In 2004 NASA's *Spitzer Space Telescope* imaged this cluster of about 130 newborn stars in a rosebud-shaped nebulosity known as NGC 7129. The star cluster and its associated nebula are located at a distance of 3,300 light-years from Earth in the constellation Cepheus. Scientists believe that our own Sun may have formed billions of years ago in a cluster similar to NGC 7129. Once radiation from new cluster stars destroys the surrounding placental material, the stars begin to slowly drift apart. *(Courtesy of NASA/JPL-Caltech/T. Megeath [Harvard-Smithsonian CfA])*

ing new insights into the birth processes of stars—an event normally hidden behind veils of cosmic dust. (*See* figure.) Outside our galaxy infrared astronomy helps astronomers and astrophysicists uncover the "cosmic engines"—thought to be galaxy collisions or black holes—powering ultraluminous infrared galaxies. Infrared astronomy also probes the birth and evolution of galaxies in the early and distant universe.

Gravitational Physics

Gravitation is the dominant long-range force in the universe. It governs the large-scale evolution of the universe and plays a major role in the violent events associated with star formation and collapse. Outer space provides the low acceleration and low-noise environment needed for the careful measurement of relativistic gravitational effects. A number of interesting experiments have been identified for a space-based experimental program in relativity and gravitational physics. Data from these experiments could substantially revise contemporary physics and the fate astrophysicists and cosmologists now postulate for the universe.

Cosmic Origins

The ultimate aim of astrophysics is to understand the origin, nature, and evolution of the universe. It has been said that the universe is not only stranger than we imagine, it is stranger than we *can* imagine! Through the creative use of modern space technology, we will continue to witness many major discoveries in astrophysics in the exciting decades ahead—each discovery helping us understand a little better the magnificent universe in which we live and the place in which we will build our extraterrestrial civilization.

See also BIG BANG; BLACK HOLES; *CHANDRA X-RAY OBSERVATORY*; *COMPTON GAMMA RAY OBSERVATORY*; COSMOLOGY; *HUBBLE SPACE TELESCOPE*; ORIGIN PROGRAM; RELATIVITY; *SPITZER SPACE TELESCOPE*.

Aten group A collection of near-Earth asteroids that cross the orbit of Earth but whose average distances from the Sun lie inside Earth's orbit. This asteroid group acquired its name from the 0.9-km-diameter asteroid Aten, discovered in 1976 by the American astronomer Eleanor Kay Helin (née Francis).

See also ASTEROID; NEAR-EARTH ASTEROID.

Athena (launch vehicle) The Lockheed-Martin Company's Athena launch vehicle program is based on that company's solid propellant rocket system experience, as well more than four decades of space launch vehicle activity. The Lockheed-Martin Company has launched more than 1,000 rocket vehicles and produced more than 2,900 solid-propellant missile systems—namely, six generations of highly reliable fleet ballistic missiles (Polaris, Poseidon, and Trident) for the U.S. Navy. The Athena program started in January 1993 and resulted in two new expendable launch vehicles: Athena I and Athena II. The Athena I launch vehicle is a two-stage solid-propellant rocket combination that can carry a payload of approximately 800 kg into low-Earth orbit (LEO). The Athena II is a three-stage, solid-propellant rocket vehicle combination that can carry a payload of about 1,900 kg into LEO. On its first operational mission the Athena I expendable launch vehicle successfully placed NASA's *Lewis* satellite into orbit from Vandenberg Air Force Base in California on August 22, 1997. The first Athena II launch took place from Cape Canaveral Air Force Station in Florida on January 6, 1998, and successfully sent NASA's *Lunar Prospector* spacecraft on its mission to study the Moon.

See also LAUNCH VEHICLE; ROCKET.

Atlas (launch vehicle) A family of versatile liquid-fuel rockets originally developed by General BERNARD SCHRIEVER of the U.S. Air Force in the late 1950s as the first operational American intercontinental ballistic missile (ICBM). Evolved and improved Atlas launch vehicles now serve many government and commercial space transportation needs. The modern Atlas rocket vehicle fleet, as developed and marketed by the Lockheed-Martin Company, included three basic families: the Atlas II (IIA and IIAS), the Atlas III (IIIA and IIIB), and the Atlas V (400 and 500 series). The Atlas II family was capable of lifting payloads ranging in mass from approximately 2,800 kg to 3,700 kg to a geostationary transfer orbit (GTO). The Atlas III family of launch vehicles can lift payloads of up to 4,500 kg to GTO, while the more powerful Atlas V family can lift payloads of up to 8,650 kg to GTO.

The family of Atlas II vehicles evolved from the Atlas I family of military missiles and cold war–era space launch

The Atlas III expendable launch vehicle made its debut at Cape Canaveral Air Force Station in Florida on May 24, 2000. The rocket's successful liftoff was powered by a new Russian RD-180 engine. ***(Courtesy of U.S. Air Force and Lockheed-Martin)***

vehicles. The transformation was brought about by the introduction of higher-thrust engines and longer propellant tanks for both stages. The Atlas III family incorporated the pressure-stabilized design of the Atlas II vehicle but used a new single-stage Atlas engine built by NPO Energomash of Russia. The RD-180 is a throttleable rocket engine that uses liquid oxygen and kerosene propellants and provides approximately 3,800 kilonewtons of thrust at sea level. (*See* figure.) The Atlas V uses the RD-180 main engine in a Common Core Booster first-stage configuration with up to five strap-on solid-propellant rockets and a Centaur upper stage. The first Atlas V vehicle successfully lifted off from Cape Canaveral Air Force Station on August 21, 2002. Later that year an important chapter in propulsion system history closed, when the last Atlas II vehicle was successfully launched from Cape Canaveral Air Force Station on December 4, 2002.

See also LAUNCH VEHICLE; ROCKET.

Atlas (moon) Atlas is a tiny inner moon of Saturn, orbiting the planet at a distance of 137,670 kilometers. This irregularly shaped moon is about 37 km by 34 km by 27 km in diameter and has an orbital period of 0.60 days. Discovered in 1980 from *Voyager I* spacecraft images, astronomers now regard Atlas as a shepherd moon within Saturn's A-ring.

See also SATURN.

atmosphere 1. Gravitationally bound gaseous envelope that forms an outer region around a planet or other celestial body. 2. (cabin) Breathable environment inside a space capsule, aerospace vehicle, spacecraft, or space station. 3. (Earth's) Life-sustaining gaseous envelope surrounding Earth. Near sea level it contains the following composition of gases (by volume): nitrogen, 78 percent; oxygen, 21 percent; argon, 0.9 percent; and carbon dioxide, 0.03 percent. There are also lesser amounts of many other gases, including water vapor and human-generated chemical pollutants. Earth's electrically neutral atmosphere is composed of four primary layers: troposphere, stratosphere, mesosphere, and thermosphere. Life occurs in the troposphere, the lowest region that extends up to about 16 km of altitude. It is also the place within which most of Earth's weather occurs. Extending from about 16 km of altitude to 48 km is the stratosphere. The vast majority (some 99 percent) of the air in Earth's atmosphere is located in these two regions. Above the stratosphere, from approximately 48 km to 85 km, is the mesosphere. In this region temperature decreases with altitude and reaches a minimum of –90°C at about 85 km (the mesopause). The uppermost layer of the atmosphere is called the thermosphere, and it extends from the mesopause out to approximately 1,000 km. In this region temperature rises with altitude as oxygen and nitrogen molecules absorb ultraviolet radiation from the Sun. This heating process creates ionized atoms and molecules and forms the layers of the ionosphere, which actually starts at about 60 km. Earth's upper atmosphere is characterized by the presence of electrically charged gases, or plasma. The boundaries of and structure within the ionosphere vary considerably according to solar activity. Finally, overlapping the ionosphere is the magnetosphere, which extends from approximately 80 km to 65,000 km on the side of Earth toward the Sun and trails out more than 300,000 km on the side of Earth away from the Sun. The geomagnetic field of Earth plays a dominant role in the behavior of charged particles in the magnetosphere.

atmospheric braking The action of slowing down an object entering the atmosphere of Earth or another planet from space by using the aerodynamic drag exerted by the atmosphere.

See also AEROBRAKING.

atmospheric compensation The physical distortion or modification of the components of an optical system to compensate for the distortion of light waves as they pass through the atmosphere and the optical system.

See also ADAPTIVE OPTICS.

atmospheric drag With respect to Earth-orbiting spacecraft, the retarding force produced on a satellite, aerospace vehicle, or spacecraft by its passage through upper regions of Earth's atmosphere. This retarding force drops off exponentially with altitude and has only a small effect on spacecraft whose perigee is higher than a few hundred kilometers. For spacecraft with lower perigee values, the cumulative effect of atmospheric drag eventually will cause them to reenter the denser regions of Earth's atmosphere (and be destroyed) unless they are provided with an onboard propulsion system that can accomplish periodic reboost.

atmospheric entry The penetration of any planetary atmosphere by an object from outer space; specifically, the penetration of Earth's atmosphere by a crewed or uncrewed space capsule, aerospace vehicle, or spacecraft. Also called *entry* and sometimes *reentry,* although typically the object is making its initial return or entry after a flight in space.

atmospheric heave In ballistic missile defense, raising a large volume of the upper atmosphere to a substantially higher altitude (i.e., hundreds of kilometers) by means of a nuclear detonation within the atmosphere. This could have several different effects on the capability of a missile defense system. For example, nuclear background radiation problems could be worsened substantially for the defense, and some directed energy weapons could be partially neutralized. On the other hand, offensive decoys could become more detectable, and offensive targeting could become more difficult.

See also BALLISTIC MISSILE DEFENSE.

atmospheric pressure The pressure (force per unit area) at any point in a planet's atmosphere due solely to the weight of the atmospheric gases above that point.

See also PASCAL.

atmospheric probe The special collection of scientific instruments (usually released by a mother spacecraft) for determining the pressure, composition, and temperature of a planet's atmosphere at different altitudes. An example is the probe released by NASA's *Galileo* spacecraft in December 1995. As it plunged into the Jovian atmosphere the probe successfully transmitted its scientific data to the *Galileo* spacecraft (the mother spacecraft) for about 58 minutes.

See also CASSINI MISSION.

atmospheric refraction Refraction resulting when a ray of radiant energy passes obliquely through an atmosphere.

atmospheric window A wavelength interval within which a planet's atmosphere is transparent, that is, easily transmits electromagnetic radiation. (*See* figure.)

See also REMOTE SENSING.

atom A tiny particle of matter (the smallest part of an element) indivisible by chemical means. It is the fundamental building block of the chemical elements. The elements, such as hydrogen (H), helium (He), carbon (C), iron (Fe), lead (Pb), and uranium (U), differ from each other because they

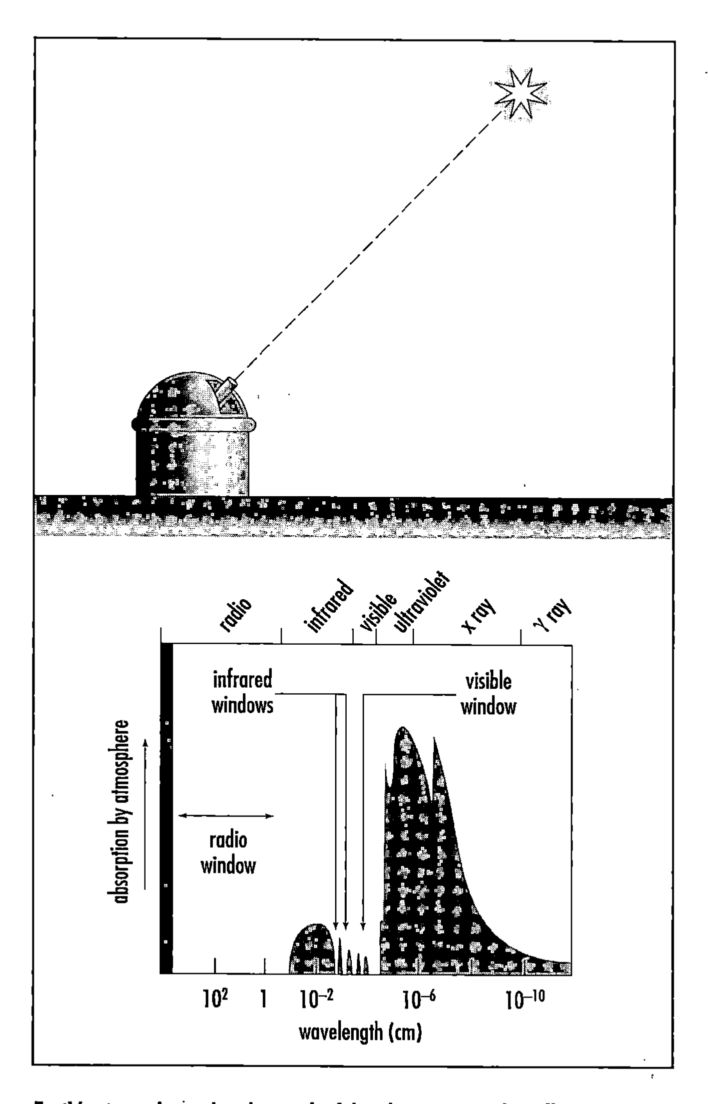

Earth's atmosphere absorbs much of the electromagnetic radiation arriving from sources in outer space. The atmosphere contains only a few wavelength bands (called atmospheric windows) that transmit most of the incident radiation within that band to a detector on the planet's surface.

consist of different types of atoms. According to (much simplified) modern atomic theory, an atom consists of a dense inner core (the nucleus) that contains protons and neutrons, and a cloud of orbiting electrons. Atoms are electrically neutral, with the number of (positively charged) protons being equal to the number of (negatively charged) electrons.

atomic clock A precise device for measuring or standardizing time that is based on periodic vibrations of certain atoms (cesium) or molecules (ammonia). Widely used in military and civilian spacecraft, as, for example, in the Global Positioning System (GPS).

atomic mass The MASS of a neutral ATOM of a particular NUCLIDE usually expressed in ATOMIC MASS UNITS (amu).

See also MASS NUMBER.

atomic mass unit (amu) One-twelfth (1/12) the mass of a neutral atom of the most abundant isotope of carbon, carbon-12.

atomic number (symbol: Z) The number of protons in the nucleus of an atom and also its positive charge. Each chemical element has its characteristic atomic number. For example, the atomic number for carbon is 6, while the atomic number for uranium is 92.

atomic weight The mass of an atom relative to other atoms. At present, the most abundant isotope of the element carbon, namely carbon-12, is assigned an atomic weight of exactly 12. As a result, 1/12 the mass of a carbon-12 atom is called one *atomic mass unit,* which is approximately the mass of one proton or one neutron. Also called *relative atomic mass.*

attached payload 1. With respect to the space shuttle, any payload that is launched in the orbiter, remains structurally and functionally attached throughout the flight, and is landed in the orbiter. 2. With respect to a space station, a payload located on the base truss outside the pressurized modules.

attenuation The decrease in intensity (strength) of an electromagnetic wave as it passes through a transmitting medium. This loss is due to absorption of the incident electromagnetic radiation (EMR) by the transmitting medium or to scattering of the EMR out of the path of the detector. Attenuation does not include the reduction in EMR wave strength due to geometric spreading as a consequence of the inverse square law. The attenuation process can be described by the equation

$$I = I_0 e^{-\alpha x}$$

where I is the FLUX (photons or particles per unit area) at a specified distance (x) into the attenuating medium, I_o is the initial flux, α is the attenuation coefficient (with units of inverse length), and e is a special mathematical function, the irrational number 2.71828. . . . The expression e^x is called the exponential function.

attitude The position of an object as defined by the inclination of its axes with respect to a frame of reference. The orientation of a space vehicle (e.g., a spacecraft or aerospace vehicle) that is either in motion or at rest, as established by the relationship between the vehicle's axes and a reference line or plane. Attitude is often expressed in terms of pitch, roll, and yaw.

attitude control system The onboard system of computers, low-thrust rockets (thrusters), and mechanical devices (such as a momentum wheel) used to keep a spacecraft stabilized during flight and to precisely point its instruments in some desired direction. Stabilization is achieved by spinning the spacecraft or by using a three-axis active approach that maintains the spacecraft in a fixed reference attitude by firing a selected combination of thrusters when necessary.

Stabilization can be achieved by spinning the spacecraft, as was done on the *Pioneer 10* and *11* spacecraft during their missions to the outer solar system. In this approach the gyroscopic action of the rotating spacecraft mass is the stabilizing mechanism. Propulsion system thrusters are fired to make any desired changes in the spacecraft's spin-stabilized attitude.

Spacecraft also can be designed for active three-axis stabilization, as was done on the *Voyager 1* and *2* spacecraft, which explored the outer solar system and beyond. In this method of stabilization, small propulsion system thrusters gently nudge the spacecraft back and forth within a deadband of allowed attitude error. Another method of achieving active three-axis stabilization is to use electrically powered reaction wheels, which are also called momentum wheels. These massive wheels are mounted in three orthogonal axes onboard the spacecraft. To rotate the spacecraft in one direction, the proper wheel is spun in the opposite direction. To rotate the vehicle back, the wheel is slowed down. Excessive momentum, which builds up in the system due to internal friction and external forces, occasionally must be removed from the system; this usually is accomplished with propulsive maneuvers.

Either general approach to spacecraft stabilization has basic advantages and disadvantages. Spin stabilized vehicles provide a continuous "sweeping motion" that is generally desirable for fields and particle instruments. However, such spacecraft may then require complicated systems to despin antennas or optical instruments that must be pointed at targets in space. Three-axis controlled spacecraft can point antennas and optical instruments precisely (without the necessity for despinning), but these craft then may have to perform rotation maneuvers to use their fields and particles science instruments properly.

See also SPACECRAFT.

***Aura* (spacecraft)** The mission of NASA's Earth-observing spacecraft *Aura* is to study Earth's ozone, air quality, and climate. Aerospace engineers and scientists designed the mission exclusively to conduct research on the composition, chemistry, and dynamics of Earth's upper and lower atmosphere using multiple instruments on a single spacecraft. A Delta II expendable launch vehicle lifted off from Vandenberg Air Force Base in California on July 15, 2004, and successfully placed the spacecraft into a polar orbit around Earth.

Aura is the third in a series of major Earth-observing spacecraft to study the environment and climate change. The first and second missions, *Terra* and *Aqua,* are designed to study the land, oceans, and Earth's radiation budget. *Aura*'s atmospheric chemistry measurements will also follow up on measurements that began with NASA's *Upper Atmospheric Research Satellite* (UARS) and continue the record of satellite ozone data collected from the total ozone mapping spectrometers (TOMS) aboard the *Nimbus-7* and other satellites.

The *Aura* spacecraft carries the following suite of instruments: high-resolution dynamics limb sounder (HIRDLS), microwave limb sounder (MLS), ozone-monitoring instrument (OMI), and troposphere emission spectrometer (TES). These instruments contain advanced technologies that aerospace engineers have developed for use on environmental satellites. Each instrument provides unique and complementary capabilities that support daily observations (on a global basis) of Earth's atmospheric ozone layer, air quality, and key climate parameters.

Managed by the NASA Goddard Space Flight Center in Greenbelt, Maryland, the *Aura* spacecraft caps off a 15-year international effort to establish the world's most comprehensive Earth-observing system. The overarching goal of the Earth-observing system is to determine the extent, causes, and regional consequences of global change. Data from the *Aura* satellite are helping scientists better forecast air quality, ozone layer recovery, and climate changes that affect health, the economy, and the environment.

See also AQUA; EARTH SYSTEM SCIENCE; GLOBAL CHANGE; *TERRA*.

aurora Visible glow in a planet's upper atmosphere (ionosphere) caused by interaction of the planet's magnetosphere and particles from the Sun (solar wind). On Earth the aurora borealis (or northern lights) and the aurora australis (or southern lights) are visible manifestations of the magnetosphere's dynamic behavior. At high latitudes disturbances in Earth's geomagnetic field accelerate trapped particles into the upper atmosphere where they excite nitrogen molecules (red emissions) and oxygen atoms (red and green emissions) (*see* figure). Auroras also occur on Jupiter, Saturn, Uranus, and Neptune. Early in the 20th century the Norwegian physicist KRISTIAN OLAF BERNHARD BIRKELAND deployed instruments during polar region expeditions and then suggested that aurora were electromagnetic in nature and were created by an interaction of Earth's magnetosphere with a flow of charged particles from the Sun. In the 1960s instruments onboard early Earth-orbiting satellites confirmed his hypothesis.

See also EARTH'S TRAPPED RADIATION BELTS; MAGNETOSPHERE.

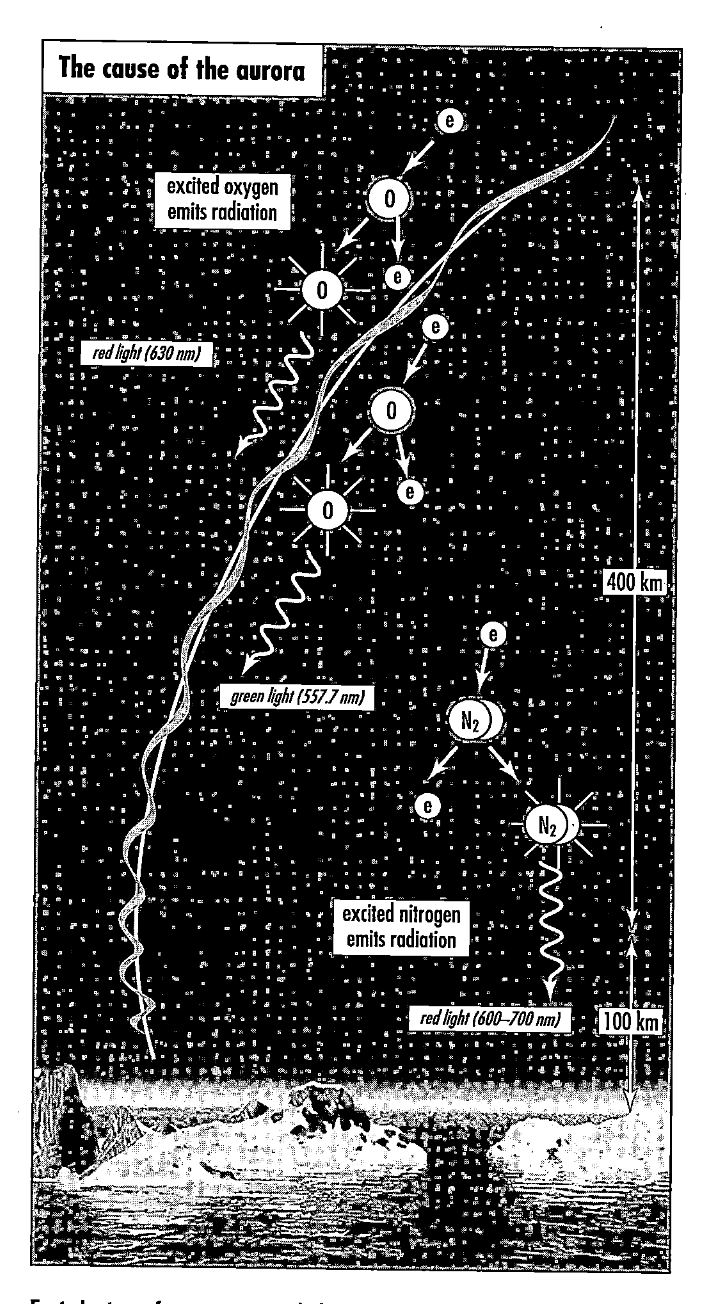

Fast electrons from space travel along the magnetic field and strike oxygen or nitrogen molecules in Earth's atmosphere. Energy from an atom or molecule excited by fast electrons is released as different colors of light. ***(Courtesy of NOAA)***

auxiliary power unit (APU) A power unit carried on a spacecraft or aerospace vehicle that supplements the main source of electric power on the craft.

auxiliary stage 1. A small propulsion unit that may be used with a payload. One or more of these units may be used to provide the additional velocity required to place a payload in the desired orbit or trajectory. 2. A propulsion system used to provide midcourse trajectory corrections, braking maneuvers, and/or orbital adjustments.

avionics The contraction of aviation and electronics. The term refers to the application of electronics to systems and equipment used in aeronautics and astronautics.

Avogadro's hypothesis The volume occupied by a mole of gas at a given pressure and temperature is the same for all gases. This hypothesis also can be expressed as: Equal volumes of all gases measured at the same temperature and pressure contain the same number of molecules. The Italian physicist Count Amedeo Avogadro (1776–1856) first proposed this hypothesis in 1811. In a strict sense, it is valid only for ideal gases.

See also IDEAL GAS; MOLE.

Avogadro's number (symbol: NA) The number of molecules per mole of any substance.

$$N_A = 6.022 \times 10^{23} \text{ molecules/mole}$$

Also called the *Avogadro constant*.

axis (plural: **axes**) Straight line about which a body rotates (axis of rotation) or along which its center of gravity moves (axis of translation). Also, one of a set of reference lines for a coordinate system, such as the x-axis, y-axis, and z-axis in Cartesian coordinates.

See also CARTESIAN COORDINATES.

azimuth The horizontal direction or bearing to a celestial body measured in degrees clockwise from north around a terrestrial observer's horizon. On Earth azimuth is 0° for an object that is due north, 90° for an object due east, 180° for an object due south, and 270° for an object due west.

See also ALTITUDE.

azimuthal mounting *See* ALTAZIMUTH MOUNTING.

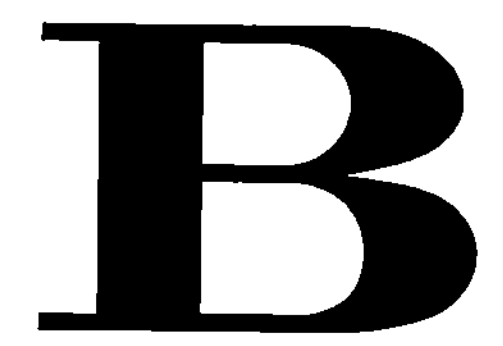

Baade, (Wilhelm Heinrich) Walter (1893–1960) German-American *Astronomer* Walter Baade was the astronomer who carefully studied Cepheid variables, enabling him to double the estimated distance, age, and scale of the universe. In 1942 he showed that the Andromeda galaxy was more than 2 million light-years away—more than twice the previously accepted distance. He also discovered that stars occupy two basic populations, or groups: Population I stars that are younger, bluish stars found in the outer regions of a galaxy and Population II stars, which are older, reddish stars found in the central regions of a galaxy. This categorization significantly advanced stellar evolution theory.

See also CEPHEID VARIABLE; POPULATION I STARS; POPULATION II STARS.

Babcock, Harold Delos (1882–1968) American *Astronomer, Physicist* Harold Delos Babcock, in collaboration with his son Horace (1912–2003), invented the solar magnetograph in 1951. This instrument allowed them and other solar physicists to make detailed (pre–space age) investigations of the Sun's magnetic field.

See also SUN.

background radiation The nuclear radiation that occurs in the natural environment, including cosmic rays and radiation from naturally radioactive elements found the air, the soil, and the human body. Sometimes referred to as natural radiation background. The term also may mean ionizing radiation (natural and human-made) that is unrelated to a particular experiment or series of measurements.

See also IONIZING RADIATION; SPACE RADIATION ENVIRONMENT.

backout The process of undoing tasks that have already been completed during the countdown of a launch vehicle, usually in reverse order.

See also COUNTDOWN; LAUNCH VEHICLE.

backup crew A crew of astronauts or cosmonauts trained to replace the prime crew, if necessary, on a particular space mission.

baffle 1. In the combustion chamber of a liquid rocket engine, an obstruction used to prevent combustion instability by maintaining uniform propellant mixtures and equalizing pressures. 2. In the fuel tank of a liquid rocket system, an obstruction used to prevent sloshing of the propellant. 3. In an optical or multispectral sensing system, a barrier or obstruction used to prevent stray (unwanted) radiation from reaching a sensitive detector element.

Baikonur Cosmodrome A major launch site for the space program of the former Soviet Union and later the Russian Federation. The complex is located just east of the Aral Sea on the barren steppes of Kazakhstan (now an independent republic). Constructed in 1955, when Kazakhstan was an integral part of the former Soviet Union, this historic cosmodrome covers more than 6,700 square kilometers and extends some 75 km from north to south and 90 km east to west. The sprawling rocket base includes numerous launch sites, nine tracking stations, five tracking control centers, and a companion 1,500-km-long rocket test range. Also known as the Tyuratam launch site during the cold war, the Soviets launched *Sputnik 1* (1957), the first artificial satellite, and cosmonaut YURI GAGARIN, the first human to travel into outer space (1961), from this location. The Russians have used Baikonur Cosmodrome for all their human-crewed launches and for the vast majority of their lunar, planetary, and geostationary Earth orbit missions. Since 1993 the Russian Federation has rented Baikonur from Kazakhstan.

See also KOROLEV, SERGEI; *SPUTNIK 1*.

Baily, Francis (1774–1844) British *Amateur Astronomer, Stockbroker* An avid amateur astronomer, Francis Baily discovered the "beads of light" phenomenon (now called BAILY'S BEADS) that occurs around the lunar disk during a total eclipse of the Sun. He made this discovery on May 15, 1836, and his work stimulated a great deal of interest in eclipses.

See also ECLIPSE; MOON.

Baily's beads An optical phenomenon that appears just before or immediately after totality in a solar eclipse, when sunlight bursts through gaps in the mountains on the Moon and a string of light beads appears along the lunar disk. The British amateur astronomer FRANCIS BAILY first described this phenomenon in 1836. His observation stimulated a great deal of interest in eclipses for the remainder of the 19th century by both professional astronomers and the general public.

ballistic flyby Unpowered flight similar to a bullet's trajectory, governed by gravity and by the body's previously acquired velocity.

See also FLYBY.

ballistic missile A missile that is propelled by rocket engines and guided only during the initial (thrust producing) phase of its flight. In the nonpowered and nonguided phase of its flight, it assumes a ballistic trajectory similar to that of an artillery shell. After thrust termination, reentry vehicles (RVs) can be released, and these RVs also follow free-falling (ballistic) trajectories toward their targets. Aerospace analysts within the U.S. Department of Defense (DOD) often classify ballistic missiles by their maximum operational ranges, using the following scale: Short-range ballistic missiles (SRBMs) are those that have a maximum operational range of about 1,100 km; medium-range ballistic missiles (MRBMs) have an operational range between 1,100 and 2,750 km; intermediate-range ballistic missiles (IRBMs) have an operational range between 2,750 and 5,500 km; and intercontinental ballistic missiles (ICBMs) have operational ranges in excess of 5,500 km. While somewhat arbitrary from an aerospace engineering perspective, this widely recognized classification scheme has proven quite useful in arms-control negotiations, ballistic missile treaty discussions, and international initiatives focused on limiting regional arms races and preventing the emergence of far-reaching ballistic missile threats from rogue nations. *Compare with* CRUISE MISSILE and GUIDED MISSILE.

See also BALLISTIC MISSILE DEFENSE.

ballistic missile defense (BMD) A defense system that is designed to protect a territory from incoming ballistic missiles in both strategic and theater tactical roles. A BMD system usually is conceived of as having several independent layers designed to destroy attacking ballistic missiles or their warheads at any or all points in their trajectories—from launch until just before impact at the targets.

The phases of a typical missile trajectory are shown in the figure below. During the boost phase the rocket engines accelerate the missile and its warhead payload through and out of the atmosphere and provide intense, highly specific observables. The booster rocket and sustainer engines of a modern ICBM (like the MINUTEMAN III) usually operate for between three and five minutes. When the boost phase ends the missile has typically reached an altitude of about 200 km.

A post-boost phase, or bus deployment phase, occurs next, during which multiple warheads (or perhaps a single warhead) and penetration aids are released from a post-boost vehicle (PVB). The post-boost vehicle, or reentry vehicle bus, is the rocket propelled final stage of an ICBM that, after booster burnout, places individual warheads and (possibly) decoys on ballistic paths toward their targets. In the midcourse phase the warheads and penetration aids travel on

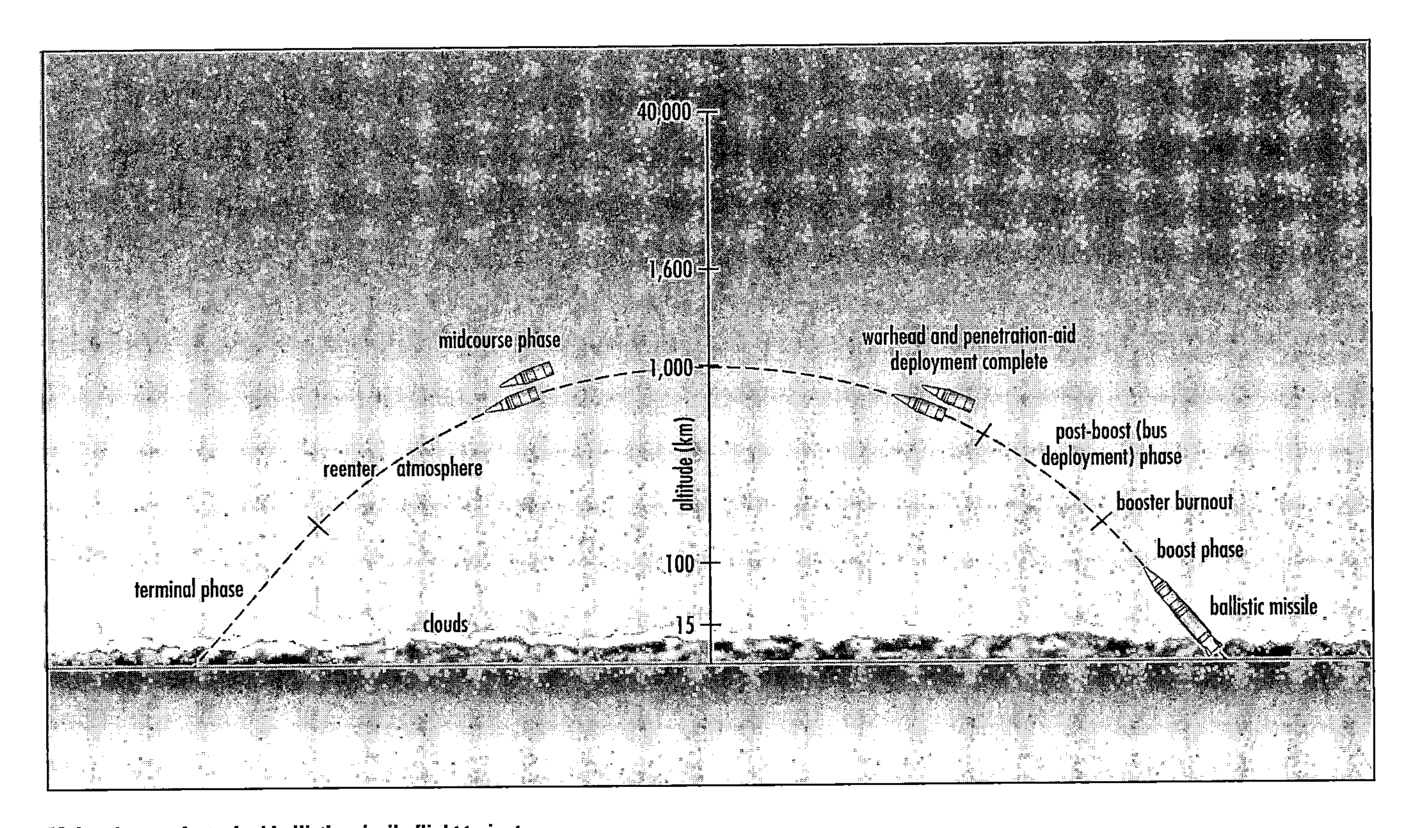

Major phases of a typical ballistic missile flight trajectory

The Ballistic Missile—A Revolution in Warfare

In the middle of the 20th century developments in space technology revolutionized warfare, transformed international politics, and changed forever our planetary civilization. One of the earliest and perhaps most significant of these developments involved the combination of two powerful World War II–era weapon systems, the American nuclear bomb and the German V-2 ballistic missile. Cold war politics encouraged a hasty marriage of these emerging military technologies. The union ultimately produced an offspring called the intercontinental ballistic missile (ICBM), a weapon system that most military experts regard as the single most influential weapon in all history. The ICBM and its technical cousin, the submarine-launched ballistic missile (SLBM), were the first weapon systems designed to attack a distant target by traveling into and through outer space. In the late 1950s the first generation of such operational "space weapons" completely transformed the nature of strategic warfare and altered the practice of geopolitics.

For the United States the development of the ICBM created a fundamental shift in national security policy. Before the ICBM the major purpose of the American military establishment was to fight and win wars. With operational nuclear weapon–armed ballistic missile forces, both the United States and the Soviet Union possessed an unstoppable weapon system capable of delivering megatons of destruction to any point on the globe within about 30 minutes. Once fired, these weapons would fly to their targets with little or no chance of being stopped. To make matters worse, national leaders could not recall a destructive ICBM strike after the missiles were launched. So from the moment on, the chief purpose of the U.S. military establishment became how to avoid strategic nuclear warfare. American and Soviet national leaders quickly realized that any major exchanged of unstoppable nuclear-armed ballistic missiles would completely destroy both adversaries and leave Earth's biosphere in total devastation.

In the era of the ICBM there are no winners, only losers—should adversaries resort to strategic nuclear conflict. For the first time in human history national leaders controlled a weapon system that could end civilization in less than a few hours. The modern ballistic missile made deterrence of nuclear war the centerpiece of American national security policy, a policy often appropriately referred to as mutually assured destruction (MAD). Some military analysts like to refer to the ICBM as an unusable weapon. For example, the United States and Russia designed their strategic nuclear forces to survive a surprise enemy first strike and still deliver a devastating second strike. Consequently, no matter which side fired their nuclear-armed ICBMs first, the other side would still be capable of firing a totally destructive retaliatory salvo, and everybody would die. In this perspective the ICBM is quite analogous to a societal suicide weapon. One cold war military analyst summarized MAD as; "He who shoots first, dies second." Such grim, gallows humor only serves to highlight the often overlooked paradox. The most destructive weapon system in history is also its most unusable weapon system. Up until now rational governments have tightly guarded the use of their nuclear-armed ICBM forces with elaborate control systems that restrict access to launch codes and nuclear weapon arming signals.

The arrival of the nuclear-armed ballistic missile is a milestone in human development. From a planetary culture perspective, this technical event brings about the end of "childhood" and the end of social innocence. Will the family of nations mature into a global society capable of avoiding strategic nuclear warfare as a means of settling political, social, and ethnic differences, or will we perish as a technically advanced but socially immature species?

Since the laws of physics that allowed the development of the nuclear-armed ICBM apply throughout the universe, it is interesting to speculate here whether other technically advancing species in the Milky Way galaxy reached the same crossroads. Did such a society survive the "ICBM technology" test, or did it perish by its own hand?

An alternate technology pathway also appeared with the arrival of the operational ICBM. Cold war politics encouraged the pioneering use of powerful ballistic missiles as space launch vehicles. So during the macabre superpower nuclear standoff that dominated geopolitics during the second half of the 20th century, the human race also acquired the very tools it needed to expand into the solar system and mature as a planetary civilization. Will tribal barbarism finally give way to the rise of a mature, conflict-free global society? The ICBM is the social and technical catalyst capable of causing great harm or positive long-term good. It is up to us to take the pathway that leads to the stars and not the one that leads to extinction.

Perhaps as a species we are beginning to go down the right road. It is interesting to observe, for example, that the threat of instant nuclear Armageddon has helped restrain those nations with announced nuclear weapon capabilities (such as the United States, the former Soviet Union, the United Kingdom, France, and the People's Republic of China) from actually using such devastating weapons in resolving a small-scale regional conflict. Because of the "no winner" standoff between unstoppable offensive ballistic missile forces against unstoppable offensive ballistic missile forces, political scientists assert that the ICBM has caused a revolution in warfare and international relations. Some military analysts even go so far as to suggest that the ICBM makes strategic warfare between rational, nuclear-armed nations impossible.

The operative term here is *rational*. Today, as nuclear weapon and ballistic missile technologies spread throughout the world, the specter of a regional nuclear conflict (such as between India and Pakistan), nuclear blackmail (as threatened by the leadership of North Korea), and nuclear terrorism (at the hands of a technically competent, financially capable subnational political group) haunt the global community. The heat of long-standing, culturally and religiously rooted regional animosities could overcome the positive example of decades of self-imposed superpower restraint on the use of nuclear-armed ballistic missiles. Today's expanding regional missile threat is encouraging some American strategic planners to revisit ballistic missile defense technologies, including concepts requiring the deployment of antimissile weapon systems in outer space.

For more than four decades the intercontinental ballistic missile has served as the backbone of U.S. nuclear deterrent forces. Throughout the cold war and up to the present day, deterring nuclear war remains the top American defense priority, but this policy produces only a meta-stable political equilibrium based on the concept of deterrence, which depends on an inherent lose-lose outcome for its efficacy. What happens to the deterrence equation if an adversary who possesses nuclear-armed ballistic missiles does not care if they lose (for whatever inconceivable reason)? Before the human race can enjoy the prosperity of a solar system civilization, we must learn how to guarantee absolute control of the powerful ballistic missile weapon systems we now possess. Our childhood has ended, and we must face future political disagreements with unprecedented maturity or perish as a result of social immaturity and foolhardiness. The expanding global presence of nuclear-armed ballistic missiles offers no other alternatives for our future global civilization.

trajectories above the atmosphere; they reenter the atmosphere in the terminal phase, where atmospheric drag affects them.

An effective ballistic missile defense system capable of engaging the missile attack all along its flight path must perform certain key functions, including: (1) promptly and reliably warning of an attack and initiating the defense; (2) continuously tracking all threatening objects from the beginning to the end of their trajectories; (3) efficiently intercepting and destroying the booster rocket or post-boost vehicle; (4) efficiently discriminating between enemy warheads and decoys through filtering of lightweight penetration aids; (5) efficiently and economically intercepting and destroying enemy warheads (reentry vehicles) in the midcourse phase of their flight; (6) intercepting and destroying the enemy warheads (reentry vehicles) at the outer reaches of the atmosphere during the terminal phase of their flight; and (7) effectively coordinating all of the defensive system's components through battle management, communications, and data processing. Battle management is the timely and accurate analysis of data on the state of a battle and the subsequent process of making decisions regarding which weapons to use and where to aim them. Subtasks include command and communication, kill assessment, maintaining knowledge of the state and positions of all elements of the defense system, and calculation of target track files. With respect to ballistic missile defense (BMD), battle management includes the set of instructions and rules and the corresponding hardware controlling the operation of a BMD system. Sensors and interceptors are allocated by the BMD system, and updated battle results are presented to the individuals in command for analysis and (as necessary) human intervention.

The figure below presents a more generalized concept for ballistic missile defense during the boost phase of the enemy's ballistic missile flight. An essential requirement is a global, full-time surveillance capability to detect a ballistic missile

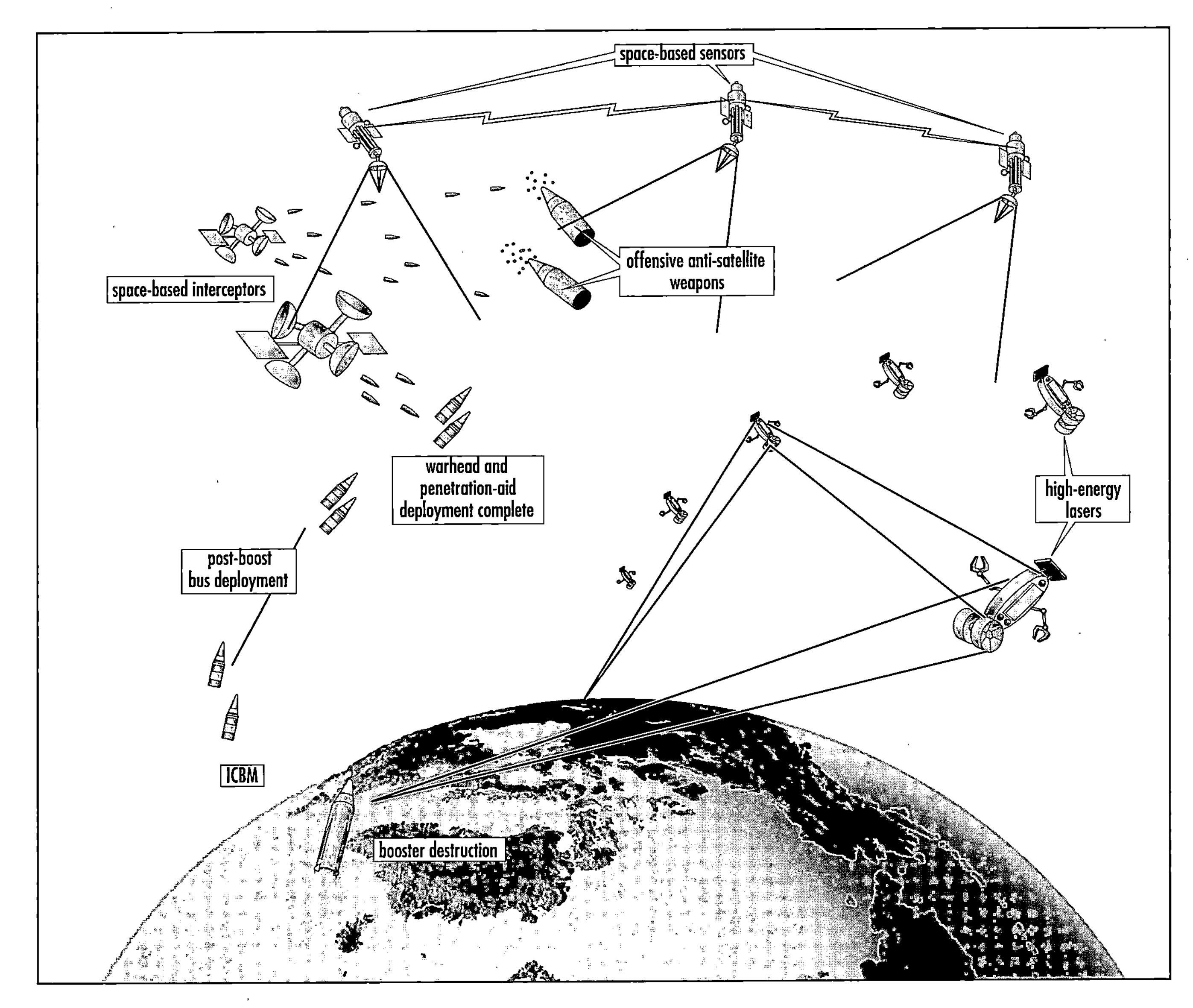

General scenario for ballistic missile defense during the boost phase of a ballistic missile's flight

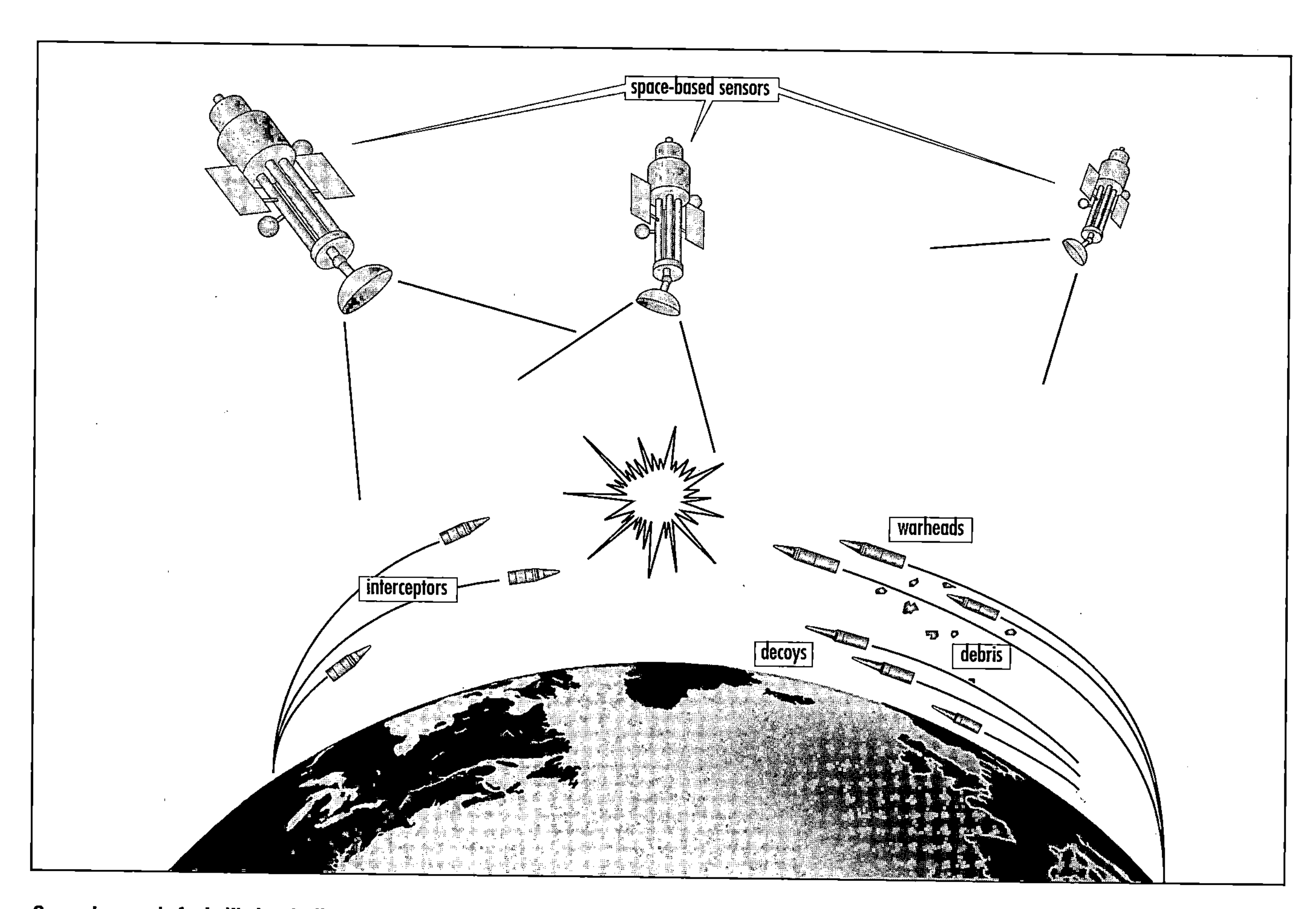

General scenario for ballistic missile defense during the midcourse phase of a ballistic missile's flight

attack, to define its destination and intensity, to determine targeted areas, and to provide data to guide boost-phase interceptors and post-boost vehicle tracking systems. Attacks may range from a single or perhaps a few missiles to a massive simultaneous launch. For every enemy booster destroyed during this phase, the number of potentially hostile objects to be identified and sorted out by the remaining elements of a multilayered (or multitiered) ballistic missile defense system will be significantly reduced. An early defensive response also will minimize the numbers of deployed penetration aids. Space-based SENSORs would be used to detect and define the ballistic missile attack. SPACE-BASED INTERCEPTORs would protect these sensor systems from enemy antisatellite (ASAT) weapons and, as a secondary mission, also would be used to attack the enemy ballistic missiles. For the boost phase defense scenario depicted in the figure, nonnuclear, direct-impact projectiles (i.e., kinetic energy weapons) are being used against the enemy's offensive weapons (i.e., ASAT weapons and warheads) while directed energy weapons (e.g., high-energy lasers) are being used to destroy the enemy's boosters early in the flight trajectory.

Similarly, the figure above illustrates a generalized ballistic missile defense scenario that might occur during the midcourse phase. Intercept outside the atmosphere during the midcourse phase of a missile attack would require the defense to cope with decoys designed to attract interceptors and to exhaust the defensive forces. However, continuing discrimination of nonthreatening objects and continuing attrition of reentry vehicles would reduce the pressure on the terminal phase missile defense system. Engagement times are longer during the midcourse phase than during other phases of a ballistic missile trajectory. The figure depicts space-based sensors that can discriminate among warheads, decoys, and debris and the interceptors that the defense has committed to the battle. The nonnuclear, direct-impact (hit-to-kill) projectiles race toward warheads that the space-based sensors have identified as credible targets.

Finally, the figure on the following page illustrates a ballistic missile defense scenario during the terminal phase of the attack. This phase is essentially the final line of defense. Threatening objects include warheads shot at but not destroyed, objects not previously detected, and decoys that have neither been discriminated against nor destroyed. The interceptors of the terminal phase defensive system must now deal with all of these threatening objects. An airborne optical adjunct (i.e., airborne sensor platform) is shown here. Surviving enemy reentry vehicles are detected in the late exoatmospheric portion of their flight with sensors

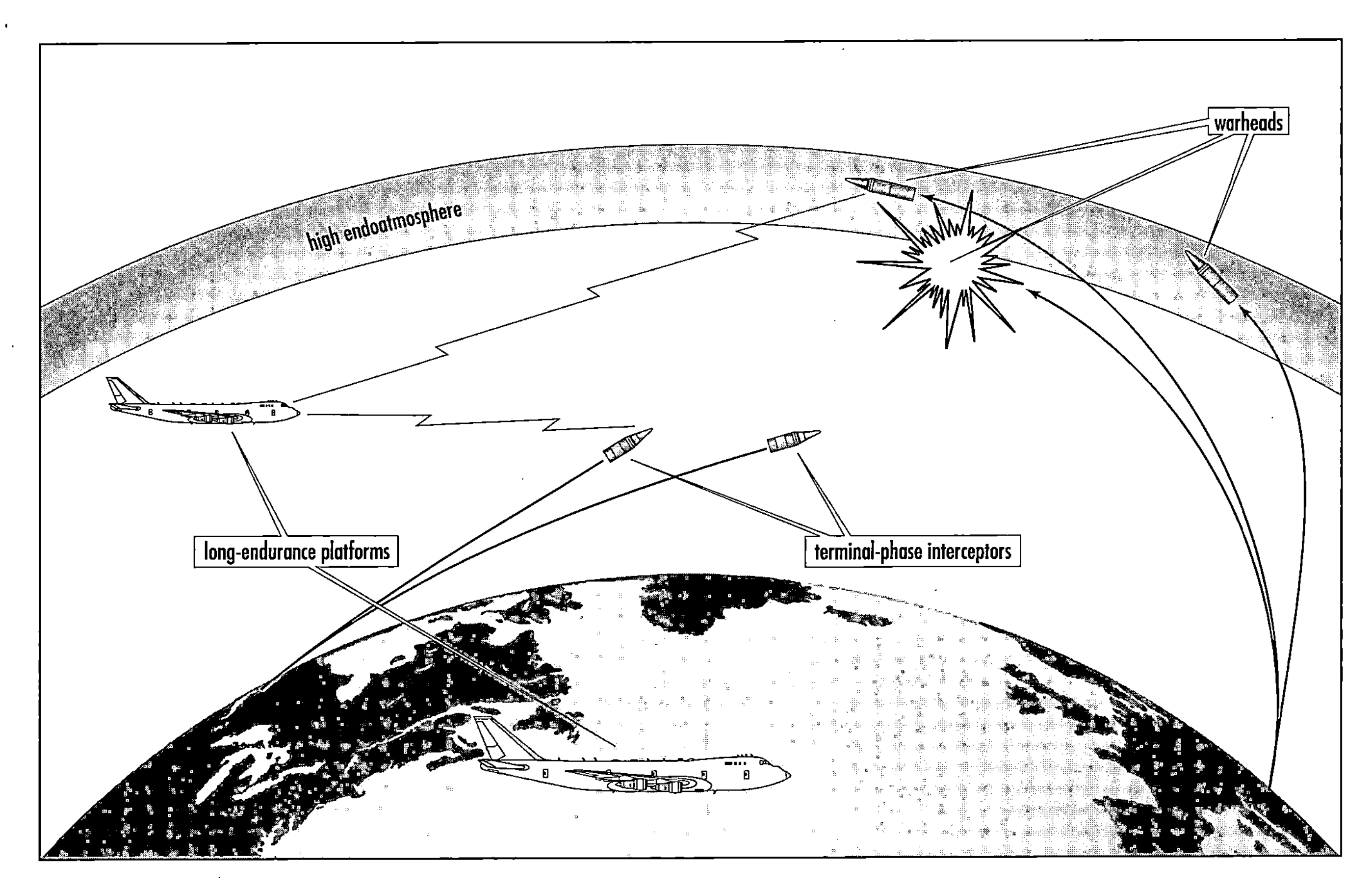

General scenario for ballistic missile defense during the terminal phase of a ballistic missile's flight

onboard the long-endurance (crewed or robotic) aerial platforms. Then terminal defense interceptors, illustrated here as nonnuclear, direct-impact projectiles, are guided to these remaining warheads and destroy them as they travel through the upper portions of the atmosphere.

Since the end of the cold war (1989), the increased proliferation of ballistic missile systems and weapons of mass destruction throughout the world has forced the United States to consider timely development of a ballistic missile defense system that can engage all classes of ballistic missile threats, including and especially the growing threat from rogue nations working in concert with terrorist groups. Consequently, in January 2003 President George W. Bush directed the U.S. Department of Defense (DOD) to begin fielding initial missile defense capabilities in the 2004–05 period to meet the near-term ballistic missile threat to the American homeland, deployed military forces, and territories of friendly and allied nations. As planned by the Missile Defense Agency, up to 20 ground-based interceptors, capable of destroying attacking intercontinental ballistic missiles during the midcourse phase of their trajectories, will be located at Fort Greely, Alaska (16 interceptors) and Vandenberg Air Force Base, California (4 interceptors). In addition, up to 20 sea-based interceptors will be deployed on existing Aegis ships. Their mission is to intercept short- and medium-range ballistic missiles in the midcourse phase of flight. These systems are being complemented by the deployment of air-transportable Patriot Advanced Capability-3 (PAC-3) systems to intercept short- and medium-range ballistic missiles.

The evolutionary American approach to missile defense includes development of the Theater High Altitude Air Defense (THAAD) system to intercept short- and medium-range missiles at high altitude and an airborne laser (ABL) aircraft to destroy a ballistic missile in the boost phase. Finally, the program includes the development and testing of space-based defenses, particularly space-based kinetic energy (hit-to-kill) interceptors and advanced target tracking satellites. However, BMD is a technically challenging and politically volatile undertaking. From a technical perspective the process is like stopping an incoming high-velocity rifle bullet with another rifle bullet—missing the threatening bullet is not a viable option.

ballistic missile submarine (SSBN) A nuclear-powered submarine armed with long-range strategic or intercontinental ballistic missiles. The U.S. Navy currently operates 18 Ohio-class/Trident ballistic missile submarines. Each nuclear-powered submarine carries 18 nuclear-armed Trident missiles.

ballistic trajectory The path an object (that does not have lifting surfaces) follows while being acted upon by only the force of gravity and any resistive aerodynamic forces of the medium through which it passes. A stone tossed into the air follows a ballistic trajectory. Similarly, after its propulsive unit stops operating, a rocket vehicle describes a ballistic trajectory.

band A range of (radio wave) frequencies. Alternatively, a closely spaced set of spectral lines that is associated with the electromagnetic radiation (EMR) characteristic of some particular atomic or molecular energy levels. In radiometry, a relatively narrow region of the electromagnetic spectrum to which a sensor element in a remote sensing system responds. A multispectral sensing system makes simultaneous measurements in a number of spectral bands. In spectroscopy, those spectral regions where atmospheric gases absorb (and emit) electromagnetic radiation; for example, the 15 micrometer (μm) carbon dioxide (CO_2) absorption band; the 6.3 micrometer (μm) water vapor (H_2O) absorption band, and the 9.6 micrometer (μm) ozone (O_3) absorption band.

bandwidth In general, the number of hertz (cycles per second) between the upper and lower limits of a frequency band. With respect to radio waves, the total range of frequency required to pass a specific modulated signal without distortion or loss of data. An ideal bandwidth allows the radio wave signal to pass under conditions of maximum amplitude modulation (AM) or frequency modulation (FM) adjustment. If the bandwidth is too narrow, it will cause a loss of data during modulation peaks. If the bandwidth is too wide, it will cause excessive noise to pass along with the signal. In frequency modulation, radio frequency (RF) signal bandwidth is determined by the frequency deviation of the signal.

bar A unit of pressure in the centimeter-gram-second (cgs) system equal to 10^5 pascal.

$$1 \text{ bar} = 10^6 \text{ dynes/cm}^2 = 10^5 \text{ newtons/m}^2 \text{ (pascals)} = 750 \text{ mm Hg}$$

The term *millibar* (symbol: mb) is encountered often in meteorology.

$$1 \text{ millibar (mb)} = 10^{-3} \text{ bar} = 100 \text{ pascals}$$

barbecue mode The slow roll of an orbiting aerospace vehicle or spacecraft to help equalize its external temperature and to promote a more favorable heat (thermal energy) balance. This maneuver is performed during certain missions. In outer space solar radiation is intense on one side of a space vehicle while the side opposite the Sun can become extremely cold.

barium star A red giant star, typically of G or K spectral type, characterized by the unusually high abundance of heavier elements, such as barium. Sometimes called a *heavy metal star.*

Barnard, Edward Emerson (1857–1923) American *Astronomer* In 1892 Edward Emerson Barnard discovered Jupiter's fifth moon, Amalthea. He also pioneered astrophotography—the use of photography in astronomy. For example, he was the first astronomer to find a comet by photographic means. An accomplished visual observer, he also discovered Barnard's star, a faint red dwarf with a large proper motion situated about six light-years from the Sun.

See also AMALTHEA.

Barnard's star A red dwarf star in the constellation Ophiuchus approximately six light-years from the Sun, making it the fourth-nearest star to the solar system. Discovered in 1916 by the American astronomer EDWARD E. BARNARD (1857–1923), it has the largest proper motion of any known star—some 10.3″ (arc-seconds) per year. Due to its large proper motion, when viewed by an observer on Earth this so-called fixed star appears to move across the night sky through a distance equivalent to the diameter of the Moon every 180 years. The absolute magnitude of Barnard's star is 13.2, and its spectral classification is M5 V.

barred spiral galaxy A type of spiral galaxy that has a bright bar or elongated ellipsoid of stars across the central regions of the galactic nucleus. A barred spiral galaxy is denoted as "SB" in EDWIN HUBBLE's classification of galaxies.

barrier cooling In rocket engineering, the use of a controlled mixture ratio near the wall of a combustion chamber to provide a film of low-temperature gases to reduce the severity of gas-side heating of the chamber.

Barringer crater *See* ARIZONA METEORITE CRATER.

barycenter The center of mass of a system of masses at which point the total mass of the system is assumed to be concentrated. In a system of two particles or two celestial bodies (that is, a binary system), the barycenter is located somewhere on a straight line connecting the geometric center of each object, but closer to the more massive object. For example, the barycenter for the Earth-Moon system is located about 4,700 km from the center of Earth—a point actually inside Earth, since our home planet has a radius of about 6,400 km.

baryon A member of the class of heavy elementary particles that contains three quarks and participates in strong interaction. Protons and neutrons, which make up the nuclei of atoms, are baryons.

See also FUNDAMENTAL FORCES IN NATURE.

basalt Fine-grained, dark-colored igneous rock composed primarily of plagioclase feldspar and pyroxene; other minerals such as olivine and ilmenite are usually present. Plagioclase feldspar is a common mineral, ranging from $NaAlSi_3O_8$ to $CaAl_2Si_2O_8$. Olivine is another mineral found in basalt, ranging from Mg_2SiO_4 to Fe_2SiO_4. Ilmenite is an opaque mineral consisting of nearly pure iron-titanium oxide ($FeTiO_3$). Basalt is the most common extrusive igneous rock, solidified from molten magma or volcanic fragments erupted on the surface of Earth, the Moon, or other solid-surface planetary bodies in the solar system. Basalt makes up the large smooth dark areas of the Moon called maria—since GALILEO GALILEI (1564–1642) and other 17th-century telescopic astronomers originally though these dark basaltic regions were lunar seas or oceans. NASA's APOLLO PROJECT MOON landing missions brought back many samples of lunar basalt.

baseline In general, any line that serves as the basis for measurement of other lines. Often in aerospace usage, a reference case, such as a *baseline design,* a *baseline schedule,* or a *baseline budget.*

basin (impact) A large, shallow, lowland area in the crust of a terrestrial planet formed by the impact of an asteroid or comet.

See also CRATER; IMPACT CRATER.

al-Battani (Latinized: Albategnius) (ca. 858–929) Arab *Astronomer, Mathematician* Al-Battani is regarded as the best and most famous astronomer of medieval Islam. As a very skilled naked eye observer he refined the sets of solar, lunar, and planetary motion data found in PTOLEMY's great work, *The Almagest,* with more accurate measurements. Centuries after his death these improved observations, as contained in his compendium, *Kitab al-Zij,* worked their way into the Renaissance and exerted influence on many western European astronomers and astrologers alike.

Al-Battani and his fellow Muslim stargazers preserved and refined the geocentric cosmology of the early Greeks. Though al-Battani was a devout follower of Ptolemy's geocentric cosmology, his precise observational data encouraged NICHOLAS COPERNICUS to pursue heliocentric cosmology—the stimulus for the great scientific revolution of the 16th and 17th centuries. Al-Battani was also an innovative mathematician who introduced the use of trigonometry in observational astronomy. In 880 he produced a major star catalog, called the *Kitab al-Zij,* and refined the length of the year to approximately 365.24 days.

Al-Battani was born in about 858 in the ancient town of Harran, located in northern Mesopotamia some 44 kilometers southeast of the modern Turkish city of Sanliufra. His full Arab name is Abu Abdullah Muhammad ibn Jabir ibn Sinan al-Raqqi al-Harrani al-Sabi al-Battani, which later European medieval scholars who wrote exclusively in Latin changed to Albategnius. His father, Jabir ibn Sinan al-Harrani, was an astronomical instrument maker and a member of the Sabian sect, a religious group of star worshippers in Harran. So al-Battani's lifelong interest in astronomy probably started at his father's knee in early childhood. Al-Battani's contributions to astronomy are best appreciated within the context of history. At this time Islamic armies were spreading through Egypt and Syria and westward along the shores of the Mediterranean Sea. Muslim conquests in the eastern Mediterranean resulted in the capture of ancient libraries, including the famous one in Alexandria, Egypt. Ruling caliphs began to recognize the value of many of these ancient manuscripts, and so they ordered Arabs scribes to translate any ancient document that came into their possession.

By fortunate circumstance, while the Roman Empire was starting to collapse in western Europe, Nestorian Christians busied themselves preserving and archiving Syrian-language translations of many early Greek books. Caliph Harun al Rashid ordered his scribes to purchase all available translations of these Greek manuscripts. His son and successor, Caliph al-Ma'mun, who ruled between 813 and 833, went a step beyond. As part of his peace treaty with the Byzantine emperor, Caliph al-Ma'mun was to receive a number of early Greek manuscripts annually. One of these tribute books turned out to be Ptolemy's great synthesis of ancient Greek astronomical learning. When translated it became widely known by its Arabic name, *The Almagest* (or "the greatness"). So the legacy of geocentric cosmology from ancient Greece narrowly survived the collapse of the Roman Empire and the ensuing Dark Ages in western Europe by taking refuge in the great astronomical observatories of medieval Islam as found in Baghdad, Damascus, and elsewhere.

But Arab astronomers did not just accept the astronomical data presented by Ptolemy. They busied themselves checking these earlier observations and often made important refinements using improved instruments, including the astrolabe and more sophisticated mathematics. History indicates that al-Battani was a far better observer than any of his contemporaries. Yet, collectively, these Arab astronomers participated in a golden era of naked eye astronomy enhanced by the influx of ancient Greek knowledge, mathematical concepts from India, and religious needs. For example, Arab astronomers refined solar methods of time keeping so Islamic clergymen would know precisely when to call the Muslim faithful to daily prayer.

Al-Battani worked mainly in ar-Raqqah and also in Antioch (both now located in modern Syria). Ar-Raqqah was an ancient Roman city along the Euphrates River just west of where it joins the Balikh River at Harran. Like other Arab astronomers, he essentially followed the writings of Ptolemy and devoted himself to refining and improving the data contained in *The Almagest.* While following this approach, he made a very important discovery concerning the motion of the aphelion point—that is, the point at which Earth is farthest from the Sun in its annual orbit. He noticed that the position at which the apparent diameter of the Sun appeared smallest was no longer located where Ptolemy said it should be with respect to the fixed stars of the ancient Greek zodiac. Al-Battani's data were precise, so he encountered a significant discrepancy with the observations of Ptolemy and early Greek astronomers. But neither al-Battani nor any other astronomer who adhered to Ptolemy's geocentric cosmology could explain the physics behind this discrepancy.

They would first need to use heliocentric cosmology to appreciate why the Sun's apparent (measured) diameter kept changing throughout the year as Earth traveled along its slightly eccentric orbit around the Sun. Second, the Sun's annual apparent journey through the signs of the zodiac corresponded to positions noted by the early Greek astronomers. But, because of the phenomenon of precession (the subtle wobbling of Earth's spin axis), this correspondence no longer took place in al-Battani's time—nor does it today. Without the benefit of a wobbling Earth model (to explain precession) and heliocentric cosmology (to explain the variation in the Sun's apparent diameter), it was clearly impossible for him, or any other follower of Ptolemy, to appreciate the true scientific significance of what he discovered.

Al-Battani's precise observations also allowed him to improve Ptolemy's measurement of the obliquity of the ecliptic—that is, the angle between Earth's equator and the plane defined by the apparent annual path of the Sun against the fixed star background. He also made careful measurements of when the vernal and autumnal equinoxes took place. These observations allowed him to determine the length of a year to be about 365.24 days. The accuracy of al-Battani's work helped the German Jesuit mathematician CHRISTOPHER CLAVIUS reform the Julian calendar and allowed Pope Gregory XIII to replace it in 1582 with the new "Gregorian" calendar that now serves much of the world as the international civil calendar.

While al-Battani was making his astronomical observations between the years 878 and 918, he became interested in some of the new mathematical concepts other Arab scholars were discovering in India. Perhaps al-Battani's greatest contribution to Arab astronomy was his introduction of the use of trigonometry, especially spherical trigonometric functions, to replace Ptolemy's geometrical methods of calculating celestial positions. This contribution gave Muslim astronomers use of some of the most complicated mathematics known in the world up to that time.

He died in 929 in Qar al-Jiss (now in modern Iraq) on a homeward journey from Baghdad, where he had presented an unfair taxation grievance from the people of ar-Raqqah to officials of the ruling caliph. His contributions to astronomy continued for many centuries after his death. Medieval astronomers became quite familiar with Albategnius, primarily through a 12th-century Latin translation of his *Kitab al-Zij*, renamed *De motu stellarum* (On stellar motion). The invention of the printing press in 1436 by Johannes Gutenberg soon made it practical to publish and widely circulate the handwritten Latin translation. This occurred in 1537 in Nuremberg, Germany—just in time for the centuries-old precise observations of al-Battani to spark interest in some of the puzzling astronomical questions that spawned the scientific revolution.

battery An electrochemical energy storage device that serves as a source of direct current or voltage, usually consisting of two or more electrolytic cells that are joined together and function as a single unit. The earliest Russian and American Earth-orbiting spacecraft (e.g., *Sputnik 1* and *Explorer 1*) depended on batteries for all their electric power. In general, however, batteries are acceptable as the sole source of electric power only on spacecraft or planetary probes with missions of very short duration—hours, perhaps day, or at the most weeks in length. Solar photovoltaic conversion is used in combination with rechargeable batteries on the vast majority of today's spacecraft. The rechargeable batteries provide electrical power during "dark times" when the solar arrays cannot view the Sun. For example, the nickel-cadmium battery

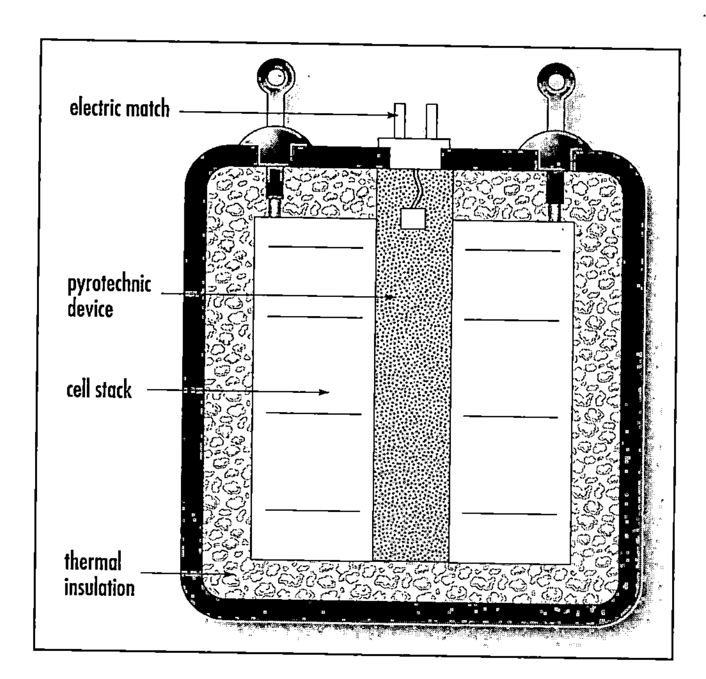

Diagram of a typical high-temperature battery with a service life of about five minutes and an operating temperature of 430°C

has been the common energy storage companion for solar cell power supply systems on many spacecraft. Specific energy densities (that is, the energy per unit mass) of about 10 watt-hours per kilogram are common at the 10 to 20 percent depths of discharge used to provide cycle life. The energy storage subsystem is usually the largest and heaviest part of a solar cell–rechargeable battery space power system.

Storable batteries can be used as the sole power supply for expendable planetary probes required to operate in harsh environments. For example, the figure shows a typical high-temperature battery with a service life of about five minutes that can operate at a temperature of 430°C. Activated when needed by a pyrotechnic device, this type of high-temperature battery could power a probe that briefly functioned in the high-temperature surface environment of Mercury or Venus. Unless extraordinary design features were built into a probe, its instruments and supporting subsystems most likely would succumb to the elevated temperature environment within a short time.

baud (rate) A unit of signaling speed. The baud rate is the number of electronic signal changes or data symbols that can be transmitted by a communications channel per second. Named after J. M. Baudot (1845–1903), a French telegraph engineer.

Bayer, Johann (1572–1625) German *Astronomer* In 1603 Johann Bayer published *Uranometria*, the first major star catalog for the entire celestial sphere. He charted more than 2,000 stars visible to the naked eye and introduced the practice of assigning Greek letters (such as α, β, and γ) to the main stars in each constellation, usually in an approximate descending order of their brightness.

beacon A light, group of lights, electronic apparatus, or other type of signaling device that guides, orients, or warns aircraft, spacecraft, or aerospace vehicles in flight. For example, a crash locator beacon emits a distinctive, characteristic signal that allows search-and-rescue (SAR) teams to locate a downed aircraft, aerospace vehicle, or aborted or reentered spacecraft of probe.

beam A narrow, well-collimated stream of particles (such as electrons or protons) or electromagnetic radiation (such as gamma ray photons) that are traveling in a single direction.

beamed energy propulsion An advanced propulsion system concept in which a remote power source (such as the Sun or a high-powered laser) supplies thermal energy to an orbital transfer vehicle (OTV). Two beamed energy concepts that have been suggested are solar thermal propulsion and laser thermal propulsion. Solar thermal propulsion makes use of the Sun and does not require an onboard energy source. The solar thermal rocket would harvest raw sunlight with concentrators or mirrors. Then the collected solar radiation would be focused to heat a propellant, such as hydrogen. Once at high temperature, the gaseous propellant would flow through a conventional converging-diverging nozzle to produce thrust. Similarly, laser thermal propulsion would use a remotely located high-power laser to beam energy to an orbital transfer vehicle to heat a propellant, such as hydrogen. The resultant high-temperature gas would then be discharged through a thrust-producing nozzle. Such advanced propulsion systems might eventually support the transport of cargo between various destinations in cislunar space and possibly even from the surface of the Moon to low lunar orbit.

beam rider A missile guided to its target by a beam of electromagnetic radiation, such as a radar beam or a laser beam.

Bean, Alan L. (1932–) American *Astronaut, U.S. Navy Officer* Astronaut Alan Bean commanded the lunar excursion module (LEM) during NASA's *Apollo 12* mission. Along with CHARLES (PETE) CONRAD, JR., he walked on the Moon's surface and deployed instruments in November 1969 as part of the second lunar landing mission of the Apollo Project. Bean also served as spacecraft commander for the *Skylab 2* space station mission in 1973. These spaceflight experiences inspired his work as a space artist.

See also APOLLO PROJECT; *SKYLAB*.

becquerel (symbol: Bq) The SI unit of radioactivity. One becquerel corresponds to one nuclear transformation (disintegration) per second. Named after the French physicist A. Henri Becquerel (1852–1908), who discovered radioactivity in 1896. This unit is related to the curie (the traditional unit for radioactivity) as follows: 1 curie (Ci) = 3.7×10^{10} becquerels (Bq).

bel (symbol: B) A logarithmic unit (n) used to express the ratio of two power levels, P_1 and P_2. Therefore, n (bels) = $\log_{10} (P_2/P_1)$. The decibel (symbol: dB) is encountered more frequently in physics, acoustics, telecommunications, and electronics, where 10 decibels = 1 bel. This unit honors the American inventor Alexander Graham Bell (1847–1922).

Compare with NEPER.

Belinda The ninth-closest moon to Uranus. Discovered as a result of the *Voyager 2* flyby in 1986, Belinda gets its name from the heroine in Alexander Pope's (1688–1744) *The Rape of the Lock,* an elegant satire about a vain young woman in English high society who has a lock of her hair stolen by an ardent young man. This small, 66-km-diameter satellite orbits Uranus every 0.6235 days at a distance of 75,260 km and with an orbital inclination of 0.03°.

Bell-Burnell, (Susan) Jocelyn (1943–) British *Astronomer* In August 1967, while still a doctoral student at Cambridge University under the supervision of Professor ANTONY HEWISH, Jocelyn Bell-Burnell discovered an unusual, repetitive radio signal that proved to be the first pulsar. She received a Ph.D. for her pioneering work on pulsars, while Hewish and a Cambridge colleague (MARTIN RYLE) eventually shared the 1974 Nobel Prize in physics for their work on pulsars. Possibly because she was a student at the time of the discovery, the Nobel Committee rather unjustly ignored her observational contribution to the discovery of the pulsar.

bell nozzle A nozzle with a circular opening for a throat and an axisymmetric contoured wall downstream of the throat that gives this type of nozzle a characteristic bell shape.

See also NOZZLE.

Belt of Orion The line of three bright stars (Alnilam, Alnitak, and Mintaka) that form the Belt of Orion, a very conspicuous constellation on the equator of the celestial sphere. The constellation honors the great hunter in Greek mythology.

bent-pipe communications An aerospace industry term (jargon) for the use of relay stations to achieve non–line-of-sight (LOS) transmission links.

Bernal, John Desmond (1910–1971) Irish *Physicist, Writer* John Desmond Bernal speculated about the colonization of space and the construction of very large spherical space settlements (now called BERNAL SPHERES) in his futuristic 1929 work *The World, the Flesh and the Devil.*

Bernal sphere Long before the space age began, the Irish physicist and writer JOHN DESMOND BERNAL predicted that the majority of the human race would someday live in "artificial globes" orbiting around the Sun. In particular, Bernal's famous 1929 work *The World, the Flesh and the Devil* conveys his bold speculation about large-scale human migration into outer space. Bernal's concept of spherical space habitats influenced both early space station designs and concepts for very large space settlements.

berthing The joining of two orbiting spacecraft using a manipulator or other mechanical device to move one into contact (or very close proximity) with the other at a selected interface. For example, NASA astronauts have used the space

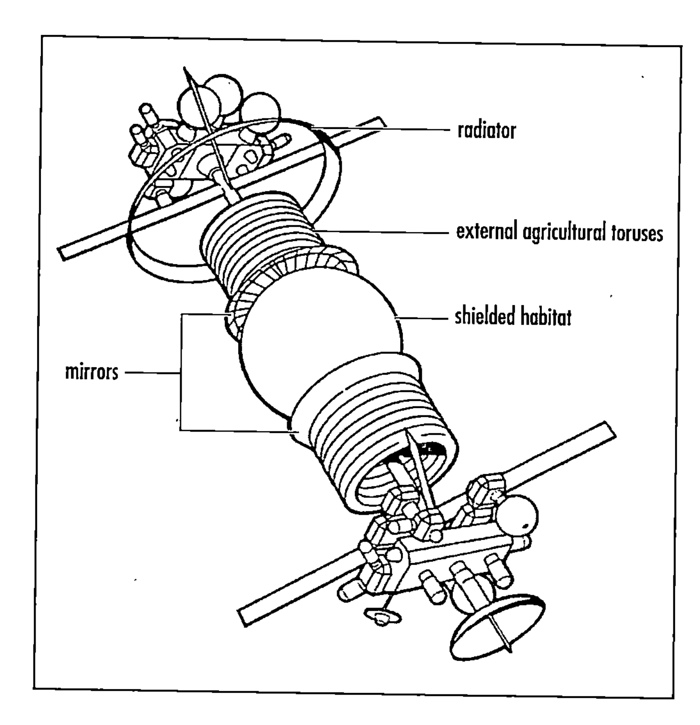

Large space settlement concept based on the Bernal sphere ***(Courtesy of NASA)***

shuttle's remote manipulator system to carefully berth a large free-flying spacecraft (such as the *Hubble Space Telescope*) onto a special support fixture located in the orbiter's PAYLOAD BAY during an on-orbit servicing and repair mission.

See also DOCKING; RENDEZVOUS.

Bessel, Friedrich Wilhelm (1784–1846) German *Astronomer, Mathematician* Friedrich Wilhelm Bessel pioneered precision astronomy and was the first to accurately measure the distance to a star (other than the Sun). In 1818 he published *Fundamenta Astronomiae,* a catalog of more than 3,000 stars. By 1833 he completed a detailed study from the Königsberg Observatory of 50,000 stars. His greatest accomplishment occurred in 1838, when he carefully observed the binary star system 61 Cygni and used its parallax (annual angular displacement) to estimate its distance at about 10.3 light-years from the Sun (current value is 11.3 light-years). His mathematical innovations and rigorous observations greatly expanded the scale of the universe and helped shift astronomical interest beyond the solar system.

beta decay Radioactivity in which an atomic nucleus spontaneously decays and emits two subatomic particles: a beta particle (β) and a neutrino (ν). In beta-minus (β^-) decay a neutron in the transforming (parent) nucleus becomes a proton, and a negative beta particle and an antineutrino are emitted. The resultant (daughter) nucleus has its atomic number (Z) increased by one (thereby changing its chemical properties), while the total atomic mass (A) remains the same as that of the parent nucleus. In beta-plus (β^+) decay a proton is converted into a neutron, and a positive beta particle (positron) is emitted along with a neutrino. The atomic number of the resultant (daughter) nucleus is decreased by one, a process that also changes its chemical properties.

beta particle (β) The negatively charged subatomic particle emitted from the atomic nucleus during the process of beta decay. It is identical to the *electron.*

See also ELECTRON; POSITRON.

Beta Regio A prominent and young upland region on Venus, situated to the north of the planet's equator. At the widest it is about 2,900 km across. This region contains a long (about 1,600-km) rift valley called Devana Chasma and also has two huge volcanoes, Rhea Mons (about 6.7 km in altitude) and Theia Mons (about 5.2 km in altitude).

See also APHRODITE TERRA; ISHTAR TERRA.

beta (β) star Within the system of stellar nomenclature introduced in 1603 by the German astronomer JOHANN BAYER, the second-brightest star in a constellation.

Betelgeuse The red supergiant star, alpha Orinois (α Ori), that marks the right shoulder in the constellation Orion, the great hunter from Greek mythology. Betelgeuse has a spectral classification of M2Ia and is a semiregular variable with a period of about 5.8 years. It has a normal apparent magnitude (visual) range between 0.3 and 0.9 (about +0.41 average), but can vary between 0.15 and 1.3. This huge star (about 500 times the diameter of the Sun) is approximately 490 light-years way from Earth. The two brightest stars in the constellation Orion were designated α Ori (Betelgeuse) and β Ori (Rigel) in order of their apparent brightness using the notation introduced by JOHANN BAYER (1572–1625) in the 17th century. However, more precise recent observations indicate that Rigel is actually brighter than Betelgeuse. Nevertheless, to avoid confusion astronomers retain the previous stellar nomenclature.

Bethe, Hans Albrecht (1906–2005) German-American *Physicist* Hans Bethe is the physicist and Nobel laureate who proposed the mechanisms by which stars generate their vast quantities for energy through the nuclear fusion of hydrogen into helium. In 1938 he worked out the sequence of nuclear fusion reactions, called the carbon cycle, that dominate energy liberation in stars more massive than the Sun. Then, working with a Cornell University colleague (Charles Critchfield), he proposed the proton-proton reaction as the nuclear fusion process for stars up to the size of the Sun. For this astrophysical work he received the 1967 Nobel Prize in physics.

Bethe was born on July 2, 1906, in Strassburg, Germany (now Strasbourg, Alsace-Lorraine, France). Starting in 1924 he studied at universities in Frankfurt and then in Munich. He received a Ph.D. in theoretical physics in July 1928 from at the University of Munich. His doctoral thesis on electron diffraction still serves as an excellent example of how a physicist should use observational data to understand the physical universe. From 1929 to 1933 he held positions as a visiting researcher or physics lecturer at various universities in Europe, including work with ENRICO FERMI at the University of Rome in 1931. With the rise of the Nazi Party in Germany,

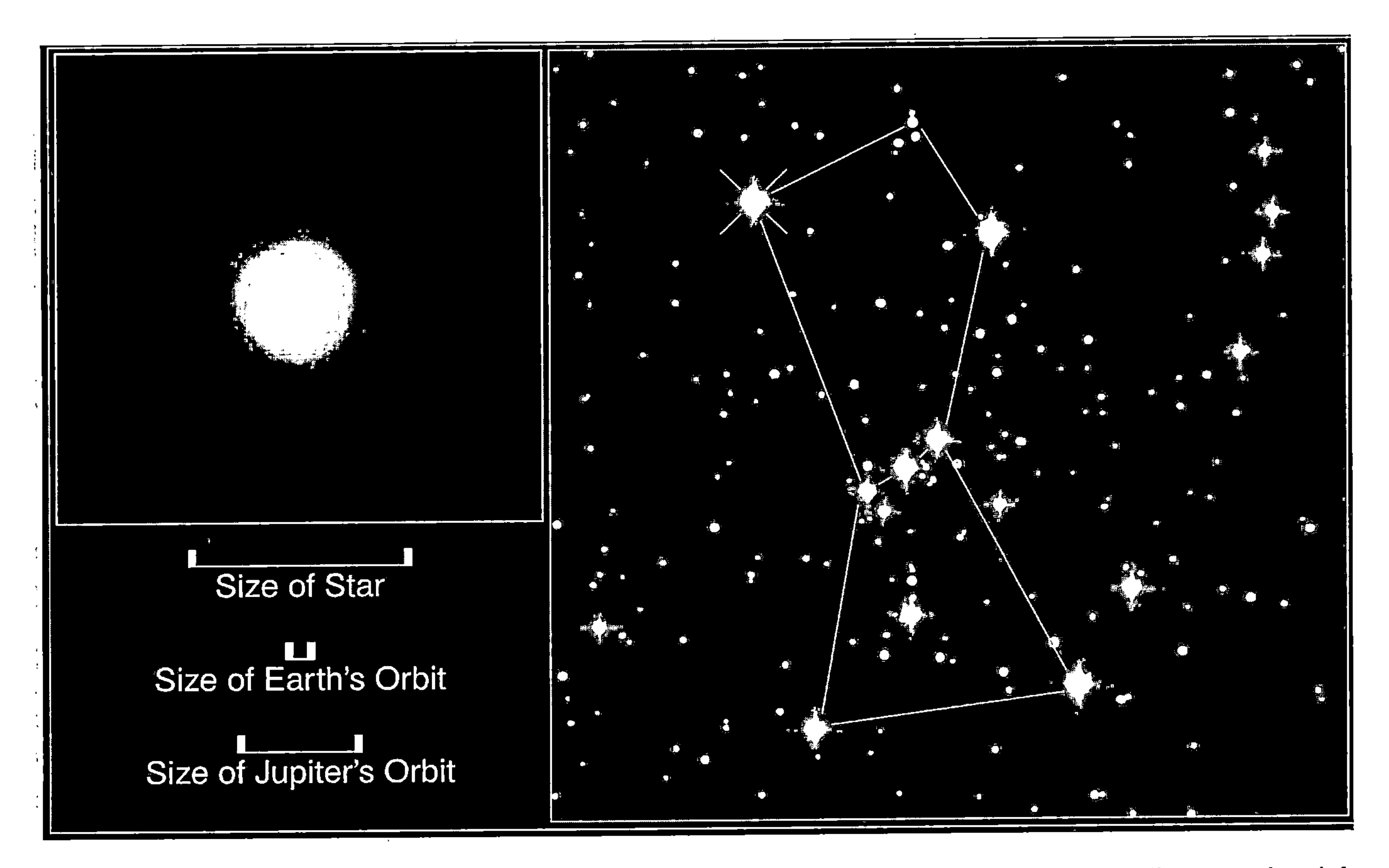

This is the first direct image of a star other than the Sun. Called Alpha Orionis, or Betelgeuse, the star is a red supergiant—a Sun-like star near the end of its life. The *Hubble Space Telescope* (HST) picture reveals a huge ultraviolet atmosphere with a mysterious hot spot on the stellar behemoth's surface. The enormous bright spot, more than ten times the diameter of Earth, is at least 2,000 kelvins hotter than the rest of the star's surface. *(Courtesy of Andrea Dupree [Harvard-Smithsonian CfA], Ronald Gilliland [STScI], NASA, and ESA)*

Bethe lost his position as a physics lecturer at the University of Tübingen. In 1933 he became a refugee from fascism. He left Germany and immigrated to the United States after spending a year (1934) working as a physicist in the United Kingdom.

In February 1935 Cornell University in Ithaca, New York, offered Bethe a position as an assistant professor of physics. By the summer of 1937 the university promoted him to the rank of full professor. Except for sabbatical leaves and an absence during World War II, he remained affiliated with Cornell as a professor of physics until 1975. At that point Bethe retired to the rank of professor emeritus. His long and very productive scientific career was primarily concerned with the theory of atomic nuclei.

In 1939 he helped solve the long-standing mystery in astrophysics and astronomy by explaining energy production processes in stars like the Sun. Bethe proposed that the Sun's energy production results from the nuclear fusion of four protons (that is, four hydrogen nuclei) into one helium-4 nucleus. The slight difference in the relative atomic masses of reactants and product of this thermonuclear reaction manifests itself as energy in the interior of stars. This hypothesis became known as the proton-proton cycle—the series of nuclear fusion reactions by which energy can be released in the dense cores of stars. Bethe received the 1967 Nobel Prize in physics for his "contributions to the theory of nuclear reactions, especially his discoveries concerning the energy production in stars."

In 1941 Bethe became an American citizen and took a leave of absence from Cornell University so he could contribute his scientific talents to the war effort of his new country. His wartime activities took him first to the Radiation Laboratory at the Massachusetts Institute of Technology, where he worked on microwave radar systems. He then went to the Los Alamos National Laboratory, New Mexico, where he played a major role in the development of the American atomic bomb under the Manhattan Project. Specifically, from 1943 to 1946 Bethe served as head of the theoretical group at Los Alamos and was responsible for providing the technical leadership needed to take nuclear physics theory and transform it into successfully functioning nuclear fission weapons.

Throughout the cold war Bethe continued to serve the defense needs of the United States, primarily as a senior scientific adviser to presidents and high-ranking government leaders. In 1952 he returned briefly to Los Alamos to provide his theoretical physics expertise as the laboratory prepared to test the first American hydrogen bomb. The following year, after the surprisingly successful atmospheric test of Andrei Sakharov's (1921–89) hydrogen bomb, Bethe provided scientific guidance to senior U.S. government officials concerning the technical significance of the emerging Soviet thermonuclear weapons program. In 1961 the U.S. Department of Energy presented him with the Enrico Fermi award in recognition of his many contributions to nuclear energy science and technology. Yet, despite the important role he played in developing the powerful arsenal of American nuclear weapons, Bethe was also a strong advocate for nuclear disarmament and the peaceful applications of nuclear energy. Bethe died on March 6, 2005, in Ithaca, New York.

Bianca The third-closest moon of Uranus, discovered in 1986 by the *Voyager* 2 spacecraft. This satellite orbits Uranus at a distance of 59,170 km and an inclination of 0.19°. It has a diameter of 42 km and an orbital period of 0.43458 day. Bianca was named after the sister of Katherine in William Shakespeare's comical play *The Taming of the Shrew.*

Biela, Wilhelm von (1782–1856) Austrian *Military Officer* Wilhelm von Biela was an Austrian military officer and amateur astronomer who "rediscovered" a short-period comet in 1826 and then calculated its orbital period (about 6.6 years). During the predicted 1846 perihelion passage of *Biela's comet,* its nucleus split in two. Following the 1852 return of Biela's *double comet,* the object disappeared, apparently disintegrating into an intense meteor shower now called the Andromedid meteors or Bielids. Last seen in 1904, the Bielids are no longer observed due to gravitational perturbation of the meteor stream. However, according to some astronomers, the Bielids may return in the early part of the 22nd century. The dynamics and disappearance of Biela's comet demonstrated to 19th-century astronomers that comets were transitory, finite objects.

See also COMET.

Biela's comet *See* BIELA, WILHELM VON.

big bang (theory) A widely accepted theory in contemporary cosmology concerning the origin of the universe. According to the big bang cosmological model, there was a very large ancient explosion called the initial singularity that started the space and time of the current universe. The universe itself has been expanding ever since, and scientists have recently observed that the rate of this expansion is increasing as well. It is presently thought that the big bang event occurred between 12 and 20 billion years ago, with 15 billion years frequently taken as the nominal value. Astrophysical observations and discoveries lend support to the big bang model, especially the discovery of the cosmic microwave background (CMB) radiation in the mid-1960s by ARNO ALLEN PENZIAS (b. 1933) and ROBERT WOODROW WILSON (b. 1936).

There are three general outcomes (or destinies) within big bang cosmology: the open universe, the closed universe, and the flat universe. In the first, the *open universe model,* scientists hypothesize that once created by this primordial explosion, the universe will continue to expand forever. The second outcome, called the *closed universe model,* assumes that the universe (under the influence of gravitational forces) will eventually stop expanding and then collapse back into another incredibly dense singularity. The endpoint of this collapse is sometimes referred to as the BIG CRUNCH. If, after the big crunch, a new cycle of explosion, cosmic expansion, and eventual collapse occurs, then the model is called the oscillating universe (or pulsating universe) model.

The third outcome, called the *flat universe model,* actually falls between the first two. This model assumes that the universe expands after the big bang up to a point and then gently halts when the forces driving the expansion and the forces of matter-induced gravitational attraction reach a state of equilibrium. Beyond this point there will be no further expansion, nor will the universe begin to contract and close itself into a compact, incredibly dense object. To make this third outcome possible, the universe must contain just the right amount of matter—a precise critical mass density described by the density parameter (symbol: Ω, or omega).

Big bang cosmologists suggest that if omega is unity (that is, $\Omega = 1$), then the universe is flat; if omega is less than unity (i.e., $\Omega < 1$), then the universe is open; and if omega is greater than unity (i.e., $\Omega > 1$), then the universe is closed. There are, of course, competing cosmological models, several of which involve the use of the COSMOLOGICAL CONSTANT (Λ, or lamda), a concept originally suggested by ALBERT EINSTEIN as he formed general relativity in 1915 and then applied it to cosmology in 1917. Briefly summarized here, the cosmological constant implies that the shape and emptiness of space gives rise to a property in the universe that somehow balances the relentless attraction of gravity. When EDWIN HUBBLE announced his observation of an expanding universe in the late 1920s, Einstein abandoned this concept and regarded it as his greatest failure. However, some modern cosmologists are again embracing the notion of a cosmological constant as part of their ongoing attempts to describe the fate of a universe that appears by more recent observations (in the late 1990s) to be expanding at an ever-increasing rate.

A very short time after the big bang explosion the concepts of space and time as we understand them today came into existence. Physicists sometimes call this incredibly tiny time interval (about 10^{-43} second) between the big bang event and the start of space and time the *Planck time* or the *Planck era.* The limitations imposed by the uncertainty principle of quantum mechanics prevent scientists from developing a model of the universe between the big bang and the end of Planck time. However, an interesting intellectual alliance has formed between astrophysicists (who study the behavior of matter and energy in the large-scale universe) and high-energy nuclear particle physicists (who investigate the behavior of matter and energy in the subatomic world).

In an important application of experimental and theoretical research from modern high-energy physics, particle physicists currently postulate that the early universe experienced a specific sequence of phenomena and phases following the big bang explosion. They generally refer to this sequence as the *standard cosmological model.* Right after the big bang, the present universe reached an incredibly high temperature—physicists suggest the unimaginably high value of about 10^{32} K. During this period, called the quantum gravity epoch, the force of gravity, the strong (nuclear) force, and a composite electroweak force all behaved as a single unified force. Scientists postulate that during this period the physics of elementary particles and the physics of space and time were one and the same. They now search for a theory of everything (TOE) to adequately describe quantum gravity.

At Planck time (about 10^{-43} second after the big bang) the force of gravity assumed its own identity. Scientists call the ensuing period the grand unified epoch. While the force of gravity functioned independently during this period, the strong nuclear force and the composite electroweak force remained together and continued to act like a single force. Today, physicists apply various grand unified theories (GUTs) in their efforts to model and explain how the strong nuclear force and the electroweak force functioned as one during this particular period in the early universe.

Then, about 10^{-35} second after the big bang the strong nuclear force separated from the electroweak force. By this time the expanding universe had "cooled" to about 10^{28} K. Physicists call the period between about 10^{-35} s and 10^{-10} s the electroweak force epoch. During this epoch the weak nuclear force and the electromagnetic force became separate entities, as the composite electroweak force disappeared. From this time forward the universe contained the four fundamental forces in nature known to physicists today—namely, the force of gravity, the strong nuclear force, the weak nuclear force, and the electromagnetic force.

Following the big bang and lasting up to about 10^{-35} s, there was no distinction between quarks and leptons. All the minute particles of matter were similar. However, during the electroweak epoch quarks and leptons became distinguishable. This transition allowed quarks and antiquarks to eventually become hadrons, such as neutrons and protons, as well as their antiparticles. At 10^{-4} s after the big bang, in the radiation-dominated era, the temperature of the universe cooled to 10^{12} K. By this time most of the hadrons disappeared because of matter-antimatter annihilations. The surviving protons and neutrons represented only a small fraction of the total number of particles in the universe, the majority of which were leptons, such as electrons, positrons, and neutrinos. However, like most of the hadrons before them, the majority of the leptons also soon disappeared as a result of matter-antimatter interactions.

At the beginning of the matter-dominated era, some three minutes (or 180 s) following the big bang, the expanding universe cooled to a temperature of 10^9 K, and small nuclei, such as helium, began to form. Later, when the expanding universe reached about 300,000 years old, the temperature dropped to 3,000 K, allowing the formation of hydrogen and helium atoms. Interstellar space is still filled with the remnants of this primordial hydrogen and helium.

Eventually, density inhomogeneities allowed the attractive force of gravity to form great clouds of hydrogen and helium. Because these clouds also experienced local density inhomogeneities, gravitational attraction formed stars and then slowly gathered groups of these stars into galaxies (about 1 billion years ago). As gravitational attraction condensed the primordial (big bang) hydrogen and helium into stars, nuclear reactions at the cores of the more massive stars created heavier nuclei up to and including iron. Supernova explosions then occurred at the end of the lives of many of these early, massive stars. These spectacular explosions produced all atomic nuclei with masses beyond iron and then scattered these heavier elements into space. The expelled "stardust" would eventually combine with interstellar gas and create new stars along with their planets, including our solar system (about 4.6 billion years ago). All things animate and inanimate here on Earth are the natural byproducts of these ancient astrophysical processes. In a very real sense, each one of us is made of stardust.

In big bang cosmology, before the galaxies and stars formed the universe was filled with hot glowing plasma that was opaque. When the expanding plasma cooled to about 10,000 K (some 10,000 years after the big bang), the era of matter began. At first, matter consisted primarily of an ionized gas of protons, electrons, and helium nuclei. Then, as this gaseous matter continued to expand, it cooled further. When it reached about 3,000 K the electrons could recombine with protons to form neutral hydrogen. Within this period of recombination (which started about 300,000 years after the big bang), the ancient fireball became transparent and continued to cool from 3,000 K to 2.7 K—the current value of the cosmic microwave background (CMB). As time passed matter condensed into galactic nebulae, which in turn gave rise to stars and eventually other celestial objects such as planets, asteroids, and comets.

Today, when scientists look deep into space with their most advanced instruments, they can see back into time only until they reach that very distant region where the early universe transitioned from an opaque gas to a transparent gas. Beyond that transition point the view of the universe is opaque, and all we can now observe is the remnant glow of that primordial hot gas. This glow was originally emitted as ultraviolet radiation (UV) but has now experienced Doppler shift to longer wavelengths by the expansion of the universe. It currently resembles emission from a cool dense gas at a temperature of only 2.73 K. Scientists, using sensitive space-based instruments (such as are found on NASA's *Cosmic Background Explorer* [COBE] spacecraft), have observed and carefully measured this fossil radiation from the ancient big bang explosion. This cosmic microwave background is the very distant spherical "wall" that surrounds and delimits the edges of the observable universe.

However, very subtle observed variations in the cosmic microwave background are challenging big bang cosmologists. They must now explain, for example, how clumpy structures of galaxies could have evolved from a previously assumed smooth (that is, uniform and homogeneous) big bang event and how large organized clusters of quasars formed. Launched on June 30, 2001, NASA's *Wilkinson Microwave Anisotropy Probe* (WMAP) spacecraft made the first detailed full-sky map of the cosmic microwave background—including data (subtle CMB fluctuations) that bring into high resolution the seeds that generated the cosmic structure scientists see today. The patterns detected by WMAP represent tiny temperature differences within an extraordinarily evenly dispersed microwave light bathing the universe, which now averages a frigid 2.73 K. These subtle fluctuations have encouraged physicists to suggest modifications in the original big bang hypothesis, and they invoke the concept of an *inflationary universe*. Scientists now postulate that just after the big bang explosion a special process (called *inflation*) occurred. During this process vacuum state fluctuations gave rise to the very rapid (exponential), nonuniform expansion of the early universe.

American physicist Alan Harvey Guth (b. 1947) was the first scientist to propose the concept of inflation. He did this in 1980 in order to help solve some of the lingering problems in cosmology that were not adequately resolved in standard big bang cosmology. As Guth and then other scientists proposed, between 10^{-35} and 10^{-33} second after the big bang, the universe actually expanded at an incredible rate—far in excess of the speed of light. In this very brief period the universe increased in size by at least a factor of 10^{30}, from an infinitesimally small subnuclear-sized dot to a region about 3 meters across. By analogy, imagine a grain of very fine sand becoming the size of the presently observable universe in 1-billionth (10^{-9}) the time it takes light to cross the nucleus of an atom. During inflation space itself expanded so rapidly that the distances between points in space increased faster than the speed of light. Scientists suggest the slight irregularities they now observe in the cosmic microwave background are evidence (the fossil remnants, or faint ghosts) of the quantum fluctuations that took place as the early universe inflated.

Astrophysicists and cosmologists have three measurable signatures that strongly support the notion that the present universe evolved from a dense, nearly featureless hot gas, just as big bang theory suggests. These scientifically measurable signatures are the expansion of the universe, the abundance of the light elements (hydrogen, helium, and lithium), and the cosmic microwave background (CMB) radiation. Edwin Hubble's 1929 observation that galaxies were generally receding (when viewed from Earth) provided scientist their first tangible clue that the big bang theory might be correct. The big bang model also suggests that hydrogen, helium, and lithium nuclei should have fused from the very energetic collisions of protons and neutrons in the first few minutes after the big bang. The universe has an incredible abundance of hydrogen and helium. Finally, within the big bang model the early universe should have been very, very hot. Scientists regard the cosmic microwave background (first detected in 1964) as the remnant heat left over from an ancient primeval fireball. But there are limitations in the basic big bang model. Perhaps most importantly, the big bang theory makes no attempt to explain how structures like stars and galaxies came into existence in the universe. Observing very subtle fluctuations in the cosmic microwave background, scientists introduced inflation theory to try to explain these puzzles while still retaining the overall framework of the hypothesis.

See also ABUNDANCE OF ELEMENTS (IN THE UNIVERSE); AGE OF THE UNIVERSE; COSMOLOGY; UNIVERSE.

big crunch Within the closed universe model of cosmology, the postulated end state that occurs after the present universe expands to its maximum physical dimensions and then collapses in on itself under the influence of gravity, eventually reaching an infinitely dense end point, or singularity. *Compare with* BIG BANG.

See also COSMOLOGY; UNIVERSE.

Big Dipper The popular name for the distinctive pattern of stars (asterism) formed by the seven brightest stars in the ancient CONSTELLATION Ursa Major (the Great Bear). The seven stars are Dubhe (α star), Merak (β star), Phecda (γ star), Megrez (δ star), Alioth (ε star), Mizar (ζ star), and Alcaid (η star). Dubhe and Merak are also called the "Pointers," or the pointing stars, because a line drawn through them points to Polaris, the North Star.

See also URSA MAJOR.

binary digit (bit) There are only two possible values (or digits) in the binary number system, namely 0 or 1. Binary notation is a common telemetry (information) encoding scheme that uses binary digits to represent numbers and symbols. For example, digital computers use a sequence of bits, such as an eight-bit long byte (*b*inary digi*t* *e*ight), to create a more complex unit of information.

See also TELEMETRY.

binary pulsar A pulsar in orbit around a companion star—possibly a white dwarf, a neutron star, a low-mass star, or a supergiant. Astronomers generally learn about the existence of the companion star by the cyclical change that occurs in the pulse period of the pulsar as the two stars orbit each other. One of the most interesting classes of binary pulsars are the *black widow pulsars*—the first of which, called PSR 1957+20, was discovered in 1988. Intense radiation from a rapidly spinning millisecond pulsar causes material to evaporate from its stellar companion. For example, PSR 1957+20 has a spin rate of 0.0016 seconds and an orbital period of 9.2 hours. The PSR 1957+20 black widow pulsar's companion now has a remaining mass of just 0.02 solar masses. In about 25 million years or so this particular millisecond pulsar will have destroyed (that is, completely evaporated) its stellar companion and thereby become a solitary millisecond pulsar. Here on Earth female black widow spiders occasionally devour their male companions after mating, but that cannibalistic behavior appears to be the exception rather than the rule. Despite this somewhat inaccurate analogy with biology, astronomers continue to refer to this interesting class of binary millisecond pulsars that destroy (evaporate) their companion stars as black widow pulsars. *Compare with* X-RAY BINARY STAR SYSTEM.

See also PULSAR.

binary star system A pair of stars that orbit about their common center of mass and are held closely together by their mutual gravitational attraction. The orbital periods of binaries range from minutes to hundreds of years. By convention, the star that is nearest the center of mass in a binary star system is called the *primary*, while the other (smaller) star of the system is called the *companion*. Binary star systems can be further classified as visual binaries, eclipsing binaries, spectroscopic binaries, and astrometric binaries. *Visual binaries* are those systems that can be resolved into two stars by an optical telescope. *Eclipsing binaries* occur when each star of the system alternately passes in front of the other, obscuring, or eclipsing, it and thereby causing their combined brightness to diminish periodically. *Spectroscopic binaries* are resolved by the Doppler shift of their characteristic spectral lines as the stars approach and then recede from the Earth while revolving about their common center of mass. Astronomers use variations in radial velocity to analyze the orbital motion of the stars in a spectroscopic binary, because these stars are generally too close to be seen separately. Frequently, the two stars in such a *closed binary* are distorted into nonspherical shapes by the action of mutual tidal forces. In a closed binary system the separation distance of the two stars might be comparable to their diameters. Astronomers subdivide closed binary star systems as detached, semidetached, or contact binaries. Semidetached and contact binaries include interacting binaries in which mass transfer occurs from the companion star to the primary star. In an *astrometric binary* one star cannot be visually observed, and its existence is inferred from the irregularities and perturbations in the motion of the visible star of the system.

An artist's rendering of a high-mass X-ray binary star system with jets. Material from a large blue giant star (on the right) is being drawn off and devoured by an invisible, cannibalistic black hole in this interacting binary system. The inflow of material from the companion normal star creates an accretion disk around the massive stellar black hole primary. Shown also are a pair of very high-speed material jets departing in opposite directions perpendicular to the plane of the accretion disk. As the inflowing material swirls in the accretion disk and spirals toward the black hole's event horizon, it heats to extremely high temperatures (millions of degrees) and emits X-rays. ***(Courtesy of NASA/CXC/SAO)***

Binary star systems are more common in the Milky Way galaxy than is generally realized. Perhaps 50 percent of all stars are contained in binary systems. The typical mean separation distance between members of a binary star system is on the order of 10 to 20 astronomical units.

The first observed binary star systems revealed their "binary characteristic" to astronomers through variations in their optical emissions. The *X-ray binary* was discovered during the 1970s by means of space-based X-RAY observations. In a typical X-ray binary system, a massive optical star is accompanied by a compact, X-ray-emitting companion that might be a neutron star or possibly even a black hole. An *interacting binary star system* is one in which mass transfer occurs, generally from a massive optical star to its compact, cannibalistic, X-ray-emitting companion. (*See* figure.) The *binary pulsar*, first observed in 1988, reveals its existence by an apparent change in the pulse period of the pulsar as it orbits its companion.

See also ASTROPHYSICS; X-RAY ASTRONOMY.

bioclean room Any enclosed area where there is control over viable and nonviable particulates in the air, with temperature, humidity, and pressure controls as required to maintain specified standards. A viable particle is a particle that will reproduce to form observable colonies when placed on a specified culture medium and incubated according to optimum environmental conditions for growth after collection. A nonviable particle will not reproduce to form observable colonies when placed on a specified culture medium and incubated according to optimum environmental conditions for growth after collection. A bioclean room is operated with emphasis on minimizing airborne viable and nonviable particle generation or concentrations to specified levels. Particle size is expressed as the apparent maximum linear dimension or diameter of the particle—usually in micrometers (μm), or microns. (A micron is 1-millionth of a meter.)

There are three clean room classes based on total (viable and nonviable) particle count, with the maximum number of particles per unit volume or horizontal surface area being specified for each class. The "cleanest," or most stringent, airborne particulate environment is called a *Class 100 Bioclean Room* (or a *Class 3.5 Room* in SI units). In this type of bioclean room, the particle count may not exceed a total of 100 particles per cubic foot (3.5 particles per liter) of a size 0.5 micrometer (μm) and larger; the viable particle count may not exceed 0.1 per cubic foot (0.0035 per liter), with an average value not to exceed 1,200 per square foot (12,900 per square

Bioclean Room Air Cleanliness Levels

Class English System (metric system)	Maximum Number of Particles per Cu Ft 0.5 Micron and Larger (per liter)	Maximum Number of Viable Particles per Cu Ft (per liter)	Average Number of Viable Particles per Sq Ft per Week (per m^2 per week)
100 (3.5)	100 (3.5)	0.1 (0.0035)	1,200 (12,900)
10,000 (350)	10,000 (350)	0.5 (0.0176)	6,000 (64,600)
100,000 (3,500)	100,000 (3,500)	2.5 (0.0884)	30,000 (323,000)

Source: NASA.

METER) per week on horizontal surfaces. A *Class 10,000 Bioclean Room* (or a *Class 350 Room* in SI units) is the next-cleanest environment, followed by a *Class 100,000 Bioclean Room* (or a *Class 3,500* Room in the SI units). The table summarizes the air cleanliness conditions for each type of clean room.

In aerospace applications bioclean rooms, or "clean rooms," as they are frequently called, are used to manufacture, assemble, test, disassemble, and repair delicate spacecraft sensor systems, electronic components, and certain mechanical subsystems. Aerospace workers wear special protective clothing, including gloves, smocks (frequently called *bunny suits*), and head and foot coverings in order to reduce the level of dust and contamination in clean rooms. A Class 10,000 Clean Room facility is typically the cleanliness level encountered in the assembly and testing of large spacecraft. If a planetary probe is being prepared for a trip to a possible life-bearing planet such as Mars, care is also taken in the bioclean room to ensure that "hitchhiking terrestrial microorganisms" are brought to the minimum population levels consistent with planetary quarantine protocols, thereby avoiding the potential of forward contamination of the target alien world. Astrobiologists will place extraterrestrial soil and rock samples from other (possibly life-bearing) worlds in the solar system in highly isolated and highly secure bioclean rooms. This will help avoid any possible back contamination of the terrestrial biosphere as a result of space travel.

See also EXTRATERRESTRIAL CONTAMINATION; ORBITING QUARANTINE FACILITY.

biogenic elements Those chemical elements generally considered by scientists to be essential for all living systems. Astrobiologists usually place primary emphasis on the elements HYDROGEN (H), carbon (C), nitrogen (N), oxygen (O), sulfur (S), and phosphorous (P). The chemical compounds of major interest are those normally associated with water (H_2O) and with organic chemistry, in which carbon (C) is bonded to itself or to other biogenic elements. Astrobiologists also include several life-essential elements associated with inorganic chemistry, such as iron (Fe), magnesium (Mg), calcium (Ca), sodium (Na), potassium (K), and chlorine (Cl) in this overall grouping, but these are often given secondary emphasis in cosmic evolution studies.

See also LIFE IN THE UNIVERSE.

biosphere The life zone of a planetary body; for example, the part of Earth inhabited by living organisms.

See also ECOSPHERE; GLOBAL CHANGE.

biotelemetry The remote measurement of life functions. Data from biosensors attached to an astronaut or cosmonaut are sent back to Earth (as telemetry) for the purposes of space crew health monitoring and evaluation by medical experts and mission managers. During a strenuous extravehicular activity (EVA), for example, medical specialists at mission control use biotelemetry to monitor an astronaut's heartbeat and respiration rate.

See also EXTRAVEHICULAR ACTIVITY; TELEMETRY.

bipropellant rocket A rocket that uses two unmixed (uncombined) liquid chemicals as its fuel and oxidizer. The two chemical propellants flow separately into the rocket's combustion chamber, where they are combined and combusted to produce high-temperature, thrust-generating gases. The combustion gases then exit the rocket system through a suitably designed nozzle.

See also ROCKET.

Birkeland, Kristian Olaf Bernhard (1867–1917) Norwegian *Physicist* Early in the 20th century Kristian Birkeland deployed instruments while he participated in polar region expeditions. After analyzing the data these instruments collected, Birkeland made the bold suggestion that aurora were electromagnetic in nature—created by an interaction of Earth's magnetosphere with a flow of charged particles from the Sun. In the 1960s instruments onboard early Earth-orbiting scientific satellites confirmed his daring yet accurate hypothesis.

See also AURORA; MAGNETOSPHERE.

bit A binary digit, the basic unit of information (either 0 or 1) in binary notation.

blackbody A perfect emitter and perfect absorber of electromagnetic radiation. All objects emit thermal radiation by virtue of their temperature. The hotter the body, the more radiant energy it emits. The Stefan-Boltzmann law states that the luminosity of a blackbody is proportional to the fourth power of the body's absolute temperature. This physical principle tells scientists that if the absolute temper-

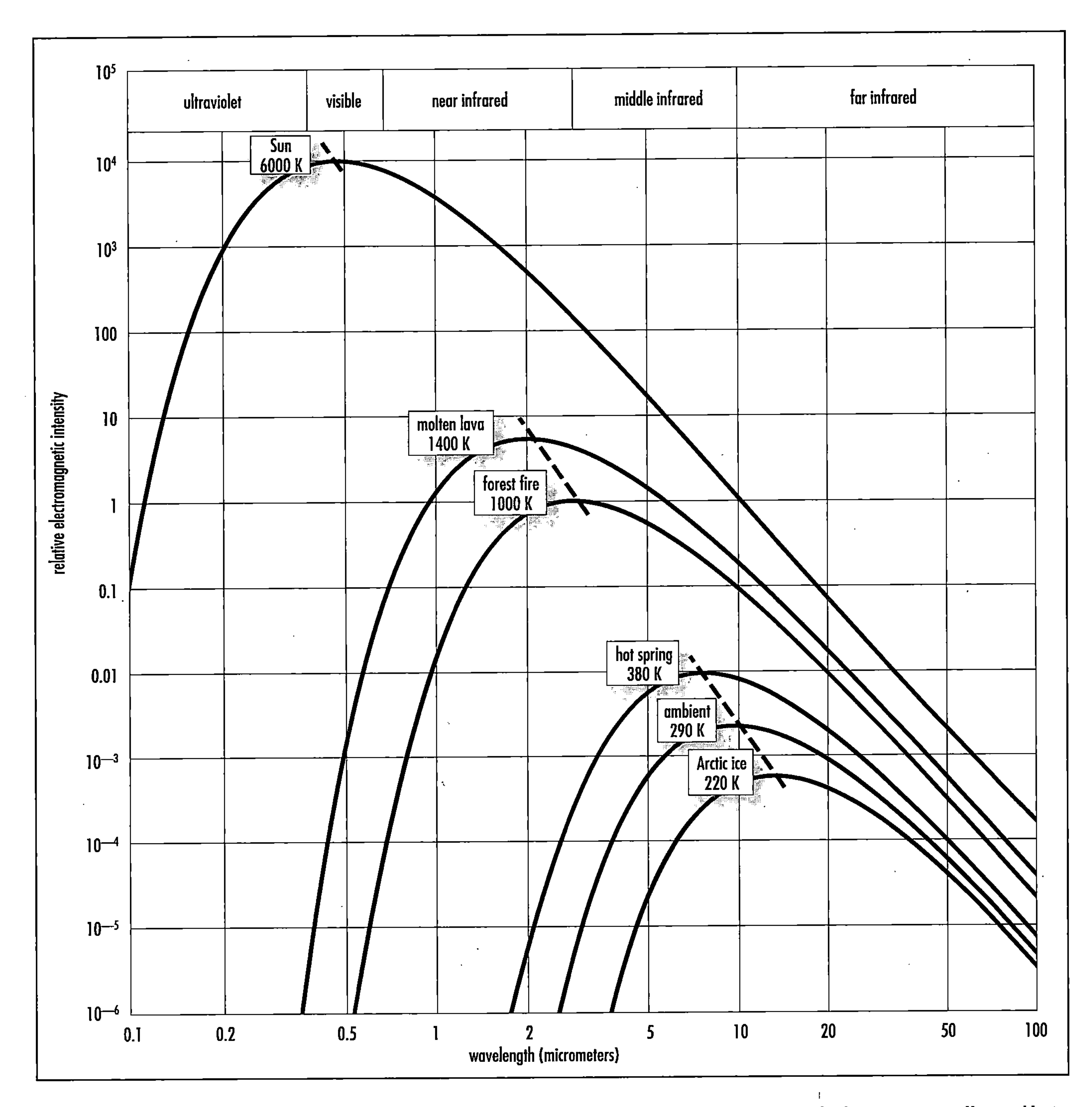

Electromagnetic radiation (thermal radiation) emitted as a function of wavelength by different natural objects at various absolute temperatures. Here, ambient temperature refers to the average temperature of Earth's land surface, assumed to behave as a blackbody radiator at 290 kelvins. ***(Courtesy of NASA)***

ature of a blackbody doubles, its luminosity will increase by a factor of 16. Specifically, the Stefan-Boltzmann law states that the rate of energy emission per unit area of a blackbody is equal to σT^4, where σ is the Stefan-Boltzmann constant (5.67×10^{-8} W/m^2 – K^4) and T is the absolute temperature.

The Sun and other stars closely approximate the thermal radiation behavior of blackbodies, so astronomers often use the Stefan-Boltzmann law to approximate the radiant energy output (or luminosity) of a stellar object. A visible star's apparent temperature is also related to its color. The coolest red stars (called stellar spectral class M stars), such as Betelgeuse, have a typical surface temperature of less than 3,500 K. However, very large hot blue stars (called spectral class B STARS), such as Rigel, have surface temperatures ranging from about 9,900 K to 28,000 K.

While the general blackbody radiation research of JOSEF STEFAN and LUDWIG BOLTZMANN proved of great importance to physicists and astronomers, scientists continued to encounter difficulty if they tried to calculate the energy radiated by a blackbody at a single, specific wavelength (λ). The German physicist Wilhelm Carl Wien (1864–1928) advanced the blackbody work of Stefan and Boltzmann by developing a relationship between the wavelength of peak emission of radiant energy by a blackbody (λ_{MAX}) and its absolute temperature (T). Wien observed that for a blackbody the product of wavelength at which maximum radiant emission occurs and the absolute temperature is a constant. Expressed in the form of an equation, Wien's displacement law is $\lambda_{MAX}T = 2.8977 \times 10^{-3}$ m·K (the Wien constant). This simple relationship provides astronomers a way of estimating the surface temperature of a star or other celestial body from observations of the thermal radiation that body emits—assuming the object behaves like a blackbody. Wien introduced this important blackbody relationship in 1893. However, he struggled in vain for the next three or so years to produce a single formula that would adequately describe the observed distribution of blackbody radiation across all wavelengths. Wien's blackbody radiation formula fit the shorter wavelengths and the peak emission quite well but failed to match the experimentally observed blackbody spectrum at longer wavelengths.

But Wien was not alone. Other late 19th-century physicists were equally frustrated in their efforts to apply classical Maxwellian physics to adequately formulate blackbody radiation. The British physicists Lord John William Rayleigh (1842–1919) and SIR JAMES HOPWOOD JEANS developed the Rayleigh-Jeans formula. Their formula accurately predicted blackbody radiation at long wavelengths but failed as the wavelengths became shorter. In fact, the Rayleigh-Jeans formulation suggested an impossible physical circumstance. As the wavelength approached zero, the radiant energy emitted by the blackbody would become infinite. Physicists call this impossible circumstance the *ultraviolet catastrophe.*

Then, in December 1900 the German physicist MAX PLANCK proposed a simple formula that correctly described the distribution of blackbody radiation over all wavelengths, but to develop this formula Planck had to introduce quantum theory—his revolutionary idea that blackbody radiation is emitted discontinuously in tiny packets, or quanta, rather than continuously, as suggested by classical physics. Planck's radiation theory agreed very well with experimental observations and overcame the ultraviolet catastrophe. Quantum theory is one of the great pillars of modern physics.

Planck's radiation law is the fundamental physical law that describes the distribution of energy radiated by a blackbody. It is an expression for the variation of monochromatic radiant flux per unit area of source as a function of wavelength (λ) of blackbody radiation at a given thermodynamic (absolute) temperature (T). Mathematically, Planck's law can be expressed as:

$$dw = [2\,\pi\,h\,c^2] / \{\lambda^5\,[e^{hc/\lambda kT} - 1]\}d\lambda$$

where dw is the radiant flux from a blackbody in the wavelength interval $d\lambda$, centered around wavelength λ, per unit area of blackbody surface at thermodynamic temperature T. In this equation, c is the speed of light, h is a constant (later named the Planck constant), and k is the Boltzmann constant. According to Planck's radiation law, the radiant energy emitted by a blackbody is a function only of the absolute temperature of the emitting object.

black dwarf The cold remains of a white dwarf star that no longer emits visible radiation or a nonradiating ball of interstellar gas that has contracted under gravitation but contains too little mass to initiate nuclear fusion.

black hole Theoretical gravitationally collapsed mass from which nothing—light, matter, or any other kind of signal—can escape. Astrophysicists now conjecture that a stellar black hole is the natural end product when a giant star dies and collapses. If the core (or central region) of the dying star has three or more solar masses left after exhausting its nuclear fuel, then no known force can prevent the core from forming the extremely deep (essentially infinite) gravitational warp in space-time called a black hole. As with the formation of a white dwarf or a neutron star, the collapsing giant star's density and gravity increase with gravitational contraction. However, in this case, because of the larger mass involved, the gravitational attraction of the collapsing star becomes too strong for even neutrons to resist, and an incredibly dense point mass, or singularity, forms. Physicists define this singularity as a point of zero radius and infinite density—in other words, it is a point where space-time is infinitely distorted. Basically, a black hole is a singularity surrounded by an event region in which the gravity is so strong that absolutely nothing can escape.

No one will ever know what goes on inside a black hole, since no light or any other radiation can escape. Nevertheless, physicists try to study the black hole by speculating about its theoretical properties and then looking for perturbations in the observable universe that provide telltale hints that a massive, invisible object matching such theoretical properties is possibly causing such perturbations. For example, here on Earth a pattern of ripples on the surface of an otherwise quiet but murky pond could indicate that a large fish is swimming just below the surface. Similarly, astrophysicists look for detectable ripples in the observable portions of the universe to support theoretical predictions about the behavior of black holes.

The very concept of a mysterious black hole exerts a strong pull on both scientific and popular imaginations. Data from space-based observatories, such as the *Chandra X-Ray Observatory* (CXO), have moved black holes from the purely theoretical realm to a dominant position in observational astrophysics. Strong evidence is accumulating that black holes not only exist, but that very large ones, called supermassive black holes (containing millions or billions of solar masses), may function like cosmic monsters that lurk at the centers of every large galaxy.

How did the idea of a black hole originate? The first person to publish a "black hole" paper was John Michell

(1724–93), a British geologist, amateur astronomer, and clergyman. His 1784 paper suggested the possibility that light (then erroneously believed to consist of tiny particles of matter subject to influence by Newton's law of gravitation) might not be able to escape from a sufficiently massive star. Michell was a competent astronomer who successfully investigated binary star system populations. He further suggested in this 18th-century paper that although no "particles of light" could escape from such a massive object, astronomers might still infer its existence by observing the gravitational influence it exerted on nearby celestial objects. The French mathematician and astronomer PIERRE-SIMON, MARQUIS DE LAPLACE introduced a similar concept in the late 1790s, when he also applied Newton's law of gravitation to a celestial body so massive that the force of gravity prevented any light particles from escaping. However, neither Laplace nor Michell used the term *black hole* to describe their postulated very massive heavenly body. In fact, the term *black hole* did not enter the lexicon of astrophysics until the American physicist John Archibald Wheeler (b. 1911) introduced it in 1967. Both of these 18th-century "black hole" speculations were on the right track but suffered from incomplete and inadequate physics.

The needed breakthrough in physics took place a little more than a century later, when ALBERT EINSTEIN introduced relativity theory. Einstein replaced the Newtonian concept of gravity as a force with the notion that gravity was associated with the distortion of space-time. As originally suggested by Einstein, the more massive an objective, the greater its ability to distort the local space and time continuum.

Shortly after Einstein introduced general relativity in 1915, the German astronomer KARL SCHWARZSCHILD discovered that Einstein's general relativity equations led to the postulated existence of a dense object into which other objects could fall but out of which no objects could ever escape. In 1916 Schwarzschild wrote the fundamental equations that describe a black hole. He also calculated the size of the event horizon, or boundary of no return, for this incredibly dense and massive object. The dimension of the event horizon now bears the name *Schwarzschild radius* in his honor.

The notion of an event horizon implies that no information about events taking place inside can ever reach outside observers. However, the event horizon is not a physical surface. Rather, it represents the start of a special region that is disconnected from normal space and time. Although scientists cannot see beyond the event horizon into a black hole, they believe that time stops there. The table presents values of the Schwarzschild radius as a function of mass.

Schwarzschild Radius as a Function of Mass

Mass of Collapsed Star (solar masses)[a]	Schwarzschild Radius (km)
1	~3
5	15
10	30
20	60
50	150

[a] One solar mass = mass of the Sun.

Inside the event horizon the escape speed exceeds the speed of light. Outside the event horizon escape is possible. It is important to remember that the event horizon is not a material surface. It is the mathematical definition of the point of no return, the point where all communication is lost with the outside world. Inside the event horizon the laws of physics as we know them do not apply. Once anything crosses this boundary, it will disappear into an infinitesimally small point—the singularity, previously mentioned. Scientists cannot observe, measure, or test a singularity, since it, too, is a mathematical definition. The little scientists currently know about black holes comes from looking at the effects they have on their surroundings *outside* the event horizon. As even more powerful space-based observatories study the universe across all portions of the electromagnetic spectrum in this century, scientists will be able to construct better theoretical models of the black hole.

The mass of the black hole determines the extent of its event horizon. The more massive the black hole, the greater the extent of its event horizon. As shown in the table, the event horizon for a black hole containing one solar mass is just three kilometers from the singularity. This is an example of a stellar black hole. In the Milky Way galaxy a supernova occurs an average of once or twice every 100 years. Since the galaxy is about 1 billion years old, scientists suggest that about 200 million supernovae have occurred, creating neutron stars and black holes. Astronomers have identified nearly a dozen stellar black hole candidates (containing between three and 20 solar masses or more) by observing how some stars appear to wobble as if nearby, yet invisible, massive companions were pulling on them. X-ray binary systems (discussed shortly) offer another way to search for candidate black holes.

In comparison to stellar black holes, the event horizon for a supermassive black hole consisting of 100 million solar masses is about 300 million kilometers from its singularity—twice the distance of the Sun from Earth. Astronomers suspect that such massive objects exist at the centers of many galaxies because they provide one of the few logical explanations for the strange and energetic events now being observed there. The suspected relationship between an active galactic nucleus (AGN) galaxy and a supermassive black hole will be discussed shortly. No one knows for sure exactly how such supermassive black holes formed. One hypothesis is that over billions of years relatively small stellar black holes (formed by supernovae) began devouring neighboring stars in the star-rich centers of large galaxies and eventually became supermassive black holes.

Once matter crosses the event horizon and falls into a black hole, only three physical properties appear to remain relevant: total mass, net electric charge, and total angular momentum. Recognizing that all black holes must have mass, physicists have proposed four basic stellar black hole models. The *Schwarzschild black hole* (first postulated in 1916) is a static black hole that has no charge and no angular momentum. The *Reissner–Nordström black hole* (introduced in 1918) has an electric charge but no angular momentum—that is, it is not spinning. In 1963 the New Zealand mathematician ROY PATRICK KERR (b. 1934) applied general relativity to describe the properties of a rapidly rotating, but uncharged, black hole. Astrophysicists think this model is the most likely "real world" black hole because the massive stars that formed them would have been rotating. One postulated feature of a rotating *Kerr black hole* is its ringlike structure (the ring singularity) that might give rise to two separate event horizons. Some daring astrophysicists have even suggested it might become possible (at least in theory) to travel through the second event horizon and emerge into a new universe or possibly a different part of this universe. The final black hole model has both charge and angular momentum. Called the *Kerr-Newman black hole,* this theoretical model appeared in 1965. However, astrophysicists currently think that rotating black holes are unlikely to have a significant electric charge, so the uncharged Kerr black hole remains the more favored "real world" candidate model for the stellar black hole.

The British astrophysicist STEPHEN WILLIAM HAWKING (b. 1942) provided additional insight into the unusual physics of stellar black holes in 1974 when he introduced the concept of *Hawking radiation,* a postulation suggesting an intimate relationship between gravity, quantum mechanics, and thermodynamics by which, under special circumstances, black holes can emit thermal radiation and eventually evaporate. Of course, this appears very unusual, since by definition nothing can escape from a black hole. The answer to this apparent paradox is found in the space around the black hole at the event horizon. Quantum mechanics suggests that pairs of virtual particles (each pair containing a particle and its antiparticle) can spontaneously pop up. According to contemporary nuclear physics, virtual particles are undetectable quantum particles that carry gravity and electromagnetic radiation, including light. Fluctuations of electromagnetic and intense gravitational fields can create pairs of virtual particles. Left to themselves, these pairs (a particle and its antiparticle) will move apart slightly and then come back together to annihilate each other on a very short timescale. However, if the pair of virtual particles is created right along a black hole's event horizon, as they move apart slightly, these particles may live long enough so that one member of the pair is pulled across the event horizon toward the black hole, while the other particle moves outward and escapes. They cannot get back together to annihilate each other.

This quantum mechanics phenomenon has profound consequences for the physics of a black hole. Without its virtual partner, the escaping particle becomes a real particle, and so it appears to an external observer that radiation is actually coming from the black hole. Physicists refer to this emission phenomenon as Hawking radiation. The virtual particle with negative energy that was captured by the black hole contributes to the reduction in mass of the singularity. If this happens over a very long period of time, the black hole will simply evaporate.

Hawking radiation gives the appearance that a black hole is emitting radiation like a blackbody with a temperature inversely proportional to its mass. While still highly speculative, Hawking's work further suggests that as virtual particles transform into real particles, the process extracts energy from the black hole's intense gravitational field. Since the transformation consumes more energy than the particles possess, it essentially contributes negative energy to the black hole. As a result, the mass of the singularity decreases, and the black hole eventually evaporates. For example, within this theoretical model, a very small *mini–black hole*—with a radius of less than 10^{-10} meter and a mass of about 10^{12} kg (that of a small asteroid)—would have a blackbody temperature of about 10^{11} K. This mini–black hole would radiate more intensely and evaporate more quickly than a more massive black hole. In contrast, a black hole of one solar mass (approximately 2×10^{30} kg) would last about 10^{66} years.

Another consequence of Hawking's postulation is the formation of a cloud of real particles and antiparticles just outside the event horizon of a black hole. These particles and antiparticles are each the surviving member of virtual particle pairs that have escaped and transformed into real particles. Their continuous annihilation outside the event horizon in the observable universe forms Hawking radiation.

Current astrophysical evidence that superdense stars, such as white dwarfs and neutron stars, really exist also supports the theoretical postulation that black holes themselves—representing the ultimate in density—must also exist. But how can scientists detect an object from which nothing, not even light, can escape? Astrophysicists think they may have found indirect ways of detecting black holes. The best currently available techniques depend upon candidate black holes being members of binary star systems. Unlike the Sun, many stars (more than 50 percent) in the Milky Way galaxy are members of a binary system. If one of the stars in a particular binary system has become a black hole, although invisible, it would betray its existence by the gravitational effects it would produce on the observable companion star. Once beyond event horizon, the black hole's gravitational influence is the same as that exerted by other objects (of equivalent mass) in the "normal" universe. So a black hole's gravitational effects on its companion would obey Newton's universal law of gravitation—that is, the mutual gravitational attraction of the two celestial objects is directly proportional to their masses and inversely proportional to the square of the distance between them.

Astrophysicists have also speculated that a substantial part of the energy of matter spiraling into a black hole is converted by collision, compression, and heating into X-rays and gamma rays that display certain spectral characteristics. X-ray and gamma radiations emanate from the material as it is pulled toward the black hole. However, once the captured material crosses the black hole's event horizon, this telltale radiation cannot escape.

Suspected black holes in binary star systems exhibit this type of prominent material capture effect. Astronomers have discovered several black hole candidates using space-based astronomical observatories (such as the *Chandra X-Ray Observatory*). One very promising candidate is called Cygnus X-1, an invisible object in the constellation Cygnus (the Swan). The notation Cygnus X-1 means that it is the

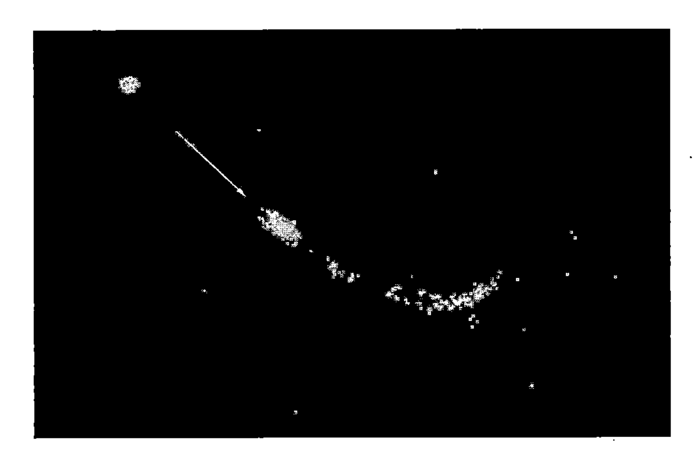

This is an artist's visualization of a doomed star being torn apart after it wandered too close to a supermassive black hole in galaxy RX J1242-11. As it neared the enormous gravity of the black hole, the star was stretched by tidal forces until it was torn apart. Before being consumed by the black hole, inflowing gas from the doomed star was heated to millions of degrees Celsius, creating one of the most extreme X-ray outbursts ever observed from the center of a galaxy. ***(Courtesy of NASA)***

first X-ray source discovered in Cygnus. X-rays from the invisible object have characteristics like those expected from materials spiraling toward a black hole. This material is apparently being pulled from the black hole's binary companion, a large star of about 30 solar masses. Based on the suspected black hole's gravitational effects on its visible stellar companion, the black hole's mass has been estimated to be about six solar masses. In time, the giant visible companion might itself collapse into a neutron star or a black hole, or else it might be completely devoured by its black hole companion. This form of stellar cannibalism would also significantly enlarge the existing black hole's event horizon.

In 1963 the Dutch-American astronomer MAARTEN SCHMIDT (b. 1929) was analyzing observations of a "star" named 3C 273. He had very confusing optical and radio data. What he and his colleagues had discovered was the quasar. Today astronomers know that the quasar is a type of active galactic nucleus (AGN) in the heart of a normal galaxy. AGN galaxies have hyperactive cores and are much brighter than normal galaxies. The AGNs emit energy equivalent to converting the entire mass of the Sun into pure energy every few years. Energy is emitted in all regions of the electromagnetic spectrum, from low-energy radio waves all the way to much higher-energy X-rays and gamma rays. Furthermore, the energy output of these AGN can vary on short time scales (hours and days), suggesting that the source is very compact. Stars by themselves, powered by nuclear fusion, cannot generate such levels of energy. Even the impressive supernovae explosions are insufficient. So astrophysicists puzzled over what physical processes could produce the power of more than 100 Milky Way galaxies and do it within a region of space only a few light-years across.

To further compound this cosmic mystery, some of these AGN galaxies have extraordinary jets of material rushing out of their cores that stretch far into space—up to 100 to 1,000 times the diameter of the galaxy. After considering all types of energetic processes, including the simultaneous explosions of thousands of supernovae, most astrophysicists now think that supermassive black holes represent the most plausible answer. Although nothing emerges from a black hole, matter falling into one can release tremendous quantities of energy just before it crosses the event horizon. For example, the region just outside the even horizon will glow in X-rays and gamma rays, the most energetic forms of electromagnetic radiation.

Matter captured by the gravity of a black hole will eventually settle into a disk around the black hole. Scientists call the inner region of swirling, superheated material an accretion disk. Stellar black holes can have accretion disks if they have a nearby companion star. Material from the companion will be drawn into orbit around the black hole, thereby forming an accretion disk. The diameter of the accretion disk depends on the mass of the black hole. The more massive the black hole, the larger the accretion disk. The accretion disk of a stellar black hole will stretch out only a few hundred or thousand kilometers from the center. However, the accretion disk of a supermassive black hole is much bigger and becomes solar system–sized.

Perhaps the most spectacular accretion disks exist in active galaxies that probably contain supermassive black holes. The *Hubble Space Telescope* (HST) has provided astronomers strong evidence for this assumption. For example, the galaxy NGC 4261 is an elliptical galaxy whose core contains an unexpected large disk of dust and gas. Astronomers think that a black hole may lurk within the central region of this galaxy. Radio observations of this galactic core have also revealed jets of material ejected from the center of this disk. This provides corroborating evidence for the existence of a very large black hole.

Exactly what creates and controls the flow of matter out these jets is still not clearly understood. It is almost as if black holes are messy eaters, consuming matter but spewing out leftovers. Most likely, the jets have something to do with the rotation and/or the magnetic fields of the black hole. Whatever the actual cause, most astronomers believe that only black holes are capable of producing such spectacular and "outrageous" behavior.

Scientists using the *Hubble Space Telescope* have also discovered a 3,700 light-year-diameter dust disk encircling a suspected 300-million-solar mass black hole in the center of the elliptical galaxy NGC 7052, located in the constellation Vulpecula about 191 million light-years from Earth. This disk is thought to be the remnant of an ancient galaxy collision, and it will be swallowed up by the giant black hole in several billion years. *Hubble Space Telescope* measurements have shown that the disk rotates rapidly at 155 km/s at a distance of 186 light-years from the center. The speed of rotation provides scientists a direct measure of the gravitational forces acting on the gas due to the presence of a suspected supermassive black hole. Though 300 million times the mass of the Sun, this suspected black hole is still only about 0.05 percent of the total mass of the NGC 7052 galaxy. The bright spot in the center of the giant dust disk is the combined light of stars that have been crowded around the black hole by its strong gravitational pull. This stellar concentration appears to match the theoretical astrophysical models that link stellar density to a central black hole's mass.

In the 1990s X-ray data from the German-American *Roentgen* Satellite (ROSAT) and the Japanese-American *Advanced Satellite for Cosmology and Astrophysics* (ASCA) suggested that a mid-mass black hole might exist in the galaxy M82. This observation was confirmed in September 2000, when astronomers compared high-resolution *Chandra X-Ray Observatory* images with optical, radio, and infrared maps of the region. Scientists now think such black holes must be the results of black hole mergers, since they are far too massive to have been formed from the death of a single star. Sometimes called the "missing link" black holes, these medium-sized black holes fill the gap in the observed (candidate) black hole masses between stellar and supermassive. The M82 "missing link" is not in the absolute center of the galaxy, where all supermassive black holes are suspected to reside, but it is fairly close to it.

blastoff The moment a rocket or aerospace vehicle rises from its launch pad under full thrust.

See also LIFTOFF.

blazar A variable extragalactic object (possibly the high-speed jet from an active galactic nucleus as viewed on end) that exhibits very dynamic, sometimes violent, behavior.

See also BL LAC OBJECT.

bl lac object (bl lacertae) A class of extragalactic objects thought to be the active centers of faint elliptical galaxies that vary considerably in brightness over very short periods of time (typically hours, days, or weeks). Scientists further speculate that a very high-speed (relativistic) jet is emerging from such an object straight toward an observer on Earth.

blockhouse (block house) A reinforced-concrete structure, often built partially underground, that provides launch crew personnel and their countdown processing and monitoring equipment protection against blast, heat, and possibly an abort explosion during the launch of a rocket. Modern launch control centers are more elaborate facilities that are usually located much farther away from launch pads than the blockhouse structures used during the 1950s and 1960s.

blowdown system In rocket engineering, a closed propellant/pressurant system that decays in ullage pressure level as the liquid propellant in a tank is consumed and ullage volume increases.

blue giant A massive, very high-luminosity star with a surface temperature of about 30,000 K that has exhausted all its hydrogen thermonuclear fuel and left the main sequence portion of the Hertzsprung-Russell diagram.

See also STAR.

blueshift When a celestial object (such as a distant galaxy) approaches an observer at high velocity, the electromagnetic radiation it emits in the visible portion of the spectrum appears shifted toward the blue (higher-frequency, shorter-wavelength) region. *Compare with* REDSHIFT.

See also DOPPLER SHIFT.

Bode, Johann Elert (1747–1826) German *Astronomer* Johann Bode was the astronomer who publicized an empirical formula that approximated the average distance to the Sun of each of the six planets known in 1772. This formula, often called BODE'S LAW, is only a convenient mathematical relationship and does not describe a physical principle or natural phenomenon. Furthermore, Bode's empirical formula was actually discovered in 1766 by JOHANN DANIEL TITIUS (1729–96). Bode popularized this relationship. As a result, he often (incorrectly) receives credit for its development. This empirical relationship is, therefore, more properly called the Titus-Bode law in recognition of the efforts of both 18th-century German astronomers. In 1801 Bode published *Uranographia,* a comprehensive listing of more than 17,000 stars and nebulas.

Bode's law An empirical mathematical relationship discovered in 1766 by JOHANN DANIEL TITIUS and popularized by JOHANN ELERT BODE in 1772. As shown in the table, the Titus-Bode formula predicts reasonably well the distances in astronomical units (AU) to the six planets known in 1772. It also indicated a "planetary gap" between Mars and Jupiter that helped lead to the discovery of the first minor planet (or asteroid), Ceres, in 1801. Uranus, discovered in 1781 by SIR WILLIAM HERSCHEL, lies about 19.2 AU from the Sun, while the Titus-Bode formula predicts 19.6 AU. This is a reasonable (though coincidental) agreement. However, the empirical relationship falls apart with respect to Neptune (discovered in 1846) and Pluto (discovered in 1930). One way to remember the Titus-Bode law is to recognize that it is obtained from the numerical sequence 0, 3, 6, 12, 24, 48, and 96 by adding four to each number in the sequence and dividing the result by 10. For example, to estimate the distance (in AU) of Mercury (represented by the first term in the numerical sequence) from the Sun, simply calculate $(0 + 4) / 10 = 0.4$. Similarly, for Jupiter (corresponding to the fifth term in the sequence),

Bode's Law

Planet	Distance from Sun (approximate AU)	Distance According to Bode-Titus Rule (AU)
Mercury	0.39	0.40
Venus	0.72	0.70
Earth	1.00	1.00
Mars	1.52	1.60
Asteroids (Main Belt)	2.2–3.3	2.80
Jupiter	5.20	5.20
Saturn	9.54	10.0
Uranus	19.2	19.6
Neptune	30.1	38.8
Pluto	39.5	77.2

the formula yields (48 + 4) / 10 = 5.2. To extend this numerical sequence to the outer planets discovered after Bode popularized formula, use 192, 384, and 768 for Uranus, Neptune, and Pluto, respectively.

boiloff The loss of a cryogenic propellant, such as liquid oxygen or liquid hydrogen, due to vaporization. This happens when the temperature of the cryogenic propellant rises slightly in the propellant tank of a rocket being prepared for launch. The longer a fully fueled rocket vehicle sits on its launch pad, the more significant the problem of boiloff becomes.

Bok, Bartholomeus Jan "Bart" (1906–1983) Dutch-American *Astronomer* The astronomer Bart Bok investigated the star-forming regions of the Milky Way galaxy in the 1930s and discovered the small, dense, cool (~10 K) dark nebulas, now called BOK GLOBULES.

Bok globules Small, cool (~10 K) interstellar clouds of gas and dust discovered by the Dutch-American astronomer BARTHOLOMEUS ("BART") JAN BOK in the 1930s. These dark nebulas are nearly circular, opaque, and relatively dense. Their masses vary between 20 and 200 solar masses, and they are between 0.5 and 2.0 light-years in diameter. Orbiting observatories with infrared telescopes, such as the *Infrared Astronomical Satellite* (IRAS), have detected protostars within some Bok globules.

bolide A brilliant meteor, especially one that explodes into fragments near the end of its trajectory in Earth's atmosphere. The initial explosion creates a sonic boom. Smaller detonations may also take place if there are several major surviving fragments. These additional fragments create their own sonic booms, resulting in a somewhat continuous thunder-like rumbling sound.

bolometer A sensitive instrument that measures the intensity of incident radiant (thermal) energy, usually by means of a thermally sensitive electrical resistor. The change in electrical resistance is an indication of the amount of radiant energy reaching the exposed material. Bolometers are especially useful in sensing radiation within the microwave and infrared portions of the electromagnetic spectrum.

Boltzmann, Ludwig (1844–1906) Austrian *Physicist* The brilliant though troubled Austrian physicist Ludwig Boltzmann developed statistical mechanics as well as key thermophysical principles that enable astronomers to better interpret a star's spectrum and its luminosity. In the 1870s and 1880s he collaborated with another Austrian physicist, JOSEF STEFAN, in discovering an important physical principle (now called the *Stefan-Boltzmann law*) that relates the total radiant energy output (luminosity) of a star to the fourth power of its absolute temperature.

Boltzmann was born on February 20, 1844, in Vienna, Austria. He studied at the University of Vienna, earning a doctoral degree in physics in 1866. His dissertation was supervised by Josef Stefan and involved the kinetic theory of gases. Following graduation he joined his academic adviser as an assistant at the university.

From 1869 to 1906 Boltzmann held a series of professorships in mathematics or physics at a number of European universities. His physical restlessness mirrored the hidden torment of his mercurial personality that would swing him suddenly from a state of intellectual contentment into mental states of deep depression. His movement from institution to institution brought him into contact with many great 19th-century scientists, including ROBERT BUNSEN, RUDOLF CLAUSIUS, and GUSTAF KIRCHHOFF.

As a physicist Boltzmann is best remembered for his invention of statistical mechanics and its pioneering application in thermodynamics. His theoretical work helped scientists connect the properties and behavior of individual atoms and molecules (viewed statistically on the microscopic level) to the bulk properties and physical behavior (such as temperature and pressure) that a substance displayed when examined on the familiar macroscopic level used in classical thermodynamics. One important development was Boltzmann's equipartition of energy principle. It states that the total energy of an atom or molecule is equally distributed, on an average basis, over its various translational kinetic energy modes. Boltzmann postulated that each energy mode had a corresponding degree of freedom. He further theorized that the average translational energy of a particle in an ideal gas was proportional to the absolute temperature of the gas. This principle provided a very important connection between the microscopic behavior of an incredibly large numbers of atoms or molecules and macroscopic physical behavior (such as temperature) that physicists could easily measure. The constant of proportionality in this relationship is now called the *Boltzmann constant,* for which physicists usually assign it the symbol k.

Boltzmann developed his kinetic theory of gases independent of the work of the great Scottish physicist JAMES CLERK MAXWELL. Their complementary activities resulted in the *Maxwell-Boltzmann distribution*—a mathematical description of the most probable velocity of a gas molecule or atom as a function of the absolute temperature of the gas. The greater the absolute temperature of the gas, the greater will be the average velocity (or kinetic energy) of individual atoms or molecules. For example, consider a container filled with oxygen (O_2) gas at a temperature of 300 K. The Maxwell-Boltzmann distribution predicts that the most probable velocity of an oxygen molecule in that container is 395 m/s.

In the late 1870s and early 1880s Boltzmann collaborated with his mentor Josef Stefan, and they developed a very important physical principle that describes the amount of thermal energy radiated per unit time by a blackbody. In physics a blackbody is defined as a perfect emitter and perfect absorber of electromagnetic radiation. All objects emit thermal radiation by virtue of their temperature. The hotter the body, the more radiant energy it emits. By 1884 Boltzmann had finished his theoretical work in support of Stefan's observations of the thermal radiation emitted by blackbody radiators at various temperatures. The result of their collaboration was the famous Stefan-Boltzmann law of thermal radiation—a physical principle of great importance to both physicists and astronomers. The Stefan-Boltzmann law states that the luminosity of a blackbody is proportional to the fourth power of the luminous body's absolute temperature. The constant of proportionality for this relationship is called the Stefan-Boltzmann constant, and scientists usually assign this constant the symbol σ, a lower-case sigma in the Greek alphabet. The Stefan-Boltzmann law tells scientists that if the absolute temperature of a blackbody doubles, its luminosity will increase by a factor of 16.

The Sun and other stars closely approximate the thermal radiation behavior of blackbodies, so astronomers often use the Stefan-Boltzmann law to approximate the radiant energy output (or luminosity) of a stellar object. A visible star's apparent temperature is also related to its color. The coolest red stars (called stellar spectral class M stars), such as Betelgeuse, have a typical surface temperature of less than 3,500 K. However, very large hot blue stars (called spectral class B stars), such as Rigel, have surface temperatures ranging from 9,900 to 28,000 K.

In the mid-1890s Boltzmann related the classical thermodynamic concept of entropy (symbol S), recently introduced by Clausius, to a probabilistic measurement of disorder. In its simplest form, Boltzmann's famous entropy equation is $S = k \ln \Omega$. Here, he boldly defined entropy (S) as a natural logarithmic function of the probability of a particular energy state (Ω). The symbol k represents Boltzmann's constant. This important equation is even engraved as an epithet on his tombstone.

Toward the end of his life Boltzmann encountered very strong academic and personal opposition to his atomistic views. Many eminent European scientists of this period could not grasp the true significance of the statistical nature of his reasoning. One of his most bitter professional and personal opponents was the Austrian physicist ERNEST MACH, who as chair of history and philosophy of science at the University of Vienna essentially forced the brilliant but depressed Boltzmann to leave that institution in 1900 and move to the University of Leipzig.

Boltzmann persistently defended his statistical approach to thermodynamics and his belief in the atomic structure of matter. However, he could not handle having to continually defend his theories against mounting opposition in certain academic circles. So while on holiday with his wife and daughter at the Bay of Duino near Trieste, Austria (now part of Italy), he hanged himself in his hotel room on October 5, 1906, as his wife and child were swimming. Was the tragic suicide of this brilliant physicist the result of a lack of professional acceptance of his work or simply the self-destructive climax of a lifelong battle with manic depression? No one can say for sure. Ironically, at the time of his death other great physicists were performing key experiments that would soon prove his statistical approach to atomic structure not only correct, but of great value to science.

Boltzmann constant (symbol: k) The physical constant describing the relationship between absolute temperature and the kinetic energy of the atoms or molecules in a perfect gas. It equals 1.380658×10^{-23} joules per kelvin (J/K) and is named after the Austrian physicist LUDWIG BOLTZMANN.

Bond, George Phillips (1825–1865) American *Astronomer* George Phillips Bond succeeded his father (WILLIAM CRANCH BOND) as the director of the Harvard College Observatory (HCO). He specialized in solar system observations, including the 1848 codiscovery (with his father) of Hyperion, a moon of Saturn. In the 1850s he pursued developments in astrophotography, demonstrating that star photographs could be used to estimate stellar magnitudes.

See also ASTROPHOTOGRAPHY.

Bond, William Cranch (1789–1859) American *Astronomer* William Cranch Bond was the founder and first director of the Harvard College Observatory (HCO). Collaborating with his son (GEORGE PHILLIPS BOND), he codiscovered Hyperion, a moon of Saturn. He also pioneered astrophotography in 1850 by taking images of Jupiter and the star Vega. Then, in 1857 he photographed the binary star system Mizar. As an important historical note, Mizar, a bright white star in the constellation Ursa Major, forms an optical double with the 4th magnitude star Alcor. In 1889 EDWARD CHARLES PICKERING discovered that Mizar was actually a spectroscopic binary—the first such discovery made in astronomy.

Bond albedo The fraction of the total amount of electromagnetic radiation (such as the total amount of light) falling upon a nonluminous spherical body that is reflected in all directions by that body. The Bond albedo is measured (or calculated) over all wavelengths and is named after the American astronomer GEORGE PHILLIPS BOND (1825–65). The Bond albedo is useful in determining the energy balance of a planetary body.

Bondi, Sir Hermann (b. 1919) Austrian-born British *Astronomer, Mathematician* In 1948 Sir Hermann Bondi collaborated with THOMAS GOLD and SIR FRED HOYLE to propose a steady-state model of the universe. Although scientists now have generally abandoned this steady-state universe hypothesis in favor of the big bang theory, Bondi's work stimulated a great deal of beneficial technical discussion within the astrophysical and astronomical communities.

See also BIG BANG; COSMOLOGY.

***Bonner Durchmusterung* (BD)** The *Bonn Survey,* an extensive four-volume star catalog containing more than 324,000 stars in the Northern Hemisphere. It was compiled and published by the German astronomer FRIEDRICH ARGELANDER (1799–1875) between 1859 and 1862.

See also CÓRDOBA DURCHMUSTERUNG.

booster engine In rocket engineering, an engine, especially a booster rocket, that adds its thrust to the thrust of a sustainer engine.

See also ROCKET.

booster rocket A solid- or liquid-propellant rocket engine that assists the main propulsion system (called the sustainer engine) of a space launch vehicle or ballistic missile during some part of its flight.

See also ROCKET.

boostglide vehicle An aerospace vehicle (part aircraft, part spacecraft) designed to fly to the limits of the sensible atmosphere, then to be boosted by rockets into outer space, returning to Earth by gliding under aerodynamic control.

See also DYNA-SOAR; X-20.

boost phase In general, that portion of the flight of a ballistic missile or launch vehicle during which the booster engine(s) and sustainer engine(s) operate. In ballistic missile defense (BMD), the first portion of a ballistic missile's trajectory during which it is being powered by its rocket engines. During this period, which typically lasts from three to five minutes for an intercontinental ballistic missile (ICBM), the missile reaches an altitude of about 200 km, whereupon powered flight ends and the ICBM begins to dispense its reentry vehicles. The other portions of the missile flight, including the midcourse phase and the reentry phase, take up the remainder of an ICBM's flight total time of about 25 to 30 minutes.

See also BALLISTIC MISSILE DEFENSE.

bootstrap Refers to a self-generating or self-sustaining process. An aerospace example is the operation of a liquid-propellant rocket engine in which, during main stage operation, the gas generator is fed by propellants from the turbopump system, and hot gases from the generator system (in turn) drive the turbopump. Of course, the operation of such a system must be started by outside power or propellant sources. However, when its operation is no longer dependent on these external sources, the rocket engine is said to be in bootstrap operation. Similarly, in computer science bootstrap refers to a sequence of commands or instructions the execution of which causes additional commands or instructions to be loaded and executed until the desired computer program is completely loaded into storage.

Borman, Frank (1928–) American *Astronaut, U.S. Air Force Officer* In December 1968 astronaut Frank Borman commanded NASA's *Apollo 8* mission, the first mission to take human beings to the vicinity of the MOON. He also served as the commander of NASA's *Gemini 7* mission in 1965.

See also APOLLO PROJECT; GEMINI PROJECT.

Boscovich, Ruggero Giuseppe (1711–1787) Croatian *Astronomer, Mathematician* The Jesuit mathematician and astronomer Ruggero Boscovich developed new methods for determining the orbits of planets and their axes of rotation. As both a priest and a multitalented scientist, Boscovich was an influential science adviser to Pope Benedict XIV. He was a strong advocate for SIR ISAAC NEWTON's universal law of gravitation and a scientific pioneer in the fields of geodesy and atomic theory.

boson An elementary particle with an integral value of spin. Photons, mesons, (postulated) gravitons, and an atomic nucleus with an even mass number (such as helium-4) are examples. The boson follows Bose-Einstein statistics and does not obey the Pauli exclusion principle in quantum mechanics. Named after the Indian physicist Satyendra Nath Bose (1894–1974).

Boss, Benjamin (1880–1970) American *Astronomer* In 1937 Benjamin Boss published the popular five-volume *Boss General Catalogue* (GC), a modern listing of more than 33,000 stars. His publication culminated the precise star position work initiated in 1912 by his father, LEWIS BOSS, when the elder Boss published the *Preliminary (Boss) General Catalogue,* containing about 6,200 stars.

Boss, Lewis (1846–1912) American *Astronomer* American astronomer Lewis Boss published the popular *Preliminary (Boss) General Catalogue* in 1912. This work was the initial version of an accurate, modern star catalog and contained about 6,200 stars. His son (BENJAMIN BOSS) completed the task by preparing a more extensive edition of the catalog in 1937.

Bouguer, Pierre (1698–1758) French *Physicist* Pierre Bouguer established the field of experimental photometry, the scientific measurement of light. In 1748 he developed the heliometer, an instrument used by astronomers to measure the diameter of the Sun, the parallax of stars, and the angular diameter of planets.

Bowen, Ira Sprague (1898–1973) American *Physicist* The physicist Ira Sprague Bowen performed detailed investigations of the light spectra from nebulas, including certain strong green spectral lines that were originally attributed by other astronomers to a hypothesized new element they called *nebulium.* In 1927 his detailed studies revealed that the mysterious green lines were actually special (forbidden) transitions of the ionized gases oxygen and nitrogen. His work supported important advances in the spectroscopic study of the Sun, other stars, and nebulas.

bow shock 1. A shock wave that is in front of a body, such as an aerospace vehicle or an airfoil. 2. The interface formed where the electrically charged solar wind encounters an obstacle in outer space, such as a planet's magnetic field or atmosphere.

See also MAGNETOSPHERE.

Bradley, James (1693–1762) British *Astronomer* James Bradley was an 18th-century British minister turned astronomer who made two important astronomical discoveries. His first accomplishment involved the discovery of the *aberration of starlight,* which he announced in 1728. This represented the first observational proof that Earth moves in space, confirming Copernican cosmology. His second discovery was that of *nutation,* a small wobbling variation in the precession of Earth's rotational axis. Bradley reported this discovery in 1748 after 19 years of careful observation. Upon the death of the legendary EDMUND HALLEY in 1742, Bradley accepted appointment as the third Astronomer Royal of England.

Bradley was born in March 1693 in Sherborne, England. As a young child he was being prepared for a life in the ministry by his uncle the Reverend James Pound. When he suffered an attack of smallpox, his uncle nursed him back to health and also introduced him to astronomy. Bradley began to demonstrate his great observational skills as he practiced amateur astronomy while studying for the clergy. However, by 1718 Bradley's talents sufficiently impressed Halley, then England's second Astronomer Royal, and he vigorously encouraged the young man to pursue astronomy as a career.

As a result of Halley's encouragement, Bradley abandoned any thoughts of an ecclesiastical career. In 1721 he accepted an academic position in astronomy at Oxford University. Throughout his astronomical career Bradley concentrated his efforts on making improvements in the precise observation of stellar positions. For example, in 1725 he started on his personal quest to become the first astronomer to successfully measure stellar parallax—the very small apparent change in the position of a fixed star when viewed by an observer on Earth as the planet reaches opposing extremes of its elliptical orbit around the Sun. However, this was a difficult measurement that managed to baffle and allude all astronomers, including Bradley, until FRIEDRICH BESSEL accomplished the task in 1838. Nevertheless, while searching for stellar parallax, Bradley accidentally discovered another very important physical phenomenon that he called the aberration of starlight.

The aberration of starlight is the tiny apparent displacement of the position of a star from its true position due to a combination of the finite velocity of light and the motion of an observer across the path of the incident starlight. For example, an observer on Earth would have Earth's orbital velocity around the Sun, approximately 30 km/s. As a result of this orbital motion effect, during the course of a year the light from a fixed star appears to move in a small ellipse around its mean position on the celestial sphere.

By a stroke of good fortune, on December 3, 1725, Bradley began observing the star Gamma Draconis (γ Dra), which was within the field of view of his large refractor telescope pointed at zenith. Bradley was forced to look at the stars overhead because he had assembled his telescope in a long chimney. Its telescopic lens was above the rooftop, and its objective (viewing lens) was far below in the unused fireplace. Bradley was actually attempting to become the first astronomer to experimentally detect stellar parallax, so he was using this relatively immobile telescope pointed at zenith to minimize atmospheric refraction. He viewed Gamma Draconis again on December 17 and was quite surprised to find that the position of the star had shifted a tiny bit. But the apparent shift was not by the amount nor in the direction expected due to stellar parallax. He carefully continued these observations over the course of an entire year and discovered that the position of Gamma Draconis actually traced out a very small ellipse.

Bradley puzzled over the Gamma Draconis anomaly for about three years, yet he still could not satisfactorily explain the cause of the displacement. Then, while sailing on the Thames River in 1728, inspiration suddenly struck as he watched a small wind-direction flag on a mast change directions. Curious, Bradley asked a sailor whether the wind had changed directions. The sailor informed him that the direction of wind had not changed, so Bradley realized that the changing position of the small flag was due to the combined motion of the boat and of the wind. This connection helped him solve the Gamma Draconis displacement mystery.

Based on the earlier attempts by the Danish astronomer OLAUS ROEMER to measure the velocity of light, Bradley correctly postulated that light must travel at a very rapid, finite velocity. He further speculated that a ray of starlight traveling from the top of his tall telescope to the bottom would have its image displaced in the objective (viewing lens) ever so slightly by Earth's orbital motion. Bradley finally recognized that the mysterious annual shift of the apparent position of Gamma Draconis, first in one direction and then in another, was due to the combined motion of starlight and Earth's annual orbital motion around the Sun. The young astronomer's discovery of the aberration of starlight was the first observational proof that Earth moves in space—confirming the Copernican hypothesis of heliocentric cosmology.

Bradley soon constructed a new telescope—one not placed in a chimney—so he could make more precise observations of many other stars. Starting in 1728 and continuing for many years, Bradley recorded the positions of many stars with great precision. His efforts improved significantly upon the work of earlier notable astronomers, including JOHN FLAMSTEED. Bradley's ability to make precise measurements led him to his second important astronomical discovery. He found that Earth's axis experienced small periodic shifts, which he called nutation after the Latin word *nutare,* meaning "to nod." The new phenomenon, characterized as an irregular wobbling superimposed on the overall precession of Earth's rotational axis, is primarily caused by the gravitational influence of the Moon as it travels along its somewhat irregular orbit around Earth. However, Bradley waited almost two decades (until 1748) to publish this discovery. The cautious scientist wanted to carefully study and confirm the very subtle shifts in stellar positions before announcing his findings.

When Halley died in 1742, Bradley received an appointment as England's third Astronomer Royal. He served with distinction in that post until his own death on July 13, 1762, in Chalford, England.

Brahe, Tycho (1546–1601) Danish *Astronomer* The quarrelsome Danish astronomer Tycho Brahe is often considered to be the greatest pretelescope astronomical observer. His precise, naked eye records of planetary motions enabled JOHANNES KEPLER to prove that all the planets move in elliptical orbits around the Sun. Brahe discovered a supernova in 1572 and reported his detailed observations in *De Nova Stella* (1573). In 1576 Danish king Frederick II provided support for him to build a world-class astronomical observatory on an island in the Baltic Sea. For two decades Brahe's Uraniborg (Castle of the Sky) was the great center for observational astronomy. Then, in 1599 German emperor Rudolf II invited him to move to Prague. Brahe became the Imperial Astronomer to the Holy Roman Emperor Rudolf II, and Kepler joined him as an assistant. However, despite the lofty title, Brahe never really resumed his own astronomical research program. Upon his death on October 14, 1601, Kepler became scientific heir to Brahe's vast collection of carefully recorded and highly accurate naked eye observations. Over the next two decades Kepler would use these data along with his own hard work and keen mathematical insights to help fan the flames of the Copernican revolution at the dawn of telescopic astronomy.

Although Brahe was undoubtedly aware of the falsity of Ptolemaic cosmology, he remained a non-Copernican and chose to advocate his own *Tychonic system* instead. The Tychonic system was a modified form of geocentric cosmology in which all the planets (except Earth) revolve around the Sun, and the Sun (along with its entire assemblage of orbiting planets) then revolved around a stationary Earth. Perhaps the reason why Brahe rejected Copernican cosmology was its inherent assumption of a moving Earth. This was a physical fact he could not prove for himself because he was unable to detect the annual parallax of the fixed stars predicted by heliocentric cosmology. That particular measurement also challenged many other astronomers in the 18th century.

Brahmagupta (ca. 598–ca. 660) Indian *Mathematician, Astronomer* The early Indian astronomer and mathematician Brahmagupta wrote the *Brahmaisiddhanta Siddhanta* around 628. This important work introduced algebraic and trigonometric methods for solving astronomical problems. His efforts and those of other Indian scientists of the period greatly influenced later Arab astronomers and mathematicians.

braking ellipses A series of ellipses, decreasing in size due to aerodynamic drag, followed by an orbiting spacecraft as it encounters and enters a planetary atmosphere.

See also AEROBRAKING.

Braun, Wernher Magnus von (1912–1977) German-American *Rocket Engineer* The brilliant rocket engineer Wernher von Braun turned the impossible dream of interplanetary space travel into a reality. He started by developing the first modern ballistic missile, the liquid-fueled V-2 rocket, for the German Army. He then assisted the United States by developing a family of ballistic missiles for the U.S. Army and later a family of powerful space launch vehicles for NASA. One of his giant Saturn V rockets successfully sent the first human explorers to the Moon's surface in July 1969 as part of the NASA's Apollo Project.

Von Braun was born on March 23, 1912, in Wirsitz, Germany (now Wyrzsk, Poland). He was the second of three sons of Baron Magnus von Braun and Baroness Emmy von Quistorp. As a young man, the science fiction novels of JULES VERNE and H. G. WELLS kindled his lifelong interest in space exploration. HERMANN J. OBERTH'S book *THE ROCKET INTO INTERPLANETARY SPACE* introduced him to rockets. In 1929 von Braun received additional inspiration from Oberth's realistic "cinematic" rocket that appeared in Fritz Lang's (1890–1976) motion picture *Die Frau im Mond* (*The Woman in the Moon*). That same year, he became a founding member of Verein für Raumschiffahrt, or VfR, the German Society for Space Travel. Through the VfR he came into contact with Oberth and other German rocket enthusiasts. He also used the VfR to carry out liquid-propellant rocket experiments near Berlin in the early 1930s.

Von Braun enrolled at the Berlin Institute of Technology in 1930. Two years later, at age 20, he received a bachelor's degree in mechanical engineering. In 1932 he entered the University of Berlin. There, as a graduate student, he received a modest rocket research grant from a German army officer (later artillery captain), Walter Dornberger (1895–1980). This effort ended in disappointment when von Braun's new rocket failed to function before an audience of military officials in 1932, but the young graduate student so impressed Dornberger that he hired him as a civilian engineer to lead the Germany Army's new rocket artillery unit. In 1934 von Braun received a Ph.D. in physics from the University of Berlin. A portion of his doctoral research involved testing small liquid-propellant rockets.

By 1934 von Braun and Dornberger had assembled a rocket development team totaling 80 engineers at Kummersdorf, a test facility located some 100 km south of Berlin. That year the Kummersdorf team successfully launched two new rockets, nicknamed Max and Moritz after German fairy tale characters. While "Max" was a modest liquid-fueled rocket, "Moritz" flew to altitudes of about 2 km and carried a gyroscopic guidance system. These successes earned the group additional research work from the German military and a larger, more remote test facility near the small coastal village of Peenemünde, on the Baltic Sea.

The Kummersdorf team moved to Peenemünde in April 1937. Soon after relocation von Braun's team began test firing the A-3 rocket, a rocket plagued with many problems. As World War II approached the German military directed von Braun to develop long-range military rockets. He accelerated design efforts for the much larger A-4 rocket. At the same time an extensive redesign of the troublesome A-3 rocket resulted in the A-5 rocket.

When World War II started in 1939 von Braun's team began launching the A-5 rocket. The results of these tests gave von Braun the technical confidence he needed to pursue final development of the A-4 rocket, the world's first long-range, liquid-fueled military ballistic missile.

The A-4 rocket had a state-of-the-art liquid-propellant engine that burned alcohol and liquid oxygen. It was approximately 14 m long and 1.66 m in diameter. These dimensions were intentionally selected to produce the largest possible mobile military rocket capable of passing through the railway tunnels in Europe. Built within these design constraints, the A-4 rocket had an operational range of up to 275 km.

Von Braun's first attempt to launch the A-4 rocket at Peenemünde occurred in September 1942 and ended in disaster. As its rocket engine came to life, the test vehicle rose from the launch pad for about one second. Then the engine's thrust suddenly tapered off, causing the rocket to slip back to Earth. As guidance fins crumpled, the doomed rocket toppled over and destroyed itself in a spectacular explosion. Success finally occurred on October 3, 1942. In its third flight test the experimental A-4 vehicle roared off the launch pad and reached an altitude of 85 km while traveling a total distance of 190 km downrange from Peenemünde. The successful flight marked the birth of the modern military ballistic missile.

Impressed, senior German military officials immediately ordered the still imperfect A-4 rocket into mass-production. By 1943 the war was turning bad for Nazi Germany, so Adolf Hitler decided to use the A-4 military rocket as a vengeance weapon against Allied population centers. He named this long-range rocket the Vergeltungwaffe-Zwei (Vengeance Weapon-Two, or V-2).

In September 1944 the German Army started launching V-2 rockets armed with high-explosive (about one metric ton) warheads against London and southern England. Another key target was the Belgian city of Antwerp, which served as a major supply port for the advancing Allied forces. More than 1,500 V-2 rockets hit England, killing about 2,500 people and causing extensive property damage. Another 1,600 V-2 rockets attacked Antwerp and other continental targets. Although modest in size and payload capacity compared to modern military rockets, von Braun's V-2 significantly advanced the state of rocketry. From a military perspective the V-2 was a formidable, unstoppable weapon that struck without warning. Fortunately, the V-2 rocket arrived too late to change the outcome of World War II in Europe.

In May 1945 von Braun, along with some 500 of his colleagues, fled westward from Peenemünde to escape the rapidly approaching Russian forces. Bringing along numerous plans, test vehicles, and important documents, he surrendered to the American forces at Reutte, Germany. Over the next few months U.S. intelligence teams, under Operation Paperclip, interrogated many German rocket personnel and sorted through boxes of captured documents.

Selected German engineers and scientists accompanied von Braun and resettled in the United States to continue their rocketry work. At the end of the war American troops also captured hundreds of intact V-2 rockets. Soon after Germany's defeat von Braun and his colleagues arrived at Fort Bliss, Texas. He helped the U.S. Army (under Project Hermes) reassemble and launch captured German V-2 rockets from the White Sands Proving Grounds in southern New Mexico. During this period von Braun also married Maria von Quistorp on March 1, 1947. Relocated German engineers and captured V-2 rockets provided a common technical heritage for the major military missiles developed by both the United States and the Soviet Union during the cold war.

In 1950 the U.S. Army moved von Braun and his team to the Redstone Arsenal, near Huntsville, Alabama. At the Redstone Arsenal von Braun supervised development of early army ballistic missiles, such as the Redstone and the Jupiter. These missiles descended directly from the technology of the V-2 rocket. As the cold war missile race heated up, the U.S. Army made von Braun chief of its ballistic weapons program.

Starting in the fall of 1952 von Braun also provided technical support for the production of a beautifully illustrated series of visionary space travel articles that appeared in *Collier* magazine. The series caught the eye of the American entertainment genius WALT DISNEY. By the mid-1950s von Braun became a nationally recognized space travel advocate through his frequent appearances on television. Along with Walt Disney, von Braun served as a host for an inspiring three-part television series on human space flight and space exploration. Thanks to von Braun's influence, when the Disneyland theme park opened in southern California in the summer of 1955, its Tomorrowland section featured a Space Station X-1 exhibit and a simulated rocket ride to the Moon. A giant 25-meter-tall needle-nosed rocket ship personally designed by WILLY LEY and von Braun also greeted visitors to the Moon mission attraction. During the mid-1950s the Disney–von Braun relationship introduced millions of Americans to the possibilities of the space age, which was actually less than two years away. In 1955 von Braun became an American citizen.

On October 4, 1957, the space age arrived when the Soviet Union launched *Sputnik 1*, the world's first artificial Earth satellite. Following the successful Soviet launches of the *Sputnik 1* and *Sputnik 2* satellites and the disastrous failure of the first American Vanguard satellite mission in late 1957, von Braun's rocket team at Huntsville was given less than 90 days to develop and launch the first American satellite. On January 31, 1958 (local time), a modified U.S. Army Jupiter C missile, called the Juno 1 launch vehicle, rumbled into orbit from Cape Canaveral Air Force Station. Under von Braun's direction this hastily converted military rocket successfully propelled *Explorer 1* into Earth orbit. The *Explorer 1* satellite was a highly accelerated joint project of the Army Ballistic Missile Agency (AMBA) and the Jet Propulsion Laboratory (JPL) in Pasadena, California. The spacecraft carried an instrument package provided by JAMES VAN ALLEN, who discovered Earth's trapped radiation belts.

In 1960 the United States government transferred von Braun's rocket development center at Huntsville from the United States Army to a newly created civilian space agency called the National Aeronautics and Space Administration (NASA). Within a year President JOHN F. KENNEDY made a bold decision to put American astronauts on the Moon within a decade. The president's decision gave von Braun the high-level mandate he needed to build giant new rockets. On July 20, 1969, two Apollo astronauts, NEIL ARMSTRONG and EDWIN (BUZZ) ALDRIN, became the first human beings to walk on the Moon. They had gotten to the lunar surface because of von Braun's flawlessly performing Saturn V launch vehicle.

As director of NASA's Marshall Space Flight Center in Huntsville, Alabama, von Braun supervised development of the Saturn family of large, powerful rocket vehicles. The Saturn I, IB, and V vehicles were well-engineered expendable rockets that successfully carried teams of American explorers to the Moon between 1968 and 1972 and performed other important human space flight missions through 1975.

Just after the first human landings on the Moon in 1969, NASA's leadership decided to move von Braun from Huntsville and assign him to a Washington, D.C., headquarters staff position in which he could "perform strategic planning" for the agency. Von Braun, now an internationally popular rocket scientist, had openly expressed his desires to press on beyond the Moon mission and send human explorers to Mars. This made him a large political liability in a civilian space agency that was now gearing down under shrinking budgets. After less than two years at NASA headquarters, von Braun decided to resign from the civilian space agency. By 1972 the rapidly declining U.S. government interest in human space exploration clearly disappointed him. He then worked as a vice-president of Fairchild Industries in Germantown, Maryland, until illness forced him to retire on December 31, 1976. He died of progressive cancer in Alexandria, Virginia, on June 16, 1977.

Von Braun's aerospace engineering skills, leadership abilities, and technical vision formed a unique bridge between the dreams of the founding fathers of astronautics, KONSTANTIN TSIOLKOVSKY, ROBERT GODDARD, and Hermann Oberth, and the first golden age of space exploration that took place from approximately 1960 to 1989. In this unique period human explorers walked on the Moon, and robot spacecraft visited all the planets in the solar system, except tiny Pluto, for the first time. Much of this was accomplished through his rocket engineering genius, which helped turn theory into powerful, well-engineered rocket vehicles that shook the ground as they thunderously broke the bonds of Earth on their journeys to other worlds.

breccia Rock consisting of angular, coarse fragments embedded in a fine-grained matrix. Generally formed by meteorite impact on planetary surfaces, breccia represents a complex mixture of crushed, shattered, and even once-melted rock all bonded together by a fine-grained material called the matrix. The breccias found by the Apollo Project astronauts in the lunar highlands are very mixed up and complicated rocks that often contain breccias within breccias. Many of the lunar highland breccia samples contain *impact melt breccias* formed in a relatively narrow span of ages, from about 3.85 to 4.0 billion years ago. As a result, some planetary geologists propose that the Moon was bombarded with exceptional intensity during this narrow time interval. However, the idea of a cataclysmic bombardment of the Moon about 4 billion years ago is not yet proven.

bremsstrahlung (German: "breaking radiation") Electromagnetic radiation emitted as energetic photons (that is, as X-rays) when a fast-moving charged particle (usually an electron) loses energy upon being accelerated and deflected by the electric field surrounding a positively charged atomic nucleus. X-rays produced in a medical X-ray machine are an example of bremsstrahlung.

brightness 1. In astronomy, a somewhat ambiguous term that is related to a celestial object's luminosity. One commonly encountered definition of brightness is the apparent intensity of light from a luminous source. An appropriate detector would measure the value of the apparent brightness in units of energy per unit time per unit area (for example, watts/m^2). Physically, the *apparent brightness* of a star is directly proportional to the star's luminosity (an intrinsic property that is sometimes referred to as the *absolute brightness*) and inversely proportional to the square of its distance to the observer. Therefore, a "bright star" is one that has a large apparent brightness. However, the ambiguity arises because a star with a large apparent brightness can have a high luminosity (that is, be a strong emitter of electromagnetic radiation), or it can be relatively close to Earth, or both. Similarly, a "faint star" is one with a low apparent brightness. It can have low luminosity (that is, be a rather weak emitter of radiation), or it can be very far from Earth, or both. Astronomers also use the term *brightness temperature*—the apparent temperature of a celestial body calculated under the basic assumption that the source emits radiation like a blackbody. 2. In ballistic missile defense (BMD), the amount of power that can be delivered on a target per unit of solid angle by a directed energy weapon.

brown dwarf A substellar (almost a star) celestial object. It has starlike material composition (i.e., mostly hydrogen with some helium) but contains too small a mass (generally between 1 percent to 8 percent of a solar mass) to allow its core to initiate thermonuclear fusion. Without such thermonuclear fusion reactions, the brown dwarf has a very low luminosity and is very difficult to detect. Today astronomers use advanced infrared (IR) imaging techniques to find these unusual degenerate stellar objects or failed stars. Some astronomers believe brown dwarfs make a significant contribution to the missing mass (or dark matter) of the universe. In 1995 the first brown dwarf candidate object (called Gliese 229B) was detected as a tiny companion orbiting the small red dwarf star Gliese 229A. Since then astronomers have found many other candidate brown dwarfs.

Bruno, Giordano (1548–1600) Italian *Philosopher, Writer* As a fiery philosopher and writer, the former Dominican monk Giordano Bruno managed to antagonize authorities throughout western Europe by adamantly supporting such politically sensitive and religiously unpopular (at the time) concepts as the heliocentric cosmology of NICHOLAS COPERNICUS, the infinite size of the universe, and the existence of intelligent life on other worlds. His self-destructive, belligerent manner eventually brought him before the Roman Inquisition. After an eight-year-long trial, an uncompromising Bruno was finally convicted of heresy and burned to death at the stake on February 17, 1600.

B star A star of spectral classification type B. The B star is massive (from 3 to 8 solar masses), hot, and blue in appearance. B stars that lie on the main sequence have surface temperatures between 9,900 K and 28,000 K, with 20,000 K considered a typical value. However, supergiant B stars exhibit surface temperatures up to 30,000 K. B stars have lifetimes on the order of about 30 million years, and their spectra primarily contain absorption lines of neutral helium and hydrogen. Spica (B 1), a blue-white B star about 220 light-years away, is an example. It is the most conspicuous star (α Vir) in the zodiac constellation Virgo. Rigel (β Ori) is another example.

See also STAR.

buffeting The beating of an aerodynamic structure or surface by unsteady flow, gusts, and the like; the irregular shaking or oscillation of a vehicle component owing to turbulent air or separated flow.

bulge of the Earth The extra extension of the equator caused by the centrifugal force of Earth's rotation, which slightly flattens the spherical shape of planet. This bulge causes the planes of satellite orbits inclined to the equator (but not polar orbits) to slowly rotate around Earth's axis.

bulkhead A dividing wall in a rocket, spacecraft, or aircraft fuselage, at right angles to the longitudinal axis of the structure, that serves to strengthen, divide, or help shape the structure. A bulkhead sometimes is designed to withstand pressure differences.

Bunsen, Robert Wilhelm (1811–1899) German *Chemist* Robert Bunsen was the innovative German chemist who collaborated with the German physicist GUSTAV KIRCHHOFF in the development of spectroscopy. Their innovative efforts entirely revolutionized astronomy by offering scientists a new and unique way to determine the chemical composition of distant celestial bodies. An accomplished experimentalist, Bunsen also made numerous contributions to the science of chemistry, including improvement of the popular laboratory gas burner that carries his name.

Bunsen was born in Göttingen, Germany, on March 31, 1811. At the age of 19 he graduated with a Ph.D. in chemistry from that city's famous university. Following graduation (in 1830) the young scholar traveled for an extensive period throughout Germany and other countries of Europe, engaging in scientific discussions and visiting many laboratories. Upon returning to Germany he lectured and conducted research at several universities until he accepted a position at the University of Heidelberg in 1852. He remained with that institution until he retired in 1889. A superb instructor and mentor, Bunsen taught many famous chemists, including the Russian Dmitri Mendeleev (1834–1907), who developed the periodic table.

In 1859 Bunsen began his historic collaboration with Kirchhoff, a very productive technical association that changed the world of chemical analysis and astronomy. While JOSEPH VON FRAUNHOFER's earlier work laid the foundations for the science of spectroscopy, Bunsen and Kirchhoff provided the key discovery that unleashed its analytical power. Their classic 1860 paper, "Chemical Analysis through Observation of the Spectrum," revealed the important fact that each element has its own characteristic spectrum—much like fingerprints help identify an individual human being. In a landmark experiment they sent a beam of sunlight through a sodium flame. With the help of their primitive spectroscope—literally a prism, a cigar box, and optical components scavenged from two otherwise abandoned telescopes—they observed two dark lines on a bright colored background just where the sodium D-lines of the Sun's spectrum occurred. Bunsen and Kirchhoff immediately concluded that gases in the sodium flame were absorbing the D-line radiation from the Sun, creating an absorption spectrum. The sodium D-line appears in emission as a bright, closely spaced double yellow line and in absorption as a dark double line against the yellow portion of the visible spectrum. The wavelengths of these double bright emission lines and dark absorption lines are identical and occur at 5889.96 angstroms (Å) and 5895.93 Å, respectively.

After some additional experimentation they realized that the Fraunhofer lines were actually absorption lines—that is, gases present in the Sun's outer atmosphere were absorbing some of the visible radiation coming from the solar interior. By comparing the solar lines with the spectra of known chemical elements, Bunsen and Kirchhoff detected a number of elements present in the Sun, the most abundant of which was hydrogen. Their pioneering discovery paved the way for other astronomers to examine the elemental composition of the Sun and other stars and made spectroscopy one of modern astronomy's most important tools.

For example, just a few years later, in 1868, the British astronomer SIR JOSEPH NORMAN LOCKYER attributed an unusual bright line that he detected in the solar spectrum to an entirely new element (then unknown on Earth). He named the new element "helium"—after *helios,* the ancient Greek word for the Sun.

Bunsen and Kirchhoff then applied their spectroscope to resolving elemental mysteries here on Earth. In 1861, for example, they discovered the fourth and fifth alkali metals, which they named cesium (from the Latin *caesium* for "sky blue") and rubidium (from the Latin *rubidus* for "darkest red"). Today scientists use spectroscopy to identify individual chemical elements from the light each emits or absorbs when heated to incandescence. Modern spectral analyses trace their heritage directly back to the pioneering work of Bunsen and Kirchhoff.

The scientific community bestowed many awards on Robert Bunsen. In 1860, for example, he received the British Royal Society's Copley Medal. He also shared the first Davy Medal with Kirchhoff in 1877 and received the Albert Medal in 1898. After a lifetime of technical accomplishment, Bunsen died peacefully in his sleep at his home in Heidelberg on August 16, 1899.

Burbidge, (Eleanor) Margaret (née Peachey) (1922–) British *Astrophysicist* In 1957 Margaret Burbidge collaborated with her husband, GEOFFREY RONALD BURBIDGE, SIR FRED HOYLE, and WILLIAM ALFRED FOWLER in the development and publication of an important scientific paper that described how *nucleosynthesis* creates heavy chemical elements from light ones in the interior of evolved stars. In 1967 she coauthored a fundamental book on quasars with her astrophysicist husband. From 1972 to 1973 she was the first woman to serve as the director of the Royal Greenwich Observatory (RGO).

Burbidge, Geoffrey Ronald (1925–) British *Astrophysicist* Geoffrey Burbidge collaborated with his astrophysicist wife, ELEANOR MARGARET BURBIDGE, SIR FRED HOYLE, and WILLIAM ALFRED FOWLER as they developed in 1957 the detailed theory explaining how nucleosynthesis creates heavier elements from lighter ones in the interior of evolved stars. In 1967 he coauthored an important book on quasars with his wife. From 1978 to 1984 he served as the director of the Kitt Peak National Observatory (KPNO) in Arizona.

burn In aerospace operations, the firing of a rocket engine. For example, the "third burn" of the space shuttle's orbital maneuvering system (OMS) engines means the third time during a particular space shuttle flight that the OMS engines had been fired. The burn time is the length of a rocket's thrusting period.

burning surface With respect to a solid-propellant rocket, the surface of the grain that is not restricted from burning at any given time during propellant combustion.

burnout The moment in time or the point in the trajectory of a chemical rocket when the combustion of propellants terminates. This usually occurs when all the propellants are consumed by the rocket engine. Rocket engineers define the *burnout velocity* as the velocity attained by the rocket vehicle at the termination of thrust—that is, at burnout. Similarly, they define *burnout angle* as the angle between the local vertical and the velocity vector at burnout.

See also ROCKET.

burn rate Literally, the rate at which a solid propellant burns, that is, the rate of recession of a burning propellant surface, perpendicular to that surface, at a specified pressure and grain temperature; with respect to solid propellant grain design, the rate at which the web decreases in thickness during rocket motor operation. The web is the minimum thickness of a solid propellant rocket grain from the initial ignition surface to the insulated case wall or to the intersection of another burning surface at the time when the burning surface experiences a major change. Also called *burning rate.*

See also ROCKET.

burn time In general, the length of a thrust-producing period for a rocket engine. For a solid-propellant rocket, the interval from attainment of a specified initial fraction of maximum thrust or pressure level to web burnout. The web is the minimum thickness of a solid propellant rocket grain from the initial surface to the insulated case wall or to the intersection of another burning surface at the time when the burning surface undergoes a major change.

See also ROCKET.

burnup 1. (aerospace technology) The vaporization and disintegration of a satellite, spacecraft, or rocket or any of their components by aerodynamic heating upon entry into a planetary atmosphere. 2. (nuclear energy technology) A measure of nuclear reactor fuel consumption. It can be expressed as (a) the percentage of nuclear fuel atoms that have undergone fission, or (b) the amount of energy produced per mass (or weight) unit of fuel in the reactor (for example, as megawattdays per metric ton [MWd/t]).

burst A sudden and short period of intense radiation emission, typically X-rays or gamma rays, from extrasolar celestial objects that are often not identified by counterpart observations in optical astronomy. Astronomers call the sources of such brief but intense emissions *bursters*—as, for example, gamma ray bursters and X-ray bursters. Within the solar system scientists have observed radio frequency (RF) bursts from Jupiter and from the Sun. Solar radio bursts are associated with solar flares.

burst disk Passive physical barrier in a fluid system that blocks the flow of fluid until ruptured by (excessive) fluid pressure.

bus deployment phase In ballistic missile defense (BMD), the portion of a missile flight during which multiple warheads and possibly decoy reentry vehicles are deployed on different paths to different targets. The warheads on a single intercontinental ballistic missile (ICBM) are carried on a platform, called the bus (or post-boost vehicle), which has small rocket motors to move it slightly from its original flight path. This phase is also called the *post-boost phase.*

byte (*b*inar*y* digi*t* *e*ight) A basic unit of information or data consisting of eight binary digits (bits). The information storage capacity of a computer system often is defined in terms of kilobytes (kb), megabytes (Mb), and even gigabytes (Gb). One kilobyte corresponds to 2^{10}, or 1,024, bytes, while 1 megabyte corresponds to 2^{20}, or 1,048,576, bytes.

Byurakan Astrophysical Observatory The observatory located at an altitude of 1.4 km on Mount Aragatz, near Yerevan, the capital of the Republic of Armenia. VIKTOR AMBARTSUMIAN founded the facility in 1946. It played a major role in the astronomy activities of the former Soviet Union.

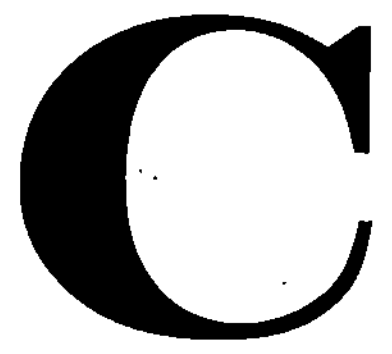

caldera A large volcanic depression more or less circular in form and much larger than the included volcanic vents. Three basic geologic processes may form a caldera: explosion, collapse, or erosion.

See also OLYMPUS MONS.

calendar A system of marking days of the year, usually devised in a way to give each date a fixed place in the cycle of seasons. Most ancient peoples, including those in Babylonia, Egypt, China, and Central America, examined the motions of the Sun, the Moon, and the planets across the sky and then invented a variety of calendars to fulfill various agricultural, religious, civil, and social needs. The basic unit of the calendar is the day, consisting of 24 hours, or 1,440 minutes, or 86,400 seconds, with each second being slightly longer in duration than the average human heartbeat. The day is defined by the motion of the Sun across the sky. Noon, the time when the Sun is at its highest, became a convenient astronomical benchmark. This is because noon corresponds to the time when the Sun is most distant from the horizon and is also exactly south or north of an observer on Earth.

Early calendar makers realized that they could conveniently define "one day" as the time from one noon to another noon. Subsequently, the sundial dial became a convenient, although coarse, early method of observing the passage of time during daylight. However, one problem encountered by early timekeepers and calendar makers in using the solar day is the fact that the time from noon to noon can vary as Earth moves in its elliptical orbit around the Sun. To resolve this problem astronomers eventually devised the concept of a *mean solar day,* which they now define as the duration of one rotation of Earth on its axis with respect to the uniform motion of the mean (average) Sun. Consequently, the length of a mean solar day is 24 hours of mean solar time. A mean solar day beginning at midnight is called a *civil day;* one beginning at noon is called an *astronomical day.*

There is also the *sidereal day* (or "star day"), which astronomers define as the time interval for a reference star to return to its original position during one full rotation of Earth. It turns out that while the definition of a sidereal day is similar to the definition of a solar day, there is one significant difference. The sidereal day uses a "fixed star" for its reference, and the solar day uses the Sun as its reference. Over a year, the Sun slowly changes its position in the sky, tracing out a full circle. If an astronomer were to keep careful count of "sidereal days" versus "solar days," he or she would discover that at the end of a year there would actually be 365.2422 solar days and 366.2422 sidereal days. In fact, use of the sidereal day provides astronomers with a period of rotation of Earth that is just 23.934 hours—a value about four minutes shy of 24 hours.

A clock-tracking mechanism designed to make a telescope follow the stars will make one full rotation per sidereal day. However, timepieces used in daily civil and social activities are based upon the mean (average) solar day. The length of a mean solar day is 24 hours of mean solar time or 24 hours 3 minutes 56.555 seconds of mean sidereal time.

The year is the time needed by Earth to make one full orbit around the Sun. At the end of each year, Earth is back to the same point in its orbit, and the Sun is back to the same apparent position in the sky (as observed from Earth). As previously mentioned, it takes Earth 365.2422 (average solar) days to complete its orbital circuit, and any calendar whose years differ from this number will gradually wander through the seasons. For example, the ancient Roman calendar had only 355 days and added a month every two or four years. But this situation was quite insufficient for the effective administration of an empire. In fact, by the time Julius Caesar became emperor of Rome, the year had slipped by three months. So in 46 B.C.E. Julius Caesar introduced a new calendar named after him, the *Julian calendar.* This calendar is similar to the one used today—having 12 months and adding a day at the end of February every fourth year (or leap year) in years whose number was divisible by the number 4. Two years following the introduction of the Julian calendar, the fifth month of the Roman year was renamed "July" in honor of Julius Caesar. Some time later (around 8 C.E.) this calendar reached its final form, including the renaming of the month following July as August in honor of Caesar Augustus.

The Julian calendar assumed a year of 365.25 days, leaving unaccounted for a small difference of 0.0078 day, or about 1/128 of a day. Therefore, the Julian calendar slips, but at a very slow rate of approximately one day every 128 years. By the year 1582 this slippage had approached two weeks, so Pope Gregory XIII sought the assistance of the German Jesuit mathematician and astronomer CHRISTOPHER CLAVIUS and others in developing a new calendar that became known as the *Gregorian calendar.* In addition to a dramatic "two week jump" (which only slowly became implemented throughout most of the world), the new Gregorian calendar introduced a number of less dramatic modifications. These changes restored March 21st as the date of the vernal equinox and took away three leap years every 40 years. Specifically, for years ending in two zeros, such as 1700, 1800, and 1900, there would not be a leap year, except when the number of the century was divisible by the number 4—such as the year 2000. Today, the Gregorian calendar is the civil calendar used in most of the world. Clavius and associates did a commendable job back in the 16th century, because the modern Gregorian calendar slips by only about one day in 20,000 years!

However, because of political and religious rivalries, many countries in the world were slow to adopt the Gregorian calendar. For example, the British Empire waited until 1752 to shift its civil calendar (from the Julian calendar), while Russia changed its civil calendar only early in the 20th century, following the 1917 Bolshevik Revolution. As a carryover of centuries-old religious rivalries, the Russian Orthodox Church still uses the Julian calendar and therefore celebrates Christmas and Easter about two weeks later than most of the Christian world.

The Moon's orbital motion has also played a significant role in calendar development. When measured with respect to certain fixed stars, the Moon's orbital period (that is, its "sidereal month") is 27.32166 days. However, the monthly lunar cycle—the period from thin crescent to half-moon, to full moon, and back to crescent—takes 29.53059 days. This is because the different shapes of the Moon as seen from Earth represent different angles of solar illumination and are determined by the position of the Sun in the sky, which changes appreciably in the course of each orbit. The average time between two successive new Moons is called the *lunar month* or the *synodic month.* This lunar-based period gave rise to the familiar division of time known as the month.

Many ancient calendars were based on the lunar month. The most successful of these is called the Metonic cycle, named after the ancient Greek astronomer METON. In about 432 B.C.E. Meton discovered that a period of 235 lunar months coincides precisely with an interval of 19 years. After each 19-year interval, the phases of the Moon start taking place on the same days of the year. Both the ancient Greek and Jewish calendars employed the Metonic cycle, and it became the main calendar of the ancient Mediterranean world until replaced by the Julian calendar in 46 B.C.E. People of the Jewish faith still use a religious calendar in which each month begins at or near the new Moon. The traditional Chinese calendar also uses an arrangement similar to the empirical relationship contained in the Metonic cycle. It is interesting to note that adding seven months in the course of 19 years keeps a calendar based on the Metonic cycle almost exactly in step with the annual seasons.

The empirical formulation in the Metonic cycle makes the length of the average year equivalent to (12 + 7/19) synodic months. Then, using the value of 29.53059 days per synodic month, we would obtain 365.2467 days per modified year, which is very close to the value of 365.2422 solar days in the *tropical year* (the time from equinox to equinox). Since the seasons recur after one tropical year, the average length of a *practical calendar year* should be as close as possible to the tropical year. The Gregorian calendar year, for example, contains an average of 365.2425 days.

Muslims use an uncorrected lunar calendar in the practice of their religion. The reason is not a lack of astronomical knowledge but rather a deliberate effort to follow a different religious schedule than those followed by other major faiths, most notably Christianity and Judaism. As a result, Islamic religious holidays slip through the seasons at a rate of about 11 days per year. For example, if the month of Ramadan, during which faithful Muslims are expected not to eat or drink from sunrise to sunset, falls in mid-winter, then some 15 years later it falls in mid-summer in the Muslim religious calendar.

The Maya Indians of Central America created an advanced civilization that peaked between 300 and 1000 C.E., although areas continued to flourish until about 1450. This civilization practiced an extensive amount of naked eye astronomy, which influenced civil, social, and religious activities. Living in the tropical climate of the Yucatán Peninsula of Mexico, Belize, and Guatemala, the Mayans developed a very complex set of calendars. Because these regions did not experience summer and winter to the same extent as do middle latitude regions (such as Europe), the Maya calendars were not strongly tied to the seasons. One Maya calendar used for civil activities was divided into 18 named partitions (loosely "months") of 20 days each, and there was one short "month" of just five days. This resulted in a calendar year of 365 days in length. According to anthropologists and archaeologists, the Maya were aware that the year was slightly longer than 365 days and even computed a value of 365.242 days for the tropical year. Other more complicated and accurate calendars were based on the precession of equinoxes. Finally, Maya astronomers were skilled astronomical observers who gave special attention to the planet Venus and accurately measured its cycles.

See also ARCHAEOLOGICAL ASTRONOMY; METONIC CYCLE; PHASES OF THE MOON; YEAR.

calendar year The time interval that is used as the basis for an annual calendar. For example, the Gregorian calendar year contains an average of 365.2425 days, so that each calendar year contains 365 days, with an extra day being added every fourth calendar year to make a leap year (which contains 366 days).

See also CALENDAR; YEAR.

Caliban A tiny moon (about 80 km in diameter) of Uranus discovered in 1997. Astronomers have very little detailed information about this satellite, and there is considerable variation in the physical data appearing in the technical literature. According to one set of NASA data (published in 2003), Caliban has an estimated mass of 8×10^{17} kg and a visual geometric albedo of 0.07. The moon travels around Uranus in a retrograde orbit at a distance of 7.2×10^6 km with a period of 579.5 days. Caliban's orbit is also characterized by an eccentricity of 0.159 and an inclination of 140.9 degrees relative to the ecliptic plane. Also called UXVI and S/1997 U1.

See also URANUS.

calibration The process of translating the signals collected by a measuring instrument (such as a telescope) into something that is scientifically useful. The calibration procedure generally involves comparing measurements made by an instrument with a standard set of reference data. This process usually removes most of the errors caused by instabilities in the instrument or in the environment through which the signal has traveled.

Callippus of Cyzicus (ca. 370–ca. 300 B.C.E.) Greek *Astronomer, Mathematician* Callippus modified the geocentric system of cosmic spheres developed by his teacher EUDOXUS OF CNIDUS. He was a skilled observer and proposed that a stationary Earth was surrounded by 34 rotating spheres upon which all celestial bodies moved. According to Callippus the Moon, Mercury, Mars, Venus, and the Sun each had five spheres, while Jupiter and Saturn each had four spheres and the fixed stars one sphere. ARISTOTLE liked this geocentric cosmology but proposed a system of 55 solid crystalline (transparent) spheres, whereas the heavenly spheres of Callippus and Eudoxus were assumed to be geometrical but not material in nature.

Callisto The second-largest moon of Jupiter and the outermost of the four Galilean satellites. This icy moon was discovered on January 7, 1610, by GALILEO GALILEI. Callisto has a mean diameter of 4,818 km, a mass of 1.077×10^{23} kg, and an estimated density of 1.85 g/cm^3 (1,850 kg/m^3)—the lowest density of the four Galilean satellites. The acceleration of gravity on the surface of Callisto is 1.24 m/s^2 (Earth has a value of 9.8 m/s^2 at sea level), and the escape velocity is 2.4 km/s. Callisto orbits Jupiter at a distance of 1.883×10^6 km with a period of 16.689 days. The moon's orbit is further characterized by an eccentricity of 0.007 and an inclination of 0.51°. Since Callisto has a rotation period of 16.689 days, it is in synchronous rotation and therefore keeps the same hemisphere facing Jupiter.

Planetary scientists believe that Callisto, sometimes called JIV, consists of equal portions of rock and water-ice.

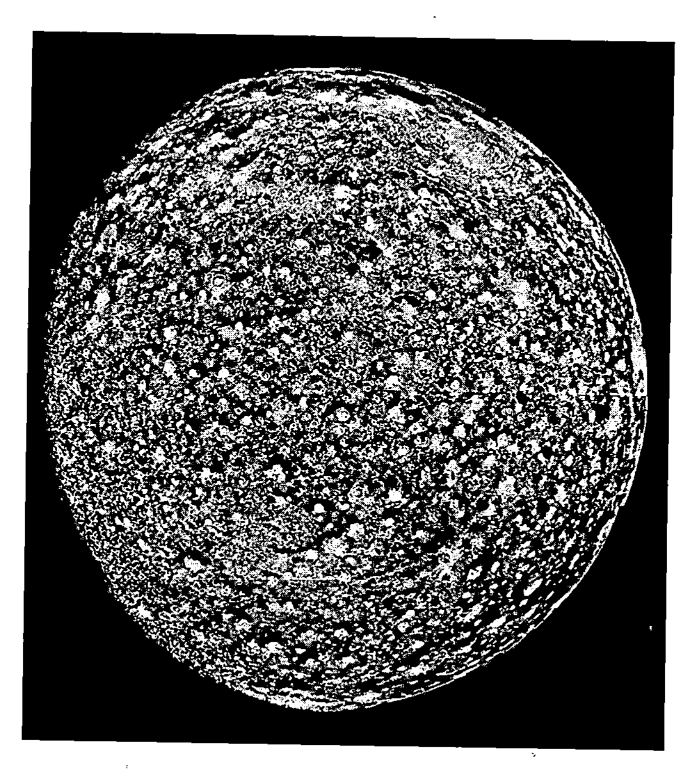

The *Voyager 2* spacecraft took this interesting nine-frame photomosaic of Callisto on July 7, 1979, at a range of 390,000 kilometers. The impact crater distribution is uniform across the moon's surface. Scientists make a special note of the very young, bright rayed craters and the giant impact structure near the limb. Many hundreds of moderate-sized impact craters also appear, a few with bright radial ray patterns. The smoothness of the limb and the lack of high topography indicate Callisto's icy composition. *(Courtesy of NASA/JPL)*

Callisto lacks mountains and is the most heavily cratered moon in the solar system. (*See* figure.) The equatorial subsurface temperature is 168 K and the surface composition is dirty ice. Callisto has a visual albedo of 0.19. The moon's crust appears very ancient, probably dating back about 4 billion years to just after the formation of the solar system.

Magnetometer data collected in the late 1990s by NASA's *Galileo* spacecraft suggest to scientists that something interesting is hidden below Callisto's icy, ancient surface—quite possibly a salty ocean. Specifically, these *Galileo* spacecraft data indicate that Callisto's magnetic field, like Europa's, is variable, which can be explained by the presence of varying electrical currents associated with Jupiter that flow near Callisto's surface. Because Callisto's atmosphere is extremely tenuous and lacks charged particles, scientists examining these magnetic field data believe this wisp of an atmosphere (primarily carbon dioxide) is not responsible for variations of the moon's magnetic field. The likely candidate currently is the possible presence of a layer of salty, melted ice beneath the surface.

See also GALILEAN SATELLITES; JUPITER; *JUPITER ICY MOONS ORBITER* (JIMO).

calorie (symbol: cal) A unit of thermal energy (heat) originally defined as the amount of energy required to raise 1 gram of water through 1°C. This energy unit (often called a *small calorie*) is related to the joule as follows: 1 cal = 4.1868 J. Scientists use the *kilocalorie* (1,000 small calories) for one *big calorie* when describing the energy content of food.

Caloris basin A very large, ringed impact basin (about 1,300 km across) on Mercury.

See also MERCURY.

Calypso A small moon of Saturn that orbits the planet at a distance of 294,660 km, the same distance as Tethys but trailing behind the larger moon by 60°. For this reason astronomers often call Calypso (as well as Telesto) a *Tethys Trojan*. Calypso has an orbital period of 1.888 days. Discovered in 1980, the tiny moon is irregularly shaped (30 km by 16 km by 16 km), has a mass of approximately 4×10^{15} kg, has a mean density of 1.0 gm/cm^3 (1,000 kg/m^3), and has a visual geometric albedo of 0.7. Astronomers sometimes refer to Calypso as SXIV or S/1980 S25.

See also LAGRANGIAN LIBRATION POINTS; SATURN; TETHYS.

Campbell, William Wallace (1862–1938) American *Astronomer* Early in the 20th century William Campbell made pioneering spectroscopic measurements of stellar motions that helped astronomers better understand the motion of our solar system within the Milky Way galaxy and the overall phenomenon of galactic rotation. He perfected the technique of photographically measuring radial velocities of stars and discovered numerous spectroscopic binaries. Campbell also provided a more accurate confirmation of ALBERT EINSTEIN's theory of general relativity when he carefully measured the subtle deflection of a beam of starlight during his solar eclipse expedition to Australia in 1922.

The astronomer was born on April 11, 1862, in Hancock County, Ohio. Campbell experienced a childhood of poverty and hardship on a farm in Ohio. He had originally enrolled in the civil engineering program at the University of Michigan. However, his encounter with SIMON NEWCOMB's work *Popular Astronomy* convinced him to become an astronomer. Upon graduation from the University of Michigan in 1886, he accepted a position in the mathematics department at the University of Colorado. After working two years in Colorado, Campbell returned to the University of Michigan in 1888 and taught astronomy there until 1891. While teaching in Colorado he also met the student who became his wife. Betsey ("Bess") Campbell proved to be a very important humanizing influence in her husband's life, since he was often domineering and inflexible—personal traits that made him extremely difficult to work with or for.

In 1890 Campbell traveled to the West from Michigan to became a summer volunteer at the Lick Observatory of the University of California, located on Mount Hamilton about 20 kilometers east of San Jose. His hard work and observing skills earned him a permanent position as an astronomer on the staff of the observatory the following year. On January 1, 1901, he became director of the Lick Observatory and ruled that facility in such an autocratic manner that his subordinates often privately called him "the tsar of Mount Hamilton." In 1923 he accepted an appointment as president of the University of California, Berkeley, but he also retained his former position as the director of the Lick Observatory.

Campbell's move to California proved to be a major turning point in his career as an astronomer. Up until then he had devoted his observational efforts to the motion of comets, but, at the Lick Observatory he brought about a revolution in astronomical spectroscopy when he attached a specially designed spectrograph to the observatory's 36-inch (91.44-centimeter) refractor telescope, then the world's largest. He used this powerful instrument to measure stellar radial (that is, line-of-sight) velocities by carefully examining the telltale Doppler shifts of stellar spectral lines. He noted that when a star's spectrum was *blueshifted,* it was approaching Earth; when *redshifted,* it was receding. He perfected his spectroscopic techniques and in 1913 published *Stellar Motions,* his classic textbook on the subject. In that same year Campbell also prepared a major catalog listing the radial velocities of more than 900 stars. He later expanded the catalog in 1928 to include 3,000 stars. His precise work greatly improved knowledge of the Sun's motion in the Milky Way and the overall rotation of the galaxy.

Campbell also used his spectroscopic skills to show that the Martian atmosphere contained very little water vapor—a finding that put him in direct conflict with other well-known astronomers at the end of the 19th century. Using the huge telescope and clear sky advantages of the Lick Observatory, Campbell, who was never diplomatic, quickly disputed the plentiful water vapor position of SIR WILLIAM HUGGINS, SIR JOSEPH LOCKYER, and other famous European astronomers. While Campbell was indeed correct, the abrupt and argumentative way in which he presented his data earned him few personal friends within the international astronomical community.

Campbell also enjoyed participating in international solar eclipse expeditions. His wife joined him on these trips, and she was often put in charge of the expedition's provisions. During the 1922 solar eclipse expedition in Australia, Campbell measured the subtle deflection of a beam of starlight just as it grazed the Sun's surface. His precise measurements helped confirm Albert Einstein's theory of general relativity and reinforced the previous, but less precise, measurements made by SIR ARTHUR EDDINGTON during the 1919 solar eclipse.

In 1930 Campbell retired as the president of the University of California, Berkeley, and as the director of the Lick Observatory. Then he and his wife moved to Washington, D.C., where he served as the president of the National Academy of Sciences from 1931 to 1935. Returning to California, he spent the last three years of his life in San Francisco. On June 14, 1938, he died by his own hand, driven to suicide by declining health and approaching total blindness.

Canadian Space Agency (CSA) The government organization that serves as the leader of the Canadian space program. CSA operates under a legislated mandate to "promote the peaceful use and development of space, to advance the knowledge of space through science, and to ensure that space science and technology provide social and economic benefits for Canadians." The Agency is currently pursuing five specific areas considered to be of critical importance to Canada: Earth and the environment, space science, human presence in space, satellite communications, and generic space technologies (including excellence in robotics).

canali The Italian word for *channels,* used in 1877 by the Italian astronomer GIOVANNI SCHIAPARELLI to describe natural surface features he observed on Mars. Subsequent pre–space age investigators, including PERCIVAL LOWELL, took the Italian word quite literally as meaning *canals* and sought additional evidence of an intelligent civilization on Mars. Since the 1960s many spacecraft have visited Mars, dispelling such popular speculations and revealing no evidence of any Martian canals constructed by intelligent beings.

See also MARS.

canard A horizontal trim and control surface on an aerodynamic vehicle.

candela (**symbol: Cd**) The SI unit of luminous intensity. It is defined as the luminous intensity in a given direction of a source that emits monochromatic light at a frequency of 540 $\times 10^{12}$ hertz and whose radiant intensity in that direction is 1/683 watt per steradian. The physical basis for this unit is the luminous intensity of a blackbody radiator at a temperature of 2,046 K (the temperature at which molten platinum solidifies).

See also LUMEN.

cannibalism 1. (galactic) The process in which a larger galaxy devours or swallows up a smaller galaxy. 2. (stellar) The process that takes place in a binary star system when a giant star devours or swallows up its nearby companion.

cannibalize The process of taking functioning parts from a nonoperating spacecraft or launch vehicle and installing these salvaged parts in another spacecraft or launch vehicle in order to make the latter operational.

Cannon, Annie Jump (1863–1941) American *Astronomer* Annie Jump Cannon worked at the Harvard College Observatory (HCO) under EDWARD CHARLES PICKERING. She was instrumental in developing a widely used system for classifying stellar spectra known as the Harvard Classification System. Cannon accomplished this by refining the alphabetical classification sequence introduced by WILLIAMINA P. FLEMING and then arranging stars into categories according to their temperatures, with type O stars being the hottest and type M stars being the coolest. Her efforts culminated in the publication of the *Henry Draper Catalogue,* a nine-volume work (completed in 1924) that contained the spectral classification of 225,300 stars.

See also SPECTRAL CLASSIFICATION.

Canopus (α Car) Formally called Alpha Carinae, this giant star is the brightest in the constellation Carina (the Keel). Excluding the Sun, it is the second-brightest star in the sky after Sirius. Classified as spectral type F0II, Canopus is about 310 light-years away.

Cape Canaveral Air Force Station (CCAFS) The region on Florida's east central coast along the Atlantic Ocean from which the U.S. Air Force and NASA have launched more than 3,000 rockets since 1950. (*See* figure.) It is the first station in the 16,000-kilometer-long Eastern Range. Together with Patrick Air Force Base (some 32 km to the south), Cape Canaveral Air Force Station forms a complex that is the center for Department of Defense launch operations on the East Coast of the United States. The adjacent NASA Kennedy Space Center serves as the spaceport for the fleet of space shuttle vehicles. CCAFS covers a 65-square-km area. Much of the land is now inhabited by large populations of animals, including deer, alligators, and wild boar. The nerve center for CCAFS and the entire Eastern Range is the range control center, from which all launches as well as the status of range resources are monitored. The range safety function also is performed in the range control center (RCC).

Launch sites include space launch complexes 40 and 41 in the Integrated Transfer Launch Facility (ITLF), where preparation and launch of powerful Titan rocket vehicles has taken place. (*See* figure.) Delta rocket vehicles are processed and launched from space complexes 17A and 17B, while

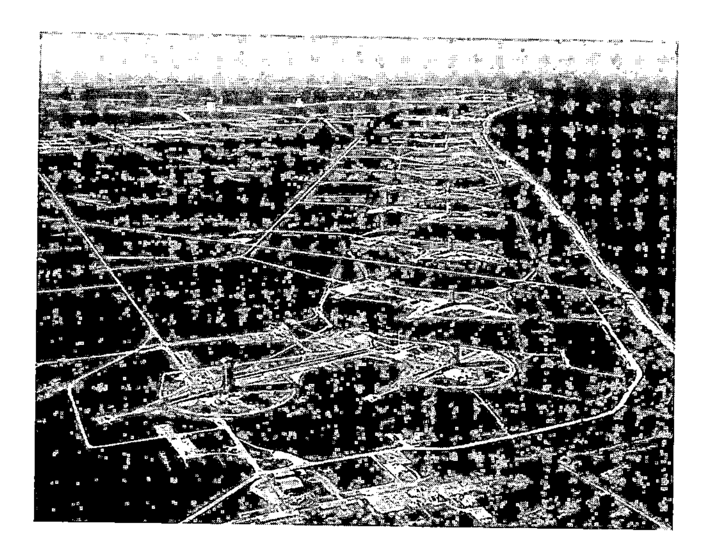

Looking north at the rocket launch complexes of Cape Canaveral Air Force Station in the early 1970s. During the ballistic missile race of the cold war, this unique collection of launch facilities, situated along the shore of the Atlantic Ocean in central Florida, became known as "missile row." The Titan rocket launch complex (called Complex 40) appears in the distance. ***(Courtesy of NASA and the U.S. Air Force)***

A *Titan 34D* space launch vehicle is shown ready for launch from Complex 40 at Cape Canaveral AFS, Florida (ca. 1982). Other portions of the U.S. Air Force's Integrated Transfer Launch Facility (ITLF) appear in the background. *(Courtesy of the U.S. Air Force and Lockheed-Martin Company)*

Atlas vehicles are launched from Complexes 36A and 36B. Space launch complex 20 is used for suborbital launches, and meteorological sounding rockets are launched from the Meteorological Rocket Launch Facility.

See also UNITED STATES AIR FORCE; VANDENBERG AIR FORCE BASE.

capital satellite A very important or very expensive satellite, as distinct from a decoy satellite or a scientific satellite of minimal national security significance.

See also ANTISATELLITE (ASAT) SPACECRAFT.

capsule 1. In general, a boxlike container, component, or unit that often is sealed. 2. A small, sealed, pressurized cabin that contains an acceptable environment for human crew, animals, or sensitive equipment during extremely high-altitude flight, space flight, or emergency escape. As these pressurized cabins became larger, such as during NASA's Apollo Project, the term *spacecraft* replaced the term *capsule* within the American aerospace program. 3. An enclosed container carried on a rocket or spacecraft that supports and safely transports a payload, experiment, or instrument intended for recovery after flight.

captive firing The firing of a rocket propulsion system at full or partial thrust while the rocket is restrained in a test stand facility. Usually aerospace engineers instrument the propulsion system to obtain test data that verify rocket design and demonstrate performance levels. Sometimes called a *holddown test.*

capture 1. In celestial mechanics, the process by which the central force field of a planet or star overcomes by gravitational attraction the velocity of a passing celestial object (such as a comet or tiny planetoid) or spacecraft and brings that body under the control of the central force field. 2. In aerospace operations, that moment during the rendezvous and docking of two spacecraft when the capture mechanism engages and locks the two orbiting vehicles together. 3. In space shuttle operations, the event of the remote manipulator system (RMS) and effector assembly making contact with and firmly attaching to a payload grappling fixture. A payload

Cape Canaveral: The American Stairway To the Stars

The fate of civilizations during pivotal moments in history is often attributed to decisive actions accomplished by a daring few persons on some hallowed piece of ground. The 20th century witnessed many important technical, social, and political events. However, none of these events are equivalent in long-term significance to the future of the human race as was the first rocket launch from a desolate patch of coastal scrubland in central Florida. On July 24,1950, Bumper 8, a modified German V-2 rocket with a WAC Corporal second stage, roared off a primitive launch pad located among the scrub palmettos of an isolated place called Cape Canaveral.

This modest technical achievement represented the first in a long procession of rocket launches from Cape Canaveral—launches that now mark a major milestone in the migration of conscious intelligence beyond the confines of our tiny planetary biosphere. The destiny of the human race would never be the same. Consider the fact that here on Earth, the last such major evolutionary unfolding occurred about 350 million years ago, when prehistoric fish, called crossopterygians, first left the ancient seas and crawled up on the land. Scientists regard these early "explorers" as the ancestors of all terrestrial animals with backbones and four limbs. Similarly, future galactic historians will note how life emerged out of Earth's ancient oceans, paused briefly on the land, then boldly ventured forth from Cape Canaveral to the stars. Humanity's bold leap toward the stars from Cape Canaveral also serves as a fitting epilogue to this region's rich and colorful past—a sometimes forgotten historical legacy replete with Native American lore, daring adventurers, sunken treasure, and hardy pioneers.

Decades after the arrival of Christopher Columbus at San Salvador (in the Bahamas) in 1492, rival European powers began to bitterly contest the ownership of Florida, primarily because it sat astride the main sea route between the Gulf of Mexico and the Old

The first rocket launch from Cape Canaveral took place on July 24, 1950. The Bumper 8 rocket, shown here lifting off, was a modified German V-2 rocket with a WAC Corporal second stage. ***(Courtesy of the U.S. Air Force)***

World. While Cape Canaveral remained a virtual wilderness with respect to European settlement, its coastal waters echoed with the sounds of cannon and muskets as pirates and privateers preyed upon Spanish treasure ships. The Ais, a warlike and sometimes cannibalistic Native American people, lived in the area and survived chiefly on seafood and indigenous vegetation. They plundered the large number of Spanish shipwrecks that came ashore near Cape Canaveral, but rarely made prisoners of any survivors. In fact, fear of the fierce Ais was the primary reason that Spanish settlers did not occupy the portion of Florida near Cape Canaveral.

By the early 18th century the Cape witnessed the arrival of English settlers and their Native American allies (who later became known as the Seminoles) from settlements in Georgia and South Carolina. This marked the beginning of a new era of expansion and conflict in the area, which continued until the end of the Second United States–Seminole War in 1842. Then, in the years following the American Civil War, small rural towns and communities sprang up along a 110-kilometer stretch of mainland, rivers, barrier islands, and sandy beaches that later became known as Brevard County, Florida. Through World War II the principal industries of Brevard County were agriculture, fishing, and tourism.

But the development of the modern ballistic missile (especially the German V-2 rocket) and the American atomic bomb during World War II soon generated military needs that placed Brevard County at the very center of the world stage. By virtue of its most prominent geographical feature, Cape Canaveral, this obscure region of central Florida became the focal point of a new era of exploration—the exploration of outer space. But this transition did not occur without a bit of luck and a lot of hard work.

As discussed here, the "Cape" is the geographic area on the east central coast of Florida (within Brevard County) that currently includes Cape Canaveral Air Force Station (CCAFS), NASA's Kennedy Space Center (KSC), the Merritt Island National Wildlife Refuge, Cape Canaveral National Seashore, and Port Canaveral. Although Cape Canaveral is possibly the most well-recognized space launch facility on the planet, just after World War II the American government almost did not select the site for long-range missile testing. The United States emerged from World War II as the world's only nuclear-armed superpower. The former Soviet Union soon challenged this dominant political and technical position. In 1946, responding to an anticipated cold war need to test military missiles over long distances, the Joint Chiefs of Staff (through their Joint Research and Development Board) established the Committee on the Long-Range Proving Ground. As the name implies, the purpose of this committee was to study possible locations for a long-range rocket proving ground. The committee considered locations in California, Florida, and Washington State. The committee's first choice was El Centro, California; Cape Canaveral earned only a second-place ranking. The committee reasoned that the El Centro site was near the aerospace industry in Southern California, and a long-range rocket testing ground down the coast of Baja California was viewed as a convenient approach in the development of new missile systems.

However, fate intervened in a very unpredictable way and changed aerospace history. President Aleman of Mexico refused to allow American missile flights over Baja California. On the other hand, the United Kingdom was willing to allow American missile flights near the Bahamas Islands (then a British colony). Therefore, the El Centro site was abandoned and Cape Canaveral selected as the nation's principal site for long-range rocket testing over water.

Cape Canaveral proved to be ideal for this task. Virtually uninhabited (in the early 1950s), the remote location let aerospace engineers and technicians inspect, fuel, and launch rockets without endangering nearby communities. Test rockets fired from the Cape were launched over open water instead of over populated land areas. In addition, the area's sunny, subtropical climate permitted year-round operations. And, as an important performance bonus, this location provided a natural ("free") velocity boost to rockets launched eastward due to the west-to-east rotation of Earth. The United States also developed Vandenberg Air Force Base in California to launch polar (north-south) orbiting spacecraft.

The international range site agreements were barely in place when a military-civilian team conducted the first rocket launch from the Cape on July 24, 1950. A tarpaper shack served as an impromptu "press site," and the launch pad service structure was composed of painter's scaffolding.

Despite the primitive facilities, Bumper 8 successfully flew. This launch inaugurated the new American rocket test range, a range that extended barely 320 kilometers at the time. The launch of Bumper 7 followed within just four days, on July 29, 1950. This second successful flight marked the end of the U.S. Army's Project Bumper, an early ballistic missile test program that started at the White Sands Proving Grounds in New Mexico and involved the use of captured German V-2 rockets. Wernher von Braun and his team of German rocket scientists, who had emigrated to the United States after World War II as part of Operation Paperclip, provided technical support for the U.S. Army's missile program. Von Braun's rocket engineering skills played a significant role in many other important launches from the Cape, including the first American satellite (*Explorer 1*), the first American to fly in space (astronaut

(continues)

Cape Canaveral: The American Stairway To the Stars *(continued)*

Shepard in the *Mercury Redstone 3* vehicle), and, of course, the incomparable Apollo lunar landing missions.

Immediately following the successful Bumper 8 and 7 launches came the "Age of the Winged Missiles." In the early 1950s the fledgling U.S. Air Force began testing the Matador, Snark, Bomarc, Mace, and many other interesting winged guided missiles at the Cape. The Air Force's most frequently launched missile, the Matador, was first flight tested at the Cape on June 20, 1951.

Learning from all-too-frequent failures was an integral part of the early missile test and development business. Erratic winged missiles often made spectacular splashes in the waters near Cape Canaveral. For example, the frequent test failures of the Snark, a particularly error-prone bird, inspired local humorists to affectionately refer to the ocean waters off Cape Canaveral as "Snark-infested" waters.

Despite numerous tests and reasonable engineering progress, the days of the early winged missiles were numbered, as more powerful ballistic missiles began to make their appearance. For instance, in August 1953 the U.S. Army launched its first Redstone rocket from Cape Canaveral. The Redstone was a medium-range, liquid-propellant ballistic missile that was a direct technical descendant of the German V-2 rocket. By the late 1950s, in response to a perceived missile gap with respect to the Soviet Union, American national security emphasis began shifting to the rapid development of a much longer-range type of ballistic missile, called the intercontinental ballistic missile (ICBM).

By the mid-1950s flight testing requirements for the Snark and Navaho winged missiles prompted the U.S. Air Force to expand the Eastern Test Range all the way down to Ascension Island, a tiny British Crown territory in the middle of the South Atlantic Ocean near the equator. With Cape Canaveral as Station One, the American long-range rocket test range extended a distance of about 8,000 km into the South Atlantic Ocean. However, the frequent testing of more powerful ballistic missiles, which eventually became the main users of the rocket range's most distant outposts, soon eclipsed the long-range testing of winged missiles. Eventually, ballistic missile tests from the Cape extended all the way past the southern tip of Africa into the Indian Ocean—a total distance of about 16,000 kilometers. To accommodate the need for timely and precise missile performance data all along flight paths, the U.S. Air Force supplemented its collection of down-range tracking sites with a fleet of specially instrumented tracking ships.

After a slow start, ballistic missile and space programs took root at the Cape and quickly dominated use of the range after 1957. The frequent flights of the winged missiles were replaced by launches of the U.S. Army's Jupiter, the U.S. Navy's Polaris, and the U.S. Air Force's Thor, Atlas, Titan, and Minuteman ballistic missiles. All this activity focused on developing powerful and accurate delivery systems for nuclear warheads—weapons of mass destruction that had grown even more lethal through improvements in American nuclear weapon engineering, especially with the arrival of the hydrogen bomb.

With the birth of the space age on October 4, 1957, the rocket testing equation became far more complicated. From a secret missile test facility in central Asia (now called the Baikonur Cosmodrome in the independent republic of Kazakhstan), the Soviet Union used its most powerful ballistic missile (called the R-7) to place the world's first artificial satellite, *Sputnik 1,* into orbit around Earth.

Rockets and missiles were no longer only part of the military nuclear arms race; they became an integral part of the game of global politics. During the cold war a superpower's international prestige was deemed directly proportional to its technical achievements in space. Caught up in this East-West competition, the Cape served as a major technology showcase for Western democracy.

In the late evening (local time) of January 31, 1958, the ground near Launch Complex 26 shook as a hastily modified U.S. Army Juno rocket roared from its pad and successfully inserted the first American satellite, called *Explorer 1,* into orbit around Earth. This effort was the culmination of a joint project of the U.S. Army Ballistic Missile Agency (ABMA) in Huntsville, Alabama, and the Jet Propulsion Laboratory (JPL) in Pasadena, California. Von Braun supervised the rocket team, while James Van Allen of the State University of Iowa provided the instruments that detected the inner of Earth's two major trapped radiation belts that now bear his name.

As the United States turned its attention to the scientific exploration of outer space, an act of the U.S. Congress formed the National Aeronautics and Space Administration (NASA)—the civilian space agency that opened its doors for business on October 1, 1958. Soon, Cape Canaveral thundered with the sound of military rockets that aerospace engineers had converted from warhead-carrying ballistic missiles to space launch vehicles. While the civilian aspects of space exploration and artificial satellite applications made headlines, military satellites—many flown or tested from Cape Canaveral—became an integral part of national defense. For example, the U.S. Air Force launched missile surveillance satellites (first Midas and later Defense Support Program satellites) to guard North America against a surprise ballistic missile attack and to ensure the efficacy of the American strategic nuclear retaliatory policy, called mutually assured destruction (MAD).

Modern post–cold war efforts by the United States and the Russian Federation in nuclear disarmament can trace their heritage to the family of Vela nuclear treaty monitoring satellites that the U.S. Air Force also launched from Cape Canaveral. The successful launch of the first pair of Vela spacecraft encouraged President John F. Kennedy to sign the Limited Test Ban Treaty (LTBT) in 1963. This treaty, verified by satellite systems, marked the end of atmospheric nuclear weapons testing by the United States, the Soviet Union, and the United Kingdom and represented the beginning of the end of the nuclear arms race that dominated the cold war.

In a magnificent wave of scientific exploration, NASA used the Cape to send progressively more sophisticated robot spacecraft to the Moon, around the Sun, and to all of the major planets in our solar system (save tiny Pluto). One epic journey, the *Voyager 2* mission, started from Complex 41 when a mighty Titan/Centaur launch vehicle ascended flawlessly into the Florida sky on August 20, 1977. This hardy, robotic explorer conducted a "Grand Tour" mission of all the giant outer planets (Jupiter, Saturn, Uranus, and Neptune). Then, like its twin (*Voyager 1*), it departed the solar system on an interstellar trajectory.

To date, four human-made objects (NASA's *Pioneer 10* and *11* spacecraft and the *Voyager 1* and *2* spacecraft) have achieved deep space trajectories that allow them to wander among the stars. Each spacecraft carries a message from Earth. *Pioneer 10* and *11* carry a special plaque, while *Voyager 1* and *2* bear a special recorded message. Thus, our first interstellar emissaries departed on their journeys through the galaxy from Cape Canaveral.

Spacecraft launched from the Cape have also revolutionized meteorology, global communications, navigation, astronomy, astrophysics, and cosmology. For example, the geostationary communications satellite (comsat), suggested in 1945 by Sir Arthur C. Clarke, became a reality in 1963 when *Syncom 2* successfully lifted off from the Cape. Today an armada of modern comsats has created a "global village" through which information travels at the speed of light. NASA's four great astronomical observatories, the *Hubble Space Telescope* (HST), the *Compton Gamma Ray Observatory* (CGRO), the *Chandra X-Ray Observatory* (CXO), and the *Spitzer Infrared Telescope Facility* (SIRTF) all reached outer space following a rocket ride from the Cape area. Data from these and many other scientific spacecraft have changed our understanding of the universe and revealed some of its most majestic and exciting phenomena.

Any celebration of the Cape's significance in human history must pay homage to the political decisions and technical efforts that allowed American astronauts to first walk on another world. Responding to President Kennedy's 1961 mandate, NASA focused the Mercury, Gemini, and Apollo Projects to fulfill his bold vision of sending American astronauts to the Moon and returning them safely to Earth in less than a decade. The Apollo Project, in particular, required the largest American rocket ever built, the colossal 111-meter-tall Saturn V rocket that was the brainchild of Wernher von Braun and a technical descendant of the V-2 rocket.

Cape Canaveral Air Force Station, which had served the nation's space program so well up to that point, was inadequate (too small) as a launch site for this monstrous vehicle. Consequently, NASA acquired scrubland and marsh for buffer space to accommodate flight of this giant rocket. (*See* figure.) NASA called the new Saturn V launch sites Complex 39A and 39B. In November 1963 this newly acquired land was named the John F. Kennedy Space Center as a tribute to the slain president. Following the Apollo Project, NASA modified Complex 39 to accommodate the space shuttle.

In the 21st century the Cape will continue to serve the needs of numerous military, scientific, and commercial customers across the United States and around the world as ventures in space grow more and more international and cooperative. Drawing upon a lesson from history, some aerospace visionaries even suggest that the Cape, as the world's premier spaceport, will eventually become the terrestrial center of solar system commerce, banking, art, and law.

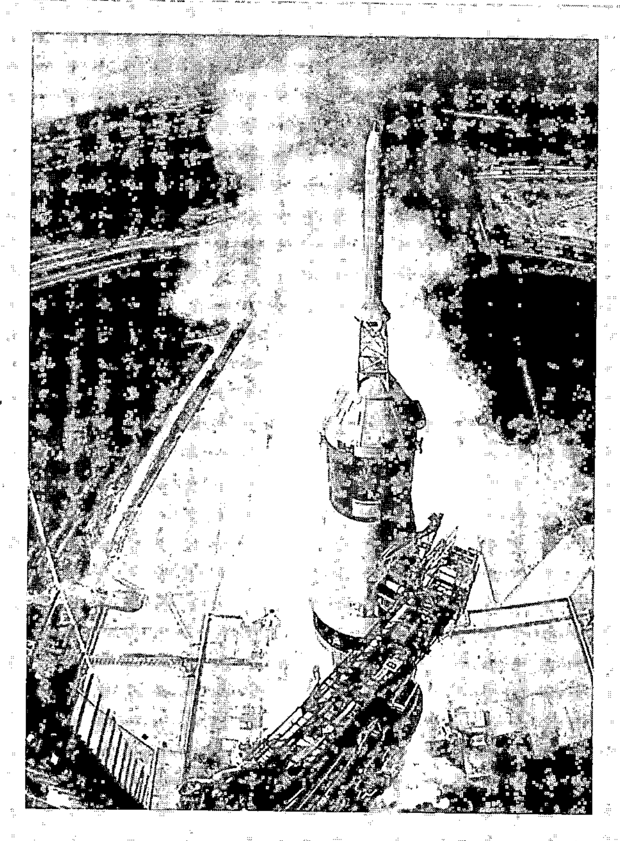

A panoramic view of the mammoth Saturn V launch vehicle as it lifted off from Complex 39 at Cape Canaveral, Florida, on July 16, 1969. Astronauts Neil Armstrong, Edwin (Buzz) Aldrin, and Michael Collins were on their way to the Moon at the start of the historic *Apollo 11* lunar landing mission. *(Courtesy of NASA)*

Access to space from the Cape is really a grand celebration of evolving human intelligence. Because of the Cape we have successfully started on the pathway to the stars. There is no turning back. For us now, it is truly the universe or nothing.

is considered captured any time it is firmly attached to the remote manipulator system. 4. In nuclear physics, a process in which an atomic or a nuclear system acquires an additional particle, such as the capture of electrons by positive ions or the capture of neutrons by nuclei.

carbon cycle 1. (astrophysics) The chain of thermonuclear fusion reactions thought to be the main energy-liberating mechanisms in stars with interior temperatures much hotter (> 16 million K) than the Sun. In this cycle hydrogen is converted to helium with large quantities of energy being released and with carbon-12 serving as a nuclear reaction catalyst. The main sequence of nuclear reactions is as follows:

$$^{12}C + {}^{1}H \rightarrow {}^{13}N + \gamma$$

$$^{13}N \rightarrow {}^{13}C + e^{+} + \nu$$

$$^{1}H + {}^{13}C \rightarrow {}^{14}N + \gamma$$

$$^{1}H + {}^{14}N \rightarrow {}^{15}O + \gamma$$

$$^{15}O \rightarrow {}^{15}N + e^{+} + \nu$$

$$^{1}H + {}^{15}N \rightarrow {}^{12}C + {}^{4}He$$

where e^+ is a positron, γ is a gamma ray, and ν is a neutrino. Since the isotope carbon-12 appears at the beginning and the end of the cycle, astrophysicists regard it as the catalyst in the overall nuclear reaction

$$4\,{}^1H + {}^{12}C \rightarrow {}^4He + {}^{12}C + 2\,e^+ + 2\,\nu + 3\,\gamma$$

Recognizing that each positron will seek a normal electron and then experience an annihilation radiation reaction that releases two gamma rays, some astrophysicists chose to express the net nuclear reaction for the overall carbon cycle as follows:

$$4\,{}^1H + {}^{12}C \rightarrow {}^4He + {}^{12}C + 2\,\nu + 7\,\gamma$$

Both reaction summaries are equivalent. Also called the *carbon-nitrogen (CN) cycle,* the *carbon-nitrogen-oxygen (CNO) cycle,* or the *Bethe-Weizsäcker cycle.* 2. (Earth system) The planetary biosphere cycle that consists of four central biochemical processes: photosynthesis, autotrophic respiration (carbon intake by green plants), aerobic oxidation, and anaerobic oxidation. Scientists believe excessive human activities involving the combustion of fossil fuels, the destruction of forests, and the conversion of wild lands to agriculture may now be causing undesirable perturbations in our planet's overall carbon cycle, thereby endangering important balances within Earth's highly interconnected biosphere.

See also GLOBAL CHANGE; GREENHOUSE EFFECT.

carbon dioxide (symbol: CO_2) A colorless, odorless, noncombustible gas present in Earth's atmosphere. Carbon dioxide is formed by the combustion of carbon and carbon compounds (e.g., fossil fuels, wood, and alcohol), by respiration, and by the gradual oxidation of organic matter in the soil. Carbon dioxide is removed from the atmosphere by green plants (during photosynthesis), and by absorption in the oceans. Since it is a greenhouse gas, the increasing presence of CO_2 in Earth's atmosphere could have significant environmental consequences, such as global warming. The pre–industrial revolution atmospheric concentration of CO_2 is estimated to have been about 275 parts per million by volume (ppmv). Environmental scientists now predict that if current trends in the rise in atmospheric CO_2 concentration continue (currently about 340 to 350 ppmv), then sometime after the middle of the 21st century the atmospheric concentration of CO_2 will reach twice its pre–industrial revolution value. This condition might severely alter Earth's environment through global warming, polar ice cap melting, and increased desertification.

See also EARTH SYSTEM SCIENCE; GLOBAL CHANGE.

carbon star A relatively cool (low-temperature) red giant or AGB star that has reached an advanced stage of evolution and displays an overabundance of carbon relative to oxygen in its outer layers. (An AGB star is one found in the asymptotic giant branch of the Hertzsprung-Russell diagram.) The extra carbon originates in the mature star's helium-burning shell and is then carried to its surface by convection. As a result, this type of star has a spectrum that features strong carbon compound bands, such as CH, CN, and C_2. Mass loss is another characteristic of the carbon star. The ejected outer atmosphere (called a *planetary nebula*) of a low-mass star carries carbon and other byproduct elements of thermonuclear burning (such as nitrogen and oxygen) into interstellar space. Once this chemically enriched material leaves the star, it mixes with the resident interstellar gas, thereby increasing the chemical diversity of the universe. Sometimes called a *C star.*

See also PLANETARY NEBULA; HERTZSPRUNG-RUSSELL DIAGRAM; STAR.

cardinal points With respect to the celestial sphere, astronomers often find it convenient to identify and use four principal (or cardinal) points on an observer's horizon. By convention, the *north point* (n) occurs where the celestial meridian crosses the horizon nearest the north celestial pole; the *south point* (s) lies diametrically opposite the north point and represents the intersection of the celestial meridian with the horizon nearest the south celestial pole. The *east point* (e) and *west point* (w) occur at the intersection of the horizon and the celestial equator. The east point lies 90° and the west point 270° in a clockwise direction from the north point.

See also CELESTIAL COORDINATES; CELESTIAL SPHERE.

cargo bay The unpressurized midpart of the space shuttle orbiter vehicle fuselage behind the cabin aft (rear) bulkhead where most payloads are carried. Its maximum usable payload envelope is 18.3 meters long and 4.6 m in diameter. Hinged doors extend the full length of the cargo bay.

See also SPACE TRANSPORTATION SYSTEM.

Carme A moon of Jupiter discovered in 1938 by the American astronomer SETH BARNES NICHOLSON (1891–1963). This lesser Jovian satellite has a diameter of about 40 km, a mass of 9×10^{16} kg, a density of 2.6 g/cm^3 (2,600 kg/m^3), and a visual geometric albedo of 0.04. It is in a retrograde orbit around Jupiter at a mean distance of 22,600,000 km, with a period of 692 days, an inclination of 164°, and an eccentricity of 0.207. Sometimes called Jupiter XI or JXI.

See also JUPITER.

Carnot cycle An idealized reversible thermodynamic cycle for a theoretical heat engine, first postulated in 1824 by the French engineer Nicolas Sadi Carnot (1746–1832). As shown in the figure, the Carnot cycle consists of two adiabatic (no energy transfer as heat) stages alternating with two isothermal (constant temperature) stages. In stage 1 an ideal working fluid experiences adiabatic compression along path ab. Since no thermal energy (heat) is allowed to enter or leave the system, the mechanical work of compression causes the temperature of the working fluid to increase from T_1 (the sink temperature) to T_2 (the source temperature). During stage 2 of the Carnot cycle the working fluid experiences isothermal expansion. While the working fluid expands in traveling along path bc, heat is transferred to it from the source at temperature T_2. Then in traveling along path cd during stage 3, the working fluid undergoes adiabatic expansion. Travel along this path essentially completes the "power stroke" of the cycle, and the work of expansion causes the temperature of the working fluid to fall from T_2 to T_1. Finally, in stage 4 the working fluid undergoes isothermal compression while traveling along path da. Mechanical work is performed on the working fluid, and an equivalent amount of heat is rejected to the sink (environment), which is at the lower temperature, T_1. The working fluid is now back to its initial state and is ready to repeat the cycle again.

The thermal efficiency (η_{th}) of the Carnot cycle is given by the expression

$$\eta_{th} = 1 - (T_1/T_2)$$

where T_1 is the sink (lower) temperature and T_2 is the source (higher) temperature expressed on an absolute scale (i.e., temperature in kelvins). The thermal efficiency of the Carnot cycle is the best possible efficiency of any heat engine operating between temperatures T_1 and T_2. This fact is sometimes referred to as the *Carnot principle*. Therefore, for maximum efficiency in a reversible heat engine, the source temperature (T_2) should be as high as possible (within the temperature limits of the materials involved), while the sink or environmental reservoir temperature (T_1) should be as low as possible. Heat engines operating on Earth must reject thermal energy to the surrounding terrestrial environment. Therefore, T_1 typically is limited to about 300 K (depending on the local environmental conditions). Heat engines operating in space have the opportunity (depending on radiator design) to reject thermal energy to the coldest regions of outer space (about 3 K). Although the Carnot cycle is a postulated, idealized cycle, it is very useful in understanding the performance limits of real heat engines that operate using the somewhat less than idealized Brayton cycle or Rankine cycle.

See also RANKINE CYCLE; SPACE NUCLEAR POWER.

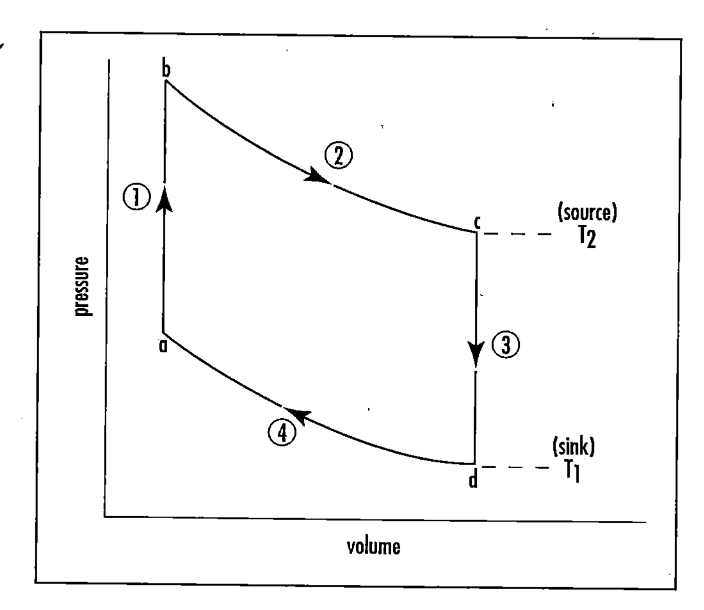

Pressure-volume relationships for an ideal working fluid undergoing the four stages of the Carnot cycle—an ideal, reversible heat engine cycle *(Drawing courtesy of the U.S. Department of Energy)*

Carpenter, Scott (1925–) American *Astronaut, U.S. Navy Officer* On May 24, 1962, Scott Carpenter flew the second American manned orbital spaceflight. His *Aurora 7* space capsule made three revolutions of Earth as part of NASA's Mercury Project.

See also MERCURY PROJECT.

carrier wave (CW) In telecommunications, an electromagnetic wave intended for modulation. This wave is transmitted at a specified frequency and amplitude. Information then is superimposed on this carrier wave by making small changes in (i.e., modulating) either its frequency or its amplitude.

See also TELECOMMUNICATIONS.

Carrington, Richard Christopher (1826–1875) British *Astronomer* Richard C. Carrington made important studies of the rotation of the Sun by carefully observing the number and positions of sunspots. Between 1853 and 1861 he discovered that sunspots at the solar equator rotated in about 25 days, while those at 45° solar latitude rotated in about 27.5 days. In 1859 (without special viewing equipment) he reported the first visual observation of a solar flare, although he thought the phenomenon he had just witnessed was the result of a large meteor falling into the Sun.

Carte du Ciel A whole-sky photographic atlas produced in the late 19th century. The effort involved 18 observatories and used standardized 0.33-meter refracting telescopes, which were especially designed in France in about 1880 to support this extensive effort in astrophotography.

Cartesian coordinates A coordinate system developed by the French mathematician Renè Descartes (1596–1650) in which locations of points in space are expressed by reference to three mutually perpendicular planes, called *coordinate planes*. The three planes intersect in straight lines called the *coordinate axes*. The distances and the axes are usually marked (x,y,z), and the *origin* is the zero point at which the three axes intersect.

case (rocket) The structural envelope for the propellant in a solid-propellant rocket motor.

See also ROCKET.

Cassegrain, Guillaume (ca. 1629–1693) French *Instrument Maker* In 1672 Guillaume Cassegrain invented a special reflecting telescope configuration. Now called the CASSEGRAIN TELESCOPE, his optical device has become the most widely used reflecting telescope in astronomy. Unfortunately, except for this important fact, very little else is known about this mysterious man who is rumored to have been possibly a priest, a teacher, an instrument maker, or a physician.

Cassegrain telescope A compound reflecting telescope in which a small convex secondary mirror reflects the convergent beam from the parabolic primary mirror through a hole in the primary mirror to an eyepiece in the back of the

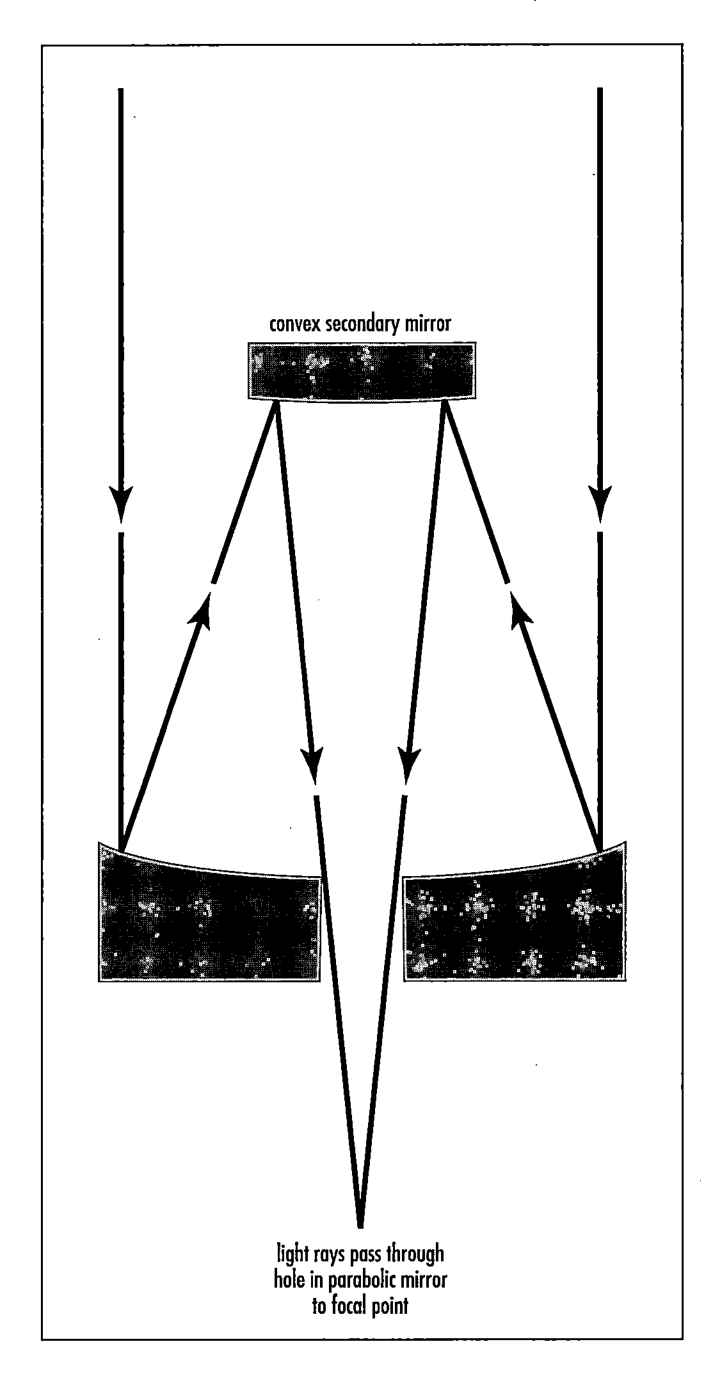

The basic components of the Cassegrain reflector telescope, the most widely used reflecting telescope in astronomy ***(Drawing courtesy of NASA)***

primary mirror. Designed by the French instrument maker GUILLAUME CASSEGRAIN in 1672.

Cassini, César-François (Cesare Francesco Cassini) (1714–1784) French *Astronomer* The French astronomer César-François Cassini, the grandson of GIOVANNI DOMENICO CASSINI, succeeded his father, JACQUES CASSINI, as director of the Paris Observatory. He started the development of the first topographical map of France, a task completed by his son, JACQUES-DOMINIQUE CASSINI. In 1761 he traveled to Vienna, Austria, to observe a transit of Venus across the Sun's disk.

Cassini, Giovanni Domenico (Jean-Dominique Cassini) (1625–1712) Italian (naturalized French) *Astronomer* In 1669 King Louis XIV of France invited the Italian astronomer Giovanni Domenico Cassini to establish and then direct the Paris Observatory. Cassini settled in France and began to use the French version of his name, Jean-Dominique Cassini. An accomplished astronomical observer, Cassini studied Mars, Jupiter, and Saturn. In 1672 he determined the parallax of Mars using observations he made in Paris and those made by JEAN RICHER in Cayenne, French Guiana. These simultaneous but geographically dispersed measurements of Mars allowed Cassini to make the first credible determination of distances in the solar system—including the all important Earth-Sun distance, which he estimated at 140 million kilometers. One astronomical unit is actually 149.6 million kilometers, so Cassini's efforts were quite good for 17th-century astronomy. In 1675 Cassini discovered that Saturn's rings are split largely into two parts by a narrow gap, since known as the *Cassini Division.* By using improved telescopes, between 1671 and 1684 he discovered four new moons of Saturn: Iapetus (1671), Rhea (1672), Dione (1684), and Tethys (1684). He was the patriarch of four generations of influential astronomers and cartographers who directed the Paris Observatory, the first national observatory of any country.

See also CASSINI MISSION; SATURN.

Cassini, Jacques (Giacomo Cassini) (1677–1756) French *Astronomer* As the son of GIOVANNI DOMENICO CASSINI, Jacques Cassini continued his father's work as director of the Paris Observatory. By carefully observing Arcturus in 1738, he became one of the first astronomers to accurately determine the proper motion of a star.

Cassini, Jacques-Dominique (1748–1845) French *Astronomer* Jacques-Dominique Cassini was the great grandson of GIOVANNI CASSINI. Maintaining a family tradition in astronomy, he succeeded his father, CÉSAR-FRANÇOIS CASSINI, as director of the Paris Observatory. He published *Voyage to California,* a discussion of an expedition to observe the 1769 transit of Venus across the Sun's disk. He also completed the topographical map of France started by his father.

Cassini mission The Cassini mission was successfully launched by a mighty Titan IV-Centaur vehicle on October 15, 1997, from Cape Canaveral Air Force Station, Florida. It is a joint NASA and European Space Agency (ESA) project to conduct detailed exploration of Saturn, its major moon Titan, and its complex system of other moons. The *Cassini* spacecraft (*see* figure) took a similar tour of the solar system as did the *Galileo* spacecraft—namely, following a Venus-Venus-Earth-Jupiter gravity assist (VVEJGA) trajectory. After a nearly seven-year journey through interplanetary space cov-

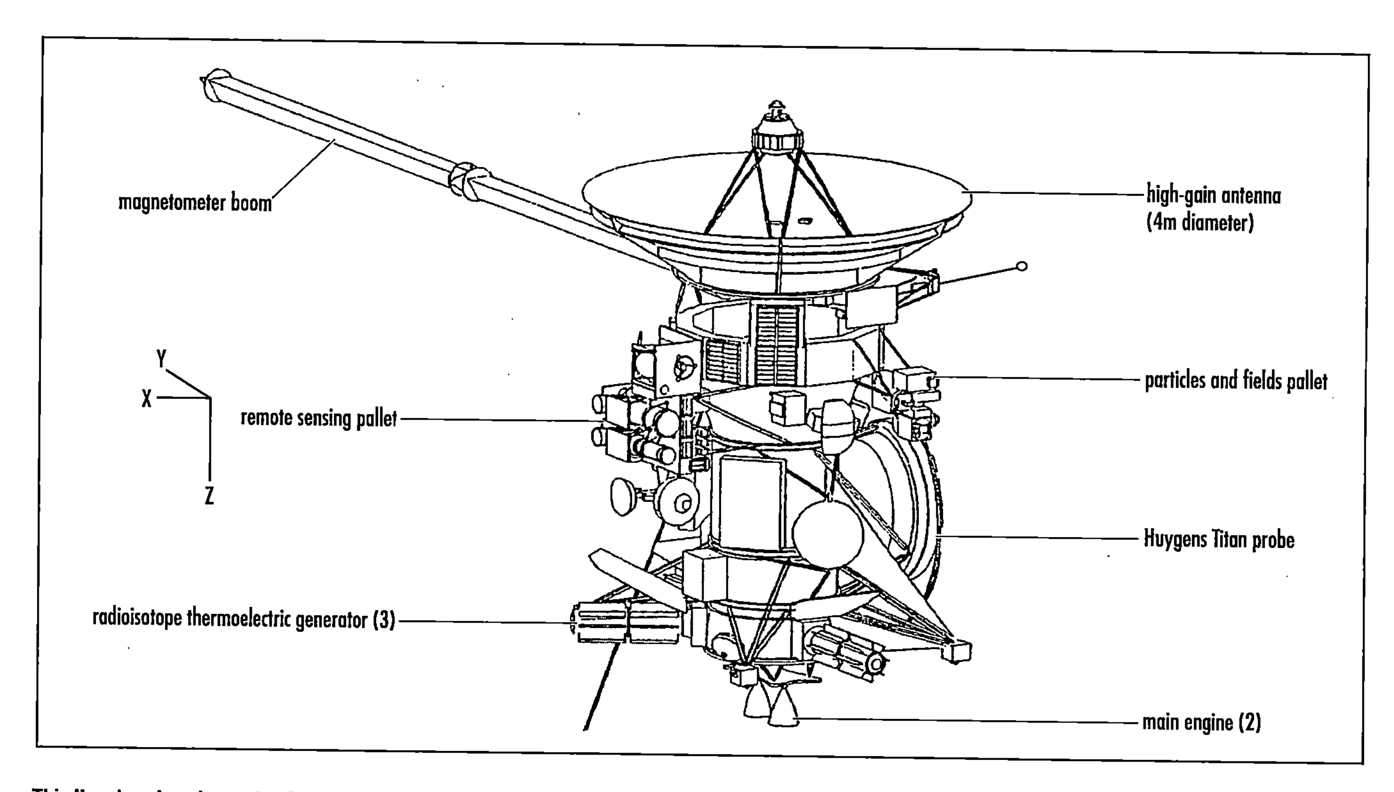

This line drawing shows the *Cassini* spacecraft in its interplanetary flight configuration. ***(Drawing courtesy of NASA)***

ering 3.5 billion kilometers, the *Cassini* spacecraft arrived at Saturn on July 1, 2004 (eastern daylight time). The spacecraft is named in honor of the Italian-born French astronomer GIOVANNI DOMENICO CASSINI, who was the first director of the Royal Observatory in Paris and conducted extensive observations of Saturn.

The most critical phase of the mission after launch was Saturn orbit insertion (SOI). When *Cassini* arrived at Saturn the sophisticated robot spacecraft fired its main engine for 96 minutes to reduce its speed and allow it to be captured as a satellite of Saturn. Passing through a gap between Saturn's F and G rings, the intrepid spacecraft successfully swung close to the planet and began the first of some six dozen orbits it will complete during its four-year primary mission.

The arrival period provided a unique opportunity to observe Saturn's rings and the planet itself, since this was the closest approach the spacecraft will make to Saturn during the entire mission. As anticipated, the *Cassini* spacecraft went right to work upon arrival and provided scientific results. Scientists examining Saturn's contorted F ring, which has baffled them since its discovery, have found one small body, possibly two, orbiting in the F ring region and a ring of material associated with Saturn's moon Atlas. *Cassini*'s close-up look at Saturn's rings revealed a small object moving near the outside edge of the F ring, interior to the orbit of Saturn's moon Pandora. This tiny object, which is about 5 km in diameter, has been provisionally assigned the name S/2004 S3. It may be a tiny moon that orbits Saturn at a distance of 141,000 km from Saturn's center. This object is located about 1,000 km from Saturn's F ring. A second object, provisionally called S/2004 S4, has also been observed in the initial imagery provided by the *Cassini* spacecraft. About the same size as S/2004 S3, this object appears to exhibit some strange dynamics that take it across the F ring.

This artist's rendering shows the *Cassini* spacecraft approaching Saturn and its magnificent rings. The glint of light behind the magnetometer boom at the bottom of the spacecraft represents the reflection of the Sun, some 1,490 million kilometers away. ***(Drawing courtesy of NASA)***

This artist's rendering shows the *Huygens* probe nearing the surface of Saturn's most intriguing moon, Titan. The *Cassini* spacecraft, which delivered the *Huygens* probe, flies overhead with its high-gain antenna pointed at the descending probe. The planet Saturn appears dimly in the background through Titan's thick atmosphere. Titan's surface may hold lakes of liquid ethane and methane, sprinkled over a thin veneer of frozen methane and ammonia. Note that the illustration contains a degree of artistic license with respect to the size of the *Cassini* spacecraft (upper right background), the sharpness of Titan's icy features, the tilt of Saturn's rings, and the visibility of the planet through Titan's thick, murky atmosphere. *(Courtesy of NASA/JPL)*

In the process of examining the F ring region, scientists also detected a previously unknown ring, now called S/2004 1R. This new ring is associated with Saturn's moon Atlas. The ring is located 138,000 km from the center of Saturn in the orbit of the moon Atlas, between the A ring and the F ring. Scientists estimate the ring has a width of 300 km.

The *Huygens* probe was carried to the Saturn system by the *Cassini* spacecraft. Bolted to the *Cassini* mother spacecraft and fed electrical power through an umbilical cable, *Huygens* rode along during the nearly seven-year journey largely in a "sleep mode." However, mission controllers awoke the *Huygens* probe about every six months for three-hour-duration instrument and engineering checkups. *Huygens* is sponsored by the European Space Agency and named after the Dutch physicist and astronomer CHRISTIAAN HUYGENS, who first described the nature of Saturn's rings and its major moon, Titan.

The *Cassini* spacecraft released the *Huygens* probe on December 24, 2004. This release activity was scheduled to occur about 20 days before the probe hit the top of Titan's atmosphere. With its umbilical cut and bolts released, *Huygens* sprang loose from its mother spacecraft and traveled on a ballistic trajectory to Titan. Spinning at about 7 revolutions per minute (rpm) for stability, the probe switched on its systems just before it reached the upper portions of Titan's atmosphere.

Two days after the probe's release, the *Cassini* orbiter spacecraft performed a deflection maneuver. This maneuver kept the orbiter from following *Huygens* into Titan's atmosphere. The deflection maneuver also established the required geometry between the probe and the orbiter for radio communications during the probe's descent. The probe descent took place on January 14, 2005.

Huygens entered Titan's atmosphere at a speed of almost 22,000 kilometers per hour. Aerospace engineers designed the probe to withstand the extreme cold of space (about –200°C) and the intense heat of atmospheric entry (a maximum temperature of more than 1,500°C is estimated). Titan is the only moon in the solar system with a sensible atmosphere. Parachutes further slowed the descent of the probe, allowing *Huygens* to conduct an intensive program of scientific observations all the way down to Titan's surface. (*See* figure.) When the probe's speed had moderated to about 1,400 kilometers per hour, its cover was pulled off by a drogue (pilot) parachute. An 8.3-meter-diameter main parachute then deployed to ensure a slow and stable descent. To limit the duration of the descent to a maximum of 2.5 hours, the main parachute was jettisoned 900 seconds after the probe entered the top of the atmosphere. A smaller, 3-meter diameter parachute was then deployed to support the probe for the remainder of the descent.

During the first part of the probe's descent, instruments onboard *Huygens* were controlled by a timer. For the final 10 to 20 kilometers of descent, a radar altimeter onboard the probe controlled its scientific instruments on the basis of altitude. Throughout the descent the probe's atmospheric structure instrument measured the physical properties of Titan's atmosphere. The gas chromatograph and mass spectrometer determined the chemical composition of the moon's atmosphere as a function of altitude. The aerosol collector and pyrolyzer instrument captured aerosol particles and heated them, sending the resulting vapor to the chromatograph for analysis.

The probe's descent imager and spectral radiometer instrument took pictures of cloud formations and of Titan's surface. This instrument also measured the visibility within Titan's atmosphere. As the surface loomed closer, the instrument switched on a bright lamp and measured the spectral reflectance of the surface. Throughout the probe's descent, the motion-induced Doppler shift of the probe's radio signal was measured by the sensitive Doppler wind experiment onboard the *Cassini* mother spacecraft. As the descending probe was shifted about by winds, the frequency of its radio signal changed slightly due to the phenomenon in physics known as Doppler shift. Scientists will use these small changes in frequency to deduce the wind speed experienced by the probe.

As *Huygens* neared impact its surface science package activated a number of sensors to measure surface properties. The probe impacted the surface of Titan at about 25 kilometers per hour. The major uncertainty was whether its landing would be a thud or a splash. If *Huygens* landed in some type of cold liquid, these instruments would measure the property of that liquid, while the probe floated for a few minutes. For example, if the *Huygens* probe landed in liquid ethane, it would not be able to return data for very long because the extremely low temperature of this liquid (about –180°C) would prevent its chemical batteries from operating. Furthermore, if liquid ethane permeated the probe's science instrument package, the radio would be badly tuned and probably not operate. Assuming, however, that *Huygens* successfully landed on a relatively soft, solid surface or else managed to "float well," the probe still could send back data from Titan's surface for only a maximum of about 30 minutes. At that point, electrical power from the probe's chemical battery was expected to run out. In addition, the line-of-sight geometry needed for radio contact with the orbiting mother spacecraft ended as the *Cassini* orbiter disappeared over the horizon.

Upon arrival at Saturn and the successful orbit insert burn (July 2004), the *Cassini* spacecraft began its extended tour of the Saturn system. This orbital tour involves at least 76 orbits around Saturn, including 52 close encounters with seven of Saturn's known moons. The *Cassini* spacecraft's orbits around Saturn are being shaped by gravity-assist flybys of Titan. Close flybys of Titan also permit high-resolution mapping of the intriguing, cloud-shrouded moon's surface. The *Cassini* orbiter spacecraft carries an instrument called the Titan imaging radar, which can see through the opaque haze covering that moon to produce vivid topographic maps of the surface.

The size of these orbits, their orientation relative to Saturn and the Sun, and their inclination to Saturn's equator are dictated by various scientific requirements. These scientific requirements include: imaging radar coverage of Titan's surface; flybys of selected icy moons, Saturn, or Titan; occultations of Saturn's rings; and crossings of the ring plane.

The *Cassini* orbiter will make at least six close, targeted flybys of selected icy moons of great scientific interest—namely, Iapetus, Enceladus, Dione, and Rhea. Images taken with *Cassini*'s high-resolution telescopic cameras during these flybys will show surface features equivalent in spatial resolution to the size of a major league baseball diamond. At least two dozen more distant flybys (at altitudes of up to 100,000 kilometers) will also be made of the major moons of Saturn other than Titan. The varying inclination of the *Cassini* spacecraft's orbits around Saturn will allow the spacecraft to conduct studies of the planet's polar regions as well as its equatorial zone.

In addition to the *Huygens* probe, Titan will be the subject of close scientific investigations by the *Cassini* orbiter. The spacecraft will execute 45 targeted, close flybys of Titan, Saturn's largest moon—some flybys as close as about 950 kilometers above the surface. Titan is the only Saturn moon large enough to enable significant gravity-assist changes in *Cassini*'s orbit. Accurate navigation and targeting of the point at which the *Cassini* orbiter flies by Titan will be used to shape the orbital tour. This mission planning approach is similar to the way the *Galileo* spacecraft used its encounters of Jupiter's large moons (the Galilean satellites) to shape its very successful scientific tour of the Jovian system.

As currently planned, the prime mission tour of the *Cassini* spacecraft will end on June 30, 2008. This date is four years after arrival at Saturn and 33 days after the last Titan flyby, which will occur on May 28, 2008. The aim point of the final flyby is being chosen (in advance) to position *Cassini* for an additional Titan flyby on July 31, 2008, providing mission controllers the opportunity to proceed with more flybys during an extended mission if resources (such as the supply of attitude control propellant) allow. Nothing in the present design of the orbital tour of the Saturn system precludes an extended mission.

The *Cassini* spacecraft, which includes the orbiter and the *Huygens* probe, is the largest and most complex interplanetary spacecraft ever built. The orbiter spacecraft alone has a dry mass of 2,125 kg. When the 320-kg *Huygens* probe and a launch vehicle adapter were attached and 3,130 kilograms of attitude control and maneuvering propellants loaded, the assembled spacecraft acquired a total launch mass of 5,712 kg. At launch the fully assembled *Cassini* spacecraft stood 6.7 meters high and 4 meters wide.

The Cassini mission involves a total of 18 science instruments, six of which are contained in the wok-shaped *Huygens* probe. This ESA-sponsored probe detached from the main orbiter spacecraft after *Cassini* arrived at its destination and then conducted its own investigations as it plunged into the atmosphere of Titan. The probe's science instruments include: the aerosol collector pyrolyzer, descent imager and spectral radiometer, Doppler wind experiment, gas chromatograph and mass spectrometer, atmospheric structure instrument, and surface science package. The orbiter spacecraft's science instruments include: a composite infrared spectrometer, imaging system, ultraviolet imaging spectrograph, visual and infrared mapping spectrometer, imaging radar, radio science, plasma spectrometer, cosmic dust analyzer, ion and neutral mass spectrometer, magnetometer, magnetospheric imaging instrument, and radio and plasma wave science. Telemetry from the spacecraft's communications antenna will also be used to make observations of the atmospheres of Titan and Saturn and to measure the gravity fields of the planet and its satellites.

Electricity to operate the *Cassini* spacecraft's instruments and computers is being provided by three long-lived radioisotope thermoelectric generators (RTGs). RTG power systems are lightweight, compact, and highly reliable. With no moving parts, an RTG provides the spacecraft electric power by directly converting the heat (thermal energy) released by the natural decay of a radioisotope (here plutonium-238, which decays by alpha particle emission) into electricity through solid-state thermoelectric conversion devices. At launch (on October 15, 1997), *Cassini*'s three RTGs were providing a total of 885 watts of electrical power from 13,200 watts of nuclear decay heat. By the end of the currently planned primary tour mission (June 30, 2008), the spacecraft's electrical power level will be approximately 633 watts. This power level is more than sufficient to support an extended exploration mission within the Saturn system should other spacecraft conditions and resources permit.

The Cassini mission (including *Huygens* probe and orbiter spacecraft) is designed to perform a detailed scientific study of Saturn, its rings, its magnetosphere, its icy satellites, and its major moon Titan. *Cassini*'s scientific investigation of the planet Saturn includes: cloud properties and atmospheric composition, winds and temperatures, internal structure and rotation, the characteristics of the ionosphere, and the origin and evolution of the planet. Scientific investigation of the Saturn ring system includes: structure and composition, dynamic processes within the rings, the interrelation of rings and satellites, and the dust and micrometeoroid environment.

Saturn's magnetosphere involves the enormous magnetic bubble surrounding the planet that is generated by its internal magnet. The magnetosphere also consists of the electrically charged and neutral particles within this magnetic bubble. Scientific investigation of Saturn's magnetosphere includes: its current configuration; particle composition, sources and sinks; dynamic processes; its interaction with the solar wind, satellites, and rings; and Titan's interaction with both the magnetosphere and the solar wind.

During the orbit tour phase of the mission (from July 1, 2004 to June 30, 2008), the *Cassini* orbiter spacecraft will perform many flyby encounters of all the known icy moons of Saturn. As a result of these numerous satellite flybys, the spacecraft's instruments will investigate: the characteristics and geologic histories of the icy satellites, the mechanisms for surface modification, surface composition and distribution, bulk composition and internal structure, and interaction of the satellites with Saturn's magnetosphere.

The moons of Saturn are diverse—ranging from the planetlike Titan to tiny, irregular objects only tens of kilometers in diameter. Scientists currently believe that all of these bodies (except for perhaps Phoebe) hold not only water ice but also other chemical components, such as methane, ammonia, and carbon dioxide. Before the advent of robotic spacecraft in space exploration, scientists believed the moons of the outer planets were relatively uninteresting and geologically dead. They assumed that (planetary) heat sources were not sufficient to have melted the mantles of these moons enough to provide a source of liquid, or even semiliquid, ice or silicate slurries. The *Voyager* and *Galileo* spacecraft have radically altered this view by revealing a wide range of geological processes on the moons of the outer planets. For example, Saturn's moon Enceladus may be feeding material into the planet's F ring, a circumstance that suggests the existence of current activity, such as geysers or volcanoes. Several of Saturn's medium-sized moons are large enough to have undergone internal melting with subsequent differentiation and resurfacing. The Cassini mission will greatly increase knowledge about Saturn's icy moons.

Two-way communication with the *Cassini* spacecraft takes place through the large dish antennas of NASA's Deep Space Network (DSN). The spacecraft transmits and receives signals in the microwave X-band using its own parabolic high-gain antenna. The orbiter spacecraft's high-gain antenna is also used for radio and radar experiments and for receiving signals from the *Huygens* probe as it plunged into Titan's atmosphere.

Because of the enormous distances involved (on average Saturn is 1.43 billion kilometers away from Earth), real-time control of the *Cassini* spacecraft is not feasible. For example, when the spacecraft arrived at Saturn on July 1, 2004, the one-way speed-of-light time from Saturn to Earth was 84 minutes. During the four-year orbital tour mission, the one-way speed-of-light time from Saturn to Earth (and vice versa) will range between 67 and 85 minutes, depending on the relative position of the two planets in their journeys around the Sun. To overcome this problem aerospace engineers have included a great deal of machine intelligence (using advanced computer hardware and software) to enable the sophisticated robot spacecraft to function with minimal human supervision.

For example, each of the *Cassini* spacecraft's science instruments is run by a microprocessor capable of controlling the instrument and formatting/packaging (packetizing) data. Ground controllers run the spacecraft at a distance by using a combination of some centralized commands to control system-level resources and some commands issued by the microprocessors of the individual science instruments. Packets of data are collected from each instrument on a schedule that can vary in the orbit tour phase of the mission. These data packets may be transmitted immediately to Earth or else stored within *Cassini*'s onboard solid-state recorders for transmission at a later time.

Because the *Cassini* spacecraft's science instruments are fixed, the entire spacecraft must be turned to point them. The spacecraft is frequently reoriented either through the use of its reaction wheels or by firing its set of small onboard thrusters. Most of the science observations are being made without real-time communications to Earth. However, the science instruments have different pointing requirements. These different requirements often conflict with each other and with the need to point the spacecraft toward Earth to transmit data home. Reprogrammable onboard software with embedded hierarchies that determine "who and what goes first" guides the onboard microprocessors and the spacecraft's main computer/clock subsystem as they resolve scheduling conflicts. Mission designers have also carefully built into the design of the orbiter tour a sufficient number of periods during which the spacecraft's high-gain antenna points toward Earth.

Mission controllers, engineering teams, and science teams monitor telemetry from the *Cassini* spacecraft and look for anomalies in real time. The flight systems operations team retrieves engineering data to determine the health, safety, and performance of the spacecraft. Members of this team also process the tracking data to determine and predict the spacecraft's trajectory. Data from *Cassini* is normally received by the Deep Space Network with one tracking pass by one antenna per day, with occasional extra coverage for special radio science experiments.

The Cassini program is an international cooperative effort involving NASA, ESA, and the Italian Space Agency (Agenzia Spaziale Italiana [ASI]), as well as several separate European academic and industrial contributors. Through this marvelous mission, some 260 scientists from 17 countries are now gaining a greatly improved understanding of Saturn, its stunning rings, its magnetosphere, Titan, and its other icy moons.

See also GALILEO PROJECT/SPACECRAFT; SATURN; SPACE NUCLEAR POWER; TITAN.

cataclysmic variable (CV) A binary star system that consists of a white dwarf primary and a normal star companion. Typically, the two stars orbit very close to each other and often have a short orbital period ranging between 1 to 10 hours. Astronomers generally refer to the white dwarf star as the primary star of the binary, while they call the normal star the secondary or companion. In a cataclysmic variable, the companion star, usually a normal star like our Sun, loses material onto the nearby white dwarf through the process of accretion. Since the white dwarf is very dense, it exerts an enormous gravitational influence on its stellar companion. As material leaves the outer regions of the companion star, it becomes intensely heated as it swirls around the accretion disk and spirals toward the surface of the white dwarf. This very hot gas can emit radiation in the ultraviolet and X-ray portions of the electromagnetic spectrum.

Astronomers classify CVs according to the properties of the observed outbursts. The two major types of CVs are *classical novae* and *dwarf novae.* Classical novae erupt once, and the amplitude of their outbursts is usually the largest observed among the cataclysmic variables. Classical nova outbursts occur when the companion's hydrogen-rich material collects on the surface of the white dwarf and suddenly undergoes thermonuclear fusion. Because a white dwarf is a essentially dense, hot cinder (the glowing remnant of a sunlike star), this cataclysmic nuclear burning occurs only in the layer of hydrogen accreted onto its surface.

A dwarf nova outburst results from temporary increases in the rate of accretion onto the surface of the nearby white dwarf. Dwarf nova outbursts may be smaller in amplitude and higher in frequency than classical novae. The first dwarf nova discovered is called U Geminorum (U Gem), and its behavior is typical of many other dwarf novae. The optical brightness of U Gem increases 100-fold every 120 days or so and then returns to the original level after a week or two.

Optical astronomers have also categorized several other types of CVs. There are *recurrent novae,* systems with eruptive outbursts that fall in between the behaviors of classical and dwarf novae. Then there are *novalike systems,* stars that have visual light spectra similar to other types of CVs but have not been observed erupting.

X-ray astronomers have observed that many CVs are weak X-ray sources. This is not surprising, since matter accreting from the companion can easily reach 100 million K or so near the surface of the white dwarf. Orbiting X-ray observatories have collected data that reveal the details of the accretion process near the primary. In the majority of CVs the flow of material from the companion to the primary proceeds by means of an accretion disk. This happens because the material leaving the secondary has angular momentum due to orbital motion in the binary system, and so accretion cannot take place in a straight line. Rather, a disklike structure forms in the plane of the binary orbit. Friction within the disk heats up the accreting material and forces this hot material to gradually spiral down onto the surface of the white dwarf. Astrophysicists speculate that the X-rays observed from CVs come from the boundary layer of the white dwarf—the region where the accretion disk hits the surface of the white dwarf. The CVs that are the strongest X-ray emitters appear to have a magnetic white dwarf as the primary.

cavitation The formation of bubbles (or cavities) in a fluid that occurs whenever the static pressure at any point in the fluid becomes less than the fluid vapor pressure. In the flow of a liquid the lowest pressure reached is the vapor pressure. If the fluid velocity is further increased, the flow condition changes and cavities are formed. The formation of these cavities (or vapor regions) alters the fluid flow path and therefore the performance of hydraulic machinery and devices, such as pumps. The collapse of these bubbles in downstream regions of high pressure creates local pressure forces that may result in pitting or deformation of any solid surface near the cavity at the time of collapse. Cavitation effects are most noticeable with high-speed hydraulic machinery, such as liquid-propellant rocket engine turbopumps.

celestial Of or pertaining to the heavens.

celestial body Any aggregation of matter in space constituting a unit for astronomical study, such as the Sun, Moon, a planet, comet, asteroid, star, nebula, and so on. Also called a *heavenly body.*

celestial coordinates In general, a mathematically logical and physically convenient system of coordinates for locating an object on the celestial sphere. A particular celestial coordinate system is usually characterized by the point chosen for observation (the origin) and a reference plane used. For example, geocentric coordinates use the center of Earth as the origin, while heliocentric coordinates use the Sun as the origin.

Geocentric equatorial coordinates use the *celestial equator* as a convenient reference plane. The *equatorial coordinate system* is the most commonly used coordinate system in astronomy. The coordinates commonly used in this system are the right ascension (RA), which is roughly the equivalent of longitude on Earth, and the declination (δ), which is roughly the equivalent of latitude on Earth. The *ecliptic coordinate system* uses the ecliptic as the fundamental reference plane. Within this system astronomers define celestial latitude (β) as the angular distance of a celestial body from 0° to 90° north (considered positive) or south (considered negative) of the ecliptic and celestial longitude (λ) as the angular distance of a celestial body from 0° to 360° measured eastward along the ecliptic to the intersection of the body's circle of celestial longitude. The vernal equinox is taken as 0°. Sometimes astronomers call these coordinates *ecliptic latitude* and *ecliptic longitude.*

In a *topocentric coordinate system* astronomers use a point on Earth's surface as the origin. A *horizontal coordinate system* represents one type of topocentric coordinate system. The horizontal system typically specifies the angular position of a celestial object relative to an observer's horizon at a given time. Often horizontal coordinates are expressed in terms of altitude (angular distance above the horizon) and azimuth (the clockwise bearing of an object from north). When astronomers use a topocentric coordinate system to locate objects in the solar system with high-precision measurements, they experience a slight difference compared to the values given by a geocentric coordinate system. However, when describing the location of more distant objects, such as a star and a galaxy, the subtle differences between the two types of geocentric coordinate systems are no longer significant.

In a *heliocentric coordinate system* the center of the Sun serves as the origin, and the ecliptic generally serves as the reference plane. Astronomers often use heliocentric latitude and heliocentric longitude to express the relative positions of celestial objects within the solar system. *Heliocentric latitude* (*b*) describes the angular position (0° to 90°) of an object north (considered positive) or south (considered negative) of the ecliptic when viewed from the center of the Sun. *Heliocentric longitude* (*l*) describes the angular position (0° to 360°) measured clockwise around the ecliptic starting at the vernal equinox.

Finally, astronomers develop galactic coordinates with reference to the plane of the Milky Way galaxy—that is, they describe the position of a celestial object in terms of its galactic latitude and longitude with respect to the galactic equator. *Galactic latitude* (*b*) gives the angular position (0° to 90°) of a celestial object north (considered positive) or south (considered negative) of the galactic equator. The *galactic equator* is the great circle on the celestial sphere that denotes the plane of the Milky Way galaxy. The galactic longitude (*l*) provides a measure of the angular position (0° to 360°) of a celestial object when measured clockwise along the galactic equator, starting from the point that marks the galactic center (or nucleus) of our galaxy (as seen from Earth), which is some 26,000 light-years away in the constellation Sagittarius. In 1959 astronomers made an international agreement to precisely define the position of zero galactic longitude (as viewed from Earth). However, recent astronomical observations suggest that the actual center of the Milky Way galaxy resides a few arc minutes away from the nominal position defined by this agreement. Because of the orientation of our solar system within the Milky Way, the galactic plane is inclined about 63° to the celestial equator.

See also CELESTIAL EQUATOR; CELESTIAL SPHERE.

celestial equator Astronomers have found it very convenient to define the celestial equator as the great circle formed by the extension of Earth's equator as it intersects with the celestial sphere. The plane of the celestial equator serves as the reference plane for measuring the right ascension (RA) and declination (δ) in the equatorial coordinate system.

See also CELESTIAL SPHERE.

celestial guidance The guidance of a missile or other aerospace vehicle by reference to celestial bodies.

See also GUIDANCE SYSTEM.

celestial-inertial guidance The process of directing the movements of an aerospace vehicle or spacecraft, especially in the selection of a flight path, by an inertial guidance system that also receives inputs from observations of celestial bodies.

See also GUIDANCE SYSTEM.

celestial latitude (symbol: β) With respect to the celestial sphere, the angular distance of a celestial body from 0° to 90° north (considered positive) or south (considered negative) of the ecliptic. Sometimes called *ecliptic latitude.*

See also CELESTIAL SPHERE.

celestial longitude (symbol: λ) With respect to the celestial sphere, the angular distance of a celestial body from 0° to 360° measured eastward along the ecliptic to the intersection of the body's circle of celestial longitude. The vernal equinox is taken as 0°. Sometimes called the *ecliptic longitude.*

See also CELESTIAL SPHERE.

celestial mechanics The scientific study of the dynamic relationships among the celestial bodies of the solar system. Analyzes the relative motions of objects under the influence of gravitational fields, such as the motion of a moon or artificial satellite in orbit around a planet.

See also KEPLERIAN ELEMENTS; KEPLER'S LAWS; ORBITS OF OBJECTS IN SPACE.

celestial navigation The process of directing an aircraft, aerospace vehicle, or spacecraft from one point to another by reference to celestial bodies of known coordinates. Celestial navigation usually refers to the process as accomplished by a human operator. The same process accomplished automatically by a machine usually is called *celestial guidance.*

See also GUIDANCE SYSTEM.

celestial sphere To create a consistent coordinate system for the heavens, early astronomers developed the very useful concept of a celestial sphere. It is an imaginary sphere of very large radius with Earth as its center and on which all observable celestial bodies are assumed projected. The rotational axis of Earth intersects the north and south poles of the celestial sphere. An extension of Earth's equatorial plane cuts the celestial sphere and forms a great circle, called the *celestial equator.* Consequently, the celestial poles and celestial equator are analogs of the corresponding constructs on the surface of Earth. Astronomers can specify the precise location of objects on the celestial sphere by giving the celestial equivalents of their latitudes and longitudes. For an observer on Earth's surface the point on the celestial sphere directly overhead is the *zenith.* Astronomers call an imaginary arc passing through the celestial poles and through the zenith the observer's *meridian.* The nadir is the direction opposite the zenith. For an observer on Earth's surface the nadir is the direction straight down to the center of the planet.

In the very widely used equatorial coordinate system the coordinates are the declination and the right ascension, and the fundamental reference circle is the celestial equator. The declination is the celestial sphere's equivalent of latitude The celestial equator has a declination of 0°, the north celestial pole a declination of +90°, and the south celestial pole a declination of –90°. Consequently, declination values expressed in positive values of degrees extend from the celestial equator to the north celestial pole, while declination values expressed in negative values of degrees extend from the celestial equator to the south celestial pole. Similiarly, the right ascension is the celestial equivalent of terrestrial longitude. Astronomers can express right ascension in degrees, but more commonly they prefer to specifiy it in hours, minutes, and seconds. They do this because as Earth rotates, the sky appears to turn 360° in 24 hours, and that corresponds to 15° per hour. As a result, an hour of right ascension is 15° of sky rotation.

Another important feature intersecting the celestial sphere is the *ecliptic plane.* When astronomers use the *ecliptic coordinate system,* they can plot the direction to any star or other celestial body in two dimensions on the inside of this large imaginary sphere by using celestial latitude and celestial longitude.

See also CELESTIAL COORDINATES.

Celsius, Anders (1701–1744) Swedish *Astronomer* Anders Celsius published a detailed account of the aurora borealis (northern lights) in 1733 and was the first scientist to associate the aurora with Earth's magnetic field. In 1742 he introduced a temperature scale (the CELSIUS TEMPERATURE SCALE) that is still widely used today. He initially selected 100° on this scale as the freezing point of water and 0° as the boiling point of water. However, after discussions with his colleagues, he soon reversed the order of these reference point temperatures.

Celsius temperature scale The widely used relative temperature scale, originally developed by ANDERS CELSIUS, in which the range between two reference points (ice at 0° and boiling water at 100°) is divided into 100 equal units or degrees.

See also ABSOLUTE TEMPERATURE.

Centaur (rocket) A powerful and versatile upper-stage rocket originally developed by the United States in the 1950s for use with the Atlas launch vehicle. Engineered by KRAFFT EHRICKE, it was the first American rocket to use liquid hydrogen as its propellant. Centaur has supported many important military and scientific missions, including the Cassini mission to Saturn (launched October 15, 1997).

Centaurs A group of unusual celestial objects residing in the outer solar system. These objects exhibit a dual asteroid/comet nature. Named after the centaurs in Greek mythology, who were half-human and half-horse beings. CHIRON is an example.

center of gravity For a rigid body, that point about which the resultant of the gravitational torques of all the particles of the body is zero. If the mass of a rigid body can be considered as concentrated at a central point, called the CENTER OF MASS, then the "weight" of this body (i.e., its response to a gravitational field) also may be considered concentrated at this point, such that the center of mass is also the center of gravity.

center of mass That point in a given body, or in a system of two or more bodies acting together in respect to another body (for example, the Earth-Moon system orbiting around the Sun), which represents the mean position of matter in the bodies or system of bodies. For a body (or system of bodies) in a uniform gravitational field, the center of mass coincides with the CENTER OF GRAVITY.

See also BARYCENTER.

centi- (symbol: c) The prefix in the SI unit system that denotes one hundredth (1/100).

central force A force that for the purposes of computation can be considered to be concentrated at one central point with its intensity at any other point being a function of the distance from the central point. For example, gravitation is considered a central force in celestial mechanics.

central processing unit (CPU) The computational and control unit of a computer—the device that functions as the "brain" of a computer system. The CPU interprets and executes instructions and transfers information within the computer. Microprocessors, which have made possible the personal computer "revolution," contain single-chip CPUs, while the CPUs in large mainframe computers and many early minicomputers contain numerous circuit boards each packed full of integrated circuits.

Centre National d'Études Spatiales (CNES) The public body responsible for all aspects of French space activity including launch vehicles and spacecraft. CNES has four main centers: Headquarters (Paris), the Launch Division at Evry, the Toulouse Space Center, and the Guiana Space Center (launch site) at Kourou, French Guiana, in South America.

See also ARIANE; EUROPEAN SPACE AGENCY.

centrifugal force A reaction force that is directed opposite to a CENTRIPETAL FORCE, such that it points out along the radius of curvature away from the center of curvature.

centripetal force The central (inward-acting) force on a body that causes it to move in a curved (circular) path. Consider a person carefully whirling a stone secured by a strong (but lightweight) string in a circular path at a constant speed. The string exerts a radial "tug" on the stone, which "tug" physicists call the centripetal force. Now as the stone keeps moving in a circle at constant speed, the stone also exerts a reaction force on the string, which is called the CENTRIFUGAL FORCE. It is equal in magnitude but opposite in direction to the centripetal force exerted by the string on the stone.

Cepheid variable A type of very bright, supergiant star that exhibits a regular pattern of changing its brightness as a function of time (typically over periods that range from 1 to 50 days). Between 1908 and 1912, while observing Cepheids in the Small Magellanic Cloud, the American astronomer HENRIETTA SWAN LEAVITT discovered the important period-luminosity relationship. Since the period of this pulsation pattern is directly related to the star's intrinsic brightness, astronomers can use Cepheid variables to determine astronomical distances.

Cepheid variables oscillate between two states: compact and expanded. In one state the star is compact, and large pressure and temperature gradients build up in the star's interior.

These large internal pressures then cause the star to expand. When the Cepheid variable reaches its expanded state, the star has a much weaker pressure gradient. Without this pressure gradient to support it against the inward pull of gravity, the star contracts and then compresses to the compact state.

The name of this interesting group of stars comes from the star Delta Cephei, the first recognized member of this class. The more luminous the Cepheid variable, the longer it takes to complete its cycle. This important period-luminosity relationship allows astronomers to use the Cepheid variables as indicators, or *standard candles,* of the distances to galaxies beyond the Milky Way. Astronomers use the term *standard candle* to describe a star that is very bright (so that it can be observed at great distances) and has a known luminosity.

Astronomers divide Cepheid variables into two general categories: *classical Cepheids* and *W Virginis stars.* Classical Cepheids are the brighter of the two categories. They are young massive (Population I) stars located in the spiral arms of galaxies. W Virginis stars, or *type II Cepheids,* are older, less massive (Population II) stars found in either the central region (nucleus) or halo of a galaxy.

Ceres The first and largest asteroid to be found. It was discovered on January 1, 1801, by the Italian astronomer GIUSEPPE PIAZZI. Ceres is about 940 km in diameter and has a mass of approximately 8.7×10^{20} kg. This asteroid has a density of 2.0 g/cm^3 (2,000 kg/m^3), an acceleration of gravity (on its surface) of 0.26 m/s^2, and an escape velocity of 0.5 km/s. Ceres has an orbital period of 4.60 years (1,680 days), an inclination of 10.6°, and an eccentricity of 0.079. The large asteroid orbits the Sun at an average distance of 414 million km (2.77 AU), with a periapsis of 381 million km (2.55 AU) and an apoapsis of 447 million km (2.98 AU). Ceres has a rotation period of 9.1 hours and an estimated mean surface temperature of –105°C. It has a low albedo (0.10) and cannot be observed with the naked eye. Since Ceres has a reflectance spectrum that suggests its composition is similar to carbonaceous chondrite meteorites, astronomers classify it as a G-type asteroid, which is a subdivision of the C-type (carbonaceous) asteroids that is distinguished spectroscopically by virtue of differences in the absorption of ultraviolet radiation. Also called Ceres-1.

See also ASTEROID.

Cernan, Eugene A. (1934–) American *Astronaut, U.S. Navy Officer* Eugene Cernan was the lunar module pilot on NASA's *Apollo 10* mission (May 16–26, 1969) that circumnavigated the Moon. He then went back to the Moon as the spacecraft commander for the *Apollo 17* lunar landing mission (December 6–19, 1972). As part of the *Apollo 17* mission he became the last human being to walk on the lunar surface in the 20th century.

See also APOLLO PROJECT.

Cerro Tololo Inter-American Observatory (CTIO) An astronomical observatory at an altitude of about 2.2-km on Cerro Tololo mountain in the Chilean Andes near La Serena. The main instrument is a 4-m reflecting telescope, and the observatory is operated by the Association of Universities for Research in Astronomy (AURA).

Chaffee, Roger B. (1935–1967) American *Astronaut, U.S. Navy Officer* Roger B. Chaffee was assigned to the crew of NASA's first crewed Apollo Project flight (called *Apollo 1*). Unfortunately, he and his crewmates (VIRGIL "GUS" GRISSOM and EDWARD H. WHITE, II) died on January 27, 1967, in a tragic launch pad accident at Cape Canaveral when an oxygen-fed flash fire swept through the interior of their Apollo space capsule. This fatal accident took place as the astronauts were performing a full-scale launch simulation test, and they could not escape from the inferno that raged within their spacecraft and consumed its interior.

chain reaction A reaction that stimulates its own repetition. In nuclear physics a fission chain reaction starts when a fissile nucleus (such as uranium-235 or plutonium-239) absorbs a neutron and splits, or "fissions," releasing additional neutrons. A fission chain reaction is self-sustaining when the number of neutrons released in a given time equals or exceeds the number of neutrons lost by escape from the system or by nonfission (parasitic) capture.

See also FISSION (NUCLEAR).

***Challenger* accident** The space shuttle *Challenger* lifted off Pad B, Complex 39, at the Kennedy Space Center (KSC) in Florida at 11:38 A.M. on January 28, 1986, on shuttle mission STS 51-L. At just under 74 seconds into the flight an explosion occurred, causing the loss of the vehicle and its entire crew, consisting of astronauts FRANCIS R. (DICK) SCOBEE (commander), MICHAEL JOHN SMITH (pilot), ELLISON S. ONIZUKA (mission specialist one), JUDITH ARLENE RESNIK (mission specialist two), RONALD ERWIN MCNAIR (mission specialist three), S. CHRISTA MCAULIFFE (payload specialist one), and GREGORY BRUCE JARVIS (payload specialist two). (*See* figure.) Christa McAuliffe was a schoolteacher from New Hampshire who was flying onboard the *Challenger* as part of NASA's Teacher-in-Space program, and Gregory Jarvis was an engineer representing the Hughes Aircraft Company. The other five were members of NASA's astronaut corps.

In response to this tragic event, President Ronald Reagan appointed an independent commission, the Presidential Commission on the Space Shuttle *Challenger* Accident. The commission was composed of people not connected with the STS 51-L mission and was charged to investigate the accident fully and to report their findings and recommendations back to the president.

The consensus of the presidential commission and participating investigative agencies was that the loss of the space shuttle *Challenger* and its crew was caused by a failure in the joint between the two lower segments of the right solid rocket booster (SRB) motor. The specific failure was the destruction of the seals (O-rings) that were intended to prevent hot gases from leaking through the joint during the propellant burn of the SRB. The commission further suggested that this joint failure was due to a faulty design that was unacceptably sensitive to a number of factors. These factors included the effects of temperature, physical dimensions, the character of materials, the effects of reusability, processing, and the reaction of the joint to dynamic loading.

The crew members of the ill-fated STS 51-L mission (left to right): payload specialist Christa McAuliffe, payload specialist Gregory Jarvis, mission specialist Judith Resnik, commander Francis R. Scobee, mission specialist Ronald E. McNair, pilot Michael J. Smith, and mission specialist Ellison Onizuka *(Courtesy of NASA)*

Hazards of Space Travel

Current experience with human performance in space is mostly for individuals operating in low-Earth orbit (LEO). However, the establishment of permanent lunar bases, three-year duration and longer expeditions to Mars, and the construction of large space settlements will require extensive human activities in space beyond the protection offered travelers in LEO by Earth's magnetosphere. The maximum continuous time spent by any single human being in space is now just several hundred days, and the people who have experienced extended periods of space travel generally represent a small number of highly trained and highly motivated individuals.

Expeditions to the *International Space Station* (ISS) are intended to expand human spaceflight experience, but the Space Shuttle *Columbia* accident (February 2003) has caused a significant curtailment in the crew size, crew rotation, and scientific and biophysical experiment schedules.

The available technical database, although limited to essentially low-Earth-orbit spaceflight, now suggests that with suitable protection people can live and work in space safely for extended periods of time and then enjoy good health after returning to Earth. Data from the three crewed *Skylab* missions, numerous long-duration *Mir* space station missions, and the first 10 expeditions to the ISS are especially pertinent to answering the important question of whether people (in small, isolated groups) can live and work together effectively in space for more than a year in a confined and relatively isolated habitat. Human factor data collected from international crew performance during extended ISS expeditions will provide a modest inkling of how best to prepare the human beings for one-year or more remote duty in establishing a lunar surface base or three-year or more remote duty on the first human expedition to Mars.

One recurring issue is the overall hazard of space travel. Three major accidents in the American space program, the *Apollo 1* mission fire (January 27, 1967), the *Challenger* accident (January 28, 1986), and the *Columbia* accident (February 1, 2003) have impressed on the national consciousness that space travel is and will remain for a significant time into the future a hazardous undertaking. Some of the major cause and effect factors related to

(continues)

Hazards of Space Travel *(continued)*

space traveler health and safety are shown below. Many of these factors require "scaling up" from current medical, safety, and occupational analyses to achieve the space technologies necessary to accommodate large groups of space travelers and permanent inhabitants. Some of these health and safety issues include: preventing launch-abort, spaceflight and space-based assembly and construction accidents; preventing failures of life-support systems; protecting space vehicles and habitats from collisions with space debris and meteoroids; protecting the crew from the ionizing radiation hazards encountered in outer space; and providing habitats and good-quality living conditions that minimize psychological stress. The biomedical effects of the substantial acceleration and deceleration forces when leaving and returning to Earth, living and working in a weightless (microgravity) environment for long periods of time, and chronic exposure to space radiation are three main factors that must be dealt with if people are to live in cislunar space (space between Earth and the Moon) and eventually populate heliocentric space.

Astronauts and cosmonauts have adapted to microgravity conditions for extended periods of time in space and have experienced maximum acceleration forces up to an equivalent of six times Earth's gravity (that is, 6 g). No acute operational problems, permanent physiological deficits, or adverse health effects on the cardiovascular or musculoskeletal systems have been observed from these experiences. However, short-term physical difficulties, such as space adaptation syndrome, or "space sickness," as well as occasional psychological problems (such as feelings of isolation and stress) and varying postflight recovery periods after long-duration missions have been encountered.

Scheduled to be phased out by the year 2010, the United States Space Transportation System, or space shuttle, can be thought of as the forerunner of more advanced "space traveler" launch vehicles. The shuttle was designed to limit acceleration/deceleration loads to a maximum of 3 g, thereby opening space travel to a larger number of individuals.

As previously mentioned, some physiological deviations have been observed in American astronauts and Russian cosmonauts during and following extended space missions, including *Skylab*, *Mir*, and the ISS. Most of these observed effects appear to be related to the adaption to microgravity conditions, with the affected physiological parameters returning to normal ranges either during the missions or shortly thereafter. No apparent persistent adverse consequences have been observed or reported to date. Nevertheless, some of these deviations could become chronic and might have important health consequences if they were to be experienced during extended missions in space, such as an approximately three-year expedition to Mars, repeated long-term tours in a space assembly facility at Lagrangian point four or five in cislunar space, or at a permanent lunar surface base with its significantly reduced (1/6th-g) gravitational environment. The physiological deviations experienced by space travelers due to microgravity (weightlessness) have usually returned to normal within a few days or weeks after return to Earth. However, bone calcium loss appears to require an extended period of recovery after a long-duration space mission.

Strategies are now being developed to overcome these physiological effects of weightlessness. An exercise regimen can be applied, and body fluid shifts can be limited by applying lower-body negative pressure. Anti–motion sickness medication is also useful for preventing temporary motion sickness or "space sickness." Proper nutrition, with mineral supplements and regular exercise, also appear to limit other observed effects. One way around this problem in the long term, of course, is to provide acceptable levels of "artificial" gravity in larger space bases and orbiting space settlements. In fact, very large space settlements will most likely offer the inhabitants a wide variety of gravity levels, ranging from microgravity up to a normal terrestrial gravity level (that is, 1 g). This multiple-gravity-level option will not only make space settlement lifestyles more diverse than those on Earth but will also prepare planetary settlers for life on their new worlds or help other space travelers gradually adjust to the "gravitational rigors" of returning to Earth.

The ionizing radiation environment encountered by workers and travelers in space is characterized primarily by fluxes of electrons, protons, and energetic atomic nuclei. In low-Earth orbit (LEO) electrons and protons are trapped by Earth's magnetic fields, forming the Van Allen belts. The amount of ionizing radiation in LEO varies with solar activity. The trapped radiation belts are of concern when space-worker crews transfer from low-Earth orbit to geostationary Earth orbit (GEO) or to lunar surface bases. In geostationary locations solar particle events (SPEs) represent a major ionizing radiation threat to space workers. Throughout cislunar space and interplanetary space (beyond the protection of the Earth's magnetosphere), space travelers are also bombarded by galactic cosmic rays. These are very energetic atomic particles consisting of protons, helium nuclei, and heavy nuclei (that is, nuclei with an atomic number greater than two). Shielding, solar flare warning systems, and excellent radiation dosimetry equipment should help prevent space travelers from experiencing ionizing radiation doses in excess of the standards established for various space missions and occupations. Lunar surface bases will require extensive radiation shielding. Because of the serious but unpredictable hazard posed by an anomalously large solar particle event (ALSPE), when lunar surface workers venture an appreciable distance away from their well-shielded base they will need ready access to a temporary radiation "storm cellar." Similarly, any human-crewed Mars expedition vehicle will also require significant amounts of radiation shielding to protect against chronic exposure to galactic cosmic rays during the long interplanetary journey. The vehicle will also have to provide additional emergency shielding provisions (possibly some type of radiation-safe room) where the crew can scramble and hide for a few hours to avoid acute radiation doses from an ALSPE.

Mars expedition personnel and lunar surface base workers might also experience a variety of psychological disorders, including the solipsism syndrome and the "shimanagashi" syndrome. The solipsism syndrome is a state of mind in which a person feels that everything is a dream and is not real. It might easily be caused in an environment (such as a small space base or confined expedition vehicle) where everything is artificial or human-made. The "shimanagashi" syndrome is a feeling of isolation in which individuals begin to feel left out, even though life may be physically comfortable. Careful design of living quarters and good communication with Earth should relieve or prevent such psychological disorders. Techniques used to maintain the psychological well-being and happiness of ISS crew members provide some valuable human factor guidelines for ensuring lunar base worker or Mars expedition crew member mental fitness.

Living and working in space in the 21st century will present some interesting challenges as well as some ever-present dangers and hazards. However, the rewards of an extraterrestrial lifestyle, for certain pioneering individuals, will more than outweigh any such personal risks.

The commission also found that the decision to launch the *Challenger* on that particular day was flawed and that this flawed decision represented a contributing cause of the accident. (Launch day for the STS 51-L mission was an unseasonably cold day in Florida.) Those who made the decision to launch were unaware of the recent history of problems concerning the O-rings and the joint. They were also unaware of the initial written recommendation of the contractor advising against launch at temperatures below 11.7°C and of the continuing opposition of the engineers at Thiokol (the manufacturer of the solid rocket motors) after the management reversed its position. Nor did the decision makers have a clear understanding of the concern at Rockwell (the main NASA shuttle contractor and builder of the orbiter vehicle) that it was not safe to launch because of the ice on the launchpad. The commission concluded that if the decision makers had known all of these facts, it is highly unlikely that they would have decided to launch the STS 51-L mission on January 28, 1986.

See also COLUMBIA ACCIDENT; SPACE TRANSPORTATION SYSTEM.

chamber pressure (symbol: P_c) The pressure of gases within the combustion chamber of a rocket engine.

See also ROCKET.

chamber volume (symbol: V_c) The volume of a rocket's combustion chamber, including the convergent portion of the nozzle up to the throat.

See also NOZZLE; ROCKET.

Chandler, Seth Carlo (1846–1913) American *Astronomer* Seth Carlo Chandler's most important contribution to science was his discovery of the irregular movements of Earth's geographical poles, a phenomenon now called the *Chandler wobble*, and the 14-month oscillation of Earth's polar axis, now called the *Chandler period*. In 1891, using a special instrument he devised for relating the positions of the certain stars to a small circle at zenith, Chandler was able to verify a cyclic variation in latitude by 0.3 arcsecond with a period of 14 months (430 days).

Chandrasekhar, Subrahmanyan (Chandra) (1910–1995) Indian-American *Astrophysicist* Subrahmanyan Chandrasekhar was a brilliant Indian-American astrophysicist who made important contributions to the theory of stellar evolution—especially the role of the white dwarf as the last stage of evolution of many stars that are about the mass of the Sun. He shared the 1983 Nobel Prize in physics with WILLIAM ALFRED FOWLER for his theoretical studies of the physical processes important to the structure and evolution of stars.

Known to the world as "Chandra" (which means "luminous" or "moon" in Sanskrit), Chandrasekhar was widely recognized as one of the foremost astrophysicists of the 20th century. He was born in Lahore, India, on October 19, 1910. Chandrasekhar trained as a physicist at Presidency College, Madras, India, and then became a Ph.D. student under Sir Ralph Howard Fowler (1889–1944) at Cambridge University in England. Early in his career he demonstrated that there was an upper limit (now called the *Chandrasekhar limit*) to the mass of a white dwarf star. In 1937 Chandrasekhar immigrated from India to the United States, where he joined the faculty of the University of Chicago. He and his wife (Lalitha Doraiswamy) became American citizens in 1953. Chandrasekhar was a popular professor at the University of Chicago, and his research interests extended to nearly all branches of theoretical astrophysics. He died in Chicago on August 21, 1995.

On July 23, 1999, NASA launched the *Advanced X-Ray Astrophysics Facility* (AXAF) as part of the STS-93 mission of the space shuttle. Upon successful orbital deployment, NASA renamed this sophisticated astronomical satellite the *Chandra X-Ray Observatory* (CXO) in his honor.

See also CHANDRASEKHAR LIMIT; *CHANDRA X-RAY OBSERVATORY*.

Chandrasekhar limit In the late 1920s the astrophysicist and Nobel laureate SUBRAHMANYAN CHANDRASEKHAR used relativity theory and quantum mechanics to show that if the mass of a degenerate star is more than about 1.4 solar masses (a maximum mass called the *Chandrasekhar limit*), it will not evolve into a white dwarf star but rather will continue to collapse under the influence of gravity and become either a neutron star or a black hole or else blow itself apart in a supernova explosion.

See also STAR.

Chandra X-Ray Observatory **(CXO)** One of NASA's four great orbiting astronomical observatories. This spacecraft was successfully launched on July 23, 1999, and deployed by the astronaut crew of the space shuttle *Columbia* during the STS-93 mission. NASA renamed this sophisticated X-ray observatory the *Chandra X-Ray Observatory* to honor the brilliant Indian-American astrophysicist and Nobel laureate SUBRAHMANYAN CHANDRASEKAR. (During its development, NASA had previously called the scientific spacecraft the *Advanced X-Ray Astrophysics Facility* [AXAF].) The Earth-orbiting astronomical facility studies some of the most interesting and puzzling X-RAY sources in the universe, including emissions from active galactic nuclei, exploding stars, neutron stars, and matter falling into black holes.

Unlike the *Hubble Space Telescope* (HST), which operates in a circular orbit that is relatively close to Earth, NASA placed the *Chandra X-Ray Observatory* in a highly elliptical (oval-shaped) orbit. To achieve its final operational orbit the CXO was first carried into low-Earth orbit by the space shuttle *Columbia* and then deployed from the orbiter's cargo bay by the astronaut crew into a 250-km-altitude orbit above Earth. Next, two firings of an attached inertial upper stage (IUS) rocket, followed by several firings of its own onboard propulsion system (following separation from the IUS rocket), placed the observatory into its highly elliptical working orbit around Earth. At its closest approach to Earth, the CXO has an altitude of about 9,600 km. At its farthest orbital distance from Earth (about 140,000 km), the observatory travels almost one-third of the way to the Moon. The CXO's working orbit is characterized by a period of 64.2 hours, an inclination of 28.5°, and an eccentricity of 0.7984.

The observatory's highly elliptical orbit carries it far outside the radiation belts that surround the planet. Exposure to the charged particle radiation found within the belts could easily upset the observatory's sensitive scientific instruments and provide faulty readings. So NASA scientists use this special elliptical orbit to keep the *Chandra X-Ray Observatory* outside the troublesome radiation belts long enough to take 55 hours of uninterrupted observations during each orbit. Of course, during the periods of interference from Earth's radiation belts (about 9 hours of each orbit), scientists do not attempt to make X-ray observations.

The *Chandra X-Ray Observatory* has three major elements. (*See* figure.) They are the spacecraft system, the telescope system, and the science instruments.

The spacecraft module contains computers, communications antennas, and data recorders to transmit and receive information between the observatory and ground stations. These onboard computers and sensors, along with human assistance from CXO's ground-based control center, command and control the space vehicle and monitor its health throughout the operational lifetime (projected as a minimum of five years).

The spacecraft module also has an onboard rocket propulsion system to move and aim the entire observatory, an aspect camera that tells the observatory its position relative to the stars, and a Sun shade and sensor arrangement that protect sensitive components from excessive sunlight. Solar arrays provide electrical power to the spacecraft and also charge three nickel-hydrogen batteries, which serve as a backup electrical power supply.

At the heart of the telescope system is the high-resolution mirror assembly (HRMA). Since high-energy X-rays would penetrate a normal mirror, aerospace engineers created special cylindrical mirrors for the *Chandra X-Ray Observatory.* The X-ray telescope consists of four nested paraboloid-hyperboloid X-ray mirror pairs, arranged in concentric cylinders within the cone. Basically, the HRMA is an assembly of tubes within tubes. Incoming X-rays graze off the highly polished mirror surfaces and are funneled to the instrument section for detection and analysis. The mirrors are slightly angled so that X-rays from sources in deep space will graze off their surfaces, much like a stone skips on a pond or lake. The function of HRMA is to accurately focus cosmic source X-rays onto the imaging instruments, which are located at the other end of the 10-meter-long telescope.

The CXO's X-ray mirrors are the largest of their kind and the smoothest ever manufactured. If the surface of the state of Colorado were as relatively smooth, Pike's Peak (elevation 4,300 meters above sea level) would be less than 2.54 centimeters tall! The largest of the eight mirrors is almost 1.20 meters in diameter and 0.91 meter long. The HRMA is contained in the cylindrical "telescope" portion of the obser-

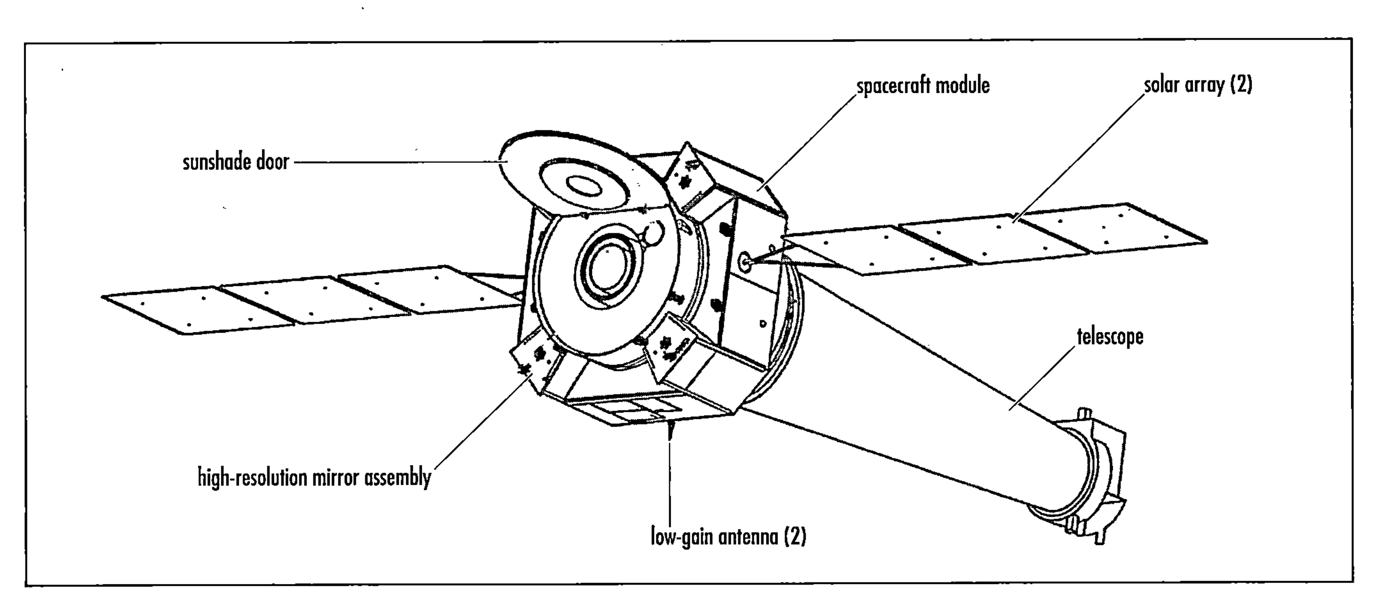

Chandra X-Ray Observatory* (CXO) *(Drawing courtesy of NASA)

vatory. The entire length of the telescope is covered with reflective multilayer insulation that assists heating elements inside the unit maintain a constant internal temperature. By maintaining a precise internal temperature, the mirrors within the telescope are not subjected to expansion or contraction, thereby ensuring greater accuracy in scientific observations at X-ray portions of the electromagnetic spectrum. With its combination of large mirror area, accurate alignment, and efficient X-ray detectors, the *Chandra X-Ray Observatory* has eight times greater resolution and is 20 to 50 times more sensitive than any previous X-ray telescope flown in space.

Within the instrument section of the observatory, two instruments at the narrow end of the telescope structure collect X-rays and study them in various ways. Each of the CXO's instruments can serve as an imager or a spectrometer. The high-resolution camera (HRC) records X-ray images, providing astrophysicists and astronomers an important look at violent, high-temperature celestial phenomena such as the death of stars and colliding galaxies.

The high-resolution camera consists of two clusters of 69 million tiny lead-oxide glass tubes. These lead-oxide tubes are only 0.127 cm long and just one-eight the thickness of a human hair. When an X-ray strikes the tubes, electrons are released. As these charged particles are accelerated down the tubes by an applied high voltage, they cause an avalanche that involves perhaps 30 million more electrons in the process. A grid of electrically charged wires at the end of the tube bundle detects this flood of particles and allows the position of the original X-ray to be precisely determined. The high-resolution camera complements the CXO's charge-coupled device imaging spectrometer (ACIS).

The observatory's imaging spectrometer is also located at the narrow end of the observatory. This detector is capable of recording not only the position, but also the energy level (or the "color"), of the incoming X-rays from cosmic sources. The imaging spectrometer consists of 10 charge-coupled device (CCD) arrays. These detectors are similar to those found in home video recorders and digital cameras but are designed to detect X-rays (as opposed to photons of visible light). Commands from the ground allow astronomers to select which of the detectors to use. The imaging spectrometer can distinguish up to 50 different energies within the range in which the observatory operates—that is, it can detect X-rays ranging from 0.09 to 10.0 kiloelectrons (keV).

In order to gain even more energy information, two screenlike instruments, called diffraction gratings, can be inserted into the path of the X-rays between the telescope and the detectors. The gratings change the path of the incident X-ray, and the CXO's X-ray cameras record its energy level ("color") and position. How much the path of an X-ray photon changes depends on its initial energy level. One grating (called the high-energy transmission grating [HETG]) concentrates on X-rays of higher and medium energies and uses the imaging spectrometer as a detector. The other grating (called the low-energy transmission grating [LETG]) disperses low-energy X-rays, and scientists use it in conjunction with the high-resolution camera.

By studying the X-ray spectra collected by the CXO and then recognizing the characteristic X-ray signatures of known elements, scientists can determine the composition of X-ray-producing objects and how these X-rays are produced. The principal science mission objectives of the *Chandra X-Ray Observatory* are to determine the nature of celestial objects from normal stars to quasars, to understand the nature of physical processes that take place in and between astronomical objects, and to help scientists study the history and evolution of the universe. In particular, the spacecraft is making observations of the cosmic X-rays from high-energy regions of the universe, such as supernova remnants, X-ray pulsars, black holes, neutron stars, and hot galactic clusters.

NASA's Marshall Space Flight Center (MSFC) has overall responsibility for the CXO mission. Under an agreement with NASA the Smithsonian Astrophysical Observatory (SAO) controls science and flight operations of the observatory from Cambridge, Massachusetts. The Smithsonian manages two electronically linked facilities, the Operations Control Center and the Science Center.

The CXO Operations Control Center is responsible for directing the observatory's mission as it orbits Earth. A control center team interacts with the CXO three times a day—receiving science and housekeeping information from its recorders. The control center team also sends new instructions to the observatory as needed. Finally, the control center team transmits scientific information from the X-ray observatory to the CXO Science Center.

The Science Center is an important resource for scientists who wish to investigate X-ray-emitting objects such as quasars and colliding galaxies. The science center team provides user support to qualified researchers throughout the scientific community. This support primarily involves the processing and archiving of the *Chandra X-Ray Observatory*'s science data.

See also ASTROPHYSICS; X-RAY ASTRONOMY.

change detection The practice in remote sensing applications of comparing two digitized images of the same scene that have been acquired at different times. By comparing the intensity differences (either graytone or natural color differences) between corresponding pixels in the time-separated images of the scene, interesting information concerning change and activity (i.e., vegetation growth, urban sprawl, migrating environmental stress in crops, etc.) can be detected quickly in a semiautomated, quasi-empirical fashion.

See also REMOTE SENSING.

channel construction Use of machined grooves in the wall of the nozzle or combustion chamber of a rocket to form coolant passages.

See also ROCKET.

chaos theory The branch of mathematical physics that studies unstable systems. First developed for use in the field of meteorology, chaos theory now finds applications in many fields involving dynamic systems whose true behaviors are very difficult to predict because they contain so many variables or unknown factors. The so-called *butterfly* *effect* provides a brief glimpse of *chaos*—or the apparently random and unpredictable behavior that sometimes governs systems that should otherwise be controlled by deterministic scientific laws. Scientists have developed and like to use laws that under perfect (ideal) conditions allow them to predict the future states of a given system from data and conditions regarding its present or past state. Unfortunately, real-world systems are often less than ideal, and a variety of discrepancies and errors creep into the best scientific laws. Nowhere is this chaotic behavior in nature more obvious to both scientists and laypersons alike than in the field of meteorology.

Weather on Earth is influenced by many variables—some obvious and some subtle and not so obvious. That is why meteorologists often describe Earth's weather system as chaotic. Weather scientists have even tried to emphasize how sensitive their forecasting equations are to the initial conditions by (somewhat humorously) stating, "when a butterfly flaps its wings in South America, its movement through the atmosphere helps determine whether a tornado forms in Kansas." Physicists also recognize that the motion of celestial objects (including planetary systems) is subject to unpredictable perturbations and chaotic behavior. For example, the random and unanticipated passage of a distant rogue star might sufficiently disturb the Oort cloud to allow one or more comets to eventually wander into the inner solar system and possibly smash into a planet. Such random acts within the apparently "clockwork" behavior of the universe introduce uncertainty and possibly enormous change in the otherwise orderly and predictable behavior of dynamic systems large and small.

See also CHAOTIC ORBIT.

chaotic orbit The orbit of a celestial body that changes in a highly unpredictable manner, usually when a small object, such as an asteroid or comet, passes close to a massive planet (such as Saturn or Jupiter) or the Sun. For example, both Uranus and Saturn influence the chaotic orbit of Chiron.

chaotic terrain A planetary surface feature (first observed on Mars in 1969) that is characterized by a jumbled, irregular assembly of fractures and blocks of rock.

charge-coupled device (CCD) An electronic (solid state) device containing a regular array of sensor elements that are sensitive to various types of electromagnetic radiation (e.g., light or X-rays) and emit electrons when exposed to such radiation. The emitted electrons are collected and the resulting charge analyzed. CCDs are used as the light-detecting component in modern television cameras and telescopes. Scientists use CCDs in spacecraft imaging systems in support of planetary exploratory or high-energy astrophysics. NASA's *Galileo* spacecraft, for example, used a solid state imagining system that contained a CCD with an 800 × 800 pixel array.

charged particle An ion; an elementary particle that carries a positive or negative electric charge, such as an electron, a proton, or an alpha particle.

charged particle detector A device that counts and/or measures the energy of charged particles; frequently found as part of the scientific instrument payload on planetary flyby and deep-space exploration missions.

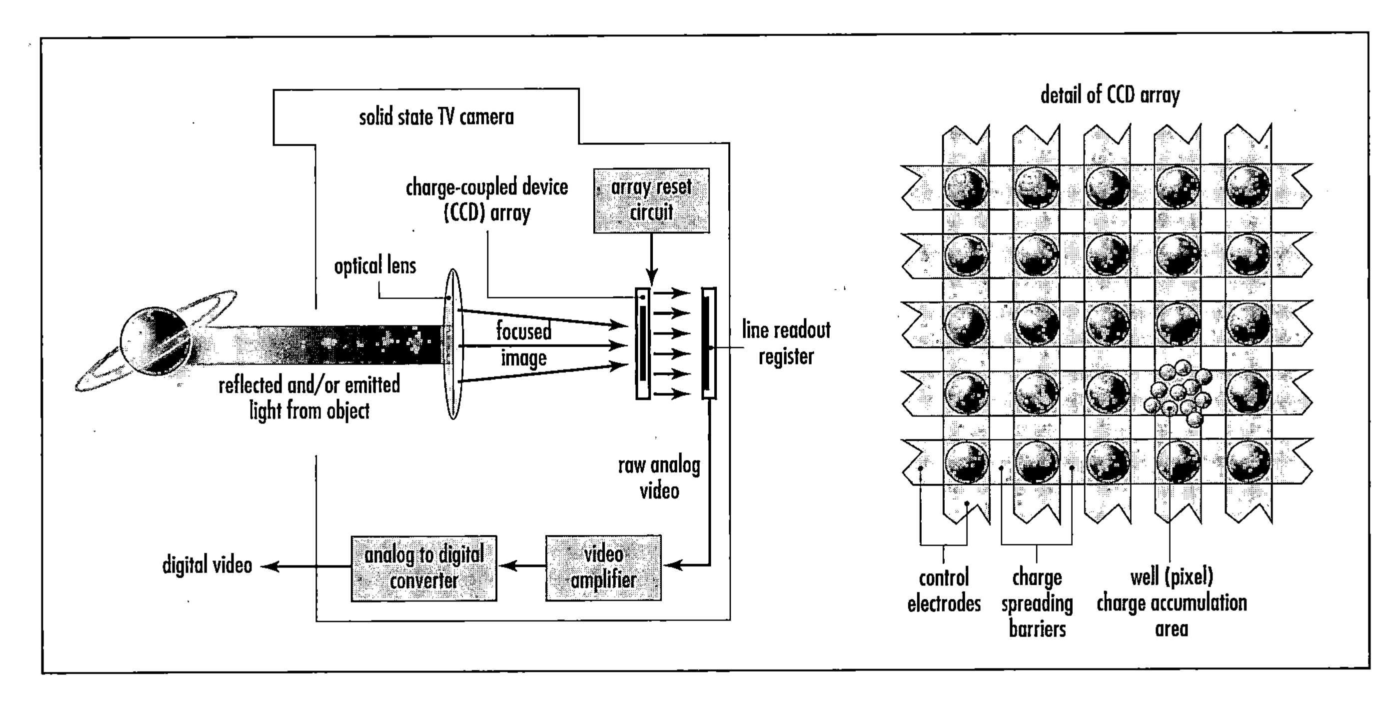

Diagram for a typical solid-state imaging system with charge-coupled device (CCD) array ***(Drawing courtesy of NASA/JPL)***

Charon The large (about 1,190-km diameter) moon of Pluto discovered in 1978 by the American astronomer James Walter Christy (b. 1938). It orbits Pluto every 6.387 days at a mean distance of 19,600 km and keeps one face permanently toward the planet.

See also PLUTO.

charring ablator An ablation material characterized by the formation of a carbonaceous layer at the heated surface that impedes thermal energy (heat) flow into the material by virtue of its insulating and radiation heat transfer characteristics.

See also ABLATION.

chaser spacecraft The spacecraft or aerospace vehicle that actively performs the key maneuvers during orbital rendezvous and docking/berthing operations. The other space vehicle serves as the target and remains essentially passive during the encounter.

chasma A canyon or deep linear feature on a planet's surface.

checkout The sequence of actions (such as functional, operational, and calibration tests) performed to determine the readiness of a spacecraft or launch vehicle to perform its intended mission.

chemical energy Energy liberated or absorbed in the process of a chemical reaction. In such a reaction energy losses or gains usually involve only the outermost electrons of the atoms or ions of the system undergoing change; here a chemical bond of some type is established or broken without disrupting the original atomic or ionic identities of the constituents. Chemical changes, according to the nature of the substance entering into the change, may be induced by thermal (thermochemical), light (photochemical), or electrical (electrochemical) energies.

See also PROPELLANT.

chemical rocket A rocket that uses the combustion of a chemical fuel in either solid or liquid form to generate thrust. The chemical fuel requires an oxidizer to support combustion.

See also PROPELLANT; ROCKET.

chilldown Cooling all or part of a cryogenic (very cold) rocket engine system from ambient (room) temperature down to cryogenic temperature by circulating cryogenic propellant (fluid) through the system prior to engine start.

See also CRYOGENIC PROPELLANT.

Chiron An unusual celestial body in the outer solar system with a chaotic orbit that lies almost entirely between the orbits of Saturn and Uranus. The American astronomer Charles Thomas Kowal (b. 1940) discovered Chiron in November 1, 1977, using a photographic plate taken several weeks earlier on October 18. This massive asteroid-sized object has a diameter that lies between 148 km and 208 km and an estimated mass between 2×10^{18} and 1×10^{19} kg. Chiron's highly elliptical orbit (eccentricity of 0.383) around the Sun has a perihelion value of 8.46 AU and an aphelion value of about 18.9 AU. Chiron's orbit is further characterized by a period of 50.7 years and an inclination of 6.94°. Chiron has a rotation period of about 5.9 hours and is the first object to be placed in the *Centaurs group*. In addition to exhibiting properties characteristic of an icy asteroid or minor planet, this intriguing object also has a detectable coma, a feature characteristic of comets. Also know as Minor Planet (2060) Chiron and Comet 95P/Chiron.

See also ASTEROID; CENTAURS; COMET.

Chladni, Ernst Florens Friedrich (1756–1827) German *Physicist* Ernst Chladni was a physicist who suggested in 1794 that meteorites were of extraterrestrial origin, possibly the debris from a planet that exploded in ancient times. In 1819 he also postulated that there might be a physical relationship between meteors and comets.

See also METEOROIDS.

choked flow A flow condition in a duct or pipe such that the flow upstream of a certain critical section (such as a nozzle or valve) cannot be increased by further reducing downstream pressure.

chondrite A stony meteorite that is characterized by the presence of small, nearly round silicate granules called *chondrules*.

See also METEORITE.

Christie, William Henry Mahoney (1845–1922) British *Astronomer* From 1881 to 1910 William Christie served as England's eighth Astronomer Royal. Under his leadership the role of the Royal Greenwich Observatory (RGO) grew in both precision astronomical measurements and in reputation as a leading institution for research. He acquired improved new telescopes for the facility and expanded its activities to include spectroscopy and astrophotography.

chromatic aberration A phenomenon that occurs in a refracting optical system because light of different wavelengths (colors) is refracted (bent) by a different amount. As a result, a simple lens will give red light a longer focal length than blue light.

See also ABERRATION.

chromosphere The reddish layer in the Sun's atmosphere located between the photosphere (the apparent solar surface) and the base of the corona. It is the source of solar prominences.

See also SUN.

chronic radiation dose A dose of ionizing radiation received either continuously or intermittently over a prolonged period of time.

See also ACUTE RADIATION SYNDROME.

Chryse Planitia A large plain on Mars characterized by many ancient channels that could have once contained

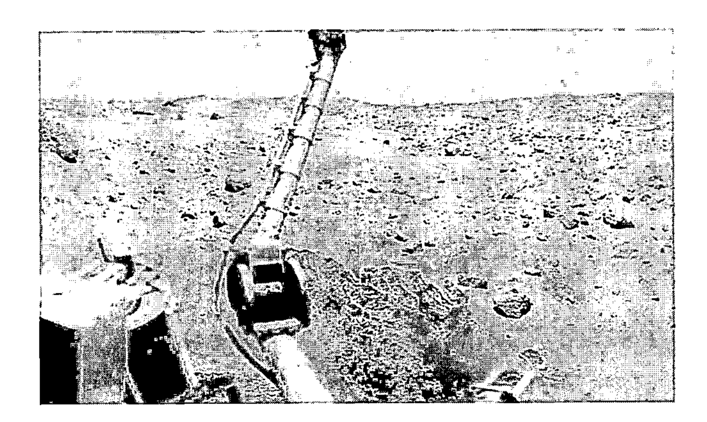

This is a close look at the relatively smooth region of Mars known as Chryse Planitia (the Plains of Gold). The *Viking 1* lander spacecraft took this image of the Martian surface on August 8, 1978. Parts of the robot spacecraft are visible in the foreground. The image shows a field of accumulated dust on the left side and a relatively smooth, rocky plain extending to the horizon, which is a few kilometers distant. Most of the rocks in the scene measure about 0.5 meter across. However, the large rock on the left, nicknamed "Big Joe" by NASA mission scientists, is about 2.5 meters across and 8 meters away from the lander. *(Courtesy of NASA/JPL)*

flowing surface water. Landing site for NASA's *Viking 1* lander spacecraft in July 1976.

chuffing Characteristic of some rockets to burn intermittently and with an irregular noise.

See also ROCKET.

chugging Form of combustion instability that occurs in a liquid-propellant rocket engine. It is characterized by a pulsing operation at a fairly low frequency.

See also ROCKET.

circadian rhythms A biological organism's day/night cycle of living; a regular change in physiological function occurring in approximately 24-hour cycles.

circularize To change an elliptical orbit into a circular one, usually by firing a rocket motor when the space vehicle is at apogee.

See also APOGEE; APOGEE MOTOR.

circular velocity At any specific distance from the primary (i.e., central) body, the orbital velocity required to maintain a constant-radius orbit. For a circular orbit around a primary body of radius R, the circular velocity v is given by the following equation, which comes from Newtonian mechanics: $v = \sqrt{(g R^2 / r)}$, where g is the acceleration of gravity (free fall) at the surface of the primary body and r is the distance between the orbiting body and the center of mass of the two-body system. If the primary body is much more massive than the satellite (secondary), then the value of r generally represents the radial distance from the center of the massive primary to the circular orbit of the satellite. Consider a satellite orbiting Earth at an altitude of 200 kilometers. The relatively simple equation presented above yields a circular velocity of approximately 7.8 km/s. More sophisticated mathematical treatments are needed to describe orbital velocities for objects traveling in elliptical orbits. However, the modest equation presented here (which is based on the assumption of a circular orbit and several other idealizations) provides a reasonably approximate value of the orbital velocity of a spacecraft moving in a circular orbit around its primary at a fixed radial distance from that primary body.

See also NEWTON'S LAW OF GRAVITATION; NEWTON'S LAWS OF MOTION; ORBITS OF OBJECTS IN SPACE.

circumlunar Around the Moon; a term generally applied to trajectories.

circumsolar space Around the Sun; heliocentric (Sun-centered) space.

circumstellar Around a star, as opposed to *interstellar*, between the stars.

cislunar Of or pertaining to phenomena, projects, or activities happening in the region of outer space between Earth and the Moon. It comes from the Latin word *cis*, meaning "on this side," and *lunar*, which means "of or pertaining to the Moon." Therefore, it means "on this side of the Moon." By convention, the outer limit of cislunar space is usually considered to be the outer limit (from the Moon) of its sphere of gravitational influence.

Clairaut, Alexis-Claude (1713–1765) French *Mathematician, Astronomer* The 18th-century French astronomer Alexis-Claude Clairaut accurately calculated the perihelion date for the 1759 return of Comet Halley. In a brilliant application of SIR ISAAC NEWTON's physical principles, Clairaut compensated for the gravitational influence of both Saturn and Jupiter on the comet's trajectory. His detailed calculations predicted a perihelion date of April 13, 1759, while the observed perihelion date for this famous comet was March 14, 1759. Prior to this activity Clairaut participated in an expedition to Lapland (northern Scandinavia) to collect geophysical evidence of the oblateness (flattened poles) of a rotating Earth.

Clark, Alvan (1804–1887) American *Optician, Telescope Maker* Alvan Clark founded the famous 19th-century telescope manufacturing company Alvan Clark & Sons in Cambridge, Massachusetts. In 1862, while testing a 0.46-meter-diameter (18-inch) objective (lens) refracting telescope, his son (ALVAN GRAHAM CLARK) pointed the new instrument at the bright star Sirius and detected its faint companion, Sirius B, a white dwarf. In 1877 the American astronomer ASAPH HALL used another of Clark's telescopes, the 0.66-meter (26-inch) refracting telescope at the U.S. Naval Observatory, to discover the two tiny moons of Mars, which he named Phobos and Deimos.

Clark, Alvan Graham (1832–1897) American *Optician, Telescope Maker* While testing a new 0.46-meter (18-inch) objective (lens) refracting telescope in 1862 for his father, ALVAN CLARK, Alvan Graham Clark pointed the instrument at the bright star Sirius and discovered its white dwarf companion, Sirius B. Previously, in 1844, FRIEDRICH WILHELM BESSEL had examined the irregular motion of Sirius and hypothesized that the star must have a dark (unseen) companion. The younger Clark's discovery of Sirius B earned him an award from the French Academy of Sciences and represented the first example of the important small dense celestial object that is the end product of stellar evolution for all but the most massive stars.

See also WHITE DWARF.

Clarke, Sir Arthur C. (1917–) British *Science Writer, Technical Visionary* The British science writer and science fiction author Sir Arthur C. Clarke is widely known for his enthusiastic support of space exploration. In 1945 he published the technical article "Extra-Terrestrial Relays," in which he predicted the development of communications satellites that operated in geostationary orbit. Then, in 1968 he worked with film director Stanley Kubrick in developing the movie version of his book *2001: A Space Odyssey.* This motion picture is still one of the most popular and realistic depictions of human spaceflight across vast interplanetary distances. Honoring his lifelong technical accomplishments, Queen Elizabeth II of England knighted Clarke in 1998.

Born in Minehead, Somerset, England, on December 16, 1917, Clarke is one of the most celebrated science fiction/space fact authors of all time. In 1936 he moved to London and joined the British Interplanetary Society (BIS), the world's longest-established organization devoted exclusively to promoting space exploration and astronautics. During World War II Clarke served as a radar instructor in the Royal Air Force (RAF). Immediately after the war he published the pioneering technical paper "Extra-Terrestrial Relays" in the October 1945 issue of *Wireless World.* Clarke proved to be a superb technical prophet, because in this paper he described the principles of satellite communications and recommended the use of geostationary orbits for a global communications system. He received many awards for this innovative recommendation, including the prestigious Marconi International Fellowship (1982). To recognize his contribution to space technology, the International Astronomical Union (IAU) named the geostationary orbit at 35,900 km altitude above Earth's surface the *Clarke orbit.*

In 1948 he received a bachelor's degree in physics and mathematics from King's College in London. Starting in 1951 Clarke discovered he could earn a living writing about space travel. He wrote factual technical books and articles about rocketry and space as well as prize-winning science fiction novels that explored the impact of space technology on the human race. In 1952 Clarke published a classic "science fact" book, *The Exploration of Space,* in which he suggested the inevitability of spaceflight. He also predicted that space exploration would occur in seven distinct phases, the last six of which involved human participation. In this pioneering book Clarke mentioned that the exploration of the Moon would serve as a stepping stone for human voyages to Mars and Venus. According to Clarke's forecasted milestones, when humans landed on Mars and Venus, the first era of interplanetary flight would come to a close. Clarke ended this book with an even bolder, far-reaching vision, namely, that humanity would eventually contact (or be contacted by) intelligent life from beyond the solar system.

The extraterrestrial contact hypothesis has served as an interesting "technical prophecy" connection between many of Clarke's science fact and science fiction works. For example, in 1953 Clarke published the immensely popular science fiction novel *Childhood's End.* This pioneering novel addressed the consequences of Earth's initial contact with a superior alien civilization that decided to help humanity "grow up." Another of Clarke's science fiction best sellers was *2001: A Space Odyssey* (1964). Four years after the book's publication film director Stanley Kubrick worked closely with Clarke to create the Academy Award–winning motion picture of the same title. Clarke subsequently published several sequels entitled *2010: Odyssey Two* (1982), *2061: Odyssey Three* (1988), and finally *3001: The Final Odyssey* (1997).

In 1962 Clarke wrote another very perceptive and delightful book, *Profiles of the Future.* In this nonfiction work he explored the impact of technology on society and how a technical visionary often succeeded or failed within a particular organization or society. To support his overall theme, Clarke introduced his three (now famous) laws of technical prophecy. Clarke's third law is especially helpful in attempting to forecast future developments in space technology far into this century and beyond. Clarke's third law of technical prophecy suggests that "Any sufficiently advanced technology is indistinguishable from magic."

Since 1956 Clarke has resided in Colombo, Sri Lanka (formerly Ceylon). A dedicated space visionary, he has consistently pleased and informed millions of people around the planet with his prize-winning works of both science fact and science fiction. Many of his more than 60 books have predicted the development and consequences of space technology. His long-term space technology projections continue to suggest some very exciting pathways for human development among the stars.

Clarke orbit A geostationary orbit; named after SIR ARTHUR C. CLARKE, who first proposed in 1945 the use of this special orbit around Earth for communications satellites.

Clausius, Rudolf Julius Emmanuel (1822–1888) German *Physicist* By introducing the concept of entropy in 1865, the German theoretical physicist Rudolf Clausius completed the development of the first comprehensive understanding of the second law of thermodynamics—a scientific feat that allowed 19th-century cosmologists to postulate an interesting future condition (or end state) of the universe they called the *heat death of the universe.*

Clausius was born on January 2, 1822, in Köslin, Prussia (now called Koszalin, in modern-day Poland). His father, the Reverend C. E. G. Clausius, served as a member of the Royal Government School Board and taught at a small private school where Clausius attended primary school. In 1840 he entered the University of Berlin as a history major but emerged in 1844 with a degree in mathematics and physics.

Following graduation Clausius taught mathematics and physics at the Frederic-Werder Gymnasium (secondary school). During this period he also pursued graduate studies and by 1848 earned a doctorate in mathematical physics from the University of Halle. Then, in 1850 he accepted a position as a professor at the Royal Artillery and Engineering School in Berlin.

As a professor Clausius published his first and what is now regarded as his most important paper discussing the mechanical theory of heat. The famous paper, entitled "Über die bewegende Kraft der Wärmer" ("On the motive force of heat"), provided the first unambiguous statement of the second law of thermodynamics. In this paper, read at the Berlin Academy on February 18, 1850, and then published in its *Annalen der Physik* (Annals of physics) later that year, Clausius built upon the heat engine theory of the French engineer and physicist Sadi Carnot (1796–1832). Clausius introduced his version of the second law of thermodynamics with a simple yet important statement: "Heat does not spontaneously flow from a colder body to a warmer body." While there is one basic statement of the first law of thermodynamics, namely the conservation of energy principle, there are quite literally several hundred different yet equally appropriate statements of the second law. Clausius's perceptive statement and insights, as expressed in 1850, leads the long and interesting parade of second law statements. This particular paper is often regarded as the foundation of classical thermodynamics.

In the 19th century many physicists, including the famous British natural philosopher BARON WILLIAM THOMSON KELVIN, were also grappling with the nature of heat and trying to develop basic mathematical relationships to describe and predict how thermal energy flows through the universe and becomes transformed into mechanical work. Clausius is generally credited as the person who most clearly described the nature, role, and importance of the second law, although others, such as Lord Kelvin, were simultaneously involved in the overall quest for understanding the mechanics of thermophysics.

Between 1850 and 1865 Clausius published a number of other papers dealing with mathematical statements of the second law. His efforts climaxed in 1865, when he introduced the concept of entropy as a thermodynamic property, which he used to describe the availability of thermal energy (heat) for performing mechanical work. In this 1865 paper Clausius considered the universe to be a closed system (that is, a system that has neither energy nor mass flowing across its boundary) and then succinctly tied together the first and second laws of thermodynamics with the following elegant statement: "The energy of the universe is constant (first law principle); the entropy of the universe strives to reach a maximum value (second law principle)."

By introducing the concept of entropy in 1865, Clausius provided scientists with a convenient mathematical way of understanding and expressing the second law of thermodynamics. In addition to its tremendous impact on classical thermodynamics, his pioneering work also had an enormous influence on 19th-century cosmology. As previously stated, Clausius assumed that the total energy of the universe (taken as a closed system) was constant, so the entropy of the universe must then strive to achieve a maximum value in accordance with the laws of thermodynamics. According to this model, the end state of the universe is one of complete temperature equilibrium, with no energy available to perform any useful work. Cosmologists refer to this condition as the heat death of the universe.

In 1855 Clausius accepted a position at the University of Zurich and in 1859 married his first wife, Adelheid Rimpam. She bore him six children but then died in 1875, while giving birth to the couple's last child. Although Clausius enjoyed teaching in Zurich, he longed for Germany and returned to his homeland in 1867 to accept a professorship at the University of Würzburg. He moved on to become a professor of physics at the University of Bonn in 1869, where he taught for the rest of his life. At almost 50 years of age, he organized some of his students into a volunteer ambulance corps for duty in the Franco-Prussian War (1870–71). Clausius was wounded in action and received the Iron Cross in 1871 for his services to the German army. He remarried in 1886, and his second wife, Sophie Stack, bore him a son. He continued to teach up until his death on August 24, 1888, in Bonn.

Despite a certain amount of nationalistically inspired international controversy over who deserved credit for developing the second law of thermodynamics, Clausius left a great legacy in theoretical physics and earned the recognition of his fellow scientists throughout Europe. In 1868 he was elected a foreign fellow of the Royal Society of London and received that society's prestigious Copley Medal in 1879. He also was awarded the Huygens Medal in 1870 by the Holland Academy of Sciences and the Poncelet Prize in 1883 by the French Academy of Sciences. The University of Würzburg bestowed an honorary doctorate on him in 1882.

Clavius, Christopher (1537–1612) German *Astronomer, Mathematician* Christopher Clavius was a Jesuit mathematician and astronomer who spearheaded the reform of the Julian calendar. His efforts enabled Pope Gregory XIII to introduce a new calendar, now called the Gregorian calendar, in 1852. Today the Gregorian calendar is widely used throughout the world as the international civil calendar.

See also CALENDAR.

clean room A controlled work environment for spacecraft and aerospace systems in which dust, temperature, and humidity are carefully controlled during the fabrication, assembly, and/or testing of critical components.

See also BIOCLEAN ROOM.

Clementine A joint project between the Department of Defense and NASA. The objective of the mission, also called the Deep Space Program Science Experiment, was to test sensors and spacecraft components under extended exposure to the space environment and to make scientific observations of the Moon and the near-Earth asteroid 1,620 Geographos. The observations included imaging at various wavelengths, including ultraviolet (UV) and infrared (IR), laser altimetry, and charged particle measurements. The *Clementine* spacecraft was launched on January 25, 1994, from Vandenburg Air Force Base in California aboard a Titan II-G rocket. After two

Earth flybys, lunar insertion was achieved on February 21, 1994. Lunar mapping took place over approximately two months in two parts. The first part consisted of a five-hour elliptical polar orbit with a perilune of about 400 kilometers at 28 degrees south latitude. After one month of mapping the orbit was rotated to a perilune of 29 degrees north latitude, where it remained for one more month. After leaving lunar orbit a malfunction in one of the onboard computers occurred on May 7, 1994. This malfunction caused a thruster to fire until it had used up all its propellant, leaving the spacecraft spinning at about 80 revolutions per minute (RPM) with no spin control. This made the planned continuation of the mission, a flyby of the asteroid 1,620 Geographos, impossible. The spacecraft remained in geocentric orbit and continued testing components until the end of mission.

climate change The long-term fluctuations in temperature, precipitation, wind, and all other aspects of Earth's weather. External processes, such as solar irradiance variations, variations of Earth's orbital parameters (e.g., eccentricity, precession, and inclination), lithosphere motions, and volcanic activity, are factors in climatic variation. Internal variations of the climate system also produce fluctuations of sufficient magnitude and variability to explain observed climate change through the feedback processes that interrelate the components of the climate system.

See also GLOBAL CHANGE.

clock An electronic circuit, often an integrated circuit that produces high-frequency timing signals. A common application is synchronization of the operations performed by a computer or microprocessor-based system. Typical clock rates in microprocessor circuits are in the megahertz range, with 1 megahertz (1 MHz) corresponding to 1 million cycles per second (10^6 cps). A *spacecraft clock* is usually part of the command and data handling subsystem. It meters the passing time during the life of the spacecraft and regulates nearly all activity within the spacecraft. It may be very simple (e.g., incrementing every second and bumping its value up by one), or it may be much more complex (with several main and subordinate fields of increasing resolution). In aerospace operations many types of commands that are uplinked to the spacecraft are set to begin execution at specific spacecraft clock counts. In downlinked telemetry spacecraft clock counts (which indicate the time a telemetry frame was created) are included with engineering and science data to facilitate processing, distribution, and analysis.

See also SPACECRAFT; TELEMETRY.

clock star A bright star generally located near the celestial equator with an accurately known position (that is, right ascension) and proper motion. Astronomers use a clock star to determine time precisely and to detect any subtle errors within observatory clocks.

closed binary *See* BINARY STAR SYSTEM.

closed ecological life support system (CELSS) A system that can provide for the maintenance of life in an isolated living chamber or facility through complete reuse of the materials available within the chamber or facility. This is accomplished, in particular, by means of a cycle in which exhaled carbon dioxide, urine, and other waste matter are converted chemically or by photosynthesis into oxygen, water, and food. On a grand (macroscopic) scale, the planet Earth itself is a closed ecological system; on a more modest scale, a "self-sufficient" space base or space settlement also would represent an example of a closed ecological system. In this case, however, the degree of "closure" for the space base or space settlement would be determined by the amount of makeup materials that had to be supplied from Earth.

Material recycling in a life support system can be based on physical and chemical processes, can be biological in nature, or can be a combination of both. Chemical and physical systems are designed more easily than are biological systems but provide little flexibility or adaptability to changing needs. A life support system based solely on physical and chemical methods also would be limited because it would still require resupply of food and some means of waste disposal. A *bioregenerative* life support system incorporates biological components in the creation, purification, and renewal of life support elements. Plants and algae are used in food production, water purification, and oxygen release. While the interactions of the biomass with the environment are very complex and dynamic, creating a fully closed ecological system—one that needs no resupply of materials (although energy can cross its boundaries, as does sunlight into Earth's biosphere)—appears possible and even essential for future permanently inhabited human bases and settlements within the solar system. Life scientists call a human-made closed ecological system that involves a combination of chemical, physical, and biological processes a *closed ecological life support system* (CELSS). This type of closed ecological system is sometimes called a *controlled ecological life support system.*

closed system 1. In thermodynamics, a system in which no transfer of mass takes place across its boundaries. *Compare with* OPEN SYSTEM. 2. A closed ecological system. *See also* CLOSED ECOLOGICAL LIFE SUPPORT SYSTEM. 3. In mathematics, a system of differential equations and supplementary conditions such that the values of all the unknowns (dependent variables) of the system are mathematically determined for all values of the independent variables (usually space and time) to which the system applies. 4. A system that constitutes a feedback loop so that the inputs and controls depend on the resulting output.

closed universe The model in cosmology that assumes the total mass of the universe is sufficiently large that one day the galaxies will slow down and stop expanding because of their mutual gravitational attraction. At that time the universe will have reached its maximum size, and then gravitation will make it slowly contract, ultimately collapsing to a single point of infinite density (sometimes called the BIG CRUNCH). Some advocates of the closed universe model also speculate that after the big crunch a new explosive expansion (that is, another BIG BANG) will occur as part of an overall pulsating (or oscillating) universe cycle. Also called *bounded universe model. Compare with* OPEN UNIVERSE.

See also ASTROPHYSICS; COSMOLOGY.

close encounter (CE) An interaction with an unidentified flying object (UFO).

See also UNIDENTIFIED FLYING OBJECT.

closest approach 1. The point in time and space when two planets or other celestial bodies are nearest to each other as they orbit about the Sun or other primary. 2. The point in time and space when a scientific spacecraft on a flyby encounter trajectory is nearest the target celestial object.

closing rate The speed at which two bodies approach each other—as, for example, two spacecraft "closing" for an orbital rendezvous and/or docking operation.

cluster (rocket) Two or more rocket motors bound together so as to function as one propulsion unit.

See also ROCKET.

cluster (stellar) A gravitationally bound collection of stars that were formed from the same giant cloud of interstellar gas.

See also GLOBULAR CLUSTER; OPEN CLUSTER.

cluster of galaxies An accumulation of galaxies that lie within a few million light-years of one another and are bound by gravitation. Galactic clusters can occur with just a few member galaxies (say 10 to 100), such as the Local Group, or they can occur in great groupings involving thousands of galaxies. At an estimated distance of 48.9 million light-years (15 megaparsecs), the Virgo Cluster is the nearest large cluster. It is an irregular cluster of galaxies in the constellation Virgo and contains approximately 2,500 galaxies.

When observed optically, these clusters of galaxies are dominated by the collective (visible) light emissions from their individual galaxies. However, when viewed in the X-ray region of the electromagnetic spectrum, most of the X-ray emissions from a particular cluster appear to come from hot gaseous material that lies between the galaxies in the cluster. Some astrophysicists speculate, therefore, that there is about as much of this hot gaseous material between the galaxies in a cluster as there is matter in the galaxies themselves.

See also GALAXY; LOCAL GROUP; VIRGO CLUSTER.

coasting flight In rocketry, the flight of a rocket between burnout or thrust cutoff of one stage and ignition of another, or between burnout and summit altitude or maximum horizontal range.

coated optics Optical elements (such as lenses, prisms, etc.) that have their surfaces covered with a thin transparent film to minimize light loss due to reflection.

coaxial injector Type of liquid rocket engine injector in which one propellant surrounds the other at each injection point.

See also ROCKET.

coelostat A device located on Earth's surface that can continuously collect light from the same area of the sky despite the apparent east-to-west rotation of the celestial sphere. Generally, the coelostat consists of a plane mirror driven by an electrical motor or mechanical clock mechanism that allows it to rotate from east to west, compensating for Earth's rotation. Driven at the proper speed, this mirror reflects the same area of the sky into the field of view of an astronomical telescope or other optical instrument kept in a stationary position.

See also HELIOSTAT.

coherence 1. The matching, in space (*transverse coherence*) or time (*temporal coherence*), of the wave structure of different parallel rays of a single frequency of electromagnetic radiation. This results in the mutual reinforcement of the energy of these different components of a larger beam. Properly designed lasers and radar systems produce coherent radiation. 2. The process of having a spacecraft telecommunications subsystem generate a downlink signal that is phase-coherent to the uplink signal it receives. The resulting spacecraft downlink signal, based on and coherent with an uplink signal, has extraordinary high-frequency stability and can be used for precisely tracking the spacecraft and for carrying out scientific experiments.

See also ELECTROMAGNETIC RADIATION; TELECOMMUNICATIONS.

coherent light The physical state in which light waves are in phase over the time period of interest. Light travels in discrete bundles of energy called photons. Each photon may be treated like an ocean wave. If all the waves are in phase, they are said to be coherent. When light is coherent, the effects of each photon build on top of the others. A laser produces coherent light and therefore can concentrate energy.

cold-flow test The thorough testing of a liquid-propellant rocket engine without actually firing (igniting) it. This type of test helps aerospace engineers verify the performance and efficiency of a propulsion system, since all aspects of propellant flow and conditioning, except combustion, are examined. Tank pressurization, propellant loading, and propellant flow into the combustion chamber (without ignition) are usually included in a cold-flow test. *Compare with* HOT FIRE TEST.

See also ROCKET.

coldsoak The exposure of a system or equipment to low temperature for a long period of time to ensure that its temperature is lowered to that of the surrounding atmosphere or operational environment.

cold war The ideological conflict between the United States and the former Soviet Union from approximately 1946 to 1989 involving rivalry, mistrust, and hostility just short of overt military action. The tearing down of the Berlin Wall in November 1989 generally is considered the (symbolic) end of the cold war period.

collimator A device for focusing or confining a beam of particles or electromagnetic radiation, such as X-ray photons.

Astronaut Eileen Marie Collins commanded the STS-114 mission of the space shuttle *Discovery,* July 26–August 9, 2005. *(Courtesy of NASA)*

Collins, Eileen Marie (1956–) American *Astronaut, U.S. Air Force Officer* Eileen Collins was the first woman to serve in the position of commander of a NASA space shuttle mission. In July 1999 she commanded the STS-93 mission, which successfully deployed the *Chandra X-Ray Observatory.* Prior to that highly successful flight, she also served (in February 1995) as the first woman pilot of the space shuttle. This milestone in human spaceflight occurred during the STS-63 mission, the first rendezvous mission of the American space shuttle with the Russian *Mir* space station.

Collins, Michael (1930–) American *Astronaut, U.S. Air Force Officer* Michael Collins served as the command module pilot during NASA's historic *Apollo 11* mission to the Moon in July 1969. He remained in lunar orbit while his fellow *Apollo 11* mission astronauts NEIL ARMSTRONG and EDWIN (BUZZ) ALDRIN became the first human beings to walk on the lunar surface. In April 1971 Collins joined the Smithsonian Institution as director of the National Air and Space Museum (NASM), where he remained for seven years.

color A quality of light that depends on its wavelength. The *special color* of emitted light corresponds to its place in the spectrum of the rainbow. *Visual light,* or *perceived color,* is the quality of light emission as recognized by the human eye. Simply stated, the human eye contains three basic types of light-sensitive cells that respond in various combinations to incoming spectral colors. For example, the color brown occurs when the eye responds to a particular combination of blue, yellow, and red light. Violet light has the shortest wavelength, while red light has the longest wavelength. All the other colors have wavelengths that lie in between.

color index Astronomers call the apparent magnitudes of stars measured by the human eye the apparent visual magnitudes. However, such apparent magnitudes can be complicated because of the range of stellar colors. In astronomy color is a measurable property of a star that is related to its absolute temperature. Considering for the moment that stars behave as blackbody radiators, a star at a temperature of 3,400 K would be a relatively cool blackbody and emit most of its radiation in the red light portion of the electromagnetic spectrum (like Betelgeuse). A star at a temperature of 5,800 K would appear yellow (like the Sun), while a star at a temperature of 25,000 K would be hot and appear bluish (like Rigel). A degree of observational confusion set in when astronomers began to use photographs of the stars to determine their apparent magnitudes. Due to evolution here on Earth, the human eye is most sensitive to the yellow-green light portion of the visual spectrum. In early stellar photographs bluish stars such as Rigel appeared brighter photographically than visually, while reddish stars such as Betelgeuse appeared brighter visually than photographically. Astronomers introduced the concept of *apparent photographic magnitude* that was later replaced by the apparent blue magnitude as measured with a specially filtered electrooptical photometer. From all this effort to standardize observations the color index emerged.

Astronomers define the *color index* as the difference in the apparent magnitude of a star at the two standard wavelengths used as a measure of the star's color and consequently its temperature. Today astronomers commonly use the UBV color index system in which the luminosity of a star is photometrically measured using different colored filters. The most common color index is *B-V,* where *B* is the apparent magnitude measured at a reference blue light wavelength (typically 440 nanometers, or 4,400 Å) and *V* is the visual apparent magnitude measured at a reference yellow-green light wavelength (typically 550 nanometers, or 5,500 Å).

By convention the *B* and *V* apparent magnitudes are assumed to have the same numerical values for white stars that have a surface temperature of approximately 9,200 K. Therefore, the value of the *B-V* color index is slightly negative for hot blue stars, is zero for white stars at 9,200 K, and becomes progressively more positive for yellow, orange, and red stars. Typical *B-V* color index values range from about –0.4 to +1.5. Astronomers also use the *U-B* color index scheme, where *U* is the apparent magnitude of a star measured with respect to a reference ultraviolet (U) wavelength (typically 365 nanometers, or 3,650 Å). A star's color index provides astronomers an efficient and consistent way of establishing the star's temperature.

color temperature (T_c) Astronomers often describe the surface temperature of a star as the color temperature, which corresponds to the temperature of an equivalent blackbody

(Planck's law) radiator whose radiant energy distribution corresponds to the measured luminosity of the star over a given range of wavelengths. However, because stars are not precise blackbody radiators, their color temperatures and effective temperatures (T_{eff}) are not generally equal. The difference is greatest for very hot stars. Astronomers define a star's *effective temperature* as the temperature the particular star would have if it were a true blackbody of the same size (i.e., surface area) radiating the same luminosity.

Columbia accident While gliding back to Earth on February 1, 2003, after a very successful 16-day scientific research mission in low-Earth orbit, NASA's space shuttle *Columbia* experienced a catastrophic reentry accident and broke apart at an altitude of about 63 kilometers over Texas. Traveling through the upper atmosphere at approximately 18 times the speed of sound, the orbiter vehicle disintegrated, taking the lives of its seven crewmembers: six American astronauts (Rick D. Husband, William C. McCool, Michael P. Anderson, Kalpana Chawla, Laurel Blair Salton Clark, and David M. Brown) and the first Israeli astronaut (Ilan Ramon). Disaster struck the STS 107 mission when *Columbia* was just 15 minutes from its landing site at the Kennedy Space Center (KSC) in Florida. Postaccident investigations indicate that a severe heating problem occurred in *Columbia*'s left wing as a result of structural damage from debris impact during launch.

coma 1. The gaseous envelope that surrounds the nucleus of a comet. 2. In an optical system, a result of spherical aberration in which a point source of light (not on axis) has a blurred, comet-shaped image.

See also ABERRATION; COMET.

combustion Burning, or rapid oxidation, accompanied by the release of energy in the form of heat and light. The combustion of hydrogen, however, emits radiation outside the visi-

The crew of the space shuttle *Columbia* who perished on February 1, 2003, when the orbiter disintegrated during STS 107 mission reentry operations. They are, left to right, front row, American astronauts Rick Husband, Kalpana Chawla, William McCool; back row, David Brown, Laurel Clark, Michael Anderson, and Israeli astronaut Ilan Ramon. *(Courtesy of NASA)*

ble spectrum, making a "hydrogen flame" essentially invisible to the naked eye.

combustion chamber The part of a rocket engine in which the combustion of chemical propellants takes place at high pressure. The combustion chamber and the diverging section of the nozzle make up a rocket's thrust chamber. Sometimes called the *firing chamber* or simply the *chamber*.

See also ROCKET.

comet A dirty ice rock consisting of dust, frozen water, and gases that orbits the Sun. As a comet approaches the inner solar system from deep space, solar radiation causes its frozen materials to vaporize (sublime), creating a coma and a long tail of dust and ions. Scientists think these icy planetesimals are the remainders of the primordial material from which the outer planets were formed billions of years ago. As confirmed by spacecraft missions, a comet's nucleus is a type of dirty ice ball consisting of frozen gases and dust. While the accompanying coma and tail may be very large, comet nuclei generally have diameters of only a few tens of kilometers or less.

As a comet approaches the Sun from the frigid regions of deep space, the Sun's radiation causes the frozen (surface) materials to sublime (vaporize). The resultant vapors form an atmosphere, or *coma,* with a diameter that may reach 100,000 kilometers. It appears that an enormous cloud of hydrogen atoms also surrounds the visible coma. This hydrogen was first detected in comets in the 1960s.

Ions produced in the coma are affected by the charged particles in the solar wind, while dust particles liberated from the comet's nucleus are impelled in a direction away from the Sun by the pressure of the solar wind. The results are the formation of the plasma (Type I) and dust (Type II) comet tails, which can extend for up to 100 million kilometers. The *Type I tail,* composed of ionized gas molecules, is straight and extends radially outward from the Sun as far as 100 million kilometers (10^8 km). The *Type II tail,* consisting of dust particles, is shorter, generally not exceeding 10 million kilometers in length. It curves in the opposite direction to the orbital movement of the comet around the Sun.

No astronomical object, other than perhaps the Sun and the Moon, has attracted more attention or interest. Since ancient times comets have been characterized as harbingers of momentous human events. William Shakespeare wrote in the play *Julius Caesar:*

> *When beggars die, there are no comets seen; but the heavens themselves blaze forth the death of princes.*

Many scientists think that comets are icy planetesimals that represent the "cosmic leftovers" when the rest of the solar system formed more than 4 billion years ago. In 1950 the Dutch astronomer JAN HENDRIK OORT suggested that most comets reside far from the Sun in a giant "cloud" (now called the *Oort cloud*). The Oort cloud is thought to extend out to the limits of the Sun's gravitational attraction, creating a giant sphere with a radius of between 50,000 and 80,000 astronomical units (AU). Billions of comets may reside in this distant region of the solar system, and their total mass is estimated to be roughly equal to the mass of the Earth. Just a few of these comets ever enter the planetary regions of the solar system, possibly through the gravitational perturbations caused by neighboring stars or other "chaotic" phenomena.

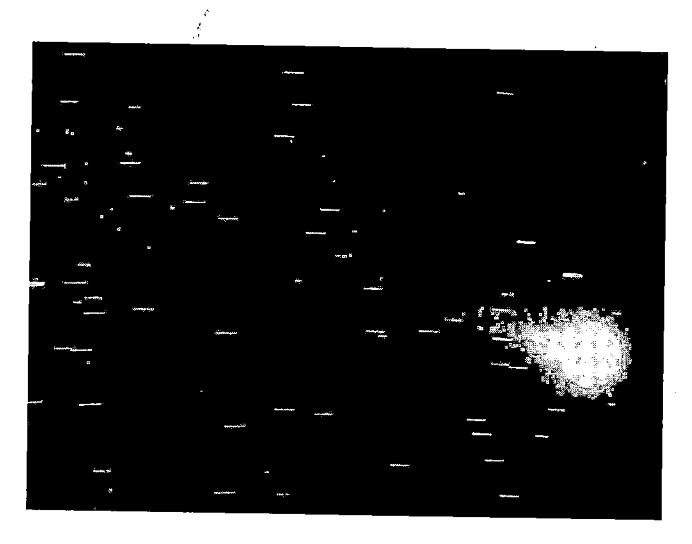

The 1965 Comet Ikeya-Seki as imaged by the U.S. Naval Observatory in Flagstaff, Arizona. Dr. Elizabeth Roemer used the observatory's 1-meter (40-inch) reflector telescope to create this photograph on October 3, 1965. The white streaks in the photograph are various fixed stars whose light was also captured by the telescope as the comet was being imaged during a 20-minute exposure period. ***(Courtesy of the U.S. Navy and NASA)***

Oort's suggestion was followed quickly by an additional hypothesis concerning the location and origin of the periodic comets we see passing through the solar system. In 1951 the Dutch-American astronomer GERARD KUIPER proposed the existence of another, somewhat nearer, region populated with cometary nuclei and icy planetesimals. Unlike the very distant Oort cloud, this region, now called the *Kuiper Belt,* lies roughly in the plane of the planets at a distance of 30 AU (Neptune's orbit) out to about 1,000 AU from the Sun. In 1992 the astronomers David C. Jewitt (b. 1958) and Jane Luu (b. 1963) discovered the first Kuiper Belt object, called QB 1992—an icy body of approximately 200 km diameter at about 44 AU from the Sun.

Once a comet approaches the planetary regions of the solar system, it is also subject to the gravitational influences of the major planets, especially Jupiter, and the comet may eventually achieve a quasi-stable orbit within the solar system. By convention, comet orbital periods are often divided into two classes: *long-period comets* (which have orbital periods in excess of 200 years) and *short-period comets* (which have periods less than 200 years). Astronomers sometimes use the term *periodic comet* for a short-period comet. In contemporary planetary defense studies related to Earth-approaching objects, the term *short-period* comet often means a comet with a period of less than 20 years.

During the space age robot spacecraft have explored several comets and greatly improved our scientific understanding of these very interesting celestial objects. Some of the most significant space missions are discussed below.

Comet Halley Exploration (1986)
As a result of the highly successful international efforts to explore Comet Halley in 1986, scientists were able to confirm their postulated dirty ice rock model of a comet's nucleus. (More recent space missions have reinforced this model.) The Halley Comet nucleus is a discrete, single "peanut-shaped" body (some 16 km by 8 km by 7.5 km). A very-low-albedo (about 0.05), dark dust layer covers the nucleus. The surface temperature of this layer was found to be 330 kelvins (57°C), somewhat higher than expected when compared to the temperature of a subliming ice surface, which would be about 190 kelvins (–83°C). The total gas production rate from the comet's nucleus at the time of the *Giotto* spacecraft's flyby encounter on March 14, 1986, was observed as 6.9×10^{29} molecules per second, of which 5.5×10^{29} molecules per second were water vapor. (The *Giotto* spacecraft was built by the European Space Agency and performed an extremely successful encounter mission with Comet Halley.) This observation means that 80 to 90 percent of Comet Halley's nucleus consists of water ice and dust. Other (parent) molecules detected in the nucleus include carbon dioxide (CO_2), ammonia (NH_3), and methane (CH_4).

Comet Shoemaker-Levy 9
Because they are only very brief visitors to the solar system, one might wonder what happens to these comets after they blaze a trail across the night sky. The chance of any particular "new" comet being captured in a short-period orbit is quite small. Therefore, most of these cosmic wanderers simply return to the Kuiper Belt or the Oort cloud, presumably to loop back into the solar system years to centuries later, or else they are ejected into interstellar space along hyperbolic orbits. Sometimes, however, a comet falls into the Sun. One such event was observed by instruments on a spacecraft in August 1979. Other comets simply break up because of gravitational (tidal) forces or possible outbursts of gases from within. When a comet's volatile materials are exhausted or when its nucleus is totally covered with nonvolatile substances, astronomers call the comet "inactive." Some space scientists believe that an inactive comet may become a "dark asteroid," such as those in the Apollo group, or else disintegrate into meteoroids. Finally, on very rare occasions, a comet may even collide with an object in the solar system, an event with the potential of causing a cosmic catastrophe.

In July 1994 20-km-diameter or so chunks of Comet Shoemaker-Levy 9 plowed into the southern hemisphere of Jupiter. These comet fragments collectively deposited the energy equivalent of about 40 million megatons of TNT. Scientists using a variety of space-based and Earth-based observatories detected plumes of hot, dark material rising higher than 1,000 km above the visible Jovian cloud tops. These plumes emerged from the holes punched in Jupiter's atmosphere by the exploding comet fragments.

Exploring and Sampling Comets
Comets are currently viewed as both extraterrestrial threats and interesting sources of extraterrestrial materials. A detailed rendezvous mission with and perhaps an automated sample return mission from a short-period comet can answer many of the questions currently puzzling comet physicists. A highly automated sampling mission, for example, would permit the collection of atomized dust grains and gases directly from the comet's coma. The simplest way to accomplish this type of space mission would be to use a high-velocity flyby technique, with the automated spacecraft passing as close to the comet's nucleus as possible. This probe would be launched on an Earth-return trajectory. Terrestrial recovery of the comet material could be accomplished by the ejection of the sample in a protective capsule designed for atmospheric entry.

Deep Space One (DS1) Mission
NASA's *Deep Space 1* (DS1) was primarily a mission to test advanced spacecraft technologies that had never been flown in space. In the process of testing its solar electric propulsion system, autonomous navigation system, advanced microelectronics and telecommunications devices, and other cutting-edge aerospace technologies, *Deep Space 1* encountered the Mars-crossing near-Earth asteroid 9,969 Braille (formerly known as 1992 KD) on July 20, 1999. Then, in September 2001 the robot spacecraft encountered Comet Borrelly and, despite the failure of the system that helped determine its orientation, returned some of the best images of a comet's nucleus.

Stardust Mission
The primary objective of NASA's Stardust mission is to fly by Comet P/Wild 2 and collect samples of dust and volatiles in the coma of the comet. The spacecraft will then return these samples to Earth for detailed study. The *Stardust* spacecraft was launched from Cape Canaveral Air Force Station in early February 1999. Following launch the spacecraft achieved a heliocentric, elliptical orbit. By midsummer 2003 it had completed its second orbit of the Sun. The spacecraft's "comet dust" collector was deployed on December 24, 2003. The encounter with Comet P/Wild 2 successfully took place in early 2004 while the spacecraft and the comet were 1.85 AU from the Sun and 2.6 AU from Earth.

The spacecraft entered the coma of Comet P/Wild 2 on December 31, 2003. The closest encounter occurred on January 2, 2004. During this encounter *Stardust* flew within 250 km of the comet's nucleus at a relative velocity of about 6.1 km/s. Six hours after the flyby, the sample collector was retracted, stowed, and sealed in the sample vault of the spacecraft's sample reentry capsule. On January 15, 2006, this capsule will separate from the main spacecraft and return to Earth. The landing is expected to take place within a 30×84 km landing ellipse at the U.S. Air Force Test and Training Range in the Utah desert at approximately 3 A.M. local time.

Contour Mission
NASA's planned Comet Nucleus Tour (CONTOUR) mission had as its primary objective close flybys of two comet nuclei with possibly a close flyby of a third known comet—or perhaps an as-yet-undiscovered comet. The two comets scheduled for visits by this spacecraft were Comet Encke and

Comet Schwassmann-Wachmann-3, and the third target was Comet d'Arrest. NASA mission planners had even hoped that a new comet would be discovered in the inner solar system between the years 2006 and 2008. In that event, they were prepared to change the trajectory of the *CONTOUR* spacecraft to rendezvous with the new comet. The CONTOUR mission's scientific objectives included imaging comet nuclei at resolutions of 4 meters, performing spectral mapping of the nuclei at resolutions of 100–200 meters, and obtaining detailed compositional data concerning gas and dust particles in the near-nucleus (coma) environment.

The *CONTOUR* spacecraft was successfully launched into space by a Delta II expendable vehicle from Cape Canaveral on July 3, 2002. Unfortunately, following a series of phasing orbits and the firing (on August 15, 2002) of the solid rocket motor needed to eject *CONTOUR* from Earth orbit and place it on a heliocentric trajectory, all contact with the spacecraft was lost.

Rosetta Mission

The European Space Agency's Rosetta mission is designed to rendezvous with Comet 67 P/Churyumov-Gerasimenko, drop a probe on the surface, study the comet from orbit, and fly past at least one asteroid en route. The *Rosetta* spacecraft was successfully launched on March 2, 2004, from Kourou, French Guiana, by an Ariane 5 rocket vehicle. Following a long interplanetary journey involving several gravity assist maneuvers, asteroid flybys, and extended spacecraft "sleep" periods, *Rosetta* will come out of hibernation in January 2014 and begin its rendezvous maneuver with Comet Churyumov-Gerasimenko in May 2014. The spacecraft will achieve orbit around the comet's nucleus in August 2014 and then deploy the *Philae* lander to an interesting site on the comet's nucleus in about November 2014. The hitchhiking *Philae* lander will touch down on the surface of the comet's nucleus at a relative velocity of 1 m/s. Following a successful landing, *Philae* will transmit data from the comet's surface to the *Rosetta* spacecraft, which will then relay these data back to scientists on Earth. The *Rosetta* spacecraft will remain in orbit about the comet past its perihelion passage (to occur in August 2015) until the planned normal end of this mission in December 2015.

Deep Impact Mission

The objectives of NASA's Deep Impact mission are to rendezvous with Comet P/Tempel 1 and launch a projectile into the comet's nucleus. Instruments on the spacecraft will observe the ejecta from the impacting projectile, much of which will represent pristine material from the interior of the comet, the crater formation process, the resulting crater, and any outgassing from the nucleus—particularly the newly exposed surface. NASA will use a Delta II rocket to launch *Deep Impact* from Cape Canaveral during a launch window that opened on December 30, 2004.

Following a successful launch, the spacecraft will transfer into a heliocentric orbit and then will rendezvous with Comet P/Tempel 1 in July 2005. As presently planned, the *Deep Impact* spacecraft will be about 880,000 km from the comet on July 3, 2005, and moving at a velocity of 10.2 km/s relative to the comet. The projectile will be released at this point, and shortly after release the flyby spacecraft will execute a maneuver to slow down (to 120 m/s) relative to the impactor. About 24 hours after release (on July 4, 2005), the impactor will strike the sunlit side of the comet's nucleus. At the relative encounter velocity of 10.2 km/s, the impactor should form a crater roughly 25 meters deep and 100 meters wide, ejecting material from the interior of the nucleus into space and vaporizing the impactor and much of the ejecta.

The flyby spacecraft will be approximately 10,000 km away at the time of impact and will begin imaging 60 seconds before impact. At 600 seconds after impact the spacecraft will be about 4,000 km from the nucleus, and observations of the crater will begin and continue until closest approach to the nucleus at a distance of about 500 km. At 961 seconds after impact imaging will end when the spacecraft reorients to cross the inner coma. At 1,270 seconds the crossing of the inner coma will be complete. At 3,000 seconds the spacecraft will begin playback of scientific data to Earth. This mission was successfully accomplished in July 2005.

Kuiper Belt and Oort Cloud Exploration

Comets are very interesting celestial objects. Robot spacecraft missions in the early decades of the 21st century have helped astronomers more fully understand their composition, physical characteristics, and dynamic behavior. More sophisticated robot probes that will reach into the Kuiper Belt are also being planned. A few daring aerospace visionaries have even suggested space exploration missions for the mid-21st century that extend all the way to the Oort cloud.

NASA's *New Horizons Pluto–Kuiper Belt Flyby* spacecraft is currently scheduled for launch in early 2006. Its reconnaissance type exploration mission will help scientists understand the interesting yet poorly understood worlds at the edge of the solar system. The first spacecraft flyby of Pluto and Charon—the frigid double planet system—will take place as early as 2015. The mission will then continue beyond Pluto and visit one or more Kuiper Belt objects by 2026. The spacecraft's long journey will help resolve some basic questions about the surface features and properties of these icy bodies as well as their geology, interior makeup, and atmospheres. The modest-sized spacecraft will have no deployable structures and will receive all its electrical power from long-lived radioisotope thermoelectric generators (RTGs) that are similar in design to those used on the *Cassini* spacecraft now touring Saturn.

Following the New Horizons mission, other advanced robot spacecraft could help determine the extent of the Kuiper Belt, verify the existence of an inner Oort cloud region (now believed to begin around 10,000 AU), help characterize the comet population of the Oort cloud, and study the depletion mechanisms for the loss of comets out of this cloud. Because of the great distances involved and the required levels of reliability and autonomy, an *Oort Cloud Probe* robot spacecraft would represent the direct technical precursor to humankind's first robot star probe.

Following the intellectual stimulus provided in 1877 by JULES VERNE in his exciting (but fictional) "Off On A Comet," an account of the imaginary interplanetary journey of Captain Hector Servadac, several space technology visionaries now boldly suggest that future human spaces explorers could someday rendezvous with and harvest the material resources of short-period comets for a variety of solar system civilization applications. Building further upon such an advanced space technology heritage, later generations of deep space explorers might even attempt to visit neighboring star systems by "riding" very-long-period comets deep into interstellar space. These human explorers and their very smart robot companions could travel in huge, fusion-propelled space habitats that could co-orbit a long-period comet and then harvest its entire reservoir of materials for propellants and life support system supplies—before finally breaking the last feeble grips of the Sun's gravitational pull and casting off into the interstellar void.

See also ASTEROID; ASTEROID DEFENSE SYSTEM; EXTRATERRESTRIAL CATASTROPHY THEORY; *DEEP IMPACT*; *DEEP SPACE ONE*; *GIOTTO* SPACECRAFT; KUIPER BELT; NEW HORIZONS PLUTO–KUIPER BELT FLYBY MISSION; OORT CLOUD; ROSETTA MISSION; STARDUST MISSION.

Comet Halley (1P/Halley) The most famous periodic comet. Named after the British astronomer EDMOND HALLEY, who successfully predicted its 1758 return. Reported since 240 B.C.E., this comet reaches perihelion approximately every 76 years. During its most recent inner solar system appearance in 1986, an international fleet of five different spacecraft, including the *Giotto* spacecraft, performed scientific investigations that supported the dirty ice rock model of a comet's nucleus. Specifically, images taken by the *Giotto* spacecraft revealed a dark, peanut-shaped nucleus some 16 km by 8 km in size with an albedo of 0.04. The comet's orbit has a perihelion distance of 0.587 astronomical unit (AU), a semimajor axis of 17.94 AU, an eccentricity of 0.967, and an inclination of 162.2°. Astronomers predict Comet Halley will next return to perihelion in 2061.

command destruct An intentional action leading to the destruction of a rocket or missile in flight. Whenever a malfunctioning vehicle's performance creates a safety hazard on or off the rocket test range, the range safety officer sends the command destruct signal to destroy it.

command guidance A guidance system wherein information (including the latest data/intelligence about a target) is transmitted to the in-flight missile from an outside source, causing the missile to traverse a directed flight path.

See also GUIDANCE SYSTEM.

command module (CM) The part of the Apollo spacecraft in which the astronauts lived and worked; it remained attached to the service module (SM) until reentry into Earth's atmosphere, when the SM was jettisoned and the command module (CM) brought the lunar astronauts back safely to the surface of Earth.

See also APOLLO PROJECT.

Common, Andrew Ainslie (1841–1903) British *Amateur Astronomer, Instrument Maker* Andrew Common was a telescope maker who pursued improvements in astrophotography in the 1880s and 1890s. He is credited with recording the first detailed photographic image of the Orion Nebula.

communications satellite An orbiting spacecraft that relays signals between two (or more) communications stations. There are two general types: the *active communications satellite*, which is a satellite that receives, regulates, and retransmits signals between stations, and the *passive communications satellite*, which simply reflects signals between stations. The active communications satellite is the type used to create today's global communication infrastructure. Communication is the most successful commercial space enterprise yet achieved. Of the hundreds of communications satellites sent into orbit over the past 40 or so years, about 150 are still operational, with newer, more capable spacecraft joining the international inventory each year.

The concept of the communications satellite can be traced back to the British space visionary and author SIR ARTHUR C. CLARKE, who in 1945 published a short article entitled "Extra-Terrestrial Relays." In this article Clarke described the use of unmanned satellites in 24-hour orbits high above the world's landmasses to distribute television programs. However, this innovative idea lay essentially unnoticed until the early 1960s, when the U.S. Department of Defense began to pursue active communication satellites (for example, the ADVENT communications satellite program, which was cancelled in 1962 due to spacecraft complexity and cost overruns), while initially NASA (under Congressional direction) was confined to research involving passive communication satellites. For example, NASA launched the *Echo 1* and 2 satellites on August 12, 1960, and January 25, 1964, respectively. These early communications satellite were large metallized balloons that served as passive reflectors of radio signals. However, because the reflected signal was so weak, such passive communications satellites proved too unattractive for commercial use. Then, in 1960 AT&T filed with the Federal Communications Commission (FCC) for permission to launch an experimental communications satellite with a view to rapidly developing and implementing an operational system. AT&T's *Telstar I* was launched from Cape Canaveral on July 10, 1962, and was followed by *Telstar II* on May 7, 1963. These privately financed spacecraft (launched with NASA's assistance on a cost reimbursable basis) were prototypes for a planned constellation of 50 medium-orbit satellites that AT&T was planning to put in place. When the U.S. government decided to give the monopoly on satellite communications to the Communications Satellite Corporation (COMSAT), which was formed as a result of the Communications Satellite Act of 1962, AT&T's satellite project was halted.

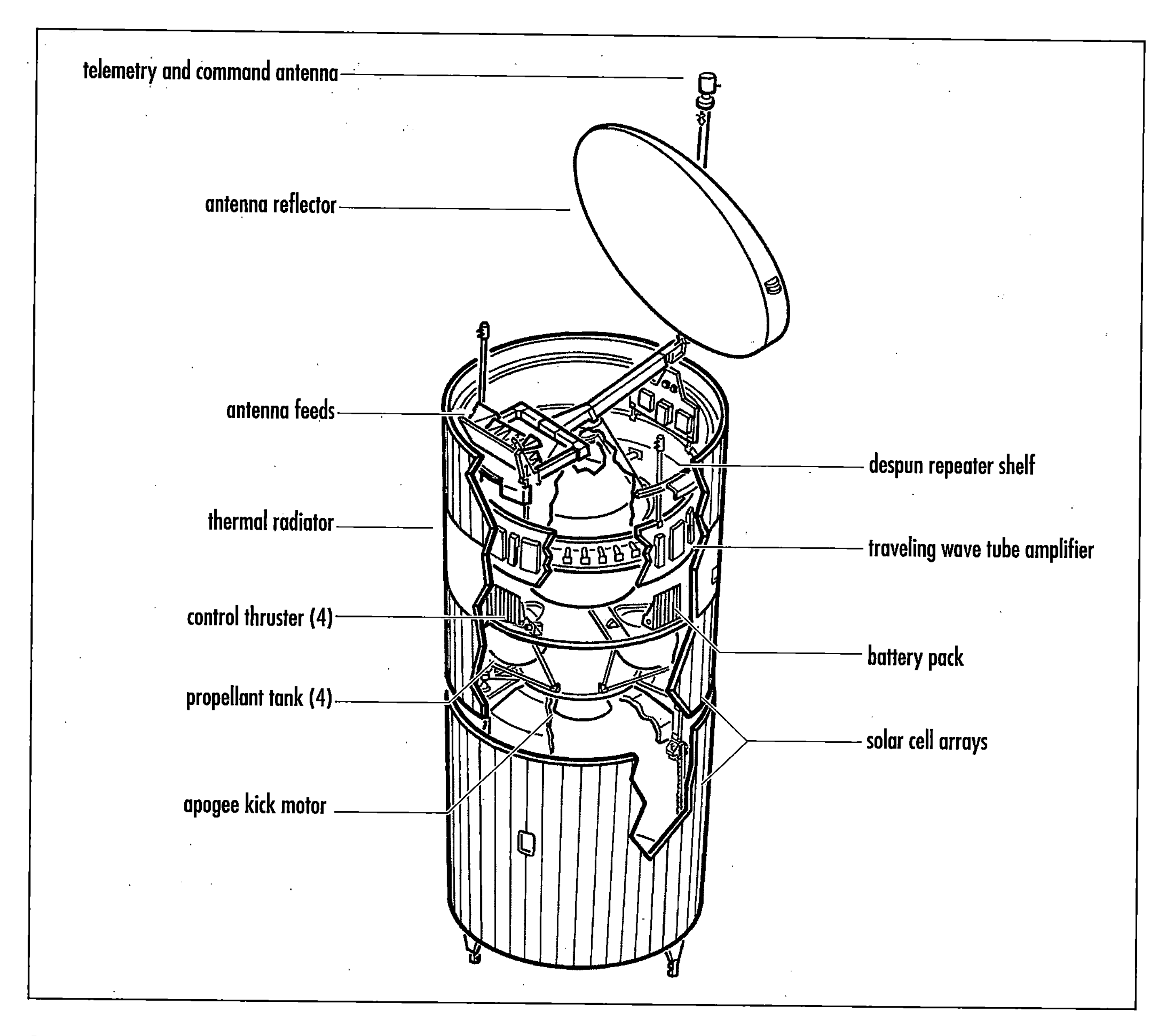

Cutaway diagram of a modern communications satellite, here the Boeing 376 spacecraft configuration. On orbit, this 2.13-meter-diameter satellite extends to a full vertical configuration (as shown) of about 6.7 meters. Numerous versions of this commercial communications satellite have been placed into service in geostationary orbit. ***(Drawing courtesy of the Boeing Company)***

Switchboards In the Sky

Satellite-based communications probably exerts more impact on the daily life of the average person than any other form of space technology. Furthermore, satellite communications is the most successful form of commercial space activity, generating billions of dollars each year in the sale of information and entertainment services and products. Our information-based "global village" is enabled by an armada of geostationary communications satellites—marvels of modern space technology that serve the citizens of planet Earth as high-capacity switchboards in the sky.

Early in the 20th century communications engineers and physicists recognized that radio waves, like other forms of electromagnetic radiation, travel in a straight line along a line-of-sight path and cannot (of themselves) bend around the curvature of

(continues)

Switchboards In the Sky *(continued)*

Earth. As a result, wireless communication is physically limited to line of sight. This means, for example, that a radio or television receiver cannot obtain broadcasts from a transmitter that lies beyond the horizon. The higher the transmitting antenna, the farther extends the line-of-sight distance available for direct wave, wireless communications. This explains why antenna towers on Earth's surface are so tall and why transmitting stations that serve large geographic areas are often located on convenient mountaintops.

Wireless communications pioneers such as the Nobel laureate Guglielmo Marconi (1874–1937) also discovered that under certain circumstances they could bounce radio waves of a certain frequency off ionized layers in Earth's atmosphere and thereby achieve short-wave radio broadcasts over long distances. However, our planet's ionosphere is a natural phenomenon that experiences many irregularities and diurnal variations, so its use is quite undependable for any continuously available, reliable, and high-capacity wireless communications system.

Then, in 1945 a creative radio engineer, Sir Arthur C. Clarke (b. 1917), recognized the technological key to establishing a dependable, worldwide wireless communications system. Why not create an incredibly tall antenna tower by putting the signal relay on a platform in space? Following his service in World War II as a radar instructor in the Royal Air Force, Clarke published his pioneering technical paper "Extra-Terrestrial Relays" in the October 1945 issue of *Wireless World*. In this visionary paper he described the principles of satellite communications and recommended the use of the geostationary orbit for a global communications system. Today, aerospace engineers often refer to geostationary orbit as the "Clarke orbit" in his honor. The modern geostationary communications satellite is a space technology application that has transformed the world of wireless communications and helped stimulate today's exciting information revolution.

The communications satellite is an Earth-orbiting spacecraft that relays signals between two or more communications stations. The space platform serves as a very-high-altitude switchboard without wires. Aerospace engineers and information specialists divide communications satellites into two general classes or types: the *active communications satellite,* which is a spacecraft that can receive, regulate, and retransmit signals between stations, and the *passive communications satellite,* which functions like a mirror and simply reflects radio frequency signals between stations. While NASA's first passive communications satellite, called *Echo 1,* helped demonstrate the use of orbiting platforms in wireless communications, hundreds of active communications satellites now serve today's global communications infrastructure.

A major advantage of geostationary orbit for the communications satellite is the fact that an Earth station (ground site) has a much easier task of tracking and antenna pointing, because the satellite is always in view at a fixed location in space. In addition, geostationary orbit provides a panoramic view of Earth that is hemispheric in extent. The same communications satellite that is in view of the United States, for example, also has good line-of-sight viewing from Canada, Mexico, Argentina, Brazil, Chile, Columbia, and Venezuela. Because of this advantage, the communications satellite has created a "global village" in which news, electronic commerce, sports, entertainment, and personal messages travel around the planet efficiently and economically at the speed of light.

With the help of communications satellites, breaking news now diffuses rapidly through political, geographic, and cultural barriers, providing millions of information-hungry people objective, or at least alternative, versions of a particular story or event. Time zones no longer represent physical or social barriers. Mobile television crews, equipped with the very latest communications satellite linkup equipment, have demonstrated an uncanny ability to pop up anywhere in the world just as an important event is taking place. Through their efforts and the global linkages provided by communications satellites, television viewers around the world will often witness significant events (pleasant and unpleasant, good and evil) in essentially real time.

This free flow of information throughout our planet is serving as a major catalyst for democracy and social improvement. As information begins to leak and then flood into politically oppressed populations along invisible lines from space, despotic governments encounter increasingly more difficult times in denying freedom to their citizens and in justifying senseless acts of aggression against their neighbors. In a very real sense, the electronic switchboard in space has become the modern tyrant's most fearsome enemy. With their unique ability to shower continuous invisible lines of information gently on the free and the oppressed alike, communications satellites serve as extraterrestrial beacons of freedom—space platforms that support the inalienable rights of human beings everywhere to pursue life, liberty, and the pursuit of individual happiness. Of course, no technology, no matter how powerful or pervasive, can instantly cure political injustices, cultural animosities, or social inequities. For example, despite news coverage, Arab slave traders are still dealing in captured black people in the Sudan, and drug cartels still exert overwhelming political influence in Latin America. However, the rapid and free flow of information is a necessary condition for creating an informed human family capable of achieving enlightened stewardship of our home planet this century.

The dramatic social transformations stimulated by the communications satellite are just beginning. Through space technology rural regions in both developed and developing countries can now enjoy almost instant access to worldwide communication services, including television, voice, facsimile, and data transmissions. The communications satellite is also supporting exponentially growing changes in health care (*telemedicine*), the workplace (*telecommuting* and *transnational outsourcing*), education (*distance learning*), electronic commerce (the open-for-business 24-hour global marketplace), and borderless banking (including the high-speed encrypted transfer of *digital money* across international boundaries). The information revolution, catalyzed to a great extent by the development of the communications satellite, is causing major social, political, and economic changes within our global civilization. If you need further proof that we live an exciting age of change, just take a glimpse at the international calling section of a current telephone directory. Such interesting and geographically remote locations as Ascension Island (area code 247), Andorra (area code 376), Christmas Island (area code 672), French Polynesia (area code 689), Greenland (area code 299), San Marino (area code 689), Wake Island (area code 808), and Antarctica (area code 672) are now just a communications satellite link away.

By the middle of 1961 NASA decided to build a medium-orbit (6,400-km-altitude) active communications satellite called *Relay.* Unlike the previous *Echo 1* and *2* satellites, which were passive, *Relay* was an active satellite (like all the communications satellites that followed). The new satellite would receive a radio signal from the ground and then retransmit (or relay) it back to another location on Earth. *Relay 1* was launched on December 13, 1962; *Relay 2* followed on January 21, 1964. The traveling wave tube (TWT) was the key power-amplifying device in the *Relay* satellite. This device, greatly improved with time, has become a basic component in modern communications satellites. In the *Relay 1* satellite the early TWT had a mass of 1.6 kg and produced a minimum radio frequency (rf) output of 11 watts at a gain of 33 decibels (dB) over the 4,050 to 4,250 MHz band with an overall efficiency of at least 21 percent.

In 1961 NASA also pursued the development of an experimental 24-hour orbit (geosynchronous) communications satellite, called Syncom. The Syncom project objective was to demonstrate the technology for synchronous-orbit communications satellites. A synchronous orbit (also called geosynchronous Earth orbit, or GEO for short) is one in which the satellite makes one orbit per day. Since the Earth rotates once a day and the satellite goes around the Earth at the same angular rate, the satellite hovers over the same area of the Earth's surface all the time. The altitude of a synchronous orbit is 35,900 km. At lower altitudes satellites orbit the Earth more than once per day. About 42% of Earth's surface is visible from a synchronous orbit. Therefore, with three properly placed satellites the entire globe is covered. Although Arthur C. Clarke proposed the idea for the synchronous communications in 1945, the first spacecraft to use a synchronous orbit, *Syncom 1,* was not launched until February 14, 1963. Unfortunately, when the apogee kick motor for circularizing the orbit was fired, the spacecraft fell silent. Telescopic observations later confirmed that the "silent" satellite was, nonetheless, in an appropriate orbit of nearly 24 hours at a 33° inclination.

A key advantage of the synchronous orbit communications satellite is that a ground station has a much easier tracking and antenna-pointing job because the satellite is always in view. For lower-altitude orbits, the spacecraft must be acquired as it comes into view above one horizon, then it must be tracked across the sky with the antenna slewing over to the opposite horizon, where the spacecraft disappears until its next pass. For continuous coverage of a satellite, ground stations would have to be distributed around the globe so that the satellite is rising over the viewing horizon of one ground station as it is setting for another station.

A particular type of synchronous orbit is the *geostationary orbit.* This is a synchronous orbit around the equator in which the satellite appears stationary over a point on Earth's surface. The *Syncom 3* spacecraft was the first to demonstrate the use of this very important type of orbit. Prior to that mission, NASA had successfully placed the *Syncom 2* spacecraft in a synchronous orbit inclined 33° to the equator—resulting in an orbit that appeared to move 33° north and 33° south in a figure-eight path over a 24-hour period as observed from the ground.

Syncom 1 was launched on February 14, 1963, into a nearly synchronous orbit but failed during the apogee motor burn. The most likely cause was determined to be a failure of the high-pressure nitrogen tank. The *Syncom* spacecraft had two separate attitude control jet propellants: nitrogen and hydrogen peroxide. One of the objectives of NASA's Syncom program was to demonstrate attitude control for antenna pointing and stationkeeping. Following improvements in its nitrogen tank design, *Syncom 2* was launched on July 26, 1963. It was a success, and the satellite transmitted data, telephone, pioneering facsimile (fax), and video signals. Finally, the NASA *Syncom 3* spacecraft was successfully launched from Cape Canaveral on August 19, 1964, into a geostationary orbit. The spacecraft had the addition of a wideband channel for television and provided "live" coverage of the 1964 Olympics, which were being held in Tokyo. Both the *Syncom 2* and *Syncom 3* spacecraft were turned over to the Department of Defense in April 1965, and they were finally retired (turned off) in April 1969. It is important to recognize that in the early 1960s just reaching synchronous orbit represented a formidable space technology challenge. A major goal of the successful Syncom program was just to show that a synchronous satellite was possible.

Syncom led to the development of the *Early Bird 1* commercial communications spacecraft as well as the Application Technology Satellite (ATS) series of NASA research satellites. By 1964 COMSAT had chosen the 24-hour-orbit (geostationary) satellite offered by the Hughes Aircraft Company for their first two commercial systems. On April 6, 1965, COMSAT's first spacecraft, *Early Bird 1,* was successfully launched from Cape Canaveral. This launch marked the beginning of the global village linked instantaneously by commercial communications satellites. By that time Earth stations for satellite-supported communications already existed in the United Kingdom, France, Germany, Italy, Brazil, and Japan. On August 20, 1964, agreements were signed that created the International Telecommunications Satellite Organization (INTELSAT). The Intelsat II series of communication satellites was a slightly more capable and longer-lived version of the *Early Bird* satellite. The Intelsat III series was the first to provide Indian Ocean coverage, thereby completing the global network.

In 1972 Telesat Canada launched the world's first domestic communications satellite, called Anik. This series of satellites served the vast Canadian continental area. On April 13, 1974, the first U.S. domestic communications satellite, Western Union's *Westar 1,* was launched. In February of 1976 COMSAT launched a new type of communications satellite, Marisat, to provide mobile services to the U.S. Navy and other maritime customers. The United Nations International Maritime Organization sponsored the establishment of the International Maritime Satellite Organization (INMARSAT) in 1979. After leasing transponders on available satellites in the 1980's, INMARSAT began to launch its own communications satellites, starting in late 1990 with *Inmarsat II F-1.*

Much of the basic technology for communications satellites existed at the beginning of the space age, but it took a period of demonstrations and experiments in the mid-1960s and then many engineering improvements to create the global fleets of communication satellites that started entering service in the 1970s. Today modern communications satellites in geostationary orbit serve as wireless switchboards in the sky for our information-hungry, highly interconnected global civilization. Rising demands for personal communication systems, such as the era of cellular telephony, are also focusing attention on systems of small communication satellites in low-Earth orbits. Such orbits would be significantly lower than the orbits used by either the Telestar or Relay satellites in the early 1960s. Typically, companies seeking to offer innovative global communications services to cellular telephone customers might place constellations of 50 to 70 small satellites in polar orbit at altitudes of about 650 km. A typical constellation might involve 10 to 11 satellites in each of several orbital planes.

The communications satellite is one of the major benefits of space technology. Numerous active communications satellites now maintain a global telecommunications infrastructure. By supporting wireless communications services around the world, these spacecraft play an important role in the information revolution.

compact body A small, very dense celestial body that represents the end product of stellar evolution: a WHITE DWARF, a NEUTRON STAR, or a BLACK HOLE.

compact galaxy A special type of galaxy first cataloged by FRITZ ZWICKY in the 1960s. Such galaxies do not fall within the conventional classification of galaxies, since their member stars are generally blue and hot and their overall brightness is almost indistinguishable from other individual stars in the sky—except when carefully viewed on photographic plates, such as those obtained from the Palomar Observatory Sky Survey (POSS).

See also GALAXY; PALOMAR OBSERVATORY SKY SURVEY.

companion body A nose cone, protective shroud, last-stage rocket, or payload separation hardware that orbits Earth along with an operational satellite or spacecraft. Companion bodies contribute significantly to a growing space (orbital) debris population in low-Earth orbit.

See also SPACE DEBRIS.

composite propellant A solid rocket propellant consisting of a fuel (typically organic resins or plastics) and an oxidizer (i.e., nitrates or perchlorates) intimately mixed in a continuous solid phase. Neither the fuel nor the oxidizer would burn without the presence of the other.

See also ROCKET.

compressible flow In fluid dynamics, flow conditions in which density changes in the fluid cannot be neglected. The opposite of INCOMPRESSIBLE FLUID.

Compton, Arthur Holly (1892–1962) American *Physicist* Arthur Holly Compton was one of the pioneers of high-energy physics. In 1927 he shared the Nobel Prize in physics for his investigation of the scattering of high-energy photons by electrons, an important phenomenon now called the *Compton effect.* His research efforts in 1923 provided the first experimental evidence that electromagnetic radiation possessed both particlelike and wavelike properties. Compton's important discovery made quantum physics credible. In the 1990s NASA named its advanced high-energy astrophysics spacecraft, the *Compton Gamma Ray Observatory* (CGRO), after him.

Compton was born on September 10, 1892, into a distinguished intellectual family in Wooster, Ohio. His father was a professor at Wooster College, and his older brother Karl studied physics and went on to become the president of the Massachusetts Institute of Technology (MIT). As a youth Arthur Compton experienced two very strong influences from his family environment: a deep sense of religious service and the noble nature of intellectual work. While he was an undergraduate at Wooster College he seriously considered becoming a Christian missionary, but his father convinced him that because of his talent and intellect, he could be of far greater service to the human race as an outstanding scientist. His older brother also helped persuade him to study science and in the process changed the course of modern physics by introducing Arthur Compton to the study of X-rays.

Compton carefully weighed his career options and then followed his family's advice by selecting a career in physics. Upon completion of his undergraduate degree at Wooster College in 1913, he joined his older brother at Princeton. He subsequently received a Master of Arts degree in 1914 and a Ph.D. degree in 1916 from Princeton University. For his doctoral research Compton studied the angular distribution of X-rays reflected from crystals. Upon graduation he married Betty McCloskey, an undergraduate classmate from Wooster College. The couple had two sons, Arthur Allen and John Joseph.

After spending a year as a physics instructor at the University of Minnesota, Compton worked for two years in Pittsburgh, Pennsylvania, as an engineering physicist with the Westinghouse Lamp Company. Then, in 1919 he received one of the first National Research Council fellowships awarded by the American government. Compton used this fellowship to study gamma ray scattering phenomena at Baron Ernest Rutherford's (1871–1937) Cavendish Laboratory in England. While working with Rutherford he verified the puzzling results obtained by other physicists—namely, that gamma rays experienced a variation in wavelength as a function of scattering angle.

The following year Compton returned to the United States to accept a position as head of the department of physics at Washington University in Saint Louis, Missouri. There, working with X-rays, he resumed his investigation of the puzzling mystery of photon scattering and wavelength change. By 1922 his experiments revealed that there definitely was a measurable shift of X-ray (photon) wavelength with scattering angle, a phenomenon now called the Compton effect. He applied special relativity and quantum mechanics to explain the results, presented in his famous paper "A Quantum Theory of the Scattering of X-rays by Light Elements," which appeared in the May 1923 issue of *The Physical Review.* In 1927 Compton shared the Nobel Prize in physics with Charles Wilson (1869–1959) for his pioneering work on the scattering of high-energy photons by electrons. (Wilson received his share of that year's prestigious award in physics for his invention of the cloud chamber.)

It was Wilson's cloud chamber that helped Compton verify the behavior of X-ray scattered recoiling electrons. Telltale cloud tracks of recoiling electrons provided Compton his corroborating evidence of the particlelike behavior of electromagnetic radiation. His precise experiments depicted the increase in wavelength of X-rays due to the scattering of the incident radiation by free electrons. Since Compton's results implied that the scattered X-ray photons had less energy than the original X-ray photons, he became the first scientist to experimentally demonstrate the particlelike quantum nature of electromagnetic waves. His book *Secondary Radiations Produced by X-Rays,* which was published in 1922, described much of this important research and his experimental procedures. The discovery of *Compton scattering* (as the Compton effect is also called) served as the technical catalyst for the acceptance and rapid development of quantum mechanics in the 1920s and 1930s.

Compton scattering is the physical principle behind many of the advanced X-ray and gamma ray detection techniques used in contemporary high-energy astrophysics. In recognition of his uniquely important contributions to modern astronomy, the National Aeronautics and Space Administration named a large orbiting high-energy astrophysics observatory, the *Compton Gamma Ray Observatory* (CGRO), in his honor. NASA's space shuttle placed this important scientific spacecraft into orbit around Earth in April 1991. Its suite of gamma ray instruments operated successfully until June 2000 and provided scientists with unique astrophysical data in the gamma ray portion of the electromagnetic spectrum—an important spectral region not observable by instruments located on Earth's surface.

In 1923 Compton became a physics professor at the University of Chicago. Once settled in at the new campus, he resumed his world-changing research with X-rays. An excellent teacher and experimenter, he wrote the 1926 textbook *X-Rays and Electrons* to summarize and propagate his pioneering research experiences. From 1930 to 1940 Compton led a worldwide scientific study to measure the intensity of cosmic rays and to determine any geographic variation in their intensity. His precise measurements showed that cosmic ray intensity actually correlated with geomagnetic latitude rather than geographic latitude. Compton's results implied that cosmic rays were very energetic charged particles interacting with Earth's magnetic field. His pre–space age efforts became a major contribution to space physics and stimulated a great deal of scientific interest in understanding the Earth's magnetosphere, an interest that gave rise to many of the early satellite payloads, including JAMES VAN ALLEN'S instruments on the American *Explorer 1* satellite.

During World War II Compton played a major role in the development and use of the American atomic bomb. He served as a senior scientific adviser and was also the director of the Manhattan Project's Metallurgical Laboratory (Met Lab) at the University of Chicago. Under Compton's leadership in the Met Lab program, the brilliant Italian-American physicist ENRICO FERMI (1901–54) was able to construct and operate the world's first nuclear reactor on December 2, 1942. This successful uranium-graphic reactor, called Chicago Pile One, became the technical ancestor for the large plutonium-production reactors built at Hanford, Washington. The Hanford reactors produced the plutonium used in the world's first atomic explosion, the Trinity device detonated in southern New Mexico on July 16, 1945, and also in the Fat Man atomic weapon dropped on Nagasaki, Japan, on August 9, 1945. Compton described his wartime role and experiences in the 1956 book *Atomic Quest—A Personal Narrative.*

Following World War II he put aside physics research and followed his family's tradition of Christian service to education by accepting the position of chancellor at Washington University in Saint Louis, Missouri. He served the university well as its chancellor until 1953 and then continued his relationship as a professor of natural philosophy until failing health forced him to retire in 1961. He died in Berkeley, California, on March 15, 1962.

Compton Gamma Ray Observatory **(CGRO)** The *Compton Gamma Ray Observatory* (CGRO) was one of the four great observatories developed by NASA to support space-based astronomy. The other three observatories in this special scientific spacecraft family are the *Chandra X-Ray Observatory* (CXO), the *Hubble Space Telescope* (HST), and the *Spitzer Space Telescope* (SST).

The CGRO was deployed successfully into low-Earth orbit (LEO) by the crew of space shuttle *Atlantis* on April 7, 1991, during shuttle mission STS-37. The observatory was then boosted to a higher circular orbit where it could accomplish its scientific mission. This large, 16,300-kg spacecraft

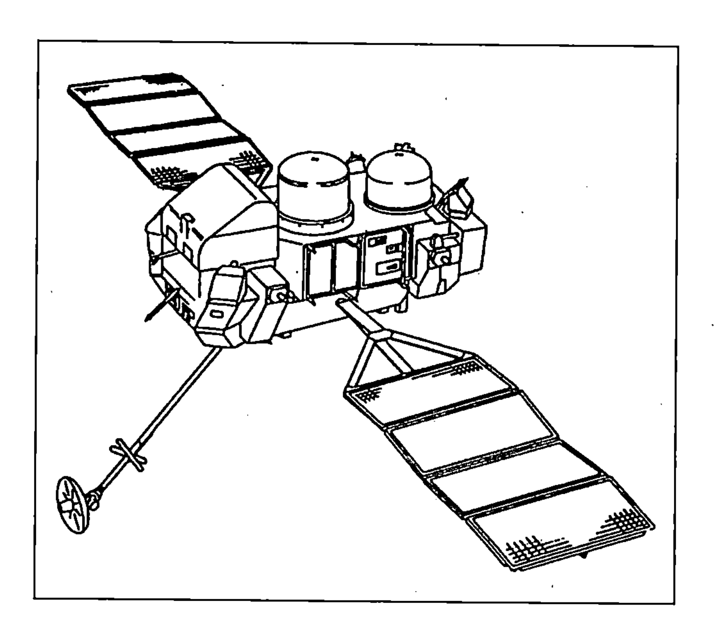

From 1991 to 2000 NASA's *Compton Gamma Ray Observatory* (CGRO) collected valuable data for scientists investigating unusual cosmic objects and phenomena that emit energetic gamma rays. *(Drawing courtesy of NASA)*

carried a variety of sensitive instruments designed to detect gamma rays over an extensive range of energies from about 30 kiloelectron volts (keV) to 30 billion electron volts (GeV). The CGRO was an extremely powerful tool for investigating some of the most puzzling astrophysical mysteries in the universe, including energetic gamma-ray bursts, pulsars, quasars, and active galaxies. NASA named this spacecraft in honor of the American physicist ARTHUR HOLLY COMPTON. At the end of its useful scientific mission, flight controllers intentionally commanded the massive spacecraft to perform a deorbit burn. This action caused the CGRO to reenter Earth's atmosphere in June 2000 and safely plunge into a remote region of the Pacific Ocean.

The CGRO carried a complement of four instruments that provided simultaneous observations, covering five decades of gamma-ray energy from 30 keV to 30 GeV. In order of increasing spectral energy coverage, they were the burst and transient source experiment (BATSE), the oriented scintillation spectrometer experiment (OSSE), the imaging Compton telescope (COMPTEL), and the energetic gamma ray experiment telescope (EGRET). For each of the instruments, an improvement in sensitivity of better than a factor of ten was realized over previous missions.

The four CGRO instruments were much larger and more sensitive than any gamma-ray telescopes previously flown in space. The large size was necessary because the number of gamma-ray interactions that can be recorded is directly related to the mass of the detector. Since the number of gamma-ray photons from celestial sources is very small compared to the number of optical photons, astrophysicists must use large instruments to detect a significant number of gamma rays in a reasonable amount of time.

An appreciation of the purpose and design of the CGRO's four instruments can be gained from understanding that above the typical energies of X-ray photons (~10 keV, or about 10,000 times the energy of optical photons) materials cannot easily refract or reflect the incoming radiation to form a picture. As a consequence, scientists had to use alternative methods to collect gamma-ray photons and thereby form images of gamma-ray sources in the sky. At gamma-ray energies astrophysicists elected to use three methods, sometimes in combination: (1) partial or total absorption of the gamma ray's energy within a high-density medium, such as a large crystal of sodium iodide, (2) collimation using heavy absorbing material to block out most of the sky and create a small field of view, and (3) at sufficiently high energies, use of the conversion process from gamma rays to electron-positron pairs in a spark chamber, which leaves a telltale directional signature of the incoming photon.

The Compton Observatory had a diverse scientific agenda that included studies of very energetic celestial phenomena: solar flares, gamma-ray bursts, pulsars, nova and supernova explosions, accreting black holes of stellar mass, quasar emission, and interactions of cosmic rays with the interstellar medium. Scientists have made many exciting discoveries using the CGRO's instruments, some previously anticipated and some completely unexpected. For example, they discovered that the all-sky map produced by EGRET was dominated by emission from interactions between cosmic rays and the interstellar gas along the plane of the Milky Way galaxy. Some point sources in this map are pulsars along the plane. At least seven pulsars are now known to emit in the gamma-ray portion of the spectrum, and five of these gamma-ray pulsars have been discovered since the CGRO was launched. One of the major discoveries made by EGRET was the class of objects known as blazars—quasars that emit the majority of their electromagnetic energy in the 30 MeV to 30 GeV portion of the spectrum. Blazars, which are at cosmological distances, have sometimes been observed to vary on time scales of days.

An all-sky map made by COMPTEL demonstrated the power of imaging in a narrow band of gamma-ray energy. This particular map revealed unexpectedly high concentrations of radioactive aluminum 26 in small regions. A COMPTEL (gamma ray) image made several interesting high-energy objects "visible," including two pulsars, a flaring black hole candidate, and a gamma-ray blazar. In another map of the galactic center region made by OSSE, the instrument's scanning observations revealed gamma-ray radiation from the annihilation of positrons and electrons in the interstellar medium. The spectrum of a solar flare recorded by OSSE gave scientists direct evidence of accelerated particles smashing into material on the Sun's surface, exciting nuclei that then radiated in gamma rays.

One of BATSE's primary objectives was the study of the mysterious phenomenon of gamma-ray bursts—brief flashes of gamma rays that occur at unpredictable locations in the sky. BATSE's all-sky map of burst positions showed that unlike galactic objects, which cluster near the plane or center of the Milky Way galaxy, these bursts come from all directions. Through the use of CGRO data astrophysicists have now established a cosmological origin for gamma-ray bursts—that is, one well beyond our galaxy. Burst light curves suggest that a chaotic phenomenon is at work; no two have ever appeared exactly the same. An average light curve for bright and dim gamma-ray bursts appears consistent with the current explanation that these bursts take place at cosmological distances: the dim ones, which presumably are farther away, are stretched more in cosmic time than are the bright ones, as the events participate in the general expansion of the universe.

The CGRO spacecraft was a three-axis stabilized, free-flying Earth satellite capable of pointing at any celestial target for a period of 14 days or more with an accuracy of 0.5°. Absolute timing was accurate to 0.1 millisecond (ms). This important orbiting laboratory had an onboard propulsion system with approximately 1,860 kg of monopropellant hydrazine for orbit maintenance. At the end of CGRO's scientific mission in June 2000, flight controllers used an intentionally reserve portion of this propellant supply to have the spacecraft successfully execute a carefully planned deorbit burn with a subsequent controlled reentry into a preselected remote area of the Pacific Ocean.

Compton scattering (Compton effect) The scattering of energetic photons (either X-ray or gamma-ray) by electrons.

In this process the electron gains energy and recoils, and the scattered photon changes direction while losing some of its energy and increasing in wavelength. First observed in 1923 by ARTHUR HOLLY COMPTON, this phenomenon demonstrated that photons have momentum and has proven very useful in gamma-ray detection.

concave lens (or mirror) A lens or mirror with an inward curvature.

Condon, Edward Uhler (1902–1974) American *Physicist* Edward Condon was a theoretical physicist who served as the director of an investigation sponsored by the U.S. Air Force (USAF) concerning unidentified flying object (UFO) sighting reports. These reports were accumulated between 1948 and 1966 under USAF Project Blue Book (and its predecessors). Condon's team at the University of Colorado investigated various cases and then wrote the report *Scientific Study of Unidentified Flying Objects*. This document, sometimes called the *Condon Report*, helped the secretary of the U.S. Air Force decide to terminate Project Blue Book in 1969. In making his decision, the secretary cited that there was no evidence to indicate that any of the sightings categorized as unidentified were extraterrestrial in origin or posed a threat to national security.

See also UNIDENTIFIED FLYING OBJECT (UFO).

conduction (thermal) The transport of heat (thermal energy) through an object by means of a temperature difference from a region of higher temperature to a region of lower temperature. For solids and liquid metals, thermal conduction is accomplished by the migration of fast-moving electrons, while atomic and molecular collisions support thermal conduction in gases and other liquids. *Compare with* CONVECTION.

Congreve, Sir William (1772–1828) British *Army Officer, Military Engineer* While a colonel of artillery, Sir William Congreve examined black powder (gunpowder) rockets captured during battles in India and then supervised the development of a series of improved British military rockets. In 1804 he wrote *A Concise Account of the Origin and Progress of the Rocket System.* During this period he also supervised the construction of a wide variety of gunpowder-fueled military rockets. His rockets ranged in mass from about 150 kg down to 8 kg.

Congreve's efforts provided the British Army with two basic types of assault rockets: the shrapnel (case-shot) rocket and the incendiary rocket. The shrapnel rocket often substituted for artillery. When this weapon flew over enemy troops, its exploding warhead showered the battlefield with rifle balls and pieces of sharp metal. Congreve filled the warhead of his incendiary rocket with a sticky, flammable material that quickly started fires when it impacted in an enemy city or in the rigging of an enemy sailing ship. His pioneering work on these early solid-propellant rockets represents an important technical step in the overall evolution of the modern military rocket.

British forces used Congreve's rockets quite effectively in large-scale bombardments during the Napoleonic Wars and the War of 1812. Perhaps the most famous application of Congreve's rockets took place in August 1814 during the British bombardment of the American Fort McHenry in the War of 1812. Throughout this attack a young American lawyer and prisoner exchange negotiator named Francis Scott Key (1780–1843) remained under guard first onboard the British ship *H.M.S. Surprise* and later on a sloop anchored behind the British battle fleet. Throughout the night Key witnessed the relentless naval bombardment of the fort, including the glowing red flames from Congreve's rockets as they wobbled skyward toward the fort from numerous small launching boats. At dawn the American flag still flew over the fort despite the massive British bombardment. Key immortalized the event and Congreve's rockets when he wrote the "rocket's red glare" phrase in a poem that became "The Star-Spangled Banner."

conic section A curve formed by the intersection of a plane and a right circular cone. Also called *conic.* The conic sections are the *ellipse,* the *parabola,* and the *hyperbola*—all curves that describe the paths of bodies moving in space. The *circle* is simply an ellipse with an eccentricity of zero.

conjunction The alignment of two bodies in the solar system so that they have the same celestial longitude as seen from Earth—that is, when they appear closest together in the

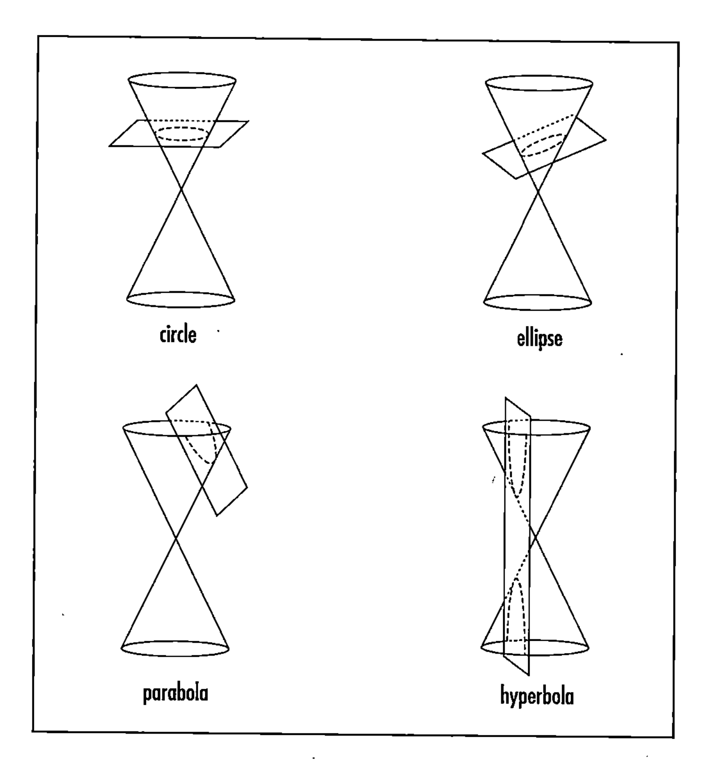

The basic conic sections

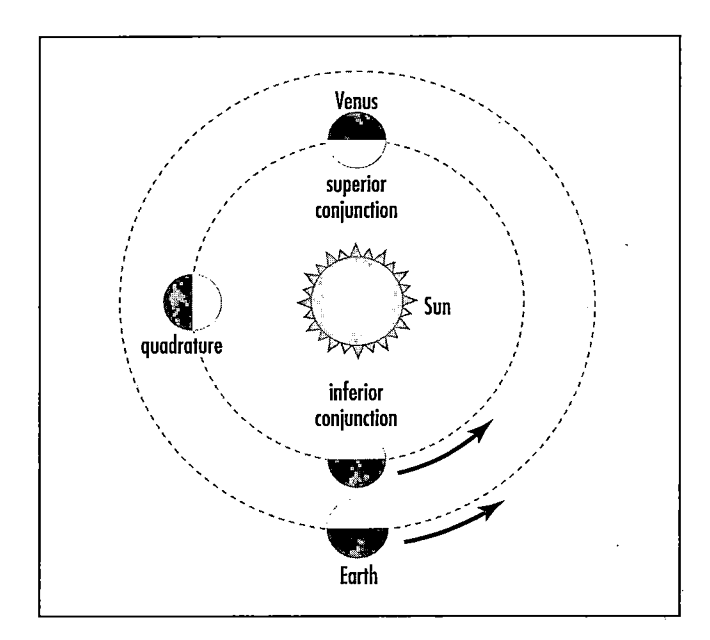

The planetary alignments of Earth and Venus within the inner solar system for inferior conjunction, superior conjunction, and quadrature

sky. For example, a superior planet forms a *superior conjunction* when the Sun lies between it and Earth. An inferior planet (either Venus or Mercury) forms an *inferior conjunction* when it lies directly between the Sun and Earth and a *superior conjunction* when it lies directly behind the Sun.

Conrad, Charles (Pete), Jr. (1930–1999) American *Astronaut, U.S. Navy Officer* Charles (Pete) Conrad served as the spacecraft commander during NASA's *Apollo 12* lunar landing mission (November 14–24, 1969). He was the third human being to walk on the surface of the Moon. As a prelude to his "Moon walk," he flew into space in 1965 as part of the *Gemini-Titan V* mission of NASA's Gemini Project and again in 1966 during the *Gemini-Titan XI* mission. Finally, a seasoned astronaut, he also served as the commander of NASA's *Skylab SL-2* mission from May 25 to June 22, 1973, making mission-saving repairs on the first American space station.

conservation of angular momentum The principle in physics that states that absolute angular momentum is a property that cannot be created or destroyed but can only be transferred from one physical system to another through the action of a net torque on the system. As a consequence the total angular momentum of an isolated physical system remains constant.

conservation of charge A principle in physics that states that for an isolated system, the total net charge is a constant; that is, charge can neither be created nor destroyed.

conservation of energy The principle in physics that states that the total energy of an isolated system remains constant if no interconversion of mass and energy takes place within the system. Also called the *first law of thermodynamics*.

See also CONSERVATION OF MASS AND ENERGY.

conservation of mass The principle in classical (Newtonian) physics that states that mass can neither be created nor destroyed but only transferred from one volume to another.

See also CONSERVATION OF MASS AND ENERGY.

conservation of mass and energy From special relativity and ALBERT EINSTEIN's famous mass-energy equivalence formula ($E = \Delta m\ c^2$), this conservation principle states that for an isolated system, the sum of the mass and energy remains constant, although interconversion of mass and energy can occur within the system.

conservation of momentum The principle in physics that states that in the absence of external forces, absolute momentum is a property that cannot be created or destroyed.

See also NEWTON'S LAWS OF MOTION.

conservation principle A principle or law in physics that states that the total quantity or magnitude of a certain physical property of a system (e.g., mass, charge, or momentum) remains unchanged even though there may be exchanges of that property between components within the system. Scientists consider such conservation principles to be universally true; so, for example, the conservation of momentum principle applies both on Earth as well as on Mars.

console A desklike array of controls, indicators, and video display devices for monitoring and controlling aerospace operations, such as the checkout, countdown, and launch of a rocket. During the critical phases of a space mission, the console becomes the central place from which to issue commands or at which to display information concerning an aerospace vehicle, a deployed payload, an system-orbiting spacecraft, or a planetary probe. The *mission control center* generally contains clusters of consoles, each assigned to specific monitoring and control tasks. Depending on the nature and duration of a particular space mission, operators will remain at their consoles continuously or only work there intermittently.

constellation (aerospace) A term used to collectively describe the number and orbital disposition of a set of satellites, such as the constellation of *Global Positioning System* (GPS) satellites.

constellation (astronomy) An easily identifiable (with the naked eye) configuration of the brightest stars in a moderately small region of the night sky. Originally there was not a single set of constellations recognized by all astronomers. Rather, early astronomers in many regions of the world often defined and named the particular collections of bright stars they observed in relatively small regions of the sky after specific heroes, events, and creatures from their ancient cultures and mythologies. The astronomical heritage of the constellations used in contemporary astronomy began about 2500 B.C.E. with the ancient stargazers of Mesopotamia. These early

The Ancient Constellations

Name/Meaning (Latin [English])	Genitive Form of Latin Name	Abbreviation	Approximate Position (Equatorial Coordinates) RA(h)	δ(°)
Andromeda (name: princess)	Andromedae	And	1	+40
Aquarius (water bearer)	Aquarii	Aqr	23	–15
Aquila (eagle)	Aquilae	Aql	20	+5
Ara (altar)	Arae	Ara	17	–55
Argo Navis (ship of Argonauts), now split into the modern constellations: Carina, Puppis, Pyxis, and Vela				
Aries (ram)	Arietis	Ari	3	+20
Auriga (charioteer)	Aurigae	Aur	6	+40
Boötes (herdsman)	Boötis	Boo	15	+30
Cancer (crab)	Cancri	Cnc	9	+20
Canis Major (great dog)	Canis Majoris	CMa	7	–20
Canis Minor (little dog)	Canis Minoris	CMi	8	+5
Capricornus (sea goat)	Capricorni	Cap	21	–20
Cassiopeia (name: queen)	Cassiopeiae	Cas	1	+60
Centaurus (centaur)	Centauri	Cen	13	–50
Cepheus (name: king)	Cephei	Cep	22	+70
Cetus (whale)	Ceti	Cet	2	–10
Corona Austrina (southern crown)	Coronae Australis	CrA	19	–40
Corona Borealis (northern crown)	Coronae Borealis	CrB	16	+30
Corvus (crow)	Corvi	Crv	12	–20
Crater (cup)	Crateris	Crt	11	–15
Cygnus (swan)	Cygni	Cyg	21	+40
Delphinus (dolphin)	Delphini	Del	21	+10
Draco (dragon)	Draconis	Dra	17	+65
Equuleus (little horse)	Equulei	Equ	21	+10
Eridanus (name: river)	Eridani	Eri	3	–20
Gemini (twins)	Geminorum	Gem	7	+20
Hercules (name: hero)	Herculis	Her	17	+30
Hydra (sea serpent; monster)	Hydrae	Hya	10	–20
Leo (lion)	Leonis	Leo	11	+15
Lepus (hare)	Leporis	Lep	6	–20
Libra (scale; balance beam)	Librae	Lib	15	–15
Lupus (wolf)	Lupi	Lup	15	–45
Lyra (lyre)	Lyrae	Lyr	19	+40
Ophiuchus (serpent bearer)	Ophiuchii	Oph	17	0
Orion (name: great hunter)	Orionis	Ori	5	0
Pegasus (name: winged horse)	Pegasi	Peg	22	+20
Perseus (name: hero)	Persei	Per	3	+45
Pisces (fishes)	Piscium	Psc	1	+15
Piscis Austrinus (southern fish)	Piscis Austrini	PsA	22	–30
Sagitta (arrow)	Sagittae	Sge	20	+10
Sagittarius (archer)	Sagittarii	Sgr	19	–25
Scorpius (scorpion)	Scorpii	Sco	17	–40
Serpens (serpent)	Serpentis	Ser	17	0
Taurus (bull)	Tauri	Tau	4	+15
Triangulum (triangle)	Trianguli	Tri	2	+30
Ursa Major (great bear)	Ursae Majoris	UMa	11	+50
Ursa Minor (little bear)	Ursae Minoria	UMi	15	+70
Virgo (virgin; maiden)	Virginis	Vir	13	0

The Modern Constellations

Name/Meaning [Latin (English)]	Genitive Form of Latin Name	Abbreviation	Approximate Position (Equatorial Coordinates)	
			RA (h)	δ (°)
Antlia (air pump)	Antliae	Ant	10	–35
Apus (bird of paradise)	Apodis	Aps	16	–75
Caelum (sculptor's chisel)	Caeli	Cae	5	–40
Camelopardalis (giraffe)	Camelopardalis	Cam	6	+70
Canes Venatici (hunting dogs)	Canum Venaticorum	CVn	13	+40
Carina (keel)*	Carinae	Car	9	–60
Chamaeleon (chameleon)	Chamaeleontis	Cha	11	–80
Circinus (compasses)	Circini	Cir	15	–60
Columba (dove)	Columbae	Col	6	–35
Coma Berenices (Berenice's hair)	Comae Berenices	Com	13	+20
Crux (southern cross)	Crucis	Cru	12	–60
Dorado (swordfish)	Doradus	Dor	5	–65
Fornax (furnace)	Fornacis	For	3	–30
Grus (crane)	Gruis	Gru	22	–45
Horologium (clock)	Horologii	Hor	3	–60
Hydrus (water snake)	Hydri	Hyi	2	–75
Indus (Indian)	Indi	Ind	21	–55
Lacerta (lizard)	Lacertae	Lac	22	+45
Leo Minor (little lion)	Leonis Minoris	LMi	10	+35
Lynx (lynx)	Lyncis	Lyn	8	+45
Mensa (table mountain)	Mensae	Men	5	–80
Microscopium (microscope)	Microscopii	Mic	21	–35
Monoceros (unicorn)	Monocerotis	Mon	7	–5
Musca (fly)	Muscae	Mus	12	–70
Norma (carpenter's square)	Normae	Nor	16	–50
Octans (octant; navigation device)	Octantis	Oct	22	–85
Pavo (peacock)	Pavonis	Pav	20	–65
Phoenix (Phoenix; mythical bird)	Phoenicis	Phe	1	–50
Pictor (painter's easel)	Pictoris	Pic	6	–55
Puppis (stern)*	Puppis	Pup	8	–40
Pyxis (nautical compass)*	Pyxidis	Pyx	9	–30
Reticulum (net)	Reticuli	Ret	4	–60
Sculptor (sculptor's workshop)	Sculptoris	Scl	0	–30
Scutum (shield)	Scuti	Sct	19	–10
Sextans (sextant)	Sextantis	Sex	10	0
Telescopium (telescope)	Telescopii	Tel	19	–50
Triangulum Australe (southern triangle)	Trianguli Australe	TrA	16	–65
Tucana (toucan)	Tucanae	Tuc	0	–65
Vela (sail)*	Velorum	Vel	9	–50
Volans (flying fish)	Volantis	Vol	8	–70
Vulpecula (fox)	Vulpeculae	Vul	20	+25

*originally part of ancient constellation Argo Navis (ship of Argonauts)

peoples used the stars to tell stories, to honor heroes and ferocious creatures—such as Orion the hunter and Ursa Major (the Great Bear)—and to remind each new human generation that the heavens were the abode of the gods. In these Mesopotamian societies (naked eye) astronomy, mythology, religion, and cultural values were closely interwoven.

Early Greek astronomers adopted the constellations they found in Mesopotamia, embellished them with their own myths and religious beliefs, and eventually created a set of 48 ancient constellations. EUDOXUS OF CNIDUS was one of the first Greeks to formally codify these ancient constellations. HIPPARCHUS reinforced his work about a century later. Finally, PTOLEMY cast the 48 ancient constellations in their present form about 150 C.E. in his great compilation of astronomical knowledge, *Syntaxis*. The early Greek astronomers were also keenly aware that these 48 constellations did not account for all the stars in the night sky. They even used a special word, *amorphotoi* (meaning "unformed"), to describe the spaces in the night sky populated by dim stars between the prominent groups of stars constituting the ancient constellations. The Ancient Constellations table contains a list of the 48 ancient Greek constellations.

Today astronomers officially recognize all but one of these ancient star patterns. The somewhat cumbersome constellation Argo Navis has now been broken up into four new constellations (as discussed shortly). By convention, when astronomers formally refer to a constellation such as Centaurus in English, they use the proper name "The Centaur." Similarly, when astronomers wish to describe celestial objects within a particular constellation they use the genitive form of the constellation's Latin name—as, for example, *Alpha Centauri* (α Cen) to describe the brightest (binary) star in the constellation Centaurus.

Astronomers in the Roman Empire were quite content to accept and use Greek celestial figures and constellations. Their primary contribution was the use of "Roman" (Latin) names for many familiar celestial objects, a tradition and heritage still followed by astronomers today. Therefore, as part of the Pax Romana, the use of the 48 ancient Greek constellations spread throughout the civilized (Western) world.

As the Roman Empire collapsed and the Dark Ages spread throughout western Europe, knowledge of Greek astronomy survived and began to flourish in Arab lands. Arab astronomers discovered Ptolemy's great work, translated it, and then renamed it *The Almagest.* By so doing, they preserved the astronomical heritage of the 48 ancient Greek constellations. Arab astronomers also refined and embellished the knowledge base of ancient Greek naked eye astronomy by providing new star names and more precise observations. Finally, the astronomical heritage of the ancient constellations returned to Europe just as people there were awakening from the Dark Ages, experiencing the Renaissance, and paving the way for the scientific revolution.

As part of the explosive interest in astronomy that occurred at the start of the scientific revolution, the German astronomer JOHANN BAYER published the important work *Uranometria* in 1603. This book was the first major star catalog for the entire celestial sphere. Bayer charted more than 2,000 stars visible to the naked eye and introduced the practice of assigning Greek letters (such as alpha α, beta β, and gamma γ) to the main stars in each constellation, usually in an approximate (descending) order of their brightness. Expanding the legacy of 48 constellations from ancient Greece, Bayer named 12 new southern hemisphere constellations: Apus, Chamaeleon, Dorado, Grus, Hydrus, Indus, Musca (originally called Apis [bee] by Bayer), Pavo, Phoenix, Triangulum Australe, Tucana, and Volans. The Modern Constellations Table describes these new constellations as well as the other modern constellations officially recognized by the International Astronomical Union since 1929.

Using the newly invented astronomical telescope the Polish-German astronomer JOHANNES HEVELIUS filled in some of the empty spaces *(amorphotoi)* in the Northern Hemisphere of the celestial sphere by identifying the following new constellations: Canes Venatici, Lacerta, Lynx, and Leo Minor. Then, in the 18th century the French astronomer Abbé NICOLAS-LOUIS DE LACAILLE described 14 new constellations he found in the Southern Hemisphere and named some of them after scientific artifacts and instruments emerging during the period. His newly identified constellations included: Antlia, Caelum, Circinus, Fornax, Horologium, Mensa, Microscopium, Norma, Octans, Pictor, Pyxis, Reticulum, Sculptor, and Telescopium. Lacaille was a precise technophile, so many of the names he carefully selected for his newly identified constellations were long and detailed. For example, Antlia (the Air Pump) honors the device invented by the British scientist Robert Boyle (1627–91), and Fornax actually means "the Laboratory Furnace." By international agreement, modern astronomers generally use shortened versions of Lacaille's original names—such as simply "Furnace" instead of "Laboratory Furnace" for Fornax. This is done for ease in technical communications; it does not represent an attempt to detract from Lacaille's important work.

Lacaille and other astronomers of his era also dismantled the cumbersome ancient constellation Argo Navis (ship of Jason and the Argonauts) and carved up the stars in this large Southern Hemisphere constellation into four smaller, more manageable ones whose names retain the original nautical theme: Carina (the Keel), Puppis (the Stern), Pyxis (the Nautical Compass), and Vela (the Sail).

Today astronomers have 88 officially recognized constellations. All of these constellations may be found (in alphabetical order) by combining the two Constellations Tables. The tables include the position of each constellation on the celestial sphere as expressed in the equatorial coordinates (right ascension [RA] and declination [δ]) that correspond to the approximate center of the constellation.

See also ZODIAC.

Constellation X-Ray Observatory (***Constellation-X***) The future NASA scientific mission that involves an array of X-ray telescopes working together in tight orbit to improve by a hundredfold how astronomers and astrophysicists observe the universe in the X-ray portion of the electromagnetic spectrum. The current plan calls for four satellites operating in unison to generate the observing power of one giant space-based X-ray telescope. (*See* figure, page 152.) The *Constellation-X* satellites will house high-resolution X-ray spectroscopy telescopes that collect high-energy X-rays produced by cataclysmic cosmic events and then interpret those event-related X-rays as spectra. Scientists regard X-ray spectra as the fingerprints of the chemicals producing the X-rays.

When observations begin (about 2010) data from *Constellation-X* will help scientists resolve many pressing issues that currently challenge their understanding of the laws of physics. For example, *Constellation-X* observations of iron spectra in the vicinity of suspected massive black holes will help astrophysicists test ALBERT EINSTEIN's theory of general relativity in an environment of extreme gravity. *Constellation-X* will also help scientists determine how black holes evolve and generate energy, thereby providing important information about the total energy content of the universe. With data from this observatory scientists will also be able to investigate galaxy formation, the evolution of the universe on large scales, the nature of dark matter, and how the universe recycles its matter and energy—for example, how heavier ele-

An artist's rendering of NASA's planned Constellation X-Ray mission, a team of powerful X-ray telescopes that orbit close to one another and work in unison to simultaneously observe the same distant objects. By combining their data, the satellites become 100 times more powerful than any previous single X-ray telescope. ***(Artist rendering courtesy of NASA)***

ments from the cores of exploding stars eventually form planets and comets and how the gas from old stars helps make new ones.

Like all X-ray telescopes, *Constellation-X* must operate in outer space because X-ray photons from cosmic phenomena do not penetrate very far into Earth's atmosphere. The scientists and engineers who designed *Constellation-X* wanted to create an X-ray observatory capable of collecting as much X-ray "light" as possible, imitating to the greatest extent possible the way giant ground-based optical telescopes such as the Keck telescope use their large optics to gather as much visible light as possible from distant celestial objects. These demanding requirements led NASA personnel to select a unique multisatellite design for *Constellation-X*. The four satellites will be of identical low-mass design, allowing each spacecraft to be launched individually or possibly in pairs. The construction of four identical spacecraft reduces the overall cost of the mission and also avoids the risk of complete mission failure should a single launch abort occur. Once successfully co-orbited, the combined capability of *Constellation-X*'s four X-ray telescopes will provide a level of sensitivity that is 100 times greater than any past or current X-ray satellite mission.

Essentially, scientists using *Constellation-X* will be able to collect more data in an hour than they can now collect in days or weeks with current space-based X-ray telescopes. Of special importance to astronomers and astrophysicists is the fact that *Constellation-X* will also allow them to discover and analyze thousands of faint X-ray-emitting sources, not just the bright sources available today. NASA's *Constellation X-Ray Observatory* promises to stimulate a revolution in X-ray astronomy and a much deeper understanding of energetic phenomena taking place throughout the universe.

See also CHANDRA X-RAY OBSERVATORY; ROSSI X-RAY TIMING EXPLORER; XMM-NEWTON SPACECRAFT; X-RAY ASTRONOMY.

continuously-crewed spacecraft A spacecraft that has accommodations for continuous habitation (human occupancy) during its mission. The *International Space Station* (ISS) is an example. Sometimes (though not preferred) called a *continuously manned spacecraft.*

continuously habitable zone (CHZ) The region around a star in which one or several planets can maintain conditions appropriate for the emergence and sustained existence of life. One important characteristic of a planet in the CHZ is that its environmental conditions support the retention of significant amounts of liquid water on the planetary surface. Sometimes called the *Goldilocks zone;* a planet in the CHZ is sometimes called a Goldilocks planet.

See also ECOSPHERE; EXOBIOLOGY; LIFE IN THE UNIVERSE.

control rocket A low-thrust rocket, such as a retrorocket or a vernier engine, used to guide, to change the attitude of, or to make small corrections in the velocity of an aerospace vehicle, spacecraft, or expendable launch vehicle.

control system (missile) A system that serves to maintain attitude stability and to correct deflections.

See also GUIDANCE SYSTEM.

control vane A movable vane used for control; especially a movable air vane or exhaust jet vane on a rocket used to control flight attitude.

convection 1. In heat transfer, mass motions within a fluid resulting in the transport and mixing of the properties of that fluid. The up-and-down drafts in a fluid heated from below in a gravitational environment. Because the density of the heated fluid is lowered, the warmer fluid rises (*natural convection*); after cooling, the density of the fluid increases, and it tends to sink. Pumps and fans promote *forced convection* when a mass of warmer fluid is driven through or across cooler surfaces or a mass of cooler fluid is driven across warmer surfaces. *Compare to* CONDUCTION and RADIATION. 2. In planetary science, atmospheric or oceanic motions that are predominately vertical and that result in the vertical transport and mixing of atmospheric or oceanic properties. Because the most striking meteorological features result if atmospheric convective motion occurs in conjunction with the rising current of air (i.e., updrafts), convection sometimes is used to imply only upward vertical motion.

converging-diverging (CD) nozzle A thrust-producing flow device for expanding and accelerating hot exhaust gases from a rocket engine. A properly designed nozzle efficiently converts the thermal energy of combustion into kinetic energy of the combustion product gases. In a supersonic converging-diverging nozzle, the hot gas upstream of the nozzle throat is at subsonic velocity (i.e., the Mach number [M] < 1), reaches sonic velocity (the speed of sound, for which $M = 1$) at the throat of the nozzle, and then expands to supersonic velocity ($M > 1$) downstream of the nozzle throat region while flowing through the diverging section of the nozzle.

See also DE LAVAL NOZZLE.

converging lens (or mirror) A lens (or mirror) that refracts (or reflects) a parallel beam of light, making it converge at a point (called the principal focus). A converging mirror is a concave mirror, while a converging lens is generally a convex lens that is thicker in the middle than at its edges. *Compare with* DIVERGING LENS.

convex lens (or mirror) A lens or mirror with an outward curvature. *Compare with* CONCAVE LENS.

Cooper, Leroy Gordon, Jr. (1927–2004) American *Astronaut, U.S. Air Force Officer* While an officer in the U.S. Air Force, Leroy Gordon Cooper was selected to become one of NASA's seven original Mercury Project astronauts. He flew the last Mercury Project mission, called *Faith-7* (or *Mercury-Atlas 9*), on May 15–16, 1963. During that mission he became the first human to perform a pilot-controlled reentry of a space capsule. In 1965 he flew again into outer space as part of the *Gemini-Titan V* mission of NASA's Gemini Project.

cooperative target A three-axis stabilized orbiting object that has signaling devices to support rendezvous and docking/capture operations by a chaser spacecraft.

co-orbital Sharing the same or very similar orbit; for example, during a rendezvous operation the chaser spacecraft and its cooperative target are said to be co-orbital.

coordinated universal time *See* UNIVERSAL TIME.

coordinate system A system that uses linear or angular quantities to designate the position of a point with respect to a selected reference position (called the origin) and an appropriate reference surface or intersecting surfaces.

See also CARTESIAN COORDINATES; CELESTIAL COORDINATES; CYLINDRICAL COORDINATES; POLAR COORDINATE SYSTEM.

Copernican system The theory of planetary motions proposed by NICHOLAS COPERNICUS in which all planets (including Earth) move in *circular* orbits around the Sun, with the planets closer to the Sun moving faster. In this system, the hypothesis of which helped trigger the scientific revolution of the 16th and 17th centuries, Earth was viewed not as an immovable object at the center of the universe (as in the geocentric Ptolemaic system) but rather as a planet orbiting the Sun between Venus and Mars. Early in the 17th century JOHANNES KEPLER showed that while Copernicus's heliocentric hypothesis was correct, the planets actually moved in (slightly) elliptical orbits around the Sun.

Copernicus, Nicholas (Nicolaus) (1473–1543) Polish *Astronomer, Church Official* This Polish astronomer and church official triggered the scientific revolution of the 17th century with his book *On the Revolution of Celestial Spheres*. Published in 1543 while Copernicus lay on his deathbed, this book overthrew the Ptolemaic system by boldly suggesting a heliocentric model for the solar system in which Earth and all the other known planets moved around the Sun. His heliocentric model (possibly derived from the long-forgotten ideas of ARISTARCHUS OF SAMOS) caused much technical, political, and social upheaval before finally displacing two millennia of Greek geocentric cosmology.

Nicholas Copernicus was born on February 19, 1473, in Torun, Poland. When Copernicus's father died in 1483, his uncle (a powerful prince-bishop) raised him and provided the young man an extensive education. From 1491 to 1494 he studied mathematics at the University of Cracow, the leading institute of learning in Poland. Then he traveled to Italy in 1496 to broaden his knowledge and remained there for about a decade. He studied medicine in Padua (from 1501 to 1505). Later, however, he grew interested in astronomy because of lectures he heard while attending the University of Bologna. It was probably there that Copernicus first became fascinated with the little-known hypothesis of Aristarchus of Samos that the Earth revolved around the Sun. He also became a doctor of canon law as a result of his studies at the University of Ferrara (1503).

The intellectual climate in late Renaissance Italy encouraged bright students to investigate new ideas. The more Copernicus studied astronomy, the more uncomfortable he became with traditional Greek astronomy and its Earth-centered model of the universe. For one thing, he considered the Ptolemaic system unnecessarily complex and incapable of predicting the positions of the planets over long periods. For another, he encountered the long-ignored but very intriguing thoughts of Aristarchus.

Around 260 B.C.E. this ancient Greek astronomer suggested that he could more easily understand the observed motions of the planets if he assumed that they (including Earth) revolved around the Sun. He further suggested that since the stars appeared motionless (except for diurnal motion due to Earth's rotation), they must be very far away. But the heliocentric hypothesis of Aristarchus was too revolutionary for the early Greeks. ARISTOTLE personally championed the then widely accepted geocentric model, and most ancient Greeks were culturally uncomfortable with the idea of a "moving" Earth. His book on the subject vanished in antiquity. The only contact Copernicus had with the heliocentric hypothesis of Aristarchus was probably through a brief mention of it in the writings of the great Greek mathematician Archimedes (ca. 287–212 B.C.E.). In any event, Copernicus's through investigation of geocentric Greek astronomy and its obvious deficiencies encouraged him to explore and validate the heliocentric hypothesis with a series of careful observations and calculations. His determined efforts in the early part of the 16th century would change science forever.

But caution about this new ideal was definitely the order of the day for Copernicus. In 1505 Copernicus returned to Poland and became a canon at his uncle's cathedral in Frombork. Despite his formal religious education, he never became a priest. He also chose not to marry. Instead, he held the lucrative church position of canon until his death. Because of this position, however, he remained very prudent about openly

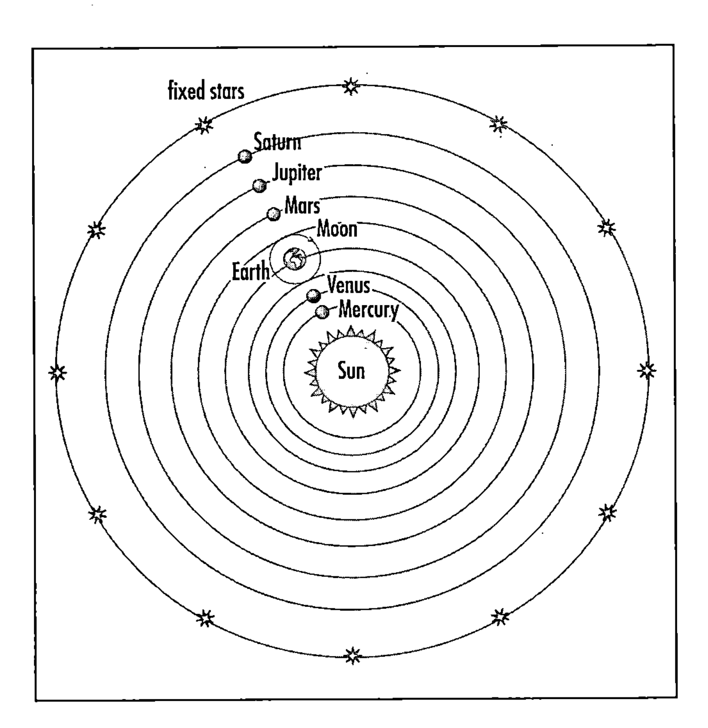

The universe according to Nicholas Copernicus. This simple heliocentric model of the solar system (using circular orbits) helped overthrow the long-standing geocentric cosmology of Aristotle and stimulated the great scientific revolution of the 17th century. ***(Drawing courtesy of NASA)***

advocating the revolutionary concept of the Sun at the center of the solar system. He reasoned quite correctly that this new model would create a direct conflict with church authorities. At that time ecclesiastical authorities regarded the geocentric model, which had been passed down from Aristotle and PTOLEMY, to be almost the equivalent of religious dogma. To openly suggest otherwise could be regarded as a form of heresy—an offense often punished by death.

Yet Copernicus enthusiastically embarked on working out all the mathematical details of his new model. From 1512 to 1529, while still performing his clerical and administrative duties as a church canon, he made careful observations of planetary motions. He found that the inability of the geocentric model to predict planetary motions quickly disappeared if he assumed that Earth and the other planets actually revolved around the Sun. But Copernicus retained some of the features of the Greek model. For example, he assumed that the planets moved in perfect circular orbits around the Sun. JOHANNES KEPLER would later correct this error by postulating that the planets move in elliptical orbits.

To avoid open conflict with church authorities, Copernicus cautiously circulated his hand-written notes to a few close friends. In 1539 the Austrian mathematician Rheticus (Georg Joachim von Lauchen [1514–76]) arrived at Frombork to study under Copernicus. He reviewed the aging astronomer's notes. Then, in a public test of the revolutionary concept, Rheticus published a summary of these notes in 1540 but was careful not to specifically mention Copernicus by name. This trial exposure of the heliocentric model actually occurred without angering church authorities. On the contrary, scientific excitement about the Copernican hypothesis spread rapidly.

As a result Copernicus finally agreed to have Rheticus supervise the publication of his complete book *De Revolutionibus Orbium Coelestium* (On the revolution of celestial spheres). To avoid any potential doctrinal problems, Copernicus dedicated the book to Pope Paul III. Unfortunately, Rheticus left the final publication steps to a Lutheran minister named Andreas Osiander. The minister, mindful that Martin Luther himself firmly opposed the new Copernican theory, added an unauthorized preface to weaken the impact of its contents. In effect, Osiander's unauthorized preface stated that Copernican theory was not being advocated as a description of the physical universe but only as a convenient way to efficiently calculate the tables of planetary motions. The book (so modified) finally appeared in 1543. Historic legend suggests that Copernicus received the first copy as he lay on his deathbed. Fortunately, Johannes Kepler discovered the unauthorized preface in 1609 and set the record straight. The cautious Polish astronomer certainly intended to change our model of the universe. Earth was definitely not the unmoving physical center of everything. It was just one of several planets following predictable pathways around their parent star.

After his death on May 24, 1543, church authorities aggressively attacked the new heliocentric model and banned Copernicus's book as heretical. His book remained on the church's official list of forbidden books until 1835. For decades after the Copernican model appeared, the church used the Inquisition to attack those who supported the new concept. Through death, Copernicus mercifully escaped the harsh public punishment inflicted on some of his later supporters. In 1600, for example, the Inquisition burned one ardent Copernican, GIORDANO BRUNO, at the stake for his beliefs. Ecclesiastical authorities condemned another Copernican, the famous Italian scientist GALILEO GALILEI, to house arrest for the remainder of his life (from about 1633 to 1642). Nevertheless, the Copernican model survived and spawned a new era in learning and wisdom. Science historians generally regard the pioneering work of Copernicus as the beginning of the scientific revolution.

Copernicus **(spacecraft)** NASA's *Copernicus* spacecraft was launched on August 21, 1972. This mission was the third in the Orbiting Astronomical Observatory (OAO) Program and the second successful spacecraft to observe the celestial sphere from above Earth's atmosphere. An ultraviolet (UV) telescope with a spectrometer measured high-resolution spectra of stars, galaxies, and planets with the main emphasis being placed on the determination of interstellar absorption lines. Three X-ray telescopes and a collimated proportional counter provided measurements of celestial X-ray sources and interstellar absorption between 1 and 100 angstroms (Å) wavelength. Also called the *Orbiting Astronomical Observatory 3* (OAO-3), its observational mission life extended from August 1972 through February 1981—some nine and a half years. NASA named this orbiting observatory in honor of the famous Polish astronomer NICHOLAS COPERNICUS.

See also ORBITING ASTRONOMICAL OBSERVATORY (OAO); ULTRAVIOLET ASTRONOMY.

Cordelia The small (about 30-km-diameter) innermost known moon of Uranus that orbits the planet at a distance of 49,770 km with a period of 0.335 days. This tiny satellite was discovered in 1986 as a result of the *Voyager 2* spacecraft encounter and subsequently named after the king's daughter in William Shakespeare's play *King Lear.* Orbiting close to Uranus, Cordelia acts as a shepherd moon for the planet's epsilon ring—as does another tiny moon, Ophelia. An inclination of 0.08° and an eccentricity of 0.00026 further characterize the orbit of Cordelia. Sometimes called Uranus VI, UVI, or S/1986 U7.

See also URANUS.

Córdoba Durchmusterung **(CD)** The *Córdoba Survey* is a massive star catalog containing approximately 614,000 stars brighter than the 10th magnitude that was compiled at the Córdoba Observatory in Argentina. This cataloging effort was finally completed in 1930 and represents an analogous extension of the *BONNER DURCHMUSTERUNG (BD) (Bonn Survey)* to the Southern Hemisphere, especially the south polar regions.

core 1. (planetary) The high-density central region of a planet. 2. (stellar) The very-high-temperature central region of a star. For main sequence stars, fusion processes within the core burn hydrogen, while for stars that have left the main sequence, nuclear fusion processes in the core involve helium and oxygen.

coriolis effect(s) 1. The physiological effects (e.g., nausea, vertigo, dizziness, etc.) felt by a person moving radially in a rotating system, such as a rotating space station. 2. The tendency for an object moving above Earth (e.g., a missile in flight) to turn to the right in the Northern Hemisphere and to the left in the Southern Hemisphere relative to Earth's surface. This effect arises because Earth rotates and is not, therefore, an inertial reference frame.

corona The outermost region of a star. The Sun's corona consists of low-density clouds of very hot gases (> 1 million K) and ionized materials.

See also SUN.

Corona **(spacecraft)** *See DISCOVERER* SPACECRAFT.

coronal hole A large region in the Sun's corona that is less dense and much cooler than the surrounding areas. The open structure of a coronal hole's magnetic field allows a constant flow of high-density plasma to stream out from the Sun. When coronal holes face in the direction of our planet, there is an increase in the intensity of the solar wind effects on Earth.

See also SPACE WEATHER; SUN.

coronal mass ejection (CME) A high-speed (10 to 1,000 km/s) ejection of matter from the Sun's corona. A CME travels through space disturbing the solar wind and giving rise to geomagnetic storms when the disturbance reaches Earth.

See also SPACE WEATHER; SUN.

cosmic Of or pertaining to the universe, especially that part outside Earth's atmosphere. This term frequently appears in the Russian (former Soviet Union) space program as the equivalent to space or astro-, such as cosmic station (versus space station) and cosmonaut (versus astronaut).

cosmic abundance of elements *See* ABUNDANCE OF ELEMENTS (IN THE UNIVERSE).

Cosmic Background Explorer **(COBE)** NASA's *Cosmic Background Explorer* (COBE) spacecraft was successfully launched from Vandenberg Air Force Base, California, by an expendable Delta rocket on November 18, 1989. The 2,270-kg spacecraft was placed in a 900-km altitude, 99°-inclination (polar) orbit, passing from pole to pole along Earth's terminator (the line between night and day on a planet or moon) to protect its heat sensitive instruments from solar radiation and to prevent the instruments from pointing directly at the Sun or Earth.

COBE's one-year space mission was to study some of the most basic questions in astrophysics and cosmology. What was the nature of the hypothesized primeval explosion (often called the *big bang*) that started the expanding universe? What started the formation of galaxies? What caused galaxies to be arranged in giant clusters with vast unbroken voids in between? Scientists have speculated for decades about the formation of the universe. The most generally accepted cosmological model is called the big bang theory of an expanding universe. The most important evidence that this gigantic explosion occurred some 15 billion years ago is the uniform diffuse cosmic microwave background (CMB) radiation that reaches Earth from every direction. This cosmic background radiation was discovered quite by accident in 1964 by ARNO ALLEN PENZIAS and ROBERT WOODROW WILSON as they were testing an antenna for satellite communications and radio astronomy. They detected a type of "static from the sky." Physicists now regard this phenomenon as the radiation remnant of the big bang event.

The COBE spacecraft carried three instruments: the far-infrared absolute spectrophotometer (FIRAS) to compare the spectrum of the cosmic microwave background radiation with a precise blackbody source, the differential microwave radiometer (DMR) to map the cosmic radiation precisely, and the diffuse infrared background experiment (DIBRE) to search for the cosmic infrared background.

The cosmic microwave background (CMB) spectrum was measured by the FIRAS instrument with a precision of 0.03 percent, and the resulting CMB temperature was found to be 2.726 ± 0.010 kelvins (K) over the wavelength range from 0.5 to 5.0 millimeters (mm). This measurement fits very well the theoretical blackbody radiation spectrum predicted by the big bang cosmological model. When the COBE spacecraft's supply of liquid helium was depleted on September 21, 1990, the FIRAS instrument (which required the liquid helium cryogen) ceased operation.

The COBE spacecraft's differential microwave radiometer (DMR) instrument was designed to search for primeval fluctuations in the brightness of the cosmic microwave background, very small temperature differences (about 1 part in 100,000) between different regions of the sky. Analysis of

DMR data suggested the presence of tiny asymmetries in the cosmic microwave background. Scientists used the existence of these asymmetries (which are actually the remnants of primordial hot and cold spots in the big bang radiation) to start explaining how the early universe eventually evolved into huge clouds of galaxies and huge empty spaces. The COBE spacecraft pioneered the study of the cosmic microwave background.

NASA's *Wilkinson Microwave Anisotropy Probe* (WMAP) was launched in June 2001 and has made a map of the temperature fluctuations of the CMB radiation with much higher resolution, sensitivity, and accuracy that COBE. The new information contained in these finer fluctuations sheds additional light on several key questions in cosmology.

See also ASTROPHYSICS; BIG BANG THEORY; COSMOLOGY; *WILKINSON MICROWAVE ANISOTROPY PROBE* (WMAP).

cosmic microwave background (**CMB**) The background of microwave radiation that permeates the universe and has a blackbody temperature of about 2.7 K. Sometimes called the *primal glow,* scientists believe it represents the remains of the ancient fireball in which the universe was created.

See also BIG BANG THEORY; COSMOLOGY; *WILKINSON MICROWAVE ANISOTROPY PROBE* (WMAP).

cosmic ray(s) Extremely energetic particles (usually bare atomic nuclei) that move through outer space at speeds just below the speed of light and bombard Earth from all directions. The existence of cosmic rays was discovered in 1912 by the Austrian-American physicist VICTOR HESS. Hydrogen nuclei (protons) make up the highest percentage of the cosmic ray population (approximately 85%), but these particles range over the entire periodic table of elements. *Galactic cosmic rays* are samples of material from outside the solar system and provide scientists direct evidence of phenomena that occur as a result of explosive processes in stars throughout the Milky Way galaxy. *Solar cosmic rays* (mostly protons and alpha particles) are ejected from the Sun during solar flare events. Solar cosmic rays are generally lower in energy than galactic cosmic rays.

See also ASTROPHYSICS.

cosmic-ray astronomy The branch of high-energy astrophysics that uses cosmic rays to provide information on the origin of the chemical elements through nucleosynthesis during stellar explosions.

Cosmic Ray Satellite-B (**COS-B**) The European Space Agency (ESA) developed the COS-B scientific satellite to study extraterrestrial gamma radiation in the energy range from about the 2.5 MeV to 5 GeV region. This spacecraft was launched by a Delta expendable rocket vehicle on August 9, 1975, from Vandenberg Air Force Base in California and successfully placed in a highly elliptical orbit around Earth. The satellite's orbit had a periapsis of 340 km, an apoapsis of 99,876 km, a period of 37.1 hours, an inclination of 90.13°, and an eccentricity of 0.881. The gamma-ray telescope onboard this satellite provided scientists with their first detailed view of the Milky Way galaxy in the gamma-ray portion of the electromagnetic spectrum. Originally projected to operate for two years, the spacecraft operated successfully for six years and eight months (from August 1975 to April 1982).

See also COMPTON GAMMA RAY OBSERVATORY.

cosmodrome In the Soviet and later Russian space program, a place that launches space vehicles and rockets. The three main cosmodromes in the Russian space program are the Plesetsk Cosmodrome, which is situated south of Archangel in the northwest corner of Russia; the Kapustin Yar Cosmodrome, on the banks of the Volga River southeast of Moscow; and the first and main cosmodrome (the site that launched *Sputnik 1*), the Baikonur Cosmodrome, located just east of the Aral Sea in Kazakstan (now an independent country). The Baikonur Cosmodrome has also been referred to as Tyuratam.

See also BAIKONUR COSMODROME.

cosmogony The study of the origin and formation of celestial objects, especially the solar system. *Compare with* COSMOLOGY.

cosmological constant (**Λ, or lambda**) A concept originally suggested by ALBERT EINSTEIN as he formed general relativity in 1915 and then applied it to cosmology in 1917. Briefly summarized here, the cosmological constant (as Einstein envisioned it) implies that the shape and emptiness of space gives rise to a property in the universe that somehow balances the relentless attraction of gravity. General relativity predicted that the universe must either expand or contract. Einstein thought the universe was static, so he added this new term to his general relativity equation to (in theory) "stop the expansion of the universe."

The core of Einstein's theory of general relativity is expressed in the famous field equation $G_{\mu\nu} = 8 \pi G T_{\mu\nu}$. This equation states that the distribution of matter and energy in the universe determines the geometry of space-time (as represented by Einstein's curvature tensor, $G_{\mu\nu}$). Here, G is Newton's universal gravitational constant and $T_{\mu\nu}$ is the stressor-energy tensor. In mathematics a tensor is an array of numbers or functions that can transform according to certain rules when there is a change in coordinates. Physicists often use tensors to describe complicated physical systems that cannot be properly handled by vectors. Einstein introduced the cosmological constant (Λ) into his field equation as a cosmic "fudge factor" that somehow would counterbalance the force of gravity on a grand scale. To come up with a static model of the universe, Einstein rewrote his general relativity equation as follows: $G_{\mu\nu} + \Lambda g_{\mu\nu} = 8 \pi G T_{\mu\nu}$. (The term $g_{\mu\nu}$ is the space-time metric tensor). By including the product $\Lambda g_{\mu\nu}$ on the left side of his general relativity field equation, Einstein implied that the cosmological constant was actually a property of space itself.

By 1922 the Russian physicist Alexander Friedmann (1888–1925) produced cosmological models that described expanding and contracting universes without resorting to a cosmological constant as found in Einstein's work. When the American astronomer EDWIN HUBBLE announced his observation of an expanding universe in 1929, Einstein formally abandoned his concept of a cosmological constant and regarded the concept as his "greatest mistake." Was the genius a bit too harsh on himself?

In the latter part of the 20th century scientists revisited the concept of a cosmological constant. Their journey began on theoretical grounds, but this intellectual excursion soon appeared to be very necessary because of some intriguing astronomical observations. In the 1960s, for example, advocates of modern field theory began to theoretically associate the notion of a cosmological constant term with the energy density of the vacuum of space. The underlying hypothesis from quantum theory is that empty space might possess a small energy density. Physicists began to speculate that a quantum level energy density arises from the very rapid appearance and then disappearance of virtual particle-antiparticle pairs. In 1967 the Russian physicist Yakov B. Zel'dovich (1914–87) made an estimate of the energy density of the quantum vacuum. His results suggested that the energy density of the vacuum of space could contribute to an immense cosmological factor. At this point theoretical physicists also recognized that if they were to successfully understand and fully appreciate how this postulated energy density might be comparable in influence to other forms of matter-energy in the universe, they would need some "new physics."

In the late 1990s two separate teams of supernova hunting astrophysicists made a startling announcement. Each team of scientists reported that their observations showed the universe was indeed expanding, but at an accelerating rate! Since 1998 scientists have gathered additional observational evidence to support the notion of cosmic acceleration. Pressed to explain what was going on and what could possibly be causing the universe to expand at a faster rate, cosmologist are beginning to cautiously embrace a refashioned cosmological constant—one that has its origins in quantum vacuum energy. Modern astronomical observations are having profound implications for particle physics and could help scientists understand the fundamental forces of nature, especially the relation between gravity and quantum mechanics.

Today the main attraction of a cosmological constant term is that it significantly improves the agreement between theory and observation. The most interesting example of this is the recent effort to measure how much the expansion of the universe has changed in the last few billion years. Within the framework of contemporary big bang cosmology, scientists previously postulated that the gravitational pull exerted by matter in the universe slows the expansion imparted by the primeval big bang explosion. Then, in the late 1990s it became practical for astronomers to accurately observe very bright rare stars called supernovas in their efforts to carefully measure how much the overall expansion of the universe has slowed over the last billion or so years. Quite surprisingly (as previously mentioned) their observations indicated that the expansion of the universe is actually speeding up, or accelerating. While many scientists understandably view these initial results with a healthy degree of caution, they are, nevertheless, willing to suggest the possibility that the universe could contain some exotic form of energy or matter that is gravitationally repulsive in its overall effect. Some physicists even suggest that the more the universe expands, the more influence this exotic form of energy or matter has.

The cosmological constant is an example of this type of gravitationally repulsive effect. However, much work remains before scientists can unravel the exciting new mystery. One thing is certain, however: The recent discovery of cosmic acceleration has forever altered cosmology in the 21st century. Now resurrected, a refashioned form of Einstein's cosmological constant could play an important role in describing the destiny of the universe.

See also ASTROPHYSICS; COSMOLOGY; RELATIVITY.

cosmological principle The hypothesis that the expanding universe is isotropic and homogeneous. In other words, there is no special location for observing the universe, and all observers anywhere in the universe would see the same recession of distant galaxies. The cosmological principle implies space curvature—that is, since there is no center of the universe, there is no outer limit or surface.

See also COSMOLOGY.

cosmology The study of the origin, evolution, and structure of the universe. Contemporary cosmology centers around the big bang hypothesis—a theory stating that about 14 billion (10^9) years ago the universe began in a great explosion and has been expanding ever since. In the open (or steady-state) model scientists postulate that the universe is infinite and will continue to expand forever. In the more widely accepted closed universe model the total mass of the universe is assumed sufficiently large to eventually stop its expansion and then make it start contracting by gravitation, leading ultimately to a big crunch. In the flat universe model the expansion gradually comes to a halt, but instead of collapsing the universe achieves an equilibrium condition with expansion forces precisely balancing the forces of gravitational contraction.

Until the late 1990s big bang cosmologist comfortably assumed that only gravitation was influencing the dynamics and destiny of the universe after the big bang explosion. Hubble's law seemed to be correct, although there was some disagreement about the value of Hubble's constant. Then, in 1998 astronomical observations of very bright supernovas indicated that the rate of expansion of the universe is actually increasing. These observations are now causing a great deal of commotion within the scientific community. The reason for this commotion is quite simple. Within the framework of contemporary big bang cosmology, scientists previously postulated that the gravitational pull exerted by matter in the universe slows the expansion imparted by the primeval big bang explosion. If the rate of expansion of the universe is actually increasing, what is counteracting the far-reaching and unrelenting force of gravity? Some scientists suggest that the universe could contain an exotic form of energy or matter that is gravitationally repulsive in its overall effect.

A resurrected and refashioned form of Einstein's cosmological constant could play an important role in describing the destiny of the universe. Before we try to estimate the impact on 21st-century cosmology of a gravitationally repulsive cosmological constant (based on quantum vacuum energy), it is important to examine briefly how cosmology evolved from prehistory to the contemporary big bang model.

Early Cosmologies

From ancient times most societies developed one or more accounts of how the world (they knew) was created. These early stories are called *creation myths*. For each of these societies their culturally based creation myth(s) attempted to explain (in very nonscientific terms) how the world started and where it was going.

In the second century C.E. PTOLEMY, assembling and synthesizing all of early Greek astronomy, published the first widely recognized cosmological model, often referred to as the *Ptolemaic system*. In this geocentric cosmology model Ptolemy codified the early Greek belief that Earth was at the center of the universe and that the visible planets (Mercury, Venus, Mars, Jupiter, and Saturn) revolved around Earth embedded on crystal spheres. The "fixed" stars appeared immutable (essentially unchangeable)—save for their gradual motions through the sky with the seasons—so they were located on a sphere beyond Saturn's sphere. While seemingly silly in light of today's scientific knowledge, this model could and did conveniently account for the motion of the planets then visible to the naked eye. Without detailed scientific data to the contrary, Ptolemy's model of the universe survived for centuries. Arab astronomers embraced and enhanced Ptolemy's work.

As a result of the efforts by Arab astronomers to preserve Greek cosmology, the notion of Earth as the center of the universe flowed back into western Europe as its people emerged from the Dark Ages and began to experience the Renaissance. Aristotle's teachings, including the unchallenged acceptance of geocentric cosmology, were integrated—almost as dogma—into a western European educational tradition and culture that drew heavily from faith-based teachings. Before the development of the scientific method (which accompanied the scientific revolution of the 17th century), even the most brilliant medieval scholars, such as Saint Thomas Aquinas (ca. 1225–74), did not feel the need to explain natural phenomena through repeatable experiments and physical laws capable of predicting the outcome of certain reactions. In a world in which people used fire (combustion) and crude gunpowder rockets but did not understand the basic principles behind thermodynamics, why should anyone challenge a comfortable Earth-centered model of the universe?

To the naked eye the heavens appeared immutable (unchanging), except for the motion of certain fixed and wandering lights, which once observed and monitored became very useful in the development of calendars for religious, cultural, and economic activities. In the majority of evolved agrarian societies, astronomically based calendars became the best friend of farmers, alerting them to the pending change of seasons and guiding them when to plant certain crops. Farming-based civilizations were well served by the Ptolemaic system, so there was no pressing economic or social need to cause intellectual mischief. For example, suppose the world did move through space at some reckless speed. What would keep people and things from falling off? A round, rotating world was difficult enough to accept, but one that also moved rapidly through the emptiness of space while spinning on its axis was a very, very unsettling concept to many people in 16th-century western Europe.

The Copernican Revolution in Cosmology

Then, a Polish church official turned astronomer named NICHOLAS COPERNICUS upset the "cosmic applecart" in 1543. His deathbed-published book, *De Revolutionibus Orbium Coelestium* (*On the Revolution of Celestial Spheres*), boldly suggested that Earth was *not* the center of the universe; rather, our planet moved around the Sun just like the other planets. *Copernican cosmology* proposed a Sun-centered (heliocentric) universe and overturned two millennia of Greek astronomy. Although church officials became very uncomfortable with Copernicus's book, nothing much scientifically happened with its important message until the marriage of the telescope and astronomy in 1610. Even the brilliant naked-eye astronomer TYCHO BRAHE died a staunch advocate of geocentric cosmology, albeit with his own refashioned version of the Ptolemaic system.

At the beginning of the scientific revolution in the early 17th century, the development of the astronomical telescope, the detailed observations of the planet Jupiter and its major moons by GALILEO GALILEI, and the emergence of JOHANN KEPLER's laws of planetary motion provided the initial observational evidence and mathematical tools necessary to validate Copernican cosmology and to dislodge Ptolemaic cosmology. Copernicus and his advocacy of heliocentric cosmology triggered the start of modern science. However, the transition from the geocentric cosmology of Aristotle to the heliocentric cosmology of Copernicus did not take place without significant turmoil—especially within the religious sectors of European societies. At the time previously homogenous European Christianity was being torn asunder by the Protestant Reformation. Any idea that appeared to challenge traditional religious teachings had the potential for causing further instability in a world already torn apart by brutal warfare over disagreements about religious doctrine. So if a person became identified as a "Copernican," it often led to a high-profile public prosecution. The trial of a "Copernican" usually resulted in an almost automatic conviction for heresy followed by imprisonment or even execution. Despite such strong political and theological opposition, the Copernican system survived and emerged to rule the cosmology for the next two centuries. The close study of the clockwork motion of the planets around the Sun and of the Moon around Earth helped inspire SIR ISAAC NEWTON when he created his universal law of gravitation (ca. 1680). Newton's concept of gravity proved valid only for bodies at rest or moving very slowly compared to the speed of light. However, these inherent limitations in Newtonian mechanics did not become apparent until the early part of the 20th century and the introduction of general relativity.

During the 18th and 19th centuries improved mathematics, better telescopes, and the hard work of many dedicated scientists pushed Newtonian mechanics and its orderly model of the universe to its limit. Of course, there were a few gnawing questions that remained beyond the grasp of classical physics and the power of the optical instruments that were available at the time. For example, how big was the universe? Up until the third decade of 20th century, most astronomers and cosmologists treated the universe and the Milky Way galaxy as one and the same. The annoying little patches of fuzzy light, called nebulas, were generally regarded as groups of stars within the Milky Way that were simply beyond the optical resolution limits of available instruments. Of course, a few scientists, such as IMMANUEL KANT, had already suggested the answer to the mystery. As discovered in the early 20th century, these fuzzy patches of light were really other galaxies—Kant's so-called *island universes.* What about the size of the universe? As EDWIN HUBBLE and other scientists eventually demonstrated, it is an incredibly large, expanding phenomenon filled with millions of galaxies rushing away from one another in all observable directions.

Big Bang Cosmology

Modern cosmology has its roots in two major theoretical developments that occurred at the beginning of the 20th century. The first is the general theory of relativity, which ALBERT EINSTEIN proposed in 1915. In it Einstein postulated how space and time can actually be influenced by strong sources of gravity. The subtle but measurable bending (warping) of a star's light as it passed behind the Sun during a 1919 solar eclipse confirmed that the gravitational force of a very massive object could indeed warp the space-time continuum.

After Einstein introduced his theory of general relativity, he as well as a number of other scientists tried to apply the new gravitational dynamics (that is, the warping of the space-time continuum) to the universe as a whole. At the time this required the scientists to make an important theoretical assumption about how the matter in the universe was distributed. The simplest hypothesis they could make was to assume that when viewed in any direction by different observers in any place, it would appear roughly the same to all of them—that is, the matter in the universe was assumed homogeneous and isotropic when averaged over very large scales. This important assumption is now called the cosmological principle, and it represents the second theoretical pillar of big bang cosmology. Space-based observatories, such as NASA's *Hubble Space Telescope,* now let astrophysicists and astronomers observe the distribution of galaxies on increasingly larger scales. The results continue to support the *cosmological principle.*

Furthermore, the cosmic microwave background (CMB)—the remnant heat from the big bang explosion—has a temperature (about 2.7 K) that is highly uniform over the entire sky. This fact strongly supports the assumption that the intensely hot primeval gas that emitted this radiation long ago was very uniformly distributed. Consequently, general relativity (linking gravity to the curvature of space-time) and the cosmological principle (the large-scale uniformity and homogeneity of the universe) form the entire theoretical basis for big bang cosmology and lead to specific predictions for observable properties of the universe.

The American astronomer Edwin Hubble provided the first important observation supporting big bang cosmology. During the 1920s he performed precise observations of diffuse nebulas and then proposed that these objects were actually independent galaxies that were moving away from us in a giant, expanding universe. Modern cosmology, based on continuously improving astrophysical observations (*observational* and *physical cosmology*) and sophisticated theoretical developments (*theoretical cosmology*), was born.

In the 1940s the Russian-American physicist GEORGE GAMOW and others proposed a method by which GEORGES LEMAÎTRE's "cosmic egg" (the initial name for the "big bang") could lead to the creation of the elements through nuclear synthesis and transformation processes in their presently observable cosmic abundances. This daring new cosmological model involved a giant explosion of an incredibly dense "point object" at the moment of creation. The so-called big bang was followed by a rapid expansion process during which matter eventually emerged. The abundance of the light elements helium and hydrogen supports big bang cosmology, because these light elements should have been formed from protons and neutrons when the universe was a few minutes old following the big bang.

Initially, the term *big bang* was sarcastically applied to this new cosmological model by rival scientists who favored a *steady-state theory* of the universe. In the steady-state cosmological model the universe is assumed to have neither a beginning nor an end, and matter is thought to be added (created) continuously to accommodate the observed expansion of the galaxies. Despite the derisive intent of the name, big bang—both the name and the cosmological model it represents—have survived and achieved general acceptability.

Astrophysical discoveries throughout the 20th century tended to support the big bang cosmology, a model stating that about 14 billion (10^9) years ago the universe began in a great explosion (sometimes called the *initial singularity*) and has been expanding ever since. Physicists define a singularity as a point of zero radius and infinite density.

The 1964 discovery of the cosmic microwave background (CMB) radiation by ARNO PENZIAS and ROBERT WILSON provided the initial observational evidence that there was, indeed, a very hot early phase in the history of the universe. More recently, space-based observatories, such as NASA's *Cosmic Background Explorer (COBE)* and the *Wilkinson Microwave Anisotropy Probe* (WMAP), have provided detailed scientific data that not only generally support the big bang cosmological model but also raise interesting questions about it. For example, big bang cosmologists had to explain how the clumpy structures of galaxies could have evolved from a previously assumed "smooth" (that is, very uniform and homogeneous) big bang event.

The *inflationary model* of the big bang attempts to correlate modern cosmology and the quantum gravitational phenomena that are believed to have been at work during the very first fleeting moments of creation. This inflationary model suggests that the very early universe expanded so rapidly that the smooth homogeneity postulated in the originally big bang model would be impossible. Although still being refined, the inflationary model appears to satisfy many of the perplexing inconsistencies that contemporary astrophysical observations had uncovered with respect to the more conventional big bang model. These refinements in big bang cosmology are expected to continue well into the 21st century as even more sophisticated space observations provide new data about the universe, its evolutionary processes, and its destiny.

Cosmology in the 21st Century

In the late 1990s two separate teams of astrophysicists examined very rare and bright stars called supernovas and then made the same startling announcement. Each team of scientists reported that the universe was expanding at an accelerating rate. Since then other scientists have gathered additional observational evidence to support the idea of cosmic acceleration. Pressed to explain what was going on and what could possibly be causing the universe to expand at a faster rate, cosmologists are now cautiously revisiting Einstein's cosmological constant but embracing a refashioned version of a gravitationally repulsive term—one that has its origins in quantum vacuum energy. Modern astronomical observations are also having profound implications for particle physics and could help scientists understand the fundamental forces of nature, especially the relation between gravity and quantum mechanics.

Today the main attraction of a cosmological constant term is that it significantly improves the agreement between theory and observation. The most interesting example of this is the recent effort to measure how much the expansion of the universe has changed in the last few billion years. Within the framework of contemporary big bang cosmology, scientists previously postulated that the gravitational pull exerted by matter in the universe slows the expansion. Now, if the universe is indeed expanding at an accelerating rate, they need to figure out what is causing this. Cosmologists must now seriously consider that the universe may contain some bizarre form of matter or energy that is, in effect, gravitationally repulsive. The concept of a cosmological constant based on quantum vacuum energy is one candidate, but much work lies ahead before big bang cosmologists can be comfortable with suggesting the ultimate destiny of the expanding universe.

The Fate of the Universe

One of the important roles of cosmology is to address the ultimate fate of the universe. In the *open* (or steady-state) model of the universe, scientists postulate that the universe is infinite and will continue to expand forever. In contrast, the *closed universe model* postulates that the total mass of universe is sufficiently large that one day it will stop expanding and begin to contract due to the mutual gravitational attraction of the galaxies. This contraction will continue relentlessly until the total mass of the universe is essentially compressed into a singularity, a process known as the big crunch. Some advocates of the closed universe model also speculate that after the big crunch there will be a new explosive expansion (that is, another big bang). This line of speculation leads to the *pulsating,* or *oscillating, universe* model—a cosmological model in which the universe appears and then disappears in an endless cycle between big bangs and big crunches.

Within big bang cosmology models whose dynamics are governed and controlled by gravity, the ultimate fate of the universe depends on the total amount of matter the universe contains. Does the universe contain enough matter to reverse its current expansion and cause closure? Astrophysical measurements of all observable luminous objects suggest that the universe contains only about 10% (or less) of the amount of matter thought needed to support the closed universe model. Where is the "missing mass," or *dark matter*? This is one important question that is perplexing modern scientists.

Cosmologists often discuss the mass of the universe in the context of the *density parameter* (symbol: Ω). It is a dimensionless parameter that expresses the ratio of the actual mean density of the universe to the critical mass density—that is, the mean density of matter within the universe (considered as a whole) that cosmologists consider necessary if gravitation is to eventually halt its expansion. One presently estimated value of this critical mass density is 8×10^{-27} kg/m^3, obtained by assuming that the Hubble constant (H_0) has a value of about 71.7 km/(s-Mpc) and that gravitation is all that is influencing the dynamic behavior of the expanding universe. If the universe does not contain sufficient mass (that is, if $\Omega < 1$), then the universe will continue to expand forever. If $\Omega > 1$, then the universe has enough mass to eventually stop its expansion and to start an inward collapse under the influence of gravitation. If the critical mass density is just right (that is, if $\Omega = 1$), then the universe is considered flat, and a state of equilibrium will eventually exist in which the outward force of expansion becomes precisely balanced by the inward force of gravitation. In the *flat universe* model the expansion of the universe gradually comes to a halt, but it does not begin to collapse.

Today many cosmologists favor the inflationary theory embellishment of the big bang hypothesis (i.e., complex explosive birth) coupled with a flat universe's ultimate fate. However, if an as yet undetected mysterious source of energy is now helping to overcome the attraction of gravity and causing the universe to expand at an accelerated rate, then its ultimate fate is more likely to be continued expansion—until all the galaxies and all the stars fly away from one another. Cosmologists will have to work hard to unlock this mystery, the answer for which may lie in quantum gravitation and new interpretations of how the very rapid appearance and then disappearance of virtual particle-antiparticle pairs gives rise to quantum vacuum energy.

Consciousness and the Universe

The history of the universe can also be viewed to follow a more or less linear time scale. This approach, sometimes called the scenario of *cosmic evolution,* links the development of the galaxies, stars, heavy elements, life, intelligence, technology, and the future. (*See* figure.) It is especially useful in *philosophical* and *theological cosmology.* Exobiologists are also interested in understanding how life, especially intelligent life and consciousness, can emerge out of the primordial matter from which the galaxies, stars, and planets evolved. This cosmological approach leads to such interesting con-

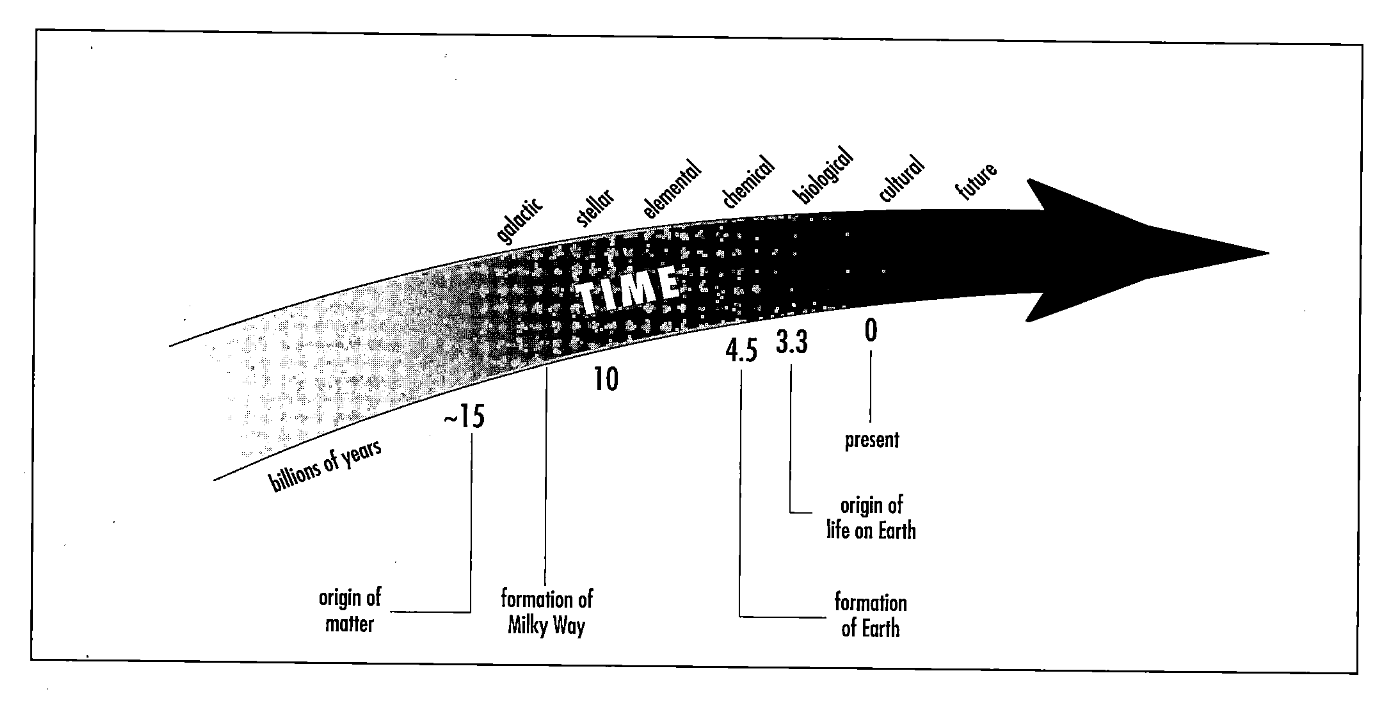

The scenario of cosmic evolution provides a grand synthesis for the long series of alterations of matter that have given rise to our galaxy, our Sun, our planet, and ourselves.

cepts as the *living universe,* the *conscious universe,* and the *thinking universe.* Is the evolution of intelligence and consciousness a normal endpoint for the development of matter everywhere in the universe? Or are human beings very unique "by-products" of the cosmic evolutionary process—perhaps the best the universe could do since the big bang? If the former question is true, then the universe should be teeming with life, including intelligent life. If the latter is true, we could be very much alone in a very big universe. If we are the only beings now capable of contemplating the universe, then perhaps it is also our destiny to venture to the stars and carry life and consciousness to places where now there is only the material potential for such.

Up until recently most scientists have avoided integrating the potential role of conscious intelligence (that is, matter that can think) into cosmological models. But what happens after conditions in the universe give rise to living matter that can think and reflect upon itself and its surroundings? The *anthropic principle* is an interesting, though highly controversial, hypothesis in contemporary cosmology that suggests that the universe evolved after the big bang in just the right way so that life, especially intelligent life, could develop on Earth. Does human intelligence (or possibly other forms of alien intelligence) have a role in the further evolution and destiny of the universe? If you thought the Copernican hypothesis livened up intellectual activities in the 16th and 17th centuries, the revival of the cosmological constant and the bold hypothesis set forth in the anthropic principle should keep cosmologists very busy for a good portion of this century.

See also ANTHROPIC PRINCIPLE; ASTROPHYSICS; BIG BANG; COSMOLOGICAL CONSTANT; UNIVERSE.

cosmonaut The title given by Russia (formerly the Soviet Union) to its space travelers. Equivalent to astronaut.

Cosmos spacecraft The general name given to a large number of Soviet and later Russian spacecraft, ranging from military satellites to scientific platforms investigating near-Earth space. *Cosmos 1* was launched in March 1962; since then well over 2,000 Cosmos satellites have been sent into outer space. Also called Kosmos.

See also RUSSIAN SPACE AGENCY.

coulomb (symbol: C) The SI unit of electric charge. The quantity of electric charge transported in one second by a current of one ampere. Named after the French physicist Charles de Coulomb (1736–1806).

countdown The precise step-by-step process that leads to the launch of a rocket or aerospace vehicle. Each countdown takes place in accordance with a specific predesignated schedule and has a detailed checklist of required events, actions, and conditions. Time is marked off (or counted down) in a reverse manner, with *zero* representing the go or launch time. A number of factors can influence the countdown process, including weather conditions, unanticipated equipment glitches or malfunctions, unauthorized people entering the launch site or its companion safety zone, and the available launch window, which is often very mission-dependent.

Cowell, Philip Herbert (1870–1949) British *Scientist* Philip Cowell was a British scientist who specialized in celestial mechanics. Of special note is his cooperation with ANDREW

C. CROMMELIN in calculating the precise time of the 1910 appearance of Comet Halley.

Crab Nebula (M1, NGC 1952) The supernova remnant of an exploding star observed in 1054 by Chinese astronomers. This "guest star," as the early Chinese astronomers called it, was so bright that people saw it in the sky during the day for almost a month. At that time the exploding star was blazing with the light of about 400 million suns. According to ancient accounts, this supernova then remained visible in the evening sky for more than a year. In 1758 CHARLES MESSIER spotted the Crab Nebula and called the interesting celestial object entry M1 in his famous catalog of nebulas and star clusters. The Crab Nebula remains a glowing mass of gas and dust located about 6,500 light-years away. It appears in the sky above the northern horn of the constellation Taurus (the Bull) and contains a pulsar that flashes optically.

See also NEBULA; SUPERNOVA.

crater 1. A bowl-shaped topographic depression with steep slopes; a volcanic orifice. Craters are formed naturally by two general processes: impact (as from an asteroid, comet, or meteoroid strike); (*see* figure below) and eruptive (as from a volcanic eruption). The large impact craters found on the terrestrial planets and on many moons throughout the solar system provide visually spectacular evidence that high-energy cosmic collisions were a common event during the early solar system and that these celestial bombardments continue up to the present. 2. The pit, depression, or cavity formed in the surface of Earth by a human-caused explosion (chemical explosives or nuclear). It may range from saucer-shaped to conical depending largely on the depth of burst. 3. The depression resulting from high-speed solid-particle impacts on a rigid material, as, for example, a meteoroid or space debris particle impact on the skin of a spacecraft.

See also EXTRATERRESTRIAL CATASTROPHE THEORY; IMPACT CRATER; SPACE DEBRIS.

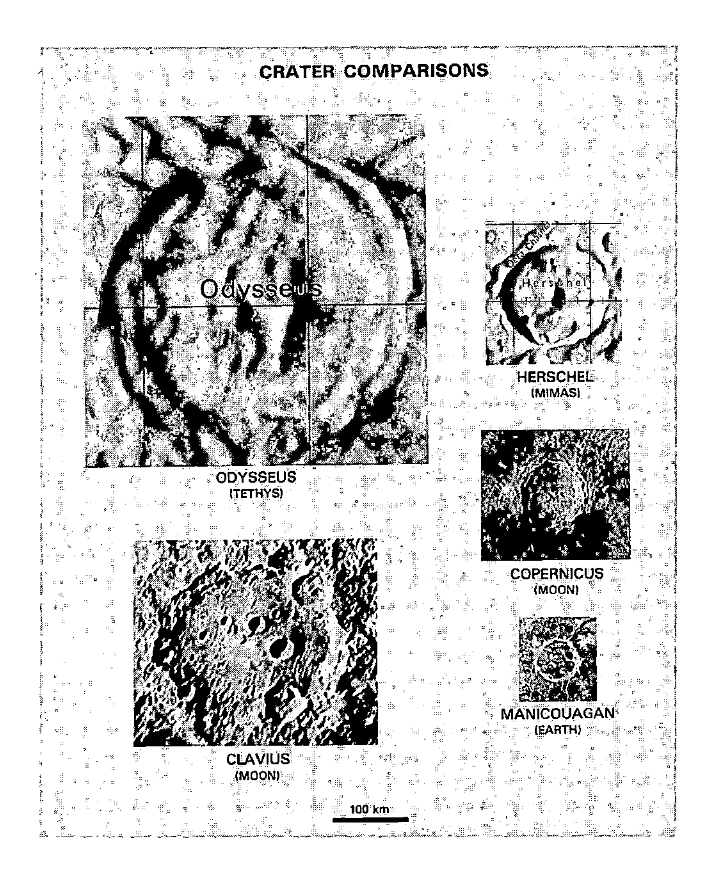

This illustration compares several large craters in the solar system. The craters Herschel and Odysseus are solitary features on Saturn's moons Mimas and Tethys, respectively. Herschel has a diameter of 140 kilometers, while Odysseus has a diameter of 430 kilometers. The craters Copernicus and Clavius are well-known features on the nearside of Earth's Moon. The Copernicus crater is 90 kilometers in diameter, while the Clavius crater has an overall diameter of about 234 kilometers. For purposes of comparison, the illustration also shows the Manicouagan structure of Canada. Scientists believe this structure to be an ancient impact scar that was formed by a cosmic collision about 210 million years ago. The ice-covered surface of the structure's circular lake is approximately 66 kilometers in diameter. *(Courtesy of NASA)*

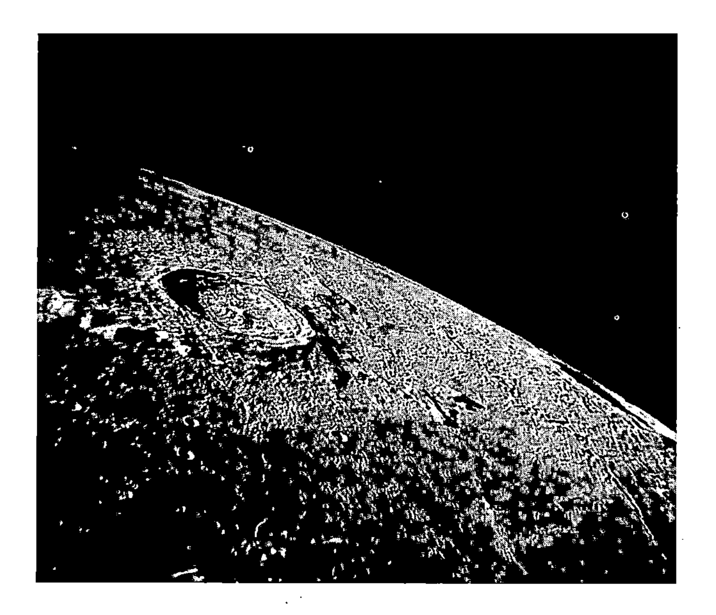

A panoramic view of the Moon's Eratosthenes Crater photographed from lunar orbit during NASA's *Apollo 17* mission in December 1972. The Copernicus Crater appears on the right near the horizon. *(Courtesy of NASA)*

crater chain A series of craters that lie approximately in a line. Planetary geologists suggest that crater chains are formed by secondary (sometimes overlapping) impact craters or by volcanic activity. Scientists postulate that the Enki Catena Crater Chain (13 craters overlapping in a line) on Ganymede was probably formed by a comet that was pulled into pieces by Jupiter's gravity as it passed too close to the giant planet. Soon after this breakup, the 13 comet fragments crashed onto Ganymede in rapid succession. In contrast, some crater chains on the Moon may be the result of volcanic activity in the form of partially collapsed lava tubes.

See also CRATER; IMPACT CRATER.

Cressida A small moon of Uranus that was discovered in 1986 as a result of the *Voyager 2* spacecraft flyby. Cressida has a diameter of about 62 km and a mass of 8×10^{17} kg. This moon travels in a circular orbit around Uranus at a distance of 61,780 km, with a period of 0.464 days and an inclination of 0.01°. Sometimes called Uranius IX.

See also URANUS.

crew-tended spacecraft A spacecraft that is visited and/or serviced by astronauts but can provide only temporary accommodations for human habitation during its overall mission. Sometimes referred to as a *man-tended spacecraft. Compare with* CONTINUOUSLY-CREWED SPACECRAFT.

Crippen, Robert L. (1937–) American *Astronaut, U.S. Navy Officer* The astronaut Robert (Bob) Crippen served as pilot and accompanied astronaut JOHN YOUNG (the spacecraft commander) on the inaugural flight of NASA's Space Transportation System (STS), or space shuttle. This first flight (designated as STS-1) took off from the Kennedy Space Center in Florida and extended from April 12 to 14, 1981. The mission involved Orbiter Vehicle (OV) 102, the *Columbia.* Crippen also served as spacecraft commander on three other space shuttle missions: STS-7 (June 18–24, 1983), STS 41-C (April 6–13, 1984), and STS 41-G (October 5–13, 1984).

critical mass density The mean density of matter within the universe (considered as a whole) that cosmologists consider necessary if gravitation is to eventually halt its expansion. One presently estimated value of the critical mass density is 8×10^{-27} kg/m³, obtained by assuming that the Hubble constant (H_0) has a value of about 71.7 km/(sMpc) and that gravitation is all that influences the dynamic behavior of the expanding universe.

See also COSMOLOGY; DENSITY PARAMETER.

critical point The highest temperature at which the liquid and vapor phases of a fluid can coexist. The temperature and pressure corresponding to this point (thermodynamic state) are called the fluid's CRITICAL TEMPERATURE and CRITICAL PRESSURE, respectively.

critical pressure 1. In thermodynamics, the pressure of a fluid at the critical point—that is, the highest pressure under which the liquid and gaseous phases of a substance can coexist. 2. In rocketry, the pressure in the nozzle throat for which the isentropic (constant entropy) mass-flow rate is at maximum.

critical speed (symbol: V^*) The flow velocity (V^*) at which the Mach number (M) is unity. It is used as a convenient reference velocity in the basic equations of high-speed, compressible fluid flow.

See also MACH NUMBER.

critical temperature The temperature above which a substance cannot exist in a liquid state, regardless of the pressure.

See also CRITICAL POINT.

critical velocity In rocketry, the speed of sound at conditions prevailing at the nozzle throat. Also called the throat velocity and critical throat velocity.

See also NOZZLE; ROCKET.

Crommelin, Andrew Claude (1865–1939) French-Irish *Astronomer* Andrew Crommelin was an astronomer who specialized in calculating the orbits of comets. He collaborated with PHILIP H. COWELL in computing and predicting the precise date of the 1910 return of Comet Halley. Based on this success, he then calculated the dates of previous appearances of this famous comet through history back to the third century B.C.E.

See also COMET HALLEY.

cross section (common symbol: σ) 1. In general, the area of a plane surface that is formed when a solid is cut; often this cut is made at right angles to the longest axis of the solid object. 2. In aerospace engineering, the physical (geometric) area presented by an aircraft, aerospace vehicle, or missile system; the frontal (or "head-on") cross section is the surface area that would be observed if the vehicle was flying directly toward an observer. 3. In aerospace operations, the apparent surface area of an aircraft, aerospace vehicle, rocket, or spacecraft that can be detected by a remote sensing system; for example, an aircraft's radar cross section. Stealth (low-observables) technology now is being used to reduce a military aircraft's radar cross section. 4. In physics, a measure of the probability that a nuclear reaction will occur. Usually measured in barns, it is the apparent (or effective) area presented by a target nucleus (or particle) to an oncoming particle or other nuclear radiation, such as a gamma-ray photon. One barn (symbol: b) is defined as an area of 1×10^{-28} m².

cross staff *See* ANCIENT ASTRONOMICAL INSTRUMENTS.

cruise missile A guided missile traveling within the atmosphere at aircraft speeds and, usually, low altitude whose trajectory is preprogrammed (*see* figure, page 164). It is capable of achieving high accuracy in striking a distant target. It is maneuverable during flight, is constantly propelled, and, therefore, does not follow a ballistic trajectory. Cruise missiles may be armed with nuclear weapons or with conventional warheads (i.e., high explosives).

See also TOMAHAWK.

cruise phase In space flight operations involving a scientific spacecraft, the cruise phase is bounded at the beginning by the launch phase and at the end by the encounter phase. After launch the scientific spacecraft is commanded to configure itself for cruise. Appendages that might have been stowed for launch (e.g., an antenna system) can now be deployed either fully or to some intermediate cruise position. Telemetry is analyzed to determine the state of health of the spacecraft and to determine how well it survived the launch. The trajectory is fine-tuned to prepare the spacecraft for its scientific encounter mission. During the cruise phase ground system upgrades can be made and appropriate tests conducted. Spacecraft flight software modifications can be implemented

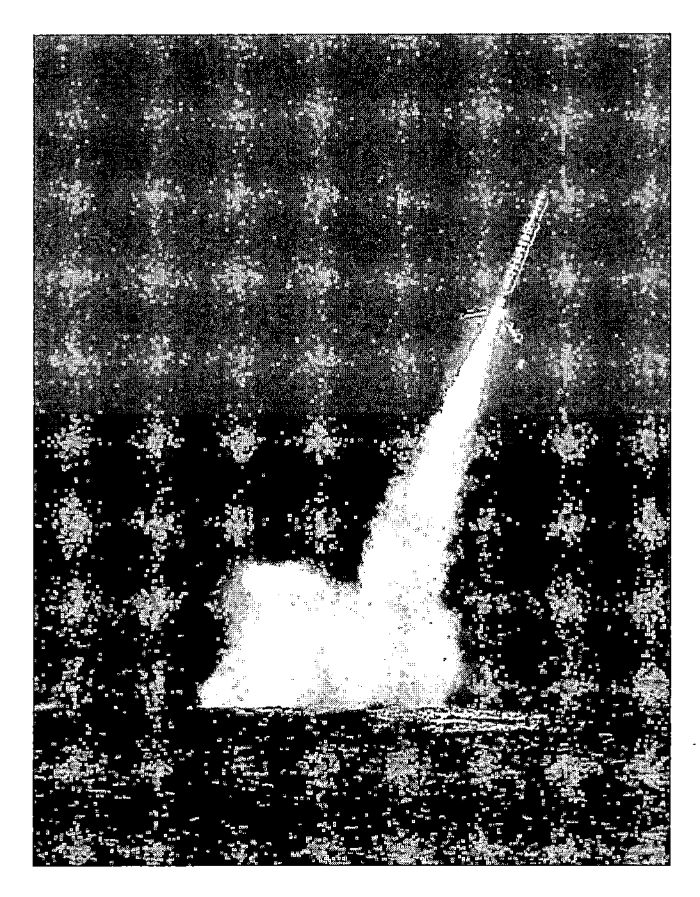

During a submerged exercise off the coast of the Bahamas in January 2003, the nuclear-powered ballistic missile submarine USS *Florida* launched a Tomahawk cruise missile to test the capabilities of the U.S. Navy's guided missile submarines. The USS *Florida* is one of four Ohio-class ballistic missile submarines (designated SSBN) being converted to guided missile submarines (designated SSGN). The SSGNs will have a capability to support and launch up to 154 Tomahawk guided missiles. *(Courtesy of the U.S. Navy)*

and also tested by the space flight operations facility. Finally, as the spacecraft nears its celestial target, scientific instruments are powered on (if necessary) and calibrated.

crust The outermost solid layer of a planet or moon.

cryogenic propellant A rocket fuel, oxidizer, or propulsion fluid that is liquid only at very low temperatures. Liquid hydrogen (LH_2) and liquid oxygen (LO_2) are examples of cryogenic propellants.

See also ROCKET.

cryogenics The branch of science dealing with very low temperatures, applications of these low-temperature environments, and methods of producing them. This science may be concerned with practical engineering problems, such as producing, transporting, and storing the metric ton quantities of liquid oxygen (LO_2) or liquid hydrogen (LH_2) needed as chemical rocket propellants, or it may help a physicist investigate some basic properties of matter at extremely low temperatures. Cryogenic researchers work with temperature environments down to one-millionth kelvin (10^{-6} K)—essentially at the physical threshold of absolute zero.

See also ABSOLUTE ZERO; CRYOGENIC TEMPERATURES.

cryogenic temperature Generally, temperatures below –150° Celsius (C) (123 kelvins [K]). Sometimes the boiling point of liquid nitrogen (LN_2), namely –195°C (77.4 K) is used as the "cryogenic temperature" threshold; in other instances the temperature associated with the boiling point of liquid helium (LHe), namely –269°C (4.2 K), is considered the threshold of the cryogenic temperature regime.

cryosphere The portion of Earth's climate system consisting of the world's ice masses and snow deposits. These include the continental ice sheets, mountain glaciers, sea ice, surface snow cover, and lake and river ice. Changes in snow cover on the land surfaces generally are seasonal and closely tied to the mechanics of atmospheric circulation. The glaciers and ice sheets are closely related to the global hydrologic cycle and to variations of sea level. Glaciers and ice sheets typically change volume and extent over periods ranging from hundreds to millions of years.

See also EARTH SYSTEM SCIENCE; GLOBAL CHANGE.

C star *See* CARBON STAR.

Cunningham, R. Walter (1932–) American *Astronaut, U.S. Marine Corps Officer* The astronaut R. Walter (Walt) Cunningham served as the lunar module (LM) pilot during NASA's *Apollo 7* mission (October 11–22, 1968). Although confined to orbit around Earth, this important mission was the first human-crewed flight of Apollo Project spacecraft hardware. Its success paved the way for the first lunar landing on July 20, 1969 by the *Apollo 11* astronauts.

See also APOLLO PROJECT.

curie (symbol: C or Ci) The traditional unit used to describe the intensity of radioactivity of a sample of material. One curie is equal to 3.7×10^{10} disintegrations per second. (This is approximately the rate of decay of one gram of radium.) Named after the Curies, Pierre and Marie, who discovered radium in 1898.

See also BECQUEREL.

current (symbol: I) The flow of electric charge through an electrical conductor. The AMPERE (symbol: A) is the SI unit of electric current.

Curtis, Heber Doust (1872–1942) American *Astronomer* Heber Curtis gained national attention in 1920 when he engaged in the "Great Debate" over the size of the universe with fellow American astronomer HARLOW SHAPLEY. Curtis supported the daring hypothesis that spiral nebulas were actually "island universes"—that is, other galaxies existing far beyond the Milky Way galaxy. This then radical position implied that the observable universe was much larger than anyone had dared to imagine. By 1924 EDWIN P. HUBBLE was able to support Curtis's hypothesis. Hubble did this by demonstrating that the great spiral Andromeda "nebula" was actually a large galaxy similar to, but well beyond, the Milky Way.

Curtis was born in Muskegon, Michigan, on June 27, 1872. As a child he went to school in Detroit. He then studied the classical languages (Latin and Greek) at the University of Michigan for five years, receiving a bachelor of arts degree in 1892 followed by a master of arts degree in 1894. After graduation Curtis moved to California to accept a position at Napa College in Latin and Greek studies. However, while teaching the classics he became interested in astronomy and volunteered as an amateur observer at the nearby Lick Observatory. In 1896 Herber Curtis became a professor of mathematics and astronomy at the College of the Pacific following a merger of that institution with Napa College.

Curtis served as a volunteer member of the Lick Observatory expedition that traveled to Thomaston, Georgia, in 1900 to observe a solar eclipse. His enjoyment of and outstanding performance during this eclipse expedition converted Curtis's amateur interests in astronomy into a lifelong profession. Following his first solar eclipse expedition he entered the University of Virginia on a fellowship and graduated in 1902 with a doctorate in astronomy. After graduation he joined the staff of the Lick Observatory as an astronomer. From 1902 to 1909 he made precise radial velocity measurements of the brighter stars in support of WALLACE CAMPBELL's observation program.

Starting in 1910 Curtis became interested in photographing and analyzing spiral nebulas. He soon became convinced that these interesting celestial objects were actually isolated independent systems of stars, or "island universes" as IMMANUEL KANT had called them in the 18th century.

Many astronomers, including Harlow Shapley, opposed Curtis's hypothesis about the extragalactic nature of spiral nebulas. On April 26, 1920, Curtis and Shapley engaged in their famous "Great Debate." They debated at the National Academy of Sciences in Washington, D.C., about the scale of the universe and the nature of the Milky Way galaxy. At the time the vast majority of astronomers considered the extent of the Milky Way galaxy synonymous with the size of the universe—that is, they thought the universe was just one big galaxy.

Neither astronomer involved in this highly publicized debate was completely correct. Curtis argued that spiral nebulas were other galaxies similar to the Milky Way—a bold hypothesis later proven to be a correct. But he also suggested that the Milky Way was small and that the Sun was near its center—both of which ideas were subsequently proven incorrect. Shapley, on the other hand, incorrectly opposed the hypothesis that spiral nebulas were other galaxies. He argued that the Milky Way galaxy was very large (much larger than it actually is) and that the Sun was far from the galactic center.

By the mid-1920s the great American astronomer Edwin Hubble helped to resolve one of the main points of controversy when he used the behavior of Cepheid variable stars to estimate the distance to the Andromeda galaxy. Hubble showed that this distance was much greater than the size of the Milky Way galaxy proposed by Shapley. So, as Curtis had suggested, the great spiral nebula in Andromeda and other spiral nebulas could not be part of the Milky Way and had to be separate galaxies. In the 1930s astronomers proved that Shapley's comments were more accurate concerning the actual size the Milky Way and the Sun's relative location within it. Therefore, when viewed from the perspective of science history, both eminent astronomers had argued their positions using partially faulty and fragmentary data. The Curtis-Shapley debate triggered a new wave of astronomical inquiry in the 1920s that allowed astronomers such as Hubble to determine the true size of the Milky Way and to recognize that it is but one among many other galaxies in an incredibly vast, expanding universe.

In 1920 Curtis became the director of the Allegheny Observatory of the University of Pittsburgh. During the next decade he improved instrumentation at the observatory and participated in four astronomical expeditions to observe solar eclipses. He then returned to the University of Michigan in 1930, where he became a professor of astronomy and the director of the university's astronomical observatories. He served in those positions until his death in Ann Arbor, Michigan, on January 9, 1942. His wife, Mary D. Rapier Curtis, his daughter, and three sons survived him.

cutoff (cut-off) 1. An act or instance of shutting something off; specifically, in rocketry, an act or instance of shutting off the propellant flow in a rocket or of stopping the combustion of the propellant. *Compare with* BURNOUT. 2. Something that shuts off or is used to shut off (e.g., a fuel cutoff valve). 3. Limiting or bounding, as, for example, in cutoff frequency.

cyborg A contraction of the expression *cyb*ernetic *org*anism. (Cybernetics is the branch of information science dealing with the control of biological, mechanical, and/or electronic systems.) While the term *cyborg* is quite common in contemporary science fiction—for example, the frightening "Borg collective" in the popular *Star Trek: The Next Generation* motion picture and television series—the concept was actually first proposed in the early 1960s by several scientists who were then exploring alternative ways of overcoming the harsh environment of space. Their suggested overall strategy was simply to adapt a human to space by developing appropriate technical devices that could be incorporated into an astronaut's body. Astronauts would become cybernetic organisms, or cyborgs! Instead of simply protecting an astronaut's body from the harsh space environment by "enclosing" it in some type of spacesuit, space capsule, or artificial habitat (the technical approach actually chosen), "cyborg approach" advocates boldly asked, Why not create "cybernetic organisms" that could function in the harsh environment of space without special protective equipment? For a variety of technical, social, and political reasons, this suggested line of research quickly ended, but the term *cyborg* has survived.

Today this term is often applied to any human being (whether on Earth, under the sea, or in outer space) using a technology-based, body-enhancing device. For example, a person with a pacemaker, hearing aid, or artificial knee can be called a cyborg. When you strap on "wearable" computer-interactive components, such as the special vision and glove devices that are used in a virtual reality system, you have actually become a "temporary" cyborg.

By further extension, the term *cyborg* is sometimes used today to describe fictional artificial humans or very sophisticated robots with near-human (or super-human) qualities. The Golem (a mythical clay creature in medieval Jewish folklore) and the Frankenstein "monster" (from Mary Shelley's classic 1818 novel) are examples of the former, while Arnold Schwarzenegger's portrayal of the Terminator (from the motion picture of the same name) is an example of the latter usage.

cycle Any repetitive series of operations or events; the complete sequence of values of a periodic quantity that occur during a period—for example, one complete wave. The period is the duration of one cycle, while the frequency is the rate of repetition of a cycle. The hertz (Hz) is the SI unit of frequency, and 1 hertz equals 1 cycle per second.

cycle life The number of times a unit may be operated (e.g., opened and closed) and still perform within acceptable limits.

Cygnus X-1 The strong X-ray source in the constellation Cygnus (the Swan) that scientists believe comes from a binary star system consisting of a supergiant star orbiting and a black hole companion. Gas drawn off the supergiant star emits X-rays as it becomes intensely heated while falling into the black hole.

See also X-RAY ASTRONOMY.

cylindrical coordinates A system of curvilinear coordinates in which the position of a point in space is determined by (a) its perpendicular distance from a given line, (b) its distance from a selected reference plane perpendicular to this line, and (c) its angular distance from a selected reference line when projected onto this plane. The coordinates thus form the elements of a cylinder and by convention are written r, θ, and z, where r is the radial distance from the cylinder's axis z and θ is the angular position from a reference line in a cylindrical cross section normal to z. Also called *polar coordinates*.

The relationships between the cylindrical coordinates (r, θ, z) and the rectangular Cartesian coordinates (x, y, z) are:

$$x = r \cos \theta; \; y = r \sin \theta; \; z = z$$

cylindrical grain A solid-propellant grain in which the internal cross section is a circle.

See also ROCKET.

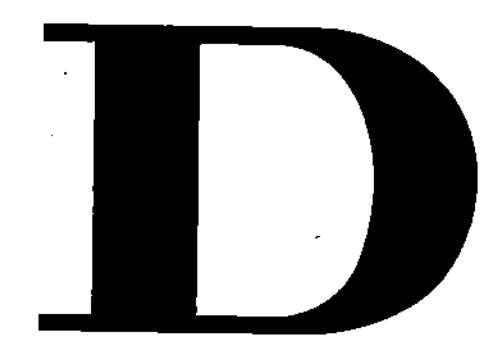

Dactyl A natural satellite of the asteroid Ida-2. NASA scientists discovered Dactyl in February 1994 as they were reviewing Galileo Project data from the spacecraft's flyby encounter with the asteroid Ida-2 on August 28, 1993. This tiny moonlet is about 1.2 km by 1.4 km by 1.6 km and orbits its parent asteroid at a distance of approximately 100 km. Dactyl is the first natural satellite of a minor planet to be discovered. The tiny moonlet has two major craters, Acmon with a diameter of 0.3 km and Celmis with a diameter 0.2 km.

See also ASTEROID; GALILEO PROJECT; IDA.

Danjon, André (1890–1967) French *Astronomer* André Danjon was noted for the development of precise astronomical instruments (such as the Danjon prismatic astrolabe), the calculation of accurate albedos of the Moon, Venus, and Mars, and studies of Earth's rotation.

See also ALBEDO; ASTROLABE.

dark energy The generic name now given by astrophysicists and cosmologists to a hypothesized unknown cosmic force field thought to be responsible for the recently observed acceleration of the expansion of the universe. In 1929 EDWIN P. HUBBLE (1889–1953) first proposed the concept of an expanding universe when he suggested that observations of Doppler-shifted wavelengths of the light from distant galaxies indicated that these galaxies were receding from Earth with speeds proportional to their distance—an empirically based postulate that became known as Hubble's law. Mathematically, this relationship is expressed as $V = H_0 \times D$, where V is the recessional velocity of a distant galaxy, H_0 is the Hubble constant, and D is its distance from Earth.

Then, in the late 1990s while scientists made systematic surveys of very distant Type I (carbon detonation) supernovas, they observed that instead of slowing down (as might be anticipated if gravity is the only significant force at work in cosmological dynamics), the rate of recession (that is, the redshift) of these very distant objects appeared to actually be increasing. It was almost as if some unknown force were neutralizing or canceling the attraction of gravity. Such startling observations proved controversial and very inconsistent with the standard "gravity only" models of an expanding universe within big bang cosmology. However, despite fierce initial resistance within the scientific community, these perplexing observations eventually became accepted. The data imply that the expansion of the universe is accelerating—a dramatic conclusion that tosses modern cosmology into as great an amount of turmoil as did Hubble's initial announcement of an expanding universe some 70 years earlier.

What could be causing this apparent acceleration of an expanding universe? Cosmologists do not yet have an acceptable answer. Some are revisiting the cosmological constant, a concept inserted by ALBERT EINSTEIN (1879–1955) into his general relativity theory to make that revolutionary theory describe a static universe. However, after introducing the concept of a mysterious force associated with empty space capable of balancing or even resisting gravitational attraction, Einstein decided to abandon the concept. In fact, he personally referred to the notion of a cosmological constant as his "greatest failure." Nevertheless, contemporary physicists are now revisiting Einstein's concept and are suggesting that there possibly is a *vacuum pressure force* (a recent name for the cosmological constant) that is inherently related to empty space and exerts its influence only on a very large scale. Consequently, this mysterious force would have been negligible during the early phases of the universe following the big bang event but might now be starting to manifest itself and serve as a major factor in cosmological dynamics—that is, the rate of expansion of the present universe. Since such a mysterious force is neither required nor explained by any of the currently known laws of physics, scientists do not yet have a clear physical interpretation of just what such a mysterious ("gravity resisting") force really means.

Today scientists grapple with various cosmological models in an effort to reconcile theory with challenging new data.

To assist themselves in this process, physicists suggest the intriguing name *dark energy* to represent the mysterious cosmic force causing the expansion of the universe to accelerate. Dark energy does not appear to be associated with either matter or radiation. What does the existence of dark energy mean? For one thing, the magnitude of the acceleration in cosmic expansion might imply that the amount of dark energy in the universe actually exceeds the total mass-energy equivalent of matter (luminous and dark) in the universe by a considerable margin. Any such (as yet unproven) dominant presence of dark energy—opposing the attractive force of gravity—then implies that the universe will continue to expand forever.

See also BIG BANG; COSMOLOGY; DARK MATTER; DENSITY PARAMETER; HUBBLE'S LAW; UNIVERSE.

dark matter Material in the universe that cannot be observed directly because it emits very little or no electromagnetic radiation, but whose gravitational effects can be measured and quantified. Dark matter was originally called *missing mass* and, as suggested by its name, was discovered only through its gravitational effects. While investigating the CLUSTER OF GALAXIES in the constellation Coma Berenices (Berenice's Hair) in 1933, the Swiss astrophysicist FRITZ ZWICKY (1898–1974) noticed that the velocities of the individual galaxies in this cluster were so high that they should have escaped from one another's gravitational attraction long ago. He concluded that the amount of matter actually present in the cluster had to be much greater than what could be accounted for by the visibly observable galaxies. In fact, from his observations Zwicky estimated that the visible matter in the cluster was only about 10 percent of the mass actually needed to gravitationally bind the galaxies together. He then focused a great deal of his scientific attention on the problem of the "missing mass" of the universe. Today astronomers refer to this mystery as the problem of dark matter. Zwicky and other scientists used the rotational speeds of individual galaxies within a cluster of galaxies (as obtained from their Doppler shifts) to provide observational evidence that most of the mass of the universe might be in the form of invisible material called dark matter.

The second direct observational evidence of the existence of dark matter came from careful radio astronomy–supported studies of the rotation rates of individual galaxies, including the Milky Way galaxy. From their rotational behavior astronomers discovered that most galaxies appear to be surrounded by a giant cloud (or galactic halo) containing matter capable of exerting gravitational influence but not emitting observable radiation. These studies also indicated that the great majority of a galaxy's mass lay in this very large halo, which is perhaps 10 times the diameter of the visible galaxy. The Milky Way galaxy is a large spiral galaxy that contains about 100 billion stars. Observations indicate that our home galaxy is surrounded by a dark matter halo that probably extends out to about 750,000 light-years. The mass of this dark matter halo appears to be about 10 times greater than the estimated mass of all the visible stars in our galaxy. However, the dark matter halo's material composition remains an astronomical mystery. Using the Milky Way galaxy as a reference, astrophysicists believe that the material contained in galactic halos could represent about 90 percent of the total mass of the universe—if current "gravity only" big bang cosmology models are correct.

It should come as no surprise that there is considerable disagreement within the scientific community as to what this dark matter really is. Two general schools of astronomical thought have emerged—one advocating MACHOs (or baryonic matter) and one advocating WIMPs (or nonbaryonic matter).

The first group assumes dark matter consists of MACHOs, or *ma*ssive *c*ompact *h*alo *o*bjects—essentially ordinary matter that astronomers have simply not yet detected. This unobserved but ordinary matter is composed of heavy particles (BARYONS), such as neutrons and protons. The brown dwarf is one candidate MACHO that could significantly contribute to resolving the missing mass problem. The brown dwarf is a substellar (almost a star) celestial body that has the material composition of a STAR but that contains too little mass to permit its CORE to initiate thermonuclear fusion. In 1995 astronomers detected the first brown dwarf candidate object—a tiny companion orbiting the small red dwarf star Gliese 229. Also very difficult to detect are low-mass white dwarfs that represent another dark matter candidate. Using sophisticated observational techniques that take advantage of GRAVITATIONAL LENSING, astronomers have collected data suggesting low-mass white dwarfs may actually make up about half of the dark matter in the universe. Finally, black holes, especially relatively low-mass primordial black holes that formed soon after the big bang, represent another MACHO candidate.

The second group of scientists speculates that dark matter consists primarily of exotic particles that they collectively refer to as WIMPs (*w*eakly *i*nteracting but *m*assive *p*article*s*). These exotic particles represent a hypothetical form of matter called nonbaryonic matter—matter that does not contain baryons (protons or neutrons). For example, if physicists determine that the neutrino actually possesses a tiny, nonzero rest mass, then these ubiquitous but weakly interacting elementary particles might carry much of the missing mass in the universe.

Some scientists suggest that the true nature of dark matter may not require an "all or nothing" characterization. To these astrophysicists it seems perfectly reasonable that dark matter might exist in several forms, including difficult to detect low-mass stellar and substellar objects (MACHOs) in the inner regions of a dark matter galactic halo as well as swarms of exotic particles (WIMPs) farther out in the galactic halo.

As presently understood, dark matter may be ordinary (but difficult to detect) baryonic matter; it may be in nonbaryonic form (such as possibly "massive" neutrinos); or perhaps it is an unexpected combination of both baryonic and nonbaryonic matter. In any event, current studies of the observable universe can account for only 10 percent of the mass needed to support gravity-only big bang cosmological models. Discovering the true nature of dark matter remains a very important and intriguing challenge for 21st-century cosmologists and astrophysicists.

See also BIG BANG THEORY; BLACK HOLES; COSMOLOGY; GALAXY; MACHO; STAR; UNIVERSE; WIMP.

dark nebula A cloud of interstellar dust and gas sufficiently dense and thick that the light from more distant stars and celestial bodies behind it is obscured. Astronomers study dark nebulas by examining their infrared radiation (IR) and radio wave emissions. The HORSEHEAD NEBULA (NGC 2024) in the constellation Orion is an example of a dark nebula.

See also INFRARED ASTRONOMY; NEBULA; RADIO ASTRONOMY.

dart configuration A configuration of an aerodynamic vehicle in which the control surfaces are at the tail of the vehicle. *Compare with* CANARD.

Darwin, Sir George Howard (1845–1912) British *Mathematical Astronomer* Sir George Darwin, second son of the famous naturalist, was a mathematical astronomer who investigated the influence of tidal phenomena on the dynamics of the Earth-Moon system. In 1879 he proposed that the Moon was formed out of material thrown from Earth's crust when Earth was a newly formed, fast-rotating young planet. His tidal theory of lunar formation remained plausible until the 1960s, when general scientific support shifted to the giant impact model of lunar formation.

See also MOON.

data 1. The plural of *datum*—a single piece of information, a fact. 2. A general term for numbers, digits, characters, and symbols that are accepted, stored, and processed by a computer. When such data become meaningful, then the data components become *information.* 3. Representation of facts, concepts, or instructions in a formalized manner suitable for communication, interpretation, or processing by humans or by automatic means. Any representations such as characters or analog quantities to which meaning is or might be assigned.

data fusion The technique in which a wide variety of data from multiple sources is retrieved and processed as a single, unified entity. A significant set of a priori databases is crucial to the effective functioning of the data fusion process. For example, this technique might be used in supporting the development of an evolving "world model" for a semi-intelligent robotic lunar rover that human beings teleoperate from a control station on Earth. Inputs from the robot vehicle's sensors would be blended or "fused" with existing lunar environment databases to support more efficient exploration strategies and surface operations. Similarly, data fusion from a variety of advanced Earth-observing spacecraft support the development and effective use of geographic information systems (GIS) in a variety of important areas, including Earth system science, meteorology, national defense, and urban planning.

See also EARTH SYSTEM SCIENCE; REMOTE SENSING; ROBOTICS IN SPACE; TELEOPERATION.

data handling subsystem The onboard computer responsible for the overall management of a spacecraft's activity. It is usually the same computer that maintains timing; interprets commands from Earth; collects, processes, and formats the telemetry data that are to be returned to Earth; and manages high-level fault protection and safing routines. Aerospace engineers frequently call this type of multifunctional spacecraft computer the *command and data handling subsystem.*

See also SPACECRAFT CLOCK; TELECOMMUNICATIONS; TELEMETRY.

data link 1. The means of connecting one location to another for the purpose of transmitting and receiving data. 2. A communications link suitable for the transmission of data. 3. Any communications channel used to transmit data from a sensor to a computer, a readout device, or a storage device.

See also DOWNLINK; TELEMETRY; UPLINK.

data relay satellite A specialized Earth-orbiting satellite that supports nearly continuous communications between other satellites or human-occupied spacecraft with controllers and researchers on Earth. NASA's constellation of Tracking and Data Relay Satellites (TDRS) is a prime example. Space systems supported by the TDRS fleet's capability to provide space-based communications include the space shuttle, the *International Space Station* (ISS), Landsat and other Earth-observing spacecraft (EOS), the *Hubble Space Telescope* (HST) and other space-based astronomical observatories, and various expendable launch vehicles (ELVs) as they ascend into space and deploy their payloads.

See also TRACKING AND DATA RELAY SATELLITE.

datum 1. A single unit of information; the singular of data. 2. Any numerical or geometrical quantity or set of such quantities that may serve as a reference or base for other quantities. When the concept is geometric, the preferred plural form is *datums,* as in the statement "Researchers used two geodetic *datums* in the analysis of the experiment."

Dawes, William Rutter (1799–1868) British *Amateur Astronomer, Physician* William Dawes made precision measurements of binary star systems. In 1850 he independently discovered the major inner C ring (or crêpe ring) of Saturn. Astronomers often call this particular Saturnian ring the crêpe ring because of its light and delicate appearance.

See also BINARY STAR SYSTEM; SATURN.

Dawn The Dawn mission, which is scheduled to launch in June 2006, is part of NASA's Discovery Program, an initiative for highly focused, rapid-development scientific spacecraft. The goal is to understand the conditions and processes during the earliest history of the solar system. To accomplish this mission the *Dawn* scientific spacecraft will investigate the structure and composition of two minor planets, 1 Ceres and 4 Vesta. These large main belt asteroids have many contrasting characteristics and appear to have remained intact since their formation more than 4.6 billion years ago.

Dawn is a mission that will rendezvous and orbit the asteroids Vesta and Ceres, two of the largest protoplanets. Ceres and Vesta reside in the extensive zone between Mars and Jupiter with many other smaller bodies in a region of the solar system called the main asteroid belt. The objectives of the mission are to characterize these two large asteroids with particular emphasis being placed on their internal structure, density, shape, size, composition, and mass. The spacecraft will also return data on surface morphology, the extent of craters, and magnetism. These measurements will help scientists determine the thermal history, size of the core, role of water in asteroid evolution, and which meteorites found on Earth came from these bodies. For both asteroids the data returned will include full surface imagery, full surface spectrometric mapping, elemental abundances, topographic profiles, gravity fields, and mapping of remnant magnetism, if any.

The top question that the mission addresses is the role of size and water in determining the evolution of the planets. Scientists consider that Ceres and Vesta are the right two celestial bodies with which to address this important question. Both asteroids are the most massive of the protoplanets in the asteroid belt—miniature planets whose growth was interrupted by the formation of Jupiter. Ceres is very primitive and wet, while Vesta is evolved and dry. Planetary scientists now suggest that Ceres may have active hydrological processes leading to seasonal polar caps of water frost. Ceres may also have a thin, permanent atmosphere—an interesting physical condition that would set it apart from the other minor planets. Vesta may have rocks more strongly magnetized than those on Mars, a discovery that would alter current ideas of how and when planetary dynamos arise.

As currently planned, the spacecraft will launch from Cape Canaveral on a Delta 7925H expendable rocket in late June 2006. After a four-year heliocentric cruise, *Dawn* will reach Vesta in late July 2010 and then go into orbit around the minor planet for 11 months. One high-orbit period at 700 km of altitude is planned, followed by a low orbit at an altitude of 120 km. Upon completion of its rendezvous and orbital reconnaissance of Vesta, the *Dawn* spacecraft will depart this minor planet in early July 2011 and fly on to Ceres. The spacecraft will reach Ceres in mid-August 2014, where it will again go into 11 months of an orbital reconnaissance mission. Both a high-orbit scientific investigation of Ceres at an altitude of 890 km and a low-orbit study at an altitude of 140 km are currently planned. The Ceres orbital reconnaissance phase of this mission will end in late July 2015. Aerospace mission planners anticipate that 288 kg of xenon will be required to reach Vesta and 89 kg to reach Ceres. The hydrazine thrusters will be used for orbit capture. Depending on the remaining supply of onboard propellants and the general health of the *Dawn* spacecraft and its complement of scientific instruments, mission controllers at the Jet Propulsion Laboratory (JPL) could elect to continue this robot spacecraft's exploration of the asteroid belt beyond July 2015.

The *Dawn* spacecraft structure is made of aluminum and is box-shaped with two solar panel wings mounted on opposite sides. A parabolic fixed (1.4-m-diameter) high-gain dish antenna is mounted on one side of the spacecraft in the same plane as the solar arrays. A medium-gain fan beam antenna is also mounted on the same side. A 5-m-long magnetometer boom extends from the top panel of the spacecraft. Also mounted on the top panel is the instrument bench holding the cameras, mapping spectrometer, laser altimeter, and star trackers. In addition, there is a gamma-ray/neutron spectrometer mounted on the top panel. The solar arrays provide 7.5 kW to drive the spacecraft and the solar electric ion propulsion system.

The *Dawn* spacecraft is the first purely scientific NASA space exploration mission to be powered by ion propulsion. The ion propulsion technology for this mission is based on the *Deep Space 1* spacecraft ion drive and uses xenon as the propellant that is ionized and accelerated by electrodes. The xenon ion engines have a thrust of 90 millinewtons (mN) and a specific impulse of 3100 s. The spacecraft maintains attitude control through the use of 12 strategically positioned 0.9 N thrust hydrazine engines. The spacecraft communicates with scientists and mission controllers on Earth by means of high- and medium-gain antennas as well as a low-gain omnidirectional antenna that uses a 135-watt traveling wave tube amplifier.

In summary, the goal of the Dawn mission is to help scientists better understand the conditions and processes that took place during the formation of our solar system. Ceres and Vesta represent two of the few large protoplanets that have not been heavily damaged by collisions with other bodies. Ceres is the largest asteroid in our solar system and Vesta is the brightest asteroid—the only one visible with the unaided eye. What makes the Dawn mission especially significant is the fact that Ceres and Vesta possess striking contrasts in composition. Planetary scientists now speculate that many of these differences arise from the conditions under which Ceres and Vesta formed during the early history of the solar system. In particular, Ceres formed wet and Vesta formed dry. Water kept Ceres cool throughout its evolution, and there is some evidence to indicate that water is still present on Ceres as either frost or vapor on the surface and possibly even liquid water under the surface. In sharp contrast, Vesta was hot, melted internally, and became volcanic early in its development. The two large protoplanets followed distinctly different evolutionary pathways. Ceres remains in its primordial state, while Vesta has evolved and changed over millions of years.

See also ASTEROID; ELECTRIC PROPULSION.

deadband In general, an intentional feature in a guidance and control system that prevents a flight path error from being corrected until that error exceeds a specified magnitude. With respect to the space shuttle Orbiter Vehicle, for example, the attitude and rate control region in which no Orbiter reaction control subsystem (RCS) or vernier engine correction forces are being generated.

See also ATTITUDE CONTROL SYSTEM; SPACE TRANSPORTATION SYSTEM; VERNIER ENGINE.

deboost A retrograde (opposite-direction) burn of one or more low-thrust rockets or an aerobraking maneuver that lowers the altitude of an orbiting spacecraft.

See also AEROBRAKING; ROCKET; THRUST.

debris Jettisoned human-made materials, discarded launch vehicle components, and derelict or nonfunctioning spacecraft in orbit around Earth.

See also SPACE DEBRIS.

decay (orbital) The gradual lessening of both the apogee and perigee of an orbiting object from its PRIMARY BODY. For example, the orbital decay process for artificial satellites and space debris often results in their ultimate fiery plunge back into the denser regions of Earth's atmosphere.

See also ORBITS OF OBJECTS IN SPACE; SPACE DEBRIS.

decay (radioactive) The spontaneous transformation of one radionuclide into a different nuclide or into a different energy state of the same nuclide. This natural process results in a decrease (with time) in the number of original radioactive atoms in a sample and involves the emission from the nucleus of alpha particles, beta particles, or gamma rays.

See also NUCLEAR RADIATION; RADIOACTIVITY; RADIOISOTOPE.

deceleration parachute A parachute attached to a craft and deployed to slow the craft, especially during landing. Also called a *drogue parachute*.

See also DROGUE PARACHUTE.

deci- (symbol: d) A prefix meaning multiplied by one-tenth (10^{-1}); for example, decibel.

See also DECIBEL; SI UNITS.

decibel (symbol: dB) 1. One-tenth of a BEL. 2. A dimensionless measure of the ratio of two POWER levels, sound intensities, or voltages. It is equal to 10 times the common logarithm (that is, the logarithm to the base 10) of the power ratio, P_2/P_1. Namely,

$$n \text{ (decibels)} = 10 \log_{10} (P_2/P_1)$$

Since the bel (B) is an exceptionally large unit, the decibel is more commonly encountered in physics and engineering.

declination (symbol: δ) For an object viewed on the celestial sphere, the angular distance north (0° to 90° positive) or south (0° to 90° negative) of the celestial equator. Declination is a coordinate used with right ascension in the equatorial coordinate system—the most commonly used coordinate system in astronomy.

See also CELESTIAL SPHERE; EQUATORIAL COORDINATE SYSTEM; KEPLERIAN ELEMENTS.

Deep Impact The objectives of NASA's Deep Impact mission are to rendezvous with Comet P/Tempel 1 and launch a projectile into the comet's nucleus. Instruments on the spacecraft will observe the ejecta from the impacting projectile, much of which will represent pristine material from the interior of the comet; the crater formation process; the resulting crater; and any outgassing from the nucleus, particularly the newly exposed surface. This project was successfully accomplished in early July 2005.

The spacecraft consists of a 350-kg cylindrical copper impactor attached to a flyby bus. The launch mass of the flyby bus and impactor is 1,010 kg. The spacecraft is a box-shaped honeycombed aluminum framework with a flat rectangular Whipple shield (a space debris shield) mounted on one side to protect components during the close approach to a comet. Mounted on the framework are one high- and one medium-resolution instrument, each of which consists of an imaging camera and an infrared spectrometer that will observe the ice and dust ejected by the comet's nucleus when the impactor strikes. Scientists believe that much of the ejected matter will be exposed to outer space for the first time in more than 4 billion years. The medium-resolution camera has a field of view (FOV) of 0.587 degrees and a resolution of 7 m/pixel at a distance of 700 km. This instrument also supports navigation and collects context images. The high-resolution camera has a FOV of 0.118 degrees and a resolution of 1.4 m/pixel at a distance 700 km. The infrared spectrometers cover the wavelength range from 1.05 to 4.8 micrometers (μm) with FOV of 0.29 degrees (in the high-resolution mode) and 1.45 degrees (in the low-resolution mode). The total flyby bus instrument payload has a mass of 90 kg.

The flyby spacecraft measures approximately 3.2 by 1.7 by 2.3 m, is three-axis stabilized, and uses an onboard hydrazine propulsion system. The flyby bus communicates with scientists on Earth via X-band (8,000 MHz) through a 1-meter-diameter parabolic dish antenna mounted on a 2-axis gimbal. Communications between the Deep Space impactor and flyby spacecraft take place via S-band. The S-band portion of the radio frequency spectrum ranges from 1,700 to 2,300 MHz. The maximum data rate will be 400 kbps. A spacecraft electric power level of 620 W at the time of the comet encounter will be provided by a 7.5-square-meter solar array and stored in a small NiH_2 battery.

The projectile is made of copper so it will be easily identifiable in the spectra after the projectile is largely vaporized and mixed in with the comet ejecta on impact. The impactor is equipped with an impactor targeting sensor, an imager that provides knowledge for autonomous control and targeting, and a cold-gas attitude control system.

NASA used a Delta II rocket to launch *Deep Impact* from Cape Canaveral on January 12, 2005. The spacecraft traveled in a heliocentric orbit and rendezvoused with Comet P/Tempel 1 in July 2005. As presently planned, *Deep Impact* will be about 880,000 km from the comet on July 3, 2005 and moving at a velocity of 10.2 km/s relative to the comet. The projectile was released at this point, and shortly after release the flyby spacecraft executed a maneuver to slow down (at 120 m/s) relative to the impactor. About 24 hours after release, on July 4, 2005, the impactor struck the sunlit side of the comet's nucleus. At the relative encounter velocity of 10.2 m/s, the impactor formed a crater roughly 25 meters deep and 100 meters wide, ejecting material from the interior of the nucleus into space and vaporizing the impactor and much of the ejecta. The flyby spacecraft was approximately 10,000 km away at the time of impact and began imaging 60 seconds before impact. At 600 seconds after impact the spacecraft was about 4000 km from the nucleus, and observations of the crater began and continued until closest approach to the nucleus, at a distance of about 500 km. At 961 seconds after impact, imaging ended when the spacecraft reoriented itself to cross the inner coma. At 1,270 seconds the crossing of the inner coma was complete, and the spacecraft

This is an artist's rendition of the *Deep Impact* flyby spacecraft releasing the impactor 24 hours before the impact event. Pictured from left to right are Comet Tempel 1, the impactor, and the flyby spacecraft. The impactor is a 370-kilogram mass with an onboard guidance system. The flyby spacecraft includes a solar panel (right), a high-gain antenna (top), a debris shield (left, background), and science instruments for high- and medium-resolution imaging, infrared spectroscopy, and optical navigation (box and cylinder, lower left). The flyby spacecraft is about 3.2 meters long, 1.7 meters wide, and 2.3 meter high. The launch payload has a mass of 1,020 kilograms. *(Courtesy of NASA/JPL/UMD)*

oriented itself to look back at the comet. At 3,000 seconds the spacecraft began playback of scientific data to Earth.

The comet and spacecraft were about 0.89 AU from Earth and 1.5 AU from the Sun during the encounter. Real-time return of selected impactor images and flyby images and spectra were returned to Earth during the encounter. Primary data return was over the first day after encounter. The mission also has a 28-day supplemental data return period. Although the most spectacular images came from the flyby spacecraft, many telescopes on Earth also collected data on the impact in the optical and other wavelengths to complement the data from the flyby spacecraft. During this mission the spacecraft ranged over a distance of 0.93 to 1.56 AU from the Sun.

See also COMETS.

deep space As a result of terminology and conventions introduced in the early part of the space age (between 1957 and 1960), deep space is often considered the region of outer space that starts at altitudes greater than 5,600 km above Earth's surface.

Deep Space Climate Observatory **(DSCOVR)** Previously called *Triana,* NASA's planned *Deep Space Climate Observatory* is the first Earth-observing mission to travel to Lagrange-1 (or L1)—the neutral gravity point between the Sun and Earth. From L1 DSCOVR will have a continuous view of the sunlit side of the Earth at a distance of 1.5 million kilometers. In order to obtain the same coverage with current Earth-observing satellites in low-Earth orbits and geostationary orbits, scientists must manipulate, calibrate, and correlate data from four or more independent spacecraft. The full view of the sunlit disk of the Earth afforded by a satellite located at L1 has tremendous potential for Earth system science. As presently envisioned, DSCOVR will be launched from Cape Canaveral in late 2008 and become the first mission to explore the untapped potential of this interesting location for the scientific observation of our home planet.

Although researchers have used spacecraft to explore many facets of Earth system science, there are still numerous unanswered questions concerning exactly how our planet functions as a complex, highly interactive system. The synoptic view of our planet from L1 will support the collection of data that could answer difficult questions facing climate researchers. For example, scientists still do not accurately know how much of the incoming radiant energy from the Sun the Earth absorbs, re-emits, and reflects. This measurement, called the planetary albedo, is vital for climate research. Unfortunately, it is difficult to make piecemeal measurements of Earth's albedo because of the influence of cloud cover, ice, snow, smoke, volcanic ash, and other regional environmental factors that cause the overall value of planetary albedo to constantly change. Scientists believe that the DSCOVR mission should provide them significantly better global measurements of planetary albedo. For example, the broadband radiometer onboard the DSCOVR spacecraft will reassess the planetary albedo every 15 minutes, making these data a useful indicator of global change. Once launched from Cape Canaveral, the DSCOVR spacecraft will make a journey of more than 1.5 million km to reach L1, from which operational location it will observe our planet and make albedo measurements for at least two years.

See also ALBEDO; EARTH'S RADIATION BUDGET; EARTH SYSTEM SCIENCE; LAGRANGIAN LIBRATION POINTS; PLANETARY ALBEDO.

Deep Space Network (DSN) The majority of NASA's scientific investigations of the solar system are accomplished through the use of robotic spacecraft. The Deep Space Network (DSN) provides the two-way communications link that guides and controls these spacecraft and brings back the spectacular planetary images and other important scientific data they collect.

The DSN consists of telecommunications complexes strategically placed on three continents, providing almost continuous contact with scientific spacecraft traveling in deep space as Earth rotates on its axis. The Deep Space Network is the largest and most sensitive scientific telecommunications system in the world. It also performs radio and radar astronomy observations in support of NASA's mission to explore the solar system and the universe. The Jet Propulsion Laboratory (JPL) in Pasadena, California, manages and operates the Deep Space Network for NASA.

In January 1958 the Jet Propulsion Laboratory established the predecessor to the DSN when, under a contract with the United States Army, the laboratory deployed portable radio tracking stations in Nigeria (Africa), Singapore (Southeast Asia), and California to receive signals from and plot the orbit of *Explorer 1,* the first American satellite to successfully orbit Earth. Later that year (on December 3, 1958), as part of the emergence of a new federal civilian space agency, JPL was transferred from U.S. Army jurisdiction to that of NASA. At the very onset of the nation's civilian space program, NASA assigned JPL responsibility for the design and execution of robotic lunar and planetary exploration programs. Shortly afterward NASA embraced the concept of the DSN as a separately managed and operated telecommunications facility that would accommodate all deep space missions. This management decision avoided the need for each space flight project to acquire and operate its own specialized space communications network.

Today the DSN features three deep space communications complexes placed approximately 120 degrees apart around the world: at Goldstone in California's Mojave Desert; near Madrid, Spain; and near Canberra, Australia. This global configuration ensures that as Earth rotates an antenna is always within sight of a given spacecraft, day and night. Each complex contains up to 10 deep space communication stations equipped with large parabolic reflector antennas.

Every deep space communications complex within the DSN has a 70-meter-diameter antenna. (*See* figure.) These antennas, the largest and most sensitive in the DSN, are capable of tracking spacecraft that are more than 16 billion (10^9) kilometers away from Earth. The 3,850-square-meter surface of the 70-meter-diameter reflectors must remain accurate within a fraction of the signal wavelength, meaning that the dimensional precision across the surface is maintained to within one centimeter. The dish and its mount have a mass of nearly 7.2 million kilograms.

There is also a 34-meter-diameter high-efficiency antenna at each complex, which incorporates more recent advances in radio frequency antenna design and mechanics. For example, the reflector surface of the 34-meter-diameter antenna is precision shaped for maximum signal-gathering capability.

The most recent additions to the DSN are several 34-meter beam waveguide antennas. On earlier DSN antennas, sensitive electronics were centrally mounted on the hard-to-reach reflector structure, making upgrades and repairs difficult. On beam waveguide antennas, however, such electronics are located in a below-ground pedestal room. An incident radio signal is then brought from the reflector to this room through a series of precision-machined radio frequency reflective mirrors. Not only does this architecture provide the advantage of easier access for maintenance and electronic equipment enhancements, but the new configuration also accommodates better thermal control of critical electronic components. Furthermore, engineers can place more electronics in the antenna to support operation at multiple frequencies. Three of these new 34-meter beam waveguide antennas have been constructed at the Goldstone, California, complex, along with one each at the Canberra and Madrid complexes.

There is also one 26-meter-diameter antenna at each complex for tracking Earth-orbiting satellites, which travel in orbits primarily 160 to 1,000 kilometers above Earth. The two-axis astronomical mount allows these antennas to point low on the horizon to acquire (pick up) fast-moving Earth-orbiting satellites as soon as they come into view. The agile 26-meter-diameter antennas can track (slew) at up to three degrees per second.

Finally, each complex also has one 11-meter-diameter antenna to support a series of international Earth-orbiting missions under the Space Very Long Baseline Interferometry Project.

All of the antennas in the DSN network communicate directly with the Deep Space Operations Center at JPL in Pasadena, California. The center staff directs and monitors operations, transmits commands, and oversees the quality of spacecraft telemetry and navigation data delivered to network users. In addition to the DSN complexes and the operations center, a ground communications facility provides communications that link the three complexes to the operations center at JPL, to space flight control centers in the United States and overseas, and to scientists around the world. Voice and data communications traffic between various locations is sent via land lines, submarine cable, microwave links, and communications satellites.

The Deep Space Network's radio link to scientific spacecraft is basically the same as other point-to-point microwave communications systems, except for the very long distances involved and the very low radio frequency signal strength received from the spacecraft. How low is "very low?" The total signal power arriving at a network antenna from a robotic spacecraft encounter among the outer planets can be *20 billion times weaker* than the power level in a modern digital wristwatch battery.

The extreme weakness of the radio frequency signals results from restrictions placed on the size, mass, and power supply of a particular spacecraft by the payload volume and mass-lifting limitations of its launch vehicle. Consequently, the design of the radio link is the result of engineering tradeoffs between spacecraft transmitter power and antenna diameter

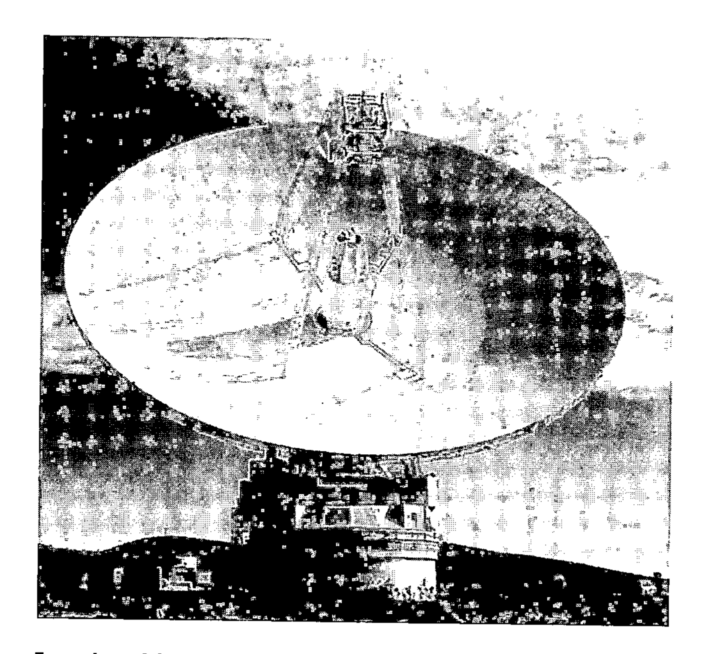

Front view of the 70-meter-diameter antenna at Goldstone, California. The Goldstone Deep Space Communications Complex, located in the Mojave Desert in California, is one of the three complexes that make up NASA's Deep Space Network (DSN). ***(Courtesy of NASA/JPL)***

and the signal sensitivity that engineers can build into the ground receiving system.

Typically, a spacecraft signal is limited to 20 watts, or about the same amount of power required to light the bulb in a refrigerator. When the spacecraft's transmitted radio signal arrives at Earth—from, for example, the neighborhood of Saturn—it is spread over an area with a diameter equal to about 1,000 Earth diameters. As a result, the ground antenna is able to receive only a very small part of the signal power, which is also degraded by background radio noise, or static.

Radio noise is radiated naturally from nearly all objects in the universe, including Earth and the Sun. Noise is also inherently generated in all electronic systems, including the DSN's own detectors. Since noise will always be amplified along with the signal, the ability of the ground receiving system to separate noise from the signal is critical. The DSN uses state-of-the-art, low-noise receivers and telemetry coding techniques to create unequalled sensitivity and efficiency.

Telemetry is basically the transmission of data to or from a spacecraft by means of radio waves or radio frequency signals. When aerospace engineers explain how a spacecraft sends messages to Earth, they describe telemetry as the process of engineering data and information about the spacecraft's own systems provided by its own scientific instruments. A typical scientific spacecraft transmits its data in binary code, using only the symbols 1 and 0. The spacecraft's telemetry system organizes and encodes these data for efficient transmission to ground stations on Earth. The ground stations have radio antennas and specialized electronic equipment to detect the individual bits, decode the data stream, and format the information for subsequent transmission to the data user (typically a team of scientists).

However, data transmission from a spacecraft can be disturbed by noise from various sources that interferes with the decoding process. If there is a high signal-to-noise ratio, the number of decoding errors will be low, but if the signal-to-noise ratio is low, then an excessive number of bit errors can occur. When a particular transmission encounters a large number of bit errors, mission controllers will often command the spacecraft's telemetry system to reduce the data transmission rate (measured in bits per second) in order to give the decoder (at the ground station) more time to determine the value of each bit.

To help solve the noise problem, a spacecraft's telemetry system might feed additional or redundant data into the data stream—which additional data are then used to detect and correct bit errors after transmission. The information theory equations used by telemetry analysts in data evaluation are sufficiently detailed to allow the detection and correction of individual and multiple bit errors. After connection, the redundant digits are eliminated from the data, leaving a valuable sequence of information for delivery to the data user.

Error detection and encoding techniques can increase the data rate many times over transmissions that are not coded for error detection. DSN coding techniques have the capability of reducing transmission errors in spacecraft science information to less than one in a million.

Telemetry is a two-way process, having a *downlink* as well as an *uplink*. Robot spacecraft use the downlink to send scientific data back to Earth, while mission controllers on Earth use the uplink to send commands, computer software, and other crucial data to the spacecraft. The uplink portion of the telecommunications process allows humans to guide spacecraft on their planned missions as well as to enhance mission objectives through such important activities as upgrading a spacecraft's onboard software while the robot explorer is traveling through interplanetary space.

Data collected by the DSN are also very important in precisely determining a spacecraft's location and trajectory. Teams of humans (called the mission navigators) use these tracking data to plan all the maneuvers necessary to ensure that a particular scientific spacecraft is properly configured and at the right place (in space) to collect its important scientific data. Tracking data produced by the DSN let mission controllers know the location of a spacecraft that is billions of kilometers away from Earth to an accuracy of just a few meters.

NASA's Deep Space Network is also a multifaceted science instrument that scientists can use to improve their knowledge of the solar system and the universe. For example, scientists use the large antennas and sensitive electronic instruments of the DSN to perform experiments in radio astronomy, radar astronomy, and radio science. The DSN antennas collect information from radio signals emitted or reflected by natural celestial sources. Such DSN-acquired radio frequency data are compiled and analyzed by scientists in a variety of disciplines, including astrophysics, radio astronomy, planetary astronomy, radar astronomy, Earth science, gravitational physics, and relativity physics.

In its role as a science instrument, the DSN provides the information needed to select landing sites for space missions; determine the composition of the atmospheres and/or the surfaces of the planets and their moons; search for BIOGENIC ELEMENTS in interstellar space; study the process of star formation; image asteroids; investigate comets, especially their nuclei and comas; search the permanently shadowed regions of the Moon and Mercury for the presence of water ice; and confirm ALBERT EINSTEIN's theory of general relativity.

The DSN radio science system also performs experiments that allow scientists to characterize the ionospheres of planets, look through the solar corona, and determine the mass of planets, moons, and asteroids. It accomplishes this by precisely measuring the small changes that take place in a spacecraft's telemetry signal as the radio waves are scattered, refracted, or absorbed by particles and gases near celestial objects within the solar system. The DSN makes its facilities available to qualified scientists as long as the research activities do not interfere with spacecraft mission support.

Another important DSN activity is *radio interferometry*—a scientific pursuit that generally involves widely separated radio telescopes on Earth but now also uses specially designed Earth-orbiting spacecraft as additional antennas. Interferometry is a technique used by radio astronomers to electronically link radio telescopes so they function as if they were a single instrument with the resolving power of a single (giant) dish with a diameter equal to the separation between the individual antenna dishes. NASA and the National Radio Astronomy Observatory (NRAO) joined with an international consortium of space agencies to support the 1997 launch of a Japanese radio telescope satellite called *HALCA*. The simultaneous use of ground radio telescopes with this orbiting radio telescope has created the largest astronomical "instrument" ever built—a radio telescope more than two-and-one-half times the diameter of Earth. The NASA-funded JPL Space Very Long Baseline Interferometry (VLBI) Project supports DSN participation in this unique radio interferometry combination, which almost triples the resolving power available to astronomers from ground-based radio telescopes. In fact, the resolving power of this radio interferometry combination is equivalent to being able to "see" a grain of rice in Toyko from Los Angeles. DSN personnel are using astronomical interferometry in cooperation with scientists in Japan to pursue such interesting science objectives as the production of high-resolution maps of quasars and the search for high radio brightness components in compact extragalactic sources.

See also ASTROPHYSICS; JET PROPULSION LABORATORY; QUASAR; RADAR ASTRONOMY; ROBOT SPACECRAFT; SPACECRAFT; TELECOMMUNICATIONS; TELEMETRY.

***Deep Space 1* (DS1)** The *Deep Space 1* (DS1) mission was the first of a series of technology demonstration probes developed by NASA's New Millennium Program. It was primarily a mission to test advanced spacecraft technologies that had never been flown in space. The space missions conducted within NASA's New Millennium Program seek to validate selected high-risk technologies on relatively inexpensive spacecraft so that future missions can use the new equipment with confidence. In the process of testing its solar electric propulsion system, autonomous navigation system, advanced microelectronics and telecommunications devices, and other cutting-edge aerospace technologies, *Deep Space 1* encountered the Mars-crossing near-Earth asteroid 9969 Braille (formerly known as 1992 KD) on July 20, 1999. In September 2001 the robot spacecraft encountered Comet Borrelly and, despite the failure of the system that helped determine its orientation, returned some of the best images of a comet's nucleus. After this encounter NASA mission controllers dubbed *Deep Space 1* "the little spacecraft that could."

As part of its mission to flight demonstrate new space technologies, the probe carried the Miniature Integrated Camera-Spectrometer (MICAS), an instrument combining two visible imaging channels with ultraviolet (UV) and infrared (IR) spectrometers. MICAS allowed scientists to study the chemical composition, geomorphology, size, spinstate, and atmosphere of the target celestial objects. *Deep Space 1* also carried the Plasma Experiment for Planetary Exploration (PEPE), an ion and electron spectrometer that measured the solar wind during cruise, the interaction of the solar wind with target bodies during encounters, and the composition of a comet's coma.

Aerospace engineers built the *Deep Space 1* spacecraft on an octagonal aluminum frame bus that measured 1.1 m by 1.1 m by 1.5 m. With instruments and systems attached, the spacecraft measured 2.5 m high, 2.1 m deep, and 1.7 m wide. The launch mass of the spacecraft was about 486.3 kg, including 31.1 kg of hydrazine and 81.5 kg of xenon gas. The probe received its electricity from a system of batteries and two solar panel "wings" attached to the sides of the frame that spanned roughly 11.75 m when deployed. The solar panels, designated SCARLET II (Solar Concentrator Arrays with Refractive Linear Element Technology), represented one of the major technology demonstration tests on the spacecraft. A cylindrical lens concentrated sunlight on a strip of photovoltaic cells (made of GaInP2/GaAs/Ge) and also protected the cells. Each solar array consisted of four 1.60-m × 1.13-m panels. At the beginning of the mission the solar array provided 2,500 watts (W) of electric power. As the spacecraft moved farther from the Sun and as the solar cells aged, the array provided less electricity to the spacecraft. Mission controllers used the spacecraft's high-gain antenna, three low-gain antennas, and a Ka-band antenna (all mounted on top of the spacecraft except for one low-gain antenna mounted on the bottom) to communicate with the probe as it traveled through deep space.

A xenon ion engine mounted in the propulsion unit on the bottom of the frame provided the spacecraft the thrust it needed to accomplish its journey. The 30-cm-diameter electric rocket engine consisted of an ionization chamber into which xenon gas was injected. Electrons emitted by a cathode traversed the discharge tube and collided with the xenon gas, stripping off electrons and creating positive ions in the process. The ions were then accelerated through a 1,280-volt grid and reached a velocity of 31.5 km/sec before ejecting from the spacecraft as an ion beam. The demonstration xenon ion engine produced 0.09 newton (N) of thrust under conditions of maximum electrical power (2,300 W) and about 0.02 N of thrust at the minimum operational electrical power level of 500 W. To neutralize electrical charge, excess electrons were collected and injected into the ion beam as it left the spacecraft. Of the 81.5 kg of xenon propellant, approximately 17 kg were consumed during the probe's primary mission.

Other technologies that were tested on this mission included a solar concentrator array, autonomous navigation, and experiments in low-power electronics, power switching, and multifunctional aerospace structures—in which electronics, cabling, and thermal control were integrated into a load-bearing element.

Deep Space 1 was launched from Pad 17-A at the Cape Canaveral Air Station at 12:08 UT (8:08 A.M. EDT) on October 24, 1998. This was the first launch of a Delta 7326-9.5 expendable rocket under NASA's Med-Lite booster program. A Delta II Lite launch vehicle has three strap-on solid-rocket boosters and a Star 37FM third stage. At 13:01 UT a successful third-stage burn placed the *Deep Space 1* probe into its solar orbit trajectory. The spacecraft then separated from the Delta II vehicle about 550 km above the Indian Ocean. About 97 minutes after launch mission controllers received telemetry from the spacecraft through the NASA Deep Space Network. This initial communication was about 13 minutes delayed from the expected time, and the reason for the delay is not known. However, all critical spacecraft systems were indicated as performing well, and the probe went about its scientific business of encountering asteroids and comets.

DS1 flew by the near-Earth asteroid 9969 Braille at 04:46 UT (12:46 A.M. EDT) on July 29, 1999, at a distance of about 26 km and a relative velocity of approximately 15.5

km/sec. Just prior to this encounter (at approximately 12:00 UT on July 28), a software problem caused the spacecraft to go into a safing mode. However, the problem was solved, and the spacecraft returned to normal operations at 18:00 UT. Up to six minor trajectory correction maneuvers were scheduled in the 48 hours prior to the flyby. The spacecraft made its final pre-encounter transmission about seven hours before its closest approach to the asteroid, after which it turned its high-gain antenna away from Earth to point the MICAS camera/spectrometer camera toward the asteroid. Unfortunately, the spacecraft experienced a target-tracking problem, and the MICAS instrument was not pointed toward the asteroid as it approached, so the probe could not obtain any close-up images or spectra. MICAS was turned off about 25 seconds before closest approach at a distance of about 350 km, and measurements were taken with the PEPE plasma instrument.

The spacecraft then turned around after the encounter to obtain images and spectra of the opposite side of the asteroid as it receded from view, but due to the target-tracking problem only two black and white images and a dozen spectra were obtained. The two images were captured at 915 and 932 seconds after closest approach from a distance of approximately 14,000 km. The spectra were taken about three minutes later. Over the next few days *Deep Space 1* transmitted these data back to Earth. The diameter of Braille is estimated at 2.2 km at its longest and 1 km at its shortest. The spectra showed this asteroid to be similar to the large minor planet Vesta. The primary mission of DS1 lasted until September 18, 1999.

By the end of 1999 the ion engine had used approximately 22 kg of xenon to impart a total delta-V of 1,300 m/s to the spacecraft. The original plan was to fly by the dormant Comet Wilson-Harrington in January 2001 and then past Comet Borrelly in September 2001, but the spacecraft's star tracker failed on November 11, 1999, so mission planners drew upon techniques developed to operate the spacecraft without the star tracker and came up with a new extended mission to fly by Comet Borrelly. As a result of these innovative actions, on September 22, 2001, *Deep Space 1* entered the coma of Comet Borrelly and successfully made its closest approach (a distance of about 2,200 km) to the nucleus at about 22:30 UT (6:30 P.M. EDT). At the time of this cometary encounter, DS1 was traveling at 16.5 km/s relative to the nucleus. The PEPE instrument was active throughout the encounter. As planned, MICAS started making measurements and imaging 80 minutes before encounter and operated until a few minutes before encounter. Both instruments successfully returned data and images from the encounter. Then mission controllers commanded the spacecraft to shut down its ion engines on December 18, 2001, at about 20:00 UT (3:00 P.M. EST). This action ended the *Deep Space 1* mission. However, aerospace engineers decided to leave the probe's radio receiver on just in case they desired to make future contact with the spacecraft. All the new space technologies flown on board DS1 were successfully tested during the primary mission.

See also COMET; DELTA (LAUNCH VEHICLE); DELTA-V.

This artist's rendering depicts NASA's New Millennium *Deep Space 1* (DS1) spacecraft approaching Comet 19P/Borrelly in September 2001. With its primary mission to serve as a space technology demonstrator—especially the flight testing of an advanced xenon electric propulsion system and 11 other advanced aerospace technologies—successfully completed by September 1999, *Deep Space 1* headed for a risky, exciting rendezvous with Comet Borrelly. On September 22, 2001, the *Deep Space 1* spacecraft entered the coma of Comet Borrelly and successfully made its closest approach, a distance of about 2,200 km, to the nucleus at about 22:30 UT (6:30 P.M. EDT). At the time of the encounter, DS1 was traveling at 16.5 km/s relative to the comet's nucleus. *(Courtesy of NASA/JPL)*

deep-space probe A spacecraft designed for exploring deep space, especially to the vicinity of the Moon and beyond. This includes lunar probes, Mars probes, outer planet probes, solar probes, and so on.

See also SPACECRAFT.

Defense Meteorological Satellite Program (DMSP) A highly successful family of weather satellites operating in polar orbit around Earth that have provided important environmental data to serve American defense and civilian needs for more than two decades. Two operational DMSP spacecraft orbit Earth at an altitude of about 832 km and scan an area 2,900 km wide. Each satellite scans the globe in approximately 12 hours. Using their primary sensor, called the Operational Linescan System (OLS), DMSP satellites take visual and infrared imagery of cloud cover. Military weather forecasters use these imagery data to detect developing weather patterns anywhere in the world. Such data are especially helpful in identifying, locating, and determining the severity of thunderstorms, hurricanes, and typhoons.

Besides the Operational Linescan System, DMSP satellites carry sensors that measure atmospheric moisture and temperature levels, X-rays, and electrons that cause auroras. These satellites also can locate and determine the intensity of auroras, which are electromagnetic phenomena that can interfere with radar system operations and long-range electromagnetic communications. Starting in December 1982 the

Block 5D-2 series spacecraft have been launched from Vandenberg Air Force Base, California. Later generations of the DMSP spacecraft contain many improvements over earlier models, including new sensors with increased capabilities and a longer life span.

Each DMSP spacecraft is placed in a sun-synchronous (polar) orbit at a nominal altitude of 832 kilometers. A sun-synchronous orbit is a special polar orbit that allows a satellite's sensors to maintain a fixed relation to the Sun, a feature that is especially useful for meteorological satellites. Each day a satellite in sun-synchronous orbit passes over a certain area on Earth at the same local time.

One way to characterize sun-synchronous orbits is by the times of day the spacecraft cross the equator. Equator crossings (called *nodes*) occur at the same local time each day, with the descending crossings occurring 12 hours (local time) from the ascending crossings. The terms "AM" and "PM" polar orbiters denote satellites with morning and afternoon equator crossings, respectively. The Block 5D-2 DMSP spacecraft measure 3.64 meters in length and are 1.21 meters in diameter. Each spacecraft has an on-orbit mass of 830 kilograms and a design life of about four years.

Starting in 2001, upgraded Block 5D-3 versions of the spacecraft with improved sensor technology became available for launch. The U.S. Air Force will maintain its support of the DMSP program until about 2010, corresponding to the projected end of life of the final DMSP satellite. Thereafter, military weather requirements will be filled by a triagency system (Department of Defense [DOD], NASA, and Department of Commerce [DOC]) called the National Polar-Orbiting Environmental Satellite System (NPOESS). In May 1994 the president directed that the Departments of Defense and Commerce converge their separate polar-orbiting weather satellite programs. Consequently, the command, control, and communications for existing DOD satellites was combined with the control for the National Oceanic and Atmospheric Administration (NOAA) weather satellites. In June 1998 the DOC assumed primary responsibility for flying both DMSP and NOAA's DMSP-cloned civilian polar orbiting weather satellites. The DOC will continue to manage both weather satellite systems, providing essential environmental sensing data to American military forces until the new converged National Polar-Orbiting Environmental Satellite System becomes operational in about 2010.

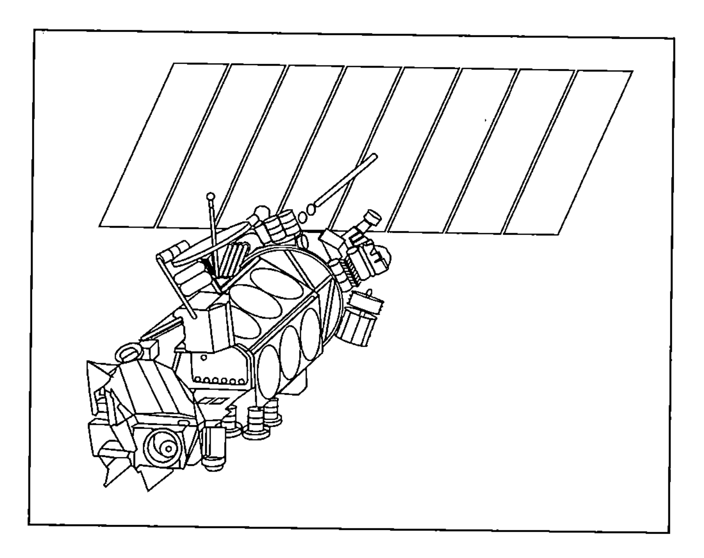

The polar-orbiting Defense Meteorological Satellite Program (DMSP) spacecraft *(Courtesy of the U.S. Air Force)*

See also NATIONAL OCEANIC AND ATMOSPHERIC ADMINISTRATION; UNITED STATES AIR FORCE; VANDENBERG AIR FORCE BASE; WEATHER SATELLITE.

Defense Satellite Communications System (DSCS) An evolving family of military satellites that provide worldwide, responsive wideband and antijam communications in support of U.S. strategic and tactical information transfer needs. Today, as part of a comprehensive plan to support globally distributed information users, the U.S. Air Force operates 13 Phase III DSCS satellites. These spacecraft orbit Earth at an altitude of approximately 35,900 km above the equator. Each Phase III satellite uses six super-high-frequency (SHF) transponder channels capable of providing worldwide secure voice and high-data-rate communications. DSCS III also carries a single-channel transponder for disseminating emergency action and force direction messages to nuclear-capable forces. The system is used for high-priority command and control communications, such as the exchange of wartime information between defense officials and battlefield commanders. This military communications satellite system can also transmit space operations and early-warning data to various defense systems and military users.

The first of the operational Phase II DSCS systems was launched in 1971. Their characteristic two-dish antennas concentrated electronic beams on small areas of Earth's surface but had limited adaptability in comparison to the newer Phase III satellites. The continuously evolving Phase III DSCS spacecraft has a mass of approximately 1,200 kilograms and a design life of about 10 years. Its rectangular body is 1.83 meters by 1.83 meters by 2.13 meters in a stowed configuration and reaches an 11.6-meter span with its solar panels deployed on orbit. The solar arrays on each satellite generate an average of 1,500 watts of electric power. The launch vehicle for the DSCS III satellite is the Atlas II rocket. The U.S. Air Force's evolved expendable launch vehicle (EELV) can also place this satellite system into orbit.

The U.S. Air Force began launching the more advanced DSCS Phase III satellites in 1982 and currently operates a constellation of 13 such satellites in geosynchronous orbit. DSCS III satellites can resist jamming and consistently have exceeded their 10-year design life. DSCS users include the National Command Authority, the White House, the Defense Information System Network (DISN), the Air Force Satellite Control Network, and the Diplomatic Telecommunications Service.

See also COMMUNICATIONS SATELLITE; UNITED STATES AIR FORCE.

Defense Support Program (DSP) The family of missile surveillance satellites operated by the U.S. Air Force since the early 1970s. Placed in geosynchronous orbit around Earth, these military surveillance satellites can detect missile launches, space launches, and nuclear detonations occurring around the world. U.S. Air Force units at Peterson Air Force Base in Colorado operate DSP satellites and report warning information via communications links to the North American Aerospace Defense Command (NORAD) and the U.S. Strategic Command early-warning centers within Cheyenne Mountain Air Force Station.

The Defense Support Program is a space-based infrared satellite system that provides global coverage. The primary (infrared) sensor of each DSP satellite supports near–real time detection and reporting of missile launches against the United States and/or allied forces, interests, and assets worldwide. Other sensors on each satellite support the near–real time detection and reporting of endoatmospheric (0–50 km), exoatmospheric (50–300 km), and deep space (> 300 km) nuclear detonations worldwide.

DSP satellites use an infrared sensor to detect heat from missile and booster plumes against Earth's background signal. The program started with the first launch of a DSP satellite in the early 1970s. Since that time DSP satellites have provided an uninterrupted early-warning capability. Developed as a strategic warning system, the DSP's effectiveness in providing tactical warning was demonstrated during the Persian Gulf conflict (from August 1990 through February 1991). During Operation Desert Storm DSP detected the launch of Iraqi Scud missiles and provided timely warning to civilian populations and United Nations coalition forces in Israel and Saudi Arabia.

Developments in the DSP have enabled it to provide accurate, reliable data in the face of tougher mission requirements, such as greater numbers of targets, smaller-signature targets, and advanced countermeasures. Through several upgrade programs DSP satellites have exceeded their design lives by some 30 percent. As the satellites' capabilities have grown, so has their mass and power levels. In the early years the DSP satellite had a mass of about 900 kilograms, and its solar paddles generated 400 watts of electric power. New-generation DSP satellites have a mass of approximately 2,270 kilograms and improved solar panels that provide more than 1,400 watts of electric power. The newer DSP satellites are about 6.7 meters in diameter and 10 meters high with the solar paddles deployed.

Other space technology developments in the program have led to techniques that have benefited other Department of Defense (DOD) military satellite systems. For example, the addition of a reaction wheel removed unwanted orbital momentum from DSP vehicles. The spinning motion of this reaction wheel serves as a countering force on the satellite's movements. This zero-sum momentum approach permits precise orbit control with a minimum expenditure of attitude control fuel. As a result of DSP experience, reaction wheels have been added to other DOD space systems, including the DEFENSE METEOROLOGICAL SATELLITE PROGRAM (DMSP), the GLOBAL POSITIONING SYSTEM (GPS), and DEFENSE SATELLITE COMMUNICATIONS SYSTEM (DSCS) satellites.

Typically, DSP satellites are placed into geosynchronous orbit by a Titan IV/Inertial Upper Stage (IUS) launch vehicle configuration. However, one DSP was successfully launched using the space shuttle during mission STS-44 in November 1991.

Numerous improvement projects have enabled DSP to provide accurate, reliable data in the face of evolving missile threats. In 1995, for example, technological advancements were made to ground processing systems, enhancing detection capability of smaller missiles to provide improved warning of attack by short-range missiles against the United States and/or allied forces overseas. Recent technological improvements in sensor design include above-the-horizon capability for full hemispheric coverage and improved resolution. An increased onboard signal-processing capability has improved clutter rejection. Enhanced reliability and survivability improvements have also been incorporated. Later in the 21st century the U.S. Air Force plans to replace DSP with a new surveillance satellite system called the Space-Based Infrared System (SBIR). For now, however, the United States and its allies depend on the constellation of DSP satellites to serve as vigilant sentinels, watching for hostile missile launches, space launches, and nuclear detonations.

See also BALLISTIC MISSILE DEFENSE; NORTH AMERICAN AEROSPACE DEFENSE COMMAND; UNITED STATES AIR FORCE.

defensive satellite weapon (DSAT) A proposed space-based weapon system that is intended to defend other orbiting satellites by destroying attacking antisatellite (ASAT) weapons.

See also ANTISATELLITE SPACECRAFT (ASAT).

degenerate matter Matter that is compressed by gravitational attraction to such high densities that the normal atomic structure is broken down and the matter no longer behaves in an ordinary way. To appreciate the nature of degenerate matter, it is helpful first to examine the characteristics of ordinary (or normal) matter in the interior of a star, such as the Sun. Scientists assume the Sun has an average density of about 1.4×10^3 kg/m^3 and is composed of a high-temperature plasma—an ionized mixture of electrons and nuclei that behave like an ideal gas of nonrelativistic classical atomic particles. In the ideal gas model temperature is proportional to pressure and increases as pressure increases. When physicists consider the pressure needed to support a star of about one solar mass against gravitational collapse while it burns its nuclear fuel, they apply the ideal gas model and obtain a core temperature of about 6×10^6 K.

However, after a star has consumed all its thermonuclear fuel, the plasma then begins to shrink under the action of its own gravity and becomes degenerate matter. In most cases the gravitational collapse of a Sunlike star's core results in the formation of a white dwarf—a compact object with a density of 10^8 kg/m^3 or more. The core of a red giant star, for example, becomes so dense that the electrons form a *degenerate gas,* characterized by a *degenerate gas pressure* that depends only on density and not temperature. In such highly compressed material very little space exists between atomic particles. Consequently, electrons, stripped of their atomic nuclei, form a degenerate electron gas whose pressure is now determined by quantum mechanics, primarily the PAULI EXCLU-

SION PRINCIPLE and the HEISENBERG UNCERTAINTY PRINCIPLE. As the density increases, the closely packed electrons begin to exert a degenerate gas pressure that becomes sufficiently large to counteract the force of gravitational contraction. Astrophysicists often describe the properties of a white dwarf in terms of the behavior of a degenerate electron gas.

For collapsing stars with masses above about 1.4 solar mass (a condition referred to as the CHANDRASEKHAR LIMIT), the electron degenerate gas pressure is insufficient to counteract gravitational attraction. As a result, gravity continues to squeeze the degenerate matter—pushing electrons into protons and forming the incredibly high-density material (about 10^{17} kg/m^3) of a neutron star. Because neutrons are subject to the laws of quantum mechanics, a neutron degenerate gas is formed whose pressure resists further gravitational collapse. However, when the mass of a collapsing star exceeds about three solar masses, even this neutron degenerate gas pressure cannot resist the relentless attraction of gravity, and the degenerate matter continues to collapse into the infinitely dense singularity of a black hole.

See also BLACK HOLE; COSMOLOGY; STAR.

degenerate star An end-of-life star whose core material has collapsed to a very-high-density condition; for example, a white dwarf or a neutron star.

See also DEGENERATE MATTER.

degree (usual symbol: °) A term that has been commonly used to express units of certain physical quantities, such as angles and temperatures. The ancient Babylonians are thought to be the first people to have subdivided the circle into 360 parts, or "degrees," thereby establishing the use of the degree in mathematics as a unit of angular measurement. When applied to the roughly spherical shape of Earth in geography and cartography, degrees are divided into 60 minutes.

See also ANGLE; ARC MINUTE; ARC SECOND; TEMPERATURE.

degrees of freedom (DOF) A mode of motion, either angular or linear, with respect to a coordinate system, independent of any other mode. A body in motion has six possible degrees of freedom, three linear (sometimes called x-, y-, and z-motion with reference to linear [axial] movements in the Cartesian coordinate system) and three angular (sometimes called pitch, yaw, and roll with reference to angular movements).

See also CARTESIAN COORDINATES; PITCH; ROLL; YAW.

Deimos The tiny, irregularly shaped (about 16-km by 12-km by 10-km) outer moon of Mars, discovered in 1877 by the American astronomer ASAPH HALL (1829–1907). Named after the god of panic (a son of Mars who accompanied his father in battle) in Greco-Roman mythology, Deimos is the smaller of the two natural Martian moons. The other moon is called Phobos (fear). Deimos orbits above the equator of the planet at a mean distance of 23,436 km. This tiny moon has an orbital period of 1.262 days (about 30 hours) and an orbital eccentricity of 0.0005. Because it makes one revolution about its shortest (10-km) axis in the same amount of time that it takes to make one orbit of Mars, astronomers say it is in a synchronous orbit—always keeping the same face toward the Red Planet. Deimos has a mass of about 1.8×10^{15} kg and an estimated mean density 1,750 kg/m^3 (or 1.75 g/cm^3). The acceleration of gravity on its surface is quite low, at only 0.005 m/s^2.

Deimos is a dark celestial body with an albedo of 0.06. Due to the way it reflects the spectrum of sunlight, this tiny moon and its companion, Phobos, appear to possess C-type surface materials, similar to that of asteroids found in the outer portions of the asteroid belt. Like Phobos, Deimos is a lumpy, heavily cratered object. However, the impact craters on Deimos are generally smaller (less than 2.5 km in diameter) and lack the ridges and grooves observed on Phobos. Ejecta deposits are not seen on Deimos, possibly because this moon's surface gravity is so low that the ejecta escaped to space. However, Deimos appears to have a thick regolith—perhaps as deep as 100 meters—formed over millennia as meteorites pulverized the moon's surface.

The most widely held hypothesis concerning the origin of Deimos and Phobos is that the moons are captured asteroids. But this popular hypothesis raises some question. For example, some scientists think that it is highly improbable for Mars to gravitationally capture two different asteroids. They suggest that perhaps a larger asteroid was initially captured and then suffered a fragmenting collision, leaving Deimos and Phobos as the postcollision products. Other scientists bypass the captured asteroid hypothesis altogether and speculate that the moons somehow evolved along with Mars. Expanded exploration of Mars this century should resolve this question.

See also MARS; PHOBOS.

De La Rue, Warren (1815–1889) British *Astronomer, Physicist* Warren De la Rue, a 19th-century British astronomer, instrument craftsperson, and physicist, made a significant advancement in solar astronomy when he developed the photoheliograph in 1858, an instrument that enabled routine photography of the Sun's outer surface. The photoheliograph is basically a telescope that records a white light image of the Sun on a photographic plate. De La Rue's first photoheliograph was installed at Kew Observatory, an observatory built by King George III for the 1769 transit of Venus that since 1871 has served as the central meteorological observatory of Great Britain. His instrument successfully photographed the Sun in March 1858, but it then took about 24 months to resolve many minor operational problems with the device. Nonetheless, in 1859 he was able to present several detailed solar photographs at a meeting of the Royal Astronomical Society (RAS).

See also ASTROPHOTOGRAPHY; SUN.

de Laval nozzle A flow device that efficiently converts the energy content of a hot, high-pressure gas into kinetic energy. Although originally developed by CARL GUSTAF DE LAVAL (1845–1913) for use in certain steam turbines, the versatile converging-diverging (CD) nozzle is now used in practically all modern rockets. The device constricts the outflow of the high-pressure (combustion) gas until it reaches sonic velocity (at the nozzle's throat) and then expands the exiting gas to very high (supersonic) velocities.

See also NOZZLE; ROCKET.

Delporte, Eugène Joseph (1882–1955) Belgian *Astronomer* In 1932 Eugène Joseph Delporte discovered the 1-km-diameter Amor asteroid. This relatively small asteroid now gives its name to the *Amor group,* a collection of near-Earth asteroids that cross the orbit of Mars but not Earth's orbit around the Sun.

See also AMOR GROUP; ASTEROID; NEAR-EARTH ASTEROID.

Delta (launch vehicle) A versatile family of American two- and three-stage liquid-propellant, expendable launch vehicles (ELVs) that uses multiple strap-on booster rockets in several configurations. Since May 1960 the Delta program has had more than 270 successful military, civil, and commercial launches. Over the years the Delta family of rocket vehicles has also accomplished many important space technology firsts. These achievements include the first international satellite, *Telstar 1* (launched in 1962); the first geosynchronous satellite, *Syncom 2* (launched in 1963); and the first commercial communications satellite, *COMSAT 1* (launched in 1965). Because of its outstanding record of launch accomplishments, the Delta rocket has earned the nickname *space workhorse vehicle.*

The evolving family of Delta rockets began in 1959, when NASA awarded a contract to Douglas Aircraft Company (now part of the Boeing Company) to produce and integrate 12 space launch vehicles. The first Delta vehicle used components from the U.S. Air Force's Thor missile for its first stage and from the U.S. Navy's Vanguard Project rocket for its second stage. On May 13, 1960, the inaugural flight of the Delta I rocket from Cape Canaveral Air Force Station successfully placed the *Echo 1* communications (relay) satellite (actually, a large, inflatable sphere) into orbit. The original Delta booster was 17 meters long and 2.4 meters in diameter. It used one Rocketdyne MB-3 Block II rocket engine (also called the LR-79-NA11 engine) that burned liquid oxygen (LOX) and kerosene to produce 667,000 newtons (N) of thrust at liftoff. Two small VERNIER ENGINES created 4,400 N of thrust each using the same propellants as the main engine. Continued improvements resulted in an evolutionary series of vehicles with increasing payload capabilities.

The Delta II is a medium-lift capacity, expendable launch vehicle used by the U.S. Air Force to launch the many satellites of the Global Positioning System (GPS). Additionally, the Delta II rocket launches civil and commercial payloads into low-Earth orbit (LEO), polar orbit, geosynchronous transfer orbit (GTO), and geostationary orbit (GEO). NASA has used the versatile Delta II vehicle to launch many important scientific missions, including the Mars Exploration Rovers *Spirit* (*see* figure) and *Opportunity.*

The Delta II launch vehicle stands a total height of 38.3 meters. In the configuration often used by the U.S. Air Force, the payload fairing—the protective shroud covering the third stage and the satellite payload—is 2.9 meters wide to accommodate GPS satellites. A 3-meter-wide stretched version fairing is also available for larger payloads. Six of

A Delta II expendable launch vehicle lifted off from Cape Canaveral Air Force Station, Florida, on June 10, 2003. This successful launch sent NASA's Mars Exploration Rover *Spirit* to the surface of the Red Planet, where, starting in January 2004, the robot rover searched for evidence of whether past environments were once wet enough on Mars to be hospitable for life. ***(Courtesy of NASA)***

the nine solid-rocket motors that ring the first stage separate after one minute of flight. The remaining three solid-rocket motors ignite and then separate after burnout about one minute later.

In the late 1990s the Boeing Company introduced the Delta III family of rockets to serve space transportation needs within the expanding commercial communications satellite market. With a payload delivery capacity of 3,800 kg delivered to a geosynchronous transfer orbit, the Delta III essentially doubled the performance of the Delta II rocket. The first stage of the Delta III is powered by a Boeing RS-27A rocket engine assisted by two vernier rocket engines that help control roll during main engine burn. The vehicle's second stage uses a Pratt and Whitney RL 10B-2 engine that burns cryogenic propellants.

On November 20, 2002, a Delta IV rocket lifted off from Cape Canaveral Air Force Station, successfully delivering the *W5 Eutelsat* commercial communications satellite to a geosynchronous transfer orbit. This event was the inaugural launch of the latest member of the Delta launch vehicle family. Delta IV rockets combine new and mature launch vehicle technologies and can lift medium to heavy payloads into space. The Delta IV vehicle also represents part of the U.S. Air Force's evolved expendable launch vehicle program. The new rocket uses the new Boeing Rocketdyne-built RS-68 liquid hydrogen and liquid oxygen main engine, an engine capable of generating 2,891,000 N of thrust. Assembled in five vehicle configurations, the Delta IV rocket family is capable of delivering payloads that range between 5,845 kg and 13,130 kg to a geosynchronous transfer orbit.

See also EVOLVED EXPENDABLE LAUNCH VEHICLE; LAUNCH VEHICLE; ROCKET.

delta star (δ) Within the system of stellar nomenclature introduced in 1603 by the German astronomer JOHANN BAYER, the fourth-brightest star in a constellation.

See also STAR.

delta-V (ΔV) Velocity change; a numerical index of the propulsive maneuverability of a spacecraft or rocket. This term often represents the maximum change in velocity that a space vehicle's propulsion system can provide. Typically described in terms of kilometers per second (km/s) or meters per second (m/s).

demodulation The process of recovering the modulating wave from a modulated carrier.

See also TELECOMMUNICATIONS.

Denning, William Frederick (1848–1931) British *Astronomer* As an avid amateur astronomer, William Denning focused his efforts on the study of meteors and the search for comets. He discovered several comets, including the periodic Comet 72P/Denning-Fujikawa, which was first observed in 1881. The writer H. G. WELLS identified Denning as a noted British meteor authority in the famous science fiction novel *The War of the Worlds*.

See also COMET; METEOR.

density (symbol: ρ) The mass of a substance per unit volume at a specified temperature. The average (or mean) density of a celestial body is its total mass divided by its total volume. Earth, for example, has an average density of 5.52 g/cm^3 (or 5,520 kg/m^3), while the Sun has an average density of 1.41 g/cm^3 (1,410 kg/m^3) and the planet Saturn, 0.69 g/cm^3 (690 kg/m^3). Scientists often use the density of water (1 g/cm^3 or 1,000 kg/m^3 at standard conditions on the surface of Earth) as a convenient reference. The density of objects encountered in astronomy and astrophysics varies widely. The density of interstellar gas is on the order of 10^{-20} kg/m^3, while the density of a neutron star is about 10^{17} kg/m^3. The *mean density of matter* in the universe (symbol: ρ_o) has a currently estimated value on the order of about 10^{-27} kg/m^3. The parameter ρ_o is important in cosmology because it helps scientists determine the dynamic behavior of the UNIVERSE; it is related to the DENSITY PARAMETER (Ω).

density parameter (symbol: Ω) A dimensionless parameter that expresses the ratio of the actual mean density of the universe to the critical mass density—that is, the mean density of matter within the universe (considered as a whole) that cosmologists consider necessary if gravitation is to eventually halt its expansion. One presently estimated value of this critical mass density is 8×10^{-27} kg/m^3, obtained by assuming that the HUBBLE CONSTANT (H_o) has a value of about 71.7 km/(s-Mpc) and that gravitation is all that is influencing the dynamic behavior of the expanding universe. If the universe does not contain sufficient mass (that is, if the $\Omega < 1$), then the universe will continue to expand forever. If $\Omega > 1$, then the universe has enough mass to eventually stop its expansion and to start an inward collapse under the influence of gravitation. If the critical mass density is just right (that is, if $\Omega = 1$), then the universe is considered flat, and a state of equilibrium eventually will exist in which the outward FORCE of expansion becomes precisely balanced by the inward force of gravitation.

See also COSMOLOGY; UNIVERSE.

de-orbit burn A retrograde (opposite-direction) rocket engine firing by which a space vehicle's velocity is reduced to less than that required to remain in orbit around a celestial body.

descending node That point in the orbit of a celestial body when it travels from north to south across a reference plane, such as the equatorial plane of the celestial sphere or the plane of the ecliptic. Also called the *southbound node*. *Compare with* ASCENDING NODE.

DeSitter, Willem (1872–1934) Dutch *Astronomer, Cosmologist* As an astronomer and mathematician, Willem DeSitter explored the cosmological consequences of ALBERT EINSTEIN's general relativity theory. In 1917 he proposed an expanding universe model. While the specific physical details of DeSitter's model proved unrealistic, his basic hypothesis served as an important stimulus to other astronomers, such as EDWIN P. HUBBLE, whose observations in the early 1920s proved the universe was indeed expanding. DeSitter's model of an expanding universe also played a formative role in modern cosmological theories that postulate an inflationary universe.

See also BIG BANG THEORY; COSMOLOGY; UNIVERSE.

Despina A small moon of Neptune (about 150 km in diameter) discovered in 1989 during the *Voyager 2* flyby encounter. Despina is Neptune's third-closest satellite, orbiting the planet at a distance of 52,530 km with an inclination of 0.07 degrees and a period of 0.33 day.

See also NEPTUNE; *VOYAGER* SPACECRAFT.

Destiny The American-built laboratory module delivered to the *International Space Station* (ISS) by the space shuttle *Atlantis* during the STS-98 mission (February 2001). Destiny is the primary research laboratory for U.S. payloads. The aluminum module is 8.5 meters long and 4.3 meters in diameter. It consists of three cylindrical sections and two end

American astronaut Susan J. Helms (left), Expedition Two flight engineer, pauses from her work to pose for a photograph while Expedition Two mission commander Russian cosmonaut Yury V. Usachev (right) speaks into a microphone aboard Destiny, the U.S. laboratory module of the *International Space Station*, in April 2001. *(Courtesy of NASA)*

cones with hatches that can be mated to other space station components. There is also a 0.509-meter (20-inch)-diameter window located on the side of the center segment. An exterior waffle pattern strengthens the module's hull, and its exterior is covered by a space debris shield blanket made of material very similar to that used in bulletproof vests worn by law enforcement personnel. A thin aluminum debris shield placed over the blanket provides additional protection against space debris and meteoroids.

NASA engineers designed the laboratory to hold sets of modular racks that can be added, removed, or replaced as necessary. The laboratory module includes a human research facility (HRF), a materials science research rack (MSRR), a microgravity science glove box (MSG), a fluids and combustion facility (FCF), a window observational research facility (WORF), and a fundamental biology habitat holding rack. The laboratory racks contain fluid and electrical connectors, video equipment, sensors, controllers, and motion dampers to support whatever experiments are housed in them. Destiny's window (the WORF), which takes up the space of one rack, is an optical gem that allows space station crewmembers to shoot very-high-quality photographs and videos of Earth's ever-changing landscape. Imagery captured from this window gives scientists the opportunity to study features such as floods, glaciers, avalanches, plankton blooms, coral reefs, urban growth, and wildfires.

See also INTERNATIONAL SPACE STATION; SPACE STATION.

destruct (missile) The deliberate action of destroying a missile or rocket vehicle after it has been launched but before it has completed its course. The range safety officer is the individual responsible for sending the command destruct whenever a missile or rocket veers off its intended (plotted) course or functions in a way so as to become a hazard.

See also ABORT; COMMAND DESTRUCT.

deuterium (usual symbol: D or $^{2}_{1}H$) A nonradioactive isotope of hydrogen whose nucleus contains one neutron and one proton. It is sometimes called *heavy hydrogen* because the deuterium nucleus is twice as heavy as that of ordinary hydrogen. The American chemist HAROLD CLAYTON UREY (1893–1981) won the 1934 Nobel Prize in chemistry for the discovery of deuterium.

See also TRITIUM.

DeVaucouleurs, Gérard Henri (1918–1995) French-American *Astronomer* Gérard deVaucouleurs performed detailed studies of distant galaxies and the Magellanic Clouds. In 1964 he published (along with his wife, Antoinnette deVaucouleur, née Piétra [1921–87]) the *Reference Catalogue of Bright Galaxies,* which revised and updated the work of earlier astronomers, including HARLOW SHAPLEY.

See also GALAXY; MAGELLANIC CLOUDS.

Dewar, Sir James (1842–1923) Scottish *Physicist, Chemist* Sir James Dewar was a Scottish scientist who specialized in the study of low-temperature phenomena and invented a special double-walled vacuum flask that now carries his name. In 1892 he started using the DEWAR FLASK to store liquefied gases at very low (cryogenic) temperatures. Then, in 1898 he successfully produced liquefied hydrogen.

Dewar flask Double-walled container with the interspace evacuated of gas (air) to prevent the contents from gaining or losing thermal energy (heat). Named for the Scottish physical chemist SIR JAMES DEWAR. Large modern versions of this device are used to store cryogenic propellants for space launch vehicles.

See also CRYOGENIC PROPELLANT; LAUNCH VEHICLE; ROCKET.

diamonds The patterns of shock waves (pressure discontinuities) often visible in a rocket engine's exhaust. These patterns resemble a series of diamond shapes placed end to end.

See also NOZZLE; ROCKET.

Dicke, Robert Henry (1916–1997) American *Physicist* In 1964 the American physicist Robert Dicke revived the hypothesis that an enormous ancient explosion, called the big bang, started the universe. He also suggested that this primordial explosion should have a detectable microwave radiation remnant. However, before he could make his own experimental observations, two other scientists (ARNO ALLEN PENZIAS and ROBERT WOODROW WILSON) detected the cosmic microwave background and provided the scientific community with direct experimental evidence supporting Dicke's revival of the big bang hypothesis.

See also BIG BANG THEORY; COSMOLOGY; UNIVERSE.

differential rotation When gaseous (non–solid surface) celestial bodies, such as the planet Jupiter and the Sun, rotate, different portions of their atmospheres can and do move at different speeds. For example, the equatorial rotation rate of Saturn is about 26 minutes faster than that in the planet's polar regions. For the Jovian planets, the phenomenon of differential rotation manifests itself in the form of large-scale wind flows that take place in each planet's frigid atmosphere. The equatorial regions of the Sun rotate about 25 percent faster than the solar polar regions. As a result, the Sun's differential rotation greatly distorts the solar magnetic field—twisting and wrapping it around the equator. This process causes any originally north–south oriented magnetic field lines to eventually realign themselves in an east–west direction. Celestial bodies with solid surfaces, such as Earth and the other terrestrial planets, do not experience differential rotation.

See also JUPITER; NEPTUNE; SATURN; SUN; TERRESTRIAL PLANETS; URANUS.

diffraction The spreading out or bending of light or other forms of electromagnetic radiation as it passes by the edge of a barrier or through an aperture, such as the light-gathering entrance of a telescope or the closely spaced parallel grooves of a DIFFRACTION GRATING. For example, when a ray of "white light" passes over a sharp opaque edge (like that of a razor blade), it is broken up into its rainbow spectrum of colors. The phenomenon of diffraction arises from the wavelike nature of electromagnetic radiation. if a scientist passes light through a single very narrow vertical slit, he or she will observe the formation of a pattern of alternating bright and dark fringes on a screen located at a suitable distance beyond the slit. The diffraction pattern on the screen occurs because the diffracted light waves interfere with one another, producing regions of reinforcement (bright fringes) and of cancellation or weakening (dark fringes). When the incident light rays are parallel, scientists call the diffraction *Fraunhofer diffraction* in honor of the German optician and physicist JOSEPH VON FRAUNHOFER (1787–1826), who studied this phenomenon in the early part of the 19th century. If the incident light rays are not parallel, scientists call the diffraction *Fresnel diffraction* in honor of the French physicist Augustin Jean Fresnel (1788–1827).

Diffraction represents a problem in astronomy because it diverts some of the light from a distant celestial object away from its intended path, slightly blurring the image produced by a telescope. The degree of blurring is a function of the wavelength (λ) of the incoming radiation and the diameter (d) of the aperture or imaging lens of the telescope. The *diffraction limit* is the ultimate limit on the resolution of a telescope. It is proportional to the ratio of the wavelength of the incoming radiation passing through the telescope's aperture or focusing lens to the diameter of that aperture or lens. The smaller the λ/d ratio, the better the resolution of the telescope. Because of diffraction, however, the telescopic image from a star or other luminous celestial object does not appear as a crisp point of light, but rather appears like a disk called the AIRY DISK. When two stars are very near each other (as viewed from Earth), their Airy disks will overlap, and diffraction produces a combined, larger blur of light that the particular telescope or optical instrument cannot separate (or resolve) into two distinct images. However, the same telescope can resolve these two stars if their Airy disks just touch.

Astronomers use the *Rayleigh criterion* to determine whether diffraction will allow a telescope to resolve two closely spaced stars (as seen from Earth). The British physicist Lord John William Rayleigh (1842–1919) recognized that the image of a star in a telescope—even an image formed with the best possible optics—consists of an Airy disk surrounded by a set of diffraction rings. He consequently postulated that a telescope will *just* resolve a point object (star) when the first dark fringe in its circular diffraction pattern falls directly on the central bright fringe in the diffraction pattern of the second (nearby) star. Astronomers call this postulation the Rayleigh criterion and use it to estimate the resolution limit of a telescope due to diffraction.

See also LENS; TELESCOPE.

diffraction grating A device containing a precise arrangement of a large number (typically up to 40,000 lines per centimeter and more) of parallel, closely spaced slits or grooves on its surface used to break up incident light or other forms of electromagnetic radiation into a spectrum by DIFFRACTION. If the diffraction grating consists of a collection of narrow slits that transmit light, scientists call the device a *transmission grating.* If the diffraction grating reflects incident light from a collection of narrow, parallel surface grooves, scientists call the device a *reflection grating.* Diffraction gratings used in optical astronomy typically contain between 100 and 1,000 narrow, parallel grooves or slits per millimeter. Since a simple diffraction grating spreads the incident light into a large number of spectra, the fraction of the radiation going into any one particular spectrum is relatively low. To overcome this problem, scientists developed the *blazed grating.* It is a reflection grating that has the grooves of its reflecting surface oriented (or pitched) at some particular angle (called the blaze angle). With the proper blaze angle this arrangement can concentrate as much as 80 to 90 percent of the incident light into one spectrum.

diffraction limit *See* DIFFRACTION.

diffraction pattern Any pattern that a telescope or other optical system produces as a result of the phenomenon of diffraction.

See also AIRY DISK.

diffuse nebula *See* HII REGION.

digit A single character or symbol in a number system. For example, the binary system has two digits, 0 and 1, while the decimal system has ten digits, ranging from 0 through 9.

digital image processing Computer processing of the digital number (DN) values assigned to each pixel in an image. For example, all pixels in a particular image with a digital number value within a certain range might be assigned a special "color" or might be changed in value some arbitrary amount to ease the process of image interpretation by a human analyst. Furthermore, two images of the same scene taken at different times or at different wavelengths might have the digital number values of corresponding pixels computer manipulated (e.g., subtracted) to bring out some special features. This is a digital image processing technique called differencing or change detection.

See also REMOTE SENSING.

digital transmission A technique in telecommunications that sends the signal in the form of one of a discrete number of codes (for example, in binary code as either 0 or 1). The information content of the signal is concerned with discrete states of this signal, such as the presence or absence of voltage or a contact in a closed or open position.

See also DEEP SPACE NETWORK; TELECOMMUNICATIONS; TELEMETRY.

digitize To express an analog measurement in discrete units, such as a series of binary digits. For example, an image or photograph can be scanned and digitized by converting lines and shading (or color) into combinations of appropriate digital values for each pixel in the image or photograph.

See also DIGITAL IMAGE PROCESSING.

Dione A satellite of Saturn discovered in 1684 by GIOVANNI CASSINI. It has a diameter of 1,120 km, a mass of approximately 1.1×10^{21} kg, and a density of about 1.4 g/cm^3. Dione is similar to the Saturnian moons Tethys (which is slightly smaller) and Rhea (which is larger). Dione orbits Saturn at a distance of 377,400 km with a period 2.74 days and an inclination of 0.02°. The moon shares this orbit with a very small moon called Helene. Since its axial rotation is the same as its orbital period, astronomers say Dione is in synchronous rotation around Saturn—meaning, like Earth's Moon, it always keeps the same hemisphere facing the parent planet. Planetary scientists consider this synchronous rotation to be the consequence of tidal action. One distinctive feature of Dione is that its surface has a nonuniform brightness, similar in hemispheric albedo variation but not quite as pronounced as the variation displayed by Iapetus. Dione's surface consists of a mixture of lightly cratered plains, moderately cratered plains, and heavily cratered terrain. The most prominent surface feature thus far discovered is a 240-km-diameter crater named Amata. Planetary scientists currently speculate that the bright streaks observed on Dione's surface may be associated with an outflow of water from the moon's interior.

See also CASSINI MISSION; SATURN.

dipole A system composed of two equal but opposite electrical charges that are separated but still relatively close together. The *dipole moment* (μ) is defined as the product of one of the charges and the distance between them. Some molecules, called *dipole molecules,* have effective centers of positive and negative charge that are permanently separated.

dipole antenna A half-wave (dipole) antenna typically consists of two straight, conducting metal rods each one-quarter of a wavelength long that are connected to an alternating voltage source. The electric field lines associated with this antenna configuration resemble those of an electric dipole. (An electric dipole consists of a pair of opposite electric charges [+q and –q] separated by a distance [d].) The dipole antenna is commonly used to transmit (or receive) radio-frequency signals below 30 megahertz (MHz).

See also TELECOMMUNICATIONS.

Dirac, Paul Adrien Maurice (1902–1984) British *Physicist* The British theoretical physicist Paul Dirac made major contributions to quantum mechanics and shared the 1933 Nobel Prize in physics. With his theory of pair production, he postulated the existence of the positron. In 1938 he proposed a link between the Hubble constant (a physical measure that describes the size and age of the universe) and the fundamental physical constants of subatomic particles. This hypothesis formed the basis of DIRAC COSMOLOGY.

See also ELEMENTARY PARTICLES; HUBBLE CONSTANT; POSITRON; QUANTUM MECHANICS.

Dirac cosmology An application of the "large numbers hypothesis" within modern cosmology that tries to relate the fundamental physical constants found in subatomic physics to the age of the universe and other large-scale cosmic characteristics. Suggested by the Nobel laureate PAUL DIRAC, it is now not generally accepted but does influence the ANTHROPIC PRINCIPLE.

See also COSMOLOGY.

direct broadcast satellite (DBS) A class of communications satellite usually placed in geostationary orbit that receives broadcast signals (such as television programs) from points of origin on Earth and then amplifies, encodes, and retransmits these signals to individual end users scattered throughout some wide area or specific region. Many households around the world now receive numerous television channels directly from space by means of small (typically less than 0.5-meter-diameter) rooftop satellite dishes that are equipped to decode DBS transmissions.

See also COMMUNICATIONS SATELLITE.

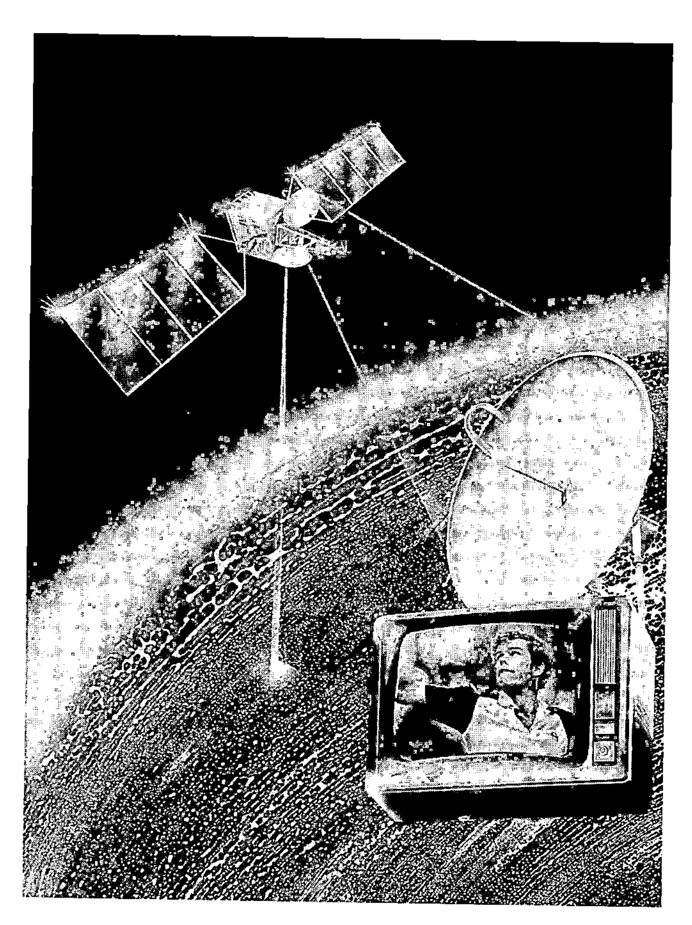

An artist's rendering of a direct broadcast satellite (DBS) and the small (typically less than 0.5-meter-diameter) rooftop satellite dish that is equipped to decode DBS transmissions, bringing numerous television channels and other signals into individual households ***(Artist's rendering courtesy of CNES)***

direct conversion The conversion of thermal energy (heat) or other forms of energy (such as sunlight) directly into electrical energy without intermediate conversion into mechanical work—that is, without the use of the moving components as found in a conventional electric generator system. The main approaches for converting heat directly into electricity include thermoelectric conversion, thermionic conversion, and magnetohydrodynamic conversion. Solar energy is directly converted into electrical energy by means of solar cells (photovoltaic conversion). Batteries and fuel cells directly convert chemical energy into electrical energy.

See also RADIOISOTOPE THERMOELECTRIC GENERATOR.

directed energy weapon (DEW) A device that uses a tightly focused beam of very intense energy, either in the form of electromagnetic radiation (e.g., light from a laser) or in the form of elementary atomic particles, to kill its target. The DEW delivers this lethal amount of energy at or near the speed of light. Also called a *speed-of-light weapon.*

See also BALLISTIC MISSILE DEFENSE; HIGH ENERGY LASER.

directional antenna Antenna that radiates or receives radio-frequency (RF) signals more efficiently in some directions than in others. A collection of antennas arranged and selectively pointed for this purpose is called a *directional antenna array.*

See also DEEP SPACE NETWORK; TELECOMMUNICATIONS; TELEMETRY.

direct motion This term now has several specialized meanings in astronomy and orbital mechanics but traces its origins to antiquity and Greek (geocentric) cosmology. First, the term means the apparent west-to-east motion of a planet or other celestial object as seen by an observer on Earth against the celestial sphere—that is, against the fixed star background. In contrast, an apparent motion from east to west is regarded as a *retrograde motion.* Second, the term describes the anticlockwise orbital motion (that is, the *direct orbit*) of a satellite around its primary body when the orbiting body is viewed from the north pole of the primary. In this context a clockwise orbital motion represents a retrograde motion. All the planets in the solar system move in direct orbits around the Sun. Third, the term refers to the anticlockwise rotation of a planet on its axis when that planet is viewed from its north pole. All the major planets in the solar system have direct rotation, except for Venus, Uranus, and Pluto, which undergo retrograde (clockwise) rotation.

direct readout The information technology capability that allows ground stations on Earth to collect and interpret the data messages (telemetry) being transmitted from satellites.

See also DEEP SPACE NETWORK; TELECOMMUNICATIONS; TELEMETRY.

Discoverer spacecraft The public (cover story) name given by the U.S. Air Force to the secret Corona photo reconnaissance satellite program. The Discoverer spacecraft series not only led to operational American reconnaissance satellites (for example, in August 1960 *Discoverer 14* successfully imaged the Soviet Union from space and then returned its film capsule from orbit), but also achieved a wide variety of space technology advances and breakthroughs—most of which remained shrouded in secrecy until the mid-1990s. The Advanced Research Projects Agency (ARPA) of the Department of Defense and the U.S. Air Force managed the Discoverer Program. The primary goal of this program was to develop a film-return photographic surveillance satellite capable of assessing how rapidly the former Soviet Union was producing long-range bombers and ballistic missiles and of locating where these nuclear-armed strategic weapon systems were being deployed. In the mid-1950s President DWIGHT D. EISENHOWER became extremely concerned about the growing threat of a surprise nuclear attack from the Soviet Union. He needed much better information about military activities inside the Soviet Union, so he decided to pursue the development of spy satellites that could take high-resolution photographs over the Sino-Soviet bloc. He viewed these spy satellites as a desperately needed replacement for the politically embarrassing U2 spy plane flights over denied Soviet territories.

The Discoverer Program was part of the publicly visible portion of the secret Corona satellite program. In addition to providing photographic information for exploitation by the American intelligence community, imagery data from the *Corona* satellites were also used to produce maps and charts for the Department of Defense and other U.S. government mapping programs. At the time, however, the true national security objectives of this important space program were not revealed to the public. Instead, the U.S. Air Force presented the various Discoverer spacecraft launches as being part of an overall research program to orbit large satellites and to test various satellite subsystems. The public news releases also described how different Discoverer spacecraft were helping to investigate the communications and environmental aspects of placing humans in space. To support this cover story, some Discoverer missions even carried biological packages that were returned to Earth from orbit. In all, 38 Discoverer satellites were launched from the beginning of the program through February 1962, when the U.S. Air Force quietly ended all public announcements concerning the Discoverer Program. However, the Discoverer satellite reconnaissance program, launched from Vandenberg Air Force Base in California, continued in secret until 1972 as Corona. From August 1960 until the 145th and final launch on May 15, 1972, the Corona (Discoverer) Program provided the leaders of the United States with imagery data collected during many important spy satellite missions. Finally, as publicly disclosed in 1995, the American intelligence community also designated the Corona spacecraft by the codename *Keyhole* (KH)—for example, personnel within the Central Intelligence Agency (CIA) called the *Corona 14* (*Discoverer 14*) spacecraft a *Keyhole 1* (or KH-1) spacecraft, and imagery products from these spy satellites could be viewed only by intelligence personnel who possessed a special high-security clearance.

The U.S. government declassified numerous Discoverer Program documents in 1995, and so this entry also contains some interesting technical details about selected missions. Without question, the collection of intelligence from Earth-orbiting spacecraft provided national leaders the crucial information they needed to help defuse cold war–era crises and to prevent even the most prickly of superpower disagreements (such as the Cuban missile crisis in 1962) from escalating into a civilization-ending nuclear war.

Discoverer 2 was launched by a Thor Agena A rocket from Vandenberg Air Force Base, California, on April 13, 1959, which placed the spacecraft into a 239-km (perigee) by 346-m (apogee) polar orbit around Earth at an inclination of 89.9 degrees and with an orbital period of 90.4 minutes. The cylindrical satellite was designed to gather spacecraft engineering data. The mission also attempted to eject an instrument package from orbit for recovery on Earth. The spacecraft was three-axis stabilized and was commanded from Earth. After 17 orbits, on April 14, 1959, a reentry vehicle was ejected. The reentry vehicle separated into two sections. The first section consisted of the protection equipment, retrorocket, and main structure; the other section was the reentry capsule itself. U.S. Air Force engineers had planned for the capsule to reenter over the vicinity of Hawaii to support a midair or ocean recovery. However, a timer malfunction caused premature capsule ejection, and it experienced reentry over Earth's north polar region. This test capsule was never recovered. The main instrumentation payload remained in orbit and carried out vehicular performance and communications tests.

The *Discoverer 2* spacecraft was 1.5 m in diameter, 5.85 m long, and had a mass after second-stage separation, including propellants, of roughly 3,800 kg. The mass excluding propellants was 743 kg, which included 111 kg for the instrumentation payload and 88 kg for the reentry vehicle. The capsule section of the reentry vehicle was 0.84 m in diameter, 0.69 m long, and held a parachute, test life-support systems, cosmic-ray film packs to determine the intensity and composition of cosmic radiation (presumably as a test for storage of future photographic film), and a tracking beacon. The capsule was designed to be recovered by a specially equipped aircraft during parachute descent but was also designed to float to permit recovery from the ocean. The main spacecraft contained a telemetry transmitter and a tracking beacon. The telemetry transmitter could send more than 100 measurements of the spacecraft performance, including 28 environmental, 34 guidance and control, 18 second-stage performance, 15 communications, and nine reentry capsule parameters. Nickel-cadmium (NiCd) batteries provided electrical power for all instruments. Orientation was provided by a cold nitrogen gas jet stream system. The spacecraft's attitude control system also included a scanner for pitch attitude and an inertial reference package for yaw and roll data.

The *Discoverer 2* mission successfully gathered data on propulsion, communications, orbital performance, and stabilization. All equipment functioned as programmed—except the timing device. Telemetry functioned until April 14, 1959, and the main tracking beacon functioned until April 21, 1959. *Discoverer 2* was the first satellite to be stabilized in orbit in all three axes, to be maneuvered on command from Earth, to separate a reentry vehicle on command, and to send its reentry vehicle back to Earth.

Discoverer 13 was an Earth-orbiting satellite designed to test spacecraft engineering techniques and to attempt deceleration, reentry through the atmosphere, and recovery from the sea of an instrument package. The U.S. Air Force launched this Discoverer series spacecraft from Vandenberg Air Force Base on August 10, 1960, using a Thor Agena A rocket vehicle. This precursor spy satellite went into a 258-km by 683-km polar orbit that had a period of 94 minutes and an inclination of approximately 83 degrees. The cylindrical Agena A stage that placed the spacecraft into orbit carried a telemetry system, a tape recorder, receivers for command signals from the ground, a horizon scanner, and a 55-kg recovery capsule, which contained biological specimens. The capsule was a bowl-shaped configuration 0.55 m in diameter and 0.68 m deep. A conical afterbody increased the total length to about 1 meter. A retrorocket, mounted at the end of the afterbody, decelerated the capsule out of orbit. An 18-kg monitoring system in the capsule reported on selected events, such as firing of the retrorocket, jettisoning of the heat shield, and others. The recovery capsule was retrieved on August 11, 1960. This achievement represented the first successful recovery of an object ejected from an orbiting satellite. The Agena upper stage rocket reentered the atmosphere and burned up on November 14, 1960.

The *Discoverer 13* spacecraft was also known as *Corona 13*—the first successful reconnaissance satellite mission. However, *Discoverer 13* carried only diagnostic equipment rather than an actual camera/film capsule payload into orbit. On August 18, 1960, a technical sibling, called *Discoverer 14* (or *Corona 14*), carried an actual camera/film capsule payload into orbit, imaged large portions of the Soviet Union, and then successfully ejected the film capsule for midair recovery over the Pacific Ocean near Hawaii. With these two successful Discoverer spacecraft missions, the era of the reconnaissance satellite began in August 1960.

See also NATIONAL RECONNAISSANCE OFFICE; RECONNAISSANCE SATELLITE; UNITED STATES AIR FORCE.

dish Aerospace jargon used to describe a parabolic radio or radar antenna, whose shape is roughly that of a large soup bowl.

See also DEEP SPACE NETWORK; RADIO ASTRONOMY; RADIO TELESCOPE.

disk 1. (astronomy) The visible surface of the Sun (or any other celestial body) seen in the sky or through a telescope. 2. (of a galaxy) Flattened, wheel-shaped region of stars, gas, and dust that lies outside the central region (nucleus) of a galaxy.

Disney, Walter (Walt) Elias (1901–1966) American *Entertainment Industry Visionary, Space Travel Advocate* Walt Disney was the American entertainment genius who popularized the concept of space travel in the early 1950s, especially through a widely acclaimed three-part television series. Because of Disney's commitment to excellence, other space visionaries such as WERNHER VON BRAUN were able to inspire millions of Americans with credible images of space exploration in the mid-1950s.

In the mid-1950s entertainment industry genius Walt Disney (left) collaborated with rocket scientist Wernher von Braun (right) in the development of a well-animated three-part television series that popularized the dream of space travel for millions of Americans. *(Courtesy of NASA)*

Born on December 5, 1901, in Chicago, Illinois, this legendary American motion picture animator and producer also introduced millions of people to the excitement of space travel. While a young boy on his family's farm near Marceline, Missouri, Disney began his cartooning career by sketching farm animals. When the family returned to Chicago in 1917, Disney attended high school, but evening art classes proved to be his real interest. Then, without graduating from high school, Disney volunteered to serve as a Red Cross ambulance driver during World War I.

In 1919 Disney began producing advertising films in Kansas City, Missouri, and eventually turned to animation. Enjoying only limited success in Kansas City, he made an eventful decision in 1923 that would change the entertainment industry forever. This decision involved a move to Hollywood, California, and the formation of a business partnership with his brother, Roy. In 1928 Walt Disney produced the first animated cartoon to use synchronized sound.

This cartoon, called *Steamboat Willie,* introduced the world to a charming new cartoon character, Mickey Mouse. Soon Minnie Mouse, Pluto, Goofy, and Donald Duck joined Mickey. The assemblage of cartoon characters soon delighted millions of fans around the world. Disney's reputation in entertainment continued to grow as he provided audiences the lighthearted, optimistic diversion they needed during the Great Depression of the 1930s.

Disney's first full-length animated feature, *Snow White and the Seven Dwarfs,* appeared in 1937 and won an Academy Award. During World War II his studios did a great deal of work for the U.S. government. Under his inspirational leadership, Disney's cartoon and motion picture studios continued to produce a wide range of award-winning cartoon shorts and innovative feature films. Following World War II he also produced *Seal Island* (1948), the first of many award-winning "true-life adventure" motion pictures. During his lifetime these consistently creative efforts earned his organization an unprecedented 30 Academy Awards.

In the early 1950s Disney started planning an entirely new form of entertainment, a family-oriented amusement complex, which he called a *theme park.* He built Disneyland in Anaheim, California, and the park had four major themes: Fantasyland, Adventureland, Frontierland, and Tomorrowland. Disney's previous cartoon and motion picture work had not ventured into the realm of future technology. Early in planning Disneyland, he recognized the power of television in promoting the new park. He also recognized the urgent need for a crowd-pleasing "future technology" to anchor Tomorrowland.

Disney turned to a longtime member of his staff, Ward Kimball (1914–2002), and asked for suggestions about Tomorrowland. Kimball mentioned that scientists were talking about the possibility of traveling in space. He showed Disney the *Collier's* magazine articles written by Wernher von Braun and other scientists that discussed space travel, space stations, missions to Mars, and the like. Disney's creative spirit recognized the opportunity. Space would become a major theme of Tomorrowland. He also developed a special television show to introduce the public to space travel, along with attractions at his new theme park.

Because he was busy developing Disneyland, Disney gave Kimball a literal "blank check" to hire the best scientists to produce a space travel television show that was both factual and entertaining. Disney frequently reviewed Kimball's progress and made creative suggestions but left Kimball in charge of the daily production activities.

On the evening of March 9, 1955, millions of television viewers across the United States tuned in to the popular Disneyland TV show. Suddenly, after the usual image of Sleeping Beauty's castle faded out, Walt Disney himself appeared on the screen. He sat on the edge of his desk and held a futuristic model rocket. Unlike the format of previous shows, Disney now personally prepared his viewers for their trip into Tomorrowland. He began this special show with a powerful (but soft-spoken) introduction that described the important influence of science in daily living. He also mentioned how things that seemed impossible today could actually become realities tomorrow. Next, he described the concept of space travel as one of humanity's oldest dreams. He concluded the piece by suggesting that recent scientific discoveries have brought people to the threshold of a new frontier—namely, the frontier of interplanetary travel. Disney's show pleased and inspired millions of viewers. In a truly magic moment for the entertainment industry, Walt Disney, supported by the world's leading rocket scientists, credibly spoke to his audience about the possibility of the impossible—travel through interplanetary space.

"Man In Space" was the first of three extremely popular space-themed Disney television shows that energized the American population toward the possibility of space travel in the mid-1950s. Each show combined careful research and factual presentation with incredibly beautiful visual displays and a splash of Disney humor for good measure. After Disney's introduction, the first show continued with a history of rocketry (featuring German-American rocket historian WILLY LEY), a discussion of the hazards of human space flight (featuring aerospace medicine expert Heinz Haber [1913–90]), and a detailed presentation by Wemher von Braun about a large, four-stage rocket that could carry six humans into space and safely return them to Earth. These space experts had previously provided technical support for the popular *Collier's* magazine series, and Disney (to his credit) spared no expense in getting their expert opinions and participation for his TV show.

"Man In Space" proved so popular with audiences that Disney rebroadcast the show on June 15 and again on September 7, 1955. One especially important person viewed the first show. President DWIGHT D. EISENHOWER liked the show so much that he personally called Disney and borrowed a copy of the show to use as a space education primer for the "brass" at the Pentagon. It is interesting to note that on June 30, 1955, President Eisenhower announced that the United States would launch an Earth-orbiting artificial satellite as part of America's participation in the upcoming International Geophysical Year (1957). Coincidence? Perhaps. Or quite possibly some of Disney's "visionary magic" worked like a much-needed catalyst in a sluggish federal bureaucracy that consistently failed to comprehend the emerging importance of space technology.

When Disneyland opened in the summer of 1955, the Tomorrowland section of the theme park featured a Space Station X-1 exhibit and a simulated rocket ride to the Moon. There was also a large (25-meter tall) needle-nosed rocket ship (designed by Willy Ley and Wemher von Braun) to greet visitors to the Moon mission attraction.

Disney's second space-themed television show, "Man and the Moon," aired on December 28, 1955. In this show von Braun enthusiastically described his wheel-shaped space station concept and how it could serve as the assembly platform for a human voyage around the Moon. Von Braun emerged from this show as the premier space travel advocate in the United States. The enthusiastic public response to the first two Disney space-themed TV shows also attracted some journalistic skepticism. Certain reporters tried to "protect" their readers by cautioning them "not to get swept away by over-enthusiasm or arm-chair speculation." Despite the cautious warnings of these skeptics, the space age arrived on October 4, 1957, when the Soviet Union launched *Sputnik 1,* the first artificial Earth satellite.

At the dawn of the space age Disney aired his third and final space-themed show, "Mars and Beyond," on December

4, 1957. Von Braun appeared only briefly in this particular show because he was busy trying to launch the first successful American satellite (*Explorer 1*). Through inputs from von Braun and his colleague, Ernst Stuhlinger, the highly animated show featured an armada of nuclear-powered interplanetary ships heading to Mars. It also contained amusing yet highly speculative cartoon-assisted discussions about the possibility of life in the solar system. With the conclusion of this episode, Disney reached millions of Americans and helped them recognize that space travel was real and no longer restricted to "Fantasyland." Through the use of television and his theme park, Disney vigorously promoted space technology in a truly imaginative, delightful way. Disney died on December 15, 1966, in Burbank, California. Five years after his death, the world-famous Disney World vacation complex opened near Orlando, Florida, about 100 kilometers west of Cape Canaveral—America's spaceport.

See also SCIENCE FICTION; VERNE, JULES; WELLS, HERBERT GEORGE (H. G.)

distortion 1. In general, the failure of a system (typically optical or electronic) to transmit or reproduce the characteristics of an input signal with exactness in the output signal. 2. An undesired changed in the dimensions or shape of a structure; for example, the *distortion* of a cryogenic propellant tank because of extreme temperature gradients.

diurnal Having a period of, occurring in, or related to a day; daily.

diverging lens (or mirror) A lens (or mirror) that refracts (or reflects) a parallel beam of light into a diverging beam. A diverging lens is generally a concave lens, while a diverging mirror is a CONVEX MIRROR. *Compare with* CONVERGING LENS.

See also LENS; MIRROR.

docking The act of physically joining two orbiting spacecraft. This is usually accomplished by independently maneuvering one spacecraft (the CHASER SPACECRAFT) into contact with the other (the TARGET SPACECRAFT) at a chosen physical interface. For spacecraft with human crews, a docking module assists in the process and often serves as a special passageway (AIRLOCK) that permits hatches to be opened and crew members to move from one spacecraft to the other without the use of a spacesuit and without losing cabin pressure.

docking module A structural element that provides a support and attachment interface between a docking mechanism and a spacecraft. For example, the special component added to the U.S. Apollo spacecraft so that it could be joined with the Soviet Soyuz spacecraft in the Apollo-Soyuz Test Project; or the component carried in the cargo bay of the U.S. space shuttle so that it could be joined with the Russian *Mir* space station.

See also APOLLO-SOYUZ TEST PROJECT; DOCKING.

doffing The act of removing wearing apparel or other apparatus, such as a spacesuit.

See also SPACESUIT.

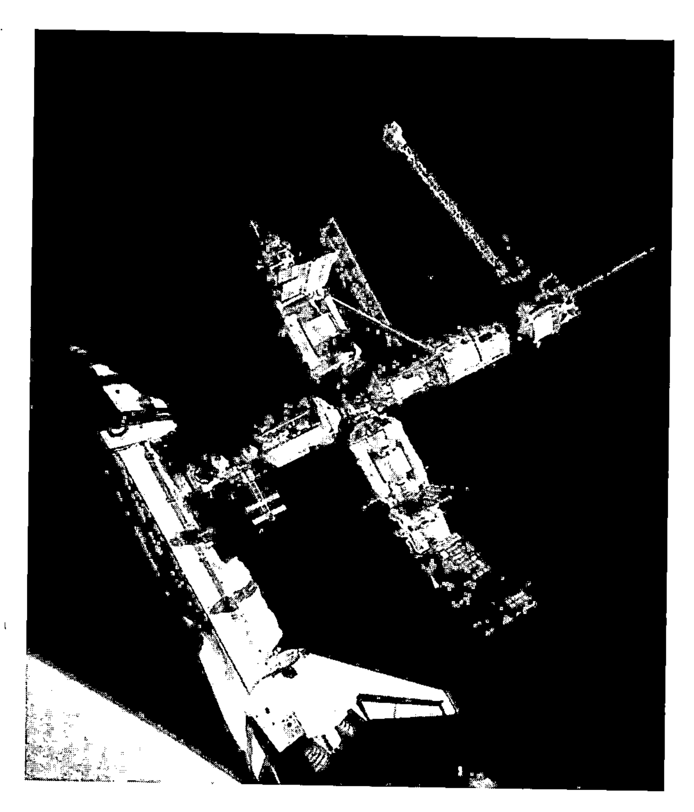

On July 4, 1995, the *Mir*-19 crew photographed this view of the space shuttle *Atlantis* still docked to Russia's *Mir* space station. Cosmonauts Anatoliy Y. Solovyev and Nikolai M. Budarin, *Mir*-19 commander and flight engineer, respectively, temporarily undocked the Soyuz spacecraft from the cluster of *Mir* elements to perform a brief fly-around. They took pictures while the STS-71 crew, with *Mir*-18's three crew members aboard, undocked *Atlantis* for the completion of this leg of the joint activities. Solovyev and Budarin had been taxied to the *Mir* space station by the STS-71 ascent trip of *Atlantis*. *(Courtesy of NASA/JSC)*

dogleg A directional turn made in a launch vehicle's ascent trajectory to produce a more favorable (orbital) inclination or to avoid passing over a populated (no-fly) region.

See also LAUNCH SITE.

Dog Star *See* SIRIUS.

Dolland, John (1706–1761) British *Optician* In the early 1750s John Dolland, a British optician, invented the first practical *heliometer*. The heliometer was the instrument previously used by astronomers to measure the diameter of the Sun or the angular separation of two stars that appear close to each other. About four years later (in 1758) he developed a composite lens for an *achromatic telescope* from two types of glass. An achromatic telescope (or lens) is one that corrects for chromatic aberration.

See also CHROMATIC ABERRATION.

Donati, Giovanni Battista (1826–1873) Italian *Astronomer* The 19th-century Italian astronomer Giovanni Donati specialized in discovering comets, including the long-period Comet Donati (C/1858), which he first sighted on June 2, 1858, and now carries his name. Donati's comet had, in addition to its major tail, two narrow extra tails. He was also the first scientist to observe the spectrum of a comet, that of Comet 55P/Tempel-Tuttle in 1864.

See also COMET.

donning The act of putting on wearing apparel or other apparatus, such as a spacesuit.

See also SPACESUIT.

Doppler, Christian Johann (1803–1853) Austrian *Physicist* In 1842 Christian Doppler published a paper that mathematically described the interesting phenomenon of how sound from a moving source changes pitch as the source approaches (frequency increases) and then goes away from (frequency decreases) an observer. This phenomenon, now called the DOPPLER SHIFT (or the Doppler effect), is widely used by astronomers to tell whether a distant celestial object is coming toward Earth (BLUESHIFT) or going away (receding) from Earth (REDSHIFT).

Doppler shift The apparent change in the observed frequency and wavelength of a source due to the relative motion of the source and an observer. If the source is approaching the observer, the observed frequency is higher, and the observed wavelength is shorter. This change to shorter wavelengths is often called the BLUESHIFT. If the source is moving away from the observer, the observed frequency will be lower, and the wavelength will be longer. This change to longer wavelengths is called the REDSHIFT. Named after CHRISTIAN DOPPLER, who discovered this phenomenon in 1842 by observing sounds.

dose In radiation protection, a general term describing the amount of energy delivered to a given volume of matter, a particular body organ, or a person (i.e., a wholebody dose) by ionizing radiation.

See also ABSORBED DOSE; IONIZING RADIATION.

dose equivalent **(symbol: H)** In radiation protection, the product of absorbed dose (D), quality factor (QF), and any other modifying factors used to characterize and evaluate the biological effects of ionizing radiation doses received by human beings (or other living creatures). The traditional unit of dose equivalent is the rem (an acronym for roentgen equivalent man). The sievert (Sv) is the SI special unit for dose equivalent. These units are related as follows:

$$100 \text{ rem} = 1 \text{ sievert (Sv)}$$

See also SI UNITS.

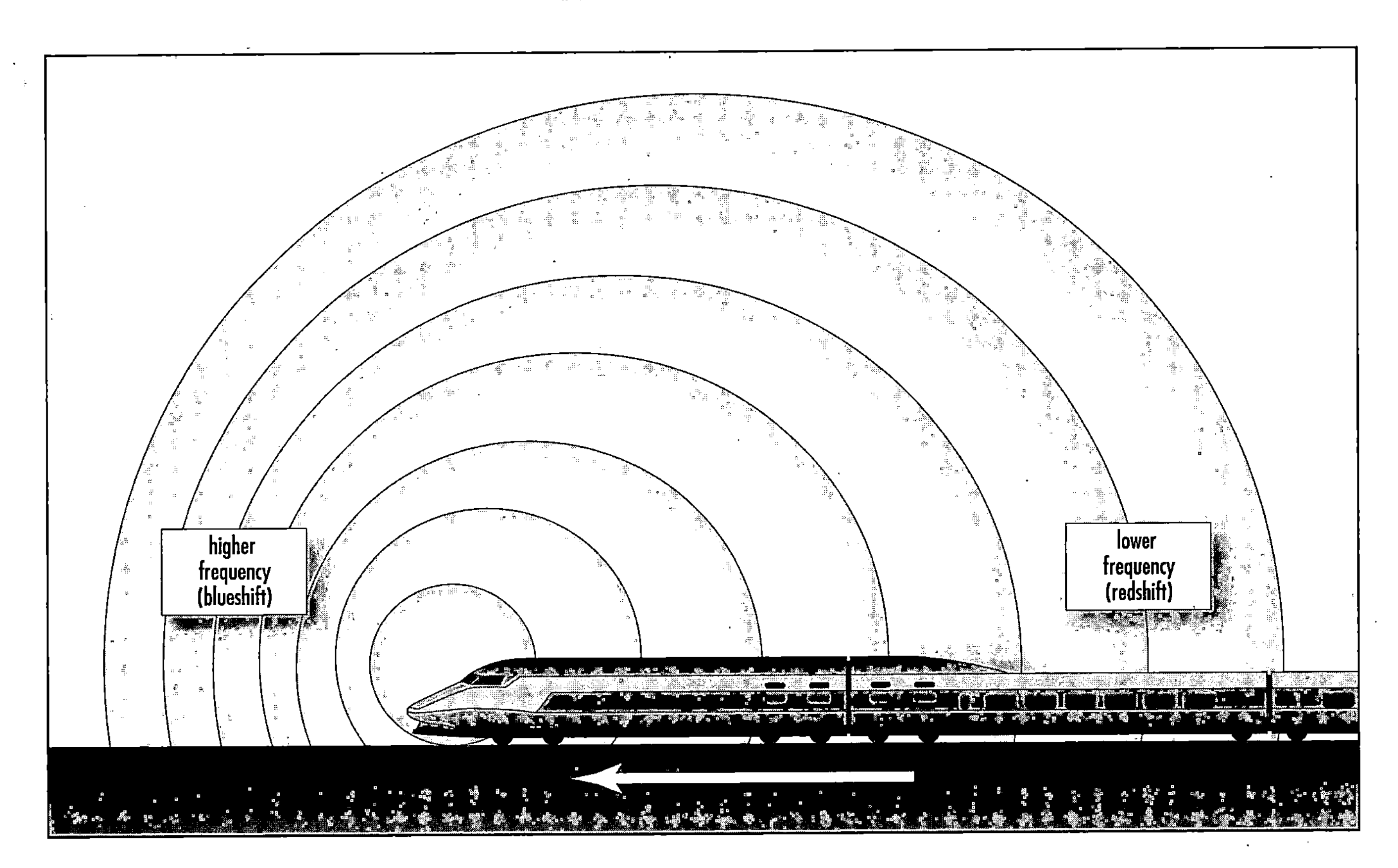

Sound from a train approaching an observer will appear to increase in frequency. Conversely, sound will appear to decrease in frequency as the train goes away (recedes) from an observer. This apparent shift in frequency due to the relative motion of an object and an observer is called the Doppler shift.

double-base propellant A solid rocket propellant using two unstable compounds, such as nitrocellulose and nitroglycerin. These unstable compounds contain enough chemically bonded oxidizer to sustain combustion.

See also ROCKET.

double star system Two stars that appear close together in the night sky. In an *optical double* the two stars have the illusory appearance of being close together when viewed by an observer on Earth, but are actually are *not* bound together by gravitational attraction; in a *physical double* the stars are gravitationally bound and orbit around their common barycenter. Astronomers generally restrict the expression *double star system* to optical doubles and use the expression *binary star system* for physical doubles.

See also BARYCENTER; BINARY STAR SYSTEM; STAR.

Douglass, Andrew Ellicott (1867–1962) American *Astronomer, Environmental Scientist* In the late 19th century, when he was 27 years old and working at the Lowell Observatory in Flagstaff, Arizona, Andrew Douglass uncovered a possible relationship between sunspot activity and Earth's climate. His link was the rate of tree growth. To confirm his suspicions he then began to record the annual rings of nearby pines and Douglas firs. By 1911 he found matching records among trees felled some 80 kilometers to the southwest of Lowell Observatory. This correlation encouraged him to begin a study of the great Sequoias and to develop the field of *dendrochronology* ("tree-dating")—a discipline that relates the rate of tree-ring development with weather effects and ultimately solar activity. He served as the director of the Steward Observatory at the University of Arizona from 1918 to 1938.

downlink The telemetry signal received at a ground station from a spacecraft or space probe.

See also DEEP SPACE NETWORK; TELECOMMUNICATIONS; TELEMETRY.

downrange A location away from a launch site but along the intended flight path (trajectory) of a missile or rocket flown from a rocket range. For example, the rocket vehicle tracking station on Ascension Island in the South Atlantic Ocean is far downrange from the launch sites at Cape Canaveral Air Force Station in Florida.

See also CAPE CANAVERAL AIR FORCE STATION.

drag (symbol: D) 1. A retarding force acting on a body in motion through a fluid, parallel to the direction of motion of the body. It is a component of the total forces acting on the body. 2. Atmospheric resistance to the orbital motion of a spacecraft. The effect of atmospheric drag is to cause the satellite's orbit to "decay," or spiral downward. A satellite of very high mass, very low cross-sectional area, and in very high orbit will not be affected greatly by drag. However, a high cross-sectional area satellite of low mass in a low-altitude orbit will be affected very strongly by drag. For Earth orbits above about 200 kilometers, a satellite's altitude decreases very slowly due to drag, but below 150 kilometers the orbit decays very rapidly. In fact, at lower orbital altitudes drag is the major space environment condition affecting satellite lifetime. For example, the abandoned U.S. *Skylab* space station came crashing back to Earth in July 1979 due to relentless action of atmospheric drag.

See also AERODYNAMIC FORCE; *SKYLAB;* SPACE DEBRIS.

Drake, Frank Donald (1930–) American *Astronomer* While working at the National Radio Astronomy Observatory (NRAO) in Green Bank, West Virginia, Frank Drake conducted the first organized attempt to detect radio wave signals from an alien intelligent civilization across interstellar distances. He performed this initial search for extraterrestrial intelligence (SETI) under an effort he called PROJECT OZMA. It led to the formulation of a speculative, semiempirical mathematical expression, popularly referred to as the *Drake equation.* The Drake equation tries to estimate the number of intelligent alien civilizations that might now be capable of communicating with one another in the Milky Way galaxy. Among his many professional accomplishments in radio astronomy, Drake was a professor of astronomy at Cornell University (1964–84), served as the director of the Arecibo Observatory in Puerto Rico, and is currently emeritus professor of astronomy and astrophysics at the University of California at Santa Cruz.

See also ARECIBO OBSERVATORY; DRAKE EQUATION; RADIO ASTRONOMY; SEARCH FOR EXTRATERRESTRIAL INTELLIGENCE.

Drake equation A probabilistic expression proposed by FRANK DRAKE in 1961 that is an interesting, though highly speculative, attempt to determine the number of advanced intelligent civilizations that might now exist in the Milky Way galaxy and be communicating (via radio waves) across interstellar distances. A basic assumption in Drake's formulation is the *principle of mediocrity*—namely, that conditions in the solar system and even on Earth are nothing particularly special but rather represent common conditions found elsewhere in the galaxy.

Just where do we look among the billions of stars in our galaxy for possible interstellar radio messages or signals from extraterrestrial civilizations? That was one of the main questions addressed by the attendees of the Green Bank Conference on Extraterrestrial Intelligent Life held in November 1961 at the National Radio Astronomy Observatory (NRAO) in Green Bank, West Virginia. One of the most significant and widely used results from this conference is the Drake equation (named after Frank Drake), which represents the first credible attempt to quantify the search for extraterrestrial intelligence (SETI). This "equation" has also been called the Sagan-Drake equation and the Green Bank equation in the SETI literature.

Although more nearly a subjective statement of probabilities than a true scientific equation, the Drake equation attempts to express the number (N) of advanced intelligent civilizations that might be communicating across interstellar distances at this time. As previously mentioned, a basic assumption inherent in this formulation is the principle of mediocrity. The Drake equation is generally expressed as

$$N = R^* f_p n_e f_l f_i f_c L$$

where

N is the number of intelligent communicating civilizations in the galaxy at present
R^* is the average rate of star formation in our galaxy (stars/year)
f_p is the fraction of stars that have planetary companions
n_e is the number of planets per planet-bearing star that have suitable ecospheres (that is, the environmental conditions necessary to support the chemical evolution of life)
f_l is the fraction of planets with suitable ecospheres on which life actually starts
f_i is the fraction of planetary life starts that eventually evolve into intelligent life-forms
f_c is the fraction of intelligent civilizations that attempt interstellar communication
L is the average lifetime (in years) of technically advanced civilizations

An inspection of the Drake equation quickly reveals that the major terms cover many disciplines and vary in technical content from numbers that are somewhat quantifiable (such as R^*) to those that are completely subjective (such as L).

For example, astrophysics can provide a reasonably approximate value for R^*. Namely, if we define R^* as the average rate of star formation over the lifetime of the Milky Way galaxy, we obtain

$$R^* = \text{number of stars in the galaxy/age of the galaxy}$$

We can then insert some typically accepted numbers for our galaxy to arrive at R^*, namely,

$$R^* = 100 \text{ billion stars/10 billion years}$$

$$R^* = 10 \text{ stars/year (approximately)}$$

Generally, the estimate for R^* used in SETI discussions is taken to fall between 1 and 20.

The rate of planet formation in conjunction with stellar evolution is currently the subject of much interest and discussion in astrophysics. Do most stars have planets? If so, then the term f_p would have a value approaching unity. On the other hand, if planet formation is rare, then f_p approaches zero. Astronomers and astrophysicists now think that planets should be a common occurrence in stellar evolutionary processes. Furthermore, advanced extrasolar planet detection techniques (involving astrometry, adaptive optics, interferometry, spectrophotometry, high-precision radial velocity measurements, and several new "terrestrial planet hunter" spacecraft) promise to provide astronomers the ability to detect Earthlike planets (should such exist) around our nearest stellar neighbors. Therefore, sometime within the next 20 years the aggressive contemporary search for extrasolar planets should provide astronomers the number of direct observations needed to establish an accurate empirical value for f_p. In typical SETI discussions, f_p is now often assumed to fall in the range between 0.4 and 1.0. The value $f_p = 0.4$ represents a more pessimistic view, while the value $f_p = 1.0$ is taken as very optimistic.

Similarly, if planet formation is a normal part of stellar evolution, we must next ask how many of these planets are actually suitable for the evolution and maintenance of life. By taking $n_e = 1.0$, we are suggesting that for each planet-bearing star system, there is at least one planet located in a suitable *continuously habitable zone* (CHZ), or *ecosphere*. This is, of course, what we see here in our own solar system. Earth is comfortably situated in the continuously habitable zone, while Mars resides on the outer (colder) edge, and Venus is situated on the inner (warmer) edge. The question of life-bearing moons around otherwise unsuitable planets was not directly addressed in the original Drake equation, but recent discussions about the existence of liquid-water oceans on several of Jupiter's moons, most notably Europa, and the possibility that such oceans might support life encourages us to consider a slight expansion of the meaning of the factor n_e. Perhaps n_e should now be taken to include the number of planets in a planet-bearing star system that lie within the habitable zone and also the number of major moons with potential life-supporting environments (liquid water, an atmosphere, etc.) around (uninhabitable) Jovian-type planets in that same star system.

We must next ask, given conditions suitable for life, how frequently does it start. One major assumption usually made (again based on the principle of mediocrity) is that wherever life can start, it will. If we invoke this assumption, then f_l equals unity. Similarly, we can also assume that once life starts, it always strives toward the evolution of intelligence, making f_i equal to 1 (or extremely close to unity).

This brings us to an even more challenging question: What fraction of intelligent extraterrestrial civilizations develop the technical means and then want to communicate with other alien civilizations? All we can do is make a very subjective guess, based on human history. The pessimists take f_c to be 0.1 or less, while the optimists insist that all advanced civilizations desire to communicate and make $f_c = 1.0$.

Let us now consider the hypothetical case of an alien scientist in a distant star system (say about 50 light-years away) who must submit numerous proposals for very modest funding to continue a detailed search for intelligent-species-generated electromagnetic signals emanating from our region of the Milky Way. Unfortunately, the Grand Scientific Collective (the leading technical organization within that alien civilization) keeps rejecting the alien scientist's proposals, proclaiming that "such proposed SETI searches are a waste of precious *zorbots* (alien units of currency) that should be used for more worthwhile research projects." The radio receivers are turned off about a year before the first detectable television signals (leaking out from Earth) pass through that star system! Consequently, although this (imagined) alien civilization might have developed the technology needed to justify use of a value of $f_c = 1$, that same alien civilization did not display any serious inclination toward SETI, thereby making $f_c \approx 0$. Extending the principle of mediocrity to the collective social behavior of intelligent alien beings (should they exist anywhere), is shortsightedness (especially among leaders) a common shortcoming throughout the galaxy?

Drake Equation Calculations

The Basic Equation: $N = R^* f_p n_e f_l f_i f_c L$

	R^*	f_p	n_e	f_l	f_i	f_c	L	N	Conclusion
Very optimistic values	20	1.0	1.0	1.0	1.0	0.5	10^6	$\sim 10^7$	The galaxy is full of intelligent life!
Your own values									
Very pessimistic values	1	0.2	1.0	1.0	0.5	0.1	100	~1	We are alone!

Source: Developed by the author.

Finally, we must also speculate on how long an advanced technology civilization lasts. If we use Earth as a model, all we can say is that (at a minimum) L is somewhere between 50 and 100 years. True high technology emerged on Earth only during the last century. Space travel, nuclear energy, computers, global telecommunications, and so on are now widely available on a planet that daily oscillates between the prospects of total destruction and a "golden age" of cultural maturity. Do most other evolving extraterrestrial civilizations follow a similar perilous pattern in which cultural maturity has to desperately race against new technologies that always threaten oblivion if they are unwisely used? Does the development of the technologies necessary for interstellar communication or perhaps interstellar travel also stimulate a self-destructive impulse in advanced civilizations, such that few (if any) survive? Or have many extraterrestrial civilizations learned to live with their evolving technologies, and do they now enjoy peaceful and prosperous "golden ages" that last for millennia to millions of years? In dealing with the Drake equation, the pessimists place very low values on L (perhaps 100 or so years), while the optimists insist that L is several thousand to several million years in duration.

Considering an appropriate value for L, it is interesting to recognize that space technology and nuclear technology also provide an intelligent species very important tools for protecting their home planet from catastrophic destruction by an impacting "killer" asteroid or comet. Although other solar systems will have cometary and asteroidal fluxes that are greater or less than those fluxes prevalent in our solar system, the threat of extinction-level planetary collisions should still be substantial. The arrival of high technology, therefore, also implies that intelligent aliens can overcome many natural hazards (including a catastrophic impact by an asteroid or comet), thereby extending the overall lifetime of the planetary civilization and increasing the value of L that we should use in the Drake equation.

Let us go back to the Drake equation and insert some "representative" values. If we take $R^* = 10$ stars/year, $f_p = 0.5$ (thereby excluding multiple-star systems), $n_e = 1$ (based on our solar system as a common model), $f_l = 1$ (invoking the principle of mediocrity), $f_i = 1$ (again invoking the principle of mediocrity) and $f_c = 0.2$ (assuming that most advanced civilizations are introverts or have no desire for space travel), then the Drake equation yields $N \approx L$. This particular result implies that the number of communicative extraterrestrial civilizations in the galaxy at present is approximately equal to the average lifetime (in Earth years) of such alien civilizations.

Now take these "results" one step further. If N is about 10 million (a very optimistic Drake equation output), then the average distance between intelligent, communicating civilizations in our galaxy is approximately 100 light-years. If N is 100,000, then these extraterrestrial civilizations on average would be about 1,000 light-years apart. But if there were only 1,000 such civilizations existing today, then they would typically be some 10,000 light-years apart. Consequently, even if the Milky Way galaxy does contain a few such civilizations, they may be just too far apart to achieve communication within the lifetimes of their respective civilizations. For example, at a distance of 10,000 light-years, it would take 20,000 years just to start an interstellar dialogue!

By now you might like to try your own hand at estimating the number of intelligent alien civilizations that could be trying to signal us today. If so, the Drake equation table has been set up just for you. Simply select (and justify to yourself) typical numbers to be used in the Drake equation, multiply all these terms together, and obtain a value for N. Very optimistic and very pessimistic values that have been used in other SETI discussions also appear in the table to help guide your own SETI efforts.

See also ASTEROID DEFENSE SYSTEM; CONTINUOUSLY HABITABLE ZONE; ECOSPHERE; EXTRATERRESTRIAL CIVILIZATIONS; FERMI PARADOX; PRINCIPLE OF MEDIOCRITY; SEARCH FOR EXTRATERRESTRIAL INTELLIGENCE (SETI).

Draper, Charles Stark (1901–1987) American *Physicist* Charles Draper was the physicist and instrumentation expert who used the principle of the gyroscope to develop inertial guidance systems for ballistic missiles, satellites, and the spacecraft used in NASA's Apollo Project.

See also APOLLO PROJECT; BALLISTIC MISSILE; GYROSCOPE; SPACECRAFT.

Draper, Henry (1837–1882) American *Physician, Amateur Astronomer* The physician and amateur astronomer Henry Draper followed his father's (JOHN WILLIAM DRAPER) example by pioneering key areas of astrophotography. In 1872 he became the first astronomer to photograph the spectrum of a star (Vega). Then, in 1880 he was the first to successfully photograph a nebula (the Orion Nebula). After he died his widow

(Mary Anna Draper, née Palmer [1839–1914]) continued his contributions to astronomy by financing publication of the now famous *Henry Draper Catalogue* of stellar spectra.

See also ASTROPHOTOGRAPHY; NEBULA; STAR.

Draper, John William (1811–1882) American *Scientist* Like his son HENRY DRAPER, the American scientist John Draper made pioneering contributions to the field of astrophotography. For example, he was the first person to photograph the Moon (1840) and then became the first to make a spectral photograph of the Sun (1844).

See also ASTROPHOTOGRAPHY.

Dreyer, Johan Ludvig Emil (John Lewis Emil Dreyer) (1852–1926) Danish *Astronomer* Johan Dreyer compiled an extensive catalog of nebulas and star clusters called the *New General Catalog (NGC) of Nebulae and Star Clusters,* which he first published in 1888. His catalog contained more than 7,800 galaxies, nebulas, and star clusters, and astronomers still refer to objects appearing in it by their NGC numbers. For example, the Orion Nebula is known as M42 (Messier Catalog number) or NGC 1976 (New General Catalog number).

See also CLUSTER (STELLAR); GALAXY; NEBULA; MESSIER CATALOG; NEW GENERAL CATALOG.

drogue parachute A small parachute used specifically to pull a larger parachute out of stowage; also a small parachute used to slow down a descending space capsule, aerospace vehicle, or high-performance airplane.

drop tower A tall tower (typically about 90 meters high or more) in which experimental packages are carefully dropped under "free-fall" conditions. A well-designed drop tower can provide low-gravity conditions for about 3 to 5 seconds—enough time to perform useful preliminary experiments simulating (briefly) the microgravity or weightlessness conditions encountered in an orbiting spacecraft. A catch tube system at the bottom of the drop tower decelerates and recovers the experimental package. Also called a *low-gravity drop tower* and a *drop tube.*

See also MICROGRAVITY.

dry emplacement A launch site that has no provision for water cooling of the pad during the launch of a rocket. *Compare with* WET EMPLACEMENT.

dry run In aerospace operations, a practice exercise or rehearsal for a specific mission task or launch operation; a practice launch without propellant loaded in the rocket vehicle.

dry weight The weight of a missile or rocket vehicle without its fuel. This term, especially appropriate for liquid-propellant rockets, sometimes is considered to include the payload.

See also LAUNCH VEHICLE; ROCKET.

Duke, Charles Moss, Jr. (1935–) American *Astronaut* Charles Duke is a U.S. Air Force officer and NASA astronaut who served as the lunar pilot on the *Apollo 16* mission to the Moon April 16–27, 1972. Along with fellow astronaut JOHN W. YOUNG (spacecraft commander), he explored the rugged lunar highlands during the fifth lunar landing mission, while their companion, astronaut THOMAS K. MATTINGLY II (command module pilot), remained in orbit around the Moon. *Apollo 16* was the first scientific expedition to inspect, survey, and sample materials and surface features in the Descartes region of the rugged lunar highlands. Duke and Young commenced their record-setting lunar surface stay of 71 hours and 14 minutes by maneuvering the lunar module, named Orion, to a landing on the rough Cayley Plains. In three subsequent excursions onto the lunar surface, they each logged 20 hours and 15 minutes in extravehicular activities involving the emplacement and activation of scientific equipment and experiments, the collection of nearly 100 kilograms of rock and soil samples, and the evaluation and use of the second lunar rover over the roughest and blockiest surface yet encountered on the Moon. As a result of his participation in the *Apollo 16* mission, Duke became the 10th human being to walk on the lunar surface.

See also APOLLO PROJECT; LUNAR EXCURSION MODULE; LUNAR ROVER.

duplexer A device that permits a single antenna system to be used for both transmitting and receiving. (Should not be confused with *diplexer,* a device that permits an antenna to be used simultaneously or separately by two transmitters.)

See also TELECOMMUNICATIONS; TELEMETRY.

dust detector A direct-sensing science instrument that measures the velocity, mass (typical range 10^{-16} g to 10^{-6} g), flight direction, charge (if any), and number of dust particles striking the instrument, carried by some spacecraft.

See also INTERPLANETARY DUST.

dwarf galaxy A small, often elliptical galaxy containing a million (10^6) to perhaps a billion (10^9) stars. The Magellanic Clouds, our nearest galactic neighbors, are examples of dwarf galaxies.

See also GALAXY; MAGELLANIC CLOUDS.

dwarf star Any star that is a main-sequence star, according to the Hertzsprung-Russell diagram. Most stars found in the galaxy, including the Sun, are of this type and are from 0.1 to about 100 SOLAR MASSes in size. However, when scientists use the term *dwarf star,* they are *not* referring to WHITE DWARFs, BROWN DWARFs, or BLACK DWARFs, which are celestial bodies that are not in the collection of main-sequence stars.

See also HERTZSPRUNG-RUSSELL DIAGRAM; MAIN-SEQUENCE STAR; STAR.

dynamic pressure (common symbol: Q or q) 1. The pressure exerted by a fluid, such as air, by virtue of its motion; it is equal to $1/2\ \rho\ V^2$, where ρ is the density and V is the velocity of the fluid. 2. The pressure exerted on a body by virtue of its motion through a fluid, as, for example, the pressure exerted on a rocket vehicle as it flies through the atmosphere. The condition of maximum dynamic pressure experienced by an ascending rocket vehicle is often called *max-Q.*

Dyna-Soar (Dynamic Soaring) An early U.S. Air Force space project from 1958 to 1963 involving a crewed boost-glide orbital vehicle that was to be sent into orbit by an expendable launch vehicle, perform its military mission, and return to Earth using wings to glide through the atmosphere during reentry (in a manner similar to NASA's space shuttle). The project was canceled in favor of the civilian (NASA) human space flight program involving the MERCURY PROJECT, GEMINI PROJECT, and APOLLO PROJECT. Also called the *X-20 Project.*

dyne (symbol: d) A unit of force in the centimeter-gram-second (c.g.s.) system equal to the force required to accelerate a one-gram mass one centimeter per second per second; that is, 1 dyne = 1 gm-cm/sec^2. Compare NEWTON.

dysbarism A general aerospace medicine term describing a variety of symptoms within the human body caused by the existence of a pressure differential between the total ambient pressure and the total pressure of dissolved and free gases within the body tissues, fluids, and cavities. For example, increased ambient pressure, as accompanies a descent from higher altitudes, might cause painful distention of the eardrums.

See also SPACE SICKNESS.

Dyson, Sir Frank Watson (1868–1939) British *Astronomer* Sir Frank Dyson participated with SIR ARTHUR STANLEY EDDINGTON on the 1919 eclipse expedition that observed the bending of a star's light by the Sun's gravitation—providing the first experimental evidence that supported ALBERT EINSTEIN's general relativity theory. He also served as England's Astronomer Royal from 1910 to 1933.

See also ASTRONOMER ROYAL; RELATIVITY.

Dyson, Freeman John (1923–) British-American *Theoretical Physicist, Mathematician* From 1957 to 1961 Freeman Dyson participated in Project Orion, a nuclear fission, pulsed rocket concept studied by the U.S. government as a means of achieving rapid interplanetary space travel. He also hypothesized the possible existence of large artificial (nonnatural) infrared radiation–emitting celestial objects that might have been constructed by very advanced alien civilizations around their parent stars to harvest all the radiant energy outputs. Today, scientists who search for signs of and signals from possible intelligent extraterrestrial civilizations refer to this type of (speculative) large-scale astronomical object as a *Dyson sphere.* A prolific writer and creative thinker, in 2003 Dyson became the president of the Space Studies Institute, a space research and advocacy organization.

See also DYSON SPHERE; PROJECT ORION.

Dyson sphere The Dyson sphere is a huge artificial biosphere created around a star by an intelligent species as part of its technological growth and expansion within an alien solar system. This giant structure would most likely be formed by a swarm of artificial habitats and miniplanets capable of intercepting essentially all the radiant energy from the parent star. The captured radiant energy would be converted for use through a variety of techniques such as living plants, direct thermal-to-electric conversion devices, photovoltaic cells, and perhaps other (as yet undiscovered) energy conversion techniques. In response to the second law of thermodynamics, waste heat and unusable radiant energy would be rejected from the "cold" side of the Dyson sphere to outer space. From our present knowledge of engineering heat transfer, the heat rejection surfaces of the Dyson sphere might be at temperatures of 200 to 300 K.

This hypothesized astroengineering project is an idea of the theoretical physicist FREEMAN JOHN DYSON. In essence, what Dyson has proposed is that advanced extraterrestrial societies, responding to Malthusian pressures, would eventually expand into their local solar system, ultimately harnessing the full extent of its energy and material resources. Just how much growth does this type of expansion represent?

We must invoke the principle of mediocrity (i.e., conditions are pretty much the same throughout the universe) and use our own solar system as a model. The energy output from our Sun—a G-spectral class star—is approximately 4×10^{26} joules per second (J/s). For all practical purposes, our Sun can be treated as a blackbody radiator at a temperature of approximately 5,800 K. The vast majority of its energy output occurs as electromagnetic radiation, predominantly in the wavelength range of 0.3 to 0.7 micrometers (μm).

As an upper limit, the available mass in the solar system for such very-large-scale (astroengineering) construction projects may be taken as the mass of the planet Jupiter, some 2×10^{27} kilograms. Contemporary energy consumption on Earth in the early 21st century amounts to about 10^{13} joules per second, which corresponds to a power level of 10 terawatts (TW). Let us now project just a 1 percent growth in terrestrial energy consumption per year. Within a mere three millennia, humankind's energy consumption needs would reach the energy output of the Sun itself! Today, several billion human beings live in a single biosphere, planet Earth, with a total mass of some 5×10^{24} kilograms. A few thousand years from now our Sun could be surrounded by a swarm of habitats containing trillions of human beings. As an exercise in the study of technology-induced social change, compare western Europe today with western Europe just two millennia ago, during the peak of the Roman Empire. What has changed, and what remains pretty much the same? Now do the same for the entire solar system, only going forward in time two or three millennia. What do you think will change (in a solar system civilization), and what will remain pretty much the same?

The Dyson sphere may therefore be taken as representing an upper limit for physical growth within our solar system. It is basically "the best we can do" from an energy and materials point of view in our particular corner of the universe. The vast majority of these human-made habitats would most probably be located in the ecosphere, or continuously habitable zone (CHZ), around our Sun—that is, ar a distance of about a one astronomical unit (AU) from our parent star. This does not preclude the possibility that other habitats, powered by nuclear fusion energy, might also be found scattered throughout the outer regions of a somewhat dismantled solar system. (These fusion-powered habitats might also become the technical precursors to the first interstellar space arks.)

Therefore, if we use our own solar system and planetary civilization as a model, we can anticipate that within a few millennia after the start of industrial development, an intelligent species might rise from the level of planetary civilization (Kardashev Type I civilization) and eventually occupy a swarm of artificial habitats that completely surround their parent star, creating a Kardashev Type II civilization. Of course, these intelligent creatures might also elect to pursue interstellar travel and galactic migration, as opposed to completing the Dyson sphere within their home star system (initiating a Kardashev Type III civilization). It was further postulated by Freeman Dyson that such advanced civilizations could be detected by the presence of thermal infrared emission (typically 8.0- to 14.0-micrometer wavelength) from very large objects in space that had dimensions of one to two astronomical units in diameter.

The Dyson sphere is certainly a grand, far-reaching concept. It is also quite interesting to realize that the space stations and space bases we build in the 21st century are, in a very real sense, the first habitats in the swarm of artificial structures that we might eventually construct as part of our solar system civilization. No other period in human history has provided the unique opportunity of constructing the first artificial habitats in our own Dyson sphere.

See also EXTRATERRESTRIAL CIVILIZATIONS; PRINCIPLE OF MEDIOCRITY; SPACE SETTLEMENT; SPACE STATION.

early warning satellite A military spacecraft that has the primary mission of detecting and reporting the launch or approach of unknown weapons or weapon systems. For example, a hostile missile launch might be detected and reported through the use of special infrared (IR) sensors onboard this type of surveillance satellite.

See also DEFENSE SUPPORT PROGRAM.

Earth The third planet from the Sun and the fifth-largest in the solar system. Our planet circles its parent star at an average distance of 149.6 million kilometers. Earth is the only planetary body in the solar system currently known to support life. The planet Earth Table presents some of the physical and dynamic properties of Earth as a planet in the solar system.

The name *Earth* comes from the Indo-European language base *er,* which produced the Germanic noun *ertho* and ultimately the German word *erde,* the Dutch *aarde,* the Scandinavian *jord,* and the English *earth.* Related word forms include the Greek *eraze,* meaning on the ground, and the Welsh *erw,* meaning a piece of land. In Greek mythology the goddess of Earth was called *Gaia,* while in Roman mythology the Earth goddess was *Tellus* (meaning fertile soil). The expression *Mother Earth* comes from the Latin expression *terra mater.* We use the word *terrestrial* to describe creatures and things related to or from planet Earth and the word *extraterrestrial* to describe creatures and things beyond or away from planet Earth. Astronomers also refer to our planet as *Terra* or *Sol III,* which means the third satellite from the Sun.

From space our planet is characterized by its blue waters and white clouds, which cover a major portion of it. Earth is surrounded by an ocean of air consisting of 78% nitrogen and 21% oxygen; the remainder is argon, neon, and other gases. The standard atmospheric pressure at sea level is 101,325 newtons per square meter. Surface temperatures range from a maximum of about 60° Celsius (°C) in desert regions along the equator to a minimum of -90°C in the frigid polar regions. In between, however, surface temperatures are generally much more benign.

Dynamic and Physical Properties of the Planet Earth

Radius	
Equatorial	6,378 km
Polar	6,357 km
Mass	5.98×10^{24} kg
Density (average)	5.52 g/cm^3
Surface area	5.1×10^{14} m^2
Volume	1.08×10^{21} m^3
Distance from the Sun (average)	1.496×10^8 km (1 AU)
Eccentricity	0.01673
Orbital period (sidereal)	365.256 days
Period of rotation (sidereal)	23.934 hours
Inclination of equator	23.45 degrees
Mean orbital velocity	29.78 km/sec
Acceleration of gravity g (sea level)	9.807 m/sec^2
Solar flux at Earth (above atmosphere)	$1,371 \pm 5$ watts/m^2
Planetary energy fluxes (approximate)	
Solar	10^{17} watts
Geothermal	2.5×10^{13} watts
Tidal friction	3.4×10^{12} watts
Human-made	
Coal-burning	3.06×10^{12} watts
Natural gas-burning	2.00×10^{12} watts
Oil-burning	3.82×10^{12} watts
Nuclear power	0.49×10^{12} watts
Hydroelectric	0.69×10^{12} watts
Total human-made	10.06×10^{12} watts
Number of natural satellites	1 (the Moon)

Planet Earth and its only natural satellite, the Moon, shown at the same scale *(Courtesy of NASA)*

Earth's rapid spin and molten nickel-iron core give rise to an extensive magnetic field. This magnetic field, together with the atmosphere, shields us from nearly all the harmful charged particles and ultraviolet radiation coming from the Sun and other cosmic sources. Furthermore, most meteors burn up in Earth's protective atmosphere before they can strike the surface. Our home planet's nearest celestial neighbor, the Moon, is its only natural satellite.

See also EARTH'S TRAPPED RADIATION BELTS; EARTH SYSTEM SCIENCE; GLOBAL CHANGE; MOON; TERRESTRIAL PLANETS.

Earth-based telescope A telescope located on the surface of Earth. *Compare with* the *HUBBLE SPACE TELESCOPE.*

Earth-crossing asteroid (ECA) An asteroid whose orbit now crosses Earth's orbit or will at some time in the future cross Earth's orbit as its orbital path evolves under the influence of perturbations from Jupiter and the other planets.

See also ASTEROID; NEAR-EARTH ASTEROID.

Earthlike planet An extrasolar planet that is located in an ecosphere and has planetary environmental conditions that resemble the terrestrial biosphere—especially a suitable atmosphere, a temperature range that permits the retention of large quantities of liquid water on the planet's surface, and a sufficient quantity of energy striking the planet's surface from the parent star. These suitable environmental conditions could permit the chemical evolution and the development of carbon-based life as we know it on earth. The planet also should have a mass somewhat greater than 0.4 Earth masses (to permit the production and retention of a breathable atmosphere) but less than about 2.4 Earth masses (to avoid excessive surface gravity conditions).

See also CONTINUOUSLY HABITABLE ZONE; ECOSPHERE; EXTRASOLAR PLANETS.

Earth Observing-1 (EO-1) One of the key responsibilities of NASA's Earth Science Office is to ensure the continuity of future Landsat data. In partial fulfillment of that responsibility, the New Millennium Program's (NMP) first *Earth Observing 1* (EO-1) flight, managed by NASA's Goddard Space Flight Center (GSFC), has validated revolutionary technologies contributing to the reduction in cost and increased capabilities for future land imaging missions. Three revolutionary land imaging instruments on EO-1 are collecting multispectral and hyperspectral scenes in coordination with the Enhanced Thematic Mapper (ETM+) on *Landsat 7*. Breakthrough space technologies in the areas of lightweight materials, high-performance integrated detector arrays, and precision spectrometers have been demonstrated in these instruments. Analysts have performed detailed comparisons of the EO-1 and *Landsat 7* images in order to validate these new instruments for follow-on Earth observation missions. Future NASA spacecraft will be an order of magnitude smaller and lighter than current versions. The EO-1 mission has also provided the on-orbit demonstration and validation of several spacecraft technologies to enable this transition. Key technology advances in communications, power, propulsion, and thermal and data storage are also part of the EO-1 mission.

EO-1 has been inserted into an orbit flying in formation with the *Landsat 7* satellite—taking a series of the same images. Comparison of these *paired scene* images support evaluation of EO-1's land imaging instruments. EO-1's smaller, less expensive, and more capable spacecraft, instruments, and technologies are setting the pace for 21st-century Earth system science (ESS) missions.

EO-1 was launched from Vandenberg Air Force Base, California, by a Delta 7320 rocket on November 21, 2000. The spacecraft was inserted into a 705-km circular, sun-synchronous (polar) orbit at a 98.7 degrees inclination—such that it is flying in formation one minute behind *Landsat 7* in the same ground track and maintaining this separation within two seconds. This close separation has enabled EO-1 to observe the same ground location (scene) through the same atmospheric region, allowing analysts to make paired scene comparisons between the two satellites. All three of the EO-1 land imaging instruments view all or subsegments of the *Landsat 7* swath. Reflected light from the ground is imaged onto the focal plane of each instrument. Each of the imaging instruments has unique filtering methods for passing light in only specific spectral bands. Analysts have selected these bands to optimize the search for specific surface features or land characteristics based on scientific or commercial applications.

For each scene, more than 20 gigabits of scene data from the Advanced Land Imager (ALI), Hyperion, and Atmospheric Corrector (AC) instruments are collected and stored on the onboard solid state data recorder at high rates. When the EO-1 spacecraft is in range of a ground station, the spacecraft automatically transmits recorded images to the ground station for temporary storage. The ground station stores the raw data on digital tapes, which are periodically sent to the Goddard Space Flight Center for processing and to the EO-1 science and technology teams for validation and research purposes.

The Advanced Land Imager is a technology verification instrument. The focal plane for the instrument is partially populated with four sensor chip assemblies (SCA) and covers 3° by 1.625°. Operating in a pushbroom fashion at an orbit of 705 km, the ALI provides Landsat-type panchromatic and multispectral bands. These bands were designed to mimic six *Landsat 7* bands with three additional bands covering the following wavelength ranges: 0.433–0.453, 0.845–0.890, and 1.20–1.30 micrometers (μm). The ALI contains wide-angle optics designed to provide a continuous 15° × 1.625° field of view for a fully populated focal plane with 30-meter resolution for the multispectral pixels and 10-meter resolution for the panchromatic pixels.

The Hyperion is a high-resolution hyperspectral imager (HSI) capable of resolving 220 spectral bands (from 0.4 to 2.5 μm) with a 30-meter spatial resolution. The instrument captures a 7.5 km × 100 km land area per image and provides detailed spectral mapping across all 220 channels with high radiometric accuracy. Hyperspectral imaging has applications in mining, geology, forestry, agriculture, and environmental management. For example, detailed classification of land assets through the use of Hyperion data support more accurate remote mineral exploration, better predictions of crop yield and assessments, and better containment mapping.

Spacecraft-derived Earth imagery is degraded by atmospheric absorption and scattering. The Atmospheric Corrector (AC) provides moderate spatial resolution hyperspectral imaging using a wedge filter technology, with spectral coverage ranging from .89 to 1.58 μm. Scientists selected these spectral bands for optimal correction of high spatial resolution images. As a result, the AC instrument provides data that directly support correction of surface imagery for atmospheric variability (primarily water vapor). The AC serves as the prototype of an important new spacecraft instrument applicable to any future scientific or commercial Earth remote sensing mission where atmospheric absorption due to water vapor or aerosols degrades surface reflectance measurements. By using the AC analysts can use measured absorption values rather than modeled absorption values. The use of measure absorption data enables environmental scientists to make more precise predictive models for various remote sensing applications. For example, new algorithms based on AC data are supporting the more accurate measurement and classification of land resources and the creation of better computer-based models for various land management applications.

See also LANDSAT REMOTE SENSING.

Earth-observing spacecraft A satellite in orbit around Earth that has a specialized collection of sensors capable of monitoring important environmental variables. This is also called an environmental satellite or a green satellite. Data from such satellites help support Earth system science.

See also *AQUA* SPACECRAFT; *EARTH OBSERVING-1*; EARTH SYSTEM SCIENCE; LANDSAT; REMOTE SENSING; *TERRA* SPACECRAFT; WEATHER SATELLITE.

Earth radiation budget **(ERB)** The Earth radiation budget is perhaps the most fundamental quantity influencing Earth's climate. The ERB components include the incoming solar radiation; the solar radiation reflected back to space by the clouds, the atmosphere, and Earth's surface; and the long-wavelength thermal radiation emitted by Earth's surface and its atmosphere. The latitudinal variations of Earth's radiation budget are the ultimate driving force for the atmospheric and oceanic circulations and the resulting planetary climate.

One of the most intriguing questions facing atmospheric scientists and climate modelers is how clouds affect climate and vice versa. Understanding these effects requires a detailed knowledge of how clouds absorb and reflect both incoming short-wavelength solar energy and outgoing long-wavelength (thermal infrared) terrestrial radiation. Scientists using satellite-derived data have discovered, for example, that clouds that form over water are very different than clouds that form over land. These differences affect the way clouds reflect sunlight back into space and how much long-wavelength thermal infrared energy from Earth the clouds absorb and re-emit.

Water vapor in the atmosphere also affects daily weather and climate, because water vapor acts like a greenhouse gas, absorbing outgoing long-wavelength radiation from Earth. Because water vapor also condenses to form clouds, an increase in water vapor in the atmosphere may also increase the amount of clouds. Scientists use a variety of instruments on Earth-observing spacecraft to try to better understand how complex natural mechanisms determine the energy balance of our planet. These efforts include NASA's Cloud and Earth's Radiant Energy System (CERES)—an effort that uses instruments on contemporary spacecraft to extend the measurements made by the previous Earth Radiation Budget Experiment (ERBE).

See also *EARTH RADIATION BUDGET SATELLITE* (ERBS); EARTH SYSTEM SCIENCE; GLOBAL CHANGE; GREENHOUSE EFFECT.

Earth Radiation Budget Satellite **(ERBS)** This spacecraft is part of NASA's three-satellite Earth Radiation Budget Experiment (ERBE), designed to investigate how energy from the Sun is absorbed and re-emitted by Earth. This process of absorption and re-radiation is one of the principal drivers of Earth's weather patterns. Observations from ERBS were also used to determine the effects of human activities (such as burning fossil fuels) and natural occurrences (such as volcanic eruptions) on Earth's radiation balance. In addition to the ERBE scanning and nonscanning instruments, the EBRS also carried the Stratospheric Aerosol Gas Experiment (SAGE II). The ERBS was the first of three ERBE platforms that would eventually carry the ERBE instruments. NASA's Goddard Space Flight Center built this Earth-observing satellite. It was launched by the space shuttle *Challenger* in October 1984 and eventually operated in a 585-km altitude with a 57° inclination orbit around Earth. The second ERBE instrument flew aboard the NOAA-9 weather satellite (launched in January 1985), and the third was flown aboard the NOAA-10 satellite (October 1986). Although the scanning instruments onboard all three ERBE satellites are no longer operational, the nonscanning instruments are still functioning.

See also EARTH RADIATION BUDGET; EARTH SYSTEM SCIENCE; NATIONAL OCEANIC AND ATMOSPHERIC ADMINISTRATION (NOAA).

Earth satellite An artificial (human-made) object placed in orbit around the planet Earth.

See also EARTH-OBSERVING SPACECRAFT; MILITARY SATELLITE; *EXPLORER 1*; *SPUTNIK 1*.

earthshine Sunlight reflected by Earth observed as a faint glowing of the dark part of the Moon when it is at or close to the new Moon phase. About five centuries ago the Italian Renaissance genius Leonardo Da Vinci (1452–1519) explained the phenomenon when he realized that both Earth and the Moon reflect sunlight. This faint illumination of the dark portion of the Moon is sometimes called the Moon's "ashen glow." The phenomenon has also given rise to the popular expression "the old Moon in the new Moon's arms." Aerospace engineers have expanded the meaning of earthshine in order to perform the sophisticated radiant energy balances required in the design of Earth-orbiting artificial satellites and spacecraft. A spacecraft or space vehicle in orbit around Earth is illuminated by both sunlight and earthshine. In this case earthshine consists of sunlight (0.4- to 0.7-micrometer wavelength radiation, or visible light) reflected by Earth and the thermal radiation (typically 10.6-micrometer wavelength infrared radiation) emitted by Earth's surface and atmosphere.

See also PHASES OF THE MOON; THERMAL CONTROL.

Earth's trapped radiation belts The magnetosphere is a region around Earth through which the solar wind cannot penetrate because of the terrestrial magnetic field. Inside the magnetosphere are two belts, or zones, of very energetic atomic particles (mainly electrons and protons) that are trapped in Earth's magnetic field hundreds of kilometers above the atmosphere. JAMES VAN ALLEN of the University of Iowa and his colleagues discovered these belts in 1958. Van Allen made the discovery using simple atomic radiation detectors placed onboard *Explorer 1*, the first American satellite.

The two major trapped radiation belts form a doughnut-shaped region around Earth from about 320 to 32,400 kilometers above the equator (depending on solar activity). Energetic protons and electrons are trapped in these belts. The inner Van Allen belt contains both energetic protons (major constituent) and electrons that were captured from the solar wind or were created in nuclear collision reactions between energetic cosmic ray particles and atoms in Earth's upper atmosphere. The outer Van Allen belt contains mostly energetic electrons that have been captured from the solar wind.

Spacecraft and space stations operating in Earth's trapped radiation belts are subject to the damaging effects of ionizing radiation from charged atomic particles. These particles include protons, electrons, alpha particles (helium nuclei), and heavier atomic nuclei. Their damaging effects include degradation of material properties and component performance, often resulting in reduced capabilities or even failure of spacecraft systems and experiments. For example, solar cells used to provide electric power for spacecraft often are severely damaged by passage through the Van Allen belts. Earth's trapped radiation belts also represent a very hazardous environment for human beings traveling in space.

Radiation damage from Earth's trapped radiation belts can be reduced significantly by designing spacecraft and space stations with proper radiation shielding. Often crew compartments and sensitive equipment can be located in regions shielded by other spacecraft equipment that is less sensitive to the influence of ionizing radiation. Radiation damage also can be limited by selecting mission orbits and trajectories that avoid long periods of operation where the radiation belts have their highest charged-particle populations. For example, for a spacecraft or space station in low-Earth orbit, this would mean avoiding the South Atlantic anomaly and, of course, the Van Allen Belts themselves.

See also IONIZING RADIATION; MAGNETOSPHERE.

Earth system science (ESS) The modern study of Earth, facilitated by space-based observations, that treats the planet as an interactive, complex system. The four major components of the Earth system are the atmosphere, the hydrosphere (which includes liquid water and ice), the biosphere (which includes all living things), and the solid Earth (especially the planet's surface and soil).

From Earth-observing satellites we can view the Earth as a whole system, observe the results of complex interactions, and begin to understand how the planet is changing. Scientists use this unique view from space to study the Earth as an integrated system. Learning more about the linkages of complex environmental processes gives scientists improved prediction capability for climate, weather, and natural hazards. Today the collection of vast quantities of data from Earth-observing satellites is causing the union of scientists from many different disciplines within a common multifaceted discipline known as Earth system science (ESS). With data derived from sophisticated spacecraft, they are striving to develop an improved understanding of the Earth system and how it responds to natural or human-induced changes. Scientists and nonscientists alike also find it convenient to personify the integrated Earth system as *Gaia*—the Mother Earth goddess found in Greco-Roman mythology.

The United States and other space-faring nations have deployed a collection of satellites with sophisticated capabilities to characterize the current state of the Earth system. In the years ahead Earth-orbiting satellites will evolve into constellations of even "smarter" satellites—spacecraft that can be reconfigured based on the changing needs of science and technology. Future-thinking scientists also envision an intelligent and integrated observation network composed of sensors deployed in vantage points from the subsurface to deep space. This *sensorweb* will provide on-demand data and analysis to a wide range of end users in a timely and cost-effective manner. By the mid-21st century an armada of advanced Earth-observing spacecraft will form a comprehensive and interconnected system that will produce many practical benefits for scientific research, national policy making, and economic growth. From an environmental information perspective the world will quite literally become *transparent.* For example, scientists and policy makers will be able to quickly assess the regional or global consequences of a particular natural or human-caused environmental hazard and launch appropriate recovery and remediation operations.

As the U.S. civilian space agency, NASA's overall goal in Earth system science is to observe, understand, and model our planet as an integrated system in order to discover how it is changing, to better predict change, and to understand the consequences for life on Earth. Scientists do so by characterizing, understanding, and predicting change in major Earth system processes and by linking their models of these processes together in an increasingly integrated way.

Earth system science involves several major activities. The first is to explore interactions among the major components of the Earth system—continents, oceans, atmosphere, ice, and life. The second is to distinguish natural from human-induced causes of change. The third is to understand and predict the consequences of change. To accomplish this, scientists are pursuing several scientific focus areas associated with the major components of the Earth system and their complex interactions. In addition to improved scientific understanding of how our planet functions, contemporary Earth system science investigations are also generating numerous beneficial applications within each focus area.

As part of Earth system science efforts, aerospace and environmental scientists are selecting those scientific questions for which space technology and remote sensing can make a defining contribution. They then assimilate new satellite, suborbital, and in-situ observations into Earth system models. A special effort is being made to develop Earth system models that incorporate observations together with process modeling to simulate linkages among the processes studied in the research focus areas—namely, (a) coupling between atmospheric composition, aerosol loading, and climate, (b) coupling between aerosol processes and the hydrological cycle, and (c) coupling between climate variability and weather.

One area of research aims to reduce the uncertainties in our understanding of the causes and consequences of global change and to enable practical options for mitigation and adaptation. Another research focus involves improvement in the duration and reliability of weather forecasts. Success here will help reduce vulnerability to natural and human-induced disasters. Still another major research dimension of Earth system science is an improved ability to understand and predict changes in the world's oceans. Scientists also seek a better understanding of the global change signals from and impacts on polar regions. Ideally, scientists will carry out their work in the next decade so that the major focus areas of Earth system science are interrelated and eventually properly integrated to develop a fully interactive and realistic representation of the Earth system.

For example, the work of scientists on characterizing, understanding, and predicting climate variability and change is centered around providing the global scale observational data sets on the higher-inertia components of the climate system (oceans and ice), their forcing functions, and their interactions with the entire Earth system. Understanding these interactions goes beyond observations. The process includes developing and maintaining a modeling capability that allows for the effective use, interpretation, and application of the data, much of which is derived from Earth-observing satellites. Here, the ultimate objective is to enable predictions of change in climate on time scales ranging from seasonal to multidecadal. Fueled by the important space-based perspective, scientists have learned much over the last several decades. Among the more recent discoveries has been that ice cover in the Arctic Ocean is shrinking, as has ice cover on land, as temperatures have warmed over the last two decades. In the Antarctic such trends are not apparent except for in a few select locations. Satellite altimetry has made a major contribution to being able to measure and monitor recent changes in global circulation and has contributed valuable insight into the net upward trend in sea level that may threaten coastal regions in the future.

The climate system is dynamic, and modeling is the only way by which scientists can effectively integrate the current knowledge of the individual components of the climate system. Through modeling studies they can estimate and project the future state of the climate system. However, they do not have a full understanding of the processes that contribute to climate variability and change. Future Earth system science work in this focus area will be to eliminate model uncertainties through better understanding of the processes.

Another research focus area addresses the distribution and cycling of carbon among the land, ocean, and atmospheric reservoirs and ecosystems as they are affected by humans, as they change due to their own biogeochemistry, and as they interact with climate variations. The goals are to quantify global productivity, biomass, carbon fluxes, and changes in land cover; document and understand how the global carbon cycle, terrestrial and marine ecosystems, and land cover and use are changing; and provide useful projections of future changes in global carbon cycling and terrestrial and marine ecosystems.

Throughout the next decade research will be needed to advance our understanding of and ability to model human–ecosystems–climate interactions so that an integrated understanding of Earth system function can be achieved. These research activities should yield knowledge of the Earth's ecosystems and carbon cycle as well as projections of carbon cycle and ecosystem responses to global environmental change.

Examples of the types of forecasts that may be possible are the outbreak and spread of harmful algal blooms, occurrence and spread of invasive exotic species, and productivity of forest and agricultural systems. This particular focus area also will contribute to the improvement of climate projections for 50 to 100 years into the future by providing key inputs for climate models. This includes projections of future atmospheric CO_2 and CH_4 concentrations and understanding of key ecosystem and carbon cycle process controls on the climate system.

Within Earth system science the goal of the scientists who investigate Earth's surface and interior structure is to assess, mitigate, and forecast the natural hazards that affect society, including such phenomena as earthquakes, landslides, coastal and interior erosion, floods, and volcanic eruptions. Satellite-based measurements are among the most practical and cost-effective techniques for producing systematic data sets over a wide range of spatial and temporal scales. Remote sensing technologies are empowering scientists to measure and understand subtle changes in the Earth's surface and interior that reflect the response of the Earth to both the internal forces that lead to volcanic eruptions, earthquakes, landslides, and sea-level change and the climatic forces that sculpt the Earth's surface. For instance, thermal infrared remote sensing image data can signal impending activity by mapping ground temperatures and variations in the composition of lava flows as well as the sulfur dioxide in volcanic plumes.

On a more fundamental level, the solid Earth science activities contribute to a better understanding of how the forces generated by the dynamism of the Earth's interior have shaped landscapes and driven the chemical differentiation of the planet. The advent of space-borne remote sensing has been vital to the move toward forecasting in the solid Earth sciences. It has created a truly comprehensive perspective for monitoring the entire Earth system.

Scientists are also studying changes in the Earth's atmospheric chemistry over time. Their research is specifically geared toward creating a better understanding of the following areas: the changes in atmospheric composition and the timescales over which they occur, the forcing functions (human-made and natural) that drive these changes, the reaction of trace components in the atmosphere to global environment change and the resulting effects on the climate, the effects of global atmospheric chemical and climate changes, and air quality.

The relationship between our atmosphere and ground emissions presents several important environmental issues. These include global ozone depletion and recovery and the effect it has on ultraviolet radiation, radioactive gasses affecting our climate, and global air quality. Earth-observing satellites such as the *Terra* spacecraft offer reliable integrated measurements of these environmental processes from the vantage point of space. Studying these areas leads to direct societal effects, such as daily air quality ratings, emissions standards, and other policies that protect planet Earth.

The first weather satellite expanded the possibilities of predicting tomorrow's weather. With new observation missions (such as *Earth Observing-1*), comprehensive data gathering, and improved modeling, meteorologists can improve prediction capabilities even more to show people how the Earth's atmosphere is changing in relation to the Earth system. Accurately predicting changes in ozone, air quality, and climate also help us understand how humans are affecting our home planet and how we can better protect it.

The weather system includes the dynamics of the atmosphere and its interaction with the oceans and land. Weather includes those local processes that occur in matters of minutes as well as the global-scale phenomena that can be predicted with a degree of success at an estimated maximum of two weeks prior. An improvement in our understanding of weather processes and phenomena is crucial to gaining a more accurate understanding of the overall Earth system. Weather is directly related to climate and to the water cycle and energy cycles that sustain life within the terrestrial biosphere.

The water cycle and the energy cycle refer to the distribution, transport, and transformation of water and energy within the Earth system. Since solar energy drives the water cycle and energy exchanges are modulated by the interaction of water with radiation, the energy cycle and the water cycle are intimately entwined.

The long-term goal of this research area within Earth system science is to allow scientists to make improved predictions of water and energy cycle processes as consequences of global change. In the past decade, for example, satellite systems have allowed scientists to quantify tropical rainfall as well as to greatly improve their ability to predict hurricanes. However, many issues remain to be resolved. In the next decade scientists will move toward balancing the water budget at global and regional spatial scales and provide global observation capability of precipitation over the diurnal cycle and important land surface quantities such as soil moisture and snow quantity at mesoscale resolution. They will also work toward providing cloud-resolving models for input into climate models, and we will gain knowledge about the major influences on variability in the water and energy cycles.

See also *AQUA* SPACECRAFT; *EARTH OBSERVING-1;* EARTH RADIATION BUDGET SATELLITE (ERBS); GLOBAL CHANGE; GREENHOUSE EFFECT; LANDSAT; REMOTE SENSING; *TERRA* SPACECRAFT; WEATHER SATELLITE.

eccentricity **(symbol: e)** A measure of the ovalness of an orbit. For example, when e = 0, the orbit is a circle; when e = 0.9, the orbit is a long, thin ellipse. (*See* figure.) The eccentricity of an ellipse can be computed by the formula:

$$e = \sqrt{[1 - (b^2/a^2)]}$$

where a is the semimajor axis and b is the semiminor axis.

See also KEPLERIAN ELEMENTS; ORBITS OF OBJECTS IN SPACE.

eccentric orbit An orbit that deviates from a circle, thus forming an ellipse.

See also ECCENTRICITY; ORBITS OF OBJECTS IN SPACE.

Echo satellite A large, inflatable passive communications-relay satellite from which radio signals from one point on Earth could be bounced off the satellite and received by another location on Earth. NASA's *Echo 1,* the world's first passive communications-relay satellite, was launched successfully from Cape Canaveral Air Force Station on August 12, 1960.

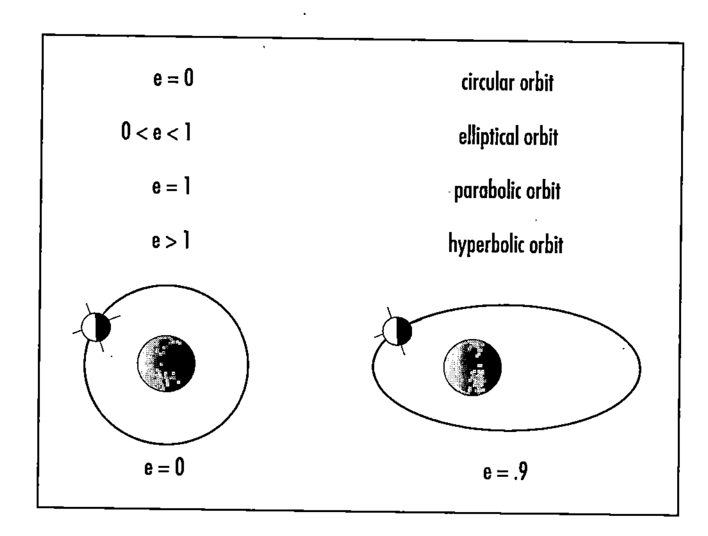

Eccentricity

eclipse 1. The reduction in visibility or the disappearance of a nonluminous body by passing into the shadow cast by another nonluminous body. 2. The apparent cutting off, wholly or partially, of the light from a luminous body by a dark (nonluminous) body coming between it and the observer.

The first type of eclipse is exemplified by a lunar eclipse, which occurs when the Moon passes through the shadow cast by Earth or when a satellite passes through the shadow cast by its planet; when a satellite passes directly behind its planet, an occultation is said to occur. The second type of eclipse is exemplified by a solar eclipse, which occurs when the Moon passes between the Sun and Earth.

eclipsing binary *See* BINARY STAR SYSTEM.

ecliptic (plane) The apparent annual path of the Sun among the stars; the intersection of the plane of Earth's orbit around the Sun with the celestial sphere. Because of the tilt in Earth's axis, the ecliptic is a great circle of the celestial sphere inclined at an angle of about 23.5° to the celestial equator.

See also CELESTIAL SPHERE; ZODIAC.

ecological system A habitable environment, either created artificially (as in a crewed space vehicle) or occurring naturally (i.e., the environment on the surface of Earth), in which human beings, animals, and/or other organisms can live in mutual relationship with one another and the environment. Under ideal circumstances the environment furnishes the sustenance for life, and the resulting waste products revert or cycle back into the environment to be used again for the continuous support of life. The interacting system of a biological community and its nonliving environmental surroundings. Also called ecosystem.

See also CLOSED ECOLOGICAL LIFE SUPPORT SYSTEM.

ecosphere The *continuously habitable zone* (CHZ) around a main sequence star of a particular luminosity in which a planet could support environmental conditions favorable to the evolution and continued existence of life. For the chemical evolution of Earthlike, carbon-based living organisms, global temperatures, and atmospheric pressure conditions must allow the retention of liquid water on the planet's surface. A viable ecosphere might lie between about 0.7 and 1.3 astronomical units from a star like the sun. However, if all the surface water has evaporated (the RUNAWAY GREENHOUSE EFFECT) or has completely frozen (the ICE CATASTROPHE), then any Earthlike planet within this ecosphere cannot sustain life.

See also CONTINUOUSLY HABITABLE ZONE; EXTRASOLAR PLANETS; LIFE IN THE UNIVERSE.

Eddington, Sir Arthur Stanley (1882–1944) British *Astronomer, Mathematician, Physicist* Sir Arthur Eddington helped create modern astrophysics. In May 1919 he led a solar eclipse expedition to Príncipe Island (West Africa) to measure the gravitational deflection of a beam of starlight as it passed close to the Sun. This effort provided early observational support for ALBERT EINSTEIN's general theory of relativity. In his 1933 publication *The Expanding Universe,* Eddington popularized the notion that the outer galaxies (spiral nebulas) were receding from one another as the universe expanded.

effector 1. In aerospace engineering, any device used to maneuver a rocket in flight, such as an aerodynamic surface, a gimbaled motor, or a control jet. 2. In robotics, the portion of a manipulator that causes the desired action, such as gripping or positioning.

See also ROBOTICS IN SPACE; ROCKET.

efficiency (symbol: η) In general, the ratio of useful output of some physical quantity to its total input. In thermodynamics, the ratio of energy output to energy input. For a heat engine, the thermal efficiency ($\eta_{thermal}$) is defined as the ratio of the work (mechanical energy) output to thermal energy (heat) input.

Ehricke, Krafft Arnold (1917–1984) German-American *Rocket Engineer* Krafft Ehricke was the visionary rocket engineer who conceived advanced propulsion systems for use in the American space program of the late 1950s and 1960s. One of his most important technical achievements was the design and development of the Centaur upper-stage rocket vehicle, the first American rocket vehicle to use liquid hydrogen (LH_2) as its propellant. The Centaur rocket vehicle made possible many important military and civilian space missions. As an inspirational space travel advocate, Ehricke's writings and lectures eloquently expounded upon the positive consequences of space technology. He anchored his far-reaching concept of an *extraterrestrial imperative* with the permanent human settlement of the Moon.

Krafft Ehricke was born in Berlin, Germany, on March 24, 1917. This was a very turbulent time because imperial Germany was locked in a devastating war with much of Europe and the United States. He grew up in the political and economic chaos of Germany's post–World War I Weimar Republic. Yet, despite the gloomy environment of a defeated and war-devastated Germany, Ehricke was able to develop his lifelong positive vision that space technology would serve as the key to improving the human condition. Following World War I a major challenge for his parents, both dentists, was that of acquiring schooling of sufficient quality to challenge their son. Unfortunately, Ehricke's frequent intellectual sparring contests with rigid Prussian schoolmasters did not help the situation and earned him a widely varying collection of grades.

By chance, at the age of 12 Ehricke saw Fritz Lang's 1929 motion picture *Die Frau im Mond* (The woman in the Moon). The Austrian filmmaker had hired the German rocket experts HERMANN OBERTH and WILLY LEY to serve as technical advisers during the production of this film. Oberth and Ley gave the film an exceptionally prophetic two-stage rocket design that startled and delighted audiences with its impressive blast-off. Ehricke viewed Lang's film at least a dozen times. Advanced in mathematics and physics for his age, he appreciated the great technical detail that Oberth and Ley had provided to make the film realistic. This motion picture served as Ehricke's introduction to the world of rockets and space travel, and he knew immediately what he wanted to do for the rest of his life. He soon discovered KONSTANTIN TSIOLKOVSKY's theoretical concept of a very efficient chemical rocket that used hydrogen and oxygen as its liquid propellants. While a teenager he also attempted to tackle Oberth's famous 1929 book *Roads to Space Travel* but struggled with some of the book's more advanced mathematics.

In the early 1930s he was still too young to participate in the German Society for Space Travel (*Verein für Raumschiffahrt,* or VfR), so he experimented in a self-constructed laboratory at home. As Adolf Hitler (1889–1945) rose to power in 1933, Ehricke, like thousands of other young Germans, became swept up in the Nazi youth movement. His free-spirited thinking, however, soon got him into difficulties and earned him an unenviable position as a conscripted laborer for the Third Reich. Just before the outbreak of World War II, he was released from the labor draft so he could attend the Technical University of Berlin. There he majored in aeronautics, the closest academic discipline to space technology. One of his professors was Hans Wilhelm Geiger (1882–1945), the noted German nuclear physicist. Geiger's lectures introduced Ehricke to the world of nuclear energy. Impressed, Ehricke would later recommend the use of nuclear power and propulsion in many of the space development scenarios he presented in the 1960s and 1970s.

Wartime conditions played havoc with Ehricke's attempt to earn a degree. While enrolled at the Technical University of Berlin, he was drafted for immediate service in the German army and sent to the western front. Wounded, he came back to Berlin to recover and resume his studies. In 1942 he obtained a degree in aeronautical engineering from the Technical University of Berlin, but while taking postgraduate courses in orbital mechanics and nuclear physics, he was again drafted into the German army, promoted to the rank of lieutenant, and ordered to serve with a panzer (tank) division on the eastern (Russian) front. But fortune played a hand, and in June 1942 the young engineer received new orders, this time reassigning him to rocket development work at Peenemünde. From 1942 to 1945 he worked there on the German army's rocket program under the overall direction of WERNHER VON BRAUN.

As a young engineer, Ehricke found himself surrounded by many other skilled engineers and technicians whose collective goal was to produce the world's first liquid-propellant ballistic missile, called the A-4 rocket. This rocket is better known as Hitler's Vengeance Weapon Two, or simply the V-2 rocket. After World War II the German V-2 rocket became the ancestor to many of the larger missiles developed by both the United States and the Soviet Union during the cold war.

Near the end of World War II Ehricke joined the majority of the German rocket scientists at Peenemünde and fled to Bavaria to escape the advancing Russian army. Swept up in Operation Paperclip along with other key German rocket personnel, Ehricke delayed accepting a contract to work on rockets in the United States by almost a year. He did this in order to locate his wife, Ingebord, who was then somewhere in war-torn Berlin. Following a long search and happy reunion, Ehricke, his wife, and their first child journeyed to the United States in December 1946 to begin a new life.

For the next five years Ehricke supported the growing U.S. Army rocket program at White Sands, New Mexico, and Huntsville, Alabama. In the early 1950s he left his position with the U.S. Army and joined the newly formed astronautics division of General Dynamics (formerly called Convair). There he worked as a rocket concept and design specialist and participated in the development of the Atlas, the irst American intercontinental ballistic missile. He became an American citizen in 1955.

Ehricke strongly advocated the use of liquid hydrogen as a rocket propellant. While at General Dynamics he recommended the development of a liquid hydrogen–liquid oxygen propellant upper-stage rocket vehicle. His recommendation became the versatile and powerful Centaur upper-stage vehicle. In 1965 he completed his work at General Dynamics as the director of the Centaur program and joined the advanced studies group at North American Aviation in Anaheim, California. From 1965 to 1968 this new position allowed him to explore pathways of space technology development across a wide spectrum of military, scientific, and industrial applications. The excitement of examining future space technologies and their impact on the human race remained with him for the rest of his life.

From 1968 to 1973 Ehricke worked as a senior scientist in the North American Rockwell space systems division in Downey, California. In this capacity he fully developed his far-ranging concepts concerning the use of space technology for the benefit of humankind. After departing Rockwell International he continued his visionary space advocacy efforts through his own consulting company, Space Global, located in La Jolla, California. As the U.S. government wound down Project Apollo in the early 1970s, Ehricke continued to champion the use of the Moon and its resources. His extraterrestrial imperative was based on the creation of a selenospheric (Moon-centered) human civilization in space. Until his death in late 1984, he spoke and wrote tirelessly about how space technology provides the human race the ability to create an unbounded "open world civilization."

Ehricke was a dedicated space visionary who not only designed advanced rocket systems (such as the Atlas-Centaur configuration) that greatly supported the first golden age of space exploration, but also addressed the important but frequently ignored social and cultural impacts of space technology. He created original art to communicate many of his ideas. He coined many interesting space technology terms. For example, Ehricke used the term *androsphere* to describe the synthesis of the terrestrial and extraterrestrial environments. The androsphere relates to our integration of Earth's biosphere—the portion of our planet that contains all the major terrestrial environmental regimes, such as the atmosphere, the hydrosphere, and the cryosphere—with the material and energy resources of the solar system. Similarly, the term *astropolis* is his concept for a large urban-like extraterrestrial facility that orbits in Earth-Moon (cislunar) space and supports the long-term use of the space environment for basic and applied research as well as for industrial development. Ehricke's *androcell* is an even bolder concept that involves a large human-made world in space, totally independent of the Earth-Moon system. These extraterrestrial city-states, with human populations of up to 100,000 or more, would offer their inhabitants the excitement of multigravity-level living at locations throughout heliocentric space.

Just weeks before his death on December 11, 1984, Ehricke served as a featured speaker at a national symposium on lunar bases and space activities for the 21st century held that October in Washington D.C. Despite being terminally ill, he traveled across the country to give a moving presentation on the importance of the Moon in creating a multiworld civilization for the human race. He ended his prophetic discussion by eloquently noting that "The Creator of our universe wanted human beings to become space travelers. We were given a Moon that was just far enough away to require the development of sophisticated space technologies, yet close enough to allow us to be successful on our first concentrated attempt."

See also APOLLO PROJECT; ATLAS LAUNCH VEHICLE; CENTAUR ROCKET; LUNAR BASES AND SETTLEMENTS; V-2 ROCKET.

Einstein, Albert (1879–1955) German-Swiss-American *Physicist* Like GALILEO GALILEI and SIR ISAAC NEWTON before him, Albert Einstein changed forever our view of the universe and how it functions. His theory of relativity (special relativity in 1905 and then general relativity in 1915) shaped modern physics. In 1921 he was awarded the Nobel Prize in physics and in 1933 escaped from Nazi Germany to the United States. Fearing nuclear weapon developments within Nazi Germany, in 1939 Einstein sent a personal letter to American president Franklin D. Roosevelt urging him to start an American nuclear weapons program—the history-changing effort later known as the Manhattan Project.

Born in Ulm, Germany, on March 14, 1879, Albert Einstein would eventually revolutionize our understanding of the universe, in much the same way that Galileo and Newton did during their own lives. He received his childhood education in both Germany and Switzerland. In 1901 he became a Swiss citizen and eventually received an appointment as a documents clerk in the Swiss patent office in Bern. This rather unchallenging job proved a blessing in disguise. It provided Einstein all the time he needed to do what he liked best—think about interesting physics problems. Because he enjoyed doing thought (*gedanken*) experiments, Einstein needed no laboratory to develop his incredibly powerful new insights into the operation of the universe.

Einstein's theory of space and time is now one of the foundations of modern physics. His famous theory of relativity falls into two general categories: the special theory of relativity, which he first proposed in 1905, and the general theory of relativity, which he presented in 1915.

The special theory of relativity deals with the laws of physics as seen by observers moving relative to one another at constant velocity—that is, by observers in nonaccelerating, or inertial, reference frames. Many experiments and scientific observations have proven the validity of Einstein's special theory of relativity.

In formulating special relativity Einstein proposed two fundamental postulates. The *First Postulate of Special Relativity* suggests that the speed of light (c) has the same value for all (inertial-reference-frame) observers, independent and regardless of the motion of the light source or the observers. The *Second Postulate of Special Relativity* states that all physical laws are the same for all observers moving at constant velocity with respect to each other. In 1905, while a clerk in the Swiss patent office, Einstein presented these bold postulates in a world-changing physics paper entitled "Zur Elektrodynamik bewegter Körper" ("On the electrodynamics of moving bodies"). The contents of this paper completely transformed 20th-century physics and astronomy.

From the theory of special relativity Einstein concluded that only a zero-rest-mass particle, like a photon, can travel at the speed of light. Another major consequence of special relativity is the equivalence of mass and energy, a relationship expressed in Einstein's famous formula $E = \Delta m\ c^2$. In this well-known equation, E is the energy equivalent of an amount of matter (Δm) that is annihilated or converted completely into pure energy, and c is the speed of light. Among many other important physical insights, this equation was the key that astronomers needed to understand energy generation in stellar interiors.

Einstein published several other important papers in 1905 and received his doctoral degree in physics as well. However, despite the growing recognition of the importance of these papers, Einstein still could not obtain a university position for another four years. By 1909 the University of Zurich finally offered him a low-paying position. Then, in 1913, through the influence of the famous German scientist MAX PLANCK, the young Einstein received a special professorship at the Kaiser Wilhelm Physical Institute in Berlin. For the first time in his life he could now pursue physics and receive an acceptable salary for his efforts.

Einstein wasted no time. In 1915 he introduced his general theory of relativity. He used this development to describe the space-time relationships of special relativity for cases where there was a strong gravitational influence, such as white dwarf stars, neutron stars, and black holes. One of Einstein's conclusions was that gravitation is not really a force between two masses (as Newtonian mechanics suggests), but rather arises as a consequence of the curvature of space-time. Einstein boldly suggested that in a four-dimensional universe (described by three spatial dimensions [x, y, and z] and time [t]), space-time became curved in the presence of matter, especially large concentrations of matter.

The fundamental postulate of general relativity states that the physical behavior inside a system in free-fall is indistinguishable from the physical behavior inside a system far removed from any gravitating matter—that is, the complete absence of a gravitational field. This very important postulate is also called *Einstein's Principle of Equivalence.*

Since its announcement in 1915, scientists have performed interesting experiments to confirm the general theory of relativity. These experiments included observation of the bending of electromagnetic radiation (starlight and radio-wave transmissions from distant spacecraft, such as NASA's Viking Project on Mars) by the Sun's immense gravitational field. Other observations included a recognition within the scientific community that the subtle perturbations (disturbances) in the orbit (at perihelion) of the planet Mercury are caused by the curvature of space-time in the vicinity of the Sun. Today scientists in the field of gravitation physics are planning new experiments to further test Einstein's relativistic theory of gravitation. Using very sensitive instruments on special spacecraft, such as NASA's *Gravity Probe B,* they hope to detect and interpret gravitational waves from cosmic sources. Such space-based research is very important because of the fundamental role of gravity in nature and its crucial influence on the evolution and destiny of the universe.

With the announcement of his general theory of relativity, Einstein's scientific reputation grew. In 1921 he received the Nobel Prize in physics for his "general contributions to physics and his discovery of the law of the photoelectric effect." At the time, his work on relativity was too sensational and "far out" for the conservative Nobel Prize committee to officially recognize. From this point on Einstein enjoyed the life of a world-famous scientific celebrity. By 1930 his best physics work was behind him. However, he continued to influence the world as a scientist-diplomat. When Adolf Hitler (1889–1945) rose to power in Germany in 1933, Einstein (a Jew) left his position at the Kaiser Wilhelm Physical Institute and sought refuge in the United States. Although a pacifist, at the urging of fellow physicists he sent an important letter in 1939 to President Franklin D. Roosevelt. In this letter Einstein urged the American president to have the United States develop an atomic bomb (based on nuclear fission) before Nazi Germany did. Because of Einstein's worldwide reputation, President Roosevelt paid special attention to his recommendation. Through the Manhattan Project in World War II, the United States developed and used the world's first atomic weapons.

In 1940 Einstein became an American citizen and also accepted a position at the Princeton Institute of Advanced Studies. He remained at this institution for the rest of his life, which ended on April 18, 1955. In his honor NASA named the second High-Energy Astronomy Observatory (HEAO-2) the *Einstein Observatory.* The spacecraft operated from 1978 to 1981, imaging the X-ray sky and making major contributions to X-ray astronomy.

See also EINSTEIN OBSERVATORY.

Einstein Observatory The second of NASA's High-Energy Astronomy Observatories, HEAO 2, renamed the *Einstein Observatory* after successful launch in November 1978 to honor the great physicist ALBERT EINSTEIN. The primary objectives of this mission were imaging and spectrographic studies of specific X-ray sources and studies of the diffuse X-ray background. The HEAO 2 spacecraft was identical to the HEAO 1 vehicle, with the addition of reaction wheels and associated electronics to enable the observatory to point its X-ray telescope at sources to an accuracy of within one min of arc.

The instrument payload had a mass of 1,450 kg. A large grazing-incidence X-ray telescope provided images of sources that were then analyzed by four interchangeable instruments mounted on a carousel arrangement that could be rotated into the focal plane of the telescope. The four instruments were a solid-state spectrometer (SSS), a focal plane crystal spectrometer (FPCS), an imaging proportional counter (IPC), and a high-resolution imaging detector (HRI). Also included in the science payload were a monitor proportional counter (MPC), which viewed the sky along the telescope axis, a broadband filter, and objective grating spectrometers that could be used in conjunction with focal plane instruments.

The major scientific objectives of the *Einstein Observatory* were to locate accurately and to examine with high spectral resolution X-ray sources in the energy range 0.2 to 4.0 keV and to perform high-spectral-sensitivity measurements with both high- and low-dispersion spectrographs. The science payload also performed high-sensitivity measurements of transient X-ray behavior. The spacecraft was a hexagonal prism 5.68 meters high and 2.67 meters in diameter. Downlink telemetry was accomplished at a data rate of 6.5 kilobits per second (kb/s) for real-time data and 128 kb/s for either of two tape recorder systems. An attitude control and determination subsystem was used to point and maneuver the spacecraft. The spacecraft also used gyroscopes, Sun sensors, and star trackers as sensing devices for pointing information and attitude determination.

An Atlas-Centaur rocket vehicle lifted off from Cape Canaveral Air Force Station on November 13, 1978 (at 05:24 UTC), and placed this scientific spacecraft into a 470-km-altitude orbit around Earth. The spacecraft's operational orbit was characterized by an inclination of 23.5° and a period of 94 minutes. The scientific mission lasted from November 1978 to April 1981. The *Einstein Observatory* was the first imaging X-ray telescope launched into space and completely changed the view of the X-ray sky. For example, the *Einstein Observatory* performed the first high-resolution spectroscopy and morphological studies of supernova remnants. Using HEAO-2 data, astronomers recognized that coronal emissions in normal stars are stronger than expected. The spacecraft resolved numerous X-ray sources in the Andromeda galaxy and the Magellanic Clouds. Furthermore, the spacecraft performed the first study of X-ray-emitting gases in galaxies and clusters of galaxies.

See also *CHANDRA X-RAY OBSERVATORY*; HIGH-ENERGY ASTRONOMY OBSERVATORY; X-RAY ASTRONOMY.

Eisele, Donn F. (1930–1987) American *Astronaut, U.S. Air Force Officer* Astronaut Donn Eisele flew onboard the *Apollo 7* mission in October 1968, the pioneering voyage in orbit around Earth of NASA's Apollo Project that ultimately landed human beings on the Moon between 1969 and 1972.

See also APOLLO PROJECT.

Eisenhower, Dwight D. (1890–1969) American *Army General, 34th President of the United States* Dwight D. Eisenhower was president of the United States between 1953 and 1961. Previously he had been a career U.S. Army officer and during World War II served as Supreme Allied Commander in Europe. Responding to the increasingly hostile cold war environment of the mid-1950s, President Eisenhower grew deeply interested in the use of space technology for national security. As a result he directed that intercontinental ballistic missiles (ICBMs) and reconnaissance satellites be developed on the highest national priority basis.

See also *DISCOVERER* SPACECRAFT; INTERCONTINENTAL BALLISTIC MISSILE; NATIONAL RECONNAISSANCE OFFICE; RECONNAISSANCE SATELLITE.

ejecta Any of a variety of rock fragments thrown out by an impact crater during its formation and subsequently deposited via ballistic trajectories onto the surrounding terrain. The deposits themselves are called ejecta blankets. It is also material thrown out of a volcano during an explosive eruption.

See also IMPACT CRATER.

ejection capsule 1. In an aircraft, aerospace vehicle, or crewed spacecraft, a detachable compartment serving as a cockpit or cabin that may be ejected as a unit and parachuted to the ground. 2. In a satellite, probe, or uncrewed spacecraft, a boxlike unit (usually containing records of observed data or experimental samples) that may be ejected and returned to Earth by a parachute or other deceleration device.

Elara A small satellite of Jupiter that is about 76 kilometers in diameter and orbits the giant planet in 259.7 days at an inclination of 28 degrees and a distance of 11,737,000 km. The American astronomer CHARLES DILLION PERRINE discovered this tiny moon in 1905. Also called Jupiter VII, the small natural satellite is 12th in order of distance from the planet. Some astronomers postulate that Himalia, Leda, Lysithea, and Elara could be the remnants of a single large asteroid that was captured by Jupiter and then broke up.

See also JUPITER.

electric propulsion The electric rocket engine is a device that converts electric power into a forward-directed force or thrust by accelerating an ionized propellant (such as mercury, cesium, argon, or xenon) to a very high exhaust velocity. The concept for an electric rocket is not new. In 1906 ROBERT GODDARD (the famous American rocket scientist) suggested that the exhaust velocity limit encountered with chemical rocket propellants might be overcome if electrically charged particles could be used as a rocket's reaction mass. This technical suggestion is often regarded as the birth of the electric propulsion concept. The basic electric propulsion system consists of three main components: (1) some type of electric thruster that accelerates the ionized propellant, (2) a suitable propellant that can be ionized and accelerated, and (3) a source of electric power. The acceleration of electrically charged particles requires a large quantity of electric power.

The needed power source may be self-contained, such as a space nuclear reactor, or it may involve the use of solar energy by photovoltaic or solar thermal conversion techniques. Electric propulsion systems using a nuclear reactor power supply are called *nuclear-electric propulsion (NEP)* systems, while those using a solar energy power supply are called *solar electric propulsion (SEP)* systems. Within the orbit of Mars both NEP and SEP systems can be considered, but well beyond Mars and especially for deep space missions to the edges of the solar system only nuclear-electric propulsion systems appear practical. This is due to the fact that the

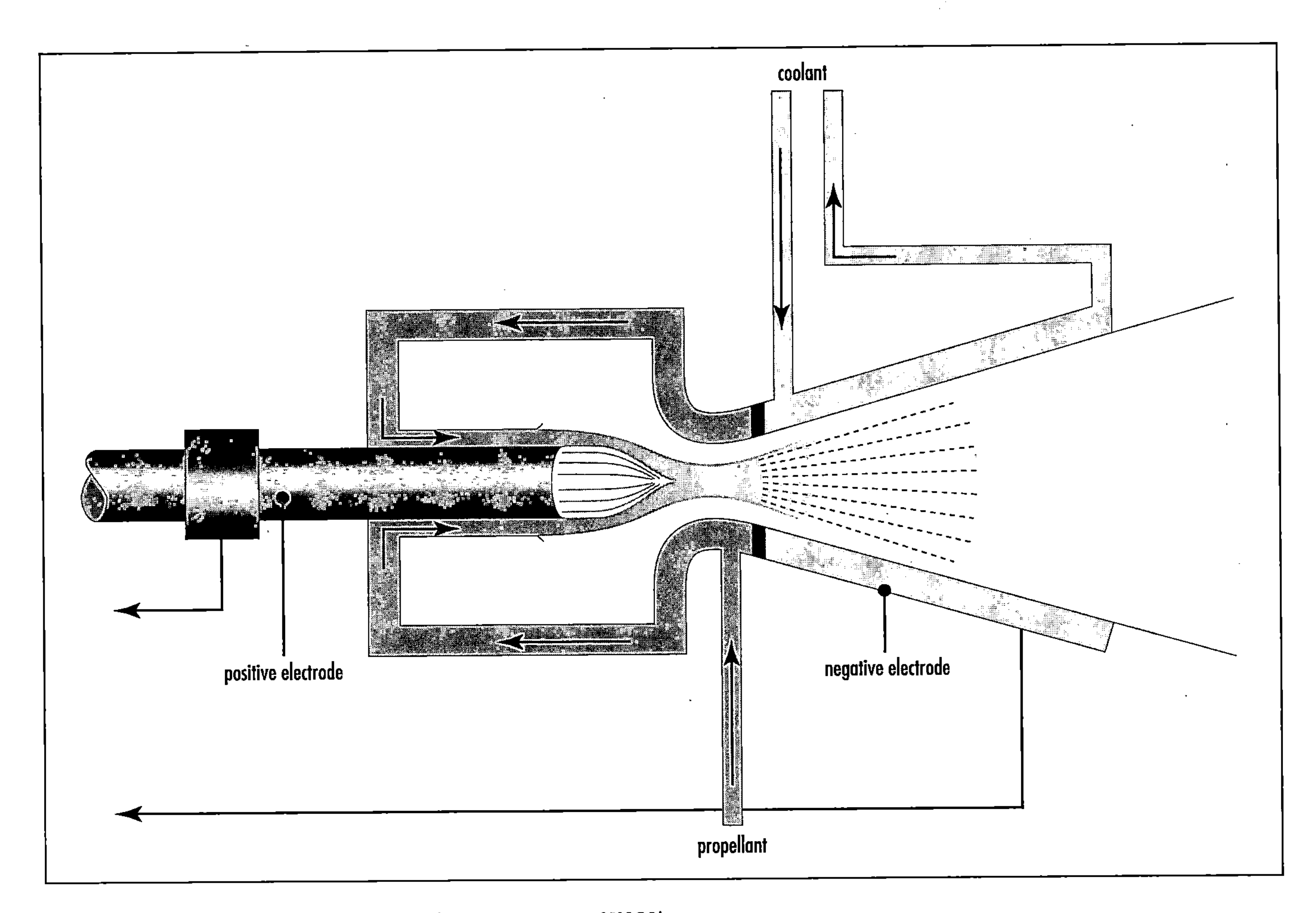

Basic components of an electrothermal rocket ***(Drawing courtesy of NASA)***

amount of solar energy available for collection and conversion falls off according to the inverse square law, that is, as 1 over the distance from the Sun squared [1/(distance)^2].

There are three general types of electric rocket engines: *electrothermal, electromagnetic,* and *electrostatic.* In the basic electrothermal rocket electric power is used to heat the propellant (e.g., ammonia) to a high temperature, and it is then expanded through a nozzle to produce thrust. (*See* figure.) Propellant heating may be accomplished by flowing the propellant gas through an electric arc (this type of electric engine is called an *arc-jet engine*) or by flowing the propellant gas over surfaces heated with electricity. Although the arc-jet engine can achieve exhaust velocities higher than those of chemical rockets, the dissociation of propellant gas molecules creates an upper limit on how much energy can be added to the propellant. In addition, other factors, such as erosion caused by the electric arc itself and material failure at high temperatures, establish further limits on the arc-jet engine. Because of these limitations, arc-jet engines appear more suitable for a role in orbital transfer vehicle propulsion and large spacecraft station keeping than as the electric propulsion system for deep space exploration missions.

The second major type of electric rocket engine is the electromagnetic engine or plasma rocket engine. (*See* figure.) In this type of engine the propellant gas is ionized to form a plasma, which is then accelerated rearward by the action of electric and magnetic fields. The *magnetoplasmadynamic (MPD)* engine can operate in either a steady state or a pulse mode. A high-power (approximately 1 megawatt-electric) steady-state MPD, using either argon or hydrogen as its propellant, is an attractive option for an electric propulsion orbital transfer vehicle (OTV).

The third major type of electric rocket engine is the electrostatic rocket engine or ion rocket engine. (*See* figure.) As in the plasma rocket engine, propellant atoms (i.e., cesium, mercury, argon, or xenon) are ionized by removing an electron from each atom. In the electrostatic engine, however, the electrons are removed entirely from the ionization region at the same rate as ions are accelerated rearward. The propellant ions are accelerated by an imposed electric field to a very high exhaust velocity. The electrons removed in the ionizer from the propellant atoms are also ejected from the spacecraft, usually by being injected into the ion exhaust beam. This helps neutralize the accumulated positive electric charge in the exhaust beam and maintains the ionizer in the electrostatic rocket at a high voltage potential.

In 1970 two 15-centimeter-diameter mercury-propellant ion thrusters were tested successfully in space onboard the *SERT 2* (*Space Electric Rocket Test-2*) spacecraft, which was placed in a 1,000-kilometer-altitude Sun-synchronous polar orbit. Each of these ion thrusters provided a maximum thrust of about 26.7 millinewton (mN). (A milli- is 10^{-3}.) The extended operation of the two thrusters demonstrated long-term ion thruster performance in the near-Earth orbital environment but also introduced the problem of "sputtering." Sputtering involves the buildup of molecular metal (in this case, mercury) contaminants on the solar arrays of the *SERT 2* spacecraft as a result of ion thruster exhaust plume contamination. Thruster experimentation was terminated in 1981 when the mercury propellant supply was exhausted.

The principal focus of the U.S. electric propulsion technology program now involves the J-series 30-cm mercury ion thruster. This type of electric rocket represents reasonably mature technology. However, because of the potential for pollution and contamination (i.e., sputtering phenomenon), mercury may not be an acceptable propellant for future heavy orbital transfer vehicle traffic operating from low-Earth orbit to destinations in cislunar space. Therefore, ion thrusters now are being developed that use argon or xenon as the propellant. Because of its potential for providing very high exhaust velocities (typically 10^5 meters per second) and very high efficiency, ion propulsion appears well-suited for interplanetary and deep-space exploration missions.

NASA's *Deep Space 1* (DS1) technology demonstration space probe was powered by two solar panel wings and an electric propulsion system. The robot spacecraft's 30-cm-diameter xenon ion engine used approximately 2,000 watts (W) of electric power to ionize the xenon gas and then accelerate these ions to about 31,500 meters per second. By ejecting these high-speed ions, this electric rocket produces some

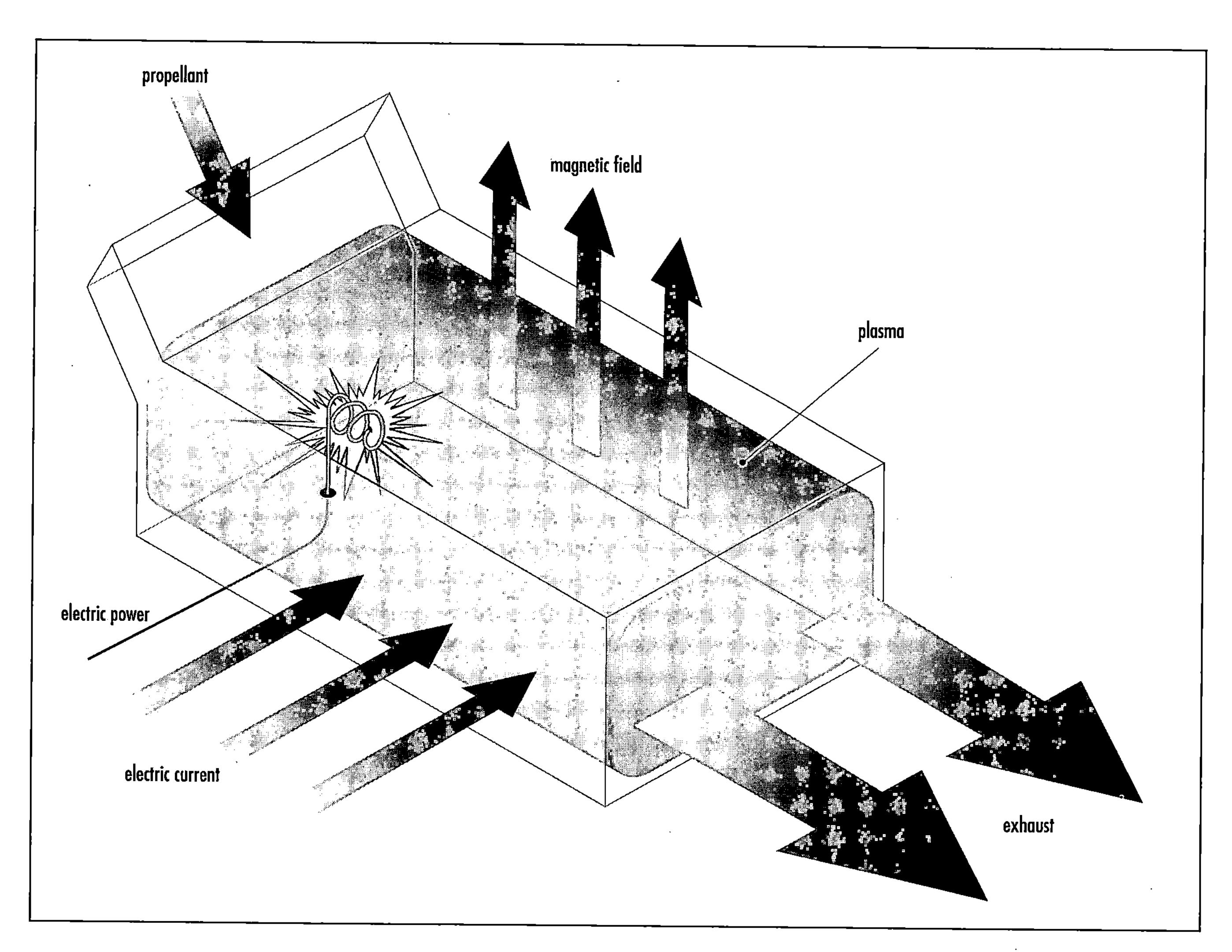

The fundamental components of an electromagnetic or plasma rocket engine ***(Drawing courtesy of NASA)***

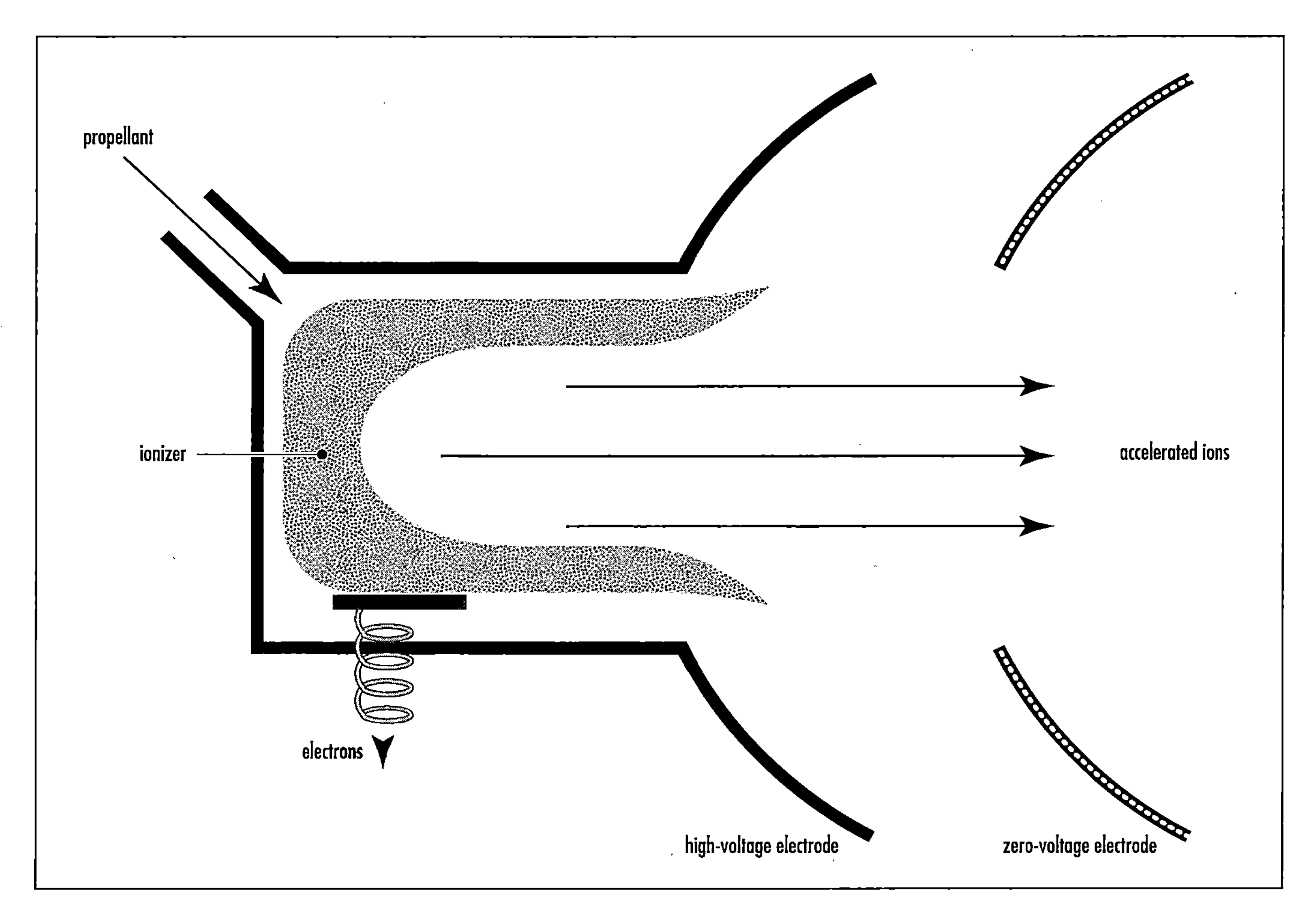

The fundamental components of an electrostatic or ion rocket engine ***(Drawing courtesy of NASA)***

0.09 newton (N) of thrust. The propellant tank contained an initial supply of 81.5 kilograms of xenon gas. This demonstration engine also produced about 0.02 N of thrust when functioning at the minimum operational electrical power level of 500 W.

Deep Space 1 was launched from Pad 17-A at the Cape Canaveral Air Force Station at 12:08 UT (8:08 A.M. EDT), on October 24, 1998, by a Delta II rocket. Once in space, the spacecraft's ion engine began to operate and allowed DS1 to perform a flyby of the near-Earth asteroid 9969 Braille at 04:46 UT (12:46 A.M. EDT) on July 29, 1999, at a distance of about 26 km and a relative velocity of approximately 15.5 km/sec. By the end of 1999 the ion engine had used approximately 22 kg of xenon to impart a total delta-V of 1,300 m/s to the spacecraft. The original plan was to fly by the dormant Comet Wilson-Harrington in January 2001 and then past Comet Borrelly in September 2001. But the spacecraft's star tracker failed on November 11, 1999, so mission planners drew upon techniques developed to operate the spacecraft without the star tracker and came up with a new extended mission to fly by Comet Borrelly. As a result of these innovative actions, on September 22, 2001, *Deep Space 1* entered the coma of Comet Borrelly and successfully made its closest approach (a distance of about 2,200 km) to the nucleus at about 22:30 UT (6:30 P.M. EDT). At the time of this cometary encounter DS1 was traveling at 16.5 km/s relative to the nucleus. Following this encounter NASA mission controllers commanded the spacecraft to shut down its ion engines on December 18, 2001, at about 20:00 UT (3:00 P.M. EST). Their action ended the *Deep Space 1* mission. All the space technologies flown on board DS1, including the 30-cm-diameter xenon ion engine, were successfully tested during the primary mission.

Space visionaries, starting with Robert Goddard, have recognized the special role electric propulsion could play in the conquest of space, namely, high-performance missions starting in a low-gravity field (such as Earth orbit or lunar orbit) and the vacuum of free space. In comparison to high-thrust, short-duration-burn chemical engines, electric propulsion systems are inherently low-thrust, high-specific-impulse rocket engines with fuel efficiencies two to 10 times greater than the propulsion efficiencies achieved using chemical propellants. Electric rockets work continuously for long periods, smoothly changing a spacecraft's trajectory. For missions to the outer solar system, the continuous acceleration provided by an electric propulsion thruster can yield shorter trip times and/or deliver higher-mass scientific payloads than those delivered by chemical rockets. NASA is now developing plans for an ambitious mission to orbit three planet-sized moons of Jupiter (Callisto, Ganymede, and Europa), which may harbor vast oceans beneath their icy surfaces. The *Jupiter Icy Moons Orbiter* (JIMO) mission would raise NASA's capability for space exploration to a revolutionary new level by pioneering the use of electric propulsion powered by a nuclear fission reactor.

See also JUPITER ICY MOONS ORBITER; NUCLEAR-ELECTRIC PROPULSION; ROCKET.

electrode A conductor (terminal) at which electricity passes from one medium into another. The positive electrode is called the *anode;* the negative electrode is called the *cathode.* In semiconductor devices, an element that performs one or more of the functions of emitting or collecting electrons or holes, or of controlling their movements by an electric field. In electron tubes, a conducting element that performs one or more of the functions of emitting, collecting, or controlling the movements of electrons or ions, usually by means of an electromagnetic field.

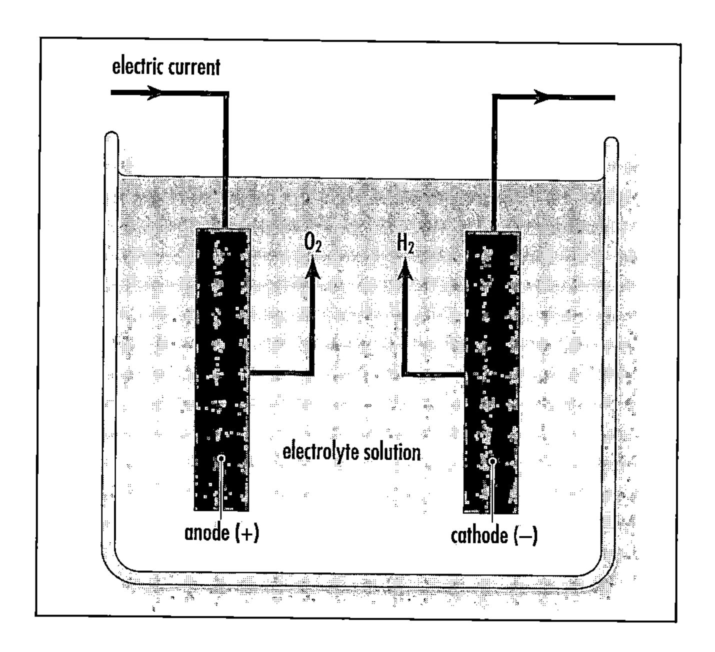

A basic electrolytic cell *(Drawing courtesy of the U.S. Department of Energy)*

electroexplosive device (EED) A pyrotechnic device in which electrically insulated terminals are in contact with, or adjacent to, a material mixture that reacts chemically (often explosively) when the required electrical energy level is discharged through the terminals. Explosive stage separation devices and missile self-destruct packages (for vehicles that have departed from an acceptable flight trajectory) are examples of EEDs used in aerospace applications.

See also EXPLOSIVE-ACTIVATED DEVICE.

electrolysis In general, a chemical reaction produced by the passage of an electric current through an electrolyte. Specifically, the process of splitting water into hydrogen and oxygen by means of a direct electric current. As shown in the figure, a basic electrolytic cell consists of two electrodes immersed in an aqueous conducting solution called an electrolyte. A source of direct current (dc) voltage is applied to the electrodes so that an electric current flows through the electrolyte from the anode (positive electrode) to the cathode (negative electrode). As a result, the water in the electrolyte solution is decomposed into hydrogen gas (H_2), which is released at the cathode, and oxygen gas (O_2), which is released at the anode. Although only the water is split, an electrolyte (for example, potassium hydroxide) is needed because water itself is a very poor conductor of electricity.

See also FUEL CELL; HYDROGEN.

electromagnetic Having both electric and magnetic properties; pertaining to magnetism produced or associated with electricity.

electromagnetic communications (EM) In aerospace, the technology involving the development and production of a variety of telecommunication equipment used for electromagnetic transmission of information over any media. The information may be analog or digital, ranging in bandwidth from a single voice or data channel to video or multiplexed channels occupying hundreds of megahertz. Included are onboard satellite communication equipment and laser communication techniques capable of automatically acquiring and tracking signals and maintaining communications through atmospheric and exoatmospheric media.

See also TELEMETRY.

electromagnetic force *See* FUNDAMENTAL FORCES IN NATURE.

electromagnetic gun (EM) A gun in which the projectile is accelerated by electromagnetic forces rather than by a chemical explosion as in a conventional gun.

electromagnetic launcher (EM) A device that can accelerate an object to high velocities using the electromotive force produced by a large current in a transverse magnetic field.

See also GUN-LAUNCH TO SPACE.

electromagnetic pulse (EMP) A large pulse of electromagnetic radiation, effectively reaching out to distances of hundreds of kilometers or more, caused by the interactions of gamma rays from a high-altitude nuclear explosion with atoms in the upper atmosphere. The resulting electromagnetic fields may couple with electrical and electronic systems to produce damaging current and voltage surges. EMP also can be created by nonnuclear means.

electromagnetic radiation (EMR) Radiation made up of oscillating electric and magnetic fields and propagated at the

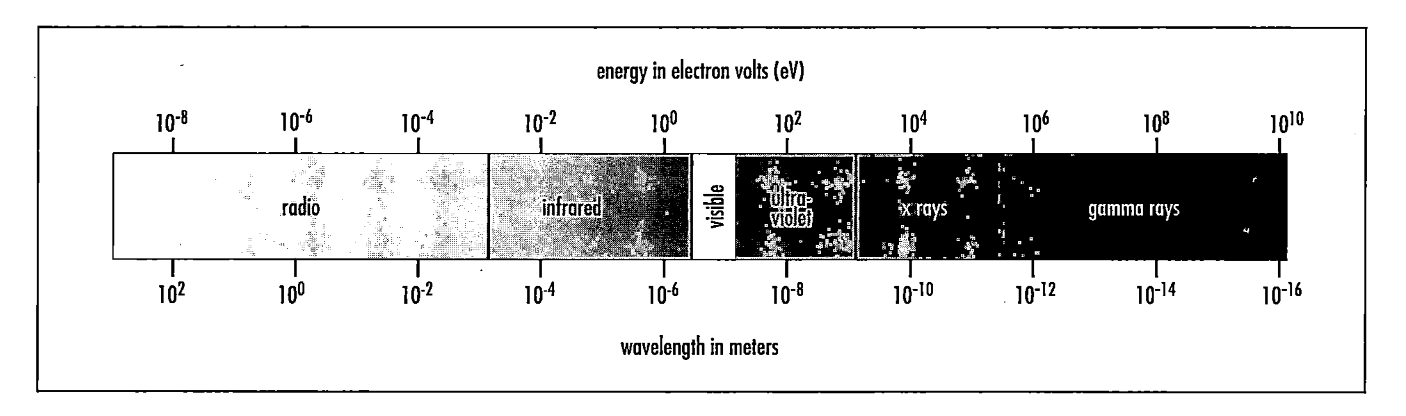

The electomagnetic (EM) spectrum

speed of light. Includes (in order of decreasing frequency) gamma radiation, X-ray, ultraviolet, visible, and infrared (IR) radiation and radar and radio waves.

See also ELECTROMAGNETIC SPECTRUM.

electromagnetic (EM) **spectrum** When sunlight passes through a prism, it throws a rainbowlike array of colors onto a surface. This display of colors is called the visible spectrum. It represents an arrangement in order of wavelength of the narrow band of electromagnetic (EM) radiation to which the human eye is sensitive.

The electromagnetic spectrum comprises the entire range of wavelengths of electromagnetic radiation, from the shortest-wavelength gamma rays to the longest-wavelength radio waves. (*See* figure.) The entire EM spectrum includes much more than meets the eye.

As shown in the figure, the names applied to the various regions of the EM spectrum are (going from shortest to longest wavelength) gamma ray, X-ray, ultraviolet (UV), visible, infrared (IR), and radio. EM radiation travels at the speed of light (i.e., about 300,000 kilometers per second) and is the basic mechanism for energy transfer through the vacuum of outer space.

One of the most interesting discoveries of 20th-century physics is the dual nature of electromagnetic radiation. Under some conditions electromagnetic radiation behaves like a wave, while under other conditions it behaves like a stream of particles, called *photons*. The tiny amount of energy carried by a photon is called a *quantum* of energy (plural: *quanta*). The word *quantum* comes to us from Latin and means "little bundle."

The shorter the wavelength, the more energy is carried by a particular form of EM radiation. All things in the universe emit, reflect, and absorb electromagnetic radiation in their own distinctive ways. The way an object does this provides scientists with special characteristics, or a signature, that can be detected by remote sensing instruments. For example, the spectrogram shows bright lines for emission or reflection and dark lines for absorption at selected EM wavelengths. Analyses of the positions and line patterns found in a spectrogram can provide information about an object's composition, surface temperature, density, age, motion, and distance.

For centuries astronomers have used spectral analyses to learn about distant extraterrestrial phenomena, but up until the space age they were limited in their view of the universe by the Earth's atmosphere, which filters out most of the EM radiation from the rest of the cosmos. In fact, ground-based astronomers are limited to just the visible portion of the EM spectrum and tiny portions of the infrared, radio, and ultraviolet regions. Space-based observatories now allow us to examine the universe in all portions of the EM spectrum. We now have examined the cosmos in the infrared, ultraviolet, X-ray, and gamma-ray portions of the EM spectrum and have made startling discoveries. We also have developed sophisticated remote sensing instruments to look back on Earth in many regions of the EM, providing powerful tools for a more careful management of the terrestrial biosphere.

See also REMOTE SENSING.

electron (symbol: e) A stable elementary particle with a unit negative electrical charge (1.602×10^{-19} coulomb) and a rest mass (m_e) of 1/1,837 that of a proton (namely, 9.109×10^{-31} kilogram). Electrons surround the positively charged nucleus and determine the chemical properties of the atom. Positively charged electrons, or positrons, also exist. The British scientist Sir J. J. Thomson (1856–1940) discovered the existence of the electron in the late 1890s.

electron volt (eV) A unit of energy equivalent to the energy gained by an electron when it experiences a potential difference of one volt. Larger multiple units of the electron volt are encountered frequently—as, for example, keV for thousand (or kilo-) electron volts (10^3 eV), MeV for million (or mega-) electron volts (10^6 eV), and GeV for billion (or giga-) electron volts (10^9 eV). One electron volt = 1.602×10^{-19} joules.

element A chemical substance that cannot be divided or decomposed into simpler substances by chemical means. A substance whose atoms all have the same number of protons (atomic number, Z) and electrons. There are 92 naturally occurring elements and more than 15 human-made, or transuranic, elements, such as plutonium (atomic number 94).

elementary particle(s) A fundamental constituent of matter. The atomic nucleus model of Ernest Rutherford (1871–1937) and James Chadwick's (1891–1974) discovery of the neutron suggested a universe consisting of three elementary particles: the proton, the neutron, and the electron. That simple model is still very useful in describing nuclear phenomena, because many of the other elementary particles are very short-lived and appear only briefly during nuclear reactions. To explain the strong forces that exist inside the nucleus between the NUCLEONs physicists developed *quantum chromodynamics* and introduced the QUARK as the basic building block of HADRONs, the class of heavy subatomic particles (including neutrons and protons) that experience this strong interaction or short-range nuclear force. Physicists call the other contemporary family of elementary particles with finite masses LEPTONs, light particles (including electrons) that participate in electromagnetic and weak interactions but not the strong nuclear force.

See also PHOTON; QUANTUM THEORY.

element set Specific information used to define and locate a particular satellite or orbiting object. Also called Keplerian elements.

See also KEPLERIAN ELEMENTS; ORBITS OF OBJECTS IN SPACE.

ellipse A smooth, oval curve accurately fitted by the orbit of a satellite around a much larger mass. Specifically, a plane curve constituting the locus of all points the sum of whose distances from two fixed points (called focuses or foci) is constant; an elongated circle. The orbits of planets, satellites, planetoids, and comets are ellipses; the center of attraction (i.e., the primary) is at one focus.

See also ORBITS OF OBJECTS IN SPACE.

elliptical galaxy A galaxy with a smooth, elliptical shape without spiral arms and having little or no interstellar gas and dust.

See also GALAXY.

elliptical orbit A noncircular Keplerian orbit.

See also ORBITS OF OBJECTS IN SPACE.

emission line A very small range of wavelengths (or frequencies) in the electromagnetic spectrum within which radiant energy is being emitted by a radiating substance. For example, the small band of wavelengths emitted by a low-density gas when it glows. The pattern of several emission lines is characteristic of the gas and is called the emission spectrum. Each radiating substance has a unique, characteristic emission spectrum.

See also ELECTROMAGNETIC SPECTRUM; REMOTE SENSING.

emission spectroscopy Analytical spectroscopic methods that use the characteristic electromagnetic radiation emitted when materials are subjected to thermal or electrical sources for purposes of identification. These thermal or electrical sources excite the molecules or atoms in the sample of material to energy levels above the ground state. As the molecules or atoms return from these higher energy states, electromagnetic radiation is emitted in discrete, characteristic wavelengths or emission lines. The pattern and intensity of these emission lines create a unique emission spectrum, which enables the analyst to identify the substance.

See also EMISSION LINE; REMOTE SENSING.

emission spectrum The distinctive pattern of emission lines produced by an element (or other material substance) when it emits electromagnetic radiation in response to heating or other types of energetic activation. Each chemical element has a unique emission spectrum.

See also REMOTE SENSING; SPECTRAL LINE; SPECTRUM.

emissivity (symbol: e or ε) The ratio of the radiant flux per unit area (sometimes called *emittance,* E) emitted by a body's surface at a specified wavelength (λ) and temperature (T) to the radiant flux per unit area emitted by a blackbody radiator at the same temperature and under the same conditions. The greatest value that an emissivity may have is unity (1)—the emissivity value for a blackbody radiator—while the least value for an emissivity is zero (0).

See also BLACKBODY; STEFAN-BOLTZMANN LAW.

empirical Derived from observation or experiment.

Encke, Johann Franz (1791–1865) German *Astronomer, Mathematician* In 1819 the German astronomer Johann Encke established the common identity of comets observed in 1786, 1795, 1805, and 1818. This faint comet has the shortest orbital period, namely 3.3 years, of any known comet. Astronomers refer to it as Comet Encke, after the mathematical astronomer who first calculated its orbit and resolved the mystery of its numerous apparitions in the era of telescope astronomy. The French astronomer Pierre François Andre Méchain (1744–1804) discovered Comet Encke in 1786. CAROLINE HERSCHEL observed it again in 1795. Similarly, the French astronomer JEAN PONS observed its passage in both 1805 and 1818. However, Encke performed the definitive calculations that proved that the comet was the same one earlier observed by other astronomers. In 1838 Encke discovered a gap near the outer edge of Saturn's A Ring, a feature now named the *Encke Division.* He was the director of the Berlin Observatory when the German astronomer JOHANN GALLE used that observatory to discover the planet Neptune on September 23, 1846.

See also COMET; NEPTUNE; SATURN.

encounter The close flyby or rendezvous of a spacecraft with a target body. The target of an encounter can be a natural celestial body (such as a planet, asteroid, or comet) or a human-made object (such as another spacecraft).

See also FLYBY; RENDEZVOUS; SPACECRAFT.

end effector The tool or "hand" at the end of a robot's arm or manipulator. This is the gripper, actuator, or mechanical device by which the robot physically grasps or acts upon objects.

See also ROBOTICS IN SPACE.

endoatmospheric Within Earth's atmosphere, generally considered to be at altitudes below 100 kilometers. *Compare with* EXOATMOSPHERIC.

See also ATMOSPHERE.

endoatmospheric interceptor In aerospace defense, an interceptor rocket that attacks incoming reentry vehicles during their terminal flight phase within Earth's atmosphere.

See also BALLISTIC MISSILE DEFENSE.

endothermic reaction In thermodynamics, a reaction in which thermal energy (heat) is absorbed from the surroundings; a reaction to which heat must be provided. *Compare with* EXOTHERMIC REACTION.

end-to-end test In aerospace operations and safety exercises, a full-up (i.e., complete) systems test of the ground and flight system's flight-termination (abort) subsystem with the destruct initiators disconnected.

See also ABORT.

energy (symbol: E) In general, the capacity to do work. Work is done when an object is moved by a force through a distance or when a moving object is accelerated. Energy can be manifested in many different forms, such as mechanical, thermal, chemical, electrical, and nuclear. The first law of thermodynamics states that energy is conserved, that is, it can neither be created nor destroyed, but simply changes form—including mass-energy transformations, in which matter is transformed into energy and vice versa according to ALBERT EINSTEIN's formula $E = \Delta m\ c^2$. The joule is the commonly encountered unit of energy in the SI unit system.

Potential energy arises from the position, state, or configuration of a body (or system). *Kinetic energy* is energy associated with a moving object. For example, in classical physics, a body of mass (m) moving at a velocity (v) has a kinetic energy of $E_{kinetic} = 1/2\ (m\ v^2)$. In thermodynamics, the concept of *internal energy* is used to provide a macroscopic description of the energy within an object due to the microscopic behavior of molecules and atoms. Similarly, in thermodynamic analyses, thermal energy (or heat) is treated as disorganized energy in transit, while work is treated as organized energy in transit. Thermodynamically, a system never stores work or heat, since these are perceived as transitory energy transfer phenomena that cease once a particular energy flow process has been completed.

Nuclear energy involves liberation of energy as a result of processes in the atomic nucleus, which often involve the transformations of small quantities of matter. *Solar energy* involves the Sun's electromagnetic energy output, while *geothermal energy* involves the thermal energy within Earth as a result of its molten core and the decay of certain radioactive nuclei in its crust.

The first law of thermodynamics of classical physics involves the conservation of energy principle. After Einstein introduced his famous mass-energy equivalence equation in 1905, modern physicists expanded this important principle and now postulate that mass and energy are conserved within the universe. One important cosmological implication is that all the energy and/or mass found in the present universe originated in the big bang event and has been changing forms ever since.

energy conversion efficiency In rocketry, the efficiency with which a nozzle converts the energy of the working substance (i.e., the propellant) into kinetic energy. Aerospace engineers commonly express this as the ratio of the kinetic energy of the jet leaving the nozzle to the kinetic energy of an ideal (hypothetical) jet leaving an ideal nozzle using the same working substance at the same initial state and under the same conditions of velocity and expansion.

See also ENERGY; ROCKET.

energy management In aerospace operations and rocketry, this term describes the monitoring of the fuel expenditure of a spacecraft or aerospace vehicle for the purposes of flight control and navigation.

energy satellite A very large space structure assembled in Earth orbit that takes advantage of the nearly continuous availability of sunlight to provide useful energy to a terrestrial power grid. As proposed most frequently, these large energy satellites would be located in geosynchronous orbit, where they would gather incoming sunlight (solar energy) and then convert it into electric energy by photovoltaic or solar-thermal conversion techniques. The electrical energy then would be transmitted to power grids on Earth or to other energy consumption locations in CISLUNAR space (e.g., a space-based manufacturing center) by beams of microwave or laser radiation.

See also SATELLITE POWER SYSTEM.

engine cut-off The specific time when a rocket engine is shut down during a flight. In the case of the space shuttle, for example, this time is referred to as MECO, or "main engine cut-off." Sometimes called *burnout.*

See also ROCKET; ROCKET ENGINE.

engine spray The part of a launch pad deluge system that cools a rocket engine and its exhaust during launch, thereby preventing thermal damage to the launch vehicle and the launch pad structure.

enthalpy (symbol: H, h) A property of a thermodynamic system often encountered in rocket engine design and performance analyses. Engineers use the following formula to define enthalpy:

$$H = U + pV$$

where H is the total enthalpy (joules), U is the total internal energy (joules), p is pressure (pascals), and V is volume (cubic meters). In many thermodynamic problems the enthalpy function does not lend itself to a specific physical interpretation, although it does have the dimension of energy (joules) or, as an intrinsic property (symbol: h), the dimensions of energy per unit mass (joules/kg). However, since U and the product (pV) appear together in many analyses, the enthalpy function is often an advantageous property to describe a process in which an energetic fluid flows through a thermodynamic system—as in the case of a heat engine or a rocket nozzle.

See also HEAT ENGINE; NOZZLE; ROCKET.

entropy (symbol: S, s) A measure of the extent to which the energy of a thermodynamic system is unavailable. In 1865 the German theoretical physicist RUDOLF CLAUSIUS developed the first comprehensive understanding of the second law of thermodynamics when he mathematically defined entropy as the thermodynamic function of state, an increase in which gives a measure of the energy of a system that has ceased to be available for work during a certain process. Clausius introduced the following important equation:

$$ds = [dq/T]_{REV}$$

where s is specific entropy (joules/kg-K), T is absolute temperature (K), and q is thermal energy (heat) input per unit mass (joules/kg) for reversible processes. Later in the 19th century the Austrian physicist LUDWIG BOLTZMANN and the American scientist Josiah Willard Gibbs (1839–1903) working independently developed a simplified statistical thermodynamics definition of entropy as a disorder or uncertainty indicator. For example, based on the work of Gibbs, entropy could now be defined as:

$$s = -k \Sigma (p_i \ln p_i)$$

where k is the Boltzmann constant and p_i is the probability of the *i*th quantum state of the system.

In the mid-20th century the American mathematician and electrical engineer Claude Shannon (1916–2001) applied the statistical concept of entropy to communication theory. He defined entropy in terms of the average information content (nonredundant) of an object. As a result of his innovative work, the concept of *information entropy* became the foundation of modern information theory.

The second law of thermodynamics is an inequality asserting that it is impossible to transfer thermal energy (heat) from a colder to a warmer system without the occurrence of other simultaneous changes in the two systems or in the environment. It follows from this important physical principle that during an adiabatic process, entropy cannot decrease. For reversible adiabatic processes, entropy remains constant, while for irreversible adiabatic processes, it increases. An equivalent formulation of the second law is that it is impossible to convert the heat of a system into work without the occurrence of other simultaneous changes in the system or its environment. This version of the second law, which requires a heat engine to have a cold sink as well as a hot source, is particularly useful in engineering applications. Another important statement of the second law is that the change in entropy (ΔS) for an isolated system is greater than or equal to zero. Mathematically,

$$(\Delta S)_{isolated} \geq 0$$

This deceptively simple mathematical inequality has profound implications in cosmology. For example, in the 19th century Clausius used the second law and his newly defined concept of entropy to describe the ultimate fate of the universe. First, he assumed that the universe was a closed system with a constant quantity of energy. He then postulated that the entropy of the universe would strive to achieve a maximum value in accordance with the laws of thermodynamics. The end state of the universe in his theoretical model is one of complete temperature equilibrium, with no energy available to perform any useful work. Physicists refer to this condition as the *heat death of the universe*. By blending the entropy concepts of Clausius and Shannon, some 21st-century cosmologists suggest that creatures with conscious intelligence (such as human beings) may have a role as the active agents of *negative entropy* within the universe. They might be able to apply information and knowledge to influence the fate of the universe and possibly prevent its ultimate heat death condition as suggested by the second law.

entry corridor The acceptable range of flight-path angles for an aerospace vehicle or space probe to enter a planetary atmosphere and safely reach the surface or a predetermined altitude in the planet's atmosphere at which parachutes or other soft-landing assistance devices would be employed. If the flight-path angle were too steep (i.e., greater than the entry corridor range), the aerospace vehicle or probe would encounter excessive aerodynamic heating as it plunged into the thickening atmosphere and burn up. If the flight path angle were too shallow (i.e., less than the entry corridor range), the vehicle or probe would bounce off the atmosphere, much like a smooth pebble skipping across a pond, and go past the target planet back into outer space. For example, a space shuttle orbiter undergoes its gliding atmospheric entry at an initial velocity of about 7.5 kilometers per second. The flight-path angle is typically –1.2° at the entry interface altitude of 122 kilometers.

See also SPACE TRANSPORTATION SYSTEM.

environmental chamber A test chamber in which temperature, pressure, fluid content and composition, humidity, noise, and movement can be controlled so as to simulate the different operational environments in which an aerospace component, subsystem, or system might be required to perform.

See also ANECHOIC CHAMBER; SHAKE-AND-BAKE TEST.

environmental satellite *See* EARTH-OBSERVING SPACECRAFT.

eolian Pertaining to, carried by, or caused by the wind. For example, eolian sand dunes on the surface of Mars.

ephemeris A collection of data about the predicted positions (or apparent positions) of a celestial object, including an artificial satellite, at various times in the future. A satellite ephemeris might contain the orbital elements of the satellite and any predicted changes.

See also KEPLERIAN ELEMENTS; ORBITS OF OBJECTS IN SPACE.

epicycle A small circle whose center moves along the circumference of a larger circle, called the deferent. Ancient astronomers such as PTOLEMY used the epicycle in an attempt to explain the motions of celestial bodies in their geocentric (nonheliocentric) models of the solar system.

epsilon (ε) star Within the system of stellar nomenclature introduced in 1603 by the German astronomer JOHANN BAYER (1572–1625), the fifth-brightest star in a constellation.

See also CONSTELLATION.

equation of state (EOS) An equation relating the thermodynamic temperature (T), pressure (p), and volume (V) for an amount of substance (n) in thermodynamic equilibrium. A large number of such equations have been developed. Of these, perhaps the simplest and most widely known is the ideal or perfect gas equation of state:

$$p V = n R T$$

where R is the universal gas constant, which has a value of 8314.5 joules per kilogram-mole per degree kelvin [J/(kmol-K)].

equations of motion A set of equations that gives information regarding the motion of a body or of a point in space as a function of time when initial position and initial velocity are known.

See also NEWTON'S LAWS OF MOTION.

equator An imaginary circle around a celestial body that is everywhere equidistant (90 degrees) from the poles and defines the boundary between the northern and southern hemispheres.

equatorial bulge The excess of a planet's equatorial diameter over its polar diameter. The increased size of the equatorial diameter is caused by centrifugal force associated with rotation about the polar axis.

equatorial coordinate system The most commonly used coordinate system in astronomy for giving the position of an object on the celestial sphere. The coordinates typically used are the RIGHT ASCENSION (RA), which is roughly the equivalent of longitude on Earth, and the DECLINATION (δ), which is the roughly equivalent of latitude on Earth. Astronomers also use the HOUR ANGLE (HA), north polar distance, and south polar distance to locate a celestial object using this coordinate system.

See also CELESTIAL SPHERE.

equatorial mounting A telescope mount that has one axis (called the polar axis) parallel to Earth's axis and the other axis (called the declination axis) at right angles to it. Astronomers attach the telescope to the declination axis and then counteract Earth's rotation by simply rotating the mount around the polar axis. As shown in the figure, the great advantage of this mount occurs if the polar axis is driven by a clock motor that turns it once each SIDEREAL DAY in the direction opposite to Earth's

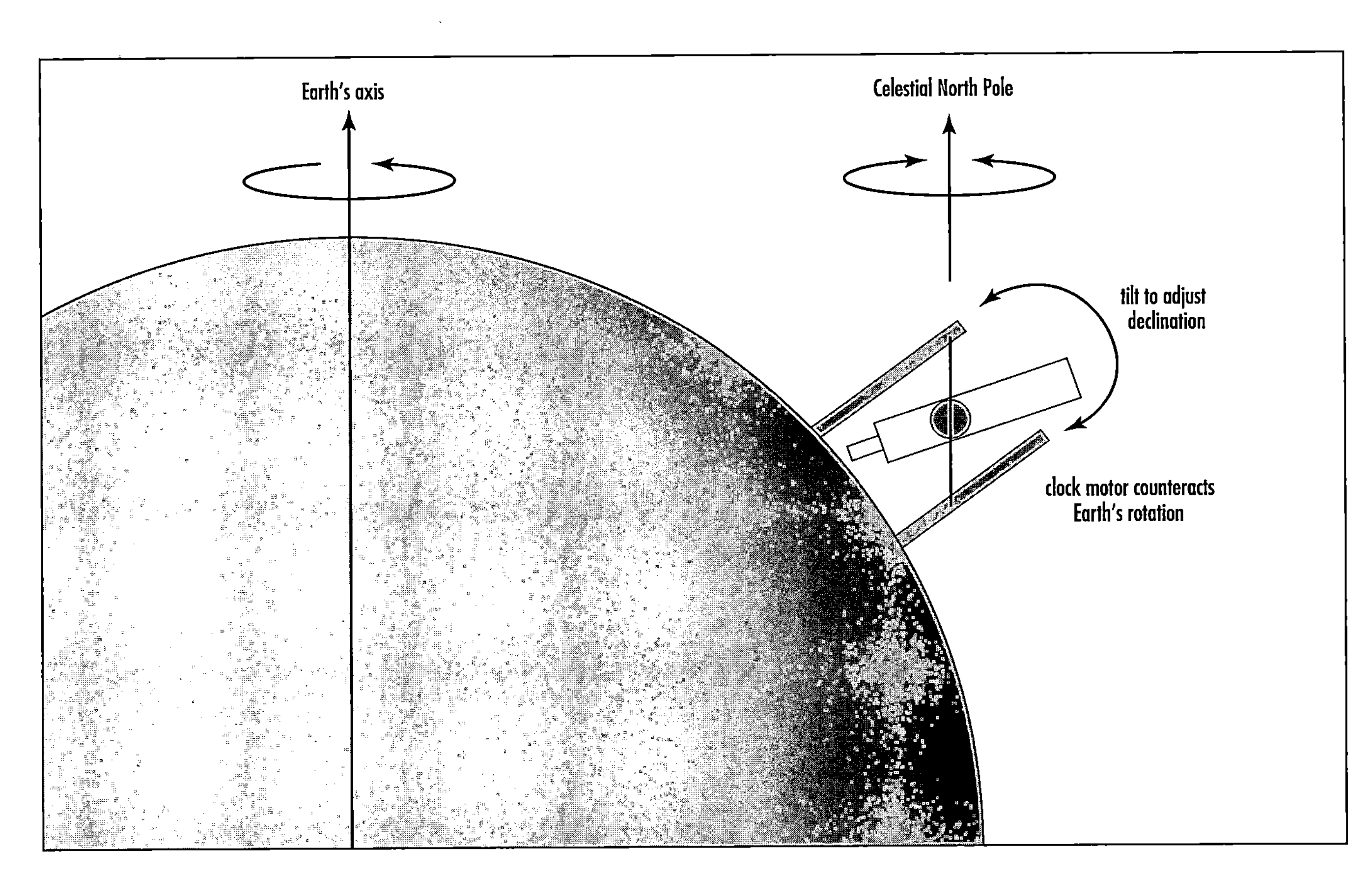

The basic equatorial mounting ***(Courtesy of NASA)***

rotation. Under such conditions any star will remain stationary in the telescope's field of view. From the mid-19th century until recently all large telescopes used equatorial mountings, which came in a variety of designs—including English mounting, fork mounting, German mounting, horseshoe mounting, Springfield mounting, and yoke mounting.

See also TELESCOPE.

equatorial orbit An orbit with an inclination of zero degrees. The plane of an equatorial orbit contains the equator of the primary body.

See also ORBITS OF OBJECTS IN SPACE.

equatorial satellite A satellite whose orbital plane coincides, or nearly coincides, with the equatorial plane of the primary body.

equinox One of two points of intersection of the ecliptic and the celestial equator that the Sun occupies when it appears to cross the celestial equator (that is, has a declination of 0°). In the Northern Hemisphere the Sun appears to go from south to north at the vernal equinox, which occurs on or about March 21st. Similarly, the Sun appears to travel from north to south at the autumnal equinox, which occurs on or about September 23. The dates are reversed in the Southern Hemisphere.

equivalence principle The fundamental postulate of ALBERT EINSTEIN's general relativity is the *principle of equivalence,* which states that the physical behavior inside a system in free-fall is indistinguishable from the physical behavior inside a system far removed from any gravitating matter (that is, the complete absence of a gravitational field).

See also RELATIVITY.

Eratosthenes of Cyrene (ca. 276–ca. 194 B.C.E.) Greek *Astronomer, Mathematician, Geographer* Eratosthenes made a remarkable attempt to measure the circumference of Earth in about 250 B.C.E. Recognizing that Earth curved, he used the difference in latitude between Alexandria and Aswan, Egypt (about 800 km apart), and the corresponding angle of the Sun at zenith at both locations during the summer solstice. Some historic interpretations of the *stadia* (an ancient unit) suggest his results were about 47,000 km or less, versus the true value of 40,000 km.

erector A vehicle used to support a rocket for transportation and to place the rocket in an upright position within a gantry.

erg The unit of energy or work in the centimeter-gram-second (c.g.s.) unit system. An erg is the work performed by a force of 1 dyne acting through a distance of 1 centimeter.

$$1 \text{ erg} = 10^{-7} \text{ joule}$$

ergometer A bicyclelike instrument used for measuring muscular work and for exercising "in place." Astronauts and cosmonauts use specially designed ergometers to exercise while on extended orbital flights.

This close-up look at Eros is a mosaic of four images taken by NASA's *NEAR-Shoemaker* spacecraft on September 5, 2000, from a distance of about 100 kilometers above the asteroid. The knobs sticking out of the surface near the top of the image surround a boulder-strewn area and are most likely remnants of ancient impact craters. *(Courtesy of NASA, JPL, and Johns Hopkins University/Applied Physics Laboratory)*

Eros The near-Earth asteroid discovered on August 13, 1898, by the German astronomer Gustav Witt (1866–1946). The French astronomer Auguste Charlois (1864–1910) independently discovered Eros on the same night that Gustav Witt did, but the German astronomer reported his findings first and so received credit for the discovery of this AMOR GROUP asteroid. Named for the god of love in Greek mythology, this inner solar system asteroid orbits the Sun at an average distance of 218.7 million kilometers. It often comes close to our planet. For example, in 1975 Eros passed within 23 million kilometers of Earth. NASA's Near-Earth Asteroid Rendezvous (NEAR) mission went into orbit around this interesting minor planet on February 14, 2000, and then touched down in the "saddle region" of Eros on February 12, 2001. The NEAR mission was the first to land a spacecraft on an asteroid and provided a great deal of scientific data. Eros is about the same age as the Sun, namely 4.5 billion years. It is an irregularly shaped asteroid 13 km by 13 km by 33 km. It has a mass of 7.2×10^{18} kilograms and a period of 1.76 years. Also called Eros-433.

See also ASTEROID; NEAR-EARTH ASTEROID; NEAR-EARTH ASTEROID RENDEZVOUS MISSION.

erosion 1. In general, the progressive loss of original material from a solid due to the mechanical interaction between that surface and a fluid or the impingement of liquid droplets or solid particles. 2. In planetary geology, the collective group of processes whereby rock material is removed and transported. The four major agents of erosion are running water, wind, glacial ice, and gravity. 3. In environmental science, the wearing away of the land surface by wind and water. On Earth erosion occurs naturally from weather or runoff, but human actions, such as land-clearing practices associated with farming, residential and industrial development, road building, and timber cutting, can intensify the process.

erosive burning The increased burning of solid propellant that results from the scouring influence of combustion products moving at high velocity over the burning surface.

See also ROCKET.

escape tower A trestle tower placed on top of a crew (space) capsule, which during liftoff connects the capsule to the escape rocket. The escape rocket is a small rocket engine attached to the leading end of an escape tower that is used to provide additional thrust to the crew capsule to obtain separation of this capsule from the expendable booster vehicle in the event of a launch pad abort or emergency. After a successful liftoff and ascent, the escape tower and escape rocket are separated from the capsule. (*See* figure.)

See also ABORT; MERCURY PROJECT.

A close-up view of the escape tower attached to astronaut John Glenn's *Friendship 7* space capsule as the two were mated to an Atlas rocket at Cape Canaveral in February 1962. Glenn became the first American to orbit Earth in a spacecraft as part of NASA's Mercury Project. *(Courtesy of NASA)*

Escape Velocity for Various Objects in the Solar System

Celestial Body	Escape Velocity (V_e) (km/sec)
Earth	11.2
Moon	2.4
Mercury	4.3
Venus	10.4
Mars	5.0
Jupiter	~ 61
Saturn	~ 36
Uranus	~ 21
Neptune	~ 24
Pluto	~ 1
Sun	~ 618

Source: Developed by the author, based on NASA and other astrophysical data.

escape velocity (symbol: V_e) The minimum velocity needed by an object to climb out of the gravity well (overcome the gravitational attraction) of a celestial body. From classical Newtonian mechanics, the escape velocity from the surface of a celestial body of mass (M) and radius (R) is given by the equation

$$V_e = \sqrt{(2\,G\,M)/R}$$

where G is the universal constant of gravitation (6.672×10^{-11} N–m^2/kg^2), M is the mass of the celestial object (kg), and R is the radius of the celestial object (meters). Here N is the newton, the basic unit of force in the SI system. The Escape Velocity Table presents the escape velocity for various objects in the solar system (or an estimated equivalent V_e for those celestial bodies that do not possess a readily identifiable solid surface, such as the giant outer planets and the Sun).

eta (η) star Within the system of stellar nomenclature introduced in 1603 by the German astronomer JOHANN BAYER (1572–1625), the seventh-brightest star in a constellation.

See also CONSTELLATION.

Eudoxus of Cnidus (ca. 400–ca. 347 B.C.E.) Greek *Astronomer, Mathematician* Eudoxus of Cnidus was an ancient Greek astronomer and mathematician who was the first to suggest a geocentric cosmology in which the Sun, Moon, planets, and fixed stars moved around Earth on a series of 27 giant geocentric spheres. CALLIPPUS OF CYZICUS, ARISTOTLE, and PTOLEMY embraced geocentric cosmology, though each offered modifications to Eudoxus's model.

See also COSMOLOGY.

Euler, Leonhard (1707–1783) Swiss *Mathematician, Physicist* Among his many scientific accomplishments, Leonhard Euler developed advanced mathematical methods for observational astronomy that supported precise predictions of the motions of the Moon, the planets, and comets. His work in optics exerted significant influence on the technical development of telescopes in the 18th and 19th centuries.

Europa The smooth, ice-covered moon of Jupiter discovered by GALILEO GALILEI in 1610 and currently thought to have a liquid water ocean beneath its frozen surface. Europa has a diameter of 3,124 kilometers and a mass of 4.84×10^{22} kilograms. It is in synchronous orbit around Jupiter at a distance of 670,900 kilometers. An eccentricity of 0.009, an inclination of 0.47°, and a period of 3.551 (Earth) days further characterize its orbit around Jupiter. The acceleration of gravity on the surface of Europa is 1.32 m^2/s, and the icy moon has an average density of 3.02 g/cm^3. Next to Mars, exobiologists currently favor Europa as a candidate world to possibly harbor some form of alien life in our solar system.

European Space Agency (ESA) An international organization that promotes the peaceful use of outer space and cooperation among the European member states in space research and applications. ESA's 15 member states are Austria, Belgium, Denmark, Finland, France, Germany, Ireland, Italy, the Netherlands, Norway, Portugal, Spain, Sweden, Switzerland, and the United Kingdom. Greece and Luxembourg are expected to become full members of ESA by December 2005. The agency also joins in cooperative ventures with other spacefaring nations such as the United States, Russia, Canada, and Japan.

ESA has its headquarters in Paris, France. The European Space Research and Technology Center (ESTEC) is situated in Noordwijk, the Netherlands, and serves as the design hub for most ESA spacecraft. The European Space Operations

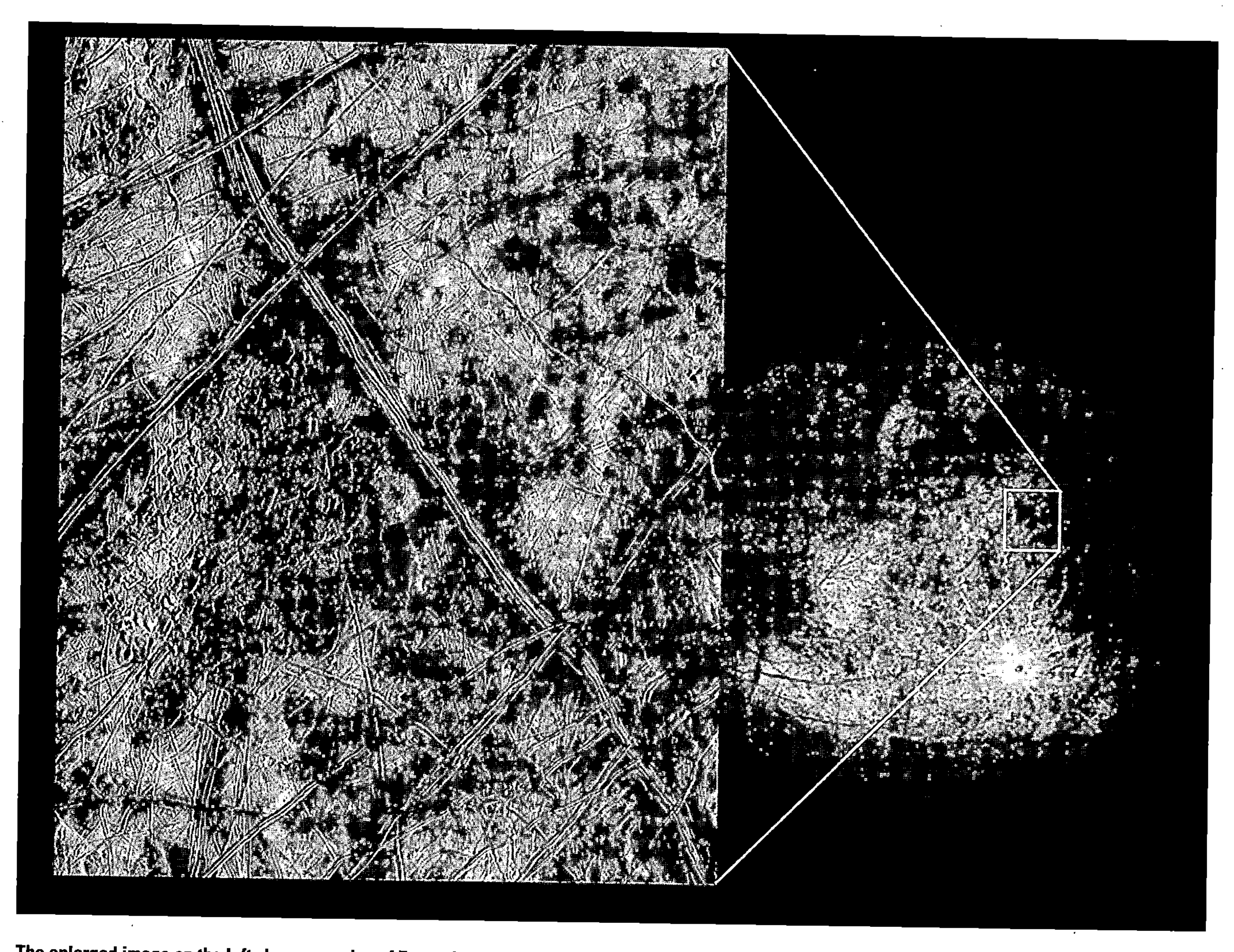

The enlarged image on the left shows a region of Europa's crust made up of blocks that planetary scientists think have broken apart and "rafted" into new positions. These features are the best geological evidence to date that Europa (seen in a global perspective on the right) may have had a subsurface ocean at some time in its past. Combined with the geological data, the presence of a magnetic field leads scientists to believe an ocean is most likely present on Europa today below its icy surface. Long, dark lines are ridges and fractures in the crust, some of which are more than 3,000 kilometers long. These images were obtained by NASA's *Galileo* spacecraft during September 1996, December 1996, and February 1997 at a distance of 677,000 kilometers. *(Courtesy of NASA/JPL)*

Center (ESOC) is located in Darmstadt, Germany, and is responsible for operating and controlling ESA satellites in orbit. The European Space Research Institute (ESRIN) is located in Frascati, Italy (near Rome), and serves as ESA's information technology center. In particular, ESRIN collects, stores, and distributes satellite-derived scientific data among the member states. The European Astronauts Center (EAC) is situated in Cologne, Germany, and trains European astronauts for future missions. ESA has liaison offices in the United States, Russia, and Belgium. In addition, ESA maintains a launch complex in Kourou, French Guiana, and tracking stations in various areas around the world.

Today ESA's involvement spans the fields of space science, Earth observation, telecommunications, space segment technologies (including orbital stations and platforms), ground infrastructures, and space transportation systems, as well as basic research in microgravity. ESA has its own corps of astronauts. They have participated on U.S. space shuttle missions involving *Spacelab,* have flown on-board the Russian *Mir* space station as part of a Euro-*Mir* mission program, and have visited the *International Space Station* (ISS).

The driving force behind the decision of European governments to coordinate and pool their efforts in joint space endeavors was quite simply economics. No individual European country could afford to individually sponsor a complete range of space projects and all the necessary infrastructure, although some countries, such as France, maintain thriving national space programs in addition to participation in ESA.

Eutelsat Eutelsat is one of the world's leading providers of satellite-based telecommunications infrastructure. The company commercializes information transfer capacity through the use of 24 communications satellites that provide coverage from the Americas to the Pacific and serve up to 90 per cent of the world population in more than 150 countries. Eutelsat's satellites offer a broad range of services, including television and radio broadcasting for the consumer public, professional video broadcasting, corporate networks, Internet services, and mobile communications. Eutelsat has also developed a suite of broadband services for local communities and businesses.

Throughout its 25 years of existence, Eutelsat has placed innovation at the center of its development. The company was the first in Europe to distribute digital satellite television, using the Digital Video Broadcasting (DVB) standard, which is now a global reference for digital broadcasting. It also played a pioneering role in the creation and promotion of two-way broadband terminals for business and local community applications.

In 2001 Eutelsat restructured as a company incorporated under French law. The firm markets its services through a network of partners that include leading telecommunications operators such as Belgacom, BT, Deutsche Telekom, France Telecom (GlobeCast), Telespazio, Xantic, and information transfer service providers, such as Hughes Network Systems. Anchor broadcast users include public and private television channels such as the BBC, Sky Italia, France Télévision, Deutsche Welle, and CNN.

Active for more than 25 years in the field of commercial satellite communications, Eutelsat owns one of the youngest fleets in the world. From geostationary orbit positions between 15° west and 70.5° east, the company provides coverage across four continents, spanning Europe, the Middle East, Africa, Asia, eastern North America, and South America. Eutelsat's system is based on 24 satellites, of which 20 are fully operated by the company. The *HOT BIRD* satellites represent Eutelsat's flagship spacecraft constellation. With five such satellites, Eutelsat is a leading global satellite service provider in terms of number of channels broadcast. For example, a total of 100 million households are now equipped to view channels broadcast by the *HOT BIRD* satellites, either through direct-to-home reception or by cable. In addition to a Ku-band payload, *HOT BIRD 6* carries a Ka-band payload, which can be combined with onboard multiplexing services for fully meshed networks using small terminals.

With a powerful beam focused on the British Isles and two steerable beams focused on Germany, *EUROBIRD 1* provides premium broadcasting capacity from 28.5° east and provides information services to more than 6 million digital homes in the United Kingdom. This satellite is also used to provide business services and channel delivery to *cable headends* (information transfer ground stations) in continental Europe. After four years operating at 13° east as *HOT BIRD™ 5,* the renamed high-power *EUROBIRD 2* satellite is now located at 25.5° east. *W1, W2, W3,* and *W5 Eutelsat* satellites combine considerable signal power delivered to the ground with wide coverage and steerable beams over Europe, Africa, the Middle East, and western Asia. *W5 Eutelsat* has made a significant contribution to the expansion of Eutelsat's coverage of the Far East through a reach to Japan and as far as Australia. *W4 Eutelsat* is configured with high-power beams for pay-TV services and for consumer Internet access in Africa and Russia. *W3A Eutelsat,* one of Eutelsat's most sophisticated satellites, combines Ku and Ka band frequencies and onboard multiplexing. This satellite was launched in March 2004 to serve markets in Europe, the Middle East, and Africa.

With coverage over Europe as far as Siberia and a spotbeam over India and neighboring regions, *SESAT 1* offers a direct connection between Europe and Asia for a wide range of telecommunications services. Similarly, with a highly flexible configuration of fixed and steerable beams, *SESAT 2* provides high-power Ku-band capacity over Europe, Africa, the Middle East, and Central Asia. *SESAT 2* can deliver telecommunications, broadband, and broadcasting services via 12 Ku-band transponders. This spacecraft is an example of the international nature of the global telecommunications industry. The satellite has a total of 24 transponders, 12 of which are referred to as *SESAT 2* and are leased to Eutelsat by the Russian Satellite Communications Company (RSCC). The remaining 12 transponders, with domestic coverage of the Russian Federation, are commercialized by the RSCC under the name *EXPRESS AM22.*

Eutelsat has grouped the three *ATLANTIC BIRD* communications satellites in a region located between 12.5° west and 5° west. Their primary mission is to offer a seamless link between the American continent, Europe, Africa, and western

Asia. The *ATLANTIC BIRD 3* satellite has also allowed Eutelsat to enter the C-band market, with 10 transponders providing full coverage of Africa.

To keep pace with the almost insatiable global demand for information services, Eutelsat deployed the first communications satellite specifically designed for 2-way broadband Internet services. Called *e-BIRD*, this satellite was launched in September 2003 and placed in geostationary orbit at 33° east to provide coverage of Europe and Turkey through four high-power beams.

Evans, Ronald E. (1933–1990) American *Astronaut, U.S. Navy Officer* Astronaut Ronald Evans served as the command module pilot during NASA's *Apollo 17* lunar landing mission (December 1972). He maintained a solo vigil in orbit around the Moon while fellow astronauts EUGENE CERNAN and HARRISON SCHMITT completed their explorations of the Taurus-Littrow landing area on the lunar surface.

evaporation The physical process by which a liquid is transformed to the gaseous state (i.e., the vapor phase) at a temperature below the boiling point of the liquid. *Compare with* SUBLIMATION.

evapotranspiration The loss (discharge) of water from Earth's surface to its atmosphere by evaporation from bodies of water or other surfaces (e.g., the soil) and by transpiration from plants.

See also EARTH SYSTEM SCIENCE.

evening star Popular, informal name for the planet Venus when it appears to observers on Earth just after sunset as a very bright object in the twilight sky.

See also VENUS.

event horizon The point of no return for a black hole; the distance from a black hole from within which nothing can escape.

See also BLACK HOLE.

evolved expendable launch vehicle (EELV) The evolved expendable launch vehicle is the space lift modernization program of the U.S. Air Force. EELV will reduce the cost of launching payloads into space by at least 25 percent over current Delta, Atlas, and Titan launch systems. Part of these savings result from the government now procuring commercial launch services and turning over responsibility for operations and maintenance of the launch complexes to contractors. This new space lift strategy reduced the government's traditional involvement in launch processing while saving a projected $6 billion in launch costs between the years 2002 and 2020. In addition, EELV improves space launch operability and standardization.

The mission statement for the EELV program is "Partner with industry to develop a national launch capability that satisfies both government and commercial payload requirements and reduces the cost of space launch by at least 25 percent." The EELV program's two primary objectives are, first, to increase the U.S. space launch industry's competitiveness in the international commercial launch services market and, second, to implement acquisition reform initiatives resulting in reduced government resources necessary to manage system development, reduced development cycle time, and deployment of commercial launch services.

There are two primary launch contractors in the EELV program, the Lockheed Martin Company and Boeing Company. The Atlas V launch vehicle resulted from the culmination of Lockheed Martin's desire to employ the best practices from both the Atlas and Titan rocket programs into an evolved and highly competitive commercial and government launch system for this century. The Atlas V launch vehicle builds on the design innovations that will be demonstrated on Atlas III and incorporates a structurally stable booster propellant tank, enhanced payload fairing options, and optional strap-on solid rocket boosters.

The Boeing Company's Delta IV family of expendable launch vehicles blends new and mature technology to lift virtually any size medium or heavy payload into space. This family of rockets consists of five vehicles based on a common booster core (CBC) first stage. Delta IV second stages are derived from the Delta III second stage, using the same RL10B-2 engine but with two sizes of expanded fuel and oxidizer tanks, depending on the model. In designing the five Delta IV configurations, the Boeing Company conducted extensive discussions with government and commercial customers concerning their present and future launch requirements. Proven technical features and processes were carried over from earlier Delta vehicles to the Delta IV family of space lift vehicles.

The first three EELV missions were all successfully launched from Cape Canaveral Air Force Station, Florida. The inaugural launch of Lockheed-Martin's Atlas V Medium Lift Vehicle took place on August 21, 2002, and placed a commercial Eutelsat payload in orbit. Three months later, on November 20, 2002, a Boeing Delta IV rocket successfully placed another commercial Eutelsat payload into orbit around Earth. Eutelsat spacecraft are commercial communications satellites that operate in geostationary orbit. The Boeing Company also accomplished the first government launch of an EELV—the Delta IV rocket that carried the *DSCS A3* military communications satellite into orbit on March 10, 2003. The first launch of the Heavy Lift Vehicle (HLV) version of the Delta IV was scheduled from Cape Canaveral in July 2004.

See also ATLAS LAUNCH VEHICLE; DEFENSE SATELLITE COMMUNICATIONS SYSTEM (DSCS); DELTA LAUNCH VEHICLE; EUTELSAT; LAUNCH VEHICLE; ROCKET.

evolved star A star near the end of its lifetime, when most of its hydrogen fuel has been exhausted; a star that has left the main sequence.

See also MAIN-SEQUENCE STAR; STAR.

excited state State of a molecule, atom, electron, or nucleus when it possesses more than its normal (i.e., ground state) energy. Excess nuclear energy often is released as a gamma ray. Excess molecular energy may appear as thermal energy (heat) or fluorescence. *Compare with* GROUND STATE.

exhaust plume Hot gas ejected from the thrust chamber of a rocket engine. The plume expands as a launch vehicle ascends through Earth's atmosphere, exposing the engine and the vehicle (especially the aft portion) to a greater thermal radiation area.

See also ROCKET.

exhaust stream The stream of gaseous, atomic, or radiant particles that emit from the nozzle of a rocket or other reaction engine.

See also ELECTRIC PROPULSION; ROCKET.

exhaust velocity The velocity (relative to the rocket) of the gases and particles that exit through the nozzle of a rocket motor.

See also ROCKET.

exit In aerospace engineering, the aft end of the divergent portion of a rocket nozzle; the plane at which the exhaust gases leave the nozzle. Also called *exit plane.*

See also NOZZLE; ROCKET.

exit pressure In aerospace engineering, the pressure of the exhaust gas as it leaves a rocket's nozzle; gas pressure at the exit plane of a nozzle.

See also NOZZLE; ROCKET.

exoatmospheric Outside Earth's atmosphere, generally considered as occurring at altitudes above 100 kilometers. *Compare with* ENDOATMOSPHERIC.

See also ATMOSPHERE.

exoatmospheric interceptor In ballistic missile defense (BMD), an interceptor rocket that destroys incoming reentry vehicles above Earth's atmosphere during the late midcourse phase of their flight path.

See also BALLISTIC MISSILE DEFENSE.

exobiology In its most general definition, exobiology is the study of the living universe. The term *astrobiology* is sometimes used for exobiology. Contemporary exobiologists are concerned with observing and using outer space to answer the following intriguing questions (among many others): Where did life come from? How did it evolve? Where is the evolution of life leading? Is the presence of life on Earth unique, or is it a common phenomenon that occurs whenever suitable stars evolve with companion planetary systems? Looking within our own solar system, exobiologists inquire: Did life evolve on Mars? If so, is it now extinct or is it extant—perhaps clinging to survival in some remote subsurface location on the Red Planet? Does Europa, one of the intriguing ice-covered moons of Jupiter, now harbor life within its suspected liquid water oceans? What about the possibility of life in suspected subsurface liquid water oceans on Jupiter's two other planet-sized icy moons, Callisto and Ganymede? What is the significance of the complex organic molecules that appear to be forming continuously in the nitrogen-rich atmosphere of Saturn's moon Titan?

This artist's rendering, entitled *20/20 Vision,* depicts a much-anticipated exobiology discovery that would have great scientific and philosophical significance in the 21st century. Shown here is a scientist-astronaut as she examines fossilized evidence of ancient life on Mars during the first human expedition to the surface of the Red Planet ca. 2025. *(Artwork courtesy of NASA; artist, Pat Rawlings)*

Exobiology is more rigorously defined as the multidisciplinary field that involves the study of extraterrestrial environments for living organisms, the recognition of evidence of the possible existence of life in these environments, and the study of any nonterrestrial lifeforms that may be encountered. Biophysicists, biochemists, and exobiologists usually define a *living organism* as a physical system that exhibits the following general characteristics: has structure (that is, it contains information), can replicate itself, and experiences few (random) changes in its information package, which supports a Darwinian type of evolution (i.e., survival of the fittest).

Observational exobiology involves the detailed investigation of other interesting celestial bodies in our solar system as well the search for extrasolar planets and the study of the organic chemistry of interstellar molecular clouds. *Exopaleontology* involves the search for the fossils or biomarkers of extinct alien life-forms in returned soil and rock samples, unusual meteorites, and in-situ by robot and/or human explorers. (*See* figure.) *Experimental exobiology* includes the evaluation of the viability of terrestrial microorganisms in space and the adaptation of living organisms to different (planetary) environments.

The challenges of exobiology can be approached from several different directions. First, pristine material samples from interesting alien worlds in our solar system can be obtained for study on Earth, as was accomplished during the Apollo Project lunar expeditions (1969–72); or such samples can be studied on the spot (in-situ) by robot explorers, as was accomplished by the robot Viking landers (1976). The returned lunar rock and soil samples have not revealed any traces of life, while the biological results of the in-situ field measurements conducted by the Viking landers have remained tantalizingly unclear. However, more recent robot spacecraft missions to the surface of Mars again raise the question of life on that planet, because their results now strongly suggest that ancient Mars was once a much wetter world.

A second major approach in exobiology involves conducting experiments in terrestrial laboratories or in laboratories in space that attempt either to simulate the primeval conditions that led to the formation of life on Earth and extrapolate these results to other planetary environments or to study the response of terrestrial organisms under environmental conditions found on alien worlds.

In 1924, Alexandr Ivanovich Oparin (1894–1980), a Russian biochemist, published *The Origins of Life On Earth,* a book in which he proposed a theory of chemical evolution that suggested that organic compounds could be produced from simple inorganic molecules and that life on Earth probably originated by this process. (The significant hypothesis of this book remained isolated within the Soviet Union, and it was not even translated into English until 1938.) A similar theory was also put forward in 1929 by JOHN BURDON HALDANE (1892–1964), a British biologist. Unfortunately, the chemical evolution of life hypothesis remained essentially dormant within the scientific community for another two decades. Then, in 1953, at the University of Chicago, American Nobel laureate HAROLD C. UREY and his former student Stanley L. Miller (b. 1930) performed what can be considered the first modern experiments in exobiology. While investigating the chemical origin of life Urey and Miller demonstrated that organic molecules could indeed be produced by irradiating a mixture of inorganic molecules. The historic Urey-Miller experiment simulated the Earth's assumed primitive atmosphere by using a gaseous mixture of methane (CH_4), ammonia (NH_3), water vapor (H_2O), and hydrogen (H_2) in a glass flask. A pool of water was kept gently boiling to promote circulation within the mixture, and an electrical discharge (simulating lightning) provided the energy needed to promote chemical reactions. Within days the mixture changed colors, indicating that more complex, organic molecules had been synthesized out of this primordial "soup" of inorganic materials.

A third general approach in exobiology involves an attempt to communicate with, or at least listen for signals from, other intelligent life-forms within our galaxy. This effort often is called the search for extraterrestrial intelligence (or SETI for short). Contemporary SETI activities throughout the world have as their principal aim to listen for evidence of extraterrestrial radio signals generated by intelligent alien civilizations.

See also APOLLO PROJECT; EUROPA; LIFE IN THE UNIVERSE; MARS; MARS EXPLORATION ROVER MISSION; *MARS PATHFINDER*; SEARCH FOR EXTRATERRESTRIAL INTELLIGENCE; VIKING PROJECT.

exoskeleton In robotics, the external supporting structure of a machine, especially an anthropomorphic device. Certain hard-shelled space suits and deep-sea diver suits sometimes are called exoskeletons. The term derives from a biological description of the rigid external covering found in certain animals, such as mollusks. The exoskeleton supports the animal's body and provides attachment points for its muscles.

See also ROBOTICS IN SPACE; SPACESUIT.

exosphere The outermost region of Earth's atmosphere. Sometimes called the region of escape.

See also ATMOSPHERE.

exotheology The organized body of opinions concerning the impact that space exploration and the possible discovery of life beyond the boundaries of Earth would have on contemporary terrestrial religions. On Earth theology involves the study of the nature of God and the relationship of human beings to God.

Throughout human history people have gazed into the night sky and pondered the nature of God. They searched for those basic religious truths and moral beliefs that define how an individual should interact with his or her Creator and toward one another. The theologies found within certain societies (especially ancient ones) sometimes involved a collection of gods. For people in such societies, the plurality of specialized gods proved useful in explaining natural phenomena as well as in codifying human behavior. For example, in the pantheism of ancient Greece, an act that displeased Zeus—the lord of the gods who ruled Earth from Mount Olympus—would cause him to hurl a powerful thunderbolt at the offending human. Thus, the fear of getting "zapped by Zeus" often helped ancient Greek societies maintain a set of moral standards. Other terrestrial religions stressed belief in a single God. Judaism, Christianity, and Islam represent the planet's three major monotheistic religions. Each has a collection of dogma and beliefs that define acceptable moral behavior by human beings and also identify individual rewards (e.g., personal salvation) and punishments (e.g., eternal damnation) for adherence or transgressions, respectively.

In the 21st century advanced space exploration missions could lead to the discovery of simple alien life-forms on other worlds within the solar system. This important scientific achievement would undoubtedly rekindle serious interest in one of the oldest philosophical questions that has puzzled a great number of people throughout history: Are we the only intelligent species in the universe? If not, what happens if contact is made with an alien intelligence? What might be our philosophical and theological relationship with such intelligent creatures?

Some space age theologians are now starting to grapple with these intriguing questions and many similar ones. For example, *should it exist,* would an alien civilization that is much older than our planetary civilization have a significantly better understanding of the universe and, by extrapolation, a clearer understanding of the nature of God? Here on Earth many brilliant scientists, such as SIR ISAAC NEWTON and ALBERT EINSTEIN, regarded their deeper personal understanding of the physical universe as an expanded perception of its Creator. Would these (hypothetical) advanced alien creatures be willing to share their deeper understanding of God? And if they do decide to share that deeper insight into the divine nature, what impact would that communication have on terrestrial religions? Exotheology involves such interesting speculations and the extrapolation of teachings found within terrestrial religions to accommodate the expanded understanding of the universe brought about by the scientific discoveries of space exploration and modern astronomy.

See also ANTHROPIC PRINCIPLE; COSMOLOGY; EXTRATERRESTRIAL CIVILIZATIONS; INTERSTELLAR COMMUNICATION AND CONTACT; SEARCH FOR EXTRATERRESTRIAL INTELLIGENCE.

exothermic reaction A chemical or physical reaction in which thermal energy (heat) is released into the surroundings. *Compare with* ENDOTHERMIC REACTION.

expandable space structure A structure that can be packaged in a small volume for launch and then erected to its full size and shape outside Earth's atmosphere.

See also LARGE SPACE STRUCTURES.

expanding universe Any model of the universe in modern cosmology that has the distance between widely separated celestial objects (e.g., distant galaxies) continuing to grow or expand with time.

See also COSMOLOGY; UNIVERSE.

expansion geometry In aerospace engineering, the contour of a rocket nozzle from the throat to the exit plane.

See also NOZZLE; ROCKET.

expansion ratio (symbol: ε) For a converging-diverging (de Laval-type) rocket nozzle, the expansion ratio (ε) is defined as the ratio of the exit plane area (A_{exit}) to the throat area (A_{throat}). Also called nozzle area expansion ratio.

$$\varepsilon = A_{exit}/A_{throat}$$

See also NOZZLE; ROCKET.

expendable launch vehicle (ELV) A ground-launched propulsion vehicle capable of placing a payload into Earth orbit or Earth-escape trajectory, whose various stages are not designed for or intended for recovery and/or reuse. The U.S. Atlas, Delta, Titan, and Scout rockets are examples of expendable launch vehicles.

See also EVOLVED EXPENDABLE LAUNCH VEHICLE; LAUNCH VEHICLE.

experimental vehicle (symbol: X) An aircraft, missile, or aerospace vehicle in the research, development, and testing portion of its life cycle and not yet approved for operational use. Often designated by the symbol X.

See also X-1; X-15; X-20; X-33; X-34.

exploding galaxies Violent, very energetic explosions centered in certain galactic nuclei where the total mass of ejected material is comparable to the mass of some 5 million average-sized sunlike stars. Jets of gas 1,000 light-years long are also typical.

See also GALAXY; JETS.

Explorer 1 The first U.S. Earth satellite. (*See* figure.) *Explorer 1* was launched successfully from Cape Canaveral Air Force Station on January 31, 1958 (local time), by the Juno I four-stage configuration of the Jupiter C launch vehicle. It was a joint project of the Army Ballistic Missile Agency (ABMA) and the Jet Propulsion Laboratory (JPL). Under the engineering supervision of WERNHER VON BRAUN, the ABMA supplied the Jupiter C launch vehicle, while JPL and WILLIAM HAYWARD PICKERING supplied the fourth-stage rocket and the satellite itself. JAMES VAN ALLEN of the State University of

Jubilant team leaders (from left to right) Pickering (JPL), Van Allen (State University of Iowa), and von Braun (ABMA) hold aloft a model of the *Explorer 1* spacecraft and its solid-fuel rocket final stage. The first U.S. Earth satellite was successfully launched from Cape Canaveral, Florida, on January 31, 1958 (local time). ***(Courtesy of NASA)***

Iowa provided the instruments, which detected the inner of Earth's two major trapped radiation belts for the first time. *Explorer 1* measured three phenomena: cosmic ray and radiation levels, the temperature in the spacecraft, and the frequency of collisions with micrometeorites. There was no provision for data storage, and therefore the satellite transmitted information continuously. Earth's trapped radiation belts, discovered by the *Explorer 1*, are called the Van Allen belts in honor of Van Allen.

See also EARTH'S TRAPPED RADIATION BELTS.

Explorer spacecraft The name *Explorer* has been used by NASA to designate a large series of scientific satellites used to "explore the unknown." Explorer spacecraft were used by NASA since 1958 to study: (1) Earth's atmosphere and ionosphere, (2) the magnetosphere and interplanetary space, (3) astronomical and astrophysical phenomena, and (4) Earth's shape, magnetic field, and surface. Many of the Explorer satellites had project names that were used before they were orbited and then were replaced by Explorer designations once they were placed in orbit. Other Explorer satellites, especially the early ones, were known before launch and after achieving orbit simply by their numerical designations. A brief (but not all-inclusive) listing will help illustrate the great variety of important scientific missions performed by these satellites: Aeronomy Explorer, Air Density Explorer, Interplanetary Monitoring Platform (IMP), Ionosphere Explorer, Meteoroid Technology Satellite (MTS), Radio Astronomy Explorer (RAE), Solar Explorer, and Small Astronomy Satellite (SAS).

Geodetic satellites (GEOS) were also called Geodetic Explorer Satellites. For example, GEOS 1 (*Explorer 29,* launched on November 6, 1965) and GEOS 2 (*Explorer 36,* launched on January 11, 1968) refined our knowledge of Earth's shape and gravity field. SAS-A, an X-ray Astronomy Explorer, became *Explorer 42* when launched on December 12, 1970, by an Italian launch crew from the San Marco platform off the coast of Kenya, Africa. Because it was launched on Kenya's Independence Day, this small spacecraft was also called *Uhuru* (the Swahili word for "freedom"). It successfully mapped the universe in X-ray wavelengths for four years, discovering X-ray pulsars and providing preliminary evidence of the existence of black holes.

See also *EXPLORER 1*; *EXTREME ULTRAVIOLET EXPLORER* (EUVE); *FAR-ULTRAVIOLET SPECTROSCOPIC EXPLORER* (FUSE); *FAST AURORAL SNAPSHOT EXPLORER* (FAST).

explosive-activated device (EAD) An electrically initiated pyrotechnic device, such as an explosive bolt. Used to separate rocket stages or to separate a spacecraft from its rocket vehicle. For example, the explosive bolt incorporates an explosive charge that can be detonated on command (usually by an electrical signal or pulse), thereby destroying the bolt and releasing pieces of aerospace equipment it was retaining. Aerospace engineers often use explosive bolts and spring-loaded mechanisms to quickly separate launch vehicle stages or to separate a spacecraft from its (upper-stage) propulsion system.

See also ELECTROEXPLOSIVE DEVICE; LAUNCH VEHICLE; UPPER STAGE.

explosive decompression A very rapid reduction of air pressure inside the pressurized portion (i.e., crew compartment) of an aircraft, aerospace vehicle, or spacecraft. For example, collision with a large piece of space debris might puncture the wall of one of the pressurized modules on a space station, causing an explosive decompression situation within that module. Air locks would activate, sealing off the stricken portion of the pressurized space habitat.

explosive valve A valve having a small explosive charge that, when detonated, provides high-pressure gas to change the valve position. Also known as a *squib valve.*

external tank (ET) The large tank that contains the propellants for the three space shuttle main engines (SSMEs) and forms the structural backbone of the space shuttle flight system in the launch configuration. At liftoff the external tank absorbs the total 28,580-kilonewton thrust loads of the three main engines and the two solid rocket boosters (SRBs). When these SRBs separate at an altitude of approximately 44 kilometers, the orbiter vehicle, with the main engines still burning, carries the ET piggyback to near orbital velocity. Then, at an altitude of approximately 113 kilometers, some 8.5 minutes into the flight, the now nearly empty tank separates from the orbiter and falls in a preplanned trajectory into a remote area of the Indian Ocean. The ET is the only major expendable element of the space shuttle.

The three main components of the ET are an oxygen tank (located in the forward position), a hydrogen tank (located in the aft position), and a collarlike intertank, which connects the two propellant tanks, houses instrumentation and processing equipment, and provides the attachment structure for the forward end of the solid rocket boosters. In June 1998 a lighter version of the external tank entered service. Made of a special lightweight material called aluminum lithium, the new tank has a mass of approximately 26,300 kilograms, about 3,400 kg less than the previous version. The reduced tank mass means heavier payloads can be carried into orbit by the space shuttle.

See also SPACE TRANSPORTATION SYSTEM.

extinction In astronomy and physics, the phenomenon that involves the reduction in the amount of light or other electromagnetic radiation received from a celestial object. It occurs due to the absorption and scattering of radiation by the intervening media through which the radiation passes. *Interstellar extinction* takes place when dust grains in space (the INTERSTELLAR MEDIUM) scatter or absorb radiation from a distant celestial object. Since the phenomenon of extinction becomes more pronounced as the wavelength of the radiation decreases, astronomers observe that starlight is reddened because blue light (which has a shorter wavelength) suffers a greater amount of extinction. *Atmospheric extinction* takes place when radiation from a celestial object passes through Earth's atmosphere. The path length the radiation travels and the density of the intervening transparent medium also contribute to extinction. One example of atmospheric extinction is the rich red-orange color of the Sun when it sets in the evening.

extragalactic Occurring, located, or originating beyond our galaxy (the Milky Way); typically, farther than 100,000 light-years distant.

extragalactic astronomy A branch of astronomy that started around 1930 that studies everything in the universe outside our galaxy, the Milky Way.

extrasolar Occurring, located, or originating outside of our solar system; as, for example, extrasolar planets.

extrasolar planets Planets that belong to a star other than the Sun. There are two general methods that can be used to detect extrasolar planets: direct, involving a search for telltale signs of a planet's infrared emissions, and indirect, involving precise observation of any perturbed motion (for example, wobbling) of the parent star or any periodic variation in the intensity or spectral properties of the parent star's light.

Growing evidence of planets around other stars is helping astronomers validate the hypothesis that planet formation is a normal part of stellar evolution. Detailed physical evidence concerning extrasolar planets—especially if scientists can determine their frequency of occurrence as a function of the type of star—would greatly assist scientists in estimating the cosmic prevalence of life. If life originates on "suitable" planets whenever it can (as many exobiologists currently hold), then knowing how abundant such suitable planets are in our galaxy would allow us to make more credible guesses about where to search for extraterrestrial intelligence and what our chances are of finding intelligent life beyond our own solar system.

Scientists have detected (through spectral light variation of the parent star) Jupiter-sized planets around Sun-like stars, such as 51 Pegasi, 70 Virginis, and 47 Ursae Majoris. Detailed computer analyses of spectrographic data have revealed that light from these stars appears redder and then bluer in an alternating periodic (sine wave) pattern. This periodic light pattern could indicate that the stars themselves are moving back and forth along the line of sight possibly due to a large (unseen) planetary object that is slightly pulling the stars away from (redder spectral data) or toward (bluer spectral data) Earth. The suspected planet around 51 Pegasi is sometimes referred to as a *hot Jupiter,* since it appears to be a large planet (about half of Jupiter's mass) that is located so close to its parent star that it completes an orbit in just a few days (approximately 4.23 days). The suspected planetary body orbiting 70 Virginis lies about .5 an astronomical unit (AU) distance from the star and has a mass approximately eight times that of Jupiter. Finally, the object orbiting 47 Ursae Majoris has an estimated mass that is 3.5 times that of Jupiter. It orbits the parent star at approximately 1 AU distance, taking about three years to complete one revolution.

To find extrasolar planets and characterize their atmospheres, scientists will use new (or planned) space-based observatories such as the *Spitzer Space Telescope* (SST), the *James Webb Space Telescope* (JWST), and the *Kepler* spacecraft. For example, in late 2003 NASA's *Spitzer Space Telescope* captured a dazzling image of a massive disc of dusty debris encircling a nearby star called Fomalhaut. Scientists consider such discs as remnants of planetary construction. Earth is thought to have been formed out of a similar disc. The *Spitzer Space Telescope* is now helping scientists identify other stellar dust clouds that might mark the sites of developing planets.

The *James Webb Space Telescope* (JWST) will be a large single telescope that is folded to fit inside its launch vehicle and cooled to low temperatures in deep space to enhance its sensitivity to faint, distant objects. Mission controllers will operate JWST in an orbit far from Earth, away from the thermal energy (heat) radiated to space by our home planet. Scheduled for launch in 2011, one of this observatory's main science goals is to determine how planetary systems form and interact. The JWST can observe evidence of the formation of planetary systems (some of which may be similar to our own solar system) by mapping the light from the clouds of dust grains orbiting stars. One star, Beta Pictoris, has such a cloud, as was discovered by the *Infrared Astronomical Satellite* (IRAS). These clouds are bright near the host stars and may be divided into rings by the gravitational influence of large planets. Scientists speculate that this dust represents material forming into planets. Around older stars such dust clouds may be the debris of material that failed to condense into planets. The JWST will have unprecedented sensitivity to observe faint dust clouds around nearby stars. The infrared wavelength range is also the best way to search for planets directly, because they are brighter relative to their central stars. For example, at visible wavelengths, Jupiter is about 100 million times fainter than the Sun, but in the infrared it is only 10,000 times fainter. A planet like Jupiter would be difficult to observe directly with any telescope on Earth, but the JWST has a chance, because it operates in space beyond the disturbing influences of the terrestrial atmosphere.

Scheduled for launch in 2006, NASA's *Kepler* spacecraft will use a unique spaceborne telescope specifically designed to search for Earthlike planets around stars beyond our solar system. This mission will allow scientists to search the galaxy for Earth-sized or even smaller planets. The extrasolar planets discovered thus far are giant planets, similar to Jupiter, and are probably composed mostly of hydrogen and helium and unlikely to harbor life. None of the planet detection methods used to date has the capability of finding Earth-sized planets—those that are 30 to 600 times less massive than Jupiter. Furthermore, none of the giant extrasolar planets discovered to date has liquid water or even a solid surface.

The Kepler Mission differs from previous ways of looking for planets, because it will look for the *transit* signature of planets. A transit occurs each time a planet crosses the line-of-sight between the planet's parent star that it is orbiting and the observer. When this happens the planet blocks some of the light from its star, resulting in a periodic dimming. This periodic signature will be used to detect the planet and to determine its size and its orbit. Three transits of a star, all with a consistent period, brightness change, and duration, will provide a robust method of detection and planet confirmation. The measured orbit of the planet and the known properties of the parent star will be used to determine if each planet discovered is in the continuously habitable zone (CHZ), that is, at the distance from its parent star where liquid water could exist on the surface of the planet. The Kepler Mission will hunt for planets using a specialized 1-meter-diameter telescope called a photometer. This instrument can measure the small changes in brightness caused by the transits. By monitoring 100,000 stars similar to our Sun for four years following the *Kepler* spacecraft's launch, scientists expect to find many hundreds of terrestrial-type planets.

NASA's Terrestrial Planet Finder (TPF) mission consists of a suite of two complementary space observatories: a visible-light coronagraph and a mid-infrared formation-flying interferometer. An interferometer consists of a collection of several (small) telescopes that function together to produce an image much sharper than would be possible with a single telescope. As currently planned, the TPF coronagraph should launch in 2014 and the TPF interferometer by 2020. The combination will detect and characterize Earthlike planets around as many as 150 stars up to 45 light-years away. The science goals of this ambitious planet-hunting mission include a survey of nearby stars in the search for terrestrial-sized planets in the continuously habitable zone (CHZ). Scientists will then perform spectroscopy on the most promising candidate extrasolar planets, looking for atmospheric signatures characteristic of habitability or even life itself.

The challenge of finding an Earth-sized (i.e., terrestrial) planet orbiting even the closest stars can be compared to finding a tiny firefly next to a blazing searchlight when both are thousands of kilometers away. Quite similarly, the infrared emissions of a parent star are a million times "brighter" than the infrared emissions of any companion planets that might orbit around it. Beyond the year 2020 data from the Terrestrial Planet Finder should allow astronomers to analyze the infrared emissions of extrasolar planets in star systems up to about 100 light-years away. They will use these data to search for signs of atmospheric gases, such as carbon dioxide, water vapor, and ozone. Together with the temperature and radius of any detected planets, these atmospheric gas data will enable scientists to determine which extrasolar planets are potentially habitable, or even whether they may be inhabited by rudimentary forms of life.

The quest for extrasolar planets is one of the most interesting areas within modern astronomy. But how can astronomers find planetary bodies around distant stars? In their search for extrasolar planets, scientists employ several important techniques. These techniques include pulsar timing, Doppler spectroscopy, astrometry, and transit photometry.

Astronomers accomplished the first widely accepted detection of extrasolar planets (namely, pulsar planets) in the early 1990s using the pulsar timing technique. Earth-mass and even smaller planets orbiting a pulsar were detected by measuring the periodic variation in the pulse arrival time. However, the planets detected are orbiting a pulsar, a "dead" star, rather than a dwarf (main-sequence) star. What is encouraging, however, about the detection is that the planets were probably formed after the supernova that resulted in the pulsar, thereby demonstrating that planet formation is probably a common rather than rare astronomical phenomenon.

Astronomers also use Doppler spectroscopy to detect the periodic velocity shift of the stellar spectrum caused by an orbiting giant planet. Scientists sometimes refer to this method as the *radial velocity method*. Using ground-based astronomical observatories, spectroscopists can measure Doppler shifts greater than about three meters per second due to the reflex motion of the parent star. This measurement sensitivity corresponds to a minimum detectable planetary mass equivalent to approximately 30 Earth masses for a planet located at 1 AU. This method can be used for main-sequence stars of spectral types mid-F through M. Stars hotter and more massive than mid-F rotate faster, pulsate, are generally more active, and have less spectral structure, thus making measurement of their Doppler shifts much more difficult. As previously mentioned, using this technique scientists have successfully detected several large (Jupiterlike) extrasolar planets.

Scientists use astrometry to look for the periodic *wobble* that a planet induces in the position of its parent star. The minimum detectable planetary mass gets smaller in inverse proportion to that planet's distance from the star. For a space-based astrometric instrument, such as the planned Space Interferometry Mission (SIM), a facility that will measure an angle as small as two micro-arc-seconds, a minimum sized planet of about 6.6 Earth masses could be detected as it travels in a one-year orbit around a one solar-mass star that is 32.61 light-years (10 parsecs) from the Earth. The SIM would also be capable of detecting a 0.4 Jupiter mass planet in a four-year orbit around a star out to a distance of 1,630 light-years (500 pc). From the ground modern telescope facilities such as the Keck Telescope can measure angles as small as 20 micro-arc-seconds, leading to a minimum detectable planetary mass in a 1 AU orbit of 66 Earth masses for a one solar-mass star at a distance of 32.6 light-years (10 pc). The limitations to this method are the distance to the star and variations in the position of the photometric center due to star spots. There are only 33 nonbinary solar-like (F, G, and K) main-sequence stars within 32.6 light-years (10 pc) of Earth. The farthest planet from its star that can be detected by this technique is limited by the time needed to observe at least one orbital period.

Finally, astronomers use the transit photometry technique to measure the periodic dimming of a star caused by a planet passing in front of it along the line of sight from the observer. Stellar variability on the time scale of a transit limits the detectable size to about half that of Earth for a one AU orbit about a one solar mass star; or, with four years of observing, transit photometry can detect Mars-sized planets in Mercurylike orbits. Mercury-sized planets can even be detected in the continuously habitable zone of K and M stars. Planets with orbital periods greater than two years are not readily detectable, since their chance of being properly aligned along the line of sight to the star becomes very small.

Giant planets in inner orbits are detectable independent of the orbit alignment based on the periodic modulation of their reflected light. The transit depth can be combined with the mass found from Doppler data to determine the density of the planet, as scientists have done for the case of a star called HD209458b. Doppler spectroscopy and astrometry measurements can be used to search for any giant planets that might also be in the systems discovered using photometry. Since the orbital inclination must be close to 90° to cause transits, there is very little uncertainty in the mass of any giant planet detected. Photometry represents the only practical method for finding Earth-sized planets in the continuously habitable zone.

See also CONTINUOUSLY HABITABLE ZONE; *JAMES WEBB SPACE TELESCOPE*; LIFE IN THE UNIVERSE; KECK OBSERVATORY; KEPLER MISSION; SPACE INTERFEROMETRY MISSION; *SPITZER SPACE TELESCOPE*; TERRESTRIAL PLANET FINDER.

extraterrestrial (ET) Refers to something that occurs in, is located in, or originates outside of the planet Earth and its atmosphere.

See also LIFE IN THE UNIVERSE.

extraterrestrial catastrophe theory For millions of years giant, thundering reptiles roamed the lands, dominated the skies, and swam in the oceans of a prehistoric Earth. Dinosaurs reigned supreme. Then, quite suddenly, some 65 million years ago they vanished. What happened to these giant creatures and to thousands of other ancient animal species?

From archaeological and geological records, we do know that some tremendous catastrophe occurred about 65 million years ago on this planet. It affected life more extensively than any war, famine, or plague in human history. In that cataclysm about 70 percent of all species then living on Earth—including, of course, the dinosaurs—disappeared within a very short period. This mass extinction is also referred to Cretaceous-Tertiary Mass Extinction event, or simply the K-T event.

In 1980 the scientists LUIS W. ALVAREZ and his son, Walter Alvarez (b. 1940), along with their colleagues at the University of California at Berkeley discovered that a pronounced increase in the amount of the element iridium in Earth's surface had occurred at precisely the time of the disappearance of the dinosaurs. First seen in a peculiar sedimentary clay area found near Gubbio, Italy, the same iridium enhancement was soon discovered in other places around the world in the thin sedimentary layer that was laid down at the time of the mass extinction. Since iridium is quite rare in Earth's crust and more abundant in the rest of the solar system, the Alvarez team postulated that a large asteroid (about 10 kilometers or more in diameter) had struck the ancient Earth. This cosmic collision would have promoted an environmental catastrophe throughout the planet. The scientists reasoned that such an asteroid would largely vaporize while passing through the Earth's atmosphere, spreading a dense cloud of dust particles, including quantities of extraterrestrial iridium atoms, uniformly around the globe.

Stimulated by the Alvarez team's postulation, many recent geological investigations have observed a global level of enhanced iridium in this thin layer (about 1 cm thick) of the Earth's lithosphere (crust) that lies between the final geological formations of the Cretaceous period (which are dinosaur fossil–rich) and the formations of the early Tertiary period (whose rocks are notably lacking in dinosaur fossils). The Alvarez hypothesis further speculated that following this asteroid impact a dense cloud of dust covered Earth for many years, obscuring the Sun, blocking photosynthesis, and destroying the very food chains upon which many ancient life-forms depended.

Despite the numerous geophysical observations of enhanced iridium levels that reinforced the Alvarez impact hypothesis, many geologists and paleontologists still preferred other explanations concerning the mass extinction that occurred about 65 million years ago. To them the impact theory of mass extinction was still a bit untidy. Where was the impact crater? This important question was answered in the early 1990s, when a 180-km-diameter ring structure, called Chicxulub, was identified from geophysical data collected in the Yucatán region of Mexico. The Chicxulub crater has been age-dated at 65 million years. Further studies have also helped confirm its impact origin. The impact of a 10-km-diameter asteroid would have created this very large crater, as well as causing enormous tidal waves, as depicted in the figure. Scientists have found evidence of tidal waves occurring about 65 million years ago around the Gulf of Mexico region.

Of course, there are still many other scientific opinions as to why the dinosaurs vanished. A popular one is that there was a gradual but relentless change in Earth's climate to which these giant reptiles and many other prehistoric animals simply could not adapt. So, one can never absolutely prove that an asteroid impact "killed the dinosaurs." Many species of dinosaurs (and smaller flora and fauna) had, in fact, become extinct over the millions of years preceding the K-T event. However, the impact of a 10-km-across asteroid would most certainly have been an immense insult to life on Earth. Locally, there would have been intense shock wave heating and fires, tremendous earthquakes, hurricane force winds, and hundreds of billions of tons of debris thrown everywhere. This debris would have created months of darkness and cooler temperatures on a global scale. There would also have been concentrated nitric acid rains worldwide. Sulfuric acid aerosols may have cooled Earth for years after the impact. Life certainly would not have been easy for those species that did survive. Fortunately, such large, extinction-level (impact) events (ELEs) are thought to occur only about once every hundred million years. It is also interesting to observe, however, that as long as those enormous reptiles roamed and dominated the Earth, mammals, including humans themselves, would have had little chance of evolving.

The possibility that an asteroid or comet will strike Earth in the future is quite real. Just look at a recent image of Mars or the Moon and ask yourself how those large impact craters were formed. A comet called Shoemaker-Levy 9 hit Jupiter in 1994. Fortunately, the probability of a *large* asteroid or comet striking Earth is quite low. For example, space scientists estimate that Earth will experience (on average) one collision with an EARTH-CROSSING ASTEROID (ECA) of 1-km-diameter size or greater every 300,000 years.

Yet, on May 22, 1989, a small ECA called 1989FC passed within 690,000 kilometers (only 0.0046 astronomical unit) of our planet. This cosmic "near-miss" occurred at less than twice the distance to the Moon. Cosmic impact specialists have estimated that if this small asteroid, presumed to be about 200 to 400 meters in diameter, had experienced a straight-on collision with Earth at a relative velocity of some 16 kilometers per second (km/s), it would have impacted with

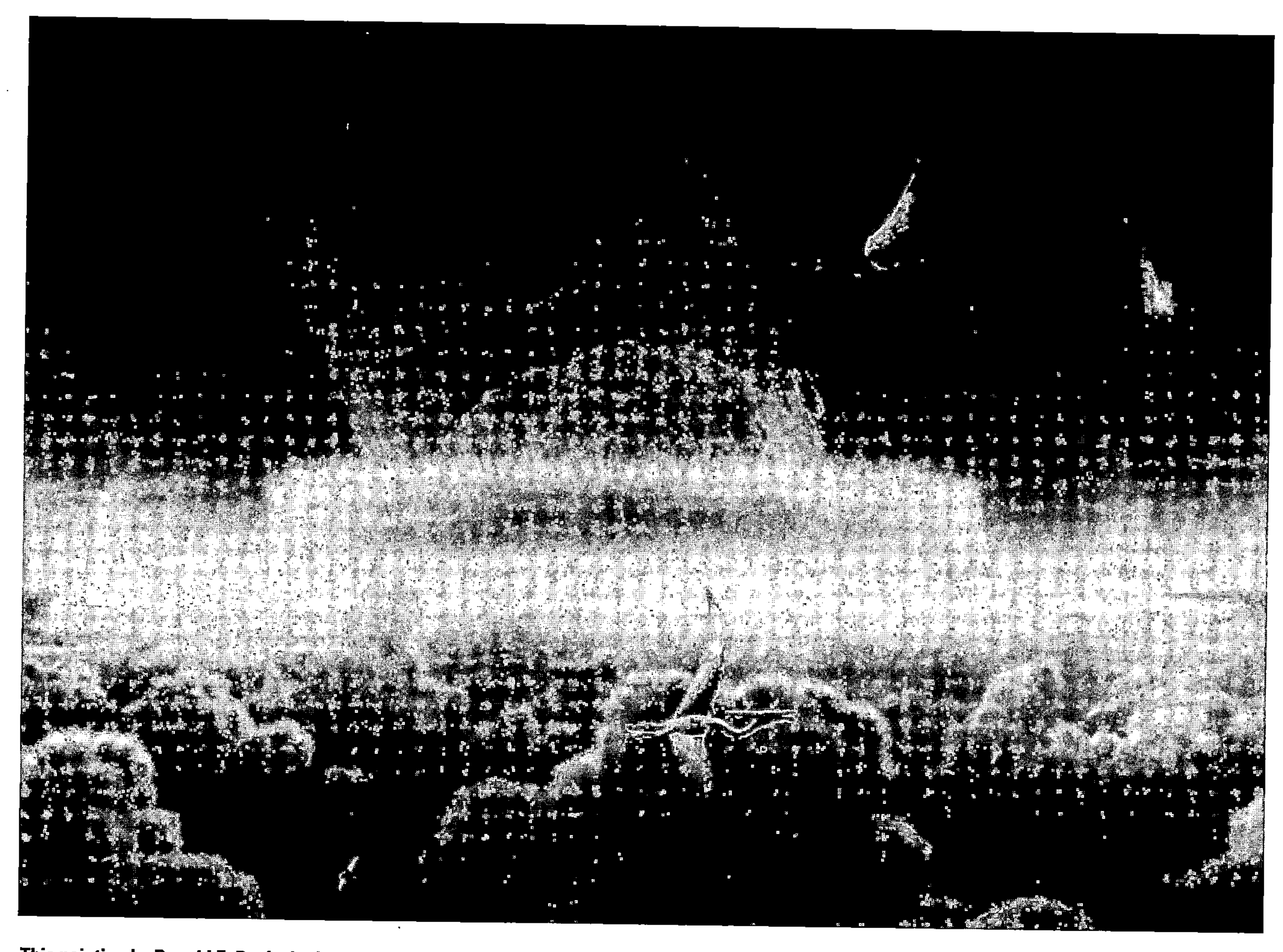

This painting by Donald E. Davis depicts an asteroid slamming into the tropical shallow seas of the sulfur-rich Yucatán Peninsula in what is today southeast Mexico. The aftermath of this immense asteroid collision, which occurred approximately 65 million years ago, is believed to have caused the extinction of the dinosaurs and many other species on Earth. The impact spewed hundreds of billions of tons of debris into the atmosphere, producing a worldwide blackout and freezing temperatures that persisted for at least a decade. Shown in this painting are pterodactyls, large flying reptiles with wingspans of up to 15 meters, gliding above local tropical clouds. ***(Courtesy of NASA/JPL)***

an explosive force of some 400 to 2,000 megatons (MT). (A megaton is the energy of an explosion that is the equivalent of 1 million tons of trinitrotoluene [TNT].) If this small asteroid had hit a terrestrial landmass, it would have formed a crater some 4 to 7 kilometers across and produced a great deal of regional (but not global) destruction.

See also ASTEROID; ASTEROID DEFENSE SYSTEM; COMET; NEMESIS; TUNGUSKA EVENT.

extraterrestrial civilizations According to some scientists, intelligent life in the universe might be thought of as experiencing three basic levels of civilization. For example, in 1964 the Russian astronomer NIKOLAI S. KARDASHEV, while examining the issue of information transmission by extraterrestrial civilizations, postulated three types of technologically developed civilizations on the basis of their energy use. A Kardashev Type I civilization would represent a planetary civilization similar to the technology level on Earth today. It would command the use of somewhere between 10^{12} and 10^{16} watts of energy—the upper limit being the amount of solar energy being intercepted by a "suitable" (or Goldilocks) planet in its orbit about the parent star.

A Kardashev Type II civilization would engage in feats of planetary engineering, emerging from its native planet through advances in space technology and extending its resource base throughout the local star system. The eventual upper limit of a Type II civilization could be taken as the creation of a Dyson sphere. A *Dyson sphere* is a postulated shell-like cluster of habitats and structures placed entirely around a star by an advanced civilization to intercept and use basically all the radiant energy from that parent star. What the physicist FREEMAN J. DYSON suggested in 1960 was that an advanced extraterrestrial civilization might eventually develop the space technologies necessary to rearrange the raw materials of all the planets in its solar system, creating a more efficient composite ecosphere around the parent star. Dyson further hypothesized that any such advanced alien civilizations might be detected by the presence of thermal infrared emissions from these artificially enclosed star systems in contrast to the normally anticipated visible radiation. Once this level of extraterrestrial civilization is achieved, the search for additional resources and the pressures of continued growth could encourage interstellar migrations. This would mark the start of a Kardashev Type III extraterrestrial civilization.

At maturity a Type III civilization would be capable of harnessing the material and energy resources of an entire galaxy (typically containing some 10^{11} to 10^{12} stars). Energy resources on the order of 10^{37} to 10^{38} watts or more would be involved.

Command of energy resources might therefore represent a key factor in the evolution of extraterrestrial civilizations. It should be noted that a Type II civilization controls about 10^{12} times the energy resources of a Type I civilization, and a Type III civilization approximately 10^{12} times as much energy as a Type II civilization.

What can we speculate about such civilizations? Starting with Earth as a model (our one and only "scientific" data point), we can presently postulate that a Type I civilization should probably exhibit the following characteristics: (1) an understanding of the laws of physics; (2) a planetary society, including a global communication network and interwoven food and materials resource networks; (3) intentional or unintentional emission of electromagnetic radiations (especially radio frequency); (4) the development of space technology and rocket propulsion–based interplanetary space travel—the tools necessary to leave the home planet; (5) (possibly) the development of nuclear energy technology, both power supplies and weapons; and (6) (possibly) a desire to search for and communicate with other intelligent life-forms in the universe. Many uncertainties, of course, are present in such characterizations. For example, given the development of space technology, will the planetary civilization decide to create a solar-system civilization? Do the planet's inhabitants develop a long-range planning perspective that supports the eventual creation of artificial habitats and structures throughout their star system? Or do the majority of Type I civilizations unfortunately destroy themselves with their own advanced technologies before they can emerge from a planetary civilization into a more stable Type II civilization? Does the exploration imperative encourage such creatures to go out from their comfortable planetary niche into an initially hostile but resource-rich star system? If this "cosmic birthing" does not occur frequently, perhaps our galaxy is indeed populated with intelligent life, but at a level of stagnant planetary (Type I) civilizations that have neither the technology nor the motivation to create an extraterrestrial civilization or even to try to communicate with any other intelligent life-forms across interstellar distances.

Assuming that an extraterrestrial civilization does, however, emerge from its native planet and create an interplanetary society, several additional characteristics would become evident. The construction of space habitats and structures (leading ultimately to a Dyson sphere around the native star) would portray feats of planetary engineering and could possibly be detected by thermal infrared emissions as incident starlight in the visible spectrum was intercepted and converted to other more useful forms of energy and the residual energy (as determined by the universal laws of thermodynamics) was rejected to space as waste heat at perhaps 300 K.

Type II civilizations might also decide to search in earnest for other forms of intelligent life beyond their star system. They would probably use portions of the electromagnetic spectrum (radio frequency and perhaps X-rays or gamma rays) as information carriers between the stars. Remembering that Type II civilizations would control 10^{12} times as much energy as Type I civilizations, such techniques as electromagnetic beacons or feats of astroengineering that yield characteristic X-ray or gamma-ray signatures may lie well within their technical capabilities. Assuming their understanding of the physical universe is far more sophisticated than ours, Type II civilizations might also use gravity waves or other physical phenomena (perhaps unknown to us now, but being sent through our own solar system at this very instant) in their effort to communicate across vast interstellar distances. Type II civilizations could also decide to make initial attempts at interstellar matter transfer. Fully autonomous robotic explorers would be sent forth on one-way scouting missions to nearby stars. Even if the mode of propulsion involved devices that achieved only a small fraction of the speed of light, Type II societies should have developed the much longer-term planning perspective and thinking horizon necessary to support such sophisticated, expensive, and lengthy missions. Type II civilizations might also use a form of panspermia (the diffusion of spores or molecular precursors through space) or even ship microscopically encoded viruses through the interstellar void, hoping that if such "seeds of life" found a suitable ecosphere in some neighboring or distant star system, they would initiate the chain of life, perhaps leading ultimately to the replication (suitably tempered by local ecological conditions) of intelligent life itself.

Finally, as the Dyson sphere was eventually completed, some of the inhabitants of this Type II civilization might respond to a cosmic wanderlust and initiate the first "peopled" interstellar missions. Complex space habitats could become *space arks* and carry portions of this civilization to neighboring star systems. However, we must ask what is the lifetime of a Type II civilization. It would appear from an extrapolation of contemporary terrestrial engineering practices that perhaps a minimum of 500 to 1,000 years would be required for even an advanced interplanetary civilization to complete a Dyson sphere.

Throughout the entire galaxy, however, if just one Type II civilization embarks on a successful interstellar migration program, then—at least in principle—it would eventually (in perhaps 10^8 to 10^9 years) sweep through the galaxy in a leapfrogging wave of colonization, establishing a Type III civilization in its wake.

This Type III civilization could eventually control the energy and material resources of up to 10^{12} stars—or the entire Milky Way galaxy! Communication or matter transfer would be accomplished by techniques that can now only be called "exotic." Perhaps directed beams of neutrinos or even (hypothesized) faster-than-light particles (such as tachyons) would serve as information carriers for this galactic society. Or they might use tunneling through black holes as their transportation network. Perhaps they might develop some kind of thought transference or telepathic skills that permitted efficient and instantaneous communications over the vast regions of interstellar space. In any event, a Type III civilization should be readily evident, since it would be galactic in extent and easily recognizable by its incredible feats of astroengineering.

In all likelihood, our galaxy at present does not contain a Type III civilization. Or else perhaps our solar system is being ignored—that is, intentionally being kept isolated—as a game preserve or zoo, as some scientists have speculated. Then again, our solar system may be simply one of the very last regions to be "filled in."

There is also another perspective. If we are indeed alone, or the most advanced civilization in our galaxy, then we now stand at the technological threshold of creating the galaxy's first Type II civilization. Should we succeed in that task, then the human race also could have the opportunity of becoming the first interstellar civilization to sweep across interstellar space, founding a Type III civilization in the Milky Way.

See also DRAKE EQUATION; DYSON SPHERE; FERMI PARADOX; INTERSTELLAR COMMUNICATION; LIFE IN THE UNIVERSE; SEARCH FOR EXTRATERRESTRIAL INTELLIGENCE (SETI); ZOO HYPOTHESIS.

extraterrestrial contamination In general, the contamination of one world by life-forms, especially microorganisms, from another world. Using the Earth and its biosphere as a reference, this planetary contamination process is called *forward contamination* if an extraterrestrial sample or the alien world itself is contaminated by contact with terrestrial organisms, and *back contamination* if alien organisms are released into the Earth's biosphere.

An alien species will usually not survive when introduced into a new ecological system, because it is unable to compete with native species that are better adapted to the environment. Once in a while, however, alien species actually thrive because the new environment is very suitable, and indigenous life-forms are unable to successfully defend themselves against these alien invaders. When this "war of biological worlds" occurs, the result might very well be a permanent disruption of the host ecosphere, with severe biological, environmental, and possibly economic consequences.

Of course, the introduction of an alien species into an ecosystem is not always undesirable. Many European and Asian vegetables and fruits, for example, have been successfully and profitably introduced into the North American environment. However, any time a new organism is released in an existing ecosystem, a finite amount of risk is also introduced.

Frequently, alien organisms that destroy resident species are microbiological life-forms. Such microorganisms may have been nonfatal in their native habitat, but once released in the new ecosystem, they become unrelenting killers of native life-forms that are not resistant to them. In past centuries on Earth entire human societies fell victim to alien organisms against which they were defenseless, as, for example, was the case with the rapid spread of diseases that were transmitted to native Polynesians and American Indians by European explorers.

But an alien organism does not have to directly infect humans to be devastating. Who can easily ignore the consequences of the potato blight fungus that swept through Europe and the British Isles in the 19th century, causing a million people to starve to death in Ireland alone?

In the space age it is obviously of extreme importance to recognize the potential hazard of extraterrestrial contamination (forward or back). Before any species is intentionally introduced into another planet's environment, we must carefully determine not only whether the organism is pathogenic (disease-causing) to any indigenous species but also whether the new organism will be able to force out native species—with destructive effects on the original ecosystem. The introduction of rabbits into the Australian continent is a classic terrestrial example of a nonpathogenic life form creating immense problems when introduced into a new ecosystem. The rabbit population in Australia simply exploded in size because of their high reproduction rate, which was essentially unchecked by native predators.

Quarantine Protocols

At the start of the space age scientists were already aware of the potential extraterrestrial contamination problem—in either direction. Quarantine protocols (procedures) were established to avoid the forward contamination of alien worlds by outbound unmanned spacecraft, as well as the back contamination of the terrestrial biosphere when lunar samples were returned to Earth as part of the Apollo Program. For example, the United States is a signatory to a 1967 international agreement, monitored by the Committee on Space Research (COSPAR) of the International Council of Scientific Unions, that establishes the requirement to avoid forward and back contamination of planetary bodies during space exploration.

A quarantine is basically a forced isolation to prevent the movement or spread of a contagious disease. Historically, quarantine was the period during which ships suspected of carrying persons or cargo (for example, produce or livestock) infected with contagious diseases were detained at their port of arrival. The length of the quarantine, generally 40 days, was considered sufficient to cover the incubation period of most highly infectious terrestrial diseases. If no symptoms appeared at the end of the quarantine, then the travelers were permitted to disembark. In modern times the term *quarantine* has obtained a new meaning, namely, that of holding a suspect organism or infected person in strict isolation until it is no longer capable of transmitting the disease. With the Apollo Project and the advent of the lunar quarantine, the term now has elements of both meanings. Of special interest in future space missions to the planets and their major moons is how we avoid the potential hazard of back contamination of Earth's environment when robot spacecraft and human explorers bring back samples for more detailed examination in laboratories on Earth.

Planetary Quarantine Program

The National Aeronautics and Space Administration (NASA) started a planetary quarantine program in the late 1950s at the beginning of the U.S. civilian space program. This quarantine program, conducted with international cooperation, was intended to prevent, or at least minimize, the possibility of contamination of alien worlds by early space probes. At that time scientists were concerned with forward contamination. In this type of extraterrestrial contamination scenario, terrestrial microorganisms, "hitchhiking" on planetary probes and landers, would spread throughout another world, destroying any native life-forms, life precursors, and perhaps even remnants of past life-forms. If forward contamination occurred, it would compromise future scientific attempts to search for and identify extraterrestrial life-forms that had arisen independently of Earth's biosphere.

A planetary quarantine protocol was therefore established. This protocol required that outbound unmanned planetary missions be designed and configured to minimize the probability of alien world contamination by terrestrial life-forms. As a design goal, these spacecraft and probes had a probability of 1 in 1,000 (1×10^{-3}) or less that they could contaminate the target celestial body with terrestrial microorganisms. Decontamination, physical isolation (for example, prelaunch quarantine), and spacecraft design techniques have all been employed to support adherence to this protocol.

One simplified formula for describing the probability of planetary contamination is

$$P(c) = m \times P(r) \times P(g)$$

where

P(c) is the probability of contamination of the target celestial body by terrestrial microorganisms
m is the microorganism burden
P(r) is the probability of release of the terrestrial microorganisms from the spacecraft hardware
P(g) is the probability of microorganism growth after release on a particular planet or celestial object

As previously stated, P(c) had a design goal value of less than or equal to 1 in 1,000. A value for the microorganism burden (m) was established by sampling an assembled spacecraft or probe. Then, through laboratory experiments scientists determined how much this microorganism burden was reduced by subsequent sterilization and decontamination treatments. A value for P(r) was obtained by placing duplicate spacecraft components in simulated planetary environments. Unfortunately, establishing a numerical value for P(g) was a bit more tricky. The technical intuition of knowledgeable exobiologists and some educated "guessing" were blended together to create an estimate for how well terrestrial microorganisms might thrive on alien worlds that had not yet been visited. Of course, today, as we keep learning more about the environments on other worlds in our solar system, we can keep refining our estimates for P(g). Just how well terrestrial life-forms grow on Mars, Venus, Europa, Titan, and a variety of other interesting celestial bodies will be the subject of future in situ (on site) laboratory experiments performed by exobiologists or their robot surrogates.

Early Mars Missions

As a point of aerospace history, the early U.S. Mars flyby missions (for example, *Mariner 4,* launched on November 28, 1964, and *Mariner 6,* launched on February 24, 1969) had P(c) values ranging from 4.5×10^{-5} to 3.0×10^{-5}. These missions achieved successful flybys of the Red Planet on July 14, 1965, and July 31, 1969, respectively. Postflight calculations indicated that there was no probability of planetary contamination as a result of these successful precursor missions.

Apollo Missions

The human-crewed U.S. Apollo Project missions to the Moon (1969–72) also stimulated a great deal of debate about forward and back contamination. Early in the 1960s scientists began speculating in earnest whether there was life on the Moon. Some of the most bitter technical exchanges during the Apollo Project concerned this particular question. If there was life, no matter how primitive or microscopic, we would want to examine it carefully and compare it with life-forms of terrestrial origin. This careful search for microscopic lunar life would, however, be very difficult and expensive because of the forward-contamination problem. For example, all equipment and materials landed on the Moon would need rigorous sterilization and decontamination procedures. There was also the glaring uncertainty about back contamination. If microscopic life did indeed exist on the Moon, did it represent a serious hazard to the terrestrial biosphere? Because of the potential extraterrestrial contamination problem, some members of the scientific community urged time-consuming and expensive quarantine procedures.

On the other side of this early 1960s contamination argument were those exobiologists who emphasized the suspected extremely harsh lunar conditions: virtually no atmosphere, probably no water, extremes of temperature ranging from 120°C at lunar noon to −150°C during the frigid lunar night, and unrelenting exposure to lethal doses of ultraviolet, charged particle, and X-ray radiations from the Sun. No life form, it was argued, could possibly exist under such extremely hostile conditions.

This line of reasoning was countered by other exobiologists, who hypothesized that trapped water and moderate temperatures below the lunar surface could sustain very primitive life-forms. And so the great extraterrestrial contamination debate raged back and forth, until finally the *Apollo 11* expedition departed on the first lunar-landing mission. As a compromise, the *Apollo 11* mission flew to the Moon with careful precautions against back contamination but with only a very limited effort to protect the Moon from forward contamination by terrestrial organisms.

The Lunar Receiving Laboratory (LRL) at the Johnson Space Center in Houston provided quarantine facilities for two years after the first lunar landing. What we learned during its operation serves as a useful starting point for planning new quarantine facilities, Earth-based or space-based. In the future these quarantine facilities will be needed to accept, handle, and test extraterrestrial materials from Mars and other solar system bodies of interest in our search for alien life-forms (present or past).

During the Apollo Project no evidence was discovered that native alien life was then present or had ever existed on the Moon. Scientists at the Lunar Receiving Laboratory performed a careful search for carbon, since terrestrial life is carbon-based. Approximately 100 to 200 parts per million of carbon were found in the lunar samples. Of this amount, only a few tens of parts per million were considered indigenous to the lunar material, while the bulk amount of carbon had been deposited by the solar wind. Exobiologists and lunar scientists have concluded that none of this carbon appears to be derived from biological activity. In fact, after the first few Apollo expeditions to the Moon, even back-contamination quarantine procedures of isolating the Apollo astronauts for a period of time were dropped.

Preventing Back Contamination

There are three fundamental approaches toward handling extraterrestrial samples to avoid back contamination. First, we could sterilize a sample while it is en route to Earth from its native world. Second, we could place it in quarantine in a remotely located, maximum confinement facility on Earth while scientists examine it closely. Finally, we could also perform a preliminary hazard analysis (called the extraterrestrial protocol tests) on the alien sample in an orbiting quarantine facility before we allowed the sample to enter the terrestrial biosphere. To be adequate, a quarantine facility must be capable of (1) containing all alien organisms present in a sample of extraterrestrial material, (2) detecting these alien organisms during protocol testing, and (3) controlling these organisms after detection until scientists could dispose of them in a safe manner.

One way to bring back an extraterrestrial sample that is free of potentially harmful alien microorganisms is to sterilize the material during its flight to Earth. However, the sterilization treatment used must be intense enough to guarantee that no life-forms as we currently know them could survive. An important concern here is also the impact the sterilization treatment might have on the scientific value of the alien world sample. For example, use of chemical sterilants would most likely result in contamination of the sample, preventing the measurement of certain soil properties. Heat could trigger violent chemical reactions within the soil sample, resulting in significant changes and the loss of important exogeological data. Finally, sterilization would also greatly reduce the biochemical information content of the sample. It is even questionable as to whether any significant exobiology data can be obtained by analyzing a heat-sterilized alien material sample. To put it simply, in their search for extraterrestrial life-forms, exobiologists want "virgin alien samples."

If we do not sterilize the alien samples en route to Earth, we have only two general ways of avoiding possible back-contamination problems. We can place the unsterilized sample of alien material in a maximum quarantine facility on Earth and then conduct detailed scientific investigations, or we can intercept and inspect the sample at an orbiting quarantine facility before allowing the material to enter Earth's biosphere.

The technology and procedures for hazardous material containment have been employed on Earth in the development of highly toxic chemical and biological warfare agents and in conducting research involving highly infectious diseases. A critical question for any quarantine system is whether the containment measures are adequate to hold known or suspected pathogens while experimentation is in progress. Since the characteristics of potential alien organisms are not presently known, we must assume that the hazard they could represent is at least equal to that of terrestrial Class IV pathogens. (A terrestrial Class IV pathogen is an organism capable of being spread very rapidly among humans; no vaccine exists to check its spread; no cure has been developed for it; and the organism produces high mortality rates in infected persons.) Judging from the large uncertainties associated with potential extraterrestrial life-forms, it is not obvious that any terrestrial quarantine facility will gain very wide acceptance by the scientific community or the general public. For example, locating such a facility and all its workers in an isolated area on Earth actually provides only a small additional measure of protection. Consider, if you will, the planetary environmental impact controversies that could rage as individuals speculated about possible ecocatastrophes. What would happen to life on Earth if alien organisms did escape and went on a deadly rampage throughout the Earth's biosphere? The alternative to this potentially explosive controversy is quite obvious: locate the quarantine facility in outer space.

Orbiting Quarantine Facility

A space-based facility provides several distinct advantages: (1) It eliminates the possibility of a sample-return spacecraft crashing and accidentally releasing its deadly cargo of alien microorganisms, (2) it guarantees that any alien organisms that might escape from confinement facilities within the orbiting complex cannot immediately enter Earth's biosphere, and (3) it ensures that all quarantine workers remain in total isolation during protocol testing (that is, during the testing procedure).

As we expand the human sphere of influence into heliocentric space, we must also remain conscious of the potential hazards of extraterrestrial contamination. Scientists, space explorers, and extraterrestrial entrepreneurs must be aware of the ecocatastrophes that might occur when "alien worlds collide"—especially on the microorganism level.

With a properly designed and operated orbiting quarantine facility, alien world materials can be tested for potential hazards. Three hypothetical results of such protocol testing are: (1) no replicating alien organisms are discovered; (2) replicating alien organisms are discovered, but they are also found not to be a threat to terrestrial life-forms, and (3) hazardous replicating alien life-forms are discovered. If potentially harmful replicating alien organisms were discovered during these protocol tests, then quarantine workers would either render the sample harmless (for example, through heat and chemical sterilization procedures); retain it under very carefully controlled conditions in the orbiting complex and perform more detailed analyses on the alien life-forms; or properly dispose of the sample before the alien life-forms could enter Earth's biosphere and infect terrestrial life-forms.

Contamination Issues in Exploring Mars
Increasing interest in Mars exploration has also prompted a new look at the planetary protection requirements for forward contamination. In 1992, for example, the Space Studies Board of the U.S. National Academy of Sciences recommended changes in the requirements for Mars landers that significantly alleviated the burden of planetary protection implementation for these missions. The board's recommendations were published in the document "Biological Contamination of Mars: Issues and Recommendations" and presented at the 29th COSPAR Assembly, which was held in 1992 in Washington, D.C. In 1994 a resolution addressing these recommendations was adopted by COSPAR at the 30th assembly; it has been incorporated into NASA's planetary protection policy. Of course, as we learn more about Mars, planetary protection requirements may change again to reflect future scientific knowledge.

These new recommendations recognize the very low probability of growth of (terrestrial) microorganisms on the Martian surface. With this assumption in mind, the forward-contamination protection policy shifts from probability-of-growth considerations to a more direct and determinable assessment of the number of microorganisms associated with any landing event. For landers that do not have life-detection instrumentation, the level of biological cleanliness required is that of the *Viking* spacecraft prior to heat sterilization. Class 100,000 clean-room assembly and component testing can accomplish this level of biological cleanliness. This is considered a very conservative approach that minimizes the chance of compromising future exploration. Landers with life-detection instruments would be required to meet *Viking* spacecraft post-sterilization levels of biological cleanliness or levels driven by the search-for-life experiment itself. Scientists recognize that the sensitivity of a life-detection instrument may impose the more severe biological cleanliness constraint on a Mars lander mission.

Included in recent changes to COSPAR's planetary protection policy is the option that an orbiter spacecraft not be required to remain in orbit around Mars for an extended time if it can meet the biological cleanliness standards of a lander without life-detection experiments. In addition, the probability of inadvertent early entry (into the Martian atmosphere) has been relaxed compared to previous requirements.

The present policy for samples returned to Earth remains directed toward containing potentially hazardous Martian material. Concerns still include a difficult-to-control pathogen capable of directly infecting human hosts (currently considered extremely unlikely) or a life form capable of upsetting the current ecosystem. Therefore, for a future Mars sample return mission (MSRM) the following backward contamination policy now applies. All samples would be enclosed in hermetically sealed containers. The contact chain between the return space vehicle and the surface of Mars must be broken in order to prevent the transfer of potentially contaminated surface material by means of the return spacecraft's exterior. The samples would be subjected to a comprehensive quarantine protocol to investigate whether harmful constituents are present. It should also be recognized that even if the sample return mission has no specific exobiological goals, the mission would still be required to meet the planetary protection sample return procedures as well as the life-detection protocols for forward contamination protection. This policy not only mitigates concern over potential contamination (forward or back), but it also prevents a hardy terrestrial microorganism "hitchhiker" from masquerading as a Martian life form.

Stardust
Stardust is the first U.S. space mission dedicated solely to the exploration of a comet and also the first robotic spacecraft mission designed to return extraterrestrial material from outside the orbit of the Moon. Launched on February 7, 1999, from Cape Canaveral Air Force Station, this spacecraft flew within 236 kilometers of Comet Wild 2 on January 2, 2004. During the close encounter *Stardust* captured thousands of particles and volatiles of cometary material. The spacecraft has also collected samples of interstellar dust, including recently discovered dust streaming into our solar system from the direction of Sagittarius. These materials are believed to consist of ancient presolar interstellar grains and nebular remnants that date back to the formation of the solar system. Scientists anticipate their analysis of such fascinating celestial specks will yield important insights into the evolution of the Sun, the planets, and possibly life itself.

In January 2006 *Stardust* returns to Earth in order to drop off its cargo of extraterrestrial material. The spacecraft will deliver these materials by precisely ejecting a specially designed 60-kilogram reentry capsule. The capsule will pass at high speed through Earth's atmosphere and then parachute to the planet's surface.

Specifically, *Stardust* is scheduled to return to Earth on January 15, 2006. Soon after the spacecraft makes its final trajectory maneuver at an altitude of about 111,000 kilometers, the spacecraft will release the sample return capsule. After the capsule has been released, the main spacecraft will perform a diversionary maneuver in order to avoid entering Earth's atmosphere. Following this maneuver, the *Stardust* spacecraft will go into orbit around the Sun.

Once ejected, the sample return capsule will enter Earth's atmosphere at a velocity of approximately 12.8 kilometers per second. Aerospace engineers have given the capsule an aerodynamic shape and center of gravity similar to that of a badminton shuttlecock. Because of this special design, the capsule will automatically orient itself with its nose down as it enters the atmosphere.

As the capsule descends, atmospheric friction on the heat shield will reduce its speed. When the capsule reaches an altitude of about 30 kilometers, it will have slowed down to about 1.4 times the speed of sound. At that point a small pyrotechnic charge will fire, releasing a drogue parachute. After descending to an altitude of about 3 kilometers, the line holding the drogue chute will be cut, allowing the drogue parachute to pull out a larger parachute that will carry the capsule to its soft landing. At touchdown, the capsule will be traveling at approximately 4.5 meters per second, or 16 kilometers per hour. About 10 minutes will elapse between the beginning of the capsule's entry into Earth's atmosphere until the main parachute is deployed.

Scientists chose the landing site at the Utah Test and Training Range near Salt Lake City because the area is a vast and desolate salt flat controlled by the U.S. Air Force in conjunction with the U.S. Army. The landing footprint for the sample return capsule will be about 30 by 84 kilometers, an ample space to allow for aerodynamic uncertainties and winds that might affect the direction the capsule travels in the atmosphere. The sample return capsule will approach the landing zone on a heading of approximately 122 degrees on a northwest to southeast trajectory. Landing will take place at about 3 A.M. (Mountain Standard Time) on January 15, 2006. A UHF radio beacon on the capsule will transmit a signal as the capsule descends to Earth, while the parachute and capsule will be tracked by radar. A helicopter will be used to fly the retrieval crew to the landing site. Given the small size and mass of the capsule, mission planners do not expect that its recovery and transportation will require extraordinary handling measures or hardware other than a specialized handling fixture to cradle the capsule during transport.

See also EXOBIOLOGY; LIFE IN THE UNIVERSE; MARS; MARS SAMPLE RETURN MISSION; STARDUST MISSION; VIKING PROJECT.

extraterrestrial life Life-forms that may have evolved independent of and now exist outside of the terrestrial biosphere.

See also EXOBIOLOGY; LIFE IN THE UNIVERSE.

extraterrestrial life chauvinisms At present we have only one scientific source of information on the emergence of life in a planetary environment—our own planet Earth. Scientists currently believe that all carbon-based terrestrial organisms descended from a common, single occurrence of the origin of life in the primeval "chemical soup" of an ancient Earth. How can we project this singular fact to the billions of unvisited worlds in the cosmos? We can only do so with great technical caution, realizing full well that our models of extraterrestrial life-forms and our estimates concerning the cosmic prevalence of life can easily become prejudiced, or chauvinistic, in their findings.

Chauvinism is defined as a strongly prejudiced belief in the superiority of one's group. Applied to speculations about extraterrestrial life, this word can take on several distinctive meanings, each heavily influencing any subsequent thought on the subject. Some of the more common forms of extraterrestrial life chauvinisms are: G-star chauvinism, planetary chauvinism, terrestrial chauvinism, chemical chauvinism, oxygen chauvinism, and carbon chauvinism. Although such heavily steeped thinking may not actually be wrong, it is important to realize that it also sets limits, intentionally or unintentionally, on contemporary speculations about life in the universe.

G-star, or solar system, chauvinism implies that life can originate only in a star system like our own—namely, a system containing a single G-spectral-class star. Planetary chauvinism assumes that extraterrestrial life has to develop independently on a particular type of planet, while terrestrial chauvinism stipulates that only "life as we know it on Earth" can originate elsewhere in the universe. Chemical chauvinism demands that extraterrestrial life be based on chemical processes, while oxygen chauvinism states that alien worlds must be considered uninhabitable if their atmospheres do not contain oxygen. Finally, carbon chauvinism asserts that extraterrestrial life-forms must be based on carbon chemistry.

These chauvinisms, singularly and collectively, impose tight restrictions on the type of planetary system that we suspect might support the rise of living systems elsewhere in the universe, possibly to the level of intelligence. If such restrictions are indeed correct, then our search for extraterrestrial intelligence is now being properly focused on Earthlike worlds around sunlike stars. If, on the other hand, life is actually quite prevalent and capable of arising in a variety of independent biological scenarios (for example, silicon-based or sulfur-based chemistry), then our contemporary efforts in modeling the cosmic prevalence of life and in trying to describe what "little green men" really look like is somewhat analogous to using the atomic theory of Democritus, the ancient Greek philosopher (ca. 460–ca. 370 B.C.E.), to help describe the inner workings of a modern nuclear fission reactor. As we continue to explore planetary bodies in our own solar system—especially Mars and the interesting larger moons of Jupiter (e.g, Europa) and Saturn (e.g., Titan)—we will be able to more effectively assess how valid these extraterrestrial life chauvinisms really are. The discovery of extinct or extant alien life on any (or all) of these neighboring worlds would suggest that the universe is probably teaming with life in a wide variety of previously unimaginable forms.

See also EXOBIOLOGY; LIFE IN THE UNIVERSE; SEARCH FOR EXTRATERRESTRIAL INTELLIGENCE (SETI).

extraterrestrial resources The resources to be found in space that could be used to help support an extended human presence and eventually become the physical basis for a thriving solar system–level civilization. These resources include unlimited solar energy, a full range of raw materials (from the Moon, asteroids, comets, Mars, numerous moons of outer planets, etc.), and an environment that is both special (e.g., access to high vacuum, microgravity, physical isolation from the terrestrial biosphere) and reasonably predictable.

extravehicular activity (EVA) Extravehicular activity, or EVA, may be defined as the activities conducted in space by an ASTRONAUT or COSMONAUT outside the protective environment of his or her spacecraft, aerospace vehicle, or space station. In the U.S. space program astronaut EDWARD H. WHITE II performed the first EVA on June 3, 1965, when he left the protective environment of his *Gemini IV* space capsule and ventured into space (while constrained by an umbilical tether). Since that historic demonstration EVA has been used successfully during a variety of American and Russian space missions to make critical repairs, perform inspections, help capture and refurbish failed satellites, clean optical surfaces, deploy equipment, and retrieve experiments. The term *EVA* (as applied to the SPACE SHUTTLE and *INTERNATIONAL*

American astronaut Jerry L. Ross, STS-88 mission specialist, is pictured during one of three space walks that were conducted on the 12-day space shuttle *Endeavour* mission in December 1998, a mission that supported the initial orbital assembly of the *International Space Station* (ISS). Astronaut James H. Newman, mission specialist, recorded this image while perched on the end of the *Endeavour's* remote manipulator system (RMS) arm. Newman can be seen reflected in Ross's helmet visor. The solar array panel for the Russian-built Zarya module for the ISS appears along the right edge of the picture. *(Courtesy of NASA)*

SPACE STATION) includes all activities for which crewmembers don their spacesuits and life support systems and then exit the orbiter vehicle's crew cabin to perform operations internal or external to the cargo bay.

Extreme Ultraviolet Explorer **(EUVE)** One of NASA's Explorer-class satellites launched from Cape Canaveral Air Force Station by a Delta II rocket on June 7, 1992 (at 16:40 UTC). The scientific satellite traveled around Earth in an approximate 525-kilometer-altitude orbit with a period of 94.8 minutes and an inclination of 28.4 degrees. It provided astronomers with a view of the relatively unexplored region of the electromagnetic spectrum—the extreme ultraviolet (EUV) region (i.e., 10 to 100 nanometers wavelength).

The science payload consisted of three grazing-incidence scanning telescopes and an extreme ultraviolet (EUV) spectrometer/deep survey instrument. The science payload was attached to a multimission modular spacecraft. For the first six months following its launch, the spacecraft performed a full-sky survey. The spacecraft also gathered important data about sources of EUV radiation within the "local bubble"—a hot, low-density region of the Milky Way galaxy (including our Sun) that is the result of a supernova explosion some 100,000 years ago. Interesting EUV sources include white dwarf stars and binary star systems in which one star is siphoning material from the outer atmosphere of its companion.

NASA extended the EUVE mission twice, but by the year 2000 operational costs and scientific merit issues led to a decision to terminate spacecraft activities. Consequently, NASA commanded the satellite's transmitters off on January 2, 2001, and then formally ended satellite operations on January 31, 2001, by placing the spacecraft in a safehold configuration. On January 30, 2002 (at approximately 11:15 pm EST), the EUVE spacecraft reentered Earth's atmosphere over central Egypt. Also called *BERKSAT* and *Explorer 67.*

See also EXPLORER SPACECRAFT.

extreme ultraviolet radiation (EUV) The region of the electromagnetic spectrum corresponding to wavelengths between 10 and 100 nanometers (100 and 1,000 angstroms).

See also ELECTROMAGNETIC SPECTRUM.

extremophile A hardy (terrestrial) microorganism that can exist under extreme environmental conditions, such as in frigid polar regions or boiling hot springs. Exobiologists speculate that similar (extraterrestrial) microorganisms might exist elsewhere in the solar system, perhaps within subsurface biological niches on Mars or in a suspected liquid water ocean beneath the frozen surface of Europa.

See also EUROPA; EXOBIOLOGY; LIFE IN THE UNIVERSE; MARS.

eyeballs in, eyeballs out Aerospace jargon derived from test pilots and used to describe the acceleration experienced by an astronaut or cosmonaut at liftoff or when the retrorockets fire. The experience at liftoff is "eyeballs in" (positive g because of vehicle acceleration), while the experience when the retrorockets fire is "eyeballs out" (negative g because of space capsule or aerospace vehicle deceleration).

eyepiece A magnifying lens that helps an observer view the image produced by a telescope.

See also LENS; TELESCOPE.

Fabricius, David (1564–1617) German *Astronomer* The clergyman David Fabricus was a skilled naked eye observer who corresponded with JOHANNES KEPLER and discovered the first variable star (Mira) in 1596. His son (JOHANNES FABRICUS) was a student in the Netherlands and in 1611 introduced his father to the telescope, an instrument recently invented in that country by HANS LIPPERSHEY. Soon father and son followed the Italian astronomer GALILEO GALILEI by making their own telescopic observations of the heavens. Unfortunately, the son's death at age 29 and the father's murder brought their early contributions to the era of telescope-based astronomy to a sudden halt.

See also TELESCOPE.

Fabricius, Johannes (1587–1616) German *Astronomer* Johannes Fabricius, the son of DAVID FABRICUS, made early telescopic observations of the heavens with his father, including the discovery of sunspots in 1611. Intrigued, he investigated the number and dynamic behavior of sunspots and wrote an early paper on the subject. However, his work is generally overshadowed by the efforts of GALILEO GALILEI and CHRISTOPH SCHEINER, who independently performed sunspot studies at about the same time.

facula (plural: faculae) A bright region of the Sun's photosphere that has a higher temperature than the surrounding region. Solar physicists note that faculae are closely related to sunspots. Faculae usually form before sunspots appear and then persist in the vicinity for many days after the sunspots have disappeared.

See also SUN.

Fahrenheit (symbol: F) A relative temperature scale with the freezing point of water (at atmospheric pressure) given a value of 32°F and the boiling point of water (at atmospheric pressure) assigned the value of 212°F.

failed star *See* BROWN DWARF.

fairing A structural component of a rocket or aerospace vehicle designed to reduce drag or air resistance by smoothing out nonstreamlined objects or sections.

fallaway section A section of a launch vehicle or rocket that is cast off and separates from the vehicle during flight, especially a section that falls back to Earth.

See also LAUNCH VEHICLE; ROCKET.

farad (symbol: F) The SI unit of electrical capacitance. It is defined as the capacitance of a capacitor whose plates have a potential difference of one volt when charged by a quantity of electricity equal to one coulomb. This unit is named after the 19th-century British scientist Michael Faraday (1791–1867), who was a pioneer in the field of electromagnetism. Since the farad is too large a unit for typical applications, submultiples, such as the microfarad (10^{-6}F), the nanofarad (10^{-9}F), and the picofarad (10^{-12}F), are encountered frequently.

Far Infrared and Submillimetre **Telescope (FIRST)** *See HERSCHEL SPACE OBSERVATORY.*

farside The side of the Moon that never faces Earth. *Compare with* NEARSIDE.

Far-Ultraviolet Spectroscopic Explorer **(FUSE)** NASA's *Far-Ultraviolet Spectroscopic Explorer* (FUSE) represented the next-generation, high-orbit, ultraviolet space observatory by examining the wavelength region of the electromagnetic spectrum ranging from about 90 to 120 nanometers (nm). This scientific spacecraft was launched from Cape Canaveral on June 24, 1999, by a Delta II rocket. The ultraviolet (UV) astronomy satellite now travels around Earth in a near-circular

An excellent view of a near-full moon as photographed by the *Apollo 16* crew during June 1972 shortly after transearth injection (TEI). Most of the lunar surface that appears in this picture is on the farside of the Moon. ***(Courtesy of NASA)***

752-km by 767-km orbit at an inclination of 25°. Through 2004 FUSE continued to operate in a satisfactory matter and provide valuable observational data. The primary objective of FUSE is to use high-resolution spectroscopy at far-ultraviolet wavelengths to study the origin and evolution of the lightest elements (hydrogen and deuterium) created shortly after the big bang and the forces and processes involved in the evolution of galaxies, stars, and planetary systems. FUSE is a part of NASA's Origins Program. The spacecraft represents a joint U.S.-Canada-France scientific project. A previous mission, the *Copernicus* spacecraft, had examined the universe in the far-ultraviolet region of the electromagnetic spectrum. However, FUSE is providing data collection with sensitivity some 10,000 times greater than the *Copernicus* spacecraft.

The FUSE satellite consists of two primary sections, the spacecraft and the science instrument. The spacecraft contains all of the elements necessary for powering and pointing the satellite, including the attitude control system, the solar panels, and communications electronics and antennas. The observatory is approximately 7.6 m long with the baffle fully deployed. The FUSE science instrument consists of four coaligned telescope mirrors (with approximately 39 cm × 35 cm clear aperture). The light from the four optical channels is dispersed by four spherical, aberration-corrected holographic diffraction gratings and recorded by two delay-line microchannel plate detectors. Two channels with SiC coatings cover the spectral range from 90.5 to 110 nm, and two channels with LiF coatings cover the spectral range from 100 to 119.5 nm.

See also COPERNICUS SPACECRAFT; ORIGINS PROGRAM; ULTRAVIOLET ASTRONOMY.

Fast Auroral Snapshot Explorer **(FAST)** NASA's *Fast Auroral Snapshot Explorer* (FAST) was launched from Vandenberg Air Force Base, California, on August 21, 1996, aboard a Pegasus XL vehicle. FAST was the second spacecraft in NASA's Small Explorer (SMEX) Program. In conjunction with other spacecraft and ground-based observations, this modestly-sized (1.8-m by 1.2-m), 187-kg-mass spacecraft collected scientific data concerning the physical processes that produce auroras, the displays of light that appear in the upper atmosphere at high latitudes.

From its highly eccentric (353-km by 4,163-km-altitude), near-polar (83°-inclination) orbit, the spacecraft investigated the plasma physics of auroral phenomena at extremely high time and spatial resolution, using its complement of particle and fields instruments. The FAST instrument set included 16 electrostatic analyzers, four electric field Langmuir probes suspended on 30-m-long wire booms, two electric field Langmuir probes on 3-m long extendable booms, a variety of magnetometers, and a time-of-flight mass spectrometer. To accomplish its scientific mission, FAST was designed as a 12-rpm, spin-stabilized spacecraft with its spin axis oriented parallel to the orbit axis. A body-mounted solar array (containing 5.6m^2 of solar cells) provided approximately 52 watts of electric power (orbit average) to the spacecraft and instruments. The spacecraft had a design life of one year.

See also AURORA.

"faster-than-light" travel The hypothesized ability to travel faster than the known physical laws of the universe will permit. In accordance with Einstein's theory of relativity, the speed of light is the ultimate speed that can be reached in the space-time continuum. The speed of light in free space is 299,793 kilometers per second. Concepts such as "hyperspace" have been introduced in science fiction to sneak around this speed-of-light barrier. Unfortunately, despite popular science fiction stories to the contrary, most scientists today feel that the speed-of-light limit is a real physical law that is not likely to change unless they can figure out a way to tunnel through (perhaps with a wormhole) or warp the space-time continuum in a clever way that permits superluminal movement through the fabric of space and time.

See also INTERSTELLAR TRAVEL; RELATIVITY; SPACE-TIME; STARSHIP; TACHYON.

fatigue 1. In aerospace engineering, a weakening or deterioration of metal or other material occurring under load, especially under repeated cyclic or continued loading. Self-explanatory compounds of this term include fatigue crack, fatigue failure, fatigue load, fatigue resistance, and fatigue test. 2. With respect to human space flight, the level of exhaustion experienced by a person after exposure to any type of physical or psychological stress, as, for example, pilot fatigue during a long mission or astronaut fatigue during a particularly demanding and lengthy extravehicular activity.

fault A fracture or zone of fractures along which the sides are displaced relative to one another. A *normal fault* is one in which rocks have been shifted vertically by extensional forces. A *reverse fault* is one in which rocks have been shifted vertically by compressional forces. A *strike-slip fault* is one in which rocks have been shifted horizontally past each other along the strike of the fault.

fault tolerance The capability of an aerospace system to function despite one or more critical failures; usually achieved by the use of redundant circuits or functions and/or reconfigurable components.

feasibility study A study to determine whether an aerospace mission plan is within the capacity of the resources and/or technologies that can be made available—for example, a NASA feasibility study involving a robotic soil sample return mission to the surface of Mars or a human-crewed expedition to the Red Planet.

feedback 1. The return of a portion of the output of a device to the input. Positive feedback adds to the input; negative feedback subtracts from the input. 2. Information concerning results or progress returned to an originating source. 3. In aeronautical (and aerospace) engineering, the transmittal of forces initiated by aerodynamic action on control surfaces or rotor blades to the cockpit controls.

femto- (symbol: f) The SI prefix for 10^{-15}. This prefix is used to designate very small quantities, such as a femtosecond (fs), which corresponds to 10^{-15} second—a very brief flash of time.

Fermi, Enrico (1901–1954) Italian-American *Physicist* Fermi helped create the nuclear age and won the 1938 Nobel Prize in physics. In addition to making numerous contributions to nuclear physics, he is also credited with the famous FERMI PARADOX—"WHERE ARE THEY?"—the popular speculative inquiry concerning the diffusion of an advanced alien civilization through the galaxy on a wave of exploration.

Fermi paradox—"Where are they?" A paradox is an apparently contradictory statement that may nevertheless be true. According to the lore of physics, the famous Fermi paradox arose one evening in 1943 during a social gathering at Los Alamos, New Mexico, when the brilliant Italian-American physicist ENRICO FERMI asked the penetrating question "Where are *they*?" "Where are who?" his startled companions replied. "Why, the extraterrestrials," responded the Nobel Prize–winning physicist, who was at the time one of the lead scientists on the top-secret Manhattan Project to build the first atomic bomb.

Fermi's line of reasoning that led to this famous inquiry has helped form the basis of much modern thinking and strategy concerning the search for extraterrestrial intelligence

(SETI). It can be summarized as follows. Our galaxy is some 10 to 15 billion (10^9) years old and contains perhaps 100 billion stars. If just one advanced civilization had arisen in this period of time and attained the technology necessary to travel between the stars, that advanced civilization could have diffused through or swept across the entire galaxy within 50 million to 100 million years—leaping from star to star, starting up other civilizations, and spreading intelligent life everywhere. But as we look around, we do not see a galaxy teeming with intelligent life, nor do we have any technically credible evidence of visitations or contact with alien civilizations, so we must conclude that perhaps no such civilization has ever arisen in the 15-billion-year history of the galaxy. Therefore, the paradox: Although we might expect to see signs of a universe filled with intelligent life (on the basis of statistics and the number of possible "life sites," given the existence of 100 billion stars in just this galaxy alone), we have seen no evidence of such. Are we, then, really alone? If we are not alone—where are *they*?

Many attempts have been made to respond to Fermi's very profound question. The "pessimists" reply that the reason we have not seen any signs of intelligent extraterrestrial civilizations is that we really are alone. Maybe we are the first technically advanced intelligent beings to rise to the level of space travel. If so, then perhaps it is our cosmic destiny to be the first species to sweep through the galaxy spreading intelligent life.

The "optimists," on the other hand, hypothesize that intelligent life exists out there somewhere and offer a variety of possible reasons why we have not "seen" these civilizations yet. A few of these proposed reasons follow. First, perhaps intelligent alien civilizations really do not want anything to do with us. As an emerging planetary civilization we may be just too belligerent, too intellectually backward, or simply below their communications horizon. Other optimists suggest that not every intelligent civilization has the desire to travel between the stars, or maybe they do not even desire to communicate by means of electromagnetic signals. Yet another response to the intriguing Fermi paradox is that *we* are actually *they*—the descendants of ancient star travelers who visited Earth millions of years ago when a wave of galactic expansion passed through this part of the galaxy.

Still another group responds to Fermi's question by declaring that intelligent aliens are out there right now but that they are keeping a safe distance, watching us either mature as a planetary civilization or else destroy ourselves. A subset of this response is the *extraterrestrial zoo hypothesis*, which speculates that we are being kept as a "zoo" or wildlife preserve by advanced alien zookeepers who have elected to monitor our activities but not be detected themselves. Naturalists and animal experts often suggest that a "perfect" zoo here on Earth would be one in which the animals being kept have no direct contact with their keepers and do not even realize they are in captivity.

Finally, other people respond to the Fermi paradox by saying that the wave of cosmic expansion has not yet reached our section of the galaxy—so we should keep looking. Within this response group are those who boldly declare that the alien visitors are now among us.

See also DRAKE EQUATION; LIFE IN THE UNIVERSE; ORIGINS PROGRAM; SEARCH FOR EXTRATERRESTRIAL INTELLIGENCE (SETI); ZOO HYPOTHESIS.

ferret satellite A military spacecraft designed for the detection, location, recording, and analyzing of electromagnetic radiation—for example, enemy radio frequency (RF) transmissions.

ferry flight An (in-the-atmosphere) flight of a space shuttle orbiter mated on top of the Boeing 747 shuttle carrier aircraft.

See also SPACE TRANSPORTATION SYSTEM (STS).

field of view (FOV) The area or solid angle that can be viewed through or scanned by an optical instrument.

See also REMOTE SENSING.

film cooling The cooling of a body or surface, such as the inner surface of a rocket's combustion chamber, by maintaining a thin fluid layer over the affected area.

filter 1. A thin layer of selective material placed in front of a detector that lets through only a selected color (e.g., a blue-light filter for a photometer), group of electromagnetic radiation wavelengths (e.g., thermal infrared band radiation between 10 and 11 micrometers wavelength), group of photon energies (e.g., gamma rays with energies between 1.0 and 1.1 MeV [million electron volts]), or a desired (often narrow)

This U.S. Department of Defense artist's rendering shows the Russian *Cosmos 389* satellite, a military spacecraft launched in December 1970 to perform electronic intelligence (ELINT) missions against American and NATO forces. *Cosmos 389* was the first in a series of "ferret satellites" used by the Soviet Union to pinpoint sources of radio and radar emissions. Russian military intelligence analysts used ELINT data obtained from such satellites to identify "hostile" air defense sites and command and control centers as part of their overall targeting and war planning efforts during the cold war. *(Artist's rendering courtesy of Department of Defense; artist, Brian W. McMullin, artwork prepared 1982)*

range of particle energies (e.g., solar protons between 2.0 and 2.2 MeV). Narrow-band filters can be quite selective, passing radiation in only a very narrow region of the electromagnetic spectrum—for example, a 427.8 nanometer (4,278 Å) wavelength narrow-band optical filter. Other radiation filters are reasonably broad—for example, a filter that passes an entire "color" in the visible spectrum (e.g., blue-light filter or a red-light filter). 2. A device that removes suspended particulate matter from a liquid or a gas. This removal process is accomplished by forcing the fluid (with the suspended matter) through a permeable material (i.e., one that has many tiny holes or pores). The permeable material retains the solid matter while permitting the passage of the fluid that has now been "filtered" to some level of purity. 3. In electronics, a device that transmits only part of the incident electromagnetic energy and thereby may change its spectral distribution. *High-pass filters* transmit energy above a certain frequency, while *low-pass filters* transmit energy below a certain frequency. *Band-pass filters* transmit energy within a certain frequency band, while *band-stop filters* transmit only energy outside a specific frequency band.

fin 1. In aeronautical and aerospace engineering, a fixed or adjustable airfoil attached longitudinally to an aircraft, missile, rocket, or similar body to provide a stabilizing effect. 2. In heat transfer, a projecting flat plate or structure that facilitates thermal energy transfer, such as a *cooling fin.*

fire arrow An early gunpowder rocket attached to a large bamboo stick; developed by the Chinese about 1,000 years ago to confuse and startle enemy troops. (*See* figure.) Modern space technology depends on the use of the rocket, a device first developed by the Chinese more than a millennium ago. Exactly when the idea of the rocket first emerged in ancient China is not clear. Historians suggest that the Chinese had formulated gunpowder by the first century C.E. However, they generally just used this explosive chemical mixture to make fireworks for festivals.

The Chinese would stuff hollowed pieces of bamboo with gunpowder and then toss the simple devices into a fire, where they soon exploded, producing the desired festive effect. Quite possibly one of these "firecracker" devices did not explode but rather shot out of the fire like a "rocket" in response to the propulsive action-reaction principle caused by the escaping hot gases from burning gunpowder. At that point the first Chinese "rocketeer" might have gotten the clever idea of attaching one of these bamboo tubes filled with gunpowder to an arrow or long stick. The famous Chinese "fire arrow" was born.

Whatever the precise technical pathway of discovery, the Battle of K'ai-fung-fu in 1232 represents the first reported use of the gunpowder-fueled rocket in warfare. During this battle Chinese troops used a barrage of rocket-propelled fire arrows to startle and defeat a band of invading Mongolian warriors. In an early attempt at passive guided missile control, the Chinese rocketeers attached a long stick to the end of the fire arrow rocket. The long stick kept the center of pressure behind the rocket's center of mass during flight. Although the addition of this long stick helped somewhat, the flight of the rocket-propelled fire arrows remained quite erratic and highly inaccurate. The heavy stick also reduced the range of these early gunpowder-fueled rockets. Despite the limitations of the fire arrow, the invading Mongol warriors quickly learned from their unpleasant experience at the Battle of K'ai-fung-fu and adopted the interesting new weapon for their own use. As a result, nomadic Mongol warriors spread rocket technology westward when they invaded portions of India, the Middle East, and Europe. By the Renaissance European armories bristled with a collection of gunpowder rockets—all technical descendants of the Chinese fire arrow.

A 13th-century Chinese warrior prepares to light a crude gunpowder rocket, known as a fire arrow. The Chinese army often used a barrage of these primitive rockets to startle and confuse enemy troops during a battle. ***(Drawing courtesy of NASA)***

See also ROCKET; ROCKETRY.

fission (nuclear) In nuclear fission the nucleus of a heavy element, such as uranium or plutonium, is bombarded by a neutron, which it absorbs. The resulting compound nucleus is unstable and soon breaks apart, or fissions, forming two lighter nuclei (called *fission products*) and releasing additional neutrons. In a properly designed nuclear reactor these fission neutrons are used to sustain the fission process in a controlled chain reaction. The nuclear fission process is accompanied by the release of a large amount of energy, typically 200 million electron volts (MeV) per reaction. Much of this energy appears as the kinetic (or motion) energy of the fission-product nuclei, which is then converted to thermal energy (or heat) as the fission products slow down in the reactor fuel material. This thermal energy is removed from the reactor core and used to generate electricity or as process heat.

Energy is released during the nuclear-fission process because the total mass of the fission products and neutrons after the reaction is less than the total mass of the original neutron and the heavy nucleus that absorbed it. From Einstein's famous mass—energy equivalence relationship, $E = mc^2$, the energy released is equal to the tiny amount of mass that has disappeared multiplied by the square of the speed of light.

Nuclear fission can occur spontaneously in heavy elements but is usually caused when these nuclei absorb neutrons. In some circumstances nuclear fission may also be induced by very energetic gamma rays (in a process called *photofission*) and by extremely energetic (billion-electron-volt-class [GeV-class]) charged particles. The most important fissionable (or fissile) materials are uranium-235, uranium-233, and plutonium-239. *Compare with* FUSION.

fixed satellite An Earth satellite that orbits from west to east at such speed as to remain constantly over a given place on Earth's equator, thereby appearing at a "fixed" location in the sky.

See also GEOSTATIONARY ORBIT.

fixed stars A term used by early astronomers to distinguish between the apparently motionless background stars and the *wandering stars* (that is, the planets). Modern astronomers now use this term to describe stars that have no detectable proper motion.

flame bucket A deep, cavelike construction built beneath a launch pad. It is open at the top to receive the hot gases from the rocket or missile positioned above it and is also open on one to three sides below, with a thick metal fourth side bent toward the open side(s) so as to deflect the exhaust gases.

See also FLAME DEFLECTOR.

flame deflector 1. In a vertical launch pad, any of the obstructions of various designs that intercept the hot exhaust zgases of the rocket engine(s), thereby deflecting them away from the launch pad structure or the ground. The flame deflector may be a relatively small device fixed to the top surface of the launch pad, or it may be a heavily constructed piece of metal mounted as a side and bottom of a flame bucket. In the latter case the flame deflector also may be perforated with numerous holes connected with a source of water. During thrust buildup and the beginning of launch, a deluge of water pours from the holes in such a deflector to keep it from melting. 2. In a captive (hot) test of a rocket engine, an elbow in the exhaust conduit or a flame bucket that can deflect the rocket's flame into the open.

Flamsteed, John (1646–1719) British *Astronomer* The astronomer John Flamsteed used both the telescope and the clock to assemble a precise star catalog. In 1675 the English king, Charles II, established the Royal Greenwich Observatory and then appointed Flamsteed its director and the first English Astronomer Royal, but Flamsteed had to run this important observatory without any significant financial assistance from the Crown. Nevertheless, he constructed instruments and collected precise astronomical data. In 1712 an impatient and somewhat devious Sir ISAAC NEWTON secured a royal command to force publication of the first volume of this important data collection without Flamsteed's consent. Newton's political maneuvering infuriated Flamsteed, who purchased and burned about 300 copies of the unauthorized book. However, an extensive three-volume set of Flamsteed's star data was published in 1725, after his death.

flare A bright eruption from the Sun's chromosphere. Solar flares may appear within minutes and fade within an hour, or else linger for days. They cover a wide range of intensity and size. Flares can cause ionospheric disturbances and telecommunications (radio) fadeouts on Earth. Solar flares also represent a major ionizing radiation hazard to astronauts or cosmonauts walking on the lunar surface or traveling through interplanetary space (i.e., traveling outside the relative protection of Earth's magnetosphere).

See also SPACE WEATHER; SUN.

flare star A cool, intrinsically faint red dwarf star that undergoes intense outbursts of energy from localized areas on its surface. Astronomers viewing such a star notice a transient but appreciable increase in its brightness. Often the brightness changes by two orders of magnitude or more in less than a few seconds and then diminishes to its normal (minimum) brightness level in a period of between 10 minutes and an hour. Astronomers generally treat the term *flare star* as synonymous with *UV Ceti star.*

flat universe A geometric model of the universe in which the laws of geometry are like those that would apply on a flat surface such as a table top.

See also COSMOLOGY; UNIVERSE.

fleet ballistic missile (FBM) An intercontinental ballistic missile usually equipped with one or more nuclear weapons carried by and launched from a submarine. The U.S. Navy equips a special class of nuclear-powered submarines, called FLEET BALLISTIC MISSILE SUBMARINES, with such nuclear warhead–carrying missiles. The Trident II (designated D-5) is the sixth generation member of the U.S. Navy's fleet ballistic missile program, which started in 1956. Since then the Polaris (A1), Polaris (A2), Polaris (A3), Poseidon (C3), and Trident I (C4) have provided a significant deterrent against nuclear aggression. At present the U.S. Navy deploys Poseidon (C3) and Trident I (C4) missiles, having retired the Polaris family of fleet ballistic missiles. The Trident II (D-5) FBM was first deployed in 1990 on the *USS Tennessee* (SSNB 734). The Trident II (D-5) FBM is a three-stage, solid-propellant, inertially guided FBM with a range of more than 7,400 kilometers. The new Trident/Ohio class fleet ballistic missile submarines each carry 24 Trident IIs. These missiles can be launched while the submarine is underwater or on the surface. Also called a *submarine-launched ballistic missile* (SLBM).

See also INTERCONTINENTAL BALLISTIC MISSILE (ICBM); ROCKET.

fleet ballistic missile submarine A nuclear-powered submarine designed to deliver ballistic missile attacks against assigned targets from either a submerged or surface condition. The U.S. Navy designates these ships as SSBN.

Fleet Satellite Communications System (FLTSATCOM) A family of military communications satellites operated in geostationary orbit. The system was designed to provide a near-global operational communications network capable of supporting the ever-increasing information transfer needs of the U.S. Navy and Air Force. The U.S. Navy began replacing and upgrading the early spacecraft in its ultrahigh-frequency (UHF) communications satellite network in the 1990s with a constellation of customized satellites, collectively referred to as the UHF Follow-On (UFO) spacecraft series. An evolving family of Atlas rockets launched these spacecraft from Cape Canaveral Air Force Station. For example, the first spacecraft in the series, called *Fleet Satellite Communications 1* (FltSatCom 1), was placed into geostationary orbit from Cape Canaveral in 1978 by an Atlas-Centaur vehicle. More recently a UFO spacecraft, designated F-10 in the series, was successfully placed into orbit on November 22, 1999, by an Atlas IIA vehicle.

See also ATLAS LAUNCH VEHICLE; COMMUNICATIONS SATELLITE.

Fleming, Williamina Paton (née Stevens) (1857–1911) Scottish-American *Astronomer* In 1881 Williamina Fleming joined EDWARD CHARLES PICKERING at the Harvard College Observatory, where she devised a classification system for stellar spectra that greatly improved the system used by PIETRO ANGELO SECCHI. Her alphabetic system became known as the Harvard Classification System and appeared in the 1890 edition of *The Henry Draper Catalogue of Stellar Spectra.*

flexible spacecraft A space vehicle (usually a space structure or rotating satellite) whose surfaces and/or appendages may be subject to and can sustain elastic flexural deformations (vibrations).

flight The movement of an object through the atmosphere or through space, sustained by aerodynamic, aerostatic, or reaction forces or by orbital speed; especially, the movement of a human-operated or human-controlled device, such as a rocket, space probe, aerospace vehicle, or aircraft, as in *space flight, high-altitude flight,* or *high-speed flight.*

flight acceptance test(s) The environmental and other tests that spacecraft, subsystems, components, and experiments scheduled for space flight must pass before launch. These tests are planned to approximate anticipated environmental and operational conditions and have the purpose of detecting flaws in material and workmanship.

flight crew In general, any personnel onboard an aerospace vehicle, space station, or interplanetary spacecraft. With respect to the space shuttle, any personnel engaged in flying the orbiter and/or managing resources onboard (e.g., the commander, pilot, mission specialist).

See also ASTRONAUT; COSMONAUT.

flight readiness firing A short-duration test operation of a rocket consisting of the complete firing of the liquid-propellant engines while the vehicle is restrained at the launch pad. This test is conducted to verify the readiness of the rocket system for a flight test or operational mission.

See also LAUNCH SITE; ROCKET.

flight test vehicle A rocket, missile, or aerospace vehicle used for conducting flight tests. These flight tests can involve either the vehicle's own capabilities or the capabilities of equipment carried onboard the vehicle.

flow work In thermodynamics, the amount of work associated with the expansion or compression of a flowing fluid.

fluid A substance that, when in static equilibrium, cannot sustain a shear stress. A *perfect* or *ideal* fluid is one that has zero viscosity—that is, it offers no resistance to shape change. Actual fluids only approximate this behavior. Physicists and engineers use this term to refer to both liquids and gases.

fluid-cooled Term aerospace engineers apply to a rocket's thrust chamber or nozzle whose walls are cooled by fluid supplied from an external source, as in regenerative cooling, transpiration cooling, and film cooling.

See also NOZZLE; ROCKET.

fluid management The isolation and separation of liquids from gas in a storage vessel that operates in a reduced- or zero-gravity environment.

See also MICROGRAVITY.

fluid mechanics The major branch of science that deals with the behavior of fluids (both gases and liquids) at rest (fluid statics) and in motion (fluid dynamics). This scientific field has many subbranches and important applications, including aerodynamics (the motion of gases, including air), hydrostatics (liquids at rest), and hydrodynamics (the motion of liquids, including water).

fluorescence Many substances can absorb energy (as, for example, from X-rays, ultraviolet light, and radioactive particles) and immediately emit this absorbed ("excitation") energy as an electromagnetic photon of visible light. This light-emitting process is called fluorescence; the emitting substances are said to be fluorescent. An *emission nebula* is a large interstellar cloud of mostly hydrogen gas that is being excited by ultraviolet radiation from extremely hot nearby stars. The giant interstellar cloud exhibits a characteristic red glow (visible light emission at 656.3 nanometers wavelength) because of the fluorescence of hydrogen atoms excited by the passage of ultraviolet photons.

flux (symbol: φ) The flow of fluid, particles, or energy through a given area within a certain time. For example, physicists define a *particle flux* as the number of particles passing through one square centimeter of a given target (normal to the beam) in one second. A *radiant flux* represents the total power (usually expressed as watts/unit area) of some form of electromagnetic radiation. Astronomers often use flux to describe the rate at which light flows. For example, the *luminous flux* is the amount of light (photons) striking a single square centimeter of a detector in one second. Finally, high-energy astrophysicists define an incident *gamma-ray flux* as the number of energetic gamma-ray photons that pass through a square centimeter of a detector in one second. In defining flux, scientists consider the target area as being normal (perpendicular) to the arriving beam of particles or electromagnetic energy.

flyby An interplanetary or deep space mission in which the *flyby spacecraft* passes close to its celestial target (e.g., a distant planet, moon, asteroid, or comet) but does not impact the target or go into orbit around it. Flyby spacecraft follow a continuous trajectory, never to be captured into a planetary orbit. Once the spacecraft has flown past its target (often after years of travel through deep space), it cannot return to recover lost data, so flyby operations often are planned years in advance of the encounter and then carefully refined and practiced in the months prior to the encounter date. Flyby operations are conveniently divided into four phases: *observatory phase, far-encounter phase, near-encounter phase,* and *post-encounter phase.*

The *observatory phase* of a flyby mission is defined as the period when the celestial target can be better resolved in the spacecraft's optical instruments than it can from Earth-based instruments. The phase generally begins a few months prior to the date of the actual planetary flyby. During this phase the spacecraft is completely involved in making observations of its target, and ground resources become completely operational in support of the upcoming encounter. The *far-encounter phase* includes the time when the full disc of the target planet (or other celestial object) can no longer fit within the field of view of the spacecraft's instruments. Observations during this phase are designed to accommodate parts of the planet (e.g., a polar region, an interesting wide-expanse cloud feature such as Jupiter's Red Spot, etc.) rather than the entire planetary disc and to take advantage of the higher resolution available.

The *near-encounter phase* includes the period of closest approach to the target. It is characterized by intensely active observations by all of the spacecraft's science experiments. This phase of the flyby mission provides scientists the opportunity to obtain the highest-resolution data about the target. Finally, the *post-encounter phase* begins when the near-encounter phase is completed and the spacecraft is receding from the target. For planetary missions this phase is characterized by day-after-day observations of a diminishing, thin crescent of the planet just encountered. It generally provides scientists an opportunity to make extensive observations of the night side of the planet. After the post-encounter phase is over, the spacecraft stops observing the target and returns to the less intense activities of its interplanetary cruise phase—a phase in which scientific instruments are powered down and navigational corrections are made to prepare the spacecraft for an encounter with another celestial object of opportunity or for a final journey of no return into deep space. Some scientific experiments, usually concerning the properties of interplanetary space, can be performed in this cruise phase.

See also *PIONEER 10, 11* SPACECRAFT; *VOYAGER* SPACECRAFT.

flywheel A massive rotating wheel that can store energy as kinetic (or motion) energy as its rate of rotation is increased; energy then can be removed from this system by decreasing the rate of rotation of the wheel. An advanced-design flywheel energy storage system has been suggested for lunar surface base applications. During the lunar daytime solar energy would be converted into electricity using Stirling-cycle heat engines. This electricity would be used for lunar base needs as well as for spinning up several very large flywheels. Excess thermal energy (heat) in the system would be transported away to a radiator by a configuration of heat pipes. During the long lunar nighttime, energy would be extracted from the flywheels and converted into electricity for the surface base.

focal length (common symbol: f) The distance between the center of a LENS or MIRROR to the focus or focal point.

folded optics Any optical system containing reflecting components for the purpose of reducing the physical length of the system or changing the path of the optical axis.

F-1 engine A powerful liquid-propellant rocket engine, five of which combined to provide thrust for the first stage of NASA's giant Saturn V launch vehicle—the rocket that sent American astronauts to the Moon in the late 1960s and early 1970s. Each F-1 engine was capable of delivering 6.7 million newtons of thrust (at sea level).

See also APOLLO PROJECT; ROCKET; SATURN LAUNCH VEHICLE.

footprint An area within which a SPACECRAFT or REENTRY VEHICLE (RV) is intended to land.

force (symbol: F) In physics, the cause of the acceleration of material objects measured by the rate of change of momentum produced on a free body. For a body of constant mass (m) traveling at a velocity (v), the momentum is (mv). Assuming a constant mass, the force (F) is given by the basic equation $F = d(mv)/dt = m[dv/dt] = ma$, where a is the acceleration. Scientists call this expression the second law of motion—a critically important relationship in classical physics first suggested by the British mathematician and physicist SIR ISAAC NEWTON in the 17th century. In his honor the newton (symbol: N) is designated as the SI unit of force. Forces between bodies occur in equal and opposite reaction pairs.

See also FUNDAMENTAL FORCES IN NATURE; NEWTON'S LAWS OF MOTION.

fossa A long, narrow, shallow (ditchlike) depression found on the surface of a planet or a moon.

Foucault, Jean-Bernard-Léon (1819–1868) French *Physicist* Jean-Bernard Foucault was a French physicist who investigated the speed of light in air and water and was the first to experimentally demonstrate that Earth rotates on its axis. He performed this experiment with a *Foucault pendulum,* a relatively large concentrated mass (about 30 kg) suspended on a very long (about 70 meters) wire.

four-dimensional Involving space-time; that is, three spatial dimensions (e.g., x, y, and z) plus the dimension of time (t).

Fowler, William Alfred (1911–1995) American *Astrophysicist* The astrophysicist William Fowler developed the widely accepted theory that nuclear processes in stars are responsible for all the heavier elements in the universe beyond hydrogen and helium, including those in the human body. He coshared the 1983 Nobel Prize in physics for his work on stellar evolution and nucleosynthesis.

Fraunhofer, Joseph von (1787–1826) German *Optician, Physicist* The German optician Joseph von Fraunhofer pioneered the important field of astronomical spectroscopy. In about 1814 he developed the prism spectrometer. Then, while using his new instrument, Fraunhofer discovered more than 500 mysterious dark lines in the Sun's spectrum—lines that now carry his name. Almost a decade later (in 1823) he observed similar (but different) lines in the spectra of other stars but left it for other scientists such as ROBERT W. BUNSEN and GUSTAV R. KIRCHHOFF to solve the mystery of the "Fraunhofer lines."

Fraunhofer was born in Straubing, Baveria, on March 6, 1787. His parents were involved in the German optical trade, primarily making decorative glass. When they died and made him an orphan at the age of 11, his guardians apprenticed him to an optician (mirror maker) in Munich. While completing this apprenticeship, Fraunhofer taught himself mathematics and physics. In 1806 he joined the Untzschneider Optical Institute, located near Munich, as an optician. He learned quickly from the master glassmakers in this company and soon started to apply his own talents—an effort that greatly improved the firm's fortunes. His lifelong contributions to the field of optics laid the foundation for German supremacy in the design and manufacture of quality optical instruments in the 19th century.

Fraunhofer's great contribution to the practice of astronomy occurred quite by accident while he was pursuing the perfection of the achromatic lens—that is, a lens capable of greatly reducing the undesirable phenomenon of chromatic aberration. As part of this quest he devised a clever way to measure the refractive indexes of optical glass by using the bright yellow emission line of the chemical element sodium, which served as his reference or standard. In 1814, in order to support these calibration efforts, he created the modern spectroscope. He accomplished this by placing a prism in front of the object glass of a surveying instrument (called a theodolite). Using this instrument he discovered (or more correctly rediscovered) that the solar spectrum contained more than 500 dark lines—a pair of which corresponded to the closely spaced bright double emission lines of sodium found in the yellow portion of the visible spectrum.

It turns out that in 1802 the British scientist William H. Wollaston (1766–1828) first observed seven dark lines in the solar spectrum, but he did not know what to make of them and speculated (incorrectly) that the mysterious dark lines were merely divisions between the colors manifested by sunlight as it passed through a prism. Working with a much better spectroscope 12 years later, Fraunhofer identified 576 dark lines in the solar spectrum and labeled the position of the most prominent lines with the letters A to K—a nomenclature physicists still use when they discuss the Fraunhofer lines. He also observed that a dark pair of his so-called D-lines in the solar spectrum corresponded to the position of a brilliant pair of yellow lines in the sodium flame he produced in his laboratory flame. However, the true significance of this discovery escaped him and, busy making improved lenses, he did not pursue the intriguing correlation any further. He also observed that the light from other stars, when passed through his spectroscope, produced spectra with dark lines that were similar to but not identical to those found in sunlight. Once again, Fraunhofer left the formal discovery of astronomical spectroscopy to other 19th-century scientists such as Robert Bunsen and Gustav Kirchhoff.

Fraunhofer was content to note the dark line phenomenon in the solar spectrum and then continue in his quest to design an achromatic objective lens, a challenging task he successfully accomplished in 1817. He did such high-quality optical work that his basic design for this type of lens is still used by opticians. Then, in 1821 he built the first diffraction grating and used the new device instead of a prism to form a spectrum and make more precise measurements of the wavelengths corresponding to the Fraunhofer lines. Again, he did not interpret these telltale dark lines as solar absorption lines. In 1859 Bunsen and Kirchhoff would finally put all the facts together and provide astronomers with the ability to assess the elemental composition of distant stars through spectroscopy. Of course, the road to their great accomplishment began with Fraunhofer's pioneering work several decades earlier.

Because Joseph Fraunhofer did not have a formal university education, the German academic community generally ignored his pioneering work in optical physics and astronomical spectroscopy. At the time it was considered improper for a "technician"—no matter how skilled—to present a scientific paper to a gathering of learned academicians. Yet 19th-century German astronomers such as FRIEDRICH BESSEL would use optical instruments made by Fraunhofer to make important discoveries. In 1823 Fraunhofer received an appointment as the director of the Physics Museum in Munich. Along with this position he also received the honorary title of "professor." He died of tuberculosis in Munich on June 7, 1826. On his tombstone is inscribed the following epithet: *"Approximavit sidera"*: "He reached for the stars.")

In his honor modern scientists refer to the dark (absorption) lines in the visible portion of the Sun's spectrum as *Fraunhofer lines.* These absorption lines are the phenomenological basis of astronomical spectroscopy and give scientists the ability to learn what stars are made of.

Fraunhofer lines A collection of more than 600 absorption (dark) lines in the spectrum of the Sun. The German optician and physicist JOSEPH VON FRAUNHOFER, for whom the lines are named, first categorized the phenomenon in 1814.

free fall The unimpeded fall of an object in a gravitational field. For example, an elevator car whose cable has snapped;

the plunge of a skydiver before his or her chute has been opened; or the fall back to Earth of a sounding rocket and its payload (after thrust termination and maximum altitude have been achieved) that is not retarded by a parachute or braking device.

See also MICROGRAVITY.

free flight Unconstrained or unassisted flight, such as the flight of a rocket after consumption of its propellant or after motor shutoff; the flight of an unguided projectile.

free-flying spacecraft (free-flyer) Any spacecraft or payload that can be detached from NASA's space shuttle or the *International Space Station* and then operate independently in orbit.

free rocket A rocket not subject to guidance or control in flight.

frequency (common symbol: f or ν) In general, the rate of repetition of a recurring or regular event; for example, the number of vibrations of a system per second or the number of cycles of a wave per second. For electromagnetic radiation, the frequency (ν) of a quantum packet of energy (e.g., a photon) is given by $\nu = E/h$, where E is the photon energy and h is the Planck constant. The SI UNIT of frequency is the hertz (Hz), which is defined as one cycle per second.

frequency modulation (FM) An information transfer technique used in telecommunications in which the frequency of the carrier wave is modulated (i.e., increased or decreased) as the signal (to be transferred) increases or decreases in value but the amplitude of the carrier wave remains constant. Specifically, angle modulation of a sine carrier wave in which the instantaneous frequency of the modulated wave differs from the carrier frequency by an amount proportional to the instantaneous value of the modulating wave.

See also TELECOMMUNICATIONS; TELEMETRY.

Fresnel lens A thin lens constructed with stepped setbacks so as to have the optical properties of a much thicker lens. The French physicist Austin Jean Fresnel (1788–1827) invented this type of compound lens in 1822. This optical device quickly found use in lighthouses around the world.

Friedman, Herbert (1916–2000) American *Astrophysicist* In 1949 the astrophysicist Herbert Friedman began using sounding rockets in astronomy, especially for studying the Sun's X-ray activity. As a scientist with the U.S. Naval Research Laboratory in Washington, D.C., his rocket-borne sensors detected X-rays from the Crab Nebula in 1964.

See also CRAB NEBULA; SOUNDING ROCKET; SUN; X-RAY ASTRONOMY.

F star A white star of spectral class F. An F star is slightly hotter and more massive than the Sun and has a surface temperature between approximately 6,000 and 7,500 kelvins.

See also STAR.

fuel 1. Any substance used to liberate thermal energy, either by chemical or nuclear reaction; a substance that might be used, for example, in a heat engine or a rocket engine. When discussing a liquid-propellant rocket engine, ordinarily the fuel (a propellant chemical such as kerosene) is distinguished from the oxidizer (e.g., liquid oxygen, or LOX) when these two are stored separately. 2. In nuclear technology, the term *fuel* applies to the fissionable (or fissile) material used to produce energy in a nuclear reactor. Uranium-235 is an example of nuclear reactor fuel.

See also ROCKET; SPACE NUCLEAR POWER.

fuel binder A continuous-phase substance that contributes the principal structural condition to solid propellant but does not contain any oxidizing element, either in solution or chemically bonded.

See also ROCKET.

fuel cell A direct conversion device that transforms chemical energy directly into electrical energy by reacting continuously supplied chemicals. In a modern fuel cell an electrochemical catalyst (such as platinum) promotes a non-

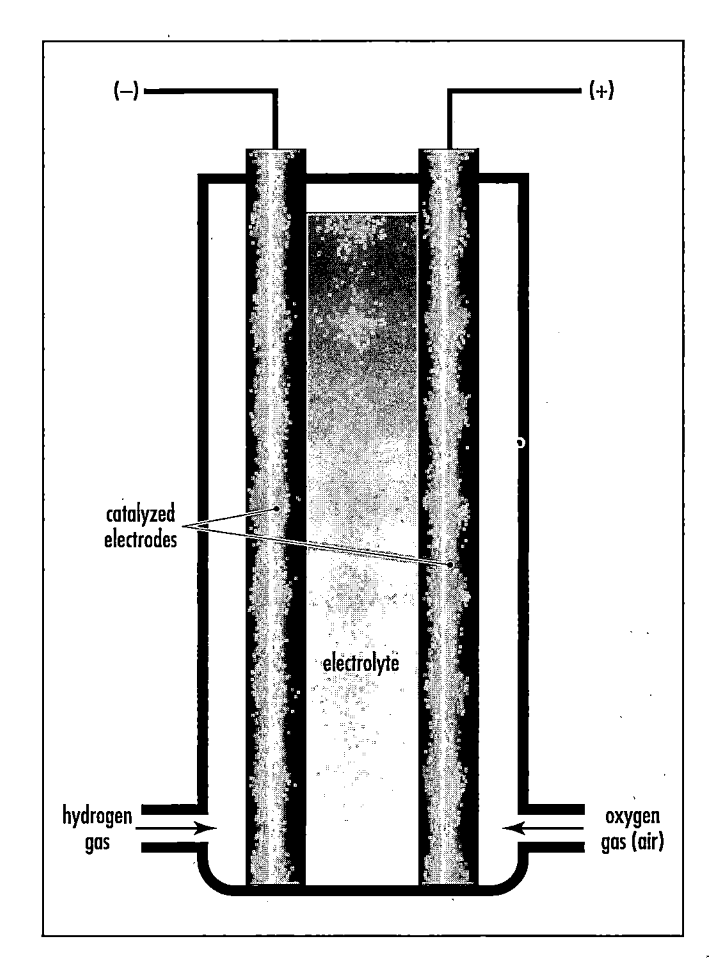

Components of a simple hydrogen-oxygen fuel cell ***(Diagram courtesy of the U.S. Department of Energy)***

combustible reaction between a fuel (such as hydrogen) and an oxidant (such as oxygen).

fuel shutoff The termination of the flow of liquid fuel into a rocket's combustion chamber, or the process of stopping the combustion (burning) of a solid rocket fuel. This term also means the event or time marking this particular action. *Compare with* CUTOFF.

full moon The phase of the Moon in which it appears as a complete circular disk in the night sky.

See also PHASES OF THE MOON.

fundamental forces in nature At present physicists recognize the existence of four fundamental forces in nature: gravity, electromagnetism, the strong force, and the weak force. Gravity and electromagnetism are part of our everyday experience. These two forces have an infinite range, which means they exert their influence over great distances. For example, the two stars in a binary star system experience mutual gravitational attraction even though they may be a light-year or more distant from each other. The other two forces, the strong force and the weak force, operate within the realm of the atomic nucleus and involve elementary particles. These forces are beyond our day-to-day experience and were essentially unknown to the physicists of the 19th century, who had a good classical understanding of the law of gravity and the fundamental principles of electromagnetism. The strong nuclear force operates at a range of about 10^{-15} m and holds the atomic nucleus together. The weak nuclear force has a range of about 10^{-17} m and is responsible for processes such as beta decay that tear nuclei and elementary particles apart. What is important to recognize is that whenever anything happens in the universe—that is, whenever an object experiences a change in motion—the event takes place because one or more of these fundamental forces is involved.

fusion (nuclear) In nuclear fusion lighter atomic nuclei are joined together, or fused, to form a heavier nucleus. For example, the fusion of deuterium ($^2{}_1D$) with tritium ($^3{}_1T$) results in the formation of a helium nucleus and a neutron. Because the total mass of the fusion products is less than the total mass of the reactants (that is, the original deuterium and tritium nuclei), a tiny amount of mass has disappeared, and the equivalent amount of energy is released in accordance with ALBERT EINSTEIN's mass-energy equivalence formula $E = \Delta mc^2$. This fusion energy then appears as the kinetic (motion) energy of the reaction products. When isotopes of elements lighter than iron fuse together, some fusion energy is liberated. However, energy must be added to any fusion reaction involving elements heavier than iron.

The Sun is our oldest source of energy, the very mainstay of all terrestrial life. The energy of the Sun and other stars comes from thermonuclear fusion reactions. Fusion reactions brought about by means of very high temperatures are called *thermonuclear reactions.* The actual temperature required to join, or fuse, two atomic nuclei depends on the nuclei and the particular fusion reaction involved. The two nuclei being joined must have enough energy to overcome the Coulomb repulsive force between like-signed (here positive and positive) electric charges. In stellar interiors fusion occurs at temperatures of tens of millions of kelvins (K). When scientists try to harness controlled thermonuclear reactions (CTRs) on Earth, they must use techniques involving reaction temperatures ranging from 50 million to 100 million kelvins.

The Fusion Table describes the major single-step thermonuclear reactions that are potentially useful in controlled fusion reactions for power and propulsion applications. Large space settlements and human-made "miniplanets" could eventually be powered by such CTR processes, and robot interstellar probes and giant space arks might use such thermonuclear reactions for both power and propulsion. Helium-3 is a rare isotope of helium. Some space visionaries

Single-Step Fusion Reactions Useful in Power and Propulsion Systems

Nomenclature	Thermonuclear Reaction	Energy Released per Reaction (MeV) [Q Value]	Threshold Plasma Temperature (keV)[a]
(D-T)	$^2_1D + ^3_1T \rightarrow ^4_2He + ^1_0n$	17.6	10
(D-D)	$^2_1D + ^2_1D \rightarrow ^3_2He + ^1_0n$	3.2	50
	$^2_1D + ^2_1D \rightarrow ^3_1T + ^1_1p$	4.0	50
(D-^{3}He)	$^2_1D + ^3_2He \rightarrow ^4_2He + ^1_1p$	18.3	100
(^{11}B-p)	$^{11}_5B + ^1_1p \rightarrow 3(^4_2He)$	8.7	300

where
D is deuterium
T is tritium
He is helium
n is neutron
p is proton
B is boron

[a]10 keV = 100 million K.

Main Thermonuclear Reactions in the Proton-Proton Cycle

$^{1}_{1}H + ^{1}_{1}H \rightarrow ^{2}_{1}D + e^{+} + \nu$

$^{2}_{1}D + ^{1}_{1}H \rightarrow ^{3}_{2}He + \gamma$

$^{3}_{2}He + ^{3}_{2}He \rightarrow ^{4}_{2}He + ^{1}_{1}H + ^{1}_{1}H$

where

$^{1}_{1}H$ is a hydrogen nucleus (that is, a proton, $^{1}_{1}p$)

$^{2}_{1}D$ is deuterium (an isotope of hydrogen)

$^{3}_{2}He$ is helium-3 (a rare isotope of helium)

$^{4}_{2}He$ is the main (stable) isotope of helium

ν is a neutrino

e^{+} is a positron

γ is a gamma ray

Major Thermonuclear Reactions in the Carbon Cycle

$^{12}_{6}C + ^{1}_{1}H \rightarrow ^{13}_{7}N + \gamma$

$^{13}_{7}N \rightarrow ^{13}_{6}C + e^{+} + \nu$ (radioactive decay)

$^{1}_{1}H + ^{13}_{6}C \rightarrow ^{14}_{7}N + \gamma$

$^{1}_{1}H + ^{14}_{7}N \rightarrow ^{15}_{8}O + \gamma$

$^{15}_{8}O \rightarrow ^{15}_{7}N + e^{+} + \nu$ (radioactive decay)

$^{1}_{1}H + ^{15}_{7}N \rightarrow ^{12}_{6}C + ^{4}_{2}He$

where

γ is a gamma ray

ν is a neutrino

e^{+} is a positron

have proposed mining Jupiter's atmosphere or the surface of the Moon for helium-3 to fuel the first interstellar probes.

At present there are immense technical difficulties preventing our effective use of controlled fusion as a terrestrial or space energy source. The key problem is that the fusion gas mixture must be heated to tens of millions of kelvins (K) and then confined for a long enough period of time for the fusion reactions to occur. For example, a deuterium-tritium (D-T) gas mixture must be heated to at least 50 million kelvins, and physicists consider this the "easiest" controlled fusion reaction for them to achieve. At 50 million kelvins any physical material used to confine the reacting fusion gases would disintegrate, and the vaporized wall materials would then "cool" the fusion gas mixture, quenching the reaction. There are three general approaches to confining these hot fusion gases, or plasmas: gravitational confinement, magnetic confinement, and inertial confinement.

Because of their large masses, the Sun and other stars are able to hold the reacting fusion gases together by gravitational confinement. Interior temperatures in stars reach tens of millions of kelvins and use complete thermonuclear fusion cycles to generate their vast quantities of energy. For main-sequence stars like or cooler than our Sun (about 10 million kelvins), the proton-proton cycle, shown in the table above, left, is believed to be the principal energy-liberating mechanism. The overall effect of the proton-proton stellar fusion cycle is the conversion of hydrogen into helium. Stars hotter than our Sun (those with interior temperatures higher than 10 million kelvins) release energy through the carbon cycle, shown in the carbon cycle table above, right. The overall effect of this cycle is again the conversion of hydrogen into helium, but this time with carbon (primarily the carbon-12 isotope) serving as a catalyst.

Here on Earth scientists attempt to achieve controlled fusion through two techniques: magnetic-confinement fusion (MCF) and inertial-confinement fusion (ICF). In magnetic confinement strong magnetic fields are employed to "bottle up," or hold, the intensely hot plasmas needed to make the various single-step fusion reactions occur (see the Fusion Table). In the inertial-confinement approach pulses of laser light, energetic electrons, or heavy ions are used to compress and heat small spherical targets of fusion material very rapidly. This rapid compression and heating of an ICF target allows the conditions supporting fusion to be reached in the interior of the pellet—before it blows itself apart.

Although there are still many difficult technical issues to be resolved before we can achieve controlled fusion, it promises to provide a limitless terrestrial energy supply. Of course, fusion also represents the energy key to the full use of the resources of our solar system and, possibly, to travel across the interstellar void.

In sharp contrast to previous and current scientific attempts at controlled nuclear fusion for power and propulsion applications, since the early 1950s nuclear weapon designers have been able to harness (however briefly) certain fusion reactions in advanced nuclear weapon systems called *thermonuclear devices*. In these types of nuclear explosives, the energy of a fission device is used to create the conditions necessary to achieve (for a brief moment) a significant number of fusion reactions of either the deuterium-tritium (D-T) or deuterium-deuterium (D-D) kind. Very powerful modern thermonuclear weapons have been designed and demonstrated with total explosive yields from a few hundred kilotons (kT) to the multimegaton (MT) range. A megaton yield device has the explosive equivalence of one million tons of the chemical high explosive trinitrotoluene (TNT). Scientists have recently suggested using such powerful thermonuclear explosive devices to deflect or destroy any future asteroid or comet that threatens to impact Earth.

See also ASTEROID DEFENSE SYSTEM; FISSION; INTERSTELLAR TRAVEL; STAR; STARSHIP; SUN.

g The symbol used for the acceleration due to gravity. For example, at sea level on Earth the acceleration due to gravity is approximately 9.8 meters per second squared—that is, "one g." This term is used as a unit of stress for bodies experiencing acceleration. When a spacecraft or aerospace vehicle is accelerated during launch, everything inside it experiences a force that may be as high as several gs.

Gagarin, Yuri A. (1934–1968) Russian *Cosmonaut, Military Officer* The Russian cosmonaut Yuri A. Gagarin became the first human being to travel in outer space. He accomplished this feat with his historic one orbit of Earth mission in the *Vostok 1* spacecraft on April 12, 1961. A popular hero of the Soviet Union, he died in an aircraft training flight near Moscow on March 27, 1968.

Gaia hypothesis The hypothesis first suggested in 1969 by the British biologist James Lovelock (b. 1919)—with the assistance of the biologist Lynn Margulis (b. 1938)—that Earth's biosphere has an important modulating effect on the terrestrial atmosphere. Because of the chemical complexity observed in the lower atmosphere, Lovelock suggested that life-forms within the terrestrial biosphere actually help control the chemical composition of the Earth's atmosphere, thereby ensuring the continuation of conditions suitable for life. Gas-exchanging microorganisms, for example, are thought to play a key role in this continuous process of environmental regulation. Without these "cooperative" interactions in which some organisms generate certain gases and carbon compounds that are subsequently removed and used by other organisms, planet Earth might also possess an excessively hot or cold planetary surface, devoid of liquid water and surrounded by an inanimate, carbon dioxide–rich atmosphere.

Gaia (also spelled *Gaea*) was the goddess of Earth in ancient Greek mythology. Lovelock used her name to represent the terrestrial biosphere—namely, the system of life on Earth, including living organisms and their required liquids, gases, and solids. Thus, the Gaia hypothesis simply states that "Gaia" (Earth's biosphere) will struggle to maintain the atmospheric conditions suitable for the survival of terrestrial life.

Human space flight began on April 12, 1961, with Yuri Gagarin's single-orbit mission. The liquid-fueled, two-stage Vostok rocket that lifted Gagarin into space was used by the Soviet Union to launch a variety of military and civilian spacecraft from the late 1950s to the 1980s. *Vostok* is Russian for "east." Human space flight in the cold war era was a very high-profile symbol of American and Soviet technological achievements and exerted considerable geopolitical influence. During the 1980s, several years after this artistic interpretation of Gagarin's historic Vostok launch was created for the U.S. Defense Intelligence Agency (DIA), the Soviets also began using Vostok rockets to place commercial satellites in orbit for other countries. *(Artist's rendering courtesy of U.S. Department of Defense/DIA; artist, Richard Terry [1978])*

If we use the Gaia hypothesis in our search for extraterrestrial life, we should look for alien worlds (e.g., extrasolar planets) that exhibit variability in atmospheric composition. Extending this hypothesis beyond the terrestrial biosphere, a planet will either be living or else it will not. The absence of chemical interactions in the lower atmosphere of an alien world could be taken as an indication of the absence of living organisms.

Although this interesting hypothesis is currently more speculation than hard, scientifically verifiable fact, it is still quite useful in developing a sense of appreciation for the complex chemical interactions that have helped to sustain life in the Earth's biosphere. These interactions among microorganisms, higher-level animals, and their mutually shared atmosphere might also have to be carefully considered in the successful development of effective closed life support systems for use on permanent space stations, lunar bases, and planetary settlements.

See also EXTRASOLAR PLANETS; EXOBIOLOGY; GLOBAL CHANGE.

Gaia mission An ambitious space-based astronomy mission of the European Space Agency (ESA) intended to create the largest and most precise three-dimensional chart of the Milky Way galaxy. By surveying more than one billion stars in our galaxy, the Gaia mission will provide astronomers new scientific details about the composition, formation, and evolution of the Milky Way. As currently planned, the Gaia mission will involve a single spacecraft carrying three optical telescopes capable of constantly sweeping the sky and recording every visible celestial object that crosses the line of sight. Astronomers anticipate that the *Gaia* spacecraft will observe each of more than 1 billion optical celestial sources about 100 times. During each observation the spacecraft will detect any changes in an object's brightness and position.

The prime mission of the *Gaia* spacecraft is to observe more than 1 billion stars, including many of the closest stars to the Sun. Since the Milky Way galaxy contains an estimated 100 billion stars, the *Gaia* spacecraft will complete an unprecedented astronomical survey of more than 1 percent of the galaxy's stellar population. During its anticipated five-year mission the *Gaia* spacecraft will log the position, brightness, and color of every optically visible celestial object that falls within its field of view. The spacecraft's repeated observations will allow astronomers to calculate the distance, speed, and direction of motion of each of the observed celestial objects. Astronomers will also be able to chart how each celestial object varies in brightness and determine whether it has any nearby companions.

For stars within a distance of approximately 150 light-years from the Sun, astronomers anticipate that the *Gaia* spacecraft will also allow them to detect the presence of Jupiter-sized planets that have orbital periods ranging between 1.5 and 9 years. The spacecraft's precise star survey will let astronomers look for the telltale tiny wobbles in a star's position caused by the gravitational pull of a Jupiter-sized planetary companion as it moves around its parent star. In our own solar system, for example, Jupiter (and to a lesser extent all the other planets) makes the Sun wobble just a bit. Astronomers now estimate that the *Gaia* spacecraft will detect between 10,000 and 50,000 extrasolar planets.

The *Gaia* spacecraft will also help astronomers detect tens of thousands of brown dwarfs ("failed stars") as they either drift through interstellar space or else orbit around a stellar companion. Astronomers will use the spacecraft to make a detailed survey of the outer regions of the solar system—detecting tens of thousands of icy celestial objects now believed resident in the Kuiper Belt. Looking beyond the Milky Way, astronomers will use the *Gaia* spacecraft to detect an anticipated 100,000 or more supernovas (exploding stars) as they take place in distant galaxies.

The *Gaia* spacecraft will be launched in June 2010 by a Soyuz-Fregat rocket. Following launch the spacecraft will travel to and then operate in an eclipse free orbit around the L2 point of the Earth-Sun system. L2 is the second Lagrangian libration point, named after its discoverer, JOSEPH-LOUIS LAGRANGE. One of the principal advantages of an L2 orbit is that it offers uninterrupted observations. Because the L2 point moves around the Sun keeping pace with Earth, the *Gaia* spacecraft will be able to observe the entire celestial sphere during the course of one year. To ensure it stays at the L2 point, the *Gaia* spacecraft will perform small station-keeping maneuvers each month.

The *Gaia* spacecraft has a planned operational lifetime of five years and will be controlled by the European Space Operations Center (ESOC) in Darmstadt, Germany. The *Gaia* spacecraft's detailed map of 1 billion stars will help astronomers improve their understanding of the Milky Way galaxy. For example, as part of the planned precision mapping of the galaxy, the spacecraft's trio of optical telescopes will detect the motion of each cataloged star in its orbit around the center of the galaxy. Much of this motion was imparted upon an individual star during its birth, so by carefully studying such motions, astronomers will actually be able to look back in time to the period when the galaxy was forming.

gain 1. An increase or amplification. 2. In electrical engineering, the ability of an electronic device to increase the magnitude of an electrical input parameter. For example, the gain of a power amplifier is defined as the ratio of the output power to the input power. Gain usually is expressed in decibels. 3. In signal processing and telecommunications, an increase in signal power in transmission from one point to another. 4. In radar technology, there are two frequently encountered general uses of this term: (a) *antenna gain,* or gain factor, which is defined as the power transmitted along the radar beam axis to that of an isotropic radiator transmitting the same total power, and (b) *receiver gain,* which is the amplification given to an incoming radar signal by the receiver.

gal (derived from: *Gal*ileo) A unit of acceleration equal to 10^{-2} meter per second squared (or 1 centimeter per second squared). The name was chosen to honor GALILEO GALILEI, the famous Italian scientist who conducted early experiments involving the acceleration of gravity. The gal and the milligal (1,000 milligal = 1 gal) often are encountered in geological survey work, in which tiny differences in the acceleration due to gravity at Earth's surface are important.

galactic Of or pertaining to a galaxy, such as the Milky Way galaxy.

galactic cannibalism A postulated model of galaxy interaction in which a more massive galaxy uses gravitation and tidal forces to pull matter away from a less massive neighboring galaxy.

See also GALAXY.

galactic cluster A diffuse collection of from 10 to perhaps several hundred stars, loosely held together by gravitational forces. Also called an *open cluster.*

See also GLOBULAR CLUSTER.

galactic coordinate system *See* CELESTIAL COORDINATES.

galactic cosmic rays (GCRs) Very energetic atomic particles that originate outside the solar system and generally are believed to come from within the Milky Way galaxy.

See also COSMIC RAY(s).

galactic disk The immense, flattened, circular region of gas and dust in a typical spiral galaxy that contains most of the galaxy's luminous stars and interstellar matter. The galactic disk bisects the galactic halo.

See also GALAXY.

galactic halo The large, roughly spherical ball of tenuous, highly ionized gas and faint old stars (that is, Population II stars and globular clusters) surrounding a galaxy. The galactic halo extends far above and far below the galactic disk. Because halo stars are so faint, astronomers generally cannot observe them.

See also DARK MATTER; GALAXY.

galactic nucleus The central region of a galaxy.

See also GALAXY.

The center of the Andromeda galaxy (M31) as imaged by the U.S. Naval Observatory in 1978 ***(Courtesy of the U.S. Navy)***

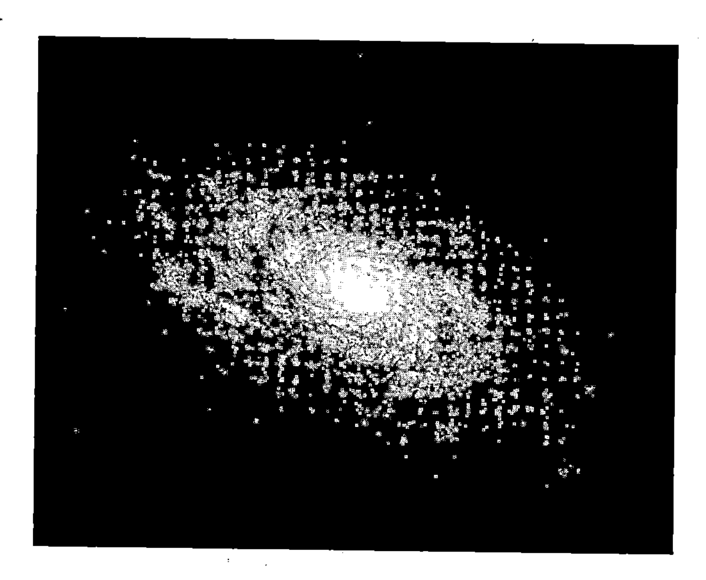

The majestic dusty spiral galaxy NGC 4414 as imaged by NASA's *Hubble Space Telescope* in a series of observations in 1995 and 1999. This galaxy is about 60 million light-years away. ***(Courtesy of NASA, Hubble Heritage Team, AURA, and STScI.)***

Galatea A small (about 160-km-diameter) moon of Neptune discovered in 1989 as a result of the *Voyager 2* spacecraft encounter. Galatea orbits Neptune at a distance of 62,000 km at an inclination of 0.05°. It has an orbital period of 0.43 day.

See also NEPTUNE.

Galaxy When capitalized, humans' home galaxy, the MILKY WAY GALAXY.

galaxy A very large accumulation of from 10^6 to 10^{12} stars. Galaxies—or "island universes," as they are sometimes called—come in a variety of shapes and sizes. They range from dwarf galaxies, such as the Magellanic Clouds, to giant spiral galaxies, such as the Andromeda galaxy. Astronomers usually classify galaxies as elliptical, spiral (or barred spiral), or irregular. Galaxies are typically tens to hundreds of thousands of light-years across, and the distance between neighboring galaxies is generally a few million light-years. For example, the Andromeda Galaxy is approximately 130,000 light-years in diameter and about 2.2 million light-years away.

In the 19th century astronomers thought the Milky Way was the only galaxy in the universe. Although the introduction of telescopes to the study of astronomy in the 17th century had opened up the universe, it still took some time for astronomers to recognize the true vastness of the universe. The telescopes revealed that the night sky was not only populated with stars but also other interesting objects that appeared like faint, patchy clouds. Astronomers called these objects *nebulas,* and they seemed to be within the Milky Way—so these objects were regarded as being relatively close patches of light.

In 1755 IMMANUEL KANT was the first to propose the nebula hypothesis. In his hypothesis he suggested that the solar system formed out of a primordial cloud of interstellar matter. Kant also introduced the term *island universe* to describe distant fuzzy patches of light. However, for the next 170 years astronomers debated the true nature of these nebulas. As telescopes became more powerful and observational techniques improved, the question arose as to whether these objects were within the Milky Way or were really communities of stars distinct from our galaxy.

It was not until the 1920s that the American astronomer EDWIN P. HUBBLE ended the debate when he observed that some of the nebulas were composed of stars. Hubble also determined the distances to these nebulas, and found that they were definitely far outside the Milky Way. His work led 20th-century astronomers and astrophysicists to recognize that the universe was expanding and filled with more than 200 billion galaxies of various sizes and shapes.

Like all galaxies, the Milky Way is held together by gravity. Gravity also holds the stars, planetary bodies, gas, and dust in orbit around the center of the galaxy. Just as planets orbit around the Sun, so do the Sun and other stars orbit around the center of the Milky Way. The same is true for other galaxies. Today astronomers even suggest that a supermassive black hole lurks at the center of each galaxy.

Galaxies come in a variety of shapes and sizes. In the 1920s Hubble was the first astronomer to investigate the morphology of galaxies. Using the 2.54-m (100 inch) Hooker reflector telescope at Mount Wilson Observatory in California, Hubble photographed numerous galaxies between 1922 and 1926. He then categorized their shapes or basic schemes as spiral, barred spiral, elliptical, irregular, and peculiar. Astronomers refer to this system as the *Hubble morphological sequence of galaxy types.*

First, Hubble noted that some galaxies, such as the Andromeda galaxy (M31), looked like disks and had arms of stars and dust that appeared in a spiral pattern. These galaxies appeared nearly uniform in brightness. Hubble also observed that in some of these types of galaxies the arms were more tightly wound. He called these *spiral galaxies.* Next, he noted that some spiral galaxies had a bright bar of gas through the center, and he called these objects *barred spiral galaxies.* Then, Hubble discovered galaxies that were slightly elliptical in shape, while others were nearly circular, such as M32, an elliptical galaxy in the constellation Andromeda. He called such objects *elliptical galaxies.* The fourth type of galaxy Hubble classified was neither spiral nor elliptical, but irregular in shape. These he referred to simply as *irregular galaxies.* The Magellanic Clouds are an example. Finally, Hubble noticed there were some galaxies that fit none of the previous descriptions. He called these *peculiar galaxies.* Centaurus A, a radio galaxy in the constellation Centaurus, is an example. Astronomers now speculate that this peculiar galaxy resulted from the merger of a spiral galaxy and an elliptical galaxy.

Today astronomers prefer to use a simpler morphology classification scheme, consisting of only two types of galaxies: spiral galaxies and elliptical galaxies. They regard barred spirals as a subclass of spirals. Irregulars may be either spiral or barred spiral. Finally, the pecular galaxy is no longer treated as a fundamentally different type of galaxy. Rather, contemporary astronomers consider peculiar galaxies as galaxies in the act of colliding; the galactic collision distorts their shapes and makes them appear "peculiar."

Galilean satellites The four largest and brightest satellites of the planet, Jupiter—namely, Io, Europa, Ganymede, and Callisto. (*See* figure.) The famous Italian scientist GALILEO GALILEI discovered and named these major Jovian moons in 1610. This astronomical discovery played an important role in Galileo's acceptance and then fiery advocacy of heliocentric cosmology. The rise of heliocentric cosmology and telescopic astronomy subsequently stimulated the emergence of the scientific revolution.

See also CALLISTO; EUROPA; GANYMEDE; IO; JUPITER.

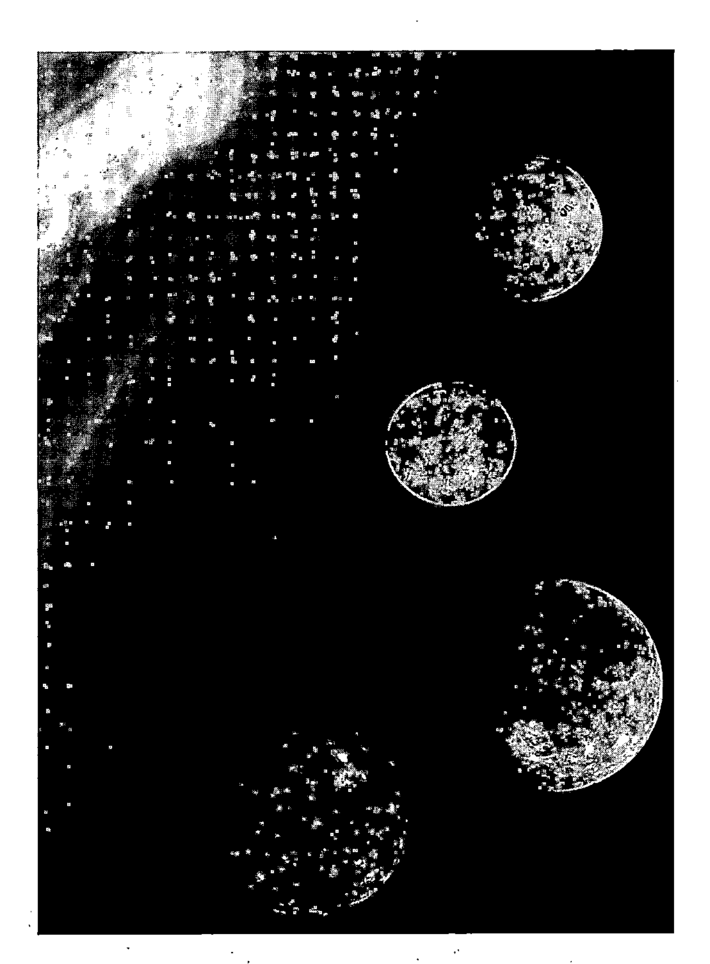

This is a scaled, composite image (scale factor: each pixel equals 15 kilometers) of the major members of the Jovian system collected by NASA's *Galileo* spacecraft during various flyby encounters in 1996 and 1997. Included in the interesting "family portrait" is the edge of Jupiter with its Great Red Spot (GRS) as well as Jupiter's four largest moons, called the Galilean satellites. From top to bottom, the moons shown are Io, Europa, Ganymede, and Callisto. *(Courtesy of NASA/JPL)*

Galilean telescope The early refracting telescope assembled by GALILEO GALILEI in about 1610. It had a converging lens as the objective and a diverging lens as the eyepiece.

See also KEPLERIAN TELESCOPE; REFRACTING TELESCOPE; TELESCOPE.

Galilei, Galileo (1564–1642) Italian *Astronomer, Mathematician, Physicist* As the first astronomer to use a telescope to view the heavens, the Italian scientist Galileo Galilei conducted early astronomical observations that helped inflame the scientific revolution of the 17th century. In 1610 he announced some of his early telescopic findings in the publication *Starry Messenger,* including the discovery of the four major moons of Jupiter (now called the Galilean satellites). Their behavior, like a miniature solar system, stimulated his enthusiastic support for the heliocentric cosmology of NICHOLAS COPERNICUS. Unfortunately, this scientific work led to a direct clash with ecclesiastical authorities, who insisted on retaining the Ptolemaic system for a number of political and social reasons. By 1632 this conflict earned the fiery Galileo an Inquisition trial at which he was found guilty of heresy (for advocating the Copernican system) and confined to house arrest for the remainder of his life.

Galileo Galilei was born in Pisa on February 15, 1564. (Scientists and astronomers commonly refer to Galileo by his first name only.) When he entered the University of Pisa in 1581, his father encouraged him to study medicine. His inquisitive mind soon became more interested in physics and mathematics than medicine. While still a medical student he attended church services one Sunday. During the sermon he noticed a chandelier swinging in the breeze and began to time its swing using his own pulse as a crude clock. When he returned home he immediately set up an experiment that revealed the pendulum principle. After just two years of study, Galileo abandoned medicine and focused on mathematics and science. This change in career pathways also changed the history of modern science.

In 1585 Galileo left the university without receiving a degree and focused his activities on the physics of solid bodies. The motion of falling objects and projectiles intrigued him. Then, in 1589 he became a mathematics professor at the University of Pisa. Galileo was a brilliant lecturer, and students came from all over Europe to attend his classrooms. This circumstance quickly angered many senior but less capable faculty members. To make matters worse, Galileo often used his tenacity, sharp wit, and biting sarcasm to win philosophical arguments at the university. This tenacious and argumentative personality earned him the nickname "the Wrangler."

In the late 16th century European professors usually taught physics (natural philosophy) as an extension of Aristotelian philosophy and not as an observational, experimental science. Through his skillful use of mathematics and innovative experiments, Galileo changed that approach. His activities constantly challenged the 2,000-year tradition of ancient Greek learning. For example, Aristotle stated that heavy objects would fall faster than lighter objects. Galileo disagreed and held the opposite view that, except for air resistance, the two objects would fall at the same rate regardless of their masses. It is not certain whether he personally performed the legendary musket ball–cannon ball drop experiment from the Leaning Tower of Pisa to prove this point. However, he did conduct a sufficient number of experiments with objects on inclined planes to upset Aristotelian physics and create the science of mechanics.

A 1640 portrait of Galileo Galilei, the fiery Italian astronomer, physicist, and mathematician who used his own version of the newly invented telescope to make detailed astronomical observations that inflamed the scientific revolution in the 17th century ***(Courtesy of NASA)***

Throughout his life Galileo was limited in his motion experiments by an inability to accurately measure small increments of time. Despite this impediment, he conducted many important experiments that produced remarkable insights into the physics of free fall and projectile motion. A century later Newton would build upon this work to create his universal law of gravitation and three laws of motion. During NASA's *Apollo 15* lunar landing mission (1971), American astronaut DAVID SCOTT released a bird's feather and a hammer. They fell on the lunar soil simultaneously, as boldly predicted more than three centuries earlier by the Italian scientist. From the surface of the Moon, astronaut Scott reported back to NASA Mission Control at the Johnson Space Center (JSC): "This proves that Mr. Galileo was correct."

By 1592 Galileo's anti-Aristotelian research and abrasive behavior had sufficiently offended his colleagues at the University of Pisa that they "invited him" to teach elsewhere.

Later that year Galileo moved to the University of Padua. This university had a more lenient policy of academic freedom, encouraged in part by the progressive Venetian government. In Padua he wrote a special treatise on mechanics to accompany his lectures. He also began teaching courses on geometry and astronomy. At the time the university's astronomy courses were primarily for medical students, who needed to learn about medical astrology.

In 1597 the German astronomer JOHANNES KEPLER provided Galileo a copy of Copernicus's book (even though the book was officially banned in Italy). Although Galileo had not previously been interested in astronomy, he discovered and immediately embraced the Copernican model. Galileo and Kepler continued to correspond until about 1610.

Between 1604 and 1605 Galileo performed his first public work involving astronomy. He observed the supernova of 1604 (in the constellation Ophiuchus) and used it to refute the cherished Aristotelian belief that the heavens were immutable (unchangeable). He delivered this challenge to Aristotle's doctrine in a series of public lectures. Unfortunately, these well-attended lectures brought him into direct conflict with the university's pro-Aristotelian philosophy professors.

In 1609 Galileo learned that a new optical instrument (a magnifying tube) had just been invented in Holland. Within six months Galileo devised his own version of the instrument. Then, in 1610 he turned this improved telescope to the heavens and started the age of telescopic astronomy. With his crude instrument he made a series of astounding discoveries, including mountains on the Moon, many new stars, and the four major moons of Jupiter—now called the *Galilean satellites* in his honor. Galileo published these important discoveries in the book *Sidereus Nuncius* (*Starry Messenger*). The book stimulated both enthusiasm and anger. Galileo used the moons of Jupiter to prove that not all heavenly bodies revolve around Earth. This provided direct observational evidence for the Copernican model, a cosmological model that Galileo now vigorously endorsed.

Unable to continue teaching "old doctrine" at the university, Galileo left Padua in 1610 and went to Florence. There he accepted an appointment as chief mathematician and philosopher to the grand duke of Tuscany, Cosimo II. He resided in Florence for the remainder of his life.

Because of *Sidereus Nuncius,* Galileo's fame spread throughout Italy and the rest of Europe. His telescopes were in demand, and he obligingly provided them to selected European astronomers, including Kepler. In 1611 he proudly took one of his telescopes to Rome and let church officials personally observe some of these amazing celestial discoveries. While in Rome he also became a member of the prestigious Academia dei Lincei (Lyncean Academy). Founded in 1603, the academy was the world's first true scientific society.

In 1613 Galileo published "Letters on Sunspots" through the academy. He used the existence and motion of sunspots to demonstrate that the Sun itself changes, again attacking Aristotle's doctrine of the immutability of the heavens. In so doing, he also openly endorsed the Copernican model. This started Galileo's long and bitter fight with ecclesiastical authorities. Above all, Galileo believed in the freedom of scientific inquiry. Late in 1615 Galileo went to Rome and publicly argued for the Copernican model. This public action angered Pope Paul V, who immediately formed a special commission to review the theory of Earth's motion. Dutifully, the (unscientific) commission concluded that the Copernican theory was contrary to Biblical teachings and possibly a form of heresy. Cardinal Robert Bellarmine (an honorable person who was later canonized) received the unenviable task of silencing the brilliant but stubborn Galileo. In late February 1616 he officially admonished Galileo to abandon his support of the Copernican hypothesis. The cardinal (under direct orders from Pope Paul V) made Galileo an offer he could not refuse. Galileo must never teach or write again about the Copernican model, or he would be tried for heresy and imprisoned or possibly executed, like GIORDANO BRUNO.

Apparently Galileo got the message—at least so it seemed for a few years. In 1623 he published *Il saggiatore* (*The Assayer*). In this book he discussed the principles of scientific research but carefully avoided support for Copernican theory. He even dedicated the book to his lifelong friend the new pope, Urban VIII. However, in 1632 Galileo pushed his luck with the new pope to the limit by publishing *Dialogue on the Two Chief World Systems.* In this masterful (but satirical) work Galileo had two people present scientific arguments to an intelligent third person concerning the Ptolemaic and Copernican worldviews. The Copernican cleverly won these lengthy arguments. Galileo represented the Ptolemaic system with an ineffective character he called Simplicio. For a variety of reasons, Pope Urban VIII regarded Simplicio as an insulting personal caricature. Within months after the book's publication, the Inquisition summoned Galileo to Rome. Under threat of execution, the aging Italian scientist publicly retracted his support for the Copernican model on June 22, 1633. The Inquisition then sentenced him to life in prison, a term that he actually served under house arrest at his villa in Arceti (near Florence). Church authorities also banned the book, but Galileo's supporters smuggled copies out of Italy, and the Copernican message again spread across Europe.

While under house arrest Galileo worked on less controversial areas of physics. He published *Discourses and Mathematical Demonstrations Relating to Two New Sciences* in 1638. In this seminal work he avoided astronomy and summarized the science of mechanics, including the very important topics of uniform acceleration, free fall, and projectile motion.

Through Galileo's pioneering work and personal sacrifice, the scientific revolution ultimately prevailed over misguided adherence to centuries of Aristotelian philosophy. Galileo never really opposed the church nor its religious teachings. He did, however, come out strongly in favor of the freedom of scientific inquiry. Blindness struck the brilliant scientist in 1638. He died while imprisoned at home on January 8, 1642. Three and a half centuries later, on October 31, 1992, Pope John Paul II formally retracted the sentence of heresy passed on him by the Inquisition.

Galileo Project/spacecraft NASA's highly successful scientific mission to Jupiter launched in October 1989. With elec-

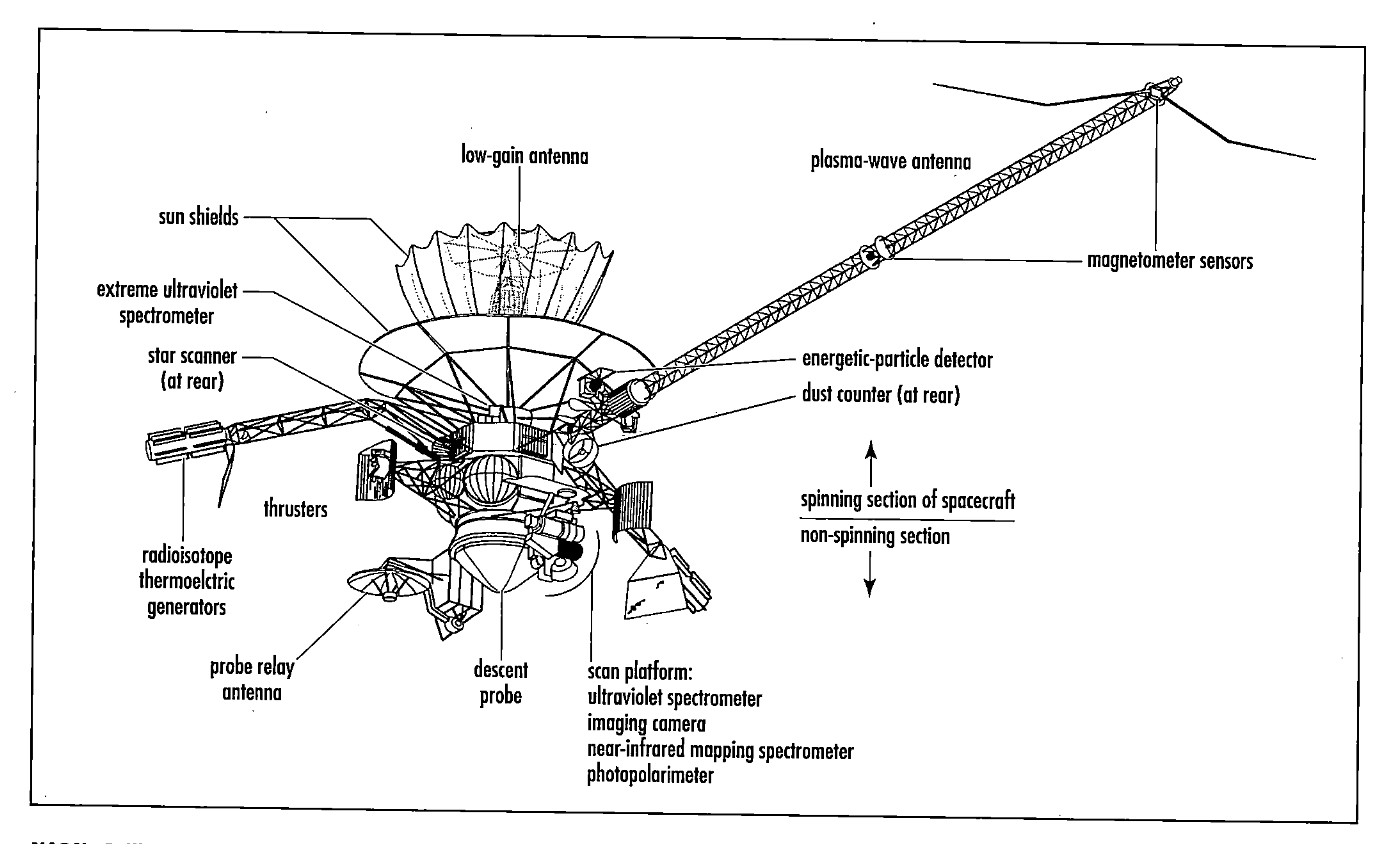

NASA's *Galileo* spacecraft *(Drawing courtesy of NASA)*

tricity supplied by two radioisotope-thermoelectric generator (RTG) units, the *Galileo* spacecraft studied the Jovian system extensively from December 1995 until February 2003. (*See* figure.) Upon arrival at the giant planet, the orbiter spacecraft, named in honor of GALILEO GALILEI, released a scientific probe into the upper portions of Jupiter's atmosphere. The spacecraft then flew in a highly elliptical orbit around Jupiter studying the planet and its moons—primarily the four Galilean satellites: Io, Europa, Ganymede, and Callisto. Right up until the end of its scientific mission, the *Galileo* spacecraft continued to deliver surprises. In December 2002 its instruments revealed that Jupiter's potato-shaped inner moon, Amalthea, had a very low density, a fact suggesting, in turn, that the tiny celestial body was probably full of holes.

Because of Amalthea's irregular shape and low density, planetary scientists speculate that this moon may have broken into many pieces that now tenuously cling together from the pull of each other's gravity. If true, the moon would possess many empty spaces throughout its interior where the fragmented pieces do not fit together well. The December 2002 flyby of Amalthea also brought the spacecraft closer to Jupiter than at any time since it began orbiting the giant planet on December 7, 1995. After more than 30 close encounters with Jupiter's four largest moons, the flyby of Amalthea became the spacecraft's last scientific adventure in the Jovian system. On February 28, 2003, the NASA flight team terminated its operation of the *Galileo* spacecraft. The directors sent a final set of commands to the far-traveling robot spacecraft, putting it on a course that resulted in its mission-ending plunge into Jupiter's atmosphere in late September 2003.

Galle, Johann Gottfried (1812–1910) German *Astronomer* By good fortune and some careful telescopic viewing, Johann Gottfried Galle became the first person to observe and properly identify the planet Neptune. Acting on a recommendation from the French astronomer URBAIN JEAN JOSEPH LEVERRIER, Galle began a special telescopic observation at the Berlin Observatory on September 23, 1846. His search quickly revealed the new planet precisely in the region of the night sky predicted by Leverrier's calculations. Improved mathematical techniques helped Galle and Leverrier convert subtle perturbations in the orbit of Uranus into one of the major discoveries of 19th-century astronomy.

Galle was born in Pabsthaus, Germany, on June 9, 1812. He received his early education at the Wittenberg Gymnasium (secondary school) and then his formal training in mathematics and physics at the University of Berlin. After graduating from the university in 1833, Galle taught mathematics in the gymnasiums at Guben and Berlin. In 1835 JOHANN FRANZ ENCKE (his former professor) invited Galle to become his assistant at the newly founded Berlin Observatory.

Galle proved to be a skilled telescopic astronomer. In 1838 he discovered the C, or "crêpe," ring of Saturn. He was also an avid "comet hunter," and during the winter of

1839–40 Galle discovered three new comets. Then, in 1846 fortune smiled on Galle and allowed him to play a major role in one of the most important moments in the history of planetary astronomy. Independently of the British astronomer JOHN COUCH ADAMS, the French astronomer and mathematician Urbain Jean Joseph Leverrier studied the perturbations in the orbit of Uranus and made his own mathematical predictions concerning the location of another planet beyond Uranus. Leverrier corresponded with Galle on September 18, 1846, and asked the German astronomer to investigate a section of the night sky to confirm his mathematical prediction that there was a planet beyond Uranus.

Upon receipt of Leverrier's letter, Galle and his assistant, HEINRICH LUDWIG D'ARREST, immediately began searching the portion of the sky recommended by the French mathematical astronomer. D'Arrest also suggested to Galle that they use Encke's most recent star chart to assist in the search for the elusive trans-Uranian planet predicted by Leverrier's computations. In less than an hour Galle detected a "star" that was not on Encke's latest chart. A good scientist, Galle waited 24 hours to confirm that this celestial object was indeed the planet Neptune. On the following night (September 24) Galle again observed the object and carefully compared its motion relative to that of the so-called fixed stars. There was no longer any question: The change in its position clearly indicated that this "wandering light" was another planet.

Galle discovered Neptune pretty much in the position predicted by Leverrier's calculations. He immediately wrote to Leverrier, informing him of the discovery and thanking him for the suggestion on where to search. The telescopic discovery of Neptune by Galle on September 23, 1846, is perhaps the greatest example of the marriage of mathematics and astronomy in the 19th century. Subtle perturbations in the calculated orbit of Uranus suggested to mathematicians and astronomers that one or more planets could lie beyond. Today astronomers jointly award credit for the combined predictive and observational discovery of Neptune to John Couch Adams, Johann Galle, and Urbain Leverrier.

In 1851 Galle accepted a position as the director of the Breslau Observatory and remained there until his retirement in 1897. He made additional contributions to astronomy by focusing his attention on determining the mean distance from Earth to the Sun—an important reference distance called the *astronomical unit* (AU). Astronomers relied on the transit of Venus across the face of the Sun to make an estimate of the astronomical unit. At the time this was a difficult measurement, since astronomers had to determine precisely the moment of first contact. However, in 1872 Galle suggested a more accurate and reliable way of measuring the scale of the solar system. He recommended that astronomers use the asteroids whose orbits come very close to Earth to measure solar parallax. Galle reasoned that the minor planets offered more precise pointlike images, and their large number provided frequent favorable oppositions. In an effort to demonstrate and refine this technique, Galle observed the asteroid Flora in 1873. Astronomers SIR DAVID GILL and SIR HAROLD SPENCER JONES greatly improved measurement of the Earth-Sun distance using Galle's suggestion.

Of the three codiscoverers of Neptune, only Johann Galle survived until 1896 to receive congratulations from the international astronomical community as it celebrated the 50th anniversary of their great achievement. He died on July 10, 1910, in Potsdam, Germany.

gamma-ray astronomy Branch of astronomy based on the detection of the energetic gamma rays associated with supernovas, exploding galaxies, quasars, pulsars, and phenomena near suspected black holes.

See also ASTROPHYSICS; *COMPTON GAMMA-RAY OBSERVATORY* (CGRO); GAMMA-RAY BURST.

gamma-ray burst (GRB) An outburst that radiates tremendous amounts of energy, equal to or greater than a supernova, in the form of gamma rays and X-rays. These bursts take place in less than a few minutes. The *gamma-ray burster,* as the intriguing phenomenon is sometimes called, represents one of the greatest astronomical mysteries.

Starting in the late 1960s, astronomers observing the heavens at very short wavelengths began detecting incredibly brief and intense bursts of gamma rays from seemingly random locations in the sky. They observed that a few times a day, the sky lights up with an incredible flash, or burst, of gamma rays. Often the burst outshines all the other sources of cosmic radiation added together. The source of the burst then disappears completely. Even more puzzling, no one was able to predict when the next burst would occur or from what direction in the sky it would come.

Following the discovery of the gamma-ray burst in the late 1960s, astronomers used the best available space-based gamma-ray detection instruments to construct a catalog of GRBs. As the number of observed GRBs grew, many theories emerged within the international astronomical community concerning their origin. Scientists also argued among themselves as to whether the GRBs were taking place in the Milky Way galaxy or in other galaxies. Unfortunately, the addition of each newly observed GRB tended to reveal little more than the fact that GRBs never repeated from the same cosmic source or location.

Then NASA's *Compton Gamma-Ray Observatory* (CGRO), launched in 1991, provided a wealth of new GRB observations. The spacecraft's burst and transient source experiment (BATSE) was capable of monitoring the sky with unprecedented sensitivity. One thing soon became clear as the catalog of GRBs observed by BATSE grew—the GRBs were in no way correlated with sources in the Milky Way galaxy. In 1997 the Italian-Dutch *BeppoSAX* satellite made a breakthrough in our basic understanding of gamma-ray bursts. Using a particularly effective combination of gamma-ray and X-ray telescopes, *BeppoSAX* detected afterglows from a few GRBs and precisely located the sources in a way that allowed other telescopes (sensitive to different portions of the electromagnetic spectrum) to promptly study the same location in the sky. This effort showed astronomers that GRBs are produced in very distant galaxies, requiring that the explosions producing the gamma-ray bursts must be incredibly powerful.

The next important breakthrough in GRB understanding occurred on January 23, 1999, when an enormously powerful event (called GRB990123) was detected. Alerted by sophisticated space-based detectors, astronomers were able to quickly observe this event at an unprecedented range of wavelengths and timing sensitivities. The event was very far away. If astronomers assumed isotropic emission of the gamma rays, this powerful GRB would have involved the release of the energy equivalent of two times the rest mass energy of a neutron star. If, on the other hand, astronomers assumed that the emitted energy was being beamed out of this distant GRB in a preferred direction (here one that just happened to point directly toward Earth), then the required energies became more reasonable and easier to explain.

Today astronomers *tenuously* suggest that gamma-ray bursts are produced by materials shooting toward Earth at nearly the speed of light. The material in question is ejected during the collision of two neutron stars or black holes. An alternate speculation is that the GRBs arise from a *hypernova*—the huge explosion hypothesized to occur when a supermassive star ends its life and collapses into a black hole. However, astronomers also recognize that many more multiwavelength, prompt observations of future GRBs will be required before they can confidently model the central engine (or engines, since there may be more than one mechanism) that produces these powerful and mysterious gamma-ray bursts.

gamma rays (symbol: γ) High-energy, very-short-wavelength packets, or quanta, of electromagnetic radiation. Gamma-ray photons are similar to X-rays except that they are usually more energetic and originate from processes and transitions within the atomic nucleus. Gamma rays typically have energies between 10,000 electron volts and 10 million electron volts (i.e., between 10 keV and 10 MeV) with correspondingly short wavelengths and high frequencies. The processes associated with gamma-ray emissions in astrophysical phenomena include (1) the decay of radioactive nuclei, (2) cosmic ray interactions, (3) curvature radiation in extremely strong magnetic fields, and (4) matter-antimatter annihilation. Gamma rays are very penetrating and are best stopped or shielded against by dense materials, such as lead and tungsten. Sometimes called *gamma radiation.*

See also GAMMA-RAY ASTRONOMY.

gamma (γ) star Within the system of stellar nomenclature introduced in 1603 by the German astronomer JOHANN BAYER (1572–1625), the third-brightest star in a constellation.

Gamow, George (Georgi Antonovich) (1904–1968) Ukrainian-American *Physicist, Cosmologist* A successful nuclear physicist, cosmologist, and astrophysicist, George Gamow was also a pioneering proponent of the big bang theory, a theory that boldly speculates that the universe began with a huge ancient explosion. As part of this hypothesis, he predicted the existence of a telltale cosmic microwave background, observational evidence for which was discovered by ARNO PENZIAS and ROBERT W. WILSON in the early 1960s.

Gamow was born on March 4, 1904, in the city of Odessa in Ukraine. His grandfather was a general in the army of the Russian czar, and his father was a teacher. On his 13th birthday Gamow received a telescope, a life-changing gift that introduced him to astronomy and helped him decide on a career in physics. Having survived the turmoil of the Russian Revolution, he entered Novorossysky University in 1922 at 18 years of age. He soon transferred to the University of Leningrad (now called the University of St. Petersburg), where he studied physics, mathematics, and cosmology. Gamow completed a Ph.D. in 1928, the same year he proposed that alpha decay could be explained by the phenomenon of the quantum mechanical tunneling of alpha particles through the nuclear potential barrier in the atomic nucleus. This innovative work represented the first successful extension of quantum mechanics to the atomic nucleus.

As a bright young nuclear physicist, he performed postdoctoral research throughout Europe—visiting at the University of Göttingen in Germany, working with Niels Bohr in Copenhagen, Denmark, and studying beside Ernest Rutherford at the Cavendish Laboratory in Cambridge, England. In 1931 he was summoned back to the Soviet Union to serve as a professor of physics at Leningrad University. However, after having lived in western Europe for several years, he found life in Stalinist Russia not to his liking. A highly creative individual, he obtained permission to attend the 1933 International Solvay Congress in Brussels, Belgium, and simply never returned to the Soviet Union. He emigrated from Europe to the United States in 1934, accepting a faculty position at George Washington University in Washington, D.C., and becoming an American citizen. In 1936 he collaborated with the Hungarian-American physicist Edward Teller (1908–2003) in developing a theory to explain beta decay—the process whereby a nucleus emits an electron.

At this point in his career, Gamow began exploring relationships between cosmology and nuclear processes. He used his knowledge of nuclear physics to interpret the processes taking place in various types of stars. In 1942, again in collaboration with Edward Teller, Gamow developed a theory about the thermonuclear reactions and internal structures within red giant stars. These efforts led Gamow to conclude that the Sun's energy results from thermonuclear reactions. He was also an ardent supporter of EDWIN HUBBLE's expanding universe theory, an astrophysical hypothesis that made him an early and strong advocate of the big bang theory.

In collaboration with RALPH ALPHER, Gamow introduced his own version of big bang cosmology in the important 1948 scientific paper "The Origin of the Chemical Elements." He based some of this work on cosmology concepts earlier suggested by GEORGES LEMAÎTRE. Starting with the explosion of Lemaître's "cosmic egg," Gamow proposed a method of thermonuclear reaction by which the various elements could emerge from a primordial mixture of nuclear particles that he called the YLEM. While later nucleosynthesis models introduced by WILLIAM FOWLER and other astrophysicists would supersede this work, Gamow's pioneering investigations provided an important starting point. Perhaps the most significant postulation to emerge from his efforts in cosmology was the prediction that the big bang event would produce a uniform cosmic microwave background—a lingering remnant at the edge of the observable universe. The discovery of this cosmic microwave background in 1964 by Arno Penzias and Robert

This historic photograph shows the construction scaffolding used to create a contemporary gantry at Cape Canaveral Air Force Station in July 1950 to help workers prepare the Bumper 8 rocket for launch. (Bumper 8, the first rocket shot from Cape Canaveral, was a captured German V-2 rocket modified by American rocket engineers.) *(Courtesy of the U.S. Air Force)*

Wilson renewed interest in Gamow's big bang cosmology and provided scientific evidence that strongly supported an ancient explosion hypothesis. Gamow's 1952 book, *Creation of the Universe,* presented a detailed discussion of his concepts in cosmology.

Just after James Watson and Francis Crick proposed their DNA model in 1953, Gamow turned his scientific interests from the physics that took place at the start of the universe to the biophysical nature of life. He published several papers on the storage and transfer of information in living cells. While his precise details were not terribly accurate by today's level of understanding in the field of genetics, he did introduce the important concept of "genetic code"—a fundamental idea confirmed by subsequent experimental investigations.

In 1956 Gamow became a faculty member at the University of Colorado in Boulder and remained at that institution for the rest of his life. He was a prolific writer who produced highly technical books, such as the *Structure of Atomic Nuclei and Nuclear Transformations* (1937), as well as very popular books about science for general audiences. Several of his most popular books were *Mr. Tompkins in Wonderland* (1936), *One, Two, and Three . . . Infinity* (1947), and *A Star Called the Sun* (1964).

Gamow was elected to the Royal Danish Academy of Sciences in 1950 and the U.S. National Academy of Sciences in 1953. The United Nations Educational, Scientific, and Cultural Organization (UNESCO) bestowed its Kalinga Prize on him in 1956 for his literary efforts that popularized modern science around the world. He died in Boulder, Colorado, on August 19, 1968.

gantry A frame that spans over something, such as an elevated platform that runs astride a work area, supported by wheels on each side. Specifically, the term is short for gantry crane or gantry scaffold.

Ganymede With a diameter of 5,262 kilometers, Ganymede is the largest moon of Jupiter and in the solar system. It was discovered and named by GALILEO GALILEI in 1610. Ganymede orbits Jupiter at a distance of 1,070,000 kilometers at an inclination of 0.21° with a period of approximately 7.16 days. This large moon has a density of just 1.94 g/cm^3 (1,940 kg/m^3), suggesting that like its Galilean satellite density twin, Callisto, Ganymede may harbor substantial amounts of ice. In 1996 instruments onboard NASA's *Galileo* spacecraft observed that Ganymede had a weak

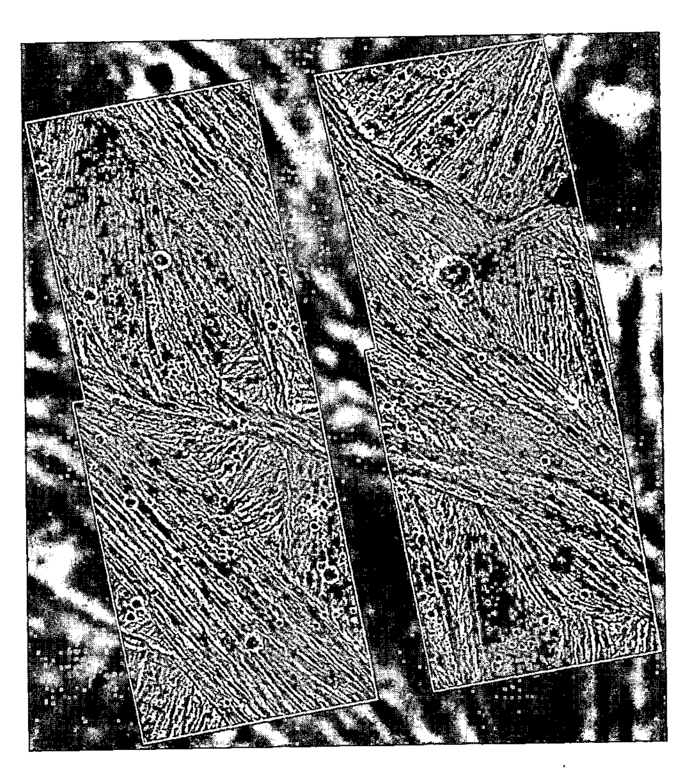

This composite image shows details of the parallel ridges and troughs that are the main surface features in the brighter regions of Ganymede. The picture consists of four high-resolution images of Ganymede's Uruk Sulcus region taken by the *Galileo* spacecraft in 1997, displayed within the context of an image of the same region made by the *Voyager 2* spacecraft in 1979. The *Galileo* spacecraft frames (with a resolution of 74 meters per pixel) unveil the fine topography of Ganymede's ice-rich surface. *(Courtesy of NASA/JPL)*

magnetosphere, making it the first moon in the solar system on which a magnetic field has been detected. In December 2000 fluctuations in magnetometer data from the *Galileo* spacecraft (similar to data fluctuations observed when the spacecraft was near Europa) encouraged some planetary scientists to suggest that Ganymede may have liquid, possibly slushy, water beneath its surface. Furthermore, high-resolution surface imagery from the *Galileo* spacecraft has provided scientists tantalizing geological evidence that Ganymede may have experienced flowing "water" lava, similar to the interesting formations observed on Europa. Analysis of these features has tended to reinforce cautious scientific speculation that Ganymede may possess a reservoir of liquid water beneath its surface. Future missions to Jupiter's icy moons will help scientists unravel this intriguing question and the one that follows immediately: "Where there is liquid water present in some abundance on another planetary body, did alien life emerge?"

See also GALILEAN SATELLITES; GALILEO PROJECT/SPACECRAFT; JUPITER; *JUPITER ICY MOONS ORBITER* (JIMO).

gas The state of matter in which the molecules are practically unrestricted by intermolecular forces so that the molecules are free to occupy any space within an enclosure.

See also IDEAL GAS.

gas cap The gas immediately in front of a body (such as a reentry vehicle or meteor) as it travels through a planetary atmosphere. This region of atmospheric gas is compressed and adiabatically heated, often to incandescence.

gas constant (symbol: R_u) The constant factor in the equation of state for an ideal (perfect) gas. The universal gas constant is

$$R_u = 8.315 \text{ joules per mole per kelvin [J/(mol-K)]}$$

The gas constant for a particular, nonideal gas is called the *specific gas constant* (symbol: R_{sp}) and is related to the universal gas constant as follows:

$$R_{sp} = R_u/M$$

where M is the molecular weight of the particular gas.

See also IDEAL GAS.

gas generator 1. An assemblage of parts similar to a small rocket engine in which a propellant is burned to provide hot exhaust gases. These hot gases are then used to (1) drive the turbine in the turbopump assembly of a rocket vehicle, (2) pressurize liquid propellants, or (3) provide thrust by exhausting through a nozzle. 2. A device used in a laboratory to generate gases, such as hydrogen and oxygen from water.

Gaspra An S-type asteroid encountered by the *Galileo* spacecraft in late October 1991. This fortuitous flyby took place as *Galileo* was traveling to Jupiter, and it provided scientists their first close-up look at an asteroid. Gaspra (about 19 km long) proved to be an irregularly shaped minor planet, rather like a potato, riddled with craters and fractures. By examining the extent of impact crater formation on its surface (which was relatively smooth compared to the surface of Ida, another main belt asteroid imaged by the *Galileo* spacecraft), planetary scientists were able to estimate that Gaspra is a rather young main belt asteroid—perhaps just 200 million years old or so. Gaspra is most likely the fragment of a much larger object that broke apart during violent ancient collisions. Also called Gaspra-951.

See also ASTEROID.

gas turbine A heat engine that uses high-temperature, high-pressure gas as a working fluid. Its operation approximates the ideal Brayton cycle, and a portion of the energy content of the gaseous working fluid is converted directly into work (i.e., rotating shaft action). Often the hot gases for operation of a gas turbine are obtained by the combustion of a fuel in air. These types of gas turbines are also called *combustion turbines*. The simplest combustion gas turbine heat engine is the open cycle system, which has three major components: the compressor, the combustion chamber (or combustor), and the gas turbine itself. The compressor and turbine usually are mounted on the same shaft, so that a portion of the mechanical ("shaft") work produced by the turbine can be used to drive the compressor.

Gauss, Carl Friedrich (1777–1855) German *Mathematician* Carl Friedrich Gauss was a brilliant German mathematician whose contributions to celestial mechanics helped many other astronomers efficiently calculate the orbits of planets, asteroids, and comets. In the early 19th century Gauss's work in orbit determination allowed astronomers to quickly rediscover the asteroid Ceres (which had become lost behind the Sun) using data from just three observations. Gauss's mathematical innovations such as the method of least squares and perturbation theory enabled the French astronomer URBAIN JEAN JOSEPH LEVERRIER to predict theoretically the position of Neptune in 1846. Urbain's effort involved an application of Gauss's celestial mechanics techniques to the subtle orbital perturbations observed for Uranus.

gauss (symbol: G or Gs) In the centimeter-gram-second (cgs) unit system, the unit of magnetic flux density. Earth's magnetic field at the planet's surface is on the average of 0.3 to 0.6 gauss.

$$1 \text{ gauss} = 10^{-4} \text{ tesla}$$

See also TESLA.

Gaussian gravitational constant (symbol: k) The constant related to JOHANN KEPLER's third law of planetary motion, which has the value of 0.01720209895. CARL FRIEDRICH GAUSS first calculated this constant for use in an astronomical unit system in which the basic unit of length is the astronomical unit (AU), the basic unit of mass is the mass of the Sun (a solar mass), and the basic unit of time is the day.

Geiger counter A nuclear radiation detection and measurement instrument. It contains a gas-filled tube containing electrodes between which there is an electrical voltage but no current flowing. When ionizing radiation passes through the tube, a short, intense pulse of current passes from the negative electrode to the positive electrode and is measured or counted. The number of pulses per second measures the intensity of the radiation. This instrument is also called a *Geiger-Müller counter* and is named for Hans Geiger and W. Müller, who developed it in the 1920s.

Geminid meteors A winter meteor shower that reaches its maximum activity on or about December 13th. An asteroid (Phaethon 3200), rather than a comet, is considered the celestial parent of this prominent meteor shower, which often produces a rough count of about 50 meteors per hour.

See also METEOR; METEOROIDS.

Gemini Project The Gemini Project (1964–66) was the second U.S. crewed space program and the beginning of sophisticated human spaceflight. It expanded and refined the scientific and technological endeavors of the Mercury Project and prepared the way for the more sophisticated Apollo Project, which carried U.S. astronauts to the lunar surface. The Gemini Project added a second crewmember and a maneuverable spacecraft. New objectives included rendezvous and docking techniques with orbiting spacecraft and extravehicular "walks in space." NASA launched the first crewed Gemini orbital flight, called the *Gemini 3* mission, from Cape Canaveral Air Force Station, Florida, on March 23, 1965. (*See* figure.) In all, 10 two-person launches occurred, successfully placing 20 astronauts in orbit and returning them safely to Earth. The Gemini Project Table summarizes the Gemini Project flights.

See also APOLLO PROJECT; MERCURY PROJECT.

Liftoff of the first two-person spacecraft flown by the United States. On March 23, 1965, a Titan rocket successfully placed astronauts Virgil I. (Gus) Grissom (command pilot) and John W. Young (pilot) aboard the *Molly Brown* spacecraft from Pad 19 at Cape Canaveral Air Force Station into low Earth orbit. The *Gemini 3* mission was the initial human-crewed flight in NASA's Gemini Project. The astronauts orbited Earth three times and then successfully splashed down in the Atlantic Ocean, where the USS *Intrepid* recovered them. *(Courtesy of NASA)*

Gemini Telescopes The Gemini Observatory consists of twin 8-meter optical/infrared telescopes located on two of the best sites on Earth for observing the universe. Together these telescopes can access the entire sky. The *Gemini South Telescope* is located at almost 2,750 meters on a mountain called Cerro Pachón, which is in the Chilean Andes. Cerro Pachón shares resources with the adjacent SOAR Telescope and the nearby telescopes of the Cerro Tololo Inter-American Observatory. The *Gemini North Telescope* is located on Hawaii's Mauna Kea in a cluster of world-class observatories at the top of a long-dormant volcano that rises almost 4,270 meters into the dry, stable air of the Pacific Ocean. The Gemini Observatory has its international headquarters in Hilo, Hawaii. Both of the Gemini Telescopes have been designed to take advantage of the latest technology and thermal controls to provide outstanding optical and infrared capabilities. By incorporating technologies such as laser guide stars, adaptive optics, and multiobject spectroscopy, astronomers in the Gemini partnership have access to some of the latest ground-based optical tools for exploring the universe. The Gemini Observatory was built and is being operated by a partnership of seven countries. These are the United States, the United Kingdom, Canada, Chile, Australia, Brazil, and Argentina. The Gemini Telescopes have been integrated with modern networking technologies to allow remote operations from control rooms at the base facilities in Hilo and La Serena, Chile. With the flexibility of remote participation and queue scheduling, qualified researchers from anywhere in the Gemini partnership nations can take advantage of the best possible match among astronomical research needs, instrument availability, and observing conditions. The Association of Universities for Research in Astronomy, Inc. (AURA) manages the Gemini Observatory under a cooperative agreement with the U.S. National Science Foundation (NSF).

general relativity ALBERT EINSTEIN's theory, introduced in 1915, that gravitation arises from the curvature of space and time—the more massive an object, the greater the curvature.

See also RELATIVITY.

Genesis Solar Wind Sample Return mission The primary mission of NASA's *Genesis* spacecraft was to collect samples of solar wind particles and return these samples of extraterrestrial material safely to Earth for detailed analysis. The mission's specific science objectives were to obtain precise solar isotopic and elemental abundances and provide a

Summary of Gemini Project Missions

Mission	Date(s) Recovery Ship	Crew	Ground Elapsed Time (GET) Hr:min	Remarks
Gemini 3 (Molly Brown)	March 23, 1965 *Intrepid (A)*	USAF Maj. Virgil I. Grissom Navy Lt. Comdr. John W. Young	4:53	3 orbits
Gemini 4	June 3–7, 1965 *Wasp (A)*	USAF Majors James A. McDivitt and Edward H. White, II	97:56	62 orbits: first U.S. EVA (White)
Gemini 5	Aug. 21–29, 1965 *Lake Champlain (A)*	USAF Lt. Col. L. Gordon Cooper. Navy Lt. Comdr. Charles Conrad, Jr.	190:55	120 orbits
Gemini 7	Dec. 4–18, 1965 *Wasp (A)*	USAF Lt. Col. Frank Borman Navy Comdr. James A. Lovell, Jr.	330:35	Longest Gemini flight; rendezvous target for *Gemini 6*; 206 orbits
Gemini 6	Dec. 15–16, 1965 *Wasp (A)*	Navy Capt. Walter M. Schirra, Jr. USAF Maj. Thomas P. Stafford	25:51	Rendezvoused within 1 ft of *Gemini 7*; 16 orbits
Gemini 8	Mar. 16, 1966 *Mason (P)*	Civilian Neil A. Armstrong USAF Maj. David R. Scott	10:41	Docked with unmanned *Agena 8*; 7 orbits
Gemini 9A	June 3–6 1966 *Wasp (A)*	USAF Lt. Col. Thomas P. Stafford Navy Lt. Comdr. Eugene A. Cernan	72:21	Rendezvoused (3) with *Agena 9*; one EVA; 44 orbits
Gemini 10	July 18–21, 1966 *Guadalcanal (A)*	Navy Comdr. John W. Young USAF Maj. Michael Collins	70:47	Docked with *Agena 10*; rendezvoused with *Agena 8*; two EVAs; 43 orbits
Gemini 11	Sept. 12–15, 1966 *Guam (A)*	Navy Comdr. Charles Conrad, Jr. Navy Lt. Comdr. Richard F. Gordon, Jr.	71:17	Docked with *Agena 11* twice; first tethered flights; two EVAs; highest altitude in Gemini program, 853 miles; 44 orbits
Gemini 12	Nov. 11–15, 1966 *Wasp (A)*	Navy Capt. James A. Lovell, Jr. USAF Maj. Edwin E. Aldrin, Jr.	94:35	Three EVAs total 5 hrs. 30 min.; 59 orbits

Note: A = Atlantic Ocean
P = Pacific Ocean
EVA = Extravehicular Activity
Source: NASA.

reservoir of solar matter for future investigation. A detailed study of captured solar wind materials would allow scientists to test various theories of solar system formation. Access to these materials would also help them resolve lingering issues about the evolution of the solar system and the composition of the ancient solar nebula.

The basic spacecraft was 2.3 meters long and 2 meters wide, and its solar array had a wingspan of 7.9 meters. At launch the spacecraft payload had a total mass of 636 kilograms, composed of the 494-kg dry spacecraft and 142 kg of onboard propellant. The scientific instruments included the solar wind collector arrays, ion concentrator, and solar wind (ion and electron) monitors. A combined solar cell array with a nickel-hydrogen storage battery provided up to 254 watts of electric power just after launch.

The mission started on August 8, 2001, when an expendable Delta II rocket successfully launched the *Genesis* spacecraft from Cape Canaveral Air Force Station, Florida. Following launch the cruise phase of the mission lasted slightly more than three months. During this period the spacecraft traveled 1.5 million kilometers from Earth to the Lagrange libration point 1 (L1). The *Genesis* spacecraft entered a *halo orbit* around the L1 point on November 16, 2001. Upon arrival the spacecraft's large thrusters fired, putting *Genesis* into a looping, elliptical orbit around the L1 point. The *Genesis* spacecraft then completed five orbits around L1; nearly 80 percent of the mission's total time was spent collecting particles from the Sun.

On December 3, 2001, *Genesis* opened its collector arrays and began accepting particles of solar wind. A total of 850 days were logged exposing the special collector arrays to the solar wind. These collector arrays are circular trays composed of palm-sized hexagonal tiles made of various high-quality materials, such as silicon, gold, sapphire, and diamondlike carbon. After the sample return capsule opened, the lid of the science canister opened as well, exposing a collector for the bulk solar wind. As long as the science canister's lid was opened, this bulk collector array was exposed to different types of solar wind that flowed past the spacecraft.

The ion and electron monitors of the *Genesis* spacecraft were located on the equipment deck outside the science canister and sample return capsule. These instruments looked for changes in the solar wind and then relayed information about these changes to the main spacecraft computer, which would then command the collector array to expose the appropriate collector. By recognizing characteristic values of temperature, velocity, density, and composition, the spacecraft's monitors were able to distinguish between three types of solar wind—fast, slow, and coronal mass ejections. The versatile robot sampling spacecraft would then fold out and extend one of three different collector arrays when a certain type of solar wind passed by.

The other dedicated science instrument of the *Genesis* spacecraft was the solar wind collector. As its name implies, this instrument would concentrate the solar wind onto a set of small collectors made of diamond, silicon carbide, and diamondlike carbon. As long as the lid of the science canister was opened, the concentrator was exposed to the solar wind throughout the collection period.

On April 1, 2004, ground controllers ordered the robot spacecraft to stow the collectors. The closeout process was completed on April 2, when the *Genesis* spacecraft closed and sealed its sample return capsule. Then, on April 22 the spacecraft began its journey back toward Earth. However, because of the position of the landing site—the U.S. Air Force's Utah Testing and Training Range in the northwestern corner of that state—and the unique geometry of the *Genesis* spacecraft's flight path, the robot sampling craft could not make a direct approach and still make a daytime landing. In order to allow the *Genesis* chase-helicopter crews an opportunity to capture the return capsule mid-air in daylight, the Genesis mission controllers designed an orbital detour toward another Lagrange point, L2, located on the other side of Earth from the Sun. After successfully completing one loop around the L2 point, the Genesis spacecraft was prepared for its return to Earth on September 8. On September 8 the spacecraft approached Earth and performed a number of key maneuvers prior to releasing the sample return capsule. Sample capsule release took place when the spacecraft flew past Earth at an altitude of about 66,000 kilometers. As planned, the *Genesis* return capsule successfully reentered Earth's atmosphere at a velocity of 11 kilometers per second over northern Oregon.

Unfortunately, during reentry on September 8, the parachute on the *Genesis* sample return capsule failed to deploy (apparently because of an improperly installed gravity switch), and the returning capsule smashed into the Utah desert at a speed of 311 km/hr. The high-speed impact crushed the sample return capsule and breeched the sample collection capsule—possibly exposing some of the collected pristine solar materials to contamination by the terrestrial environment.

However, mission scientists worked diligently to recover as many samples as possible from the spacecraft wreckage and then to ship the recovered materials in early October to the Johnson Space Center in Houston, Texas, for evaluation and analysis. One of the cornerstones of the recovery process was the discovery that the gold foil collector was undamaged by the hard landing. Another postimpact milestone was the recovery of the *Genesis* spacecraft's four separate segments of the concentrator target. Designed to measure the isotopic ratios of oxygen and nitrogen, the segments contained within their structure the samples that are the mission's most important science goals.

The United States is a signatory to the Outer Space Treaty of 1966. This document states in part that exploration of the Moon and other celestial bodies shall be conducted "so as to avoid their harmful contamination and also adverse changes in the environment of the Earth resulting from the introduction of extraterrestrial matter." The Genesis sample consists of atoms collected from the Sun. NASA's planetary protection officer has categorized the Genesis mission as a

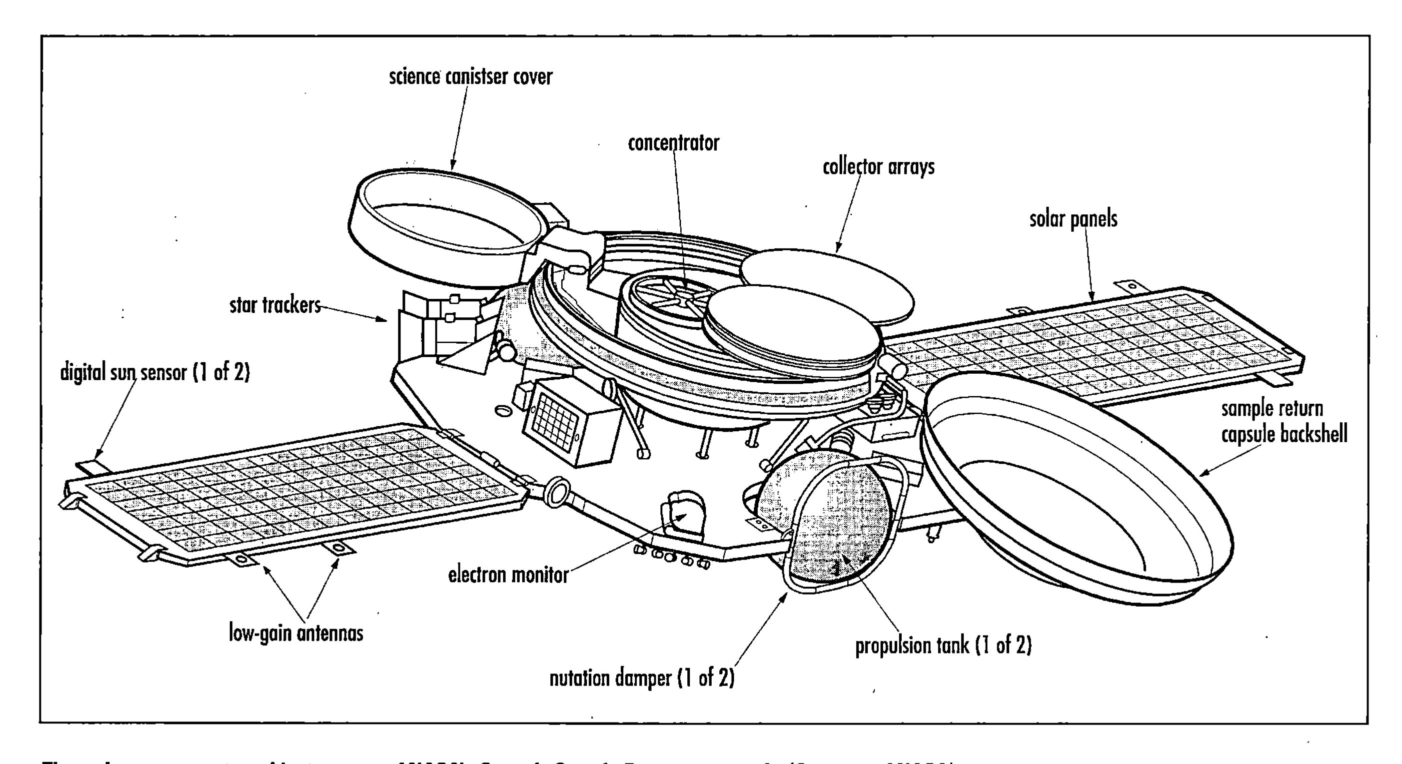

The major components and instruments of NASA's *Genesis Sample Return* spacecraft *(Courtesy of NASA)*

mission "safe for unrestricted Earth return." This declaration means that exobiologists and other safety experts have concluded that there is no chance of extraterrestrial biological contamination during sample collection at the L1 point. The U.S. National Research Council's Space Studies Board has also concurred on a planetary protection designation of unrestricted Earth return for the Genesis mission. The board determined that the sample had no potential for containing life. Consequently, there is no significant issue of extraterrestrial contamination of planet Earth due to the aborted sample return operation of the Genesis capsule. However, the issue of planetary contamination remains of concern when robot-collected sample capsules return from potentially life-bearing celestial bodies, such as Mars and Europa.

geo- A prefix meaning the planet Earth, as in geology and geophysics.

geocentric Relative to Earth as a center; measured from the center of Earth.

geocentric coordinate system *See* CELESTIAL COORDINATES.

geocentric cosmology A cosmology that places planet Earth at the center of the universe. Ptolemaic cosmology is the most familiar form of geocentric cosmology, although many ancient peoples held similar worldviews.

See also COSMOLOGY; PTOLEMAIC SYSTEM.

geodesy In general, the science of determining the exact size and shape of bodies in our solar system and of the distribution of mass within these bodies. Specifically, the science that deals mathematically with the size and shape of Earth, with Earth's external gravity field, and with surveys of such precision that the overall size and shape of Earth must be taken into consideration.

geographic information system (GIS) A computer-assisted system that acquires, stores, manipulates, compares, and displays geographic data, often including multispectral sensing data sets from Earth-observing satellites.

See also REMOTE SENSING.

geoid The figure of Earth as defined by the geopotential surface that most nearly coincides with mean sea level over the entire surface of Earth.

geomagnetic storm Sudden worldwide fluctuations in Earth's magnetic field, associated with solar flare–generated shock waves that propagate from the Sun to Earth. Geomagnetic storms may significantly alter ionospheric current densities over large portions of the middle- and high-latitude regions of Earth, impacting high-frequency (HF) systems. During a solar storm large numbers of particles are dumped into the high-altitude atmosphere. These particles produce heating effects (especially near the auroral zone), which then alter the near-Earth-space atmospheric densities from about 100 kilometers out to about 1,000 kilometers, impacting the operation of spacecraft orbiting Earth at altitudes below 1,000 kilometers. Polar-orbiting spacecraft can be especially impacted, as, for example, by an increase in atmospheric drag.

See also AURORA; IONOSPHERE; SPACE WEATHER.

geophysics The branch of science dealing with Earth and its environment—that is, the solid Earth itself, the hydrosphere (oceans and seas), the atmosphere, and (by extension) near-Earth space. Classically, geophysics has been concerned primarily with physical phenomena occurring at and below the surface of Earth. This traditional emphasis gave rise to such companion disciplines as geology, physical oceanography, seismology, hydrology, and geodesy. With the advent of the space age the trend has been to extend the scope of geophysics to include meteorology, geomagnetism, astrophysics, and other sciences concerned with the physical nature of the universe.

See also EARTH SYSTEM SCIENCE.

geoprobe A rocket vehicle designed to explore outer space near Earth at a distance of more than 6,440 kilometers (km) from Earth's surface. By convention, rocket vehicles operating lower than 6,440 km are called sounding rockets.

See also SOUNDING ROCKET.

geosphere The solid (lithosphere) and liquid (hydrosphere) portions of Earth. Above the geosphere lies the atmosphere. At the interface between these two regions is found almost all of the biosphere, or zone of life.

See also EARTH SYSTEM SCIENCE.

Geostationary Operational Environmental Satellite (GOES) GOES weather satellites maintain orbital positions over the same Earth location along the equator at about 35,900 kilometers, giving them the ability to make continuous observations of weather patterns over and near the United States. Operated by the U.S. National Oceanic and Atmospheric Administration (NOAA), these satellites provide both visible light and infrared images of cloud patterns as well as "soundings," or indirect measurements, of the temperature and humidity throughout the atmosphere. NOAA has been operating GOES satellites since 1974. Data from these spacecraft provide input for the forecasting responsibilities of the National Weather Service. Among other applications, the GOES data assist in monitoring storms and provide advance warning of emerging severe weather. The geostationary vantage point of GOES satellites permits the observation of large-scale weather events, which is required for forecasting small-scale events.

See also WEATHER SATELLITE.

geostationary orbit (GEO) A satellite in a circular orbit around Earth at an altitude of 35,900 kilometers above the equator that travels around the planet at the same rate as Earth spins on its axis. Communications satellites, environmental satellites, and surveillance satellites use this important orbit. If the spacecraft's orbit is circular and lies in the equatorial plane (to an observer on Earth), the spacecraft appears stationary over a given point on Earth's surface. If the spacecraft's orbit is inclined to the equatorial plane (when observed from Earth), the spacecraft traces out a figure-eight path every 24 hours.

See also GEOSYNCHRONOUS ORBIT; SYNCHRONOUS SATELLITE.

geosynchronous orbit (GEO) An orbit in which a satellite completes one revolution at the same rate as the Earth spins, namely, 23 hours, 56 minutes, and 4.1 seconds. A satellite placed in such an orbit (at approximately 35,900 kilometers above the equator) revolves around Earth once per day.

See also GEOSTATIONARY ORBIT; SYNCHRONOUS SATELLITE.

g-force An inertial force usually expressed in multiples of terrestrial gravity.

Giacconi, Riccardo (1931–) Italian-American *Astrophysicist* Using special instruments carried into space on sounding rockets and satellites, Riccardo Giacconi helped establish the exciting new field of X-ray astronomy. This astrophysical adventure began quite by accident in June 1962, when Giacconi's team of scientists placed a specially designed instrument package on a sounding rocket. As the rocket climbed above Earth's sensible atmosphere above White Sands, New Mexico, the instruments unexpectedly detected the first cosmic X-ray source—that is, a source of X-rays from beyond the solar system. Subsequently named Scorpius X-1, this "X-ray star" is the brightest of all nontransient X-ray sources ever discovered. He shared the 2002 Nobel Prize in physics for his pioneering contributions to astrophysics.

Giacconi was born in Milan, Italy, on October 6, 1931. He earned a Ph.D. degree in cosmic-ray physics at the University of Milan in 1956. Upon graduation he immigrated to the United States to accept a position as a research associate at Indiana University in Bloomington. In 1959 he moved to the Boston area to join American Science and Engineering, a small scientific research firm established by scientists from the Massachusetts Institute of Technology (MIT). Giacconi began his pioneering activities in X-ray astronomy at American Science and Engineering.

Collaborating with BRUNO ROSSI at MIT and others researchers in the early 1960s, Giaconni designed novel X-ray detection instruments for use on sounding rockets and eventually spacecraft. In June 1962 one of Giacconi's sounding rocket experiments proved especially significant. His team placed an instrument package on a sounding rocket in an attempt to perform a brief search for possible solar-induced X-ray emissions from the lunar surface. To the team's great surprise, the instruments accidentally detected the first known X-ray source outside the solar system, a star called Scorpius X-1. Previous rocket-borne experiments by personnel from the U.S. Naval Research Laboratory that started in 1949 had detected X-rays from the Sun, but Scorpius X-1, the so-called "X-ray star," was the first source of X-rays known to exist beyond the solar system. It is the brightest and most persistent nontransient X-ray source in the sky. This fortuitous discovery is often regarded as the beginning of X-ray astronomy, an important field within high-energy astrophysics.

Because Earth's atmosphere absorbs X-rays, instruments to detect and observe X-rays produced by astronomical phenomena must be placed in space high above the sensible atmosphere. Sounding rockets provide a short duration (typically just a few minutes) way of searching for cosmic X-ray sources, while specially instrumented orbiting observatories provide scientists a much longer period of time to conduct all-sky surveys and detailed investigations of interesting X-ray sources. Astrophysicists investigate X-ray emissions because they provide a unique insight into some of the most violent and energetic processes taking place in the universe.

In the mid-1960s, as an executive vice president and senior scientist at American Science and Engineering, Giacconi headed the scientific team that constructed the first imaging X-ray telescope. In this novel instrument incoming X-rays graze, or ricochet off, precisely designed and shaped surfaces and then collect at a specific place, called the focal point. The first such device flew into space on a sounding rocket in 1965 and captured images of X-ray hot spots in the upper atmosphere of the Sun. NASA used greatly improved versions of this type of instrument on the *Chandra X-Ray Observatory* (CXO).

The following analogy illustrates the significance of Giacconi's efforts in X-ray astronomy and also serves as a graphic testament to the rapid rate of progress in astronomy brought on by the space age. By historic coincidence, Giacconi's first relatively crude X-ray telescope was approximately the same length and diameter as the first astronomical (optical) telescope used by GALILEO GALILEI in 1610. Over a period of about four centuries, optical telescopes improved in sensitivity by 100 million times as their technology matured from Galileo's first telescope to the capability of NASA's *Hubble Space Telescope* (HST). About 40 years after Giacconi tested the first X-ray telescope on a sounding rocket, NASA's magnificent *Chandra X-Ray Observatory* provided scientists a leap in measurement sensitivity of about 100 million. Much like Galileo's optical telescope revolutionized observational astronomy in the 17th century, Giacconi's X-ray telescope triggered a modern revolution in high-energy astrophysics in the 20th century.

Giacconi functioned well as both a skilled astrophysicist and as a successful scientific manager—capable of overseeing the execution of large, complex scientific projects. Along these lines he envisioned and then supervised development of two important orbiting X-ray observatories, the *Uhuru* satellite and the *Einstein Observatory.*

In the late 1960s his group at American Science and Engineering supported NASA in the design, construction, and operation of the *Uhuru* satellite, the first orbiting observatory dedicated to making surveys of the X-ray sky. *Uhuru,* also known as the *Small Astronomical Satellite 1,* was launched from the San Marco platform off the coast of Kenya, Africa, on December 12, 1970. By coincidence, December 12th marked seventh anniversary of Kenyan independence, so NASA called this satellite *Uhuru,* which is the Swahili word for "freedom." The well-designed spacecraft operated for more than three years and detected 339 X-ray sources, many of which turned out to be due to matter from companion stars being pulled at extremely high velocity into suspected black holes or super-dense neutron stars. With this successful spacecraft Giacconi's team firmly established X-ray astronomy as an exciting new astronomical field.

In 1973 Giacconi moved his X-ray astronomy team to the Harvard-Smithsonian Center for Astrophysics, headquartered in Cambridge, Massachusetts. There he supervised the development of a new satellite for NASA, called the *High-Energy Astronomy Observatory 2* (HEAO-2). Following a successful launch in November 1978, NASA renamed the spacecraft the *Einstein X-Ray Observatory* in honor of the brilliant physicist ALBERT EINSTEIN. This spacecraft was the first orbiting observatory to use the grazing incidence X-ray telescope (a concept initially proposed by Giacconi in 1960) to produce detailed images of cosmic X-ray sources. The *Einstein Observatory* operated until April 1981 and produced more than 7,000 images of various extended X-ray sources, such as clusters of galaxies and supernova remnants.

Giacconi's display of superior scientific management skills during the *Uhuru* and *Einstein Observatory* projects secured his 1981 appointment as the first director of the Space Telescope Science Institute. The Space Telescope Science Institute is the astronomical research center on the Johns Hopkins Homewood campus in Maryland responsible for operating NASA's *Hubble Space Telescope* (HST).

Then, from 1993 to 1999, Giacconi served as the director general of the European Southern Observatory (ESO) in Garching, Germany. This was his first scientific leadership role involving ground-based astronomical facilities. Before leaving this European intergovernmental organization he successfully saw the "first light" for ESO's new Very Large Telescope in the Southern Hemisphere on Cerro Paranal in Chile. In July 1999 Giacconi returned to the United States to become president and chief executive officer of Associated Universities, Inc., the Washington, D.C.–headquartered not-for-profit scientific management corporation that operates the National Radio Astronomy Observatory in cooperation with the National Science Foundation.

In October 2002 the Nobel committee awarded Giacconi half of that year's Nobel Prize in physics for his pioneering contributions to astrophysics, an effort that led to the discovery of cosmic X-ray sources. A renowned scientist and manager, Giacconi has received numerous other awards, including the Bruce Gold Medal from the Astronomical Society of the Pacific (1981) and the Gold Medal of the Royal Astronomical Society in London (1982). He is the author of more than 200 scientific publications that deal with the X-ray universe, ranging from black hole candidates to distant clusters of galaxies. He currently resides with his wife in Chevy Chase, Maryland.

giant-impact model The hypothesis that the Moon originated when a Mars-sized object struck a young Earth with a glancing blow. The giant (oblique) impact released material that formed an accretion disk around Earth out of which the Moon formed.

See also MOON.

giant molecular cloud (GMC) Massive clouds of gas in interstellar space composed primarily of molecules of hydrogen (H_2) and dust. GMCs can contain enough mass to make several million stars like the Sun and are often the sites of star formation.

See also STAR.

This interesting photographic montage helps answer the frequently asked question in astronomy: How big is a giant planet? This illustration uses properly scaled, spacecraft-derived images to compare Earth (12,756 km diameter) with the gaseous giant worlds Jupiter (142,982 km diameter) and Saturn (120,540 km diameter). ***(Courtesy of NASA)***

giant planets In our solar system, the large, gaseous outer planets: Jupiter, Saturn, Uranus, and Neptune.

See also SOLAR SYSTEM.

giant star A star near the end of its life that has swollen in size, such as a BLUE GIANT or a red giant.

gibbous A term used to describe a phase of a moon or planet when more than half, but not all, of the illuminated disk can be viewed by the observer.

See also PHASES OF THE MOON.

Gibson, Edward G. (1936–) American *Astronaut, Solar Physicist* From November 16, 1973, to February 8, 1974, the American astronaut Edward G. Gibson served as the science pilot on NASA's *Skylab 4* mission. This was the third and final mission to Skylab—the first American space station placed in low-Earth orbit. As a solar physicist, Gibson used the Apollo Telescope Mount (ATM) console during the record-setting 84-day mission to make direct scientific observations of the Sun (unobstructed by Earth's atmosphere), to monitor interesting transient solar phenomena, and to closely

American astronaut and solar physicist Edward G. Gibson, shown here working at the Apollo Telescope Mount (ATM) console during NASA's *Skylab 4* mission *(Photograph taken onboard* Skylab *on December 5, 1973, during a live television broadcast to Earth) (Courtesy of NASA/JSC)*

follow other transient events, such as the passage of Comet Kohoutek.

giga- (symbol: G) A prefix meaning multiplied by 10^9.

Gill, Sir David (1843–1914) British *Astronomer* Sir David Gill was a Scottish watchmaker turned astronomer who passionately pursued the measurement of solar parallax. His careful observations, often involving travel with his wife (Lady Isabel Gill) to remote regions of the Earth, established a value of the astronomical unit that served as the international reference until 1968. As an example, in 1877 the couple lived for six months on tiny Ascension Island in the middle of the South Atlantic so Gill could estimate a value for solar parallax by observing Mars during its closest approach. The couple also resided in South Africa for 28 years so he could serve as Her Majesty's Royal Astronomer in charge of the Cape of Good Hope Observatory.

Gill was born in Aberdeen, Scotland, on June 12, 1843. From 1858 to 1860 he studied at the Marischal College of the University of Aberdeen but left without a degree because of family circumstances. As the eldest surviving son, he abandoned college and trained to be a watchmaker so he could take over his aging father's business. However, while at Marischal College Gill had the brilliant Scottish physicist JAMES CLERK MAXWELL as his professor of natural philosophy. Maxwell's lectures greatly inspired Gill, and the young watchmaker would ultimately pursue a career in science.

By good fortune, his activities in precision timekeeping led him directly into a career in astronomy. While operating his watch and clock making business, Gill developed a great facility with precision instruments. As a hobby, he restored an old, abandoned transit telescope in order to provide the city of Aberdeen accurate time. He also purchased a 50-centimeter-diameter reflecting telescope and modified the device to take photographs. Gill's excellent photograph of the Moon (taken on May 18, 1869) soon caught the attention of the young Lord Lindsay, who was trying to persuade his father (the earl of Crawford) to build him a private astronomical observatory. After brief negotiations Gill abandoned the family watch making business in 1872 to become the director of Lindsay's private astronomical observatory near Aberdeen. As a competent instrument maker and self-educated observational astronomer, Gill set about the task of designing, equipping, and operating Lord Lindsay's private observatory at Dun Echt, Scotland.

His years as a privately employed astronomer at Dun Echt brought Gill in contact with the most prominent astronomers in Europe. The work, especially his experience with precision measurements using the heliometer, also prepared him well for the next step in his career as a professional astronomer. In the 19th century astronomers used this (now obsolete) instrument to measure the parallax of stars and the angular diameter of planets. David Gill became a world-recognized expert in the use of the heliometer, an instrument that he relentlessly applied with great enthusiasm in pursuit of his lifelong quest to precisely measure solar parallax.

Astronomers define solar parallax as the angle subtended by Earth's equatorial radius when our planet is observed from the center of the Sun—that is, at a distance of 1 astronomical unit. The astronomical unit (AU) is the basic reference distance in solar system astronomy. By international agreement, the solar parallax is approximately 8.794 arc seconds.

Sponsored by Lord Lindsay, Gill participated in an astronomical expedition to Mauritius in 1874 to observe the transit of Venus. He took 50 precision chronometers on the long sea voyage around the southern tip of Africa. Mauritius is a small inhabited island in the Indian Ocean east of Madagascar. Gill focused on providing precise timing for the other astronomers. A planetary transit takes place when one celestial object moves across the face of another celestial object of larger diameter, such as Mercury or Venus moving across the face of the Sun as viewed by an observer from Earth. The 1874 transit of Venus provided Gill an opportunity to refine the elusive value of the astronomical unit. However, he soon discovered firsthand the difficulty of using the transit of Venus to determine solar parallax. Upon magnification the image of the planetary disk became fuzzy and insufficiently sharp. As a result, it was very difficult for astronomers to measure the precise moment of first contact, even with all Gill's precision chronometers available. In 1876 Gill departed on friendly terms from his position as director of Lord Lindsay's observatory.

Along with his wife, Isabel, Gill then conducted his own privately organized expedition to Ascension Island in 1877 to measure solar parallax. Sir David and Lady Gill made their arduous journey to this tiny island in the middle of the South Atlantic Ocean in order to precisely measure Mars during its closest approach to Earth. Gill used the distance from the Royal Greenwich Observatory to Ascension Island as a baseline. After six months on Ascension, he was able to obtain reasonable results.

However, Gill was still not satisfied with the precision of his work, so he began to follow JOHANN GALLE's suggestion of using asteroids with their pointlike masses to make measurements of solar parallax. He knew this was the right approach because he had already made a preliminary (but unsuccessful) effort by observing the asteroid Juno during the Mauritius expedition.

In 1889 Gill organized an international effort that successfully made precise measurements of the solar parallax by observing the motion of three asteroids (Iris, Victoria, and Sappho). From these observations Gill calculated a value of the astronomical unit that remained the international standard for almost a century. His value of 8.80 arc seconds for solar parallax was replaced in 1968 by a value of 8.794 arc seconds, a value obtained by radar observations of solar system distances. Between 1930 and 1931 SIR HAROLD SPENCER JONES followed Gill's work and used observations of the asteroid Eros to make a refined estimate of solar parallax. After a decade of careful calculation Jones obtained a value of 8.790 arc seconds.

The astronomical expeditions of 1874 and 1877 brought Gill recognition within the British astronomical community. In 1879 he was appointed Her Majesty's Royal Astronomer at the Cape of Good Hope Observatory in South Africa. He served with great distinction as royal astronomer at this facility until his retirement in 1907. Early in this appointment he accidentally became a proponent for astrophotography when he successfully photographed the great comet of 1882 and then noticed the clarity of the stars in the background in his comet photographs. He upgraded and improved the observatory so it could play a major role in photographing the Southern Hemisphere sky in support of such major international efforts as the Carte du Ciel project. He collaborated with JACOBUS KAPTEYN in the creation of the *Cape Photographic Durchmusterung,* an enormous star catalog published in 1904. The *Cape Durchmusterung* contained more than 400,000 Southern Hemisphere stars whose positions and magnitudes were determined from photographs taken at the Cape of Good Hope Observatory. Gill was knighted in 1900.

Sir David and Lady Gill returned to Great Britain from South Africa in 1907 and retired in London. There he busied himself by completing the book *History and Description of the Cape Observatory,* published in 1913. He died in London on January 24, 1914. He is considered one of the great observational astronomers of the period.

gimbal 1. A device on which an engine or other object may be mounted that has two mutually perpendicular and intersecting axes of rotation and therefore provides free angular movement in two directions. 2. In a gyroscope, a support that provides the spin axis with a degree of freedom. 3. To move a reaction engine about on a gimbal so as to obtain pitching and yawing correction moments. 4. To mount something on a gimbal.

gimbaled motor A rocket engine mounted on a gimbal.

Giotto di Bondone (1266–1337) Italian *Artist* Giotto di Bondone was an artist of the Florentine school who apparently witnessed the 1301 passage of Comet Halley and was then inspired to include the first "scientific" representation of this famous comet in his fresco *Adoration of the Magi,* now found in the Scrovengi Chapel in Padua, Italy. The European Space Agency (ESA) named its *Giotto* spacecraft in his honor.

***Giotto* (spacecraft)** The European Space Agency (ESA)'s scientific spacecraft launched on July 2, 1985, from the agency's Kourou, French Guiana, launch site by an Ariane 1 rocket. The Giotto mission was designed to encounter and investigate Comet Halley during its 1986 return to the inner solar system. Following this successful encounter, the *Giotto* spacecraft studied Comet Grigg-Skjellerup during an extended mission in 1992.

The major objectives of the Giotto mission were to (1) obtain color photographs of Comet Halley's nucleus, (2) determine the elemental and isotopic composition of volatile components in the cometary coma, particularly parent molecules, (3) characterize the physical and chemical processes that occur in the cometary atmosphere and ionosphere, (4) determine the elemental and isotopic composition of dust particles, (5) measure the total gas production rate and dust flux and size/mass distribution and derive the dust-to-gas ratio, and (6) investigate the macroscopic systems of plasma flows resulting from the cometary–solar wind interaction.

The *Giotto* spacecraft encountered Comet Halley on March 13, 1986, at a distance of 0.89 AU from the Sun and 0.98 AU from Earth and an angle of 107° from the comet-Sun line. A design goal of this mission was to have the spacecraft come within 500 km of the comet's nucleus at closest encounter. The actual closest approach distance was measured at 596 km. To protect itself during the encounter, the *Giotto* spacecraft had a dust shield designed to withstand impacts of particles up to 0.1 g. The scientific payload consisted of 10 hardware experiments: a narrow-angle camera, three mass spectrometers for neutrals, ions, and dust, various dust detectors, a photopolarimeter, and a set of plasma experiments. All experiments performed well and produced an enormous quantity of important scientific results. Perhaps the most significant accomplishment was the clear identification of the comet's nucleus.

At 14 seconds before closest approach the *Giotto* spacecraft was hit by a "large" dust particle. The impact caused a significant shift (about 0.9°) of the spacecraft's angular momentum vector. Following this "bump," scientific data were received intermittently for the next 32 minutes. Some experiment sensors suffered damage during this 32-minute interval. Other experiments (the camera baffle and deflecting mirror, the dust detector sensors on the front sheet of the bumper shield, and most experiment apertures) were exposed to dust particles regardless of the accident and also suffered damage. Many of the sensors survived the encounter with little or no damage. Other scientific instruments did experience damage.

Giotto's cameras recorded many images and gave scientists a unique opportunity to determine the shape and material composition of Comet Halley's nucleus, an irregular, peanut-shaped object somewhat larger (about 15 km by 10 km) than they had estimated. The nucleus was dark and surrounded by a cloud of dust. Though damaged by multiple particle impacts, *Giotto* successfully conducted the Halley encounter. Upon completion of the encounter, ESA mission controllers put the spacecraft in a hibernation (or "quiet") mode as it continued to travel in deep space. In February 1990 ESA controllers reactivated the hibernating spacecraft for a new task—to observe Comet Grigg-Skjellerup, a short-period comet with an orbital period of 5.09 years as it experienced perihelion in July 1992.

During the *Giotto* extended mission (GEM) the hardy and far-traveled spacecraft successfully encountered Comet Grigg-Skjellerup on July 10, 1992. The closest approach was approximately 200 km. At the time of the encounter the heliocentric distance of the spacecraft was 1.01 AU, and the geocentric distance, 1.43 AU. ESA controllers had switched on the scientific payload on the previous evening (July 9). Eight experiments were operated and provided a surprising amount of interesting data. For example, at 12 hours before the closest approach the plasma analyzer detected the first presence of cometary ions about 600,000 km from the nucleus. On July 23 ESA controllers terminated the Comet Grigg-Skjellerup encounter operation. The again hibernating *Giotto* spacecraft flew by Earth on July 1, 1999, at a closest approach of about 219,000 km. During this encounter the spacecraft was moving at about 3.5 km/sec relative to Earth.

The *Giotto* spacecraft was named after the Italian painter GIOTTO DI BONDONE, who apparently witnessed the 1301 passage of Comet Halley. The Renaissance artist then included the first scientific representation of this comet in his famous fresco *Adoration of the Magi,* which can be found in the Scrovegni Chapel in Padua, Italy.

See also COMET.

Glenn, John Herschel, Jr. (1921–) American *Astronaut, Military Officer, U.S. Senator* John H. Glenn, Jr., an American astronaut, U.S. Marine Corps officer, and U.S. senator, was the first American to orbit Earth in a spacecraft. He accomplished this historic feat on February 20, 1962, as part of NASA's Mercury Project. Launched from Cape Canaveral by an Atlas rocket, Glenn flew into space aboard the *Friendship 7* Mercury space capsule and made three orbits of Earth. His mission lasted about five hours. More than three and a half decades later he became the oldest human being to travel in space when he joined the space shuttle *Discovery* crew on the nine-day-duration STS-95 orbital mission, a mission that lasted from October 29 to November 7, 1998.

glide 1. A controlled descent by a heavier-than-air aeronautical or aerospace vehicle under little or no engine thrust in which forward motion is maintained by gravity and vertical descent is maintained by lift forces. 2. A descending flight path of a glide, such as a shallow glide. 3. To descend in a glide.

glide path The flight path (seen from the side) of an aeronautical or aerospace vehicle in a glide.

global change Earth's environment has been subject to great change over eons. Many of these changes have occurred quite slowly, requiring numerous millennia to achieve their full effect. However, other global changes have occurred relatively rapidly over time periods as short as a few decades or less. These global changes appear in response to such phenomena as the migration of continents, the building and erosion of mountains, changes in the Sun's energy output, variations in Earth's orbital parameters, the reorganization of oceans, and even the catastrophic impact of a large asteroid or comet. Such natural phenomena lead to planetary changes on local, regional, and global scales, including a succession of warm and cool climate epochs, new distributions of tropical rain forests and rich grasslands, the appearance and disappearance of large deserts and marshlands, the advances and retreats of great ice sheets (glaciers), the rise and fall of ocean and lake levels, and even the extinction of vast numbers of species. The last great mass extinction (on a global basis) appears to have occurred some 65 million years ago, possibly due to the impact of a large asteroid. The peak of the most recent period of glaciation generally is considered to have occurred about 18,000 years ago, when average global temperatures were about 5°C cooler than today.

Although such global changes are the inevitable results of major natural forces currently beyond human control, it is also apparent to scientists that humans have become a powerful agent for environmental change. For example, the chemistry of Earth's atmosphere has been altered significantly by both the agricultural and industrial revolutions. The erosion

On February 20, 1962, Mercury Project astronaut John H. Glenn, Jr., became the first American to orbit Earth in a spacecraft. Decades later he returned to space as a payload specialist onboard NASA's STS-95 shuttle mission, which took place from October 29 to November 7, 1998. *(Courtesy of NASA)*

of the continents and sedimentation of rivers and shorelines have been influenced dramatically by agricultural and construction practices. The production and release of toxic chemicals have affected the health and natural distributions of biotic populations. The ever-expanding human need for water resources has affected the patterns of natural water exchange that take place in the hydrological cycle (the oceans, surface and ground water, clouds, etc.). One example is the enhanced evaporation rate from large human-engineered reservoirs compared to the smaller natural evaporation rate from wild, unregulated rivers. This increased evaporation rate from large engineered reservoirs demonstrates how human activities can change natural processes significantly on a local and even regional scale. Such environmental changes may not always be beneficial and often may not have been anticipated at the start of the project or activity. As the world population grows and human civilization undergoes further technological development this century, the role of our planet's most influential animal species as an agent of environmental change undoubtedly will expand.

Over the past four decades scientists have accumulated technical evidence that indicates that ongoing environmental changes are the result of complex interactions among a number of natural and human-related systems. For example, the changes in Earth's climate now are considered to involve not only wind patterns and atmospheric cloud populations but also the interactive effects of the biosphere and ocean currents, human influences on atmospheric chemistry, Earth's orbital parameters, the reflective properties of our planetary system (Earth's albedo), and the distribution of water among the atmosphere, hydrosphere, and cryosphere (polar ice). The aggregate of these interactive linkages among our planet's major natural and human-made systems that appear to affect the environment has become known as global change.

The governments of many nations, including the United States, have started to address the long-term issues associated with global change. Over the last decade preliminary results from global observation programs (many involving space-based systems) have stimulated a new set of concerns that the dramatic rise of industrial and agricultural activities during the 19th and 20th centuries may be adversely affecting the overall Earth system. Today the enlightened use of Earth and its resources has become an important political and scientific issue.

The global changes that may affect both human well-being and the quality of life on this planet include global climate warming, sea-level change, ozone depletion, deforestation, desertification, drought, and a reduction in biodiversity. Although complex phenomena in themselves, these individual global change concerns cannot be fully understood and addressed unless they are studied collectively in an integrated, multidisciplinary fashion. An effective and well-coordinated international research program, which includes use of advanced environmental observation satellite systems, will be required to significantly improve our scientific knowledge of natural and human-induced changes in the global environment and their regional effects.

See also EARTH-OBSERVING SPACECRAFT; EARTH SYSTEM SCIENCE; GREENHOUSE EFFECT; REMOTE SENSING.

Global Positioning System (GPS) NAVSTAR GPS is a space-based, radio positioning system nominally consisting of a constellation of 24 Earth-orbiting satellites that provide navigation and timing information to military and civilian users worldwide. The GPS consists of three major segments: *space, control,* and *user.* The *space* segment consists of 24 operational satellites in six orbital planes (four satellites in each plane). The satellites operate in circular 20,200-km orbits at an inclination of 55° and with a 12-hour period. The spacecrafts' positions are, therefore, the same at the same sidereal time each day—that is, the satellites appear four minutes earlier each day. The *control* segment of the GPS consists of five monitor stations (Hawaii, Kwajalein, Ascension Island, Diego Garcia, Colorado Springs), three ground antennas, (Ascension Island, Diego Garcia, Kwajalein), and a Master Control Station (MCS) located at Schriever Air Force Base in Colorado. The monitor stations passively track all satellites in view, accumulating ranging data. This information is processed at the MCS to determine satellite orbits and to update each satellite's navigation message. Updated information is transmitted to each satellite via the ground antennas. The *user* segment consists of antennas and receiver-processors that provide positioning, velocity, and precise timing to the user.

GPS satellites orbit Earth every 12 hours emitting continuous radio wave signals on two different L-band frequencies. The GPS constellation is designed and operated as a 24-satellite system consisting of six planes with a minimum of four satellites per plane. With proper equipment, users can receive these signals and use them to calculate location, time, and velocity. The signals provided by GPS are so accurate that time can be calculated to within a millionth of a second, velocity within a fraction of a kilometer per hour, and location within a few meters or less. GPS receivers have been developed by both military and civilian manufacturers for use in aircraft, ships, and land vehicles. Hand-held systems are also available for use by individuals in the field.

GPS provides 24-hour navigation services including extremely accurate, three-dimensional information (latitude, longitude, and altitude), velocity, and precise time; a worldwide common grid that is easily converted to any local grid; passive, all-weather operations; continuous real-time information; support to an unlimited number of military users and areas; and support to civilian users at a slightly less accurate level.

The Global Positioning System has significantly enhanced functions such as mapping, aerial refueling and rendezvous, geodetic surveys, and search and rescue operations. Many military applications were put to the test in the early 1990s during the U.S. involvement in Operations Desert Shield and Desert Storm. Allied troops relied heavily on GPS data to navigate the featureless regions of the Saudi Arabian desert. Forward air controllers, pilots, tank drivers, and support personnel all used the system so successfully that American defense officials cited GPS as a key to the Desert Storm victory. During Operations Enduring Freedom, Noble Eagle, and Iraqi Freedom (OIF) the space-based navigation system's contributions increased significantly and allowed the delivery of GPS-guided munitions with pinpoint precision and a minimum of collateral damage. One U.S. Air Force estimate concluded that almost one-fourth of the total 29,199 bombs and missiles coalition forces released against Iraqi targets were guided by GPS signals.

There are four generations of the GPS satellite, designated as the Block I, the Block II/IIA, the Block IIR, and the Block IIF. Block I satellites were used to test the principles of the Global Positioning System and to demonstrate the efficacy of space-based navigation. Lessons learned from the operation of the 11 Block I satellites were incorporated into later design blocks.

Block II and IIA satellites make up the first operational constellation. With a mass of 1,670 kilograms, each GPS Block IIA satellite operates in a specially designated circular nearly 20,200-km orbit. Delta II expendable launch vehicles have lifted these satellites into their characteristic 12-hour orbits from Cape Canaveral Air Force Station in Florida. GPS Block IIA satellites have a design lifetime of approximately seven and a half years.

Block IIR satellites represent a dramatic improvement over the satellites in the previous design blocks. These satellites have the ability to determine their own positions by performing intersatellite ranging with other Block IIR spacecraft. Each Block IIR spacecraft has a mass of approximately 2,035 kilograms and a design lifetime of 10 years. The Block IIR satellites are replacing Block II and IIA satellites as the latter reach the end of their service lifetimes. On November 6, 2004, for example, a Delta II rocket successfully launched the GPS IIR-13 satellite into orbit from Space Launch Complex 17B at Cape Canaveral.

The 1,705-kilogram Block IIF spacecraft represents the fourth generation of GPS navigation satellite. With a design lifetime of 11 years (minimum), these spacecraft serve as sustainment vehicles, ensuring the vitality of the GPS constellation for the next two decades. The U.S. Air Force intends to use either Delta IV or Atlas V expendable launch vehicles to place Block IIF satellites into orbit.

The U.S. Air Force Space Command (AFSC) formally declared the GPS satellite constellation as having met the requirement for *full operational capability* (FOC) as of April 27, 1995. Requirements included 24 operational satellites (Block II/IIA) functioning in their assigned orbits and successful testing completed for operational military functionality.

Plans to upgrade the current GPS will reduce vulnerabilities to jamming and respond to national policy that encourages widespread civilian use of this important navigation system without degrading its military utility. For example, upgrades for the satellites and their ground control segment include civil signals and new military signals transmitted at higher power levels. Projected improvements in military end-user equipment will also minimize the effect of adversarial jamming and protect efficient use of the GPS in time of war by American and friendly forces.

The NAVSTAR Global Positioning System is a multiservice program within the U.S. Department of Defense. The U.S. Air Force serves as the designated executive service for management of the system. The system is operated and controlled by the 50th Space Wing, located at Schriever Air Force Base, Colorado. The GPS Master Control Station, operated by the 50th Space Wing, is responsible for monitoring and controlling the GPS satellite constellation. The GPS-dedicated ground system consists of monitor stations and ground antennas located around the world. The monitor stations use GPS receivers to passively track the navigation signals on all satellites. Information from the monitor stations is then processed at the Master Control Station and used to update the navigation messages from the satellites. The Master Control Station crew sends updated navigation information to GPS satellites through ground antennas using an S-band signal. The ground antennas are also used to transmit commands to satellites and to receive state-of-health data (telemetry).

See also UNITED STATES AIR FORCE.

globular cluster Compact cluster of from tens to hundreds, thousands, or possibly even 1 million stars. Astronomers believe that all the stars in a cluster share a common origin. Roughly spherical in shape, a globular cluster is a gravitationally bound system of older (Population II) stars, stars that are older than the Sun and have very low metal content. Because they date back to the birth of the Milky Way galaxy

This is a composite infrared image (at four wavelengths: 3.6 micrometers [μm], 4.5 μm, 5.8 μm, and 8.0 μm) taken by NASA's *Spitzer Space Telescope* on April 21, 2004. It shows a previously unknown globular cluster that lies in the dusty galactic plane about 9,000 light-years away. The inset (upper right) is a visible light image of the same region in the constellation Aquila and shows only a dark patch of sky where this large globular cluster is located. To help appreciate the important role space-based infrared telescopes play in modern astronomy, as viewed from Earth, this new cluster's apparent size is comparable to a grain of rice held at arm's length. *(Courtesy of NASA/JPL-Caltech/ H. Kobulnicky, University of Wyoming; inset courtesy of Digitized Sky Survey, California University of Technology, Pasedena, California)*

some 13 billion years ago, astrophysicists view these ancient bundles of stars as "cosmic fossils." Astronomers use globular clusters as tools for studying the age and formation of the Milky Way.

There are about 150 known globular clusters sprinkled around the center of our galaxy like seeds in a pumpkin. In 2004 NASA's *Spitzer Space Telescope* found a previously unknown globular cluster fairly close by (about 9,000 light-years away) within the dusty plane of the Milky Way. Most globular clusters (like M13, the Great Cluster in Hercules) orbit around the center of our galaxy well above the dust enshrouded galactic plane. However, NASA's new infrared telescope was able to "see" into the dusty galactic plane and detect this new globular cluster. (*See* figure.) The cluster is located in the constellation Aquila and has an estimated mass equivalent of about 300,000 Suns.

See also GALACTIC CLUSTER; GREAT CLUSTER IN HERCULES.

Glonass Name given to Russian global positioning satellites used in a constellation to provide navigation support.

See also GLOBAL POSITIONING SYSTEM (GPS).

gluon The hypothetical subnuclear particle that holds together quarks via the strong force.

See also ELEMENTARY PARTICLE(S); FUNDAMENTAL FORCES IN NATURE.

gnomon The part of a sundial that casts a shadow. Also, a device used by ancient astronomers to measure the altitude of the Sun in order to determine the time of day or estimate the time of year.

See also ANCIENT ASTRONOMICAL INSTRUMENTS.

Goddard, Robert Hutchings (1882–1945) American *Physicist, Rocket Engineer* Robert H. Goddard was a brilliant but reclusive American physicist who promoted rocket science and cofounded the field of astronautics in the early part of the 20th century—independently of KONSTANTIN E. TSIOLKOVSKY and HERMANN J. OBERTH. However, unlike Tsiolkovsky and Oberth, who were primarily theorists, Goddard performed many "hands-on" pioneering rocket experiments. Included in his long list of accomplishments is the successful launch of the world's first liquid-propellant rocket on March 16, 1926. Goddard's lifelong efforts resulted in more than 214 U.S. patents, almost all rocket related. Today Goddard is widely recognized as the "Father of American Rocketry."

Goddard was born in Worcester, Massachusetts, on October 5, 1882. When he was a very young child the family moved to Boston but then in 1898 returned to Worcester. His father was a machine shop superintendent, so visits to his father's shop helped nurture young Goddard's fascination with gears, levers, and all types of mechanical gadgets. As a child he was very sickly and suffered from pulmonary tuberculosis. Forced to avoid strenuous youthful games and sports, Goddard retreated to more cerebral pursuits such as reading, daydreaming, and keeping a detailed personal diary.

In the 1890s Goddard became very interested in space travel. He was especially influenced by the classic JULES VERNE science fiction novel *From the Earth to the Moon* and by the H. G. WELLS story *The War of the Worlds,* an exciting tale of alien invaders from Mars that appeared as a daily serial feature in a Boston newspaper. Many years later Goddard sent a personal letter to H. G. Wells explaining what a career stimulus and source of inspiration that particular story was to him in his youth. The British author wrote a letter back to Goddard acknowledging the kind remarks.

Starting in childhood, Goddard began to keep a diary in which he meticulously recorded his thoughts and his experiences—the successes as well as the failures. For example, one day during his early teens an explosive mixture of hydrogen and oxygen shattered the windows of his improvised home laboratory. Without much emotion he dutifully noted in his diary "such a mixture must be handled with great care." Later, his wife, Esther, helped him carefully document the results of his rocket experiments.

Goddard's youthful diary describes a very special event in his life that happened shortly after his 17th birthday. On October 19, 1899, he climbed an old cherry tree in his backyard with the initial intention of pruning it, but because it was such a pleasant autumn day, he remained up in the tree for hours, thinking about his future and whether he wanted to devote his life to rocketry. His thoughts even wandered to developing a device that could reach Mars. So October 19th became Goddard's personal holiday, or special "Anniversary Day"—the day on which he committed his life to rocketry and spaceflight. Of that special life-changing day, Goddard later wrote: "I was a different boy when I descended the tree from when I ascended, for existence at last seemed very purposive."

Because of his frequent absence from school due to illness, he did not graduate from high school until 1904, when he was in his early 20s. At high school graduation he concluded his oration with the following words: "It is difficult to say what is impossible, for the dream of yesterday, is the hope of today, and the reality of tomorrow." This is Goddard's most frequently remembered quotation.

When he was an undergraduate student most people, including many supposed scientific experts who should have known better, mistakenly believed that a rocket could not operate in the vacuum of space because "it needed an atmosphere to push on." Goddard knew better. He understood the theory of reaction engines and would later use novel experimental techniques to demonstrate the fallacy of this common and very incorrect hypothesis about a rocket's inability to function in outer space. By the time he died the large liquid-propellant rocket was not only a technical reality, but also scientifically recognized as the technical means for achieving another 20th-century impossible dream—interplanetary travel.

Goddard enrolled at Worcester Polytechnic Institute in 1904. In his freshman year he wrote a very interesting essay in response to a professor's assignment about transportation systems people might use "far" in the future—the professor's target year was 1950. Goddard's paper described a high-speed vacuum tube railway system that would take a specially designed commuter train from New York to Boston in only 10 minutes! His proposed train used electromagnetic levitation and accelerated continuously for the first half of the trip

Dr. Robert H. Goddard, the reclusive genius who is widely acknowledged as the "father of American rocketry," is shown with a steel combustion chamber and rocket nozzle in this 1915 photograph. ***(Courtesy of NASA)***

and then decelerated continuously at the same rate during the second half of the trip.

In 1907 he achieved his first public notoriety as a "rocket man" when he ignited a black powder (gunpowder) rocket in the basement of the physics building. As clouds of gray smoke filled the building, school officials became immediately "interested" in Goddard's work. Much to their credit, the tolerant academic officials at Worcester Polytechnic Institute did not expel him for the incident. Instead, he went on to graduate with a bachelor of science degree in 1908 and then remained at the school as an instructor of physics. He began graduate studies in physics at Clark University in 1908. His enrollment began a long relationship with the institution, both as a student and later as a faculty member. In 1910 he earned a master's degree, followed the next year by a Ph.D. in physics. After receiving his doctorate he accepted a research fellowship at Princeton University from 1912 to 1913. However, he suffered a near fatal relapse of pulmonary tuberculosis in 1913, an attack that incapacitated him for many months. After recovery Goddard joined the faculty at Clark University the following year as an instructor of physics. By 1920 he became a full professor of physics and also the director of the physical laboratories at Clark. He retained these positions until he retired in August 1943.

Clark University provided Goddard the small laboratory, supportive environment, and occasional, very modest amount of institutional funding necessary to begin serious experimentation with rockets. (*See* figure.) Following his return to Clark University Goddard met his future wife, Esther Christine Kisk. Beginning in 1918, she started helping him carefully compile the notes and reports of his many experiments. In time their professional relationship turned pleasantly personal, and they were married on June 21, 1924.

Although the couple remained childless, Esther Kisk Goddard created a home environment that provided her husband with a continuous flow of encouragement and support. She cheered with him when an experiment worked; she was also there to cushion the blows of experimental failures and to buffer him from the harsh remarks of uninformed detractors. She was the publicly transparent but essential partner in "Goddard Inc."—a chronically underfunded, federally ignored, but nevertheless incredibly productive husband-wife rocket research team who achieved many marvelous engineering milestones in liquid-propellant rocket technology in the 1920s and 1930s.

As a professor at Clark University, Goddard responded to his childhood fascination with rockets in a methodical, scientifically rigorous way. He recognized the rocket was the enabling technology for space travel. But unlike Tsiolkovsky and Oberth, who also appreciated the significance of the rocket, Goddard went well beyond a theoretical investigation of these interesting reaction devices. He enthusiastically designed new rockets, invented special equipment to analyze their performance, and then flight-tested his experimental vehicles to transform theory into real-world action. In 1914 Goddard obtained the first two of his numerous rocketry related U.S. patents. One was for a rocket that used liquid fuel, the other for a two-stage solid-fuel rocket.

In 1915 Goddard performed a series of cleverly designed experiments in his laboratory at Clark University that clearly demonstrated that a rocket engine generates thrust in a vacuum. Goddard's work provided indisputable experimental evidence showing space travel was indeed possible, yet his innovative research was simply too far ahead of its time. His academic colleagues generally did not understand or appreciate the significance of the rocket experiments, and Goddard continued to encounter great difficulty in gathering any financial support to continue. Since no one else thought rocket physics was promising, he became disheartened and even thought very seriously about abandoning his lifelong quest.

But somehow Goddard persevered, and in September 1916 he sent a description of his experiments along with a modest request for funding to the Smithsonian Institution. Upon reviewing Goddard's proposal, officials at the Smithsonian found his rocket experiments to be "sound and ingenious." One key point in Goddard's favor was his well-expressed desire to develop the liquid-propellant rocket as a means of taking instruments into the high-altitude regions of Earth's atmosphere, regions that were well beyond the reach of weather balloons. This early marriage between rocketry and meteorology proved very important, because this relationship provided officials in private funding institutions a tangible reason for supporting Goddard's proposed rocket work.

On January 5, 1917, the Smithsonian Institution informed Goddard that he would receive a $5,000 grant from that institution's Hodgkins Fund for Atmospheric Research. Not only did the Smithsonian provide additional funding to Goddard over the years, but equally important the organization published Goddard's two classic monographs on rocket propulsion, the first in 1919 and the second in 1936.

In 1919 Goddard summarized his early rocket theory work in the classic report "A Method of Reaching Extreme Altitudes." It appeared in the *Smithsonian Miscellaneous Collections,* Volume 71, Number 2 in late 1919 and then as "Smithsonian Monograph Report Number 2540" in January 1920. This was the first American scientific work that carefully discussed all the fundamental principles of rocket propulsion. Goddard described the results of his experiments with solid-propellant rockets and even included a final chapter on how the rocket might be used to get a modest payload to the Moon. He suggested carrying a payload of explosive powder that would flash upon impact and signal observers on Earth. Unfortunately, nontechnical newspaper reporters missed the true significance of this great treatise on rocket propulsion and decided instead to sensationalize his suggestion about reaching the Moon with a rocket. The press gave him such unflattering nicknames as "Moony" and the "Moon man." For example, the headlines of the January 12, 1920, edition of the *New York Times* boldly proclaimed "Believes Rocket Can Reach Moon." The accompanying article proceeded to soundly criticize Goddard for not realizing that a rocket cannot operate in a vacuum. Goddard received publicity, but his new found "fame" was all derogatory.

Offended by such inappropriate and uninformed notoriety, Goddard chose to work in seclusion for the rest of his life. He avoided further controversy by publishing as little as possible and refusing to grant interviews with the members of the press. These understandable actions caused much of his brilliant work in rocketry to go unrecognized during his lifetime. The German rocket scientists from Peenemünde who emigrated to the United States after World War II expressed amazement at the extent of Goddard's inventions, many of which they had painfully duplicated. They also wondered aloud why the U.S. government had chosen to totally ignore this brilliant man. Independent of Tsiolkovsky and Oberth, Goddard recognized that the liquid-propellant rocket was the key to interplanetary travel. He then devoted his professional life to inventing and improving liquid-propellant rocket technology. Some of his most important contributions to rocket science are briefly summarized below.

In 1912 Goddard first explored on a theoretical and mathematical basis the practicality of using the rocket to reach high altitudes and even attain a sufficiently high velocity (called the *escape velocity*) to leave Earth and get to the Moon. He obtained patents in 1914 for the liquid-propellant rocket and for the concept of a two-stage rocket.

In 1915 he used a clever static test experiment in a vacuum chamber to show that the rocket would indeed function in outer space. As World War I drew to a close, Goddard demonstrated several rockets to American military officials, including the prototype for the famous *bazooka rocket* used in World War II. However, with the war almost over, these U.S. government officials simply chose to ignore Goddard and the value of his rocket research.

Undiscouraged by a lack of government interest, he turned his creative energy to the development of the liquid-propellant rocket and performed a series of important experiments at Clark University between 1921 and 1926. One test was particularly noteworthy. On December 6, 1925, Goddard performed a successful captive (that is, a static, or nonflying) test of a liquid-fueled rocket in the annex of the physics building. During this test Goddard's rocket generated enough thrust to lift its own weight. Encouraged by the static test, Goddard proceeded to construct and flight-test the world's first liquid-fueled rocket.

On March 26, 1926, Goddard made space technology history by successfully launching the world's first liquid-propellant rocket. Even though this rocket was quite a primitive device, it is nonetheless the technical ancestor of all modern liquid-propellant rockets. The rocket engine itself was just 1.2 meters tall and 15.2 centimeters in diameter. Gasoline and liquid oxygen served as its propellants. Only four people witnessed this great historic event. The publicity-shy physicist took his wife, Esther, and two loyal staff members from Clark University, P. M. Roope and Henry Sachs, to a snow-covered field at his aunt Effie Ward's farm in Auburn, Massachusetts. (Note that although Goddard refers to her as "Auntie Effie" in his diary, she was actually a more distant relative.)

Goddard assembled a metal frame structure that resembled a child's play-gym to support the rocket prior to launch. His wife served as the team's official photographer. The launch procedure was quite simple, since there was no countdown. Goddard carefully opened the fuel valves. Then, with the aid of a long pole, Sachs applied a blowtorch to ignite the engine. With a great roar, the tiny rocket vehicle jumped from the metal support structure, flew in an arc, and then landed unceremoniously some 56 meters away in a frozen cabbage patch. In all, this historic rocket rose to a height of about 12 meters, and its engine burned for only two and a half seconds. Despite its great technical significance, the world would not learn about Goddard's great achievement for some time.

With modest funding from the Smithsonian, Goddard continued his liquid-propellant rocket work. In July 1929 he launched a larger and louder rocket near Worcester. It successfully carried an instrument payload (a barometer and a camera)—the first liquid-propellant rocket to do so. However, it also created a major disturbance. As a consequence the local authorities ordered him to cease his rocket flight experiments in Massachusetts.

Fortunately, the aviation pioneer Charles Lindbergh (1902–74) became interested in Goddard's work and helped him secure additional funding from the philanthropist Daniel Guggenheim. A timely grant of $50,000 allowed Goddard to set up a rocket test station in a remote part of the New Mexico desert near Roswell. There, undisturbed and well out of the public view, Goddard conducted a series of important liquid-propellant rocket experiments in the 1930s. These experiments led to Goddard's second important Smithsonian monograph, entitled "Liquid-Propellant Rocket Development," which was published in 1936. On March 26, 1937, Goddard launched a liquid-propellant rocket nicknamed "Nell" (as were all his flying rockets) that flew for 22.3 seconds and reached a maximum altitude of about 2.7 kilometers. This was the highest altitude ever obtained by one of Goddard's rockets.

Goddard retained only a small technical staff to assist him in New Mexico and avoided contact with the rest of the scientific community, yet his secretive, reclusive nature did not impair the innovative quality of his work. In 1932 he pioneered the use both of vanes in the rocket engine's exhaust and a gyroscopic control device to help guide a rocket during flight. Then, in 1935 he became the first person to launch a liquid-propellant rocket that attained a velocity greater than the speed of sound. Finally, in 1937 Goddard successfully launched a liquid-propellant rocket that had its motor pivoted on gimbals so it could respond to the guidance signals from its onboard gyroscopic control mechanism. He also pioneered the use of the converging-diverging (de Laval) nozzle in rocketry, tested the first pulse jet engine, constructed the first turbopumps for use in a liquid-propellant rocket, and developed the first liquid-propellant rocket cluster.

Over his lifetime he registered 214 patents on various rockets and their components. Unfortunately, throughout his life the U.S. government essentially ignored Goddard's pioneering rocket work, even during World War II. So he finally closed his rocket facility near Roswell, New Mexico, and moved to Annapolis, Maryland. There he participated in a modest project with the U.S. Navy involving the use of solid-propellant rockets to assist seaplane takeoff. As a lifelong "solitary" researcher, Goddard found the social demands of

Dr. Robert H. Goddard observes a rocket test from his launch control shack site in the New Mexican desert near Roswell. While standing by the firing control panel, Goddard could fire, release, or stop testing his rocket if conditions became unsatisfactory. The firing, releasing, and stopping keys are shown on the control panel. In this 1930s-era photograph Goddard is using a small telescope to view the launch of one of his pioneering liquid-propellant rockets, which is located in the launch tower in the distance. ***(Courtesy of NASA)***

the team developing jet-assisted takeoff (JATO) for the government not particularly to his liking.

Just before his death he was able to inspect the twisted remains of a German V-2 rocket that had fallen into American hands and was taken back to the United States for analysis. As he performed a technical autopsy on the enemy rocket, a military intelligence analyst inquired: "Dr. Goddard, isn't this just like your rockets?" Without showing any emotion, Goddard calmly replied: "Seems to be." America's visionary "rocket man" died of throat cancer on August 10, 1945, in Annapolis, Maryland. Goddard did not see the dawn of the space age.

In 1951 Esther Goddard and the Guggenheim Foundation, which helped fund much of Goddard's rocket research, filed a joint lawsuit against the U.S. government for infringement on his patents. In 1960 the government settled this lawsuit and in the process officially acknowledged Goddard's great contributions to modern rocketry. That June the National Aeronautics and Space Administration and the Department of Defense jointly awarded his estate the sum of $1 million for the use of his patents. NASA's Goddard Space Flight Center (GSFC) in Greenbelt, Maryland, now honors his memory by carrying his name.

Goddard Space Flight Center (GSFC) NASA's Goddard Space Flight Center is located within the city of Greenbelt, Maryland, approximately 10 kilometers northeast of Washington, D. C. This NASA field center is a major U.S. laboratory for developing and operating unmanned scientific spacecraft. GSFC manages many of NASA's Earth observation, astronomy, and space physics missions. GSFC includes several other properties, most significantly the Wallops Flight Facility (WFF) near Chincoteague, Virginia. NASA named this important aerospace facility in honor of ROBERT H. GODDARD, the "father" of American rocketry and one of the cofounders of astronautics in the 20th century.

The GSFC Greenbelt Facility encompasses 514 ha (1,270 acres) and, in addition to the main site, maintains the adjacent Magnetic Test Facility and Propulsion Research site and outlying sites, including the Antenna Performance Measuring Range and the Optical Tracking and Ground Plane Facilities. NASA has ownership of 454 ha (1,121 acres) of land at Greenbelt. The remaining 60 ha (149 acres) are the outlying sites and are held by revocable lease from the U.S. Department of Agriculture (USDA).

Under the leadership of its director, Goddard Space Flight Center is managed by a system of directorates. The *Office of the Director* provides overall management and coordinates control over the center's diversified activities. *Earth Sciences* conducts scientific studies in the Earth sciences leading to a better understanding of processes affecting global change and the distribution of natural resources through research, development, and application of space technologies. *Space Sciences* plans, organizes, directs, and evaluates a broad spectrum of scientific research both theoretical and experimental in the study of space phenomena. This directorate also provides scientific counsel to other directorates that are working on space science projects. *Applied Engineering and Technology* provides expertise for science conceptualization, end-to-end mission development, and space communications support. This directorate develops advanced technologies to meet current and future science needs. *Flight Programs and Projects* plans, organizes, and directs the management of the center's major flight projects, new start studies, international projects, and the small- and medium-class expendable launch vehicles. *Suborbital and Special Orbital Projects* is responsible for the overall management, operation, and support of NASA's sounding rocket and balloon programs and the conduct of aeronautical research. This function is located at the Wallops Flight Facility, Wallops Island, Virginia.

See also EARTH SYSTEM SCIENCE; NATIONAL AERONAUTICS AND SPACE ADMINISTRATION (NASA); SOUNDING ROCKET; WALLOPS FLIGHT FACILITY.

Gold, Thomas (1920–2004) Austrian-American *Astronomer* In 1948 the astronomer Thomas Gold collaborated with SIR FRED HOYLE and SIR HERMANN BONDI and proposed a steady-state universe model in cosmology. This model, based on the perfect cosmological principle, has now been largely abandoned in favor of the big bang theory. In the late 1960s Gold suggested that the periodic signals from pulsars represented emissions from rapidly spinning neutron stars, a theoretical hypothesis consistent with subsequent observational data.

See also BIG BANG; COSMOLOGY; PULSAR.

Goldilocks zone *See* CONTINUOUSLY HABITABLE ZONE.

Goodricke, John (1764–1786) British *Astronomer* Though deaf and mute, John Goodricke made a significant contribution to 18th-century astronomy when he observed that certain variable stars, such as Algol, were periodic in nature. Prior to his death from pneumonia at age 21, he suggested that this periodic behavior might be due to a dark, invisible companion regularly passing in front of (that is, eclipsing) its visible companion, the phenomenon modern astronomers call an *eclipsing binary.*

See also ALGOL; BINARY STAR SYSTEM; VARIABLE STAR.

Gordon, Richard F., Jr. (1929–) American *Astronaut, U.S. Navy Officer* Astronaut Richard F. Gordon, Jr., occupied the command module pilot seat during NASA's *Apollo 12* mission to the Moon, which took place from November 14 to 24, 1969. While Gordon orbited overhead in the Apollo command module spacecraft, his companions, astronauts ALAN L. BEAN and CHARLES (PETE) CONRAD, JR., explored the Moon's surface during the second lunar landing mission of the Apollo Project.

See also APOLLO PROJECT.

Gorizont spacecraft An extended series of Russian geostationary communications satellites that have provided telephone, telegraph, and facsimile communications services in addition to relaying television and radio broadcasts. *Gorizont 1,* the first satellite in this evolving family of spacecraft, was launched into a high-Earth orbit (HEO) from the Baikonur Cosmodrome (aka Tyuratam) on December 19,

1978. Almost two decades later *Gorizont 32* was sent into geostationary orbit (GEO) from the Baikonur Cosmodrome on May 25, 1996. In addition to supporting telecommunications services across Russia, Gorizont spacecraft also support maritime and international communications.

See also COMMUNICATIONS SATELLITE; MOLNIYA SATELLITE; RUSSIAN SPACE AGENCY.

Gould, Benjamin Apthorp (1824–1896) American *Astronomer* Benjamin Apthorp Gould founded the *Astronomical Journal* in 1849 and produced an extensive 19th-century star catalog for the Southern Hemisphere sky. In 1868 he was invited by the government of Argentina to establish and direct a national observatory at Córdoba. This gave him the opportunity to make detailed astronomical observations, which resulted in a star catalog comparable to the star catalog produced by FRIEDRICH WILHELM AUGUST ARGELANDER for the Northern Hemisphere sky.

GOX Gaseous oxygen. *Compare with* LOX.

grain 1. In rocketry, the integral piece of molded or extruded solid propellant that constitutes both fuel and oxidizer in a solid rocket motor and is shaped to produce, when burned, a specified performance versus time relation. 2. In photography, a small particle of metallic silver remaining in a photographic emulsion after development and fixing. In the agglomerate, these grains form the dark areas of a photographic image. Excessive graininess reduces the quality of a photographic image, especially when the image is enlarged or magnified. 3. In materials science, an individual crystal in a polycrystalline metal or alloy.

gram (symbol: g) The fundamental unit of mass in the centimeter-gram-second (CGS) unit system. One gram is equal to one-thousandth (1/1,000) of a kilogram.

***Granat* (satellite)** Russian spacecraft dedicated to an X-ray and gamma-ray astronomy mission in collaboration with other European countries. Launched on December 1, 1989, the satellite operated in orbit around Earth until November 1998. *Granat* allowed scientists to study the spectra and time variability of black hole candidates and provided a very deep imaging of the galactic center region.

grand unified theories (GUTs) Theories in contemporary physics that attempt to describe the behavior of the single force that results from unification of the strong, weak, and electromagnetic forces in the early universe.

See also BIG BANG THEORY; COSMOLOGY; FUNDAMENTAL FORCES IN NATURE.

granules The small bright features found in the photosphere of the Sun. Also called granulation.

See also SUN.

gravitation The acceleration produced by the mutual attraction of two masses, directed along the line joining their centers of mass, and of magnitude inversely proportional to the square of the distance between the two centers of mass. In classical physics the force of attraction (F_g) between two masses (m and M) is given by SIR ISAAC NEWTON's law of gravitation:

$$F_g = [G\ m\ M] / r^2$$

where G is the gravitational constant (6.6726×10^{-11} N m^2kg^{-2}) and r is the distance between the two masses.

According to Newton's theory of gravity (which he proposed in about 1687), all masses "pull" on one another with an invisible force called "gravity." This force is an inherent property of matter and is directly proportional to an object's mass. In our solar system the Sun reaches out across enormous distances and "pulls" smaller masses, such as planets, comets, and asteroids, into orbit around it using its force of gravity (or gravitation).

In the early 20th century ALBERT EINSTEIN discovered a contradiction between Newton's theory of gravity and his own theory of special relativity, which he proposed in 1905. In special relativity the speed of light is assumed to be the speed limit of all energy in the universe. No matter what kind of energy it is, it cannot propagate or transmit across the universe any faster than 299,792 km/sec. Yet Newton's theory assumed that the Sun instantaneously transmits its force of gravity to the planets at a speed much faster than the speed of light. Einstein began to wonder whether gravity was unique in its ability to fly across the universe. He also explored another possibility: Perhaps masses react with one another for a different reason.

In 1915 Einstein published his general relativity theory, a theory that transformed space from the classic Newtonian concept of vast emptiness with nothing but the invisible force of gravity to rule the motion of matter to an ephemeral fabric of space-time (a space-time continuum), which enwraps and grips matter and directs its course through the universe. Einstein postulated that this space-time fabric spans the entire universe and is intimately connected to all matter and energy within it.

At this point one may wonder how this dramatic change in thinking explains the motion of the planets and the orbits of the Moon and artificial satellites around Earth. Theoretically, when a mass sits in the space-time fabric, it will deform the fabric itself, changing the shape of space and altering the passage of time around it. At this point in the discussion, it may be helpful to imagine the space-time fabric as a sturdy but flexible sheet of rubber stretched out to form a plane upon which an object, such as a baseball, is placed. The ball causes the sheet to bend. Therefore, in Einstein's general relativity a mass causes the fabric of space-time to bend in a similar manner. The more massive the object, the deeper is the dip in the fabric of space-time.

In the case of the Sun, for example, Einstein reasoned that because of its mass the space-time fabric would curve around it, creating a "dip" in space-time. As the planets (as well as asteroids and comets) travel across the space-time fabric, they would respond to this Sun-caused dip and travel around the massive object by following the curvature in space-time. As long as they never slow down, the planets would maintain regular orbits around the Sun, neither spiraling in toward it nor flying off into interstellar space.

gravitational collapse The unimpeded contraction of any mass caused by its own gravity.

gravitational constant (symbol: G) The coefficient of proportionality in SIR ISAAC NEWTON's universal law of gravitation:

$$G = 6.672 \times 10^{-11} \text{ newton} - \text{meter}^2 - \text{kilogram}^{-2} \ (\text{N m}^2\text{kg}^{-2})$$

gravitational field Field extending out in all directions that is created by an object with mass. In classical (Newtonian) physics the strength of this field determines how that object's mass can influence other objects with mass. The strength of the gravitation field follows the inverse square law, meaning it decreases as the square of the distance from the source.

See also GRAVITATION.

gravitational instability A condition whereby an object's (inward-pulling) gravitational potential energy exceeds its (outward-pushing) thermal energy, thereby causing the object to collapse under the influence of its own mass.

See also STAR; SUPERNOVA.

gravitational lensing The bending of light from a distant celestial object by a massive foreground object.

In 1990 the European Space Agency's Faint Object Camera onboard NASA's *Hubble Space Telescope* provided astronomers with a detailed image of the gravitational lens G2237 + 0305, also called the Einstein cross. This photograph clearly depicts the interesting optical illusion caused by gravitational lensing. Shown are four images of the same very distant quasar that has been multiple-imaged by a relatively nearby galaxy (central image), which acts as a gravitational lens. The angular separation between the upper and lower images is 1.6 arc seconds. *(Courtesy of NASA and ESA.)*

gravitational redshift A prediction of ALBERT EINSTEIN's general relativity theory implying that photons will lose energy as they escape the gravitational field of a massive object. Because a photon's energy is proportional to its frequency, a photon that loses energy escaping a gravitational field will also suffer a decrease in its frequency—that is, the photon will redshift in wavelength.

gravitational wave In modern physics, the gravitational analog of an electromagnetic wave whereby gravitational radiation is emitted at the speed of light from any mass that undergoes rapid acceleration.

graviton The hypothetical elementary particle in quantum gravity that modern physicists speculate plays a role similar to that of the photon in quantum electrodynamics (QED). Although experiments have not yet provided direct physical evidence of the existence of the graviton, ALBERT EINSTEIN predicted the existence of a quantum (or particle) of gravitational energy as early as 1915, when he started formulating general relativity theory.

See also RELATIVITY.

gravity (gravitational force) In general, the attraction of a celestial object for any nearby mass. Specifically, the downward force imparted by Earth on a mass near Earth or on its surface. The greater the mass of the object, the greater the force of gravity. Since Earth is rotating, the force observed as "gravity" is actually the result of the force of gravitation and the centrifugal force arising from its rotation.

See also GRAVITATION.

gravity anomaly A region on a celestial body where the local force of gravity is lower or higher than expected. If the celestial object is assumed to have a uniform density throughout, then we would expect the gravity on its surface to have the same value everywhere.

See also MASCON.

gravity assist The change in a spacecraft's direction and velocity achieved by a carefully calculated flyby through a planet's gravitational field. The change in direction and velocity is achieved without the use of supplementary propulsive energy.

See also ORBITS OF OBJECTS IN SPACE; *VOYAGER* SPACECRAFT.

Gravity Probe B A NASA spacecraft launched on April 20, 2004, by a Delta II rocket from Vandenberg Air Force Base, California. This scientific spacecraft, developed for NASA by Stanford University, is essentially a relativity gyroscope experiment designed to test two extraordinary, currently unverified predictions of ALBERT EINSTEIN's general relativity theory. *Gravity Probe B*'s experiment is very precisely investigating tiny changes in the axis of spin of four gyroscopes contained in a satellite that orbits at an altitude of 640 km directly over Earth's poles. So free are these special gyroscopes from disturbance that they provide an almost perfect space-time reference system. Specifically, they are allowing scientists to measure how the fabric of space-time is being warped by the presence of Earth. Even more profoundly, *Gravity Probe B* is also allowing scientists to measure how Earth's rotation drags space-time around with it. Confirmation of these anticipated general relativity effects, though small for an object the mass of Earth, would have far-reaching implications for the nature of matter and the structure of the universe. In early September 2004 NASA reported that *Gravity Probe B* had successfully completed the necessary calibrations following ascent into space and was now entering the science phase of its mission.

See also RELATIVITY (THEORY).

gray (symbol: Gy) In radiation protection, the SI unit of absorbed dose of ionizing radiation. One gray (1 Gy) is equal to the absorption of 1 joule of energy per kilogram of matter. This unit is named in honor of L. H. Gray (1905–65), a British radiologist. The gray is related to the traditional radiation dose unit system as follows:

$$1 \text{ Gy} = 100 \text{ rad}$$

gray body A body that absorbs some constant fraction, between zero and one, of all electromagnetic radiation incident upon it. The fraction of electromagnetic radiation absorbed is called the absorptivity (symbol: a) and is independent of wavelength for a gray body. Many objects fit the gray body model, which represents a surface of absorptive characteristics intermediate between those of a white body and a blackbody. In radiation transport theory, a white body is a hypothetical body whose surface absorbs no electromagnetic radiation at any wavelength, that is, it is a perfect reflector. Since the white body exhibits zero absorptivity at all wavelengths, it is an idealization that is exactly opposite to that of the blackbody. The gray body is sometimes called *grey body*.

See also BLACKBODY.

Great Bear *See* URSA MAJOR.

Great Cluster in Hercules The magnificent globular cluster in the constellation Hercules, discovered in 1714 by SIR EDMUND HALLEY. Also called M13 and NGC 6205.

See also GLOBULAR CLUSTER.

Great Dark Spot (GDS) A large, dark, oval-shaped feature in the clouds of Neptune, discovered in 1989 by NASA's *Voyager 2* spacecraft.

See also NEPTUNE; *VOYAGER* SPACECRAFT.

The magnificent globular cluster M13 in the constellation Hercules. Astronomers also call this gravitationally bound bundle of ancient stars NGC 6205 and the Great Star Cluster in Hercules. ***(Courtesy of the U.S. Naval Observatory)***

Great Observatories Program (NASA) Each portion of the electromagnetic spectrum (that is, radio waves, infrared radiation, visible light, ultraviolet radiation, X-rays, and gamma rays) brings astronomers and astrophysicists unique information about the universe and the objects within it. For example, certain radio frequency (RF) signals help scientists characterize cold molecular clouds, and the cosmic microwave background (CMB) represents the "fossil" radiation from the big bang. The infrared (IR) portion of the spectrum provides signals that let astronomers observe nonvisible objects such as near-stars (brown dwarfs) and relatively cool stars. Infrared radiation also helps scientists peek inside dust-shrouded stellar nurseries (where new stars are forming) and unveil optically opaque regions at the core of the Milky Way galaxy. Ultraviolet (UV) radiation provides astrophysicists special information about very hot stars and quasars, while visible light helps observational astronomers characterize planets, main sequence stars, nebulas, and galaxies. Finally, the collection of X-rays and gamma rays by space-based observatories brings scientists unique information about high-energy phenomena, such as supernovas, neutron stars, and black holes—whose presence is inferred by intensely energetic radiation emitted from extremely hot material as it swirls in an accretion disk before crossing the particular black hole's event horizon.

Scientists recognized that they could greatly improve their understanding of the universe if they could observe all portions of the electromagnetic spectrum. As the technology for space-based astronomy matured toward the end of the 20th century, NASA created the Great Observatories Program. This important program involved a series of four highly sophisticated space-based astronomical observatories, each carefully designed with state-of-the-art equipment to gather "light" from a particular portion (or portions) of the electromagnetic spectrum. (*See* figure.)

NASA initially assigned each Great Observatory a developmental name and then renamed the orbiting astronomical facility to honor a famous scientist. The first Great Observatory was the *Space Telescope* (ST), which became the *Hubble Space Telescope* (HST). It was launched by a space shuttle in 1990 and then refurbished on-orbit through a series of subsequent shuttle missions. With constantly upgraded instruments and improved optics, this long-term space-based observatory is designed to gather light in the visible, ultraviolet, and near-infrared portions of the spectrum. This Great Observatory honors the American astronomer EDWIN POWELL HUBBLE. Although NASA has now cancelled plans for a future refurbishment mission by the space shuttle, the HST could operate for several more years until being replaced by the *James Webb Space Telescope* (JWST) around 2010.

The second Great Observatory was the *Gamma Ray Observatory* (GRO), which NASA renamed the *Compton Gamma Ray Observatory* (CGRO) following its launch by a space shuttle in 1991. Designed to observe high-energy gamma rays, this observatory collected valuable scientific information from 1991 to 1999 about some of the most violent processes in the universe. NASA renamed the observatory to honor the American physicist and Nobel laureate ARTHUR HOLLY COMPTON. The CGRO's scientific mission officially ended in 1999. The following year NASA mission managers commanded the massive spacecraft to perform a controlled deorbit burn. This operation resulted in a safe reentry in June 2000 and a harmless impact in a remote part of the Pacific Ocean.

NASA originally called the third observatory in this series the *Advanced X-Ray Astrophysics Facility* (AXAF).

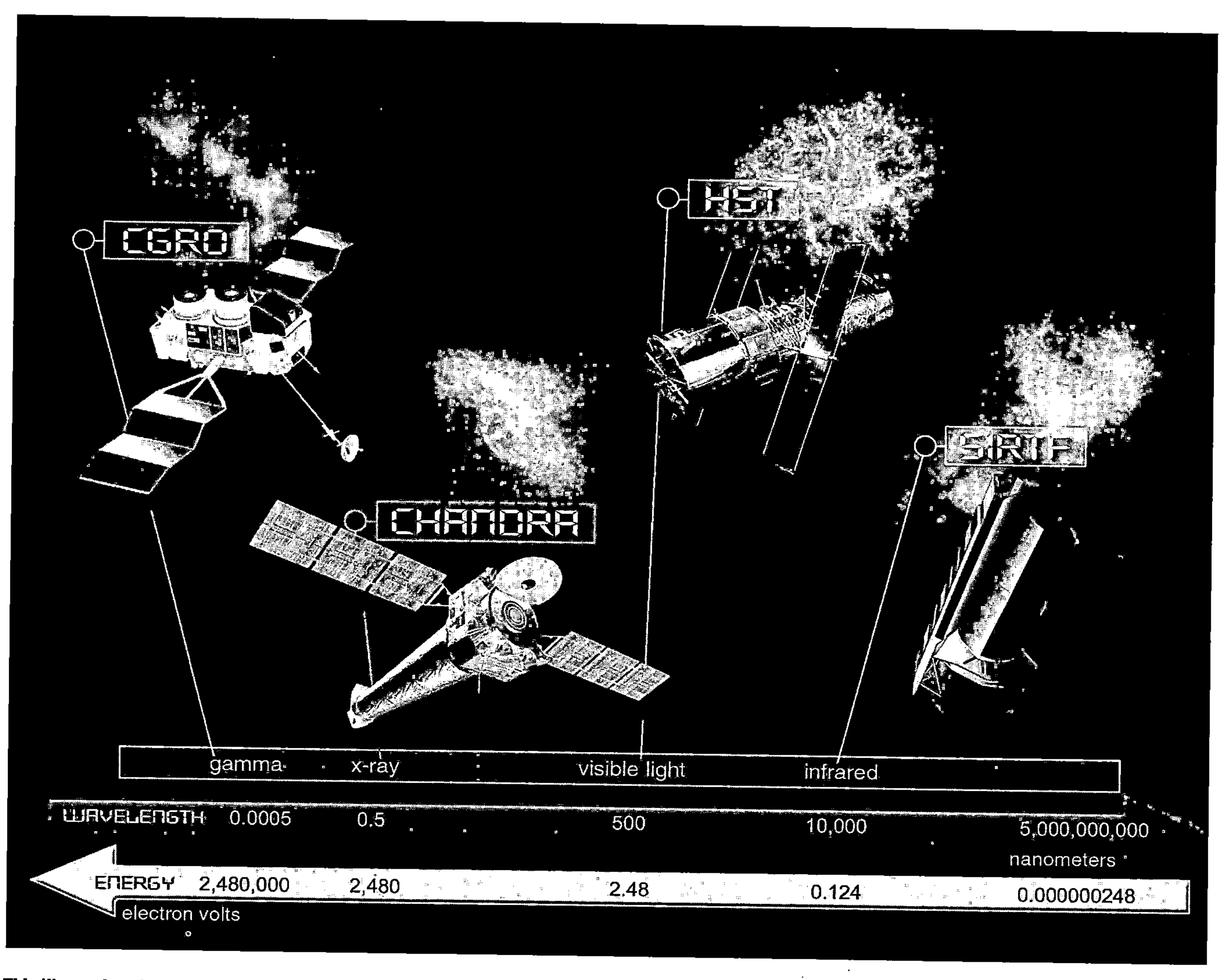

This illustration shows each of NASA's Great Observatories and the region of the electromagnetic spectrum from which the particular space-based astronomy facility collects scientific data. From left to right (in order of decreasing photon energy and increasing wavelength): the *Compton Gamma Ray Observatory* (CGRO), the *Chandra X-Ray Observatory* (CXO), the *Hubble Space Telescope* (HST), and the *Space Infrared Telescope Facility* (SIRTF), now called the *Spitzer Space Telescope* (SST). *(Courtesy of NASA)*

Renamed the *Chandra X-Ray Observatory* (CXO) to honor the Indian-American astrophysicist and Nobel laureate SUBRAHMANYAN CHANDRASEKHAR, the observatory was placed into a highly elliptical orbit around Earth in 1999. The CXO is examining X-ray emissions from a variety of energetic cosmic phenomena, including supernovas and the accretion disks around suspected black holes, and should operate until at least 2009.

The fourth and final member of NASA's Great Observatories Program is the *Space Infrared Telescope Facility* (SIRTF). NASA launched this observatory in 2003 and renamed it the *Spitzer Space Telescope* (SST) to honor the American astrophysicist LYMAN SPITZER, JR. The sophisticated infrared observatory is now providing scientists a fresh vantage point from which to study processes that have until now remained mostly in the dark, such as the formation of galaxies, stars, and planets. The SST also serves as an important technical bridge to NASA's Origins Program, an ongoing attempt to scientifically address such fundamental questions as "Where did we come from?" and "Are we alone?"

See also ASTROPHYSICS; *CHANDRA X-RAY OBSERVATORY; COMPTON GAMMA RAY OBSERVATORY; HUBBLE SPACE TELESCOPE;* ORIGINS PROGRAM; *SPITZER SPACE TELESCOPE.*

Great Red Spot (GRS) A distinctive oval-shaped feature in the southern hemisphere clouds of Jupiter. Observed by telescopic astronomers since the 17th century, the GRS was also noted in 1831 by amateur German astronomer SAMUEL HEINRICH SCHWABE.

See also JUPITER.

Greek alphabet

alpha	Α, α	xi	Ξ, ξ
beta	Β, β	omicron	Ο, ο
gamma	Γ, γ	pi	Π, π
delta	Δ, δ	rho	Ρ, ρ
epsilon	Ε, ε	sigma	Σ, σ, ς (at end of word)
zeta	Ζ, ζ		
eta	Η, η	tau	Τ, τ
theta	Θ, θ	upsilon	Υ, υ
iota	Ι, ι	phi	Φ, φ
kappa	Κ, κ	chi	Χ, χ
lambda	Λ, λ	psi	Ψ, ψ
mu	Μ, μ	omega	Ω, ω
nu	Ν, ν		

Green Bank Telescope (GBT) On November 15, 1988, the original 100-meter Green Bank radio telescope collapsed due to a sudden failure of a key structural element. This unexpected loss of the National Radio Astronomy Observatory (NRAO)'s major facility resulted in the construction of a replacement project called the Robert C. Byrd Green Bank Telescope (GBT). NRAO operates the Green Bank Telescope, which is the world's largest fully steerable single aperture antenna. In addition to the GBT, there are several other radio telescopes at the Green Bank site in West Virginia. The GBT is generally described as a 100-meter telescope, but the actual dimensions of the surface are 100 by 110 meters. The overall structure of the GBT is a wheel-and-track design that allows the telescope to view the entire sky above 5 degrees elevation. The track, 64 m in diameter, is level to within a few thousandths of a centimeter in order to provide precise pointing of the structure while bearing 7,300 metric tons of moving mass. The GBT is of an unusual design. Unlike conventional telescopes, which have a series of supports in the middle of the surface, the GBT's aperture is unblocked so that incoming radio frequency radiation meets the surface directly.

See also RADIO ASTRONOMY.

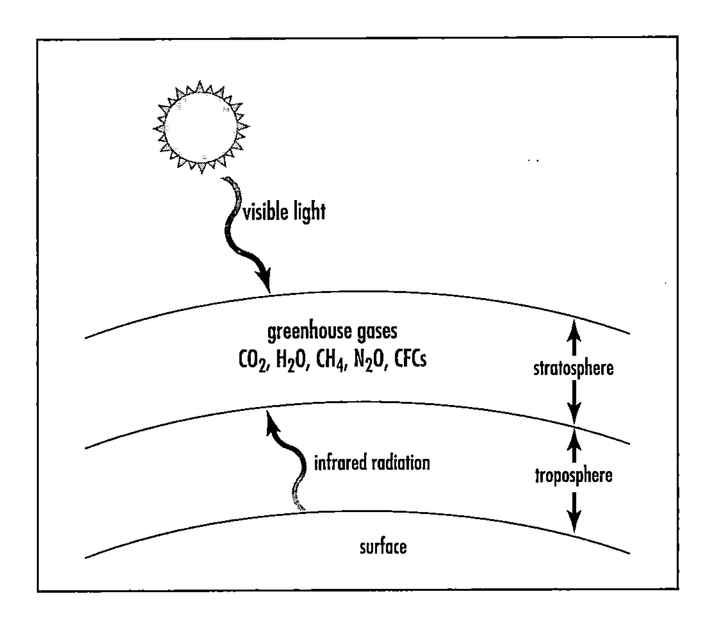

The greenhouse effect in Earth's atmosphere

greenhouse effect The general warming of the lower layers of a planet's atmosphere caused by the presence of "greenhouse gases," such as water vapor (H_2O), carbon dioxide (CO_2), and methane (CH_4). On Earth the greenhouse effect might occur because our atmosphere is relatively transparent to visible light from the Sun (typically 0.3 to 0.7 micrometer wavelength) but is essentially opaque to the longer-wavelength (typically 10.6 micrometer) thermal infrared radiation emitted by the planet's surface. Because of the presence of greenhouse gases in our atmosphere, such as carbon dioxide, water vapor, methane, nitrous oxide (NO_2) and human-made chlorofluorocarbons (CFCs), this outgoing thermal radiation from Earth's surface could be blocked from escaping to space, and the absorbed thermal energy could cause a rise in the temperature of the lower atmosphere. Therefore, as the presence of greenhouse gases increases in Earth's atmosphere, more outgoing thermal radiation is trapped, and a global warming trend occurs.

Scientists around the world are concerned that human activities, such as increased burning of vast amounts of fossil fuels, are increasing the presence of greenhouse gases in our atmosphere and upsetting the overall planetary energy balance. These scientists are further concerned that we may be creating the conditions of a *runaway greenhouse,* as appears to have occurred in the past on the planet Venus. Such an effect is a planetary climatic extreme in which all of the surface water has evaporated from the surface of a life-bearing or potentially life-bearing planet. Planetary scientists believe that the current Venusian atmosphere allows sunlight to reach the planet's surface, but its thick clouds and rich carbon dioxide content prevent surface heat from being radiated back to space. This condition has led to the evaporation of all surface water on Venus and has produced the present infernolike surface temperatures of approximately 485°C, a temperature hot enough to melt lead.

See also GLOBAL CHANGE; VENUS.

"greening the galaxy" A visionary term used to describe the diffusion of human beings, their technology, and their culture through interstellar space, first to neighboring star systems and then eventually across the entire Milky Way galaxy.

See also EXTRATERRESTRIAL CIVILIZATIONS; FERMI PARADOX.

green satellite An Earth-orbiting satellite that collects environmental data.

See also *AQUA* SPACECRAFT; *EARTH OBSERVING-1;* EARTH-OBSERVING SPACECRAFT; *TERRA* SPACECRAFT.

Greenwich mean time (GMT) Mean solar time at the meridian of Greenwich, England, used as a basis for standard time throughout the world. Normally expressed in four numerals, 0001 through 2400. Also called universal time, Z-time, and zulu time.

Greenwich Observatory *See* ROYAL GREENWICH OBSERVATORY.

Gregorian calendar A more precise version of the Julian calendar that was devised by CHRISTOPHER CLAVIUS and introduced by Pope Gregory XIII in 1582. The changes restored March 21st as the vernal equinox. It is now the civil calendar used in most of the world.

See also CALENDAR.

Gregorian telescope A compound reflecting telescope consisting of a small concave ellipsoidal secondary mirror and a concave paraboloidal primary mirror. The Scottish mathematician JAMES GREGORY published this design in 1663. This type of reflecting telescope proved popular with 18th-century astronomers but eventually gave way to the more compact and now very widely used Cassegrain reflecting telescope.

See also CASSEGRAIN TELESCOPE; REFLECTING TELESCOPE; TELESCOPE.

Gregory, James (1638–1675) Scottish *Mathematician, Astronomer* James Gregory was a Scottish mathematician who described the components and operation of a two-mirror reflecting telescope (a design later called the Gregorian telescope) in his 1663 publication, *The Advance of Optics.*

Grissom, Virgil (Gus) I. (1926–1967) American *Astronaut, U.S. Air Force Officer* Virgil (Gus) I. Grissiom was the second American to travel in outer space, a feat he accomplished during NASA's suborbital Mercury Project *Liberty Bell 7* flight on July 21, 1961. In March 1965 he served as the command pilot during the first crewed Gemini Project orbital mission. Grissom, along with his fellow Apollo Project astronauts EDWARD H. WHITE, II and ROGER B. CHAFFEE, died on January 27, 1967, at Cape Canaveral when a flash fire consumed their Apollo spacecraft during a mission simulation and training test at the launch pad.

gross liftoff weight (GLOW) The total weight of an aircraft, aerospace vehicle, or rocket, as loaded; specifically, the total weight with full crew, full tanks, and payload.

Ground-based Electro-Optical Deep Space Surveillance (GEODSS) There are more than 9,000 known objects in orbit around Earth. These objects range from active satellites to pieces of space debris, or space junk, such as expended upper-stage vehicles, fragments from exploded rockets, and even missing tools and cameras from astronaut-performed extra vehicular activity. The Space Control Center (SCC) at Cheyenne Mountain Air Force Station supports the U.S. Strategic Command (USSTRATCOM) missions of space surveillance and the protection of American assets in outer space. The SCC maintains a current computerized catalog of all orbiting human-made objects, charts present positions, plots future orbital paths, and forecasts the times and general locations for significant human-made objects reentering Earth's atmosphere. The center currently tracks more than 9,000 human-made objects in orbit around Earth, of which approximately 20% are functioning payloads or satellites. The SCC receives data continuously from the Space Surveillance Network (SSN) operated by the U.S. Air Force. These data include information about objects considered to be in deep space—that is, at an altitude of more than 4,800 kilometers.

Four Ground-based Electro-Optical Deep Space Surveillance (GEODSS) sites around the world play a major role in helping the SCC track human-made objects orbiting Earth. GEODSS is the successor to the Baker-Nunn camera system developed in the mid-1950s. The GEODSS system performs its space object tracking mission by bringing together telescopes, low-light-level television, and modern computers. Each operational GEODSS site has at least two main telescopes and one auxiliary telescope. Operational GEODSS sites are located at Socorro, New Mexico (Detachment 1 of the Air Force 18th Space Surveillance Squadron headquartered at Edwards Air Force Base); at Diego Garcia, British Indian Ocean Territories (Detachment 2); at Maui, Hawaii (Detachment 3, collocated on top of Mount Haleakala with the U.S. Air Force Maui Space Surveillance System); and at Morón Air Base, Spain (Detachment 4).

The GEODSS telescopes move across the sky at the same rate that the stars appear to move. This keeps the distant stars in the same positions in the fields of view. As the telescopes slowly move, the GEODSS cameras take very rapid electronic snapshots of the fields of view. Computers then take these snapshot images and overlay them on each other.

Star images, which essentially remain fixed, are erased electronically. Human-made space objects circling Earth, however, do not remain fixed, and their movements show up as tiny streaks that can be viewed on a console screen. Computers then measure these streaks and use the data to determine the orbital positions of the human-made objects. The GEODSS system can track objects as small as a basketball orbiting Earth at an altitude of more than 32,200 kilometers.

See also SPACE DEBRIS.

ground elapsed time (GET) The time since launch. For example, at two hours GET, the space shuttle crew will open the orbiter's cargo bay doors and prepare to deploy a payload on orbit.

ground receiving station A facility on the surface of Earth that records data transmitted by an Earth-observing satellite.

ground state The state of a nucleus, atom, or molecule at its lowest (stable) energy level. *Compare with* EXCITED STATE.

ground support equipment (GSE) Any nonflight (i.e., ground-based) equipment used for launch, checkout, or in-flight support of an aerospace vehicle, expendable rocket, spacecraft, or payload. More specifically, GSE consists of nonflight equipment, devices, and implements that are required to inspect, test, adjust, calibrate, appraise, gauge, measure, repair, overhaul, assemble, transport, safeguard, record, store, or otherwise function in support of a rocket, space vehicle, or the like, either in the research and development phase or in the operational phase. In general, GSE is not considered to include land and buildings but may include equipment needed to support another item of GSE.

groundtrack The path followed by a spacecraft over Earth's surface.

ground truth In remote sensing, measurements made on the ground to support, confirm, or help calibrate observations made from aerial or space platforms. Typical ground truth data include weather, soil conditions and types, vegetation types and conditions, and surface temperatures. Best results are obtained when these ground truth measurements are performed simultaneously with the airborne or spaceborne sensor measurements. Also referred to as ground data and ground information.

See also REMOTE SENSING.

Early flight (July 18, 1951) of the U.S. Air Force's Matador surface-to-surface guided missile at Cape Canaveral Air Force Station, Florida. Historic Cape Canaveral Lighthouse appears in the lower left portion of the photograph. ***(Courtesy of United States Air Force)***

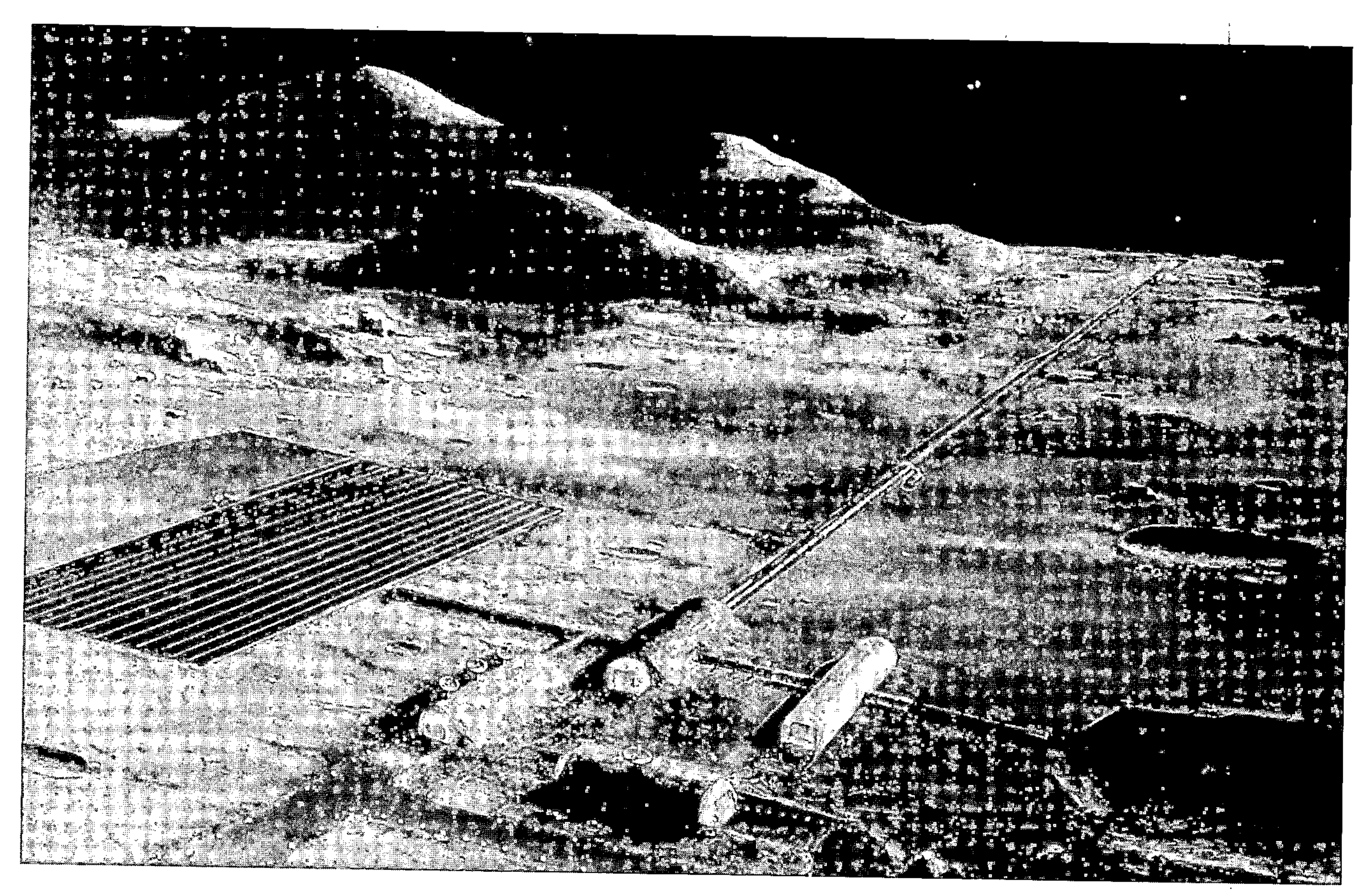

An artist's rendering that shows a future lunar base with a kilometers-long electromagnetic cannon (mass driver) capable of shooting four-kg mass packages of lunar material from the surface of the Moon to a point in deep space, where the "payload packages" are collected and processed for use in orbiting space settlements and industrial complexes *(Courtesy of NASA/JSC)*

G stars Stars of spectral type G; yellow stars that have surface temperatures ranging from 5,000 to 6,000 kelvins. The Sun is a G star.

See also STAR; SUN.

g-suit (or **G-suit**) A suit that exerts pressure on the abdomen and lower parts of the body to prevent or retard the collection of blood below the chest under conditions of positive acceleration.

See also G.

guest star The name ancient Chinese astronomers of the Sung dynasty gave to the spectacular Crab nebula supernova, which appeared in 1054 C.E.

See also CRAB NEBULA; SUPERNOVA.

Guiana Space Center *See* KOUROU LAUNCH COMPLEX.

guidance system A system that evaluates flight information; correlates it with target or destination data; determines the desired flight path of the missile, spacecraft, or aerospace vehicle; and communicates the necessary commands to the vehicle's flight control system.

guided missile (GM) An unmanned, self-propelled vehicle moving above the surface of Earth whose trajectory or course is capable of being controlled while in flight. An air-to-air guided missile (AAGM) is an air-launched vehicle for use against aerial targets. An air-to-surface guided missile (ASGM) is an air-launched missile for use against surface targets. A surface-to-air guided missile (SAGM) is a surface-launched guided missile for use against targets in the air. Finally, a surface-to-surface guided missile (SSGM) is a surface-launched missile for use against surface targets.

See also BALLISTIC MISSILE; CRUISE MISSILE.

guide vane A control surface that may be moved into or against a rocket's jetstream; used to change the direction of the jet flow for thrust vector control.

See also ROCKET.

gun-launch to space (GLTS) An advanced launch concept involving the use of a long and powerful electromagnetic launcher to hurl small satellites and payloads into orbit. One recent concept suggests the development of a hypervelocity coil-gun launcher to place 100-kg payloads into Earth orbit at altitudes ranging from 200 to 500 kilometers. This coil-gun launcher would accelerate the specially designed (and acceleration-hardened) payload package to an initial velocity of about 6 kilometers per second through a long evacuated tube. The payload package would consist of the payload itself (e.g., a satellite or bulk cargo), a solid-propellant orbital insertion rocket, a guidance system, and an aeroshell. The specially designed payload package would penetrate easily through the atmosphere and be protected from atmospheric heating by its aeroshell. Once out of the sensible atmosphere, the aeroshell would be discarded, and the solid-propellant rocket would be fired to provide the final velocity increment necessary for orbital insertion and circularization. Payloads launched (or perhaps more correctly, "shot") into orbit by such electromagnetic guns would experience peak accelerations ranging from hundreds to thousands of g's (one g is the normal acceleration due to gravity at Earth's surface).

GLTS has also been suggested for use on the lunar surface, where the absence of an atmosphere and the Moon's reduced gravitational field (about one-sixth that experienced on Earth's surface) might make this launch approach attractive for hurling raw or refined materials into low-lunar orbit. (*See* figure, page 283.) Launch concepts involving this approach are sometimes referred to as the Jules Verne approach to orbit—in recognition of the giant gun used by JULES VERNE to send his explorers on a voyage around the Moon in the famous science fiction story *From the Earth to the Moon.*

See also LAUNCH VEHICLE; SPACEPORT.

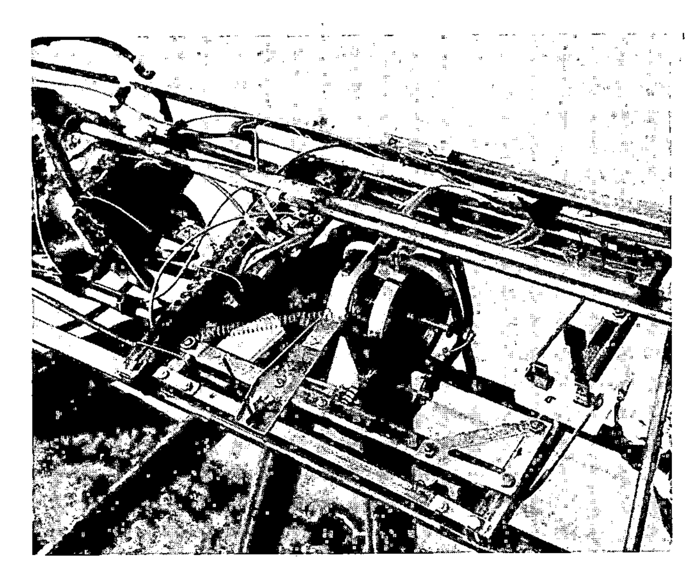

A close-up look at the gyroscope guidance system designed and constructed by Dr. Robert Goddard for his 1930s-era liquid-propellant rocket system ***(Courtesy of NASA)***

gyroscope A device that uses the angular momentum of a spinning mass (rotor) to sense angular motion of its base about one or two axes orthogonal (mutually perpendicular) to the spin axis. Also called a *gyro.*

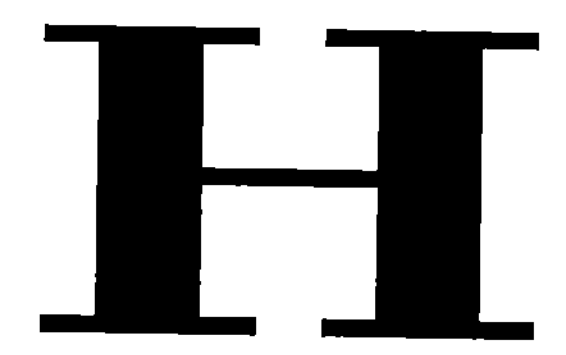

habitable payload A payload with a pressurized compartment suitable for supporting a crewperson in a shirtsleeve environment.

habitable zone *See* CONTINUOUSLY HABITABLE ZONE (CHZ).

Hadley Rille A long, ancient lava channel on the Moon that was the landing site for the *Apollo 15* mission during NASA's Apollo Project.

See also APOLLO PROJECT; MOON.

hadron Any of the class of subatomic particles that interact by means of the strong (nuclear) force, including protons and neutrons.

See also ELEMENTARY PARTICLE(S); FUNDAMENTAL FORCES IN NATURE.

Haise, Fred Wallace, Jr. (1933–) American *Astronaut* Astronaut Fred W. Haise, Jr. served as the lunar excursion module (LEM) pilot for NASA's ill-fated *Apollo 13* mission to the Moon from April 11 to 17, 1970. Instead of landing as planned in the Fra Mauro region, he helped his fellow astronauts, JAMES A. LOVELL, JR., and JOHN LEONARD "JACK" SWIGERT, JR., maneuver their explosion-crippled Apollo Project spacecraft on a life-saving emergency trajectory around the Moon and back to safety on Earth. As the LEM pilot, Haise was instrumental in converting that two-person lunar lander spacecraft into a three-person space lifeboat. In 1977 he participated as a pilot-astronaut during the space shuttle's approach and landing tests conducted at Edwards Air Force Base in California. These nonpowered gliding tests of the space shuttle *Enterprise* served as an important preparation for the shuttle's space flight program, which began in 1981.

See also APOLLO PROJECT; SPACE TRANSPORTATION SYSTEM.

Haldane, J(ohn) B(urdon) S(anderson) (1892–1964) British *Geneticist, Science Writer* The British geneticist J. B. S. Haldane was a multitalented scientist and a skilled writer whose works popularized science. He pioneered several important research pathways in evolution and population genetics. An avowed Marxist, he left the United Kingdom in 1957 and went to India, where he became the director of that country's Genetics and Biometry Laboratory in Orissa. He also became a citizen of India and died there in 1964.

Hale, George Ellery (1868–1938) American *Astronomer* The American astronomer George Ellery Hale pioneered the field of modern astrophysics. In the late 1880s he invented the *spectroheliograph,* an instrument that allowed astronomers to photograph the Sun at a particular wavelength or spectral line. He also was responsible for establishing several major observatories in the United States, including the Yerkes Observatory (Wisconsin), the Mount Wilson Observatory (California), and the Palomar Observatory (California).

See also ASTROPHYSICS.

half-life 1. (radioactive) The time in which half the atoms of a particular radioactive isotope disintegrate to another nuclear form. Measured half-lives vary from millionths of a second to billions of years. The half-life ($T_{1/2}$) is given by the expression:

$$T_{1/2} = (\ln 2)/\lambda = 0.69315/\lambda$$

where λ is the decay constant for the particular radioactive isotope and ln 2 is the natural (Napierian) logarithm of the number 2 with a numerical value of approximately 0.69315. 2. (effective) The time required for a radionuclide contained in a biological system, such as a human being or an animal, to reduce its activity by half as a combined result of radioactive decay and biological elimination. 3. (biological) The time required for the elimination of one-half of the total amount of a toxic substance contained in a biological system by natural processes. 4. (environmental) The time required for a pollutant to lose half its effect on the environment. For example, the half-life of the insecticide DDT in the environment is about 15 years.

Hall, Asaph (1829–1907) American *Astronomer* In 1877, while he was a staff member at the U.S. Naval Observatory in Washington, D.C., Asaph Hall discovered the two small moons of Mars. He called the larger, innermost moon "Phobos" (fear), and the smaller, outermost moon "Deimos" (terrified flight) after the attendants of Mars (Ares), the god of war in Roman (Greek) mythology.

Asaph Hall was born in Goshen, Connecticut, on October 15, 1829. His father died when he was just 13 years of age, so he had to leave school in order to support his family as a carpenter's apprentice. In the early 1850s Hall and his wife, Angelina, were working as schoolteachers in Ohio when he decided to become a professional astronomer. Since the difficult days of his childhood, he had always entertained a deep interest in astronomy and taught himself the subject as best he could. So in 1857, with only a little formal training (approximately one year) at the University of Michigan and a good deal of encouragement from his wife, he approached WILLIAM CRANCH BOND, the director of the Harvard College Observatory, for a position as an assistant astronomer. Recalling his own struggle to become an astronomer, William Bond admired Hall's spunk. He hired Hall as an assistant researcher to support his son, GEORGE PHILLIPS BOND, who also worked at the observatory. Hall's starting salary was just three dollars a week, but that did not deter him from pursuing his dream. His experience at the Harvard College Observatory polished Hall's skills as a professional astronomer. By 1863 he joined the U.S. Naval Observatory in Washington, D.C., and eventually took charge of its new 66-centimeter (26-inch) refractor telescope. During this period the observatory and its instruments were under the overall direction of SIMON NEWCOMB, who was actually more interested in the mathematical aspects of astronomy than in performing personal observations of the heavens.

In December 1876 Hall was observing the moons of Saturn when he noticed a white spot on the generally featureless, butterscotch-colored globe of the ringed planet. He carefully observed this spot and used it to estimate the rate of Saturn's rotation to considerable accuracy. Hall worked out the period of rotation to be about 10.23 hours, a value comparable to the value reported by Sir WILLIAM HERSCHEL in 1794. The contemporary value for the period of rotation of Saturn's equatorial region is 10.56 hours.

In the summer of 1877 Mars was in opposition and approached within 56 million kilometers of Earth. Hall was curious whether the Red Planet had any natural satellites. Other astronomers had previously searched for Martian moons and were unsuccessful. However, rather curiously, the 18th-century writer Jonathan Swift mentioned two Martian satellites in his famous 1726 novel *Gulliver's Travels*. On the night of August 10, 1877, Hall made his first attempt with the powerful Naval Observatory telescope, but viewing conditions along the Potomac River were horrible, and Mars produced a glare in the telescope that made searching for any satellites very difficult. He returned home very frustrated, but his wife provided encouragement and told him to keep trying.

On the following evening Hall detected a suspicious object near the planet just before fog rolled in from the Potomac River and enveloped the Naval Observatory. It was not until the evening of August 16 that Hall again observed a faint starlike object near Mars, which proved to be its tiny outermost moon, Deimos. He showed the interesting object to an assistant and instructed him to keep quiet about the discovery. Hall wanted to confirm his observations before other astronomers could take credit for his finding. On the evening of August 17, as Hall waited for Deimos to reappear in the telescope, he suddenly observed the larger, inner moon of Mars, which he called Phobos. By August 18 Hall's excitement and observing log notes had informally leaked news of his discovery to the other astronomers at the Naval Observatory.

That evening the observatory was packed with eager individuals, including Simon Newcomb, each trying to snatch a piece of the "astronomical glory" resulting from Hall's great discovery. For example, Newcomb improperly implied in a subsequent newspaper story that his calculations helped Hall realize that the two new objects were satellites orbiting Mars. This attempt at "sharing" Hall's discovery caused a great deal of personal friction between the two astronomers that lasted for decades. Imitating Hall's work, astronomers at other observatories soon claimed they had discovered a third and fourth moon of Mars—hasty "discoveries" that proved not only incorrect but based on proclaimed orbital motion data in clear violation of JOHANNES KEPLER's laws of planetary motion.

When the dust settled, astronomers around the world confirmed that Mars possessed just two tiny moons (Phobos and Deimos), and Hall received full and sole credit for the discovery. For example, the Royal Astronomical Society in London awarded him its prestigious Gold Medal in 1877.

Hall remained at the Naval Observatory until his retirement in 1891. He focused his attention on planetary astronomy, determining stellar parallax, and investigating the orbital mechanics of binary star systems, such as the visual binary 61 Cygnus. Following retirement from the Naval Observatory, Hall became a professor of astronomy at Harvard from 1896 to 1901. He died in Annapolis, Maryland, on November 22, 1907.

See also MARS; U.S. NAVAL OBSERVATORY.

Halley, Sir Edmond (Edmund) (1656–1742) British *Mathematician, Astronomer* The British mathematician and astronomer Sir Edmond Halley encouraged Sir Isaac Newton to write the *Principia*, one of the most important scientific books ever written. Halley personally financed Newton's famous publication. In his own book *A Synopsis of the Astronomy of Comets*, which appeared in 1705, Halley suggested that the comet he observed in 1682 was actually the same comet that had appeared in 1531. He then used Newton's laws of motion (as contained in the *Principia*) to boldly predict that this comet would return again in about 1758. It did (after his death), and Comet Halley carries his name because of the accuracy of his prediction and discovery. He succeeded JOHN FLAMSTEED as the second Astronomer Royal of England (1720–42).

Halley's comet *See* COMET HALLEY.

halo orbit A circular or elliptical orbit in which a spacecraft remains in the vicinity of a Lagrangian libration point.

See also LAGRANGIAN LIBRATION POINT.

hangfire A faulty condition in the ignition system of a rocket engine.

hard landing A relatively high-velocity impact landing of a "lander" spacecraft or probe on the surface of a planet or moon that destroys all equipment, except perhaps a very rugged instrument package or payload container. This hard landing could be intentional, as occurred during the early U.S. Ranger missions to the lunar surface, or unintentional, as when a retrorocket system fails and the "lander" spacecraft strikes the planetary surface at an unplanned high speed.

hardware 1. A generic term dealing with physical items, as distinguished from ideas or "paper designs"; includes aerospace equipment, tools, instruments, devices, components, parts, assemblies, and subassemblies. Satellites and rockets often are referred to as space hardware. The term also can be used in aerospace engineering to describe the stage of development, as in the passage of a device or component from the design stage into the hardware stage as a finished object. 2. In computer science and data automation, the physical equipment or devices that make up a computer or one of its peripheral components (e.g., a printer), as opposed to the software, which operates and controls the computer system.

HARM missile The AGM-88 High-Speed Antiradiation Missile (HARM) is an air-to-surface tactical missile designed to seek out and destroy enemy radar-equipped air defense systems. An antiradiation missile passively homes in on enemy electronic signal emitters, especially those associated with radar sites used to direct antiaircraft guns and surface-to-air missiles. The AGM-88 can detect, attack, and destroy a target with minimum aircrew input. Used extensively by the U.S. Air Force and Navy.

Harpoon missile The AGM-84D is an all-weather, over-the-horizon, antiship missile. The Harpoon's active radar guidance, warhead design, and low-level, sea-skimming cruise

Emergency crew personnel examine a crumpled NASA X-15 rocket plane after a hard landing on the dry lakebed at Edwards Air Force Base in California in 1963. The pilot, Scott Crossfield, survived the rough landing, but the X-15 rocket plane suffered a broken airframe. ***(Courtesy of NASA/Dryden Flight Research Center)***

trajectory assure high survivability and effectiveness. This missile is capable of being launched from surface ships, submarines, or (without its booster component) from aircraft. Originally developed for the U.S. Navy to serve as its basic antiship missile for fleetwide use, the AGM-84D also has been adapted for use on the U.S. Air Force's B-52G bombers.

Harvard classification system A method of classifying stars by their spectral characteristics (such as O, B, A, F, G, K, and M) in order of decreasing surface temperature. The system was developed by WILLIAMINA PATON FLEMING and others while working for EDWARD CHARLES PICKERING at the Harvard College Observatory in the 1880s. In the 1940s the Morgan-Keenan (M-K) classification system replaced the Harvard classification system. The M-K system retained the basic Harvard classification system's nomenclature and sequence of stellar spectral types (namely, O, B, A, F, G, K, and M) but included a more precise definition (based on observations) of each spectra type.

See also SPECTRAL CLASSIFICATION; STAR.

Harvard College Observatory (HCO) Founded in 1839, the Harvard College Observatory (HCO) is the astronomical observatory of Harvard University, which is located in Cambridge, Massachusetts. Starting in the late 19th century, EDWARD CHARLES PICKERING supervised the production of the *Henry Draper Catalogue* and related star classification efforts by ANNIE JUMP CANNON, WILLIAMINA PATON FLEMING, HENRIETTA LEAVITT, and other innovative American astronomers who worked at HCO. Their efforts led to the classification of 225,000 stars according to their spectra as defined by the Harvard classification system.

See also SMITHSONIAN ASTROPHYSICAL OBSERVATORY (SAO).

Harvard-Smithsonian Center for Astrophysics (CfA) *See* SMITHSONIAN ASTROPHYSICAL OBSERVATORY (SAO).

hatch A door in the pressure hull of a spacecraft, aerospace vehicle, or space station. The hatch is sealed tightly to prevent the cabin or pressurized module atmosphere from escaping to the vacuum of space outside the pressure hull.

Hawk A surface-to-air guided missile developed by the U.S. Army in the late 1950s and continually upgraded since its first field deployment in 1960. Designed to provide air defense protection, the Hawk system is employed as a platoon consisting of an acquisition radar, a tracking radar, an identification friend-or-foe (IFF) system, and up to six launchers with three missiles each. The Hawk missile is highly reliable, accurate, and lethal. During the 1991 Persian Gulf War the Hawk downed several enemy aircraft.

Hawking, Sir Stephen William (1942–) British *Astrophysicist, Cosmologist* Despite a disabling neurological disorder, the brilliant British scientist Stephen W. Hawking has made major theoretical contributions to the study of general relativity and quantum mechanics, especially by providing an insight into the unusual physics of black holes and mathematical support for the big bang theory. In 1988 he published *A Brief History of Time,* a popular book that introduced millions of readers to modern cosmology. His concept of Hawking radiation suggests an intimate relationship between gravity, quantum mechanics, and thermodynamics. Hawking travels around the world giving public lectures and attending scientific conferences and symposia. He has been awarded 12 honorary degrees and is a member of the Royal Society and the U.S. National Academy of Sciences.

See also BLACK HOLE.

Hawking radiation A theory (currently under discussion and revision) proposed in 1974 by the British astrophysicist STEPHEN W. HAWKING in which he suggests that due to a combination of the properties of quantum mechanics and gravity, under certain conditions black holes can seem to emit radiation.

***Hayabusa* (spacecraft)** On May 9, 2003, the Japanese Aerospace Exploration Agency (JAXA) successfully launched its space engineering spacecraft called *Hayabusa* on an asteroid sample return mission. An M-5 solid fuel rocket booster lifted the spacecraft into space from the Kagoshima launch center in Japan. Previously called MUSES-C, the spacecraft has been flying smoothly in a heliocentric orbit using its ion engine.

Hayabusa successfully completed a gravity assist maneuver as it flew past Earth on May 19, 2004, and it is now on a new elliptical trajectory toward the asteroid 25143 Itokawa. This asteroid is a 0.3-km by 0.7-km near-Earth asteroid (NEA) named in honor of the late Dr. Hideo Itokawa, the "father" of Japan's space development program. *Hayabusa* (which means "falcon" in Japanese) is now traveling through interplanetary space using an ion engine. In October 2005 the robot spacecraft will arrive at the asteroid Itokawa and begin to travel with it. After observing the asteroid from a distance of about 20 kilometers, the spacecraft will then move closer to the asteroid and perform a series of soft landings in order to collect from 1 to 10 grams of surface samples at three landing zones. *Hayabusa* will then bring the samples back to Earth for analysis. As currently planned, the spacecraft's sample return capsule will descend through Earth's atmosphere in October 2007 and be recovered by scientists on Earth, who will then analyze the collected materials.

See also ASTEROID; JAPANESE AEROSPACE EXPLORATION AGENCY.

head (headrise) The increase in fluid pressure supplied by a pump; the difference between pressure at the pump inlet and pressure at the pump discharge. Fluid pressure sometimes is expressed as an equivalent height of a fluid column, such as 960 millimeters of mercury, or else as pascals (newtons per meter-squared [N/m^2]).

head-up display A display of flight, navigation, attack, or other information superimposed on a pilot's forward field of view.

health physics The branch of science that is concerned with the recognition, evaluation, and control of health hazards from ionizing radiation.

See also SPACE RADIATION ENVIRONMENT.

heat Energy transferred by a thermal process. Heat (or thermal energy) can be measured in terms of the mechanical units of energy, such as the joule (J), or in terms of the amount of energy required to produce a definite thermal change in some substance, as, for example, the energy required to raise the temperature of a unit mass of water at some initial temperature (e.g., calorie), 1 joule = 0.239 calorie.

See also THERMODYNAMICS.

heat balance 1. The equilibrium that exists on average between the radiation received by a planet and its atmosphere from the Sun and that emitted by the planet and its atmosphere. 2. The equilibrium that is known to exist when all sources of heat (thermal energy) gain and loss for a given region or object are accounted for. In general, this balance includes conductive, convective, evaporative, and other terms as well as a radiation term.

heat capacity (usual symbol: C) In general, a measure of the increase in energy content of an object or system per degree of temperature rise. This thermodynamic property describes the ability of a substance to store the energy that has been delivered to it as heat (i.e., thermal energy flow). The specific heat capacity (C) is related to the total heat capacity (C) as follows:

$$c = C/\rho$$

where c has the units (joules/kg-K), C has the units (joules/m^3- K), and ρ is the density of the substance (kg/m^3). The specific heat capacity (c), sometimes shortened to specific heat, has very special significance and meaning in classical thermodynamics. For example, the specific heat at constant volume (c_v) is defined as:

$$c_v = (\partial u/\partial T)_v$$

where u is internal energy (joules/kg) and T is temperature (kelvins [K]). The expression $(\delta u/\delta T)_v$ means the partial derivative of internal energy (u) with respect to temperature (T) holding volume (v) constant. Similarly, the specific heat at constant pressure (c_p) is defined as:

$$c_p = (\partial h/\partial T)_p$$

where h is enthaply (joules/kg) and T is temperature (K). The terms c_v and c_p are very important thermodynamic derivative functions.

See also THERMODYNAMICS.

Heat Capacity Mapping Mission (HCMM) NASA's Heat Capacity Mapping Mission, or HCMM, was launched on April 26, 1978, and operated successfully between April 1978 and September 1980 in a near-polar orbit at 620 kilometers of altitude. This Earth-observing spacecraft was the first NASA research effort directed mainly toward observations of the thermal state of Earth's land surface by a satellite. The HCMM sensor measured reflected solar radiation and thermal emission from the surface with a spatial resolution of 600 meters. The satellite was placed in an orbit that permitted it to survey thermal conditions during midday and at night. HCMM data have been used to produce temperature difference and apparent thermal inertia images for selected areas within much of North America, Europe, North Africa, and Australia. These data can be used by scientists for rock-type discrimination, soil-moisture detection, assessment of vegetation states, thermal current monitoring in water bodies, urban heat-island assessments, and other environmental studies of Earth.

See also EARTH SYSTEM SCIENCE; GLOBAL CHANGE.

heat death of the universe A possible ultimate fate of the universe suggested in the 19th century by the German theoretical physicist RUDOLF JULIUS CLAUSIUS. As he evaluated the consequences of the second law of thermodynamics on a grand scale, he concluded that the universe would end (die) in a condition of maximum entropy in which there was no energy available for useful work.

See also COSMOLOGY; THERMODYNAMICS; UNIVERSE.

heat engine A thermodynamic system that receives energy in the form of heat and that, in the performance of an energy transformation on a working fluid, does work. Heat engines function in cycles. An ideal heat engine works in accordance with the Carnot cycle, while practical heat engines use thermodynamic cycles such as Brayton, Rankine, and Stirling. The steam engine, which helped create the Industrial Revolution, is a heat engine. Gas turbines and automobile engines are also heat engines.

See also CARNOT CYCLE; THERMODYNAMICS.

heater blanket An electrical heater, usually in strip or sheet form, that is wrapped around all or a portion of a cryogenic component (e.g., a valve or actuator) to prevent the temperature within the component from falling below a stated operating minimum. In aerospace engineering applications, the heater blanket represents an active form of spacecraft thermal control.

See also THERMAL CONTROL.

heat pipe A very-high-efficiency heat (thermal energy) transport device. The basic heat pipe is a closed system that contains a working fluid that transports thermal energy by means of phase change (i.e., through the use of the latent heat of vaporization). An external thermal energy (heat) source is applied to the outside of the evaporator section of the heat pipe. This heat addition causes the working fluid contained in a wick to undergo a phase change from liquid to vapor. The vapor leaves the wick, flows through the central region of the heat pipe, and eventually arrives at the condenser region. At the condenser section of the heat pipe, the vapor also undergoes phase change—this time, however, going from a gas to a liquid. The condensation process releases heat, which then is removed through the external portion of the evaporator section. The wick (which lines the interior surface of the entire heat pipe) uses capillary action to return the liquid to the evaporator section. Once there, the liquid can again evaporate, continuing the overall heat transport process. In aerospace engineering applications, the heat pipe represents a passive form of spacecraft thermal control.

See also THERMAL CONTROL.

heat shield 1. In general, any device that protects something from unwanted thermal energy (heat). 2. In aerospace engineering, the special structure used to protect a reentry body or aerospace vehicle from aerodynamic heating on

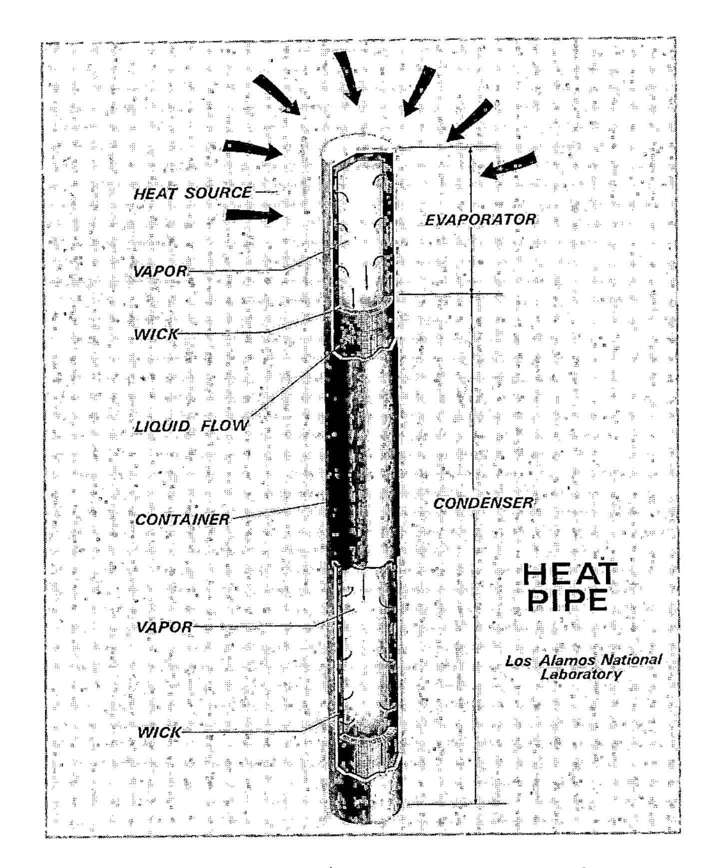

The components and operation of a basic heat pipe, a very-high-efficiency heat transfer device often used in spacecraft thermal control systems *(Courtesy of the U.S. Department of Energy and Los Alamos National Laboratory)*

flight through a planet's atmosphere, or the protective structure used to block unwanted heat from entering into a spacecraft, its components, or its sensing instruments.

heat sink In thermodynamics, a material, region, or environment in which thermal energy (heat) is stored or to which heat is rejected from a heat engine. The term also is used to describe a material capable of absorbing large quantities of heat, such as the special thermal protection materials used on reentry bodies and aerospace vehicles.

heat-sink chamber In rocketry, a combustion chamber in which the heat capacity of the chamber wall limits wall temperature. This approach is effective only for short-duration firings.

heat soak The increase in the temperature of rocket engine components after firing has ceased. This increase results from heat transfer through contiguous parts of the engine when no active cooling is present.

heat transfer The transfer or exchange of thermal energy (heat) by conduction, convection, or radiation within an object and between an object and its surroundings. Thermal energy also is transferred when a working fluid experiences phase change, such as evaporation or condensation.

heavy hydrogen An expression for deuterium (D), the nonradioactive isotope of hydrogen with one proton and one neutron in its nucleus. The American chemist and Nobel laureate HAROLD CLAYTON UREY discovered deuterium in the early 1930s.

heavy-lift launch vehicle (HLLV) A conceptual large-capacity, space lift vehicle capable of carrying tons of cargo into low-Earth orbit (LEO) at substantially less cost than today's expendable launch vehicles (ELVs).

heavy metal star *See* BARIUM STAR.

hectare A unit of area equal to 10,000 square meters; 1 hectare = 2.4711 acres.

hecto- (symbol: h) A prefix in the SI unit system meaning multiplied by 10^2.

Heisenberg uncertainty principle The important physical principle that states that it is impossible to simultaneously know the momentum and precise position of a subatomic particle. This quantum mechanics principle also implies that a physicist cannot precisely know both energy and time on a microscopic scale. In 1927 the German physicist Werner Heisenberg (1901–76) suggested that the uncertainties of position (Δx) and momentum (Δp) of a subatomic particle are related by the following inequality:

$$\Delta x \, \Delta p \geq h/4\pi,$$

where h is Planck's constant.

See also ELEMENTARY PARTICLE(S); QUANTUM MECHANICS.

Helene A small (approximately 40-km-diameter), irregularly shaped moon of Saturn that co-orbits the planet at a distance of 377,400 km along with the larger moon Dione. Helene (like Dione) has an orbital period of 2.74 days. The astronomers P. Laques and J. Lecacheux, using ground-based observations, discovered Helene in 1980. Astronomers sometimes refer to this tiny moon as *Dione-B,* because it precedes the larger moon by 60° and is located at the leading Lagrangian point in their shared orbit around Saturn. Also called Saturn XII.

See also SATURN.

heliocentric Relative to the Sun as a center, as in heliocentric orbit or heliocentric space.

heliocentric coordinate system *See* CELESTIAL COORDINATES.

heliocentric cosmology A model of the solar system with the Sun at the center. Sometimes called Copernican cosmology.

See also COPERNICUS, NICHOLAS; COSMOLOGY.

heliometer A useful but now obsolete 19th-century optical instrument that was used by astronomers to measure the angular separation between stars that appeared close together in the night sky or to measure the Sun's diameter. The device is basically a modified refracting telescope, the objective lens of which has been divided in half. This arrangement allowed astronomers to superimpose the opposite limbs of the Sun or two star images and then arrive at an angular measurement from the movement and separation of the divided lenses and their angular position.

heliopause The boundary in deep space thought to be roughly circular or perhaps teardrop-shaped marking the edge of the Sun's influence. It occurs perhaps 100 astronomical units (AU) from the Sun.

See also SUN.

heliosphere The space within the boundary of the heliopause, containing the Sun and the solar system.

heliostat A mirrorlike device arranged to follow the Sun as it moves through the sky and to reflect the Sun's rays on a stationary collector or receiver.

helium (symbol: He) A noble gas, the second-most abundant element in the universe. Natural helium is mostly the isotope helium-4, which contains two protons and two neutrons in the nucleus. Helium-3 is a rare isotope of helium containing two protons and one neutron in the nucleus. Helium was initially discovered in the Sun's spectrum in 1868 by the British physicist SIR JOSEPH LOCKYER almost four decades before it was found on Earth.

Hellas (plural: Hellades) A quantity of information first proposed by the physicist Philip Morrison. It corresponds to 10^{10} bits of information—more or less the amount of information we know about ancient Greece. *Hellas* [ΕΛΛΑΣ] is the Greek name for Greece. In considering interstellar communication with other intelligent civilizations, we would hope to send and receive something on the order of 100 Hellades of information or more at each contact.

See also INTERSTELLAR COMMUNICATION AND CONTACT.

Hellfire (AGM-114) The Hellfire is a family of air-to-ground missiles developed by the U.S. Army to provide heavy antiarmor capability for attack helicopters. The first three generations of Hellfire missiles used a laser seeker, while the fourth generation missile, called *Longbow Hellfire,* employs a radio frequency (RF) or radar seeker. As part of the war on terrorism, Hellfire missiles have been carried by and successfully fired from the Predator unmanned aerial vehicle (UAV). The U.S. Navy and Marine Corps use versions of the Hellfire missile (namely, the AGM-114B/K/M models) as air-to-air weapons against hostile helicopters or slow-moving fixed-wing aircraft.

Henderson, Thomas (1798–1844) Scottish *Astronomer* In the 1830s the Scottish astronomer Thomas Henderson made careful parallax measurements of Alpha Centauri from the Cape of Good Hope Observatory, South Africa. As a result of these observations, he reported in 1839 that this star (actually a triple star system) was a distance of less than four light-years away, making it the closest known star, excluding the Sun. Henderson's estimate was quite reasonable, since the precise value of the speed of light and, therefore, the distance equivalence of a light-year was not precisely known within the scientific community until ALBERT MICHELSON's efforts later in the 19th century.

See also ALPHA CENTAURI; SPEED OF LIGHT.

henry (symbol: H) The SI unit of inductance (L). Inductance relates to the production of an electromotive force (E) in a conductor when there is a change in the magnetic flux (Φ) in that conductor. The induced electromotive force (E) is proportional to the time rate of change of the current (dI/dt), namely:

$$E = -L\ (dI/dt)$$

where the inductance (L) serves as a proportionality constant and depends on the geometric design of the circuit.

One henry (H) is defined as the inductance occurring in a closed electric circuit in which an electromotive force (or emf) of 1 volt is produced when the current (I) in the circuit is varied uniformly at the rate of 1 ampere per second. This unit has been named in honor of the American physicist Joseph Henry (1797–1878).

Henry Draper (HD) Catalog *See* DRAPER, HENRY.

Heraclides of Pontus (ca. 388–315 B.C.E.) Greek *Astronomer, Philosopher* The early Greek astronomer and philosopher Heraclides of Pontus was the first person to suggest that our planet had a daily rotation on its axis from west to east. While believing Earth to be the center of the universe, he boldly speculated that Mercury and Venus might travel in orbits around the Sun, since neither planet was ever observed far from it. Unfortunately, his revolutionary idea about planets traveling around the Sun clashed with geocentric Greek cosmology, so it remained essentially unnoticed until rediscovered by NICHOLAS COPERNICUS in the 16th century.

Hero of Alexandria (ca. 20–ca. 80 C.E.) Greek *Mathematician, Engineer* The Greek engineer Hero of Alexandria invented many clever mechanical devices, including the *aeolipile,* a spinning, steam-powered spherical apparatus that demonstrated the action-reaction principle by which all rocket engines work.

Herschel, Caroline (1750–1848) German-born, British *Astronomer* Caroline Herschel was the first notable woman astronomer. She assisted her famous brother, the British astronomer SIR (FREDERICK) WILLIAM HERSCHEL, and was a skilled observer who personally discovered eight comets between 1786 and 1797. In 1783 she observed the Andromeda Galaxy, which she referred to as a nebula, consistent with practices of 18th century astronomers.

Herschel, Sir (Frederick) William (1738–1822) German-born, British *Astronomer* The British astronomer Sir William Herschel discovered Uranus in 1781, the first new planet since the start of ancient astronomy and the first found through the use of the telescope. A skilled observer, he located more than 2,500 nebulas and star clusters as well as more than 800 binary star systems. His study of the distribution of

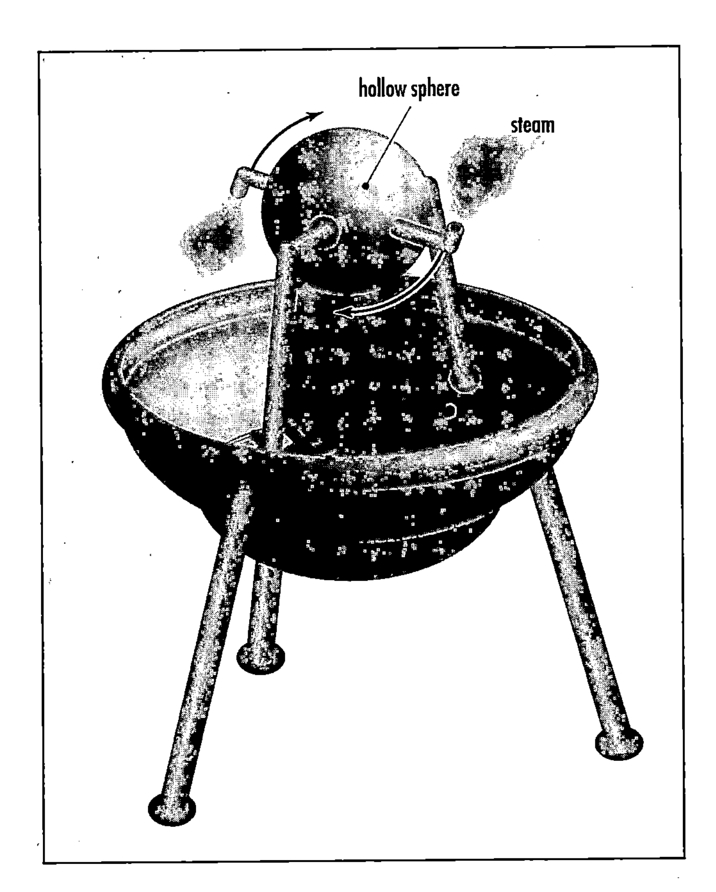

In about 60 C.E. the Greek engineer and mathematician Hero of Alexandria created the *aeoliphile* (shown here), a toylike device that demonstrates the action-reaction principle, which is the basis of operation of all rocket engines.

the observed stars established a basic understanding of the form of the Milky Way galaxy. In 1800, while investigating the energy content of sunlight (incoming solar radiation) with the help of a prism and a thermometer, he discovered the existence of (thermal) infrared radiation, which lies just beyond red light in a longer wavelength portion of the electromagnetic spectrum.

Herschel, Sir John (Frederick William) (1792–1871) British *Astronomer* As the son of SIR (FREDERICK) WILLIAM HERSCHEL, the British astronomer Sir John Herschel continued his famous father's investigations of binary star systems and nebulas. Between about 1833 and 1838 he performed an extensive survey of the night sky in the Southern Hemisphere from an observatory at the Cape of Good Hope, South Africa.

Herschel Space Observatory (HSO) The European Space Agency's *Herschel Space Observatory,* also called the *Far Infrared and Submillimetre Telescope* (FIRST), is due for launch in 2007 from Kourou, French Guiana, by an Ariane 5 expendable launch vehicle (ELV). The main science goal of this observatory will be to study how the first galaxies and stars formed and evolved. HSO will measure the chemical and dynamic evolution of galaxies and stars. It will be capable of probing deeply into dust-enshrouded star-forming regions to observe protostars, planetary systems in the making, and the remnants of planet formation. The spacecraft is named in honor of the great German-born British astronomer SIR WILLIAM HERSCHEL, who discovered the planet Uranus in 1781 and the existence of infrared radiation in 1800.

The giant infrared observatory will carry three major scientific instruments: the Heterodyne Instrument for FIRST (HIFI), a high-resolution spectrometer; the Photoconductor Array Camera and Spectrometer (PACS); and the Spectral and Photometric Imaging Receiver (SPIRE), which contains cameras and imaging spectrometers. These instruments operate at cryogenic temperatures and cover the 60- to 670-micrometer (μm) wavelength interval of the electromagnetic spectrum. The large space-based telescope's primary mirror is 3.5 m in diameter. The National Aeronautics and Space Administration (NASA) is participating in the mission by providing the telescope and contributing to the HIFI and SPIRE instruments.

As now planned, the *Herschel Space Observatory* will be launched on an Ariane 5 rocket along with the *Planck* spacecraft. However, the two satellites will separate shortly after launch and proceed to different operational orbits. HSO will go into an operational orbit at the L2 Lagrangian point, an optimal vantage point for far-infrared (IR) and submillimeter astronomical observation. This location in heliocentric space is well isolated from Earth's strong far-IR emission and provides good sky visibility because the Sun and Earth both lie in approximately the same direction.

See also INFRARED ASTRONOMY; *PLANCK* SPACECRAFT.

hertz (symbol: Hz) The SI unit of frequency. One hertz is equal to 1 cycle per second. Named in honor of the German physicist HEINRICH RUDOLF HERTZ.

Hertz, Heinrich Rudolf (1857–1894) German *Physicist* In 1888 Heinrich Hertz produced and detected radio waves for the first time. He also demonstrated that this form of electromagnetic radiation, like light, propagates at the speed of light. His discoveries form the basis of both the global telecommunications industry (including communications satellites) and radio astronomy. The hertz (Hz) is the SI unit of frequency named in his honor.

Hertz was born on February 22, 1857, in Hamburg, Germany, into a prosperous and cultured family. Following a year of military service from 1876 to 1877, he entered the University of Munich to study engineering. However, after just one year he found engineering not to his liking and began to pursue a life of scientific investigation as a physicist in academia. Consequently, in 1878 he transferred to the University of Berlin and started studying physics with the famous German scientist Herman von Helmholtz (1821–94) as his mentor. Hertz graduated magna cum laude with a Ph.D. degree in physics in 1880. Following graduation he continued working at the University of Berlin as an assistant to Helmholtz for the next three years.

He left Berlin in 1883 to work as a physicist at the University of Kiel. There, following suggestions from his mentor, Hertz began investigating the validity of the electromagnetic theory recently proposed by the Scottish physicist JAMES CLERK MAXWELL. As a professor of physics at the Karlsruhe Polytechnic from 1885 to 1889, Hertz finally gained access to the equipment he needed to perform the famous experiments that demonstrated the existence of electromagnetic waves and verified Maxwell's equations. During this period Hertz not only produced electromagnetic (radio frequency) waves in the laboratory but also measured their wavelength and velocity. Of great importance to modern physics and the fields of telecommunications and radio astronomy, Hertz showed that his newly identified radio waves propagated at the speed of light, as predicted by Maxwell's theory of electromagnetism. He also discovered that radio waves were simply another form of electromagnetic radiation similar to visible light and infrared radiation save for their longer wavelengths and lower frequencies. Hertz's experiments verified Maxwell's electromagnetic theory and set the stage for others, such as Guglielmo Marconi (1874–1937), to use the newly discovered "radio waves" to transform the world of communications in the 20th century.

In 1887, while experimenting with ultraviolet radiation, Hertz observed that incident ultraviolet radiation was releasing electrons from the surface of a metal. Unfortunately, he did not recognize the significance of this phenomenon, nor did he pursue further investigation of the photoelectric effect. In 1905 ALBERT EINSTEIN wrote a famous paper describing this effect, linking it to MAX PLANCK's idea of photons as quantum packets of electromagnetic energy. Einstein earned the 1921 Nobel Prize in physics for his work on the photoelectric effect.

Hertz performed his most famous experiment in 1888 with an electric circuit in which he oscillated the flow of current between two metal balls separated by an air gap. He observed that each time the electric potential reached a peak in one direction or the other, a spark would jump across the gap. Hertz applied Maxwell's electromagnetic theory to the situation and determined that the oscillating spark should generate a very long electromagnetic wave that traveled at the speed of light. He also used a simple loop of wire with a small air gap at one end to detect the presence of electromagnetic waves produced by his oscillating spark circuit. With this pioneering experiment, Hertz produced and detected *Hertzian waves,* later called *radiotelegraphy waves* by Marconi and then simply radio waves. By establishing that Hertzian waves were electromagnetic in nature, the young German physicist extended human knowledge about the electromagnetic spectrum, validated Maxwell's electromagnetic theory, and identified the fundamental principles for wireless communications.

In 1889 Hertz accepted a professorship at the University of Bonn. There he used cathode ray tubes to investigate the physics of electric discharges in rarified gases, again just missing another important discovery, the discovery of X-rays, which was accomplished by the German physicist WILHELM CONRAD ROENTGEN at Würzburg in 1895.

Hertz was an excellent physicist whose pioneering research with electromagnetic waves gave physics a solid foundation upon which others could build. His major publications included *Electric Waves* (1890) and *Principles of Mechanics* (1894). He suffered from lingering ill health due to blood poisoning and died as a young man (in his late 30s) on January 1, 1894, in Bonn, Germany. The international scientific community named the basic unit of frequency the hertz (symbol Hz) in his honor.

Hertzsprung, Ejnar (1873–1967) Danish *Astronomer* At the start of the 20th century the Danish astronomer Ejnar Hertzsprung made one of the most important contributions to modern astronomy when he showed in 1905 how the luminosity of a star is related to its color (or spectrum). His work, independent of the American astronomer HENRY NORRIS RUSSELL, contributed to one of the great observational syntheses in astrophysics, the famous Hertzsprung-Russell (HR) diagram, an essential tool for anyone seeking to understand stellar evolution.

Hertzsprung was born on October 8, 1873, in Frederiksberg, near Copenhagen. His father had graduated with a master of science degree in astronomy but for financial reasons worked as a civil servant in the department of finance within the Danish government. Consequently, Severin Hertzsprung encouraged his son to enjoy astronomy as an amateur. He also admonished him not to pursue astronomy as a profession because the field presented very few opportunities for financial security. Hertzsprung responded to his father's well-intended advice and studied chemical engineering at Copenhagen Polytechnic Institute. Upon graduating in 1898 he accepted employment as a chemical engineer in St. Petersburg, Russia, and remained there for two years. He then journeyed to the University of Leipzig, where he studied photochemistry for several months under the German physical chemist Wilhelm Ostwald (1853–1932). The thorough understanding of photochemical processes he acquired from Ostwald allowed Hertzsprung to make his most important contribution to modern astronomy.

Hertzsprung became one of the great observational astronomers of the 20th century despite the fact he never received any formal academic training in astronomy. When he returned to Denmark in 1902 the chemical engineer revived his lifelong interest in astronomy by becoming a private astronomer at the Urania and University Observatories in Copenhagen. During this period he taught himself a great deal about astronomy and used the telescopes available at the two small observatories to make detailed photographic observations of the light from stars.

Hertzsprung's earliest and perhaps most important contribution to astronomy started with the publication of two papers, both entitled "Zur Strahlung der Sterne" ("On the radiation of the stars") in a relatively obscure German scientific photography journal. These papers appeared in 1905 and 1907, and their existence was unknown to the American astronomer Henry Norris Russell, who would soon publish similar observations in 1913. Publication of these two seminal papers marks the beginning of the independent development of Hertzsprung-Russell (HR) diagrams, the famous tool in modern astronomy and astrophysics that graphically portrays the evolutionary processes of visible stars. Its creation is equally credited to both astronomers.

Succinctly stated, Hertzsprung's two papers presented his insightful interpretation of the stellar photography data that revealed the existence of a relationship between the color of a star and its respective brightness. He stated that his photographic data also suggested the existence of both giant and dwarf stars. Hertzsprung then developed these and many other new ideas over the course of his long career as an astronomer.

Between 1905 and 1913 Hertzsprung and Russell independently observed and reported that any large sample of stars, when analyzed statistically using a two-dimensional plot of magnitude (or luminosity) versus spectral type (color or temperature), form well defined groups, or bands. Most stars in the sample will lie along an extensive central band, called the main sequence, that extends from the upper-left corner to the lower-right corner of the HR diagram. Giant and super giant stars appear in the upper-right portion of the HR diagram, while the white dwarf stars populate the lower-left region. Modern astronomers use several convenient forms of the HR diagram to support their visual observations and to describe where a particular type fits in the overall process of stellar evolution. For example, our parent star, the Sun, is a representative yellow dwarf (main sequence) star. Previously astronomers used the term *dwarf star* to describe any star lying on the main sequence of the HR diagram. Today the term *main-sequence star* is preferred to avoid possible confusion with white dwarf stars, the extremely dense final evolutionary phase of most low-mass stars.

While Hertzsprung was still in Copenhagen, he also started making detailed photographic investigations of star clusters. In 1906 he apparently constructed his first color versus magnitude precursor diagrams based on photographic observations of the Pleiades and subsequently published such data along with companion data for the Hyades in 1911. This activity represents an important milestone in the evolution of the Hertzsprung-Russell diagram. The Pleidaes is a prominent open star cluster in the constellation Taurus about 410 light-years distant. The Hyades is an open cluster of about 200 stars in the constellation Taurus some 150 light-years away. Astronomers use both clusters as astronomical yardsticks, comparing the brightness of their stars with the brightness of stars in other clusters. Hertzsprung would actually spend the next two decades making detailed observations of the Pleidaes, a favorite astronomical object used frequently by other astronomers to compare the performance of their telescopes in the Carte du Ciel astrophotography program that began in 1887.

In 1909 the German astronomer KARL SCHWARZSCHILD invited Hertzsprung to visit him in Göttingen, Germany. Schwarzschild quickly recognized that Hertzsprung, despite his lack of formal training in astronomy, possessed exceptional talents in the field and gave Hertzsprung a staff position at the Göttingen Observatory. Later that year Schwarzschild became the director of the Astrophysical Observatory in Potsdam, Germany, and he offered Hertzsprung an appointment as a senior astronomer.

Hertzsprung proved to be a patient, exacting observer who was always willing to do much of the tedious work himself. While at the Potsdam Observatory he investigated the Cepheid stars in the Small Magellanic Cloud. He used the periodicity relationship for Cepheid variables announced in 1912 by HENRIETTA LEAVITT to help him calculate intergalactic distances. While the currently estimated distance (195,000 light-years away) to the Small Magellanic Cloud is about six times larger than the value of 32,600 light-years Hertzsprung estimated in 1913, his work still had great importance. It introduced innovative methods for measuring incredibly large distances and presented an astronomical distance that was significantly larger than any previously known distance in the universe. This "enlarged view" of the universe encouraged other astronomers, such as HEBER CURTIS and HARLOW SHAPLEY, to vigorously debate its true extent.

Hertzsprung left the Potsdam Observatory in 1919 and accepted an appointment as associate director of the Leiden Observatory in the Netherlands. He became the director of the observatory in 1935 and served in that position until he retired in 1944. Upon retirement he returned home to Denmark. He remained active in astronomy, primarily by examining numerous photographs of binary stars and extracting precise position data. The international astronomical community recognized his contributions to astronomy through several prestigious awards. He received the Gold Medal of the Royal Astronomical Society in 1929 and the Bruce Gold Medal from the Astrophysical Society of the Pacific in 1937. After a full life dedicated to progress in observational astronomy, Hertzsprung, the chemical engineer turned astronomer, died at the age of 94 on October 21, 1967, in Roskilde, Denmark. The Hertzsprung-Russell diagram permanently honors his important role in understanding the life cycle of stars.

Hertzsprung-Russell (H-R) diagram A useful graphic depiction of the different types of STARS arranged according to their SPECTRAL CLASSIFICATION and LUMINOSITY. Named in honor of the Danish astronomer EJNAR HERTZSPRUNG (1873–1967) and the American astronomer HENRY NORRIS RUSSELL (1877–1957), who developed the diagram independently of each other.

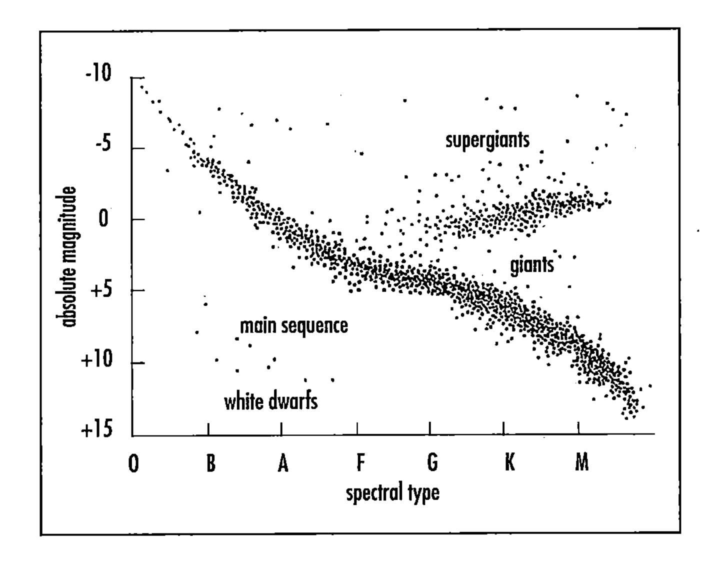

The Hertzsprung-Russell (H-R) diagram

Hess, Victor Francis (1883–1964) Austrian-American *Physicist* As a result of ionizing radiation measurements made during perilous high-altitude balloon flights between 1911 and 1913, Victor Hess discovered the existence of cosmic rays that continually bombard Earth from space. Detection of this very important astrophysical phenomenon earned him a coshare of the 1936 Nobel Prize in physics. Cosmic rays are very energetic nuclear particles that arrive at Earth from all over the Milky Way galaxy. Scientists used his discovery to turn Earth's atmosphere into a giant natural laboratory, a clever research approach that opened the door to many new discoveries in high-energy nuclear physics.

Hess was born in Waldstein Castle, near Peggau in Styria, Austria, on June 24, 1883. His father, Vinzens Hess, was a royal forester in the service of Prince Öttinger-Wallerstein. Hess completed his entire education in Graz, Austria, attending secondary school (the gymnasium) from 1893 to 1901 and then the University of Graz from 1901 to 1905 as an undergraduate. He then continued on at the university as a graduate student in physics and received a Ph.D. degree in 1910. For approximately a decade after earning his doctorate, Hess investigated various aspects of radioactivity while working as a staff member at the Institute of Radium Research of the Viennese Academy of Sciences.

Between 1911 and 1913 Hess performed the pioneering research that gained him a coshare of the 1936 Nobel Prize in physics. He published this important work in the *Proceedings of the Viennese Academy of Sciences*. In the years 1909 and 1910 other scientists had used electroscopes (an early nuclear radiation detection instrument) to compare the level of ionizing radiation in high places, such as the top of the Eiffel Tower in Paris and during balloon ascents into the atmosphere. These early studies gave a vague yet interesting indication that the level of ionizing radiation at higher altitudes was actually greater than the level detected on Earth's surface. These somewhat indefinite readings puzzled scientists. As they operated their radiation detectors farther away from Earth's surface and the sources of natural radioactivity within the planet's crust, the scientists expected the observed radiation levels to simply decrease as a function of altitude. They had not anticipated the existence of energetic nuclear particles arriving from outer space.

Hess attacked the interesting mystery by first making considerable improvements in radiation detection instrumentation and then taking these improved instruments on a number of daring balloon ascents in 1911 and 1912 to heights up to 5.3 kilometers. To investigate solar influence he made balloon flights both during the daytime and at night, when the aerial operations were much more hazardous. The results were similar, as they were in 1912 when he made a set of balloon flight measurements during a total solar eclipse. His careful, systematic measurements revealed that there was a decrease in ionization up to about an altitude of one kilometer, but beyond that height the level of ionizing radiation increased considerably, so that at an altitude of 5 kilometers it had twice the intensity as at sea level. Hess completed analysis of his measurements in 1913 and published his results in the *Proceedings of the Viennese Academy of Sciences*. Carefully examining these measurements, he concluded that there was an extremely penetrating radiation, an *ultra radiation* (as he initially called it), entering Earth's atmosphere from outer space. Hess had discovered *cosmic rays*, although the term was actually coined in 1925 by the American physicist Robert Milikan (1868–1953).

In 1919 Hess received the Lieben Prize from the Viennese Academy of Sciences for his discovery of "ultra radiation." The following year he received an appointment as a professor of experimental physics at the University of Graz. From 1921 to 1923 he took a leave of absence from his university position to work as, first, the director of the research laboratory of the United States Radium Company in New Jersey and then as a consultant to the Bureau of Mines of the U.S. Department of the Interior in Washington, D.C.

Returning to Austria in 1923, Hess resumed his position as a physics professor at the University of Graz. He moved to the University of Innsbruck in 1931 and became the director of its newly established Institute of Radiology. As part of his activities at Innsbruck, he also founded a research station on Mount Hafelekar (at 2.3 kilometers altitude) to observe and study cosmic rays. In 1932 the Carl Zeiss Institute in Jena awarded him the Abbe Memorial Prize and the Abbe Medal. That same year Hess became a corresponding member of the Academy of Sciences in Vienna.

The greatest acknowledgment of the importance of his pioneering research came in 1936, when Hess shared that year's Nobel Prize in physics for his discovery of cosmic rays. An excellent experimental physicist, Hess had gathered important radiation data during daytime as well as more perilous nocturnal balloon flights and then, after careful analysis, realized that his instruments had provided him the initial scientific evidence for the existence of extraterrestrial sources of very energetic atomic particles. Hess's fascinating discovery established a major branch of high-energy astrophysics and helped change our understanding of the universe and some of its most violent, high-energy phenomena.

Cosmic rays are very energetic nuclear particles that carry information to Earth from all over the galaxy. Hess's work also turned Earth's atmosphere into a giant natural laboratory and opened the door for other important discoveries in high-energy physics. When a primary cosmic ray particle hits the nucleus of an atmospheric atom, the result is an enormous number of interesting secondary particles. This cosmic ray "shower" of secondary particles gives nuclear scientists a detailed look at the consequences of energetic nuclear reactions. Hess shared the 1936 Nobel Prize with the American physicist Carl David Anderson (1905–91), who used energetic cosmic ray interactions to discover the positron.

In 1938 the Nazis came to power in Austria, and Hess, a Roman Catholic with a Jewish wife, was immediately dismissed from his university position. The couple fled to the United States by way of Switzerland. Later that year Hess accepted a position at Fordham University in the Bronx, New York, as a professor of physics. He became an American citizen in 1944 and retired from Fordham University in 1956.

Victor Hess wrote more than 60 technical papers and published several books, including *The Electrical Conductivity of the Atmosphere and Its Causes* (1928) and *The Ionization Balance of the Atmosphere* (1933). He died on

December 17, 1964, in Mount Vernon, New York. Hess's discovery of cosmic rays opened a new area within astrophysics and also provided scientists with a new source of very energetic nuclear particles that led to many other important breakthroughs in physics in the 1920s, 1930s, and beyond.

Hevelius, Johannes (1611–1687) Polish-German *Astronomer* In 1647 the Polish-German astronomer Johannes Hevelius published a lunar atlas called *Selenographica*—the first detailed description of the nearside of the Moon. A skilled observer, he used his large personally constructed observatory named Stellaburgum, which was located in Danzig (now Gdańsk, Poland), to develop a comprehensive catalog of more than 1,500 stars and to discover four comets.

Hewish, Antony (1924–) British *Astronomer* The British astronomer and Nobel laureate Antony Hewish collaborated with SIR MARTIN RYLE in the development of radio wave–based astrophysics. His efforts included the discovery of the pulsar, for which he shared the 1974 Nobel Prize in physics with Ryle. During a survey of galactic radio waves in August 1967, his graduate student JOCELYN (SUSAN) BELL-Bernal was actually the first person to notice the repetitive signals from this pulsar, but the 1974 Nobel awards committee inexplicably overlooked her contributions. However, while giving his Nobel lecture, Hewish publicly acknowledged his student's (Bell's) contributions in the discovery of the pulsar.

See also PULSAR.

high-Earth orbit (HEO) An orbit around Earth at an altitude greater than 5,600 kilometers.

High-Energy Astronomy Observatory (HEAO) A series of three NASA spacecraft placed in Earth orbit. *HEAO-1* was launched in August 1977, *HEAO-2* in November 1978, and *HEAO-3* in September 1979. These orbiting observatories supported X-ray astronomy and gamma-ray astronomy. After launch NASA renamed *HEAO-2* the *Einstein Observatory* in honor of the famous German-Swiss-American physicist ALBERT EINSTEIN.

high-energy laser (HEL) A space-based laser capable of generating high-energy beams (i.e., 20 kilowatts or greater average power; 1 kilojoule or more per pulse) in the infrared, visible, or ultraviolet portion of the electromagnetic spectrum. When projected to a target, this beam would accomplish damage ranging from degradation to destruction.

highlands Oldest exposed areas on the surface of the Moon; extensively cratered and chemically distinct from the maria.

See also MOON.

high-power microwave (HPM) A proposed space-based weapon system that would be capable of destroying not only targets in space but also airborne and ground targets. This type of weapon system typically would involve peak powers of 100 megawatts or more and a single pulse energy of 100 joules or more. Target damage would result from electronic upset or burnout.

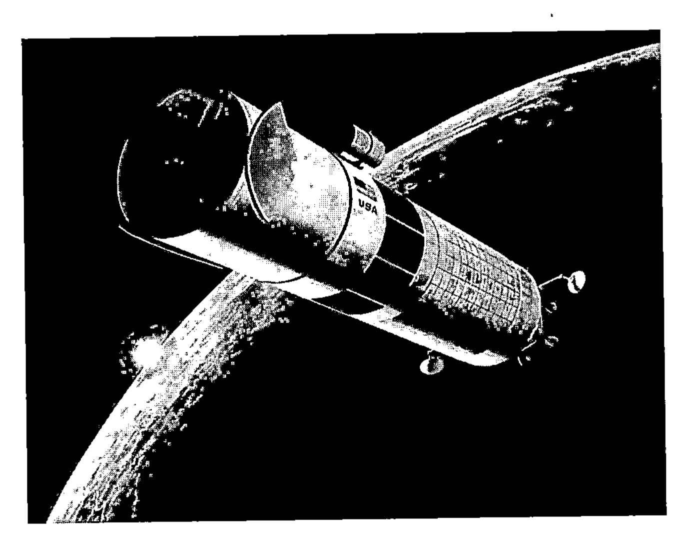

Artist's rendering of a space-based, high-energy laser system (ca. 2020) that would intercept enemy ballistic missiles in the boost phase, before they can deploy their warheads and decoys ***(Courtesy of the U.S. Department of Defense)***

Hill, George William (1838–1914) American *Astronomer, Mathematician* The American mathematical astronomer George William Hill made fundamental contributions to celestial mechanics, including pioneering mathematical methods to precisely calculate the Moon's orbit with perturbations from Jupiter and Saturn.

Himalia The small (about 180-km-diameter) moon of Jupiter discovered in 1904 by the American astronomer CHARLES DILLION PERRRINE. Himalia travels around the giant planet at a distance of 11,470,000 km with an orbital period of 251 days. Also called Jupiter VI.

See also JUPITER.

Hipparchus of Nicaea (ca. 190–ca. 120 B.C.E.) Greek *Astronomer, Mathematician* Science historians generally regard the early Greek astronomer and mathematician Hipparchus of Nicaea as the greatest ancient astronomer. Although he embraced geocentric cosmology, he carefully studied the Sun's annual motion to determine the length of a year, and his results had an accuracy of about six minutes. His observational legacy includes a star catalog completed in about 129 B.C.E. that contained 850 fixed stars. He also divided naked eye observations into six magnitudes, ranging from the faintest (or least visible to the naked eye) to the brightest observable celestial bodies in the night sky.

***Hipparcos* (spacecraft)** The European Space Agency (ESA) launched the *Hipparcos* spacecraft in August 1989. *Hipparcos* was a pioneering space experiment dedicated to the precise measurement of the positions, parallaxes, and proper motions of the stars. The goal was to measure the five astrometric parameters of some 120,000 primary program stars (an effort called the Hipparcos experiment) to a precision of some 2 to 4 milliarcseconds over a planned spacecraft lifetime of about 2.5 years. In addition, *Hipparcos* measured the astrometric and two-color photometric properties of some 400,000 additional stars (an effort called the *Tycho experiment*) to a somewhat lower astrometric precision. Once these mission goals were achieved, ESA flight controllers terminated communications with *Hipparcos* in August 1993. In June 1997 ESA published the final products of the *Hipparcos* mission as the *Hipparcos Catalog* and the *Tycho Catalog*.

The *Hipparcos* spacecraft was the first European space-based astrometry mission to measure the positions, distances, motions, brightness, and colors of stars. The *Hipparcos* spacecraft pinpointed more than 100,000 stars 200 times more accurately than ever before. Because astrometry has served as the bedrock science for the study of the universe since ancient times, this successful mission represented a major leap forward that affected every branch of astronomy. The name *Hipparcos* is an acronym for *Hi*gh *P*recision *Par*allax *Co*llecting *S*atellite; the word also closely resembles the name of the greatest naked eye astrometrist in ancient Greece, HIPPARCHUS OF NICAEA. The other great astrometrist honored by this mission is the famous 16th-century Danish naked eye observer TYCHO BRAHE.

Hohmann, Walter (1880–1945) German *Engineer* The German engineer Walter Hohmann wrote the 1925 book *The Attainability of Celestial Bodies.* In this work he described the mathematical principles that govern space vehicle motion, including the most efficient (that is, minimum energy) orbit transfer path between two orbits in the same geometric plane. This widely used (but time-consuming) orbit transfer technique is now called the *Hohmann transfer orbit* in his honor.

Hohmann transfer orbit The most efficient orbit transfer path between two coplanar circular orbits. The maneuver consists of two impulsive high-thrust burns (or firings) of a spacecraft's propulsion system. The first burn is designed to change the original circular orbit to an elliptical orbit whose perigee is tangent with the lower-altitude circular orbit and whose apogee is tangent with the higher-altitude circular orbit. After coasting for half of the elliptical transfer orbit (that is, the Hohmann transfer orbit) and when tangent with the destination circular orbit, the second impulsive high-thrust burn is performed by the spacecraft's onboard propulsion system (sometimes called an *apogee motor*). This circularizes the spacecraft's orbit at the desired new altitude. The technique can also be used to lower the altitude of a satellite from one circular orbit to another circular orbit (of less altitude). In this case two impulsive retrofirings are required. The first retrofire (that is, a rocket firing with the thrust directed opposite to the direction of travel) places the spacecraft on an elliptical transfer orbit. The second retrofire then takes place at perigee of the elliptical transfer orbit and circularizes the spacecraft's orbit at the desired lower altitude. Named after WALTER HOHMANN, the German engineer who proposed this orbital transfer technique in 1925.

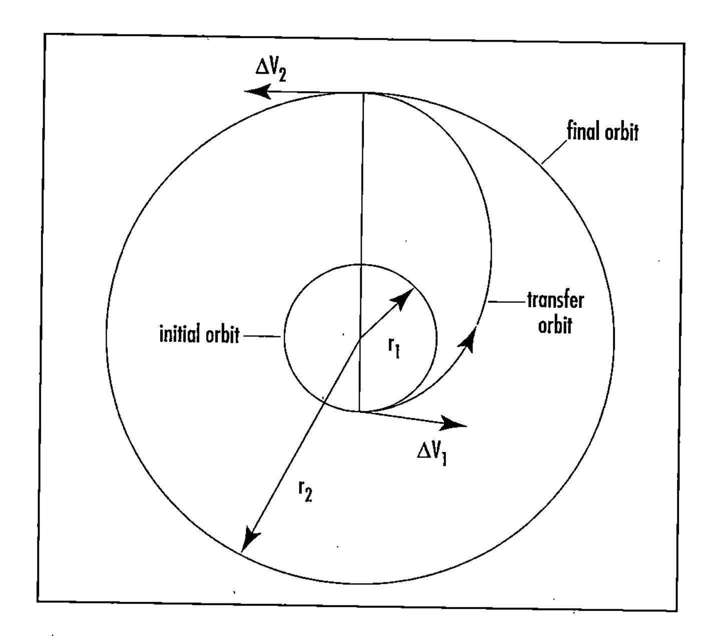

The Hohmann transfer for orbit raising from a lower altitude circular orbit to a higher altitude circular orbit

hold In rocketry, to halt the sequence of events during a countdown until an impediment has been removed so that the countdown to launch can be resumed; for example, "T minus 25 and holding because of a sudden change in weather conditions."

holddown test The testing of some system or subsystem in a rocket while the rocket is firing but restrained in a test stand.

homing guidance A guidance system by which a missile steers itself toward a target by means of a self-contained mechanism that is activated by some distinguishing characteristics of the target. The process is called active homing guidance if the target information is received in response to signals transmitted by the missile (e.g., reflected radar beams when the search radar is carried by the missile); semiactive homing guidance if the missile's homing system follows a reflected signal from the target and the source of the reflected signal is other than the missile or the target (e.g., a missile following laser light reflected from the target when the laser beam [also called a target designator beam] is placed or "painted" on the target by the attacking aircraft platform); and passive guidance if the missile follows a signal emitted by the target itself (e.g., an infrared signature associated with an exhaust plume).

HI region (**H⁰ region**) A diffuse region of neutral, predominantly atomic hydrogen in interstellar space. Neutral hydrogen emits radio radiation at 1,420.4 megahertz (MHz), corresponding to a wavelength of approximately 21 centimeters. However, the temperature of the region (approximately 100 kelvins) is too low for optical emission.

See also HII REGION; NEBULA.

Honest John An early surface-to-surface missile deployed by the U.S. Army. It has a range of about 50 kilometers.

hook Aerospace industry jargon for a design feature to accommodate the addition or upgrade of computer software at some future time; as, for example, a new space platform that has been hooked and scarred to accommodate a telerobotic laboratory capability.

horizon The line marking the apparent junction of Earth and sky.

horizon mission methodology (HMM) A systematic methodology developed within NASA for identifying and evaluating innovative aerospace technology concepts that offer revolutionary, breakthrough-type capabilities for late 21st-century and early 22nd-century space missions and for assessing the potential mission impact of these advanced technologies. The methodology is based on the concept of the *horizon mission* (HM), which is defined as a hypothetical space mission having performance requirements that cannot be met by extrapolating currently known space technologies. Horizon missions include an interstellar probe, an unpiloted star probe (USP), human-tended planetary stations in the Jupiter or Saturn systems, and an unknown spacecraft (or alien artifact) investigation.

In the intellectually stimulating process of examining advanced space technologies, candidate horizon missions are bounded on one side (the lower extreme) by the planned or proposed 21st-century missions that can (in principle) be accomplished by extrapolating current space technologies. These *extrapolated-technology missions* (ETMs) include a human expedition to Mars and a permanent lunar surface base. Such missions appear achievable (from a technology perspective, at least) between the years 2025 and 2050. The other boundary (the upper limit) of this process is formed by those truly *over-the-horizon missions* (OHMs) whose scale is so vast or whose driving motivation is so far removed from current cultural and political aspirations that it is difficult to firmly discuss their real technology requirements. An example of an over-the-horizon mission is the self-replicating interstellar robot spacecraft. This very smart (perhaps 22nd-century technology) spacecraft would travel to a nearby star system, explore it, and extract the resources necessary to both refurbish itself and make several copies of itself. The parent craft and its mechanical progeny would then depart on separate routes to other star systems and repeat the exploration and replication cycle. By using just one successful self-replicating machine (SRM), the sponsoring civilization could (in principle) trigger a wave of robot exploration throughout the entire Milky Way galaxy.

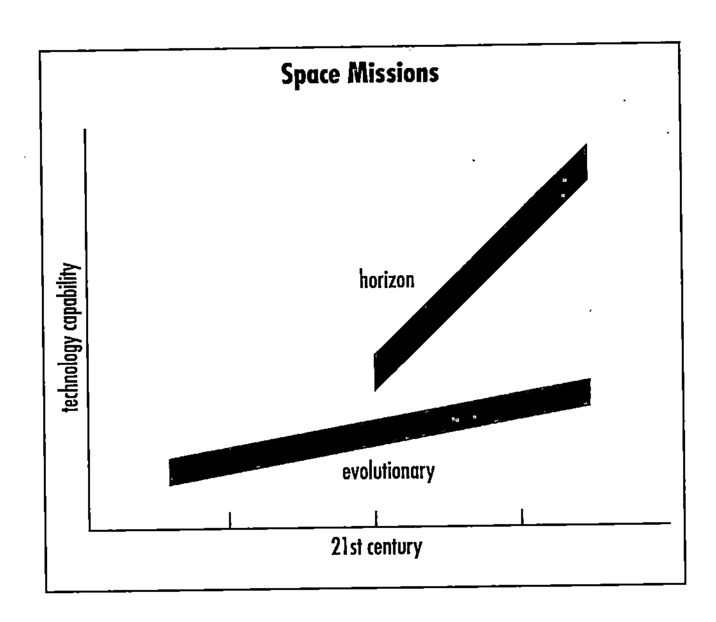

The realm of horizon mission methodology, a creative tool for examining hypothetical mid- to late 21st-century space missions that have performance requirements that cannot be met by extrapolating current space technologies ***(Courtesy of NASA)***

Lying between these two boundaries (i.e., the extrapolated-technology missions and the over-the-horizon technology missions) are the *horizon missions* (HMs). Horizon missions could occur in the latter part of the 21st century (i.e., after 2050), when the space imperatives of a solar system civilization may provide the cultural and political stimulus needed to pursue the requisite advanced space technologies.

Horizon mission thinking helps aerospace strategic planners create a collective vision for space missions in the second half of this century and beyond. A discussion of just a few of these exciting missions will help demonstrate the overall methodology involved. The essence of the horizon mission methodology is simply to define a future mission capability and then to "look backward from the future to the present" to identify the functional, operational, and technological capabilities needed to enable that particular horizon mission capability. Because horizon missions are intentionally chosen to be beyond any projected space assets or extrapolated technology levels, their functional requirements can be established independent of current technology availability. This approach fosters creative, "breakthrough" technical thinking and avoids the trap of limiting future options to what can be postulated only from extrapolations (no matter how bold) of current technologies.

A brief historical anecdote will help illustrate an important point about "breakthrough" technical thinking. Consider that you are back at the very beginning of the 20th century (ca. 1901). Imagine yourself as a transportation systems engineer who is charged with conceiving a future transportation system that is capable of crossing the Atlantic Ocean (say from New York to London) in less than two days by the year 1950. Like most system designers in the early 1900s, you would probably vigorously pursue this "futuristic" design problem by looking at ways of improving the performance of the ocean-going steamship, perhaps through improved hull design, or through better power plants, or even through the use of exotic new materials (such as aluminum). You might even boldly suggest a hydrofoil approach. But, surrounded by 1901 technology in the "golden age of steam," would you have dared to suggest the airplane as a practical solution to this problem? Yet, by 1950 most travelers were crossing the Atlantic Ocean in less than 10 hours via propeller-driven aircraft, and by 1960 commercial jet liners reduced this transit time even further. Revolutionary technical "breakthroughs" usually emerge from nonlinear thinking, not linear extrapolations of contemporary technology levels.

We will now briefly examine four very interesting horizon missions. Remember, however, that a reader in the year 2101 (probably a technical historian) will almost certainly look at the next few paragraphs and wonder: How could *that author* have been so *reserved* in projecting 21st-century space technology?

The first horizon mission is the *Interstellar Probe* (IP), a probe whose main objective would be the dedicated scientific investigation of the near interstellar region out to a distance of about 200 astronomical units (AUs) from our solar system. This first-generation "extrasolar" robot probe would investigate the heliosphere, determine the location of the heliopause, and record the composition of the interstellar medium. (Note: as you read this paragraph, NASA's *Pioneer 10* and *11* and *Voyager 1* and *2* spacecraft are departing our solar system. These spacecraft were actually designed for planetary exploration, not exploration of the interstellar medium. However, because of the interstellar messages they carry, we can consider them to be the "zeroth" generation of extrasolar probe spacecraft.) The Interstellar Probe might use a close solar flyby to obtain a trajectory in the same direction as the Sun's motion through the interstellar medium. It would travel the distance of 200 AU in about 30 to 35 years. Some of the technology challenges associated with this particular mission are propulsion systems, thermal protection systems (for the close solar flyby), appropriate detectors for in-situ measurements of the interstellar medium, and long-lived materials and spacecraft systems (possibly even self-repairing).

The *Unpiloted Star Probe* (UPS) is a challenging 100-year-duration mission to a nearby star system, perhaps the Alpha Centauri triple star system (approximately 4.3 light-years away) or possibly Barnard's star (about 6 light-years away). A sophisticated robot probe would conduct interstellar research starting from the outer boundaries of our solar system through the interstellar medium and then into the heliosphere of another star system. Within the alien star system the robot probe should also be capable of searching for planets and possibly even detecting signs of life (e.g., the presence and variation of bioindicative atmospheric gases, artificial electromagnetic emissions, etc.) This horizon mission presents several quite formidable technology challenges, including propulsion; very-long-lived, autonomous operations; instrumentation to effectively search the "unknown"; and data return over a period of 100 years and over a distance of up to 10 light-years.

Another horizon mission involves the development of advanced space habitats, called *Planetary Stations.* These human-tended stations would be placed in the Jupiter or Saturn systems for 20-year research missions. Such stations could orbit the giant planets themselves, orbit a particular moon, or be located on the surface of a very interesting moon (e.g., Jupiter's moon Europa or Saturn's moon Titan). These planetary stations would serve as long-term observation platforms and laboratories for scientific research. They would also function as a sortie base for human excursions to the numerous moons in each planetary system. During these sortie missions resource assessment evaluations would be conducted, and, if appropriate, pilot plants for resource extraction might be established. Some of the technology challenges associated with this horizon mission include space radiation protection, regenerative life support systems for the permanent stations, reliable excursion vehicles, long-lived power supplies (including rechargeable units), mobile life support systems and extravehicular activity (EVA) systems for diverse sortie operations, and advanced instrumentation for in-situ science and resource evaluation.

The final horizon mission that we will discuss is called the *Unknown Spacecraft Investigation.* In the late 21st century widespread exploration and human presence in space could result in the serendipitous discovery of artifacts or even a derelict spacecraft of non-Earth origin. Such a discovery would unquestionably trigger the need for a detailed investigation of the object to determine its purpose, capabilities, and enabling technologies. Investigative protocols (including careful consideration of planetary contamination issues) and special instrumentation would be required to accomplish this mission. There might even be an attempt to activate the alien craft or artifact (if either appears to be in a dormant but potentially functional condition). A variety of technical challenges are associated with this type of "response" horizon mission, including the sensing and spectral analysis of an intelligently crafted object of unknown origin and purpose; nondestructive diagnostic instruments and procedures for sampling and evaluation; advanced artificial intelligence to assist in the diagnosis of identified characteristics, patterns, and anomalies; and possibly advanced code-breaking to decipher an alien language or any strange signals (should the object emit such upon stimulation or in response to our attempts at activation or dismantling).

Horrocks, Jeremiah (1619–1641) British *Astronomer* The 17th-century British astronomer Jeremiah Horrocks was the first person to predict and then record a transit of Venus across the Sun. This particular transit occurred on November 24, 1639.

Horsehead Nebula (NGC 2024) A dark nebula in the constellation Orion that has the shape of a horse's head.
See also NEBULA.

hot fire test A liquid-fuel propulsion system test conducted by actually firing the rocket engine(s) (usually for a short period of time) with the rocket vehicle secured to the launch pad by holddown bolts. *Compare with* COLD-FLOW TEST.

hot Jupiter A Jupiter-sized planet in another solar system that orbits close enough to its parent star to have a high surface temperature.

Hound Dog An early U.S. Air Force air-to-surface missile with a range of about 800 kilometers. It was externally carried under the wing of long-range strategic bombers, such as the B-52.

hour (symbol: h) A unit of time equal to 3,600 seconds or 60 minutes.
See also TIME.

hour angle (HA) The angle (measured in a westward direction along the celestial equator) between an observer's meridian and the hour circle of a celestial object. Generally, the hour angle is expressed from 0 to 24 hours in hours, minutes, and seconds.

"housekeeping" The collection of routine tasks that must be performed to keep an aerospace vehicle or spacecraft functioning properly. For example, the astronaut crew or ground support team performing frequent status checks on and making minor adjustments to the life support and waste management systems of a human-occupied space vehicle; or the mission control team performing regular telemetry management checks and trajectory adjustments on an interplanetary spacecraft as it cruises toward its planetary target.

Hoyle, Sir Fred (1915–2001) British *Astrophysicist, Science Writer* In 1948 the British astrophysicist and well-known science writer Sir Fred Hoyle collaborated with SIR HERMAN BONDI and THOMAS GOLD to develop and promote the steady universe model. It was Hoyle who first coined the expression *big bang theory*—essentially as a derogatory remark about this competitive theory in cosmology. The term stuck, and big bang theory has all but displaced the steady state universe model in contemporary astrophysics. He also collaborated with GEOFFREY RONALD BURBIDGE, (ELEANOR) MARGARET BURBIDGE, and WILLIAM ALFRED FOWLER in the development of the theory of nucleosynthesis. Though often controversial, he promoted astronomy to the general public with several very popular books. He was also an accomplished science fiction writer.

H-II launch vehicle An advanced expendable booster sponsored by the Japanese Aerospace Exploration Agency (formerly called National Space Development Agency of Japan, or NASDA). The H-II is a two-stage launch vehicle capable of delivering a 2,000-kg payload into geostationary orbit. The H-II is equipped with high-performance rocket engines using liquid hydrogen and liquid oxygen. The thrust of the first stage is supplemented by two large solid rocket boosters. The first stage uses a large liquid oxygen–liquid hydrogen rocket engine called the LE-7, which employs a staged-combustion cycle. The second-stage engine uses a liquid oxygen–liquid hydrogen engine called the LE-5A, which employs a hydrogen bleed cycle. The first launch of the H-II vehicle from the Tanegashima Space Center, Japan, occurred on February 4, 1994; it successfully placed several experimental payloads into their planned orbits.

See also LAUNCH VEHICLE; JAPANESE AEROSPACE EXPLORATION AGENCY (JAXA).

HII region (H^+ region) A region in interstellar space consisting mainly of ionized hydrogen existing mostly in discrete clouds. The ionized hydrogen of HII regions emits radio waves by thermal emission and recombination-line emission, in comparison to the 21-centimeter-wavelength radio-wave emission of neutral hydrogen in HI regions.

See also HI REGION; NEBULA.

Hubble, Edwin Powell (1889–1953) American *Astronomer* The American astronomer Edwin Powell Hubble revolutionized our understanding of the universe by proving that galaxies existed beyond the Milky Way galaxy. In the 1920s he made important observational discoveries that completely changed how scientists view the universe. This revolution in cosmology started in 1923, when Hubble used a Cephid variable to estimate the distance to the Andromeda galaxy. His results immediately suggested that such spiral nebulas were actually large, distant, independent stellar systems, or island universes. Next, he introduced a classification scheme in 1925 for such "nebulas" (galaxies), calling them either elliptical, spiral, or irregular. This scheme is still quite popular in astronomy. In 1929 Hubble announced that in an expanding universe the other galaxies are receding from us with speeds proportional to their distance, a postulate now known as Hubble's Law. Hubble's concept of an expanding universe filled with numerous galaxies forms the basis of modern observational cosmology.

Hubble was born in Marshfield, Missouri, on November 20, 1889. Early in his life he studied law at the University of Chicago and then as a Rhodes scholar at Oxford from 1910 to 1913. However, in 1913 he made a fortunate decision for the world of science by abandoning law and studying astronomy instead. From 1914 to 1917 Hubble held a research position at the Yerkes Observatory of the University of Chicago. There, on the shores of Lake Geneva at Williams Bay, Wisconsin, Hubble started studying interesting nebulas. By 1917 he concluded that the spiral-shaped ones (we now call these galaxies) were quite different from the diffuse nebulas (which are actually giant clouds of dust and gas).

Following military service in World War I, he joined the staff at the Carnegie Institute's Mount Wilson Observatory, located in the San Gabriel Mountains northwest of Los Angeles. When Hubble arrived in 1919, this observatory's 2.54-meter (100-inch) telescope was the world's biggest instrument. Except for other scientific work during World War II, Hubble remained affiliated with this observatory for the remainder of his life. Once at Mount Wilson, Hubble used its large instrument to resume his careful investigation of nebulas.

In 1923 Hubble discovered a Cepheid variable star in the Andromeda "nebula," now known to astronomers as the Andromeda galaxy, or M31. A Cepheid variable is one of a group of important very bright supergiant stars that pulsate periodically in brightness. By carefully studying this particular Cepheid variable in M31, Hubble was able to conclude that it was very far away and belonged to a separate collection of stars far beyond the Milky Way galaxy. This important discovery provided the first tangible observational evidence that galaxies existed beyond the Milky Way. Through Hubble's pioneering efforts, the size of the known universe expanded by incredible proportions.

Hubble continued to study other galaxies and in 1925 introduced the well-known classification system of spiral galaxies, barred spiral galaxies, elliptical galaxies, and irregular galaxies. Then, in 1929 Hubble investigated the recession velocities of galaxies (i.e., the rate at which galaxies are moving apart) and their distances away. He discovered that the more distant galaxies were receding faster than the galaxies closer to us. This very important discovery revealed that the universe was expanding. Hubble's work provided the first observational evidence that supported big bang cosmology. Today the concept of the universe expanding at a uniform, steady rate is codified in a simple mathematical relationship called Hubble's law in his honor.

Hubble's law describes the expansion of the universe as a linear, steady rate. As Hubble initially observed and subsequent astronomical studies confirmed, the apparent recession velocity (v) of galaxies is proportional to their distance (r) from an observer. The proportionality constant is H_0, the Hubble constant. Currently proposed values for H_0 fall between 50 and 90 kilometers per second per megaparsec (km s^{-1} Mpc^{-1}). The inverse of the Hubble constant ($1/H_0$) is called the *Hubble time*. Astronomers use the value of the Hubble time as one measure of the age of the universe. If H_0 has a value of 50 km s^{-1} Mpc^{-1}, then the universe is about 20 billion years old (which appears far too old by recent astrophysical estimates). An H_0 value of 80 km s^{-1} Mpc^{-1} suggests a much younger universe, ranging in age between 8 and 12 billion years old. Despite space age investigations of receding galaxies, there is still much debate within the astrophysical community as to the proper value of the Hubble constant. Cosmologists and astrophysicists using several techniques currently suggest that the universe is about 13.7 billion years old. Astrophysical observations of distant supernovas made in the late 1990s suggest that rate of expansion of the universe is actually increasing. This will influence the linear expansion rate hypothesis inherent in Hubble's law and will also influence age-of-the-universe estimates based on the Hubble time. However, these recent observations, while very significant and important, take very little away from Hubble's brilliant pioneering work, which introduced the modern cosmological model of an enormously large, expanding universe filled with billions of galaxies.

Hubble, like many astronomical pioneers before him, created a revolution in our understanding of the physical universe. Through his dedicated observational efforts, we learned about the existence of galaxies beyond the Milky Way and the fact that the universe appears to be expanding. Quite fittingly, NASA named the *HUBBLE SPACE TELESCOPE* after him. This powerful orbiting astronomical observatory continues the tradition of extragalactic investigation started by Hubble, who died on September 28, 1953, in San Marino, California.

Hubble constant (symbol: H_0) The constant within HUBBLE'S LAW proposed in 1929 by the American astronomer EDWIN POWELL HUBBLE that establishes an empirical relationship between the distance to a galaxy and its velocity of recession due to the expansion of the universe. A value of 70 kilometers per second per megaparsec km s^{-1} ± 7 Mpc^{-1} is currently favored by astrophysicists and astronomers.

See also UNIVERSE.

Hubble's law The hypothesis that the redshifts of distant galaxies are directly proportional to their distances from Earth. The American astronomer EDWIN POWELL HUBBLE first proposed this relationship in 1929. It can be expressed as $V = H_0 D$, where V is the recessional velocity of a distant galaxy, H_0 is the Hubble constant, and D is its distance from Earth. Assuming expansion at a steady state, H_0 is the constant of proportionality in the relationship between the relative recessional velocity (V) of a distant galaxy or a very remote extragalactic object and its distance (D) from Earth. Unfortunately, there is still much debate and controversy among astrophysicists and cosmologists as to what the value of the Hubble constant really is (based on measurements of distant receding galaxies). An often-encountered value is 70 kilometers per second per megaparsec (km s^{-1} Mpc^{-1}), although values ranging from 50 to 90 km s^{-1} Mpc^{-1} can be found in the technical literature.

The inverse of the Hubble constant ($1/H_o$ = *Hubble time*) has the dimension of time, and astronomers consider it one measure of the age of the universe, especially if a constant expansion rate is assumed. For example, if H_o is given a value of 80 km s^{-1} Mpc^{-1}, this would suggest that the universe is between 8 and 12 billion (10^9) years old. However, if H_o is given a value of 50 km s^{-1} Mpc^{-1}, then the universe has an estimated age of about 20 billion years. The smaller the value assigned to the Hubble constant, the older the estimated age of the universe, that is, the time that has passed since the primordial big bang. Astronomers and astrophysicists currently think the universe is between 12 and 14 billion years old, with 13.7 billion years as a popular consensus value. Astrophysical observations of distant supernovas made in the late 1990s now suggest that the rate of expansion of the universe is actually increasing. These observations, once confirmed and widely accepted within the astrophysical community, will cause a reexamination of Hubble's law and its inherent assumption of a linear rate of expansion since the big bang. Of course, nothing is being taken away from the brilliant work of Edwin Hubble and his pioneering advocacy of the modern cosmological model based on an expanding universe. Scientific progress is made through refinements and improvement in models of the physical universe, so Hubble's law, which has served the astronomical community well for more than 70 years, will undergo some form of evolutionary refinement and modification to embody the latest discoveries in observational astronomy.

See also ASTROPHYSICS; COSMOLOGY; UNIVERSE.

Hubble Space Telescope **(HST)** The *Hubble Space Telescope* (HST) is a cooperative program of the European Space Agency (ESA) and NASA to operate a long-lived space-based optical observatory for the benefit of the international astronomical community. The orbiting facility is named for the American astronomer EDWIN P. HUBBLE, who revolutionized our knowledge of the size, structure, and makeup of the universe through his pioneering observations in the first half of the 20th century. The HST is being used by astronomers and space scientists to observe the visible universe to distances never before obtained and to investigate a wide variety of interesting astronomical phenomena. In 1996, for example, HST observations revealed the existence of approximately 50 billion (10^9) more galaxies than scientists previously thought existed!

The HST is 13.1 meters long and has a diameter of 4.27 meters. This space-based observatory is designed to provide detailed observational coverage of the visible, near-infrared, and ultraviolet portions of the electromagnetic (EM) spectrum. The HST power supply system consists of two large solar panels (unfurled on orbit), batteries, and power-conditioning equipment. The 11,000 kilogram free-flying astronomical observatory was initially placed into a 600-kilometer low-Earth orbit (LEO) during the STS-31 space shuttle *Discovery* mission on April 25, 1990.

The years since the launch of HST have been momentous, including the discovery of a spherical aberration in the telescope optical system, a flaw that severely threatened the space-based telescope's usefulness. A practical solution was found, however, and the STS-61 space shuttle *Endeavour* mission (December 1993) successfully accomplished the first on-orbit servicing and repair of the telescope. The effects of the spherical aberration were overcome, and the HST was restored to full functionality. During this mission the astronauts changed out the original Wide Field/Planetary Camera (WF/PC1) and replaced it with the WF/PC2. The relay mirrors in the WF/PC2 are spherically aberrated to correct for the spherically aberrated primary mirror of the observatory. *Hubble*'s primary mirror is two micrometers (2 μm) too flat at the edge, so the corrective optics within WF/PC2 are too high by that same amount. In addition, a corrective optics package called the Corrective Optics Space Telescope Axial Replacement, or COSTAR, replaced the High-Speed Photometer during the STS-61 servicing mission. COSTAR is designed to optically correct the effect of the primary mirror's aberration on the telescope's three other scientific instruments: the Faint Object Camera (FOC), built by ESA, the Faint Object Spectrograph (FOS), and the Goddard High Resolution Spectrograph (GHRS).

In February 1997 a second servicing operation (HST SM-02) was successfully accomplished by the STS-82 space shuttle *Discovery* mission. As part of this HST servicing mission, astronauts installed the Space Telescope Imaging Spectrograph (STIS) and the Near-Infrared Camera and Multi-Object Spectrometer (NICMOS). These instruments replaced the Goddard High-Resolution Spectrograph (GHRS) and the Faint Object Spectrograph (FOS), respectively. The STIS observes the universe in four spectral bands that extend from the ultraviolet through the visible and into the near-infrared. Astronomers use the STIS to analyze the temperature, composition, motion, and other important properties of celestial objects. Operating at near-infrared wavelengths, the NICMOS is letting astronomers observe the dusty cores of active galactic nuclei (AGN) and investigate interesting protoplanetary disks around stars.

On November 13, 1999, flight controllers placed the *Hubble Space Telescope* in a "safe-mode" after the failure of a fourth gyroscope. In this safe mode the HST could not observe celestial targets, but the observatory's overall safety was preserved. This protective mode allowed ground control of the telescope, but with only two gyroscopes working, *Hubble* could not point with the precision necessary for scientific observations. Flight controllers on the ground also closed the telescope's aperture door in order to protect the optics and

NASA's *Hubble Space Telescope* (HST) being unberthed and carefully lifted up out of the payload bay of the space shuttle *Discovery* and then placed into the sunlight by the shuttle's robot arm in February 1997. This event took place during the STS-82 mission, which NASA also calls the second HST servicing mission (HTS SM-02). *(Courtesy of NASA/JSC)*

then aligned the spacecraft to make sure that *Hubble*'s solar panels would receive adequate illumination from the Sun.

On December 19, 1999, seven astronauts boarded the space shuttle *Discovery* and departed the Kennedy Space Center to make an on-orbit "house call" to the stricken observatory. When the third of the HST's gyroscopes failed (the space-based telescope needs at least three to point with scientific accuracy), NASA made a decision to split the third HST servicing mission into two parts, called SM3A and SM3B. Six days and three extravehicular activities (EVAs) later, the STS-103 mission crew had replaced worn or outdated equipment and performed critical maintenance during Servicing Mission 3A. The most important task was to replace the gyroscopes that accurately point the telescope at celestial targets. The astronauts also installed an advanced central computer, a digital data recorder, battery improvement kits, and new outer layers of thermal protection material. After the *Discovery* and its crew returned safely to Earth (on December 27), the HST flight controllers turned on the orbiting instrument. Because of this successful repair mission, the *Hubble Space Telescope* was as good as new and returned to its astronomical duties.

The second part of the third HST servicing mission took place in March 2002. NASA launched the space shuttle *Columbia* so its crew of seven astronauts could rendezvous with the HST and perform a series of upgrades. The STS-109 mission is also known as servicing mission 3B. During this mission, which started on March 1, 2002, the astronauts performed five EVAs, or space walks. Their primary task was to install a new science instrument called the Advanced Camera for Surveys (ACS). The ACS brought the nearly 12-year-old orbiting telescope into the 21st century. With its wide field of view, sharp image quality, and enhanced sensitivity, ACS doubled the HST's field of view and collects data 10 times faster than the Wide Field and Planetary Camera 2, the HST's earlier surveying instrument. During this servicing mission the astronauts upgraded the telescope's electric power system. They replaced the spacecraft's four eight-year-old solar panels with smaller rigid ones that produce 30 percent more electricity. During the final extravehicular activity the STS-109 astronauts installed a new cooling system for the Near Infrared Camera and Multi-Object Spectrometer (NICMOS), which became inactive when its 100-kg block of nitrogen ice coolant became depleted. The new refrigeration system chills NICMOS's infrared detectors to below –193°C. Finally, the shuttle astronauts replaced one of the four reaction wheel assemblies that make up the HST's pointing control system.

On January 16, 2004, NASA administrator Sean O'Keefe announced his decision to cancel servicing mission 4, the last scheduled flight of the space shuttle to the *Hubble Space Telescope.* Astronauts on this servicing mission (planned for 2006) would have performed maintenance work and installed new instruments. New shuttle flight safety guidelines, following the loss of the *Columbia* (on February 1, 2003), weighed heavily in the administrator's decision. NASA is currently reviewing options on how to extend the *Hubble*'s lifetime beyond 2007, when critical components on the telescope will require replacement or repair. NASA has a replacement for the *Hubble Space Telescope,* a new space-based observatory called the *James Webb Space Telescope,* now scheduled for launch in August 2011.

Although HST operates around the clock, not all of its time is spent observing the universe. Each orbit lasts about 95 minutes, with time allocated for "housekeeping" functions and for observations. Housekeeping functions include turning the telescope to acquire a new target or to avoid the Sun or the Moon, switching communications antennas and data transmission modes, receiving command loads and downloading data, calibrating, and similar activities. Responsibility for conducting and coordinating the science operations of the *Hubble Space Telescope* rests with the Space Telescope Science Institute (STScI) on the Johns Hopkins University Homewood campus in Baltimore, Maryland.

Because of HST's location above the Earth's atmosphere, its science instruments can produce high-resolution images of astronomical objects. Ground-based telescopes, influenced by atmospheric effects, can seldom provide resolutions better than 1.0 arc-second, except perhaps momentarily under the very best observing conditions. The *Hubble Space Telescope*'s resolution is about 10 times better, or 0.1 arc-second.

Here are just a few of the exciting discoveries provided by the *Hubble Space Telescope:* (1) It gave astronomers their first detailed view of the shapes of 300 ancient galaxies located in a cluster 5 billion light-years away; (2) it provided astronomers their best look yet at the workings of a black hole "engine" in the core of the giant elliptical galaxy NGC4261, located 45 million light-years away in the constellation Virgo; (3) HST's detailed images of newly forming stars helped confirm more than a century of scientific hypothesis and conjecture on how a solar system begins; and (4) HST gave astronomers their earliest look at a rapidly ballooning bubble of gas blasted off a star (Nova Cygni, which erupted on February 19, 1992).

Pushing the limits of its powerful vision, the HST has also uncovered the oldest burned-out stars in the Milky Way galaxy. Detection of these extremely old, dim white dwarfs has provided astronomers a completely independent indication of the age of the universe, an assessment that does not rely on measurements of the universe's expansion. The ancient white dwarfs, as imaged by *Hubble* in 2001, appear to be between 12 and 13 billion years old. Finding these "oldest" stars puts astronomers well within arm's reach of calculating the absolute age of the universe, since the faintest and coolest white dwarfs within globular clusters can yield a globular cluster's age and globular clusters are among the oldest clusters of stars in the universe.

Finally, from September 2003 to January 2004, the *Hubble Space Telescope* performed the Hubble Ultra Deep Field (HUDF) observation, the deepest portrait of the visible universe ever achieved by humankind. The HUDF revealed the first galaxies to emerge from the so-called dark ages—the time shortly after the big bang event when the first stars reheated the cold universe. This historic portrait of the early universe involved two separate long-exposure images captured by the HST's Advanced Camera for Surveys (ACS) and the Near-Infrared Camera and Multi-object Spectrometer (NICMOS).

Imagery from the HST has revolutionized our view of the visible universe. Refurbished by shuttle servicing missions, this amazing facility should continue to provide astrophysicists and astronomers with incredibly interesting data for several more years.

Huggins, Margaret Lindsay (née Murray) (1848–1915) British *Astronomer* The British astronomer Lady Margaret Lindsay Huggins collaborated with her husband (SIR WILLIAM HUGGINS) and collected pioneering spectra of celestial objects, including the Orion nebula.

Huggins, Sir William (1824–1910) British *Astronomer, Spectroscopist* The British astronomer Sir William Huggins helped revolutionize the observation of celestial objects by performing early spectroscopy measurements and then comparing the observed stellar spectra with laboratory spectra. In 1868 he collaboratively made the first measurement of a star's Doppler shift and in 1881 (independently of HENRY DRAPER) also made a pioneering photograph of the spectrum of a comet.

Hulse, Russell A. (1950–) American *Radio Astronomer* In 1974, while performing research using the Arecibo Observatory in Puerto Rico, radio astronomer Russell A. Hulse (then a doctoral student at the University of Massachusetts) codiscovered an unusual binary star system with his professor, JOSEPH H. TAYLOR, JR. Their discovery was the first binary pulsar, called PSR 1913 + 16. They detected a pair of neutron stars in a close binary system, and one neutron star was observable as a pulsar here on Earth. This unique binary system provided scientists with a very special deep space laboratory for investigating the modern theory of gravity and the validity of ALBERT EINSTEIN's general theory of relativity. Because their discovery had such great significance, both Hulse and Taylor shared the 1993 Nobel Prize in physics.

human factors engineering The branch of engineering involved in the design, development, testing, and construction of devices, equipment, and artificial living environments to the anthropometric, physiological, and/or psychological requirements of the human beings who will use them. In aerospace, human factors engineering is involved in such diverse areas as the design of a functional microgravity toilet (for both male and female crewpersons), the creation of crew living and sleeping quarters that are both efficient (from a size perspective) and yet provide a suitable level of aesthetics and privacy, and the development of crew workstations that can accommodate a variety of mission needs and still provide immediate attention to hazardous or emergency situations (e.g., a warning light that "jumps out" immediately and can be recognized easily amid 100 or so other lights and indicators).

Humason, Milton Lassell (1891–1972) American *Astronomer* In the late 1920s the American astronomer Milton Lassell Humason assisted EDWIN POWELL HUBBLE during Hubble's discovery of the recession of the distant galaxies and the expansion of the universe.

humidity The amount of water vapor in the air; the "wetness" of the atmosphere. The relative humidity represents the ratio of the amount of water vapor present in the air at a given temperature to the greatest amount of water vapor possible for that temperature. For example, a 10% relative humidity is very dry air, while a 90% relative humidity is very moist air.

Huygens, Christiaan (1629–1695) Dutch *Astronomer, Physicist, Mathematician* In 1655 the Dutch astronomer Christiaan Huygens discovered Titan, the largest moon of Saturn. Then, in about 1659, using a personally constructed telescope, he was able to discern the true, thin, disklike shape of Saturn's rings. A creative and productive individual, he also constructed and patented the first pendulum clock (about 1656) and founded the wave theory of light. In his honor the European Space Agency named the Titan atmospheric probe portion of the Cassini mission the *Huygens* probe.

See also CASSINI MISSION; TITAN.

***Huygens* probe** A scientific probe sponsored by the European Space Agency (ESA) and named after the Dutch astronomer CHRISTIAAN HUYGENS, who discovered Titan, SATURN's largest moon, in 1655. The probe was deployed into the atmosphere of Titan in January 2005 from NASA's *Cassini* spacecraft, which functioned as the probe's mother spacecraft. The scientific objectives of the *Huygens* probe were to determine the physical characteristics (density, pressure, temperature, etc.) of Titan's atmosphere as a function of height; measure the abundance of atmospheric constituents; investigate the atmosphere's chemistry and photochemistry; characterize the meteorology of Titan; and examine the physical state, topography, and composition of the moon's surface.

See also CASSINI MISSION; TITAN.

Hyades An open cluster of about 200 stars in the constellation Taurus some 150 light-years away.

See also STAR CLUSTER.

hydrazine (symbol: N_2H_4) A toxic, colorless liquid that often is used as a rocket propellant because it reacts violently with many oxidizers. It is spontaneously ignitable with concentrated hydrogen peroxide (H_2O_2) and nitric acid. When decomposed by a suitable catalyst, hydrazine is also a good monopropellant and can be used in simple, small rocket engines, such as those for spacecraft attitude control.

hydrogen (symbol: H) A colorless, odorless gas that is the most abundant chemical element in the universe. Hydrogen occurs as molecular hydrogen (H_2), atomic hydrogen (H), and ionized hydrogen (that is, broken down into a proton and its companion electron). Hydrogen has three isotopic forms, protium (ordinary hydrogen), deuterium (heavy hydrogen), and tritium (radioactive hydrogen). Liquid hydrogen (LH_2) is an excellent high-performance cryogenic chemical propellant for rocket engines, especially those using liquid oxygen (LO_2) as the oxidizer.

hydrosphere The water on Earth's surface (including oceans, seas, rivers, lakes, ice caps, and glaciers) considered as an interactive system.

See also EARTH SYSTEM SCIENCE.

hyperbaric chamber A chamber used to induce an increase in ambient pressure such as would occur in descending below sea level in a water or air environment. It is the only type of chamber suitable for use in the treatment of decompression sickness.

hyperbola An open curve with two branches, all points of which have a constant difference in distance from two fixed points called focuses (or foci).

hyperbolic orbit An orbit in the shape of a hyperbola. All interplanetary flyby spacecraft follow hyperbolic orbits, both for Earth departure and again upon arrival at the target planet.

See also CONIC SECTION.

hyperbolic velocity A velocity sufficient to allow escape from a planetary system or the solar system; for example, all interplanetary spacecraft missions follow hyperbolic orbits, both for Earth departure and then again upon arrival at the target planet. Comets, if not gravitationally captive to the Sun, can have hyperbolic velocities; their trajectories would be escape trajectories or hyperbolas.

See also ORBITS OF OBJECTS IN SPACE.

hypergolic fuel (hypergol) A rocket fuel that spontaneously ignites when brought into contact with an oxidizing agent (oxidizer); for example, aniline ($C_6H_5NH_2$) mixed with red fuming nitric acid (85% HNO_3 and 15% N_2O_4) produces spontaneous combustion.

Hyperion An irregularly-shaped (roughly 350-km by 200-km) smaller moon of Saturn that orbits the planet at a distance of 1,481,100 km with a period of 21.28 days, an inclination of 0.43°, and an eccentricity of 0.104. Discovered in 1848 by WILLIAM CRANCH BOND and GEORGE PHILLIPS BOND. Two days after the father-son team of American astronomers discovered Hyperion, the British amateur astronomer WILLIAM LASSELL independently discovered the moon.

See also SATURN.

hypernova A newly discovered type of supernova that some astrophysicists suggest may be connected to the mysterious and very energetic phenomenon called gamma-ray bursts (GRBs). In 2003 astronomers fortuitously observed an energetic gamma-ray burst, called GRB 030329, in the X-ray, optical, and radio wavelengths.

See also GAMMA-RAY BURST.

hyperoxia An aerospace medicine term describing a condition in which the total oxygen content of the body is increased above that normally existing at sea level.

hypersonic Of or pertaining to speeds equal to or in excess of five times the speed of sound (i.e., $\geq$ Mach 5).

hypersonic glider An unpowered aerospace vehicle designed to fly and maneuver at hypersonic speeds upon reentering a planetary atmosphere.

hyperspace A concept of convenience developed in science fiction to make "faster-than-light" travel appear credible. Hyperspace is frequently described as a special dimension or property of the universe in which physical things are much closer together than they are in the normal space-time continuum.

In a typical science fiction story the crewmembers of a spaceship simply switch into *hyperspace drive,* and distances to objects in the "normal" universe are considerably shortened. When the spaceship emerges out of hyperspace, the crew is where they wanted to be essentially instantly. Although this concept violates the speed-of-light barrier predicted by Albert Einstein's special relativity theory, it is nevertheless quite popular in modern science fiction.

See also "FASTER-THAN-LIGHT" TRAVEL; WORMHOLE.

hypervelocity Extremely high velocity. The term is applied by physicists to speeds approaching the speed of light, but in aerospace engineering applications, the term generally implies speeds of the order of satellite or spacecraft speed and greater, that is, 5 to 10 kilometers per second and greater.

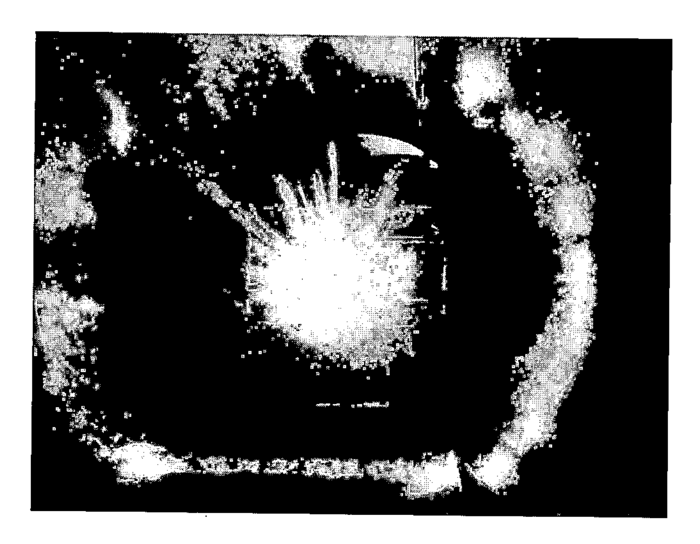

This photograph shows the "energy flash" when a hypervelocity particle traveling at speeds up to 7.5 km/s impacts a solid surface. Starting in 1963, NASA scientists at the Ames Research Center (ARC) performed this and similar tests to simulate what would happen when a small piece of space debris hits a spacecraft in orbit around Earth. ***(Courtesy of NASA/ARC)***

hypervelocity gun (HVG) A gun that can accelerate projectiles to 5 kilometers per second or more; for example, an electromagnetic gun.

See also GUN-LAUNCH TO SPACE.

hypervelocity impact A collision between two objects that takes place at a very high relative velocity—typically at a speed in excess of 5 kilometers per second. A spacecraft colliding with a piece of space debris or an asteroid striking a planet are examples.

hypobaric chamber A chamber used to induce a decrease in ambient pressure such as would occur in ascending altitude. This type of chamber is used primarily for flight training purposes and for aerospace experiments. Also called an *altitude chamber.*

hypothesis A scientific theory proposed to explain a set of data or observations; can be used as a basis for further investigation and testing.

hypoxia An aerospace medicine term describing an oxygen deficiency in the blood, cells, or tissues of the body in such degree as to cause psychological and physiological disturbances.

hysteresis 1. Any of several effects resembling a kind of internal friction, accompanied by the generation of thermal energy (heat) within the substance affected. *Magnetic hysteresis* occurs when a ferromagnet is subjected to a varying magnetic intensity. *Electric hysteresis* occurs when a dielectric is subjected to a varying electric intensity. *Elastic hysteresis* is the internal friction in an elastic solid subjected to varying stress. 2. The delay of an indicator in registering a change in a parameter being measured.

HZE particles The most potentially damaging cosmic rays, with high atomic number (Z) and high kinetic energy (E). Typically, HZE particles are atomic nuclei with Z greater than 6 and E greater than 100 million electron volts (100 MeV). When these extremely energetic particles pass through a substance, they deposit a large amount of energy along their tracks. This deposited energy ionizes the atoms of the material and disrupts molecular bonds.

See also COSMIC RAYS; SPACE RADIATION ENVIRONMENT.

I

Iapetus A mid-sized (1,440-km-diameter) moon of Saturn that orbits the ringed planet at a distance of 3,561,300 km. Iapetus has an orbital period of 79.33 days and is in synchronous rotation around Saturn. With a density of only 1.1 g/cm^3 (1,100 kg/m^3), Iapetus must be composed almost entirely of water ice. Discovered by Giovanni Cassini in 1671, this moon has an interesting appearance than makes it unique in the solar system. The leading hemisphere of Iapetus is dark, with an albedo of between 0.03 and 0.05, while its trailing hemisphere is almost as bright as the Jovian moon Europa, with an albedo of 0.5. Some planetary scientists suggest that the dark material on Iapetus's leading surface is a thin layer of swept-up dust that came from another Saturnian moon, Phoebe, as a result of numerous micrometeorite impacts on that moon. However, the color of the leading half of Iapetus and that of Phoebe do not quite match. So other scientists suggest that some active process within Iapetus is responsible. Perhaps it is a layer of dark organic material that formed within Iapetus and then oozed to the surface. However, the puzzle is further compounded by the fact that the dividing line between the two sides is inexplicably sharp. The *Cassini* spacecraft and its planned four-year scientific investigation of the Saturnian system should help resolve this interesting question and shed new light on this strange moon of Saturn. Finally, almost all Saturn's moons except for Iapetus and Phoebe are very nearly in the plane of Saturn's equator. Iapetus has an inclination of 14.7°, and Phoebe has an inclination of 175.3°.

See also SATURN.

ice catastrophe A planetary climatic extreme in which all the liquid water on the surface of a life-bearing or potentially life-bearing planet has become frozen or completely glaciated.

See also EARTH SYSTEM SCIENCE; GLOBAL CHANGE.

Ida On August 28, 1993, Ida became the first asteroid to be observed up close by a spacecraft as NASA's *GALILEO* SPACECRAFT sped by on the way to Jupiter and took pictures of it. These images revealed not only that Ida has a cratered surface, but also that it has its own small moon, called Dactyl, which is about 1.6 km by 1.2 km in diameter and orbits some 90 km away from the asteroid. Ida is a potato-shaped main belt asteroid some 58 km long by 23 km wide. This asteroid is about the same age as the Sun, namely some 4.5 billion years old. Ida travels around the Sun in 4.84 years at an average distance of 429,150,000 km. Planetary scientists believe that Dactyl is probably made from pieces of Ida that broke off during an ancient impact.

ideal gas The pressure (p), volume (V), and temperature (T) behavior of many gases at low pressures and moderate temperatures is approximated quite well by the ideal (or perfect) gas equation of state, which is

$$p\,V = N\,R_u T$$

where N is the number of moles of gas and R_u is the universal gas constant.

$$R_u = 8314.5 \text{ joules/kg-mole-K}$$

This very useful relationship is based on the experimental work originally conducted by Robert Boyle (1627–91) (*Boyle's law*), Jacques Charles (1746–1823) (*Charles's law*) and Joseph Louis Gay-Lussac (1778–1850) (*Gay-Lussac's law*). In the ideal gas approximation scientists assume that there are no forces exerted between the molecules of the gas and that these molecules occupy negligible space in the containing region. The ideal gas equation above and its many equivalent forms enjoy widespread application in thermodynamics, as, for example, in describing the performance of an ideal rocket.

See also IDEAL ROCKET.

ideal nozzle The nozzle of an ideal rocket or a nozzle designed according to the ideal gas laws.

ideal rocket A theoretical rocket postulated for design and performance parameters that are then corrected (i.e., modified) in aerospace engineering practice. The ideal rocket model often assumes one or more of the following conditions: a homogeneous and invariant propellant; exhaust gases that behave in accordance with the ideal (perfect) gas laws; the absence of friction between the exhaust gases and the nozzle walls; the absence of heat transfer across the rocket engine and/or nozzle wall(s); an axially directed velocity of all exhaust gases; a uniform gas velocity across every section normal to the nozzle axis; and chemical equilibrium established in the combustion chamber and maintained in the nozzle.

See also ROCKET.

igniter A device used to begin combustion, such as a squib used to ignite the fuel in a solid-propellant rocket. The igniter typically contains a specially arranged charge of a ready-burning composition (such as blackpowder) and is used to amplify the initiation of a primer.

image A graphic representation of a physical object or scene formed by a mirror, lens, or electro-optical recording device. In remote sensing, the term *image* generally refers to a representation acquired by nonphotographic methods.

See also REMOTE SENSING.

image compensation Movement intentionally imparted to a photographic film system or electro-optical imagery system at such a rate as to compensate for the forward motion of an air or space vehicle when observing objects on the ground. Without image compensation the forward motion of the air or space vehicle would cause blurring and degradation of the image produced by the system.

image converter An electro-optical device capable of changing the spectral (i.e., wavelength) characteristics of a radiant image. The conversion of an infrared image to a visible image, an X-ray image to a visible image, and a gamma-ray image to a visible image are all examples of image conversion performed by this type of device.

image processing A general term describing the contemporary technology used for acquiring, transferring, analyzing, displaying, and applying digital imagery data. Often this image processing can take place in real time or near real time. It is especially useful in defense applications that involve the use of mobile sensors for real-time target acquisition and guidance, the processing and display of large complex data sets, timely data transmission and compression techniques, and the three-dimensional presentation of battlefield information. The remote sensing of Earth and other planetary objects also has been greatly enhanced by advanced image processing techniques. These techniques (often conducted on today's personal computers) enable remote sensing analysts to compare, interpret, archive, and recall vast quantities of "signature" information emanating from objects on Earth or other planetary bodies. Modern image processing supports applications in such important areas as crop forecasting, rangeland and forest management, land use planning, mineral and petroleum exploration, mapmaking, water quality evaluation, and disaster assessment.

See also REMOTE SENSING.

imagery Collectively, the representations of objects produced by image-forming devices such as electro-optical sensor systems and photographic systems.

imaging instruments Optical imaging from scientific spacecraft is performed by two families of detectors, vidicons and the newer charge-coupled devices (CCDs). Although the detector technology differs, in each case an image of the target celestial object is focused by a telescope onto the detector, where it is converted to digital data. Color imaging requires three exposures of the same target through three different color filters selected from a filter wheel. Ground processing combines data from the three black-and-white images, reconstructing the original color by using three values for each picture element (pixel).

A vidicon is a vacuum tube resembling a small cathode ray tube (CRT). An electron beam is swept across a phosphor coating on the glass where the image is focused, and its electrical potential varies slightly in proportion to the levels of light it encounters. This varying potential becomes the basis of the video signal produced. *Viking*, *Voyager*, and many earlier NASA spacecraft used vidicon-based imaging systems to send back spectacular images of Mars (*Viking*) and the outer planets, Jupiter, Saturn, Uranus, and Neptune (*Voyager*).

The newer charge-coupled device (CCD) imaging system is typically a large-scale integrated circuit that has a two-dimensional array of hundreds of thousands of charge-isolated wells, each representing a pixel. Light falling on a well is absorbed by a photoconductive substrate (e.g., silicon) and releases a quantity of electrons proportional to the intensity of the incident light. The CCD then detects and stores an accumulated electrical charge representing the light level on each well. These charges subsequently are read out for conversion to digital data. CCDs are much more sensitive to light over a wider portion of the electromagnetic spectrum than are vidicon tubes; they are also less massive and require less energy to operate. In addition, they interface more easily with digital circuitry, simplifying (to some extent) onboard data processing and transmission back to Earth. The *Galileo* spacecraft's solid state imaging (SSI) instrument contained a CCD with an 800 by 800 pixel array.

See also CHARGE-COUPLED DEVICE (CCD); GALILEO PROJECT/SPACECRAFT; VIKING PROJECT; *VOYAGER* SPACECRAFT.

Imbrium Basin An ancient impact crater on the Moon's nearside. Planetary scientists suggest that this basin formed about 3.9 billion years ago as a result of an enormous impact on the Moon. About 1,300 km across, the impact that formed this basin came close to splitting the young Moon apart.

See also MOON.

impact 1. A single forceful collision of one mass in motion with a second mass that may be either in motion or at rest; for example, the impact of a meteoroid traveling at high velocity

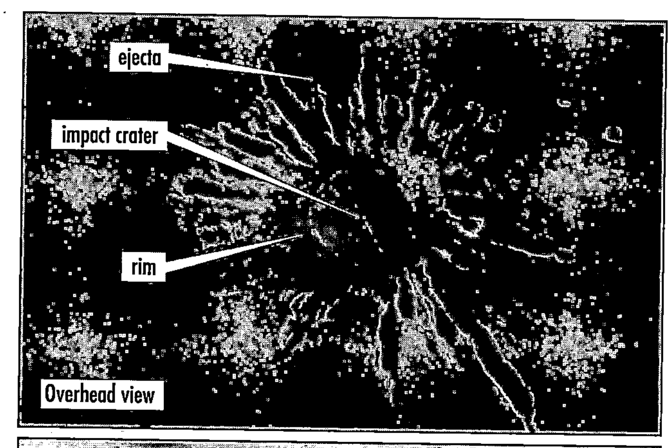

Ideal example of a small, fresh impact crater

with the surface of the Moon. 2. Specifically in aerospace operations, the action or event of an object, such as a rocket or a space probe, striking the surface of a planet or natural satellite; the time of this event, as in from launch to impact. 3. To strike an object or surface, as in the kinetic energy weapon impacted its target at a velocity of 5 kilometers per second or the missile impacted 20 minutes after launch.

impact crater The crater or basin formed on the surface of a planetary body as a result of a high-speed impact of a meteoroid, asteroid, or comet.

See also CRATER; EXTRATERRESTRIAL CATASTROPHE THEORY.

impact line With respect to missile range safety, an imaginary line on the outside of the destruct line and running parallel to it, which defines the outer limits of impact for fragments from a rocket or missile destroyed under command destruct procedures. Also called impact limit line.

See also ABORT.

impactor 1. A meteoroid, asteroid, or comet that strikes a planetary surface at high velocity, creating an impact crater. 2. A hypervelocity projectile fired by a kinetic energy weapon (KEW) system that strikes its target. 3. Piece of space debris that strikes an aerospace vehicle or spacecraft at hypervelocity.

impedance (symbol: Z) A quantity describing the total opposition to current flow (both resistance [R] and reactance [X]) in an alternating current (ac) circuit. For an ac circuit, the impedance can be expressed as:

$$Z^2 = R^2 + X^2$$

where Z is the impedance, R is the resistance, and X is the reactance, all expressed in ohms (Ω).

impeller A device that imparts motion to a fluid. For example, in a centrifugal compressor the impeller is a rotary disk that, faced on one or both sides with radial vanes, accelerates the incoming fluid outward into a diffuser.

impinging-stream injector In a liquid-propellant rocket engine, a device that injects the fuel and oxidizer into the combustion chamber in such a manner that the streams of fluid intersect each other.

See also ROCKET.

implosion The rapid inward collapse of the walls of an evacuated system or device as a result of the failure of the walls to sustain the ambient pressure. An implosion also can occur when a sudden sharp inward pressure force is exerted uniformly on the surface of an object or composite structure, causing it to crush and collapse inward. A submarine diving below the "crush depth" of its hull design will experience an implosion. The *Galileo* probe plunging into the depths of the Jovian atmosphere in December 1995 also experienced an implosion at the end of its brief but successful scientific mission. If the application of external pressure is sufficiently rapid and intense, even a solid metal object will experience some degree of "squeezing" and compression as its volume is decreased and its density increased during an implosion.

See also GALILEO PROJECT/SPACECRAFT.

impulse (symbol: I) In general, a mechanical "jolt" delivered to an object that represents the total change in momentum the object experiences. Physically, the thrust force (F) integrated over the period of time (t_1 to t_2) it is applied. The impulse theorem is expressed mathematically as:

$$\text{Impulse (I)} = \int_1^2 F\,dt = \text{change of momentum of object,}$$

where F is the time-dependent thrust force applied from t_1 to t_2. If the thrust force is constant (or can be treated as approximately constant) during the period, then the impulse (I) becomes simply:

$$I = F\,\Delta t \text{ (the impulse approximation)}$$

where Δt is the total time increment. Impulse (I) has the units of newton-second (N-s) in the SI system. Also called the total impulse, I_{total}. *Compare with but do not confuse with* SPECIFIC IMPULSE (I_{sp}).

impulse intensity Mechanical impulse per unit area. The SI unit of impulse intensity is the pascal-second (Pa-s). A conventionally used unit of impulse intensity is the tap, which is 1 dyne-second per square centimeter.

$$1 \text{ tap} = 0.1 \text{ Pa-s}$$

incandescence Emission of light due to high temperature of the emitting material, as, for example, the very hot filament in an incandescent lamp. Any other emission of light is called luminescence.

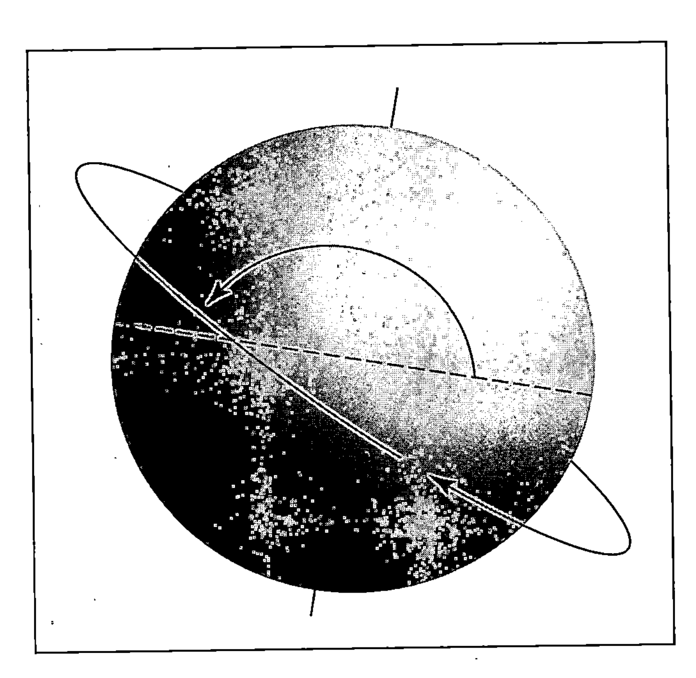

Inclination (i) is one of the six Keplerian elements used to describe the orbit of an object around its primary (central body). The inclination is the angle of the orbital plane with respect to the equatorial plane of the central body. ***(Diagram courtesy of NASA)***

inclination (symbol: i) One of the six Keplerian (orbital) elements, inclination describes the angle of an object's orbital plane with respect to the central body's equator. For Earth-orbiting objects, the orbital plane always goes through the center of Earth, but it can tilt at any angle relative to the equator. By general agreement, inclination is the angle between Earth's equatorial plane and the object's orbital plane measured counterclockwise at the ascending node.

See also KEPLERIAN ELEMENTS; ORBITS OF OBJECTS IN SPACE.

incompressible fluid A fluid for which the density (ρ) is assumed constant.

independent system In aerospace engineering, a system not influenced by other systems. For example, an independent circuit would not require other circuits to be functioning properly in order to perform its task. This "independent" characteristic requires that the circuit be powered from an independent power supply, be controlled from a single source, and (in terms of redundancy) be tolerant of all credible failure modes in the corresponding redundant circuit or system. To make a system truly independent, the aerospace engineer also must exclude interface items such as mounting brackets and connectors that can create a common failure.

Indian Space Research Organization (ISRO) The organization that conducts the major portions of the national space program of India, including the development of launch vehicles, satellites, and sounding rockets. Under its government-assigned charter, ISRO has produced the *Indian National Satellite* (INSAT) for communications, television broadcasting, meteorology, and disaster warning and the *Indian Remote Sensing* (IRS) satellite for resource monitoring and management. ISRO's launch vehicle program started with the development of the SLV-3, first successfully flown on July 18, 1980. This effort was followed by the development and operational use of the Augmented Satellite Launch Vehicle (ASLV) in 1992 and the Polar Satellite Launch Vehicle (PSLV) in 1997. The Geosynchronous Satellite Launch Vehicle (GSLV) is under development.

induction heating The heating of an electrically conducting material using a varying electromagnetic field to induce eddy currents within the material. This may be an undesirable effect in electric power generation and distribution equipment but a desirable effect in materials processing (e.g., an induction heating furnace).

inert In aerospace operations, not containing an explosive or other hazardous material such that checkout operations can be conducted in a less restrictive manner and so that anomalies can occur in a safe configuration.

inert gas Any one of six "noble" gases: helium (He), neon (Ne), argon (Ar), krypton (Kr), xenon (Xe), and radon (Rn), all of which are almost completely chemically inactive. All of these gases are found in Earth's atmosphere in small amounts.

inertia The resistance of a body to a change in its state of motion. Mass (m) is an inherent property of a body that helps quantify inertia. SIR ISAAC NEWTON's first law of motion sometimes is called the "law of inertia." Namely, a body in motion will remain in motion unless acted upon by an external force, and a body at rest will remain at rest unless acted upon by an external force.

See also NEWTON'S LAWS OF MOTION.

inertial guidance system A guidance system designed to project a missile or rocket over a predetermined path, wherein the path of the vehicle is adjusted after launching by (inertial) devices wholly within the missile or rocket and independent of outside information. The inertial guidance system measures and converts accelerations experienced to distance traveled in a certain direction.

See also GUIDANCE SYSTEM.

inertial orbit An orbit that conforms with JOHANNES KEPLER's laws of celestial motion. Kepler's laws describe the behavior of natural celestial bodies and spacecraft that are not under any type of propulsive power.

See also KEPLER'S LAWS; ORBITS OF OBJECTS IN SPACE.

inertial upper stage (IUS) A versatile orbital transfer vehicle (OTV) developed by the U.S. Air Force and manufactured by the Boeing Company. The IUS is a two-stage payload delivery system that travels from low-Earth orbit (LEO) to higher altitude destinations, such as geosynchronous orbit.

The versatile upper-stage vehicle is compatible with both the space shuttle and the Titan IV expendable launch vehicle. The IUS consists of two high-performance, solid-propellant rocket motors, an interstage section, and supporting guidance, navigation, and communications equipment. The first-stage rocket motor typically contains about 9,700 kg of solid propellant and generates a thrust of about 185,000 newtons. The second-stage solid rocket motor contains about 2,720 kg of propellant and generates some 78,300 newtons of thrust. The nozzle of the second stage has an extendable exit cone for increased system performance.

In July 1999 an IUS system propelled NASA's *Chandra X-Ray Observatory* from low-Earth orbit (LEO) into its operational highly elliptical orbit around Earth that reaches one-third of the distance to the Moon. In addition, the IUS helped send the *Galileo* spacecraft on its journey to Jupiter, the *Magellan* spacecraft to Venus, and the *Ulysses* spacecraft on its extended journey to explore the polar regions of the Sun. The U.S. Air Force has used IUS rockets to send early-warning satellites, such as the large Defense Support Program spacecraft, from low-Earth orbit to their operational locations in geosynchronous orbit.

"infective theory of life" The belief that some primitive form of life—perhaps selected, hardy bacteria or bioengineered microorganisms—was placed on an ancient Earth by members of a technically advanced extraterrestrial civilization. This planting, or "infecting," of simple microscopic life on a then-lifeless planet could have been intentional (that is, *directed panspermia*) or accidental—for example, through the arrival of a biologically contaminated space probe or even from space garbage left behind by ancient extraterrestrial visitors.

See also LIFE IN THE UNIVERSE; PANSPERMIA.

inferior planet(s) Planets that have orbits that lie inside Earth's orbit around the Sun—namely, Mercury and Venus.

infinity **(symbol: ∞)** A quantity beyond measurable limits.

inflation **(inflationary universe)** The inflation theory proposes a period of extremely rapid (exponential) expansion of the universe very shortly after the big bang, during which time the energy density of the universe was dominated by a cosmological constant term that later decayed to produce the matter and radiation that fill the present universe. The inflation theory links important ideas in modern physics, such as symmetry breaking and phase transitions, to cosmology.

See also BIG BANG (THEORY); COSMOLOGICAL CONSTANT; COSMOLOGY.

in-flight phase The flight of a missile or rocket from launch to detonation or impact; the flight of a spacecraft from launch to the time of planetary flyby, encounter and orbit, or impact.

in-flight start An engine ignition sequence after take-off and during flight. This term includes starts both within and above the sensible atmosphere.

Infrared Astronomical Satellite **(IRAS)** The *Infrared Astronomical Satellite,* which was launched in January 1983, was the first extensive scientific effort to explore the universe in the infrared (IR) portion of the electromagnetic spectrum. IRAS was an international effort involving the United States, the United Kingdom, and the Netherlands. By the time IRAS ceased operations in November 1983, this successful space-based infrared telescope had completed the first all-sky survey in a wide range of IR wavelengths with a sensitivity 100 to 1,000 times greater than any previous telescope.

See also *SPITZER SPACE TELESCOPE.*

infrared astronomy (IR) The branch of astronomy dealing with infrared (IR) radiation from celestial objects. Most celestial objects emit some quantity of infrared radiation. However, when a star is not quite hot enough to shine in the visible portion of the electromagnetic spectrum, it emits the bulk of its energy in the infrared. IR astronomy, consequently, involves the study of relatively cool celestial objects, such as interstellar clouds of dust and gas (typically about 100 kelvins [K]) and stars with surface temperatures below about 6,000 K.

Many interstellar dust and gas molecules emit characteristic infrared signatures that astronomers use to study chemical processes occurring in interstellar space. This same interstellar dust also prevents astronomers from viewing visible light coming from the center of the Milky Way galaxy. However, IR radiation from the galactic nucleus is absorbed not as severely as radiation in the visible portion of the electromagnetic spectrum, and IR astronomy enables scientists to study the dense core of the Milky Way. Infrared astronomy also allows astrophysicists to observe stars as they are being formed (they call these objects *protostars*) in giant clouds of dust and gas (called *nebulas*), long before their thermonuclear furnaces have ignited and they have "turned on" their visible emission.

Unfortunately, water and carbon dioxide in Earth's atmosphere absorb most of the interesting IR radiation arriving from celestial objects. Earth-based astronomers can use only a few narrow IR spectral bands or windows in observing the universe, and even these IR windows are distorted by "sky noise" (undesirable IR radiation from atmospheric molecules). With the arrival of the space age, however, astronomers have placed sophisticated IR telescopes (such as the *Infrared Astronomical Satellite*) in space, above the limiting and disturbing effects of Earth's atmosphere, and have produced comprehensive catalogs and maps of significant IR sources in the observable universe.

See also ASTROPHYSICS; *INFRARED ASTRONOMICAL SATELLITE; SPITZER SPACE TELESCOPE.*

infrared imagery (IR) That imagery produced as a result of electro-optically sensing electromagnetic radiations emitted or reflected from a given object or target surface in the infrared portion of the electromagnetic spectrum, which extends from approximately 0.72 micrometers (μm) (near-infrared) to about 1,000 micrometers (μm) (far-infrared). *Thermal imagery* is infrared imagery associated with the 8- to 14-micrometer (μm) thermal infrared region. Infrared imagery *is not* the same as photographs produced using infrared film.

See also ELECTROMAGNETIC SPECTRUM; IMAGE CONVERTER.

infrared photography (IR) Photography employing an optical system and direct image recording on "infrared film" that is sensitive to near-infrared wavelengths—about 1.0 micrometer (μm). Infrared photography *is not* the same as infrared imagery.

infrared radiation (IR) That portion of the electromagnetic (EM) spectrum lying between the optical (visible) and radio wavelengths. It is generally considered to span three decades of the EM spectrum, from 1 micrometer (μm) to 1,000 micrometers (μm) wavelength. The British-German astronomer SIR WILLIAM HERSCHEL is credited with the discovery of infrared radiation.

See also ELECTROMAGNETIC SPECTRUM.

infrared radiometer (IR) A telescope-based instrument that measures the intensity of thermal (infrared) energy radiated by targets. This type of instrument often is found as part of the instrument package on a scientific spacecraft. For example, by filling the IR radiometer's field of view completely with the disc of a target planet and measuring its total thermal energy output, the planet's thermal energy balance can be computed, revealing to space scientists the ratio of solar heating to the planet's internal heating.

infrared sensor (IR) A sensor (often placed on a military satellite) that detects the characteristic infrared (IR) radiation from a missile plume, a reentry vehicle, or a "cold body" target in space. Depending on the detector materials and instrument design (e.g., actively cooled versus passively cooled), IR sensors respond to different portions of the IR region of the electromagnetic spectrum and, therefore, support different applications in defense and space science.

See also BALLISTIC MISSILE DEFENSE; DEFENSE SUPPORT PROGRAM; INFRARED ASTRONOMY.

infrared telescope (IR) A telescope specifically designed to observe the universe and celestial objects within it in the infrared (IR) portion of the electromagnetic spectrum. Astronomers place ground-based infrared telescopes at the summit of very high mountains to help the instrument avoid most of Earth's interfering atmosphere. Astronomers also use aircraft (flying observatories) to carry infrared telescopes above most of Earth's atmosphere for short periods (hours) of observation. However, the most sensitive and important infrared telescopes are space-based IR telescopes, such as the *Spitzer Space Telescope*. Modern astronomers enjoy the unique data collection benefits of sophisticated space-based IR telescopes. They can continuously view the universe across the entire IR portion of the spectrum completely unhampered by our planet's limiting atmosphere.

See also AIRBORNE ASTRONOMY; KECK OBSERVATORY; *SPITZER SPACE TELESCOPE*.

inhibit device An electromechanical device that prevents a hazardous event from occurring. This device has direct control—that is, it is not simply a device monitoring a potentially hazardous situation, nor is it in indirect control of some device experiencing the hazardous circumstances. All inhibit devices are independent from one another and are verifiable. A temperature limit switch that shuts down a device or system before a potentially hazardous temperature condition is reached (i.e., at some safe preset temperature limit) is an example of an inhibit device.

inhibitor In general, anything that inhibits; specifically, a substance bonded, taped, or dip-dried onto a solid propellant to restrict the burning surface and to give direction to the burning process.

initiator The part of the solid-rocket igniter that converts a mechanical, electrical, or chemical input stimulus to an energy output that ignites the energy release system. Also called an initiation system.

injection 1. The introduction of fuel and oxidizer in the combustion chamber of a liquid-propellant rocket engine. 2. The time following launch when nongravitational forces (e.g., thrust, lift, and drag) become negligible in their effect on the trajectory of a rocket or aerospace vehicle. 3. The process of putting a spacecraft up to escape velocity or a satellite into orbit around Earth.

injection cooling The method of reducing heat transfer to a body by mass transfer cooling, which is accomplished by injecting a fluid into the local flow field through openings in the surface of the body.

injector A device that propels (injects) fuel and/or oxidizer into a combustion chamber of a liquid-propellant rocket engine. The injector atomizes and mixes the propellants so that they can be combusted more easily and completely. Numerous types of injectors have been designed, including splash plate injectors, shower head (nonimpinging) injectors, and spray-type injectors.

See also ROCKET.

inner planets The terrestrial planets: Mercury, Venus, Earth, and Mars. These planets all have orbits around the Sun that lie inside the main asteroid belt.

See also EARTH; MARS; MERCURY; VENUS.

Innes, Robert Thorburn Ayton (1861–1933) Scottish *Astronomer* In 1915 the Scottish astronomer Robert Thorburn Ayton Innes discovered Proxima Centauri. Except for the Sun, this star is the closest one to Earth (about 4.2 light-years away) and is the faintest member of the Alpha Centauri triple star system.

See also ALPHA CENTAURI.

insertion The process of putting an artificial satellite, aerospace vehicle, or spacecraft into orbit.

insolation The amount of solar energy that falls on a planetary body, usually expressed as radiant flux per unit area per unit time. The *solar constant*, for example, is the total amount of the Sun's radiant energy that normally crosses perpendicular to a unit area at the top of Earth's atmosphere. At one astronomical unit distance from the Sun, the solar constant has a value of 1,371 ± 5 watts per square meter. The

American astronomer SAMUEL P. LANGLEY pioneered solar insolation measurements.

instrument 1. (verb) To provide a rocket, aerospace vehicle, spacecraft, or component with instrumentation. 2. (noun) A device that detects, measures, records, or otherwise provides data about certain quantities (e.g., propellant supply available), activities (e.g., vehicle velocity), or environmental conditions (e.g., space radiation environment or surface temperature on a planet).

instrumentation 1. The installation and use of electronic, gyroscopic, and other instruments for the purpose of detecting, measuring, recording, telemetering, processing, or analyzing different values or quantities as encountered in the flight of a rocket, aerospace vehicle, or spacecraft. 2. The collection of such instruments in a rocket, aerospace vehicle, or spacecraft, as, for example, the instrumentation package for a scientific spacecraft. 3. The specialized branch of aerospace engineering concerned with the design, composition, arrangement, and operation of such instruments.

intact abort An abort of an aerospace vehicle mission in which the crew, payload, and vehicle are returned to the landing site.

See also ABORT; ABORT MODES; SPACE TRANSPORTATION SYSTEM.

integral tank A fuel or oxidizer tank built within the normal contours of an aircraft or rocket vehicle that uses the skin of the vehicle as a wall of the tank.

integrated circuit (IC) Electronic circuits, including transistors, resistors, capacitors, and their interconnections, fabricated on a single small piece of semiconductor material (chip). Categories of integrated circuits such as LSI (large-scale integration) and VLSI (very-large-scale integration) refer to the level of integration, which denotes the number of transistors on a chip.

integration The collection of activities leading to the compatible assembly of payload and launch vehicle into the desired final (flight) configuration.

intensity (symbol: I or *I*) The radiant flux emitted from a point source in a given direction per unit solid angle; typically measured as watts per steradian.

interacting binary *See* BINARY STAR SYSTEM.

intercontinental ballistic missile (ICBM) A ballistic missile with a range capability in excess of 5,500 kilometers.

See also ATLAS (LAUNCH VEHICLE); MINUTEMAN (LGM-30); TITAN (LAUNCH VEHICLE).

interface 1. In general, a common boundary, whether material or nonmaterial, between two parts of a system. 2. In engineering, a mechanical, electrical, or operational common boundary between two elements of an aerospace system, subsystem, or component. 3. In aerospace operations, the point

On November 28, 1958, the Atlas rocket vehicle became the first operational intercontinental ballistic missile (ICBM) developed by the United States during the cold war. Shown here is an Atlas-D booster departing Complex 12 at Cape Canaveral Air Force Station (ca. mid-1960s) as part of Project Fire, a reentry vehicle test program supporting NASA's Apollo Project. ***(Courtesy of NASA)***

An artist's rendering of the Soviet mobile SS-25 (NATO designation: Sickle) intercontinental ballistic missile (ICBM). The deployment of the SS-25 mobile ICBM in the 1980s made Soviet land-based nuclear forces harder to locate and destroy. As shown in this 1986 painting, the missile and support equipment were mounted on massive off-road vehicles that supported rapid dispersal of Soviet strategic nuclear forces during the cold war. The SS-25 carried a single nuclear warhead and was about the same size as the U.S. Air Force Minuteman ICBM. ***(Courtesy of the U.S. Department of Defense/DIA; artist, Edward L. Cooper [1986])***

or area where a relationship exists between two or more parts, systems, programs, persons, or procedures wherein physical and functional compatibility is required. 4. In fluid mechanics, a surface separating two fluids across which there is a dissimilarity of some fluid property such as density or velocity, or some derivative of these properties. The equations of motion do not apply at the interface but are replaced by the boundary conditions.

interference 1. Extraneous signals, noises, and similar disturbing phenomena that hinder proper reception of the desired signal in electronic, optical, or audio equipment. 2. In physics, the mutual effect of the meeting of two or more traveling waves or vibrations of any kind. Sometimes called wave interference.

interferometer An instrument that achieves high angular resolution by combining signals from at least two widely separated telescopes (*optical interferometer*) or a widely separated antenna array (*radio interferometer*). Radio interferometers are one of the basic instruments of radio astronomy. In principle, the interferometer produces and measures interference fringes from two or more coherent wave trains from the same source. These instruments are used to measure wavelengths, to measure the angular width of sources, to determine the angular position of sources (as in satellite tracking), and for many other scientific purposes.

See also VERY LARGE ARRAY (VLA).

intergalactic Between or among the galaxies. Although no place in the universe is truly "empty," the space between clusters of galaxies comes very close. These intergalactic regions are thought to contain less than an one atom in every 10 cubic meters.

intermediate-range ballistic missile (IRBM) A ballistic missile with a range capability from about 1,000 to 5,500 kilometers.

internal energy (symbol: u) A thermodynamic property interpretable through statistical mechanics as a measure of the microscopic energy modes (i.e., molecular activity) of a system. This property appears in a first law of thermodynamics energy balance for a closed system as:

$$du = dq - dw$$

where du is the increment of specific internal energy, dq is the increment of thermal energy (heat) added to the system, and dw is the increment of work done by the system per unit mass.

See also THERMODYNAMICS.

International Astronautical Federation (IAF) Founded in 1951, the International Astronautical Federation (IAF) provides a global forum for the exchange of information on and the promotion of space activities. Each year the IAF organizes an International Astronautical Congress in cooperation with the International Academy of Astronautics (IAA) and the International Institute of Space Law (IISL). The IAF also organizes specialized workshops and is managing an international Earth observations data networking project in cooperation with the European Space Agency (ESA).

U.S. Army personnel prepare to launch a Redstone intermediate-range ballistic missile (IRBM) at the White Sands Missile Range in the early 1950s. A direct technical descendent of the German V-2 military rocket, the Redstone rocket also was pressed into service as a space launch vehicle, helping to successfully place the first American satellite, *Explorer 1*, into orbit in 1958 and then propelling the Project Mercury astronauts Alan B. Shepard, Jr. and Virgil "Gus" I. Grissom on the first two successful American manned space missions (both suborbital) in 1961. *(Courtesy of the U.S. Army)*

International Astronomical Union (IAU) Founded in 1919 in Brussels, Belgium, and now headquartered in Paris, France, the International Astronomical Union (IAU) is the governing body for the science and professional practice of astronomy around the world. The IAU's mission is to promote and safeguard the science of astronomy in all its aspects through international cooperation. Its individual members are professional astronomers all over the world—at the Ph.D. level or beyond—and active in professional research and education in astronomy. However, the IAU also maintains cordial relations with organizations that include amateur astronomers in their membership. National IAU members are generally those countries with a significant level of professional astronomy. With more than 9,000 individual members and 65 national members worldwide, the IAU plays a pivotal role in promoting and coordinating worldwide cooperation in astronomy. The IAU also serves as the internationally rec-

In December 1998 the space shuttle *Endeavour* supported the first on-orbit assembly mission of the *International Space Station* (ISS). As shown here, NASA astronauts James H. Newman (left) and Jerry L. Ross (right) worked between the Zarya and Unity (foreground) modules of the ISS during an extravehicular activity conducted as part of the STS-88 mission. Newman was tethered to the Unity module, while Ross was anchored at the feet to a mobile foot restraint device, which was mounted to the end of the shuttle's robot arm (called the remote manipulator system, or RMS). *(Courtesy of NASA/JSC)*

ognized authority for assigning designations to celestial bodies and any surface features on them.

The scientific and educational activities of the IAU are organized by its 12 scientific divisions and, through them, its 37 more specialized commissions, which cover the full spectrum of astronomy, along with its 86 working and program groups. The long-term policy of the IAU is defined by a general assembly and implemented by an executive committee, while the IAU officers direct day-to-day operations. The focal point of its activities is the permanent IAU Secretariat, located at the Institut d'Astrophysique of Paris, France.

International Space Station (ISS) A major human space flight project headed by NASA. Russia, Canada, Europe, Japan, and Brazil are also contributing key elements to this large, modular space station in low Earth orbit that represents a permanent human outpost in outer space for microgravity research and advanced space technology demonstrations. On-orbit assembly began in December 1998 (*see* figure) with completion originally anticipated by 2004. (*See* figure.) How-

In December 2001 the crew of the space shuttle *Endeavour* photographed the *International Space Station* (ISS) against the blackness of outer space and Earth's horizon. The STS-108 crew snapped this panoramic image as the orbiter vehicle was separating from the station. *(Courtesy of NASA)*

Vital Statistics for the *International Space Station* (as of 06/30/05)

ISS:	Major Elements:
Zarya:	launched Nov. 20, 1998
Unity:	attached Dec. 8, 1998
Zvezda:	attached July 25, 2000
Z1 Truss:	attached Oct. 14, 2000
Soyuz:	docked Apr. 16, 2005
Progress:	docked Jun. 18, 2005
P6 Integrated Truss:	attached Dec. 3, 2000
Destiny:	attached Feb. 10, 2001
Canadarm2:	attached April 22, 2001
Joint Airlock:	attached July 15, 2001
Pirs:	attached Sept. 16, 2001
S0 Truss:	attached April 11, 2002
S1 Truss:	attached Oct. 10, 2002
P1 Truss:	attached Nov. 26, 2002
Weight:	178,594 kg (393,733 lbs.)
Habitable Volume:	425 cubic meters (15,000 cubic feet)
Surface Area (solar arrays):	892 square meters (9,600 square feet)
Dimensions:	
Width:	73 meters (240 feet) across solar arrays
Length:	44.5 meters (146 feet) from Destiny Lab to Zvezda 52 meters (170 feet) with a Progress resupply vessel docked
Height:	27.5 meters (90 feet)

Source: NASA.

ever, the space shuttle *Columbia* accident that took place on February 1, 2003, killing the seven astronaut crewmembers and destroying the orbiter vehicle, has exerted a major impact on the ISS schedule. The Vital Statistics Table and the Expedition Crews Table provide a summary of ISS construction and crew activities through June 30, 2005.

See also SPACE STATION.

International System of Units *See* SI UNIT(S).

International Telecommunications Satellite (INTELSAT) The international organization involved in the development of a family of highly successful commercial communications satellites of the same name.

See also COMMUNICATIONS SATELLITE.

***International Ultraviolet Explorer* (IUE)** A highly successful scientific spacecraft launched in January 1978. Operated jointly by NASA and the European Space Agency (ESA), the IUE has helped astronomers from around the world obtain access to the ultraviolet (UV) radiation of celestial objects in unique ways not available by other means. The spacecraft contains a 0.45-m-aperture telescope solely for spectroscopy in the wavelength range from 115 to 325 nanometers (nm) (1,150 to 3,250 angstroms [Å]). IUE data help support fundamental studies of comets and their evaporation rates when they approach the Sun and of the mechanisms driving the stellar winds that make many stars lose a significant fraction of their mass (before they die slowly as white dwarfs or suddenly in supernova explosions). This long-lived international spacecraft also has assisted astrophysicists in their search to understand the ways by which black holes possibly power the turbulent and violent nuclei of active galaxies.

See also ULTRAVIOLET ASTRONOMY.

Internet An enormous global computer network that links many government agencies, research laboratories, universities, private companies, and individuals. This worldwide computer network had its origins in the "ARPA-net"—a small experimental computer network established by the Advanced Research Project Agency (ARPA) of the U.S. Department of Defense in the 1970s to permit rapid communication among universities, laboratories, and military project offices. Appendix II in the back matter of this book identifies several useful Internet sources for additional information on astronomy, astrophysics, and space technology

interplanetary Between the planets; within the solar system.

interplanetary dust (IPD) Tiny particles of matter (typically less than 100 micrometers in diameter) that exist in space within the confines of our solar system. By convention, the term applies to all solid bodies ranging in size from submicrometer diameter to tens of centimeters in diameter, with corresponding masses ranging from 10^{-17} gram to approximately 10 kilograms. Near Earth the IPD flux is taken as approximately 10^{-13} to 10^{-12} gram per square meter per second (g/m^2-s). Space scientists have made rough estimates that Earth collects about 10,000 metric tons of IPD per year. They also estimate that the entire IPD "cloud" in the solar system has a total mass of between 10^{+16} and 10^{+17} kg.

interstage section A section of a missile or rocket that lies between stages.

See also STAGING.

interstellar Between or among the stars.

interstellar communication and contact Several methods of achieving contact with (postulated) intelligent extraterrestrial life-forms have been suggested. These methods include: (1) interstellar travel by means of starships, leading to physical contact between different civilizations; (2) indirect contact through the use of robot (i.e., uncrewed) interstellar probes; (3) serendipitous contact, such as finding a derelict alien starship or probe drifting in the main asteroid belt; (4) interstellar communication involving the transmission and reception of electromagnetic signals; and (5) very "exotic" techniques involving information transfer through the modulation of

Expedition Crews to the *International Space Station* (as of 06/30/05)

Expedition One	Launch: 10/31/00 Land: 03/18/01 Time: 140 days, 23 hours, 28 minutes Crew: Commander William Shepherd, Soyuz Commander Yuri Gidzenko, Flight Engineer Sergei Krikalev
Expedition Two	Launch: 03/08/01 Land: 08/22/01 Time: 167 days, 6 hours, 41 minutes Crew: Commander Yury Usachev, Flight Engineer Susan Helms, Flight Engineer James Voss
Expedition Three	Launch: 08/10/01 Land: 12/17/01 Time: 128 days, 20 hours, 45 minutes Crew: Commander Frank Culbertson, Soyuz Commander Vladimir Dezhurov, Flight Engineer Mikhail Tyurin
Expedition Four	Launch: 12/05/01 Land: 06/19/02 Time: 195 days, 19 hours, 39 minutes Crew: Commander Yury Onufrienko, Flight Engineer Dan Bursch, Flight Engineer Carl Walz
Expedition Five	Launch: 06/06/02 Land: 12/07/02 Time: 184 days, 22 hours, 14 minutes Crew: Commander Valery Korzun, NASA ISS Science Officer Peggy Whitson, Flight Engineer Sergei Treschev
Expedition Six	Launch: 11/23/02 Land: 05/03/03 Time: 161 days, 19 hours, 17 minutes Crew: Commander Ken Bowersox, Flight Engineer Nikolai Budarin, NASA ISS Science Officer Don Pettit
Expedition Seven	Launch: 04/25/03 Land: 10/27/03 Time: 184 days, 21 hours, 47 minutes Crew: Commander Yuri Malenchenko, Flight Engineer Ed Lu
Expedition Eight	Launch: 10/18/03 Land: 04/29/04 Time: 194 days, 18 hours, 35 minutes Crew: Commander Michael Foale, Flight Engineer Alexander Kaleri, Flight Engineer (ESA) Pedro Duque* ** ESA astronaut Duque launched with Expedition 8 crew on Soyuz TMA-3 spacecraft and returned with Expedition 7 crew on Soyuz TMA-2 spacecraft.*
Expedition Nine	Launch: 04/18/04 Land: 10/23/04 Crew: Commander Gennady Padalka, Flight Engineer Mike Fincke, Flight Engineer (ESA) André Kuipers* ** ESA astronaut Kuipers launched with Expedition 9 crew on Soyuz TMA-4 spacecraft and returned with Expedition 8 crew on Soyuz TMA-3 spacecraft.*
Expedition Ten	Launch: 10/13/04 Land: 04/24/05 Time: 192 days, 19 hours, 2 minutes Crew: Commander Leroy Chiao, Flight Engineer Salizhan Sharipov, Flight Engineer Yuri Shargin* ** Cosmonaut Shargin launched with Expedition 10 crew on Soyuz TMA-5 spacecraft and returned with Expedition 9 crew on Soyuz TMA-4 spacecraft.*
Expedition Eleven	Launch: 04/14/05 Land: 10/07/05 (planned) Crew: Commander Sergei Krikalev, Flight Engineer John Phillips, Flight Engineer (ESA) Roberto Vittori* ** ESA astronaut Vittori launched with Expedition 11 crew on Soyuz TMA-6 and returned with Expedition 10 crew on Soyuz TMA-5.*

Source: NASA.

gravitons, neutrinos, or streams of tachyons; the use of some form of telepathy; and perhaps even matter transport by means of distortions in the space-time continuum that help "beat" the speed-of-light barrier that now underlies our understanding of the physical universe.

See also EXTRATERRESTRIAL CIVILIZATIONS; LIFE IN THE UNIVERSE; SEARCH FOR EXTRATERRESTRIAL INTELLIGENCE (SETI).

interstellar medium (ISM) The gas and dust particles that are found between the stars in our galaxy, the Milky Way. Up until about three decades ago, the interstellar medium was considered to be an uninteresting void. Today, through advances in astronomy (especially infrared and radio astronomy), we know that the interstellar medium contains a rich and interesting variety of atoms and molecules as well as a population of fine-grained dust particles. More than 100 interstellar molecules have been discovered to date, including many organic molecules considered essential in the development of life. Interstellar dust (which "reddens" the visible light from the stars behind it because of its preferential scattering of shorter-wavelength photons) is considered to consist of very fine silicate particles, typically 0.1 micrometers (μm) in diameter. These interstellar "sands" sometimes may have an irregularly shaped coating of water ice, ammonia ice, or solidified carbon dioxide.

interstellar probe A highly automated interstellar spacecraft sent from our solar system to explore other star systems. Most likely this type of probe would make use of very smart machine systems capable of operating autonomously for decades or centuries. Once the robot probe arrived at a new star system, it would begin a detailed exploration procedure. The target star system would be scanned for possible life-bearing planets, and if any were detected, they would become the object of more intense scientific investigations. Data collected by the "mother" interstellar probe and any miniprobes (deployed to explore individual objects of interest within the new star system) would be transmitted back to Earth. There, after light-years of travel, the signals would be intercepted and analyzed by scientists, and interesting discoveries and information would enrich our understanding of the universe.

Robot interstellar probes also might be designed to carry specially engineered microorganisms, spores, and bacteria. If a probe encountered ecologically suitable planets on which life had not yet evolved, then it could "seed" such barren but potentially fertile worlds with primitive life-forms or at least life precursors. In that way human beings (through their robot probes) would not only be exploring neighboring star systems but would be participating in the spreading of life itself through some portion of the Milky Way galaxy.

This artist's rendering shows humanity's first robotic interstellar probe departing the solar system (ca. 2075) on an epic journey of scientific exploration. ***(Courtesy of NASA)***

interstellar travel In general, matter transport between star systems in a galaxy; specifically, travel beyond our own solar system. The "matter" transported may be (1) a robot interstellar probe on an exploration mission, (2) an automated spacecraft that carries a summary of the cultural and technical heritage of a civilization (e.g., the plaques on the *Pioneer 10* and *11* spacecraft and the "recorded message" on the *Voyager 1* and *2* spacecraft), (3) a "crewed" starship on a long-term round-trip voyage of scientific exploration, or even (4) a giant "interstellar ark" that is designed to transport a portion of the human race on a one-way mission beyond the solar system in search of new suitable planetary systems to explore and inhabit.

See also EXTRATERRESTRIAL CIVILIZATIONS; STARSHIP.

intravehicular activity (IVA) Astronaut or cosmonaut activities performed inside an orbiting spacecraft or aerospace vehicle. *Compare with* EXTRAVEHICULAR ACTIVITY.

inverse-square law A relation between physical quantities of the form x proportional to $1/y^2$, where y is usually a distance and x terms are of two kinds, forces and fluxes. SIR ISAAC NEWTON's law of gravitation is an example of the inverse-square law. This relationship is

$$F = [Gm_1\ m_2]/r^2$$

where F is the gravitational force between two objects having masses, m_1 and m_2, G is the gravitational constant, and r is the distance between these two masses. Note that the force of gravitational attraction decreases as the inverse of the distance squared (i.e., as $1/r^2$).

inviscid fluid A hypothesized "perfect" fluid that has a zero coefficient of viscosity. Physically, this means that shear stresses are absent despite the occurrence of shearing deformations in the fluid. The inviscid fluid also glides past solid boundaries without sticking. No real fluids are inviscid. In fact, since real fluids are viscous, they stick to solid boundaries during flow processes, creating thin boundary layers where shear forces are significant. However, the inviscid (perfect) fluid approximation provides a useful model that approximates the behavior of real fluids in many flow situations.

in vitro Literally, "in glass." A life sciences term describing biological tests using cells from an organism but occurring outside that living organism in a laboratory apparatus, such as a glass test tube.

in vivo Literally, "in a living organism (body)." A life sciences term describing biological experiments that take place within (or using) a living organism, such as a test animal or human volunteer. Astronauts often participate in an interesting variety of in vivo life science experiments while on extended flights in microgravity.

Io The pizza-colored volcanic Galilean satellite of Jupiter with a diameter of 3,630 km. Io travels in synchronous rotation around Jupiter at a distance of 422,000 km with a period of 1.77 days. Io was discovered on January 7, 1610, by GALILEO GALILEI. Because of the enormous tidal forces exerted by Jupiter, the tormented surface of Io is constantly changing and renewing itself. Even though Io keeps the same hemisphere pointing toward Jupiter, Europa and Ganymede tug on Io enough to perturb its orbit and create large tidal bulges. In fact, this gravitational tug-of-war causes the surface of Io to bulge up and down by as much as 100 m. Io is the most volcanically active body in the solar system. (*See* figure.) Volcanic plumes spew sulfur and other materials up to 300 kilometers above the Galilean moon's surface, giving Io a surface that is radically different from any other moon in the solar system. Io's interesting and unusual surface contains lakes of molten sulfur, (nonvolcanic) mountains, active volcanoes with extensive lava flows, and numerous calderas, some several kilometers deep. The moon also has a thin atmosphere consisting primarily of sulfur dioxide (SO_2).

See also GALILEAN SATELLITES; JUPITER.

This image of Io represents a historic moment in space exploration because it provided scientists with their first evidence of active volcanism on another celestial body. It was taken on March 4, 1979, when the *Voyager 1* spacecraft was at a distance of 499,000 kilometers. The dark, fountainlike feature near the limb of the pizza-colored Jovian moon is the plumelike structure of ejected material from an ongoing volcanic eruption. The ejected material rises more than 100 km above the moon's surface and follows a ballistic trajectory, producing the characteristic domelike shape of the cloud top. During its flyby encounter *Voyager 1* photographed a total of eight erupting volcanoes, identifying Io as the most geologically active object in the entire solar system. *(Courtesy of NASA/JPL)*

ion An atom or molecule that has lost or (more rarely) gained one or more electrons. By this ionization process, the atom or molecule becomes electrically charged.

ion engine An electrostatic rocket engine in which a propellant (e.g., cesium, mercury, argon, or xenon) is ionized and the propellant ions are accelerated by an imposed electric field to very high exhaust velocity.

See also ELECTRIC PROPULSION.

ionization The process of producing ions by the removal of electrons from, or the addition of electrons to, atoms or molecules. High temperatures, electrical discharges, and nuclear radiations can cause ionization.

ionizing radiation Any type of nuclear radiation that displaces electrons from atoms or molecules, thereby producing ions within the irradiated material. Examples include alpha (α) radiation, beta (β) radiation, gamma (γ) radiation, protons, neutrons, and X-rays.

See also ALPHA PARTICLE; BETA PARTICLE; GAMMA RAYS; SPACE RADIATION ENVIRONMENT.

ionosphere That portion of Earth's upper atmosphere extending from about 50 to 1,000 km, in which ions and free electrons exist in sufficient quantity to reflect radio waves.

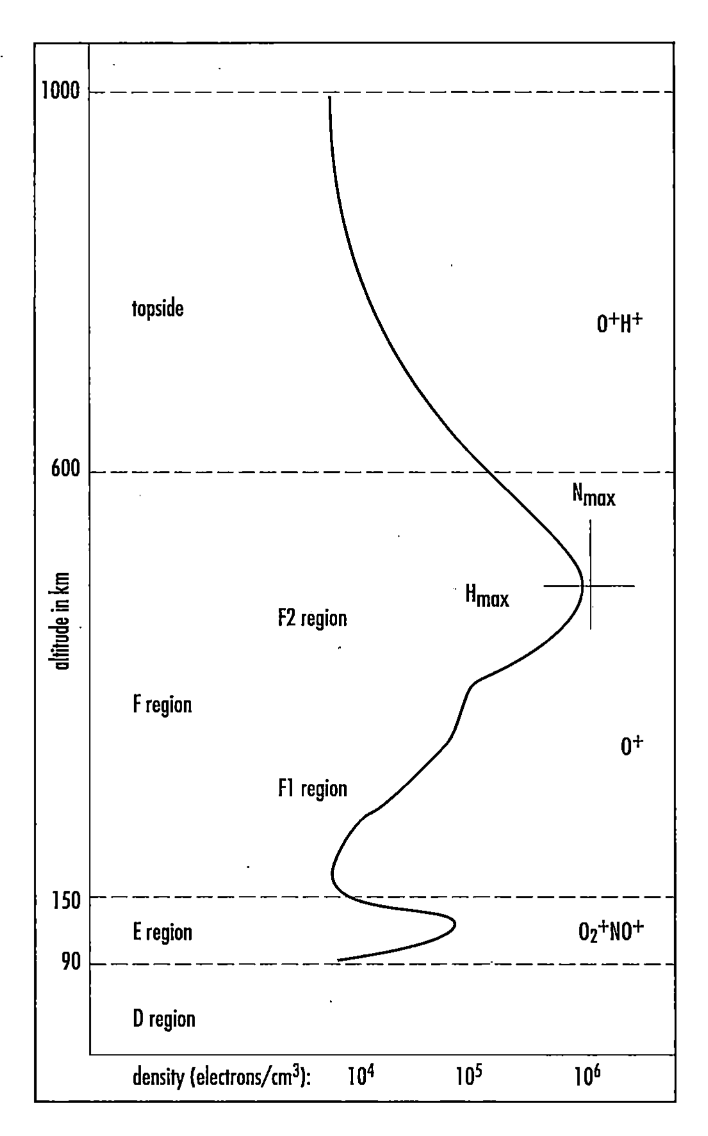

This diagram shows the various layers of Earth's ionosphere and their predominant ion populations as a function of altitude. As shown, the ion density varies considerably. ***(Courtesy of NOAA)***

ionospheric storm Disturbances of Earth's ionosphere (caused by solar flare activity), resulting in anomalous variations in its characteristics and effects on radio wave propagation. In general, ionospheric storms can persist for several days and cover large portions of the globe. Two types of storms of particular interest are the *polar-cap absorption* (PCA) event and the geomagnetically induced storm. As the name implies, the polar-cap absorption (PCA) event occurs at high latitudes and is associated with the arrival of very energetic solar flare protons. Radio communications blackouts usually occur as a result of a PCA event. The geomagnetically induced ionospheric storm occurs along with periods of special auroral activity. Unlike PCA events, the geomagnetic storms are more intense at night. Finally, the *sudden ionospheric disturbance* (SID) appears within a few minutes of the occurrence of strong solar flares and causes a fade-out of long-distance radio communications on the sunlit part of Earth.

irradiance (symbol: E) The total radiant flux (electromagnetic radiation of all wavelengths) received on a unit area of a given surface; usually measured in watts per meter squared (W/m^2).

irregular galaxy A galaxy with a poorly defined structure or shape.

See also GALAXY.

irreversible process In thermodynamics, a process that cannot return both the system and the surroundings to their original conditions. Four common causes of irreversibility in a thermodynamic process are friction, heat transfer through a finite temperature difference, the mixing of two different substances, and unrestrained expansion of a fluid. The exhausting of combustion gases through a rocket engine's nozzle is an irreversible process.

Irwin, James Benson (1930–1991) American *Astronaut, U.S. Air Force Officer* Astronaut James Benson Irwin served as the lunar excursion module (LEM) pilot during NASA's *Apollo 15* mission to the Hadley Rille region of the Moon. As part of this Apollo Project mission, which took place from July 26 to August 7, 1971, Irwin became the eighth person to walk on the lunar surface.

See also APOLLO PROJECT.

Ishtar Terra A very large highland plateau in the northern hemisphere of Venus, about 5,000 km long and 600 km wide.

island universe(s) Term introduced in the 18th century by the German philosopher IMMANUEL KANT to describe other galaxies, which at the time were fuzzy patches of light called by the general term *nebulas*.

See also GALAXY; NEBULA.

isothermal process In thermodynamics, any process or change of state of a system that takes place at constant temperature.

See also THERMODYNAMICS.

isotope One of two or more atoms with the same atomic number (Z) (i.e., the same chemical element) but with different atomic weights. An equivalent statement is that the nuclei of isotopes have the same number of protons but different numbers of neutrons. Therefore, carbon-12 ($^{12}_{6}C$), carbon-13 ($^{13}_{6}C$), and carbon-14 ($^{14}_{6}C$), are all isotopes of the element carbon. The subscripts denote their common atomic number (i.e., Z = 6), while the superscripts denote their differing atomic mass numbers (i.e., A = 12, 13, and 14, respectively), or approximate atomic weights. Isotopes usually have very nearly the same chemical properties but different physical and nuclear properties. For example, the isotope carbon-14 is radioactive, while the isotopes carbon-12 and carbon-13 are both stable (nonradioactive).

isotope power system A device that uses the thermal energy deposited by the absorption of alpha or beta particles from a radioisotope (radioactive material) as the heat source to produce electric power by direct energy conversion or dynamic conversion (thermodynamic cycle) processes. (Note: Radioisotopes that emit gamma rays are not normally considered suitable for this application because of the shielding requirements.) A variety of radioisotope thermoelectric generators (RTGs), powered by the alpha-emitting radioisotope plutonium-238 ($^{238}_{94}Pu$), has been used very successfully to provide continuous long-term electric power to spacecraft such as *Pioneer 10* and *11*, *Voyager 1* and *2*, *Ulysses*, *Galileo*, and the *Viking 1* and *2* lander spacecraft on Mars.

See also CASSINI MISSION; GALILEO PROJECT/SPACECRAFT; *PIONEER 10, 11;* SPACE NUCLEAR POWER; ULYSSES MISSION; VIKING PROJECT; VOYAGER SPACECRAFT.

isotropic Having uniform properties in all directions.

isotropic radiator An energy source that radiates (emits) particles or photons uniformly in all directions.

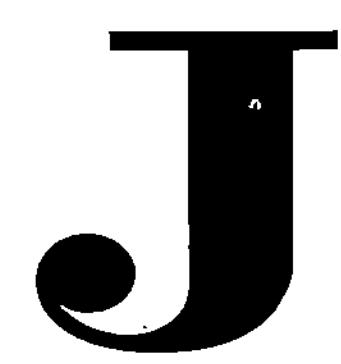

James Webb Space Telescope (JWST) NASA's *James Webb Space Telescope,* previously called the *Next Generation Space Telescope* (NGST), is scheduled for launch in 2010. An expendable launch vehicle will send the spacecraft on a three-month journey to its operational location, about 1.5 million kilometers from Earth in orbit around the second Lagrangian libration point (or L2) of the Earth-Sun system. This distant location will provide an important advantage for the spacecraft's infrared radiation (IR) imaging and spectroscopy instruments. At that location a single shield on just one side of the observatory will protect the sensitive telescope from unwanted thermal radiation from both Earth and the Sun. As a result, the spacecraft's infrared sensors will be able to function at a temperature of 50 kelvins (K) without the need for complicated refrigeration equipment.

The telescope is being designed to detect electromagnetic radiation whose wavelength lies in the range from 0.6 to 28 micrometers (μm) but will have an optimum performance in the 1 to 5 micrometer (μm) region. The JWST will be able to collect infrared signals from celestial objects that are much fainter than are those now being studied with very large ground-based infrared telescopes (such as the Keck Observatory) or the current generation of space-based infrared telescopes (such as the *Spitzer Space Telescope*). With a primary mirror diameter of at least 6 meters, the JWST will make infrared measurements that are comparable in spatial resolution (that is, image sharpness) to images currently collected in the visible light portion of the spectrum by the *Hubble Space Telescope.*

In September 2002 NASA officials renamed the NGST to honor JAMES E. WEBB, the civilian space agency's administrator from February 1961 to October 1968 during the development of the Apollo Project. Webb also initiated many other important space science programs within the fledgling agency. NASA is developing the JWST to observe the faint infrared signals from the first stars and galaxies in the universe. Data from the orbiting observatory should help scientists better respond to lingering fundamental questions about the universe's origin and ultimate fate. One important astrophysical mystery involves the nature and role of dark matter.

jansky (symbol: Jy) A unit used to describe the strength of an incoming electromagnetic wave signal. The jansky frequently is used in radio and infrared astronomy. It is named after the American radio engineer KARL G. JANSKY, who discovered extraterrestrial radio wave sources in the 1930s, a discovery generally regarded as the birth of radio astronomy.

$$1 \text{ jansky (Jy)} = 10^{-26} \text{ watts per meter squared per hertz } [\text{W/(m}^2\text{-Hz)}]$$

See also RADIO ASTRONOMY.

Jansky, Karl Guthe (1905–1950) American *Radio Engineer* In 1932 the American radio engineer Karl Guthe Jansky started the field of radio astronomy by initially detecting and then identifying interstellar radio waves from the direction of the constellation Sagittarius, a direction corresponding to the central region of the Milky Way galaxy. In his honor the unit describing the strength of an extraterrestrial radio wave signal is now called the jansky (Jy).

See also RADIO ASTRONOMY.

Janssen, Pierre-Jules-César (1824–1907) French *Astronomer* The French astronomer Pierre-Jules-César Janssen was a scientific pioneer in the field of solar physics. His precise measurements of the Sun's spectral lines in 1868 enabled SIR JOSEPH NORMAN LOCKYER to conclude the presence of a previously unknown element (which Lockyer called helium). At the end of the 19th century Janssen focused his efforts on the new field of astrophotography and created an extensive collection of more than 6,000 photographic images of the Sun, which he published as a solar atlas in 1904.

See also ASTROPHOTOGRAPHY; SUN.

Janus A small moon of Saturn that orbits at a distance of 151,470 km from the planet. Janus is an irregularly shaped object, 220 km by 160 km, that travels around Saturn with an orbital period of 0.695 days. This moon was discovered in 1966 by the French astronomer Audouin Charles Dollfus (b. 1924). As confirmed in 1980 by data received from the *Voyager 1* spacecraft flyby of Saturn, Janus essentially co-orbits Saturn with another small moon called Epimetheus. Also called Saturn X and 1980 S1.

See also SATURN.

Japanese Aerospace Exploration Agency (JAXA) On October 1, 2003, the Japanese Aerospace Exploration Agency (JAXA) was formed by the merger of the three agencies that had previously promoted aeronautical and space activities within Japan, namely, the Institute of Space and Astronautical Science (ISAS), the National Aerospace Laboratory (NAL), and the National Space Development Agency of Japan (NASDA). Prior to the merger the three aeronautical and space agencies operated individually in their respective areas. Specifically, ISAS was the agency responsible for space and planetary research, NAL was responsible for next generation aviation and space research development, and NASDA was responsible for the development of large-size rockets (such as the H-IIA) and satellites and participation in the *International Space Station.* By forming this merger, the Japanese government intended to create a new agency comparable in scope, responsibility, and activity to the American National Aeronautics and Space Administration (NASA) and the European Space Agency (ESA).

JAXA is exclusively engaged in the peaceful applications of space technology and exploration. Some of the current major operations and activities of JAXA include (1) academic research related to space science in cooperation with the universities, (2) basic research and development involving space technology and aeronautical technology, (3) development of satellites and launch vehicles, (4) launch operations and spacecraft tracking and monitoring operations, and (5) sharing research and development results on a national and international basis.

Jarvis, Gregory B. (1944–1986) American *Astronaut* On January 28, 1986, astronaut Gregory B. Jarvis died in the explosion of the space shuttle *Challenger* while serving as a civilian payload specialist on the ill-fated STS 51-L mission. At the time of this tragic accident Jarvis was an electrical engineer for the Hughes Aircraft Company.

See also *CHALLENGER* ACCIDENT.

Jeans, Sir James Hopwood (1877–1946) British *Astronomer, Physicist* The British astronomer and physicist Sir James Hopwood Jeans proposed various theories of stellar evolution. In 1928 he suggested that matter was being continuously created in the universe. This hypothesis led other scientists to establish a steady-state universe model in cosmology. From 1928 onward Jeans focused on writing popular books about astronomy. His efforts did much to popularize astronomy just at the time big bang cosmology and the notion of an expanding universe began to emerge.

jet(s) 1. A strong, well-defined stream of fluid either issuing from an orifice or moving in a contracted duct, such as the jet of combustion gases coming out of a reaction engine or the jet in the test section in a wind tunnel. 2. A tube or nozzle through which fluid passes or emerges in a "jet," such as the jet in a liquid-propellant rocket motor's injector. 3. A jet engine. 4. A jet aircraft. 5. In astronomy, a narrow, bright feature associated with certain types of celestial objects.

jet propulsion Reaction propulsion in which the propulsion unit obtains oxygen from the air, as distinguished from rocket propulsion, in which the unit carries its own oxidizer (i.e., oxygen-producing material). In connection with aircraft propulsion, the term refers to a turbine jet unit (usually burning hydrocarbon fuel) that discharges hot gas through a tail pipe and a nozzle, producing a thrust that propels the aircraft.

See also ROCKET PROPULSION.

Jet Propulsion Laboratory (JPL) NASA's Jet Propulsion Laboratory (JPL) is located near Pasadena, California, approximately 32 kilometers northeast of Los Angeles. JPL is a government-owned facility operated by the California Institute of Technology under a NASA contract. In addition to the Pasadena site, JPL operates the worldwide Deep Space Network (DSN), including a DSN station at Goldstone, California.

This laboratory is engaged in activities associated with deep space automated scientific missions, such as subsystem engineering, instrument development, and data reduction and analysis required by deep space flight. The *Cassini* spacecraft's exploration of Saturn is a current JPL mission.

See also NATIONAL AERONAUTICS AND SPACE ADMINISTRATION (NASA).

jetstream In aerospace engineering, a jet issuing from an orifice into a much more slowly moving medium, such as the stream of combustion products ejected from a reaction engine. This term also is used in meteorology, but appears as two words, namely *jet stream.*

See also JET STREAM.

jet stream In meteorology, a strong band of wind or winds in the upper troposphere or in the stratosphere moving in a general direction from west to east and often reaching velocities of hundreds of kilometers an hour.

See also JETSTREAM.

jettison To discard. For example, when the propellant in the booster stage of a multistage rocket vehicle is used up, the now-useless booster stage is jettisoned—that is, it is separated from the upper stage(s) of the launch vehicle and allowed to fall back to Earth, usually impacting in a remote ocean area.

Johnson Space Center (JSC) NASA's Lyndon B. Johnson Space Center (JSC) is located about 32 kilometers southeast of downtown Houston, Texas. This center was established in September 1961 as NASA's primary center for (1) the design, development, and testing of spacecraft and associated systems for human space flight; (2) the selection and training of astronauts; (3) the planning and supervision of crewed space missions; and (4) extensive participation in the medical, engineering, and scientific experiments aboard human-crewed space flights. The Mission Control Center (MCC) at JSC is the central facility from which all U.S. crewed space flights are monitored.

Jones, Sir Harold Spencer (1890–1960) British *Astronomer* Sir Harold Spencer Jones was a British astronomer who made precise measurements of the astronomical unit—that is, the average distance from Earth to the Sun. From 1933 to 1955 he served as the tenth Astronomer Royal of England. He also wrote a number of popular books on astronomy, including *Life on Other Worlds* (1940).

joule (symbol: J) The unit of energy or work in the International System of Units. It is defined as the work done (or its energy equivalent) when the point of application of a force of one newton moves a distance of one meter in the direction of the force.

$$1 \text{ joule (J)} = 1 \text{ newton-meter (N-m)} = 1 \text{ watt-second (W-s)} = 10^7 \text{ erg} = 0.2388 \text{ calories}$$

This unit is named in honor of the British physicist James Prescott Joule (1818–89).

See also SI UNITS.

Jovian planet One of the major, or giant, planets in the solar system that is characterized by a large total mass, low average density, and an abundance of the lighter elements (especially hydrogen and helium). The Jovian planets are Jupiter, Saturn, Uranus, and Neptune.

See also JUPITER; NEPTUNE; SATURN; URANUS.

JP-4 A liquid hydrocarbon fuel for jet and rocket engines, the chief ingredient of which is kerosene.

Julian calendar The 12-month (approximately 365-day) calendar introduced by the Roman emperor Julius Caesar in 46 B.C.E.

See also CALENDAR; GREGORIAN CALENDAR.

Juliet A small (about 85-km-diameter) moon of Uranus discovered in 1986 as a result of the *Voyager 2* spacecraft encounter. This moon orbits Uranus at a distance of 64,350 km with a period of 0.49 days. Named after the heroine in William Shakespeare's famous tragic play *Romeo and Juliet.* Also called Uranus XI and S/1986 U2.

See also URANUS; VOYAGER SPACECRAFT.

Juno (asteroid) A large minor planet discovered in 1804 by the German astronomer Karl Ludwig Harding (1765–1834). Juno is a type-S asteroid with a diameter of 240 kilometers and a rotation period of 7.2 hours. Its orbital period is 4.36 years. Juno's orbit around the Sun is further characterized by an aphelion of 3.36 astronomical units (AU), a perihelion of 1.98 AU, an inclination of almost 13°, and an eccentricity of 0.26. Juno was the third minor planet discovered in the main asteroid belt and is sometimes called Juno-3 or even Asteroid-3.

See also ASTEROID.

Juno (launch vehicle) The Juno I and Juno II were early NASA expendable launch vehicles adapted from existing U.S. Army missiles. These rocket vehicles were named after the ancient Roman goddess Juno, queen of the gods and wife of Jupiter, king of the Gods. The Juno I, a four-stage configuration of the Jupiter C vehicle, orbited the first U.S. satellite, *Explorer 1,* on January 31, 1958. The Juno V was the early designation of the launch vehicle that eventually became the Saturn I rocket.

See also EXPLORER 1; JUPITER C; LAUNCH VEHICLE.

Jupiter The largest planet in our solar system, with more than twice the mass of all the other planets and their moons combined and a diameter of approximately 143,000 km. It is the fifth planet from the Sun and is separated from the four terrestrial planets by the main asteroid belt. The giant planet, named after Jupiter, king of the gods in Roman mythology (Zeus in Greek mythology), rotates at a dizzying pace—about once every 9 hours and 55 minutes. It takes Jupiter almost 12 Earth-years to complete a journey around the Sun. Its mean distance from the Sun is about 5.2 astronomical units (AU), or 7.78×10^8 km. The Properties Table provides a summary of the physical and dynamic characteristics of this giant planet.

The planet is distinguished by bands of colored clouds that change their appearance over time. One distinctive feature, the Great Red Spot, is a huge oval-shaped atmospheric storm that has persisted for at least three centuries. (*See* fig-

Physical and Dynamic Properties of Jupiter

Diameter (equatorial)	142,982 km
Mass	1.9×10^{27} kg
Density (mean)	1.32 g/cm^3
Surface gravity (equatorial)	23.1 m/s^2
Escape velocity	59.5 km/s
Albedo (visual geometric)	0.52
Atmosphere	Hydrogen (~89%), helium (~11%), also ammonia, methane, and water
Natural satellites	63
Rings	3 (1 main, 2 minor)
Period of rotation (a "Jovian day")	0.413 day
Average distance from Sun	7.78×10^8 km (5.20 AU) [43.25 light-min]
Eccentricity	0.048
Period of revolution around Sun (a "Jovian year")	11.86 years
Mean orbital velocity	13.1 km/s
Magnetosphere	Yes (intense)
Radiation belts	Yes (intense)
Mean atmospheric temperature (at cloud tops)	~129 K
Solar flux at planet (at top of atmosphere)	50.6 W/m^2 (at 5.2 AU)

Source: NASA.

The planet Jupiter as imaged by the *Voyager 1* spacecraft on January 9, 1979. This planetary portrait, taken when the spacecraft was still 54 million kilometers away from its closest encounter with the gaseous giant, is dominated by Jupiter's Great Red Spot, a persistent and vast atmospheric storm. *(Courtesy of NASA/JPL)*

ure.) In July 1994 20 large fragments from Comet Shoemaker-Levy crashed into the clouds of Jupiter, producing gigantic disturbances in the Jovian cloud system. Scientists estimated that the largest of the comet's fragments (about 3 to 5 km in diameter) produced a blast in the Jovian atmosphere equivalent perhaps to as much as 6 million megatons of TNT.

Jupiter's atmosphere was explored in situ by the *Galileo* spacecraft's atmospheric probe on December 7, 1995. The Jovian atmosphere is composed primarily of hydrogen and helium in approximately "stellar composition" abundances—that is, about 89% hydrogen (H) and 11% helium (He). In fact, Jupiter is sometimes called a near star. If it had been only about 100 times larger, nuclear burning could have started in its core, and Jupiter would have become a star to rival the Sun itself. Its interesting complement of natural satellites even resembles a miniature solar system.

Jupiter has 16 named (known) natural satellites of significant size. The four largest of Jupiter's moons are often called the Galilean satellites. They are Io, Europa, Ganymede, and Callisto. These moons were discovered in 1610 by the Italian astronomer GALILEO GALILEI. Very little actually was known about the Jovian moons until the *Pioneer 10* and *11* and *Voyager 1* and *2* spacecraft encountered Jupiter between 1973 and 1979. These flybys provided a great deal of valuable information and initial imagery, including the discovery of three new moons (Metis, Adrastea, and Thebe), active volcanism on Io, and a possible liquid-water ocean beneath the smooth icy surface of Europa. Then, the highly elliptical orbit of NASA's *Galileo* spacecraft provided the opportunity for several closeup inspections of Jupiter's major moons and greatly improved the current knowledge base for the Jovian system. The table on page 326 provides selected physical and dynamic data for the moons of Jupiter.

See also GALILEAN SATELLITES; GALILEO PROJECT/SPACECRAFT; *PIONEER 10, 11* SPACECRAFT; VOYAGER SPACECRAFT.

Jupiter C A modified version of the Redstone ballistic missile developed by the U.S. Army and a direct descendant of the V-2 rocket developed in Germany during World War II. The Juno I, a four-stage configuration of the Jupiter C rocket, orbited the first American satellite, *Explorer 1,* on January 31, 1958.

See also EXPLORER 1; JUNO (LAUNCH VEHICLE).

***Jupiter Icy Moons Orbiter* (JIMO)** NASA's proposed *Jupiter Icy Moons Orbiter* (JIMO) mission involves an advanced technology spacecraft that would orbit three of

Late in the evening on January 31, 1958 (local time), this four-stage configuration of the Jupiter C rocket blasted off from Cape Canaveral Air Force Station and successfully placed the *Explorer 1* spacecraft, the first United States satellite, into orbit around Earth. *(Courtesy of NASA)*

Properties of the Moons of Jupiter

Moon	Diameter (km)	Semimajor Axis of Orbit (km)	Period of Rotation (days)
Sinope	30 (approx)	23,700,000	758 (retrograde)
Pasiphae	40 (approx)	23,500,000	735 (retrograde)
Carme	30 (approx)	22,600,000	692 (retrograde)
Ananke	20 (approx)	21,200,000	631 (retrograde)
Elara	80 (approx)	11,740,000	260.1
Lysithea	20 (approx)	11,710,000	260
Himalia	180 (approx)	11,470,000	251
Leda	16 (approx)	11,110,000	240
Callisto[a]	4,820	1,880,000	16.70
Ganymede[a]	5,262	1,070,000	7.16
Europa[a]	3,130	670,900	3.55
Io[a]	3,640	422,000	1.77
Thebe (1979J2)	100 (approx)	222,000	0.675
Amalthea	270 (approx)	181,300	0.498
Adrastea (1979J1)	40 (approx)	129,000	0.298
Metis (1979J3)	40 (approx)	127,900	0.295

[a] Galilean satellite.
Source: NASA.

Jupiter's most intriguing moons, Callisto, Ganymede, and Europa. All three planet-sized moons may have liquid water oceans beneath their icy surfaces. Following up on the historic discoveries made by the *Galileo* spacecraft, the JIMO mission would make detailed studies of the makeup, history, and potential for sustaining life of each of these three large icy moons. The mission's proposed science goals include scouting for potential life on the moons, investigating the origin and evolution of the moons, exploring the radiation environment around each moon, and determining how frequently each moon is battered by space debris.

The JIMO spacecraft would pioneer the use of electric propulsion powered by a nuclear fission reactor. Contemporary electric propulsion technology, successfully tested on the NASA's *Deep Space One* (DS 1) spacecraft, would allow the planned JIMO spacecraft to orbit three different moons during a single mission. Current spacecraft, such as *Cassini,* have enough onboard propulsive thrust capability (upon arrival at a target planet) to orbit that single planet and then use various orbits to flyby any moons or other objects of interest, such as ring systems. In contrast, the JIMO's proposed nuclear-electric propulsion system would have the necessary long-term thrust capability to gently maneuver through the Jovian system and allow the spacecraft to successfully orbit each of the three icy moons of interest.

See also JUPITER; NUCLEAR-ELECTRIC PROPULSION SYSTEM (NEP); SPACE NUCLEAR POWER.

This artist's rendering depicts NASA's proposed Project Prometheus nuclear reactor–powered, ion-propelled spacecraft entering the Jovian system ca. 2015. The *Jupiter Icy Moons Orbiter* (JIMO) mission would perform detailed scientific studies of Callisto, Ganymede, and Europa (in that order), searching for liquid water oceans beneath their surfaces. Europa is of special interest to the scientific community because its suspected ocean of liquid water could contain alien life-forms. *(Courtesy of NASA/JPL)*

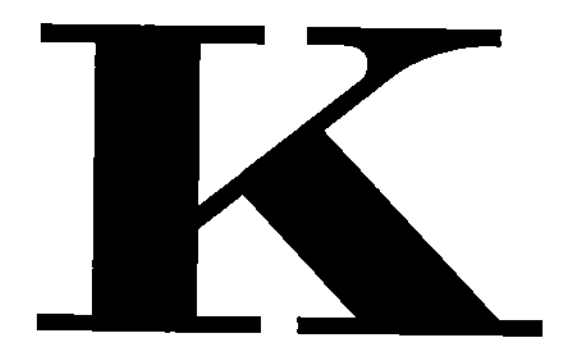

Kant, Immanuel (1724–1804) German *Philosopher* The German philosopher Immanuel Kant was the first astronomer to propose the nebula hypothesis. He did this in the 1755 book *General History of Nature and Theory of the Heavens.* In the nebula hypothesis Kant suggested that the solar system formed out of a primordial cloud of interstellar matter. Kant also introduced the term *island universes* to describe distant collections of stars—now called galaxies. A truly brilliant thinker, his works in metaphysics and philosophy had a great influence on Western thinking in the 18th century and beyond.

Although Kant is primarily regarded as a great philosopher and not an astronomer or physicist, his early work addressed Newtonian cosmology and contained several important concepts that amazingly anticipated future developments in astronomy and astrophysics. First, his *Theory of the Heavens* introduced the nebular hypothesis of the formation of the solar system—namely that the planets formed from a cloud of primordial interstellar material that slowly collected and began swirling around a protosun under the influence of gravitational attraction, and eventually portions of this rotating disc condensed into individual planetary "clumps." PIERRE-SIMON, MARQUIS DE LAPLACE independently introduced a similar version of Kant's nebular hypothesis in 1796. Second, Kant suggested that the Milky Way galaxy was a lens-shaped collection of many stars that orbited around a common center, similar to the way the rings of Saturn orbited around the gaseous giant planet. Third, he speculated that certain distant spiral nebulas were actually other island universes, a term he coined to describe the modern concept of other galaxies. Fourth, Kant suggested that tidal friction was slowing down Earth's rotation just a bit, a bold scientific speculation confirmed by precise measurements in the 20th century. Finally, he also suggested the tides on Earth are caused by the Moon and that this action is responsible for the fact that the Moon is locked in a synchronous orbit around Earth, always keeping the same "face" (called the nearside) presented to observers on our planet. For a person who had just completed his university degree and was about to embark on a long and productive academic career best known for its important philosophical contributions to Western civilization, these were remarkable hypotheses.

Kapteyn, Jacobus Cornelius (1851–1922) Dutch *Astronomer* Toward the end of the 19th century Jacobus C. Kapteyn made significant contributions to the emerging new field of astrophotography—that is, taking analysis-quality photographs of celestial objects. He also pioneered the use of statistical methods to evaluate stellar distributions and the radial and proper motions of stars. Between 1886 and 1900 Kapteyn carefully scrutinized numerous photographic plates collected by SIR DAVID GILL at the Cape of Good Hope Royal Observatory. He then published these data as an extensive, three-volume catalog called the *Cape Photographic Durchmusterung,* which contained the positions and photographic magnitudes of more than 450,000 Southern Hemisphere stars.

Kapteyn was born on January 19, 1851, at Barneveld in the Netherlands. From 1869 to 1875 he studied physics at the University of Utrecht. After graduating with a Ph.D. Kapteyn accepted a position as a member of the observing staff at the Leiden Observatory. During his studies he had not given any special attention to astronomy, but this employment opportunity convinced him that studying stellar populations and the structure of the universe was how he wanted to spend his professional life. In 1878 Kapteyn received an appointment as the chair of astronomy and theoretical mechanics at the University of Groningen. Since the university did not have its own observatory, Kapteyn decided to establish a special astronomical center dedicated to the detailed analysis of photographic data collected by other astronomers. The year following this brilliant career maneuver, he married Catharina Karlshoven, with whom he had three children. In the mid-1880s Kapteyn interacted with Sir David Gill, Royal Astronomer at the Cape of Good Hope

Observatory. The two astronomers developed a cordial relationship, and Kapteyn volunteered to make an extensive catalogue of the stars in the Southern Hemisphere using the photographs taken by Gill at the Cape Observatory. In 1886 Kapteyn's collaboration with Gill began to get underway.

Underestimating the true immensity of the task he had just volunteered for, Kapteyn worked for more than a decade in two small rooms at the University of Groningen. During his labor-intense personal crusade in photographic plate analysis, he enjoyed support from just one permanent assistant, several part-time assistants, and some occasional convict labor from the local prison. The end product of Kapteyn's massive personal effort was the *Annals of the Cape Observatory* (also called the *Cape Photographic Durchmusterung*), an immense three-volume work published between 1896 and 1900 that contained the position and photographic magnitudes of 454,875 Southern Hemisphere stars. His pioneering effort led to the creation of the astronomical laboratory at the University of Groningen and to the establishment of the highly productive "Dutch school" of 20th-century astronomers.

In 1904 Kapteyn noticed that the proper motion of the stars in the Sun's neighborhood did not occur randomly, but rather moved in two different, opposite streams. However, he did not recognize the full significance of his discovery. Years later one of his students, JAN HENDRIK OORT, demonstrated that the star-streaming phenomenon first detected by Kapteyn was observational evidence that the Milky Way galaxy was rotating.

Along with statistical astronomy, Kapteyn was interested in the general structure of the galaxy. In 1906 he attempted to organize a cooperative international effort in which astronomers would determine the population of different magnitude stars in 200 selected areas of the sky and also measure the radial velocities and proper motions of these stars. The radial velocity of a star is a measure of its velocity along the line of sight with Earth. Astronomers use the Doppler shift of a star's spectrum to indicate whether it is approaching Earth (blueshifted spectrum) or receding from Earth (redshifted spectrum). The proper motion of a star is the gradual change in the position of a star due to its motion relative to the Sun. Astronomers define proper motion as the apparent angular motion per year of a star as observed on the celestial sphere—that is, the change in its position in a direction that is perpendicular to the line of sight. Kapteyn gathered enormous quantities of data, but World War I interrupted this multinational cooperative effort.

Nevertheless, Kapteyn examined the data that were collected but overlooked the starlight-absorbing effect of interstellar dust. So in the tradition of WILLIAM HERSCHEL, he postulated that the Milky Way must be lens-shaped, with the Sun located close to its center. The American astronomer HARLOW SHAPLEY disagreed with the Dutch astronomer's erroneous model and in 1918 conclusively demonstrated that the Milky Way galaxy was far larger than Kapteyn had estimated and that the Sun actually resided far away from the galactic center.

Kapteyn's model of the Milky Way suffered when applied to the galactic plane because he lacked sufficient knowledge about the consequences of interstellar absorption, not because he was careless in his evaluation of observational data. On the contrary, he was a very skilled and careful analyst who helped found the important field of statistical astronomy. He died in Amsterdam, Netherlands, on June 18, 1922. Perhaps his greatest overall contribution to astronomical sciences was the creation of the Dutch school of astronomy, an institution that provided many great astronomers who continued his legacy of excellence in data analysis throughout the 20th century.

Kapustin Yar A minor Russian launch complex located on the banks of the Volga River near Volgograd at approximately 48.4° north latitude and 45.8° east longitude. This complex originally was built to support the early Soviet ballistic missile test program; its first missile launch occurred in 1947. Since the launch of *Cosmos 1* in 1962, the Kapustin Yar site has conducted relatively few (less than 100) space launches in comparison to the other two Russian launch complexes, Baikonur Cosmodrome (now in Kazakhstan) and Plesetsk Cosmodrome (south of Archangel in the northwest corner of Russia), which is sometimes called the world's busiest spaceport. However, Kapustin Yar has been used for small and intermediate size payloads, including those launched for or in cooperation with other nations. For example, on April 19, 1975, with 50 scientists and technicians from India in attendance, the Indian satellite *Aryabhata* was sent into orbit from this complex.

See also ARYABHATA; BAIKONUR COSMODROME; PLESETSK COSMODROME.

Kardashev, Nikolai Semenovich (1932–) Russian *Astronomer* While considering factors that could influence the search for extraterrestrial intelligence, the Russian astronomer Nikolai S. Kardashev suggested that there were three possible types (or levels) of extraterrestrial civilizations. He based this characterization on how each type of civilization might be capable of manipulating and harnessing energy resources at the planetary, solar system, and galactic scale. The more energy a particular civilization controlled, the more powerful the signals they might beam through interstellar space. This interesting line of speculation became known within the SETI community as the Kardashev civilizations.

See also KARDASHEV CIVILIZATIONS; SEARCH FOR EXTRATERRESTRIAL INTELLIGENCE (SETI).

Kardashev civilizations The Russian astronomer NIKOLAI S. KARDASHEV, in describing the possible technology levels of various alien civilizations around distant stars, distinguished three types, or levels, of extraterrestrial civilizations. He used the civilization's overall ability to manipulate and harness energy resources as the prime comparative figure of merit. A *Kardashev Type I civilization* would be capable of harnessing the total energy capacity of its home planet; a *Type II civilization*, the energy output of its parent star; while a *Type III civilization* would be capable of using and manipulating the energy output of its entire galaxy.

See also EXTRATERRESTRIAL CIVILIZATIONS; SEARCH FOR EXTRATERRESTRIAL INTELLIGENCE (SETI).

Kármán, Theodore von (1881–1963) Hungarian-American *Mathematician, Engineer* Theodore von Kármán was a gifted aeronautical research engineer who pioneered the application of advanced mathematics in aerodynamics and astronautics. In 1944 he cofounded Caltech's Jet Propulsion Laboratory and initiated research on both solid- and liquid-propellant rockets. This laboratory would eventually create some of NASA's most successful deep space exploration spacecraft. In 1963 he received the first National Medal of Science from President JOHN F. KENNEDY.

Keck Observatory Located near the 4,200-meter high summit of Mauna Kea, Hawaii, the W. M. Keck Observatory possesses two of the world's largest optical/infrared reflector telescopes, each with a 10-meter-diameter primary mirror. Keck I started astronomical observations in May 1993, and its twin, Keck II, in October 1996. The pair of telescopes operate as an optical interferometer.

Keeler, John Edward (1857–1900) American *Astronomer* In 1881 the American astronomer John Edward Keeler assisted SAMUEL PIERPONT LANGLEY in making measurements of solar radiation from Mount Whitney, California. By studying the spectral composition of the rings of Saturn, Keeler showed that the rings were composed of collections of discrete particles, as was earlier theorized by JAMES CLERK MAXWELL.

keep-out zone A volume of space around a space asset (e.g., a military surveillance satellite) that is declared to be forbidden (i.e., "off-limits") to parties who are not owners or operators of the asset. Enforcement of such a zone is intended to protect space assets against attack, especially by space mines. Keep-out zones can be negotiated or unilaterally declared. The right to defend such a zone by force and the legality of unilaterally declared zones under the terms of the current Outer Space Treaty remain to be determined in space law.

kelvin (symbol: K) The International System (SI) unit of thermodynamic temperature. One kelvin is defined as the fraction 1/273.16 of the thermodynamic temperature of the triple point of water. Absolute zero of temperature has a temperature of 0K. This unit is named after the Scottish physicist WILLIAM THOMSON, BARON KELVIN (1824–1907).

See also SI UNITS.

Kelvin, William Thomson, Baron (aka Sir William Thomson) (1824–1907) Scottish *Physicist, Engineer, Mathematician* The multitalented William Thomson, Lord Kelvin made major contributions to science including a better understanding of the electromagnetic theory of light and advances in thermodynamics. For example, he developed the absolute temperature scale that now bears his name. In his honor, the SI unit of absolute temperature is now called the kelvin (K).

Kennedy, John Fitzgerald (1917–1963) American *Politician, 35th President of the United States* In May 1961 President John F. Kennedy boldly proposed that NASA send astronauts to the Moon to demonstrate American space technology superiority over the Soviet Union during the cold war. Shot by an assassin on November 22, 1963, Kennedy did not live to see the triumphant Apollo Project lunar landings (1969–72), a magnificent technical accomplishment that his vision and leadership set in motion almost a decade earlier

Kennedy was born on May 29, 1917, in Brookline, Massachusetts. He graduated from Harvard University in 1940 and then served in the U.S. Navy as a commissioned officer during World War II. Following the war Kennedy became the Democratic representative from the 11th Massachusetts Congressional District and served his district in the House of Representatives from 1946 to 1952. He ran for the U.S. Senate in 1952 and defeated the Republican incumbent, Henry Cabot Lodge, Jr. In the 1960 presidential election, Kennedy narrowly defeated his Republican opponent, Richard M. Nixon, and became the 35th president of the United States.

During his brief term in office (January 20, 1961, to November 22, 1963), President Kennedy had to continuously deal with conflicts involving the Soviet Union, led by an aggressive premier, NIKITA KHRUSHCHEV. Kennedy's challenges included clashes over Cuba and Berlin, as well as a growing world community perception that the United States had lost its technical superiority to the Soviet Union. During Kennedy's presidency the Soviet premier constantly flaunted his nation's space technology accomplishments as an illustration of the superiority of Soviet communism over Western capitalism. President Kennedy worked hard to maintain a balance between American and Soviet spheres of influence in global politics. While not a space technology enthusiast per se, Kennedy recognized that its civilian space technology achievements were giving the Soviet Union greater influence in global politics. Driven by political circumstances early in his presidency, Kennedy took steps to respond to this challenge.

In spring 1961 Kennedy needed something special to restore America's global image as leader of the free world. Space technology was the new, highly visible arena for Soviet-American competition. On April 12, 1961, the Soviet Union launched the first human into orbit around Earth, cosmonaut YURI GAGARIN. The American response was a modest, suborbital Mercury Project flight by astronaut ALAN SHEPARD on May 6, 1961. Soviet premier Khrushchev continued to flaunt his nation's space technology superiority, forcing the young president to make a bold and decisive decision.

During May 1961 Kennedy consulted with his advisers and reviewed many space achievement options with his vice president, Lyndon B. Johnson, who headed the National Aeronautics and Space Council. After much thought Kennedy boldly selected the Moon landing project. Kennedy did so not to promote space science or to satisfy a personal, long-term space exploration vision, but because this mission was a truly daring project that would symbolize American strength and technical superiority in head-to-head cold war competition with the Soviet Union. During a special message to the U.S. Congress on May 25, 1961, in which he discussed urgent national needs, President Kennedy announced the Moon landing mission with these immortal words:

> I believe that this nation should commit itself to achieving the goal, before this decade is out, of landing a man on the Moon and returning him safely to Earth. No single space project in this period will be more impressive

On May 25, 1961, President John F. Kennedy delivered his historic space message to a joint session of the U.S. Congress in which he declared: "I believe this nation should commit itself to achieving the goal, before the decade is out, of landing a man on the Moon and returning him safely to Earth." Appearing in the background are (left) Vice President Lyndon B. Johnson and (right) Speaker of the House Sam T. Rayburn. ***(Courtesy of NASA)***

> to mankind, or more important for the long-range exploration of space; and none will be so difficult or expensive to accomplish.

This speech gave NASA the mandate to expand and accelerate its Mercury Project activities and configure itself to accomplish the "impossible" through the Apollo Project. When Kennedy made his decision, the United States had not yet successfully placed a human being in orbit around Earth. Kennedy's mandate galvanized the American space program and marshaled incredible levels of technical and fiscal resources. Science historians often compare NASA's Apollo Project to the Manhattan Project (World War II atomic bomb program) or the construction of the Panama Canal in extent, complexity, and national expense. On July 20, 1969, two Apollo astronauts, NEIL A. ARMSTRONG and EDWIN E. "BUZZ" ALDRIN, stepped on the lunar surface and successfully fulfilled Kennedy's bold initiative. Sadly, the young president who launched the most daring space exploration project of the cold war did not personally witness its triumphant conclusion. An assassin's bullet took his life in Dallas, Texas, on November 22, 1963. NASA's Kennedy Space Center, site of Launch Complex 39, from which humans left Earth to explore the Moon, bears his name.

Often forgotten in the glare of the Moon landing announcement are several other important space technology initiatives that President Kennedy called for in his historic "Urgent National Needs" speech on May 25, 1961. President Kennedy accelerated development of the Rover nuclear rocket program as a means of preparing for more ambitious space exploration missions beyond the Moon. (Due to a dramatic change in space program priorities, the Nixon administration cancelled this program in 1972.)

Kennedy also requested additional funding to accelerate the use of communication satellites to expand worldwide communications and the use of satellites for worldwide weather observation. Both of these initiatives quickly evolved into major areas of space technology that now serve the global community.

President John F. Kennedy, in responding to the Soviet space technology challenge, satisfied a major dream of space pioneers through the ages. Through his bold and decisive leadership, human beings traveled through interplanetary space and walked on another world for the first time in history.

Kennedy Space Center (KSC) NASA's John F. Kennedy Space Center (KSC) is located on the east coast of Florida 241 km south of Jacksonville and approximately 80 km east of Orlando. KSC is immediately north and west of Cape Canaveral Air Force Station (CCAFS). The center is about 55 km long and varies in width from 8 km to 16 km. The total land and water occupied by the installation is 56,817 hectares. Of this total area, 34,007 hectares are NASA-owned. The remainder of the area is owned by the state of Florida. This large area, with adjoining bodies of water, provides the buffer space necessary to protect the nearby civilian communities during space vehicle launches. Agreements have been made with the U.S. Department of the Interior supporting the use of the nonoperational (buffer) area as a wildlife refuge and national seashore. The complex honors JOHN FITZGERALD KENNEDY, who as 35th president of the United States committed the nation to the Apollo Project.

KSC was established in the early 1960s to serve as the launch site for the Apollo-Saturn V lunar landing missions.

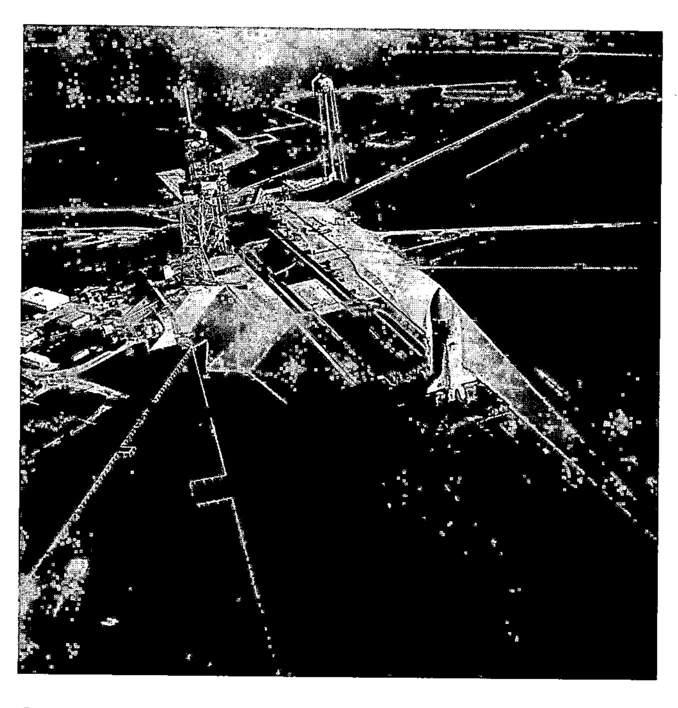

A panoramic early-morning view of Complex 39 (Pad A) at the Kennedy Space Center in Florida as an assembled space shuttle vehicle nears the completion of its nearly six-kilometer-long slow rollout journey from the Vehicle Assembly Building (VAB) to the launch pad. Shown here is the *Challenger* on November 30, 1982. *(Courtesy of NASA)*

After the Apollo Project ended in 1972, Launch Complex 39 was used to support both the *Skylab* program (early nonpermanent U.S. space station) and then the Apollo-Soyuz Test Project (an international rendezvous and docking demonstration involving spacecraft from the United States and the Soviet Union).

The Kennedy Space Center now serves as the primary center within NASA for the test, checkout, and launch of space vehicles. KSC responsibility includes the launching of crewed (space shuttle) vehicles at Launch Complex (LC) 39 and NASA uncrewed, expendable launch vehicles (such as the Delta II rocket) at both nearby Cape Canaveral Air Force Station and at Vandenberg Air Force Base in California.

The assembly, checkout, and launch of the space shuttle vehicles and their payloads take place at the center. Weather conditions permitting, the orbiter vehicle lands at the Kennedy Space Center (after an orbital mission) and undergoes "turn-around," or processing, between flights. The Vehicle Assembly Building (VAB) and the two shuttle launch pads at LC 39 may be the best-known structures at KSC, but other facilities also play critical roles in prelaunch processing of payloads and elements of the space shuttle system. Some buildings, such as the 160-meter-tall VAB, were originally designed for the Apollo Project in the 1960s and then altered to accommodate the space shuttle. Other facilities, such as the Orbiter Processing Facility (OPF) high bays, were designed and built exclusively for the space shuttle program.

See also APOLLO PROJECT; LAUNCH VEHICLE; NATIONAL AERONAUTICS AND SPACE ADMINISTRATION (NASA); SPACE TRANSPORTATION SYSTEM (STS).

Kepler, Johannes (Johann Kepler) (1571–1630) German *Astronomer, Mathematician* The German astronomer Johannes Kepler developed three laws of planetary motion that described the elliptical orbits of the planets around the Sun and provided the empirical basis for the acceptance of NICHOLAS COPERNICUS's heliocentric hypothesis. Kepler's laws gave astronomy its modern, mathematical foundation. His publication *De Stella Nova* (The new star) described the supernova in the constellation Ophiuchus that he first observed (with the naked eye) on October 9, 1604.

Kepler was born on December 27, 1571, in Württemberg, Germany. A sickly child, he pursued a religious education at the University of Tübingen in the hopes of becoming a Lutheran minister. He graduated in 1588 and went on to complete a master's degree in 1591. During this period of religious study, his skill in mathematics also emerged. Consequently, by 1594 he abandoned his plans for the ministry and became a mathematics instructor at the University of Graz in Austria. While pursuing mathematical connections with astronomy, he encountered the new Copernican (heliocentric) model and embraced it.

Throughout his life as an accomplished astronomer and mathematician, Kepler also displayed a strong interest in mysticism. He extracted many of these mystical notions from ancient Greek astronomy, such as the "music of the celestial spheres" originally suggested by PYTHAGORAS. A common practice for many 17th-century astronomers, he dabbled in astrology. He often cast horoscopes for important benefactors, such as the Holy Roman Emperor Rudolf II and Duke (Imperial General) Albrecht von Wallenstein.

In 1596 he published *Mysterium Cosmographicum* (The cosmographic mystery), an intriguing work in which he tried (unsuccessfully) to analytically relate the five basic geometric solids (from Greek mathematics) to the distances of the six known planets from the Sun. The work attracted the attention of TYCHO BRAHE, a famous pretelescope astronomer. In 1600 the elderly Danish astronomer invited Kepler to join him as his assistant in Prague. When Brahe died in 1601, Kepler succeeded him as the imperial mathematician to Rudolf II.

In 1604 Kepler wrote the book *De Stella Nova* (The new star) in which he described a supernova in the constellation Ophiuchus that he first observed on October 9, 1604. This supernova (radio source 3C 358) is called KEPLER'S STAR.

From 1604 until 1609 Kepler's main interest involved a detailed study of Mars. The movement of Mars could not be explained unless he assumed the orbit was an ellipse, with the Sun located at one focus. This assumption produced a major advance in our understanding of the solar system and provided observational evidence of the validity of the Copernican model. Kepler recognized that the other planets also followed elliptical orbits around the Sun. He published this discovery in 1609 in the book *Astronomia Nova* (New astronomy). The book, dedicated to Rudolf II, confirmed the Copernican model and permanently shattered 2,000 years of geocentric Greek astronomy. Astronomers now call Kepler's announcement that the orbits of the planets are ellipses with the Sun as a common focus *Kepler's first law of planetary motion.* Possibly because of his powerful benefactors, Kepler was never officially attacked for supporting Copernican cosmology.

When he published *De Harmonica Mundi* (Concerning the harmonies of the world) in 1619, Kepler continued his great work involving the orbital dynamics of the planets. Although this book extensively reflected Kepler's fascination with mysticism, it also provided a very significant insight that connected the mean distances of the planets from the Sun with their orbital periods. This discovery became known as *Kepler's third law of planetary motion.*

Between 1618 and 1621 Kepler summarized all his planetary studies in the publication *Epitome Astronomica Copernicanae* (Epitome of Copernican astronomy). This work contained *Kepler's second law of planetary motion.* As a point of scientific history, Kepler actually based his second law (the law of equal areas) on a mistaken physical assumption that the Sun exerted a strong magnetic influence on all the planets. Later in the century SIR ISAAC NEWTON (through his universal law of gravitation) provided the "right physical explanation" (within classical physics) for the planetary motion correctly described by Kepler's second law.

Kepler's three laws of planetary motion are (1) the planets move in elliptical orbits, with the Sun as a common focus; (2) as a planet orbits the Sun, the radial line joining the planet to the Sun sweeps out equal areas within the ellipse in equal times; and (3) the square of the orbital period (P) of a planet is proportional to the cube of its mean distance (a) from the Sun—that is, there is a fixed ratio between the time it takes a planet to complete an orbit around the Sun and the size of the orbit. Astronomers express this ratio as P^2/a^3, where a is the semimajor axis of the ellipse and P is the period of revolution around the Sun.

In 1627 Kepler's *Rudolphine Tables* (named after his benefactor, Emperor Rudolf, and dedicated to Tycho Brahe) provided astronomers detailed planetary position data. The tables remained in use until the 18th century. Kepler was a skilled mathematician, and he used the logarithm (newly invented by the Scottish mathematician John Napier [1550–1617]) to help perform the extensive calculations. This was the first important application of the logarithm.

Kepler also worked in the field of optics. Prior to 1610 Galileo and Kepler communicated with each other, although they never met. According to one historic anecdote (ca. 1610), Kepler refused to believe that Jupiter had four moons and behaved like a miniature solar system unless he personally observed them. A Galilean telescope somehow arrived at his doorstep. Kepler promptly used the device and immediately described the four major Jovian moons as *satellites,* a term he derived from the Latin word *satelles,* meaning people who escort or loiter around a powerful person. Kepler's laws are now used extensively in the space age to describe the motion of artificial (human-made) satellites. In 1611 Kepler improved the design of Galileo's original telescope by introducing two convex lenses in place of the one convex lens and one concave lens arrangement used by the Italian astronomer.

Before his death in 1630 Kepler wrote a very interesting novel called *Somnium* (The dream). It is a story about an Icelandic astronomer who travels to the Moon. While the tale contains demons and witches (who help get the hero to the Moon's surface in a dream state), Kepler's description of the lunar surface is quite accurate. Consequently, many historians treat this story, published after Kepler's death in 1634, as the first genuine piece of science fiction.

Kepler fathered 13 children and constantly battled financial difficulties. He worked in a part of Europe (Germany) torn by religious unrest and continual warfare (Thirty Years' War). He died of fever on November 15, 1630, in Regensburg, Bavaria, while searching for new funds from the government officials there.

Keplerian elements The Keplerian elements, or orbital elements, are the six parameters that uniquely specify the position and path of a satellite (natural or human-made) in its orbit as a function of time. These elements and their characteristics are described in the figure. Named after the German astronomer JOHANNES KEPLER.

See also ORBITS OF OBJECTS IN SPACE.

Keplerian telescope In 1609 the Italian astronomer GALILEO GALILEI learned about a new optical instrument, the telescope, which had just been invented in the Netherlands (Holland) by HANS LIPPERSHEY. Within six months Galileo devised his own version of the instrument, a simple refracting telescope, but one three times more powerful than earlier models. Galileo then turned his more powerful instrument to the heavens and started the age of telescopic astronomy. In 1611 the German astronomer JOHANNES KEPLER improved

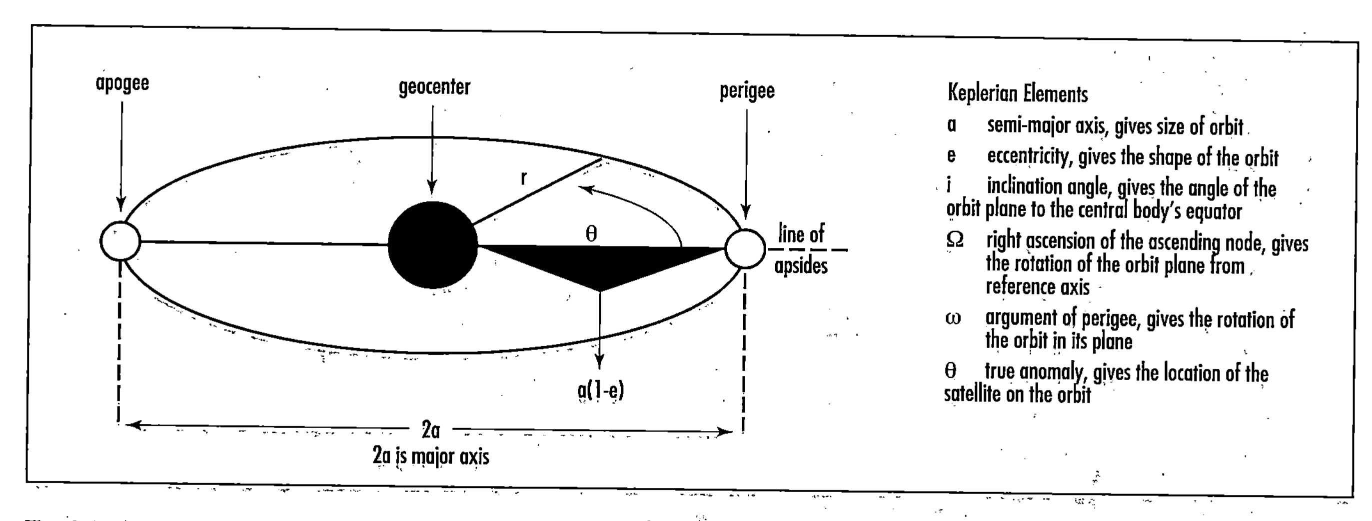

The six Keplerian elements (or classical orbital elements) that uniquely define the position of a satellite

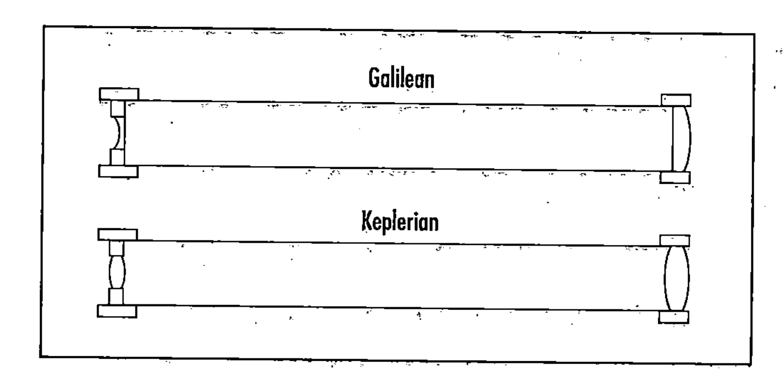

This drawing compares the two refracting telescopes that ushered in the age of telescopic astronomy in the early 17th century. The Galilean telescope, designed by Galileo Galilei in 1609, has a converging lens as the objective and a diverging lens as its eyepiece. The Keplerian telescope, designed by Johannes Kepler in 1611, uses simple convex lenses for both the objective lens and the eyepiece. ***(Courtesy of NASA)***

the design of Galileo's original telescope by introducing two convex lenses in place of the one convex lens and one concave lens arrangement used by the Italian astronomer. The basic Galilean and Keplerian telescopes are compared in the figure.

See also REFRACTING TELESCOPE; TELESCOPE.

Kepler mission Scheduled for launch in 2006, NASA's *Kepler* mission spacecraft will carry a unique space-based telescope specifically designed to search for earthlike planets around stars beyond the solar system. As of 2004, astronomers had discovered more than 90 extrasolar planets. However, these discoveries, made using ground-based telescopes, all involved giant planets similar to Jupiter. Such giant extrasolar planets are most likely to be gaseous worlds composed primarily of hydrogen and helium and unlikely to harbor life. None of the detection methods used to date has had the capability of finding Earthlike planets—tiny objects that are between 30 and 600 times less massive than Jupiter.

The *Kepler* spacecraft's specialized one-meter-diameter telescope will look for the transit signature of extrasolar planets. A transit occurs each time a planet crosses the line of sight (LOS) between that planet's parent star and a distant observer. As the orbiting extrasolar planet blocks some of the light from its parent star, the Kepler spacecraft's sensitive photometer will record a periodic dimming of the incident starlight. Astronomers will then use such periodic signature data to detect the presence of the planet and to determine its size and orbit. Three transits of a star, all with consistent period, brightness changes, and duration, will provide a robust method for detecting and confirming the presence of extrasolar planets around neighboring stars. Scientists can use the measured orbit of the planet and the known properties of the parent star to determine if a newly discovered extrasolar planet lies in the continuously habitable zone (CHZ), that is, at a distance from the parent star where liquid water could exist on the planet's surface. NASA named this cornerstone mission in the search for extraterrestrial life after the famous German astronomer and mathematician JOHANNES KEPLER.

See also EXTRASOLAR PLANETS.

Kepler's laws Three empirical laws describing the motion of the planets in their orbits around the Sun, first formulated by the German astronomer JOHANNES KEPLER on the basis of the detailed observations of the Danish astronomer Tycho Brahe. These laws are (1) the orbits of the planets are ellipses, with the Sun at a common focus; (2) as a planet moves in its orbit, the line joining the planet and the Sun sweeps over equal areas in equal intervals of time (sometimes called the *law of equal areas*); and (3) the square of the orbital period (P) of any planet is proportional to the cube of its mean distance (a) from the Sun (i.e., the semimajor axis a for the elliptical orbit). An empirical statement of Kepler's third law is that for any planet, P^2/a^3 = a constant. About a century after Kepler's initial formulation of these laws of planetary motion, SIR ISAAC NEWTON introduced his law of gravitation and three laws of motion, which provided the mathematical basis

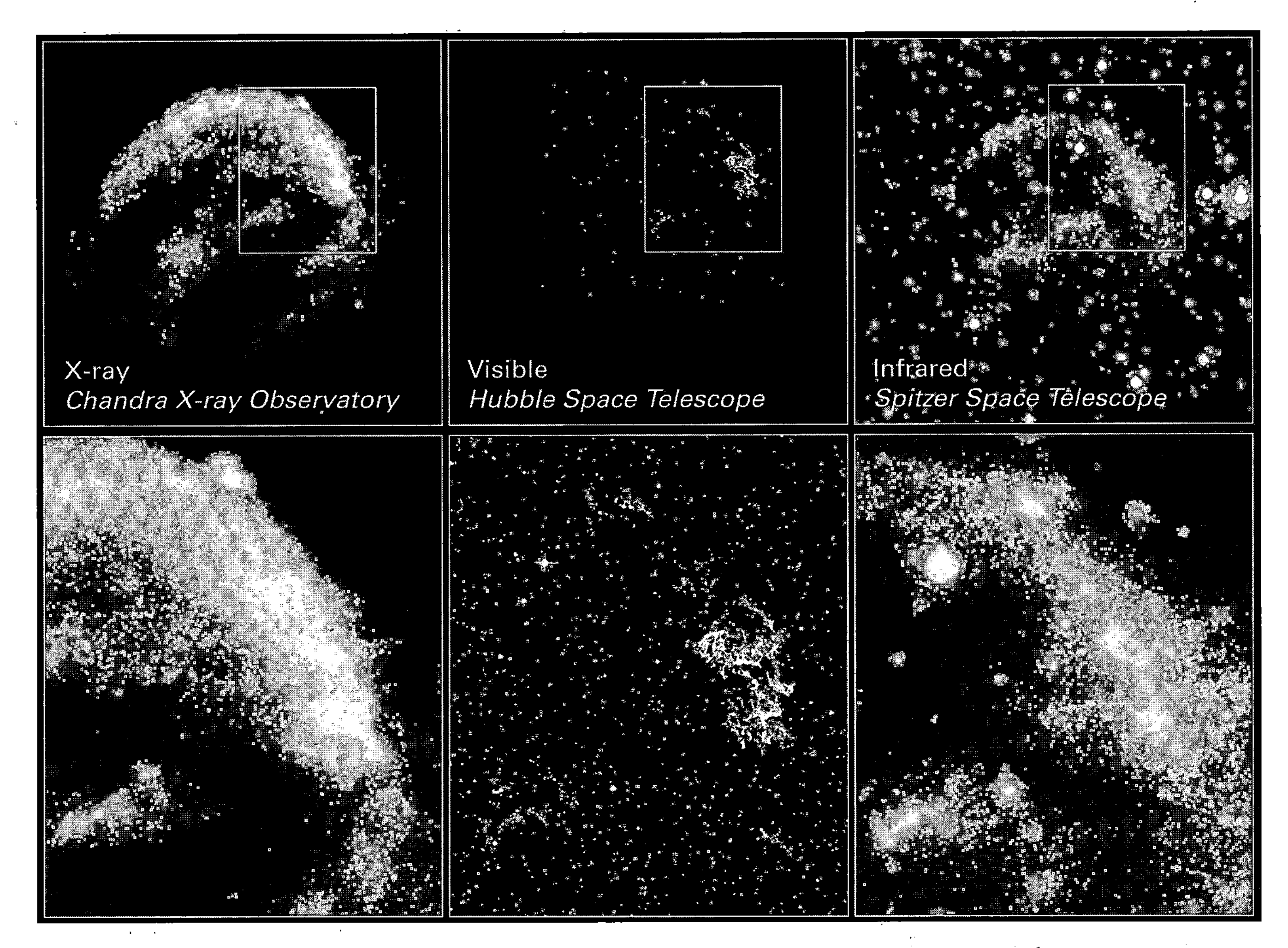

This illustration provides views of Kepler's supernova remnant in the X-ray, visible light, and infrared portions of the electromagnetic spectrum as collected by NASA's *Chandra X-ray Observatory* (CXO), *Hubble Space Telescope* (HST), and *Spitzer Space Telescope* (SST), respectively. The human eye can see neither the X-ray nor infrared regions of the spectrum, so astronomers use special space-based telescopes and then process the data collected to create various color-coded images that make these "invisible radiations" visible and, therefore, more interpretable by the human mind. The lower portion of this composite provides close-up views of the supernova remnant in different portions of the spectrum, each view corresponding to the region outlined in the upper portion of the composite image. ***(Courtesy of NASA/ESA/Johns Hopkins University)***

for a more complete physical understanding of the motion of planets and satellites.

Kepler's star (SN 1604) On October 9, 1604, the German astronomer JOHANNES KEPLER first observed the supernova SN 1604 in the constellation Ophiuchus. He described his observations in the book *De Stella Nova* (The new star). This supernova remained visible to the naked eye for about a year and was studied by astronomers in Europe, Korea, and China. By examining these 17th-century records, modern astronomers recognized Supernova 1604 as a Type 1a supernova. Astronomers now study Kepler's Supernova Remnant using a variety of space-based instruments, including the *Hubble Space Telescope,* the *Chandra X-Ray Observatory,* and the *Spitzer Space Telescope.*

See also SUPERNOVA.

kerosene A liquid hydrocarbon fuel used in certain rocket and jet engines. Also spelled kerosine.

Kerr, Roy Patrick (1934–) New Zealander *Mathematician* In 1963 the New Zealander mathematician Roy Patrick Kerr applied general relativity to describe the properties of a rapidly rotating but uncharged black hole. Astrophysicists think this model is the most likely "real-world" black hole because the massive stars that formed them would have been rotating. One postulated feature of a rotating Kerr black hole is its ringlike structure (the ring singularity) that might give rise to two separate event horizons. Some daring astrophysicists have even suggested it might become possible (at least in theory) to travel through the second event horizon and emerge into a new universe or possibly a different part of this universe. Another theoretical black hole model, called the

Kerr-Newman black hole, appeared in 1965. In this model the black hole has both charge and angular momentum. However, astrophysicists currently think that rotating black holes are unlikely to have a significant electric charge, so the uncharged Kerr black hole remains the more favored real-world candidate model for the stellar black hole.

See also BLACK HOLE.

Kerr black hole A rapidly rotating, but uncharged, black hole, first proposed by the New Zealander mathematician ROY PATRICK KERR in 1963.

See also BLACK HOLE.

keV A unit of energy corresponding to 1,000 electron volts; keV means kiloelectron volts.

See also ELECTRON VOLT; SI UNITS.

Khrushchev, Nikita S. (1894–1971) Ukraine-born Russian *Politician, Soviet Premier* The provocative premier of the Soviet Union who used early Russian space exploration achievements to imply the superiority of Soviet communism over the United States (and capitalism) during the cold war. With his permission and encouragement, the Russian rocket engineer SERGEI KOROLEV used a powerful military rocket as a space launch vehicle to successfully place the world's first artificial satellite, called *Sputnik 1,* into orbit around Earth on October 4, 1957. Khrushchev also allowed Korolev to use a military rocket to launch the first human into orbit around Earth, cosmonaut YURI A. GAGARIN, on April 12, 1961. Despite these Soviet space accomplishments, failing domestic economic programs and a major loss of political prestige during the Cuban missile crisis (October 1962) eventually forced Khrushchev from office in October 1964. Yet it was Khrushchev's aggressive use of space exploration as a tool of international politics that encouraged American president JOHN F. KENNEDY to initiate the Apollo Project, ultimately giving all humanity one of its greatest technical triumphs.

kilo- (symbol: k) An SI unit prefix meaning that a basic space/time/mass unit is multiplied by 1,000; as, for example, a kilogram (kg) or a kilometer (km). Note, however, that in computer technology and digital data processing, kilo designated with a capital "K" refers to a precise value of 1,024 (which corresponds in binary notation to 2^{10}). Therefore, a kilobyte (abbreviated KB or K-byte) is actually 1,024 bytes; similarly, a kilobit (Kb or K-bit) stands for 1,024 bits.

See also SI UNITS.

kiloparsec (kpc) A distance of 1,000 parsecs, or approximately 3,261.6 light-years.

kiloton (symbol: kT) A unit of energy defined as 10^{12} calories (4.2×10^{12} joules). This unit is used to describe the energy released in a nuclear detonation. It is approximately the amount of energy that would be released by the explosion of 1,000 metric tons (1 kiloton) of the high explosive trinitrotoluene (TNT).

See also YIELD.

kinetic energy (symbol: KE or E_{KE}) The energy that a body possess as a result of its motion, defined (in classical, or Newtonian, physics) as one-half the product of its mass (m) and the square of its speed (v), that is, $E_{KE} = 1/2\ m\ v^2$.

kinetic energy interceptor An interceptor that uses a nonexplosive projectile moving at very high speed (e.g., several kilometers per second) to destroy a target on impact. The projectile may include homing sensors and onboard rockets to improve its accuracy, or it may follow a preset trajectory (as with a projectile launched from a high-velocity gas or electromagnetic gun).

See also BALLISTIC MISSILE DEFENSE (BMD).

kinetic energy weapon (KEW) A weapon that uses kinetic energy, or the energy of motion, to destroy, or "kill," an object. Kinetic energy weapon projectiles strike their targets at very high velocities (typically more than 5 kilometers per second) and achieve a great deal of impact damage, which destroys the target. Typical KEW projectiles include spherical or shaped plastic pellets (fired by high-velocity gas guns) or tiny metal rods, pellets, or shaped shells (fired by electromagnetic railgun systems.) Even tiny grains of sand, explosively dispersed into a cloud directly in the path of a high-velocity reentry vehicle, can achieve a "kinetic energy" kill as the hostile reentry vehicle (and its warhead) is quite literally "sandblasted" out of existence.

See also BALLISTIC MISSILE DEFENSE.

kinetic kill vehicle (KKV) In space defense, a rocket that homes in on its target and then destroys it either by striking it directly or by showering it with fragments of material from a specially designed exploding device.

See also BALLISTIC MISSILE DEFENSE.

kinetic theory A physical theory, initially developed in the 19th century, that describes the "macroscopic" physical properties of matter (e.g., pressure and temperature) in terms of the "microscopic" behavior (i.e., motions and interactions) of the constituent atoms and molecules. For example, the simple kinetic theory of a gas explains pressure in terms of a statistical treatment of the continuous impacts of gas molecules (or atoms) on the walls of the container—that is, the pressure is described as being proportional to the average translational kinetic energy of the system of particles per unit volume.

Kirchhoff, Gustav Robert (1824–1887) German *Physicist, Spectroscopist* This gifted German physicist collaborated with ROBERT BUNSEN in developing the fundamental principles of spectroscopy. While investigating the phenomenon of blackbody radiation, Kirchhoff applied spectroscopy to study the chemical composition of the Sun, especially the production of the Fraunhofer lines in the solar spectrum. His pioneering work contributed to the development of astronomical spectroscopy, one of the major tools in modern astronomy.

Kirchhoff was born on March 12, 1824, in Königsberg, Prussia (now Kaliningrad, Russia). He was a student of CARL

FRIEDRICH GAUSS at the University of Königsberg and graduated in 1847. While still a student, Kirchhoff extended the work of Georg Simon Ohm (1787–1854) by introducing his own set of physical laws (now called Kirchhoff's laws) to describe the network relationship among current, voltage, and resistance in electrical circuits. Following graduation Kirchhoff taught as an unsalaried lecturer (that is, as a privatdozent) at the University of Berlin and remained there for approximately three years before joining the University of Breslau as a physics professor. Four years later he accepted a more prestigious appointment as a professor of physics at the University of Heidelberg and remained with that institution until 1875. At Heidelberg he collaborated with Robert Bunsen in a series of innovative experiments that significantly changed observational astronomy through the introduction of astronomical spectroscopy.

Prior to his innovative work in spectroscopy, Kirchhoff produced a theoretical calculation in 1857 at the University of Heidelberg demonstrating the physical principle that an alternating electric current flowing through a zero-resistance conductor would flow through the circuit at the speed of light. His work became a key step for the Scottish physicist JAMES CLERK MAXWELL during his formulation of electromagnetic wave theory.

Kirchhoff's most significant contributions to astrophysics and astronomy were in the field of spectroscopy. JOSEPH VON FRAUNHOFER's work had established the technical foundations for the science of spectroscopy, but undiscovered was the fact that each chemical element had its own characteristic spectrum. That critical leap of knowledge was necessary before physicists and astronomers could solve the puzzling mystery of the Fraunhofer lines and make spectroscopy an indispensable tool in both observational astronomy and numerous other applications. In 1859 Kirchhoff achieved that giant step in knowledge while working in collaboration with the German chemist Robert W. Bunsen at the University of Heidelberg. In his breakthrough experiment Kirchhoff sent sunlight through a sodium flame and with the primitive spectroscope he and Bunsen had constructed observed two dark lines on a bright background just where the Fraunhofer D-lines in the solar spectrum occurred. He and Bunsen immediately concluded that the gases in the sodium flame were absorbing the D-line radiation from the Sun, producing an absorption spectrum.

After additional experiments Kirchhoff also realized that all the other Fraunhofer lines were actually absorption lines—that is, gases in the Sun's outer atmosphere were absorbing some of the visible radiation coming from the solar interior, thereby creating these dark lines, or "holes," in the solar spectrum. By comparing solar spectral lines with the spectra of known elements, Kirchhoff and Bunsen detected a number of elements present in the Sun, with hydrogen being the most abundant. This classic set of experiments, performed in Bunsen's laboratory in Heidelberg with a primitive spectroscope assembled from salvaged telescope parts, gave rise to the entire field of spectroscopy, including astronomical spectroscopy.

Bunsen and Kirchhoff then applied their spectroscope to resolving elemental mysteries on Earth. In 1861, for example, they discovered the fourth and fifth alkali metals, which they named cesium (from the Latin *caesium* for "sky blue") and rubidium (from the Latin *rubidus* for "darkest red"). Today scientists use spectroscopy to identify individual chemical elements from the light each emits or absorbs when heated to incandescence. Modern spectral analyses trace their heritage directly back to the pioneering work of Bunsen and Kirchhoff. Astronomers and astrophysicists use spectra from celestial objects in a number of important applications, including composition evaluation, stellar classification, and radial velocity determination.

In 1875 Kirchhoff left Heidelberg because the cumulative effect of a crippling injury he sustained in an earlier accident now prevented him from performing experimental research. Confined to a wheelchair or crutches, he accepted an appointment to the chair of mathematical physics at the University of Berlin. In the new, less physically demanding position, he pursued numerous topics in theoretical physics for the remainder of his life. During this period he made significant contributions to the field of radiation heat transfer. For example, he discovered that the emissive power of a body compared to the emissive power of a blackbody at the same temperature is equal to the absorptivity of the body. His work in blackbody radiation was fundamental in the development of quantum theory by MAX PLANCK.

Recognizing his great contributions to physics and astronomy, the British Royal Society elected Kirchhoff a fellow in 1875. Unfortunately, failing health forced him to retire prematurely in 1886 from his academic position at the University of Berlin. He died in Berlin on October 17, 1887, and some of his scientific work appeared posthumously. While at the University of Berlin he prepared his most comprehensive publication, *Lectures on Mathematical Physics,* a four-volume effort that appeared between 1876 and 1894. His collaborative experiments with Bunsen and his insightful interpretation of their results started a new era in astronomy.

Kirchhoff's law (of radiation) The physical law of radiation heat transport states that at a given temperature, the ratio of the emissivity (ε) to the absorptivity (α) for a given wavelength is the same for all bodies and is equal to the emissivity of an ideal blackbody at that temperature and wavelength. Developed by the 19th century German physicist GUSTAV ROBERT KIRCHHOFF.

See also BLACKBODY.

Kirkwood, Daniel (1814–1895) American *Astronomer, Mathematician* In 1857 the American mathematical astronomer Daniel Kirkwood explained the uneven distribution and gaps in the main belt asteroid population as being the result of orbital resonances with the planet Jupiter. Today these gaps are called the *Kirkwood gaps* in recognition of his work.

See also ASTEROID; KIRKWOOD GAPS.

Kirkwood gaps Gaps, or "holes," in the main asteroid belt between Mars and Jupiter where essentially no asteroids are located. These gaps initially were explained in 1857 by the American astronomer DANIEL KIRKWOOD as being the result of complex orbital resonances with Jupiter. Specifically, the gaps correspond to the absence of asteroids with orbital periods that are simple fractions (e.g., 1/2, 2/5, 1/3, 1/4, etc.) of Jupiter's orbital period. Consider, for example, an asteroid that orbits the Sun in exactly half the time it takes Jupiter to orbit the Sun. (This example corresponds to a 2:1 orbital resonance.) The minor planet in this particular orbital resonance would then experience a regular, periodic pull (i.e., gravitational tug) from Jupiter at the same point in every other orbit. The cumulative effect of these recurring gravitational pulls is to deflect the asteroid into a chaotic (elongated) orbit that could then cross the orbit of Earth or Mars.

See also ASTEROID.

Kitt Peak National Observatory (KPNO) Part of the National Optical Astronomy Observatory (NOAO), Kitt Peak supports the most diverse collection of astronomical observatories on Earth for nighttime optical and infrared astronomy and daytime study of the Sun. Founded in 1958, KPNO operates three major nighttime telescopes, shares site responsibilities with the National Solar Observatory for the McMath-Pierce solar telescope complex, and hosts the facilities of consortia that operate 19 optical telescopes and two radio telescopes. Kitt Peak is located 90 km southwest of Tucson, Arizona, and has a visitor center open daily to the public. Kitt Peak, nestled in the Quinlan Mountains of the Sonoran Desert, comprises 81 hectares of the nearly 1.2-million hectare Tohono O'odham Nation. This land is leased by NOAO from the Tohono O'odham under a perpetual agreement that is valid for as long as scientific research facilities are maintained at the site.

Researchers have used various KPNO facilities to make major contributions to astronomy. For example, the study of spiral galaxy rotation curves by astronomers at KPNO provided the first indication of dark matter in the universe. Dark matter may dominate over ordinary matter in regulating the dynamics of galaxies and the entire universe. By using radio emissions as a selection criterion, very-high-redshift galaxies were also discovered. The systematic study of these galaxies, with redshifts greater than 5 and dating back to very early epochs, has resulted in a new understanding of the rate of galaxy formation.

klystron A high-powered electron tube that converts direct current (d.c.) electric energy into radio-frequency waves (microwaves) by alternately speeding up and slowing down the electrons (i.e., by electron beam velocity oscillation).

knot (symbol: kn) A unit of speed originating in the late 16th-century sailing industry that represents a speed equal to 1 nautical mile per hour.

$$1 \text{ knot (kn)} = 1 \text{ nautical mile/hr} = 1.1508 \text{ statute miles/hr} = 1.852 \text{ kilometers/hr}$$

Komarov, Vladimir M. (1927–1967) Russian *Cosmonaut, Soviet Air Force Officer* Cosmonaut Vladimir M. Komarov was the first person to make two different trips into outer space. He also became the first person to die while engaged in space travel. On April 23, 1967, he ascended into orbit around Earth onboard a new Soviet spacecraft, called *Soyuz 1.* The flight encountered many difficulties, and Komarov finally had to execute an emergency reentry maneuver on the 18th orbit (April 24). During the final stage of his reentry over the Kazakh Republic, the recovery parachute became entangled, causing his spacecraft to impact the ground at high speed. He died instantly and was given a hero's state funeral.

Korolev, Sergei (1907–1966) Ukraine-born Russian *Rocket Engineer* The Russian (Ukraine-born) rocket engineer Sergei Korolev was the driving technical force behind the initial intercontinental ballistic missile (ICBM) program and the early outer space exploration projects of the Soviet Union. In 1954 he started work on the first Soviet ICBM, the R-7. This powerful rocket system was capable of carrying a massive payload across continental distances. As part of cold war politics Soviet premier NIKITA KHRUSHCHEV allowed Korolev to use this military rocket to place the first artificial satellite, named *SPUTNIK 1,* into orbit around Earth on October 4, 1957. This event is now generally regarded as the beginning of the space age.

Korolev was born on January 12, 1907, in Zhitomir, Ukraine—at the time part of czarist Russia. (Korolev's birth date sometimes appears as December 30, 1906—a date corresponding to an obsolete czarist calendar system that is no longer used in Russia.) As a young boy Korolev obtained his first ideas about space travel in the inspirational books of KONSTANTIN TSIOLKOVSKY. After discovering the rocket Korolev decided on a career in engineering. He entered Kiev Polytechnic Institute in 1924 and two years later transferred to Moscow's Bauman High Technical School, where he studied aeronautical engineering under such famous Russian aircraft designers as Andrey Tupolev. Korolev graduated as an aeronautical engineer in 1929.

He began to champion rocket propulsion in 1931, when he helped organize the Moscow Group for the Investigation of Reactive Motion (GIRD) (Gruppa isutcheniya reaktvnovo dvisheniya). Like its German counterpart, the Verein für Raumschiffahrt, or VfR (Society for Space Travel), this Russian technical society began testing liquid-propellant rockets of increasing size. In 1933 GIRD-Moscow merged with a similar group from St. Petersburg (then called Leningrad), and together they formed the Reaction Propulsion Scientific Research Institute (RNII). Korolev was very active in this new organization and encouraged its members to develop and launch a series of rocket-propelled missiles and gliders during the mid-1930s. The crowning achievement of Korolev's early aeronautical engineering efforts was his creation of the RP-318, Russia's first rocket-propelled aircraft.

In 1934 the Soviet Ministry of Defense published Korolev's book *Rocket Flight into the Stratosphere.* Between 1936 and 1938 he supervised a series of rocket engine tests and winged-rocket flights within RNII. However, Soviet dictator Joseph Stalin (1879–1953) was eliminating many intellectuals through a series of brutal purges. Despite his technical brilliance, Korolev along with most of the staff at RNII found themselves imprisoned in 1938. During World

War II Korolev remained in a scientific labor camp. His particular prison design bureau, called *sharashka* TsKB-29, worked on jet-assisted take-off (JATO) systems for aircraft.

Once freed from the labor camp after the war, Korolev resumed his work on rockets. He accepted an initial appointment as chief constructor for the development of a long-range ballistic missile. At this point in his life Korolev essentially disappeared from public view, and all his rocket and space activities remained a tightly guarded state secret. For many years after his death in 1966, the Soviet government publicly referred to him only as the "Chief Designer of Carrier Rockets and Spacecraft."

Following World War II Joseph Stalin became more preoccupied with developing an atomic bomb than with exploiting captured German V-2 rocket technology, but he apparently sanctioned some work on long-range rockets and also approved the construction of a ballistic missile test range at Kapustin Yar. It was this approval that allowed Korolev to form a group of rocket experts to examine captured German rocket hardware and to resume the Russian rocket research program. In late October 1947 Korolev's group successfully test fired a captured German V-2 rocket from the new launch site at Kapustin Yar, near the city of Volgograd. By 1949 Korolev had developed a new rocket, called the Pobeda (Victory-class) ballistic missile. He used Russian-modified GermanV-2 rockets and Pobeda rockets to send instruments and animals into the upper atmosphere. In May 1949 one of his modified V-2 rockets lifted a 120-kg payload to an altitude of 110 km.

As cold war tensions mounted between the Soviet Union and the United States in 1954, Korolev began work on the first Russian intercontinental ballistic missile (ICBM). Soviet leaders were focusing their nation's defense resources on developing very powerful rockets to carry the country's much heavier, less design-efficient nuclear weapons. Responding to this emphasis, Korolev designed an ICBM called the R-7. This powerful rocket was capable of carrying a 5,000-kg payload more than 5,000 km.

With the death of Stalin in 1953, a new leader, Nikita Khrushchev, decided to use Russian technological accomplishments to emphasize the superiority of Soviet communism over Western capitalism. Under Khrushchev Korolev received permission to send some of his powerful military missiles into the heavens on missions of space exploration as long as such space missions also had high-profile political benefits.

In summer 1955 construction began on a secret launch complex in a remote area of Kazakhstan north of a town called Tyuratam. This Central Asian site is now called the Baikonur Cosmodrome and lies within the political boundaries of the Republic of Kazakhstan. In August and September 1957 Korolev successfully launched the first Russian intercontinental ballistic missile (the R-7) on long-range demonstration flights from this location. Encouraged by the success of these flight tests, Khrushchev allowed Korolev to use an R-7 military missile as a space launch vehicle in order to beat the United States into outer space with the first artificial satellite.

On October 4, 1957, a mighty R-7 rocket roared from its secret launch pad at Tyuratam, placing *Sputnik 1* into orbit around Earth. Korolev, the "anonymous" engineering genius, had propelled the Soviet Union into the world spotlight and started the space age. To Khrushchev's delight, a supposedly technically inferior nation won a major psychological victory against the United States. With the success of *Sputnik 1,* space technology became a key instrument of cold war politics and superpower competition.

The provocative and boisterous Khrushchev immediately demanded additional high-visibility space successes from Korolev. The rocket engineer responded on November 3, 1957, by placing a much larger satellite into orbit. *Sputnik 2* carried the first living space traveler into orbit around Earth. The passenger was a dog named Laika. Korolev continued to press his nation's more powerful booster advantage by developing the *Vostok* one-person spacecraft to support human space flight. On April 12, 1961, another of Korolev's powerful military rockets placed the *Vostok 1* spacecraft, carrying cosmonaut YURI GAGARIN, into orbit around Earth. Gagarin's brief flight (only one orbital revolution) took place just before the United States sent its first astronaut, ALAN B. SHEPARD, JR., on a suborbital mission from Cape Canaveral, Florida, on May 5, 1961.

Using Korolev's powerful rockets like so many trump cards, Khrushchev continued to scoop up a great deal of international prestige for the Soviet Union, which had just become the first nation to place a person into orbit around Earth. This particular Russian space accomplishment forced American president JOHN F. KENNEDY into a daring response. In May 1961 Kennedy announced his bold plan to land American astronauts on the Moon in less than a decade. From the perspective of history Korolev's space achievements became the political catalyst by which WERNHER VON BRAUN received a mandate to build more powerful rockets for the United States. Starting in July 1969, von Braun's new rockets would carry American explorers to the surface of the Moon, a technical accomplishment that soundly defeated the Soviet Union in the so-called great space race of the 1960s.

Following the success of the *Sputnik* satellites, Korolev started using his powerful rockets to propel large Soviet spacecraft to the Moon, Mars, and Venus. One of these spacecraft, called *Lunik 3,* took the first images of the Moon's farside in October 1959. These initial images of the Moon's long-hidden hemisphere excited both astronomers and the general public. Appropriately, one of the largest features on the Moon's farside now bears Korolev's name. He also planned a series of interesting follow-on *Soyuz* spacecraft for future space projects involving multiple human crews. However, he was not allowed to pursue these developments in a logical fashion. Premier Khrushchev kept demanding other space mission "firsts" from Korolev's team.

For example, the Soviet leader wanted to fly a three-person crew in space before the United States could complete the first flight of its new two-person Gemini Project space capsule. (The first crewed Gemini flight occurred on March 23, 1965.) Korolev recognized how dangerous it would be to convert his one-person *Vostok* spacecraft design into an internally stripped-down "three-seater," but he also remembered the painful time he spent in a political prison camp for perceived disloyalty to Stalin's regime.

So, despite strong opposition from his design engineers, Korolev removed the *Vostok* spacecraft's single ejection seat and replaced it with three couches. Without an ejection seat, the cosmonaut crew could not leave the space capsule during the final stages of reentry descent, as was done during earlier successful *Vostok* missions. To accommodate the demands of this mission, Korolev came up with several clever ideas, including a new retrorocket system and a larger reentry parachute. To quell the vehement safety objections from his design team, he also made them an offer they could not refuse. If they could design this modified three-person spacecraft in time to beat the Americans, one of the engineers would be allowed to participate in the flight as a cosmonaut.

On October 12, 1964, a powerful military booster sent the "improvised" *Voskhod 1* space capsule into a 170-km by 409-km orbit around Earth. *Voshkod* is the Russian word for "sunrise." The flight lasted just one day. The retrofitted *Vostok* spacecraft carried three cosmonauts without spacesuits under extremely cramped conditions. Cosmonaut VLADIMIR M. KOMAROV commanded the flight; medical expert Boris Yegorov and design engineer Konstantin Feoktistov accompanied him. Korolev won an extremely high-risk technical gamble. His engineering skill and luck not only satisfied Khrushchev's insatiable appetite for politically oriented space accomplishments, but he also kept his promise of a ride in space to one of his engineers.

Khrushchev responded to Kennedy's "Moon race" challenge by ordering Korolev's Experimental Design Bureau No. 1 (OKB-1) to accelerate their plans for a very large booster. As originally envisioned by Korolev, the Russian N-1 rocket was to be a very large and powerful booster for placing extremely heavy payloads in Earth orbit, including space stations, nuclear rocket upper stages, and various military payloads, but after Korolev's untimely death in 1966, the highly secret, monstrous N-1 rocket became the responsibility of another rocket design group, suffered four catastrophic failures between 1969 and1972, and then was quietly cancelled.

From 1962 to 1964 Khrushchev's continued political use of space technology seriously diverted Korolev's creative energies from much more important projects such as new boosters, the *Soyuz* spacecraft, a human Moon-landing mission, and a space station. His design team was just beginning to recover from Khrushchev's constant interruptions when disaster struck. On January 14, 1966, Korolev died during a botched routine surgery at a hospital in Moscow. He was only 58 years old. Some of Korolev's contributions to space technology include the powerful legendary R-7 rocket (1956), the first artificial satellite (1957), pioneering lunar spacecraft missions (1959), the first human space flight (1961), a spacecraft mission to Mars (1962), and the first space walk (1965). Even after his death, the Soviet government chose to hide Korolev's identity by publicly referring to him only as the "Chief Designer of Carrier Rockets and Spacecraft." Despite this official anonymity, Chief Designer and Academician Korolev is now properly recognized as the rocket engineer who started the space age.

Kourou launch complex (aka Guiana Space Center) In 1964 the French government selected Kourou, French Guiana, as the site from which to launch its satellites. After the European Space Agency (ESA) was formed in 1975, the French government offered to share the Guiana Space Center (GSC), or Centre Spatial Guyanais (CSG), with the ESA. For its part, the ESA approved funding to upgrade the launch facilities at Kourou to prepare the spaceport for the family of Ariane launch vehicles then under development. Since that time, the ESA has continued to fund two-thirds of the spaceport's annual budget in order to finance the operations and facilities needed to accommodate an evolving family of European space launch vehicles, especially the Ariane family of rockets.

Kourou lies at approximately 5° north latitude and 52.4° west longitude, just 500 km above the equator on the Atlantic Ocean in coastal French Guiana. This location makes the space complex ideally situated for launches into geostationary transfer orbit. Thanks to its favorable geographical position in the northeast corner of South America, Europe's spaceport can also support a wide range of missions from due east (e.g., geostationary transfer orbit) to north

An Ariane 5 expendable launch vehicle lifts off from ELA-3 at the Guiana Space Center in Kourou, French Guiana, on December 10, 1999. This first operational flight of the Ariane 5 launch vehicle successfully carried the European Space Agency's X-Ray Multi-Mirror Mission (XMM) observatory into orbit. ***(Courtesy of ESA/CNES/Arianespace-Service Optique CSG)***

A panoramic view of the Ariane 5 launch vehicle complex (EL-3) at Europe's spaceport, the Guiana Space Center in Kourou, French Guiana (January 2004) ***(Courtesy of ESA-S. Corvaja)***

(e.g., polar orbit). In fact, Kourou is so well placed that with just one spaceport all possible European space missions can be launched with minimum risk. Tropical forests largely cover French Guiana, and the launch site itself has no significant natural risks from either hurricanes or earthquakes. In addition to clients for space transportation services from countries throughout Europe, Kourou provides launch services to aerospace industry clients from the United States, Japan, Canada, Brazil, and India.

See also ARIANE; EUROPEAN SPACE AGENCY; LAUNCH VEHICLE.

Krikalev, Sergei (1958–) Russian *Cosmonaut* In February 1994 cosmonaut Sergei Krikalev flew on the first joint U.S.-Russian space shuttle *Discovery* mission (STS-60). A little more than four years later he flew as a crewmember on the space shuttle *Endeavour* during the STS-88 mission, which was the inaugural assembly mission (December 1998) for the *International Space Station* (ISS). This veteran Russian space traveler also served as a member of the ISS Expedition-1 crew, which launched from the Baikonur Cosmodrome (Kazakhstan) on a Russian Soyuz expendable launch vehicle on October 31, 2000. The space station's first crew docked with the orbiting international facility (still under assembly) on November 2. Krikalev and his ISS Expedition-1 crewmates then departed the ISS on March 18, 2001, riding as Earth-bound passengers on the space shuttle (STS-102 mission), which landed at NASA's Kennedy Space Center on March 21, 2001.

See also INTERNATIONAL SPACE STATION.

K star An orange colored star of spectral class K with a surface temperature between 3,500 and 5,000 K. Aldebaran is an example.

See also STAR.

Kuiper, Gerard Peter (Gerrit Pieter) (1905–1973) Dutch-American *Astronomer* In 1951 Gerard Kuiper, a Dutch-American planetary astronomer, boldly postulated the presence of thousands of icy planetesimals in an extended region at the edge of the solar system, beyond the orbit of Pluto. Today this region of frigid, icy objects has been detected and is called the Kuiper Belt in his honor. In 1944 Kuiper discovered that Saturn's largest moon (Titan) had an atmosphere. He continued to arouse interest in planetary astronomy by discovering Miranda, the fifth-largest moon of Uranus, in 1948 and Nereid, the outermost moon of Neptune, in 1949. Transitioning to space age astronomy, he served with distinction as a scientific adviser to the U.S. National Aeronautics and Space Administration (NASA) on the early lunar and planetary missions in the 1960s, especially NASA's Ranger Project and Surveyor Project.

Kuiper was born on December 7, 1905, in Harenkarspel, Netherlands. When he was young two factors influenced Kuiper's interest in astronomy. First, his father gave him a gift of a small telescope that he used to great advantage because of his exceptional visual acuity. Second, he was drawn to astronomy by the cosmological and philosophical writings of the great French mathematician René Descartes (1596–1650). While enrolling at the University of Leiden in September 1924, Kuiper made the acquaintance of a fellow student, BART J. BOK, who would remain a lifelong friend. He completed a bachelor of science degree in 1927 and immediately pursued graduate studies. One of Kuiper's professors at the University of Leiden was EJNAR HERTZSPUNG, under whom he did his doctoral thesis on the subject of binary stars. After receiving a doctoral degree in 1933, Kuiper traveled to the United States under a fellowship and conducted postdoctoral research on binary stars at the Lick Observatory, near San Jose, California. In August 1935 he left California and spent a year at the Harvard College Observatory. While there he met and later married (in June 1936) Sarah Parker Fuller, whose family had donated the property on which the Harvard Oak Ridge Observatory stood. The couple had two children: a son and a daughter. In 1936 Kuiper joined the Yerkes Observatory of the University of Chicago and became a naturalized American citizen the following year. He joined other members of the Yerkes staff who worked with the University of Texas in planning and developing the McDonald Observatory near Fort Davis, Texas. The 2.1-meter reflector of this important astronomical facility was dedicated and began operation in 1939. At the University of Chicago Kuiper progressed up the academic ranks, becoming a full professor of astronomy in 1943.

In an important 1941 technical paper Kuiper introduced the concept of "contact binaries," a term describing close, mass-exchanging binary stars characterized by accretion disks. During World War II he took a leave of absence from the University of Chicago to support various defense-related projects. His duties included special technical service missions as a member of the U.S. War Department's ALSOS Mission, which assessed the state of science in Nazi Germany during the closing days of the war. In one of his most interesting adventures, he helped rescue MAX PLANCK from the advancing Soviet armies. Kuiper, an astronomer now turned commando, commandeered a military vehicle and driver, dashed across war-torn Germany from the American lines to Planck's location in eastern Germany, and then spirited the aging German physicist and his wife away to a safe location in the western part of Germany near Göttingen.

While performing his wartime service, Kuiper still managed to make a major contribution to modern planetary astronomy, the area within which he would become the recognized world leader. Taking a short break from his wartime research activities in the winter of 1943–44, Kuiper conducted an opportunistic spectroscopic study of the giant planets Jupiter and Saturn and their major moons at the McDonald Observatory. To his great surprise, early in 1944 he observed methane on the largest Saturnian moon, Titan, making the moon the only known satellite in the solar system with an atmosphere (at that time). This fortuitous discovery steered Kuiper into his very productive work in solar system astronomy, especially the study of planetary atmospheres.

Through spectroscopy he discovered in 1948 that the Martian atmosphere contained carbon dioxide. In 1948 he discovered the fifth moon of Uranus, which he named Miranda, and then Neptune's second moon, Nereid, in 1949. He went on to postulate the existence of a region of icy planetesimals and minor planets beyond the orbit of Pluto. The first members of this region, now called the Kuiper Belt in his honor, were detected in the early 1990s. Kuiper left the University of Chicago and joined the University of Arizona in 1960. At Arizona he founded and directed the Lunar and Planetary Laboratory, making major contributions to planetary astronomy at the dawn of the space age. He produced major lunar atlases for both the U.S. Air Force and NASA. His research on the Moon in the 1950s and 1960s provided strong support for the impact theory of crater formation. Prior to the space age, most astronomers held that the craters on the Moon had been formed by volcanic activity.

Kuiper promoted a multidisciplinary approach to the study of the solar system, and this emphasis stimulated the creation of a new astronomical discipline, called planetary science. He assisted in the selection of the superior ground-based observatory site on Mauna Kea on the island of Hawaii. Because of his reputation as an outstanding planetary astronomer, he served as a scientific adviser on many of NASA's 1960s and 1970s space missions to the Moon and the inner planets. For example, Kuiper was the chief scientist for NASA's Ranger Project (1961–65), which sent robot spacecraft equipped with television cameras crashing into the lunar surface. Later he helped identify suitable landing sites for NASA's Surveyor robot lander spacecraft as well as the Apollo Project's human landing missions.

He served as principal author or general editor on several major works in modern astronomy including *The Solar System,* a four-volume series published between 1953 and 1963; *Stars and Stellar Systems,* a nine-volume series appearing in 1960; the famous *Photographic Atlas of the Moon* (1959); the *Orthographic Atlas of the Moon* (1961); the *Rectified Lunar Atlas* (1963); and the *Consolidated Lunar Atlas* (1967).

Kuiper was also a pioneer in applying infrared technology to astronomy and played a very influential role in the development of airborne infrared astronomy in the 1960s and early 1970s. Starting in 1967 Kuiper used NASA's Convair 990 jet aircraft equipped with a telescope system capable of performing infrared spectroscopy. As the aircraft flew above 12,200 meters (about 40,000 feet) altitude, Kuiper and his research assistants performed trend-setting infrared spectroscopy measurements of the Sun, planets, and stars—discovering many interesting phenomena that could not be observed by ground-based astronomers because of the blocking influence of the denser portions of Earth's atmosphere. Kuiper died on December 24, 1973, while on a trip to Mexico City with his wife and several family friends. As one tribute to his numerous contributions to astronomy, in 1975 NASA named its newest airborne infrared observatory, a specially outfitted C-141 jet transport aircraft, the *Gerard P. Kuiper Airborne Observatory* (KAO). This unique airborne astronomical facility operated out of Moffett Field, California, for more than two decades until retired from service in October 1995. No astronomer did more in the 20th century to revitalize planetary astronomy. Celestial objects carry his name or continue his legacy of discovery across the entire solar system—from a specially named crater on Mercury, to Miranda around Uranus, to Nereid around Neptune, to an entire cluster of icy minor planets beyond Pluto.

Kuiper Airborne Observatory (KAO) *See* AIRBORNE ASTRONOMY.

Kuiper Belt A region in the outer solar system beyond the orbit of Neptune that contains millions of icy planetesimals (small solid objects). These icy objects range in size from tiny particles to Plutonian-sized planetary bodies. The Dutch-American astronomer GERARD PETER KUIPER first suggested the existence of this disk-shaped reservoir of icy objects in 1951.

See also OORT CLOUD.

laboratory hypothesis A variation of the zoo hypothesis response to the Fermi paradox. This particular hypothesis postulates that the reason we cannot detect or interact with technically advanced extraterrestrial civilizations in the Milky Way galaxy is because they have set the solar system up as a "perfect" laboratory. These hypothesized extraterrestrial experimenters want to observe and study us but do not want us to be aware of or be influenced by their observations.

See also FERMI PARADOX; ZOO HYPOTHESIS.

Lacaille, Nicolas-Louis de (1713–1762) French *Astronomer* Between 1751 and 1753 the French astronomer Nicolas Louis de Lacaille mapped the positions of nearly 10,000 stars in the Southern Hemisphere from the Cape of Good Hope, South Africa. His effort was summarized in the book *Star Catalog of the Southern Sky,* published in 1763, the year after he died.

Lagrange, Comte Joseph-Louis (Giuseppe Luigi Lagrangia) (1736–1813) Italian-French *Mathematician, Celestial Mechanics Expert* The 18th-century mathematician Joseph-Louis Lagrange made significant contributions to celestial mechanics. About 1772 he identified certain special regions in outer space, now called the Lagrangian libration points, that mark the five equilibrium points for a small celestial object moving under the gravitational influence of two much larger bodies. Other astronomers used his discovery of the Lagrangian libration points to find new objects in the solar system, such as the Trojan group of asteroids. His influential book *Analytical Mechanics* represented an elegant compendium of the mathematical principles that described the motion of heavenly bodies.

Lagrange was born on January 25, 1736, in Turin, Italy. His father, Giuseppe Francesco Lodovico Lagrangia, was a prosperous public official in the service of the king of Savoy but impoverished his family through unwise financial investments and speculations. Lagrange, a gentle individual with a brilliant mathematical mind, responded to his childhood poverty by quipping: "If I had been rich, I probably would not have devoted myself to mathematics." Although given an Italian family name at baptism, throughout his life Lagrange preferred to emphasize his French ancestry, derived from his father's side of the family. He would, therefore, sign his name "Luigi Lagrange" or "Lodovico Lagrange," combining an Italian first name with a French family name.

Lagrange studied at the College of Turin, enjoying instruction in classical Latin and ignoring Greek geometry, but after reading EDMOND HALLEY's treatise on the application of algebra in optics, Lagrange changed his earlier career plans to become a lawyer and embraced the study of higher mathematics. In 1755 he was appointed professor of mathematics at the Royal Artillery School in Turin. At the time Lagrange was just 19 years old and already exchanging impressive mathematical notes on his calculus of variations with the famous Swiss mathematician LEONHARD EULER. In 1757 Lagrange became a founding member of a local scientific group that eventually evolved into the Royal Academy of Science of Turin. He published many of his elegant papers on the calculus of variations in the society's journal, *Mélanges de Turin.*

By the early 1760s Lagrange had earned a reputation as one of Europe's most gifted mathematicians, yet he was a humble and gentle person, who did not seek fame or position. He just wanted to pursue the study of mathematics with a modest amount of financial security. In 1764 Lagrange received a prize from the Paris Academy of Sciences for his brilliant mathematical paper on the libration of the Moon. The Moon's libration is the phenomenon by which 59 percent of the lunar surface is visible to an observer on Earth over a period of 30 years. This results from a complicated collection of minor perturbations in the Moon's orbit as it travels around Earth. Soon after that award he won another prize from the Paris Academy in 1776 for his theory on the motion of Jupiter's moons.

When Euler left Berlin in 1766 and returned to St. Petersburg, he recommended to King Frederick II of Prussia that Lagrange serve as his replacement. Therefore, in November 1766 Frederick II invited Lagrange to succeed Euler as mathematical director of the Berlin Academy of Science within the Prussian Academy of Sciences. Just under a year from his arrival in Berlin, Lagrange married Vittoria Conti, a cousin who had lived for an extended time with his family in Turin. The couple had no children. For two decades Lagrange worked in Berlin and during this period produced a steady number of top-quality, award-winning mathematical papers. He corresponded frequently with PIERRE-SIMON LAPLACE, a contemporary French mathematician living in Paris. Lagrange repeated winning prizes from the Paris Academy of Sciences for his mathematics in 1772, 1774, and 1780. He shared the 1772 Paris Academy prize with Leonhard Euler for his superb work on the challenging three-body problem in celestial mechanics—an effort that involved the existence and location of the Lagrangian libration points. These five points (usually designated as L_1, L_2, L_3, L_4, and L_5) are the locations in outer space where a small object can experience a stable orbit in spite of the force of gravity exerted by two much more massive celestial bodies when they all orbit about a common center of mass. He won the 1774 prize for another brilliant paper on the motion of the Moon. His 1780 award-winning effort discussed the perturbation of cometary orbits by the planets.

However, during his twenty years in Berlin his health began to fail. His wife also suffered from poor health and died in 1783 after an extended illness. Her death plunged Lagrange into a state of depression. After the death of Frederick II Lagrange departed from Berlin to accept an invitation from French king Louis XVI. In May 1787 Lagrange became a member of the Academy of Sciences in Paris, a position he retained for the rest of his career despite all the turmoil induced by the French Revolution. Upon his arrival in Paris King Louis XVI offered Lagrange apartments in the Louvre from which comfortable surroundings he published in 1788 his elegant synthesis of mechanics entitled *Mécanique analytique* (Analytical mechanics). Lagrange had actually written this book in Berlin but published it in Paris.

Despite his friendship with King Louis XVI, he survived the turmoil of the French Revolution that began in 1789. In May 1790 Lagrange became a member of and eventually chaired the committee within the Paris Academy of Sciences that standardized measurements and created the international system (that is, the SI system) of units currently used throughout the world.

In 1792 Lagrange married his second wife, a much younger woman named Renée Le Monnier, the daughter of one of his astronomer colleagues at the Paris Academy. Following the Reign of Terror (1793) Lagrange became the first professor of analysis at the famous École Polytechnique, founded in March 1794. Though brilliant, the aging mathematician was not an accomplished lecturer, and much of what he discussed sailed clearly over the heads of his audience of inattentive college-aged students. Despite his shortcomings as a lecturer, the notes from his calculus lectures were collected and published as the *Theory of Analytic Functions* (1797) and *Lessons on the Calculus of Functions* (1804), the first textbooks on the mathematics of real analytical functions.

In 1791 the Royal Society in London made him a fellow. Napoleon, the emperor of France, bestowed many honors upon the aging mathematician, making him a senator, a member of the Legion of Honor, and a Count of the Empire (*comte de l'Empire*). Despite the lavish political attention, Lagrange remained a quiet academician who preferred to keep absorbed in his thoughts about mathematics. He died in Paris on April 10, 1813. His numerous works on celestial mechanics prepared the way for 19th-century mathematical astronomers to make discoveries using the subtle perturbations and irregularities in the motions of solar system bodies.

Lagrangian libration point(s) In celestial mechanics, one of five points in outer space (called L_1, L_2, L_3, L_4, and L_5) where a small object can experience a stable orbit in spite of the force of gravity exerted by two much more massive celestial bodies when they orbit about a common center of mass. JOSEPH LOUIS LAGRANGE, an Italian-French mathematician, calculated the existence and location of these points in 1772. Three of the points (L_1, L_2, and L_3) lie on the line joining the center of mass of the two large bodies. A small object placed at any of these points is actually in unstable equilibrium. Any slight displacement of its position results in the object's rapid departure from the point due to the gravitational influence of the more massive bodies. However, the fourth and fifth libration points (L_4 and L_5) are locations of stable equilibrium. The Trojan group of asteroids can be found in such stable Lagrangian points 60° ahead of (L_4) and 60 degrees behind (L_5) Jupiter's orbit around the Sun. Space visionaries have suggested placing large space settlements in the L_4 and L_5 libration points of the Earth-Moon system.

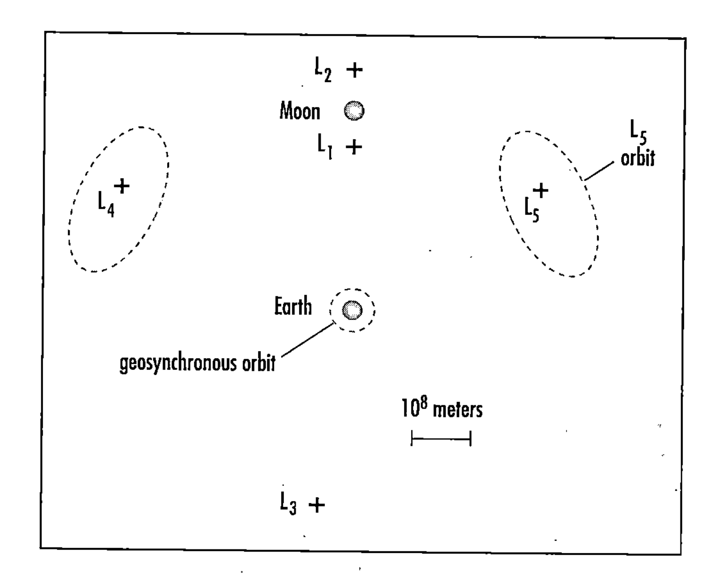

The five Lagrangian libration points in the Earth-Moon system

Lalande, Joseph-Jérôme Le Français de (1732–1807) French *Astronomer* In 1751 the French astronomer Joseph-Jérôme Le Français de Lalande collaborated with NICOLAS-LOUIS DE LACAILLE in measuring the distance to the Moon. He also published an extensive catalog in 1801 containing about 47,000 stars, including a "star" that actually was the planet Neptune, which was later discovered in 1846 by the German astronomer JOHANN GOTTFRIED GALLE.

lambert (symbol: L) A unit of luminance defined as equal to 1 LUMEN of (light) flux emitted per square centimeter of a perfectly diffuse surface. This unit is named in honor of the German mathematician Johann H. Lambert (1728–77).

laminar boundary layer In fluid flow, the layer next to a fixed boundary layer. The fluid velocity is zero at the boundary, but the molecular viscous stress is large because the velocity gradient normal to the wall is large. The equations describing the flow in the laminar boundary layer are the Navier-Stokes equations containing only inertia and molecular viscous terms.

laminar flow In fluid flow, a smooth flow in which there is no crossflow of fluid particles occurring between adjacent stream lines. The flow velocity profile for laminar flow in circular pipes is parabolic in shape with a maximum flow in the center of the pipe and a minimum flow at the pipe walls. *Compare with* TURBULENT FLOW.

Landau, Lev Davidovich (1908–1968) Russian *Physicist* In the early 1930s the Russian physicist Lev Davidovich Landau speculated about the possible existence of a neutron star. Later in his career he received the 1962 Nobel Prize in physics for his pioneering work on condensed matter, especially involving the properties of helium.

lander (spacecraft) A spacecraft designed to reach the surface of a planet and capable of surviving long enough to telemeter data back to Earth. The U.S. *Viking 1* and *2* landers on Mars and *Surveyor* landers on the Moon are examples of highly successful lander missions. The Russian *Venera* landers were able to briefly survive the extremely harsh surface environment of Venus, carried out chemical composition analyses of Venusian rocks, and relayed images of the surface.

Landsat A family of versatile NASA-developed Earth-observing satellites that have pioneered numerous applications of multispectral sensing. The first spacecraft in this series, *Landsat-1* (originally called the *Earth Resources Technology Satellite,* or ERTS-1), was launched successfully in July 1972 and quite literally changed the way scientists and nonscientists alike could study Earth from the vantage point of outer space. *Landsat-1* was the first civilian spacecraft to collect relatively high-resolution images of the planet's land surfaces and did so simultaneously in several important wavelength bands of the electromagnetic spectrum. NASA made these information-packed, visible to near-infrared images available to researchers around the globe, who quickly applied the new data to a wide variety of important disciplines, including agriculture, water resource evaluation, forestry, urban planning, and pollution monitoring. *Landsat-2* (launched in January 1975) and *Landsat-3* (launched in March 1978) were similar in design and capability to *Landsat-1.*

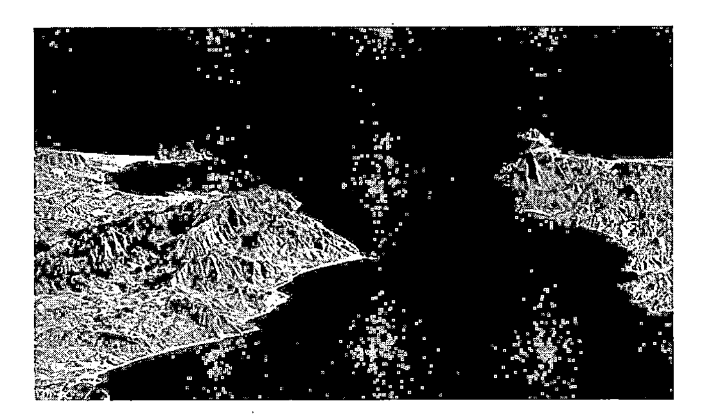

This interesting perspective view of the Strait of Gibraltar was generated by draping a Landsat satellite image acquired on July 6, 1987, over an elevation model produced by the Shuttle Radar Topography Mission (SRTM) in February 2000. The scene looks eastward with Europe (Spain) on the left and Africa (Morocco) on the right. The famous Rock of Gibraltar, administered by Great Britain, is the peninsula in the back left portion of the scene. The strategic strait is only about 13 kilometers wide. The image has a computer-generated three-times vertical exaggeration to enhance topographical expression. ***(Courtesy of NASA/JPL/National Imagery and Mapping Agency [NIMA])***

In July 1982 NASA introduced its second-generation civilian remote sensing spacecraft with the successful launch of *Landsat-4.* That satellite had both an improved multispectral scanner (MSS) and a new thematic mapper (TM) instrument. *Landsat-5,* carrying a similar complement of advanced instruments, was placed successfully into polar orbit in March 1984. Unfortunately, an improved *Landsat-6* spacecraft failed to achieve orbit in October 1993.

On April 15, 1999, a Boeing Delta II expendable launch vehicle lifted off from Vandenberg Air Force Base in California and successfully placed the *Landsat-7* spacecraft into a 705-kilometer-altitude, sun-synchronous (polar) orbit. This operational orbit allows the spacecraft to completely image Earth every 16 days. *Landsat-7* carries an enhanced thematic mapper plus (ETM+)—an improved remote sensing instrument with eight bands sensitive to different wavelengths of visible and infrared radiation. *Landsat-7* sustains the overall objectives of the Landsat program by providing continuous, comprehensive coverage of Earth's land surfaces, thereby extending the unique collection of environmental and global change data that began in 1972. Because of the program's continuous global coverage at the 30-meter spatial resolution, scientists can use Landsat data in global change research, in regional environmental change studies, and for many other civil and commercial purposes.

The 1992 Land Remote Sensing Act identified data continuity as the fundamental goal of the Landsat program. *Landsat-7* was developed as a triagency program among

NASA, the National Oceanic and Atmospheric Administration (NOAA), and the Department of the Interior's U.S. Geological Survey (USGS)—the agency responsible for receiving, processing, archiving, and distributing the data. Landsat data are now being applied in many interesting areas, including precision agriculture, cartography, geographic information systems, water management, flood and hurricane damage assessment, environmental monitoring and protection, global change research, rangeland management, urban planning, and geology.

Langley, Samuel Pierpont (1834–1906) American *Astronomer, Aeronautics Pioneer* Samuel Pierpont Langley believed that all life and activity on Earth was made possible by solar radiation. In 1878 he invented the bolometer, an instrument sensitive to incident electromagnetic radiation and especially suitable for measuring the amount of incoming infrared (IR) radiation. Modern versions of Langley's instrument are placed on Earth-observing spacecraft to provide regional and global measurements of the planet's radiant energy budget. He also led an expedition to the top of Mount Whitney, California, in 1881 to analyze the infrared component of the Sun's spectrum. Although Langley is remembered today primarily as an aeronautical pioneer, he was also an astronomer and the first American scientist to perceive astrophysics as a distinct field. Furthermore, Langley's invention of the bolometer and his careful measurement of infrared radiation from the Sun foreshadowed modern concerns about climate change and the search for links between solar and terrestrial phenomena.

See also EARTH RADIATION BUDGET; GLOBAL CHANGE.

Laplace, Pierre-Simon, marquis de (1749–1827) French *Mathematician, Astronomer, Celestial Mechanics Expert* Called the "French Newton," Pierre Laplace's work in celestial mechanics established the foundations of 19th-century mathematical astronomy. Laplace extended SIR ISAAC NEWTON's gravitational theory and provided a more complete mechanical interpretation of the solar system, including the subtle perturbations of planetary motions. His mathematical formulations supported the discoveries of Uranus (18th century), Neptune (19th century), and Pluto (20th century). In 1796, apparently independently of IMMANUEL KANT, Laplace introduced his own version of the nebula hypothesis, suggesting that the Sun and the planets condensed from a primeval interstellar cloud of gas.

Laplace was born on March 23, 1749, at Beaumont-en-Auge in Normandy, France. The son of a moderately prosperous farmer, he attended a Benedictine priory school in Beaumont, a preparatory school for future members of the clergy or military. At the age of 16 Laplace enrolled in the University of Caen, where he studied theology in response to his father's wishes. However, after two years at the university Laplace discovered his great mathematical abilities and abandoned any plans for an ecclesiastical career. At 18 he left the university without receiving a degree and went to live in Paris. There Laplace met Jean D'Alembert (1717–83) and greatly impressed the French mathematician and philosopher. D'Alembert secured an appointment for Laplace at the prestigious École Miltaire. By age 19 Laplace had become the professor of mathematics at the school.

In 1773 he began to make his important contributions to mathematical astronomy. Working independently of but cooperatively (through correspondence) with JOSEPH-LOUIS LAGRANGE, who then resided in Berlin, Laplace applied and refined Newton's law of gravitation to bodies throughout the solar system. Careful observations of the six known planets and the Moon indicated subtle irregularities in their orbital motions. This raised the very puzzling question: How could the solar system remain stable? Newton, for all his great scientific contributions, considered this particular problem far too complicated to be treated with mathematics. He simply concluded that "divine intervention" took place on occasion to keep the entire system in equilibrium.

Laplace, however, was determined to solve this challenging problem with his own excellent skills in mathematics. He immediately tackled the first of several challenging issues. In the 18th century astronomers could not explain why Saturn's orbit seemed to be continually expanding, while Jupiter's orbit seemed to be getting smaller. Laplace was able to mathematically demonstrate that while there certainly were measured irregularities in the orbital motions of these giant planets, such anomalies were not cumulative, but rather periodic and self-correcting. His important discovery represented a mathematically and physically sound conclusion that the solar system was actually an inherently stable system.

As a result of this brilliant work, Laplace became recognized as one of France's great mathematicians. The other great French mathematician of the time was Lagrange. In 1773 Laplace became an associate member of the French Academy of Sciences and then a full member in 1785. However, as his fame grew he more frequently acted the part of a political opportunist than a great scientist, tending to "forget" his humble roots and rarely acknowledging his many benefactors and collaborators. Good fortune came to Laplace in 1784, when he became an examiner at the Royal Artillery Corps. He evaluated young cadets who would soon serve in the French army. While fulfilling this duty in 1785, he had the opportunity to grade the performance and pass a 16-year-old cadet named Napoleon Bonaparte. The examiner position actually allowed Laplace to make direct contact with many of the powerful men who would run France over its next four politically turbulent decades—including the monarchy of Louis XVI, the revolution, the rise and fall of Napoleon, and the restoration of the monarchy under Louis XVIII. While other great scientists, such as Antoine Lavoisier (1743–94), would quite literally lose their heads around him, Lagrange maintained an uncanny ability to correctly change his views to suit the changing political events.

In 1787 Lagrange left Berlin and joined Laplace in Paris. Despite a rivalry between them, each of the great mathematical geniuses benefited from their constant exchange of ideas. They both served on the commission within the French Academy of Sciences that developed the metric system, although Laplace was eventually discharged from the commission during the Reign of Terror that took place in 1793 as a by-product of the French Revolution. Laplace married Marie-Charlotte de Coutry de Romanges in May 1788. His wife was 20 years younger than Laplace and bore him two children. Just before the Reign of Terror began, Laplace departed Paris with his wife and two children, and they did not return until after July 1794.

Laplace published *The System of the World* (*Exposition du système du monde*) in 1796. This book was a five-volume popular (that is, nonmathematical) treatment of astronomy and celestial mechanics. From a historic perspective this book also served as the precursor for his more detailed mathematical work on the subject. Laplace used the book to introduce his version of the nebular hypothesis, in which he suggested that the Sun and planets formed from the cooling and gravitational contraction of a large, flattened, and slowly rotating gaseous nebula. Apparently, Laplace was unaware that Kant had proposed a similar hypothesis about 40 years earlier. Laplace wrote this book in such exquisite prose that in 1816 he won admission to the French Academy, an honor bestowed only rarely upon an astronomer or mathematician. In 1817 he served as president of this distinguished and exclusive literary organization.

Between 1799 and 1825 Laplace summed up gravitational theory and completed Newton's pioneering work in a monumental five-volume work entitled *Celestial Mechanics* (*Mécanique céleste*). Laplace provided a complete mechanical interpretation of the solar system and the application of the law of gravitation. Central to this work is Laplace's great discovery of the invariability of the mean motions of the planets. With this finding he demonstrated the inherent stability of the solar system and gained recognition throughout his country and the rest of Europe as the "French Newton."

Laplace maintained an extraordinarily high scientific output despite the tremendous political changes taking place around him. He always demonstrated political adroitness and "changed sides" whenever it was to his advantage. Under Napoleon he became a member and then chancellor of the French senate, received the Legion of Honor in 1805, and became a Count (comte) of the Empire in 1806. Yet, when Napoleon fell from power, Laplace quickly supported restoration of the monarchy and was rewarded by King Louis XVIII, who named him a marquis in 1817.

Laplace made scientific contributions beyond his great work in celestial mechanics. He provided the mathematical foundation for probability theory in two important books. In 1812 he published his *Analytic Theory of Probability* (*Théorie analytique des probabilitiés*), in which he presented many of the mathematical tools he invented to apply probability theory to games of chance and events in nature, including astronomy and collisions by comets with planets. In addition, he wrote a popular discussion of probability theory entitled *A Philosophical Essay on Probability* (*Essai philosophique sur les probabilités*), which appeared in 1814.

Laplace explored other areas of physical science and mathematics between 1805 and 1820. These areas included heat transfer, capillary action, optics, the behavior of elastic fluids, and the velocity of sound. However, as other physical theories began to emerge with their intelligent young champions, Laplace's dominant position in French science came to an end. He died in Paris on March 5, 1827. Throughout France and Europe Laplace was highly regarded as one of the greatest scientists of all time. He held membership in many foreign societies, including the British Royal Society, which elected him a fellow in 1789.

Large Magellanic Cloud (LMC) An irregular galaxy about 20,000 light-years in diameter and approximately 160,000 light-years from Earth.

See also MAGELLANIC CLOUDS.

large space structures The building of very large and complicated structures in space will be one of the hallmarks of space activities in the 21st century. The capability to supervise deployment and construction operations on orbit is a crucial factor in the effective use of space. Space planners and visionaries already have identified three major categories of large space structures: (1) very large surfaces (i.e., 100s of square meters to perhaps a few square kilometers in area) to provide power (solar energy collection), to reject thermal energy (from a multimegawatt power-level space nuclear reactor), or to reflect sunlight (illumination of Earth's surface from space); (2) very large antennas (50-meter to 100-meter-diameter and more) to receive and transmit energy over the radio-frequency portion of the electromagnetic spectrum; and (3) large space platforms (perhaps 200 meters or more in dimension) that provide the docking, assembly, checkout, and resupply facilities to service spacecraft and orbital transfer vehicles or that accommodate the highly automated "factory" facilities for space-based manufacturing operations.

In the beginning these large space structures will be deployed on-orbit—that is, erected or inflated from some stowed launch configuration. However, there is a limit to the size of such deployable and erectable systems. Larger, more sophisticated space structures (i.e., those with dimensions exceeding 25 meters or so) will require assembly on orbit, most likely by teams of space workers supported by fleets of robotic spacecraft that can intelligently assist in their fabrication, assembly, and maintenance.

Larissa A small moon of Neptune (about 200 km in diameter) discovered in July 1989 as a result of the *Voyager 2* encounter. Larissa orbits Neptune at a distance of 73,550 km with a period of 0.555 days.

See also NEPTUNE.

laser A device capable of producing an intense beam of coherent light. The term is an acronym derived from the expression "light amplification by the stimulated emission of radiation." In a laser the beam of light is amplified when photons (quanta of light) strike excited atoms or molecules. These atoms or molecules are thereby stimulated to emit new photons (in a cascade, or chain reaction) that have the same wavelength and are moving in phase and in the same direction as the original photon. Modern lasers come in a wide variety of wavelengths and power levels. Very-high-energy lasers, for example, can destroy a target by heating, melting, or vaporizing the object. Other types of lasers (operating at much lower power levels) are used in such diverse areas as medicine, communications, environmental monitoring, and geophysics research (i.e., to study movement of Earth's crust).

See also BALLISTIC MISSILE DEFENSE.

laser designator A device that emits a low-power beam of laser light that is used to mark a specific place or object. A weapon system equipped with a special tracking sensor then detects this reflected laser light and can then home in on the designated target with great accuracy.

Laser Geodynamics Satellite (LAGEOS) A series of passive, spherical satellites launched by NASA and the Italian Space Agency (Agenzia Spaziale Italiana [ASI]) and dedicated exclusively to satellite laser ranging (SLR) technologies and demonstrations. These SLR activities include measurements of global tectonic plate motion, regional crustal deformation near plate boundaries, Earth's gravity field, and the orientation of Earth's polar axis and spin rate. The *LAGEOS-1* satellite was launched in 1976 and the *LAGEOS-2* in 1992. Each satellite was placed into a nearly circular, 5,800-km orbit but with a different inclination (110° for *LAGEOS-1* and 52° for *LAGEOS-2*.) The small, spherical satellites are just 60 centimeters in diameter but have a mass of 405 kg. This compact, dense design makes their orbits as stable as possible, thereby accommodating satellite laser ranging measurements of submillimeter precision. The massive, compact spacecraft have a design life of 50 years.

The LAGEOS satellite looks like a dimpled golf ball with a surface covered by 426 nearly equally spaced retroreflectors. Each retroreflector has a flat, circular front face with a prism-shaped back. These retroreflectors are three-dimensional prisms that reflect laser light directly back to its source. A timing signal starts when the laser beam leaves the ground station and continues until the pulse, reflected from one of the spacecraft's retroreflectors, returns to the ground station. Since the speed of light is constant, the distance between the ground station and the satellite can be determined precisely. This process is called *satellite laser ranging*. Scientists use this technique to accurately measure movements of Earth's surface, centimeters per year in certain locations. For example, *LAGEOS-1* data have shown that the island of Maui (in Hawaii) is moving toward Japan at a rate of approximately 7 centimeters per year and away from South America at 8 centimeters per year.

See also EARTH SYSTEM SCIENCE.

laser guided weapon A weapon that uses a seeker to detect laser energy reflected from a laser-marked or designated target and through signal processing provides guidance commands to a control system that guides the weapon to the point from which the laser energy is being reflected.

laser propulsion system A proposed form of advanced propulsion in which a high-energy laser beam is used to heat a propellant (working fluid) to extremely high temperatures and then the heated propellant is expanded through a nozzle to produce thrust. If the high-energy laser is located away from the space vehicle—as, for example, on Earth, on the Moon, or perhaps at a different orbital position in space—the laser energy used to heat the propellant would be beamed into the space vehicle's thrust chamber. This concept offers the advantage that the space vehicle being propelled need not carry a heavy onboard power supply. A fleet of laser-propelled orbital transfer vehicles operating between low-Earth orbit and lunar orbit has been suggested as part of a lunar base transportation system. This propulsion concept also has been suggested for deep space probes on one-way missions to the edge of the solar system and beyond. In this case the laser energy ("light energy") might first be gathered by a large optical collector and then converted into electric energy, which would be used to operate some type of advanced electric propulsion system.

See also BEAMED ENERGY PROPULSION.

laser radar An active remote sensing technique analogous to radar but that uses laser light rather than radio- or microwave radiation to illuminate a target. Also called ladar or lidar.

See also LIDAR.

laser rangefinder A device that uses laser energy for determining the distance from the device to a place or object. A short pulse of laser light usually is placed on the target, and the time required for the reflected laser signal to return to the device is used to determine the range.

Lassell, William (1799–1880) British *Astronomer* William Lassell was a wealthy British brewer and amateur astronomer who discovered Triton, the largest moon of Neptune, on October 10, 1846, just 17 days after JOHANN GOTTFRIED GALLE discovered the planet. During other observations of the solar system in the 19th century, Lassell codiscovered the eighth known Saturnian moon, Hyperion (1848), and then discovered two moons of Uranus, Ariel and Umbriel, in 1851.

latch A device that fastens one object or part to another but is subject to ready release on demand so the objects or parts can be separated. For example, a sounding rocket can be held on its launcher by a latch or several latches and then quickly released after ignition and proper thrust development.

latent heat The amount of thermal energy (heat) added to or removed from a substance to produce an isothermal (constant temperature) change in phase. There are three basic types of latent heat. The *latent heat of fusion* is the amount of thermal energy added or removed from a unit mass of substance to cause an isothermal change in phase between liquid and solid (e.g., freezing water or melting an ice cube). The *latent heat of vaporization* is the amount of thermal energy added to or removed from a unit mass of substance to cause an isothermal change of phase between liquid and gas (e.g., evaporating water or condensing steam). The latent heat of vaporization is sometimes called the *latent heat of condensation,* when the change in phase is from gas to liquid. Finally, the *latent heat of sublimation* is the amount of thermal energy added to or removed from a unit mass of substance to cause an isothermal change in phase from solid to vapor (e.g., a block of "dry ice" disappearing in a cloud of carbon dioxide vapor).

In the technical literature of thermodynamics, the latent heat is often treated as the enthalpy of vaporization, condensation, melting, solidification (fusion), or sublimation, as appropriate. The term *latent heat* is used to describe only the amount of thermal energy (heat) added to or removed from a substance that causes a change in phase but no change in temperature; the term *sensible heat* is used to describe energy transfer as heat that produces a temperature change. This somewhat confusing nomenclature has its roots in the historical evolution of the science of thermodynamics.

See also THERMODYNAMICS.

lateral 1. Of or pertaining to the side; directed or moving toward the side. 2. Of or pertaining to the lateral axis; directed, moving, or located along, or parallel to, the lateral axis.

Latin space designations The language of the Roman Empire (from more than 2 millennia ago) is used today by space scientists to identify places and features found on other worlds. This is done by international agreement to avoid confusion or contemporary language favoritism in the naming of newly discovered features on the planets and their moons. As in biology and botany, Latin is treated as a "neutral" language in space science. Two of the more common Latin terms that are encountered in space science and planetary geology are *mare, maria* (pl.), meaning "sea," such as Mare Tranquillitatis, the Sea of Tranquility, site of the first human landing (*Apollo 11* mission) on the Moon, and *mons, monte* (pl.), meaning "mountain," such as Olympus Mons, the largest volcano on Mars. As an interesting historical footnote, when the Italian scientist GALILEO GALILEI first explored the Moon with his newly developed telescope (ca. 1610), he thought the dark (lava flow) regions he saw were bodies of water and mistakenly called them *maria* (or seas). Other 17th-century astronomers made similar mistakes while attempting to identify and name different features on the Moon's surface by comparing them to their assumed terrestrial counterparts.

However, despite the obvious inaccuracies, modern space scientists have preserved this nomenclature tradition. Some of the other more commonly encountered Latin language–derived space designations are presented here. *Catena* (pl. *catenae*), meaning "chain," identifies a line (or chain) of craters (e.g., Tithoniae Catena on Mars). *Chasma* (pl. *chasmata*) describes a long, linear canyon or narrow gorge (e.g., the Artemis Chasmata on Venus). *Dorsum* (pl. *dorsa*), meaning "back," identifies a ridge on a planetary surface (e.g., Schiaparelli Dorsum on Mercury). *Fossa* (pl. *fossae*), meaning "ditch," describes a long, linear depression (e.g., Erythraea Fossa on Mars). *Lacus* (pl. *lacus*), meaning "lake," classifies a small (often irregular) dark patch on the surface of a moon or planet (e.g., Lacus Somniorum on the Moon). *Linea* (pl. *lineae*), meaning "thread," describes an elongated, linear feature or region (e.g., Hippolyta Linea on the surface of Venus). *Palus* (pl. *paludes*), meaning "marsh" or "swamp," identifies a mottled area (an area covered with spots and streaks of different shades and colors) on a planetary surface (e.g., Oxia Palus on Mars). *Patera* (pl. *paterae*), meaning "bowl" or "shallow dish," distinguishes an irregular crater such as one occurring at or near the summit of a mountain or ancient volcano (e.g., Apollinaris Patera on Mars). *Planitia* (pl. *planitiae*) describes a large, low plain on the surface of a planet or moon (e.g., Utopia Planitia, the vast, sloping plain on Mars that contains the *Viking 2* lander site). *Planum* (pl. *plana*) characterizes a high, relatively flat plateau (e.g., Lakshmi Planum, a high volcanic plateau on Venus). *Regio* (pl. *regiones*), meaning "area" or "region," is traditionally applied to any feature on a planetary surface that is not clearly defined or understood (e.g., Galileo Regio, a large dark area about 3,000 km across on the Jovian moon Ganymede). *Rima* (pl. *rimae*), meaning "crack," describes a long, narrow furrow (or well-defined fissure) on the surface of a planet or moon (e.g., Rima Hadley [or Hadley Rille] on the Moon, which was the site of the *Apollo 15* crewed landing). *Rupes* (pl. *rupes*), meaning "rock" or "cliff," identifies a linear feature on a planetary surface that coincides with an abrupt topographical change, such as a cliff (e.g., Vesta Rupes on Venus). *Sulcus* (pl. *sulci*), meaning "groove," characterizes a planetary surface that consists of an intricate network of parallel ridges and linear depressions (e.g., Sulci Gordii on Mars). *Terra* (pl. *terrae*), meaning "land" or "territory," describes a large area of highland terrain that corresponds to the size of a terrestrial continent (e.g., Ishtar Terra on Venus). *Unda* (pl. *undae*), meaning "wave," is commonly used in plural form (*undae*) to identify (sand) dunes on a planetary surface (e.g., Ningal Undae on Venus). *Vallis* (pl. *valles*) identifies a large or long valley on a planetary surface (e.g., Valles Marineris, the vast system of Martian canyons discovered by NASA's *Mariner 9* spacecraft in 1971).

latitude The angle between a perpendicular at a location and the equatorial plane of Earth. Latitude is measured in degrees north or south of the equator, with the equator being defined as 0°, the North Pole as 90° north, and the South Pole as 90° south latitude.

launch 1. (noun) a. The action taken in launching a rocket or aerospace vehicle from a planetary surface. b. The transition from static repose to dynamic flight of a missile or rocket. c. The action of sending forth a rocket, probe, or other object from a moving vehicle such as an aircraft, aerospace vehicle, or spacecraft. 2. (verb) a. To send off a rocket or missile under its own propulsive power. b. To send off a missile, rocket, or aircraft by means of a catapult, as in the case of the German V-1 cruise missile (World War II era), or by means of inertial force, as in the release of a bomb from a flying aircraft. c. To give a payload or space probe an added boost for flight into space just before separation from its launch vehicle.

launch aircraft The aircraft used to air launch a missile, rocket, human-crewed experimental aircraft, air-launched space booster, or remotely piloted vehicle (RPV). Launch aircraft, such as the NASA/Air Force B-52s used during the X-15 hypersonic flight program, often are called "motherships."

See also PEGASUS LAUNCH VEHICLE; X-15.

launch azimuth The initial compass heading of a powered vehicle at launch. The term often is applied to space launch vehicles.

launch complex In general, the site, facilities, and equipment used to launch a rocket, missile, or aerospace vehicle.

See also LAUNCH SITE.

launch configuration The assembled combination of strap-on booster rockets, primary launch vehicle, spacecraft/payload, and upper stages (if appropriate) that must be lifted off the ground at launch.

launch crew The group of engineers, technicians, managers, and safety personnel who prepare and launch a rocket, missile, or aerospace vehicle.

launching angle The angle between a horizontal plane and the longitudinal axis of a rocket, missile, or aerospace vehicle being launched.

launchpad (*also* **launch pad**) In general, the load-bearing base or platform from which a rocket, missile, or aerospace vehicle is launched. Often simply called "the pad."

launch site A defined area from which an aerospace or rocket vehicle is launched, either operationally or for test purposes. Major launch sites in the United States include the Eastern Range at Cape Canaveral Air Force Station/Kennedy Space Center on the central east coast of Florida, the Western Range at Vandenberg Air Force Base on the west central coast of California, and NASA's Wallops Flight Facility on the east coast of Virginia. Major Russian launch sites include Plesetsk, Kapustin Yar, and Tyuratam (also called the Baikonur Cosmodrome), which is located in Kazakhstan. The European Space Agency (ESA) has a major launch site in Kourou, French Guiana, on the northeast coast of South America.

launch time The time at which an aircraft, missile, rocket, or aerospace vehicle is scheduled for flight.

launch under attack (LUA) Execution by National Command Authorities (e.g., the president) of Single Integrated Operational Plan (SIOP) forces subsequent to tactical warning of strategic nuclear attack against the United States and prior to first impact of an enemy's nuclear warheads.

launch vehicle (LV) An expendable (ELV) or reusable (RLV) ROCKET-propelled vehicle that provides sufficient thrust to place a spacecraft into orbit around Earth or to send a payload on an interplanetary trajectory to another celestial body. Sometimes called booster or space lift vehicle. (*See* the launch vehicle table.)

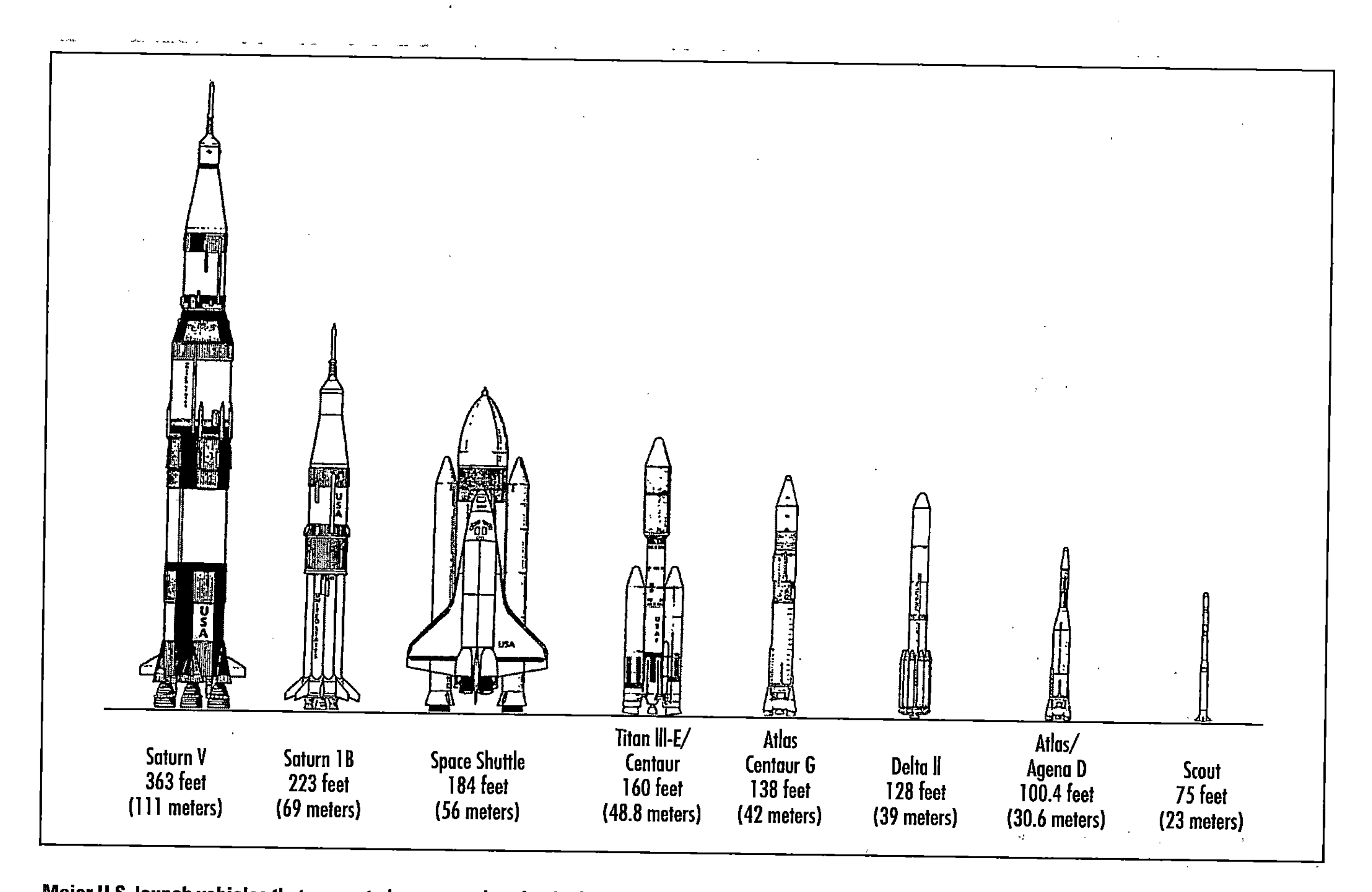

Major U.S. launch vehicles that supported space exploration in the 20th century *(Courtesy of NASA)*

Characteristics of Some of the World's Launch Vehicles

Country	Launch Vehicle	Stages	First Launch	Performance
China	Long March 2 (CZ-2C)	2 hypergolic, optional solid upper stage	1975	3,175 kg to LEO
	Long March 2F	2 hypergolic, 4 hypergolic strap-on rockets	1992	8,800 kg to LEO
	Long March 3	2 hypergolic, 1 cryogenic	1984	5,000 kg to LEO
	Long March 3A	2 hypergolic, 1 cryogenic	1994	8,500 kg to LEO
	Long March 4	3 hypergolic	1988	4,000 kg to LEO
Europe (ESA/ France)	Ariane 40	2 hypergolic, 1 cryogenic	1990	4,625 kg to LEO
	Ariane 42P	2 hypergolic, 1 cryogenic, 2 strap-on solid rockets	1990	6,025 kg to LEO
	Ariane 42L	2 hypergolic, 1 cryogenic 2 hypergolic strap-on rockets	1993	3,550 kg to GTO
	Ariane 5	2 large solid boosters, cryogenic core, hypergolic upper stage	1996	18,000 kg to LEO, 6,800 kg to GTO
India	Polar Space Launch Vehicle (PSLV)	2 solid stages, 2 hypergolic, 6 strap-on solid rockets	1993	3,000 kg to LEO
Israel	Shavit	3 solid-rocket stages	1988	160 kg to LEO
Japan	M-3SII	3 solid-rocket stages, 2 strap-on solid rockets	1985	770 kg to LEO
	H-2	2 cryogenic, 2 strap-on solid rockets	1994	10,000 kg to LEO, 4,000 kg to GTO
Russia	Soyuz	2 cryogenic, 4 cryogenic strap-on rockets	1963	6,900 kg to LEO
	Rokot	3 hypergolic	1994	1,850 kg to LEO
	Tsyklon	3 hypergolic	1977	3,625 kg to LEO
	Proton (D-I)	3 hypergolic	1968	20,950 kg to LEO
	Energia	cryogenic core, 4 cryogenic strap-on rockets, optional cryogenic upper stages	1987	105,200 kg to LEO
USA	Atlas I	1-1/2 cryogenic lower stage, 1 cryogenic upper stage	1990	5,580 kg to LEO, 2,250 to GTO
	Atlas II	1-1/2 cryogenic lower stage, 1 cryogenic upper stage	1991	6,530 kg to LEO, 2,800 kg to GTO
	Atlas IIAS	1-1/2 cryogenic lower stage, 1 cryogenic upper stage, 4 strap-on solid rockets	1993	8,640 kg to LEO, 4,000 kg to GTO
	Delta II	1 cryogenic, 1 hypergolic, 1 solid stage, 9 strap-on solid rockets	1990	5,050 kg to LEO, 1,820 kg to GTO
	Athena I	2 solid stages	1995	815 kg to LEO
	Pegasus (aircraft-launched)	3 solid stages	1990	290 kg to LEO
	Space shuttle	2 large solid-rocket boosters, cryogenic core	1981	25,000 kg to LEO
	Taurus	4 solid stages	1994	1,300 kg to LEO
	Titan 4	2 hypergolic stages, 2 large strap-on solid rockets, variety of upper stages	1989	18,100 kg to LEO

LEO, low Earth orbit; GTO, geostationary transfer orbit.
Source: NASA, DoD, OTA (U.S. Congress), and others.

launch window An interval of time during which a launch may be made to satisfy some mission objective. Usually a short period of time each day for a certain number of days. An interplanetary launch window generally is constrained within a number of weeks each year by the location of Earth in its orbit around the Sun, in order to permit the launch vehicle to use Earth's orbital motion for its trajectory as well as to time the spacecraft to arrive at its destination when the target planet is in the correct position. A launch that involves a rendezvous in Earth orbit with another spacecraft or platform must be timed precisely to account for the orbital motion of the "target" space object, creating launch windows of just a few minutes each day. Similarly, an interplanetary launch window is further constrained to just a few hours each appropriate day in order to take full advantage of Earth's rotational motion.

Laval, Carl Gustaf Patrik de (1845–1913) Swedish *Engineer, Inventor* Carl de Laval developed the converging-diverging nozzle—a flow device that he applied to steam turbines. Today the DE LAVAL NOZZLE has also become an integral part of most modern rocket engines.

See also NOZZLE; ROCKET.

layered defense An approach to ballistic missile defense (BMD) consisting of several relatively independent layers of missile defense technologies and systems designed to operate against different portions of the trajectory of an enemy's ballistic missile. For example, there could be a boost-phase (first layer) defense, with any remaining targets being passed on to succeeding layers (i.e., mid-course, terminal) of the defense system.

See also BALLISTIC MISSILE DEFENSE.

Leavitt, Henrietta Swan (1868–1921) American *Astronomer* In 1902 the American astronomer Henrietta Swan Leavitt joined the staff at Harvard College Observatory (HCO) and within a decade (about 1912) discovered the period-luminosity relationship for Cepheid variable stars. Her important discovery allowed other astronomers, such as HARLOW SHAPLEY, to make more accurate estimates of distances in the Milky Way galaxy and beyond. During her career she found more than 2,400 variable stars. Her efforts helped change people's knowledge of the size of the universe.

See also CEPHEID VARIABLE.

Leda A small outer moon of Jupiter discovered in 1974 by the American astronomer Charles T. Kowal (b. 1940). Leda has a diameter of 16 km and orbits Jupiter at a distance of 11,110,000 km with a period of 240 days. Also called Jupiter XIII.

See also JUPITER.

Lemaître, Abbé Georges-Édouard (1894–1966) Belgian *Astrophysicist, Cosmologist* Georges Lemaître was an innovative cosmologist who suggested in 1927 that a violent explosion might have started an expanding universe. He based this hypothesis on his interpretation of ALBERT EINSTEIN's general relativity theory and on EDWIN POWELL HUBBLE's contemporary observation of galactic redshifts, an observational indication that the universe was indeed expanding. Other physicists, such as GEORGE GAMOW, built on Lemaître's pioneering work and developed it into the widely accepted big bang theory of modern cosmology. Central to Lemaître's model is the idea of an initial superdense primeval atom, his "cosmic egg," which started the universe in a colossal ancient explosion.

Lemaître was born on July 17, 1894, in Charleroi, Belgium. Before World War I he studied civil engineering at the University of Louvain. At the outbreak of war in 1914 he volunteered for service in the Belgian Army and earned the Belgian Croix de Guerre for his combat activities as an officer in the artillery corps. Following the war he returned to the University of Louvain, where he pursued additional studies in physics and mathematics. In the early 1920s he also responded to another vocational calling and became an ordained priest in the Roman Catholic Church. After ordination (1923) Abbé Lemaître pursued a career devoted to science, especially cosmology, in which he would make one of the major contributions in the 20th century.

Taking advantage of an advanced studies scholarship from the Belgian government, Lemaître traveled to England to study from 1923 to 1924 with SIR ARTHUR EDDINGTON at the solar physics laboratory of the University of Cambridge. He continued his travels and came to the United States, where he studied at the Massachusetts Institute of Technology (MIT) from 1925 to 1927, earning a Ph.D. in physics. While at MIT Lemaître became familiar with the expanding universe concepts just being developed by the American astronomers HARLOW SHAPLEY and Edwin P. Hubble.

In 1927 Lemaître returned to Belgium and joined the faculty of the University of Louvain as a professor of astrophysics, a position he retained for the rest of his life. That year he published his first major paper related to cosmology. Unaware of similar work by the Russian mathematician Alexander Friedmann (1888–1925), Lemaître blended Einstein's general relativity and Hubble's early work on expanding galaxies to reach the conclusion that an expanding universe model is the only appropriate explanation for observed redshifts of distant galaxies. Lemaître suggested the distant galaxies serve as "test particles" that clearly demonstrate the universe is in a state of expansion. Unfortunately, Lemaître's important insight attracted little attention within the astronomical community, mainly because he published this paper in a rather obscure scientific journal in Belgium.

However, in 1931 Sir Arthur Eddington discovered Lemaître's paper and with his permission had the paper translated into English and then published in the *Monthly Notices of the Royal Astronomical Society.* Sir Eddington also provided a lengthy commentary to accompany the translation of Lemaître's paper. By this time Hubble had formally announced his famous law (Hubble's law) that related the distance of galaxies to their observed redshift, so Lemaître's paper, previously ignored, now created quite a stir in the astrophysical community. He was invited to come to London to lecture about his expanding universe concept. During this visit he not only presented an extensive account of his original theory of the expanding universe, he also introduced his idea about a "primitive atom," sometimes called Lemaître's cosmic egg. He was busy thinking not only about how to show the universe was expanding, but also about what caused the expansion and how this great process began.

Lemaître cleverly reasoned that if the galaxies are now everywhere expanding, then in the past they must have been much closer together. In his mind he essentially ran time backward to see what conditions were as the galaxies came closer and closer together in the early universe. He hypothesized that eventually a point would be reached when all matter resided in a super dense "primal atom" (or "cosmic egg"). He further suggested that some instability within this primal atom would result in an enormous explosion that would start the universe expanding. Big bang cosmology results directly from his concepts, although it took several decades and other scientists, such as George Gamow, ARNO PENZIAS, and ROBERT W. WILSON, to fully develop and establish the big bang theory as the currently preferred cosmological model.

In addition to the important 1927 big bang scientific note, his other significant papers include "Discussion on the Evolution of the Universe" (1933) and "Hypothesis of the Primeval Atom" (1946). His contributions to astrophysics and modern cosmology were widely recognized. In 1934 he received the Prix Francqui (Francqui Award) directly from the hands of Albert Einstein, who had personally nominated Lemaître for this prestigious award. In 1941 the Royal Belgian Academy of Sciences and Fine Art voted him membership. In 1936 he became a member of the Pontifical Academy of Sciences and then presided over this special papal scientific assembly as its president from 1960 until his death. The Belgian government bestowed its highest award for scientific achievement upon him in 1950, and the British Royal Astronomical Society awarded him the society's first Eddington Medal in 1953. He died at Louvain, Belgium, on June 20, 1966. Abbé Lemaître lived long enough to witness the detection by Penzias and Wilson of the microwave remnants of the big bang explosion, an event he had boldly hypothesized almost four decades earlier.

American astronaut Thomas P. Stafford (left) examines packages of space food with Russian cosmonaut Alexei P. Leonov (right) in the Soyuz orbital module trainer at the NASA Johnson Space Center in Houston, Texas. Both space travelers were preparing for their participation as spacecraft commanders during the first international spacecraft rendezvous and docking mission, called the Apollo-Soyuz Test Project (ASTP), which took place in 1975. *(Courtesy of NASA)*

lens A curved piece of glass, quartz, plastic, or other transparent material that has been ground, polished, and shaped accurately to focus light from a distant object so as to form an image of that object.

Leonov, Alexei Arkhipovich (1934–) Russian *Cosmonaut* On March 18, 1965, cosmonaut Alexei Arkhipovich Leonov became the first person to perform a space walk, a tethered extravehicular activity (EVA) outside his Earth-orbiting *Voskhod* 2 spacecraft. In July 1975 he served as the Russian spacecraft commander during the Apollo-Soyuz Test Project (ASTP), the first international rendezvous and docking mission of crewed spacecraft.

lepton One of a class of light elementary particles that have small or negligible masses, including electrons, positrons, neutrinos, antineutrinos, muons, and antimuons.

See also ELEMENTARY PARTICLE(s).

Leverrier, Urbain-Jean-Joseph (1811–1877) French *Astronomer, Mathematician, Orbital Mechanics Expert* Urbain Leverrier was a skilled celestial mechanics practitioner who mathematically predicted in 1846 (independently of JOHN COUCH ADAMS) the possible location of an eighth, as yet undetected, planet in the outer regions of the solar system. Leverrier provided his calculations to JOHANN GOTTFRIED GALLE at the Berlin Observatory. Leverrier's correspondence allowed the German astronomer to quickly discover Neptune by telescopic observation on September 23, 1846. Despite his great success with mathematical astronomy in the discovery of Neptune, Leverrier failed by similarly suggesting that observed perturbations in the orbit of Mercury were due to another undiscovered planet that he called Vulcan. Many 19th-century astronomers, including SAMUEL HEINRICH SCHWABE, searched without success for Leverrier's Vulcan inside the orbit of Mercury. In the early part of the 20th century ALBERT EINSTEIN's general relativity theory explained the perturbations of Mercury without resorting to Leverrier's hypothetical, nonexistent planet Vulcan. Today, the only planet Vulcan that can be found resides in science fiction, most notably in the popular *Star Trek* series.

Urbain-Jean-Joseph Leverrier was born on March 11, 1811, in Saint-Lô, France. His father, a local government official, made a great financial sacrifice so that his son could receive a good education at the prestigious École Polytechnique. Upon graduation Leverrier began his professional life as a chemist. He investigated the nature of certain chemical compounds under the supervision of the French chemist Joseph Gay-Lussac (1778–1850), who was a professor of chemistry at the École Polytechnique. In 1836 Leverrier accepted an appointment as a lecturer in astronomy at the same institution. Consequently, with neither previous personal interest in nor extensive formal training in astronomy, Leverrier suddenly found himself embarking on an astronomically oriented academic career. The process all came about rather quickly, when the opportunity for academic promotion at the École Polytechnique presented itself and Leverrier seized the moment.

As he settled into this new academic position, Leverrier began investigating lingering issues in celestial mechanics. He decided to focus his attention on continuing the work of PIERRE-SIMON, MARQUIS DE LAPLACE in mathematical astronomy and demonstrate with even more precision the inherent stability of the solar system. Following the suggestion given him in 1845 by DOMINIQUE FRANÇOIS ARAGO, he began investigating the subtle perturbations in the orbital motion of Uranus, the outer planet discovered by Sir WILLIAM HERSCHEL some 50 years earlier.

In 1846 Leverrier, following the suggestions of other astronomers (such as FRIEDRICH WILHELM BESSEL), hypothesized that an undiscovered planet lay beyond the orbit of Uranus. He then examined the perturbations in the orbit of Uranus and used the law of gravitation to calculate the approximate size and location of this as yet undetected outer planet. Unknown to Leverrier, a British astronomer named John Couch Adams was making similar calculations. However, during this period SIR GEORGE BIDDELL AIRY, Britain's Astronomer Royal, basically ignored Couch's calculations and correspondence about the possible location of a new outer planet.

In mid-September 1846 Leverrier sent a detailed letter to the German astronomer Johann G. Galle at the Berlin Observatory. Leverrier's correspondence told Galle where to look in the night sky for a planet beyond Uranus. Upon receipt of Leverrier's letter, Galle responded quickly. On the evening of September 23, 1846, Galle found Neptune after about an hour of searching. The French-German team of mathematical and observational astronomers had beaten the British astronomical establishment to one of that century's greatest discoveries. In France Leverrier experienced the celebrity that often accompanies a widely recognized scientific accomplishment. For example, he received appointment to a specially created chair of astronomy at the University of Paris, and the French government made him an officer in the Legion of Honor. Leverrier also received international credit for mathematical discovery of Neptune. The Royal Society of London awarded him its Copley Medal in 1846 and elected him a fellow in 1847. As a point of scientific justice, John Couch Adams received the Royal Society's Copley Medal in 1848 for his "discovery" of Neptune.

Bristling with success, fame, and good fortune, Leverrier pressed his luck a second time in 1846 by applying mathematical astronomy to explain minor perturbations in the solar system. This time he looked inward at the puzzling behavior in the orbital motion of Mercury, the innermost known planet. Unfortunately, Leverrier failed quite miserably with a similar hypothesis that the observed perturbations in Mercury were due to the presence of an inner undiscovered planet (one he called Vulcan) or possibly even a belt of asteroids orbiting closer to the Sun. (Some contemporary astronomers have revived Leverrier's speculation about the possible presence of asteroids, called *Vulcanoids,* traveling around the Sun within the orbit of Mercury.) Many 19th-century observational astronomers, including Schwabe, searched in vain for Leverrier's Vulcan. Einstein's general relativity theory in the early part of the 20th century explained the perturbations of Mercury without resorting to Leverrier's hypothetical, nonexistent planet Vulcan.

In 1854 Leverrier succeeded Dominique F. Arago as the director of the Paris Observatory. The French king Louis XIV had founded the Paris Observatory in 1667 and appointed GIOVANNI DOMENICO CASSINI its first director. This observatory served as the national observatory of France and was the first such national observatory established in the era of telescopic astronomy. Leverrier focused his efforts from 1847 until his death on producing more accurate planetary data tables and on creating a set of standard references for use by astronomers. Unfortunately, Leverrier was an extremely unpopular director who ran the Paris Observatory like a tyrant. By 1870 his coworkers had had enough, and he was replaced as director. However, he was reinstated to the directorship in 1873, when his successor, Charles-Eugene Delaunay (1816–72), died suddenly. This time, however, Leverrier served as director under the very watchful authority of a directing council.

Leverrier died in Paris on September 23, 1877. His role in celestial mechanics and the predictive discovery of Neptune represents one of the greatest triumphs of mathematical astronomy.

Ley, Willy (1906–1969) German-American *Engineer* The German-American engineer and technical writer Willy Ley promoted interplanetary space travel in the United States following World War II, especially by writing the popular book *Rockets, Missiles and Space Travel.*

libration A real or apparent oscillatory motion, particularly the apparent oscillation of the Moon. Because of libration, more than half (about 59%) of the Moon's surface actually can be seen over time by an observer on Earth despite the fact that it is in synchronous motion around our planet (i.e., keeping the same side always facing Earth).

libration points The unique positions of gravitational balance measured from a celestial body and its satellite at which points smaller objects can orbit in stable equilibrium with the other two bodies.

See also LAGRANGIAN LIBRATION POINTS.

Lick Observatory An astronomical observatory located at an altitude of 1,280 meters on Mount Hamilton near San Jose, California. In 1888, when endowed by the philanthropist James Lick (1796–1876) to the University of California, the major instrument was a powerful 91-cm-diameter refracting telescope. The observatory's principal instrument today is a 3-m reflecting telescope.

lidar An active remote sensing technique analogous to radar but that uses laser light rather than radio- or microwave radiation to illuminate a target. The laser light is reflected off a target and then detected. Using time-of-flight techniques to analyze the returned beam, information on both the distance to and velocity of the target can be obtained. The term is an acronym for light detection and ranging. Sometimes referred to as ladar (laser detection and ranging).

life cycle All the phases through which an item, component, or system passes from the time engineers envision and initially develop it until the time it is either consumed in use (for example, expended during a mission) or disposed of as excess to known requirements (for example, a flight spare that is not needed because of the success of the original system). NASA engineers generally use the following life cycle phases: Pre-Phase A (conceptual study), Phase A (preliminary analysis), Phase B (definition), Phase C/D (design and development), and the operations phase.

life in the universe Any search for life in the universe requires that we develop and agree upon a basic definition of what life is. For example, according to contemporary exobiologists and biophysicists, life (in general) can be defined as a living system that exhibits the following three basic characteristics: (1) it is structured and contains information, (2) it is able to replicate itself, and (3) it experiences few random changes in its "information package"—which random changes when they do occur enable the living system to evolve in a Darwinian context (i.e., survival of the fittest).

The history of life in the universe can be explored in the context of a grand, synthesizing scenario called *cosmic evolution.* This sweeping scenario links the development of galaxies, stars, planets, life, intelligence, and technology and then speculates on where the ever-increasing complexity of matter is leading. The emergence of conscious matter (especially conscious intelligence), the subsequent ability of a portion of the universe to reflect upon itself, and the destiny of this intelligent consciousness are topics often associated with contemporary discussions of cosmic evolution. One interesting speculation is the *anthropic principle*—namely, was the universe designed for life, especially the emergence of human life?

The cosmic evolution scenario is not without scientific basis. The occurrence of organic compounds in interstellar clouds, in the atmospheres of the giant planets of the outer solar system, and in comets and meteorites suggests the existence of a chain of astrophysical processes that links the chemistry of interstellar clouds with the prebiotic evolution of organic matter in the solar system and on early Earth. There is also compelling evidence that cellular life existed on Earth some 3.56 billion years ago (3.56 Gy). This implies that the cellular ancestors of contemporary terrestrial life emerged rather quickly on a geological time scale. These ancient creatures may have also survived the effects of large impacts from comets and asteroids in those ancient, chaotic times when the solar system was evolving.

The figure summarizes some of the factors believed important in the evolution of complex life. These include (A) endogenous factors stemming from physical-chemical properties of Earth and those of *eukaryotic* organisms; (B) factors associated with properties of the Sun and of Earth's position with respect to the Sun; (C) factors originating within the

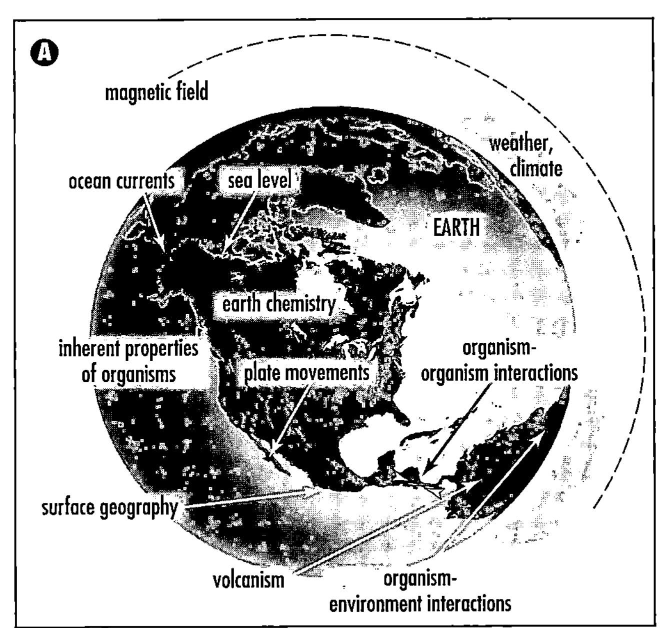

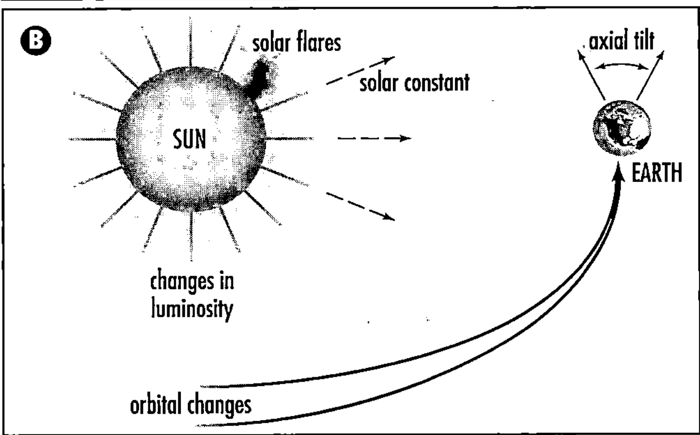

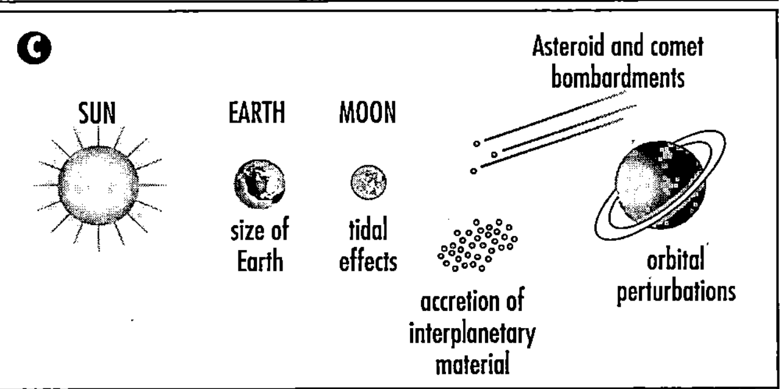

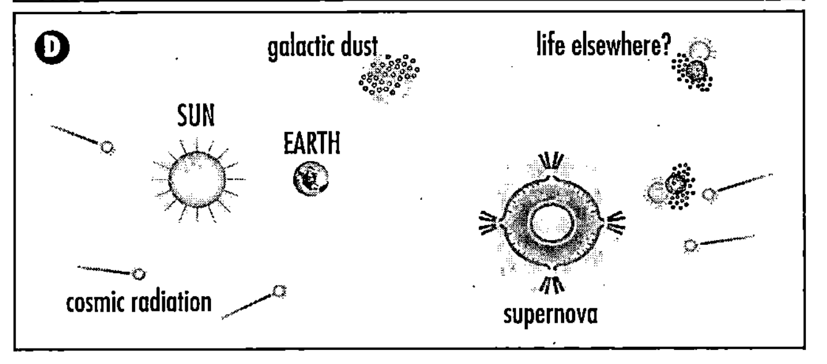

Factors considered important in the evolution of complex life ***(Courtesy of NASA)***

solar system, including Earth as a representative planet; and (D) factors originating in space far from our solar system.

The word *eukaryotic* refers to cells whose internal construction is complex, consisting of organelles (e.g., nucleus, mitochondria, etc.), chromosomes, and other structures. All higher terrestrial organisms are built of eukaryotic cells, as are many single-celled organisms (called *protists*). The evolution of complex life apparently had to await the evolution of eukaryotic cells, an event that is believed to have occurred on Earth about 1 billion years ago. A *eukaryote* is an organism built of eukaryotic cells.

And where does all this lead in the cosmic evolution scenario? We should first recognize that all living things are extremely interesting pieces of matter. Life-forms that have achieved intelligence and have developed technology are especially interesting and valuable in the cosmic evolution of the universe. Intelligent creatures with technology, including human beings, can exercise conscious control over matter in progressively more effective ways as the level of their technology grows. Ancient cave dwellers used fire to provide light and heat. Modern humans harness solar energy, control falling water, and split atomic nuclei to provide energy for light, warmth, industry, and entertainment. People later in this century will most likely "join atomic nuclei" (controlled fusion) to provide energy for light, warmth, industrial applications, and entertainment here on Earth as well as for interplanetary power and propulsion systems for emerging human settlements on the Moon and Mars. The trend should be obvious. Some scientists even speculate that if technologically advanced civilizations throughout the Milky Way galaxy can learn to live with the awesome powers unleashed in such advanced technologies, then it may be the overall destiny of advanced intelligent life-forms (including human beings) to ultimately exercise (beneficial) control over all the matter and energy in the universe.

According to modern scientific theory living organisms arose naturally on the primitive Earth through a lengthy process of chemical evolution of organic matter. This process began with the synthesis of simple organic compounds from inorganic precursors in the atmosphere; continued in the oceans, where these compounds were transformed into increasingly more complex organic substances; and then culminated with the emergence of organic microstructures that had the capability of rudimentary self-replication and other biochemical functions.

Human interest in the origins of life extends back deep into antiquity. Throughout history each society's "creation myth" seemed to reflect that particular people's view of the extent of the universe and their place within it. Today, in the space age, the scope of those early perceptions has expanded well beyond the reaches of our solar system to the stars, vast interstellar clouds, and numerous galaxies that populate the seemingly limitless expanse of outer space. Just as the concept of *biological evolution* implies that all living organisms have arisen by divergence from a common ancestry, so, too, the concept of *cosmic evolution* implies that all matter in our solar system has a common origin. Following this line of reasoning, scientists now postulate that life may be viewed as the product of countless changes in the form of primordial stellar matter, changes brought about by the interactive processes of astrophysical, cosmochemical, geological, and biological evolution.

If we use the even larger context of cosmic evolution, we can further conclude that the chain of events that led to the origins of life here on Earth extends well beyond planetary history: to the origin of the solar system itself, to processes occurring in ancient interstellar clouds that spawned stars like our Sun, and ultimately to the very birth within these stars (through nucleosynthesis and other processes) of the elements that make up living organisms—the *biogenic elements.* The biogenic elements are those generally judged to be essential for all living systems. Scientists currently place primary emphasis on the elements hydrogen (H), carbon (C), nitrogen (N), oxygen (O), sulfur (S), and phosphorous (P). The compounds of major interest are those normally associated with water and with organic chemistry, in which carbon is bonded to itself or to other biogenic elements. The essentially "universal" presence of these compounds throughout interstellar space gives exobiologists the scientific basis for forming the important contemporary hypothesis that the *origin of life is inevitable throughout the cosmos wherever these compounds occur and suitable planetary conditions exist.* Present-day understanding of life on Earth leads modern scientists to the conclusion that life originates on planets and that the overall process of biological evolution is subject to the often chaotic processes associated with planetary and solar system evolution (e.g., the random impact of a comet on a planetary body or the unpredictable breakup of a small moon).

Scientists now use four major epochs in describing the evolution of living systems and their chemical precursors. These are:

1. The cosmic evolution of biogenic compounds—an extended period corresponding to the growth in complexity of the biogenic elements from nucleosynthesis in stars, to interstellar molecules, to organic compounds in comets and asteroids.
2. Prebiotic evolution—a period corresponding to the development (in planetary environments) of the chemistry of life from simple components of atmospheres, oceans, and crustal rocks, to complex chemical precursors, to initial cellular life-forms.
3. The early evolution of life—a period of biological evolution from the first living organisms to the development of multicellular species.
4. The evolution of advanced life—a period characterized by the emergence of progressively more advanced life-forms, climaxing perhaps with the development of intelligent beings capable of communicating, using technology, and exploring and understanding the universe within which they live.

As scientists unravel the details of this process for the chemical evolution of terrestrial life, we should also ask ourselves another very intriguing question: If it happened here, did it or could it happen elsewhere? In other words, what are the prospects of finding extraterrestrial life—in this solar system or perhaps on Earthlike planets around distant stars?

According to the principle of mediocrity (frequently used by exobiologists), there is nothing "special" about the solar system or planet Earth. Within this speculation, therefore, if similar conditions have existed or are now present on "suitable" planets around alien suns, the chemical evolution of life will also occur.

Contemporary planetary formation theory strongly suggests that objects similar in mass and composition to Earth may exist in many planetary systems. In order to ascertain whether any of these extrasolar planets may be life-sustaining, we will need to use advanced space-based systems, such as NASA's planned TERRESTRIAL PLANET FINDER (TPF), to accomplish direct detection of terrestrial (i.e., small and rocky) planetary companions to other stars, as well as to investigate the composition of their atmospheres. Liquid water is a basic requirement for life as we know it, and it is the key indicator that will be used by scientists to determine whether planets revolving around other stars may indeed be life-sustaining.

Perhaps an even more demanding question is: Does alien life (once started) develop to a level of intelligence? If we speculate that alien life does evolve to some level of intelligence, then we must also ask: Do intelligent alien life-forms acquire advanced technologies and learn to live with these vast powers over nature?

Exobiologists postulate that under suitable environmental conditions, life, including intelligent life, should arise on planets around alien stars. This is an artist's rendering of an intelligent reptilelike creature. Some scientists suggest that on Earth this type of very smart warm-blooded dinosaur might have eventually evolved had not a massive extinction event occurred some 65 million years ago, displacing the dinosaurs and allowing mammals, including *Homo sapiens*, to emerge. *(Artist's rendering courtesy of the Department of Energy/Los Alamos National Laboratory)*

In sharp contrast to a universe full of emerging intelligent creatures, other scientists suggest that life itself is a very rare phenomenon and that we here on Earth are the only life-forms anywhere in the galaxy to have acquired conscious intelligence and to have developed (potentially self-destructive) advanced technologies. Just think about the powerful implications of this latter conjecture. Are we the best the universe has been able to produce in more than 14 billion years or so of cosmic evolution? If so, every human being is something very special.

Our preliminary search on the Moon and on Mars for extant (living) extraterrestrial life in the solar system has to date been unsuccessful. However, the recent detailed study of a Martian meteorite by NASA scientists has renewed excitement and speculation about the possibility of microbial life on Mars, past or perhaps even present in some isolated biological niche.

Microbial life-forms on Earth are found in acid-rich hot springs, alkaline-rich soda lakes, and saturated salt beds. Additionally, microbial life has been found in the Antarctic living in rocks and at the bottoms of perennially ice-covered lakes. Life is found in deep-sea hydrothermal vents at temperatures of up to 120°C (393 kelvins). Bacteria have even been discovered in deep (1-km or deeper) subsurface ecosystems that derive their energy from basalt weathering. Some microorganisms can survive ultraviolet radiation, while others can tolerate extreme starvation, low nutrient levels, and low water activity. Remarkably, spore-forming bacteria have been revived from the stomachs of wasps entombed in amber that was between 25 and 40 million years old. Clearly, life is diverse, tenacious, and adaptable to extreme environments.

Results from the biology experiments onboard NASA's *Viking 1* and 2 lander spacecraft suggest that extant life is absent in surface environments on Mars. However, life could be present in deep subsurface environments where liquid water may exist. Furthermore, although the present surface of Mars appears very inhospitable to life as we know it, recent space missions to the Red Planet have provided scientists good evidence that the Martian surface environment was more Earthlike early in its history (some 3.5 to 4.0 billion years ago), with a warmer climate and liquid water at or near the surface. We know that life originated very quickly on the early Earth (perhaps within a few hundred million years), and so it seems quite reasonable to assume that life could have emerged on Mars during a similar early window of opportunity when liquid water was present at the surface.

Of course, the final verdict concerning life on Mars (past or present) will not be properly resolved until more detailed investigations of the Red Planet take place. Perhaps later this century a terrestrial explorer (robot or human) will stumble upon a remote exobiological niche in some deep Martian canyon, or possibly a team of astronaut-miners, searching for certain ores on Mars, will uncover the fossilized remains of a tiny ancient creature that roamed the surface of the Red Planet in more hospitable environmental eras. Speculation, yes—but not without reason.

The giant outer planets and their constellations of intriguing moons also present some tantalizing possibilities. Who cannot get excited about the possible existence of an ocean of liquid water beneath the Jovian moon Europa and the (remote) chance that this extraterrestrial ocean might contain communities of alien life-forms clustered around hydrothermal vents? Furthermore, data from NASA's *GALILEO* SPACECRAFT suggest that Callisto and Ganymede may also have liquid water deposits beneath their icy surfaces.

All we can say now with any degree of certainty is that our overall understanding of the cosmic prevalence of life will be significantly influenced by the exobiological discoveries (pro and con) that will occur in the next few decades. In addition to looking for extraterrestrial life on other worlds in our solar system, exobiologists can also search for life-related molecules in space in order to determine the cosmic nature of prebiotic chemical synthesis.

Recent discoveries, for example, show that comets appear to represent a unique repository of information about chemical evolution and organic synthesis at the very outset of the solar system. (For example, after reviewing Comet Halley encounter data, space scientists suggest that comets have remained unchanged since the formation of the solar system.) Exobiologists now have evidence that the organic molecules considered to be the molecular precursors to those essential for life are prevalent in comets. These discoveries have provided further support for the hypothesis that the chemical evolution of life has occurred and is now occurring widely throughout the Milky Way galaxy. Some scientists even suggest that comets have played a significant role in the chemical evolution of life on Earth. They hypothesize that significant quantities of important life-precursor molecules could have been deposited in an ancient terrestrial atmosphere by cometary collisions.

Meteoroids are solid chunks of extraterrestrial matter. As such, they represent another source of interesting information about the occurrence of prebiotic chemistry beyond the Earth. In 1969, for example, meteorite analysis provided the first convincing proof of the existence of extraterrestrial amino acids. (Amino acids are a group of molecules necessary for life.) Since that time a large amount of information has been gathered showing that many more of the molecules considered necessary for life are also present in meteorites. As a result of this line of investigation, it now seems clear to exobiologists that the chemistry of life is not unique to Earth. Future work in this area should greatly help our understanding of the conditions and processes that existed during the formation of the solar system. These studies should also provide clues concerning the relations between the origin of the solar system and the origin of life.

The basic question, "Is life—especially intelligent life—unique to the Earth?," lies at the very core of our concept of self and where we fit in the cosmic scheme of things. If life is extremely rare, then we have a truly serious obligation to the entire (as yet "unborn") universe to carefully preserve the precious biological heritage that has taken more than 4 billion years to evolve on this tiny planet. If, on the other hand, life (including intelligent life) is abundant throughout the galaxy, then we should eagerly seek to learn of its existence and ultimately become part of a galactic family of conscious, intelligent creatures. Fermi's famous paradoxical question "Where are they?" takes on special significance this century.

See also AMINO ACID; ANTHROPIC PRINCIPLE; DRAKE EQUATION; EXOBIOLOGY; EXTRATERRESTRIAL CIVILIZATIONS; EXTRASOLAR PLANETS; FERMI PARADOX; JUPITER; MARS; MARTIAN METEORITES; VIKING PROJECT.

life support system (LSS) The system that maintains life throughout the aerospace flight environment, including (as appropriate) travel in outer space, activities on the surface of another world (for example, the lunar surface), and ascent and descent through Earth's atmosphere. The LSS must reliably satisfy the human crew's daily needs for clean air, potable water, food, and effective waste removal.

lift 1. That component (commonly used symbol, L) of the total aerodynamic force acting upon a body perpendicular to the undisturbed airflow relative to the body. 2. To lift off, or to take off in a vertical ascent. Said of a rocket or an aerostat.

See also AERODYNAMIC FORCES.

liftoff The action of a rocket or aerospace vehicle as it separates from the launch pad in a vertical ascent. In aerospace operations, this term applies only to vertical ascent, while takeoff applies to ascent at any angle.

light Electromagnetic radiation, ranging from about 0.4 to 0.7 micrometers (μm) in wavelength, to which the human eye is sensitive. Light helps us form our visual awareness of the universe and its contents.

See also ELECTROMAGNETIC SPECTRUM.

light curve A plot used by astronomers to illustrate how the light output of a star (or other optically variable astronomical object) changes with time.

light flash A momentary flash of light seen by astronauts in space, even with their eyes closed. Scientists believe that there are probably at least three causes for these light flashes. First, energetic cosmic rays passing through the eye's "detector" (the retina) ionize a few atoms or molecules, resulting in a signal in the optic nerve. Second, extremely energetic HZE particles can produce Cerenkov radiation in the eyeball. (Cerenkov radiation is the bluish light emitted by a particle traveling very near the speed of light when it enters a medium in which the velocity of light is less than the particle's speed.) Finally, alpha particles from nuclear collisions caused by very energetic Van Allen belt protons can produce ionization in the retina, again triggering a signal in the optic nerve. Astronauts and cosmonauts have reported seeing these flashes in a variety of sizes and shapes.

See also COSMIC RAYS; HZE PARTICLES.

light-gathering power (LGP) The ability of a telescope or other optical instrument to collect light.

light-minute (lm) A unit of length equal to the distance traveled by a beam of light (or any electromagnetic wave) in the vacuum of outer space in one minute. Since the speed of light (c) is 299,792.5 kilometers per second (km/s) in free space, a light-minute corresponds to a distance of approximately 18 million kilometers.

lightning protection Aerospace launch complexes and the launch vehicles themselves must be protected from lightning strikes. At any given instant, more than 2,000 thunderstorms are taking place throughout the world. All of these combine to produce about 100 lightning flashes per second.

On November 14, 1969, an out-of-the-blue lightning bolt struck and temporarily disabled the electrical systems on the *Apollo 12* spacecraft onboard a Saturn V launch vehicle about 1.6 kilometers above NASA's Kennedy Space Center (KSC) on the way to the Moon. On March 26, 1987, a NASA Atlas/Centaur vehicle and its payload were lost when the vehicle was hit by lightning while climbing up through the atmosphere after launch from Cape Canaveral Air Force Station (CCAFS). The lightning strike caused the rocket's computer to upset and issue an extreme yaw command that led to the vehicle's breakup in flight.

Lightning tends to strike the highest points in any given area, and aerospace safety engineers take special care to protect tall structures at a launch complex from these high-voltage bolts. The U.S. Lightning Protection Code for structures requires a pathway, or conductor, that will lead a lightning bolt safely to the ground. Additional protection is provided by circuit breakers, fuses, and electrical surge arrestors.

NASA's Kennedy Space Center operates extensive lightning protection and detection systems in order to keep workers, the space shuttle vehicles, launchpads, and processing facilities safe from harm. While the NASA lightning protection system is exclusively on KSC property, the lightning detection system incorporates equipment and personnel both at the KSC and CCAFS, which is located just east of the NASA complex. The Launch Pad Lightning Warning System (LPLWS) and the Lightning Surveillance System (LSS) provide the data necessary for evaluating weather conditions that can lead to the issuance of lightning warnings for the entire KSC/CCAFS launch complex. When the threat of lightning is present (typically within 8 km of critical facilities), a lightning policy is put into effect. This lightning policy restricts certain outdoor activities and hazardous operations, such as the start of launch vehicle fueling.

light pollution Human-made nocturnal sources of light, such as street lights and signs, that limit or interfere with sensitive astronomical observations in a region on Earth. Natural sources of optical pollution include the light of the Moon (especially during the full Moon), zodiacal light, and airglow.

See also AIRGLOW; ZODIACAL LIGHT.

light-second (ls) A unit of length equal to the distance traveled by a beam of light (or any electromagnetic wave) in the vacuum of outer space in one second. Since the speed of light (c) in free space is 299,792.5 kilometers per second (km/s), a light-second corresponds to a distance of approximately 300,000 kilometers.

light water Ordinary water (H_2O), as distinguished from heavy water (D_2O), where D is deuterium.

light-year (ly) The distance light (or any other form of electromagnetic radiation) can travel in one year. One light-year is equal to a distance of approximately 9.46×10^{12} kilometers, or 63,240 astronomical units (AU).

limb The visible outer edge or observable rim of the disk of a celestial body—for example, the "limb" of Earth, the Sun, or the Moon.

limit switch A mechanical device that can be used to determine the physical position of equipment. For example, an extension on a valve shaft mechanically trips a limit switch as it moves from open to shut or shut to open. The limit switch gives "on/off" output that corresponds to valve position. Normally, limit switches are used to provide full open or full shut indications. Many limit switches are of the pushbutton variety. When the valve extension comes in contact with the limit switch, the switch depresses to complete, or turn on, the electrical circuit. As the valve extension moves away from the limit switch, spring pressure opens the switch, turning off the circuit.

linear accelerator A long straight tube (or series of tubes) in which charged particles, such as electrons or protons, gain in energy by the action of oscillating electromagnetic fields.

See also ACCELERATOR.

linear energy transfer (LET) The energy lost (ΔE) per unit path length (Δx) by ionizing radiation as it passes through matter. Often it is used as a measure of the relative ability of different types of ionizing radiation to produce a potentially harmful effect on living matter; that is, the higher the LET value, the greater the potential for biological effect. $LET = \Delta E/\Delta x$ = energy deposited/unit path length; often expressed in units of keV/micrometer.

Lindblad, Bertil (1895–1965) Swedish *Astronomer* The Swedish astronomer Bertil Lindblad studied the dynamics of stellar motions and suggested in the mid-1920s that the Milky Way galaxy was rotating. His pioneering work contributed to the overall theory of galactic motion and structure.

line of apsides The line connecting the two points of an orbit that are nearest and farthest from the center of attraction, such as the perigee and apogee of a satellite in orbit around Earth or the perihelion and aphelion of a planet around the Sun; the major axis of any elliptical orbit and extending indefinitely in both directions.

See also ORBITS OF OBJECTS IN SPACE.

line of flight (LOF) The line in air or space along which an aircraft, rocket, aerospace vehicle, or spacecraft travels. This line can be either the intended (i.e., planned) line of movement or the actual line of movement of the vehicle or craft.

line of force A line indicating the direction in which a force acts.

line of nodes The straight line created by the intersection of a satellite's orbital plane and a reference plane (usually the equatorial plane) of the planet or object being orbited.

line of sight (LOS) The straight line between a sensor or the eye of an observer and the object or point being observed. Sometimes called the optical path.

liner 1. With respect to solid-propellant rocket motors, the thin layer of material used to bond the solid propellant to the motor case; this material also can provide thermal insulation. 2. With respect to a rocket nozzle, the thin layer of ablative material placed on the nozzle wall (especially in the throat area) to help reduce heat transfer from the hot exhaust gases to the nozzle wall structure.

line spectrum (pl: spectra) The discontinuous lines produced when bound electrons go from a higher energy level to a lower energy level in an excited atom, spontaneously emitting electromagnetic radiation. This radiation is essentially emitted at a single frequency, which is determined by the transition, or "jump," in energy. Each different "jump" in energy level, therefore, has its own characteristic frequency. The overall collection of radiation emitted by the excited atom is called the line spectra. Since these line spectra are characteristic of the atom, they can be used for identification purposes.

link In telecommunications, a general term used to indicate the existence of communications pathways and/or facilities between two points. In referring to communications between a ground station and a spacecraft or satellite, the term UPLINK describes communications from the ground site to the spacecraft, while the term DOWNLINK describes communications from the spacecraft to the ground site.

See also TELECOMMUNICATIONS; TELEMETRY.

Lippershey, Hans (Jan Lippersheim) (ca. 1570–1619) Dutch *Optician* In 1608 the Dutch optician Hans Lippershey invented and patented a simple, two-lens telescope. Once the basic concept for this new optical instrument began circulating throughout western Europe, creative persons such as GALILEO GALILEI and JOHANNES KEPLER quickly embraced the telescope's role in observational astronomy and made improved devices of their own design.

See also TELESCOPE.

liquid A phase of matter between a solid and a gas. A substance in a state in which the individual particles move freely with relation to one another and take the shape of the container but do not expand to fill the container. *Compare with* FLUID.

liquid-fuel rocket A rocket vehicle that uses chemical propellants in liquid form for both the fuel and the oxidizer.

See also ROCKET.

liquid hydrogen (LH_2) A liquid propellant used as the fuel with liquid oxygen serving as the oxidizer in high-performance cryogenic rocket engines. Hydrogen remains a liquid only at very low (cryogenic) temperatures, typically about 20 kelvins (K) (–253°C) or less. This cryogenic temperature requirement creates significant propellant storage and handling problems. However, the performance advantage experienced with the use of liquid hydrogen/liquid oxygen in high-energy space boosters often offsets these cryogenic handling problems.

See also LIQUID PROPELLANT.

liquid oxygen (LOX or LO_2) A cryogenic liquid propellant requiring storage at temperatures below 90 kelvins (K) (–183°C). Often used as the oxidizer with RP-1 (a kerosene-type) fuel in many expendable boosters or with liquid hydrogen (LH_2) fuel in high-performance cryogenic propellant rocket engines.

See also LIQUID PROPELLANT.

liquid propellant Any liquid combustible fed to the combustion chamber of a rocket engine. The petroleum used as a rocket fuel is a type of highly refined kerosene, called RP-1 (for refined petroleum-1). Usually it is burned with liquid oxygen (LOX) in rocket engines to provide thrust.

Cryogenic propellants are very cold liquid propellants. The most commonly used are LOX, which serves as an oxidizer, and liquid hydrogen (LH_2), which serves as the fuel. Cooling and compressing gaseous hydrogen and oxygen into liquids vastly increases their density, making it possible to store them (as cryogenic liquids) in smaller tanks. Unfortunately, the tendency of cryogenic propellants to return to their gaseous form unless kept supercool makes them difficult to store for long periods of time. Therefore, cryogenic propellants are less satisfactory for use in strategic nuclear missiles or tactical military rockets, which must be kept launch-ready for months to years at a time. However, cryogenic propellants are favored for use in space boosters because of the relatively high thrust achieved per unit mass of chemical propellant consumed. For example, the space shuttle's main engines burn LH_2 and LOX.

Hypergolic propellants are liquid fuels and oxidizers that ignite on contact with each other and need no ignition source. This easy start and restart capability makes these liquid propellants attractive for both crewed and uncrewed spacecraft maneuvering systems. Another positive feature of hypergolic propellants is the fact that, unlike cryogentic propellants, these liquids do not have to be stored at extremely low temperatures. One favored hypergolic combination is monomethyl hydrazine (MMH) as the fuel and nitrogen tetroxide (N_2O_4) as the oxidizer. However, both of these fluids are highly toxic and must be handled under the most stringent safety conditions. Hypergolic propellants are used in the core liquid-propellant stages of the Titan family of launch vehicles.

Some of the desirable properties of a liquid rocket propellant are high specific gravity (to minimize tank space), low freezing point (to permit cold-weather operation), stability (to minimize chemical deterioration during storage), favorable thermophysical properties (e.g., a high boiling point and high thermal conductivity—especially valuable in support of regenerative cooling of the combustion chamber and nozzle throat), and low vapor pressure (to support more efficient propellant pumping operations). Some of the physical hazards encountered in the use of liquid propellants include toxicity, corrosion, ignitability (when exposed to air or heat), and explosive potential (when the propellant deteriorates, is shocked, or is excessively heated).

See also ROCKET.

liquid-propellant rocket engine A rocket engine that uses chemical propellants in liquid form for both the fuel and the oxidizer.

liter (symbol: l or L) A unit of volume in the metric (SI) system. Defined as the volume of 1 kilogram of pure water at standard (atmospheric) pressure and a temperature of 4°C. Also spelled *litre.* 1 liter = 0.2642 gallons = 1.000028 cubic decimeters (dm^3).

See also SI UNITS.

lithium hydroxide (LiOH) A white crystalline compound used for removing carbon dioxide from a closed atmosphere, such as that found on a crewed spacecraft, aerospace vehicle, or space station. Spacesuit life support systems also use lithium hydroxide canisters to purge the suit's closed atmosphere of the carbon dioxide exhaled by the astronaut occupant.

little green men (LGM) A popular expression (originating in science fiction literature) for extraterrestrial beings, presumably intelligent.

live testing The testing of a rocket, missile, or aerospace vehicle by actually launching it. *Compare* STATIC TESTING.

Liwei, Yang (1965–) Chinese *Taikonaut (Astronaut), Chinese Army Officer* On October 15, 2003, Yang Liwei became the first taikonaut (astronaut) from the People's Republic of China. He rode onboard the *Shenzhou 5* spacecraft as a Long March 2F rocket carried it into orbit from the Jiuquan Satellite Launch Center. After Liwei made 14 orbits of Earth, the *Shenzhou 5* spacecraft made a successful reentry and soft landing on October 16. He was then safely recovered in the Chinese portion of Inner Mongolia.

lobe An element of a beam of focused electromagnetic (e.g., radio frequency, or RF) energy. Lobes define surfaces of equal power density at varying distances and directions from the radiating antenna. Their configuration is governed by two factors: (1) the geometrical properties of the antenna reflector and feed system and (2) the mutual interference between the direct and reflected rays for an antenna situated above a reflecting surface. In addition to the major lobes of an antenna system, there exist side lobes (or minor lobes) that result from the unavoidable finite size of the reflector. They exist at appreciable angles from the axis of the beam, and, while objectionable, they normally contain much less energy than in the major lobe.

Local Group A small cluster of about 30 galaxies, of which the Milky Way (our galaxy) and the Andromeda galaxy are dominant members.

See also CLUSTER OF GALAXIES; GALAXY.

local time Time adjusted for location around Earth or other planets in time zones. *Compare with* UNIVERSAL TIME.

local vertical The direction in which the force of gravity acts at a particular point on the surface of Earth (or other planet).

Lockheed Martin Launch Vehicle (LMLV) The original name given to the family of small- to medium-payload-capacity commercial launch vehicles developed by the Lockheed Martin Company in the 1990s. Renamed the Athena Program, these rockets combined several decades of launch vehicle and solid-propellant rocket experience. The Athena I (formerly called the LMLV-1) is a two-stage solid-propellant rocket capable of placing about 815 kg into low Earth orbit (LEO). The first stage of the Athena I is a large solid rocket motor called the Thiokol Castor 120 motor and is derived from the first stage of the U.S. Air Force's Peacekeeper intercontinental ballistic missile (ICBM). The Athena II (formerly called the LMLV-2) is a three-stage solid-propellant rocket vehicle that can lift about 1,810 kg into LEO. The first launch of the Athena II took place on January 6, 1998, from Cape Canaveral Air Force Station (CCAFS), Florida. On that date the Athena II sent NASA's *Lunar Prospector* spacecraft on its 19-month scientific mission to study the Moon. On January 26, 1999, an Athena I vehicle successfully launched the Republic of China's *ROCSAT-1* satellite from CCAFS.

See also LAUNCH VEHICLE.

lock on Signifies that a tracking or target-seeking system is continuously and automatically tracking a target in one or more coordinates (e.g., range, bearing, elevation).

Lockyer, Sir Joseph Norman (1836–1920) British *Physicist, Astronomer, Spectroscopist* Helium is the second-most-abundant element in the universe, yet its existence was totally unknown prior to the late 19th century. In 1868 the British physicist Sir Joseph Norman Lockyer collaborated with the French astronomer PIERRE JULES JANSSEN and discovered the element through spectroscopic studies of solar prominences. However, it was not until 1895 that the Scottish chemist Sir William Ramsay (1852–1916) finally detected the presence of this elusive noble gas within Earth's atmosphere. Lockyer also founded the prestigious scientific periodical *Nature* and pioneered the field of archaeological astronomy.

Lockyer was born on May 17, 1836, in Rugby, England. Following a traditional (classical) postsecondary education outside Great Britain on the European continent, Lockyer began a career as a civil servant in the War Office of the British government. While serving Her Majesty's government in this capacity from approximately 1857 to 1869, he also became an avid amateur astronomer, enthusiastically pursuing this scientific "hobby" every moment he could spare. Following observation of a solar eclipse in 1858, he constructed a private observatory at his home. In 1861, while working at the War Office in London, he married and settled in Wimbledon. Lockyer made several important astronomical discoveries in the 1860s and turned to astronomy and science on a full-time basis at the end of that decade.

While still an amateur astronomer in the 1860s, Lockyer decided to apply the emerging field of spectroscopy to study the Sun. This decision enabled Lockyer to join two other early astronomical spectroscopists, Pierre Jules Janssen and SIR WILLIAM HUGGINS, in extending the pioneering work of ROBERT BUNSEN and GUSTAV KIRCHHOFF. Lockyer's efforts directly contributed to the development of modern astrophysics. In 1866 he started spectroscopic observation of sunspots. He observed Doppler shifts in their spectral lines and suggested that strong convective currents of gas existed in the outer regions of the Sun, a region he named the chromosphere. By 1868 he devised a clever method of acquiring the spectra of the solar prominences without waiting for an eclipse of the Sun to take place. A prominence is a cooler, cloudlike feature found in the Sun's corona. They often appear around the Sun's limb during total solar eclipses. Lockyer discovered that he could observe the spectra of prominences without eclipse conditions if he passed light from the very edge of the Sun through a prism.

Almost simultaneously, but independently of Lockyer, the French astronomer Pierre Jules Janssen also developed this method of obtaining the spectra of solar prominences. Janssen and Lockyer reported their identical discoveries at the same meeting of the French Academy of Sciences. Officials of the academy wisely awarded (10 years later) both scientists a special "joint medallion" to honor their simultaneous and independent contribution to solar physics.

In 1868 Janssen used his newly developed spectroscopic technique to investigate prominences during a solar eclipse expedition to Guntur, India. Janssen reported an unknown bright orange line in the spectral data he collected. Lockyer reviewed Janssen's report and then compared the position of the reported unknown orange line to the spectral lines of all the known chemical elements. When he could not find any correlation, Lockyer concluded that this line was the spectral signature of a yet undiscovered element—an element he named helium after Helios, the sun god in Greek mythology. Because spectroscopy was still in its infancy, most scientists waited for almost three decades before accepting Lockyer's identification of a new element in the Sun. Wide scientific acceptance of Lockyer's discovery of helium took place only after the Scottish chemist Sir William Ramsay (1852–1916) detected the presence of this elusive noble gas within Earth's atmosphere in 1895.

In 1869 Lockyer won election as a fellow to the Royal Society. He decided to abandon his civil service career so he could pursue astronomy and science on a full-time basis. As a result, he founded the internationally acclaimed scientific journal *Nature* in 1869 and then served as its editor for almost 50 years. Between 1870 and 1905 Lockyer remained an active solar astronomer and personally participated in eight solar eclipse expeditions.

In 1873 Lockyer turned his attention to stellar spectroscopy and introduced his theory of atomic and molecular dissociation in an attempt to explain puzzling green lines in the spectra of certain nebulas. He disagreed with fellow British spectroscopist Sir William Huggins, who suggested that these particular green lines came from an unknown element, which he called "nebulium." Instead, Lockyer hypothesized that the unfamiliar spectral lines might actually be from terrestrial compounds that had dissociated into unusual combinations of simpler substances, making their spectral signatures unrecognizable. The American astrophysicist IRA SPRAGUE BOWEN solved the mystery of "nebulium" in the 1920s, when he demonstrated that the green emission lines from certain nebulas were actually caused by forbidden electron transitions taking place in oxygen and nitrogen under various excited states of ionization. Lockyer's theory of dissociation was definitely on the right track.

Lockyer was an inspiring lecturer. In 1881 he developed an interesting new course in astrophysics at the Royal College of Science (today part of the Imperial College). By 1885 he received an academic promotion to professor of astrophysics, making him the world's first professor in this discipline. The Royal College also constructed the Solar Physics Observatory at Kensington to support his research and teaching activities. Lockyer served as the observatory's director until approximately 1913, when the facility moved to a new location at Cambridge University.

During travel to Greece and Egypt in the early 1890s, Lockyer noticed how many ancient temples had their foundations aligned along an east-west axis, a consistent alignment that suggested to him some astronomical significance with respect to the rising and setting Sun. To pursue this interesting hypothesis, Lockyer then visited Karnack, one of the great temples of ancient Egypt. He discussed the hypothesis in his 1894 book *The Dawn of Astronomy*. This book is often regarded as the beginning of archaeological astronomy, the scientific investigation of the astronomical significance of ancient structures and sites. As part of this effort, Lockyer studied Stonehenge, an ancient site located in south England. However, he could not accurately determine the site's construction date. As a result, he could not confidently project the solar calendar back to a sufficiently precise moment in history that would reveal how the curious circular ring of large vertical stones topped by capstones might be connected to some astronomical practice of the ancient Britons. Lockyer's visionary work clearly anticipated the results of modern studies of Stonehenge, results that suggest the site could have served as an ancient astronomical calendar around 2000 B.C.E.

A prolific writer, his best-known published works include *Studies in Spectrum Analysis* (1872), *The Chemistry of the Sun* (1887), and *The Sun's Place in Nature* (1897). In 1874 the Royal Society of London recognized his important achievements in solar physics and astronomical spectroscopy and awarded him its Rumford Medal.

In 1897 Lockyer joined the order of "knight commander of the Bath (KCB)," a royal honor bestowed upon him by Britain's Queen Victoria for his discovery of helium and lifetime contributions to science. After the Solar Physics Observatory at the Royal College (in Kensington) transferred to Cambridge University, Lockyer remained active in astronomy by constructing his own private observatory in 1912 in Sidmouth, Devonshire. This facility, called the Norman Lockyer Observatory, currently operates under the supervision of an organization of amateur astronomers. Lockyer, the British civil servant who entered astronomy in the 1850s as a hobby and became one of the great pioneers in astrophysics and stellar spectroscopy, died on August 16, 1920, at Salcombe Regis, Devonshire.

logarithm The power (p) to which a fixed number (b), called the base (usually 10 or e [2.71828 . . .]), must be raised to produce the number (n) to which the logarithm corresponds. Any number (n) can be written in the form $n = b^p$. The term "p" is then the logarithm to the base "b" of the number "n"; that is, $p = \log_b n$. *Common logarithms* have 10 as the base and usually are written log or $\log_{10}$. *Natural logarithms* (also called Napierian logarithms) have e as the base and often are denoted ln or $\log_e$. The irrational number e is defined as the limit as x tends to infinity (∞) of the expression $(1 + 1/x)^x$. An antilogarithm (or inverse logarithm) is the value of the number corresponding to a given logarithm. Using the previous nomenclature, if "p" is the logarithm, then "n" is the antilogarithm.

logarithmic scale A scale in which the line segments that are of equal length are those representing multiples of 10.

logistics carriers Carriers that deliver payloads to a permanent space station (or other platform/depot in orbit) and do not require a pressurized environment.

logistics module In aerospace operations, a self-contained, pressurized-environment unit (module) designed to deliver payloads to a permanent space station or other orbiting platform/depot.

Long Duration Exposure Facility **(LDEF)** NASA's large (about the size of a bus), free-flying spacecraft that exposed numerous trays of experiments to the space environment during an extended 69-month mission in orbit around Earth from April 1984 to January 1990. LDEF gathered information on the possible consequences of the space radiation environment, atomic oxygen, meteoroids, spacecraft contamination, and space debris on aerospace hardware and spacecraft components.

longeron The main longitudinal member of a fuselage or nacelle (a streamlined enclosure).

longitude 1. Great circles that pass through both the north and south poles. Also called *meridians.* 2. Angular distance along a primary great circle from the adopted reference point; the angle between a reference plane through the polar axis (often called the prime meridian) and the second plane through that axis. On Earth longitude is measured eastward and westward through the prime meridian through 180°.

longitudinal axis The fore-and-aft line through the center of gravity of a craft.

Long March launch vehicle (LM) A series of launch vehicles developed by China. The Long March 2 (LM-2), or Changzheng-2C (CZ-2C), has two hypergolic stages and an optional solid rocket upper stage. It can place approximately 3,175 kg into low Earth orbit (LEO). This booster was first launched in 1975. The Long March-3 (LM-3) vehicle has two hypergolic stages (essentially the Long March-2 vehicle) and a cryogenic third stage. The Long March-3 vehicle was first launched in 1984 and is capable of placing 5,000-kg payload into LEO. The Long March-3A (LM-3A) vehicle has two hypergolic stages and a cryogenic upper stage. It was first launched in 1994 and can place 8,500 kg into LEO. The Long March-4 (LM-4) vehicle was first launched in 1988. It consists of three hypergolic stages and can place a 4,000-kg payload into LEO. Finally, the Long March-2E vehicle (LM-2E, or CZ-2E) has two hypergolic stages and four hypergolic strap-on rockets. It can place 8,800 kg into LEO and was first launched in 1992.

See also LAUNCH VEHICLE.

long-period comet A comet with an orbital period around the sun greater than 200 years. *Compare with* SHORT-PERIOD COMET.

look angle(s) The elevation and azimuth at which a particular satellite is predicted to be found at a specified time. Look angles are used for satellite tracking and data acquisition to minimize the amount of searching needed to acquire the satellite in the telescopic field of view or the antenna beam.

look-back time The time in the past at which the light astronomers now receive from a distant celestial object was emitted.

Lovell, Sir Alfred Charles Bernard (1913–) British *Radio Astronomer* The British radio astronomer Sir Alfred Charles Bernard Lovell founded and directed the famous Jodrell Bank Experimental Station between 1951 and 1981. It was this radio telescope facility that recorded the beeping signals from *Sputnik 1* and other early Russian spacecraft at the beginning of the space age.

Lovell, James Arthur, Jr. (1928–) American *Astronaut, U.S. Navy Officer* Astronaut James A. Lovell, Jr. was the commander of NASA's near-tragic *Apollo 13* mission to the Moon in 1970. His leadership and space travel experience were instrumental in the safe return to Earth of all three *Apollo 13* astronauts. Prior to that in-flight aborted lunar landing mission, Lovell successfully flew into orbit around Earth with astronaut FRANK BORMAN during the *Gemini 7* mission (1965) and with astronaut EDWIN (BUZZ) ALDRIN on the *Gemini 12* mission (1966). Together with astronauts Frank Borman and WILLIAM A. ANDERS, Lovell also participated in the historic *Apollo 8* mission (December 1968), the first human flight to the vicinity of the Moon.

low Earth orbit (LEO) An orbit, usually close to a circle, just beyond Earth's appreciable atmosphere at a height (typically 300 to 400 km or more) sufficient to prevent the orbit from decaying rapidly because of normal conditions of atmospheric drag.

Lowell, Percival (1855–1916) American *Astronomer* Late in the 19th century Percival Lowell established a private astronomical observatory (called the Lowell Observatory) near Flagstaff, Arizona, primarily to support his personal interest in Mars and his aggressive search for signs of an intelligent civilization there. Driving Lowell was his misinterpretation of GIOVANNI SCHIAPARELLI's use of *canali* in an 1877 technical report in which the Italian astronomer discussed his telescopic observations of the Martian surface. Lowell took this report as early observational evidence of large, water-bearing canals built by intelligent beings. Lowell then wrote books, such as *Mars and Its Canals* (1906) and *Mars As the Abode of Life* (1908), to communicate his Martian civilization theory to the public. While his nonscientific (but popular) interpretation of observed surface features on Mars proved quite inaccurate, his astronomical instincts were correct for another part of the solar system. Based on perturbations in the orbit of Neptune, Lowell predicted in 1905 the existence of a planet-sized, trans-Neptunian object. In 1930 CLYDE TOMBAUGH, working at the Lowell Observatory, discovered Lowell's Planet X and called the tiny planet Pluto.

Percival Lowell was born on March 13, 1855, in Boston, Massachusetts, into an independently wealthy aristocratic family. His brother (Abbott) became president of Harvard University and his sister (Amy) became an accomplished poet. Following graduation with honors from Harvard University in 1876, Lowell devoted his time to business and to traveling throughout the Far East. Based on his experiences between 1883 and 1895, Lowell published several books about the Far East, including *Chosön* (1886), *The Soul of the Far East* (1888), *Noto* (1891), and *Occult Japan* (1895).

He was not especially attracted to astronomy until later in life, when he discovered an English translation of Schiaparelli's 1877 Mars observation report that included the Italian word *canali.* As originally intended by Schiaparelli, the word *canali* in Italian simply meant channels. In the early 1890s Lowell unfortunately became erroneously inspired by the thought of "canals" on Mars—that is, artificially constructed structures, the (supposed) presence of which he then extended to imply the existence of an advanced alien civilization. From this point on Lowell decided to become an astronomer and dedicate his time and wealth to a detailed study of Mars.

Unlike other observational astronomers, however, Lowell was independently wealthy and already had a general idea of what he was searching for—namely, evidence of an advanced civilization on the Red Planet. To support this quest he spared no expense and sought the assistance of professional astronomers. For example, WILLIAM HENRY PICKERING from the Harvard College Observatory and his assistant ANDREW DOUGLASS helped Lowell find an excellent "seeing" site upon which to build the private observatory for the study of Mars. Lowell used his wealth to construct this private observatory near Flagstaff, Arizona. The facility, called the Lowell Observatory, was opened in 1894 and housed a top-quality 61-centimeter (24-inch) refractor telescope that allowed Lowell perform some excellent planetary astronomy. However, his beliefs tended to anticipate the things he reported, such as oases and seasonal changes in vegetation. Other astronomers, such as Douglass, labeled these blurred features simply indistinguishable natural markings. As Lowell more aggressively embellished his Mars observations, Douglass began to question Lowell's interpretation of these data. Perturbed by Douglass's scientific challenge, Lowell simply fired him in 1901 and then hired VESTO M. SLIPHER to fill the vacancy.

In 1902 the Massachusetts Institute of Technology gave Lowell an appointment as a nonresident astronomer. He was definitely an accomplished observer but often could not resist the temptation to greatly stretch his interpretations of generally fuzzy and optically distorted surface features on Mars into observational evidence of the presence of artifacts from an advanced civilization. With such books as *Mars and Its Canals,* published in 1906, Lowell became popular with the general public, which drew excitement from his speculative (but scientifically unproven) theory of an intelligent alien civilization on Mars struggling to distribute water from the planet's polar regions with a series of elaborate giant canals. While most planetary astronomers shied away from such unfounded speculation, science fiction writers flocked to Lowell's alien civilization hypothesis, a premise that survived in various forms until the dawn of the space age. On July 14, 1965, NASA's *Mariner 4* spacecraft flew past the Red Planet and returned images of its surface that shattered all previous speculations and romantic myths about a series of large canals built by a race of ancient Martians.

Since the *Mariner 4* encounter with Mars, an armada of other robot spacecraft has also studied Mars in great detail. No cities, no canals, and no intelligent creatures have been found on the Red Planet. What has been discovered, however, is an interesting "halfway" world. Part of the Martian surface is ancient, like the surfaces of the Moon and Mercury, while part is more evolved and Earthlike. In this century robot spacecraft and eventually human explorers will continue Lowell's quest for Martians, but this time they will hunt for tiny microorganisms *possibly* living in sheltered biological niches or else *possible* fossilized evidence of ancient Martian life-forms from the time when the Red Planet was a milder, kinder, and wetter world.

While Lowell's quest for signs of intelligent life on Mars may have lacked scientific rigor by a considerable margin, his astronomical instincts about "Planet X"—his name for a suspected icy world lurking beyond the orbit of Neptune—proved technically correct. In 1905 Lowell began to make detailed studies of the subtle perturbations in Neptune's orbit and predicted the existence of a planet-sized trans-Neptunian object. He then initiated an almost decade-long telescopic search but failed to find this elusive object. In 1914, near the end of his life, he published the negative results of his search for Planet X and bequeathed the task to some future astronomer. Lowell died in Flagstaff, Arizona, on November 12, 1916.

In 1930 a farm boy turned amateur astronomer named Clyde Tombaugh fulfilled Lowell's quest. Hired by Vesto Slipher to work as an observer at the Lowell Observatory and to search for Planet X, Tombaugh made use of the blinking comparator technique to find the planet Pluto on February 18, 1930. The Lowell Observatory still functions today as a major private astronomical observatory, a fitting tribute to its founder, the wealthy Bostonian aristocrat who had a passion for Martians that ultimately resulted in the discovery of a new planet. An often forgotten milestone in astronomy took place at the Lowell Observatory in 1912, when Vesto Slipher made early measurements of the Doppler shift of distant nebulas. He found many to be receding from Earth at a high rate of speed. Slipher's work provided EDWIN P. HUBBLE the basic direction he needed to discover the expansion of galaxies.

Lowell Observatory The observatory in Flagstaff, Arizona, at an altitude of 2,210 m founded by Percival Lowell. He established this facility as a private astronomical observatory in 1894, primarily to support his personal interest in Mars and his aggressive search for signs of an intelligent civilization.

low-gain antenna (LGA) A spacecraft's low-gain antenna (LGA) provides wide-angle coverage at the expense of gain. Coverage is nearly omnidirectional, except for areas that may be shadowed by the spacecraft body. For interplanetary space missions, especially in the inner solar system, low-gain antennas often are designed to be usable for relatively low data rates, as long as the spacecraft is within relatively close "interplanetary range" (typically a few astronomical units) and the Earth-to-spacecraft transmitter is powerful enough. For example, a LGA can be used on a mission to Venus, but a spacecraft exploring Saturn or Uranus must use a high-gain antenna (HGA) because of the distances involved. An LGA sometimes is mounted on top of a HGA's subreflector.

See also TELECOMMUNICATIONS; TELEMETRY.

LOX Liquid oxygen at a temperature of 90 kelvins (K) (–183°C) or lower. A common propellant (the oxidizer) for liquid rocket engines. Encountered in a variety of aerospace uses, such as LOX tank, LOX valve, or loxing (i.e., to fill a rocket vehicle's propellant tank with liquid oxygen).

LOX-hydrogen engine A rocket engine that uses liquid hydrogen as the fuel and liquid oxygen (LOX) as the oxidizer, such as the space shuttle main engines (SSMEs).

See also ROCKET.

lubrication (in space) Because of the very low pressure conditions encountered in outer space, conventional lubricating oils and greases evaporate very rapidly. Even soft metals (such as copper, lead, tin, and cadmium) that are often used in bearing materials on Earth will evaporate at significant rates in space. Responding to the harsh demands on materials imposed by the space environment, aerospace engineers use two general lubrication techniques in space vehicle design and operation: thick-film lubrication and thin-film lubrication. In thick-film lubrication (also known as hydrodynamic or hydrostatic lubrication), the lubricant remains sufficiently viscous during operation so that the moving surfaces do not come into physical contact with each other. Thin-film lubrication takes place whenever the film of lubricant between two moving surfaces is squeezed out so that surfaces actually come into physical contact with each other (on a microscopic scale). Aerospace engineers lubricate the moving components of a spacecraft or satellite with dry films, liquids, metallic coatings, special greases, or combinations of these materials. Liquid lubricants are often used on space vehicles with missions of a year or more duration.

See also SPACE TRIBOLOGY.

Lucid, Shannon W. (1943–) American *Astronaut* Astronaut Shannon W. Lucid served as a NASA mission specialist on four separate space shuttle flights: STS-51G (June 1985), STS-34 (October 1989), STS-43 (August 1991) and STS-58 (October 1993). She was the cosmonaut engineer 2 onboard Russia's *Mir* space station (1996)—joining the Russian crew from the space shuttle *Atlantis* (STS-76 mission) in March and returning to Earth after 188 days in orbit as a passenger on the space shuttle *Atlantis* (STS-79 mission).

lumen (symbol: lm) The SI unit of luminous flux. It is defined as the luminous flux emitted by a uniform point source with an intensity of one candela in a solid angle of one steradian.

See also CANDELA; SI UNITS.

luminosity (symbol: L) The rate at which a star or other luminous object emits energy, usually in the form of electromagnetic radiation. The luminosity of the Sun is approximately 4×10^{26} watts.

See also STEFAN-BOLTZMANN LAW.

luminous intensity Luminous energy per unit of time per unit of solid angle; the intensity (i.e., flux per unit solid angle) of visible radiation weighted to take into account the variable response of the human eye as a function of the wavelength of light; usually expressed in candelas. Sometimes called luminous flux density.

lumped mass Concept in engineering analysis wherein a mass is treated as if it were concentrated at a point.

Luna A series of Russian spacecraft sent to the Moon in the 1960s and 1970s. For example, *Luna 9,* a 1,581-kg spacecraft, was launched on January 31, 1966, and successfully soft-landed on the Ocean of Storms on February 3, 1966. This spacecraft transmitted several medium-resolution photographs of the lunar surface before its batteries failed four days after landing. The lander spacecraft also returned data on radiation levels at the landing site. *Luna 12* was an orbiter spacecraft that was launched on October 22, 1966, successfully orbited the Moon, and returned television pictures of the surface. *Luna 16,* launched on September 12, 1970, was the first successful automated (robotic) sample-return mission to the lunar surface. (Of course, the U.S. *Apollo 11* and *12* lunar landing missions in July 1969 and November 1969, respectively, already had returned lunar samples collected by astronauts who walked on the Moon's surface.) After landing on the Sea of Fertility, this robot spacecraft deployed a drill that bore 35 centimeters into the surface. The lunar soil sample, which had a mass of about 0.1 kg, was transferred automatically to a return vehicle that then left the lunar surface and landed in the Soviet Union on September 24, 1970. *Luna 17* introduced the first robot roving vehicle, called *Lunokhod 1,* on the lunar surface. The spacecraft successfully touched down on the Sea of Rains and deployed the sophisticated *Lunokhod 1* rover. This eight-wheel vehicle was radio-controlled from Earth. The rover covered 10.5 kilometers during a surface exploration mission

This is an artist's conception of a lunar habitat being assembled out of components delivered to the Moon's surface by a series of automated cargo flights ca. 2020. Pressurized rovers, logistics modules, and a space suit maintenance and storage module combine to provide the living and working quarters for the lunar astronauts. *(Courtesy of NASA; artist, John Frassanito and Associates [1993])*

that lasted 10.5 months. The rover's cameras transmitted more than 20,000 images of the Moon's surface, and instruments of the vehicle analyzed properties of the lunar soil at many hundreds of locations. *Luna 20* (launched February 14, 1972) and *Luna 24* (launched August 9, 1976) were also successful robot soil sample return missions. *Luna 21,* launched in January 1973, successfully deployed another robot rover, *Lunokhod 2,* in the Le Monnier crater in the Sea of Tranquility. This 840 kg rover vehicle traveled about 37 kilometers during its four-month-long surface exploration mission. Numerous photographs were taken and surface experiments conducted by this robot rover under radio control by Russian scientists and technicians on Earth.

See also MOON.

lunar Of or pertaining to the Moon.

lunar bases and settlements When human beings return to the Moon, it will not be for a brief moment of scientific inquiry, as occurred in NASA's Apollo Project, but rather as permanent inhabitants of a new world. They will build bases from which to completely explore the lunar surface, establish science and technology laboratories that take advantage of the special properties of the lunar environment, and harvest the Moon's resources (including the suspected deposits of lunar ice in the polar regions) in support of humanity's extraterrestrial expansion.

A lunar base is a permanently inhabited complex on the surface of the Moon. In the first permanent lunar base camp, a team of from 10 up to perhaps 100 lunar workers will set about the task of fully investigating the Moon. (*See* figure.) The word *permanent* here means that the facility will always be occupied by human beings, but individuals probably will serve tours of from one to three years before returning to Earth. Some workers at the base will enjoy being on another world. Some will begin to experience isolation-related psychological problems, similar to the difficulties often experienced by members of scientific teams who "winter-over" in

Antarctic research stations. Still other workers will experience injuries or even fatal accidents while working at or around the lunar base.

For the most part, however, the pioneering lunar base inhabitants will take advantage of the Moon as a science-in-space platform and perform the fundamental engineering studies needed to confirm and define the specific roles the Moon will play in the full development of space later this century and in centuries beyond. For example, the discovery of frozen volatiles (including water) in the perpetually frozen recesses of the Moon's polar regions could change lunar base logistics strategies and accelerate development of a large lunar settlement of up to 10,000 or more inhabitants. Many lunar base applications have been proposed. Some of these concepts include (1) a lunar scientific laboratory complex, (2) a lunar industrial complex to support space-based manufacturing, (3) an astrophysical observatory for solar system and deep space surveillance, (4) a fueling station for orbital transfer vehicles that travel through cislunar space, and (5) a training site and assembly point for the first human expedition to Mars. Social and political scientists suggest that a permanent lunar base could also become the site of innovative political, social, and cultural developments, essentially rejuvenating our concept of who we are as intelligent beings and boldly demonstrating our ability to beneficially apply advanced technology in support of the positive aspects of human destiny. Another interesting suggestion for a permanent lunar base is its use as a field operations center for the rapid response portion of a planetary defense system that protects Earth from threatening asteroids or comets.

As lunar activities expand, the original lunar base could grow into an early settlement of about 1,000 more or less permanent residents. Then, as the lunar industrial complex develops further and lunar raw materials, food, and manufactured products start to support space commerce throughout cislunar space, the lunar settlement itself will expand to a population of around 10,000. At that point the original settlement might spawn several new settlements, each taking advantage of some special location or resource deposit elsewhere on the lunar surface.

In the 22nd century this collection of permanent human settlements on the Moon could continue to grow, reaching a combined population of about 500,000 persons and attaining a social and economic "critical mass" that supports true self-

Surface operations at a permanent lunar base ca. 2025 ***(Artist rendering courtesy of NASA; Pat Rawlings, artist [The Deal, 1997])***

This artist's rendering suggests how lunar astronauts might respond to a medical emergency ca. 2025. In the scenario illustrated, an antenna installer has fallen over a 30-meter escarpment and fractured his right leg. Two other lunar workers respond to the emergency using their *medivac hopper* (parked left foreground) to promptly reach the accident scene. While communicating with the lunar base (distant background), the pair of first responders use a variety of scanning devices to assess the severity of the worker's injury. They decide to use an inflatable splint to brace the semiconscious worker's leg and then use the medivac hopper to evacuate him to the lunar base. ***(Artist rendering courtesy of NASA; artist, Pat Rawlings [Moon 911, 1992])***

sufficiency. This moment of self-sufficiency for the lunar civilization will also be a historic moment in human history. From that time on the human race will exist in two distinct and separate biological niches—we will be *terran* and *nonterran* (or extraterrestrial).

With the rise of a self-sufficient, autonomous lunar civilization, future generations will have a choice of worlds on which to live and prosper. Of course, such a major social development will most likely produce its share of cultural backlash in both worlds. Citizens of the 22nd century may start seeing personal ground vehicles with such bumper-sticker slogans such as, "This is my world—love it or leave it!," "Terran go home," or even "Protect terrestrial jobs—ban lunar outsourcing!"

The vast majority of lunar base development studies include the use of the Moon as a platform from which to conduct science in space. Scientific facilities on the Moon will take advantage of its unique environment to support platforms for astronomical, solar, and space science observations. The unique environmental characteristics of the lunar surface include low gravity (one-sixth that of the Earth), high vacuum (about 10^{-12} torr [a torr is a unit of pressure equal to 1/760 of the pressure of Earth's atmosphere at sea level]), seismic stability, low temperatures (especially in permanently shadowed polar regions), and a low radio noise environment on the Moon's farside.

Astronomy from the lunar surface offers the distinct advantages of a low radio noise environment and a stable platform in a low-gravity environment. The farside of the Moon is permanently shielded from direct terrestrial radio emissions. As future radio telescope designs approach their ultimate (theoretical) performance limits, this uniquely quiet lunar environment may be the only location in all cislunar space where sensitive radio wave detection instruments can be used to full advantage, both in radio astronomy and in our search for extraterrestrial intelligence (SETI). In fact, radio astronomy, including extensive SETI efforts, may represent one of the main "lunar industries" late in this century. In a sense, SETI performed by lunar-based scientists will be extraterrestrials searching for other extraterrestrials.

The Moon also provides a solid, seismically stable, low-gravity, high-vacuum platform for conducting precise interferometric and astrometric observations. For example, the availability of ultrahigh-resolution (microarcsecond) optical, infrared, and radio observatories will allow astronomers to carefully search for extrasolar planets encircling nearby stars.

A lunar scientific base also will provide life scientists with a unique opportunity to extensively study biological processes in reduced gravity (1/6 g) and in low magnetic fields. Genetic engineers can conduct their experiments in comfortable facilities that are nevertheless physically isolated from the Earth's biosphere. Exobiologists can experiment with new types of plants and microorganisms under a variety of simulated alien-world conditions. Genetically engineered "lunar plants," grown in special greenhouse facilities, could become a major food source while also supplementing the regeneration of a breathable atmosphere for the various lunar habitats.

The true impetus for large, permanent lunar settlements will most likely arise from the desire for economic gain, a time-honored stimulus that has driven much technical, social, and economic development on Earth. The ability to create useful products from native lunar materials will have a controlling influence on the overall rate of growth of the lunar civilization. Some early lunar products can now easily be identified. Lunar ice, especially when refined into pure water or dissociated into the important chemicals hydrogen (H_2) and oxygen (O_2) represents the Moon's most important resource. Other important early lunar products include (1) oxygen (extracted from lunar soils) for use as a propellant by orbital transfer vehicles traveling throughout cislunar space, (2) raw (i.e., bulk, minimally processed) lunar soil and rock materials for space radiation shielding, and (3) refined ceramic and metal products to support the construction of large structures and habitats in space.

The initial lunar base can be used to demonstrate industrial applications of native Moon resources and to operate small pilot factories that provide selected raw and finished products for use both on the Moon and in Earth orbit. Despite the actual distances involved, the cost of shipping a kilogram of "stuff" from the surface of the Moon to various locations in cislunar space may prove much cheaper than shipping the same "stuff" from the surface of Earth.

The Moon has large supplies of silicon, iron, aluminum, calcium, magnesium, titanium and oxygen. Lunar soil and rock can be melted to make glass in the form of fibers, slabs, tubes, and rods. Sintering (a process whereby a substance is formed into a coherent mass by heating, but without melting) can produce lunar bricks and ceramic products. Iron metal can be melted and cast or converted to specially shaped forms using powder metallurgy. These lunar products would find a ready market as shielding materials, in habitat construction, in the development of large space facilities, and in electric power generation and transmission systems.

Lunar mining operations and factories can be expanded to meet growing demands for lunar products throughout cislunar space. With the rise of lunar agriculture (accomplished in special enclosed facilities), the Moon may even become our "extraterrestrial breadbasket," providing the majority of all food products consumed by humanity's extraterrestrial citizens.

One interesting space commerce scenario involves an extensive lunar surface mining operation that provides the required quantities of materials in a preprocessed condition to a giant space manufacturing complex located at Lagrangian libration point 4 or 5 (L_4 or L_5). These exported lunar materials would consist primarily of oxygen, silicon, aluminum, iron, magnesium, and calcium locked into a great variety of complex chemical compounds. It is often suggested by space visionaries that the Moon will become the chief source of materials for space-based industries in the latter part of this century.

Numerous other tangible and intangible advantages of lunar settlements will accrue as a natural part of their creation and evolutionary development. For example, the high-technology discoveries originating in a complex of unique lunar laboratories could be channeled directly into appropriate economic and technical sectors on Earth, as "frontier" ideas, techniques, products, and so on. The permanent presence of people on another world (a world that looms large in the night sky) will continuously suggest an *open world philosophy* and a sense of cosmic destiny to the vast majority of humans who remain behind on the home planet. The human generation that decides to venture into cislunar space and to create permanent lunar settlements will long be admired not only for its great technical and intellectual achievements but also for its innovative cultural accomplishments. Finally, it is not too remote to speculate that the descendants of the first lunar settlers will become first the interplanetary, then the interstellar, portion of the human race. The Moon can be viewed as humanity's stepping stone to the universe.

lunar crater A depression, usually circular, on the surface of the Moon. It frequently occurs with a raised rim called a *ringwall.* Lunar craters range in size up to 250 kilometers in diameter. The largest lunar craters are sometimes called "walled plains." The smaller crater, 15 to 30 kilometers across, often are called "craterlets," while the very smallest, just a few hundred meters across, are called "beads." Many lunar craters have been named after famous people, usually astronomers.

See also IMPACT CRATER; MOON.

lunar day The period of time associated with one complete orbit of the Moon about Earth. It is equal to 27.322 Earth days. The lunar day is also equal in length to the sidereal month.

lunar eclipse The phenomenon observed when the Moon enters the shadow of Earth. A lunar eclipse is called *penumbral* if the Moon enters only the penumbra of Earth, *partial* if the Moon enters the umbra without being totally immersed, and *total* if the Moon is entirely immersed in the umbra.

lunar excursion module (LEM) The lander spacecraft used to bring the Apollo astronauts to the surface of the Moon. The lunar excursion module (LEM) lifted off from Earth enclosed in a compartment of the Saturn V launch vehicle, below the command-service module (CSM) that housed the astronauts. Once the craft was on its way to the Moon, the CSM pulled the LEM from its storage area. In lunar orbit the two Apollo astronauts who were to explore the lunar surface climbed into the LEM, undocked, and "flew" the lander spacecraft to the Moon's surface. The descent stage of the LEM contained a main, centrally located retrorocket engine that accomplished the soft landing. Upon completion of the surface expedition, the crew lifted off from the lunar surface using the top segment of the LEM, which was powered by another, smaller rocket. The "four-legged" bottom portion of the LEM remained on the Moon. After reaching lunar orbit with the now-smaller LEM spacecraft, the crew rendezvoused with and transferred back into the Apollo CSM. The ascent stage of the LEM was then jettisoned and placed in a decaying orbit that sent it crashing into the lunar surface to avoid possible interference with the Apollo CSM or a future mission. During the aborted *Apollo 13* mission, the LEM served as a lifeboat for all three Apollo astronauts, providing them the emergency life support that helped get them safely home after the CSM suffered a catastrophic explosion on the way to the Moon.

The LEM is also referred to as the lunar module (LM).

See also APOLLO PROJECT.

lunar gravity The force imparted by the Moon to a mass that is at rest relative to the lunar surface. It is approximately one-sixth the acceleration experienced by a mass at rest on the surface (sea level) of Earth. On the Moon's surface the acceleration due to gravity is about 1.62 meters per second per second (m/s^2).

lunar highlands The light-colored, heavily cratered, mountainous part of the lunar landscape.

See also MOON.

lunar maria Vast expanses of low-albedo (that is, comparatively dark) ancient volcanic lava flows found mostly on the nearside surface of the Moon.

lunar module (LM) *See* LUNAR EXCURSION MODULE (LEM).

lunar orbit Orbit of a spacecraft around the Moon.

Lunar Orbiter Five Lunar Orbiter missions were launched by NASA in 1966 and 1967 with the purpose of mapping the Moon's surface prior to the landings by the Apollo astronauts (which occurred from 1969 to 1972). All five missions were highly successful; 99% of the lunar surface was photographed with 60-meter spatial resolution or better. The first three Lunar Orbiter missions were dedicated to imaging 20 potential Apollo landing sites that had been preselected based on observations from Earth. The fourth and fifth missions were committed to broader scientific objectives and were flown in high-altitude polar orbits around the Moon. *Lunar Orbiter 4* photographed the entire nearside and 95% of the farside, and *Lunar Orbiter 5* completed the farside coverage and acquired medium- (20-meter) and high- (2-meter) resolution images of 36 preselected areas. These probes were sent into orbit around the Moon to gather information and then purposely crashed at the end of each mission to prevent possible interference with future projects.

See also APOLLO PROJECT; MOON.

lunar probe A probe for exploring and reporting conditions on or about the Moon.

Apollo 11* astronaut Edwin "Buzz" Aldrin descends the ladder of the lunar excursion module (LEM) and becomes the second human being to walk on the Moon on July 20, 1969. *(Courtesy of NASA)

Lunar Prospector A NASA Discovery Program spacecraft designed for a low-altitude polar orbit investigation of the Moon. The spacecraft was launched successfully from Cape Canaveral Air Force Station, Florida, by a Lockheed Athena II vehicle (formerly called the Lockheed Martin Launch Vehicle) on January 6, 1998. After swinging into orbit around the Moon on January 11, the *Lunar Prospector* used its complement of instruments to perform a detailed study of surface composition. The spacecraft also searched for resources, especially suspected deposits of water ice in the permanently shadowed regions of the lunar poles. The 126-kg spacecraft carried a gamma-ray spectrometer, a neutron spectrometer, a magnetometer, an electron reflectometer, an alpha particle spectrometer, and a Doppler gravity experiment.

The data from this mission complemented the detailed imagery data from the Clementine mission. Data from the *Lunar Prospector*'s neutron spectrometer hinted at the presence of significant amounts of water ice at the lunar poles. While still subject to confirmation, the presence of large quantities of water ice at the lunar poles would make the Moon a valuable supply depot for any future human settlement of the Moon and regions beyond.

After a highly successful 19-month scientific mapping mission, flight controllers decided to turn the spacecraft's originally planned end-of-life crash into the lunar surface into an impact experiment that might possibly confirm the presence of water ice on the Moon. Therefore, as its supply of attitude control fuel neared exhaustion, the spacecraft was directed to crash into a crater near the Moon's south pole on July 31, 1999. Observers from Earth attempted to detect signs of water in the impact plume, but no such signal was found. However, this impromptu impact experiment should be regarded only as a long shot opportunity and not a carefully designed scientific procedure. In contrast, the *Lunar Prospector*'s scientific data have allowed scientists to construct a detailed map of the Moon's surface composition. These data have also greatly improved knowledge of the origin, evolution, and current inventory of lunar resources.

lunar rover(s) Crewed or automated (robot) vehicles used to help explore the Moon's surface. The Lunar Rover Vehicle (LRV), shown in the figure, also was called a space buggy and the Moon car. It was used by American astronauts during the *Apollo 15, 16,* and *17* expeditions to the Moon. This vehicle was designed to climb over steep slopes, go over rocks, and move easily over sandlike lunar surfaces. It was able to carry more than twice its own mass (about 210 kilograms) in passengers, scientific instruments, and lunar material samples. This electric-powered (battery) vehicle could travel about 16 kilometers per hour on level ground. The vehicle's power came from two 36-volt silver zinc batteries that drove independent 1/4-horsepower electric motors in each wheel. Apollo astronauts used their space buggies to explore well beyond their initial lunar landing sites. With these vehicles they were able to gather Moon rocks and travel much farther and quicker across the lunar surface than if they had had to explore on foot. For example, during the *Apollo 17* expedition the lunar rover traveled 19 kilometers on just one of its three excursions.

Automated, or robot, rovers also can be used to explore the lunar surface. For example, during the Russian *Luna 17* mission to the Moon in 1970, the "mother" spacecraft soft-landed on the lunar surface in the Sea of Rains and deployed the *Lunokhod 1* robot rover vehicle. Controlled from Earth by radio signals, this eight-wheeled lunar rover vehicle traveled for months across the lunar surface, transmitting more than 20,000 television images of the surface and performing more than 500 lunar soil tests at various locations. The Russian *Luna 21* mission to the Moon in January 1973 successfully deployed another robot rover, called *Lunokhod 2.*

See also APOLLO PROJECT; LUNA; MOON.

lunar satellite A human-made object that is placed in orbit for one or more revolutions around the Moon. For example, the American *Lunar Orbiter* spacecraft was a lunar satellite.

See also LUNAR ORBITER.

Lundmark, Knut Emil (1889–1958) Swedish *Astronomer* The Swedish astronomer Knut Emil Lundmark investigated galactic phenomena and suggested that the term *supernova* be applied to the brightest novas observed in other galaxies.

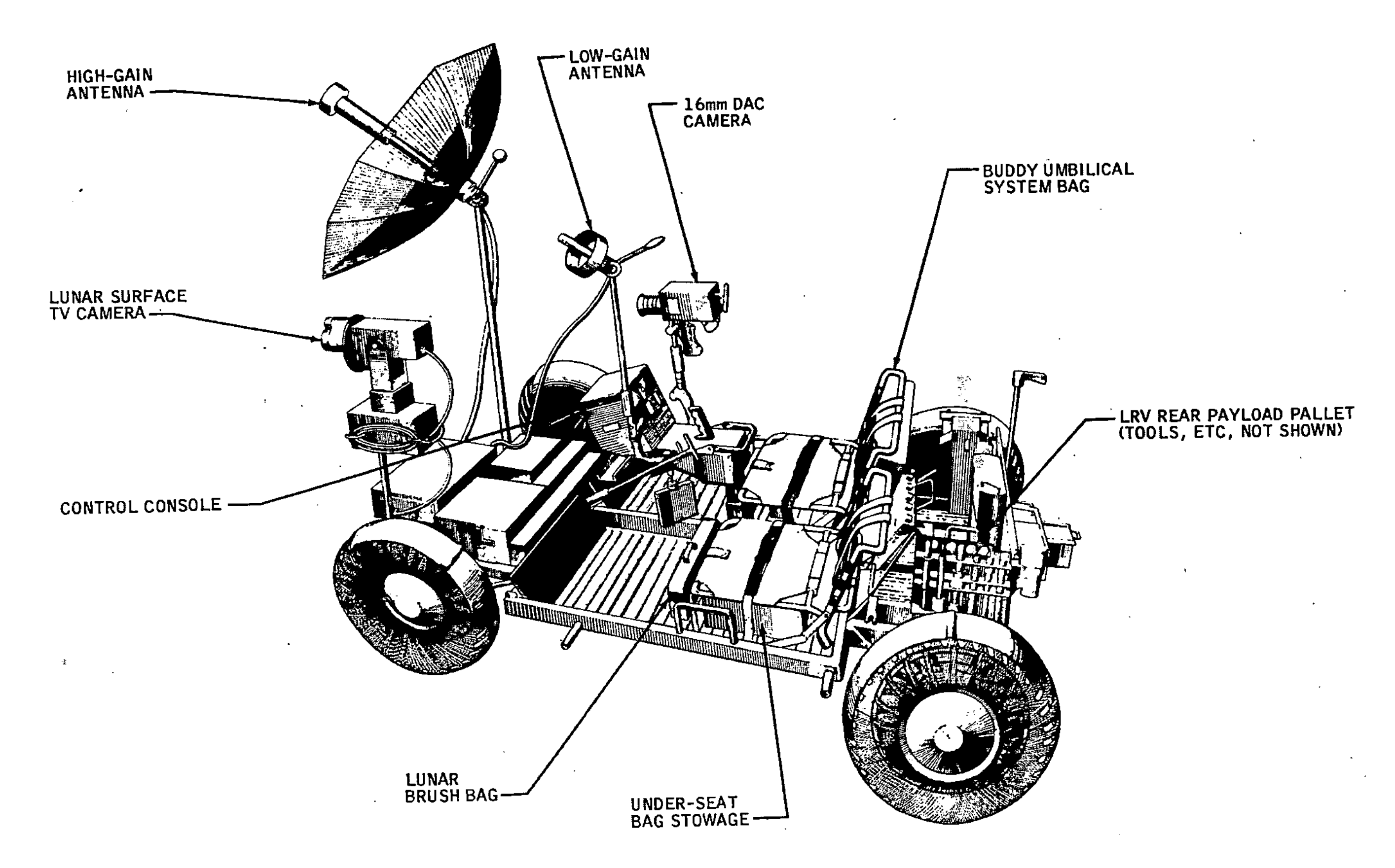

The Lunar Rover Vehicle (LRV) was used by the *Apollo 15, 16,* and *17* mission astronauts to explore the Moon's surface a significant distance away from their particular landing sites. *(Drawing courtesy of NASA)*

Lunokhod A Russian eight-wheeled robot vehicle controlled by radio signals from Earth and used to conduct surface exploration of the lunar surface.

See also LUNA; LUNAR ROVER(S).

lux (symbol: lx) The SI unit of illuminance. It is defined as 1 lumen per square meter.

See also LUMEN; SI UNITS.

Luyten, Willem Jacob (1899–1994) Dutch-American *Astronomer* The Dutch-American astronomer Willem Jacob Luyten specialized in the detection and identification of white dwarf stars.

Lyot, Bernard Ferdinand (1897–1952) French *Astronomer* The French astronomer Bernard Ferdinand Lyot invented the coronagraph in 1930. This instrument allowed astronomers to study the corona of the Sun in the absence of a solar eclipse, thereby greatly advancing the field of solar physics.

Lysithea A moon of Jupiter discovered in 1938 by the American astronomer SETH BARNES NICHOLSON. This small moon has a diameter of about 20 kilometers and orbits Jupiter at a distance of 11,710,000 km with a period of 260 days. Also called Jupiter X.

See also JUPITER.

M

Mace An improved version of the U.S. Air Force's Matador missile (MGM-1C), differing primarily in its improved guidance system, longer range, low-level attack capability, and higher-yield warhead. The Mace was designated as the MGM-13. This early "winged missile" was guided by a self-contained radar guidance system (MGM-13A) or by an inertial guidance system (MG-13B).

Mach, Ernst (1838–1916) Austrian *Physicist* The Austrian physicist Ernst Mach investigated the phenomena associated with high-speed flow. In recognition of his efforts, the number expressing the ratio of the speed of a body or of a point on a body with respect to the surrounding air or other fluid or the speed of a flow to the speed of sound in the medium is called the Mach number (M).

Mach number (symbol: M) A number expressing the ratio of the speed of a body or of a point on a body with respect to the surrounding air or other fluid or the speed of a flow to the speed of sound in the medium. It is named after the Austrian scientist ERNST MACH.

If the Mach number is less than one (i.e., $M < 1$), the flow is called *subsonic,* and local disturbances can propagate ahead of the flow. If the Mach number is equal to unity (i.e., $M = 1$), the flow is called *sonic.* If the Mach number is greater than one (i.e., $M > 1$), the flow is called *supersonic,* and disturbances cannot propagate ahead of the flow, with the general result that oblique shock waves form (i.e., shock waves inclined in the direction of flow). The formation and characteristics of such oblique shock waves involve complicated compressible flow phenomena. Sometimes the oblique shock wave is attached to the surface of the body, but this usually occurs only when the supersonic stream encounters a sharp, narrow-angled object. Most times, however, the oblique shock wave takes the form of a rounded, detached shock that appears in front of a more blunt-shaped object moving at supersonic speed through a compressible fluid medium.

MACHO The *massive compact halo object* hypothesized to populate the outer regions (or halos) of galaxies, accounting for most of the dark matter. Current theories suggest that MACHOs might be low-luminosity stars, Jovian-sized planets, or possibly black holes.

See also WIMP.

Magellan, Ferdinand (1480–1521) Portuguese *Explorer* In 1519 the Portuguese explorer Ferdinand Magellan became the first person to record the two dwarf galaxies visible to the naked eye in the Southern Hemisphere. Astronomers call these nearby galaxies the Magellanic Clouds in his honor.

Magellanic Clouds The two dwarf, irregularly shaped neighboring galaxies that are closest to our Milky Way galaxy. The Large Magellanic Cloud (LMC) is about 160,000 light-years away, and the Small Magellanic Cloud (SMC) is approximately 180,000 light-years away. Both can be seen with the naked eye in the Southern Hemisphere. Their presence was first recorded in 1519 by the Portuguese explorer FERDINAND MAGELLAN, after whom they are named.

See also GALAXY.

Magellan mission A NASA solar system exploration mission to the planet Venus. On May 4, 1989, the 3,550-kilogram *Magellan* spacecraft was delivered to Earth orbit by the space shuttle *Atlantis* during the STS 30 mission and then sent on an interplanetary trajectory to the cloud-shrouded planet by a solid-fueled inertial upper-stage (IUS) rocket system. *Magellan* was the first interplanetary spacecraft to be launched by the space shuttle. On August 10, 1990, *Magellan* (named for the famous 16th-century Portuguese explorer FERDINAND MAGELLAN) was inserted into orbit around Venus and began initial operations of its very successful radar mapping mission.

Magellan used a sophisticated imaging radar system to make the most detailed map of Venus ever captured during its four years in orbit around Earth's "sister planet" from 1990 to 1994. After concluding its radar mapping mission, *Magellan* made global maps of the Venusian gravity field. During this phase of the mission, the spacecraft did not use its radar mapper but instead transmitted a constant radio signal back to Earth. When it passed over an area on Venus with higher than normal gravity, the spacecraft would speed up slightly in its orbit. This movement then would cause the frequency of *Magellan*'s radio signal to change very slightly due to the Doppler effect. Because of the ability of the radio receivers in the NASA Deep Space Network to measure radio frequencies extremely accurately, scientists were able to construct a very detailed gravity map of Venus. In fact, during this phase of its mission, the spacecraft provided high-resolution gravity data for about 95% of the planet's surface. Flight controllers also tested a new maneuvering technique called aerobraking, a technique that uses a planet's atmosphere to slow or steer a spacecraft.

The craters revealed by *Magellan*'s detailed radar images suggested to planetary scientists that the Venusian surface is relatively young, perhaps "recently" resurfaced or modified about 500 million years ago by widespread volcanic eruptions. The planet's current harsh environment has persisted at least since then. No surface features were detected that suggested the presence of oceans or lakes at any time in the planet's past. Furthermore, scientists found no evidence of plate tectonics, that is, the movements of huge crustal masses.

Magellan's mission ended with a dramatic plunge through the dense atmosphere to the planet's surface. This was the first time an operating planetary spacecraft had ever been crashed intentionally. Contact was lost with the spacecraft on October 12, 1994, at 10:02 Universal Time (3:02 A.M. Pacific Daylight Time). The purpose of this last maneuver was to gather data on the Venusian atmosphere before the spacecraft ceased functioning during its fiery descent. Although much of the *Magellan* is believed to have been vaporized by atmospheric heating during this final plunge, some sections may have survived and hit the planet's surface intact.

See also VENUS.

magma Molten rock material beneath the surface of a planet or moon. It may be ejected to the surface by volcanic activity.

magnet A substance with the property of attracting certain other substances. Some metals, such as cobalt, iron, and nickel, can be magnetized to attract other magnetic metals. A magnet has a magnetic field, which is a region around it where there are forces exerted on other magnetic materials. Earth has a magnetic field, as do several other planets. An electric current flowing through a conductor also creates a magnetic field.

magnetars The class of neutron stars that have magnetic fields hundreds of times more intense than typical neutron stars.

See also NEUTRON STAR.

magnetic field strength **(symbol: H)** The strength of the magnetic force on a unit magnetic pole in a region of space affected by other magnets or electric currents. The magnetic field strength is a vector quantity that is measured in amperes per meter.

magnetic flux **(symbol: φ_M)** A measure of the total size of a magnetic field. The flux through any area in the medium surrounding a magnet is equal to the integral of the magnetic flux density (B) over the area (A). That is,

$$\varphi_M = \int B \cdot d\,A$$

The weber (Wb) is the SI unit of magnetic flux.

magnetic flux density **(symbol: B)** The magnetic flux that passes through a unit area of a magnetic field in a direction at right angles to the magnetic force. It is a vector quantity. The SI unit of magnetic flux density is the tesla (T), which is equal to one weber of magnetic flux per square meter.

magnetic star A star with a strong magnetic field, usually characterized by measurements of the Zeeman effect splitting of its spectral lines.

See also ZEEMAN EFFECT.

magnetic storm A sudden, worldwide disturbance of Earth's magnetic field caused by solar disturbances, such as a solar flare. The onset of a magnetic storm may occur in an hour, while the gradual return to normal conditions may take several days.

magnetometer An instrument for measuring the strength and sometimes the direction of a magnetic field. Magnetometers often are used to detect and measure the interplanetary and solar magnetic fields in the vicinity of a scientific spacecraft. These direct-sensing instruments typically detect the strength of magnetic fields in three dimensions (i.e., in the x-, y-, and z-planes). As a magnetometer sweeps an arc through a magnetic field when the spacecraft rotates, an electrical signal is produced proportional to the magnitude and structure of the field.

magnetoplasmadynamic thruster (MPD) An advanced electric propulsion device capable of operating with a wide range of propellants in both pulsed and steady-state modes. MPD thrusters are well suited for orbit transfer and spacecraft maneuvering applications. In an MPD thruster device the current flowing from the cathode to the anode sets up a ring-shaped magnetic field. This magnetic field then pushes against the plasma in the arc. As propellant, such as argon, flows through the arc plasma, it is ionized and forced away by the magnetic field. A thrusting force therefore is created by the interaction of an electrical current and a magnetic field.

See also ELECTRIC PROPULSION.

magnetosphere The region around a planet in which charged atomic particles are influenced by the planet's own magnetic field rather than by the magnetic field of the Sun as

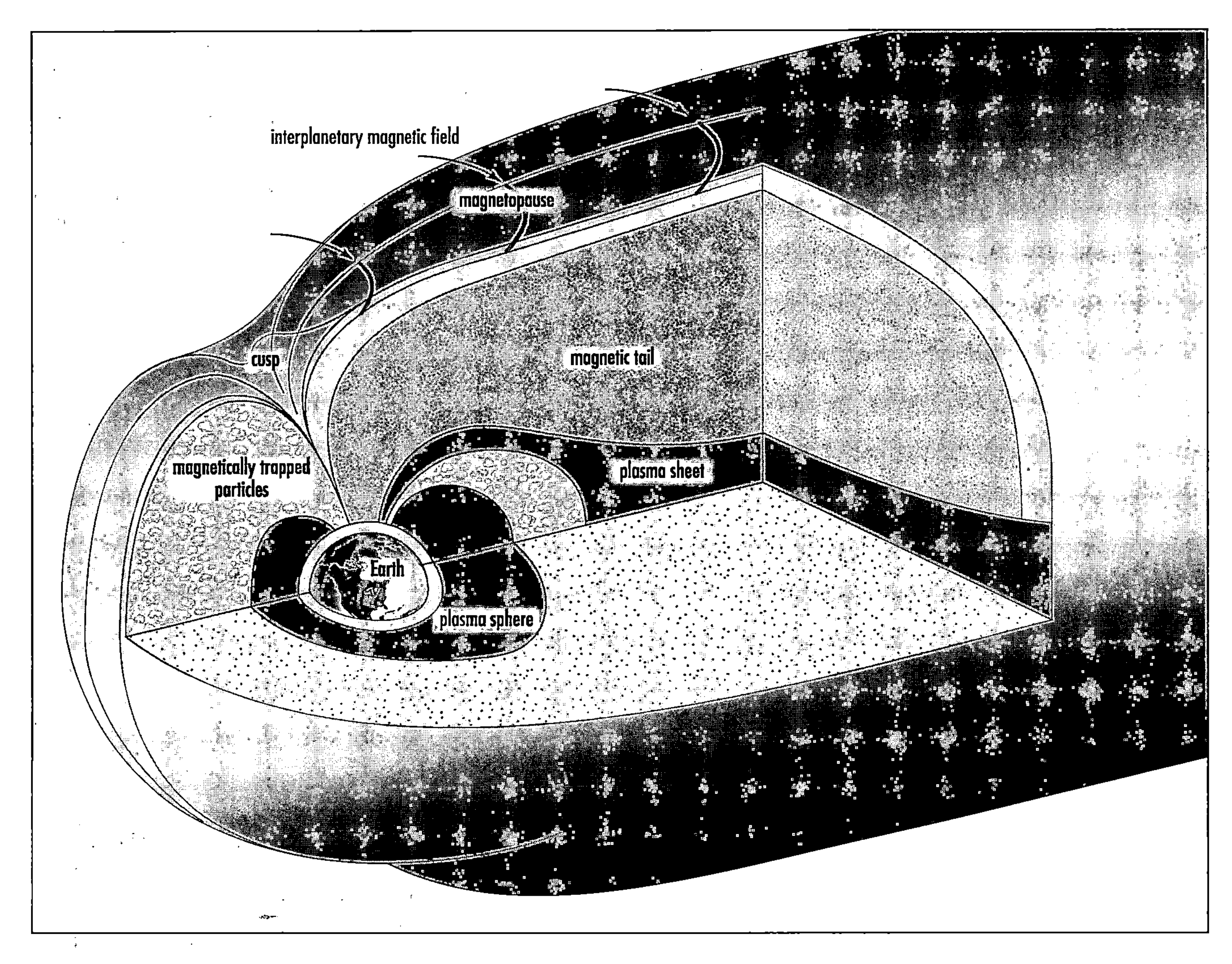

A synoptic view of Earth's magnetosphere

projected by the solar wind. Because Earth has its own magnetic field, the interaction of Earth's magnetic field and the solar wind results in this very dynamic and complicated region surrounding Earth. As shown in the figure, studies by spacecraft and probes have now mapped much of the region of magnetic field structures and streams of trapped particles around Earth. The *solar wind,* a plasma of electrically charged particles (mostly protons and electrons) that flows at speeds of 1 million kilometers per hour or more from the Sun, shapes Earth's magnetosphere into a teardrop, with a long magnetic tail (called the *magnetotail*) stretching out opposite the Sun.

Earth and other planets of the solar system exist in the *heliosphere,* the region of space dominated by the magnetic influence of the Sun. Interplanetary space is not empty but filled with the solar wind. The geomagnetic field of Earth presents an obstacle to the solar wind, behaving much like a rock in a swiftly flowing stream of water. A shock wave, called the *bow shock,* forms on the sunward side of Earth and deflects the flow of the solar wind. The bow shock slows down, heats, and compresses the solar wind, which then flows around the geomagnetic field, creating Earth's magnetosphere. The steady pressure of the solar wind compresses the otherwise spherical field lines of Earth's magnetic field on the sunward side at about 15 Earth radii, or some 100,000 kilometers, a distance still inside the Moon's orbit around Earth. On the night side of Earth away from the Sun, the solar wind pulls the geomagnetic field lines out to form a long magnetic tail, the magnetotail. The magnetotail is believed to extend for hundreds of Earth radii, although it is not known precisely how far it actually extends into space away from Earth.

The outermost boundary of Earth's magnetosphere is called the *magnetopause.* Some solar wind particles do pass through the magnetopause and become trapped in the inner magnetosphere. Some of those trapped particles then travel down through the *polar cusps* at the North and South Poles and into the uppermost portions of Earth's atmosphere. These trapped solar wind particles then have enough energy to trigger the AURORA, which are also called the Northern (*aurora borealis*) and Southern (*aurora australis*) Lights because they occur in circles around the North and South Poles. These spectacular aurora are just one dramatic manifestation of the many connections among the Sun, the solar wind, and Earth's magnetosphere and atmosphere.

magnitude A number, now measured on a logarithmic scale, used to indicate the relative brightness of a celestial body. The smaller the magnitude number, the greater the brightness. When ancient astronomers studied the heavens they observed objects of varying brightness and color and decided to group them according to their relative brightness. The 20 brightest stars of the night sky were called *stars of the first magnitude.* Other stars were called 2nd, 3rd, 4th, 5th, and 6th magnitude stars according to their relative brightness. The 6th magnitudes stars are the faintest stars visible to the unaided (i.e., naked) human eye under the most favorable observing conditions. By convention the star Vega (Alpha Lyra) was defined as having a magnitude of zero. Even brighter objects, such as the star Sirius and the planets Mars and Jupiter, thus acquired *negative magnitude* values. In 1856 the British astronomer NORMAN ROBERT POGSON proposed a more precise logarithmic magnitude in which a difference of five magnitudes represented a relative brightness ratio of 100 to 1. Consequently, with Pogson's proposed scale two stars differing by five magnitudes would differ in brightness by a factor of 100. Today the Pogson magnitude scale is used almost universally in astronomy.

See also STAR.

main-belt asteroid An asteroid located in the main asteroid belt, which occurs between Mars and Jupiter. This term sometimes is limited to asteroids found in the most populous portion of about 2.2 to 3.3 astronomical units (AUs) from the Sun.

See also ASTEROID.

main-sequence star A star in the prime of its life that shines with a constant luminosity achieved by steadily converting hydrogen into helium through thermonuclear fusion in its core.

See also STAR.

main stage 1. In a multistage rocket vehicle, the stage that develops the greatest amount of thrust, with or without booster engines. 2. In a single-stage rocket vehicle powered by one or more engines, the period when "full thrust" (i.e., at or above 90% of the rated thrust) is attained. 3. A sustainer engine that is considered as a stage after booster engines have fallen away, as in the main stage of the Atlas launch vehicle.

See also ROCKET.

make safe One or more actions necessary to prevent or interrupt complete function of a system. Among the necessary actions are (1) install safety devices such as pins or locks; (2) disconnect hoses, linkages, or batteries; (3) bleed (i.e., drain) fluids from accumulators and reservoirs; (4) remove explosive devices such as initiators, fuses, and detonators; and (5) intervene by welding components in place or fixing their movement with lockwires. Often used to describe the "disarming" or "disabling" of a weapon or rocket system.

maneuvering reentry vehicle (MaRV) A reentry vehicle that can maneuver in the late-midcourse or terminal phase of its flight, either to enhance its accuracy or to avoid antiballistic missiles. Maneuvers within the atmosphere usually can be accomplished by aerodynamic means, while maneuvers in space are accomplished by small rockets.

See also BALLISTIC MISSILE DEFENSE.

maneuver pad Data or information on spacecraft attitude, thrust values, event times, and the like that are transmitted in advance of a maneuver.

manipulators Mechanical devices used for handling objects, frequently involving remote operations (i.e., teleoperation) and/or hazardous substances or environmental conditions. That portion of a robot system that is capable of grasping or handling. A manipulator often has a versatile end effector (i.e., a special tool or "grasping element" installed at the end of the manipulator) that can respond to a

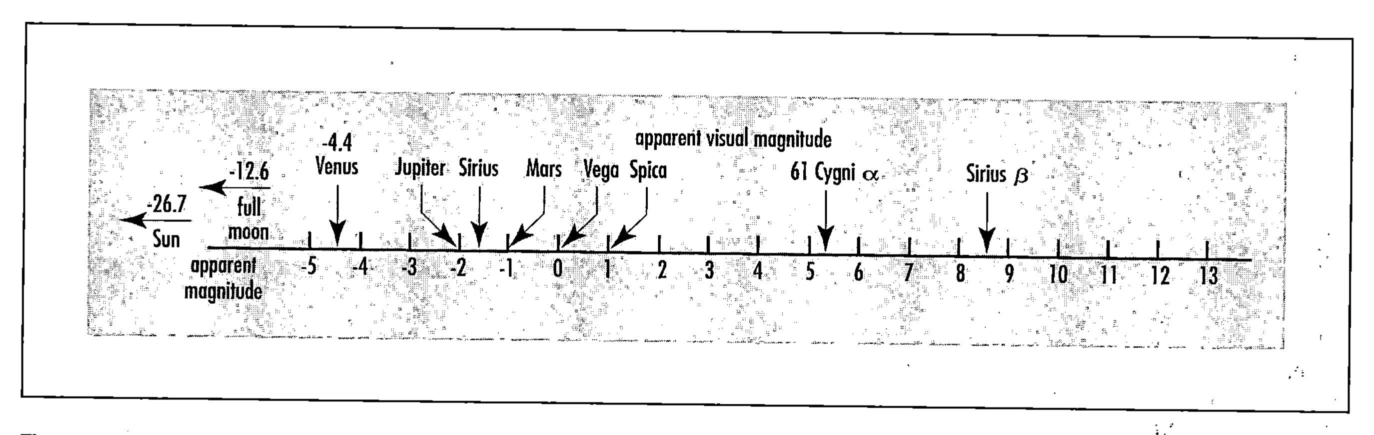

The apparent visual magnitude of several celestial objects

variety of different handling requirements. For example, the U.S. space shuttle has a very useful manipulator called the remote manipulator system (RMS).

man-machine interface The boundary where human and machine characteristics and capabilities are joined in order to obtain optimum operating conditions and maximum efficiency of the combined man-machine system. A joystick and a control panel are examples of man-machine interfaces. Also called human-machine interface.

manned An aerospace vehicle or system that is occupied by one or more persons, male or female. The terms "crewed," "human," or "personed" are preferred today in the aerospace literature. For example, a "manned mission to Mars" should be called a "human mission to Mars."

manned vehicle An older aerospace term describing a rocket or spacecraft that carried one or more human beings, male or female. Used to distinguish that craft from a robot (i.e., pilotless) aircraft, a ballistic missile, or an automated (and uncrewed) satellite or planetary probe. The expression *crewed vehicle* or *personed vehicle* is now preferred.

man-rated A launch vehicle, spacecraft, aerospace system, or component considered safe and reliable enough to be used by human crew members. The term *human-rated* is now preferred.

maria (singular: mare) Latin word for "seas." It was originally used by GALILEO GALILEI to describe the large, dark, ancient lava flows on the lunar surface, thought by early astronomers to be bodies of water on the Moon. Despite the physical inaccuracy, this term is still used by astronomers.

See also LATIN SPACE DESIGNATIONS.

Mariner spacecraft A series of NASA planetary exploration spacecraft that performed reconnaissance flyby and orbital missions to Mercury, Mars, and Venus in the 1960s and 1970s. *Mariner 2* encountered Venus at a distance of about 41,000 kilometers on December 14, 1962. It provided important scientific data about conditions in interplanetary space (e.g., solar wind, interplanetary magnetic field, cosmic rays, etc.) and on the planet Venus. *Mariner 4* flew by Mars on July 14, 1965, and returned images of the planet's surface. The closest encounter distance for its mission was 9,846 km from the Martian surface. The *Mariner 5* spacecraft passed within 4,000 km of Venus on October 19, 1967. Its instruments successfully measured both interplanetary and Venusian magnetic fields, charged particles, and plasmas as well as certain properties of the Venusian atmosphere. *Mariner 6* passed within 3,431 km of the Martian surface on July 31, 1969. The spacecraft's instruments took images of the planet's surface and measured ultraviolet and infrared emissions of the Martian atmosphere. The *Mariner 7* spacecraft was identical to the *Mariner 6* spacecraft. *Mariner 7* passed within 3,430 km of Mars on August 5, 1969, and acquired images of the planet's surface. Emissions of the Martian atmosphere also were measured. *Mariner 9* arrived at and orbited Mars on November 14, 1971. This spacecraft gathered data on the composition, density, pressure, and temperature of the Martian atmosphere as well as performed studies of the Martian surface. After depleting its supply of attitude control gas, the spacecraft was turned off on October 27, 1972. *Mariner 10* was the seventh successful launch in the Mariner series. (*Mariner 1* and *Mariner 8* experienced launch failures, while *Mariner 3* ceased transmissions nine hours after launch and entered solar orbit.) The *Mariner 10* spacecraft was the first to use the gravitational pull of a planet (Venus) to reach another planet (Mercury). It passed Venus on February 5, 1974, at a distance of 4,200 km. The spacecraft then crossed the orbit of Mercury at a distance of 704 km from the surface on March 29, 1974. A second encounter with Mercury occurred on September 21, 1974, at an altitude of about 47,000 km. A third and final Mercury encounter occurred on March 16, 1975, when the spacecraft passed the planet at an altitude of 327 km. Many images of the planet's surface were acquired during these flybys, and magnetic field measurements were performed. When the supply of attitude control gas became depleted on March 24, 1975, this highly successful mission was terminated.

See also MARS; MERCURY; VENUS.

Mars The fourth planet in the solar system, with an equatorial diameter of 6,794 kilometers. (*See* figure.) Throughout human history Mars, the Red Planet, has been at the center of astronomical thought. The ancient Babylonians, for example, followed the motions of this wandering red light across the night sky and named it after Nergal, their god of war. In time, the Romans, also honoring their own god of war, gave the planet its present name. The presence of an atmosphere, polar caps, and changing patterns of light and dark on the surface caused many pre–space age astronomers and scientists to consider Mars an Earthlike planet, the possible abode of extraterrestrial life. In fact, when actor Orson Welles broadcast a radio drama in 1938 based on H. G. WELLS's science fiction classic *War of the Worlds*, enough people believed the report of invading Martians to create a near panic in some areas.

Over the past three decades, however, sophisticated robot spacecraft—flybys, orbiters, and landers—have shattered these romantic myths of a race of ancient Martians struggling to bring water to the more productive regions of a dying world. Spacecraft-derived data have shown instead that the Red Planet is actually a "halfway" world. Part of the Martian surface is ancient, like the surfaces of the Moon and Mercury, while part is more evolved and Earthlike. Contemporary information about Mars is presented in the table.

In August and September 1975 two *Viking* spacecraft were launched on a mission to help answer the question: Is there life on Mars? Each *Viking* spacecraft consisted of an orbiter and lander. While scientists did not expect these spacecraft to discover Martian cities bustling with intelligent life, the exobiology experiments on the lander were designed to find evidence of primitive life-forms, past or present. Unfortunately, the results sent back by the two robot landers were teasingly inconclusive.

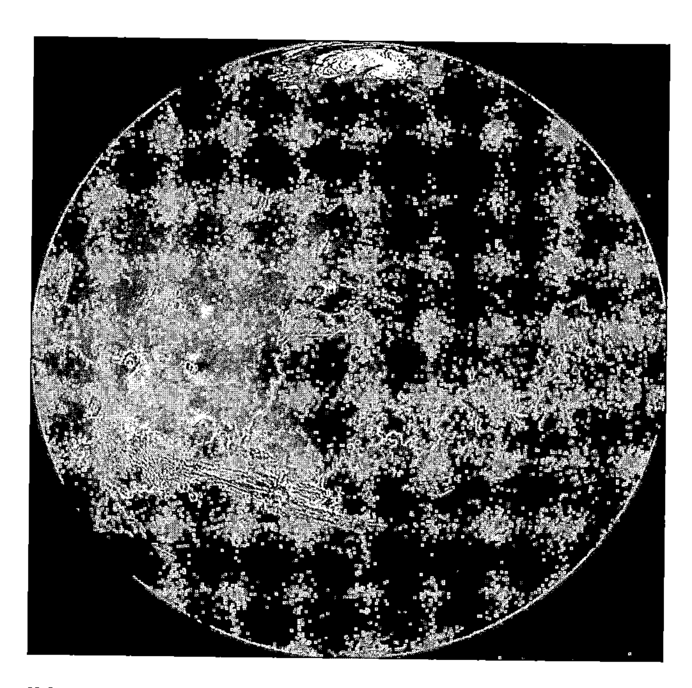

U.S. Geological Survey scientists processed approximately 1,000 *Viking Orbiter 1* spacecraft images of Mars to produce this global view of the Red Planet. The northern polar cap is visible in this projection at the top of the image as well as the great equatorial canyon system (Valles Marineris) below center and four huge Tharsis volcanoes and several smaller ones at the left. Scientists also note the heavy impact cratering of the highlands (bottom and right portions of this mosaic) and the younger, less heavily cratered terrain elsewhere. *(Courtesy of NASA/JPL and USGS)*

The Viking Project was the first mission to successfully soft-land a robot spacecraft on another planet (excluding Earth's Moon). All four *Viking* spacecraft (two orbiters and two landers) exceeded by considerable margins their design goal lifetime of 90 days. The spacecraft were launched in 1975 and began to operate around or on the Red Planet in 1976. When the *Viking 1* lander touched down on the Plain of Chryse on July 20, 1976, it found a bleak landscape. Several weeks later its twin, the *Viking 2* lander, set down on the Plain of Utopia and discovered a more gentle, rolling landscape. *The Viking 2* orbiter spacecraft ceased operation in July 1978; the *Lander 2* fell silent in April 1980; the *Viking 1* orbiter managed at least partial operation until August 1980; the *Viking 1* lander spacecraft made its final transmission on November 11, 1982. NASA officially ended the Viking mission to Mars on May 21, 1983.

As a result of these interplanetary missions, we now know that Martian weather changes very little. For example, the highest atmospheric temperature recorded by either *Viking* lander was –21°C (midsummer at the *Viking 1* site), while the lowest recorded temperature was –124°C (at the more northerly *Viking 2* site during winter).

The atmosphere of Mars was found to be primarily carbon dioxide (CO_2). Nitrogen, argon, and oxygen are present in small percentages, along with trace amounts of neon, xenon, and krypton. The Martian atmosphere contains only a trace of water, about 1/1000th as much as is found in Earth's atmosphere. But even this tiny amount can condense and form clouds that ride high in the Martian atmosphere or form patches of morning fog in valleys. There is also evidence that Mars had a much denser atmosphere in the past, one capable of permitting liquid water to flow on the planet's surface. Physical features resembling riverbeds, canyons and gorges, shorelines, and even islands hint that large rivers and maybe even small seas once existed on the Red Planet.

Mars has two small, irregularly shaped moons called Phobos ("fear") and Deimos ("terror"). These natural satellites were discovered in 1877 by the American astronomer ASAPH HALL. They both have ancient, cratered surfaces with some indication of regoliths to depths of possibly 5 meters or more. The physical properties of these two moons are presented in the second table. It is hypothesized that these moons actually may be asteroids "captured" by Mars.

Scientists believe that at least 12 unusual meteorites found on Earth actually are pieces of Mars that were blasted off the Red Planet by ancient meteoroid impact collisions. One particular Martian meteorite, called ALH84001, has stimulated a great deal of interest in the possibility of life on Mars. In the summer of 1996 a NASA research team at the Johnson Space Center (JSC) announced that they had found evidence in ALH84001 that "strongly suggests primitive life may have existed on Mars more than 3.6 billion years ago." The NASA team found the first organic molecules thought to be of Martian origin, several mineral features characteristic

Physical and Dynamic Data for Mars

Diameter (equatorial)	6,794 km
Mass	6.42×10^{23} kg
Density (mean)	3.9 g/cm^3
Surface gravity	3.73 m/sec^2
Escape velocity	5.0 km/sec
Albedo atmosphere	0.15
(main components by volume)	
Carbon dioxide (CO_2)	95.32%
Nitrogen (N_2)	2.7%
Argon (Ar)	1.6%
Oxygen (O_2)	0.13%
Carbon monoxide (CO)	0.07%
Water vapor (H_2O)	0.03%[a]
Natural satellites	2 (Phobos and Deimos)
Period of rotation (a Martian day)	1.026 days
Average distance from Sun	2.28×10^8 km (1.523 AU)
Eccentricity	0.093
Period of revolution around Sun (a Martian year)	687 days
Mean orbital velocity	24.1 km/sec
Solar flux at planet (at top of atmosphere)	590 W/m^2 (at 1.52 AU)

[a] variable
Source: Based on NASA data.

Physical and Dynamic Properties of the Martian Moons, Phobos and Deimos

Property	Phobos	Deimos
Characteristic dimensions (both are irregularly shaped)		
Longest dimension	27 km	15 km
Intermediate dimension	21 km	12 km
Shortest dimension	19 km	11 km
Mass	10.8×10^{15} kg	1.8×10^{15} kg
Density	1.9 g/cm^3	1.8 g/cm^3
Albedo	0.06	0.07
Surface gravity	1 cm/sec^2	0.5 cm/sec^2
Rotation	Synchronous	Synchronous
Semimajor axis of orbit	9,378 km	23,436
Eccentricity	0.018	0.0008
Sidereal period (approx.)	0.319 days	1.262 days

Source: Based on NASA data.

of biological activity, and possibly microfossils (i.e., very tiny fossils) of primitive, bacterialike organisms inside this ancient Martian rock that fell to Earth as a meteorite.

Stimulated by the exciting possibility of life on Mars, NASA and other organizations have launched (or will soon launch) a variety of robot spacecraft to accomplish more focused scientific investigations of the Red Planet. Starting in 1996, some of these missions have proven highly successful, while others have ended in disappointing failures.

This new wave of exploration started on November 7, 1996, when NASA launched the *Mars Global Surveyor* (MGS) from Cape Canaveral Air Force Station, Florida. The spacecraft arrived at Mars on September 12, 1997, an event representing the first successful mission to the Red Planet in two decades. After spending a year and a half carefully trimming its orbit from a looping ellipse to a more useful circular track around the planet, the MGS began its mapping mission in March 1999. Using its high-resolution camera, the MGS observed the planet from its low-altitude, nearly polar orbit over the course of an entire Martian year, the equivalent of nearly two Earth years. At the conclusion of its primary scientific mission on January 31, 2001, the spacecraft entered an extended mission phase. The MGS has successfully studied the entire Martian surface, atmosphere, and interior, returning an enormous amount of valuable scientific data. Among its most significant scientific contributions so far are high-resolution images of gullies and debris flow features that invitingly suggest there may be current sources of liquid water, similar to an aquifer, at or near the surface of the planet. These findings are shaping and guiding upcoming robot missions to Mars.

NASA launched the *Mars Pathfinder* mission to the Red Planet on December 4, 1996, using a Delta II expendable launch vehicle. The mission, formerly called the *Mars Environmental Survey* (or MESUR) *Pathfinder*, had as its primary objective the demonstration of innovative, low-cost technology for delivering an instrumented lander and free-ranging robotic rover to the Martian surface. The *Mars Pathfinder* not only accomplished that important objective, but it also returned an unprecedented amount of data and operated well beyond its anticipated design life. From *Mars Pathfinder*'s innovative airbag bounce and roll landing on July 4, 1997, until its final data transmission on September 27, the lander-minirover team returned numerous images of the Ares Vallis landing site and useful chemical analyses of proximate rocks and soil deposits. Data from this successful mission have suggested that ancient Mars was once warm and wet, further stimulating the intriguing question of whether life could have emerged on that planet when liquid water flowed on its surface and its atmosphere was significantly thicker.

However, the exhilaration generated by these two successful missions was quickly dampened by two glaring failures. On December 11, 1998, NASA launched the *Mars Climate Orbiter* (MCO). This spacecraft was to serve as both an interplanetary weather satellite and a data relay satellite for another mission, called the *Mars Polar Lander* (MPL). The MCO also carried two science instruments, an atmospheric sounder and a color imager. However, just as the spacecraft arrived at the Red Planet on September 23, 1999, all contact was lost with it. NASA engineers have concluded that because of human error in programming the final trajectory, the spacecraft most probably attempted to enter orbit too deep in the planet's atmosphere and consequently burned up.

NASA used another Delta II expendable launch vehicle to send the MPL to the Red Planet on January 3, 1999. The MPL was an ambitious mission to land a robot spacecraft on the frigid Martian terrain near the edge of the planet's southern polar cap. Two small penetrator probes (called *Deep Space 2*) piggybacked with the lander spacecraft on the trip to Mars. After an uneventful interplanetary journey the MPL and its companion *Deep Space 2* experiments were mysteriously lost when the spacecraft arrived at the planet on December 3, 1999.

Undaunted by these disappointing sequential failures, NASA officials sent the *2001 Mars Odyssey* mission to the Red Planet on April 7, 2001. The scientific instruments onboard the orbiter spacecraft were designed to determine the composition of the planet's surface, to detect water and shallow buried ice, and to study the ionizing radiation environment in the vicinity of Mars. The spacecraft arrived at the planet on October 24, 2001, and successfully entered orbit around it. After executing a series of aerobrake maneuvers that properly trimmed it into a near-circular polar orbit around Mars, the spacecraft began to make scientific measurements in January 2002. The orbiting spacecraft continued to collect scientific data until the end of its primary scientific mission in late summer 2004. After its primary mission the spacecraft functioned as a communications relay satellite until about October 2005, supporting information transfer from the *Mars Exploration Rover* (MER) Mission back to scientists on Earth.

In summer 2003 NASA launched identical Mars Exploration rovers that were to operate on the surface of the Red Planet during 2004. *Spirit* (MER-A) was launched by a Delta II rocket from Cape Canaveral Air Force Station on June 10,

2003, and successfully landed on Mars on January 4, 2004. *Opportunity* (MER-B) was launched from Cape Canaveral Air Force Station on July 7, 2003, by a Delta II rocket and successfully landed on the surface of Mars on January 25, 2004. Both successful landings resembled the airbag bounce and roll arrival demonstrated during the *Mars Pathfinder* mission. Following arrival on the surface of the Red Planet, each rover drove off and began its surface exploration mission in decidedly different locations on Mars.

In 2003 NASA also participated in a mission called *Mars Express,* sponsored by the European Space Agency (ESA) and the Italian Space Agency. Launched in June 2003, the *Mars Express* spacecraft arrived at the Red Planet in December 2003. Following successful arrival maneuvers, the spacecraft's scientific instruments began to study the atmosphere and surface of Mars from a polar orbit. The main objective of the *Mars Express* is to search from orbit for suspected subsurface water locations. The spacecraft also delivered a small lander spacecraft to more closely investigate the most suitable candidate site. The small lander was named *Beagle 2* in honor of the famous ship in which the British naturalist Charles Darwin (1809–82) made his great voyage of scientific discovery. After coming to rest on the surface of Mars, *Beagle 2* was to have performed exobiology and geochemistry research. The *Beagle 2* was scheduled to land on December 25, 2003. However, European Space Agency ground controllers have been unable to communicate with the probe. Despite the problems with *Beagle 2,* the *Mars Express* spacecraft has functioned well in orbit around the planet and accomplished its main mission of global high-resolution photogeology and mineralogical mapping of the Martian surface.

On August 12, 2005 NASA successfully launched a powerful new scientific spacecraft called the *Mars Reconnaissance Orbiter* (MRO). This mission to the Red Planet will focus on scrutinizing those candidate water-bearing locations previously identified by the *Mars Global Surveyor* and *2001 Mars Odyssey* missions. The MRO will be capable of measuring thousands of Martian landscapes at a resolution of between 0.2 and 0.3 meters. By way of comparison, this spacecraft's imaging capability will be good enough to detect and identify rocks on the surface of Mars that are as small as beach balls. Such high-resolution imagery data will help bridge the gap between detailed localized surface observations accomplished by landers and rovers and the synoptic global measurements made by orbiting spacecraft.

Possibly as early as 2009, NASA plans to develop and launch a long-duration, long-range mobile science laboratory. This effort will demonstrate the efficacy of developing and deploying truly "smart" landers—advanced robotic systems that are capable of autonomous operation, including hazard avoidance and navigation around obstacles to reach very promising but difficult-to-reach scientific sites. NASA also proposes to create a new family of small scout missions. These missions would involve airborne robotic vehicles (such as a Mars airplane) and special miniature surface landers or penetrator probes, possibly delivered to interesting locations around Mars by the airborne robotic vehicles. These scout missions, beginning in about 2007, would greatly increase the number of interesting sites studied and set the stage for even more sophisticated robotic explorations in the second decade of this century.

As presently planned, NASA intends to launch its first Mars Sample Return mission (MSRM) in about 2014. Future advances in robot spacecraft technology will enhance and accelerate the search for possible deposits of subsurface water and life (existent or extinct) on the Red Planet. Two decades of intensely focused scientific missions by robot spacecraft will not only greatly increase scientific knowledge about Mars, but the effort will set the stage for the first Mars expedition by human explorers.

Mars airplane A low-mass, unpiloted (robot) aircraft that carries experiment packages or performs detailed reconnaissance operations on Mars. For example, the Mars airplane could be used to deploy a network of science stations, such as seismometers or meteorology stations, at selected Martian sites with an accuracy of a few kilometers. If designed with a payload capacity of about 50 kilograms, this flying platform could also perform high-resolution imagery or conduct detailed geochemical surveys of candidate exobiological sites. It would be capable of flying at altitudes of 500 m to 15 km, with corresponding ranges of 25 km to 6,700 km.

Two basic design approaches have been considered. In the first the airplane is designed as a one-way, "disposable" flying platform. After descending into the Martian atmosphere, it automatically deploys and performs aerial surveys, atmospheric soundings, and so on and then crashes when its fuel supply is exhausted. Or else it can be equipped with a small, variable-thrust rocket motor, so that it may soft-land and take off again. The latter design approach gives the robot aircraft an operational ability to make in-situ measurements and to gather samples at a variety of distant sites. The soil specimens can even be returned by the airplane to a robot lander–ascent vehicle spacecraft at some primary surface site.

Mars balloon A specially designed balloon package (or *aerobot*) that could be deployed into the Martian atmosphere and then used to explore the surface. During the Martian daytime the balloon would become buoyant enough due to solar heating to lift its instrumented guide rope off the surface. Then at night the balloon would sink when cooled, and the instrumented guide rope would then come in contact with the Martian surface, allowing various surface science measurements to be made. A typical balloon exploration system might operate for 10 to 50 sols (that is, from 10 to 50 Martian days) and provide surface and (in situ) atmospheric data from many different surface sites. Data would be relayed back to Earth via a Mars orbiting spacecraft.

One proposed NASA mission, called the Mars Geoscience Aerobots mission, involves high-spatial-resolution (less than 100 meters per pixel or less) spectral mapping of the Martian surface from aerobot platforms. One or more aerobots would be deployed at an altitude between 4 and 6 kilometers and operate for up to 50 days. Onboard instruments would perform high-resolution mineralogy and geochemistry measurements in support of future exobiological sample return missions.

Mars base For automated Mars missions the spacecraft and robotic surface rovers generally will be small and self-contained. For human expeditions to the surface of the Red Planet, however, two major requirements must be satisfied: life support (habitation) and surface transportation (mobility). Habitats, power supplies, and life support systems will tend to be more complex in a permanent Martian surface base that must sustain human beings for years at a time. Surface mobility systems will also grow in complexity and sophistication as early Martian explorers and settlers travel 10s to 100s of kilometers from their base camp. At a relatively early time in any Martian surface base program, the use of Martian resources to support the base must be tested vigorously and then quickly integrated in the development of an eventually self-sustaining surface infrastructure.

In one possible scenario the initial Martian habitats will resemble standardized lunar base (or space station) pressurized modules and will be transported from cislunar space to Mars in prefabricated condition by interplanetary nuclear-electric propulsion (NEP) cargo ships. These modules will be configured and connected as needed on the surface of Mars and then covered with a meter or so thickness of Martian soil for protection against the lethal effects of solar flare radiation or continuous exposure to cosmic rays on the planet's surface. (Unlike Earth's atmosphere, the very thin Martian atmosphere does not shield very well against ionizing radiations from space.)

See also MARS.

Mars Climate Orbiter **(MCO)** Originally part of the *Mars Surveyor '98* mission, NASA launched the *Mars Climate Orbiter* on December 11, 1998, from Cape Canaveral Air Force Station, Florida, using a Delta II expendable launch vehicle. The mission of this orbiter spacecraft was to circle Mars and serve as both an interplanetary weather satellite and a communications satellite, relaying data back to Earth from the other part of the *Mars Surveyor '98* mission, a lander called the *Mars Polar Lander* (MPL). The MCO also carried two science instruments: an atmospheric sounder and a color imager. Unfortunately, when it arrived at the Red Planet all contact with the spacecraft was lost on September 23, 1999. NASA engineers believe that the MCO burned up in the Martian atmosphere due to a fatal error in its arrival trajectory. This human-induced computational error caused the spacecraft to enter too deeply into the planet's atmosphere and encounter destructive aerodynamic heating.

Mars crewed expedition The Mars crewed expedition will be the first visit to Mars by human beings and will most likely occur before the mid-part of this century. Current concepts suggest a 600- to 1,000-day duration mission that most likely will start from Earth orbit, possibly powered by a nuclear thermal rocket. (*See* figure.) A total crew size of up to 15 astronauts is anticipated. After hundreds of days of travel through interplanetary space, the first Martian explorers will have about 30 days allocated for surface excursion activities on the Red Planet.

The commitment to a human expedition to Mars, perhaps as early as the third decade of this century, is an ambitious undertaking that will require extended political and social commitment for several decades. One nation, or several nations in a cooperative venture, must be willing to make a lasting statement about the value of human space exploration in our future civilization. A successful crewed mission to Mars will establish a new frontier both scientifically and philosophically. This frontier spirit is further amplified if the first crewed mission to Mars is viewed as a precursor to human settlement of the Red Planet.

This artist's rendering shows the first human expedition nearing Mars ca. 2030. Upon arrival the mission's primary propulsion system, a nuclear thermal rocket (NTR), fires to insert the space vehicle into the proper parking orbit around the Red Planet. Nuclear propulsion technology can shorten interplanetary trip times and/or deliver more payload mass to the planet for the same initial Earth-orbit (or lunar orbit) departure mass than can chemical propulsion technology. ***(Courtesy of NASA; artist, Pat Rawlings [1995])***

Exactly what happens after the first human expedition to Mars is, of course, open to wide speculation at present. People on Earth could simply marvel at "another outstanding space exploration first" and then settle back to their more pressing terrestrial pursuits. This pattern unfortunately followed the spectacular Moon landing missions (1969–72) of NASA's Apollo Project. On the other hand, if this first human expedition to Mars were widely recognized and accepted as the precursor to our permanent occupancy of heliocentric space, then Mars would truly become the central object of greatly expanded human space activities, perhaps complementing the rise of a self-sufficient lunar civilization.

Outside the Earth-Moon system, Mars is the most hospitable body in the solar system for humans and is currently the only practical candidate for human exploration and settlement in the early decades of this century. Mars also offers the opportunity for in-situ resource utilization (ISRU). With

ISRU initiatives the planet can provide air for the astronauts to breathe and fuel for their surface rovers and return vehicle. In fact, ISRU is an integral part of the many recent Mars expedition scenarios. In one recent NASA study, for example, it was suggested that a Mars ascent vehicle (MAV) (for crew departure from the surface of planet), critical supplies, an unoccupied habitat, and an ISRU extraction facility be prepositioned on the surface of the Red Planet before the human crew ever leaves Earth.

Of course, the logistics of a crewed mission to Mars are very complex, and many factors (including ISRU) must be considered before a team of human explorers sets out for the Red Planet with acceptable levels of risk and reasonable hope of returning safely to Earth. The establishment of a permanent lunar base is now being viewed by NASA long-range planners as a necessary step before human explorers are sent to Mars. The overall space technology performance and human factors experience gained from extended lunar surface operations should provide Mars mission planners the ability to reduce the risk inherent in a long-duration interplanetary journey.

A human expedition to Mars will most likely involve an interplanetary voyage that takes between 600 and 1,000 days (depending on the particular propulsion system and mission scenario selected). Some of the other important factors that must be carefully considered include the overall objectives of the expedition, the selection of the transit vehicles and their trajectories, the desired stay-time on the surface of Mars, the primary site to be visited, the required resources and equipment, and crew health and safety throughout the extended journey. Due to the nature of interplanetary travel, there is no quick return to Earth nor even the possibility of supplementary help from Earth should the unexpected happen. Once the crew departs from the Earth-Moon system and heads for Mars, they must be totally self-sufficient and flexible enough to adapt to new situations.

Mars Exploration Rover mission (MER) In summer 2003 NASA launched identical twin Mars rovers that were to operate on the surface of the Red Planet during 2004. (*See* figure on page 382.) *Spirit* (MER-A) was launched by a Delta II rocket from Cape Canaveral Air Force Station on June 10, 2003, and successfully landed on Mars on January 4, 2004. *Opportunity* (MER-B) was launched from Cape Canaveral Air Force Station on July 7, 2003, by a Delta II rocket and successfully landed on the surface of Mars on January 25, 2004. Both successful landings resembled the airbag bounce and roll arrival demonstrated during the *Mars Pathfinder* mission. Following arrival on the surface of the Red Planet, each rover drove off and began its surface exploration mission in decidedly different locations on Mars. *Spirit* (MER-A) landed in Gusev Crater, which is roughly 15° south of the Martian equator. NASA mission planners selected Gusev Crater because it had the appearance of a crater lakebed.

An artist's rendering of a mid-21st-century Mars base near Pavonis Mons, a large shield volcano on the Martian equator that overlooks the ancient water-eroded canyon in which the base is located. The base infrastructure shown here includes a habitation module, a power module, central base work facilities, a greenhouse, a launch and landing complex, and even a robotic Martian airplane. In the foreground human explorers have taken their surface rover vehicle to an interesting spot, where one of the team members has just made the discovery of the century, a large fossil of an ancient Martian creature. ***(Artwork courtesy of NASA/JSC; artist, Pat Rawlings [1985])***

A permanent research station on the surface of Mars ca. 2040 ***(Artist's rendering courtesy of NASA)***

Mars Exploration Rover (MER) on the surface of the Red Planet in 2004 ***(Artist's rendering courtesy of NASA)***

Opportunity (MER-B) landed at Terra Meridiani (*See* figure at lower right.) Terra Meridiania is also known as the "hematite site" because this location displayed evidence of coarse-grained hematite, an iron-rich mineral that typically forms in water. Among this mission's principal scientific goals is the search for and characterization of a wide range of rocks and soils that hold clues to past water activity on Mars. At the end of 2004 both rovers functioned extremely well and scampered across Mars far beyond the primary mission goal of 90 days.

With much greater mobility than the *Mars Pathfinder* minirover, each of these powerful new robot explorers could successfully travel up to 100 meters per Martian day across the surface of the planet. Each rover carried a complement of sophisticated instruments that allowed it to search for evidence that liquid water was present on the surface of Mars in ancient times. Each rover visited a different region of the planet. Immediately after landing each rover performed reconnaissance of the particular landing site by taking panoramic (360°) visible (color) and infrared images. Then, using images and spectra taken daily by the rovers, NASA scientists at the Jet Propulsion Laboratory used telecommunications and teleoperations to supervise the overall scientific program. With intermittent human guidance, the pair of mechanical explorers functioned like "robot prospectors," examining particular rocks and soil targets and evaluating composition and texture at the microscopic level.

Each rover had a set of five instruments with which to analyze rocks and soil samples. The instruments included a

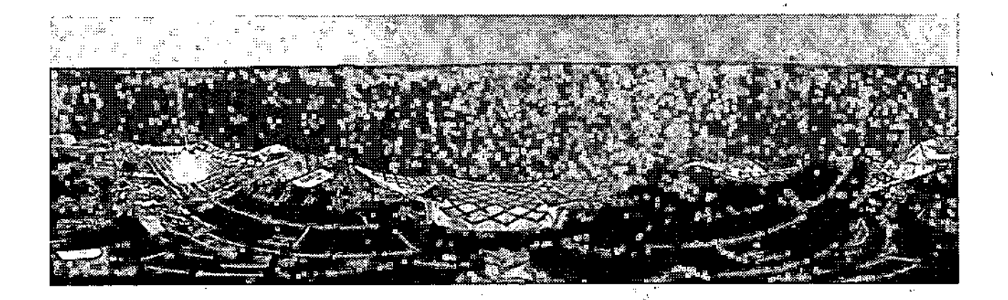

This 360° panorama is one of the first images beamed back to Earth from the Mars Exploration Rover *Opportunity* shortly after it successfully touched down at Meridiani Planum on January 25, 2004. The image was captured by the robot rover's navigation camera. ***(Courtesy of NASA/JPL)***

panoramic camera (Pancam), a miniature thermal emission SPECTROMETER (Mini-TES), a Mössbauer spectrometer (MB), an alpha particle X-ray spectrometer (APXS), magnets, and a microscopic imager (MI). There was also a special rock abrasion tool (or RAT) that allowed each rover to expose fresh rock surfaces for additional study of interesting targets.

Each rover had a mass of 185 kilograms and a range of up to 100 meters per sol (Martian day). Surface operations lasted beyond the goal of 90 sols. Communication to Earth was accomplished primarily by means of Mars-orbiting spacecraft, such as the *MARS ODYSSEY 2001*, serving as data relays.

Mars Express A mission to MARS launched in June 2003 and developed by the European Space Agency (ESA) and the Italian Space Agency. After the 1,042-kilogram spacecraft arrived at the Red Planet in December 2003, its scientific instruments began to study the atmosphere and surface of Mars from a polar orbit. The main objective of the *Mars Express* mission is to search from orbit for suspected subsurface water locations. The spacecraft also delivered a small lander spacecraft to more closely investigate the most suitable candidate site. The small lander was named *Beagle 2* in honor of the famous ship in which the British naturalist Charles Darwin (1809–82) made his great voyage of scientific discovery. After coming to rest on the surface of Mars, *Beagle 2* was to have performed exobiology and geochemistry research. *Beagle 2* was scheduled to land on December 25, 2003. However, European Space Agency ground controllers have been unable to communicate with the probe. Despite the problems with *Beagle 2*, the *Mars Express* spacecraft has functioned very well in orbit around the planet and accomplished its main mission of global high-resolution photogeology and mineralogical mapping of the Martian surface. In August 2004 the *Mars Express* also sent images back to Earth from NASA's *Opportunity* (MER-B) surface rover as part of an international interplanetary networking demonstration.

***Mars Global Surveyor* (MGS)** NASA launched the Mars Global Surveyor mission from Cape Canaveral Air Force Station, Florida, on November 7, 1996, using a Delta II expendable launch vehicle. The safe arrival of this robot spacecraft at Mars on September 12, 1997, represented the first successful mission to the Red Planet in two decades. MGS was designed as a rapid, low-cost recovery of the *MARS OBSERVER* mission (MO) objectives. After a year and a half trimming its orbit from a looping ellipse to a circular track around the planet, the spacecraft began its primary mapping mission in March 1999. Using a high-resolution camera, the MGS spacecraft observed the planet from a low-altitude, nearly polar orbit over the course of one complete Martian year, the equivalent of nearly two Earth years. Completing its primary mission on January 31, 2001, the spacecraft entered an extended mission phase.

The MGS science instruments included a high-resolution camera, a thermal emission spectrometer, a laser altimeter, and a magnetometer/electron reflectometer. With these instruments, the spacecraft successfully studied the entire Martian surface, atmosphere, and interior, returning an enormous amount of valuable scientific data in the process. Among the key scientific findings of this mission so far are high-resolution images of gullies and debris flow features that suggest there may be current sources of liquid water, similar to an aquifer, at or near the surface of the planet. Magnetometer readings indicate the Martian magnetic field is not globally generated in the planet's core but appears to be localized in particular areas of the crust. Data from the spacecraft's laser altimeter have provided the first three-dimensional views of the northern ice cap on Mars. Finally, new temperature data and close-up images of the Martian moon Phobos suggest that its surface consists of a powdery material at least 1 meter thick, most likely the result of millions of years of meteoroid impacts.

Marshall Space Flight Center (MSFC) A major NASA complex in Huntsville, Alabama. It was the center where WERNHER VON BRAUN and his team of German rocket scientists developed a series of highly successful rockets, including the giant Saturn V vehicle that sent human expeditions to the lunar surface.

See also NATIONAL AERONAUTICS AND SPACE ADMINISTRATION.

***Mars Observer* mission** NASA's *Mars Observer* (MO), the first of the Observer series of planetary missions, was designed to study the geoscience of Mars. The primary science objectives for the mission were to (1) determine the global elemental and mineralogical character of the surface; (2) define globally the topography and gravitational field of the planet; (3) establish the nature of the Martian magnetic field; (4) determine the temporal and spatial distribution, abundance, sources, and sinks of volatiles (substances that readily evaporate) and dust over a seasonal cycle; and (5) explore the structure and circulation of the Martian atmosphere. The 1,018-kilogram spacecraft was launched successfully on September 25, 1992. Unfortunately, for unknown reasons contact with the *Mars Observer* was lost on August 22, 1993, just three days before scheduled orbit insertion around Mars. Contact with the spacecraft was not reestablished, and it is not known whether this spacecraft was able to follow its automatic programming and go into Mars orbit or if it flew by Mars and is now in a heliocentric orbit. Although none of the primary objectives of the mission were achieved, cruise mode (i.e., interplanetary) data were collected up to the loss of contact.

See also MARS; *MARS GLOBAL SURVEYOR*.

***Mars Odyssey 2001* mission** NASA launched the *Mars Odyssey 2001* mission to the Red Planet from Cape Canaveral Air Force Station on April 7, 2001. The robot spacecraft, previously called the *Mars Surveyor 2001 Orbiter*, was designed to determine the composition of the planet's surface, to detect water and shallow buried ice, and to study the ionizing radiation environment in the vicinity of Mars. The spacecraft arrived at the planet on October 24, 2001, successfully entered orbit, and then performed a series of aerobrake maneuvers to trim itself into a polar orbit around Mars for

This artist's rendering shows NASA's *Mars Odyssey 2001* spacecraft starting its science mission around Mars in January 2002. ***(Artist's rendering courtesy of NASA/JPL)***

scientific data collection. The scientific mission began in January 2002.

Mars Odyssey has three primary science instruments: the thermal emission imaging system (THEMIS), a gamma-ray spectrometer (GRS), and the Mars radiation environment experiment (MARIE). THEMIS is examining the surface distribution of minerals on Mars, especially those that can form only in the presence of water. The GRS is determining the presence of 20 chemical elements on the surface of Mars, including shallow, subsurface pockets of hydrogen that act as a proxy for determining the amount and distribution of possible water ice on the planet. Finally, MARIE is analyzing the Martian radiation environment in a preliminary effort to determine the potential hazard to future human explorers. The spacecraft collected scientific data until the nominal end of its primary scientific mission in July 2004. From that point it began to function as an orbiting communications relay and will continue to do so until about October 2005. It is currently supporting information transfer between the Mars Exploration Rover (MER) 2003 mission and scientists on Earth. NASA selected the name *Mars Odyssey 2001* for this important spacecraft as a tribute to the vision and spirit of space exploration embodied in the science fact and science fiction works of the famous British writer SIR ARTHUR C. CLARKE.

Mars Pathfinder NASA launched the *Mars Pathfinder* mission to the Red Planet using a Delta II expendable launch vehicle on December 4, 1996. This mission, previously called the *Mars Environmental Survey* (or MESUR) *Pathfinder,* had the primary objective of demonstrating innovative technology for delivering an instrumented lander and free-ranging robotic rover to the Martian surface. The *Mars Pathfinder* not only accomplished this primary mission but also returned an unprecedented amount of data, operating well beyond the anticipated design life.

Mars Pathfinder used an innovative landing method that involved a direct entry into the Martian atmosphere assisted by a parachute to slow its descent through the planet's atmosphere and then a system of large airbags to cushion the impact of landing. From its airbag-protected bounce and roll landing on July 4, 1997, until the final data transmission on September 27, the robotic lander/rover team returned numerous close-up images of Mars and chemical analyses of various rocks and soils found in the vicinity of the landing site.

The landing site was at 19.33 N, 33.55 W, in the Ares Vallis region of Mars, a large outwash plain near Chryse Planitia (the Plains of Gold), where the *Viking 1* lander had successfully touched down on July 20, 1976. Planetary geologists speculate that this region is one of the largest outflow channels on Mars, the result of a huge ancient flood that occurred over a short period of time and flowed into the Martian northern lowlands.

The lander, renamed by NASA the *Carl Sagan Memorial Station,* first transmitted engineering and science data collected during atmospheric entry and landing. The American astronomer CARL SAGAN popularized astronomy and astrophysics and wrote extensively about the possibility of extraterrestrial life. Then the lander's imaging system (which was on a pop-up mast) obtained views of the rover and the immediate surroundings. These images were transmitted back to Earth to assist the human flight team in planning the robot rover's operations on the surface of Mars. After some initial maneuvers to clear an airbag out of the way, the lander deployed the ramps for the rover. The 10.6-kilogram minirover had been stowed against one of the lander's petals. Once commanded from Earth, the tiny robot explorer came to life and rolled onto the Martian surface. Following rover deployment, the bulk of the lander's remaining tasks were to support the rover by imaging rover operations and relaying data from the rover back to Earth. Solar cells on the lander's three petals, in combination with rechargeable batteries, powered the lander, which also was equipped with a meteorology station.

The rover, renamed *Sojourner* (after the American civil rights crusader Sojourner Truth), was a six-wheeled vehicle that was teleoperated (that is, driven over great distances by remote control) by personnel at the Jet Propulsion Laboratory (JPL) in Pasadena, California. The rover's human controllers used images obtained by both the rover and the lander systems. Teleoperation at interplanetary distances required that the rover be capable of some semiautonomous operation, since the time delay of the signals averaged between 10 and 15 minutes depending on the relative positions of Earth and Mars.

For example, the rover had a hazard avoidance system, and surface movement was performed very slowly. The small rover was 280 millimeters high, 630 millimeters long, and 480 millimeters wide, with a ground clearance of 130 millimeters. While stowed in the lander, the rover had a height of just 180 millimeters. However, after deployment on the Martian surface, the rover extended to its full height and rolled down a deployment ramp. The relatively far-traveling little rover received its supply of electrical energy from its 0.2-square-meter array of solar cells. Several nonrechargeable batteries provided backup power.

The rover was equipped with a black-and-white imaging system. This system provided views of the lander, the surrounding Martian terrain, and even the rover's own wheel tracks that helped scientists estimate soil properties. An alpha particle X-ray spectrometer (APXS) onboard the rover was used to assess the composition of Martian rocks and soil.

Both the lander and the rover outlived their design lives—the lander by nearly three times and the rover by 12 times. Data from this very successful lander-rover surface mission suggest that ancient Mars was once warm and wet, stimulating further scientific and popular interest in the intriguing question of whether life could have emerged on the planet when it had liquid water on the surface and a thicker atmosphere.

Mars penetrator *See* PENETRATOR.

Mars Polar Lander **(MPL)** Originally designated as the lander portion of the *Mars Surveyor '98* mission, NASA launched the *Mars Polar Lander* spacecraft from Cape Canaveral Air Force Station, Florida, on January 3, 1999, using a Delta II expendable launch vehicle. MPL was an ambitious mission to land a robot spacecraft on the frigid surface of Mars near the edge of the planet's southern polar cap. Two small penetrator probes (called *Deep Space 2*) piggybacked with the lander spacecraft on the trip to Mars. After an uneventful interplanetary journey, all contact with the MPL and the *Deep Space 2* experiments was lost as the spacecraft arrived at the planet on December 3, 1999. The missing lander was equipped with cameras, a robotic arm, and instruments to measure the composition of the Martian soil. The two tiny penetrators were to be released as the lander spacecraft approached Mars and then follow independent ballistic trajectories, impacting on the surface and plunging below it in search of water ice.

The exact fate of the lander and its two tiny microprobes remains a mystery. Some NASA engineers believe that the MPL might have tumbled down into a steep canyon, while others speculate the MPL may have experienced too rough a landing and become disassembled. A third hypothesis suggests the MPL may have suffered a fatal failure during its descent through the Martian atmosphere. No firm conclusions could be drawn because the NASA mission controllers were completely unable to communicate with the missing lander or either of its planetary penetrators.

Mars Reconnaissance Orbiter **(MRO)** NASA launched the *Mars Reconnaissance Orbiter* (MRO) on August 12, 2005, from Cape Canaveral Air Force Station using an Atlas III expendable booster. This spacecraft will make high-resolution measurements of the surface from orbit. (*See* figure.) It will be equipped with a stereo imaging camera (HiRISE) with resolution much better than one meter and a visible/near-infrared spectrometer (CRISM) to study the surface composition. Also on board will be an infrared radiometer, an accelerometer, and a shallow subsurface sounding radar (SHARAD) provided by the Italian Space Agency to search for underground water. Tracking of the orbiter will give information on the gravity field of Mars. The primary objectives

This artist's rendering shows NASA's *Mars Reconnaissance Orbiter* spacecraft taking extremely high-resolution images of the planet's surface and using its sounder to investigate scientifically interesting areas for possible subsurface water ca. 2006. *(Artist's rendering courtesy of NASA/JPL)*

of the mission will be to look for evidence of past or present water, to study the weather and climate, and to identify landing sites for future missions.

MRO has a mass of approximately 1,975 kg. It is scheduled to reach Mars in March 2006 and will undergo four to six months of aerobraking to lower the initially highly elliptical orbit to a 250-by-320-km polar science orbit. Science observations, including detailed studies of selected target regions, will take place over one Martian year (roughly two Earth years), after which NASA will use the orbiter spacecraft as a communications relay link for future missions.

See also MARS.

Mars Sample Return mission (MSRM) The purpose of a Mars Sample Return mission (MSRM) is, as the name implies, to use a combination of robot spacecraft and lander systems to collect soil and rock samples from Mars and then return them to Earth for detailed laboratory analysis. A wide variety of options for this type of mission are being explored. For example, one or several small robot rover vehicles could be carried and deployed by the lander vehicle. These rovers (under the control of operators on Earth) would travel away from the original landing site and collect a wide range of rock and soil samples for return to Earth. Another option is to design a nonstationary, or mobile, lander that could travel (again guided by controllers on Earth) to various surface locations and collect interesting specimens. After the soil collection mission was completed, the upper portion of the lander vehicle would lift off from the Martian surface and rendezvous in orbit with a special "carrier" spacecraft. (*See* figure.) This automated rendezvous/return "carrier" spacecraft would remove the soil sample canisters from the ascent portion of the lander vehicle and then depart Mars orbit on a trajectory that would bring the samples back to Earth. After an interplanetary journey of about one year, this automated "carrier" spacecraft, with its precious cargo of Martian soil and rocks, would achieve orbit around Earth.

To avoid any potential problems of extraterrestrial contamination of Earth's biosphere by alien microorganisms that might possibly be contained in the Martian soil and rocks, the sample canisters might first be analyzed in a special human-tended orbiting quarantine facility. An alternate return mission scenario would be to bypass an Earth-orbiting quarantine process altogether and use a direct reentry vehicle operation to bring the encapsulated Martian soil samples to Earth.

Whatever sample return mission profile ultimately is selected, contemporary analysis of Martian meteorites (that have fallen to Earth) has stimulated a great scientific interest in obtaining well-documented and well-controlled "virgin" samples of Martian soil and rocks. Carefully analyzed in laboratories on Earth, these samples will provide a wealth of important and unique information about the Red Planet. These samples might even provide further clarification of the most intriguing question of all: Is there (or, at least, has there been) life on Mars? A successful Mars Sample Return mission is also considered a significant and necessary step toward eventual human expeditions to Mars in the 21st century.

See also EXTRATERRESTRIAL CONTAMINATION; MARS; MARS EXPLORATION ROVER MISSION; *MARS PATHFINDER;* MARTIAN METEORITES.

Mars surface rover(s) Automated robot rovers and human-crewed mobility systems used to satisfy a number of surface exploration objectives on Mars in this century.

See also MARS EXPLORATION ROVER MISSION; *MARS PATHFINDER.*

Martian Of or relating to the planet Mars.

Martian meteorites Scientists now believe that at least 12 unusual meteorites are pieces of Mars that were blasted off the Red Planet by meteoroid impact collisions. These interesting meteorites were previously called SNC meteorites, after the first three samples discovered (namely, Shergotty, Nakhla, and Chassigny) but are now generally referred to as Martian meteorites. The Chassigny meteorite was discovered in Chassigny, France, on October 3, 1815. It established the name of the chassignite type subgroup of the SNC meteorites. Similarly, the Shergotty meteorite fell on Shergotty, India, on August 25, 1865, and provides the name of the shergottite type subgroup of SNC meteorites. Finally, the Nakhla meteorite was found in Nakhla, Egypt, on June 28, 1911, and establishes the name for the nakhlite type subgroup of SNC meteorites.

All 12 known SNC meteorites are igneous rocks crystallized from molten lava in the crust of the parent planetary body. The Martian meteorites discovered so far on Earth represent five different types of igneous rocks, ranging from simple plagioclase-pyroxene basalts to almost monomineralic

Artist's rendering of a Mars Sample Return Mission (MSRM). The sample return spacecraft is shown departing the surface of the Red Planet after soil and rock samples previously gathered by robot rovers have been stored on board in a specially sealed capsule. To support planetary protection protocols, once in a rendezvous orbit around Mars the sample return spacecraft would use a mechanical device to transfer the sealed capsule of Martian soil samples to an orbiting Earth-return spacecraft ("mother ship") that would then take the samples to Earth. ***(Artist's rendering courtesy of NASA/JPL; artist, Pat Rawlings)***

Martian Meteorites

Name	Classification	Mass (kg)	Find/Fall	Year
Shergotty	S-basalt (pyx-plag)	4.00	fall	1865
Zagami	S-basalt	18.00	fall	1962
EETA 79001	S-basalt	7.90	find-A	1980
QUE94201	S-basalt	0.012	find-A	1995
ALHA77005	S-lherzolite (ol-pyx)	0.48	find-A	1978
LEW88516	S-lherzolite	0.013	find-A	1991
Y793605	S-lherzolite	0.018	find-A	1995
Nakhla	N-clinopyroxenite	40.00	fall	1911
Lafayette	N-clinopyroxenite	0.80	find	1931
Gov. Valadares	N-clinopyroxenite	0.16	find	1958
Chassigny	C-dunite (olivine)	4.00	fall	1815
ALH84001	orthopyroxenite	1.90	find-A	1993

Classification: S = shergottite, N = nakhlte, C = chassignite. ALH84001 is none of these. find-A designates Antarctic meteorites (all recent finds). Year is recovery date for non-Antarctic meteorites, and date of Martian classification for Antarctic meteorites.
Source: NASA.

cumulates of pyroxene or olivine. These Martian meteorites are summarized in the table.

The only natural process capable of launching Martian rocks to Earth is meteoroid impact. To be ejected from Mars, a rock must reach a velocity of 5 kilometers per second or more. During a large meteoroid impact on the surface of Mars, the kinetic energy of the incoming cosmic "projectile" causes shock deformation, heating, melting, and vaporization as well as crater excavation and ejection of target material. The impact and shock environment of such a collision provide scientists with an explanation as to why the Martian meteorites are all igneous rocks. Martian sedimentary rocks and soil would not be consolidated sufficiently to survive the impact as intact rocks and then wander through space for millions of years and eventually land on Earth as meteorites.

One particular Martian meteorite, called ALH84001, has stimulated a great deal of interest in the possibility of life on Mars. In the summer of 1996 a NASA research team at the Johnson Space Center (JSC) announced that they had found evidence in ALH84001 that "strongly suggests primitive life may have existed on Mars more than 3.6 billion years ago." The NASA research team found the first organic molecules thought to be of Martian origin, several mineral features characteristic of biological activity, and possibly microscopic fossils of primitive, bacterialike organisms. (*See* figure.) While the NASA research team did not claim that they had conclusively proved life existed on Mars, they did believe that "they have found quite reasonable evidence of past life on Mars."

Martian meteorite ALH84001 is a 1.9-kilogram, potato-sized igneous rock that has been age-dated to about 4.5 billion years, the period when the planet Mars formed. This rock is believed to have originated underneath the Martian surface and to have been extensively fractured by impacts as meteorites bombarded the planet during the early history of the solar system. Between 3.6 and 4.0 billion years ago, Mars is believed to have been a warmer and wetter world. Water is thought to have penetrated fractures in the subsurface rock, possibly forming an underground water system. Since the water was saturated with carbon dioxide from the Martian atmosphere, carbonate materials were deposited in the fractures. The NASA research team estimates that this rock from Mars entered Earth's atmosphere about 13,000 years ago and fell in Antarctica as a meteorite. ALH84001 was discovered in 1984 in the Allan Hills ice field of Antarctica by an annual expedition of the National Science Foundation's Antarctic Meteorite Program. It was preserved for study at the NASA JSC Meteorite Processing Laboratory, but its possible Martian origin was not fully recognized until 1993. It is the oldest of the Martian meteorites yet discovered.

See also MARS; METEORITE; METEOROIDS.

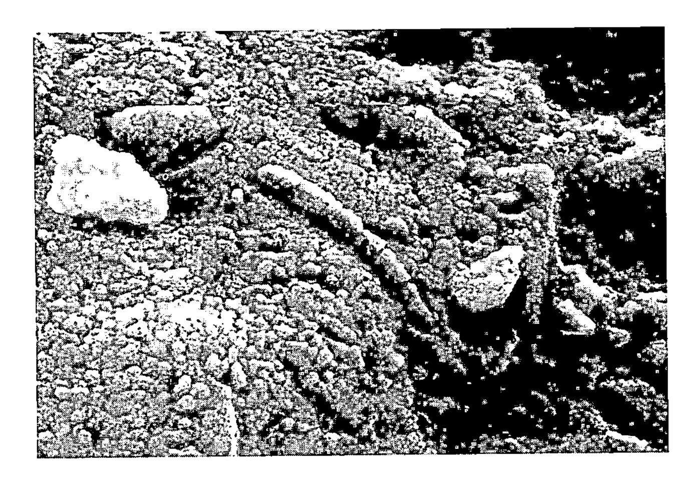

A close-up view of a piece of the Martian meteorite ALH84001, which some scientists postulate contains microscopic fossils that provide evidence for the existence of life on Mars (photo taken October 10, 1996) ***(Courtesy of NASA/JSC)***

mascon A term meaning "mass concentration." An area of mass concentration, or high density, within a celestial body, usually near the surface. In 1968 data from five U.S. lunar orbiter spacecraft indicated that regions of high density, or "mascons," existed under circular maria (i.e., the extensive dark areas) on the Moon. The Moon's gravitational attraction is somewhat higher over such mass concentrations, and their presence perturbs (causes variations in) the orbits of spacecraft around the Moon.

See also MOON.

Maskelyne, Nevil (1732–1811) British *Astronomer* The British astronomer Nevil Maskelyne served as the fifth Astronomer Royal (1765–1811) and made significant contributions to the use of astronomical observations in navigation. In 1769 he studied a transit of Venus and used data from this event to calculate the Earth-Sun distance (that is, the astronomical unit) to within 1 percent of its modern value.

mass (symbol: m) The amount of material present in an object. This fundamental unit describes "how much" material makes up an object. The SI unit for mass is the kilogram (kg), while it is the pound-mass (lbm) in the traditional, or engineering, unit system. The terms *mass* and *weight* are very often confused in the traditional, or engineering, unit system because they have similar names: the pound-mass (lbm) and the pound-force (lbf). However, these terms are not the same, since "weight" is a derived unit that describes the action of the local force of gravity on the "mass" of an object. An object of 1 kilogram mass on Earth will also have a 1 kilogram mass on the surface of Mars. However, the "weight" of this 1-kilogram-mass object will be different on the surface of each planet, since the acceleration of gravity is different on each planet (i.e., about 9.8 meters per second per second on Earth's surface versus 3.7 meters per second per second on Mars).

mass driver An electromagnetic device currently under study and development that can accelerate payloads (nonliving and nonfragile) to a very high terminal velocity. Small, magnetically levitated vehicles, sometimes called "buckets," would be used to carry the payloads. These buckets would contain superconducting coils and be accelerated by pulsed magnetic fields along a linear trace or guideway. When these buckets reached an appropriate terminal velocity (e.g., several kilometers per second), they would release their payloads and then be decelerated for reuse. Mass drivers have been suggested as a way "shooting" lunar ores into orbit around the Moon for collection and subsequent use by space-based manufacturing facilities.

See also GUN-LAUNCH TO SPACE.

mass-energy equation (mass-energy equivalence) The statement developed by ALBERT EINSTEIN that "the mass of a body is a measure of its energy content," as an extension of his 1905 special theory of relativity. The statement subsequently was verified experimentally by measurements of mass and energy in nuclear reactions. This equation is written as $E = m c^2$ and illustrates that when the energy of a body changes by an amount E, the mass, m, of the body will change by an amount equal to E/c^2. (The factor c^2, the square of the speed of light in a vacuum, may be regarded as the conversion factor relating units of mass and energy.) This famous equation predicted the possibility of releasing enormous amounts of energy in a nuclear chain reaction (e.g., as found in a nuclear explosion) by the conversion of some of the mass in the atomic nucleus into energy. This equation is sometimes called the Einstein equation.

See also RELATIVITY.

mass number (symbol: A) The number of nucleons (i.e., the number of protons and neutrons) in an atomic nucleus. It is the nearest whole number to an atom's atomic weight. For example, the mass number of the isotope uranium-235 is 235.

mass spectrometer An instrument used to measure the relative atomic masses and relative abundances of isotopes. A sample (usually gaseous) is ionized, and the resultant stream of charged particles is accelerated into a high-vacuum region where electric and magnetic fields deflect the particles and focus them on a detector. A *mass spectrum* (i.e., a series of lines related to mass/charge values) then is created. This characteristic pattern of lines helps scientists identify different molecules.

mass transfer cooling Heat transfer or thermal control accomplished primarily by the use of an amount of mass to cool a surface or region. There are three general categories of mass transfer cooling: (1) transpiration cooling, (2) film cooling, and (3) ablation cooling. Aerospace designers have long recognized mass transfer to the boundary layer as an effective technique for significantly altering the adverse aerodynamic heating phenomena associated with a body in hypersonic flight through a planetary atmosphere. Similarly, rocket engineers often use some form of mass transfer cooling to prevent the overheating or destruction of thermally sensitive rocket nozzle components.

Matador An early medium-range, surface-to-surface winged missile developed by the U.S. Air Force in the 1950s.

materials processing in space (MPS) Materials processing is the science by which ordinary and comparatively inexpensive raw materials are made into useful crystals, chemicals, metals, ceramics, and countless other manufactured products. Modern materials processing on Earth has taken us into the space age and opened up the microgravity environment of Earth orbit. The benefits of extended periods of "weightlessness" promise to bring new and unique opportunities for the science of materials processing. In the microgravity environment of an orbiting spacecraft, scientists can use materials processing procedures that are all but impossible on Earth.

In orbit materials processing can be accomplished without the effects of gravity, which on Earth causes materials of different densities and temperatures to separate and deform under the influence of their own masses. However, when scientists refer to an orbiting object as being "weightless," they do not literally mean there is an absence of gravity. Rather, they are referring to the microgravity conditions, or the absence of relative motion between objects in a free-falling environment, as experienced in an Earth-orbiting spacecraft. These useful "free-fall" conditions can be obtained only briefly on Earth using drop towers or "zero-gravity" aircraft. Extended periods of microgravity can be achieved only on an orbiting spacecraft, such as the space shuttle orbiter, a space station, or a crew-tended free-flying platform.

Hydrostatic pressure places a strain on materials during solidification processes on Earth. Certain crystals are sufficiently dense and delicate that they are subject to strain under the influence of their own weight during growth. Such strain-induced deformations in crystals degrade their overall performance. In microgravity heat-treated, melted, and resolidified crystals and alloys can be developed free of such deformations.

Containerless processing in microgravity eliminates the problems of container contamination and wall effects. Often these are the greatest source of impurities and imperfections when a molten material is formed on Earth, but in space a material can be melted, manipulated, and shaped free of contact with a container wall or crucible by acoustic, electromagnetic, or electrostatic fields. In microgravity the surface tension of the molten material helps hold it together, while on Earth this cohesive force is overpowered by gravity.

Over the next few decades space-based materials processing research will emphasize both scientific and commercial goals. Potential space-manufactured products include special crystals, metals, ceramics, glasses, and biological materials. Processes will include containerless processing and fluid and chemical transport. As research in these areas progresses, specialized new materials and manufactured products could become available in this century for use in space as well as on Earth.

See also DROP TOWER; INTERNATIONAL SPACE STATION; MICROGRAVITY; ZERO-GRAVITY AIRCRAFT.

mating The act of fitting together two major components of a system, such as the mating of a launch vehicle and a spacecraft. Also, the physical joining of co-orbiting spacecraft either through a docking or a berthing process.

matter-antimatter propulsion Spacecraft propulsion by use of matter-antimatter annihilation reactions.

Mattingly, Thomas K., II (1936–) American *Astronaut, U.S. Navy Officer* Astronaut Thomas K. Mattingly II served as the spacecraft pilot during NASA's *Apollo 16* Moon landing mission (April 1972). He orbited the Moon while astronauts JOHN W. YOUNG and CHARLES MOSS DUKE, JR. descended to the surface in the lunar excursion module (LEM). Mattingly also commanded the space shuttle on the STS-4 (1982) and STS-51C (1985) Earth orbital missions.

Maverick (AGM-65) A tactical air-to-surface guided missile designed for close air support, interdiction, and defense suppression missions. It provides standoff capability and high probability of strike against a wide range of tactical targets, including armor, air defenses, ships, transportation equipment, and fuel storage facilities. Maverick A and B models have an electro-optical television guidance system; the Maverick D and G models have an imaging infrared guidance system. The U.S. Air Force accepted the first AGM-65A Maverick missile in August 1972. AGM-65 missiles were used successfully by F-16 and A-10 aircraft in 1991 to attack armored targets in the Persian Gulf during Operation Desert Storm.

max-Q An aerospace term describing the condition of maximum dynamic pressure—the point in the flight of a launch vehicle when it experiences the most severe aerodynamic forces as it rises up through Earth's atmosphere toward outer space.

Maxwell, James Clerk (1831–1879) Scottish *Physicist* James Clerk Maxwell was a brilliant theoretical physicist who made many important contributions to science, including the fundamental concept of electromagnetic radiation. In 1857 he also made a major contribution to astronomy by correctly predicting from theoretical principles alone that the rings of Saturn must consist of numerous small particles.

Maxwell Montes A mountain range on Venus located in Ishtar Terra containing the highest peak (11-km altitude) on the planet. It is named after the Scottish theoretical physicist JAMES CLERK MAXWELL.

McAuliffe, S. Christa Corrigan (1948–1986) American *Astronaut, School Teacher* S. Christa Corrigan McAuliffe was a gifted school teacher who participated in NASA's "Teacher in Space" program in the 1980s and was invited to became an astronaut. Unfortunately, she died along with the rest of the STS-51L mission crew when the space shuttle *Challenger* exploded during launch ascent on January 28, 1986.

See also *CHALLENGER* ACCIDENT.

McNair, Ronald E. (1950–1986) American *Astronaut* In February 1984 the astronaut Ronald E. McNair successfully traveled in outer space as a NASA mission specialist during the STS-41B mission of the space shuttle *Challenger.* However, on the STS-51L mission he lost his life along with the rest of the *Challenger*'s crew when the space shuttle exploded during launch ascent on January 28, 1986.

See also *CHALLENGER* ACCIDENT.

mean solar day The duration of one rotation of Earth on its axis with respect to the uniform motion of the mean Sun. The length of the mean solar day is 24 hours of mean solar time or 24 hours 3 minutes 56.555 seconds of mean sidereal time. A mean solar day beginning at midnight is called a civil day; one beginning at noon is called an astronomical day.

mean solar time Time based on Earth's rate of rotation, which originally was assumed to be constant.

See also MEAN SOLAR DAY.

mean time between failures (MTBF) The average time between the failure of elements in a system composed of many elements.

mean time to repair (MTTR) In a multielement system, the average time required to repair the system in the event of a failure.

medium-range ballistic missile (MRBM) A ballistic missile with a range capability from about 1,100 km to 2,780 km.

mega- (symbol: M) A prefix in the SI unit system meaning multiplied by 1 million (10^6), such as, for example, megahertz (MHz), meaning 1 million hertz.

megaparsec (Mpc) A million parsecs; a distance of approximately 3,260,000 light-years.

See also LIGHT-YEAR; PARSEC.

megaton (MT) A measure of the explosive yield of a nuclear weapon that is equivalent to the explosion of 1 million tons of trinitrotoluene (TNT), a chemical high explosive. By definition, 1 MT = 10^{15} calories or 4.2×10^{15} joules. Here, a ton represents 1,000 kilograms of TNT.

Mercury The innermost planet in the solar system, orbiting the Sun at approximately 0.4 astronomical unit (AU). This planet, named for the messenger god of Roman mythology, is a scorched, primordial world that is only 40 percent larger in diameter than Earth's Moon. NASA's *Mariner 10* spacecraft provided the first close-up views of Mercury. This spacecraft was launched from Cape Canaveral in November 1973. After traveling almost five months, including a flyby of Venus, this spacecraft passed within 805 kilometers of Mercury on March 29, 1974. *Mariner 10* then looped around the Sun and made another rendezvous with Mercury on September 21, 1974. This encounter process was repeated a third time on March 16, 1975, before the control gas used to orient the

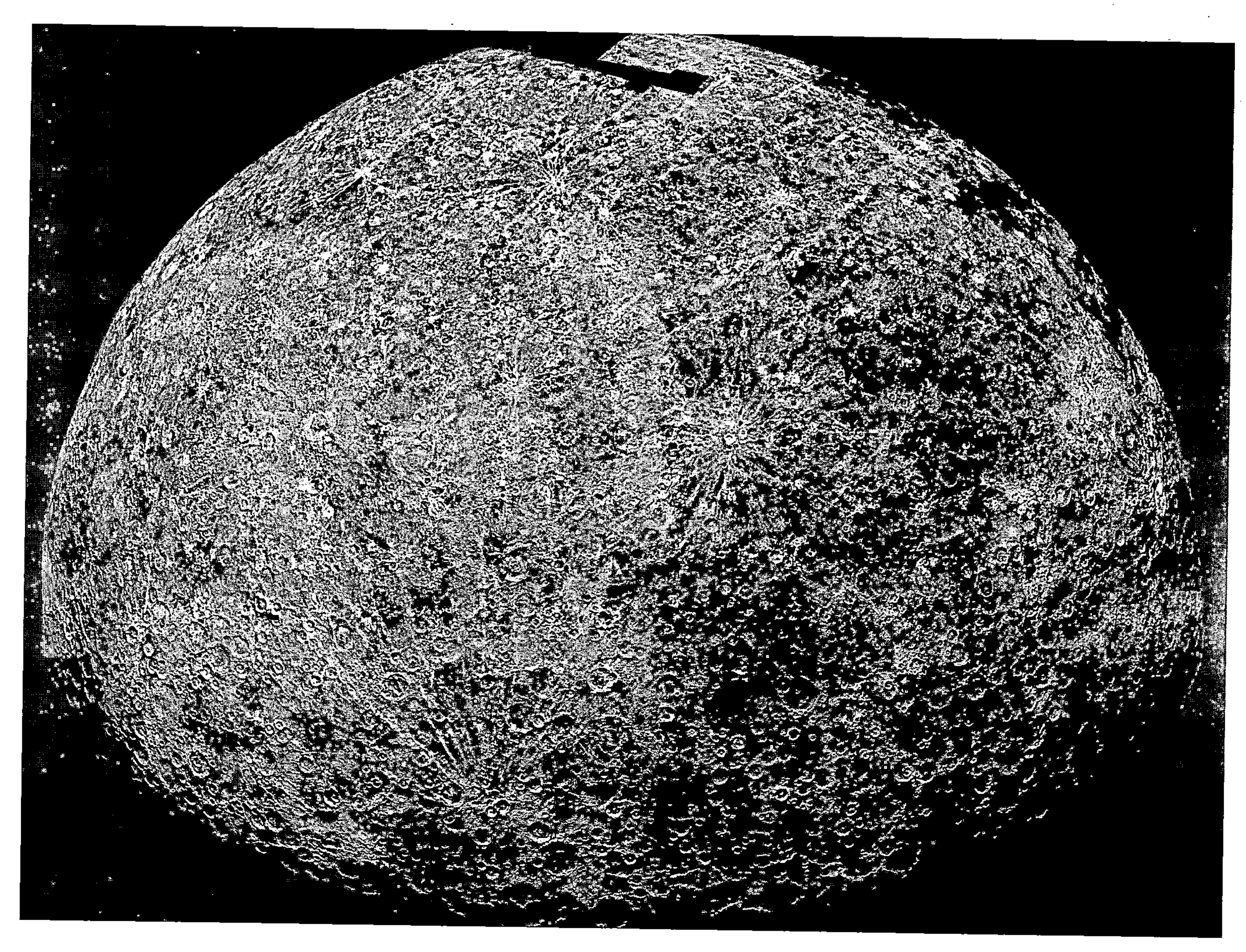

A mosaic image of Mercury (southern hemisphere) created from more than 200 high-resolution photographs collected by NASA's *Mariner 10* spacecraft during its second encounter with the planet in September 1974 *(Courtesy of NASA/JSC)*

Physical and Dynamic Properties of the Planet Mercury

Radius (mean equatorial)	2,439 km
Mass	3.30×10^{23} kg
Mean density	5.44 g/cm³
Acceleration of gravity (at the surface)	3.70 m/sec²
Escape velocity	4.25 km/sec
Normal albedo (averaged over visible spectrum)	0.125
Surface temperature extremes	100 K to 700 K
Atmosphere	negligible
Number of natural satellites	none
Flux of solar radiation	
Aphelion	6,290 W/m²
Perihelion	14,490 W/m²
Semimajor axis	5.79×10^{7} km (0.387 AU)
Perihelion distance	4.60×10^{7} km (0.308 AU)
Aphelion distance	6.98×10^{7} km (0.467 AU)
Eccentricity	0.20563
Orbital inclination	7.004 degrees
Mean orbital velocity	47.87 km/sec
Sidereal day (a Mercurean "day")	58.646 Earth days
Sidereal year (a Mercurean "year")	87.969 Earth days

Source: NASA.

spacecraft was exhausted. This triple flyby of the planet Mercury by *Mariner 10* is sometimes referred to as Mercury I, II, and III in the technical literature.

The images of Mercury transmitted back to Earth by *Mariner 10* revealed an ancient, heavily cratered world that closely resembled Earth's Moon. Unlike the Moon, however, huge cliffs called lobate scarps crisscross Mercury. These great cliffs apparently were formed when Mercury's interior cooled and shrank, compressing the planet's crust. The cliffs are as high as 2 km and as long as 1,500 km.

To the surprise of scientists, instruments onboard *Mariner 10* discovered that Mercury has a weak magnetic field. It also has a wisp of an atmosphere a trillionth of the density of Earth's atmosphere and made up mainly of traces of helium, hydrogen, sodium, and potassium.

Temperatures on the sunlit side of Mercury reach approximately 700 kelvins (K) (427°C), a temperature that exceeds the melting point of lead; on the dark side temperatures plunge to a frigid 100 kelvins (K) (–173°C). Quite literally, Mercury is a world seared with intolerable heat in the daytime and frozen at night.

In the late 1960s scientists on Earth bounced radar signals off the surface of Mercury. Analysis of the scattered radar signals indicated that the planet rotates slowly on its axis, with a period of about 59 days. Consequently, the "days" and "nights" on this planet are quite long by terrestrial standards, with one Mercurian day equal to 59 Earth days. It takes the planet approximately 88 days to orbit around the Sun.

Mercury's surface features include large regions of gently rolling hills and numerous impact craters like those found on the Moon. Many of these craters are surrounded by blankets of ejecta (material thrown out at the time of a meteoroid impact) and secondary craters that were created when chunks of ejected material fell back down to the planet's surface. Because Mercury has a higher gravitational attraction than the Moon, these secondary craters are not spread as widely from each primary craters as occurs on the Moon. One major surface feature discovered by *Mariner 10* is a large impact basin called Caloris, which is about 1,300 km in diameter. Scientists now believe that Mercury has a large iron-rich core, the source of its weak but detectable magnetic field. The table presents some contemporary physical, and dynamic property data about the Sun's closest planetary companion.

Because Mercury lies deep in the Sun's gravity well, detailed exploration with sophisticated orbiters and landers requires the development of advanced planetary spacecraft that take advantage of intricate "gravity assist" maneuvers involving both Earth and Venus. On August 3, 2004, NASA launched the *Messenger* spacecraft to Mercury from Cape Canaveral Air Force Station. *Messenger* will enter orbit around Mercury in March 2011.

See also MARINER SPACECRAFT; *MESSENGER* SPACECRAFT.

Mercury Project America's pioneering project to put a human being into orbit. The series of six suborbital and orbital flights was designed to demonstrate that human beings could withstand the high acceleration of a rocket launching, a prolonged period of weightlessness, and then a period of high deceleration during reentry.

The Mercury Project became an official program of NASA on October 7, 1958. Seven astronauts were chosen in April 1959 after a nationwide call for jet pilot volunteers. The one-person Mercury spacecraft was designed and built with a maximum orbiting mass of about 1,452 kilograms. Shaped somewhat like a bell, this small spacecraft was about 189 centimeters wide across the bottom and about 2.7 meters tall. The astronaut escape tower added another 5.2 meters, for an overall length of approximately 8 meters at launch. Two boosters were chosen: the U.S. Army Redstone with its 346,944 newtons thrust for the suborbital flights and the U.S. Air Force Atlas with its 1,601,280 newtons thrust for the orbital missions.

On May 5, 1961, the astronaut Alan B. Shepard, Jr. was launched from Complex 5 at Cape Canaveral Air Force Station (CCAFS) by a Redstone booster on the first U.S. crewed space flight. (*See* figure.) His brief suborbital mission lasted just 15 minutes, and his *Freedom 7* Mercury capsule traveled 186.7 kilometers into space. On July 21, 1961, another Redstone booster hurled the astronaut Virgil I. "Gus" Grissom through the second and last suborbital flight in the *Liberty Bell 7* Mercury capsule.

Following these two successful suborbital missions, NASA then advanced the project to the Mercury-Atlas series of orbital missions. Another space milestone was reached on

NASA's Mercury Project began with the Mercury Redstone 3 (MR3) mission from Cape Canaveral Air Force Station, Florida, on May 5, 1961. A U.S. Army Redstone rocket successfully carried astronaut Alan B. Shepard, Jr. onboard the *Freedom 7* space capsule for a 15-minute suborbital flight, ending with a downrange splashdown and recovery in the Atlantic Ocean. By flying this rocket-propelled ballistic trajectory, Shepard became the first American to travel in outer space. *(Courtesy of NASA/JSC)*

February 20, 1962, when the astronaut John H. Glenn, Jr. became the first American in orbit, circling the Earth three times in the *Friendship 7* spacecraft.

On May 24, 1962, astronaut M. Scott Carpenter completed another three-orbit flight in the *Aurora 7* spacecraft. Astronaut Walter M. Schirra, Jr. doubled the flight time in space and orbited six times, landing the *Sigma 7* Mercury capsule in a Pacific Ocean recovery area. All previous Mercury Project landings had been in the Atlantic Ocean. Finally, on May 15–16, 1963, astronaut L. Gordon Cooper, Jr. completed a 22-orbit mission in the *Faith 7* spacecraft, triumphantly concluding the program and paving the way for the Gemini Project and ultimately the Apollo Project, which took American astronauts to the lunar surface. The table summarizes the Mercury Project missions.

See also APOLLO PROJECT; GEMINI PROJECT.

meridian A great circle that passes through both the North and South Poles; also called a *line of longitude.*

meson A class of very short-lived elementary nuclear particles. As a subdivision of hadrons, mesons consist of quark and antiquark pairs and have masses that lie between the mass of an electron and the mass of a proton. Mesons exist with positive, negative, and zero charge. They are a major component of secondary cosmic ray showers.

See also ELEMENTARY PARTICLE(S).

mesosphere The region of Earth's atmosphere above the stratosphere that is characterized by temperature decreasing with height. The top of this layer, called the mesopause, occurs between 80 and 85 kilometers altitude. The mesopause is the coldest region of the entire atmosphere and has a temperature of approximately 180 kelvins (–93°C).

See also ATMOSPHERE.

***Messenger* spacecraft** NASA used a Delta II rocket to launch the *Messenger* spacecraft to the planet Mercury from

Mercury Project Missions (1961–1963)

Mission	Date(s)/Recovery Ship, Ocean	Astronaut	Mission Duration	Remarks
Mercury Redstone 3 *(Freedom 7)*	May 5, 1961 *Lake Champlain,* Atlantic	Navy Comdr. Alan B. Shepard, Jr.	0:15:22	suborbital
Mercury Redstone 4 *(Liberty Bell 7)*	July 21, 1961 *Randolph,* Atlantic	USAF Maj. Virgil I. Grissom	0:15:37	suborbital
Mercury Atlas 6 *(Friendship 7)*	Feb. 20, 1962 *Noa,* Atlantic	Marine Lt. Col. John H. Glenn	4:55:23	3 orbits
Mercury Atlas 7 *(Aurora 7)*	May 24, 1962 *Pierce,* Atlantic	Navy Lt. Comdr. Scott Carpenter	4:56:05	3 orbits
Mercury Atlas 8 *(Sigma 7)*	Oct. 3, 1962 *Kearsarge,* Pacific	Navy Comdr. Walter M. Schirra, Jr.	9:13:11	6 orbits
Mercury Atlas 9 *(Faith 7)*	May 15–16, 1963 *Kearsarge,* Pacific	USAF Maj. L. Gordon Cooper	34:19:49	22 orbits

Note: Names in parentheses in the first column are those given to Mercury spacecraft.
Source: NASA.

Cape Canaveral Air Force Station on August 3, 2004. This mission is actually called the *Mercury Surface, Space Environment. Geochemisty, and Ranging* mission (MESSENGER). The spacecraft's instruments will support a scientific investigation of Mercury and help scientists understand some of the forces that shaped the solar system's innermost planet. Following flybys of three planets (Earth once, Venus twice, and Mercury three times), *Messenger* will finally ease into an orbit around Mercury in March 2011 and begin its scientific studies from a working orbit.

See also MERCURY.

Messier, Charles (1730–1817) French *Astronomer* Charles Messier, the first French astronomer to spot Comet Halley during its anticipated return in 1758, was an avid "comet hunter." He personally discovered at least 13 new comets and assisted in the codiscovery of six or more others. In 1758 he began compiling a list of nebulas and star clusters that eventually became the famous noncomet Messier Catalog. He assembled this list of "unmoving" fuzzy nebulas and star clusters primarily as a tool for astronomers engaged in comet searches. False alarms were time consuming and often caused by diffuse celestial objects, such as nebulas, star clusters, and galaxies, that were easily mistaken by 18th-century astronomers for distant new comets.

Messier was born in Badonviller, Lorraine, France, on June 26, 1730. His father died when Messier was a young boy, and the family was thrust into poverty. The appearance of a multitailed comet and the solar eclipse of 1748 stimulated his childhood interest in astronomy. Messier went to Paris in 1751 and became a clerk for Joseph Nicolas Delisle (1688–1768), who was then the astronomer of the French Navy. Messier's ability to carefully record data attracted Delisle's attention, and soon the young clerk was cataloging observations made at the Hotel de Cluny, including the transit of Mercury in May 1753. Delisle, like many other astronomers, anticipated the return of Comet Halley sometime in 1758 and made his own calculations in an attempt to be the first to detect its latest visit.

Delisle assigned the task of searching for Comet Halley to his young clerk. Messier searched diligently throughout 1758 but mostly in the wrong location, because Delisle's calculations were in error. However, in August of that year Messier discovered and tracked a different comet for several months. Then, in early November he detected a distant fuzzy object that he thought was the anticipated Comet Halley. However, it was not. To Messier's frustration the object did not move, so he realized the fuzzy patch had to be a nebula. On September 12, 1758, he carefully noted its position in his observation log. This particular "fuzzy patch" was actually the Crab Nebula, a fuzzy celestial object previously discovered by another astronomer.

Cometography, that is "comet hunting," was one of the great triumphs of 18th-century astronomy. Messier enjoyed the thrill and excitement that accompanied a successful quest for a new "hairy star"—the early Greek name (κομετες]) for a comet. The young clerk-astronomer also realized that fuzzy nebulas and star clusters caused time consuming false alarms. He found a solution—prepare a list of noncomet objects, such as nebulas and star clusters. In the 18th century astronomers used the word *nebula* for any fuzzy, blurry, luminous celestial object that could not be sufficiently resolved by available telescopes. The fuzzy object he mistook for a comet in late August 1758 eventually became object number one (M1) in the Messier Catalogue—namely, the Crab Nebula in the constellation Taurus.

Despite false alarms caused by nebulas and his discovery of another comet, Messier kept vigorously searching for Comet Halley throughout 1758. Then, on the evening of December 25, an amateur German astronomer named Georg Palitzch became the first observer to catch a glimpse of the greatly anticipated comet as it returned to the inner portion of the solar system. About a month later, on the evening of January 21, 1759, Messier succeeded in viewing this comet and became the first French astronomer to have done so during the comet's 1758–59 journey through perihelion. His work earned him recognition at the highest levels. For example, the French king, Louis XV, fondly referred to Messier as "the comet ferret."

Following the passage of Comet Halley, Messier continued to search for new comets and to assemble his noncomet list of nebulas and star clusters that could cause false alarms for comet-chasing astronomers. By 1764 this list contained 40 objects, 39 of which he personally verified through his own observations. Within the resolution limits of the telescopes of his day, Messier's nonmoving fuzzy objects included nebulas, star clusters, and distant galaxies. By 1771 he had completed the first version of his famous noncomet list and published it as the *Catalogue of Nebulae and of Star Clusters* in 1774. This initial catalog contained 45 Messier objects. He completed the final version of his noncomet list of celestial objects in 1781. It contained 103 objects (seven additional were added later) and was published in 1784. Messier designated each of the entries in the list by a number prefixed by the letter "M." For example, Messier object M1 represented the Crab Nebula, and object M31, the Andromeda galaxy. His designations persist to the present day, although the Messier designations have generally been superseded by the NGC designations presented in the *New General Catalogue (NGC) of Nebulae and Clusters of Stars* published in 1888 by the Danish astronomer Johan Dreyer (1852–1926).

In 1764 Messier was elected a foreign member of the Royal Society of London. He became a member of the Paris Academy of Sciences in 1770. That year Messier also became the astronomer of the French navy, officially assuming the position formerly held by Delisle, who had retired six years earlier. However, Messier lost this position during the French Revolution, and his observatory at the Hotel de Cluny fell into general disrepair. In 1806 Napoleon Bonaparte (1769–1821) awarded Messier the cross of the Legion of Honor, possibly because Messier had openly and rather unscientifically suggested in a technical paper that the appearance of the great comet of 1769 coincided with Napoleon's birth. This politically inspired act of astrology is the last known attempt by an otherwise knowledgeable astronomer to tie the appearance of a comet to a significant

Selected Messier Objects and Their New General Catalog (NGC) Equivalent Numbers

Messier Number	NGC* IC	Description of Celestial Object
M1	1952	Crab Nebula in constellation Taurus (Tau)
M2	7089	Globular cluster in constellation Aquarius (Aqr)
M3	5272	Globular cluster in constellation Canes Venatici (CVn)
M4	6121	Globular cluster in constellation Scorpius (Sco)
M5	5904	Globular cluster in constellation Serpens (Ser)
M8	6523	Lagoon Nebula in constellation Sagittarius (Sgr)
M13	6205	Hercules Cluster (globular cluster in constellation Hercules [Her])
M17	6618	Omega Nebula in constellation Sagittarius (Sgr)
M20	6514	Trifid Nebula in constellation Sagittarius (Sgr)
M27	6853	Dumbbell Nebula (planetary nebula) in constellation Vulpecula (Vul)
M31	224	Andromeda galaxy (spiral galaxy in constellation Andromeda [And])
M42	1976	Orion Nebula in constellation Orion (Ori)
M57	6720	Ring Nebula in constellarion Lyra (Lyr)
M58	4579	Spiral galaxy in constellation Virgo (Vir)
M87	4486	Giant elliptical galaxy in constellation Virgo (Vir)
M104	4594	Sombrero galaxy in constellation Virgo (Vir)
M110	205	Elliptical galaxy in constellation Andromeda (And)

* NGC = *New General Catalog of Nebulae and Clusters of Stars* published by Danish astronomer Johan Dreyer in 1888; IC = *Index Catalog* which Dreyer published as supplements to NGC in 1895 and 1908.

historic event on Earth. Most likely, the aging Messier, with his life and observatory now in complete disrepair, was willing to tarnish his scientific reputation to regain some of the political and financial advantage he had enjoyed just prior to the French Revolution. Messier died in Paris on April 11, 1817. Most of the comets he so eagerly sought have been forgotten, but his famous noncomet list of celestial objects serves as a permanent tribute to one of the 18th century's most successful "comet hunters."

Messier Catalog A compilation of bright (but fuzzy) celestial objects, such as nebulas and galaxies, developed by CHARLES MESSIER in the late 18th century to assist his search for new comets. Astronomers still use Messier object nomenclature—for example, the Orion Nebula is M42. See the Messier Objects table.

meteor The luminous phenomena that occurs when a meteoroid enters Earth's atmosphere. Sometimes called a shooting star.

See also METEOROIDS.

meteorite Metallic or stony material that has passed through the atmosphere and reached Earth's surface. Meteorites are of extraterrestrial origin. Known sources of meteorites include asteroids and comets; some even originated on the Moon and on Mars. Presumably, meteorites from the Moon and Mars were ejected by impact-cratering events. The term *meteorite* also is applied to meteoroids that land on planetary bodies other than Earth.

See also MARTIAN METEORITES; METEOROIDS.

meteoroids An all-encompassing term that refers to solid objects found in space, ranging in diameter from micrometers to kilometers and in mass from less than 10^{-12} gram to more than 10^{+16} grams. If these pieces of extraterrestrial material are less than 1 gram, they often are called micrometeoroids. When objects of more than approximately 10^{-6} gram reach Earth's atmosphere, they are heated to incandescence (i.e., they glow with heat) and produce the visible effect popularly called a meteor. If some of the original meteoroid survives its glowing plunge into Earth's atmosphere, the remaining unvaporized chunk of space matter is then called a meteorite.

Scientists currently think that meteoroids originate primarily from asteroids and comets that have perihelia (the portions of their orbits nearest the Sun) near or inside Earth's orbit around the Sun. The parent celestial objects are assumed to have been broken down into a collection of smaller bodies by numerous collisions. Recently formed meteoroids tend to remain concentrated along the orbital path of their parent body. These "stream meteoroids" produce the well-known meteor showers that can be seen at certain dates from Earth.

Meteoroids generally are classified by composition as stony meteorites (chondrites), irons, and stony-irons. Of the meteorites that fall on Earth, stony meteorites make up about 93%, irons about 5.5 percent, and stony-irons about 1.5%. Space scientists estimate that about 10^{+7} kilograms (or 10,000 metric tons) of "cosmic rocks" now fall on our planet annually.

Some of the meteorites found on Earth are thought to have their origins on the Moon and on Mars. It is now believed that these lunar and Martian meteorites were ejected

Time Between Meteoroid Collisions for a Space Shuttle Orbiter in Low Earth Orbit (300 km Altitude)

Minimum Meteoroid Mass (g)	Estimated Time Between Collisions (yr)
10	350,000
1	25,000
0.1	1,800
0.01	130

Source: Based on NASA data.

by ancient impact-cratering events. If an asteroid or comet of sufficient mass and velocity hits the surface of the Moon, a small fraction of the material ejected by the impact collision could depart from the Moon's surface with velocities greater than its escape velocity (2.4 kilometers per second [km/sec]). A fraction of that ejected material eventually would reach Earth's surface, with Moon-to-Earth transit times ranging from under 1 million years to upward of 100 million years. Similarly, a very energetic asteroid or comet impact collision on Mars (which has an escape velocity of 5.0 km/sec) could be the source of the interesting Martian meteorites found recently in Antarctica.

Micrometeoroids and small meteoroids large enough to damage an Earth-orbiting spacecraft are considered "somewhat rare" by space scientists. For example, the table presents a contemporary estimate for the time between collisions between an object the size of a space shuttle orbiter in low Earth orbit and a meteoroid of mass greater than a given meteoroid mass.

On a much larger "collision scale," meteoroid impacts now are considered by planetary scientists to have played a basic role in the evolution of planetary surfaces in the early history of the solar system. Although still dramatically evident in the cratered surfaces found on other planets and moons, here on Earth this stage of surface evolution has essentially been lost due to later crustal recycling and weathering processes.

See also ASTEROID; COMET; MARTIAN METEORITES.

meteorological rocket A rocket designed for routine upper-air observation (as opposed to scientific research), especially in that portion of Earth's atmosphere inaccessible to weather balloons, namely above 30,500 meters altitude.

meteorological satellite An Earth-observing spacecraft that senses some or most of the atmospheric phenomena (e.g., wind and clouds) related to weather conditions on a local, regional, or hemispheric scale.

See also WEATHER SATELLITE.

meteorology The branch of science dealing with phenomena of the atmosphere. This includes not only the physics, chemistry, and dynamics of the atmosphere but also many of the direct effects of the atmosphere on Earth's surface, the oceans, and life in general. Meteorology is concerned with actual weather conditions, while climatology deals

Will It Rain on My Parade?

Before the space age observations were basically limited to areas relatively close to Earth's surface, with vast gaps over oceans and sparsely populated regions. Meteorologists could only dream of having a synoptic view of our entire planet. In the absence of sensors on Earth-orbiting satellites, scientists were severely limited in their ability to observe Earth's atmosphere. Most of their measurements were made from below and a few from within the atmosphere, but none from above. Their plight was clearly reflected in the opening statement of a 1952 U.S. Weather Bureau pamphlet concerning the "future" of weather forecasting. This pamphlet began with a then-fanciful technical wish, wondering "if it were possible for a person to rise by plane or rocket to a height where he could see the entire country from the Atlantic to the Pacific."

Of course, a number of pre–space age meteorologists recognized the exciting promise that the Earth-orbiting satellite meant to their field. When available, weather satellites would routinely provide the detailed view of Earth they so desperately needed to make more accurate forecasts. Even more important was the prospect that a system of operational weather satellites could help reduce the number of lethal surprises from the atmosphere. A few bold visionaries even speculated (quite correctly) that cameras on Earth-orbiting platforms could detect hurricane-generating disturbances long before these highly destructive killer storms matured and threatened life and property. With sophisticated instruments on weather satellites, meteorologists might even be able to detect dangerous thunderstorms hidden by frontal clouds and provide warning to communities in their paths. Finally, "weather eyes" in space offered the promise of greatly improved routine forecasting (especially three- to five-day predictions)—a service that would certainly improve the quality of life for almost everyone.

Then, on April 1, 1960, NASA launched the *Television Infrared Observation Satellite* (TIROS-1) and the dreams and fondest wishes of many meteorologists became a reality. TIROS-1 was the first satellite capable of imaging clouds from space. Operating in a mid-latitude (approximately 44° inclination) orbit, the trailblazing spacecraft quickly proved that properly instrumented satellites could indeed observe terrestrial weather patterns. This successful launch represented the beginning of satellite-based meteorology and opened the door to a deeper knowledge of terrestrial weather and the natural forces that control and affect it.

TIROS-1 carried a television camera and during its 78-day operational lifetime transmitted approximately 23,000 cloud photographs, more than half of which proved very useful to meteorologists. TIROS-1 also stimulated unprecedented levels of interagency cooperation within the federal government. In particular, the success of this satellite initiated a long-term, interagency development effort that produced an outstanding (civilian) operational meteorological satellite system. Within this arrangement NASA performed the necessary space technology research and development efforts, while the U.S. Department of Commerce (through the auspices of the National Oceanic and Atmospheric Administration [NOAA]) managed and operated the emerging national system of weather satellites. Over the years the results of that cooperative, interagency arrangement have provided the citizens of the United States with the world's most advanced weather forecast system. As a purposeful example of the peaceful applications of outer space, the United States (through NOAA) now makes weather satellite information available to other federal agencies, to other countries, and to the private sector.

Once proven feasible in 1960, the art and science of space-based meteorological observations quickly expanded and evolved. Atmospheric scientists and aerospace engineers soon developed more sophisticated sensors that were capable of providing improved environmental data of great assistance in weather forecasting. In the early days of satellite-based meteorology, the terms *environmental satellite, meteorological satellite,* and *weather satellite* were often used interchangeably. More recently the term *environmental satellite* has also acquired a more specialized meaning within the fields of Earth system science and global change research.

In 1964 NASA replaced the very successful family of TIROS spacecraft with a new family of advanced weather satellite, called Nimbus—after the Latin word for cloud. Among the numerous space technology advances introduced by the Nimbus spacecraft was the fact that they all flew in near-polar, sun-synchronous orbits around Earth. This operational orbit enabled meteorologists to piece spacecraft data together into mosaic images of the entire globe. As remarkable as the development of these civilian low-altitude, polar-orbiting weather satellites was in the 1960s, this achievement represented only half of the solution to high-payoff space-based meteorology. During this period the first generation of polar-orbiting, military weather spacecraft, called the Defense Meteorological Satellite Program (DMSP), also began to appear. At first highly classified, these low-altitude military weather satellites eventually emerged from under the cloak of secrecy and then played a significant role in national weather forecasting for both defense-related and civil activities.

Meteorologists recognized that to completely serve the information needs of the global weather forecasting community, they would have to develop and deploy operational geostationary weather satellites that were capable of providing good-quality hemispheric views on a continuous basis. Responding to this need, NASA launched the first *Applications Technology Satellite* (ATS-1) in 1966. The ATS-1 spacecraft operated in geostationary orbit over a Pacific Ocean equatorial point at about 150° west longitude. In December of that year this spacecraft's spin-scan cloud camera began transmitting nearly continuous photographic coverage of most of the Pacific Basin. For the next few years this very successful technology demonstration spacecraft provided synoptic cloud photographs and became an important component of weather analysis and forecast activities for this data-sparse ocean area.

NASA launched the ATS-3 spacecraft in November 1967, and the technology demonstration satellite exerted a similar influence on meteorology. From its particular vantage point in geostationary orbit, the advanced multicolor spin-scan camera on ATS-3 had a field of view that covered much of the North and South Atlantic Oceans, all of South America, the vast majority of North America, and even the western edges of Europe and Africa.

Scientists from NASA and NOAA used data from both the ATS-1 and ATS-3 satellites to pioneer important new weather analysis techniques. In particular, these geostationary spacecraft not only provided meteorologists with cloud system and wind field data on a hemispheric scale, but they also allowed atmospheric scientists to observe small-scale weather events on an almost continuous basis. This represented a major breakthrough in meteorology. By repeating their photographs at about 27-minute intervals, the ATS spacecraft demonstrated that geostationary weather satellites could watch a thunderstorm develop from cumulus clouds. This significantly improved the early detection of severe weather. ATS data also became a routine part of the information flowing into the National Hurricane Center in Florida. For example, in August 1969 data from the ATS-3 spacecraft helped forecasters track Hurricane Camille and provided accurate and timely warning about the threatened region along the Gulf Coast of the United States.

The meteorology-related accomplishments of NASA's ATS spacecraft established the technical foundation for an important family of weather satellites known as the Geostationary Operational Satellites (GOES). NOAA currently uses GOES spacecraft to provide a complete line of forecasting services throughout the United States and around the world. When NASA launched GOES-1 for NOAA in October 1975, the field of space-based meteorology achieved complete operational maturity.

In May 1994 a presidential directive instructed the Departments of Defense (DOD) and Commerce to converge their separate low-altitude, polar orbiting weather satellite programs. Consequently, the command, control, and communications for existing DOD satellites were combined with the control for NOAA weather satellites. In June 1998 the Commerce Department assumed primary responsibility for flying both the military's DMSP and NOAA's DMSP-cloned civilian polar orbiting weather satellites. The Department of Commerce will continue to manage both weather satellite systems, providing essential environmental sensing data to American military forces until the new converged National Polar-Orbiting Environmental Satellite System (NPOESS) becomes operational in about 2010.

The current NOAA operational environmental satellite system consists of two basic types of weather satellites: Geostationary Operational Satellites (GOES) for short-range warning and "nowcasting" and Polar Orbiting Environmental Satellites (POES) for longer-term forecasting. Data from both types of weather satellites are needed to support a comprehensive global weather monitoring system.

Geostationary weather satellites provide the kind of continuous monitoring needed for intensive data analysis. Because they operate above a fixed spot on Earth's surface and are far enough away to enjoy a full-disk view of our planet's surface, the GOES spacecraft (and similar geostationary meteorological satellites launched by other nations) provide a constant vigil for the atmospheric "triggers" of severe weather conditions, such as tornadoes,

(continues)

Will It Rain on My Parade? *(continued)*

flash floods, hailstorms, and hurricanes. When these dangerous weather conditions develop, the GOES spacecraft monitor the storms and track their movements. Meteorologists also use GOES imagery to estimate rainfall during thunderstorms and hurricanes for flash-flood warnings. Finally, they also use weather satellite imagery to estimate snowfall accumulations and the overall extent of snow cover. Satellite-derived snowfall data help meteorologists issue winter storm warnings and spring melt advisories.

Each of NOAA's low-altitude, polar-orbiting weather satellites monitors the entire surface of Earth, tracking environmental variables and providing high-resolution cloud images and atmospheric data. The primary mission of these NOAA spacecraft is to detect and track meteorological patterns that affect the weather and climate of the United States. To fulfill their mission, the spacecraft carry instruments that collect visible and infrared radiometer data—which data are then used for imaging purposes, radiation measurements, and temperature profiles. Ultraviolet radiation sensors on each polar-orbiting spacecraft monitor ozone levels in the atmosphere and help scientists monitor the ozone hole over Antarctica. Typically, each polar-orbiting spacecraft performs more than 16,000 environmental measurements as it travels around the world. Meteorologists use these data to support forecasting models, especially pertaining to remote ocean areas, where conventional data are lacking.

Today the weather satellite is an indispensable part of modern meteorology. Satellite-derived meteorological data has become an integral part of our daily lines. For example, most television weather persons include a few of the latest satellite cloud images to support their daily forecasts. Professional meteorologists use weather satellites to observe and measure a wide range of atmospheric properties and processes in their continuing efforts to provide ever more accurate forecasts and severe weather warnings. Imaging instruments provide detailed visible and near-infrared pictures of clouds and cloud motions, as well as an indication of sea surface temperature. Atmospheric sounders collect data in several infrared and microwave bands. When processed, these sounder data provide useful profiles of moisture and temperature as a function of altitude. Radar altimeters, scatterometers, and synthetic aperture radar (SAR) imagery systems measure ocean currents, sea-surface winds, and the structure of snow and ice cover.

Nowhere have operational weather satellites had a greater social impact than in the early detection and continuous tracking of tropical cyclones—the hurricanes of the Atlantic Ocean and the typhoons of the Pacific Ocean. Few things in nature can compare to the destructive force and environmental fury of a hurricane. Often called the greatest storm on Earth, a hurricane is capable of annihilating coastal areas with sustained winds of 250 kilometers per hour or more, storm surges, and intense amounts of rainfall. Scientists estimate that during its life cycle a major hurricane can expend as much energy as 10,000 nuclear bombs. Today, because of weather satellites, meteorologists can provide people who live in at-risk coastal regions timely warning about the pending arrival of a killer storm. Witness the unprecedented massive evacuation of millions of people in Florida and nearby states in 2004, as four powerful and lethal hurricanes (named Charley, Frances, Ivan, and Jeanne) pounded the region within the space of just a few weeks.

---

with average weather conditions and long-term weather patterns.

See also WEATHER SATELLITE.

meter (symbol: m) The fundamental SI UNIT of length. 1 meter = 3.281 feet. Also spelled metre (British spelling).

Metis A small (40-km-diameter) inner moon of Jupiter discovered in 1979 as a result of the *Voyager 1* encounter. Metis orbits Jupiter at a distance of 127,900 km with an orbital period of 0.295 day. Metis is the innermost known moon of Jupiter and together with Adrastea orbits within Jupiter's main ring. Also called Jupiter XVI and 1979 J3.

See also JUPITER.

Meton (of Athens) (ca. 460– ? B.C.E.) Greek *Astronomer* Meton discovered around 432 B.C.E. that a period of 235 lunar months coincides with precisely an interval of 19 years. After each 19-year interval, the phases of the Moon start taking place on the same days of the year. Both the ancient Greek and Jewish calendars used the Metonic cycle, and it became the main calendar of the ancient Mediterranean world until replaced by the Julian calendar in 46 B.C.E.

Metonic calendar (cycle) Named for the ancient astronomer METON (OF ATHENS), it is based on the Moon and counts each cycle of the phases of the Moon as one month (*one lunation*). After a period of 19 years, the lunations will occur on the same days of the year.

See also CALENDAR.

metric system The international system (SI) of weights and measures based on the meter as the fundamental unit of length, the kilogram as the fundamental unit of mass, and the second as the fundamental unit of time. Also called the mks system.

See also SI UNITS.

metrology The science of dimensional measurement; sometimes includes the science of weighing.

MeV An abbreviation for 1 million electron volts, a common energy unit encountered in the study of nuclear reactions. 1 MeV = 10^6 eV.

Michelson, Albert Abraham (1852–1931) German-American *Physicist* How fast does light travel? Physicists have pondered that challenging question since GALILEO GALILEI's time. In the 1880s Albert Michelson provided the first very accurate answer—a velocity just below 300,000 kilometers per second. He received the 1907 Nobel Prize in physics for his innovative optical and precision measurements.

Albert Michelson was born in Strelno, Prussia (now Strzelno, Poland), on December 19, 1852. When he was just two years old his family immigrated to the United States, settling first in Virginia City, Nevada, and then in San Francisco, California, where he received his early education. In 1869 Michelson received an appointment from President Grant to the U.S. Naval Academy in Annapolis, Maryland. More skilled as a scientist than as a seaman, he graduated from Annapolis in 1873 and received his commission as an ensign in the U.S. Navy. Following two years of sea duty, Michelson became an instructor in physics and chemistry at the academy. In 1879 he received an assignment to the Navy's Nautical Almanac Office in Washington, D.C. There he met and worked with Simon Newcomb, who among many other things was attempting to measure the velocity of light. After about a year at the Nautical Almanac Office, Michelson took a leave of absence to study advanced optics in Europe. For example, in Germany he visited the University of Berlin and the University of Heidelberg. He also studied at the College de France and the École Polytechnique in Paris. In 1881 he resigned from the U.S. Navy and then returned to the United States to accept an appointment as a professor of physics at the Case School in Applied Sciences in Cleveland, Ohio. Michelson had conducted some preliminary experiments in measuring the speed of light in 1881 at the University of Berlin. Once settled in at Case, he refined his experimental techniques and obtained a value of 299,853 kilometers per second. This value remained a standard within physics and astronomy for more than two decades and changed only when Michelson himself improved the value in the 1920s.

In the early 1880s, with financial support from the Scottish-American inventor Alexander Graham Bell (1847–1922), Michelson constructed a precision optical interferometer. An interferometer is an instrument that splits a beam of light into two paths and then reunites the two separate beams. Should either beam experience travel across different distances or at different velocities (due to passage through different media), the reunited beam will appear out of phase and produce a distinctive arrangement of dark and light bands, called an interference pattern. Using an interferometer in 1887, Michelson collaborated with the American physicist William Morley (1838–1923) to perform one of the most important "failed" experiments ever undertaken in science.

Now generally referred to as the "Michelson-Morley experiment," this important experiment used an interferometer to test whether light traveling in the same direction as Earth through space moves more slowly than light traveling at right angles to Earth's motion. Their "failure" to observe velocity differences in the perpendicular beams of light dispelled the prevailing concept that light traveled through the universe using some sort of invisible cosmic ether as the medium. Michelson had reasoned that when rejoined in the interferometer the two beams of light should be out of phase due to Earth's motion through the postulated ether. But their very careful measurements "failed" to detect any interfering influence of the hypothetical, all-pervading ether. Of course, the real reason their classic experiment "failed" is now very obvious—there is no cosmic ether. Michelson's work was exact and precise and provided a very correct, albeit null, result. The absence of ether gave ALBERT EINSTEIN important empirical evidence upon which he constructed special relativity theory in 1905. The Michelson-Morley experiment provided the first direct evidence that light travels at a constant speed in space, the very premise upon which all of special relativity is built.

From 1889 to 1892 Michelson served as a professor of physics at Clark University in Worcester, Massachusetts. He then left Clark University to accept a position in physics at the newly created University of Chicago. He remained affiliated with this university until his retirement in 1929. In 1907 Michelson became the first American scientist to receive the Nobel Prize in physics. He received this distinguished international award because of his excellence in the development and application of precision optical instruments.

During World War I he rejoined the U.S. Navy. Following this wartime service Michelson returned to the University of Chicago and turned his attention to astronomy. Using a sophisticated improvement of his optical interferometer, he measured the diameter of the star Betelgeuse in 1920. His achievement represented the first accurate determination of the size of a star, excluding the Sun. In 1923 he resumed his life-long quest to improve the measurement of the velocity of light. This time he used a 35-kilometer-long pathway between two mountain peaks in California that he had carefully surveyed to an accuracy of less than 2.5 centimeters. With a specially designed, revolving, eight-sided mirror, he measured the velocity of light as 299,798 kilometers per second. After retiring from the University of Chicago in 1929, Michelson joined the staff of the Mount Wilson Observatory in California. In the early 1930s he bounced light beams back and forth in an evacuated tube to produce an extended 16-kilometer pathway to measure optical velocity in a vacuum. The final result, announced in 1933 after his death, was a velocity of 299,794 kilometers per second. To appreciate the precision of Michelson's work, the currently accepted value of the velocity of light in a vacuum is 299,792.5 kilometers per second.

Michelson died in Pasadena, California, on May 9, 1931. His most notable scientific works include *Velocity of Light* (1902), *Light Waves and Their Uses* (1899–1903), and *Studies In Optics* (1927). In addition to being the first American scientist to win a Nobel Prize in physics, Michelson received numerous other awards and international recognition for his contributions to physics. He became a member in 1888 and served as president (1923–27) of the American National Academy of Sciences. The Royal Society of London made him a foreign fellow in 1902 and presented him its Copley Medal in 1907. He also received the Draper Medal from the American National Academy of Sciences in 1912, the Franklin Medal from the Franklin Institute in Philadelphia in 1923, and the Gold Medal of the Royal Astronomical Society in 1923. His greatest scientific accomplishment was the accurate measurement of the velocity of light, a physical "yardstick" of the universe and an important constant throughout all of modern physics.

micro- (symbol: μ) A prefix in the SI UNIT system meaning divided by 1 million; for example, a micrometer (mm) is 10^{-6} meter. The term also is used as a prefix to indicate something is very small, as in micrometeoroid and micromachine.

microgravity Because the inertial trajectory of a spacecraft (e.g., the space shuttle orbiter) compensates for the force of Earth's gravity, an orbiting spacecraft and all its contents approach a state of free fall. In this state of free fall all objects inside the spacecraft appear "weightless."

It is important to understand how this condition of weightlessness, or the apparent lack of gravity, develops. SIR ISAAC NEWTON's law of gravitation states that any two objects have a gravitational attraction for each other that is proportional to their masses and inversely proportional to the square of the distance between their centers of mass. It is also interesting to recognize that a spacecraft orbiting Earth at an altitude of 400 kilometers is only 6 percent farther away from the center of Earth than it would be if it were on Earth's surface. Using Newton's law, we find that the gravitational attraction at this particular altitude is only 12 percent less than the attraction of gravity at the surface of Earth. In other words, an Earth-orbiting spacecraft and all its contents are very much under the influence of Earth's gravity! The phenomenon of weightlessness occurs because the orbiting spacecraft and its contents are in a continual state of free fall.

The figure describes the different orbital paths a falling object may take when "dropped" from a point above Earth's sensible atmosphere. With no tangential-velocity component, an object would fall straight down (trajectory 1) in this simplified demonstration. As the object receives an increasing tangential-velocity component, it still "falls" toward Earth under the influence of terrestrial gravitational attraction, but the tangential-velocity component now gives the object a trajectory that is a segment of an ellipse. As shown in trajectories 2 and 3 in the figure, as the object receives a larger tangential velocity, the point where it finally hits Earth moves farther and farther away from the release point. If we keep increasing this velocity component, the object eventually "misses Earth" completely (trajectory 4). As the tangential velocity is increased further, the object's trajectory takes the form of a

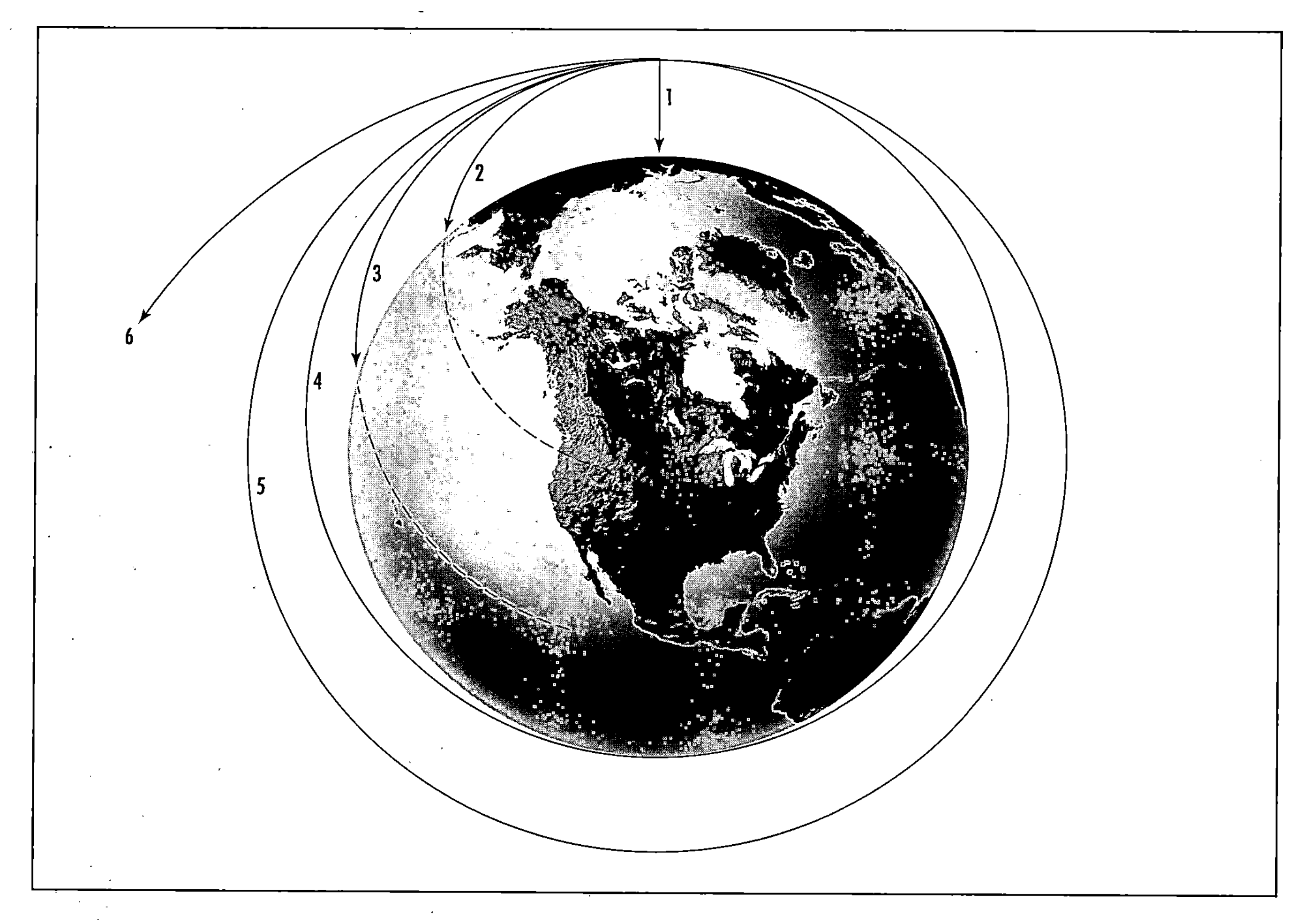

Various orbital paths of a falling body around Earth

circle (trajectory 5) and then a larger ellipse, with the release point representing the point of closest approach to Earth (or "perigee"). Finally, when the initial tangential-velocity component is about 41 percent greater than that needed to achieve a circular orbit, the object follows a parabolic, or escape, trajectory and will never return (trajectory 6).

ALBERT EINSTEIN's principle of equivalence states that the physical behavior inside a system in free fall is identical to that inside a system far removed from other matter that could exert a gravitational influence. Therefore, the terms *zero gravity* (also called "zero g") and *weightlessness* often are used to describe a free-falling system in orbit.

Sometimes people ask what is the difference between mass and weight. Why do we say, for example, "weightlessness" and not "masslessness"? Mass is the physical substance of an object—it has the same value everywhere. Weight, on the other hand, is the product of an object's mass and the local acceleration of gravity (in accordance with Newton's second law of motion, $F = ma$). For example, you would weigh about one-sixth as much on the Moon as on Earth, but your mass remains the same in both places.

A "zero-gravity" environment is really an ideal situation that can never be totally achieved in an orbiting spacecraft. The venting of gases from the space vehicle, the minute drag exerted by a very thin terrestrial atmosphere at orbital altitude, and even crew motions create nearly imperceptible forces on people and object alike. These tiny forces are collectively called "microgravity." In a microgravity environment astronauts and their equipment are almost, but not entirely, weightless.

Microgravity represents an intriguing experience for space travelers. However, life in microgravity is not necessarily easier than life on Earth. For example, the caloric (food-intake) requirements for people living in microgravity are the same as those on Earth. Living in microgravity also calls for special design technology. A beverage in an open container, for instance, will cling to the inner or outer walls and, if shaken, will leave the container as free-floating droplets or fluid globs. Such free-floating droplets are not merely an inconvenience. They can annoy crew members, and they represent a definite hazard to equipment, especially sensitive electronic devices and computers.

Therefore, water usually is served in microgravity through a specially designed dispenser unit that can be turned on or off by squeezing and releasing a trigger. Other beverages, such as orange juice, typically are served in sealed containers through which a plastic straw can be inserted. When the beverage is not being sipped, the straw is simply clamped shut.

Microgravity living also calls for special considerations in handling solid foods. Crumbly foods are provided only in bite-sized pieces to avoid having crumbs floating around the space cabin. Gravies, sauces, and dressings have a viscosity (stickiness) that generally prevents them from simply lifting off food trays and floating away. Typical space food trays are equipped with magnets, clamps, and double-adhesive tape to hold metal, plastic, and other utensils. Astronauts are provided with forks and spoons. However, they must learn to eat without sudden starts and stops if they expect the solid food to stay on their eating utensils.

Personal hygiene is also a bit challenging in microgravity. For example, space shuttle astronauts must take sponge baths rather than showers or regular baths. Because water adheres to the skin in microgravity, perspiration also can be annoying, especially during strenuous activities. Waste elimination in microgravity represents another challenging design problem. Special toilet facilities are needed that help keep an astronaut in place (i.e., prevent drifting). The waste products themselves are flushed away by a flow of air and a mechanical "chopper-type" device.

Sleeping in microgravity is another interesting experience. For example, shuttle and space station astronauts can sleep either horizontally or vertically while in orbit. Their fireproof sleeping bags attach to rigid padded boards for support, but the astronauts themselves quite literally sleep "floating in air."

Working in microgravity requires the use of special tools (e.g., torqueless wrenches), handholds, and foot restraints. These devices are needed to balance or neutralize reaction forces. If these devices were not available, an astronaut might find him- or herself helplessly rotating around a "work piece."

Exposure to microgravity also causes a variety of physiological (bodily) changes. For example, space travelers appear to have "smaller eyes" because their faces become puffy. They also get rosy cheeks and distended veins in their foreheads and necks. They may even be a little bit taller than they are on Earth, because their body masses no longer "weigh down" their spines. Leg muscles shrink, and anthropometric (measurable postural) changes also occur. Astronauts tend to move in a slight crouch, with head and arms forward.

Some space travelers suffer from a temporary condition resembling motion sickness. This condition is called space sickness or space adaptation syndrome. In addition, sinuses become congested, leading to a condition similar to a cold.

Many of these microgravity-induced physiological effects appear to be caused by fluid shifts from the lower to the upper portions of the body. So much fluid goes to the head that the brain may be fooled into thinking that the body has too much water. This can result in an increased production of urine.

Extended stays in microgravity tend to shrink the heart, decrease production of red blood cells, and increase production of white blood cells. A process called resorption occurs. This is the leaching of vital minerals and other chemicals (e.g., calcium, phosphorous, potassium, and nitrogen) from the bones and muscles into the body fluids that are then expelled as urine. Such mineral and chemical losses can have adverse physiological and psychological effects. In addition, prolonged exposure to a microgravity environment might cause bone loss and a reduced rate of bone tissue formation.

While a relatively brief stay (from seven to 70 days) in microgravity may prove a nondetrimental experience for most space travelers, long-duration (i.e., one to several years) missions such as a human expedition to Mars could require the use of artificial gravity (created through the slow rotation of the living modules of the spacecraft) to avoid any serious health effects that might arise from such prolonged exposure to a microgravity environment. While cruising to Mars, this artificial gravity environment also would help condition the astronauts for activities on the Martian surface, where they will once again experience the "tug" of a planet's gravity.

Besides providing an interesting new dimension for human experience, the microgravity environment of an orbiting space system offers the ability to create new and improved materials that cannot be made on Earth. Although microgravity can be simulated on Earth using drop towers, special airplane trajectories, and sounding rocket flights, these techniques are only short-duration simulations (lasting only seconds to minutes) that are frequently "contaminated" by vibrations and other undesirable effects. However, the long-term microgravity environment found in orbit provides an entirely new dimension for materials science research, life science research, and even manufacturing of specialized products.

See also DROP TOWER; MATERIALS PROCESSING IN SPACE; ZERO-GRAVITY AIRCRAFT.

micrometeoroids Tiny particles of meteoritic dust in space ranging in size from about 1 to 200 or more micrometers (microns) in diameter.

micrometer 1. An SI UNIT of length equal to one-millionth (10^{-6}) of a meter; also called a micron. 1 μm = 10^{-6} m. 2. An instrument or gauge for making very precise linear measurements (e.g., thicknesses and small diameters) in which the displacements measured correspond to the travel of a screw of accurately known pitch.

micron (symbol: μm) An SI UNIT of length equal to one-millionth (10^{-6}) of a meter. Also called a micrometer.

microorganism A tiny plant or animal, especially a protozoan or a bacterium.

See also EXTRATERRESTRIAL CONTAMINATION.

microsecond (symbol: μs) A unit of time equal to one-millionth (10^{-6}) of a second.

microwave (radiation) A comparatively short-wavelength electromagnetic (EM) wave in the radio-frequency portion of the EM spectrum. The term *microwave* usually is applied to those EM wavelengths that are measured in centimeters, approximately 30 centimeters to 1 millimeter (with corresponding frequencies of 1 gigahertz [GHz] to 300 GHz).

See also ELECTROMAGNETIC SPECTRUM.

***Midori* spacecraft** *See* ADVANCED EARTH OBSERVATION SATELLITE.

Mie scattering Any scattering produced by spherical particles without special regard to the comparative size of radiation wavelength and particle diameter.

milestone An important event or decision point in a program or plan. The term originates from the use of stone markers set up on roadsides to indicate the distance in miles to a given point. Milestone charts are used extensively in aerospace programs and planning activities.

military satellite (MILSAT) A satellite used for military or defense purposes such as missile surveillance, navigation, and intelligence gathering.

Milky Way galaxy Our home galaxy. The immense band of stars stretching across the night sky represents our "inside view" of the Milky Way. Classified as a spiral galaxy, the Milky Way is characterized by the following general features: a spherical central bulge at the galactic nucleus; a thin disk of stars, dust, and gas formed in a beautiful, extensive pattern of spiral arms; and a halo defined by an essentially spherical distribution of globular clusters. This disk is between 2,000 and 3,000 light-years thick and is some 100,000 light-years in diameter. It contains primarily younger, very luminous, metal-rich stars (called Population I stars) as well as gas and dust.

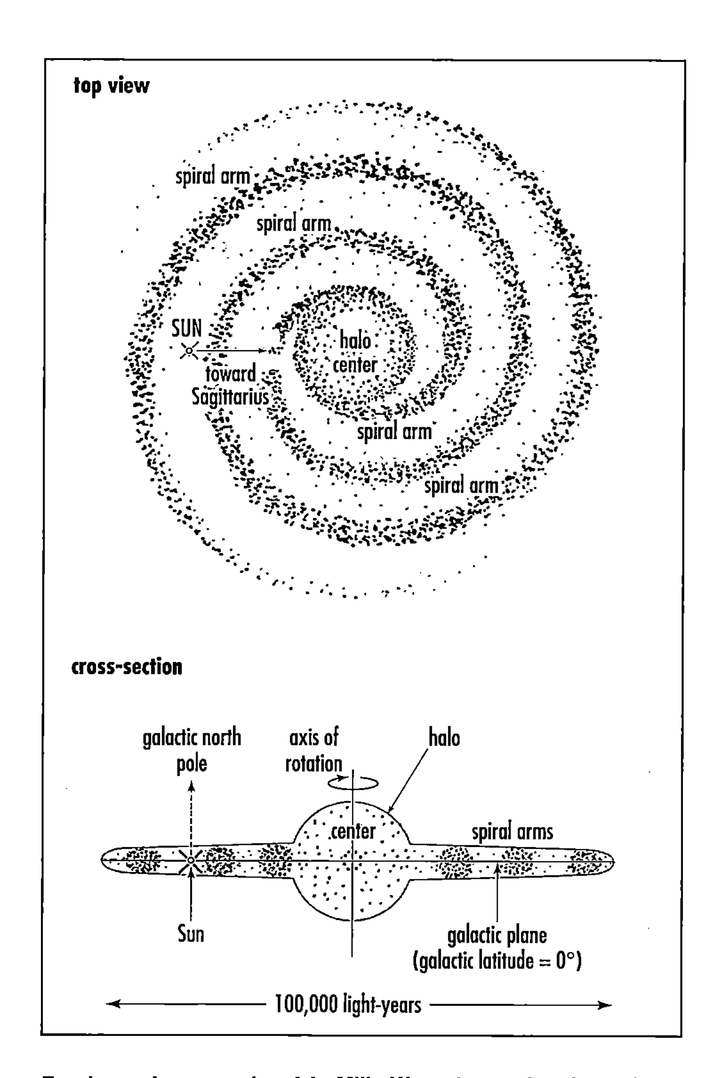

Top view and cross section of the Milky Way galaxy, a gigantic rotating disk of stars, gas, and dust about 100,000 light-years in diameter and about 3,000 light-years thick, with a bulge, or halo, around the center

Most of the stars found in the halo are older, metal-poor stars (called Population II stars), while the galactic nucleus appears to contain a mixed population of older and younger stars. Some astrophysicists now speculate that a massive black hole containing millions of "devoured" solar masses may lie at the center of many galaxies, including our own. Current estimates suggest that our galaxy contains between 200 and 600 billion solar masses. (A solar mass is a unit used for comparing stellar masses, with one solar mass being equal to the mass of our Sun.) Our solar system is located about 30,000 light-years from the center of the galaxy.

See also BLACK HOLE; GALAXY; STAR.

milli- **(symbol: m)** The SI unit system prefix meaning multiplied by 1/1000 (10^{-3}). For example, a millivolt (mV) is 0.001 volt, a millimeter (mm) is 0.001 meter, and a millisecond (msec) is 0.001 second.

millibar **(symbol: mbar or mb)** A unit of pressure equal to 0.001 bar (i.e., 10^{-3} bar) or 1,000 dynes per square centimeter. The millibar is used as a unit of measure of atmospheric pressure, with a standard atmosphere being equal to about 1,013 millibars, or 29.92 inches (760 mm) of mercury. 1 mbar = 100 newtons/m^2 = 1,000 dynes/cm^2.

See also BAR; DYNE.

millimeter **(symbol: mm)** One-thousandth (1/1000, 10^{-3}) of a meter. 1 mm = 0.001 m = 0.1 cm = 0.03937 in.

millimeter astronomy The branch of radio astronomy that studies electromagnetic radiation from cosmic sources (typically interstellar matter as found in giant molecular clouds) at wavelengths ranging from 1 to 10 millimeters.

See also SUBMILLIMETER ASTRONOMY.

millisecond **(symbol: msec or ms)** One-thousandth (1/1000, 10^{-3}) of a second; 1 msec = 0.001 sec = 1,000 msec.

Milne, Edward Arthur (1896–1950) English *Astrophysicist* Studied the dynamic processes and energy transfer mechanisms in stellar atmospheres.

Milstar The Milstar (Military Strategic and Tactical Relay) System is an advanced military COMMUNICATIONS SATELLITE that provides the Department of Defense (DOD), the National Command Authorities (NCA), and the armed forces of the United States worldwide assured, survivable communications. Designed to penetrate enemy jammers and overcome the disruptive effects of NUCLEAR DETONATIONS, the Milstar is the most robust and reliable satellite communications system ever deployed by the DOD. The program originated near the end of the COLD WAR with the primary objective of creating a secure, nuclear survivable, space-based communication system for the National Command Authorities. Today, in the age of information warfare, the system supports global high-priority defense communications needs with a constellation of improved satellites (Milstar II) that provide an exceptionally low probability of interception or detection by hostile forces.

The Milstar satellite has a MASS of approximately 4,540 kilograms and a design life of approximately 10 years. The operational Milstar satellite constellation consists of four spacecraft positioned around EARTH in GEOSYNCHRONOUS ORBITS. Each Milstar satellite serves as a smart switchboard in space directing message traffic from terminal to terminal anywhere on Earth. Since the satellite actually processes the communications signal and can link with other Milstar satellites through crosslinks, the requirement for ground controlled switching is significantly reduced. In fact, the satellite establishes, maintains, reconfigures, and disassembles required communications circuits as directed by the users. Milstar terminals provide encrypted voice, data, teletype, or facsimile communications. A key goal of the contemporary Milstar system is to provide interoperable communications among users of U.S. Army, Navy, and Air Force Milstar terminals. Timely and protected communications among different units of the American armed forces is essential if such combat forces are to bring about the rapid and successful conclusion of a high-intensity modern conflict.

The first Milstar satellite was launched on February 7, 1994, by a TITAN IV rocket from CAPE CANAVERAL AIR FORCE STATION (CCAFS). On January 15, 2001, another Titan IV EXPENDABLE LAUNCH VEHICLE successfully lifted off from CCAFS with a Milstar II communications satellite as its PAYLOAD. There are now three operational Milstar satellites in orbit; another satellite was lost due to BOOSTER rocket failure, and two additional satellites are in production.

Mimas An intermediate-sized moon of Saturn with a diameter of 390 km. Mimas orbits Saturn at a distance of 185,520 km and with a period of rotation of 0.942 days.

See also SATURN.

mini- An abbreviation for "miniature."

miniature homing vehicle **(MHV)** An air-launched, direct-ascent (i.e., "pop-up"), kinetic energy kill, antisatellite (ASAT) weapon.

Minkowski, Rudolph Leo (1895–1976) German-American *Astrophysicist* Used spectroscopy to examine nebulas (especially the Crab Nebula) and supernova remnants. In the mid-1950s he collaborated with WALTER BAADE in the optical identification of several extragalactic radio wave sources.

minor planet *See* ASTEROID.

minute 1. A unit of time equal to the 60th part of an hour; that is, 60 minutes = 1 hour. 2. A unit of angular measurement such that 60 minutes (60″) equal 1° of arc.

Minuteman **(LGM-30)** A three-stage solid-propellant ballistic missile that is guided to its target by an all-inertial guidance and control system. These strategic missiles are equipped with nuclear warheads and designed for deployment in hardened and dispersed underground silos. The LGM-30 Minuteman intercontinental ballistic missile (ICBM) is an element of the U.S. strategic deterrent force. The "L" in LGM stands for silo-configuration; "G" means surface attack; and "M" means guided missile.

The Minuteman weapon system was conceived in the late 1950s and deployed in the mid-1960s. Minuteman was a revolutionary concept and an extraordinary technical achievement. Both the missile and basing components incorporated significant advances beyond the relatively slow-reacting, liquid-fueled, remotely controlled ICBMs of the previous generation of missiles (such as the Atlas and the Titan). From the beginning, Minuteman missiles have provided a quick-reacting, inertially guided, highly survivable component of America's nuclear triad. Minuteman's maintenance concept capitalizes on high reliability and a "remove-and-replace" approach to achieve a near 100% alert rate.

Through state-of-the-art improvements, the Minuteman system has evolved over three decades to meet new challenges and assume new missions. Modernization programs have resulted in new versions of the missile, expanded targeting options, and significantly improved accuracy. For example, when the Minuteman I became operational in October 1962, it had a single-target capability. The Minuteman II became operational in October 1965. While looking similar to the Minuteman I, the Minuteman II had greater range and targeting capability. Finally, the Minuteman III became operational in June 1970. This missile, with its improved third stage and the postboost vehicle, can deliver multiple independently targetable reentry vehicles and their penetration aids onto multiple targets. More than 500 Minuteman III's are currently deployed at bases in the United States.

Miranda (Uranus V) A small moon of Uranus discovered on February 16, 1948, by the Dutch-American astronomer GERARD PETER KUIPER. Miranda has a diameter of 470 km and travels in synchronous orbit around Uranus at a distance of 129,800 km with a period of 1.414 days. The *Voyager 2* spacecraft encounter of January 1986 provided scientists incredibly detailed images of the icy moon with an unusual, cobbled together surface unlike any other in the solar system. Planetary scientists think the small moon was shattered in ancient times (possibly as many as five times), and Miranda appears to have reassembled itself each time from the shattered remnants.

mirror A surface capable of reflecting most of the LIGHT (that is, ELECTROMAGNETIC RADIATION in the visible portion of the SPECTRUM) falling upon it.

mirror matter A popular name for antimatter, which is the "mirror image" of ordinary matter. For example, an antielectron (also called a positron) has a positive charge, while an ordinary electron has a negative charge.

See also ANTIMATTER.

***Mir* space station** *Mir* ("peace" in Russian) was a third-generation Russian SPACE STATION of modular design that was assembled in ORBIT around a core MODULE launched in February 1986. Although used extensively by many COSMONAUTs and guest researchers (including American ASTRONAUTs), the massive station was eventually abandoned because of economics and the construction of the *INTERNATIONAL SPACE STATION* that started in 1998. *Mir* was safely deorbited into a remote area of the Pacific Ocean in March 2001.

missile Any object thrown, dropped, fired, launched, or otherwise projected with the purpose of striking a target. Short for ballistic missile or guided missile. *Missile* should not be used loosely as a synonym for rocket or launch vehicle.

Missile Defense Agency (MDA) An organization within the U.S. Department of Defense with the assigned mission to develop, test, and prepare for deployment a MISSILE defense system. MDA is integrating advanced interceptors, land-, sea-, air- and space-based SENSORs, and battle management command and control systems in order to design and demonstrate a layered missile defense system that can respond to and engage all classes and ranges of ballistic missile threats. The contemporary missile defense systems being developed and tested by MDA are primarily based on hit-to-kill vehicle technology and sophisticated, data-fusing battle management systems.

See also BALLISTIC MISSILE DEFENSE.

missilry The art and science of designing, developing, building, launching, directing, and sometimes guiding a rocket-propelled missile; any phase or aspect of this art or science. Also called missilery.

missing mass *See* DARK MATTER.

mission 1. In aerospace operations, the performance of a set of investigations or operations in space to achieve program goals. For example, the *Voyager 2* "Grand Tour" mission to the outer planets. 2. In military operations, a duty assigned to an individual or unit; a task. Also, the dispatching of one or more aircraft to accomplish a particular task. For example, a search-and-destroy mission.

mission control center (MCC) In general, the operational headquarters of a space mission. Specifically, the NASA facility responsible for providing total support for all phases of a crewed space flight (e.g., space shuttle flight)—prelaunch, ascent, on-orbit activities, reentry, and landing. It provides systems monitoring and contingency support, maintains communications with the crew and onboard systems, performs flight data collection, and coordinates flight operations. NASA's MCC is located at the Johnson Space Center in Houston, Texas.

mission specialist The space shuttle crew member (and career astronaut) responsible for coordinating payload–Space Transportation System (STS) interaction and, during the payload operation phase of a shuttle flight, directing the allocation of STS and crew resources to accomplish the combined payload objectives.

Mission to Planet Earth (MTPE) *See* EARTH-OBSERVING SPACECRAFT; EARTH SYSTEM SCIENCE.

Mitchell, Edgar Dean (1930–) American *Astronaut, U.S. Navy officer* Served as the lunar excursion module (LEM) pilot during the *Apollo 14* Moon landing mission (January–February 1971). Although this was his first NASA space mission, Mitchell immediately became a member of the exclusive "Moon walkers club" when he accompanied Alan B. Shepard as they explored the Fra Mauro region of the lunar surface.

Mitchell, Maria (1818–1889) American *Astronomer* Achieved international recognition for discovering a comet in 1847. As the first professionally recognized American woman astronomer, she worked as a "computer" for the Nautical Almanac Office of the U.S. government from about 1849 to 1868, carefully performing (by hand) precise calculations for the motions of the planet Venus. From 1865 until her death she served as a professor of astronomy at the newly opened Vassar College in Poughkeepsie, New York, and also directed the school's observatory.

mixture ratio In liquid-propellant rockets, the mass flow rate of oxidizer to the combustion chamber divided by the mass flow rate of fuel to the combustion chamber.

See also ROCKET.

MKS system The international system of units based on the meter (length), the kilogram (mass), and the second (time) as fundamental units of measure.

See also SI UNITS.

mobile launch platform (MLP) The structure on which the elements of the space shuttle are stacked in the Vehicle Assembly Building (VAB) and then moved to the launch pad (Complex 39) at the NASA Kennedy Space Center in Florida.

See also SPACE TRANSPORTATION SYSTEM.

mobility aids Handrails and footrails that help crew members move about an orbiting spacecraft or space station.

mock-up A full-size replica or dummy of something, such as a spacecraft, often made of some substitute material, such as wood, and sometimes incorporating actual functioning pieces of equipment, such as engines or power supplies. Mock-ups are used to study construction procedures, to examine equipment interfaces, and to train personnel.

modeling A scientific investigative technique that uses a mathematical or physical representation of a system or theory. This representation, or "model," accounts for all, or at least some, of the known properties of the system or the characteristics of the theory. Models are used frequently to test the effects of changes of system components on the overall performance of the system or the effects of variation of critical parameters on the behavior of the theory.

modulation The process of modifying a radio frequency (RF) signal by shifting its phase, frequency, or amplitude to carry information. The respective processes are called phase modulation (PM), frequency modulation (FM), and amplitude modulation (AM).

module 1. A self-contained unit of a launch vehicle or spacecraft that serves as a building block for the overall structure. It is common in aerospace practice to refer to the module by its primary function, for example, the Apollo spacecraft "command module" that was used in the Apollo Project to the Moon. 2. A pressurized, crewed laboratory suitable for conducting science, applications, and technology activities; for example, the spacelab module in the Space Transportation System. 3. A one-package assembly of functionally related parts, usually a "plug-in" unit arranged to function as a system or subsystem; a "black box."

mole (symbol: mol) The SI unit of the amount of a substance. It is defined as the amount of substance that contains as many elementary units as there are atoms in 0.012 kilograms of carbon-12, a quantity known as Avogadro's number (N_A). (The Avogadro number, N_A, has a value of about 6.022×10^{23} molecules/mole.) These elementary units may be atoms, molecules, ions, or radicals, and are specified. For example, often it is convenient to express the amount of an "ideal" gas in a given volume in terms of the number of moles (n). As previously stated, a mole of any substance is that mass of the substance that contains a specific number of atoms or molecules. Therefore, the number of moles of a substance can be related to its mass (m) (in grams) by the equation:

$$n = m/M_W$$

where M_W is the molecular weight of the substance (expressed as grams per mole, or g/mol). For the case of oxygen (O_2), which has a molecular weight (M_W) of 32 g/mol, the mass of 1 mole of oxygen is 32.0 grams; a half mole of oxygen would have a mass of 16.0 grams, and so on.

molecule A group of atoms held together by chemical forces. The atoms in the molecule may be identical, as in hydrogen (H_2), or different, as in water (H_2O) and carbon dioxide (CO_2). A molecule is the smallest unit of matter that can exist by itself and retain all its chemical properties.

Molniya launch vehicle A Russian launch vehicle that is descended from the first Russian intercontinental ballistic missile (ICBM). This vehicle, first used in 1961, consists of three cryogenic stages and four cryogenic strap-on motors. It can place a 1,590-kilogram payload into a highly elliptical (Molniya) orbit. This vehicle was used to launch communication satellites (of the same name, i.e., "Molniya" satellites), as well as many of the Russian interplanetary missions in the 1960s.

Molniya orbit A highly elliptical 12-hour orbit that places the apogee (about 40,000 kilometers) of a spacecraft over the Northern Hemisphere and the perigee (about 500 km) of the spacecraft over the Southern Hemisphere. Developed and used by the Russians for their communications satellites (called Molniya satellites). A satellite in a Molniya orbit spends the bulk of its time (i.e., apogee) above the horizon in view of the high northern latitudes and very little of its time (i.e., perigee) over southern latitudes.

Molniya satellite One of a family of Russian communications spacecraft that operate in highly elliptical orbits, called Molniya orbits, so that the spacecraft spends most of its time above the horizon in the Northern Hemisphere.

moment (symbol: M) A tendency to cause rotation about a point or axis, such as of a control surface about its hinge or of an airplane about its center of gravity; the measure of this tendency, equal to the product of the force and the perpendicular distance between the point or axis of rotation and the line of action of the force.

moment of inertia (symbol: I) For a massive body made up of many particles or "point masses" (m_i), the moment of inertia (I) about an axis is defined as the sum (Σ) of all the products formed by multiplying each point mass or particle (m_i) by the square of its distance $(r_i)^2$ from the line or axis of rotation; that is, $I = \Sigma_i\, m_i\, (r_i)^2$. The moment of inertia can be considered as the analog in rotational dynamics of mass in linear dynamics.

momentum (linear) The linear momentum (p) of a particle of mass (m) is given by the equation $p = m \cdot v$, where v is the particle's velocity. Newton's second law of motion states that the time rate of change of momentum of a particle is equal to the resultant force (F) on the particle, namely $F = dp/dt$.

monopropellant A liquid rocket propellant consisting of a single chemical substance (such as hydrazine) that decomposes exothermally and produces a heated exhaust jet without the use of a second chemical substance. Often used in attitude control systems on spacecraft and aerospace vehicles.

This photographic montage compares Earth's Moon on the same scale to the six medium-sized moons of Saturn. While our Moon (3,476-km diameter) consists of rock, the six satellites of Saturn shown here are composed predominantly of ice. They range in size from Mimas with a diameter of 390 km to Rhea with a diameter of 1,530 km. *(Courtesy of NASA; photograph of Earth's Moon courtesy of Lick Observatory)*

Moon The Moon is Earth's only natural satellite and closest celestial neighbor. (*See* figure.) While life on Earth is made possible by the Sun, it is also regulated by the periodic motions of the Moon. For example, the months of our year are measured by the regular motions of the Moon around Earth, and the tides rise and fall because of the gravitational tug-of-war between Earth and the Moon. Throughout history the Moon has had a significant influence on human culture, art, and literature. Even in the space age it has proved to be a major technical stimulus. It was just far enough away to represent a real technical challenge to reach it, yet it was close enough to allow success on the first concentrated effort. Starting in 1959 with the U.S. *Pioneer 4* and the Russian *Luna 1* lunar flyby missions, a variety of American and Russian missions have been sent to and around the Moon. The most exciting of these missions were the APOLLO PROJECT's human expeditions to the Moon from 1968 to 1972.

In 1994 the *Clementine* spacecraft, which was developed and flown by the Ballistic Missile Defense Organization of the U.S. Department of Defense as a demonstration of certain advanced space technologies, spent 70 days in lunar orbit mapping the Moon's surface. Subsequent analysis of the *Clementine* data offered tantalizing hints that water ice might be present in some of the permanently shadowed regions at the Moon's poles. NASA launched the *LUNAR PROSPECTOR* mission in January 1998 to perform a detailed study of the Moon's surface composition and to hunt for signs of the suspected deposits of water ice. Data from the *Lunar Prospector* mission strongly suggested the presence of water ice in the lunar polar regions, although the results require additional confirmation. Water on the Moon (as trapped surface ice in the permanently shadowed regions of the Moon's poles) would be an extremely valuable resource that would open up many exciting possibilities for future development of LUNAR BASES AND SETTLEMENTS.

From evidence gathered by the early uncrewed lunar missions (such as Ranger, Surveyor, and the LUNAR ORBITER spacecraft) and by the Apollo missions, lunar scientists have learned a great deal more about the Moon and have been able to construct a geological history dating back to its infancy. The table provides selected physical and dynamic properties of the Moon.

Because the Moon does not have any oceans or other free-flowing water and lacks a sensible atmosphere, appreciable erosion, or "weathering," has not occurred there. In fact, the Moon is actually a "museum world." The primitive materials that lay on its surface for billions of years are still in an excellent state of preservation. Scientists believe that the Moon was formed more than 4 billion years ago and then differentiated quite early, perhaps only 100 million years later. Tectonic activity ceased eons ago on the Moon. The lunar crust and mantle are quite thick, extending inward to more than 800 kilometers (480 miles). However, the deep interior of the Moon is still unknown. It may contain a small iron core at its center, and there is some evidence that the lunar interior may be hot and even partially molten. Moonquakes have been measured within the lithosphere and interior, most being the result of gravitational stresses. Chemically, Earth and the Moon are quite similar, although compared to Earth, the Moon is depleted of more easily vaporized materials. The lunar surface consists of highlands composed of alumina-rich rocks that formed from a globe-encircling molten sea and maria made up of volcanic melts that surfaced about 3.5 billion years ago. However, despite all we have learned in the past three decades about our nearest celestial neighbor, lunar exploration really has only just started. Several puzzling mysteries still remain, including the origin of the Moon itself.

Recently a new lunar origin theory has been suggested, a cataclysmic birth of the Moon. Scientists who support this theory suggest that near the end of Earth's accretion from the primordial solar nebula materials (i.e., after its core was formed, but while Earth was still in a molten state), a Mars-sized celestial object (called an "impactor") hit Earth at an oblique angle. This ancient explosive collision sent vaporized impactor and molten Earth material into Earth orbit, and the Moon then formed from these materials.

Physical and Astrophysical Properties of the Moon

Diameter (equatorial)	3,476 km
Mass	7.350×10^{22} kg
Mass (Earth's mass = 1.0)	0.0123
Average density	3.34 g/cm^3
Mean distance from Earth (center-to-center)	384,400 km
Surface gravity (equatorial)	1.62 m/sec^2
Escape velocity	2.38 km/sec
Orbital eccentricity (mean)	0.0549
Inclination of orbital plane (to ecliptic)	5° 09′
Sidereal month (rotation period)	27.322 days
Albedo (mean)	0.07
Mean visual magnitude (at full)	–12.7
Surface area	37.9×10^6 km^2
Volume	2.20×10^{10} km^3
Atmospheric density (at night on surface)	2×10^5 molecules/cm^3
Surface temperature	102 K–384 K

Source: NASA.

As previously mentioned, the surface of the Moon has two major regions with distinctive geological features and evolutionary histories. First are the relatively smooth, dark areas that Galileo originally called "maria" (because he thought they were seas or oceans). Second are the densely cratered, rugged highlands (uplands), which Galileo called "terrae." The highlands occupy about 83 percent of the Moon's surface and generally have a higher elevation (as much as 5 kilometers above the Moon's mean radius.) In other places the maria lie about 5 kilometers below the mean radius and are concentrated on the "near side" of the Moon (i.e., on the side of the Moon always facing Earth).

The main external geological process that modifies the surface of the Moon is meteoroid impact. Craters range in size from very tiny pits only micrometers in diameter to gigantic basins hundreds of kilometers across.

The surface of the Moon is strongly brecciated, or fragmented. This mantle of weakly coherent debris is called regolith. It consists of shocked fragments of rocks, minerals, and special pieces of glass formed by meteoroid impact. Regolith thickness is quite variable and depends on the age of the bedrock beneath and on the proximity of craters and their ejecta blankets. Generally, the maria are covered by 3 to 16 meters of regolith, while the older highlands have developed a "lunar soil" at least 10 meters thick.

moon A small natural body that orbits a larger one; a natural satellite.

Morgan, William Wilson (1906–1994) American *Astronomer* Morgan performed detailed investigations of large bluish-white stars (that is, O-type and B-type spectral classification) in the Milky Way galaxy and used these data in 1951 to infer the structure of the two nearest galactic spiral arms. Previous to that effort, Morgan collaborated with Philip Childs Keenan (1908–2002) at Yerkes Observatory and published the *Atlas of Stellar Spectra* in 1943. This work introduced the new, more sophisticated spectral classification system called the *Morgan-Keenan (MK) classification system.*

Morgan-Keenan (MK) classification A refined and more precise stellar classification system introduced by WILLIAM WILSON MORGAN and Philip Child Keenan (1908–2002) in 1943. This scheme is based on stellar spectra work performed at the Yerkes Observatory. Morgan and Keenan kept the basic Harvard classification system (namely the sequence of stellar spectral types: O, B, A, F, G, K, M) but then provided a more precise numerical division (0 to 9) of each spectral type. They also included a range of luminosity classes (I through V) to distinguish supergiant stars from giant stars, from main-sequence stars, and so on. This is the spectral classification system used in modern astronomy. Within the Morgan-Keen (MK) classification system the Sun, which was classified as a star of G spectral type in the Harvard system, now is called a star of G2V spectral type in the MK system.

See also STAR.

Morley, Edward Williams (1838–1923) American *Chemist* Morley collaborated with Albert Michelson in conducting the classic "ether-drift" experiment of 1887, which led to the collapse of the commonly held (but erroneous) concept that light needed this invisible ether medium to propagate through outer space.

morning star *See* VENUS.

mother spacecraft Exploration SPACECRAFT that carries and deploys one or several ATMOSPHERIC PROBES and ROVER or LANDER SPACECRAFT when arriving at a target PLANET. The mother spacecraft then relays its data back to EARTH and may orbit the planet to perform its own scientific mission. NASA's GALILEO PROJECT spacecraft to JUPITER and the CASSINI MISSION spacecraft to SATURN are examples.

Mount Wilson Observatory The observatory founded in 1904 on Mount Wilson in the San Gabriel Mountains near Los Angeles by GEORGE ELLERY HALE. The observatory was one of the original scientific enterprises of the Carnegie Institution of Washington, D.C. At an altitude of 1,742 meters, Hale originally selected this site to perform observations of the Sun. In 1917 the observatory opened its main nighttime astronomical instrument, a 2.5-meter (100-in) reflector telescope called the Hooker telescope after the American businessman John D. Hooker (1837–1910), who funded the telescope's large mirror. During the first half of the 20th century Mount Wilson was successively home to the world's two largest telescopes, as well as the most powerful solar observatories. Using these world-class facilities, scientists revolutionized astronomy, astrophysics, and cosmology.

For example, the American astronomer EDWIN POWELL HUBBLE made two of his most important discoveries using the Hooker telescope in the 1920s. The first was his discovery that nebulas were really distant galaxies. The second was Hubble's discovery that these galaxies were moving away from each other and that the entire universe was expanding.

Since 1986 the Mount Wilson Institute (MWI) has operated the observatory under an agreement with the Carnegie Institution. The addition of an adaptive optics system keeps the Hooker telescope in active service to the astronomical community. The observatory's two solar telescope towers continue to make daily data collections, maintaining a solar astronomy effort that represents the world's longest continuous observational record of the Sun.

M star A cool, reddish-colored star of spectral type M. M stars typically have surface temperatures of less than 3,500 K. Betelgeuse is an example.

muon (mu-meson) An elementary particle classed as a lepton with 207 times the mass of an electron. It may have a single positive or negative charge.

See also ELEMENTARY PARTICLE(s).

multiple independently targetable reentry vehicle (MIRV) A package of two or more reentry vehicles that can be carried by a single ballistic missile and guided to separate targets. MIRVed ballistic missiles use a warhead-dispensing mechanism called a postboost vehicle to target and release the reentry vehicles.

multispectral sensing A method of using many different bands of the electromagnetic spectrum (e.g., the visible, near-infrared, and thermal infrared bands) to sense a target. When several bands are used to sense a target, deceptive measures become much less effective.

See also REMOTE SENSING.

multistage rocket A vehicle having two or more rocket units, each firing after the one in back of it has exhausted its propellant. Normally, each unit, or stage, is jettisoned after completing its firing. Also called a multiple-stage rocket or, infrequently, a step rocket.

See also ROCKET.

mutual assured destruction (MAD) The strategic situation that existed during the cold war in which either superpower (i.e., either the United States or the Soviet Union) could inflict massive nuclear destruction on the other, no matter which side struck first.

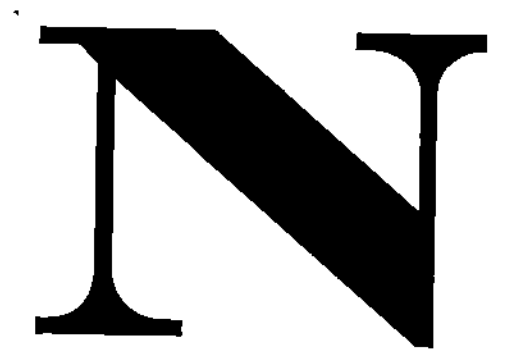

nadir 1. The direction from a spacecraft directly down toward the center of a planet. Opposite of the zenith. 2. That point on the celestial sphere directly beneath the observer and directly opposite the zenith.

Naiad A small irregularly shaped moon of Neptune discovered in August 1989 as a result of the *Voyager 2* encounter. Naiad has a diameter of approximately 60 km and orbits Neptune at a distance of 48,230 km with a period of 0.294 day. Also called Neptune III and S/1989 N6.

See also NEPTUNE.

naked eye The normal human eye unaided by any optical instrument, such as a telescope. The use of corrective lenses (glasses) or contact lenses that restore an individual's normal vision are included in the concept of naked eye observing.

naked eye astronomy The practice of astronomy without the assistance of an optical telescope. All astronomical observations before the early 17th century were "naked eye" studies of the heavens. However, naked eye astronomers often used ancient astronomical instruments, such as the astrolabe, to locate objects and track celestial objects. The Danish astronomer TYCHO BRAHE is considered the last and greatest naked eye astronomer.

See also ANCIENT ASTRONOMICAL INSTRUMENTS.

nano- (symbol: n) A prefix in the SI unit system meaning multiplied by 10^{-9}.

nanometer (nm) A billionth of a meter (i.e., 10^{-9} meter).

nanosecond (ns) A billionth of a second (i.e., 10^{-9} second).

National Aeronautics and Space Administration (NASA) The civilian space agency of the United States, created in 1958 by an act of Congress (the National Aeronautics and Space Act of 1958). NASA belongs to the executive branch of the federal government. Its overall mission is to plan, direct, and conduct civilian (including scientific) aeronautical and space activities for peaceful purposes. This mission is carried out at NASA headquarters in Washington, D.C., supported by centers and field facilities throughout the United States. Current NASA centers and facilities include the Ames Research Center (Moffett Field, California), the Dryden Flight Research Center (Edwards AFB, California), the Glenn Research Center (Lewis Field, Cleveland, Ohio), the Goddard Institute for Space Studies (New York City, New York), the Goddard Space Flight Center (Greenbelt, Maryland), the Independent Verification and Validation Facility (Fairmont, West Virginia), the contractor-operated Jet Propulsion Laboratory (Pasadena, California), the Johnson Space Center (Houston, Texas), the Kennedy Space Center (Cape Canaveral, Florida), the Langley Research Center (Hampton, Virginia), the Marshall Space Flight Center (Huntsville, Alabama), the Stennis Space Center (Mississippi), the Wallops Flight Facility (Wallops Island, Virginia), and the White Sands Test Facility (White Sands, New Mexico).

NASA headquarters in Washington, D.C., exercises management over the space flight centers, research centers, and other installations that make up the civilian space agency. Responsibilities include general determination of programs and projects, establishment of policies and procedures, and evaluation and review of all phases of the NASA aerospace program. NASA's human space flight program placed astronauts on the Moon during the Apollo Project and currently involves the Space Transportation System (space shuttle) and the *International Space Station.* Contemporary NASA programs dedicated to the exploration of the universe include the *Hubble Space Telescope,* the *Chandra X-Ray Observatory,* and the *Spitzer Space Telescope.* NASA's scientific investigation of the solar system continues with such projects as the Cassini mission to Saturn, the *Mars Odyssey 2001* spacecraft, the Mars Exploration Rover (MER) 2003 mission, the *Mars Reconnaissance Orbiter,* and the *Messenger* spacecraft.

For more than four decades NASA has also sponsored Earth monitoring from space with the *Aqua, Aura,* and *Terra* spacecraft now providing leading-edge collections of environmental data in support of Earth system science activities. Since its inception NASA has also supported many important space technology applications including the development of weather satellites, communications satellites, and the civilian application of multispectral sensing from satellites, such as Landsat.

There are several pressing issues facing NASA. One of the most critical current (2005) problems involves the return of the space shuttle to a safe and reliable flight status. The shuttle fleet (*Atlantis, Discovery,* and *Endeavour*) was grounded after the *Columbia* accident on February 1, 2003. Until the space shuttle can once again perform orbital missions, all logistics for the *International Space Station* and the rotation of astronaut and cosmonaut crews depends exclusively on the use of Russian launch vehicles. NASA must also address the very pressing issue of how to best replace the aging Space Transportation System (i.e., space shuttle fleet). An advanced rocket propulsion system, capable of safely and more inexpensively carrying astronauts and payloads into orbit, is the key to the efficient long-term operation of the *International Space Station,* as well as future human space flight missions—such as a return to the Moon and an expedition to Mars. Furthermore, NASA officials must also examine how to perform another servicing mission to the *Hubble Space Telescope* sometime before the year 2007. On a slightly more distant planning horizon, NASA must construct and successfully launch the *James Webb Space Telescope* in 2011 as a replacement for the *Hubble Space Telescope.* Finally, senior NASA planners are attempting to respond to President George W. Bush's long-term vision for the nation's space program, a vision he announced on January 14, 2004. In that vision President Bush committed the United States to a long-term human and robotic program to explore the solar system, starting with a return to the Moon that will ultimately enable future exploration of Mars and other destinations.

National Air and Space Museum (NASM) The National Air and Space Museum of the Smithsonian Institution maintains the largest collection of historic aircraft and spacecraft in the world. Located on the National Mall in Washington, D.C., the museum offers millions of visitors each year hundreds of professionally displayed space exploration and rocketry artifacts. Highlighting the exhibits is the *Apollo 11* command module and a lunar rock that guests can actually touch. Other displays include astronaut JOHN H. GLENN, JR.'s *Friendship 7* space capsule, a *Corona KH-4B* reconnaissance satellite camera and film-return capsule, a U.S. Air Force and Minuteman III ICBM, a V-2 rocket, rockets developed by SIR WILLIAM CONGREVE and ROBERT H. GODDARD, and a Russian SS-20 ICBM. Spacecraft displays (accurate replicas) include *Pioneer 10, Voyager 2,* and *Sputnik 1.*

National Oceanic and Atmospheric Administration (NOAA) In 1970 the National Oceanic and Atmospheric Administration (NOAA) was established as an agency within the U.S. Department of Commerce (DOC) to ensure the safety of the general public from atmospheric phenomena and to provide the public with an understanding of Earth's environment and resources. NOAA conducts research and gathers data about the global oceans, the atmosphere, outer space, and the Sun through five major organizations: the National Weather Service, the National Ocean Service, the National Marine Fisheries Service, the National Environmental Satellite, Data, and Information Service, and NOAA Research. NOAA research and operational activities are supported by the seventh uniformed service of the U.S. government. The NOAA Corps contains the commissioned personnel who operate NOAA's ships and fly its aircraft.

The National Environmental Satellite, Data, and Information Service (NESDIS) is responsible for the daily operation of American environmental satellites, such as the Geosynchronous Operational Environmental Satellite (GOES). The prime customer for environmental satellite data is the National Weather Service, but NOAA also distributes these data to many other users within and outside the government. NOAA's operational environmental satellite system is composed of geostationary operational environmental satellites for short-range warning and "nowcasting" and polar orbiting environmental satellites (POES) for long-term forecasting. Both types of satellites are necessary to provide a complete global weather monitoring system. The satellites also carry search and rescue (SAR) capabilities for locating people lost in remote regions on land (including victims of aircraft crashes) or stranded at sea as a result of maritime disasters and accidents.

As a result of a presidential directive in May 1994, NESDIS now operates the spacecraft in the Department of Defense's Defense Meteorological Satellite Program (DMSP). The executive order combined the U.S. military and civilian operational meteorological satellite systems into a single national system capable of satisfying both civil and national security requirements for space-based remotely sensed environmental data. As part of this merger, a triagency (NOAA, NASA, and DOD) effort is underway to develop and deploy the National Polar-Orbiting Operational Environmental Satellite System (NPOESS) starting in about 2010. In addition to operating satellites, NESDIS also manages global databases for meteorology, oceanography, solid earth geophysics, and solar-terrestrial sciences.

See also EARTH SYSTEM SCIENCE; GLOBAL CHANGE; WEATHER SATELLITE.

National Polar-Orbiting Operational Environmental Satellite System (NPOESS) The planned U.S. system of advanced polar-orbiting environmental satellites that converges existing American polar-orbiting meteorological satellite systems—namely the Defense Meteorological Satellite Program (DMSP) and the Department of Commerce's Polar-Orbiting Environmental Satellite (POES)—into a single national program. Responding to a 1994 presidential directive, a triagency team of Department of Defense, Department of Commerce (NOAA), and NASA personnel are currently developing this new system of polar-orbiting environmental satellites. When deployed (starting in about 2010), NPOESS will continue to monitor global environmental conditions and to collect and disseminate data related to weather, the atmosphere, the oceans, various land masses, and the near-Earth space environment. The global and regional environmental imagery and specialized environmental data from NPOESS will support the peacetime and wartime missions of the Department of Defense as well as civil mission requirements of organizations such as the National Weather Service within the National Oceanic and Atmospheric Administration (NOAA). In particular, NPOESS will use instruments that sense surface and atmospheric radiation in the visible, infrared, and microwave bands of the electromagnetic spectrum, monitor important parameters of the space environment, and measure distinct environmental parameters such as soil moisture, cloud levels, sea ice, and ionospheric scintillation.

See also WEATHER SATELLITE.

National Radio Astronomy Observatory (NRAO) A collection of government-owned radio astronomy facilities throughout the United States, including the radio telescope in Green Bank, West Virginia.

See also GREEN BANK TELESCOPE (GBT); RADIO ASTRONOMY.

National Reconnaissance Office (NRO) The National Reconnaissance Office (NRO) is the national program to meet the needs of the U.S. government through spaceborne reconnaissance. The NRO is an agency of the Department of Defense (DOD) and receives its budget through that portion of the National Foreign Intelligence Program (NFIP) known as the National Reconnaissance Program (NRP), which is approved both by the secretary of defense and the director of central intelligence (DCI). The existence of the NRO was declassified by the deputy secretary of defense, as recommended by the DCI, on September 18, 1992.

The mission of the NRO is to ensure that the United States has the technology and spaceborne assets needed to acquire intelligence worldwide. This mission is accomplished through research, development, acquisition, and operation of the nation's intelligence satellites. The NRO's assets collect intelligence to support such functions as intelligence and warning, monitoring of arms control agreements, military operations and exercises, and monitoring of natural disasters and environmental issues.

The director of the NRO is appointed by the president and confirmed by the Congress as the assistant secretary of the air force for space. The secretary of defense has the responsibility, which is exercised in concert with the DCI, for the management and operation of the NRO. The DCI establishes the collection priorities and requirements for the collection of satellite data. The NRO is staffed by personnel from the Central Intelligence Agency (CIA), the military services, and civilian DOD personnel.

The Corona program was the first satellite photo reconnaissance program developed by the United States. Corona was a product of the cold war. Soviet nuclear weapons coupled with the long-range bomber and missile delivery systems created the possibility of a devastating surprise attack—a "nuclear Pearl Harbor"—and President DWIGHT D. EISENHOWER demanded better intelligence. But the Iron Curtain and the sheer size of the Soviet landmass made traditional intelligence collection methods marginally useful, at best. The U-2 reconnaissance aircraft served admirably for four years collecting data over the Soviet Union, but its effectiveness was known to be limited by advancements in Soviet air defenses. That, coupled with the urgency for a national space program stimulated by *SPUTNIK 1* (the first satellite launched by the Soviet Union on October 4, 1957), gave birth to the Corona satellite program.

The U.S. Air Force, under a program with the innocuous designation Weapon System 117L Advanced Reconnaissance System, had been developing a family of satellites since 1956 in connection with development of the ballistic missile, which would serve as the satellite launch vehicle. But priority for the satellite program had been low, and funding was limited.

Following *Sputnik 1* all that changed. The most promising near-term project was split out from Weapon System 117L in early 1958 and named Corona. The program was provided increased priority and funding and designated for streamlined management under a joint USAF/CIA team. That team was told to take a space booster, a spacecraft, a reentry vehicle, a camera, photographic film, and a control network, all untested, and to make them work together as a system quickly. However, strong leadership, coupled with unwavering White House support, would be needed (and provided) to take this pioneering reconnaissance satellite program through many early setbacks, and there were numerous setbacks. All but one of the first 13 Corona launches (numbered 0–12) resulted in failure of one type or another, but each failure was investigated, identified, and corrected as the team moved closer to an operational photo reconnaissance capability.

Then, on August 10, 1960, *Corona XIII* was launched with diagnostic equipment rather than an actual cameral payload. A day later the film capsule reentered and was recovered in the Pacific Ocean. On August 18 *Corona XIV*, with a camera and film payload, was launched in a perfect mission ending with a successful midair capsule recovery. The film retrieved from *Corona XIV* was processed quickly and exploited by the intelligence community. This film capsule provided more coverage than all previous U-2 flights over the Soviet Union combined. The United States now had the space-based "eyes" so necessary to protect its people and keep the peace throughout the cold war. The 145th and final Corona launch took place on May 15, 1972.

From a technical perspective Corona was the first space program to recover an object from orbit and the first to deliver photo reconnaissance information from a satellite. It would go on to be the first program to use multiple reentry vehicles, pass the 100 mission mark, and produce stereoscopic space imagery. Its most remarkable technological advance, however, was the improvement in its ground resolution from an initial 7.6- to 12.2-meter capability to an ultimate 1.82-meter resolution.

See also DISCOVERER SPACECRAFT.

National Solar Observatory (NSO) The mission of the U.S. National Solar Observatory (NSO) is to advance knowledge of the Sun, both as an astronomical object and as the dominant external influence on Earth, by providing forefront observational opportunities to the research community. The mission includes the operation of cutting-edge facilities, the continued development of advanced instrumentation both in-house and through partnerships, conducting solar research, and educational and public outreach. The National Solar Observatory has two observing facility locations: Sacramento Peak, New Mexico, and Kitt Peak, Arizona. The 0.76-m Dunn Solar Telescope, located on Sacramento Peak at an altitude of 2,804 meters, is a premier facility for high-resolution solar physics. The evacuated light path eliminates the loss of image clarity due to distortions from the air. NSO has pioneered solar adaptive optics and high-resolution, ground-based solar physics. The McMath-Pierce Solar Telescope on Kitt Peak, at an altitude of 2,096 meters, is currently the largest unobstructed-aperture optical telescope in the world, with a diameter of 1.5 meters. It is capable of panchromatic, flux-limiting studies of the Sun. In particular, it is the only solar telescope in the world on which investigations in the relatively unexplored infrared domain beyond 2.5 microns are routinely accomplished.

See also KITT PEAK NATIONAL OBSERVATORY (KPNO); SOLAR TELESCOPE; SUN.

National Space Development Agency of Japan (NASDA) Up until the formation of the Japanese Aerospace Exploration Agency (JAXA) on October 1, 2003, Japan's large rocket vehicle and space development activities were primarily implemented by the National Space Development Agency. On that date, the government merged the three agencies that had previously promoted aeronautical and space activities within Japan, namely, the Institute of Space and Astronautical Science (ISAS), the National Aerospace Laboratory (NAL), and the National Space Development Agency of Japan (NASDA). Save for the agency name change, many of NASDA's functions and facilities (as described below) remain the same.

NASDA was established on October 1, 1969, and the organization's activities are limited to the peaceful uses of outer space. The agency is primarily engaged in research and development involving satellites and launch vehicles for practical uses, launch and tracking operations for Japanese satellites, and promoting the development of civilian remote sensing technologies and materials processing in space (MPS).

NASDA's H-II rocket is the centerpiece of Japan's space program. It is the first fully domestically manufactured Japanese rocket vehicle and has been designed to send 2,000-kilogram-class missions to geostationary orbit. The first stage of this modern expendable launch vehicle is a high-performance liquid oxygen–liquid hydrogen engine called the LE-7. The vehicle's second stage uses the LE-5A liquid hydrogen–liquid oxygen engine. It is a reignitable engine that offers higher performance and reliability. H-II vehicle Number 1 was successfully launched on February 4, 1994, but vehicle Number 8 suffered a launch failure on November 15, 1999, providing a checkered pattern of success and failure that testifies to the difficulties of developing truly advanced high-performance rocket systems using cryogenic propellants. The H-II also uses two strap-on solid-propellant booster rockets.

The main JAXA (formerly NASDA) facilities include the Tsukuba Space Center (Ibaraki), Kakuda Propulsion Center, and Tanegashima Space Center. The Tanegashima Space Center is located on the southeastern portion of Tanegashima Island, Kagoshima. This center includes the Takesake Range for small rockets, the Osaki Range for the H-II launch vehicle, and the Yoshinobu Launch Complex for the H-II expendable vehicle.

JAXA is participating in the *International Space Station* (ISS) and has also sponsored Japanese astronauts on flights onboard the U.S. space shuttle. For example, astronaut Takao Doi served as a mission specialist during the STS-87 mission (November 20, 1997, through December 5, 1997) and successfully performed two extra vehicular activities (EVAs).

See also H-II LAUNCH VEHICLE; JAPANESE AEROSPACE EXPLORATION AGENCY (JAXA).

NATO III A constellation of military communications satellites launched from Cape Canaveral Air Force Station, Florida, between April 1976 and November 1984. These satellites, designated A, B, C, and D, have provided communications for North Atlantic Treaty Organization (NATO) officials in Belgium, Canada, Denmark, the United Kingdom (England), Germany, as well as Greece, Iceland, and Italy. The United States, the Netherlands, Norway, Portugal, and Turkey also have used the satellite communications system. NATO III simultaneously accommodated hundreds of NATO users and provided voice and facsimile services. The 2.13-meter-diameter, 2.74-meter-high cylinder-shaped, spinning spacecraft are located in geostationary orbit. Each satellite has three "horn" antennas mounted on a platform that spines in the opposite direction of the body of the spacecraft, enabling the antennas to point at the same place on Earth constantly.

See also COMMUNICATIONS SATELLITE.

natural satellite A moon in orbit around a planet (such as Phobos around Mars), as opposed to a human-made spacecraft in orbit around the planet (such as the *Mars Odyssey 2001* spacecraft in orbit around Mars); also a small asteroid in orbit around a larger (parent) asteroid (such as Dactyl around Ida). The German astronomer JOHANNES KEPLER introduced the use of the word *satellite* after observing the Galilean moons of Jupiter through a telescope in the early 17th century.

navigation The navigation of a spacecraft has two main aspects. The first is *orbit determination*—a task involving the knowledge and prediction of the spacecraft's position and velocity. The second aspect of spacecraft navigation is *flight path control*—a task involving firing a spacecraft's onboard propulsion systems (such as a retrorocket motor or tiny attitude control rockets) to alter the spacecraft's velocity.

Navigating a robot spacecraft in deep space is a challenging operation. For example, no single measurement directly provides mission controllers information about the lateral motion of a spacecraft as it travels on a mission deep in the solar system. Aerospace engineers define *lateral motion* as any motion except motion directly toward or way from Earth (which is called *radial motion*). Spacecraft flight controllers use the measurements of the Doppler shift of telemetry (particularly a coherent downlink carrier) to obtain the radial component of a spacecraft's velocity relative to Earth. Spacecraft controllers add a uniquely coded ranging pulse to an uplink communication with a spacecraft and record the transmission time. When the spacecraft receives this special ranging pulse, it returns a similarly coded pulse on its downlink transmission. Engineers know how long it takes the spacecraft's onboard electronics to "turn" the ranging pulse around (for example, the *Cassini* spacecraft takes 420 nanoseconds [ns] ± 9 ns). There are other known and measured (calibrated) delays in the overall transmission process, so when the return pulse is received back on Earth—at NASA's Deep Space Network, for example—spacecraft controllers can then calculate how far (radial distance) the spacecraft is away from Earth. Mission controllers also use angular quantities to express a spacecraft's position in the sky.

Spacecraft that carry imaging instruments can use them to perform *optical navigation.* For example, they can observe the destination (target) planet or moon against a known background starfield. Mission controllers will often carefully plan and uplink appropriate optical navigation images as part of a planetary encounter command sequence. When the spacecraft collects optical navigation images, it immediately downlinks (transmits) these images to the human navigation team at mission control. The mission controllers immediately process the optical imagery and use these data to obtain precise information about the spacecraft's trajectory as it approaches its celestial target.

Once a spacecraft's solar or planetary orbital parameters are known, these data are compared to the planned mission data. If there are discrepancies, mission controllers plan for and then have the spacecraft execute an appropriate *trajectory correction maneuver* (TCM). Similarly, small changes in a spacecraft's orbit around a planet may become necessary to support the scientific mission. In that case the mission controllers plan for and instruct the spacecraft to execute an appropriate *orbit trim maneuver* (OTM). This generally involves having the spacecraft fire some of its small-thrust, attitude-control rockets. Trajectory correction and orbit trim maneuvers use up a spacecraft's onboard propellant supply, which is often a very carefully managed, mission-limiting consumable.

navigation satellite A spacecraft placed into a well-known, stable orbit around earth that transmits precisely timed radio wave signals useful in determining locations on land, at sea, or in the air. Such satellites are deployed as part of an interactive constellation.

See also GLOBAL POSITIONING SYSTEM; TRANSIT (SPACECRAFT).

Navstar The Navstar Global Positioning System is a constellation of advanced radio navigation satellites developed for the U.S. Department of Defense and operated by the U.S. Air Force. In addition to serving defense navigation needs, this system also supports many civilian applications.

See also GLOBAL POSITIONING SYSTEM (GPS).

n-body problem In celestial mechanics, the challenging mathematical problem involving the determination of the trajectories on *n*-objects (treated as point masses) under the assumption that the only interaction between these n-point masses is gravitational attraction. Within classical physics, the starting basis for such calculations is Sir ISAAC NEWTON's law of gravitation. Mathematical astronomers have achieved solutions for the two-body problem. They have also obtained restricted solutions (for certain special cases) of the three-body problem. One example of a restricted three-body problem involves determining the perturbing influence of the planet Jupiter on the orbital motion of a main-belt asteroid. Mathematical astronomers regard a determination of the orbits of stars within a cluster as an n-body problem that defies analytical solution.

near-Earth asteroid (NEA) An inner solar system asteroid whose orbit around the Sun brings it close to Earth, perhaps even posing a collision threat in the future.

See also ASTEROID; ASTEROID DEFENSE SYSTEM; EARTH-CROSSING ASTEROID.

Near Earth Asteroid Rendezvous mission (NEAR) NASA's Near Earth Asteroid Rendezvous (NEAR) mission was launched on February 17, 1996, from Cape Canaveral Air Force Station by a Delta II expendable launch vehicle. It was the first of NASA's Discovery missions, a series of small-scale spacecraft designed to proceed from development to flight in less than three years for a cost of less than $150 million. The NEAR spacecraft was equipped with an X-ray/gamma-ray spectrometer, a near-infrared imaging spectrograph, a multispectral camera fitted with a charge-coupled device (CCD) imaging detector, a laser altimeter, and a magnetometer. The primary goal of this mission was to rendezvous with and achieve orbit around the near-earth asteroid Eros-433 (also referred to as 433 Eros and sometimes just simply as Eros).

Eros-433 is an irregularly shaped S-class asteroid about 13 by 13 by 33 km in size. This asteroid, the first-near Earth asteroid to be found, was discovered on August 13, 1898, by the German astronomer Gustav Witt (1866–1946). In Greek mythology Eros (Roman name: Cupid) was the son of Hermes (Roman name: Mercury) and Aphrodite (Roman name: Venus) and served as the god of love.

As a member of the Amor group of asteroids, Eros has an orbit that crosses the orbital path of Mars but does not intersect the orbital path of Earth around the Sun. The asteroid follows a slightly elliptical orbit, circling the Sun in 1.76 years at an inclination of 10.8° to the ecliptic. Eros-433 has a perihelion of 1.13 astronomical units (AU) and an aphelion of 1.78 AU. The closest approach of Eros to Earth in the 20th century occurred on January 23, 1975, when the asteroid came within 0.15 AU (about 22 million km) of our home planet.

After launch and departure from Earth orbit, NEAR entered the first part of its cruise phase. It spent most of this phase in a minimal activity (hibernation) state that ended a few days before the successful flyby of the asteroid Mathilde-253 on June 27, 1997. During that encounter the spacecraft flew within 1,200 km of Mathilde-253 at a relative velocity of 9.93 km/s. Imagery and other scientific data were collected.

On July 3, 1997, the NEAR spacecraft executed its first major deep space maneuver, a two-part propulsive burn of its main 450-newton (hydrazine/nitrogen tetroxide-fueled) thruster. This maneuver successfully decreased the spacecraft's velocity by 279 meters per second (m/s) and lowered perihelion from 0.99 AU to 0.95 AU. Then on January 23, 1998, the spacecraft performed an Earth gravity assist flyby—a critical maneuver that altered its orbital inclination from 0.5° to 10.2° and its aphelion distance from 2.17 AU to 1.77 AU. This gravity assist maneuver gave the NEAR spacecraft orbital parameters that nearly matched those of the target asteroid, Eros-433.

The original mission plan was to rendezvous with and achieve orbit around Eros-433 in January 1999 and then to study the asteroid for approximately one year. However, a software problem caused an abort of the first encounter burn, and NASA revised the mission plan to include a flyby of Eros-433 on December 23, 1998, and then an encounter and orbit on February 14, 2000. The radius of the spacecraft's orbit around Eros-433 was brought down in stages to a 50-by-50-km orbit on April 30, 2000, and decreased to a 35-by-35-km orbit on July 14, 2000. The orbit was then raised over the succeeding months to 200 by 200 km and next slowly decreased and altered to a 35-by-35-km retrograde orbit around the asteroid on December 13, 2000. The mission ended with a touchdown in the "saddle" region of Eros on February 12, 2001. NASA renamed the spacecraft NEAR-*Shoemaker* in honor of American astronomer and geologist Eugene M. Shoemaker (1928–97) following his untimely death in an automobile accident on July 18, 1997.

See also EROS.

near infrared That portion of the electromagnetic spectrum involving the shorter wavelengths in the infrared region; generally considered to extend from just beyond the visible red portion of the spectrum (about 0.7 micrometer wavelength) out to about 3.0 micrometers wavelength.

See also ELECTROMAGNETIC SPECTRUM; INFRARED (IR) RADIATION.

nearside The side of the Moon that always faces Earth.

See also MOON.

nebula (plural: nebulas or nebulae) A cloud of interstellar gas or dust. It can be seen as either a dark hole against a brighter background (called a *dark nebula*) or as a luminous patch of light (called a *bright nebula*).

nebula hypothesis The hypothesis that the Sun and planets formed from a large rotating cloud of interstellar dust and gas (a primordial nebula) due to the collapsing influence of gravitational attraction. The German philosopher and astronomer IMMANUEL KANT initially suggested the concept in 1755, and the French mathematician and astronomer PIERRE SIMON DE MARQUIS LAPLACE (apparently) independently of Kant introduced his own version of the nebula hypothesis in 1796.

See also SOLAR SYSTEM.

Nemesis A postulated dark stellar companion to the Sun whose close passage every 26 million years is thought to be responsible for the cycle of mass extinctions that seem also to have occurred on Earth at 26-million-year intervals. This "death star" companion has been named for the Greek goddess of retributive justice, or vengeance. If it really does exist, it might be a white dwarf, a rogue star that was captured by the Sun, or a tiny but gravitationally influential neutron star. The passage of such a death star through the Oort Cloud (a postulated swarm of comets surrounding the solar system) could lead to a massive shower of comets into the solar system. One or several of these "perturbed" comets hitting Earth then would trigger massive extinctions and catastrophic environmental changes within a very short period of time.

neper (symbol: N or N_p) A natural logarithmic unit (x) used to express the ratio of two power levels, P_1(input) and P_2(output), such that x (nepers) = $1/2 \ln (P_1/P_2)$. The unit is named after John Napier (1550–1617), the Scottish mathematician who developed natural logarithms (symbol: ln). This unit is often encountered in telecommunications engineering. 1 neper = 8.686 decibels.

See also BEL.

Neptune The outermost of the Jovian planets and the first planet to be discovered using theoretical predictions. Neptune's discovery was made by the German astronomer JOHANN GOTTFRIED GALLE at the Berlin Observatory in 1846. This discovery was based on independent orbital perturbation (disturbance) analyses by the French astronomer URBAIN JEAN JOSEPH LEVERRIER and the British scientist JOHN COUCH ADAMS. It is considered to be one of the triumphs of 19th-century theoretical astronomy.

Because of its great distance from Earth, little was known about this majestic blue giant planet until the *Voyager 2* spacecraft swept through the Neptunian system on August 25, 1989. Neptune's characteristic blue color comes from the selective absorption of red light by the methane (CH_4) found in its atmosphere, an atmosphere consisting primarily of hydrogen (more than 89 percent) and helium (about 11 percent) with minor amounts of methane, ammonia ice, and water ice. At the time of the *Voyager 2* encounter, Neptune's most prominent "surface" feature was called the Great Dark Spot (GDS), which was somewhat analogous in relative size and scale to Jupiter's Red Spot. However, unlike Jupiter's Red Spot, which has been observed for at least 300 years, Neptune's GDS, which was located in its southern hemisphere in 1989, had disappeared by June of 1994, when the *Hubble Space Telescope* looked for it. Then a few months later a nearly identical spot appeared in Neptune's northern hemisphere. Neptune is an extraordinarily dynamic planet that continues to surprise space scientists. The *Voyager 2* encounter also revealed the existence of six additional satel-

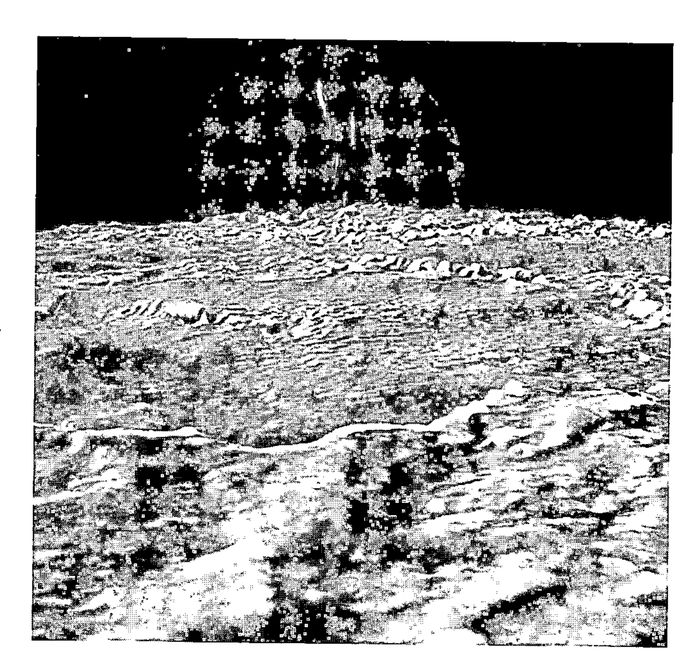

This composite view shows Neptune on Triton's horizon. Neptune's south pole is to the left, and the planet's large anticyclonic storm system, called the Great Dark Spot (GDS), appears in the planet's southern hemisphere. The foreground consists of a computer-generated view of Triton's maria as they would appear from a point approximately 45 km above the moon's surface. In creating this interesting three-dimensional view from a *Voyager 2* spacecraft image, scientists exaggerated vertical distances roughly 30-fold. The actual range of the relief is about 1 km. *(Courtesy of NASA/JPL and U.S. Geological Survey)*

Physical and Dynamic Properties of Neptune

Diameter (equatorial)	49,532 km
Mass	1.02×10^{26} kg
Density (mean)	1.64 g/cm^3
Surface gravity	11 m/sec^2 (approx.)
Escape velocity	23.5 km/sec (approx.)
Albedo (visual geometric)	0.4 (approx.)
Atmosphere	Hydrogen (~80%), helium (~18.5%) methane (~1.5%)
Temperature (blackbody)	33.3 K
Natural satellites	13
Rings	6 (Galle, LeVerrier, Lassell, Arago, unnamed, Adams [arcs])
Period of rotation (a Neptunian day)	0.6715 day (16 hr 7 min)
Average distance from Sun	4.5×10^9 km (30.06 AU)
Eccentricity	0.0086
Period of revolution around Sun (a Neptunian year)	165 yr
Mean orbital velocity	5.48 km/sec
Magnetic field	Yes (strong, complex; tilted 50° to planet's axis of rotation)
Radiation belts	Yes (complex structure)
Solar flux at planet (at top of atmosphere)	1.5 W/m^2 (at 30 AU)

Source: NASA.

lites and an interesting ring system. The Properties of Neptune Table provides contemporary physical and dynamic data for Neptune.

Triton, Neptune's largest moon, is one of the most interesting and coldest objects (about 35 kelvins [K] surface temperature) yet discovered in the solar system. Because of its inclined retrograde orbit, density (2.0g/cm^3 [2,000 kg/m^3]), rock and ice composition, and frost-covered (frozen nitrogen) surface, space scientists consider Triton to be a "first cousin" to the planet Pluto. Triton shows remarkable geological history; *Voyager* 2 images have revealed active geyserlike eruptions spewing invisible nitrogen gas and dark dust several kilometers into space.

See also PLUTO; URANUS; VOYAGER SPACECRAFT.

Nereid A moon of Neptune discovered by the Dutch-American astronomer GERARD PETER KUIPER in 1949. Nereid is relatively small, with an average diameter of 340 km. The moon is very distant from Neptune and takes a little over 360 days to make one orbit. Nereid's orbit has an inclination of 27.6° and an eccentricity of 0.751, giving the moon the honor of being the most eccentric of any known in the solar system. Nereid travels around Neptune at an average radial distance of 5,513,400 km. However, because of its high eccentricity, the moon's distance from Neptune ranges from about 1,353,600 km (periapsis) to 9,623,700 km (apoapsis). Also called Neptune II.

See also NEPTUNE.

NERVA (**N**uclear **E**ngine for **R**ocket **V**ehicle **A**pplication) The acronym applied to a series of nuclear reactors developed as part of the overall U.S. nuclear rocket development program (e.g., the ROVER Program) from 1959 to 1973.

See also NUCLEAR ROCKET; SPACE NUCLEAR PROPULSION.

Neumann, John von (1903–1957) Hungarian-born German-American *Mathematician* John von Neumann was a brilliant mathematician who made significant contributions to nuclear physics, game theory, and computer science in the 1940s. He was selected by the U.S. Air Force in 1953 to chair a special panel of experts who evaluated the American strategic ballistic missile program in the face of an anticipated Soviet ballistic missile threat. As a result of the recommendations from von Neumann's panel, President DWIGHT D. EISENHOWER assigned strategic ballistic missile development the highest national priority. General BERNARD SCHRIEVER was then tasked with creating an operational Atlas intercontinental ballistic missile (ICBM) as quickly as possible.

neutral atmosphere That portion of a planetary atmosphere consisting of neutral (uncharged) atoms or molecules, as opposed to electrically charged ions.

See also ATMOSPHERE.

neutral buoyancy simulator Orbiting objects in space, being "weightless," behave a little like neutral buoyancy objects in water on Earth. NASA astronauts have trained for certain on-orbit tasks by working with and moving space hardware while underwater in a large tank or submersion facility. These activities help simulate the actual mission activities to be performed during weightlessness.

neutral burning With respect to a solid-propellant rocket, a condition of propellant burning such that the burning surface or area remains constant, thereby producing a constant pressure and thrust.

neutrino (**symbol:** ν) An elusive uncharged elementary particle with no (or possibly extremely little) mass, produced in many nuclear reactions, including the process of beta decay. Neutrinos interact very weakly with matter and are consequently quite difficult to detect. For example, neutrinos from the Sun usually pass right through Earth without interacting.

See also ELEMENTARY PARTICLE(S).

neutrino astronomy The branch of modern astronomy that investigates neutrinos produced in various celestial phenomena involving nuclear reactions. These include supernova explosions, very-high-energy collisions between cosmic rays and nuclei in the upper atmosphere, and the energy-liberating thermonuclear reactions taking place in stellar cores.

Solar physicists estimate that the total integrated flux of all solar neutrinos reaching Earth is about 65 billion per square centimeter per second. Despite this impressive number, the neutrinos interact so weakly with matter that the probability of detecting any one of them is minuscule. Another way of describing the neutrino detection problem is that an atom presents an extremely thin target to the neutrino. Therefore, to have any hope of "catching a neutrino" in a detector, a scientist has to be willing to wait an extremely long period of time, or else build an enormously large detector that contains a huge number of target atoms.

In the late 1960s a pioneering neutrino astronomy experiment began about 1,500-meters underground in the Homestake Gold Mine in South Dakota. The experiment used a large tank of perchloroethylene (C_2Cl_4), a common dry cleaning fluid, to try to snare a few of the very elusive solar neutrinos. Electron neutrinos (ν_e) from the Sun manifest themselves through the inverse beta decay reaction on chlorine nuclei:

$$\nu_e + {}^{37}Cl \rightarrow {}^{37}Ar + e^-$$

In this reaction a neutron in the parent chlorine-37 (^{37}Cl) nucleus is transformed into a proton to yield the daughter argon-37 (^{37}Ar) along with the emission of an electron (e^-). Nuclear physicists know that the solar neutrino requires at least 0.814 MeV of energy to drive this reaction. Once produced, the unstable argon-37 atoms (half-life of approximately 35 days) eventually decay by recapturing an orbital electron and becoming once again chlorine-37. The emission of characteristic X-rays accompanies the argon-37 decay, but these telltale X-rays can be detected only after the argon-37 atoms have been extracted from the giant chlorine tank. Scientists hunting solar neutrinos bubble helium gas through the large tank of perchloroethylene (C_2Cl_4) to remove any of the unstable argon-37 atoms. The extracted argon-37 atoms are then passed through another detector (outside the chlorine tank) that searches for the telltale X-rays. This simple-sounding process is actually an enormously challenging one from a logistics perspective. Only a single atom of argon-37 is produced every two days in the large 615-ton detector, which contains approximately 2×10^{30} chlorine nuclei, yet, exhaustive tests of the argon-37 extraction process have proven not only efficient but also quite reliable. After more than three decades of operation in the Homestake Gold Mine, this radiochemical approach to neutrino detection experiment has measured an average neutrino flux that falls a factor of 3 below the predictions of the standard solar model used by astrophysicists to describe thermonuclear reaction processes in the Sun's interior. With a solar neutrino detection threshold of 0.814 MeV, the Homestake Mine chlorine experiment has consistently provided an observed solar neutrino rate of 2.5 ± 0.2 SNU versus a theoretically predicted rate of 9 ± 1 SNU. (One SNU is defined as 10^{-36} captures per target atom per second.)

Similarly, Russian and American scientists have also devised an alternate neutrino detection scheme, called the gallium experiment (or GALLEX). In this radiochemical technique, gallium is the neutrino target (specifically the isotope gallium-71) and germanium-71 is the byproduct of the reaction:

$$\nu_e + {}^{71}Ga \rightarrow {}^{71}Ge + e^-$$

Of special importance in neutrino astronomy is the fact that for this reaction the neutrino needs a threshold energy of only 0.2332 MeV. Solar neutrino rates observed with this radiochemical technique are 74 ± 8 SNU versus a theoretically predicted value (for neutrinos above 0.234 MeV) of 132 ± 7 SNU.

In Japan scientists have constructed a very different type of neutrino detector 1 kilometer underground in the Kamiokande Mine in the Japanese Alps. The Kamiokande detector is a huge cylindrical tank filled with 3,000 tons of purified water (H_2O) that functions as an enormous Cerenkov radiation detector. Cerenkov radiation is electromagnetic radiation, usually bluish light, emitted by a charged particle moving through a transparent medium at a speed greater than the speed of light in that material. This phenomenon was discovered by the Russian physicist Pavel Cerenkov (1904–90). In the Kamiokande detector neutrinos are detected in real time after they undergo elastic scattering reactions with electrons in the water. The neutrino imparts energy to the electron, which then streaks through the water at relativistic speeds. The Cerenkov radiation (bluish light) from these relativistic electrons is then detected by an array of photomultiplier tubes, which surround the water tank. Scientists then use computers to reconstruct the light flash event. With sophisticated analyses, they are able to determine the energy and the direction of the neutrino that caused the reaction.

The Kamiokande detector thus functions as a *neutrino telescope*. It not only observes the presence of elusive particle, but it also provides scientists a general sense of the direction from which the neutrino came. For example, in the search for solar neutrinos the Kamiokande detector has provided scientists unique experimental proof that the neutrinos they were observing did indeed come from the Sun. The Kamiokande detector has the capability of detecting neutrinos with energies greater than 7 MeV. This facility has measured a solar neutrino flux of $2.9 \pm 0.4 \times 10^6$ neutrinos per cm^2-s versus a theoretical predicted value of $5.7 \pm 0.8 \times 10^6$ neutrinos per cm^2-s.

These interesting solar neutrino data have created what modern astronomers refer to as the "missing" solar neutrino problem. Solar physicists remain puzzled over the consistently observed disparity between theoretical neutrino flux predictions for the Sun and experimentally measured values obtained from different techniques at different neutrino energy thresholds.

Another major milestone in neutrino astronomy involved Supernova 1987A. Approximately 20 hours before astronomers detected this supernova in the optical portion of the electromagnetic spectrum, a short (about 13-second) pulse of neutrinos was observed by both the American Homestake Mine (chlorine tank) neutrino detection facility in South Dakota and the Japanese Kamiokande underground neutrino detection facility. Physicists have suggested that these extragalactic neutrinos were emitted as a dying giant star was collapsing in the Large Magellanic Cloud—just prior to the optically observable supernova explosion. This event marks the first time astronomers have received information about a cosmic event beyond the solar system by means of particulate radiation that is outside the electromagnetic spectrum. Neutrino astronomy is in its infancy, but the ability to observe these elusive particles and even sense the direction (and therefore object or phenomenon) from which they come provides 21st-century astronomers a very important new way to study the universe.

neutron (symbol: **n**) An uncharged elementary particle with a mass slightly greater than that of a proton. It is found in the nucleus of every atom heavier than ordinary hydrogen. A free neutron is unstable, with a half-life of about 10 minutes, and decays into an electron, a proton, and a neutrino. Neutrons sustain the fission chain reaction in a nuclear reactor and support the supercritical reaction in a fission-based nuclear weapon.

neutron star A very small (typically 20 to 30 kilometers in diameter), superdense stellar object—the gravitationally collapsed core of a massive star that has undergone a supernova explosion. Astrophysicists believe that pulsars are rapidly spinning young neutron stars that possess intense magnetic fields.

See also PULSAR; STAR.

Newcomb, Simon (1835–1909) Canadian-American *Mathematical Astronomer* Simon Newcomb was one of the 19th century's leading mathematical astronomers. While working for the Nautical Almanac Office of the United States Naval Observatory, he prepared extremely accurate tables that predicted the positions of solar system bodies. Before the arrival of navigation satellites, sailors found their way at sea using an accurate knowledge of the positions of natural celestial objects, such as the Sun, Moon, and planets. The more accurate the available nautical tables, the more precisely sea captains could chart their voyages. In 1860 Newcomb also presented a trend-busting astronomical paper in which he correctly speculated that the main-belt asteroids did not originate from the disintegration of a single ancient planet, as was then commonly assumed.

Newcomb was born on March 12, 1835, in Wallace, Nova Scotia, Canada. As the son of an itinerant teacher, he received little formal schooling, and what education he did experience in childhood took place privately. In 1853 Newcomb moved to the United States and, much like his father, worked as a teacher in Maryland between 1854 and 1856. While near the libraries in Washington, D.C., he began to study mathematics extensively on his own. By 1857 Newcomb's mathematical aptitude caught the attention of the American physicist Joseph Henry (1797–1878), who helped him find employment as a "computer" (that is, a person who does precise astronomical calculations) at the United States Nautical Office, then located at Harvard University in Cambridge, Massachusetts. In 1858 Newcomb graduated from Harvard University with a bachelor of science degree and continued for three years afterward as a graduate student.

In 1861 Newcomb received an appointment as a professor of mathematics for the United States Navy and was assigned to duty at the U.S. Naval Observatory in Washington, D.C. A year earlier he had made his initial mark of mathematical astronomy when he presented a paper that demonstrated that the orbits of the asteroids (minor planets) did not diverge from a single point. His careful calculations dismantled the then popular hypothesis in solar system astronomy that the main belt of minor planets between Mars and Jupiter were the remnants of a single, ancient planet that tore itself apart. Unfortunately, Newcomb's interesting paper was quickly overshadowed by the start of the American Civil War (in April 1861) and the pressing needs of the U.S. Navy for more accurate nautical charts.

At the U.S. Naval Observatory in Washington, D.C., Newcomb focused all his mathematical talents and energies on preparing the most accurate nautical charts ever made (prior to the era of electronic computers). One of his first major responsibilities was to negotiate the contract for the Naval Observatory's new 0.66-meter (26-inch) telescope, which had recently been authorized by the U.S. Congress. Newcomb also planned the tower and dome and supervised construction. Later in his career Newcomb assisted in the development of the Lick Observatory in California.

Although primarily a mathematical astronomer rather than a skilled observer, Newcomb participated in a number of astronomical expeditions while serving at the U.S. Naval Observatory. For example, he traveled to Gibraltar in 1870–71 to observe an eclipse of the Sun and to the Cape of Good Hope at the southern tip of Africa in 1882 to observe a transit of Venus. In 1877 he received promotion to senior professor of mathematics in the U.S. Navy (a position with the equivalent naval rank of captain) and was put in charge of the office of the important publication the *American Ephemeris and Nautical Almanac*. Newcomb had a number of military and civilian assistants to help him produce the *Nautical Almanac,* including ASAPH HALL, the American astronomer who discovered the two tiny moons of Mars in 1877 while a staff member at the U.S. Naval Observatory.

By the 1890s, as American naval power reached around the globe, Newcomb supported its needs for improved navigation by producing the highest-quality nautical almanac yet developed. To accomplish this, he supervised a large staff of astronomical computers (that is, human beings) and enjoyed control of the largest observatory budget in the world. To support the production of a high-quality *Nautical Almanac,* Newcomb pursued, produced, and promoted a new and more accurate system of astronomical constants that became the world standard by 1896. He retired from his position as superintendent of the American Nautical Almanac Office in 1897 and received promotion to the rank of rear admiral in 1905.

Newcomb was also a professor of mathematics and astronomy at Johns Hopkins University between 1884 and 1893. He served as editor of the *American Journal of Mathematics* for many years and as president of the American Mathematical Society from 1897 to 1898. He was a founding member and the first president (from 1899 to 1905) of the American Astronomical Society.

He wrote numerous technical papers, including "An Investigation of the Orbit of Uranus, with General Table of its Motion" (1874) and "Measurement of the Velocity of Light" (1884), a paper that brought him in contact with a brilliant young scientist named ALBERT MICHELSON. A gifted writer, Newcomb published books on various subjects for different levels of audiences. His books included *Popular Astronomy* (1877), *Principles of Political Economy* (1886), and *Calculus* (1887).

As a gifted mathematical astronomer, Newcomb received many honors and won election to many distinguished societies. For example, in 1874 he received the Gold Medal of the Royal Astronomical Society in Great Britain. He also became a fellow of the Royal Society of London in 1877 and received that society's Copley Medal in 1890. The Astronomical Society of the Pacific made him the first Bruce Gold Medallist in 1898. He died in Washington, D.C., on July 11, 1909, and was buried with full military honors at Arlington Cemetery. To commemorate his great contributions to the United States in the field of mathematical astronomy and nautical navigation, the U.S. Navy named a surveying ship the *USS Simon Newcomb.*

New General Catalog of Nebulae and Clusters of Stars (NGC) The extensive catalog of nebulas and star clusters prepared by the Danish astronomer JOHAN LUDVIG EMIL DREYER, first published in 1888 and then expanded with supplements in 1895 and 1908. Dreyer listed these "fuzzy" celestial objects by their right ascension values and provided a brief description of the objects' appearance. The *New General Catalog* is a revised and enlarged version of the *General Catalog of Nebulae and Clusters of Stars* published in 1864 by SIR JOHN FREDERICK WILLIAM HERSCHEL. NGC numbers are used extensively in modern astronomy. Bright celestial objects appear in both the (more extensive) NGC and the (limited entry) *Messier Catalog.* The Orion Nebula, for example, is called NGC 1976 in the *New General Catalog* and M 42 in the Messier Catalog.

See also MESSIER CATALOG.

New Horizons Pluto–Kuiper Belt Flyby mission Originally conceived as the Pluto Fast Flyby (PFF), NASA's New Horizons Pluto–Kuiper Belt Flyby mission is planned for launch in 2006. This reconnaissance-type exploration mission will help scientists understand the interesting yet poorly understood worlds at the edge of the solar system. The first spacecraft flyby of Pluto and Charon—the frigid double planet system—could take place as early as 2015. (*See* figure.) The mission will then continue beyond Pluto and visit one or more Kuiper Belt objects (of opportunity) by 2026. The spacecraft's long journey will help resolve some basic questions about the surface features and properties of these icy bodies as well as their geology, interior makeup, and atmospheres.

With respect to the Pluto-Charon system, some of the major scientific objectives include the characterization of the global geology and geomorphology of Pluto and Charon, the mapping of the composition of Pluto's surface, and the determination of the composition and structure of Pluto's transitory atmosphere. It is intended that the spacecraft reach Pluto before the tenuous Plutonian atmosphere can refreeze onto the surface as the planet recedes from its 1989 perihelion. Studies of the double-planet system will actually begin some 12 to 18 months before the spacecraft's closest approach to Pluto in about 2015. The modest-sized spacecraft will have no deployable structures and will receive all its electric power from long-lived radioisotope thermoelectric generators (RTGs) that are similar in design to those used on the *Cassini* spacecraft now orbiting Saturn.

This important mission will complete the initial scientific reconnaissance of the solar system with robot spacecraft. At present Pluto is the most poorly understood planet in the solar system. As some scientists speculate, the tiny planet may even be considered the largest member of the family of primitive icy objects that reside in the Kuiper Belt. In addition to the first close-up view of Pluto's surface and atmosphere, the spacecraft will obtain gross physical and chemical surface properties of Pluto, Charon, and (possibly) several Kuiper Belt objects. The celestial mechanics opportunity to launch this mission to Pluto by way of a gravity assist from Jupiter opens in January 2006.

new moon The Moon at conjunction, when little or none of it is visible to an observer on Earth because the illuminated side is away from him or her.

See also MOON; PHASES OF THE MOON.

newton (symbol: N) The SI unit of force, which honors SIR ISAAC NEWTON. It is defined as the force that provides a 1 kilogram mass with an acceleration of 1 meter per second per second.

$$1\ \text{N} = 1\ \text{kg-m/s}^2$$

Newton, Sir Isaac (1642–1727) British *Physicist, Astronomer, Mathematician* Sir Isaac Newton was the brilliant though introverted British astrophysicist and mathematician whose law of gravitation, three laws of motion, development of the calculus, and design of a new type of reflecting telescope make him one of the greatest scientific minds in human history. Through the patient encouragement and financial support of the British mathematician SIR EDMUND HALLEY, Newton published his great work, *The Principia* (or, *Mathematical Principles of Natural Philosophy*), in

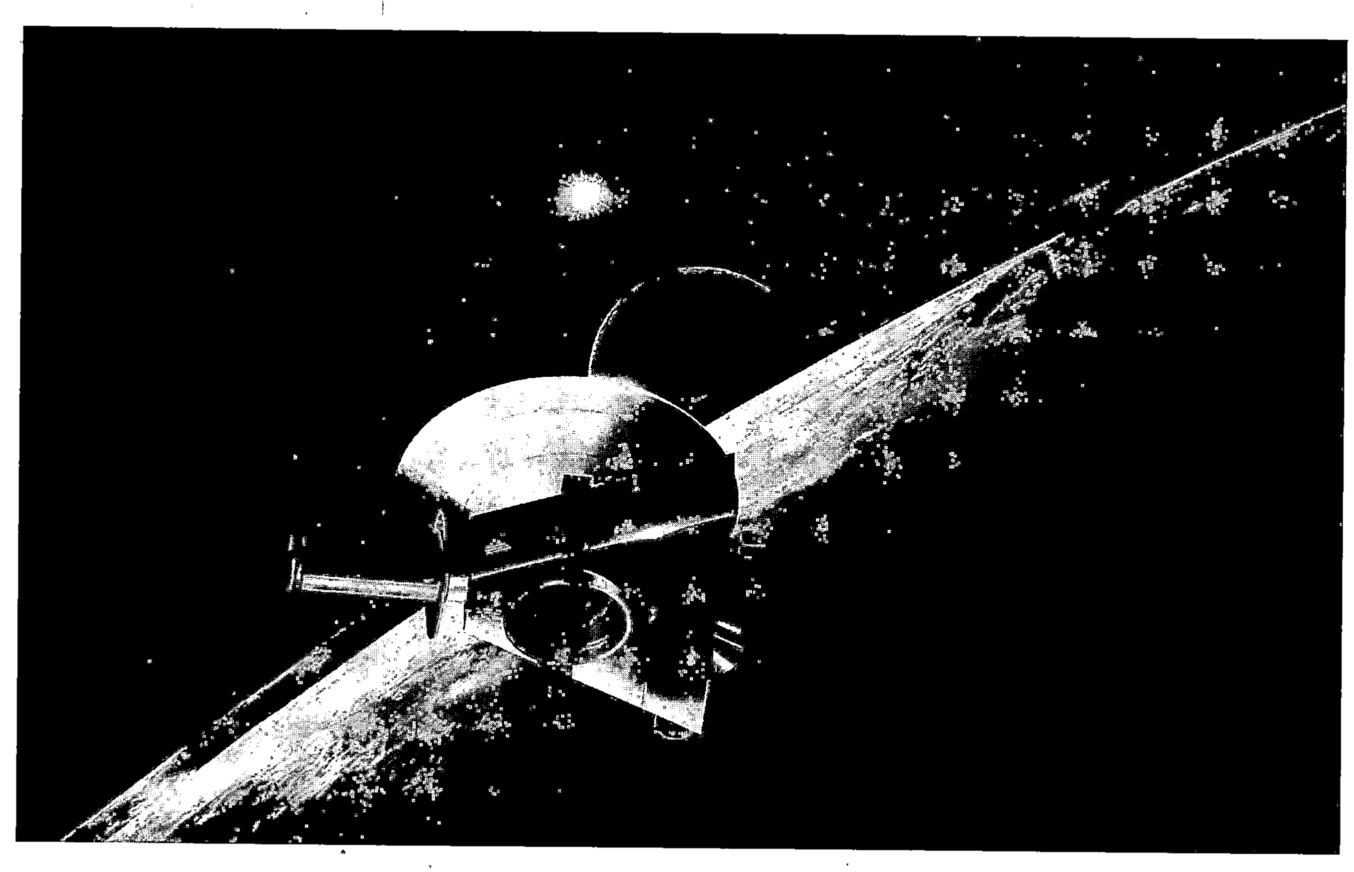

This is an artist's rendering of NASA's *New Horizons* spacecraft during its planned encounter with the tiny planet Pluto (foreground) and its relatively large moon, Charon, possibly as early as July 2016. A long-lived, plutonium-238 fueled, radioisotope thermoelectric generator (RTG) subsystem (cylinder on lower-left portion of the spacecraft) provides electric power as it flies past these distant icy worlds billions of kilometers from the Sun. As depicted here, one of the spacecraft's most prominent design features is a 2.1-meter-diameter dish antenna through which it communicates with scientists on Earth from as far as 7.5 billion kilometers away. Following its encounter with the Pluto-Charon planetary system, the spacecraft is hoped to encounter one or several icy planetoid targets of exploration in the Kuiper Belt. *(Artwork courtesy of NASA/JPL)*

1687. This monumental book transformed the practice of physical science and completed the scientific revolution started by NICHOLAS COPERNICUS, JOHANNES KEPLER, and GALILEO GALILEI.

Newton was born prematurely in Woolsthorpe, Lincolnshire, on December 25, 1642 (using the former Julian calendar, or on January 4, 1643, under the current Gregorian calendar). His father had died before Newton's birth, and this contributed to a very unhappy childhood. In order to remarry, his mother placed her three-year-old son in the care of his grandmother. Separation from his mother and other childhood stresses are believed to have significantly contributed to his very unusual adult personality. Throughout his life Newton would not tolerate criticism, remained hopelessly absent-minded, and often tottered on the verge of emotional collapse. British historians claim that Newton laughed only once or twice in his entire life, yet he is still considered by many to be the greatest human intellect who ever lived. His brilliant work in physics, astronomy, and mathematics combined the discoveries of Copernicus, Kepler, and Galileo. Newton's universal law of gravitation and his three laws of motion fulfilled the scientific revolution and dominated science for the next two centuries.

In 1653 his stepfather died, and Newton's mother returned to the farm at Woolsthorpe and removed her son from school so he could practice farming. Fortunately for science, Newton failed miserably as a farmer and in June 1661 departed Woolsthorpe for Cambridge University. He graduated without any particular honors or distinction from Cambridge with a bachelor's degree in 1665. Following graduation Newton returned to his mother's farm in Woolsthorpe to avoid the plague that had broken out in London. For the next two years he pondered mathematics and physics at home, and this self-imposed exile laid the foundation for his brilliant contributions. By Newton's own account, one day on the farm he saw an apple fall to the ground and began to wonder if the same force that pulled on the apple also kept the Moon in its place. At this point heliocentric cosmology as expressed in the works of Copernicus, Galileo, and Kepler was becoming widely accepted (except where banned on political or religious grounds), but the mechanism for planetary motion around the Sun remained unexplained.

By 1667 the plague epidemic subsided, and Newton returned to Cambridge as a minor fellow at Trinity College. The following year he received a master of arts degree and became a senior fellow. In about 1668 he constructed the first working reflecting telescope, a device that now carries his name. The Newtonian telescope uses a parabolic mirror to collect light. The primary mirror then reflects the collected light by means of an internal secondary mirror to an external focal point at the side of the telescope's tube. This new telescope design earned Newton a great deal of professional acclaim, including eventual membership in the Royal Society

In 1669 Isaac Barrow, Newton's former mathematics professor, resigned his position so that the young Newton could succeed him as Lucasian Professor of Mathematics. This position provided Newton the time to collect his notes and properly publish his work, a task he was always tardy to perform.

Shortly after his election to the Royal Society in 1671, Newton published his first paper in that society's transactions. While an undergraduate, Newton had used a prism to refract a beam of white light into its primary colors (red, orange, yellow, green, blue, and violet.) Newton reported this important discovery to the Royal Society. However, Newton's pioneering work involving the physics of light was immediately attacked by Robert Hooke (1635–1703), an influential member of the society.

This was the first in a lifelong series of bitter disputes between Hooke and Newton. Newton only skirmished lightly then quietly retreated. This was Newton's lifelong pattern of avoiding direct conflict. When he became famous later in his life, Newton would start a controversy, withdraw, and then secretly manipulate others, who would then carry the brunt of the battle against Newton's adversary. For example, Newton's famous conflict with the German mathematician Gottfried Leibniz (1646–1716) over the invention of calculus followed precisely such a pattern. Through Newton's clever manipulation the calculus controversy even took on nationalistic proportions as carefully coached pro-Newton British mathematicians bitterly argued against Leibniz and his supporting group of German mathematicians.

In August 1684 Edmund Halley made a historic trip to visit Newton at Woolsthorpe and convinced the reclusive genius to address the following puzzle about planetary motion: What type of curve does a planet describe in its orbit around the Sun, assuming an inverse square law of attraction? To Halley's delight, Newton immediately responded, "An ellipse." Halley pressed on and asked Newton how he knew the answer to this important question. Newton nonchalantly informed Halley that he had already done the calculations years ago (in 1666) while living at the family farm to avoid plague-ravaged London. Then, the absent-minded Newton tried to find the old calculations that solved one of the major scientific questions of the day. Of course, he could not find them but promised to send Halley another set as soon as he could.

To partially fulfill his promise, Newton sent Halley his *De Motu Corporum* (1684). In this document Newton demonstrated that the force of gravity between two bodies is directly proportional to the product of their masses and inversely proportional to the square of the distance between them (Physicists now call this relationship Newton's *universal law of gravitation*). Halley was astounded and begged Newton to carefully document all of his work on gravitation and orbital mechanics. Through the patient encouragement and financial support of Halley, Newton published his great work, *Philosophiae Naturalis Principia Mathematica* (Mathematical principles of natural philosophy) in 1687. In the *Principia* Newton gave the world his famous three laws of motion and the universal law of gravitation. This monumental work transformed physical science and completed the scientific revolution. Many consider the *Principia* the greatest scientific accomplishment of the human mind.

In 1705 Halley, who was also a proficient astronomer, used the contents of the *Principia* to properly identify (as the same object) the comet observed in the years 1531, 1607, and 1682. He also predicted that this comet would return in 1758. Halley died in 1742, so he did not personally witness the predictive success of his celestial mechanic skills. The comet did return as he predicted, and so other scientists posthumously honored him by calling the periodic comet Comet Halley.

For all his brilliance, Newton was also extremely fragile. After completing the *Principia,* he drifted away from physics and astronomy and eventually suffered a serious nervous disorder in 1693. Upon recovery he left Cambridge in 1696 to assume a government post in London as warden (then later master) of the Royal Mint. During his years in London Newton enjoyed power and worldly success. Robert Hooke, his lifelong scientific antagonist, died in 1703. The following year the Royal Society elected Newton its president. Unrivaled, he won annual reelection to this position until his death. However, Newton was so bitter about his quarrels with Hooke that he waited until 1704 to publish his other major work, *Opticks*. Queen Anne knighted him in 1705.

Although his most innovative years were now clearly behind him, Newton at this point in his life continued to exert great influence on the course of modern science. He used his position as president of the Royal Society to exercise autocratic (almost tyrannical) control over the careers of many younger scientists. Even late in life he could not tolerate controversy, but now, as society president, he skillfully maneuvered younger scientists to fight his intellectual battles. In this manner he continued to rule the scientific landscape until his death in London on March 20, 1727.

Newtonian mechanics The system of mechanics based on SIR ISAAC NEWTON's laws of motion in which mass and energy are considered as separate, conservative mechanical properties, in contrast to their treatment in relativistic mechanics, in which mass and energy are treated as equivalent (from ALBERT EINSTEIN's mass-energy equivalency formula: $E = \Delta mc^2$).

See also NEWTON'S LAWS OF MOTION.

Newtonian telescope A reflecting telescope in which a small plane mirror reflects the convergence beam from the objective (primary mirror) to an eyepiece at the side of the

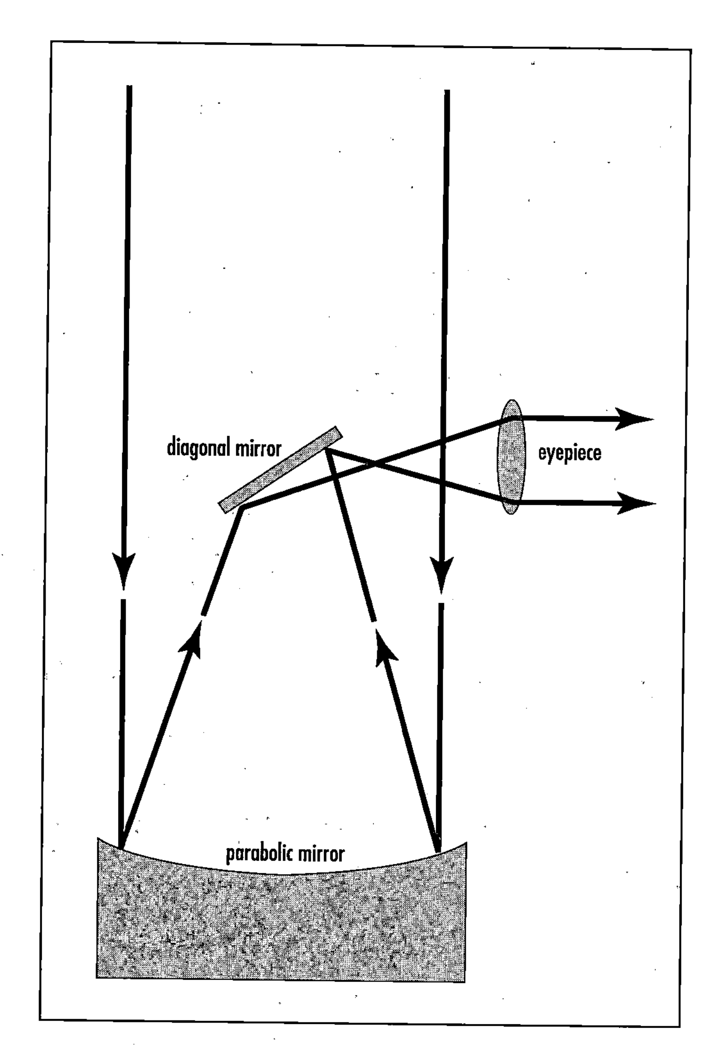

The basic components and fundamental design of the reflecting telescope invented by Sir Isaac Newton in about 1668 *(Drawing courtesy of NASA)*

telescope. After the second reflection the rays travel approximately perpendicular to the longitudinal axis of the telescope. (*See* figure.) SIR ISAAC NEWTON constructed the first reflecting telescope in about 1668. As a result, astronomers now call this type of reflecting telescope a Newtonian telescope in recognition of his accomplishment.

See also REFLECTING TELESCOPE; TELESCOPE.

Newton's law of gravitation Every particle of matter in the universe attracts every other particle with a force (F), acting along the line joining the two particles, proportional to the product of the masses (m_1 and m_2) of the particles and inversely proportional to the square of the distance (r) between the particles, or

$$F = [G\, m_1\, m_2]/r^2$$

where F is the force of gravity (newtons), m_1 and m_2 are the masses of the (attracting) particles (kg), r is the distance between the particles (m_1 and m_2), and G is the universal gravitational constant.

$$G = 6.6732\ (\pm 0.003) \times 10^{-11}\ \text{N m}^2/\text{kg}^2 \text{ (in SI units)}$$

This important physical law was introduced by Sir Isaac Newton in his paper *De Motu Corporum* (1684) and then more fully elaborated upon in the *Principia,* published in 1687.

Newton's laws of motion A set of three fundamental postulates that form the basis of the mechanics of rigid bodies. These laws were formulated in about 1685 by the brilliant British scientist and mathematician SIR ISAAC NEWTON as he was studying the motion of the planets around the Sun. Newton described this work in the book *Mathematical Principles of Natural Philosophy,* which is often referred to simply as Newton's *Principia.*

Newton's first law is concerned with the principle of inertia and states that if a body in motion is not acted upon by an external force, its momentum remains constant. This law also can be called the *law of conservation of momentum.*

The second law states that the rate of change of momentum of a body is proportional to the force acting upon the body and is in the direction of the applied force. A familiar statement of this law is the equation F = m a, where F is the vector sum of applied forces, m is the mass, and a is the vector acceleration of the body.

Newton's third law is the principle of action and reaction. It states that for every force acting upon a body, there is a corresponding force of the same magnitude exerted by the body in the opposite direction.

Next-Generation Space Telescope (NGST) *See* JAMES WEBB SPACE TELESCOPE (JWST).

Nicholson, Seth Barnes (1891–1963) American *Astronomer* The American astronomer Seth Barnes Nicholson discovered the minor Jovian moons Lysithea (discovered in 1938), Ananke (1951), Carme (1938), and Sinope (1914). He also performed important research on sunspots.

nickel-cadmium batteries (nicad) Long-lived, rechargeable batteries frequently used in spacecraft applications as a secondary source of power. For example, in a typical Earth-orbiting spacecraft solar cell arrays capture sunlight and provide electric power to operate the spacecraft's electrical systems and also to charge the nickel-cadmium batteries. Then, when the spacecraft is shadowed by Earth and is not illuminated by the Sun, spacecraft electric power is provided by the nickel-cadmium batteries. This cycle continues many times during the lifetime of the spacecraft. While the nicad batteries have a low energy density (typically 15 to 30 watt-hours per kg), they have a good deep-discharge tolerance and long cycle life, making their use nearly standard in contemporary spacecraft operations.

Nike Hercules A U.S. Army air defense surface-to-air guided missile system that provides nuclear or conventional, medium- to high-altitude air defense coverage against manned bombers and air-breathing missiles. The system is designed to operate in either a mobile or fixed-site configuration and has a capability of performing surface-to-surface missions. Also designated as the MIM-14. The army developed two other surface-to-air missiles in the Nike missile family, the Nike Ajax and the Nike Zeus.

Nimbus satellites A series of second-generation weather satellites created by NASA and the National Oceanographic and Atmospheric Administration (NOAA) as a follow-up to the Tiros satellites, which were the first polar-orbiting meteorological satellites. The Nimbus satellites were more complex than the Television and Infrared Observation Satellites (TIROS). For example, they carried advanced television cloud-mapping cameras and an infrared radiometer that permitted the collection of cloud images at night for the first time. From 1964 to 1978 seven Nimbus spacecraft were placed in polar orbit from Vandenberg Air Force Base, California, creating a 24-hour-per-day capability to observe weather conditions on the planet.

See also WEATHER SATELLITE.

NOAA *See* NATIONAL OCEANIC AND ATMOSPHERIC ADMINISTRATION.

node 1. In orbital mechanics, one of the two points of intersection of the orbit of a planet, planetoid, or comet with the ecliptic, or of the orbit of a satellite with the plane of the orbit of its primary. By convention, that point at which the body crosses to the north side of the reference plane is called the *ascending node;* the other node is called the *descending node.* The line connecting these two nodes is called the *line of nodes.* 2. In physics, a point, line, or surface in a standing wave where some characteristic of the wave field has essentially zero amplitude. By comparison, a position of maximum amplitude in a standing wave is referred to as an antinode. 3. In communications, a network junction or connection point (e.g., a terminal).

nominal In the context of aerospace operations and activities, a word meaning "within prescribed or acceptable limits." For example, the Mars mission is now on a "nominal" interplanetary trajectory; or the pressure in the combustion chamber is "nominal."

noncoherent communication Communications mode wherein a spacecraft generates its downlink frequency independently of any uplink frequency.

See also TELECOMMUNICATIONS.

nondestructive testing Testing to detect internal and concealed defects in materials and components using techniques that do not damage or destroy the items being tested. Aerospace engineers and technicians frequently use X-rays, gamma rays, and neutron irradiation as well as ultrasonics to accomplish nondestructive testing.

nonflight unit An aerospace unit that is not intended for flight usage. Also called *test article* or a *hangar queen.*

nonimpinging injector An injector used in liquid rocket engines that employs parallel streams of propellant usually emerging normal to the face of the injector. In this type of injector propellant mixing usually is obtained by turbulence and diffusion. The World War II–era German V-2 rocket used a nonimpinging injector.

See also INJECTOR; ROCKET.

nonionizing electromagnetic radiation Radiation that does not change the structure of atoms but does heat tissue and may cause harmful biological effects. Microwaves, radio waves, and low-frequency electromagnetic fields from high-voltage transmission lines are considered to be nonionizing radiation.

normal 1. (in aerospace operations) Equivalent to usual, regular, rational, or standard conditions. 2. (in mathematics) Perpendicular. A line is *normal* to another line or a plane when it is perpendicular to it. A line is *normal* to a curve or curved surface when it is perpendicular to the tangent line or plane at the point of tangency.

North American Aerospace Defense Command (NORAD) The North American Aerospace Defense Command is a binational (U.S. and Canadian) organization charged with the missions of aerospace warning and aerospace control of North America. Aerospace warning includes the monitoring of human-made objects in space and the detection, validation, and warning of an attack against North America whether by aircraft, missiles, or space vehicles. Aerospace control includes ensuring air sovereignty and air defense of the airspace of Canada and the United States. To achieve these mission objectives NORAD relies on mutual support arrangements with other commands, including United States Strategic Command (USSTRATCOM). The commander of NORAD is responsible to both the president of the United States and the prime minister of Canada.

The headquarters for the NORAD commander is located at Peterson AFB, Colorado, and NORAD's command and control center is a short distance away at Cheyenne Mountain Air Station (CMAS). Deep within the underground complex at Cheyenne Mountain are all the facilities needed by NORAD personnel to collect and coordinate data from a worldwide system of sensors. Data from these sensors provide an accurate picture of any emerging aerospace threat to North America. NORAD has three subordinate regional headquarters located at Elmendorf AFB, Alaska, Canadian Forces Base, Winnipeg, Manitoba, and Tyndall AFB, Florida. These regional headquarters receive direction from the NORAD commander and then take appropriate actions to control air operations within their respective geographic areas of responsibility.

As part of the aerospace warning mission, NORAD's commander must provide an integrated tactical warning and attack assessment (called an ITW/AA) concerning a particular aerospace attack on North America to the governments of both Canada and the United States. A portion of the necessary threat information comes from warning systems controlled directly by NORAD, while other attack assessment information comes to NORAD from other commands under various mission support agreements.

NORAD's aerospace control mission includes detecting and responding to any air-breathing threat to North America. NORAD uses a network of ground-based radar systems and fighter aircraft to detect, intercept, and (if necessary) engage any air-breathing threat to the continent. The supersonic aircraft presently used by NORAD consist of U.S. F-15 and F-16 fighters and Canadian CF-18 fighters. NORAD also assists in the detection and monitoring of aircraft suspected of illegal drug trafficking. This information is passed on to civilian law enforcement agencies to help combat the flow of illegal drugs by air into the United States and Canada. Responding to the threat of terrorism, NORAD provides its military response capabilities to civil authorities and homeland defense organizations to counter domestic airspace threats.

nose cone The cone-shaped leading edge of a rocket vehicle. Designed to protect and contain a warhead or payloads, such as satellites, instruments, animals, or auxiliary equipment. The outer surface and structure of a nose cone is built to withstand the high temperatures associated with the aerodynamic heating and vibrations (buffeting) that results from flight through the atmosphere.

See also REENTRY VEHICLE; ROCKET.

nova (plural: novas or novae) A star that exhibits a sudden and exceptional brightness, usually of a temporary nature, and then returns to its former luminosity. A supernova is a much brighter explosion of a giant star during which a large fraction of its mass is blown outward into space. The term derives from the Latin word *nova,* which means "new."

See also CATACLYSMIC VARIABLE; STAR; SUPERNOVA.

nozzle In a rocket engine, the carefully shaped aft portion of the thrust chamber that controls the expansion of the exhaust gas so that the thermal energy of combustion is converted effectively into kinetic energy of the combustion product gases, thereby propelling the rocket vehicle. The nozzle is a major component of a rocket engine, having a significant influence on the overall engine performance and representing a large fraction of the engine structure. The design of the nozzle consists of solving simultaneously two different problems: defining the shape of the wall that forms the expansion surface and delineating the nozzle structure and hydraulic system. In general, the nozzle shape is selected to maximize performance within the constraints placed on the rocket system. The nozzle structure of a large rocket must provide strength and rigidity to a system in which mass is at a high premium and the loads are not readily predictable. The maximum loads on the nozzle structure often occur during the start transients (i.e., initial rapid rise in pressure and fluid flow rates) before full flow is established.

The nozzles used on liquid rocket engines now are inherently very efficient components, since they are the product of many years of engineering development. (*See* figure.) With increasing area ratio (i.e., exit area to throat area), the nozzle becomes a proportionally larger part of the rocket engine. Because there is relatively little to be gained in trying to increase the efficiencies of conventional bell and conical nozzles, recent

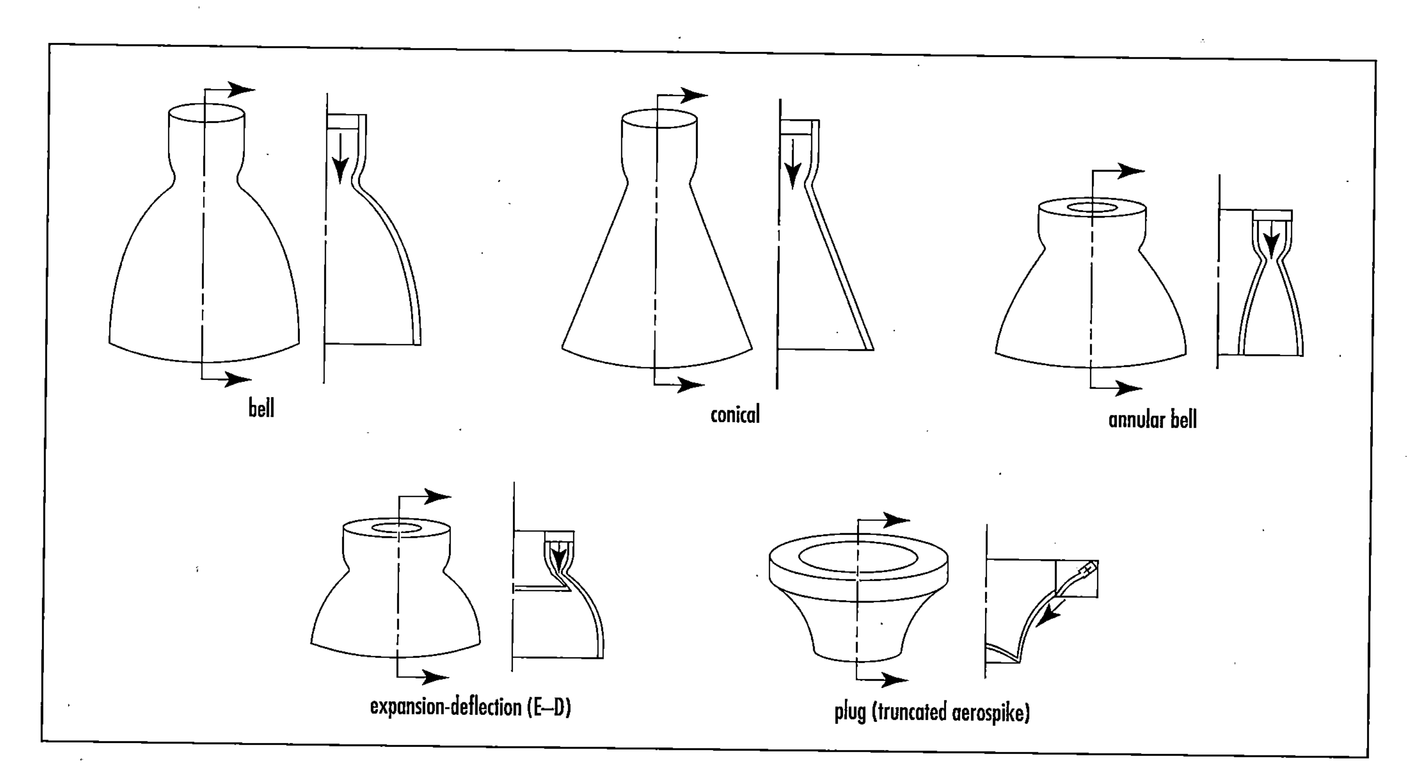

Basic types of nozzles used in liquid propellant rocket engines

aerospace engineering activities in nozzle development have focused on obtaining the same nozzle efficiency from a shorter package through the use of short bell nozzles and annular nozzles, such as expansion-deflection (E-D) and plug (aerospike) nozzles.

An *aerospike nozzle* is an annular nozzle that allows the combustion gas to expand from one surface—a centerbody spike—to ambient pressure. A *bell nozzle* is a nozzle with a circular opening for a throat and an axisymmetric contoured wall downstream of the throat that gives the nozzle a characteristic bell shape. The *expansion-deflection,* or E-D, *nozzle* is a nozzle that has an annular throat that discharges exhaust gas with a radial outward component. A *plug nozzle* is an annular nozzle that discharges exhaust gas with a radial inward component—a truncated aerospike.

The nozzle configuration best suited to a particular application depends on a variety of factors, including the altitude regime in which the nozzle will be used, the diversity of stages in which the same nozzle will be employed, and limitations imposed by available development time and funding. The propellant combination, chamber pressure, mixture ratio, and thrust level for a rocket engine generally are determined on the basis of factors other than the nozzle configuration. These parameters will have only a minor effect on the selection of a nozzle configuration but will have a significant effect on the selection of the nozzle cooling method.

The throat of the nozzle is the region of transition from subsonic to supersonic flow. For typical rocket nozzles, local mass flux and, consequently, the rate of heat transfer to the wall are highest in this area of the nozzle. The shape of the nozzle wall in the vicinity of the throat dictates the distribution of exhaust-gas flow across the nozzle at the throat. The nozzle wall geometry immediately upstream of the throat determines the distribution of gas properties at the throat. The expansion geometry extends from the throat to the nozzle exit. The function of this part of the nozzle is to accelerate the exhaust gases to a high velocity in a short distance while providing near-ideal performance. Both engine length and propellant specific impulse strongly influence the payload capability of a rocket vehicle. Therefore, aerospace engineers give considerable attention to the design of the nozzle expansion geometry in order to obtain the maximum performance from a nozzle length consistent with optimum vehicle payload.

The *area ratio* of a nozzle is defined as the ratio of the geometric flow area of the nozzle exit to the geometric flow area of the nozzle throat. This important nozzle parameter is also called the *expansion area ratio.* An ideal nozzle is a nozzle that provides theoretically perfect performance for the given area ratio when analyzed on the basis of one-dimensional point-source flow. The *aerodynamic throat area* of a nozzle is the effective flow area of the throat, which is less than the geometric flow area because the flow is not uniform. A *nozzle extension* is a nozzle structure that has been added to the main nozzle in order to increase the expansion area ratio or to provide a change in nozzle construction. Finally, *overexpansion* is a condition that occurs when a nozzle expands the exhaust gas to an ambient pressure that is higher than that for which the nozzle was originally designed.

See also ROCKET.

nozzle extension Nozzle structure that is added to the main nozzle in order to increase the expansion area ratio or to provide a change in nozzle construction.

See also NOZZLE.

nuclear detonation A nuclear explosion resulting from fission or fusion reactions in the nuclear materials contained in a nuclear weapon.

nuclear-electric propulsion system (NEP) A propulsion system in which a nuclear reactor is used to produce the electricity needed to operate the electric propulsion engine(s). Unlike the solar-electric propulsion (SEP) system, the nuclear electric propulsion (NEP) system can operate anywhere in the solar system and even beyond, since the NEP system's performance is independent of its position relative to the Sun. Furthermore, it can provide shorter trip times and greater payload capacity than any of the advanced chemical propulsion technologies that might be used in the next two decades or so for detailed exploration of the outer planets, especially Saturn, Uranus, Neptune, Pluto, and their respective moons. Closer to Earth, a NEP orbital transfer vehicle (OTV) can serve effectively in cislunar space, gently transporting large cargoes and structures from low Earth orbit to geosynchronous orbit or to a logistics depot in lunar orbit that supports the needs of an expanding lunar settlement. Finally, advanced nuclear electric propulsion systems can become the main space-based transport vehicles for both human explorers and their equipment as permanent surface bases are established on Mars in the mid-21st century.

See also ELECTRIC PROPULSION; *JUPITER ICY MOONS ORBITER* (JIMO); SPACE NUCLEAR POWER; SPACE NUCLEAR PROPULSION.

nuclear energy Energy released as a result of interactions involving atomic nuclei; that is, reactions in which there is a rearrangement of the constituents (i.e., protons and neutrons) of the nucleus. Any nuclear process in which the total mass of the products is less than the mass before interaction is accompanied by the release of energy. The amount of energy released is related to the decrease in (nuclear) mass by ALBERT EINSTEIN's famous mass-energy equivalence formula:

$$E = mc^2$$

where E is the energy released, m is the *decrease* in mass, and c is the speed of light. The disappearance of 1 atomic mass unit (i.e., the mass of a proton or a neutron) results in the release of about 931 million electron volts (MeV) of energy.

See also FISSION (NUCLEAR); FUSION (NUCLEAR).

nuclear explosion A catastrophic, destructive event in which the primary cause of damage to human society and/or the environment is a direct result of the rapid (explosive) release of large amounts of energy from nuclear reactions (fission and/or fusion).

nuclear fission The splitting of a heavy atomic nucleus into two smaller masses accompanied by the release of neutrons, gamma rays, and about 200 million electron volts (MeV) of energy.

See also FISSION (NUCLEAR).

nuclear fusion The nuclear reaction (process) whereby several smaller nuclei combine to form a more massive nucleus, accompanied by the release of a large amount of energy (which represents the difference in nuclear masses between the sum of the smaller particles and the resultant combined, or "fused," nucleus). Nuclear fusion is the source of energy in stars, including the Sun. For example, the nuclear fusion processes now taking place within the Sun convert approximately 5 million metric tons of mass into energy every second.

See also FUSION (NUCLEAR).

nuclear radiation Particles (e.g., alpha particles, beta particles, and neutrons) and very energetic electromagnetic radiation (e.g., gamma-ray photons) emitted from atomic nuclei during various nuclear reaction processes, including radioactive decay, fission, and fusion. Nuclear radiations are ionizing radiations.

nuclear reaction A reaction involving a change in an atomic nucleus, such as neutron capture, radioactive decay, fission, or fusion. A nuclear reaction is distinct from and far more energetic than a chemical reaction, which is limited to changes in the electron structure surrounding the nucleus.

See also FISSION (NUCLEAR); FUSION (NUCLEAR).

nuclear reactor A device in which a fission chain reaction can be initiated, maintained, and controlled. Its essential component is the core, which contains the fissile fuel. Depending on the type of reactor, its purpose, and its design, the reactor also can contain a moderator and a reflector and shielding, coolant, and control mechanisms.

See also FISSION (NUCLEAR); NUCLEAR ROCKET; SPACE NUCLEAR POWER; SPACE NUCLEAR PROPULSION.

nuclear rocket In general, a rocket vehicle that derives its propulsive thrust from nuclear energy sources, primarily fission. There are two general classes of nuclear rockets, the *nuclear-thermal rocket* (NTR) and the *nuclear-electric propulsion* (NEP) system. The nuclear thermal rocket uses a nuclear reactor to heat the propellant (generally hydrogen) to extremely high temperatures before it is expelled through a rocket nozzle. (*See* figure.) The nuclear electric propulsion system uses a nuclear reactor to produce electric power, which in turn is used to operate an electric propulsion system.

See also NUCLEAR-ELECTRIC PROPULSION SYSTEM (NEP); SPACE NUCLEAR PROPULSION.

Nuclear Rocket Development Station (NRDS) Major test facilities for the U.S. nuclear rocket program (from 1959 to 1973) were located at the Nuclear Rocket Development Station (NRDS) at the Nevada Test Site (NTS) in southern Nevada. Both nuclear reactors and complete nuclear rocket engine assemblies were tested (in place) at the NRDS, which had three major test areas: Test cells A and C and Engine Test Stand Number One (ETS-1). The reactor test facilities were designed to test a reactor in an upward-firing position, while the engine test facility tested nuclear rocket engines in a downward-firing mode. During a nuclear rocket test at the NRDS, the surrounding desert basin (called Jackass Flats

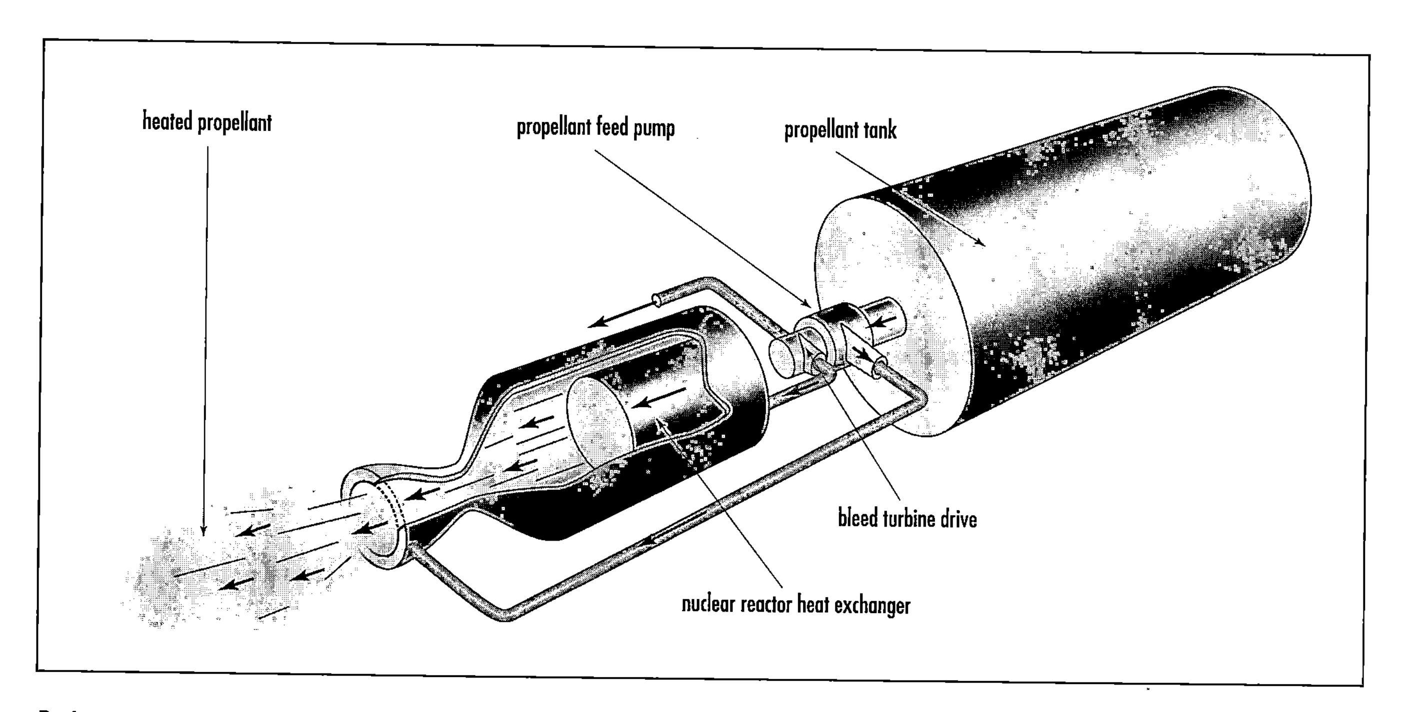

Basic components and principle of operation of a nuclear thermal rocket. The propellant tank carries liquid hydrogen at cryogenic temperatures. This single propellant is heated to extremely high temperatures as it passes through the nuclear reactor, and then very hot gaseous hydrogen is expelled out the rocket's nozzle to produce thrust.

after some of the indigenous wildlife) literally became an inferno as the very hot hydrogen gas spontaneously ignited upon contact with the air and burned (with atmospheric oxygen) to form water.

See also NUCLEAR ROCKET; SPACE NUCLEAR PROPULSION.

nuclear weapon A device in which the explosion results from the energy released by reactions involving atomic nuclei, either fission, or fusion, or both. The basic fission weapon can either be a gun-assembly type weapon or an implosion type weapon, while a thermonuclear weapon involves the use of the energy of a fission weapon to ignite the desired thermonuclear reactions.

nucleon The common name for a major constituent particle of the atomic nucleus. It is applied to protons and neutrons. The number of nucleons in a nucleus represents the atomic mass number (A) of that nucleus. For example, the isotope uranium-235 has 235 nucleons in its nucleus (92 protons and 143 neutrons). This isotope also is said to have an atomic mass number of 235 (i.e., A = 235).

nucleosynthesis The production of heavier chemical elements from the fusion or joining of lighter chemical elements (e.g., hydrogen or helium nuclei) in thermonuclear reactions in stellar interiors.

See also CARBON CYCLE; FUSION (NUCLEAR); PROTON-PROTON CHAIN REACTION; STAR.

nucleus (plural: nuclei) 1. (atomic) The small, positively charged central region of an atom that contains essentially all of its mass. All nuclei contain both protons and neutrons except the nucleus of ordinary (light) hydrogen, which consists of a single proton. The number of protons in the nucleus determines what element an atom is, while the total number of protons and neutrons determines a particular atom's nuclear properties (such as radioactivity, the ability to fission, etc.). Isotopes of a given chemical element have the same number of protons in their nuclei but different numbers of neutrons, resulting in different atomic masses—as, for example, uranium-235 and uranium-238. 2. (cometary) The small (few-kilometer-diameter), permanent, solid ice and rock central body of a comet. 3. (galactic) The central region of a galaxy a few light-years in diameter characterized by a dense cluster of stars or possibly even the hidden presence of a massive black hole.

See also ACTIVE GALACTIC NUCLEUS.

nuclide A general term applicable to all atomic (isotopic) forms of all the elements; nuclides are distinguished by their atomic number, relative mass number (atomic mass), and energy state. *Compare with* ISOTOPE.

nutation A small, irregular wobbling variation in the precession of Earth's rotational axis. The British astronomer JAMES BRADLEY reported this phenomenon in 1748 after 19 years of careful observation. Starting in 1728 and continuing for many years, Bradley recorded the positions of many stars with great precision. His efforts improved significantly upon the work of earlier notable astronomers, including JOHN FLAMSTEED. Bradley's ability to make precise measurements led him to this important astronomical discovery. He found that Earth's axis experienced small periodic shifts, which he called nutation after the Latin word *nutare,* meaning to nod. This phenomenon, characterized as an irregular wobbling superimposed on the overall precession of Earth's rotational axis, is primarily caused by the gravitational influence of the Moon as it travels along its somewhat irregular orbit around Earth. Bradley waited almost two decades (until 1748) to publish his discovery. The cautious scientist wanted to carefully study and confirm the very subtle shifts in stellar positions before announcing his findings.

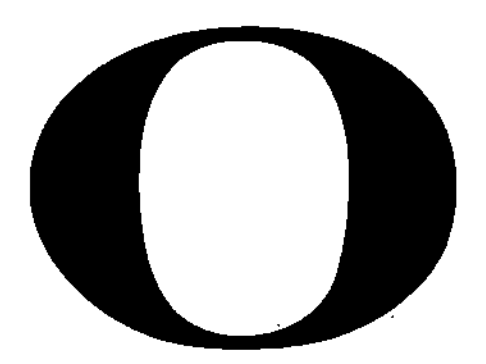

Oberon A large, heavily cratered moon of Uranus discovered in 1787 by SIR WILLIAM HERSCHEL. Oberon has a diameter of 1,520 km and travels in a synchronous orbit around Uranus at a distance of 582,600 km with a period of 13.46 days. The *Voyager 2* flyby (January 1986) revealed an ancient surface with several large impact craters and bright rays similar to those observed on the Jovian moon Callisto. Oberon has a density of 1.6 g/cm^3 (1,600 kg/m^3) and a visual albedo of 0.24. Also known as Uranus IV.

Oberth, Hermann Julius (1894–1989) Romanian-German *Mathematician, Physicist, (Theoretical) Rocket Engineer* Transylvanian-born German physicist Hermann Julius Oberth was a theoretician who helped establish the field of astronautics early in the 20th century. He originally worked independently of KONSTANTIN TSIOLKOVSKY and ROBERT GODDARD in his advocacy of rockets for space travel but later discovered and acknowledged their prior efforts. Unlike Tsiolkovsky and Goddard, however, Oberth made human beings an integral part of the space travel vision. His inspirational 1923 publication *The Rocket Into Interplanetary Space* provided a comprehensive discussion of all the major aspects of space travel, and his 1929 award-winning book *Roads to Space Travel* popularized the concept of space travel for technical and nontechnical readers alike. Although he remained a theorist and was generally uncomfortable with actually "bending metal," his technical publications and inspiring lectures exerted a tremendous career influence on many young Germans, including the legendary WERNHER VON BRAUN.

Oberth was born on June 25, 1894, in the town of Hermannstadt in a German enclave within the Transylvanian region of Romania (then part of the Austro-Hungarian Empire). His parents were members of the historically German-speaking community of the region. His father, Julius, was a prominent physician and tried to influence his son to a career in medicine. But Oberth was a free-spirited thinker who preferred to ponder the future while challenging the "politically correct" ideas held by the current scientific establishment.

At 11, Oberth discovered the works of JULES VERNE, especially *De la terre à la lune* (From the Earth to the Moon). Through his famous 1865 novel, Verne was first to provide young readers like Oberth a somewhat credible account of a human voyage to the Moon. Verne's travelers are blasted on a journey around the Moon in a special hollowed-out capsule fired from a large cannon. Verne correctly located the cannon at a low-latitude site on the west coast of Florida called "Tampa Town." By coincidence, this fictitious site is about 120 kilometers west of Launch Complex 39 at the NASA Kennedy Space Center from which NASA's Apollo astronauts actually left for journeys to the Moon between 1968 and 1972. Oberth read this novel many times and then, while remaining excited about space travel, questioned the story's technical efficacy. He soon discovered that the acceleration down the barrel of this huge cannon would have crushed the intrepid explorers and that the capsule itself would have burned up in Earth's atmosphere. But Verne's story made space travel appear technically possible, and this important idea thoroughly intrigued Oberth. After identifying the technical limitations in Verne's fictional approach, he started searching for a more practical way to travel into space. That search quickly led him to the rocket.

In 1909, at the age of 15, Oberth completed his first plan for a rocket to carry several people. His design for a liquid-propellant rocket that burned hydrogen and oxygen came three years later. He graduated from the gymnasium (high school) in June 1912, receiving an award in mathematics. The next year he entered the University of Munich with the intention of studying medicine, but World War I interrupted his studies. As a soldier he was wounded and transferred to a military medical unit. To the great disappointment of his father, Oberth's three years of duty in this medical unit reinforced his decision not to study medicine.

Hermann J. Oberth was one of the cofounders of astronautics. Throughout his life he vigorously promoted the concept of space travel. Unlike Robert Goddard and Konstantin Tsiolkovsky, the other founding fathers of astronautics, Oberth lived to see the arrival of the space age and human spaceflight, including the Apollo Project lunar landings. ***(Courtesy of NASA)***

During World War I Oberth tried to interest the Imperial German War Ministry in developing a long-range military rocket. In 1917 he submitted his specific proposal for a large liquid-fueled rocket. He received a very abrupt response for his efforts. Certain Prussian armaments "experts" within the ministry quickly rejected his plan and reminded him that their experience clearly showed that military rockets could not fly farther than 7 kilometers. Of course, these officials were hung up on their own limited experience with contemporary military rockets that used inefficient black powder for propellant. They totally missed Oberth's breakthrough idea involving a better-performing liquid-fueled rocket.

Undaunted, Oberth returned to the university environment and continued to investigate the theoretical problems of rocketry. In 1918 he married Mathilde Hummel and a year later went back to the University of Munich to study physics. He studied at the University of Göttingen (1920–21), followed by the University of Heidelberg, before becoming certified as a secondary school mathematics and physics teacher in 1923. At this point in his life, he was unaware of the contemporary rocket theory work by Konstantin Tsiolkovsky in Russia and Robert H. Goddard in the United States. In 1922 he presented a doctoral dissertation on rocketry to the faculty at the University of Heidelberg. Unfortunately, the university committee rejected his dissertation.

Still inspired by space travel, he revised this work and published it in 1923 as *Die Rakete zu den Planetenräumen* (The rocket into interplanetary space). This modest-sized book provided a thorough discussion of the major aspects of space travel, and its contents inspired many young German scientists and engineers to explore rocketry, including Wernher von Braun. Oberth worked as a teacher and writer in the 1920s. He discovered and acknowledged the rocketry work of Goddard and Tsiolkovsky in the mid-1920s and became the organizing principle around which the practical application of rocketry developed in Germany. He served as a leading member of Verein für Raumschiffahrt (VfR), the Society for Space Travel. Members of this technical society conducted critical experiments in rocketry in the late 1920s and early 1930s, until the German army absorbed their efforts and established a large military rocket program.

Fritz Lang (1890–1976), the popular Austrian filmmaker, hired Oberth as a technical adviser during the production of the 1929-released film *Die Frau im Mond* (The woman in the Moon). Collaborating with WILLY LEY, Oberth provided the film an exceptionally prophetic two-stage cinematic rocket that startled and delighted audiences with its impressive blast-off. Lang, ever the showman, also wanted Oberth to build and launch a real rocket as part of a publicity stunt to promote the new film. Unfortunately, with little engineering experience, Oberth accepted Lang's challenge but was soon overwhelmed by the arduous task of turning theory into practical hardware. Shocked and injured by a nearly fatal liquid-propellant explosion, Oberth abandoned Lang's publicity rocket and retreated to the comfort and safety of writing and teaching mathematics in his native Romania.

Oberth was a much better theorist and technical writer than a nuts-and-bolts rocket engineer. In 1929 Oberth expanded his ideas concerning the rocket for space travel and human space flight in the award-winning book entitled *Wege zur Raumschiffahrt* (Roads to space travel). The Astronomical Society of France gave Oberth this award. His book helped popularize the concept of space travel for both technical and nontechnical audiences. As the newly elected president of the VfR, Oberth used some of the book's prize money to fund rocket engine research within the society. Young engineers such as Wernher von Braun had a chance to experiment with liquid-propellant engines, including one of Oberth's own concepts, the *Kegeldüse* (conic) engine design. In this visionary book Oberth also anticipated the development of electric rockets and ion propulsion systems.

Throughout the 1930s Oberth continued to work on liquid-propellant rocket concepts and on the idea of human space flight. In 1938 he joined the faculty at the Technical University of Vienna. There he participated briefly in a rocket-related project for the German air force. In 1940 he became a

German citizen. The following year he joined von Braun's rocket development team at Peenemünde.

But Oberth worked only briefly with von Braun's military rocket team at Peenemünde and in 1943 transferred to another location to work on solid-propellant antiaircraft rockets. At the end of the war Allied forces captured him and placed him in an internment camp. Upon release he left a devastated Germany and sought rocket-related employment as a writer and lecturer in Switzerland and Italy. In 1955 he joined von Braun's team of German rocket scientists at the U.S. Army's Redstone Arsenal. He worked there for several years before returning to Germany in 1958 and retiring.

Of the three founding fathers of astronautics, only Oberth lived to see some of his pioneering visions come true. These visions included the dawn of the space age (1957), human space flight (1961), the first human landing on the Moon (1969), the first space station (1971), and the first flight of a reusable launch vehicle, NASA's space shuttle (1982). He died in Nuremberg, Germany, on December 29, 1989. Oberth studied the theoretical problems of rocketry and outlined the technology needed for people to live and work in space. The last paragraph of his 1954 book, *Man Into Space,* addresses the important question: Why space travel? His eloquently philosophical response is "This is the goal: To make available for life every place where life is possible. To make inhabitable all worlds as yet uninhabitable, and all life purposeful."

objective The main light-gathering lens or mirror of a telescope. It is sometimes called the primary lens or primary mirror of a telescope.

See also REFLECTING TELESCOPE; REFRACTING TELESCOPE; TELESCOPE.

objective prism The narrow-angle prism (wedge of optical-quality glass) that astronomers place in front of the objective lens or primary mirror of a telescope in order to disperse the incoming light from each star (or celestial object) in the field of view of the instrument into small, low-resolution spectra. This device allows astronomers to simultaneously record the spectra of all the stars (or celestial objects) in the field of view on a single digital or photographic image. A transmission diffraction grating serves the same purpose.

See also TELESCOPE; TRANSMISSION GRATING.

oblateness The degree of flattening of the poles of a planet or star caused by its own rotation. The polar diameter of a rotating planet or star is less than its equatorial diameter. The greater the oblateness, the more extensive is this condition of *polar flattening.* Astronomers calculate the oblateness of a planet or star by subtracting the polar diameter of the object from its equatorial diameter and dividing the result by the equatorial diameter. This produces a dimensionless number that characterizes how far a particular rotating celestial object departs from being a perfect sphere. A perfect sphere has an oblateness of zero, since the polar diameter is equal to the equatorial diameter. The most oblate planet in the solar system is Saturn. This magnificent ringed planet has an equatorial diameter of 120,540 km (at the cloud tops) and a polar diameter of 108,730 km. These diameters yield an oblateness of 0.098 for Saturn, which is rapidly rotating. By way of comparison, Earth has an equatorial diameter of 12,756 km and a polar diameter of 12,714 km, resulting in an oblateness of just 0.0033.

See also OBLATE SPHEROID.

oblate spheroid A sphere flattened such that its polar diameter is smaller than its equatorial diameter.

obliquity of the ecliptic (**symbol:** ε) The angle between Earth's equator and the ecliptic. This angle is equivalent to the tilt of Earth's axis (which varies slightly with time due to nutation and precession). At present the tilt of Earth's axis is approximately 23°26′. This value is decreasing slowly at approximately 50″ per century.

observable universe In general, the portion of the entire universe that can be seen from Earth. Originally, in early 20th-century astronomy, the portion of the universe that could be detected and studied by the light emitted. In the context of 21st-century astronomy, the portion of the universe that is detectable across all portions of the electromagnetic spectrum (including gamma-ray, X-ray, ultraviolet, visible light, infrared, microwave, and radio frequency), as well as through the study of energetic cosmic rays and neutrinos. *Compare with* DARK MATTER.

observatory The place (or facility) from which astronomical observations are made. For example, the Keck Observatory is a ground-based observatory, while the *Hubble Space Telescope* is a space-based (Earth-orbiting) observatory.

observed In astronomy and navigation, pertaining to a value that has been measured, in contrast to one that has been computed.

occult, occulting The disappearance of one celestial object behind another. For example, a solar eclipse is an occulting of the Sun by the Moon; that is, the Moon comes between Earth and the Sun, temporarily blocking the Sun's light and darkening regions of Earth.

ocean remote sensing Covering about 70 percent of Earth's surface, the oceans are central to the continued existence of life on our planet. The oceans are where life first appeared on Earth. Today the largest creatures on the planet (whales) and the smallest creatures (bacteria and viruses) live in the global oceans. Important processes between the atmosphere and the oceans are linked. Oceans store energy. The wind roughens the ocean surface, and the ocean surface, in turn, extracts more energy from the wind and puts it into wave motion, currents, and mixing. When ocean currents change, they cause changes in global weather patterns and can cause droughts, floods, and storms.

However, our knowledge of the global oceans is still limited. Ships, coastlines, and islands provide places from which we can observe, sample, and study small portions of the oceans. But from Earth's surface we can look at only a very small part of the global ocean. Satellites orbiting Earth can

Important Data Provided by Ocean Remote Sensing from Space

Sensor	Data	Science Question	Application
Ocean-color sensor	Ocean color	Phytoplankton concentration, ocean currents; ocean surface temperature; pollution and sedimentation	Fishing productivity, ship routing; monitoring coastal pollution
Scatterometer	Wind speed; wind direction	Wave structure; currents; wind patterns	Ocean waves; ship routing; currents; ship, platform safety
Altimeter	Altitude of ocean surface; wave height; wind speed	El Niño onset and structure	Wave and current forecasting
Microwave imager	Surface wind speed; ice edge; precipitation	Thickness, extent of ice cover, internal stress of ice; ice growth and ablation rates	Navigation information; shiprouting; wave and surf forecasting
Microwave	Sea-surface temperature	Ocean-air interactions	Weather forecasting

Source: U.S. Congress, Office of Technology Assessment, 1994.

survey an entire ocean in less than an hour. Sensors on these satellites can "look" at clouds to study the weather or at the sea surface to measure its temperature, wave heights, and the direction of the waves. Some satellite sensors use microwave (radar) wavelengths to look through the clouds at the sea surface. One other important characteristic that we can see from space is the color of the ocean. Changes in the color of ocean water over time or across a distance on the surface provide additional valuable information.

When we look at the ocean from space, we see many shades of blue. Using instruments that are more sensitive than the human eye, scientists can measure the subtle array of colors displayed by the global ocean. To skilled remote-sensing analysts and oceanographers, different ocean colors can reveal the presence and concentration of phytoplankton, sediments, and dissolved organic chemicals. Phytoplankton are small, single-celled ocean plants, smaller than the size of a pinhead. These tiny plants contain the chemical chlorophyll. Plants use chlorophyll to convert sunlight into food using a process called photosynthesis. Because different types of phytoplankton have different concentrations of chlorophyll, they appear as different colors to sensitive satellite instruments. Therefore, looking at the color of an area of the ocean allows scientists to estimate the amount and general type of phytoplankton in that area, and such information tells them about the health and chemistry of that portion of the ocean. Comparing images taken at different periods reveals any changes and trends in the health of the ocean that have occurred over time.

Why are phytoplankton so important? Besides acting as the first link in a typical oceanic food chain, phytoplankton are a critical part of ocean chemistry. Carbon dioxide in the atmosphere is in balance with carbon dioxide (CO_2) in the ocean. During photosynthesis, phytoplankton remove carbon dioxide from sea water and release oxygen as a by-product. This allows the oceans to absorb additional carbon dioxide from the atmosphere. If fewer phytoplankton existed, atmospheric carbon dioxide would increase.

Phytoplankton also affect carbon dioxide levels when they die. Phytoplankton, like plants on land, are composed of substances that contain carbon. Dead phytoplankton can sink to the ocean floor. The carbon in the phytoplankton is soon covered by other material sinking to the ocean bottom. In this way the oceans serve as a reservoir, that is, a place to dispose of, global carbon, which otherwise would accumulate in the atmosphere as carbon dioxide.

Observation of the oceans from space with a variety of special instruments provides oceanographers and environmental scientists with very important data. The Remote Sensing Table summarizes some of the data that these satellite sensors can provide.

See also AQUA SPACECRAFT; REMOTE SENSING.

oersted (symbol: Oe) The unit of magnetic field strength in the centimeter-gram-second (c.g.s.) system of units.

$$1 \text{ oersted} = 79.58 \text{ amperes/meter}$$

This unit is named in honor of Hans Christian Oersted (1777–1851), a Danish physicist who was first to demonstrate the relationship between electricity and magnetism.

ogive The tapered or curved front of a missile.

ohm (symbol: Ω) The SI unit of electrical resistance. It is defined as the resistance (R) between two points on a conductor produced by a current flow (I) of one ampere when there is a constant voltage difference (potential) (V) of one volt between these points. From Ohm's law, the resistance in a conductor is related to the voltage and the current by the equation $R = V/I$, so that 1 ohm (W) of resistance = 1 volt per ampere. The unit and physical law are named in honor of the German physicist George Simon Ohm (1787–1854).

See also SI UNITS.

Ohm's law Statement that the current (I) in a linear constant current electric circuit is inversely proportional to the resistance (R) of the circuit and directly proportional to the voltage difference (V) (i.e., the electromotive force) in the circuit. Expressed as an equation, V = I R. First postulated by the German physicist George S. Ohm (1787–1854).

See also OHM.

Olbers, Heinrich Wilhelm Matthäus (1758–1840) German *Astronomer (amateur)* Sometimes a person is primarily remembered for asking an interesting but really tough question that baffled the best scientific minds of the time. The German astronomer Heinrich Wilhelm Olbers was such a person. To his credit as an observational astronomer, he discovered the main-belt asteroids Pallas in 1802 and Vesta in 1807. But today he is best remembered for formulating the interesting philosophical question now known as *Olbers's paradox*. In about 1826 he challenged his fellow astronomers by posing the following seemingly innocent question: Why isn't the night sky with its infinite number of stars not as bright as the surface of the Sun? As discussed shortly, astronomers did not develop a satisfactory response to Olbers's paradox until the middle of the 20th century.

Olbers was born on October 11, 1758, at Ardbergen (near Bremen), Germany. The son of a Lutheran minister, he entered the University of Göttingen in 1777 to study medicine but also took courses in mathematics and astronomy. By 1781 he became qualified to practice medicine and established a practice in Bremen. By day he served his community as a competent physician, and in the evenings he pursued his hobby of astronomy. He converted the upper floor of his home into an astronomical observatory and used several telescopes to conduct regular observations.

Cometography (or "comet hunting") was a favorite pursuit of 18th-century astronomers and Olbers, a dedicated amateur astronomer, eagerly engaged in this quest. In 1796 he discovered a comet and then used a new mathematical technique he had developed to compute its orbit. His technique, often referred to as *Olbers's method*, proved computationally efficient and was soon adopted by many other astronomers.

At the end of the 18th century astronomers began turning their attention to the interesting gap between Mars and Jupiter. At the time, popularization of the Titius-Bode law—actually just an empirical statement of relative planetary distances from the Sun—caused astronomers to speculate that a planet should have been in this gap. Responding to this wave of interest, the Italian astronomer GIUSEPPE PIAZZI discovered the first and largest of the minor planets, which he named Ceres, on January 1, 1801. Unfortunately, Piazzi soon lost track of Ceres. However, based on Piazzi's few observations the brilliant German mathematician CARL FRIEDRICH GAUSS was able to calculate its orbit.

At the start of the 19th century Olbers continued to successfully practice medicine in Bremen, but he also became widely recognized as a skilled astronomer. Exactly a year after Piazzi discovered Ceres, Olbers relocated the large minor planet on January 1, 1802. While observing Ceres, Olbers also discovered another minor planet, Pallas. He then continued looking in the gap and found the third main-belt asteroid to be discovered, Vesta, in 1807.

In 1811, anticipating the discovery of solar radiation pressure, Olbers suggested that a comet's tail points away from the Sun during perihelion passage because the material ejected by the comet's nucleus is influenced (pushed) by the Sun. He remained an active observational astronomer and discovered four more comets, including one in 1815 that now bears his name.

Despite these accomplishments, Olbers is best remembered for popularizing the very interesting question, "Why is the night sky dark?" He presented a technical paper in 1826 that attempted to solve this interesting puzzle. Earlier astronomers in the 17th and 18th centuries were certainly aware of the problem posed by this question. For example, the basic question is sometimes traced back to the writings of the 17th-century astronomer JOHANNES KEPLER or to a 1744 publication by the Swiss astronomer Jean Phillipe Loys de Chéseaux. However, Olbers's paper represents the most enduring formulation of the question and so today we acknowledge his definitive role by calling the question *Olbers's paradox*. Within the perspective of 19th-century cosmology, the basic problem can be summarized as follows: If the universe is infinite, unchanging, homogeneous, and therefore filled with stars (in the early 20th century the "galaxies" were included in the statement), the entire night sky should be filled with light and as bright as daytime. Of course, Olbers and all other astronomers saw that the night sky was obviously not "bright as the day"—so what was to account for this inconsistency, or paradox, between observation and theory? Olbers attempted to provide his own explanation by suggesting that the dust and gas in interstellar space blocked much of the light traveling to Earth from distant stars. Unfortunately, he was wrong in this particular suggestion, because within his cosmological model such absorbing gas would have heated up to incandescent temperatures and glowed.

To solve this apparent contraction, Olbers needed to discard the prevalent infinite, static, homogeneous model of the universe and replace that model of the universe with the big bang model of cosmology with its expanding, finite, and nonhomogeneous universe. Modern astronomers now provide a reasonable explanation of Olbers's paradox with the assistance of an expanding, finite universe model. Some suggest that as the universe expands the very distant stars (and galaxies) become obscure due to extensive redshift, a phenomenon that weakens their apparent light and makes the night sky dark. Other astrophysicists declare that the observed redshift to lower energies is not a sufficient cause to darken the sky. They suggest that the finite extent of the physical, observable universe and its changing, time-dependent nature provide the reason the night sky is truly dark despite the faint amounts of starlight reaching Earth. Today modern cosmology tells us the universe is dynamic and not statically filled with unchanging stars and galaxies. Finally, as is inherently implied within big bang cosmology, since the universe is finite in extent and the speed of light is assumed constant, there has not even been a sufficient amount of time for the light from those galaxies beyond a certain range (called the observable universe) to reach Earth yet.

Olbers died in Bremen, Germany, on March 2, 1840. The chief astronomical legacy of this successful German

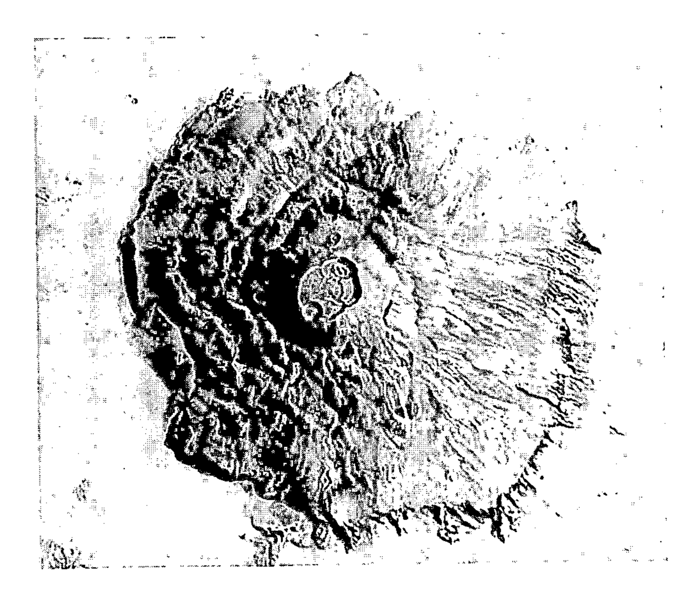

The huge Martian volcano Olympus Mons is the largest known shield volcano in the solar system. ***(Courtesy of NASA)***

physician and accomplished amateur astronomer is Olbers's paradox, a coherent response to which helps scientists highlight some of the most important aspects and implications of modern cosmology.

Olbers's paradox *See* HEINRICH WILHELM OLBERS.

Olympus Mons The largest known single volcano in the solar system. This huge mountain is located on Mars at 18° north, 133° west and is a shield volcano some 650 km wide, rising 26 km above the surrounding plains. At its top is a complex caldera 80 km across. Lava flows may be traced down its sides and over a 4-km-high scarp at its base.

See also MARS.

omni- A prefix meaning "all," as in omnidirectional.

omnidirectional antenna An antenna that radiates or receives radio frequency (RF) signals with about the same efficiency in or from all directions. *Compare with* DIRECTIONAL ANTENNA.

See also TELECOMMUNICATIONS; TELEMETRY.

one-g The downward acceleration of gravity at Earth's surface; at sea level, one-g corresponds to an acceleration of 9.8 meters per second per second (m/s^2).

See also G.

one-way communications Communications mode consisting only of the downlink received from a spacecraft.

See also TELECOMMUNICATIONS.

one-way light time (OWLT) Elapsed time (in units of light-seconds or light-minutes) for a radio signal to travel (one way) between Earth and a spacecraft or solar system body.

Onizuka, Ellison S. (1946–1986) American *Astronaut, U.S. Air Force Officer* Astronaut Ellison S. Onizuka successfully served as a NASA mission specialist on the successful Department of Defense–sponsored STS-51C space shuttle *Discovery* mission in January 1985. Then, on January 28, 1986, he lost his life when the space shuttle *Challenger* exploded during launch ascent at the start of the fatal STS-51L mission. The *Challenger* accident claimed the lives of all seven astronauts onboard and destroyed the vehicle.

See also CHALLENGER ACCIDENT.

Oort, Jan Hendrik (1900–1992) Dutch *Astronomer* An accomplished astronomer, Jan Oort made pioneering studies of the dimensions and structure of the Milky Way galaxy in the 1920s. However, he is now most frequently remembered for the interesting hypothesis he extended in 1950. At that time he postulated that a large swarm of comets circles the Sun at a great distance—somewhere between 50,000 and 100,000 astronomical units. Today astronomers refer to this very distant postulated reservoir of icy bodies as the Oort cloud in his honor.

Oort was born in the township of Franeker, Friesland Province, in the Netherlands on April 28, 1900. At 17 years of age he entered the University of Groningen, where he studied stellar dynamics under JACOBUS C. KAPTEYN. He completed his doctoral course work in 1921 and remained at the university as a research assistant for Kapteyn for about one year. He then performed astrometry-related research at Yale University between 1922 and 1924. Upon returning to Holland in 1924, he accepted an appointment as a staff member at Leiden Observatory and remained affiliated with the University of Leiden in some capacity for the remainder of his life.

Following completion of all the requirements for his doctoral degree in 1926, Oort began making important contributions to our understanding of the structure of the cosmos. Over the span of seven decades of productive intellectual activity, Oort make significant contributions to the rapidly emerging body of astronomical knowledge. At the start of his astronomical career the universe was considered to be contained within just the Milky Way galaxy, a vast collection of billions of stars bound within some rather poorly defined borders. When he published his last paper in 1992, Oort was describing some of the interesting characteristics of an expanding universe now considered to contain billions of galaxies, with space and time dimensions extending in all directions to the limits of observation.

In 1927 he revisited Kapetyn's star streaming data while also building on the recent work of the Swedish astronomer BERTIL LINDBLAD. By this approach Oort produced a classic paper that demonstrated that the differential systematic motions of stars in the solar neighborhood could be explained in terms of a rotating galaxy hypothesis. The paper, entitled "Observational Evidence Confirming Lindblad's Hypothesis of a Rotation of the Galactic System," served as Oort's platform for introducing his concept of differential galactic rotation. *Differential rotation* occurs when different parts of a gravitationally bound gaseous system rotate at different speeds. This implies that the various components of a rotating galaxy share in the overall rotation around the common center to varying degrees. Oort continued to pursue evidence for differential galactic rotation and was able to establish the mathematical theory of galactic structure. Starting in 1935, Oort served the University of Leiden as both an associate professor of astronomy and a joint director of the observatory. In 1945, at the end of World War II, he received a promotion to full professor and director of its observatory. He continued his university duties until retirement in 1970. However, after retirement from the university he continued to work regularly at the Leiden Observatory and produced technical papers until just before his death on November 5, 1992.

Despite the research-limited conditions that characterized German-occupied Holland during World War II, Oort saw the important galactic research potential of the newly emerging field of radio astronomy. He therefore strongly encouraged his graduate student, Hendrik van de Hulst (1918–2000), to investigate whether hydrogen clouds might emit a useful radio frequency signal. In 1944 van de Hulst was able to theoretically predict that hydrogen should have a characteristic radio wave emission at 21 centimeters wavelength. After Holland recovered from World War II and the University of Leiden returned to its normal academic functions, Oort was able to form a Dutch team, including van de Hulst, that discovered in 1951 the predicted 21-centimeter wavelength radio frequency emission from neutral hydrogen in outer space. By measuring the distribution of this telltale radiation, Oort and his colleagues mapped the location of hydrogen gas clouds throughout the Milky Way galaxy. Then they used this application of radio astronomy to find the large-scale spiral structure of the galaxy, the location of the galactic center, and the characteristic motions of large interstellar clouds of hydrogen.

Throughout his career as an astronomer, comets remained one of Oort's favorite topics. He supervised research by a doctoral student who was investigating the origin of comets and the statistical distribution of the major axes of their elliptical orbits. Going beyond where this student's research ended, Oort decided to investigate the consequences that passing rogue stars might have on cometary orbits. As early as 1932 the Estonian astronomer ERNEST JULIUS ÖPIK had suggested the long-period comets that occasionally passed through the inner solar system might reside in some gravitationally bound cloud at a great distance from the Sun. In 1950 Oort revived and extended the concept of a heliocentric reservoir of icy bodies in orbit around the Sun. This reservoir is now called the *Oort cloud* to commemorate his work. In refining this comet reservoir model, Oort first suggested that in the early solar system these icy bodies formed at a distance of less than 30 astronomical units (AU) from the Sun—that is, within the orbit of Neptune. But then they diffused outward as a result of gravitational interaction with the giant gaseous planets such as Jupiter and Saturn. According to Oort's hypothesis, these deflected comets then collected in a swarm (or cloud) that extended between 50,000 and 80,000 AU from the sun. At this extreme range of distances (about 25 percent of the way to the nearest star, Proxima Centuri), perturbations by stars and gas clouds helped shaped their heliocentric orbits so that only an occasional comet strays out of the cloud on a visit back through the inner solar system. Oort's theory has become a generally accepted model for where very-long-period comets come from.

He received numerous awards for his theory of galactic structure. These include the Bruce Medal of the Astronomical Society of the Pacific in 1942 and the Gold Medal of the Royal Astronomical Society in 1946. However, it is the Oort cloud—his comet cloud hypothesis—that gives his work a permanent legacy in astronomy.

Oort cloud A large reservoir, or "cloud," of icy bodies theorized to orbit the Sun at a distance of between 50,000 and 80,000 astronomical units (AU)—that is, out to the limits of the Sun's gravitational attraction. First proposed in 1950 by the Dutch astronomer JAN HENDRIK OORT. His hypothesis has become the generally accepted model in astronomy for where very-long-period comets come from.

See also COMET.

opacity The degree to which light is prevented from passing through an object or substance. Opacity is the opposite of transparency. As an object's opacity increases, the amount of light passing through the object decreases. Clean, clear air is glasslike and generally transparent, but most clouds found in Earth's atmosphere are opaque.

Oparin, Alexandr Ivanovich (1894–1980) Russian *Biochemist* Long before the space age the Russian biochemist Alexandr I. Oparin explored the biochemical origin of life. For example, in the 1920s Oparin and the British scientist J. B. S. HALDANE suggested that a reducing atmosphere (one without oxygen) and high doses of ultraviolet radiation might produce organic molecules. The *Oparin-Haldane hypothesis* implied that life (on Earth) slowly emerged from an ancient biochemical soup. However, their speculation generally went unnoticed until 1953. In that year Nobel laureate HAROLD UREY and his student Stanley Miller performed a classic exobiology experiment (now called the *Urey-Miller experiment*) in which gaseous mixtures simulating Earth's primitive atmosphere were subjected to energy sources such as ultraviolet radiation and lightning discharges. To Urey and Miller's great surprise, life-forming organic compounds, called amino acids, appeared. Oparin conducted a variety of biochemical experiments that investigated life-forming processes by means of molecular accumulations. He also wrote *The Origin of Life on Earth* (1936) and *The Chemical Origin of Life* (1964).

See also EXOBIOLOGY; LIFE IN THE UNIVERSE.

open cluster An open cluster consists of numerous young stars that formed at the same time within a large cloud of interstellar dust and gas. Open clusters are located in the spiral arms or disks of galaxies. The Hyades is an example of an open cluster. Also known as a galactic cluster.

See also GALACTIC CLUSTER; HYADES.

open loop A control system operating without feedback or perhaps with only partial feedback. An electrical or mechanical system in which the response of the output to an input is preset; there is no feedback of the output for comparison and corrective adjustment.

See also FEEDBACK.

open system In thermodynamics, a system defined by a control volume in space across whose boundaries energy and mass can flow. *Compare with* CLOSED SYSTEM.

open universe The open, or unbounded, universe model in cosmology assumes that there is not enough matter in the universe to halt completely (by gravitational attraction) the currently observed expansion of the galaxies. Therefore, the galaxies will continue to move away from one another, and the expansion of the universe (which started with the big bang) will continue forever.

See also BIG BANG (THEORY); COSMOLOGY; UNIVERSE.

operating life The maximum operating time (or number of cycles) that an item can accrue before replacement or refurbishment without risk of degradation of performance beyond acceptable limits.

operational missile A missile that has been accepted by the armed services for tactical and/or strategic application.

operations planning Performing those tasks that must be done to ensure that the launch vehicle systems and ground-based flight control operations support flight objectives. Consumables analyses and the preparation of flight rules are a part of operations planning.

Ophelia A small moon of Uranus discovered as a result of the *Voyager 2* flyby in January 1986. Ophelia is 30 km in diameter and orbits Uranus at a distance of 53,790 km with a period of 0.376 day. This moon is named after the daughter of Polonius in William Shakespeare's tragedy *Hamlet*. Astronomers call Ophelia a *shepherd moon* because together with Cordelia (another small Uranian moon) this tiny moon helps to shape and tend Uranus's outermost (epsilon) ring. It is also known as Uranus VII and S/1986 U8.

Öpik, Ernest Julius (1893–1985) Estonian *Astronomer* With many astronomical interests, the Estonian astronomer Ernest Julius Öpik chose to emphasize the study of comets and meteoroids. In 1932, after examining the perturbations of cometary orbits, he suggested that there might be a huge population of cometary nuclei at a great distance from the Sun (perhaps at about 60,000 astronomical units). He further speculated that the close passage of a rogue star might perturb this reservoir of icy celestial bodies, causing a few new comets to approach the inner solar system. The Dutch astronomer JAN OORT revived and expanded this idea in 1950. Öpik also developed a theory of ablation to describe how meteoroids burn up and disintegrate in Earth's atmosphere due to aerodynamic heating.

***Opportunity* rover** *See* MARS EXPLORATION ROVER MISSION (MER).

opposition In astronomy, the alignment of a superior planet (e.g., Mars, Jupiter, etc.) with the Sun so that they appear to an observer on Earth to be in opposite parts of the sky; that is, the planetary object has a celestial longitude differing from that of the Sun by 180 degrees. More simply put, these bodies lie in a straight line with Earth in the middle. *Compare with* CONJUNCTION.

optical astronomy The branch of astronomy that uses radiation in the visible light portion of the electromagnetic spectrum (nominally 0.4 to 0.7 micrometers [μm] in wavelength) to study celestial objects and cosmic phenomena. Naked eye astronomy is inherently optical astronomy because the human eye is senstitive to visible light, and light helps human beings form a visual awareness of the universe and its contents. Telescopic astronomy is also optical astronomy if the instrument is sensitive to sources of visible light.

See also ELECTROMAGNETIC SPECTRUM; TELESCOPE.

optical double Two stars that appear close together in the night sky. In an *optical double* the two stars have the illusionary appearance of being close together when viewed by an observer on Earth, but are actually *not* bound together by gravitational attraction; in a *physical double* the stars are gravitationally bound and orbit around their common barycenter. Astronomers generally restrict the expression *double star system* to optical doubles and use the expression *binary star system* for physical doubles.

optical interferometer In the early 1880s, with financial support from the Scottish-American inventor Alexander Graham Bell (1847–1922), the Nobel laurate ALBERT MICHELSON constructed a precision optical interferometer (sometimes called the *Michelson interferometer*). An optical interferometer is an instrument that splits a beam of light into two paths and then reunites the two separate beams. Should either beam experience travel across different distances or at different velocities (due to passage through different media), the reunited beam will appear out of phase and produce a distinctive arrangement of dark and light bands, called an interference pattern. In principle, therefore, the optical interferometer produces and measures interference fringes from two or more coherent wave trains from the same optical radiation (light) source.

optoelectronic device A device that combines optical (light) and electronic technologies, such as a fiber optics communications system.

orbit 1. In physics, the region occupied by an electron as it moves around the nucleus of an atom. 2. In astronomy and space science, the path followed by a satellite around an astronomical body, such as Earth or the Moon. When a body is moving around a primary body such as Earth under the influence of gravitational force alone, the path it takes is called its orbit. If a spacecraft is traveling along a closed path with respect to the primary body, its orbit will be a circle or an ellipse. Perfectly circular orbits are not achieved in practice. However, the ellipse approaches a circle when the eccentricity becomes small (i.e., approaches zero). The planets, for example, have the Sun as their primary body and follow nearly circular orbits. When a satellite makes a full trip around its primary body it is said to complete a revolution, and the time required is called its *period* or the *period of revolution.*

See also ORBITS OF OBJECTS IN SPACE.

orbital elements A set of six parameters used to specify a Keplerian orbit and the position of a satellite in such an orbit at a particular time. These six parameters are semimajor axis (a), which gives the size of the orbit; eccentricity (e), which gives the shape of the orbit; inclination angle (i), which gives the angle of the orbit plane with respect to the central (primary) body's equator; right ascension of the ascending node (Ω), which gives the rotation of the orbit plane from the reference axis; argument of perigee (ω), which is the angle from the line of the ascending node to perigee point measured along the orbit in the direction of the satellite's motion; and true anomaly (θ), which gives the location of the satellite on the orbit.

See also KEPLERIAN ELEMENTS.

orbital injection The process of providing a space vehicle with sufficient velocity to establish an orbit.

orbital maneuvering system (OMS) Two orbital maneuvering engines, located in external pods on each side of the aft fuselage of the space shuttle orbiter structure, that provide thrust for orbit insertion, orbit change, orbit transfer, rendezvous and docking, and deorbit. Each pod contains a high-pressure helium storage bottle, a tank, pressurization regulators and controls, a fuel tank, an oxidizer tank, and a pressure-fed regeneratively cooled rocket engine. Each OMS engine develops a vacuum thrust of 27 kilonewtons and uses a hypergolic propellant combination of nitrogen tetroxide and monomethyl hydrazine. These propellants are burned at a nominal oxidizer-to-fuel mixture ratio of 1.65 and a chamber pressure of 860 kilonewtons per square meter (kN/m^2). The system can provide a velocity change of 305 meters per second when the orbiter carries a payload of 29,500 kilograms.

See also SPACE TRANSPORTATION SYSTEM (STS).

orbital mechanics The scientific study of the dynamic relationships among the celestial bodies of the solar system and the motion of human-made objects in outer space, such as spacecraft and satellites. Classical orbital mechanics (sometimes called *celestial mechanics*) involves the mathematical tools and physical principles for analyzing the relative motions of objects under the influence of gravitational fields, such as the motion of a moon or artificial satellite in orbit around a planet. Space age orbital mechanics (sometimes referred to as *astrodynamics*) incorporates the mathematical tools and physical laws necessary to treat the motion of human-made objects under the influence of gravity and various thrust-producing rocket engines.

See also KEPLERIAN ELEMENTS; KEPLER'S LAWS OF PLANETARY MOTION; ORBITS OF OBJECTS IN SPACE.

orbital period The interval between successive passages of a satellite or spacecraft through the same point in its orbit. Often called period.

See also ORBITS OF OBJECTS IN SPACE.

orbital plane The imaginary plane that contains the orbit of a satellite and passes through the center of the primary (i.e., the celestial body being orbited). The angle of inclination (θ) is defined as the angle between Earth's equatorial plane and the orbital plane of the satellite.

See also ORBITS OF OBJECTS IN SPACE.

orbital transfer vehicle (OTV) A propulsion system used to transfer a payload from one orbital location to another—as, for example, from low Earth orbit (LEO) to geostationary Earth orbit (GEO). Orbital transfer vehicles can be expendable or reusable; many involve chemical, nuclear, or electric propulsion systems. An expendable orbital transfer vehicle frequently is referred to as an upper-stage unit, while a reusable OTV is sometimes called a space tug. OTVs can be designed to move people and cargo between different destinations in cislunar space.

See also UPPER STAGE.

orbital velocity The average velocity at which a satellite, spacecraft, or other orbiting body travels around its primary.

orbit determination The process of describing the past, present, or predicted (i.e., future) position of a satellite or other orbiting body in terms of its orbital parameters.

orbiter (spacecraft) A spacecraft specially designed to travel through interplanetary space, achieve a stable orbit around a target planet (or other celestial body), and conduct a program of detailed scientific investigation.

Orbiter (space shuttle) The winged aerospace vehicle portion of NASA's space shuttle. It carries astronauts and payload into orbit and returns from outer space by gliding and landing like an airplane. The operational Orbiter Vehicle (OV) fleet includes *Discovery* (OV-103), *Atlantis* (OV-104), and *Endeavour* (OV-105). The *Challenger* (OV-99) was lost in a launch accident on January 28, 1986, that claimed the lives of all seven crewmembers, and the *Columbia* (OV-102) was destroyed during a reentry accident on February 1, 2003, that claimed the lives of all seven crewmembers.

orbit inclination The angle between an orbital plane and Earth's equatorial plane.

See also ORBITS OF OBJECTS IN SPACE.

Orbiting Astronomical Observatory (OAO) A series of large astronomical observatories developed by NASA in the late 1960s to significantly broaden our understanding of the universe. The first successful large observatory placed in Earth orbit was the *Orbiting Astronomical Observatory 2* (OAO-2), nicknamed *Stargazer,* which was launched on December 7, 1968. In its first 30 days of operation OAO-2 collected more than 20 times the celestial ultraviolet (UV) data than had been acquired in the previous 15 years of sounding rocket launches. *Stargazer* also observed Nova Serpentis for 60 days after its outburst in 1970. These observations confirmed that mass loss by the nova was consistent with theory. NASA's *Orbiting Astronomical Observatory 3*

(OAO-3), named *Copernicus* in honor of the famous Polish astronomer NICHOLAS COPERNICUS, was launched successfully on August 21, 1972. This satellite provided much new data on stellar temperatures, chemical compositions, and other properties. It also gathered data on the black hole candidate Cygnus X-1, so named because it was the first X-ray source discovered in the constellation Cygnus.

Orbiting Geophysical Observatory (OGO) A series of six NASA scientific spacecraft placed in Earth orbit between 1961 and 1965. At the beginning of the U.S. civilian space program, data from the Orbiting Geophysical Observatory (OGO) spacecraft made significant contributions to an initial understanding of the near-Earth space environment and Sun-Earth interrelationships. For example, they provided the first evidence of a region of low-energy electrons enveloping the high-energy Van Allen radiation belt region.

See also EARTH'S TRAPPED RADIATION BELTS.

orbiting laser power station A large orbiting system that collects incoming solar radiation and concentrates it to operate a direct solar-pumped laser. The orbiting facility then directs the beam from this solar-pumped laser to a suitable collection facility (e.g., on the lunar surface), where photovoltaic converters tuned to the laser wavelength produce electric power with an efficiency of perhaps 50%. In one scheme for the proposed orbiting laser power station, a nearly parabolic solar collector with a radius of about 300 meters captures sunlight and directs it in a line focus onto a 10-meter-long laser, with an average concentration of several thousand solar constants. (A solar constant is the total amount of the Sun's radiant energy that crosses normal to a unit area at the top of Earth's atmosphere. At one astronomical unit from the Sun a solar constant is about 1,371 ± 5 watts per meter squared [W/m^2].)

An organic iodide gas lasant flows through the laser, circulated by a turbine-compressor combination. (A lasant is a chemical substance that supports stimulated light emission in a laser.) The hot lasant then is cooled and purified at the radiator. New lasant is added from the supply tanks to make up for the small amount of lasant lost in each pass through the laser. Power from the laser then is spread and focused by a combination of transmission mirrors to provide a 1-meter-diameter spot at distances up to 10,000 kilometers away or more.

Orbiting Quarantine Facility (OQF) A proposed Earth-orbiting laboratory in which soil and rock samples from other worlds—for example, Martian soil and rock specimens—could first be tested and examined for potentially harmful alien microorganisms before the specimens are allowed to enter Earth's biosphere. (*See* figure.) A space-based quarantine facility provides several distinct advantages: (1) it eliminates the possibility of a sample-return spacecraft's crashing and accidentally releasing its potentially deadly cargo of alien microorganisms; (2) it guarantees that any alien organisms that might escape from the orbiting laboratory's confinement facilities cannot immediately enter Earth's biosphere; and (3) it ensures that all quarantine workers remain in total isolation during *protocol testing* of the alien soil and rock samples.

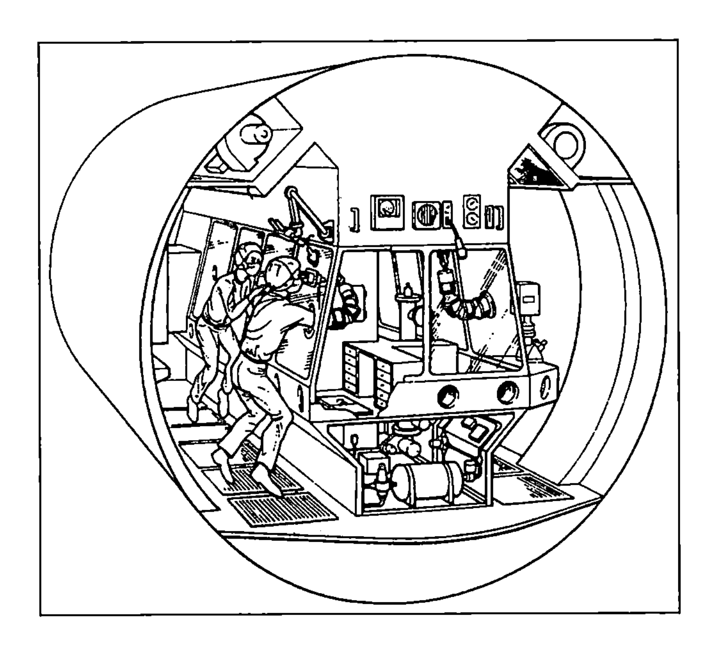

Exobiologists in an Orbiting Quarantine Facility (OQF) test for potentially harmful alien microorganisms that might be contained in an extraterrestrial soil sample. ***(Drawing courtesy of NASA)***

Three hypothetical results of such protocol testing are (1) no replicating alien organisms are discovered; (2) replicating alien organisms are discovered, but they are found not to be a threat to terrestrial life-forms; or (3) hazardous replicating alien life-forms are discovered. If potentially harmful replicating alien organisms were discovered during these protocol tests, then orbiting quarantine facility workers would render the samples harmless (e.g., through heat and chemical sterilization procedures); retain them under very carefully controlled conditions in the orbiting complex and perform more detailed studies on the alien life-forms; or properly dispose of the samples before the alien life-forms could enter Earth's biosphere and infect terrestrial life-forms.

See also EXTRATERRESTRIAL CONTAMINATION.

Orbiting Solar Observatory (OSO) A series of eight NASA Earth-orbiting scientific satellites that were used to study the Sun from space, with emphasis on its electromagnetic radiation emissions in the range from ultraviolet to gamma rays. These scientific observatories were launched between 1962 and 1975. They acquired an enormous quantity of important solar data during an 11-year solar cycle, when solar activity went from low to high and then back to low again.

See also SUN.

orbits of objects in space We must know about the science and mechanics of orbits to launch, control, and track spacecraft and to predict the motion of objects in space. An *orbit* is the path in space along which an object moves around a primary body. Common examples of orbits include Earth's path around its celestial primary (the Sun) and the Moon's path around Earth (its primary body). A single orbit is a complete path around a primary as viewed from space. It differs from a revolution. A single revolution is accomplished whenever an orbiting object passes over the primary's longitude or latitude from which it started. For example, the space shuttle *Discovery* completed a revolution whenever it passed over approximately 80° west longitude on Earth. However, while *Discovery* was orbiting from west to east around the globe, Earth itself was also rotating from west to east. Consequently, *Discovery*'s period of time for one revolution actually was longer than its orbital period. (*See* figure.) If, on the other hand, *Discovery* were orbiting from east to west (not a practical flight path from a propulsion-economy standpoint), then because of Earth's west-to-east spin its period of revolution would be shorter than its orbital period. An east-to-west orbit is called a *retrograde orbit* around Earth; a west-to-east orbit is called a *posigrade orbit.* If *Discovery* were traveling in a north-south orbit, or *polar orbit,* it would complete a period of revolution whenever it passed over the latitude from which it started. Its orbital period would be about the same as the revolution period, but not identical, because Earth actually wobbles slightly north and south.

Other terms are used to describe orbital motion. The *apoapsis* is the farthest distance in an orbit from the primary; the *periapsis,* the shortest. For orbits around planet Earth, the comparable terms are *apogee* and *perigee.* The *line of apsides* is the line connecting the two points of an orbit that are nearest and farthest from the center of attraction, such as the perigee and apogee of a satellite in orbit around Earth. For objects orbiting the Sun, *aphelion* describes the point on an orbit farthest from the Sun; *perihelion,* the point nearest to the Sun.

Another term frequently encountered is the *orbital plane.* An Earth satellite's orbital plane can be visualized by thinking of its orbit as the outer edge of a giant flat plate that cuts Earth in half.

Inclination is another orbital parameter. This term refers to the number of degrees the orbit is inclined away from the equator. The inclination also indicates how far north and south a spacecraft will travel in its orbit around Earth. If, for example, a spacecraft has an inclination of 56°, it will travel around Earth as far north as 56° north latitude and as far south as 56° south latitude. Because of Earth's rotation it will not, however, pass over the same areas of Earth on each orbit. A spacecraft in a polar orbit has an inclination of about 90°. As such, this spacecraft orbits Earth traveling alternately in north and south directions. A polar-orbiting satellite eventually passes over the entire Earth because Earth is rotating from west to east beneath it. NASA's *Terra* spacecraft is an example of a spacecraft whose cameras and multispectral sensors observe the entire Earth from a nearly polar orbit, providing valuable information about the terrestrial environment and resource base.

A satellite in an equatorial orbit around Earth has zero inclination. The Intelsat communications satellites are examples of satellites in equatorial orbits. By their placement into near-circular equatorial orbits at just the right distance above Earth, these spacecraft can be made essentially to "stand still" over a point on Earth's equator. Such satellites are called *geostationary.* They are in *synchronous orbits,* meaning they take as long to complete an orbit around Earth as it takes for Earth to complete one rotation about its axis (i.e., approximately 24 hours). A satellite at the same "synchronous" altitude but in an inclined orbit also may be called synchronous. While this

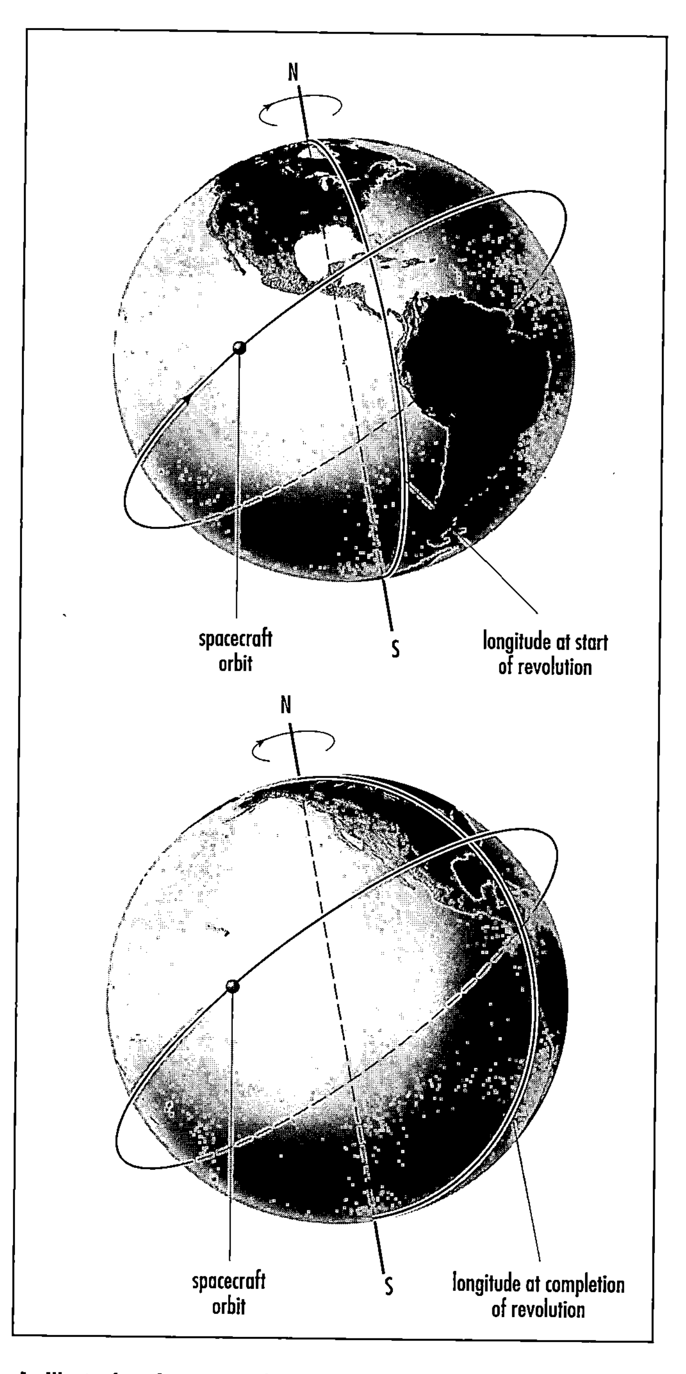

An illustration of a spacecraft's west-to-east orbit around Earth and how Earth's west-to-east rotation moves longitude ahead. As shown here, the period of one revolution can be longer than the orbital period.

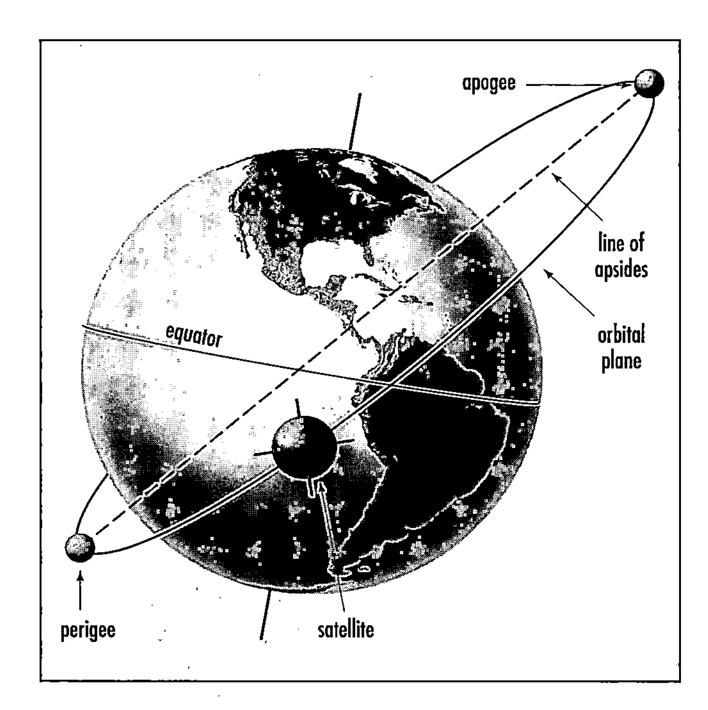

The terms *apogee* and *perigee* described in terms of a satellite's orbit around Earth

particular spacecraft would not move much east and west, it would move north and south over Earth to the latitudes indicated by its inclination. The terrestrial ground track of such a spacecraft resembles an elongated figure eight, with the crossover point on the equator.

All orbits are elliptical, in accordance with JOHANNES KEPLER's first law of planetary motion (described shortly). However, a spacecraft generally is considered to be in a circular orbit if it is in an orbit that is nearly circular. A spacecraft is taken to be in an elliptical orbit when its apogee and perigee differ substantially.

Two sets of scientific laws govern the motions of both celestial objects and human-made spacecraft. One is Newton's law of gravitation; the other, Kepler's laws of planetary motion.

The brilliant English scientist and mathematician SIR ISAAC NEWTON observed the following physical principles:

1. All bodies attract each other with what we call gravitational attraction. This applies to the largest celestial objects and to the smallest particles of matter.
2. The strength of one object's gravitational pull upon another is a function of its mass—that is, the amount of matter present.
3. The closer two bodies are to each other, the greater their mutual attraction. These observations can be stated mathematically as

$$F = [G\ m_1\ m_2] / r^2$$

where F is the gravitational force acting along the line joining the two bodies (N); m_1 and m_2 are the masses (in kilograms) of body one and body two, respectively; r is the distance between the two bodies (m); and G is the universal gravitational constant (6.6732×10^{-11} newton-meter^2/kilogram^2).

Specifically, Newton's law of gravitation states that two bodies attract each other in proportion to the product of their masses and inversely as the square of the distance between them. This physical principle is very important in launching spacecraft and guiding them to their operational locations in space and frequently is used by astronomers to estimate the masses of celestial objects. For example, Newton's law of gravitation tells us that for a spacecraft to stay in orbit, its velocity (and therefore its kinetic energy) must balance the gravitational attraction of the primary object being orbited. Consequently, a satellite needs more velocity in low than in high orbit. For example, a spacecraft with an orbital altitude of 250 km will have an orbital speed of about 28,000 km per hour.

Our Moon, on the other hand, which is about 442,170 km from Earth, has an orbital velocity of approximately 3,660 km per hour. Of course, to boost a payload from the surface of Earth to a high-altitude (versus low-attitude) orbit requires the expenditure of more energy, since we are in effect lifting the object farther out of Earth's "gravity well."

Following the thrusting period any spacecraft launched into orbit moves in accordance with the same laws of motion that govern the motions of the planets around our Sun and the motion of the Moon around Earth. The three laws that describe these planetary motions, first formulated by the German astronomer Johannes Kepler, may be stated as follows:

1. Each planet revolves around the Sun in an orbit that is an ellipse, with the Sun as it focus, or primary body.
2. The radius vector—such as the line from the center of the Sun to the center of a planet, from the center of Earth to the center of the Moon, or from the center of Earth to the center (of gravity) of an orbiting spacecraft—sweeps out equal areas in equal periods of time.
3. The square of a planet's orbital period is equal to the cube of its mean distance from the Sun. We can generalize this last statement and extend it to spacecraft in orbit about Earth by saying that a spacecraft's orbital period increases with its mean distance from the planet.

In formulating his first law of planetary motion, Kepler recognized that purely circular orbits did not really exist. Rather, only elliptical ones were found in nature, being determined by gravitational perturbations (disturbances) and other factors. Gravitational attractions, according to Newton's law of gravitation, extend to infinity, although these forces weaken with distance (as an inverse square law phenomenon) and eventually become impossible to detect. However, spacecraft orbiting Earth, while influenced primarily by the gravitational attraction of Earth (and anomalies in Earth's gravitational field), also are influenced by the Moon and the Sun and possibly other celestial objects, such as the planet Jupiter.

Kepler's third law of planetary motion states that the greater a body's mean orbital altitude, the longer it will take for it to go around its primary. Let us take this principle and apply it to a rendezvous maneuver between a space shuttle orbiter and a satellite in low Earth orbit (LEO). To catch up with and retrieve an uncrewed spacecraft in the same orbit, the space shuttle first must be decelerated. This action causes the orbiter vehicle to descend to a lower orbit. In this lower orbit the shuttle's velocity would increase. When properly positioned in the vicinity of the target satellite, the orbiter then would be accelerated, raising its orbit and matching orbital velocities for the rendezvous maneuver with the target spacecraft.

Another very interesting and useful orbital phenomenon is the Earth satellite that appears to "stand still" in space with respect to a point on Earth's equator. Such satellites were first envisioned by the British scientist and writer SIR ARTHUR C. CLARKE in a 1945 essay in *Wireless World.* Clarke described a system in which satellites carrying telephone and television transmitters would circle Earth at an orbital altitude of approximately 35,580 km above the equator. Such spacecraft move around Earth at the same rate that the Earth rotates on its axis. Therefore, they neither rise nor set in the sky like the planets and the Moon but rather always appear to be at the same longitude, synchronized with Earth's motion. At the equator Earth rotates about 1,600 km per hour. Satellites placed in this type of orbit are called *geostationary* (or *geosynchronous*) spacecraft.

It is interesting to note here that the spectacular Voyager missions to Jupiter, Saturn, and beyond used a "gravity assist" technique to help speed them up and shorten their travel time. How can a spacecraft be speeded up while traveling past a planet? A spacecraft increases in speed as it approaches a planet (due to gravitational attraction), but the gravity of the planet also should slow it down as it begins to move away again. So where does this increase in speed really come from?

There are three basic possibilities for a spacecraft trajectory when it encounters a planet. (*See* figure.) The first possible trajectory involves a direct hit, or hard landing. This is an *impact trajectory.* (See trajectory a.) The second type of trajectory is an *orbital-capture trajectory.* The spacecraft is simply "captured" by the gravitational field of the planet and enters orbit around it. (See trajectories b and c.) Depending on its precise speed and altitude (and other parameters), the spacecraft can enter this captured orbit from either the leading or trailing edge of the planet. In the third type of trajectory, a *flyby trajectory,* the spacecraft remains far enough away from the planet to avoid capture but passes close enough to be strongly affected by its gravity. In this case the speed of the spacecraft will be increased if it approaches from the trailing side of the planet (see trajectory d) and diminished if it approaches from the leading side (see trajectory e). In addition to changes in speed, the direction of the spacecraft's motion also changes.

Thus, the increase in speed of the spacecraft actually comes from a decrease in speed of the planet itself. In effect, the spacecraft is being "pulled" by the planet. Of course, this is a greatly simplified discussion of complex encounter phenomena. A full account of spacecraft trajectories must consider the speed and actual trajectory of the spacecraft and planet, how close the spacecraft will come to the planet, and the size (mass) and speed of the planet in order to make even a simple calculation.

Perhaps an even better understanding of gravity assist can be obtained if we use vectors in a more mathematical explanation. The way in which speed is added to the flyby spacecraft during close encounters with the planet Jupiter is shown in the figure. During the time that spacecraft, such as *Voyager 1* and *2,* were near Jupiter, the heliocentric (Sun-centered) path they followed in their motion with respect to Jupiter closely approximated a hyperbola.

The heliocentric velocity of the spacecraft is the vector sum of the orbital velocity of Jupiter (V_j) and the velocity of the spacecraft with respect to Jupiter (i.e., tangent to its trajectory—the hyperbola). The spacecraft moves toward Jupiter along an asymptote, approaching from the approximate direction of the Sun and with asymptotic velocity (V_a). The heliocentric arrival velocity (V_1) is then computed by vector addition:

$$V_1 = V_j + V_a$$

The spacecraft departs Jupiter in a new direction, determined by the amount of bending that is caused by the effects of the

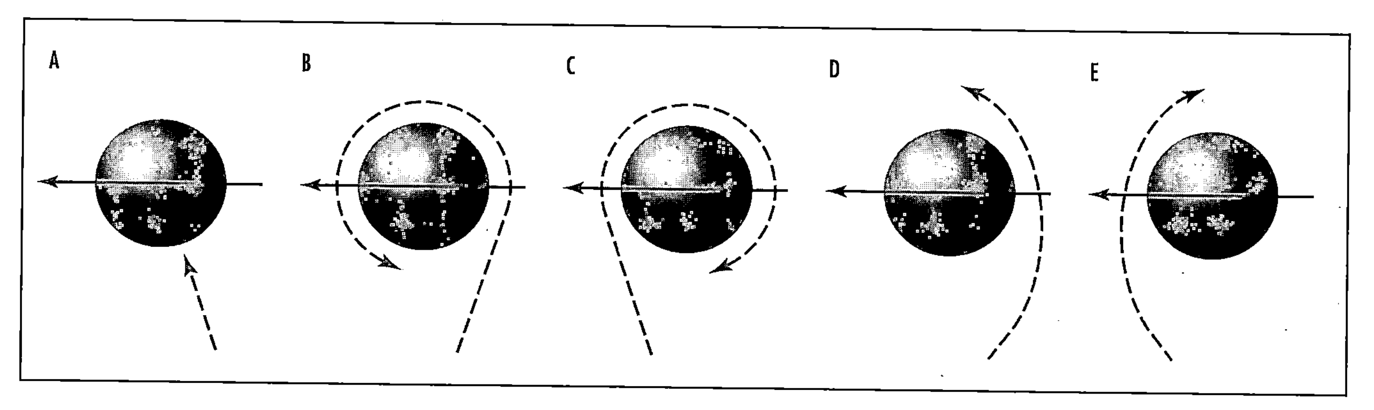

Possible trajectories of a spacecraft encountering a planet

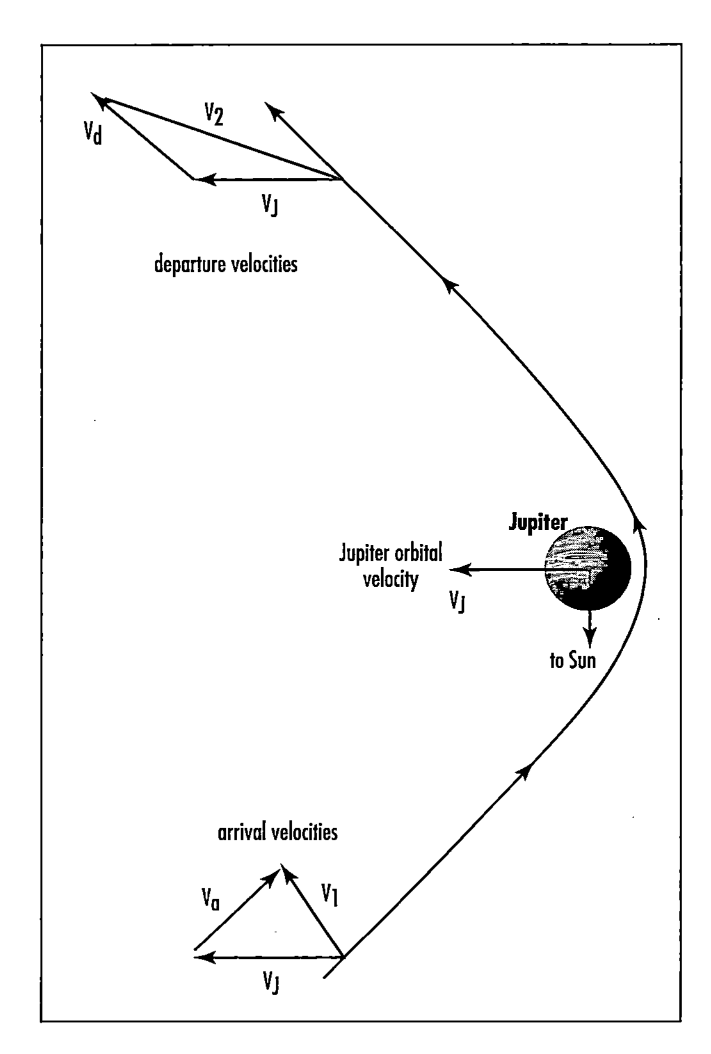

Velocity change of a spacecraft during a Jupiter flyby

gravitational attraction of Jupiter's mass on the mass of the spacecraft. The asymptotic departure speed (V_d) on the hyperbola is equal to the arrival speed. Thus, the length of V_a equals the length of V_d. For the heliocentric departure velocity, $V_2 = V_j + V_d$. This vector sum is also depicted in the velocity figure.

During the relatively short period of time that the spacecraft is near Jupiter, the orbital velocity of Jupiter (V_j) changes very little, and we assume that V_j is equal to a constant. The vector sums in the velocity figure illustrate that the deflection, or bending, of the spacecraft's trajectory caused by Jupiter's gravity results in an increase in the speed of the spacecraft along its hyperbolic path, as measured relative to the Sun. This increase in velocity reduces the total flight time necessary to reach Saturn and points beyond. This "indirect" type of deep space mission to the outer planets saves two or three years of flight time when compared to "direct-trajectory" missions, which do not take advantage of gravity assist.

Of course, while the spacecraft gains speed during its Jovian encounter, Jupiter loses some of its speed. However, because of the extreme difference in their masses, the change in Jupiter's velocity is negligible.

orbit trim maneuvers Small changes in a spacecraft's orbit around a planet for the purpose of adjusting an instrument's field-of-view footprint, improving sensitivity of a gravity field survey, or preventing too much orbital decay. To make a change increasing the altitude of periapsis ("perigee" for an Earth-orbiting satellite), an orbit trim maneuver (OTM) would be designed to increase the spacecraft's velocity when it is at apoapsis ("apogee" for an Earth-orbiting satellite). To decrease the apoapsis altitude, an OTM would be performed at periapsis, reducing the spacecraft's velocity. Slight changes in the orbital plane's orientation also can be made with orbit trim maneuvers. However, the magnitude of such changes is necessarily small due to the limited amount of "maneuver" propellant typically carried by a spacecraft.

See also NAVIGATION.

order of magnitude A factor of 10; a value expressed to the nearest power of 10—for example, a cluster containing 9,450 STARs has approximately 10,000 stars in an order of magnitude estimate.

Origins Program (NASA) Thousands of years ago on a small rocky planet orbiting a modest star in an ordinary spiral galaxy, our remote ancestors looked up and wondered about their place between Earth and sky. At the start of a new millennium, we ask the same profound questions: How did the universe begin and evolve? How did we get here? Where are we going? Are we alone?

Today, after only the blink of an eye in cosmic time, scientists are beginning to answer these questions. Space probes and space-based and ground-based astronomical observatories have played a central role in this process of discovery. NASA's Origins Program involves a series of space missions intended to help scientists answer these age-old astronomical questions. NASA's Origins-themed missions include the *Spitzer Space Telescope* (SST) (previously called the *Space Infrared Telescope Facility*), the *James Webb Space Telescope* (JWST) (previously called the *Next Generation Space Telescope*), the *Terrestrial Planet Finder* (TPF), and the *Space Interferometry Mission* (SIM). Earth-based observations using the Keck Telescopes on Mauna Kea, Hawaii, are also part of the current program.

The central principle of the Origins Program mission architecture has been that each major mission builds on the scientific and technological legacy of previous missions while providing new capabilities for the future. In this way the complex challenges of the theme can be achieved with reasonable cost and acceptable risk. For example, the techniques of interferometry developed for the Keck Interferometer and the *Space Interferometry Mission,* along with the infrared detector technology from the *Spitzer Space Telescope* and the large optics technology needed for the *James Webb Space Telescope,* will enable the *Terrestrial Planet Finder* mission to search out and characterize habitable planets.

The *Kepler* mission in NASA's Discovery Program will also provide valuable planetary system statistics and exemplifies the kind of alternate approaches the overall Origins Program is attempting to embrace. The dynamic state of this emergent scientific field (that is, extrasolar planet hunting) suggests strongly that the actions undertaken to achieve the Origins Program goals must remain flexible and must adapt to and make use of evolving technical and scientific knowledge and capability.

Support for Origins missions is provided by two key science centers: the Michelson Science Center (MSC) and the Space Telescope Science Institute (STScI). MSC is a science operations and analysis service sponsored by the Origins Program and operated by the California Institute of Technology. The MSC facilitates timely and successful execution of projects pursuing the discovery and characterization of extrasolar planetary systems and terrestrial planets. STScI is operated by the Association of Universities for Research in Astronomy, Inc., for NASA under contract with NASA's Goddard Space Flight Center in Greenbelt, Maryland.

Even as scientists and engineers work to develop the missions for this decade and the next, they must also begin to envision where their explorations will lead them afterward, since developing the needed advanced space technologies can easily take a decade or more before they are ready to be applied. For example, beyond the Terrestrial Planet Finder (TPF) mission, scientific attention should turn to detailed studies of any indications of life found on the planets that TPF discovers. This will require a still more capable spectroscopic mission, a *Life Finder* (LF) mission, which will probe the infrared spectrum with great sensitivity and resolution. Anticipated follow-on space exploration missions, therefore, call for advanced investigations in galaxy and planetary system formation and cosmology that require a high-resolution interferometry telescope such as a proposed 8-meter space-based telescope called the *Single Aperture Far-Infrared Observatory* (SAFIR).

In a decades-long program of organized investigation, the suggested *Single Aperture Far-Infrared Observatory* (SAFIR) might launch and operate between TPF and LF, carrying out its own science program while leading to the 25-meter spaced-based telescopes needed for the LF mission. The technology developed for such a mission might also be used as a building block for a kilometer-baseline interferometer used at far-infrared wavelengths for cosmological studies. Investigations in the distribution of matter in the universe (including dark matter) will require a large-scale UV/optical observatory that will build on the technology developments of the *James Webb Space Telescope* (JWST) and the *Space Interferometry Mission* (SIM) and pave the way for more challenging UV/optical telescopes of the future. The astronomical search for scientific answers to the key questions that form the theme of the Origins Program provides an inspiring and far-reaching vision that challenges both scientists and engineers.

One of the most exciting parts of the Origins Program is exploring the diversity of other worlds and searching for those that might harbor life. During the past three decades scientists have used both ground- and space-based facilities to look inside the cosmic nurseries where stars and planets are born. Parallel studies conducted in the solar system with planetary probes and of meteorites have revealed clues to the processes that shaped the early evolution of our own planetary system. An overarching goal of space science in this century is to connect what we observe elsewhere in the universe with objects and phenomena in our own solar system. Scientists now have strong evidence, based on the telltale wobbles measured for nearly 100 nearby stars, that they are orbited by otherwise unseen planets. One remarkable star, Upsilon Andromedae, shows evidence of three giant planet companions. For another, HD209458, astronomers have observed the periodic decrease in its brightness due to the transit of one of its planets across the stellar disk and have thereby been able to measure the planet's radius and mass.

However, these newly discovered planetary systems are quite unlike our own solar system. The masses of the extrasolar planets span a broad range from one-eighth to more than ten Jupiter masses. Many of these planets (often called HOT JUPITERS) are surprisingly close to their parent stars, and the majority are in eccentric orbits. Extreme proximity and eccentricity are two characteristics not seen in the giant planets of our solar system. Although planetary systems like our own are only now becoming detectable, the lack of a close analog to our own solar system and the striking variety of the detected systems raises a fascinating question: Is our solar system of a rare (or even unique) type?

In tandem, astronomers have now identified the basic stages of star formation. The process begins in the dense cores of cold gas clouds (so-called molecular clouds) that are on the verge of gravitational collapse. It continues with the formation of protostars, infant stellar objects with gas-rich, dusty circumstellar disks that evolve into adolescent "main-sequence" stars. Tenuous disks of ice and dust that remain after most of the gas disk has dispersed often surround these more mature stars. It is in the context of these last stages of star formation that planets are born. Scientists have now found many extrasolar planets. Most are unlike those in our own solar system. But might there be near-twins of our solar system as well? Are there Earthlike planets? What are their characteristics? Could they support life? Do some actually show signs of past or present life?

See also EXTRASOLAR PLANETS; *JAMES WEBB SPACE TELESCOPE;* KEPLER MISSION; LIFE IN THE UNIVERSE; SPACE INTERFEROMETRY MISSION; *SPITZER SPACE TELESCOPE;* TERRESTRIAL PLANET FINDER.

Orion (Ori) A very conspicuous constellation on the celestial equator, named after the great hunter in Greek mythology. Prominent stars such as Betelgeuse and Rigel help outline Orion, and a line of three bright stars (Alnilam, Alnitak, and Mintaka) form the belt of Orion.

See also CONSTELLATION (ASTRONOMY).

Orion Nebula A bright nebula about 1,500 light-years away in the constellation Orion. Astronomers also refer to the Orion Nebula as M42 or NGC 1976.

See also NEBULA.

orrery A mechanical model of the solar system that shows the movement of the planets in their relative orbital periods around the Sun. This device was popular in the 18th century and obtained its name from its inventor, Charles Boyle, the fourth Earl of Cork and Orrery (1676–1731). Also called a planetarium (pl: planetaria).

oscillating universe A closed universe in which gravitational collapse is followed by a new wave of expansion.

See also COSMOLOGY.

oscillatory combustion Unstable combustion in a rocket engine or motor, characterized by pressure oscillations.

See also ROCKET.

O stars Very hot, large blue stars of spectral type O with surface temperatures ranging from 28,000 K to 40,000 K. These stars have the shortest lifetimes, of approximately 3 to 6 million years.

See also STAR.

outer planets The planets in the solar system with orbits greater than the orbit of Mars. These are Jupiter, Saturn, Uranus, Neptune, and Pluto. Of these outer planets, all except Pluto also are called the giant planets.

outer space A general term referring to any region beyond Earth's atmospheric envelope. By informal international agreement, outer space usually is considered to begin at between 100 and 200 km altitude. Within the U.S. aerospace program, persons who have traveled beyond about 80 km altitude often are recognized as space travelers or astronauts.

outgassing Release of gas from a material when it is exposed to an ambient pressure lower than the vapor pressure of the gas. Generally refers to the gradual release of gas from enclosed surfaces when an enclosure is vacuum pumped or to the gradual release of gas from a spacecraft's surfaces and components when they are first exposed to the vacuum conditions of outer space following launch. Outgassing presents a problem to spacecraft designers because the released vapor might recondense on optical surfaces (instruments) or other special spacecraft surfaces where these unwanted material depositions can then degrade performance of the component or sensor. Aerospace engineers try to avoid such outgassing problems by carefully selecting the materials used in the spacecraft. In addition, certain (outgassing prone) components can be placed through a lengthy thermal vacuum ("bake-out") treatment to remove unwanted substances prior to final installation and launch.

overexpanding nozzle A nozzle in which the working fluid is expanded to a lower pressure than the external pressure; that is, an overexpanding nozzle has an exit area larger than the optimum exit area for the particular external pressure environment. For example, a nozzle designed to operate in space (where the ambient or external pressure is essentially zero) would overexpand the rocket's exhaust gases if that particular rocket engine nozzle configuration were used in Earth's atmosphere, where the ambient pressure is greater than the vacuum conditions found in space.

See also NOZZLE.

overpressure The pressure resulting from the blast wave of an explosion. It is referred to as "positive" when it exceeds atmospheric pressure and "negative" during the passage of the wave when resulting pressures are less than atmospheric pressure.

oxidizer Material whose main function is to supply oxygen or other oxidizing materials for deflagration (burning) or a solid propellant for combustion of a liquid fuel.

See also PROPELLANT.

Ozma project *See* PROJECT OZMA.

Paaliaq A small moon of Saturn discovered by ground-based observations in 2000. Paaliaq has a diameter of 19 km and travels around Saturn at a distance of 15,199,000 km with an orbital period of 687 days. Also known as Saturn XX and S/2000 S 2.

See also SATURN.

pad The platform from which a rocket vehicle is launched.

See also LAUNCHPAD.

pad deluge Water sprayed on certain launch pads during the launch of a rocket so as to reduce the temperatures of critical parts of the pad or the rocket.

pair production The gamma-ray interaction process in which a high-energy (>1.02 million electron volts [MeV]) photon is annihilated in the vicinity of the nucleus of an absorbing atom with the simultaneous formation of a negative and positive electron (positron) pair.

Pallas The large (about 540 kilometers in diameter) main-belt asteroid discovered in 1802 by the German astronomer HEINRICH WILHELM OLBERS. Pallas was the second asteroid found. Pallas travels in an elliptical heliocentric orbit characterized by a perihelion of 2.12 astronomical units (AU), an aphelion of 3.42 astronomical units, and an inclination of 34.8°. It rotates every 7.8 hours and orbits around the Sun with a period of 4.62 years.

See also ASTEROID.

Palomar Observatory A major astronomical observatory located at an altitude of 1,710 m on Mount Palomar, California (northern San Diego County). The observatory is owned and operated by the California Institute of Technology and is home to the historic 5-m (200-inch) Hale Telescope, which between 1948 and 1976 was the world's largest optical telescope. This telescope is named after the American astronomer GEORGE ELLERY HALE, who obtained funding for the telescope in 1928 and selected the site on Mount Palomar in 1934. Although construction of the Hale Telescope began in the mid-1930s, work was not completed until late 1947 because of delays caused by World War II. Since 1948 the Palomar Observatory, which also includes the 1.24-m (48-inch) Oschin Telescope (a Schmidt-type reflecting telescope) for wide-field viewing, has engaged in a variety of important astronomical studies, including the Palomar Observatory Sky Survey (POSS).

Palomar Observatory Sky Survey (POSS) A detailed photographic star atlas of the northern sky made by the 1.24-m (48-inch) Oschin Telescope at the Palomar Observatory in the mid-1950s in collaboration with the National Geographic Society. The original survey was completed in 1958 and consisted of 935 pairs of photographic plates, one plate sensitive to red light and one plate sensitive to blue light. A second such survey was conducted in the 1980s, and a third survey is in progress.

See also PALOMAR OBSERVATORY.

Pan A small (20-km diameter) moon of Saturn discovered in 1990 from images previously taken in 1981 by the *Voyager 2* spacecraft and then rediscovered in 2004 in images taken by the *Cassini* spacecraft. Pan travels around Saturn within the Encke Division in Saturn's A Ring at a distance of 133,580 km with an orbital period of 0.575 day. Also called Saturn XVIII and S/1981 S13.

See also SATURN.

Pandora A satellite of Saturn discovered in 1980 as a result of the *Voyager 1* encounter. The irregularly shaped moon (about 110-km by 70-km diameter) travels around Saturn at a distance of 141,700 km (between Saturn's F and G Rings) with an orbital period of 0.629 day. Also called Saturn XVII or S/1980 S26.

See also SATURN.

panspermia The general hypothesis that microorganisms, spores, or bacteria attached to tiny particles of matter have diffused through space, eventually encountering a suitable planet and initiating the rise of life there. The word itself means "all-seeding."

In the 19th century the Scottish scientist WILLIAM THOMSON KELVIN suggested that life may have arrived on Earth from outer space, perhaps carried inside meteorites. Then in 1908, with the publication of his book *Worlds in the Making,* the Swedish chemist and Nobel laureate SVANTE ARRHENIUS put forward the idea that is now generally regarded as the *panspermia hypothesis.* Arrhenius said that life really did not start on Earth but rather was "seeded" by means of extraterrestrial spores (seedlike germs), bacteria, or microorganisms. According to his hypothesis, these microorganisms, spores, or bacteria originated elsewhere in the Milky Way galaxy (on a planet in another star system where conditions were more favorable for the chemical evolution of life) and then wandered through space attached to tiny bits of cosmic matter that moved under the influence of stellar radiation pressure.

The greatest difficulty most scientists have today with Arrhenius's original panspermia concept is simply the question of how these "life-seeds" could wander through interstellar space for up to several billion years, receive extremely severe radiation doses from cosmic rays, and still be "vital" when they eventually encounter a solar system that contains suitable planets. Even on a solar system scale, the survival of such microorganisms, spores, or bacteria would be difficult. For example, "life seeds" wandering from the vicinity of Earth to Mars would be exposed to both ultraviolet radiation from the Sun and ionizing radiation in the form of solar-flare particles and cosmic rays. The interplanetary migration of spores might take several hundred thousand years in the airless, hostile environmental conditions of outer space. Francis Crick and Leslie Orgel attempted to resolve this difficulty by proposing the *directed-panspermia hypothesis.* Feeling that the overall concept of panspermia was too interesting to abandon entirely, in the early 1970s they suggested that an ancient, intelligent alien race could have constructed suitable interstellar robot spacecraft, loaded these vehicles with an appropriate cargo of microorganisms, spores, or bacteria, and then proceeded to "seed the galaxy" with life, or at least the precursors of life. This "life-seed" cargo would have been protected during the long interstellar journey and then released into suitable planetary atmospheres or oceans when favorable planets were encountered by the robot starships.

Why would an extraterrestrial civilization undertake this type of project? It might first have tried to communicate with other races across the interstellar void. Then, when this failed, it could have convinced itself that it was *alone.* At this point in its civilization, driven by some form of "missionary zeal" to "green" (or perhaps "blue") the Milky Way galaxy with life as it knew it, the alien race might have initiated a sophisticated directed panspermia program. Smart robot spacecraft containing well-protected spores, microorganisms, or bacteria were launched into the interstellar void to seek new "life sites" in neighboring star systems. This effort might have been part of an advanced technology demonstration program—a form of planetary engineering on an interstellar scale. These life-seeding robot spacecraft may also have been the precursors of an ambitious colonization wave that never came—or perhaps is just now getting underway!

In their directed panspermia discussions, Crick and Orgel identified what they called the *theorem of detailed cosmic reversibility.* This theorem suggests that if we can now contaminate other worlds in this solar system with microorganisms hitchhiking on terrestrial spacecraft, then it is also reasonable to assume that an advanced intelligent extraterrestrial civilization could have used its robot spacecraft to contaminate or seed our world with spores, microorganisms, or bacteria sometime in the very distant past.

Others have suggested that life on Earth might have evolved as a result of microorganisms inadvertently left here by ancient astronauts themselves. It is most amusing to speculate that we may be here today because ancient space travelers were "litterbugs," scattering their garbage on a then-lifeless planet. This line of speculation is sometimes called the *extraterrestrial garbage theory* of the origin of terrestrial life.

SIR FRED HOYLE and N. C. Wickramasinghe have also explored the issue of directed panspermia and the origin of life on Earth. In several publications they argued convincingly that the biological composition of living things on Earth has been and will continue to be radically influenced by the arrival of "pristine genes" from space. They further suggested that the arrival of these cosmic microorganisms and the resultant complexity of terrestrial life is not a random process, but one carried out under the influence of a greater cosmic intelligence.

This brings up another interesting point. As people on Earth develop the technology necessary to send smart machines and humans to other worlds in our solar system (and eventually even to other star systems), should we initiate a program of directed panspermia? If we became convinced that we might really be alone in the galaxy, then strong intellectual and biological imperatives might urge us to "green the galaxy," or to seed life as we know it where there is now none. Perhaps late in the 21st century robot interstellar explorers will be sent from our solar system not only to search for extraterrestrial life but also to plant life on potentially suitable extrasolar planets when no life is found. This may be one of our higher cosmic callings, to be the first intelligent species to rise to a level of technology that permits the expansion of life itself within the galaxy. Of course, our directed panspermia effort might be only the next link in a cosmic chain of events that was started eons ago by a long-since extinct alien civilization. Perhaps millions of years from now on an Earthlike planet around a distant Sunlike star other intelligent beings will start wondering whether life on their world started spontaneously or was seeded there by an ancient civilization (in this case, *we humans*) that has long since disappeared from view in the galaxy. While the panspermia and directed panspermia hypotheses do not address how life originally started somewhere in the galaxy, they certainly provide some intriguing concepts regarding how, once started, life might "get around."

See also EXTRATERRESTRIAL CIVILIZATIONS; LIFE IN THE UNIVERSE.

parabolic Pertaining to or shaped like a parabola. A parabola is a conic section with an eccentricity (e) equal to unity (i.e., e = 1). If a parabola is symmetrical about the x-axis and has its vertex at the origin, then it is defined by the equation $y^2 = 4ax$, where a is the distance from the vertex to the focus and 4a is the length of the latus rectum.
See also CONIC SECTION.

parabolic orbit An orbit shaped like a parabola; the orbit representing the least eccentricity for escape from an attracting body.
See also ORBITS OF OBJECTS IN SPACE.

parabolic reflector A reflecting surface having a cross section along the axis in the shape of a parabola. Incoming parallel rays striking the reflector are brought to a focus point, or if the source of the rays is located at the focus, the rays are reflected as a parallel beam. Also called parabolic effect.
See also CONIC SECTION; PARABOLIC.

parallax In general, the difference in the apparent position of an object when viewed from two different points. Specifically, the angular displacement in the apparent position of a celestial body when observed from two widely separated points (e.g., when Earth is at the extremities of its orbit around the Sun). This difference is very small and is expressed as an angle, usually in arc seconds (where 1 arc second = 1/3,600 of a degree).
See also ANNUAL PARALLAX; ASTROMETRY; PARSEC.

Paris Observatory At the invitation of King Louis XIV of France, GIOVANNI DOMENICO CASSINI founded this astronomical observatory in 1667 and then directed its operation. The Paris Observatory was Europe's first national observatory.

parking orbit The temporary (but stable) orbit of a spacecraft around a celestial body. It is used for the assembly and/or transfer of equipment or to wait for conditions favorable for departure from that orbit.

parsec (symbol: pc) A unit of distance frequently encountered in astronomical studies. The parsec is defined as a parallax shift of one second of arc. The term itself is a shortened form of "parallax second." The parsec is the extraterrestrial distance at which the main radius of Earth's orbit (one astronomical unit [AU] by definition) subtends an angle of one arc-second. It is therefore also the distance at which a star would exhibit an annual parallax of one arc-second.

1 parsec = 3.2616 light-years (or 206,265 AU)

The kiloparsec (kpc) represents a distance of 1,000 parsecs (or 3,261.6 light-years), and the megaparsec (Mpc), a distance of 1 million parsecs (or about 3.262 million light-years).
See also ANNUAL PARALLAX.

particle A minute constituent of matter, generally one with a measurable mass. An elementary atomic particle such a proton, neutron, electron (beta particle), or alpha particle.
See also ELEMENTARY PARTICLE(S).

particle accelerator A device that accelerates charged nuclear and subnuclear particles by means of changing electromagnetic fields. Accelerators impart energies of millions of electron volts (MeV) and higher to particles such as protons or electrons. Accelerators used in nuclear physics research have achieved energy levels of more than 10^{12} electron volts. Accelerators useful in a ballistic missile defense (BMD) role, such as neutral particle beam (NPB) accelerators and free electron laser (FEL) devices, need only achieve particle energies of between 10^8 and 10^9 electron volts but generally require much higher beam currents.

pascal (symbol: Pa) The SI unit of pressure. It is defined as the pressure that results from a force of one newton (N) acting uniformly over an area of 1 square meter.

$$1 \text{ pascal (Pa)} = 1 \text{ N/m}^2$$

This unit is named after the French scientist Blaise Pascal (1623–62).

Pasiphaë An outer satellite of Jupiter discovered in 1908 by the British astronomer Philibert J. Melotte (1880–1961). Pasiphaë is about 40 km in diameter and travels in a retrograde orbit around Jupiter at a distance of 23,500,000 km with a period of 735 days. Its orbit is further characterized by an inclination of 145° and an eccentricity of 0.378. Also known as Jupiter VIII.
See also JUPITER.

Pasithee A very small moon of Jupiter (about 2 km in diameter) discovered in December 2001 as a result of ground-based observations at the University of Hawaii at Mauna Kea. Pasithee travels in a retrograde orbit around Jupiter at an average distance of 23,096,000 km with an inclination of approximately 165°. Also known as Jupiter XXXVIII and S/2001 J6.
See also JUPITER.

pass 1. A single circuit of Earth (or a planet) by a spacecraft or satellite. Passes start at the time the satellite crosses the equator from the Southern Hemisphere into the Northern Hemisphere (the ascending node). 2. The period of time a satellite or spacecraft is within telemetry range of a ground station.

passive Containing no power sources to augment output power or signal, such as a passive electrical network or a passive reflector. Applied to a device that draws all its power from the input signal. A dormant device or system, that is, one that is not active.

passive defense Measures taken to reduce the probability and to minimize the effects of damage caused by hostile action without the intention of taking the initiative (e.g., short of launching a preemptive first strike).

passive homing guidance A system of homing guidance wherein the receiver in the missile uses radiation from the target. *Compare with* ACTIVE HOMING GUIDANCE.
See also GUIDANCE SYSTEM.

passive satellite A satellite that does not transmit a signal. This type of "silent" human-made space object could be a dormant (but still functional) replacement satellite, a retired or decommissioned satellite, a satellite that has failed prematurely (and is now a piece of "space junk"), or possibly even a silent stalking space mine.

passive sensor A sensor that detect radiation naturally emitted by (e.g., infrared) or reflected (e.g., sunlight) from a target.

passive target A "target" orbital payload or satellite that is stabilized along its three axes and is detected, acquired, and tracked by means of electromagnetic energy reflected from the target's surface (or skin).

path 1. Of a satellite or spacecraft, the projection of the orbital plane on Earth's surface. Since Earth is turning under the satellite, the path of a single orbital pass will not be a closed curve. The terms *path* and *track* are used interchangeably. On a cylindrical map projection, a satellite's path is a sine-shaped curve. 2. Of an aircraft or aerospace vehicle, the flight path. 3. Of a meteor, the projection of the trajectory on the celestial sphere as seen by an observer.

See also GROUNDTRACK; ORBITS OF OBJECTS IN SPACE.

Patriot The U.S. Army's Patriot missile system provides high- and medium-altitude defense against aircraft and tactical ballistic missiles. The combat element of the Patriot missile system is the fire unit that consists of a radar set, an engagement control station (ECS), an equipment power plant (EPP), an antenna mast group (AMG), and eight remotely located launchers. The single phased-array radar provides the following tactical functions: airspace surveillance, target detection and tracking, and missile guidance. The ECS provides the human interface for command and control operations. Each firing battery launcher contains four ready-to-fire missiles sealed in canisters that serve a dual purpose as shipping containers and launch tubes. The Patriot's fast reaction capability, high firepower, ability to track up to 50 targets simultaneously with a maximum range of 68.5 kilometers, and ability to operate in a severe electronic countermeasures (ECM) environment are features not available in previous air defense systems. The U.S. Army received the Patriot in 1985, and the system gained notoriety during the Persian Gulf War of 1991 as the "Scud killer."

The Patriot Advanced Capability-3 (PAC-3) upgrade program is currently being managed by the Missile Defense Agency (MDA) with the Department of Defense. PAC-3 is a terminal defense system being built upon the previous Patriot and missile defense infrastructure. The PAC-3 missile is a high-velocity hit-to-kill vehicle that represents the latest generation of Patriot missile. It is being developed to provide an increased capability against tactical ballistic missiles, cruise missiles, and hostile aircraft. The PAC-3 missile provides the range, accuracy, and lethality to effectively defend against theater ballistic missiles (TBMs) that may be carrying either conventional high-explosive (HE), biological, chemical, or nuclear warheads. Unlike previous Patriot missiles, the PAC-3 uses an active radar seeker and closed-loop guidance to directly hit the target.

With the PAC-3 missile and ground system, threat missiles can be engaged and destroyed at higher altitudes and greater ranges with better lethality. The system can operate despite electronic countermeasures. Due to the active seeker and closed-loop guidance, a greater number of interceptors can be controlled at one time than with earlier Patriot missiles.

See also BALLISTIC MISSILE DEFENSE; THEATER HIGH ALTITUDE AREA DEFENSE.

Pauli exclusion principle A principle in quantum mechanics that states that no two fermions (particles such as quarks, protons, neutrons, and electrons) may exist in the same quantum state. This physical principle was initially proposed in the mid-1920s by the German physicist Wolfgang Pauli (1900–58). The Pauli exclusion principle implies, for example, that no two electrons in a given atom can have the same quantum number values for energy, spin, and angular momentum.

See also ELEMENTARY PARTICLE(S); QUANTUM MECHANICS.

payload 1. Originally, the revenue-producing portion of an aircraft's load, such as passengers, cargo, and mail. By extension, the term *payload* has become applied to that which a rocket, aerospace vehicle, or spacecraft carries over and above what is necessary for the operation of the vehicle during flight. 2. With respect to the Space Transportation System (or space shuttle), the total complement of specific instruments, space equipment, support hardware, and consumables carried in the orbiter vehicle (especially in the cargo bay) to accomplish a discrete activity in space but not included as part of the basic orbiter payload support (e.g., the remote manipulator system, or RMS). 3. With respect to a space station, an aggregate of instruments and software for the performance of specific scientific or applications investigations, or for commercial activities. Payloads may be located inside the pressurized modules of the space station, attached to the space station structure, or attached to a platform, or they may be free-flyers. 4. With respect to a military missile, the payload generally consists of a warhead, its protective container, and activating devices.

payload assist module (PAM) A family of commercially developed upper-stage rocket vehicles intended for use with NASA's space shuttle or with expendable launch vehicles, such as the Delta rocket.

payload bay The 4.57-meter-diameter by 18.3-meter-long enclosed volume within the space shuttle orbiter vehicle designed to carry payloads, upper-stage vehicles with attached payloads, payload support equipment, and associated mounting hardware. Also called the cargo bay.

See also SPACE TRANSPORTATION SYSTEM.

payload buildup The process by which the instrumentation (e.g., sensors, detectors, etc.) and equipment, including the necessary mechanical and electronic subassemblies, are combined into a complete operational package capable of achieving the objectives of a particular mission.

payload carrier With respect to the Space Transportation System (STS), one of the major classes of standard payload carriers certified for use with the space shuttle to support payload operations. Payload carriers are identified as habitable modules (e.g., spacelabs) and attached but uninhabitable modules and equipment (e.g., pallets, free-flying systems, satellites, and propulsive upper stages).

See also SPACE TRANSPORTATION SYSTEM.

payload changeout room An environmentally controlled room on a movable support structure (including a manipulator system) at the space shuttle launch pad (Complex 39 A/B at the Kennedy Space Center). It is used for inserting payloads vertically into the orbiter vehicle's cargo bay at the launch pad and supports payload transfer between the transport canister and the cargo bay.

See also SPACE TRANSPORTATION SYSTEM.

payload integration The compatible installation of a complete payload package into the spacecraft and space launch vehicle.

payload retrieval mission A space shuttle mission in which an orbiting payload is captured, secured in the cargo bay, and then returned to Earth.

See also SPACE TRANSPORTATION SYSTEM.

payload servicing mission A space shuttle mission in which an orbiting payload is given inspection, maintenance, modification, and/or repair. This type of mission generally involves rendezvous and capture operations as well as extravehicular activity (EVA), during which shuttle astronauts perform repair and servicing activities while outside the orbiter vehicle's pressure cabin. Sometimes called payload repair mission.

See also *HUBBLE SPACE TELESCOPE*; SPACE TRANSPORTATION SYSTEM.

payload specialist A noncareer astronaut who flies as a space shuttle passenger and is responsible for achieving the payload and experiment objectives. He or she is the onboard scientific expert in charge of payload and experiment operations. The payload specialist has a detailed knowledge of the payload instruments and their subsystems, operations, requirements, objectives, and supporting equipment. As such, he or she is either the principal investigator (PI) conducting experiments while in orbit or the direct representative of the principal investigator. Of course, the payload specialist also must be knowledgeable about certain basic orbiter systems, such as food and hygiene accommodations, life support systems, hatches, tunnels, and caution and warning systems.

Payne-Gaposchkin, Cecilia Helena (1900–1979) British-American *Astronomer* In 1925 Cecilia Helena Payne-Gaposchkin proposed that the element hydrogen was the most abundant element in stars and in the universe at large. While collaborating with her husband, Russian-American astronomer Sergei Gaposchkin (1898–1984), in 1938 she published an extensive catalog of variable stars.

See also ABUNDANCE OF ELEMENTS (IN THE UNIVERSE).

Peacekeeper (LGM-118A) The Peacekeeper (also called LGM-118A) is the newest U.S. intercontinental ballistic missile (ICBM). Its deployment filled a key goal of the strategic modernization program and increased the strength and credibility of the ground-based leg of the U.S. strategic (nuclear) triad. The Peacekeeper is capable of delivering 10 independently targeted warheads with greater accuracy than any other ballistic missile. It is a three-stage rocket ICBM system consisting of three major sections: the boost system, the post-boost vehicle system, and the reentry system.

The boost system consists of four rocket stages that launch the missile into space. These rocket stages are mounted atop one another and fire successively. Each of the first three stages exhausts its solid-propellant materials through a single moveable nozzle that guides the missile along its flight path. The fourth-stage post-boost vehicle is made up of a maneuvering rocket and a guidance and control system. The reentry vehicle system consists of the deployment module containing up to 10 cone-shaped reentry vehicles and a protective shroud.

The air force achieved initial operational capability of 10 deployed Peacekeeper missiles at F. E. Warren Air Force Base, Wyoming, in December 1986. Full operational capability was achieved in December 1988 with the establishment of a squadron of 50 missiles. However, with the end of the cold war the United States has begun to revise its strategic nuclear policy and has agreed to eliminate the multiple reentry vehicle Peacekeeper ICBMs by the year 2007 as part of the Strategic Arms Reduction Treaty II.

See also UNITED STATES STRATEGIC COMMAND.

Pegasus (launch vehicle) The Pegasus air-launched space booster is produced by Orbital Sciences Corporation and Hercules Aerospace Company to provide small satellite users with a cost-effective, flexible, and reliable method for placing payloads into low Earth orbit (LEO). Launching a Pegasus rocket from an airplane flying at an altitude of approximately 12 kilometers—the same altitude most commercial jet airliners fly—reduces the amount of thrust needed to overcome Earth's gravity by 10 to 15 percent.

See also LAUNCH VEHICLE.

Peltier effect The production or absorption of thermal energy (heat) that occurs at the junction of two dissimilar metals or semiconductors when an electric current flows through the junction. If thermal energy is generated by the flow of current in one direction, it will be absorbed when the current flow is reversed. This phenomenon, discovered by the French physicist J. C. A. Peltier (1785–1845), can be used in direct conversion cooling applications (e.g., infrared sensor arrays), especially for small volumes or regions where regular refrigeration and cooling techniques are impractical.

See also DIRECT CONVERSION.

penetration aid In general, techniques and/or devices employed by offensive aerospace weapon systems to increase the probability of penetration of enemy defenses. With respect to ballistic missiles, a device or group of devices that accompanies a reentry vehicle (RV) during its flight to misdirect

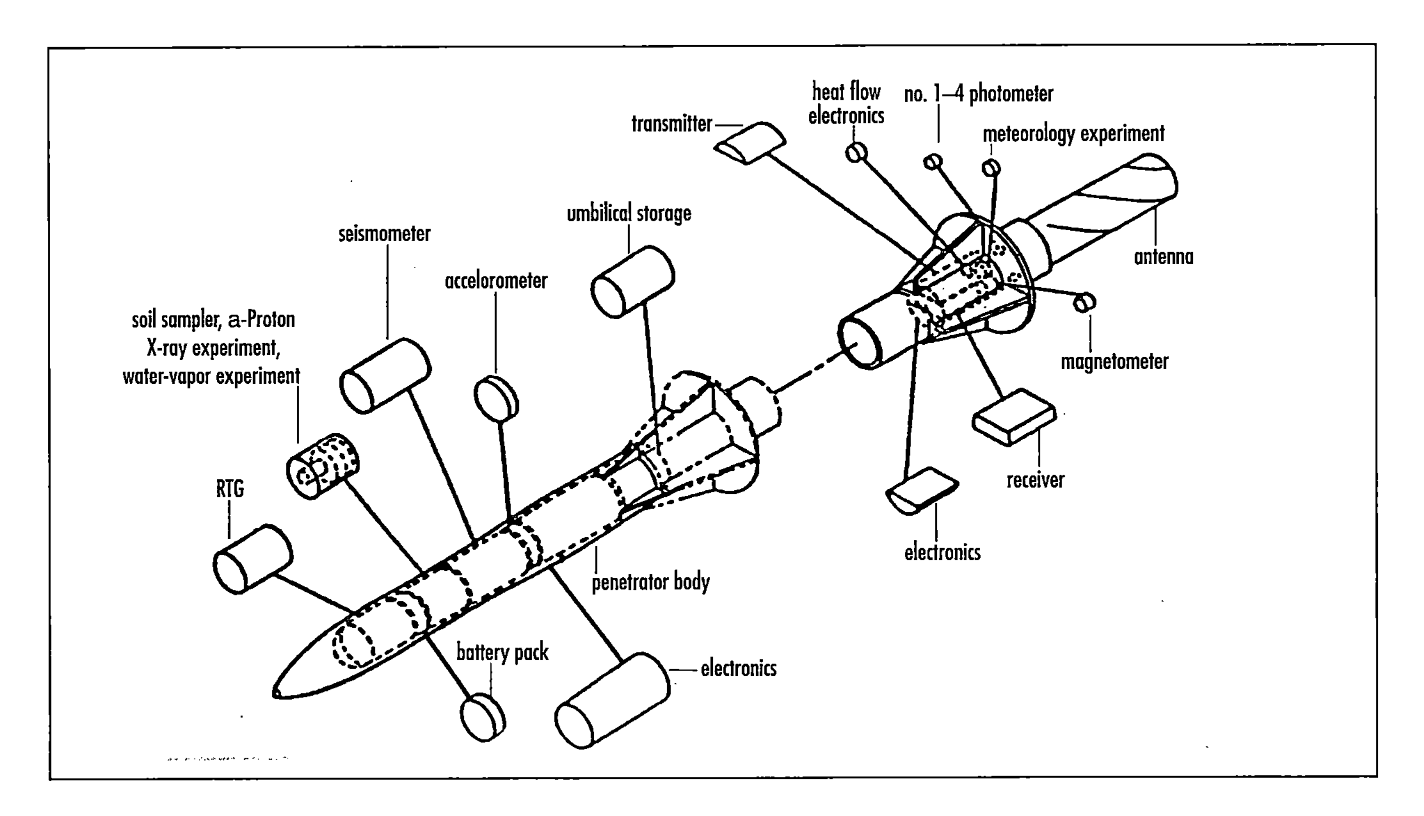

Components of a typical planetary penetrator ***(Drawing courtesy of NASA)***

defenses and thereby allow the RV to reach its target and detonate its warhead.

See also BALLISTIC MISSILE DEFENSE.

penetrator Space scientists have concluded that experiments performed from a network of penetrators can provide essential facts needed to begin understanding the evolution, history, and nature of a planetary body such as Mars. The scientific measurements performed by penetrators might include seismic, meteorological, and local site characterization studies involving heat flow, soil moisture content, and geochemistry. A typical penetrator system consists of four major subassemblies: (1) the launch tube, (2) the deployment motor, (3) the decelerator (usually a two-stage device), and (4) the penetrator itself. (*See* figure.) The launch tube attaches to the host spacecraft and houses the penetrator, deployment motor, and two-stage decelerator. The deployment motor is based on well-proven solid-rocket motor technology and provides the required deorbit velocity. If the planetary body has an atmosphere, the two-stage decelerator includes a furlable umbrella heat shield for the first stage of hypersonic deceleration. The penetrator itself is a steel device, shaped like a rocket, with a blunt ogive (curved) nose and conical flared body. The aftbody (or afterbody) of the penetrator remains at the planet's surface, with the forebody penetrating the subsurface material. Communications with scientists on Earth from the surface and subsurface penetrator sites is accomplished by means of an orbiting mothership, which interrogates each penetrator at regular intervals.

A network of penetrators can be used to study planets with solid surfaces (e.g., Mercury, Venus, Mars, and Pluto) as well as many of the interesting moons in the solar system (e.g., Titan, Io, Callisto, Ganymede, Triton, and Charon). The Jovian moon Europa is an especially interesting target because of the possibility of a liquid water ocean underneath its smooth, icy surface. A network of penetrators also can be deployed on Earth's own Moon to conduct a variety of scientific experiments, including in-situ evaluation and assessment of surface deposits of frozen volatiles now suspected to occur in certain permanently shadowed polar regions.

Penzias, Arno Allen (1933–) German-American *Physicist, Radio Astronomer* Sometimes the world's greatest scientific discoveries take place when least expected. While collaborating in the mid-1960s with ROBERT WOODROW WILSON, a fellow radio astronomer at Bell Laboratories, Arno Penzias went about the task of examining natural sources of extraterrestrial radio wave noise that might interfere with transmissions from communication satellites. To his great surprise, he and Wilson quite by accident stumbled upon the "Holy Grail" of modern cosmology—the cosmic microwave background. This all-pervading microwave background radiation resides at the very edge of the observable universe and is considered the cooled remnant (about 3K) of the ancient big bang explosion. They carefully confirmed the data that rocked the world of physics and gave cosmologists the first empirical evidence pointing to a very hot, explosive phase at the beginning of the universe. Penzias and Wilson shared the 1978 Nobel Prize in physics for this most important discovery.

Penzias was born in Munich, Germany, on April 26, 1933. His Jewish family was one of the last to successfully flee Nazi Germany before the outbreak of World War II and escaped almost certain death in one of Hitler's concentration camps. In 1939 his parents placed him and his younger brother on a special refugee train to Great Britain and then traveled there under separate exit visas to join their sons. Once reunited, the family sailed for the United States in December and arrived in New York City in January 1940. Penzias followed the education route taken by many thousands of upwardly mobile immigrants. He used hard work and the New York public school system as his pathway to a better life. In 1954 he graduated from the City College of New York (CCNY) with a bachelor of science degree in physics.

Following graduation Penzias married and joined the U.S. Army, where he served in the Signal Corps for two years. He then enrolled at Columbia University in 1956 and graduated from that institution with a master of arts degree in 1958 and a Ph.D. in 1962. His technical experience in the Army Signal Corps allowed him to obtain a research assistantship in the Columbia Radiation Laboratory. There he met and studied under Charles Townes (b. 1915), an American physicist and Nobel laureate who developed the first operational maser (microwave amplification by stimulated emission of radiation), the forerunner to the laser. For a doctoral research project Townes assigned Penzias the challenging task of building a maser amplifier suitable for use in a radio astronomy experiment of his own choosing.

Upon completion of his thesis work in 1961, Penzias sought temporary employment at the Bell Laboratories in Holmdel, New Jersey. To his surprise, the director of the Radio Research Laboratory offered him a permanent position. Penzias quickly accepted the radio astronomer position and subsequently remained a member of Bell Laboratories in various technical and management capacities until his retirement in May 1998.

In 1963 Penzias met another radio astronomer, Robert Woodrow Wilson, who had recently been hired to work at Bell Laboratories after graduating from Caltech. The two scientists embarked on the task of using a special 6-meter-diameter horn-reflector antenna that had just been constructed at Bell Laboratories to serve as an ultra–low noise receiver for signals bounced from NASA's *Echo-1* satellite, the world's first passive communications satellite experiment, launched in August 1960. At the time this ultrasensitive 6-meter horn-reflector was no longer needed for satellite work at the laboratory, so both Penzias and Wilson began to modify the instrument for radio astronomy. Once modified with traveling-wave maser amplifiers, their modest-sized horn reflector became the most sensitive radio telescope in the world (at the time). Both radio astronomers were eager to extend portions of their doctoral research with this newly converted, ultrasensitive radio telescope. Their initial objective was to measure the intensity of several interesting extraterrestrial radio sources at a radio frequency wavelength of 7.53 centimeters, a task with potential value in both the development of satellite-based telecommunications and radio astronomy.

On May 20, 1964, the two radio astronomers made a historic measurement that later proved to be the very first measurement that clearly indicated the presence of the cosmic microwave background, the remnant "cold light" from the dawn of creation. At the time neither was sure of why this strange background "noise" at around 3 K temperature kept showing up in all the measurements. As excellent scientists do, they checked every potential source for this unexplained signal. They ruled out equipment error (including antenna noise due to faulty joints), the lingering effects of previous nuclear tests in outer space, the presence of human-made radio frequency signals (including New York City), sources within the Milky Way galaxy, and even the possible echoing effects of pigeon droppings from a pair of birds that made a portion of the antenna their home. Nothing explained this persistent, omnidirectional, uniform microwave signal at 7 centimeters wavelength, corresponding to a blackbody temperature of approximately 3K. By early 1965 Penzias and Wilson had completed their initial data collection with the 6-meter horn-reflector antenna but were still no closer to unraveling the true physical nature of this persistent signal. Their careful analysis indicated the signal was definitely "real"—but what was it?

Then the mystery unraveled quickly when Penzias had a meeting with another physicist in spring 1965. During their discussions Penzias casually mentioned some recent radio astronomy work and the interesting noiselike signal he and Wilson had collected. The physicist suggested that Penzias contact ROBERT HENRY DICKE and members of his group at Princeton University, who were exploring the physics of the universe and something called the "big bang" hypothesis. Contact was made with Dicke, and he visited Bell Laboratories. After Dicke reviewed their precise work, he immediately realized that Penzias and Wilson had totally by accident beaten all other astrophysicists and cosmologists, including him, in being the first to detect evidence of the hypothesized microwave remnant of the big bang explosion.

Recognizing the scientific importance of this work, they agreed to publish side-by-side letters in the *Astrophysical Journal*. As a result, Volume 142 of the *Astrophysical Journal* contains two historic letters that appear in side-by-side fashion. In one letter Dicke and his team at Princeton University discuss the cosmological implications of the cosmic microwave background; in the other letter Penzias and Wilson from Bell Laboratories present their cosmic microwave background measurements without elaborating (in that particular letter) on the cosmological significance of the data.

Once the Penzias and Wilson data began to circulate within the astronomical community, its true significance quickly emerged. Scientists revisited the earlier big bang hypothesis work of GEORGES LEMAÎTRE and GEORGE GAMOW, elaborated upon it, and engaged in a series of corroborating measurements. For providing the first definitive evidence of the cosmic microwave background, Penzias and Wilson shared the 1978 Nobel Prize in physics with the Russian physicist Pytor Kapitsa, who received his award for unrelated achievements in low-temperature physics.

While responding to the research needs and priorities of Bell Laboratories in the 1960s and 1970s, Penzias and Wilson also continued to participate in pioneering radio astronomy research. For example, in 1973, while investigating molecules in interstellar space, they discovered the presence of the deuterated molecular species DCN and were then able to use this molecular species to trace the distribution of deuterium in the Milky Way galaxy.

In 1972 Penzias became head of the Radio Physics Research Department at Bell Laboratories, and early in 1979 his management responsibilities increased when he received a promotion to head the laboratory's Communications Sciences Research Division. At the end of 1981 he received another promotion and became vice president of research as the Bell System experienced a major transformation. While discharging his responsibilities in this new executive position, Penzias began to concentrate on the creation and effective use of new technologies. In 1989 he published a book, *Ideas and Information,* that discussed the impact of computers and other new technologies on society. In May 1998 Penzias retired from his position as chief scientist at Lucent/Bell Laboratories. He now resides with his wife in California.

Best known as the radio astronomer who codiscovered the cosmic microwave background, Penzias received many awards and honors in addition to the 1978 Nobel Prize. Some of his other awards include the Henry Draper Medal of the American National Academy of Sciences (1977) and the Herschel Medal of the Royal Astronomical Society (1977). He also received membership in many distinguished organization, including the National Academy of Sciences, the American Academy of Arts and Sciences, and the American Physical Society.

perfect cosmological principle The postulation that at all times the universe appears the same to all observers.

See also COSMOLOGY.

perfect fluid In simplifying assumptions made as part of preliminary engineering analyses, a fluid chiefly characterized by a lack of viscosity and, usually, by incompressibility. Also called an ideal fluid or an inviscid fluid.

perfect gas A gas that obeys the following equation of state: $pv = RT$, where p is pressure, v is specific volume, T is absolute temperature, and R is the gas constant.

See also IDEAL GAS.

perforation With respect to a solid-propellant rocket, the central cavity or hole down the center of a propellant grain. This perforation can take different shapes, ranging from a right circular cylinder to a complex star pattern. The geometry and dimensions of the perforation affect the solid-propellant burn rate and, consequently, the thrust delivered by the rocket as a function of time.

See also ROCKET.

peri- A prefix meaning near, as in PERIGEE.

periapsis The point in an orbit closest to the body being orbited.

See also ORBITS OF OBJECTS IN SPACE.

periastron That point of the orbit of one member of a binary star system at which the stars are nearest to each other. The opposite of APASTRON.

See also BINARY STAR SYSTEM.

pericynthian That point in the trajectory of a vehicle that is closest to the Moon.

perigee In general, the point at which a satellite's orbit is closest to the primary (central body); the minimum altitude attained by an Earth-orbiting object.

See also ORBITS OF OBJECTS IN SPACE.

perigee propulsion A programmed thrust technique for escape from a planet by an orbiting space vehicle. The technique involves the intermittent application of thrust at perigee (when the object's orbital velocity is high) and coasting periods.

perihelion The place in a solar orbit that is nearest the Sun, when the Sun is the center of attraction. *Compare with* APHELION.

perilune In an orbit around the Moon, the point in the orbit nearest the lunar surface.

period (common symbol: T) 1. In general, the time required for one full cycle of a regularly repeated series of events. 2. The time taken by a satellite to travel once around its orbit. Also called the orbital period. 3. The time between two successive swings of a pendulum or between two successive crests in a wave, such as a radio wave. 4. The time required for one oscillation; consequently, the reciprocal of frequency. 5. In nuclear engineering, the time required for the power level of a reactor to change by the factor *e* (2.718, the base of natural logarithms). Also called the reactor period.

periodic comet A comet with a period of less than 200 years. Also called a short period comet.

See also COMET.

permanently crewed capability (PCC) The capability to operate a space station (or a planetary surface base) with a human crew on board, 24 hours a day, 365 days a year. Sometimes called permanently manned capability (PMC).

Perrine, Charles Dillion (1867–1951) American *Astronomer* The American astronomer Charles Perrine worked on the staff of the LICK OBSERVATORY from 1893 to 1909 and then served as the director of the Argentine National Observatory (in South America) from 1909 to 1936. Between 1900 and 1908 he participated in four eclipse expeditions and also directed the 1901 eclipse expedition to Sumatra (now part of modern Indonesia) sponsored by the Lick Observatory. In 1904 he discovered HIMALIA, the sixth known moon of Jupiter, and then in early 1905 discovered ELARA, the seventh known moon of Jupiter.

Pershing (MGM-31A) A U.S. Army mobile surface-to-surface inertially guided missile of a two-stage solid-propellant type. It possesses a nuclear warhead capability and is designed to support ground forces by attacking long-range ground targets.

perturbation 1. Any departure introduced into an assumed steady-state system, or a small departure from the nominal path, such as a specified trajectory. 2. In astronomy, a disturbance in the regular motion of a celestial body resulting from a force that is additional to that which causes the regular motion, specifically a gravitational force. For example, the orbit of an asteroid or comet is strongly *perturbed* when the small celestial object passes near a major planetary body such as Jupiter.

phase 1. The fractional part of a periodic quantity through which the particular value of an independent variable has advanced as measured from an arbitrary reference. 2. The physical state of a substance, such as solid, liquid, or gas. 3. The extent that the disk of the Moon or a planet, as seen from Earth, is illuminated by the Sun.

phase change The physical process that occurs when a substance changes between the three basic phases (or states) of matter: solid, liquid, and vapor. Sublimation is a change of phase directly from the solid phase to the vapor phase. Vaporization is a change of phase from the liquid phase to the vapor (gaseous) phase. Condensation is the change of phase from the vapor phase to the liquid phase (or from the vapor phase directly to the solid phase). Fusion is the change of phase from the liquid phase to the solid phase, while melting is a change of phase from the solid phase to the liquid phase. Phase change is a thermodynamic process involving the addition or removal of heat (thermal energy) from the substance. The temperature at which these processes occur is dependent on the pressure.

phase modulation (PM) A type of modulation in which the relative phase of the carrier wave is modified or varied in accordance to the amplitude of the signal. Specifically, a form of angle modulation in which the angle of a sine-wave carrier is caused to depart from the carrier angle by an amount proportional to the instantaneous value of the modulating wave.

See also AMPLITUDE MODULATION; FREQUENCY MODULATION.

phases of the moon The changing illuminated appearance of the nearside of the Moon to an observer on Earth. The major phases include the new Moon (not illuminated), first quarter, full Moon (totally illuminated), and last (third) quarter.

Phobos The larger of the two irregularly shaped natural moons of Mars, discovered in 1877 by the American astronomer ASAPH HALL. Named after the god of fear (a son of Mars who accompanied his father in battle) in Greco-Roman mythology. The other moon is called DEIMOS (panic). With a major axis of 26 km and a minor axis of 18 km, Phobos orbits above the Martian equator at a mean distance of 9,377 km from the center of the planet. Phobos has an orbital period of 0.3189 days (about 7.65 hours) and an orbital eccentricity of 0.015. It is in a synchronous orbit, always keeping the same face toward the Red Planet. This moon has a mass of about 1.08×10^{16} kg and an estimated mean density 1,900 kg/m^3 (or 1.90 g/cm^3). The acceleration of gravity on its surface is quite low, at only 0.01 m/s^2.

Phobos is a dark celestial body, with an albedo of 0.06. Due to the way it reflects the spectrum of sunlight, this moon and its companion moon, Deimos, appear to possess C-type surface materials, similar to that of asteroids found in the outer portions of the asteroid belt. The surface of Phobos is cratered and covered with regolith—perhaps as deep as 100 meters—formed over millennia as meteorites pulverized the moon's surface. The largest crater on Phobos is called Stickney and has a diameter of about 10 kilometers. The second-largest crater is called Hall and has a diameter of approximately 6 kilometers.

The most widely held hypothesis concerning the origin of Phobos and Deimos is that the tiny moons are captured asteroids, but this popular hypothesis raises some questions. For example, some scientists think that it is highly improbable for Mars to gravitationally capture two different asteroids. They suggest that perhaps a larger asteroid was initially captured and then suffered a fragmenting collision, leaving Deimos and Phobos as the postcollision products. Other scientists bypass the captured asteroid hypothesis altogether and speculate that the tiny moons somehow evolved along with Mars. Expanded exploration of Mars this century should resolve this question. Recent measurements made by Mars-orbiting robot spacecraft indicate that the orbit of Phobos is contracting because of tidal forces. Some estimates now suggest that Phobos may decay from orbit within about 100 million years. Also known as Mars I.

See also DEIMOS; MARS.

Phoebe The outermost known moon of Saturn, discovered in 1898 by the American astronomer William Henry Pickering. Phoebe has a diameter of 220 km and travels in a retrograde

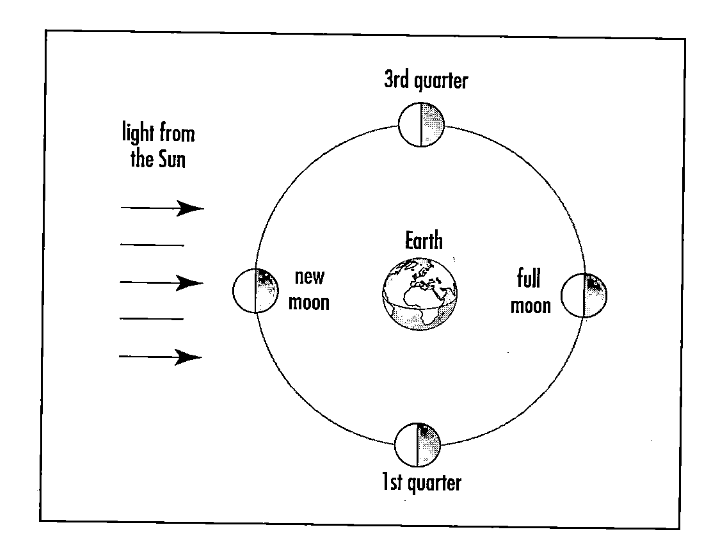

Phases of the Moon

orbit around Saturn at a distance of 12,952,000 km with a period of 550.5 days. Also known as Saturn IX.

See also SATURN.

Phoenix (AIM-54A) A long-range air-to-air missile with electronic guidance and homing developed by the U.S. Navy. Usually carried in clusters of up to six missiles on an aircraft such as the Tomcat, this missile is an airborne weapons control system with multiple-target-handling capabilities. It is used to kill multiple air targets with conventional warheads. Near-simultaneous launch is possible against up to six targets in all weather and heavy jamming environments. This missile is powered by a solid-propellant rocket motor. It is 3.9 meters long, 38.1 centimeters wide, and has a wing span of 0.9 meters. It has a range in excess of 184 kilometers. Its conventional warhead has a mass of 60.75 kg, while the entire missile has a total mass of 460.8 kg.

photoelectric effect The emission of an electron from a surface as the surface absorbs a photon of electromagnetic radiation. Electrons so emitted are called photoelectrons. The effectiveness of the process depends on the surface metal involved and the wavelength of the radiant energy to which it is exposed. For example, cesium will emit photoelectrons when exposed to visible radiation. The energy of the photoelectron produced is equal to the energy of the incident photon ($h\nu$) minus the amount of work (Φ) needed to raise the electron to a sufficient energy level to free it from the surface. This relationship is expressed in ALBERT EINSTEIN's photoelectric effect equation:

$$E_{photoelectron} = h\nu - \Phi$$

where h is the Planck constant, ν is the frequency of the incident photon, and Φ is the work function of the solid (metal) surface. In 1905 Einstein published an important physics paper that described the photoelectric effect.

photography *See* ASTROPHOTOGRAPHY.

photoionization The ionization of an atom or molecule caused by collisions with an energetic photon.

photometer An instrument that measures light intensity and the brightness of celestial objects, such as stars.

photon According to quantum theory, the elementary bundle or packet of electromagnetic radiation, such as a photon of light. Photons have no mass and travel at the speed of light. The energy (E) of the photon is equal to the product of the frequency (ν) of the electromagnetic radiation and Planck's constant (h):

$$E = h\nu$$

where h is equal to 6.626×10^{-34} joule-sec and ν is the frequency (hertz).

See also ELEMENTARY PARTICLE(S).

photon engine A conceptual reaction engine in which thrust would be obtained by emitting photons, such as light rays. Although the thrust from such an engine would be very small, it would be applied continuously in the vacuum of outer space (if a suitable power source became available) and eventually might achieve significant speeds, perhaps even supporting interstellar flight.

See also LASER PROPULSION SYSTEM.

photo reconnaissance system *See* RECONNAISSANCE SATELLITE.

photosphere The intensely bright, visible portion (surface) of the Sun.

See also SUN.

photosynthesis The process by which green plants manufacture oxygen and carbohydrates out of carbon dioxide (CO_2) and water (H_2O) with chlorophyll serving as the catalyst and sunlight as the radiant energy source. Photosynthesis is dependent on favorable temperature and moisture conditions as well as on the atmospheric carbon dioxide concentration. Without photosynthesis there would be no replenishment of our planet's fundamental food supply, and our atmosphere's supply of oxygen would disappear.

See also EARTH SYSTEM SCIENCE.

photovoltaic conversion A form of direct conversion in which a photovoltaic material converts incoming photons of visible light directly into electricity. The SOLAR CELL is an example.

photovoltaic material A substance (semiconductor material) that converts sunlight directly into electricity using the photovoltaic effect. This effect involves the production of a voltage difference across a p-n junction, resulting from the absorption of a photon of energy. The p-n junction is the region at which two dissimilar types of semiconductor materials (i.e., materials of opposite polarity) meet. The voltage difference in the semiconductor material is caused by the internal drift of holes and electrons.

Crystalline silicon (Si) and gallium arsenide (GaAs) are the photovoltaic materials often used in spacecraft power system applications. The typical silicon *solar cell* delivers about 11.5 percent solar energy-to-electric energy conversion efficiency, while the newer but more expensive gallium arsenide cell promises to provide about 18 percent or more conversion efficiency. Solar panels and solar arrays are constructed in a variety of configurations from large numbers of individual solar cells that are made of these photovoltaic materials. The series-parallel pattern of electrical connections between the cells determines the total output voltage of the solar panel or array.

On orbit, prolonged exposure to sunlight causes the performance of these photovoltaic materials to degrade, typically by 1 or 2 percent per year. Exposure to radiation (both the permanently trapped radiation belts surrounding Earth and transitory solar flare radiations) can cause even more rapid degradation in the performance of photovoltaic materials. Solar cells constructed from gallium arsenide are more radiation-resistant than silicon solar cells.

See also SOLAR CELL.

Piazzi, Giuseppe (1746–1826) Italian *Astronomer* In the late 18th century this intellectually gifted Italian monk was a professor of mathematics at the academy of Palermo and also served as the director of the new astronomical observatory constructed in Sicily in 1787. Giuseppe Piazzi was a skilled observational astronomer and discovered the first asteroid on New Year's Day in 1801. Following prevailing astronomical traditions, he named the newly found celestial object Ceres after the patron goddess of agriculture in Roman mythology.

Piazzi was born at Ponte in Valtellina, Italy, on July 16, 1746. Early in his life he decided to enter the Theatine Order, a religious order for men founded in southern Italy in the 16th century. Upon completing his initial religious training, Piazzi accepted the order's strict vow of poverty. Then, as a newly ordained monk, he pursued advanced studies at colleges in Milan, Turin, Genoa, and Rome. Piazzi emerged as a professor of theology and a professor of mathematics. In the late 1770s he taught briefly at the university on the island of Malta and then returned to a college post in Rome to serve with distinction as a professor of dogmatic theology. An academic colleague in Rome was another monk named Luigi Chiaramonti, later elected Pope Pius VII (1800–23).

In 1780 his religious order sent Piazzi to the academy in Palermo to assume the chair of higher mathematics. While working at this academy in Sicily, Piazzi obtained a grant from the viceroy of Sicily, Prince Caramanico, to construct an observatory. As part of the development of this new observatory, he traveled to Paris in 1787 to study astronomy with JOSEPH-JEROME LALANDE and then on to Great Britain the following year to work with England's fifth Astronomer Royal, NEVIL MASKELYNE. As a result of his travels, Piazzi gathered new astronomical equipment for the observatory in Palermo. Upon returning to Sicily he began to set up the new equipment on top of a tower in the royal palace between 1789 and 1790.

All was soon ready, and Piazzi began to make astronomical observations in May 1791. He published his first astronomical reports in 1792. He primary goal was to improve the stellar position data in existing star catalogs. Piazzi paid particular attention to compensating for errors induced by such subtle phenomena as the aberration of starlight, discovered in 1728 by the British astronomer JAMES BRADLEY. The monk-astronomer was a very skilled observer and was soon able to publish a refined positional list of approximately 6,800 stars in 1803. He followed this initial effort with the publication of a second star catalog containing about 7,600 stars in 1814. Both of these publications were well received within the astronomical community and honored by prizes from the French Academy.

However, Piazzi is today best remembered as the person who discovered the first asteroid (or minor planet). He made this important discovery on January 1, 1801, from the observatory that he founded and directed in Palermo, Sicily. To please his royal benefactors while still maintaining astronomical traditions, he called the new celestial object Ceres after the ancient Roman goddess of agriculture, who was once widely worshiped on the island of Sicily.

Ceres (asteroid 1) is the largest asteroid, with a diameter of about 935 kilometers. However, because of its very low albedo (about 10 percent), Ceres cannot be observed by the naked eye. After just three good telescopic observations in early 1801, Piazzi lost the asteroid in the Sun's glare, yet Piazzi's carefully recorded positions enabled the German mathematician CARL FRIEDRICH GAUSS to accurately predict the asteroid's orbit. Gauss's calculations allowed another early asteroid hunter, HEINRICH OLBERS, to relocate the minor planet on January 1, 1802. Olbers then went on to find the asteroids Pallas (1802) and Vesta (1807).

When Piazzi died in Naples, Italy, on July 22, 1826, only four asteroids were known to exist in the gap between Mars and Jupiter, but his pioneering discovery of Ceres in 1801 stimulated renewed attention on this interesting region of the solar system—a gap that should contain a missing planet according to the Titus-Bode law. In 1923 astronomers named the 1,000th asteroid discovered Piazzia to honor the Italian astronomer's important achievement.

Picard, Jean (1620–1682) French *Astronomer* In 1671 the French astronomer and priest Jean Picard made a careful measurement of Earth's circumference using the length of a degree on the meridian at Paris. His effort provided the first major improvement in that important measurement since the efforts of the early Greek astronomer and mathematician ERATOSTHENES OF CYRENE. The value obtained by Picard came quite close to the modern value.

See also EARTH.

Pickering, Edward Charles (1846–1919) American *Astronomer, Physicist* Starting in 1877, Edward Charles Pickering dominated American astronomy for the last quarter of the 19th century. He served as the director of the Harvard College Observatory for more than four decades. In this capacity he supervised the production of the *Henry Draper Catalog* (by ANNIE J. CANNON, WILLIAMINA FLEMING, HENRIETTA LEAVITT, and other astronomical "computers"). This important work listed more than 225,000 stars according to their spectra, as defined by the newly introduced Harvard Classification System. He also vigorously promoted the exciting new field of astrophotography and in 1889 discovered the first spectroscopic binary star system, two stars so visually close together that they can be distinguished only by the Doppler shift of their spectral lines.

Pickering was born on July 19, 1846, in Boston, Massachusetts. After graduating from Harvard University in 1865, he taught physics for approximately 10 years at the Massachusetts Institute of Technology (MIT). To assist in his physics lectures, he constructed the first instructional physics laboratory in the United States. Then in 1876, at just 30 years of age, he became a professor of astronomy and the director of the Harvard College Observatory. He remained in this position for 42 years and used it to effectively shape the course of American astronomy through the last two decades of 19th century and the first two decades of the 20th century. One important innovation was Pickering's decision to hire many very-well-qualified women to work as observational astronomers and "astronomical computers" at the Harvard Observatory. With the funds provided by Henry Draper's widow, Ann Palmer Draper (1839–1914), Pickering employed

Annie J. Cannon, Williamina P. Fleming, Antonia Maury, Henrietta S. Leavitt, and others to work at the observatory. He then supervised these talented women as they produced the contents of the *Henry Draper Catalog.* This catalog, published in 1918, was primarily compiled by Cannon and presented the spectra classifications of some 225,000 stars based on Cannon's "Harvard classification system," in which the stars were ordered in a sequence that corresponded to the strength of their hydrogen absorption lines.

In 1884 Pickering produced a new catalog, called the *Harvard Photometry.* This work presented the photometric brightness of 4,260 stars using the North Star (Polaris) as a reference. Pickering revised and expanded this effort in 1908. He was an early advocate of astrophotography and in 1903 published the first photographic map of the entire sky. To achieve his goals in astrophotography he also sought the assistance of his younger brother, WILLIAM HENRY PICKERING, who established the Harvard College Observatory's southern station in Peru. Independently of the German astronomer HERMANN CARL VOGEL (1841–1907), Pickering discovered the first spectroscopic binary star system (Mizar) in 1889.

He was a dominant figure in American astronomy and a great source of encouragement to amateur astronomers. He founded the American Association of Variable Star Observers and received a large number of awards for his contributions to astronomy. These awards include the Henry Draper Medal of the National Academy of Sciences (1888), the Rumford Prize of the American Academy of Arts and Sciences (1891), the Gold Medal of the Royal Astronomical Society (awarded both in 1886 and 1901), and the Bruce Gold Medal of the Astronomical Society of the Pacific (1908). He died in Cambridge, Massachusetts, on February 3, 1919.

Pickering, William Hayward (1910–2004) New Zealander–American *physicist, Engineer* The engineering physicist William Hayward Pickering directed the Jet Propulsion Laboratory (JPL) operated for NASA by Caltech from 1954 to 1976. He supervised the development of many of the highly successful U.S. planetary exploration spacecraft. His leadership efforts included the first American satellite (*EXPLORER 1*), Mariner spacecraft to Venus and Mars, Ranger and Surveyor spacecraft to the Moon, the *Viking 1* and 2 spacecraft (landers and orbiters) to Mars, and the incredible *Voyager 1* and 2 spacecraft missions to the outer planets and beyond.

Pickering, William Henry (1858–1938) American *Astronomer* The American astronomer William Henry Pickering was the younger brother of EDWARD CHARLES PICKERING. At his brother Edward's request, William Pickering established an auxiliary astronomical observatory for the Harvard College Observatory at Arequipa, Peru, in 1891. He then used this observatory to discover the ninth moon of Saturn (called Phoebe) in 1898. He published a photographic atlas of the Moon in 1903 using data collected at Harvard's astronomical station on the island of Jamaica. Unfortunately, he became overzealous in interpreting these lunar photographs and reported seeing evidence of vegetation and frost. Influenced by such "mistaken evidence," he suspected that there might be life-forms on the Moon. He also conducted a photographic search for a planet beyond Neptune, and he may have even captured an image of Pluto in 1919 but failed to recognize it.

pico- (symbol: p) A prefix in the SI unit system meaning multiplied by 10^{-12}; for example, a *picosecond* is 10^{-12} second.

See also SI UNITS.

piggyback experiment An experiment that rides along with the primary experiment on a space-available basis without interfering with the mission of the primary experiment.

pilot 1. In general, the person who handles the controls of an aircraft, aerospace vehicle, or spacecraft from within the craft and in so doing guides or controls it in three-dimensional flight. 2. An electromechanical system designed to exercise control functions in an aircraft, aerospace vehicle, or spacecraft; for example, the *automatic pilot.* 3. (verb) To operate, control, or guide an aircraft, aerospace vehicle, or spacecraft from within the vehicle so as to move in three-dimensional flight through the air or space. 4. With respect to the U.S. Space Transportation System (STS), the second in command of a space shuttle flight. He or she assists the commander as required in conducting all phases of the orbiter vehicle flight. The pilot has such authority and responsibilities as are delegated to him or her by the commander. The pilot normally operates the remote manipulator system (RMS) at the orbiter's payload handling station, using the RMS to deploy, release, or capture payloads. The pilot also is generally the second crewperson (behind the mission specialist) for extravehicular activity (EVA).

See also SPACE TRANSPORTATION SYSTEM.

***Pioneer* plaque** On June 13, 1983, the *Pioneer 10* spacecraft became the first human-made object to leave the solar system. In an initial attempt at interstellar communication, the *Pioneer 10* spacecraft and its sistercraft (*Pioneer 11*) were equipped with identical special plaques (*see* figure on page 456). The plaque is intended to show any intelligent alien civilization that might detect and intercept either spacecraft millions of years from now when the spacecraft was launched, from where it was launched, and by what type of intelligent beings it was built. The plaque's design is engraved into a gold-anodized aluminum plate 152 millimeters by 229 millimeters. The plate is approximately 1.27 millimeters thick. It is attached to the Pioneer spacecrafts' antenna support struts in a position that helps shield it from erosion by interstellar dust.

Numbers have been superimposed on the plaque illustrated in the figure to assist in this discussion. At the far right the bracketing bars (1) show the height of the woman compared to the Pioneer spacecraft. The drawing at the top left of the plaque (2) is a schematic of the hyperfine transition of neutral atomic hydrogen used here as a universal "yardstick" that provides a basic unit of both time and space (length) throughout the Milky Way galaxy. This figure illustrates a reverse in the direction of the spin of the electron in a hydrogen atom. The transition depicted emits a characteristic radio wave with an approximately 21-centimeter wavelength. Therefore, by providing this drawing we are telling any tech-

Reaching Beyond the Solar System

One of the fundamental characteristics of human nature is our desire to communicate. In recent years we have begun to respond to a deep cosmic yearning to reach beyond our own solar system to other star systems hoping not only that someone or something is out there but that "it" will eventually "hear us" and perhaps even return a message.

Because of the vast distances between even nearby stars, when we say "interstellar communication" we are not talking about communication in "real time." On Earth radio wave communication does not involve a very perceptible time lag—that is, messages and responses are normally received immediately after they are sent. In contrast, electromagnetic waves take many light-years to cross the interstellar void from star system to star system. So our initial attempts at interstellar communication have actually been more like putting a message in a bottle and tossing it into the "cosmic sea" or else placing a message in a type of interstellar time capsule for some future generation of human or alien beings to discover.

Positive, active attempts to "communicate" with any alien civilizations that might exist among the stars are often referred to as CETI, an acronym for "communication with extraterrestrial intelligence." If, on the other hand, we passively scan the skies for signs of an advanced extraterrestrial civilization or patiently listen for their radio wave messages, we call the process the "search for extraterrestrial intelligence," or SETI.

Starting in 1960, there have been several serious SETI efforts, the vast majority of which involved listening to selected portions of the microwave spectrum in hopes of detecting structured electromagnetic signals indicative of the existence of some intelligent extraterrestrial civilization among the stars. To date, unfortunately, none of these efforts have provided any positive evidence that such "intelligent" (that is, coherent and artificially produced versus naturally occurring) radio signals exist. However, SETI researchers have examined only a small portion of the billion of stars in the Milky Way galaxy. Furthermore, they have generally listened only to rather narrow portions of the electromagnetic spectrum. Optimistic investigators, therefore, suggest "the absence of evidence is not necessarily the evidence of absence." Detractors counter by suggesting that SETI is really an activity without a subject. Such detractors also regard the SETI researcher as a modern Don Quixote who is tilting at cosmic windmills.

However, our galaxy is a vast place, and it is only within the past few decades that the people of Earth have enjoyed radio, television, and communication technologies sophisticated enough to cross the threshold of some minimal "interstellar communication technology horizon." A little more than a century ago, for example, Earth could have been quite literally "bombarded" with many alien radio wave signals, but no one here had the technology capable of receiving or interpreting such hypothetical signals. After all, even the universe's most colossal natural signal, the remnant microwave background from the big bang, was detected and recognized only in the mid-1960s, yet this interesting signal is detectable (as static) by any common television set connected to a receiving antenna.

Early in the space age a few scientists did boldly attempt to communicate "over time" with alien civilizations. They did this by including carefully prepared messages on several deep space explorations scheduled to travel beyond the solar system and also by intentionally transmitting a very powerful radio message to a special group of stars using the world's largest radio telescope, the Arecibo Observatory.

The three most important deliberate attempts at interstellar communication (from Earth to the stars) to date are the special message plaques placed on both the *Pioneer 10* and *11* spacecraft, the "Sounds of Earth" recordings included on the *Voyager 1* and *2* spacecraft, and the famous Arecibo interstellar message transmitted from Puerto Rico by the radio telescope on November 16, 1974. The four spacecraft mentioned are now on separate and distinct interstellar trajectories that will take them deep into the galaxy over the next few million years. The special Arecibo message is traveling at the speed of light toward the Great Cluster in Hercules (called Messier 13), which lies about 25,000 light-years away from Earth. The globular cluster M13 contains about 300,000 stars within a radius of approximately 18 light-years.

It may come as bit of a surprise that the people of Earth have also been *unintentionally* leaking radio frequency signals into the galaxy since the middle of the 20th century. Imagine the impact that some of our early television shows could have on an alien civilization capable of intercepting and reconstructing the television signals. Some of the earliest of these broadcast signals are now about 60 light-years into the galaxy. Who knows what is looking at the *Ed Sullivan Show* during which Elvis Presley made his initial television appearance? Should these speculative circumstances occur, a group of alien scientists, philosophers, and religious leaders might right now be engaged in a very heated debate concerning the true significance of the special message from Earth that proclaims "You Ain't Nothin But A Hound Dog!"

nically knowledgeable alien civilization finding it that we have chosen 21 centimeters as a basic length in the message. While extraterrestrial civilizations will certainly have different names and defining dimensions for their basic system of physical units, the wavelength size associated with the hydrogen radio-wave emission will still be the same throughout the galaxy. Science and commonly observable physical phenomena represent a general galactic language—at least for starters.

The horizontal and vertical ticks (3) represent the number 8 in binary form. It is hoped the alien beings pondering this plaque will eventually realize that the hydrogen wavelength (21 centimeters) multiplied by the binary number representing 8 (indicated alongside the woman's silhouette) describes her overall height, namely, 8×21 centimeters = 168 centimeters, or approximately 5 feet 5 inches tall. Both human figures are intended to represent the intelligent beings that built the Pioneer spacecraft. The man's hand is raised as a gesture of goodwill. These human silhouettes were carefully selected and drawn to maintain ethnic neutrality. Furthermore, no attempt was made to explain terrestrial "sex" to an

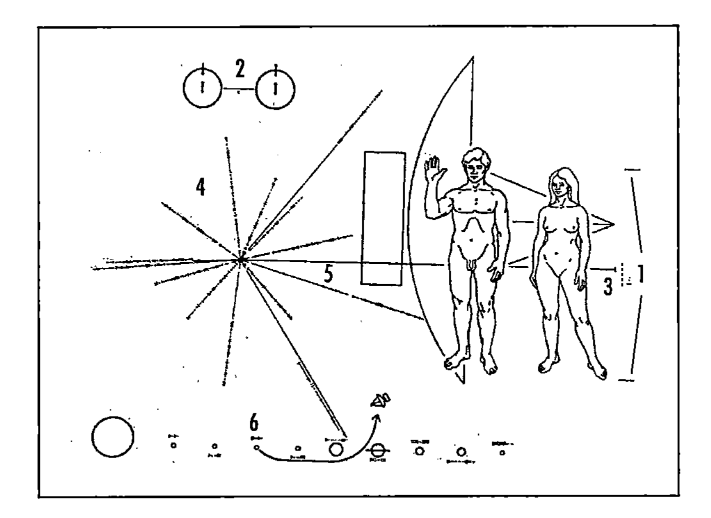

Annotated *Pioneer* plaque *(Drawing courtesy of NASA)*

alien culture—that is, the plaque makes no specific effort to explain the potentially "mysterious" differences between the man and woman depicted.

The radial pattern (4) should help alien scientists locate our solar system within the Milky Way galaxy. The solid bars indicate distance, with the long horizontal bar (5) with no binary notation on it representing the distance from the Sun to the galactic center, while the shorter solid bars denote directions and distances to 14 pulsars from our Sun. The binary digits following these pulsar lines represent the periods of the pulsars. From the basic time unit established by the use of the hydrogen-atom transition, an intelligent alien civilization should be able to deduce that all times indicated are about 0.1 second—the typical period of pulsars. Since pulsar periods appear to be slowing down at well-defined rates, the pulsars serve as a form of galactic clock. Alien scientists should be able to search their astrophysical records and identify the star system from which the *Pioneer* spacecraft originated and approximately when they were launched, even if each spacecraft is not found for hundreds of millions of years. Consequently, through the use of this pulsar map we have attempted to locate ourselves both in galactic space and in time.

As a further aid to identifying the *Pioneers'* origin, a diagram of our solar system (6) is also included on the plaque. The binary digits accompanying each planet indicate the relative distance of that planet from the Sun. The *Pioneers'* trajectory is shown starting from the third planet (Earth), which has been offset slightly above the others. As a final clue to the terrestrial origin of the *Pioneer* spacecraft, their antenna are depicted pointing back to Earth. This message was designed by FRANK DRAKE and the late CARL SAGAN, and the artwork was prepared by Linda Salzman Sagan.

When the *Pioneer 10* spacecraft sped past Jupiter in December 1973, it acquired sufficient kinetic energy (through the gravity-assist technique) to carry it completely out of the solar system. The table describes some of the "near" star encounters that *Pioneer 10* will undergo in the next 860,000 years.

The *Pioneer 11* spacecraft carries an identical message. After that spacecraft's encounter with Saturn, it also acquired sufficient velocity to escape the solar system, but in almost the opposite direction to *Pioneer* 10. In fact, *Pioneer 11* is departing in the same general direction in which our solar system is moving through space. As some scientists have philosophically noted, the Pioneer plaque represents at least one *intellectual cave painting*—a mark of humanity—that might survive not only all the caves on Earth, but also the solar system itself.

See also PIONEER 10, 11 SPACECRAFT.

Pioneer spacecraft NASA's solar-orbiting Pioneer spacecraft have contributed enormous amounts of data concerning the solar wind, solar magnetic field, cosmic radiation, micrometeoroids, and other phenomena of interplanetary space. *Pioneer 4*, launched March 3, 1959, was the first U.S. spacecraft to go into orbit around the Sun. It provided excel-

Near-Star Encounters Predicted for the *Pioneer 10* Spacecraft

Star No.	Name	Information
1	Proxima Centauri	Red dwarf "flare" star. Closest approach is 6.38 light-years after 26,135 years
2	Ross 248	Red dwarf star. Closest approach is 3.27 light-years after 32,610 years
3	Lambda Serpens	Sun-type star. Closest approach is 6.9 light-years after 173,227 years
4	G 96	Red dwarf star. Closest approach is 6.3 light-years after 219,532 years
5	Altair	Fast-rotating white star (1.5 times the size of the Sun and 9 times brighter). Closest approach is 6.38 light-years after 227,075 years
6	G 181	Red dwarf star. Closest approach is 5.5 light-years after 292,472 years
7	G 638	Red dwarf star. Closest approach is 9.13 light-years after 351,333 years
8	D + 19 5036	Sun-type star. Closest approach is 4.9 light-years after 423,291 years
9	G 172.1	Sun-type star. Closest approach is 7.8 light-years after 847,919 years
10	D + 25 1496	Sun-type star. Closest approach is 4.1 light-years after 862,075 years

Source: NASA/Kennedy Space Center.

lent radiation data. *Pioneer 5,* launched March 11, 1960, confirmed the existence of interplanetary magnetic fields and helped explain how solar flares trigger magnetic storms and the northern and southern lights (auroras) on Earth. *Pioneers 6* through *9* were launched between December 1965 and November 1968. These spacecraft provided large quantities of valuable data concerning the solar wind, magnetic and electrical fields, and cosmic rays in interplanetary space. Data from these spacecraft helped scientists draw a new picture of the Sun as the dominant phenomenon of interplanetary space.

See also PIONEER 10, 11 SPACECRAFT; PIONEER VENUS MISSION.

***Pioneer 10, 11* spacecraft** The *Pioneer 10* and *11* spacecraft, as their names imply, are true deep-space explorers, the first human-made objects to navigate the main asteroid belt, the first spacecraft to encounter Jupiter and its fierce radiation belts, the first to encounter Saturn, and the first to leave the solar system. These spacecraft also investigated magnetic fields, cosmic rays, the solar wind, and the interplanetary dust concentrations as they flew through interplanetary space.

At Jupiter and Saturn scientists used the spacecraft to investigate the giant planets and their interesting complement of moons in four main ways: (1) by measuring particles, fields, and radiation; (2) by spin-scan imaging the planets and some of their moons; (3) by accurately observing the paths of the spacecraft and measuring the gravitational forces of the planets and their major satellites acting on them; and (4) by observing changes in the frequency of the S-band radio signal before and after occultation (the temporary "disappearance" of the spacecraft caused by their passage behind these celestial bodies) to study the structures of their ionospheres and atmospheres.

The *Pioneer 10* spacecraft was launched from Cape Canaveral Air Force Station, Florida by an Atlas-Centaur

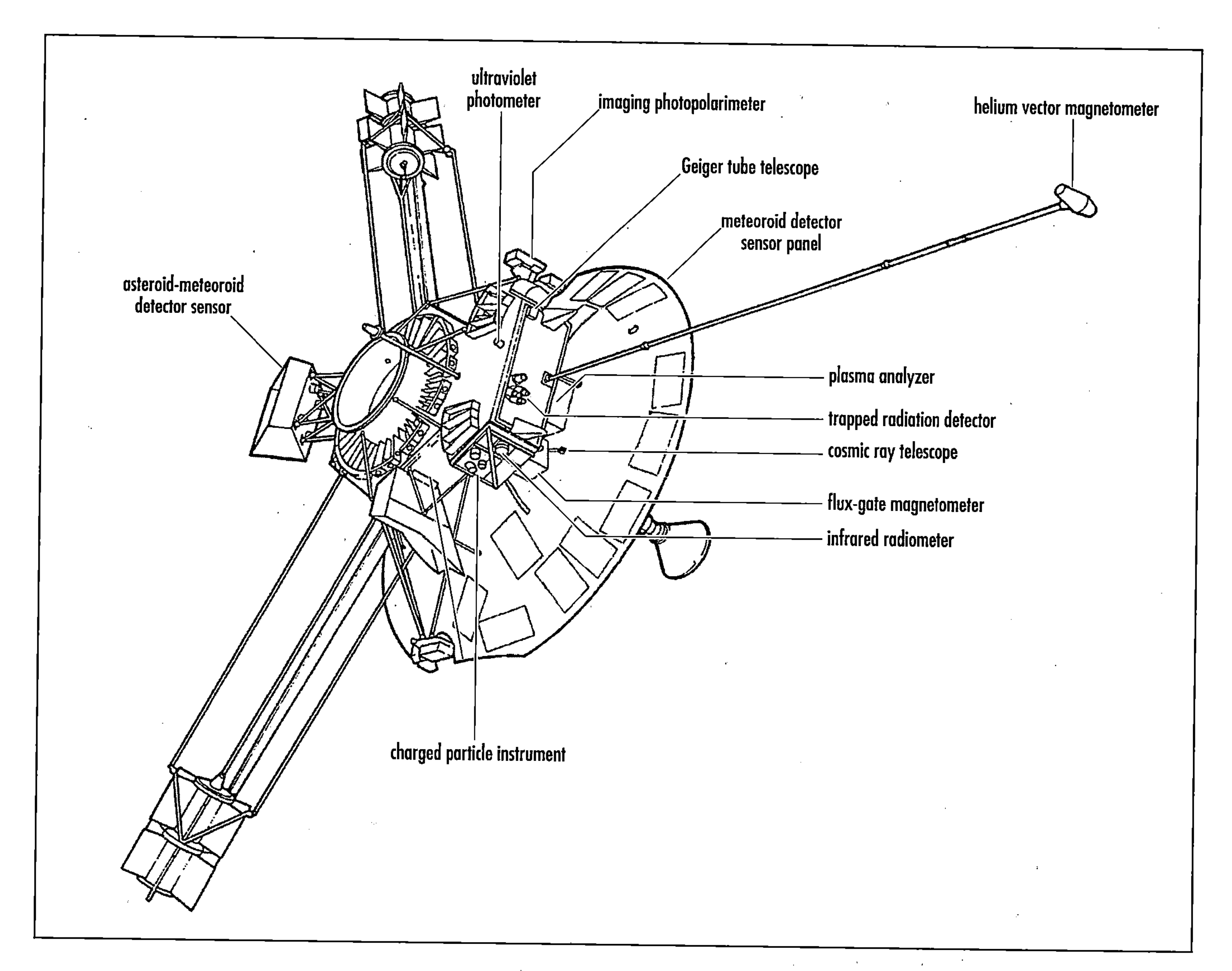

The *Pioneer 10* and *11* spacecraft with their complement of scientific instruments. The spacecraft's electric power was provided by a long-lived radioisotope-thermoelectric generator (RTG). ***(Drawing courtesy of NASA)***

rocket on March 2, 1972. It became the first spacecraft to cross the main asteroid belt and the first to make close-range observations of the Jovian system. Sweeping past Jupiter on December 3, 1973 (its closest approach to the giant planet), it discovered no solid surface under the thick layer of clouds enveloping the giant planet, an indication that Jupiter is a liquid hydrogen planet. *Pioneer 10* also explored the giant Jovian magnetosphere, made close-up pictures of the intriguing Red Spot, and observed at relatively close range the Galilean satellites Io, Europa, Ganymede, and Callisto. When *Pioneer 10* flew past Jupiter it acquired sufficient kinetic energy to carry it completely out of the solar system.

Departing Jupiter, *Pioneer 10* continued to map the *heliosphere* (the Sun's giant magnetic bubble, or field, drawn out from it by the action of the solar wind). Then, on June 13, 1983, *Pioneer 10* crossed the orbit of Neptune, which at the time was (and until 1999 was) the planet farthest out from the Sun. This unusual circumstance was due to the eccentricity in Pluto's orbit, which had taken the icy planet inside the orbit of Neptune. This historic date marked the first passage of a human-made object beyond the known planetary boundary of the solar system. Beyond this solar system boundary, *Pioneer 10* measured the extent of the heliosphere as it traveled through "interstellar" space. Along with its sister ship, *Pioneer 11,* the spacecraft helped scientists investigate the deep space environment.

The *Pioneer 10* spacecraft is heading generally toward the red star Aldebaran. It is more than 68 light-years away from Aldebaran, and the journey will require about 2 million years to complete. Budgetary constraints forced the termination of routine tracking and project data processing operations on March 31, 1997. However, occasional tracking of *Pioneer 10* continued beyond that date. The last successful data acquisition from *Pioneer 10* by NASA's Deep Space Network (DSN) occurred on March 3, 2002—30 years after launch—and again on April 27, 2002. The spacecraft signal was last detected on January 23, 2003, after an uplink message was transmitted to turn off the remaining operational experiment, the Geiger Tube Telescope. However, no downlink data signal was achieved, and by early February no signal at all was detected. NASA personnel concluded that the spacecraft's RTG power supply had finally fallen below the level needed to operate the onboard transmitter. Consequently, no further attempts were made to communicate with *Pioneer 10*.

The *Pioneer 11* spacecraft was launched on April 5, 1973, and swept by Jupiter at an encounter distance of only 43,000 km on December 2, 1974. It provided additional detailed data and pictures of Jupiter and its moons, including the first views of Jupiter's polar regions. Then, on September 1, 1979, *Pioneer 11* flew by Saturn, demonstrating a safe flight path through the rings for the more sophisticated *Voyager 1* and *2* spacecraft to follow. *Pioneer 11* (by then officially renamed *Pioneer Saturn*) provided the first close-up observations of Saturn and its rings, satellites, magnetic field, radiation belts, and atmosphere. It found no solid surface on Saturn but discovered at least one additional satellite and ring. After rushing past Saturn, *Pioneer 11* also headed out of the solar system toward the distant stars.

The *Pioneer 11* spacecraft operated on a backup transmitter from launch. Instrument power sharing began in February 1985 due to declining RTG power output. Science operations and daily telemetry ceased on September 30, 1995, when the RTG power level became insufficient to operate any of the spacecraft's instruments. All contact with *Pioneer 11* ceased at the end of 1995. At that time the spacecraft was 44.7 astronomical units (AU) away from the Sun and traveling through interstellar space at a speed of about 2.5 AU per year.

Both Pioneer spacecraft carry a special message (the *Pioneer* plaque) for any intelligent alien civilization that might find them wandering through the interstellar void millions of years from now. This message is an illustration engraved on an anodized aluminum plaque. The plaque depicts the location of Earth and the solar system, a man and a woman, and other points of science and astrophysics that should be decipherable by a technically intelligent civilization.

See also JUPITER; *PIONEER* PLAQUE; SATURN.

Pioneer Venus mission The Pioneer Venus mission consisted of two separate spacecraft launched by the United States to the planet Venus in 1978. The *Pioneer Venus Orbiter* spacecraft (also called *Pioneer 12*) was a 553-kg spacecraft that contained a 45-kg payload of scientific instruments. It was launched on May 20, 1978, and placed into a highly eccentric orbit around Venus on December 4, 1978. For 14 years (1978–92) the *Pioneer Venus Orbiter* spacecraft gathered a wealth of scientific data about the atmosphere and ionosphere of Venus and their interactions with the solar wind as well as details about the planet's surface. Then in October 1992 this spacecraft made an intended, final entry into the Venusian atmosphere, gathering data up to its final fiery plunge and dramatically ending the operations portion of the Pioneer Venus mission.

The *Pioneer Venus Multiprobe* spacecraft (also called *Pioneer 13*) consisted of a basic bus spacecraft, a large probe, and three identical small probes. The *Pioneer Venus Multiprobe* spacecraft was launched on August 8, 1978, and separated about three weeks before entry into the Venusian atmosphere. The four now-separated probes and their spacecraft bus successfully entered the Venusian atmosphere at widely dispersed locations on December 9, 1978, and returned important scientific data as they plunged toward the planet's surface. Although the probes were not designed to survive landing, one hardy probe did and transmitted data for about an hour after impact.

The *Pioneer Venus Orbiter* and *Multiprobe* spacecraft provided a wealth of scientific data about Venus and its surface, atmosphere, and interaction with the solar wind. For example, the orbiter spacecraft made an extensive radar map covering about 90 percent of the Venusian surface. Using its radar to look through the dense Venusian clouds, this spacecraft revealed that the planet's surface was mostly gentle, rolling plains with two prominent plateaus, Ishtar Terra and Aphrodite Terra. This highly successful, two-spacecraft mission also provided important groundwork for NASA's subsequent Magellan mission to Venus.

See also MAGELLAN MISSION; VENUS.

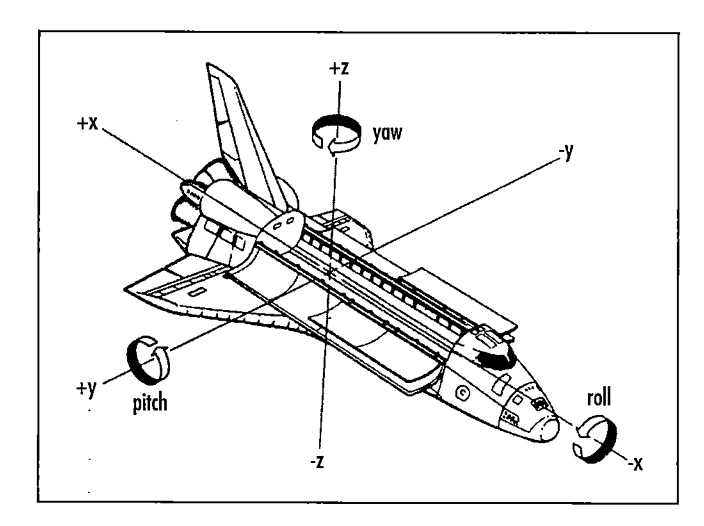

Pitch, roll, and yaw ***(Drawing courtesy of NASA)***

pitch 1. In acoustics, the "highness" or "lowness" of a sound. Pitch depends mainly on the frequency of the sound stimulus but also on the pressure and waveform of the stimulus. 2. In aerospace technology, the rotation (angular motion) of an aircraft, aerospace vehicle, or spacecraft about its lateral axis. (*See* figure.) 3. In engineering, the distance between corresponding points on adjacent teeth of a gear or on adjacent blades on a turbine wheel, as measured along a prescribed arc called the *pitchline.*

pitch angle The angle between an aircraft's or missile's longitudinal axis and Earth's horizontal plane. Also called inclination angle.

pitchover The programmed turn from the vertical that a rocket or launch vehicle (under power) takes as it describes an arc and points in a direction other than vertical.

pixel Contraction for picture element; the smallest unit of information on a screen or in an image; the more pixels, the higher the potential resolution of the video screen or image.

plage A bright patch in the Sun's chromosphere.
See also SUN.

Planck, Max Karl (1858–1947) German *Physicist* Modern physics, with its wonderful new approach to viewing both the most minute regions of inner space and the farthest regions of outer space, had two brilliant cofounders, Max Planck and ALBERT EINSTEIN. Many other intelligent people have built upon the great pillars of 20th-century physics, quantum theory and relativity, but to these two giants of scientific achievement go the distinctive honor of leading the way. Max Planck was the gentle, cultured German physicist who introduced quantum theory in 1900, his powerful new theory concerned with the transport of electromagnetic radiation in discrete energy packets or quanta. Planck received the 1918 Nobel Prize in physics for this important world-changing scientific accomplishment.

Planck was born in Kiel, Germany, on April 23, 1858. His father was a professor of law at the University of Kiel and gave his son a deep sense of integrity, fairness, and the value of intellectual achievement, important traits that characterized the behavior of this scientist through all phases of his life. When Planck was a young boy his family moved to Munich, and that city is where he received his early education. A gifted student who could easily have become a great pianist, Planck studied physics at the Universities of Munich and Berlin. In Berlin he had the opportunity to directly interact with such famous scientists as GUSTAV KIRCHHOFF and RUDOLF CLAUSIUS. The former introduced Planck to sophisticated classical interpretations of blackbody radiation, while the latter challenged him with the profound significance of the second law of thermodynamics and the elusive concept of entropy.

In 1879 Planck received his doctoral degree in physics from the University of Munich. From 1880 to 1885 he remained in Munich as a privatdozent (unsalaried lecturer) in physics at the university. He became an associate professor in physics at the University of Kiel in 1885 and remained in that position until 1889, when he succeeded Kirchhoff as professor of physics at the University of Berlin. His promotion to associate professor in 1885 provided Planck the income to marry his first wife, Marie Merck, a childhood friend from Kiel with whom he lived happily until her death in 1909. The couple had two sons and twin daughters. In 1910 Planck married his second wife, Marga von Hösling. He remained a professor of physics at the University of Berlin until his retirement in 1926. Good fortune and success would smile upon Planck's career as a physicist in Berlin, but tragedy would stalk his personal life.

While teaching at the University of Berlin at the end of the 19th century, Planck began to address the very puzzling problem involved with the emission of energy by a blackbody radiator as a function of its temperature. A decade earlier JOSEF STEFAN had performed important heat transfer experiments concerning blackbody radiators. However, classical physics could not adequately explain his experimental observations. To make matters more puzzling, classical electromagnetic theory incorrectly predicted that a blackbody radiator should emit an infinite amount of thermal energy at very high frequencies—a paradoxical condition referred to as the "ultraviolet catastrophe." Early in 1900 the British scientists Lord Rayleigh (1842–1919) and SIR JAMES JEANS introduced their mathematical formula (called the Rayleigh-Jeans formula) to describe the energy emitted by a blackbody radiator as a function of wavelength. While their formula adequately predicted behavior at long wavelengths (for example, at radio frequencies), their model failed completely in trying to predict blackbody behavior at shorter wavelengths (higher frequencies), the portion of the electromagnetic spectrum of great importance in astronomy, since stars approximate high-temperature blackbody radiators.

Planck solved the problem when he developed a bold new formula that successfully described the behavior of a blackbody radiator over all portions of the electromagnetic spectrum. To reach his successful formula Planck assumed that the atoms of the blackbody emitted their radiation only in discrete, individual energy packets, which he called *quanta.*

In a classic paper that he published in late 1900, Planck presented his new blackbody radiation formula. He included the revolutionary idea that the energy for a blackbody resonator at a frequency (ν) is simply the product h ν, where h is a universal constant, now called *Planck's constant.* This 1900 paper, published in *Annalen der Physik,* contained Planck's most important work and represented a major turning point in the history of physics. The introduction of quantum theory had profound implications on all modern physics, from the way scientists treated subatomic phenomena to the way they modeled the behavior of the universe on cosmic scales.

However, Planck himself was a reluctant revolutionary. For years he felt that he had only created the quantum postulate as a "convenient means" of explaining his blackbody radiation formula. However, other physicists were quick to seize on Planck's quantum postulate and then to go forth and complete his revolutionary movement, displacing classical physics with modern physics. For example, Albert Einstein used Planck's quantum postulate to explain the photoelectric effect in 1905, and Niels Bohr (1885–1962) applied quantum mechanics in 1913 to create his world-changing model of the hydrogen atom.

Planck received the 1918 Nobel Prize in physics in recognition of his epoch-making investigations into quantum theory. He also published two important books, *Thermodynamics* (1897) and *The Theory of Heat Radiation* (1906), that summarized his major efforts in the physics of blackbody radiation. He maintained a strong and well-respected reputation as a physicist even after his retirement from the University of Berlin in 1926. That same year the British Royal Society honored his contributions to physics by electing him a foreign member and awarding him its prestigious Copley Medal.

But while Planck climbed to the pinnacle of professional success, his personal life was marked with deep tragedy. At the time Planck won the Nobel Prize in physics, his oldest son, Karl, died in combat in World War I, and his twin daughters, Margarete and Emma, died about a year apart in childbirth. Then, in the mid-1930s, when Adolf Hitler seized power in Germany, Planck, in his capacity as the elder statesperson for the German scientific community, bravely praised Einstein and other German-Jewish physicists in open defiance of the ongoing Nazi persecutions. Planck even met personally with Hitler to try to stop the senseless attacks against Jewish scientists, but Hitler simply flew into a tirade and ignored the pleas from the aging Nobel laureate. So in 1937, as a final protest, Planck resigned as the president of the Kaiser Wilhelm Institute, the leadership position in German science that he had proudly held with great distinction since 1930. In his honor that institution is now called the Max Planck Institute.

During the closing days of World War II his second son, Erwin, was brutally tortured and then executed by the Nazi gestapo for his role in the unsuccessful 1944 assassination attempt against Hitler. Just weeks before the war ended, Planck's home in Berlin was completely destroyed by Allied bombing. Then, in the very last days of the war, American troops launched a daring rescue across war-torn Germany to keep Planck and his second wife, Marga, from being captured by the advancing Russian Army. The Dutch-American astronomer GERARD KUIPER was a participant in this military action that allowed Planck to live the remainder of his life in the relative safety of the Allied-occupied portion of a divided Germany. On October 3, 1947, at the age of 89, the brilliant physicist who ushered in a revolution in physics with his quantum theory died peacefully in Göttingen, Germany.

Planck's constant (symbol: h) A fundamental constant in modern physics that is equal to the ratio of the energy (E) of a quantum of energy to the frequency (ν) of this quantum (or photon); that is,

$$h = E/\nu = 6.626 \times 10^{-34} \text{ joule-second}$$

This unit is named in honor of the German physicist MAX PLANCK, who first developed quantum theory.

Planck's radiation law The fundamental physical law that describes the distribution of energy radiated by a blackbody. Introducing the concept of the *quantum* (or photon) as a small unit of energy transfer, the German physicist Max Planck first proposed this law in 1900. It is an expression for the variation of monochromatic radiant flux per unit area of source as a function of wavelength (λ) of blackbody radiation at a given thermodynamic temperature (T). Mathematically, Planck's law can be expressed as:

$$dw = [2\pi hc^2]/\{\lambda^5 [e^{hc/\lambda kT} - 1]\} d\lambda$$

where dw is the radiant flux from a blackbody in the wavelength interval dλ, centered around wavelength λ, per unit area of blackbody surface at thermodynamic temperature T. In this equation c is the velocity of light, h is a constant (later named the Planck constant), and k is the Boltzmann constant.

***Planck* (spacecraft)** A European Space Agency (ESA) scientific mission planned for launch in 2007 that will examine with the highest instrumentation accuracy to date the first light that filled the universe after the big bang explosion. Today this ancient explosion's remnant is called the cosmic microwave background (CMB). Detailed CMB data from the *Planck* spacecraft, named after the German physicist MAX PLANCK, will help astrophysicists and cosmologists better model the early universe and further test the validity of big bang cosmology with inflation (i.e., the inflationary universe model).

Planck time The incredibly short period of time (1.3×10^{-43} s) associated with the Planck era, the first cosmic epoch when (as physicists now hypothesize) *quantum gravity* controlled the universe immediately following the big bang explosion.

See also BIG BANG; COSMOLOGY; QUANTUM GRAVITY.

planet A nonluminous celestial body that orbits around the Sun or some other star. There are nine such large objects, or "major planets," in the solar system and numerous "minor planets," or asteroids. The distinction between a planet and a satellite may not always be clear-cut, except for the fact that a satellite orbits around a planet. For example, our Moon is nearly the size of the planet Mercury and is very large in comparison to its parent planet, Earth. In some cases Earth and the Moon can be treated almost as a "double-planet system;" the same is true for icy Pluto and its large satellite, or moon, Charon.

The largest planet is Jupiter, which has more mass than all the other planets combined. Mercury is the planet nearest the Sun, while (on average) Pluto is the farthest away. At perihelion (the point in an orbit at which a celestial body is nearest the Sun), Pluto actually is closer to the Sun than Neptune. Saturn is the least dense planet in the solar system. If we could find some giant cosmic swimming pool, Saturn would float, since it is less dense than water. Seven of the nine planets have satellites, or moons, some of which are larger than the planet Mercury.

See also ASTEROID; EARTH; JUPITER; MARS; MERCURY; NEPTUNE; PLUTO; SATURN; URANUS; VENUS.

planetarium Originally a mechanical device, such as an orrery, that could depict the relative motions of the known planets in the solar system. In modern times, a theaterlike domed facility in which various celestial objects (such as are found in Earth's night skies) can be projected upon the interior ceiling, simulating views of the heavens past, present, and future. Themed "sky shows" held at planetaria provide excellent ways to gain familiarity with the constellations in the night sky and the relative motion of the planets in the solar system, as well as other interesting topics in astronomy.

planetary Of or pertaining to a planet.

planetary albedo The fraction of incident solar radiation that is reflected by a planet (and its atmosphere) and returned to space. For example, Earth has a planetary albedo of

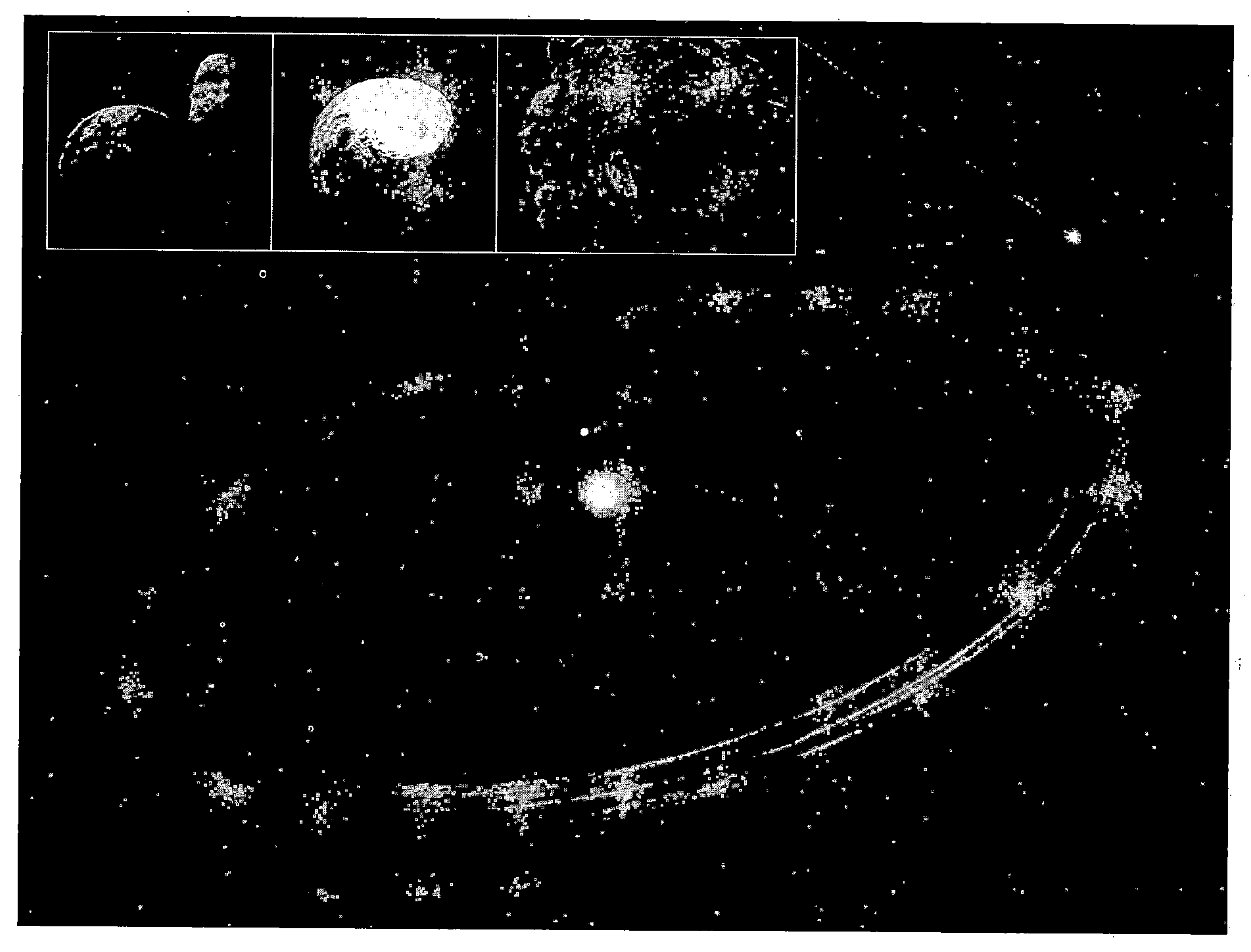

This artist's rendering depicts a young star encircled by full-sized planets and rings of dust beyond. These rings, or debris discs, arise when embryonic planets smash into one another. (The sequence of events for one of these catastrophic planetary body collisions appears in the upper-left insert.) The illustration shows how planetary systems arise out of massive collisions between rocky bodies. NASA's *Spitzer Space Telescope* is collecting data that indicate that these catastrophes continue to occur around stars when they are as old as 100 million years and even after they have developed full-sized planets. *(Courtesy of NASA/JPL-Caltech)*

approximately 0.3—that is, it reflects about 30 percent of the incident solar radiation back to space, mostly by backscatter from clouds in the atmosphere.

planetary engineering Planetary engineering, or *terraforming*, as it is sometimes called, is the large-scale modification or manipulation of the environment of a planet to make it more suitable for human habitation. In the case of Mars, for example, human settlers would probably seek to make its atmosphere more dense and breathable by adding more oxygen. Early "Martians" would most likely also attempt to alter the planet's harsh temperatures and modify them to fit a more terrestrial thermal pattern. Venus represents an even larger challenge to the planetary engineer. Its current atmospheric pressure would have to be significantly reduced, its infernolike surface temperatures greatly diminished, the excessive amounts of carbon dioxide in its atmosphere reduced, and perhaps the biggest task of all, its rotation rate would have to be increased to shorten the length of the solar day.

It should now be obvious that when we discuss planetary engineering projects, we are speaking of truly large, long-term projects. Typical time estimates for the total terraforming of a planet such as Mars or Venus range from centuries to a few millennia. However, we can also develop ecologically suitable enclaves, or niches, especially on the Moon or Mars. Such localized planetary modification efforts could probably be accomplished within a few decades of project initiation.

Just what are the "tools" of planetary engineering? The planetary pioneers in the next century will need at least the following if they are to convert presently inhospitable worlds into new ecospheres that permit human habitation with little or no personal life support equipment: First, and perhaps the most often overlooked, human ingenuity; second, a thorough knowledge of the physical processes of the particular planet or moon undergoing terraforming (especially the existence and location of environmental pressure points at which small modifications of the local energy or material balance can cause global environmental effects); third, the ability to manipulate large quantities of energy; fourth, the ability to manipulate the surface or material composition of the planet; and fifth, the ability to move large quantities of extraterrestrial materials (for example, small asteroids, comets, or water-ice shipments from the Saturnian rings) to any desired locations within heliocentric space.

One frequently suggested approach to planetary engineering is the use of biological techniques and agents to manipulate alien worlds into more desirable ecospheres. For example, scientists have proposed seeding the Venusian atmosphere with special microorganisms (such as genetically engineered algae) capable of converting excess carbon dioxide into free oxygen and combined carbon. This biological technique would provide not only a potentially more breathable Venusian atmosphere, it would also help to lower the currently intolerable surface temperatures by reducing the runaway greenhouse effect.

Other individuals have suggested the use of special vegetation (such as genetically engineered lichen, small plants, or scrubs) to help modify the polar regions on Mars. The use of specially engineered survivable plants would reduce the albedo of these frigid regions by darkening the surface, thereby allowing more incident sunlight to be captured. In time, an increased amount of solar energy absorption would elevate global temperatures and cause melting of the long-frozen volatiles, including water. This would raise the atmospheric pressure on Mars and possibly cause a greenhouse effect. With the polar caps melted, large quantities of liquid water would be available for transport to other regions of the planet. Perhaps one of the more interesting Martian projects late in the next century will be to construct a series of large irrigation canals.

Of course, there are other alternatives to help melt the Martian polar caps. The Martian settlers could decide to construct giant mirrors in orbit above the Red Planet. These mirrors would be used to concentrate and focus raw sunlight directly on the polar regions. Other scientists have suggested dismantling one of the Martian moons (Phobos or Deimos) or perhaps a small dark asteroid and then using its dust to physically darken the polar regions. This action would again lower the albedo and increase the absorption of incident sunlight.

Another approach to terraforming Mars would be to use nonbiological replicating systems (that is, *self-replicating robot systems*). These self-replicating machines would probably be able to survive more hostile environmental conditions than genetically engineered microorganisms or plants. To examine the scope and magnitude of this type of planetary engineering effort, we first assume that the Martian crust is mainly silicone dioxide (SiO_2) and then that a general-purpose 100-ton self-replicating system (SRS) "seed machine" can make a replica of itself on Mars in just one year. This SRS unit would initially make other units like itself using native Martian raw materials. In the next phase of the planetary engineering project, these SRS units would be used to reduce SiO_2 into oxygen that is then released into the Martian atmosphere. In just 36 years from the arrival of the "seed machine," a silicon dioxide reduction capability would be available that could release up to 220,000 tons per second of pure oxygen into the thin atmosphere of the Red Planet. In only 60 years of operation, this array of SRS units would have produced and liberated 4×10^{17} kg of oxygen into the Martian environment. Assuming negligible leakage through the Martian exosphere, this is enough "free" oxygen to create a 0.1 bar pressure breathable atmosphere across the entire planet. This pressure level is roughly equivalent to the terrestrial atmosphere at an altitude of 3,000 meters.

What would be the environmental impact of all these mining operations on Mars? Scientists estimate that the total amount of material that would have to be excavated to terraform Mars is on the order of 10^{18} kilograms of silicon dioxide. This is enough soil to fill a surface depression 1 km deep and about 600 km in diameter. This is approximately the size of the crater Edom near the Martian equator. The Martians might easily rationalize, "Just one small hole for Mars, but a new ecosphere for humankind!"

Asteroids also could play an interesting role in planetary engineering scenarios. People have suggested crashing one or two "small" asteroids into depressed areas on Mars (such as the Hellas Basin) to instantly deepen and enlarge the depression. The goal would be to create individual or multiple (connected) instant depressions about 10 km deep and 100 km across. These human-made impact craters would be deep enough to trap a denser atmosphere, allowing a small ecological enclave, or niche, to develop. Environmental conditions in such enclaves could range from typical polar conditions to perhaps something almost balmy.

Others have suggested crashing an asteroid into Venus to help increase its spin. If the asteroid hit the Venusian surface at just the right angle and speed, it could conceivably help speed up the planet's rotation rate, greatly assisting any overall planetary engineering project. Unfortunately, if the asteroid were too small or too slow it would have little or no effect, while if it were too large or hit too fast it could possibly shatter the planet.

It has also been proposed that several large-yield nuclear devices be used to disintegrate one or more small asteroids that had previously been maneuvered into orbits around Venus. This would create a giant dust and debris cloud that would encircle the planet and reduce the amount of incoming sunlight. This, in turn, would lower surface temperatures on Venus and allow the rocks to cool sufficiently to start absorbing carbon dioxide from the dense Venusian atmosphere. Finally, other scientists have suggested mining the rings of Saturn for frozen volatiles, especially water-ice, and then transporting these back for use on Mars, the Moon, or Venus for large-scale planetary engineering projects.

planetary nebula (plural: planetary nebulas or nebulae) The large, bright, usually symmetric cloud of gas surrounding a highly evolved (often dying) star. This shell of gas expands at a velocity on the order of tens of kilometers per second. Despite the name (based on early astronomical observations), planetary nebulas are not associated with planets nor are they associated with interstellar matter. Rather, they represent the expanding shell of hot gas ejected by mature stars (e.g., red giants) at the end of their active lives. In about 5 billion years from now our Sun will go through its "red giant" phase and develop a planetary nebula. *Compare with* NEBULA.

See also STAR.

planetary orbital operations Planetary spacecraft can engage in two general categories of orbital operations: exploration of a planetary system and mapping of a planet. Exploring a planetary system includes making observations of the target planet and its system of satellites and rings. A mapping mission generally is concerned with acquiring large amounts of data about the planet's surface features. An orbit of low inclination at the target planet usually is well suited to a planetary system exploration mission, since it provides repeated exposure to satellites (moons) orbiting with the equatorial plane as well as adequate coverage of the planet and its magnetosphere. However, an orbit of high inclination is much better suited for a mapping mission, because the target planet will fully rotate below the spacecraft's orbit, thereby providing eventual exposure to every portion of the planet's surface. During either type of mission the orbiting spacecraft is involved in an extended encounter period with the target planet and requires continuous (or nearly continuous) support from the flight team members at mission control on Earth. *Galileo* is an example of a planetary system exploration mission, while *Magellan* is an example of a planetary mapping mission.

See also CASSINI MISSION; GALILEO PROJECT/SPACECRAFT; MAGELLAN MISSION.

planetary orbit insertion The same type of highly precise navigation and course correction procedures used in flyby missions are applied during the cruise phase for a planetary orbiter mission. This process then places the spacecraft at precisely the correct location at the correct time to enter into an orbit about the target planet. However, orbit insertion requires not only the precise position and timing of a flyby mission but also a controlled deceleration. As the spacecraft's trajectory is bent by the planet's gravity, the command sequence aboard the spacecraft fires its retroengine(s) at the proper moment and for the proper duration. Once this retroburn (or retrofiring) has been completed successfully, the spacecraft is captured into orbit by its target planet. If the retroburn fails (or is improperly sequenced), the spacecraft will continue to fly past the planet. It is quite common for this retroburn to occur on the farside of a planet as viewed from Earth, requiring this portion of the orbit insertion sequence to occur essentially automatically (based on onboard commands) and without any interaction with the flight controllers on Earth.

planetary probe An instrument-containing spacecraft deployed in the atmosphere or on the surface of a planetary body in order to obtain in situ scientific data and environmental information about the celestial body.

planetary quarantine protocols(s) *See* EXTRATERRESTRIAL CONTAMINATION.

planetesimals Small celestial objects in the solar system, such as asteroids, comets, and moons.

planet fall The landing of a spacecraft or space vehicle on a planet.

planetoid An asteroid, or minor planet.

See also ASTEROID.

planet X The name used by the American astronomer PERCIVAL LOWELL to describe the planet he speculated existed beyond the orbit of Neptune. Lowell searched for Planet X but could not find it. However, his quest for Planet X ultimately allowed CLYDE TOMBAUGH to discover the planet Pluto using facilities at the Lowell Observatory in Flagstaff, Arizona.

planisphere *See* ANCIENT ASTRONOMICAL INSTRUMENTS.

Plaskett, John Stanley (1865–1941) Canadian *Astronomer* During his career the Canadian astronomer John Stanley Plaskett made numerous radial velocity measurements, which enabled him to discover many spectroscopic binary star systems. One of these binary systems, now called Plaskett's star, is the most massive binary star system yet discovered in the observable universe. Current astronomical measurements suggest that Plaskett's star is a binary star system consisting of two blue supergiant stars, each of which has a mass on the order of 50 solar masses.

plasma An electrically neutral gaseous mixture of positive and negative ions. Sometimes called the "fourth state of matter," since a plasma behaves quite differently from solids, liquids, or gases.

plasma engine A reaction (rocket) engine using electromagnetically accelerated plasma as propellant.

See also ELECTRIC PROPULSION.

plasma instruments Plasma detectors on a scientific spacecraft measure the density, composition, temperature, velocity, and three-dimensional distribution of plasmas that exist in interplanetary space and within planetary magnetospheres. Plasma detectors are typically sensitive to solar and planetary plasmas, and they can observe the solar wind and its interactions with a planetary system.

See also MAGNETOSPHERE; SOLAR WIND.

plasma rocket A rocket engine in which the ejection of plasma generates thrust; a rocket using a plasma engine.

See also ELECTRIC PROPULSION.

platform 1. An uncrewed (unmanned) orbital element of a space station program that provides standard support services to payloads not attached to the space station. 2. With respect to ballistic missile defense, a satellite in space that is used as a carrier for weapons, sensors, or both.

See also BALLISTIC MISSILE DEFENSE.

Pleiades A prominent open star cluster in the constellation Taurus about 410 light-years distant.

See also STAR CLUSTER.

Plesetsk Cosmodrome The northern Russian launch site about 300 km south of Archangel that supports a wide variety of military space launches, ballistic missile testing, and space lift services for scientific or civilian spacecraft requiring a polar orbit.

plug nozzle An annular nozzle that discharges exhaust gas with a radial inward component; a truncated aerospike.

See also NOZZLE.

plume 1. In aerospace, the hot, bright exhaust gases from a rocket. 2. In environmental studies, a visible or measurable discharge of a contaminant from a given point of origin; can be visible or thermal in water, or visible in the air (e.g., a plume of smoke).

plutino (little Pluto) Any of numerous, small (about 100 km in diameter), icy celestial bodies that occupy the inner portions of the Kuiper Belt and whose orbital motion resonance with Neptune resembles that of Pluto—namely, each icy object completes two orbits around the sun in the time Neptune takes to complete three orbits.

See also TRANS-NEPTUNIAN OBJECT.

Pluto Pluto, the smallest planet in the solar system, has remained a mystery since its discovery by astronomer CLYDE TOMBAUGH in 1930. Pluto is the only planet not yet viewed close up by spacecraft, and given its great distance from the Sun and tiny size, study of the planet continues to challenge and extend the skills of planetary astronomers. In fact, most of what scientists know about Pluto has been learned since the late 1970s. Such basic characteristics as the planet's radius and mass were virtually unknown before the discovery of Pluto's moon, Charon, in 1978. Since then observations and speculations about the Pluto-Charon system—it is now considered a "double planet" system—have progressed steadily to a point that many of the key questions remaining about the system must await the close-up observation of a flyby space mission, such as NASA's planned *New Horizons Pluto–Kuiper Belt Flyby* mission.

For example, there is a strong variation in brightness, or albedo, as Pluto rotates, but planetary scientists do not know if what they are observing is a system of varying terrains, areas of different composition, or both. Scientists know there is a dynamic, largely nitrogen and methane atmosphere around Pluto that waxes (grows) and wanes (diminishes) with the planet's elliptical orbit around the Sun, but they still need to understand how the Plutonian atmosphere arises, persists, is again deposited on the surface, and how some of it escapes into space.

Telescopic studies (both Earth-based telescopes and NASA's Earth-orbiting *Hubble Space Telescope*) indicate that Pluto and Charon are very different bodies. Pluto is mostly rock, while Charon appears more icy. How and when the two bodies in this interesting double-planet system could have evolved so differently is another key question that awaits data from a close-up observation. Data about Pluto and Charon, as gathered using ground-based and Earth-orbiting observatories, have helped improve our fundamental understanding of these planetary bodies. The most recent of these data are presented in Table 1 for Pluto and Table 2 for Charon.

Because of Pluto's highly eccentric orbit, its distance from the Sun varies from about 4.4 to 7.4 billion kilometers, or some 29.5 to 49.2 astronomical units (AU). For most of its orbit Pluto is the outermost of the planets. However, from 1979 it actually orbited closer to the Sun than Neptune. This condition remained until 1999, when Pluto again became the outermost planet in the solar system. Upon its discovery in 1930, Clyde Tombaugh (keeping with astronomical tradition) named the planet Pluto after the god of the underworld in Roman mythology. Incidentally, this discovery was not a real astronomical surprise, since astronomers at the turn of the century had predicted Pluto's existence from perturbations (disturbances) observed in the orbits of both Uranus and Neptune.

However, the discovery of Charon by James W. Christy of the U.S. Naval Observatory in 1978 triggered a revolution in our understanding of Pluto. Charon was the boatman in Greek and Roman mythology who ferried the dead across the

Dynamic Properties and Physical Data for Pluto

Diameter	2,290 km
Mass	1.25×10^{22} kg
Mean density	2.05 g/cm^3
Albedo (visual)	0.3
Surface temperature (average)	~40K
Surface gravity	0.4–0.6 m/s^2
Escape velocity	1.1 km/s
Atmosphere (a transient phenomenon)	Nitrogen (N_2) and methane (CH_4)
Period of rotation	6.387 days
Inclination of axis (of rotation)	119.6°
Orbital period *(around Sun)*	248 years (90,591 days)
Orbit inclination	17.15°
Eccentricity of orbit	0.2482
Mean orbital velocity	4.75 km/s
Distance from Sun	
Aphelion	7.38×10^9 km (49.2 AU) (409.2 light-min)
Perihelion	4.43×10^9 km (29.5 AU) (245.3 light-min)
Mean distance	5.91×10^9 km (39.5 AU) (328.5 light-min)
Solar flux (at 30 AU ~perihelion)	1.5 W/m^2
Number of known natural satellites	1

(Note: Some of these data are speculative.)
Source: NASA.

River Styx to the underworld, ruled by Pluto ("Hades" in Greek myth). This large moon has a diameter of about 1,270 km, compared to tiny Pluto's diameter of 2,290 km. These similar sizes encourage many planetary scientists to regard the pair as the only true double-planet system in our solar system. Charon circles Pluto at a distance of 19,405 km. Its orbit is gravitationally synchronized with Pluto's rotation period (approximately 6.4 days) so that both the planet and its moon keep the same hemisphere facing each other at all times.

Planetary scientists currently believe that Pluto possesses a very thin atmosphere that contains nitrogen (N_2) and methane (CH_4). Pluto's atmosphere is unique in the solar system in that it undergoes a formation-and-decay cycle each orbit around the Sun. The atmosphere begins to form several decades before perihelion (the planet's closet approach to the Sun) and then slowly collapses and freezes out decades later as the planet's orbit takes it farther and farther away from the Sun to the frigid outer extremes of the solar system. In September 1989 Pluto experienced perihelion. Several decades from now its thin atmosphere will freeze out and collapse, leaving a fresh layer of nitrogen and methane snow on the planet's surface.

Dynamic Properties and Physical Data for Charon

Diameter	1,186 km
Mass	1.7×10^{21} kg
Mean density	1.9 g/cm^3
Surface gravity	0.2 m/s^2
Escape velocity	.58 m/s
Albedo (visual)	0.375
Mean distance from Pluto	19,600 km
Sidereal period (gravitational synchronized orbit)	6.387 days
Orbital inclination	98.8°
Orbital eccentricity	0

(Note: Some of these data are speculative.)
Source: NASA.

Pluto–Kuiper Belt Mission Express *See* NEW HORIZONS PLUTO–KUIPER BELT FLYBY MISSION.

pod An enclosure, housing, or detachable container of some kind; for example, an instrument pod or an engine pod.

pogo A term developed within the aerospace industry to describe the longitudinal (vertical) dynamic oscillations (vibrations) generated by the interaction of a launch vehicle's structural dynamics with the propellant and engine combustion process. The name appears to have originated from the similarity of this bumping phenomenon to the up-and-down bumping motion of a pogo stick.

pogo suppressor A device within a liquid rocket engine that absorbs any vibration of the vehicle's structure during the engine combustion process.

See also POGO.

Pogson, Norman Robert (1829–1891) British *Astronomer* In 1856 the British astronomer Norman Robert Pogson suggested the use of a new mathematically based scale for describing stellar magnitudes. First, he verified SIR WILLIAM HERSCHEL's discovery that a first-magnitude star is approximately 100 times brighter than a sixth-magnitude star—the limit of naked eye observing as originally proposed by the early Greek astronomer HIPPARCHUS and later refined by PTOLEMY. Pogson then suggested that because an interval of five magnitudes corresponds to a factor of 100 in brightness, a one-magnitude difference in brightness should correspond to the fifth root of 100, or 2.512. Pogson's scale is now universally used in modern astronomy.

Pogson scale *See* POGSON, NORMAN ROBERT.

Poincaré, (Jules) Henri (1854–1912) French *Mathematician* Among his many accomplishments, the French mathematician attacked celestial mechanics's challenging *three-body problem* and *n-body problem,* providing astronomers at least partial solutions that supported significant advances in orbital motion prediction. His pioneering mathematical work paved the way for the development of more efficient computational solutions based on the use of digital computers.

point of impact The point at which a projectile, bomb, or reentry vehicle impacts or is expected to impact.

point of no return 1. The point along the launch track of an aerospace vehicle or space vehicle from which it can no longer return to its launch base. 2. For a human surface expedition on Mars, the point at which the team can no longer return to its surface base with the remaining supplies (e.g., air, food, water, or fuel). 3. For an aircraft, a point along its flight track beyond which its endurance will not permit return to its own or some other associated base with its remaining fuel supply.

poise (symbol: P) Unit of dynamic viscosity in the centimeter-gram-second (cgs) unit system. It is defined as the tangential force per unit area (e.g., dynes/cm^2) required to maintain a unit difference in velocity (e.g., 1 cm/sec) between two parallel plates in a liquid that are separated by a unit distance (e.g., 1 cm).

$$1 \text{ poise} = (1 \text{ dyne-second})/(\text{centimeter})^2 = 0.1 \text{ (newton-second)}/(\text{meter})^2$$

The unit is named after the French scientist Jean Louis Poiseuille (1799–1869). The centipoise, or 0.01 poise, is encountered often. For example, the dynamic viscosity of water at 20°C is approximately 1 centipoise.

polar coordinate system A coordinate system in which a point (P) that is defined as P(x, y) in two-dimensional Cartesian coordinates is now represented as P(r, θ), where $x = r \cos \theta$ and $y = r \sin \theta$. Physically, in the polar coordinate system r is the radial distance in the x-y plane from the origin (O) to point P, and θ is the angle formed between the x-axis and the radial vector (i.e., the line of length r from the origin O to point P).

In three dimensions, the Cartesian coordinate system point P(x,y,z) becomes P(r,θ,z) in cylindrical polar coordinates—where the point P (r,θ,z) is now regarded as lying on the surface of a cylinder. The terms r and θ are as previously defined, while z represents the height above (or below) the x-y plane.

See also CARTESIAN COORDINATES; CELESTIAL COORDINATES; CYLINDRICAL COORDINATES.

Polaris (North Star) A creamy white-colored F spectral class supergiant star that is about 820 light-years away. The North Star is also known as Alpha Ursae Minoris (αUMi), meaning it is the brightest (or alpha) star in the constellation Ursa Minor. Since Polaris lies very close to the north celestial pole, astronomers treat it as the current pole star.

Polaris (UGM-27) An underwater- or surface-launched, surface-to-surface, solid-propellant ballistic missile with inertial guidance and nuclear warhead. The Polaris (version UGM-27C) has a range of 4,600 km.

See also FLEET BALLISTIC MISSILE.

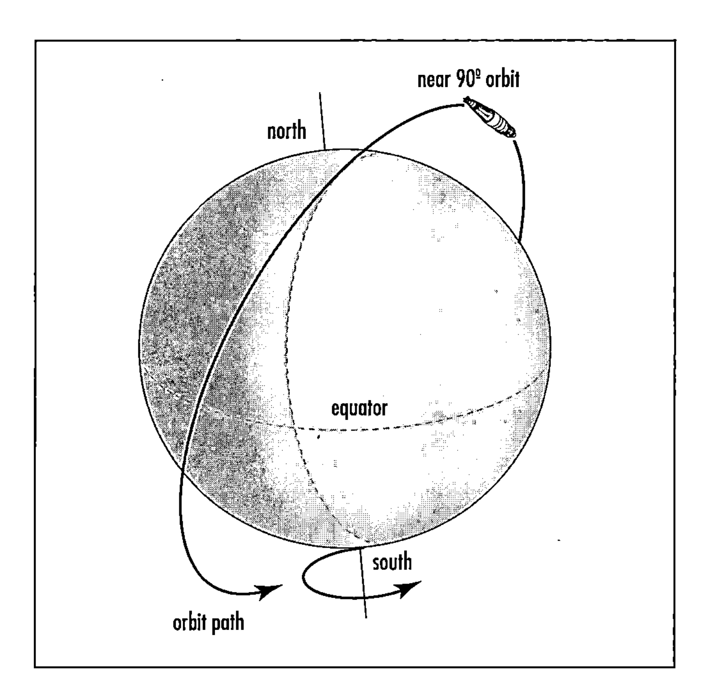

A spacecraft in polar orbit around Earth (or other planet)

polarization A distinct orientation of the wave motion and direction of travel of electromagnetic radiation, including light. Along with brightness and color, polarization is a special quality of light. It represents a condition in which the planes of vibration of the various rays in a beam of light are at least partially (if not completely) aligned.

polar orbit An orbit around a planet (or primary body) that passes over or near its poles; an orbit with an inclination of about 90°.

Polar-Orbiting Operational Environmental Satellite (POES) The National Oceanic and Atmospheric Administration's (NOAA's) Polar-Orbiting Operational Environmental Satellites follow orbits that pass close to the North and South Poles as Earth rotates beneath them. Orbiting at an altitude of about 840 kilometers, these satellites provide continuous global coverage of the state of Earth's atmosphere, including such essential information as atmospheric temperature, humidity, cloud cover, ozone concentration, and Earth's energy (radiation) budget. They also provide important surface data, such as sea ice and sea surface temperature and snow and ice coverings.

See also NATIONAL POLAR-ORBITING OPERATIONAL ENVIRONMENTAL SATELLITE SYSTEM (NPOESS).

polar orbiting platform (POP) An uncrewed (unmanned) spacecraft in polar or near-polar inclination that is operated from ground stations but dependent on a permanent space station program to provide services for its complement of payloads.

See also SPACE STATION.

Polar Space Launch Vehicle (PSLV) A launch vehicle developed by the Indian Space Research Organization (ISRO). It consists of two solid stages, two hypergolic stages, and six strap-on solid-rocket motors. This vehicle can place up to 3,000 kg into low Earth orbit (LEO). It was first launched in 1993.

See also LAUNCH VEHICLE.

poles The poles for a rotating celestial body are located at the ends (usually called north and south) of the body's axis of rotation.

Pond, John (1767–1836) British *Astronomer* The British astronomer John Pond was appointed England's sixth Astronomer Royal in 1811. He served in this position until 1835, during which time he upgraded the Royal Greenwich Observatory (RGO) with new instruments. In 1833 he published a very accurate star catalog.

Pons, Jean (1761–1831) French *Astronomer* The French astronomer Jean Pons was an avid comet hunter who between 1801 and 1827 discovered (or codiscovered) 37 comets, a number that represents a personal observing record in astronomy and about 75 percent of all comets discovered within that period.

popping Sudden, short-duration surges of pressure in a liquid-propellant rocket's combustion chamber.

population I stars Hot, luminous young stars, including those like the Sun, that reside in the disk of a spiral galaxy and are higher in heavy element content (about 2 percent abundance) than older Population II stars.

population II stars Older stars that are lower in heavy element content than Population I stars and reside in globular clusters as well as in the halo of a galaxy—that is, the distant spherical region that surrounds a galaxy.

port 1. An opening; a place to access an aerospace system through which energy, data, or fluids may be supplied to or withdrawn from the system. For example, an observation port, a refueling port, or a communications port. 2. The left-hand side of an aerospace vehicle as a person looks from the rear of the craft to the nose.

Portia A small moon of Uranus (about 110 km in diameter) discovered in 1986 as a result of the *Voyager 2* encounter. Portia travels around Uranus at an average distance of 66,100 km with an orbital period of 0.51 day. Also called Uranus XII or S/1986 U1.

See also URANUS.

posigrade rocket An auxiliary rocket that fires in the direction in which the vehicle is pointed, used, for example, in separating two stages of a vehicle. A posigrade course correction adds to the spacecraft's speed, in contrast to a *retrograde correction,* which slows it down.

See also ORBITS OF OBJECTS IN SPACE.

positron (symbol: β^+) An elementary particle with the mass of an electron but charged positively. It is the "antielectron." A positron is emitted in some types of radioactive decay and also is formed in pair production by the interaction of high-energy gamma rays with matter.

See also ANTIMATTER; ELEMENTARY PARTICLE(s).

post-boost phase The phase of a missile trajectory, after the booster's stages have finished firing, in which the various reentry vehicles (RVs) are independently placed on ballistic trajectories toward their targets. In addition, penetration aids can be dispensed from the post-boost vehicle during this phase, which typically lasts between three and five minutes.

See also BALLISTIC MISSILE DEFENSE.

post-boost vehicle (PBV) The portion of a ballistic missile payload that carries the multiple warheads and has maneuvering capability to place each warhead on its final trajectory to a target. Also referred to as a bus.

See also BALLISTIC MISSILE DEFENSE.

potential energy (PE) Energy possessed by an object by virtue of its position in a gravity field. In general, the potential energy (PE) of an object can be expressed as

$$PE = m\,g\,z$$

where m is the mass of the object, g is the acceleration due to gravity, and z is the height of the object above some reference position or datum.

power (symbol: P) 1. In physics and engineering, the rate with respect to time at which work is done or energy is transformed or transferred to another location. In the SI unit system, power typically is measured in multiples of the watt, which is defined as one joule (of energy) per second. 2. In mathematics, the number of times a quantity is multiplied by itself; for example, b^3 means that b has been raised to the third power, or simply $b^3 = b \cdot b \cdot b$.

power plant In the context of aerospace engineering, the complete installation of an engine (or engines) with supporting subsystems (e.g., cooling subsystem, ignition subsystem) that generates the motive power for a self-propelled vehicle or craft, such as an aircraft, rocket, or aerospace vehicle.

preburners Some liquid-hydrogen–fueled rocket engines, such as the space shuttle main engines (SSMEs), have fuel and oxidizer preburners that provide hydrogen-rich hot gases at approximately 1,030 kelvins (K) (757°C). These gases then are used to drive the fuel and oxidizer high-pressure turbopumps.

precession The gradual periodic change in the direction of the axis of rotation of a spinning body due to the application of an external force (torque).

precession of equinoxes The slow westward motion of the equinox points across the ecliptic relative to the stars of the zodiac caused by the slight wobbling of Earth about its axis of rotation.

See also EQUINOX; ZODIAC.

precombustion chamber In a liquid-propellant rocket engine, a chamber in which the propellants are ignited and from which the burning mixture expands torchlike to ignite the mixture in the main combustion chamber.

preflight In aerospace operations, something that occurs before launch vehicle liftoff.

preset guidance A technique of missile control wherein a predetermined flight path is set into the control mechanism and cannot be adjusted after launching.

See also GUIDANCE SYSTEM.

pressurant Gas that provides ullage pressure in a propellant tank.

pressure (symbol: p) A thermodynamic property that two systems have in common when they are in mechanical equilibrium. Pressure can be defined as the normal component of force per unit area exerted by a fluid on a boundary. According to the Pascal principle, the pressure at any point in a fluid at rest is the same in all directions. Blaise Pascal (1623–62) was the French scientist who discovered this hydrostatic principle in the 1650s. The fundamental principle of hydrostatics is

$$dp/dz = -\rho g$$

where dp is the change in pressure corresponding to a change in height or depth (dz), ρ is the fluid density, and g is the acceleration due to gravity. Assuming that the density (ρ) of the fluid is constant, we can compute the hydrostatic pressure of a liquid from the following relationship:

$$p(z) = \rho g\, z$$

where p is the pressure and z is the height or depth with respect to some reference or datum, such as sea level.

In aerostatics, density varies (decreases) with height in the atmosphere, and, therefore, pressure cannot be evaluated as a function of height until the density is specified in some manner. The ideal gas law often can be used to describe the thermodynamic properties of a gas, such as air. Using this relationship

$$p = \rho\, R\, T$$

where p is the pressure of the gas, ρ is the density, T is the absolute temperature, and R is the specific gas constant.

The pascal (named in honor of Blaise Pascal) is the SI unit of pressure. It is defined as

$$1 \text{ pascal} = 1 \text{ newton}/(\text{meter})^2$$

pressure fed In rocketry, a propulsion system in which tank ullage pressure expels the propellants from the tanks into the combustion chamber of the engine.

pressure-laden sequence In rocketry, a method to accomplish fail-safe liquid-propellant engine starts by sequencing the operation of rocket engine control valves; the sequencing is achieved by vent mechanisms on the control system or propellant feed system, or both, that are triggered by pressure changes.

pressure regulator A pressure control valve that varies the volumetric flow rate through itself in response to a downstream pressure signal so as to maintain the downstream pressure nearly constant.

pressure suit In general, a garment designed to provide pressure upon the body so that the respiratory and circulatory functions may continue normally, or nearly so, under low-pressure conditions such as occur at high altitudes or in space without benefit of a pressurized module or crew compartment. 1. (partial) A skintight suit that does not completely enclose the body but that is capable of exerting pressure on the major portion of the body in order to counteract an increased intrapulmonary oxygen pressure. 2. (full) A suit that completely encloses the body and in which a gas pressure, sufficiently above ambient pressure for maintenance of respiratory and circulatory functions, may be sustained.

pressurization For a launch vehicle prior to liftoff, the sequence of operations that increases the propellant tank ullage pressure to the desired level just before the main sequence of propellant flow and engine firing.

pressurization system The set of fluid system components that provides and maintains a controlled gas pressure in the ullage space of the launch vehicle propellant tanks.

pressurized habitable environment Any module or enclosure in space in which an astronaut may perform activities in a "shirt-sleeve" environment.

prestage A step in the action of igniting a large liquid-propellant rocket taken prior to the ignition of the full flow. It involves igniting a partial flow of propellants into the thrust chamber.

primary body The celestial body about which a satellite, moon, or other object orbits or from which it is escaping or toward which it is falling. For example, the primary body of Earth is the Sun; the primary body of the Moon is Earth.

primary mirror The principal light-gathering mirror of a reflecting telescope.

See also REFLECTING TELESCOPE; TELESCOPE.

primitive atmosphere The atmosphere of a planet or other celestial object as it existed in the early stages of its formation. For example, the primitive atmosphere of Earth some 3 billion years ago is thought to have consisted of water vapor, carbon dioxide, methane, and ammonia.

principal investigator (PI) The research scientist who is responsible for a space experiment and for reporting the results.

principle of equivalence *See* EQUIVALENCE PRINCIPLE.

principle of mediocrity A general assumption or speculation often used in discussions concerning the nature and probability of extraterrestrial life. It assumes that things are pretty much the same all over—that is, it assumes that there is nothing special about Earth or our solar system. By invoking this hypothesis, we are guessing that other parts of the universe are pretty much as they are here. This philosophical position allows us to then take the things we know about Earth, the chemical evolution of life that occurred here, and the facts we are learning about other objects in our solar system and extrapolate these to develop concepts of what may be occurring on alien worlds around distant suns.

The simple premise of the principle of mediocrity is very often employed as the fundamental starting point for contemporary speculations about the cosmic prevalence of life. If Earth is indeed nothing special, then perhaps a million worlds in the Milky Way galaxy (which is one of billions of galaxies) not only are suitable for the origin of life but have witnessed its chemical evolution in their primeval oceans and are now (or at least were) habitats for a myriad of interesting living creatures. Some of these living systems may also have risen to a level of intelligence by which they are at this very moment gazing up into the heavens of their own world and wondering if they, too, are alone.

If, on the other hand, Earth and its delicate biosphere really are something special, then life—especially intelligent life capable of comprehending its own existence and contemplating its role in the cosmic scheme of things—may be a rare, very precious jewel in a vast, lifeless cosmos. In this latter case the principle of mediocrity would be most inappropriate to use in estimating the probability that extraterrestrial life exists elsewhere in the universe.

Today we cannot pass final judgment on the validity of the principle of mediocrity. We must, at an absolute minimum, wait until human and robot explorers have made more detailed investigations of the interesting objects in our own solar system. Celestial objects of particular interest to exobiologists include the planet Mars and certain moons of the giant outer planets Jupiter and Saturn. Once we have explored these alien worlds in depth, scientists will have a much more accurate technical basis for suggesting that we are either "something special" or else "nothing special," as the principle of mediocrity implies.

See also EXOBIOLOGY; LIFE IN THE UNIVERSE; SEARCH FOR EXTRATERRESTRIAL INTELLIGENCE.

prism A block (often triangular) of transparent material that disperses an incoming beam of white light into the visible spectrum of rainbow colors—that is, red, orange, yellow, green, blue, indigo, and violet, in order of decreasing wavelength.

See also ELECTROMAGNETIC SPECTRUM; LIGHT.

probe 1. Any device inserted in an environment for the purpose of obtaining information about the environment. 2. An instrumented spacecraft or vehicle moving through the upper atmosphere or outer space or landing on another celestial body in order to obtain information about the specific environment, such as, for example, a deep space probe, a lunar probe, or a Jovian atmosphere probe. 3. A slender device (e.g., a pitot tube) projected into a moving fluid for measurement purposes. 4. A ground-based set of sensors that could be launched rapidly into space on warning of attack and then function as tracking and acquisition sensors to support weapon allocation and firing by ballistic missile defense (BMD) weapons against enemy intercontinental ballistic missiles (ICBMs) and reentry vehicles (RVs). 5. An extended structure on one spacecraft that is inserted into a compatible receptacle (called a *drogue*) on another spacecraft, creating a linkage (*docking*) of the two spacecraft.

Prognoz spacecraft A series of scientific Russian spacecraft placed in highly elliptical Earth orbits for the purpose of investigating the Sun, the effects of the solar wind on Earth's magnetosphere, and solar flares. The first spacecraft in this series, *Prognoz 1,* was launched from Tyuratam (Baikonour Cosmodrome) on April 14, 1972. This 845-kg spacecraft was placed in a highly elliptical Earth orbit that had a perigee of 950 km and an apogee of 200,000 km, an inclination of 65°, and a period of 4.04 days.

prograde orbit An orbit having an inclination of between 0 and 90°.

See also ORBITS OF OBJECTS IN SPACE.

program In an aerospace context, a major activity that involves the human resources, material, funding, and scheduling necessary to achieve desired space exploration or space technology goals. For example, NASA's Origins Program, the International Space Station Program, and the Hubble Space Telescope Program.

progressive burning The condition in which the burning area of a solid-propellant rocket grain increases with time, thereby increasing pressure and thrust as the burning progresses.

See also ROCKET.

Progress spacecraft An uncrewed supply spacecraft configured to perform automated rendezvous and docking with orbiting Russian space stations, such as the *Salyut* and the *Mir.* After the Russian cosmonauts have unloaded the *Progress* cargo ship, it separates from the space station, deorbits, and then burns up in the upper regions of Earth's atmosphere. Since the space shuttle *Columbia* accident (February 1, 2003), Progress spacecraft have also played an essential role in supplying the crews onboard the *International Space Station* (ISS). For example, on December 25, 2004, the *Progress M-51* spacecraft docked with the ISS and brought an emergency supply of food, ending a shortage that had alarmed NASA and Russian Space Agency officials. The *Progress M-51* lifted off from the Baikonour Cosmodrome on December 24, 2004, and successfully delivered 2.5 tons of important equipment and supplies for the Expedition 10 crew

(American astronaut Leroy Chiao and Russian cosmonaut Salizhan Sharipov), including more than 200 kg of food. Had the *Progress M-51* spacecraft not successfully docked with the ISS, space officials in both countries were seriously considering ordering the Expedition 10 crew to evacuate the space station. This is the first time in four decades of human space flight that a shortage of food almost caused the evacuation of a space station.

See also INTERNATIONAL SPACE STATION.

project A planned undertaking of something to be accomplished, produced, or constructed, having a finite beginning and a finite ending; for example, the Mercury Project or the Apollo-Soyuz Test Project.

Project Cyclops A very large array of dish antennas proposed for use in a detailed search of the radio frequency spectrum (especially the 18- to 21-centimeter wavelength "water hole") for interstellar signals from intelligent alien civilizations. The engineering details of this SETI (search for extraterrestrial intelligence) configuration were derived in a special summer institute design study sponsored by NASA at Stanford University in 1971. The stated objective of Project Cyclops was to assess what would be required in hardware, human resources, time, and funding to mount a realistic effort using present (or near-term future) state-of-the-art techniques aimed at detecting the existence of extraterrestrial (extra-solar system) intelligent life.

Named for the one-eyed giants found in Greek mythology, the proposed Cyclops Project would use as its "eye" a large array of individually steerable 100-meter-diameter parabolic dish antennas. These Cyclops antennas would be arranged in a hexagonal matrix, so that each antenna unit would be equidistant from all its neighbors. A 292-meter separation distance between antenna dish centers would help avoid shadowing. In the Project Cyclops concept an array of about 1,000 of these antennas would be used to simultaneously collect and evaluate radio signals falling on them from a target star system. The entire Cyclops array would function like a single giant radio antenna some 30 to 60 square kilometers in size.

Project Cyclops can be regarded as one of the foundational studies in the search for extraterrestrial intelligence. Its results, based on the pioneering efforts of such individuals as FRANK DRAKE, Philip Morrison, John Billingham, and the late Bernard Oliver, established the technical framework for subsequent SETI activities. Project Cyclops also reaffirmed the interstellar microwave window, the 18- to 21-centimeter-wavelength "water hole," as perhaps the most suitable part of the electromagnetic spectrum for interstellar civilizations to communicate with one another.

See also DRAKE EQUATION; FERMI PARADOX; SEARCH FOR EXTRATERRESTRIAL INTELLIGENCE; WATER HOLE.

Project Daedalus The name given to an extensive study of interstellar space exploration conducted from 1973 to 1978 by a team of scientists and engineers under the auspices of the British Interplanetary Society. This hallmark effort examined the feasibility of performing a simple interstellar mission using only contemporary technologies and reasonable extrapolations of imaginable near-term capabilities.

In mythology Daedulus was the grand architect of King Minos's labyrinth for the Minotaur on the island of Crete. Daedalus also showed the Greek hero Theseus, who slew the Minotaur, how to escape from the labyrinth. An enraged King Minos imprisoned both Daedalus and his son, Icarus. Undaunted, Daedalus, a brilliant engineer, fashioned two pairs of wings out of wax, wood, and leather. Before their aerial escape from a prison tower, Daedalus cautioned his son not to fly too high, so that the Sun would not melt the wax and cause the wings to disassemble. They made good their escape from King Minos's Crete, but while over the sea, Icarus, an impetuous teenager, ignored his father's warnings and soared high into the air. Daedalus (who reached Sicily safely) watched his young son, wings collapsed, tumble to his death in the sea below.

The proposed Daedalus spaceship structure, communications systems, and much of the payload were designed entirely within the parameters of 20th-century technology. Other components, such as the advanced machine intelligence flight controller and onboard computers for in-flight repair, required artificial intelligence capabilities expected to be available in the mid-21st century. The propulsion system, perhaps the most challenging aspect of any interstellar mission, was designed as a nuclear-powered, pulsed-fusion rocket engine that burned an exotic thermonuclear fuel mixture of deuterium and helium-3 (a rare isotope of helium). This pulsed-fusion system was believed capable of propelling the robot interstellar probe to velocities in excess of 12 percent of the speed of light (that is, above 0.12 c). The best source of helium-3 was considered to be the planet Jupiter, and one of the major technologies that had to be developed for Project Daedalus was an ability to mine the Jovian atmosphere for helium-3. This mining operation might be achieved by using "aerostat" extraction facilities (floating balloon-type factories).

The Project Daedalus team suggested that this ambitious interstellar flyby (one-way) mission might possibly be undertaken at the end of the 21st century—when the successful development of civilization had generated the necessary wealth, technology base, and exploratory zeal. The target selected for this first interstellar probe was Barnard's star, a red dwarf (spectral type M) about 5.9 light-years away in the constellation Ophiuchus.

The Daedalus spaceship would be assembled in cislunar space (partially fueled with deuterium from Earth) and then ferried to an orbit around Jupiter, where it could be fully fueled with the helium-3 propellant that had been mined from the Jovian atmosphere. These thermonuclear fuels would then be prepared as pellets, or "targets," for use in the ship's two-stage pulsed-fusion power plant. Once fueled and readied for its epic interstellar voyage, somewhere around the orbit of Callisto the ship's mighty pulsed-fusion first-stage engine would come alive. This first-stage pulsed-fusion unit would continue to operate for about two years. At first-stage shutdown the vessel would be traveling at about 7 percent of the speed of light (0.07 c).

The expended first-stage engine and fuel tanks would be jettisoned in interstellar space, and the second-stage pulsed-fusion engine would ignite. The second stage would also operate in the pulsed-fusion mode for about two years. Then it, too, would fall silent, and the giant robot spacecraft, with its cargo of sophisticated remote sensing equipment and nuclear fission–powered probe ships, would be traveling at about 12 percent of the speed of light (0.12 c). It would take the Daedalus spaceship about 47 years of coasting after second-stage shutdown to encounter Barnard's star.

In this scenario, when the Daedalus interstellar probe was about 3 light-years away from its objective (about 25 years of elapsed mission time), smart computers onboard would initiate long-range optical and radio astronomy observations. A special effort would be made to locate and identify any extrasolar planets that might exist in the Barnardian system.

Of course, traveling at 12 percent of the speed of light, Daedalus would have only a very brief passage through the target star system. This would amount to a few days of "close-range" observation of Barnard's star itself and only "minutes" of observation of any planets or other interesting objects by the robot mother spacecraft.

However, several years before the Daedalus mother spacecraft passed through the Barnardian system, it would launch its complement of nuclear-powered probes (also traveling at 12 percent of the speed of light initially). These probe ships, individually targeted to objects of potential interest by computers onboard the robot mother spacecraft, would fly ahead and act as data-gathering scouts. A complement of 18 of these scout craft, or small robotic probes, was considered appropriate in the Project Daedalus study.

Then, as the main Daedalus spaceship flashed through the Barnardian system, it would gather data from its own onboard instruments as well as information telemetered to it by the numerous probes. Over the next day or so it would transmit all these mission data back toward our solar system, where team scientists would patiently wait the approximately six years it would take for these information-laden electromagnetic waves, traveling at light speed, to cross the interstellar void. Its mission completed, the Daedalus mother spaceship without its probes would continue on a one-way journey into the darkness of the interstellar void, to be discovered perhaps millennia later by an advanced alien race, which might puzzle over humankind's first attempt at the direct exploration of another star system.

The main conclusions that can be drawn from the Project Daedalus study might be summarized as follows: (1) exploration missions to other star systems are, in principle, technically feasible; (2) a great deal could be learned about the origin, extent, and physics of the Milky Way galaxy, as well as the formation and evolution of stellar and planetary systems, by missions of this type; (3) the prerequisite interplanetary and initial interstellar space system technologies necessary to successfully conduct this class of mission also contribute significantly to humankind's search for extraterrestrial intelligence (for example, smart robot probes and interstellar communications); (4) a long-range societal commitment on the order of a century would be required to achieve such a project; and (5) the prospects for interstellar flight by human beings do not appear very promising using current or foreseeable 21st-century technologies.

The Project Daedalus study also identified three key technology advances that would be needed to make even a robot interstellar mission possible. These are (1) the development of controlled nuclear fusion, especially the use of the deuterium–helium-3 thermonuclear reaction; (2) advanced machine intelligence; and (3) the ability to extract helium-3 in large quantities from the Jovian atmosphere.

Although the choice of Barnard's star as the target for the first interstellar mission was somewhat arbitrary, if future human generations can build such an interstellar robot spaceship and successfully explore the Barnardian system, then with modest technology improvements, all star systems within 10 to 12 light-years of Earth become potential targets for a more ambitious program of (robotic) interstellar exploration.

See also BARNARD'S STAR; EXTRASOLAR PLANETS; FUSION; HORIZON MISSION METHODOLOGY; INTERSTELLAR PROBE(S); INTERSTELLAR TRAVEL; STARSHIP.

Project Orion Project Orion was the name given to a nuclear fission pulsed rocket concept studied by agencies of the U.S. government in the early 1960s. A human-crewed interplanetary spaceship would be propelled by exploding a series of nuclear fission devices behind it. A giant pusher plate mounted on large shock absorbers would receive the energy pulse from each successive nuclear detonation, and the spaceship configuration would be propelled forward by Newton's action-reaction principle.

In theory this concept is capable of achieving specific impulse values ranging from 2,000 to 6,000 seconds, depending on the size of the pusher plate. Specific impulse is a performance index for rocket propellants. It is defined as the thrust produced by the propellant divided by the mass flow rate. As a point of comparison, the very best chemical rockets have specific impulse values ranging from 450 to about 500 seconds.

A crewed Orion spaceship would move rapidly throughout interplanetary space at a steady acceleration of perhaps 0.5 g (one-half the acceleration of gravity on Earth's surface). Typically, a 1- to 10-kiloton fission device would be exploded every second or so close behind the giant pusher plate. A kiloton is the energy of a nuclear explosion that is equivalent to the detonation of 1,000 tons of TNT (trinitrotluene, a chemical high explosive). A few thousand such detonations would be needed to propel a crew of 20 astronauts to Mars or the moons of Jupiter.

Work by the United States on this nuclear fission pulsed rocket concept came to an end in the mid-1960s as a result of the Limited Test Ban Treaty of 1963. This treaty prohibited the signatory nations (which included the United States) from testing nuclear devices in Earth's atmosphere, underwater, or in outer space.

Advanced versions of the original Orion concept have emerged. In these new spaceship concepts, externally detonated nuclear fission devices have been replaced by many small, controlled thermonuclear fusion explosions taking place inside a specially constructed thrust chamber. These small thermonuclear explosions would occur in an inertial confinement fusion (ICF) process in which many powerful laser, electron, or ion beams simultaneously impinged on a tiny fusion

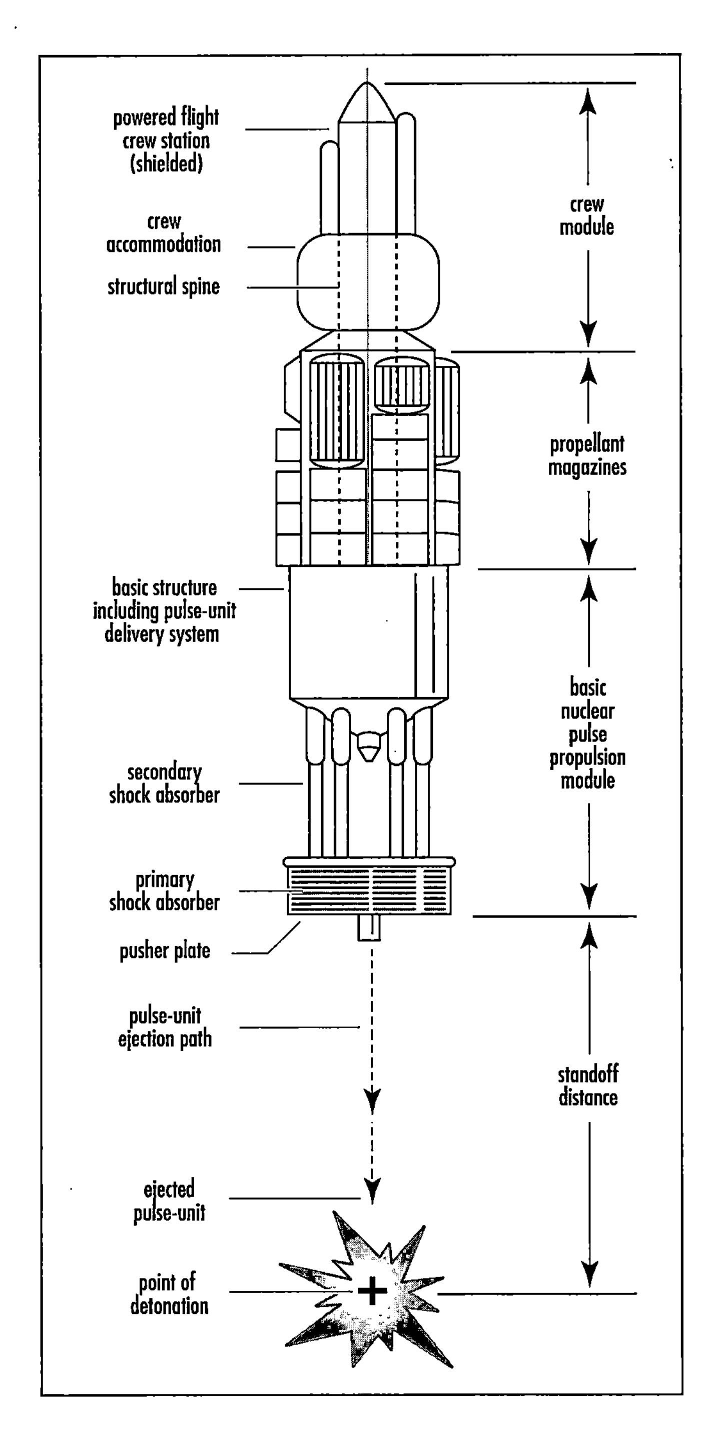

Major elements of the original Project Orion nuclear-pulsed spaceship concept ***(Drawing courtesy of the U.S. Department of Energy)***

pellet. Each miniature thermonuclear explosion would have an explosive yield equivalent to a few tons of TNT. The expanding shell of very hot, ionized gas from the thermonuclear explosion would be directed into a thrust-producing exhaust stream. Such pulsed nuclear fusion spaceships, when developed, could open up our entire solar system to exploration by human crews. For example, Earth-to-Neptune travel would take less than 15 days at a steady, comfortable constant acceleration of 1 g. Pulsed fusion systems also represent a possible propulsion system for interstellar travel.

See also FISSION (NUCLEAR); FUSION (NUCLEAR); PROJECT DAEDALUS.

Project Ozma The first attempt to detect interstellar radio signals from an intelligent extraterrestrial civilization. It was conducted by FRANK DRAKE at the National Radio Astronomy Observatory (NRAO) in Green Bank, West Virginia, in 1960. Drake derived the name for this effort from the queen of the imaginary land of Oz, since in his own words Oz was "a place very far away, difficult to reach, and populated by exotic beings."

A frequency of 1,420 megahertz (MHz) was selected for this initial search, the frequency of the 21-centimeter interstellar hydrogen line. Since this is a radio frequency at which most emerging technical civilizations would first use narrow-bandwidth, high-sensitivity radio telescopes, scientists reasoned that this would also be the frequency that more advanced alien civilizations would use in trying to signal emerging civilizations across the interstellar void.

In 1960 the 29.5-meter-diameter Green Bank radio telescope was aimed at two Sunlike stars about 11 light-years away, Tau Ceti and Epsilon Eridani. Patiently, Frank Drake and his Project Ozma team listened for intelligent signals, but after more than 150 hours of listening, no evidence of strong radio signals from intelligent extraterrestrial civilizations was obtained. Project Ozma is generally considered the first serious attempt to listen for intelligent interstellar radio signals in our search for extraterrestrial intelligence (SETI).

See also DRAKE EQUATION; EXTRATERRESTIAL CIVILIZATIONS; SEARCH FOR EXTRATERRESTRIAL INTELLIGENCE.

Prometheus A small moon of Saturn discovered in 1980 as a result of the *Voyager 1* spacecraft flyby. Prometheus is an irregularly shaped moon approximately 140 by 80 km that travels around Saturn at a distance of 139,350 km with a period of 0.613 day. Also called Saturn XVI or S/1980 S27. Together with Pandora, Prometheus acts as a shepherd moon for Saturn's F Ring.

See also SATURN.

prominence A cloud of cooler plasma extending high above the Sun's visible surface, rising above the photosphere into the corona.

See also SUN.

propellant In general, a material, such as a fuel, an oxidizer, an additive, a catalyst, or any compound or mixture of these, carried in a rocket vehicle that releases energy during combustion and thus provides thrust to the vehicle. Propellants commonly are in either liquid or solid form. Modern launch vehicles use three types of liquid propellants: petroleum-based, cryogenic, and hypergolic.

The petroleum used as a rocket fuel is a type of kerosene similar to the compound burned in heaters and lanterns. However, the rocket petroleum is highly refined and is called

RP-1 (Refined Petroleum). It usually is burned with liquid oxygen (LOX) to produce thrust. However, the RP-1/LOX propellant combination delivers considerably less *specific impulse* (I_{sp}) than that delivered by cryogenic propellant combinations.

The specific impulse is an important parameter in describing the performance of a rocket engine and the propellant combination being used. It is defined as:

$$I_{sp} = \text{(thrust force)}/\text{(mass flow rate of propellant)}$$

In the SI unit system, specific impulse has the units of newtons/(kilogram/second), which then reduces to meters/second. In the U.S. standard (or engineering) unit system, this important propellant performance parameter has the units of pounds-force [lb_f] per pounds-mass [lb_m] per second ($lb_f/[lb_m/s]$), which often is simplified to just "second(s)," since 1 lb_f = 1 lb_m at sea level on Earth. This is a somewhat confusing situation, but as long as the basic definition is used consistently within a system of units, the specific impulse (expressed as $lb_f/(lb_m/s)$, "seconds," or m/s will describe the relative performance capabilities of different types of rocket propellant combinations.

The most important cryogenic propellants are liquid oxygen (LOX), which serves as the oxidizer, and liquid hydrogen (LH_2), which serves as the fuel. The word *cryogenic* is a derivative of the Greek word *kruos* (κρυοζ) meaning "very cold." Oxygen remains in a liquid state at temperatures of –183° Celsius (°C) and below. Hydrogen remains in a liquid state at temperatures of –253°C and below.

In gaseous form oxygen and hydrogen have such low densities that extremely large tanks would be required to store them onboard a rocket or launch vehicle. But cooling and compressing these propellants into liquids greatly increases their densities, making it possible to store large quantities in much-smaller-volume tanks. However, the tendency of liquid cryogenic propellants to return to gaseous form unless kept very cold makes them difficult to store over long periods of time. Therefore, cryogenic propellants generally are not considered suitable for military rockets, which must be kept launch-ready for months at a time.

However, the high-performance capability of the LH_2/LOX combination makes this low-temperature storage problem acceptable when the launch vehicle reaction time and propellant storability issues are not too critical. The LOX/LH_2 combination of cryogenic propellants has been used for a variety of U.S. rockets including the RL-10 engines on the Centaur upper stage, the J-2 engines used on the Saturn V rocket vehicle's second and third stages, and the space shuttle's main engines (SSMEs).

Using the simplified U.S. customary unit of specific impulse for comparison purposes, the Centaur RL-10 engine has a specific impulse of 444 seconds, the J-2 engines of the Saturn V upper stages had specific impulse ratings of 425 seconds, and the space shuttle main engines have a specific impulse rating of 455 seconds. In contrast, the giant cluster of F-1 engines in the Saturn V vehicle first stage burned LOX and kerosene and had a specific impulse rating of just 260 seconds. This same propellant combination was used by the booster stages of the Atlas/Centaur rocket combination and yielded a specific impulse rating of 258 seconds in the original Atlas rocket's booster engine and 220 seconds in its sustainer engine.

Hypergolic propellants are fuels and oxidizers that ignite on contact with one another and, therefore, need no ignition source. This easy start and restart capability makes them attractive for both crewed and uncrewed spacecraft maneuvering systems. Another advantage of hypergolic propellants is the fact that they do not have the extremely low storage temperature requirements needed for cryogenic propellants.

Typically, monomethyl hydrazine (MMH) is used as the hypergolic fuel and nitrogen tetroxide (N_2O_4) as the oxidizer. Hydrazine is a clear nitrogen and hydrogen compound with a "fishy" smell; it is similar to ammonia. Nitrogen tetroxide is a reddish fluid that has a pungent, sweetish smell. Both of these fluids are highly toxic and must be handled under the most stringent safety precautions.

Hypergolic propellants are used in the core liquid-propellant stages of the Titan family of launch vehicles and on the second stage of the Delta launch vehicle. The space shuttle orbiter uses hypergolic propellants (or "hypergols") in its orbital maneuvering system (OMS), which is used for orbital insertion, major orbital maneuvers, and deorbit operations. The orbiter's reaction control system (RCS) also uses hypergols to accomplish attitude control.

The solid-propellant rocket motor is the oldest and simplest of all forms of rocketry, dating back to the ancient Chinese. In its simplest form the solid-propellant rocket is just a casing (usually steel) that is filled with a mixture of chemicals (fuel and oxidizer) in solid form that burn at a rapid rate after ignition, expelling hot gases from the nozzle to produce thrust. Solid-propellant rocket motors do not require turbopumps or complex propellant feed systems. A simple squib device at the top of the motor directs a high-temperature flame along the surface of the propellant grain, igniting it instantaneously.

Generally, solid propellants are stable and storable. Unlike a liquid-propellant rocket engine, however, a solid-propellant motor cannot be shut down. Once ignited, it will burn until all the propellant is exhausted.

Solid-propellant rockets have a variety of uses in space operations. Small solids often power the final stage of a launch vehicle or are attached to payload elements to boost satellites and spacecraft to higher orbits.

Medium-sized solid-propellant rocket motors are used in special upper-stage systems that are designed specifically to boost a payload from low Earth orbit (LEO) to geosynchronous Earth orbit (GEO) or to place a spacecraft on an interplanetary trajectory. The Payload Assist Module (PAM) and the Inertial Upper Stage (IUS) are examples of upper-stage systems.

The Scout launch vehicle is a four-stage rocket used to place small satellites into orbit. All four stages are solid-propellant rockets.

Finally, liquid-propellant launch vehicles such as the Titan, Delta, and space shuttle use solid-propellant rockets to provide added thrust at liftoff. The space shuttle, for example, uses the largest solid rocket motors ever built and flown. Each reusable solid rocket booster (SRB) contains 453,600 kilograms of propellant in the form of a hard, rubbery substance with a consistency like that of a pencil eraser. A solid-propellant rocket motor always contains its own oxygen supply. The oxidizer in the shuttle's solid rocket booster is ammonium perchlorate, which forms 69.93 percent of the mixture. The fuel is a form of powdered aluminum (16 percent), with an iron oxidizer powder (0.07 percent) as a catalyst. The binder that holds the mixture together is polybutadiene acrylic acid acrylonitrile (12.04 percent). In addition, the mixture contains an epoxy-curing agent (1.96 percent). The binder and epoxy also burn as fuel, adding thrust. The specific impulse of the shuttle's solid rocket booster is 242 seconds at sea level.

propellant mass fraction (symbol: ζ) In rocketry, the ratio of the propellant mass (m_p) to the total initial mass (m_o) of the vehicle before operation, including propellant load, payload, and structure.

$$\zeta = m_p/m_o$$

proper motion (symbol: μ) The apparent angular displacement of a star with respect to the celestial sphere. Barnard's star has the largest known proper motion.
See also BARNARD'S STAR.

proportional counter A nuclear radiation detection instrument in which an electronic detection system receives pulses that are proportional to the number of ions formed in a gas-filled tube by ionizing radiation.
See also GEIGER COUNTER; SPACE RADIATION ENVIRONMENT.

propulsion system The launch vehicle or space vehicle system that includes the rocket engines, propellant tanks, fluid lines, and all associated equipment necessary to provide the propulsive force as specified for the vehicle. For example, typical launch vehicles, expendable or reusable, depend on chemical propulsion systems (i.e., solid-propellant and/or liquid-propellant rockets) to take a payload from the surface of Earth into orbit (generally low Earth orbit). Upper-stage chemical rockets also serve as the propulsion systems used to take certain payloads from low Earth orbit to geosynchronous Earth orbit or to place a spacecraft on an interplanetary trajectory. However, nuclear propulsion systems (thermal or electric) also might be used to take a payload or spacecraft from Earth orbit to Mars, to the outer reaches of the solar system, and even beyond. Solar-electric propulsion systems can be used to take nonpriority payloads from Earth orbit to various locations in the inner solar system. Finally, very exotic propulsion systems based on nuclear fusion reactions or even matter-antimatter annihilation reactions have been suggested (stretching our current knowledge of physics and engineering) as the propulsive systems necessary for interstellar travel.
See also ELECTRIC PROPULSION; LAUNCH VEHICLE; ROCKET; SPACE NUCLEAR PROPULSION; UPPER STAGE.

Prospero A small (dimensions currently uncertain) outer moon of Uranus discovered in July 1999 by ground-based telescopic observations. Prospero is named after the magician who is master of the island in William Shakespeare's comedy *The Tempest*. This moon travels in a retrograde orbit around Uranus at an average distance of 16,665,000 km with a period of 1,953 days and an inclination of 146°. Also known as Uranus XVIII or S/1999 U3.
See also URANUS.

Proteus An irregularly shaped moon (436 by 416 by 402 km) of Neptune with an orbital period of 1.12 days discovered in 1989 as a result of the *Voyager 2* spacecraft flyby. Proteus travels around Neptune at a distance of 117,650 km. This moon is the second-largest in the Neptune system and has a dark, low-albedo, crater-covered surface. Also called Neptune VIII or S/1989 N1.
See also NEPTUNE.

protium The most abundant isotope of hydrogen; ordinary hydrogen consisting of one proton in the nucleus surrounded by one electron.
See also DEUTERIUM; TRITIUM.

protogalaxy A galaxy at the early stages of its evolution.
See also GALAXY.

proton (symbol: p) A stable elementary nuclear particle with a single positive charge and a mass of about 1.672×10^{-27} kilograms, which is about 1,836 times the mass of an electron. A single proton makes up the nucleus of an ordinary (or light) hydrogen atom. Protons are also constituents of all other nuclei. The atomic number (Z) of an atom is equal to the number of protons in its nucleus.

Proton A Russian liquid-propellant expendable launch vehicle. The four-stage configuration of this vehicle often is used for interplanetary missions. It consists of three hypergolic stages and one cryogenic stage and is capable of placing 4,790 kg into geostationary transfer orbit (GTO), that is, an elliptical orbit with an apogee of 35,400 km and a perigee of 725 km. The Proton D-l vehicle consists of three hypergolic stages and can place up to 20,950 kg into low Earth orbit. The three-stage Proton vehicle configuration was used to place the *Salyut* and *Mir* space station modules into orbit. A Proton rocket vehicle also placed the *Zarya* module—the first component of the *International Space Station*—into orbit in November 1998.
See also LAUNCH VEHICLE.

proton-proton chain reaction The series of thermonuclear fusion reactions in stellar interiors by which four hydrogen nuclei are fused into a helium nucleus. This is the main energy liberation mechanism in stars such as the Sun.
See also FUSION (NUCLEAR).

proton storm The burst, or flux, of protons sent into space by a solar flare.
See also SPACE WEATHER; SUN.

protoplanet Any of a star's planets as such planets emerge during the process of accretion (accumulation), in which planetesimals collide and coalesce into large objects.
See also PLANET.

A Russian Proton rocket lifts off from the Baikonur Cosmodrome in Kazahkstan on November 20, 1998. This launch successfully placed the Zarya module, the first component of the *International Space Station* (ISS), into low Earth orbit. ***(Courtesy of NASA)***

protostar A star in the making. Specifically, the stage in a young star's evolution after it has separated from a gas cloud but prior to its collapsing sufficiently to support thermonuclear reactions.

See also STAR.

protosun A sun as it emerges in the formation of a solar system.

See also SUN.

prototype 1. Of any mechanical device, a production model suitable for complete evaluation of mechanical and electrical form, design, and performance. 2. The first of a series of similar devices. 3. A physical standard to which replicas are compared. 4. A spacecraft or component that has passed or is undergoing tests, qualifying the design for fabrication of complete flight units or components thereof. 5. A model suitable for evaluation of design, performance, and production potential.

Proxima Centauri The closest star to the Sun and the third member of the Alpha Centauri triple star system. It is some 4.2 light-years away.

See also ALPHA CENTAURI.

Ptolemaic system The ancient Greek model of an Earth-centered (geocentric) universe as described and embellished by PTOLEMY in the *Syntaxis* (written around 150 C.E.) This Earth-centered model of the universe lingered in Western civilization as a result of Ptolemy's great compendium of early Greek astronomical knowledge, which was preserved during the Dark Ages by Arab astronomers (in about 820 C.E.) as *The Almagest* and then revived by European astronomers during the Renaissance. *Compare with* COPERNICAN SYSTEM.

See also ARISTOTLE; EUDOXUS OF CNIDUS.

Ptolemy (Claudius Ptolemaeus) (ca. 100–ca. 170 C.E.) Greek *Astronomer* The last of the early Greek astronomers, Ptolemy lived in Alexandria, Egypt, and wrote (in about 150 C.E.) *Syntaxis,* a compendium of astronomical and mathematical knowledge from all the great ancient Greek philosophers and astronomers. His book preserved the Greek geocentric cosmology in which Earth was considered the unmoving center of the universe while the wandering planets and fixed stars revolved around it. Arab astronomers translated Ptolemy's book in about 820 C.E., renaming it *The Almagest* (The greatest). The Ptolemaic system remained essentially unchallenged in Western thinking until NICHOLAS COPERNICUS and the start of the scientific revolution in the 16th century.

Puck A small moon of Uranus discovered in 1985 as a result of long-distance imagery taken by *Voyager 2* as the spacecraft was approaching the planet for the January 1986 flyby encounter. Puck is named after the mischievous sprite (spirit) in William Shakespeare's comedy *A Midsummer Night's Dream.* This moon has a diameter of 154 km and travels around Uranus at a distance of 86,000 km with an orbital period of 0.762 day. Also known as Uranus XV and S/1985 U1.

See also URANUS.

pulsar A celestial object thought to be a young, rapidly rotating neutron star that emits radiation in the form of rapid pulses with a characteristic pulse period and duration. The graduate student JOCELYN BELL-BURNELL and her academic adviser ANTONY HEWISH detected the first pulsar in August 1967. This unusual celestial object emitted radio waves in a pulsating rhythm. Because of the structure and repetition in its radio signal, their initial inclination was to consider the possibility that this repetitive signal was from an intelligent extraterrestrial civilization. However, subsequent careful investigation and the discovery of another radio wave pulsar that December quickly dispelled the "little green man" hypothesis and showed the scientists that the unusual signal was being emitted by an interesting newly discovered natural phenomenon, the pulsar.

See also NEUTRON STAR.

pulsating variable star A star whose luminosity varies in a predictable, periodic way.

See also CEPHEID VARIABLE.

pulse The variation of a quantity whose value is normally constant; this variation is characterized by a rise and a decay and has a finite duration. This definition is so broad that it covers almost any transient phenomenon, such as the short burst of electromagnetic energy associated with a radar pulse. The only features common to all pulses are rise, finite duration, and decay (fall). The rise, duration, and decay must be of a quantity that is constant (not necessarily zero) for some time before the pulse and must have the same constant value for some time afterward.

pulse code modulation (PCM) The transmission of information by controlling the amplitude, position, or duration of a series of pulses. An analog signal (e.g., an image, music, a voice, etc.) is broken up into a digital signal (i.e., binary code) and then transmitted in this series of pulses via telephone line or radio waves.

pulsejet A jet engine containing neither compressor nor turbine in which combustion takes place intermittently, producing thrust by a series of "miniexplosions," or combustion pulses. This type of engine is equipped with valves or shutters in the front that open and shut to take in air for the periodic combustion pulses. The World War II German V-1 "buzz-bomb" contained a pulsejet engine.

pump A machine for transferring mechanical energy from an external source to the fluid flowing through it. The increased energy is used to lift the fluid, to increase its pressure, or to increase its rate of flow.

pump-fed Term used to describe a propulsion system that incorporates a pump that delivers propellant to the combustion chamber at a pressure greater than the tank ullage pressure.

See also ROCKET.

purge To rid a line or tank of residual fluid, especially of fuel or oxide in the tanks or lines of a rocket after a test firing or simulated test firing.

pyrogen A small rocket motor used to ignite a larger rocket motor.

pyrolysis Chemical decomposition of a substance due to the application of intense heat (thermal energy).

pyrometer An instrument for the remote (noncontact) measurement of temperatures. This term is generally applied to instruments that measure temperatures above 600° Celsius.

pyrophoric fuel A fuel that ignites spontaneously in air. *Compare with* HYPERGOLIC FUEL (HYPERGOL).

pyrotechnic As an adjective, a term used to describe a mixture of chemicals that, when ignited, is capable of reacting exothermically to produce light, heat, smoke, sound, or gas and also may be used to introduce a delay into an explosive train because of its known burning time. The term includes propellants and explosives; for example, pyrotechnic device and pyrotechnic delay.

pyrotechnics As a noun, igniters (other than pyrogens) in which solid explosives or energetic propellantlike chemical formulations are used as the heat-producing materials.

pyrotechnic subsystems Space system designers use electrically initiated pyrotechnic devices to open certain valves, ignite solid rocket motors, explode bolts to separate or jettison hardware, and to deploy spacecraft appendages. Pyrotechnic subsystems receive their electrical power from a bank of capacitors that are charged from the spacecraft's main power bus a few minutes prior to the planned detonation (activation) of a particular pyrotechnic device.

Pythagoras (ca. 580–ca. 500 B.C.E.) Greek *philosopher, mathematician* Pythagoras was an early Greek philosopher and mathematician who taught that Earth was a perfect sphere at the center of the universe and that the planets and stars moved in circles around it. His thoughts significantly influenced Greek cosmology and subsequently Western thinking for centuries. However, they arose primarily from his mystical belief in the circle as a perfect form rather than from careful observation. Although Nicholas Copernicus successfully challenged geocentric cosmology in the 16th century and replaced it with a heliocentric (Sun-centered) model of the solar system, it was JOHANNES KEPLER who displaced the concept of circular orbits for the planets early in the 17th century. Students remember Pythagoras today when they have to struggle with the famous "Pythagorean theorem" in various mathematics courses.

q In aerospace engineering, the symbol commonly used for dynamic pressure. For example, after liftoff the launch vehicle encountered "max q" —that is, maximum dynamic pressure—at 45 seconds into the flight.

quadrant An early astronomical instrument that incorporated a 90° graduated arc for making angular measurements in astronomy and navigation. Ancient astronomers used the quadrant to measure the altitudes and angular separations of stars. The basic instrument consisted of a quarter circle (90° graduated arc) along with a swiveling arm and sighting mechanism. A plumb bob suspended from the quadrant's center provided vertical alignment. The sighting mechanism, mounted on the moveable arm, allowed an astronomer to make observations of the stars and planets as they crossed the meridian. The last great Greek astronomer, PTOLEMY, mentions three basic instruments in his writings. These ancient instruments he identified are the quadrant, the triquetrum, and the astrolabe. Arab astronomers improved upon the quadrant, and it remained the most important astronomical instrument until the invention of the telescope in western Europe early in the 17th century.

See also ANCIENT ASTRONOMICAL INSTRUMENTS.

quantization The fact that electromagnetic radiation (including light) and matter behave in a discontinuous manner and manifest themselves in the form of tiny "packets" of energy called *quanta* (singular: quantum).

quantum (*plural:* **quanta**) In modern physics, a discrete bundle of energy possessed by a photon.

See also QUANTUM THEORY.

quantum chromodynamics (QCD) A quantum field theory (gauge theory) that describes the strong force interactions among quarks and antiquarks as these minute, subnuclear-sized elementary particles exchange gluons. The strong force is one of the four fundamental forces in nature, and physicists also refer to it as the nuclear force or the strong nuclear force. It is the force that binds quarks together to form baryons (3 quarks) and mesons (2 quarks). For example, QCD addresses how the nucleons of everyday matter, neutrons and protons, consist of the quark combinations *uud* and *udd*, respectively. Here, the symbol *u* represents a single up quark, while the symbol *d* represents a single down quark. The force that holds nucleons together to form an atomic nucleus is envisioned as a residual interaction between the quarks inside each nucleon.

See also ELEMENTARY PARTICLE(S); FUNDAMENTAL FORCES IN NATURE; QUARK.

quantum cosmology Theories of modern cosmology that address the physics of the very early universe by combining quantum mechanics and gravity. Ultimately, quantum cosmology requires a viable theory of quantum gravity. In an interesting application of experimental and theoretical research from modern high-energy physics, particle physicists suggest that the early universe experienced a specific sequence of phenomena and phases following the big bang explosion. They generally refer to this sequence as the *standard cosmological model.* Right after the big bang the present universe reached an incredibly high temperature—physicists offer the unimaginable value of about 10^{32} K. During this period, called the *quantum gravity epoch,* the force of gravity, the strong nuclear force, and a composite electroweak force all behaved as a single unified force.

At *Planck time* (about 10^{-43} second after the big bang), the force of gravity assumed its own identity. Scientists call the ensuing period the *grand unified epoch.* While the force of gravity functioned independently during this period, the strong nuclear force and the composite electroweak force remained together and continued to act like a single force. Today physicists apply various grand unified theories (GUTs) in their efforts to model and explain how the strong nuclear force and the electroweak force functioned as one during this period in the early universe.

Then, about 10^{-35} s after the big bang, the strong nuclear force separated from the electroweak force. By this time the expanding universe had "cooled" to about 10^{28} K. Physicists call the period between about 10^{-35} s and 10^{-10} s the *electroweak epoch*. During this epoch the weak nuclear force and the electromagnetic force became separate entities as the composite electroweak force disappeared. From this time forward the universe contained the four basic forces known to physicists today—namely, the force of gravity, the strong nuclear force, the weak nuclear force, and the electromagnetic force.

Following the big bang and lasting up to about 10^{-35} s there was no distinction between quarks and leptons. All the minute particles of matter were similar. However, during the electroweak epoch quarks and leptons became distinguishable. This transition allowed quarks and antiquarks to eventually become hadrons, such as neutrons and protons, as well as their antiparticles. At 10^{-4} s after the big bang, in the radiation-dominated era, the temperature of the universe cooled to 10^{12} K. By this time most of the hadrons disappeared because of matter-antimatter annihilations. The surviving protons and neutrons represented only a small fraction of the total number of particles in the universe, the majority of which were leptons, such as electrons, positrons, and neutrinos. However, like most of the hadrons before them, the majority of the leptons also soon disappeared as a result of matter-antimatter interactions.

At the beginning of the matter-dominated era, about 200 seconds (or some three minutes) following the big bang, the expanding universe cooled to a temperature of 10^9 K, and small nuclei, such as helium, began to form. Later, when the expanding universe reached an age of about 500,000 years (or some 10^{13} seconds old), the temperature dropped to 3,000 K, allowing the formation of hydrogen and helium atoms. Interstellar space is still filled with the remnants of this primordial hydrogen and helium.

Eventually, density inhomogeneities allowed the attractive force of gravity to form great clouds of hydrogen and helium. Because these clouds also experienced local density inhomogeneities, gravitational attraction formed stars and then slowly gathered groups of these stars into galaxies. As gravitational attraction condensed the primordial (big bang) hydrogen and helium into stars, nuclear reactions at the cores of the more massive stars created heavier nuclei up to and including iron. Supernova explosions then occurred at the end of the lives of many of these early, massive stars. These spectacular explosions produced all atomic nuclei with masses beyond iron and then scattered the elements into space. The expelled "stardust" would later combine with interstellar gas and eventually create new stars and their planets. For example, all things animate and inanimate here are on Earth are the natural by-products of these ancient astrophysical processes.

quantum electrodynamics (QED) The branch of modern physics, independently copioneered by Richard Feynman (1918–88), Julian Schwinger (1918–94), and Sinitiro Tomonaga (1906–79), that combines quantum mechanics with the classical electromagnetic theory of the 19th-century Scottish physicist JAMES CLERK MAXWELL. QED provides a quantum mechanical description of how electromagnetic radiation interacts with charged matter.

quantum gravity A hypothetical theory of gravity that represents the successful merger of general relativity and quantum mechanics. Although no such theory yet exists, there is a need for this theory to help physicists address problems involving the very early universe—that is, from the big bang up to Planck time ($\sim 10^{-43}$ s)—and the behavior of black holes. The graviton is the hypothetical particle in quantum gravity that plays a role similar to that of the photon in quantum electrodynamics (QED).

See also GRAVITATION.

quantum mechanics The physical theory that emerged from MAX PLANCK's original quantum theory and developed into wave mechanics, matrix mechanics, and relativistic quantum mechanics in the 1920s and 1930s. Within the realm of quantum mechanics, the Heisenberg uncertainty principle and the Pauli exclusion principle provide a framework that dictates how particles behave at the atomic and subatomic levels.

See also HEISENBERG UNCERTAINTY PRINCIPLE; PAULI EXCLUSION PRINCIPLE.

quantum theory The theory in modern physics, first stated in 1900 by the German physicist MAX PLANCK, that all electromagnetic radiation is emitted and absorbed in "quanta," or discrete energy packets, versus continuously. According to quantum theory, the energy of a photon (E) is related to its frequency (ν) by the equation $E = h\,\nu$, where E is the photon energy (joules), ν is the frequency (hertz), and h is a constant called Planck's constant (6.626×10^{-34} joule-seconds). In 1905 ALBERT EINSTEIN used quantum theory to explain the photoelectric effect. Einstein's Nobel-prize-winning work assumed that light propagated in quanta, or photons. In 1913 the Danish physicist Niels Bohr (1885–1962) combined quantum theory with Ernest Rutherford's (1871–1937) nuclear atom hypothesis. Bohr's work resolved the difficulties with the Rutherford atomic model and also served as a major intellectual catalyst promoting the emergence of quantum mechanics in the 1920s.

See also QUANTUM MECHANICS.

Quaoar A large object in the outer solar system first observed in June 2002. Quaoar is an icy world with a diameter of about 1,250 km, making it about half the size of Pluto. Located in the Kuiper Belt some 1.6 billion km beyond Pluto and about 6.4 billion km away from Earth, Quaoar takes 285 years to go around the Sun and travels in a nearly circular orbit with an eccentricity of just 0.04. The name Quaoar (pronounced kwah-o-wahr) comes from the creation mythology of the Tongva, a Native American people who inhabited the Los Angeles, California, area before the arrival of European explorers and settlers.

Like the planet Pluto, Quaoar dwells in the Kuiper Belt, an icy debris field of cometlike bodies extending 5 billion kilometers beyond the orbit of Neptune. However, while astronomers generally treat Pluto as both a planet and a member of the Kuiper Belt, they regard Quaoar as simply a Kuiper Belt object (KBO). In addition, despite its size, astronomers do not consider Quaoar to be the long-sought, hypothesized tenth planet. Nevertheless, it remains an intriguing and impressive new world, most likely consisting of equal portions of rock and various ices, including water ice, methane ice (frozen natural gas), methanol ice (alcohol ice), carbon dioxide ice (dry ice), and carbon monoxide ice. Measurements made at the Keck Telescope indicate the presence of water ice on Quaoar.

See also KUIPER BELT; PLUTO; SEDNA.

quark Any of six extremely tiny elementary particles (diameter less than 10^{-18} m) independently postulated in 1963 by the physicists M. Gell-Mann (b. 1929) and G. Zweig (b. 1937) as the basic building blocks of matter residing inside hadrons, such as neutrons and protons. The six types, or "flavors," of quarks are the up (u), down (d), strange (s), charmed (c), top, (t), and bottom (b). Physicists postulate that all quarks have spin 1/2. Another interesting feature of quarks is that they have fractional electrical charges. For example, the up quark carries an electrical charge of +2/3 e, while the down quark carries one of –1/3 e. The symbol e represents the basic charge of an electron. Quarks possess another property called color, for which physicists assign three possibilities: blue, green, and red. Corresponding to each quark, there is an antiquark with oppositely signed electric charge (namely, +1/3 e or –2/3 e), and antiblue, antigreen, and antired colors. Note that physicists selected the name of the characteristics and rather whimsically. The quark quality of color is important in quantum chromodynamics (QCD) but has nothing to do with the visible portion of the electromagnetic spectrum. As presently assumed in the STANDARD MODEL of nuclear physics, hadrons (baryons and mesons) are made of doublet and triplet combinations of quarks. The neutron, for example, consists of one up (u) and two down (d) quarks, while a proton consists of 2 up (u) and one down (d) quarks. Physicists think exchanging gluons (massless particles serving as the carriers of the strong nuclear force) hold quarks together. Today some astrophysicists suggest that deep within neutron stars are states of condensed matter best characterizes as a soup of free quarks.

See also QUANTUM CHROMODYNAMICS (QCD).

quasars Mysterious objects with high redshifts (i.e., traveling away from Earth at great speed) that appear almost like stars but are far more distant than any individual star we can now observe. These unusual objects were first discovered in the 1960s with radio telescopes, and they were called *quasi-stellar radio sources*, or *quasars*, for short. Quasars emit tremendous quantities of energy from very small volumes. These puzzling objects emit various portions of their energy in the form of radio waves, visible light, ultraviolet radiation, X-rays, and even gamma rays. A quasar that is relatively quiet in the radio frequency portion of the electromagnetic spectrum is often called a *quasi-stellar object* (QSO) or sometimes a *quasi-stellar galaxy* (QSG). Some of the most distant quasars yet observed are so far away that they are receding at more than 90 percent of the speed of light.

As bright, concentrated radiation sources, quasars are thought to be the nuclei of active galaxies. The optical brightness of some quasars has been observed to change by a factor of two in about a week, with detectable changes occurring in just one day. Therefore, astrophysicists now speculate that such quasars cannot be much larger than about one light-day across (a distance about twice the dimension of our solar system), because a light source cannot change brightness significantly in less time than it takes for light itself to travel.

The problem facing astrophysicists is to explain how a quasar can generate more energy than is possessed by an entire galaxy and generate this energy in so small a region of space. In fact, quasars are such peculiar astronomical objects they can radiate as much energy per second as is radiated by a thousand or more galaxies from a region that has a diameter about one-millionth of the host galaxy. To scientists, the powerhouse of a quasar seems comparable to a small pocket flashlight that is capable of producing as much light as all the houses and businesses in a sprawling metropolis the size of the Los Angeles basin in southern California. Quasars are intense sources of X-rays as well as visible light. They represent the most powerful type of X-ray source yet discovered in the universe.

A pair of quasars called Q2345+007 A, B was imaged by the *CHANDRA X-RAY OBSERVATORY* in mid-2000. Close scrutiny of this X-ray image revealed that the supposed "twin quasars" were neither a mirage nor identical twins. After careful analysis scientists now regard this unusual quasar pair not as an illusion, but rather two different quasars created by merging galaxies. When galaxies collide the flow of gas onto the central supermassive black holes of each of the galaxies can be enhanced, resulting in two quasars. The X-ray "light" from this quasar pair started its journey toward Earth some 11 billion years ago. At that time galaxies were about three times closer together than they are now, so collisions between them were much more likely to happen.

Astrophysicists postulate that the power of a quasar depends on the mass of its central supermassive black hole and the rate at which it swallows matter. Almost all galaxies, including our own Milky Way galaxy, have monsters lurking in the middle—that is, most galaxies are believed to contain supermassive black holes at their centers. Quasars represent the extreme cases, where large quantities of gas pour into the black hole so rapidly that the energy output appears to be a thousand times greater than that of the galaxy itself. Astronomers call a galaxy with a somewhat less active supermassive black hole an active galaxy and its centrally located black hole an active galactic nucleus, or AGN. The Milky Way galaxy and the neighboring Andromeda galaxy are examples of normal galaxies in which the supermassive black hole lurking in the center has very little gas to capture.

X-rays from quasars and AGNs are produced when infalling matter is heated to temperatures of millions of degrees as it swirls toward the supermassive black hole at the galactic center. However, not all the matter in the gravitational whirlpool is doomed to fall into the black hole. In many quasars and AGNs a portion of the swirling gas escapes as a hot wind that is blown away from the accretion disk at speeds as high as a tenth of the speed of light.

Even more dramatic are the high-energy jets that radio and X-ray observations reveal exploding away from some supermassive black holes. These jets move at nearly the speed of light in tight beams that blast out of the galaxy and travel hundreds of thousands of light-years.

quasi-stellar object (QSO) *See* QUASAR.

Quickbird A high-resolution commercial imagery satellite system launched from Vandenberg Air Force Base, California, on October 18, 2001, and placed into a 450-km-altitude, sun-synchronous orbit at an inclination of 98° by a Boeing Delta 2 expendable launch vehicle. The satellite's imaging system can collect 0.61-meter-resolution panchromatic and 2.5-meter-resolution multispectral stereoscopic data. High-resolution imagery data from this commercial satellite constructed by Ball Aerospace and Technologies Corporation contribute to mapping, agricultural and urban planning, weather research, and military surveillance. Also called *Quickbird 2* because an earlier satellite, called *Quickbird 1,* suffered a launch failure in November 2000.

See also REMOTE SENSING.

QuikSCAT **(Quick Scatterometer)** An American oceanographic satellite sponsored by NASA. A Titan II expendable launch vehicle (ELV) successfully placed the satellite into an 804-by-806-km polar orbit with an inclination of 98.6° from Vandenberg Air Force Base, California, on June 20, 1999 (UTC). This Earth-observing spacecraft provides climatologists, meteorologists, and oceanographers with daily detailed snapshots of the winds swirling above the world's oceans. It measures ocean winds and directions by monitoring wind-induced ripples by means of a microwave scatterometer. The winds play a major role in every aspect of weather on Earth. For example, they directly affect the turbulent exchanges of heat, moisture, and greenhouse gases between Earth's atmosphere and the oceans.

QuikSCAT carries a state-of-the-art radar instrument known as "SeaWinds." This scatterometer operates by transmitting high-frequency microwave pulses at a frequency of 13.4 gigahertz (GHz) to the ocean surface and measuring the backscattered, or echoed, radar pulse bounced back to the satellite. The instrument senses ripples caused by winds near the ocean's surface. Using these data, scientists can then compute the speed and direction of the winds. QuikSCAT acquires hundreds of times more observations of surface wind velocity each day than can ships and buoys. The satellite provides continuous, accurate, and high-resolution measurements of both wind speeds and directions regardless of local weather conditions. The SeaWinds scatterometer uses a rotating dish ANTENNA with two spot beams that sweep in a circular pattern. The antenna radiates microwave pulses across broad regions of Earth's surface. The instrument collects data over ocean, land, and ice in a continuous 1,800-km-wide band. QuikSCAT makes approximately 400,000 measurements and covers approximately 90 percent of Earth's surface each day.

See also EARTH SYSTEM SCIENCE; OCEAN REMOTE SENSING.

quiet Sun The collection of solar phenomena and features, including the photosphere, the solar spectrum, and the chromosphere, that are always present. *Compare with* ACTIVE SUN.

rad The traditional unit in radiation protection and nuclear technology for absorbed dose of ionizing radiation. A dose of 1 rad means the absorption of 100 ergs of ionizing radiation energy per gram of absorbing material (or 0.01 joule per kilogram in SI units). The term is an acronym derived from *r*adiation *a*bsorbed *d*ose.

See also ABSORBED DOSE; SPACE RADIATION ENVIRONMENT.

radar An active form of remote sensing generally used to detect objects in the atmosphere and space by transmitting electromagnetic waves (e.g., radio or microwaves) and sensing the waves reflected by the object. The reflected waves (called "returns" or "echoes") provide information on the distance to the object (or target) and the range and velocity of the object (or target). The reflected waves also can provide information about the shape of the object (or target). The term is an acronym for *radio detection and ranging.*

Satellite-borne radar systems (called SYNTHETIC APERTURE RADAR, or SAR systems) can be used to study the features of a planet's surface when conditions (e.g., clouds or nighttime) obscure the surface from observation in the visible portion of the electromagnetic spectrum.

See also RADAR ASTRONOMY; RADAR IMAGING; *RADARSAT;* REMOTE SENSING.

radar altimeter An active instrument carried onboard an aircraft, aerospace vehicle, or spacecraft used for measuring the distance (or altitude) of the vehicle or craft above the surface of a planet. An accurate determination of altitude is obtained by carefully timing the travel of a radar pulse down to the surface and back.

radar astronomy The use of radar by astronomers to study objects in our solar system, such as the Moon, planets, asteroids, and even planetary ring systems. For example, a powerful radar telescope, such as the Arecibo Observatory, can hurl a radar signal through the "opaque" Venusian clouds (some 80 km thick) and then analyze the faint return signal to obtain detailed information for the preparation of high-resolution surface maps. Radar astronomers can precisely measure distances to celestial objects, estimate rotation rates, and also develop unique maps of surface features, even when the actual physical surface is obscured from view by thick layers of clouds.

See also ARECIBO OBSERVATORY; MAGELLAN MISSION.

radar imaging An active remote sensing technique in which a radar antenna first emits a pulse of microwaves that illuminates an area on the ground (called a footprint). Any of the microwave pulse that is reflected by the surface back in the direction of the imaging system is then received and recorded by the radar antenna. An image of the ground thus is made as the radar antenna alternately transmits and receives pulses at particular microwave wavelengths and polarizations. Generally, a radar imaging system operates in the wavelength range of one centimeter to one meter, which corresponds to a frequency range of about 300 megahertz (MHz) to 30 gigahertz (GHz). The emitted pulses can be polarized in a single vertical or horizontal plane. About 1,500 high-power pulses per second are transmitted toward the target or surface area to be imaged, with each pulse having a pulse width (i.e., pulse duration) of between 10 to 50 microseconds (μs). The pulse typically involves a small band of frequencies, centered on the operating frequency selected for the radar. The bandwidths used in imaging radar systems generally range from 10 to 200 MHz.

At the surface of Earth (or other planetary body, such as cloud-enshrouded Venus), the energy of this radar pulse is scattered in all directions, with some reflected back toward the antenna. The roughness of the surface affects radar backscatter. Surfaces whose roughness is much less than the radar wavelength scatter in the specular direction. Rougher surfaces scatter more energy in all directions, including the direction back to the receiving antenna.

As the radar imaging system moves along its flight path, the surface area illuminated by the radar also moves along the surface in a swath, building the image as a result of the radar platform's motion. The length of the radar antenna determines the resolution in the azimuth (i.e., along-track) direction of the image. The longer the antenna, the finer the spatial resolution in this dimension. The term *synthetic aperture radar* (SAR) refers to a technique used to synthesize a very long antenna by electronically combining the reflected signals received by the radar as it moves along its flight track.

As the radar imaging system moves, a pulse is transmitted at each position and the return signals (or echoes) are recorded. Because the radar system is moving relative to the ground, the returned signals (or echoes) are Doppler-shifted. This Doppler shift is negative as the radar system approaches a target and positive as it moves away from a target. When these Doppler-shifted frequencies are compared to a reference frequency, many of the returned signals can be focused on a single point, effectively increasing the length of the antenna that is imaging the particular point. This focusing operation, often referred to as *SAR processing,* is done rapidly through the use of high-speed digital computers and requires a very precise knowledge of the relative motion between the radar platform and the objects (surface) being imaged.

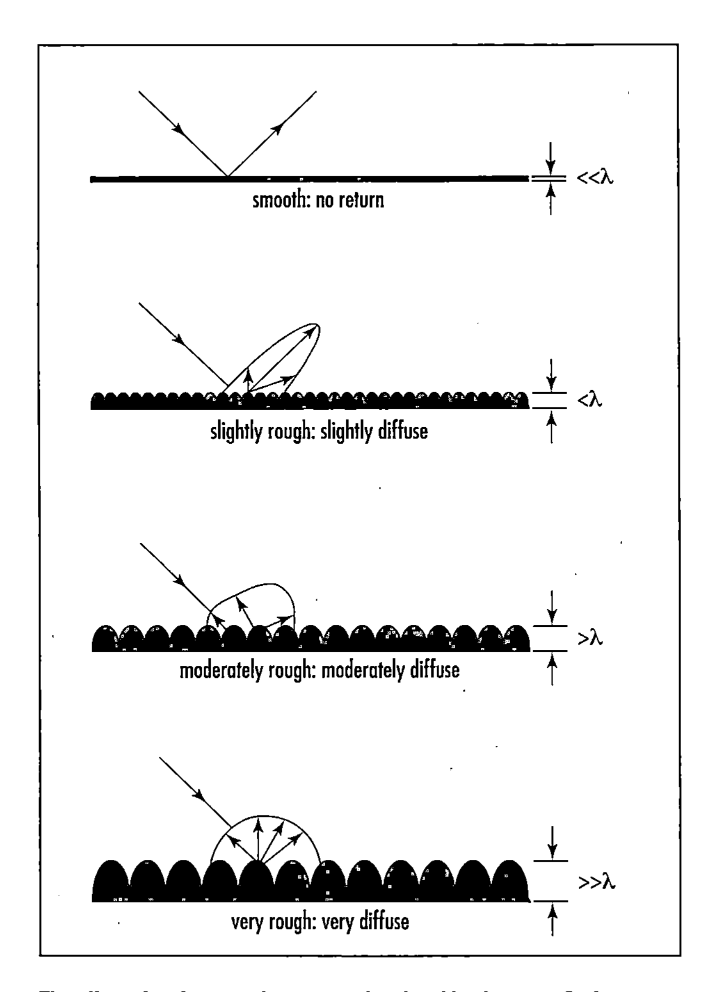

The effect of surface roughness on radar signal backscatter. Surfaces whose roughness is much less than the radar wavelength λ scatter in the specular direction. Rougher surfaces scatter more radar signal energy in all directions, including the direction back to the receiving antenna. *(Courtesy of NASA)*

The synthetic aperture radar is now a well-developed technology that can be used to generate high-resolution radar images. The SAR imaging system is a unique remote sensing tool. Since it provides its own source of illumination, it can image at any time of the day or night, independent of the level of sunlight available. Because its radio-frequency wavelengths are much longer than those of visible light or infrared radiation, the SAR imaging system can penetrate through clouds and dusty conditions, imaging surfaces that are obscured to observation by optical instruments.

Radar images are composed of many picture elements, or PIXELS. Each pixel in a radar image represents the radar backscatter from that area of the surface. Typically, dark areas in a radar image represent low amounts of backscatter (i.e., very little radar energy being returned from the surface), while bright areas indicate a large amount of backscatter (i.e., a great deal of radar energy being returned from the surface).

A variety of conditions determine the amount of radar signal backscatter from a particular target area. These conditions include the geometric dimensions and surface roughness of the scattering objects in the target area, the moisture content of the target area, the angle of observation, and the wavelength of the radar system. As a general rule, the greater the amount of backscatter from an area (i.e., the brighter it appears in an image), the rougher the surface being imaged. Therefore, flat surfaces that reflect little or no radar (microwave) energy back toward the SAR imaging system usually appear dark or black in the radar image. In general, vegetation is moderately rough (with respect to most radar imaging system wavelengths) and consequently appears as gray or light gray in a radar image. Natural and human-made surfaces that are inclined toward the imaging radar system will experience a higher level of backscatter (i.e., appear brighter in the radar image) than similar surfaces that slope away from the radar system. Some areas in a target scene (e.g., the back slope of mountains) are shadowed and do not receive any radar illumination. These shadowed areas also will appear dark or black in the image. Urban areas provide interesting radar image results. When city streets are lined up in such a manner that the incoming radar pulses can bounce off the streets and then bounce off nearby buildings (a "double-bounce") directly back toward the radar system, the streets appear very bright (i.e., white) in a radar image. However, open roads and highways are generally physically flat surfaces that reflect very little radar signal, so often they appear dark in a radar image. Buildings that do not line up with an incoming radar pulse so as to reflect the pulse straight back to the imaging system appear light gray in a radar image because buildings behave like very rough (diffuse) surfaces.

The amount of signal backscattered also depends on the electrical properties of the target and its water content. For

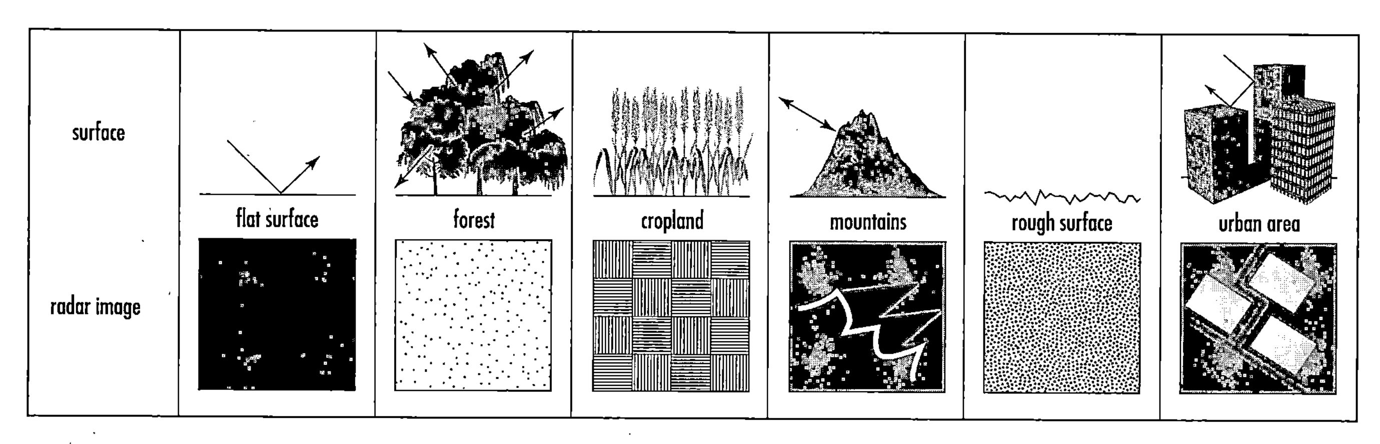

How different types of surfaces can appear in a radar image ***(Computer-generated artwork courtesy of NASA/JPL)***

example, a wetter object will appear bright, while a drier version of the same object will appear dark. A smooth body of water is the exception to this general rule. Since a smooth body of water behaves like a flat surface, it backscatters very little signal and will appear dark in a radar image.

Spacecraft-carried radar imaging systems have been used to explore and map the cloud-enshrouded surface of Venus. Earth-orbiting radar systems, such as NASA's Spaceborne Imaging Radar-C/X-band Synthetic Aperture Radar (SIR-C/X-SAR), which is carried into space in the space shuttle's cargo bay, support a variety of contemporary Earth resource observation and monitoring programs.

See also MAGELLAN MISSION; PIONEER VENUS MISSION; *RADARSAT*; REMOTE SENSING.

Radarsat Canada's first remote sensing satellite, *Radarsat* was placed in an 800-km polar orbit by a U.S. Delta II expendable launch vehicle on November 4, 1995, from Vandenberg Air Force Base in California. The principal instrument onboard this spacecraft is an advanced synthetic aperture radar (SAR), which produces high-resolution surface images of Earth despite clouds and darkness. The *Radarsat*'s SAR is an active sensor that transmits and receives a microwave signal (variations of C-band) that is sensitive to the moisture content of vegetation and soil and supports crop assessments in Canada and around the world. SAR imagery can provide important geological information for mineral and petroleum exploration. The instrument also delineates ice cover and its extent (e.g., first-year ice versus heavier multiyear ice), thereby permitting the identification of navigable Arctic sea routes. *Radarsat* was developed under the management of the Canadian Space Agency (CSA) in cooperation with the United States National Aeronautics and Space Administration (NASA), the National Oceanic and Atmospheric Administration (NOAA), the provincial governments of Canada, and the private sector. The United States supplied the launch vehicle for the 2,700-kg satellite in exchange for 15 percent of its viewing time. Canada has 51 percent of the observation time. Radarsat International is marketing the rest of the observation time. The spacecraft has exceeded its five-year life expectancy.

See also RADAR IMAGING; REMOTE SENSING.

radial-burning In rocketry, a solid-propellant grain that burns in the radial direction, either outwardly (e.g., an internal-burning grain) or inwardly (e.g., an internal-external burning tube or rod and tube grain).

See also ROCKET.

radial velocity (symbol: v_r) The component of a celestial object's velocity (speed and direction) as measured along an observer's line of sight. The Doppler shift in the spectrum of a celestial object helps astronomer's characterize the object's radial velocity. If the object is receding, the spectrum is redshifted, and astronomers assign a positive value to the radial velocity. If, on the other hand, the celestial object is approaching the observer, its spectrum is blueshifted, and astronomers assign a negative value to the radial velocity. The majority of stars in the Milky Way galaxy have radial velocities that lie between +40 km/s and –40 km/s. For example, the star Alpha Centauri has a radial velocity of –20 km/s, meaning it is approaching us and that its spectral lines are slightly blueshifted (about 0.0067 percent).

See also DOPPLER SHIFT.

radian A unit of angle. One radian is the angle subtended at the center of a circle by an arc equal in length to a radius of the circle.

$$1 \text{ radian} = 360°/(2\pi) = 57.2958°$$

radiance (symbol: L_e or L) For a point source of radiant energy, radiance is defined as the radiant intensity (I_e) in a specified direction per unit projected area; namely, $L_e = [dI_e/dA] \cos \theta$, where A is the area and θ is the angle between the surface and the specified direction. Radiance is expressed in watts per steradian per square meter (W sr^{-1} m^2).

radiant flux (symbol: Φ_e or Φ) The total power (i.e., energy per unit time) emitted or received by a body in the form of electromagnetic radiation. Generally expressed in watts (W).

radiant heat transfer The transfer of thermal energy (heat) by electromagnetic radiation that arises due to the temperature of a body. Most energy transfer of this type is in the infrared portion of the electromagnetic spectrum. However, if the emitting object has a high enough temperature, it also will radiate in the visible spectrum and beyond (e.g., the Sun). The term *thermal radiation* often is used to distinguish this form of electromagnetic radiation from other forms, such as radio waves, light, X-rays, and gamma rays. Unlike convection and conduction, radiant heat transfer takes place in and through a vacuum.

radiant intensity (symbol: I_e or I) The radiant flux (Φ_e) emitted by a point source per unit solid angle; measured in watts per steradian (W sr^{-1}).

radiation The propagation of energy by electromagnetic waves (photons) or streams of energetic nuclear particles and the energy so propagated. NUCLEAR RADIATION generally is emitted from atomic nuclei (as a result of various nuclear reactions) in the form of alpha particles, beta particles, neutrons, protons, and/or gamma rays.

radiation belt(s) The region(s) in a planet's magnetosphere where there is a high density of trapped atomic particles from the solar wind.

See also EARTH'S TRAPPED RADIATION BELTS.

radiation cooling In rocketry, the process of cooling a combustion chamber or nozzle in which thermal energy losses by radiation heat transfer from the outer combustion chamber or nozzle wall (to the surroundings) balance the heat gained (on the inner wall) from the hot combustion gases. As a result of radiation cooling, the combustion chamber or nozzle operates in a state of thermal equilibrium. To accommodate such heat transfer processes, radiatively cooled rocket engine components often are made of refractory metals or graphite.

radiation hardness The ability of electronic components and systems to function in high-intensity nuclear (ionizing) radiation environments. Spacecraft passing through or operating in a planet's trapped radiation belts would experience high-intensity nuclear radiation environments. Similarly, spacecraft exposed to solar flare conditions can experience high-intensity radiation environments depending on their distance from the Sun and the relative strength of the solar flare event. Techniques for increasing the radiation hardness of electronic components and systems include use of semiconductors that are less susceptible to ionizing radiation upset, shielding, reduction in size of the equipment and components, and redundancy.

See also SPACE RADIATION ENVIRONMENT.

radiation heat transfer *See* RADIANT HEAT TRANSFER.

radiation laws The collection of empirical and theoretical laws that describe radiative transport phenomena. The four fundamental laws that together provide a fundamental description of the behavior of blackbody radiation are (1) *Kirchhoff's law of radiation,* which describes the relationship between emission and absorption at any given wavelength and a given temperature; (2) *Planck's radiation law,* which describes the variation of intensity of a blackbody radiator at a given temperatures as a function of wavelength; (3) the *Stefan-Boltzmann law,* which relates the time rate of radiant energy emission from a blackbody to the fourth power of its absolute temperature; and (4) *Wien's displacement law,* which relates the wavelength of maximum intensity emitted by a blackbody to its absolute temperature.

See also BLACKBODY; KIRCHHOFF'S LAW (OF RADIATION); PLANCK'S RADIATION LAW; STEFAN-BOLTZMANN LAW; WIEN'S DISPLACEMENT LAW.

radiation pressure The tiny force that light or other forms of electromagnetic radiation exert per unit surface area of an illuminated object. Physicists define pressure as the force per unit area.

See also SOLAR SAIL.

radiation sickness A potentially fatal illness resulting from excessive exposure to ionizing radiation.

See also ACUTE RADIATION SYNDROME.

radiator 1. Any source of radiant energy. 2. In aerospace engineering, a device that rejects waste heat from a spacecraft or satellite to outer space by radiant heat transfer processes. Radiator design depends on both operating temperature and the amount of thermal energy to be rejected. The amount of waste heat that can be radiated to space by a given surface area is determined by the Stefan-Boltzmann law and is proportional to the fourth power of the radiating surface temperature. The surface area and mass of the thermal radiator are very sensitive to the heat rejection temperature. Higher heat rejection temperatures correspond to smaller radiator areas and, therefore, lower radiator masses. For radioisotope thermoelectric generator (RTG) power systems, for example, radiator temperatures are typically about 575 kelvins (K). Radiators can be fixed or deployable. Flat plate and cruciform configurations often are used. The radiator can be of solid, all-metal construction, or else contain embedded coolant tubes and passageways to assist in the transport of thermal energy to all portions of the radiator surface. However, radiators with embedded coolant tubes and channels must be sufficiently thick ("armored") to protect against meteoroid impact damage and the subsequent loss of coolant. Heat pipe radiator configurations also have been considered.

See also STEFAN-BOLTZMANN LAW; SPACE NUCLEAR POWER.

radioactivity The spontaneous decay or disintegration of an unstable atomic nucleus, usually accompanied by the emission of ionizing radiation, such as alpha particles, beta particles, and gamma rays. The radioactivity, often shortened to just *activity,* of natural and human-made (artificial) radioisotopes decreases exponentially with time, governed by the fundamental relationship

$$N = N_0 \, e^{-\lambda t}$$

where N is the number of radionuclides (of a particular radioisotope) at time (t), N_0 is the number of radionuclides of that particular radioisotope at the start of the count (i.e., at t = 0), and λ is the decay constant of the radioisotope.

The decay constant (λ) is related to the half-life ($T_{1/2}$) of the radioisotope by the equation

$$\lambda = (\ln 2)/T_{1/2} = 0.69315/T_{1/2}$$

The half-lives for different radioisotopes vary widely in value, from as short as about 10^{-8} seconds to as long as 10^{10} years and more. The longer the half-life of the radioisotope, the more slowly it undergoes radioactive decay. For example, the natural radioisotope uranium-238 has a half-life of 4.5×10^9 years and therefore is only slightly radioactive (emitting an alpha particle when it undergoes decay).

Radioisotopes that do not normally occur in nature but are made in nuclear reactors or accelerators are called *artificial radioactivity* or *human-made radioactivity.* Plutonium-239, with a half-life of 24,400 years, is an example of artificial radioactivity.

radio astronomy The branch of astronomy that collects and evaluates radio signals from extraterrestrial sources. Radio astronomy is a relatively young branch of astronomy. It was started in the 1930s, when KARL JANSKY, an American radio engineer, detected the first extraterrestrial radio signals. Until Jansky's discovery, astronomers had used only the visible portion of the electromagnetic spectrum to view the universe. The detailed observation of cosmic radio sources is difficult, however, because these sources shed so little energy on Earth. However, starting in the mid-1940s with the pioneering work of the British astronomer SIR ALFRED CHARLES BERNARD LOVELL at the United Kingdom's Nuffield Radio Astronomy Laboratories at Jodrell Bank, the radio telescope has been used to discover some extraterrestrial radio sources so unusual that their very existence had not even been imagined or predicted by scientists.

One of the strangest of these cosmic radio sources is the pulsar, a collapsed giant star that has become a neutron star and now emits pulsating radio signals as it spins. When the first pulsar was detected in 1967, it created quite a stir in the scientific community. Because of the regularity of its signal, scientists thought they had detected the first interstellar signals from an intelligent alien civilization.

Another interesting celestial object is the quasar, or quasi-stellar radio source. Discovered in 1964, quasars now are considered to be entire galaxies in which a very small part (perhaps only a few light-days across) releases enormous amounts of energy—equivalent to the total annihilation of millions of stars. Quasars are the most distant known objects in the universe; some of them are receding from us at more than 90 percent of the speed of light.

See also ARECIBO OBSERVATORY; PULSAR; QUASAR.

radio brightness *See* RADIO SOURCE CATALOG.

radio frequency (RF) In general, a frequency at which electromagnetic radiation is useful for communication purposes; specifically, a frequency above 10,000 hertz and below 3×10^{11} hertz. One hertz is defined as 1 cycle per second.

See also ELECTROMAGNETIC SPECTRUM.

radio galaxy A galaxy, often exhibiting a dumbbell-shaped structure, that produces very strong signals at radio wavelengths. Cygnus A, one of the closest bright radio galaxies, is about 650 million light-years away.

See also RADIO ASTRONOMY.

radio interferometer *See* VERY LARGE ARRAY.

radioisotope A radioactive isotope. An unstable isotope of an element that spontaneously decays or disintegrates, emitting nuclear radiation (e.g., an alpha particle, beta particle, or gamma ray). More than 1,300 natural and artificial radioisotopes have been identified.

radioisotope thermoelectric generator (RTG) A space power system in which thermal energy (heat) deposited by the absorption of alpha particles from a radioisotope source (generally plutonium-238) is converted directly into electricity. Radioisotope thermoelectric generators, or RTGs, have been used in space missions for which long life, high reliability, operation independent of the distance or orientation to the Sun, and operation in severe environments (e.g., lunar night, Martian dust storms) are critical. The first RTG used by the United States in space was the SNAP 3B design for the U.S. Navy's *Transit 4A* and *4B* satellites, which were launched successfully into orbit around Earth in June and November 1961, respectively. SNAP is an acronym that stands for Systems for Nuclear Auxiliary Power. SNAP-27 units provided electric power to instruments left on the lunar surface by the Apollo astronauts. SNAP-19 units provided electric power to the two highly successful Viking lander spacecraft that touched down on Mars in 1976. The *Pioneer 10* and *11* spacecraft that visited Jupiter and Saturn (*Pioneer 11* only) also were powered by SNAP-19 units. The *Voyager 1* and *2* spacecraft that explored the outer regions of our solar system, including flybys of Jupiter, Saturn, Uranus (*Voyager 2* only) and Neptune (*Voyager 2* only) were powered by an RTG unit called the Multihundred Watt (MHW) Generator. Finally, an RTG unit called the General Purpose Heat Source (GPHS) is providing electric power to more recent spacecraft, such as *Galileo, Ulysses,* and *Cassini.*

The RTGs currently designed for space missions (i.e., the GPHS) contain several kilograms of an isotopic mixture of plutonium (primarily the isotope plutonium-238) in the form of an oxide pressed into a ceramic pellet. The pellets are arranged in a converter housing and function as a heat source to generate the electricity provided by the RTG. The radioactive decay of plutonium-238, which has a half-life of 87.7 years, produces heat, some of which is converted into electricity by an array of thermocouples made of silicon-germanium junctions. Waste heat then is radiated into space from an array of metal fins.

Plutonium, like all radioactive substances and many non-radioactive materials, can be a health hazard under certain circumstances, especially if inadvertently released into the

Role of the Radioisotope Thermoelectric Generator (RTG) in Space Exploration

In the early years of the U.S. space program, lightweight batteries, fuel cells, and solar panels provided electric power for space missions. Then, as these missions became more ambitious and complex, their power needs increased significantly. Aerospace engineers examined other technical options to meet the challenging demands for more spacecraft electrical power, especially on deep space missions beyond the orbit of Mars. One selected option was the use of nuclear energy in the form of the radioisotope thermoelectric generator, or RTG.

Starting in the 1960s, cooperative efforts among several federal agencies, most notably NASA and the Department of Energy (DOE), resulted in RTG-powered spacecraft that explored the outer planets of the solar system in an unprecedented wave of scientific discovery. These trailblazing missions produced an enormous quantity of scientific information about the history and composition of the solar system. However, none of these important space exploration achievements would have been possible without the RTG. Nuclear power played a key role in establishing the United States as the world leader in outer planetary exploration and space science.

Since 1961 there have been more than 20 missions launched by either NASA or the U.S. Department of Defense (DOD) that incorporated RTGs developed by the DOE. These nuclear energy systems satisfied some or all of the electrical power needs of these missions. Some of the early American space activities that involved the use of an RTG include the Apollo lunar surface experiment packages (ALSEPs), the *Pioneer 10* and *11* spacecraft, the *Viking 1* and *2* robot landers on Mars, and the far-traveling *Voyager 1* and *2* spacecraft. More recent RTG-powered missions include the *Galileo* spacecraft to Jupiter (launched in 1989), the *Ulysses* spacecraft to explore the polar regions of the Sun (launched in 1990), and the Cassini mission to Saturn (launched in 1997). Other planned RTG-powered missions include the New Horizons spacecraft that will travel to Pluto and then go beyond into the Kuiper Belt of icy planetoids at the distant fringes of the solar system.

On March 2, 1972 (local time), NASA initiated an incredible new era of deep space exploration by launching the RTG-powered *Pioneer 10* spacecraft. *Pioneer 10* became the first spacecraft to transit the main asteroid belt and to encounter Jupiter. On its historic flyby *Pioneer 10* investigated the giant planet's magnetosphere, observed its four major satellites (the Galilean satellites Io, Europa, Ganymede, and Callisto), and then continued on its famous interstellar journey, becoming the first human-made object to cross the planetary boundary of the solar system. This space exploration milestone happened on June 13, 1983. NASA successfully launched its twin, the *Pioneer 11* spacecraft, on April 5, 1973. *Pioneer 11* swept by Jupiter and its intense radiation belts at an encounter distance of only 43,000 km on December 2, 1974. Then, renamed *Pioneer Saturn,* the spacecraft flew on to Saturn, becoming the first spacecraft to make close-up observations of this beautiful gaseous giant planet, its majestic ring system, and several of its many moons. Following these successful planetary encounters, both Pioneer spacecraft continued to send back scientific data about the nature of the interplanetary environment in very deep space. The effort continued for many years, and both spacecraft eventually passed beyond the recognized planetary boundary of the solar system. Each spacecraft now follows a different trajectory into interstellar space. Continuous and reliable electrical power from each spacecraft's RTGs made the *Pioneer 10* and *11* missions possible.

When it comes to journeys of deep space exploration enabled by the RTG, NASA's *Voyager 2* resides in a class by itself. Once every 176 years the giant outer planets (Jupiter, Saturn, Uranus, and Neptune) align themselves in a special orbital pattern such that a spacecraft launched at just the right time from Earth to Jupiter can visit the other three planets on the same mission by means of the gravity assist technique. NASA scientists called this special multiple outer planet encounter mission the "Grand Tour." They took advantage of this unique celestial alignment in 1977 and launched two sophisticated nuclear-powered robotic spacecraft called *Voyager 1* and *2.* The *Voyager 2* spacecraft lifted off from Cape Canaveral on August 20, 1977, and successfully encountered Jupiter (closest approach) on July 9, 1979. The far-traveling robot spacecraft then used gravity assist maneuvers to successfully encounter Saturn (August 25, 1981), Uranus (January 24, 1986), and Neptune (August 25, 1989). Since January 1, 1990, both RTG-powered Voyager spacecraft are participating in the Voyager Interstellar Mission (VIM). The spacecraft are now examining the properties of deep space and searching for the heliopause, the boundary that marks the end of the Sun's influence on the interplanetary medium and the true beginning of interstellar space.

The Italian scientist Galileo Galilei helped create the modern scientific method and was the first to apply the telescope in astronomy. It was only appropriate that NASA name one its most successful scientific spacecraft in his honor. The *Galileo* spacecraft lifted into space in October 1989 aboard the space shuttle *Atlantis.* An upper-stage rocket burn followed by several clever gravity assist maneuvers within the inner solar system (called the VEEGA maneuver for Venus-Earth-Earth-gravity-assist) eventually delivered the sophisticated RTG-powered robot spacecraft to Jupiter in late 1995. What ensued was a magnificent period of detailed scientific investigation of the Jovian system unparalleled in the history of planetary astronomy. The effectiveness of the *Galileo* spacecraft's instruments depended not only on the continuous supply of electrical energy received from its two GPHS-RTGs (a total of about 570 watts-electric at the start of the mission), but also the availability of a long-lasting reliable supply of small quantities (about 1 watt from each unit) of thermal energy provided by 120 radioisotope heater units (RHUs) strategically located throughout the spacecraft and its hitchhiking atmospheric probe.

The primary mission at Jupiter began when the *Galileo* spacecraft entered orbit in December 1995 and its descent probe, which had been released five months earlier, dove into the giant planet's atmosphere. Its primary mission included a 23-month, 11-orbit tour of the Jovian system, including 10 close encounters of the four major moons discovered by Galileo Galilei in 1610. Although the primary mission was successfully completed in December 1997, NASA decided to extend the *Galileo* spacecraft's mission three times before its intentional impact into Jupiter's atmosphere on September 21, 2003. As a result of these extended missions, the *Galileo* spacecraft experienced 34 encounters with Jupiter's major moons—11 with Europa, eight with Callisto, eight with Ganymede, seven with Io—and even one with the tiny innermost moon, Amalthea. NASA mission controllers purposely put the spacecraft on a collision course with Jupiter to eliminate any chance of an unwanted impact between the spacecraft and Jupiter's moon Europa, which astrobiologists suspect has a subsurface ocean that could harbor some form of extraterrestrial life.

The nuclear-powered Galileo mission produced a string of discoveries while circling the solar system's largest planet, Jupiter, 34 times. For example, Galileo was the first mission to measure Jupiter's atmosphere directly with a descent probe and was the first to conduct long-term observations of the Jovian system from orbit. It found evidence of subsurface liquid layers of saltwater on Europa, Ganymede, and Callisto. The spacecraft also examined a diversity of volcanic activity on Io. While traveling to Jupiter from the inner solar system, Galileo became the first spacecraft to fly by an asteroid and the first to discovery a tiny moon (Dactyl) orbiting an asteroid (Ida). From launch (1989) to Jovian impact (2003), the spacecraft traveled more than 4.6 billion kilometers.

Despite some contemporary controversy regarding the overall risks to Earth's biosphere from the use of space nuclear power in the form of the RTG, these long-duration electric power supplies have enabled a variety of important space exploration missions in the 20th century and promise to continue enabling spectacular missions in the outer portions of the solar system in this century. For example, the nuclear-powered *Cassini* spacecraft arrived at Saturn in summer 2004, and another nuclear-powered spacecraft, the New Horizons mission, will visit Pluto and its moon Charon sometime before 2015.

terrestrial environment. Consequently, RTGs fueled by plutonium are designed under rigorous safety standards with the specific goal that these devices be capable of surviving all credible launch accident environments without releasing their plutonium content.

Prior to launch, a nuclear-powered spacecraft undergoes a thorough safety analysis and review by various agencies of the government, including the Department of Energy. The results of that safety analysis and review also are evaluated by an independent panel of experts. Both these safety reviews are then sent to the White House, where the staff makes a final evaluation concerning the overall risk presented by the mission. Finally, presidential approval is required before a spacecraft carrying an RTG can be launched by the United States.

See also SPACE NUCLEAR POWER; THERMOELECTRIC CONVERSION.

radiometer An instrument for detecting and measuring radiant energy, especially infrared radiation.

See also REMOTE SENSING.

radionuclide A radioactive isotope characterized according to its atomic mass (A) and atomic number (Z). Radionuclides experience spontaneous decay in accordance with their characteristic half-life and can be either naturally occurring or human-made.

radio source catalog A compilation of radio sources that astronomers have detected in a particular survey, usually with the radio frequency and limiting flux density specified. The *Third Cambridge Catalog (3C)* is perhaps the most recognized example. Started in the early days of radio astronomy, this catalog provides radio astronomers the position of each source in the sky (typically using right ascension and declination data) as well as an indication of its *radio brightness*—that is, the spectral flux density per unit solid angle.

See also RADIO ASTRONOMY.

radio telescope A large metallic device, generally parabolic (dish-shaped), that collects radio wave signals from extraterrestrial objects (e.g., active galaxies and pulsars) or from distant spacecraft and focuses these radio signals onto a sensitive radio-frequency (RF) receiver.

See also ARECIBO OBSERVATORY; DEEP SPACE NETWORK; RADIO ASTRONOMY; VERY LARGE ARRAY.

radio waves Electromagnetic waves of wavelengths between about 1 millimeter (0.001 meter) and several thousand kilometers, and corresponding frequencies between 300 gigahertz and a few kilohertz. The higher frequencies are used for spacecraft communications.

A large parabolic radio telescope, or "dish," as it is often called

See also ELECTROMAGNETIC SPECTRUM; HERTZ, HEINRICH RUDOLF; TELECOMMUNICATIONS.

rail gun A device that uses the electromagnetic force experienced by a moving current in a transverse magnetic field to accelerate very small objects to very high velocities.

See also GUN-LAUNCH TO SPACE.

ramjet A reaction (jet-propulsion) engine containing neither compressor nor turbine that depends for its operation on the air compression accomplished by the forward motion of the vehicle. The ramjet has a specially shaped tube or duct open at both ends into which fuel is fed at a controlled rate. The air needed for combustion is shoved, or "rammed," into the duct and compressed by the forward motion of the vehicle and engine assembly. This rammed air passes through a diffuser and is mixed with fuel and burned. The combustion products then are expanded in a nozzle. A ramjet cannot operate under static conditions. The duct geometry depends on whether the ramjet operates at subsonic or supersonic speeds.

range 1. The distance between any given point and an object or target. 2. Extent or distance limiting the operation or action of something, such as the range of a surface-to-air missile. 3. An area in and over which rockets, missiles, or aerospace vehicles are fired for testing.

Ranger Project The Ranger spacecraft were the first U.S. robot spacecraft sent toward the Moon in the early 1960s to pave the way for the Apollo Project's human landings at the end of that decade. The Rangers were a series of fully attitude-controlled spacecraft designed to photograph the lunar surface at close range before impacting. *Ranger 1* was launched on August 23, 1961, and set the stage for the rest of the Ranger missions by testing spacecraft navigational performance. *Ranger 2* through *9* were launched from November 1961 through March 1965. (*See* Table) All of the early Ranger missions (*Ranger 1* through *6*) suffered setbacks of one type or another. Finally, *Ranger 7, 8,* and *9* succeeded with flights that returned many thousands of images (before impact) and greatly advanced scientific knowledge about the lunar surface.

See also APOLLO PROJECT; MOON; SURVEYOR PROJECT.

Rankine cycle A fundamental thermodynamic heat engine cycle in which a working fluid is converted from a liquid to a vapor by the addition of heat (in a boiler), and the hot vapor then is used to spin a turbine, thereby providing a work output. This work output often is used to rotate a generator and produce electric power. After passing through the turbine, the vapor is cooled in a condenser (rejecting heat to the surroundings). While being cooled, the working fluid changes phase (from vapor to liquid) and then passes through a pump to start the cycle again.

See also CARNOT CYCLE; HEAT ENGINE; THERMODYNAMICS.

Rayleigh criterion *See* DIFFRACTION.

Rayleigh scattering Selective scattering process caused by spherical particles (e.g., gas molecules) whose characteristic dimensions are about one-tenth (or less) of the wavelength (λ) of the radiation being scattered. Rayleigh scattering is inversely proportional to the fourth power of the wavelength—that is, this scattering process goes as λ^{-4}. For the scattering of visible light, this means that blue light (λ = 0.4 μm) is scattered much more severely by Earth's atmosphere than longer-wavelength red light (λ = 0.7 μm). For a clear daytime atmosphere this selective scattering causes the sky to appear blue to an observer who is not along the direction of illumination. However, at sunrise and sunset sunlight must pass through a much thicker column of the atmosphere, and the preferential scattering of blue light is so complete that essentially only red, orange, and yellow light reaches an observer on Earth. Named after the British physicist Lord Rayleigh (1842–1919).

Another selective scattering process is called *Mie scattering*. It involves scattering of radiation by particles whose characteristic size is between one-tenth (0.1) and 10 times the wavelength of the incident radiation. Smoke, fumes, and haze in Earth's atmosphere result in this type of scattering, which is inversely proportional to the wavelength of the radiation (i.e., the process goes as λ^{-1}).

Nonselective scattering is caused by particles (such as are found in clouds and fog) with characteristic dimensions of more than about 10 times the wavelength of the radiation being scattered. These larger particles scatter all wavelengths of light equally well, so that clouds and fog appear white

Ranger Mission Summary (1961–1965)

Mission	Launch Date	Objective	Result(s)
Ranger 1	08/23/61	Lunar mission prototype	Launch failure
Ranger 2	11/18/61	Lunar mission prototype	Launch failure
Ranger 3	01/26/62	Lunar (impact) probe	Spacecraft failed; missed Moon
Ranger 4	04/23/62	Lunar (impact) probe	Spacecraft failed; impact farside
Ranger 5	10/18/62	Lunar (impact) probe	Spacecraft failed; missed Moon
Ranger 6	01/30/64	Lunar (impact) probe	Successful impact; cameras failed
Ranger 7	07/28/64	Lunar (impact) probe	Successful impact; numerous images
Ranger 8	02/17/65	Lunar (impact) probe	Successful impact; numerous images
Ranger 9	03/21/65	Lunar (impact) probe	Successful impact; numerous images

Source: NASA data.

even though they contain particles of water, which is actually colorless.

See also MIE SCATTERING.

rays (lunar) Bright streaks extending across the surface from young impact craters on the Moon; also observed on Mercury and on several of the large moons of the outer planets.

reaction 1. In chemistry, a chemical change that occurs when two or more substances are mixed, usually in solution; chemical activity between substances. 2. In physics, the equal-in-magnitude but opposite-in-direction force that occurs in response to the action of some other force in accordance with SIR ISAAC NEWTON's third law of motion (i.e., the action-reaction principle).

See also NEWTON'S LAWS OF MOTION.

reaction-control jets Small propulsion units on a spacecraft or aerospace vehicle that are used to rotate the craft or to accelerate it in a specific direction.

reaction control system (RCS) The collection of thrusters on the space shuttle orbiter vehicle that provides attitude control and three-axis translation during orbit insertion, on-orbit operations, and reentry.

See also SPACE TRANSPORTATION SYSTEM.

reaction engine An engine that develops thrust by its physical reaction to the ejection of a substance (including possibly photons and nuclear radiations) from it; commonly, the reaction engine ejects a stream of hot gases created by combusting a propellant within the engine.

A reaction engine operates in accordance with SIR ISAAC NEWTON's third law of motion (i.e., the action-reaction principle). Both rocket engines and jet engines are reaction engines. Sometimes called a reaction motor.

See also NEWTON'S LAWS OF MOTION; ROCKET.

reaction time In human factors engineering, the interval between an input signal (physiological) or a stimulus (psychophysiological) and a person's response to it. Normally, it takes a human being at least three-tenths of a second (0.3 s) to begin to respond to a crisis or emergency situation (e.g., applying the brakes of an automobile). Training and special alarms and signals can reduce this reaction time a bit, while inexperience, panic, or fear can significantly lengthen it.

readout 1. (verb) The action of a spacecraft's transmitter sending data that either are being instantaneously acquired or else extracted from storage (often by playing back a magnetic tape upon which the data have been recorded previously). 2. The data transmitted by the action described in sense 1.

readout station A recording (or receiving) station at which the data-carrying radio-frequency signals transmitted by a spacecraft or space probe are acquired and initially processed.

See also TELECOMMUNICATIONS; TELEMETRY.

real time Time in which reporting on or recording events is simultaneous with the events; essentially, "as it happens."

real-time data Data presented in usable form at essentially the same time an event occurs.

Reber, Grote (1911–2002) American *Radio Engineer* In 1937 the American radio engineer Grote Reber built the world's first radio telescope in his backyard and for several years thereafter was the world's only practicing radio astronomer. His steerable, approximately 10-meter-diameter parabolic dish antenna detected many radio sources, including the radio wave signals in the constellation Sagittarius from the direction of the center of the Milky Way galaxy, confirming KARL GUTHE JANSKY's discovery in the early 1930s.

reconnaissance satellite An Earth-orbiting military satellite intended to perform reconnaissance missions against enemy nations and potential adversaries. In the 1960s the United States developed and flew its first generation of photo reconnaissance satellites, called the Corona, Argon, and Lanyard systems.

In 1958 during the cold war, President Dwight D. Eisenhower approved a program that would answer questions about Soviet missile capability and replace risky U-2 reconnaissance flights over Soviet territory. The Central Intelligence Agency (CIA) and the U.S. Air Force would jointly develop satellites to photograph denied areas from space. The program had both a secret mission and a secret name—Corona.

The CIA and the USAF developed this first-generation space program with great speed and tight secrecy. In August 1960, after numerous unsuccessful attempts, a Corona satellite was launched successfully, and its film capsule was recovered from space. During the next 12 years Corona satellites ushered in a new era of technical intelligence and a new era of space "firsts" that also contributed to advancements in other areas of the national space program. For example, the Corona program provided valuable experience on how to recover objects from orbit, experience that was later used to recover astronauts from orbit during NASA's Mercury, Gemini, and Apollo Projects. Corona also provided a fast and relatively inexpensive way to map Earth from space.

But the most important contribution of the Corona system to national security came from the intelligence it provided. During the cold war Corona looked through the Iron Curtain and helped lay the groundwork for important disarmament agreements. For example, with reconnaissance satellites the United States could verify reductions in missiles without on-site inspections. (*See* figure on page 491.) Satellite imagery gave U.S. leaders the confidence to enter into negotiations and to sign important arms control agreements with the Soviet Union. Successor programs continued to monitor intercontinental ballistic missile (ICBM) sites and verify strategic arms agreements and the Nuclear Nonproliferation Treaty.

The Corona program operated from August 1960 to May 1972 and collected both intelligence and mapping imagery. Argon was a mapping system that used the organizational framework of Corona and achieved seven successful

Space Technology and National Security

In the mid-20th century the development of reconnaissance satellites, surveillance satellites, and other information-related, Earth-orbiting military spacecraft significantly transformed the practice of national security and the conduct of military operations. From the launching of the very first successful American reconnaissance satellite in 1960, "spying from space" produced enormous impacts on how the United States government conducted peacekeeping and war fighting. Recognizing the immense value of the unobstructed view of Earth provided by the "high ground" of outer space, defense leaders immediately made space technology an integral part of projecting national power and protecting national assets. However, most of these military space activities were performed in secret, so only civilian space accomplishments of the 1960s and 1970s made the headlines.

What is the basic difference between a reconnaissance satellite and a surveillance satellite? Reconnaissance systems employ their sensors to search for specific types of denied information of value to intelligence analysts. Surveillance systems use their special instruments to monitor Earth, its atmosphere, and near-Earth space for hostile events or activities (generally military in nature) that could threaten the interests of the United States or its allies. Such hostile actions include the launching of a surprise ballistic missile attack or the clandestine testing of a nuclear weapon. Nuclear surveillance satellites, for example, have special sensors that look for characteristic signals from a nuclear detonation conducted secretly in Earth's atmosphere or in outer space by a signatory nation trying to violate its participation in a nuclear test ban treaty.

Founded in 1960, the National Reconnaissance Office (NRO) is the national program that meets the needs of the U.S. government through space-based reconnaissance. The NRO is an agency of the Department of Defense (DOD) and is staffed by personnel from the Central Intelligence Agency (CIA), the military services, and civilian defense personnel. NRO satellites collect data in support of such functions as intelligence and warning, monitoring of arms control agreements, military operations and exercises, and monitoring of natural disasters and environmental issues.

The launch of *Sputnik 1* in October 1957 provided the political stimulus to accelerate the development of the first American photo reconnaissance satellite, called Corona. Under strict secrecy in 1958, the most promising portion of an innocuously named U.S. Air Force advanced reconnaissance program, called Weapon System 117L, was separated from that program and designated the Corona Program. Placed under the management of a joint USAF/CIA team, Corona enjoyed increased priority and funding. Despite numerous early launch failures, President Dwight D. Eisenhower provided his unwavering support for this photoreconnaissance satellite. His patience was rewarded on August 18, 1960, when the *Corona 14* spacecraft was successfully launched from Vandenberg AFB, California. A day later the spacecraft ejected a film capsule that reentered Earth's atmosphere and was recovered in midair by a military aircraft flying over the Pacific Ocean near Hawaii. The end of this perfect mission marked the beginning of overhead photoreconnaissance from space.

The film capsule retrieved from *Corona 14* was quickly processed and analyzed by the American intelligence community. This single, very-short-duration satellite mission provided more imagery coverage than all previous high-risk U-2 spy plane flights over the Soviet Union combined. The photoreconnaissance satellite gave the United States the space-based "eyes" it needed to protect its people and maintain the peace throughout the cold war. The 145th and final Corona launch took place on May 15, 1972. It is not an overstatement to suggest that space technology helped avoid World War III. During the politically tense years of the cold war the United States and the Soviet Union were engaged in an aggressive nuclear arms race in which both superpowers maintained nuclear-armed ballistic missiles on hair-trigger alert. Military satellites provided the vital information needed by national leaders to follow a sane course of action in the otherwise insane world of mutually assured destruction (MAD).

From a space technology perspective, Corona was the first space program to recover an object from orbit and the first to deliver photoreconnaissance information from a satellite. It would go on to be the first program to use multiple reentry vehicles, pass the 100-mission mark, and produce stereoscopic overhead imagery from space. The most remarkable technological advance, however, was the improvement in the ground resolution of its imagery from an initial 7.6-meter to 12.2-meter capability to an eventual 1.82-meter capability.

However, almost all of these important space technology advances and applications went unnoticed by the American public. While the reconnaissance satellite transformed the art and practice of technical intelligence collection, it did so under the cloak of secrecy. From its formation in 1960, the National Reconnaissance Office operated in a highly classified environment. It was only in the early 1990s that even the existence of this organization and its extremely important role in successfully applying space technology to collect intelligence data (imagery and signals) over denied areas in support of national security was made public by presidential decision.

In 1966 the U.S. Air Force started development of an important family of surveillance satellites, now known as the Defense Support Program (DSP). These early-warning satellites operate in geostationary orbit at an altitude of about 35,900 km above the equator. Missile surveillance spacecraft use special infrared detectors to continuously scan the planet's surface (both land and water) for the telltale hot exhaust plume signature that characterizes the launch of a ballistic missile. Since American missile warning satellites first became operational in the early 1970s, these spacecraft have provided the national leadership with an uninterrupted, 24-hour-per-day, worldwide ballistic missile surveillance capability. Continuously improved versions of the DSP satellite have served as the cornerstone of the national early warning program. By eliminating the possibility of a surprise enemy ballistic missile attack, this important family of military spacecraft enabled the fundamental national policy of strategic nuclear deterrence.

In the post–cold war era these silent sentinels still stand guard, always ready to alert national authorities about a hostile ballistic missile attack. Now their missile surveillance mission has expanded to include shorter-range ballistic missiles launched by rogue nations during regional conflicts. Later this century a new generation of space-based infrared surveillance systems will help American military leaders satisfy the following four critical defense missions: missile warning, missile defense, technical intelligence, and battle space characterization, including battle damage assessment (BDA).

Three other broad classes of military satellite systems have significantly influenced the use of information in support of national defense. These military spacecraft have also pioneered closely related civilian space technology applications. Included in the group are polar-orbiting military meteorological satellites, the special constellation of navigation satellites called the Global Positioning System (GPS), and the family of advanced military communication satellites.

Space technology has greatly influenced the practice of national defense and the conduct of modern warfare. Today space systems are the primary source of warning of an impending attack and have the ability to fully characterize that attack. Highly capable reconnaissance satellites also monitor arms control agreements and continuously gather technical data that allows intelligence analysts to evaluate the world situation and avoid military or political surprises. Unfortunately, even the best levels of technical intelligence cannot predict irrational human behavior as might be displayed by a rogue political leader or a terrorist group of religious fanatics who choose to lurk in the shadows and prey on innocent victims. No level of advanced space-based technical intelligence can peer into such twisted minds. Technical intelligence collected from satellite systems remains important in the case of unconventional warfare and counterterrorism operations, but it does not have the same exceptionally high value in battle space dominance that it provides during more conventional conflicts.

Today military planners often speak in terms of a battle space that extends beyond the traditional land, sea, and air battlefields of the 20th century and includes the application of various military spacecraft operating far above. The concept of a transparent battle space refers to the state of information superiority that allows the American commander to "see" everything within the battle space—all enemy activity with complete accountability of friendly forces. Modern space technology is an important force multiplier because military satellites support the surgical application of military resources with maximum impact on hostile forces and minimal risk to friendly forces.

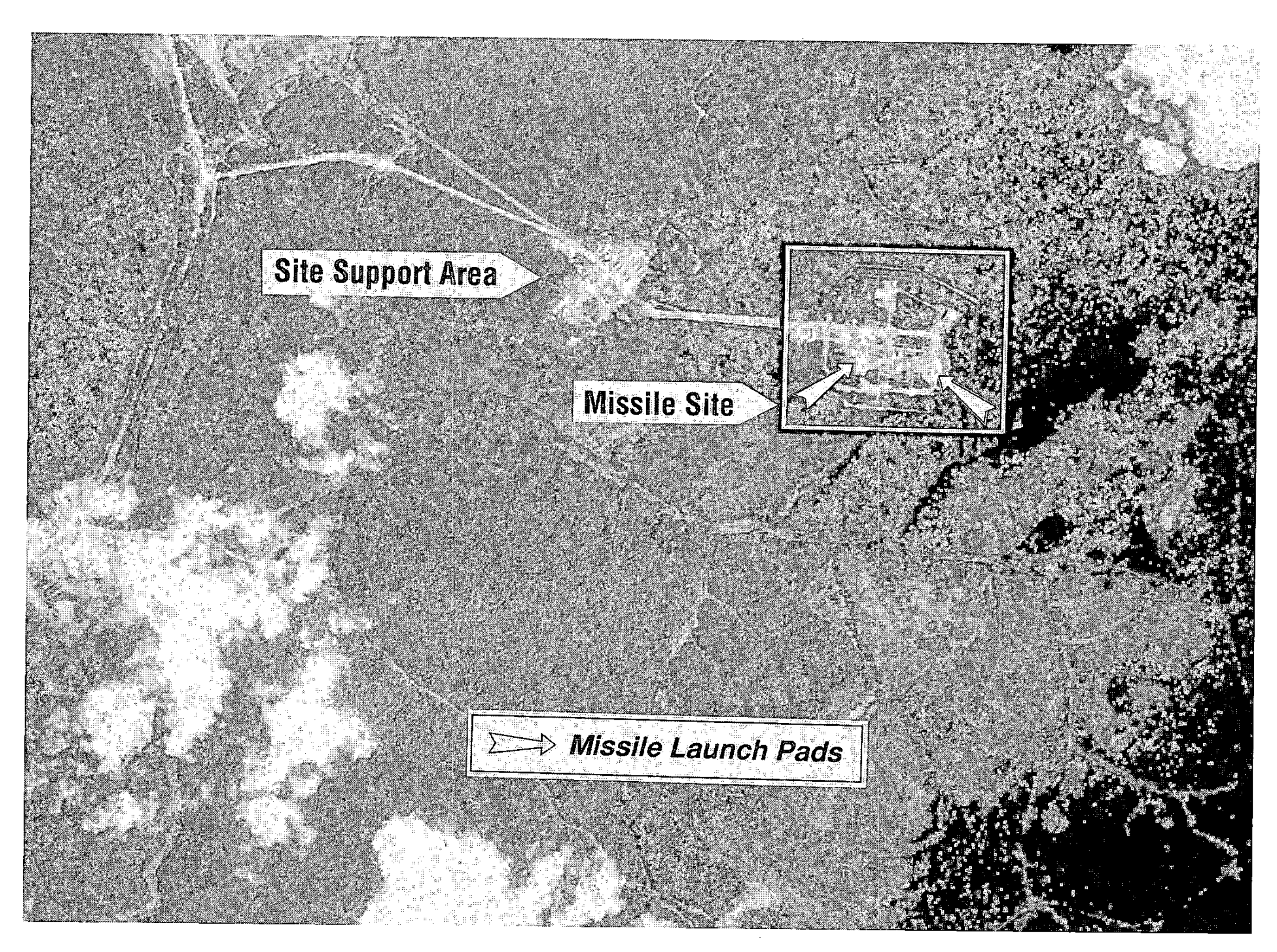

This is a declassified reconnaissance satellite photograph of the Yurya ICBM complex in the Soviet Union. The image was captured by a United States Corona spy satellite in June 1962. While exploiting this photograph, American intelligence analysts identified the construction of an SS-7 missile launch site, as noted in the processed image. Reconnaissance satellites provided critical information about the status of the Soviet missile program and helped maintain global stability during some of the most dangerous portions of the cold war and its civilization-threatening nuclear arms race. ***(Photograph courtesy of National Reconnaissance Office [NRO])***

missions of 12 from May 1962 to August 1964. Finally, Lanyard was an attempt to gain higher-resolution imagery. It flew one successful mission in 1963.

Early imagery collections were driven, for the most part, by the pressures of the cold war and the needs of the United States to confirm suspected developments in Soviet strategic arms capabilities. Worldwide photographic coverage also was used to produce maps and charts for the Department of Defense and other U.S. government mapping programs.

A presidential executive order dated February 24, 1995, declassified more than 800,000 images collected by these early photoreconnaissance systems from 1960 to 1972. The historic declassified imagery (some with a resolution of 1.82 meters) is being made available to assist environmental scientists to improve their understanding of global environmental processes and to develop a baseline in the 1960s for assessing important environmental changes.

See also *DISCOVERER* SPACECRAFT; NATIONAL RECONNAISSANCE OFFICE.

recovery The procedure or action that occurs when the whole of a satellite, spacecraft, scientific instrument package, or component of a rocket vehicle is recovered after a launch or mission; the result of this procedure.

rectifier A device for converting alternating current (a.c.) into direct current (d.c.); usually accomplished by permitting current flow in one direction only.

red dwarf (star) Reddish main-sequence stars (spectral type K and M) that are relatively cool (about 4000 K surface temperature) and have low mass (typically 0.5 solar masses or less). These faint, low-luminosity stars are inconspicuous, yet they represent the most common type of star in the universe and the longest lived. Barnard's star is an example.

See also BARNARD'S STAR; STAR.

Redeye (FIM-43) A U.S. Army, lightweight, person-portable, shoulder-fired air defense artillery weapon for low-altitude air defense of forward combat area troops.

red fuming nitric acid (RFNA) Concentrated nitric acid (HNO_3) in which nitrogen dioxide (NO_2) has been dissolved. This toxic and corrosive mixture typically contains between 5 and 20 percent nitrogen dioxide. It is used as an oxidizer in liquid-propellant rockets.

red giant (star) A large, cool star with a surface temperature of about 2,500 K and a diameter 10 to 100 times that of the Sun. This type of highly luminous star is at the end of its evolutionary life, departing the main sequence after exhausting the hydrogen in its core. A red giant is often a variable star. Some 5 billion years from now, the Sun will evolve into a massive red giant.

See also STAR; VARIABLE STAR.

redline 1. (noun) Term denoting a critical value for a parameter or a condition that, if exceeded, threatens the integrity of a system, the performance of a vehicle, or the success of a mission. 2. (verb) To establish a critical value as described in sense 1.

Red Planet The planet Mars—so named because of its distinctive reddish soil.

See also MARS.

redshift The apparent increase in the wavelength of a light source caused by the receding motion of the source. The Doppler effect shift of the visible spectra of distant galaxies toward red light (i.e., longer wavelength) indicates that these galaxies are receding. The greater redshift observed in more distant galaxies has been interpreted to mean that the universe is expanding. *Compare with* BLUESHIFT.

Redstone An early liquid-propellant, medium-range, surface-to-surface missile used by the U.S. Army. The Redstone served as the first stage of the Jupiter-C vehicle that launched the first American satellite, *Explorer 1,* on January 31, 1958. It also served as the launch vehicle for Mercury Project astronaut ALAN SHEPARD's suborbital flight on May 5, 1961, in the Mercury *Freedom 7* capsule, the first American human space mission.

See also EXPLORER 1; MERCURY PROJECT.

reentry The return of objects, originally launched from Earth, back into the sensible (measurable) atmosphere; the action involved in this event. The major types of reentry are ballistic, gliding, and skip. To perform a safe controlled reentry a spacecraft or aerospace vehicle must be capable of achieving a controlled dissipation of its kinetic and potential energies. The kinetic and potential energy values are determined by the returning object's velocity and altitude at the atmospheric reentry interface. Successful reentry, culminating in a safe landing on the surface of Earth (land or water), requires a very carefully designed and maintained flight trajectory.

Random or uncontrolled reentry, as might occur with a derelict satellite or piece of space debris, normally results in excessive aerodynamic heating and the burnup of the object in the upper atmosphere. On certain occasions, however, natural or human-made objects have undergone uncontrolled reentry and actually survived the fiery plunge through the atmosphere, impacting on Earth. Also called entry in the aerospace literature.

reentry body That part of a rocket or space vehicle that reenters Earth's atmosphere after flight above the sensible (measurable) atmosphere.

reentry corridor The narrow region or pathway, established primarily by velocity and altitude, along which a spacecraft or aerospace vehicle must travel in order to return safely to Earth's surface. The design of the vehicle and the operational circumstances (e.g., normal end-of-mission return to Earth, emergency reentry, etc.) also contribute to the dimensions of the reentry corridor.

reentry nose cone A nose cone designed especially for reentry, usually consisting of one or more chambers protected by a specially designed outer surface that serves as both an aerodynamic surface and a heat shield.

reentry phase That portion of the trajectory of a ballistic missile or aerospace vehicle when there is a significant interaction of the vehicle and Earth's sensible (measurable) atmosphere following flight in outer space.

See also BALLISTIC MISSILE DEFENSE.

reentry trajectory The trajectory or pathway followed by an aerospace vehicle, spacecraft, or other object during reentry. If the object is unguided after passing the reentry interface at the top of the sensible (measurable) atmosphere, its trajectory will be essentially ballistic. With controlled aerodynamic maneuvering and guidance during reentry, however, the object will follow a glide or skip trajectory as it descends to the surface of Earth.

reentry vehicle (RV) 1. In general, that part of a space vehicle designed to reenter Earth's atmosphere in the terminal portion of its trajectory. 2. In the context of BALLISTIC MISSILE DEFENSE (BMD), that part of a ballistic missile (or post-boost vehicle) that carries the nuclear warhead to its target. The reentry vehicle is designed to enter Earth's atmosphere in the terminal portion of its trajectory and proceed to its target. The reentry vehicle is designed to survive rapid heating during high-velocity flight through the atmosphere and to protect its warhead until the nuclear weapon can detonate at the target.

reentry window The area at the limits of Earth's sensible (measurable) atmosphere through which an aerospace vehicle or spacecraft in a given trajectory should pass to accomplish a successful reentry.

Rees, Sir Martin John (1942–) British *Astrophysicist* Sir Martin John Rees was among the first astrophysicists to suggest that enormous black holes could power quasars. His investigation of the distribution of quasars helped discredit steady state cosmology. He has also contributed to the theories of galaxy formation and studied the nature of the so-called dark matter of the universe. Dark matter is believed to be present throughout the universe but cannot be observed directly because it emits little or no electromagnetic radiation. Contemporary cosmological models suggest this "missing mass" controls the ultimate fate of the universe. In 1995 Sir Rees accepted an appointment to serve Great Britain as the 15th Astronomer Royal.

Rees was born on June 23, 1942, in York, England. He received his formal education at Cambridge University. Rees graduated from Trinity College in 1963 with a bachelor of arts degree in mathematics and then continued on for his graduate work, completing a doctor of philosophy degree in 1967. Before becoming a professor at Sussex University in 1972, he held various postdoctoral positions in both the United Kingdom and the United States. In 1973 Rees became the Plumian Professor of Astronomy and Experimental Philosophy at Cambridge and remained in that position until 1991. During this period he also served two separate terms as director of the Institute of Astronomy at Cambridge. These terms ran from 1977 to 1982 and from 1987 to 1991. Among his many professional affiliations, Rees became a fellow of the Royal Society of London in 1979, a foreign associate of the U.S. National Academy of Sciences in 1982, and a member of the Pontifical Academy of Sciences in 1990. He is currently a Royal Society professor and a fellow of King's College at Cambridge University. He is also a visiting professor at the Imperial College in London.

In 1992 he became president of the Royal Astronomical Society. That year Queen Elizabeth II conferred knighthood upon him. By royal decree in 1995, Sir Martin Rees became Great Britain's Astronomer Royal, the 15th distinguished astronomer to hold this position since King Charles II created the post in 1675.

As an eminent astrophysicist, he has modeled quasars and studied their distribution, important work that helped discredit the steady state universe model in cosmology. He was one of the first astrophysicists to postulate that enormous black holes might power these powerful and puzzling compact extragalactic objects that were first discovered in 1963 by MAARTEN SCHMIDT. Rees maintains a strong research interest in many areas of contemporary astrophysics, including gamma-ray bursts, black hole formation, the mystery of dark matter, and anthropic cosmology. One of today's most interesting cosmological issues involves the so-called ANTHROPIC PRINCIPLE, the somewhat controversial premise that suggests the universe evolved in just the right way after the big bang event to allow for the emergence of life, especially intelligent human life.

One of Rees's greatest talents is his ability to effectively communicate the complex topics of modern astrophysics to both technical and nontechnical audiences. He has written more than 500 technical articles, mainly on cosmology and astrophysics, and several books, including *Our Cosmic Habitat* (2001), *Just Six Numbers* (2000), *New Perspectives in Astrophysical Cosmology* (2000), *Cosmic Coincidences* (1989), and *Gravity's Fatal Attraction: Black Holes in the Universe* (1995). Rees has received numerous awards for his contributions to modern astrophysics, including the Gold Medal of the Royal Astronomical Society (1987), the Bruce Gold Medal from the Astronomical Society of the Pacific, the Science Writing Award from the American Institute of Physics (1996), the Bower Science Medal of the Franklin Institute (1998), and the Rossi Prize of the American Astronomical Society (2000).

reflectance (symbol: ρ) In radiation heat transfer, the reflectance (ñ) of a body is defined as the ratio of the incident radiant energy reflected by the body to the total radiant energy falling on the body. The reflecting surface of the body may be specular (mirrorlike) or diffuse. For the special case of an ideal blackbody, all the radiant energy incident upon this blackbody is absorbed and none is reflected, regardless of wavelength or direction. Therefore, the reflectance for a blackbody has a value of zero, that is, $\rho_{blackbody} = 0$. All other "real-world" solid objects have a reflectance of greater than (or perhaps approximately equal to) zero. Also called *reflectivity. Compare with* ABSORPTANCE and TRANSMITTANCE.

reflecting telescope A telescope that collects and focuses light from distant objects by means of a mirror, called the PRIMARY MIRROR. The first reflecting telescope was introduced

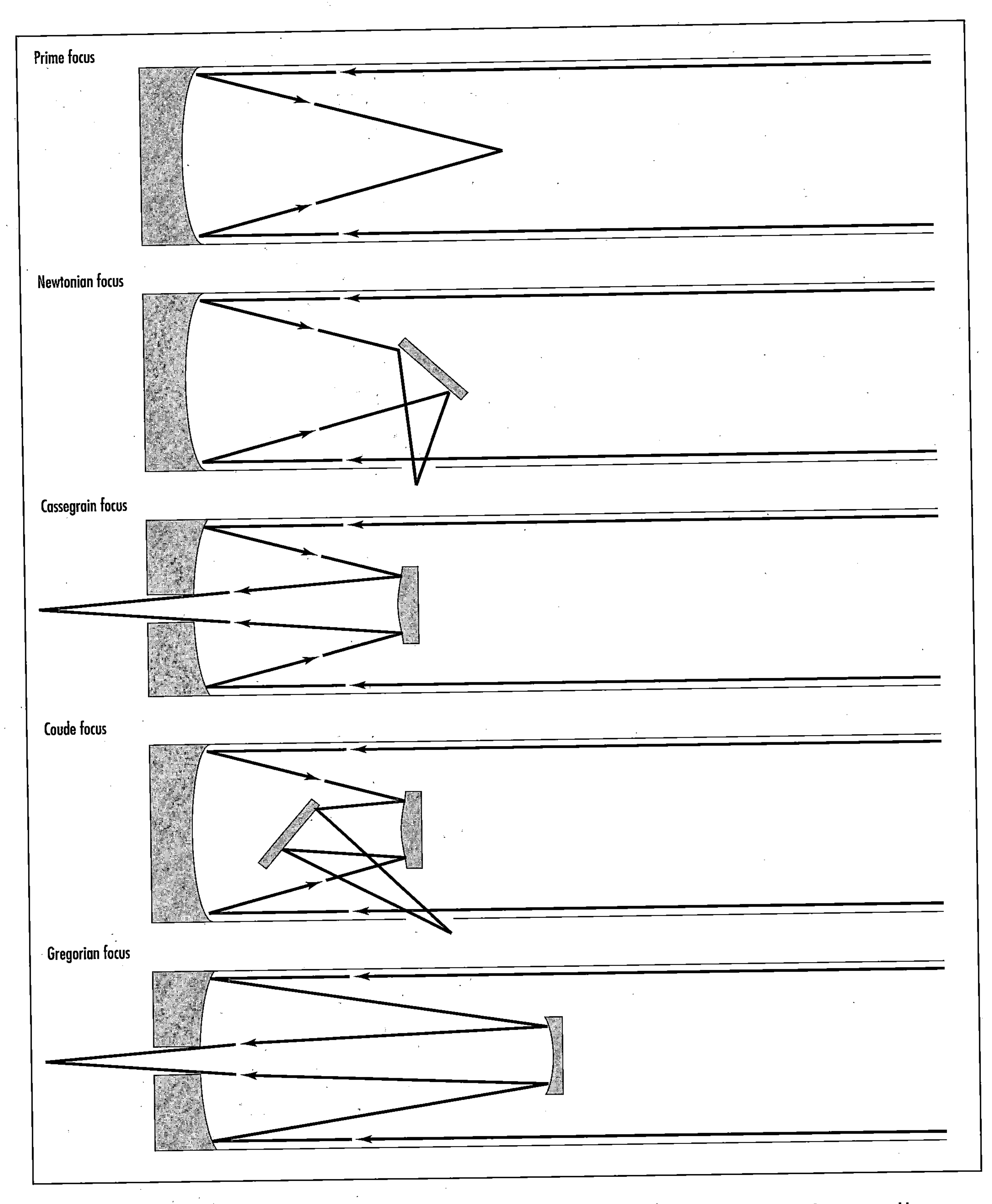

This illustration provides a basic comparison of the general design features and components of the major types of reflecting telescopes used in astronomy. As shown here, the light from distant celestial objects enters each device from the right. ***(Diagram courtesy of NASA)***

by SIR ISAAC NEWTON in about 1670. His design is sometimes referred to as a *Newtonian reflector.*

See also TELESCOPE.

reflection The return of all or part of a beam of light when it encounters the interface (boundary) between two different media, such as air and water. A mirrorlike surface reflects most of the light falling upon it.

refracting telescope A telescope that collects and focuses light from distant objects by means of a *primary lens* (*objective*) or system of lenses. (*See* figure.) In 1609 GALILEO GALILEI began improving the first refracting telescope invented by the Dutch optician HANS LIPPERSHEY and then applied his own instrument design (called the GALILEAN TELESCOPE) to astronomical observations. It was this refracting Galilean telescope that helped launch the scientific revolution of the 17th century. Also called a *refractor.*

See also TELESCOPE.

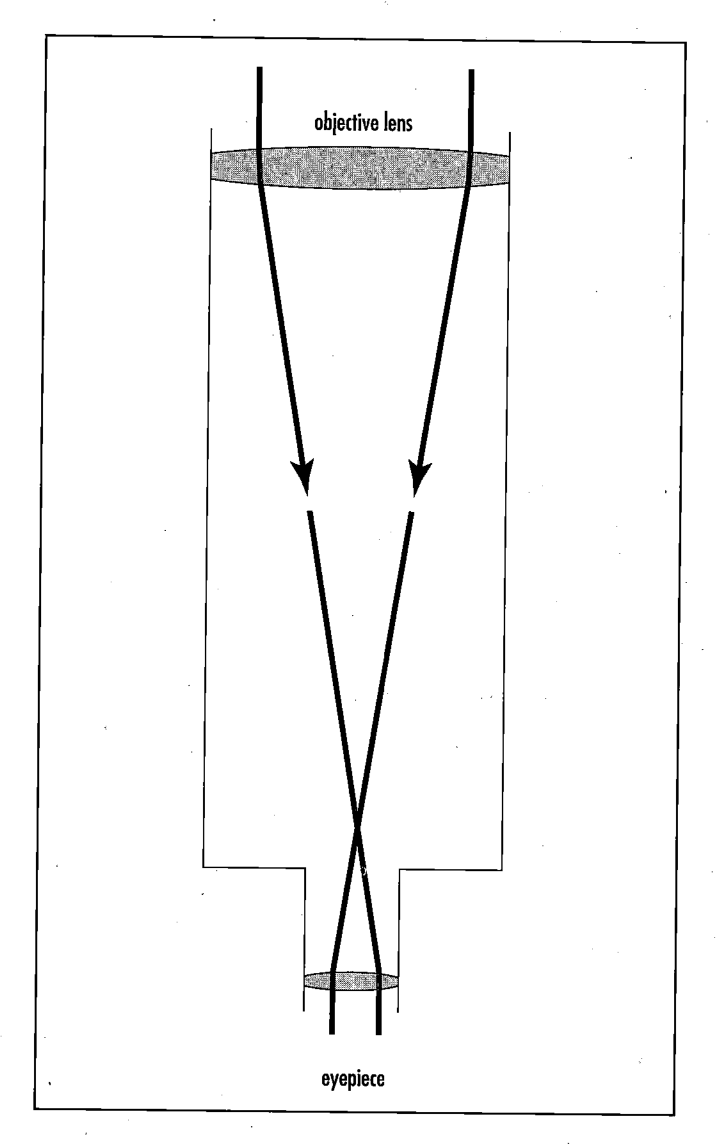

This illustration shows the design features and components of the basic refracting telescope. Light from a distant celestial object enters the telescope from the right. In 1609 Galileo Galilei designed his own simple refracting telescope, now called the *Galilean telescope,* and with this modest optical device the brilliant Italian scientist started the age of telescopic astronomy. *(Diagram courtesy of NASA)*

refraction The deflection or bending of electromagnetic waves when these waves pass from one type of transparent medium into another. For example, visible light passing through a prism at a suitable angle is dispersed into a continuum of colors. This happens because of refraction. When visible light waves cross an interface between two transparent media of different densities (e.g., from air into glass) at an angle other than 90°, the light waves are bent (or refracted). Different wavelengths of visible light are bent different amounts; this causes them to be dispersed into a continuum of colors.

The *index of refraction* is defined as the ratio of the speed of light in a vacuum to the speed of light in the transparent substance of the observed medium. Each type of transparent substance lowers the speed of light slightly and by a different amount; low-density air has a different index of refraction from high-density air and from water, glass, carbon dioxide (gas), and so on.

regenerative cooling A common approach to cooling large liquid-propellant rocket engines and nozzles, especially those that must operate for an appreciable period of time. In this cooling technique one of the liquid propellants (e.g., liquid oxygen) is first sent through special cooling passages in the thrust chamber and nozzle walls before being injected into the combustion chamber.

See also ROCKET.

regenerative life support system (RLSS) A controlled ecological life support system in which biological and physiochemical subsystems produce plants for food and process solid, liquid, and gaseous wastes for reuse in the system. (*See* figure on page 496.) Varying degrees of closure are possible. As the amount of recycling of the consumable materials necessary for life to acceptable standards increases, the quantity of makeup materials that must be supplied to the RLSS of a space habitat, planetary outpost, or even starship decreases. The ideal RLSS requires no resupply of any consumable material, since it would be capable of recycling everything. By definition, the ideal RLSS would have achieved complete closure and would be totally self-contained, including all the energy resources needed to support the recycling processes. In reality, most contemporary RLSS concepts for space habitats involve some amount of resupply and usually need a flow of energy across their boundaries from external sources (especially solar energy). On a grand scale, Earth's biosphere can be thought of as a natural regenerative life support system with almost complete closure save for the flow of life-sustaining solar energy.

regenerator A device used in a thermodynamic process for capturing and returning to the process thermal energy (heat) that otherwise would be lost. The use of a regenerator helps increase the thermodynamic efficiency of a heat engine cycle.

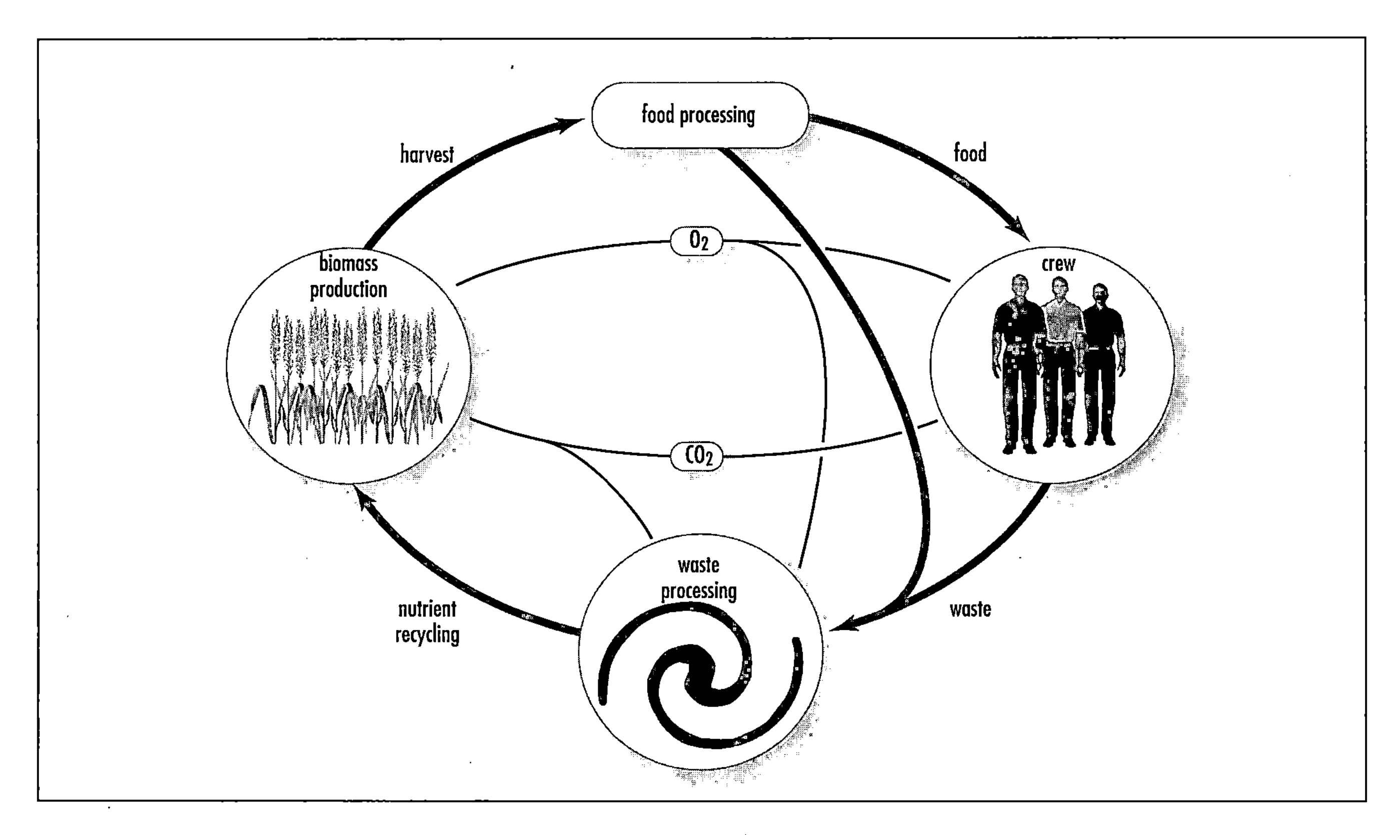

The basic components of a regenerative life support system ***(Drawing courtesy of NASA)***

regolith The lunar regolith is the unconsolidated mass of surface debris that overlies the Moon's bedrock. This blanket of pulverized lunar dust and soil was created by eons of meteoric and cometary impacts. The "fines" are the fraction of the lunar regolith containing particles that are less than 1 millimeter in diameter.

See also MOON.

regressive burning With respect to a solid-propellant rocket, the condition in which the thrust, pressure, or burning surface decreases with time or with the amount of web burned.

See also WEB.

Reinmuth, Karl (1892–1979) German *Astronomer* Karl Reinmuth was a German astronomer who discovered the 1.4-km-diameter asteroid that came within 10 million km (about 0.07 astronomical unit) of Earth in 1932. Called Apollo, this asteroid now gives its name to the Apollo group of near-Earth asteroids (NEAs) whose orbits cross Earth's orbit around the Sun.

See also APOLLO GROUP; ASTEROID; ASTEROID DEFENSE SYSTEM; NEAR-EARTH ASTEROID.

relative atomic mass (symbol: A) The total number of nucleons (that is, both protons and neutrons) in the nucleus of an atom. Also called the ATOMIC MASS or sometimes *atomic mass number.* For example, the relative atomic mass of the isotope carbon-12 is 12.

See also ATOMIC MASS UNIT.

relative state In aerospace operations, the position and motion of one spacecraft relative to another.

relativistic In general, pertaining to an object (including nuclear particles) moving at speeds that are an appreciable fraction of the speed of light (c).

relativity (theory) The theory of space and time developed by ALBERT EINSTEIN that has become one of the foundations of modern physics. Einstein's theory of relativity often is discussed in two general categories: the special theory of relativity, which he first proposed in 1905, and the general theory of relativity, which he presented in 1915.

The special theory of relativity is concerned with the laws of physics as seen by observers moving relative to one another at constant velocity—that is, by observers in nonaccelerating or inertial reference frames. Special relativity has been well demonstrated and verified by many types of experiments and observations.

Einstein proposed two fundamental postulates in formulating special relativity: (1) *First postulate of special relativity:* The speed of light (c) has the same value for all (inertial-reference-frame) observers, regardless and independent of the motion of the light source or the observers. (2) *Second postulate of special relativity:* All physical laws are the same for all observers moving at constant velocity with respect to one another.

The first postulate appears contrary to our everyday "Newtonian mechanics" experience, yet the principles of special relativity have been more than adequately validated in experiments. Using special relativity, scientists can now predict the space-time behavior of objects traveling at speeds from essentially zero up to those approaching that of light itself. At lower velocities the predictions of special relativity become identical with classical Newtonian mechanics. However, when we deal with objects moving close to the speed of light, we must use relativistic mechanics. What are some of the consequences of the theory of special relativity?

The first interesting relativistic effect is called *time dilation*. Simply stated—with respect to a stationary observer/clock—time moves more slowly on a moving clock system. This unusual relationship is described by the equation:

$$\Delta t = (1/\beta)\ \Delta T_p$$

where Δt is called the time dilation (the apparent slowing down of time on a moving clock relative to a stationary clock and observer) and ΔT_p is the "proper time" interval as measured by an observer and clock on the moving system.

$$\beta = \sqrt{[1 - (v^2/c^2)]}$$

where v is the velocity of the object and c is the velocity of light.

Let us now explore the time-dilation effect with respect to a postulated starship flight from our solar system. We start with twin brothers, Astro and Cosmo, who are both astronauts and are currently 25 years of age. Astro is selected for a special 40-year-duration starship mission, while Cosmo is selected for the ground control team. This particular starship, the latest in the fleet, is capable of cruising at 99 percent of the speed of light (0.99 c) and can quickly reach this cruising speed. During this mission Cosmo, the twin who stayed behind on Earth, has aged 40 years. (We are taking Earth as our fixed or stationary reference frame "relative" to the starship.) But Astro, who has been onboard the starship cruising the galaxy at 99 percent of the speed of light for the last 40 Earth-years, has aged just 5.64 Earth-years! Therefore, when he returns to Earth from the starship mission, he is a little over 30 years old, while his twin brother, Cosmo, is now 65 and retired in Florida. Obviously, starship travel (if we can overcome some extremely challenging technical barriers) also presents some very interesting social problems.

The time-dilation effects associated with near-light-speed travel are real, and they have been observed and measured in a variety of modern experiments. All physical processes (chemical reactions, biological processes, nuclear-decay phenomena, and so on) appear to slow down when in motion relative to a "fixed" or stationary observer/clock.

Another interesting effect of relativistic travel is *length contraction*. We first define an object's proper length (L_p) as its length measured in a reference frame in which the object is at rest. Then, the length of the object when it is moving (L)—as measured by a stationary observer—is always smaller, or contracted. The relativistic length contraction is given by

$$L = \beta\ (L_p)$$

This apparent shortening, or contraction, of a rapidly moving object is seen by an external observer (in a different inertial reference frame) only in the object's direction of motion. In the case of a starship traveling at near-light speeds, to observers on Earth this vessel would appear to shorten, or contract, in the direction of flight. If an alien starship were 1 kilometer long (at rest) and entered our solar system at an encounter velocity of 90 percent of the speed of light (0.9 c), then a terrestrial observer would see a starship that appeared to be about 435 meters long. The aliens on board and all their instruments (including tape measures) would look contracted to external observers but would not appear any shorter to those on board the ship (that is, to observers within the moving reference frame). If this alien starship were really "burning rubber" at a velocity of 99 percent of the speed of light (0.99 c), then its apparent contracted length to an observer on Earth would be about 141 meters. If, however, this vessel were just a "slow" interstellar freighter that was lumbering along at only 10 percent of the speed of light (0.1 c), then it would appear about 995 meters long to an observer on Earth.

Special relativity also influences the field of dynamics. Although the rest mass (m_o) of a body is invariant (does not change), its "relative" mass increases as the speed of the object increases with respect to an observer in another fixed or inertial reference frame. An object's relative mass is given by

$$m = (1/\beta)\ m_o$$

This simple equation has far-reaching consequences. As an object approaches the speed of light, its mass becomes infinite. Since things cannot have infinite masses, physicists conclude that material objects cannot reach the speed of light. This is basically the *speed-of-light barrier*, which appears to limit the speed at which interstellar travel can occur. From the theory of special relativity, scientists now conclude that only a "zero-rest-mass" particle, such as a PHOTON, can travel at the speed of light. There is one other major consequence of special relativity that has greatly affected our daily lives—the equivalence of mass and energy from Einstein's very famous formula:

$$E = \Delta m\ c^2$$

where E is the energy equivalent of an amount of matter (Δm) that is annihilated, or converted completely into pure energy, and c is the speed of light.

This simple yet powerful equation explains where all the energy in nuclear fission or nuclear fusion comes from. The complete annihilation of just one gram of matter releases about 9×10^{13} joules of energy.

In 1915 Einstein introduced his general theory of relativity. He used this development to describe the space-time relationships developed in special relativity for cases in which there was a strong gravitational influence such as white dwarf stars, neutron stars, and black holes. One of Einstein's conclusions was that gravitation is not really a force between two masses (as postulated in Newtonian mechanics) but rather arises as a consequence of the curvature of space-time. In our four-dimensional universe (x, y, z, and time), space-time becomes curved in the presence of matter, especially very massive, compact objects.

The fundamental postulate of general relativity is also called Einstein's *principle of equivalence*: The physical behavior inside a system in free fall is indistinguishable from the physical behavior inside a system far removed from any gravitating matter, that is, in the complete absence of a gravitational field.

Several experiments have been performed to confirm the general theory of relativity. These experiments have included observation of the bending of electromagnetic radiation (starlight and radio wave transmissions from various spacecraft missions, such as NASA's *Viking* spacecraft on Mars) by the Sun's immense gravitational field and recognizing the subtle perturbations (disturbances) in the orbit (at perihelion—the point of closest approach to the Sun) of the planet Mercury as caused by the curvature of space-time in the vicinity of the Sun. While some scientists do not think that these experiments have conclusively demonstrated the validity of general relativity, additional astronomical observations investigating phenomena such as neutron stars and black holes with more powerful space-based observatories and special experimental spacecraft, such as NASA's *Gravity Probe B,* continue to provide evidence in support of general relativity.

See also ASTROPHYSICS; BLACK HOLE; COSMOLOGICAL CONSTANT; *GRAVITY PROBE B;* SPACE-TIME.

reliability The probability of specified performance of a piece of equipment or system under stated conditions for a given period of time.

rem In radiation protection and nuclear technology, the traditional unit for dose equivalent (symbol: H). The dose equivalent in rem is the product of the absorbed dose (symbol: D) in rad and the quality factor (QF) as well as any other modifying factors considered necessary to characterize and evaluate the biological effects of ionizing radiation doses received by human beings or other living creatures. The term is an acronym derived from the expression *r*oentgen *e*quivalent *m*an. The rem is related to the SIEVERT, the SI unit of dose equivalent, as follows: 100 rem = 1 sievert.

See also DOSE EQUIVALENT; SPACE RADIATION ENVIRONMENT.

remaining body That part of an expendable rocket vehicle that remains after the separation of a fallaway section or companion body. In a multistage rocket, the remaining body diminishes in size as each section or part is cast away and successively becomes a different body.

remote control Control of an operation from a distance, especially by means of telemetry and electronics; a controlling switch, level, or other device used in this type of control, as in remote control arming switch.

See also TELEOPERATION.

remotely piloted vehicle (RPV) An aircraft or aerospace vehicle whose pilot does not fly on board but rather controls it at a distance (i.e., remotely) using a telecommunications link from a crewed aircraft, aerospace vehicle, or ground station. RPVs often are used on extremely hazardous missions or on long-duration missions involving extended loitering and surveillance activities. Also called an unmanned (or uncrewed) aerial vehicle (UAV).

remote manipulator system (RMS) The Canadian-built, 15.2-meter-long articulating arm that is remotely controlled from the aft flight deck of the space shuttle orbiter. The elbow and waist movements of the RMS permit payloads to be grappled for deployment out of the cargo bay or to be retrieved and secured in the cargo bay for on-orbit servicing or return to Earth. Because the RMS can be operated from the shirt-sleeve environment of the orbiter cabin, an extravehicular activity (EVA) often is not required to perform certain orbital operations.

See also SPACE TRANSPORTATION SYSTEM.

remote sensing The sensing of an object, event, or phenomenon without having the sensor in direct contact with the object being studied. Information transfer from the object to the sensor is accomplished through the use of the electromagnetic spectrum. Remote sensing can be used to study Earth in detail from space or to study other objects in the solar system, generally using flyby and orbiter spacecraft. Modern remote sensing technology uses many different portions of the electromagnetic spectrum, not just the visible portion we see with our eyes. As a result, often very different and very interesting "images" are created by these new remote sensing systems.

For example, the figure on page 501 is a radar image showing the volcanic island of Réunion, about 700 kilometers east of Madagascar in the southwest Indian Ocean. The southern half of the island is dominated by the active volcano Piton de la Fournaise. This is one of the world's most active volcanoes, with more than 100 eruptions in the last 300 years. The latest activity occurred in the vicinity of Dolomieu Crater, shown in the lower center of the image within a horseshoe-shaped collapse zone. The radar illumination is from the left side of the image and dramatically emphasizes the precipitous cliffs at the edges of the central canyons of the island. These canyons are remnants from the collapse of formerly active parts of the volcanoes that built the island. This image was acquired by the Spaceborne Imaging Radar-C/X-Band Synthetic Aperture Radar (SIR-C/X-SAR) flown aboard the space shuttle *Endeavour* on October 5, 1994. The SIR-C/X-SAR is part of the Earth system science program of the National Aeronautics and Space Administration (NASA). The radars illuminate Earth with microwaves, allowing detailed observations at any time regardless of weather or sunlight conditions. SIR-C/X-SAR, a joint mission of the German (DARA), Italian (ASI), and American (NASA) space agencies, uses three microwave wavelengths: L-band (24 cm), C-band (6 cm), and X-band (3 cm). The international scientific community is using these multifrequency radar imagery data to better understand the global environment and how it is changing.

Earth receives and is heated by energy in the form of electromagnetic radiation from the Sun. Some of this incoming solar radiation is reflected by the atmosphere, while most penetrates the atmosphere and subsequently is reradiated by atmospheric gas molecules, clouds, and the surface of Earth itself (including, e.g., oceans, mountains, plains, forests, ice sheets, and urbanized areas). All remote sensing systems (including those used to observe Earth from space) can be

Space-Age Archaeology

One unusual but very interesting application of space technology is in the field of archaeology. Archaeology is essentially the scientific study of the past through careful analysis of objects left behind. To study ancient civilizations, for example, archaeologists often explore crumbling ruins and monuments, excavate tombs, and sift through the soil at prehistoric campsites for bone fragments and other interesting "refuse items." Other important clues in deciphering the history and culture of early peoples are found along trading routes, at ancient springs and water holes, and even in the garbage dumps outside long-abandoned cities. At a promising dig archaeologists will carefully excavate and then patiently sift through a ton or so of soil in search of telltale broken bits of pottery and other items, including discarded food fragments. Digging is an integral part of archaeologists' fieldwork, so shovels and picks remain their primary investigative tool and provide the practical means by which these researchers can carefully search a promising archaeological site. Sometimes archaeologists use mesh screens to sift through mounds of rubble and isolate interesting objects, such as tiny shards of pottery, broken tools, or an occasional piece of bronze or copper. They then use a fine brush to gingerly dust and clean any precious finds.

However, modern archaeologists also employ space age technology to help them discover and unravel the past. Their high technology assistance includes satellite-based wireless communications from the world's most remote regions, the Global Position Satellite (GPS) System for precision navigation to and accurate location of isolated dig sites, and satellite-based remote sensing data to help find promising new dig sites. Weather satellites also support their field activities, and satellite-aided search and rescue is always available should the archaeological team encounter transportation difficulties while traveling to and from a site.

Experienced archaeologists know that the first critical step in reconstructing the past is the ability to successfully locate interesting new places to dig. One of the newest and most promising ways that modern archaeologists search for clues of promising sites is with space-based remote sensing. Remote sensing involves the sensing of an object, location, or phenomenon without having the sensor in direct contact with the object or location being studied. Telltale information about the object or site is transferred to the sensor by portions of the electromagnetic spectrum. Modern satellite-based remote sensing technology uses many portions of the electromagnetic spectrum, not just the very narrow visible portion we see with our eyes. So archaeologists take advantage of sophisticated imaging instruments on Earth-orbiting satellites to gain a unique regional perspective about a target area. Computer-assisted image analysis performed in a comfortable office environment allows them to detect and measure subtle environmental signatures that would otherwise be undetectable by a group of explorers on the ground. One of the big advantages of satellite-based remote sensing is that it extends human perceptual abilities. Another advantage is the ability to explore large portions of Earth's surface, including very inaccessible regions, efficiently and without risk.

The high-resolution (typically one to two meters ground spot size) images of Earth routinely collected by the current family of Earth-observing spacecraft provide today's archaeologists with an enormous amount of complementary data about promising sites. Before the scientists ever leave the comfort of their university or research institution, they have a fairly good idea about the physical characteristics of the candidate site and its surrounding region. Much like a military campaign, archaeologists use satellite data to construct high-resolution digital maps and three-dimensional renderings of a site, plan entrance and exit routes, and organize logistics operations for their expeditions to very remote, physically inhospitable regions. This saves time and money and reduces the overall risks of traveling into "uncharted territories."

In 1992 fairly low spatial resolution Landsat satellite imagery and data from NASA's space shuttle imaging radar system helped scientists locate the remains of the lost ancient city of Ubar (founded in about 2,800 B.C.E.) on the Arabian Peninsula in a remote part of the modern country of Oman. In ancient times Ubar served as the trailhead city in the southern Arabian Peninsula for caravans of frankincense-laden camels as they departed across the desert to ancient Greece, Rome, or Mesopotamia. The key that helped scientists locate this lost fortress city was the clever combination of visible and near-infrared images of the region taken by Landsat and radar images acquired during space shuttle imaging missions. Careful examination of combinations of these images revealed a regional network of tracks, some made by camel caravans that dated back more than two millennia. Radar images often reveal subsurface features that are not obvious in visible imagery or even to observers standing on the ground. The detection of networks of such subsurface features often leads to interesting archaeological finds.

For example, radar images from another space shuttle mission allowed archaeologists and other researchers to discover a previously unknown section of the ancient Cambodian city of Angkor. Since radar imagery can see through tree canopies, dense vegetation, clouds, and the dark of night, images of Angkor were able to show a network of irrigation canals north of the main temple area. The presence of these canals told researchers that the ancient city, which once housed more than a million people, extended farther north than previously thought. Modern archaeologists regard Angkor as one of the world's great archaeological ruins. However, the huge complex now largely lies hidden beneath dense jungle growth. The combination of dense jungle growth and sporadic political turbulence makes long-term research access to the site difficult. So space-based remote sensing serves as a valuable tool for archaeologists who wish to examine this important ancient city in Southeast Asia and help document how it grew, flourished, and then died over an 800-year period.

Today high-resolution visible, infrared, and radar satellite imagery provide archaeologists, adventurers, and researchers with a relatively inexpensive and safe "armchair" alternative for conducting exploratory probes of physically inaccessible or politically unstable regions of the world. Turkey's Mount Ararat, the reputed legendary resting place of Noah's Ark, represents just such an example. Because of political disturbances and wartime conditions

(continues)

Space-Age Archaeology *(continued)*

involving Kurdish guerrilla bands, the government of Turkey generally prohibits expeditions to this remote site. However, high-resolution commercial satellite imagery provides archaeologists tantalizing (but unconfirmed) hints of what could be the partially exposed portions of an alleged human-made structure embedded in the mountain's ice cap. Today's archaeologists still use picks, shovels, brushes, and wire mesh screens to unlock the past, but they also carry powerful laptop computers crammed with the latest satellite imagery of the promising site they are about to "dig."

This perspective view shows Mount Ararat in easternmost Turkey, which has been the site of several searches for the remains of Noah's Ark. The main peak, known as Great Ararat, is the tallest peak in Turkey, rising to 5,165 meters. This southerly, near-horizontal view also shows on the left the conically shaped peak known as Little Ararat. Both peaks are relatively young volcanoes from a geological standpoint, but their activity during historical times is uncertain. This interesting image was generated from a Landsat multispectral image acquired on August 31, 1989, draped over an elevation model produced by the Space Shuttle Radar Topography Mission flown by *Endeavour* in February 2000. *(Courtesy of NASA/JPL/NIMA)*

divided into two general classes: PASSIVE SENSORS and ACTIVE SENSORS. Passive sensors observe reflected solar radiation (or emissions characteristic of and produced by the target itself), while active sensors (such as a radar system) provide their own illumination on the target. Both passive and active remote sensing systems can be used to obtain images of the target or scene or else simply to collect and measure the total amount of energy (within a certain portion of the spectrum) in the field of view.

Passive sensors collect reflected or emitted radiation. Types of passive sensors include *imaging radiometers* and *atmospheric sounders.* Imaging radiometers sense the visible, near-infrared, thermal-infrared, or ultraviolet wavelength regions and provide an image of the object or scene being viewed. Atmospheric sounders collect the radiant energy emitted by atmospheric constituents, such as water vapor or carbon dioxide, at infrared and microwave wavelengths. These remotely sensed data are then used to infer temperature and humidity throughout the atmosphere.

Active sensors provide their own illumination (radiation) on the target and then collect the radiation reflected back by the object. Active remote sensing systems include *imaging radar, scatterometers,* RADAR ALTIMETERS, and *lidar altimeters.* An imaging radar emits pulses of microwave radiation from a radar transmitter and collects the scattered radiation to generate an image. (Look again at the radar image on page 501 for an example of a detailed radar image collected from space.) Scatterometers emit microwave radiation and sense the amount of energy scattered back from the surface over a wide field of view. These types of instruments are used to measure surface wind speeds and direction and to determine cloud content. Radar altimeters emit a narrow pulse of microwave energy toward the surface and accurately time the return pulse reflected from the surface, thereby providing a precise measurement of the distance (altitude) above the surface. Similarly, lidar altimeters emit a narrow pulse of laser light (visible or infrared) toward the surface and time the return pulse reflected from the surface.

Today remote sensing of Earth from space provides scientific, military, governmental, industrial, and individual users with the capacity to gather data to perform a variety of important tasks. These tasks include (1) simultaneously observing key elements of an interactive Earth system; (2) monitoring clouds, atmospheric temperature, rainfall, wind speed, and wind direction; (3) monitoring ocean surface temperatures and ocean currents; (4) tracking anthropogenic and natural changes to the environment and climate; (5) viewing remote or difficult-to-access terrain; (6) providing synoptic views of large portions of Earth's surface without being hindered by political boundaries or natural barriers; (7) allowing repetitive coverage of the same area over comparable viewing conditions to support change detection and long-term environmental monitoring; (8) identifying unique surface features (especially with the assistance of multispectral imagery); and (9) performing terrain analysis and measuring moisture levels in soil and plants.

Space-based remote sensing is a key technology in our intelligent stewardship of planet Earth, its resources, and its interwoven biosphere in this century and beyond. Monitoring of the weather and climate supports accurate weather forecasting and identifies trends in the global climate. Monitoring of the land surface assists in global change research, the management of known natural resources, the exploration for new resources (e.g., oil, gas, and minerals), detailed mapping, urban planning, agriculture, forest management, water resource assessment, and national security. Monitoring of the oceans helps determine such properties as ocean productivity, the extent of ice cover, sea-surface winds and waves, ocean currents and circulation, and ocean-surface temperatures. These types of ocean data have particular value to scientists as well as to the fishing and shipping industries.

In 1999 two successful satellite launches dramatically changed the civilian (that is, publicly accessible) use of remote sensing of Earth from space. Since September 1999 and the launch of Space Imaging's *Ikonos* satellite, excellent quality, high-resolution (one-meter or better) imagery from space has become commercially available from a private firm. Prior to that launch, there was a clear distinction between tightly controlled, high-resolution military imaging (photo reconnaissance) satellites and civilian (government-owned and -sponsored) Earth observing satellites (such as the Landsat family of spacecraft) that collected openly available but lower-resolution multispectral images of Earth. A presidential directive in March 1994 (called PDD-23) blurred that long-standing distinction within the U.S. government.

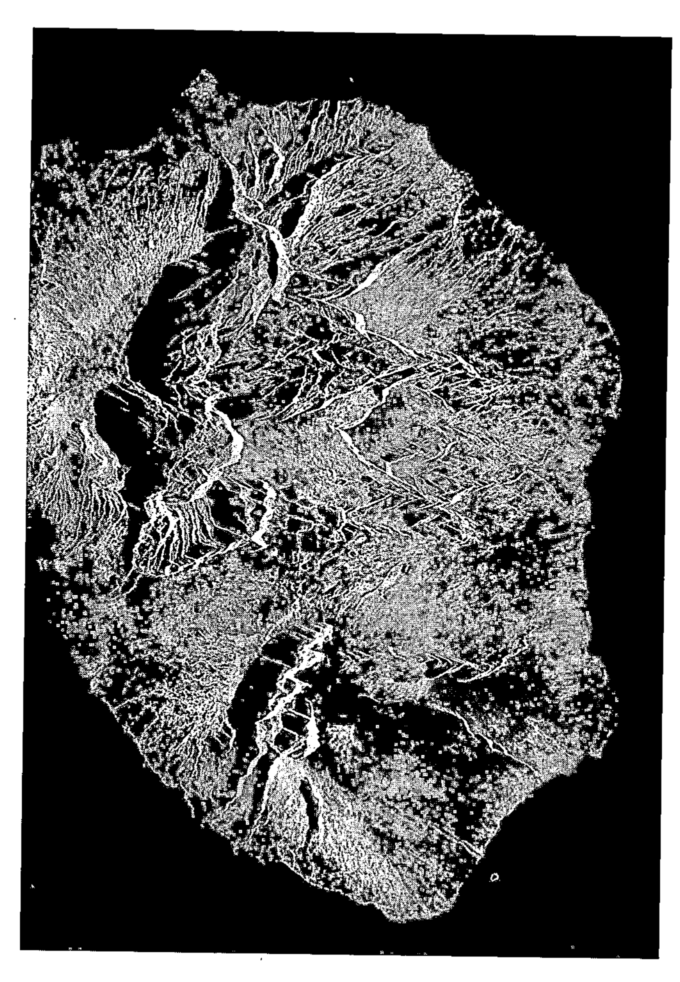

A radar image (C and X band) of the volcanic island of Réunion, which lies about 700 kilometers east of Madagascar in the southwest Indian Ocean. This image was acquired on October 5, 1994, by NASA's Spaceborne Imaging Radar-C/X Band Synthetic Aperture Radar (SIR-C/X-SAR) system, flown onboard the space shuttle *Endeavour.* *(Courtesy of NASA/JPL)*

The launch of another nonmilitary Earth-observing satellite, called *Terra,* also helped to accelerate the growth of the orbiting information revolution, an ongoing process sometimes referred to as the transparent globe. In December 1999 NASA successfully placed the TERRA SPACECRAFT into orbit around Earth. *Terra* carries a payload of five state-of-the-art sensors that are simultaneously collecting data about Earth's atmosphere, lands, oceans, and solar energy balance.

NASA further accelerated the use of remote sensing in support of Earth system science with the successful launch of the *AQUA* SPACECRAFT on May 4, 2002 and the *AURA* SPACECRAFT on July 15, 2004. This sophisticated Earth-observing spacecraft is a technical sibling to the *Terra* spacecraft. The *Aqua* spacecraft carries six state-of-the-art remote sensing instruments that simultaneously collect data about the role and movement of water in the Earth system and a suite of scientific instruments that are investigating Earth's ozone, air quality, and climate. The first and second missions, *Terra* and *Aqua,* are designed to study the land, oceans, and Earth's radiation budget. *Aura's* atmospheric chemistry measurements are following up on measurements that began with NASA's *UPPER ATMOSPHERE RESEARCH SATELLITE* (UARS) and also continuing the record of satellite ozone data collected from the total ozone mapping spectrometers (TOMS) aboard the *Nimbus 7* and other satellites.

See also EARTH SYSTEM SCIENCE; GLOBAL CHANGE.

rendezvous The close approach of two or more spacecraft in the same orbit so that docking can take place. These objects meet at a preplanned location and time with essentially zero relative velocity. A rendezvous would be involved, for example, in the construction, servicing, or resupply of a space station or when the space shuttle orbiter performs on-orbit repair or servicing of a satellite. The term is also applied to a space mission, such as the Near Earth Asteroid Rendezvous (NEAR) mission, in which a scientific spacecraft is maneuvered so as to fly alongside a target body, such as a comet or asteroid, at zero relative velocity.

See also NEAR EARTH ASTEROID RENDEZVOUS (NEAR) MISSION.

repressurization Sequence of operations during a rocket vehicle or spacecraft flight that uses an onboard pressurant supply to restore the ullage pressure to the desired level after an engine burn period.

resilience The property of a material that enables it to return to its original shape and size after deformation. For example, the resilience of a sealing material is the property that makes it possible for a seal to maintain sealing pressure despite wear, misalignment, or out-of-round conditions. This term is also applied in aerospace operations to describe the relative hardiness or "robustness" of a robot spacecraft or crew-occupied space vehicle that can suffer several significant component or subsystem degradations or failures and still function (through hardware and software "workarounds" and automated fault isolation procedures) at a performance level sufficient to continue and/or complete the mission.

resistance (symbol: R) 1. Electrical resistance (R) is defined as the ratio of the voltage (or potential difference) (V) across a conductor to the current (I) flowing through it. In accordance with Ohm's law, $R = V/I$. The SI unit of resistance is the ohm (Ω), where 1 ohm = 1 volt per ampere. 2. Mechanical resistance is the opposition by frictional effects to forces tending to produce motion. 3. Biological resistance is the ability of plants and animals to withstand poor environmental conditions and/or attacks by chemicals or disease. This ability may be inborn or developed.

Resnik, Judith A. (1949–1986) American *Astronaut* NASA astronaut Judith A. Resnik successfully traveled into orbit around Earth as a mission specialist during the space shuttle *Discovery*'s STS 41-D mission (1984). Then, while attempting her next space flight, she lost her life as the space shuttle *Challenger* exploded during launch ascent on January 28, 1986, at the start of the STS 51-L mission. This fatal accident claimed the lives of all seven astronauts onboard and destroyed the *Challenger* space vehicle.

See also *CHALLENGER* ACCIDENT.

resolution 1. In general, a measurement of the smallest detail that can be distinguished by a sensor system under specific conditions. 2. The degree to which fine details in an image or photograph can be seen as separated or resolved. *Spatial resolution* often is expressed in terms of the most closely spaced line-pairs per unit distance that can be distinguished. For example, when the resolution is said to be 10 line-pairs per millimeter, this means that a standard pattern of black-and-white lines whose line plus space width is 0.1 millimeter is barely resolved by an optical system, finer patterns are not resolved, and coarser patterns are more clearly resolved.

See also REMOTE SENSING.

resolving power In general, the finest detail an optical instrument can provide. With respect to remote sensing, a measure of the ability of individual components or the entire remote sensing system to define (and therefore "resolve") closely spaced targets.

See also AIRY DISK.

restart In aerospace operations, the act of firing a spacecraft's or space vehicle's "restartable" rocket engine after a previous thrust period (powered flight) and a coast phase in a parking or transfer orbit.

rest mass (symbol: m_0) The mass that an object has when it is at (absolute) rest. From ALBERT EINSTEIN's special relativity theory, the mass of an object increases when it is in motion according to the formula

$$m = \frac{m_0}{\sqrt{1-[v^2/c^2]}}$$

where m is its mass in motion, m_0 is its mass at rest, v is the speed of the object, and c is the speed of light. This is especially significant at speeds approaching a considerable fraction of the speed of light. Newtonian physics, in contrast, makes no distinction between rest mass (m_0) and motion mass (m). This is acceptable in classical physics as long as the object under consideration is traveling at speeds well below the speed of light—that is, when $v < c$.

See also RELATIVITY (THEORY).

restricted propellant A solid propellant that has only a portion of its surface exposed for burning; the other propellant surfaces are covered by an inhibitor.

restricted surface The surface of a solid-propellant grain that is prevented from burning through the use of inhibitors.

retrieval With respect to the space shuttle, the process of using the remote manipulator system (RMS) and/or other handling aids to return a captured payload to a stowed or berthed position. No payload is considered retrieved until it is fully stowed for safe return or berthed for repair and maintenance.

See also REMOTE MANIPULATOR SYSTEM; SPACE TRANSPORTATION SYSTEM.

retrofire To ignite a retrorocket.

retrofit The modification of or addition to a spacecraft, aerospace vehicle, or expendable launch vehicle after it has become operational.

retrograde In a reverse or backward direction.

retrograde motion Motion in an orbit opposite to the usual orbital direction of celestial bodies within a given system. Specifically, for a satellite, motion in a direction opposite to the direction of rotation of the primary.

retrograde orbit An orbit having an inclination of more than 90°. In a retrograde orbit the spacecraft or satellite has a motion that is opposite in direction to the rotation of the primary body (such as Earth). *Compare with* PROGRADE ORBIT.

See also ORBITS OF OBJECTS IN SPACE.

retroreflection Reflection in which the reflected rays return along paths parallel to those of their corresponding incident rays.

retroreflector A mirrorlike instrument, usually a corner reflector design, that returns light or other electromagnetic radiation (e.g., an infrared laser beam) in the direction from which it comes.

See also LASER GEODYNAMICS SATELLITE.

retrorocket A small rocket engine on a satellite, spacecraft, or aerospace vehicle used to produce a retarding thrust or force that opposes the object's forward motion. This action reduces the system's velocity.

reusable launch vehicle (RLV) A launch vehicle that incorporates simple, fully reusable designs for airline-type operations using advanced technology and innovative operational techniques.

See also X-33.

reverse thrust Thrust applied to a moving object in a direction to oppose the object's motion; used to decelerate a spacecraft or aerospace vehicle.

reversible process In thermodynamics, a process that goes in either the forward or the reverse direction without violating the second law of thermodynamics; a constant entropy process. For example, the entropy of a control mass experiencing a reversible, adiabatic process will not change. Such a constant entropy process is also called an isentropic process. However, the "reversible process" is just an idealization used in engineering analyses to help approximate the behavior of certain real world systems. Some of the more commonly encountered reversible process idealizations are frictionless motion, current flow without resistant, and frictionless pulleys. In contrast, irreversible processes, such as molecular diffusion, a spontaneous chemical reaction, or motion with friction all produce entropy.

See also ENTROPY; THERMODYNAMICS.

revetment In aerospace operations and safety, a wall of concrete, earth, or sandbags installed for the protection of launch crew personnel and equipment against the blast and flying shrapnel from an exploding rocket during a launch abort near the pad.

revolution 1. Orbital motion of a celestial body or spacecraft about a center of gravitational attraction, such as the Sun or a planet, as distinct from rotation around an internal axis. For example, Earth revolves around the Sun annually and rotates daily about its axis. 2. One complete cycle of the movement of a spacecraft or a celestial object around its primary.

See also ORBITS OF OBJECTS IN SPACE.

Rhea A major satellite of Saturn, similar to but somewhat larger than Dione, discovered in 1672 by GIOVANNI DOMENICO CASSINI. Rhea has a diameter of 1,530 km and orbits Saturn at a distance of 527,040 km with an orbital period of 4.52 days. With a density of 1.2 g/cm^3 (1,200 kg/m^3), Rhea probably consists mostly of water ice. The leading edge of this Saturnian moon is uniformly bright and heavily cratered, while its trailing edge has shown few visible craters (based on *Voyager* spacecraft imagery analysis). The *Cassini* spacecraft is scheduled to flyby Rhea on November 26, 2005 at an altitude of 500 km. This close encounter should provide a large amount of interesting new information about Rhea, including one issue puzzling planetary scientists: Why are Dione and Rhea so similar with respect to their surface features, even though Rhea is almost twice as massive? Also called *Saturn V.*

See also SATURN.

Richer, Jean (1630–1696) French *Astronomer* In 1671 the French astronomer Jean Richer led a scientific expedition to Cayenne, French Guiana. This expedition made careful astronomical measurements of Mars at the same time GIOVANNI DOMENICO CASSINI was performing similar observations in France. These simultaneous observations supported the first adequate parallax of Mars and enabled Cassini to calculate the distance from Earth to Mars. This work led to a significant improvement in astronomical knowledge concerning the dimensions of the solar system. In fact, Cassini's values closely approached modern values. On this expedition Richer also made careful geophysical observations with a pendulum. These observations eventually allowed SIR ISAAC NEWTON to discover the oblateness of Earth.

See also OBLATENESS; PARALLAX; SOLAR SYSTEM.

Ride, Sally K. (1951–) American *Astronaut* NASA astronaut Sally K. Ride was the first American woman to travel in outer space. She did this as a crewmember (mission specialist position) on the STS-7 mission of the space shuttle *Challenger,* which lifted off from the Kennedy Space Center on June 18, 1983. She flew again in space as a crewmember on the STS 41-G mission of the *Challenger* (1984). Then in January 1986 Ride was selected to serve as a member of the presidential commission that investigated the *Challenger* accident of January 28, 1986.

rift valley A depression in a planet's surface due to crustal mass separation.

Rigel A very conspicuous, massive blue-white supergiant star of spectral type B. Rigel is also called Beta Orionis (B Ori). If Rigel is about 1,400 light-years away, as some astronomers estimate, then this star would have a luminosity on the order of 150,000 times the luminosity of the Sun.

See also STAR.

right ascension *See* CELESTIAL COORDINATES; KEPLERIAN ELEMENTS.

rille A deep, narrow depression on the lunar surface that cuts across all other types of topographical features on the Moon. From the German word *Rille,* meaning groove. Planetary geologists also use this term to describe similar features on other solar system bodies.

ring (planetary) A disk of matter that encircles a planet. Such rings usually contain ice and dust particles ranging in size from microscopic fragments up to chunks that are tens of meters in diameter.

ringed world A planet with a ring or set of rings encircling it. In the solar system Jupiter, Saturn, Uranus, and Neptune all have ring systems of varying degrees of composition and complexity. This indicates that ring systems may be a common feature of Jupiter-sized planets circling other stars.

Astronomers speculate that there are three general ways such ring systems are formed around a planet. The first is meteoroidal bombardment of a large body, so that the fragments inside the planet's ROCHE LIMITS form a ring. The second is condensation of material inside the Roche limit when the planet was forming. This trapped nebular material can neither join the parent planet nor form a moon or satellite. Finally, ring systems can also form when a satellite or other celestial object (for example, a comet) comes within the planet's Roche limit and gets torn apart by tidal forces.

The Roche limit, named after the French mathematician, EDOUARD ROCHE, who formulated this relationship in the 19th century, is the critical distance from the center of a massive celestial object (or planet) within which tidal forces will tear apart a moon or satellite. If we assume that (1) both celestial objects have the same density and (2) that the moon in question is held together only by gravitational attraction of its matter, then the Roche limit is typically about 1.2 times the diameter of the parent planet or primary celestial body.

See also JUPITER; NEPTUNE; SATURN; URANUS.

ring seal Piston-ring type of seal that assumes its sealing position under the pressure of the fluid to be sealed.

robotics in space Robotics is basically the science and technology of designing, building, and programming robots. Robotic devices, or robots as they are usually called, are primarily "smart machines" with manipulators that can be programmed to do a variety of manual or human labor tasks automatically. A robot, therefore, is simply a machine that does mechanical, routine tasks on human command. The expression *robot* is attributed to Czech writer Karel Čapek, who wrote the play *R.U.R.* (*Rossum's Universal Robots*). This play first appeared in English in 1923 and is a satire on the mechanization of civilization. The word *robot* is derived from *robata,* a Czech word meaning "compulsory labor" or "servitude."

A typical robot consists of one or more manipulators (arms), end effectors (hands), a controller, a power supply, and possibly an array of sensors to provide information about the environment in which the robot must operate. Because most modern robots are used in industrial applications, their classification is based on these industrial functions. Terrestrial robots frequently are divided into the following classes: nonservo (or pick-and-place), servo, programmable, computerized, sensory, and assembly robots.

The *nonservo robot* is the simplest type. It picks up an object and places it at another location. The robot's freedom of movement usually is limited to two or three directions.

The *servo robot* represents several categories of industrial robots. This type of robot has servo-mechanisms for the manipulator and end effector to enable it to change direction in midair (or midstroke) without having to trip or trigger a mechanical limit switch. Five to seven directions of motion are common, depending on the number of joints in the manipulator.

The *programmable robot* is essentially a servo robot that is driven by a programmable controller. This controller memorizes (stores) a sequence of movements and then repeats these movements and actions continuously. Often engineers program this type of robot by "walking" the manipulator and end effector through the desired movements.

The *computerized robot* is simply a servo robot run by computer. This kind of robot is programmed by instructions fed into the controller electronically. These "smart robots" may even have the ability to improve upon their basic work instructions.

The *sensory robot* is a computerized robot with one or more artificial senses to observe and record its environment and to feed information back to the controller. The artificial senses most frequently employed are sight (robot or computer vision) and touch.

Finally, the *assembly robot* is a computerized robot, generally with sensors, that is designed for assembly line and manufacturing tasks, both on Earth and eventually in space.

In industry robots are designed mainly for manipulation purposes. The actions that can be produced by the end effector or hand include (1) motion (from point to point along a desired trajectory or along a contoured surface), (2) a change in orientation, and (3) rotation.

Nonservo robots are capable of point-to-point motions. For each desired motion the manipulator moves at full speed until the limits of its travel are reached. As a result, nonservo robots often are called "limit sequence," "bang-bang," or "pick-and-place" robots. When nonservo robots reach the end of a particular motion, a mechanical stop or limit switch is tripped, stopping the particular movement.

Servo robots are also capable of point-to-point motions, but their manipulators move with controlled variable velocities and trajectories. Servo robot motions are controlled without the use of stop or limit switches.

Four different types of manipulator arms have been developed to accomplish robot motions. These are the rectangular, cylindrical, spherical, and anthropomorphic (articulated or jointed arm). Each of these manipulator arm designs features two or more degrees of freedom (DOFs), a term that refers to the directions in which a robot's manipulator arm is able to move. For example, simple straight line or linear movement represents one DOF. If the manipulator arm is to follow a two-dimensional curved path, it needs two degrees of freedom: up and down and right and left. Of course, more complicated motions will require many degrees of freedom. To locate an end effector at any point and to orient this effector in a particular work volume require six DOFs. If the manipulator arm needs to avoid obstacles or other equipment, even more degrees of freedom will be required. For each DOF, one linear or rotary joint is needed. Robot designers sometimes combine two or more of these four basic manipulator arm configurations to increase the versatility of a particular robot's manipulator.

Actuators are used to move a robot's manipulator joints. Three basic types of actuators are used in contemporary robots: pneumatic, hydraulic, and electrical. Pneumatic actuators employ a pressurized gas to move the manipulator joint. When the gas is propelled by a pump through a tube to a particular joint, it triggers or actuates movement. Pneumatic actuators are inexpensive and simple, but their movement is not precise. Therefore, this kind of actuator usually is found in nonservo or pick-and-place robots. Hydraulic actuators are quite common and capable of producing a large amount of power. The main disadvantages of hydraulic actuators are their accompanying apparatus (pumps and storage tanks) and problems with fluid leaks. Electrical actuators provide smoother movements, can be controlled very accurately, and are very reliable. However, these actuators cannot deliver as much power as hydraulic actuators of comparable mass. Nevertheless, for modest power actuator functions, electrical actuators often are preferred.

Many industrial robots are fixed in place or move along rails and guideways. Some terrestrial robots are built into wheeled carts, while others use their end effectors to grasp handholds and pull themselves along. Advanced robots use articulated manipulators as legs to achieve a walking motion.

A robot's end effector (hand or gripping device) generally is attached to the end of the manipulator arm. Typical functions of this end effector include grasping, pushing and pulling, twisting, using tools, performing insertions, and various types of assembly activities. End effectors can be mechanical, vacuum, or magnetically operated and can use a snare device or have some other unusual design feature. The final design of the end effector is determined by the shapes of the objects that the robot must grasp. Usually most end effectors are some type of gripping or clamping device.

Robots can be controlled in a wide variety of ways, from simple limit switches tripped by the manipulator arm to sophisticated computerized remote sensing systems that provide machine vision, touch, and hearing. In the case of computer-controlled robots, the motions of the manipulator and end effector are programmed, that is, the robot "memorizes" what it is supposed to do. Sensor devices on the manipulator help to establish the proximity of the end effector to the object to be manipulated and feed information back to the computer controller concerning any modifications needed in the manipulator's trajectory.

Another interesting type of robot system, the *field robot,* has become practical recently. A field robot is a robot that operates in unpredictable, unstructured environments, typically outdoors (on Earth), and often operates autonomously or by teleoperation over a large workspace (typically a square kilometer or more). For example, in surveying a potentially dangerous site, a human operator will stay at a safe distance away in a protected work environment and control by cable or radio frequency link the field robot, which then actually operates in the hazardous environment. These terrestrial field robots can be considered "technological first cousins" to the more sophisticated teleoperated planetary rovers that roam the Moon, Mars, and other planetary bodies.

Robotic systems have played and will continue to play a major role in the exploration and development of the solar system. Early American space robots included NASA's *Surveyor* spacecraft, which soft-landed on the lunar surface starting in 1966, operating soil scoops and preparing the way for the Apollo astronauts (*see* figure), and NASA's *Viking* lander spacecraft, which explored and examined the Martian surface for signs of microbial life-forms starting in 1976. The Soviet Lunokhod remotely controlled robot moon rovers

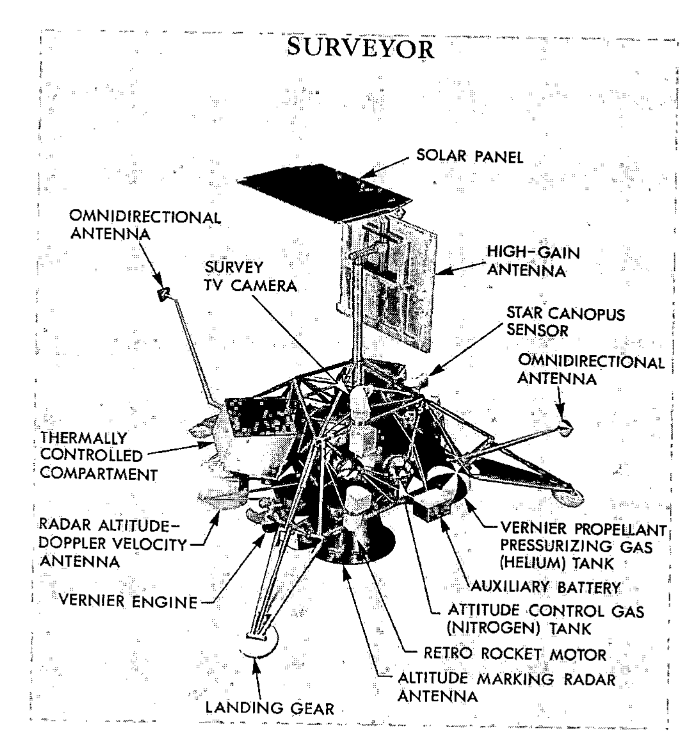

NASA's *Surveyor* spacecraft performed robotic exploration of the lunar surface from 1966 to 1968 in preparation for the landings by the Apollo astronauts (1969–72). ***(Courtesy of NASA)***

roamed across the lunar surface in the early 1970s, conducting numerous experiments and soil property investigations.

The space shuttle orbiter's remote manipulator system (RMS) was first flown in 1981. This versatile "robot arm" was designed to handle spacecraft deployment and retrieval operations, as well as to permit the assembly of large structures (e.g., a permanent space station). It is installed in the space shuttle orbiter's port (left) side cargo bay door hinges and is operated by a "shirt-sleeved" astronaut from inside the crew cabin. The RMS was designed and built by the National Research Council of Canada. It is a highly sophisticated robotic device that is similar to a human arm. The 15-meter-long RMS features a shoulder, wrist, and hand, although its "hand" does not look like a human hand. The skeleton of this mechanical arm is made of lightweight graphite composite materials. Covering the skeleton are skin layers consisting of thermal blankets. The muscles driving the joints are electric actuators (motors). Built-in sensors act like nerves and sense joint positions and rotation rates.

The RMS includes two closed-circuit television cameras, one at the wrist and one at the elbow. These cameras allow an astronaut who is operating the RMS from the orbiter's aft flight deck to see critical points along the arm and the target toward which the arm is moving.

The INTERNATIONAL SPACE STATION (ISS), currently under development in low Earth orbit, when fully operational, will have several robots to help astronauts and cosmonauts complete their tasks in space. Japan is developing a remote manipulator system for the Japanese Experiment Module (JEM). The European Space Agency and the Russian Space Agency are developing the European robotic arm. Finally, building upon space shuttle RMS technology and experience, Canada and the United States are developing the space station's mobile servicing system (MSS). The MSS is the most complex robotic system on the ISS. It consists of the space station remote manipulator system (SSRMS), the mobile remote servicer base system (MBS), the special purpose dexterous manipulator (SPDM), and the mobile transporter (MT). The MSS is controlled by an astronaut or cosmonaut at one of two robotics workstations inside the ISS. The primary functions of the MSS robotic system on the ISS are to assist in the assembly of the main elements of the station, to handle large payloads, to exchange orbital replacement units, to support astronaut extravehicular activities, to assist in station maintenance, and to provide transportation around the station outside the pressurized habitable environment.

Expanding on its previous space robot experience, NASA is now pursuing "intelligent teleoperation" technology as an interim step toward developing the technology needed for truly autonomous space robots. The goals of NASA's current space telerobotics program are to develop, integrate, and demonstrate the science and technology of remote telerobotics, leading to increases in operational capability, safety, cost effectiveness, and probability of success of NASA missions. Space telerobotics technology requirements can be characterized by the need for manual and automated control, nonrepetitive tasks, time delay between operator and manipulator, flexible manipulators with complex dynamics, novel locomotion, operations in the space environment, and the ability to recover from unplanned events. There are three specific areas of focus: on-orbit servicing, science payload maintenance, and exploration robotics.

The on-orbit servicing telerobotics program is concerned with the development of space robotics for eventual application to on-orbit satellite servicing by both free-flying and platform attached servicing robots. Relevant technologies include virtual reality telepresence, advanced display technologies, proximity sensing for perception technologies, and robotic flaw detection. Potential mission applications include repair of free-flying small satellites, ground-based control of robotic servicers, and servicing of external space platform payloads (including payloads externally mounted on the *International Space Station*).

The science payload maintenance telerobotics program is intended to mature technologies for robotics that will be used inside pressurized living space to maintain and service payloads. Once developed, this telerobotic capability will off-load the requirements of intensive astronaut maintenance of these payloads and permit the operation of the payloads during periods when the astronauts may not be present. Relevant technologies include lightweight manipulators, redundant robotic safety systems, and self-deploying systems. One particular mission application involves intravehicular activity (IVA) robots for the ISS.

The exploration robotics program involves the development of robot systems for surface exploration of the Moon and Mars, including robotic reconnaissance and surveying systems as precursors to eventual human missions. During such surface exploration missions these robots will explore potential landing sites and areas of scientific interest, deploy scientific instruments, gather samples for in-situ analysis or possible return to Earth, acquire and transmit video imagery, and provide the images needed to generate "virtual environments" of the lunar and Martian surfaces. The robotic systems for these operations will need high levels of local autonomy, including the ability to perform local navigation, identify areas of potential scientific interest, regulate on-board resources, and schedule activities—all with limited ground command intervention.

For example, NASA's planned *Phoenix Mars Scout* will land in icy soils near the north polar permanent ice cap of the Red Planet and explore the history of water in these soils and any associated rocks. (*See* figure.) This sophisticated space robot will serve as NASA's first exploration of a potential modern habitat on Mars and open the door to a renewed search for carbon-bearing compounds, last attempted with the Viking lander spacecraft missions in the 1970s. The *Phoenix* spacecraft is currently in development and will launch in August 2007. The robot explorer will land in May 2008 at a candidate site in the Martian polar region previously identified by the *Mars Odyssey* orbiter spacecraft as having high concentrations of ice just beneath the top layer of soil. The *Phoenix* spacecraft's stereo color camera and a weather station will study the surrounding environment, while its other instruments will check excavated soil samples for water, organic chemicals, and conditions that could indicate whether the site was ever hospitable to life. Of special interest to exobiologists, the spacecraft's microscopes will reveal features as small as one one-thousandth the width of a human hair.

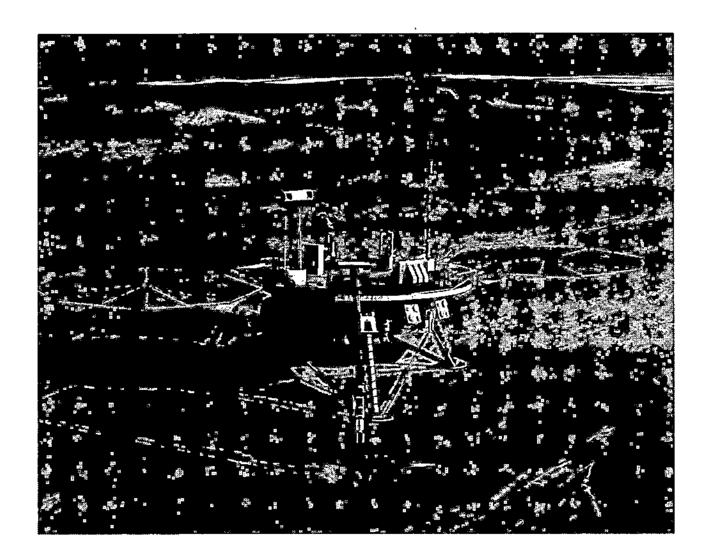

This artist's rendering shows NASA's proposed "scout-class" robot lander spacecraft, called *Phoenix,* deployed on the surface of Mars ca. 2008. The spacecraft would use its robotic arm to dig into a spot in the water-ice–rich northern polar region of Mars for clues on the planet's history of water. The robot explorer would also search for environments suitable for microscopic organisms (microbes). *(Courtesy of NASA)*

In this artist's rendering NASA's planned Mars Science Laboratory (MSL) travels near a canyon on the Red Planet in 2010. With a greater range than any previous Mars rover, the MSL will analyze dozens of samples scooped up from the soil and cored from rocks at scientifically interesting locations on the planet. One of the primary objectives of this sophisticated robot explorer is to investigate the past or present ability of Mars to support life. ***(Artwork courtesy of NASA/JPL)***

The science rover/remote geologist robot represents a 20-kilogram future microrover that can autonomously traverse many kilometers on the surface of Mars and perform scientist-directed experiments and then return relevant data back to Earth. NASA's proposed Mars Science Laboratory (MSL) is an example of a sophisticated extraterrestrial field robot.

An *aerobot* is an autonomous robotic aerovehicle (e.g., a free-flying balloon or a specially designed "aeroplane") that is capable of flying in the atmospheres of Venus, Mars, Titan, or the outer planets. For Martian or Venusian aerobots the robotic system would be capable of one or more of the following activities: autonomous state determination, periodic altitude variations, altitude control and the ability to follow a designated flight path within a planetary atmosphere using prevailing planetary winds, and landing at a designated surface location.

Future exploration of interplanetary small bodies, such as comets and asteroids, will require telerobotic technology developments in a variety of areas. Landing and surface operations in the very low-gravity environment of small interplanetary bodies (where the acceleration due to gravity is typically 10^{-4} to 10^{-2} meter per second-squared) is an extremely challenging problem. The robotic comet or asteroid explorer must have mechanisms and autonomous control algorithms to perform landing, anchoring, surface/subsurface sampling, and sample manipulation for a complement of scientific instruments. A robotic lander might use, for example, crushable material on the underside of a base plate designed to absorb almost all of the landing kinetic energy. An anchoring, or attachment, system then would be used to secure the lander and compensate for the reaction forces and moments generated by the sample acquisition mechanisms. For example, the European Space Agency's *Rosetta* robotic spacecraft is now traveling through interplanetary space and is scheduled to rendezvous with Comet 67P/Churyumov-Gerasimenko, drop a probe on the surface of the comet's nucleus, and study the comet from orbit.

Recent advances in microtechnology and mobile robotics have made it possible to consider the creation and use of extremely small automated and remote-controlled vehicles, called nanorovers, in planetary surface exploration missions. A nanorover is a robotic vehicle with a mass of between 10 and 50 grams. One or several of these tiny robots could be used to survey areas around a lander and to look for a particular substance, such as water ice or microfossils. The nanorover would then communicate its scientific findings back to Earth via the lander spacecraft (possibly in conjunction with an orbiting mother spacecraft).

Eventually, space robots will achieve higher levels of artificial intelligence, autonomy, and dexterity, so that servicing and exploration operations will become less and less dependent on a human operator being present in the control loop.

These robots will be capable of interpreting very high-level command structures and executing commands without human intervention. Erroneous command structures, incomplete task operations, and the resolution of differences between the robot's built-in "world model" and the real-world environment it is encountering will be handled autonomously. This is especially important as more sophisticated robots are sent deeper into the outer solar system and telecommunications time delays of minutes become hours. This higher level of autonomy will also be very important in the development and operation of permanent lunar or Martian surface bases, where smart machines will become our permanent partners in the development of the space frontier.

These increasingly more sophisticated space robots will have working lifetimes of decades with little or no maintenance. Some space planners envision robots capable of repairing themselves or other robots—again with little or no direct human supervision. The brilliant Hungarian-American mathematician JOHN VON NEUMANN was the first person to seriously explore the concept of the mechanical SELF-REPLICATING SYSTEM (SRS)—that is, a robot smart and dexterous enough to make copies of itself. From von Neumann's work and the more recent work of other investigators, five broad classes of SRS behavior have been suggested:

1. *Production.* The generation of useful output from useful input. In the production process the machine remains unchanged. Production is a simple behavior demonstrated by all working machines, including SRS devices.
2. *Replication.* The complete manufacture of a physical copy of the original machine by the machine itself.
3. *Growth.* An increase in the mass of the original machine by its own actions while still retaining the integrity of its original design. For example, the machine might add an additional set of storage compartments in which to keep a larger supply of parts or constituent materials.
4. *Evolution.* An increase in the complexity of the machine's function or structure. This is accomplished by additions or deletions to existing subsystems or by changing the characteristics of these subsystems.
5. *Repair.* Any operation performed by a machine on itself that helps reconstruct, reconfigure, or replace existing subsystems but does not change the SRS unit population, the original unit mass, or its functional complexity.

In theory, such replicating systems can be designed to exhibit any or all of these machine behaviors. When such machines are actually built (perhaps in the mid- to late 21st century), a particular SRS unit most likely will emphasize just one or several kinds of machine behavior, even if it were capable of exhibiting all of them. For example, a particular SRS unit might be the fully autonomous, general-purpose, self-replicating lunar factory that first makes a sufficient number of copies of itself and then sets about harvesting lunar resources and converting these resources into products needed to support a permanently inhabited (by humans) lunar settlement.

Once we have developed the sophisticated robotic devices needed for the detailed investigation of the outer regions of the solar system, the next step becomes quite obvious. Sometime late in the 21st century, humankind will build and launch its first fully automated robot explorer to a nearby star system. This interstellar explorer will be capable of searching for extrasolar planets around other suns, targeting any suitable planets for detailed investigation, and then initiating the search for extraterrestrial life. Light-years away, terrestrial scientists will wait for faint distant radio signals by which the robot starship describes any new worlds it has encountered, perhaps shedding light on the greatest cosmic mystery of all: Does life exist elsewhere in the universe?

The early development of SRS technology for use on Earth and in space should trigger an *era of superautomation* that will transform most terrestrial industries and lay the foundation for efficient space-based industries. One interesting machine is called the *Santa Claus machine,* originally suggested and named by the American physicist Theodore Taylor (1925–2004). In this particular version of an SRS unit, a fully automatic mining, refining, and manufacturing facility gathers scoopfuls of terrestrial or extraterrestrial materials. It then processes these raw materials by means of a giant mass spectrograph that has huge superconducting magnets. The material is converted into an ionized atomic beam and sorted into stockpiles of basic elements, atom by atom. To manufacture any item, the Santa Claus machine selects the necessary materials from its stockpile, vaporizes them, and injects them into a mold that changes the materials into the desired item. Instructions for manufacturing, including directions on adapting new processes and replication, are stored in a giant computer within the Santa Claus machine. If the product demands became excessive, the Santa Claus machine would simply reproduce itself.

SRS units might be used in very large space construction projects (such as lunar mining operations) to facilitate and accelerate the exploitation of extraterrestrial resources and to make possible feats of planetary engineering. For example, we could deploy a seed SRS unit on Mars as a prelude to permanent human habitation. This machine would use local Martian resources to automatically manufacture a large number of robot explorer vehicles. This armada of vehicles would be disbursed over the surface of the Red Planet searching for the minerals and frozen volatiles needed in the establishment of a Martian civilization. In just a few years a population of some 1,000 to 10,000 smart machines could scurry across the planet, completely exploring its entire surface and preparing the way for permanent human settlements.

Replicating systems would also make possible large-scale interplanetary mining operations. Extraterrestrial materials could be discovered, mapped, and mined using teams of surface and subsurface prospector robots that were manufactured in large quantities in an SRS factory complex. Raw materials could be mined by hundreds of machines and then sent wherever they were needed in heliocentric space.

Atmospheric mining stations could be set up at many interesting and profitable locations throughout the solar system. For example, Jupiter and Saturn could have their atmospheres mined for hydrogen, helium (including the very valuable isotope helium-3), and hydrocarbons using aerostats. Cloud-enshrouded Venus might be mined for carbon dioxide, Europa for water, and Titan for hydrocarbons. Intercepting and mining comets with fleets of robot spacecraft might also yield large quantities of useful volatiles. Similar mechanized space armadas could mine water ice from Saturn's ring system. All of these smart robot devices would be mass-produced by seed SRS units. Extensive mining operations in the main asteroid belt would yield large quantities of heavy metals. Using extraterrestrial materials, these replicating machines could, in principle, manufacture huge mining or processing plants or even ground-to-orbit or interplanetary vehicles. This large-scale manipulation of the solar system's resources would occur in a very short period of time, perhaps within one or two decades of the initial introduction of replicating machine technology.

From the viewpoint of our solar system civilization, perhaps the most exciting consequence of the self-replicating system is that it would provide a technological pathway for organizing potentially infinite quantities of matter. Large reservoirs of extraterrestrial matter might be gathered and organized to create an ever-widening human presence throughout the solar system. Self-replicating space stations, space settlements, and domed cities on certain alien worlds of the solar system would provide a diversity of environmental niches never before experienced in the history of the human race.

The SRS unit would provide such a large amplification of matter manipulating capability that it is possible even now to start seriously considering planetary engineering (or terraforming) strategies for the Moon, Mars, Venus, and certain other alien worlds. In time, advanced self-replicating systems could be used in the 22nd century by humankind's solar system civilization to perform incredible feats of astroengineering. The harnessing of the total radiant energy output of our parent star through the robot-assisted construction of a Dyson sphere is an exciting example of the large-scale astroengineering projects that might be undertaken.

Advanced SRS technology also appears to be the key to human exploration and expansion beyond the very confines of the solar system. Although such interstellar missions may today appear highly speculative, and indeed they certainly require technologies that exceed contemporary or even projected levels in many areas, a consideration of possible interstellar applications is actually quite an exciting and useful mental exercise. It illustrates immediately the fantastic power and virtually limitless potential of the SRS concept.

It appears likely that before humans move out across the interstellar void, smart robot probes will be sent ahead as scouts. Interstellar distances are so large and search volumes so vast that self-replicating probes represent a highly desirable, if not totally essential, approach to surveying other star systems for suitable extrasolar planets and for extraterrestrial life. One study on galactic exploration suggests that search patterns beyond the 100 nearest stars would most likely be optimized by the use of SRS probes. In fact, reproductive probes might permit the direct reconnaissance of the nearest 1 million stars in about 10,000 years and the entire Milky Way galaxy in less than 1 million years—starting with a total investment by the human race of just one self-replicating interstellar robot spacecraft.

Of course, the problems in keeping track of, controlling, and assimilating all the data sent back to the home star system by an exponentially growing number of robot probes are simply staggering. We might avoid some of these problems by sending only very smart machines capable of greatly distilling the information gathered and transmitting only the most significant quantities of data, suitably abstracted, back to Earth. We might also set up some kind of command and control hierarchy, in which each robot probe only communicated with its parent. Thus, a chain of "ancestral repeater stations" could be used to control the flow of messages and exploration reports. Imagine the exciting chain reaction that might occur as one or two of the leading probes encountered an intelligent alien race. If the alien race proved hostile, an interstellar alarm would be issued, taking light-years to ripple across interstellar space, repeater station by repeater station, until Earth received notification. Would future citizens of Earth retaliate and send more sophisticated, possibly predator robot probes to that area of the galaxy or would we elect to position warning beacons all around the area, signaling any other robot probes to swing clear of the alien hazard?

In time, as first predicted early in the 20th century by ROBERT H. GODDARD, giant space arks representing an advanced level of synthesis between human crew and robot "crew" will depart from the solar system and plunge into the interstellar medium. Upon reaching another star system that contained suitable planetary resources, the space ark itself could undergo replication. The human passengers (perhaps several generations of humans beyond the initial crew that departed the solar system) would then redistribute themselves between the parent space ark, offspring space arks, and any suitable extrasolar planets (if found). Consequently, the original space ark would serve as a self-replicating "Noah's Ark" for the human race and any terrestrial life-forms included on the giant mobile habitat. This dispersal of conscious intelligence (that is, intelligent human life) to a variety of ecological niches among the stars would ensure that not even disaster on a cosmic scale, such as the death of our Sun, could threaten the complete destruction of humanity and all our accomplishments. The self-replicating space ark would enable human beings to literally "green the galaxy" with a propagating wave of consciousness and life (as we know it). From a millennial perspective, this is perhaps the grandest role for robotics in space.

See also ARTIFICIAL INTELLIGENCE; ASTROENGINEERING; CASSINI MISSION; DYSON SPHERE; LUNOKHOD; MARS EXPLORATION ROVER (MER) MISSION; *MARS PATHFINDER;* MARS SAMPLE RETURN MISSION; PLANETARY ENGINEERING; TELEOPERATION; TELEPRESENCE; THOUSAND ASTRONOMICAL UNIT (TAU) MISSION; VIKING PROJECT.

robot spacecraft A semiautomated or fully automated spacecraft capable of executing its primary exploration MISSION with minimal or no human supervision.

See also ROBOTICS IN SPACE.

Roche, Edouard Albert (1820–1883) French *Mathematician* Edouard Albert Roche was a French mathematician who calculated how close a natural satellite can orbit around its parent planet (primary body) before the influence of gravity creates tidal forces that rip the moon apart, creating a ring of debris. Assuming the planet and its satellite have the same density, he calculated that this limit (called the *Roche limit*) occurs when the moon is at a distance of approximately 2.45 times the radius of the planet or less.

Roche limit As postulated by the French mathematician EDOUARD ALBERT ROCHE in the 19th century, the smallest distance from a planet at which gravitational forces can hold together a natural satellite or moon that has the same average density as the primary body. If the moon's orbit falls within the Roche limit, it will be torn apart by tidal forces. The rings of Saturn, for example, occupy the region within Saturn's Roche limit.

If we assume that (1) both celestial objects have the same density and (2) that the moon in question is held together only by gravitational attraction of its matter, then the Roche limit is typically about 1.225 times the diameter (or about 2.45 times the radius) of the parent planet or primary celestial body. Also known as the tidal stability limit.

See also RINGED WORLD.

rocket In general, a completely self-contained projectile, pyrotechnic device, or flying vehicle propelled by a reaction (rocket) engine. Since it carries all of its propellant, a rocket vehicle can function in the vacuum of outer space and represents the key to space travel. Rockets obey SIR ISAAC NEWTON's third law of motion, which states, "For every action there is an equal and opposite reaction." Rockets can be classified by the energy source used by the reaction engine to accelerate the ejected matter that creates the vehicle's thrust, such as, for example, chemical rocket, nuclear rocket, and electric rocket. Chemical rockets, in turn, often are divided into two general subclasses, solid-propellant rockets and liquid-propellant rockets.

A *solid-propellant chemical rocket* is the simplest type of rocket. It can trace its technical heritage all the way back to the gunpowder-fueled "fire arrow" rockets of ancient China (ca. 1045 C.E.). Those in use today generally consist of a solid propellant (i.e., fuel and oxidizer compound) with the following associated hardware: case, nozzle, insulation, igniter, and stabilizers.

Solid propellants, commonly referred to as the "grain," are basically a chemical mixture or compound containing a fuel and oxidizer that burn (combust) to produce very hot gases at high pressure. The important feature is that these propellants are self-contained and can burn without the introduction of outside oxygen sources (such as air from Earth's atmosphere). Consequently, the solid-propellant rocket and its technical sibling, the liquid-propellant rocket (which is discussed shortly), can operate in outer space. In fact, a rocket vehicle performs best in space, as was originally postulated in 1919 by the brilliant American rocket scientist ROBERT H. GODDARD. Goddard is often called the father of American rocketry and is one of the three cofounders of astronautics in the early 20th century.

Solid propellants often are divided into three basic classes: monopropellants, double-base, and composites. Monopropellants are energetic compounds such as nitroglycerin or nitrocellulose. Both of these compounds contain fuel (carbon and hydrogen) and oxidizer (oxygen). Monopropellants are rarely used in modern rockets. Double-base propellants are mixtures of monopropellants, such as nitroglycerin and nitrocellulose. The nitrocellulose adds physical strength to the grain, while the nitroglycerin is a high-performance, fast-burning propellant. Usually double-base propellants are mixed together with additives that improve the burning characteristics of the grain. The mixture becomes a puttylike material that is loaded into the rocket case.

Composite solid propellants are formed from mixtures of two or more unlike compounds that by themselves do not make good propellants. Usually one compound serves as the fuel and the other as the oxidizer. For example, the propellants used in the solid rocket boosters (SRBs) of the space

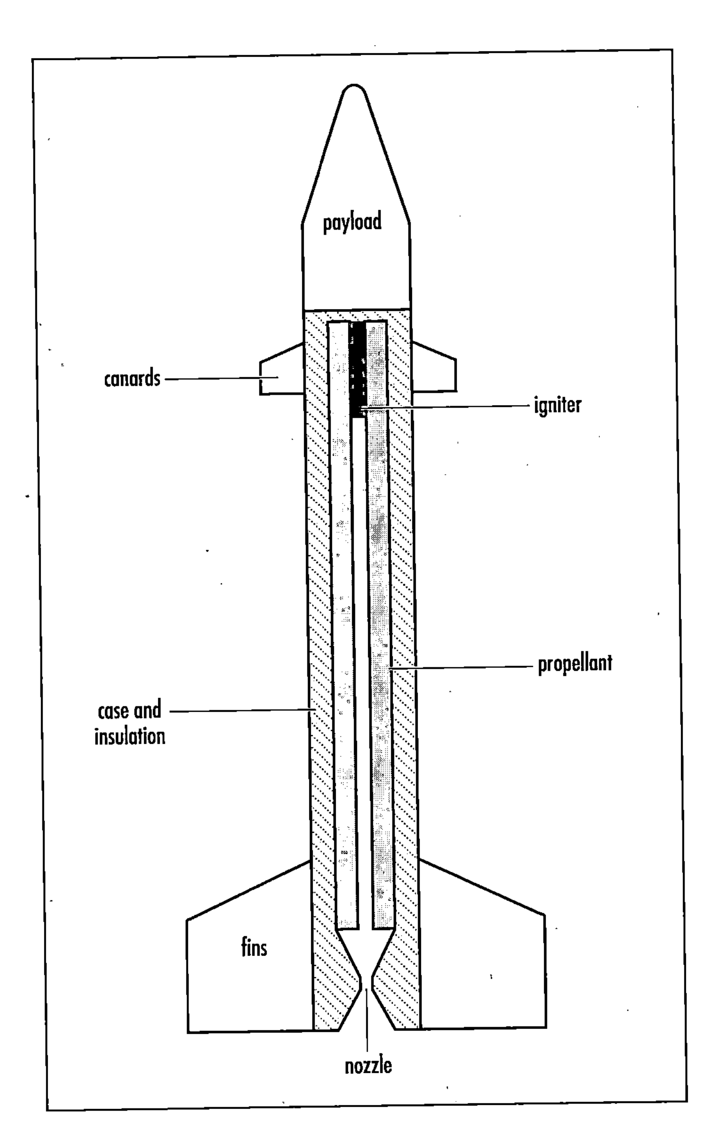

The basic components of a solid-propellant rocket

shuttle fall into this category. The propellant type used in this case is known as PBAN, which means polybutadiene–acrylic acid–acrylonitrile terpolymer. This somewhat exotic-sounding chemical compound is used as a binder for ammonium perchlorate (oxidizer), powdered aluminum (fuel), and iron oxide (an additive). The cured propellant looks and feels like a hard rubber eraser.

The thrust produced by the propellants is determined by the combustive nature of the chemicals used and by the shape of their exposed burning surfaces. A solid propellant will burn at any point that is exposed to heat or hot gases of the right temperature. Grains usually are designed to be either end-burning or internal-burning.

End-burning grains burn the slowest of any grain design. The propellant is ignited close to the nozzle, and the burning proceeds the length of the propellant load. The area of the burning surface is always at a minimum. While the thrust produced by end-burning is lower than for other grain designs, the thrust is sustained over longer periods.

Much more massive thrusts are produced by internal-burning. In this design the grain is perforated by one or more hollow cores that extend the length of the case. With an internal-burning grain the burning surface of the exposed cores is much larger than the surface exposed in an end-burning grain. The entire surfaces of the cores are ignited at the same time, and burning proceeds from the inside out. To increase the surface available for burning, a core may be shaped into a cruciform or star design.

By varying the geometry of the core design, the thrust produced by a large internal-burning grain also can be customized as a function of time to accommodate specific mission needs. For example, the massive solid rocket boosters (SRBs) used by the space shuttle feature a single core that has an 11-point star design in the forward section. At 65 seconds into the launch, the star points are burned away, and thrust temporarily diminishes. This coincides with the passage of the space shuttle vehicle through the sound barrier. Buffeting occurs during this passage, and the reduced SRB thrust helps alleviate strain on the vehicle.

The rocket case is the pressure- and load-carrying structure that encloses the solid propellant. Cases are usually cylindrical, but some are spherical in shape. The case is an inert part of the rocket, and its mass is an important factor in determining how much payload the rocket can carry and how far it can travel. Efficient, high-performance rockets require that the casing be constructed of the lightest materials possible. Alloys of steel and titanium often are used for solid rocket casings. Upper-stage vehicles may use thin metal shells that are wound with fiberglass for extra strength.

Unless protected by insulation, the solid rocket motor case will lose strength rapidly and burst or burn through. Therefore, to protect the casing insulation is bonded to the inside wall of the case before the propellant is loaded. The thickness of this insulation is determined by its thermal properties and how long the casing will be exposed to the high-pressure, very hot combustion gases. A frequently used insulation for solid rockets is an asbestos-filled rubber compound that is thermally bonded to the casing wall.

During the combustion process the resulting high-temperature and high-pressure gases exit the rocket through a narrow opening called the nozzle that is located at the lower end of the motor. The most efficient nozzles are convergent-divergent designs. During operation, the exhaust gas velocity in the convergent portion of the nozzle is subsonic. The gas velocity increases to sonic speed at the throat and then to supersonic speeds as it flows through and exits the divergent portion of the nozzle. The narrowest part of the nozzle is the throat. Escaping gases flow through this constricted region with relatively high velocity and temperature. Excessive heat transfer to the nozzle wall at the throat is a great problem. Often thermal protection for the throat consists of a liner that either withstands these high temperatures (for a brief period of operation) or else ablates (i.e., intentionally erodes, carrying heat away). Large solid rocket motors, such as the space shuttle's solid rocket booster, generally rely on ablative materials to protect the nozzle's throat. Smaller solid rocket motors, such as might be used in an air-to-air combat missile or in a short-range surface-to-surface military missile, often use high-temperature-resistant materials to protect the nozzle. These temperature-resistant materials (possibly augmented by a thin layer of heat-resistant liner) can protect the nozzle's throat sufficiently, since the rocket's burn period is quite short (typically a few seconds).

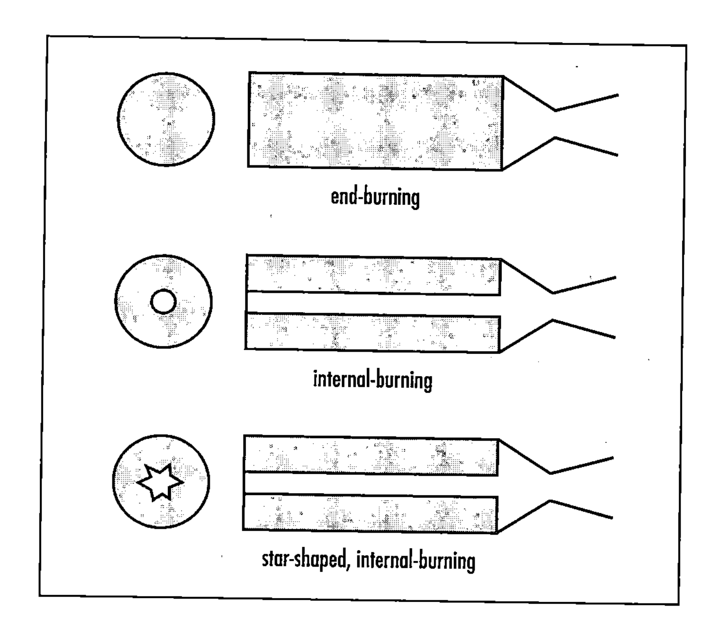

Typical solid-rocket motor grain designs

To ignite the propellants of a solid rocket, the grain surface must be saturated with hot gases. The igniter is usually a rocket motor itself, but much smaller in size. The igniter can be placed inside the upper end of the hollow core, at the lower end of the core, or even completely outside the solid rocket motor. In the latter case the exhaust of the igniter is directed into the nozzle of the larger solid rocket motor. An electrical circuit with a hot-wire resistor or an exploding bridgewire starts the igniter. The initial part of the ignition sequence begins with fast-burning pellets that, in turn, fire up the main igniter propellants.

Active directional control of solid-propellant rockets in flight generally is accomplished by one of two basic approaches. First, fins and possibly canards (small fins on the front end of the casing) can be mounted on the rocket's exterior. During flight in the atmosphere, these structures tilt to steer the rocket in much the same way that a rudder operates on a boat. Canards have an opposite effect on the directional changes of the rocket from the effect produced by the (tail) fins. Second, directional control also may be accomplished by using a gimbaled nozzle. Slight changes in the direction of the exhaust gases are accomplished by moving the nozzle from side to side. Large solid rocket motors, such as the space shuttle's solid rocket boosters, use the gimbaled-nozzle approach to help steer the vehicle.

Compared to liquid-propellant rocket systems, solid-propellant rockets offer the advantage of simplicity and reliability. With the exception of stability controls, solid-propellant rockets have no moving parts. When loaded with propellant and erected on a launch pad (or placed in an underground strategic missile silo or inside a launch tube in a ballistic missile submarine), solid rockets stand ready for firing at a moment's notice. In contrast, liquid-propellant rocket systems require extensive prelaunch preparations.

Solid rockets generally have an additional advantage in that a greater portion of their total mass can consist of propellants. Liquid-propellant rocket systems require fluid feed lines, pumps, and tanks, all adding additional (inert) mass to the vehicle. In fact, aerospace engineers often describe the performance of a rocket vehicle in terms of the propellant mass fraction. The *propellant mass fraction* is defined as the rocket vehicle's propellant mass divided by its total mass (i.e., propellant, payload, structure, and supporting systems mass). A well-designed, "ideal" rocket vehicle should have a propellant mass fraction of between 91 and 93 percent (i.e., 0.91 to 0.93). This means that 91 to 93 percent of the rocket's total mass is propellant. Large liquid-propellant rocket vehicles generally achieve a mass fraction of about 90 percent or less.

The principal disadvantage of solid-propellant rockets involves the burning characteristics of the propellants themselves. Solid propellants are generally less energetic (i.e., deliver less thrust per unit mass consumed) than do the best liquid propellants. Also, once ignited, solid-propellant motors burn rapidly and are extremely difficult to throttle or extinguish. In contrast, liquid-propellant rocket engines can be started and stopped at will.

Today solid-propellant rockets are used for strategic nuclear missiles (e.g., the U.S. Air Force's Minuteman), for tactical military missiles (e.g., Sidewinder), for small expendable launch vehicles (e.g., Scout), and as strap-on solid boosters for a variety of liquid-propellant launch vehicles, including the reusable space shuttle and the expendable Titan-IV. Solid-rocket motors also are used in small sounding rockets and in many types of upper-stage vehicles, such as the Inertial Upper Stage (IUS) system.

The American physicist Robert H. Goddard invented the liquid-propellant rocket engine in 1926. So important were his contributions to the field of rocketry that it is appropriate to state that "Every liquid propellant rocket is essentially a Goddard rocket."

The figure describes the major components of a typical liquid-propellant chemical rocket. Here, the propellants (liquid hydrogen for the fuel and liquid oxygen for the oxidizer) are pumped to the combustion chamber, where they begin to react. The liquid fuel often is passed through the tubular walls of the combustion chamber and nozzle to help cool them and prevent high-temperature degradation of their surfaces.

Liquid-propellant rockets have three principal components in their propulsion system: propellant tanks, the rocket engine's combustion chamber and nozzle assembly, and turbopumps. The propellant tanks are load-bearing structures that contain the liquid propellants. There is a separate tank for the fuel and for the oxidizer. The combustion chamber is the region into which the liquid propellants are pumped, vaporized, and reacted (combusted), creating the hot exhaust gases, which then expand through the nozzle generating thrust. The turbopumps are fluid-flow machinery that deliver the propellants from the tanks to the combustion chamber at high pressure and sufficient flow rate. (In some liquid-propellant rockets the turbopumps are eliminated by using an "over-pressure" in the propellant tanks to force the propellants into the combustion chamber.) A simplified liquid-propellant rocket engine is illustrated in the figure.

Propellant tanks serve to store one or two propellants until needed in the combustion chamber. Depending on the

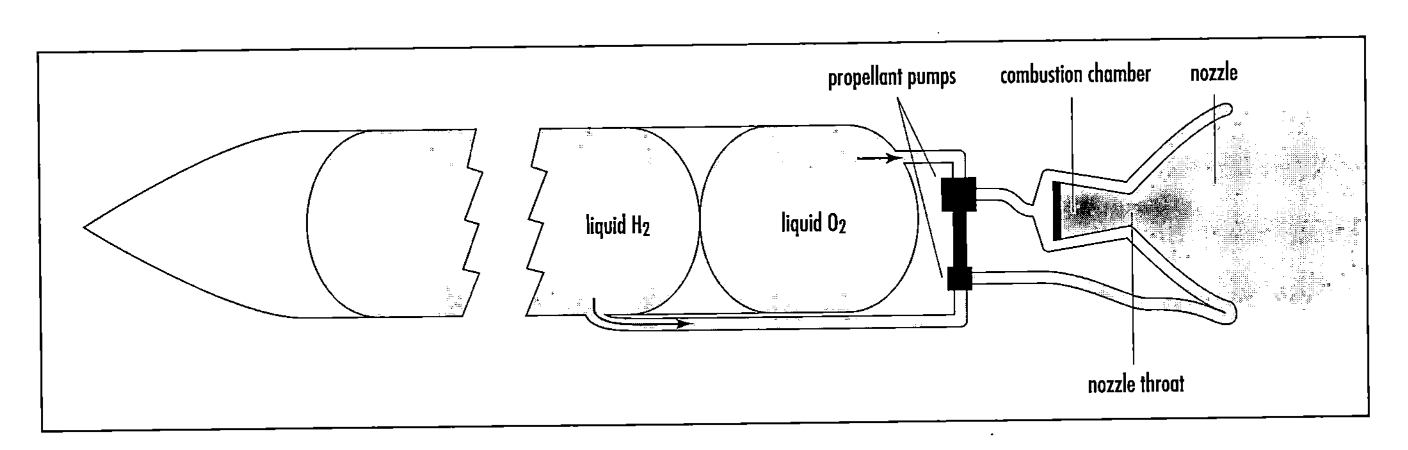

Basic hardware associated with a bipropellant (here LH_2 and LO_2) liquid rocket

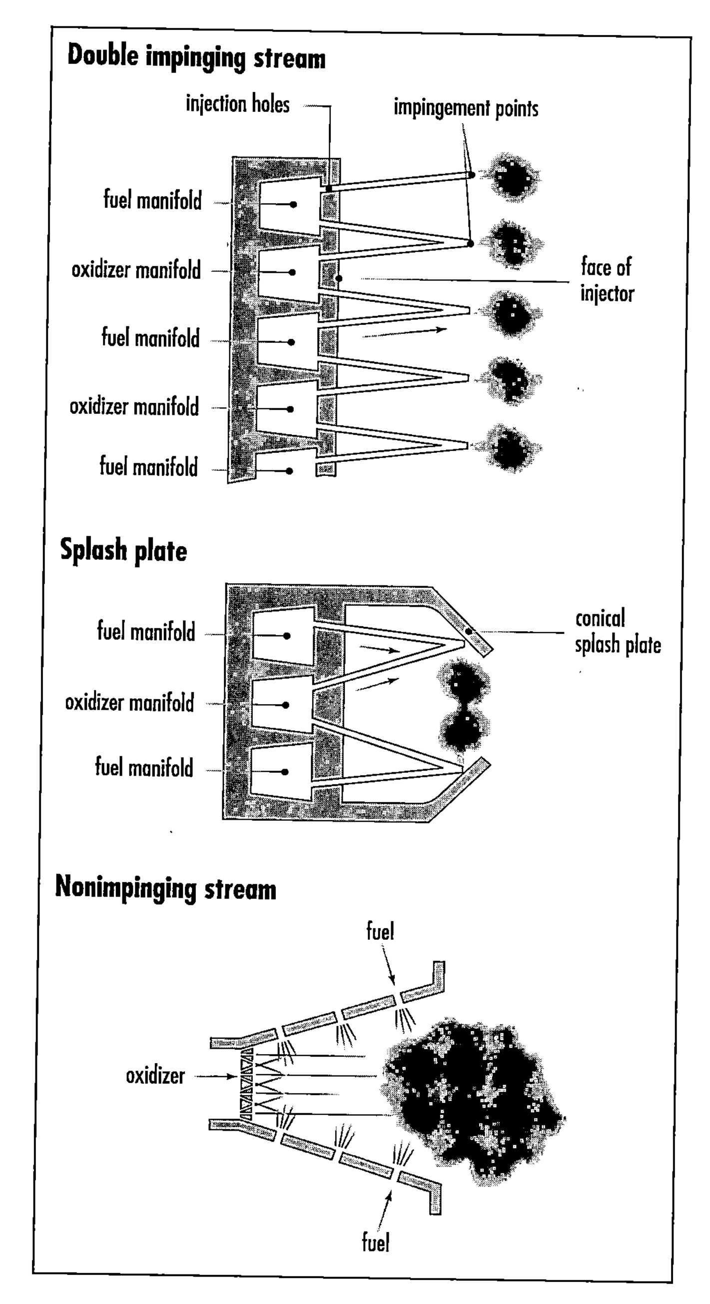

Some typical injector designs for liquid-propellant rocket engines *(Courtesy of NASA)*

type of liquid propellants used, the tank may be nothing more than a low-pressure envelope, or it may be a pressure vessel capable of containing propellants under high pressure. In the case of cryogenic (extremely low-temperature) propellants, the tank has to be an extremely well-insulated structure to prevent the very cold liquids from boiling away.

As with all rocket vehicle components, the mass of the propellant tanks is an important design factor. Aerospace engineers fully recognize that the lighter they can make the propellant tanks, the more payload that can be carried by the rocket or the greater its range. Many liquid-propellant tanks are made of very thin metal or are thin metal sheaths wrapped with high-strength fibers and cements. These tanks are stabilized by the internal pressure of their contents, much the same way the wall of a balloon gains strength from the gas inside (at least up to a certain level of internal pressure). However, very large propellant tanks and tanks that contain cryogenic propellants require additional strengthening or layers. Structural rings and ribs are used to strengthen tank walls, giving the tanks the appearance of an aircraft frame. With cryogenic propellants extensive insulation is needed to keep the propellants in their liquefied form. Unfortunately, even with the best available insulation, cryogenic propellants are difficult to store for long period of time and eventually will boil off (i.e., vaporize). For this reason cryogenic propellants are not used in liquid-propellant military rockets, which must be stored for months at a time in a launch-ready condition.

Turbopumps provide the required flow of propellants from the low-pressure propellant tanks to the high-pressure combustion chamber. Power to operate the turbopumps is produced by combusting a fraction of the propellants in a preburner. Expanding gases from the burning propellants drive one or more turbines, which in turn drive the turbopumps. After passing through the turbines, these exhaust gases are either directed out of the rocket through a nozzle or are injected, along with liquid oxygen, into the combustion chamber for more complete burning.

The combustion chamber of a liquid-propellant rocket is a bottle-shaped container with openings at opposite ends. The openings at the top inject propellants into the chamber. Each opening consists of a small nozzle that injects either fuel or oxidizer. The main purpose of the injectors is to mix the propellants to ensure smooth and complete combustion and to avoid detonations. Combustion chamber injectors come in many designs, and one liquid-propellant engine may have hundreds of injectors. (*See* figure.)

After the propellants have entered the combustion chamber, they must be ignited. Hypergolic propellant combinations ignite on contact, but other propellants need a starter device, such as a spark plug. Once combustion has started, the thermal energy released continues the process.

The opening at the opposite (lower) end of the combustion chamber is the throat, or narrowest part of the nozzle. Combustion of the propellants builds up gas pressure inside the chamber, which then exhausts through this nozzle. By the time the gas leaves the exit cone (widest part of the nozzle), it achieves supersonic velocity and imparts forward thrust to the rocket vehicle.

Because of the high temperatures produced by propellant combustion, the chamber and nozzle must be cooled. For example, the combustion chambers of the space shuttle's main engines (i.e., the SSMEs) reach 3,590 kelvins (K) (3,317° Celsius) during firing. All surfaces of the combustion chamber and nozzle need to be protected from the eroding effects of the high-temperature, high-pressure gases.

Two general approaches can be taken to cool the combustion chamber and nozzle. One approach is identical to the cooling approach taken with many solid-propellant rocket nozzles. The surface of the nozzle is covered with an ablative material that sacrificially erodes when exposed to the high-temperature gas stream. This intentional material erosion process keeps the surface underneath cool, since the ablated material carries away a large amount of thermal energy. However, this cooling approach adds extra mass to a liquid-propellant engine, which in turn reduces payload and range capability of the rocket vehicle. Therefore, ablative cooling is used only when the liquid-propellant engine is small or when a simplified engine design is more important than high performance. In considering such technical choices, an aerospace engineer is making "design tradeoffs."

The second method of cooling is called *regenerative cooling*. A complex plumbing arrangement inside the combustion chamber and nozzle walls circulates the fuel (in the case of the space shuttle's main engines, very cold [cryogenic] liquid hydrogen fuel) before it is sent through the preburner and into the combustion chamber. This circulating fuel then absorbs some of the thermal energy entering the combustion chamber and nozzle walls, providing a level of cooling. Although more complicated than ablative cooling, regenerative cooling reduces the mass of large rocket engines and improves flight performance.

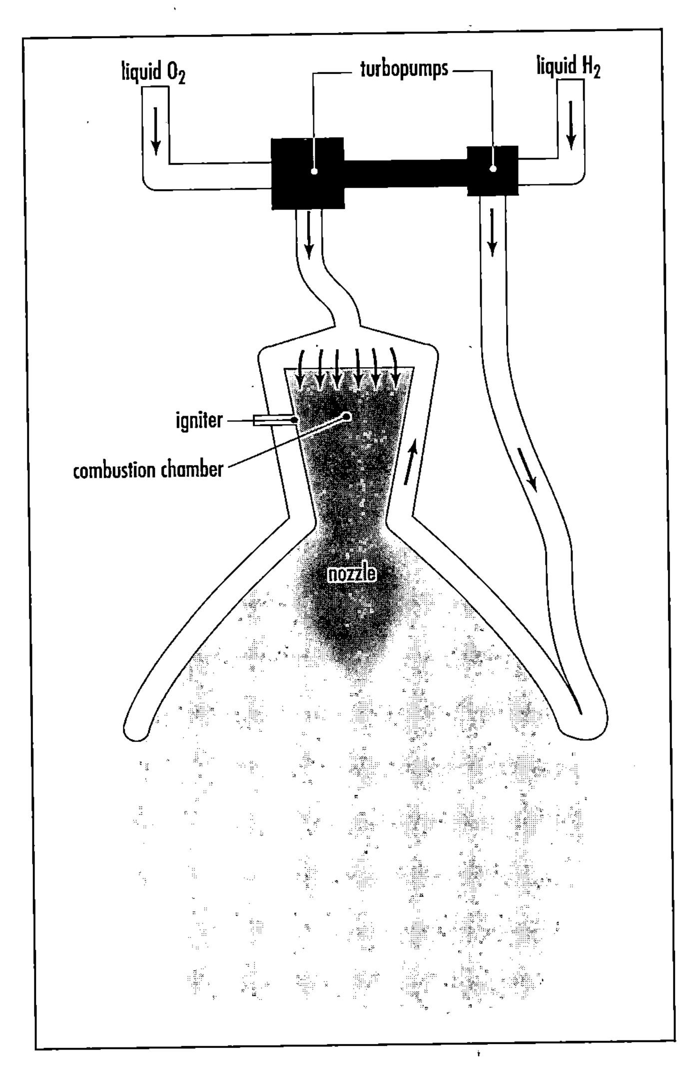

A simplified liquid-propellant rocket engine that employs regenerative cooling of the combustion chamber and nozzle walls

Propellants for liquid rockets generally fall into two categories: monopropellants and bipropellants. Monopropellants consist of a fuel and an oxidizing agent stored in one container. They can be two premixed chemicals, such as alcohol and hydrogen peroxide, or a homogeneous chemical, such as nitromethane. Another chemical, hydrazine, becomes a monopropellant when first brought into contact with a catalyst. The catalyst initiates a reaction that produces heat and gases from the chemical decomposition of the hydrazine.

Bipropellants have the fuel and oxidizer separate from each other until they are mixed in the combustion chamber. Commonly used bipropellant combinations include liquid oxygen (LO_2) and kerosene, liquid oxygen and liquid hydrogen (LH_2), and monomethylhydrazine (MMH) and nitrogen tetroxide (N_2O_4). The last bipropellant combination, MMH and N_2O_4, is hypergolic, meaning these two propellants ignite spontaneously when brought into contact with each other. Hypergolic propellants are especially useful for attitude control rockets for which frequent firings and high reliability are required.

Aerospace engineers must consider many factors in selecting bipropellant combinations for a particular rocket system. For example, LH_2 and N_2O_4 make a good combination based on propellant performance, but their widely divergent storage temperatures (cryogenic and room temperature, respectively) require the use of large quantities of thermal insulation between the two tanks, adding considerable mass to the rocket vehicle. Another important factor is the toxicity of the chemicals used. MMH and N_2O_4 are both highly toxic. Rocket vehicles that use this propellant combination require special propellant handling and prelaunch preparation.

See also ELECTRIC PROPULSION; LAUNCH VEHICLE; PROPELLANT; ROCKETRY; SPACE NUCLEAR PROPULSION.

rocket engine 1. In general, a reaction engine that contains within itself (or carries along with itself) all the substances (i.e., matter and energy sources) necessary for its operation. A rocket engine does not require the intake of any outside substance and is therefore capable of operating in outer space. The rocket engine converts energy from some suitable source and uses this energy to eject stored matter at high velocity. The ejected matter is called the rocket engine's propellant. Rocket engines often are classified according to the type of energy sources used to manipulate the ejected matter, such as, for example, chemical rocket engines, nuclear rocket engines, and electric rocket engines. Very commonly, however, the term *rocket engine* is used to refer to chemical propulsion system engines as a class. Chemical propulsion system engines are further classified by the type of propellant: liquid-propellant rocket engine or solid-propellant rocket engine. 2. Specifically, that portion of a rocket's chemical propulsion system in which combustible materials (propellants) are supplied to a chamber and burned under specified conditions. The thermal energy released in the chemical combustion reactions is converted into the kinetic energy of hot combustion gases that are ejected through a nozzle, creating the thrust (reaction force) that then propels the vehicle to which the rocket engine is attached. In common aerospace practice the term *rocket engine* usually is applied to a device (machine) that burns liquid propellants and therefore requires a rather complex system of propellant tanks, ducts, pumps, flow control devices, and so on. The term *rocket motor* is customarily applied to a device (machine) that burns solid propellants and therefore is relatively simple, normally requiring just the solid-propellant grain within a case, an igniter, and a nozzle.

See also ELECTRIC PROPULSION; ROCKET; SPACE NUCLEAR PROPULSION.

rocket propellant 1. In general, the matter ejected by a rocket (reaction) engine. 2. Any agent used for consumption or combustion in a rocket engine and from which the rocket derives its thrust, such as a chemical fuel, oxidizer, catalyst, or any compound or mixture of these. Chemical rocket propellants can be either liquid or solid. For example, liquid hydrogen (fuel) and liquid oxygen (oxidizer) often are used as the cryogenic liquid propellants in modern high-performance launch vehicles. Nuclear thermal rockets use a monopropellant, such as liquid hydrogen, which is heated to extremely high temperatures by a nuclear reactor; electric rockets use a readily ionizable element such as cesium, mercury, argon, or xenon as the propellant, which is accelerated and ejected at extremely high velocities by electromagnetic processes.

See also ELECTRIC PROPULSION; PROPELLANT; ROCKET; SPACE NUCLEAR PROPULSION.

rocket propulsion Reaction propulsion by means of a rocket engine. Rocket propulsion differs from "jet propulsion" in that a rocket propulsion system is completely self-contained (i.e., it carries all the substances necessary for its operation, such as chemical fuel and oxidizer), while a jet propulsion system needs external materials to achieve reactive propulsion (e.g., air from the atmosphere to support the chemical combustion processes). Rocket propulsion works both within and outside Earth's atmosphere; jet propulsion works only up to a certain altitude within Earth's atmosphere.

See also JET PROPULSION; ROCKET; X-15.

rocket ramjet A ramjet engine that has a rocket mounted within the ramjet duct. This rocket is used to bring the ramjet up to the necessary operating speed.

See also RAMJET.

rocketry The art or science of making rockets; the branch of science that deals with rockets, including theory, research, development, experimentation, and application.

According to certain historical records, the Chinese were the first to use gunpowder rockets, which they called "fire arrows," in military applications. In the Battle of K'ai-fung-fu (1232 C.E.) these rocket fire-arrows helped the Chinese repel Mongol invaders. About 1300 the rocket migrated to Europe, where over the next few centuries (especially during the Renaissance) it ended up in most arsenals as a primitive bombardment weapon. However, artillery improvements eventually made the cannon more effective in battle, and military rockets remained in the background.

Rocketry legend suggests that around 1500, a Chinese official named Wan-Hu (also spelled Wan-Hoo) conceived of the idea of flying through the air in a rocket-propelled chair-kite assembly. Serving as his own test pilot, he vanished in a bright flash during the initial flight of this rocket-propelled device.

The rocket also found military application in the Middle East and India. For example, in the late 1700s Rajah Hyder Ali, prince of Mysore, India, used iron-cased stick rockets to defeat a British military unit. Profiting from their adverse rocket experience in India, the British, led by SIR WILLIAM CONGREVE, improved the design of these Indian rockets and developed a series of more efficient bombardment rockets that ranged in mass from about 8 to 136 kilograms. Perhaps the most famous application of Congreve's rockets was the British bombardment of the American Fort McHenry in the War of 1812. This rocket attack is now immortalized in the "rockets' red glare" phrase of "The Star-Spangled Banner."

An American named William Hale attempted to improve the inherently low accuracy of 19th-century military rockets through a technique called spin stabilization. U.S. Army units used bombardment rockets during the Mexican-American War (e.g., the siege of Veracruz in March 1847) and during the U.S. Civil War. However, improvements in artillery technology again outpaced developments in military rocketry; by the beginning of the 20th century rockets remained more a matter of polite speculation than a widely accepted technology.

The Russian schoolteacher KONSTANTIN TSIOLKOVSKY wrote a series of articles about the theory of rocketry and space flight at the turn of the century. Among other things, his visionary works suggested the necessity for liquid-propellant rockets, the very devices that the American physicist ROBERT H. GODDARD would soon develop. (Because of the geopolitical circumstances in czarist Russia, Goddard and many other scientists outside Russia were unaware of Tsiolkovsky's work.) Today Tsiolkovsky is regarded as the father of Russian rocketry and one of the founders of modern rocketry. His tombstone bears the prophetic inscription: "Mankind will not remain tied to the Earth forever!"

Similarly, the brilliant physicist Robert H. Goddard is regarded as the father of American rocketry and the developer of the practical modern rocket. He started working with rockets by testing solid-fuel models. In 1917, when the United States entered World War I, Goddard worked to perfect rockets as weapons. One of his designs became the technical forerunner of the bazooka, a tube-launched antitank rocket. Goddard's device (about 45.7 cm long and 2.54 cm in diameter) was tested in 1918, but the war ended before it could be used against enemy tanks.

In 1919 Goddard published the important technical paper "Method of Reaching Extreme Altitudes," in which he concluded that the rocket actually would work better in the vacuum of outer space than in Earth's atmosphere. At the time Goddard's "radical" (but correct) suggestion cut sharply against the popular (but wrong) belief that a rocket needed air to "push against." He also suggested that a multistage rocket could reach very high altitude and even attain sufficient velocity to "escape from Earth." Unfortunately, the press scoffed at his

Robert H. Goddard and the liquid-oxygen–gasoline rocket in the launch frame from which he fired it on March 16, 1926, in a snow-covered field at Auburn, Massachusetts. The event represents the birth of the liquid-propellant rocket. ***(Courtesy of NASA)***

ideas, and the general public failed to appreciate the great scientific merit of this paper. Despite this adverse publicity, Goddard continued to experiment with rockets, but now he intentionally avoided publicity. On March 16, 1926, he launched the world's first liquid-fueled rocket from a snow-covered field at his aunt Effie Goddard's farm in Auburn, Massachusetts. (*See* figure.) This simple gasoline-and-liquid-oxygen–fueled device rose to a height of just 12 meters and landed about 56 meters away in a frozen field. Regardless of its initial "range," Goddard's liquid-propellant rocket successfully flew, and the world would never be quite the same. The technical progeny of this simple rocket have taken human beings into Earth orbit and to the surface of the Moon. They also have sent sophisticated space probes throughout the solar system and beyond.

After this initial success Goddard flew other rockets in rural Massachusetts, at least until they started crashing into neighbors' pastures. After the local fire marshal declared that his rockets were a fire hazard, Goddard terminated his New England test program. The famous aviator Charles Lindbergh came to Goddard's rescue and helped him receive a grant from the Guggenheim Foundation. With this grant Goddard moved to sparsely populated Roswell, New Mexico, where he could experiment without disturbing anyone. At his Roswell test complex Goddard developed the first gyro-controlled rocket guidance system. He also flew rockets faster than the speed of sound and at altitudes up to 2,300 meters. Yet, despite his numerous technical accomplishments in rocketry, the U.S. government never developed an interest in his work. In fact, only during World War II did he receive any government funding, and that was for him to design small rockets to help aircraft take off from navy carriers. By the time he died in 1945, Goddard held more than 200 patents in rocketry. It is essentially impossible to design, construct, or launch a modern liquid-propellant rocket without using some idea or device that originated from Goddard's pioneering work in rocketry.

While Goddard worked essentially unnoticed in the United States, a parallel group of "rocketeers" thrived in Germany, centered originally around the Society for Space Travel. In 1923 HERMANN J. OBERTH published a highly prophetic book entitled *The Rocket into Interplanetary Space*. In this important work he proved that flight beyond the atmosphere was possible. One of the many readers inspired by this book was a brilliant young teenager named WERNHER VON BRAUN. Oberth published another book in 1929 entitled *The Road to Space Travel*. In this work he proposed liquid-propellant rockets, multistage rockets, space navigation, and guided reentry systems. From 1939 to 1945 Oberth, along with other German rocket scientists (including von Braun), worked in the military rocket program. Under the leadership of von Braun this program produced a number of experimental designs, the most famous of which was the large liquid-fueled A-4 rocket. The German military gave this rocket its more sinister and well-recognized name, V-2, for "vengeance weapon number two." It was the largest rocket vehicle at the time, being about 14 meters long and 1.7 meters in diameter and developing some 249,000 newtons of thrust.

At the end of World War II the majority of the German rocket development team from the Peenemünde rocket test site, led by von Braun, surrendered to the Americans. This team of German rocket scientists and captured V-2 rocket components were sent to White Sands Missile Range (WSMR), New Mexico, to initiate an American military rocket program. The first reassembled V-2 rocket was launched by a combined American-German team on April 16, 1946. A V-2, assembled and launched on this range, was America's first rocket to carry a heavy payload to high altitude. Another V-2 rocket set the first high-altitude and velocity record for a single-stage missile, and still another V-2 rocket was the first large missile to be controlled in flight.

Stimulated by the cold war, the American and German scientists at White Sands worked to develop a variety of missiles and rockets including the Corporal, Redstone, Nike, Aerobee, and, Atlas. The need for more room to fire rockets of longer range became evident in the late 1940s. In 1949 the Joint Long Range Proving Ground was established at a remote, deserted location on Florida's eastern coast known as Cape Canaveral. On July 24, 1950, a two-stage Bumper rocket became the first rocket vehicle to be launched from this now world-recognized location. The Bumper-8 rocket vehicle consisted of a V-2 first stage and a WAC-Corporal rocket second stage.

The missile race and the space (Moon) race of the cold war era triggered tremendous developments in rocketry during the second half of the 20th century. On October 4, 1957, the Russian rocket engineer SERGEI KOROLEV received permission to use a powerful military rocket to launch *SPUTNIK 1*, the first artificial satellite to orbit Earth. The United States quickly responded on January 31, 1958, by launching *EXPLORER 1*, the first American satellite. The U.S. Army Ballistic Missile Agency (including von Braun's team of German rocket scientists then located at the Redstone Arsenal in Huntsville, Alabama) modified a Redstone-derived booster into a four-stage launch vehicle configuration, called the Juno I. (This Juno I launch vehicle was the satellite-launching version of the Army's Jupiter C rocket.) The Jet Propulsion Laboratory (JPL) was responsible for the fourth stage, which included America's first satellite.

Another descendant of the V-2 rocket, the mighty Saturn V launch vehicle, carried U.S. astronauts to the surface of the Moon between 1969 and 1972. The first flight of the space shuttle on April 12, 1981, opened the era of aerospace vehicles and interest in reusable space transportation systems.

With the end of the cold war in 1989, missile and space confrontation races have been replaced by cooperative programs that are truly international in perspective. For example, during the construction and operation of the *INTERNATIONAL SPACE STATION*, both American and Russian rockets have carried people and equipment. In this century advanced engineering versions of the "dream machines" advocated by such space travel visionaries as Tsiolkovsky, Oberth, Goddard, Korolev, and von Braun at the beginning of the 20th century will serve as the space launch vehicles that will help fulfill our destiny among the stars.

See also CAPE CANAVERAL AIR FORCE STATION; LAUNCH VEHICLE; ROCKET; SPACE TRANSPORTATION SYSTEM; V-2 ROCKET.

rocketsonde A rocket-borne instrument package for the measurement and transmission of upper-air meteorological data (up to about 76 kilometers altitude), especially that portion of the atmosphere inaccessible to radiosonde techniques. A meteorological rocket.

rocket thrust The thrust of a rocket engine, usually expressed in SI units as newtons (N) or in the American engineering unit system as pounds-force. One pound-force is equal to 4.448 newtons.

rocket vehicle A vehicle propelled by a rocket engine.
See also ROCKET.

rockoon A high-altitude sounding system consisting of a small solid-propellant rocket carried aloft by a large plastic balloon. The rocket is fired near the maximum altitude of the balloon flight. It is a relatively mobile sounding system and has been used extensively from ships.

Roemer, Olaus Christensen (aka Ole Römer) (1644–1710) Danish *Astronomer* While working at the Paris Observatory with GIOVANNI DOMENICO CASSINI, the Danish astronomer Olaus Christensen Roemer noticed time discrepancies for successive predicted eclipses of the moons of Jupiter and correctly concluded that light must have a finite velocity. He then attempted to calculate the speed of light in 1675. However, his numerical results (about 227,000 km/s) were lower than the currently accepted value (299,792.458 km/s), mainly because of the inaccuracies in solar system distances at that time.

roentgen A unit of exposure to ionizing radiation, especially X-rays or gamma rays. The roentgen is defined as the quantity of X-rays or gamma rays required to produce a charge of 2.58×10^{-4} coulomb in a kilogram of dry air under standard conditions. The unit is named after WILHELM CONRAD ROENTGEN, the German scientist who discovered X-rays in 1895.

Roentgen, Wilhelm Conrad (also spelled Röntgen) (1845–1923) German *Physicist* The German physicist Wilhelm Conrad Roentgen started world-changing revolutions in both physics and medicine when he discovered X-rays in 1895. His achievement led to a better understanding of the interaction of matter and energy, and the X-rays themselves and X-ray sensor technology spawned many important applications in such diverse disciplines as modern medicine, astrophysics, and astronomy.

Roentgen was born on March 27, 1845, in Lennep, Prussia (now Remscheid, Germany). A variety of circumstances led him to enroll at the polytechnic in Zurich, Switzerland, in 1865. Roentgen graduated in 1868 as a mechanical engineer and then completed a Ph.D. in 1869 at the University of Zurich. Upon graduation he pursued an academic career. He became a physics professor at various universities in Germany, lecturing and also performing research on piezoelectricity, heat conduction in crystals, the thermodynamics of gases, and the capillary action of fluids. His affiliations included the Universities of Strasbourg (1876–79), Giessen (1879–88), Würzburg (1888–1900), and Munich (1900–20).

His great discovery happened at the University of Würzburg on the evening of November 8, 1895. Roentgen, like many other late 19th-century physicists, was investigating luminescence phenomena associated with cathode ray tubes. This device, also called a *Crookes tube* after its inventor, Sir William Crookes (1832–1919), consisted of an evacuated glass tube containing two electrodes, a cathode and an anode. Electrons emitted by the cathode often missed the anode and struck the glass wall of the tube, causing it to glow, or fluoresce. On that particular evening Roentgen decided to place a partially evacuated discharge tube inside a thick black cardboard carton. As he darkened the room and operated the light-shielded tube, he suddenly noticed that a paper plate covered on one side with barium platinocyanide began to fluoresce, even though it was located some 2 meters from the discharge tube. He concluded that the phenomenon causing the plate to glow was some new form of penetrating radiation originating within the opaque paper-enshrouded discharge tube. He called this unknown radiation "X-rays," because x was the traditional algebraic symbol for an unknown quantity.

During subsequent experiments Roentgen discovered that objects of different thicknesses, when placed in the path of these mysterious X-rays, demonstrated variable transparency when he recorded the interaction of the X-rays on a photograph plate. As part of his investigations, Roentgen even held his wife's hand over a photograph plate and produced the first "medical X-ray" of the human body. When he developed this particular X-ray–exposed photograph plate, he observed an interior image of his wife's hand. This first "roentgenogram" contained dark shadows cast by the bones within her hand and by the wedding ring she was wearing. These dark shadows were surrounded by a less darkened (penumbral) shadow corresponding to the fleshy portions of her hand. Roentgen formally announced this important discovery on December 28, 1895, a discovery that instantly revolutionized both the field of physics and the field of medicine.

Although the precise physical nature of X-rays as very-short-wavelength, high-energy photons of electromagnetic radiation was not recognized until about 1912, the field of physics and the medical profession immediately embraced Roentgen's discovery. Scientists often consider this event the beginning of modern physics. In 1896 the American inventor Thomas A. Edison (1847–1931) developed the first practical fluoroscope, a noninvasive device that used X-rays to allow a physician to observe how internal organs of the body function within a living patient. At the time, however, no one recognized the potential health hazards associated with exposure to excessive quantities of X-rays or other forms of ionizing radiation, so Roentgen, his assistant, and many early X-ray technicians would eventually exhibit the symptoms of and suffer from acute radiation syndrome.

Unfortunately, Roentgen had to endure several bitter personal attacks by jealous rival scientists who had themselves overlooked the very phenomena that he observed during his experiments. Eventually, Roentgen received numerous honors for his discovery of X-rays, including the first Nobel Prize ever awarded in physics in 1901. Both a dedicated scientist and a humanitarian, Roentgen elected not to patent his discovery so the world could freely benefit from his work, and he even donated the money he received for his Nobel Prize to the university. Four years after the death of his wife, Roentgen died on February 10, 1923, in Munich, Germany. The cause of his death was carcinoma of the intestine, a condition most likely promoted by chronic exposure to ionizing radiation resulting from his extensive experimentation with the X-rays he discovered. At the time of his death the scientist was nearly bankrupt due to the hyperinflationary economic conditions in post–World War I Germany.

See also *ROENTGEN* SATELLITE (ROSAT); X-RAY ASTRONOMY.

***Roentgen* Satellite** (ROSAT) An Earth-orbiting X-ray observatory successfully launched by an expendable Delta II rocket on June 1, 1990, and placed into a 580-km-altitude, 53°-inclination orbit. This mission represents a cooperative effort by the United States (NASA), the United Kingdom, and Germany. Germany constructed the spacecraft and main X-ray telescope, the United Kingdom contributed a wide-field camera (WFC), and NASA provided the high-resolution imager (HRI), the launch vehicle, and launch services. Germany also performed the mission operations, including data collection and initial processing. During its first six months in orbit, ROSAT was dedicated to an all-sky survey and detected some 60,000 X-ray and extreme ultraviolet (EUV) sources. The spacecraft's primary mission (all-sky X-ray source survey) was completed by February 1992. The spacecraft was named in honor of WILHELM CONRAD ROENTGEN, the German scientist who discovered X-rays in 1895.

See also X-RAY ASTRONOMY.

rogue star A wandering star that passes close to a solar system, disrupting the celestial bodies in the system and triggering cosmic catastrophes on life-bearing planets.

See also NEMESIS.

roll The rotational or oscillatory movement of an aircraft, aerospace vehicle, or missile about its longitudinal (lengthwise) axis.

See also PITCH; YAW.

rollout That portion of the landing of an aircraft or aerospace vehicle following touchdown.

Roosa, Stuart Allen (1933–1994) American *Astronaut, U.S. Air Force Officer* Astronaut Stuart Allen Roosa served as the Apollo spacecraft pilot during the *Apollo 14* Moon-landing mission (1971). He remained in lunar orbit as fellow Apollo Project astronauts ALAN B. SHEPARD, JR. and EDGAR DEAN MITCHELL explored the lunar surface in the Fra Mauro region.

See also APOLLO PROJECT.

RORSAT A family of ocean reconnaissance satellites operated in low Earth orbit by the Soviet Union during the cold war to track the larger surface ships (such as aircraft carriers) of the U.S. Navy. The radar systems of these particular military surveillance spacecraft received their electrical power from a space nuclear reactor. The *Cosmos 469* spacecraft, launched on December 25, 1971, is believed to be the first of these ocean surveillance spacecraft to be powered by a BES-5 space nuclear reactor. Reports released after the cold war have suggested that the Russian BES-5 was a small, compact 100 kW-thermal (about 10 kW-electrical) class nuclear reactor that used enriched uranium-235 as its fuel.

On January 24, 1978, *Cosmos 954,* a Soviet nuclear-powered ocean surveillance satellite, crashed in the Northwest Territories of Canada. However, the reentry and crash of *Cosmos 954* was not really a surprise. Launched on September 18, 1977, this satellite began behaving oddly within the first few weeks of its ocean surveillance mission. The computers at the North American Air Defense Command (NORAD) responsible for tracking all space objects near Earth predicted that this misbehaving Russian spacecraft would decay from its low-altitude orbit and reenter Earth's atmosphere by April 1978. Because of the potential radiation hazard, the Soviet government also admitted to the world that its nuclear-powered *r*adar *o*cean *r*econnaissance *sat*ellite (or RORSAT) called *Cosmos 954* was out of control.

According to reports released by the Canadian government (Nuclear Emergency Preparedness and Response Division), *Cosmos 954*'s reentry and crash scattered a large amount of radioactivity over a 124,000-square-kilometer area in northern Canada, stretching southward from Great Slave Lake into northern Alberta and Saskatchewan. The clean-up operation, called Operation Morning Light, was a coordinated effort between the United States and Canada. These clean-up efforts in frozen wilderness areas across northern Canada continued into October 1978. The Canadian Nuclear Safety Commission (formerly called the Atomic Energy Control Board) has estimated that in addition to numerous pieces of contaminated structural materials from *Cosmos 954,* about 0.1 percent of the spacecraft's nuclear reactor was recovered on the ground. This supports the opinion of many aerospace nuclear experts who have suggested that most of *Cosmos 954*'s highly radioactive reactor core had dispersed in Earth's upper atmosphere during reentry. The *Cosmos 954* incident turned world opinion against the use of nuclear power sources in space, especially the use of nuclear reactors on spacecraft in low Earth orbit.

Apparently, the last RORSAT mission was *Cosmos 1932,* launched on March 14, 1988. After this mission the Soviet leader Mikhail Gorbachev canceled the program.

See also SPACE NUCLEAR POWER.

Rosalind A small moon of Uranus discovered in 1986 as a result of the *Voyager 2* spacecraft flyby. Rosalind is named after the daughter of the duke who was banished in William Shakepeare's comedy *As You Like It.* The moon has a diameter of 54 km and orbits Uranus at an average distance of 69,930 km with an orbital period of 0.558 day. Also known as Uranus XIII or S/1986 U4.

See also URANUS.

Rosetta mission Rosetta is a European Space Agency (ESA) robot spacecraft mission to rendezvous with Comet 67P/Churyumov-Gerasimenko, drop a probe on the surface of the comet's nucleus, study the comet from orbit, and fly by at least one asteroid enroute to the target comet. The major goals of this mission are to study the origin of comets and the relationship between cometary and interstellar material and its implications with regard to the origin of the solar system.

An Ariane 5 rocket successfully launched the *Rosetta* spacecraft on March 2, 2004 (at 07:17 UT), from the Kourou launch complex in French Guiana. As of December 31, 2004, *Rosetta* continued in the interplanetary cruise phase of its mission, which it entered on October 17. *Rosetta* has a complex trajectory, including three Earth and one Mars gravity assist maneuvers, before finally reaching the target comet in late 2014. Upon arrival at Comet 67P/Churyumov-Gerasimenko, the spacecraft will study the comet remotely as well as by means of a sophisticated instrument probe that will land on the surface of the nucleus. The *Rosetta* spacecraft carries the following science instruments: an imager, infrared and ultraviolet spectrometers, a plasma package, a magnetometer, particle analysis instruments, and a radio frequency sounder to study any subsurface layering of materials in the comet's nucleus. Instruments carried by the lander probe include an imager, a magnetometer, and an alpha particle/proton/x-ray spectrometer to determine the chemical composition of materials on the surface of the comet's nucleus. The nominal end of the Rosetta mission will occur in 2015, following the perihelion passage of Comet 67P/Churyumov-Gerasimenko. On its 10-year journey to this comet, ESA mission planners hope the spacecraft will experience at least one asteroid flyby.

Comet 67P/Churyumov-Gerasimenko was discovered in 1969 by astronomers Klim Churyumov and Svetlana Gerasimenko from Kiev, Ukraine, as they were conducting a survey of comets at the Alma-Ata Astrophysical Institute. This comet has been observed from Earth on six approaches to the Sun—1969 (discovery), 1976, 1982, 1989, 1996, and 2002. These observations have revealed that the comet has a small nucleus (about 3 km by 5 km) that rotates with a period of approximately 12 hours. The comet travels around the Sun in a highly eccentric (0.632) orbit, characterized by an orbital period of 6.57 years, a perihelion distance from the Sun of 1.29 astronomical units (AU), and an aphelion distance from the Sun of 5.74 AU. The orbit of Comet 67P/Churyumov-Gerasimenko also has an inclination of 7.12°.

The *Rosetta* spacecraft was originally going to rendezvous with and examine Comet 46 P/Wirtanen. However, due to postponement of the previously planned launch date, the original target comet was dropped, and a new target comet, Comet 67P/Churyumov-Gerasimenko, was selected.

See also COMET.

Rosse, third earl of (aka William Parsons) (1800–1867) Irish *Astronomer* The third earl of Rosse (aka William Parsons) was an astronomer who used his personal wealth to construct a massive 1.8-meter-diameter reflecting telescope on his family castle grounds in Ireland. Although located in a geographic region poorly suited for observation, the earl engaged in astronomy as a hobby and used his giant telescope (the largest in the world at the time) to detect and study "fuzzy spiral nebulas," many of which were later identified as spiral galaxies. In 1848 he gave the Crab Nebula its name and also made detailed observations of the Orion Nebula.

Rossi, Bruno Benedetto (1905–1993) Italian-American *Physicist, Astronomer, Space Scientist* Attracted to high-energy astronomy, Bruno Rossi investigated the fundamental nature of cosmic rays in the 1930s. Using special instruments carried into outer space by sounding rockets, he collaborated with RICCARDO GIACCONI and other scientists in 1962 and discovered X-ray sources outside the solar system. In the first decade of the space age, Rossi made many important contributions to the emerging fields of X-ray astronomy and space physics. To honor these important contributions, NASA renamed the X-ray astronomy satellite successfully launched in December 1995 the *Rossi X-Ray Timing Explorer* (RXTE).

Rossi was born on April 13, 1905, in Venice, Italy. The son of an electrical engineer, he began his college studies at the University of Padua and received a doctorate in physics from the University of Bologna in 1927. He began his scientific career at the University of Florence and then became the chair of physics at the University of Padua, serving in that post from 1932 to 1938. However, in 1938 the fascist regime of Benito Mussolini suddenly dismissed Rossi from his position at the university. That year fascist leaders went about purging the major Italian universities of many "dangerous" intellectuals who might challenge Italy's totalitarian government and question Mussolini's alliance with Nazi Germany. Like many other brilliant European physicists in the 1930s, Rossi became a political refugee from fascism. Together with his new bride, Nora Lombroso, he departed Italy in 1938.

The couple arrived in the United States in 1939 after short stays in Denmark and the United Kingdom. Rossi eventually joined the faculty at Cornell University in 1940 and remained with that university as an associate professor until 1943. In spring 1943 Rossi had his official status as an "enemy alien" changed to "cleared to top secret," so he joined the many other refugee nuclear physicists at the Los Alamos National Laboratory in New Mexico. Allowed to support the wartime needs of his newly adopted country, Rossi collaborated with many other gifted refugee scientists from Europe as they developed the atomic bomb under the top-secret program called the Manhattan Project. Rossi used all his skills in radiation detection instrumentation to provide his colleague, the great nuclear physicist ENRICO FERMI, an ultrafast measurement of the exponential growth of the chain reaction in the world's first plutonium bomb (called Trinity) that was tested near Alamogordo, New Mexico, on July 16, 1945. In a one-microsecond oscilloscope trace Rossi's instrument captured the rising intensity of gamma rays from the bomb's supercritical chain reaction—marking the precise historical moment before and after the age of nuclear weapons.

Following World War II Rossi left Los Alamos in 1946 to become a professor of physics at the Massachusetts Institute of Technology (MIT). In 1966 Rossi became an institute professor, an academic rank at MIT reserved for scholars of great distinction. Upon retirement in 1971 the university honored his great accomplishments by bestowing on him the distinguished academic rank of institute professor emeritus.

Early in his career Rossi's experimental investigations of cosmic rays and their interactions with matter helped establish the foundation of modern high-energy particle physics. Cosmic rays are extremely energetic nuclear particles that enter Earth's atmosphere from outer space at velocities approaching that of light. When cosmic rays collide with atoms in the upper atmosphere, they produce cascades of numerous short-lived subatomic particles, such as mesons. Rossi carefully measured the nuclear particles associated with such cosmic ray showers and effectively turned Earth into one giant nuclear physics laboratory. Rossi began his detailed study of cosmic rays in 1929, when only a few scientists had an interest in them or realized their great importance. That year, to support his cosmic ray experiments, he invented the first electronic circuit for recording the simultaneous occurrence of three or more electrical pulses. This circuit is now widely known as the *Rossi coincidence circuit.* It not only became one of the fundamental electronic devices used in high-energy nuclear physics research, it also represented the first electronic AND circuit—a basic element in modern digital computers. While at the University of Florence Rossi demonstrated in 1930 that cosmic rays were extremely energetic, positively charged nuclear particles that could pass through more than a meter of lead. Through years of research Rossi helped remove much of the mystery surrounding the *Höhenstrahlung* ("radiation from above") first detected by VICTOR HESS in 1911–12.

By the mid-1950s large particle accelerators had replaced cosmic rays in much of nuclear particle physics research, so Rossi used the arrival of the space age (late 1957) to became a pioneer in two new fields within observational astrophysics, space plasma physics and X-ray astronomy. In 1958 he focused his attention on the potential value of direct measurements of ionized interplanetary gases by space probes and Earth-orbiting satellites. He and his colleagues constructed a detector (called the *MIT plasma cup*) that flew into space onboard NASA's *Explorer 10* satellite. Launched in 1961, this instrument discovered the *magnetopause,* the outermost boundary of the magnetosphere beyond which Earth's magnetic field loses its dominance.

Then, in 1962 Rossi collaborated with Riccardo Giacconi (then at American Science and Engineering, Inc.) and launched a sounding rocket from White Sands, New Mexico, with an early grazing incidence X-ray mirror as its payload. With funding from the U.S. Air Force, the scientists mainly hoped to observe any X-rays scattered from the Moon's surface as a result of interactions with energetic atomic particles from the solar wind. To their great surprise, the rocket's payload detected the first X-ray source from beyond the solar system—Scorpius X-1, the brightest and most persistent X-ray source in the sky. Rossi's fortuitous discovery of this intense cosmic X-ray source marked the beginning of extrasolar (cosmic) X-ray astronomy.

Other astronomers performed optical observations of Scorpius X-1 and found a binary star system consisting of a visually observable ordinary dwarf star and a suspected neutron star. In this type of so-called X-RAY BINARY star system, matter drawn from the normal star becomes intensely heated and emits X-rays as it falls to the surface of its neutron star companion.

Rossi was a member of many important scientific societies, including the National Academy of Sciences and the American Academy of Arts and Sciences. A very accomplished writer, Rossi's books included *High Energy Particles* (1952), *Cosmic Rays* (1964), *Introduction to the Physics of Space* (coauthored with Stanislaw Olbert in 1970), and the insightful autobiography *Moments in the Life of a Scientist* (1990). He received numerous awards, including the Gold Medal of the Italian Physical Society (1970), the Cresson Medal from the Franklin Institute, Philadelphia (1974), the Rumford Award of the American Academy of Arts and Sciences (1976), and the U.S. National Science Medal (1985). Rossi's scientific genius laid many of the foundations of high-energy physics and astrophysics. He died at home in Cambridge, Massachusetts, on November 21, 1993. As a tribute to his contributions to X-ray astronomy and space science, NASA named its 1995 X-ray astronomy satellite the *Rossi X-Ray Timing Explorer* (RXTE).

Rossi X-Ray Timing Explorer (RXTE) A NASA spacecraft designed to study the temporal and broadband spectral phenomena associated with stellar and galactic systems containing compact X-ray–emitting objects. The X-ray energy range observed by this spacecraft extended from 2 to 200 kiloelectron volts (keV), and the time scales monitored varied from microseconds to years.

The 3,200-kg spacecraft carried three special instruments to measure X-ray emissions from deep space—the proportional counter array (PCA), the high-energy X-ray timing experiment (HEXTE), and the all-sky monitor (ASM). The PCA and the HEXTE worked together to form a large X-ray observatory that was sensitive to X-rays from 2 to 200 keV. The ASM instrument observed the long-term behavior of X-ray sources and also served as a sentinel, monitoring the sky and enabling the spacecraft to swing rapidly to observe targets of opportunity with its other two instruments. Working together, these instruments gathered data about interesting X-ray emissions from the vicinity of black holes and from neutron stars and white dwarfs, along with telltale energetic electromagnetic radiation from exploding stars and active galactic nuclei.

NASA successfully launched this spacecraft into a 580-km-altitude circular orbit around Earth on December 30, 1995, with an expendable Delta II rocket from Cape Canaveral Air Force Station, Florida. Following launch NASA renamed the spacecraft the *Rossi X-Ray Timing Explorer* (RXTE) in honor of BRUNO B. ROSSI, the distinguished Italian-American physicist who helped pioneer the field of X-ray astronomy. This spacecraft is also referred to as *Explorer 69* and the *NASA X-Ray Timing Explorer.*

See also X-RAY ASTRONOMY.

rotate To turn about an internal axis. Said especially of celestial bodies; for example, Earth rotates, or spins, on its axis once every 24 hours (approximately).

rotating service structure (RSS) An environmentally controlled facility at a launch pad that is used for inserting payloads vertically into the space shuttle orbiter's cargo bay.

See also SPACE TRANSPORTATION SYSTEM.

rotation 1. The turning of an object (especially a celestial body) about an axis within the body, such as the daily rotation of Earth. 2. One turn (i.e., a rotation) of a body about an internal axis. For example, Earth has a sidereal rotation period of 23.9345 hours.

See also ORBITS OF OBJECTS IN SPACE.

round off To adjust or delete less significant digits from a number and possibly apply some rule of correction to the part retained.

rover A crewed or robotic vehicle used to explore a planetary surface.

See also LUNAR ROVER(S); MARS EXPLORATION ROVER MISSION (MER); MARS SURFACE ROVER(S).

Rover Program The name given to the overall U.S. nuclear rocket development program conducted from 1959 to 1973. It was envisioned that nuclear (thermal) rockets, placed in Earth orbit by the giant Saturn V launch vehicle, would propel a human expedition to Mars sometime in the 1980s and safely return the astronauts to Earth.

See also NUCLEAR ROCKET; SPACE NUCLEAR PROPULSION.

Royal Greenwich Observatory (RGO) The famous British observatory founded by King Charles II in 1675 at Greenwich in London. The original structure is now a museum, while the RGO is presently located at Cambridge and serves as the principal astronomical institute of the United Kingdom. Until 1971 the Astronomer Royal also served as the RGO director, but these positions are now separate appointments.

RP-1 Rocket propellant number one (RP-1)—a common hydrocarbon-based, liquid-propellant rocket fuel that is essentially a refined kerosene mixture. It is relatively inexpensive and generally burned with liquid oxygen (LOX).

See also PROPELLANT; ROCKET.

rubber-base propellant A solid-propellant mixture in which the oxygen supply is obtained from a perchlorate and the fuel is provided by a synthetic rubber compound.

See also PROPELLANT; ROCKET.

rumble In rocketry, a form of combustion instability, especially in a liquid-propellant rocket engine, characterized by a low-pitched, low-frequency rumbling noise.

runaway greenhouse An environmental catastrophe during which the greenhouse effect produces excessively high global temperatures that cause all the liquid (surface) water on a life-bearing planet to evaporate permanently. *Compare with* ICE CATASTROPHE.

See also GLOBAL CHANGE; GREENHOUSE EFFECT.

Russell, Henry Norris (1877–1957) American *Astronomer, Astrophysicist* Henry Norris Russell was one of the most influential astronomers in the first half of the 20th century. As a student, professor, and observatory director, he worked nearly 60 years at Princeton University, a truly productive period that also included vigorous retirement activities as professor emeritus and observatory director emeritus. Primarily a theoretical astronomer, he made significant contributions in spectroscopy and astrophysics. Independent of EJNAR HERTZSPRUNG, Russell investigated the relationship between absolute stellar magnitude and a star's spectral class. By 1913 their independent but complementary efforts resulted in the development of the famous Hertzsprung-Russell (HR) diagram—a diagram of fundamental importance to all modern astronomers who wish to understand the theory of stellar evolution. Russell also performed pioneering studies of eclipsing binaries and made preliminary estimates of the relative abundance of elements in the universe. Often called "the dean of American astronomers," he served the astronomical community well as a splendid teacher, writer, and research adviser.

Russell was born on October 25, 1877, in Oyster Bay, New York. The son of a Presbyterian minister, he received his first introduction to astronomy at age five, when his parents showed him the 1882 transit of Venus across the Sun's disk.

He completed his undergraduate education at Princeton University in 1897, graduating with the highest academic honors ever awarded by that institution, namely insigni cum laude (with extraordinary honor). Russell remained at Princeton for his doctoral studies in astronomy, again graduating with distinction (summa cum laude) in 1900.

Following several years of postdoctoral research at Cambridge University, he returned to Princeton in 1905 to accept an appointment as an instructor in astronomy. He remained affiliated with Princeton for the remainder of his life. He became a full professor in 1911 and the following year became director of the Princeton Observatory. He remained in these positions until his retirement in 1947. Starting in 1921 Russell also held an additional appointment as a research associate of the Mount Wilson Observatory in the San Gabriel Mountains just north of Los Angeles, California. In 1927 he received an appointment to the newly endowed C. A. Young Research Professorship at Princeton, a special honor bestowed on him by his undergraduate classmates from the class of 1897.

Starting with his initial work at Cambridge University on the determination of stellar distances, Russell began to assemble data from different classes of stars. He noticed these data related spectral type and absolute magnitude and soon concluded (independent of Hertzsprung) that there were actually two general types of stars, giants and dwarfs. By the early 1910s their independent but complementary efforts resulted in the development of the famous Hertzsprung-Russell (HR) diagram.

The Hertzsprung-Russell diagram is actually just an innovative graph that depicts the brightness and temperature characteristics of stars. However, since its introduction by Russell at a technical meeting in 1913, it has become one of the most important tools in modern astrophysics. So-called dwarf stars, including our Sun, lie on the main sequence of the HR diagram, the well-populated region that runs from the top left to lower right portions of this graph. Giant and supergiant stars lie in the upper right portion of the diagram above the main sequence band. Finally, huddled down in the lower left portion of the HR diagram below the main sequence band are the extremely dense, fading cores of burned-out stars known collectively as white dwarfs. This term is sometimes misleading because it actually applies to a variety of compact, fading stars that have experienced gravitational collapse at the ends of their life cycles.

Following his pioneering work in stellar evaluation, Russell engaged in equally significant work involving eclipsing binaries. An eclipsing binary is a binary star that has its orbital plane positioned with respect to Earth such that each component star is totally or partially eclipsed by the other star every orbital period. With his graduate student HARLOW SHAPLEY, Russell analyzed the light from such stars to estimate stellar masses. Later he collaborated with another assistant, Charlotte Emma Moore Sitterly (1898–1990), in using statistical methods to determine the masses of thousands of binary stars.

Before the discovery of nuclear fusion, Russell tried to explain stellar evolution in terms of gravitational contraction and continual shrinkage and used the Hertzsprung-Russell diagram as a flowchart. When HANS BETHE and other physicists began to associate nuclear fusion processes with stellar life cycles, Russell abandoned his contraction theory of stellar evolution. However, the basic information he presented in the HR diagram remained very useful.

In the late 1920s Russell performed a detailed analysis of the Sun's spectrum, showing hydrogen was a major constituent. He noted the presence of other elements as well as their relative abundances. Extending this work to other stars, he postulated that most stars exhibit a similar general combination of relative elemental abundances (dominated by hydrogen and helium) that became known as the "Russell mixture." This work reached the same general conclusion that was previously suggested in 1925 by the British astronomer CECILIA HELENA PAYNE-GAPOSCHKIN.

Russell was an accomplished teacher, and his excellent two-volume textbook, *Astronomy,* jointly written with Raymond Dugan and John Stewart, appeared in 1926–27. It quickly became a standard text in astronomy curricula at universities around the world. His *Solar System and Its Origin* (1935) served as a pioneering guide for future research in astronomy and astrophysics. Even after his retirement from Princeton in 1947, Russell remained a dominant force in American astronomy. Honored with appointments as both an emeritus professor and an emeritus observatory director, he pursued interesting areas in astrophysics for the remainder of his life.

He received recognition from many organizations, including the Gold Medal of the Royal Astronomical Society (1921), the Henry Draper Medal from the National Academy of Sciences (1922), the Rumford Prize from the American Academy of Arts and Sciences (1925), and the Bruce Gold Medal from the Astronomical Society of the Pacific (1925). He died in Princeton, New Jersey, on February 18, 1957.

Russian Space Agency (RKA) An agency of the Russian government formed after the breakup of the former Soviet Union. The acronym RKA stands for Rosaviacosmos in Russian. This agency uses the technology and launch sites that belonged to the former Soviet space program and provides centralized control of Russia's civilian space program, including all crewed and uncrewed nonmilitary space flights. The most significant current program is the agency's partnership with and active participation in the *INTERNATIONAL SPACE STATION* (ISS) program. The organizational predecessors to RKA directed the former Soviet civilian space program during the cold war and sponsored a high-profile space technology competition with the United States, including the launch of the world's first artificial satellite (*SPUTNIK 1*) and the race to place human explorers on the Moon. The table provides a chronological summary of selected Russian (and former Soviet Union) space achievements.

The RKA employs only a few hundred people and contracts out much of the work. The prime contractor used by the RKA is the S. P. Korolev RSC Energia, a leader in the Russian rocket and space industry. Established in 1946, S. P. Korolev Rocket and Space Corporation (and its predecessor organizations) became a pioneer in practically all areas of rocket and space technology. Today it serves as the prime contractor for crewed space stations, crewed spacecraft, and space systems constructed to support human space activities.

Chronological Summary of Selected Russian Space Achievements

1948	Launch of the R-1 ballistic missile (developed by Sergei P. Korolev).
1957	Launch of the R-7 intercontinental ballistic missile.
1957	Launch of the first artificial Earth satellite (*Sputnik 1*).
1959	The interplanetary probe *Luna-1* accomplishes the first flyby of the Moon.
1959	*Luna-2* impacts on the Moon's surface carrying national symbols of the USSR.
1959	Interplanetary probe *Luna-3* takes first pictures of Moon's farside.
1961	First human to orbit around Earth (Cosmonaut Yuri Gagarin onboard *Vostok-1* spacecraft).
1963	First woman in space (Cosmonaut Valentina Tereshkova onboard *Vostok-6* spacecraft).
1965	First extravehicular activity (space walk) performed by Cosmonaut Alexei Leonov.
1965	Launch of interplanetary probe *Venera-3*, which, in 1966 became the first spacecraft to reach the surface of Venus.
1966	*Luna-9* makes first soft landing on the Moon and transmits images from the surface.
1966	*Luna-10* becomes the first artificial satellite of the Moon.
1967	First automatic docking and undocking of two unmanned spacecraft.
1969	First docking of crewed spacecraft (*Soyuz-4* and *Soyuz-5*) and crew transfer from one spacecraft to the other in outer space.
1969	First formation flying of three manned spacecraft, *Soyuz-6*, *Soyuz-7* and *Soyuz-8*, during which they maneuvered relative to each other while ground facilities provided simultaneous support for the three spacecraft.
1971	Launch of the first long-duration crewed space station (*Salyut*).
1975	First international on-orbit docking of spacecraft (*Soyuz-19* and Apollo spacecraft).
1978	Flight of the first space logistics vehicle (*Progress-1*), which resupplied propellant to the propulsion system of the *Salyut-6* space station.
1978	Beginning of participation by international crews onboard *Salyut-6* space station.
1984	First space walk of a female cosmonaut (S.E. Savitskaya).
1986	Beginning of deployment of the *Mir* space station.
1995	Cosmonaut Valeri Polyakov sets an orbital flight endurance record by continuously staying in space for 437 days 17 hours 58 minutes.
1995	Start of the international space shuttle–*Mir* rendezvous and docking missions.
1996	American astronaut Shannon Lucid sets female space travel endurance record by continuously staying in space for 188 days 4 hours (including 183 days 23 hours onboard *Mir*).
1998	Launch of the functional and cargo module *Zarya*—the first module of the *International Space Station*.
1999	Cosmonaut Sergei Avdeev sets a cumulative space travel endurance record of 747 days 14 hours 12 minutes (accrued over three missions).
2000	Successful in-orbit docking of *Zvezda* Service Module with the *Zarya-Unity* module, making the space station ready for operation by human crews.
2000	*Soyuz TM-31* spacecraft delivered the first crew to the International Space Station (Russian cosmonauts Yu P. Gidzenko and S. K. Krikalev and American astronaut W. Shepherd); this was the start of the permanently crewed mode of ISS operations.
2001	Decommissioned *Mir* space station safely deorbited into a remote area of the Pacific Ocean.
2001	*Soyuz TM-32* spacecraft delivers the first visiting crew to the *International Space Station* (cosmonauts Yu. M. Baturin, T. A. Musabaev, accompanied by the first space tourist, an American businessman named Dennis Tito); the event inaugurates a new line of commercial space activities, space tourism.
2003	Following the space shuttle *Columbia* accident, Soyuz spacecraft missions provide access to the *International Space Station* and support the rotation of astronaut and cosmonaut crews.

Source: S. P. Korolev RSC Energia

The military counterpart of the RKA is the Military Space Forces (VKS). The VKS controls Russia's Plesetsk Cosmodrome, one of the world's busiest launch facilities. The RKA and VKS share control of the Baikonur Cosmodrome in Kazakhstan (now an independent republic). At this launch facility the RKA reimburses the VKS for the wages of many of the flight controllers during civilian launches. The RKA and VKS also share control of the Gagarin Cosmonaut Training Center, located at Star City (Zvezdny Gorodok) near Moscow.

Ryle, Sir Martin (1918–1984) British *Radio Astronomer* Sir Martin Ryle established a center for radio astronomy at Cambridge University after World War II and developed in about 1960 the technique of aperture synthesis. He shared the 1974 Nobel Prize in physics with ANTONY HEWISH for the discovery of the pulsar and their collaborative research activities in radio astronomy, including the pioneering application of aperture synthesis. Ryle also served as England's 12th Astronomer Royal from 1972 to 1982.

safe/arm system (S/A) Mechanism in a solid-propellant rocket motor igniter that, when in the "safe" condition, physically prevents the initiating charge from propagating to the energy release system, which would then ignite the main propellant charge. It is an aerospace safety device intended to prevent an accidental or premature ignition of a solid rocket motor.

Safeguard An early, midcourse-, and terminal-phase ballistic missile defense system deployed by the United States in North Dakota in 1975. It was dismantled in 1976 because of its limited cost effectiveness.

See also BALLISTIC MISSILE DEFENSE.

safety analysis The determination of potential sources of danger and recommended resolutions in a timely manner. A safety analysis addresses those conditions found in either the hardware-software systems, the human-machine interface, or the human-environment relationship (or combinations thereof) that could cause the injury or death of aerospace personnel, damage to or loss of the aerospace system, injury or loss of life to the public, or harm to the environment.

safety critical 1. (noun) Aerospace facility, support, test, and flight systems containing: (1) pressurized vessels, lines, and components; (2) propellants, including cryogenic propellants; (3) hydraulics and pneumatics; (4) high voltages; (5) ionizing and nonionizing radiation sources; (6) ordnance and explosive devices or devices used for ordnance and explosive checkout; (7) flammable, toxic, cryogenic, or reactive elements or compounds; (8) high temperatures; (9) electrical equipment that operates in the area where flammable fluids or solids are located; (10) equipment used for handling aerospace program hardware; and (11) equipment used by personnel for walking or as work platforms. 2. (adjective) Describes something (e.g., a function, a piece of equipment, or a set of data) that may affect the safety of aerospace ground and flight personnel, the aerospace system or vehicle, payloads, the general public, or the environment.

Sagan, Carl Edward (1934–1996) American *Astronomer, Science Writer* Carl Sagan investigated the origin of life on Earth and the possibility of extraterrestrial life. In the early 1960s he suggested that a runaway greenhouse effect could be operating on Venus. In the 1970s and beyond he used his collection of popular books and a television series, *Cosmos,* to effectively communicate science, especially astronomy, astrophysics, and cosmology, to the general public.

SAINT An early U.S. satellite inspector system designed to demonstrate the feasibility of intercepting, inspecting, and reporting on the characteristics of satellites in orbit.

See also ANTISATELLITE (ASAT) SPACECRAFT.

Salyut A series of Earth-orbiting space stations launched by the Soviet Union (now Russian Federation) in the 1970s and early 1980s to support a variety of civilian and military missions. The word *salyut* means "salute." *Salyut 1* to *5* often are regarded as first-generation crewed spacecraft in this series, while *Salyut 6* and *7* are considered to be an evolved, second-generation design. One major design difference is the fact that the first generation had just one docking port, while the second generation had two docking ports.

Salyut 1 was launched into Earth orbit on April 19, 1971. It functioned in a civilian research capacity as an observation platform, gathering data in the fields of astronomy, Earth resources monitoring, and meteorology. On June 29 the three cosmonauts who had just spent 24 days aboard the space station died during reentry operations. Cosmonauts Georgi Dobrovolsky, Vladislav Volkov, and Victor Patseyev suffocated when a vent valve on their *Soyuz 11* spacecraft opened and air rushed out of the capsule as they separated from the space station. After the capsule touched down, startled recovery crews found all three cosmonauts dead. The *Salyut 1* station decayed from orbit on October 11, 1971.

Salyut 2, considered the first "military mission" Salyut station, was launched on April 3, 1973. However, before a cosmonaut crew could be sent to this station it suffered a catastrophic explosion on April 14. This explosion tore away the space station's solar panels, telecommunications equipment, and docking apparatus. The derelict space station then tumbled out of orbit and on about May 28 burned up in Earth's atmosphere. *Salyut 3,* the first operational military Salyut, was launched into a low (219-km by 270-km) orbit in late June 1974. Apparently the cosmonaut crews performed military photoreconnaissance operations, although little information has been released about this spacecraft or the other military space station, *Salyut 5.*

Salyut 4, launched on December 26, 1974, was similar to *Salyut 1.* Its crews performed civilian missions including astronomical, Earth resources and biomedical observations, and materials processing experiments.

Salyut 5, the last station in this series dedicated primarily to military activities, was launched on June 22, 1976. In addition to military mission equipment, this station also had materials processing equipment.

Salyut 6 was launched on September 29, 1977. Its design represented an evolution of the earlier generation of civilian Salyut spacecraft. This station functioned successfully from 1977 to 1981, more than twice its design lifetime of 18 months. The addition of a second docking port gave this second-generation space station much greater flexibility. The use of the Progress supply ship also improved orbital operations. Civilian research activities included microgravity materials processing, Earth resource photography, and astronomical observations. The last cosmonaut crew left the *Salyut 6* on April 25, 1977, and the station reentered the atmosphere and was destroyed in July 1982.

Salyut 7 was launched on April 19, 1982, and, like the *Salyut 6* crew visit program, enjoyed a pattern of progressively longer flights by the resident cosmonaut crew. However, in early 1985, when it was not occupied by a crew, this station suffered several major technical problems. In June 1985 a cosmonaut crew (*Soyuz T-13* mission) was sent to repair and reoccupy the derelict space station. Repairs enabled it to perform one additional long-duration flight. However, the final *Salyut 7* crew suddenly departed in their spacecraft on November 11, 1985, terminating the mission early because of a serious illness involving one cosmonaut.

Building on the experience of the Salyut space stations, the Russian launched their third-generation space station, called *Mir,* on February 19, 1986. In March a crew from the *Mir* space station flew to the *Salyut 7,* visited for about six weeks, and then returned to the *Mir.* After this "space station" to "space station" visit, the *Salyut 7* was placed in a parking orbit for possible reuse. However, the uncrewed space station experienced a more rapid orbital decay than anticipated, was permanently abandoned, and reentered Earth's atmosphere in February 1991.

See also *MIR*; SPACE STATION.

Sänger, Eugen (1905–1964) Austrian *Astronautics Pioneer* In the 1920s and 1930s the Austrian astronautics pioneer Eugen Sänger envisioned rocket planes and reusable space transportation systems. The arrival of NASA's Space Transportation System, or space shuttle, about a half century later partially fulfilled some of his far-reaching technical ideas for future of rocket-propelled vehicles. During World War II Sänger directed a rocket research program for the German Air Force and focused his attention on a long-range, winged rocket bomber that could travel intercontinental distances by skipping in and out of Earth's atmosphere.

satellite A smaller (secondary) body in orbit around a larger (primary) body. Earth is a natural satellite of the Sun, and the Moon is a natural satellite of Earth. Human-made spacecraft placed in orbit around Earth are called artificial satellites, or more commonly just satellites.

satellite laser ranging (SLR) In satellite laser ranging (SLR), ground-based stations transmit short, intense laser pulses to a retroreflector-equipped satellite, such as the Laser Geodynamics Satellite (LAGEOS). The roundtrip time of flight of the laser pulse is measured precisely and corrected for atmospheric delay to obtain a geometric range. Transmitting pulses of light to these retroflector-equipped spacecraft with a global network of laser ranging stations (both fixed and mobile) allows scientists to determine both the precise orbit of a satellite and the position of the individual ground stations. By monitoring the carefully measured position of these ground stations over time (e.g., several months to years), researchers can deduce the motion of the ground site locations due to plate tectonics or other geodynamic processes, such as subsidence. In fact, a global network of more than 30 SLR stations already has provided a basic framework for determining plate motion, confirming the anticipated motion for most plates.

The theory of plate tectonics tries to explain how the continents arrived at their current positions and to predict where the continents will be in the future. Scientists theorize that up to about 200 million years ago, one giant continent existed where the Atlantic Ocean is today. It is called *Pangaea,* meaning "all lands." Then about 180 million years ago Pangaea started to break up into several continents. The plates carrying the continents drifted away from one another, and a low layer of rock formed between the plates. At present the plates comprise the solid outer 100 kilometers of Earth. These plates move slowly, typically not faster than about 15 centimeters a year. For example, North America and Europe are moving apart at a rate of about 3 cm per year. Although the accumulated motion of the plates is slow when averaged over time, the effects of their short-term drastic movements (such as during earthquakes) can be catastrophic. Plates may bump into one another, spread apart, or move horizontally past one another, in the process causing earthquakes, building mountains, or triggering volcanoes.

Scientists have used satellite laser ranging techniques for the past two decades to study the motions of Earth. These geodynamic studies include measurements of global tectonic plate motions, regional crustal deformation near plate boundaries,

Earth's gravity field, and the orientation of its polar axis and its spin rate. In the future scientists will also "invert" the traditional SLR system by placing the laser ranging hardware onboard a satellite and then pulsing selected retroreflecting objects on Earth's surface from space. This "inverted" SLR technology, sometimes called a *laser altimeter system* (LAS), can be used to measure ice sheet topography and temporal changes, cloud heights, planetary boundary heights, aerosol vertical structure, and land and water topography. A laser altimeter system also can be used to perform similar laser ranging measurements on other planets, such as Mars.

See also LASER GEODYNAMICS SATELLITE.

Satellite Power System (SPS) A very large space structure constructed in Earth orbit that takes advantage of the nearly continuous availability of sunlight to provide useful energy to a terrestrial power grid. The original SPS concept, called the Solar Power Satellite, was first presented in 1968 by Peter Glaser. The basic concept behind the SPS is quite simple: each SPS unit is placed in geosynchronous orbit above the Earth's equator, where it experiences sunlight more than 99 percent of the time. These large orbiting space structures then gather the incoming sunlight for use on Earth in one of three general ways: microwave transmission, laser transmission, or mirror transmission.

In the fundamental microwave transmission SPS concept, solar radiation would be collected by the orbiting SPS and converted into radio frequency (RF) or microwave energy. This energy would then be precisely beamed to a receiving antenna on Earth, where it would be converted into electricity. The process of beaming energy using the radio frequency portion of the electromagnetic spectrum is often referred to as *wireless power transmission* (WPT). In the laser transmission SPS concept, solar radiation would be converted into infrared laser radiation, which would then be beamed down to special receiving facilities on the Earth for the production of electricity. Finally, in the mirror transmission SPS concept, very large (kilometer dimensions) orbiting mirrors would be used to reflect raw (unconverted) sunlight directly to terrestrial energy conversion facilities 24 hours a day.

In the microwave transmission SPS concept described, incoming sunlight would be converted into electrical power on the giant space platform by either photovoltaic (solar cell) or heat engine (thermodynamic cycle turbogenerator) techniques. This electric power would then be converted at high efficiency into microwave energy. The microwave energy, in turn, would be focused into a beam and aimed precisely at special ground stations. The ground station receiving antenna (also called a *rectenna*) would reconvert the microwave energy into electricity for distribution in a terrestrial power grid.

Because of its potential to relieve long-term national and global energy shortages, the Satellite Power System concept was studied extensively in the 1970s after it was first introduced and revisited again in the 1990s. The original series of studies embraced a sweeping vision of extraterrestrial development and suggested that the giant SPS units could be constructed using extraterrestrial materials gathered from the Moon and Earth-crossing asteroids, with manufacturing and construction activities being accomplished by thousands of space workers who would reside with their families in large, permanent space settlements at selected locations in cislunar space. For example, a permanent space settlement might be developed at Lagrangian libration point 5, popularly referred to as L_5.

Other early SPS studies considered the development and construction of SPS units using only terrestrial materials. These materials were to be placed in low Earth orbit (LEO) by a fleet of special heavy-lift launch vehicles (HLLV). The main emphasis was to significantly reduce the cost of launching large quantities of mass into low Earth orbit. Human-assisted SPS construction work would be accomplished in LEO, perhaps at the site of a permanent space station or space construction base. Assembled SPS sections would then be ferried to geosynchronous Earth orbit (GEO) by a fleet of orbital transfer vehicles (OTVs). At GEO a crew of space workers would complete final assembly and prepare the SPS unit for operation.

In 1980 the U.S. Department of Energy (DOE) and National Aeronautics and Space Administration (NASA) defined an SPS reference system to serve as the basis for conducting initial environmental, societal, and comparative assessments; alternative concept trade-off studies; and supporting critical technology investigations. This SPS reference system was, of course, not an optimum or necessarily preferred system design. It does, however, represent one potentially plausible approach for achieving SPS concept goals and has been cited extensively in the technical literature.

The main technical characteristics of this reference SPS system are summarized in the SPS table. The proposed configuration would provide 5 gigawatts (GW) of electric power at the terrestrial grid interface. (A gigawatt is 10^9 watts.) In the reference scenario 60 such SPS units would be placed in geostationary orbit and thus provide some 300 gigawatts of electric power for use on Earth. It was optimistically estimated that only about six months would be required to construct each SPS unit, once the appropriate space infrastructure had been developed. Despite the exciting visions of permanent human settlements in space, lunar mining bases, and harvesting the Sun's energy to provide plentiful "nonpolluting" power to an energy-hungry world, the original SPS concept remained just that—a concept. No single government or even consortium of nations exhibited the desire to commit the resources to develop the necessary space infrastructure.

More recently, however, careful examination of global energy needs in the post-2020 period and concerns about the rising levels of carbon dioxide (CO_2) emissions due to an anticipated increase in fossil fuel combustion have encouraged scientists and strategic planners to revisit the SPS concept. The major purpose of this "new look" effort (conducted by NASA in the 1990s) was to determine whether a contemporary solar power satellite and its associated terrestrial support systems could be defined that would deliver energy into terrestrial electrical power grids at prices equal to or below ground alternatives in a variety of terrestrial markets projected beyond 2020. This "new" SPS concept must also be implemented without major environmental consequences. Of parallel importance is the constraint that the "new" SPS design concept must involve a feasible system that could be

SPS Reference

System Characteristics

General capability (utility interface)
300 GW—total
5 GW—single unit

Number of units: 60
Design life: 30 years
Deployment rate: 2 units/year

Satellite

Overall dimensions: 10 x 5 x 0.5 km
Structural material: graphite composite

Satellite mass: 35–50 x 10^6 kg
Geostationary orbit: 35,800 km

Energy Conversion System

Photovoltaic solar cells; silicon or gallium aluminum arsenide

Power Transmission and Reception

DC-RF conversion: klystron
Transmission antenna diameter: 1 km
Frequency: 2.45 GHz
Rectenna dimensions (at 35° latitude)
Active area: 10 x 13 km
Including exclusion area: 12 x 15.8 km

Rectenna construction time: ≈ 2 years
Rectenna peak power density: 23 mW/cm^2
Power density at rectenna edge: 1 mW/cm^2
Power density at exclusion edge: 0.1 mW/cm^2
Active, retrodirective array control system with pilot beam reference

Space Transportation System

Earth-to-LEO—Cargo: vertical takeoff, winged two-stage (425-metric-ton payload)
Personnel: modified shuttle
LEO-to-GEO—Cargo: electric orbital transfer vehicle
Personnel: two-stage liquid oxygen/liquid hydrogen

Space Construction

Construction staging base—LEO: 480 km
Final construction—GEO: 35,800 km
Satellite construction time: 6 months

Construction crew: 600
System maintenance crew: 240

Source: NASA/Department of Energy.

developed at only a fraction of the initial capital investment projected for the SPS Reference System of the 1970s. The original SPS concept required that more than $250 billion (1996 equivalent dollars) be invested before the first commercial kilowatt-hour could be delivered.

In response to the NASA-sponsored "new look" effort, several interesting concepts for a contemporary SPS have emerged. These concepts generally emphasize the use of advanced power conversion technologies, modularity, extensive use of robotic systems for assembly, and minimal use of human space workers. More modest power output goals for each orbiting platform have also appeared. We will examine one of the most interesting, a concept called the *Sun tower,* a large tethered-satellite system employing highly modularized power generation.

The total Sun tower concept involves a constellation of medium-scale, gravity-gradient-stabilized, RF-transmitting space solar power (SSP) systems. Each satellite resembles a large Earth-pointing "sunflower." Carrying this analogy further, the Sun-pointing "leaves" on the stalk are the modular solar collector units. Each advanced photovoltaic (sunlight-to-electrical) power conversion unit is between 50 and 100 meters in diameter and produces approximately one megawatt of electric power. The Earth-pointing "face" of the flower is the RF transmitter array. About 200 megawatts (MW) of RF power is transmitted from the space platform, which travels around Earth in an initial sun-synchronous orbit of about 1,000-kilometers altitude. Multiple satellites would be required to maintain constant power generation at a particular ground site. The nominal ground receiver is a 4-km-diameter site that has direct electrical feed into the power utilities interface. The Sun tower concept is thought to be achievable with a projected cost to first power on the order of between $8 and $15 billion (1996 equivalent dollars).

It is perhaps too early to validate the SPS concept (original or "new look") or to totally dismiss it. What can be stated at this time, however, is that the controlled beaming to Earth of solar energy (either as raw concentrated sunlight or as converted microwaves or laser radiation) represents an interesting alternative energy pathway in this century and beyond. This "space-based" energy economy could also serve as a very powerful stimulus toward the creation of an extraterrestrial civilization. However, any SPS concept also involves potential impacts on the terrestrial environment. Some of these environmental impacts are comparable in type and magnitude to those arising from other large-scale terrestrial energy technologies; others are unique to the SPS concept. Some of these unique environmental and health impacts are potential adverse effects on Earth's upper atmosphere from launch vehicle effluents and from energy beaming (that is, microwave heating of the ionosphere); potential hazards to terrestrial life-forms from nonionizing radiation (microwave or infrared laser); electromagnetic interference with other spacecraft, terrestrial communications, and astronomy; and the potential hazards to space workers, especially exposure to ionizing radiation doses well beyond currently accepted industrial criteria. These issues will have to be favorably resolved if the SPS concept is to emerge as a major pathway in humanity's creative use of the resources of outer space.

See also SPACE RESOURCES; SPACE SETTLEMENT.

satellite reconnaissance The use of specially designed Earth-orbiting spacecraft to collect intelligence data. Photoreconnaissance satellites are an example.

See also NATIONAL RECONNAISSANCE OFFICE; RECONNAISSANCE SATELLITE.

saturated liquid In thermodynamics, a substance that exists as a liquid at the saturation temperature and pressure of a mixture of vapor and liquid.

See also THERMODYNAMICS.

saturated vapor In thermodynamics, a substance that exists as a vapor at the saturation temperature and pressure of a mixture of vapor and liquid.

See also THERMODYNAMICS.

saturation A condition in which a mixture of vapor and liquid can exist together at a given temperature and pressure. The vapor dome on a traditional pressure (p)–specific volume (v) plot for a simple compressible substance is bounded by the saturated liquid line on the left and the saturated vapor line on the right. The point where the two lines meet at the peak of this dome-shaped region is called the critical point. It represents the highest pressure and temperature at which distinct liquid and vapor (gas) phases of a substance can coexist.

See also THERMODYNAMICS.

Saturn (launch vehicle) Family of expendable launch vehicles developed by NASA (WERNHER VON BRAUN's team at the Marshall Space Flight Center in Huntsville, Alabama) to support the Apollo Project. The Saturn 1B was used initially to launch Apollo lunar spacecraft into Earth orbit to help the astronauts train for the crewed flights to the Moon. The first launch of a Saturn 1B vehicle with an unmanned Apollo spacecraft took place in February 1966. A Saturn 1B vehicle launched the first crewed Apollo flight, *Apollo 7,* on October 11, 1968. After completion of the Apollo Project, the Saturn 1B was used to launch the three crews (three astronauts in each crew) for the *Skylab* space station in 1973. Then in 1975 a Saturn 1B launched the American astronaut crew for the Apollo-Soyuz Test Project, a joint American-Soviet docking mission. With an Apollo spacecraft on top, the Saturn 1B vehicle was approximately 69 meters tall. This expendable launch vehicle developed 7.1 million newtons of thrust at liftoff.

The Saturn V rocket, America's most powerful staged rocket, successfully carried out the ambitious task of sending astronauts to the Moon during the Apollo Project. The first launch of the Saturn V vehicle, the unmanned *Apollo 4* mission, occurred on November 9, 1967. The first crewed flight of the Saturn V vehicle, the *Apollo 8* mission, was launched in December 1968. This historic mission was the first human flight to the Moon. The three astronauts aboard *Apollo 8* circled the Moon (but did not land) and then returned safely to Earth. On July 16, 1969, a Saturn V vehicle sent the *Apollo 11* spacecraft and its crew on the first lunar landing mission. The last crewed mission of the Saturn V vehicle occurred on December 7, 1972, when the *Apollo 17* mission lifted off on the final human expedition to the Moon in the 20th century. The Saturn V vehicle flew its last mission on May 14, 1973, when it lifted the unmanned *SKYLAB* space station into Earth orbit. *Skylab* later was occupied by three different astronaut crews for a total period of 171 days.

All three stages of the Saturn V vehicle used liquid oxygen (LO_2) as the oxidizer. The fuel for the first stage was kerosene, while the fuel for the upper two stages was liquid hydrogen (LH_2). The Saturn V vehicle, with the Apollo spacecraft and its small emergency escape rocket on top, stood 111 meters tall and developed 34.5 million newtons of thrust at liftoff.

See also APOLLO PROJECT; LAUNCH VEHICLE.

Saturn (planet) Saturn is the sixth planet from the Sun and to many the most beautiful celestial object in the solar system. To the naked eye the planet is yellowish in color. This planet is named after the elder god and powerful Titan of Roman mythology (Cronus in Greek mythology), who ruled supreme until he was dethroned by his son Jupiter (Zeus to the ancient Greeks).

Composed mainly of hydrogen and helium, Saturn (with an average density of just 0.7 grams per cubic centimeter) is so light that it would float on water if there were some cosmic ocean large enough to hold it. The planet takes about 29.5 Earth years to complete a single orbit around the Sun, but a Saturnian day is only approximately 10 hours and 30 minutes long.

The first telescopic observations of the planet were made in 1610 by GALILEO GALILEI. The existence of its magnificent ring system was not known until the Dutch astronomer CHRISTIAAN HUYGENS, using a better-resolution telescope, properly identified the ring system in 1655. Galileo had seen the rings but mistook them for large moons on either side of the planet. Huygens is also credited with the discovery in 1655 of Saturn's largest moon, Titan.

Astronomers had very little information about Saturn, its rings, and its constellation of moons until the *Pioneer 11* spacecraft (September 1, 1979), *Voyager 1* spacecraft (November 2, 1980), and *Voyager 2* spacecraft (August 26, 1981) encountered the planet. (*See* figure.) These spacecraft encounters revolutionized our understanding of Saturn and have provided the bulk of the current known information about this interesting giant planet, its beautiful ring system, and its large complement of moons (30 named moons [6 major, 24 minor] discovered and identified at present). The planet now is being observed on occasion by the *HUBBLE SPACE TELESCOPE* and is now being visited for a detailed scientific mission by NASA's *Cassini* spacecraft. The *Cassini* spacecraft's planned four-year investigation of the Saturn system will greatly expand the body of scientific knowledge about this magnificent ringed planet and its family of intriguing moons.

Saturn is a giant planet, second in size only to Jupiter. Like Jupiter, it has a stellar-type composition, rapid rotation, strong magnetic field, and intrinsic internal heat source. Saturn has a diameter of approximately 120,540 kilometers at its equator, but 10 percent less at the poles because of its rapid rotation. Saturn has a mass of approximately 5.68×10^{26} kilograms, the lowest of any planet in the solar system; this indicates that much of Saturn is in a gaseous state. The Saturn table lists contemporary data for the planet.

Although the other giant planets—Jupiter, Uranus, and Neptune—all have rings, Saturn has an unrivaled complex yet delicate ring system of billions of icy particles whirling around the planet in an orderly fashion. The main ring areas stretch from about 7,000 km above Saturn's atmosphere out to the F Ring, a total span of about 74,000 km. Within this vast region the icy particles generally are organized into ringlets, each typically less than 100 km wide. Beyond the F Ring lies the G and E rings, with the latter extending some 180,000 to 480,000 km from the planet's center. The complex ring system

This photographic montage of the Saturn system was prepared from a collection of images taken by NASA's *Voyager 1* spacecraft during its flyby encounter in November 1980. The artistically assembled collection of images shows Dione in the foreground, Saturn rising behind, Tethys and Mimas in the lower right, Titan in the far upper right, and Enceladus and Rhea off Saturn's rings to the left. *(Courtesy of NASA/JPL)*

Physical and Dynamic Properties of the Planet Saturn

Diameter (equatorial)	120,540 km
Mass	5.68×10^{26} kg
Density	0.69 g/cm^3
Surface gravity (equatorial)	9.0 m/s^2 (approx)
Escape velocity	35.5 km/s (approx)
Albedo (visual geometric)	0.5
Atmosphere	Hydrogen (89%), helium (11%), small amounts of methane (CH_4), ammonia (NH_3), and ethane (C_2H_6); water-ice aerosols
Natural satellites	33
Rings (thousands)	Complex system
Period of rotation (a Saturnian day)	0.44 days (approx)
Average distance from the Sun	14.27×10^8 km (9.539 AU) [79.33 light-min]
Eccentricity	0.056
Period of revolution around the Sun (a Saturnian year)	29.46 years
Mean orbital velocity	9.6 km/s
Solar flux at planet (at top of clouds)	15.1 watts/m^2 (at 9.54 AU)
Magnetosphere	Yes (strong)
Temperature (blackbody)	77 K

Source: NASA.

contains a variety of interesting physical features, including kinky rings, clump rings, resonances, spokes, shepherding moons (moons that keep the icy particles in an organized structure), and probably additional as-yet-undiscovered moonlets.

Scientists think that the Saturnian rings resulted from one of three basic processes: a small moon venturing too close to the planet and ultimately getting torn apart by large gravitationally induced tidal forces, some of the planet's primordial material failing to coalesce into a moon, or collisions among several larger objects that orbited the planet.

Saturn has 30 named moons (data related to the most significant of these satellites appears in the tables). Saturn has many small (as yet unnamed) satellites, and the number of these small moons will in all likelihood continue to grow as planetary scientists examine the *Cassini* spacecraft's high-resolution images of the planet and its complex ring system. Saturn's moons form a diverse and remarkable constellation of celestial objects. The largest satellite, Titan, is in a class by itself because of its size and dense atmosphere. The *Cassini* spacecraft and its companion HUYGENS PROBE should help resolve many of the most pressing questions about this intriguing cloud-shrouded moon. The six other major satellites (Iapetus, Rhea, Dione, Tethys, Enceladus, and Mimas) are somewhat similar; all are of intermediate size (some 400 to 1,500 km in diameter) and consist mainly of water ice. Saturn also has many smaller moons, ranging in size from the irregularly shaped Hyperion (about 350 km by 200 km across) to tiny Pan (about 20 km in diameter). Hyperion orbits Saturn with a random, chaotic tumbling motion—an orbital condition perhaps indicative of an ancient shattering collision. Phoebe, the outermost moon of Saturn, orbits in a retrograde direction (opposite the direction of the other moons) in a plane much closer to the ecliptic than to Saturn's equatorial plane.

Titan is the largest and perhaps most interesting of Saturn's satellites. It is the second-largest moon in the solar system and the only one known to have a dense atmosphere. The atmospheric chemistry now taking place on Titan may be similar to those processes that occurred in Earth's atmosphere several billion years ago.

Larger in size than the planet Mercury, Titan's density appears to be about twice that of water ice. Scientists believe, therefore, that it may be composed of nearly equal amounts of rock and ice. Prior to the arrival of the *Cassini* spacecraft in 2004, Titan's surface was hidden from view by a dense, optically thick photochemical haze whose main layer is about 300 km above its surface. Titan's atmosphere is mostly nitrogen, and planetary scientists have speculated that carbon-nitrogen compounds may exist on Titan because of the great abundance of both nitrogen and hydrocarbons. The Cassini mission is helping scientists answer long-standing questions about the nature of Titan's surface.

Since it arrived at Saturn in July 2004, the *Cassini* spacecraft has made many new and exciting discoveries. These new findings include wandering and rubble-pile moons, new and clumpy Saturn rings, splintering storms, and a dynamic magnetosphere. Another discovery involved an interesting tiny moon (about 5 km in diameter) named Polydeuces. Polydeuces is a companion, or Trojan, moon of Dione. Saturn is the only planet known to have moons with companion Trojan moons.

Following its release from the *Cassini* mother spacecraft on December 25, 2004, the *Huygens* probe reached Titan's outer atmosphere after cruising for 20 days. On January 14, 2005, starting at an altitude of about 1,270 km, the probe then successfully descended through Titan's hazy cloud lay-

Physical Data for the Larger Moons of Saturn

Moon	Diameter (km)	Mass (kg)	Density (g/cm3)	Albedo (visual geometric)
Iapetus	1,440	1.6×10^{21}	1.0	0.05–0.5
Titan	5,150	1.35×10^{23}	1.9	0.2
Rhea	1,530	2.3×10^{21}	1.2	0.7
Dione	1,120	1.05×10^{21}	1.4	0.7
Tethys	1,050	6.2×10^{20}	1.0	0.9
Enceladus	500	7.3×10^{19}	1.1	~1.0
Mimas	390	3.8×10^{19}	1.2	0.5

Source: NASA.

Physical and Dynamic Properties of the More Significant Moons of Saturn

Moon	Diameter (km)	Semimajor Axis of Orbit (km)	Period of Rotation (days)
Phoebe	220 (approx.)	12,952,000	550.5 (retrograde)
Iapetus	1,440	3,561,300	79.33
Hyperion	350 × 200	1,481,000	21.28
Titan	5,150	1,221,850	15.95
Rhea	1,530	527,040	4.52
Helene	40 (approx.)	377,400	2.74
Dione	1,120	377,400	2.74
Calypso	30 (approx.)	294,660	1.89
Tethys	1,050	294,660	1.89
Telesto	25 (approx.)	294,660	1.89
Enceladus	500	238,020	1.37
Mimas	390	185,520	0.942
Janus	220 × 160	151,470	0.695
Epimetheus	140 × 100	151,420	0.694
Pandora	110 × 70	141,700	0.629
Prometheus	140 × 80	139,350	0.613
Atlas	40 (approx.)	137,640	0.602
Pan	20 (approx.)	133,580	0.575

Source: NASA.

ers. *Huygens* came to rest on Titan's frozen surface and continued transmitting from the surface for several hours. During the almost four-hour duration descent and on the ground, the probe collected some 350 pictures, which revealed a landscape apparently modeled by erosion with drainage channels, shore-like features, and even pebble-shaped objects on the surface. Titan's surface resembles wet sand or clay with a thin solid crust. The composition appears to be mainly a mix of dirty water ice and hydrocarbon ice. The temperature measured by the probe at ground level was about –180° C. Sampling of Titan's atmosphere from an altitude of 160 km to ground level revealed a uniform mix of methane and nitrogen in the stratosphere. *Huygens* also detected clouds of methane at about 20 km altitude and methane and ethane fog near the surface.

See also CASSINI MISSION; *PIONEER 10, 11* SPACECRAFT; *VOYAGER* SPACECRAFT.

scalar Any physical quantity whose field can be described by a single numerical value at each point in space. A scalar quantity is distinguished from a vector quantity by the fact that a scalar quantity possesses only magnitude, while a vector quantity possesses both magnitude and direction.

scar Aerospace jargon for design features to accommodate the addition or upgrade of hardware at some future time.

scarp A cliff produced by erosion or faulting.

scattering 1. (particle) A process that changes a particle's trajectory. Scattering is caused by particle collisions with atoms, nuclei, and other particles or by interactions with fields of magnetic force. If the scattered particle's internal energy (as contrasted with its kinetic energy) is unchanged by the collision, *elastic scattering* prevails; if there is a change in the internal energy, the process is called *inelastic scattering*. 2. (photon) Scattering also may be viewed as the process by which small particles suspended in a medium of a different refractive index diffuse a portion of the incident radiation in all directions. In scattering, no energy transformation results, only a change in the spatial distribution of the radiation. Along with absorption, scattering is a major cause of the attenuation of radiation by the atmosphere. Scattering varies as a function of the ratio of the particle diameter to the wavelength of the radiation. When this ratio is less than about one-tenth, Rayleigh scattering occurs, in which the scattering process varies inversely as the fourth power of the wavelength. At larger values of the ratio of particle diameter to wavelength, the scattering varies in a complex fashion described by Mie scattering theory; at a ratio on the order of 10, the laws of geometric optics begin to apply.

See also MIE SCATTERING; RAYLEIGH SCATTERING.

Scheiner, Christoph (1573–1650) German *Mathematician, Astronomer* The Jesuit priest Christoph Scheiner was a mathematician and astronomer who, independently of GALILEO GALILEI, designed and used his own telescope to observe the Sun and discovered sunspots in 1611. Attempting to preserve ARISTOTLE's hypothesis of the immutable (unchanging) heavens, Scheiner interpreted the sunspots as being small satellites that encircled the Sun as opposed to changing, moving features of the Sun itself. This interpretation stirred up a great controversy with Galileo, who vigorously endorsed the Copernican revolution in astronomy.

See also SUNSPOT.

Schiaparelli, Giovanni Virginio (1835–1910) Italian *Astronomer* In the 1870s the Italian astronomer Giovanni Virginio Schiaparelli carefully observed Mars and made a detailed map of its surface, including some straight markings that he described as *canali,* meaning "channels" in Italian. Unfortunately, when his reference to these linear features was mistakenly translated into English as "canals," some other astronomers (such as the American astronomer PERCIVAL LOWELL) completely misunderstood Schiaparrelli's meaning and launched a frantic observational search for canals as the presumed artifacts of an intelligent alien civilization. Schiaparelli also worked on the relationship between meteor showers and the passage and/or disintegration of comets.

Schirra, Walter M. (1923–) American *Astronaut, U.S. Navy Officer* Walter (Wally) M. Schirra was selected as one of NASA's original seven Mercury Project astronauts. Of this group of initial American astronauts, he is the only one to have traveled into space as a participant in all three of NASA's 1960s human space flight programs—namely, the Mercury Project, the Gemini Project, and the Apollo Project.

See also APOLLO PROJECT; GEMINI PROJECT; MERCURY PROJECT.

schlieren (German: streaks, striae) 1. Regions of different density in a fluid, especially as shown by special apparatus. 2. Pertaining to a method or apparatus for visualizing or photographing regions of varying density in a fluid field.

See also SCHLIEREN PHOTOGRAPHY.

schlieren photography A method of photography for flow patterns that takes advantage of the fact that light passing through a density gradient in a gas is refracted as if it were passing through a prism. Schlieren photography allows the visualization of density changes, and therefore shock waves, in fluid flow. Schlieren techniques have been used for decades in laboratory wind tunnels to visualize supersonic flow around special scale-model aircraft, missiles, and aerospace vehicles.

See also WIND TUNNEL.

Schmidt, Maarten (1929–) Dutch-American *Astronomer* In 1963 the Dutch-American astronomer Maarten Schmidt discovered the first quasar. This interesting celestial object caught his attention because its spectrum had an enormous redshift, indicating that the very distant object was traveling away from Earth at more than 15 percent of the speed of light.

See also QUASAR.

Schmitt, Harrison H. (1935–) American *Astronaut, Geologist, U.S. Senator* The geologist and astronaut Harrison H. Schmitt was a member of NASA's *Apollo 17* lunar landing mission in December 1972. He and EUGENE A. CERNAN became the last two human beings to walk on the Moon in the 20th century. As the first geologist to personally visit another world, Schmitt explored the Taurus-Littrow region of the lunar surface with a great deal of professional enthusiasm.

Schriever, Bernard (1913–2005) American *U.S. Air Force Officer, Engineer* General Bernard Schriever supervised the rapid development of the Atlas, Thor, and Titan ballistic missiles during the cold war. He created and applied an innovative systems engineering approach that saved a great deal of development time. His rockets not only served the United States in the area of national defense, they also supported the space launch vehicle needs of NASA and the civilian space exploration community.

See also ATLAS (LAUNCH VEHICLE); THOR (ROCKET); TITAN (LAUNCH VEHICLE).

Schwabe, Samuel Heinrich (1789–1875) German *Pharmacist, Amateur Astronomer* As an avid amateur astronomer, the German pharmacist S. Heinrich Schwabe made systematic observations of the Sun for many years in his search for a hypothetical planet (called Vulcan) that 19th-century astronomers assumed traveled around the Sun inside the orbit of Mercury. However, instead of this fictitious planet, Schwabe discovered that sunspots have a cycle of about ten years. He announced his findings in 1843 and eventually received recognition for this discovery in the 1860s.

Schwarzschild, Karl (1873–1916) German *Astronomer* The German astronomer Karl Schwarzschild applied ALBERT EINSTEIN's general relativity theory to very-high-density celestial objects and point masses (singularities). In 1916 he introduced the concept that has become known as the Schwarzschild radius—the zone (event horizon) around a super-dense gravitationally collapsing star from which nothing, not even light, can escape. His insightful theoretical work marked the start of black hole astrophysics.

See also ASTROPHYSICS; BLACK HOLE; RELATIVITY.

Schwarzschild black hole An uncharged black hole that does not rotate; the German astronomer KARL SCHWARZSCHILD hypothesized this basic model of a black hole in 1916.

See also BLACK HOLE.

Schwarzschild radius The radius of the event horizon of a black hole. Named after the German astronomer KARL SCHWARZSCHILD, who applied relativity theory to very-high-density objects and point masses (singularities).

See also BLACK HOLE.

science fiction A form of fiction in which technical developments and scientific discoveries represent an important part of the plot or story background. Frequently, science fiction involves the prediction of future possibilities based on new scientific discoveries or technical breakthroughs. Some of the most popular science fiction predictions waiting to happen are interstellar travel, contact with extraterrestrial civilizations, the development of exotic propulsion or communication devices that would permit us to break the speed-of-light barrier, travel forward or backward in time, and very smart machines and robots.

According to the well-known writer Isaac Asimov (1920–92), one very important aspect of science fiction is not just its ability to predict a particular technical breakthrough, but rather its ability to predict change itself through technology. Change plays a very important role in modern life. People responsible for societal planning must consider not only how things are now, but how they will (or at least might be) in the upcoming decades. Gifted science fiction writers, such as JULES VERNE, H. G. WELLS, Isaac Asimov, and SIR ARTHUR C. CLARKE, are also skilled technical prophets who help many people peek at tomorrow before it arrives.

For example, the famous French writer Jules Verne wrote *De la Terre à la Lune (From the Earth to the Moon)* in 1865, an account of a human voyage to the Moon from a Floridian launch site near a place Verne called "Tampa Town." A little more than 100 years later, directly across the state from the modern city of Tampa, the once isolated regions of the east central Florida coast shook to the mighty roar of a Saturn V rocket. The crew of NASA's *Apollo 11* mission had embarked from Earth, and people were to walk for the first time on the lunar surface.

science payload The complement of scientific instruments on a spacecraft, including both remote sensing and direct sensing devices that together cover large portions of the electromagnetic spectrum, large ranges in particle energies, or a detailed set of environmental measurements.

scientific airlock An opening in a crewed spacecraft or space station from which experiment and research equipment can be extended outside (into space) while the interior of the vehicle retains its atmospheric integrity (i.e., remains pressurized).

scientific notation A method of expressing powers of 10 that greatly simplifies writing large numbers. In scientific notation a number expressed in a positive power of 10 means the decimal point moves to the right (e.g., 3×10^6 = 3,000,000); a number expressed in a negative power of 10 means that the decimal point moves to the left (e.g., 3×10^{-6} = 0.000003).

scintillation counter An instrument that detects and measures ionizing radiation by counting light flashes (scintillations) caused by radiation interacting with certain materials (such as phosphors).

See also IONIZING RADIATION; SPACE RADIATION ENVIRONMENT.

Scobee, Dick (1939–1986) American *Astronaut, U.S. Air Force Officer* The astronaut Francis Richard (Dick) R. Scobee successfully flew into space as the pilot of the space shuttle *Challenger* during the STS-41C mission in April 1984. Then as shuttle commander he perished along with the rest of the crew on January 28, 1986, when the *Challenger* exploded during launch ascent at the start of the ill-fated STS-51L mission.

See also *CHALLENGER* ACCIDENT.

Scorpius X-1 The first cosmic (extrasolar) X-ray source discovered and the brightest and most persistent X-ray source in the sky. Scientists who were flying a sounding rocket experiment from the White Sands Missile Range in New Mexico unexpectedly detected this object in June 1962. They had designed their rocket-borne X-ray detection payload to search for X-ray sources within the solar system, especially possible X-rays from the lunar surface. Subsequent investigations suggest that Scorpius X-1 is a low-mass X-ray binary with an orbital period of 18.9 hours.

See also X-RAY ASTRONOMY; X-RAY BINARY STAR SYSTEM.

Scott, David R. (1932–) American *Astronaut, U.S. Air Force Officer* Astronaut David R. Scott participated in NASA's Gemini and Apollo Projects. He traveled to the Moon during the *Apollo 15* mission (July–August 1971) and, along with astronaut JAMES BENSON IRWIN, walked on its surface. They were also the first pair of Moon-walking Apollo astronauts to use the lunar rover vehicle, an electric battery-powered vehicle to assist in their exploration of the Moon near the landing site.

See also APOLLO PROJECT; GEMINI PROJECT; LUNAR ROVER.

Scout The four-stage, solid-propellant expendable launch vehicle developed and used by NASA to place small payloads into Earth orbit and to place probes on suborbital trajectories. The first Scout (Solid Controlled Orbital Utility Test) vehicle was launched on July 1, 1960, from the Mark 1 Launcher at the NASA Goddard Space Flight Center's Wallops Flight Facility, Wallops Island, Virginia. The Scout's first-stage motor was based on an earlier version of the U.S. Navy's Polaris missile motor; the second-stage solid-propellant motor was derived from the U.S. Army's Sergeant surface-to-surface missile; and the third- and fourth-stage motors were adapted by NASA from the USN's Vanguard missile.The standard Scout vehicle is a slender, all solid-propellant, four-stage booster system, approximately 23 m in length with a launch mass of 21,500 kg. This expendable vehicle is capable of placing a 186-kg-mass payload into a 560-km orbit around Earth.

See also LAUNCH VEHICLE; ROCKET.

screaming A form of combustion instability, especially in a liquid-propellant rocket engine, of relatively high frequency and characterized by a high-pitched noise.

See also ROCKET.

screeching A form of combustion instability, especially in an afterburner, of relatively high frequency and characterized by a harsh, shrill noise.

See also ROCKET.

scrub To cancel or postpone a rocket firing, either before or during the countdown.

See also ABORT; COUNTDOWN.

Scud A family of mobile, short-range tactical ballistic missiles developed by the Soviet Union during the cold war. Scud guided missiles are single-stage, liquid-propellant rockets about 11.25 meters long, except for the Scud-A, which has a length of 10.25 meters. The Scud A through D models were designed to carry various warheads ranging from conventional high-explosive (HE) weapons to nuclear, chemical, or biological weapons. The Scud-B, first deployed in 1965, had a maximum range of about 300 km, while the Scud-D, first deployed in the 1980s, had a maximum range of about 700 km. During the cold war the Soviet Union exported numerous nonnuclear warhead models of the Scud to Warsaw Pact and non–Warsaw Pact nations, such as Iraq.

Iraqi Scud missiles earned some degree of notoriety during the Persian Gulf War of 1991. All the Iraqi variants of the Soviet Scud-B guided missile used kerosene as the fuel and some form of red fuming nitric acid as the oxidizer. With a maximum range of about 300 km, the 1991-era Iraqi Scuds exhibited notoriously poor accuracy—the farther they flew the more inaccurate they became. Relatively unsophisticated gyroscopes guided the missile and only during the powered phase of the flight, a period generally lasting about 80 seconds. Once the Iraqi Scud's engine shut down, the entire spent missile body with warhead attached followed a simple ballistic trajectory to the target area. Upon reentry into the sensible atmosphere, the dynamically awkward missile body–warhead combination became unstable and often disintegrated before impacting within the intended target area.

sealed cabin The crew-occupied volume of an aircraft, aerospace vehicle, or spacecraft characterized by walls that do not allow any gaseous exchange between the cabin (inner) atmosphere and its surroundings and containing its own mechanisms for maintenance of the cabin atmosphere.

sea-level engine A rocket engine designed to operate at sea level, that is, the exhaust gases achieve complete expansion at sea-level ambient pressure.

See also ROCKET; ROCKET ENGINE.

search and rescue satellite The basic COSPAS-SARSAT system consists of radio beacons that transmit emergency signals during distress situations, instruments onboard satellites (usually weather satellites) in geostationary and low Earth orbits that can detect the distress signals transmitted by the radio beacons, ground receiving stations that receive and process the satellite downlink signals used to initiate distress alerts, mission control centers (MCCs) that receive the distress alerts from the ground receiving stations and then forward the distress alerts to appropriate rescue coordination centers (RCCs), and search and rescue points of contact (SPOC). The acronym COSPAS stands for Cosmicheskaya Systyema Poiska Avariynich Sudov, which translated from Russian means "space system for the detection of vessels in distress." The acronym SARSAT means "Search and Rescue Satellite Aided Tracking." The U.S. National Oceanic and Atmospheric Administration (NOAA) uses this acronym to identify the American portion of this international program, which it operates in conjunction with its meteorological satellites.

See also BEACON; NATIONAL OCEANIC AND ATMOSPHERIC ADMINISTRATION (NOAA); WEATHER SATELLITE.

Space-Age Guardian Angels

In the Old Testament book of *Tobias,* Archangel Raphael cares for the young Tobias during a perilous biblical journey on behalf of his father. Today spacecraft in the COSPAS-SARSAT system serve the human race much like a constellation of orbiting electronic guardian angels, locating people who are lost or in distress throughout the world. Specifically, the COSPAS-SARSAT program assists search and rescue (SAR) activities on a worldwide basis by providing accurate, timely, and reliable distress alert location data to the international community on a nondiscriminatory basis. The acronym COSPAS stands for Cosmicheskaya Systyema Poiska Avariynich Sudov, which translated from Russian means "space system for the detection of vessels in distress." The acronym SARSAT means "Search and Rescue Satellite Aided Tracking." The U.S. National Oceanic and Atmospheric Administration (NOAA) uses this acronym to identify the American portion of this international program, which it operates in conjunction with its meteorological satellites.

The basic COSPAS-SARSAT system consists of radio beacons that transmit emergency signals during distress situations, instruments onboard satellites (usually weather satellites) in geostationary and low Earth orbits that can detect the distress signals transmitted by the radio beacons, ground receiving stations that receive and process the satellite downlink signals used to initiate distress alerts, mission control centers (MCCs) that receive the distress alerts from the ground receiving stations and then forward the distress alerts to appropriate rescue coordination centers (RCCs), and search and rescue points of contact (SPOC).

Three basic types of distress radio beacons are in use: the emergency locator transmitter (ELT) for aviation use, the emergency position-indicating radio beacons (EPIRBs) for maritime use, and the personal locator beacon (PLB) for land use and personal applications that are neither aviation or maritime. Distress beacons operating at 406 megahertz (MHz) and 121.5 MHz are compatible with the COSPAS-SARSAT system. However, the operational capabilities of the system are significantly different for the two beacon frequencies, and the more accurate 406-MHz beacon frequency will be used exclusively starting in February 2009. The ground receiving stations that support the COSPAS-SARSAT system are referred to as local users terminals, or LUTs.

The COSPAS-SARSAT satellite constellation consists of search and rescue (SAR) satellites in low earth orbit (LEOSAR) and geostationary orbit (GEOSTAR). The nominal system configuration in low Earth orbit contains four satellites, two COSPAS spacecraft and two SARSAT spacecraft. Russia supplies the two COSPAS satellites placed in near-polar orbits at 1,000 km altitude. The Russian spacecraft are equipped with search and rescue instrumentation at 121.5 MHz and 406 MHz. The United States also supplies two low Earth orbit spacecraft to the international system. These SARSATs are actually two NOAA meteorological satellites that operate in sun-synchronous, near-polar orbits at an altitude of about 850 km. Reflecting the international nature of this important program, the American spacecraft are equipped with 121.5 MHz and 406 MHz search and rescue instrumentation supplied by Canada and France.

The current geostationary orbit search and rescue satellite constellation consists of two NOAA weather satellites provided by the United States (called GOES East and GOES West) and one satellite provided by India (called INSAT). Collectively, these three satellites provide continuous Earth coverage. The acronym GOES stands for Geostationary Orbiting Environmental Satellite.

The idea of satellite-aided search and rescue traces its origins to a tragic accident that took place in 1970, when a plane carrying two U.S. congressmen crashed in a remote region of Alaska. Despite a massive search and rescue effort, no trace of the missing aircraft or its passengers has ever been found. In reaction to this tragedy, the U.S. Congress mandated that all aircraft operated in the United States carry an Emergency Locator Transmitter (ELT). This device was designed to automatically activate after a crash and transmit a homing signal. Since space technology was still in its infancy, the frequency chosen for ELT transmissions was 121.5 MHz, the frequency used by international aircraft for distress signals. This system worked, but it had many technical limitations. After several years these limitations began to outweigh the benefits. In addition, space technology had improved to the point that a satellite-aided search and rescue system had become practical. The space-based system would operate on a frequency (406 MHz) reserved exclusively for emergency radio beacons, it would have a digital signal that uniquely identified each registered beacon, and it would provide global search and rescue coverage.

The original SARSAT system emerged in the 1970s as a result of a joint effort by the United States, Canada, and France. NASA developed the satellite system for the United States, but once the system was functional, its operation became the responsibility of NOAA, where it remains today. The Soviet Union developed a similar system called COSPAS. The four founding nations (the United States, Canada, France, and the Soviet Union) banded together in 1979 to form COSPAS-SARSAT. In 1982 the first search and rescue–equipped satellite was launched, and by 1984 the system was declared fully operational.

Since 1982 almost 17,000 lives have been saved worldwide with the assistance of COSPAS-SARSAT, including more than 4,600 lives in the United States (through January 2004). The COSPAS-SARSAT organization also grew. The original four member nations have now been joined by 29 other nations that operate 45 ground stations and 23 mission control centers or serve as search and rescue points of contact. The COSPAS-SARSAT system continues to serve as a model of international cooperation. Even during the politically tense cold war period of the 1980s, the United States and the Soviet Union were able to put aside their ideological differences to tackle some of the tough technology questions, whose successful resolution ultimately made the COSPAS-SARSAT system a global reality.

Today travelers with registered radio beacons are accompanied and protected by their COSPAS-SARSAT electronic guardian angels anywhere they venture on Earth, whether at sea, on land, or through the air. Satellite-aided search and rescue is one of the marvelous benefits of modern space technology.

search for extraterrestrial intelligence (SETI) The major aim of SETI programs is to listen for evidence of radio frequency (microwave) signals generated by intelligent extraterrestrial civilizations. This search is an attempt to answer an important philosophical question: Are we alone in the universe? The classic paper by Giuseppe Cocconi and Philip Morrison entitled "Searching for Interstellar Communications" (*Nature,* 1959) often is regarded as the start of modern SETI. With the arrival of the space age the entire subject of extraterrestrial intelligence (ETI) has left the realm of science fiction and now is regarded as a scientifically respectable (although currently speculative) field of endeavor.

The current understanding of stellar formation leads scientists to think that planets are normal and frequent companions of most stars. As interstellar clouds of dust and gas condense to form stars, they appear to leave behind clumps of material that form into planets. The Milky Way galaxy contains at least 100 billion to 200 billion stars. This line of reasoning has been reinforced by the detection of extrasolar planets.

Current theories on the origin and chemical evolution of life indicate that it probably is not unique to Earth but may be common and widespread throughout the galaxy. Further, some scientists believe that life on alien worlds could have developed intelligence, curiosity, and the technology necessary to build the devices needed to transmit and receive electromagnetic signals across the interstellar void. For example, an intelligent alien civilization might, like humans, radiate electromagnetic energy into space. This can happen unintentionally, as a result of planetary-level radio-frequency communications, or intentionally, through the beaming of structured radio signals out into the Milky Way galaxy in the hope some other intelligent species can intercept and interpret these signals against the natural electromagnetic radiation background of space.

SETI observations may be performed using radio telescopes on Earth, in space, or even (some day) on the far side of the Moon. Each location has distinct advantages and disadvantages.

Until recently only very narrow portions of the electromagnetic spectrum have been examined for "artifact signals" (i.e., those generated by intelligent alien civilizations). Human-made radio and television signals—the kind radio astronomers reject as clutter and interference—actually are similar to the signals SETI researchers are hunting for.

The sky is full of radio waves. In addition to the electromagnetic signals we generate as part of our technical civilization (e.g., radio, TV, radar, etc.), the sky also contains natural radio wave emissions from such celestial objects as the Sun, the planet Jupiter, radio galaxies, pulsars, and quasars. Even interstellar space is characterized by a constant, detectable radio-noise spectrum.

Consequences of Interstellar Contact

The discovery of planets around other stars has renewed scientific speculation about the possibility of intelligent life elsewhere in the Milky Way galaxy. Should such intelligent life exist, would they also be interested in searching for and contacting other intelligent beings such as us? Just what would happen if we made contact with an extraterrestrial civilization? No one on Earth can really say for sure. However, this contact will very probably be one of the most momentous events in all human history.

The postulated interstellar contact could be direct or indirect. Direct contact might involve a visit to Earth by a starship from an advanced stellar civilization. It could also take the form of the discovery of an alien probe, artifact, or derelict spaceship in the outer regions of our solar system. Some space travel experts have suggested that the hydrogen- and helium-rich giant outer planets might serve as convenient "fueling stations" for passing interstellar spaceships from other worlds. Indirect contact via radio frequency communication represents a more probable contact pathway (at least from a contemporary terrestrial viewpoint). The consequences of a successful search for extraterrestrial intelligence (SETI) would be nothing short of extraordinary. Were we to locate and identify but a single extraterrestrial signal, humankind would know immediately one great truth: We are not alone. We might also learn that the universe is teeming with life.

The overall impact of this postulated interstellar contact will depend on the circumstances surrounding the event. If it happens by accident or after only a few years of searching, the news, once verified, would surely startle most citizens of this world. If, however,intelligent alien signals were detected only after an extended effort lasting generations and involving extensive search facilities, the terrestrial impact of the event might be less overwhelming.

The reception and decoding of a radio frequency (or other type of) signal from an extraterrestrial civilization in the depths of space offers the promise of practical and philosophical benefits for all humanity. Response to that signal, however, would involve a potential planetary risk. If we do intercept an alien signal, we can decide (as a planetary society) to respond. We may also choose not to respond. If we are suspicious of the motives of the alien culture that sent the message, we are under no obligation to answer. There would be no practical way for them to realize that their signal had been intercepted, decoded, and understood by the intelligent inhabitants of a tiny world called Earth.

Optimists emphasize the friendly nature of such an interstellar contact and anticipate large technical gains for our planetary society, including the reception of information and knowledge of extraordinary value. They imagine that there will be numerous scientific and technological benefits from such contacts. However, because of the long round-trip times associated with speed-of-light–limited interstellar communications (perhaps decades or centuries), any information exchange will most likely be in the form of semi-independent transmissions. Each transmission burst would contain a bundle of significant facts about the sending society, such as planetary data, its life-forms, its age, its history, its philosophies and beliefs, and whether it has successfully contacted other alien cultures. An interstellar dialogue with questions asked and answered in somewhat rapid succession would not be as practical. Therefore, over the period of a century or more, the people of Earth might receive a wealth of information at a gradual enough rate to assemble a comprehensive picture of the alien civilization without inducing severe culture shock here at home.

Some scientists feel that if we successfully establish interstellar contact, we would probably not be the first planetary civilization to have accomplished this feat. In fact, they speculate that interstellar communications may have been going on since the first intelligent civilizations evolved in our galaxy about 4 or 5 billion years ago. One of the most exciting consequences of this type of interstellar conversation would be the accumulation by all participants of an enormous body of information and knowledge that has been passed down from alien race to alien race since the beginning of the galaxy's communicative phase. Included in this vast body of knowledge, something we might call the "galactic heritage," could be the entire natural and social history of numerous species and planetary civilizations. Also included might be an extensive compendium of astrophysical data that extend back countless millennia. Access to such a collection of scientific data would allow the best and brightest minds on Earth to develop accurate new insights into the origin and destiny of the universe.

However, interstellar contact should lead to far more than merely an exchange of scientific knowledge. Humankind would discover other social forms and structures, probably better capable of self-preservation and genetic evolution. We would also discover new forms of beauty and become aware of different endeavors that promote richer, more rewarding lives. Such contacts might also lead to the development of branches of art or science that simply cannot be undertaken by just one planetary civilization. Rather, they might require joint multiple-civilization participation across interstellar distances. Most significant, perhaps, is the fact that interstellar contact and communication would represent the end of the cultural isolation of the human race. The contact optimists further speculate that our own civilization would be invited to enter a sophisticated "cosmic community" as mature "adults" who are proud of our human heritage, rather than remaining isolated, with a destructive tendency to annihilate one another in childish planetary rivalries. Indeed, perhaps the very survival of the human race ultimately depends on finding ourselves cast in a larger cosmic role, a role far greater in significance than any human being can now imagine.

We should also speculate about the possible risks that could accompany confirming our existence to an alien culture that is most likely far more advanced and powerful than our own. Contact pessimists suggest such risks range from planetary annihilation to the humiliation of the human race. For discussion purposes, these risks can be divided into four general categories: invasion, exploitation, subversion, and cultural shock.

The invasion of Earth is a popular and recurring theme in science fiction. By actively sending out signals into the cosmic void or responding to intelligent signals we have detected and decoded, we would be revealing our existence and announcing the fact that Earth is a habitable planet. Within this contact scenario, our planet might be invaded by vastly superior beings who are set on conquering the galaxy. Another major interstellar contact hazard is exploitation. Human beings could appear to be a very primitive form of conscious life, perhaps at the level of an experimental animal or an unusual pet.

Another interstellar contact hazard is that of subversion. This appears to be a more plausible and subtle form of contact risk because it can occur with an exchange of signals. Here, an advanced alien race—under the guise of teaching and helping us join a cosmic community—might actually trick us into building devices that would allow "them" to conquer "us." The alien civilization would not necessarily have to make direct contact, since their interstellar "Trojan horse" might arrive via radio frequency signals. Since computer worms, viruses, and Trojan horses are a constant Internet threat here on Earth, why not imagine a similar, but more sophisticated, interstellar contact threat?

The last general contact risk is massive cultural shock. Some individuals have expressed concerns that even mere contact with a vastly superior extraterrestrial race could prove extremely damaging to human psyches despite the best intentions of the alien race. Terrestrial cultures, philosophies, and religions that now place human beings at the very center of creation would have to be "modified and expanded" to compensate for the confirmed existence of other far superior intelligent beings. We would now have to "share the universe" with someone or something better and more powerful than we are. As the dominant species on this planet, we must seriously consider whether the majority of human beings could accept this new role.

Advances in space technology are encouraging many people to consider the possibility of the existence of intelligent species elsewhere in the universe. At this happens, we must also keep asking ourselves a more fundamental question: Are human beings generally prepared for the positive identification of such a fact? Will contact with intelligent aliens open up a golden age on Earth or initiate devastating cultural regression?

However, the choice of initiating interstellar contact may no longer really be ours. In addition to the radio and television broadcasts that are leaking out into the galaxy at the speed of light, the powerful radio/radar telescope at the Arecibo Observatory was used to beam an interstellar message of friendship to the fringes of the Milky Way galaxy on November 16, 1974. We have, therefore, already announced our presence to the galaxy and should not be too surprised if someone or something eventually answers.

And just what would a radio-frequency signal from an intelligent extraterrestrial civilization look like? The figure presents a spectrogram that shows a simulated "artifact signal" from outside the solar system. This particular signal was sent by the *Pioneer 10* spacecraft from beyond the orbit of Neptune and was received by a Deep Space Network (DSN) radio telescope at Goldstone, California, using a 65,000-channel spectrum analyzer. The three signal components are quite visible as spikes above the always-present background radio noise. The center spike appearing in the figure has a transmitted signal power of approximately 1 watt, about half the power of a miniature Christmas tree light. SETI scientists are looking for a radio-frequency signal that might appear this clearly, or for one that may actually be quite difficult to distinguish from the background radio noise. To search through this myriad of radio-frequency signals, SETI scientists have developed state-of-the-art spectrum analyzers that can sample millions of frequency channels simultaneously and identify candidate "artifact signals" automatically for further observation and analysis.

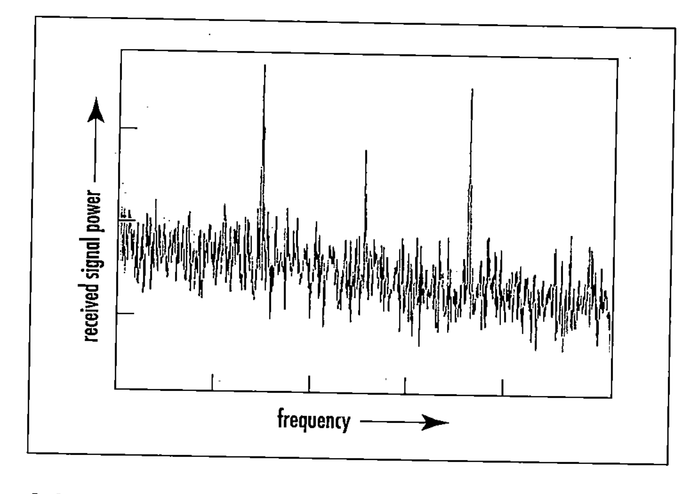

A simulated signal from extraterrestrial civilization using the *Pioneer 10* spacecraft transmitting an artifact signal from beyond the orbit of Neptune *(Courtesy of NASA)*

In October 1992 NASA started a planned decade-long SETI program called the High Resolution Microwave Survey (HRMS). The objective of HRMS was to search other solar systems for microwave signals using radio telescopes at the National Science Foundation's Arecibo Observatory in Puerto Rico, NASA's Goldstone Deep Space Communications Complex in California, and other locations. Coupled with these telescopes were HRMSs—dedicated high-speed digital data processing systems that contained off-the-shelf hardware and specially designed software.

The search proceeded in two different modes: a targeted search and an all-sky survey. The targeted search focused on about 1,000 nearby stars that resembled our Sun. In a somewhat less sensitive search mode, the all-sky survey was planned to search the entire celestial sphere for unusual radio signals. However, severe budget constraints and refocused national objectives resulted in the premature termination of the NASA HRMS program in 1993—after just one year of observation. Since then, while NASA has remained deeply interested in searching for life within our solar system, specific SETI projects no longer receive government funding.

Today privately funded organizations and foundations (such as the SETI Institute in Mountain View, California, a nonprofit corporation that focuses on research and educational projects relating to the search for extraterrestrial life) are conducting surveys of the heavens in search of radio signals from intelligent alien civilizations.

If an alien signal is ever detected and decoded, then the people of Earth will face another challenging question: Do we answer? For the present time, SETI scientists are content to listen passively for "artifact signals" that might arrive across the interstellar void.

See also ARECIBO OBSERVATORY; INTERSTELLAR COMMUNICATION AND CONTACT; LIFE IN THE UNIVERSE.

Seasat A NASA Earth observation spacecraft designed to demonstrate techniques for global monitoring of oceanographic phenomena and features, to provide oceanographic data, and to determine key features of an operational ocean dynamics system. *Seasat-A* was launched on June 26, 1978, by an expendable Atlas-Agena launch vehicle from Vandenberg Air Force Base, California. It was placed into a near-circular (~800-km-altitude) polar orbit. The major difference between *Seasat-A* and previous Earth observation satellites was the use of active and passive microwave sensors to achieve an all-weather capability. After 106 days of returning data, contact with *Seasat-A* was lost when a short circuit drained all power from its batteries.

See also AQUA SPACECRAFT; OCEAN REMOTE SENSING; REMOTE SENSING.

Sea Sparrow (RIM-7M) The U.S. Navy's Sea Sparrow missile, designated RIM-7, is a radar-guided air-to-air missile with a high-explosive (HE) warhead. It has a cylindrical body with four wings (canards) at mid-body and four tail fins. A solid-propellant rocket motor powers this versatile all-weather, all-altitude operational weapon system that can attack high-performance aircraft and missiles from any direction. The navy also uses the Sea Sparrow aboard ships as a surface-to-air antimissile defense system.

See also SPARROW.

Secchi, Pietro Angelo (1818–1878) Italian *Astronomer* The Jesuit priest Pietro Angelo Secchi was the first scientist to systematically apply spectroscopy to astronomy. By 1863 he completed the first major spectroscopic survey of the stars and published a catalog containing the spectra of more than 4,000 stars. After examining these data, he proposed in 1867 that such stellar spectra could be divided into four basic classes—the Secchi classification system. His pioneering work eventually evolved into the Harvard classification system and to an improved understanding of stellar evolution. He also pursued advances in astrophotography by photographing solar eclipses to assist in the study of solar phenomena such as prominences.

See also SPECTRAL CLASSIFICATION; STAR.

second (symbol: s) 1. The SI unit of time, now defined as the duration of 9,192,631,770 periods of radiation corresponding to the transition between two hyperfine levels of the ground state of the cesium-133 atom. Previously, this unit of time had been based on astronomical observations. 2. A unit of angle (symbol: ″) equal to 1/3600 of a degree or 1/60 of a minute of angle.

secondary crater A crater formed when a large chunk of material from a primary-impact crater strikes the surrounding planetary surface.

See also IMPACT CRATER.

secondary radiation Electromagnetic radiation (photons) or ionizing particulate radiation that results from the absorption of other (primary) ionizing radiation in matter. For example, the ionizing occurring within a spacecraft when high-energy cosmic rays impact the wall of a spacecraft and remove electrons or nuclei from the matter in the wall. These removed electrons or nuclei are called secondary radiation.

See also SPACE RADIATION ENVIRONMENT.

second law of thermodynamics An inequality asserting that it is impossible to transfer thermal energy (heat) from a colder to a warmer system without the occurrence of other simultaneous changes in the two systems or in the environment. It follows from this important physical principle that during an adiabatic process, entropy cannot decrease. For reversible adiabatic processes, entropy remains constant, while for irreversible adiabatic processes it increases.

An equivalent formulation of the law is that it is impossible to convert the heat of a system into work without the occurrence of other simultaneous changes in the system or its environment. This version of the second law, which requires a heat engine to have a cold sink as well as a hot source, is particularly useful in engineering applications. Another important statement of the second law is that the change in entropy (ΔS) for an isolated system is greater than or equal to zero. Mathematically, $(\Delta S)_{isolated} \geq 0$.

See also ENTROPY; THERMODYNAMICS.

section One of the cross-sectional parts that a rocket vehicle is divided into, each adjoining another at one or both of its ends. Usually described by a designating word, such as nose section, aft section, center section, tail section, thrust section, and propellant tank section.

Sedna Sedna is the most distant object known to be orbiting the Sun. Named for the Inuit goddess of the ocean, Sedna is a mysterious planetlike body three times farther from Earth than Pluto. At present the frigid planetoid is 13 billion kilometers away. Some astronomers suggest that the discovery of Sedna represents the first detection of an icy body in the long-hypothesized Oort cloud. In 1950 the Dutch astronomer JAN HENDRIK OORT proposed the existence of a very distant repository of small icy bodies that supplies the long-period comets, which show up on occasion in the inner solar system.

Sedna is approximately three-fourths the size of Pluto and has a reddish color. In fact, after Mars, it is the second-reddest object in the solar system. With an estimated diameter of between 1,300 to 1,750 kilometers, Sedna represents the largest solar system object discovered since CLYDE WILLIAM TOMBAUGH detected Pluto in 1930. The NASA-funded researchers who discovered Sedna on November 14, 2003, used the 1.2-meter (48 in) Samuel Oschin Telescope at Caltech's Palomar Observatory near San Diego, California. They were Michael Brown (California Institute of Technology), Chad Trujillo (Gemini Observatory, Hawaii), and David Rabinowitz (Yale University). Sedna is extremely far from the Sun in the coldest known region of the solar system, where temperatures never rise above –240°C (that is, a frigid 33 K).

The mysterious object travels in a highly elliptical orbit unlike anything previously observed in the solar system. At aphelion Sedna is 130 billion kilometers from the Sun. Sedna will continue to come closer to Earth and become brighter over the next 72 years, after which it starts its long journey to the farthest reaches of the solar system. With a 10,500-year orbital period, the last time Sedna was this close to Earth our planet was just starting to come out of the most recent ice age.

Seebeck effect The establishment of an electric potential difference that tends to produce a flow of current in a circuit of two dissimilar metals, the junctions of which are at different temperatures. It is the physical phenomenon involved in the operation of a thermocouple and is named for the German scientist Thomas Seebeck (1770–1831), who first observed it in 1822.

See also THERMOCOUPLE; THERMOELECTRIC CONVERSION.

Selenian Of or relating to Earth's Moon. Once a permanent lunar base is established, a resident of the Moon.

See also LUNAR BASES AND SETTLEMENTS; MOON.

selenocentric 1. Relating to the center of Earth's Moon; referring to the Moon as a center. 2. Orbiting about the Moon as a central body.

selenodesy That branch of science (applied mathematics) that determines, by observation and measurement, the exact positions of points and the figures and areas of large portions of the Moon's surface, or the shape and size of the Moon.

selenoid A satellite of Earth's Moon; a spacecraft in orbit around the Moon, such as, for example, NASA's *Lunar Orbiter* spacecraft.

See also LUNAR ORBITER.

selenography The branch of astronomy that deals with Earth's Moon and its surface, composition, motion, and the like. Selene is Greek for "Moon."

self-cooled Term applied to a combustion chamber or nozzle in which temperature is controlled or limited by methods that do not involve flow within the wall of coolant supplied from an external source.

self-pressurization Increase of ullage pressure by vaporization or boil-off of contained fluid without the aid of additional pressurant.

See also ULLAGE.

self-replicating system (SRS) An advanced robotic device. A single SRS unit is a machine system that contains all the elements required to maintain itself, to manufacture desired products, and even (as the name implies) to reproduce itself.

The Hungarian-American mathematician JOHN VON NEUMANN was the first person to seriously consider the concept of self-replicating machine systems. During and following World War II he became interested in the study of automatic replication as part of his wide-ranging interests in

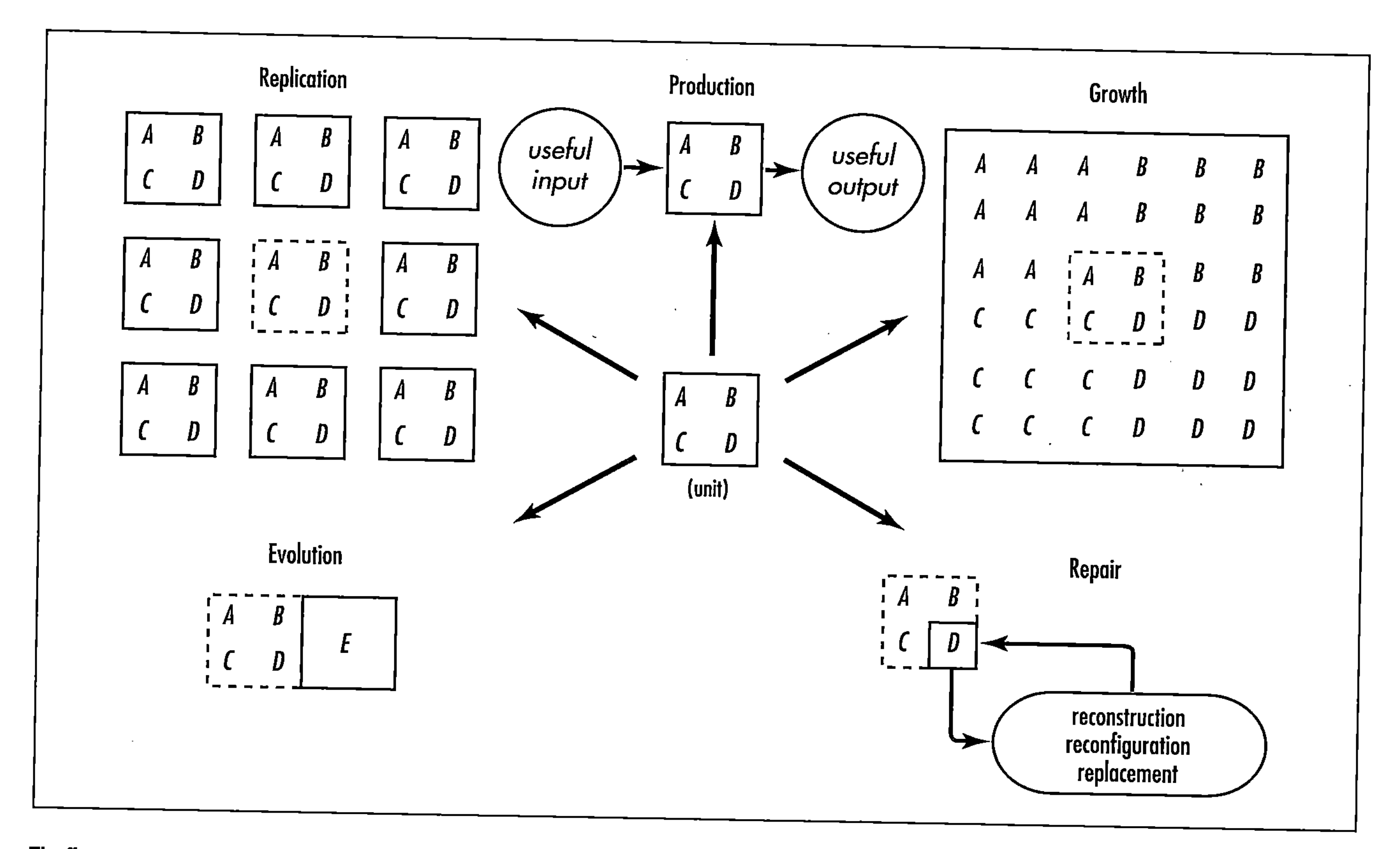

The five general classes of self-replicating system (SRS) behavior: production, replication, growth, repair, and evolution ***(Drawing courtesy of NASA)***

complicated machines. From von Neumann's initial work and the more recent work of other investigators, five general classes of SRS behavior have been defined: production, replication, growth, repair, and evolution.

The issue of closure (total self-sufficiency) is one of the fundamental problems in designing self-replicating systems. In an arbitrary SRS unit there are three basic requirements necessary to achieve total closure: (1) matter closure, (2) energy closure, and (3) information closure. If the machine is only partially self-replicating, then it is said that only partial closure of the system has occurred. In this case some essential matter, energy, or information must be provided from external sources, or else the machine will fail to reproduce itself.

The self-replicating system could be used to assist in planetary engineering projects in which a seed machine would be sent to a target planet, such as Mars. The SRS unit would make a sufficient number of copies of itself using Martian resources and then set about some production task, such as manufacturing oxygen to make the planet's atmosphere more breathable for future human settlers. Similarly, a seed SRS unit could be sent into interstellar space and trigger a wave of galactic exploration, stopping to repair itself or make copies of itself in the various solar systems it encounters on its cosmic journey.

See also ROBOTICS IN SPACE.

semiactive homing guidance A system of homing guidance in which the receiver in the missile uses radiation reflected from a target that has been illuminated by a source outside the missile. For example, a pilot in an aircraft can use a laser designator to mark, or "paint," a particular target. Some of this laser light then is reflected back to the missile, which follows it to the target.

See also GUIDANCE SYSTEM.

semimajor axis One-half the major axis of an ellipse. For a planet this corresponds to its average orbital distance from the Sun.

See also KEPLER'S LAWS (OF PLANETARY MOTION).

sensible atmosphere That portion of a planet's atmosphere that offers resistance to a body passing through it.

See also ATMOSPHERE.

sensible heat The thermal energy (heat) that, when added to or removed from a thermodynamic system, causes a change in temperature. *Compare with* LATENT HEAT.

sensor In general, a device that detects and/or measures certain types of physically observable phenomena. More specifically, that part of an electronic instrument that detects electromagnetic radiations (or other characteristic emissions, such as nuclear particles) from a target or object at some distance away and then converts these incident radiations (or particles) into a quantity (i.e., an internal electronic signal) that is amplified, measured (quantified), displayed, and/or recorded by another part of the instrument. A passive sensor uses characteristic emissions from the object or target as its input signal. In contrast, an active sensor (such as a radar) places a burst of electromagnetic energy on the object or target being observed and then uses the reflected signal as its input.

See also REMOTE SENSING.

Sentinel A former antiballistic missile (ABM) system that was designed for light-area defense against a low-level ballistic missile attack on the United States. It was developed into the Safeguard system in the late 1960s.

See also BALLISTIC MISSILE DEFENSE; SAFEGUARD.

separation 1. The action of a fallaway section or companion body as it casts off from the remaining body of a vehicle, or the action of the remaining body as it leaves a fallaway section behind it. 2. The moment at which this action occurs.

separation velocity The velocity at which a rocket, aerospace vehicle, or spacecraft is moving when some part or section is separated from it; specifically, the velocity of a space probe or satellite at the time of separation from the launch vehicle.

Sergeant (MGM-29A) A U.S. Army mobile, inertially guided, solid-propellant, surface-to-surface missile with nuclear capability, designed to attack targets up to a range of about 140 kilometers.

service module (SM) The large part of the Apollo spacecraft that contained support equipment; it was attached to the command module (CM) until just before the CM (carrying three astronauts) reentered Earth's atmosphere at the end of a mission.

See also APOLLO PROJECT.

Setebos A minor moon of Uranus that orbits the planet at a distance of about 18 million kilometers. Discovered in 1999, the moon is in retrograde orbit with a period of approximately 2,200 days. Originally called S/1999 U1, the moon now has the name of the character who enslaves Ariel in William Shakespeare's play *The Tempest.* Astronomers have very little information about this distant Uranian moon.

See also URANUS.

Seyfert, Carl Keenan (1911–1960) American *Astronomer* In 1943 the American astronomer Carl K. Seyfert discovered a special group of spiral galaxies now known as Seyfert galaxies.

See also GALAXY; SEYFERT GALAXY.

Seyfert galaxy A type of spiral galaxy with a very bright galactic nucleus, first observed in 1943 by the American astronomer CARL K. SEYFERT.

See also ACTIVE GALAXIES; GALAXY.

sextant An instrument used by navigators to measure the altitude of certain celestial objects above the horizon. The basic device contains a graduated arc of 60° with a sighting device and a movable arm.

See also ANCIENT ASTRONOMICAL INSTRUMENTS.

shake-and-bake test A series of prelaunch tests performed on a spacecraft to simulate the launch vibrations and thermal environment (i.e., temperature extremes) it will experience during the mission. A typical test plan might involve the use of enormous speakers to blast the spacecraft with acoustic vibrations similar to those it will encounter during launch. As part of a typical shake-and-bake test plan, aerospace engineers also will place the spacecraft in a special environmental chamber and subject it to the extreme hot and cold temperatures it will experience during space flight.

Shapley, Harlow (1885–1972) American *Astronomer* In about 1914 the American astronomer Harlow Shapley used a detailed study of variable stars (especially Cepheid variables) to establish more accurate dimensions for the Milky Way galaxy and to discover that the Sun was actually two-thirds of the way out in the rim of this spiral galaxy—some 30,000 light-years from its center. Up until then astronomers thought the Sun was located near the center of the galaxy. In 1921 Shapley engaged in a public debate with HEBER CURTIS concerning the nature of distant spiral nebulas, which Shapley originally believed were either part of the Milky Way galaxy or very close neighbors. He also studied the Magellanic Clouds and clusters of galaxies.

Shenzhou 5 On October 15, 2003, the People's Republic of China became the third nation—following the former Soviet Union and the United States—to place a human being in orbit around Earth using a national launch vehicle. On that date a Chinese Long March 2F rocket lifted off from the Jiuquan Satellite Launch Center and placed the *Shenzhou 5* spacecraft with TAIKONAUT YANG LIWEI onboard into orbit around Earth. *Taikonaut* is a suggested Chinese equivalent to astronaut and cosmonaut. *Taikong* is the Chinese word for space or cosmos, so the prefix *taiko-* assumes the same concept and significance as that of *astro-* and *cosmo-* to form the words astronaut and cosmonaut. After 14 orbits around Earth the spacecraft reentered the atmosphere on October 16, 2003, and Yang Liwei was safely recovered in the Chinese portion of Inner Mongolia.

Shenzhou 6 On October 12, 2005, the People's Republic of China successfully launched its second human-crewed mission from the Jiuquan Satellite Launch Center. A Long March 2F rocket blasted off with a pair of TAIKONAUTS on board this spacecraft that has a general design similar to the Russian Soyuz spacecraft but with significant modifications. The *Shenzhou 6* had a reentry capsule, an orbital module, and a propulsion module. Taikonauts Fei Junlong and Nie Haisheng sat in the reentry module during takeoff and during the reentry/landing portion of the mission. During nearly five days (115.5 hours) in space, they took turns entering the orbital module that contained life-support equipment and experiments. After 76 orbits, their reentry capsule safely returned to Earth on October 16, making a soft, parachute-assisted landing in northern Inner Mongolia.

See also SHENZHOU 5.

Shepard, Alan B., Jr. (1923–1998) American *Astronaut, U.S. Navy Officer* Selected as one of the original seven Mercury Project astronauts, Alan B. Shepard, Jr., became the first American to travel in outer space. He accomplished this important space technology milestone on May 5, 1961, when

Mercury Project astronaut Alan B. Shepard, Jr., the first American to travel in outer space (suborbital flight, May 5, 1961) ***(Courtesy of NASA)***

he rode inside the *Freedom 7* space capsule as it was lifted off from Cape Canaveral Air Force Station, Florida, by a Redstone rocket. (*See* figure.) The suborbital Mercury Redstone 3 mission hurled Shepard on a ballistic trajectory downrange from Cape Canaveral. After about 15 minutes his tiny space capsule splashed down in the Atlantic Ocean some 450 kilometers from the launch site. U.S. Navy personnel recovered him and the *Freedom 7* space capsule from the ocean. Later in his astronaut career Shepard made a second, much longer journey into space. In February 1971 he served as the commander of NASA's *Apollo 14* lunar landing mission. Together with astronaut EDGAR DEAN MITCHELL, Shepard explored the Moon's Fra Mauro region.

See also APOLLO PROJECT; MERCURY PROJECT.

shepherd moon A small inner moon (or pair of moons) that shapes and forms a particular ring around a ringed planet. For example, the shepherds Ophelia and Cordelia tend the Epsilon Ring of Uranus.

shield A quantity of material used to reduce the passage of ionizing radiation. The term *shielding* refers to the arrangement of shields used for a particular nuclear radiation protection circumstance.

See also SPACE RADIATION ENVIRONMENT.

shield volcano A wide, gently sloping volcano formed by the gradual outflow of molten rock; many occur on Venus.

Shillelagh (MGM-51) A U.S. Army missile system that is mounted on a main battle tank or assault reconnaissance vehicle for use against enemy armor, troops, and field fortifications.

***shimanagashi* syndrome** Will terrestrial immigrants to extraterrestrial communities suffer from the *shimanagashi* syndrome? During Japan's feudal period political offenders were often exiled on small islands. This form of punishment was known as *shimanagashi*. Today in many modern prisons one can find segregation or isolation units in which inmates who are considered chronic troublemakers are confined for a period of time. Similarly, but to a lesser degree, mainlanders who spend a few years on an isolated island or island chain might feel a strange sense of isolation even though the island environment (such as is experienced in Hawaii) includes large cities and many modern conveniences. These mainland visitors start feeling left out and even intellectually crippled despite the fact that life might be physically very comfortable there. The term *island fever* is often used to describe this situation.

Early extraterrestrial communities will be relatively small and physically isolated from Earth. However, electronic communications, including the transmission of books, journals, and contemporary literature, could avoid or minimize such feelings of isolation. As the actual numbers of extraterrestrial settlements grow, physical travel between them could also reduce the sense of physical isolation.

See also SPACE SETTLEMENT.

shirt-sleeve environment A space station module or spacecraft cabin in which the atmosphere is similar to that found on the surface of Earth, that is, it does not require a pressure suit.

shock diamonds The diamond-shaped (or wedge-shaped) shock waves that sometimes become distinctly visible in the exhaust of a rocket vehicle during launch ascent.

shock front 1. A shock wave regarded as the forward surface of a fluid region having characteristics different from those of the region ahead of it. 2. The boundary between the pressure disturbance created by an explosion (in air, water, or earth) and the ambient atmosphere, water, or earth.

shock wave 1. A surface of discontinuity (i.e., of abrupt changes in conditions) set up in a supersonic field of flow, through which the fluid undergoes a finite decrease in velocity accompanied by a marked increase in pressure, density, temperature, and entropy. Sometimes called a shock. 2. A pressure pulse in air, water, or earth propagated from an explosion. It has two distinct phases: in the first, or positive, phase the pressure rises sharply to a peak, then subsides to the normal pressure of the surrounding medium; in the second, or negative, phase the pressure falls below that of the medium, then returns to normal. A shock wave in air is also called a *blast wave*.

short-period comet A comet with an orbital period of less than 200 years.

short-range attack missile (SRAM or AGM-69) A supersonic U.S. Air Force air-to-surface missile that can be armed with a nuclear warhead and can be launched from the B-52 and the B-1B aircraft. The missile's range, speed, and accuracy allow the carrier aircraft to "stand off" from its intended targets and launch this missile while still outside enemy defenses. The SRAM is used by bombers to neutralize enemy air defenses and to strike heavily defended targets.

short-range ballistic missile A ballistic missile with a range capability up to about 1,100 kilometers.

shot A colloquial aerospace term used to describe the act or instance of firing a rocket, especially from Earth's surface; for example, the launch crew prepared for the next shot of the space shuttle.

Shrike (AGM-45) An air-launched, anti-radiation missile used by the U.S. Navy to home in on and destroy hostile antiaircraft radars. The missile uses a solid-propellant rocket motor, is 3.05 meters long, and has a mass of 177 kilograms. It employs a passive radar homing guidance system and contains a conventional warhead.

shroud 1. In general, the protective outer covering for a payload that is being transported into space by a launch vehicle. Once the launch vehicle has left Earth's sensible atmosphere, the shroud is discarded. 2. In the context of ballistic missile defense (BMD), a shroud would be used with a reentry vehicle (RV) or a post-boost vehicle (PBV) to confuse a missile defense system. For example, the RV shroud is a thin envelope that would enclose a reentry vehicle, interfering with the infrared radiation it normally would emit; a PBV shroud is a loose conical device that would be positioned behind a PBV to conceal the deployment of reentry vehicles and decoys.

shuttle Commonly encountered, shortened name for NASA's space shuttle vehicle, or the U.S. Space Transportation System.

shuttle glow A phenomenon first observed on early space shuttle missions in which a visible glow caused by the optical excitation of the tenuous residual atmosphere of atomic oxygen and molecular nitrogen found between 250 and 300 km altitude is seen around the leading edges of the vehicle. As the space shuttle or other spacecraft flies through these thin gases at orbital velocities (about 8 km per second), the impact of these gases actually gives rise to two phenomena: Some portions of the space vehicle glow while in orbit, while other portions of the space vehicle can erode away slowly due to a chemical interaction with the highly reactive residual atmosphere (i.e., atomic oxygen chemically attacks the surfaces of the orbiting spacecraft).

sideband 1. Either of the two frequency bands on both sides of the carrier frequency within which fall the frequencies of the wave produced by the process of modulation. 2. The wave components lying within such a band.

sidereal Of or pertaining to the stars.

sidereal day The duration of one rotation of Earth on its axis with respect to the vernal equinox. It is measured by successive transits of the vernal equinox over the upper branch of a meridian. Because of the precession of the equinoxes, the sidereal day so defined is slightly less than the period of rotation with respect to the stars, but the difference is less than 0.01 second. The length of the mean sidereal day is 24 hours of sidereal time or 23 hours 56 minutes 4.091 seconds of mean solar time.

sidereal month The average amount of time the moon takes to complete one orbital revolution around Earth when using the fixed stars as a reference; approximately 27.32 days.

sidereal period The period of time required by a celestial body to complete one revolution around another celestial body with respect to the fixed stars.

sidereal time Time measured by the rotation of Earth with respect to the stars, which are considered "fixed" in position. Sidereal time may be designated as local or Greenwich, depending on whether the local meridian or the Greenwich meridian is used as the reference.

sidereal year The period of one apparent revolution of Earth around the Sun with respect to a referenced ("fixed") star. A sidereal year averages 365.25636 days in duration. Because of the precession of the equinoxes, this is about 20 minutes longer than a tropical year. The tropical year is defined as the average period of time between successive passages of the Sun across the vernal equinox.

See also YEAR.

Sidewinder (AIM-9) A supersonic, heat-seeking, air-to-air missile carried by U.S. Navy, Marine, and Air Force fighter aircraft. The Sidewinder has the following main components: an infrared homing guidance section, an active optical target detector, a high-explosive warhead, and a solid-propellant rocket motor. The heat-seeking infrared guidance system enables the missile to home on a target aircraft's engine exhaust. This infrared seeker also permits the pilot to launch the missile and then leave the area or take evasive action while the missile guides itself to the target. The AIM-9A, a prototype of Sidewinder, was fired first successfully in September 1953. Since then the design of this missile has evolved continuously such that it has become the most widely used air-to-air missile by the United States and its Western allies.

siemens (symbol: S) The SI unit of electrical conductance. It is defined as the conductance of an electrical circuit or element that has a resistance of 1 ohm. In the past this unit sometimes was called the mho or reciprocal ohm. The unit is named in honor of the German scientist Ernst Werner von Siemens (1816–92).

sievert (symbol: Sv) The special SI unit for dose equivalent (H). In radiation protection, the dose equivalent in sieverts (Sv) is the product of the absorbed dose (D) in grays (Gy) and the radiation weighting factor (w_R) (formerly the quality factor [QF]) as well as any other modifying factors considered necessary to characterize and evaluate the biological effects of the ionizing radiation received by human beings or other living creatures. The sievert is related to the traditional dose equivalent unit (rem) as follows: 1 sievert = 100 rem.

See also DOSE EQUIVALENT; SPACE RADIATION ENVIRONMENT.

signal 1. A visible, audible, or other indication used to convey information. 2. Information to be transmitted over a communications system. 3. Any carrier of information; as opposed to noise. 4. In electronics, any transmitted electrical impulse. The variation of amplitude, frequency, and waveform are used to convey information.

signal-to-noise ratio (SNR) The ratio of the amplitude of the desired signal to the amplitude of noise signals at a given point in time. The higher the signal-to-noise ratio, the less interference with reception of the desired signal.

signature In general, the set of characteristics by which an object or target can be identified; for example, the distinctive type of radiation emitted or reflected by a target that can be used to recognize it. A target's signature might include a characteristic infrared emission, an unusual pattern of motion, or a distinctive radar return.

simulation 1. In general, the art of replicating relevant portions of the "real-world" environment to test equipment, train mission personnel, and prepare for emergencies. Simulations can involve the use of physical mass and energy replicants, high-fidelity (i.e., reasonably close to the original) hardware, and supporting software. With the incredible growth in computer and display technologies, computer-based simulations are assuming an ever-increasing role as aerospace design tools and astronaut training aids. In fact, today's computer-based simulations often are referred to as "reality in a box." Virtual reality systems are extremely versatile simulation tools that can be operated to test equipment and operating procedures and to train personnel safely using a variety of highly interactive mission scenarios. 2. In a military context, the art of making a decoy look like a more valuable strategic target.

See also VIRTUAL REALITY.

single-event upset (SEU) A bit flip (i.e., a "0" is changed to a "1," or vice versa) in a digital microelectronic circuit. The single-event upset (SEU) is caused by the passage of high-energy ionizing radiation through the silicon material of which the semiconductors are made. These space radiation–induced bit flips (or SEUs) can damage data stored in memory, corrupt operating software, cause a central processing unit (CPU) to write over critical data tables or to halt, and even trigger an unplanned event involving some piece of computer-controlled hardware (e.g., fire a thruster) that then severely impacts a mission. Aerospace engineers deal with the SEU problem in a variety of ways, including the use of additional shielding around sensitive spacecraft electronics components, the careful selection of electronic components (i.e., using more radiant-resistant parts), the use of multiple-redundancy memory units and "polling" electronics, and the regular resetting of a spacecraft's onboard computers.

See also SPACE RADIATION ENVIRONMENT.

single failure point (SFP) A single element of hardware the failure of which would lead directly to loss of life, vehicle, or mission. When safety conditions dictate that abort be initiated when a redundant element fails, that element also is considered a single failure point.

single-stage rocket A rocket vehicle provided with a single rocket propulsion system.

See also ROCKET.

single-stage-to-orbit vehicle (SSTO) Since the early 1960s aerospace engineers have envisioned building reusable launch vehicles because they offer the potential of relative operational simplicity and reduced costs compared with expendable launch vehicles. Until recently, however, the necessary technologies were not available. Now, thanks to aerospace technology improvements, many engineers believe that it is technically feasible to design and produce a single-stage-to-orbit (SSTO) launch vehicle with sufficient payload capacity to meet most government and commercial space transportation requirements.

Despite these recent technology advances, the development of an SSTO space transportation system involves considerable risk. For example, because an SSTO launch vehicle will have no expendable components, it will need to carry more fuel than would be necessary if it were shedding mass by dropping (expended) stages during launch. Achieving the propellant mass fraction necessary to reach orbit with a useful payload will require numerous technological advances that improve fuel efficiency and lower structural mass without compromising structural integrity. In addition, completely reusable launch vehicles are technologically much more difficult to achieve because components must be capable of resisting deterioration and surviving multiple launches and reentries.

The difficulties of a horizontal takeoff and horizontal landing vehicle are known from the National Aerospace Plane (NASP) Program. Other candidate SSTO vehicle configurations include vertical takeoff and vertical landing, vertical takeoff and horizontal landing (winged body), and vertical takeoff and horizontal landing (lifting body). NASA's former (now cancelled) X-33 program was focused on SSTO vehicle development using a vertical-takeoff and horizontal-landing (lifting body) approach.

singularity The hypothetical central point in a black hole at which the curvature of space and time becomes infinite; a theoretical point that has infinite density and zero volume. The principle of cosmic censorship—a theorem of black hole physics—states that an event horizon always conceals a singularity, preventing that singularity from communicating its existence to observers elsewhere in the universe.

See also BLACK HOLE.

Sinope The small outermost known moon of Jupiter discovered in 1914 by the American astronomer SETH BARNES NICHOLSON. Sinope has a diameter of 36 km, a mass of 7.8×10^{16} kg, and orbits Jupiter at a distance of 23,700,000 km. Also called Jupiter IX.

See also JUPITER.

Sirius (Dog Star) The alpha star in the constellation Canis Major (α CMa) and the brightest star in the night sky. Astronomers also call Sirius the Dog Star and classify it as an A1V star (that is, a main sequence dwarf star). It is about 8.6 light-years away. Sirius has a white dwarf companion called Sirius B (or the Pup).

See also STAR.

Fundamental SI Units

Name (and Symbol)	Physical Quantity Represented
meter (m)	length
kilogram (kg)	mass
second (s)	time
ampere (A)	electric current
candela (cd)	luminous intensity
kelvin (K)	thermodynamic temperature
mole (mol)	amount of substance

SI unit(s) The Système International d'Unités (International System of Units, or SI units) is the internationally agreed-upon system of coherent units that is now in use throughout the world for scientific and engineering purposes. The metric (or SI) system was developed originally in France in the late 18th century. Its basic units for length, mass, and time—the meter (m), the kilogram (kg), and the second (s)—were based on natural standards. The modern SI units are still based on natural standards and international agreement, but these standards now are ones that can be measured with greater precision than the previous natural standards. For example, the *meter* (m) (British spelling: metre) is the basic SI unit of length. Currently it is defined as being equal to 1,650,763.73 wavelengths in a vacuum of the orange-red line of the spectrum of krypton-86. (Previously, the meter had been defined as one ten-millionth the distance from the equator to the North Pole along the meridian nearest Paris.)

Derived and Supplementary SI Units

Name (and symbol)	Physical Quantity Represented
becquerel (Bq)	radioactivity
coulomb (C)	electric charge
farad (F)	electric capacitance
gray (Gy)	absorbed dose of ionizing radiation
henry (H)	inductance
hertz (Hz)	frequency
joule (J)	energy
lumen (lm)	luminous flux
lux (lx)	illuminance
newton (N)	force
ohm (Ω)	electric resistance
pascal (Pa)	pressure
radian (rad)	plane angle
siemens (S)	electric conductance
sievert (Sv)	ionizing radiation dose equivalent
steradian (sr)	solid angle
tesla (T)	magnetic flux density
volt (V)	electric potential difference
watt (W)	power
weber (Wb)	magnetic flux

Recommended SI Unit Prefixes

Submultiple	Prefix	Symbol	Multiple	Prefix	Symbol
10^{-1}	deci-	d	10^{1}	deca-	da
10^{-2}	centi-	c	10^{2}	hecto-	h
10^{-3}	milli-	m	10^{3}	kilo-	k
10^{-6}	micro-	μ	10^{6}	mega-	M
10^{-9}	nano-	n	10^{9}	giga-	G
10^{-12}	pico-	p	10^{12}	tera-	T
10^{-15}	femto-	f	10^{15}	peta-	P
10^{-18}	atto-	a	10^{18}	exa-	E
10^{-21}	zepto-	z	10^{21}	zetta-	Z
10^{-24}	yocto-	y	10^{24}	yotta-	Y

Metric measurements play an important part in the global aerospace industry. Unfortunately, a good deal of American aerospace activity still involves the use of another set of units, the English system of measurement (sometimes referred to as the American Engineering System of units). One major advantage of the metric system over the English system is that the metric system uses the number 10 as a base. Therefore, multiples or submultiples of SI units are reached by multiplying or dividing by 10.

The metric system uses the kilogram as the basic unit of mass. In contrast, the English system has a unit called the pound-mass (lbm), which is a *unit of mass,* and another unit, the pound-force (lbf), which is a *unit of force.* Furthermore, these two "pound" units are related by an arbitrary (pre–space age) definition within the English system of measurement that states that a 1 pound-mass object "weighs" 1 pound-force at sea level on Earth. Novices and professionals alike all too often forget that this "pound" equivalency is valid only at sea level on the surface of Earth. Therefore, this particular English unit arrangement can cause much confusion in working aerospace engineering problems. For example, in the metric system a kilogram is a kilogram, whether on Earth or on Mars. However, in the English system, if an object "weighs" one pound on Earth, it only "weighs" 0.38 pound on Mars. Of course, from the basic physical concept of mass, one pound-mass (lbm) on Earth is still one pound-mass

International System (SI) Units and Their Conversion Factors

Quantity	Name of Unit	Symbol	Conversion Factor
distance	meter	m	1 km = 0.621 mi. 1 m = 3.28 ft. 1 cm = 0.394 in. 1 mm = 0.039 in. 1 μm = 3.9×10^{-5} in. = 10^{4} Å 1 nm = 10 Å
mass	kilogram	kg	1 tonne = 1.102 tons 1 kg = 2.20 lb. 1 g = 0.0022 lb. = 0.035 oz. 1 mg = 2.20×10^{-6} lb. = 3.5×10^{-5} oz.
time	second	s	1 yr. = 3.156×10^{7} s 1 day = 8.64×10^{4} s 1 hr. = 3,600 s
temperature	kelvin	K	273 K = 0°C = 32°F 373 K = 100°C = 212°F
area	square meter	m^2	1 m^2 = 10^4 cm^2 = 10.8 $ft.^2$
volume	cubic meter	m^3	1 m^3 = 10^6 cm^3 = 35 $ft.^3$
frequency	hertz	Hz	1 Hz = 1 cycle/s 1 kHz = 1,000 cycles/s 1 MHz = 10^6 cycles/s
density	kilogram per cubic meter	kg/m^3	1 kg/m^3 = 0.001 g/cm^3 1 g/cm^3 = density of water
speed, velocity	meter per second	m/s	1 m/s = 3.28 ft./s 1 km/s = 2,240 mi./hr.
force	newton	N	1 N = 10^5 dynes = 0.224 lbf
pressure	newton per square meter	N/m^2	1 N/m^2 = 1.45×10^{-4} lb./$in.^2$
energy	joule	J	1 J = 0.239 cal; 1 J = 10^7 erg
photon energy	electronvolt	eV	1 eV = 1.60×10^{-19} J
power	watt	W	1 W = 1 J/s
atomic mass	atomic mass unit	amu	1 amu = 1.66×10^{-27} kg
wavelength of light	angstrom	Å	1 Å = 0.1 nm = 10^{-10} m
acceleration of gravity	g	g	1 g = 9.8 m/s^2

Source: NASA.

Common Metric/English Conversion Factors (for Space Technology Activities)

	Multiply	By	To Obtain
length	inches	2.54	centimeters
	centimeters	0.3937	inches
	feet	0.3048	meters
	meters	3.281	feet
	miles	1.6093	kilometers
	kilometers	0.6214	miles
	kilometers	0.54	nautical miles
	nautical miles	1.852	kilometers
	kilometers	3281	feet
	feet	0.0003048	kilometers
weight and mass	ounces	28.350	grams
	grams	0.0353	ounces
	pounds	0.4536	kilograms
	kilograms	2.205	pounds
	tons	0.9072	metric tons
	metric tons	1.102	tons
liquid measure	fluid ounces	0.0296	liters
	gallons	3.7854	liters
	liters	0.2642	gallons
	liters	33.8140	fluid ounces
temperature	degrees Fahrenheit plus 459.67	0.5555	kelvins
	degrees Celsius plus 273.16	1.0	kelvins
	kelvins	1.80	degrees Fahrenheit minus 459.67
	kelvins	1.0	degrees Celsius minus 273.16
	degrees Fahrenheit minus 32	0.5555	degrees Celsius
	degrees Celsius	1.80	degrees Fahrenheit plus 32
thrust (force)	pounds force	4.448	newtons
	newtons	0.225	pounds force
pressure	millimeters mercury	133.32	pascals (newtons per square meter)
	pounds per square inch	6.895	kilopascals (1,000 pascals)
	pascals	0.0075	millimeters mercury at 0°C
	kilopascals	0.1450	pounds per square inch

Source: NASA.

(lbm) on Mars. What changes the object's "weight" (as calculated from Newton's second law of motion: force = mass = acceleration) is the difference in the local acceleration due to gravity. On Earth the sea-level acceleration due to gravity (g) is equal to 9.8 meters per second per second (m/s^2) (or about 32.2 feet per second per second [ft/s^2]), while on the surface of Mars the local acceleration due to gravity (g_{Mars}) is only 3.73 meters per second per second (m/s^2) (or about 12.2 feet per second per second [ft/s^2]). Confusing? That is why the metric system is used extensively throughout this book and why the companion English units (on rare occasions for clarity or historic continuity) are included in parentheses. On such occasions special effort has also been made to identify the pound-mass and the pound-force as separate and distinct units within the English system of measurements. It is also helpful to recognize that the metric companion unit to the pound-mass is the kilogram (kg), namely 1 kg = 2.205 pounds-mass, while the metric companion unit to the pound-force is the newton (N), namely 1 N = 0.2248 pound-force.

The tables provide a useful summary of contemporary SI units and their conversion factors, the recommended SI unit prefixes, special units used in astronomical investigations, and common metric to English conversion factors.

Special Units for Astronomical Investigations

Astronomical unit (AU): The mean distance from Earth to the Sun—approximately 1.495979×10^{11} m

Light-year (ly): The distance light travels in 1 year's time—approximately 9.46055×10^{15} m

Parsec (pa): The parallax shift of 1 second of arc (3.26 light-years)—approximately 3.085768×10^{16} m

Speed of light (c): 2.9979×10^{8} m/s

Source: NASA.

Skylab (SL) The first U.S. space station, which was placed in orbit in 1973 by a two-stage configuration of the Saturn V expendable launch vehicle and then visited by three astronaut crews who worked on scientific experiments in space for a total of approximately 172 days, with the last crew spending 84 days in Earth orbit. *Skylab* (SL) was composed of five major parts: the Apollo telescope mount (ATM), the multiple docking adapter (MDA), the airlock module (AM), the instrument unit (IU), and the orbital workshop (OWS), which included the living and working quarters. The ATM was a solar observatory, and it provided attitude control and experiment pointing for the rest of the cluster. The retrieval and installation of film used in the ATM was accomplished by the astronauts during extravehicular activity (EVA). The MDA served as a dock for the modified Apollo spacecraft that taxied the crews to and from the space station. The AM was located between the docking port (MDA) and the living and working quarters and contained controls and instrumentation. The IU, which was used only during launch and the initial phases of operation, provided guidance and sequencing functions for the initial deployment of the ATM, its solar arrays, and the like. The OWS was a modified Saturn IV-B stage that had been converted into a "two-story" space laboratory with living quarters for a crew of three. This orbital laboratory was capable of unmanned, in-orbit storage, reactivation, and reuse.

There were four launches in the Skylab Program from Complex 39 at the Kennedy Space Center. The first launch was on May 14, 1973. A two-stage Saturn V rocket placed the unmanned 90-metric-ton *Skylab* space station in an initial 435-km orbit around Earth. As the rocket accelerated past 7,620 meters, atmospheric drag began clawing at *Skylab*'s meteoroid/sun shield. This cylindrical metal shield was designed to protect the orbital workshop from tiny particles and the Sun's scorching heat. At 63 seconds after launch the shield ripped away from the spacecraft, trailing an aluminum strap that caught on one of the unopened solar wings. The shield became tethered to the laboratory while at the same time prying the opposite solar wing partly open. Minutes later, as the booster rocket staged, the partially deployed solar wing and meteoroid/Sun shield were flung into space. With the loss of the meteoroid/sun shield, temperatures inside *Skylab* soared, rendering the space station uninhabitable and threatening the food, medicine, and film stored onboard. The Apollo telescope mount (ATM), the major piece of scientific equipment, did deploy properly, however, an action that included the successful unfolding of its four solar panels.

The countdown for the launch of the first *Skylab* crew was halted. NASA engineers worked quickly to devise a solar parasol to cover the workshop and to find a way to free the remaining stuck solar wing. On May 25, 1973, astronauts CHARLES "PETE" CONRAD, JR., Joseph P. Kerwin, and Paul J. Weitz were launched by a Saturn 1B rocket toward *Skylab*.

After repairing *Skylab*'s broken docking mechanism, which had refused to latch, the astronauts entered the space station and erected a mylar solar parasol through a space access hatch. It shaded part of the area where the protective meteoroid/sun shield had been ripped away. Temperatures within the spacecraft immediately began dropping, and *Skylab* soon became habitable without space suits. Even so, the many experiments on board demanded far more electric power than the four ATM solar arrays could generate. Skylab could fulfill its scientific mission only if the first crew freed the remaining crippled solar wing. Using equipment that resembled long-handled pruning shears and a prybar, the astronauts pulled the stuck solar wing free. The space station was now ready to meet its scientific mission objectives.

The duration of the first crewed mission was 28 days. The second astronaut crew, ALAN BEAN, Jack Lousma, and Owen Garriott, was launched on July 28, 1973; mission duration on the space station was approximately 59 days and 11 hours. The third Skylab crew, Gerald Carr, William Pogue, and EDWARD GIBSON, was launched November 16, 1973; mission duration was a little more than 84 days. Saturn IB rockets launched all three crews in modified Apollo spacecraft, which also served as their return-to-Earth vehicle. The third and final manned Skylab mission ended with splashdown in the Pacific Ocean on February 8, 1974. (See the table.)

Skylab Mission Summary (1973–1974)

Mission	Dates	Crew	Mission Duration	Remarks
Skylab 1	Launched 05/14/73	Unmanned	Re-entered atmosphere 07/11/79	90-metric ton space station visited by three astronaut crews
Skylab 2	05/25/73 to 06/22/73	Charles Conrad, Jr. Paul J. Weitz Joseph P. Kerwin (M.D.)	28 days 49 min	Repaired Skylab; 392 hrs experiments; 3 EVAs
Skylab 3	07/28/73 to 09/25/73	Alan L. Bean Jack R. Lousma Owen K. Garriott (Ph.D.)	59 days 11 hrs	Performed maintenance; 1,081 hrs experiments; 3 EVAs
Skylab 4	11/16/73 to 02/08/74	Gerald P. Carr William R. Pogue Edward G. Gibson (Ph.D.)	84 days 1 hr	Observed Comet Kohoutek; 4 EVAs; 1,563 hrs experiments

After the last astronaut crew departed the space station in February 1974, it orbited Earth as an abandoned derelict. Unable to maintain its original altitude, the station finally reentered the atmosphere on July 11, 1979, during orbit 34,981. While most of the station was burned up during reentry, some pieces survived and impacted in remote areas of the Indian Ocean and sparsely inhabited portions of Australia.

See also SPACE STATION.

Slayton, Deke, Jr. (1924–1993) American *Astronaut* NASA selected Deke Slayton, Jr., as one of the original seven Mercury Project astronauts. During astronaut training a previously undetected (minor) medical issue developed, which kept Slayton from flying on a Mercury Project or Gemini Project mission in the 1960s. However, he persisted in achieving his career goal and traveled in space in July 1975 as one of three American astronauts who took part in the Apollo-Soyuz Test Project (ASTP), the first cooperative international rendezvous and docking mission.

See also APOLLO-SOYUZ TEST PROJECT (ASTP).

Slipher, Earl Carl (1883–1964) American *Astronomer* The American astronomer Earl Carl Slipher contributed to advances in astrophotography, especially innovative techniques to produce high-quality images of Mars, Jupiter, and Saturn. His was the brother of VESTO M. SLIPHER, another American astronomer.

See also ASTROPHOTOGRAPHY.

Slipher, Vesto Melvin (1875–1969) American *Astronomer* In 1912 the American astronomer Vesto Melvin Slipher began spectroscopic studies of the light from so-called spiral nebulas (now recognized as galaxies) and observed Doppler shift (that is, redshift) phenomena. These pioneering observations suggested that the distant objects were receding from Earth at very high speed. Slipher's work served as the foundation upon which EDWIN POWELL HUBBLE and other astronomers developed the concept of an expanding universe. He was the brother of EARL CARL SLIPHER, another American astronomer.

See also COSMOLOGY.

sloshing The back-and-forth movement of a liquid rocket propellant in its tank(s), creating problems of stability and control in the vehicle. Aerospace engineers often use antislosh baffles in a rocket vehicle's propellant tanks to avoid this problem.

Small Magellanic Cloud (SMC) An irregular galaxy about 9,000 light-years in diameter and 180,000 light-years from Earth.

See also GALAXY; MAGELLANIC CLOUDS.

Smith, Michael J. (1945–1986) American *Astronaut, U.S. Navy Officer* NASA astronaut Michael J. Smith served as the pilot on the STS 51-L mission of the space shuttle *Challenger,* an ill-fated mission on January 28, 1986, in which the launch vehicle exploded shortly after liftoff from the Kennedy Space Center in Florida. About 74 seconds after launch, a fatal accident claimed the lives of Smith and the other six *Challenger* crewmembers.

See also CHALLENGER ACCIDENT.

Smithsonian Astrophysical Observatory (SAO) The Smithsonian Astrophysical Observatory (SAO) is a research bureau of the Smithsonian Institution, which was created by the U.S. Congress in 1846. The SAO was founded in 1890 by SAMUEL PIERPONT LANGLEY, the Smithsonian Institution's third secretary, primarily for studies of the Sun. In 1955 the SAO moved from Washington, D.C., to Cambridge, Massachusetts, to affiliate with the Harvard College Observatory and to expand its staff, facilities, and, most important, its scientific scope. FRED L. WHIPPLE, the first director of the SAO in this new era, accepted a national challenge to create a worldwide satellite-tracking network. His decision helped establish the SAO as a leading center for space science research. In 1973 the ties between the Smithsonian and Harvard were strengthened and formalized by the creation of the joint Harvard-Smithsonian Center for Astrophysics (CfA).

Today the SAO is part of one of the world's largest and most diverse astrophysical institutions. The CfA has pioneered the development of orbiting observatories and large ground-based telescopes, the application of computers to astrophysical problems, and the integration of laboratory measurements, theoretical astrophysics, and observations across the electromagnetic spectrum. Scientists of the CfA are engaged in a broad program of research in astronomy, astrophysics, and Earth and space sciences. This research is organized by divisions: atomic and molecular physics, high-energy astrophysics, optical and infrared astronomy, planetary sciences, radio and geoastronomy, solar and stellar physics, and theoretical astrophysics. Observational data are gathered by instruments onboard rockets, balloons, and spacecraft as well as by ground-based telescopes at the Fred Lawrence Whipple Observatory in Arizona and the Oak Ridge Observatory in Massachusetts and by a millimeter-wave radio telescope in Cambridge. Current major initiatives involve the creation of a submillimeter telescope array in Hawaii, the conversion of the Multiple Mirror Telescope in Arizona to a single-mirror instrument 6.5 meters in diameter, and preparations for the launch of a variety of space experiments. Scientific staff will also have access to two twin 6.5-meter optical telescopes, Magellan I and II, now under construction in Chile by the Carnegie Institution and its partners, including Harvard College Observatory.

The Submillimeter Array (SMA) in Hawaii is a unique astronomical instrument designed to have the highest feasible resolution in the last waveband of the electromagnetic spectrum to be explored from Earth. This *submillimeter window,* whose wavelengths range from about 0.3 to 1.0 millimeter, is well-suited for studies of the structure and motions of the matter that forms stars; of the spiral structure of galaxies, as outlined by their giant molecular clouds; and of quasars and active galactic nuclei. The array consists initially of eight movable reflectors, each 6 meters in diameter, located near the 4,000-meter summit of Mauna Kea, an extinct volcano on the island of Hawaii. Planning for the array started in 1984. The Institute of Astronomy and Astrophysics of Taiwan's Academia Sinica is SAO's partner in the project.

SNAP An acronym for Systems for Nuclear Auxiliary Power. The SNAP program was created by the U.S. Atomic Energy Commission (forerunner of the U.S. Department of Energy) to develop small auxiliary nuclear power sources for specialized outer space, remote land, and sea applications. Two general approaches were pursued. The first used the thermal energy from the decay of a radioisotope source to produce electricity directly by thermoelectric or thermionic methods. The second approach used the thermal energy liberated in a small nuclear reactor to produce electricity by either direct conversion methods or dynamic conversion methods.

See also SPACE NUCLEAR POWER.

Snark A surface-to-surface guided missile developed by the U.S. Air Force in the early 1950s. This now obsolete winged missile had a range of about 8,000 km and was capable of carrying a nuclear warhead.

SNC meteorites *See* MARTIAN METEORITES.

soft landing The act of landing on the surface of a planet without damage to any portion of the space vehicle or payload, except possibly the landing gear. NASA's *Surveyor* spacecraft and *Viking* lander spacecraft were designed for soft landings on the Moon and Mars, respectively.

See also SURVEYOR PROJECT; VIKING PROJECT.

software The programs (that is, sets of instructions and algorithms) and data used to operate a digital computer. *Compare with* HARDWARE.

Sol The Sun.

sol A Martian day (about 24 hours 37 minutes 23 seconds in duration); 7 sols equal approximately 7.2 Earth days.

solar 1. Of or pertaining to the Sun or caused by the Sun, such as solar radiation and solar atmospheric tide. 2. Relative to the Sun as a datum or reference, such as solar time.

solar activity Any type of variation in the appearance or energy output of the Sun.

Solar and Heliospheric Observatory **(SOHO)** The primary scientific aims of the European Space Agency's *Solar and Heliospheric Observatory* are to investigate the physical processes that form and heat the Sun's corona, maintain it, and give rise to the expanding solar wind and to study the interior structure of the Sun. SOHO is part of the International Solar-Terrestrial Physics Program (ISTP) and involves NASA participation. The 1,350-kilogram (on-orbit dry mass) spacecraft was launched on December 2, 1995, and placed in a halo orbit at the Earth-Sun Lagrangian libration point 1 (L1) to obtain uninterrupted sunlight. It had a two-year design life, but onboard consumables are sufficient for an extra four years of operation. The spacecraft carries a complement of 12 scientific instruments.

In April 1998 SOHO successfully completed its nominal two-year mission to study the Sun's atmosphere, surface, and interior. The major science highlights of this mission include the detection of rivers of plasma beneath the surface of the Sun and the initial detection of solar flare–induced solar quakes. In addition, the SOHO spacecraft discovered more than 50 sungrazing comets. Then on June 24, 1998, during routine housekeeping and maintenance operations, contact was lost with the SOHO spacecraft. After several anxious weeks ground controllers were able to reestablish contact on August 3. They then proceeded to recommission various defunct subsystems and to perform an orbit correction maneuver that brought SOHO back to a normal operational mode with all its scientific instruments properly functioning by November 4, 1998.

See also EUROPEAN SPACE AGENCY; LAGRANGIAN LIBRATION POINTS; SUN.

solar cell Solar cells are direct energy conversion devices that have been used for more than three decades to provide electric power for spacecraft. In a direct energy conversion (DEC) device, electricity is produced directly from the primary energy source without the need for thermodynamic power conversion cycles involving the heat engine principle and the circulation of a working fluid. A solar cell, or "photovoltaic system," turns sunlight directly into electricity. The solar cell has no moving parts to wear out and produces no noise, fumes, or other polluting waste products. However, the space environment, especially trapped radiation belts and the energetic particles released in solar flares, can damage solar cells used on spacecraft and reduce their useful lifetime.

See also PHOTOVOLTAIC MATERIAL; SATELLITE POWER SYSTEM.

solar constant The total amount of the Sun's radiant energy that normally crosses perpendicular to a unit area at the top of Earth's atmosphere (i.e., at 1 astronomical unit from the Sun). The currently used value of the solar constant is 1,371 ± 5 watts per square meter. The spectral distribution of the Sun's radiant energy resembles that of a blackbody radiator with an effective temperature of 5,800 K. This means that most of the Sun's radiant energy lies in the visible portion of the electromagnetic spectrum, with a peak value near 0.45 micrometer (μm).

See also SUN.

solar cycle The periodic (semiregular) change in the number of sunspots. It is the interval between successive minima and is about 11.1 years (unless the magnetic polarities of Northern and Southern Hemisphere sunspots are considered; then the 11-year interval is actually part of a 22-year magnetic cycle).

See also SUN; SUNSPOT.

solar day *See* YEAR.

solar electric propulsion (SEP) A low-thrust propulsion system in which the electricity required to power the ion engines (or other type of electric rocket engines) is generated either by a solar-thermal conversion system or by a solar-photovoltaic conversion system.

See also ELECTRIC PROPULSION.

solar energy Energy from the Sun; radiant energy in the form of sunlight.

See also SOLAR CONSTANT; SOLAR RADIATION.

solar flare A highly concentrated explosive release of energy within the solar atmosphere. It appears as a sudden, short-lived brightening of a localized area within the Sun's chromosphere. The electromagnetic radiation output from a solar flare ranges from radio to X-ray frequencies. Energetic nuclear particles, primarily electrons, protons, and a few alpha particles, also are ejected during a solar flare.

See also SPACE RADIATION ENVIRONMENT; SUN.

solar mass The mass of the Sun, namely 1.99×10^{30} kilograms; it is commonly used as a unit in comparing stellar masses.

See also SUN.

***Solar Maximum* mission (SMM)** A NASA scientific satellite designed to provide coordinated observations of solar activity, especially solar flares, during a period of maximum solar activity. The 2,315-kg spacecraft was placed in orbit on February 14, 1980. Its scientific payload consisted of seven instruments specifically selected to study the short-wavelength and coronal manifestations of solar flares. Data were obtained on the storage and release of flare energy, particle acceleration, formation of hot plasma, and mass ejection.

In April 1984 space shuttle mission STS 41-C involved on-orbit repair of the SMM spacecraft. During this shuttle mission the astronauts rendezvoused with the spacecraft while performing an extravehicular activity (EVA) from the orbiter and successfully repaired it. The SMM spacecraft then continued to collect data until November 24, 1989. It reentered Earth's atmosphere on December 2, 1989.

solar nebula The cloud of dust and gas from which the Sun, the planets, and other minor bodies of the solar system are postulated to have formed (condensed).

See also SUN.

solar panel A winglike set of solar cells used by a spacecraft to convert sunlight directly into electric power; also called a solar array.

See also SOLAR CELL; SOLAR PHOTOVOLTAIC CONVERSION.

solar photovoltaic conversion The direct conversion of sunlight (solar energy) into electrical energy by means of the photovoltaic effect. A single photovoltaic (PV) converter cell is called a solar cell, while a combination of cells designed to increase the electric power output is called a solar array or a solar panel.

Since 1958 solar cells have been used to provide electric power for a wide variety of spacecraft. The typical spacecraft solar cell is made of a combination of n-type (*negative*) and p-type (*positive*) semiconductor materials, generally silicon. When this combination of materials is exposed to sunlight, some of the incident electromagnetic radiation removes bound electrons from the semiconductor material atoms, thereby producing free electrons. A hole (positive charge) is left at each location from which a bound electron has been removed. Consequently, an equal number of free electrons and holes are formed. An electrical barrier at the p-n junction causes the newly created free electrons near the barrier to migrate deeper into the n-type material and the matching holes to migrate farther into the p-type material.

If electrical contacts are made with the n- and p-type materials and these contacts are connected through an external load (conductor), the free electrons will flow from the n-type material to the p-type material. Upon reaching the p-type material the free electrons will enter existing holes and once again become bound electrons. The flow of free electrons through the external conductor represents an electric current that will continue as long as more free electrons and holes are being created by exposure of the solar cell to sunlight. This is the general principle of solar photovoltaic conversion.

See also PHOTOVOLTAIC MATERIAL; SOLAR CELL.

solar power satellite (SPS) *See* SATELLITE POWER SYSTEM.

solar radiation The total electromagnetic radiation emitted by the Sun. The Sun is considered to radiate as a blackbody at a temperature of about 5,770 kelvins (K). Approximately 99.9% of this radiated energy lies within the wavelength interval from 0.15 to 4.0 micrometers (μm), with some 50% lying within the visible portion of the electromagnetic spectrum (namely 0.4 to 0.7 μm) and most of the remaining radiated energy in the near-infrared portion of the spectrum.

See also ELECTROMAGNETIC SPECTRUM; SUN.

solar sail A proposed method of space transportation that uses solar radiation pressure to gently push a giant gossamer structure and its payload through interplanetary space. As presently envisioned, the solar sail would use a large quantity of very thin reflective material to produce a net reaction force by reflecting incident sunlight. Because solar radiation pressure is very weak and decreases as the square of the distance from the Sun, enormous sails—perhaps 100,000 to 200,000 square meters—would be needed to achieve useful accelerations and payload transport.

The main advantage of the solar sail would be its long-duration operation as an interplanetary transportation system. Unlike rocket propulsion systems that must expel their onboard supply of propellants to generate thrust, solar sails would have operating times limited only by the effective lifetimes in space of the sail materials. The solar photons that would do the "pushing" constantly pour in from the Sun and are essentially "free." This makes the concept of solar sailing particularly interesting for cases in which we must ship large numbers of nonpriority payloads through interplanetary space, such as, for example, a shipment of special robotic exploration vehicles from Earth to Mars.

However, because the large reflective solar sail could not generate a force opposite to the direction of the incident solar radiation flux, its maneuverability would be limited. This lack of maneuverability along with long transit times represent the major disadvantages of the solar sail as a space transportation system.

solar storm A disturbance in the space environment triggered by an intense solar flare (or flares) that produces bursts of electromagnetic radiation and charged particles. A solar storm can adversely affect human space operations, cause damage to orbiting spacecraft (especially geostationary satellites that are outside Earth's magnetosphere), and affect Earth's upper atmosphere and magnetic field. Solar storms are associated with unexpected surges in solar activity.

Periods of intense solar activity present several major problems for orbiting spacecraft. First, a solar storm heats Earth's atmosphere, causing it to expand. This atmospheric expansion increases the drag experienced by low-Earth-orbiting (LEO) satellites, forcing them to use additional altitude control propellant and shortening their operating lifetimes. Second, during a solar storm more energetic charged particles bombard spacecraft, causing electronic upsets and damage to sensors and electronic equipment. Finally, a solar storm involves bursts of electromagnetic radiation that upset Earth's ionosphere and interfere with radio wave communications. Also called a geomagnetic storm.

See also SOLAR FLARE; SPACE WEATHER; SUN.

solar system The Sun and the collection of celestial objects that are bound to it gravitationally. These celestial objects include the nine major planets, more than 60 known moons, more than 2,000 minor planets or asteroids, and a very large number of comets. Except for the comets, all of these celestial objects orbit around the Sun in the same direction, and their orbits lie close to the plane defined by Earth's own orbit and the Sun's equator.

The nine major planets can be divided into two general categories: (1) the terrestrial, or Earthlike, planets, consisting of Mercury, Venus, Earth and Mars; and (2) the outer, or Jovian, planets, consisting of the gaseous giants Jupiter, Saturn, Uranus, and Neptune. Tiny Pluto is currently regarded as a "frozen snowball" in a class by itself. Because of the size of its moon, Charon, some astronomers like to consider Pluto and Charon as forming a double planet system.

As a group the terrestrial planets are dense, solid bodies with relatively shallow or no atmospheres. In contrast, the Jovian planets are believed to contain modest-sized rock cores surrounded by concentric layers of frozen hydrogen, liquid hydrogen, and gaseous hydrogen. Their atmospheres also contain such gases as helium, methane, and ammonia. Astronomers also use the term *solar system* to refer to any star and its gravitationally bound collection of planets, asteroids, and comets.

See also ASTEROID; COMET; EARTH; JUPITER; MARS; MERCURY; NEPTUNE; PLUTO; SATURN; SUN; URANUS; VENUS.

solar telescope A telescope, either reflecting or refracting, designed specifically to observe the nearest star, the Sun. Solar telescopes such as the McMath Pierce Solar Telescope at the Kitt Peak National Observatory in Arizona are generally located on high mountains in desert regions to minimize the influence of cloud cover and atmospheric turbulence. At present the Sun is the only star whose disk can be adequately resolved and studied by telescope. Because of the large apparent diameter of the Sun's disk, solar telescopes are large structures with their lenses or mirrors having long focal lengths in order to form suitable large images. The solar telescope is generally constructed in a fixed position, and a heliostat is then used to bring in the sunlight. The intense heat of solar radiation often requires special cooling procedures for both the instruments and the facilities.

When the McMath Pierce Solar Telescope was dedicated on March 15, 1960, it was the world's largest and most sophisticated telescope for studying the Sun. The telescope incorporates a tower rising nearly 30 meters above the ground, from which a shaft slants approximately 60 meters to the ground, where a tunnel continues an additional 90 meters into the mountain. Nearly 140 10.3-m by 2.4-m copper panels make up the outer skin of the telescope. Through these panels are circulated 13,600 liters of antifreeze and an incredible 57,000 liters of water. A refrigerating plant located about 60 meters away pumps the chilled mixture to the telescope through underground pipes. In this facility incoming sunlight is captured and reflected by a 2-meter-diameter heliostat located 30 meters above the ground at the top of the tower, and the beam travels 150 meters as it is focused by first a 1.6-meter-diameter concave mirror and then a flat mirror before being sent on to various measuring instruments in the observation room.

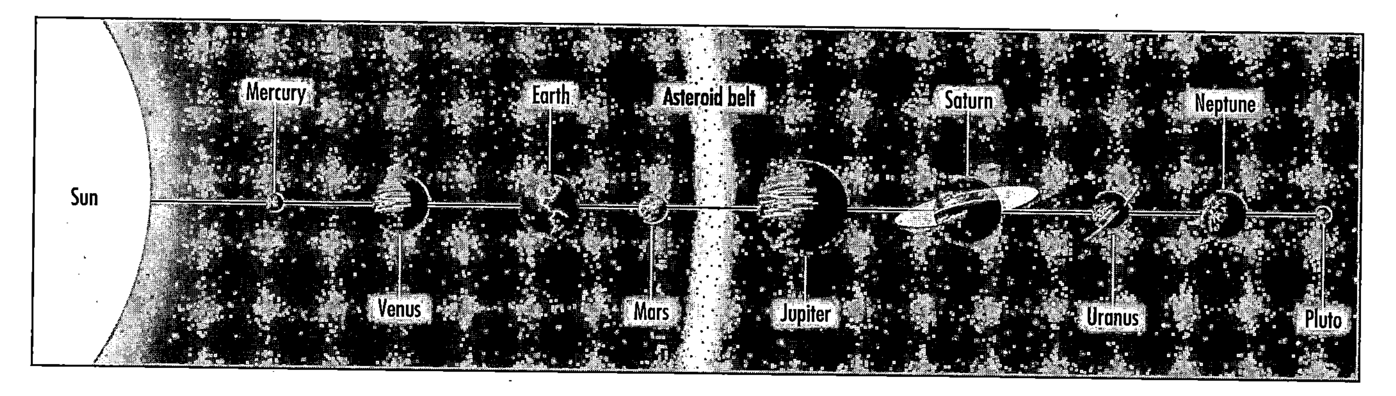

The major components of our solar system (not to scale) *(Drawing courtesy of NASA)*

The Advanced Technology Solar Telescope (ATST), representing a collaboration of 22 institutions from the solar physics community, is being proposed for development at Haleakala, Hawaii. Considered to be the next generation ground-based telescope, the ATST would have a 4-meter aperture, integrated adaptive optics, low scattered light, infrared coverage, and state-of-the-art postfocus instrumentation.

solar thermal conversion The conversion of sunlight (solar energy) into electricity by means of a thermodynamic cycle involving a heat engine. In general, incoming solar energy is concentrated and focused to heat a working fluid. The solar-heated working fluid undergoes a series of changes in its thermodynamic state during which mechanical energy (work) is extracted from the hot fluid. This mechanical work then is used to spin a generator to produce electricity. Thermodynamic cycles that could be used in solar thermal conversion include the Brayton, Rankine, and Stirling cycles.

See also HEAT ENGINE; THERMODYNAMICS.

solar wind The variable stream of electrons, protons, and various atomic nuclei (e.g., alpha particles) that flows continuously outward from the Sun into interplanetary space. The solar wind has typical speeds of a few hundred kilometers per second.

See also MAGNETOSPHERE; SUN.

solenoid Helical coil of insulated wire that, when conducting electricity, generates a magnetic field that actuates a movable core.

solid A state of matter characterized by a three-dimensional regularity of structure. When a solid substance is heated beyond a certain temperature, called the melting point, the forces between its atoms or molecules can no longer support the characteristic lattice structure, causing it to break down as the solid material transforms into a liquid or (more rarely) transforms directly into a vapor (sublimation).

solid angle (symbol: Ω) Three-dimensional angle formed by the vertex of a cone; that portion of the whole of space about a given point, bounded by a conical surface with its vertex at that point and measured by the area cut by the bounding surface from the surface of a sphere of unit radius centered at that point. The STERADIAN (sr) is the SI unit of solid angle.

solid Earth The lithosphere portion of the Earth system, including this planet's core, mantle, crust, and all surface rocks and unconsolidated rock fragments.

See also EARTH SYSTEM SCIENCE.

solid lubricant A dry film lubricant.

See also SPACE TRIBOLOGY.

solid propellant A rocket propellant in solid form, usually containing both fuel and oxidizer combined or mixed, and formed into a monolithic (not powdered or granulated) grain.

See also PROPELLANT; ROCKET.

solid-propellant rocket A rocket propelled by a chemical mixture or compound of fuel and oxidizer in solid form. Also called a solid rocket.

See also PROPELLANT; ROCKET.

solid rocket booster (SRB) Two solid rocket boosters (SRBs) operate in parallel to augment the thrust of the space shuttle main engines (SSMEs) from the launch pad through the first two minutes of powered flight. These boosters also assist in guiding the entire vehicle during the initial ascent. Following separation they are recovered (after a parachute-assisted splashdown in the ocean) for refurbishment and reuse. In addition to its basic component, the solid rocket motor (SRM), each booster contains several subsystems: structural, thrust vector control (TVC), separation, recovery, and electrical and instrumentation.

The heart of the solid rocket booster is the solid rocket motor, the first ever developed for use on a crewed launch vehicle. The huge solid rocket motor is composed of a segmented motor case loaded with solid propellants, an ignition system, a movable nozzle, and the necessary instrumentation and integration hardware.

Each SRB develops about 11,790 kilonewtons at liftoff and contains more than 500,000 kilograms of propellant. The type of propellant used in the solid rocket motor is known as PBAN, which stands for polybutadiene–acrylic acid–acrylonitrile terpolymer. In addition to the PBAN (which serves as the binder), the propellant consists of approximately 70 percent ammonium perchlorate (the oxidizer), 16 percent powdered aluminum (the fuel), and a trace of iron oxide (a catalyst) to control the burning rate. The cured solid propellant looks and feels like a hard rubber eraser.

See also PROPELLANT; ROCKET; SPACE TRANSPORTATION SYSTEM.

solid-state device A device that uses the electric, magnetic, and photonic properties of solid materials, mainly semiconductors. It contains no moving parts and depends on the internal movement of charge carriers (i.e., electrons and "positive" holes) for its operation.

solipsism syndrome A psychological disorder that could happen to the inhabitants of large space bases or space settlements. It is basically a state of mind in which a person feels that everything is a dream and not real. The whole of life becomes a long dream from which the individual can never awaken. A person with this syndrome feels very lonely and detached and eventually becomes apathetic and indifferent. This syndrome might easily occur in a space habitat environment where everything is artificial or human-made. To prevent or alleviate the tendency toward solipsism syndrome in space habitats, we would use large-geometry interior designs (e.g., have something beyond the obvious horizon); place some things beyond the control or reach of the inhabitants' manipulation (e.g., an occasional rainy day weather pattern variation or small animals that have freedom of movement); and provide living and growing objects such as vegetation, animals, and children.

See also SPACE SETTLEMENT.

solstice The two times of the year when the Sun's position in the sky is the most distant from the celestial equator. For the Northern Hemisphere, the summer solstice (longest day) occurs about June 21 and the winter solstice (shortest day) about December 21.

sonic 1. In aerodynamics, of or pertaining to the speed of sound; that which moves at acoustic velocity, as in sonic flow; designed to operate or perform at the speed of sound, as in a *sonic leading edge*. 2. Of or pertaining to sound, as in *sonic amplifier.* In this sense, *acoustic* is preferred to *sonic.*

sonic boom A thunderlike noise caused by a shock wave that emanates from an aircraft, aerospace vehicle, or flying object that is traveling at or above sonic velocity in Earth's atmosphere. As supersonic objects travel through the air, the air molecules are pushed aside with great force, and this action forms a shock wave. The bigger and heavier the supersonic object, the more the air is displaced and the stronger the resultant shock waves. Several factors can influence sonic booms—the mass, size, and shape of the supersonic vehicle or object; its altitude, attitude, and flight path; and local weather or atmospheric conditions.

The shock wave forms a cone of pressurized air molecules that moves outward and rearward in all directions and extends to the ground. As the cone spreads across the landscape along the flight path, it creates a continuous sonic boom along the full width of its base.

sounding rocket A rocket, usually with a solid-propellant motor, used to carry scientific instruments on parabolic trajectories into the upper regions of Earth's sensible atmosphere (i.e., beyond the reach of aircraft and scientific balloons) and into near-Earth space. A sounding rocket basically is divided into two major components, the solid rocket motor and the payload. Payloads typically include the experiment package, the nose cone, a telemetry system, an attitude control system, a radar tracking beacon, the firing despin module, and the recovery section. Many sounding rocket payloads are recovered for refurbishment and reuse.

Sounding rockets fly vertical trajectories from 48 km to over 1,285 km in altitude. The flight normally lasts less than 30 minutes. Sounding rockets come in a wide variety of sizes and types. For example, NASA's sounding rocket fleet ranges from the single-stage Super Arcas, which stands 3 meters high and 11 centimeters in diameter, to the single-stage Aries, which stands 11 meters high and 1.11 meter in diameter. The Super Arcas has been used by NASA since 1962 for carrying meteorological measuring devices. It can carry a 4.5-kilogram payload to an altitude of 91 km. The four-stage Black Brant XII, the tallest in NASA's fleet, is 20 meters high.

All of NASA's sounding rockets use solid-propellant rocket motors. Furthermore, all of these rocket vehicles are unguided except the Aries and those that use the S-19 Boost Guidance System. During flight all sounding rockets except the Aries are imparted with a spinning motion to reduce potential dispersion of the flight trajectory due to vehicle misalignments. NASA conducts between 30 and 35 suborbital sounding rocket flights per year, providing low-cost, quick-

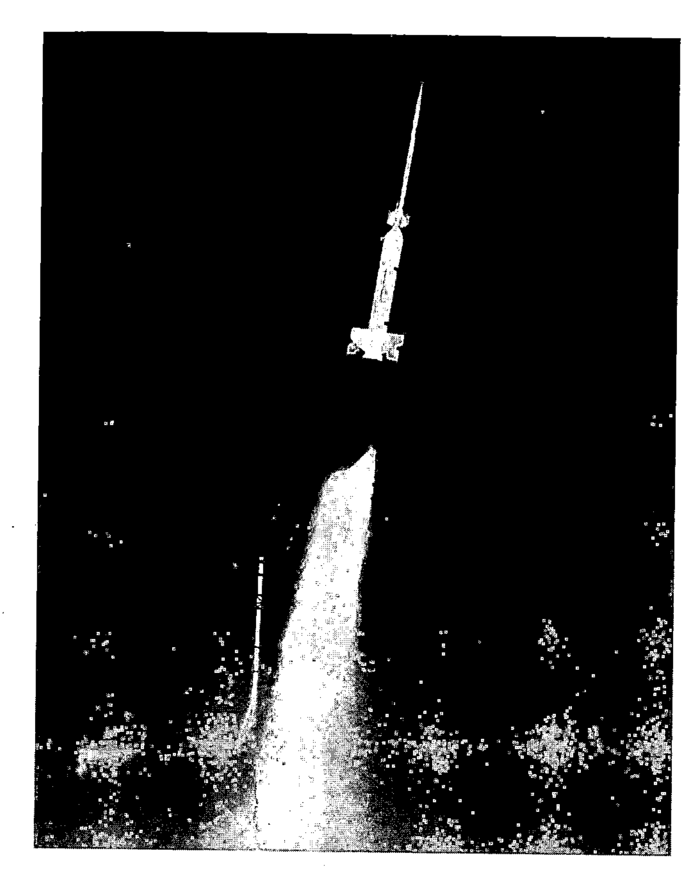

A Nike-Apache sounding rocket blasts off from NASA's Wallops Flight Facility, Wallops Island, Virginia, in 1965. In its seven-minute, ballistic trajectory flight over the Atlantic Ocean, this rocket's scientific payload, designed to make measurements within the ionosphere, reached a maximum altitude of approximately 180 kilometers. ***(Courtesy of NASA)***

response flight opportunities to scientists in the areas of physics and astronomy, microgravity materials processing research, and instrument development. Most of these sounding rocket launches take place at NASA's Wallops Flight Facility, Wallops Island, Virginia.

See also ROCKET; WALLOPS FLIGHT FACILITY.

South Atlantic anomaly (SAA) A region of Earth's trapped radiation particle zone that dips close to the planet in the South Atlantic Ocean southeast of the Brazilian coast. This region represents the most significant source of ionizing radiation for space travelers in low-inclination, low-altitude orbits around Earth.

See also EARTH'S TRAPPED RADIATION BELTS; SPACE RADIATION ENVIRONMENT.

Soyuz launch vehicle The "workhorse" Soviet (and later Russian) launch vehicle that was first used in 1963. With its two cryogenic stages and four cryogenic strap-ons, this vehicle is capable of placing up to 6,900 kg into low Earth orbit (LEO). At present it is the most frequently flown launch vehicle in the world. Since 1964 the Soyuz rocket has been used to launch every Russian human crewed space mission.

See also LAUNCH VEHICLE.

Soyuz spacecraft An evolutionary family of crewed spacecraft that have been used by the Soviet Union and later the Russian Federation on a wide variety of space missions. The word *soyuz* is Russian for "union." *Soyuz 1* was launched in April 1967. Unfortunately, upon reentry a parachute failed to open properly, and the spacecraft was destroyed on impact with its occupant, cosmonaut VLADIMIR M. KOMAROV. The second Russian space tragedy occurred at the end of the *Soyuz 11* mission (June 1971), when a valve malfunctioned as the spacecraft was separating from the *Salyut 1* space station, allowing all the air to escape from the crew compartment. This particular early version of the *Soyuz* spacecraft did not have sufficient room for the crew to wear their pressure suits during reentry; consequently, the three cosmonauts, Georgi Dobrovolsky, Victor Patseyev, and Vladislav Volkov, suffocated during the reentry operation. They were found dead by the Russian recovery team after touchdown. In July 1975 the *Soyuz 19* spacecraft was used successfully by cosmonauts ALEXEI LEONOV and Valeri Kubasov in the Apollo-Soyuz Test Project, an international rendezvous and docking mission. The next major variant of this versatile spacecraft, called the *Soyuz-T* ("T" standing for transport), was first flown in December 1979. The *Soyuz-TM* is a modernized version of the *Soyuz-T*. It was flown in May 1986 and has been used to ferry crew and supplies to orbiting stations such as the *Mir* and the *INTERNATIONAL SPACE STATION*.

The *Soyuz TMA-1* is a Russian automatic passenger spacecraft designed for launch by a Soyuz launch vehicle from the Baikonur Cosmodrome. Following launch the spacecraft proceeds in an automated fashion to rendezvous and dock with the *International Space Station* (ISS). The *Soyuz TMA-1* is a larger craft with a more comfortable interior than the previous *Soyuz TM* models. After docking the spacecraft remains parked at the ISS, serving as an emergency escape spacecraft until it is relieved by the arrival of another *Soyuz* spacecraft. For example, in late October 2002 a *Soyuz TMA-1* was launched from the Baikonur Cosmodrome and successfully carried three cosmonauts (two Russian and one Belgian) to the ISS. The *Soyuz TMA-1* automatically docked with the ISS. After 10 days of microgravity research the three visiting cosmonauts departed from the ISS using a previously parked *Soyuz TM-34* spacecraft. The *Soyuz TMA-1* spacecraft that carried them into space remained behind as a lifeboat for the permanent crew of the ISS.

The *Soyuz TMA-1* spacecraft approaches the Pirs docking compartment on the *International Space Station* (ISS) on November 1, 2002. The spacecraft is carrying the *Soyuz 5* taxi crew: Russian commander Sergei Zalyotin, Belgian flight engineer Frank DeWinne (representing the European Space Agency), and Russian flight engineer Yuri V. Lonchakov, the three of whom conducted an eight-day visit to the station. Among other engineered improvements, the *Soyuz TMA* spacecraft has upgraded computers, a new cockpit control panel, and improved avionics. *(Courtesy of NASA)*

See also APOLLO-SOYUZ TEST PROJECT; *MIR* SPACE STATION; SALYUT.

space 1. Specifically, the part of the universe lying outside the limits of Earth's atmosphere. By informal international agreement, outer space is usually considered to begin at between 100 and 200 kilometers altitude. Within the U.S. aerospace program persons who have traveled beyond 80 km altitude often are recognized as space travelers or astronauts. 2. More generally, the volume in which all celestial bodies, including Earth, move.

space base A large, permanently inhabited space facility located in orbit around a celestial body or on its surface. The space base would serve as a center of human operations in some particular region of the solar system, supporting exploration, scientific missions, and extraterrestrial resource applications. An orbiting facility also could serve as a space transportation hub, a robot repair and maintenance facility, a space construction site, and even a recreation and medical services center for space travelers. With from 10 to perhaps 200 occupants, the space base would have a much larger human population than a space station. Modular construction, use of extraterrestrial materials for radiation shielding, development of a closed regenerative life support system, and large solar or nuclear power supplies (e.g., 50 to 100 kilowatts-electric or more) would be some characteristic design features. For an orbiting facility, artificial gravity would be provided by rotating the crew habitats and certain work areas. For surface bases on the Moon and Mars, the human inhabitants would experience the natural gravity of that particular celestial object.

See also LUNAR BASES AND SETTLEMENTS; MARS BASE; SPACE SETTLEMENT; SPACE STATION.

space-based astronomy Astronomical observations conducted by human-crewed or robotic spacecraft above Earth's atmosphere; for example, the excellent cosmic observations performed by the *Hubble Space Telescope* (a human-tended but remotely operated Earth-orbiting instrument platform) and the solar observations made by the crew of the U.S. *Skylab* space station (1973–74).

See also *CHANDRA X-RAY OBSERVATORY*; COSMIC BACKGROUND EXPLORER; *HUBBLE SPACE TELESCOPE*; *SKYLAB*; *SPITZER SPACE TELESCOPE*.

space-based interceptor (SBI) A kinetic energy kill rocket vehicle that is based in space.

See also BALLISTIC MISSILE DEFENSE.

spaceborne imaging radar In general, a synthetic aperture radar (SAR) system that is placed on a spacecraft. The radar is an active remote sensing instrument that transmits pulses of microwave energy toward the planet's surface and measures the amount of energy that is reflected back to the radar antenna. An SAR system therefore provides its own illumination and can produce images of the planet's surface during the day or night and through cloud cover. These types of radar systems have been used to image the cloud-enshrouded surface of Venus and to help scientists study planet Earth more effectively.

For example, the Spaceborne Imaging Radar-C/X Band Synthetic Aperture Radar (SIR-C/X-SAR) was flown in the space shuttle orbiter's cargo bay during the STS-59 mission (April 1994). This radar system, a joint project of NASA, the German Space Agency (DARA), and the Italian Space Agency (ASI), collected an enormous amount of radar imagery data in support of NASA's Earth system science and other planetary study efforts.

See also MAGELLAN MISSION; RADAR IMAGING; REMOTE SENSING.

space capsule A container used for conducting an experiment or operation in space, usually involving a human being or other living organisms. The Mercury Project spacecraft were small space capsules designed to carry a single American astronaut into space and return him safely to Earth. As American human-rated spacecraft became a little larger during the Gemini and Apollo projects, the term *space capsule* generally was abandoned in favor of *spacecraft*.

See also MERCURY PROJECT.

space colony An early term used to describe a large, permanent space habitat and industrial complex occupied by up to 10,000 persons. Currently the term *space settlement* is preferred.

See also SPACE SETTLEMENT.

space commerce The commercial sector of space operations and activities. At least six major areas are now associated with the field commonly referred to as space commerce. These areas are (1) space transportation, (2) satellite communications, (3) satellite-based positioning and navigational services, (4) satellite remote sensing (including geographic information system support), (5) space-based industrial facilities, and (6) materials research and processing in space. In the past few years wealthy individuals have interacted with the Russian Space Agency and privately funded their own commercial trips (as passengers and "guest cosmonauts") into outer space, so *space tourism* might also be considered a fledgling area of space commerce. Specifically, in April 2001 the American businessperson Dennis Tito blasted off onboard a Russian Soyuz rocket bound for the *INTERNATIONAL SPACE STATION*. This trip, at a cost of $20 million, and his one-week stay in space qualified him as the world's first space tourist. In April 2002 Mark Shuttleworth, a South African businessperson, rode another Russian Soyuz rocket bound for the ISS and became the world's second space tourist.

space construction Large structures in space, such as modular space stations, global communication and information services platforms, and satellite power systems (SPSs), will all require on-orbit assembly operations by space construction workers. Space construction requires protection of the workforce and some materials from the hard vacuum, intense sunlight, and natural radiation environment encountered above Earth's protective atmosphere.

Outer space, however, is also an environment that in many ways is ideal for the construction process. First, because of the absence of significant gravitational force (i.e., the microgravity experienced by the free-fall condition of orbiting objects), the structural loads are quite small, even minute. Structural members may therefore be much lighter than terrestrial structures of the same span and stiffness. Second, the absence of gravitational forces greatly facilitates the movement of material and equipment. On Earth the movement of materials during a construction operation absorbs a large portion of the total work effort expended by construction personnel and their machines. Third, the absence of an atmosphere with its accompanying wind loads, inclement weather, and unpredictable changes permits space work to be planned accurately and executed readily without environmental interruptions (except perhaps due to solar flares, which would increase the radiation hazard).

Automated fabrication is considered to be a key requirement for viable space construction activities. Work in space will require the close interaction of astronauts (i.e., space workers) and very smart machines. For example, the advanced maneuverable spacesuit will be a versatile, self-contained, life-supporting backpack with gaseous-nitrogen-propelled jet thrusters that will enable a space worker to travel back and forth to various space construction locations. The automated beam builder will be a machine designed for fabricating "building-block" structural beams in space. Combined with a space structure fabrication system, the beam builder could allow space workers to manufacture and assemble structures in low Earth orbit (LEO) using the *INTERNATIONAL SPACE STATION* (ISS) as an early "construction camp."

Eventually, as the demand for more sophisticated space construction and assembly efforts grows, permanent space construction bases could be established in LEO and elsewhere in cislunar space. Remote astronaut workstations could be mounted on large manipulator arms attached to the space station or space base. These "open cherry pickers" would have a convenient tool and parts bin, a swing-away control and display panel, and lights for general and point illumination. The closed version of this cherry picker would involve a pressurized human-occupied remote workstation (space construction module) that contains life support equipment and controls and displays for operating dexterous manipulators. The table on page 556 lists some of the typical space construction equipment that could be used by the middle of this century.

See also LARGE SPACE STRUCTURES; SATELLITE POWER SYSTEM.

Typical Space Construction Equipment and Supporting Systems

Advanced maneuverable space suits
Automated beam builder
Space structure fabrication system
Remote astronaut work stations
 "Closed cherry picker"
 "Open cherry picker"
 Free-flyer work station
Manned orbital transfer vehicle (MOTV)
Cargo orbital transfer vehicle
Advanced launch vehicles (for personnel and cargo)

spacecraft In general, a human-occupied or uncrewed platform that is designed to be placed into an orbit about Earth or into a trajectory to another celestial body. The spacecraft is essentially a combination of hardware that forms a space platform. It provides structure, thermal control, wiring, and subsystem functions, such as attitude control, command, data handling, and power. Spacecraft come in all shapes and sizes, each tailored to meet the needs of a specific mission in space. For example, spacecraft designed and constructed to acquire scientific data are specialized systems intended to function in a specific hostile environment. The complexity of these scientific spacecraft varies greatly. Often they are categorized according to the missions they are intended to fly.

Scientific spacecraft include flyby spacecraft, orbiter spacecraft, atmospheric probe spacecraft, atmospheric balloon packages, lander spacecraft, surface penetrator spacecraft, and surface rover spacecraft.

Flyby spacecraft follow a continuous trajectory and are not captured into a planetary orbit. These spacecraft have the capability to use their onboard instruments to observe passing celestial targets (e.g., a planet, a moon, or an asteroid), even compensating for the target's apparent motion in an optical instrument's field of view. They must be able to transmit data at high rates back to Earth and also must be capable of storing data onboard for those periods when their antennas are not pointing toward Earth. They must be capable of surviving in a powered-down cruise mode for many years of travel through interplanetary space and then of bringing all their sensing systems to focus rapidly on the target object during an encounter period that may last for only a few crucial hours or minutes. NASA's *Pioneer 10* and *11* and *Voyager 1* and *2* are examples of highly successful flyby scientific spacecraft. NASA used the flyby spacecraft during the initial, or reconnaissance, phase of solar system exploration.

An *orbiter spacecraft* is designed to travel to a distant planet and then orbit around that planet. This type of scientific spacecraft must possess a substantial propulsive capability to decelerate at just the right moment in order to achieve a proper orbit insertion. Aerospace engineers design an orbiter spacecraft recognizing the fact that solar occultations will occur frequently as it orbits the target planet. During these periods of occultation, the spacecraft is shadowed by the planet, cutting off solar array production of electric power and introducing extreme variations in the spacecraft's thermal environment. Generally, a rechargeable battery system augments solar electric power. Active thermal control techniques (e.g., the use of tiny electric-powered heaters) are used to complement traditional passive thermal control design features. The periodic solar occultations also interrupt uplink and downlink communications with Earth, making onboard data storage a necessity. NASA used orbiter spacecraft as part of the second, in-depth study phase of solar system exploration. The *Lunar Orbiter, Magellan, Galileo,* and *Cassini* spacecraft are examples of successful scientific orbiters.

Some scientific exploration missions involve the use of one or more smaller, instrumented spacecraft, called *atmospheric probe spacecraft.* These probes separate from the main spacecraft prior to closest approach to a planet in order to study the planet's gaseous atmosphere as they descend through it. Usually an atmospheric probe spacecraft is deployed from its "mother spacecraft" (i.e., the main spacecraft) by the release of springs or other devices that simply separate it from the mother spacecraft without making a significant modification of the probe's trajectory. Following probe release the mother spacecraft usually executes a trajectory correction maneuver to prevent its own atmospheric entry and to help the main spacecraft continue on with its flyby or orbiter mission activities. NASA's Pioneer Venus (four probes), Galileo (one probe), and Cassini (*Huygens* probe) missions involved the deployment of a probe or probes into the target planetary body's atmosphere (i.e., Venus, Jupiter, and Saturn's moon Titan, respectively). An aeroshell protects the atmospheric probe spacecraft from the intense heat caused by atmospheric friction during entry. At some point in the descent trajectory, the aeroshell is jettisoned and a parachute then is used to slow the probe's descent sufficiently that it can perform its scientific observations. Data usually are telemetered from the atmospheric probe to the mother spacecraft, which then either relays the data back to Earth in real time or else records the data for later transmission to Earth.

An *atmospheric balloon package* is designed for suspension from a buoyant gas-filled bag that can float and travel under the influence of the winds in a planetary atmosphere. Tracking of the balloon package's progress across the face of the target planet will yield data about the general circulation patterns of the planet's atmosphere. A balloon package needs a power supply and a telecommunications system to relay data and support tracking. It also can be equipped with a variety of scientific instruments to measure the planetary atmosphere's composition, temperature, pressure, and density.

Lander spacecraft are designed to reach the surface of a planet and survive at least long enough to transmit back to Earth useful scientific data, such as imagery of the landing site, measurement of the local environmental conditions, and an initial examination of soil composition. For example, the Russian Venera lander spacecraft have made brief scientific investigations of the infernolike Venusian surface. In contrast, NASA's Surveyor lander craft extensively explored the lunar surface at several landing sites in preparation for the human Apollo Project landing missions, while NASA's *Viking 1* and

2 lander craft investigated the surface conditions of Mars at two separate sites for many months.

A *surface penetrator spacecraft* is designed to enter the solid body of a planet, an asteroid, or a comet. It must survive a high-velocity impact and then transmit subsurface information back to an orbiting mother spacecraft.

Finally, a *surface rover spacecraft* is carried to the surface of a planet, soft-landed, and then deployed. The rover can either be semiautonomous or fully controlled (through teleoperation) by scientists on Earth. Once deployed on the surface, the electrically powered rover can wander a certain distance away from the landing site and take images and perform soil analyses. Data then are telemetered back to Earth by one of several techniques: via the lander spacecraft, via an orbiting mother spacecraft, or (depending on the size of the rover) directly from the rover vehicle. The Soviet Union deployed two highly successful robot surface rovers (called *Lunokhod 1* and *2*) on the Moon in the 1970s. NASA plans to explore the surface of Mars with a variety of small mobile robots over the next decade. The *Spirit* and *Opportunity* Mars Exploration Rovers (2003) are just the first of many robot rovers that will scamper across the Red Planet.

spacecraft charging In orbit or in deep space, spacecraft and space vehicles can develop an electric potential up to tens of thousands of volts relative to the ambient outer space plasma (the solar wind). Large potential differences (called *differential charging*) also can occur on the space vehicle. One of the consequences is electrical discharge, or arcing, a phenomenon that can damage space vehicle surface structures and electronic systems. Many factors contribute to this complex problem, including the spacecraft configuration, the materials from which the spacecraft is made, whether the spacecraft is operating in sunlight or shadow, the altitude at which the spacecraft is performing its mission, and environmental conditions, such as the flux of high-energy solar particles and the level of magnetic storm activity.

Wherever possible, spacecraft designers use conducting surfaces and provide adequate grounding techniques. These design procedures can significantly reduce differential charging, which is generally a more serious problem than the development of a high spacecraft-to-space (plasma) electrical potential.

spacecraft clock Generally, the timing component within the spacecraft's command and data-handling subsystem; it meters the passing time during the life of the spacecraft and regulates nearly all activity within the spacecraft.

See also CLOCK.

spacecraft drag A space vehicle or spacecraft operating at an altitude below a few thousand kilometers will encounter a significant number of atmospheric particles (i.e., the residual atmosphere) during each orbit of Earth. These encounters result in drag, or friction, on the spacecraft, causing it gradually to "slow down" and lose altitude unless some onboard propellant is expended to overcome this drag and maintain the original orbital altitude. If the density of the residual atmosphere at the space vehicle's altitude increases, so will the drag on the vehicle. Any mechanism that can heat Earth's atmosphere (e.g., a geomagnetic storm) will create density changes in the upper atmosphere that can alter a spacecraft's orbit rapidly and significantly. When the residual upper atmosphere is heated by these solar disturbances, it expands outward and makes its presence felt at even higher (than normal) altitudes.

The significance and severity of spacecraft drag was demonstrated clearly by the rapid and premature demise of the abandoned U.S. *Skylab* space station. Atmospheric heating during a period of maximum solar activity caused the space station's drag to increase considerably, a situation that then resulted in a much more rapid rate of loss of orbital altitude than had been projected. As a result, the 90-metric-ton station (last occupied by an astronaut crew in 1974) experienced a fiery reentry in July 1979, years before its originally projected demise. In fact, NASA had been considering using an early space shuttle mission to "rescue" *Skylab* by providing a reboost to higher altitude. However, the first shuttle mission (STS-1) was not flown until April 1981, almost two years after *Skylab* became a "front-page" victim of spacecraft drag.

To avoid similar problems of large derelict space objects making unplanned or erratic reentries due to increasing levels of atmospheric drag, Russian spacecraft controllers successfully deorbited the abandoned *MIR* SPACE STATION in March 2001. Any parts of this massive space station complex that survived reentry splashed down harmlessly in a remote area of the Pacific Ocean. Similarly, at the end of its useful scientific mission NASA spacecraft controllers intentionally commanded the massive (16,300-kg) *COMPTON GAMMA RAY OBSERVATORY* to perform a carefully planned deorbit maneuver. This caused the derelict spacecraft to reenter Earth's atmosphere in June 2000, and any surviving pieces crashed safely in a remote part of the Pacific Ocean.

See also SKYLAB.

spacecraft navigation In general, navigating a spacecraft involves measuring its radial distance and velocity, the angular distance to the spacecraft, and its velocity in the plane of the sky. From these data a mathematical model can be developed and maintained that describes the history of the spacecraft's location in three-dimensional space as a function of time. Any necessary corrections to a spacecraft's trajectory or orbit then can be identified based on this mathematical model. A spacecraft's navigational history often is reconstructed and incorporated in its scientific observations during a planetary encounter.

Within NASA the art of spacecraft navigation draws on tracking data, which include measurements of the Doppler shift of the downlink carrier and the pointing angles of antennas in the Deep Space Network (DSN). Such navigational data differ from the telemetry data associated with scientific instruments and spacecraft state-of-health sensors.

The distance to a spacecraft can be measured in the following manner. A uniquely coded ranging pulse can be added to the uplink communications sent to a spacecraft. The transmission time of this specially coded pulse is recorded. When the spacecraft receives the ranging pulse, it then returns the special pulse on its downlink communications transmission.

The time it takes the spacecraft to turn the pulse around within its onboard electronics package is known from prelaunch calibration testing. Thus, when the specially coded downlink pulse is received at one or several DSN sites, the true travel time elapsed can be determined and the spacecraft's distance from Earth can then be calculated. When several DSN sites have received this coded downlink transmission, the angular position of the spacecraft also can be determined using triangulation techniques (since the angles at which the DSN antennas point are recorded with an accuracy of thousandths of a degree). Similarly, a spacecraft's velocity component can be determined by a careful measurement of an induced Doppler shift in the downlink frequency when a two-way coherent communications mode is being used.

Spacecraft that are equipped with imaging instruments can employ these instruments to perform optical navigation. The imaging instrument observes the spacecraft's target (destination) planet against a known background star field. These images are called *optical navigation* (OPNAV) images, and their interpretation provides a very precise data set that is useful for refining knowledge of a spacecrafts's trajectory.

Once a spacecraft's solar or planetary orbital parameters are known, they are compared to the parameters desired for the flight. A *trajectory correction maneuver* (TCM) then can be planned and executed to correct any discrepancy between a spacecraft's actual and desired/planned trajectory. A TCM typically involves a spacecraft velocity change (ΔV) on the order of just meters to tens of meters per second because of the limited propellant supply carried by a spacecraft.

Similarly, small changes in a spacecraft's orbit around a planet might be needed to adjust an instrument's field-of-view footprint, to improve sensitivity of a gravity field survey, or to prevent too much orbital decay (i.e., loss of altitude). These orbit changes are called *orbit trim maneuvers* (OTMs), and generally they are carried out in the same manner as TCMs. For example, to make a change that increases the altitude of perigee, an OTM would be performed that increased the spacecraft's velocity when the craft was at apogee. Finally, slight changes in a spacecraft's orbital plane orientation also can be accomplished using OTMs. However, the magnitude of the orbital plane change often is necessarily small due to the limited amount of propellant a spacecraft carries for these navigational maneuvers.

space debris Space junk or derelict human-made space objects in orbit around Earth. Space debris represents a hazard to astronauts, spacecraft, and large space facilities such as space stations.

Since the start of the space age in 1957, the natural meteoroid environment has been a design consideration for spacecraft. Meteoroids are part of the interplanetary environment and sweep through Earth orbital space at an average speed of about 20 kilometers (km) per second. Space science data indicate that at any one moment, a total of approximately 200 kg of meteoroid mass is within some 2,000 km of Earth's surface, the region of space (called low Earth orbit, or LEO) most frequently used. The majority of this mass is found in meteoroids about 0.01 cm diameter. However, lesser amounts of this total mass occur in meteoroid sizes both smaller and larger than 0.01 cm. The natural meteoroid flux varies in time as Earth travels around the Sun.

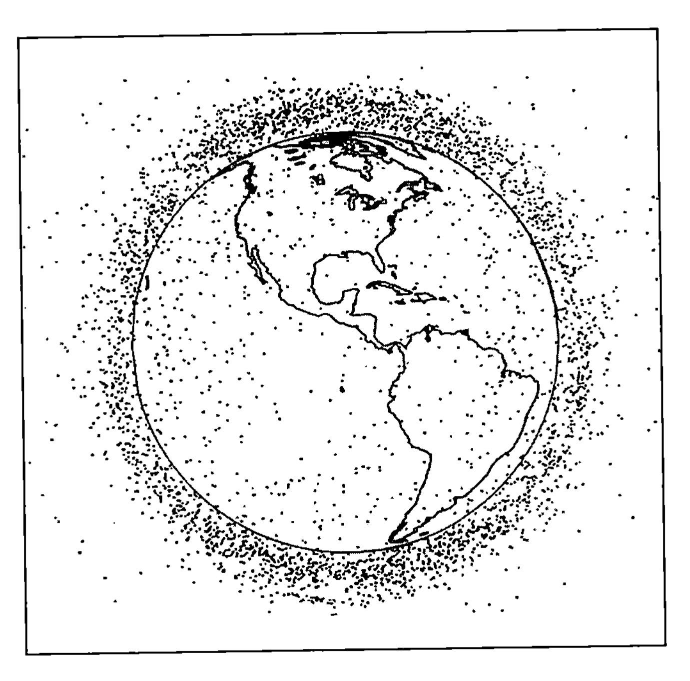

Space debris is a growing problem that will require more and more attention as space activities increase in this century. The drawing is a "snapshot" representation showing all objects larger than 10 cm in diameter (about the size of a baseball) that were found in low Earth orbit (LEO) on May 30, 1987, by the U.S. Space Command (now part of U.S. Strategic Command). Although the debris objects are not drawn to scale on this map, the illustration provides a meaningful depiction of the space debris population, especially as found in LEO. ***(Drawing courtesy of the U.S. Air Force and NASA)***

Human-made space debris is also called *orbital debris* and differs from natural meteoroids because it remains in Earth orbit during its lifetime and is not a transient phenomenon such as the meteoroid showers that occur as the Earth travels through interplanetary space around the Sun. The estimated mass of human-made objects orbiting the Earth within about 2,000 km of its surface is about 3 million kg, or about 15,000 times more mass than that represented by the natural meteoroid environment. These human-made objects are for the most part in high-inclination orbits and pass one another at an average relative velocity of 10 km per second. Most of this mass is contained in more than 3,000 spent rocket stages, inactive satellites, and a comparatively few active satellites. A lesser amount of space debris mass (some 40,000 kg) is distributed in more than 4,000 smaller-sized orbiting objects currently being tracked by space surveillance systems. The majority of these smaller space debris objects are the byproducts of more than 130 on-orbit fragmentations (satellite breakup events). Recent studies indicate a total mass of at least 1,000 kg for orbital debris sizes of 1 cm or smaller, and about 300 kg for orbital debris smaller than 0.1 cm. The explosion or fragmentation of a large space object also has the potential to produce a large number of smaller objects, objects too small to be detected by contemporary ground-based space surveillance systems. Consequently, this orbital debris environment is now considered more hazardous than the natural meteoroid environment to spacecraft operating in Earth orbit below an altitude of 2,000 km.

Two general types of orbital debris are of concern: (1) large objects (greater than 10 cm in diameter) whose population, while small in absolute terms, is large relative to the population of similar masses in the natural meteoroid environment, and (2) a much greater number of smaller objects (less than 10 cm in diameter), whose size distribution approximates the natural meteoroid population and whose numbers add to the "natural debris" environment in those size ranges. The interaction of these two general classes of space debris objects, combined with their long residence time in orbit, create further concern that new collisions producing additional fragments and causing the total space debris population to grow are inevitable.

An orbiting object loses energy through frictional encounters with the upper limits of Earth's atmosphere and as a result of other orbit-perturbing forces (e.g., gravitational influences). Over time the object falls into progressively lower orbits and eventually makes a final plunge toward Earth. Once an object enters the sensible atmosphere, atmospheric drag slows it down rapidly and causes it either to burn up completely or to fall through the atmosphere and impact on Earth's surface or in its oceans. A *decayed satellite* (or piece of orbital debris) is one that reenters Earth's atmosphere under the influence of natural forces and phenomena (e.g., atmospheric drag). Space vehicles and satellites that are intentionally removed from orbit are said to have been "deorbited."

One of the most celebrated reentries of a large human-made object occurred on July 11, 1979, when the then-decommissioned and abandoned first American space station, called *Skylab*, came plunging back to Earth over Australia and the Indian Ocean, a somewhat spectacular reentry event that occurred without harm to life or property. In fact, although human-made objects reenter from orbit on the average of more than one per day, only a very small percentage of these reentry events result in debris surviving to reach Earth's surface. The aerodynamic forces and heating associated with reentry processes usually break up and vaporize most incoming space debris.

Solar activity greatly affects the natural decay of Earth-orbiting objects. High levels of solar activity heat Earth's upper atmosphere, causing it to expand farther into space and to reduce the orbital lifetimes of space objects found at somewhat higher altitudes in the LEO regime. However, at about 600 km of altitude, the atmospheric density is sufficiently low, and solar activity–induced atmospheric density increases do not noticeably affect the debris population lifetimes. This solar cycle–based natural cleansing process for space debris in LEO is extremely slow and by itself cannot offset the current rate of human-made space debris generation.

The effects of orbital debris impacts on spacecraft and space facilities depend on the velocity and mass of the debris. For debris sizes less than approximately 0.01 cm in diameter, surface pitting and erosion are the primary effects. Over a long period of time the cumulative effect of individual particles colliding with a satellite could become significant because the number of such small debris particles is very large in LEO.

For debris larger than about 0.1 cm in diameter, the possibility of structural damage to a satellite or space facility becomes an important consideration. For example, a 0.3-cm-diameter sphere of aluminum traveling at 10 kilometers per second has about the same kinetic energy as a bowling ball traveling at 100 kilometers per hour. Aerospace engineers anticipate significant structural damage to a satellite or space facility if such an impact occurs.

Space system engineers, therefore, find it helpful to distinguish three space debris size ranges in designing spacecraft. These are debris sizes 0.01 cm diameter and below, which produce surface erosion; debris sizes ranging from 0.01 to 1.0 cm diameter, which produce significant impact damage that can be quite serious; and (3) space debris objects greater than 1.0 cm diameter, which can readily produce catastrophic damage in a satellite or space facility.

Today only about 5 percent of the cataloged objects in Earth orbit are active, operational spacecraft. The remainder of these human-made space objects represent various types of orbital debris. Space debris often is divided into four general categories: (1) operational debris (about 12 percent), objects intentionally discarded during satellite delivery or satellite operations (this category includes lens caps, separation and packing devices, spin-up mechanisms, payload shrouds, empty propellant tanks, and a few objects discarded or "lost" during extravehicular activities [EVAs] by astronauts or cosmonauts); (2) spent and intact rocket bodies (14 percent); (3) inactive (decommissioned or dead) payloads (20 percent); and (4) fragmentation (on-orbit space object breakup) (49 percent).

Aerospace engineers consider the growing space debris problem when they design new spacecraft. In an attempt to make these new spacecraft as "litter-free" as possible, new spacecraft are designed with provisions for retrieval or removal at the end of their useful operations. Telerobotic space debris collection systems also have been proposed.

See also METEOROIDS.

space defense All defensive measures designed to destroy attacking enemy vehicles (including missiles) while they are in space or to nullify or reduce the effectiveness of such an attack from (or through) space.

Space Detection and Tracking System (SPADATS) A network of space surveillance sensors operated by the U.S. Air Force. This ground-based network is capable of detecting and tracking space vehicles and objects in orbit around Earth. The orbital characteristics of such objects are reported to a central control facility.

See also UNITED STATES AIR FORCE.

space food Eating is a basic survival need that an astronaut or cosmonaut must accomplish in order to make space a suitable place to work and accomplish mission objectives. The space food (meals) he or she consumes must be nutritious, safe, lightweight, easily prepared, and convenient to use and require little storage space, need no refrigeration, and be psychologically acceptable (especially for crews on long-duration missions).

Since the beginning of human space flight in the early 1960s, eating in space has become more natural and "Earthlike" while better meeting the other criteria. Space flight feeding has progressed from squeezing pastelike foods from "toothpaste" tubes to eating a "sit-down" dinner complete with normal utensils, except for the addition of scissors, which today's astronauts use to cut open packages.

In general, there are five basic approaches to preparing food for use in space: rehydratable food, intermediate-moisture food, thermostabilized food, irradiated food, and natural-form food.

Rehydratable food has been dehydrated by a technique such as freeze drying. In the space shuttle program, for example, foods are dehydrated to meet launch vehicle mass and volume restrictions. They are rehydrated later in orbit when they are ready to be eaten. Water used for rehydration comes from the orbiter vehicle's fuel cells, which produce electricity by combining hydrogen and oxygen; water is the resultant by-product. More than 100 different food items, such as cereals, spaghetti, scrambled eggs, and strawberries, go through this dehydration-rehydration process. For example, when a strawberry is freeze-dried, it remains full size in outline, with its color, texture, and quality intact. The astronaut can then rehydrate the strawberry with either saliva (mouth moisture) as it is chewed or by adding water to a package.

A total of 20 varieties of drinks, including tea and coffee, also are dehydrated for use in space travel, but pure orange juice and whole milk cannot be included. If water is added to dehydrated orange juice, orange "rocks" form and do not rehydrate. Dehydrated whole milk does not dissolve properly upon rehydration. It forms lumps and has a disagreeable taste, so skim milk must be used. In the 1960s General Foods Corporation developed a synthetic orange juice product called Tang that could be used in place of orange juice.

Intermediate-moisture food is partially dehydrated food, such as dried apricots, dried pears, and dried peaches. Thermostabilized food is cooked at moderate temperatures to destroy bacteria and then sealed in cans or aluminum pouches. This type of space food includes tuna, canned fruit in heavy syrup, and ground beef. *Irradiated food* is preserved by exposure to ionizing radiation. Various types of meat and bread are processed in this manner. Finally, *natural-form food* is low in moisture and taken into space in much the same form as is found on Earth. Peanut butter, nuts, graham crackers, gum, and hard candies are examples. Salt and pepper are packaged in liquid form because crystals would float around the crew cabin and could cause eye irritation or contaminate equipment.

All food in space must be packaged in individual serving portions that allow easy manipulation in the microgravity environment of an orbiting spacecraft. These packages can be off-the-shelf thermostabilized cans, flexible pouches, or semirigid containers.

In the space shuttle program the variety of food carried into orbit is sufficiently broad that crew members can enjoy a six-day menu cycle. A typical dinner might consist of a shrimp cocktail, steak, broccoli, rice, fruit cocktail, chocolate pudding, and grape drink.

When the *INTERNATIONAL SPACE STATION* (ISS) becomes fully operational (sometime after 2006 because of the *COLUMBIA* ACCIDENT), the station's permanent crew of three could eventually grow in number to a maximum of seven persons. Their food and other supplies will be replenished at regular intervals. ISS residents are now using an extension of the joint U.S.-Russian food system that was developed during the initial shuttle-*Mir* phase of the ISS program. ISS crewmembers have a menu cycle of eight days, meaning the menu repeats every eight days. Half of the food system is American and half is Russian. However, plans are also to include the foods of other ISS partner countries, including Europe, Japan and Canada. The packaging system for the daily menu food is based on single-service, disposable containers. Single-service containers eliminate the need for a dishwasher, and the disposal approach is quite literally out of this world.

Since electrical power for the ISS is generated by solar panels rather than fuel cells (as on the space shuttle), there is no extra water generated onboard the station. Water is recycled from cabin air, but not enough for significant use in the food system. Consequently, the percentage of shuttle-era rehydratable foods is being decreased and the percentage of thermostabilized foods increased over time.

Generally, the American portion of the current ISS food system is similar to the shuttle food system. It uses the same basic types of foods—thermostabilized, rehydratable, natural form, and irradiated—and the same packaging methods and materials. As on the shuttle, beverages on the ISS are in powdered form. The water temperature is different on the station; unlike the shuttle, there is no chilled water. Crewmembers have only ambient, warm, and hot water available to them.

Space station crewmembers usually eat breakfast and dinner together. They use the food preparation area in the Russian Zvezda service module to prepare their meals. The module has a fold-down table designed to accommodate three astronauts or cosmonauts eating together under microgravity conditions. Used food packaging materials are bagged and placed along with other trash in a Progress supply vehicle. Then the robot spacecraft assumes a secondary mission as an extraterrestrial garbage truck. It is jettisoned from the ISS and burns up upon reentry into Earth's atmosphere. Garbage management is a major problem, especially when regularly scheduled resupply missions are interrupted and delayed.

Long-duration human missions beyond Earth orbit (where resupply will be difficult or impossible) will require food supplies capable of extended storage. The menu cycle will have to be greatly expanded to support crew morale and nutritional well-being. A permanent lunar base could have a highly automated greenhouse to provide fresh vegetables and fruits. (Cosmonauts and astronauts have performed several modest greenhouse experiments on the ISS.) A rotating "space greenhouse" to provide an appropriate level of artificial gravity for plant development might accompany human expeditions to Mars and beyond. A permanent Martian surface base most likely would include an "agricultural facility" as part of its closed environment life support system (CELSS).

Space Infrared Telescope Facility **(SIRTF)** *See* SPITZER SPACE TELESCOPE.

Space Interferometry Mission (SIM) A NASA Origins Program–themed mission scheduled for launch in 2009 and continuing in operation through 2019. The Space Interferometry Mission (SIM) will be the first spacecraft whose primary mission is to perform astrometry by long-baseline interferometry. Astrometry involves the precise measurement of the positions and motions of objects in the sky. Free from the distortions and noise of Earth's atmosphere and with a maximum baseline of 10 meters, SIM will enable astronomers for the first time to measure the positions and motions of stars with microarcsecond accuracy. This represents a capability of detecting the small reflex motion of stars induced by orbiting planets a few times the mass of Earth. SIM will conduct a detailed survey for planetary companions to stars in the solar neighborhood and will give scientists a more complete picture of the architecture of planetary systems around a representative sample of different stellar types. SIM will determine the positions and distances of stars throughout the Milky Way hundreds of times more accurately than any previous program and create a stellar reference grid providing a visible light astronomical reference frame with unprecedented precision. Against this reference frame SIM will measure the internal dynamics of our galaxy and the dynamics in our local group of galaxies, measure the photometric and astrometric effects of condensed dark matter in our galactic halo, and calibrate the brightness of several classes of astronomical "standard candles." The Michelson Science Center (MSC) is responsible for developing and operating the science operations system (SOS) for SIM, including program solicitation, user interface and consultation, data infrastructure, and (jointly with the Jet Propulsion Laboratory [JPL]) science operations for SIM.

See also ORIGINS PROGRAM; STANDARD CANDLE (ASTRONOMICAL).

space junk A popular aerospace industry expression for space debris or orbital debris.

See also SPACE DEBRIS.

Spacelab (SL) An orbiting laboratory facility delivered into space and sustained while in orbit within the huge cargo bay of the space shuttle orbiter. Developed by the European Space Agency (ESA) in cooperation with NASA, Spacelab featured several interchangeable elements that were arranged in various configurations to meet the particular needs of a given flight. The major elements were a habitable module (short or long configuration) and pallets. Inside the pressurized habitable research module, astronaut scientists (payload specialists) worked in a shirt-sleeve environment and performed a variety of experiments while orbiting Earth in a microgravity environment. Several platforms (called pallets) could also be placed in the orbiter's cargo bay behind the habitable module. Any instruments and experiments mounted on these pallets were exposed directly to the space environment when the shuttle's cargo bay doors were opened after the aerospace vehicle achieved orbit around Earth. A train of pallets could also be flown without the concurrent use of the habitable module.

The first Spacelab mission (called STS-9/*Spacelab 1*) was launched in November 1983. It was a highly successful joint NASA and ESA mission consisting of both the habitable module and an exposed instrument platform. The final Spacelab mission (called STS-55/*Spacelab D-2*) was launched in April 1993. It was the second flight of the German Spacelab configuration and continued microgravity research that had started with the first German Spacelab mission (STS-61A/*Spacelab D-1*) flown in October 1985. NASA, other ESA countries, and Japan contributed some of the 90 experiments conducted during the *Spacelab D-2* mission.

space launch vehicle (SLV) An expendable or reusable rocket-propelled vehicle used to lift a payload or spacecraft from the surface of Earth and to place it in orbit around the planet or on an interplanetary trajectory.

See also LAUNCH VEHICLE; SPACE TRANSPORTATION SYSTEM.

space law Space law is basically the code of international law that governs the use and control of outer space by different nations on Earth. Four major international agreements, conventions, or treaties help govern space activities: (1) Treaty on Principles Governing the Activities of States in the Exploration and Use of Outer Space, Including the Moon and Other Celestial Bodies (1967), which is also called the Outer Space Treaty; (2) Agreement on the Rescue of Astronauts, the Return of Astronauts and the Return of Objects Launched into Outer Space (1968); (3) Convention on International Liability for Damage Caused by Space Objects (1972); and (4) Convention on Registration of Objects Launched into Outer Space (1975). A fifth major treaty, called The United Nations Moon Treaty, or Moon Treaty, was adopted by the U.N. General Assembly on December 5, 1979, and entered into force on July 11, 1984, although many nations, especially the major powers, have yet to ratify it.

The Moon Treaty is based to a considerable extent on the 1967 Outer Space Treaty and is considered to represent a meaningful advance in international law dealing with outer space. It contains obligations of both immediate and long-term application to such matters as the safeguarding of human life on celestial bodies, promotion of scientific investigations, exchange of information about and derived from activities on celestial bodies, and enhancement of opportunities and conditions for evaluation, research, and exploitation of the natural resources of celestial bodies. Perhaps the recent discovery of (suspected) ice deposits in the Moon's polar regions by NASA's *LUNAR PROSPECTOR* spacecraft may rekindle international debate regarding natural resources found on alien worlds. Could such deposits (if real) be used commercially? Or should they be preserved for their scientific value as part of the common heritage of mankind (CHM) theme emphasized in the Moon Treaty? The provisions of the Moon Treaty also apply to other celestial bodies within the solar system (other than Earth) and to orbits around the Moon. Two other international conventions address (in part) the use of space nuclear power systems: (1) Convention on Early

Notification of a Nuclear Accident (1986) and (2) Convention on Assistance in the Case of a Nuclear Accident or Radiological Emergency (1987).

Few human undertakings have stimulated so great a degree of legal scrutiny on an international level as has the development of modern space technology. Perhaps this is because space activities involve technologies that do not respect national (terrestrial) boundaries and therefore place new stresses on traditional legal principles. In fact, these traditional legal principles, which are based on the rights and powers of territorial sovereignty, are often in conflict with the most efficient application of new space systems. In order to resolve such complicated and complex space age legal issues, both the technologically developed and developing nations of Earth have been forced to rely even more on international cooperation.

The United Nations Committee on the Peaceful Uses of Outer Space (COPUOS) has been and continues to be the main architect of international space law; it was established by resolution of the U.N. General Assembly in 1953 to study the problems associated with the arrival of the space age. COPUOS is made up of two subcommittees, one of which studies the technical and scientific aspects and the other the legal aspects of space activities. Some contemporary topics for space law discussion include (1) remote sensing, (2) direct broadcast communications satellites, (3) the use of nuclear power sources in space, (4) the delimitation of outer space (i.e., where does space begin and national "airspace" end from a legal point of view), (5) military activities in space, (6) space debris, (7) "Who speaks for Earth?" if we ever receive a radio signal or some other form of direct contact from an intelligent alien species; and even (8) "How do we treat an alien visitor?"—a particularly delicate legal question if the alien creature's initial actions are apparently belligerent.

space medicine The branch of aerospace medicine concerned specifically with the health of persons who make, or expect to make, flights beyond Earth's sensible (measurable) atmosphere into outer space.

space mine A hypothetical military satellite with an explosive charge, either nuclear or nonnuclear, that is designed to position itself within lethal range of a target satellite and then detonate at a preprogrammed condition or time, upon direct command, or when it is attacked.

space nuclear power (SNP) Through the cooperative efforts of the U.S. Department of Energy (DOE), formerly called the Atomic Energy Commission, and NASA, the United States has used nuclear energy in its space program to provide electrical power for many missions, including science stations on the Moon, extensive exploration missions to the outer planets—Jupiter, Saturn, Uranus, Neptune, and beyond—and even to search for life on the surface of Mars.

For example, when the *Apollo 12* mission astronauts departed from the lunar surface on their return trip to Earth (November 1969), they left behind a nuclear-powered science station that sent information back to scientists on Earth for several years. That science station as well as similar stations left on the Moon by the *Apollo 14* through *17* missions operated on electrical power supplied by plutonium-238–fueled radioisotope thermoelectric generators (RTGs). In fact, since 1961 nuclear power systems have helped assure the success of many space missions, including the *Pioneer 10* and *11* missions to Jupiter and Saturn; the *Viking 1* and *2* landers on Mars; the spectacular *Voyager 1* and *2* missions to Jupiter, Saturn, Uranus, Neptune, and beyond; the *Ulysses* mission to the Sun's polar regions; the *Galileo* mission to Jupiter, and the *Cassini* mission to Saturn (*see* table).

Energy supplies that are reliable, transportable, and abundant represent a very important technology in the development of an extraterrestrial civilization. Space nuclear power systems can play an ever-expanding role in supporting the advanced space exploration and settlement missions of this century, including permanent lunar and Martian surface bases. Even more ambitious space activities, such as asteroid movement and mining, planetary engineering, and human expeditions to the outer regions of the solar system, will require compact energy systems in advanced space nuclear reactor designs at the megawatt and gigawatt levels.

Space nuclear power supplies offer several distinct advantages over the more traditional solar and chemical space power systems. These advantages include compact size, modest mass requirements, very long operating lifetimes, the ability to operate in extremely hostile environments (e.g., intense trapped radiation belts, the surface of Mars, the moons of the outer planets, and even interstellar space), and the ability to operate independently of the distance from the Sun or orientation to the Sun.

Space nuclear power systems use the thermal energy or heat released by nuclear processes. These processes include the spontaneous but predictable decay of radioisotopes, the controlled splitting, or fissioning, of heavy atomic nuclei (such as fissile uranium-235) in a self-sustained neutron chain reaction, and eventually the joining together, or fusing, of light atomic nuclei such as deuterium and tritium in a controlled thermonuclear reaction. This "nuclear" heat then can be converted directly or through a variety of thermodynamic (heat-engine) cycles into electric power. Until controlled thermonuclear fusion capabilities are achieved, space nuclear power applications will be based on the use of either radioisotope decay or nuclear fission reactors.

The radioisotope thermoelectric generator consists of two main functional components: the thermoelectric converter and the nuclear heat source. The radioisotope plutonium-238 has been used as the heat source in all U.S. space missions involving radioisotope power supplies. Plutonium-238 has a half-life of about 87.7 years and therefore supports a long operational life. (The half-life is the time required for one-half the number of unstable nuclei present at a given time to undergo radioactive decay.) In the nuclear decay process plutonium-238 emits primarily alpha radiation that has very low penetrating power. Consequently, only a small amount of shielding is required to protect the spacecraft from its nuclear emissions. A thermoelectric converter uses the "thermocouple principle" to directly convert a portion of the nuclear (decay) heat into electricity.

A nuclear reactor also can be used to provide electric power in space. The Soviet and, later, Russian space program

Summary of Space Nuclear Power Systems Launched by the United States (1961–1997)

Power Source	Spacecraft	Mission Type	Launch Date	Status
SNAP-3A	*Transit 4A*	Navigational	June 29, 1961	Successfully achieved orbit
SNAP-3A	*Transit 4B*	Navigational	November 15, 1961	Successfully achieved orbit
SNAP-9A	*Transit-5BN-1*	Navigational	September 28, 1963	Successfully achieved orbit
SNAP-9A	*Transit-5BN-2*	Navigational	December 5, 1963	Successfully achieved orbit
SNAP-9A	*Transit-5BN-3*	Navigational	April 21, 1964	Mission aborted: burned up on reentry
SNAP-10A	*Snapshot*	Experimental	April 3, 1965	Successfully achieved orbit (reactor)
SNAP-19B2	*Nimbus-B-1*	Meteorological	May 18, 1968	Mission aborted: heat source retrieved
SNAP-19B3	*Nimbus III*	Meteorological	April 14, 1969	Successfully achieved orbit
SNAP-27	*Apollo 12*	Lunar	November 14, 1969	Successfully placed on lunar surface
SNAP-27	*Apollo 13*	Lunar	April 11, 1970	Mission aborted on way to Moon. Heat source returned to South Pacific Ocean
SNAP-27	*Apollo 14*	Lunar	January 31, 1971	Successfully placed on lunar surface
SNAP-27	*Apollo 15*	Lunar	July 26, 1971	Successfully placed on Lunar surface
SNAP-19	*Pioneer 10*	Planetary	March 2, 1972	Successfully operated to Jupiter and beyond; in interstellar space
SNAP-27	*Apollo 16*	Lunar	April 16, 1972	Successfully placed on lunar surface
Transit-RTG	*"Transit" (Triad-01-IX)*	Navigational	September 2, 1972	Successfully achieved orbit
SNAP-27	*Apollo 17*	Lunar	December 7, 1972	Successfully placed on lunar surface
SNAP-19	*Pioneer 11*	Planetary	April 5, 1973	Successfully operated to Jupiter, Saturn, and beyond
SNAP-19	*Viking 1*	Mars	August 20, 1975	Successfully landed on Mars
SNAP-19	*Viking 2*	Mars	September 9, 1975	Successfully landed on Mars
MHW	*LES 8/9*	Communications	March 14, 1976	Successfully achieved orbit
MHW	*Voyager 2*	Planetary	August 20, 1977	Successfully operated to Jupiter, Saturn, Uranus, Neptune, and beyond
MHW	*Voyager 1*	Planetary	September 5, 1977	Successfully operated to Jupiter, Saturn, and beyond
GPHS-RTG	*Galileo*	Planetary	October 18, 1989	Successfully sent on interplanetary trajectory to Jupiter (1996 arrival)
GPHS-RTG	*Ulysses*	Solar-Polar	October 6, 1990	Successfully sent on interplanetary trajectory to explore polar regions of Sun (1994–95)
GPHS-RTG	*Cassini*	Planetary	October 15, 1997	Successfully sent on interplanetary trajectory to Saturn (July 2004 arrival)

Source: NASA; Department of Energy.

has flown several space nuclear reactors (most recently a system called Topaz). The United States has flown only one space nuclear reactor, an experimental system called the SNAP-10A, which was launched and operated on-orbit in 1965. The objective of the SNAP-10A program was to develop a space nuclear reactor power unit capable of producing a minimum of 500 watts-electric (W_e) for a period of one year while operating in space. The SNAP-10A reactor was a small zirconium hydride (ZrH) thermal reactor fueled by uranium-235. The SNAP-10A orbital test was successful, although the mission was prematurely (and safely) terminated on-orbit by the failure of an electronic component outside the reactor.

Since the United States first used nuclear power in space, great emphasis has been placed on the safety of people and the protection of the terrestrial environment. A continuing major objective in any new space nuclear power program is to avoid undue risks. In the case of radioisotope power supplies, this means designing the system to contain the radioisotope fuel under all normal and potential accident conditions. For space nuclear reactors, such as the SNAP-10A and more advanced systems, this means launching the reactor in a "cold" (nonoperating) configuration and starting up the reactor only after a safe, stable Earth orbit or interplanetary trajectory has been achieved.

See also FISSION (NUCLEAR); FUSION (NUCLEAR); *JUPITER ICY MOONS ORBITER*; NUCLEAR-ELECTRIC PROPULSION SYSTEM; RADIOISOTOPE THERMOELECTRIC GENERATOR; SPACE NUCLEAR PROPULSION.

space nuclear propulsion Nuclear fission reactors can be used in two basic ways to propel a space vehicle: (1) to generate electric power for an electric propulsion unit, and (2) as a thermal energy, or heat, source to raise a propellant (working material) to extremely high temperatures, for subsequent expulsion out a nozzle. In the second application the system is often called a NUCLEAR ROCKET.

In a nuclear rocket chemical combustion is not required. Instead, a single propellant, usually hydrogen, is heated by the energy released in the nuclear fission process, which occurs in a controlled manner in the reactor's core. Conventional rockets, in which chemical fuels are burned, have severe limitations in the specific impulse a given propellant combination can produce. These limitations are imposed by the relatively high molecular weight of the chemical combustion products. At attainable combustion chamber temperatures, the best chemical rockets are limited to specific impulse values of about 4,300 meters per second. Nuclear rocket systems using fission reactions, fusion reactions, and even possibly matter-antimatter annihilation reactions (the "photon rocket") have been proposed because of their much greater propulsion performance capabilities.

Engineering developments will be needed in this century to permit the use of advanced fission reactor systems, such as the gaseous-core reactor rocket or even fusion-powered systems. However, the solid-core nuclear reactor rocket is within a test-flight demonstration of engineering reality. In this nuclear rocket concept hydrogen propellant is heated to extremely high temperatures while passing through flow channels within the solid-fuel elements of a compact nuclear reactor system that uses uranium-235 as the fuel. The high-temperature gaseous hydrogen then expands through a nozzle to produce propulsive thrust. From the mid-1950s until the early 1970s, the United States conducted a nuclear rocket program called Project Rover. The primary objective of the project was to develop a nuclear rocket for a human-crewed mission to Mars. Unfortunately, despite the technical success of this nuclear rocket program, mission emphasis within the overall American space program changed, and the nuclear rocket and the human-crewed Mars mission planning were discontinued in 1973.

See also FISSION (NUCLEAR); FUSION (NUCLEAR); ELECTRIC PROPULSION; NERVA; NUCLEAR-ELECTRIC PROPULSION SYSTEM; ROVER PROGRAM; SPACE NUCLEAR POWER; SPECIFIC IMPULSE; STARSHIP.

space physics The branch of science that investigates the magnetic and electrical phenomena that occur in outer space, in the upper atmosphere of planets, and on the Sun. Space physicists use balloons, rockets, satellites, and deep-space probes to study many of these phenomena in situ (i.e., in their place of origin). The scope of space physics ranges from investigating the generation and transport of energy from the Sun to searching past the orbit of Pluto for a magnetic boundary (called the *heliopause*) separating our solar system from the rest of the Milky Way galaxy. Phenomena such as auroras, trapped radiation belts, the solar wind, solar flares, and sunspots fall within the realm of space physics.

See also ASTRONOMY; ASTROPHYSICS; COSMOLOGY.

space platform An uncrewed (unmanned), free-flying orbital platform that is dedicated to a specific mission, such as commercial space activities (e.g., materials processing in space) or scientific research (e.g., the *Hubble Space Telescope*). This platform can be deployed, serviced, and retrieved by the space shuttle. In concept, platform servicing can be accomplished by astronauts and cosmonauts from the *INTERNATIONAL SPACE STATION*—if a future space platform orbits near the station or can be delivered to the vicinity of the space station by a space tug.

See also HUBBLE SPACE TELESCOPE; SPACE TUG.

spaceport A spaceport is both a doorway to outer space from the surface of a planet and a port of entry from outer space to a planet's surface. At a spaceport one finds the sophisticated facilities required for the assembly, testing, launching, and (in the case of reusable aerospace vehicles) landing and postflight refurbishment of space launch vehicles. Typical operations performed at a spaceport include the assembly of space vehicles; preflight preparation of space launch vehicles and their payloads; testing and checkout of space vehicles, spacecraft, and support equipment; coordination of launch vehicle tracking and data-acquisition requirements; countdown and launch operations; and (in the case of reusable space vehicles) landing operations and refurbishment. A great variety of technical and administrative activities also are needed to support the operation of a spaceport. These include design engineering, safety and security, quality assurance, cryogenic fluids management, toxic and hazardous materials handling, maintenance, logistics, computer operations, communications, and documentation.

Expendable (one-time use) space launch vehicles now can be found at spaceport facilities around the globe. NASA's Kennedy Space Center in Florida is the spaceport for the (partially) reusable aerospace vehicle, the space shuttle. In this century highly automated spaceports also will appear on the lunar and Martian surfaces to support permanent human bases on these alien worlds.

See also BAIKONUR COSMODROME; CAPE CANAVERAL AIR FORCE STATION; KENNEDY SPACE CENTER; KOUROU LAUNCH COMPLEX.

space radiation environment One of the major concerns associated with the development of a permanent human presence in outer space is the ionizing radiation environment, both natural and human-made. The natural portion of the space radiation environment consists primarily of Earth's trapped radiation belts (also called the Van Allen belts), solar particle events (SPEs), and galactic cosmic rays (GCRs). (*See* table.) Ionizing radiation sources associated with human activities can include space nuclear power systems (fission reactors and radioisotope), the detonation of nuclear explosives in the upper portion of Earth's atmosphere or in outer space (activities currently banned by international treaty), space-based particle accelerators, and radioisotopes used for calibration and scientific activity.

Earth's trapped radiation environment is most intense at altitudes ranging from 1,000 km to 30,000 km. Peak intensities occur at about 4,000 km and 22,000 km. Below approximately 10,000 km altitude, most trapped particles are relatively low-energy electrons (typically, a few million electron volts [MeV]) and protons. In fact, below about 500 km altitude, only the trapped protons and their secondary nucle-

Components of the Natural Space Radiation Environment

Galactic Cosmic Rays

Typically 85% protons, 13% alpha particles, 2% heavier nuclei
Integrated yearly fluence
 1 x 10^8 protons/cm^2 (approximately)
Integrated yearly radiation dose:
 4 to 10 rads (approximately)

Geomagnetically Trapped Radiation

Primarily electrons and protons
Radiation dose depends on orbital altitude
Manned flights below 300 km altitude avoid Van Allen belts

Solar-Particle Events

Occur sporadically; not predictable
Energetic protons and alpha particles
Solar-flare events may last for hours to days
Dose very dependent on orbital altitude and amount of shielding

ar interaction products represent a chronic ionizing radiation hazard.

Trapped electrons collide with atoms in the outer skin of a spacecraft, creating penetrating X-rays and gamma rays (called secondary radiations) that can cause tissue damage. Trapped energetic protons can penetrate several grams of material (typically 1- to 2-centimeters-thick shields of aluminum are required to stop them), causing ionization of atoms as they terminate their passage in a series of nuclear collisions. Most human-occupied spacecraft missions in low Earth orbit (LEO) are restricted to altitudes below 500 km and inclinations below about 60° to avoid prolonged (chronic) exposure to this type of radiation. For orbits below 500 km altitude and inclinations less than about 60° the predominant part of an astronaut's overall radiation exposure will be due to trapped protons from the South Atlantic anomaly (SAA). The SAA is a region of Earth's inner radiation belts that dips close to the planet over the southern Atlantic Ocean southeast of the Brazilian coast. Passage through the SAA generally represents the most significant source of chronic natural space radiation for space travelers in LEO. Earth's geomagnetic field generally protects astronauts and spacecraft in LEO from cosmic ray and solar flare particles.

However, spacecraft in highly elliptical orbits around Earth will pass through the Van Allen belts each day. Furthermore, those spacecraft with a high-apogee altitude (say, greater than 30,000 km) also will experience long exposures to galactic cosmic rays and solar flare environments. Similarly, astronauts traveling through interplanetary space to the Moon or Mars will be exposed to both a continuous galactic cosmic ray environment and a potential solar flare environment (i.e., a solar particle event, or SPE).

A solar flare is a bright eruption from the Sun's chromosphere that may appear within minutes and then fade within an hour. Solar flares cover a wide range of intensity and size. They eject high-energy protons that represent a serious hazard to astronauts traveling beyond LEO. The SPEs associated with solar flares can last one to two days. Anomalously large solar particle events (ALSPEs), the most intense variety, can deliver potentially lethal doses of energetic particles—even behind modest spacecraft shielding (e.g., 1 to 2 grams per centimeter squared of aluminum). The majority of SPE particles are energetic protons, but heavier nuclei also are present.

Galactic cosmic rays (GCRs) originate outside the solar system. GCR particles are the most energetic of the three general types of natural ionizing radiation in space and contain all elements from atomic numbers 1 to 92. Specifically, galactic cosmic rays have the following general composition: protons (82–85 percent), alpha particles (12–14 percent), and highly ionizing heavy nuclei (1–2 percent), such as carbon, oxygen, neon, magnesium, silicon, and iron. The ions that are heavier than helium have been given the name HZE particles, meaning high atomic number (Z) and high energy (E). Iron (Fe) ions appear to contribute substantially to the overall HZE population. Galactic cosmic rays range in energy from 10s of MeV to 100s of GeV (a GeV is one billion [10^9] electron volts) and are very difficult to shield against. In particular, HZE particles produce high-dose ionization tracks and kill living cells as they travel through tissue.

An effective space radiation protection program for astronauts and cosmonauts on extended missions in interplanetary space or on the lunar or Martian surface should include sufficient permanent shielding (of the spacecraft or surface base habitat modules), adequate active dosimetry, the availability of "solar storm shelters" (zones of increased shielding protection) on crewed spacecraft or on the planetary surface, and an effective solar particle event warning system that monitors the Sun for ALSPEs.

The ionizing radiation environment found in space also can harm sensitive electronic equipment (e.g., a single-event upset) and spacecraft materials. Design precautions, operational planning, localized shielding, device redundancy, and computer-memory "voting" procedures are techniques used by aerospace engineers to overcome or offset space radiation–induced problems that can occur in a spacecraft, especially one operating beyond LEO.

See also EARTH'S TRAPPED RADIATION BELTS; RADIATION SICKNESS; SINGLE-EVENT UPSET.

space resources Generally, when people think about outer space, visions of vast emptiness, devoid of anything useful, come to their minds. However, space is really a new frontier that is rich with resources, including unlimited solar energy, a full range of raw materials, and an environment that is both special (e.g., high vacuum, microgravity, physical isolation from the terrestrial biosphere) and reasonably predictable.

Since the start of the space age preliminary investigations of the Moon, Mars, several asteroids and comets, and meteorites have provided tantalizing hints about the rich mineral potential of the extraterrestrial environment. For example, the U.S. Apollo Project expeditions to the lunar surface established that the average lunar soil contains more than 90 percent of the material needed to construct a complicated space industrial facility. The soil in the lunar highlands is rich in anorthosite, a mineral suitable for the extraction of aluminum, silicon, and oxygen. Other lunar soils have been found to contain ore-bearing granules of ferrous metals such as iron, nickel, titanium, and chromium. Iron can be concentrated from the lunar soil (called regolith) before the raw material is even refined simply by sweeping magnets over regolith to gather the iron granules scattered within.

Remote sensing data of the lunar surface obtained by the *CLEMENTINE* spacecraft and *LUNAR PROSPECTOR* spacecraft have encouraged some scientists to suggest that water ice may be trapped in perpetually shaded polar regions. If this speculation proves true, then "ice mines" on the Moon could provide both oxygen and hydrogen, vital resources for permanent lunar settlements and space industrial facilities. The Moon would be able both to export chemical propellants for propulsion systems and to resupply materials for life support systems.

Its vast mineral resource potential, frozen volatile reservoirs, and strategic location will make Mars a critical "supply depot" for human expansion into the mineral-rich asteroid belt and to the giant outer planets and their fascinating collections of resource-laden moons. Smart robot explorers will assist the first human settlers on Mars, enabling these "Martians" to assess quickly and efficiently the full resource potential of their new world. As these early settlements mature, they will become economically self-sufficient by exporting propellants, life support system consumables, food, raw materials, and manufactured products to feed the next wave of human expansion to the outer regions of the solar system. Cargo spacecraft also will travel between cislunar space and Mars carrying specialty items to eager consumer markets of both civilizations.

The asteroids, especially Earth-crossing asteroids, represent another interesting category of space resources. Recent space missions and analysis of meteorites (many of which scientists believe originate from broken-up asteroids) indicate that carbonaceous (C-type) asteroids may contain up to 10 percent water, 6 percent carbon, significant amounts of sulfur, and useful amounts of nitrogen. S-class asteroids, which are common near the inner edge of the main asteroid belt and among the Earth-crossing asteroids, may contain up to 30 percent free metals (alloys of iron, nickel, and cobalt, along with high concentrations of precious metals). E-class asteroids may be rich sources of titanium, magnesium, manganese, and other metals. Finally, chondrite asteroids, which are found among the Earth-crossing population, are believed to contain accessible amounts of nickel, perhaps more concentrated than the richest deposits found on Earth.

Using smart machines, possibly including self-replicating systems, space settlers in the latter portions of this century will be able to manipulate large quantities of extraterrestrial matter and move it to wherever it is needed in the solar system. Many of these space resources will be used as the feedstock for the orbiting and planetary surface base industries that will form the basis of interplanetary trade and commerce. For example, atmospheric ("aerostat") mining stations could be set up around Jupiter and Saturn extracting such materials as hydrogen and helium—especially helium-3, an isotope of great potential value in nuclear fusion research and applications. Similarly, Venus could be "mined" for the carbon dioxide in its atmosphere, Europa for water, and Titan for hydrocarbons. Large fleets of robot spacecraft might even be used to gather chunks of water ice from Saturn's ring system while a sister fleet of robot vehicles extracts metals from the main asteroid belt. Even the nuclei of selected comets could be intercepted and mined for frozen volatiles, including water ice.

Suggested Physiological Design Criteria for a Space Settlement

Pseudogravity	0.95 g
Average rotation rate	≤1 rpm
Radiation exposure for the general population	≤0.5 rem/yr (5 mSv/yr)
Magnetic field intensity	≤100 μT
Temperature	23° ± 8°C
Atmospheric composition	
pO_2	22.7 ± 9 kPa (170 ± 70 mm Hg)
p (Inert gas; most likely N_2)	26.7 kPa < pN_2 < 78.9 kPa (200 < pN_2 < 590 mm Hg)
pCO_2	<0.4 kPa (<3 mm Hg)
pH_2O	1.00 ± 0.33 kPa (7.5 ± 2.5 mm Hg)

Source: NASA.

space settlement A large extraterrestrial habitat where from 1,000 to perhaps 10,000 people would live, work, and play while supporting space industrialization activity, such as the operation of a large space manufacturing complex or the construction of satellite power systems. One possible design is a spherical space settlement called the Bernal sphere. This giant spherical habitat would be approximately 2 kilometers in circumference. Up to 10,000 people would live in residences along the inner surface of the large sphere. Rotation of the settlement at about 1.9 revolutions per minute (RPM) would provide Earth-like gravity levels at the sphere's equator, but there would be essentially microgravity conditions at the poles. Because of the short distances between locations in the equatorial residential zone, passenger vehicles would not

Suggested Quantitative Environmental Design Criteria for a Space Settlement

Population: men, women, children	10,000
Community and residential, projected area per person, m^2	47
Agriculture, projected area per person, m^2	20
Community and residential, volume per person, m^3	823
Agriculture, volume per person, m^3	915

Source: NASA.

Suggested Qualitative Environmental Design Criteria for a Space Settlement

Long lines of sight
Larger overhead clearance
Noncontrollable, unpredictable parts of the environment, for example, plants, animals, children, weather
External views of large natural objects
Parts of interior out of sight of others
Natural light
Contact with the external environment
Availability of privacy
Good internal communications
Capability of physically isolating segments of the habitat from each other
Modular construction
 of the habitat
 of the structures within the habitat
Flexible internal organization
Details of interior design left to inhabitants

Source: NASA.

be necessary. Instead, the space settlers would travel on foot or perhaps by bicycle. The climb from the residential equatorial area up to the sphere's poles would take about 20 minutes and would lead the hiker past small villages, each at progressively lower levels of artificial gravity. A corridor at the axis would permit residents to float safely in microgravity out to exterior facilities, such as observatories, docking ports, and industrial and agricultural areas. Ringed areas above and below the main sphere in this type of space settlement would be the external agricultural toruses. The three tables present some of NASA's suggested physiological design criteria, quantitative environmental design criteria, and qualitative environmental design criteria for a future space settlement.

See also BERNAL SPHERE; SATELLITE POWER SYSTEM.

spaceship In general, any crewed vehicle or craft capable of traveling in outer space; specifically, a crewed space vehicle capable of performing an interplanetary (and eventually an interstellar) journey.

space shuttle Space shuttles (i.e., the orbiter vehicles) are the main element of the U.S. Space Transportation System (STS) and are used for space research and applications. The shuttles are the first vehicles capable of being launched into space and returning to Earth on a routine basis.

Space shuttles are used as orbiting laboratories in which scientists and mission specialists conduct a wide variety of experiments. Astronaut crews aboard space shuttles place satellites into orbit. They also rendezvous with satellites to carry out repairs and/or return a malfunctioning spacecraft to Earth for refurbishment. Space shuttles have also been used as part of the *INTERNATIONAL SPACE STATION* (ISS) program. In Phase I of this program, for example, orbiter vehicles such as the *Atlantis* (OV-104) performed important rendezvous and docking missions with the Russian *Mir* space station. Up until the *Columbia* accident of February 1, 2003, space shuttles were also actively involved in ISS activities such as on-orbit assembly, crew rotation, and resupply.

Space shuttles are true aerospace vehicles. They leave Earth and its atmosphere under rocket power provided by three liquid-fueled main engines and two solid-propellant boosters attached to an external liquid-propellant tank. After their missions in orbit end, the orbiter vehicles streak back through the atmosphere and are maneuvered to land like an airplane. The shuttles, however, are without power during reentry, and they land on runways much like a glider.

The operational space shuttle fleet consists of the *Discovery* (OV-103), *Atlantis* (OV-104), and *Endeavour* (OV-105). The *Enterprise* (OV-101) served as a test vehicle but never flew in space. This vehicle is now an exhibit at the National Air and Space Museum (NASM). The *Challenger* (OV-99) was lost along with its crew on January 28, 1986, during the first two minutes of the STS 51-L mission, and the *Columbia* (OV-102) and its crew were lost on February 1, 2003, at the end of the STS 107 mission while returning to Earth.

See also SPACE TRANSPORTATION SYSTEM.

space shuttle main engine (SSME) Each of three main liquid-propellant rocket engines on an orbiter vehicle that is capable of producing a thrust of 1.67 million newtons at sea level and about 2.09 million newtons in the vacuum of outer space. These engines burn for approximately eight minutes during launch ascent and together consume about 242,000 liters of cryogenic propellants (liquid hydrogen and liquid oxygen) per minute when all three operate at full power. The space shuttle's large external tank provides the liquid hydrogen and liquid oxygen for these engines and is then discarded after main engine cutoff (MECO).

See also SPACE TRANSPORTATION System.

space sickness A form of motion sickness experienced by about 50 percent of astronauts and cosmonauts when they encounter the microgravity ("weightless") environment of an orbiting spacecraft a few minutes or hours after launch. Space sickness symptoms include nausea, vomiting, and general malaise. This condition is generally only temporary, typically lasting no more than a day or so, and can be treated but not prevented with medications. When a person enters an extended period of microgravity, the fluids shift from the lower part of the body to the upper part, causing many physical changes to occur. For example, a person's eyes will appear smaller because the face becomes fuller (i.e., puffy or swollen in appearance), and the waistline shrinks a few centimeters. Also referred to as *space adaptation syndrome.*

space simulator A device or facility that simulates some condition or conditions found in space. It is used for testing equipment and in training programs.

space station A space station is an orbiting space system that is designed to accommodate long-term human habitation in space. The concept of people living and working in artificial habitats in outer space appeared in 19th-century science fiction literature in stories such as Edward Everett Hale's "Brick Moon" (1869) and JULES VERNE's "Off on a Comet" (1878).

At the beginning of the 20th century KONSTANTIN TSIOLKOVSKY provided the technical underpinnings for this concept with his truly visionary writings about the use of orbiting stations as a springboard for exploring the cosmos. Tsiolkovsky, the father of Russian astronautics, provided a more technical introduction to the space station concept in his 1895 work *Dreams of Earth and Heaven, Nature and Man.* He greatly expanded on the idea of a space station in his 1903 work *The Rocket into Cosmic Space.* In this technical classic Tsiolkovsky described all the essential ingredients needed for a crewed space station including the use of solar energy, the use of rotation to provide artificial gravity, and the use of a closed ecological system complete with "space greenhouse."

Throughout the first half of the 20th century the space station concept continued to evolve technically. For example, the German scientist HERMANN OBERTH described the potential applications of a space station in his classic treatise *The Rocket to Interplanetary Space* (German title, *Das Rakete zu den Planetenraumen*) (1923). The suggested applications included the use of a space station as an astronomical observatory, an Earth-monitoring facility, and a scientific research platform. In 1928 an Austrian named Herman Potocnik (pen name Hermann Noordung) introduced the concept of a rotating, wheel-shaped space station. Noordung called his design "Wohnrad" ("Living Wheel"). Another Austrian, Guido von Pirquet, wrote many technical papers on space flight, including the use of a space station as a refueling node for space tugs. In the late 1920s and early 1930s von Pirquet also suggested the use of multiple space stations at different locations in cislunar space. After World War II WERNHER VON BRAUN (with the help of space artist Chesley Bonestell) popularized the concept of a wheel-shaped space station in the United States.

Created in 1958, NASA became the forum for the American space station debate. How long should such an orbiting facility last? What was its primary function? How many crew? What orbital altitude and inclination? Should it be built in space or on the ground and then deployed in space? In 1960 space station advocates from every part of the fledgling space industry gathered in Los Angeles for a Manned Space Station Symposium, where they agreed that the space station was a logical goal but disagreed on what it was, where it should be located, and how it should be built.

Then in 1961 President JOHN F. KENNEDY decided that the Moon was a worthy target of the American spirit and her-

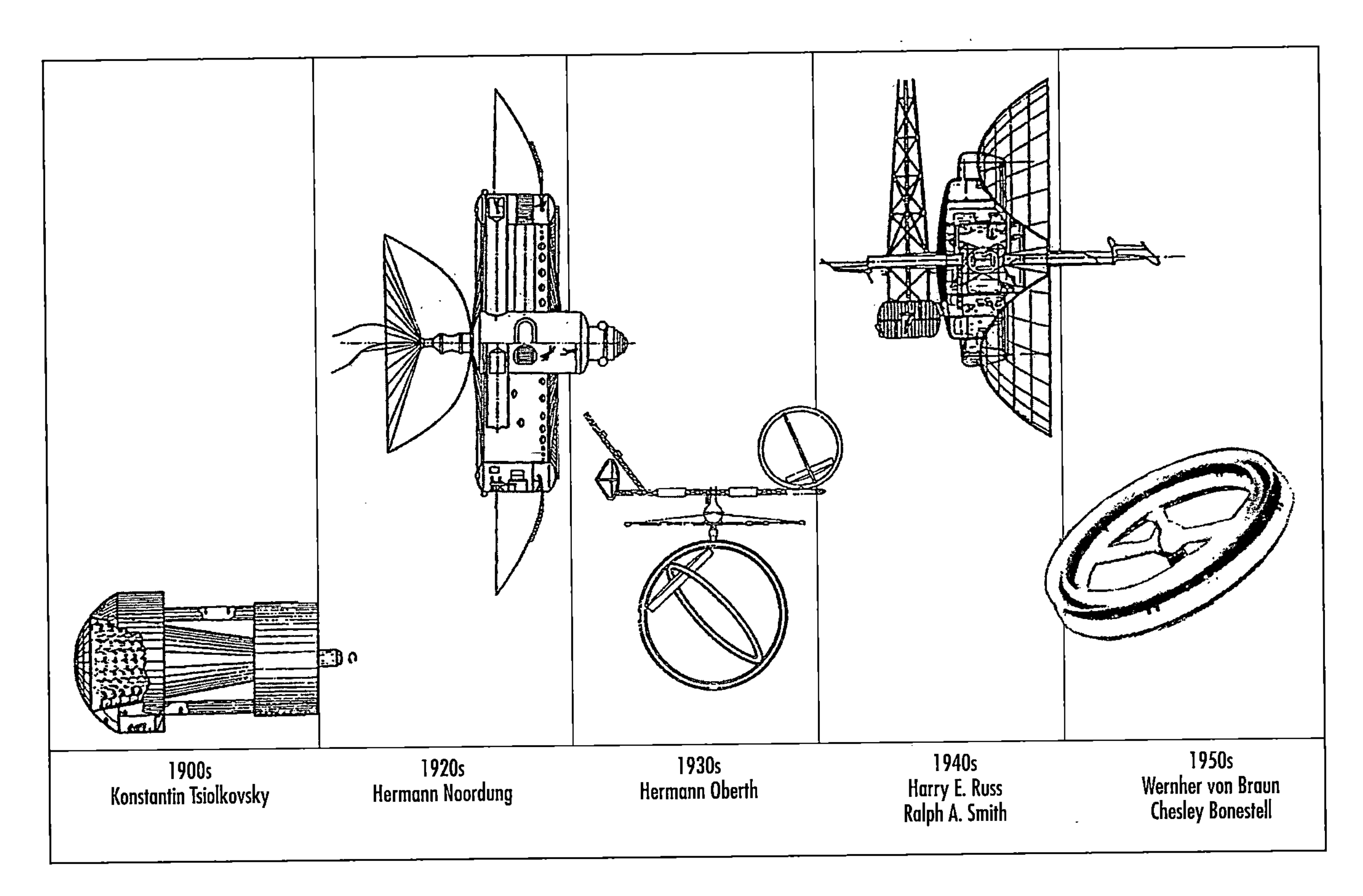

Early space station concepts ***(Drawing courtesy of NASA)***

Russian Space Station Experience

Name	Type	Launched	Remarks
First-Generation Space Stations (1964–1977)			
Salyut-1	Civilian	1971	First space station
Unnamed	Civilian	1972	Failure
Salyut-2	Military	1973	First Almaz station; failure
Cosmos 557	Civilian	1973	Failure
Salyut-3	Military	1974–1975	Almaz station
Salyut-4	Civilian	1974–1977	
Salyut-5	Military	1976–1977	Last Almaz station
Second-Generation Space Stations (1977–1985)			
Salyut-6	Civilian	1977–1982	Highly successful
Salyut-7	Civilian	1982–1991	Last staffed in 1986
Third-Generation Space Stations (1986–1999)			
Mir	Civilian	1986–1999	First permanent space station

Source: NASA.

itage. A lunar landing mission had a definite advantage over a space station: Everyone could agree on the definition of landing on the Moon, but few could agree on the definition of a space station. However, this disagreement was actually beneficial. It forced space station designers and advocates to think about what they could do, the cost of design, and what was necessary to make the project a success. What were the true requirements for a space station? How could they best be met? The space station requirements review process started informally within NASA in 1963 and has continued up to the present day. For more than four decades NASA planners and officials have asked the scientific, engineering, and business communities over and over again: What would you want? What do you need? As answers flowed in NASA developed a variety of space station concepts to help satisfy these projected requirements. The *International Space Station* (ISS) is the latest in this evolving series of space station concepts.

Even before the Apollo Project had landed men successfully on the Moon, NASA engineers and scientists were considering the next giant step in the U.S. crewed spaceflight program. That next step became the simultaneous development of two complementary space technology capabilities. One was a safe, reliable transportation system that could provide routine access to space. The other was an orbital space station where human beings could live and work in space. The space station would serve as a base camp from which other, more advanced space technology developments could be initiated. This long-range strategy set the stage for two of the most significant American space activities carried out in the 1970s and 1980s: SKYLAB and the space shuttle (or U.S. Space Transportation System).

On May 14, 1973, the United States launched its first space station, called *Skylab*. The massive space station was launched in one piece by a powerful Saturn V booster left over from the Apollo Project. *Skylab* demonstrated that people could function in space for periods of up to 12 weeks and, with proper exercise, could return to Earth with no ill effects.

In particular, the flight of *Skylab* proved that human beings could operate very effectively in a prolonged microgravity environment and that it was not essential to provide artificial gravity for people to live and work in space (at least for periods up to about six months). Long-duration flights by Russian cosmonauts and American astronauts on *Mir* and the ISS have reinforced and extended these findings. The *Skylab* astronauts accomplished a wide range of emergency repairs on station equipment, including freeing a stuck solar panel array (a task that saved the entire mission), replacing rate gyros, and repairing a malfunctioning antenna. On two separate occasions the crew installed portable sun shields to replace the original equipment that was lost when *Skylab* was launched. These on-orbit activities clearly demonstrated the unique and valuable role people have in space.

Unfortunately, because of schedule pressures and and budget limitations, NASA engineers did not design *Skylab* for a permanent presence in space. For example, the facility was not designed to be routinely serviced on orbit, although the *Skylab* crews were able to perform certain repair functions. *Skylab* was not equipped to maintain its own orbit, a design deficiency that eventually caused its fiery demise on July 11, 1979, over the Indian Ocean and portions of western Australia. Finally, the first American space station was not designed for evolutionary growth and therefore was subject to rapid technological obsolescence. Future space station programs, including the ISS, should take these shortcomings into account and then effectively apply the highly successful *Skylab* program experience toward develop of permanent, evolutionary, and modular space stations.

While the United States was concentrating on the Apollo Moon landing project, the Soviet Union (now the Russian Federation) embarked on an ambitious space station program. As early as 1962 Russian engineers described a space station comprised of modules launched separately and brought together in orbit. The world's first space station, called *Salyut-1*, was launched on April 19, 1971, by a Proton booster. (The Russian word *salyut* means "salute.") The first generation of Russian space stations had one docking port and could not be resupplied or refueled. The stations were launched uncrewed and later occupied by crews. Two types of early Russian space stations existed: Almaz military stations and Salyut civilian stations. During the cold war, to confuse Western observers the Russians referred to both kinds of station as Salyut (*see* table on page 569).

The Almaz military station program was the first approved. When proposed in 1964, it had three parts: the Almaz military surveillance space station, transport logistics spacecraft for delivering military-cosmonauts and cargo, and Proton rockets for launching both. All of these spacecraft were built, but none was actually used as originally planned.

Russian engineers completed several Almaz space station hulls by 1970. Russian leaders then ordered that the hulls be transferred to a crash program to launch a civilian space station. Work on the transport logistics spacecraft was deferred, and the *Soyuz* spacecraft originally built for the Russian manned Moon program was reapplied to ferry crews to the space stations.

Unfortunately, the early first-generation Russian space stations were plagued by failures. For example, the crew of *Soyuz 10,* the first spacecraft sent to *Salyut-1,* was unable to enter the station because of a docking mechanism problem. The *Soyuz 11* crew lived aboard *Salyut-1* for three weeks but died during the return to Earth because the air escaped from their spacecraft. Then three first-generation stations failed to reach orbit or broke up in orbit before the cosmonaut crews could reach them. The second failed space station, called *Salyut-2,* was the first Almaz military space station to fly.

However, the Russians recovered rapidly from these failures. *Salyut-3, Salyut-4,* and *Salyut-5* supported a total of five crews. In addition to military surveillance and scientific and industrial experiments, the cosmonauts performed engineering tests to help develop the second-generation stations.

The second-generation Russian space station was introduced with the launch on September 29, 1977, and successful operation of the *Salyut-6* station. Several important design improvements appeared on this station, including the addition of a second docking port and the use of an automated Progress resupply spacecraft (a space "freighter" derived from the *Soyuz* spacecraft.)

With the second-generation stations, the Russian space station program evolved from short-duration to long-duration stays. Like the first-generation stations, they were launched uncrewed, and their crews arrived later in a *Soyuz* spacecraft. Second-generation Russian stations had two docking ports. This permitted refueling and resupply by Progress spacecraft, which docked automatically at the aft port. After docking cosmonauts on the station opened the aft port and unloaded the space freighter. Transfer of fuel to the station was accomplished automatically under supervision from ground controllers.

The availability of a second, docking port also meant long-duration crews could receive visitors. Visiting crews often included cosmonaut researchers from the Soviet bloc countries or countries that were politically sympathetic to the Soviet Union. For example, the Czech cosmonaut Vladimir Remek visited the *Salyut-6* station in 1978 and became the first space traveler not from either the United States or Russia.

These visiting crews helped relieve the monotony that can accompany long stays in space. They often traded their *Soyuz* spacecraft for the one already docked at the station, because the *Soyuz* spacecraft had only a limited lifetime in orbit. The spacecraft's lifetime was gradually extended from 60 to 90 days for the early Soyuz ferry to more than 180 days for the *Soyuz-TM*. By way of comparison, the Soyuz TMA crew transfer (and escape) vehicle used with the *International Space Station* has a lifetime of more than a year.

The *Salyut-6* station received 16 cosmonaut crews, including six long-duration crews. The longest stay duration for a *Salyut-6* crew was 185 days. The first *Salyut-6* long-duration crew stayed in orbit for 96 days, surpassing the 84-day space endurance record that had been established in 1974 by the last *Skylab* crew. The *Salyut-6* hosted cosmonauts from Hungary, Poland, Romania, Cuba, Mongolia, Vietnam, (East) Germany, and Czechoslovakia. A total of 12 Progress freighter spacecraft delivered more than 20 tons of equipment, supplies, and fuel. An experimental transport logistics spacecraft called *Cosmos 1267* docked with *Salyut-6* in 1982. The transport logistics spacecraft was originally designed for the Almaz program. *Cosmos 1267* demonstrated that a large module could dock automatically with a space station, a major space technology step toward the multimodular *Mir* space station and the *International Space Station.* The last cosmonaut crew left the *Salyut-6* station on April 25, 1977. The station reentered Earth's atmosphere and was destroyed in July 1982.

The *Salyut-7* space station was launched on April 19, 1982, and was a near twin of the *Salyut-6* station. It was home to 10 cosmonaut crews, including six long-duration crews. The longest crew stay time was 237 days. Guest cosmonauts from France and India worked aboard the station, as did cosmonaut Svetlana Savitskaya, who flew aboard the *Soyuz-T-7/Salyut-7* mission and became the first Russian female space traveler since VALENTINA TERESHKOVA in 1963. Savitskaya also became the first woman to "walk in space" (i.e., perform an extravehicular activity) during the *Soyuz-T-12/Salyut 7* mission in 1984. Unlike the *Salyut-6* station, however, the *Salyut-7* station suffered some major technical problems. In early 1985, for example, Russian ground controllers lost contact with the then unoccupied station. In July 1985 a special crew aboard the *Soyuz-T-13* spacecraft docked with the derelict space station and made emergency repairs that extended its lifetime for another long-duration mission. The *Salyut-7* station finally was abandoned in 1986; it reentered Earth's atmosphere over Argentina in 1991.

During its lifetime on orbit, 13 Progress spacecraft delivered more than 25 tons of equipment, supplies, and fuel to *Salyut-7.* Two experimental transport logistics spacecraft, called *Cosmos 1443* and *Cosmos 1686,* docked with the station. *Cosmos 1686* was a transitional vehicle, a transport logistics spacecraft that had been redesigned to serve as an experimental space station module.

In February 1986 the Soviets introduced a third-generation space station, the *Mir* ("peace") station. Design improvements included more extensive automation, more spacious crew accommodations for resident cosmonauts (and later American astronauts), and the addition of a multiport docking adapter at one end of the station. In a very real sense, *Mir* represented the world's first "permanent" space station. When docked with the *Progress-M* and *Soyuz-TM* spacecraft, this station measured more than 32.6 meters long and was about 27.4 meters wide across its modules. The orbital complex consisted of the *Mir* core module and a variety of additional modules, including the Kvant (quantum), the Kvant 2, and the Kristall modules.

The *Mir* core resembled the *Salyut-7* station but had six ports instead of two. The fore and aft ports were used primarily for docking, while the four radial ports that were located in a node at the station's front were used for berthing large modules. When launched in 1986, the core had a mass of about 20 tons. The Kvant module was added to the *Mir* core's aft port in 1987. This small, 11-ton (11,050-kg) module contained astrophysics instruments and life support and attitude control equipment. Although Kvant blocked the core module's aft port, it had its own aft port that then served as the station's aft port.

The Russians added the 18.5-ton Kvant 2 module in 1989. They based the design of this module on the transport logistics spacecraft originally intended for the Almaz military space station program of the early 1970s. The purpose of Kvant 2 was to provide biological research data, Earth observation data, and extravehicular activity (EVA) capability. Kvant 2 carried an EVA airlock, two solar arrays, and life support equipment.

The 19.6-ton Kristall module was added in 1990. It carried scientific equipment, retractable solar arrays, and a docking node equipped with a special androgynous interface docking mechanism designed to receive spacecraft with masses of up to 100 tons. This docking unit (originally developed for the former Russian space shuttle Buran) was attached to the docking module (DM) that was used by the American space shuttle orbiter vehicles to link up with *Mir* during Phase I of the *International Space Station* (ISS) program.

Three more modules, all carrying American equipment, were also added to the *Mir* complex as part of the ISS program. These were the Spektr module, the Priroda ("nature") module, and the docking module. The Spetkr module, carrying scientific equipment and solar arrays, was severely damaged on June 25, 1997, when a Progress resupply spacecraft collided with it during practice docking operations.

The Priroda module carried microgravity research and Earth observation equipment. This 19.7-ton module was the last module added to the *Mir* complex (April 1996). Unlike the other modules, however, Priroda had none of its own solar power arrays and depended on other portions of the *Mir* complex for electric power.

The docking module was delivered by the space shuttle *Atlantis* during the STS-74 mission (November 1995) and berthed at Kristall's androgynous docking port. The Russian-built docking module became a permanent extension on *Mir* and provided better clearances for space shuttle orbiter–*Mir* linkups.

From 1986 the *Mir* space station served as a major part of the Russian space program. However, because of active participation in the ISS program and severe budget constraints, the Russian Federation decommissioned *Mir* and abandoned the station in 1999. For safety reasons Russian spacecraft controllers successfully deorbited the large space station in March 2001 and intentionally crashed any surviving remnants in a remote area of the Pacific Ocean.

In January 1984, as part of his state of the union address, President Ronald Reagan called for a space station program that would include participation by U.S.-allied countries. With this presidential mandate NASA established a Space Station Program Office in April 1984 and requested proposals from the American aerospace industry. By March 1986 the baseline design was the dual keel configuration, a rectangular framework with a truss across the middle for holding the station's living and working modules and solar arrays.

Japan, Canada, and the European Space Agency (ESA) each signed a bilateral memorandum of understanding in the spring of 1985 with the United States, agreeing to participate in the space station project. In 1987 the station's dual keel configuration was revised to compensate for a reduced space shuttle flight rate in the wake of the CHALLENGER ACCIDENT of January 28, 1986. The revised baseline had a single truss with the built-in option to upgrade to the dual keel design. The need for a space station "lifeboat," called the assured crew return vehicle, was also identified.

In 1988 President Reagan named the space station *Freedom*. With each annual budget cycle, *Freedom*'s design underwent modifications as the U.S. Congress called for reductions in its cost. The truss was shortened, and the U.S. habitation and laboratory modules were reduced in size. The truss was to be launched in sections with subsystems already in place. Despite these redesign efforts, NASA and its contractors were able to produce a substantial amount of hardware. In 1992 the United States agreed to purchase Russian Soyuz spacecraft to serve as *Freedom*'s lifeboat. This action presaged the greatly increased space cooperation between the United States and Russia that followed the cold war. Another important activity, the space shuttle–*Mir* program (later called Phase I of the ISS program), was also started in 1992. The shuttle-*Mir* program used existing assets (primarily U.S. space shuttle orbiter vehicles and the Russian *Mir* space station) to provide the joint operational experience and to perform the joint research that would eventually lead to the successful construction and operation of the ISS.

In 1993 President Bill Clinton called for *Freedom* to be redesigned once again to reduce costs and to include more international involvement. The White House staff selected a design option that was called *Space Station Alpha*, a downsized configuration that would use about 75 percent of the hardware designs originally intended for *Freedom*. After the Russians agreed to supply major hardware elements (many of which were originally intended for a planned *Mir 2* space station), *Space Station Alpha* became officially known as the *International Space Station*, or ISS.

The *International Space Station* program is divided into three basic phases. Phase I (an expansion of the shuttle-*Mir* docking mission program) provided U.S. and Russian aerospace engineers, flight controllers, and cosmonauts and astronauts the valuable experience needed to cooperatively assemble and build the ISS. Phase I officially began in 1995 and involved more than two years of continuous stays by a total of seven American astronauts aboard the *Mir* space station and nine shuttle-Mir docking missions. This phase of the ISS program ended in June 1998 with the successful completion of the STS-91 mission, a mission in which the space shuttle *Discovery* docked with *Mir* and "downloaded" (that is, returned to Earth) astronaut Andrew Thomas, the last American occupant of the Russian space station.

Phases II and III involve the in-orbit assembly of the station's components—Phase II the core of the ISS and Phase III its various scientific modules. ISS program milestones for 2003 and beyond have been delayed by NASA's space shuttle return-to-flight activities following the *COLUMBIA* ACCIDENT of February 1, 2003.

A very historic moment in aerospace history took place near the end of the 20th century. On December 10, 1998, STS-88 shuttle mission commander Robert Cabana and Russian cosmonaut and mission specialist SERGEI KRIKALEV swung open the hatch between the shuttle *Endeavour* and the first element of the ISS. With this action the STS-88 astronauts completed the first steps in the orbital construction of the ISS. In late November 1998 a Russian Proton rocket had successfully placed the NASA-owned, Russian-built *Zarya* ("sunrise") control module into a perfect parking orbit. A few days later, in early December, the shuttle *Endeavour* carried the American-built *Unity* connecting module into orbit for rendezvous with *Zarya*. Astronauts Jerry Ross and James Newman then performed three arduous extravehicular activities (totaling 21 hours and 22 minutes) to complete the initial assembly of the space station. When their space walking efforts were completed, Cabana and Krikalev were quite literally able to open the door of an important new space station era. As predicted by Konstantin Tsiolkovsky at the start of the 20th century, humankind will soon leave the "cradle of Earth" and build permanent outposts in space on the road to the stars.

See also INTERNATIONAL SPACE STATION.

spacesuit Outer space is a very hostile environment. If astronauts and cosmonauts are to survive there, they must take part of Earth's environment with them. Air to breathe, acceptable ambient pressures, and moderate temperatures have to be contained in a shell surrounding the space traveler. This can be accomplished by providing a very large enclosed structure or habitat or, on an individual basis, by encasing the astronaut in a protective flexible capsule called a spacesuit.

Spacesuits used on NASA missions from the Mercury Project up through the Apollo-Soyuz Test Project provided effective protection for American astronauts. However, certain design problems handicapped the suits. These suits were custom-fitted garments. In some suit models more than 70 different measurements had to be taken of the astronaut in order to manufacture the spacesuit to the proper fit. As a result, the spacesuit could be worn by only one astronaut on only one mission. These early spacesuits were stiff, and even simple motions such as grasping objects quickly drained an astronaut's strength. Even donning the suit was an exhausting process that at times lasted more than an hour and required the help of an assistant.

For example, the Mercury Project spacesuit was a modified version of a U.S. Navy high-altitude jet aircraft pressure suit. It consisted of an inner layer of Neoprene-coated nylon fabric and an outer layer of aluminized nylon. Joint mobility at the elbows and knees was provided by simple break lines sewn into the suit, but even with these break lines it was difficult for the wearer to bend arms or legs against the force of a pressurized suit. As an elbow or knee joint was bent, the suit joints folded in on themselves, reducing internal volume and increasing pressure. The Mercury spacesuit was worn "soft," or unpressurized, and served only as a backup for possible spacecraft cabin pressure loss, an event that never happened.

NASA spacesuit designers then followed the U.S. Air Force approach toward greater suit mobility when they developed the spacesuit for the two-man Gemini Project spacecraft. Instead of the fabric-type joints used in the Mercury suit, the Gemini spacesuit had a combination of a pressure bladder and a link-net restraint layer that made the whole suit flexible when pressurized.

The gas-tight, human-shaped pressure bladder was made of Neoprene-coated nylon and covered by load-bearing link-net woven from Dacron and Teflon cords. The net layer, being slightly smaller than the pressure bladder, reduced the stiffness of the suit when pressurized and served as a type of structural shell. Improved arm and shoulder mobility resulted from the multilayer design of the Gemini suit.

Walking on the Moon's surface presented a new set of problems to spacesuit designers. Not only did the spacesuits for the "Moonwalkers" have to offer protection from jagged rocks and the intense heat of the lunar day, but the suits also had to be flexible enough to permit stooping and bending as the Apollo astronauts gathered samples from the Moon and used the lunar rover vehicle for transportation over the surface of the Moon.

The additional hazard of micrometeoroids that constantly pelt the lunar surface from deep space was met with an outer protective layer on the Apollo spacesuit. A backpack portable life support system provided oxygen for breathing, suit pressurization, and ventilation for Moonwalks lasting up to seven hours.

Apollo spacesuit mobility was improved over earlier suits by use of bellowslike molded rubber joints at the shoulders, elbows, hips, and knees. Modifications to the suit waist for the *Apollo 15* through *17* missions provided flexibility and made it easier for astronauts to sit on the lunar rover vehicle.

From the skin out, the Apollo A7LB spacesuit began with an astronaut-worn liquid-cooling garment, similar to a pair of longjohns with a network of spaghetti-like tubing sewn onto the fabric. Cool water, circulating through the tubing, transferred metabolic heat from the astronaut's body to the backpack, where it was then radiated away to space. Next came a comfort and donning improvement layer of lightweight nylon, followed by a gas-tight pressure bladder of Neoprene-coated nylon and bellowslike molded joint components, a nylon restraint layer to prevent the bladder from ballooning, a lightweight thermal superinsulation of alternating

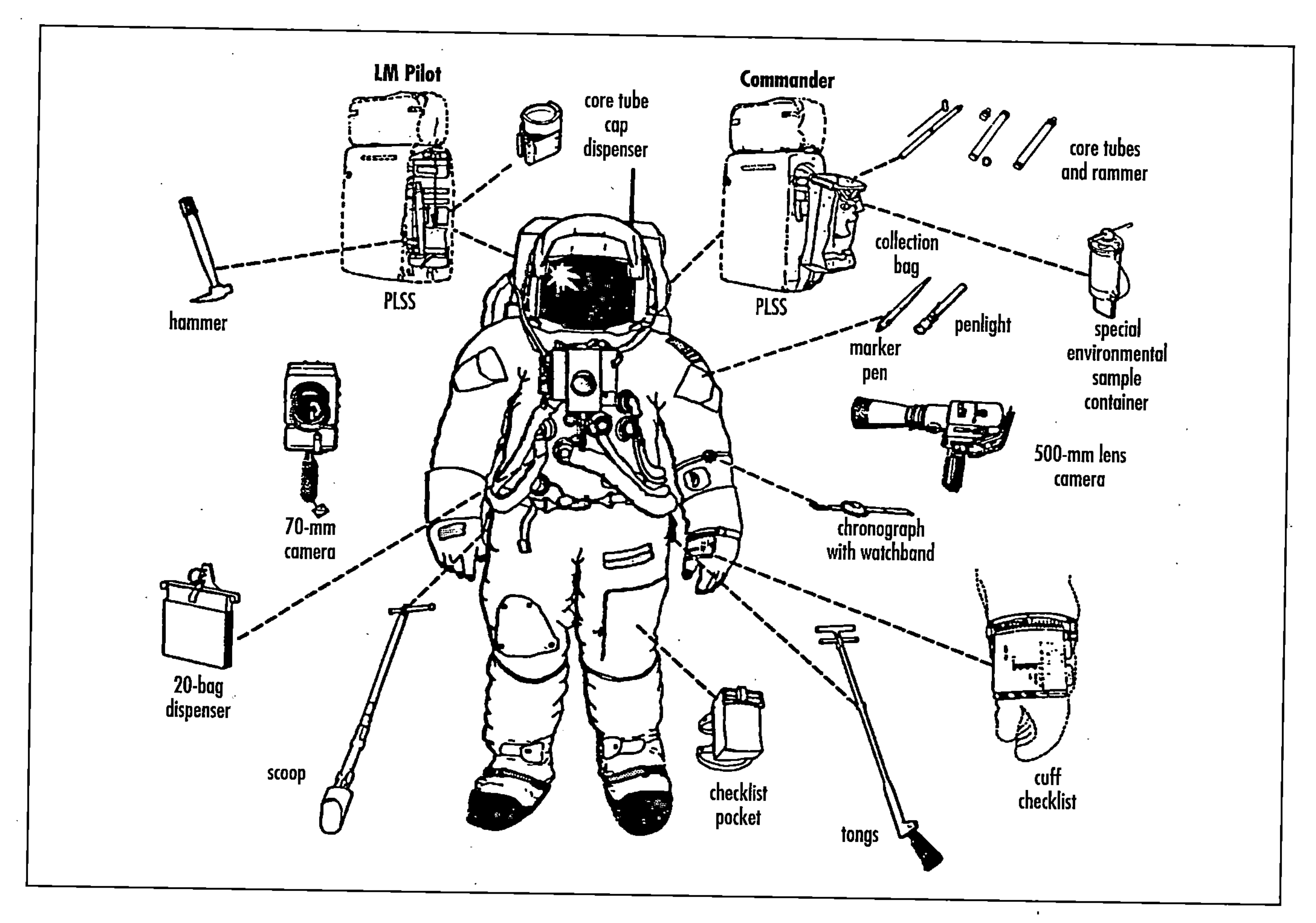

The Apollo astronaut spacesuit and its supporting equipment ***(Drawing courtesy of NASA)***

layers of thin Kapton and glass-fiber cloth, several layers of Mylar and spacer material, and, finally, protective outer layers of Teflon-coated glass-fiber Beta cloth.

Apollo space helmets were formed from high-strength polycarbonate and were attached to the spacesuit by a pressure-sealing neck ring. Unlike Mercury Project and Gemini Project helmets, which were closely fitted and moved with the astronaut's head, the Apollo Project helmet was fixed, and the astronaut's head was free to move within it. While walking on the Moon the Apollo crew wore an outer visor assembly over the polycarbonate helmet to shield against eye-damaging ultraviolet radiation and to maintain head and face thermal comfort.

Lunar gloves and boots completed the Apollo spacesuit. Both were designed for the rigors of exploring; the gloves also could adjust sensitive instruments. The gloves consisted of integral structural restraint and pressure bladders, molded from casts of the crewperson's hands and covered by multilayered superinsulation for thermal and abrasion protection. Thumbs and fingertips were molded of silicone rubber to permit a degree of sensitivity and "feel." Pressure sealing disconnects, similar to the helmet-to-suit connection, attached the gloves to the spacesuit arms.

The lunar boot was actually an overshoe that the Apollo astronaut slipped on over the integral pressure boot of the spacesuit. The outer layer of the lunar boot was made from metal-woven fabric, except for the ribbed silicone rubber sole; the tongue area was made from Teflon-coated glass-fiber cloth. The boot inner layers were made from Teflon-coated glass-fiber cloth followed by 25 alternating layers of Kapton film and glass-fiber cloth to form an efficient, lightweight thermal insulation. Modified versions of the Apollo spacesuit were also used during the *Skylab* Program (1973–74) and the Apollo-Soyuz Test Project (1975).

A new spacesuit was developed for shuttle-era astronauts that provided many improvements in comfort, convenience, and mobility over previous models. This suit, which is worn outside the orbiter during extravehicular activity (EVA), is modular and features many interchangeable parts. Torso, pants, arms, and gloves come in several different sizes and can be assembled for each mission in the proper combination to suit individual male and female astronauts. The design approach is cost-effective because the suits are reusable and not custom-fitted.

The shuttle spacesuit is called the *extravehicular mobility unit* (EMU) and consists of three main parts: liner, pressure vessel, and primary life support system (PLSS). These components are supplemented by a drink bag, communications set, helmet, and visor assembly.

Containment of body wastes is a significant problem in spacesuit design. In the shuttle-era EMU, the PLSS handles odors, carbon dioxide, and the containment of gases in the suit's atmosphere. The PLSS is a two-part system consisting of a backpack unit and a control and display unit located on the suit chest. A separate unit is required for urine relief. Two different urine-relief systems have been designed to accommodate both male and female astronauts. Because of the short time durations for extravehicular activities, fecal containment is considered unnecessary.

The *manned maneuvering unit* (MMU) is a one-person, nitrogen-propelled backpack that latches to the EMU spacesuit's PLSS. Using rotational and translational hand controllers, the astronaut can fly with precision in or around the orbiter's cargo bay or to nearby free-flying payloads or structures and can reach many otherwise inaccessible areas outside the orbiter vehicle. Astronauts wearing MMUs have deployed, serviced, repaired, and retrieved satellite payloads.

The MMU has been called "the world's smallest reusable spacecraft." The MMU propellant (noncontaminating gaseous nitrogen stored under pressure) can be recharged from the orbiter vehicle. The reliability of the unit is guaranteed with a dual parallel system rather than a backup redundant system. In the event of a failure in one parallel system, the system would be shut down and the remaining system would be used to return the MMU to the orbiter's cargo bay. The MMU includes a 35-mm still camera that is operated by the astronaut while working in space.

Shuttle-era spacesuits are pressurized at 29.6 kilopascals, while the shuttle cabin pressure is maintained at 101 kilopascals. Because the gas in the suit is 100 percent oxygen (instead of 20 percent oxygen as is found in Earth's atmosphere), the person in the spacesuit actually has more oxygen to breathe than is available at an altitude of 3,000 meters or even at sea level without the spacesuit. However, prior to leaving the orbiter to perform tasks in space, an astronaut has to spend several hours breathing pure oxygen. This procedure (called "prebreathing") is necessary to remove nitrogen dissolved in body fluids and thereby prevent its release as

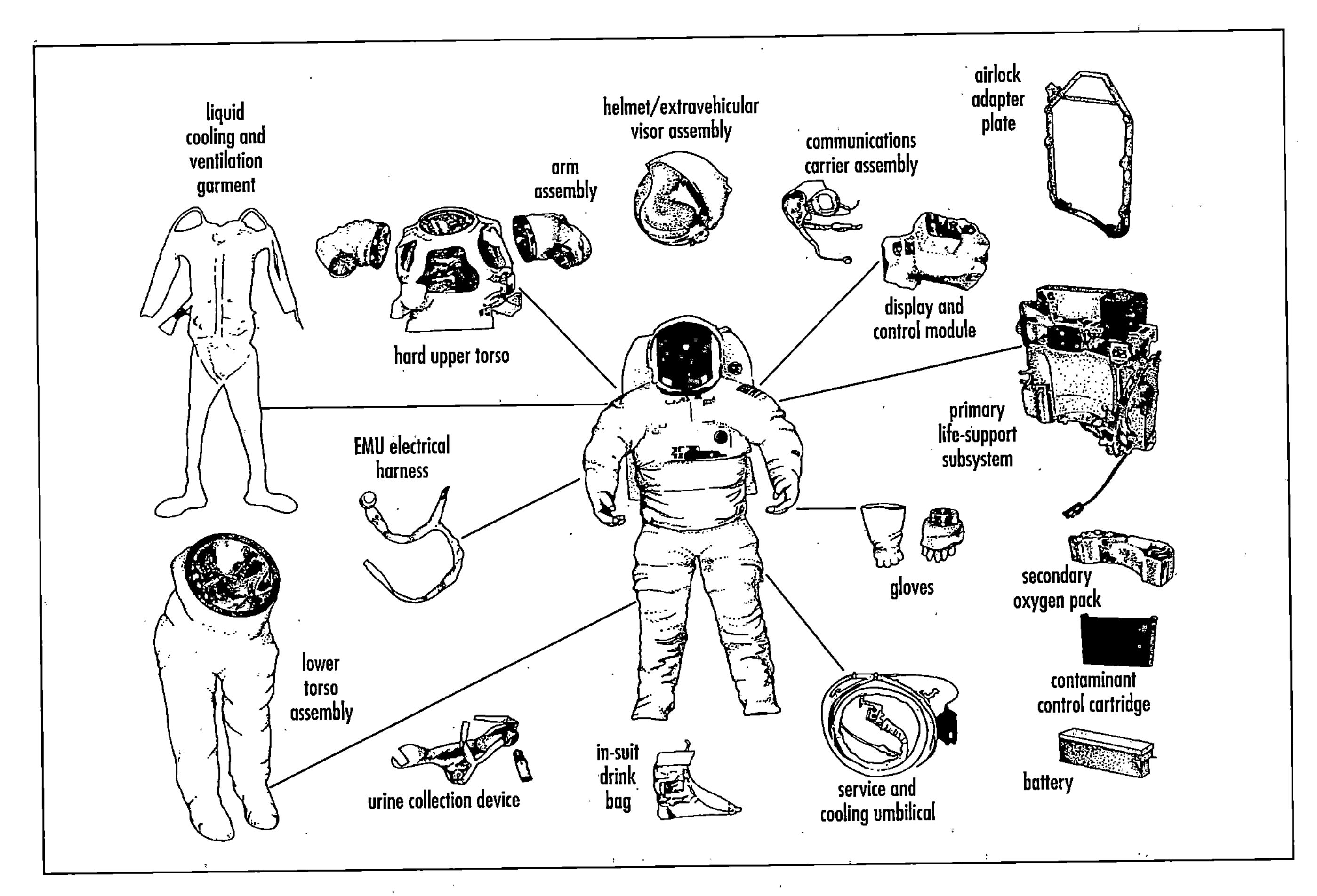

The components of the shuttle-era Extravehicular Mobility Unit (EMU) ***(Drawing courtesy of NASA)***

gas bubbles when pressure is reduced, a condition commonly referred to as "the bends."

In addition to new space walking tools and philosophies for astronaut-assisted assembly of the *International Space Station* (ISS), American astronaut space walkers have an enhanced spacesuit. The shuttle spacesuit (the extravehicular mobility unit) was originally designed for sizing and maintenance between flights by skilled specialists on Earth. Such maintenance and refurbishment activities would prove difficult, if not impossible, for astronauts aboard the station.

The shuttle spacesuit has been improved for use on the ISS. It can now be stored in orbit and is certified for up to 25 extravehicular activities before it must be returned to Earth for refurbishment. The spacesuit can be adjusted in flight to fit different astronauts and can be easily cleaned and refurbished between EVAs onboard the station. The modified spacesuit has easily replaceable internal parts, reusable carbon dioxide removal cartridges, metal sizing rings that accommodate in-flight suit adjustments to fit different crewmembers, new gloves with enhanced dexterity, and a new radio with more channels to allow up to five people to communicate with one another simultaneously.

Due to orbital motion induced periods of darkness and component caused shadowing, assembly work on the space station is frequently performed at much colder temperatures than those encountered during most of the space shuttle mission EVAs. Unlike the shuttle, the ISS cannot be turned to provide an optimum amount of sunlight to moderate temperatures during an extravehicular activity, so a variety of other enhancements now make the shuttle spacesuit more compatible for use aboard the space station. Warmth enhancements include fingertip heaters and the ability to shut off the spacesuit's cooling system. To assist assembly work in shadowed environments, the spacesuit has new helmet-mounted floodlights and spotlights. There is also a jet-pack "life jacket" to allow an accidentally untethered astronaut to fly back to the station in an emergency. In 1994, as part of the STS-64 mission, astronaut Mark Lee performed an EVA during which he tested a new mobility system called the Simplified Aid for EVA Rescue (SAFER). This system is similar to but smaller and simpler than the MMU.

Before the arrival of the joint airlock module (called Quest) as part of the STS-104 mission to the ISS in July 2001, space walks conducted from the space station could use only Russian spacesuits unless the space shuttle was present. The facilities of the Zvezda service module limited space station–based EVAs to only those with Russian Orlan spacesuits. The Quest module, now attached to the ISS, gives the station's occupants the capability to conduct EVAs with either Russian- or American-designed spacesuits. Prebreathing protocols and spacesuit design differences no longer limit EVA activities by astronauts and cosmonauts on the space station.

See also APOLLO PROJECT; *INTERNATIONAL SPACE STATION*.

space system Generally, a system consisting of launch vehicle(s), spacecraft, and ground support equipment (GSE).

Space Telescope Science Institute (STScI) Responsibility for conducting and coordinating the science operations of the *Hubble Space Telescope* (HST) rests with the Space Telescope Science Institute (STScI) on the Johns Hopkins University Homewood Campus in Baltimore, Maryland. STScI is operated for NASA by the Association of Universities for Research in Astronomy, Inc. (AURA).

See also HUBBLE SPACE TELESCOPE.

space-time (or spacetime) The synthesis of three spatial dimensions and a fourth dimension, time. ALBERT EINSTEIN used the concept of space-time in both his special and general relativity theories. For example, general relativity relates the curvature of space-time to gravitation and its influence on the positions and motions of particles of matter in the universe

See also RELATIVITY.

Spacetrack A global system of radar, optical, and radiometric sensors linked to a computation and analysis center in the North American Aerospace Defense Command (NORAD) combat operations center. The Spacetrack mission is detection, tracking, and cataloging of all human-made objects in orbit around Earth. It is the U.S. Air Force portion of the North American Aerospace Defense Command Space Detection and Tracking system (SPADATS).

See also NORTH AMERICAN AEROSPACE DEFENSE COMMAND.

Space Transportation System (STS) NASA's name for the overall space shuttle program, including intergovernmental agency requirements and international and joint projects. The major components of the space shuttle system are the winged orbiter vehicle (often referred to as the shuttle or the space shuttle); the three space shuttle main engines (SSMEs); the giant external tank (ET), which feeds liquid hydrogen fuel and liquid oxygen oxidizer to the shuttle's three main engines; and the two solid-rocket boosters (SRBs).

The orbiter is the only part of the space shuttle system that has a name in addition to a part number. The first orbiter built was the *Enterprise* (OV-101), which was designed for flight tests in the atmosphere rather than operations in space. It is now at the Smithsonian Museum at Dulles Airport outside Washington, D.C. Five operational orbiters were constructed (listed in order of completion): *Columbia* (OV-102), *Challenger* (OV-99), *Discovery* (OV-103), *Atlantis* (OV-104), and *Endeavour* (OV-105). The *Columbia* and its crew were lost in a reentry accident on February 1, 2003. The *Challenger* and its crew were lost in a launch accident on January 28, 1986.

Shuttles are launched from either Pad 39A or 39B at the Kennedy Space Center, Florida. Depending on the requirements of a particular mission, a space shuttle can carry about 22,680 kg of payload into low Earth orbit (LEO). An assembled shuttle vehicle has a mass of about 2.04 million kg at liftoff.

The two solid-rocket boosters (SRBs) are each 45.4 m high and 3.7 m in diameter. Each has a mass of about 590,000 kg. Their solid propellant consists of a mixture of powdered aluminum (fuel), ammonium perchlorate (oxidizer), and a trace of iron oxide to control the burning rate. The solid mixture is held together with a polymer binder. Each

Liftoff of the space shuttle *Discovery* and its five-person crew from Pad 39-B at the Kennedy Space Center on September 29, 1988. This was the start of the successful STS-26 mission, the return-to-flight mission that followed the 1986 *Challenger* accident. *(Courtesy of NASA)*

booster produces a thrust of approximately 13.8 million newtons for the first few seconds after ignition. The thrust then gradually declines for the remainder of the two-minute burn to avoid overstressing the flight vehicle. Together with the three main liquid-propellant engines on the orbiter, the shuttle vehicle produces a total thrust of 32.5 million newtons at liftoff.

Typically, the SRBs burn until the shuttle flight vehicle reaches an altitude of about 45 km and a speed of 4,970 km per hour. Then they separate and fall back into the Atlantic Ocean to be retrieved, refurbished, and prepared for another flight. After the solid-rocket boosters are jettisoned, the orbiter's three main engines, fed by the huge external tank, continue to burn and provide thrust for another six minutes before they too are shut down at MECO (main engine cutoff). At this point the external tank is jettisoned and falls back to Earth, disintegrating in the atmosphere with any surviving pieces falling into remote ocean waters.

The huge external tank is 47 m long and 8.4 m in diameter. At launch it has a total mass of about 760,250 kg. The two inner propellant tanks contain a maximum of 1,458,400 liters of liquid hydrogen (LH_2) and 542,650 liters of liquid oxygen (LO_2). The external tank is the only major shuttle flight vehicle component that is expended on each launch.

The winged orbiter vehicle is both the heart and the brains of America's Space Transportation System. About the same size and mass as a commercial DC-9 aircraft, the orbiter contains the pressurized crew compartment (which can normally carry up to eight crew members), the huge cargo bay (which is 18.3 m long and 4.57 m in diameter), and the three main engines mounted on its aft end. The orbiter vehicle itself is 37 m long and 17 m high and has a wingspan of 24 m. Since each of the operational vehicles varies slightly in construction, an orbiter generally has an empty mass of between 76,000 and 79,000 kg.

Each of the three main engines on an orbiter vehicle is capable of producing a thrust of 1.668 million newtons at sea level and 2.09 million newtons in the vacuum of space. These engines burn for approximately eight minutes during launch ascent and together consume about 242,250 liters of cryogenic propellants each minute when all three operate at full power.

An orbiter vehicle also has two smaller orbital maneuvering system (OMS) engines that operate only in space. These engines burn nitrogen tetroxide as the oxidizer and monomethyl hydrazine as the fuel. These propellants are supplied from onboard tanks carried in the two pods at the upper rear portion of the vehicle. The OMS engines are used for major maneuvers in orbit and to slow the orbiter vehicle for reentry at the end of its mission in space. On most missions the orbiter enters an elliptical orbit, then coasts around Earth to the opposite side. The OMS engines then fire just long enough to stabilize and circularize the orbit. On some missions the OMS engines also are fired soon after the external tank separates to place the orbiter vehicle at a desired altitude for the second OMS burn that then circularizes the orbit. Later OMS engine burns can raise or adjust the orbit to satisfy the needs of a particular mission. A shuttle flight can last from a few days to more than a week or two.

After deploying the payload spacecraft (some of which can have attached upper stages to take them to higher-altitude operational orbits, such as a geostationary orbit), operating the onboard scientific instrument (e.g., Spacelab), making scientific observations of Earth or the heavens, or performing other aerospace activities, the orbiter vehicle reenters Earth's atmosphere and lands. This landing usually occurs at either the Kennedy Space Center in Florida or at Edwards Air Force Base in California (depending on weather conditions at the landing sites). Unlike prior manned spacecraft, which followed a ballistic trajectory, the orbiter (now operating like an unpowered glider) has a cross-range capability (i.e., it can move to the right or left off the straight line of its reentry path) of about 2,000 km. The landing speed is between 340 and 365 km per hour. After touchdown and rollout, the orbiter vehicle immediately is "safed" by a ground crew with special equipment. This safing operation is also the first step in preparing the orbiter for its next mission in space.

The orbiter's crew cabin has three levels. The uppermost is the flight deck, where the commander and pilot control the mission. The middeck is where the galley, toilet, sleep stations, and storage and experiment lockers are found. Also located in the middeck are the side hatch for passage to and from the orbiter vehicle before launch and after landing and the airlock hatch into the cargo bay and to outer space to support on-orbit extravehicular activities (EVAs). Below the middeck floor is a utility area for air and water tanks.

The orbiter's large cargo bay is adaptable to numerous tasks. It can carry satellites, large space platforms such as the *LONG-DURATION EXPOSURE FACILITY* (LDEF), and even an entire scientific laboratory, such as the European Space Agency's Spacelab to and from low Earth orbit. It also serves as a workstation for astronauts to repair satellites, a foundation from which to erect space structures, and a place to store and hold spacecraft that have been retrieved from orbit for return to Earth.

Mounted on the port (left) side of the orbiter's cargo bay behind the crew quarters is the remote manipulator system (RMS), which was developed and funded by the Canadian government. The RMS is a robot arm and hand with three joints similar to those found in a human being's shoulder, elbow, and wrist. There are two television cameras mounted on the RMS near the "elbow" and "wrist." These cameras provide visual information for the astronauts who are operating the RMS from the aft station on the orbiter's flight deck. The RMS is about 15 m in length and can move anything from astronauts to satellites to and from the cargo bay as well as to different points in nearby outer space.

The table provides a brief summary of all the space shuttle flights from 1981 up to the fatal *COLUMBIA* ACCIDENT on February 1, 2003. On that day, about 15 minutes before the anticipated end of a highly successful microgravity research mission, the *Columbia* disintegrated over the western United States as it glided at high speed toward the Kennedy Space Center landing site. The cause of the fatal accident remains the subject of technical discussion and debate. However, extensive postaccident investigations suggest that some type of very intense heating took place inside *Columbia*'s left wing. Some type of launch ascent debris (such as insulation foam falling away from the external tank) appears to have impacted the left wing and compromised its ability to withstand the severe thermal conditions of reentry. The accident claimed the first operational orbiter vehicle and its seven crewmembers. Despite the tragic loss of both the *Challenger* in 1986 (during the STS 51-L mission) and the *Columbia* in 2003 (during the STS-107 mission), the space shuttle has had more than 100 successful missions and has played an important role in the American space program.

NASA plans to upgrade the remaining Orbiter vehicles—namely *Discovery, Atlantis,* and *Endeavour*—and return them to flight. The space shuttle will then continue to serve as the centerpiece space launch vehicle for the American human space flight program for the next decade or so, after which NASA hopes to develop and deploy an alternative form of space transportation for human crewed missions. Until that planned, new, human-rated space launch system is ready,

NASA Space Shuttle Launches (1981–2003)

Year	Launches
1981	STS-1, STS-2
1982	STS-3, STS-4, STS-5
1983	STS-6, STS-7, STS-8, STS-9
1984	41-B, 41-C, 41-D, 41-G, 51-A
1985	51-C, 51-D, 51-B, 51-G, 51-F, 51-I, 51-J, 61-A, 61-B
1986	61-C, 51-L (*Challenger* accident)
1987	No Launches
1988	STS-26, STS-27
1989	STS-29, STS-30, STS-28, STS-34, STS-33
1990	STS-32, STS-36, STS-31, STS-41, STS-38, STS-35
1991	STS-37, STS-39, STS-40, STS-43, STS-48, STS-44
1992	STS-42, STS-45, STS-49, STS-50, STS-46, STS-47, STS-52, STS-53
1993	STS-54, STS-56, STS-55, STS-57, STS-51, STS-58, STS-61
1994	STS-60, STS-62, STS-59, STS-65, STS-64, STS-68, STS-66
1995	STS-63, STS-67, STS-71, STS-70, STS-69, STS-73, STS-74
1996	STS-72, STS-75, STS-76, STS-77, STS-78, STS-79, STS-80
1997	STS-81, STS-82, STS-83, STS-84, STS-94, STS-85, STS-86, STS-87
1998	STS-89, STS-90, STS-91, STS-95, STS-88
1999	STS-96, STS-93, STS-103
2000	STS-99, STS-101, STS-106, STS-92, STS-97
2001	STS-98, STS-102, STS-100, STS-104, STS-105, STS-108
2002	STS-109, STS-110, STS-111, STS-112, STS-113
2003	STS-107 (*Columbia* accident)
2004	No launches
2005	STS-114

Source: NASA (as of December 31, 2005).

however, the space shuttle remains essential for the successful completion and operation of the *International Space Station.*

space tribology The branch of aerospace engineering that deals with friction, lubrication, and the behavior of lubricants under the harsh environmental conditions encountered during space flight. Space tribology is actually a combination of scientific disciplines: engineering, physics, chemistry, and metallurgy, all focused on and concerned with the contact of surfaces in the vacuum of outer space. The space environment often imposes special demands on the moving parts of a spacecraft. Factors such as high vacuum (i.e., very low pressure), radiation, thermal stress, and microgravity all must be considered in developing suitable approaches to reducing friction between moving components of a spacecraft. For example, new liquid lubricants, such as perfluorinated fluids, have been developed with extremely low vapor pressure and high viscosity index. Physical vapor deposition techniques also have been used to produce thin, highly adherent films of solid lubricant on metal components. Future space missions with extended lifetimes (e.g., five to 10 years) and intermittent periods of operation (as might occur in a multiple asteroid encounter mission) are placing new demands on the field of space tribology. Lubricants possessing extended lifetimes and increased reliability must be developed to support these new space mission requirements.

See also LUBRICATION (IN SPACE).

space tug A proposed but as yet undeveloped space-based reusable upper-stage vehicle that is capable of delivering, retrieving, and servicing payloads in orbits and trajectories beyond the planned operating altitude of the *International Space Station.* The space tug would be "hangared" in space, refurbished on orbit, and reused many times.

space vehicle In general, a vehicle or craft capable of traveling through outer space. This vehicle can be occupied by a human crew or be uncrewed and can be capable of returning to Earth or sent on a one-way (expendable) mission. A space vehicle capable of operating both in space and in Earth's atmosphere is called an AEROSPACE VEHICLE (such as, for example, the space shuttle orbiter vehicle).

See also SPACECRAFT; SPACE TRANSPORTATION SYSTEM.

space walk Aerospace jargon used to describe an extravehicular activity (EVA).

See also EXTRAVEHICULAR ACTIVITY.

space warfare 1. The conduct of offensive and/or defensive operations in outer space, usually resulting in the destruction or negation of an adversary's military space systems; for example, the use of antisatellite (ASAT) systems against capital military space assets. 2. The use of military space systems (e.g., surveillance satellites and secure communications satellites) to support friendly forces during hostile actions in Earth's atmosphere, on land, or at sea. 3. The use of military space systems to support ballistic missile defense operations.

See also ANTISATELLITE SPACECRAFT WEAPON; BALLISTIC MISSILE DEFENSE.

space weather The closest star, the Sun, looks serene at a distance of about 150 million kilometers away, but it is actually a seething nuclear cauldron that churns, boils, and often violently erupts. Parts of the Sun's surface and atmosphere are constantly being blown into space, where they become the solar wind. Made up of hot charged particles, this wind streams out from the Sun and flows through the solar system, bumping and buffeting any objects it encounters. Traveling at more than a million kilometers per hour, the solar wind takes about three to four days to reach Earth. When it arrives at Earth the solar wind interacts with our planet's magnetic field, generating millions of amps of electric current. It blows Earth's magnetic field into a tear-shaped region called the magnetosphere. Collectively, the eruptions from the Sun, the disturbances in the solar wind, and the stretching and twisting of Earth's magnetosphere are referred to by the term *space weather.* Quite similar to terrestrial weather, space weather can also be calm and mild or completely wild and dangerous.

Adverse space weather conditions triggered by solar eruptions not only affect astronauts and spacecraft but also activities and equipment on Earth, including terrestrial power lines, communications, and navigation. For example, in space coronal mass ejections (CMEs) and solar flares can damage the sensitive electronic systems of satellites or trigger phantom commands in the computers responsible for operating various spacecraft. Even astronauts are at risk if they venture beyond the radiation-shielded portions of their space vehicles. On Earth space weather can interfere with radar. During a solar-induced magnetic storm electric currents can surge through Earth's surface and sometimes disrupt terrestrial power lines. In 1989, for example, one such surge produced a cascade of broken circuits at Canada's Hydro-Quebec electric power company, causing the entire grid to collapse in less than 90 seconds. All over Quebec the lights went out. Some Canadians even went without electric power and heat for an entire month because of this severe space weather episode.

Today an armada of satellites from the United States (primarily sponsored by NASA and NOAA), Europe, Japan, and Russia help scientists around the world monitor and forecast space weather. Much of this effort has been focused through the International Solar Terrestrial Physics (ISTP) Program. Near solar maximum, more frequent episodes of "inclement" space weather are generally anticipated. Therefore, as the Sun approaches the maximum of its cycle, space weather forecasters and space scientists more closely monitor the space environment for signs of potentially stormy relationships between Earth and its parent star.

See also MAGNETOSPHERE; SUN.

Sparrow (AIM-7) The Sparrow is a radar-guided air-to-air missile with a high-explosive warhead. The versatile missile has all-weather, all-altitude operational capability and can attack high-performance aircraft and missiles from any direction. It is a widely deployed missile used by the forces of the United States and the North Atlantic Treaty Organization (NATO).

The missile has five major sections: radome, radar guidance, warhead, flight control (autopilot plus hydraulic control system), and solid-propellant rocket motor. It has a cylindrical body with four wings at midbody and four tail fins. Although the external dimensions of the Sparrow have remained relatively unchanged from model to model, the internal components of newer missiles represent major improvements with vastly increased capabilities.

The AIM-7F joined U.S. Air Force inventory in 1976 as the primary medium-range air-to-air missile for the F-15 Eagle. The AIM-7M, the only current operational version, entered service in 1982. It has improved reliability and performance over earlier models at low altitudes and in electronic countermeasure environments. It also has a significantly more lethal warhead. Today the F-4 Phantom, F-15 Eagle, and F-16 Fighting Falcon fighter aircraft carry the AIM-7M Sparrow. U.S. and NATO navies operate a surface-to-air version of this missile called the RIM-7F/M Sea Sparrow.

See also SEA SPARROW.

Spartan 1. A nuclear-armed long-range (mid-course–intercept) surface-to-air guided missile that had been deployed by the United States as part of the former Safeguard ballistic missile defense system in 1975. It was designed to intercept strategic ballistic missile reentry vehicles in space. 2. A free-flying unmanned space platform developed by NASA to support various scientific studies. A Spartan space platform is launched aboard the space shuttle and deployed from the orbiter vehicle, where it then performs a preprogrammed mission on orbit. Scientific data are collected during each mission using a tape recorder and (in many cases) film cameras. After the Spartan platform is deployed from the orbiter, there is no command and control capability. At the end of its mission the platform is retrieved by the orbiter, stowed in the cargo bay, and returned to Earth for data recovery. The platform then is refurbished and made available for a future mission. Power during the deployed phase of the mission is provided by onboard batteries, while platform attitude control is accomplished with pneumatic gas jets.

See also BALLISTIC MISSILE DEFENSE; SAFEGUARD; SPACE TRANSPORTATION SYSTEM.

special relativity Theory introduced in 1905 by the German-Swiss-American physicist ALBERT EINSTEIN.

See also RELATIVITY (THEORY).

specific heat capacity (symbol: c) In thermodynamics, a measure of the heat capacity of a substance per unit mass. It is the quantity of thermal energy (heat) required to raise the temperature of a unit mass of a substance by 1 degree. In the SI unit system, specific heat capacity has the units joules per kilogram-kelvin (J kg^{-1} K^{-1}).

For a simple compressible substance, the specific heat at constant volume (c_v) can be defined as $c_v = (\delta/\delta T)_v$, where u is internal energy and T is temperature; while the specific heat at constant pressure (c_p) can be defined as $c_p = (\delta h/\delta T)_p$, where h is enthalpy.

specific impulse (symbol: I_{sp}) An important performance index for rocket propellants. It is defined as the thrust (or thrust force) produced by propellant combustion divided by the propellant mass flow rate. Expressed as an equation, the specific impulse is

$$I_{sp} = \text{thrust} / \text{mass flow rate} = F_{thrust} / (\dot{m})$$

where F_{thrust} is the thrust and ($\dot{m}$) is the mass flow rate of propellant. It is also important to understand the units associated with specific impulse. In the SI unit system, thrust is expressed in newtons and mass flow rate in kilograms per second. Since 1 newton equals 1 kilogram-meter per second-squared (i.e., 1 N = 1 kg-m/s^2), the specific impulse in SI units becomes

$$I_{sp} = \text{newtons} / (\text{kg/sec}) = \text{meter/second (m/s)}$$

In the English (or traditional engineering) system of units, thrust is expressed in pounds-force (lbf), while the mass flow rate of propellant is in pounds-mass per second (lbm/s). Since (by definition), 1 pound-force is equal to 1 pound-mass at sea level on the surface of Earth, aerospace engineers often use the following simplification, which is, strictly speaking, valid only at sea level on Earth:

$$I_{sp} = \text{lbf/(lbm/s)} = \text{seconds (s)}$$

This apparently convenient oversimplification by rocket engineers in using the traditional system of units to express specific impulse often leads to a great deal of "unit confusion" for people both within and outside the aerospace profession. That is why SI units have been emphasized and are used almost exclusively throughout this book. However, the reader is also cautioned that many older aerospace documents, especially NASA documents that discuss rocket engine performance, use the oversimplified unit "seconds" rather than the more precise "meter/second" for specific impulse. This difference in the choice of unit systems causes what appear to be major differences in the stated values of specific impulse for various types of rocket engines.

See also PROPELLANT; ROCKET; ROCKET ENGINE.

specific volume (symbol: v) Volume per unit mass of a substance; the reciprocal of density.

spectral classification The system in which stars are given a designation. It consists of a letter and a number according to their spectral lines, which correspond roughly to surface temperature. Astronomers classify stars as O (hottest), B, A, F, G, K, and M (coolest). The numbers represent subdivisions within each major class. The Sun is a G2 star. M stars are numerous but very dim, while O and B stars are very bright but rare. Sometimes referred to as the Harvard classification because astronomers at the Harvard College Observatory (HCO) introduced the system there in the 1890s.

See also MORGAN-KEENAN (MR) CLASSIFICATION; STAR.

spectral line A bright (or dark) line found in the spectrum of some radiant source. Bright lines indicate emission, and dark lines indicate absorption.

spectrogram The photographic image of a spectrum.

spectrometer An optical instrument that splits (disperses) incoming visible light or other electromagnetic radiation from a celestial object into a spectrum by diffraction and then measures the relative amplitudes of the different wavelengths. In EMISSION SPECTROSCOPY, scientists use these data to infer the material composition and other properties of the objects that emitted the light. Using *absorption spectroscopy*, scientists can infer the composition of the intervening medium that absorbed specific wavelengths of light as the radiation passed through the medium. A spectrometer is a very useful remote sensing tool for scientists who wish to study planetary atmospheres. Scientific spacecraft often carry infrared radiation or ultraviolet radiation spectrometers.

spectroscopic binary *See* BINARY STAR SYSTEM.

spectroscopy The study of spectral lines from different atoms and molecules. Astronomers use emission spectroscopy to infer the material composition of the objects that emitted the light and absorption spectroscopy to infer the composition of the intervening medium.

spectrum 1. In physics, any series of energies arranged according to frequency (or wavelength). 2. The electromagnetic spectrum; the sequence of electromagnetic waves from gamma rays (high frequency) to radio waves (low frequency). 3. The series of images produced when a beam of radiant energy is dispersed.

specular reflection Reflection in which the reflected radiation is not diffused; reflection as if from a mirror. The angle between the normal (perpendicular) to the surface and the incident beam is equal to the angle between the normal to the surface and the reflected beam. Any surface irregularities on a specular reflector must be small compared to the wavelength of the incident radiation.

speed of light (symbol: c) The speed of propagation of electromagnetic radiation (including light) through a perfect vacuum. Scientists consider the speed of light as a universal constant equal to 299,792.458 kilometers per second.

speed of sound The speed at which sound travels in a given medium under specified conditions. The speed of sound at sea level in the international standard atmosphere is 1,215 kilometers per hour. Sometimes called the acoustic velocity.

See also ATMOSPHERE.

spherical coordinates A system of coordinates defining a point on a sphere by its angular distances from a primary great circle and from a reference secondary great circle, such as latitude and longitude.

See also CELESTIAL COORDINATES; COORDINATE SYSTEM.

spicule A small solar storm that expels narrow (about 1,000-km-diameter) jets of hot gas into the Sun's lower atmosphere. These jets ascend from the lower chromosphere with velocities of up to 20 to 30 km/s and travel several thousand kilometers into the inner corona. Spicules last for 300 to 600 seconds before falling back and fading away.

See also SUN.

spin-off A general term used to describe benefits that have derived from the (secondary) application of aerospace technologies to terrestrial problems and needs. For example, a special coating material designed to protect a reentry vehicle is also applied to fire-fighting equipment to make it more heat resistant. The manufacturer of the fire-fighting equipment improved the terrestrial product by "spinning-off" aerospace technology.

spin rocket A small rocket that imparts spin to a large rocket vehicle or spacecraft.

spin stabilization Directional stability of a missile or spacecraft obtained by the action of gyroscopic forces that result from spinning the body about its axis of symmetry.

spiral galaxy A galaxy with spiral arms, such as the Milky Way galaxy and the Andromeda galaxy.

See also GALAXY.

***Spirit* rover** *See* MARS EXPLORATION ROVER (MER) MISSION.

Spitzer, Lyman, Jr. (1914–1997) American *Astrophysicist* In 1946, more than a decade before the launch of the first artificial satellite, the American astrophysicist Lyman Spitzer, Jr., proposed the development of a large space-based observatory that could operate unhindered by distortions in Earth's atmosphere. His vision ultimately became NASA's *Hubble Space Telescope,* which was launched in 1990. Spitzer was a renowned astrophysicist who made major contributions in the areas of stellar dynamics, plasma physics, thermonuclear fusion, and space-based astronomy. NASA launched the *Space Infrared Telescope Facility* (SIRTF) in 2003 and renamed this sophisticated new space-based infrared telescope the *Spitzer Space Telescope* (SST) in his honor.

Spitzer was born on June 26, 1914, in Toledo, Ohio. He attended Yale University, where he earned a bachelor's degree in physics in 1935. Following a year at Cambridge University, he entered Princeton University, where he earned a master's degree in 1937 and then a doctorate in astrophysics in 1938. His mentor and doctoral adviser was the famous American astronomer HENRY NORRIS RUSSELL. After receiving his Ph.D. Spitzer spent a year as a postdoctoral fellow at Harvard University, after which he joined the faculty of Yale University (1939).

During World War II Spitzer performed underwater sound research, working with a team that led to the development of sonar. When the war was over he returned for a brief time to teach at Yale University. In 1946, more than a decade before the first artificial satellite (*SPUTNIK 1*) was launched into space and 12 years before NASA was created, Spitzer proposed the pioneering concept of placing an astronomical observatory in space, where it could observe the universe over a wide range of wavelengths and not have to deal with the blurring (or absorbing) effects of Earth's atmosphere. He further proposed that a space-based telescope would be able to collect much clearer images of very distant objects in comparison to any ground-based telescope. To support these views, Spitzer wrote "Astronomical Advantages of an Extra-Terrestrial Observatory." In this visionary paper he enumerated the advantages of putting a telescope in space. He then invested a considerable amount of his time over the next five decades to lobby for a telescope in space, both with members of the U.S. Congress and with fellow scientists. His efforts were instrumental in the development of the *Hubble Space Telescope* (HST).

In 1947 Spitzer received an appointment as chairman of Princeton University's astrophysical sciences department. Accepting this appointment, he succeeded his doctoral adviser, Henry Norris Russell. Spitzer also became the director of Princeton's observatory. While at Princeton he made many contributions to the field of astrophysics. For example, he thoroughly investigated interstellar dust grains and magnetic fields as well as the motions of star clusters and their evolution. He also studied regions of star formation and was among the first astrophysicists to suggest that bright stars in spiral galaxies formed recently from the gas and dust there. Finally, he accurately predicted the existence of a hot galactic halo surrounding the Milky Way galaxy.

In 1951 Spitzer founded the Princeton Plasma Physics Laboratory, originally called Project Matterhorn by the U.S. Atomic Energy Commission. This laboratory became Princeton University's pioneering program in controlled thermonuclear research. Spitzer promoted efforts to harness nuclear fusion as a clean source of energy and remained the laboratory's director until 1967. In 1952 Spitzer became the Charles A. Young Professor of Astronomy at Princeton. He retained that prestigious title for the rest of his life.

From 1960 to 1962 Spitzer served as president of the American Astronomical Society. As the fledgling U.S. space program emerged in the 1960s, Spitzer's visionary idea for space-based astronomy finally began to look more promising. In 1962 he led a program to design an observatory that would orbit the Earth and study the ultraviolet (UV) light from space. (Earth's atmosphere normally blocks ultraviolet light, so scientists cannot study UV emissions from cosmic sources using ground-based facilities.) This proposed observatory eventually became NASA's *Copernicus* spacecraft, which operated successfully between 1972 and 1981.

In 1965 the National Academy of Sciences (NAS) established a committee to define the scientific objectives for a proposed large space telescope, and the academy selected Spitzer to chair this committee. At the time many astronomers did not support the idea of a large space-based telescope. They were concerned, for example, that the cost of an orbiting astronomical facility would reduce the government's financial support for ground-based astronomy. Spitzer invested a great personal effort to convince members of the scientific community, as well of the U.S. Congress, that placing a large telescope into space had great scientific value. In 1968 the first step in making Spitzer's vision of putting a large telescope in space came true. That year NASA launched its highly successful *Orbiting Astronomical Observatory* (OAO).

Between 1973 and 1975 Spitzer was awarded several prestigious honors, including the Catherine Wolfe Bruce Gold Medal (1973), the NAS Henry Draper Medal (1974), and the first James Clerk Maxwell Prize for Plasma Physics of the American Physical Society (1975).

Through the early 1970s Spitzer continued to lobby NASA and the U.S. Congress for the development of a large space telescope. Finally, in 1975 NASA, along with the European Space Agency, began development of what would eventually become the *Hubble Space Telescope.* In 1977, due in large part to Spitzer's continuous efforts, the U.S. Congress approved funding for the construction of NASA's *Space Telescope* (ST), an orbiting facility eventually named the *Hubble Space Telescope* in honor of the great American astronomer EDWIN P. HUBBLE. In 1990, more than 50 years after Spitzer first proposed placing a large telescope into space, NASA used the space shuttle to successfully deploy the *Hubble Space Telescope* in orbit around Earth. Refurbished several times through on-orbit servicing missions by the space shuttle, the *Hubble Space Telescope* still provides scientists stunning images of the universe and still produces amazing new discoveries.

At the age of 82, Spitzer passed away on March 31, 1997, in Princeton, New Jersey. NASA launched a new space telescope on August 25, 2003. This space-based observatory consists of a large and lightweight telescope and three cryogenically cooled science instruments capable of studying the universe at near- to far-infrared wavelengths. Incorporating state-of-the-art infrared detector arrays and launched into an innovative Earth-trailing solar orbit, the observatory is orders of magnitude more capable than any previous space-based infrared telescope. NASA calls this new facility the *Spitzer Space Telescope* in honor of Spitzer and his vision of and contributions to space-based astronomy.

See also GREAT OBSERVATORY PROGRAM; *HUBBLE SPACE TELESCOPE*; *SPITZER SPACE TELESCOPE*.

Spitzer Space Telescope (SST) The final mission in NASA's Great Observatories Program, a family of four orbiting observatories each studying the universe in a different portion of the electromagnetic spectrum. The *Spitzer Space Telescope* (SST), previously called the *Space Infrared Telescope Facility* (SIRTF), consists of a 0.85-meter-diameter telescope and three cryogenically cooled science instruments. NASA renamed this space-based infrared telescope to honor the American astronomer LYMAN SPITZER, JR. The SST represents the most powerful and sensitive infrared telescope ever launched. The orbiting facility obtains images and spectra of celestial objects at infrared radiation wavelengths between 3 and 180 micrometers (μm), an important spectral region of observation mostly unavailable to ground-based telescopes because of the blocking influence of Earth's atmosphere. Following a successful launch on August 25, 2003, from Cape Canaveral Air Force Station by an expendable Delta rocket, the SST traveled to an Earth-trailing heliocentric orbit that allowed the telescope to cool rapidly with a minimum expenditure of onboard cryogenic coolant. With a projected mission lifetime of at least 2.5 years, the SST has taken its place alongside NASA's other great orbiting astronomical observatories and is collecting high-resolution infrared data that help scientists better understand how galaxies, stars, and planets form and develop. Other missions in this program include the *Hubble Space Telescope* (HST), the *Compton Gamma-Ray Observatory* (CGRO), and the *Chandra X-Ray Observatory* (CXO).

See also GREAT OBSERVATORY PROGRAM; INFRARED ASTRONOMY.

splashdown That portion of a human space mission in which the space capsule (reentry craft) containing the crew lands in the ocean—quite literally "splashing down." The astronauts then are recovered by a team of helicopters, aircraft, and/or surface ships. This term was used during NASA's Mercury, Gemini, Apollo, *Skylab,* and Apollo-Soyuz projects (1961–75). In the space shuttle era the orbiter vehicle returns to Earth by landing much like an aircraft. Consequently, the orbiter vehicle and its crew are said to "touch down."

The term *splashdown* also can be applied to an uncrewed recoverable space capsule or spacecraft that has been deorbited intentionally and lands in the ocean. Once again, if the space capsule or spacecraft lands on the surface of Earth (under controlled circumstances), the event is described as touching down or making a soft landing. An uncontrolled or destructive landing on Earth's surface is called an impact, hard landing, or crash landing depending on the circumstances. For example, a piece of space debris is said to "impact"; a space capsule with an undeployed or ripped parachute will either "hard land" or "crash land." Similar terms apply to probes or spacecraft landing on other (solid surface) celestial bodies.

SPOT A family of Earth-observing satellites developed by the French Space Agency (CNES). The spacecraft's name is an acronym for Systeme Probatoire d'Observation de la Terre. *SPOT 1* was launched by an Ariane rocket from Kourou, French Guiana, on February 22, 1986, and successfully placed into an 825-km-altitude, circular polar orbit. The *SPOT 2* spacecraft was launched by an Ariane 4 rocket from Kourou on January 22, 1990, and placed in an orbit exactly opposite to the *SPOT 1* spacecraft, providing more frequent (every 13 days) repeat visits to scenes on Earth. More recent additions to this family of remote sensing spacecraft include *SPOT 3* (launched in September 1993), *SPOT 4* (launched in March 1998), and *SPOT 5* (launched on May 4, 2002). *SPOT 5* takes a variety of relatively high-resolution (for example, 2.5-m panchromatic and 5-m multispectral) images of Earth that are sold commercially to users around the globe through the company SpotImage.

Sprint A high-acceleration, short-range, nuclear-armed, surface-to-air guided missile that was deployed in 1975 as part of the former Safeguard ballistic missile defense system. The Sprint system was designed to intercept strategic ballistic reentry vehicles in the atmosphere during the terminal phase of their flight.

See also BALLISTIC MISSILE DEFENSE; SAFEGUARD.

Sputnik 1 On October 4, 1957, the Soviet Union (now the Russian Federation) launched *Sputnik 1,* the first human-made object to be placed in orbit around Earth. A 29-meter-tall Russian A-1 rocket boosted this artificial satellite into an approximate 230-kilometer by 950-kilometer orbit. The simple 83.5-kg-mass spacecraft was essentially a hollow sphere made of steel that contained batteries and a radio transmitter to which were attached four whip antennas. As it orbited Earth *Sputnik 1* provided scientists with information on temperatures and electron densities in Earth's upper atmosphere. It reentered the atmosphere and burned up on January 4, 1958.

The name *Sputnik* means "fellow traveler." Launched during the cold war under the guidance of the Soviet aerospace visionary SERGEI KOROLEV, *Sputnik 1* represented a technological surprise that sent tremors through the United States and its allies. This launch is often regarded as the birth date of the modern space age. It also marks the beginning of a heated "space race" between the United States and the Soviet Union, a space technology competition that culminated in July 1969 with an American astronaut becoming the first human to walk on the surface of the Moon.

spy satellite Popular name for a reconnaissance satellite.

See also RECONNAISSANCE SATELLITE.

squib A small pyrotechnic device that may be used to fire the igniter in a rocket.

See also ROCKET.

S star A red giant star of spectral type S that is similar to a spectral type M star but distinguished through spectroscopy by dominant molecular bands of zirconium oxide (ZrO) rather than titanium oxide (TiO). About half of the S stars are irregular long-period variable stars. Sometimes called a zirconium star.

See also STAR.

stacking Assembling the coolant tubes of a liquid propellant rocket engine's thrust chamber vertically on a mandrel that simulates the chamber-nozzle contour. This procedure facilitates fitting and adjusting the tubes to the required contour prior to brazing.

See also ROCKET.

Stafford, Thomas P. (1930–) American *Astronaut, U.S. Air Force Officer* In December 1965 the astronaut Thomas P. Stafford traveled into space for the first time as the pilot of the *Gemini 6* mission, during NASA's Gemini Project. He returned to space in June 1966 as the spacecraft commander during the *Gemini 9* mission. In May 1969 he served as the commander of the *Apollo 10* lunar mission. The *Apollo 10* crew traveled to the Moon and demonstrated all the operational steps for a landing mission while in lunar orbit except the actual physical landing on the lunar surface. That mission and historic moment were assigned by NASA to the crew of the *Apollo 11* mission.

In 1975 Stafford served as the U.S. spacecraft commander during the Apollo-Soyuz Test Project (ASTP), the first international orbital rendezvous and docking mission. Despite the political stresses of the cold war, he trained with his Russian counterpart, cosmonaut ALEXEI LEONOV, and together both of these experienced space travelers supervised the historic first meeting in space between American astronauts and Soviet cosmonauts.

Upon completing his assignment with the NASA astronaut corps, he returned to the U.S. Air Force and commanded many important aerospace technology developments in the late 1970s. On March 15, 1978, he was promoted to the rank of lieutenant general. General Stafford retired from the air force in November 1979.

In June 1990 Vice President Quayle and Admiral Richard Truly (then NASA administrator) asked General Stafford to serve as chair for a team of experts to independently advise NASA how to carry out President George H.W. Bush's vision of returning to the Moon, this time to stay, and then go on to explore Mars. General Stafford skillfully assembled teams of experts from the Department of Defense, the Department of Energy, and NASA and completed the study, entitled *America at the Threshold.* This report provided U.S. government officials a definitive technical roadmap spanning the next three decades of the U.S. human space flight program. Emphasis was placed on a human return to the Moon as a pathway for the human exploration of Mars and beyond. Early in this century President George W. Bush has proposed the development of Moon bases and human expeditions to Mars as worthy long-range space program goals for the United States.

stage 1. In rocketry, an element of a missile or launch vehicle system that usually separates from the vehicle at burnout or engine cutoff. In multistage rockets, the stages are numbered chronologically in the order of burning (i.e., first stage, second stage, etc.). 2. In mechanical engineering, a set of rotor blades and stator vanes in a turbine or in an axial-flow compressor. 3. In fluid mechanics and thermodynamics, a step or process through which a fluid passes, especially in compression or expansion.

stage-and-a-half A liquid-rocket propulsion unit of which only part falls away from the rocket vehicle during flight, as in the case of booster rockets falling away to leave the sustainer engine to consume the remaining fuel.

staged combustion Rocket engine cycle in which propellants are partially burned in a preburner prior to being burned in the main combustion chamber.

staging 1. In rocketry, the process during the flight of a multistage launch vehicle or missile during which a spent stage separates from the remaining vehicle and is free to decelerate or follow its own trajectory. 2. In mechanical engineering, the use of two or more stages in a turbine or pump.

stagnation point The point in a flow field about a body immersed in a flowing fluid at which the fluid particles have zero velocity with respect to the body. In thermodynamics, *stagnation* is a condition in which a flowing fluid is brought to rest isentropically (i.e., with no change in entropy). Engineers call the pressure that a flowing fluid would attain if it were brought to rest isentropically the *stagnation pressure.* The *stagnation region* is the region in the vicinity of a stagnation point in a flow field about an immersed body where the fluid velocity is negligible. Finally, the temperature that a flowing fluid would attain if the fluid were brought to rest with no change in entropy is called the *stagnation temperature.*

standard candle (astronomical) A celestial object whose properties allow astronomers to measure large distances through space. The absolute brightness of a standard candle can be determined without a measurement of its apparent brightness. Comparing the absolute brightness of a standard candle to its apparent brightness therefore allows astronomers to measure its distance. For example, the distinct variations of Cepheid variable stars in other galaxies indicate their absolute brightness. By accurately measuring the apparent brightness of these variable stars, astronomers can precisely determine the distance to the galaxy in which they reside. Astronomers also find Type Ia supernovas very useful as a class of standard candles.

Standard Missile (SM) A U.S. Navy surface-to-air and surface-to-surface missile that is mounted on surface ships. The Standard Missile is produced in two major types: the SM-1 MR/SM-2 Medium Range (MR) and the SM-2 Extended Range (ER). It is one of the most reliable in the U.S. Navy's inventory and is used against enemy missiles, aircraft, and ships. The SM joined the fleet more than a decade ago. It replaced the Terrier and Tartar Missiles and now is part of the weapons inventory of more than 100 navy ships. The SM-2 (MR) is a medium-range defense weapon for Ticonderoga-class AEGIS cruisers, Arleigh Burke class AEGIS destroyers, and California- and Virginia-class nuclear cruisers. Oliver Hazard Perry–class frigates use the SM-1 MR.

The SM-1/SM-2 Medium Range (MR) missiles have a dual-thrust solid-propellant rocket, a semiactive radar homing guidance system, and a proximity fuse, high-explosive warhead. The SM-2 Extended Range (ER) missile has a two-stage, solid-fuel rocket, an inertial/semiactive radar homing guidance system, and a proximity fuse, high-explosive warhead. The SM-1 MR missile has a range of between 27 and 37 km; the SM-2 MR missile, between 74 and 167 km; and the SM-2 ER missile, between 120 and 185 km.

standard model Contemporary model used by nuclear physicists to explain how elementary particles interact to form matter. Within the standard model physicists assume that matter is composed of molecules, such as carbon dioxide (CO_2). Each molecule, in turn, contains atoms, the smallest identifiable physical unit of a chemical element. Each atom consists of a tiny, very dense and massive, positively charged central region (nucleus) surrounded by a cloud of negatively charged electrons. The electrons in this cloud arrange themselves according to the allowed energy states described by quantum mechanics. The nucleus of each atom (except for the simplest atom, hydrogen [H]) consists of combinations of nucleons called protons (positively charged) and neutrons (uncharged). Within the nucleus the resident nucleons form complex relationships and assume a variety of complicated subnuclear energy states based on proton and neutron populations and how these particles exchange quarks. Quarks are extremely tiny (less than 10^{-18} meter radius) particles found within protons, neutrons, and other very-short-lived elementary particles.

See also ELEMENTARY PARTICLES; FUNDAMENTAL FORCES IN NATURE.

star A star is essentially a self-luminous ball of very hot gas that generates energy through thermonuclear fusion reactions that take place in its core. Stars may be classified as either "normal" or "abnormal." Normal stars, such as the SUN, shine steadily. These stars exhibit a variety of colors: red, orange, yellow, blue, and white. Most stars are smaller than the Sun, and many stars also resemble it. However, a few stars are much larger than the Sun. In addition, astronomers have observed several types of abnormal stars including giants, dwarfs, and a variety of variable stars.

Most stars can be put into one of several general spectral types called O, B, A, F, G, K, and M. (*See* the table below.) This classification is a sequence established in order of decreasing surface temperature. The table at the bottom of page 585 identifies the 10 brightest stars in Earth's night sky. Of course, this excludes the closet star to Earth, our parent star, the Sun. The spectral classification stars listed in this table not only contain the basic letters (O, B, A, F, etc.) previously mentioned but also contain some additional information that astronomers find helpful in describing stars. Astronomers have found it useful, for example, to further subdivide each lettered spectral classification into 10 subdivisions, denoted by the numbers 0 to 9. By convention within modern astronomy, the hotter the star, the lower this number. Astronomers classify the Sun as a G2 star. This means the Sun is a bit hotter than a G3 star and a bit cooler than a G1 star. Betelgeuse is an M2 star, and Vega is an A0 star.

Astronomers have also found it helpful to categorize stars by luminosity class. The table on luminosity contains the standard stellar luminosity classes. This classification scheme is based on a single spectral property—the width of a star's spectral lines. From astrophysics scientists know that spectral line width is sensitive to the density conditions in a star's photosphere. In turn, astronomers can correlate a star's atmospheric density to its luminosity. The use of luminosity classes allows astronomers to distinguish giants from main sequence stars, supergiants from giants, and so forth.

Our parent star, the Sun, is approximately 1.4 million kilometers in diameter and has an effective surface temperature of about 5,800 K. The Sun, like other stars, is a giant nuclear furnace in which the temperature, pressure, and density are sufficient to cause light nuclei to join together, or "fuse." For example, deep inside the solar interior, hydrogen, which makes up 90 percent of the Sun's mass, is fused into helium atoms, releasing large amounts of energy that eventu-

Stellar Spectral Classes

Type	Description	Typical Surface Temperatures (K)	Remarks/Examples
O	very hot, large blue stars (hottest)	28,000–40,000	ultraviolet stars; very short lifetimes (3–6 million years)
B	large, hot blue stars	11,000–28,000	Rigel
A	blue-white, white stars	7,500–11,000	Vega, Sirius, Altair
F	white stars	6,000–7,500	Canopis, Polaris
G	yellow stars	5,000–6,000	the Sun
K	orange-red stars	3,500–5,000	Arcturus, Aldebaran
M	red stars (coolest)	<3,500	Antares, Betelgeuse

Source: NASA.

Stellar Luminosity Classes

Luminosity Class	Type of Stars
Ia	Bright Supergiants
Ib	Supergiants
II	Bright Giants
III	Giants
IV	Subgiants
V	Main-Sequence Stars (Dwarfs)
VI	Subdwarfs
VII	White Dwarfs

ally works its way to the surface and then is radiated throughout the solar system. The Sun is currently in a state of balance, or equilibrium, between two competing forces: gravity (which wants to pull all its mass inward) and the radiation pressure and hot gas pressure resulting from the thermonuclear reactions (which push outward).

Many stars in the galaxy appear to have companions with which they are gravitationally bound in binary, triple, or even larger systems. The table lists the 10 stars (excluding the Sun) nearest Earth. As shown in the table, several of these nearby stars have companions. Compared to other stars throughout the Milky Way galaxy, the Sun is slightly unusual. It does not have a known stellar companion. However, the existence of a very distant, massive, dark companion called Nemesis has been postulated by some astrophysicists in an attempt to explain an apparent "cosmic catastrophe cycle" that occurred on Earth about 65 million years ago.

Astrophysicists have discovered what appears to be the life cycle of stars. Stars originate with the condensation of enormous clouds of cosmic dust and hydrogen gas, called nebulas. Gravity is the dominant force behind the birth of a star. According to Newton's universal law of gravitation, all bodies attract one another in proportion to their masses and distance apart. The dust and gas particles found in these huge interstellar clouds attract one another and gradually draw closer together. Eventually enough of these particles join together to form a central clump that is sufficiently massive to bind all the other parts of the cloud by gravitation. At this point the edges of the cloud start to collapse inward, separating it from the remaining dust and gas in the region.

Initially the cloud contracts rapidly because the thermal energy release related to contraction is radiated outward easily. However, when the cloud grows smaller and more dense, the heat released at the center cannot escape to the outer surface immediately. This causes a rapid rise in internal temperature, slowing down but not stopping the relentless gravitational contraction.

The actual birth of a star occurs when its interior becomes so dense and its temperature so high that thermonuclear fusion occurs. The heat released in thermonuclear fusion reactions is greater than that released through gravitational contraction, and fusion becomes the star's primary energy-producing mechanism. Gases heated by nuclear fusion at the cloud's center begin to rise, counterbalancing the inward pull of gravity on the outer layers. The star stops collapsing and reaches a state of equilibrium between outward and inward forces. At this point the star has become what astronomers and astrophysicists call a "main sequence star." Like the Sun, it will remain in this state of equilibrium for billions of years, until all the hydrogen fuel in its core has been converted into helium.

How long a star remains on the main sequence, burning hydrogen for its fuel, depends mostly on its mass. The Sun has an estimated main sequence lifetime of about 10 billion years, of which approximately 5 billion years have now passed. Larger stars burn their fuels faster and at much higher temperatures. These stars, therefore, have short main sequence lifetimes, sometimes as little as 1 million years. In comparison, the "red dwarf stars," which typically have less than one-tenth the mass of the Sun, burn up so slowly that trillions of years must elapse before their hydrogen supply is exhausted. When a star has used up its hydrogen fuel, it leaves the "normal" state, or departs the main sequence. This

The Ten Nearest Stars (excluding the Sun)

Name	Distance (pc)	Radial Velocity (km/s)	Transverse Velocity (km/s)	Spectral Type I	Spectral Type II*
Proxima Centauri	1.30	–16	23.8	M5	
Alpha Centauri	1.35	–22	23.2	G2V	K1V
Barnard's Star	1.82	–108	89.7	M5V	
Wolf 359 (CN Leo)	2.38	+13	53.0	M8V	
Lalande 21185 (HD 95735)	2.52	–84	57.1	M2V	
UV Ceti	2.58	+30	41.1	M6V	M6V
Sirius	2.64	–8	16.7	A1V	white dwarf
Ross 154	2.90	–4	9.9	M5V	
Ross 248	3.18	–81	23.8	M6V	
Epsilon Eridani	3.22	+16	15.3	K2V	

* Columns I and II correspond to individual members of a binary-star system. One parsec (pc) = 3.2616 light years (ly).

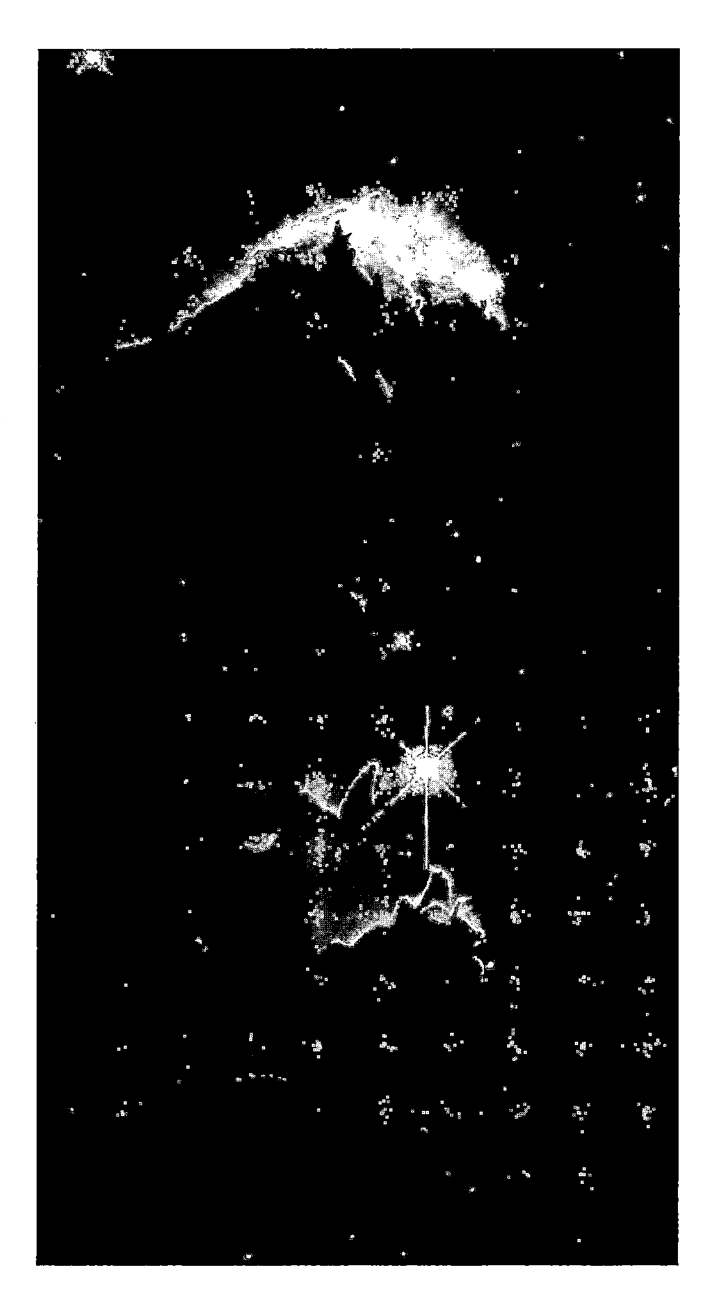

This spectacular image, taken by NASA's *Hubble Space Telescope* in 1995, shows a "star birth" cloud in M16, an open cluster in the constellation Serpens (Ser). Such dense clouds of gas, called *giant molecular clouds* (GMCs), are located primarily in the spiral arms of galaxies. Astronomers believe the GMC is the birthplace of stars. A new star is born when gravitational attraction causes a denser region in the cloud to form into a *protostar.* The protostar continues to contract until its compressed central core initiates hydrogen and helium burning. At that point it becomes a main sequence star. *(Courtesy of NASA)*

happens when the core of the star has been converted from hydrogen to helium by thermonuclear reactions.

When the hydrogen fuel in the core of a main sequence star has been consumed, the core starts to collapse. At the same time, the hydrogen fusion process moves outward from the core into the surrounding regions. There the process of converting hydrogen into helium continues, releasing radiant energy. As this burning process moves into the outer regions, the star's atmosphere expands greatly, and it becomes a "red giant." The term *giant* is quite appropriate. If we put a red giant where the Sun is now, the innermost planet, Mercury, would be engulfed by it; similarly, if we put a larger "red supergiant" there, this supergiant would extend past the orbit of Mars.

As the star's nuclear evolution continues, it may become a "variable star," pulsating in size and brightness over periods of several months to years. The visual brightness of such an "abnormal" star might now change by a factor of 100, while its total energy output might vary by only a factor of 2 or 3.

As an abnormal star grows, its contracting core may become so hot that it ignites and burns nuclear fuels other than hydrogen, beginning with the helium created in millions to perhaps billions of years of main sequence burning. The subsequent behavior of such a star is complex, but in general it can be characterized as a continuing series of gravitational contractions and new nuclear reaction ignitions. Each new series of fusion reactions produces a succession of heavier elements in addition to releasing large quantities of energy. For example, the burning of helium produces carbon, the burning of carbon produces oxygen, and so forth.

Finally, when nuclear burning no longer releases enough radiant energy to support the giant star, it collapses, and its dense central core becomes either a compact white dwarf or a tiny neutron star. This collapse also may trigger an explosion of the star's outer layers, which displays itself as a supernova. In exceptional cases with very massive stars, the core (or perhaps even the entire star) might become a black hole.

When a star like the Sun has burned all the nuclear fuels available, it collapses under its own gravity until the collective resistance of the electrons within it finally stops the contraction process. The "dead star" has become a "white dwarf" and may now be about the size of Earth. Its atoms are packed so tightly together that a fragment the size of a sugar cube would have a mass of thousands of kilograms. The white dwarf then cools for perhaps several billion years, going from white, to yellow, to red, and finally becoming a cold, dark sphere sometimes called a "black dwarf." (Note that a white dwarf does not experience nuclear burning; rather, its light comes from a thin gaseous atmosphere that gradually dissipates its heat to space.) Astrophysicists estimate that there are more than 10 billion white dwarf stars in the Milky Way galaxy alone, many of which have now become black dwarfs. This fate appears to be awaiting our own Sun and most other stars in the galaxy.

However, when a star with a mass of about 1.5 to 3 times the mass of the Sun undergoes collapse, it will contract even further and end up as a "neutron star," with a diameter of perhaps only 20 kilometers. In neutron stars intense gravitational forces drive electrons into atomic nuclei, forcing them to combine with protons and transforming this combination into neutrons. Atomic nuclei are, therefore, obliterated in this process, and only the collective resistance of neutrons to compression halts the collapse. At this point the star's matter is so dense that each cubic centimeter has a mass of several billion tons.

For stars that end their life having more than a few solar masses, even the resistance of neutrons is not enough to stop the unyielding gravitational collapse. In death such massive stars may ultimately become black holes, incredibly dense point masses, or singularities, that are surrounded by a literal "black region" in which gravitational attraction is so strong that nothing, not even light itself, can escape.

Currently, many scientists relate the astronomical phenomena called supernovas and pulsars with neutron stars and their evolution. The final collapse of a giant star to the neutron stage may give rise to the physical conditions that cause its outer portions to explode, creating a supernova. This type of cosmic explosion releases so much energy that its debris products will temporarily outshine all the ordinary stars in the galaxy.

A regular "nova" (the Latin word for "new," the plural of which is novas or novae) occurs more frequently and is far less violent and spectacular. One common class, called recurring novas, is due to the nuclear ignition of gas being drawn from a companion star to the surface of a white dwarf. Such binary star systems are quite common; sometimes the stars will have orbits that regularly bring them close enough for one to draw off gas from the other.

When a supernova occurs at the end of a massive star's life, the violent explosion fills vast regions of space with matter that may radiate for hundreds or even thousands of years. The debris created by a supernova explosion eventually will cool into dust and gas, become part of a giant interstellar cloud, and perhaps once again be condensed into a star or planet. Most of the heavier elements found on Earth are thought to have originated in supernovas because the normal thermonuclear fusion processes cannot produce such heavy elements. The violent power of a supernova explosion can, however, combine lighter elements into the heaviest elements found in nature (e.g., lead, thorium, and uranium). Consequently, both the Sun and its planets were most likely enriched by infusions of material hurled into the interstellar void by ancient supernova explosions.

Pulsars, first detected by radio astronomers in 1967, are sources of very accurately spaced bursts, or pulses, of radio signals. These radio signals are so regular, in fact, that the scientists who made the first detections were startled into thinking that they might have intercepted a radio signal from an intelligent alien civilization.

The pulsar, named because its radio-wave signature regularly turns on and off, or pulses, is considered to be a rapidly spinning neutron star. One pulsar is located in the center of the Crab Nebula, where a giant cloud of gas is still glowing from a supernova explosion that occurred in the year 1054 C.E., a spectacular celestial event observed and recorded by ancient Chinese astronomers. The discovery of this pulsar allowed scientists to understand both pulsars and supernovas.

In a supernova explosion a massive star is literally destroyed in an instant, but the explosive debris lingers and briefly outshines everything in the galaxy. In addition to scattering material all over interstellar space, supernova explosions leave behind a dense collapsed core made of neutrons. This neutron star, with an immense magnetic field, spins many times a second, emitting beams of radio waves, X-rays, and other radiations. These radiations may be focused by the pulsar's powerful magnetic field and sweep through space much like a revolving lighthouse beacon. The neutron star, the end product of a violent supernova explosion, becomes a pulsar.

Astrophysicists must develop new theories to explain how pulsars can create intense radio waves, visible light, X-rays, and gamma rays all at the same time. Orbiting X-ray observatories have detected X-ray pulsars that are believed to be caused by a neutron star pulling gaseous matter from a normal companion star in a binary star system. As gas is sucked away from the normal companion to the surface of the neutron star, the gravitational attraction of the neutron star heats up the gas to millions of kelvins, and this causes the gas to emit X-rays.

The advent of the space age and the use of powerful orbiting observatories, such as the *Hubble Space Telescope* (HST), the *Chandra X-Ray Observatory,* (CXO), and the *Spitzer Space Telescope* (SST), to view the universe as never before possible has greatly increased our knowledge about the many different types of stellar phenomena. Most exciting of all, perhaps, is the fact that this process of astrophysical discovery has really only just begun.

See also ASTROPHYSICS; BLACK HOLE; *CHANDRA X-RAY OBSERVATORY;* FUSION; *HUBBLE SPACE TELESCOPE; SPITZER SPACE TELESCOPE;* SUN.

starburst galaxy A galaxy that is experiencing a period of intense star-forming activity. While this burst of activity can last for 10 million years or more, on a cosmic scale it is really a brief period compared to the 10-billion-year-or-so overall lifetime of a typical galaxy. During a starburst stars form at a rate that is 10s and even 100s of times greater than the normal rate of star formation in a galaxy. The burst occurs over a region that is a few thousand light-years in diameter. The most popular theory in astrophysics is that the starburst is triggered by a close encounter or even collision with another galaxy. The collision of galaxies sends shock waves rushing through the affected galaxies. These shock waves (manifested as bursts of radiation, gravitational field distortions, etc.) then push on the giant molecular clouds (GMCs) of dust and gas, causing the GMCs to form a few hundred new stars. Many of the newly formed stars are very massive and very bright, making a starburst galaxy among the most luminous types of galaxies. The massive new stars use up their nuclear fuel very quickly and then explode as supernovas. The exploding supernovas send out bursts of radiation and material, a process that helps stimulate the formation of other new stars, which in turn experience a supernova. This star formation chain reaction can propagate through the central region of a galaxy, where most of the interstellar gas is located. The starburst ends when most of this interstellar gas is used up or blown away.

See also GALAXY; STAR.

star cluster A group of stars, numbering from a few to perhaps thousands, that were formed from a common gas cloud and are now bound together by their mutual gravitational attraction.

Stardust mission The primary objective of NASA's Discovery class Stardust mission is to fly by the Comet Wild 2 and collect samples of dust and volatiles in the coma of the comet. NASA launched the *Stardust* spacecraft from Cape Canaveral Air Force Station, Florida, on February 7, 1999, using an expendable Delta II rocket. Following launch, the spacecraft successfully achieved an elliptical, heliocentric orbit. By mid-summer 2003 it had completed its second orbit of the Sun. The spacecraft then successfully flew by the nucleus of Comet Wild 2 on January 2, 2004. When *Stardust* flew past the comet's nucleus it did so at an approximate relative velocity of 6.1 kilometers per second. At closest approach during this close encounter, the spacecraft came within 250 km of the comet's nucleus and returned images of the nucleus. The spacecraft's dust monitor data indicated that many particle samples were collected. *Stardust* then traveled on a trajectory that brought it near Earth in early 2006. The comet material samples that were collected had been stowed and sealed in the special sample storage vault of the reentry capsule carried onboard the *Stardust* spacecraft. As the spacecraft flew past Earth in mid-January 2006, it ejected the sample capsule. The sample capsule descended through Earth's atmosphere and was successfully recovered in the Utah desert on January 15, 2006.

See also AEROGEL; COMET.

star grain A hollow, solid-propellant rocket grain with the cross section of the hole having a multipointed star shape.

See also ROCKET.

star probe A specially designed and instrumented probe spacecraft that is capable of approaching within 1 million kilometers or so of the Sun's surface (photosphere). This close encounter with the Sun, the nearest star, would provide scientists with their first in-situ measurements of the physical conditions in the corona (the Sun's outer atmosphere). The challenging mission will require advanced space technologies (including propulsion, laser communications, thermal protection, etc.) and might be flown about 2030.

See also SUN.

starship A starship is a space vehicle capable of traveling the great distances between star systems. Even the closest stars in the Milky Way galaxy are often light-years apart. By convention, the word *starship* is used here to describe interstellar spaceships capable of carrying intelligent beings to other star systems, while robot interstellar spaceships are called interstellar probes.

What are the performance requirements for a starship? First, and perhaps most important, the vessel should be capable of traveling at a significant fraction of the speed of light (c). A velocity of 10 percent of the speed of the light (0.1c) is often considered the lowest acceptable speed for a starship, while cruising speeds of 0.9 c and beyond are considered highly desirable. This *optic velocity* cruising capability is necessary to keep interstellar voyages to reasonable lengths of time, both for the home civilization and for the starship crew.

Consider, for example, a trip to the nearest star system, Alpha Centauri, a triple star system about 4.23 light-years away. At a cruising speed of 0.1 c, it would take about 43 years just to get there and another 43 years to return. The time dilation effects of travel at these "relatively low" relativistic speeds would not help too much either, since a ship's clock would register the passage of about 42.8 years versus a terrestrial elapsed time of 43 years. In other words, the crew would age about 43 years during the journey to Alpha Centauri. If we started with 20-year-old crewmembers departing from the outer regions of the solar system in the year 2100 at a constant cruising speed of 0.1 c, they would be approximately 63 years old when they reached the Alpha Centauri star system some 43 years later in 2143. The return journey would be even more dismal. Any surviving crewmembers would be 106 years old when the ship returned to the solar system in the year 2186. Most, if not all, of the crew would probably have died of old age or boredom, and that is for just a journey to the nearest star.

A starship should also provide a comfortable living environment for the crew and passengers (in the case of an interstellar ark). Living in a relatively small, isolated, and confined habitat for a few decades to perhaps a few centuries could certainly overstress even the most psychologically adaptable individuals and their progeny. One common technique used in science fiction to avoid this crew stress problem is to have all or most of the crew placed in some form of "suspended animation" while the vehicle travels through the interstellar void, tended by a ship's company of smart robots.

Any properly designed starship must also provide an adequate amount of radiation protection for the crew, passengers, and sensitive electronic equipment. Interstellar space is permeated with galactic cosmic rays. Nuclear radiation leakage from an advanced thermonuclear fusion engine or a matter-antimatter engine (photon rocket) must also be prevented from entering the crew compartment. In addition, the crew will have to be protected from nuclear radiation showers produced when a starship's hull, traveling at near light speed, slams into interstellar molecules, dust, or gas. For example, a single proton (which we can assume is "stationary") being hit by a starship moving at 90 percent of the speed of light (0.9c) would appear to those on board like a one gigaelectron volt (GeV) proton being accelerated at them. Imagine traveling for years at the beam output end of a very-high-energy particle accelerator. Without proper deflectors or shielding, survival in the crew compartment from such nuclear radiation doses would be doubtful.

To truly function as a starship, the vessel must be able to cruise at will light-years from its home star system. The starship must also be able to accelerate to significant fractions of the speed of light, cruise at these near-optic velocities, and then decelerate to explore a new star system or to investigate a derelict alien spaceship found adrift in the depths of interstellar space.

We will not discuss the obvious difficulties of navigating through interstellar space at near-light velocities. It will be sufficient just to mention that when you look forward at near light speeds everything is *blueshifted;* while when you look aft (backwards) things appear *redshifted.* The starship and its

Characteristics of Possible Starship Propulsion Systems

Pulsed Nuclear Fission System (Project Orion)

Principle of Operation: Series of nuclear fission explosions are detonated at regular intervals behind the vehicle; special giant pusher plate absorbs and reflects pulse of radiation from each atomic blast; system moves forward in series of pulses.

Performance Characteristics: Very low efficiency in converting propellant (explosive device) mass into pure energy for propulsion; limited to number of nuclear explosives that can be carried on board; radiation hazards to crew (needs heavy shielding); probably limited to a maximal speed of about 0.01 to 0.10 the speed of light.

Potential Applications: Most useful for interplanetary transport (especially for rapid movement to far reaches of solar system); not suitable for a starship; very limited application for an interstellar robot probe; possible use for a very slow, huge interstellar ark (several centuries of flight time). Interplanetary version could be built in a decade or so; limited interstellar version by end of 21st century.

Pulsed Nuclear Fusion System (Project Daedalus)

Principle of Operation: Thermonuclear burn of tiny deuterium/helium-3 pellets in special chamber (using laser or particle beam inertially confined fusion techniques); very energetic fusion reaction products exit chamber to produce forward thrust.

Performance Characteristics: Uses energetic single-step fusion reaction; thermonuclear propellant carried on board vessel; maximal speed of about 0.12 c considered possible.

Potential Application: Not suitable for starship; possible use for robot interstellar probe (flyby) mission or slow interstellar ark (centuries of flight time). Limited system might be built by end of 21st century (interstellar probe).

Interstellar Ramjet

Principle of Operation: First proposed by R. Bussard; after vehicle has an initial acceleration to near-light speed, its giant scoop (thousands of square kilometers in area) collects interstellar hydrogen, which then fuels a proton-proton thermonuclear cycle or perhaps the carbon cycle (both of which are found in stars); thermonuclear reaction products exit vehicle and provide forward thrust.

Performance Characteristics: In principle, not limited by amount of propellant that can be carried; however, construction of light-mass giant scoop is major technical difficulty; in concept, cruising speeds of from 0.1 c up to 0.9 c might be obtained.

Potential Applications: Starship; interstellar robot probe; giant space ark. Would require many major technological breakthroughs—several centuries away, if ever.

Photon Rocket

Principle of Operation: Uses matter and antimatter as propellant; equal amounts are combined and annihilate each other, releasing an equivalent amount of energy in form of hard nuclear (gamma) radiation; these gamma rays are collected and emitted in a collimated beam out the back of vessel, providing a forward thrust.

Performance Characteristics: The best (theoretical) propulsion system our understanding of physics will permit; cruising speeds from 0.1 c to 0.99 c.

Potential Applications: Starship; interstellar probes (including self-replicating machines); large space arks; many major technological barriers must be overcome—centuries away, if ever.

crew must be able to find their way from one location in the Milky Way galaxy to another on their own.

What appears to be the major engineering technology needed to make a starship a real part of humans' extraterrestrial civilization is an effective propulsion system. Interstellar class propulsion technology is the key to the galaxy for any emerging civilization that has mastered space flight within and to the limits of its own solar system. Despite the tremendous engineering difficulties associated with the development of a starship propulsion system, several concepts have been proposed. These include the pulsed nuclear fission engine (Project Orion concept), the pulsed nuclear fusion concept (Project Daedalus study), the interstellar nuclear ramjet, and the photon rocket. These systems are briefly described in the table along with their potential advantages and disadvantages.

Unfortunately, in terms of our current understanding of the laws of physics, all known phenomena and mechanisms that might be used to power a starship are either not energetic enough or simply entirely beyond today's technology level and even the technology levels anticipated for several tomorrows. Perhaps major breakthroughs will occur in our understanding of the physical laws of the universe—breakthroughs that provide insight into more intense energy sources or ways around the speed of light barrier now imposed by ALBERT EINSTEIN's theory of relativity. But until such new insights occur (if ever), human travel to another star system on board a starship must remain in the realm of future dreams.

See also INTERSTELLAR PROBE; INTERSTELLAR TRAVEL; PROJECT DAEDALUS; PROJECT ORION; RELATIVITY (THEORY); SPACE NUCLEAR PROPULSION.

star tracker An instrument on a missile or spacecraft that locks onto a star (or pattern of "fixed" stars) and provides reference information to the missile or spacecraft's guidance system and/or attitude control system during flight.

See also GUIDANCE SYSTEM.

state of the art The level to which technology and science have been developed in any given discipline or industry at some designated cutoff time.

See also HORIZON MISSION METHODOLOGY.

static firing The firing of a rocket engine in a hold-down position. Usually done to measure thrust and to accomplish other performance tests.

static testing The testing of a rocket or other device in a stationary or hold-down position, either to verify structural design criteria, structural integrity, and the effects of limit loads or to measure the thrust of a rocket engine.

stationary orbit An orbit in which a satellite revolves about the primary at the same angular rate at which the primary rotates on its axis. To an observer on the surface of the primary, the satellite would appear to be stationary over a point on the primary.

See also GEOSTATIONARY ORBIT; SYNCHRONOUS SATELLITE.

stationkeeping The sequence of maneuvers that maintains a space vehicle or spacecraft in a predetermined orbit.

steady flow A flow whose velocity vector components at any point in the fluid do not vary with time.

steady state Condition of a physical system in which parameters of importance (fluid velocity, temperature, pressure, etc.) do not vary significantly with time; in particular, the condition or state of rocket engine operation in which mass, momentum, volume, and pressure of the combustion products in the thrust chamber do not vary significantly with time.

steady-state universe A model in cosmology based on the perfect cosmological principle suggesting that the universe looks the same to all observers at all times.

See also COSMOLOGY; UNIVERSE.

Stefan, Josef (1835–1893) Austrian *Physicist* In about 1879 the Austrian physicist Josef Stefan experimentally demonstrated that the energy radiated per unit time by a blackbody is proportional to the fourth power of the body's absolute temperature. In 1884 another Austrian physicist, LUDWIG BOLTZMANN, provided the theoretical foundations for this relationship. They collaborated on the formulation of the *Stefan-Boltzmann law*, a physical principle of great importance to astronomers and astrophysicists.

Stefan-Boltzmann law One of the basic laws of radiation heat transfer. This law states that the amount of energy radiated per unit time from a unit surface area of an ideal blackbody radiator is proportional to the fourth power of the absolute temperature of the blackbody. It can be expressed as

$$E = \varepsilon \sigma T^4$$

where E is the emittance per unit area (watts per square meter per second [$Wm^{-2}s^{-1}$]) of the blackbody, ε is the emissivity ($\varepsilon = 1$ for a blackbody), σ is the Stefan-Boltzmann constant (5.67×10^{-8} watts per square meter per kelvin to the fourth power [$Wm^{-2}\,K^{-4}$]), and T is the absolute temperature (kelvins) of the blackbody.

stellar Of or pertaining to the stars.

stellar evolution The different phases in the lifetime of a star, from its formation out of interstellar gas and dust to the time after its nuclear fuel is exhausted.

stellar guidance A system wherein a guided missile may follow a predetermined course with reference primarily to the relative position of the missile and certain preselected celestial bodies.

See also GUIDANCE SYSTEM.

stellar spectrum (plural: spectra) *See* SPECTRAL CLASSIFICATION.

stellar wind The flux of electromagnetic radiation and energetic particles that streams outward from a star. The stellar wind associated with the Sun is called the solar wind.

See also SOLAR WIND.

Stephano A lesser moon of Uranus discovered in 1999 and named after the ship's butler in William Shakespeare's play *The Tempest*. Astronomers have little knowledge about Stephano, and the data presented here are regarded as best estimates. This moon has a diameter of about 20 km and is in a retrograde orbit around Uranus at a distance of 7,942,000 km. Stephano's retrograde orbit is characterized by a period of 676.5 days and an inclination of 144°. Also called S/199 U2 and UXX.

See also URANUS.

step rocket A multistage rocket.

steradian (symbol: sr) The supplementary unit of solid angle in the international system (SI). One steradian is equal to the solid angle subtended at the center of a sphere by an area of surface equal to the square of the radius. The surface of a sphere subtends a solid angle of 4π steradians about its center.

Stinger A shoulder-fired infrared missile that homes in on the heat (thermal energy) emitted by either jet- or propeller-driven fixed-wing aircraft or helicopters. The Stinger system employs a proportional navigation system that allows it to fly an intercept course to the target. Once the missile has traveled a safe distance from the gunner, its main engine ignites and propels it to the target.

Stirling cycle A thermodynamic cycle for a heat engine in which thermal energy (heat) is added at constant volume, followed by isothermal expansion with heat addition. The heat then is rejected at constant volume, followed by isothermal compression with heat rejection. If a regenerator is used so that the heat rejected during the constant volume process is recovered during heat addition at constant volume, then the thermodynamic efficiency of the Stirling cycle approaches (in the limit) the efficiency of the Carnot cycle. The basic principle of the Stirling engine was first proposed and patented in 1816 by Robert Stirling, a Scottish minister.

See also CARNOT CYCLE; HEAT ENGINE; THERMODYNAMICS.

stoichiometric combustion The burning of fuel and oxidizer in precisely the right proportions required for complete reaction with no excess of either reactant.

Stonehenge A circular ring of large vertical stones topped by capstones located in southern England. Some astronomers and archaeologists suggest that this structure was built by ancient Britons between 3000 B.C.E. and 1000 B.C.E. for use as an astronomical calendar.

See also ARCHAEOLOGICAL ASTRONOMY.

storable propellant Rocket propellant (usually liquid) capable of being stored for prolonged periods of time without special temperature or pressure controls.

stowing In aerospace operations, the process of placing a payload in a retained, or attached, position in the space shuttle orbiter's cargo bay for ascent to or return from orbit.

strain In engineering, the change in the shape or volume of an object due to applied forces. There are three basic types of strain: longitudinal, volume, and shear. Longitudinal (or tensile) strain is the change in length per unit length, such as occurs, for example, with the stretching of a wire. Volume (or bulk) strain involves a change in volume per unit volume, such as occurs, for example, when an object is totally immersed in a liquid and experiences a hydrostatic pressure. Finally, shear strain is the angular deformation of an object without a change in its volume. Shear occurs, for example, when a rectangular block of metal is strained or distorted in such a way that two opposite faces become parallelograms, while the other two opposite faces do not change their shape.

See also STRESS.

Stratospheric Observatory for Infrared Astronomy (SOFIA) *See* AIRBORNE ASTRONOMY.

stress In engineering, a force per unit area on an object that causes it to deform (i.e., experience strain). Stress can be viewed as either the system of external forces applied to deform an object or the system of internal "opposite" forces (a function of the material composition of the object) by which the object resists this deformation. The three basic types of stress are compressive (or tensile) stress, hydrostatic pressure, and shear stress.

See also STRAIN.

stretchout An action whereby the time for completing an aerospace activity, especially a contract, is extended beyond the time originally programmed or contracted for. Cost overruns, unanticipated technical delays, and budget cuts are frequent reasons why a stretchout is needed.

stringer A slender, lightweight, lengthwise fill-in structural member in a rocket body. A stringer reinforces and gives shape to the rocket vehicle's skin.

string theory A contemporary theory in physics now under extensive investigation and intense debate. In this theory physicists suggest that particles and forces are really unified, since they are all considered to be particular vibrational aspects (or modes) of a single fundamental submicroscopic physical object called the *string*. Thus, within string theory the various elementary particles are simply different modes of vibration of very tiny strings. How small is one of these hypothesized strings? The physical dimension that characterizes the realm of string theory is called the *Planck length*—a length on the order of 10^{-35} m. How did physicists arrive at this incredibly tiny dimension? The Planck length is simply the distance a photon of light travels in PLANCK TIME, a very small period of time on the order of 10^{-43} s. It is on the very tiny scale of the Planck length that physicists now anticipate resolving the unification of quantum mechanics and general relativity. The mathematical details of string theory are extremely complex, and the physical phenomena string theory implies are quite bizarre, at the very least. For example, theoretical physicists suggest that strings must vibrate in no fewer than 11 space-time dimensions, versus the four dimensions (x-, y-, z-, and time) usually associated with the space-time of general relativity.

See also ELEMENTARY PARTICLE; FUNDAMENTAL FORCES IN NATURE; SPACE-TIME.

Struve, Friedrich Georg Wilhelm (1793–1864) German-Russian *Astronomer* The German-Russian astronomer Friedrich Georg Wilhelm Struve set up the Pulkovo Observatory near St. Petersburg, Russia, in the mid-1830s and spent many years investigating and cataloging binary star systems. In about 1839 he made one of the early attempts to quantify interstellar distances by measuring the parallax of the star Vega. He was the patriarch of the line of Struve astronomers, an astronomical dynasty that extended for four generations and included his son, OTTO WILHELM STRUVE, and great-grandson, OTTO STRUVE.

Struve, Otto (1897–1963) Russian-American *Astronomer* The Russian-American astronomer Otto Struve was the great-grandson of FRIEDRICH GEORG WILHELM STRUVE and the last in a long line of Struve astronomers. Otto Struve investigated stellar evolution and was a strong proponent for the existence of extrasolar planets, especially around Sunlike stars. As the childless son of Ludwig Struve (1858–1920), his death in 1963 ended the famous four-generation Struve family dynasty in astronomy.

Struve, Otto Wilhelm (1819–1905) German-Russian *Astronomer* As the son of FRIEDRICH GEORG WILHELM STRUVE, Otto Wilhelm Struve succeeded his father as director of the Pulkovo Observatory and continued the study of binary star systems, adding some 500 new binary systems to the list. He had two sons who were also astronomers. Hermann Struve (1854–1920) became director of the Berlin Observatory, and Ludwig Struve (1858–1920) became a professor of astronomy at Kharkov University, Ukraine.

Subaru Telescope The Subaru Telescope is an optical-infrared telescope located at the 4,200-m-altitude summit of Mauna Kea on the island of Hawaii. This telescope is an important part of the National Astronomical Observatory of Japan (NAOJ). The Subaru Telescope represents a new generation of ground-based astronomical telescopes not only because of the size of its mirror, with an effective aperture of 8.2 meters, but also because of the various revolutionary technologies used to achieve an outstanding observational performance. Some of the telescope's special features include an active support system that maintains an exceptionally high mirror surface accuracy, a new dome design that suppresses local atmospheric turbulence, an extremely accurate tracking mechanism that uses magnetic driving systems, seven observational instruments installed at the four foci, and an auto-exchanger system to use the observational instruments effectively. These sophisticated systems have been used and fine-tuned since the Subaru Telescope experienced first light in January 1999 and began scientific observations.

subgiant A star that has exhausted the hydrogen in its core and is now evolving into a giant star.

See also STAR.

sublimation In thermodynamics, the direct transition of a material from the solid phase to the vapor phase, and vice versa, without passing through the liquid phase.

submarine-launched ballistic missile (SLBM) *See* FLEET BALLISTIC MISSILE.

submarine rocket (subroc) Submerged, submarine-launched, surface-to-surface rocket with nuclear depth charge or homing torpedo payload, primarily used in antisubmarine applications. Also designated by the U.S. Navy as UUM-44A.

submillimeter astronomy The branch of astronomy that involves millimeter and submillimeter electromagnetic radiation, so named because its wavelength ranges from 0.3 to 1.7 millimeter. The main celestial source of millimeter and submillimeter radiation is cold interstellar material. The material consists of gas, dust, and small rocklike bodies. It is also the substance out of which stars and planets are eventually formed. So detecting submillimeter emissions from interstellar material plays an important role in studying the birth and death of stars. When stars are born out of dense interstellar clouds, their first visible light is trapped within them, so submillimeter instruments are important in modern astronomy.

For example, located near the 4,000-meter summit of Mauna Kea, an extinct volcano on the island of Hawaii, the Submillimeter Array (SMA) is a unique astronomical instrument. It is designed to have the highest feasible resolution in the last waveband of the electromagnetic spectrum to be explored from ground-based observatories on Earth. The facility's submillimeter window (from about 0.3 to 1.0 millimeter) is well suited for studies of the structure and motions of the matter that forms stars; of the spiral structure of galaxies, as outlined by their giant molecular clouds; and of quasars and active galactic nuclei. The initial array consists of eight movable reflectors, each 6 meters in diameter. Planning for the array started in 1984. The Institute of Astronomy and Astrophysics of Taiwan's Academia Sinica is the Harvard-Smithsonian Center for Astrophysics (CfA) partner in the SMA project.

See also ELECTROMAGNETIC SPECTRUM; SMITHSONIAN ASTROPHYSICAL OBSERVATORY.

subsatellite point Intersection with Earth's (or a planet's) surface of the local vertical passing through an orbiting satellite.

subsonic Of or pertaining to speeds less than the speed of sound.

See also SPEED OF SOUND.

al-Sufi, Abd al-Rahman (903–986) Arab *Astronomer* The great Arab astronomer al-Sufi developed a star catalog in about 964 based on PTOLEMY's *Almagest.* He incorporated some of his own observations in this important work, including the first astronomical reference to an extragalactic object—the Andromeda galaxy (M31), which he reported as a fuzzy nebula.

Sullivan, Kathryn D. (1951–) American *Astronaut* NASA astronaut Kathryn D. Sullivan became the first American woman to perform an extravehicular activity (EVA). Her historic space walk occurred in October 1984, during the STS 41-G mission of the space shuttle *Challenger.* She traveled again in outer space as part of the STS-31 mission (1990), during which the space shuttle *Discovery* crew deployed the *HUBBLE SPACE TELESCOPE* (HST) and as part of the STS-45 mission (1992), during which space shuttle *Atlantis* performed Earth-observing experiments.

Sun The Sun is our parent star and the massive, luminous celestial object about which all other bodies in the solar system revolve. It provides the light and warmth upon which almost all terrestrial life depends. Its gravitational field determines the movement of the planets and other celestial bodies (e.g., comets). The Sun is a main sequence star of spectral type G2V. Like all main sequence stars, the Sun derives its abundant energy output from thermonuclear fusion reactions involving the conversion of hydrogen to helium and heavier nuclei. Photons associated with these exothermic (energy-releasing) fusion reactions diffuse outward from the Sun's core until they reach the convective envelope. Another by-product of the thermonuclear fusion reactions is a flux of neutrinos that freely escape from the Sun.

At the center of the Sun is the core, where energy is released in thermonuclear reactions. Surrounding the core are concentric shells called the radiative zone, the convective envelope (which occurs at approximately 0.8 of the Sun's radius), the photosphere (the layer from which visible radiation emerges), the chromosphere, and, finally, the corona (the Sun's outer atmosphere). Energy is transported outward through the convective envelope by convective (mixing) motions that are organized into cells. The Sun's lower, or inner, atmosphere, the photosphere, is the region from which energy is radiated directly into space. Solar radiation approximates a Planck distribution (blackbody source) with an effective temperature of 5,800 K. The table provides a summary of the physical properties of the Sun.

The chromosphere, which extends for a few thousand kilometers above the photosphere, has a maximum temperature of approximately 10,000 K. The corona, which extends several solar radii above the chromosphere, has temperatures of more than 1 million K. These regions emit electromagnetic (EM) radiation in the ultraviolet (UV), extreme ultraviolet (EUV), and X-ray portions of the spectrum. This shorter-wavelength EM radiation, although representing a relatively small portion of the Sun's total energy output, still plays a dominant role in forming planetary ionospheres and in photochemistry reactions occurring in planetary atmospheres.

Since the Sun's outer atmosphere is heated, it expands into the surrounding interplanetary medium. This continuous outflow of plasma is called the solar wind. It consists of protons, electrons, and alpha particles as well as small quantities of heavier ions. Typical particle velocities in the solar wind fall between 300 and 400 kilometers per second, but these velocities may reach as high as 1,000 kilometers per second.

Although the total energy output of the Sun is remarkably steady, its surface displays many types of irregularities.

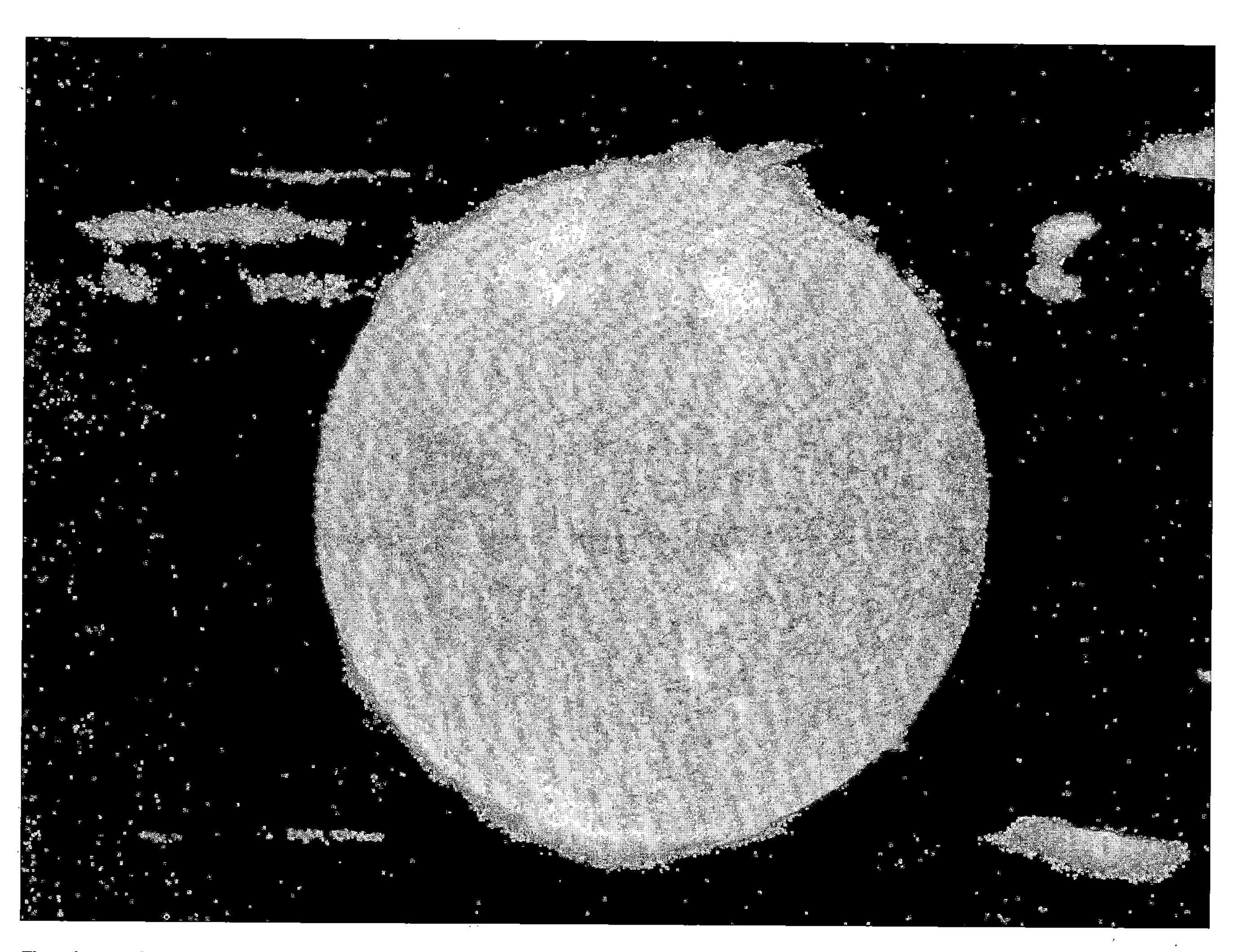

The solar eruption of June 10, 1973, as seen in this spectroheliogram obtained during NASA's Skylab mission. At the top of this image a great eruption can be observed extending more than one-third of a solar radius from the Sun's surface. In the picture solar north is to the right, and east is up. The wavelength scale (150 to 650 angstroms, or 15 to 65 nanometers) increases to the left. ***(Photograph courtesy of NASA)***

Physical and Dynamic Properties of the Sun

Diameter	1.39×10^6 km
Mass	1.99×10^{30} kg
Distance from the Earth (average)	1.496×10^8 km [1 AU] (8.3 light min)
Luminosity	3.9×10^{26} watts
Density (average)	1.41 g/cm^3
Equivalent blackbody temperature	5,800 kelvins (K)
Central temperature (approx.)	15,000,000 kelvins (K)
Rotation period (varies with latitude zones)	27 days (approx.)
Radiant energy output per unit surface area	6.4×10^7 W/m^2
Solar cycle (total cycle of polarity reversals of Sun's magnetic field)	22 years
Sunspot cycle	11 years (approx.)
Solar constant (at 1 AU)	1371 ± 5 W/m^2

Source: NASA.

These include sunspots, faculae, plages (bright areas), filaments, prominences, and flares. All are believed ultimately to be the result of interactions between ionized gases in the solar atmosphere and the Sun's magnetic field. Most solar activity follows the SUNSPOT CYCLE. The number of sunspots varies, with a period of about 11 years. However, this approximately 11-year sunspot cycle is only one aspect of a more general 22-year *solar cycle* that corresponds to a reversal of the polarity patterns of the Sun's magnetic field.

Sunspots were originally observed by GALILEO GALILEI in 1610. They are less bright than the adjacent portions of the Sun's surface because they are not as hot. A typical sunspot temperature might be 4,500 K compared to the photosphere's temperature of about 5,800K. Sunspots appear to be made up of gases boiling up from the Sun's interior. A small sunspot may be about the size of Earth, while larger ones could hold several hundred or even thousands of Earth-sized planets. Extra-bright solar regions, called plages, often overlie sunspots. The number and size of sunspots appear to rise and fall through a fundamental 11-year cycle (or in an overall 22-year cycle, if polarity reversals in the Sun's magnetic field are considered). The greatest number occurs in years when the Sun's magnetic field is the most severely twisted (called sunspot maximum). Solar physicists think that sunspot migration causes the Sun's magnetic field to reverse its direction. It then takes another 22 years for the Sun's magnetic field to return to its original configuration.

A SOLAR FLARE is the sudden release of tremendous energy and material from the Sun. A flare may last minutes or hours, and it usually occurs in complex magnetic regions near sunspots. Exactly how or why enormous amounts of energy are liberated in solar flares is still unknown, but scientists think the process is associated with electrical currents generated by changing magnetic fields. The maximum number of solar flares appears to accompany the increased activity of the sunspot cycle. As a flare erupts, it discharges a large quantity of material outward from the Sun. This violent eruption also sends shock waves through the SOLAR WIND.

Data from space-based solar observatories have indicated that PROMINENCES (condensed streams of ionized hydrogen atoms) appear to spring from sunspots. Their looping shape suggests that these prominences are controlled by strong magnetic fields. About 100 times as dense as the solar corona, prominences can rise at speeds of hundreds of kilometers per second. Sometimes the upper end of a prominence curves back to the Sun's surface, forming a "bridge" of hot glowing gas hundreds of thousands of kilometers long. On other occasions the material in the prominence jets out and becomes part of the solar wind.

High-energy particles are released into heliocentric space by solar events, including very large solar flares called *anomalously large solar particle events* (ALSPEs). Because of their close association with infrequent large flares, these bursts of energetic particles are also relatively infrequent. However, solar flares, especially ALSPEs, represent a potential hazard to astronauts traveling in interplanetary space and working on the surfaces of the Moon and Mars.

See also FUSION; SPACE RADIATION ENVIRONMENT; STAR; ULYSSES MISSION; *YOHKOH* SPACECRAFT.

sundial An ancient astronomical instrument that uses the position of the shadow cast by the Sun to measure the passage of time during the day.

See also ANCIENT ASTRONOMICAL INSTRUMENTS.

sungrazer One of a group of long-period comets that have very small perihelion distances. Sometimes called *Kreutz sungrazers* after the German astronomer Heinrich Cral Kreutz (1854–1907), who studied them in 1888. With a most precarious astronomical existence, some sungrazer comets pass through the solar corona, while others venture too close to the Sun's surface and are torn apart.

sunlike stars Yellow main sequence stars with surface temperatures of 5,000 to 6,000 K; spectral type G stars.

See also STAR; SUN.

sunspot A relatively dark, sharply defined region on the Sun's surface marked by an umbra (darkest region) that is approximately 2,000 kelvins (K) cooler than the effective photospheric temperature of the Sun (i.e., 5,800 K). The umbra is surrounded by a less dark but sharply bounded region called the penumbra. The average sunspot diameter is about 3,700 km, but this diameter can range up to 245,000 km. Most sunspots are found in groups of two or more, but they can occur singly.

See also SUN.

sunspot cycle The approximately 11-year cycle in the variation of the number of sunspots. A reversal in the Sun's magnetic polarity also occurs with each successive sunspot cycle, creating a 22-year solar magnetic cycle.

sun-synchronous orbit A space-based sensor's view of Earth will depend on the characteristics of its orbit and the sensor's field of view (FOV). A sun-synchronous orbit is a special polar orbit that enables a spacecraft's sensor to maintain a fixed relation to the Sun. This feature is particularly important for environmental satellites (i.e., weather satellites) and multispectral imagery remote sensing satellites (such as the Landsat family of spacecraft). Each day a spacecraft in a sun-synchronous orbit around Earth passes over a certain area on the planet's surface at the same local time. One way to characterize sun-synchronous orbits is by the time the spacecraft cross the equator. These equator crossings (called NODES) occur at the same local time each day, with the descending (north-to-south) crossings occurring 12 hours (local time) from the ascending (south-to-north) crossings. In aerospace operations, the terms "A.M. polar orbiter" and "P.M. polar orbiter" are used to describe sun-synchronous satellites with morning and afternoon equator crossings, respectively.

An A.M. polar orbiter permits viewing of the land surface with adequate illumination, but before solar heating and the daily cloud buildup occurs. Such a "morning platform" also provides an illumination angle that highlights geological features. A P.M. polar orbiter provides an opportunity to study the role of developed clouds in Earth's weather and climate and provides a view of the land surface after it has experienced a good deal of solar heating.

A typical morning meteorological platform might orbit at an altitude of 810 km at an inclination of 98.86° and would have a period of 101 minutes. This morning platform would have its equatorial crossing time at approximately 0730 (local time). Similarly, an early-afternoon P.M. polar orbiter might orbit the Earth at an altitude of 850 km at an inclination of 98.70° and would have a period of 102 minutes. This early-afternoon platform would cross the equator at approximately 1330 (local time). Each satellite (i.e., the morning and afternoon platform) views the same portion of Earth twice each day. Therefore, the pair would provide environmental data collections with approximately six-hour gaps between each collection. In the United States the afternoon platform is usually considered the primary meteorological mission, while the morning platform is considered to provide supplementary and backup coverage.

See also ORBITS OF OBJECTS IN SPACE; WEATHER SATELLITE.

supercluster A cluster of galactic clusters—a large-scale physical feature of the universe that represents an enormously large grouping of matter. Superclusters are between 50 and 100 megaparsecs (Mpc) in size and contain the mass of perhaps 10^{15} to 10^{16} solar masses.

See also GALACTIC CLUSTER; UNIVERSE.

supergiant The largest and brightest type of star, with a luminosity of 10,000 to 100,000 times that of the Sun.

See also STAR.

superior planets Planets that have orbits around the Sun that lie outside Earth's orbit. These planets include Mars, Jupiter, Saturn, Uranus, Neptune, and Pluto.

superluminal With a speed greater than the speed of light, 3×10^8 meters per second (m/s).

supernova A catastrophic stellar explosion occurring near the end of a star's life in which the star collapses and explodes, manufacturing (by nuclear transmutation) the heavy elements that it spews out into space. During a supernova explosion the brightness of a star increases by a factor of several million times in a matter of days. Supernovas are fascinating celestial phenomena that are also extremely important for understanding the Milky Way galaxy. For example, supernovas heat up the interstellar medium, create and distribute heavy elements throughout the galaxy, and accelerate cosmic rays. Astrophysicists divide supernovas into two basic physical types, Type Ia and Type II.

The *Type Ia supernova* involves the sudden explosion of a white dwarf star in a binary star system. A white dwarf is the evolutionary endpoint for stars with masses up to about 5 times the mass of the Sun. The remaining white dwarf has a mass of about 1.4 times the mass of the Sun and is about the size of Earth. A white dwarf in a binary star system will draw material off its stellar companion if the two stars orbit close to each other. Mass capture happens because the white dwarf is a very dense object and exerts a very strong gravitational pull. Should the in-falling matter from a suitable companion star, such as a red giant, cause the white dwarf to exceed 1.4 solar masses (the Chandrasekhar limit), the white dwarf begins to experience gravitational collapse and creates internal conditions sufficiently energetic to support the thermonuclear fusion of elements such as carbon. The carbon and other elements that make up the white dwarf begin to fuse uncontrollably, resulting in an enormous thermonuclear explosion that involves and consumes the entire star. Astronomers sometimes call this type of supernova a *carbon-detonation supernova.*

The *Type II supernova* is also referred to as a *massive star supernova.* Stars that are five times or more massive than the Sun end their lives in a most spectacular way. The process starts when there is no longer enough fuel for the fusion process to occur in the core. When a star is functioning normally and in equilibrium, fusion takes place in its core, and the energy liberated by the thermonuclear reactions produces an outward pressure that combats the inward gravitational attraction of the star's great mass. But when a massive star that is going to become a supernova begins to die, it first swells into a red supergiant, at least on the outside. In the interior its core begins shrinking. As the star's core shrinks, the material becomes hotter and denser. Under these extremely heated and compressed conditions, a new series of thermonuclear reactions take place involving elements all the way up to, but not including, iron. Energy released from these fusion reactions temporarily halts the further collapse of the core.

However, this pause in the gravity-driven implosion of the core is only temporary. When the compressed core contains essentially only iron (Fe), all fusion ceases. (From nuclear physics scientists know that iron nuclei are extremely stable and cannot fuse into higher atomic number elements.) At this point the dying star begins the final phase of gravitational collapse. In less than a second the core temperature rises to more than 100 billion degrees as the iron nuclei are quite literally crushed together by the unrelenting influence of gravitational attraction. Then this "atom crushing" collapse abruptly halts due to the buildup of neutron pressure in the degenerate core material. The collapsing core recoils as it encounters this neutron pressure, and a rebound shock wave travels outward through the overlying material. The passage of this intense shock causes numerous nuclear reactions in the overlying outer material of the red supergiant, and the end result is a gigantic explosion. The shock wave also propels remnants of the original overlying material and any elements synthesized by various nuclear reactions out into space. Astronomers call the material that is exploded out into space a *supernova remnant* (SNR).

All that remains of the original star is a small, superdense core composed almost entirely of neutrons—a neutron star. Because of the sequence of physical processes just mentioned, astrophysicists sometimes call a Type II supernova an *implosion-explosion supernova.* If the original star was very massive (perhaps 15 or more solar masses) even neutrons are not able to withstand the relentless collapse of the core's degenerate matter, and a black hole forms.

The two different types of supernovas were originally classified on the basis of the presence or absence of hydrogen spectral lines. In the 1930s astrophysicists such as FRITZ ZWICKY observed that Type Ia supernovas do not exhibit hydrogen lines, while Type II supernovas do. In general, this early classification scheme based on spectral line observation agrees with the two basic explosive processes just described. Massive stars have atmospheres consisting of mostly hydrogen, while white dwarfs stars are bare, glowing, cinderlike stellar cores rich in elements such as carbon (the by-product of earlier hydrogen and helium fusion).

However, if the original dying star was so massive that its strong stellar wind had already blown off the hydrogen from its atmosphere by the time of the supernova explosion, then it, too, will not exhibit hydrogen spectral lines. In a somewhat confusing choice, astrophysicists often call this particular type of supernova a *Type 1b supernova*—despite the fact that such supernovas are technically within the Type II class of "massive star" supernovas. The discrepancy between the modern classification of supernovas (based on a true physical difference in how supernovas explode) and the historic classification of supernovas (based on early spectral line observations) shows how classification schemes used in science to describe new physical phenomena can change over time as scientists better understand the universe.

See also DEGENERATE MATTER; NEUTRON STAR; STARS.

Supernova 1987A (SN 1987A) A spectacular Type II supernova event observed in the Large Magellanic Cloud (LMC) starting on February 24, 1987. This was the first supernova bright enough to be seen by the naked eye (for observers in the Southern Hemisphere) since the supernova of 1604, which is also called Kepler's star. Because this event took place relatively close to Earth, scientists from around the world used a variety of space-based, balloon-borne, and ground-based instruments to examine the explosive death of a red supergiant star. For example, almost a day (some 20 hours) before SN 1987A was detected optically, giant underground neutrino detectors in both Japan and the United States simultaneously recorded a brief (13-second) burst of neutrinos from the supernova. This successful data collection was a major milestone in NEUTRINO ASTRONOMY and astrophysics because it marked the first time scientists collected information about a particular celestial object beyond the solar system that did not involve radiation within the electromagnetic spectrum. Balloon-borne and space-based gamma-ray detectors provided confirmation that heavier elements (such as cobalt) were also produced in this supernova. SN 1987A remained visible to the naked eye until the end of 1987, but the study of this interesting event did not end there. Today, scientists are using sophisticated space-based observatories such as the *CHANDRA X-RAY OBSERVATORY* to carefully analyze the expanding debris (*supernova remnant*) from this gigantic explosion.

See also NEUTRINO ASTRONOMY; SUPERNOVA.

supersonic Of or pertaining to speed in excess of the speed of sound.

See also SPEED OF SOUND.

supersonic nozzle A converging-diverging nozzle designed to accelerate a fluid to supersonic speed.

See also NOZZLE.

surface penetrator spacecraft A spacecraft probe designed to enter the surface of a celestial body, such as a comet or asteroid. The penetrator is capable of surviving a high-velocity impact and then making in-situ measurements of the penetrated surface. Data are sent back to a mother spacecraft for retransmission to scientists on Earth.

See also DEEP IMPACT; PENETRATOR.

surface rover spacecraft An electrically powered robot vehicle designed to explore a planetary surface. Depending on the size of the rover and its level of sophistication, this type of mobile craft is capable of semiautonomous to fully autonomous operation. A rover can perform a wide variety of exploratory functions, including the acquisition of multispectral imagery, soil sampling and analysis, and rock inspection and collection. Data are transmitted back to Earth either directly by the rover vehicle or via a lander spacecraft or orbiting mother spacecraft.

See also LUNAR ROVER(S); MARS EXPLORATION ROVER MISSION; *MARS PATHFINDER;* MARS SURFACE ROVER(S).

surface tension The tendency of a liquid that has a large cohesive force to keep its surface as small as possible, forming spherical drops. Surface tension arises from intermolecular forces and is manifested in such phenomena as the absorption of liquids by porous surfaces, the rise of water (and other fluids) in a capillary tube, and the ability of liquids to "wet" a surface.

See also HEAT PIPE.

surveillance In general, the systematic observation of places, persons, or things usually accomplished by optical, infrared, radar, or radiometric sensors. These sensors can be placed on a satellite to observe Earth or other regions of space (called *space-based surveillance*); installed on an aircraft platform (called *aerial surveillance*) to observe outer space (i.e., airborne astronomy), the surrounding atmosphere, or Earth's surface; or else situated on fixed or mobile terrestrial platforms to monitor the atmosphere and/or outer space (called *ground-based surveillance*). Surveillance includes tactical observations, strategic warning, and meteorological and environmental assessments.

See also DEFENSE SUPPORT PROGRAM.

surveillance satellite An Earth-orbiting military satellite that watches regions of the planet for hostile military activities, such as ballistic missile launches and nuclear weapon detonations.

See also DEFENSE SUPPORT PROGRAM.

Surveyor Project NASA's highly successful Surveyor Project began in 1960. It consisted of seven unmanned lander spacecraft that were launched between May 1966 and January 1968 as a precursor to the human expeditions to the lunar surface in the Apollo Project. These robot lander craft were used to develop soft-landing techniques, to survey potential Apollo mission landing sites, and to improve scientific understanding of the Moon.

The *Surveyor 1* spacecraft was launched on May 30, 1966, and soft-landed in the Ocean of Storms region of the Moon. It found that the load-bearing strength of the lunar soil was more than adequate to support the Apollo Project lander spacecraft (called the lunar excursion module, or LEM). This contradicted the then-prevalent hypothesis that

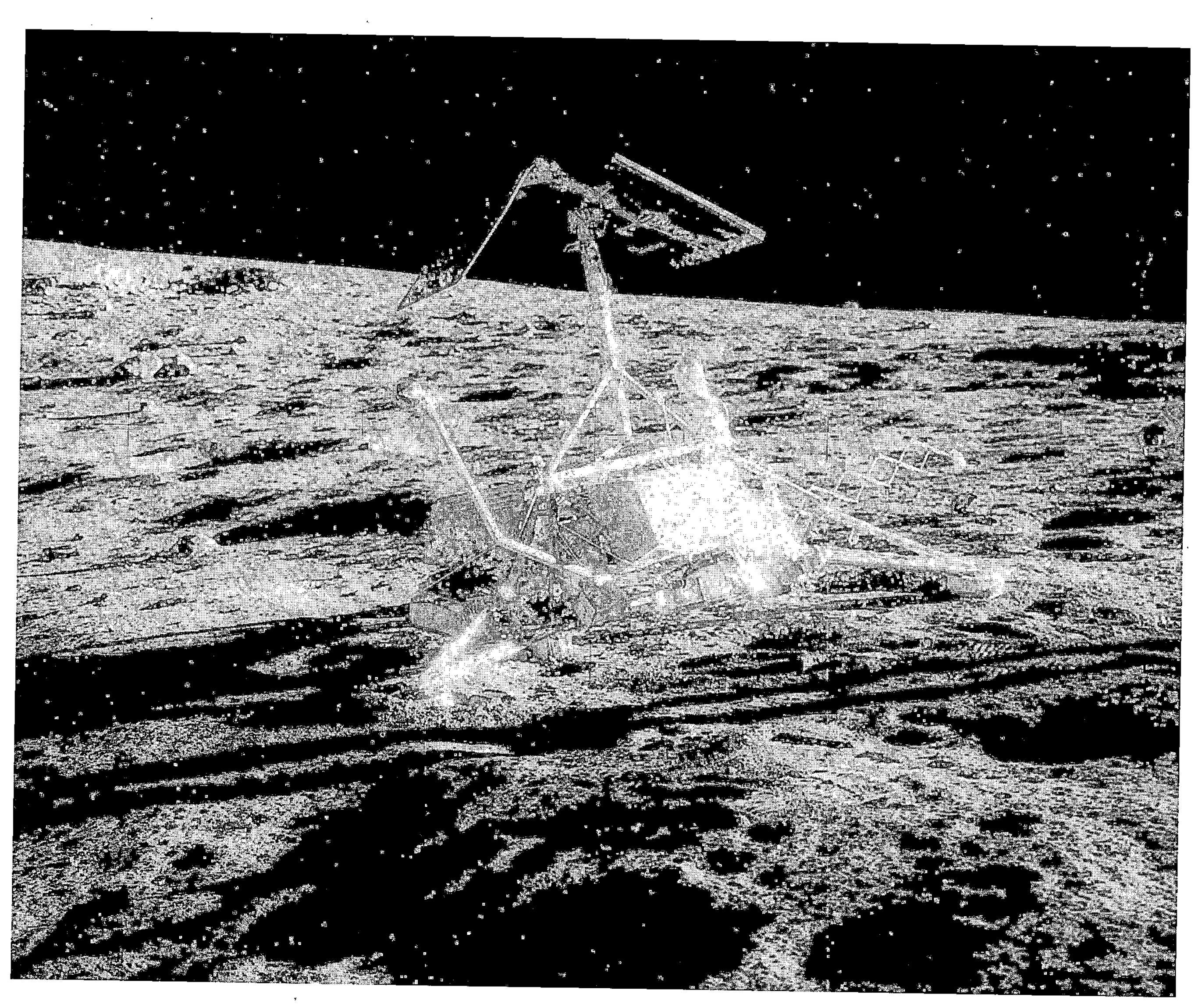

NASA's *Surveyor 3* spacecraft on the surface of the Moon. From 1966 to 1968 *Surveyor* spacecraft performed robotic explorations of the lunar surface in preparation for the human landings by the Apollo astronauts. This photograph was taken by two of the Apollo 12 mission astronauts, Charles Conrad Jr. and Alan L. Bean, during their excursions on the lunar surface in late November 1969. *(Courtesy of NASA)*

the LEM might sink out of sight in the fine lunar dust. The *Surveyor 1* spacecraft also telecast many pictures from the lunar surface.

The *Surveyor 3* spacecraft was launched on April 17, 1967, and soft-landed on the side of a small crater in another region of the Ocean of Storms. This robot spacecraft used a shovel attached to a mechanical "arm" to dig a trench and discovered that the load-bearing strength of the lunar soil increased with depth. It also transmitted many pictures from the lunar surface.

The *Surveyor 5* spacecraft was launched on September 8, 1967, and soft-landed in the Sea of Tranquility. An alpha particle–scattering device onboard this craft examined the chemical composition of the lunar soil and revealed a similarity to basalt on Earth.

The *Surveyor 6* was launched on November 7, 1967, and soft-landed in the Sinus Medii (Central Bay) region of the Moon. In addition to performing soil analysis experiments and taking many images of the lunar surface, this spacecraft also performed an extremely critical "hop experiment." NASA engineers on Earth remotely fired the *Surveyor*'s vernier rockets to launch it briefly above the lunar surface. The spacecraft's launch did not create a dust cloud and resulted only in shallow cratering. This important demonstration indicated that the Apollo astronauts could safely lift off from the lunar surface with their rocket-propelled craft (upper portion of the LEM) when their surface exploration mission was completed.

Finally, the *Surveyor 7* spacecraft was launched on January 7, 1968, and landed in a highland area of the Moon near the crater Tycho. Its alpha particle–scattering device showed that the lunar highlands contained less iron than the soil found in the mare regions (lunar plains). Numerous images of the lunar surface also were returned.

Despite the fact that the *Surveyor 2* and *4* spacecraft crashed on the Moon (rather than soft-landed and functioned), the overall Surveyor Project was extremely successful and paved the way for the human-crewed Apollo surface expeditions that occurred between 1969 and 1972.

See also APOLLO PROJECT; LANDER (SPACECRAFT); MOON; ROBOTICS IN SPACE.

sustainer engine A rocket engine that maintains the velocity of a rocket vehicle once the vehicle has achieved its intended (programmed) velocity by means of a more powerful booster engine or engines (which is/are usually jettisoned). A sustainer engine also is used to provide the small amount of thrust needed to maintain the speed of a spacecraft or orbital glider that has dipped into the atmosphere at perigee.

***Swift* (spacecraft)** NASA's *Swift* spacecraft lifted off aboard a Boeing Delta II rocket from pad 17-A at Cape Canaveral Air Force Station, Florida, on November 20, 2004. *Swift* has successfully begun its mission to study gamma-ray bursts and to identify their origins. *Swift* is a first-of-its-kind multiwavelength space-based observatory dedicated to the study of gamma-ray burst (GRB) science. Its three instruments will work together to observe GRBs and afterglows in the gamma-ray, X-ray, ultraviolet, and optical wavebands. The main mission objectives for *Swift* are to determine the origin of gamma-ray bursts, classify gamma-ray bursts and search for new types, determine how the GRB blast wave evolves and interacts with the surroundings, use gamma-ray bursts to study the early universe, and perform the first sensitive hard X-ray survey of the sky. This represents a swath of the electromagnetic spectrum more than a million times wider than what the *HUBBLE SPACE TELESCOPE* detects. During its planned two-year mission *Swift* is expected to observe more than 200 gamma-ray bursts, the most comprehensive study of GRB afterglows to date.

See also GAMMA-RAY BURST.

Swigert, John Leonard "Jack," Jr. (1931–1982) American *Astronaut* In April 1970 the astronaut Jack Swigert, Jr., flew as part of NASA's *Apollo 13* mission to the Moon. Some 55 hours into the translunar flight, an explosion within the Apollo spacecraft service module forced the crew to reconfigure the lunar excursion module (LEM) into a lifeboat. This life-threatening in-space emergency forced the crew to abort any attempt at landing in the programmed Fra Mauro area of the Moon. Although a space flight rookie, Swigert's skillful response to this crisis helped fellow astronauts JAMES ARTHUR LOVELL, JR., and FRED WALLACE HAISE, JR., survive the perilous journey and return to Earth safely.

See also APOLLO PROJECT; LUNAR EXCURSION MODULE (LEM).

Sycorax A lesser moon of Uranus discovered in 1997. Sycorax has a diameter of about 160 km and is in a retrograde orbit around the planet at an average distance of 12,175,000 km. The moon's retrograde orbit is characterized by a period of 1,284 days, an inclination of 153°, and an eccentricity of 0.509. Also called S/1997 U2 or UXVII.

See also URANUS.

synchronous orbit An orbit over the equator in which the orbital speed of a spacecraft or satellite matches exactly the rotation of Earth on its axis, so that the spacecraft or satellite appears to stay over the same location on Earth's surface.

See also GEOSYNCHRONOUS ORBIT.

synchronous rotation In astronomy, the rotation of a natural satellite (moon) about its primary in which the orbital period is equal to the period of rotation of the satellite about its own axis. As a consequence of this condition, the satellite (moon) always presents the same side (face) to the parent body. The Moon, for example, is in synchronous rotation around Earth. (However, libration allows slightly more than one hemisphere to be seen from Earth.) Sometimes called captured rotation.

synchronous satellite An equatorial west-to-east satellite orbiting Earth at an altitude of approximately 35,900 kilometers; at this altitude the satellite makes one revolution in 24 hours, synchronous with Earth's rotation.

See also GEOSYNCHRONOUS ORBIT; GEOSTATIONARY EARTH ORBIT.

synchrotron radiation Electromagnetic radiation (light, ultraviolet, or X-ray) produced by very energetic electrons as they spiral around lines of force in a magnetic field.

Syncom A family of NASA spacecraft launched in the 1960s to demonstrate the technologies needed to operate commercial communications satellites in geostationary Earth orbit. *Syncom 1* was launched in February 1963. After a successful launch and injection into a highly elliptical orbit, contact was lost with the spacecraft during the firing of its built-in orbit circularization rocket. *Syncom 2* was launched in July 1963 and provided important operational experience for communications satellites operating at synchronous orbit. *Syncom 3* was launched in August 1964 and successfully provided live coverage of the 1964 Olympic games from Japan.

See also COMMUNICATIONS SATELLITE.

synodic period The time required for a celestial body to return to the same apparent position relative to the Sun, taking Earth's motion into account; the time between consecutive conjunctions of two planets when viewed from Earth.

See also CONJUNCTION.

synthetic aperture radar (SAR) A radar system that correlates the echoes of signals emitted at different points along a satellite's orbit or an aircraft's flight path. For example, a SAR system on a spacecraft illuminates its target to the side of its direction of movement and travels a known distance in orbit while the reflected, phase shift–coded pulses are returned and collected. With extensive computer processing, this procedure provides the basis for synthesizing an antenna (aperture) on the order of kilometers in size. The highest resolution achievable by such a system is theoretically equivalent to that of a single large antenna as wide as the distance between the most widely spaced points along the orbit that are used for transmitting positions.

See also MAGELLAN MISSION; RADAR IMAGING; *RADARSAT.*

system integration process The process of uniting the parts (components) of a spacecraft, launch vehicle, or space platform into a complete and functioning aerospace system.

tachyon A hypothetical faster-than-light subatomic particle. Although no experimental evidence for the tachyon has yet been discovered, the existence of such a particle is not in conflict with the theory of relativity. Since the tachyon would exist only at speeds above the speed of light, it would have a positive, real mass. Within the general construct of relativity theory, when a tachyon lost energy, it would accelerate. Conversely, the faster a tachyon traveled, the less energy it would have. If the tachyon really exists, advanced alien civilizations might use it in some way to achieve more rapid interstellar communications.

See also RELATIVITY (THEORY).

taikonaut The suggested Chinese equivalent to astronaut and cosmonaut. *Taikong* is the Chinese word for space or cosmos, so the prefix "taiko-" assumes the same concept and significance as the use of "astro-" and "cosmo-" to form the words astronaut and cosmonaut.

See also *SHENZHOU 5*.

tail 1. The rear surfaces of an aircraft, aerospace vehicle, missile, or rocket. 2. The rear portion of a body, such as of an aerospace vehicle or rocket. 3. Short for the tail of a comet.

tailoring Modification of a basic solid rocket propellant by adjustment of the propellant properties to meet the requirements of a specific rocket motor.

takeoff 1. The ascent of a rocket vehicle as it departs from the launch pad at any angle. 2. The action of an aircraft as it becomes airborne. Compare this term to LIFTOFF, which refers only to the vertical ascent of a rocket or missile from its launch pad.

tandem launch The launching of two or more spacecraft or satellites using a single launch vehicle.

Tanegashima Space Center *See* JAPANESE AEROSPACE EXPLORATION AGENCY.

tank 1. A container incorporated into the structure of a liquid-propellant rocket from which a liquid propellant or propellants are fed into the combustion chamber(s). 2. A container incorporated into the structure of a nuclear rocket from which a monopropellant, such as liquid hydrogen, is fed into the nuclear reactor. 3. A ground-based or space-based container for the storage of liquid hydrogen, liquid oxygen, or other liquid propellants until they are transferred to a rocket's tanks or some other receptacle.

tankage The aggregate of the tanks carried by a liquid-propellant rocket or a nuclear rocket.

tap A unit of impulse intensity, defined as one dyne-second per square centimeter. 1 tap = 0.1 pascal-second.

target 1. Any object, destination, point, and so on toward which something is directed. 2. An object that reflects a sufficient amount of radiated signal to produce a return, or echo, signal on detection equipment. 3. The cooperative (usually passive) or noncooperative partner in a space rendezvous operation. 4. In military operations, a geographic area, complex, or installation planned for capture or destruction by military forces. 5. In intelligence usage, a country, area, installation, or person against which intelligence operations are directed. 6. In radar, (1) generally, any discrete object that reflects or retransmits energy back to the radar equipment; (2) specifically, an object of radar search or surveillance. 7. In nuclear physics, a material subjected to particle bombardment (such as in an accelerator) or neutron irradiation (such as in a nuclear reactor) in order to induce a nuclear reaction.

target acquisition The detection, identification, and location of a target in sufficient detail to permit the effective employment of weapons.

target spacecraft The nonmaneuvering spacecraft in rendezvous and proximity operations.

Taylor, Joseph H., Jr. (1941–) American *Radio Astronomer* In 1974, while performing research using the Arecibo Observatory in Puerto Rico, American radio astronomer Joseph H. Taylor, Jr., codiscovered the first binary pulsar, called PSR 1913 + 16, with his then doctoral student RUSSELL A. HULSE. Their discovery—a pair of neutron stars in a close binary system with one neutron star observable as a pulsar here on Earth—provided scientists with a very special natural deep space laboratory for investigating the modern theory of gravity and the validity of ALBERT EINSTEIN's general theory of relativity. Because their discovery had such great significance both Taylor and Hulse shared the 1993 Nobel Prize in physics.

telecommunications 1. In general, any transmission, emission, or reception of signs, signals, writings, images, sounds, or information of any nature by wire, radio, visual, or other electromagnetic or electro-optical systems. 2. In aerospace, the flow of data and information (usually by radio signals) between a spacecraft and an Earth-based communications system. A spacecraft has only a limited amount of power available to transmit a signal that sometimes must travel across millions or even billions of kilometers of space before reaching Earth. For example, an interplanetary spacecraft might have a transmitter that has no more than 20 watts of radiating power. One aerospace engineering approach is to concentrate all available power into a narrow radio beam and then to send this narrow beam in just one direction, instead of broadcasting the radio signal in all directions. Often this is accomplished by using a parabolic dish antenna on the order of 1 to 5 meters in diameter. However, even when these concentrated radio signals reach Earth, they have very small power levels. The other portion of the telecommunications solution is to use special, large-diameter radio receivers on Earth, such as are found in NASA's Deep Space Network. These sophisticated radio antennas are capable of detecting the very-low-power signals from distant spacecraft.

In telecommunications, the radio signal transmitted to a spacecraft is called the UPLINK. The transmission from the spacecraft to Earth is called the DOWNLINK. Uplink or downlink communications may consist of a pure radio-frequency (RF) tone called a carrier, or these carriers may be modified to carry information in each direction. Commands transmitted to a spacecraft sometimes are referred to as an *upload.* Communications with a spacecraft involving only a downlink are called ONE-WAY COMMUNICATIONS. When an uplink signal is being received by the spacecraft at the same time that a downlink signal is being received on Earth, the telecommunications mode is often referred to as *two-way.*

Spacecraft carrier signals usually are modulated by shifting each waveform's phase slightly at a given rate. One scheme is to modulate the carrier with a frequency, for example, near 1 megahertz (1 MHz). This 1-MHz modulation is then called a *subcarrier.* The subcarrier is modulated to carry individual phase shifts that are designated to represent binary ones (1s) and zeros (0s)—the spacecraft's telemetry data. The amount of phase shift used in modulating data onto the subcarrier is referred to as the *modulation index* and is measured in degrees. This same type of communications scheme is also on the uplink. Binary digital data modulated onto the uplink are called *command data.* They are received by the spacecraft and either acted upon immediately or stored for future use or execution. Data modulated onto the downlink are called TELEMETRY and include science data from the spacecraft's instruments and spacecraft state-of-health data from sensors within the various onboard subsystems (e.g., power, propulsion, thermal control, etc.).

Demodulation is the process of detecting the subcarrier and processing it separately from the carrier, detecting the individual binary phase shifts, and registering them as digital data for further processing. The device used for this is called a *modem,* which is short for *mo*dulator/*dem*odulator. These same processes of modulation and demodulation are commonly used with Earth-based computer systems and facsimile (fax) machines to transmit data back and forth over a telephone line. For example, if you have used a personal computer to "chat" over the Internet, your modem used a familiar audio frequency carrier that the telephone system could handle.

The dish-shaped *high-gain antenna* (HGA) is the type of spacecraft antenna mainly used for communications with Earth. The amount of gain achieved by an antenna refers to the amount of incoming radio signal power it can collect and focus into the spacecraft's receiving subsystems. In the frequency ranges used by spacecraft, the high-gain antenna incorporates a large parabolic reflector. Such an antenna may be fixed to the spacecraft bus or steerable. The larger the collecting area of the high-gain antenna, the higher the gain and the higher the data rate it will support. However, the higher the gain, the more highly directional the antenna becomes. Therefore, when a spacecraft uses a high-gain antenna, the antenna must be pointed within a fraction of a degree of Earth for communications to occur. Once this accurate antenna pointing is achieved, communications can take place at a high rate over the highly focused radio signal.

The *LOW-GAIN ANTENNA* (LGA) provides wide-angle coverage at the expense of gain. Coverage is nearly omnidirectional except for areas that may be shadowed by the spacecraft structure. The low-gain antenna is designed for relatively low data rates. It is useful as long as the spacecraft is relatively close to Earth (e.g., within a few astronomical units). Sometimes a spacecraft is given two low-gain antennas to provide full omnidirectional coverage, since the second LGA will avoid the spacecraft structure "blind spots" experienced by the first LGA. The low-gain antenna can be mounted on top of the high-gain antenna's subreflector.

The *medium-gain antenna* (MGA) represents a compromise in spacecraft engineering. It provides more gain than the low-gain antenna and has wider-angle antenna-pointing accuracy requirements (typically 20° to 30°) than the high-gain antenna.

See also DEEP SPACE NETWORK.

telemetry 1. The process of making measurements at one point and transmitting the data to a distant location for evaluation and use. 2. Data modulated onto a spacecraft's communications downlink, including science data from the spacecraft's instruments and subsystem state-of-health data.

See also TELECOMMUNICATIONS.

teleoperation The process by which a human worker, usually in a safe and comfortable environment, operates a versatile robot system that is at a distant, often hazardous location. Communications can be accomplished with radio signals, laser beams, or even "cable" if the distance is not too great and the deployed cable does not interfere with operations or safety. For example, an astronaut onboard the space shuttle (or the *INTERNATIONAL SPACE STATION*) might "teleoperate" at a safe distance using radio signals a free-flying space robot retrieval system. In one space operations scenario the teleoperated space robot rendezvouses with a damaged, potentially hazardous spacecraft. After initial rendezvous the human operator uses instruments on the robot system to inspect the target space object. Then, if it appears that repairs can be performed safely, the human worker uses the space robot to capture it and bring it closer to the shuttle (or space station). Otherwise, again through teleoperations, the human operator uses the retrieval robot system to move the derelict spacecraft to a designated parking (storage) orbit or else to put the hunk of "space junk" on a reentry trajectory that will cause it to plunge into Earth's upper atmosphere and burn up harmlessly.

See also ROBOTICS IN SPACE.

teleportation A concept used in science fiction to describe the instantaneous movement of material objects to other locations in the universe.

See also SCIENCE FICTION.

telepresence The use of telecommunications, interactive displays, and a collection of sensor systems on a robot (which is at some distant location) to provide the human operator a sense of being present where the robot system actually is located. Depending on the level of sophistication in the operator's workplace as well as on the robot system, this telepresence experience can vary from a simple "steroscopic" view of the scene to a complete virtual reality activity in which sight, sound, touch, and motion are provided. Telepresence actually combines the technologies of virtual reality with robotics. Some day in the not too distant future, human controllers on Earth, wearing sensor-laden bodysuits and three-dimensional viewer helmets, will use telepresence actually to "walk" and "work" on the Moon and other planetary bodies through their robot surrogates. Also called virtual residency.

See also ROBOTICS IN SPACE; TELEOPERATION; VIRTUAL REALITY.

telescience A mode of scientific activity in which a distributed set of users (investigators) can interact directly with their instruments, whether in space or ground facilities, with databases, data handling, and processing facilities, and with each other.

telescope In general, a device that collects electromagnetic radiation from a distant object so as to form an image of the object or to permit the radiation to be analyzed. Optical (or astronomical) telescopes can be divided into two main classes, *refracting telescopes* and *reflecting telescopes*. The oldest optical telescope is the refracting telescope, first constructed in Holland in 1608 and then developed for astronomical use by GALILEO GALILEI in 1609. In its simplest form the refracting telescope (or *refractor*) uses a converging lens, often called the primary lens, to gather and focus incoming light onto an eyepiece, which then magnifies the image. Galileo used a diverging lens in the eyepiece of his GALILEAN TELESCOPE, while JOHANNES KEPLER created an improved telescope in about 1611 by replacing the concave (negative) lens in the eyepiece with a convex (positive) lens. In the reflecting telescope, invented by SIR ISAAC NEWTON around 1668, a primary concave mirror located at the bottom of the telescope tube is used to collect and focus incoming light back up to a small secondary mirror, usually at a 45° angle to the incident light beams. This secondary mirror then reflects the incident light into an eyepiece mounted on the side of the telescope tube, which magnifies the image. A variety of optical telescopes consisting of various combinations of mirrors and lenses have since been developed, including the Cassegrain telescope and Schmidt telescope. An optical telescope often is described in terms of its resolving power, aperture, and light-gathering power.

In addition to optical telescopes, Earth-based observatories use radio telescopes and (in limited locations such as high mountaintops) INFRARED TELESCOPES. However, space-based observatories unhampered by the attenuation of Earth's atmosphere also include infrared telescopes, *gamma-ray telescopes, X-ray telescopes,* and *ultraviolet telescopes.*

See also CASSEGRAIN TELESCOPE; *CHANDRA X-RAY OBSERVATORY; COMPTON GAMMA-RAY OBSERVATORY; HUBBLE SPACE TELESCOPE; INFRARED ASTRONOMICAL SATELLITE;* KEPLERIAN TELESCOPE; HANS LIPPERSHEY; REFLECTING TELESCOPE; REFRACTING TELESCOPE; ZUCCHI, NICCOLO.

Telesto A small satellite of SATURN that was discovered in 1980 as a result of ground-based observations. Telesto orbits Saturn at a distance of 294,660 km and shares this orbit with Tethys and Calypso. Telesto is an irregularly shaped moon that is about an average of 25 km in diameter. It has an orbital period of 1.89 days. The two smaller moons, Telesto and Calypso, orbit in the Lagrangian points of the larger moon, Tethys. This means that Telesto is 60° ahead and Calypso 60° behind Tethys in their co-orbital travels around Saturn.

See also SATURN.

Television and Infrared Observation Satellite (TIROS) The TIROS series of weather satellites carried special television cameras that viewed Earth's cloud cover from a 725-km-altitude orbit. The images telemetered back to Earth provided meteorologists with a new tool, a *nephanalysis,* or cloud chart. On April 1, 1960, *TIROS 1,* the first true weather satellite, was launched into a near-equatorial orbit. By 1965 nine more TIROS satellites were launched. These spacecraft had progressively longer operational times and carried infrared radiometers to study Earth's heat distribution. Several were placed in polar orbits to increase cloud picture coverage. *TIROS 8* had the first Automatic Picture Transmission (APT) equipment. *TIROS 9* and *10* were test satellites of improved spacecraft configurations for the Tiros Operational

Satellite (TOS) system. Operational use of the TOS satellites began in 1966. They were placed in sun-synchronous (polar) orbits so they could pass over the same location on Earth's surface at exactly the same time each day. This orbit enabled meteorologists to view local cloud changes on a 24-hour basis. Several Improved TOS Satellites (ITOS) were launched in the 1970s. The ITOS spacecraft served as workhorses for meteorologists of the National Oceanographic and Atmospheric Administration (NOAA), which was responsible for their operation.

See also SUN-SYNCHRONOUS ORBIT; WEATHER SATELLITE.

temperature (symbol: T) A thermodynamic property that determines the direction of heat (thermal energy) flow. From the laws of thermodynamics, when two objects or systems are brought together, heat naturally will flow from regions of higher temperature to regions of lower temperature. In statistical mechanics, temperature can be considered as a "macroscopic" measurement of the overall kinetic energy of the individual atoms and molecules of a substance or body.

See also ABSOLUTE TEMPERATURE.

Tenth Planet Some astronomers have speculated that the currently estimated mass of the Pluto-Charon system is too small to account for subtle perturbations observed in the orbits of Uranus and Neptune. Although the specific source of these perturbations is not presently known, one hypothesis that has occasionally surfaced is that there is a massive object (perhaps a few times more massive than Earth) circling the Sun far beyond the orbit of Neptune. The speculated Tenth Planet has not been detected, and today the Tenth Planet hypothesis is not widely accepted as an adequate explanation of the perturbation data. Instead, astronomers now view the large icy objects within the Kuiper Belt, such as Quaoar, as better candidates.

See also QUAOAR.

tera- (symbol: T) A prefix in the International System of Units meaning multiplied by 10^{12}.

Tereshkova, Valentina (1937–) Russian *Cosmonaut* The cosmonaut Valentina Tereshkova holds the honor of being the first woman to travel in outer space. She accomplished this feat on June 16, 1963, by riding the *Vostok 6* spacecraft into orbit. During this historic mission she completed 48 orbits of Earth. Upon her return she received the Order of Lenin and was made a hero of the Soviet Union.

terminal 1. A point at which any element in a circuit may be connected directly to one or more other elements. 2. Pertaining to a final condition or the last division of something, such as terminal guidance or terminal ballistics. 3. In computer operations, an input/output (I/O) device (or station) usually consisting of a video screen or printer for output and a keyboard for input.

terminal guidance 1. The guidance applied to a missile between midcourse guidance and its arrival in the vicinity of the target. 2. With respect to an interplanetary spacecraft, the guidance and navigation performed during its approach to the target planet, which is accomplished by observing the angular position and motion and possibly the apparent size of the target body.

See also GUIDANCE SYSTEM.

terminal phase The final phase of a ballistic missile trajectory, lasting about a minute or less, during which warheads reenter the atmosphere and detonate at their targets.

See also BALLISTIC MISSILE DEFENSE.

terminator The boundary line separating the illuminated (i.e., sunlit) and dark portions of a nonluminous celestial body, such as the Moon.

Terra A Latin word meaning "Earth."

terrae The highland regions of the Moon. Typically, the terrae are heavily cratered with large, old-appearing craters on a rugged surface with high albedo.

See also LATIN SPACE DESIGNATIONS; MOON.

terraforming The proposed large-scale modification or manipulation of the environment of a planet, such as Mars or Venus, to make it more suitable for human habitation. Also called planetary engineering.

terran Of or relating to planet Earth; a native of planet Earth.

***Terra* (spacecraft)** The first in a new family of sophisticated NASA Earth-observing spacecraft successfully placed into polar orbit on December 18, 1999, from Vandenberg Air Force Base, California. The five sensors aboard *Terra* are designed to enable scientists to simultaneously examine the world's climate system. The instruments comprehensively observe and measure the changes of Earth's landscapes, in its oceans, and within the lower atmosphere. One of the main objectives is to determine how life on Earth affects and is affected by changes within the climate system, with a special emphasis on better understanding the global carbon cycle. Formerly called EOS-AM1, this morning equator-crossing platform has a suite of sensors designed to study the diurnal properties of cloud and aerosol radiative fluxes. Another cluster of instruments on the spacecraft is addressing issues related to air-land exchanges of energy, carbon, and water. Other spacecraft, including *Aqua* and *Aura,* have joined *Terra* as part of NASA's comprehensive Earth system science effort.

See also *AQUA* SPACECRAFT; *AURA* SPACECRAFT; EARTH-OBSERVING SPACECRAFT; EARTH SYSTEM SCIENCE.

terrestrial Of or relating to planet Earth; an inhabitant of planet Earth.

terrestrial environment Earth's land area, including its human-made and natural surface and subsurface features and its interfaces and interactions with the atmosphere and the oceans.

See also EARTH.

Terrestrial Planet Finder (TPF) A planned NASA mission (launch date during or after 2010) whose goal is to take "family portraits" of other planetary systems—that is, to image neighboring stars and any planets around them. In the search for Earthlike (terrestrial) planets, the bright light from parent stars needs to be cancelled so scientists can observe any dim planets that might orbit these stars. One concept of the Terrestrial Planet Finder (TPF) involves a 100-meter-long spacecraft that carries several precisely located but widely separated telescopes, all functioning together as an optical interferometer. Light collected from these telescopes would be cleverly combined so the long spacecraft would act like a giant telescope. Using the TPF scientists would search for and study the atmospheres of candidate extrasolar planets to see if any these planets could support life. Some exobiologists even hope that the TPF would allow them to catch a spectroscopic glimpse of whether life already exists on an alien planet. Seasonal variations in the level of certain gases in a planet's atmosphere—such as water vapor, carbon dioxide, ozone, and methane—would suggest to exobiologists that some type of biological activity, such as photosynthesis, could be taking place on an extrasolar planet within the ecosphere of a distant star. Should an even longer interferometer be needed to accomplish these searches, then several spacecraft flying in precise formation and each carrying a telescope might also be used to form a giant "virtual" interferometer.

See also ECOSPHERE; EXTRASOLAR PLANETS; LIFE IN THE UNIVERSE.

terrestrial planets In addition to Earth itself, the terrestrial (or inner) planets include Mercury, Venus, and Mars. These planets are similar in their general properties and characteristics to the Earth; that is, they are small, relatively high-density bodies composed of metals and silicates with shallow (or no) atmospheres as compared to the gaseous outer planets.

See also EARTH; MARS; MERCURY; VENUS.

Mariner: The Family of NASA Spacecraft that Changed Our Understanding of the Inner Solar System

Since the start of the space age robot spacecraft have played a very important role in the exploration of the solar system. Between 1962 and late 1973 NASA's Jet Propulsion Laboratory designed and constructed 10 spacecraft named Mariner. Their purpose was to explore the inner solar system, visiting the terrestrial planets Venus, Mars, and Mercury for the first time and then returning to Venus and Mars for close-up observation. The terrestrial planets are similar in their general properties and characteristics to Earth—that is, they are small, relatively high-density bodies composed of metals an silicates with shallow (or negligible) atmospheres as compared to the gaseous giant planets of the outer solar system. In a series of trail-blazing missions these early robot explorers helped to change many long-held, even romantic, perceptions about Earth's nearest planetary neighbors and in so doing greatly revised our scientific understanding of the terrestrial worlds of the inner solar system.

The 1962 *Mariner 2* flyby of Venus was the first successful interplanetary mission achieved by any nation. The 1965 *Mariner 4* flyby of Mars was the first successful attempt by a robot spacecraft to collect close-up photographs of another planet. *Mariner 9,* the next-to-last spacecraft in the series, became the first spacecraft to orbit another planet when it went into orbit around Mars in 1971 and began its one-year-duration mapping and measurement mission. The final mission in the series, *Mariner 10,* flew past Venus and then went on to Mercury, performing a total of three flybys of the innermost planet. After more than four decades of space exploration, the *Mariner 10* encounters with Mercury remain the only close-up visits to this elusive and hard to reach planet.

For untold millennia human beings gazed up at the night sky wondering, speculating, and observing. Ancient astronomers noticed that certain points of light appeared to move relative to the "fixed" lights. The early Greek astronomers called these particularly noticeable five objects *planetes,* a word that means "wanders." Only Mercury, Venus, Mars, Jupiter, and Saturn are observable to the naked eye. Before the arrival of each successful Mariner spacecraft, there was much speculation in both the scientific literature and science fiction about the nature of these interesting celestial objects, especially about how "Earthlike" the so-called terrestrial planets really were. Then, in a little more than a decade, this modest family of relatively small robot explorers transformed our understanding of the inner solar system, shattering many popular myths in the process.

The early Greek astronomers considered Venus both the evening star Hesperos and the morning star Phosphoros. Venus received its current name from the Romans when they honored their goddess of love and beauty. In the not too distant past it was quite popular to imagine Venus as literally Earth's planetary twin. Pre–space age scientists reasoned that since Venus had a diameter, density, and gravity that were only slightly less than those of Earth, the cloud-enshrouded planet must be physically similar. They were encouraged in these speculations by the presence of an obviously dense atmosphere and the fact that Venus was a little nearer to the Sun. Stimulated by such scientific speculations, romantic visions of a neighboring planet with oceans, tropical forests, giant reptiles, and even primitive humans made frequent appearances in popular science fiction stories during the first half of the 20th century.

However, on December 14, 1962, the *Mariner 2* spacecraft passed Venus at a range of 34,762 km and became the first spacecraft of any nation to accomplish a fully successful interplanetary mission. During a 42-minute scan of the planet *Mariner 2* gathered data about the Venusian atmosphere and surface before continuing on to a heliocentric orbit. The *Mariner 2* flyby provided the first direct scientific evidence that Venus was not a "tropical twin" of Earth. Since that historic encounter, visits to Venus by numerous other American and Russian spacecraft have completely dispelled any pre–space age romantic fantasies that this cloud-enshrouded planet was a prehistoric world that mirrored a younger Earth. Except for a few physical similarities involving size and surface gravity, Earth and Venus are very different worlds. For example, the surface temperature on Venus approaches 500°C, making its surface the hottest in the solar system, even hotter than Mercury. In addition, the atmospheric pressure on the Venusian surface is more than 90 times that of Earth. The planet has no surface water, and its dense, hostile atmosphere, with sulfuric acid clouds and an overabundance of carbon dioxide (about 96 percent), represents a runaway greenhouse of disastrous proportions.

Throughout human history the Red Planet, Mars, has remained at the center of astronomical and philosophical interest. The ancient Babylonians followed the motions of this wandering red light across the night sky and named it Nergal after their god of war. The ancient Greeks called the fourth planet from the Sun Ares for their god of war. Similarly, the Romans honored their god of war by giving the planet its present name. The presence of an atmosphere, polar caps, and changing patterns of light and dark on the surface caused many pre–space age astronomers and scientists to consider Mars an "Earthlike planet"—the possible abode of extraterrestrial life. In fact, when Orson Welles broadcast a radio drama in 1938 based on H. G. Wells's science-fiction classic *War of the Worlds,* enough people believed the report of invading Martians to create a near panic in some parts of the eastern United States. However, on July 14, 1965, the *Mariner 4* spacecraft flew past Mars, collecting the first close-up photographs of another planet. Over the past four decades an armada of sophisticated robot spacecraft—flybys, orbiters, landers, and rovers—have continued to shatter the romantic, pre–space age myths of a race of ancient Martians struggling to bring water to the more productive regions of a dying world. Instead, spacecraft-derived data have shown that the Red Planet is actually a "halfway" world. The *Mariner 5, 6,* and *7* spacecraft provided photographs of Mars, while the *Mariner 9* spacecraft became the first artificial satellite of Mars. Data from these and many other missions suggest that part of the Martian surface is ancient, like the surfaces of the Moon and Mercury, while part is more evolved and Earthlike, including tantalizing hints that liquid water flowed freely on Mars in the past. While the early Mariner spacecraft photographs did not show ancient canals and cities, they greatly increased scientific knowledge about the planet. Today much more sophisticated orbiter, lander, and rover spacecraft continue the trail-blazing Mariner spacecraft efforts to study Mars and to assist in the search for life (extinct or existent). In 2004, for example, two NASA robot spacecraft, called *Mars Spirit Rover* and *Mars Opportunity Rover,* simultaneously conducted detailed surface investigations of different regions of the Red Planet.

Two years after *Mariner 9*'s pioneering orbital investigations of Mars, the last Mariner took off from Cape Canaveral and performed the first successful two-planet exploration by a single spacecraft. With the scorched inner planet Mercury as its ultimate target, *Mariner 10* was the first spacecraft to demonstrate the use of the gravity assist technique, swinging by the planet Venus in such a precise way as to bend its trajectory and place the spacecraft on a flight path to the innermost planet, Mercury. Before moving on to the airless, small, and heavily cratered Mercury, *Mariner 10* used instrumentation equipped with a near-ultraviolet filter to produce the first clear images of the Venusian chevron cloud. The gravity assist maneuver enabled *Mariner 10* to return at six-month intervals for three mapping passes over the planet.

Before the *Mariner 10* mission in 1974, Mercury's surface features were little more than a blur to Earth-bound observers. *Mariner 10* experienced the first of its three flyby encounters of Mercury on March 29, 1974, at a distance of about 700 km. Its high-resolution photographs, covering about 50 percent of the planet's surface, gave scientists their first-ever close-up look at the innermost world of the solar system. They discovered that the surface of Mercury closely resembles our Moon. Impact craters cover the majority of the planet, but unlike the Moon, Mercury's cratered upland regions are covered with large areas of smooth plains. *Mariner 10* images showed that the most distinguishing features on Mercury's surface are scarps, or long cliffs. These wind across Mercury's surface for tens to hundreds of kilometers and range from 100 meters to more than 1.5 kilometers in height. The largest surface feature imaged by the *Mariner 10* mission was the Caloris basin. Resembling a gigantic bull's-eye, the Caloris basin is a multiringed impact basin some 1,340 km across, almost one-quarter of the full diameter of the planet.

Space technology and the practice of planetary science have come a long way since the Mariner spacecraft first explored the inner solar system. This marvelous family of robot spacecraft introduced such important new space technology developments as the three-axis stabilized spacecraft and the use of the gravity assist technique. Each of the Mariner projects was designed to have two spacecraft launched on separate rockets, just in case there were difficulties with the then nearly untried launch vehicles that were rapidly being adapted for space exploration missions from duty as military missiles. Although *Mariner 1, Mariner 3,* and *Mariner 8* were lost during launch, their backups proved successful. Save for the three launch accidents noted, no Mariner spacecraft were lost in later flight to their destination planets or before completing their scientific missions.

Terrier (RIM-2) A U.S. Navy surface-to-air missile with a solid-fuel rocket motor. It is equipped with radar beam rider or homing guidance and a nuclear or nonnuclear warhead.

Terrier land weapon system A surface-to-air missile system using the Terrier RIM-2B and Terrier RIM-2C missiles with ground-launching and guidance equipment developed specifically for land operations. This equipment is a lighter and land-mobile version of the U.S. Navy Terrier system.

tesla (symbol: T) The SI unit of magnetic flux density. It is defined as one weber of magnetic flux per square meter. The unit is named in honor of Nikola Tesla (1870–1943), a Croatian-American electrical engineer and inventor.

$$1 \text{ tesla} = 1 \text{ weber}/(\text{meter})^2 = 10^4 \text{ gauss}$$

test bed 1. A base, mount, or frame within or upon which a piece of equipment, such as an engine or sensor system, is secured for testing. 2. A flying test bed.

test chamber A place or room that has special characteristics where a person or device is subjected to experiment, such as an altitude chamber, wind tunnel, or acoustic chamber.

test firing The firing of a rocket engine, either live or static, with the purpose of making controlled observations of the engine or of an engine component.

test flight A flight to make controlled observations of the operation or performance of an aircraft, aerospace vehicle, rocket, or missile.

Tethys A major moon of Saturn discovered by GIOVANNI DOMENICO CASSINI in 1684. Tethys has a diameter of 1,050 km and orbits Saturn at a distance of 294,660 km with a period of 1.89 days. With its relatively low density (about 1.2 gm/cm^3, or 1,200 kg/m^3), this moon appears similar to Dione and Rhea in that they are all most likely composed primarily of water ice. A giant (about 400-km-diameter) impact crater, called Odysseus, is one dominant surface feature; another is a huge canyon system called the Ithaca Chasma that extends from near the moon's north pole to its equator and then on to the moon's south. Tethys co-orbits with two smaller moons—Telesto, which occupies the leading Lagarangian point 60° in front of Tethys, and Calypso, which occupies the trailing Lagarangian point 60° behind Tethys's.

See also SATURN.

Thalassa A small (about 100-km average diameter), irregularly shaped moon of Neptune discovered in 1989 as a result of the *Voyager 2* encounter. Thalassa orbits Neptune at a distance of 50,075 km with a period of 0.311 days.

See also NEPTUNE.

theater ballistic missile (TBM) A ballistic missile with limited range, typically less than 400 kilometers, that is capable of carrying either a conventional high-explosive, biological, chemical, or nuclear warhead.

See also BALLISTIC MISSILE DEFENSE; THEATER DEFENSE.

theater defense Defense against nuclear weapons on a regional level (e.g., Europe, Japan, Israel) rather than at the strategic level (i.e., globally or involving the United States and Russia).

Theater High Altitude Area Defense (THAAD) The Theater High Altitude Area Defense (THAAD) system is one element of the Missile Defense Agency's terminal defense system against hostile ballistic missiles launched by rogue nations in politically unstable regions of the world. The THAAD system represents a land-based upper-tier ballistic missile defense (BMD) system that will engage short- and medium-range ballistic missiles, including those armed with weapons of mass destruction (WMD). THAAD is the only BMD system designed for intercepts inside and outside Earth's atmosphere and has rapid mobility to defend against ballistic missile threats anywhere in the world. The THAAD system consists of four major components: truck-mounted launchers, interceptors, a radar system, and a battle management/command and control system. The kinetic energy interceptor consists of a single-stage booster and a kinetic kill vehicle that destroys its target through the force of kinetic energy upon impact. The first successful body-to-body intercept of a ballistic missile target was achieved on June 10, 1999.

See also BALLISTIC MISSILE DEFENSE.

Thebe A small (about 100-km-diameter) inner moon of Jupiter that orbits the giant planet at a distance of 222,000-km with a period of 0.675 days. Discovered in 1979 as a result of the *Voyager 1* encounter. Also called Jupiter XIV and S/1979 J2.

See also JUPITER.

theorem of detailed cosmic reversibility A premise developed by Francis Crick and Leslie Orgel in support of their directed panspermia hypothesis. This theorem states that if we can now contaminate another world in our solar system with terrestrial microorganisms, then it is also reasonable to assume that an intelligent alien civilization could have developed the advanced technologies needed to "infect," or seed, the early prebiotic Earth with spores, microorganisms, or bacteria.

See also EXTRATERRESTRIAL CONTAMINATION; PANSPERMIA.

thermal conductivity (symbol: k) An intrinsic physical property of a substance, describing its ability to conduct heat (thermal energy) as a consequence of molecular motion. Typical units for thermal conductivity are joules/(second-meter-kelvin).

thermal control Regulation of the temperature of a space vehicle is a complex problem because extreme temperatures are encountered during a typical space mission. In the vacuum environment of outer space radiative heat transfer is the only natural means to transfer thermal energy (heat) into and out of a space vehicle. In some special circumstances a gaseous or liquid working fluid can be "dumped" from the space vehicle to provide a temporary solution to a transient heat load, but this is an extreme exception rather than the general design approach. The overall thermal energy balance for a space vehicle near a planetary body is determined by several factors: (1) thermal energy sources within the space vehicle itself; (2) direct solar radiation (the Sun has a characteristic blackbody temperature of about 5,770 kelvins (K); (3) direct thermal (infrared) radiation from the planet (e.g., Earth has an average surface temperature of about 288 K; (4) indirect (reflected) solar radiation from the planetary body (e.g., Earth reflection); and (5) vehicle emission, that is, radiation from the surface of the space vehicle to the low-temperature sink of outer space (deep space has a temperature of about 3 K).

Under these conditions thermally isolated portions of a space vehicle in orbit around Earth could encounter temperature variations from about 200 K, during Earth-shadowed or darkness periods, to 350 K, while operating in direct sunlight. Spacecraft materials and components can experience thermal fatigue due to repeated temperature cycling during such extremes. Consequently, aerospace engineers need to exercise great care in providing the proper thermal control for a spacecraft. Radiative heat transport is the principal mechanism for heat flow into and out of the spacecraft, while conduction heat transfer generally controls the flow of heat within the spacecraft.

There are two major approaches to spacecraft thermal control, passive and active. Passive thermal control techniques include the use of special paints and coatings, insulation blankets, radiating fins, Sun shields, heat pipes, and careful selection of the space vehicle's overall geometry (i.e., both the external and internal placement of temperature-sensitive components). Active thermal control techniques include the use of heaters (including small radioisotope sources) and coolers, louvers, shutters, and the closed-loop pumping of cryogenic materials.

An open-loop flow (or "overboard dump") of a rapidly heated working fluid might be used to satisfy a one-time or occasional special mission requirement to remove a large amount of thermal energy in a short period of time. Similarly, a sacrificial ablative surface could be used to handle a singular, large transitory external heat load, but these transitory (essentially one-shot) thermal control approaches are the exception rather than the aerospace engineering norm.

For interplanetary spacecraft aerospace engineers often use passive thermal control techniques such as surface coatings, paint, and insulation blankets to provide an acceptable thermal environment throughout the mission. Components painted black will radiate more efficiently. Surfaces covered with white paint or white thermal blankets will reflect sunlight effectively and protect the spacecraft from excessive solar heating. Gold (i.e., gold-foil surfaces) and quartz mirror tiles also are used on the surfaces of special components. Active heating can be used to keep components within tolerable temperature limits. Resistive electric heaters, controlled either autonomously or via command, can be applied to special components to keep them above a certain minimum allowable temperature during the mission. Similarly, radioisotope heat sources generally containing a small quantity of plutonium-238 can be installed where necessary to provide "at-risk" components with a small, essentially permanent supply of thermal energy.

thermal cycling Exposure of a component to alternating levels of relatively high and low temperatures.

thermal equilibrium A condition that exists when energy transfer as heat between two thermodynamic systems (e.g., system 1 and system 2) is possible but none occurs. We say that system 1 and system 2 are in thermal equilibrium, and that they have the same temperature.

See also ZEROTH LAW OF THERMODYNAMICS.

thermal infrared The infrared region (IR) of the electromagnetic spectrum extending from about 3 to 14 micrometers (μm) wavelength. The thermal IR band most frequently used in remote sensing of Earth (from space) is the 8- to 14-micrometer-wavelength band. This band corresponds to windows in the atmospheric absorption bands, and, since Earth has an average surface temperature of about 288 kelvins (K), its thermal emission peaks near 10 micrometers wavelength.

thermal protection system A system designed to protect a space vehicle from undesirable heating. Usually this is accomplished by using a system of special materials that can reject, absorb, or reradiate the unwanted thermal energy.

thermal radiation In general, the electromagnetic radiation emitted by any object as a consequence of its temperature. Thermal radiation ranges in wavelength from the longest infrared wavelengths to the shortest ultraviolet wavelengths and includes the optical (or visible) region of the spectrum.

See also PLANCK'S RADIATION LAW.

thermionic conversion The direct conversion of heat (thermal energy) into electricity by evaporating electrons from a hot metal surface and condensing them on a cooler surface. The energy required to remove an electron completely from a metal is called the *work function,* which varies with the nature of the metal and its surface condition. The basic thermionic converter consists of two metals (or electrodes) with different work functions sealed in an evacuated chamber. The electrode with the larger work function is maintained at a higher temperature than the one with the smaller work function.

The emitter (or higher-temperature electrode) evaporates electrons, thereby acquiring a positive charge. The collector (or lower-temperature electrode) collects these evaporated, or emitted, electrons and becomes negatively charged. As a result, a voltage develops between the two electrodes, and a direct electric current will flow in an external circuit (or load) connecting them. The voltage is determined mainly by the difference in the work functions of the electrode materials.

However, the emission of electrons from the higher-temperature electrode can be inhibited as a result of an accumulation of electrons in its vicinity. This phenomenon is called the *space charge effect,* and its impact can be reduced by introducing a small quantity of cesium metal into the evacuated thermionic converter chamber that contains the electrodes.

To achieve a substantial electron emission rate per unit area of emitter, the emitter temperature in a thermionic converter containing cesium must be at least 1,000°C (1,273 K). Thermionic conversion has been suggested for space power generation, especially when used in conjunction with a space nuclear reactor as the heat source.

See also DIRECT CONVERSION.

thermistor A semiconductor electronic device that uses the temperature-dependent change of resistivity of the substance. The thermistor has a very large negative temperature coefficient of resistance; that is, the electrical resistance decreases as the temperature increases. It can be used for temperature measurements or electronic circuit control.

thermocouple A device consisting essentially of two conductors made of different metals, joined at both ends, producing a loop in which an electric current will flow when there is a difference in temperature between the two junctions. The amount of current that will flow in an attached circuit is dependent on the temperature difference between the measurement (hot) and reference (cold) junction, the characteristics of the two different metals used, and the characteristics of the attached circuit. Depending on the different metals chosen, a thermocouple can be used as a thermometer over a certain temperature range.

thermodynamics The branch of science that treats the relationships between thermal energy (heat) and mechanical energy. The field of thermodynamics in both physics and engineering involves the study of systems. A thermodynamic system is simply a collection of matter and space with its boundaries defined in such a way that energy transfer (as work and heat) across the boundaries can be identified and understood easily. The surroundings represent everything else that is not included in the thermodynamic system being studied.

Thermodynamic systems usually are placed in one of three groups: closed systems, open systems, and isolated systems. A CLOSED SYSTEM is a system for which only energy (but not matter) can cross the boundaries. An OPEN SYSTEM can experience both matter and energy transfer across its boundaries. An *isolated system* can experience neither matter nor energy transfer across its boundaries. A *control volume* is a fixed region in space that is defined and studied as a thermodynamic system. Often the control volume is used to help in the analysis of open systems.

STEADY STATE refers to a condition in which the properties at any given point within the thermodynamic system are constant over time. Neither mass nor energy accumulate or deplete in a steady state system.

A *thermodynamic process* is the succession of physical states that a system passes through. Thermodynamic processes often can described by one of the following terms: cyclic, reversible, irreversible, adiabatic, or isentropic. In a *cyclic process* the system experiences a series of states and ultimately returns to its original state. The Carnot cycle and the Brayton cycle are examples. A REVERSIBLE PROCESS can proceed in either direction and results in no change in the thermodynamic system or its surroundings. An IRREVERSIBLE PROCESS can proceed in only one direction; if the process were reversed, there would be a permanent (i.e., irreversible) change made to the system or its surroundings. An ADIABATIC PROCESS is one in which there is no flow of heat across the boundaries of the thermodynamic system. An *isentropic process* is one in which the entropy of the thermodynamic system remains constant, or unchanged. An isentropic process is actually a reversible adiabatic process.

Thermodynamics is an elegant branch of physics based on two fundamental laws. The first law of thermodynamics states *Energy can neither be created nor destroyed, but only altered in form.* This concept often is referred to as the CONSERVATION OF ENERGY principle. In its simplest formulation involving a thermodynamic system, this law can be expressed as

$$\Sigma \text{ energy}_{in} - \Sigma \text{ energy}_{out} = \Delta \text{ energy}_{storage}$$

This expression often is referred to as an *energy balance* in which the energy flows into a system and out of a system are balanced with respect to any change in energy stored in the system. First-law energy balances are very useful in engineering thermodynamics.

The second law of thermodynamics involves the concept of entropy and is used to determine the possibility of certain processes and to establish the maximum efficiency of allowed processes. There are many statements of the second law. The statement attributed to the German physicist RUDOLPH CLAUSIUS, which was made in 1850, is considered one of the earliest: *It is impossible to construct a device that operates in a cycle and produces no effect other than the removal of heat from a body at one temperature and the absorption of an equal quantity of heat by a body at a higher temperature.*

The second law of thermodynamics demonstrates that the maximum possible efficiency of a system (e.g., heat engine) is the Carnot efficiency, which is expressed as

$$\eta_{max} = \eta_{Carnot} = (T_H - T_L)/T_H$$

where T_H is the high (absolute) temperature and T_L is the low (absolute) temperature associated with the system—that is, the temperature at which heat is added to the system (T_H) and the temperature at which heat is rejected from the system (T_L).

Another way of expressing the second law is to look at the possible changes in entropy of an isolated system. From the second law of thermodynamics, we know that the entropy of an isolated system can never decrease, it can only remain constant or increase—that is,

$$\Delta \text{ entropy}_{isolated} \geq 0$$

Thermodynamics plays a very important role in aerospace engineering, especially in defining the limits of performance of space power and propulsion systems. Thermodynamics also plays an important role in physics, such as in helping scientists describe the flow of energy on a planet, in a solar system, throughout a galaxy, or in the universe taken as a whole. If we treat the universe as a closed system, then according to the first law of thermodynamics, all the energy and matter in the universe was present immediately after the big bang—no more has been added or taken away; rather, it has merely been changing form.

See also HEAT DEATH OF UNIVERSE.

thermoelectric conversion The direct conversion of thermal energy (heat) into electrical energy based on the *Seebeck thermoelectric effect.* In 1821 the Russian-German scientist Thomas Johann Seebeck (1770–1831) was the first person to observe that if two different metals are joined at two locations that are then kept at different temperatures, an electric current will flow in a loop. The combination of two materials capable of directly producing electricity as a result of a temperature difference is called a *thermocouple.*

If the thermocouple materials A and B are joined at the hot junction and the other ends are kept cold, an electromotive force is generated between the cold ends. A direct current also will flow in a circuit (or load) connected between these cold ends. This electric current will continue to flow as long as heat is supplied at the hot junction and removed from the cold ends. For a given thermocouple (using either different metals or semiconductor materials), the voltage and electric power output are increased by increasing the temperature difference between the hot and cold junctions. The practical performance limits of a particular thermoelectric conversion device generally are set by the nature of the thermocouple materials and the (high) temperature of the available heat source. Often, such as in space power applications, several thermocouples are connected in series to increase both voltage and power.

See also RADIOISOTOPE THERMOELECTRIC GENERATOR; SPACE NUCLEAR POWER; THERMOCOUPLE.

thermonuclear reaction *See* FUSION.

thermonuclear weapon A nuclear weapon in which very high temperatures from a nuclear fission explosion are used to bring about the fusion of light nuclei, such as deuterium (D) and tritium (T), which are isotopes of hydrogen, with the accompanying release of enormous quantities of energy. Sometimes referred to as a hydrogen bomb.

Third Cambridge Catalog **(3C)** *See* RADIO-SOURCE CATALOG.

third law of thermodynamics Based on the work of Hermann Nernst (1864–1941), MAX PLANCK, LUDWIG BOLTZMANN, Gilbert Lewis (1875–1946), ALBERT EINSTEIN, and other physicists in the early 1900s, the third law of thermodynamics can be stated as: *The entropy of any pure substance in thermodynamic equilibrium approaches zero as the absolute temperature approaches zero.* The third law is important in that it furnishes a basis for calculating the absolute entropies of substances, either elements or compounds; these data then can be used in analyzing chemical reactions.

See also THERMODYNAMICS.

Thomson, William *See* KELVIN, WILLIAM THOMSON, BARON.

Thor An early intermediate-range ballistic missile (IRBM) developed by the U.S. Air Force. This one-stage, liquid-propellant rocket had a range of more than 2,700 km. It was first launched on January 25, 1957, and is now retired from service. Originally developed as a nuclear weapon delivery system, this rocket also served the U.S. Air Force and NASA as a space launch vehicle, especially when combined with a variety of upper-stage vehicles, such as the Agena.

See also AGENA.

Thousand Astronomical Unit mission (TAU) A conceptual mid-21st century NASA mission involving an advanced-technology robot spacecraft that would travel a 50-year journey into very deep space, a distance of about 1,000 astronomical units (some 160 billion kilometers) away from Earth. The *Thousand Astronomical Unit* (TAU) spacecraft would feature an advanced multimegawatt nuclear reactor, ion propulsion, and a laser (optical) communications system. Initially, the TAU spacecraft would be directed for an encounter with Pluto and its large moon, Charon, followed by passage through the heliopause and perhaps even reaching the inner Oort cloud, the hypothetical region where long-period comets are thought to originate, at the end of its long mission. This advanced robot spacecraft would investigate low-energy cosmic rays, low-frequency radio waves, interstellar gases, and deep space phenomena. It would also perform high-precision astrometry and measurements of distances between stars.

See also OORT CLOUD; ROBOTICS IN SPACE; SPACE NUCLEAR PROPULSION.

three-body problem The problem in classical celestial mechanics that treats the motion of a small body, usually of negligible mass, relative to and under the gravitational influence of two other finite point masses.

three-dimensional grain configuration In rocketry, a solid-propellant grain whose surface is described by three-dimensional analytical geometry (i.e., one that considers end effects).

throat In rocketry, the portion of a convergent/divergent nozzle at which the cross-sectional area is at a minimum. This is the region in which exhaust gases from the combustion chamber transition from subsonic flow to supersonic flow.

See also NOZZLE; ROCKET.

throttling The varying of the thrust of a rocket engine during powered flight by some technique. Adjusting the propellant flow rate (e.g., tightening of fuel lines), changing of thrust chamber pressure, pulsed thrust, and variation of nozzle expansion are methods to achieve throttling.

thrust (symbol: T) 1. In general, the forward force provided by a reaction motor. 2. In rocketry, the fundamental thrust equation is

$$T = \dot{m}\, V_e + (P_e - P_a)\, A_e$$

where T is the thrust (force), $\dot{m}$ is the mass flow rate of propellant, V_e is the exhaust velocity at the nozzle exit, P_e is the exhaust pressure at the nozzle exit, P_a is the ambient pressure, and A_e is exit area of the nozzle. For a well-designed nozzle in which the exhaust gases are properly and completely expanded, $P_e \approx P_a$, so that the thrust becomes approximately represented by the term: $T \approx \dot{m}\; Ve$.

thrust barrel A structure in a rocket vehicle designed to accept the thrust load from two or more engines. Also called the thrust structure.

thrust chamber The place in a chemical rocket engine where the propellants are burned (or combusted) to produce the hot, high-pressure gases that are then accelerated and exhausted by the nozzle, thereby creating forward thrust. The combustion chamber and nozzle assembly.

See also ROCKET.

thrust terminator A device for ending the thrust in a rocket engine, either through propellant cutoff (in the case of a liquid-propellant rocket engine) or through the diversion of the flow of exhaust gases from the nozzle (in the case of a solid-propellant rocket).

thrust-time profile A plot of thrust level versus time for the duration of firing of a rocket engine or engines.

thrust vector control (TVC) The intentional change in the direction of the thrust of a rocket engine to guide the vehicle. Thrust vector control can be achieved by using such devices as jet vanes and rings, which deflect the exhaust gas flow as it leaves the rocket's nozzle, or by using gimbaled nozzles and nozzle fluid injection techniques. However, interrupting the nozzle exhaust flow will introduce some degradation in the rocket engine's performance. Therefore, the gimbaled nozzle approach to thrust vector control often is the technique that imposes the minimum engine performance degradation.

tidal force A force that arises in a system of one or more bodies because of a difference in gravitation. One dramatic example of the effects of tidal forces in the solar system is the rings around Saturn. These rings consist of many small particles. The particles in the rings cannot clump together to form a larger body via accretion and mutual gravitational attraction because of the strong tidal forces exerted by Saturn. Farther away from the planet, however, the tidal forces are not strong enough to rip large objects apart, and as a result Saturn has a diverse collection of moons. Some very small moons, called shepherd moons, even orbit within the outer rings. Tidal forces can bring about the destruction of a comet that passes too close to the Sun or a natural satellite that orbits too closely to a massive planet. This destruction occurs when the smaller orbiting body crosses the Roche limit, and the tidal forces across the smaller body exceed the cohesive forces holding the smaller body together.

On Earth we experience examples of tidal forces on a daily basis. The tides of the oceans are the result of the combined tidal forces of the Moon and the Sun. In the center of a distant galaxy, if a rogue star wanders too close to a supermassive black hole, the black hole's tidal forces will rip the doomed star apart. Most of the gaseous debris from the star escapes the black hole, but a small amount of material is captured by the black hole's immense gravitational pull, forming a rotating disk of gas. Close to the black hole, the swirling disk of in-falling material becomes intensely hot and begins to glow in X-rays, just before it crosses the event horizon.

See also ACCRETION; GRAVITATION; ROCHE LIMIT.

tidal theory of lunar formation *See* DARWIN, SIR GEORGE HOWARD.

time (symbol: t) In general, time is defined as the duration between two instants. The SI unit of time is the second, which now is precisely defined using an atomic standard. Time also may be designated as solar time, lunar time, or sidereal time depending on the astronomical reference used. For example, *solar time* is measured by successive intervals between transits of the Sun across the meridian. *Lunar time* involves phases of Earth's Moon. The *lunar month,* for example, represents the time taken by the Moon to complete one revolution around Earth, as measured from new Moon to new Moon. The lunar month is also called the *synodic month* and is 29.5306 days long. SIDEREAL TIME is measured by successive transits of the vernal equinox across the local meridian. On Earth time can be local or *universal time* (UT), which is related to the Greenwich meridian.

See also CALENDAR; UNIVERSAL TIME.

time dilation *See* RELATIVITY (THEORY).

time line In aerospace operations involving human crews, the planned schedule for astronauts during their space mission.

time tic Markings on telemetry records to indicate time intervals.

time tick A time signal consisting of one or more short audible sounds or beats.

Titan (launch vehicle) The family of U.S. Air Force launch vehicles that was started in 1955. The Titan I missile was the first American two-stage intercontinental ballistic missile (ICBM) and the first underground silo–based ICBM. The Titan I vehicle provided many structural and propulsion techniques that were later incorporated into the Titan II vehicle. Years later the Titan IV evolved from the Titan III family and is similar to the Titan 34D vehicle.

The Titan II was a liquid-propellant, two-stage, intercontinental ballistic missile that was guided to its target by an all-inertial guidance and control system. This missile, designated as LGM-25C, was equipped with a nuclear warhead and designed for deployment in hardened and dispersed underground silos. The U.S. Air Force built more than 140 Titan ICBMs; at one time they served as the foundation of America's nuclear deterrent force during the cold war. The Titan II vehicles also were used as space launch vehicles in NASA's Gemini Project in the mid-1960s. Deactivation of the Titan II ICBM force began in July 1982. The last missile was taken from its silo at Little Rock Air Force Base, Arkansas, on June 23, 1987. The deactivated Titan II missiles then were placed in storage at Norton Air Force Base, California. Some of these retired ICBMs have found new uses as space launch vehicles.

The Titan II space launch vehicle is a modified Titan II ICBM that is designed to provide low- to medium-mass launch capability into polar low Earth orbit (LEO). The modified Titan II is capable of lifting a 1,900-kg mass into polar LEO. With the addition of two strap-on solid rocket motors (i.e., the graphite epoxy motors) to the first stage, the payload capability to polar LEO is increased to 3,530 kg .

The versatile Titan III vehicles supported a variety of defense and civilian space launch needs. The Titan IIIA vehicle was developed to test the integrity of the inertially guided three-stage Titan IIIC liquid-propulsion system core vehicle. The Titan IIIB consists of the first two stages of the core Titan III vehicle (without the two strap-on solid-rocket motors) with an Agena vehicle used as the third stage. The Titan IIIC consists of the Titan IIIA core vehicle with two solid-rocket motors (sometimes called "Stage 0") attached on opposites sides of the liquid-propellant core vehicle. This con-

On October 15, 1997, a Titan IV/Centaur rocket, the United States's most powerful expendable rocket vehicle combination, successfully lifted off from Cape Canaveral Air Force Station, Florida, and started the *Cassini* spacecraft (with attached *Huygens* probe) on a seven-year journey through the solar system. On July 1, 2004, the two-story tall *Cassini* spacecraft reached the magnificent ringed planet and began its four-year mission of intensive scientific investigation within the Saturnian system. *(Courtesy of NASA and the U.S. Air Force)*

figuration was developed for space launches from Complex 40 at Cape Canaveral Air Force Station (AFS), Florida. The Titan IIID vehicle, launched from Vandenberg Air Force Base, California, is essentially a Titan IIIC configuration without the Transtage and is radio-guided during launch. The Titan IIIE configuration was developed for NASA and is launched from Complex 41 at Cape Canaveral AFS. The Titan IIIE is basically a standard Titan IIID with a Centaur vehicle used as the third stage. Finally, the Titan 34D configuration uses a stretched core vehicle in conjunction with larger solid-rocket motors to increase booster performance.

The Titan IV is the newest and largest expendable space booster used by the U.S. Air Force. It was developed to launch the nation's largest, high-priority, high-value "shuttle-class" defense payloads. This "heavy-lift" vehicle is flexible in that it can be launched with several optional upper stages (such as the Centaur and the Inertial Upper Stage) for greater and more varied spacelift capability. The Titan IV vehicle's first stage consists of an LR87 liquid-propellant rocket that features structurally independent tanks for its fuel (aerozine 50) and oxidizer (nitrogen tetroxide). This design minimizes the hazard of the two propellants mixing if a leak should develop in either tank. Additionally, these liquid propellants can be stored at normal temperature and pressure, eliminating launch pad delays (such as are often encountered with the boil-off and refilling of cryogenic propellants) and giving the Titan IV vehicle the capability to meet critical defense program and planetary mission launch windows.

The Titan IV vehicle can have a length of up to 61.2 meters and carry up to 17,550 kg into a 144-kilometer (km)-altitude orbit when launched from Cape Canaveral AFS and up to 13,950 kg into a 160-km-altitude polar orbit when launched from Vandenberg AFB.

The Titan IVB rocket is the most recent and largest unmanned space booster used by the USAF. It is a heavy-lift space launch vehicle designed to carry government payloads, such as the Defense Support Program surveillance satellite and National Reconnaissance Office (NRO) satellites, into space from either Cape Canaveral Air Force Station, Florida, or Vandenberg Air Force Base, California. The Titan IVB can place a 21,670-kilogram payload into low Earth orbit or more than 5,760 kilograms into geosynchronous orbit. The first Titan IVB rocket was successfully flown from Cape Canaveral Air Force Station on February 23, 1997. A powerful Titan IV-Centaur configuration successfully sent NASA's *Cassini* spacecraft to Saturn from Cape Canaveral Air Force Station on October 15, 1997.

An important chapter in American aerospace history came to an end on April 29, 2005. Late that evening, the U.S. Air Force successfully launched the last Titan IVB rocket from Complex 40 at Cape Canaveral Air Force Station. This rocket's payload was a government spy satellite.

Titan (moon) The largest Saturnian moon, approximately 5,150 km in diameter, was discovered by CHRISTIAAN HUYGENS in 1655. It has a sidereal period of 15.945 days and orbits at a mean distance of 1,221,850 km from the planet. Titan is the second-largest moon in the solar system and the only one known to have a dense atmosphere. The atmospheric chemistry presently taking place on Titan appears similar to those processes that occurred in Earth's atmosphere several billion years ago.

Larger in size than the planet Mercury, Titan has a density that appears to be about twice that of water ice. Scientists believe, therefore, that it may be composed of nearly equal amounts of rock and ice. Titan's surface is hidden from the normal view of spacecraft cameras by a dense, optically thick photochemical haze whose main layer is about 300 km above the moon's surface. Measurements prior to the July 2004 arrival of the *Cassini* spacecraft with its hitchhiking companion *HUYGENS PROBE* indicated that Titan's atmosphere is mostly nitrogen. The existence of carbon-nitrogen compounds on Titan was thought possible because of the great abundance of both nitrogen and hydrocarbons.

What does the surface of Titan look like? Prior to the Cassini mission scientists speculated that there must be large quantities of methane on the surface, enough perhaps to form methane rivers or even a methane sea. The temperature on the surface is about 91 kelvins (–182°C), which is close enough to the temperature at which methane can exist as a liquid or solid under the atmospheric pressure near the surface. Some researchers have speculated that Titan is like Earth but colder, with methane playing the role that water plays on our planet.

This analogy has stimulated scientific visions of methane-filled seas near Titan's equator and frozen methane ice caps in the moon's polar regions. Titan's surface might also experience a constant rain of organic compounds from the upper atmosphere, perhaps creating up to a 100-meter-thick layer of tarlike materials.

At 1.25 billion km from Earth, after a 7-year journey through the solar system, the ESA's *Huygens* probe separated from the *Cassini* orbiter spacecraft to enter a ballistic trajectory toward Titan in order to dive into its atmosphere on January 14, 2005. This was the first human-made object to explore in situ this unique environment, whose chemistry is assumed to be very similar to that of the early Earth just before life began 3.8 billion years ago. The *Cassini* orbiter, carrying *Huygens* on its flank, entered an orbit around Saturn on July 1, 2004, and began to investigate the ringed planet and its moons for a mission that will last at least four years. The first distant flyby of Titan took place on July 2–3, 2004. It provided data on Titan's atmosphere that were confirmed by the data obtained during the first close flyby on October 26, 2004, at an altitude of 1,174 km. These data were used to validate the entry conditions of the *Huygens* probe. A second close flyby of Titan by *Cassini-Huygens* at an altitude of 1,200 km occurred on December 13 and provided additional data to further validate the entry conditions of the *Huygens* probe. On December 17 the *Cassini* orbiter was placed on a controlled collision course with Titan in order to release *Huygens* on the proper trajectory, and on December 21 all systems were set up for separation and the *Huygens* timers set to wake the probe a few hours before its arrival at Titan.

The *Huygens* probe separated on the morning of December 25. Since the *Cassini* orbiter spacecraft had to achieve precise pointing for the release, there was no real-time telemetry available until it turned back its main antenna toward Earth and beamed the recorded data of the release. It took more than an hour (about 67 min) for the signals to reach Earth.

After release, *Huygens* moved away from the *Cassini* mother spacecraft at a speed of about 35 cm per second and, to keep on track, spun on its axis, making about 7 revolutions a minute. *Huygens* did not communicate with *Cassini* for the whole period until after deployment of the main parachute following entry into Titan's atmosphere. On December 28, 2004, *Cassini* maneuvered off its collision course with Titan and resumed its scientific mission, while also preparing to function as the mother spacecraft to receive and relay *Huygens* data. As *Huygens* plunged into Titan's atmosphere, *Cassini* recorded the probe's data signals and then transmitted them back to Earth at a later time.

Preliminary results from *Cassini-Huygens* indicate that Titan has a surface shaped largely by Earth-like processes of tectonics, erosion, winds, and perhaps volcanism. Titan has liquid methane on its cold surface. Among the new discoveries is what may be a long river, roughly 1,500 kilometers long. Scientists have also concluded that winds on Titan blow a lot faster than the moon rotates—a phenomenon called super-rotation. Since most of the cloud activity observed on Titan by the *Cassini* spacecraft has occurred over the moon's south pole, scientists think this may be where the cycle of methane rain, channel carving, runoff, and evaporation is most active. This hypothesis could explain the presence of extensive channel-like features in the region.

Following its release from the *Cassini* mother spacecraft on December 25, 2004, the *Huygens* probe traveled for 20 days before it reached Titan's outer atmosphere. On January 14, 2005, the probe successfully descended through Titan's hazy cloud layers and came to rest on Titan's frozen surface, where it continued transmitting from the surface for several hours. During its almost four-hour descent and on the ground, *Huygens* collected some 350 pictures, which revealed a landscape apparently modeled by erosion with drainage channels, shore-like features and even pebble-shaped objects on the surface. Where the probe landed, Titan's surface resembles wet sand or clay with a thin solid crust. The composition appears to be mainly a mixture of dirty water ice and hydrocarbon ice. The temperature measured by the probe at ground level was about –180°C. Sampling of Titan's atmosphere from an altitude of 160 km to ground level revealed a uniform mix of methane and nitrogen in the stratosphere. *Huygens* also detected clouds of methane at about 20 km altitude and methane and ethane fog near the surface.

Images collected during the *Cassini* spacecraft's close flybys of Titan show dark, curving linear patterns in various regions, but mostly concentrated near the moon's south pole. Scientists suggest that these large curved and linear patterns may also be channels cut by liquid methane—similar to the small (a few kilometers long) channels observed by *Huygens*. During a close flyby (within 1,027 kilometers of moon's surface) of Titan on April 16, 2005, the *Cassini* spacecraft found that the outer layer of the moon's thick, hazy atmosphere is brimming with complex hydrocarbons. Hydrocarbons containing as many as seven carbon atoms were observed, as well as nitrogen-containing hydrocarbons (nitriles).

These dramatic events marked the first attempt ever to unveil the mysteries of Titan using a robot orbiter spacecraft and an atmospheric probe. Titan is an intriguing, distant world that may hold some important clues to the early days of the evolution of life on our planet.

See also CASSINI MISSION; SATURN.

Titania SIR WILLIAM HERSCHEL discovered Titania in 1787. With a diameter of 1,580 km, it is the largest known moon of Uranus (also discovered by Sir William Herschel). Titania orbits Uranus at an average distance of 435,800 km with a period of 8,706 days and is in synchronous rotation with Uranus. The moon is named after the Queen of the Fairies, who is also Oberon's wife, in William Shakespeare's comedy *A Midsummer-Night's Dream.* Its estimated average density of 1.7 g/cm^3 (1,700 kg/m^3) suggests a composition of about 40 to 50 percent water ice, with rock making up the remaining material. Also called Uranus III.

See also URANUS.

Titius, Johann Daniel (1729–1796) German *Astronomer* In 1766 the German astronomer Johann Daniel Titius was the first to notice an empirical relationship describing the distances of the six known planets from the Sun. Another German astronomer, JOHANNES BODE, later popularized this empirical relationship, which has now become alternately called *Bode's law* or the *Titius-Bode law.*

See also BODE'S LAW.

Titov, Gherman S. (1935–2000) Russian *Cosmonaut* In August 1961 the Russian cosmonaut Gherman S. Titov became the second person to travel in orbit around Earth. His *Vostok 2* spacecraft made 17 orbits of the planet, during which he became the first of many space travelers to experience space sickness.

See also SPACE SICKNESS.

Tomahawk Long-range subsonic cruise missile used by the U.S. Navy for land attack and for antisurface warfare. Tomahawk is an all-weather submarine- or ship-launched antiship or land-attack cruise missile. After launch a solid-propellant rocket engine propels the missile until a small turbofan engine takes over for the cruise portion of the flight. Radar detection of the cruise missile is difficult because of its small cross section and low-altitude flight. Similarly, infrared detection is difficult because the turbofan emits little heat.

The antiship variant of Tomahawk uses a combined active radar seeker and passive system to seek out, engage, and destroy a hostile ship at long range. Its modified Harpoon cruise missile guidance system permits the Tomahawk to be launched and fly at low altitudes in the general direction of an enemy warship to avoid radar detection. Then, at a programmed distance, the missile begins an active radar search to seek out, acquire, and hit the target ship.

The land-attack version has inertial and terrain contour matching (TERCOM) guidance. The TERCOM guidance system uses a stored map reference to compare with the actual terrain to help the missile determine its position. If necessary, a course correction is made to place the missile on course to the target.

The basic Tomahawk is 5.56 m long and has a mass of 1,192 kg, not including the booster. It has a diameter of 51.81 cm and a wing span (when deployed) of 2.67 m. This missile is subsonic and cruises at about 880 km per hour. It can carry a conventional or nuclear warhead. In the land-attack conventional warhead configuration, it has a range of 1,100 km, while in the land-attack nuclear warhead configuration, it has a range of 2,480 km. In the antiship role it has a range of more than 460 km. This missile was first deployed in 1983.

Tombaugh, Clyde William (1906–1997) American *Astronomer* On February 18, 1930, this American "farm boy–astronomer" joined a very select astronomy club when he discovered the planet Pluto. He accomplished this much-anticipated astronomical feat while working as a junior astronomer at the Lowell Observatory in Flagstaff, Arizona. Success came through very hard work, perseverance, and Tombaugh's skilled use of the blinking comparator, an innovative approach to astrophotography based on the difference in photographic images taken a few days apart. His discovery of the elusive trans-Neptunian planet also helped fulfill the dreams of the observatory's founder, PERCIVAL LOWELL, who had predicted the existence of a "Planet X" decades earlier.

Tombaugh was born on February 4, 1906, on a farm near Streator, Illinois. When he was in secondary school his family moved to the small farming community of Burdett in the western portion of Kansas. When Tombaugh graduated from Burdett High School in 1925, he had to abandon any immediate hope of attending college because crop failures had recently impoverished his family. However, while growing up on the farm he acquired a very strong interest in amateur astronomy from his father. To complement this interest, he also taught himself mathematics and physics. In 1927 he constructed several homemade telescopes, including a 23-centimeter (9-inch) reflector built from discarded farm machinery, auto parts, and handmade lenses and mirrors. Using this reflector telescope under the dark skies of western Kansas, he spent evenings observing and made detailed drawings of Mars and Jupiter. Searching for advice and comments about his observations, Tombaugh submitted his drawings in 1928 to the staff of professional planetary astronomers at the Lowell Observatory in Flagstaff. To his great surprise, the staff members at the Lowell Observatory were very impressed with his detailed drawings, which revealed his great talent as a careful and skilled astronomical observer.

Consequently, despite the fact that Tombaugh lacked a formal education in astronomy, VESTO SLIPHER, the director of the Lowell Observatory, hired him in 1929 as a junior astronomer. Slipher placed Tombaugh on a rendezvous trajectory with astronomical history when he assigned the young farm boy–astronomer to the monumental task of searching the night sky for Planet X. As early as 1905 the American astronomer Percival Lowell had noticed subtle perturbations in the orbits of Neptune and Uranus—perturbation he attributed to the gravitational tug of some undetected planet farther out in the solar system. After several more years of investigation, Lowell boldly predicted in 1915 the existence of this Planet X. He began to personally conduct a detailed photographic search in candidate sections of the sky using his newly built observatory in Flagstaff. Unfortunately, Lowell died in 1916 without finding his mysterious Planet X.

Using a 0.33-meter (13-inch) photographic telescope at the Lowell Observatory, Tombaugh embarked on a systematic search for the long-sought Planet X. He worked through the nights in a cold, unheated dome, making pairs of exposures of portions of the sky with time intervals of two to six days. He then carefully examined these astrographs (star photographs) under a special device called a blink-comparator in the hope of detecting the small shift in position of one very special faint point of light among hundreds of thousands of points of light. This tiny shift could be a sign of the long sought distant planet among a field of stars. On the nights of

January 23 and 29, 1930, Tombaugh made two such photographs of the region of the sky near the star Delta Geminorum. Then, on the afternoon of February 18, 1930, he triumphed in his struggle to find the ninth major planet. While comparing these photographic plates under the blink-comparator, Tombaugh detected the telltale shift of a faint starlike object. Slipher and other astronomers at the observatory reviewed the results and rejoiced in Tombaugh's discovery. However, they urged caution before publicly announcing this discovery. Like any competent scientist, Tombaugh took time to confirm his important discovery. The discovery of Pluto was confirmed with subsequent photographic measurements and announced to the world on March 13, 1930.

With this announcement Tombaugh joined a very exclusive group of astronomers who observed and then named major planets in the solar system. Of the many thousands of astronomers in history who have searched the heavens with telescopes, only the following accompany Tombaugh in this very elite "planet-finders" club: Sir WILLIAM HERSCHEL, who discovered Uranus in 1781, and JOHANN GALLE, who discovered Neptune in 1846. In keeping with astronomical tradition, Tombaugh had the right as the planet's discoverer to give the distant celestial body a name. He chose Pluto, god of darkness and the underworld in Roman mythology (Hades in Greek mythology).

Pluto is a tiny, frigid world, very different from the gaseous giant planets that occupy the outer solar system. For that reason there is some contemporary debate within the astronomical community as to whether Pluto is really a "major planet" or should simply be regarded as an icy moon that escaped from Neptune or perhaps as a large Kuiper Belt object. As more and more trans-Neptunian objects (TNOs) are discovered beyond the orbit of Pluto, this debate will most likely intensify. But for now Pluto is regarded as the ninth major planetary body in the solar system, and Tombaugh's discovery represents a marvelous accomplishment in 20th-century observational astronomy.

After discovering Pluto Tombaugh remained with the Lowell Observatory for the next 13 years. During this period he also entered the University of Kansas on a scholarship in 1932 to pursue the undergraduate education he was forced to delay because of financial constraints. In 1934, while attending the university, he met and married Patricia Edson of Kansas City. The couple remained married for more than 60 years and raised two children. While observing during the summers at the Lowell Observatory, Tombaugh earned a bachelor of science degree in astronomy in 1936 and then a master of arts degree in astronomy in 1939. He frequently told the story of his perplexed astronomy professor who did not want a "planet discoverer" in his basic astronomy course.

Upon graduation he returned to the Lowell Observatory and continued a rigorous program of sky watching that resulted in his meticulous cataloging of more than 30,000 celestial objects, including hundreds of variable stars, thousands of new asteroids, and two comets. He also engaged in a search for possible small natural satellites encircling Earth. However, save for the Moon, he could not find any natural satellites of Earth that were large enough or bright enough to be detected by means of photography. Since the start of the space age in 1957, human-made (artificial) Earth-orbiting objects have been routinely imaged and catalogued by the U.S. Air Force.

During World War II Tombaugh taught navigation to military personnel from 1943 to 1945 at Arizona State College (now Northern Arizona University). Because of funding reductions following World War II, the Lowell Observatory did not rehire him as an astronomer, so he returned to work for the military at the White Sands Missile Range in Las Cruces, New Mexico. There he supervised the development and installation of the optical instrumentation used during the testing of ballistic missiles, including WERNHER VON BRAUN's V-2 rockets that were captured in Germany by the U.S. Army at the close of World War II.

Tombaugh left his position at the White Sands Missile Range in 1955 and joined the faculty at New Mexico State University in Las Cruces. He helped this university establish a planetary astronomy program and remained an active observational astronomer. Upon retirement from the university in 1973, he and his wife went on extensive lecture tours throughout the United States and Canada to raise money for scholarships in astronomy at New Mexico State University. Several weeks before his 91st birthday, Tombaugh died on January 17, 1997, in Las Cruces, New Mexico.

Tombaugh was recognized around the world as the discoverer of Pluto. He became a professor emeritus in 1973 at New Mexico State University and published several books, including *The Search for Small Natural Earth Satellites* (1959) and *Out of the Darkness: The Planet Pluto* (1980), coauthored with Patrick Moore.

See also KUIPER BELT; PLUTO.

topside sounder A spacecraft instrument that is designed to measure ion concentration in the ionosphere from above Earth's atmosphere.

See also REMOTE SENSING.

torque (symbol: τ) In physics, the moment of a force about an axis; the product of a force and the distance of its line of action from the axis.

torr A unit of pressure named in honor of the Italian physicist Evangelista Torricelli (1608–47). One standard atmosphere (on Earth) is equal to 760 torr; or 1 torr = 1 millimeter of mercury = 133.32 pascals.

total impulse (symbol: I_T) The integral of the thrust force (T) over an interval of time, t:

$$I_T = \int T\, dt$$

See also SPECIFIC IMPULSE.

touchdown The (moment of) landing of an aerospace vehicle or spacecraft on the solid surface of a planet or moon. *Compare* with SPLASHDOWN.

TOW missile The U.S. TOW (Tube-launched Optically Tracked, Wire Command–Link Guided) missile is a long-range, heavy antitank system designed to attack and defeat armored vehicles and other targets, such as field fortifications. The TOW 2A missile is 14.91 cm in diameter and 128.02 cm long; the TOW 2B missile is 14.9 cm in diameter and 121.9 cm long. The maximum effective range is about 3.75 kilometers. The basic TOW weapon system was fielded in 1970. This system is designed to attack and defeat tanks and other armored vehicles. It is used primarily in antitank warfare and is a command to line-of-sight, wire-guided weapon. The system will operate in all weather conditions and on an obscured battlefield. The TOW 2 launcher is the most recent launcher upgrade. The TOW 2B missile incorporates a new fly-over, shoot-down technology. The TOW missile can be mounted on the Bradley Fighting Vehicle System (BFVS), the Improved TOW Vehicle (ITV), the High-Mobility Multipurpose Wheeled Vehicle (HMMWV), and the AH-1S Cobra Helicopter. The TOW missile is also in use by more than 40 nations besides the United States as their primary heavy antiarmor weapons system.

track 1. (noun) The actual path or line of movement of an aircraft, rocket, aerospace vehicle, and the like over the surface of Earth. It is the projection of the vehicle's flight path on the planet's surface. 2. (verb) To observe or plot the path of something moving, such as an aircraft or rocket, by one means or another, such as by telescope or by radar; said of persons or of electronic equipment, such as, for example, "The radar tracked the satellite while it passed overhead."

tracking 1. The process of following the movement of a satellite, rocket, or aerospace vehicle. Tracking usually is accomplished with optical, infrared, radar, or radio systems. 2. In ballistic missile defense (BMD), the monitoring of the course of a moving target. Ballistic objects may have their tracks predicted by a defensive system using several observations and physical laws.

Tracking and Data Relay Satellite (TDRS) NASA's Tracking and Data Relay Satellite (TDRS) network provides almost full-time coverage not only for the space shuttle but also for up to 24 other orbiting spacecraft simultaneously. Services provided include communications, tracking, telemetry, and data acquisition.

The TDRS satellites operate in geosynchronous orbits at an altitude of 35,888 kilometers above Earth. From this high-altitude perspective, TDRS satellites look down on an orbiting space shuttle or other spacecraft in low- and medium-altitude orbits. For most of their orbits around Earth, such spacecraft remain in sight of one or more TDRS satellites. In the past spacecraft could communicate with Earth only when they were in view of a ground tracking station, typically less than 15 percent of the time during each orbit. The full TDRS constellation enables other spacecraft to communicate with Earth for about 85 to 100 percent of their orbits, depending on their specific operating altitudes. All space shuttle missions and nearly all other NASA spacecraft in Earth orbit require TDRS support capabilities for mission success.

Each TDRS is a three-axis-stabilized satellite measuring 17.4 m across when its solar panels are fully deployed. The satellite has a mass of 2,268 kg. The spacecraft's design uses three modules. The equipment module, forming the base of the satellite's central hexagon, houses the subsystems that actually control and operate the satellite. Attached solar power arrays generate more than 1,700 watts of electrical power. When a TDRS is in the shadow of Earth, nickel-cadmium batteries supply electric power.

The communications payload module is the middle portion of the hexagon and includes the electronic equipment that regulates the flow of transmissions between the satellite's antennas and other communications functions.

Finally, the antenna module is a platform atop the hexagon holding seven antenna systems. TDRS has uplink, or forward, channels that receive transmissions (radio signals) from the ground, amplify them, and retransmit them to user spacecraft. The downlink, or return, channels receive the user spacecraft's transmissions, amplify them, and retransmit them to the ground terminal. The prime transmission link between Earth and a TDRS is a 2-meter dish antenna attached by a boom to the central hexagon. This parabolic reflector operates in the Ku-band (12–14 gigahertz [GHz]) frequency to relay transmissions to and from the ground terminal. All spacecraft telemetry data downlinked by a TDRS is channeled through a highly automated ground station complex at White Sands, New Mexico. This location was chosen for its clear line of sight to the satellites and dry climate. The latter minimizes rain degradation of radio signal transmission.

The TDRS spacecraft with its attached upper stage is launched aboard the space shuttle from the Kennedy Space Center in Florida and deployed from the orbiter's payload bay, nominally about six hours into the mission. To boost the TDRS into a geosynchronous orbit, the Inertial Upper Stage (IUS) vehicle fires twice. The upper-stage vehicle's first-stage motor fires about an hour after deployment, placing the attached TDRS into an elliptical geotransfer orbit. The first stage then separates. The IUS second-stage motor fires about 12.5 hours into the mission, circularizing the orbit and shifting the flight path so that the satellite is moving above the equator. The IUS second stage and the TDRS then separate at about 13 hours after launch. Once in geosynchronous orbit the satellite's appendages, including the solar panels and parabolic antennas, are deployed. About 24 hours after launch the satellite is ready for ground controllers to begin checkout procedures. Initially the spacecraft is positioned at an intermediate location (in geosynchronous orbit) for checkout and testing, then it is moved to its final operational location (in geosynchronous orbit).

In June 2000 NASA launched the first of three newly designed and improved TDRS spacecraft (called TDRS-H, -I, and -J) from Cape Canaveral Air Force Station to replenish the existing on-orbit fleet of spacecraft. The TDRS-H was joined in geosynchronous orbit by the TDRS-I when an Atlas IIA expendable launch vehicle successfully lifted that new satellite into space on September 30, 2002. The improved TDRS satellites retain and augment two large antennas that move smoothly to track satellites orbiting below, providing high-data-rate communications for the *INTERNATIONAL SPACE STATION*, the *HUBBLE SPACE TELESCOPE*, and other Earth-orbiting astronomy spacecraft.

trajectory In general, the path traced by any object or body moving as a result of an externally applied force, considered in three dimensions. Trajectory sometimes is used to mean flight path or orbit, but "orbit" usually means a closed path, and "trajectory" means a path that is not closed.

trajectory correction maneuver (TCM) Once a spacecraft's solar or planetary orbital parameters are known, spacecraft controllers on Earth can compare these values to those actually desired for the mission. To correct any discrepancy, a trajectory correction maneuver (TCM) can be planned and executed. This process involves computing the direction and magnitude of the vector required to correct to the desired trajectory. An appropriate time then is determined for making the change. The spacecraft is commanded to rotate to the attitude (in three-dimensional space) computed for implementing the change, and its thrusters are fired for the required amount of time. TCMs generally involved a velocity change (or "delta-V") on the order of meters per second or tens of meters per second.

transceiver A combination of transmitter and receiver in a single housing, with some components being used by both units.

transducer General term for any device that converts one form of energy (usually in some type of signal) to another form of energy. For example, a microphone is an electroacoustic transducer in which sound waves (acoustic signals) are converted into corresponding electrical signals that then can be amplified, recorded, or transmitted to a remote location. The photocell and thermocouple are also transducers, converting light and heat, respectively, into electrical signals.

transfer orbit In interplanetary travel, an elliptical trajectory tangent to the orbits of both the departure planet and target planet (or moon).

See also HOHMANN TRANSFER ORBIT.

transient The condition of a physical system in which the parameters of importance (e.g., temperature, pressure, fluid velocity, etc.) vary significantly with time; in particular, the condition or state of a rocket engine's operation in which the mass, momentum, volume, and pressure of the combustion products within the thrust chamber vary significantly with time.

transient period The interval from start or ignition to the time when steady-state conditions are reached, as in a rocket engine.

transit (planetary) The passage of one celestial body in front of another larger-diameter celestial body, such as Venus across the face of the Sun. From Earth observers can witness only transits of Mercury and Venus. There are about 13 transits of Mercury every century, but transits of Venus are much rarer events. In fact, only seven such events have occurred since the invention of the astronomical telescope. These transits took place in 1631, 1639, 1761, 1769, 1874, 1882, and most recently on June 8, 2004. If you missed personally observing the 2004 transit, mark your astronomical calendar, because the next transit of Venus takes place on June 6, 2012.

Astronomers use *contacts* to characterize the principal events that occur during a transit. During one of the rare transits of Venus, for example, the event begins with contact I, which is the instant the planet's disk is externally tangent to the Sun. The entire disk of Venus is first seen at contact II, when the planet is internally tangent to the Sun. During the next several hours, Venus gradually traverses the solar disk at a relative angular rate of approximately 4 arc-minutes per hour. At contact III the planet reaches the opposite limb and is once again internally tangent to the Sun. The transit ends at contact IV, when the planet is externally tangent to the Sun. Contacts I and II define the phase of the transit called *ingress,* while astronomers refer to contacts III and IV as the *egress phase,* or simply the *egress.*

From celestial mechanics, transits of Venus are possible only in early December and early June, when Venus's nodes pass across the Sun. If Venus reaches inferior conjunction at this time, a transit occurs. As you may have noticed from the list of the historic transits of Venus, these transits show a clear pattern of recurrence and take place at intervals of 8, 121.5, 8, and 105.5 years. The next pair of Venus transits will occur over a century from now on December 11, 2117, and December 8, 2125.

So it is probably best to make plans to observe the upcoming June 6, 2012, transit of Venus. The entire 2012 transit (that is, all four contacts) will be visible from northwestern North America, Hawaii, the western Pacific, northern Asia, Japan, Korea, eastern China, the Philippines, eastern Australia, and New Zealand. Unfortunately, no portion of the 2012 transit will be visible from Portugal or southern Spain, western Africa, or the southeastern two-thirds of South America. For the parts of the world not previously mentioned, the Sun will be setting or rising while the transit is in progress, so it will not be possible to observe the complete event.

See also VENUS.

***Transit* (spacecraft)** A U.S. Navy navigational satellite system that has provided passive, all-weather, worldwide position information to surface ships (military and commercial) and submarines for more than 30 years.

transit time The length of time required by a rocket, missile, or spacecraft to reach its intended target or destination.

translation Movement in a straight line without rotation.

translunar Of or pertaining to space beyond the Moon's orbit around Earth. *Compare with* CISLUNAR.

transmission grating A diffraction grating in which incoming signal energy is resolved into spectral components upon transmission through the grating.

See also DIFFRACTION GRATING.

transmittance (**transmissivity**) In radiation heat transfer, the transmittance (commonly used symbol: τ) for a body is defined as the ratio of the incident radiant energy transmitted through the body to the total radiant energy falling upon the body. The body may be transparent to the incident radiant energy (i.e., permitting some or all of the radiant energy to travel through the body), or else it may be opaque and therefore inhibit or prevent the transmission of incident radiant energy. An ideal blackbody is perfectly opaque. Therefore, this blackbody has a transmittance of zero, that is, $\tau_{blackbody} = 0$. *Compare with* ABSORPTANCE and REFLECTANCE.

transmitter A device for the generation of signals of any type and form that are to be transmitted. For example, in radio and radar, a transmitter is that portion of the equipment that includes electronic circuits designed to generate, amplify, and shape the radio frequency (RF) energy that is delivered to the antenna, where it is radiated out into space. In aerospace applications, the spacecraft's transmitter generates a tone at a single designated frequency, typically in the S-band (about 2 gigahertz) or X-band (about 5 gigahertz) region of the electromagnetic spectrum. This tone is called the *carrier*. The carrier then can be sent from the spacecraft to Earth as it is, or it can be modulated within the transmitter with a data-carrying subcarrier. The signal generated by the spacecraft's transmitter is passed to a power amplifier, where its power typically is boosted several tens of watts. The microwave-band power amplifier can be a solid-state amplifier (SSA) or a traveling wave tube (TWT) amplifier. The output of the power amplifier then is conducted through wave guides to the spacecraft's antenna, often a high-gain antenna (HGA).

See also TELECOMMUNICATIONS.

Trans-Neptunian object (TNO) Any of the numerous small, icy celestial bodies that lie in the outer fringes of the solar system beyond Neptune, the gaseous giant planet that is about 30 astronomical units from the Sun. Astronomers suspect that there are perhaps tens of thousands of these frozen objects with diameters in excess of 100 kilometers. The distant icy world Quaoar, with a diameter of about 1,250 km, is an example. TNOs include plutinos and Kuiper Belt objects.

See also KUIPER BELT; PLUTINO; QUAOAR.

transpiration cooling A form of mass transfer cooling that involves controlled injection of a fluid mass through a porous surface. This process basically is limited by the maximum rate at which the coolant material can be pumped through the surface.

transponder A combined receiver and transmitter whose function is to transmit signals automatically when triggered by an appropriate interrogating signal.

trap A part of a solid-propellant rocket engine used to prevent the loss of unburned propellant through the nozzle.

See also ROCKET.

Triana *See* DEEP SPACE CLIMATE OBSERVATORY.

Trident missile (D-5) The U.S. Navy's Trident II (D-5) missile is the main armament aboard the Ohio-class fleet ballistic missile submarines (SSBNs), which now have replaced the aging fleet ballistic missile submarines built in the 1960s. The Trident-II (D-5) three-stage, solid-propellant rocket is the latest in a line of fleet ballistic missiles that began with the Polaris (A-1) missile. The Trident-II missile incorporates many state-of-the-art advances in rocketry and electronics, giving this submarine-launched fleet ballistic missile greater range, payload capacity, and accuracy than its predecessors.

See also FLEET BALLISTIC MISSILE.

triple point In thermodynamics, the temperature of a mixture of a substance that contains the liquid, gaseous, and solid phases of the material in thermal equilibrium at standard pressure. For example, by international agreement, the temperature for the triple point of water has been set at 273.16 kelvins (K), creating an easily reproducible standard for a one-point absolute temperature scale.

triquetrum *See* ANCIENT ASTRONOMICAL INSTRUMENTS.

tritium (**symbol: T or** $^3{}_1H$) The radioisotope of hydrogen with two neutrons and one proton in the nucleus. It has a half-life of 12.3 years.

Triton The largest moon of Neptune. It has a diameter of approximately 2,700 km and an average surface temperature of just 35 K, making it one of the coldest objects yet discovered in the solar system. Triton was discovered by the British astronomer WILLIAM LASSELL in 1846, barely a month after the planet Neptune itself was discovered by the German astronomer JOHANN GALLE. This large moon is in a unique synchronous retrograde orbit around Neptune at a distance of 354,760 km. The orbit is characterized by a period of 5.877 days and an inclination of 157.4°. Triton is the only large moon in the solar system that rotates in a direction opposite to the direction of rotation of its planet, while keeping the same hemisphere toward the planet. Triton has a tenuous atmosphere consisting of primarily nitrogen with small amounts of methane and traces of carbon monoxide.

See also NEPTUNE.

Trojan group A group of asteroids that lies near the two Lagrangian points in Jupiter's orbit around the Sun. Achilles was the first asteroid in the group to be identified (in 1906); many subsequently discovered members of this group have been named in honor of the heroes, both Greek and Trojan, of the Trojan War.

See also ASTEROID.

tropical year *See* YEAR.

truncation error In computations, the error resulting from the use of only a finite number of terms of an infinite series or from the approximation of operations in the infinitesimal calculus by operations in the calculus of finite differences.

Tsiolkovsky, Konstantin Eduardovich (1857–1935) Russian *Schoolteacher, Space Travel Pioneer* The Russian schoolteacher Konstantin Eduardovich Tsiolkovsky is one of the three founding fathers of astronautics—the other two technical visionaries being ROBERT GODDARD and HERMANN OBERTH. At the beginning of the 20th century Tsiolkovsky worked independently of the other two individuals, but the three shared and promoted the important common vision of using rockets for interplanetary travel.

Tsiolkovsky, a nearly deaf Russian schoolteacher, was a theoretical rocket expert and space travel pioneer light-years ahead of his time. This brilliant schoolteacher lived a simple life in isolated, rural towns within Czarist Russia. Yet, despite his isolation from the mainstream of scientific activity, he somehow wrote with such uncanny accuracy about modern rockets and space that he cofounded the field of astronautics. Primarily a theorist, he never constructed any of the rockets he proposed in his prophetic books. His 1895 book *Dreams of Earth and Sky* included the concept of an artificial satellite orbiting Earth. Many of the most important principles of astronautics appeared in his seminal 1903 work *Exploration of Space by Reactive Devices*. This book linked the use of the rocket to space travel and suggested the use of a high-performance liquid hydrogen and liquid oxygen rocket engine. Tsiolkovsky's 1924 work *Cosmic Rocket Trains* introduced the concept of the multistage rocket. His books inspired many future Russian cosmonauts, space scientists, and rocket engineers, including SERGEI KOROLEV, whose powerful rockets helped fulfill Tsiolkovsky's predictions.

The Russian space flight pioneer was born on September 17, 1857, in the village of Izhevskoye, in the Ryazan Province of Russia. His father, Eduard Ignatyevich Tsiolkovsky, was a Polish noble by birth living in exile working as a provincial forestry official. His mother, Mariya Yumasheva, was Russian and Tartar. At the age of nine a near-fatal attack of scarlet fever left Tsiolkovsky almost totally deaf. With his loss of hearing, he adjusted to a lonely, isolated childhood in which books became his friends. He also learned to educate himself and in the process acquired a high degree of self-reliance.

At age 16 Tsiolkovsky ventured to Moscow, where he studied mathematics, astronomy, mechanics, and physics. He used an ear trumpet to listen to lectures and struggled with a meager allowance of just a few kopecks (pennies) each week for food. Three years later Tsiolkovsky returned home. He soon passed the schoolteacher's examination and began his teaching career at a rural school in Borovsk, located about 100 kilometers from Moscow. In Borovsk he met and married his wife, Varvara Sokolovaya. He remained a young provincial schoolteacher in Borovsk for more than a decade. Then in 1892 Tsiolkovsky moved to another teaching post in Kaluga, where he remained until he retired in 1920.

As he began his teaching career in rural Russia, Tsiolkovsky also turned his fertile mind to science, especially concepts about rockets and space travel. Despite his severe hearing impairment, Tsiolkovsky's tenacity and self-reliance allowed him to become an effective teacher and also to make significant contributions to the fields of aeronautics and astronautics.

However, teaching in rural Russian villages in the late 19th century physically isolated Tsiolkovsky from the mainstream of scientific activities, both in his native country and elsewhere in the world. In 1881, for example, he independently worked out the kinetic theory of gases and then proudly submitted a manuscript concerning this original effort to the Russian Physico-Chemical Society. Unfortunately, Dmitri Mendeleyev (1834–1907), the famous chemist who developed the periodic table of the elements, had to inform Tsiolkovsky that the particular theory had already been developed a decade earlier. However, the originality and quality of Tsiolkovsky's paper sufficiently impressed Mendeleyev and the other reviewers, so they invited the rural schoolteacher to become a member of the society.

While teaching in Borovsk, Tsiolkovsky used his own meager funds to construct the first wind tunnel in Russia. He did this so he could experiment with airflow over various streamlined bodies. He also began making models of gas-filled metal-skinned dirigibles. His interest in aeronautics served as a stimulus for his more visionary work involving the theory of rockets and their role in space travel. As early as 1883 he accurately described the weightlessness conditions of space in an article entitled "Free Space." In his 1895 book, *Dreams of Earth and Sky,* Tsiolkovsky discussed the concept of an artificial satellite orbiting Earth. By 1898 he correctly linked the rocket to space travel and concluded that the rocket would have to be a liquid-fueled chemical rocket in order to achieve the necessary escape velocity. The escape velocity is the minimum velocity an object must acquire in order to overcome the gravitational attraction of a large celestial body, such as planet Earth. To completely escape from Earth's gravity, for example, a spacecraft needs to reach a minimum velocity of approximately 11 kilometers per second.

Many of the fundamental principles of astronautics were described in his seminal work, *Exploration of Space by Reactive Devices*. This important theoretical treatise showed that space travel was possible using the rocket. Another pioneering concept found in the book was a design for a liquid-propellant rocket that used liquid hydrogen and liquid oxygen. Tsiolkovsky delayed publishing the important document until 1903. One possible reason for the delay is the fact that Iganty, Tsiolkovsky's son, had committed suicide in 1902.

Because Tsiolkovsky was a village teacher in an isolated rural area of prerevolutionary Russia, his important work in aeronautics and astronautics went essentially unnoticed by the world scientific community. Furthermore, in those days few people in Russia cared about space travel, so he never received significant government funding to pursue any type of practical demonstration of his innovative concepts. His suggestions included the spacesuit, space stations, multistage rockets, large habitats in space, the use of solar energy, and closed life support systems. In the first two decades of the 20th century, things seemed to go from bad to worse in this space travel visionary's life. For example, in 1908 an overflowing river flooded his home and destroyed many of his notes and scientific materials. Undaunted, he salvaged what he could, rebuilt, and pressed on with teaching and writing about space.

Cosmic Consequences of Space Exploration

The space age has been an era of discovery and scientific achievement without equal in human history. This essay provides a perspective on how space exploration helps scientists answer some of humankind's most important philosophical questions: Who are we? Where did we come from? Where are we going? Are we alone in this vast universe? Future space missions will define the cosmic philosophy of an emerging solar system civilization. As the detailed exploration of distant worlds leads to presently unimaginable discoveries, the human race will continue its search for life beyond Earth. In so doing, space exploration allows us to learn more about the human role as an intelligent species in a vast and beautiful universe.

Very early in the 20th century the legendary Russian spaceflight pioneer Konstantin Eduardovich Tsiolkovsky (1857–1935) boldly predicted interplanetary travel and the eventual migration of human beings from planet Earth into the solar system. Engraved on his tombstone in Kaluga, Russia, are the prophetic words: "Mankind will not remain tied to earth forever." To commemorate the centennial of Tsiolkovsky's birth, the Soviet Union launched *Sputnik 1*, the world's first artificial Earth-orbiting satellite. This technical achievement took place on October 4, 1957, a date now generally considered the birthday of the space age.

The launch of *Sputnik 1* took place during the cold war (1946–89), a period of intense ideological conflict and technical competition between the United States and the Soviet Union. Nowhere did this rivalry assume such public manifestation as in the so-called space race, especially the undeclared but highly visible Moon race. Space exploration achievements became directly associated on the world political stage with superpower prestige. Ultimately, and perhaps by political accident, this competitive environment helped inaugurate the central space exploration event that marks the beginning of humanity's solar system civilization. The Apollo Project grew out of President John F. Kennedy's (1917–63) visionary lunar lending mandate in response to the early Soviet space exploration challenge.

This inspiring event occurred on July 20, 1969, when two American astronauts, Neil A. Armstrong and Edwin E. "Buzz" Aldrin, Jr., became the first human beings to walk on another world. The Apollo lunar expeditions (1968–72) provided historians with a very definitive milestone by which to divide all human history—namely, the time before and after we demonstrated the ability to become a multiple-planet species. It is reasonable to treat the Moon as Earth's nearest "planetary" neighbor because astronomers and planetary scientists often regarded the Earth-Moon system as a double planet system. The footprints of the 12 humans who walked on the lunar surface during the Apollo Project now serve as a collective permanent beacon that challenges future generations to follow and seek their destiny beyond the boundaries of Earth.

Far-traveling robotic exploring machines also contributed to the first golden age of space exploration. In a magnificent initial wave of scientific exploration spanning the 1960s, 1970s, and 1980s, an armada of robotic spacecraft departed Cape Canaveral, Florida, and traveled to the Moon, around the Sun, and to all the planets in the solar system save for tiny Pluto. One particularly epic journey, that of the *Voyager 2* spacecraft, clearly defined this rich era of discovery, an intellectually electrifying period within which scientists learned more about the solar system than in all previous human history.

The legendary journey of the *Voyager 2* spacecraft started from Complex 41, when a mighty Titan/Centaur expendable rocket vehicle ascended flawlessly into the heavens on August 20, 1977. This robotic exploring machine successfully performed its famous "grand tour" mission by sweeping past all the gaseous giant outer planets (Jupiter encountered on July 9, 1979; Saturn on August 25, 1981; Uranus on January 24, 1986; and Neptune on August 25, 1989). After its encounter with Neptune scientists placed *Voyager 2* on an interstellar trajectory, and it continues on this journey beyond the solar system at a speed of about 3.1 astronomical units per year.

As we examine the significance of space exploration on human culture and philosophy, we should remember that four human-made objects (NASA's *Pioneer 10* and *11* spacecraft and the *Voyager 1* and *2* spacecraft) now travel on trajectories sufficient to carry them into the interstellar void. Each of these spacecraft carries a message from Earth. *Pioneer 10* and *11* bear a specially designed plaque that features the stylized images of a man and woman, as well as simply coded reference astronomical data. The *Voyager 1* and *2* spacecraft each carry a digitally recorded message that includes a variety of sounds and images from Earth. These are the human race's first interstellar emissaries. Perhaps a thousand millennia from now (star date 1,002,004, if you prefer), an advanced alien civilization will discover one of these artifacts of space exploration, decode its message, and learn about our "ancient" 20th-century civilization and its first attempts to come of age in the Milky Way galaxy.

Contemporary space exploration now provides tantalizing hints about the possibilities of life (extinct or existent) on Mars or within suspected liquid water oceans beneath Europa, Callisto, and Ganymede, three of the four satellites of Jupiter discovered by Galileo at the dawn of telescopic astronomy. Somewhat closer to home, continuous streams of quality data from sophisticated Earth-orbiting astronomical observatories, such as the refurbished *Hubble Space Telescope* (HST) and the *Chandra X-Ray Observatory* (CXO), provide and surprise astronomers and cosmologists with an ever-improving view of the universe across space and time. Each day scientists confirm that the universe is not just a beautiful, violent, and strange place, but a place that is stranger than the human mind could ever imagine.

The discovery of extraterrestrial life, extinct or existing, no matter how humble in form, will force a major revision in the anthrophic principle, that comfortable premise by which human beings have tacitly assumed over the centuries that planet Earth and (by extrapolation) the universe were created primarily for their benefit. The discovery of life beyond Earth will shatter such a chauvinistic terrestrial viewpoint and encourage a wholesale reevaluation of key philosophical questions, such as "Who are we as a species?" and "What is our role in the cosmic scheme of things?" Now the subject of extraterrestrial life resides in the nebulous buffer zone between science fiction and highly speculative

(continues)

Cosmic Consequences of Space Exploration *(continued)*

science, but through space exploration Martian or Europan life could easily become science fact. For example, in March 2004 the *Mars Opportunity Rover* discovered strong geological evidence that the Meridiani Planum region was once soaked with liquid water. Astrobiologists suggest that wherever liquid water is available, life can and should emerge.

Absurd? Perhaps. But less than 500 years ago a bold Polish church official, Nicholas Copernicus, and a fiery Italian physicist, Galileo Galilei, helped trigger the first scientific revolution, an event that completely revised the traditional and long-cherished view about Earth's central role in the physical scheme of things. Likewise in this century, as we use space technology to investigate interesting locations in the solar system, we must be prepared for the arrival of a similar technical and philosophical revolution. For some, demonstrating that life abounds in the universe will represent a logical and exciting pathway for science and space exploration; for others, such a discovery will create cultural and religious turmoil, as did the Copernican revolution.

Space exploration offers human beings the universe as both a destination and a destiny. Some future galactic historians will look back and note how life emerged out of Earth's ancient seas, paused briefly on the land, and then boldly ventured forth into the solar system and beyond.

Following the Russian Revolution of 1917, the new Soviet government grew interested in rocketry and rediscovered Tsiolkovsky's amazing work. The Soviet government honored him for his previous achievements in aeronautics and astronautics and encouraged him to continue his pioneering research. He received membership in the Soviet Academy of Sciences in 1919, and the government granted him a pension for life in 1921 in recognition of his overall teaching and scientific contributions. Tsiolkovsky used the free time of retirement to continue to make significant contributions to astronautics. For example, in his 1924 book, *Cosmic Rocket Trains,* he recognized that on its own a single-stage rocket would not be powerful enough to escape Earth's gravity, so he developed the concept of a staged rocket, which he called a rocket train. He died in Kaluga on September 19, 1935. His epitaph conveys the important message "Mankind will not remain tied to Earth forever."

Tsiolkovsky's visionary writings inspired many future Russian cosmonauts and aerospace engineers, including Sergei Korolev. As part of the Soviet Union's celebration of the centennial of Tsiolkovsky's birth, Korolev received permission to use a powerful Russian military rockets to launch *Sputnik 1,* the world's first artificial satellite, on October 4, 1957, a date generally regarded as the birth of the space age.

Tsyklon launch vehicle A medium-capacity Russian-Ukranian launch vehicle capable of placing a 3,600-kilogram payload into low Earth orbit (LEO). First launched in 1977, this vehicle consists of three hypergolic-propellant stages. The Tsyklon rocket is currently being manufactured by NPO Yuzhnoye, Ukraine.

See also LAUNCH VEHICLE.

T-time In aerospace operations, any specific time (minus or plus) that is referenced to "launch time," or "zero," at the end of a countdown. T-time is used to refer to times during a countdown sequence that is intended to result in the ignition of a rocket propulsion unit or units to launch a missile or rocket vehicle. For example, one will hear the phrase "T-minus 20 seconds and counting" during a countdown sequence. This refers to the point in the launch sequence that occurs 20 seconds before the rocket engines are ignited.

See also COUNTDOWN.

tube-wall construction Use of parallel metal tubes that carry coolant to or from the combustion chamber or nozzle wall.

tumble 1. To rotate end over end—said of a rocket, of an ejection capsule, or of a spent propulsion stage that has been jettisoned. 2. Of a gyroscope, to precess suddenly and to an extreme extent as a result of exceeding its operating limits of pitch.

Tunguska event A violent explosion that occurred in a remote part of Siberia (Russia) on the morning of June 30, 1908. One hypothesis is that the wide-area (about 80 kilometers in diameter) destructive event was caused by the entrance into Earth's atmosphere of a relatively small, extinct cometary nucleus about 60 meters in diameter or a volatile near-Earth asteroid with a composition similar to a carbonaceous chondrite meteorite. Most of the kinetic energy of the cometary or asteroidal fragments was probably dissipated through an explosive disruption of the atmosphere some 8 to 9 kilometers above the surface of the devastated Siberian forest region.

Subsequent investigations decades after the event failed to find an impact crater, although many square-kilometers of forest were laid flat by the explosive event. The trees in this remote Siberian forest were mostly knocked to the ground out to distances of about 20 km from the end point of the fireball trajectory, and some were snapped off or knocked over at distances as great as 40 km. The energy released is estimated to have been equivalent to the detonation of a thermonuclear weapon with a yield between 12 and 20 megatons.

Circumstantial evidence further suggests that fires were ignited up to 15 km from the endpoint of an intense burst of radiant energy. The combined environmental effects were quite similar to those expected from a large-yield nuclear detonation at a similar altitude, except, of course, that there were no accompanying bursts of neutrons or gamma rays nor any lingering radioactivity. Should a Tunguska-like event occur today over a densely populated area, the resulting 10 to 20 megaton "airburst" would flatten buildings over an area some 40 km in diameter and would ignite exposed flammable materials near the center of the devastated area.

Because of the unusual nature of this destructive event, several of the original investigators even speculated that it was caused by the explosion of an alien spacecraft. However, no firm technical evidence has ever been gathered to support this unfounded hypothesis.

See also ASTEROID; ASTEROID DEFENSE SYSTEM; COMET; EXTRATERRESTRIAL CATASTROPHE THEORY; NEAR-EARTH ASTEROID.

turbine A machine that converts the energy of a fluid stream into mechanical energy of rotation. The working fluid used to drive a turbine can be gaseous or liquid. For example, a highly compressed gas drives an expansion turbine, hot gas drives a gas turbine, steam (or other vapor) drives a steam (or vapor) turbine, water drives a hydraulic turbine, and wind spins a wind turbine (or windmill). Generally (except for wind and water turbines), a turbine consists of two sets of curved blades (or vanes) along side each other. One set of vanes is fixed, and the other set of vanes can move. The moving vanes are spaced around the circumference of a cylinder or rotor, which can rotate about a central shaft. The fixed set of vanes (the stator) often is attached to the inside casing that encloses the rotor, or moving portion of the turbine.

In order to make efficient use of the energy of the working fluid, gas and steam turbines often have a series of successive stages. Each stage consists of a set of fixed (stator) and moving (rotor) blades. The pressure of the working fluid decreases as it passes from stage to stage. The overall diameter of each successive stage of the turbine therefore can be increased to maintain a constant torque (or rotational effect) as the working fluid expands and loses pressure and energy.

turbopump system In a liquid-propellant rocket, the assembly of components (e.g., propellant pumps, turbine[s], power source, etc.) designed to raise the pressure of the propellants received from the vehicle tanks and deliver them to the main thrust chamber at specified pressures and flow rates.

See also ROCKET.

turbulence A state of fluid flow in which the instantaneous velocities exhibit irregular and apparently random fluctuations so that, in practice, only statistical properties can be recognized and subjected to analysis. These fluctuations often constitute major deformations of the flow and are capable of transporting momentum, energy, and suspended matter at rates far in excess of the rate of transport by the molecular processes of diffusion and conduction in a nonturbulent or laminar flow.

turbulence ring In a liquid-propellant rocket engine, the circumferential protuberance in the gas-side wall of a combustion chamber intended to generate turbulent flow and thereby enhance the mixing of burning gases.

See also ROCKET.

turbulent flow Fluid flow in which the velocity at a given point fluctuates randomly and irregularly in both magnitude and direction. The opposite of LAMINAR FLOW.

turnoff point The point on the Hertzsprung-Russell (HR) diagram at which stars of a particular globular cluster leave the main sequence and enter a more evolved state. Astronomers reason that since all the stars of a cluster formed at about the same time from a parent gas cloud, then the most massive of these stars will reach the turnoff point first. Astronomers subsequently use the location of the turnoff points to estimate the age of a particular globular cluster.

See also GLOBULAR CLUSTER; HERTZSPRUNG-RUSSELL (HR) DIAGRAM.

twilight The time following sunset and preceding sunrise when the Earth's sky is partially illuminated by the setting or rising Sun. The position of the center of the Sun below the horizon defines the three different periods of twilight: *civil twilight* (less than 6° below the horizon), *nautical twilight* (between 6° to 12° below the horizon), and *astronomical twilight* (between 12° and 18° below the horizon).

two-body problem The problem in classical celestial mechanics that addresses the relative motion of two point masses under their mutual gravitational attraction.

two-dimensional grain configuration In rocketry, the solid-propellant grain whose burning surface is described by two-dimensional analytical geometry (i.e., the cross section is independent of length).

See also ROCKET.

two-phase flow Simultaneous flow of gases and solid particles (e.g., condensed metal oxides); or the simultaneous flow of liquids and gases (vapors).

Tychonic system Despite the fact that he was Europe's most skilled naked eye astronomer, the 16th-century Danish astronomer Tycho Brahe bitterly resisted accepting or even considering the heliocentric hypothesis of NICHOLAS COPERNICUS. Instead, Brahe chose to concoct his own elaborate geocentric model of the solar system, an erroneous model that became known as the Tychonic system. He advocated that all the planets except Earth revolved around the Sun and that the Sun and its entire assemblage of planets revolved around a stationary Earth. Few, if any, late 16th-century astronomers paid serious attention to the Tychonic system.

Tycho's star A supernova (Type 1) that appeared in the constellation Cassiopeia in November 1572. The Danish astronomer TYCHO BRAHE discovered and carefully studied this supernova, which remained visible to the naked eye for about 18 months. In 1573 he reported his observations in *De Nova Stella* (The new star).

See also SUPERNOVA.

***Uhuru* (spacecraft)** NASA's *Uhuru* satellite was the first Earth-orbiting mission dedicated entirely to celestial X-ray astronomy. The spacecraft was launched by a Scout rocket on December 12, 1970 from the San Marco platform off the coast of Kenya, Africa. The satellite's launch date also coincided with the seventh anniversary of Kenyan independence, so in recognition of this, NASA renamed the *Small Astronomical Satellite 1* (SAS-1) *Uhuru,* the Swahili word for "freedom." The mission operated for more than two years and ended in March 1973. *Uhuru* performed the first comprehensive and uniform all-sky survey of cosmic X-ray sources. Instruments (two sets of proportional counters) onboard the spacecraft detected 339 X-ray sources of astronomical interest, such as X-ray binaries, supernova remnants, and Seyfert galaxies. Also called *Explorer 42*.

See also X-RAY ASTRONOMY.

ullage The amount that a container, such as a fuel tank, lacks of being full.

ullage pressure Pressure in the ullage space of a container, either supplied or self-generated.

ultra- A prefix meaning "surpassing a specified limit, range, or scope" or "beyond."

ultrahigh frequency (UHF) A radio frequency in the range 0.3 gigahertz (GHz) to 3.0 GHz.

ultrasonic Of or pertaining to frequencies above those that affect the human ear, that is, acoustic waves at frequencies greater than approximately 20,000 hertz.

ultraviolet (UV) astronomy Astronomy based on the ultraviolet (10- to 400-nanometer wavelength) portion of the electromagnetic (EM) spectrum. Because of the strong absorption of UV radiation by Earth's atmosphere, ultraviolet astronomy must be performed using high-altitude balloons, rocket probes, and orbiting observatories. Ultraviolet data gathered from spacecraft are extremely useful in investigating interstellar and intergalactic phenomena. Observations in the ultraviolet wavelengths have shown, for example, that the very-low-density material that can be found in the interstellar medium is quite similar throughout the Milky Way galaxy but that its distribution is far from homogeneous. In fact, UV data have led some astrophysicists to postulate that low-density cavities, or "bubbles," in interstellar space are caused by supernova explosions and are filled with gases that are much hotter than the surrounding interstellar medium. Ultraviolet data gathered from space-based observatories have revealed that some stars blow off material in irregular bursts and not in a steady flow, as was originally thought. Astrophysicists also find that ultraviolet data are of considerable use when they want to study many of the phenomena that occur in distant galaxies.

See also ASTROPHYSICS.

ultraviolet (UV) radiation That portion of the electromagnetic spectrum that lies beyond visible (violet) light and is longer in wavelength than X-rays. Generally taken as electromagnetic radiation with wavelengths between 400 nanometers (just past violet light in the visible spectrum) and about 10 nanometers (the extreme ultraviolet cutoff and the beginning of X-rays).

See also ELECTROMAGNETIC SPECTRUM.

Ulugh Beg (Muhammed Targai) (1394–1449) Mongol *Astronomer, Mathematician* The astronomer and mathematician Ulugh Beg (aka Muhammed Targai) was a 15th-century Mongol prince and the grandson of the great Mongol conqueror Tamerlane (aka Timur). In about 1420 he began developing a great observatory in the city of Samarkand (modern-day Uzbekistan). Using large instruments, Ulugh Beg and his staff carefully observed the positions of the Sun,

Moon, planets, and almost 1,000 fixed stars and published an accurate star catalog in 1437. His star catalog, called the *Zij-i Sultani,* represented the first improved listing of stellar positions and magnitudes since PTOLEMY. Following the death of his father, who was king, in 1447, Ulugh Beg briefly served as ruler over the region. In 1449 he was assassinated by his own son, 'Abd al-Latif. With this brutal act of politically motivated patricide also died the golden age of Mongol astronomy and the greatest astronomer of the time.

Ulysses mission An international space project to study the poles of the Sun and the interstellar environment above and below these solar poles. The Ulysses mission is named for the legendary Greek hero in Homer's epic saga of the Trojan War, who wandered into many previously unexplored areas on his return home. It is a survey mission designed to examine the properties of the solar wind, the structure of the Sun–solar wind interface, the heliospheric magnetic field, solar radio bursts and plasma waves, solar and galactic cosmic rays, and the interplanetary/interstellar neutral gas and dust environment—all as a function of solar latitude. The *Ulysses* spacecraft was built by Dornier Systems of Germany for the European Space Agency (ESA), which is also responsible for in-space operations of the Ulysses mission.

NASA provided launch support using the space shuttle *Discovery* and an upper-stage configuration consisting of a two-stage inertial upper-stage (IUS) rocket and a PAM-S (Payload Assist Module) configuration. In addition, the United States, through the Department of Energy, provided the radioisotope thermoelectric generator (RTG) that supplies electric power to this spacecraft. *Ulysses* is tracked and its scientific data collected by NASA's Deep Space Network (DSN). Spacecraft monitoring and control as well as data reduction and analysis is performed at NASA's Jet Propulsion Laboratory (JPL) by a joint ESA-JPL team.

Ulysses is the first spacecraft to travel out of the ecliptic plane in order to study the unexplored region of space above the Sun's poles. To reach the necessary high solar latitudes *Ulysses* was initially aimed close to Jupiter so that the giant planet's large gravitational field would accelerate the spacecraft out of the ecliptic plane to high latitudes. The gravity assist encounter with Jupiter occurred on February 8, 1992. After the Jupiter encounter *Ulysses* traveled to higher latitudes, with maximum southern latitude of 80.2° being achieved on September 13, 1994 (South Polar Pass 1).

Since *Ulysses* was the first spacecraft to explore the third dimension of space over the poles of the Sun, space scientists experienced some surprising discoveries. For example, they learned that two clearly separate and distinct solar wind regimes exist, with fast wind emerging from the solar poles. Scientists also were surprised to observe how cosmic rays make their way into the solar system from galaxies beyond the Milky Way galaxy. The magnetic field of the Sun over its poles turns out to be very different from that previously expected—based on observations from Earth. Finally, *Ulysses* detected a beam of particles from interstellar space that was penetrating the solar system at a velocity of about 80,000 kilometers per hour, or about 22.22 km/s.

Ulysses then traveled through high northern latitudes during June through September 1995 (North Polar Pass 1). The spacecraft's high-latitude observations of the Sun occurred during the minimum portion of the 11-year solar cycle.

In order to fully understand the Sun, however, scientists also wanted to study our parent star at near maximum activity conditions of the 11-year cycle. The extended mission of the far-traveling, nuclear-powered scientific spacecraft provided the opportunity. During solar maximum conditions, *Ulysses* achieved high southern latitudes between September 2000 and January 2001 (South Polar Pass 2) and then traveled through high northern latitudes between September 2001 and December 2001 (North Polar Pass 2). Now well into its extended mission, *Ulysses* continues to send back valuable scientific information on the inner workings of the Sun, especially concerning its magnetic field and how that magnetic field influences the solar system.

This mission was originally called the International Solar Polar Mission (ISPM). The original mission planned for two spacecraft, one built by NASA and the other by ESA. However, NASA canceled its spacecraft part of the original mission in 1981 and instead provided launch and tracking support for the single spacecraft built by ESA.

See also SUN.

umbilical An electrical or fluid servicing line between the ground or a tower and an upright rocket vehicle before launch. Also called umbilical cord.

umbra 1. The central (darkest) region of the shadow cast by an eclipsing body. The dark umbra is surrounded by a less dark outer region of shadow called the *penumbra.* 2. The central part of a sunspot, which is the coolest and darkest region.

See also ECLIPSE; SUNSPOT.

Umbriel The mid-sized (1,170-km-diameter) moon of Uranus discovered by the British amateur astronomer WILLIAM LASSELL in 1851. Named after a character in Alexander Pope's satirical poem *The Rape of the Lock,* Umbriel is in synchronous rotation around Uranus at a distance of 266,000 km with a period of 4.144 days. Like the other large Uranian moons, Umbriel appears to consist of a mixture of about 40 to 50 percent water ice with the rest rock. Images from the *Voyager 2* flyby of January 1986 show Umbriel has an older, heavily cratered surface, which appears to have remained stable since its formation. One interesting surface feature that puzzles scientists is the fact that Umbriel is very dark, having only about half of the visual albedo (namely, 0.18) of Ariel (0.34), which is the brightest of the large Uranian moons.

See also URANUS.

UMR injector Injector that produces a uniform mixture ratio (UMR) in the combustion chamber of a liquid-propellant rocket engine and thus a combustion region with relatively uniform temperature distribution.

See also ROCKET.

underexpansion A condition in the operation of a rocket nozzle in which the exit area of the nozzle is insufficient to permit proper expansion of the propellant gas. Consequently, the exiting gas is at a pressure greater than the ambient pressure, and this leads to the formation of external expansion waves.

unidentified flying object (UFO) A flying object apparently seen in terrestrial skies by an observer who cannot determine its nature. The vast majority of such UFO sightings can, in fact, be explained by known phenomena. However, these phenomena may be beyond the knowledge or experience of the person making the observation. Common phenomena that have given rise to UFO reports include artificial Earth satellites, aircraft, high-altitude weather balloons, certain types of clouds, and even the planet Venus.

There are, nonetheless, some reported sightings that cannot be fully explained on the basis of the data available (which may be insufficient or too scientifically unreliable) or on the basis of comparison with known phenomena. It is the investigation of these relatively few UFO sighting cases that has given rise, since the end of World War II, to the *UFO hypothesis*. This popular, though technically unfounded, hypothesis speculates that these unidentified flying objects are under the control of extraterrestrial beings who are surveying and visiting the Earth.

Modern interest in UFOs appears to have begun with a sighting report made by a private pilot named Kenneth Arnold. In June 1947 he reported seeing a mysterious formation of shining disks in the daytime near Mount Rainier in the state of Washington. When newspaper reporters heard of his account of "shining saucer-like disks," the popular term *flying saucer* was born.

In 1948 the U.S. Air Force (USAF) began to investigate these UFO reports. Project Sign was the name given by the air force to its initial study of UFO phenomena. In the late 1940s Project Sign was replaced by Project Grudge, which in turn became the more familiar Project Blue Book. Under Project Blue Book the U.S. Air Force investigated many UFO reports from 1952 to 1969. Then, on December 17, 1969, the secretary of the USAF announced the termination of Project Blue Book.

The USAF decision to discontinue UFO investigations was based on the following circumstances: (1) an evaluation of a report prepared by the University of Colorado and entitled "Scientific Study of Unidentified Flying Objects" (this report is also often called the Condon report after its principal author, EDWARD UHLER CONDON); (2) a review of this University of Colorado report by the National Academy of Sciences; (3) previous UFO studies; and (4) U.S. Air Force experience from nearly two decades of UFO report investigations.

As a result of these investigations and studies and of experience gained from UFO reports since 1948, the conclusions of the U.S. Air Force were: (1) No UFO reported, investigated, and evaluated by the USAF ever gave any indication of threatening national security; (2) there was no evidence submitted to or discovered by the USAF that sightings categorized as "unidentified" represent technological developments or principles beyond the range of present-day scientific knowledge; and (3) there was no evidence to indicate that the sightings categorized as "unidentified" were extraterrestrial vehicles.

With the termination of Project Blue Book, the U.S. Air Force regulation establishing and controlling the program for investigating and analyzing UFOs was rescinded. All documentation regarding Project Blue Book investigations was then transferred to the Modern Military Branch, National Archives and Records Service, 8th Street and Pennsylvania Avenue, N.W., Washington, D.C. 20408. This material is presently available for public review and analysis. If you wish to review these files personally, you need simply to obtain a researcher's permit from the National Archives and Records Service. Of a total of 12,618 sightings reported to Project Blue Book, 701 remained "unidentified" when the USAF ended the project. Since the termination of Project Blue, nothing has occurred that has caused the USAF or any other federal agency to support a resumption of UFO investigations. Today reports of unidentified objects entering North American air space are still of interest to the military—primarily as part of its overall defense surveillance program and support of heightened national antiterrorism activities—but beyond these national defense–related missions, the U.S. Air Force no longer investigates reports of UFO sightings.

Over the past half century the subject of UFOs has evoked strong opinions and emotions. For some people the belief in or study of UFOs has assumed almost the dimensions of a religious quest. Other individuals remain nonbelievers, or at least very skeptical, concerning the existence of alien beings and elusive vehicles that never quite seem to manifest themselves to scientific investigators or competent government authorities. Regardless of one's conviction, nowhere has the debate about UFOs in the United States been more spirited than over the events that unfolded near the city of Roswell, New Mexico, in the summer of 1947. This event, popularly known as the Roswell incident, has become widely celebrated as a UFO encounter. Numerous witnesses, including former military personnel and respectable members of the local community, have come forward with stories of humanoid beings, alien technologies, and government coverups that have caused even the most skeptical observer to pause and consider the reported circumstances. Inevitably, over the years these tales have spawned countless articles, books, and motion pictures concerning visitors from outer space who crashed in the New Mexico desert.

As a result of increasing interest and political pressure concerning the Roswell incident, in February 1994 the U.S. Air Force was informed that the General Accounting Office (GAO), an investigative agency of Congress, planned to conduct a formal audit in order to determine the facts regarding the reported crash of a UFO in 1947 at Roswell, New Mexico. The GAO's investigative task actually involved numerous federal agencies, but the focus was on the U.S. Air Force, the agency most often accused of hiding information and records concerning the Roswell incident. The GAO research team conducted an extensive search of U.S. Air Force archives, record centers, and scientific facilities. Seeking information that might help explain peculiar tales of odd wreckage and alien bodies, the researchers reviewed a large number of documents concerning a variety of events, including aircraft crashes, errant missile tests (from White Sands, New Mexico), and nuclear mishaps.

This extensive research effort revealed that the Roswell incident was not even considered a UFO event until the

1978–80 time frame. Prior to that, the incident was generally dismissed because officials in the U.S. Army Air Force (predecessor to the U.S. Air Force) had originally identified the debris recovered as being that of a weather balloon. The GAO research effort located no records at existing air force offices that indicated any cover-up by the USAF or that provided any indication of the recovery of an alien spacecraft or its occupants. However, records were located and investigated concerning a then–top secret balloon project known as Project Mogul attempted to monitor Soviet nuclear tests. Comparison of all information developed or obtained during this effort indicated that the material recovered near Roswell was consistent with a balloon device and most likely from one of the Project Mogul balloons that had not previously been recovered. This government response to contemporary inquiries concerning the Roswell incident is described in an extensive report released in 1995 by Headquarters USAF, entitled "The Roswell Report: Fact versus Fiction in the New Mexico Desert." While the National Aeronautics and Space Administration (NASA) is the current federal agency focal point for answering public inquiries to the White House concerning UFOs, the civilian space agency is not engaged in any research program involving these UFO phenomena or sightings, nor is any other agency of the U.S. government.

One interesting result that emerged from Project Blue Book is a scheme, developed by Dr. J. Allen Hynek, to classify, or categorize, UFO sighting reports. The table describes the six levels of classification that have been used. A type-A UFO report generally involves seeing bright lights in the night sky. These sightings usually turn out to be a planet (typically Venus), a satellite, an airplane, or meteors. A type-B UFO report often involves the daytime observation of shining disks (that is, flying saucers) or cigar-shaped metal objects. This type of sighting usually ends up as a weather balloon, a blimp or lighter-than-air ship, or even a deliberate prank or hoax. A type-C UFO report involves unknown images appearing on a radar screen. These signatures might linger, be tracked for a few moments, or simply appear and then quickly disappear—often to the amazement and frustration of the scope operator. These radar visuals frequently turn out to be something like swarms of insects, flocks of birds, unannounced aircraft, and perhaps the unusual phenomena radar operators call "angels." To radar operators, angels are anomalous radar wave propagation phenomena.

Close encounters of the first kind (visual sighting of a UFO at moderate to close range) represent the type-D UFO reports. Typically, the observer reports something unusual in the sky that "resembles an alien spacecraft." In the type-E UFO report, not only does the observer claim to have seen the alien spaceship but also reports the discovery of some physical evidence in the terrestrial biosphere (such as scorched ground, radioactivity, mutilated animals, etc.) that is associated with the alien craft's visit. This type of sighting has been named a close encounter of the second kind. Finally, in the type-F UFO report, which is also called a close encounter of the third kind, the observer claims to have seen and sometimes to have been contacted by alien visitors. Extraterrestrial contact stories range from simple sightings of "UFOnauts," to communication with them (usually telepathic), to cases of kidnapping and then release of the terrestrial observer. There are even some reported stories in which a terran was kidnapped and then seduced by an alien visitor, a challenging task of romantic compatibility even for an advanced star-faring species!

Despite numerous stories about such UFO encounters, not a single shred of scientifically credible, indisputable evidence has yet to be acquired. If we were to judge these reports on some arbitrary proof scale, the second table might be used as a guide for helping us determine what type of data or testimony we will need to convince ourselves that the "little green men" (LGM) have arrived in their flying saucer. Unfortunately, we do not have any convincing data to support categories 1 to 3 in the table. Instead, all we have are large quantities of eyewitness accounts of various UFO encounters (category-4 items). Even the most sincere human testimony changes over time and is often subject to wide variation and contradiction. The scientific method puts very little weight on human testimony in validating a hypothesis.

UFO Report Classifications

A. Noctural (nighttime) light
B. Diurnal (daytime) disk
C. Radar contact (radar visual [RV])
D. Visual sighting of alien craft at modest to close range (also called *close encounter of the first kind* [CE I])
E. Visual sighting of alien craft plus discovery of (hard) physical evidence of craft's interaction with terrestrial environment (also called *close encounter of the second kind* [CE II])
F. Visual sighting of aliens themselves, including possible physical contact (also called *close encounter of the third kind* [CE III])

Source: Derived from work of Dr. J. Allen Hynek and Project Blue Book.

Proposed "Proof Scale" to Establish Existence of UFOs

Highest Value[a]

(1) The alien visitors themselves or the alien spaceship
(2) Irrefutable physical evidence of a visit by aliens or the passage of their spaceship
(3) Indisputable photograph of an alien spacecraft or one of its occupants
(4) Human eyewitness reports

Lowest Value

[a] From a standpoint of the scientific method and validation of the UFO hypothesis with "hard" technical data

Source: Based on work of Dr. J. Allen Hynek and Project Blue Book.

Even from a more philosophical point of view, it is very difficult to logically accept the UFO hypothesis. Although intelligent life may certainly have evolved elsewhere in the universe, the UFO encounters reported to date hardly reflect the logical exploration patterns and encounter sequences we might anticipate from an advanced, star-faring alien civilization.

In terms of our current understanding of the laws of physics, interstellar travel appears to be an extremely challenging, if not technically impossible, undertaking. Any alien race that developed the advanced technologies necessary to travel across vast interstellar distances would most certainly be capable of developing sophisticated remote sensing technologies. With these remote sensing technologies they could study the Earth essentially undetected—unless, of course, they wanted to be detected. And if they wanted to make contact, they could most surely observe where the Earth's population centers are and land in places where they could communicate with competent authorities. It is insulting not only to their intelligence but to our own human intelligence as well to think that these alien visitors would repeatedly contact only people in remote, isolated areas, scare the dickens out of them, and then lift off into the sky. Why not land in the middle of the Rose Bowl during a football game or near the site of an international meeting of astronomers and astrophysicists? And why conduct only short, momentary intrusions into the terrestrial biosphere? After all, the Viking landers we sent to Mars gathered data for years. It is hard to imagine that an advanced culture would make the tremendous resource investment to send a robot probe or even arrive here themselves and then only flicker through an encounter with just a very few beings on this planet. Are we that uninteresting? If that is the case, then why are there so many reported visits? From a simple exercise of logic, the UFO hypothesis just does not make sense—terrestrial or extraterrestrial!

Hundreds of UFO reports have been made since the late 1940s. Again, why are we so interesting? Are we at a galactic crossroads? Are the outer planets of our solar system an "interstellar truck stop" where alien starships pull in and refuel? (Some people have already proposed this hypothesis.) Let us play a simplified interstellar traveler game to see if so many reported visits are realistic, even if we are very interesting. First, we assume that our galaxy of more than 100 billion stars contains about 100,000 different star-faring alien civilizations that are more or less uniformly dispersed. (This is a very optimistic number according to the Drake equation and to scientists who have speculated about the likelihood of Kardashev Type II civilizations.) Then each of these civilizations has, in principle, 1 million other star systems to visit without interfering with any other civilization. (Yes, the Milky Way galaxy is a really big place!) What do you think the odds are of two of these civilizations both visiting our solar system and each only casually exploring the planet Earth during the last five decades? The only logical conclusion that can be drawn is that the UFO encounter reports are not credible indications of extraterrestrial visitations. Yet, despite the lack of scientific evidence, UFO-related Web sites are among the most popular and frequently visited on the Internet.

See also ANCIENT ASTRONAUT THEORY; DRAKE EQUATION; INTERSTELLAR COMMUNICATION AND CONTACT.

unit The unit defines a measurement of a physical quantity, such as length, mass, or time. There are two common unit systems in use in aerospace applications today, the International System of Units (SI), which is based on the meter-kilogram-second (mks) set of fundamental units, and the American standard system of units, which is based on the foot–pound-mass–second (fps) set of fundamental units. In the standard system of units, 1 pound-mass (lbm) is defined as equaling 1 pound-force (lbf) on the surface of Earth at sea level. Derived units, such as energy, power, and force, are based on combinations of the fundamental units in accordance with physical laws, such as Newton's laws of motion. The International System of Units is used by scientists and engineers around the world and is the preferred choice in astronomy and space exploration. However, the American standard system of units is often encountered in the practice of aerospace engineering in the United States and occasionally in the field of observational astronomy, especially in describing the lens or mirror size of a telescope, such as the phrase "a 70-inch reflector."

See also SI UNITS.

United States Air Force Starting with the nuclear missile race during the cold war, the United States Air Force, through various commands and organizations, has served as the primary agent for the space defense needs of the United States. For example, all military satellites have been launched from Cape Canaveral Air Force Station (CCAFS), Florida, or Vandenberg Air Force Base (VAFB), California, rocket ranges owned and operated by the U.S. Air Force. The end of the cold war (circa 1989) and the dramatic changes in national defense needs at the start of this century have placed additional emphasis on access to space and the use of military space systems for intelligence gathering, surveillance, secure and dependable information exchange, and the protection and enhancement of American fight forces deployed around the world.

The Air Force Space Command (AFSPC) was created on September 1, 1982. Headquartered at Peterson AFB, Colorado, it is the major command that directly discharges much of the U.S. Air Force responsibilities for national defense with respect to space systems and land-based intercontinental ballistic missiles (ICBMs). Through its subordinate organization at Vandenberg AFB, for example, the AFSPC provides space war fighting forces to the U.S. Strategic Command and warning support to the North American Air Defense Command (NORAD). AFSPC also exercises overall responsibility for the development, acquisition, operation, and maintenance of land-based ICBMs. The AFSPC's Space Warfare Center at Schriever AFB, Colorado, plays a major role in fully integrating space systems into the operational components of today's air force. The Space Warfare Center's force enhancement mission examines ways to use space systems to support fighters in the areas of navigation, weather, intelligence, communications, and theater ballistic missile warning and also how these apply to theater operations. Finally, AFSPC's Space and Missile Center at Los Angeles AFB, California, designs and acquires all U.S. Air Force and most Department of Defense space systems. It oversees launches, completes on-orbit checkouts, and then turns the various space systems over to the appropriate user agencies.

United States Space Command (USSPACECOM) A unified command of the Department of Defense with headquarters at Peterson Air Force Base, Colorado Springs, Colorado. USSPACECOM was activated on September 23, 1985, to consolidate and operate military assets affecting U.S. activities in space. The command was composed of three components, the U.S. Air Force, Naval, and Army Space Commands. USSPACECOM had three primary missions, space operations, warning, and planning for ballistic missile defense (BMD). The commander in chief of USSPACECOM was directly responsible to the president through the secretary of defense and the Joint Chiefs of Staff. USSPACECOM exercised operational command of assigned U.S. military space assets through the air force, navy, and army component space commands. Merged into the U.S. Strategic Command (USSTRATCOM) on October 1, 2002.

United States Strategic Command (USSTRATCOM) The United States Strategic Command (USSTRATCOM) was formed on October 1, 2002, with the merger of the United States Space Command (USSPACECOM) and USSTRATCOM, the strategic forces organization within the Department of Defense. The merger was part of an ongoing initiative within the Department of Defense to transform the American military into a 21st-century fighting machine. The blending of the two previously existing organizations improved combat effectiveness and accelerated the process of information collection and assessment needed for strategic decision making. As a result, the merged command is now responsible for both early warning of and defense against missile attacks as well as long-range strategic attacks.

The newly created United States Strategic Command is headquartered at Offutt Air Force Base, Nebraska, the well-known headquarters of the U.S. Air Force's Strategic Air Command during the cold war. USSTRATCOM is one of nine unified commands within the Department of Defense. The command serves as the command and control center for U.S. strategic (nuclear) forces and controls military space operations, computer network operations, information operations, strategic warning and intelligence, and global strategic planning. Finally, the command is charged with deterring and defending against the proliferation of weapons of mass destruction.

USSTRATCOM coordinates the use of the Department of Defense's military space forces in providing missile warning, communications, navigation, weather, imagery, and signals intelligence. It provides space support to deployed American military forces worldwide and also defends the information technology structure within the Department of Defense.

The command's space missions consist of space support, force enhancement, space control, and force application. All military satellites are launched from Cape Canaveral Air Force Station, Florida, or Vandenberg Air Force Base, California. Once the satellites reach their final orbits and begin operating, U.S. Army, Navy, and/or Air Force personnel track and "fly" the spacecraft and operate their specialized payloads through a worldwide network of ground stations. USSTRATCOM has several military space components—the Army Space Command (ARSPACE) in Arlington, Virginia; the Naval Space Command (NAVSPACE) in Dahlgren, Virginia; and Space Air Force (SPACEAF) at Vandenberg Air Force Base in California.

Unity The first U.S.-built component of the *International Space Station* (ISS); a six-sided connecting module and passageway (node), Unity was the primary cargo of the space shuttle *Endeavour* during the STS-88 mission in early December 1998. Once delivered into orbit, astronauts mated Unity to the Russian-built Zarya module, delivered earlier into orbit by a Russian Proton rocket that lifted off from the Baikonour Cosmodrome.

See also INTERNATIONAL SPACE STATION.

universal time (UT) The worldwide civil time standard, equivalent to Greenwich mean time (GMT).

universal time coordinated (UTC) The worldwide scientific standard of timekeeping, based on carefully maintained atomic clocks. It is kept accurate to within microseconds. The addition (or subtraction) of "leap seconds" as necessary at two opportunities every year keeps UTC in step with Earth's rotation. Its reference point is Greenwich, England. When it is midnight there on Earth's prime meridian, it is midnight (00:00:00.000000) UTC, often referred to as "all balls" in aerospace jargon.

universe Everything that came into being at the moment of the big bang, and everything that has evolved from that initial mass of energy; everything that we can (in principle) observe. All energy (radiation), all matter, and the space that contains them.

From ancient times people have looked up at the night sky and wondered how many celestial objects existed and how far away they were. The fortuitous union of modern space technology and astronomy allows us to look deeper into the universe than ever before.

Cosmology is the scientific study of the large-scale properties of the universe as a whole. It endeavors to apply the scientific method to understand the origin, evolution, and ultimate fate (destiny) of the entire universe. Today the big bang model is a broadly accepted theory for the origin and evolution of our universe. It postulates that 12 to 14 billion years ago, the portion of the universe we can see today was only a few millimeters across. It has since expanded from this hot dense state into the vast and much cooler cosmos we currently inhabit. We now see remnants of this original hot dense matter as the very cold cosmic microwave background radiation that still pervades the universe and is observable by microwave detection instruments as a uniform glow across the entire sky.

Presented with a variety of interesting new astrophysical data in the latter portion of the 20th century (much of these data acquired by space-based observatories), astrophysicists and cosmologists are now responding to some very intriguing questions concerning the future of the universe. Is the universe going to continue expanding forever—possibly at an accelerating rate? If not, when will the expansion stop? What role do dark matter and dark energy play in the fate of the universe? Space-based instruments will continue to provide interesting pieces of the cosmic puzzle, and 21st-century scientists will be better prepared to respond to such questions and the ultimate question: Where is the universe going and when will it end (if ever)?

See also ASTROPHYSICS; BIG BANG THEORY; COSMOLOGY.

This view of nearly 10,000 galaxies is called the Hubble Ultra Deep Field (HUDF), the deepest portrait of the visible universe ever achieved by humankind. The million-second-long exposure taken by the *Hubble Space Telescope* (HST) began on September 24, 2003, and continued through January 16, 2004. The HUDF reveals the first galaxies to emerge from the so-called dark ages, the time shortly after the big bang when the first stars reheated the cold, dark universe. This historic new view is actually two separate images taken by the HST's Advance Camera for Surveys (ACS) and Near-Infrared Camera and Multi-Object Spectrometer (NICMOS). Both images reveal galaxies that are too faint to be seen by ground-based telescopes or even by the HST's previous faraway looks, called the Hubble Deep Fields (HDFs), which were taken in 1995. ***(Courtesy of NASA, ESA, S. Beckwith [STScI] and the HUDF Team)***

unmanned Without human crew; *unpersoned* or *uncrewed* are more contemporary terms.

unmanned aerial vehicle (UAV) A robot aircraft flown and controlled through teleoperation by a distant human operator. Also called a remotely piloted vehicle (RPV).

upcomer Nozzle tube in which coolant flows in a direction opposite to that of the exhaust gas flow.

uplink The telemetry signal sent from a ground station to a spacecraft or planetary probe.

See also TELECOMMUNICATIONS.

uplink data Information that is passed from a ground station on Earth to a spacecraft, space probe, or space platform.

See also TELECOMMUNICATIONS.

upper atmosphere The general term applied to outer layers of Earth's atmosphere above the troposphere. It includes the stratosphere, mesosphere, thermosphere, and exosphere. Sounding rockets and Earth-observing satellites often are used to gather scientific data about this region.

See also ATMOSPHERE; *UPPER ATMOSPHERE RESEARCH SATELLITE*.

Upper Atmosphere Research Satellite **(UARS)** NASA satellite launched on September 12, 1991, from the space shuttle *Discovery* during the STS-48 mission. UARS was released by the astronaut crew into a 585-km-altitude, 57°-inclination orbit. UARS studied the physical and chemical processes of Earth's stratosphere, mesosphere, and lower thermosphere. The spacecraft had a mission lifetime of three years.

upper stage In general, the second, third, or later stage in a multistage rocket. Often getting into low Earth orbit (LEO) is only part of the effort against the planet's gravitational forces for geosynchronous satellites and interplanetary spacecraft. Once propelled into LEO by a reusable aerospace vehicle, such as the space shuttle, or an expendable launch vehicle, such as the Delta or Titan, these payloads depend on an upper-stage vehicle to boost them on the next phase of their journey. Upper-stage vehicles carried by the space shuttle are solid-propellant rocket vehicles that provide the extra thrust, or "kick," needed to move spacecraft into higher orbits or interplanetary trajectories. Solid-propellant or liquid-propellant rocket vehicles are used as upper stages with expendable launch vehicles.

The upper-stage vehicle, its payload, and any required supporting hardware (e.g., a special cradle attached to the shuttle's cargo bay or an adapter mounted to the first or second stage of an expendable launch vehicle) are carried into LEO by the launch vehicle. Once in LEO, the upper-stage vehicle and its attached payload separate from the launch vehicle and fire the rocket engine (or engines) necessary to deliver the payload to its final destination.

See also AGENA; CENTAUR; INERTIAL UPPER STAGE; PAYLOAD ASSIST MODULE.

Uranus Unknown to ancient astronomers, the planet Uranus was discovered by SIR WILLIAM HERSCHEL in 1781. Initially called Georgium Sidus (George's star, after England's King George III) and Herschel (after its discoverer), in the 19th century the seventh planet from the Sun was finally named Uranus after the ancient Greek god of the sky and father of the Titan Cronos (Saturn in Roman mythology).

At nearly 3 billion kilometers from the Sun, Uranus is too distant from Earth to permit telescopic imaging of its features by ground-based telescopes. Because of the methane in its upper atmosphere, the planet appears as only a blue-green disk or blob in the most powerful of terrestrial telescopes. On January 24, 1986, a revolution took place in our understanding and knowledge about this planet, as the *Voyager 2*

Selected Physical and Dynamic Properties of the Planet Uranus

Diameter (equatorial)	51,120 km
Mass (estimated)	8.7×10^{25} kg
"Surface" gravity	8.69 m/s^2
Mean density (estimated)	1.3 g/cm^3
Albedo (visual)	0.66
Temperature (blackbody)	58 kelvins (K)
Magnetic field	Yes, intermediate strength (field titled 60° with respect to axis of rotation)
Atmosphere	Hydrogen (~83%) helium (~15%) methane (~2%)
"Surface" features	Bland and featureless (except for some discrete methane clouds)
Escape velocity	21.3 km/s
Radiation belts	Yes (intensity similar to those at Saturn)
Rotation period	17.24 hours
Eccentricity	0.047
Mean orbital velocity	6.8 km/s
Sidereal year (a Uranian year)	84 years
Inclination of planet's equator to its orbit around the Sun	97.9°
Number of (known) natural satellites (>10 km diameter)	26
Rings	Yes (11)
Average distance from Sun	2.871×10^9 km (19.19 AU) [159.4 light-min]
Solar flux at average distance from Sun	3.7 W/m^2 (approx.)

Source: Based on NASA data.

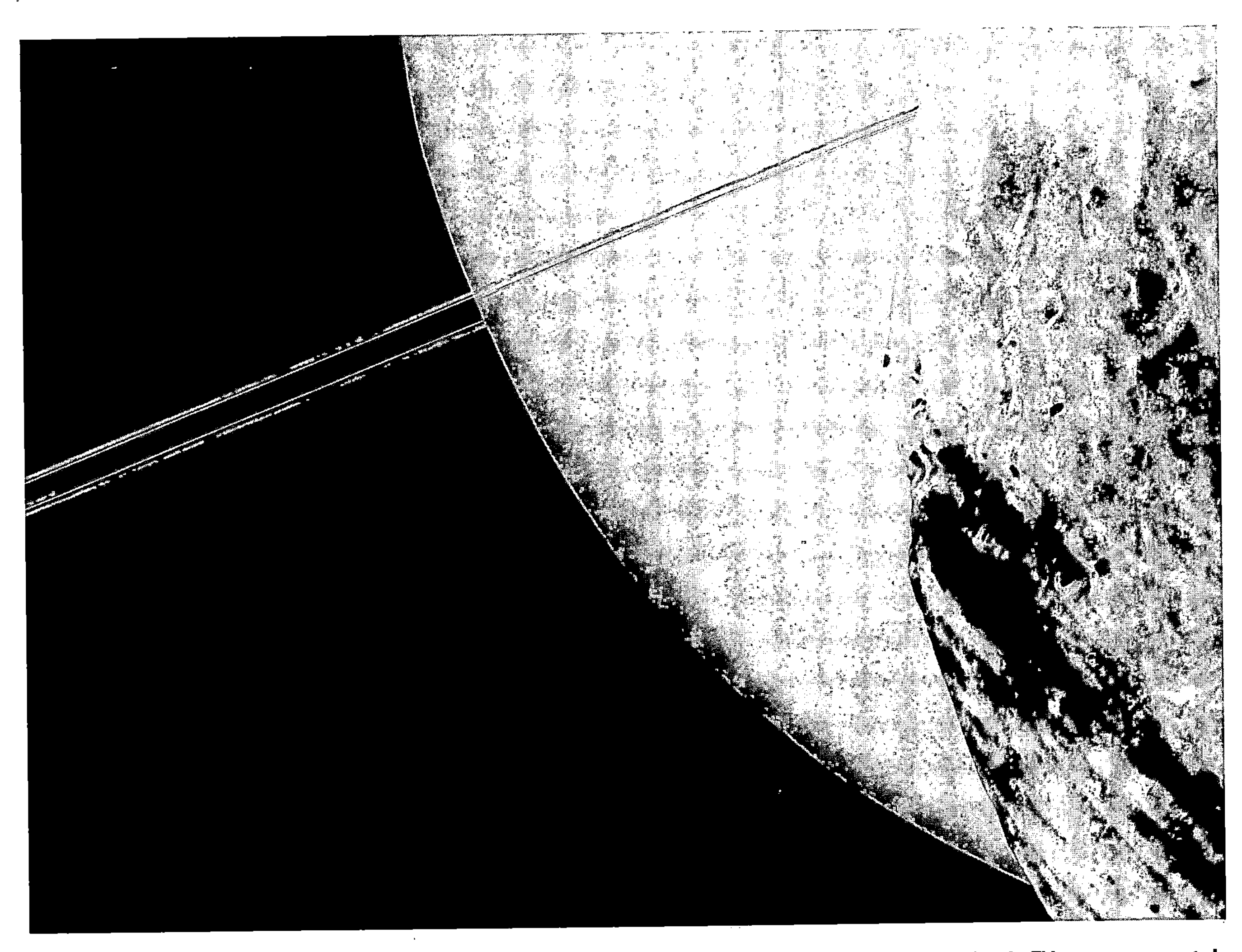

Stunning view of the rings and cloud tops of Uranus as seen at a distance of 105,000 kilometers from the Uranian moon Miranda. This scene was created from a montage of *Voyager 2* spacecraft encounter imagery. *(Courtesy of NASA/JPL)*

encountered the Uranian system at a relative velocity of more than 14 kilometers per second. What we know about Uranus today is largely the result of that spectacular encounter.

Uranus has one particularly interesting property—its axis of rotation lies in the plane of its orbit rather than vertical to the orbital plane, as occurs with the other planets. Because of this curious situation, Uranus moves around the Sun like a barrel rolling along on its side rather than like a top spinning on its end. In other words, Uranus is tipped over on its side, with its orbiting moons and rings creating the appearance of a giant bull's-eye. The northern and southern polar regions are alternatively exposed to sunlight or to the

Selected Physical and Dynamic Property Data for Major Uranian Moons

Name	Diameter (km)	Period (day)	Distance from Center of Uranus (km)	Visual Albedo	Average Density (g/cm³)
Miranda	470	1.414	129,800	0.27	1.3
Ariel	1,160	2.520	191,200	0.34	1.7
Umbriel	1,170	4.144	266,000	0.18	1.5
Titania	1,580	8.706	435,800	0.27	1.7
Oberon	1,520	13.463	582,600	0.24	1.6

Source: Based on NASA data.

darkness of space during the planet's 84-year-long orbit around the Sun. At its closest approach *Voyager 2* came within 81,500 kilometers of the Uranian cloudtops. The spacecraft telemetered back to Earth thousands of spectacular images and large quantities of other scientific data about the planet and its moons, rings, atmosphere, and interior.

The upper atmosphere of Uranus consists mainly of hydrogen (H_2) (approximately 83 percent) and helium (He) (approximately 15 percent) with small amounts of methane (CH_4) (about 2 percent), water vapor (H_2O), and ammonia (NH_3). The methane in the upper atmosphere of Uranus and its preferential absorption of red light gives the planet its overall blue-green color.

The table on page 629 presents selected physical and dynamic property data for Uranus, while the second table describes some of the physical features of the major Uranian moons. The large Uranian moons appear to be about 50 percent water ice, 20 percent carbon- and nitrogen-based materials, and 30 percent rock. Their surfaces, almost uniformly dark gray in color, display varying degrees of geological history. Very ancient, heavily cratered surfaces are apparent on some of the moons, while others show strong evidence of internal geological activity. Miranda, the innermost of the five large Uranian moons, is considered by scientists to be one of the strangest bodies yet observed in the solar system. *Voyager 2* images revealed an unusual world consisting of huge fault canyons as deep as 20 kilometers, terraced layers, and a mixture of old and young surfaces.

See also VOYAGER SPACECRAFT.

Urey, Harold Clayton (1893–1981) American *Chemist, Early Exobiologist* Nobel laureate Harold Clayton Urey investigated the possible origins of life on Earth from a chemical perspective and summarized some of his interesting concepts in the 1952 book *The Planets: Their Origin and Development*. In 1953, together with his student Stanley Miller, he performed a classic exobiology experiment now called the Urey-Miller experiment in which gaseous mixtures, simulating Earth's primitive atmosphere, were subjected to energy sources such as ultraviolet radiation and lightning discharges. To their great surprise life-forming organic compounds, called amino acids, appeared. Prior to this important and interesting exobiology experiment, Urey had previously been awarded the 1934 Nobel Prize in chemistry for his discovery of deuterium. He also played a key research role in the American atomic bomb development program, called the Manhattan Project, during World War II.

See also EXOBIOLOGY; LIFE IN THE UNIVERSE.

Ursa Major (UMa) Better known as the Great Bear constellation, Ursa Major is a very large and conspicuous constellation in the northern sky whose seven brightest stars form a very familiar shape (asterism) known as the *Big Dipper*. The "Pointers," Dubhe and Merak, are the two stars in the Big Dipper that point to Polaris and the North Pole.

See also CONSTELLATION (ASTRONOMY).

U.S. Geological Survey (astrogeology) The mission of the U.S. Geological Survey's (USGS) Astrogeology Research Program is to establish and maintain geoscientific and technical expertise in planetary science and remote sensing to perform the following tasks: scientifically study and map extraterrestrial bodies, plan and conduct planetary exploration missions, and explore and develop new technologies in data processing and analysis, archiving, and distribution.

The USGS Astrogeology Research Program has a rich history of participation in space exploration efforts and planetary mapping, starting in 1963 when the Flagstaff Field Center was established to provide lunar geological mapping and assist in training astronauts destined for the Moon. Throughout the years the program has participated in processing and analyzing data from various missions to the planetary bodies in our solar system, assisting in finding potential landing sites for exploration vehicles, mapping our neighboring planets and their moons, and conducting research to better understand the origins, evolution, and geological processes operating on these bodies.

There were a number of important factors in choosing Flagstaff, Arizona, as a location for the field center. For instance, the area provides excellent atmospheric conditions for astronomical observations of the Moon. A telescope was built here specifically to support a USGS program of lunar geological mapping in addition to the topographical maps of the Moon that were being made at Flagstaff's Lowell Observatory. Another important factor was Flagstaff's location near volcanic craters, Meteor Crater (also called the Arizona Meteorite Crater and the Ballinger Crater), and the Grand Canyon, which provide natural laboratories for field studies and astronaut training on terrains similar to the surface of the Moon.

The Astrogeology Research Program was the first occupant of the Flagstaff Field Center. The campus now has five buildings housing more than 150 science, research, technical, and administrative staff members from several USGS and Department of the Interior programs.

Eugene Merle (Gene) Shoemaker (1928–97) founded the USGS Astrogeology Research Program in 1961 and was its first chief scientist. He established the Flagstaff Field Center in 1963 and retired from the USGS in 1993. He remained on emeritus status with the USGS and maintained an affiliation with the Lowell Observatory until his death in a car accident in Australia in 1997.

Shoemaker was involved in the NASA's robot spacecraft Ranger and Surveyor Project and continued with the manned Apollo Project. He culminated his lunar studies in 1994 with new data on the Moon from the Department of Defense's *CLEMENTINE* project, for which he served as the science team leader. He also collaborated closely with his wife, Carolyn (b. 1929), a planetary astronomer. They codiscovered Comet Shoemaker-Levy (fragments of which impacted Jupiter in July 1994) along with colleague David Howard Levy (b. 1948).

Starting in 1963, the Astrogeology Research Program played an important role in training astronauts destined to explore the lunar surface and in supporting the testing of equipment for both manned and unmanned missions. As part of the astronauts' training, USGS and NASA geoscientists gave lectures and field trips during the 1960s and early 1970s to teach astronauts the basics of terrestrial and lunar geology. Field trips included excursions into the Grand Canyon to demonstrate the development of geological structure over time; to Lowell Observatory (Flagstaff) and Kitt Peak National Observatory (Tucson); to Meteor Crater east of

Flagstaff; and to Sunset Crater cinder cone and nearby lava flows in the Flagstaff area. This training was essential in giving astronauts the skills and understanding to make observations about what they would see on the lunar surface and to collect samples for later study on Earth.

The volcanic fields around Flagstaff have proven particularly useful in testing equipment and training astronauts. Cameras planned for use in NASA's Surveyor Project were tested on the Bonito Flow in Sunset Crater National Monument because the lava flow appeared to be similar to flows on the lunar surface. A field of artificial impact craters was created in the Cinder Lakes volcanic field near Flagstaff to create a surface similar to the proposed first manned American landing site on the Moon.

HARRISON H. SCHMITT joined the astrogeology team as a geologist at the Flagstaff Field Center in 1964, having recently earned a doctoral degree from Harvard University. In addition to assisting in the geological mapping of the Moon, he led the Lunar Field Geological Methods Project. When NASA announced a special recruitment for scientist-astronauts in late 1964, Schmitt applied. Out of more than 1,000 applicants, six were chosen. Of those six, Joe Kerwin, Owen Garriott, and EDWARD G. GIBSON would fly in the Skylab missions in 1973 and 1974, and Schmitt would go to the Moon on the *Apollo 17* mission.

Just one year after joining the USGS, Schmitt was transferred to NASA and began his pilot training at Williams Air Force Base in Arizona. On December 7, 1972, *Apollo 17* was launched, carrying Schmitt, RONALD E. EVANS, and mission commander EUGENE A. CERNAN. On December 11 the lunar excursion module (LEM) landed at Taurus-Littrow on the Moon, and four hours later Cernan and Schmitt became the 11th and 12th men to walk on the Moon. The pair did a total of three EVAs (extravehicular activities) before their departure on December 14, collecting rock and soil samples, taking photographs, setting up equipment, and making observations. During the mission Schmitt and Cernan discovered orange soil, a surprising find that created a great deal of excitement in the scientific community. Schmitt was the only geologist to go to the Moon. *Apollo 17* was the final human mission in a four-year exploration of the Moon. The crew of *Apollo 17* left a plaque that reads

> *"Here man completed his first exploration of the Moon, December 1972 A.D. May the spirit of peace in which we came be reflected in the lives of all mankind."*

See also APOLLO PROJECT; ARIZONA METEORITE CRATER; KITT PEAK NATIONAL OBSERVATORY; LOWELL OBSERVATORY.

U.S. Naval Observatory (USNO) Performs an essential scientific role for the United States, the U.S. Navy, and the Department of Defense. Its astrometry mission includes determining the positions and motions of the Earth, Sun, Moon, planets, stars, and other celestial objects; providing astronomical data; determining precise time; measuring the Earth's rotation; and maintaining the Master Clock for the United States. Founded in 1830 as the Depot of Charts and Instruments, the Naval Observatory is one of the oldest scientific agencies in the country. Observatory astronomers formulate the theories and conduct the relevant research necessary to improve these mission goals. This astronomical and timing data, essential for accurate navigation and the support of communications on Earth and in space, is vital to the U.S. Navy and Department of Defense. These data are also used extensively by other agencies of the government and the public at large.

The largest telescope located on the observatory grounds in Washington, D.C., is the historic 0.66-m (26-inch) refractor. Acquired in 1873, it was the world's largest refracting telescope until 1883. The American astronomer ASAPH HALL used this telescope in 1877 to discover the two tiny moons of Mars. Today this telescope is used mainly for determining the orbital motions and masses of double stars using a special camera known as a speckle interferometer and for planetary satellite observations.

The USNO's largest optical telescope is located at the Flagstaff Station in Arizona. It is the 1.55-meter astrometric reflector, used to obtain distances of faint objects and to measure the brightness and colors of stars both photographically and with charge-coupled device (CCD) technology. In 1978 photographic plates taken with this telescope led to the discovery of a large moon called Charon circling the planet Pluto. This is the largest optical telescope operated by the U.S. Navy. It was designed to produce extremely accurate astrometric measurements in small fields, and has been used to measure parallaxes and therefore distances for faint stars. Since 1964 more than 1,000 of the world's most accurate stellar distances and proper motions have been measured with this telescope. In recent years this telescope has also served as a test-bed for the development of state-of-the-art near-infrared detectors.

The USNO also uses radio telescopes at various locations to determine astronomical time and the orientation of the Earth in space. Finally, the USNO's newest instrument, under development at the Lowell Observatory's Anderson Mesa Station near Flagstaff, is the Navy Prototype Optical Interferometer (NPOI), which will provide unprecedented ground-based astrometry data and images.

Utopia Planitia The smooth Martian plain on which NASA's *Viking 2* lander successfully touched down on September 3, 1976. The location of the *Viking 2* landing site is 47.96° north latitude and 225.77° west longitude.

See also MARS; VIKING PROJECT.

UV Ceti star *See* FLARE STAR.

vacuum The absence of gas or a region in which there is a very low gas pressure. This is a relative term. For example, a *soft vacuum* (or *low vacuum*) has a pressure of about 0.01 pascal (i.e., 10^{-2} pascal); a *hard vacuum* (or *high vacuum*) typically has a pressure between 10^{-2} and 10^{-7} pascal; while pressures below 10^{-7} pascal are referred to as an *ultrahard* (or *ultrahigh*) vacuum.

Valles Marineris An extensive canyon system on Mars near the planet's equator discovered in 1971 by NASA's *Mariner 9* spacecraft.

Van Allen, James Alfred (1914–2006) American *Physicist, Space Scientist* James Van Allen, a pioneering space scientist, placed the Iowa cosmic ray experiment on the first American satellite, called *EXPLORER 1,* and with this instrument discovered the inner portion of Earth's trapped radiation belts in early 1958. Today space scientists call this distinctive zone of magnetically trapped atomic particles around Earth the Van Allen radiation belts in his honor.

Van Allen was born in Mount Pleasant, Iowa, on September 7, 1914. He graduated from Iowa Wesleyan College in 1935. During his sophomore year he made his first measurements of cosmic ray intensities. After graduation he attended the University of Iowa, where he earned a master's degree in 1936 and completed a Ph.D. in physics in 1939. From 1939 to 1942 Van Allen worked at the Carnegie Institution of Washington as a physics research fellow in the department of terrestrial magnetism. As a Carnegie fellow, he received valuable cross-training in geomagnetism, cosmic rays, nuclear physics, solar-terrestrial physics, and ionosphere physics. All of this scientific cross-fertilization prepared him well for a leading role as the premier American space scientist at the beginning of the space age.

In 1942 Van Allen transferred to the Applied Physics Laboratory at Johns Hopkins University to work on rugged vacuum tubes. That fall he received a wartime commission in the U.S. Navy, where he then served as an ordnance specialist for the remainder of World War II. One of his primary contributions was the development of an effective radio proximity fuse, one that detonated an explosive shell when the ordnance came near its target. Following combat duty in the Pacific, he returned to the laboratory at John Hopkins University in Maryland. There on an afternoon in March 1945 he quite literally ran into his future wife, Abigail, in a minor traffic accident. Six months later both drivers were married, and their fortuitous traffic encounter eventually yielded five children and a house full of grandchildren.

In the postwar research environment Van Allen began applying his wartime engineering experience to miniaturize leading-edge rugged electronic equipment. He used this "small but tuff" electronic equipment in conjunction with his pioneering rocket and satellite scientific instrument payloads. By spring 1946 the USN transferred then Lieutenant Commander Van Allen to its inactive reserve, and he resumed his war-interrupted research work at the Johns Hopkins University. He remained at the Applied Physics Laboratory until 1950, when he returned to the University of Iowa as head of the physics department. While at Johns Hopkins Van Allen started to perform a series of preliminary space science experiments that anticipated his great discovery at the dawn of the American space program. He began to design and construct rugged, miniaturized instruments to collect geophysical data at the edge of space using rides on captured German V-2 rockets, Aerobee sounding rockets, and even rockets launched from high-altitude balloons (*rockoons*). One of his prime interests was the measurement of the intensity of cosmic rays and any other energetic particles arriving at the top of Earth's atmosphere from outer space. He carried these research interests back to the University of Iowa and over the years established an internationally recognized space physics program.

On October 4, 1957, the Soviet Union shocked the world by sending the first artificial satellite, called *SPUTNIK 1,* into orbit around Earth. In addition to starting the space age, this Russian satellite forced the United States into a hotly contested space technology race. In a desperate attempt to gather political influence, both cold war adversaries started pumping large quantities of money into the construction of military and scientific (civilian) space systems.

Fortune often favors the prepared, and as a gifted scientist, Van Allen was well prepared for the great opportunity that suddenly came his way. With the dramatic failure of the Vanguard rocket vehicle in early December 1957, senior officials in the American government made an emergency decision to launch the country's first satellite with a military rocket. A payload that was rugged enough to fly on a rocket was needed at once. By good fortune and preparedness, Van Allen's Iowa cosmic-ray experiment was available and quickly selected to become the principal component of the payload on *Explorer 1*. He responded to this opportunity by providing a rugged single Geiger-Muller tube to the Jet Propulsion Laboratory, which was under contract with the U.S. Army to construct the upper-stage spacecraft portion of *Explorer 1*.

Van Allen's scientific instrument was sturdy enough to ride into space on a rocket and then collect interesting geophysical data that was sent back to Earth by the host spacecraft. All was ready on January 31, 1958. A U.S. Army Jupiter C rocket, hastily pressed into service as a launch vehicle in an improvised configuration cobbled together by WERNHER VON BRAUN, rumbled into orbit from Cape Canaveral, Florida. Soon the first American satellite, *Explorer I,* traveled in orbit around Earth, restoring national pride and announcing a most interesting scientific bonus. As *Explorer I* glided through space, Van Allen's ionizing radiation detector, primarily designed to measure cosmic-ray intensity, jumped off scale. A great surprise to Van Allen and other scientists, his instrument had unexpectedly detected the inner portion of Earth's trapped radiation belts.

The *Explorer 3* spacecraft, launched into orbit on March 26, 1958, carried an augmented version of the Iowa cosmic-ray instrument. That spacecraft also harvested an enormous quantity of radiation data, confirming the presence of trapped energetic charged particle belts (mainly electrons and protons) within Earth's magnetosphere. These belts are now called the Van Allen radiation belts in his honor. The Van Allen radiation belts served as a major research focus for many other early scientific satellites. Soon an armada of spacecraft probed the region around Earth, defining the extent, shape, and composition of the planet's trapped radiation belts. Van Allen became a scientific celebrity who eventually met eight different American presidents.

In the 1960s and 1970s he assisted the National Aeronautics and Space Administration (NASA) in planning, designing, and operating energetic particle instruments on planetary exploration spacecraft to Mars, Venus, Jupiter, and Saturn. Since *Explorer 1* Van Allen and his team of accomplished researchers and graduate students at the University of Iowa have actively participated in many other pioneering space exploration missions. For example, members of this group published major papers dealing with Jupiter's intensely powerful magnetosphere, the discovery and preliminary survey of Saturn's magnetosphere, and the energetic particles population in interplanetary space.

Van Allen retired from the department of physics and astronomy in 1985. However, as the Carver Professor of Physics emeritus, he pored over interesting space physics data in his office in the campus building appropriately named "Van Allen Hall." His space physics efforts not only thrust the University of Iowa into international prominence, but he also served his nation as a truly inspirational scientific hero.

Van Allen radiation belts A doughnut-shaped zone of high-intensity particulate radiation around Earth from an altitude of about 320 to 32,400 km above the magnetic equator. The radiation of the Van Allen belts is composed of protons and electrons temporarily trapped in Earth's magnetic field. Spacecraft and their occupants and sensitive equipment orbiting within the belts or passing through them must be protected against this ionizing radiation. The existence of the belts was first confirmed by instruments placed on EXPLORER *1*, the first U.S. Earth satellite, by JAMES VAN ALLEN (b. 1914), after whom the region is now named.

See also EARTH'S TRAPPED RADIATION BELTS.

Vandenberg Air Force Base (VAFB) Vandenberg Air Force Base is located 89 km north of Santa Barbara near Lompoc, California. It is the site of all military, NASA, and commercial space launches accomplished on the West Coast of the United States. The base, named in honor of General Hoyt S. Vandenberg (air force chief of staff from 1948 to 1953), also provides launch facilities for the testing of intercontinental ballistic missiles. The first missile was launched from Vandenberg in 1958, and the world's first polar-orbiting satellite was launched from there aboard a Thor/Agena launch vehicle in 1959.

See also CAPE CANAVERAL AIR FORCE STATION; UNITED STATES AIR FORCE.

Vanguard Project An early U.S. space project involving a new three-stage rocket (Vanguard rocket) and a series of scientific spacecraft (Vanguard satellites.) The U.S. Navy successfully launched the first satellite in this project, called *Vanguard 1,* on March 17, 1958. This tiny Earth-orbiting satellite provided information leading in 1959 to the identification of the slight but significant pear shape of the Earth. An earlier widely publicized launch attempt had ended in dramatic failure when the Vanguard rocket exploded just as it lifted off the launch pad. The *Vanguard 2* satellite was launched on February 17, 1959. It transmitted the world's first picture of Earth's cloud cover (as observed from a satellite), but a wobble in the spacecraft due to an inadvertent bump by its launch vehicle resulted in less than satisfactory picture quality. Finally, the *Vanguard 3* satellite was launched on September 18, 1959. This early spacecraft deepened our understanding of the space environment, including the Van Allen radiation belt and micrometeoroids.

vapor The gaseous phase of a substance; in thermodynamics, this term often is used interchangeably with gas.

vaporization In thermodynamics, the transition of a material from the liquid phase to the gaseous (or vapor) phase, generally as a result of heating or pressure change.

vapor turbine A turbine in which part of the thermal energy (heat) supplied by a vapor is converted into mechanical work of rotation. The steam turbine is a common type of vapor turbine. Sometimes called a condensing turbine.
See also TURBINE.

variable-area exhaust nozzle An exhaust nozzle on a jet engine or a rocket engine that has an exhaust opening that can be varied in area by means of some mechanical device.
See also NOZZLE.

variable star A star that does not shine steadily but whose brightness (luminosity) changes over a short period of time.
See also CEPHEID VARIABLE; STAR.

Vatican Observatory The Vatican Observatory has its headquarters at the papal residence in Castel Gandolfo, Italy, outside Rome. Its dependent research center, the Vatican Observatory Research Group (VORG), is hosted by Steward Observatory at the University of Arizona, Tucson. The Vatican Observatory Research Group operates the 1.8-m Alice P. Lennon Telescope (reflector) with its Thomas J. Bannan Astrophysics Facility, known together as the Vatican Advanced Technology Telescope, at the Mount Graham International Observatory in southeastern Arizona.

vector Any physical quantity, such as force, velocity, or acceleration, that has both magnitude and direction at each point in space, as opposed to a scalar, which has magnitude only.

vector steering A steering method for rockets and spacecraft in which one or more thrust chambers are gimbal-mounted ("gimbaled") so that the direction of the thrust (i.e., the thrust vector) may be tilted in relation to the vehicle's center of gravity to produce a turning movement.

***Vega* (spacecraft)** Twin Russian spacecraft that were launched in December 1984 and performed a probe and flyby mission to Venus and then continued on for an encounter with Comet Halley. These spacecraft were modified versions of the Venera spacecraft. In June 1985 the *Vega 1* and *2* spacecraft flew by Venus. Each spacecraft dropped off a probe package consisting of an instrumented balloon and a lander and then proceeded, using a gravity assist from Venus, to encounter Comet Halley. *Vega 1* flew past Comet Halley on March 6, 1986, coming within 9,000 km of its nucleus. During its closest approach the *Vega 1* spacecraft acquired hundreds of images of the comet as well as other important scientific data. Similarly, *Vega 2* encountered Comet Halley on March 9, 1986, and also returned many images and important data. The Russians quickly provided these images and data to scientists at the European Space Agency (ESA) so they could make last-minute adjustments in the trajectory of the *Giotto* spacecraft prior to its encounter with Comet Halley on March 14, 1986. The *Vega* spacecraft are now in orbit around the Sun.
See also COMET; *GIOTTO* SPACECRAFT; VENERA PROBES AND SPACECRAFT; VENUS.

Vega (star) A conspicuous white star, Alpha Lyrae (α Lyr), in the constellation Lyra. Vega is about 25 light-years (7.8 pc) away and has the spectral classification A0V.
See also CONSTELLATION; STAR.

vehicle In general, an aerospace structure, machine, or device (e.g., a rocket) that is designed to carry a payload through the atmosphere and/or space; more specifically, a rocket vehicle.
See also LAUNCH VEHICLE.

vehicle tank Tank that serves both as a primary integral structure of a rocket vehicle (or spacecraft) and as a container of pressurized propellants.

Vela spacecraft A family of research and development spacecraft launched by the United States in the 1960s and early 1970s to detect nuclear detonations in the atmosphere down to Earth's surface or in outer space at distances of more than 160 million km. These spacecraft were jointly developed by the U.S. Department of Defense and the U.S. Atomic Energy Commission (now the Department of Energy) and were placed in pairs 180° apart in very-high-altitude (about 115,000-km) orbits around Earth. The first pair of Vela spacecraft, called *Vela 1A* and *Vela 1B,* were launched successfully on October 17, 1963. The last pair of these highly successful, 26-sided (polyhedron-shaped) spacecraft, called *Vela 6A* and *Vela 6B,* were launched successfully on April 8, 1970. It is interesting to note that the United States, the Soviet Union, and the United Kingdom signed the Limited Nuclear Test Ban Treaty in October 1963. This treaty prohibits the signatories from testing nuclear weapons in Earth's atmosphere, underwater, or in outer space. In addition to supporting important U.S. government nuclear test monitoring objectives, the *Vela* satellites also supported a modest revolution in astrophysics. Between 1969 and 1972 the *Vela* satellites detected 16 very short bursts of gamma-ray photons with energies of 0.2 to 1.5 million electron volts. These mysterious cosmic gamma-ray bursts lasted from less than a tenth of a second to about 30 seconds. Although the Vela instruments were not designed primarily for astrophysical research, simultaneous observations by several spacecraft started astrophysicists on their contemporary hunt for "gamma-ray bursters."
See also GAMMA-RAY ASTRONOMY; GAMMA-RAY BURST.

velocity (symbol: v) A vector quantity that describes the rate of change of position. Velocity has both magnitude (speed) and direction, and it is expressed in terms of units of length per unit of time (e.g., meters per second).

velocity of light (symbol: c) *See* SPEED OF LIGHT.

Venera probes and spacecraft A family of mostly successful robotic space missions flown by the Soviet Union to the planet Venus between 1961 and 1984. These missions included orbiters, landers, and atmospheric probes. In October 1967, for example, the *Venera 4* spacecraft placed a landing capsule/probe into the Venusian atmosphere, collecting data that indicated that the planet's atmosphere was from 90

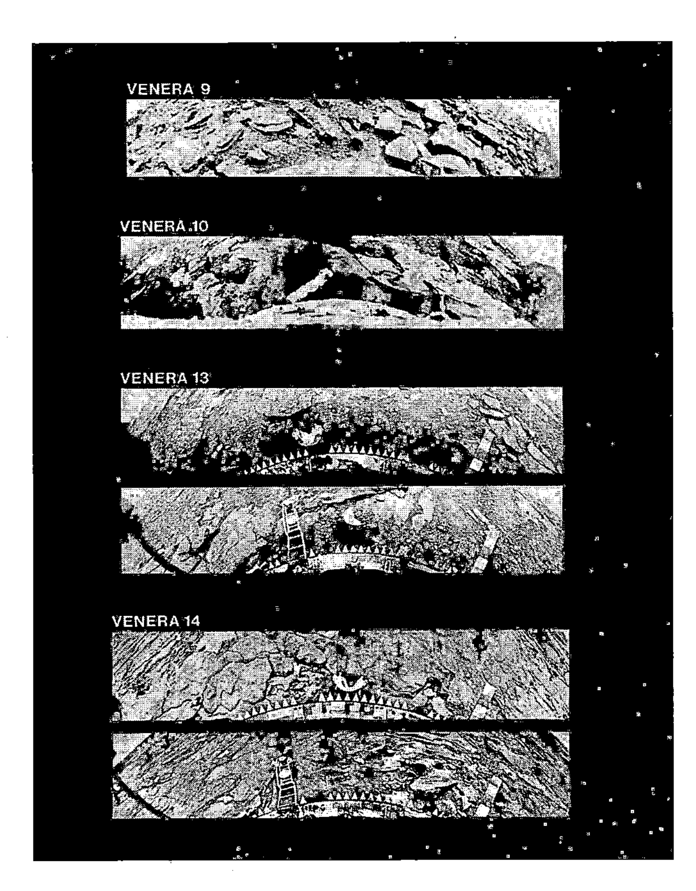

Wide-angle views of the surface of Venus from the Russian *Venera 9, 10, 13,* and *14* lander spacecraft *(Photograph courtesy of NASA)*

to 95 percent carbon dioxide (CO_2). This 380-kg probe descended by parachute for about 94 minutes, when data transmissions ceased at an altitude of some 25 km. In December 1970 the 495-kg atmospheric probe from the *Venera 7* spacecraft reached the surface of Venus and transmitted data for about 20 minutes. The *Venera 8* spacecraft launched a probe into the Venusian atmosphere in July 1972. This probe, with an improved communications system, successfully landed on the surface and survived for about 50 minutes in the infernolike conditions found on the surface of Venus.

In October 1975 the *Venera 9* and *10* spacecraft sent probes to the surface of Venus that landed successfully and transmitted the first black-and-white images of the planet's rock-strewn surface. *Venera 11* and *12,* launched in 1978, were also lander probe/flyby missions with improved sensors.

In the 1980s the Soviets launched four more sophisticated Venera spacecraft to Venus. The *Venera 13* and *14* spacecraft were 5,000-kg flyby/lander configurations sent to the planet at the end of 1981. *Venera 13* landed on Venus on March 1, 1982, and its identical companion, *Venera 14,* touched down on March 5, 1982. These landers returned black-and-white and color images of the Venusian surface. They also performed the first soil analysis of the Venusian surface.

The final pair of *Venera* spacecraft, *Venera 15* and *16,* were launched in June 1983. These spacecraft did not carry landers but rather performed mapping missions of the planet's cloud-enshrouded surface using their synthetic aperture radar (SAR) systems. The spacecraft orbited for about one year and produced detailed radar images of the planet's surface at between 1 and 2 kilometers resolution.

See also VENUS.

vengeance weapon 2 (Vergeltungwaffe 2) *See* V-2 ROCKET.

Venus Venus is the second planet from the Sun. Because the planet appears to observers on Earth as either an evening or a morning star, it is often called the Evening Star or the Morning Star. Venus is named after the Roman goddess of love and beauty. Among the planets in our solar system, it is the only one named after a female mythological deity. It is also called an inferior planet because it revolves around the Sun within the orbit of Earth. The planet maintains an average distance of about 0.723 astronomical unit (AU) (about 108 million km) from the Sun. The first table provides physical and dynamic data for Earth's nearest planetary neighbor. At closest approach Venus is approximately 42 million km from Earth.

Physical and Dynamic Properties of Venus

Diameter (equatorial)	12,100 km
Mass	4.87×10^{24} kg
Density (mean)	5.25 g/cm^3
Surface gravity	8.88 m/sec^2
Escape velocity	10.4 km/sec
Albedo (over visible spectrum)	0.7–0.8
Surface temperature (approx.)	750 K (477°C)
Atmospheric pressure (at surface)	9600 kPa (~1,400 psi)
Atmosphere	CO_2 (96.4%),
(main components)	N_2 (3.4%)
(minor components)	Sulfur dioxide (150 ppm), argon (70 ppm), water vapor (20 ppm)
Surface wind speeds	0.3–1.0 m/sec
Surface materials	Basaltic rock and altered materials
Magnetic field	Negligible
Radiation belts	None
Number of natural satellites	None
Average distance from Sun	1.082×10^8 km (0.723 AU)
Solar flux (at top of atmosphere)	2,620 W/m^2
Rotation period (a Venusian "day")	243 days (retrograde)
Eccentricity	0.007
Mean orbital velocity	35.0 km/sec
Sidereal year (period of one revolution around Sun)	224.7 days
Earth-to-Venus distances	
Maximum	2.59×10^8 km (1.73 AU)
Minimum	0.42×10^8 km (0.28 AU)

Source: NASA.

Spacecraft Exploration of Venus (1961–89)

Spacecraft	Country	Launch Date	Comments
Venera 1	USSR	Feb. 12, 1961	Passed Venus at 100,000 km May 1961 but lost radio contact Feb. 27, 1961.
Mariner 2	USA	Aug. 27, 1962	First successful interplanetary probe; passed Venus Dec. 14, 1962, at 35,000 km.
Venera 2	USSR	Nov. 12, 1965	Passed Venus Feb. 27, 1966, at 24,000 km but communications failed.
Venera 3	USSR	Nov. 16, 1965	Impacted on Venus Mar. 1, 1966, but communications failed earlier.
Venera 4	USSR	June 12, 1967	Probed Venusian atmosphere; flyby spacecraft and descent module.
Mariner 5	USA	June 14, 1967	Venus flyby Oct. 19, 1967, within 3,400 km.
Venera 5	USSR	Jan. 5, 1969	Descent probe entered atmosphere May 16, 1969.
Venera 6	USSR	Jan. 10, 1969	Descent module entered atmosphere May 17, 1969.
Venera 7	USSR	Aug. 17, 1970	Descent module soft-landed on Venus Dec. 15, 1970.
Venera 8	USSR	Mar. 27, 1972	Descent module soft-landed on Venus July 22, 1972.
Mariner 10	USA	Nov. 3, 1973	Flyby investigation of Venus Feb. 5, 1974, at 5,800 km; *Mariner 10* continued on to Mercury.
Venera 9	USSR	June 8, 1975	Orbiter and descent module arrived at Venus Oct. 22, 1975; descent module soft-landed and returned picture; orbiter circled planet at 1,545 km.
Venera 10	USSR	June 14, 1975	Orbiter and descent module arrived at Venus Oct. 25, 1975; descent module soft-landed and returned picture; orbiter circled planet at 1,665 km.
Pioneer Venus			
Orbiter	USA	May 20, 1978	Orbited Venus Dec. 4, 1978; radar mapping mission.
Multiprobe	USA	Aug. 8, 1978	Three small probes, 1 large probe, and main bus entered atmosphere Dec. 9, 1978.
Venera 11	USSR	Sept. 9, 1978	Descent module soft-landed; flyby vehicle passed planet at 35,000 km Dec. 25, 1978.
Venera 12	USSR	Sept. 14, 1978	Descent module soft-landed; flyby vehicle passed planet at 35,000 km Dec. 21, 1978.
Venera 13	USSR	Oct. 30, 1981	Orbiter and descent module; descent module soft-landed Mar. 3, 1982, and returned color picture.
Venera 14	USSR	Nov. 4, 1981	Orbiter and descent module; descent module soft-landed Mar. 5, 1982, and returned color picture.
Venera 15	USSR	June 2, 1983	Orbiter; radar mapping mission.
Venera 16	USSR	June 6, 1983	Orbiter; radar mapping mission.
Vega 1	USSR	Dec. 15, 1984	Venus flyby spacecraft (on way to Comet Halley encounter); Venus lander and instrumented balloon in Venusian atmosphere.
Vega 2	USSR	Dec. 21, 1984	Venus flyby spacecraft (on way to Comet Halley encounter); Venus lander (automated soil sampling); instrumented balloon in Venusian atmosphere.
Magellan	USA	May 4, 1989	Orbiter; high-resolution radar mapping mission.

Source: NASA.

Since the 1960s visits by numerous American and Russian spacecraft have dispelled the pre–space age romantic fantasies that this cloud-enshrouded planet was a prehistoric world that mirrored a younger Earth. Except for a few physical similarities of size and gravity, Earth and Venus are very different worlds. For example, the surface temperature on Venus approaches 500° Celsius (773 kelvins), its atmospheric pressure is more than 90 times that of Earth, it has no surface water, and its dense, hostile atmosphere with sulfuric acid clouds and an overabundance of carbon dioxide (about 96 percent) represents a runaway greenhouse of disastrous proportions. The second table provides a brief summary of the American and Russian spacecraft that have successfully explored Venus from 1961 to 1989.

Cloud-penetrating synthetic aperture radar (SAR) systems carried by the American Pioneer Venus orbiter and *Magellan* spacecraft and the Russian *Venera 15* and *16* spacecraft have provided a detailed characterization of the previously unobservable Venusian surface. These radar imagery data, especially the high-resolution images collected by the *Magellan* spacecraft, are now challenging planetary scientists to explain some of the interesting findings.

Scientists generally divide the surface of Venus into three classes of terrain: highlands (or tesserae) (10 percent of surface), rolling uplands (70 percent), and lowland plains (20 percent). One very prominent highland region is called Ishtar Terra after Ishtar, the ancient Babylonian goddess of love. This region is about the size of Australia and stands several kilometers above the average planetary radius. Ishtar Terra contains the highest peaks yet discovered on Venus, including Maxwell Montes, which is often regarded as the single most impressive topographical feature on the planet.

The relatively fresh appearance and small number of impact craters suggest to planetary scientists that the surface as a whole is young by geological standards. It also appears that Venus experienced a dramatic resurfacing event some 300 to 500 million years ago, but whether this resurfacing phenomenon, which covered most of the planet with lava, was caused by a brief chain of catastrophic events or by the influences of low-level volcanism operating over longer periods of time is currently the subject of scientific investigation and debate. The fractured Venusian highlands, such as Ishtar Terra and Aphrodite Terra, represent the planet's older surface material not covered by younger flows of lava. Maat Mons is the largest volcano on Venus, and it appears to have experienced a recent surge of volcanic activity. Volcanic features unique to the planet are the *coronae,* large (often hundreds of kilometers across), heavily fractured circular regions that sometimes are surrounded by a trench. The channels, or *canali,* on Venus are long, riverlike surface features where lava once flowed.

Despite all the interesting facts that have been revealed about Venus by the American and Soviet space missions, equally interesting questions also have been raised by scientists as they examined these new data. Questions such as "What was Venus like before the resurfacing event?" await resolution in this century, when even more rugged probes and landers again visit Earth's nearest planetary neighbor.

See also MAGELLAN MISSION; PIONEER VENUS MISSION; VENERA PROBES AND SPACECRAFT.

vernal equinox The spring equinox, which occurs on or about March 21 in the Northern Hemisphere.

See also EQUINOX.

Verne, Jules (1828–1905) French *Writer* The French writer and technical visionary Jules Verne created modern science fiction and its dream of space travel with his classic 1865 novel *From the Earth to the Moon,* an acknowledged source of youthful inspiration for KONSTANTIN TSIOLKOVSKY, ROBERT H. GODDARD, HERMANN J. OBERTH, WERNHER VON BRAUN, and other space technology pioneers.

vernier engine A rocket engine of small thrust used primarily to obtain a fine adjustment in the velocity and trajectory or in the attitude of a rocket or aerospace vehicle.

Very Large Array (VLA) The Very Large Array (VLA) is a spatially extended radiotelescope facility at Socorro, New Mexico. It consists of 27 antennas, each 25 meters in diameter, that are configured in a giant "Y" arrangement on railroad tracks over a 20-kilometer distance. The VLA is operated by the National Radio Astronomy Observatory and sponsored by the National Science Foundation.

The VLA has four major antenna configurations: A array, with a maximum antenna separation of 36 km; B array, with a maximum antenna separation of 10 km; C array, with a maximum antenna separation of 3.6 km; and D array, with a maximum antenna separation of 1 km. The operating resolution of the VLA is determined by the size of the array. At its highest the facility has a resolution of 0.04 arc-seconds. This corresponds, for example, to an ability to "see" a 43-gigahertz (GHz) radio frequency source the size of a golf ball at a distance of 150 km. The facility collects the faint radio waves emitted by a variety of interesting celestial objects and produces radio images of these objects with as much clarity and resolution as the photographs from some of the world's largest optical telescopes.

The technique of focusing and combining the signals from a distributed array of smaller telescopes to simulate the resolution of a single, much larger telescope is called APERTURE SYNTHESIS. In astronomy, a radio telescope is used to measure the intensity of the weak, staticlike cosmic radio waves coming from some particular direction in the universe. *Sensitivity* is defined as the radio telescope's ability to detect these very weak radio signals, while RESOLUTION is defined as the telescope's ability to locate the source of these signals. The sensitivity of a distributed array of telescopes, such as the VLA, is proportional to the sum of the collecting areas of all the individual elements, while the array's resolution is determined by the distance (baseline) over which the array elements can be spread. Each of the 25-m-diameter dish antennas in the VLA was specially designed with aluminum panels formed into a parabolic surface accurate to 0.5 millimeter, a design condition that enables the antennas to focus radio signals as short as 1 cm wavelength.

In the VLA each antenna collects incoming radio signals and sends them to a central location, where they are combined. The sensitive radio receivers of the VLA can be tuned to wavelengths of 90 cm (P band), 20 cm (L band), 6 cm (C band), 3.6 cm (X band), 2 cm (U band), 1.3 cm (K band), and 0.7 cm (Q band). The corresponding range of frequencies extends from 0.30–0.34 gigahertz (GHz) (P band) up to 40–50 GHz (Q band). These sensitive radio frequency receivers are cooled to low temperature (typically about 18 kelvins) to reduce internally generated noise, which tends to mask the very weak radio signals from space. Once received, the incoming cosmic radio signals are amplified several million times and sent to the VLA's control building by means of a waveguide.

The data collected by the VLA are conveniently stored so that an astronomer can evaluate a radio image days or even years after the actual observation was made. Different astronomical observations require different resolution capabilities. For example, a single highly detailed radio image might involve more than 40 hours of observation, while a crude, low-resolution radio image (i.e., a "thumbnail sketch" radio signal) of a particular source may require only 10 minutes of observing time. Radio astronomers might use the high-resolution radio image to explore the inner core of an interesting radio galaxy. In contrast, they would use the low-resolution image, which provides only the faint overall radio emission features of a galaxy, during an initial search for interesting sources.

The VLA is used to produce radio images with as much detail as those made by an optical telescope. To accomplish this the VLA's 27 dish-shaped antennas are arranged in a giant Y-pattern, with the southeast and southwest arms of the Y-pattern each 21 km long and the north arm 19 km long. The resolution of this radio telescope array is varied by changing the separation and spacing of its 27 antenna elements. The VLA is generally found in one of four standard array configurations. In the smallest antenna dispersion configuration (D Array, the low-resolution configuration), the 27 individual antennas are clustered together and form an equivalent radio antenna with a baseline of just 1 km. In the largest antenna dispersion configuration (A array, the high-resolution configuration), the individual antennas stretch out in a giant Y-pattern that produces a maximum baseline of 36 km.

Each of the 235-ton VLA dish antennas is carried along the Y-pattern array arms by a special transporter that moves on two parallel sets of railroad tracks. Generally, it takes about two hours to move an antenna from one station (pedestal) to another and about a week is needed to reconfigure the entire VLA array. Since its startup in 1981 the high-resolution and high-sensitivity radio images produced by the Very Large Array have made it one of the world's leading radio telescope facilities.

See also RADIO ASTRONOMY.

Very Large Telescope (VLT) The European Southern Observatory (ESO) optical and infrared Very Large Telescope consists of four 8-m telescopes that can work independently or in combined mode. In the combined mode the VLT provides the total light collecting power of a 16-m single telescope, making it the largest optical telescope in the world. The four 8-m telescopes supplemented with 3 auxiliary 1-m telescopes can also be used as an interferometer, providing high angular resolution imaging. The useful wavelength range extends from the near-ultraviolet to 25 micrometers (μm) in the infrared. The Paranal Observatory is located on Cerro Paranal in the Atacama Desert of northern Chile, a region considered to be one of the driest areas on Earth. Cerro Paranal is a 2,635-m-high mountain about 120 km south of the town of Antofagasta and 12 km inland from the Pacific Coast.

Vesta The third-largest main belt asteroid (576-km-diameter) discovered in 1807 by HEINRICH WILHELM OLBERS. It orbits the Sun at an average distance of 2.362 AU with a period of 3.63 years. At perihelion Vesta is 2.15 AU from the Sun, while at aphelion it is 2.57 AU distant. Vesta has a rotation rate of 5.3 hours. It is the brightest of the main belt asteroids, with an albedo of 0.23. Also called Vesta-4.

See also ASTEROID.

Viking Project The Viking Project was the culmination of an initial series of American missions to explore Mars in the 1960s and 1970s. Viking was designed to orbit Mars and to land and operate on the surface of the Red Planet. Two identical spacecraft, each consisting of a lander and an orbiter, were built.

The orbiters carried the following scientific instruments:

1. A pair of cameras with 1,500-millimeter focal length that performed a systematic search for landing sites, then looked at and mapped almost 100 percent of the Martian surface. Cameras onboard the *Viking 1* and *Viking 2* orbiters took more than 51,000 photographs of Mars.
2. A Mars atmospheric water detector that mapped the Martian atmosphere for water vapor and tracked seasonal changes in the amount of water vapor.
3. An infrared thermal mapper that measured the temperatures of the surface, polar caps, and clouds; it also mapped seasonal changes. In addition, although the Viking orbiter

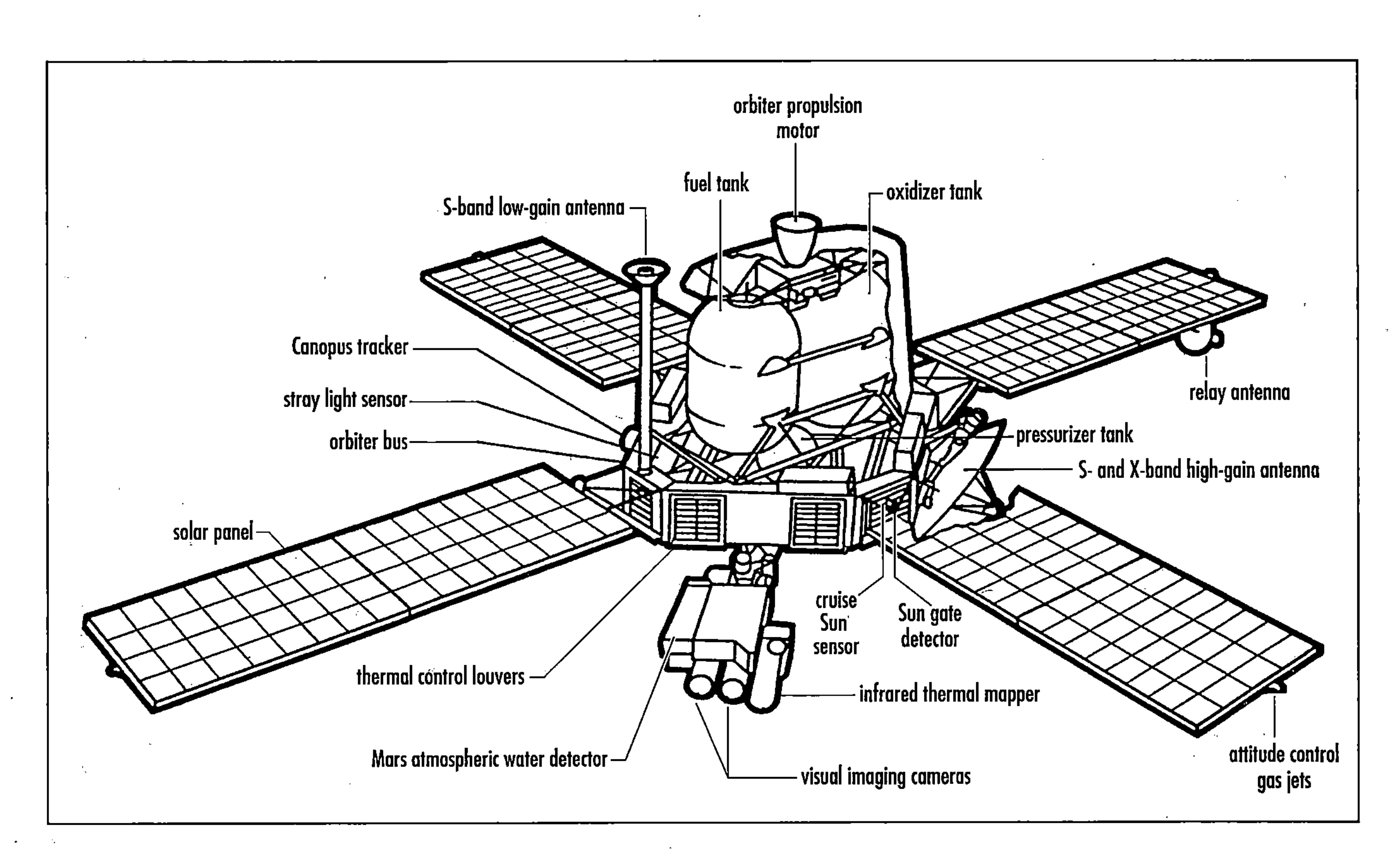

NASA's Viking orbiter spacecraft and its complement of instruments ***(Drawing courtesy of NASA)***

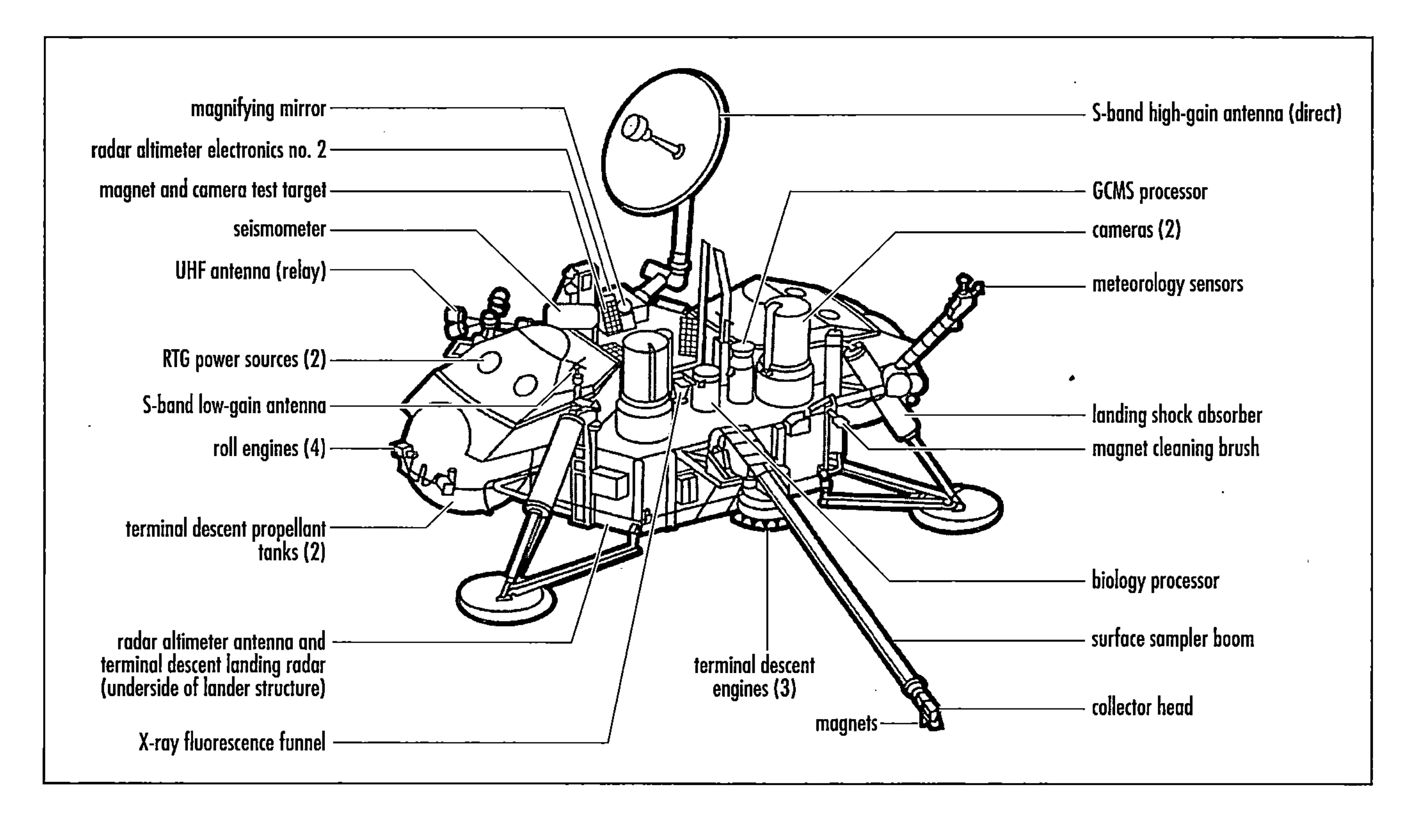

NASA's Viking lander spacecraft and its complement of instruments *(Drawing courtesy of NASA)*

radios were not considered scientific instruments, they were used as such. By measuring the distortion of radio signals as these signals traveled from the Viking orbiter spacecraft to Earth, scientists were able to measure the density of the Martian atmosphere.

The Viking landers carried the following instruments:

1. The biology instrument, consisting of three separate experiments designed to detect evidence of microbial life in the Martian soil.
2. A gas chromatograph/mass spectrometer (GCMS) that searched the Martian soil for complex organic molecules.
3. An X-ray fluorescence spectrometer that analyzed samples of the Martian soil to determine its elemental composition.
4. A meteorology instrument that measured air temperature and wind speed and direction at the landing sites.
5. A pair of slow-scan cameras that were mounted about 1 meter apart on the top of each lander. These cameras provided black-and-white, color, and stereo photographs of the Martian surface.
6. A seismometer had been designed to record any "Marsquakes" that might occur on the Red Planet. Such information would have helped planetary scientists determine the nature of Mars's internal structure. Unfortunately, the seismometer on Lander 1 did not function after landing and the instrument on Lander 2 observed no clear signs of internal tectonic activity.
7. An upper-atmosphere mass spectrometer that conducted its primary measurements as each lander plunged through the Martian atmosphere on its way to the landing site. This instrument made the lander's first important scientific discovery, the presence of nitrogen in the Martian atmosphere.
8. A retarding potential analyzer that measured the Martian ionosphere, again during entry operations.
9. Accelerometers, a stagnation pressure instrument, and a recovery temperature instrument that helped determine the structure of the lower Martian atmosphere as the landers approached the surface.
10. A surface sampler boom that employed its collector head to scoop up small quantities of Martian soil to feed the biology, organic chemistry, and inorganic chemistry instruments. It also provided clues to the soil's physical properties. Magnets attached to the sampler, for example, provided information on the soil's iron content.
11. The lander radios also were used to conduct scientific experiments. Physicists were able to refine their estimates of Mars's orbit by measuring the time for radio signals to travel between Mars and Earth. The great accuracy of these radio-wave measurements also allowed scientists to confirm portions of ALBERT EINSTEIN's general theory of relativity.

Both Viking missions were launched from Cape Canaveral, Florida. *Viking 1* was launched on August 20, 1975, and *Viking 2* on September 9, 1975. The landers were sterilized before launch to prevent contamination of Mars by terrestrial microorganisms. These spacecraft spent nearly a year in transit to the Red Planet. *Viking 1* achieved Mars orbit on June 19, 1976, and *Viking 2* began orbiting Mars on August 7, 1976. The *Viking 1* lander accomplished the first soft landing on Mars on July 20, 1976, on the western slope of Chryse Planitia (the Plains of Gold) at 22.46° north latitude and 48.01° west longitude. The *Viking 2* lander touched down successfully on September 3, 1976, at Utopia Planitia at 47.96° north latitude and 225.77° west longitude.

The Viking mission was planned to continue for 90 days after landing. Each orbiter and lander, however, operated far beyond its design lifetime. For example, the *Viking 1* orbiter exceeded four years of active flight operations in orbit around Mars. The Viking Project's primary mission ended on November 15, 1976, just 11 days before Mars passed behind the Sun (an astronomical event called a superior conjunction). After conjunction, in mid-December 1976, telemetry and command operations were reestablished and extended mission operations began.

The *Viking 2* orbiter mission ended on July 25, 1978, due to exhaustion of attitude-control system gas. The *Viking 1* orbiter spacecraft also began to run low on attitude-control system gas, but through careful planning it was possible to continue collecting scientific data (at a reduced level) for another two years. Finally, with its attitude-control gas supply exhausted, the *Viking 1* orbiter's electrical power was commanded off on August 7, 1980.

The last data from the *Viking 2* lander were received on April 11, 1980. The *Viking 1* lander made its final transmission on November 11, 1982. After more than six months of effort to regain contact with the *Viking 1* lander, the Viking mission came to an end on May 23, 1983.

With the single exception of the seismic instruments, the entire complement of scientific instruments of the Viking Project acquired far more data about Mars than was ever anticipated. The primary objective of the landers was to determine whether (microbial) life currently exists on Mars. The evidence provided by the landers is still subject to debate, although most scientists feel these results are strongly indicative that life does not now exist on Mars. However, recent analyses of Martian meteorites have renewed interest in this very important question, and Mars is once again the target of intense scientific investigation by even more sophisticated scientific spacecraft.

See also *MARINER* SPACECRAFT; MARS EXPLORATION ROVER MISSION; *MARS EXPRESS; MARS GLOBAL SURVEYOR; MARS ODYSSEY 2001* MISSION; *MARS PATHFINDER; MARS RECONNAISSANCE ORBITER;* MARS SAMPLE RETURN MISSION; MARTIAN METEORITES.

Virgo Cluster The nearest large cluster of galaxies, The Virgo Cluster, an irregular large cluster of about 2,000 galaxies, is also at the center of the local supercluster.

See also CLUSTER OF GALAXIES; SUPERCLUSTER.

virtual particles A mental construct or scientific language of convenience invented by physicists so they can more meaningfully discuss how elementary particles interact in quantum field theory. When modern physicists talk about elementary particle interactions, they use Feynman diagrams. These diagrams, developed by the Nobel laureate Richard P. Feynman (1918–88), are a shorthand for a quantum field theory calculation that gives the probability of the process. Feynman diagrams have lines that represent mathematical expressions, but each line can also be viewed as representing a particle. However, in the intermediate stages of a process the lines represent particles that can never be observed. These particles do not have the required Einsteinian relationship between their energy, momentum, and mass. As a consequence, physicists call them virtual particles. The important thing to realize is that the virtual particle represents an intermediate stage and not the final stage in the interaction process.

See also ELEMENTARY PARTICLE(S).

virtual reality (VR) A computer-generated artificial reality that captures and displays in varying degrees of detail the essence or effect of a physical reality (i.e., a "real-world" scene, event, or process) being modeled or studied. With the aid of a data glove, headphones, and/or head-mounted stereoscopic display, a person is projected into the three-dimensional world created by the computer.

A virtual reality system generally has several integral parts. There is always a computerized description (i.e., the "database") of the scene or event to be studied or manipulated. It can be a physical place, such as a planet's surface made from digitized images sent back by robot space probes. It can even be more abstract, such as a description of the ozone levels at various heights in Earth's atmosphere or the astrophysical processes occurring inside a pulsar or a black hole.

VR systems also use a special helmet or headset ("goggles") to supply the sights and sounds of the artificial, computer-generated environment. Video displays are coordinated to produce a three-dimensional effect. Headphones make sounds appears to come from any direction. Special sensors track head motions, so that the visual and audio images shift in response.

Most VR systems also include a glove with special electronic sensors. The "data glove" lets a person interact with the virtual world through hand gestures. He or she can move or touch objects in the computer-generated visual display, and these objects then respond as they would in the physical world. Advanced versions of such gloves also provide artificial "tactile sensations" so that an object "feels like the real thing" being touched or manipulated (e.g., smooth or rough, hard or soft, cold or warm, light or heavy, flexible or stiff, etc.).

The field of virtual reality is quite new, and rapid advances should be anticipated over the next decade as computer techniques, visual displays, and sensory feedback systems (e.g., advanced data gloves) continue to improve in their ability to project and model the real world. VR systems have many potential roles in aerospace technology. For example, sophisticated virtual reality systems will let scientists "walk on another world" while working safely here on Earth. Future mission planners will identify the best routes (based on safety, resource consumption, and mission objectives) for both robots and humans to explore the surfaces of the Moon and Mars before the new missions are even launched. Astronauts will use VR training systems regularly to try out space maintenance and repair tasks and perfect their skills long before they lift off on an actual mission. Aerospace engineers will use VR systems as an indispensable design tool to fully examine and test new aerospace hardware long before any "metal is bent" in building even a prototype model of the item.

viscosity A measure of the internal friction or flow resistance of a fluid when it is subjected to shear stress. The dynamic viscosity is defined as the force that must be applied per unit area to permit adjacent layers of fluid to move with unit velocity relative to each other. The *dynamic viscosity* is sometimes expressed in poise (centimeter-gram-second [cgs] system) or in pascal-seconds (SI unit system). One poise is equal to 0.1 newton-second per square meter (i.e., 1 poise = 10^{-1} N s m^{-2}). The *kinematic viscosity* is defined as the dynamic viscosity divided by the fluid's density. The kinematic viscosity can be expressed in stokes (centimeter-gram-second [cgs] system) or in square meters per second (SI unit system). One stoke is equal to 10^{-4} m^2 s^{-1}. In general, the viscosity of a liquid usually decreases as the temperature is increased; the viscosity of a gas increases as the temperature increases.

See also SI UNIT(S).

viscous fluid A fluid whose molecular viscosity is sufficiently large to make the viscous forces a significant part of the total force field in the fluid.

visual binary *See* BINARY STAR SYSTEM.

Vogel, Hermann Carl (1841–1907) German *Astronomer* The German astronomer Hermann Carl Vogel performed spectroscopic analyses of stars that supported the study of stellar evolution and used Doppler shift measurements to obtain their radial velocities. In the course of this work in the late 1880s, he detected the first spectroscopic binary star systems.

See also BINARY STAR SYSTEM.

void In rocketry, an air bubble in a cured propellant grain or in a rocket motor insulation.

volcano A vent in the crust of a planet or moon from which molten lava, gases, and other pyroclastic materials flow.

volt (symbol: V) The SI unit of electric potential difference and electromotive force. A volt is equal to the difference of electric potential between two points of a conductor carrying a constant current of 1 ampere when the power dissipated between these points equals 1 watt. This unit is named after the Italian scientist Count Alessandro Volta (1745–1827), who performed pioneering work involving electricity and electric cells.

volume (symbol: V) The space occupied by a solid object or a mass of fluid (liquid or confined gas).

Voskhod An early Russian three-person spacecraft that evolved from the Vostok spacecraft. *Voskhod 1* was launched on October 12, 1964, and carried the first three-person crew into space. The cosmonauts Vladimir Komarov, Konstantin Feoktistov, and Boris Yegorov flew on a one-day Earth orbital mission. *Voskhod 2* was launched on March 18, 1965, and carried a crew of two cosmonauts, including ALEXEI LEONOV, who performed the world's first "space walk" (about 10 minutes in duration) during the orbital mission. Voskhod means "sunrise" in Russian.

Vostok The first Russian manned spacecraft. This spacecraft was occupied by a single cosmonaut and consisted of a spherical cabin (about 2.3 m in diameter) that was attached to a biconical instrument module. *Vostok 1* was launched on April 12, 1961, carrying the cosmonaut YURI GAGARIN, the first human to fly in space. Gagarin's flight made one orbit of Earth and lasted about 108 minutes.

***Voyager* record** The *Voyager 1* and 2 spacecraft, launched during the summer of 1977, will eventually cross the heliopause and leave our solar system. (The heliopause is that boundary in deep space that marks the edge of the Sun's influence.) As these spacecraft wander through the Milky Way galaxy over the next million or so years, each has the

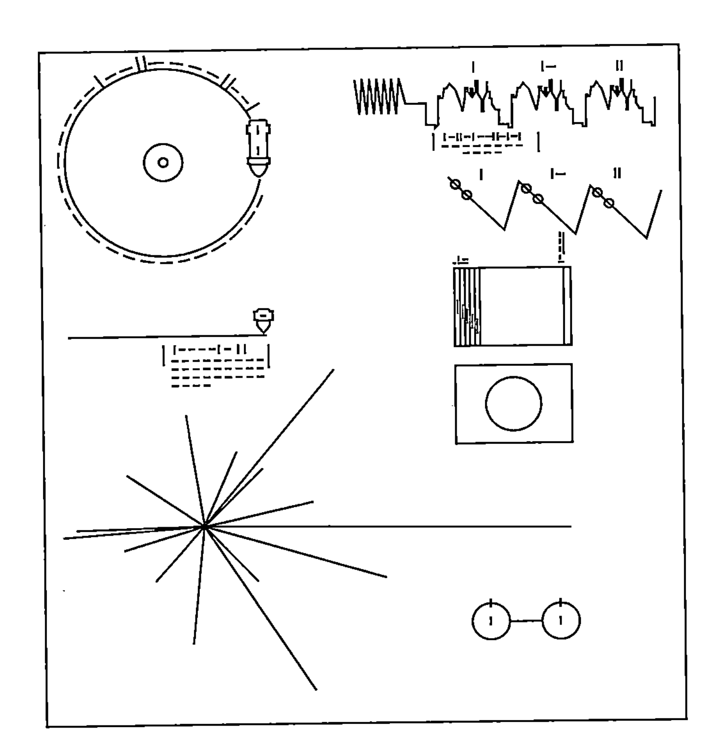

The set of instructions to any alien civilization that might find the *Voyager 1* or *2* spacecraft explaining how to operate the *Voyager* record and where the spacecraft and message came from *(Drawing courtesy of NASA)*

potential of serving as an interstellar ambassador, since each carries a special message from Earth on the chance that an intelligent alien race might eventually find one of the spacecraft floating among the stars.

The *Voyager* spacecrafts' interstellar message is a phonograph record called "The Sounds of Earth." Electronically imprinted on it are words, photographs, music, and illustrations that will tell an extraterrestrial civilization about our planet. Included are greetings in more than 50 languages, music from various cultures and periods, and a variety of natural terrestrial sounds such as the wind, the surf, and different animals. The *Voyager* record also includes a special message from former president Jimmy Carter. The late CARL SAGAN described in detail the full content of this phonograph message to the stars in his delightful book *Murmurs of Earth*. (*See* table.)

Each record is made of copper with gold plating and is encased in an aluminum shield that also carries instructions on how to play it. Look at the figure without reading beyond this paragraph. Can you decipher the instructions we have given to alien civilizations?

In the upper left is a drawing of the phonograph record and the stylus carried with it. Written around it in binary notation is the correct time for one rotation of the record, 3.6 seconds. Here, the time units are 0.70 billionths of a second, the time period associated with a fundamental transition of the hydrogen atom. The drawing further indicates that the record should be played from the outside in. Below this drawing is a side view of the record and stylus, with a binary number giving the time needed to play one side of the record (approximately one hour).

The information provided in the upper-right portion of the instructions is intended to show how pictures (images) are to be constructed from the recorded signals. The upper-right drawing illustrates the typical wave form that occurs at the start of a picture. Picture lines 1, 2, and 3 are given in binary numbers, and the duration of one of the picture "lines" is also noted (about 8 milliseconds). The drawing immediately below shows how these lines are to be drawn vertically, with a staggered interlace to give the correct picture rendition. Immediately below this is a drawing of an entire picture raster, showing that there are 512 vertical lines in a complete picture. Then, immediately below this is a replica of the first picture on the record. This should allow extraterrestrial recipients to verify that they have properly decoded the terrestrial pictures. A circle was selected for this first picture to guarantee that any aliens who find the message use the correct aspect ratio in picture reconstruction.

Finally, the drawing at the bottom of the protective aluminum shield is that of the same pulsar map drawn on the *Pioneer 10* and *11* plaques. (These spacecraft are also headed on interstellar trajectories.) The map shows the location of our solar system with respect to 14 pulsars, whose precise periods are also given. The small drawing with two circles in the lower-right-hand corner is a representation of the hydrogen atom in its two lowest states, with a connecting line and digit, 1. This indicates that the time interval associated with the transition from one state to the other is to be used as the fundamental time scale, both for the times given on the protective aluminum shield and in the decoded pictures.

A Partial List of the Contents of the *Voyager* Record, *The Sounds of Earth*

A. Sounds of Earth

Whale, volcanoes, rain, surf, cricket frogs, birds, hyena, elephant, chimpanzee, wild dog, laughter, fire, tools, Morse code, train whistle, *Saturn V* rocket liftoff, kiss, baby

B. Music

Bach: *Brandenberg Concerto #2*, 1st movement; Zaire: "Pygmy Girls" initiation song; Mexico: mariachi band playing "El Cascabel"; Chuck Berry: "Johnny B. Goode"; Navajo: night chant; Louis Armstrong: "Melancholy Blues"; China (zither) "Flowering Streams"; Mozart: Queen of the Night aria (*The Magic Flute*); Beethoven: Symphony #5, 1st movement

C. Greetings (in 55 Languages)

(example) English: "Hello from the children of planet Earth"

D. Pictures (Digital Data to Be Reconstructed into Images)

Calibration circle, solar location map, the Sun, Mercury, Mars, Jupiter, Earth, fetus, birth, nursing mother, group of children, sequoia (giant tree), snowflake, seashell, dolphins, eagle, Great Wall of China, Taj Mahal, UN Building, Golden Gate Bridge, radio telescope (Arecibo), Titan Centaur launch, astronaut in space

See also INTERSTELLAR COMMUNICATION AND CONTACT; *PIONEER* PLAQUE; *PIONEER 10, 11* SPACECRAFT; Voyager SPACECRAFT.

Voyager spacecraft Once every 176 years the giant outer planets Jupiter, Saturn, Uranus, and Neptune align in such a pattern that a spacecraft launched from Earth to Jupiter at just the right time can be able to visit the other three planets on the same mission using a technique called gravity assist. NASA space scientists named this multiple giant planet encounter mission the "Grand Tour" and took advantage of a unique celestial alignment opportunity in 1977 by launching two sophisticated spacecraft called *Voyager 1* and *2*.

Each Voyager spacecraft had a mass of 825 kg and carried a complement of scientific instruments to investigate the outer planets and their many moons and intriguing ring systems. These instruments, provided electric power by a long-lived nuclear system called a radioisotope thermoelectric generator (RTG), recorded spectacular close-up images of the giant outer planets and their interesting moon systems, explored complex ring systems, and measured properties of the interplanetary medium.

Taking advantage of the 1977 Grand Tour launch window, the *Voyager 2* spacecraft lifted off from Cape Canaveral, Florida, on August 20, 1977, onboard a Titan-Centaur rocket. (NASA called the first Voyager spacecraft launched *Voyager 2* because the second Voyager spacecraft to be launched eventually would overtake it and become *Voyager 1*.) *Voyager 1* was launched on September 5, 1977. This spacecraft followed the same trajectory as its *Voyager 2* twin and overtook its sister ship just after entering the asteroid belt in mid-December 1977.

Voyager 1 made its closest approach to Jupiter on March 5, 1979, and then used Jupiter's gravity to swing itself to Saturn. On November 12, 1980, *Voyager 1* successfully encountered the Saturnian system and then was flung out of the ecliptic plane on an interstellar trajectory. The *Voyager 2* spacecraft successfully encountered the Jovian system on July 9, 1979 (closest approach), and then used the gravity assist technique to follow *Voyager 1* to Saturn. On August 25, 1981, *Voyager 2* encountered Saturn and then went on to successfully encounter both Uranus (January 24, 1986) and Neptune (August 25, 1989). Space scientists consider the end of *Voyager 2*'s encounter with the Neptunian system as the end of a truly extraordinary epoch in planetary exploration. In the first 12 years after they were launched from Cape Canaveral, these incredible spacecraft contributed more to our understanding of the giant outer planets of our solar system than was accomplished in more than three millennia of Earth-based observations. Following its encounter with the Neptunian system, *Voyager 2* also was placed on an interstellar trajectory and, like its *Voyager 1* twin, now continues to travel away from the Sun.

As the influence of the Sun's magnetic field and solar wind grow weaker, both *Voyager* spacecraft eventually will pass out of the heliosphere and into the interstellar medium. Through NASA's Voyager Interstellar Mission (VIM) (which began officially on January 1, 1990), the two Voyager spacecraft will continue to be tracked on their outward journey. The two major objectives of the VIM are an investigation of the interplanetary and interstellar media and a characterization of the interaction between the two and a continuation of the successful Voyager program of ultraviolet astronomy. During the VIM the spacecraft will search for the heliopause, the outermost extent of the solar wind, beyond which lies interstellar space. It is hoped that one Voyager spacecraft will still be functioning when it penetrates the heliopause and will provide scientists with the first true sampling of the interstellar environment. Barring a catastrophic failure on board either spacecraft, their nuclear power systems should provide useful levels of electric power until at least 2015.

Since both Voyager spacecraft eventually would journey beyond the solar system, their designers placed a special interstellar message (a recording entitled "The Sounds of Earth") on each in the hope that perhaps millions of years from now some intelligent alien race will find either spacecraft drifting quietly through the interstellar void. If they are able to decipher the instructions for using this record, they will learn about our contemporary terrestrial civilization and

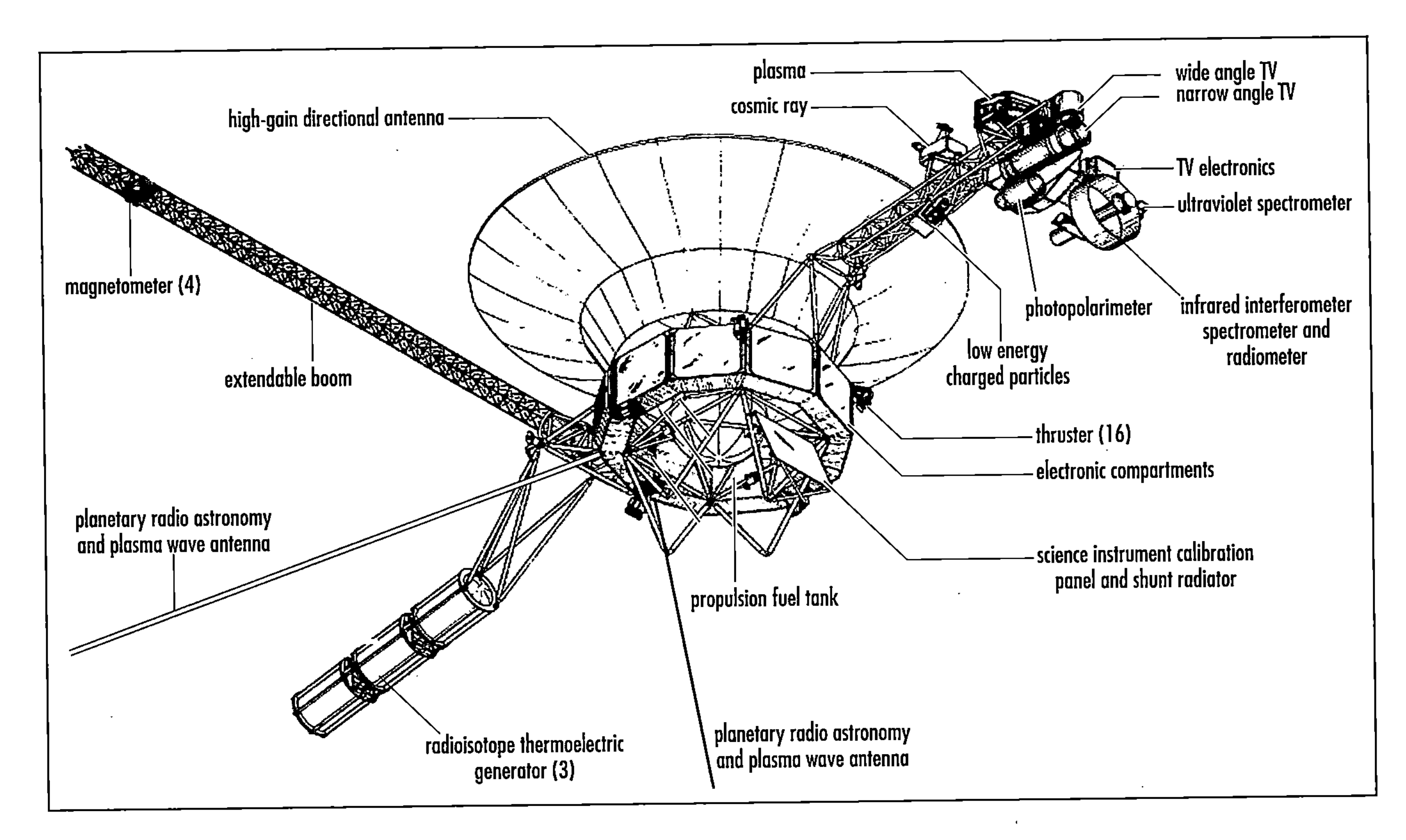

The *Voyager* spacecraft and its complement of sophisticated instruments *(Drawing courtesy of NASA)*

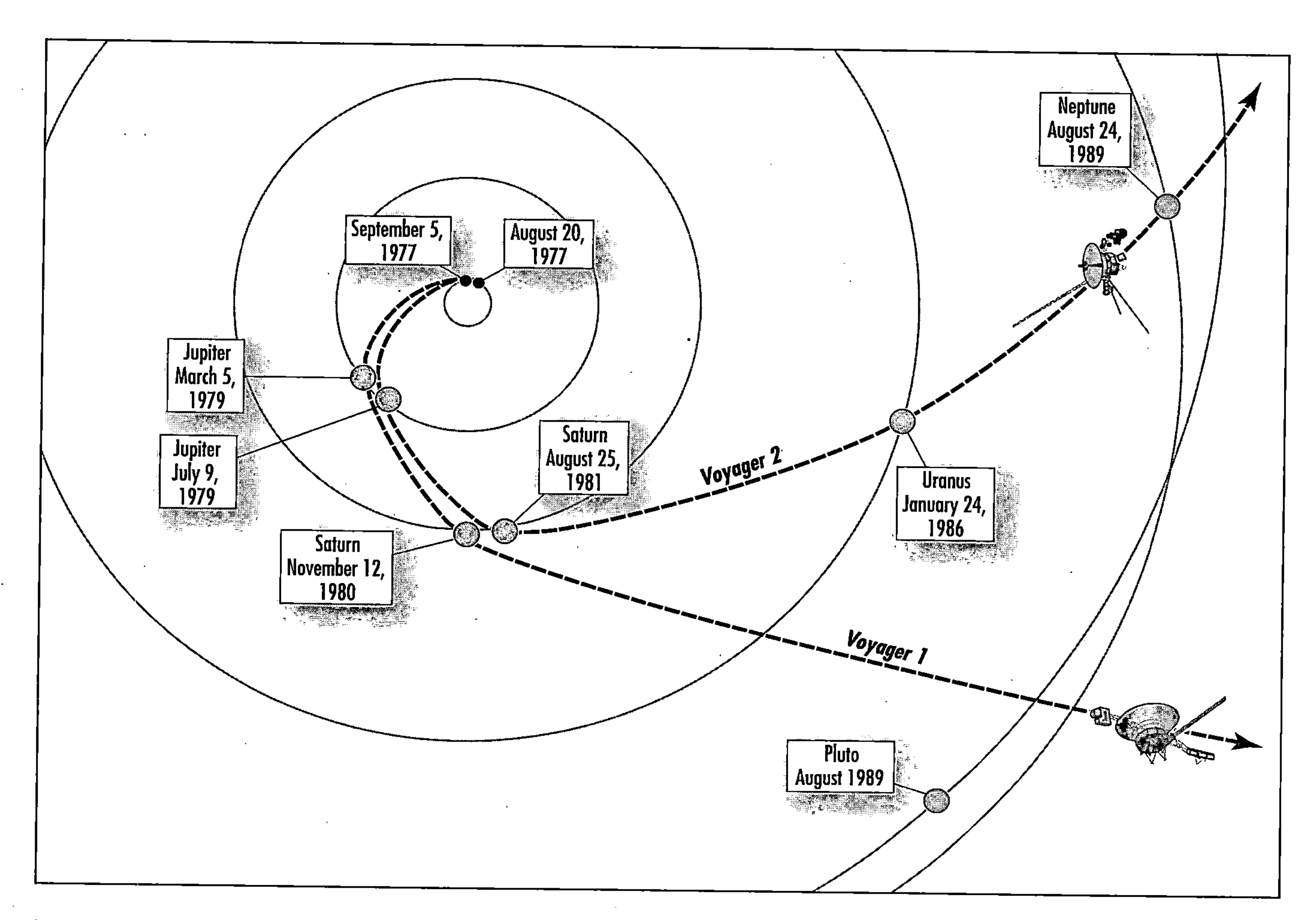

Planetary encounters of the *Voyager 1* and *2* spacecraft (1977–89) *(Drawing courtesy of NASA)*

"Nearby" Stars That Will Be Encountered by *Voyager 2* in the Next Million Years

	Year of Closest Approach	*Voyager 2*-to-Star Distance (Light-years)	Sun-to-*Voyager 2* Distance (Light-years)	Sun-to-Star Distance (Light-years)
Barnard's Star	8,571	4.03	0.42	3.80
Proxima Centauri	20,319	3.21	1.00	3.59
Alpha Centauri	20,629	3.47	1.02	3.89
Lalande 21185	23,274	4.65	1.15	4.74
Ross 248	40,176	1.65	1.99	3.26
DM-36 13940	44,492	5.57	2.20	7.39
AC + 79 3888	46,330	2.77	2.29	3.76
Ross 154	129,084	5.75	6.39	8.83
DM + 15 3364	129,704	3.44	6.42	6.02
Sirius	296,036	4.32	14.64	16.58
DM-5 4426	318,543	3.92	15.76	12.66
44 Ophiuchi	442,385	6.72	21.88	21.55
DM + 27 1311	957,963	6.62	47.38	47.59

Source: NASA/JPL.

the men and women who sent both *Voyagers* on its stellar journey. The table on page 645 identifies the "nearby" stars that *Voyager 2* will encounter in the next million years.

See also JUPITER; NEPTUNE; ORBITS OF OBJECTS IN SPACE; SATURN; URANUS.

V-2 rocket The German V-2 rocket of World War II was the "grandfather" of many of the large missiles flown by the United States. Encouraged by the findings made by the American rocket scientist ROBERT H. GODDARD in the 1920s and 1930s, a team of German rocket scientists developed this weapon system at Peenemünde, Germany, and terrorized Allied populations of continental Europe and Great Britain until the end of the war. The German program began in early 1940; the first V-2 rocket was launched on July 6, 1942. In the closing days of World War II, the United States captured many German rocket scientists, including WERNHER VON BRAUN, as part of Operation Paper Clip and shipped these scientists plus numerous captured V-2 rocket components to the United States. These V-2 rockets were then assembled and tested at the White Sands Missile Range in New Mexico as part of an emerging U.S. missile program.

The V-2 rocket itself had a liquid-propellant engine that burned alcohol and liquid oxygen. It was about 14 meters long and 1.66 meters in diameter. After World War II the United States (using American and captured German rocket personnel) assembled and tested 67 V-2 rockets in New Mexico between 1946 and 1952. This experience provided the technical expertise for many important American rockets, including the Redstone, Jupiter, and Saturn vehicles.

A reassembled German V-2 rocket lifting off its launch pad at the U.S. Army's White Sands Missile Range in southern New Mexico ca. 1947 *(Courtesy of the U.S. Army)*

Vulcan A planet that some 19th-century astronomers believed existed in an extremely hot orbit between Mercury and the Sun. Named after the Roman god of fire and metalworking, Vulcan's existence was postulated to account for gravitational perturbations observed in the orbit of Mercury. Modern astronomical observations have failed to reveal this celestial object, and ALBERT EINSTEIN's theory of relativity enabled 20th-century astronomers to account for the observed irregularities in Mercury's orbit. As a result, the planet Vulcan, created out of theoretical necessity by 19th-century astronomers, has now quietly disappeared in contemporary discussions about our solar system.

See also MERCURY; RELATIVITY (THEORY); SOLAR SYSTEM.

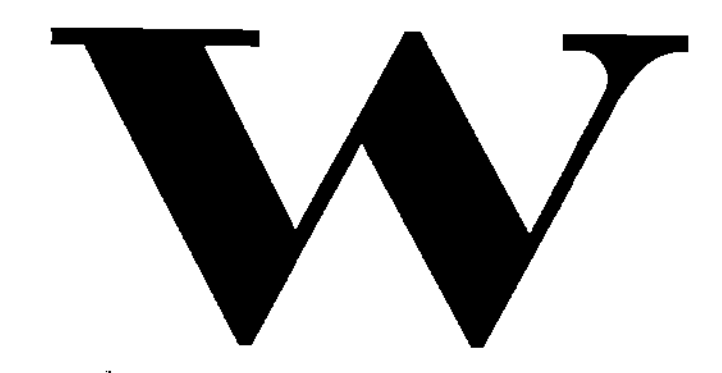

Wallops Flight Facility (WFF) The NASA Goddard Space Flight Center's Wallops Flight Facility is located on Virginia's Eastern Shore. The facility was originally established in 1945 by the National Advisory Committee for Aeronautics (NACA) as a center for aeronautic research. Wallops, one of the oldest launch sites in the world, is now NASA's primary facility for management and implementation of suborbital research programs. Built to conduct aeronautical research using rocket-propelled vehicles, Wallops launched its first rocket on July 4, 1945. Since then the facility has fulfilled its basic mission with the launch of more than 14,000 rockets, primarily sounding rockets. Over the years the WFF launch range has grown to include six launch pads, assembly facilities, and state-of-the-art instrumentation. In addition, the facility's mobile launch facilities enable NASA scientists and engineers to launch rockets around the world. The science dictates the launch site; WFF personnel provide the expertise and instrumentation to launch, track, acquire data, and recover the payload. After consolidating with NASA's Goddard Space Flight Center (GSFC) in 1982, the Wallops Flight Facility became NASA's primary facility for suborbital programs. The facility's customer base for sounding rocket services diversified to include more commercial users for the WFF research airport, tracking facilities, and launch range.

See also GODDARD SPACE FLIGHT CENTER (GSFC); SOUNDING ROCKET.

warhead The portion of a missile or rocket that contains either a nuclear or thermonuclear weapon system, a high explosive system, chemical or biological agents, or harmful materials intended to inflict damage on an enemy.

water hole A term used in the search for extraterrestrial intelligence (SETI) to describe a narrow portion of the electromagnetic spectrum that appears especially appropriate for interstellar communications between emerging and advanced civilizations. This band lies in the radio frequency (RF) part of the spectrum between 1,420 megahertz (MHz) frequency (21.1 cm wavelength) and 1,660 megahertz (MHz) frequency (18 cm wavelength).

Hydrogen (H) is abundant throughout interstellar space. When hydrogen experiences a "spin-flip" transition due to an atomic collision, it emits a characteristic 1,420 megahertz frequency (or 21.1 cm wavelength) radio wave. Any intelligent race within the Milky Way galaxy that has risen to the technological level of radio astronomy will eventually detect these natural emissions. Similarly, there is another grouping of characteristic spectral emissions centered near the 1,660 megahertz frequency (18 cm wavelength) that are associated with hydroxyl (OH) radicals. As we know from elementary chemistry, $H + OH = H_2O$, so we have, as suggested by SETI investigators, two radio wave emission signposts associated with the dissociation products of water that "beckon all waterbased life to search for its kind at the age-old meeting place of all species: the water hole."

Is this high regard for the 1,420 to 1,660 megahertz frequency band reasonable, or simply a case of terrestrial chauvinism? Many exobiologists (astrobiologists) currently feel that if other life exists in the universe, it will most likely be carbon-based life, and water is essential for carbon-based life as we know it. In addition, for purely technical reasons, if we were to scan all the decades of the electromagnetic spectrum in search of a suitable frequency at which to send or receive interstellar messages, we would arrive at the narrow microwave region between 1 and 10 gigahertz (GHz) as the most suitable candidate for conducting interstellar communication. The two characteristic emissions of dissociated water, namely 1,420 megahertz for H and 1,660 megahertz for OH, are situated within an apparently optimum radio frequency communications band.

On the basis of this type of reasoning, the water hole has been favored by scientists engaged in SETI projects involving the reception and analysis of radio signals. They generally consider that this portion of the electromagnetic spectrum represents a logical starting place for us to listen for interstellar signals from other intelligent civilizations.

See also SEARCH FOR EXTRATERRESTRIAL INTELLIGENCE.

watt (symbol: W) The SI unit of power (i.e., work per unit time). A watt represents 1 joule (J) of energy per second. In electrical engineering, 1 watt corresponds to the product of 1 ampere (A) times 1 volt (V). This represents the rate of electric energy dissipation in a circuit in which a current of 1 ampere is flowing through a voltage difference of 1 volt. This unit is named in honor of James Watt (1736–1819), the Scottish engineer who developed the steam engine.

See also HEAT ENGINE; THERMODYNAMICS.

wave A periodic disturbance that is propagated in a medium in such a manner that at any point in the medium, the quantity serving as a measure of the disturbance is a function of time, while at any instant the displacement at a point is a function of the position of the point. At each spatial point there is an oscillation. The number of oscillations that occur per unit time is the FREQUENCY (symbol: ν). The distance between one wave crest to the next wave crest (or one trough to the next trough) is called the *wavelength* (symbol: λ).

See also WAVELENGTH.

wavelength (symbol: λ) In general, the mean distance between maxima (or minima) of a periodic pattern. Specifically, the least distance between particles moving in the same phase of oscillation in a wave disturbance. The wavelength is measured along the direction of propagation of the wave, usually from the midpoint of a crest (or trough) to the midpoint of the next crest (or trough). The wavelength (λ) is related to the frequency (ν) and phase speed (c) (i.e., speed of propagation of the wave disturbance) by the simple formula $\lambda = c/\nu$. The reciprocal of the wavelength is called the *wave number.*

weakly interacting massive particle *See* WIMP.

weak nuclear force *See* FUNDAMENTAL FORCES IN NATURE.

weapons of mass destruction (WMD) In defense and arms control usage, weapons that can cause a high order of destruction and kill many people. These types of weapons include nuclear, chemical, biological, and radiological. The term generally excludes the means of transporting or propelling the weapon when such means is a separable and divisible part of the weapon (e.g., a guided missile).

weather satellite One of the first applications of data and images supplied by Earth-orbiting satellites was to improve the understanding and prediction of weather. The weather satellite (also known as the METEOROLOGICAL SATELLITE or the *environmental satellite*) is an uncrewed spacecraft that carries a variety of sensors. Two types of weather satellites generally are encountered, geostationary and polar-orbiting weather satellites, which are named for their types of orbits.

Today weather satellites are used to observe and measure a wide range of atmospheric properties and processes to support increasingly more sophisticated weather warning and forecasting activities. Imaging instruments provide detailed pictures of clouds and cloud motions as well as measurements of sea-surface temperatures. Sounders collect data in several infrared or microwave spectral bands that are processed to provide profiles of temperature and moisture as a function of altitude. Radar altimeters, scatterometers, and imagers (i.e., synthetic aperture radar, or SAR) can measure ocean currents, sea-surface winds, and the structure of snow and ice cover. Some weather satellites are even equipped with a search and rescue satellite-aided tracking (SARSAT) system that is used on a global basis to help locate people who are lost and who have the appropriate emergency transmitters. SARSAT-equipped weather satellites can immediately receive and relay distress signals, increasing the probability of a prompt, successful rescue mission.

In the United States several federal agencies have charters for monitoring and forecasting weather. The National Weather Service (NWS) of the National Oceanic and Atmospheric Administration (NOAA) has the primary responsibility for providing severe storm and flood warnings as well as short- and medium-range weather forecasts. The Federal Aviation Administration (FAA) provides specialized forecasts and warnings for aircraft. The Defense Meteorological Satellite Program (DMSP) within the Department of Defense (DOD) formerly supported the specialized needs of the military and intelligence communities, which emphasize global capabilities to monitor clouds and visibility in support of combat and reconnaissance activities and to monitor sea-surface conditions in support of naval operations. In response to a 1994 presidential directive, NOAA has assumed operational responsibility for the DMSP.

Global change research strives to monitor and understand the processes of natural and anthropogenic (people-caused) changes in Earth's physical, biological, and human environments. Weather satellites support this research by providing measurements of stratospheric ozone and ozone-depleting chemicals; by providing long-term scientific records of Earth's climate; by monitoring Earth's radiation balance and the concentrations of greenhouse gases and aerosols; by monitoring ocean temperatures, currents, and biological productivity; by monitoring the volume of ice sheets and glaciers; and by monitoring land use and vegetation. These variables provide important information concerning the complex processes and interactions of global environment change, including climate change.

See also DEFENSE METEOROLOGICAL SATELLITE PROGRAM; GEOSTATIONARY OPERATIONAL ENVIRONMENTAL SATELLITE; GLOBAL CHANGE; POLAR-ORBITING OPERATIONAL ENVIRONMENTAL SATELLITE.

web The minimum thickness of a solid-propellant grain from the initial ignition surface to the insulated case wall or to the intersection of another burning surface at the time when the burning surface undergoes a major change; for an end-burning grain, the length of the grain.

See also ROCKET; SOLID-PROPELLANT ROCKET.

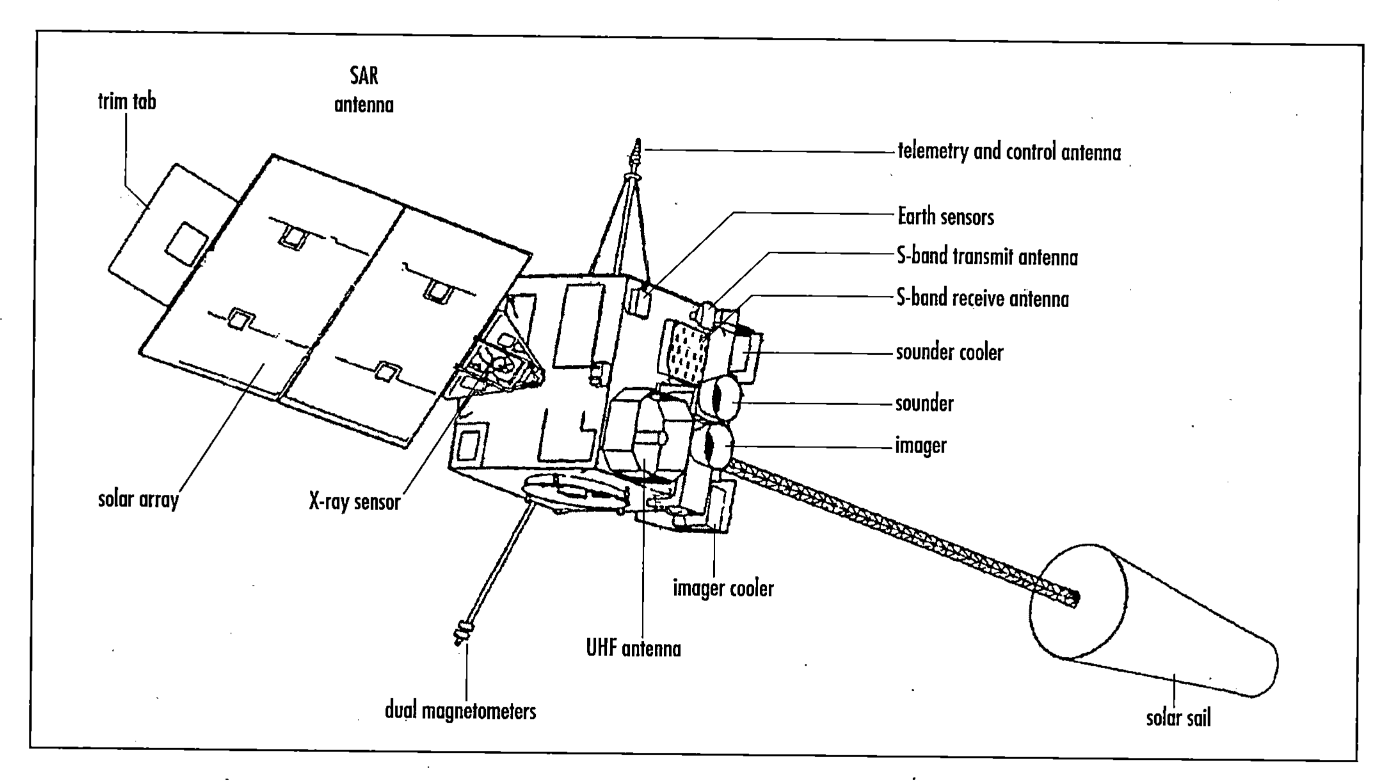

The major components and features of a family of modern weather satellites, NOAA's Geostationary Operational Environmental Satellites (GOES I-M). Note that the "SAR antenna" represents a special search-and-rescue (SAR) mission feature that enables this type of satellite to receive and then relay distress signals from people who are lost on Earth. ***(Drawing courtesy of NASA and NOAA)***

Webb, James E. (1906–1992) American *NASA Administrator* James E. Webb was the second administrator of the National Aeronautics and Space Administration (NASA), the American civilian space agency formally established on October 1, 1958, under the National Aeronautics and Space Act of 1958. As the leader of NASA, he responded to President JOHN F. KENNEDY's vision and was the management force behind the daring and spectacular Apollo Project that placed American astronauts on the lunar surface and returned them safely to Earth.

Webb was born on October 7, 1906, in Tally Ho, North Carolina. Starting in 1932, he enjoyed a long and diversified career in and out of public service, including an appointment as under secretary of state, U.S. Department of State, by President Harry S. Truman in 1949. On February 14, 1961, Webb became the second administrator of NASA. Under his direction the fledgling civilian agency undertook one of the most impressive projects in history, the goal of landing an American on the Moon before the end of the decade through the execution of the Apollo Project. For seven years after President Kennedy's May 25, 1961, lunar landing announcement through October 1968, James Webb successfully politicked and maneuvered for NASA in Washington. As a longtime Washington insider, he was a master at bureaucratic politics. In the end, through a variety of methods Webb built a network of political liaisons that brought continued support for and resources to accomplish the first human Moon landing on the bold schedule President Kennedy had announced.

Unfortunately, Webb was also NASA's leader when tragedy struck the Apollo Project. On January 27, 1967, Apollo-Saturn (AS) 204 was on the launch pad at Kennedy Space Center, Florida, moving through simulation tests when a flash fire within the space capsule killed the three astronauts aboard, VIRGIL "GUS" I. GRISSOM, EDWARD H. WHITE, II, and ROGER B. CHAFFEE. Under his guidance NASA set out to discover the details of the tragedy, corrected the problems, and got the Apollo Project back on schedule. Webb deflected much of the backlash over the Apollo tragedy from both NASA as an agency and from the administration of President Lyndon Johnson. While he was personally tarred with the disaster, the space agency's image and popular support was largely undamaged. He left NASA in October 1968, just as the Apollo Project was approaching its successful completion. After retiring from NASA, Webb remained in Washington, D.C., serving on several advisory boards, including as a regent of the Smithsonian Institution. He died on March 27, 1992, and is buried in Arlington National Cemetery. To honor his leadership accomplishments in both human space flight and space science, NASA has named the successor to the *HUBBLE SPACE TELESCOPE* the *JAMES WEBB SPACE TELESCOPE* in his honor.

weber (symbol: Wb) The SI unit of magnetic flux. It is defined as the flux that produces an electromotive force of 1 volt when, linking a circuit of 1 turn, the flux is reduced to 0 at a uniform rate in 1 second. Named in honor of Wilhelm Weber, a German physicist (1804–91) who studied magnetism. 1 weber = 10^8 maxwells.

weight (symbol: w) 1. The force with which a body is attracted toward Earth or another planetary body by gravity. 2. The product of the mass (m) of a body and the gravitational acceleration (g) acting on the body, w = mg. For example, a 1-kilogram mass on the surface of Earth would experience a downward force, or "weight," of approximately 9.8 newtons. In a dynamic situation, the weight can be a multiple of that under resting conditions. Weight varies on other planets in accordance with their value of gravitational acceleration.

weightlessness The condition of free fall, or zero-g, in which objects inside an orbiting, unaccelerated spacecraft are "weightless"—even though the objects and the spacecraft are still under the influence of the celestial object's gravity; the condition in which no acceleration, whether of gravity or other force, can be detected by an observer within the system in question.

See also g; MICROGRAVITY.

Weizsäcker, Carl Friedrich Baron (1912–) German *Theoretical Physicist* In 1938, the German theoretical physicist, Baron Carl Friedrich Weizsäcker (independent of the work performed by HANS ALBRECHT BETHE) suggested that stars derive their energy from a chain of thermonuclear fusion reactions, primarily by joining hydrogen into helium in a process called the carbon cycle. Reviving (in part) some of the 18th-century work of IMMANUEL KANT and PIERRE-SIMON, MARQUIS DE LAPLACE, in 1944 Weizsäcker developed a 20th-century version of the NEBULA HYPOTHESIS in an attempt to explain how the solar system might have formed from an ancient cloud of interstellar gas and dust. One of the consequences of his modern version of this hypothesis is that stars with planets are a normal part of stellar evolution.

Wells, Herbert George (H. G.) (1866–1946) British *Science fiction Writer* At the beginning of the 20th century the British writer H. G. Wells inspired many future astronautical pioneers with his exciting fictional works that popularized the idea of space travel and life on other worlds. For example, in 1897 he wrote *War of the Worlds,* the classic tale about extraterrestrial invaders from Mars.

Wells was born on September 21, 1866, in Bromley, Kent, England. In 1874 a childhood accident forced him to recuperate with a broken leg. The prolonged convalescence encouraged him to become an ardent reader, and this period of intensive self-learning served him well. He went on to become an accomplished author of both science fiction and more traditional novels.

He settled in London in 1891 and began to write extensively on educational matters. His career as a science fiction writer started in 1895 with the publication of the incredibly popular book *The Time Machine.* At the turn of the century he focused his attention on space travel and the consequences of alien contact. Between 1897 and 1898 *The War of the Worlds* appeared as a magazine serial and then a book. Wells followed this very popular space invasion story with *The First Men in the Moon,* which appeared in 1901. Like JULES VERNE, Wells did not link the rocket to space travel, but his stories did excite the imagination. *The War of the Worlds* was the classic tale of an invasion of Earth from space. In his original story hostile Martians land in 19th-century England and prove to be unstoppable, conquering villains until they themselves are destroyed by tiny terrestrial microorganisms.

In writing this story Wells was probably influenced by the then popular (but incorrect) assumption that supposedly observed Martian "canals" were artifacts of a dying civilization on the Red Planet. This was a very fashionable hypothesis in late 19th-century astronomy. The "canal craze" started quite innocently in 1877, when the Italian astronomer GIOVANNI SCHIAPARELLI reported linear features he observed on the surface of Mars as *canali,* the Italian word for "channels." Schiaparelli's accurate astronomical observations became misinterpreted when translated as "canals" in English. Consequently, other notable astronomers such as the American PERCIVAL LOWELL began to enthusiastically search for and soon "discover" other surface features on the Red Planet that resembled signs of an intelligent Martian civilization.

H. G. Wells cleverly solved (or more than likely ignored) the technical aspects of space travel in his 1901 novel *The First Men in the Moon.* He did this by creating "cavorite," a fictitious antigravity substance. This story inspired many young readers to think about space travel. However, space age missions to the Moon have now completely vanquished the delightful (though incorrect) products of this writer's fertile imagination, including giant moon caves, a variety of lunar vegetation, and even bipedal Selenites.

However, in many of his other fictional works Wells was often able to correctly anticipate advances in technology. This earned him the status of a technical prophet. For example, he foresaw the military use of the airplane in his 1908 work *The War in the Air* and foretold of the splitting of the atom in his 1914 novel *The World Set Free.*

Following his period of successful fantasy and science fiction writing, Wells focused on social issues and the problems associated with emerging technologies. For example, in his 1933 novel *The Shape of Things to Come,* he warned about the problems facing Western civilization. In 1935 Alexander Korda produced a dramatic movie version of this futuristic tale. The movie closes with a memorable philosophical discussion on technological pathways for the human race. Sweeping an arm, as if to embrace the entire universe, one of the main characters asks his colleague: "Can it really be our destiny to conquer all this?" As the scene fades out, his companion replies: "The choice is simple. It is the whole universe or nothing. Which shall it be?"

The famous novelist and visionary died in London on August 13, 1946. He had lived through the horrors of two world wars and witnessed the emergence of many powerful new technologies, except space technology. His last book, *Mind at the End of Its Tether,* appeared in 1945. In this work

Wells expressed a growing pessimism about humanity's future prospects.

See also SCIENCE FICTION.

wet emplacement A launch pad that provides a deluge of water for cooling the flame bucket and other equipment during the launch of a missile or aerospace vehicle. *Compare with* DRY EMPLACEMENT and FLAME DEFLECTOR.

Whipple, Fred Lawrence (1906–2004) American *Astronomer* In the period 1949–50 the American astronomer Fred Lawrence Whipple proposed the "dirty snowball" hypothesis for comet nuclei. When the European Space Agency's *GIOTTO* spacecraft encountered Comet Halley at close range in March 1986, the imagery collected by the spacecraft confirmed Whipple's hypothesis. In the 1940s Whipple introduced the concept of the *Whipple shield,* a sacrificial bumper usually made of aluminum that is placed in front of a spacecraft to absorb the initial impact of a small meteoroid collision. The Whipple shield absorbs the threatening high-speed projectile, creating a debris cloud that contains many smaller, less lethal bumper and projectile fragments. The full force of the debris cloud is then diluted over a larger area on the rear wall of the Whipple shield. Depending on the design characteristics of the shield and the size of the projectile, the shield should prevent perforation of a spacecraft's hull (wall) under projected mission scenarios. The Whipple shield was the first spacecraft shield ever implemented, and aerospace engineers still use this basic concept to protect modern spacecraft against anticipated meteoroid and space debris hazards.

See also COMET; METEOROIDS; SPACE DEBRIS.

White, Edward H., II (1930–1967) American *Astronaut, U.S. Air Force Officer* On June 7, 1965, Edward H. White, II performed the first extravehicular activity (EVA) from an orbiting American spacecraft. This important space technology milestone took place during the *Gemini 4* mission of NASA's Gemini Project. White later died in the fatal flash fire that consumed the interior of the Apollo Project spacecraft during a training test on the launch pad at Cape Canaveral on January 26, 1967. Apollo astronauts VIRGIL "GUS" I. GRISSOM and ROGER B. CHAFFEE also perished in this deadly accident.

See also APOLLO PROJECT; GEMINI PROJECT.

white dwarf A compact, dense star nearing the end of its evolutionary life. A white dwarf is a star that has exhausted the thermonuclear fuel (hydrogen and helium) in its interior, causing nuclear burning to cease. When a star of 1 solar mass or less exhausts its nuclear fuel, it collapses under gravity into a very dense object about the size of Earth. Initially, the surface temperature of the white dwarf is high, typically 10,000 K or more. However, as cooling by thermal radiation continues, the object becomes fainter and fainter until it finally degenerates into a nonvisible stellar core derelict called a BLACK DWARF.

See also STAR.

white hole A highly speculative concept defining a region where matter spontaneously enters into the universe, the opposite (time-reverse) phenomenon of an object falling into a black hole. Some scientists cautiously suggest that the wormhole may serve as a passageway or connection between a black hole and a white hole. No such object or phenomenon has yet been observed.

See also BLACK HOLE; WORMHOLE.

white room A clean, dust-free room that is used for the assembly, calibration, and (if necessary) repair of delicate spacecraft components and devices, such as gyros and sensor systems.

See also BIOCLEAN ROOM.

White Sands Missile Range (WSMR) White Sands Missile Range is a multiservice test range whose main function is the support of missile development and test programs for the U.S. Army, Navy, Air Force, NASA, and other government agencies and private firms. WSMR is under operational control of the U.S. Army Test and Evaluation Command (TECOM), Aberdeen Proving Ground, Maryland. The missile range is located in the Tularosa Basin of south-central New Mexico. The range boundaries extend almost 160 km north to south by about 65 km east to west.

During World War II the U.S. Army Ordnance Corps recognized the possibilities of rocket warfare and sponsored research and development in methods of missile guidance. The missile range was established on July 9, 1945, as White Sands Proving Ground (the name was later changed to White Sands Missile Range) to serve as America's testing range for the development of rocket technology and missile weapons. Following World War II captured German V-2 rockets were assembled and tested at White Sands. Other rockets such as Nike, Corporal, Lance, and Viking also experienced developmental testing at the range. Because it was the United States' first major rocket launch facility, Launch Complex 33 (LC-33) at WSMR is designated a National Historic Landmark. It is also notable that the world's first atomic bomb, a test device called the Trinity shot, was exploded on July 16, 1945, near the northern boundary of the WSMR complex.

During the early part of the space shuttle program, the astronauts Jack Lousma and C. Gordon Fullerton landed the orbiter *Columbia* at the Northrup Strip in the Alkali Flats area of the WSMR to successfully end the STS-3 mission (March 30, 1982). Because heavy rains had made the primary shuttle landing site at Edwards Air Force Base, California, unacceptably wet, NASA chose the alternate site to support *Columbia*'s return to Earth. After this event the Northrup Strip was renamed the White Sands Space Harbor. Although STS-3 has been the only shuttle mission to land at the range, WSMR continues to support ongoing shuttle missions, including support as a contingency or end-of-mission (EOM) landing site.

Wien's displacement law In radiation transport, the statement that the wavelength of maximum radiation intensity (λ_{max}) for a blackbody is inversely proportional to the absolute temperature of the radiating blackbody:

$$\lambda_{max}\, T = b \text{ (a constant)}$$

If the absolute temperature (T) is in kelvins and the maximum wavelength (λ_{max}) is in centimeters, then the constant (b) has

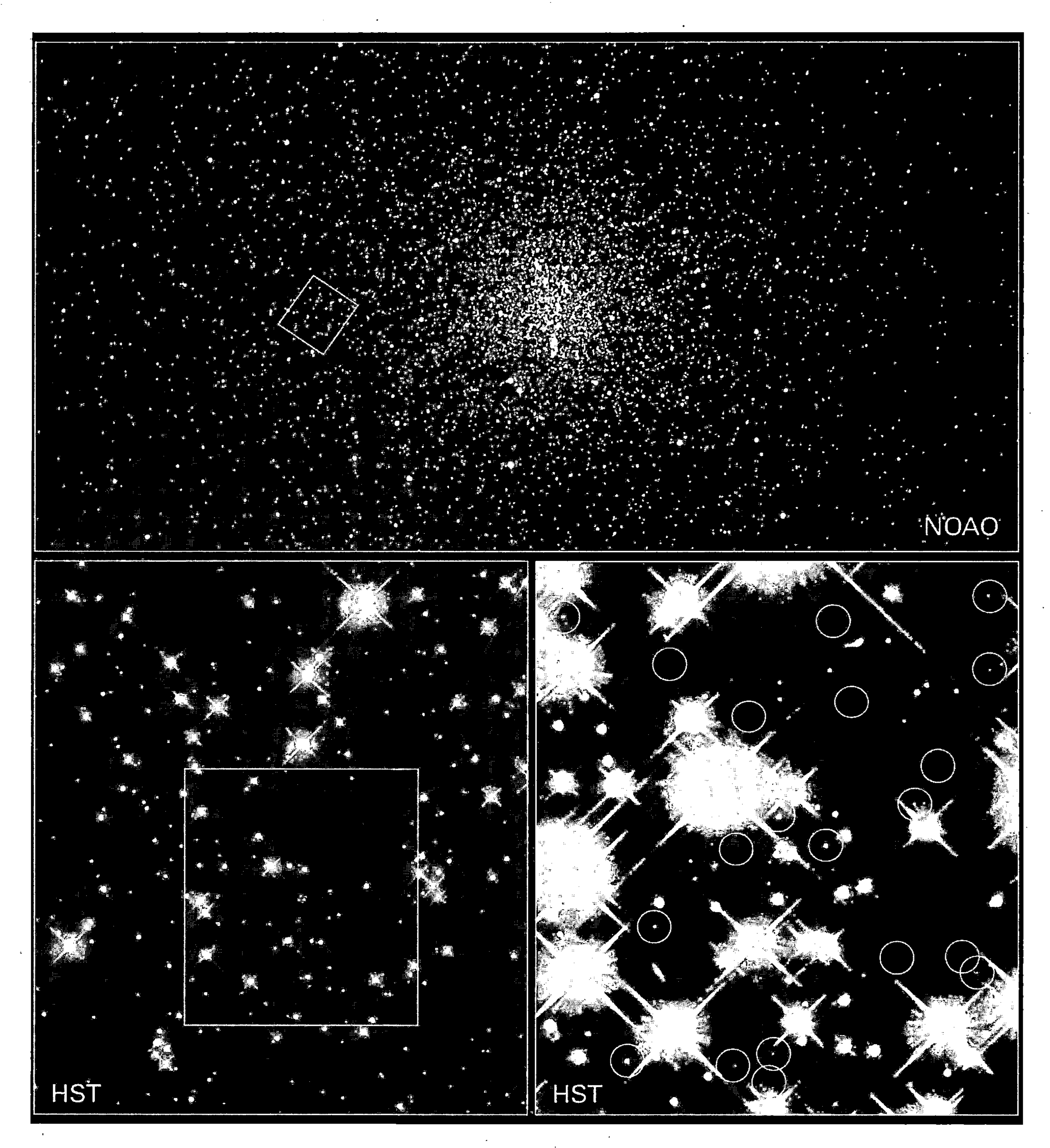

Peering deep inside a globular cluster of several hundred thousand stars called M4, NASA's *Hubble Space Telescope* (HST) uncovered the oldest known burned-out stars in the Milky Way galaxy. These small dying stars are white dwarfs, and by observing them astronomers obtained an additional perspective on the age of the universe. The top panel is a panoramic view of the entire M4 globular cluster as photographed by the 0.9-meter-diameter ground-based telescope at the Kitt Peak National Observatory in March 1995. The box to the left is the small region observed by the HST from January through April 2001. An even smaller, one-light-year-across HST sampling of this region appears at the bottom right. In this image astronomer-added circles pinpoint a number of faint white dwarfs. *(HST photos courtesy of NASA and H. Richer [University of British Columbia]; upper panel courtesy of NOAO/AURA/NSF)*

the value 0.2898 cm-K. Building upon the work of LUDWIG BOLTZMANN, the German physicist Wilhelm Carl Wien (1864–1928) presented this relationship in 1893.

"wilco" Aerospace language for "I will comply."

Wildt, Rupert (1905–1976) German-American *Astronomer* In the 1930s the German-American astronomer Rupert Wildt performed detailed spectroscopic studies of the atmospheres of Jupiter and Saturn. His investigations suggested the presence of methane and ammonia.

Wilkinson Microwave Anisotropy Probe **(WMAP)** NASA's WMAP mission is designed to determine the geometry, content, and evolution of the universe through a high-resolution full-sky map of the temperature anisotropy of the cosmic microwave background (CMB). The choice of the spacecraft's orbit, sky-scanning strategy, and instrument design were driven by the science goals of the mission. The CMB sky map data products derived from the WMAP observations have 45 times the sensitivity and 33 times the angular resolution of NASA's *Cosmic Background Explorer* (COBE) mission.

In 1992 NASA's COBE satellite detected tiny fluctuations, or anisotropy, in the cosmic microwave background. This spacecraft found, for example, that one part of the sky has a cosmic microwave background radiation temperature of 2.7251 K, while another part of the sky has a temperature of 2.7249 K. These fluctuations are related to fluctuations in the density of matter in the early universe and therefore carry information about the initial conditions for the formation of cosmic structures such as galaxies, clusters, and voids. The COBE spacecraft had an angular resolution of 7° across the sky—this is 14 times larger than the Moon's apparent size. Such limitations in angular resolution made COBE sensitive only to broad fluctuations in the CMB of large size.

NASA launched the *Wilkinson Microwave Anisotropy Probe* (WMAP) in June 2001, and this spacecraft has been making sky maps of temperature fluctuations of the CMB radiation with much higher resolution, sensitivity, and accuracy than COBE. The new information contained in WMAP's finer study of CMB fluctuations is shedding light on several very important questions in cosmology.

WMAP was launched on June 30, 2001, from Cape Canaveral Air Force Station, Florida, by a Delta II rocket and placed into a controlled Lissajous orbit about the second Sun-Earth Lagrange point (L2), a distant orbit that is four times farther than the Moon and 1.5 million kilometers from Earth. *Lissajous orbits* are the natural motion of a satellite around a collinear libration point in a two-body system and require the expenditure of less momentum change for station keeping than do halo orbits, in which the satellite follows a simple circular or elliptical path around the libration point.

The WMAP spacecraft is expected to collect high-quality science data involving faint anisotropy, or variations, in the temperature of the CMB for at least four years. The WMPA spacecraft is mapping the cosmic microwave background at five frequencies: 23 gigahertz (GHz) (K-band), 33 GHz (Ka-band), 41 GHz (Q-band), 61 GHz (V-band), and 94 GHz (W-band). The cosmic microwave background radiation that WMAP is observing is the oldest light in the universe and has been traveling across the universe for about 14 billion years.

See also BIG BANG; *COSMIC BACKGROUND EXPLORER* (COBE); COSMOLOGY.

Wilson, Robert Woodrow (1936–) American *Physicist, Radio Astronomer* Robert Wilson collaborated with ARNO PENZIAS at Bell Laboratories in the mid-1960s and made the most important discovery in 20th-century cosmology. They detected the cosmic microwave background radiation and provided the first empirical evidence to support the big bang hypothesis. For this milestone achievement Wilson and Penzias shared the 1978 Nobel Prize in physics.

Wilson was born in Houston, Texas, on January 10, 1936. His childhood experiences involving visits to the company shop with his father, a chemist working for the oil industry, provided Wilson his lifelong interest in electronics and machinery. In 1957 he received an undergraduate degree in physics with honors from Rice University and then enrolled for graduate work in physics at the California Institute of Technology (Caltech). At the time Wilson had not decided about what specific area of physics he wanted to pursue for his doctoral research. However, fortuitous meetings and discussions on campus led him to become involved in interesting work at Caltech's Owens Valley Radio Observatory, then undergoing development in the Sierra Nevada Mountains near Bishop, California. Today this facility, located about 400 km north of Los Angeles, is the largest university-operated radio observatory in the world. For Wilson graduate research in radio astronomy provided a nice mixture of electronics and physics. His decision to pursue this area of research placed him on a highly productive career path in radio astronomy that included a share of the Nobel Prize in physics in 1978 for the most important discovery in cosmology in the 20th century. However, in 1958 he briefly delayed his entry into radio astronomy by returning to Houston so he could court and then marry his wife, Elizabeth Rhoads Sawin.

In 1959 Wilson took his first astronomy courses at Caltech and began working at the Owens Valley Radio Observatory during breaks in the academic calendar. He completed the last part of his thesis research under the supervision of MAARTEN SCHMIDT, who was exploring quasars at the time. Following graduation with a doctoral degree, Wilson stayed on at Caltech as a postdoctoral researcher and completed several other projects he had worked on. In 1963 he joined Bell Laboratories and soon met Arno Penzias, the only other radio astronomer at the laboratory. Between 1963 and 1965 they collaborated on applying a special 6-m-diameter antenna to investigate the problem of radio noise from the sky. At this time Bell Laboratories had a general interest in detecting and resolving any cosmic radio noise problems that might adversely affect the operation of early communication satellites. Expecting radio noise to be less severe at shorter wavelengths, Wilson and Penzias were surprised to discover that the sky was uniformly "bright" in the microwave region at 7.3 centimeters wavelength. Discussions with ROBERT H. DICKE in the spring of 1965 indicated that Wilson and Penzias had stumbled upon the first direct evidence of the cosmic microwave background, the "Holy Grail" of big bang cosmology.

Wilson and Penzias estimated the detected cosmic "radio noise" to be about 3 K. Measurements performed by NASA's *Cosmic Background Explorer* (COBE) spacecraft between 1989 and 1990 indicated that the cosmic microwave background closely approximates a blackbody radiator at a temperature of 2.735 K. Wilson and Penzias clearly hit the scientific jackpot. For this great (though somewhat serendipitous) discovery, Wilson shared one half of the 1978 Nobel Prize in physics with Penzias. The other half of the award that year went to the Russian physicist Pyotr Kapitsa for unrelated work in low-temperature physics.

In the late 1960s and early 1970s Wilson and Penzias used techniques in radio astronomy at millimeter wavelengths to investigate molecular species in interstellar space. They were excited to find large quantities of carbon monoxide (CO) behind the Orion Nebula and soon found that this molecule was rather widely distributed throughout the Milky Way galaxy. From 1976 to 1990 Wilson served as the head of the Radio Physics Research Department at Bell Laboratories. He was elected to the U.S. National Academy of Sciences in 1979. In addition to the 1978 Nobel Prize in physics, Wilson also received the Herschel Medal from the Royal Astronomical Society (1977) and the Henry Draper Medal from the American National Academy of Sciences (1977). Wilson's most important contribution to astronomy and astrophysics remains his codiscovery of the cosmic microwave background, the lingering remnant of the big bang explosion that started the universe.

See also ASTROPHYSICS; BIG BANG (THEORY); COSMOLOGY; RADIO ASTRONOMY.

WIMP A hypothetical particle, called the *weakly interacting massive particle,* thought by some scientists to pervade the universe as the hard-to-observe dark matter.

See also DARK MATTER; ELEMENTARY PARTICLE(S); FUNDAMENTAL FORCES IN NATURE; MACHO.

window In general, a gap in a linear continuum. An ATMOSPHERIC WINDOW, for example, is a range of wavelengths in the electromagnetic spectrum to which the atmosphere is transparent; a LAUNCH WINDOW is the time interval during which conditions are favorable for launching an aerospace vehicle or spacecraft on a specific mission.

wind tunnel A ground facility that supports aerodynamic testing of aircraft, missiles, propulsion systems, or their components under simulated flight conditions. The concept of a wind tunnel is quite simple. Instead of flying the object (or a precise scale model of the object) through the air at the desired test speed and attitude, it is supported in the test attitude within the wind tunnel, and the air flows past the object at the test speed. Wind tunnels exist in a wide variety of shapes and sizes. One wind tunnel design parameter of particular interest is the velocity of airflow, usually given in free-stream Mach number (M) range. (The Mach number is the ratio of the free stream velocity to the local velocity of sound.) The table provides a general classification of wind tunnels according to the speed of the airflow in the test section. The test section in a wind tunnel usually includes windows and appropriate illumination for viewing the test object. Today no aircraft, spacecraft, or space launch or reentry vehicle is built or committed to flight until after its design and components have been thoroughly tested in wind tunnels.

Classification of Wind Tunnels (According to Airflow Speed in Test Section)

Tunnel Class	Free-stream Mach-number[a] Range
Low speed	0 to 0.5
High speed	0.5 to 0.9
Transonic	0.7 to 1.4
Supersonic	1.4 to 5.0
Hypersonic	>5.0

[a] $\text{Mach number} = \frac{\text{Local stream velocity}}{\text{Local velocity of sound}}$

Source: NASA.

See also MACH NUMBER.

wing The aerodynamic lifting surface that provides conventional lift and control for an aircraft, winged missile, or aerospace vehicle.

See also AIRFOIL.

Wolf, Johann Rudolph (1816–1893) Swiss *Astronomer, Mathematician* The Swiss astronomer and mathematician Johann Rudolph Wolf counted sunspots and sunspot groups to confirm SAMUEL HEINRICH SCHWABE's work and discovered that the period of the sunspot cycle was approximately 11 years. He was also one of several observers who noticed the relationship between sunspot number and solar activity. He became director of the Bern Observatory in 1847. In 1849 Wolf formalized his system for describing solar activity by counting sunspots and sunspot groups, a system formerly referred to as *Wolf's sunspot number* or the *Wolf number* but now called the *relative sunspot number.*

Wolf, Maximilian (Max) Franz Joseph Cornelius (1863–1932) German *Astronomer* Max Wolf was an asteroid hunter who pioneered the use of astrophotography to help him in his search for these elusive solar system bodies. He discovered more than 200 minor planets, including asteroid Achilles in 1906, the first of the Trojan group of asteroids. This interesting group of minor planets moves around the Sun in Jupiter's orbit, but at the Lagrangian libration points 60° ahead of and 60° behind the giant planet.

Wolf was born on June 21, 1863, in Heidelberg, Germany, the city where he spent almost his entire lifetime. His father was a respected physician, and the family was comfortably wealthy. When Wolf was still a young boy, his father encouraged him to develop an interest in science and astronomy by building a private observatory adjacent to the family's residence. His father's investment was a good one. Wolf completed a doctoral degree in mathematical studies at the University of Heidelberg in 1888 and then went to the University of Stockholm for two years of postdoctoral work.

Wolf returned to Heidelberg in 1890 and began lecturing in astronomy at the university as a privat-docent (unpaid instructor). In 1893 Wolf received a dual appointment from the University of Heidelberg to serve as a special professor of astrophysics and to direct the new observatory being constructed at nearby Königsstuhl. In 1902 he became the chair of astronomy at the university. He also became the director of the Königsstuhl Observatory. Wolf retained these two positions for the remainder of his life.

He performed important spectroscopic work concerning nebulas and published 16 "Lists of Nebulae," which contained a total of approximately 6,000 celestial objects. Wolf is best remembered as being the first astronomer to use photography to find asteroids. In 1891 he demonstrated the ability to photograph a large region of the sky with a specially configured telescope assembly that followed the fixed stars precisely as Earth rotated. With this astrophotography arrangement, the stars appeared as points, while the minor planets Wolf was hunting appeared as short streaks on his photographic plates. On September 11, 1898, he discovered his first asteroid using this photographic technique and named the celestial object (also called asteroid 323) Brucia to honor the American philanthropist Catherine Wolfe Bruce. She had donated money for Wolf's new telescope at the Königsstuhl Observatory. Catherine Wolfe Bruce also established the *Bruce Gold Medal* as an award presented annually to an astronomer in commemoration of his or her lifetime achievements.

By the end of that year Wolf had discovered nine other minor planets using astrophotography. Over the next three decades this technique allowed him to personally discover more than 200 new asteroids, including Achilles-588, the first member of the Trojan group of asteroids. Wolf found and named Achilles in 1906, noting that it orbited the Sun at L_4, the Lagrangian libration point 60° ahead of Jupiter. The Trojan group is the collection of minor planets found at the two Lagrangian libration points lying on Jupiter's orbital path around the Sun. The name comes from the fact that many of these asteroids were named after the mythical heroes of the Trojan War.

Wolf was an avid asteroid hunter and a pioneer in astrophotography. In recognition of his contributions to astronomy he received the Gold Medal of the Royal Astronomical Society (1914) and the Bruce Gold Medal from the Astronomical Society of the Pacific (1930). He died in Heidelberg on October 3, 1932.

Wolf-Rayet star (WR) A very massive (more than 10 solar masses) hot, luminous (spectral type O) star in a late stage of evolution. The outer hydrogen-rich envelope of a WR star has been driven off by radiation pressure and strong stellar winds, exposing its hot helium core. This type of very hot star (up to 90,000 K) was first observed in 1867 by the French astronomers Charles Joseph Wolf (1827–1918) and Georges Antoine Pons Rayet (1839–1906). Since their discovery more than 300 WR stars have been identified in the Milky Way galaxy.

See also STAR.

Woolley, Sir Richard van der Riet (1906–1986) British *Astronomer* Sir Richard van der Riet Woolley investigated stellar and solar atmospheres, globular clusters, and stellar evolution. From 1956 to 1971 he served as Great Britain's 11th Astronomer Royal.

Worden, Alfred Merrill (1932–) American *Astronaut, U.S. Air Force Officer* The astronaut Alfred Merrill Worden served as the command module pilot during NASA's *Apollo 15* lunar landing mission (July–August 1971). While fellow Apollo astronauts DAVID R. SCOTT and JAMES BENSON IRWIN explored the Moon's surface in the Hadley Rille area, Worden orbited overhead in the Apollo Project spacecraft.

See also APOLLO PROJECT.

work (symbol: W) In physics and thermodynamics, work (W) is defined as the energy (E) expended by a force (F) acting though a distance (s). Mathematically,

$$\text{Work} = \int F \cdot d\, s'$$

In the SI unit system, work is expressed in joules (J). When a force of 1 newton (N) moves through a distance of 1 meter, 1 joule of work is performed, that is,

$$1 \text{ joule} = 1 \text{ newton-meter}$$

working fluid A fluid (gas or liquid) used as the medium for the transfer of energy from one part of a system to another part.

See also CARNOT CYCLE.

wormhole Some scientists speculate that matter falling into a black hole may actually survive. They suggest that under very special circumstances such matter might be conducted by means of passageways, called *wormholes,* to emerge in another place or time in this universe or perhaps in another universe. These hypothetical wormholes are considered to be distortions or holes in the space-time continuum. In forming this concept scientists are speculating that black holes can play "relativistic tricks" with space and time. If wormholes do exit, then in principle (at least) a person might be able to use one to travel faster than light, visiting distant parts of the universe or possibly traveling though time as well as through space.

See also BLACK HOLE; RELATIVITY (THEORY).

W Virginis Star *See* CEPHEID VARIABLE.

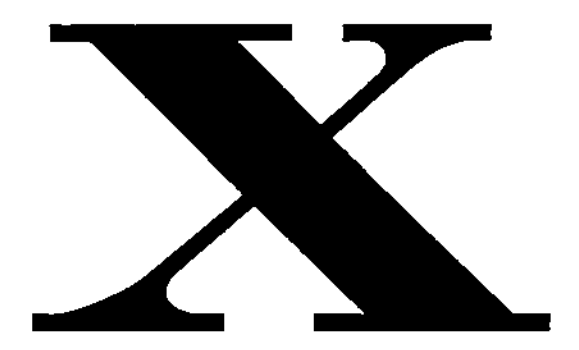

X In aerospace engineering, a prefix used to designate an experimental missile, aerospace vehicle, or rocket. The X-15 experimental rocket plane is an example.

See also X-15.

X-1 A rocket-powered research aircraft patterned on the lines of a 50-caliber machine-gun bullet that was the first human-crewed vehicle to fly faster than the speed of sound. The speed of sound in air varies with altitude. On October 14, 1947, the *Bell X-1,* named "Glamorous Glennis" and piloted by Captain Charles "Chuck" Yeager, was carried aloft under the bomb bay of a B-29 bomber and then released. The pilot ignited the aircraft's rocket engine, climbed, and accelerated, reaching Mach 1.06, or 1,127 kilometers per hour, as it flew over Edwards Air Force Base in California at an altitude of 13.1 km. At this altitude the speed of sound (or Mach 1.00) is 1,078.7 km per hour. The rocket-powered experimental aircraft, having used up all of its propellant, then glided to a landing on its tricycle gear at Muroc Dry Lake in the Mojave Desert.

See also X-15.

X-15 The North American X-15 was a rocket-powered experimental aircraft 15.24 meters (m) long with a wingspan of 6.71 m. It was a missile-shaped vehicle with an unusual wedge-shaped vertical tail, thin stubby wings, and unique fairings that extended along the side of the fuselage. The X-15 had an empty mass of 6,340 kilograms and a launch mass of 15,420 kg. The vehicle's pilot-controlled rocket engine was capable of developing 253,500 newtons of thrust.

The X-15 research aircraft helped bridge the gap between human flight within the atmosphere and human flight in space. It was developed and flown in the 1960s to provide in-flight information and data on aerodynamics, structures, flight controls, and the physiological aspects of high-speed, high-altitude flight. For flight in the dense air of the lower ("aircraft-usable") portions of the atmosphere, the X-15 employed conventional aerodynamic controls. However, for flight in the thin upper portions of Earth's atmospheric envelope, the X-15 used a ballistic control system. Eight hydrogen peroxide–fueled thruster rockets, located on the nose of the aircraft, provided pitch and yaw control.

Because of its large fuel consumption, the X-15 was air launched from a B-52 aircraft (i.e., a "mothership") at an altitude of about 13,700 m and a speed of about 805 kilometers per hour. Then the pilot ignited the rocket engine, which provided thrust for the first 80 to 120 seconds of flight, depending on the type of mission being flown. The remainder of the normal 10- to 11-minute flight was powerless and ended with a 322-km per hour glide landing at Edwards Air Force Base in California. Generally, one of two types of X-15

NASA's X-15 rocket plane makes a textbook landing on the dry lakebed at Edwards Air Force Base, California, in 1961. ***(Courtesy of NASA/Dryden Flight Research Center)***

flight profiles was used: a high-altitude flight plan that called for the pilot to maintain a steep rate of climb, or a speed profile that called for the pilot to push over and maintain a level altitude.

First flown in 1959, the three X-15 aircraft made a total of 199 flights. The X-15 flew more than six times the speed of sound and reached a maximum altitude of 107.8 km and a maximum speed of 7,273 km per hour. The final flight occurred on October 24, 1968. It is interesting to note that Apollo astronaut NEIL ARMSTRONG (the first human to walk on the Moon) was one of the pilots who flew the X-15 aircraft.

X-20 An early U.S. Air Force space plane concept whose mission was to explore human flight in the hypersonic and orbital regimes. However, the proposed vehicle did not have a significant payload capacity, and therefore it would not have been suitable in space logistics missions (i.e., in "payload-hauling" missions). The X-20, or DYNA-SOAR, space plane was to be placed into orbit by a Titan rocket booster. When its orbital mission was completed, the military pilot was to control this sleek glider through atmospheric reentry and then land it on a runway, much like a conventional jet fighter. Despite its innovative nature, the project was canceled in December 1963, before the prototype vehicle was ever built.

X-33 NASA's X-33 reusable launch vehicle (RLV) prototype program attempted to simultaneously demonstrate several major advancements in space launch vehicle technology that would increase safety and reliability while lowering the cost of placing a payload into low earth orbit by an order of one magnitude or so—that is, from about $20,000 per kilogram to perhaps as little as $2,000 per kilogram. The X-33 vehicle was a half-scale prototype of the conceptual SINGLE-STAGE-TO-ORBIT (SSTO) RLV, called VentureStar, being pursued by the Lockheed-Martin Company as a commercial development. However, after a series of disappointing technical setbacks and serious cost overruns, NASA abruptly cancelled the X-33 program in March 2001.

X-34 The vehicle being developed under NASA's X-34 program served as a suborbital flying laboratory for several space technology advancements applicable to future reusable launch vehicles. The program involved a partnership between NASA and the Orbital Sciences Corporation of Dulles, Virginia. Despite success in X-34 vehicle development and testing, NASA decided to cancel the program early in 2001 due to funding constraints and other factors.

X-37 NASA's X-37 vehicle is an advanced technology demonstrator that is helping define the future of reusable space transportation systems. It is a cooperative program involving NASA, the U.S. Air Force, and the Boeing Company. The X-37 is a reusable launch vehicle designed to operate in both the orbital and reentry phases of space flight. The robotic space plane can be ferried into orbit by the space shuttle or by an expendable launch vehicle. The program is demonstrating new airframe, avionics, and operational technologies relevant to future spacecraft and launch vehicle designs. The shape of the X-37 vehicle is a 120 percent scale derivative of the U.S. Air Force X-40A space vehicle, also designed by the Boeing Company.

***XMM-Newton* (spacecraft)** The European Space Agency's large X-ray observatory capable of sensitive X-ray spectroscopy measurements. *XMM-Newton* derives its name from its X-ray Multi-Mirror (XMM) design and honors SIR ISAAC NEWTON. The *XMM-Newton* spacecraft is a three-axis stabilized X-ray astronomy satellite with a pointing accuracy of one arc second. *XMM-Newton* carries three very advanced X-ray telescopes that are sensitive to the spectral range from 0.1 keV to 12 keV. Each of these X-ray telescopes contains 58 high-precision concentric mirrors, carefully nested to provide the most collecting area possible. The spacecraft—the largest scientific satellite ever constructed in Europe—was launched on December 10, 1999, by an Ariane 5 rocket from Kourou, French Guiana. Following launch the *XMM-Newton* spacecraft achieved a 48-hour highly elliptical orbit around Earth with an apogee of 114,000 km and a perigee of 7,000 km. The European Space Operations Center (ESOC) at Darmstadt, Germany, controls the spacecraft, which has an expected maximum operational lifetime of ten years. Also referred to as the *High Throughput Spectroscopy* mission and the *X-ray Multi-Mirror (XMM)* mission.

See also EUROPEAN SPACE AGENCY; X-RAY ASTRONOMY.

X-ray A penetrating form of electromagnetic radiation of very short wavelength (approximately 0.01 to 10 nanometers, or 0.1 to 100 angstroms) and high photon energy (approximately 100 electron volts to some 100 kiloelectron volts). X-rays are emitted when either the inner orbital electrons of an excited atom return to their normal energy states (these photons are called *characteristic X-rays*) or when a fast-moving charged particle (generally an electron) loses energy in the form of photons upon being accelerated and deflected by the electric field surrounding the nucleus of a high atomic number element (this process is called BREMSSTRAHLUNG, or "braking radiation"). Unlike gamma rays, X-rays are nonnuclear in origin.

See also ASTROPHYSICS; *CHANDRA X-RAY OBSERVATORY*; X-RAY ASTRONOMY.

X-ray astronomy X-ray astronomy is the most advanced of the three general disciplines associated with high-energy astrophysics, namely, X-ray, gamma-ray, and cosmic ray astronomy. Since Earth's atmosphere absorbs most of the X-rays coming from celestial phenomena, astronomers must use high-altitude balloon platforms, sounding rockets, or orbiting spacecraft to study these interesting emissions, which are usually associated with very energetic, violent processes occurring in the universe. X-ray emissions carry detailed information about the temperature, density, age, and other physical conditions of the celestial objects that have produced them. The observation of X-ray emissions has been very valuable in the study of high-energy events, such as mass transfer in binary star systems, the interaction of supernova remnants with interstellar gas, and the functioning of quasars.

The first solar X-ray measurements were accomplished by rocket-borne instruments in 1949. Some 13 years later RICCARDO GIACCONI and BRUNO B. ROSSI, along with their colleagues, detected the first nonsolar source of cosmic X-rays, called Scorpius X-1. They made the important though unanticipated discovery in June 1962 as a result of a sounding rocket flight. This event is often considered the start of X-ray astronomy. During the next eight years instruments launched on rockets and balloons detected several dozen bright X-ray sources in the Milky Way galaxy and a few sources in other galaxies. The excitement in X-ray astronomy grew when scientific spacecraft became available. Satellites allowed scientists to place complex instruments above Earth's atmosphere for extended periods of time. As a result, over 35 years, orbiting observatories have provided a greatly improved understanding of the energetic, often violent phenomena that are now associated with X-ray emissions from astrophysical objects.

In December 1970 NASA launched *Explorer 42* (or the *Small Astronomical Satellite [SAS]-1*), the first spacecraft devoted entirely to X-ray astronomy. This satellite, renamed *Uhuru* (the Swahili word for "freedom"), was lifted into Earth orbit from San Marco, a rocket launch platform off the coast of Kenya on the east coast of Africa. Successfully functioning until April 1973, the scientific spacecraft performed the first survey of the X-ray sky. In addition to detecting more than 300 X-ray sources, *Uhuru* also provided data about X-ray binaries and diffuse X-ray emission from galactic clusters. Since then increasingly more sophisticated spacecraft have performed the observation of celestial X-ray emissions.

For example, in November 1978 NASA successfully launched the second *HIGH ENERGY ASTROPHYSICAL OBSERVATORY* (HEAO-2), also called the *EINSTEIN OBSERVATORY* in honor of the physicist ALBERT EINSTEIN. This massive 3,130-kg satellite contained a large *grazing-incidence X-ray telescope* that provided the first comprehensive images of the X-ray sky. Until then scientists studied cosmic X-ray sources mostly by determining their positions, measuring their X-ray spectra, and monitoring changes in their X-ray brightness over time. With HEAO-2 it became possible to routinely produce images of cosmic X-ray sources rather than simply locate their positions. This breakthrough in observation was made possible by the grazing-incidence X-ray telescope.

An X-ray is a very energetic packet (photon) of electromagnetic energy that cannot be reflected or refracted by glass mirrors and lenses the way photons of visible light are focused in traditional optical telescopes. However, if an X-ray arrives almost parallel to a surface—that is, if the X-ray arrives at a grazing incidence—then the energetic photon can actually be reflected in a useful manner. The way the incident X-ray is reflected by this special surface depends on the atomic structure of the material and on the wavelength (energy level) of the X-ray. Using this grazing incidence technique, scientists can arrange special materials to help "focus" incident X-rays over a limited range of energies onto an array of detection instruments. The *Einstein Observatory* (HEAO-2) was the first such imaging X-ray telescope to be deployed in Earth orbit. The scientific objectives of HEAO-2 were to locate accurately and examine X-ray sources in the 0.2 to 4.0 kiloelectron volt (keV) energy range, to perform high-spectral-sensitivity spectroscopy, and to perform high-sensitivity measurements of transient X-ray sources. Operating successfully until 1981, HEAO-2 provided astronomers with X-ray images of such extended optical objects as supernova remnants, normal galaxies, clusters of galaxies, and active galactic nuclei. Among the *Einstein Observatory*'s most unexpected discoveries was the finding that all stars, from the coolest to the very hottest, emit significant amounts of X-rays.

Thousands of cosmic X-ray sources became known due to observations by NASA's *Einstein Observatory* and the *European X-Ray Observatory* satellite (EXOSAT), launched by the European Space Agency (ESA) in May 1983. As a result of these important discoveries in X-ray astronomy, astronomers now recognize that a significant fraction of the radiation emitted by virtually every type of interesting astrophysical object emerges as X-rays.

In June 1990 the joint German-U.S.-U.K. *ROENTGEN SATELLITE* (ROSAT) was placed into orbit around Earth. This orbiting extreme ultraviolet (EUV) and soft (low-energy) X-ray observatory was named in honor of WILHELM CONRAD ROENTGEN, the German physicist who discovered X-rays. Some of the major scientific objectives of the very successful ROSAT mission were to study X-ray emission from stars of all spectral types, to detect and map X-ray emission from galactic supernova remnants, to perform studies of various active galaxy sources, and to perform a detailed EUV survey of the local interstellar medium. More recent orbiting X-ray observatories, such as NASA's *CHANDRA X-RAY OBSERVATORY* (CXO) and the European Space Agency's *XMM-Newton* Observatory, have been providing scientists higher spectral and angular resolution data and greater detection sensitivity. These sophisticated orbiting X-ray astronomy facilities are helping scientists understand many previously unresolved mysteries in high-energy astrophysics.

Formerly called the *Advanced X-Ray Astrophysics Facility* (AXAF), NASA launched the *Chandra X-Ray Observatory* in July 1999 and named the space-based observatory after the Indian-American astrophysicist SUBRAHMANYAN CHANDRASEKAR. This Earth-orbiting astrophysical facility has the capability to study some of the most interesting and puzzling X-ray sources in the universe, including emissions from active galactic nuclei, exploding stars, neutron stars, and matter falling into black holes. The European Space Agency's space observatory is also making significant contributions to contemporary X-ray astronomy. Launched on December 10, 1999, from Kourou, French Guiana, by an Ariane 5 rocket, *XMM-Newton* contains three very advanced X-ray telescopes. Previously called the *X-Ray Multi-Mirror* satellite, *XMM-Newton* is the largest science satellite ever built in Europe, a very high sensitive astrophysical instrument that can detect millions of X-ray sources.

Space-based observatories, such as the *Chandra X-Ray Observatory* and *XMM-Newton* as well as NASA's planned *Constellation X-Ray Observatory* provide modern astrophysicists with important X-ray emission data they need to better understand stellar structure and evolution, including binary star systems, supernova remnants, pulsars, and black hole candidates; large-scale galactic phenomena, including the interstellar medium itself and soft X-ray emissions of local galaxies; the nature of active galaxies, including the spectral characteristics and time variation of emissions from the central regions of such galaxies; and rich clusters of galaxies, including their associated X-ray emissions. A bit closer to home, X-ray emission data are also helping space scientists monitor violent and dangerous solar flares as they occur on the Sun.

See also ASTROPHYSICS; *CHANDRA X-RAY OBSERVATORY; CONSTELLATION X-RAY OBSERVATORY; XMM-NEWTON.*

X-ray binary star system The most-often-encountered type of luminous galactic X-ray source. It is a close binary star system in which material from a large, normal star flows under gravitational forces onto a compact stellar companion, such as a neutron star or a black hole (for the most luminous X-ray sources) or perhaps a white dwarf (for less luminous sources).

See also BINARY STAR SYSTEM; X-RAY ASTRONOMY.

X-ray burster X-ray source that radiates thousands of times more energy than the Sun in short bursts that last only seconds. Astrophysicists suggest that a neutron star in a close binary star system accretes matter onto its surface from a companion until temperatures reach the level needed for hydrogen fusion to take place. The result is a sudden period of rapid nuclear burning and the subsequent release of an enormous quantity of energy. The gaseous matter pulled away from the companion star generally forms an accretion disk around the neutron star. Spiraling matter in the inner portions of the accretion disk becomes extremely hot, releasing a steady stream of X-rays. As the inward flowing gas builds up on the neutron star's surface, the temperature of the gas rises due to the pressure of overlying material. Eventually, the temperature of the captured material reaches thermonuclear fusion conditions. The result is a rapid period of thermonuclear hydrogen burning, a phenomenon that releases an enormous amount of energy, characterized by a burst of X-rays. Hours then pass as fresh new accreted material accumulates on the neutron star's surface and conditions approach the production of another rapid X-ray burst.

Sometimes not all of the in-falling material from the stellar companion makes it to the surface of the nearby neutron star. In such cases a portion of the in-falling gaseous material is shot out into interstellar space at enormously high speeds in narrow jets. For example, astronomers have observed an interesting celestial object called SS 433. This compact object expels the equivalent of one Earth mass of material each year in two narrow jets that travel in opposite directions roughly perpendicular to an accretion disk. When these jets, traveling at about 80,000 km/s, or about 25 percent of the speed of light, interact with the interstellar medium, they emit radiation in the radio frequency portion of the electromagnetic spectrum.

See also NEUTRON STAR.

X-Ray Multi-Mirror Mission (XMM) *See XMM-NEWTON* SPACECRAFT.

X-Ray Timing Explorer (XTE) *See ROSSI X-RAY TIMING EXPLORER.*

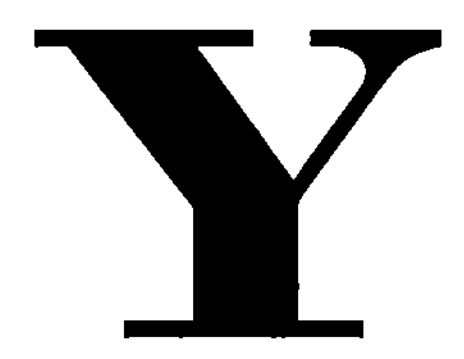

Y In aerospace technology and industry, the symbol used for prototype. For example, aerospace engineers will usually designate the prototype version of a guided missile (GM) as YGM. Prototype vehicles are manufactured in limited quantities and are generally used to support operational and preproduction tests.

Yagi antenna A highly directional, linear antenna named after Hidetsuga Yagi (1886–1976). The device generally consists of three or more half-wave elements—a dipole (the driven element), a parallel reflector, and one or more closely spaced director elements in front of the dipole. When the Yagi antenna is used for reception (as, for example, in a radio telescope), the arrangement focuses the incoming radio frequency (RF) signal on the dipole. When the Yagi antenna is used for transmission, the directors reinforce the output of the dipole.

yardangs In planetary geology, streamlined hills oriented parallel to the prevailing winds and produced by wind erosion of rock or soft sediments.

Yarkovsky effect A subtle sunlight-induced thrusting effect that some scientists suggest could push asteroids away from their natural orbits in the main belt between Mars and Jupiter and into the inner portions of the solar system. This effect is produced when sunlight warms an asteroid as it spins. The hot surfaces of the asteroid radiate more thermal energy than cooler surfaces of the asteroid. This tiny difference in radiant energy emission produces a very small anisotropic reaction force that can gradually propel a minor planet from its original orbital path. The actual magnitude of this orbital shift varies from asteroid to asteroid and depends very much on its thermal and rotational properties. The phenomenon bears the name of the Russian engineer Ivan O. Yarkovsky (1844–1902), who studied how small rotating particles can experience a change in their orbital momentum due to the absorption and subsequent (differential) emission of solar radiation.

yaw The rotation or oscillation of an aircraft, missile, or aerospace vehicle about its vertical axis so as to cause the longitudinal axis of the vehicle to deviate from the flight line or heading in its horizontal plane.

See also PITCH; ROLL.

year A period of one revolution of Earth around the Sun. The choice of a reference point determines the exact length of a year. The CALENDAR YEAR (based on the Gregorian calendar) is made up of an average of 365.2425 mean solar days. For convenience, this calendar is taken as three successive years of 365 days followed by one "leap" year consisting of 366 days. A *solar day* is defined as the time interval between two successive passages of the Sun across the meridian, while a MEAN SOLAR DAY of 24 hours is considered the average value of the solar day for one year. Scientists and astronomers also use other more specialized definitions. For example, the period of revolution from perihelion to perihelion (point of Earth's closest approach to the Sun) is called the *anomalistic year* and corresponds to approximately 365.259 mean solar days. The *tropical year* is defined as the average period of time between successive passages of the Sun across the vernal equinox (that is, return of the Sun to the first point in Aries) and corresponds to approximately 365.2422 mean solar days. Finally, the SIDEREAL YEAR is based on the return passage of certain fixed stars and is approximately 365.25636 mean solar days. Note that the positions of the so-called fixed stars gradually change because of the phenomenon of precession.

See also CALENDAR; GREGORIAN CALENDAR; EQUINOX; PRECESSION.

Yerkes Observatory An astronomical observatory of the University of Chicago located about 330 m above sea level at Williams Bay, Wisconsin. The facility houses the world's largest refracting telescope, a 102-cm (40-inch)-diameter device completed in 1897. The American astronomer GEORGE ELLERY HALE (1868–1938) established this observatory with a grant from Charles Tyson Yerkes (1837–1905), a philanthropic transportation tycoon. In 1967 the facility added a 1-m reflecting telescope and then in 1994 modified this instrument with the addition of adaptive optics. Currently Yerkes astronomers are measuring the velocities of the most distant star clusters within the Milky Way to more accurately determine the mass of our galaxy. They are also making spectroscopic measurements of lithium abundances to answer questions about stellar structure and the chemical evolution of the Milky Way. Finally, Yerkes astronomers continue to contribute to airborne astronomy by developing one of the instruments for NASA's new *Strategic Observatory For Infrared Astronomy* (SOFIA).

yield The total energy released in a nuclear explosion. It is usually expressed in equivalent tons of TNT (the quantity of trinitrotoluene required to produce a corresponding amount of energy). Within the U.S. Department of Defense, low yield is generally considered to be less than 20 kilotons; low-intermediate yield, from 20 to 200 kilotons; intermediate yield, from 200 kilotons to 1 megaton. There is no standardized term to describe yields from 1 megaton upward.

ylem In cosmology and astrophysics, the postulated form of high-density primordial matter, consisting mainly of neutrons at a density of about 10^{16} kg/m^3 that existed in the ancient fireball after the big bang event but before the formation of the chemical elements.

YM A prefix often used in aerospace engineering to designate a prototype missile.

***Yohkoh* (spacecraft) (Sunbeam)** A 390-kg-mass solar X-ray observation satellite launched by the Japanese Institute of Space and Astronautical Sciences (ISAS) on August 30, 1991. The main objective of this satellite is to study the high-energy radiations from solar flares (i.e., hard and soft X-rays and energetic neutrons), as well as quiet Sun structures and pre–solar flare conditions. *Yohkoh* means "sunbeam" in Japanese. It is a three-axis stabilized observatory-type satellite in a nearly circular orbit around Earth carrying four instruments, two imagers and two spectrometers. The imaging instruments (a hard X-ray telescope [20–80 keV energy range] and a soft X-ray telescope [0.1–4.0 keV energy range]) have almost full-Sun fields of view to avoid missing any flares on the visible disk of the Sun. This was a cooperative mission of Japan, the United States, and the United Kingdom. For example, *Yohkoh*'s soft X-ray telescope (SXT) was developed for NASA by the Lockheed Palo Alto Research Laboratory, in partnership with the National Astronomical Observatory of Japan and the Institute for Astronomy of the University of Tokyo.

By the end of its mission in December 2001, the *Yohkoh* spacecraft had completed a decade of observing solar X-rays over the entire sunspot cycle. One unexpected result of the *Yohkoh* investigation was to show that the Sun's corona is much more active than scientists previously thought. In addition, the corona within active regions (that is, the sites of solar flares) was found to be expanding, in some cases almost continuously. Such expanding active regions apparently contribute to mass loss from the Sun and other stars.

See also SUN.

Astronaut John W. Young is a six-time space flight veteran. He appears here in a 1992 photograph along with models of all the NASA spacecraft he has flown. As a member of the exclusive "Moon Walkers" club, he is also one of the 12 human beings who walked on the lunar surface in the 20th century. *(Courtesy of NASA/JSC)*

Young, John W. (1930–) American *Astronaut, U.S. Navy Officer* The astronaut John W. Young, Jr. has traveled into space on six separate NASA missions, starting with the first human-crewed Gemini Project flight (*Gemini 3*), on which he was accompanied by the astronaut VIRGIL "GUS" GRISSOM on March 23, 1965. After the *Gemini 10* flight (July 1966), he traveled to the Moon as part of the *Apollo 10* mission (May 1969), a key rehearsal mission that orbited the Moon and completed all appropriate Apollo Project operations just short of landing on the lunar surface. Then in April 1972 he served as the commander of the *Apollo 16* lunar landing mission and walked on the Moon's surface along with astronaut CHARLES MOSS DUKE, JR. As a tribute to his astronaut skills, NASA selected Young to command the space shuttle *Columbia* on the inaugural mission (STS-1) of the U.S. Space Transportation System (STS) in April 1981. He returned to space for the sixth time as the commander of the space shuttle *Columbia* during the STS-9 mission (1983), a mission that served as the inaugural flight of the European Space Agency's (ESA) SPACELAB.

Young, Thomas (1773–1829) British *Physicist, Physician* In 1803 the British physicist and physician Thomas Young performed an important experiment that demonstrated the wave nature of light. Young illuminated a pair of parallel narrow slits with a source of monochromatic light and observed an emerging pattern of light and dark fringes. This interference phenomenon revealed that light behaved like a wave. In Young's famous experiment each narrow slit acts like a new source of light, spreading light into what would otherwise be dark areas. Physicists call this phenomenon DIFFRACTION. Light waves emanating from each narrow slit also interfere with one another. When such waves add together in *constructive interference,* they create bright fringes. When the crest of one emerging wave falls into the trough of the other emerging wave, the waves cancel and create a dark fringe. Physicists call this behavior *destructive interference.*

young Population I stars A class of stars found in the disk of a spiral galaxy, especially in the spiral arms, that have a metal abundance similar to that of the Sun. Such stars represent the youngest stars in the spiral galaxy.

See also POPULATION I STARS.

young stellar object (YSO) A general term used to describe any celestial object in an early stage of star formation, ranging from a protostar to a pre–main sequence star.

See also STAR.

Z The symbol for *atomic number,* or the number of protons in the nucleus of an atom of a given element. For example, Z = 92 is the atomic number for the element uranium (U); a neutral atom of uranium has 92 protons in its nucleus and 92 electrons in orbit around the nucleus.

Zarya (Dawn) The 31,100-kg-mass Russian-built and American financed module that was the first-launched of numerous modules that make up the *International Space Station* (ISS), a large habitable spacecraft being assembled in low Earth orbit (LEO). The ISS travels around Earth in an approximately circular orbit at an altitude of 386 km with an inclination of 52.6°. The first assembly step of the ISS occurred in late November and early December 1998. During a space shuttle–supported orbital assembly operation, astronauts linked Zarya, the initial control module, together with Unity, the American six-port habitable connection module. Zarya is also known as the Functional Cargo Block, or FGM when the Russian equivalent acronym is transliterated. Because Zarya was the first-orbited element of the ISS, its international spacecraft identification (1998-067A) also serves as the spacecraft identification for the *International Space Station.*

See also INTERNATIONAL SPACE STATION.

Zeeman effect The splitting of atomic spectral lines in the presence of an external magnetic field. The Dutch physicist Pieter Zeeman (1865–1943) discovered this phenomenon in 1896–97 and shared the 1902 Nobel Prize for physics for his work. Astronomers use the Zeeman effect to study of the magnetism of the Sun and other stars. Fundamental atomic absorption and emission phenomena change slightly under the influence of a star's magnetic field. Specifically, there is a slight splitting of spectral lines (the Zeeman effect), resulting in overall line broadening. As an approximate rule of thumb, the greater the magnetic field the more extensive the broadening of spectral lines.

Zenit (launch vehicle) A two-stage liquid-propellant Russian space launch vehicle (SLV) developed within Ukraine in the early 1980s by NPO Yuzhnoye. The first stage of this rocket uses an RD-171 booster engine, burns liquid oxygen and kerosene, and is capable of producing a thrust of 7.25 million newtons (N). The second stage uses an RD-120 sustainer engine along with an RD-8 vernier engine, burns liquid oxygen and kerosene, and develops approximately 912,000 N of thrust. The Zenit features automated fueling and launch processing. Since its introduction in 1985 the Zenit served as a medium-capacity launch vehicle with one Russian military space program as its primary user. Following the end of the cold war Zenit has enjoyed commercial application in an innovative international maritime-based launch services program called Sea Launch.

Sea Launch was formed in April 1995 as a commercial partnership between the Boeing Commercial Space Company (Seattle, Washington), Kvaerner (Oslo, Norway), RSC-Energia (Moscow, Russia), and SDO Yuzhnoye/PO Yuzhmash (Dnepropetrovsk, Ukraine). Basically, this international consortium launches commercial satellites from a platform at sea. The Zenit-3SL rocket makes up the first two stages of the Sea Launch™ rocket, and an Energia-produced Block DM-SL third stage completes the vehicle. An assembled launch vehicle (including payload unit) has an overall length of approximately 60 m with an average diameter of 3.9 m. All three stages of this marine-launched rocket use liquid oxygen and kerosene as the propellants. The two-stage Zenit-3SL rocket with a Block DM-SL third stage can send a 6,000-kg-mass payload into a geosynchronous transfer orbit (GTO).

Long Beach, California, serves as the home port for the Sea Launch assembly and command ship (ACS), a modified roll-on, roll-off cargo vessel. A modified self-propelled ocean oil-drilling platform serves as the rocket's marine launch platform. Upon arrival at the marine launch site, the launch platform is positioned, ballasted to its launch depth, and oriented to minimize wind and wave effects. About 27 hours before liftoff, the Zenit rocket is rolled out of its environmentally protected hangar on the assembly and command ship and automatically erected on the sea-going launch pad. Propellant loading begins approximately 180 minutes before launch.

On March 27, 1999, the Zenit-3SL carried a demonstration payload into orbit, validating the overall marine launch concept and inaugurating the Sea Launch system. The first commercial launch took place on October 9, 1999. Several successful commercial launches of the Zenit-3SL rocket have taken place since then, including that of the *EchoStarIX/Telstar 13* communications satellite on August 7, 2003.

See also LAUNCH VEHICLE.

zenith The point on the celestial sphere vertically overhead. Compare with NADIR, the point 180° from the zenith.

zenith distance (symbol: z or ζ) The vertical arc measured straight downward from the zenith to the altitude of a celestial object; expressed mathematically, the zenith distance becomes $z = 90° - \text{altitude}$. Also called the coaltitude.

zero-age main sequence (ZAMS) The region on the Hertzsprung-Russell (H-R) diagram at which new stars are just beginning to experience nuclear burning in their cores. Astronomers refer to the main sequence line predicted by theory as the zero-age main sequence line, or ZAMS. They consider that when its core achieves a stable state and starts fusing hydrogen into helium, a protostar or prestellar object has reached the main sequence and is at age zero.

See also HERTZSPRUNG-RUSSELL DIAGRAM; STAR.

zero-g The condition of continuous free fall and apparent weightlessness experienced by passengers and objects in an orbiting spacecraft.

See also MICROGRAVITY.

zero-gravity aircraft (zero-g) An aircraft that flies a special parabolic trajectory to create low-gravity conditions (typically 0.01 g) for short periods of time (10 to 30 seconds), where one g represents the acceleration of gravity on the surface of Earth (namely, 9.8 meters per second per second). For example, a modified KC-135 aircraft can simulate up to 40 periods of low gravity for 25-second intervals during one flight. This aircraft accommodated a variety of experiments, and NASA often used it to support crew training and to refine space flight experiment equipment and techniques. (*See* figure.) The KC-135, like other "zero-gravity" research aircraft, obtains simulated weightlessness by flying a parabolic trajectory. In general, the plane climbs rapidly at a 45° angle (pull-up), slows as it traces a parabola (pushover), and then descends at a 45° angle (pull-out). The forces associated with acceleration and deceleration produce twice the normal gravity during the pull-up and pull-out legs of the flight, while the brief pushover at the top of the parabola produces less than 1 percent of Earth's sea-level gravity.

See also MICROGRAVITY.

zero-length launching A technique in which the first motion of a missile or aircraft removes it from the launcher.

zeroth law of thermodynamics Two systems, each in thermal equilibrium with a third, are in thermal equilibrium with each other. This statement is frequently called the zeroth law of thermodynamics and is actually an implicit part of the concept of temperature. It is important in the field of thermometry and in the establishment of empirical temperature scales.

See also ABSOLUTE TEMPERATURE; CELSIUS TEMPERATURE SCALE; THERMODYNAMICS.

zeta star (ζ) Within the system of stellar nomenclature introduced in 1603 by the German astronomer JOHANN BAYER (1572–1625), the sixth-brightest star in a constellation.

zodiac From the ancient Greek language meaning "circle of figures." Early astronomers described the band in the sky about 9° on each side of the ecliptic, which they divided into 30° intervals, each representing a sign of the zodiac. Within their geocentric cosmology the Sun appeared to enter a different constellation of the zodiac each month, so the signs of the zodiac helped them mark the annual revolution of Earth around the Sun. These 12 constellations (or signs) are Aries (ram), Taurus (bull), Gemini (twins), Cancer (crab), Leo (lion), Virgo (maiden), Libra (scales), Scorpius (scorpion), Sagittarius (archer), Capricornus (goat), Aquarius (water-bearer), and Pisces (fish). Although the signs of the zodiac originally (more than 2,500 years ago) corresponded in position to the 12 constellations just named, because of the phenomenon of precession the zodiacal signs do not presently coincide with these constellations. For example, when people today say the Sun enters Aries at the vernal equinox, it has in fact shifted forward (from ancient times) and is now actually in the constellation Pisces.

See also CONSTELLATION (ASTRONOMY); ECLIPTIC; EQUINOX; PRECESSION.

zodiacal light A faint cone of light extending upward from the horizon in the direction of the ecliptic. Zodiacal light is best seen on moonless nights from the tropical latitudes for a few hours after sunset or before sunrise. The phenomenon is due to the scattering of sunlight by tiny pieces of interplanetary dust (IPD) that orbit the Sun in the plane of the ecliptic.

See also INTERPLANETARY DUST.

Zond (Probe) A family of Soviet spacecraft that explored the Moon, Mars, Venus, and the interplanetary medium in the 1960s. *Zond 1*, a partial success, was launched toward Venus by the Soviet Union on April 2, 1964. However, the 890-kg-mass spacecraft stopped transmitting about two months before it flew past Venus on July 14, 1964, at a distance of 100,000 km. It then entered a heliocentric orbit.

Zond 2 was an 890-kg-mass automatic interplanetary station launched from a parking orbit around Earth toward Mars on November 30, 1964. Russian scientists used this planetary probe to test space-borne systems and to conduct scientific investigations. *Zond 2* carried a descent craft and a suite of instruments similar to those carried by the Russian *Mars 1* spacecraft, including a magnetometer, television pho-

Canadian Astronaut Marc Garneau experiences a brief period of weightlessness during a flight aboard a KC-135 "zero-gravity" aircraft in July 1984. NASA has used such zero-g training flights to help prepare astronauts for space shuttle missions and to evaluate equipment and experiments scheduled for use in future missions. ***(Courtesy of NASA)***

tographic equipment, a spectro-reflectometer, ionizing radiation sensors, and a micrometeoroid instrument. The *Zond 2* also had six experimental, low-thrust, electric propulsion devices (plasma ion engines). Russian scientists placed *Zond 2* on a long curving trajectory toward Mars to minimize the relative velocity upon arrival. Despite the failure of one of the spacecraft's two solar panels, the spacecraft controllers were able to successfully test the ion engines for about 10 days (from December 8 to 18, 1964). Then, after a mid-course maneuver in early May 1965, all communications with the *Zond 2* were lost. The spacecraft flew past Mars on August 6, 1965, at a distance of 1,500 km and a relative speed of 5.62 km/s. Because of the communications failure, however, no scientific data were sent back to Earth during this relatively close flyby.

The 960-kg-mass *Zond 3* was launched as a test interplanetary probe on July 18, 1965. About 33 hours later (on July 20, 1965) it flew past the Moon at a closest approach distance of 9,200 km. As it encountered the Moon the spacecraft obtained 25 good-quality photographs of the lunar farside. It then continued to explore interplanetary space and entered a heliocentric orbit.

Some Western intelligence analysts regarded the remainder of the Zond missions (namely *Zond 4, 5, 6, 7,* and *8*) as possible human-crew precursor missions, flown by the Soviet Union during the cold war in competitive response to the American Apollo Project and the so-called race to the Moon in the late 1960s. Even though the larger Zond capsule used in these later missions could not support a lunar landing, the space capsule's true (but undeclared) purpose appears to have been the first human circumnavigation flight around the Moon. (As a historical footnote, three Apollo Project astronauts flew around the Moon in late December 1968 during the *Apollo 8* mission and achieved this important space exploration milestone for the United States.)

Soviet space officials used a Proton rocket to launch the 5,140-kg-mass *Zond 4* on March 2, 1968. The larger and heavier spacecraft departed on a ballistic trajectory into deep space, out to a distance of about 300,000 km from Earth. The officially announced purpose of this mission was to explore circumterrestrial space and to flight test new equipment. In order to avoid any trajectory complications from the Moon's gravity, Russian mission controllers launched the spacecraft in a direction away from the Moon. *Zond 4* was a cylindrical space capsule approximately 4.5 m in length and about 2.5 m in diameter. Two solar panels spanning a total of 9 m when deployed were attached on opposite sides of the capsule's body. *Zond 4* apparently served as the uncrewed testbed and precursor for a future cosmonaut-occupied lunar mission spacecraft. *Zond 4*'s return to Earth was supposed to involve a skip reentry maneuver. However, an apparent attitude control system malfunction led to a reentry flight path with too steep an angle of attack, so as the spacecraft entered Earth's atmosphere at high speed over West Africa, Russian controllers (on a tracking ship in the Atlantic Ocean) activated a self-destruct mechanism, and *Zond 4* exploded over the Gulf of Guinea at an altitude of about 10 km.

The 5,375-kg-mass *Zond 5* space capsule was also a precursor to a future human-crewed lunar mission. Launched on September 14, 1968, at 21:42 (UTC), the spacecraft flew around the Moon on September 18, 1968, coming as close as 1,950 km to the lunar surface. *Zond 5* was the first human-made object to circumnavigate the Moon and return to Earth. According to Soviet space officials, this spacecraft carried a biological payload consisting of an assortment of living things, such as turtles, worms, and plant seeds. A minor malfunction at the end of the mission caused the reentry capsule to splash down in the Indian Ocean, where the biological payload was successfully recovered.

Soviet space officials launched the 5,375-kg-mass *Zond 6* on a lunar flyby mission from a parent satellite in parking orbit around Earth on November 10, 1968. The spacecraft carried a scientific payload, including photographic equipment. *Zond 6* flew around the Moon on November 14, 1968, at a minimum distance of 2,420 km and collected photographs of the lunar nearside and farside. On November 17, 1968, the spacecraft made a controlled reentry and landed in the Soviet Union. Some accounts suggest that the returning space capsule experienced operational difficulties at the end of the flight and crashed near its launch site at the Baikonur Cosmodrome.

The *Zond 7* mission took place on August 7, 1969, after American Apollo Project astronauts had already circumnavigated the Moon (in December 1968) and walked on the lunar surface for the first time (in July 1969). As a result, any cold war political advantage of performing a relatively high-risk lunar flyby mission with a cosmonaut inside the Zond space capsule disappeared. The uncrewed *Zond 7* flew past the Moon on August 11, 1969, at a distance of 1985 km, obtaining color photographs. The space capsule then looped around the Moon and returned to Earth, reentering Earth's atmosphere on August 14, 1969, and achieving a soft landing in the Soviet Union.

The Soviets launched the 5,375-kg *Zond 8* toward the Moon on October 20, 1970. As a lunar circumnavigation mission, the spacecraft flew past the Moon on October 24, 1970, at a distance of 1,110 km and obtained both black-and-white and color photographs of the lunar surface. After returning to Earth *Zond 8* reentered the atmosphere and splashed down in the Indian Ocean on October 27, 1970. This was the last Zond mission.

zone of avoidance The region around the plane of the Milky Way galaxy where other galaxies are not seen because of dust absorption.

zoo hypothesis One response to the Fermi paradox. It assumes that intelligent, very technically advanced species do exist in the Milky Way galaxy but that we cannot detect or interact with them because they have set the solar system aside as a perfect zoo or wildlife preserve. The reasoning behind this hypothesis goes something like this. Technically advanced beings exert a great deal of control over their environment. For example, on Earth human beings have a far greater influence on the biosphere than all of the other creatures that co-inhabit the planet. Occasionally, we decide not to exert this influence and set aside certain regions as specially designed zoos, wildlife sanctuaries, or wilderness areas. In these designated places other species can develop naturally with little or no human interaction. In fact, the perfect wildlife sanctuary or zoo is one set up so that the species within are not even aware of the presence or existence of their zookeeper. Applying this terrestrial analogy, the zoo hypothesis suggests that human beings cannot detect their "extraterrestrial zookeepers." These postulated interstellar–traveling beings have used their advanced technologies to cleverly set aside the solar system as a perfect zoo in which human beings can develop "naturally" as a species without any awareness of or direct interaction with them.

See also FERMI PARADOX.

Zucchi, Niccolo (1586–1670) Italian *Astronomer, Instrument Maker* The Jesuit priest Niccolo Zucchi participated in the development of telescopic astronomy by making telescopes in the early 17th century. About 1616 he designed and constructed one of the earliest known reflecting telescopes. He later described the principles of this telescope in his 1652 book *Optica Philosophia.* It used an ocular lens to observe the image produced by reflection from a concave metal mirror. Zucchi's ability to view the heavens with his original device was probably seriously hampered by the presence of obstructions in the light path.

This is actually a common problem in the design of any type of reflecting telescope. Many refinements soon followed his initial idea. For example, the Scottish mathematician JAMES GREGORY described a more sophisticated reflecting telescope design in his 1663 work *Optica Promota.* However, credit for the first practical (that is, "working) reflecting telescope goes to the British physicist and mathematician SIR ISAAC NEWTON. Newton introduced his device in 1668. Modern astronomers often speak of a NEWTONIAN TELESCOPE and even a GREGORIAN TELESCOPE, but no one ever mentions a "Zucchian reflector"—mainly because Zucchi's fundamental idea needed some additional work, and he never provided the required effort due to other interests and duties.

However, while Zucchi was serving as the papal legate to the court of Ferdinand II of Austria (ca. 1627), a fellow Jesuit suggested that he provide JOHANNES KEPLER with one of his telescopes. Zucchi obliged and delivered the instrument as a special gift from the Jesuit order. Despite the fact that Kepler's poor eyesight made him less than a skilled observer, he appreciated the gift, and the two astronomers became friends.

The contact with Kepler may have rekindled Zucchi's interest in astronomy, because on May 17, 1630, he became the first person to report observing two colored belts (bands) and spots on Jupiter. Colorful bands do characterize the upper portions of the atmosphere of the gaseous giant planet, and Jupiter also has a large distinctive red spot, called the Great Red Spot. While a professor at the Jesuit College in Rome, Zucchi reported seeing spots on the planet Mars in 1640. Some astronomers question whether Zucchi's early telescopes could have actually detected such characteristic marks on the surface of the Red Planet. They suggest instead that he might merely have misinterpreted an instrument-caused distortion in the image of Mars.

In any event, Zucchi's report most likely encouraged the Dutch astronomer CHRISTIAAN HUYGENS to use his more powerful new telescope to carefully examine Mars in 1659. The result was an instant success. On November 28, 1659, Huygens recorded seeing a large dark spot on the surface of Mars—probably Syrtis Major, the most prominent dark marking on the planet. This triangular-shaped dark surface feature is about 1,200 km long and 1,000 km wide.

Zucchi lived during an interesting yet extremely turbulent period, when both scientific and religious revolutions were sweeping through Europe. As an astronomer and instrument builder his work helped to feed the flames of the Copernican revolution, yet he was also mindful of the physical flames that burned GIORDANO BRUNO to death for heresy and of the extreme difficulties GALILEO GALILEI encountered when each individual attempted to advocate heliocentric cosmology. So more than likely Father Zucchi chose to make his scientific contributions discretely while walking an apparently conservative political path. He was a teacher and preacher who was quite familiar with the traditional (Aristotelian) aspects of natural philosophy, but he was also a skilled instrument maker and scientist whose pioneering efforts helped establish telescopic astronomy during its tumultuous birthing process. As a testament to his wide-ranging scientific interests and skills, he was also the first person to recognize the phenomenon of bioluminescence. He presented his hypothesis in the section of *Optica Philosophia* that reported his experiments involving the Bolognian stone, an interesting type of "glowing" rock. Toward the end of his life he served as the director of the Jesuit house in Rome, where he died on May 21, 1670. In 1935 the International Astronomical Union (IAU) named a lunar crater Zucchius in his honor. Crater Zucchius is 64 km in diameter and can be found at lunar latitude 61.4° south and longitude 50.3° west.

Zulu time **(symbol: Z)** Time based on Greenwich mean time (GMT).

See also GREENWICH MEAN TIME; UNIVERSAL TIME.

Zvezda **(star)** The Russian service module for the *International Space Station* (ISS). The 20-ton module has three docking hatches and 14 windows. Launched by a Proton rocket from the Baikonur Cosmodrome at 04:56 on July 12, 2000 (UTC), the module automatically docked with the Zarya module of the orbiting ISS complex on July 26, 2000, at 0:0145 (UTC). Prior to docking with the ISS, Zvezda bore the international spacecraft designation 2000-037A. Once attached and functional, the module became an integral part of the ISS and began to serve as the living quarters for the astronaut and cosmonaut crews during the on-orbit assembly phase. The first ISS crew, called the Expedition One crew, began to occupy the ISS and live in Zvezda on November 2, 2000. American astronaut Bill Shepherd and Russian cosmonauts Yuri Gidzenko and SERGEI KRIKALEV made the Zvezda module their extraterrestrial home until March 14, 2001. At that point the Expedition Two crew, consisting of cosmonaut Yuri Usachev and astronauts Susan Helms and Jim Voss, replaced them as Zvezda's occupants. In addition to supporting human habitation in space, this module also provides electrical power distribution, data processing, flight control, and on-orbit propulsion for the space station complex.

See also *INTERNATIONAL SPACE STATION;* PROTON LAUNCH VEHICLE.

Zwicky, Fritz (1898–1974) Swiss *Astrophysicist* Shortly after EDWIN HUBBLE introduced the concept of an expanding universe in the late 1920s, the Swiss astrophysicist Fritz Zwicky pioneered the early search for the dark matter—that is, the missing mass of the universe. Then in 1934, while collaborating with WALTER BAADE, Zwicky remained at the frontiers of astrophysics by postulating the creation of a neutron star as a result of a supernova explosion. He was a truly visionary physicist, but also one with an extremely eccentric and often caustic personality. In the 1960s Zwicky published an extensive catalog of galaxies and clusters of galaxies.

Zwicky was born on February 14, 1898, in Varna, Bulgaria. A lifelong Swiss citizen, he spent his childhood in Switzerland, where he also received a Ph.D. in physics from the Federal Institute of Technology in Zürich in 1922. After graduation he went to the United States to work as a postdoctoral fellow at the California Institute of Technology (Caltech). Claiming he enjoyed the mountains near Pasadena, Zwicky remained affiliated with that institution and stayed in California for the rest of his life. Despite living and working for almost five decades in the United States, the eccentric Zwicky retained his Swiss citizenship and never chose to apply for American citizenship.

Zwicky had a rather dark side to his technical genius. He was a very abrasive individual who was fond of using rough language and intimidating colleagues and students alike. Only a few extremely tolerant individuals, such as Walter Baade, could, or would, overlook his eccentric and difficult personality in order to encounter the bursts of genius that occasionally could be found among his characteristically caustic comments.

In 1927 he became a professor of theoretical physics at Caltech and served in that academic position until he received an appointment as a professor in astrophysics in 1942. Upon his retirement in 1972, Caltech made him a professor emeritus in recognition of his lifelong service to the university and to astrophysics. While at the California Institute of Technology Zwicky frequently interacted with many of the world's best astronomers, such as Baade, who came to southern California to use the Mount Wilson Observatory, located in the nearby San Gabriel Mountains. Between 1943 and 1946 Zwicky committed much of his time directing research at the Aerojet Corporation in Azusa, California, in support of jet engine development and other defense-related activities.

Despite his socially difficult personality, Zwicky was a gifted astrophysicist who helped establish the framework of modern cosmology. In 1933, while investigating the Coma Cluster, a rich cluster of galaxies in the constellation Coma Berenices (Berenice's Hair), Zwicky observed that the velocities of the individual galaxies in this cluster were so high that they should definitely have escaped from one another's gravitational attraction long ago. Never afraid of controversy, he quickly came to the pioneering conclusion that the amount of matter actually in that cluster had to be much greater than what could be accounted for by the visible galaxies alone. Visual observation indicated the presence of an amount of matter that he estimated to be only about 10 percent of the mass needed to gravitationally bind the galaxies in the cluster. He then focused a great deal of his attention on the problem of this "missing mass," or as it is now more commonly called, the problem of DARK MATTER.

To support his investigation of dark matter, Zwicky and his collaborators started to use the 0.457-meter (18-inch) Schmidt telescope at the Palomar Observatory near San Diego in 1936 to extensively study and then catalog numerous clusters of galaxies. Between 1961 and 1968 Zwicky published the results of this intense effort in a six-volume work called *Catalog of Galaxies and of Clusters of Galaxies.* Often simply referred to as the *Zwicky Catalog,* this important work contained about 10,000 clusters, classified as either compact, medium compact, or open, and approximately 31,000 individual galaxies.

While investigating bright novas with Walter Baade in 1934, Zwicky coined the term SUPERNOVA to describe the family of novas that appeared to be far more energetic and cataclysmic than any of the more frequently encountered novas. The classical nova involves a sudden and unpredictable increase in the brightness (perhaps up to ten orders of magnitude) of a binary star system. In a typical spiral galaxy such as the Milky Way, astrophysicists expect about 25 such eruptions annually. As Zwicky noted, the supernova was a much more violent and rare stellar event. A star undergoing a supernova explosion temporarily shines with a brightness that is more than a hundred times more luminous than an ordinary nova. The supernova is a very special astronomical event and occurs in a galaxy such as the Milky Way only about two or three times per century.

From 1937 to 1941 Zwicky performed an extensive search for such events beyond the Milky Way. He personally discovered 18 extragalactic supernovas. Up until then only about 12 such events had been recorded in the entire history of astronomy.

Since the pioneering search by Zwicky and Baade, other astrophysicists have found more than 800 extragalactic supernovas. Despite their brilliance, not all supernovas are visible to naked eye observers on Earth. The most famous naked eye supernovas include the very bright new star reported by Chinese and Korean astronomers in 1054, Tycho's star, appearing in 1572, Kepler's star, appearing in 1604, and supernova 1987A, the most recent event, appearing in the Large Magellanic Cloud (LMC) on February 23, 1987.

Zwicky recognized that supernovas were incredibly interesting and violent celestial events. During a supernova explosion the dying star might become more than 1 billion times brighter than the Sun. In 1932 the British physicist Sir James Chadwick (1891–1974) performed a key nuclear reaction experiment and discovered the neutron. Zwicky quickly recognized the astronomical significance of Chadwick's discovery and boldly presented his visionary "neutron star" hypothesis in 1934. Specifically, Zwicky proposed that a neutron star might be formed as a by-product of a supernova explosion. He reasoned that the violent explosion and subsequent gravitational collapse of the dying star's core should create a compact object of such incredible density that all its electrons and protons would become so closely packed together that they transformed into neutrons.

However, for about three decades other scientists chose to ignore Zwicky's daring hypothesis. Then in 1968 astronomers confirmed the presence of the first suspected neutron star, a young optical pulsar in the cosmic debris of the great supernova event of 1054. The discovery proved that the eccentric Zwicky was clearly years ahead of his contemporaries in suggesting a linkage between supernovas and neutron stars.

Zwicky also published the creative book *Morphological Cosmology* in 1957. He received the Gold Medal from the Royal Astronomical Society in 1973 for his contributions to modern cosmology. After living for five decades in the United States, he died in Pasadena, California, on February 8, 1974.

APPENDIX I

FURTHER READING

Angelo, Joseph A., Jr. *The Dictionary of Space Technology.* Rev. ed. New York: Facts On File, Inc., 2004.

———. *Encyclopedia of Space Exploration.* New York: Facts On File, Inc., 2000.

———, and Irving W. Ginsberg, eds. *Earth Observations and Global Change Decision Making, 1989: A National Partnership.* Malabar, Fla.: Krieger Publishing, 1990.

Brown, Robert A., ed. *Endeavour Views the Earth.* New York: Cambridge University Press, 1996.

Burrows, William E. *The Infinite Journey: Eyewitness Accounts of NASA and the Age of Space.* New York: Discovery Book, 2000.

Chaisson, Eric, and Steave McMillian. *Astronomy Today.* 5th ed. Upper Saddle River, N.J.: Pearson Prentice Hall, 2005.

Cole, Michael D. *International Space Station. A Space Mission.* Springfield, N.J.: Enslow Publishers, 1999.

Collins, Michael. *Carrying the Fire.* New York: Cooper Square Publishers, 2001.

Consolmagno, Guy J., et al. *Turn Left at Orion: A Hundred Night Objects to See in a Small Telescope—And How to Find Them.* New York: Cambridge University Press, 2000.

Damon, Thomas D. *Introduction to Space: The Science of Spaceflight.* 3rd ed. Malabar, Fla.: Krieger Publishing Co., 2000.

Dickinson, Terence. *The Universe and Beyond.* 3rd ed. Willowdater, Ont.: Firefly Books, Ltd., 1999.

Heppenheimer, Thomas A. *Countdown: A History of Space Flight.* New York: John Wiley, 1997.

Kluger, Jeffrey. *Journey Beyond Selene: Remarkable Expeditions Past Our Moon and to the Ends of the Solar System.* New York: Simon & Schuster, 1999.

Kraemer, Robert S. *Beyond the Moon: A Golden Age of Planetary Exploration, 1971–1978.* Smithsonian History of Aviation and Spaceflight Series. Washington, D.C.: Smithsonian Institution Press, 2000.

Lewis, John S. *Rain of Iron and Ice: The Very Real Threat of Comet and Asteroid Bombardment.* Reading, Mass.: Addison-Wesley, 1996.

Logsdon, John M. *Together in Orbit: The Origins of International Participation in the Space Station.* NASA History Division Monographs in Aerospace History 11. Washington, D.C.: Office of Policy and Plans, November 1998.

Matloff, Gregory L. *The Urban Astronomer: A Practical Guide for Observers in Cities and Suburbs.* New York: John Wiley, 1991.

Neal, Valerie, Cathleen S. Lewis, and Frank H. Winter. *Spaceflight: A Smithsonian Guide.* New York: Macmillan, 1995.

Pebbles, Curtis L. *The Corona Project: America's First Spy Satellites.* Annapolis, Md.: Naval Institute Press, 1997.

Seeds, Michael A. *Horizons: Exploring the Universe.* 6th ed. Pacific Grove, Calif.: Brooks/Cole Publishing, 1999.

Sutton, George Paul. *Rocket Propulsion Elements.* 7th ed. New York: John Wiley, 2000.

Todd, Deborah, and Joseph A. Angelo, Jr. *A to Z of Scientists in Space and Astronomy.* New York: Facts On File, Inc., 2005.

APPENDIX II

EXPLORING CYBERSPACE

In recent years numerous Web sites dealing with astronomy, astrophysics, cosmology, space exploration, and the search for life beyond Earth have appeared on the Internet. Visits to such sites can provide information about the status of ongoing missions, such as NASA's *Cassini* spacecraft as it explores the Saturn system. This book can serve as an important companion as you explore a new Web site and encounter an unfamiliar person, technology phrase, or physical concept not fully discussed within the particular site. To help enrich the content of this book and make your astronomy and space technology–related travels in cyberspace more enjoyable and productive, the following is a selected list of Web sites that are recommended for your viewing. From these sites you will be able to link to many other astronomy or space-related locations on the Internet. This is obviously just a partial list of the many astronomy and space-related Web sites now available. Every effort has been made to ensure the accuracy of the information provided. However, due to the dynamic nature of the Internet, changes might occur, and any inconvenience you may experience is regretted.

Selected Organizational Home Pages

The European Space Agency (ESA) is an international organization whose task is to provide for and promote, exclusively for peaceful purposes, cooperation among European states in space research and technology and their applications. URL: www.esrin.esa.it. Accessed on April 12, 2005.

The National Aeronautics and Space Administration (NASA) is the civilian space agency of the United States government and was created in 1958 by an act of Congress. NASA's overall mission is to plan, direct, and conduct American civilian (including scientific) aeronautical and space activities for peaceful purposes. URL: www.nasa.gov. Accessed on April 12, 2005.

The National Oceanic and Atmospheric Administration (NOAA) was established in 1970 as an agency within the U.S. Department of Commerce to ensure the safety of the general public from atmospheric phenomena and to provide the public with an understanding of Earth's environment and resources. URL: www.noaa.gov. Accessed April 12, 2005.

The National Reconnaissance Office (NRO) is the organization within the Department of Defense that designs, builds, and operates U.S. reconnaissance satellites. URL: www.nro.gov. Accessed on April 12, 2005.

The United States Air Force (USAF) serves as the primary agent for the space defense needs of the United States. All military satellites are launched from Cape Canaveral Air Force Station, Florida, or Vandenberg Air Force Base, California. URL: www.af.mil. Accessed on April 12, 2005.

The United States Strategic Command (USSTRATCOM) is the strategic forces organization within the Department of Defense that commands and controls U.S. nuclear forces and military space operations. URL: www.stratcom.mil. Accessed on April 14, 2005.

Selected NASA Centers

The Ames Research Center (ARC), Mountain View, California, is NASA's primary center for exobiology, information technology, and aeronautics. URL: www.arc.nasa.gov. Accessed on April 12, 2005.

The Dryden Flight Research Center (DFRC), Edwards, California, is NASA's center for atmospheric flight operations and aeronautical flight research. URL: www.dfrc.nasa.gov. Accessed on April 12, 2005.

The Glenn Research Center (GRC), Edwards, California, develops aerospace propulsion, power, and communications technology for NASA. URL: www.grc.nasa.gov. Accessed on April 12, 2005.

The Goddard Space Flight Center (GSFC), Greenbelt, Maryland, has a diverse range of responsibilities within NASA including Earth system science, astrophysics, and operation of the *Hubble Space Telescope* and other Earth-orbiting spacecraft. URL: www.nasa.gov/goddard. Accessed on April 12, 2005.

The Jet Propulsion Laboratory (JPL), Pasadena, California, is a government-owned facility operated for NASA by Caltech. JPL manages and operates NASA's deep space scientific missions, as well as the NASA's Deep Space Network, which communicates with solar system exploration spacecraft. URL: www.jpl.nasa.gov. Accessed on April 12, 2005.

The Johnson Space Center (JSC), Houston, Texas, is NASA's primary center for design, development, and testing of spacecraft and associated systems for human space flight, including astronaut selection and training. URL: www.jsc.nasa.gov. Accessed on April 12, 2005.

The Kennedy Space Center (KSC), Florida, is the NASA center responsible for ground turnaround and support operations, prelaunch checkout, and launch of the space shuttle. This center is also responsible for NASA launch facilities at Vandenberg Air Force Base, California. URL: www.ksc.nasa.gov. Accessed on April 12, 2005.

Langley Research Center (LaRC), Hampton, Virginia, is NASA's center for structures and materials, as well as hypersonic flight research and aircraft safety. URL: www.larc.nasa.gov. Accessed on April 12, 2005.

The Marshall Space Flight Center (MSFC), Huntsville, Alabama, serves as NASA's main research center for space propulsion, including contemporary rocket engine development as well as advanced space transportation system concepts. URL: www.msfc.nasa.gov. Accessed on April 12, 2005.

The Stennis Space Center (SSC), Mississippi, is the main NASA center for large rocket engine testing, including space shuttle engines as well as future generations of space launch vehicles. URL: www.ssc.nasa.gov. Accessed on April 14, 2005.

The Wallops Flight Facility (WFF), Wallops Island, Virginia, manages NASA's suborbital sounding rocket program and scientific balloon flights to Earth's upper atmosphere. URL: www.wff.nasa.gov. Accessed on April 14, 2005.

The White Sands Test Facility (WSTF), White Sands, New Mexico, supports the space shuttle and space station programs by performing tests on and evaluating potentially hazardous materials, space flight components, and rocket propulsion systems. URL: www.wstf.nasa.gov. Accessed on April 12, 2005.

Selected Space Missions

The Cassini Mission is an ongoing scientific exploration of the planet Saturn. URL: saturn.jpl.nasa.gov. Accessed on April 14, 2005.

The *Chandra X-Ray Observatory* (CXO) is a space-based astronomical observatory that is part of NASA's Great Observatory program. CXO observes the universe in the X-ray portion of the electromagnetic spectrum. URL: www.chandra.harvard.edu. Accessed on April 14, 2005.

Exploration of Mars by numerous contemporary and previous flyby, orbiter, and lander robotic spacecraft. URL: mars.jpl.nasa.gov. Accessed on April 14, 2005.

The National Space Science Data Center (NSSDC) provides a worldwide compilation of space missions and scientific spacecraft. URL: nssdc.gsfc.nasa.gov/planetary. Accessed on April 14, 2005.

Voyager (Deep Space/Interstellar) updates the status of NASA's *Voyager 1* and *2* spacecraft as they travel beyond the solar system. URL: voyager.jpl.nasa.gov. Accessed on April 14, 2005.

Other Interesting Astronomy and Space Sites

The Arecibo Observatory in the tropical jungle of Puerto Rico is the world's largest radio/radar telescope. URL: www.naic.edu. Accessed on April 14, 2005.

Astrogeology (USGS) describes the USGS Astrogeology Research Program, which has a rich history of participation in space exploration efforts and planetary mapping. URL: planetarynames.wr.usgs.gov. Accessed on April 14, 2005.

The *Hubble Space Telescope* (HST) is an orbiting NASA Great Observatory that is studying the universe primarily in the visible portions of the electromagnetic spectrum. URL: hubblesite.org. Accessed on April 14, 2005.

NASA's Deep Space Network (DSN) is a global network of antennas that provide telecommunications support to distant interplanetary spacecraft and probes. URL: deepspace.jpl.nasa.gov/dsn. Accessed on April 14, 2005.

NASA's Space Science News provides contemporary information about ongoing space science activities. URL: science.nasa.gov. Accessed on April 14, 2005.

The National Air and Space Museum (NASM) of the Smithsonian Institution in Washington, D.C., maintains the largest collection of historic aircraft and spacecraft in the world. URL: www.nasm.si.edu. Accessed on April 14, 2005.

Planetary Photojournal is a NASA/JPL-sponsored Web site that provides an extensive collection of images of celestial objects within and beyond the solar system, historic and contemporary spacecraft used in space exploration, and advanced aerospace technologies. URL: photojournal.jpl.nasa.gov. Accessed on April 14, 2005.

The Planetary Society is the nonprofit organization founded in 1980 by Carl Sagan and other scientists that encourages all spacefaring nations to explore other worlds. URL: planetary.org. Accessed on April 14, 2005.

The Search for Extraterrestrial Intelligence (SETI) Project at UC Berkeley is a Web site that involves contemporary activities in the search for extraterrestrial intelligence, especially a radio SETI project that lets anyone with a computer and an Internet connection participate. URL: www.setiathome.ssl.berkeley.edu. Accessed on April 14, 2005.

Solar System Exploration is a NASA-sponsored and maintained web site that presents the latest events, discoveries, and missions involving the exploration of the solar system. URL: solarsystem.nasa.gov. Accessed on April 14, 2005.

Space Flight History is a gateway Web site sponsored and maintained by the NASA Johnson Space Center. It provides access to a wide variety of interesting data and historic reports dealing with primarily U.S. human space flight. URL: www11.jsc.nasa.gov/history. Accessed on April 14, 2005.

Space Flight Information (NASA) is a NASA-maintained and sponsored gateway Web site that provides the latest information about human space flight activities, including the *International Space Station* (ISS) and the space shuttle. URL: spaceflight.nasa.gov/home. Accessed on April 14, 2005.

APPENDIX III

A CHRONOLOGY OF SPACE AND ASTRONOMY

This appendix presents some of the most interesting events that occurred in the development of space and astronomy as a science and the development and application of modern space technology. Many of the older dates are historic estimates; the more recent dates are known more accurately, although even some of these dates are subject to discussion and debate. For continuity with many existing NASA historical documents, some of the most memorable American space technology dates are presented with respect to local time versus universal time (UT). For example, the first U.S. Earth-orbiting satellite, *Explorer 1,* was launched in the late evening on January 31, 1958 (local time) from Cape Canaveral AFS, Florida. Most NASA histories use January 31, 1958 (local time) as the date of this famous event. However, some recent space publications report the event as February 1, 1958 (UT). Both statements are correct based on the time reference used. Obviously, the use of universal time is appropriate in reporting scientific experiments and performing operational space missions. However, there is an extensive tradition within many aerospace documents that involves the use of local time to report important events. Apparent calendar discrepancies between local time and universal time can cause confusion. In this chronology, unless otherwise specified, all historical dates referenced are the most traditionally accepted dates, which are generally those based on the local time when and where the event occurred. Exceptions to this will be noted when appropriate.

ca. 3000 B.C.E. (to perhaps 1000 B.C.E.)

Stonehenge erected on the Salisbury Plain of southern England (its possible use: ancient astronomical calendar for prediction of summer solstice).

ca. 1300 B.C.E.

Egyptian astronomers recognize all the planets visible to the naked eye (Mercury, Venus, Mars, Jupiter, and Saturn), and they also identify more than 40 star patterns, or constellations.

ca. 500 B.C.E.

Babylonians devise zodiac, which is later adopted and embellished by Greeks and used by other early peoples.

ca. 432 B.C.E.

The early Greek astronomer Meton discovers that 235 lunar months make up about 19 years.

ca. 375 B.C.E.

The early Greek mathematician and astronomer Eudoxus of Cnidos starts codifying the ancient constellations from tales of Greek mythology.

ca. 366 B.C.E.

The ancient Greek astronomer and mathematician Eudoxus of Cnidos constructs a naked eye astronomical observatory. Eudoxus also postulates that Earth is the center of the universe and develops a system of 27 concentric spheres (or shells) to explain and predict the motion of the visible planets (Mercury, Venus, Mars, Jupiter, and Saturn), the Moon, and the Sun. His geocentric model represents the first attempt at a coherent theory of cosmology. With Aristotle's endorsement this early Greek cosmological model eventually leads to the dominance of the Ptolemic system in Western culture for almost two millennia.

ca. 350 B.C.E.

Greek astronomer Heraclides of Pontus suggests Earth spins daily on an axis of rotation and teaches that the planets Mercury and Venus move around the Sun.

In his work *Concerning the Heavens,* the great Greek philosopher Aristotle suggests Earth is a sphere and makes an attempt to estimate its circumference. He also endorses and modifies the geocentric model of Eudoxus of Cnidus. Because of Aristotle's intellectual prestige, his concentric crystal sphere model of the heavens (cosmology) with Earth at the center of the universe essentially remains unchallenged in Western civilization until Nicholas Copernicus in the 16th century.

ca. 275 B.C.E.

The Greek astronomer Aristarchus of Samos suggests an astronomical model of the universe (solar system) that anticipates the modern heliocentric theory proposed by Nicholas Copernicus. However, these "correct" thoughts that Aristarchus presented in his work *On the Size and Distances of the Sun and the Moon* are essentially ignored in favor of the geocentric model of the universe proposed by Eudoxus of Cnidus and endorsed by Aristotle.

ca. 225 B.C.E.

The Greek astronomer Eratosthenes of Cyrene uses mathematical techniques to estimate the circumference of Earth.

ca. 129 B.C.E.

The Greek astronomer Hipparchus of Nicaea completes a catalog of 850 stars that remains important until the 17th century.

0 C.E.

According to Christian tradition, the Star of Bethlehem guides three wise men (thought to be Middle Eastern astronomers) to the nativity of Jesus.

ca. 60 C.E.

The Greek engineer and mathematician Hero of Alexandria creates the aeoliphil, a toylike device that demonstrates the action-reaction principle that is the basis of operation of all rocket engines.

ca. 150 C.E.

The Greek astronomer Ptolemy writes *Syntaxis* (later called *The Almagest* by Arab astronomers and scholars), an important book that summarizes all the astronomical knowledge of the ancient astronomers, including the geocentric model of the universe that dominates Western science for more than one and a half millennia.

820

Arab astronomers and mathematicians establish a school of astronomy in Baghdad and translate Ptolemy's work into Arabic, after which it became known as the *Great Work,* or *al-Majisti* (*The Almagest* by medieval scholars).

850

The Chinese begin to use gunpowder for festive fireworks, including a rocketlike device.

964

Arab astronomer Abd Al-Rahman Al-Sufi publishes his *Book of the Fixed Stars,* a star catalog with Arabic star names; includes earliest known reference to the Andromeda galaxy.

1232

The Chinese army uses fire arrows (crude gunpowder rockets on long sticks) to repel Mongol invaders at the battle of Kai-fung-fu. This is the first reported use of the rocket in warfare.

1280–90

The Arab historian Al-Hasan al Rammah writes *The Book of Fighting on Horseback and War Strategies,* in which he gives instructions for making both gunpowder and rockets.

1379

Rockets appear in western Europe; used in the siege of Chioggia (near Venice), Italy.

1420

The Italian military engineer Joanes de Fontana writes *Book of War Machines,* a speculative work that suggests military applications of gunpowder rockets, including a rocket-propelled battering ram and a rocket-propelled torpedo.

1424

The Mongolian ruler and astronomer Beg Ulugh constructs a great observatory at Samarkand.

1429

The French army uses gunpowder rockets to defend the city of Orléans. During this period arsenals throughout Europe begin to test various types of gunpowder rockets as an alternative to early cannons.

ca. 1500

According to early rocketry lore, a Chinese official named Wan-Hu (or Wan-Hoo) attempts to use an innovative rocket-propelled kite assembly to fly through the air. As he sat in the pilot's chair, his servants lit the assembly's 47 gunpowder (black powder) rockets. Unfortunately, this early rocket test pilot disappeared in a bright flash and explosion.

1519

The Portuguese explorer Ferdinand Magellan becomes the first person to record observing the two irregular dwarf galaxies visible to the naked eye in the Southern Hemisphere. These celestial objects now bear his name as the Magellanic Clouds.

1543

The Polish church official and astronomer Nicholas Copernicus changes history and initiates the scientific revolution with his book *On the Revolution of Celestial Orbs* (*De revolutionibus orbium coelestium*). This important book, published while Copernicus lay on his deathbed, proposed a Sun-centered (heliocentric) model of the universe in contrast to the long-standing Earth-centered (geocentric) model advocated by Ptolemy and many of the early Greek astronomers.

1572

The Danish astronomer Tycho Brahe discovers a supernova that appears in the constellation Cassiopeia. He describes his precise (pretelescope) observations in *De Nova Stella* (1573). The dynamic nature of this "brilliant" star causes Brahe and other astronomers to question the long-cherished hypothesis of Aristotle concerning the unchanging nature of the heavens.

1576

With the generous support of the Danish king Frederick II, Tycho Brahe starts construction of a great naked eye astronomical observatory on the island of Hven in the Baltic Sea. For the next two decades Brahe's Uraniborg ("Castle of the Sky") serves as the world's center for astronomy. The precise data from this observatory eventually allow Johannes Kepler to develop his important laws of planetary motion.

1577

The observation of a comet's elongated trajectory causes Tycho Brahe to question Aristotle's geocentric crystal spheres model of planetary motion. He proposes his own "Tychonic system," a modified version of Ptolemaic cosmology in which all the planets (except Earth) revolve around the Sun, and the Sun with its entire assembly of planets and comets revolves around a stationary Earth.

1601

Following the death of Tycho Brahe, the German emperor Rudolf II appoints Johannes Kepler to succeed him as the imperial mathematician in Prague. Kepler thus acquires Brahe's collection of precise pretelescopic astronomical observations and uses them to help develop his three laws of planetary motion.

1602–04

The German astronomer David Fabricus makes precise pretelescope observations of the orbit of Mars. His data greatly assist Johannes Kepler in the development of Kepler's three laws of planetary motion.

1608

The Dutch optician Hans Lippershey develops a crude telescope.

1609

The German astronomer Johannes Kepler publishes *New Astronomy*, in which he modifies Nicholas Copernicus's model of the universe by announcing that the planets have elliptical orbits rather than circular ones. Kepler's laws of planetary motion help put an end to more than 2,000 years of geocentric Greek astronomy.

1610

On January 7, 1610, Galileo Galilei uses his telescope to gaze at Jupiter and discovers the giant planet's four major moons (Callisto, Europa, Io, and Ganymede). He proclaims this and other astronomical observations in his book *Starry Messenger* (*Sidereus Nuncius*). Discovery of these four Jovian moons encourages Galileo to advocate the heliocentric theory of Nicholas Copernicus and brings him into direct conflict with church authorities.

1619

Johannes Kepler publishes *The Harmony of the World*, in which he presents his third law of planetary motion.

1639

On November 24 (Julian calendar) the British astronomer Jeremiah Horrocks makes the first observation of a transit of Venus.

1642

Galileo Galilei dies while under house arrest near Florence, Italy, for his clashes with church authorities concerning the heliocentric theory of Nicholas Copernicus.

1647

The Polish-German astronomer Johannes Hevelius publishes his work *Selenographia,* in which he provides a detailed description of features on the surface (nearside) of the Moon.

1655

The Dutch astronomer Christiaan Huygens discovers Titan, the largest moon of Saturn.

1668

Giovanni Domenico Cassini publishes tables that describe the motions of Jupiter's major satellites.

Sir Isaac Newton constructs the first reflecting telescope.

1675

Giovanni Domenico Cassini discovers the division of Saturn's rings, a feature that now carries his name.

The British king Charles II establishes the Royal Greenwich Observatory and appoints John Flamsteed as its director and also the first Astronomer Royal.

1680

The Russian czar Peter the Great sets up a facility to manufacture rockets in Moscow. The facility later moves to St. Petersburg and provides the czarist army with a variety of gunpowder rockets for bombardment, signaling, and nocturnal battlefield illumination.

1687

Financed and encouraged by Sir Edmond Halley, Sir Isaac Newton publishes his great work, *The Principia* (*Philosophiae naturalis principia mathimatica*). This book provides the mathematical foundations for understanding the motion of almost everything in the universe, including the orbital motion of planets and the trajectories of rocket-propelled vehicles.

1728

The British astronomer James Bradley discovers the optical phenomenon of aberration of starlight. He explains this phenomenon as a combination of Earth's motion around the Sun and light having a finite speed.

1740

The Swedish astronomer Anders Celsius assumes responsibility for the construction and management of the Uppsala Observatory. In 1742 he suggests a temperature scale based on 0° as the boiling point of water and 100° as the melting point of ice. Following his death in 1744 his colleagues at the Uppsala Observatory continue to use this new temperature scale but invert it. Their action creates the familiar Celsius temperature scale, with 0° as the melting point of ice and 100° as the boiling point of water.

1748

The French physicist Pierre Bouguer invents the heliometer, an instrument that measures the light of the Sun and other luminous bodies.

1755

The German philosopher Immanuel Kant introduces the nebula hypothesis, suggesting that the solar system may have formed out of an ancient cloud of interstellar matter. He also uses the term *island universes* to describe distant disklike collections of stars.

1758

The English optician John Dolland and his son make the first achromatic (that is, without chromatic aberration) telescope.

1766

The German astronomers Johann Elbert Bode and Johann Daniel Titius discover an empirical rule (eventually called Bode's law) that describes the apparently proportional distances of the six known planets from the Sun. Their work stimulates the search by other astronomers to find a "missing" planet between Mars and Jupiter, eventually leading to the discovery of the main belt asteroids.

1772

The Italian-French mathematician Joseph Lewis Lagrange describes the points in the plane of two objects in orbit around their common center of gravity at which their combined forces of gravitation are zero. Today such points in space are called Lagrangian libration points.

1780s

The Indian ruler Hyder Ally of Mysore creates a rocket corps within his army. Hyder's son, Tippo Sultan, successfully uses rockets against the British in a series of battles in India between 1782 and 1799.

1781

The German-born British astronomer Sir (Frederick) William Herschel discovers Uranus, the first planet to be found through the use of the telescope. Herschel originally called this new planet Georgium Sidus (George's Star) to honor the British king, George III. However, astronomers from around the world insisted that a more traditional name from Greco-Roman mythology be used. Herschel bowed to peer pressure and eventually chose the name Uranus, as suggested by the German astronomer Johann Elert Bode.

1782

The British astronomer John Goodricke is the first to recognize that the periodic behavior of the variable star Algol (Beta Persei) is actually that of an eclipsing binary star system.

1786

The German-born British astronomer Caroline Lucretia Herschel becomes the world's first (recognized) woman astronomer. While working with her brother Sir (Frederick) William Herschel, she discovers eight comets during the period 1786–97.

1796

The French mathematician and astronomer Pierre-Simon, marquis de Laplace proposes the nebular hypothesis of planetary formation, suggesting that the solar system originated from a massive cloud of gas. Over time the forces of gravitation helped the center of this cloud collapse to form the Sun and smaller remnant clumps of matter to form the planets.

1801

On January 1 the Italian astronomer Giuseppe Piazzi discovers the first asteroid (minor planet), which he names Ceres.

1802

The German astronomer Heinrich Wilhelm Olbers discovers the second asteroid, Pallas, and then continues his search, finding the brightest minor planet, Vesta, in 1807.

1804

Sir William Congreve writes *A Concise Account of the Origin and Progress of the Rocket System* and documents the British military's experience in India. He then starts the development of a series of British military black-powder rockets.

1807

The British use about 25,000 of Sir William Congreve's improved military black-powder rockets to bombard Copenhagen, Denmark, during the Napoleonic Wars.

1809

The brilliant German mathematician, astronomer, and physicist Carl Friedrich Gauss publishes a major work on celestial mechanics that revolutionizes the calculation of perturbations in planetary orbits. His work paves the way for other 19th-century astronomers to mathematically anticipate and then discover Neptune in 1846 using perturbations in the orbit of Uranus.

1812

British forces use Sir William Congreve's military rockets against American troops during the War of 1812. The British rocket bombardment of Fort William McHenry inspires Francis Scott Key to add the "rocket's red glare" verse in the "Star Spangled Banner."

1814–17

The German physicist Joseph von Fraunhofer develops the prism spectrometer into a precision instrument. He subsequently discovers the dark lines in the Sun's spectrum that now bear his name.

1840

The American astronomers Henry Draper and John William Draper (his son) photograph the Moon and start the field of astrophotography.

1842

The Austrian physicist Christian Johann Doppler describes the Doppler shift in a scientific paper. This phenomenon is experimentally verified in 1845 through an interesting experiment in which a locomotive pulls an open railroad car carrying several trumpeters.

1846

On September 23 the German astronomer Johann Gottfried Galle discovers Neptune in the location theoretically predicted and calculated by Urbain Jean Joseph Leverrier.

1847

The American astronomer Maria Mitchell makes the first telescopic discovery of a comet.

1850

The American astronomer William Cranch Bond helps expand the field of astrophotography by successfully photographing the planet Jupiter.

1851

The French physicist Jean-Bernard Léon Foucault uses a 65-meter-long pendulum to demonstrate Earth's rotation to a large audience in Paris.

The British astronomer William Lassell discovers the two satellites of Uranus, named Ariel and Umbriel.

1852

The French physicist Jean-Bernard Léon Foucault constructs the first gyroscope.

1864

The Italian astronomer Giovanni Battista Donati collects the spectrum of Comet Tempel.

1865

The French science fiction writer Jules Verne publishes his famous story *From the Earth to the Moon* (*De la terre a la lune*). This story interests many people in the concept of space travel, including young readers who go on to become the founders of astronautics: Robert Hutchings Goddard, Hermann J. Oberth, and Konstantin Eduardovich Tsiolkovsky.

1866

The American astronomer Daniel Kirkwood explains the unequal distribution of asteroids found in the main belt between Mars and Jupiter as a celestial mechanics resonance phenomenon involving the gravitational influence of Jupiter. In his honor these empty spaces, or gaps, are now called the Kirkwood gaps.

1868

During a total eclipse of the Sun, the French astronomer Pierre Jules César Janssen makes important spectroscopic observations that reveal the gaseous nature of solar prominences.

The British astronomer Sir William Huggins becomes the first to observe the gaseous spectrum of a nova.

While observing the Sun the British astronomer Sir Joseph Norman Lockyer postulates the existence of an unknown element that he names helium (meaning "the Sun's element"), but it is not until 1895 that helium is detected on Earth.

1869

The American clergyman and writer Edward Everett Hale publishes *The Brick Moon,* a story that is the first fictional account of a human-crewed space station.

1874

The Scottish astronomer Sir David Gill makes the first of several very precise measurements of the distance from Earth to the Sun while on an expedition to Mauritius (Indian Ocean). He then visits Ascension Island (South Atlantic Ocean) in 1877 to perform similar measurements.

1877

While a staff member at the U.S. Naval Observatory in Washington, D.C., the American astronomer Asaph Hall discovers and names the two tiny Martian moons, Deimos and Phobos.

1881

The American astronomer and aeronautical engineer Samuel P. Langley leads an expedition to Mount Whitney in the Sierra Nevada Mountains to examine how incoming solar radiation is absorbed by Earth's atmosphere.

1897

The British author H. G. Wells writes the science fiction story *War of the Worlds,* a classic tale about extraterrestrial invaders from Mars.

1903

The Russian technical visionary Konstantin Eduardovich Tsiolkovsky becomes the first person to link the rocket and space travel when he publishes *Exploration of Space with Reactive Devices.*

1906

The American astronomer George Ellery Hale sets up the first tower telescope for solar research on Mount Wilson in California.

1918

The American physicist Robert Hutchings Goddard writes *The Ultimate Migration,* a far-reaching technology piece within which he postulates the use of an atomic-powered space ark to carry human beings away from a dying Sun. Fearing ridicule, however, Goddard hides the visionary manuscript, and it remains unpublished until November 1972, many years after his death in 1945.

1919

The American rocket pioneer Robert Hutchings Goddard publishes the Smithsonian monograph *A Method of Reaching Extreme Altitudes.* This important work presents all the fundamental principles of modern rocketry. Unfortunately, members of the press completely miss the true significance of his technical contribution and decide to sensationalize his comments about possibly reaching the Moon with a small rocket-propelled package. For such "wild fantasy," newspaper reporters dubbed Goddard with the unflattering title "Moon man."

On May 29 a British expedition led by the astronomer Sir Arthur Eddington observes a total solar eclipse and provides scientific verification of Albert Einstein's general relativity theory by detecting the slight deflection of a beam of light from a star near the edge of the "darkened" Sun. This important experiment helps usher in a new era in humans understanding of the physical universe.

1922

The American astronomer William Wallace Campbell measures the deflection of a beam of starlight that just skims the Sun's edge during solar eclipse. This work provides additional scientific evidence in favor of Albert Einstein's general relativity theory.

1923

Independently of Robert Hutchings Goddard and Konstantin Eduardovich Tsiolkovsky, the German space travel visionary Hermann J. Oberth publishes the inspiring book *The Rocket into Planetary Space (Die Rakete zu den Planetenräumen).*

The American astronomer Edwin Powell Hubble uses Cepheid variable stars as astronomical distance indicators and shows that the great spiral "nebula" in the constellation Andromeda (now called the Andromeda galaxy) is a celestial object well beyond the Milky Way galaxy. As a result of Hubble's work, astronomers begin to postulate that an expanding universe contains many such galaxies, or "island universes."

1924

The German engineer Walter Hohmann writes *The Attainability of Celestial Bodies (Die Erreichbarkeit der Himmelskörper),* an important work that details the mathematical principles of rocket and spacecraft motion. He includes a description of the most efficient (that is, minimum energy) orbit transfer path between two coplanar orbits, a frequently used space operations maneuver now called the Hohmann transfer orbit.

1926

On March 16 in a snow-covered farm field in Auburn, Massachusetts, the American physicist Robert Hutchings Goddard makes space technology history by successfully firing the world's first liquid-propellant rocket. Although his primitive gasoline (fuel) and liquid oxygen (oxidizer) device burned for only two and a half seconds and landed about 60 m away, it represents the technical ancestor of all modern liquid-propellant rocket engines.

In April the first issue of *Amazing Stories* appears. The publication becomes the world's first magazine dedicated exclusively to science fiction. Through science fact and fiction the modern rocket and space travel become firmly connected. As a result of this union, the visionary dream for many people in the 1930s (and beyond) becomes that of interplanetary travel.

1927

The Belgian astrophysicist and cosmologist Georges Édouard Lemaître publishes a major paper introducing a cosmology model that features an expanding universe. His work stimulates development of the big bang theory.

1929

The American astronomer Edwin Powell Hubble announces that his measurements of galactic redshift values indicate that other galaxies are receding from the Milky Way galaxy with speeds that increase in proportion to their distance from humans' home galaxy. This observation becomes known as Hubble's law.

The British scientist and writer John Desmond Bernal speculates about large human settlements in outer space later called Bernal spheres in his book *The World, the Flesh, and the Devil.*

The German space travel visionary Hermann J. Oberth writes the award-winning book *Roads to Space Travel* (*Wege zur Raumschiffahrt*) that helps popularize the notion of space travel among nontechnical audiences.

1930

On 18 February the American astronomer Clyde Tombaugh discovers the planet Pluto.

1932

In December, the American radio engineer Karl Jansky announces his discovery of a stellar radio source in the constellation of Sagittarius. His paper is regarded as the birth of radio astronomy.

Following the discovery of the neutron by the English physicist Sir James Chadwick, the Russian physicist Lev D. Landau suggests the existence of neutron stars.

1933

P. E. Cleator founds the British Interplanetary Society (BIS), which becomes one of the world's most respected space travel advocacy organizations.

1935

Konstantin Tsiolkovsky publishes his last book, *On the Moon,* in which he strongly advocates the spaceship as a means of lunar and interplanetary travel.

1936

P. E. Cleator, founder of the British Interplanetary Society, writes *Rockets Through Space,* the first serious treatment of astronautics in the United Kingdom. However, several established British scientific publications ridicule his book as the premature speculation of an unscientific imagination.

1937

The American radio engineer Grote Reber builds the first radio telescope, a modest device located in his backyard. For several years following this construction, Reber is the world's only radio astronomer.

1939–45

Throughout World War II combating nations use rockets and guided missiles of all sizes and shapes. Of these, the most significant with respect to space exploration is the development of the liquid-propellant V-2 rocket by the German Army at Peenemünde under Wernher von Braun.

1942

On October 3 the German A-4 rocket (later renamed Vengeance Weapon Two, or V-2 rocket) completes its first successful flight from the Peenemünde test site on the Baltic Sea. This is the birth date of the modern military ballistic missile.

1944

In September the German Army begins a ballistic missile offensive by launching hundreds of unstoppable V-2 rockets (each carrying a one-ton high explosive warhead) against London and southern England.

1945

Recognizing the war is lost, the German rocket scientist Wernher von Braun and key members of his staff surrender to American forces near Reutte, Germany, in early May. Within months U.S. intelligence teams under Operation Paperclip interrogate German rocket personnel and sort through carloads of captured documents and equipment. Many of these German scientists and engineers join von Braun in the United States to continue their rocket work. Hundreds of captured V-2 rockets are also disassembled and shipped back to the United States.

Similarly, on May 5 the Soviet Army captures the German rocket facility at Peenemünde and hauls away any remaining equipment and personnel. In the closing days of the war in Europe, captured German rocket technology and personnel helps set the stage for the great missile and space race of the cold war.

On July 16 the United States explodes the world's first nuclear weapon. The test shot, code named Trinity, occurs in a remote portion of southern New Mexico and changes the face of warfare forever. As part of the cold war confrontation between the United States and the Soviet Union, the nuclear-armed ballistic missile will become the most powerful weapon ever developed by the human race.

In October a then obscure British engineer and writer named Sir Arthur C. Clarke suggests the use of satellites at geostationary orbit to support global communications. His article, "Extra-Terrestrial Relays" in *Wireless World,* represents the birth of the communications satellite concept, an application of space technology that actively supports the information revolution.

1946

On April 16 the U.S. Army launches the first American-adapted, captured German V-2 rocket from the White Sands Proving Ground in southern New Mexico.

Between July and August the Russian rocket engineer Sergei Korolev develops a stretched-out version of the German V-2 rocket. As part of his engineering improvements Korolev increases the rocket engine's thrust and lengthens the vehicle's propellant tanks.

1947

On October 30 Russian rocket engineers successfully launch a modified German V-2 rocket from a desert launch site near a place called Kapustin Yar. This rocket impacts about 320 km downrange from the launch site.

1948

The September issue of the *Journal of the British Interplanetary Society (JBIS)* starts a four-part series of technical papers by L. R. Shepherd and A. V. Cleaver that explores the feasibility of applying nuclear energy to space travel, including the concepts of nuclear-electric propulsion and the nuclear rocket.

1949

On August 29 the Soviet Union detonates its first nuclear weapon at a secret test site in the Kazakh desert. Code named First Lightning (Pervaya Molniya), the successful test breaks the nuclear weapon monopoly enjoyed by the United States. It plunges the world into a massive nuclear arms race that includes the accelerated development of strategic ballistic missiles capable of traveling thousands of kilometers. Because they are well behind the United States in nuclear weapons technology, the leaders of the Soviet Union decide to develop powerful, high-thrust rockets to carry their heavier, more primitively designed nuclear weapons. That decision gives the Soviet Union a major launch vehicle advantage when both superpowers decide to race into outer space (starting in 1957) as part of a global demonstration of national power.

1950

On July 24 the United States successfully launches a modified German V-2 rocket with an American-designed WAC Corporal second-stage rocket from the U.S. Air Force's newly established Long-Range Proving Ground at Cape Canaveral, Florida. The hybrid multistage rocket (called Bumper 8) inaugurates the incredible sequence of military missile and space vehicle launches to take place from Cape Canaveral, the world's most famous launch site.

In November the British technical visionary Sir Arthur C. Clarke publishes "Electromagnetic Launching As a Major Contribution to Space-Flight." Clarke's article suggests mining the Moon and launching the mined lunar material into outer space with an electromagnetic catapult.

1951

Cinema audiences are shocked by the science fiction movie *The Day the Earth Stood Still.* This classic story involves the arrival of a powerful, humanlike extraterrestrial and his robot companion who come to warn the governments of the world about the foolish nature of their nuclear arms race. It is the first major science fiction story to portray powerful space aliens as friendly, intelligent creatures who come to help Earth.

The Dutch-American astronomer Gerard Peter Kuiper suggests the existence of a large population of small icy planetesimals beyond the orbit of Pluto, a collection of frozen celestial bodies now known as the Kuiper Belt.

1952

Collier's magazine helps stimulate a surge of American interest in space travel by publishing a beautifully illustrated series of technical articles written by space experts such as Wernher von Braun and Willy Ley. The first of the famous eight-part series appears on March 22 and is boldly titled "Man Will Conquer Space Soon." The magazine also hires the most influential space artist, Chesley Bonestell, to provide stunning color illustrations. Subsequent articles in the series introduce millions of American readers to the concept of a space station, a mission to the Moon, and an expedition to Mars.

Wernher von Braun publishes *The Mars Project* (*Das Marsprojekt*), the first serious technical study regarding a human-crewed expedition to Mars. His visionary proposal involves a convoy of 10 spaceships with a total combined crew of 70 astronauts to explore the Red Planet for about one year and then return to Earth.

1953

In August, the Soviet Union detonates its first thermonuclear weapon, a hydrogen bomb. This is a technological feat that intensifies the superpower nuclear arms race and increases emphasis on the emerging role of strategic nuclear-armed ballistic missiles.

In October the U.S. Air Force forms a special panel of experts headed by John von Neumann to evaluate the American strategic ballistic missile program. In 1954 this panel recommends a major reorganization of the American ballistic missile effort.

1954

Following the recommendations of John von Neumann, President Dwight D. Eisenhower gives strategic ballistic missile development the highest national priority. The cold war missile race explodes on the world stage as the fear of a strategic ballistic missile gap sweeps through the American government. Cape Canaveral becomes the famous proving ground for such important ballistic missiles as the Thor, Atlas, Titan, Minuteman, and Polaris. Once developed, many of these powerful military ballistic missiles will also serve the United States as space launch vehicles. The U.S. Air Force general Bernard Schriever oversees the time-critical development of the Atlas ballistic missile, an astonishing feat of engineering and technical management.

1955

Walt Disney, an American entertainment visionary, promotes space travel by producing an inspiring three-part television series that includes appearances by noted space experts such as Wernher von Braun. The first episode, "Man In Space," airs on March 9 and popularizes the dream of space travel for millions of American television viewers. This show, along with its companion episodes, "Man and the Moon" and "Mars and Beyond," make von Braun famous and the term *rocket scientist* a popular household phrase.

1957

On October 4, the Russian rocket scientist Sergei Korolev, with permission from Soviet premier Nikita S. Khrushchev, uses a powerful military rocket to successfully place *Sputnik 1,* the world's first artificial satellite, into orbit around Earth. News of the Soviet success sends a political and technical shockwave across the United States. The launch of *Sputnik 1* marks the beginning of the space age. It also is the start of the great space race of the cold war, a period when people measured national strength and global prestige by accomplishments (or failures) in outer space.

On November 3 the Soviet Union launches *Sputnik 2,* the world's second artificial satellite. It is a massive spacecraft (for the time) that carries a live dog named Laika, which is euthanized at the end of the mission.

The highly publicized attempt by the United States to launch its first satellite with a newly designed civilian rocket ends in complete disaster on December 6. The Vanguard rocket explodes after rising only a few centimeters above its launch pad at Cape Canaveral. Soviet successes with *Sputnik 1* and *Sputnik 2* and the dramatic failure of the Vanguard rocket heighten American anxiety. The exploration and use of outer space becomes a highly visible instrument of cold war politics.

1958

On January 31 the United States successfully launches *Explorer 1,* the first American satellite in orbit around Earth. A hastily formed team from the U.S. Army Ballistic Missile Agency (ABMA) and Caltech's Jet Propulsion Laboratory (JPL), led by Wernher von Braun, accomplishes what amounts to a national prestige rescue mission. The team uses a military ballistic missile as the launch vehicle. With instruments supplied by James Van Allen of the State University of Iowa, *Explorer 1* discovers Earth's trapped radiation belts, now called the Van Allen radiation belts in his honor.

The National Aeronautics and Space Administration (NASA) becomes the official civilian space agency for the United States government on October 1. On October 7 the newly created NASA announces the start of the Mercury Project, a pioneering program to put the first American astronauts into orbit around Earth.

In mid-December an entire Atlas rocket lifts off from Cape Canaveral and goes into orbit around Earth. The missile's payload compartment carries Project Score (Signal Communications Orbit Relay Experiment), a prerecorded Christmas season message from President Dwight D. Eisenhower. This is the first time the human voice is broadcast back to Earth from outer space.

1959

On January 2 the Soviet Union sends a 360-kg spacecraft called *Lunik 1* toward the Moon. Although it misses hitting the Moon by between 5,000 and 7,000 km, it is the first human-made object to escape Earth's gravity and go into orbit around the Sun.

In mid-September the Soviet Union launches *Lunik 2.* The 390-kg spacecraft successfully impacts on the Moon and becomes the first human-made object to land (crash) on another world. *Lunik 2* carries Soviet emblems and banners to the lunar surface.

The September issue of *Nature* contains an article by Philip Morrison and G. Cocconi entitled "Searching for Interstellar Communications," marking the start of SETI (search for extraterrestrial intelligence).

On October 4 the Soviet Union sends *Lunik 3* on a mission around the Moon. The spacecraft successfully circumnavigates the Moon and takes the first images of the lunar farside. Because of the synchronous rotation of the Moon around Earth, only the nearside of the lunar surface is visible to observers on Earth.

1960

The United States launches the *Pioneer 5* spacecraft on March 11 into orbit around the Sun. The modest-sized (42-kg) spherical American space probe reports conditions in interplanetary space between Earth and Venus over a distance of about 37 million km.

On April 1 NASA successfully launches the world's first meteorological satellite, called TIROS (Television and Infrared Observation Satellite). The images of cloud patterns from outer space create a revolution in weather forecasting.

The U.S. Navy places the world's first experimental navigation satellite, called *Transit 1B,* into orbit on April 13. The spacecraft serves as a space-based beacon providing radio signals that allow military users to determine their locations at sea more precisely.

On May 24 the U.S. Air Force launches a MIDAS (Missile Defense Alarm System) satellite from Cape Canaveral. This event inaugurates an important American program of special military surveillance satellites intended to detect enemy missile launches by observing the characteristic infrared (heat) signature of a rocket's exhaust plume. Essentially unknown to the general public for decades because of the classified nature of their mission, the emerging family of missile surveillance satellites provides U.S. government authorities with a reliable early warning system concerning a surprise enemy (Soviet) ICBM attack. Surveillance satellites help support the national policy of strategic nuclear deterrence throughout the cold war and prevent an accidental nuclear conflict.

The U.S. Air Force successfully launches the *Discoverer 13* spacecraft from Vandenberg Air Force Base on August 10. This spacecraft is actually part of a highly classified air force and Central Intelligence Agency (CIA) reconnaissance satellite program called Corona. Started under special executive order by President Dwight D. Eisenhower, the joint agency spy satellite program begins to provide important photographic images of denied areas of the world from outer space. On August 18 *Discoverer 14* (also called *Corona XIV*) provides the U.S. intelligence community its first satellite-acquired images of the Soviet Union. The era of satellite reconnaissance is born. Data collected by the spy satellites of the National Reconnaissance Office (NRO) contribute significantly to U.S. national security and help preserve global stability during many politically troubled times.

On August 12 NASA successfully launches the *Echo 1* experimental spacecraft. This large (30.5-m-diameter), inflatable, metalized balloon becomes the world's first passive communications satellite. At the dawn of space-based telecommunications, engineers bounce radio signals off the large inflated satellite between the United States and the United Kingdom.

The Soviet Union launches *Sputnik 5* into orbit around Earth. This large spacecraft is actually a test vehicle for the new Vostok spacecraft that will soon carry cosmonauts into outer space. *Sputnik 5* carries two dogs, Strelka and Belka. When the spacecraft's recovery capsule functions properly the next day, these two dogs become the first living creatures to return to Earth successfully from an orbital flight.

1961

On January 31 NASA launches a Redstone rocket with a Mercury Project space capsule on a suborbital flight from Cape Canaveral. The passenger astrochimp, Ham, is safely recovered down range in the Atlantic Ocean after reaching an altitude of 250 km. This successful primate space mission is a key step in sending American astronauts safely into outer space.

The Soviet Union achieves a major space exploration milestone by successfully launching the first human being into orbit around Earth. Cosmonaut Yuri Gagarin travels into outer space in the *Vostok 1* spacecraft and becomes the first person to observe Earth directly from an orbiting space vehicle.

On May 5 NASA uses a Redstone rocket to send astronaut Alan B. Shephard, Jr. on his historic 15-minute suborbital flight into outer space from Cape Canaveral. Riding inside the Mercury Project *Freedom 7* space capsule, Shepard reaches an altitude of 186 km and becomes the first American to travel in space.

President John F. Kennedy addresses a joint session of the U.S. Congress on May 25. In an inspiring speech concerning many urgent national needs, the newly elected president creates a major space challenge for the United States when he declares, "I believe that this nation should commit itself to achieving the goal, before this decade is out, of landing a man on the Moon and returning him safely to Earth." Because of his visionary leadership, when American astronauts Neil A. Armstrong and Edwin E. "Buzz" Aldrin, Jr. step onto the lunar surface for the first time on July 20, 1969, the United States is recognized around the world as the undisputed winner of the cold war space race.

On June 29 the United States launches the *Transit 4A* navigation satellite into orbit around Earth. The spacecraft uses a radioisotope thermoelectric generator (RTG) to provide supplementary electric power. The mission represents the first successful use of a nuclear power supply in space.

On July 21 the astronaut Virgil "Gus" I. Grissom becomes the second American to travel in outer space. A Redstone rocket successfully hurls his NASA Mercury Project *Liberty Bell 7* space capsule on a 15-minute suborbital flight from Cape Canaveral.

The Soviet Union launches the *Vostok 2* spacecraft into orbit on August 6. It carries cosmonaut Gherman S. Titov, the second person to orbit Earth in a spacecraft successfully. About 10 hours into the flight, Titov also becomes the first of many space travelers to suffer from the temporary discomfort of space sickness, or space adaptation syndrome (SAS).

1962

On February 20 the astronaut John Herschel Glenn, Jr. becomes the first American to orbit Earth in a spacecraft. An Atlas rocket launches the NASA Mercury Project *Friendship 7* space capsule from Cape Canaveral. After completing three orbits Glenn's capsule safely splashes down in the Atlantic Ocean.

NASA launches *Telstar 1* on July 10, the world's first commercially constructed and funded active communications satellite. Despite its relatively low operational orbit (about 950 km by 5,600 km), the pioneering American Telephone and Telegraph (AT&T) satellite triggers a revolution in international television broadcasting and telecommunications services.

In late August NASA sends the *Mariner 2* spacecraft to Venus from Cape Canaveral. *Mariner 2* passes within 35,000 km of the planet on December 14, 1962, thereby becoming the world's first successful interplanetary space probe. The spacecraft observes very high surface temperatures (~430°C). These data shatter pre–space age visions about Venus being a lush, tropical planetary twin of Earth.

During October the placement of nuclear-armed Soviet offensive ballistic missiles in Fidel Castro's Cuba precipitates the Cuban Missile Crisis. This dangerous superpower confrontation brings the world perilously close to nuclear warfare. Fortunately, the crisis dissolves when Premier Nikita S. Khrushchev withdraws the Soviet ballistic missiles after much skillful political maneuvering by President John F. Kennedy and his national security advisers.

1963

The Soviet cosmonaut Valentina Tereskova becomes the first woman to travel in outer space. On June 16 she ascends into space onboard the *Vostok 6* spacecraft and then returns to Earth after a flight of almost three days. While in orbit she flies the *Vostok 6* within 5 km of the *Vostok 5* spacecraft, piloted by cosmonaut Valery Bykovskiy. During their proximity flight the two cosmonauts communicate with each other by radio.

On July 26 NASA successfully launches the *Syncom 2* satellite from Cape Canaveral. This spacecraft is the first communications satellite to operate in high-altitude, synchronous (figure eight) orbit and helps fulfill the vision of Sir Arthur C. Clarke. About a year later *Syncom 3* will achieve a true geosynchronous orbit above the equator. Both of these experimental NASA satellites clearly demonstrate the feasibility and great value of placing communications satellites into geostationary orbits. The age of instantaneous global communications is born.

In October President John F. Kennedy signs the Limited Test Ban Treaty for the United States, and the important new treaty enters force on October 10. Within a week (on October 16) the U.S. Air Force successfully launches the first pair of Vela nuclear detonation detection satellites from Cape Canaveral. These spacecraft orbit Earth at a very high altitude and continuously monitor the planet and outer space for nuclear detonations in violation of the Test Ban Treaty. From that time on American spacecraft carrying nuclear detonation detection instruments continuously provide the United States with a reliable technical ability to monitor Earth's atmosphere and outer space for nuclear treaty violations.

1964

Following a series of heartbreaking failures, NASA successfully launches the *Ranger 7* spacecraft to the Moon from Cape Canaveral on July 24. About 68 hours after liftoff the robot probe transmits more than 4,000 high-resolution television images before crashing into the lunar surface in a region known as the Sea of Clouds. The *Ranger 7, 8,* and *9* spacecraft help prepare the way for the lunar landing missions by the Apollo Project astronauts (1969–72).

In August the International Telecommunications Satellite Organization (INTELSAT) is formed, its mission being to develop a global communications satellite system.

On October 12 the Soviet Union launches the first three-person crew into outer space when the specially-configured *Voskhod 1* spacecraft is used to carry cosmonauts Vladimir Komarov, Boris Yegorov, and Konstantin Feoktistov into orbit around Earth.

On November 28 NASA's *Mariner 4* spacecraft departs Cape Canaveral on its historic journey as the first spacecraft from Earth to visit Mars. It successfully encounters the Red Planet on July 14, 1965, at a flyby distance of about 9,800 km. *Mariner 4*'s close-up images reveal a barren, desertlike world and quickly dispel any pre–space age notions about the existence of ancient Martian cities or a giant network of artificial canals.

1965

The German-American physicist Arno Allen Penzias and the American physicist Robert Woodrow Wilson detect the faint cosmic microwave background considered by cosmologists to be the lingering signal from the big bang explosion that began the universe.

On March 18 the Soviet Union launches the *Voskhod 2* spacecraft carrying cosmonauts Pavel Belyayev and Alekesey Leonov. During this mission Leonov becomes the first human to leave the confines of an orbiting spacecraft and perform an extravehicular activity (EVA). While tethered he conducts this historic 10-minute space walk. Then he encounters some significant difficulties when he tries to get back into the *Voskhod*'s airlock with a bloated and clumbersome spacesuit. Their reentry proves equally challenging, as they land in an isolated portion of a snowy forest with only wolves to greet them. Rescuers arrive the next day, and the entire cosmonaut rescue group departs the improvised campsite on skis.

A Titan II rocket carries astronauts Virgil "Gus" I. Grissom and John W. Young into orbit on March 23 from Cape Canaveral inside a two-person Gemini Project spacecraft. NASA's *Gemini 3* flight is the first crewed mission for the new spacecraft and marks the beginning of more sophisticated space activities by American crews in preparation for the Apollo Project lunar missions.

On April 6 NASA places the *Intelsat-1 (Early Bird)* communications satellite into orbit from Cape Canaveral. It is the first commercial communications satellite placed in geosynchronous orbit.

The Soviet Union launches its first communications satellite on April 23. Designed to facilitate telecommunications across locations at high northern latitudes, the *Molniya 1A* spacecraft uses a special, highly elliptical, 12-hour orbit (about 500 km by 12,000 km). This type of orbit is now called a Molniya orbit.

On June 3 NASA launches the *Gemini 4* mission from Cape Canaveral with astronauts James A. McDivitt and Edward H. White, II as the crew. During this mission, astronaut White conducts the first American space walk and spends about 21 minutes on a tether outside the spacecraft.

In December NASA expands the scope of the Gemini Project activities. *Gemini 7* lifts off on December 4 carrying astronauts Frank Borman and James A. Lovell, Jr. into space on an almost 14-day mission. On December 15 they are joined in orbit by the *Gemini 6* spacecraft carrying astronauts Walter M. Schirra and Thomas P. Stafford. Once in orbit, the *Gemini 6* spacecraft comes within 2 meters of the *Gemini 7* "target spacecraft," thereby accomplishing the first successful orbital rendezvous operation.

1966

The Soviet Union sends the *Luna 9* spacecraft to the Moon on January 31. The 100-kg-mass spherical spacecraft soft lands in the Ocean of Storms region on February 3, rolls to a stop, opens four petal-like covers, and then transmits the first panoramic television images from the Moon's surface.

On March 16 NASA launches the *Gemini 8* mission from Cape Canaveral using a Titan II rocket. Astronauts Neil A. Armstrong and David R. Scott guide their spacecraft to a Gemini Agena target vehicle (GATV), accomplishing the first successful rendezvous and docking operation between a crewed chaser spacecraft and an uncrewed target vehicle. However, after an initial period of stable flight, the docked spacecraft begin to tumble erratically. Only quick corrective action by the astronauts prevents a major space disaster. They make an emergency reentry and are recovered in a contingency landing zone in the Pacific Ocean.

The Soviet Union launches the *Luna 10* to the Moon on March 31. This massive 1,500-kg spacecraft becomes the first human-made object to achieve orbit around the Moon.

On May 30 NASA sends the *Surveyor 1* lander spacecraft to the Moon. The versatile robot spacecraft successfully makes a soft landing on June 1 in the Ocean of Storms. It then transmits more than 10,000 images from the lunar surface and performs numerous soil mechanics experiments in preparation for the Apollo Project human landing missions.

In mid-August NASA sends the *Lunar Orbiter 1* spacecraft to the Moon from Cape Canaveral. It is the first of five successful missions to collect detailed images of the Moon from lunar orbit. At the end of each mapping mission the orbiter spacecraft is intentionally crashed into the Moon to prevent interference with future orbital activities.

On September 12 NASA launches the *Gemini 11* mission from Cape Canaveral. The Gemini Project spacecraft with astronauts Charles (Pete) Conrad, Jr. and Richard F. Gordon, Jr. quickly accomplishes rendezvous and docking with the Agena target vehicle. The astronauts then use the Agena's restartable rocket engine to propel themselves in a docked configuration to a record-setting altitude of 1,370 km, the highest ever flown by an Earth-orbiting human-crewed spacecraft.

1967

On January 27 disaster strikes NASA's Apollo Project. While inside their *Apollo 1* spacecraft during a training exercise on Launch Pad 34 at Cape Canaveral, astronauts Virgil "Gus" I. Grissom, Edward H. White II, and Roger B. Chaffee are killed when a flash fire sweeps through their spacecraft. The Moon landing program is delayed by 18 months while major design and safety changes are made in the Apollo Project spacecraft.

On April 23 tragedy strikes the Russian space program when the Soviets launch cosmonaut Vladimir Komarov in the new *Soyuz* ("union") spacecraft. Following an orbital mission plagued with difficulties, Komarov is killed on April 24 during reentry operations when the spacecraft's parachute fails to deploy properly and the vehicle impacts the ground at high speed.

On October 27 and 30 the Soviet Union launches two uncrewed Soyuz-type spacecraft, called *Cosmos 186* and *Cosmos 188*. On October 30 the orbiting spacecraft accomplish the first automatic rendezvous and docking operation. The craft then separate and are recovered on October 31 and November 2.

1968

NASA launches the *Apollo 7* mission into orbit around Earth. The astronauts Walter M. Schirra, Don F. Eisele, and R. Walter Cummingham perform a variety of orbital operations with the redesigned Apollo Project spacecraft.

On December 21 NASA's *Apollo 8* spacecraft (Command and Service Modules only) departs Launch Complex 39 at the Kennedy Space Center during the first flight of the mighty Saturn V launch vehicle with a human crew as part of the payload. Astronauts Frank Borman, James Arthur Lovell, Jr., and William A. Anders become the first people to leave Earth's gravitational influence. They go into orbit around the Moon and capture images of an incredibly beautiful Earth "rising" above the starkly barren lunar horizon, pictures that inspire millions and stimulate an emerging environmental movement. After 10 orbits around the Moon the first lunar astronauts return safely to Earth on December 27.

1969

In a full "dress-rehearsal" for the first Moon landing, NASA's *Apollo 10* mission departs the Kennedy Space Center on May 18. The astronauts Eugene A. Cernan, John W. Young, and Thomas P. Stafford successfully demonstrate the complete Apollo Project mission profile and evaluate the performance of the lunar excursion module (LEM) down to within 15 km of the lunar surface.

The entire world watches as NASA's *Apollo 11* mission leaves for the Moon on July 16 from the Kennedy Space Center. The astronauts Neil A. Armstrong, Michael Collins, and Edwin E. "Buzz" Aldrin, Jr. make a long-held dream of humanity a reality. On July 20 Neil Armstrong cautiously descends the steps of the lunar excursion module's (LEMs) ladder and steps on the lunar surface, exclaiming "One small step for a man, one giant leap for mankind!" He and Buzz Aldrin become the first two people to walk on another world. Many people regard the Apollo Project lunar landings as the greatest technical accomplishment in all human history.

On November 14 NASA's *Apollo 12* mission lifts off from the Kennedy Space Center. The astronauts Charles (Pete) Conrad, Jr., Richard F. Gordon, Jr., and Alan L. Bean continue the scientific objectives of the Apollo Project. Conrad and Bean become the third and fourth "Moon walkers," collecting samples from a larger area and deploying the Apollo Lunar Surface Experiment Package (ALSEP).

1970

NASA's *Apollo 13* mission leaves for the Moon on April 11. Suddenly, on April 13 a life-threatening explosion occurs in the Service Module portion of the Apollo spacecraft. Astronauts James A. Lovell, Jr., John Leonard Swigert, and Fred Wallace Haise, Jr. must use their lunar excursion module (LEM) as a lifeboat. While an anxious world waits and listens, the crew skillfully maneuvers their disabled spacecraft around the Moon. With critical supplies running low, they limp back to Earth on a free-return trajectory. At just the right moment on April 17 they abandon the LEM *Aquarius* and board the Apollo Project spacecraft (Command Module) for a successful atmospheric reentry and recovery in the Pacific Ocean.

The former Soviet Union launches its *Venera 7* mission to Venus on August 17. When the spacecraft arrives on December 15, a probe is parachuted into the dense Venusian atmosphere. Subsequent analysis of the data from the instrumented capsule confirms that it has landed on the surface. It records an ambient temperature of approximately 475°C and an atmospheric pressure that is 90 times the pressure found at sea level on Earth. This represents the first successful transmission of data from the surface of another planet.

On September 12 the Soviet Union sends the *Luna 16* robot spacecraft to the Moon. Russian engineers use teleoperation of the spacecraft's drill to collect about 100 grams of lunar dust, which is then placed in a sample return canister. On September 21 the sample return canister leaves the lunar surface on a trajectory back to Earth. Three days later it lands in Russia. *Luna 16* is the first robot spacecraft to return a sample of material from another world. In similar missions *Luna 20* (February 1972) and *Luna 24* (August 1976) return automatically collected lunar samples, while *Luna 18* (September 1971) and *Luna 23* (October 1974) land on the Moon but fail to return samples for various technical reasons.

On November 10 the Soviet Union launches another interesting robot mission to the Moon. *Luna 17* lands in the Sea of Rains. On command from Earth, a 750-kg robot rover called *Lunakhod 1* rolls down an extended ramp and begins exploring the lunar surface. Russian engineers use teleoperation to control this eight-wheeled rover as it travels for months across the lunar surface, transmitting more than 20,000 television images and performing more than 500 soil tests at various locations. The mission represents the first successful use of a mobile, remotely controlled (teleoperated) robot vehicle to explore another planetary body. The *Luna 23* mission in January 1973 successfully deploys another rover, *Lunakhod 2*. However, the technical significance of these machine missions is all but ignored in the global glare of NASA's Apollo Project and its incredible human triumphs.

1971

NASA sends the *Apollo 14* mission to the Moon on January 31. While astronaut Stuart Allen Roosa orbits overhead in the Apollo spacecraft, astronauts Alan B. Shepard, Jr. and Edgar Dean Mitchell descend to the lunar surface. After departing the lunar excursion module (LEM) they become the fifth and sixth Moon walkers.

On April 19 the Soviet Union launches the first space station, called *Salyut 1*. It remains initially uncrewed because the three-cosmonaut crew of the *Soyuz 10* mission launched on April 22 attempts to dock with the station but cannot go on board.

At the end of May NASA launches the *Mariner 9* spacecraft to Mars with an Atlas-Centaur rocket vehicle. The spacecraft successfully enters orbit around the planet on November 13, 1971, and provides numerous images of the Martian surface.

The second major Russian space tragedy occurs in late June. The fatal accident takes place as the crew separates from *Salyut 1* to return to Earth in their *Soyuz 11* spacecraft after spending 22 days onboard the space station. While not wearing pressure suits for reentry, the cosmonauts Victor Patseyev, Vladislav Volkov, and Georgi Dobrovolsky suffocate when a pressure valve malfunctions and the air rushes out of their *Soyuz 11* spacecraft. On June 30 a startled cosmonaut recovery team on Earth finds all three men dead inside their spacecraft.

On July 26 NASA launches the *Apollo 15* mission to the Moon. The astronaut James Benson Irwin remains in the Apollo spacecraft orbiting the Moon while astronauts David R. Scott and Alfred Merrill Worden become the seventh and eighth Moon walkers. They also are the first persons to drive a motor vehicle (electric powered) on another world using their lunar rover to scoot across the surface.

1972

In early January President Richard M. Nixon approves NASA's space shuttle program. This decision shapes the major portion of NASA's program for the next three decades.

On March 2 an Atlas-Centaur launch vehicle successfully sends NASA's *Pioneer 10* spacecraft from Cape Canaveral on its historic mission. This far-traveling robot spacecraft becomes the first to transit the main belt asteroids, the first to encounter Jupiter (December 3, 1973) and by crossing the orbit of Neptune on June 13, 1983, which at the time was the farthest planet from the Sun, the first human-made object ever to leave the planetary boundaries of the solar system. On an interstellar trajectory *Pioneer 10* (and its twin, *Pioneer 11*) carries a special plaque greeting any intelligent alien civilization that might find it drifting through interstellar space millions of years from now.

On April 16 NASA launches the *Apollo 16* mission, the fifth human landing mission to the Moon. While the astronaut Thomas K. Mattingly II orbits the Moon in the Apollo spacecraft, the astronauts John W. Young and Charles Moss Duke, Jr. become the ninth and tenth Moon walkers. They also use the battery-powered lunar rover to travel across the Moon's surface in the Descartes lunar highlands.

NASA places a new type of satellite, called the *E*arth *R*esources *T*echnology *S*atellite-1 (ERTS-1), into a sun-synchronous orbit from Vandenberg Air Force Base on July 23. Later renamed *Landsat-1,* it is the first civilian spacecraft to provide relatively high-resolution multispectral images of Earth's surface, creating a revolution in the way we look at our home planet. Over next three decades a technically evolving family of Landsat spacecraft helps scientists study the Earth as an interconnected system, and this gives rise to the important new field of Earth system science.

On December 7 NASA's *Apollo 17* mission, the last expedition to the Moon in the 20th century, departs from the Kennedy Space Center propelled by a Saturn V rocket. While the astronaut Ronald E. Evans remains in lunar orbit, fellow astronauts Eugene A. Cernan and Harrison H. Schmitt become the 11th and 12th members of the exclusive Moon walkers club. Using a lunar rover they explore the Taurus-Littrow region. Their safe return to Earth on December 19 brings to a close one of the epic periods of human exploration.

1973

In early April, propelled by an Atlas-Centaur rocket, NASA's *Pioneer 11* spacecraft departs on an interplanetary journey from Cape Canaveral. The spacecraft encounters

Jupiter on December 2, 1974, and then uses a gravity assist maneuver to establish a flyby trajectory to Saturn. It is the first spacecraft to view Saturn at close range (closest encounter on September 1, 1979) and then follows a path into interstellar space.

On May 14 NASA launches *Skylab,* the first American space station. A Saturn V rocket is used to place the entire large facility into orbit in a single launch. The first crew of three American astronauts arrives on May 25 and makes the emergency repairs necessary to save the station, which suffered damage during the launch ascent. The astronauts Charles (Pete) Conrad, Jr., Paul J. Weitz, and Joseph P. Kerwin stay onboard for 28 days. They are replaced by the astronauts Alan L. Bean, Jack R. Lousma, and Owen K. Garriott, who arrive on July 28 and live in space for about 59 days. The final *Skylab* crew, astronauts Gerald P. Carr, William R. Pogue, and Edward G. Gibson, arrive on November 11 and reside in the station until February 8, 1974, setting a space endurance record for the time of 84 days. NASA then abandons *Skylab.*

In early November NASA launches the *Mariner 10* spacecraft from Cape Canaveral. It encounters Venus on February 5, 1974, and uses a gravity assist maneuver to become the first spacecraft to investigate Mercury at close range.

1974

On May 30 NASA launches the Applications Technology Satellite-6 (ATS-6) to demonstrate the use of a large antenna structure on a geostationary orbit communications satellite to transmit good-quality television signals to small, inexpensive ground receivers.

1975

In June the Soviet Union sends twin *Venera 9* and *10* spacecraft to Venus. *Venera 9,* launched on June 8, goes into orbit around Venus on October 22. It releases a lander that reaches the surface and transmits the first television images of the planet's infernolike landscape. *Venera 10,* launched June 14, follows a similar mission profile.

In July the United States and the Soviet Union conduct the first cooperative international rendezvous and docking mission, called the Apollo-Soyuz Test Project (ASTP). On July 15 the Russians launch the *Soyuz 19* spacecraft with the cosmonauts Alexei A. Leonov and Valerie N. Kubasov on board. Several hours later NASA launches the *Apollo 18* spacecraft with the astronauts Thomas P. Stafford, Vance D. Brand, and Deke Slayton, Jr. on board.

In late August and early September NASA launches the twin *Viking 1* (August 20) and *Viking 2* (September 9) orbiter/lander combination spacecraft to the Red Planet from Cape Canaveral. Arriving at Mars in 1976, all Viking Project spacecraft (two landers and two orbiters) perform exceptionally well, but the detailed search for microscopic alien lifeforms on Mars remains inconclusive.

On October 16 NASA launches the first Geostationary Operational Environmental Satellite (GOES-1) for the U.S. National Oceanic and Atmospheric Administration (NOAA). This spacecraft is the first in a long series of operational meteorological satellites that continuously monitor weather conditions on a hemispheric scale, providing warnings about hurricanes and other severe weather patterns.

1976

NASA launches the first Laser Geodynamics Satellite (LAGEOS) into a precise orbit around Earth from Vandenberg Air Force Base. The heavy but small (60-cm-diameter), golf ball–shaped spacecraft has a surface that is completely covered with mirrorlike retroreflectors. The joint NASA–Italian Space Agency project demonstrates the use of ground-to-satellite laser ranging systems in the study of solid Earth dynamics, an important part of Earth system science.

1977

On August 20 NASA sends the *Voyager 2* spacecraft from Cape Canaveral on an epic grand tour mission during which it encounters all four giant planets and then departs the solar system on an interstellar trajectory. Using a gravity assist maneuver, *Voyager 2* visits Jupiter on July 9, 1979, Saturn on August 25, 1981, Uranus on January 24, 1986, and Neptune on August 25, 1989. The resilient far-traveling robot spacecraft and its twin, *Voyager 1,* also carry a special interstellar message from Earth, a digital record entitled *The Sounds of Earth.*

On September 5 NASA sends the *Voyager 1* spacecraft from Cape Canaveral on its fast-trajectory journey to Jupiter (March 5, 1979), Saturn (March 12, 1980), and beyond the solar system.

In late September the Soviet Union launches the *Salyut 6* space station, a second-generation design with several important improvements, including an additional docking port and the use of automated Progress resupply spacecraft.

NASA launches *Meteosat-1* from Cape Canaveral on November 22 for the European Space Agency (ESA). Upon reaching geostationary orbit, it becomes Europe's first weather satellite.

1978

In May the British Interplanetary Society releases its Project Daedalus report, a conceptual study about a one-way robot spacecraft mission to Barnard's star at the end of the 21st century.

NASA successfully launches the Pioneer–Venus Orbiter spacecraft (*Pioneer 12*) from Cape Canaveral on May 20. After arriving on December 4, it becomes the first American spacecraft to orbit Venus. It uses its radar mapping system from 1978 to 1992 to image extensively the hidden surface of the cloud-enshrouded planet.

The American astronomer James Christy discovers Pluto's large moon, Charon, on June 22.

On August 8 NASA launches the Pioneer-Venus Multiprobe (*Pioneer 13*), which encounters the planet on December 9 and releases four probes into the Venusian atmosphere.

During a visit to NASA's Kennedy Space Center on October 1, President Jimmy Carter publicly mentions that American reconnaissance satellites have made immense contributions to international security.

1979

When unable to maintain its orbit, the abandoned NASA *Skylab* becomes a dangerous orbiting derelict that eventually decays in a dramatic fiery plunge through Earth's atmosphere on July 11, a reentry that leaves debris fragments scattered over remote regions of western Australia.

On December 24 the European Space Agency (ESA) successfully launches the first Ariane 1 rocket from the Guiana Space Center in Kourou, French Guiana.

1980

India's Space Research Organization (ISRO) successfully places a modest 35-kg test satellite called *Rohini* into low Earth orbit on July 1. The launch vehicle is a four-stage, solid-propellant rocket manufactured in India. The SLV-3 (Standard Launch Vehicle-3) gives India independent national access to outer space.

1981

On April 12 NASA launches the space shuttle *Columbia* on its maiden orbital flight from Complex 39-A at the Kennedy Space Center. The astronauts John W. Young and Robert L. Crippen thoroughly test the new aerospace vehicle. Upon reentry it becomes the first spacecraft to return to Earth by gliding through the atmosphere and landing like an airplane. Unlike all previous one-time use space vehicles, *Columbia* is prepared for another mission in outer space.

In autumn the Soviet Union sends the *Venera 13* and *14* spacecraft to Venus. *Venera 13* departs on October 30, and its lander touches down on the Venusian surface on March 1, 1982. Similarly, *Venera 14* lifts off from Earth on November 4, 1981, and its capsule lands on Venus on March 5, 1982. Both hardy robot landers successfully return color images of the infernolike surface of Venus and perform the first soil sampling experiments on that planet.

1982

On November 11 NASA launches the space shuttle *Columbia* with a crew of four astronauts on the first operational flight (called STS-5) of the U.S. Space Transportation System.

1983

An expendable Delta rocket places the *Infrared Astronomy Satellite* (IRAS) into a polar orbit from Vandenberg Air Force Base on January 25. The international scientific spacecraft (United States–United Kingdom–Netherlands) completes the first comprehensive all-sky infrared radiation (IR) survey of the universe.

The first flight of the *Challenger* occurs on April 4, when NASA launches the STS-6 space shuttle mission. During the mission, the astronauts Donald Peterson and Story Musgrave put on their spacesuits and perform the first extravehicular activity (EVA) from an orbiting shuttle.

On June 18 NASA launches the space shuttle *Challenger* (STS-7 mission) with the astronaut Sally K. Ride, the first American woman to travel in outer space.

In late August the space shuttle *Challenger* (STS-8 mission) flies into space with an astronaut crew of five, including Guion S. Bluford, Jr., first African American to orbit Earth.

1984

In his State of the Union address on January 25, President Ronald Reagan calls for a permanent American space station. However, his vision must wait until December 1998, when a combined astronaut and cosmonaut crew assembles the first two components of the *International Space Station* as part of the STS-88 space shuttle mission.

The Soviet Union launches the *Soyuz T-12* spacecraft on July 17. The spacecraft carries three cosmonauts, including Svetlana Savistskaya. While the *Soyuz T-12* docks with the *Salyut-7* space station, cosmonaut Savistskaya performs a series of experiments in space during an extravehicular activity (EVA) and becomes the first female space walker.

During the STS 41-G space shuttle mission launched on October 5, the astronaut Kathyrn D. Sullivan becomes the first American woman to perform an extravehicular activity (EVA).

1985

On April 12 NASA's space shuttle *Discovery* carries U.S. Senator "Jake" Garn into orbit as a member of the astronaut crew of the STS 51-D mission.

1986

On January 24 NASA's *Voyager 2* spacecraft encounters Uranus.

On January 28 the space shuttle *Challenger* lifts off from the NASA Kennedy Space Center on its final voyage. At just under 74 seconds into the STS 51-L mission, a deadly explosion occurs, killing the crew and destroying the vehicle. Led by President Ronald Reagan, the United States mourns seven astronauts lost in the *Challenger* accident.

In late February the Soviet Union launches the first segment of a third-generation space station, the modular orbiting complex called *Mir* (peace).

In March an international armada of spacecraft encounters Comet Halley. The European Space Agency's *Giotto* spacecraft makes and survives the most hazardous flyby, streaking within 610 km of the comet's nucleus on March 14 at a relative velocity of 68 km/s.

1987

In late February Supernova 1987A is observed in photographic images of the Large Magellanic Cloud by Canadian astronomer Ian Shelton. It is the first supernova visible to the naked eye since the one discovered by Johannes Kepler in 1604.

1988

On September 19 Israel uses a Shavit ("comet") three-stage rocket to place the country's first satellite, called *Ofeq 1,* into an unusual east-to-west orbit, one that is opposite to the direction of Earth's rotation but necessary because of launch safety restrictions.

As the *Discovery* successfully lifts off on September 29 for the STS-26 mission, NASA returns the space shuttle to service following a 32-month hiatus after the *Challenger* accident.

1989

During the STS-30 mission in early May, the space shuttle *Atlantis* deploys NASA's *Magellan* spacecraft and sends it on a trajectory to Venus.

On August 25 the *Voyager 2* spacecraft encounters Neptune.

During the STS-34 mission in mid-October, the space shuttle *Atlantis* deploys NASA's *Galileo* spacecraft and sends it on its long interplanetary journey to Jupiter.

In mid-November NASA launches the *Cosmic Background Explorer* (COBE) into a polar orbit from Vandenberg Air Force Base. The spacecraft carefully measures the cosmic microwave background and helps scientists answer questions about the big bang explosion.

1990

NASA officially begins the Voyager Interstellar Mission (VIM) on January 1. This is an extended mission in which both Voyager spacecraft search for the heliopause.

During the STS-31 mission in late April, the astronaut crew of the space shuttle *Discovery* deploys NASA's *Hubble Space Telescope* (HST) into orbit around Earth. Subsequent shuttle missions in 1993, 1997, 1999, and 2002 will repair design flaws and perform maintenance on this orbiting optical observatory.

1991

In early April NASA uses the space shuttle *Atlantis* to deploy the *Compton Gamma-Ray Observatory* (CGO), a major Earth-orbiting observatory that explores the universe in the gamma-ray portion of the electromagnetic spectrum.

While on its way to Jupiter, NASA's *Galileo* spacecraft passes within 1,600 km of the asteroid 951 Gaspra. The encounter provides the first close-up images of a main belt asteroid. Gaspra is a type-S (silicaceous) asteroid about 19 by 12 by 11 km in size.

1992

On February 11 the National Space Development Agency of Japan (NASDA) successfully launches that country's first Earth-observing spacecraft from the Tanegashima Space Center using a Japanese-manufactured H-1 rocket.

On September 25 NASA successfully launches the *Mars Observer* (MO) spacecraft from Cape Canaveral. For unknown reasons all contact with the spacecraft is lost in late August 1993, just a day or so before it was to go into orbit around Mars.

1993

While coasting to Jupiter NASA's *Galileo* spacecraft encounters Ida, a main belt asteroid, on August 28 at a distance of 2,400 km. *Galileo*'s imagery reveals that Ida has a tiny satellite of its own, named Dactyl.

In early December during the STS-61 mission, the astronaut crew of the space shuttle *Endeavour* perform a complicated on-orbit repair of NASA's *Hubble Space Telescope,* thereby restoring the orbiting observatory to its planned scientific capabilities.

1994

In late January a joint Department of Defense and NASA advanced technology demonstration spacecraft called *Clementine* lifts off for the Moon from Vandenberg Air Force Base. Some of the spacecraft's data suggest that the Moon may actually possess significant quantities of water ice in its permanently shadowed polar regions.

On February 3 the space shuttle *Discovery* lifts off from the NASA Kennedy Space Center. The six-person crew of the STS-60 mission includes cosmonaut Sergei Krikalev, the first Russian to travel into outer space using an American launch vehicle.

In March the U.S. Air Force successfully launches the final satellite in the initial constellation of the Global Positioning System (GPS), and the navigation satellite system becomes fully operational. GPS revolutionizes navigation on land, at sea, and in the air for numerous military and civilian users.

1995

In February during NASA's STS-63 mission, the space shuttle *Discovery* approaches (encounters) the Russian *Mir* space station as a prelude to the development of the *International Space Station* (ISS). The astronaut Eileen Marie Collins serves as the first female shuttle pilot.

On March 14 the Russians launch the *Soyuz TM-21* spacecraft to the *Mir* space station from the Baikonur Cosmodrome. The crew of three includes the American astronaut Norman Thagard, the first American to travel into outer space on a Russian rocket and the first to stay on the *Mir* space station. The *Soyuz TM-21* cosmonauts also relieve the previous *Mir* crew, including the cosmonaut Valeriy Polyakov, who returns to Earth on March 22 after setting a world record for remaining in space for 438 days.

In late June NASA's space shuttle *Atlantis* docks with the Russian *Mir* space station for the first time. During this shuttle mission (STS-71) *Atlantis* delivers the *Mir 19* crew (the cosmonauts Anatoly Solovyev and Nikolai Budarin) to the Russian space station and then returns the *Mir 18* crew back to Earth, including the American astronaut Norman Thagard, who has spent 115 days in space onboard the *Mir*. The shuttle-*Mir* docking program is the first phase of the *International Space Station* (ISS). A total of nine shuttle-*Mir* docking missions will occur between 1995 and 1998.

In early December NASA's *Galileo* spacecraft arrives at Jupiter and starts its multiyear scientific mission by successfully deploying a probe into the Jovian atmosphere.

1996

In late March the space shuttle *Atlantis* delivers the astronaut Shannon Lucid to *Mir*, who becomes a cosmonaut engineer and researcher and the first American woman to live on the Russian *Mir* space station.

In the summer a NASA research team from the Johnson Space Center announces that they have found evidence in a Martian meteorite called ALH84001 that "strongly suggests primitive life may have existed on Mars more than 3.6 billion years ago." The Martian microfossil hypothesis touches off a great deal of technical debate.

In mid-September the space shuttle *Atlantis* docks with the Russian *Mir* space station and returns astronaut Shannon Lucid to Earth after she spends 188 days onboard the *Mir*, setting a new U.S. and world space flight record for a woman.

NASA launches the *Mars Global Surveyor* (MGS) on November 7 and then the *Mars Pathfinder* on December 4 from Cape Canaveral.

1997

In February the astronaut crew of the space shuttle *Discovery* (STS-82 mission) successfully accomplishes the second *Hubble Space Telescope* (HST) servicing mission.

1998

In early January NASA sends the *Lunar Prospector* to the Moon from Cape Canaveral. Data from this orbiter spacecraft reinforce previous hints that the Moon's polar regions may contain large reserves of water ice in a mixture of frozen dust lying at the frigid bottoms of some permanently shadowed craters.

On October 29 the space shuttle *Discovery* (STS-95 mission) lifts off, and its crew includes the astronaut U.S. Senator John Herschel Glenn, Jr., who returns to outer space after 36 years and becomes the oldest person, at age 77, to experience space flight in the 20th century.

In early December the space shuttle *Endeavour* ascends from the NASA Kennedy Space Center on the first assembly mission of the *International Space Station* (ISS). During the STS-88 shuttle mission, *Endeavour* performs a rendezvous with the previously launched Russian-built Zarya (sunrise) module. An international crew connects this module with the American-built Unity module carried in the shuttle's cargo bay.

1999

In July the astronaut Eileen Marie Collins serves as the first female space shuttle commander (STS-93 mission) as the *Columbia* carries NASA's *Chandra X-Ray Observatory* (CXO) into orbit.

After a successful launch from Cape Canaveral (on December 11, 1998) and an uneventful interplanetary trip, all contact with the *Mars Climate Orbiter* (MCO) is lost as it approaches the Red Planet on September 23. A trajectory calculation error most likely caused the orbiter spacecraft to approach the Martian atmosphere too steeply and burn up.

On December 3 NASA loses all contact with the *Mars Polar Lander* mission just prior to arrival at Mars. Successfully launched from Cape Canaveral on January 3, it is the second NASA mission to fail upon arrival at the Red Planet in 1999.

2000

Aware of the growing space debris problem, NASA spacecraft controllers intentionally deorbit the massive *Compton Gamma-Ray Observatory* (CGRO) on June 4 at the end of its useful mission. Their action assures that any pieces surviving atmospheric reentry fall harmlessly into a remote part of the Pacific Ocean.

2001

NASA launches the *Mars Odyssey 2001* mission to the Red Planet in early April. The spacecraft successfully orbits the planet in October.

On March 23 Russian officials safely deorbit the decommissioned *Mir* space station, which has become a large space derelict. Their action assures that any components surviving atmospheric reentry fall harmlessly into a remote part of the Pacific Ocean.

2002

On March 1 the space shuttle *Columbia* with a crew of seven astronauts takes off from the Kennedy Space Center and successfully accomplishes Servicing Mission 3B to the *Hubble Space Telescope.*

On May 4 NASA successfully launches its *Aqua* satellite from Vandenberg Air Force Base. This sophisticated Earth-observing spacecraft joins the *Terra* spacecraft in performing Earth system science studies.

On October 1 the U.S. Department of Defense forms the U.S. Strategic Command as the control center for all American strategic (nuclear) forces. USSTRATCOM also conducts military space operations, strategic warning and intelligence assessment, and global strategic planning.

2003

On February 1, while gliding back to Earth after a successful 16-day scientific research mission (STS-107), the space shuttle *Columbia* experiences a catastrophic reentry accident at an altitude of about 63 km over the western United States. Traveling at 18 times the speed of sound, the orbiter vehicle disintegrates, taking the lives of all seven crew members: six American astronauts (Rick Husband, William McCool, Michael Anderson, Kalpana Chawla, Laurel Clark, and David Brown) and the first Israeli astronaut (Ilan Ramon).

On June 2 the European Space Agency launches the *Mars Express Orbiter* along with its companion lander, *Beagle 2,* from Kourou, French Guiana. The *Beagle 2* was released from the *Mars Express Orbiter* on December 19, 2003. The *Mars Express* arrived successfully on December 25, 2003. The *Beagle 2* was also scheduled to land on December 25, 2003. However, ground controllers were unable to communicate with the probe.

NASA's Mars Exploration Rover *Spirit* is launched by a Delta II rocket to the Red Planet on June 10. *Spirit,* also known as MER-A, arrives safely on Mars on January 3, 2004, and begins its teleoperated surface exploration mission under the supervision of mission controllers at the NASA Jet Propulsion Laboratory.

NASA launches the second Mars Exploration Rover (MER), called *Opportunity,* using a Delta II rocket launch, which lifts off from Cape Canaveral Air Force Station on July 7, 2003. *Opportunity,* also called MER-B, successfully lands on Mars on January 24, 2004, and starts its teleoperated surface exploration mission under the supervision of mission controllers at the NASA Jet Propulsion Laboratory.

2004

On July 1 NASA's *Cassini* spacecraft arrives at Saturn and begins its four-year mission of detailed scientific investigation.

2005

On January 14 the *Huygens* probe descends through the atmosphere of Titan and lands on the moon's surface where the ambient temperature is measured as –180°C.

On April 29 the U.S. Air Force launches the last Titan IVB rocket from Complex 40 at Cape Canaveral Air Force Station.

NASA successfully launched the space shuttle *Discovery* on the STS-114 mission on July 26 from the Kennedy Space Center in Florida. After docking with the *International Space Station,* the *Discovery* returned to Earth and landed at Edwards AFB, California, on August 9.

On August 12, NASA launched the *Mars Reconnaissance Orbiter* from Cape Canaveral AFS, Florida.

The Expedition 12 crew (Commander William McArthur and Flight Engineer Valery Tokarev) arrived at the *International Space Station* on October 3 and replaced the Expedition 11 crew.

The People's Republic of China successfully launched its second human space flight mission, called *Shenzhou 6,* on October 12. Two taikonauts, Fei Junlong and Nie Haisheng, traveled in space for almost five days and made 76 orbits of Earth before returning safely to Earth, making a soft, parachute-assisted landing in northern Inner Mongolia.

2006

NASA launched the *New Horizons* spacecraft from Cape Canaveral on January 19 and successfully sent this robot probe on its long one-way mission to conduct a scientific encounter with the Pluto system (in 2015) and then to explore portions of the Kuiper belt that lies beyond.

Follow-up observations by NASA's *Hubble Space Telescope,* reported on February 22, have confirmed the presence of two new moons around the distant planet Pluto. The HST first discovered the two new moons, now called Hydra (S/2005 P 1) and Nix (S/2005 P 2), in May 2005.

On March 10, NASA's *Mars Reconnaissance Orbiter* successfully arrived at Mars.

The Expedition 13 crew (Commander Pavel Vinogradov and Flight Engineer Jeff Williams) arrived at the *International Space Station* on April 1 and replaced the Expedition 12 crew. Joining them for several days before returning back to Earth with the Expedition 12 crew was Brazil's first astronaut Marcos Pontes.

On August 24, the members of the International Astronomical Union (IAU) decided (by vote) to demote Pluto from its traditional status as one of the nine major planets and place the object into a new class, called a *dwarf planet.* The IAU decision now leaves the solar system with eight major planets and three dwarf planets: Pluto (which serves as the prototype dwarf planet), Ceres (the largest asteroid), and the large, distance Kuiper belt object identified as 2003 UB313 (nicknamed Xena). Astronomers anticipate the discovery of other dwarf planets in the distant parts of the solar system.

APPENDIX IV

Planetary Data

	Mercury	Venus	Earth	Moon	Mars	Jupiter	Saturn	Uranus	Neptune	Pluto
Mass (10^{24}kg)	0.330	4.87	5.97	0.073	0.642	1,899	568	86.8	102	0.0125
Diameter (km)	4,879	12,104	12,756	3,475	6,794	142,984	120,536	51,118	49,528	2,390
Density (kg/m^3)	5,427	5,243	5,515	3,340	3,933	1,326	687	1,270	1,638	1,750
Gravity (m/s^2)	3.7	8.9	9.8	1.6	3.7	23.1	9.0	8.7	11.0	0.6
Escape Velocity (km/s)	4.3	10.4	11.2	2.4	5.0	59.5	35.5	21.3	23.5	1.1
Rotation Period (hours)	1,407.6	–5,832.5	23.9	655.7	24.6	9.9	10.7	–17.2	16.1	–153.3
Length of Day (hours)	4,222.6	2,802.0	24.0	708.7	24.7	9.9	10.7	17.2	16.1	153.3
Distance from Sun (10^6 km)	57.9	108.2	149.6	0.384	227.9	778.6	1,433.5	2,872.5	4,495.1	5,870.0
Perihelion (10^6 km)	46.0	107.5	147.1	0.363	206.6	740.5	1,352.6	2,741.3	4,444.5	4,435.0
Aphelion (10^6 km)	69.8	108.9	152.1	0.406	249.2	816.6	1,514.5	3,003.6	4,545.7	7,304.3
Orbital Period (days)	88.0	224.7	365.2	27.3	687.0	4,331	10,747	30,589	59,800	90,588
Orbital Velocity (km/s)	47.9	35.0	29.8	1.0	24.1	13.1	9.7	6.8	5.4	4.7
Orbital Inclination (degrees)	7.0	3.4	0.0	5.1	1.9	1.3	2.5	0.8	1.8	17.2
Orbital Eccentricity	0.205	0.007	0.017	0.055	0.094	0.049	0.057	0.046	0.011	0.244
Axial Tilt (degrees)	0.01	177.4	23.5	6.7	25.2	3.1	26.7	97.8	28.3	122.5
Mean Temperature (°C)	167	464	15	–20	–65	–110	–140	–195	–200	–225
Surface Pressure (bars)	0	92	1	0	0.01	Unknown	Unknown	Unknown	Unknown	0
Number of Moons	0	0	1	0	2	63	33	26	13	1
Ring System?	No	No	No	No	No	Yes	Yes	Yes	Yes	No
Global Magnetic Field?	Yes	No	Yes	No	No	Yes	Yes	Yes	Yes	Unknown

Explanatory Notes:

Aphelion (10^6 km)	The point in a body's orbit furthest from the Sun, in 10^6 kilometers.
Mean orbital velocity (km/s)	The average speed of the body in orbit, in kilometers/second.
Orbital inclination (deg)	The inclination of the orbit to the ecliptic, in degrees. For satellites, this is with respect to the planet's equator.
Orbital eccentricity	A measure of the circularity of the orbit, equal to (aphelion – perihelion distance) / (2 × semi-major axis). For a circular orbit, eccentricity = 0. Dimensionless.
Sidereal rotation period (hrs)	The time for one rotation of the body on its axis relative to the fixed stars, in hours. A minus sign indicates retrograde rotation.
Length of day (hrs)	The average time in hours for the Sun to move from the noon position in the sky at a point on the equator back to the same position; on Earth this defines a 24-hour day.
Obliquity to orbit (deg)	The tilt of the body's equator relative to the body's orbital plane, in degrees.

Atmospheres

Surface Pressure: Atmospheric pressure at the surface, in bars, millibars (mb = 10^{-3} bar), or picobars (10^{-12} bar).
Surface Density: Atmospheric density at the surface in kilograms/meters^3.
Average Temperature: Mean temperature of the body over the entire surface in Kelvin.

Related Definitions

Astronomical Unit (AU) — The mean distance from the Sun to the Earth = 149,597,900 km.
Bar — A measure of pressure or stress. 1 bar = 10^5 Pascal (Pa) = 10^5kg m^{-1} s^{-2}.
Ecliptic — An imaginary plane defined by the Earth's orbit.
Equinox — The point in a body's orbit when the sub-solar point is exactly on the equator.
Gravitational Constant — Relates gravitational force to mass; = 6.6726×10^{-11} meters^3 kilograms^{-1} seconds^{-2}.
Opposition — An orbital configuration in which two bodies are on exact opposite sides of the Sun or are on the same side of the Sun forming a line with the Sun (neglecting inclination).
Phase Angle — The angle between the Earth and Sun as seen from the body.

Source: NASA/NSSDC Planetary Fact Sheet, December 2004.

APPENDIX V

HOW PLANETS, MOONS, ASTEROIDS, COMETS, AND INTERESTING CELESTIAL OBJECTS AND PLANETARY SURFACE FEATURES ARE NAMED

Ever wonder how a mountain on Mars got its name? This appendix contains a variety of interesting information involving solar system astronomy as extracted from the Astrogeology Research Program of the U.S. Geological Survey (USGS). Created by an act of Congress in 1879, the USGS has evolved over the ensuing years, matching its talents and knowledge to the progress of science and technology, and now stands as the sole science agency for the Department of the Interior.

Today, when images are first obtained of the surface of a planet or satellite, a theme for naming features is chosen, and a few important features are named, usually by members of the appropriate International Astronomical Union (IAU) task group. Later, as higher-resolution images and maps become available, additional features are named at the request of investigators mapping or describing specific surfaces, features, or geological formations. Anyone may suggest that a task group consider a specific name. If the members of the IAU task group agree that the name is appropriate, it can be retained for use on a provisional basis. Sometimes there is a request from a member of the scientific community that a specific planetary feature be named. Following acceptance during a review by various committees within the IAU, the names successfully reviewed by an IAU task group then earn provisional approval and may be used on maps and in publications as long as the provisional status is clearly stated. Provisional names are then presented for adoption to the IAU's general assembly, which meets triennially. A name is not considered to be official—that is "adopted"—until the general assembly has given its approval. Approved names are listed in the transactions of the IAU.

The IAU was founded in Brussels, Belgium, in 1919, so its formal "naming" procedures and protocols have really influenced astronomy starting only in the early 20th century. Before then, a rather informal tradition within the astronomical community made naming a new planet, moon, or asteroid the privilege and task of the discoverer. However, since there was no "formal" international process, not all initially selected names gained acceptance. Occasionally, peer pressure from the scientific community forced the discoverer to have a "change of heart."

The naming of the planet Uranus is an interesting example. Unknown to ancient astronomers, Uranus was discovered by Sir William Herschel in 1781. Initially he called his discovery Georgium Sidus (George's star), after England's King George III. But astronomers throughout Europe protested his "politically motivated" choice. Then some of his fellow astronomers began calling the planet Herschel in his honor, but that choice proved equally unacceptable. Finally, the German astronomer Johann Bode stepped into the fracas and called Herschel's new planet Uranus, after the ancient Greek god of the sky and father of the Titan Cronos (Saturn in Roman mythology). Astronomers around the world were much more comfortable with this more "traditional" choice.

The table involves planetary nomenclature, planet and satellite names, and (where known) the people credited with the discovery. The source of these materials is the *Gazetteer of Planetary Nomenclature* from the USGS Astrogeology Research Program.

Planet and Satellite Names and Discoverers

Mercury

Body	Description	Date of Discovery	Discovery Location	Discoverer
Mercury	Named Mercurius by the Romans because it appears to move so swiftly.			

Venus

Body	Description	Date of Discovery	Discovery Location	Discoverer
Venus	Roman name for the goddess of love. This planet was considered to be the brightest and most beautiful planet or star in the heavens. Other civilizations have named it for their god or goddess of love or war.			

Earth System

Body	Description	Date of Discovery	Discovery Location	Discoverer
Earth	The name Earth comes from the Indo-European base "er," which produced the Germanic noun "ertho," and ultimately German "erde," Dutch "aarde," Scandinavian "jord," and English "earth." Related forms include Greek "eraze," meaning "on the ground," and Welsh "erw," meaning "a piece of land."			
Moon	Every civilization has had a name for the satellite of Earth that is known, in English, as the Moon. The Moon is known as Luna in Italian, Latin, and Spanish, as Lune in French, as Mond in German, and as Selene in Greek.			

Martian System

Body	Description	Date of Discovery	Discovery Location	Discoverer
Mars	Named by the Romans for their god of war because of its red, bloodlike color. Other civilizations also named this planet from this attribute; for example, the Egyptians named it "Her Desher," meaning "the red one."			
Phobos	Inner satellite of Mars. Named for one of the horses that drew Mars' chariot; also called an "attendant" or "son" of Mars, according to chapter 15, line 119 of Homer's "Iliad." This Greek word means "flight."	August 17, 1877	Washington	A. Hall
Deimos	This outer Martian satellite was named for one of the horses that drew Mars' chariot; also called an "attendant" or "son" of Mars, according to chapter 15, line 119 of Homer's "Iliad." Deimos means "fear" in Greek.	August 11, 1877	Washington	A. Hall

Asteroids and Their Satellites Visited by Spacecraft

Body	Description	Date of Discovery	Discovery Location	Discoverer
Dactyl (1993 (243)1)	Named for a group of mythological beings who lived on Mount Ida, where the infant Zeus was hidden and raised (according to some accounts) by the nymph Ida.	August 28, 1993		*Galileo* imaging and infrared science teams
Eros	Named for the Greek god of love.	August 13, 1898	Berlin	C. G. Witt
Gaspra	Named for a resort on the Crimean Peninsula.	July 30, 1916	Simeis	G. Neujmin
Ida	Named for a nymph who raised the infant Zeus. Ida is also the name of a mountain on the island of Crete, the location of the cave where Zeus was reared.	September 29, 1884	Vienna	J. Palisa

Body	Description	Date of Discovery	Discovery Location	Discoverer
Mathilde	The name was suggested by a staff member of the Paris Observatory who first computed an orbit for Mathilde. The name is thought to honor the wife of the vice director of the Paris Observatory at that time.	November 12, 1885	Vienna	J. Palisa

Jovian System

Satellites in the Jovian system are named for Zeus/Jupiter's lovers and descendants. Names of outer satellites with a prograde orbit generally end with the letter "a" (although an "o" ending has been reserved for some unusual cases), and names of satellites with a retrograde orbit end with an "e."

Body	Description	Date of Discovery	Discovery Location	Discoverer
Jupiter	The largest and most massive of the planets was named Zeus by the Greeks and Jupiter by the Romans; he was the most important deity in both pantheons.			
Io (JI)	Io, the daughter of Inachus, was changed by Jupiter into a cow to protect her from Hera's jealous wrath. But Hera recognized Io and sent a gadfly to torment her. Io, maddened by the fly, wandered throughout the Mediterranean region.	January 8, 1610	Padua	Galileo (Simon Marius probably made an inde pendent discovery of the Galilean satellites at about the same time that Galileo did, and he may have unwittingly sighted them up to a month earlier, but the priority must go to Galileo because he published his discovery first.)
Europa (JII)	Beautiful daughter of Agenor, king of Tyre, she was seduced by Jupiter, who had assumed the shape of a white bull. When Europa climbed on his back he swam with her to Crete, where she bore several children, including Minos.	January 8, 1610	Padua	Galileo (who evidently observed the combined image of Io and Europa the previous night)
Ganymede (JIII)	Beautiful young boy who was carried to Olympus by Jupiter disguised as an eagle. Ganymede then became the cupbearer of the Olympian gods.	January 7, 1610	Padua	Galileo
Callisto (JIV)	Beautiful daughter of Lycaon, she was seduced by Jupiter, who changed her into a bear to protect her from Hera's jealousy.	January 7, 1610	Padua	Galileo
Amalthea (JV)	A naiad who nursed the new-born Jupiter. She had as a favorite animal a goat which is said by some authors to have nourished Jupiter. The name was suggested by Flammarion.	September 9, 1892	Mt. Hamilton	E. E. Barnard
Himalia (JVI)	A Rhodian nymph who bore three sons of Zeus.	December 4, 1904	Mt. Hamilton	C. D. Perrine
Elara (JVII)	Daughter of King Orchomenus, a paramour of Zeus, and by him the mother of the giant Tityus.	January 3, 1905	Mt. Hamilton	C. D. Perrine
Pasiphaë (JVIII)	Wife of Minos, king of Crete. Zeus made approaches to her as a bull (taurus). She then gave birth to the Minotaur.	January 27, 1908	Greenwich	P. J. Melotte
Sinope (JIX)	Daughter of the river god Asopus. Zeus desired to make love to her. Instead of this he granted perpetual virginity, after he had been deceived by his own promises. (In the same way, she also fooled Apollo.)	July 21, 1914	Mt. Hamilton	S. B. Nicholson

(continues)

Jovian System (continued)

Body	Description	Date of Discovery	Discovery Location	Discoverer
Lysithea (JX)	Daughter of Kadmos, also named Semele, mother of Dionysos by Zeus. According to others, she was the daughter of Evenus and mother of Helenus by Jupiter.	July 6, 1938	Mt. Wilson	S. B. Nicholson
Carme (JXI)	A nymph and attendant of Artemis; mother, by Zeus, of Britomartis.	July 30, 1938	Mt. Wilson	S. B. Nicholson
Ananke (JXII)	Goddess of fate and necessity, mother of Adrastea by Zeus.	September 28, 1951	Mt. Wilson	S. B. Nicholson
Leda (JXIII)	Seduced by Zeus in the form of a swan, she was the mother of Pollux and Helen.	September 11, 1974	Palomar	C. T. Kowal
Thebe (JXIV, 1979 J2)	An Egyptian king's daughter, granddaughter of Io, mother of Aigyptos by Zeus. The Egyptian city of Thebes was named after her.	March 5, 1979	*Voyager 1*	S. Synnott and the Voyager team
Adrastea (JXV, 1979 J1)	A nymph of Crete to whose care Rhea entrusted the infant Zeus.	July, 1979	*Voyager 2*	D. Jewitt and E. Danielson
Metis (JXVI, 1979 J3)	First wife of Zeus. He swallowed her when she became pregnant; Athena was subsequently born from the forehead of Zeus.	March 4, 1979	*Voyager 1*	S. Synnott and the Voyager team
Callirrhoe (JXVII, 1999 J1)	Daughter of the river god Achelous and stepdaughter of Jupiter.	October 19, 1999	Spacewatch	J. V. Scotti, T. B. Spahr, R. S. McMillan, J. A. Larson, J. Montani, A. E. Gleason, and T. Gehrels
Themisto (JXVIII, 1975 J1, 2000 J1	Daughter of the Arcadian river god Inachus, mother of Ister by Zeus.	September 30, 1975, rediscovered November 21, 2000	Palomar, rediscovered at Mauna Kea	C. T. Kowal and E. Roemer (1975), and S. S. Sheppard, D. C. Jewitt, Y. R. Fernandez, G. Magnier, M. Holman, B. G. Marsden, and G. V. Williams (2000)
Megaclite (JXIX, 2000 J8)	Daughter of Macareus, who with Zeus gave birth to Thebe and Locrus.	November 25, 2000	Mauna Kea	S. S. Sheppard, D. C. Jewitt, Y. R. Fernandez, and G. Magnier
Taygete (JXX, 2000 J9)	Daughter of Atlas, one of the Pleiades, mother of Lakedaimon by Zeus.	November 25, 2000	Mauna Kea	S. S. Sheppard, D. C. Jewitt, Y. R. Fernandez, and G. Magnier
Chaldene (JXXI, 2000 J10)	Bore the son Solymos with Zeus.	November 25, 2000	Mauna Kea	S. S. Sheppard, D. C. Jewitt, Y. R. Fernandez, and G. Magnier
Harpalyke (JXXII, 2000 J5)	Daughter and wife of Clymenus. In revenge for this incestuous relationship, she killed the son she bore him, cooked the corpse, and served it to Clymenus. She was transformed into the night bird called Chalkis, and Clymenus hanged himself. Some say that she was transformed into that bird because she had intercourse with Zeus.	November 23, 2000	Mauna Kea	S. S. Sheppard, D. C. Jewitt, Y. R. Fernandez, and G. Magnier
Kalyke (JXXIII, 2000 J2)	Nymph who bore the handsome son Endymion with Zeus.	November 23, 2000	Mauna Kea	S. S. Sheppard, D. C. Jewitt, Y. R. Fernandez, and G. Magnier

Body	Description	Date of Discovery	Discovery Location	Discoverer
Iocaste (JXXIV, 2000 J3)	Wife of Laius, King of Thebes, and mother of Oedipus. After Laius was killed, Iocaste unknowingly married her own son Oedipus. When she learned that her husband was her son, she killed herself. Some say she was the mother of Agamedes by Jupiter.	November 23, 2000	Mauna Kea	S. S. Sheppard, D. C. Jewitt, Y. R. Fernandez, and G. Magnier
Erinome (JXXV, 2000 J4)	Daughter of Celes, compelled by Venus to fall in love with Jupiter.	November 23, 2000	Mauna Kea	S. S. Sheppard, D. C. Jewitt, Y. R. Fernandez, and G. Magnier
Isonoe (JXXVI, 2000 J6)	A Danaid, bore with Zeus the son Orchomenos.	November 23, 2000	Mauna Kea	S. S. Sheppard, D. C. Jewitt, Y. R. Fernandez, and G. Magnier
Praxidike (JXXVII, 2000 J7)	Goddess of punishment, mother of Klesios by Zeus.	November 23, 2000	Mauna Kea	S. S. Sheppard, D. C. Jewitt, Y. R. Fernandez, and G. Magnier
Autonoe (JXXVIII, 2001, J1)	Mother of the Graces by Jupiter according to some authors.	December 10, 2001	Mauna Kea	S. S. Sheppard, D. C. Jewitt, and J. Kleyna
Thyone (JXXIX, 2001 J2)	Semele, mother of Dionysos by Zeus. She received the name of Thyone in Hades by Dionysos before he ascended up with her from there to heaven.	December 11, 2001	Mauna Kea	S. S. Sheppard, D. C. Jewitt, and J. Kleyna
Hermippe (JXXX, 2001 J3)	Consort of Zeus and mother of Orchomenos by him.	December 9, 2001	Mauna Kea	S. S. Sheppard, D. C. Jewitt, and J. Kleyna
Aitne (JXXXI, 2001 J11)	A Sicilian nymph, conquest of Zeus.	December 11, 2001	Mauna Kea	S. S. Sheppard, D. C. Jewitt, and J. Kleyna
Eurydome (JXXXII, 2001 J4)	Mother of the Graces by Zeus, according to some authors.	December 9, 2001	Mauna Kea	S. S. Sheppard, D. C. Jewitt, and J. Kleyna
Euanthe (JXXXIII, 2001 J7)	The mother of the Graces by Zeus, according to some authors.	December 11, 2001	Mauna Kea	S. S. Sheppard, D. C. Jewitt, and J. Kleyna
Euporie (JXXXIV, 2001 J10)	One of the Horae, a daughter of Jupiter and Themis.	December 11, 2001	Mauna Kea	S. S. Sheppard, D. C. Jewitt, and J. Kleyna
Orthosie (JXXXV, 2001 J9)	One of the Horae, a daughter of Jupiter and Themis.	December 11, 2001	Mauna Kea	S. S. Sheppard, D. C. Jewitt, and J. Kleyna
Sponde (JXXXVI, 2001 J5)	One of the Horae (Seasons), daughter of Jupiter.	December 9, 2001	Mauna Kea	S. S. Sheppard, D. C. Jewitt, and J. Kleyna
Kale (JXXXVII, 2001 J8)	One of the Graces, a daughter of Zeus, husband of Hephaestos.	December 9, 2001	Mauna Kea	S. S. Sheppard, D. C. Jewitt, and J. Kleyna
Pasithee (JXXXVIII, 2001 J6)	One of the Graces, a daughter of Zeus.	December 11, 2001	Mauna Kea	S. S. Sheppard, D. C. Jewitt, and J. Kleyna

Saturnian System

Satellites in the Saturnian system are named for Greco-Roman Titans, descendants of the Titans, the Roman god of the beginning, and giants from Greco-Roman and other mythologies. Gallic, Inuit, and Norse names identify three different orbit inclination groups, where inclinations are measured with respect to the ecliptic, not Saturn's equator or orbit. Retrograde satellites (those with an inclination of 90 to 180 degrees) are named for Norse giants (except for Phoebe, which was discovered long ago and is the largest). Prograde satellites with an orbit inclination of around 36 degrees are named for Gallic giants, and prograde satellites with an inclination of around 48 degrees are named for Inuit giants.

Body	Description	Date of Discovery	Discovery Location	Discoverer
Saturn	Roman name for the Greek Kronos, father of Zeus/Jupiter. Other civilizations have given different names to Saturn, which is the farthest planet from Earth that can be observed by the naked human eye. Most of its satellites were named for Titans who, according to Greek mythology, were brothers and sisters of Saturn.			
Mimas (SI)	Named by Herschel's son John in the early 19th century for a Giant felled by Hephaestus (or Ares) in the war between the Titans and Olympian gods.	July 18, 1789	Slough	W. Herschel
Enceladus (SII)	Named by Herschel's son John for the Giant Enceladus. Enceladus was crushed by Athene in the battle between the Olympian gods and the Titans. Earth piled on top of him became the island of Sicily.	August 28, 1789	Slough	W. Herschel
Tethys (SIII)	Cassini wished to name Tethys and the other three satellites that he discovered (Dione, Rhea, and Iapetus) for Louis XIV. However, the names used today for these satellites were applied in the early 19th century by John Herschel, who named them for Titans and Titanesses, brothers and sisters of Saturn. Tethys was the wife of Oceanus and mother of all rivers and Oceanids.	March 21, 1684	Paris	G. D. Cassini
Dione (SIV)	Dione was the sister of Kronos and mother (by Zeus) of Aphrodite.	March 21, 1684	Paris	G. D. Cassini
Rhea (SV)	A Titaness, mother of Zeus by Cronos.	December 23, 1672	Paris	G. D. Cassini
Titan (SVI)	Named by Huygens, who first called it "Luna Saturni."	March 25, 1655	The Hague	C. Huygens
Hyperion (SVII)	Named by Lassell for one of the Titans.	September 16, 1848	Cambridge, Mass.	W. C. Bond and G. P. Bond; independently discovered September 18, 1848 at Liverpool by W. Lassell
Iapetus (SVIII)	Named by John Herschel for one of the Titans.	October 25, 1671	Paris	G. D. Cassini
Phoebe (SIX)	Named by Pickering for one of the Titanesses.	August 16, 1898	Arequipa	W. H. Pickering
Janus (SX, 1980 S1)	First reported (though with an incorrect orbital period) and named by A. Dollfus from observations in December 1966, this satellite was finally confirmed in 1980. It was proven to have a twin, Epimetheus, sharing the same orbit but never actually meeting. It is named for the Roman god of the beginning. The two-faced god could look forward and backward at the same time.	December 15, 1966 (Dollfus), February 19, 1980 (Pascu)	Pic du Midi (Dollfus), Washington (Pascu)	A. Dollfus (1966), D. Pascu (1980)
Epimetheus (SXI, 1980 S3)	First suspected by J. Fountain and S. Larson as confusing the detection of Janus. They assigned the correct orbital period, and the satellite was finally confirmed in 1980. Named for the son of the Titan Iapetus. In contrast with his far-sighted brother Prometheus, he "subsequently realized" that he was in the wrong.	1977 (Fountain and Larson), February 26, 1980 (Cruikshank)	Tucson (Fountain and Larson), Mauna Kea (Cruikshank)	J. Fountain and S. Larson (1977), D. Cruikshank (1980)
Helene (SXII, 1980 S6)	A granddaughter of Cronos, her beauty triggered the Trojan War.	March 1, 1980	Pic du Midi	P. Laques and J. Lecacheux
Telesto (SXIII, 1980 S13)	Daughter of the Titans Oceanus and Tethys.	April 8, 1980	Tucson	B. A. Smith, H. Reitsema, S. M. Larson, and J. Fountain

Body	Description	Date of Discovery	Discovery Location	Discoverer
Calypso (SXIV, 1980 S25)	Daughter of the Titans Oceanus and Tethys and paramour of Odysseus.	March 13, 1980	Flagstaff	D. Pascu, P. K. Seidelmann, W. Baum, and D. Currie
Atlas (SXV, 1980 S28)	A Titan; he held the heavens on his shoulders.	October 1980	*Voyager 2*	R. Terrile and the Voyager team
Prometheus (SXVI, 1980 S27)	Son of the Titan Iapetus, brother of Atlas and Epimetheus, he gave many gifts to humanity, including fire.	October 1980	*Voyager 2*	S. Collins and the Voyager team
Pandora (SXVII, 1980 S26)	Made of clay by Hephaestus at the request of Zeus. She married Epimetheus and opened the box that loosed a host of plagues upon humanity.	October 1980	*Voyager 2*	S. Collins and the Voyager team
Pan (SXVIII, 1981 S13)	Son of the Titans Cronos and Rhea. He was the half human, half goat god of pastoralism. Discovered orbiting in the Encke Gap in Saturn's A ring.	August 1981	*Voyager 2*	M. R. Showalter
Ymir (SXIX, 2000 S1)	Ymir is the primordial Norse giant and the progenitor of the race of frost giants.	August 7, 2000	La Silla	B. Gladman, J. Kavelaars, J.-M. Petit, H. Scholl, M. Holman, B. G. Marsden, P. Nicholson, and J. A. Burns
Paaliaq (SXX, 2000 S2)	Named for an Inuit giant.	August 7, 2000	La Silla	B. Gladman, J, Kavelaars, J.-M. Petit, H. Scholl, M. Holman, B. G. Marsden, P. Nicholson, and J. A. Burns
Tarvos (SXXI, 2000 S4)	Named for a Gallic giant.	September 23, 2000	Mauna Kea	B. Gladman, J. Kavelaars, J.-M. Petit, H. Scholl, M. Holman, B. G. Marsden, P. Nicholson, and J. A. Burns
Ijiraq (SXXII, 2000 S6)	Named for an Inuit giant.	September 23, 2000	Mauna Kea	B. Gladman, J. Kavelaars, J.-M. Petit, H. Scholl, M. Holman, B. G. Marsden, P. Nicholson, and J. A. Burns
Suttungr (SXXIII, 2000 S12)	Named for a Norse giant who kindled flames that destroyed the world.	September 23, 2000	Mauna Kea	B. Gladman, J. Kavelaars, J.-M. Petit, H. Scholl, M. Holman, B. G. Marsden, P. Nicholson, and J. A. Burns
Kiviuq (SXXIV, 2000 S5)	Named for an Inuit giant.	August 7, 2000	La Silla	B. Gladman, J. Kavelaars, J.-M. Petit, H. Scholl, M. Holman, B. G. Marsden, P. Nicholson, and J. A. Burns

(continues)

Saturnian System (continued)

Body	Description	Date of Discovery	Discovery Location	Discoverer
Mundilfari (SXXV, 2000 S9)	Named for a Norse giant.	September 23, 2000	Mauna Kea	B. Gladman, J. Kavelaars, J.-M. Petit, H. Scholl, M. Holman, B. G. Marsden, P. Nicholson, and J. A. Burns
Albiorix (SXXVI, 2000 S11)	Named for a Gallic giant who was considered to be the king of the world.	November 9, 2000	Mt. Hopkins	M. Holman
Skathi (SXXVII, 2000 S8)	Named for a Norse giantess.	September 23, 2000	Mauna Kea	B. Gladman, J. Kavelaars, J.-M. Petit, H. Scholl, M. Holman, B. G. Marsden, P. Nicholson, and J. A. Burns
Erriapo (SXXVIII, 2000 S10)	Named for a Gallic giant.	September 23, 2000	Mauna Kea	B. Gladman, J. Kavelaars, J.-M. Petit, H. Scholl, M. Holman, B. G. Marsden, P. Nicholson, and J. A. Burns
Siarnaq (SXXIX, 2000 S3)	Named for an Inuit giant.	September 23, 2000	Mauna Kea	B. Gladman, J. Kavelaars, J.-M. Petit, H. Scholl, M. Holman, B. G. Marsden, P. Nicholson, and J. A. Burns
Thrymr (SXXX, 2 000 S7)	Named for a Norse giant.	September 23, 2000	Mauna Kea	B. Gladman, J. Kavelaars, J.-M. Petit, H. Scholl, M. Holman, B. G. Marsden, P. Nicholson, and J. A. Burns

Uranian System

Satellites in the Uranian system are named for characters from Shakespeare's plays and from Pope's "Rape of the Lock."

Body	Description	Date of Discovery	Discovery Location	Discoverer
Uranus	Several astronomers, including Flamsteed and Le Monnier, had observed Uranus earlier but had recorded it as a fixed star. Herschel tried unsuccessfully to name his discovery "Georgian Sidus" after George III; the planet was named by Johann Bode in 1781 for the father of Saturn.	March 13, 1781	Bath	W. Herschel
Ariel (UI)	Named by John Herschel for a sylph in Pope's "Rape of the Lock."	October 24, 1851	Liverpool	W. Lassell
Umbriel (UII)	Umbriel was named by John Herschel for a malevolent spirit in Pope's "Rape of the Lock."	October 24, 1851	Liverpool	W. Lassell
Titania (UIII)	Named by Herschel's son John in early 19th century for the queen of the fairies in Shakespeare's "A Midsummer Night's Dream."	January 11, 1787	Slough	W. Herschel

Body	Description	Date of Discovery	Discovery Location	Discoverer
Oberon (UIV)	Named by Herschel's son John in the early 19th century for the king of the fairies in Shakespeare's "A Midsummer-Night's Dream."	January 11, 1787	Slough	W. Herschel
Miranda (UV)	Named by Kuiper for the heroine of Shakespeare's "The Tempest."	February 16, 1948	Fort Davis	G. P. Kuiper
Cordelia (UVI, 1986 U7)	Daughter of Lear in Shakespeare's "King Lear."	January 20, 1986	*Voyager 2*	R. Terrile and the *Voyager 2* team
Ophelia (UVII, 1986 U8)	Daughter of Polonius, fiance of Hamlet in Shakespeare's "Hamlet, Prince of Denmark."	January 20, 1986	*Voyager 2*	R. Terrile and the *Voyager 2* team
Bianca (UVIII, 1986 U9)	Daughter of Baptista, sister of Kate, in Shakespeare's "Taming of the Shrew."	January 23, 1986	*Voyager 2*	*Voyager 2* team
Cressida (UIX, 1986 U3)	Title character in Shakespeare's "Troilus and Cressida."	January 9, 1986	*Voyager 2*	S. Synnott and the *Voyager 2* team
Desdemona (UX, 1986 U6)	Wife of Othello in Shakespeare's "Othello, the Moor of Venice."	January 13, 1986	*Voyager 2*	S. Synnott and the *Voyager 2* team
Juliet (UXI, 1986 U2)	Heroine of Shakespeare's "Romeo and Juliet."	January 3, 1986	*Voyager 2*	S. Synnott and the *Voyager 2* team
Portia (UXII, 1986 U1)	Wife of Brutus in Shakespeare's "Julius Caesar."	January 3, 1986	*Voyager 2*	S. Synnott and the *Voyager 2* team
Rosalind (UXIII, 1986 U4)	Daughter of the banished duke in Shakespeare's "As You Like It."	January 13, 1986	*Voyager 2*	S. Synnott and the *Voyager 2* team
Belinda (UXIV, 1986 U5)	Character in Pope's "Rape of the Lock."	January 13, 1986	*Voyager 2*	S. Synnott and the *Voyager 2* team
Puck (UXV, 1985 U1)	Mischievous spirit in Shakespeare's "A Midsummer-Night's Dream."	December 30, 1985	*Voyager 2*	S. Synnott and the *Voyager 2* team
Caliban (UXVI, 1997 U1)	Named for the grotesque, brutish slave in Shakespeare's "The Tempest."	September 6, 1997	Palomar	B. Gladman, P. Nicholson, J. A. Burns, and J. Kavelaars
Sycorax (UXVII, 1997 U2)	Named for Caliban's mother in Shakespeare's "The Tempest."	September 6, 1997	Palomar	P. Nicholson, B. Gladman, J. Burns, and J. Kavelaars
Prospero (UXVIII, 1999 U3)	Named for the rightful Duke of Milan in "The Tempest."	July 18, 1999	Mauna Kea	M. Holman, J. Kavelaars, B. Gladman, J.-M. Petit, and H. Scholl
Setebos (UXIX, 1999 U1)	Setebos was a new-world (South American) deity's name that Shakespeare popularized as Sycorax's god in "The Tempest."	July 18, 1999	Mauna Kea	J. Kavelaars, B. Gladman, M. Holman, J.-M. Petit, and H. Scholl
Stephano (UXX, 1999 U2)	Named for a drunken butler in "The Tempest."	July 18, 1999	Mauna Kea	B. Gladman, M. Holman, J. Kavelaars, J.-M. Petit, and H. Scholl
Trinculo (UXXI, 2001 U1)	A jester in Shakespeare's "The Tempest."	August 13, 2001	Cerro Tololo	M. Holman, J. J. Kavelaars, and D. Milisavljevic

Neptunian System

Satellites in the Neptunian system are named for characters from Greek or Roman mythology associated with Neptune or Poseidon or the oceans.

Body	Description	Date of Discovery	Discovery Location	Discoverer
Neptune	Neptune was "predicted" by John Couch Adams and Urbain Le Verrier who, independently, were able to account for the irregularities in the motion of Uranus by correctly predicting the orbital elements of a trans-Uranian body. Using the predicted parameters of Le Verrier (Adams never published his predictions), Johann Galle observed the planet in 1846. Galle wanted to name the planet for Le Verrier, but that was not acceptable to the international astronomical community. Instead, this planet is named for the Roman god of the sea.	September 23, 1846	Berlin	J. G. Galle
Triton (NI)	Triton is named for the sea-god son of Poseidon (Neptune) and Amphitrite. The first suggestion of the name Triton has been attributed to the French astronomer Camille Flammarion.	October 10, 1846	Liverpool	W. Lassell
Nereid (NII)	The Nereids were the 50 daughters of Nereus and Doris and were attendants of Neptune.	May 1, 1949	Fort Davis	G. P. Kuiper
Naiad (NIII, 1989 N6)	The name of a group of Greek water nymphs who were guardians of lakes, fountains, springs, and rivers.	August 1989	*Voyager 2*	*Voyager 2* team
Thalassa (NIV, 1989 N5)	Greek sea goddess. Mother of Aphrodite in some legends; others say she bore the Telchines.	August 1989	*Voyager 2*	R. Terrile and the *Voyager 2* team
Despina (NV, 1989 N3)	Daughter of Poseidon (Neptune) and Demeter.	July 1989	*Voyager 2*	S. Synnott and the *Voyager 2* team
Galatea (NVI, 1989 N4)	One of the Nereids, attendants of Poseidon.	July 1989	*Voyager 2*	S. Synnott and the *Voyager 2* team
Larissa (NVII, 1989 N2)	A lover of Poseidon. After the discovery by *Voyager 2,* it was established that an occultation of a star by this satellite had been fortuitously observed in 1981 by H. Reitsema, W. Hubbard, L. Lebofsky, and D. J. Tholen.	July 1989	*Voyager 2*	*Voyager 2* team
Proteus (NVIII, 1989 N1)	Greek sea god, son of Oceanus and Tethys.	June 1989	*Voyager 2*	S. Synnott and the *Voyager 2* team

Plutonian System

Body	Description	Date of Discovery	Discovery Location	Discoverer
Pluto	Pluto was discovered at Lowell Observatory in Flagstaff, Arizona, during a systematic search for a trans-Neptune planet predicted by Percival Lowell and William H. Pickering. Named after the Greek god of the underworld who was able to render himself invisible.	January 23, 1930	Flagstaff	C. W. Tombaugh
Charon (P1, 1978 P1)	Named after the mythological boatman who ferried souls across the river Styx to Pluto for judgment.	April 13, 1978	Flagstaff	J. W. Christy

Ring and Ring Gap Nomenclature

System	Name	Distance (km) from Planet's Center	Radial Width (km)
Rings of Jupiter	1979 J1R("Halo")	100,000–122,800	22,800
	1979 J2R ("Main")	122,800–129,200	6,400
	1979 J3R ("Gossamer")	129,200–214,200	85,000
Rings of Saturn	D	67,000–74,500	7,500
	C	74,500–92,000	17,500
	Columbo gap	77,800	100
	Maxwell gap	87,500	270
	B	92,000–117,500	25,500
	Cassini division	117,500–122,200	4,700
	Huygens gap	117,680	285–440
	A	122,200–136,800	14,600
	Encke division	133,570	325
	Keeler gap	136,530	35
	F	140,210	30–500
	G	165,800–173,800	8,000
	E	180,000–480,000	300,000
Rings of Uranus	1986 U2R	38,000	2,500?
	6	41,840	1–3
	5	42,230	2–3
	4	42,580	2–3
	Alpha	44,720	7–12
	Beta	45,670	7–12
	Eta	47,190	0–2
	Gamma	47,630	1–4
	Delta	48,290	3–9
	1986 U1R	50,020	1–2
	Epsilon	51,140	20–100
Rings and ring arcs of Neptune	1989 N3R Galle	41,900	15
	1989 N2R Leverrier	53,200	15
	Lassell	55,400	–
	Arago	57,600	–
	1989 N1R Adams	62,930	< 50
	Liberté ("Leading" arc)	62,900	–
	Egalité ("Equidistant" arc)	62,900	–
	Fraternité ("Following" arc)	62,900	–
	Courage (arc)	62,900	–

Categories for Naming Features on Planets and Satellites

Mercury

Craters	Famous deceased artists, musicians, painters, authors
Montes	Caloris, from Latin word for "hot"
Planitiae	Names for Mercury (either planet or god) in various languages
Rupes	Ships of discovery or scientific expeditions
Valles	Radio telescope facilities

Venus

Astra	Goddesses, miscellaneous
Chasmata	Goddesses of hunt; moon goddesses
Colles	Sea goddesses
Coronae	Fertility and earth goddesses
Craters	Over 20 km; famous women; under 20 km, common female first names
Dorsa	Sky goddesses
Farra	Water goddesses
Fluctus	Goddesses, miscellaneous
Fossae	Goddesses of war
Labyrinthi	Goddesses, miscellaneous
Lineae	Goddesses of war
Montes	Goddesses, miscellaneous (also one radar scientist)
Paterae	Famous women
Plana	Goddesses of prosperity
Planitiae	Mythological heroines
Regiones	Giantesses and Titanesses (also two Greek alphanumeric)
Rupes	Goddesses of hearth and home
Terrae	Goddesses of love
Tesserae	Goddesses of fate and fortune
Tholi	Goddesses, miscellaneous
Undae	Desert goddesses
Valles	Word for planet Venus in various world languages (400 km and longer); river goddesses (less than 400 km in length)

The Moon

Craters, Catenae, Dorsa, Rimae	Craters: famous deceased scientists, scholars, artists and explorers; other features named from nearby craters
Lacus, Maria, Paludes, Sinus	Latin terms describing weather and other abstract concepts
Montes	Terrestrial mountain ranges or nearby craters
Rupes	Name of nearby mountain ranges (terrestrial names)
Valles	Name of nearby features

Mars and Martian Satellites

Mars	
Large craters	Deceased scientists who have contributed to the study of Mars; writers and others who have contributed to the lore of Mars
Small craters	Villages of the world with a population of less than 100,000
Large valles	Name for Mars/star in various languages
Small valles	Classical or modern names of rivers
Other features	From nearest named albedo feature on Schiaparelli or Antoniadi maps
Deimos	Authors who wrote about Martian satellites
Phobos	Scientists involved with the discovery, dynamics, or properties of the Martian satellites

Jovian Satellites

Amalthea	People and places associated with the Amalthea myth
Thebe	People and places associated with the Thebe myth
Io	
Active eruptive centers	Fire, sun, thunder gods, and heroes
Catenae	Sun gods
Fluctus	Name derived from nearby named feature, or fire, sun, thunder, volcano gods, goddesses and heroes, mythical blacksmiths
Mensae	People associated with Io myth, derived from nearby feature, or people from Dante's "Inferno"
Montes	Places associated with Io myth, derived from nearby feature, or places from Dante's "Inferno"
Paterae	Fire, sun, thunder, volcano gods, heroes, goddesses, mythical blacksmiths
Plana	Places associated with Io myth, derived from nearby feature, or places from Dante's "Inferno"
Regiones	Places associated with Io myth, derived from nearby feature, or places from Dante's "Inferno"
Tholi	Places associated with Io myth, derived from nearby feature, or places from Dante's "Inferno"
Europa	
Chaos	Places associated with Celtic myths
Craters	Celtic gods and heroes
Flexus	Places associated with the Europa myth, or Celtic stone rows
Large ringed features	Celtic stone circles
Lenticulae	Celtic gods and heroes
Lineae	People associated with the Europa myth, or Celtic stone rows

Maculae	Places associated with the Europa myth
Regiones	Places associated with Celtic myths
Ganymede	
Catenae	Gods and heroes of ancient Fertile Crescent people
Craters	Gods and heroes of ancient Fertile Crescent people
Faculae	Places associated with Egyptian myths
Fossae	Gods (or principals) of ancient Fertile Crescent people
Paterae	Dry wadis (channels) of the Fertile Crescent region
Regiones	Astronomers who discovered Jovian satellites
Sulci	Places associated with myths of ancient people
Callisto	
Catenae	Mythological places in high latitudes
Craters	Heroes and heroines from northern myths
Large ringed features	Homes of the gods and of heroes

Satellites of Saturn

Janus	People from myth of Castor and Pollux (twins)
Epimetheus	People from myth of Castor and Pollux (twins)
Mimas	People and places from Malory's "Le Morte Darthur" legends (Baines translation)
Enceladus	People and places from Burton's "Arabian Nights"
Tethys	People and places from Homer's "Odyssey"
Dione	People and places from Virgil's "Aeneid"
Rhea	People and places from creation myths (with Asian emphasis)
Titan	
Craters (and Lakes if present)	Lakes from all continents on Earth
Fluvial channels	Rivers from all continents on Earth
Major bright albedo features	Sacred or enchanted places from legends, myths, stories, and poems of cultures from around the world
Major dark albedo features	Legendary/mythical primordial seas or enchanted waters from world cultures
Other features	Deities of happiness, peace, and harmony from world cultures
Hyperion	Sun and Moon deities
Iapetus	People and places from Sayers' translation of "Chanson de Roland"
Phoebe	
Craters	People associated with Phoebe, people from the "Argonautica" by Apollonius Rhodius and Valerius Flaccus
Other features	Islands of the Greek archipelagos, places from the "Argonautica"

Satellites of Uranus

Puck	Mischievous (Pucklike) spirits (class)
Miranda	Characters and places from Shakespeare's plays
Ariel	Light spirits (individual and class)
Umbriel	Dark spirits (individual)
Titania	Female Shakespearean characters and places
Oberon	Shakespearean tragic heroes and places
Small Satellites	Heroines from Shakespeare and Pope

Satellites of Neptune

Proteus	Water-related spirits, gods, goddesses (excluding Greek and Roman names)
Triton	Aquatic names, excluding Roman and Greek. Possible categories include worldwide aquatic spirits, famous terrestrial fountains or fountain locations, terrestrial aquatic features, famous terrestrial geysers or geyser locations, terrestrial islands
Nereid	Individual nereids
Small Satellites	Gods ad goddesses associated with Neptune/Poseidon mythology or generic mythological aquatic beings

Pluto

Pluto	Underworld deities

Asteroids

Ida	
Craters	Caverns and grottos of the world
Dorsa	Galileo project participants
Regiones	Discoverer of Ida and places associated with the discoverer
Dactyl	
Craters	Idaean dactyls
Gaspra	
Craters	Spas of the world
Regiones	Discoverer of Gaspra, and Galileo project participants
Mathilde	
Craters	Coal fields and basins of the world
Eros	
Craters	Mythological and legendary names of an erotic nature
Dorsa	Scientists who have contributed to the exploration and study of Eros
Regiones	Discoverers of Eros

INDEX

Note: Page numbers in **boldface** indicate main entries.

B

C

E

H

I

J

K

L

M

N

O

P

S

T

W

X